FOREIGN WORD ROOTS, PREFIXES, SUFFIXES, AND COMBINING FORMS

ach entry starts with the commonly used form or forms of the prefix, suffix, or combining form followed by the word oot (shown in italics) and its English translation. One example is also given to illustrate the use of each entry.

, *a-*, without: avascular

-, *ab*, from: abduct

c, *-akos*, pertaining to: cardiac

r-, *akron*, extremity: acromegaly

l-, *ad*, to, toward: adduct

len-, **adeno-**, *adenos*, gland: adenoid

ip-, *adipos*, fat: adipocytes

er-, *aeros*, air: aerobic metabolism

-, *ad*, toward: afferent

l, *-alis*, pertaining to: brachial

b-, *albicans*, white: albino

lgia, *algos*, pain: neuralgia

lo-, *allos*, other: allograft

na-, *ana*, up, back: anaphase

ndro-, *andros*, male: androgen

ngio-, *angeion*, vessel: angiogram

nte-, *ante*, before: antebrachial

nti-, **ant-**, *anti*, against: antibiotic

po-, *apo*, from: apocrine

rachn-, *arachne*, spider: arachnoid

rter-, *arteria*, artery: arterial

rthro-, *arthros*, joint: arthroscopy

sis, **-asia**, *-asis*, state, condition: homeostasis

stro-, *aster*, star: astrocyte

tel-, *ateles*, imperfect: atelectasis

ur-, *auris*, ear: auricle

uto-, *auto*, self: autonomic

aro-, *baros*, pressure: baroreceptor

i-, *bi-*, two: bifurcate

io-, *bios*, life: biology

last-, **-blast**, *blastos*, precursor: blastocyst

rachi-, *brachium*, arm: brachiocephalic

rachy-, *brachys*, short: brachydactyly

rady-, *bradys*, slow: bradycardia

ronch-, *bronchus*, windpipe, airway: bronchial

arcin-, *karkinos*, cancer: carcinoma

ardi, **cardio-**, **-cardia**, *kardia*, heart: cardiac

ele-, *kele*, tumor, hernia, or swelling: blastocele

entesis, *kentesis*, puncture: thoracocentesis

ephal-, *cephalos*, head: brachiocephalic

erebr-, *cerebrum*, brain: cerebral hemispheres

erebro-, *cerebros*, brain: cerebrospinal fluid

ervic-, *cervicis*, neck: cervical vertebrae

hole-, *chole*, bile: cholecystitis

chondrion, *chondrion*, granule: mitochondrion

hondro-, *chondros*, cartilage: chondrocyte

hrom-, **chromo-**, *chroma*, color: chromatin

ircum-, *circum*, around: circumduction

clast, *klastos*, broken: osteoclast

oel-, **-coel**, *koila*, cavity: coelom

olo-, *kolon*, colon: colonoscopy

ontra-, *contra*, against: contralateral

orp-, *corpus*, body: corpuscle

ortic-, *cortex*, rind or bark: corticospinal

ost-, *costa*, rib: costal

ranio-, *cranium*, skull: craniosacral

ribr-, *cribrum*, sieve: cribriform

rine, *krinein*, to separate: endocrine

ut-, *cutis*, skin: cutaneous

cyan-, *kyanos*, blue: cyanosis

cyst-, **-cyst**, *kystis*, sac: blastocyst

cyt-, **cyto-**, kyton, a hollow cell: cytology

de-, *de*, from, away: deactivation

dendr-, *dendron*, tree: dendrite

dent-, *dentes*, teeth: dentition

derm-, *derma*, skin: dermatome

desmo-, *desmos*, band: desmosome

di-, *dis*, twice: disaccharide

dia-, *dia*, through: diameter

digit-, *digit*, a finger or toe: digital

dipl-, *diploos*, double: diploid

dis-, apart, away from: disability

diure-, *diourein*, to urinate: diuresis

dys-, *dys-*, painful: dysmenorrhea

-ectasis, *ektasis*, expansion: atelectasis

ecto-, *ektos*, outside: ectoderm

-ectomy, *ektome*, excision: appendectomy

ef-, *ex*, away from: efferent

emmetro-, *emmetros*, in proper measure: emmetropia

encephalo-, *enkephalos*, brain: encephalitis

end-, **endo-**, *endos*, inside: endometrium

entero-, *enteron*, intestine: enteric

epi-, *epi*, on: epimysium

erythema-, *erythema*, flushed (skin): erythematosis

erythro-, *erythros*, red: erythrocyte

ex-, *ex*, out, away from: exocytosis

extra-, *exter*, outside of, beyond, in addition: extracellular

ferr-, *ferrum*, iron: transferrin

fil-, *filum*, thread: filament

-form, *-formis*, shape: fusiform

gastr-, *gaster*, stomach: gastrointestinal

-gen, **-genic**, *gennan*, to produce: mutagen

genicula-, *geniculum*, kneelike structure: geniculates

genio-, *geneion*, chin: geniohyoid

gest-, *gesto*, to bear: gestation

glosso-, **-glossus**, *glossus*, tongue: hypoglossal

glyco-, *glykys*, sugar: glycogen

-gram, *gramma*, record: myogram

gran-, *granulum*, grain: granulocyte

-graph, **-graphia**, *graphein*, to write, record: electroencephalograph

gyne-, **gyno-**, *gynaikos*, woman: gynecologist

hem-, **hemato-**, *haima*, blood: hemopoiesis

hemi-, *hemi-*, half: hemisphere

hepato-, *hepaticus*, liver: hepatocyte

hetero-, *heteros*, other: heterozygous

histo-, *histos*, tissue: histology

holo-, *holos*, entire: holocrine

homeo-, **homo-**, *homos*, same: homeostasis

hyal-, **hyalo-**, *hyalos*, glass: hyaline

hydro-, *hydros*, water: hydrolysis

hyo-, *hyoeides*, U-shaped: hyoid bone

hyper-, *hyper*, above: hyperpolarization

hypo-, *hypo*, under: hypothyroid

hyster-, *hystera*, uterus: hysterectomy

-ia, *-ia*, state or condition: insomnia

idi-, *idios*, one's own: idiopathic

ile-, *ileum:* ileocolic sphincter
ili-, ilio-, *ilium:* iliac
in-, *in-,* in, within, or denoting negative effect: inactivate
infra-, *infra,* beneath: infraorbital
inter-, *inter,* between: interventricular
intra-, *intra,* within: intracapsular
ipsi-, *ipse,* itself: ipsilateral
iso-, *isos,* equal: isotonic
-itis, *-itis,* inflammation: dermatitis
karyo-, *karyon,* body: megakaryocyte
kerato-, *keros,* horn: keratin
kino-, -kinin, *kinein,* to move: bradykinin
lact-, lacto-, -lactin, *lac,* milk: prolactin
lapar-, *lapara,* flank or loins: laparoscopy
-lemma, *lemma,* husk: plasmalemma
leuko-, *leukos,* white: leukocyte
liga-, *ligare,* to bind together: ligase
lip-, lipo-, *lipos,* fat: lipoid
lith-, *lithos,* stone: cholelithiasis
lyso-, -lysis, -lyze, *lysis,* dissolution: hydrolysis
macr-, *makros,* large: macrophage
mal-, *mal,* abnormal: malabsorption
mamilla-, *mamilla,* little breast: mamillary
mast-, masto-, *mastos,* breast: mastoid
mega-, *megas,* big: megakaryocyte
melan-, *melas,* black: melanocyte
men-, *men,* month: menstrual
mero-, *meros,* part: merocrine
meso-, *mesos,* middle: mesoderm
meta-, *meta,* after, beyond: metaphase
micr-, *mikros,* small: microscope
mono-, *monos,* single: monocyte
morpho-, *morphe,* form: morphology
multi-, *multus,* much, many: multicellular
-mural, *murus,* wall: intramural
myelo-, *myelos,* marrow: myeloblast
myo-, *mys,* muscle: myofilament
narc-, *narkoun,* to numb or deaden: narcotics
nas-, *nasus,* nose: nasolacrimal duct
natri-, *natrium,* sodium: natriuretic
necr-, *nekros,* corpse: necrosis
nephr-, *nephros,* kidney: nephron
neur-, neuro-, *neuron,* nerve: neuromuscular
oculo-, *oculus,* eye: oculomotor
odont-, *odontos,* tooth: odontoid process
-oid, *eidos,* form, resemblance: odontoid process
oligo-, *oligos,* little, few: oligopeptide
-ology, *logos,* the study of: physiology
-oma, *-oma,* swelling: carcinoma
onco-, *onkos,* mass, tumor: oncology
oo-, *oon,* egg: oocyte
ophthalm-, *ophthalmos,* eye: ophthalmic nerve
-opia, *ops,* eye: myopia
orb-, *orbita,* a circle: orbicularis oris
orchi-, *orchis,* testis: orchiectomy
orth-, *orthos,* correct, straight: orthopedist
-osis, *-osis,* state, condition: neurosis
osteon, osteo-, *os,* bone: osteocyte
oto-, *otikos,* ear: otolith
para-, *para,* beyond: paraplegia
patho-, -path, -pathy, *pathos,* disease:
    pathology
pedia-, *paidos,* child: pediatrician
per-, *per,* through, throughout: percutaneous
peri-, *peri,* around: perineurium
phag-, *phagein,* to eat: phagocyte

-phasia, *phasis,* speech: aphasia
-phil, -philia, *philus,* love: hydrophilic
phleb-, *phleps,* a vein: phlebitis
-phobe, -phobia, *phobos,* fear: hydrophobic
phot-, *phos,* light: photoreceptor
-phylaxis, *phylax,* a guard: prophylaxis
physio-, *physis,* nature: physiology
-plasia, *plasis,* formation: dysplasia
platy-, *platys,* flat: platysma
-plegia, *plege,* a blow, paralysis: paraplegia
-plexy, *plessein,* to strike: apoplexy
pneum-, *pneuma,* air: pneumotaxic center
podo-, podo, *podon,* foot: podocyte
-poiesis, *poiesis,* making: hemopoiesis
poly-, *polys,* many: polysaccharide
post-, *post,* after: postanal
pre-, *prae,* before: precapillary sphincter
presby-, *presbys,* old: presbyopia
pro-, *pro,* before: prophase
proct-, *proktos,* anus: proctology
pterygo-, *pteryx,* wing: pterygoid
pulmo-, *pulmo,* lung: pulmonary
pulp-, *pulpa,* flesh: pulpitis
pyel-, *pyelos,* trough or pelvis: pyelitis
quadr-, *quadrans,* one quarter: quadriplegia
re-, *re-,* back, again: reinfection
retro-, *retro,* backward: retroperitoneal
rhin-, *rhis,* nose: rhinitis
-rrhage, *rhegnymi,* to burst forth: hemorrhage
-rrhea, *rhein,* flow, discharge: amenorrhea
sarco-, *sarkos,* flesh: sarcomere
scler-, sclero-, *skleros,* hard: sclera
-scope, *skopeo,* to view: colonoscope
-sect, *sectio,* to cut: transect
semi-, *semis,* half: semitendinosus
-septic, *septikos,* putrid: antiseptic
-sis, *-sis,* state or condition: metastasis
som-, -some, *soma,* body: somatic
spino-, *spina,* spine, vertebral column:
    spinothalamic pathway
-stalsis, *staltikos,* contractile: peristalsis
sten-, *stenos,* a narrowing: stenosis
-stomy, *stoma,* mouth, opening: colostomy
stylo-, *stylus,* stake, pole: styloid process
sub-, *sub,* below; subcutaneous
super-, *super,* above or beyond: superficial
supra-, *supra,* on the upper side: supraspinous fossa
syn-, *syn,* together: synthesis
tachy-, *tachys,* swift: tachycardia
telo-, *telos,* end: telophase
tetra-, *tettares,* four: tetralogy of Fallot
therm-, thermo-, *therme,* heat: thermoregulation
thorac-, *thorax,* chest: thoracentesis
thromb-, *thrombos,* clot: thrombocyte
-tomy, *temnein,* to cut: appendectomy
tox-, *toxikon,* poison: toxemia
trans-, *trans,* through: transudate
tri-, *tres,* three: trimester
-tropic, *trope,* turning: adrenocorticotropic
tropho-, *trophe,* nutrition: trophoblast
-trophy, *trophikos,* nourishing: atrophy
tropo-, *tropikos,* turning: troponin
uni-, *unus,* one: unicellular
uro-, -uria, *ouron,* urine: glycosuria
vas-, *vas,* vessel: vascular
zyg-, *zygotos,* yoked: zygote

Free Student Aid.

Log on.

Explore.

Succeed.

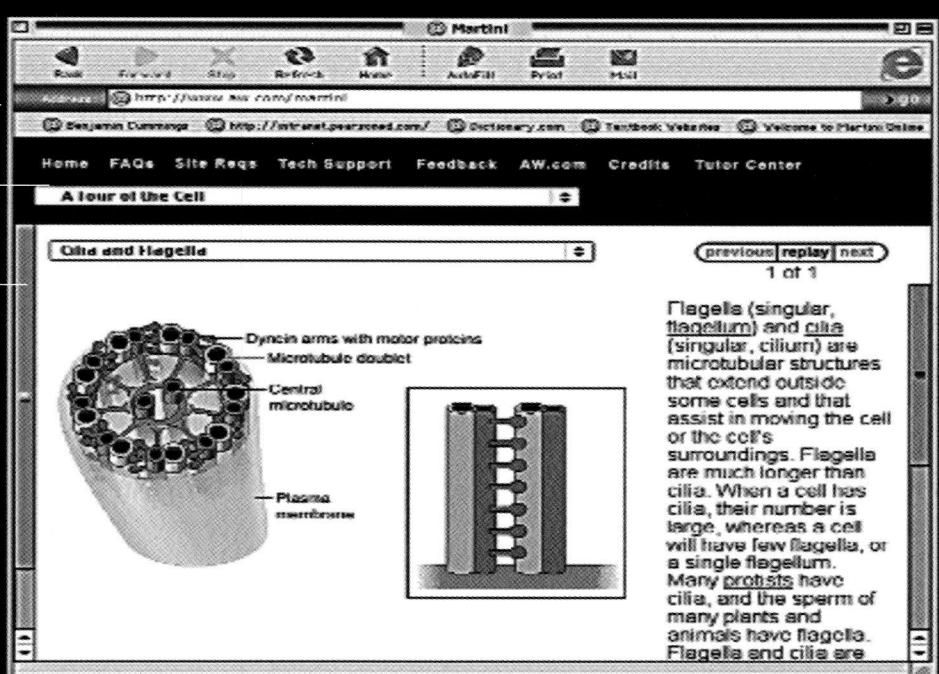

To help you succeed in Anatomy and Physiology, your professor has arranged for you to enjoy access to great media resources, including the *InterActive Physiology®* CD-ROM and the **Martini Online** Web site. You'll find that these resources that accompany your textbook will enhance your course materials.

Got technical questions?
For technical support, please visit www.aw.com/techsupport to complete an online form requesting assistance. Technical support is available Monday - Friday 8 a.m. to 5 p.m. CST.

What your system needs to use Martini Online:

WINDOWS
250 MHz Intel Pentium processor or greater
Windows 95, 98, NT4, 2000, XP
64 MB RAM installed
800 X 600 screen resolution
4x CD-ROM drive
Browser: Internet Explorer 5.0 or higher or Netscape Communicator 4.7
Plug-Ins: Shockwave Player 8, Flash Player 5, QuickTime 5

MACINTOSH
233 MHz PowerPC
OS 8.6 or higher
64 MB RAM available
800 x 600 screen resolution, thousands of colors
4x CD-ROM Drive
Browser: Internet Explorer 5.0 or higher or Netscape Communicator 4.7
Plug-Ins: Shockwave Player 8, Flash Player 5, QuickTime 5

Here's your personal ticket to success:

How to log on to www.aw.com/martini

1. Go to **www.aw.com/martini**.
2. Click **Fundamentals of Anatomy & Physiology, Sixth Edition**.
3. Click "Register."
4. Enter your pre-assigned access code exactly as it appears below.
5. Complete online registration form to create your own personal Login Name and Password.
6. Once your personal Login Name and Password are confirmed by email, go back to www.aw.com/martini, click **Fundamentals of Anatomy & Physiology, Sixth Edition,** type in your new Login Name and Password, and click "Enter."

Your Access Code is:

Record your new Login Name and Password on the back of this card.

Cut out this card and keep it handy. It's your ticket to valuable information.

Important: Please read the License Agreement located on the login screen before using the Martini Online Web site or *InterActive Physiology®* CD-ROM. By using the Web site or CD-ROM, you indicate that you have read, understood and accepted the terms of this agreement.

SIXTH EDITION

Fundamentals of
Anatomy
&
Physiology

FREDERIC H. MARTINI, Ph.D.

with

WILLIAM C. OBER, M.D.
ART COORDINATOR AND ILLUSTRATOR

CLAIRE W. GARRISON, R.N.
ILLUSTRATOR

KATHLEEN WELCH, M.D.
CLINICAL CONSULTANT

RALPH T. HUTCHINGS
BIOMEDICAL PHOTOGRAPHER

KATHLEEN IRELAND, Ph.D.
CONTRIBUTOR AND MEDIA CONSULTANT

Benjamin
Cummings

SAN FRANCISCO   BOSTON   NEW YORK
CAPETOWN   HONG KONG   LONDON   MADRID   MEXICO CITY
MONTREAL   MUNICH   PARIS   SINGAPORE   SYDNEY   TOKYO   TORONTO

Prentice Hall Staff

Editor-in-Chief, Life and Geosciences: *Sheri L. Snavely*
Senior Acquisitions Editor: *Halee Dinsey*
Senior Development Editor: *Dan Schiller*
Production Editor: *Shari Toron*
Project Manager: *Crissy Dudonis*
Editorial Assistant: *Susan Zeigler*
Editor-in-Chief, Development: *Carol Trueheart*
Assistant Vice President of Production & Manufacturing: *David W. Riccardi*
Executive Managing Editor: *Kathleen Schiaparelli*
Assistant Managing Editor, Science: *Beth Sweeten*
Formatting Manager: *Jim Sullivan*
Electronic Production Specialist: *Karen Noferi*
Electronic Page Makeup: *Karen Noferi, Karen Stephens, Joanne Del Ben, William Johnson, Clara Bartunek, Judith Wilkens, Richard Foster, Vicki Croghan, Prepare*
Manufacturing Manager: *Trudy Pisciotti*
Assistant Manufacturing Manager: *Michael Bell*
Photo Researcher: *Yvonne Gerin*

Director, Creative Services: *Paul Belfanti*
Director of Design: *Carole Anson*
Art Director: *Jonathan Boylan*
Interior Design: *Anne Flanagan*
Cover Designer: *Anne DeMarinis*
Left Front Cover Photograph and Title Page: *Johner/Amana America, Inc. Photonica IMA USA, Inc.*
Right Front Cover Photograph: *Students from The Ailey School. Photograph by Beatriz Schiller.*
Left Back Cover Photograph: *Mike Powell/Getty Images, Inc.*
Right Back Cover Photograph: *Ailey II Company Members. Photograph by Roy Volkman.*
Art Manager: *Patricia Burns*
Art Editor: *Adam Velthaus, Ronda Whitson*
Assistant Managing Editor, Science Supplements: *Dinah Thong*
Copy Editor: *Brian Baker*
Proofreader: *Michael Rossa*

Notice: Our knowledge in clinical sciences is constantly changing. The author and the publisher of this volume have taken care that the information contained herein is accurate and compatible with the standards generally accepted at the time of publication. Nevertheless, it is difficult to ensure that all information given is entirely accurate for all circumstances. The author and the publisher disclaim any liability, loss, or damage incurred as a consequence, directly or indirectly, of the use and application of any of the contents of this volume.

Printed in the United States of America

10 9 8 7 6 5 4 3 2 1

ISBN 0-13-061568-4 (college edition)

ISBN 0-13-111158-2 (school edition)

Library of Congress Cataloging-in-Publication Data

Martini, Frederic.
  Fundamentals of anatomy & physiology / Frederic H. Martini, with
William C. Ober, art coordinator and illustrator; Claire W. Garrison,
illustrator; Kathleen Welch, clinical consultant; Ralph T. Hutchings,
biomedical photographer; Kathleen Ireland, contributor and media
consultant.-- 6th ed.
      p. cm.
Includes index.
  ISBN 0-13-061568-4
  1. Human physiology. 2. Human anatomy. I. Title: Fundamentals of
anatomy and physiology. II. Title.
QP34.5 .M27 2004
612—dc21   2002152645

To my family, whose support makes this project possible, and to my readers, whose thanks and suggestions for improvement are deeply appreciated.

TEXT AND ILLUSTRATION TEAM

Dr. Frederic (Ric) Martini
Author

Dr. Martini received his Ph.D. from Cornell University in comparative and functional anatomy. His publications include journal articles, technical reports, magazine articles, and a book for naturalists about the biology and geology of tropical islands. He is the coauthor of six other undergraduate texts on anatomy or anatomy and physiology. He is currently on the faculty of the University of Hawaii and remains affiliated with the Shoals Marine Laboratory, a joint venture between Cornell University and the University of New Hampshire. Dr. Martini is a member of the Human Anatomy and Physiology Society, the American Physiological Society, and the American Association of Anatomists. He is also a member of the American Association for the Advancement of Science, the National Association of Biology Teachers, the Society for College Science Teachers, the Society for Integrative and Comparative Biology, the Western Society of Naturalists, and the International Society of Vertebrate Morphologists.

William C. Ober, M.D.
Art Coordinator and Illustrator

Dr. Ober received his undergraduate degree from Washington and Lee University and his M.D. from the University of Virginia in Charlottesville. While in medical school, he also studied in the Department of Art as Applied to Medicine at Johns Hopkins University. Dr. Ober is currently on the faculty of the University of Virginia in the Department of Sports Medicine. He is also part of the Core Faculty at the Shoals Marine Laboratory.

Claire W. Garrison, R.N.
Illustrator

Ms. Garrison practiced pediatric and obstetric nursing for nearly 20 years before turning to medical illustration as a full-time career. She has been Dr. Ober's associate since 1986, following a five-year apprenticeship. Ms. Garrison is also a Core Faculty member at the Shoals Marine Laboratory.

Dr. Kathleen Welch
Clinical Consultant

Dr. Welch received her M.D. from the University of Washington in Seattle and did her residency at the University of North Carolina in Chapel Hill. For two years, she served as Director of Maternal and Child Health at the LBJ Tropical Medical Center in American Samoa and subsequently was a member of the Department of Family Practice at the Kaiser Permanente Clinic in Lahaina, Hawaii. She has been in private practice since 1987. Dr. Welch is a Fellow of the American Academy of Family Practice and a member of the Hawaii Medical Association and the Human Anatomy and Physiology Society.

Ralph T. Hutchings
Biomedical Photographer

Mr. Hutchings was associated with the Royal College of Surgeons for 20 years. An engineer by training, he has focused for years on photographing the structure of the human body. The result has been a series of color atlases, including the *Color Atlas of Human Anatomy*, the *Color Atlas of Surface Anatomy*, and *The Human Skeleton* (all published by Mosby-Yearbook Publishing). For his anatomical portrayal of the human body, the International Photographers Association has chosen Mr. Hutchings as the best photographer of humans in the twentieth century. He lives in North London, where he tries to balance the demands of his photographic assignments with his hobbies of early motor cars and airplanes.

Dr. Kathleen Anne Ireland
Contributor and Media Consultant

Dr. Ireland currently teaches biology at Seabury Hall, a private college preparatory school on the island of Maui. She has taught at state universities, community colleges, and private college preparatory schools for over 15 years. Dr. Ireland has an extensive background in teaching traditional as well as television and Internet courses. In the early 1990s she began one of the first completely on-line Internet A&P courses, offered through Iowa State University—a course that became the center of her doctoral studies. Dr. Ireland also has many years of experience in agricultural genetics, medical physiology, and marine biology. "Dr. Kate," as her students call her, continues her research on the integration of technology into the science curriculum in both secondary and post-secondary education. An active member of HAPS, AAAS, HSTA and MSET, she is also a dedicated runner, swim team coach, and mother of two sons, Greg and Marc.

BRIEF CONTENTS

CONTENTS

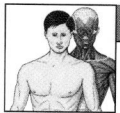
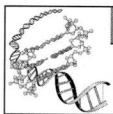

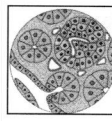

SYSTEMS OVERVIEW 146

Unit 2 Covering, Support, and Movement

CHAPTER 5

THE INTEGUMENTARY SYSTEM 154

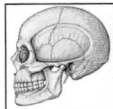

CHAPTER 6

OSSEOUS TISSUE AND SKELETAL STRUCTURE 182

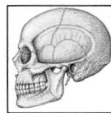

CHAPTER 7

THE AXIAL SKELETON 208

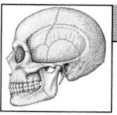
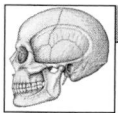
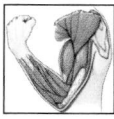
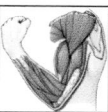

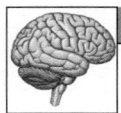
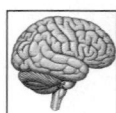
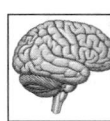
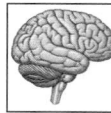

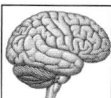
CHAPTER 16

NEURAL INTEGRATION II: THE AUTONOMIC NERVOUS SYSTEM AND HIGHER ORDER FUNCTIONS 532

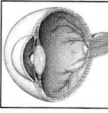
CHAPTER 17

THE SPECIAL SENSES 564

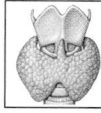
CHAPTER 18

THE ENDOCRINE SYSTEM 604

Unit 4 Fluids and Transport

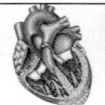

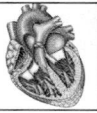

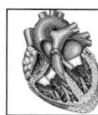

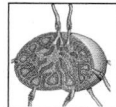

Unit 5 Environmental Exchange

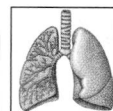

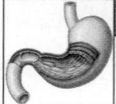

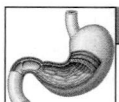

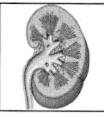
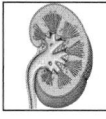
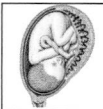

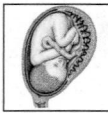

☤ CLINICAL TOPICS IN THE TEXT

Topics in **boldface** are covered in boxed Clinical Discussions.
Many more clinical topics are included in the *Applications Manual* that accompanies this text.

PREFACE

To the Instructor

You will find many changes in this edition, starting right here in the Preface. You may or may not be familiar with earlier editions of this book, so I will focus on things that I'd want to know about if I were considering a text for use in one of my classes.

WHAT SETS THIS TEXT APART FROM THE OTHERS?

I have often heard people say, "all A & P books are the same." Well, certainly the available texts are about the same size and length, and they all have a similar organization, feature colorful illustrations, and include an assortment of supplements. But they differ widely in the way they organize information and present complex information. I will cite four specifics that set this text apart from all others on the market:

☐ THE ART PROGRAM

Most texts are illustrated by a dozen or more different artists and studios, all working in relative isolation. There are no common conventions, no consistent presentation or theme. This is the only text that has all of the art done by a pair of medical illustrators, Bill Ober, M.D., and Claire Garrison, R.N. (see the biographical notes on p. iv). Bill and Claire have worked with me since the start of the first edition illustration program in 1986. They know the material, they know the text, and as they work they know how they will illustrate related topics in later chapters. As a result, the art flows as smoothly as the text.

☐ THE PRESENTATION STYLE OF THE NARRATIVE

I believe that it is important to organize material so that it can easily be studied, understood, and reviewed for exams. This is not a novel that will be read for fun. It is a resource that will help students to accomplish a specific task—mastering the material —as efficiently as possible. The narrative has been structured accordingly, with special attention paid to those features that have been shown to aid in the actual learning process:

- Overviews that help students to see the big picture
- Navigator figures that provide a roadmap through complex discussions
- Lists that make it easier to organize information and focus on key facts and concepts
- Tables that summarize information concisely, in a format that makes it easy to memorize when memorization is required.

☐ THE PEDAGOGICAL FRAMEWORK

This focus on the real needs of real students is reflected in many pedagogical devices used throughout the text:

- Students can check their progress periodically by answering short Concept Check questions at the ends of major sections.
- When material relates to topics presented earlier, link icons ∞ p. 000 with specific page numbers are provided to facilitate review.
- The end-of-chapter Study Outlines are unusually comprehensive.
- The end-of-chapter questions are organized in increasing levels of difficulty and sophistication, so that students can test their mastery first of facts and terminology, then of concepts, and finally of critical thinking and clinical applications.

☐ THE ANCILLARIES PROVIDED WITH THE TEXT

Today, no single text, no matter how good or how comprehensive, can provide all the resources that students should be able to draw on for help in mastering this demanding subject. Accordingly, this book has been designed as the central component of a fully integrated learning system that includes both print and media components. The complete package includes five additional components, all of which are bundled with the text:

- *InterActive Physiology® CD-ROM*, which uses animations, tutorials, and quizzes to teach the most difficult physiological concepts.
- *3-D Anatomy CD-ROM*, featuring rotatable, layered visualizations derived from the Visible Human Project of the National Library of Medicine, "Hot Spot" roll-over labeling, and three-dimensional Fly-Throughs of specific anatomical structures.
- An *Applications Manual*, organized to parallel the text, that includes discussions of clinically important disorders for each of the body's organ systems, together with information about the relevant diagnostic techniques, laboratory tests, and treatment options.
- *Martini's Atlas of the Human Body*, containing 120 anatomical photographs, a scanning atlas with 51 images produced by a variety of modern radiological techniques, and 21 embryology summaries.
- A passcode to the associated website, *Martini Online*, where students will find a variety of interactive self-quizzes for each chapter and an online e-book version of *Fundamentals of Anatomy & Physiology*.

All of these components are completely integrated with the text through icons and specific references that appear both within the chapters and in the Chapter Reviews. More complete descriptions appear below in the "Supplements" section.

WHAT'S DIFFERENT ABOUT THE SIXTH EDITION?

The basic chapter sequence and organization of the text remains largely unchanged. However, you will find several new organizational features within the narrative, as well as refinements to (and more generous use of) features that have been highly praised by students and instructors:

■ In response to feedback from users of the text, I have nearly doubled the number of **Navigator figures** to help students keep track of their progress through difficult physiological material. In this edition, Navigators will be found in Chapters 10 (muscle physiology), 12 (neurophysiology), 20 (cardiac physiology and cardiac output), 21 (cardiovascular physiology), 22 (immunity), 23 (respiratory physiology), 25 (cellular metabolism), and 26 (renal physiology). A comparable pattern of presentation—overview followed by blocks of text and art, accompanied by strategically positioned summary tables—has been used throughout, even in chapters that do not contain Navigators.

■ A **Systems Overview** section, which provides basic information about the major body systems and their components, has been inserted after the introductory material on the chemical, cellular, and tissue levels but before the systems are treated individually. This unique feature is designed to help prepare students for the in-depth chapters to follow.

■ **System in Perspective** sections, one per body system, which:
   ◆ summarize the components and functions of each organ system
   ◆ discuss the integration between each system and other systems
   ◆ introduce key clinical patterns in the disorders characteristic of each system
   ◆ provide Media Connections exercises that draw on the *InterActive Physiology*® CD-ROM, and relevant websites.

■ The **Media Connections** were developed by Kathleen Ireland, Ph.D. Kate has extensive research and teaching experience in anatomy and physiology, and she specializes in media use and instructional design. (Her brief biography appears on p. v.) It is our hope that this feature will enable instructors to capitalize on the wealth of Web-based content, as it provides students with alternative learning opportunities.

■ **Concept check questions**, which provide an opportunity for review of important material within a chapter, are now accompanied by **references to relevant exercises** on the *InterActive Physiology*® CD-ROM, the *3-D Anatomy CD-ROM*, and the *Companion Website*.

■ **Summary sections** that provide an overview linked to a comprehensive table and, in many cases, an associated illustration, have been added in key locations.

   Clinical material in the new edition has been reorganized to integrate it more organically with the text and increase its pedagogical value.

   Short *Clinical Notes*, embedded in the text but identified with a caduceus icon, deal with pathologies that shed particular light on the principles of normal physiological function.

   Longer, boxed *Clinical Discussions* deal with topics of particular medical or social importance (such as cancer and AIDS) that require more extended exploration.

   The *Applications Manual* packaged with the text contains an abundance of supplementary clinical material. It includes authoritative discussions of a broad range of diseases (both those mentioned in the text and many others), together with extensive information about the relevant laboratory tests, diagnostic procedures, and treatments. Organized to parallel the text, system by system and chapter by chapter, the *Applications Manual* is designed to be used as a supplement to the text, as a reference, or as a resource for those students who want to learn more about clinical topics.

   Perhaps the most far-reaching change to the narrative has been the reorganization and rescripting of the chapters dealing with the nervous system. For many years I have been troubled by the tendency to treat this system as a collection of parts or functions; the isolation of the autonomic nervous system is one particularly obvious example of this approach. In this edition, the presentation of the nervous system follows functional patterns and stresses the integration of information. The new sequence of topics is:

■ Chapter 12: neural tissue and neurophysiology
■ Chapter 13: the spinal cord, the spinal nerves, and the wiring of spinal reflexes
■ Chapter 14: the brain (with structures considered in inferior-to-superior sequence), the cranial nerves, and the wiring of cranial reflexes
■ Chapter 15: somatic and visceral sensory pathways and somatic motor pathways (the SNS)
■ Chapter 16: visceral motor pathways (the ANS) and higher order functions
■ Chapter 17: the special senses

   The process of revising and rewriting these key chapters resulted in extensive changes to the art program as well. In addition to the beautiful illustrations produced by Bill Ober and Claire Garrison, I have been lucky enough to work with Ralph Hutchings, a biomedical photographer. Ralph was recently cited as one of the best photographers of the twentieth century for his anatomical portrayals of the human body. His photos, many of them new to this edition, appear throughout the textbook, the media, and the new companion *Atlas*.

   Dramatic improvements in the illustration program for this edition include the following:

■ Chapters 10, 12, 13, 14, 18, 20, 21, 22, 26, and 28 (among others): New anatomical art and new physiological diagrams have been added to accompany a revised coverage of neurophysiology, endocrine physiology, cardiovascular function, immune function, and renal function.

■ *Fundamentals of Anatomy & Physiology* pioneered the use of step lists in text and art. In this edition, we have increased the number of such figures while continuing to ensure that the steps in the text are unambiguously tied to the related art.

■ Linking the art and photos in the text with new photographs in the companion *Atlas* and the anatomical simulations on the *3-D Anatomy CD-ROM* will give students a better understanding of the three-dimensional structure of the body.

THE SUPPLEMENTS PACKAGE

The supplements team consists of talented A & P instructors, most of whom have been using this textbook since its first or second edition. Their efforts are coordinated by the supplements editors and multimedia editors at Prentice Hall and at Benjamin Cummings. During the development of the Sixth Edition, the supplements team has worked with me as an extra team of reviewers. In return, I have provided additional suggestions and comments while their revisions were under way.

□ FOR THE INSTRUCTOR

Lecture Presentation Box (0-13-046169-5)

Benjamin Cummings and I have tried to simplify the task of preparing lectures for this course. The Lecture Presentation Box contains three sets of materials for each chapter. The first set contains traditional Instructor's Manual materials, including detailed lecture notes and an organizational grid of visual resources by chapter. This first set also contains suggested demonstrations, analogies, answers to end-of-chapter questions, and mnemonic activities designed to engage students' interest and to provide some additional flavor for your lectures. The second set contains student handouts, including crossword puzzles and Quiz Art (line art without labels) for each chapter. The third set contains all the transparency acetates, organized by chapter. Also included is a *Lecture Presentation Notebook* to help you organize your lecture material.

Instructor's Manual (0-13-046409-0)

This useful resource includes a wealth of materials to help you prepare and organize your lectures, such as lecture notes, vocabulary aids, applications, and classroom demonstrations. To help you organize all the visual resources provided in the supplements package, this manual also includes a detailed grid that correlates each section of the textbook with the visual aids that support that section. At a glance you can determine which transparencies, animations, and CD-ROM images are available for your lecture. New to this edition are Media Connections Instructor Notes, designed to ensure successful and meaningful use of the Media Connections activities found in chapters that precede System in Perspective sections.

Transparencies (0-13-046172-5)

The transparency set has been significantly improved and expanded. More than 800 acetates are included, with art and labels enlarged for use in large lecture halls. Where possible, complex figures have been divided into separate transparencies for improved clarity and teaching effectiveness. The selection of acetates and the quality of each image have been assessed by faculty reviewers to ensure that we offer the best possible presentation.

Printed Test Bank and Computerized Test Bank
(Print format: 0-13-046173-3)
(Software: Cross-Platform CD-ROM 0-13-046168-7)

The Test Bank has been thoroughly revised and reorganized for easier selection of questions for every objective in the textbook. A test bank of more than 3000 questions organized around the three-level learning system (as seen in the textbook and *Study Guide*) will help you design a variety of tests and quizzes.

Instructor's Resource CD-ROM (0-13-046171-7)

This cross-platform CD-ROM includes all the illustrations from the textbook, which can be edited and customized for lecture presentation or testing purposes. These images are available for easy use in PowerPoint or on the Web in three formats: images with labels and leaders, images with leaders only, and images with no labels or leaders. In addition, the CD includes 3-D animations plus all 160 photographic images (with and without labels) from *Martini's Atlas of the Human Body*—creating a powerful teaching tool.

Martini Online (www.aw.com/martini)

The Companion Website provides a wealth of materials to help enhance courses, including the Syllabus tool, which enables instructors to easily put their syllabuses online. The website also includes Instructor Resources, which offers instructors the ability to download media and print supplements. And the site allows students to link from a quiz question to the corresponding eBook by clicking on a "hint" button.

The new website has been designed to make it easier to incorporate Web-based activities into the core curriculum. One of the key advances is the inclusion of chapter-based Media Connections. Media Connections provide specific, time-constrained exercises with clearly stated objectives that relate the Web-based activities to the chapter content. These exercises have been designed (1) to enable instructors to capitalize on the wealth of Web-based content and (2) to provide alternative learning opportunities. To help you implement these Media Connections in your course, the *Instructor's Manual* offers instructions for their use.

The website also includes clinical case studies organized into steps that give you the opportunity to practice and apply your knowledge of anatomy and physiology in a practical and realistic way. These case studies, derived from actual case histories, support problem-based learning.

In addition to the basic tools available on the Companion Website for this textbook, Benjamin Cummings provides rich and thorough online course material in major course management systems, including **Blackboard**, **CourseCompass** (Pearson's nationally hosted version of **Blackboard**) and **WebCT**.

I hope that as the year passes, you will use the Companion Website or my email address, listed at the end of the Preface, to contact me with suggestions for future improvements, new illustrations, or news of interesting research developments.

Distance is no longer a limiting factor in our "global A & P village," and I hope that even more instructors and students will participate in the development of future editions.

☐ FOR THE STUDENT

Applications Manual (0-13-031117-0)

Each new copy of the text comes packaged with an *Applications Manual*, which is written in collaboration with Kathleen Welch, M.D., my wife, who is also the clinical consultant for the text. This unique supplement provides access to interesting and relevant clinical and diagnostic information, and it presents that information in a framework that helps students develop problem-solving skills. The book begins with an overview of the way diseases are diagnosed, and this leads to a discussion of the scientific method and to the use of logic and reasoning to evaluate health-related statements and claims. It then provides background information about chemical, cellular, and molecular disorders and techniques. Finally, it presents the major categories of disorders and important diagnostic techniques for each body system. Although it can be read separately from the textbook, each major topic in the *Applications Manual* is cross-referenced to specific pages in the textbook.

The *Applications Manual* is designed to demonstrate the relevance of key concepts presented in the textbook. By placing the material in a separate volume, topics can be covered in greater depth—or skipped—without interrupting the flow of information in the textbook. Few instructors will require their students to read the entire *Applications Manual*. The *Manual* is available so that each student can refer to segments of particular importance to his or her career plans or personal or family health. In subsequent courses, the *Applications Manual* becomes an invaluable reference for students, linking the topics covered in advanced courses with the basic anatomy and physiology covered in students' introductory-level courses.

InterActive Physiology CD-ROM (0-8053-5999-0)
3-D Anatomy CD-ROM (0-321-17028-8)

Two powerful media resources accompany the Sixth Edition: the *InterActive Physiology® CD-ROM* and the new *3-D Anatomy CD-ROM*.

Inside the back cover of the textbook, students will find the award-winning **InterActive Physiology® CD-ROM**. Complete with animations, tutorials, and quizzes, IP is referenced throughout the text in appropriate Concept Check boxes, end-of-chapter Study Outlines, and end-of-body-system Media Connections. IP offers an indispensable resource for enhancing physiological topics covered within the text.

The second media resource, bound in *Martini's Atlas of the Human Body*, is the **3-D Anatomy CD-ROM**. The anatomical renderings and animations presented on the *3-D Anatomy CD-ROM* were developed with data from the Visible Human Project, sponsored by the National Library of Medicine. Students and instructors can rotate images, providing powerful views of internal structures.

Martini's Atlas of the Human Body (0-13-146123-0)

This new, spiral-bound *Atlas* contains a wealth of information that augments the figures in the text. Most prominently, it includes 160 anatomical plates, 53 of them new to this edition, as well as 53 high-quality radiological images produced by CT, MRI, and contrast X-ray techniques. Each image is cross-referenced to figures in *Fundamentals of Anatomy & Physiology* through icons that appear in the figure captions. In addition, the *Atlas* contains 36 pages of full-color Embryology Summaries, dealing with the developmental origins of each body system, and is bundled with the new *3-D Anatomy CD-ROM*.

Study Guide (0-13-046407-4)

This very popular *Study Guide*, written by Charles Seiger, is an excellent way to review basic facts and concepts as well as to develop problem-solving skills. A variety of questions, including labeling and concept mapping, are keyed to every learning objective in the textbook and are organized around the same three-level learning system.

Video Tutor for Anatomy and Physiology (0-13-751843-9)

This videotape offers high-quality tutorials for anyone with access to a VCR. Each segment includes a professionally narrated animation that walks you through the most difficult physiological concepts and offers self-check questions that allow you to test your understanding of the material. Segments of the Video Tutor include membrane transport, protein synthesis, muscle contraction, action potential propagation, vision, auditory function, heart function, urine formation, and the immune response.

☐ FOR THE LABORATORY

Benjamin Cummings publishes a variety of laboratory manuals to meet the diverse needs of anatomy and physiology labs. Please see your Benjamin Cummings sales representative for more details. Here is a list of those manuals:

Laboratory Textbook of Anatomy and Physiology, Second Edition, by Michael G. Wood. This full-color, 700-page lab manual has been designed to complement *Fundamentals of Anatomy and Physiology*. It features cat dissections. (0-13-019694-0)

Laboratory Exercises in Anatomy and Physiology with Cat Dissections, Seventh Edition, by Gerard J. Tortora and Robert J. Amitrano. This full-color manual features cat dissections and graphics. (0-13-047791-5)

Anatomy and Physiology Laboratory Manual, Sixth Edition, by Gerard J. Tortora and Robert J. Amitrano. This full-color manual offers a variety of laboratory exercises with minimal mammalian dissections. (0-13-089670-5)

ACKNOWLEDGMENTS

This textbook is not the product of any single individual; rather, it represents a group effort, and the members of the group deserve to be acknowledged. Foremost on my thank-you list are the faculty and reviewers throughout the world who offered suggestions that helped guide me through the revision process. Their interest in the subject, concern for the accuracy and method of presentation, and experience with students of widely varying abilities and backgrounds made the revision process an educational experience for me. To them, I express my sincere thanks and best wishes:

Shylaja Akkaraju, *College of DuPage*
John Aliff, *Georgia Perimeter College*
Steven Amdur, *Nassau Community College*
Karen Apel, *University of Wisconsin–Milwaukee*
Steven Bassett, *Southeast Community College*
Greg Bohm, *Hillsborough Community College*
Mark Bolke, *Clark College*
Leslie Carlson, *Iowa State University*
Catherine Carter, *Georgia Perimeter College*
Lucia Cepriano, *SUNY Farmingdale*
Ana Christensen, *Lamar University*
Darrell Davies, *Kalamazoo Valley Community College*
Mary Dawson, *Kingsborough Community College*
Danielle Desroches, *William Paterson University*
Eric Dewar, *Lansing Community College*
Paul Emerick, *Monroe Community College*
Ralph Fregosi, *University of Arizona*
Paul Garcia, *Houston Community College*
Michelle Geremia, *Quinnipiac University*
Linda Gingerich, *St. Petersburg College*
Susan Grigsby, *Houston Community College*
Cecil Hampton, *Jefferson College*
Timothy Henry, *University of Texas at Arlington*
Cynthia Herbrandson, *Kellogg Community College*
Jacqueline Homan, *South Plains College*
James Horowitz, *Palm Beach Community College*
George A. Jacob, *Xavier University*
Renu Jain, *Houston Community College*
Kelly Johnson, *University of Kansas*
Carissa Krane, *University of Dayton*
Cris Martin, *University of Maryland*
Robert McDonough, *Georgia Perimeter College*
Eddie McNack, *Houston Community College*
Judy Megaw, *Indian River Community College*
Beta Meyer, *Mount San Antonio College*
Leslie Miller, *Iowa State*
Mahtas Moussavi, *Houston Community College*
Judy Nath, *Lourdes College*
Auguste Nioupin, *Houston Community College*
Joyce Ono, *California State University–Fullerton*
Betsy Ott, *Tyler Junior College*
Beverly Perry, *Houston Community College*
Ed Pivorum, *Clemson University*

Dan Porter, *Amarillo College*
Linda Powell, *Community College of Philadelphia*
Anil Rao, *Metropolitan State College*
Jackie Reynolds, *Richland College*
Chris Riegle, *Irvine Valley College*
Todd Rimkus, *Marymount University*
Nidia Romer, *Miami Dade Community College*
Judith Shardo, *University of South Alabama*
Marilyn Shopper, *Johnson County Community College*
Milton Shult, *Houston Community College*
Carl Shuster, *Amarillo College*
Sharon Simpson, *Broward Community College*
Sandy Stewart, *Vincennes University*
Dennis Strete, *Houston Community College*
Kris Stuempfle, *Gettysburg College*
P. Swaroop, *Houston Community College*
Richard Symmons, *California State University–Hayward*
Robert Tallitsch, *Augustana College*
Mark Taylor, *Baylor University*
Diane Tice, *SUNY Morrisville*
Connie Vinton-Schoepske, *Iowa Central Community College*
Jyoti Wagle, *Houston Community College*
Cheryl Watson, *Central Connecticut State University*
Mary Weis, *Collin County Community College*
Mary Pat Wenderoth, *University of Washington*
Shirley Whitescarver, *Lexington Community College*
Vernon Wiersema, *Houston Community College*

After the initial drafts were completed, four dedicated instructors functioned as technical editors, assisting with the review and correction of page proofs:

Kathleen Andersen, *University of Iowa*
Gillian Bice, *Michigan State University*
Kelly J. Johnson, *University of Kansas*
Marilyn Shannon, *Indiana University, Purdue University*

I extend special thanks to them not only for their work in reviewing pages, but also for providing suggestions during the manuscript stage. They have been immensely helpful to me during the revision of this edition.

Focus groups and casual meetings with students around the world helped me concentrate on the needs of individual students. I first undertook this project in 1983 to address the needs of my own students, and their feedback continues to be very important to me. I also thank the many instructors and students who took the time to contact me by phone or e-mail with specific suggestions, kudos, questions, or problems.

Over time, a textbook evolves. As with organisms, each evolutionary step builds on a preexisting framework. Thus, I also thank the individuals who helped with the development of previous editions. That list includes the following:

Maxine A'Hearn, *Prince George's Community College*
Ahmed Naguy Ali, *Alexandria, Egypt*
Paul Anderson, *Massachusetts Bay Community College*
Timothy Alan Ballard, *University of North Carolina–Wilmington*

Debra Joan Barnes, Contra Costa College
Edwin Bartholomew, Lahainaluna High School
CeCe Barto, Tomball College
Steven Bassett, Southeast Community College
Robert Bauman, Jr., Amarillo College
Dean Beckwith, Illinois Central College
Doris Benfer, Montgomery Community College
Michèle Bertholf, Metropolitan State College and Front Range
  Community College
Latsy Best, Palm Beach Community College–North
Charles Biggers, University of Memphis
Michael Bonnert, University of Toronto
Cynthia Bottrell, Scott Community College
Spencer R. Bowers, Oakton Community College
Mimi Bres, Prince George's Community College
Alan Bretag, University of South Australia School of Pharmacy
C. David Bridges, Purdue University
Sandra Bruner, Polk Community College
Gene Carella, Niagara County Community College
Robert M. Carey, University of Arizona
Wayne Carley, Lamar University
Lucia Cepriano, SUNY Farmingdale
William M. Chamberlain, Indiana State University
William D. Chapple, University of Connecticut
Beng Cheah, University of Newcastle
Anthony Chee, Houston Community College
Suzette Chopin, Texas A & M University at Corpus Christi
Chin Moi Chow, Cumberland College of Health Sciences,
  University of Sydney
O. D. Cockrum, Texas State Technical College
Kim Cooper, Arizona State University
Richard Coppings, Chattanooga State College
William F. Crowley, Harvard Medical School
Grant Dahmer, University of Arizona
Charles Daniels, Kapiolani Community College
Brent DeMars, Lakeland Community College
Charles Dick, Pasco–Hernando Community College
Gerald R. Dotson, Front Range Community College
Ellen Dupre, Indian River Community College
John Dziak, Community College of Allegheny County
Lee Famiano, Cuyahoga Community College
Lee Farello, Niagara County Community College
Marion Fintel, Jefferson State Community College
Kathleen A. Flickinger, Maui Community College
Ruby Fogg, New Hampshire Tech
Mildred Fowler, Tidewater Community College
Sharon Fowler, Dutchess Community College
Denise Friedman, Hudson Valley Community College
Ann Funkhouser, University of the Pacific
Mildred Galliher, Cochise College
Anthony J. Gaudin, Ivy Tech State College
Lori K. Garrett, Danville Area Community College
Jeff Gerst, North Dakota State University
Louis Giacinti, Milwaukee Area Technical College
Delaine Gilcrease, Mesa Community College

Freda Glaser, University of Maryland–Baltimore County
Kathleen M. Gorczyca, North Shore Community College
Bonnie Gordon, Memphis State University
Mac E. Hadley, University of Arizona
William Hairston, Harrisburg Area Community College
Cecil Hampton, Jefferson College
Ernest Harber, San Antonio College
John P. Harley, Eastern Kentucky University
Ann Harmer, Orange Coast Community College
Linden Haynes, Hinds Community College
Ruth Lanier Hays, Clemson University
Mary Healey, Springfield College
Jean Helgeson, Collin County Community College
Vickie S. Hennessy, Sinclair Community College
Donna Hoel, Stark Community College
Elvis J. Holt, Purdue University
Jacqueline A. Homan, South Plains College
James Horwitz, Palm Beach Community College
Beth Howard, Rutgers University
Yvette Huet-Hudson, University of North Carolina–Charlotte
Angie Huxley, Pima Community College
Aaron James, Gateway Community College
Desiree Jett, Essex County College
Drusilla Jolly, Forsyth Technical Community College
David Kalichstein, Ocean County College
Eileen Kalmar, St. John Fisher College
George Karleskint, St. Louis Community College
Nancy G. Kincaid, Troy State University Montgomery
C. Ward Kischer, University of Arizona
Frank Kitakis, Wayne County Community College
William Kleinelp, Middlesex County College
Michael Kokkinn, University of South Australia School of
  Pharmacy
Bob Kucera, University of Newcastle
Jerry K. Lindsey, Tarrant County Junior College
Mary Lockwood, University of New Hampshire
Susan Lustick, San Jacinto College North
Greg Maravellas, Bristol Community College
Dan Mark, Penn Valley Community College
Jane Marks, Scottsdale Community College
William Mautz, University of Hawaii at Hilo
Alice Gerke McAfee, University of Toledo Technical College
Thomas W. McCort, Cuyahoga Community College
Mike McCusker, Eastern College
Bob McDonough, Georgia Perimeter College
Ruth McFarland, Mt. Hood Community College
Paul McGrath, University of Newcastle
Michael P. McKinley, Glendale Community College
Roberta Meehan, University of North Carolina
Richard F. Meginniss, College of Lake County
Ann Miller, Middlesex Community College
Alice Mills, Middle Tennessee State University
Melvin Mills, Scottsdale Community College
Ron Mobley, Wake Technical Community College
Rose Morgan, Minot State University

Aubrey Morris, Pensacola Junior College
Robert L. Moskowitz, Community College of Philadelphia
Richard Mostardi, Akron University
Ann Murphy, Sydney University
Elizabeth Naugle, Lane Community College
Martha Newsome, Tomball College
Bill Nicholson, University of Arizona
Richard Northrup, Delta College
Erik Nyholm, Umea University
Claire R. Oakley, Rocky Mountain College
John M. Olson, Villanova University
Betty Orr, Sinclair Community College
Betsy Ott, Tyler Community College
Mary Theresa Ortiz, Kinsborough Community College
Michael Palladino, Monmouth University
David L. Parker, Northern Virginia Community College
Mark Paulissen, Slippery Rock University
Brian K. Paulsen, California University of Pennsylvania
Lois Peck, Philadelphia College of Pharmacy
Philip Penner, Borough of Manhattan Community College
Ingrid Persson, Umea University
Clifford Pohl, Duquesne University
Robert Pollack, Nassau Community College
Robert L. Preston, Illinois State University
Gary Quick, Paradise Valley Community College
Anil Rao, Metro State College of Denver
Joel Reicherter, SUNY Farmingdale
Peta Reid, University of Sydney, Faculty of Nursing
Jean Rigden, Scottsdale Community College
John M. Ripper, Butler County Community College
Carolyn J. Rivard, Fanshawe College
Carolyn C. Robertson, Tarrant County Junior College
Kevin Ryan, Stark Technical College
Kristi Sather-Smith, Hinds Junior College
Frank Schwartz, Cuyahoga Community College
Charles Seiger, Atlantic Community College
Mark Shoop, Macon College
Sherrie Shupe, Delaware Technical and Community College
P. George Simone, Eastern Michigan University
Robert A. Sinclair, San Antonio College
Tom Smeaton, University of South Australia School of
    Pharmacy
David S. Smith, San Antonio College
Jeffery Smith, Delgado Community College
Philip Sokolove, University of Maryland–Baltimore County
Michael Soules, University of Washington
Thomas S. Spurgeon, Colorado State University
Sandra Stewart, Vincennes University
Ann Stoeckmann, Pennsylvania State University–Worthington
    Scranton
Jenna Sullivan, University of Arizona
Eric Sun, Macon College
Richard Symmons, California State University at Hayward
Dennis Taylor, Hiram College
Kathy Taylor, University of Arizona
Jay Templin, Widener University

Diane G. Tice, SUNY Morrisville
Caryl Tickner, Stark State Technical College
Marge Torode, Cumberland College of Health Sciences,
    University of Sydney
Lucia Tranel, St. Louis Community College and St. Louis
    College
Steve Trautwein, Southeast Missouri State University
Pat Turner, Howard Community College
Kent M. Van De Graaff, Brigham Young University
Sheila Van Holst, University of Sydney, Faculty of Nursing
Michael Vennig, University of South Australia School of
    Pharmacy
Jane Wallace, Chattanooga State Technical Community College
Eva Weinreb, Community College of Philadelphia
Debra A. Wellner, Wichita State University
Rosamund Wendt, Community College of Philadelphia
J. Wilkinson, University of Sydney, School of Biological Sciences
Stephen Williams, Glendale Community College
Bruce Wingerd, San Diego State University
Eric Wise, University of California at Santa Barbara
Michael G. Wood, Del Mar College
Jamie Young, Chattanooga State Technical Community College
Nancy L. Young, Seattle Pacific University

The accuracy and currency of the clinical material in this edition and in the *Applications Manual* in large part reflect the work of my wife, Kathleen Welch, M.D. I also owe thanks to Kate Ireland for her creative efforts on the Media Connections and for generating new media components incorporated into the Companion Website.

Virtually without exception, reviewers stressed the importance of accurate, integrated, and visually attractive illustrations in aiding the students to understand essential material. The revision of the art program was directed by Bill Ober, M.D., and Claire Garrison, R.N. Their suggestions about topics of clinical importance, presentation sequence, and revisions to the proposed art were of incalculable value to me and to the project.

Many of the text's illustrations include color photographs or micrographs collected from a variety of sources. The striking anatomical photos in the text, in *Martini's Atlas of the Human Body*, and the Wood lab manual were the work of Ralph Hutchings, whose efforts on this project are deeply appreciated. Much of the work in tracking down micrographs and other visual materials was performed by Yvonne Gerin; I greatly value her help. Dr. Eugene C. Wasson, III, and the staff of Maui Radiology Consultants, Inc., provided valuable assistance in the selection and printing of the CT and MRI scans that are included in the new *Atlas*.

I also express my appreciation to the editors and support staff both at Prentice Hall, with whom I started this revision, and Benjamin Cummings, with whom I finished it. First on the list is Shari Toron, the Production Editor for all of my projects. She somehow managed to handle every crisis, and kept things moving in the right direction, all the while maintaining her poise and sense of humor. Her support, hard work, and patience are deeply appreciated.

I owe special thanks to Halee Dinsey, Senior Acquisitions Editor at Prentice Hall, for her extraordinary creativity and dedication. Her vision helped shape this book in countless ways. Thanks are also due to her Editorial Assistant, Susan Zeigler, and to Crissy Dudonis, who assisted with the package and supplements.

Daniel Schiller, Senior Development Editor at Prentice Hall, played a vital role in fashioning the Sixth Edition. I could not have survived this process without him, and his unfailing attention to detail and quality made a tremendous difference in the final package.

This book would not exist without the extraordinary dedication of Jonathan Boylan, art director, who solved so many problems under pressure with unfailing good cheer; Jim Sullivan, Manager of Page Formatting; and Karen Noferi, who, together with her team of formatters, combined craftsmanship and speed in astonishing measure. Their efforts, far above and beyond the call of duty, are deeply appreciated.

I would also like to express my gratitude to Paul Corey, President of the Engineering, Science, and Math Division (ESM), of Prentice Hall, for his support of this project; Carol Trueheart, Editor-in-Chief of Development; David Riccardi, Assistant Vice President of Production and Manufacturing; Kathleen Schiaparelli, Executive Managing Editor, and Beth Sweeten, Assistant Managing Editor, for providing the extraordinary resources (human as well as technological) required to produce a book of this size and complexity; Paul Belfanti, Director of Creative Services, Carole Anson, Director of Design, Patricia Burns, Managing Editor for AV Assets and Production, and her talented team at Artworks, and Adam Velthaus, AV Production Editor, for their outstanding work on the design, the art program, and much else; Mike Rossa, for his eagle-eyed checking of page proofs; and Fran Daniele of PREPARE, for her work on the *Atlas* and the *Applications Manual*, which are texts in their own right. On the Benjamin Cummings side, I'd like to thank Linda Davis, President, Daryl Fox, VP and Publisher, Lauren Fogel, Executive Producer, Lauren Harp, Executive Marketing Manager, and my support team of Vicki Gundrum, Sr. Project Editor, Kim Neumann, Associate Online Project Editor, Ziki Dekel, Associate Project Editor, and Michael Roney, Editorial Assistant. It's good to have friends.

No one person could expect to produce a flawless textbook of this scope and complexity. Any errors or oversights are strictly my own rather than those of the reviewers, artists, or editors. To help improve future editions, I encourage you to send any pertinent information, suggestions, or comments about the organization or content of this textbook to me directly, using the e-mail address below. I will deeply appreciate any and all comments and suggestions and will carefully consider them in the preparation of the Seventh Edition.

Frederic H. Martini

Frederic H. Martini
Haiku, Hawaii
martini@maui.net

To the Student

HOW TO GET THE MOST OUT OF THIS PACKAGE

You probably have several reasons for taking this course. Like most people, you probably have many questions about your own body and about the origins of health problems or diseases. Perhaps you also need this course to further your career plans. If you are like most College students, you have two short-term goals in mind: (1) to learn and understand the material, and (2) to be able to demonstrate your mastery to the satisfaction of the instructor. I've tried to build features into the text that will help you reach both of these goals. In many respects, I've assembled this text like an Owner's Manual for the human body—it not only has the important information, but is also designed to help you find that information quickly. As a result, you may find that it is very different from the texts you've used for other classes. In this section, I've tried to give you some advice on how to use the features and unique organization of this text to your advantage. If you charge off into the text without reading further, you may miss some features that could help you succeed.

☐ WATCH THE BIG PICTURE

- Each chapter begins with an outline. Use it to get a sense of the scope and organization of the chapter as a whole.
- The major sections are numbered so that you can more easily keep track of your position within the chapter as you proceed.
- Each of these sections begins with a list of objectives. They will give you a preview of the main points that you can expect to learn as you work through the material.
- Many of the chapters include Navigator figures, which function as a kind of roadmap. Use them to help keep track of where you are in a complex discussion without getting lost in the details.

☐ LEARN THE TERMINOLOGY

- You will also find that there are a lot of new words to learn. The faster you can learn the terminology, the easier it will be for you to master the concepts. Throughout the text, the most important terms appear in **boldface** print so they are easy to spot. In addition, any time there's a chance of confusion, I've included a pronunciation guide. Accented syllables are in capital letters, and for the vowels:

ā as in tray	a as in track
ē as in tree	e as in help
ī as in spine	i as in ink
ō as in bone	o as in Tom
ū as in use	u as in run
oo as in moon	oy as in boy

☐ ALWAYS TRY TO ORGANIZE IMPORTANT INFORMATION

You've got to have a study plan and a method of organizing your notes. Don't try to memorize everything–organize the material, focus on the key points, and look for common themes and patterns.

- Take advantage of the many lists in the text, as well as the abundance of tables. Both of these formats summarize important information in a way that will make it easier for you to grasp and memorize.
- The material in each chapter builds upon the material presented in earlier chapters. Often you will want to review the relevant discussions. To make this easier, I've inserted a symbol with a page reference: ∞ p. 000. To help you proceed directly to the proper page, you will find that the page numbers appear in bold within a color band in the upper margin of each page. (The colors indicate the body system, in case you just need to find a general topic rather than a specific page.)

☐ KEEP AN EYE ON THE ART

I've provided a lot of visual support in this package–because, whether you are considering processes or structures, you need to be able to visualize what's being described in the narrative.

There are a LOT of illustrations in the text, and I work closely with the artist, an M.D., who reviews the text before creating the images. Because the text and art are developed together, you will find yourself moving back and forth between the text and the art as you read the chapter. At each spot in the narrative where you are sent to the art, there's a red dot that serves as a place holder–use it to help you find your way back to the point in the text that you left.

Many of the figures show dynamic processes underway in a series of steps; these same steps are also clearly identified and explained in the narrative. Breaking a complex process into stages is one way to make it easier to follow and understand.

Some figures are accompanied by an icon, ATLAS, which directs you to another view of the same structure in ***Martini's Atlas of the Human Body*** packaged with your text. You'll find that looking at a single anatomical feature from different perspectives or visualized using different techniques is a powerful way of enhancing your understanding of what you are seeing.

☐ PACE YOURSELF

It is usually a very bad idea to read a chapter in this textbook as if it were a chapter in a novel–from beginning to end, with no breaks. It's much more efficient to pace yourself and take a few pauses along the way to test your understanding of the material presented thus far. You are building your knowledge base, and there's no use reading the *next* section until you understand what was covered in *this* section.

- At the end of each major section, you will find a few short review questions. Take a moment to answer them. If you find them easy to answer, you can go on to the next section. If you find them difficult to answer, go back and reread the material. It's like building a house–it's not wise to start working in the second floor if the first floor is shaky.

■ Many of these questions are also accompanied by references to tutorials or three-dimensional anatomical views on the *InterActive Physiology® CD-ROM* packaged with your text or the *3-D Anatomy CD-ROM* that accompanies your *Atlas*. Don't hesitate to take advantage of these resources if you feel that you need to reinforce or deepen your understanding of the material you have just covered.

☐ CONSIDER WHAT THE INFORMATION *MEANS*

You probably would not be taking this course if you did not have at least some interest in health and medicine. The text and its ancillaries contain a wealth of clinical material, organized so that you can easily focus on precisely what is relevant to your own or your instructor's concerns:

■ Short **Clinical Notes** embedded in the text deal with many kinds of pathologies. These topics have been selected to help you understand the principles of normal function by showing what happens when something goes wrong.

■ Topics of great general importance or that require more extended treatment are considered in boxed **Clinical Discussions**.

■ Information about a variety of diseases not discussed in the text, as well as additional details about many that are, can be found in the *Applications Manual* that came with your text, together with information about the diagnosis and treatment of these conditions. The *Applications Manual* is organized in exactly the same way as the text to make it as easy as possible for you to consult. Use it as a supplement to the text, as a reference, or just to satisfy your individual curiosity. Look for the blue icon, **AM**, that will direct you to relevant topics in the *Manual*.

☐ PUT IT TOGETHER

The human body is an integrated whole, not a collection of isolated systems, so it's important that you be able to see information pertaining to specific systems in a larger context.

■ For each body system, you'll find a two-page "System in Perspective" section that summarizes the components and functions of the system, displays its interactions with every other body system, outlines relevant clinical aspects of the system, and provides media-based exercises to deepen your insight or open fresh perspectives. These sections can be an important part of your study, because they not only review the material in an entire chapter (or series of chapters) but integrate the contents of the entire book.

☐ ALWAYS REVIEW THE CHAPTER AFTER YOU COMPLETE IT

It might take you a week to work through a chapter, and by the time you reach the end, you may not have a clear perspective on the presentation as a whole. That's the time to review the end-of-chapter-material.

■ The *Study Outline* at the end of each chapter gives you an opportunity to review all the chapter's key information and concepts. You can also use it as a jumping-off point for the CD-based tutorials and visualizations referenced within the chapter, because those references are repeated here.

■ The *Review Questions* are organized so that you can assess your mastery of the chapter on three levels. You can test your grasp of facts and terminology (Level 1), your understanding of concepts (Level 2), and your ability to apply the material to realistic clinical situations (Level 3). If you need additional practice, there are more self-assessment quizzes in various formats on the **Companion Website** (www.aw.com/martini).

I wrote this book to give you the information you need as clearly as possible. If you find certain sections especially difficult, if you don't find information that you or your instructor would really like to cover, or if you have other comments or suggestions for future editions, please let me know using the e-mail address below.

Good luck and best wishes,
Ric (martini@maui.net)

UNIT 1   CHAPTER   1   2   3   4

CLINICAL DISCUSSIONS
- The Visible Human Project   8
- Homeostasis and Disease   14

AN INTRODUCTION TO ANATOMY AND PHYSIOLOGY

We have all had questions regarding the functioning of the human body at one time or another. Most people wonder, "Why do I have a fever?" or "Why am I so achy?" when stricken by the flu, and there's usually a close relative—a child, a sibling, a parent, or a grandparent—with some medical problem that nobody in the family understands. At some point when you needed some basic information on the human body, you may have wished that you'd been born with an owner's manual. This textbook, you will find, is designed to take the place of that missing owner's manual.

1–1 INTRODUCTION: STUDYING THE HUMAN BODY

Objectives

- Describe the basic functions of organisms.
- Define anatomy and physiology, and describe various specialties of each discipline.

This course will serve as an introduction to the inner workings of your body, providing a wealth of information about both its structure and its functioning. Many of the students who use this book are preparing for careers in health-related fields—but regardless of your career choice, you will find the information within these pages to be relevant to your future. You do, after all, live in a human body! Being human, you most likely exhibit the psychological characteristics of a typical human, one of the most basic of which is a seemingly insatiable curiosity—and few subjects arouse so much curiosity as our own bodies. The study of anatomy and physiology, which you are now beginning, will provide answers to questions regarding the functioning of your body in both health and disease.

Within the confines of anatomy and physiology, several scientific disciplines converge. Most broadly, since we are living organisms, the study of the human body involves many aspects of biological science. Learning how the body functions, and how it can continue to function effectively even as conditions around it and within it change, requires a basic understanding of chemistry. Much of the movement of the body, as well as the movement of materials within it, can best be understood in terms of the principles of physics. There are times when mathematical models must be used to represent anatomical relationships or physiological processes. In turn, the health sciences, including pharmacology, medicine, psychology, and kinesiology, are all rooted in the study of anatomy and physiology.

◻ THE CHARACTERISTICS OF LIVING THINGS

One way to gain insight into how life—specifically, your life—is maintained within the confines of your body is to consider the nature of life in general. Our world contains an enormous diversity of living organisms that vary widely in appearance and lifestyle. One aim of biology—the science of life—is to discover the unity and patterns that underlie this diversity, and so shed light on what we have in common with other living things. Biologists have found that all living things share certain basic characteristics, including the following:

- **Organization:** Every organism has a characteristic pattern of organization which differs from that of inanimate objects. Each organism has a discrete boundary separating it from its environment, and conditions within the organism are often very different from those in the outside world. Life continues only as long as the organism can maintain its physical integrity, and can keep its internal organization and internal environment within tolerable limits. We will look more closely at the organization of living things and how they regulate internal conditions in later sections.

- **Responsiveness:** Organisms respond to changes in their immediate environment; this property is also called *irritability*. You move your hand away from a hot stove; your dog barks at approaching strangers; fish are scared by loud noises; and amoebas glide toward their prey. Organisms also make longer term changes as they adjust to their environments. For example, an animal may either grow a heavier coat as winter approaches or migrate to a warmer climate. The potential to make such adjustments is termed *adaptability*.

- **Growth and Differentiation:** Over a lifetime, organisms grow larger, increasing in size by increasing the size or number of their cells. In multicellular organisms, the individual cells become specialized to perform particular functions. This specialization is called *differentiation*. Growth and differentiation often produce changes in form and function. For example, the body proportions and functional capabilities of an adult human are quite different from those of an infant.

- **Reproduction:** Organisms reproduce, creating subsequent generations of similar organisms.

- **Movement:** Organisms are capable of producing movement, which may be internal (transporting food, blood, or other materials inside the body) or external (moving through the environment).

- **Metabolism and Excretion:** Organisms rely on complex chemical reactions to provide the energy for responsiveness, growth, reproduction, and movement. They must also synthesize complex chemicals, such as proteins. *Metabolism* refers to all the chemical operations under way in the body. Many normal metabolic operations require the absorption of certain materials from the environment. These materials, called *nutrients*, are used for growth and maintenance; many nutrients also are important as energy sources. To generate energy efficiently, most cells require oxygen, an atmospheric gas. *Respiration* is the absorption, transport, and use of oxygen by cells. Metabolic operations often generate unneeded or potentially harmful waste products that must be removed from body fluids through the process of *excretion*.

Several additional functions can be distinguished when you consider animals as complex as fish, cats, or humans. Whereas very small organisms can absorb materials across exposed surfaces, creatures larger than a few millimeters seldom absorb nutrients directly from their environment. For example, humans cannot absorb steaks, apples, or ice cream without processing them first. That processing, called *digestion*, occurs in specialized areas of the body where complex foods are broken down into simpler components that can be absorbed easily. Respiration and excretion are also more complicated in large organisms than in small ones. Humans have specialized structures responsible for gas exchange (lungs) and for waste elimination (kidneys). Finally, because absorption, respiration, and excretion are performed in different regions of the body, humans require an internal transportation system, or *cardiovascular* system.

All of these adaptations to larger body size are instances of a very basic biological principle that you will encounter again and again in the study of anatomy and physiology: *Form follows function*. That is, the characteristics of particular structures can best be understood in the light of what they do. Every part of the human body plays some role in carrying on the processes that keep us alive, and every part has developed in ways that help it to play that role.

When we compare the particular adaptations of human beings with those of other creatures, we find another important principle: Humans resemble other organisms in many respects. Figure 1–1● highlights various similarities among birds, fishes, and humans. Although each animal is unique in some respects, there are obvious structural and functional similarities. Animals can be classified according to their shared characteristics, and birds, fish, and humans are members of a group called the *vertebrates*. The shared characteristics and organizational patterns provide useful clues about how these animals have evolved over time. Many of the complex structures and functions of the human body discussed in this text have distant evolutionary origins.

◻ THE SCIENCES OF ANATOMY AND PHYSIOLOGY

The word *anatomy* has Greek origins, as do many other anatomical terms and phrases. A literal translation would be "a cutting open." **Anatomy** is the study of internal and external structures of the body and the physical relationships among body parts. For example, someone studying anatomy might examine how a particular muscle attaches to the skeleton. **Physiology**, another word adopted from Greek, is the study of how organisms perform their vital functions. Someone studying physiology might consider how a muscle contracts or what forces a contracting muscle exerts on the skeleton.

Anatomy and physiology are closely integrated both theoretically and practically. Anatomical information provides clues about functions, and physiological mechanisms can be explained only in terms of the underlying anatomy. This observation leads to a very important concept: *All specific functions are performed by specific structures.*

Anatomists and physiologists approach the relationship between structure and function from different perspectives.

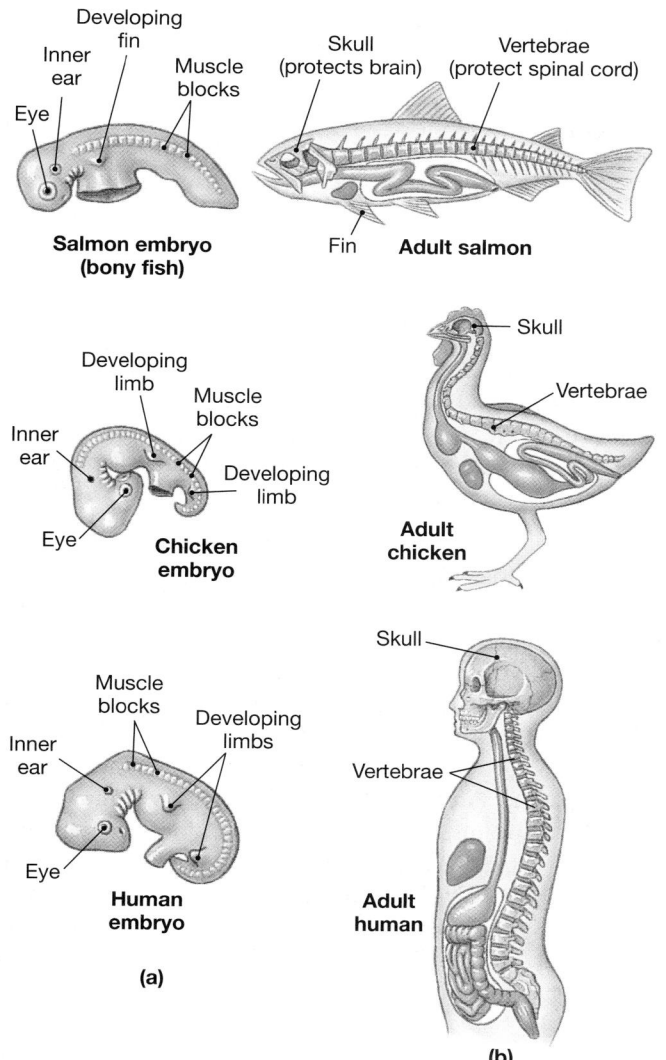

The link between structure and function is always present, but not always understood. For example, the anatomy of the heart was clearly described in the 15th century, but almost 200 years passed before the pumping action of the heart was demonstrated.

This text will introduce anatomical structures and the physiological processes that make human life possible. The information will enable you to understand important mechanisms of disease and will help you make intelligent decisions about personal health.

□ ANATOMY

How you look often determines what you see; you get a very different view of your neighborhood from a satellite photo than you get standing in your front yard. Your method of observation has an equally dramatic effect on your understanding of the structure of the human body. Anatomy can be divided into *microscopic anatomy* and *gross (macroscopic) anatomy* on the basis of the degree of structural detail under consideration. Other anatomical specialties focus on specific processes, such as respiration, or medical applications, such as *surgical anatomy*, which deals with landmarks on the body that are useful during surgical procedures.

Anatomy is a living, dynamic field. Despite a long history of observation and dissection, new information and interpretations occur frequently. As recently as 1996, researchers working on the Visible Human database (see the box on p. 8) described a facial muscle that had previously been overlooked.

Gross Anatomy

Gross anatomy, or *macroscopic anatomy,* involves the examination of relatively large structures and features usually visible with the unaided eye. There are many ways to approach gross anatomy:

- *Surface anatomy* is the study of general form and superficial markings.
- *Regional anatomy* focuses on the anatomical organization of specific areas of the body, such as the head, neck, or trunk. Many advanced courses in anatomy stress a regional approach, because it emphasizes the spatial relationships between structures already familiar to students.
- *Systemic anatomy* is the study of the structure of organ systems, such as the *skeletal system* or the *muscular system.* **Organ systems** are groups of organs that function together in a coordinated manner. For example, the heart, blood, and blood vessels form the *cardiovascular system,* which distributes oxygen and nutrients throughout the body. Introductory texts such as this present systemic anatomy, because that approach clarifies functional relationships among the component organs. The human body has 11 organ systems, and we shall introduce them later in the chapter.
- *Developmental anatomy* deals with the changes in form that occur during the period between conception and physical maturity. Because developmental anatomy considers anatomical structures over such a broad range of sizes (from a single cell to an adult human), the techniques of developmental anatomists are similar to those used in gross anatomy and in microscopic

●FIGURE 1–1
Comparative Anatomy. Humans are classified as vertebrates, a group that includes animals as different in adult appearance as fish, birds, and people. All vertebrates share a basic pattern of anatomical organization which differs from that of other animals. The similarities are often most apparent in comparing embryos at comparable stages of development (a) rather than in comparing adult vertebrates (b).

To understand the difference, consider a simple nonbiological analogy. Suppose that an anatomist and a physiologist were asked to examine and report on a pickup truck. The anatomist might begin by measuring and photographing the various parts of the truck and, if possible, taking it apart and putting it back together. The anatomist could then explain its key structural relationships—for example, how the movement of the pistons in the engine cylinders causes the driveshaft to rotate and how the transmission then conveys this motion to the wheels. The physiologist also would note the relationships among the truck's components, but his or her primary focus would be on functional characteristics, such as the amount of power that the engine could generate, the amount of force transmitted to the wheels in different gears, and so on.

anatomy. The most extensive structural changes occur during the first two months of development. The study of these early developmental processes is called *embryology* (em-brē-OL-o-jē).

Several anatomical specialties are important in a clinical setting. Examples include *medical anatomy* (anatomical features that change during illness), *radiographic anatomy* (anatomical structures seen by using specialized imaging techniques, discussed later in the chapter), and *surgical anatomy* (anatomical landmarks important in surgery).

Microscopic Anatomy

Microscopic anatomy deals with structures that cannot be seen without magnification. The boundaries of microscopic anatomy are established by the limits of the equipment used. With a light microscope, you can see basic details of cell structure; with an electron microscope, you can see individual molecules that are only a few nanometers across. Consequently, as technology has increased our ability to resolve smaller and smaller structures, the amount of information available about microscopic anatomy has increased. As a result, microscopic anatomy is a very active area of research. We do not yet know all there is to know about anatomy! As we proceed through the text, we shall be considering details at all levels, from macroscopic to microscopic. (Readers unfamiliar with the terms used to describe measurements and weights should consult the reference tables in Appendix I.)

Microscopic anatomy includes two major subdivisions; cytology and histology. **Cytology** (sī-TOL-o-jē) is the analysis of the structure of individual **cells**, the simplest units of life. Cells are composed of chemical substances in various combinations, and our lives depend on the chemical processes occurring in the trillions of cells in the body. For this reason, we consider basic chemistry (Chapter 2) before we examine cell structure (Chapter 3).

Histology (his-TOL-o-jē) is the examination of **tissues**—groups of specialized cells and cell products that work together to perform specific functions (Chapter 4). Tissues combine to form **organs**, such as the heart, kidney, liver, or brain. Many organs are easily examined without a microscope, so at the organ level we cross the boundary from microscopic anatomy to gross anatomy.

▢ PHYSIOLOGY

As we noted earlier, physiology is the study of the function of anatomical structures; **human physiology** is the study of the functions of the human body. These functions are complex and much more difficult to examine than most anatomical structures. As a result, there are even more specialties in physiology than in anatomy:

- *Cell physiology*, the study of the functions of cells, is the cornerstone of human physiology. Cell physiology deals with events at the chemical and molecular levels: both chemical processes within cells and chemical interactions between cells. Chapters

2–4 focus on the chemical structure, internal organization, and control mechanisms of cells and tissues.

- *Special physiology* is the study of the physiology of specific organs. For example, *cardiac physiology* is the study of heart function.

- *Systemic physiology* includes all aspects of the functioning of specific organ systems. Cardiovascular physiology, respiratory physiology, and reproductive physiology are examples of systemic physiology.

- *Pathological physiology* is the study of the effects of diseases on organ or system functions. (*Pathos* is the Greek word for "disease.") Modern medicine depends on an understanding of both normal physiology and pathological physiology. You will find extensive information on clinically important topics in subsequent chapters and in the *Applications Manual* that accompanies this text. **AM** Disease, Pathology, and Diagnosis

Physicians normally use a combination of anatomical, physiological, and psychological information when they evaluate patients. When a patient presents symptoms to a physician, the physician will look at the structures affected (gross anatomy), perhaps take a fluid or tissue sample (microscopic anatomy), and ask questions to determine what alterations from normal functioning the patient is experiencing. Think back to your last trip to a doctor's office. Not only did the attending physician examine your body, noting any anatomical abnormalities, but he or she also evaluated your physiological processes by asking questions, observing your movements, listening to your body sounds, taking your temperature, and perhaps obtaining chemical analyses of fluids such as blood or urine. In evaluating all these observations to reach a diagnosis, physicians rely on a logical framework, developed by scientific investigators over the centuries, known as the *scientific method*. **AM** The Scientific Method

---

1–2 LEVELS OF ORGANIZATION

Objectives

- Identify the major levels of organization in organisms, from the simplest to the most complex.
- Identify the organ systems of the human body and the major components of each system.

Our study of the human body will begin with an overview of chemistry (Chapter 2) and microscopic anatomy (Chapters 3 and 4). We then proceed to the anatomy and physiology of each organ system. When considering events from the microscopic to the macroscopic scale, we shall examine several interdependent levels of organization. Figure 1–2● presents an example of the relationships among these various levels of organization:

- **The Chemical, or Molecular, Level.** Atoms, the smallest stable units of matter, can combine to form molecules with complex shapes. Even at this simplest level, form follows function: The specialized shape of a molecule determines its function.

●FIGURE 1–2
Levels of Organization. Interacting atoms
form molecules that combine in the protein fibers
of a heart muscle cell. Such cells interlock, creating
heart muscle tissue, which constitutes most of the
walls of the heart, a three-dimensional organ. The
heart is one component of the cardiovascular
system, which also includes the blood and blood
vessels. The combined organ systems form an
organism—a human.

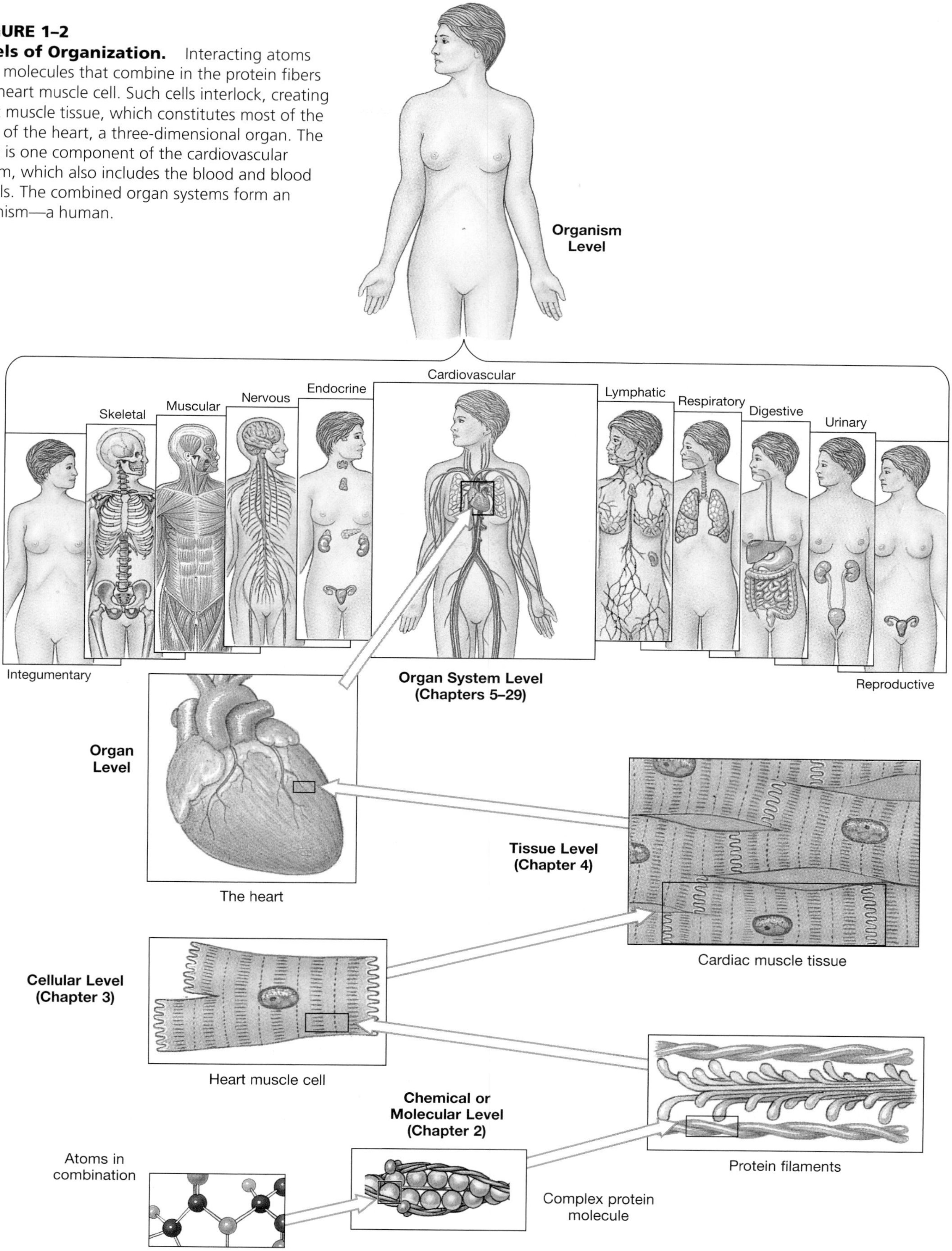

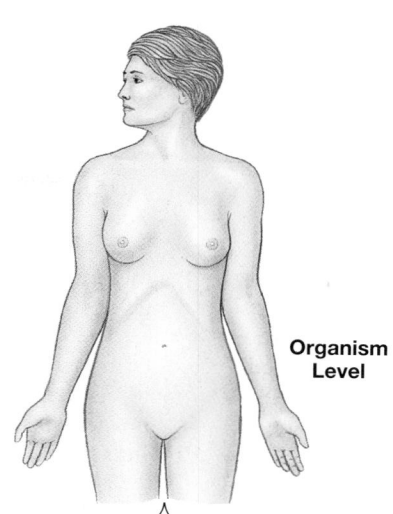

**Organism
Level**

Integumentary   Skeletal   Muscular   Nervous   Endocrine   Cardiovascular   Lymphatic   Respiratory   Digestive   Urinary   Reproductive

**Organ System Level
(Chapters 5–29)**

**Organ
Level**

The heart

**Tissue Level
(Chapter 4)**

Cardiac muscle tissue

**Cellular Level
(Chapter 3)**

Heart muscle cell

**Chemical or
Molecular Level
(Chapter 2)**

Atoms in
combination

Complex protein
molecule

Protein filaments

The Visible Human Project

In 1993, Joseph Paul Jernigan, a career criminal, was executed in Texas for a particularly brutal murder. Neither he nor the state of Texas could have predicted the impact his death would have on medical research and education. Before his execution, Jernigan had donated his body to the State Anatomical Board of Texas. Because of his age (39), his size (1.76m [5′10″] and 90.4 kg [199 lb]), and his excellent medical history prior to his death, his body was selected for the Visible Human project. This project, funded by the U.S. National Library of Medicine, was designed to create accurate computerized versions of the human body. The data sets, or "visible humans," could then be studied and manipulated in ways that would be impossible using real bodies.

Data sets of a virtual male and a virtual female were planned; Joseph Paul Jernigan became the basis for the former. The data set was generated by Dr. Victor Spitzer and colleagues at the University of Colorado. After serial CT and MRI scans of the body were taken, it was frozen and then sectioned at 1-mm intervals. As each slice was generated, high-resolution digital images and 70-mm color photographs were taken. Image generation and processing were completed in 1995. The data set of a female was generated somewhat later, from a 59-year-old woman who died of natural causes and who had donated her body to the Anatomy Board of Maryland. The processing method was similar, except the sections were taken at 0.33 mm, giving roughly three times the resolution of the male data set.

The Visible Human data sets include an impressive volume of information, combining X rays, MRI, and CT scans with digitized images of cross sections through the body. The digitized images, in computerized format, can be accessed through the Internet at the NLM/NIH Website, http://www.nlm.nih.gov.

The real value in these data is that the three-dimensional relationships among anatomical structures can be visualized and explored from multiple perspectives. CT scans and MRI images can often be directly correlated with the appearance of the body by matching the scans or MRI images with the sectional images. The data have subsequently been used to generate a variety of enhanced images and interactive educational projects. You will find images based on the Visible Human project scattered throughout this book, and they form the basis for the anatomical animations found on the **Anatomy CD-ROM** that accompanies the Atlas.

- **The Cellular Level.** Molecules can interact to form *organelles*, such as the protein filaments found in muscle cells. Each type of organelle has specific functions. For example, interactions among protein filaments produce the contractions of muscle cells in the heart. Cells are the smallest living units in the body; organelles are their structural and functional components.

- **The Tissue Level.** A *tissue* is a group of cells working together to perform one or more specific functions. Heart muscle cells, or cardiac muscle cells (*cardium*, heart), interact with other types of cells and with extracellular materials to form *cardiac muscle tissue.*

- **The Organ Level.** *Organs* consist of two or more tissues working in combination to perform several functions. Layers of cardiac muscle tissue, in combination with *connective tissue*, another type of tissue, form the bulk of the wall of the *heart*, a hollow, three-dimensional organ.

- **The Organ System Level.** Organs interact in *organ systems*. Each time it contracts, the heart pushes blood into a network of blood vessels. Together, the heart, blood, and blood vessels form the cardiovascular system, one of 11 organ systems in the body.

- **The Organism Level.** All organ systems of the body work together to maintain life and health. This brings us to the highest level of organization, that of the *organism*—in this case, a human.

The organization at each level determines not only the characteristics, but also the functions, of higher levels. For example, the arrangement of atoms and molecules at the chemical level creates the protein filaments that, at the cellular level, give cardiac muscle cells the ability to contract powerfully. At the tissue level, these cells are linked, forming cardiac muscle tissue. The structure of the tissue ensures that the contractions are coordinated, producing a heartbeat. When that beat occurs, the internal anatomy of the heart, an organ, enables it to function as a pump. The heart is filled with blood and connected to the blood vessels, and the pumping action circulates blood through the vessels of the cardiovascular system. Through interactions with the respiratory, digestive, urinary, and other systems, the cardiovascular system performs a variety of functions essential to the survival of the organism.

Something that affects a system will ultimately affect each of the system's components. For example, the heart cannot pump blood effectively after massive blood loss. If the heart cannot pump and blood cannot flow, oxygen and nutrients cannot be distributed. Very soon, the cardiac muscle tissue begins to break down as individual muscle cells die from oxygen and nutrient starvation. These changes will not be restricted to the cardiovascular system; all cells, tissues, and organs in the body will be damaged.

Figure 1–3● (pp. 9–10) introduces the 11 organ systems in the human body. These organ systems are interdependent, interconnected, and occupy a relatively small space. The cells, tissues, organs, and organ systems of the body coexist in a shared environment, like the inhabitants of a large city. Just as city dwellers breathe the city air

THE INTEGUMENTARY SYSTEM

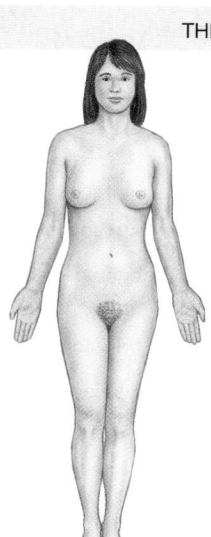

Major Organs:
- Skin
- Hair
- Sweat glands
- Nails

Functions:
- Protects against environmental hazards
- Helps regulate body temperature
- Provides sensory information

THE NERVOUS SYSTEM

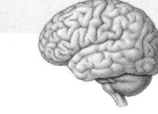

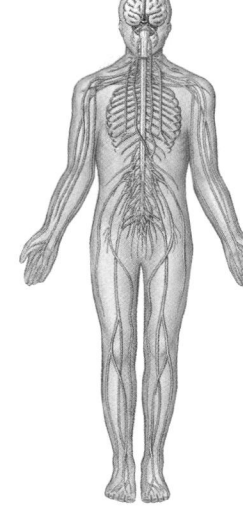

Major Organs:
- Brain
- Spinal cord
- Peripheral nerves
- Sense organs

Functions:
- Directs immediate responses to stimuli
- Coordinates or moderates activities of other organ systems
- Provides and interprets sensory information about external conditions

THE SKELETAL SYSTEM

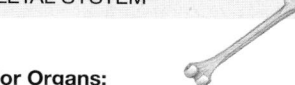

Major Organs:
- Bones
- Cartilages
- Associated ligaments
- Bone marrow

Functions:
- Provides support and protection for other tissues
- Stores calcium and other minerals
- Forms blood cells

THE ENDOCRINE SYSTEM

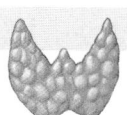

Major Organs:
- Pituitary gland
- Thyroid gland
- Pancreas
- Adrenal glands
- Gonads (testes and ovaries)
- Endocrine tissues in other systems

Functions:
- Directs long-term changes in the activities of other organ systems
- Adjusts metabolic activity and energy use by the body
- Controls many structural and functional changes during development

THE MUSCULAR SYSTEM

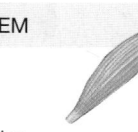

Major Organs:
- Skeletal muscles and associated tendons and aponeuroses (tendinous sheets)

Functions:
- Provides movement
- Provides protection and support for other tissues
- Generates heat that maintains body temperature

THE CARDIOVASCULAR SYSTEM

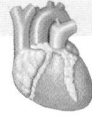

Major Organs:
- Heart
- Blood
- Blood vessels

Functions:
- Distributes blood cells, water, and dissolved materials, including nutrients, waste products, oxygen, and carbon dioxide
- Distributes heat and assists in control of body temperature

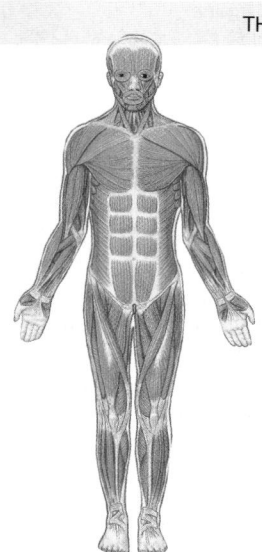

•**FIGURE 1–3**
An Introduction to Organ Systems

THE LYMPHATIC SYSTEM

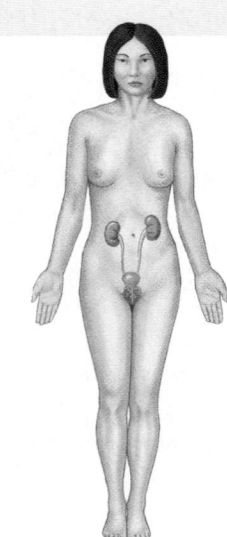

Major Organs:
- Spleen
- Thymus
- Lymphatic vessels
- Lymph nodes
- Tonsils

Functions:
- Defends against infection and disease
- Returns tissue fluids to the bloodstream

THE URINARY SYSTEM

Major Organs:
- Kidneys
- Ureters
- Urinary bladder
- Urethra

Functions:
- Excretes waste products from the blood
- Controls water balance by regulating volume of urine produced
- Stores urine prior to voluntary elimination
- Regulates blood ion concentrations and pH

THE RESPIRATORY SYSTEM

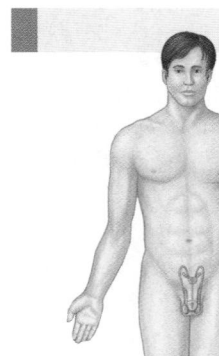

Major Organs:
- Nasal cavities
- Sinuses
- Larynx
- Trachea
- Bronchi
- Lungs
- Alveoli

Functions:
- Delivers air to alveoli (sites in lungs where gas exchange occurs)
- Provides oxygen to bloodstream
- Removes carbon dioxide from bloodstream
- Produces sounds for communication

THE MALE REPRODUCTIVE SYSTEM

Major Organs:
- Testes
- Epididymis
- Ductus deferens
- Seminal vesicles
- Prostate gland
- Penis
- Scrotum

Functions:
- Produces male sex cells (sperm) and hormones

THE DIGESTIVE SYSTEM

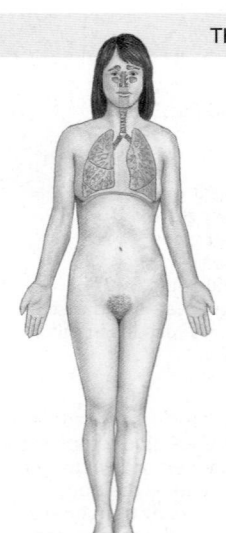

Major Organs:
- Teeth
- Tongue
- Pharynx
- Esophagus
- Stomach
- Small intestine
- Large intestine
- Liver
- Gallbladder
- Pancreas

Functions:
- Processes and digests food
- Absorbs and conserves water
- Absorbs nutrients (ions, water, and the breakdown products of dietary sugars, proteins, and fats)
- Stores energy reserves

THE FEMALE REPRODUCTIVE SYSTEM

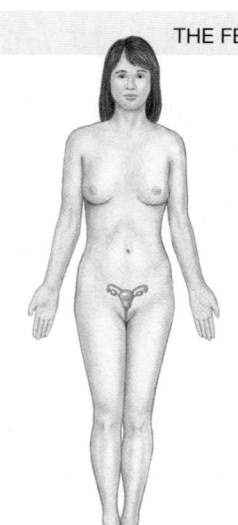

Major Organs:
- Ovaries
- Uterine tubes
- Uterus
- Vagina
- Labia
- Clitoris
- Mammary glands

Functions:
- Produces female sex cells (oocytes) and hormones
- Supports developing embryo from conception to delivery
- Provides milk to nourish newborn infant

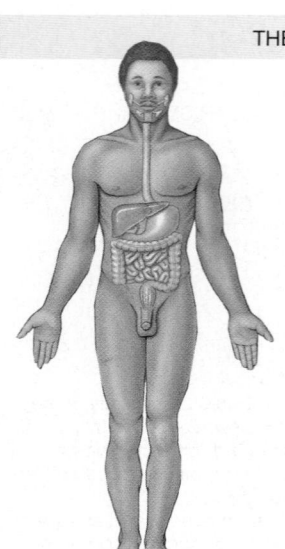

●FIGURE 1–3
An Introduction to Organ Systems (*continued*)

and drink the water provided by the local water company, cells in the human body absorb oxygen and nutrients from the fluids that surround them. If a city is blanketed in smog or its water supply is contaminated, the inhabitants will become ill. Similarly, if the body fluid composition becomes abnormal, cells will be injured or destroyed. Suppose the temperature or salt content of the blood changes. The effect on the heart could range from a minor adjustment (heart muscle tissue contracts more often, so the heart rate goes up) to a total disaster (the heart stops beating, so the individual dies).

Various physiological mechanisms act to prevent damaging changes in the composition of body fluids and the environment inside our cells. **Homeostasis** (*homeo*, unchanging + *stasis*, standing) refers to the existence of a stable internal environment. To survive, every organism must maintain homeostasis.

✓ Which characteristics of life do white blood cells exhibit when they migrate to the site of an injury in response to chemicals released from damaged cells?

✓ At which level of organization does a histologist investigate structures?

✓ A researcher studies the factors that cause heart failure. What might this person's specialty be called?

Answers start on page Q-1

To view a sample of the animations created from the Visible Human Data set, visit the **Anatomy CD-ROM:** Skeletal System. How many of these bones can you identify already?

1–3 HOMEOSTASIS

Objectives

- Explain the concept of homeostasis and its significance for organisms.
- Describe how positive feedback and negative feedback are involved in homeostatic regulation.

Homeostasis is absolutely vital; a failure to maintain it soon leads to illness or even death. The principle of homeostasis is the central theme of this text and the foundation of all modern physiology. Homeostatic regulation is the adjustment of physiological systems to preserve homeostasis. An understanding of homeostatic regulation will help you make accurate predictions about the body's responses to both normal and abnormal conditions.

Two general mechanisms are involved in homeostatic regulation: autoregulation and extrinsic regulation:

1. **Autoregulation**, or *intrinsic regulation*, occurs when the activities of a cell, a tissue, an organ, or an organ system adjust automatically in response to some environmental change. For example, when oxygen levels decline in a tissue, the cells release chemicals that dilate local blood vessels. This dilation increases the rate of blood flow and provides more oxygen to the region.

2. **Extrinsic regulation** results from the activities of the nervous system or endocrine system, two organ systems that

control or adjust the activities of many systems simultaneously. For example, when you exercise, your nervous system issues commands that increase your heart rate so that blood will circulate faster. Your nervous system also reduces blood flow to less active organs, such as the digestive tract. The oxygen in circulating blood is thus saved for the active muscles.

In general, the nervous system directs rapid, short-term, and very specific responses. When you accidentally set your hand on a hot stove, the heat produces a painful, localized disturbance of homeostasis. Your nervous system responds by ordering the contraction of specific muscles that will pull your hand away from the stove. The effects last only as long as the neural activity continues, usually a matter of seconds.

In contrast, the endocrine system releases chemical messengers, called *hormones*, that affect tissues and organs throughout the body. The responses may not be immediately apparent, but when the effects appear, they may persist for days or weeks. Examples of endocrine function include the long-term regulation of blood volume and composition and the adjustment of organ system function during starvation. The endocrine system also plays a major role in growth and development: it is responsible for the changes that take place in your body as you mature and age.

Regardless of the system involved, homeostatic regulation always attempts to keep the characteristics of the internal environment within certain limits. The regulatory mechanism consists of three parts: (1) a **receptor**, a sensor that is sensitive to a particular environmental change, or *stimulus*; (2) a **control center**, or *integration center*, which receives and processes the information supplied by the receptor; and (3) an **effector**, a cell or organ that responds to the commands of the control center and whose activity either opposes or enhances the stimulus. You are probably already familiar with comparable regulatory mechanisms, such as the thermostat in your house or apartment (Figure 1–4●).

A thermostat is a control center that receives information about room temperature from an internal or remote thermometer (a receptor). The dial on the thermostat establishes the *set point*, or desired value, which in this case is the temperature you select. In our example, the set point is 22°C (about 72°F). The function of the thermostat is to keep room temperature within acceptable limits, usually within a degree or so of the set point. In summer, the thermostat accomplishes this function by controlling an air conditioner (an effector). When the temperature at the thermometer rises above the acceptable range, the thermostat turns on the air conditioner, which then cools the room (Figure 1–4a●). When the temperature at the thermometer reaches the set point, the thermostat turns off the air conditioner (Figure 1–4b●). The control is not precise; the room is large, and the thermostat is located on just one wall. Over time, the temperature in the center of the room oscillates around the set point.

☐ NEGATIVE FEEDBACK

The essential feature of temperature control by thermostat can be summarized very simply: A variation outside the desired range triggers an automatic response that corrects the situation. This

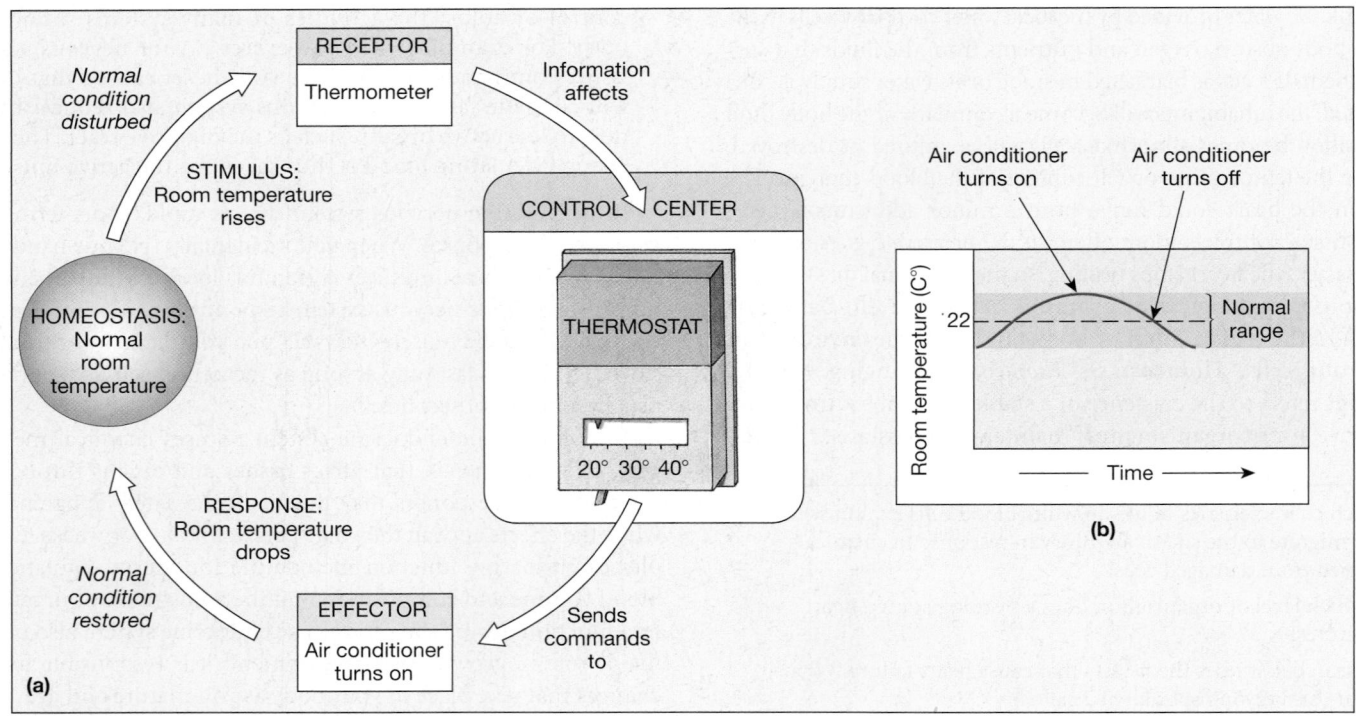

●**FIGURE 1–4**
The Control of Room Temperature. **(a)** A thermostat turns on an air conditioner (or a heater) to keep room temperature near the desired set point. Whether room temperature rises or falls, the thermostat (a control center) triggers an effector response that restores normal temperature. When room temperature rises, the thermostat turns on the air conditioner, and the temperature returns to normal. **(b)** With this regulatory system, room temperature oscillates around the set point.

method of homeostatic regulation is called *negative feedback*, because an effector activated by the control center opposes, or *negates*, the original stimulus. Negative feedback thus tends to minimize change, keeping variation in key systems within limits that are compatible with our long-term survival.

Most homeostatic regulatory mechanisms involve negative feedback. Consider the control of body temperature, a process called *thermoregulation*. In thermoregulation, the relationship between heat loss, which occurs primarily at the body surface, and heat production, which occurs in all active tissues, is altered.

The control center for thermoregulation is in the *hypothalamus*, a region of the brain (Figure 1–5a●). This control center receives information from two sets of temperature receptors, one in the skin and the other inside the hypothalamus. At the normal set point, body temperature, as measured with an oral thermometer, will be approximately 37°C (98.6°F). If body temperature rises above 37.2°C, activity in the control center targets two effectors: (1) muscle tissue in the walls of blood vessels supplying the skin and (2) sweat glands. The muscle tissue relaxes and the blood vessels dilate, increasing blood flow through vessels near the body surface; the sweat glands accelerate their secretion. The skin then acts like a radiator by losing heat to the environment, and the evaporation of sweat speeds the process. As body temperature returns to normal, the thermoregulatory control center becomes less active. Superficial blood flow and sweat gland activity then decrease to previous levels.

Negative feedback is the primary mechanism of homeostatic regulation, and it provides long-term control over internal conditions and systems. Homeostatic mechanisms using negative feedback normally ignore minor variations, and they maintain a normal *range* rather than a fixed value. In the previous example, body temperature oscillated around the set-point temperature (Figure 1–5b●). The regulatory process itself is dynamic, because the set point may vary with changing environments or differing activity levels. For example, when you are asleep, your thermoregulatory set point lowers, whereas when you work outside on a hot day (or when you have a fever), it climbs. Thus, body temperature can vary from moment to moment or from day to day for any individual, due to (1) small oscillations around the set point or (2) changes in the set point. Comparable variations occur in all other physiological characteristics.

The variability among individuals is even greater than that within an individual. Each of us has homeostatic set points determined by genetic factors, age, gender, general health, and environmental conditions. It is therefore impractical to define "normal" homeostatic conditions very precisely. By convention, physiological values are reported either as averages—average values obtained by sampling a large number of individuals—or as a range that includes 95 percent or more of the sample population. For instance, 5 percent of healthy adults have resting body temperatures that are below 36.7°C or above 37.2°C. These temperatures are perfectly normal for them, and the variations have no clinical significance. Physicians must keep this variability in mind when they review lab reports or clinical

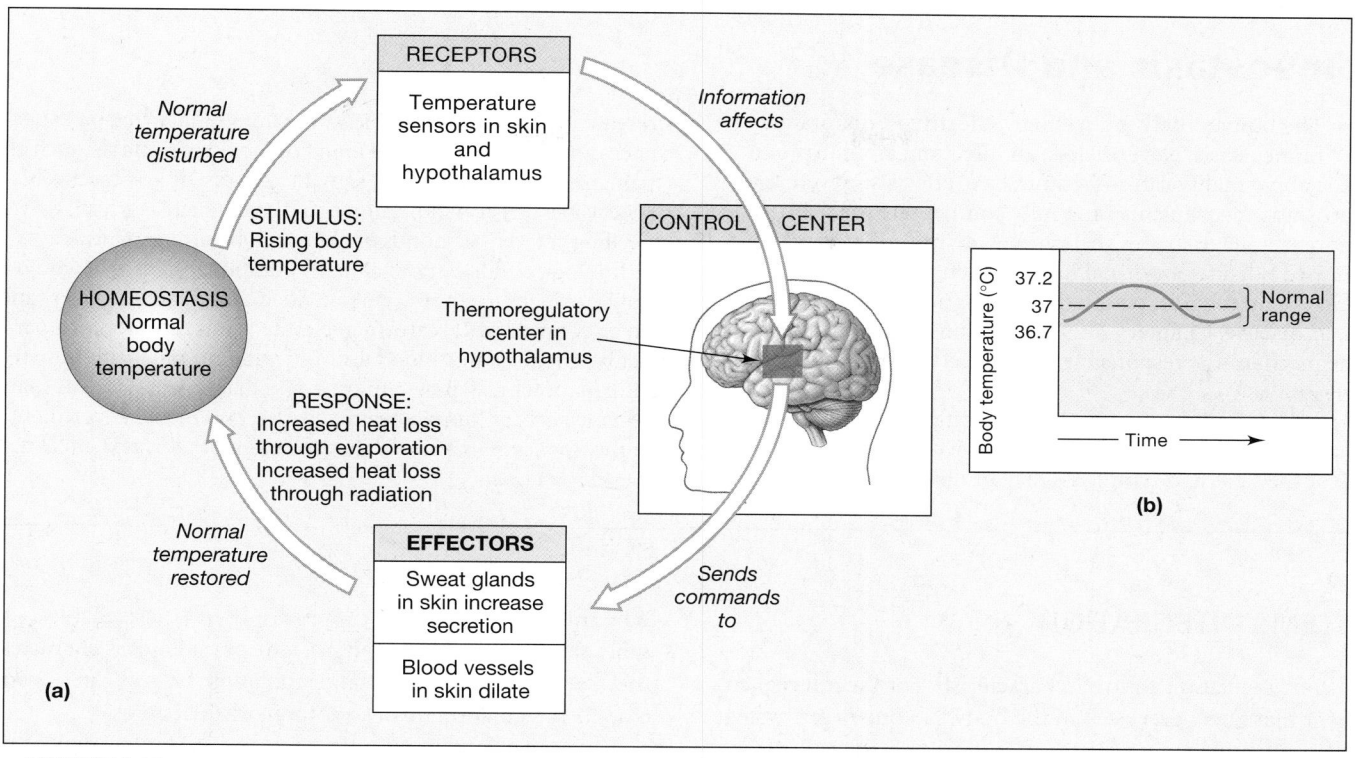

●FIGURE 1–5
Negative Feedback: The Control of Body Temperature.   In negative feedback, a stimulus produces a response that opposes or eliminates the stimulus. **(a)** Events comparable to those shown in Figure 1–3 occur in the regulation of body temperature. A control center in the brain functions as a thermostat with a set point of 37°C. If body temperature exceeds 37.2°C, heat loss is increased through enhanced blood flow to the skin and increased sweating. **(b)** The thermoregulatory center keeps body temperature oscillating within an acceptable range, usually between 36.7 and 37.2°C.

discussions, because unusual values—even those outside the normal range—may represent individual variation rather than disease.

☐ POSITIVE FEEDBACK

In *positive feedback*, an initial stimulus produces a response that exaggerates or enhances the change in the original conditions rather than opposing it. Suppose that the thermostat in Figure 1–4a● is connected to a heater rather than to an air conditioner. Now, when room temperature rises, the thermostat turns on the heater, causing a further rise in room temperature. Room temperature will continue to increase until someone switches off the thermostat, turns off the heater, or intervenes in some other way. This kind of escalating cycle is often called a *positive feedback loop*.

In the body, positive feedback loops are often incorporated into control mechanisms in which a potentially dangerous or stressful process must be completed quickly. For example, the immediate danger from a severe cut is the loss of blood, which can lower blood pressure and reduce the efficiency of the heart. The body's response is diagrammed in Figure 1–6●. Damage to the blood vessel wall releases chemicals that begin the process of blood clotting. As clotting gets under way, each step releases chemicals that accelerate the process. This escalating process is a positive feedback loop that ends with the formation of a blood clot, which patches the vessel wall and stops the bleeding. Blood clotting will be examined more closely in Chapter 19. Labor and delivery, another example of positive feedback in action, will be discussed in Chapter 29.

●FIGURE 1–6
Positive Feedback: Blood Clotting.   Positive feedback loops are important in accelerating processes that must proceed to completion rapidly. In this example, positive feedback accelerates the clotting process until a blood clot forms and stops the bleeding.

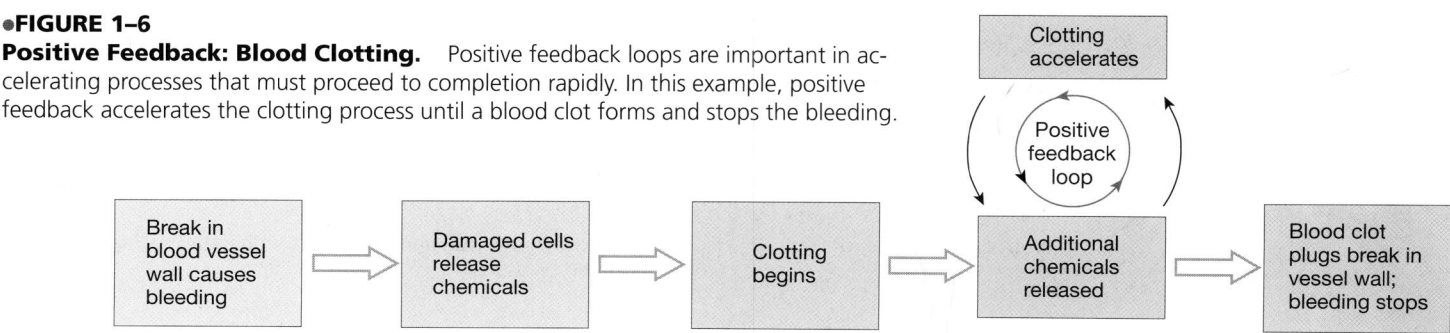

Homeostasis and Disease

The human body is amazingly effective in maintaining homeostasis. Nevertheless, an infection, an injury, or a genetic abnormality can sometimes have effects so severe that homeostatic mechanisms can't fully compensate for them. One or more characteristics of the internal environment may then be pushed outside of normal limits. When this happens, organ systems begin to malfunction, producing a state known as illness, or **disease**. Chapters 2–29 devote considerable attention to the mechanisms responsible for a variety of human diseases. **AM** Homeosstasis and Disease

An understanding of normal homeostatic mechanisms usually enables one to draw conclusions about what might be responsible for the signs and symptoms that are characteristic of many

diseases. **Symptoms** are subjective—things that a person experiences and describes but that aren't otherwise detectable, such as pain, nausea, and anxiety. A **sign**, by contrast, is an objectively observable physical indication of a disease, such as a rash, a swelling, a fever, or sounds of abnormal breathing. Nowadays, technological aids can reveal many additional signs that would not be evident to a physician's unaided senses: an unusual shape on an X ray or MRI scan, or an elevated concentration of a particular chemical in a blood test. You will find much additional information on signs, symptoms, and diagnosis, as well as many other aspects of human health, disease, and medical treatment, in the *Applications Manual* that accompanies this text. **AM** The Diagnosis of Disease

☐ SYSTEMS INTEGRATION

Homeostatic regulation controls characteristics of the internal environment that affect every cell in the body. No one organ system has total control over any of these characteristics; they all require the coordinated efforts of multiple organ systems. In later chapters, we shall explore the functions of each organ system and see

how the systems interact to preserve homeostasis. Table 1–1 presents several examples of important physiological characteristics that are subject to homeostatic regulation. As you can see, in each case such regulation involves several organ systems.

Each organ system interacts with and is, in turn, dependent on other organ systems. However, it is much easier for introductory students to learn the basics of anatomy and physiology one system

TABLE 1–1 SYSTEMS INVOLVED IN HOMEOSTATIC REGULATION

Internal Characteristic	Primary Systems	Functions
Body temperature	Integumentary system	Heat loss
	Muscular system	Heat production
	Cardiovascular system	Heat distribution
	Nervous system	Coordination of blood flow, heat production, and heat loss
Body fluid composition		
Nutrient concentration	Digestive system	Nutrient absorption, storage, and release
	Cardiovascular system	Nutrient distribution
	Urinary system	Control of nutrient loss in the urine
Oxygen, carbon dioxide levels	Respiratory system	Absorption of oxygen, elimination of carbon dioxide
	Cardiovascular system	Internal transport of oxygen and carbon dioxide
Body fluid volume	Urinary system	Elimination or conservation of water from the blood
	Digestive system	Absorption of water; loss of water in feces
	Integumentary system	Loss of water through perspiration
	Cardiovascular system	Distribution of water
Waste product concentration	Urinary system	Elimination of waste products from the blood
	Cardiovascular system	Transport of waste products to sites of excretion
Blood pressure	Cardiovascular system	Pressure generated by the heart moves blood through blood vessels
	Nervous system and endocrine system	Adjustments in heart rate and blood vessel diameter can raise or lower blood pressure

at a time. Although Chapters 5–29 will be organized around individual systems, you should not lose sight of the fact that these systems all work together. The Systems in Perspective sections in later chapters will help reinforce this message. There are 11 of these sections; each provides an overview of one system's functions and summarizes its functional relationships with other systems.

✔ Why is homeostatic regulation important to humans?

✔ What happens to the body when homeostasis breaks down?

✔ Why is positive feedback helpful in blood clotting, but unsuitable for the regulation of body temperature?

Answers start on page Q-1

1–4 A FRAME OF REFERENCE FOR ANATOMICAL STUDIES

Objectives

■ Use anatomical terms to describe body sections, body regions, and relative positions.

■ Identify the major body cavities and their subdivisions.

Early anatomists faced serious problems in communication. Stating that a bump is "on the back" does not give very precise information about its location. So anatomists created maps of the human body. Landmarks are prominent anatomical structures, distances are measured in centimeters or inches, and specialized directional terms are used. In effect, anatomy uses a special language that must be learned almost at the start of your study.

A familiarity with Latin and Greek word roots and patterns makes anatomical terms more understandable. As new terms are introduced, notes on pronunciation and relevant word roots will be provided. The front endpapers contain additional information on roots, prefixes, suffixes, and combining forms.

Latin and Greek terms are not the only ones that have been imported into the anatomical vocabulary over the centuries, and the vocabulary continues to expand. Many anatomical structures and clinical conditions were initially named after either the discoverer or, in the case of diseases, the most famous victim. Over the last 100 years, most of these commemorative names, or *eponyms*, have been replaced by more precise terms. Appendix V lists the most important eponyms and related historical details.

☐ SUPERFICIAL ANATOMY

A familiarity with anatomical landmarks, regions, and directional references will make subsequent chapters more understandable. As you encounter new terms, create your own mental maps from the information provided in the accompanying anatomical illustrations.

Anatomical Landmarks

Important anatomical landmarks are presented in Figure 1–7●. Anatomical terms are given in boldface type, common names in plain type, and anatomical adjectives in parentheses. Understanding the terms and their etymological origins will help you remember the location of a particular structure as well as its name. For example, *brachium* refers to the arm; later we will consider the *brachialis muscle* and the *brachial artery*, which are (as their names suggest) in the arm.

Standard anatomical illustrations show the human form in the **anatomical position**. When the body is in this position, the hands are at the sides with the palms facing forward. Figure 1–8a● shows an individual in the anatomical position as seen from the front (an *anterior view*), Figure 1–7b● from the back (a *posterior view*). Unless otherwise noted, all descriptions in this text refer to the body in the anatomical position. A person lying down in the anatomical position is said to be **supine** (soo-PĪN) when face up and **prone** when face down.

Anatomical Regions

Major regions of the body are listed in Table 1–2 and shown in Figure 1–8● (p. 16). To describe a general area of interest or injury, anatomists and clinicians often need broader terms as

TABLE 1–2 REGIONS OF THE HUMAN BODY (SEE FIGURE 1–7)	
Structure	Region
Cephalon (head)	Cephalic region
Cervicis (neck)	Cervical region
Thoracis (thorax or chest)	Thoracic region
Brachium (arm)	Brachial region
Antebrachium (forearm)	Antebrachial region
Carpus (wrist)	Carpal region
Manus (hand)	Manual region
Abdomen	Abdominal region
Lumbus (loin)	Lumbar region
Gluteus (buttock)	Gluteal region
Pelvis	Pelvic region
Pubis (anterior pelvis)	Pubic region
Inguen (groin)	Inguinal region
Femur (thigh)	Femoral region
Crus (anterior leg)	Crural region
Sura (calf)	Sural region
Tarsus (ankle)	Tarsal region
Pes (foot)	Pedal region
Planta (sole)	Plantar region

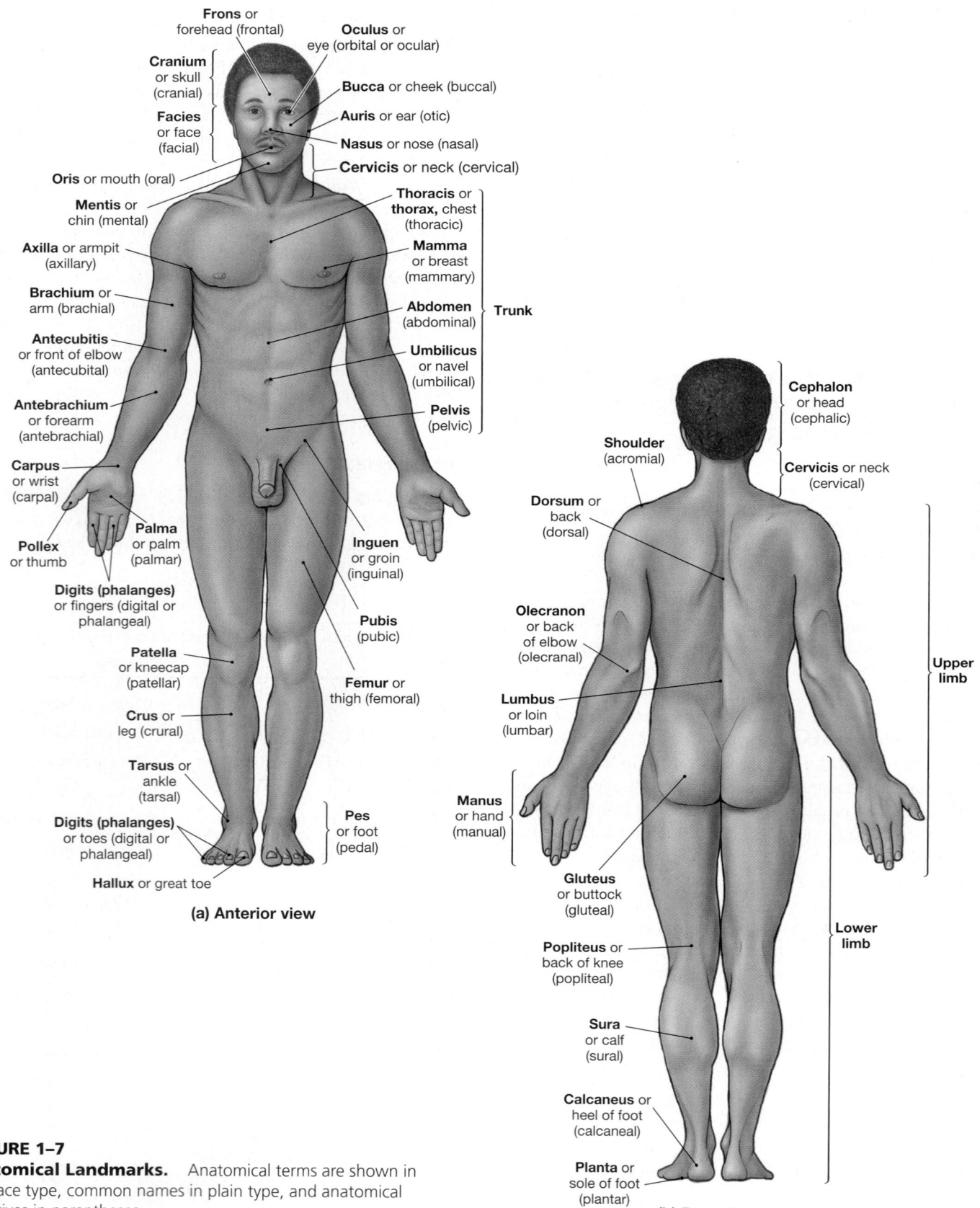

Frons or forehead (frontal)
Oculus or eye (orbital or ocular)
Cranium or skull (cranial)
Bucca or cheek (buccal)
Auris or ear (otic)
Facies or face (facial)
Nasus or nose (nasal)
Cervicis or neck (cervical)
Oris or mouth (oral)
Thoracis or **thorax,** chest (thoracic)
Mentis or chin (mental)
Mamma or breast (mammary)
Axilla or armpit (axillary)
Brachium or arm (brachial)
Abdomen (abdominal) — Trunk
Antecubitis or front of elbow (antecubital)
Umbilicus or navel (umbilical)
Antebrachium or forearm (antebrachial)
Pelvis (pelvic)
Carpus or wrist (carpal)
Palma or palm (palmar)
Pollex or thumb
Inguen or groin (inguinal)
Digits (phalanges) or fingers (digital or phalangeal)
Pubis (pubic)
Patella or kneecap (patellar)
Femur or thigh (femoral)
Crus or leg (crural)
Tarsus or ankle (tarsal)
Digits (phalanges) or toes (digital or phalangeal)
Pes or foot (pedal)
Hallux or great toe

(a) Anterior view

Cephalon or head (cephalic)
Shoulder (acromial)
Cervicis or neck (cervical)
Dorsum or back (dorsal)
Olecranon or back of elbow (olecranal)
Lumbus or loin (lumbar)
Manus or hand (manual)
Upper limb
Gluteus or buttock (gluteal)
Popliteus or back of knee (popliteal)
Sura or calf (sural)
Lower limb
Calcaneus or heel of foot (calcaneal)
Planta or sole of foot (plantar)

(b) Posterior view

●**FIGURE 1–7**
Anatomical Landmarks. Anatomical terms are shown in boldface type, common names in plain type, and anatomical adjectives in parentheses.

well as specific landmarks. Two methods are used, both concerned with mapping the surface of the abdomen and pelvis. Clinicians refer to four **abdominopelvic quadrants**. The area is divided into four segments by using a pair of imaginary lines that intersect at the umbilicus (navel). This simple method, shown in Figure 1–8a●, provides useful references for the description of aches, pains, and injuries. The location can help the physician determine the possible cause; for example, tenderness in the right lower quadrant (RLQ) is a symptom of appendicitis, whereas tenderness in the right upper quadrant (RUQ) may indicate gallbladder or liver problems.

Anatomists like to use more precise terms to describe the location and orientation of internal organs. They recognize nine **abdominopelvic regions** (Figure 1–8b●). Figure 1–8c● shows the relationships among quadrants, regions, and internal organs.

Anatomical Directions

Figure 1–9● and Table 1–3 show the principal directional terms and examples of their use. There are many different terms, and some can be used interchangeably. For example, *anterior* refers to the front of the body when viewed in the anatomical position; in humans, this term is equivalent to *ventral*, which refers to the belly. Before you read further, analyze the table in detail, and practice using these terms. If you are familiar with the basic vocabulary, the descriptions in subsequent chapters will be easier to follow. When reading anatomical descriptions, you will find it useful to remember that the terms *left* and *right* always refer to the *left* and *right* sides of the *subject*, not of the observer.

☐ SECTIONAL ANATOMY

A presentation in sectional view is sometimes the only way to illustrate the relationships among the parts of a three-dimensional object. An understanding of sectional views has become increasingly important since the development of electronic imaging techniques that enable us to see inside the living body without resorting to surgery. Although these views are sometimes difficult to interpret, it is worth your spending the time required to fully understand what they show. Once you are able to recognize sectional views, you will have a good mental

●**FIGURE 1–8**
Abdominopelvic Quadrants and Regions.
(a) Abdominopelvic quadrants divide the area into four sections. These terms, or their abbreviations, are most often used in clinical discussions. **(b)** Abdominopelvic regions divide the same area into nine sections, providing more precise regional descriptions. **(c)** Overlapping quadrants and regions and the relationship between superficial anatomical landmarks and underlying organs.

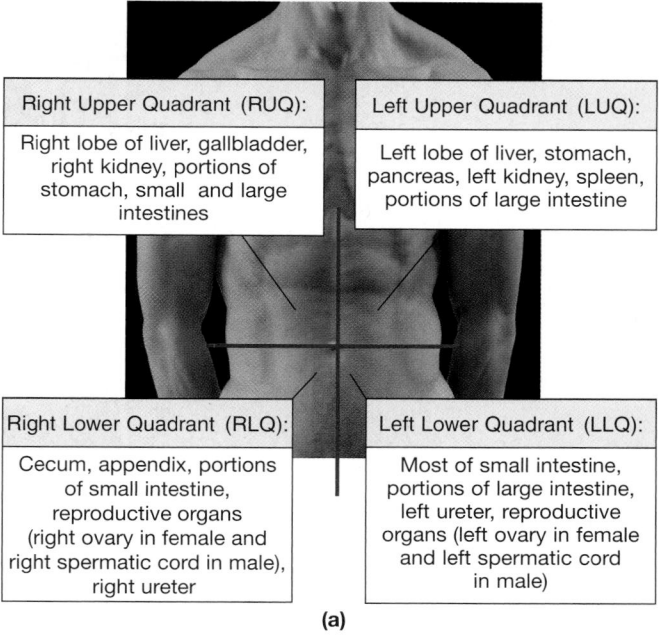

Right Upper Quadrant (RUQ): Right lobe of liver, gallbladder, right kidney, portions of stomach, small and large intestines

Left Upper Quadrant (LUQ): Left lobe of liver, stomach, pancreas, left kidney, spleen, portions of large intestine

Right Lower Quadrant (RLQ): Cecum, appendix, portions of small intestine, reproductive organs (right ovary in female and right spermatic cord in male), right ureter

Left Lower Quadrant (LLQ): Most of small intestine, portions of large intestine, left ureter, reproductive organs (left ovary in female and left spermatic cord in male)

(a)

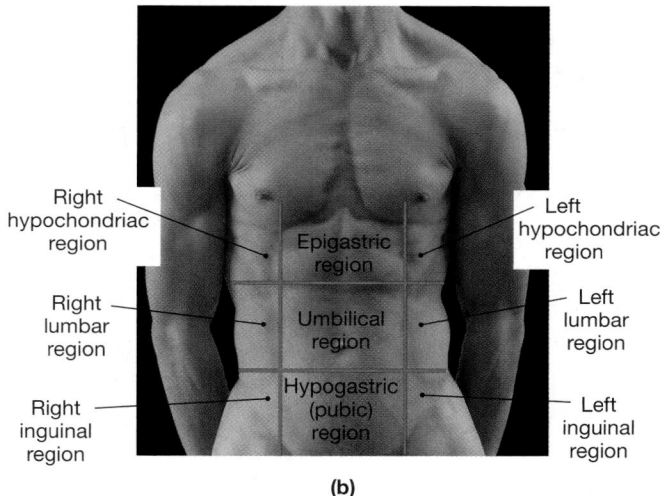

Right hypochondriac region
Epigastric region
Left hypochondriac region
Right lumbar region
Umbilical region
Left lumbar region
Right inguinal region
Hypogastric (pubic) region
Left inguinal region

(b)

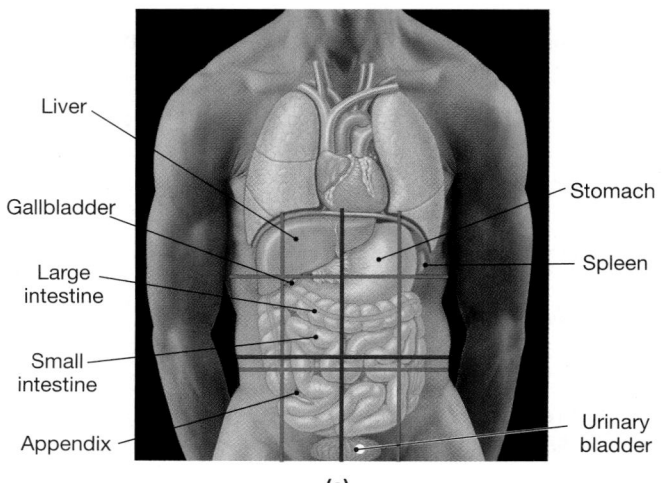

Liver
Gallbladder
Large intestine
Small intestine
Appendix
Stomach
Spleen
Urinary bladder

(c)

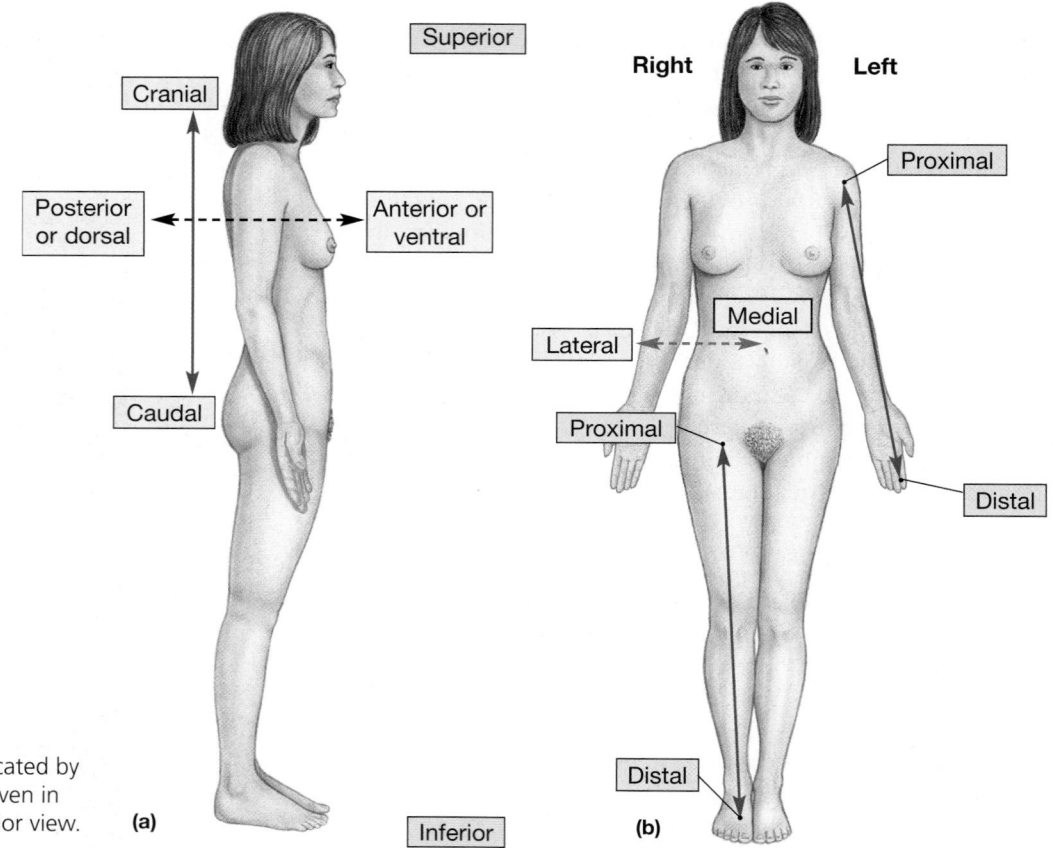

●**FIGURE 1–9**
Directional References. Important directional terms used in this text are indicated by arrows; definitions and descriptions are given in Table 1–3. **(a)** A lateral view. **(b)** An anterior view.

TABLE 1–3 DIRECTIONAL TERMS (SEE FIGURE 1–9)		
Term	**Region or Reference**	**Example**
Anterior	The front; before	The navel is on the *anterior* surface of the trunk.
Ventral	The belly side (equivalent to anterior when referring to human body)	The navel is on the *ventral* surface.
Posterior	The back; behind	The shoulder blade is located *posterior* to the rib cage.
Dorsal	The back (equivalent to posterior when referring to human body)	The *dorsal* body cavity encloses the brain and spinal cord.
Cranial or cephalic	The head	The *cranial*, or *cephalic*, border of the pelvis is on the side toward the head rather than toward the thigh.
Superior	Above; at a higher level (in human body, toward the head)	In humans, the cranial border of the pelvis is *superior* to the thigh.
Caudal	The tail (coccyx in humans)	The hips are *caudal* to the waist.
Inferior	Below; at a lower level	The knees are *inferior* to the hips.
Medial	Toward the body's longitudinal axis; toward the midsagittal plane	The *medial* surfaces of the thighs may be in contact; moving medially from the arm across the chest surface brings you to the sternum.
Lateral	Away from the body's longitudinal axis; away from the midsagittal plane	The thigh articulates with the *lateral* surface of the pelvis; moving laterally from the nose brings you to the cheeks.
Proximal	Toward an attached base	The thigh is *proximal* to the foot; moving proximally from the wrist brings you to the elbow.
Distal	Away from an attached base	The fingers are *distal* to the wrist; moving distally from the elbow brings you to the wrist.
Superficial	At, near, or relatively close to the body surface	The skin is *superficial* to underlying structures.
Deep	Farther from the body surface	The bone of the thigh is *deep* to the surrounding skeletal muscles.

model for studying the anatomy and physiology of a particular region or system. Radiologists and other medical professionals responsible for interpreting medical scans spend much of their time analyzing sectional views of the body.

Planes and Sections

Any slice through a three-dimensional object can be described with reference to three **sectional planes**, as indicated in Figure 1–10● and Table 1–4. The **transverse plane** lies at right angles to the long axis of the body, dividing it into *superior* and *inferior* sections. A cut in this plane is called a **transverse section**, or *cross section*. The **frontal plane**, or *coronal plane*, and the **sagittal plane** are parallel to the long axis of the body. The frontal plane extends from side to side, dividing the body into *anterior* and *posterior* sections. The sagittal plane extends from front to back, dividing the body into left and right sections. A cut that passes along the midline and divides the body into left and right halves is a *midsagittal section*, or *median section*; a cut

parallel to the midsagittal line is a *parasagittal section*. The atlas that accompanies this text contains images of sections taken through the body in various planes. You will be referred to these images later in the text, for comparison with specific illustrations in figures.

□ BODY CAVITIES

Many vital organs are suspended in internal chambers called *body cavities*. These cavities have two essential functions: (1) They protect delicate organs, such as the brain and spinal cord, from accidental shocks and cushion them from the thumps and bumps that occur when we walk, jump, or run; and (2) they permit significant changes in the size and shape of internal organs. For example, because they are inside body cavities, the lungs, heart, stomach, intestines, urinary bladder, and many other organs can expand and contract without distorting surrounding tissues or disrupting the activities of nearby organs.

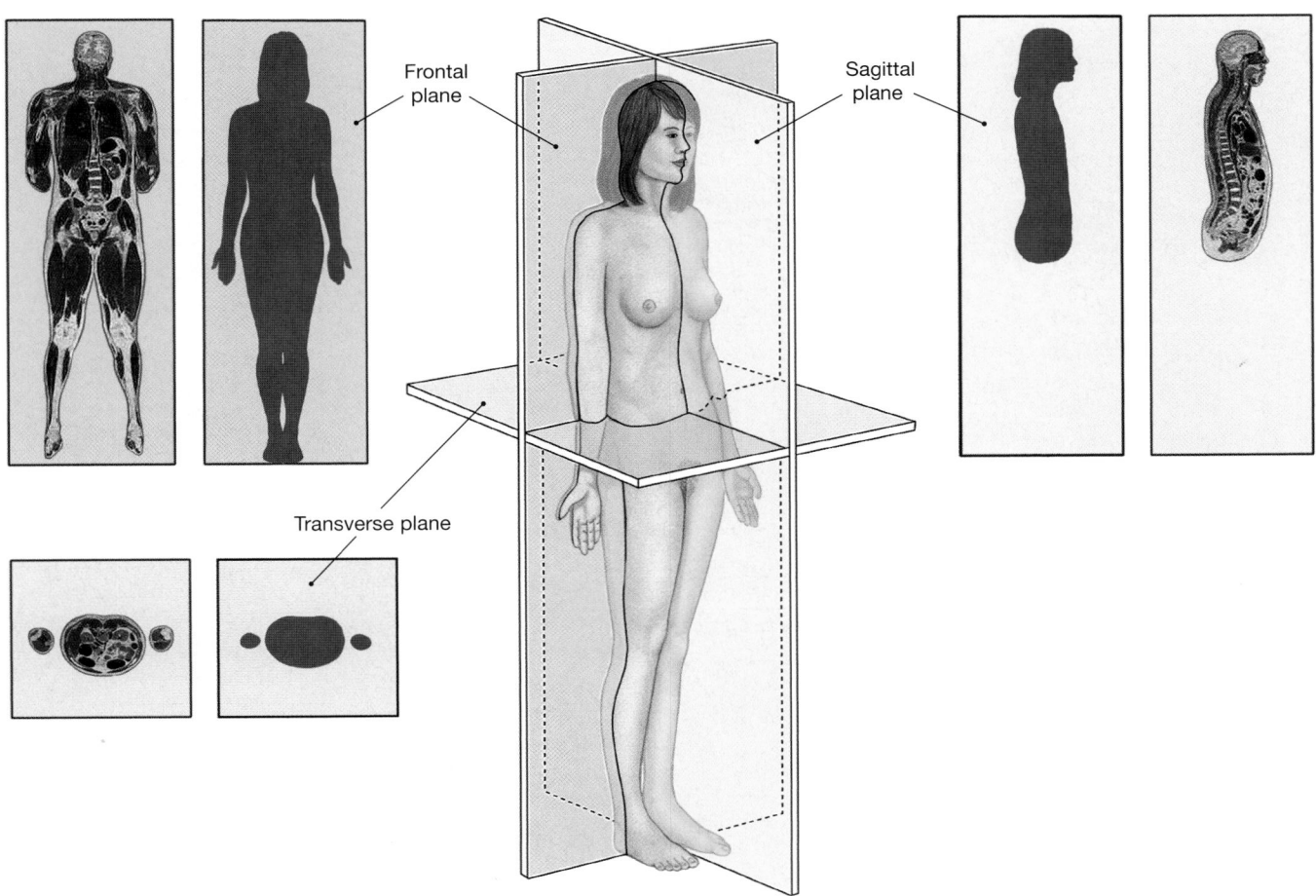

Frontal plane

Sagittal plane

Transverse plane

●**FIGURE 1–10**
Planes of Section. The three primary planes of section, which are defined and described in Table 1–4. The photos of sectional images were generated by the Anatomy Explorer/Digital Cadaver software, based on the Visible Human data set.

TABLE 1–4 Terms That Indicate Planes of Section (see Figure 1–10)

Orientation of Plane	Plane	Directional Reference	Description
Perpendicular to long axis	Transverse or horizontal	Transversely or horizontally	A *transverse*, or *horizontal*, *section* separates superior and inferior portions of the body.
Parallel to long axis	Sagittal	Sagittally	A *sagittal section* separates right and left portions. You examine a sagittal section, but you section sagittally.
	Midsagittal		In a *midsagittal section*, the plane passes through the midline, dividing the body in half and separating the right and left sides.
	Parasagittal		A *parasagittal section* misses the midline, separating right and left portions of unequal size.
	Frontal or coronal	Frontally or coronally	A *frontal*, or *coronal*, *section* separates anterior and posterior portions of the body; coronal usually refers to sections passing through the skull.

The term *dorsal body cavity* is sometimes used to refer to the fluid-filled space that surrounds the brain and spinal cord. As such, it includes the **cranial cavity**, which contains the brain, and the **spinal cavity**, which contains the spinal cord. The much larger **ventral body cavity**, or *coelom* (SĒ-lōm; *koila*, cavity), contains organs of the respiratory, cardiovascular, digestive, urinary, and reproductive systems (Figures 1–11, 1–12●).

The ventral body cavity appears early in embryological development. As internal organs grow, their relative positions change and the ventral body cavity is gradually subdivided. The **diaphragm**. (DĪ-ah-fram), a flat muscular sheet, divides the ventral body cavity into a superior **thoracic cavity**, bounded by the chest wall, and an inferior **abdominopelvic cavity**, enclosed by the abdominal wall and by the bones and muscles of the pelvis.

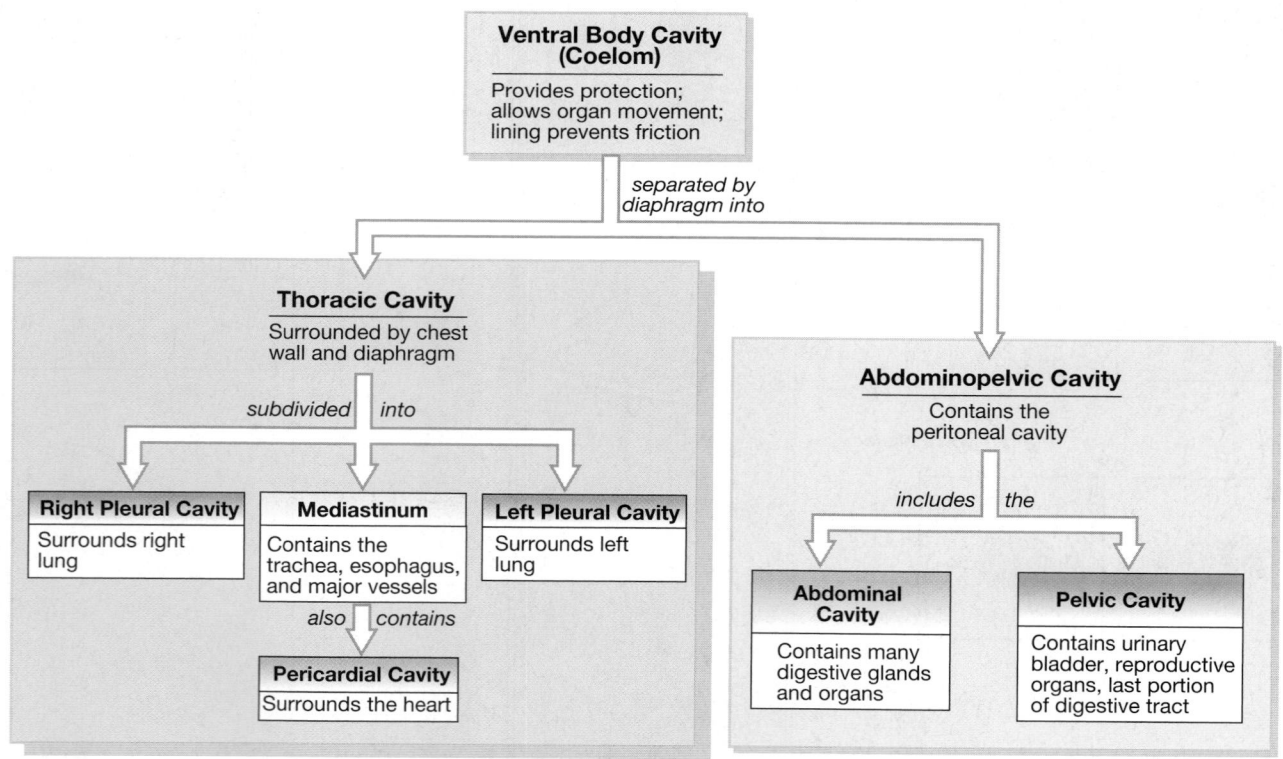

●FIGURE 1–11
Relationships among the Subdivisions of the Ventral Body Cavity

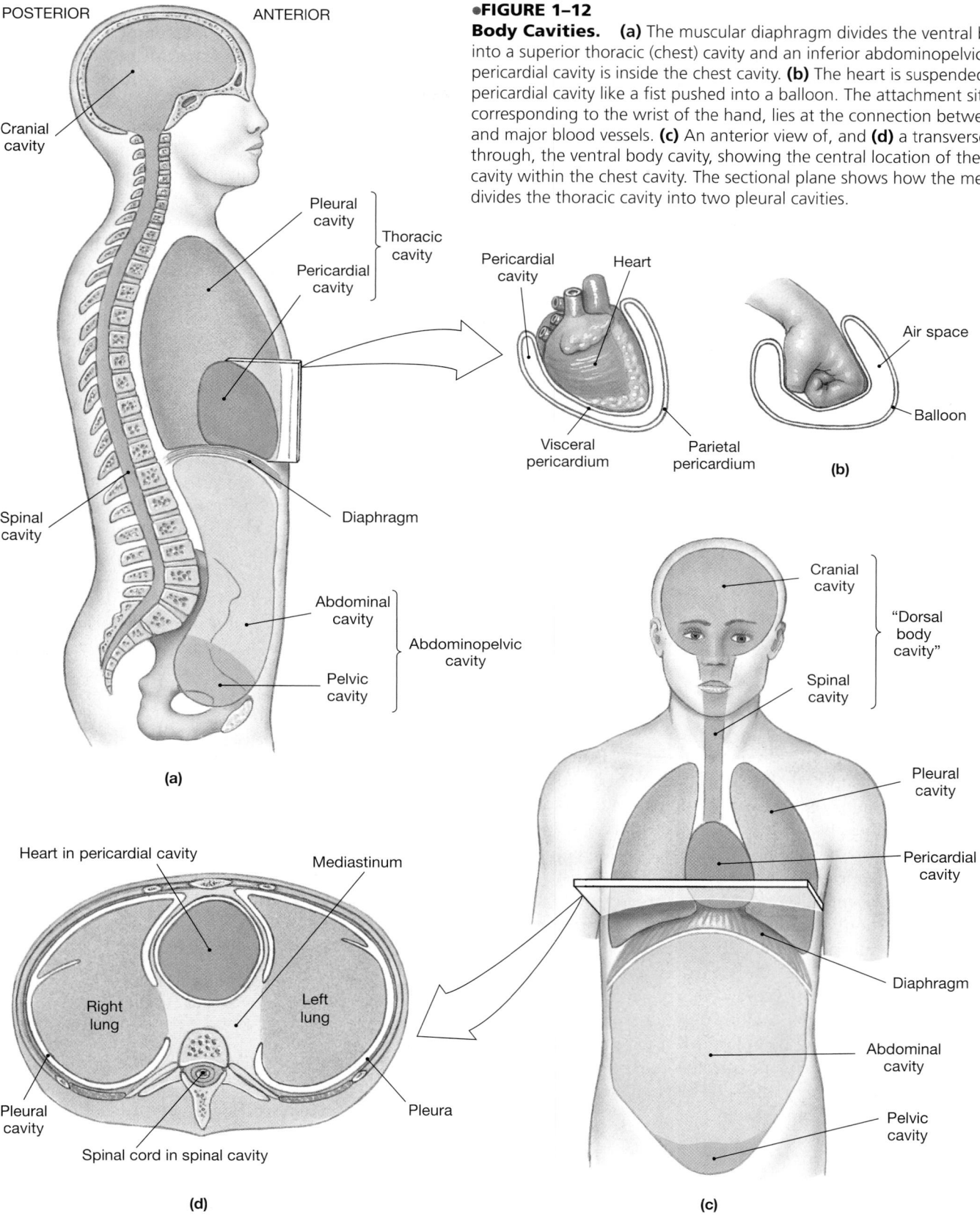

POSTERIOR   ANTERIOR

Cranial cavity

Spinal cavity

Pleural cavity

Pericardial cavity

Thoracic cavity

Diaphragm

Abdominal cavity

Pelvic cavity

Abdominopelvic cavity

(a)

●FIGURE 1–12

Body Cavities.   **(a)** The muscular diaphragm divides the ventral body cavity into a superior thoracic (chest) cavity and an inferior abdominopelvic cavity. The pericardial cavity is inside the chest cavity. **(b)** The heart is suspended within the pericardial cavity like a fist pushed into a balloon. The attachment site, corresponding to the wrist of the hand, lies at the connection between the heart and major blood vessels. **(c)** An anterior view of, and **(d)** a transverse section through, the ventral body cavity, showing the central location of the pericardial cavity within the chest cavity. The sectional plane shows how the mediastinum divides the thoracic cavity into two pleural cavities.

Pericardial cavity   Heart

Visceral pericardium   Parietal pericardium

Air space

Balloon

(b)

Cranial cavity

Spinal cavity

"Dorsal body cavity"

Pleural cavity

Pericardial cavity

Diaphragm

Abdominal cavity

Pelvic cavity

(c)

Heart in pericardial cavity   Mediastinum

Right lung   Left lung

Pleural cavity

Spinal cord in spinal cavity

Pleura

(d)

Many of the organs in the thoracic and abdominopelvic cavities change size and shape as they perform their functions. For example, the lungs inflate and deflate as you breathe, and your stomach swells at each meal and shrinks between meals. These organs are surrounded by moist internal spaces that permit expansion and limited movement, but prevent friction. The thoracic cavity is subdivided into three separate spaces, whereas the abdominopelvic cavity contains a single extensive chamber. The internal organs that are partially or completely enclosed by these cavities are called **viscera** (VIS-e-ruh). A delicate layer called a *serous membrane* lines the walls of these internal cavities and covers the surfaces of the enclosed viscera. Serous membranes secrete a watery fluid that coats the opposing surfaces and reduces friction.

The Thoracic Cavity

The **thoracic cavity** contains the lungs and heart; associated organs of the respiratory, cardiovascular, and lymphatic systems; the inferior portions of the esophagus; and the thymus. The boundaries of the thoracic cavity are established by the muscles and bones of the chest wall and the diaphragm (Figure 1–12a●). The thoracic cavity is subdivided into the left and right **pleural cavities**, separated by the **mediastinum** (mē-dē-as-TĪ-num or mē-dē-AS-ti-num) (Figure 1–12a,c,d●). Each pleural cavity, which contains a lung, is lined by a shiny, slippery serous membrane that reduces friction as the lung expands and recoils during respiration. The serous membrane lining a pleural cavity is called a *pleura* (PLOO-rah). The *visceral pleura* covers the outer surfaces of a lung, and the *parietal pleura* covers the opposing mediastinal surface and the inner body wall.

The mediastinum consists of a mass of connective tissue that surrounds, stabilizes, and supports the esophagus, trachea, and thymus, as well as the major blood vessels that originate or end at the heart. The mediastinum also contains the **pericardial cavity**, a small chamber that surrounds the heart. The relationship between the heart and the pericardial cavity resembles that of a fist pushing into a balloon (Figure 1–12b●). The wrist corresponds to the *base* (attached portion) of the heart, and the balloon corresponds to the serous membrane that lines the pericardial cavity. The serous membrane covering the heart is called the *pericardium* (*peri-*, around + *kardia*, heart). The layer covering the heart is the *visceral pericardium*, and the opposing surface is the *parietal pericardium*. During each beat, the heart changes in size and shape. The pericardial cavity permits these changes, and the slippery pericardial lining prevents friction between the heart and adjacent structures in the thoracic cavity.

The Abdominopelvic Cavity

The abdominopelvic cavity extends from the diaphragm to the pelvis. It is subdivided into a superior **abdominal cavity** and an inferior **pelvic cavity** (Figure 1–12a,c●). The abdominopelvic cavity contains the *peritoneal* (per-i-tō-NĒ-al) *cavity*, a chamber lined by a serous membrane known as the *peritoneum* (per-i-tō-NĒ-um). The *parietal peritoneum* lines the inner surface of the body wall. A narrow space containing a small amount of fluid separates the parietal peritoneum from the *visceral peritoneum*, which covers the enclosed organs. You are probably already aware of the movements of the organs in this cavity. Who has not had at least one embarrassing moment when the contraction of a digestive organ produced a movement of liquid or gas and a gurgling or rumbling sound? The peritoneum allows the organs of the digestive system to slide across one another without damage to themselves or the walls of the cavity.

The abdominal cavity extends from the inferior surface of the diaphragm to the level of the superior margins of the pelvis. This cavity contains the liver, stomach, spleen, small intestine, and most of the large intestine. (The positions of most of these organs are shown in Figure 1–8c●, p.17.) The organs are partially or completely enclosed by the peritoneal cavity, much as the heart or lungs are enclosed by the pericardial or pleural cavities, respectively. A few organs, such as the kidneys and pancreas, lie between the peritoneal lining and the muscular wall of the abdominal cavity. Those organs are said to be *retroperitoneal* (*retro*, behind).

The pelvic cavity is the portion of the ventral body cavity inferior to the abdominal cavity. The bones of the pelvis form the walls of the pelvic cavity, and a layer of muscle forms its floor. The pelvic cavity contains the last portion of the large intestine, the urinary bladder, and various reproductive organs. For example, the pelvic cavity of females contains the ovaries, uterine tubes, and uterus; in males, it contains the prostate gland and seminal vesicles. The pelvic cavity contains the inferior portion of the peritoneal cavity. The peritoneum covers the ovaries and the uterus in females, as well as the superior portion of the urinary bladder in both sexes.

✓ Which type of section would separate the two eyes?

✓ If a surgeon makes an incision just inferior to the diaphragm, which body cavity will be opened?

Answers start on page Q-1

 Practice locating structures in the abdominal and thoracic cavities by viewing the **Anatomy CD-ROM:** Digestive System Dissection and Cardiovascular System/Heart Dissection.

This chapter provided an overview of the locations and functions of the major components of each organ system. It also introduced the anatomical vocabulary needed for you to follow more detailed anatomical descriptions in later chapters. Many of the figures in those chapters contain images produced by the procedures outlined in Figures 1–13 through 1–15● (see "Sectional Anatomy and Clinical Technology," pp. 23–25), which summarize the most common methods of visualizing anatomical structures in living individuals.

Sectional Anatomy and Clinical Technology | focus

Radiological procedures include (1) scanning techniques that involve the use of beams of radiation, such as X rays, to create a photographic or computer-generated image of internal structures and (2) methods that involve the administration of radioactive materials.

Physicians who specialize in the performance of these procedures and the analysis of the resulting radiological images are called *radiologists*. Radiological procedures can provide detailed information about internal systems. Figures 1–13 through 1–15● compare the views provided by several techniques radiologists and other clinicians use. The figures show *X rays, computerized to-*

mography (CT) scans, magnetic resonance imaging (MRI) scans, ultrasound images, spiral-CT scans, digital subtraction angiography (DSA) images, and *positron emission tomography (PET) scans.* These clinical technologies and more specialized MRI procedures are described further in the *Applications Manual*; many examples are included as figures later in the text. **AM** The Diagnosis of Disease

Whenever you see anatomical diagrams or clinical procedures that present cross-sectional views of the body, remember that the sections are oriented as though the observer is standing at the feet of the subject and looking toward the head.

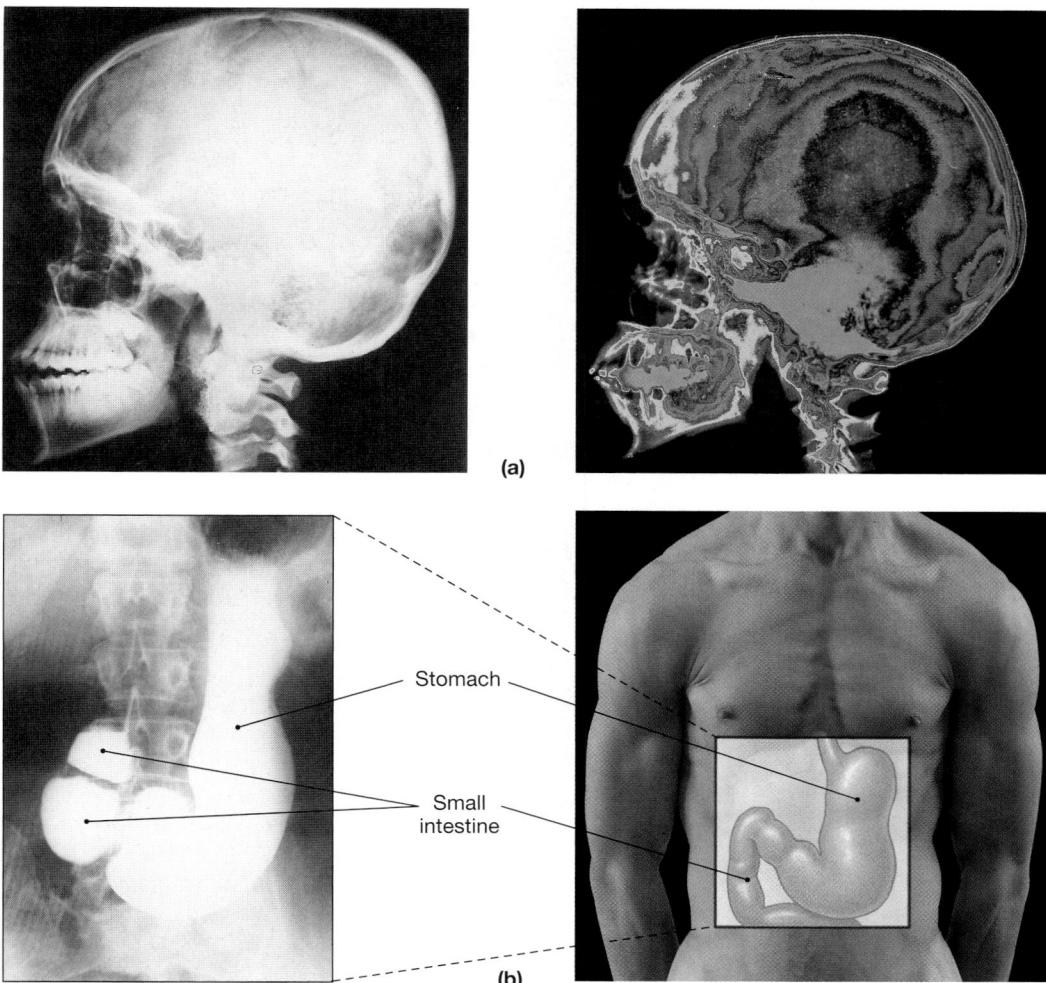

(a)

Stomach

Small intestine

(b)

●**FIGURE 1–13**
X Rays   **(a)** X rays of the skull, taken from the left side. **X rays** are a form of high-energy radiation that can penetrate tissues. In the most familiar procedure, a beam of X rays travels through the body and strikes a photographic plate. Not all of the projected X rays reach the film; some are absorbed or deflected as they pass through the body. The resistance to X-ray penetration is called **radiodensity**. Radiodensity increases in the following sequence: air, fat, liver, blood, muscle, bone. The usual result is an image with radiodense tissues, such as bone, appearing in white, and less dense tissues in shades of gray to black. (The image on the right was color enhanced.) **(b)** A **barium contrast X ray** of the upper digestive tract. Such an X ray is produced by introducing a radiodense material into the body to provide sharp outlines and contrast and to check the distribution of fluids or the movements of internal organs. In this instance, the patient swallowed a solution of barium, an element that is very dense. The contours of the stomach and intestinal linings are clearly outlined against the white of the barium solution.

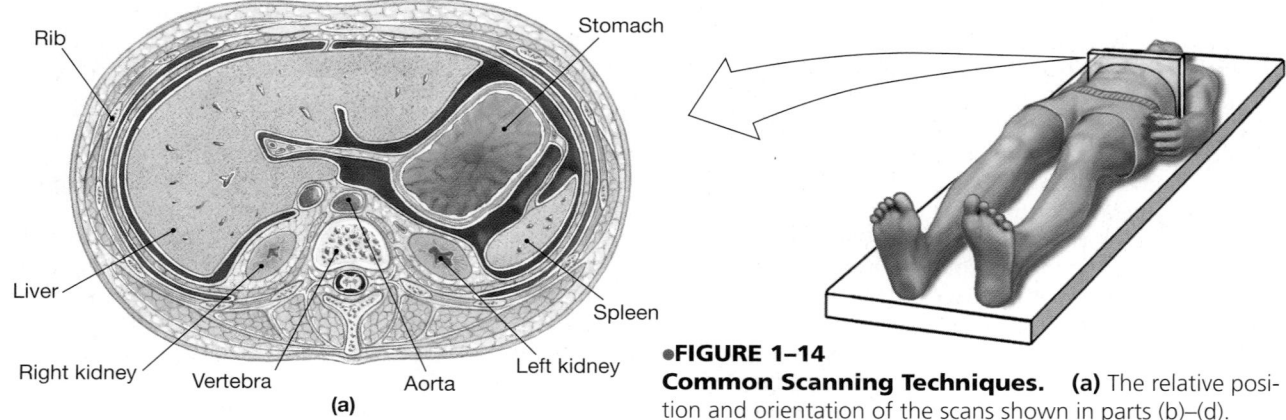

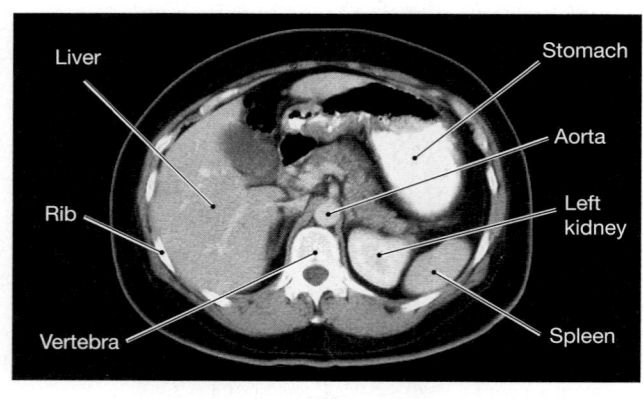

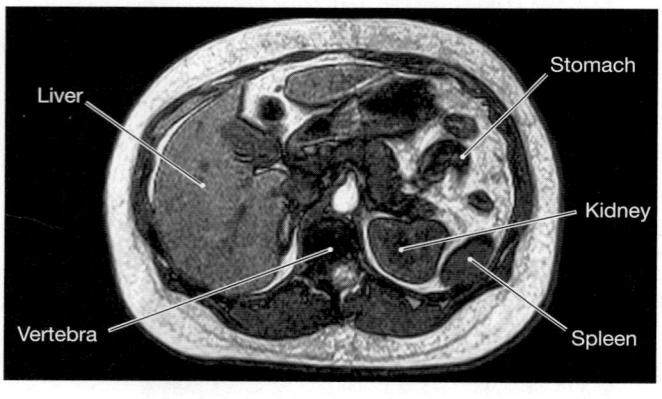

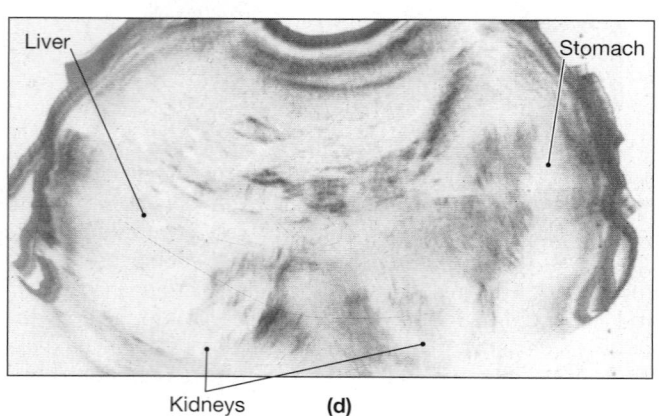

●**FIGURE 1–14**
Common Scanning Techniques.   **(a)** The relative position and orientation of the scans shown in parts (b)–(d).

(b) A CT scan of the abdomen. **C**omputerized **t**omography (**CT**), formerly called **c**omputerized **a**xial **t**omography (**CAT**), uses computers to reconstruct sectional views. A single X-ray source rotates around the body, and the X-ray beam strikes a sensor monitored by the computer. The source completes one revolution around the body every few seconds; it then moves a short distance and repeats the process. The result is usually displayed as a sectional view in black and white, but it can be colorized for visual effect. CT scans show three-dimensional relationships and soft-tissue structure more clearly than do standard X rays.

(c) An MRI scan of the abdomen. **M**agnetic **r**esonance **i**maging (**MRI**) surrounds part or all of the body with a magnetic field about 3000 times as strong as that of Earth. The MRI field affects protons within atomic nuclei throughout the body. The protons line up along the magnetic lines of force like compass needles in Earth's magnetic field. When struck by a radio wave of a certain frequency, a proton will absorb energy. When the wave pulse ends, that energy is released and the source of the radiation is detected. Each element differs in terms of the radio frequency required to affect its protons.

(d) An ultrasound scan of the abdomen. In **ultrasound** procedures, a small transmitter contacting the skin broadcasts a brief, narrow burst of high-frequency sound and then picks up the echoes. The sound waves are reflected by internal structures. An **echogram**, or ultrasound picture, can be assembled from the pattern of echoes. These images lack the clarity of those produced by other procedures, but no adverse effects have been reported, and fetal development can be monitored without a significant risk of birth defects. Special methods of transmission and processing permit analysis of the beating heart, without the complications that can accompany injections of a dye. Note the differences in detail among this image, the CT scan, and the MRI image.

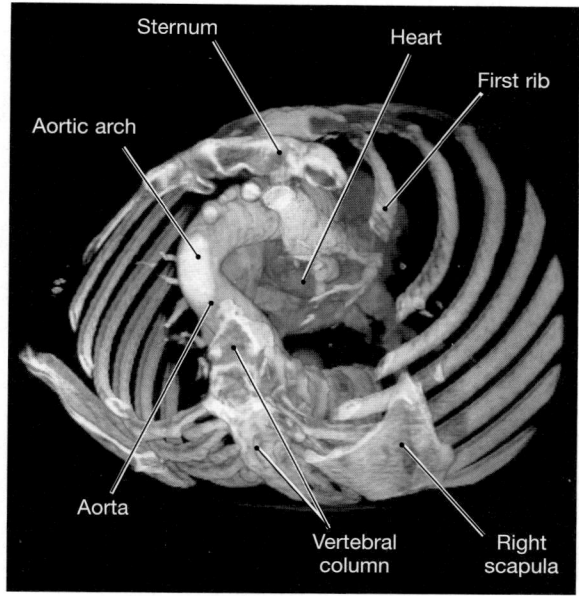

(a)

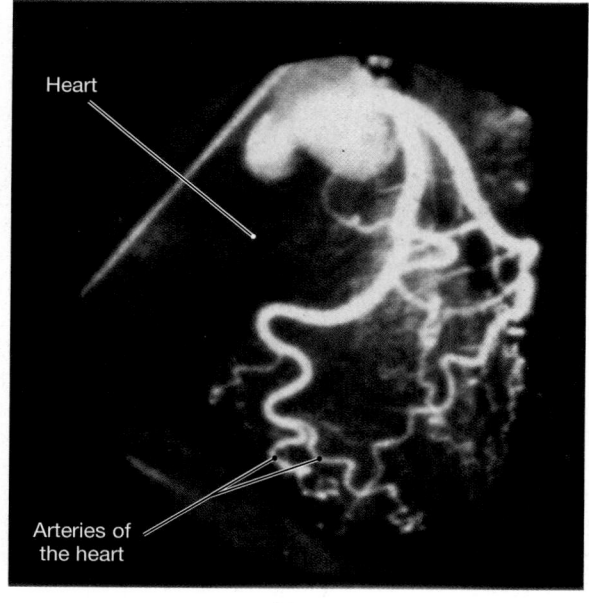

(b)

(c)

●**FIGURE 1–15**
Special Scanning Methods.   **(a)** A **spiral-CT** scan of the chest. Such an image is created by special processing of CT data to permit rapid three-dimensional visualization of internal organs. Spiral-CT scans are becoming increasingly important in clinical settings. **(b)** **D**igital **s**ubtraction **a**ngiography (**DSA**) is used to monitor blood flow through specific organs, such as the brain, heart, lungs, or kidneys. X rays are taken before and after a radiopaque dye is administered, and a computer "subtracts" details common to both images. The result is a high-contrast image showing the distribution of the dye. **(c)** **P**ositron-**e**mission **t**omography (**PET**) scans rely on the administration of radioactive isotopes that are later detected by gamma-ray detectors and interpreted by computers.

Chapter Review

SELECTED CLINICAL TERMINOLOGY

Terms Discussed in This Chapter

abdominopelvic quadrant: One of four divisions of the anterior abdominal surface. *(p. 17)*

abdominopelvic region: One of nine divisions of the anterior abdominal surface. *(p. 17)*

CT, CAT (computerized [axial] tomography): An imaging technique that uses X rays to reconstruct the body's three-dimensional structure. *(p. 23)*

disease: A malfunction of organs or organ systems resulting from a failure of homeostatic regulation. *(p. 14)*

DSA (digital subtraction angiography): A technique used to monitor blood flow through specific organs, such as the brain, heart, lungs, or kidneys. X rays are taken before and after a radiopaque dye is administered, and a computer "subtracts" details common to both images. The result is a high-contrast image showing the distribution of the dye. *(p. 23)*

embryology: The study of structural changes during the first two months of development. *(p. 6)*

histology (his-TOL-o-jē): The study of tissues. *(p. 6)*

MRI (magnetic resonance imaging): An imaging technique that employs a magnetic field and radio waves to portray subtle structural differences. *(p. 23)*

PET scan (positron emission tomography): An imaging technique that shows the chemical functioning, as well as the structure, of an organ. *(p. 23)*

radiologist: A physician who specializes in performing and analyzing radiological procedures. *(p. 23)*

spiral-CT: A method of processing computerized tomography data to provide rapid, three-dimensional images of internal organs. *(p. 23)*

ultrasound: An imaging technique that uses brief bursts of high-frequency sound reflected by internal structures. *(p. 23)*

X rays: High-energy radiation that can penetrate living tissues. *(p. 23)*

AM Additional Terms Discussed in the *Applications Manual*

auscultation
diagnosis
infection
inspection
palpation
pathology
percussion

STUDY OUTLINE

1–1 INTRODUCTION: STUDYING THE HUMAN BODY 3

THE CHARACTERISTICS OF LIVING THINGS 4

1. **Biology** is the study of life. One of its goals is to discover the unity and patterns that underlie the diversity of organisms. *(Figure 1–1)*

2. All living things have certain common characteristics, including **organization, responsiveness, growth** and **differentiation, reproduction, movement,** and **metabolism** and **excretion**.

THE SCIENCES OF ANATOMY AND PHYSIOLOGY 4

3. **Anatomy** is the study of internal and external structures and the physical relationships among body parts. **Physiology** is the study of how living organisms perform vital functions. All specific functions are performed by specific structures.

ANATOMY 5

4. In *gross (macroscopic) anatomy*, we consider features that are visible without a microscope. This field includes *surface anatomy* (general form and superficial markings); *regional anatomy* (superficial and internal features in a specific area of the body); and *systemic anatomy* (structure of major organ systems). In *developmental anatomy*, we examine the changes in form that occur between conception and physical maturity. In *embryology*, we study developmental processes that occur during the first two months of development.

5. The boundaries of *microscopic anatomy* are established by the equipment used. In **cytology**, we analyze the internal structure of individual cells. In histology, we examine **tissues**, groups of cells that have specific functional roles. Tissues combine to form **organs**, anatomical units with multiple functions. Organs combine to form **organ systems**, groups of organs that function together.

PHYSIOLOGY 6

6. **Human physiology** is the study of the functions of the human body. It is based on *cell physiology*, the study of the functions of cells. In *special physiology*, we study the physiology of specific organs. In *systemic physiology*, we consider all aspects of the function of specific organ systems. In *pathological physiology*, we study the effects of diseases on organ or system functions.

1–2 LEVELS OF ORGANIZATION 6

1. Anatomical structures and physiological mechanisms are arranged in a series of interacting levels of organization. *(Figure 1–2)*

2. The 11 organ systems of the body are the integumentary, skeletal, muscular, nervous, endocrine, cardiovascular, lymphatic, respiratory, digestive, urinary, and reproductive systems. *(Figure 1–3)*

3. **Homeostasis** is the presence of a stable environment within the body.

Identifying bones: **Anatomy CD-ROM:** Skeletal System.

1–3 HOMEOSTASIS   11

1. Physiological systems preserve homeostasis through **homeostatic regulation.**
2. **Autoregulation** occurs when the activities of a cell, a tissue, an organ, or an organ system change automatically in response to an environmental change. **Extrinsic regulation** results from the activities of the nervous or endocrine system.
3. Homeostatic regulation usually involves a **receptor** that is sensitive to a particular stimulus, a **control center** that receives and processes the information from the receptor, and an **effector** whose activities are regulated by the control center and whose actions have a direct or indirect effect on the same stimulus.

NEGATIVE FEEDBACK   11

4. **Negative feedback** is a corrective mechanism involving an action that directly opposes a variation from normal limits. *(Figure 1–5)*

POSITIVE FEEDBACK   13

5. In **positive feedback**, a stimulus produces a response that exaggerates the stimulus, creating a *positive feedback loop*. *(Figure 1–6)*

SYSTEMS INTEGRATION   14

6. No one organ system has total control over the internal environmental characteristics; all systems work in concert. *(Table 1–1)*

1–4 A FRAME OF REFERENCE FOR ANATOMICAL STUDIES   15

SUPERFICIAL ANATOMY   15

1. Standard anatomical illustrations show the body in the **anatomical position**. If the figure is shown lying down, it can be either **supine** (face up) or **prone** (face down). *(Figure 1–7)*
2. **Abdominopelvic quadrants** and **abdominopelvic regions** represent two approaches to describing anatomical regions of the body. *(Figure 1–8; Table 1–2)*

3. The use of special directional terms provides clarity for the description of anatomical structures. *(Figure 1–9; Table 1–3)*

SECTIONAL ANATOMY   17

4. The three **sectional planes** (**frontal**, or *coronal*, **plane**; **sagittal plane**; and **transverse plane**) describe relationships among the parts of the three-dimensional human body. *(Figure 1–10; Table 1–4)*

BODY CAVITIES   19

5. *Body cavities* protect delicate organs and permit changes in the size and shape of internal organs. The term *dorsal body cavity* refers to the **cranial cavity** (enclosing the brain) and the **spinal cavity** (surrounding the spinal cord). The **ventral body cavity**, or *coelom*, surrounds developing respiratory, cardiovascular, digestive, urinary, and reproductive organs. *(Figures 1–11, 1–12)*
6. The **diaphragm** divides the ventral body cavity into the (superior) **thoracic** and (inferior) **abdominopelvic cavities**. The thoracic cavity consists of two **pleural cavities** (each containing a lung), separated by the **mediastinum**. Within the mediastinum is the **pericardial cavity**, which contains the heart. The abdominopelvic cavity consists of the **abdominal cavity** and the **pelvic cavity** and contains the *peritoneal cavity*, an internal chamber lined by the *peritoneum*, a *serous membrane*. *(Figure 1–12)*

 Identifying structures: **Anatomy CD-ROM:** Digestive System Dissection and Cardiovascular System/Heart Dissection.

FOCUS: Sectional Anatomy and Clinical Technology   23

7. Important *radiological procedures* (which can provide detailed information about internal systems) include **CT**, **MRI**, **ultrasound**, **X rays**, **spiral-CT**, and **DSA**. *(Figures 1–13, 1–14, 1–15)*

REVIEW QUESTIONS

More assessment questions are available to you on the Companion Website. You will find Matching, Multiple Choice, True/False, and other quizzes to help further your understanding of the material covered in this chapter. To access the site, go to www.aw.com/martini.

LEVEL 1 Reviewing Facts and Terms

Match each numbered item with the most closely related lettered item. Use letters for answers in the spaces provided.

_____ 1. cytology	a. study of tissues
_____ 2. physiology	b. constant internal environment
_____ 3. histology	c. face up
_____ 4. metabolism	d. study of functions
_____ 5. homeostasis	e. positive feedback
_____ 6. muscle	f. organ system
_____ 7. heart	g. study of cells
_____ 8. endocrine	h. negative feedback
_____ 9. temperature regulation	i. brain and spinal cord
_____ 10. labor and delivery	j. all chemical activity in body
_____ 11. supine	k. thoracic and abdominopelvic
_____ 12. prone	l. tissue
_____ 13. ventral body cavity	m. serous membrane
_____ 14. dorsal body cavity	n. organ
_____ 15. pericardium	o. face down

16. The process by which an organism increases the size or number of its cells is
    (a) reproduction
    (b) adaptation
    (c) growth
    (d) metabolism

17. The following is a list of six levels of organization that make up the human body:

    (1) tissue   (2) cell
    (3) organ   (4) molecule
    (5) organism   (6) organ system

    The correct order, from the smallest to the largest level, is
    (a) 2, 4, 1, 3, 6, 5
    (b) 4, 2, 1, 3, 6, 5
    (c) 4, 2, 1, 6, 3, 5
    (d) 4, 2, 3, 1, 6, 5
    (e) 2, 1, 4, 3, 5, 6

18. The study of internal and external structures and the physical relationships among body parts is
    (a) histology   (b) anatomy
    (c) physiology   (d) embryology

19. The specialist who attempts to determine the physical and chemical processes responsible for vital functions is a
    (a) physiologist
    (b) physician
    (c) pathologist
    (d) cytologist

20. The relative stability of an organism's internal environment is
    (a) homeopathy
    (b) uniformity
    (c) equilibrium
    (d) homeostasis

21. Which of the following is a characteristic of complex living organisms such as humans, but not simpler life-forms, such as bacteria or single-celled protozoans?
    (a) digestion
    (b) organization
    (c) metabolism
    (d) responsiveness

22. Crisis management by directing rapid, short-term, specific responses is a function of the
    (a) endocrine system
    (b) nervous system
    (c) hormones
    (d) autoregulatory system

23. The breakdown and absorption of nutrients is performed by the organs of the
    (a) excretory system
    (b) circulatory system
    (c) respiratory system
    (d) digestive system

24. When a variation outside normal limits triggers a response that restores the normal condition, the regulatory mechanism involves
    (a) negative feedback
    (b) positive feedback
    (c) compensation
    (d) adaptation

25. In positive feedback, the initial stimulus produces a response that
    (a) suppresses the stimulus
    (b) has no effect on the stimulus
    (c) interferes with the completion of the process
    (d) exaggerates the stimulus

26. Failure of homeostatic regulation in the body results in
    (a) autoregulation   (b) extrinsic regulation
    (c) disease   (d) positive feedback

27. The terms that apply to the front of the body when it is in the anatomical position are
    (a) posterior, dorsal   (b) back, front
    (c) medial, lateral   (d) anterior, ventral

28. A plane through the body that passes perpendicular to the long axis of the body and divides the body into a superior and an inferior section is a
    (a) sagittal section
    (b) transverse section
    (c) coronal section
    (d) frontal section

29. Gross anatomy includes all the following subspecialties, except
    (a) surface anatomy
    (b) developmental anatomy
    (c) histological anatomy
    (d) radiographic anatomy

30. The diaphragm divides the ventral body cavity into a superior _____ cavity and an inferior _____ cavity.
    (a) pleural, pericardial
    (b) abdominal, pelvic
    (c) thoracic, abdominopelvic
    (d) cranial, thoracic

31. The mediastinum is the region between the
    (a) lungs and heart
    (b) two pleural cavities
    (c) chest and abdomen
    (d) heart and pericardium

LEVEL 2 Reviewing Concepts

32. What basic functions do all living things perform?

33. (a) Define *anatomy*.
    (b) Define *physiology*.

34. Why is it important to understand the different levels of organization in the human body?

35. What is the role of serous membranes in the body?

36. What is homeostatic regulation, and what is its physiological importance?

37. What distinguishes autoregulation from extrinsic regulation?

38. What necessary components are involved in homeostatic regulation? What is the function of each?

39. How does negative feedback differ from positive feedback?

40. Describe the position of the body when it is in the anatomical position.

41. Name the two upper abdominal quadrants and list the organs that lie in each.

42. As a surgeon, you perform an invasive (surgical) procedure that requires you to cut through the pleura. What body cavity are you operating on, and what specific organ are you exposing?

43. In which body cavity would each of the following organs or organ systems be?
    (a) brain, spinal cord
    (b) cardiovascular, digestive, and urinary systems
    (c) heart, lungs
    (d) stomach, intestines

44. What differentiates sectional anatomy from superficial anatomy?

LEVEL 3 Critical Thinking and Clinical Applications

45. The hormone *calcitonin* is released from the thyroid gland in response to increased levels of calcium ions in the blood. If this hormone is controlled by negative feedback, what effect would calcitonin have on blood calcium levels?

46. During exercise, blood flow to skeletal muscles increases. The initial response that increases blood flow is automatic and independent of the nervous and endocrine systems. Which type of homeostatic regulation is this? Why?

47. A chemical imbalance in a heart muscle cell can cause the heart to cease pumping blood; the cessation of blood flow will in turn cause other tissues and organs to cease functioning. This observation supports the view that
(a) all organisms are composed of cells
(b) all levels of organization within an organism are interdependent
(c) chemical molecules make up cells
(d) all cells are independent of each other
(e) congenital defects can be life threatening

UNIT 1   CHAPTER 1 **2** 3 4

CLINICAL DISCUSSION
- Fatty Acids and Health   50

THE CHEMICAL LEVEL OF ORGANIZATION

Water is so common, and seems so ordinary, that it is often taken for granted. Yet all living things, including humans, are totally dependent on water. You can survive for weeks without food, but for only a few days without water. The basic reason may surprise you: In large part, you *are* water. Water makes up almost two-thirds of your body weight.

Chemically, water is a simple substance: two atoms of hydrogen joined to one atom of oxygen. Yet, when these three atoms are linked by chemical bonds, they produce a substance with many unusual properties. In this chapter, you will learn how atoms are bound together to form molecules, the building blocks of cells. You will also meet the larger molecules that form the structural framework of the body and that enable our cells to grow, divide, and perform other essential functions.

2–1 ATOMS, MOLECULES, AND BONDS

Objectives

- Describe an atom and how atomic structure affects interactions between atoms.
- Compare the ways in which atoms combine to form molecules and compounds.

Our study of the human body begins at the chemical level of organization. *Chemistry* is the science that deals with the structure of *matter*, defined as anything that takes up space and has mass. *Mass* is a physical property that determines the weight of an object in Earth's gravitational field. For our purposes, the mass of an object is the same as its weight. If we were to ride on the space shuttle, however, we would find that the two are not always equivalent: In orbit we would be weightless, but our mass would remain unchanged.

The smallest stable units of matter are called **atoms**. Air, elephants, oranges, oceans, rocks, and people are all composed of atoms in varying combinations. The unique characteristics of each object, living or nonliving, result from the types of atoms involved and the ways those atoms combine and interact. You will need to become familiar with the chemical principles governing those interactions in order to understand the anatomy and physiology of the cells, tissues, organs, and organ systems of the human body.

Atoms are composed of **subatomic particles**. Although many different subatomic particles exist, only three are important for understanding the chemical properties of matter: *protons, neutrons*, and *electrons*. Protons and neutrons are similar in size and mass, but **protons** (p^+) bear a positive electrical charge, whereas **neutrons** (n or n^0) are electrically *neutral*, or uncharged. **Electrons** (e^-) are much lighter than protons—only 1/1836th as massive—and bear a negative electrical charge. The mass of an atom is therefore determined primarily by the number of protons and neutrons in the nucleus. The mass of a large object, such as your body, is the sum of the masses of all the component atoms.

☐ ATOMIC STRUCTURE

Atoms normally contain equal numbers of protons and electrons. The number of protons in an atom is known as its **atomic number**. *Hydrogen* (H) is the simplest atom, with an atomic number of 1. Thus, an atom of hydrogen contains one proton, and it contains one electron as well. The proton is located in the center of the atom and forms the nucleus. Hydrogen atoms seldom contain neutrons, but when neutrons are present, they are also located in the nucleus. All atoms other than hydrogen have neutrons as well as protons in their nuclei.

The electron orbits the nucleus at high speed, forming a spherical **electron cloud**. For convenience, we often illustrate atomic structure in the simplified form shown in Figure 2–1a●. In this two-dimensional representation, the electrons occupy a circular **electron shell**. One reason an electron tends to remain in its electron shell is that the negatively charged electron is attracted to the positively charged proton. The attraction between opposite electrical charges is an example of an *electrical force.* As you will see in later chapters, electrical forces are involved in many physiological processes.

The dimensions of the electron cloud determine the overall size of the atom. To get an idea of the scale involved, consider that if the nucleus were the size of a tennis ball, the electron cloud would have a radius of 10 km (roughly 6 miles!). In reality, atoms are so small that atomic measurements are most conveniently reported in *nanometers* (NA-nō-mē-terz) (nm). One nanometer is 10^{-9} meter (0.000000001 m). The very largest atoms approach 0.5 nm in diameter (0.0000000005 m, or 0.00000002 in.).

Elements and Isotopes

Atoms can be classified on the basis of their atomic number into groups called **elements**. Each element includes all the atoms that have the same number of protons, and thus the same atomic number. Only 92 elements exist in nature, although about two dozen additional elements have been created through nuclear reactions in research laboratories. Every element has a chemical symbol, an abbreviation recognized by scientists everywhere. Most of the symbols are easily connected with the English names of the elements (O for oxygen, N for nitrogen, C for carbon, and so on), but a few are abbreviations of their Latin names. For example, the symbol for sodium, Na, comes from the Latin word *natrium*. (Appendix II, the *periodic table*, gives the chemical symbols and atomic numbers of each element.)

Because atomic nuclei are unaltered by ordinary chemical processes, elements cannot be changed or broken down into simpler substances in chemical reactions. Thus, an atom of carbon always remains an atom of carbon, regardless of the chemical events in which it may take part.

The relative contributions of the 13 most abundant elements in the human body to the total body weight are shown in Table 2–1. The human body also contains atoms of another 13 elements—called *trace elements*—that are present in such small amounts that percentage values are not shown.

The atoms of a single element can differ in terms of the number of neutrons in the nucleus. For example, although most hydrogen nuclei consist of a single proton, 0.015 percent also contain one neutron, and a very small percentage contain two neutrons. Atoms whose nuclei contain the same number of protons, but different numbers of neutrons, are called **isotopes**. Different isotopes of an element have essentially identical chemical properties, and so are indistinguishable except on the basis of mass. The **mass number**—the total number of protons plus neutrons in the nucleus—is used to designate a particular isotope. Thus, the three isotopes of hydrogen are hydrogen-1, or ^{1}H, with one proton and one electron (Figure 2–1a●); hydrogen-2, or ^{2}H, also known as *deuterium*, with one proton, one

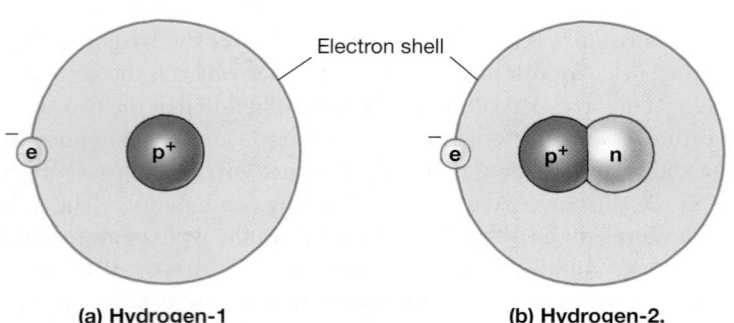
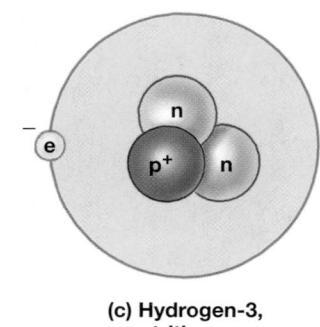

| (a) Hydrogen-1 (electron-shell model) | (b) Hydrogen-2, deuterium | (c) Hydrogen-3, tritium |

●**FIGURE 2–1**
Hydrogen Atoms. **(a)** A two-dimensional model depicting the electron in an electron shell around the nucleus lets us visualize the components of the atom. A typical hydrogen nucleus contains a proton and no neutrons. **(b)** A deuterium (^{2}H) nucleus contains a proton and a neutron. **(c)** A tritium (^{3}H) nucleus contains a pair of neutrons in addition to the proton.

electron, and one neutron (Figure 2–1b●); and hydrogen-3, or ^{3}H, also known as *tritium*, with one proton, one electron, and two neutrons (Figure 2–1c●).

TABLE 2–1 PRINCIPAL ELEMENTS IN THE HUMAN BODY	
Element **(% of total body weight)**	**Significance**
Oxygen (65)	A component of water and other compounds; gaseous form is essential for respiration
Carbon (18.6)	Found in all organic molecules
Hydrogen (9.7)	A component of water and most other compounds in the body
Nitrogen (3.2)	Found in proteins, nucleic acids, and other organic compounds
Calcium (1.8)	Found in bones and teeth; important for membrane function, nerve impulses, muscle contraction, and blood clotting
Phosphorus (1.0)	Found in bones and teeth, nucleic acids, and high-energy compounds
Potassium (0.4)	Important for proper membrane function, nerve impulses, and muscle contraction
Sodium (0.2)	Important for membrane function, nerve impulses, and muscle contraction
Chlorine (0.2)	Important for membrane function and water absorption
Magnesium (0.06)	A cofactor for several enzymes
Sulfur (0.04)	Found in many proteins
Iron (0.007)	Essential for oxygen transport and energy capture
Iodine (0.0002)	A component of hormones of the thyroid gland
Trace elements: silicon (Si), fluorine (F), copper (Cu), manganese (Mn), zinc (Zn), selenium (Se), cobalt (Co), molybdenum (Mo), cadmium (Cd), chromium (Cr), tin (Sn), aluminum (Al), and boron (B)	Some function as cofactors; the functions of many trace elements are poorly understood

The nuclei of some isotopes spontaneously emit subatomic particles or radiation in measurable amounts. Such isotopes are called **radioisotopes**, and the emission process is called *radioactive decay*. Strongly radioactive isotopes are dangerous, because the emissions can break molecules apart, destroy cells, and otherwise damage living tissues. Weakly radioactive isotopes are sometimes used in diagnostic procedures to monitor the structural or functional characteristics of internal organs.

Radioisotopes differ in how rapidly they decay. The decay rate of a radioisotope is commonly expressed in terms of its **half-life**: the time required for half of a given amount of the isotope to decay. The half-lives of radioisotopes range from fractions of a second to billions of years. **AM** Medical Use of Radioisotopes

Atomic Weights

A typical atom of *oxygen*, which has an atomic number of 8, contains eight protons, but it also contains eight neutrons. The mass number of this isotope is therefore 16. The mass numbers of other isotopes of oxygen depend on the number of neutrons present. Mass numbers are useful because they tell us the number of subatomic particles in the nuclei of different atoms. However, they do not tell us the *actual* mass of the atoms. For example, they do not take into account the masses of the electrons or the slight difference between the mass of a proton and that of a neutron. The actual mass of an atom is known as its **atomic weight**.

The unit used to express the atomic weight is the **dalton** (also known as the *atomic mass unit*, or *amu*). One dalton is very close to the weight of a single proton. Thus, the atomic weight of the most common isotope of hydrogen is very close to 1, and that of the most common isotope of oxygen is very close to 16.

The atomic weight of an element is an average mass number that reflects the proportions of different isotopes. For example, the atomic number of hydrogen is 1, but the atomic weight of hydrogen is 1.0079, primarily because some hydrogen atoms (0.015 percent) have a mass number of 2, and even fewer have a mass number of 3. The atomic weight of an element is usually very close to the mass number of the most common isotope of that element. The atomic weights of the elements are included in Appendix II.

Atoms participate in chemical reactions in fixed numerical ratios. To form water, for example, exactly two atoms of hydrogen combine with each atom of oxygen. But individual atoms are far too small and too numerous to be counted. To overcome this difficulty, chemists use a unit called the *mole*. For any element, a **mole** (abbreviated *mol*) is a quantity with a weight in grams equal to that elements atomic weight. The reason that the mole is so useful is that one mole of a given element always contains the same number of atoms as one mole of any other element. That number (called *Avogadro's number*) is 6.023×10^{23}, or about 600 billion trillion. Expressing relationships in moles rather than in grams makes it much easier to keep track of the relative numbers of atoms in chemical samples and processes. For example, if a report stated that a sample contains 0.5 mol of hydrogen atoms and 0.5 mol of oxygen atoms, you would know immediately that they were present in equal numbers. That would not be so evident if the report stated that there were 0.505 g of hydrogen atoms and 8.00 g of oxygen atoms. Most chemical analysis and clinical laboratory tests report data in moles (mol), millimoles (mmol—1/1000 mol, or 10^{-3} mol), or micromoles (μmol—1/1,000,000 mol, or 10^{-6} mol). (Additional information on metric weights and measures can be found in Appendix I.)

Electrons and Energy Levels

Atoms are electrically neutral; every positively charged proton is balanced by a negatively charged electron. Thus, each increase in the atomic number is accompanied by a comparable increase in the number of electrons orbiting the nucleus. Within the electron cloud, electrons occupy an orderly series of **energy levels**. Although the electrons in an energy level may travel in complex orbits around the nucleus, for our purposes the orbits can be diagrammed as a series of concentric electron shells. The first electron shell (the one closest to the nucleus) corresponds to the lowest energy level.

Each energy level is limited in the number of electrons it can hold. The first energy level can hold at most two electrons, while the next two levels can each hold up to eight electrons. The electrons in an atom occupy successive shells in an orderly manner: The first energy level is filled before any electrons enter the second, and the second energy level is filled before any electrons enter the third.

The outermost energy level forms the "surface" of the atom. The number of electrons in this level determines the chemical properties of the element. Atoms with unfilled energy levels are unstable—that is, they will react with other atoms, usually in ways that give them full outer electron shells. In contrast, atoms with a filled outermost energy level are stable and therefore do not readily react with other atoms.

As indicated in Figure 2–2●, a hydrogen atom has one electron in the first energy level, and the level is thus unfilled. A hydrogen atom readily reacts with other hydrogen atoms or with the atoms of other elements. A *helium* atom has two electrons in its first energy level. Because its outer energy level is filled, a helium atom is very stable; it will not ordinarily react with other atoms. A *lithium* atom has three electrons. Its first energy level can hold only two of them, so lithium has a single electron in a second, unfilled energy level. As you would expect, lithium is extremely reactive. The second energy level is filled in a *neon* atom, with an atomic number of 10. Neon atoms, like helium atoms, are thus very stable. The atoms that are most important to biological systems are *un*stable, because those are the atoms that interact to form larger structures. The elements most important to the human body are listed in Table 2–1. (Further information on these and other elements is given in Appendix II.)

☐ CHEMICAL BONDS

Elements that do not readily participate in chemical processes are said to be *inert*. Helium, neon, and argon have filled outermost energy levels. These elements are called *inert gases*, because their atoms neither react with one another nor combine with atoms of

other elements. Elements with unfilled outermost energy levels, such as hydrogen and lithium, are called *reactive*, because they readily interact or combine with other atoms. In doing so, these atoms achieve stability by gaining, losing, or sharing electrons to fill their outermost energy level. The interactions often involve the formation of **chemical bonds**, which hold the participating atoms together once the reaction has ended. In the sections that follow, we will consider three basic types of chemical bonds: *ionic bonds, covalent bonds*, and *hydrogen bonds*.

When chemical bonding occurs, the result is the creation of new chemical entities that we call *molecules* and *compounds*. The term **molecule** refers to any chemical structure consisting of atoms held together by covalent bonds. A **compound** is any chemical substance made up of atoms of two or more elements, regardless of the type of bond joining them. The two categories overlap, but they aren't the same. Not all molecules are compounds, because some molecules consist of atoms of only one element. (Two oxygen atoms, for example, can be joined by a covalent bond to form a molecule of oxygen.) And not all compounds consist of molecules, because some compounds, such as ordinary salt (sodium chloride) are held together by ionic rather than covalent bonds. Many substances, however, belong to both categories. Water is a compound because it contains two different

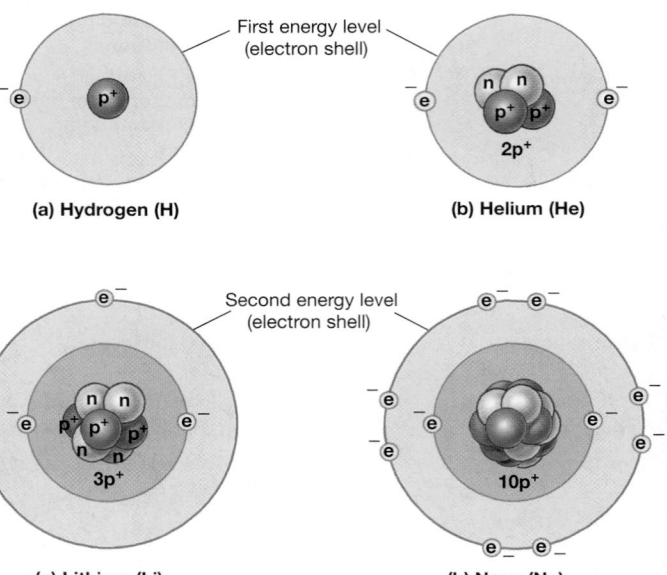

(a) Hydrogen (H)   **(b) Helium (He)**

(c) Lithium (Li)   **(b) Neon (Ne)**

●**FIGURE 2–2**
Atoms and Energy Levels. **(a)** A typical hydrogen atom has one proton and one electron. The electron orbiting the nucleus occupies the first energy level, diagrammed as an electron shell. **(b)** An atom of helium has two protons, two neutrons, and two electrons. The two electrons orbit in the same energy level. **(c)** The first energy level can hold only two electrons. In a lithium atom, with three protons, four neutrons, and three electrons, the third electron occupies a second energy level. **(d)** The second level can hold up to eight electrons. A neon atom has 10 protons, 10 neutrons, and 10 electrons; thus, both the first and second energy levels are filled.

elements—hydrogen and oxygen—and it consists of molecules, because the hydrogen and oxygen atoms are held together by co-valent bonds. As we shall see in subsequent sections, most biolog-ically important compounds, from water to DNA, are molecular.

Regardless of the type of bonding involved, a chemical com-pound has properties that can be quite different from those of its components. For example, a mixture of hydrogen and oxygen is highly flammable, but the chemical combination of hydrogen and oxygen atoms produces water, a compound used to put out fires.

The human body consists of countless molecules and com-pounds, so it is a challenge to describe these substances and their varied interactions. Fortunately, chemists rely on a standardized system of *chemical notation*. The rules of this system are given in "FOCUS: Chemical Notation" on p. 38, and you will find them useful in this chapter and in later chapters.

Ionic Bonds

As the name implies, ionic bonds form between ions. **Ions** are atoms or molecules that carry an electric charge, either positive or negative. Ions with a positive charge ($+$) are called **cations** (KAT-ī-onz); ions with a negative charge ($-$) are called **anions** (AN-ī-onz). **Ionic bonds** are chemical bonds created by the elec-trical attraction between anions and cations.

Ions have an unequal number of protons and electrons. Atoms become ions by losing or gaining electrons. If we assign a value of $+1$ to the charge on a proton, the charge on an electron is -1. If the number of protons is equal to the number of elec-trons, an atom is electrically neutral. An atom that loses an elec-tron becomes a cation with a charge of $+1$, because it then has one proton that lacks a corresponding electron. Losing a second electron would give the cation a charge of $+2$. Adding an extra electron to a neutral atom produces an anion with a charge of -1; adding a second electron gives the anion a charge of -2.

In the formation of an ionic bond,

- one atom—the *electron donor*—loses one or more electrons and becomes a cation, with a positive ($+$) charge.

- another atom—the *electron acceptor*—gains those same elec-trons and becomes an anion, with a negative ($-$) charge.

- attraction between the opposite charges then draws the two ions together.

The formation of an ionic bond is illustrated in Figure 2–3a●. The sodium atom diagrammed in STEP 1 has an atomic number of 11, so this atom normally contains 11 protons and 11 elec-trons. (Because neutrons are electrically neutral, their presence has no effect on the formation of ions or ionic bonds.) Electrons fill the first and second energy levels, and a single electron occu-pies the outermost level. Losing that one electron would give the sodium atom a full outermost energy level—the second level—and would produce a **sodium ion**, with a charge of $+1$. (The chemical shorthand for sodium abbreviates this ion as Na^+.) But a sodium atom cannot simply throw away the electron: The elec-tron must be donated to an electron acceptor. A chlorine atom has seven electrons in its outermost energy level, so it needs only

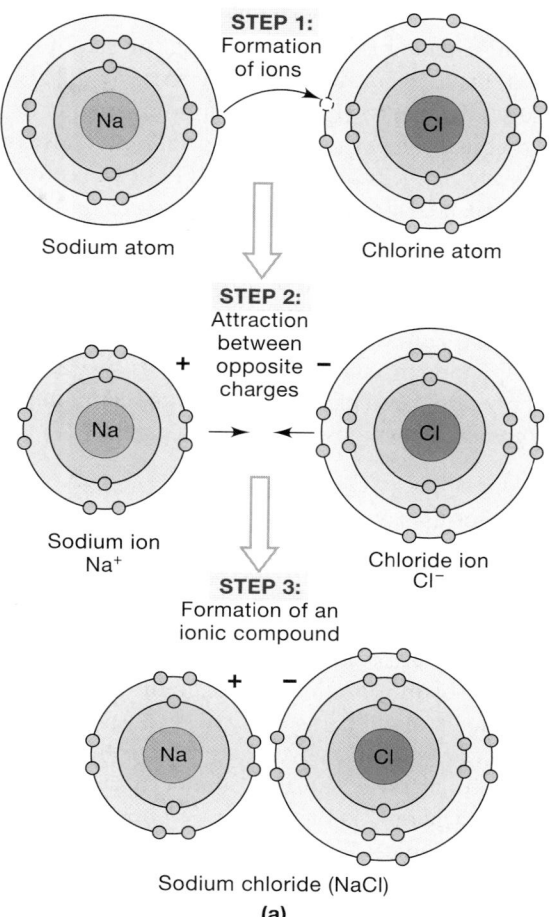

STEP 1: Formation of ions

Sodium atom   Chlorine atom

STEP 2: Attraction between opposite charges

Sodium ion Na^+ Chloride ion Cl^-

STEP 3: Formation of an ionic compound

Sodium chloride (NaCl)

(a)

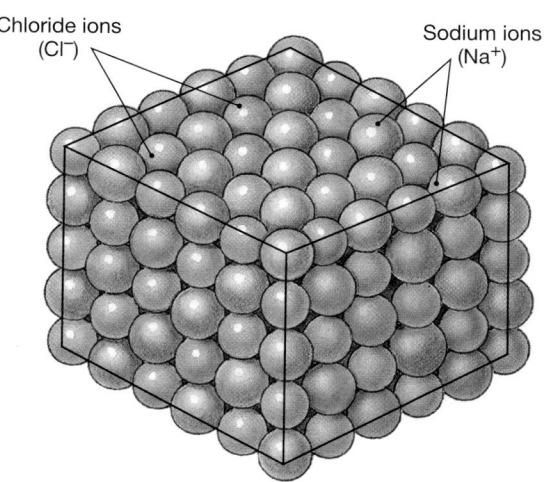

Chloride ions (Cl^-)   Sodium ions (Na^+)

(b) Sodium chloride crystal

●**FIGURE 2–3**
Ionic Bonding. **(a) STEP 1:** A sodium (Na) atom loses an elec-tron, which is accepted by a chlorine (Cl) atom. **STEP 2:** Because the sodium (Na^+) and chloride (Cl^-) ions have opposite charges, they are attracted to one another. **STEP 3:** The association of sodium and chloride ions forms the ionic compound sodium chloride. **(b)** Large numbers of sodium and chloride ions form a crystal of sodium chlo-ride (table salt).

one electron to achieve stability. A sodium atom can provide the extra electron. In the process (STEP 2), the chlorine atom becomes a **chloride ion** (Cl^-), with a charge of -1.

Both atoms have now become stable ions with filled outermost energy levels. But the two ions do not move apart after the electron transfer, because the positively charged sodium ion is attracted to the negatively charged chloride ion (STEP 3). The combination of oppositely charged ions forms an *ionic compound*—in this case **sodium chloride**, otherwise known as table salt. Large numbers of sodium and chloride ions interact to form highly structured crystals, held together only by the electrical attraction of oppositely charged ions (Figure 2–3b●). Although sodium chloride and other ionic compounds are common in body fluids, they are not present as intact crystals. When placed in water, ionic compounds dissolve, and the component anions and cations separate.

Covalent Bonds

Some atoms can complete their outer electron shells not by gaining or losing electrons, but by sharing electrons with other atoms. Such sharing creates **covalent** (KŌ-vā-lent) **bonds** between the atoms involved.

Individual hydrogen atoms, as diagrammed in Figure 2–2a●, do not exist in nature. Instead, we find hydrogen molecules (Figure 2–4a●). Molecular hydrogen consists of a pair of hydrogen atoms. In chemical shorthand, molecular hydrogen is indicated by H_2, where H is the chemical symbol for hydrogen and the subscript 2 indicates the number of atoms. Molecular hydrogen is a gas that is present in the atmosphere in very small quantities. The two hydrogen atoms share their electrons, and each electron whirls around both nuclei. The sharing of one pair of electrons creates a **single covalent bond** (—).

Oxygen, with an atomic number of 8, has two electrons in its first energy level and six in its second. The oxygen atoms diagrammed in Figure 2–4b● attain a stable electron configuration by sharing two pairs of electrons, thereby forming a **double covalent bond**. In a structural formula, a double covalent bond is represented by two lines (=). Molecular oxygen (O_2) is an atmospheric gas that is very important to most organisms. Our cells would die without a relatively constant supply of oxygen.

In our bodies, chemical processes that consume oxygen generally also produce **carbon dioxide** (CO_2) as a waste product. The oxygen atoms in a carbon dioxide molecule form double covalent bonds with the carbon atom (Figure 2–4c●).

A triple covalent bond, such as the one joining two nitrogen molecules (N_2), is indicated by three lines (≡). Molecular nitrogen accounts for roughly 79 percent of our planet's atmosphere, but our cells ignore it completely. In fact, deep-sea divers live for long periods while breathing artificial air that does not contain nitrogen. (We will discuss the reasons for eliminating nitrogen under these conditions in Chapter 23.)

Covalent bonds usually form molecules that complete the outer energy levels of the atoms involved. An ion or molecule that contains unpaired electrons in its outermost energy level is called a *free radical*. Free radicals are highly reactive. Almost as

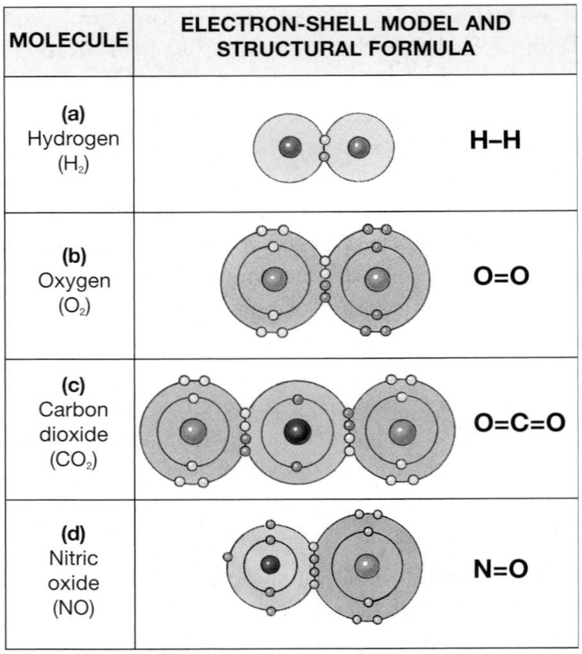

MOLECULE	ELECTRON-SHELL MODEL AND STRUCTURAL FORMULA
(a) Hydrogen (H_2)	H–H
(b) Oxygen (O_2)	O=O
(c) Carbon dioxide (CO_2)	O=C=O
(d) Nitric oxide (NO)	N=O

●**FIGURE 2–4**
Covalent Bonds. **(a)** In a hydrogen molecule, two hydrogen atoms share electrons such that each atom has a filled outermost electron shell. This sharing creates a single covalent bond. **(b)** In an oxygen molecule, two oxygen atoms share two pairs of electrons. The result is a double covalent bond. **(c)** In a carbon dioxide molecule, a central carbon atom forms double covalent bonds with two oxygen atoms. **(d)** A nitric oxide molecule is held together by a double covalent bond, but the outer electron shell of the nitrogen atom requires an additional electron to be complete. Thus, nitric oxide is a free radical, which reacts readily with another atom or molecule.

fast as it forms, a free radical enters additional reactions that are typically destructive. For example, free radicals can damage or destroy vital compounds, such as proteins. Free radicals sometimes form in the course of normal metabolism, but cells have several methods of removing or inactivating them. However, *nitric oxide* (NO) (Figure 2–4d●) is a free radical that has important functions in the body. It is involved in chemical communication in the nervous system, in the control of blood vessel diameter, in blood clotting, and in the defense against bacteria and other pathogens. It has been suggested that the cumulative damage produced by free radicals inside and outside our cells is a major factor in the aging process.

NONPOLAR COVALENT BONDS Covalent bonds are very strong, because the shared electrons hold the atoms together. In typical covalent bonds the atoms remain electrically neutral, because each shared electron spends just as much time "at home" as away. (If you and a friend were tossing a pair of baseballs back and forth as fast as you could, on average, each of you would have just one baseball.) Many covalent bonds involve an equal sharing of electrons. Such bonds, which occur, for instance, between two atoms of the same type, are called **nonpolar covalent bonds**. Nonpolar covalent bonds, especially those between carbon atoms, create the stable framework of the large molecules that make up most of the structural components of the human body.

POLAR COVALENT BONDS Covalent bonds involving different types of atoms may instead involve an unequal sharing of electrons, because the elements differ in how strongly they attract electrons. An unequal sharing of electrons creates a **polar covalent bond**. For example, in a molecule of water (Figure 2–5●), an oxygen atom forms covalent bonds with two hydrogen atoms. The oxygen nucleus has a much stronger attraction for the shared electrons than the hydrogen atoms do. As a result, the electrons spend more time orbiting the oxygen nucleus than orbiting the hydrogen nuclei. Because it has two extra electrons most of the time, the oxygen atom develops a slight (partial) negative charge, indicated by δ^-. At the same time, each hydrogen atom develops a slight (partial) positive charge, δ^+, because its electron is away much of the time. (Suppose you and a friend were tossing a pair of baseballs back and forth, but one of you returned them back as fast as possible, and the other held onto them for a while before throwing them back. One of you would now, on average, have more than one baseball, and the other would have fewer than one.) The unequal sharing of electrons makes polar covalent bonds somewhat weaker than nonpolar covalent bonds. Polar covalent bonds often create *polar molecules*—molecules that have positive and negative ends. Polar molecules have very interesting properties; we will consider the characteristics of the most important polar molecule in the body, water, in a later section.

Hydrogen Bonds

Covalent and ionic bonds tie atoms together to form molecules and compounds. There are also comparatively weak forces that act between adjacent molecules and even between atoms within a large molecule. The most important of these weak attractive forces is the **hydrogen bond**. A hydrogen bond is the attraction between a δ^+ on the hydrogen atom of a polar covalent bond and a δ^- on an oxygen or nitrogen atom of another polar covalent bond. The polar covalent bond containing the oxygen or nitrogen atom can be in a different molecule from, or in the same molecule as, the hydrogen atom.

Hydrogen bonds are too weak to create molecules, but they can change molecular shapes or pull molecules together. For example, hydrogen bonding occurs between water molecules (Figure 2–6●). At the water surface, the attraction between molecules slows the rate of evaporation and creates the phenomenon known as surface tension. **Surface tension** acts as a barrier that keeps small objects from entering the water. For example, it allows insects to walk across the surface of a pond or puddle. Surface tension in a layer of tears keeps small objects such as dust particles from touching the surface of the eye. At the cellular level, hydrogen bonds affect the shapes and properties of complex molecules, such as proteins and nucleic acids (including DNA), and they may also determine the three-dimensional relationships between molecules.

States of Matter

Matter in our environment exists in one of three states: solid, liquid, or gaseous. *Solids* maintain their volume and shape at ordinary temperatures and pressures. A lump of granite, a brick, and a textbook are solid objects. *Liquids* have a constant volume, but no fixed shape. The shape of a liquid is determined by the shape of its container. Water, coffee, and soda are liquids. A *gas* has neither a

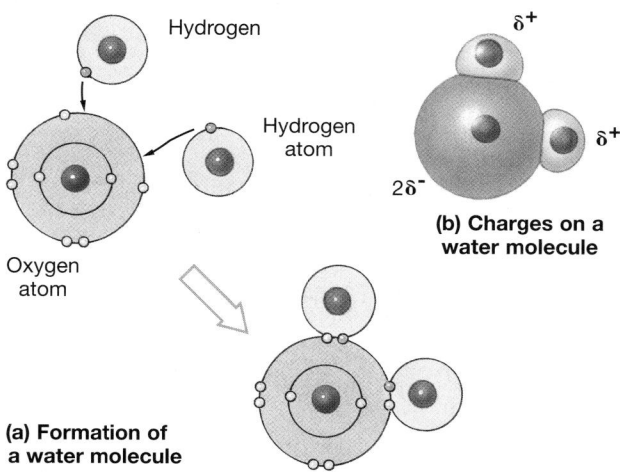

●**FIGURE 2–5**
Polar Covalent Bonds and the Structure of Water.   **(a)** In forming a water molecule, an oxygen atom completes its outermost energy level by sharing electrons with a pair of hydrogen atoms. The sharing is unequal, because the oxygen atom holds the electrons more tightly than do the hydrogen atoms. **(b)** Because the oxygen atom has two extra electrons much of the time, it develops a slight negative charge, and the hydrogen atoms become weakly positive. The bonds in a water molecule are polar covalent bonds.

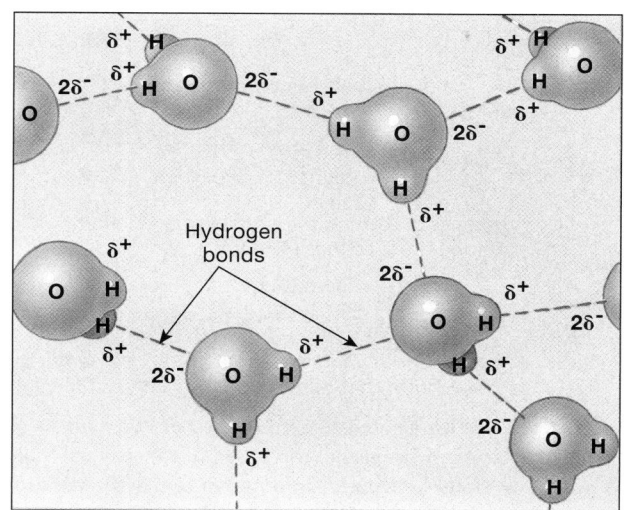

●**FIGURE 2–6**
Hydrogen Bonds.   The hydrogen atoms of a water molecule have a slight positive charge, and the oxygen atom has a slight negative charge. (*See Figure 2-5b.*) The attraction between a hydrogen atom of one water molecule and the oxygen atom of another is a hydrogen bond.

constant volume nor a fixed shape. Gases can be compressed or expanded, and they will fill a container of any shape. The air of our atmosphere is the gas with which we are most familiar.

Whether a particular substance is a solid, a liquid, or a gas depends on the degree of interaction among its atoms or molecules. For example, hydrogen molecules have little attraction for one another; in our environment, molecular hydrogen exists as a gas. Water is the only substance that occurs as a solid (ice), a liquid (water), and a gas (water vapor) at temperatures compatible with life. Water exists in the liquid state over a broad range of temperatures primarily because of hydrogen bonding among the water molecules. We will talk more about the unusual properties of water in a later section.

Molecular Weights

The **molecular weight** of a molecule is the sum of the atomic weights of its component atoms. It follows from the definition of the mole given previously that the molecular weight of a molecule in grams is equal to the weight of one mole of molecules. Molecular weights are important because you cannot handle individual molecules; nor could you easily count the billions of molecules involved in chemical reactions in the body. Using molecular weights, you can calculate the quantities of reactants needed to perform a specific re-

action and determine the amount of product generated. For example, suppose you want to create water from hydrogen and oxygen according to the equation

$$2H_2 + O_2 \rightarrow 2H_2O$$

The first step in performing such an experiment would be to calculate the molecular weights involved. The atomic weight of hydrogen is close to 1.0, so one hydrogen molecule (H_2) has a molecular weight near 2.0. Oxygen has an atomic weight of about 16, so the molecular weight of an oxygen molecule (O_2) is roughly 32. To perform the experiment, you would combine 4 g of hydrogen with 32 g of oxygen to produce 36 g of water. You could also work with ounces, pounds, or tons, as long as the proportions remained the same.

✔ Both oxygen and neon are gases at room temperature. Oxygen combines readily with other elements, but neon does not. Why?

✔ How is it possible for two samples of hydrogen to contain the same number of atoms, yet have different weights?

✔ Which kind of bond holds atoms in a water molecule together? What attracts water molecules to one another?

Answers start on page Q-1

---

focus | Chemical Notation

Before we can consider the specific compounds that occur in the human body, we must be able to describe chemical compounds and reactions effectively. The use of sentences to describe chemical structures and events often leads to confusion. A simple form of "chemical shorthand" makes communication much more efficient. The chemical shorthand we shall use is known as **chemical notation**. Chemical notation enables us to describe complex events briefly and precisely; its rules are summarized in Table 2–2.

TABLE 2–2 RULES OF CHEMICAL NOTATION

1. The symbol of an element indicates one atom of that element:

   H = one atom of hydrogen
   O = one atom of oxygen

2. A number preceding the symbol of an element indicates more than one atom of that element:

   2H = two atoms of hydrogen
   2O = two atoms of oxygen

3. A subscript following the symbol of an element indicates a molecule with that number of atoms of that element:

   H_2 = hydrogen molecule, composed of two hydrogen atoms
   O_2 = oxygen molecule, composed of two oxygen atoms
   H_2O = water molecule, composed of two hydrogen atoms and one oxygen atom

4. In a description of a chemical reaction, the participants at the start of the reaction are called reactants, and the reaction generates one or more products. An arrow indicates the direction of the reaction, from reactants (usually on the left) to products (usually on the right). In the following reaction,

   two atoms of hydrogen combine with one atom of oxygen to produce a single molecule of water:

   $$2H + O \rightarrow H_2O$$

5. A superscript plus or minus sign following the symbol of an element indicates an ion. A single plus sign indicates a cation with a charge of +1. (The original atom has lost one electron.) A single minus sign indicates an anion with a charge of −1. (The original atom has gained one electron.) If more than one electron has been lost or gained, the charge on the ion is indicated by a number preceding the plus or minus sign:

   Na^+ = sodium ion (the sodium atom has lost one electron)
   Cl^- = chloride ion (the chlorine atom has gained one electron)
   Ca^{2+} = calcium ion (the calcium atom has lost two electrons)

6. Chemical reactions neither create nor destroy atoms; they merely rearrange atoms into new combinations. Therefore, the numbers of atoms of each element must always be the same on both sides of the equation for a chemical reaction. When this is the case, the equation is balanced:

   Unbalanced: $H_2 + O_2 \rightarrow H_2O$
   Balanced: $2H_2 + O_2 \rightarrow 2H_2O$

2–2 CHEMICAL REACTIONS

Objectives

- Use chemical notation to symbolize chemical reactions.
- Distinguish among the major types of chemical reactions that are important for studying physiology.
- Describe the crucial role of enzymes in metabolism.

Cells remain alive and functional by controlling chemical reactions. In a **chemical reaction**, new chemical bonds form between atoms, or existing bonds between atoms are broken. These changes occur as atoms in the reacting substances, or **reactants**, are rearranged to form different substances, or **products**. (See "FOCUS: Chemical Notation," p. 38.)

In effect, each cell is a chemical factory. For example, growth, maintenance and repair, secretion, and contraction all involve complex chemical reactions. Cells use chemical reactions to provide the energy needed to maintain homeostasis and to perform essential functions. All of the reactions underway in the cells and tissues of the body at any given moment constitute its **metabolism** (me-TAB-ō-lizm).

☐ BASIC ENERGY CONCEPTS

An understanding of some basic relationships between matter and energy is essential for any discussion of chemical reactions. **Work** is movement or a change in the physical structure of matter. **Energy** is the capacity to perform work; movement or physical change cannot occur unless energy is provided. The two major types of energy are kinetic energy and potential energy:

1. **Kinetic energy** is the energy of motion—energy that is doing work. When you fall off a ladder, it is kinetic energy that does the damage.
2. **Potential energy** is stored energy—energy that has the potential to do work. It may derive from an object's position (a book sitting on a high shelf or you standing on a ladder) or from its physical or chemical structure (a stretched spring or a charged battery).

Kinetic energy must be used in lifting the book to the shelf, in climbing the ladder, in stretching the spring, or in charging the battery. The resulting potential energy is converted back into kinetic energy when the book falls (or you fall), the spring recoils, or the battery discharges. The kinetic energy can then be used to perform work. For example, the chemical potential energy stored in gasoline is converted to kinetic energy in the engine of your car. The gas is ignited in the cylinders, driving the pistons up and down and thus ultimately providing kinetic energy to the wheels.

Energy cannot be lost; it can only be converted from one form to another. A conversion between potential energy and kinetic energy is never 100 percent efficient. Each time an energy exchange occurs, some of the energy is released in the form of heat. *Heat* is an increase in random molecular motion; the temperature of an object is proportional to the average kinetic energy of its molecules. When energy is converted to heat, some useable energy is always lost, because heat can never be completely converted to work or any other form of energy. Moreover, cells have no effective way of using heat to perform work.

Cells perform work as they synthesize complex molecules and move materials into, out of, and within the cell. Some cells are specialized for motion or contraction. For example, a skeletal muscle at rest contains potential energy in the form of the positions of protein filaments and the covalent bonds between molecules inside the cell. When a muscle contracts, it performs work; potential energy is converted into kinetic energy, and heat is released. The amount of heat is proportional to the amount of work done. As a result, when you exercise, your body temperature rises.

☐ TYPES OF REACTIONS

Three types of chemical reactions are important to the study of physiology: (1) decomposition reactions, (2) synthesis reactions, and (3) exchange reactions.

Decomposition

Decomposition is a reaction that breaks a molecule into smaller fragments. You could diagram a typical *decomposition reaction* as

$$AB \rightarrow A + B$$

Decomposition reactions occur outside cells as well as inside them. For example, a typical meal contains molecules of fats, sugars, and proteins that are too large and too complex to be absorbed and used by your body. Decomposition reactions in the digestive tract break these molecules down into smaller fragments before absorption begins.

Decomposition reactions involving water are important in the breakdown of complex molecules in the body. In **hydrolysis** (hī-DROL-i-sis; *hydro-*, water + *lysis*, dissolution), one of the bonds in a complex molecule is broken, and the components of a water molecule (H and OH) are added to the resulting fragments:

$$A-B-C-D-E + H_2O \rightarrow A-B-C-H + HO-D-E$$

Collectively, the decomposition reactions of complex molecules within the body's cells and tissues are referred to as **catabolism** (ka-TAB-ō-lizm; *katabole*, a throwing down). When a covalent bond—a form of potential energy—is broken, it releases kinetic energy that can perform work. By harnessing the energy released in this way, cells perform vital functions such as growth, movement, and reproduction.

Synthesis

Synthesis (SIN-the-sis) is the opposite of decomposition. A *synthesis reaction* assembles larger molecules from smaller components. Simple synthetic reactions could be diagrammed as

$$A + B \rightarrow AB$$

Synthesis reactions may involve individual atoms or the combination of molecules to form even larger products. The formation of water from hydrogen and oxygen molecules is a synthesis reaction. Synthesis always involves the formation of new chemical bonds, whether the reactants are atoms or molecules.

Dehydration synthesis, or *condensation*, is the formation of a complex molecule by the removal of water:

$$A–B–C–H + HO–D–E \rightarrow A–B–C–D–E + H_2O$$

Dehydration synthesis is therefore the opposite of hydrolysis. We shall encounter examples of both reactions in later sections.

Collectively, the synthesis of new compounds within the body's cells and tissues is known as **anabolism** (a-NAB-ō-lizm; *anabole*, a throwing upward). Because it takes energy to create a chemical bond, anabolism is usually an "uphill" process. Cells must balance their energy budgets, with catabolism providing the energy to support anabolism as well as other vital functions.

Exchange

In an **exchange reaction**, parts of the reacting molecules are shuffled around. The following equation represents a simple exchange reaction:

$$AB + CD \rightarrow AD + CB$$

This equation has two reactants and two products. Although the reactants and products contain the same components (A, B, C, and D), those components are present in different combinations. In an exchange reaction, the reactant molecules AB and CD must break apart (a decomposition) before they can interact with each other to form AD and CB (a synthesis).

▢ REVERSIBLE REACTIONS

Chemical reactions are (at least theoretically) reversible, so if $A + B \rightarrow AB$, then $AB \rightarrow A + B$. Many important biological reactions are freely reversible. Such reactions can be represented as an equation:

$$A + B \rightleftharpoons AB$$

This equation reminds you that, in a sense, two reactions are occurring simultaneously, one a synthesis ($A + B \rightarrow AB$) and the other a decomposition ($AB \rightarrow A + B$). At **equilibrium** (ē-kwi-LIB-rē-um), the rates at which the two equations proceed are in balance. As fast as one molecule of AB forms, another degrades into $A + B$.

The result of a disturbance in the equilibrium condition can be predicted. In our example, the rate at which the synthesis reaction proceeds is directly proportional to the frequency of encounters between A and B. The frequency of encounters depends on the degree of crowding: You are much more likely to run into another person in a crowded room than in a room that is almost empty. The addition of more AB molecules will increase the rate of conversion of AB to A and B. The amounts of A and B will then increase, leading to an increase in the rate of

the reverse reaction—the formation of AB from A and B. Eventually, an equilibrium is again established.

Not all chemical reactions, however, are easily reversed. The requirements for two paired reactions may differ, so at any time and place the overall reaction will proceed chiefly in one direction only. For example, the synthesis reaction may occur only when A and B molecules are heated; the decomposition reaction may occur only when AB molecules are placed in water. In that case, the reaction would be written as

$$A + B \underset{H_2O}{\overset{heat}{\rightleftharpoons}} AB$$

▢ ENZYMES, ENERGY, AND CHEMICAL REACTIONS

Most chemical reactions do not occur spontaneously, or they occur so slowly that they would be of little value to cells. Before a reaction can proceed, enough energy must be provided to activate the reactants. The amount of energy required to start a reaction is called the **activation energy**. Although many reactions can be activated by changes in temperature or acidity, such changes are deadly to cells. For example, every day your cells break down complex sugars as part of your normal metabolism. Yet, to break down a complex sugar in a laboratory, you must boil it in an acidic solution. Your cells don't have that option; temperatures that high and solutions that corrosive would immediately destroy living tissues. Instead, your cells use special proteins called *enzymes* to perform most of the complex synthesis and decomposition reactions in your body.

Enzymes promote chemical reactions by lowering the activation energy requirements (Figure 2–7●). In doing so, they make it possible for chemical reactions, such as the breakdown of sugars, to proceed under conditions compatible with life. Enzymes belong to a class of substances called **catalysts** (KAT-uh-lists; *katalysis*, dissolution), compounds that accelerate chemical reactions without

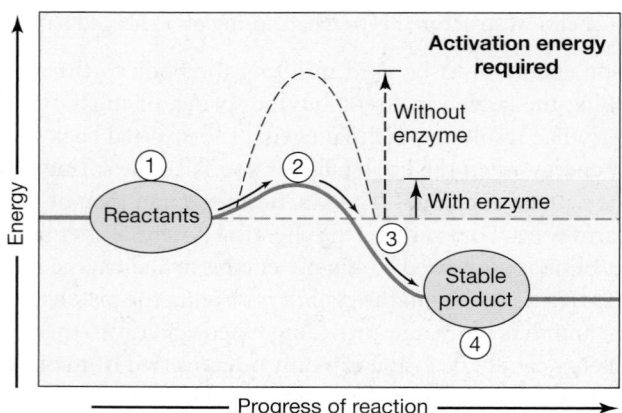

●**FIGURE 2–7**
Enzymes and Activation Energy. Enzymes lower the activation energy requirements, so a reaction can occur readily, in order from 1–4, under conditions in the body.

themselves being permanently changed or consumed. A cell makes an enzyme molecule to promote a specific reaction. Enzymatic reactions, which are reversible, can be written as

$$A + B \underset{}{\overset{enzyme}{\rightleftharpoons}} AB$$

Although the presence of an appropriate enzyme can accelerate a reaction, an enzyme affects only the rate of the reaction, not the direction of the reaction or the products that will be formed. An enzyme cannot bring about a reaction that would otherwise be impossible. Enzymatic reactions are generally reversible, and they proceed until an equilibrium becomes established.

The complex reactions that support life proceed in a series of interlocking steps, and each step is controlled by a specific enzyme. Such a reaction sequence is called a *pathway*. A synthetic pathway can be diagrammed as

$$A \underset{Step\ 1}{\overset{enzyme\ 1}{\rightleftharpoons}} B \underset{Step\ 2}{\overset{enzyme\ 2}{\rightleftharpoons}} C \underset{Step\ 3}{\overset{enzyme\ 3}{\rightleftharpoons}} \quad \text{and so on}$$

In many cases, the steps in the synthetic pathway differ from those in the decomposition pathway, and separate enzymes are often involved.

It takes activation energy to start a chemical reaction, but once it has begun, the reaction as a whole may absorb or release energy, generally in the form of heat, as it proceeds to completion. If the amount of energy released is greater than the activation energy needed to start the process, there will be a net release of energy. Reactions that release energy are said to be **exergonic** (*exo-*, outside). If more energy is required to begin the reaction than is released as it proceeds, the reaction as a whole will absorb energy. Such reactions are called **endergonic** (*endo-*, inside). Exergonic reactions are relatively common in the body; they are responsible for generating the heat that maintains your body temperature.

✓ In cells, glucose, a six-carbon molecule, is converted into two three-carbon molecules by a reaction that releases energy. How would you classify this reaction?

✓ Why are enzymes needed in our cells?

Answers start on page Q-1

2-3 INORGANIC COMPOUNDS

Objectives

- Distinguish between organic and inorganic compounds.
- Explain how the chemical properties of water make life possible.
- Discuss the importance of pH and the role of buffers in body fluids.
- Describe the physiological roles of inorganic compounds.

Although the human body is very complex, it contains relatively few elements, as indicated in Table 2-1. But knowing the identity and quantity of each element in the body will not help you understand the body any more than studying the alphabet will help you understand this textbook. In this book, just 26 letters are combined to form many different words, and those words are combined in various ways to form many different sentences, each with a specific meaning. Similarly, in our bodies, only about 26 elements combine to form thousands of different chemical compounds. As we saw in Chapter 1, these compounds make up the living cells that constitute the framework of the body and carry on all its life processes. So it is impossible to understand the structure and functioning of the human body without learning about the major classes of chemical compounds.

The rest of this chapter focuses on nutrients and metabolites. **Nutrients** are the essential elements and molecules normally obtained from the diet. **Metabolites** (me-TAB-ō-līts; *metabole*, change) include all the molecules synthesized or broken down by chemical reactions inside our bodies. Nutrients and metabolites can be broadly categorized as inorganic or organic. **Inorganic compounds** generally do not contain carbon and hydrogen atoms as their primary structural ingredients, whereas carbon and hydrogen *always* form the basis for **organic compounds**.

The most important inorganic compounds in the body are (1) *carbon dioxide*, a by-product of cell metabolism; (2) *oxygen*, an atmospheric gas required in important metabolic reactions; (3) *water*, which accounts for most of our body weight; and (4) *inorganic acids, bases,* and *salts*—compounds held together partially or completely by ionic bonds. In this section, we shall be concerned primarily with water, its properties, and how those properties establish the conditions necessary for life. Most of the other inorganic molecules and compounds in the body exist in association with water, the primary component of our body fluids. For example, carbon dioxide and oxygen are gas molecules that occur both in the atmosphere and in body fluids; all the inorganic acids, bases, and salts we will discuss are dissolved in body fluids.

☐ WATER AND ITS PROPERTIES

Water, H_2O, is the single most important constituent of the body, accounting for up to two-thirds of total body weight. A change in the body's water content can have fatal consequences because virtually all physiological systems will be affected.

Although water is familiar to everyone, it has some highly unusual properties. These properties are a direct result of the hydrogen bonding that occurs between adjacent water molecules:

1. *Solubility.* A remarkable number of inorganic and organic molecules will *dissolve* (break up) in water, creating a solution. Every **solution** is a uniform mixture of two or more substances—whether in solid, liquid, or gaseous form. The medium in which other atoms, ions, or molecules are dispersed is called the **solvent**; the dispersed substances are the **solutes**. In *aqueous solutions*, water is the solvent. The solvent properties of water are so important that we shall consider them in detail in the next section.

2. *Reactivity.* In our bodies, chemical reactions occur in water, and water molecules are also participants in some reactions. Hydrolysis and dehydration synthesis are two examples noted earlier in the chapter.

3. *High Heat Capacity.* *Heat capacity* is the ability to absorb and retain heat. Water has an unusually high heat capacity, because water molecules in the liquid state are attracted to one another through hydrogen bonding. Important consequences of this attraction include the following:

   • The temperature must be quite high before individual molecules have enough energy to break free and become water vapor, a gas. Consequently, water stays in the liquid state over a broad range of environmental temperatures, and the freezing and boiling points of water are far apart.

   • Water carries a great deal of heat away with it when it finally does change from a liquid to a gas. This feature accounts for the cooling effect of perspiration on the skin.

   • An unusually large amount of heat energy is required to change the temperature of 1 g of water by 1°C. Thus, once a quantity of water has reached a particular temperature, it will change temperature only slowly. This property is called *thermal inertia.* Because water accounts for up to two-thirds of the weight of the human body, thermal inertia helps stabilize body temperature. Similarly, water's high heat capacity allows the blood plasma to transport and redistribute large amounts of heat as it circulates within the body. For example, heat absorbed as the blood flows through active muscles will be released when the blood reaches vessels near the relatively cool surface of the body.

4. *Lubrication.* Water is an effective lubricant. There is little friction between water molecules. Thus, if two opposing surfaces are separated by even a thin layer of water, friction between those surfaces will be greatly reduced. (That is why driving on wet roads can be tricky; your tires may start sliding on a layer of water rather than maintaining contact with the road.) With-

in joints such as the knee, an aqueous solution prevents friction between the opposing surfaces. Similarly, a small amount of fluid in the ventral body cavities prevents friction between internal organs, such as the heart or lungs, and the body wall. ⊃⊃ p. 22

Aqueous Solutions

Its chemical structure makes water an unusually effective solvent. The bonds in a water molecule are oriented such that the hydrogen atoms are relatively close together (Figure 2–8a●). As a result, the water molecule has positive and negative poles. A water molecule is therefore called a *polar molecule*, or a *dipole*.

Many inorganic compounds are held together partially or completely by ionic bonds. In water, these compounds undergo *ionization* (ī-on-i-ZĀ-shun), or *dissociation* (di-sō-sē-Ā-shun). In this process (Figure 2–8b●), ionic bonds are broken as the individual ions interact with the positive or negative ends of polar water molecules. The result is a mixture of cations and anions surrounded by water molecules. The water molecules around each ion form a *hydration sphere.*

An aqueous solution containing anions and cations will conduct an electrical current. Cations (+) move toward the negative side, or negative *terminal*, and anions (−) move toward the positive terminal. Electrical forces across cell membranes affect the functioning of all cells, and small electrical currents carried by ions are essential to muscle contraction and nerve function. Chapters 10 and 12 will discuss these processes in more detail.

ELECTROLYTES AND BODY FLUIDS Soluble inorganic molecules whose ions will conduct an electrical current in solution are called **electrolytes** (e-LEK-trō-līts). Sodium chloride is an electrolyte. The dissociation of electrolytes in blood and other body fluids releases sodium ions (Na^+), potassium ions (K^+), calcium

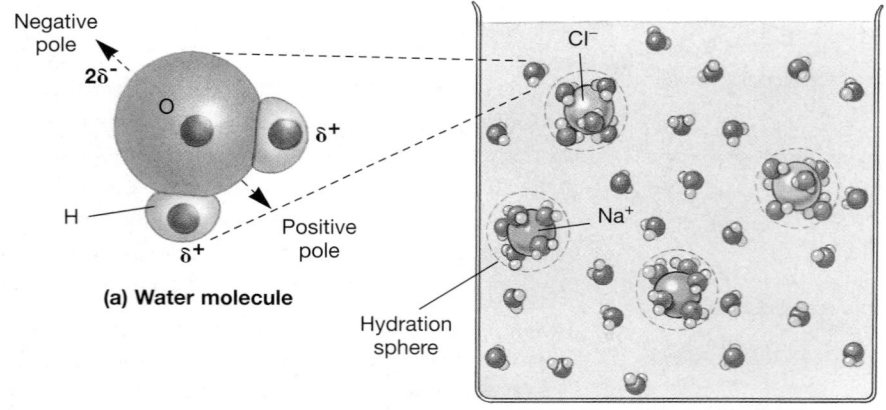
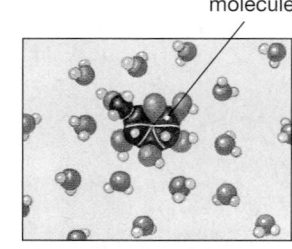

●**FIGURE 2–8**
Water Molecules and Solutions. **(a)** In a water molecule, oxygen forms polar covalent bonds with two hydrogen atoms. Because both hydrogen atoms are at one end, the molecule has an uneven distribution of charges, creating positive and negative poles. **(b)** Ionic compounds, such as sodium chloride, dissociate in water as the polar water molecules break the ionic bonds. Each ion in solution is surrounded by water molecules, creating hydration spheres. **(c)** Hydration spheres also form around an organic molecule containing polar covalent bonds. If the molecule is small, as is glucose, it will be carried into solution.

TABLE 2–3 IMPORTANT ELECTROLYTES THAT DISSOCIATE IN BODY FLUIDS

Electrolyte		Ions Released
NaCl (sodium chloride)	$\rightarrow$	$Na^+ + Cl^-$
KCl (potassium chloride)	$\rightarrow$	$K^+ + Cl^-$
CaCl$_2$ (calcium chloride)	$\rightarrow$	$Ca^{2+} + 2\ Cl^-$
NaHCO$_3$ (sodium bicarbonate)	$\rightarrow$	$Na^+ + HCO_3^-$
MgCl$_2$ (magnesium chloride)	$\rightarrow$	$Mg^{2+} + 2\ Cl^-$
Na$_2$HPO$_4$ (disodium phosphate)	$\rightarrow$	$2\ Na^+ + HPO_4^{2-}$
Na$_2$SO$_4$ (sodium sulfate)	$\rightarrow$	$2\ Na^+ + SO_4^{2-}$

ions (Ca^{2+}), and chloride ions (Cl^-). Table 2–3 lists important electrolytes and the ions released by their dissociation.

Changes in the concentrations of electrolytes in body fluids will disturb almost every vital function. For example, declining potassium levels will lead to a general muscular paralysis, and rising concentrations will cause weak and irregular heartbeats. The concentrations of ions in body fluids are carefully regulated, primarily by the coordination of activities at the kidneys (ion excretion), the digestive tract (ion absorption), and the skeletal system (ion storage or release).

HYDROPHILIC AND HYDROPHOBIC COMPOUNDS Some organic molecules contain polar covalent bonds, which also attract water molecules. The hydration spheres that form may then carry these molecules into solution (Figure 2–8c•). Molecules such as *glucose*, an important soluble sugar, that interact readily with water molecules in this way are called **hydrophilic** (hī-drō-FI-lik; *hydro-*, water + *philos*, loving).

Many other organic molecules have few polar covalent bonds or none at all. Such molecules are said to be *nonpolar*. When nonpolar molecules are exposed to water, hydration spheres do not form and the molecules do not dissolve. Molecules that do not readily interact with water are called **hydrophobic** (hī-drō-FŌ-bik; *hydro-*, water + *phobos*, fear). Among the most familiar such molecules are fats and oils of all kinds. Body fat deposits, for example, consist of large, hydrophobic droplets trapped in the watery interior of cells. Gasoline, heating oil, and diesel fuel are hydrophobic molecules not found in the body; accidentally discharged into lakes or oceans, they form tenacious oil slicks instead of dissolving.

SOLUTE CONCENTRATIONS The **concentration** of a substance is the amount of that substance present in a specified volume of solvent. Physiologists and clinicians often monitor inorganic and organic solute concentrations in body fluids such as blood or urine. Each solute has a normal range of values (see Appendix IV), and variations outside this range may indicate disease. Many solutes are participants in biochemical reactions, and as we noted earlier, the concentrations of reactants and products in a chemical reaction directly affect reaction rates.

Solute concentrations can be expressed in several ways, and we shall introduce two methods here. In the first method, we ex-

press the number of solute atoms, molecules, or ions in a specific volume of solution. Values are reported in moles per liter (mol/l, or M) or millimoles per liter (mmol/l, or mM). A concentration expressed in these units is referred to as the *molarity* of the solution. As we discussed earlier in the chapter, a mole is a quantity of any substance having a weight in grams equal to the atomic or molecular weight of that substance. Physiological concentrations today are most often reported in millimoles per liter.

You can report concentrations in moles or millimoles only when you know the molecular weight of the ion or molecule in question. When the chemical structure is unknown or when you are dealing with a complex mixture of materials, concentration is expressed in terms of the weight of material dissolved in a unit volume of solution. Values are most often given in milligrams (mg) or grams (g) per deciliter (dl, or 100 ml). This is the method used, for example, in reporting the concentration of plasma proteins in a blood sample. **AM** Solutions and Concentrations

Colloids and Suspensions

Body fluids may contain large and complex organic molecules, such as proteins and protein complexes, that are held in solution by their association with water molecules (Figure 2–8c•). A solution containing dispersed proteins or other large molecules is called a **colloid**. Liquid Jell-O™ is a familiar viscous (thick) colloid.

The particles or molecules in a colloid will remain in solution indefinitely. A **suspension** contains even larger particles that will, if undisturbed, settle out of solution due to the force of gravity. Stirring beach sand into a bucket of water creates a temporary suspension that will last only until the sand settles to the bottom. Whole blood is another temporary suspension, because the blood cells are suspended in the blood plasma. If clotting is prevented, the cells in a blood sample will gradually settle to the bottom of the container. Measuring that settling rate, or "sedimentation rate," is a common laboratory test.

Hydrogen Ions in Body Fluids

A hydrogen atom involved in a chemical bond or participating in a chemical reaction can easily lose its electron, to become a hydrogen ion, H^+. Hydrogen ions are extremely reactive in solution. In excessive numbers, they will break chemical bonds, change the shapes of complex molecules, and generally disrupt cell and tissue functions. As a result, the concentration of hydrogen ions in body fluids must be regulated precisely.

A few hydrogen ions are normally present even in a sample of pure water, because some of the water molecules dissociate spontaneously, releasing cations and anions. The dissociation of water is a reversible reaction that can be represented as

$$H_2O \rightleftharpoons H^+ + OH^-$$

The dissociation of one water molecule yields a hydrogen ion and a *hydroxide* (hī-DROK-sīd) *ion*, OH^-.

Very few water molecules ionize in pure water, and the number of hydrogen and hydroxide ions is small. The quantities are usually reported in moles, making it easy to keep track of the relative numbers of hydrogen and hydroxide ions. One liter of pure water contains about 0.0000001 mol of hydrogen ions and an equal number of hydroxide ions. In other words, the concentration of hydrogen ions in a solution of pure water is 0.0000001 mol per liter. This can be written as

$$[H^+] = 1 \times 10^{-7} \text{ mol/l}$$

The brackets around the H^+ signify "the concentration of," another example of chemical notation.

pH The hydrogen ion concentration in body fluids is so important to physiological processes that a special shorthand is used to express it. The **pH** of a solution is defined as the negative logarithm of the hydrogen ion concentration in moles per liter. Thus, instead of using the equation $[H^+] = 1 \times 10^{-7}$ mol/l, we say that the pH of pure water is $-(-7)$, or 7. Using pH values saves space, but always remember that the pH number is an *exponent* and that the pH scale is logarithmic. For instance, a pH of 6 ($[H^+] = 1 \times 10^{-6}$, or 0.000001) means that the concentration of hydrogen ions is *10 times as great* as it is at a pH of 7 ($[H^+] = 1 \times 10^{-7}$, or 0.0000001). For common liquids, the pH scale, included in Figure 2–9●, ranges from 0 to 14.

Although pure water has a pH of 7, solutions display a wide range of pH values, depending on the nature of the solutes involved:

- A solution with a pH of 7 is said to be **neutral**, because it contains equal numbers of hydrogen and hydroxide ions.
- A solution with a pH below 7 is **acidic** (a-SI-dik), meaning that it contains more hydrogen ions than hydroxide ions.
- A pH above 7 is **basic**, or *alkaline* (AL-kuh-lin), meaning that it has more hydroxide ions than hydrogen ions.

The pH of blood normally ranges from 7.35 to 7.45. Abnormal fluctuations in pH can damage cells and tissues by breaking chemical bonds, changing the shapes of proteins, and altering cellular functions. *Acidosis* is an abnormal physiological state caused by low blood pH (below 7.35); a pH below 7 can produce coma. *Alkalosis* results from an abnormally high pH (above 7.45); a blood pH above 7.8 generally causes uncontrollable and sustained skeletal muscle contractions.

☐ INORGANIC ACIDS AND BASES

The body contains both inorganic and organic *acids* and *bases* that may cause acidosis or alkalosis, respectively. An **acid** is any solute that dissociates in solution and releases hydrogen ions, thereby lowering the pH. Because a hydrogen atom that loses its electron consists solely of a proton, hydrogen ions are often referred to simply as protons, and acids as *proton donors*.

A *strong acid* dissociates completely in solution, and the reaction occurs essentially in one direction only. *Hydrochloric acid* (HCl) is a representative strong acid; in water, it ionizes as

$$HCl \rightarrow H^+ + Cl^-$$

The stomach produces this powerful acid to assist in the breakdown of food. Hardware stores sell HCl under the name muriatic acid, for cleaning sidewalks and swimming pools.

A **base** is a solute that removes hydrogen ions from a solution and thereby acts as a *proton acceptor*. In solution, many bases release a hydroxide ion (OH^-). Hydroxide ions have a strong affinity for hydrogen ions and react quickly with them to form water molecules. A *strong base* dissociates completely in solution. *Sodium hydroxide*, NaOH, is a strong base; in solution, it releases sodium ions and hydroxide ions:

$$NaOH \rightarrow Na^+ + OH^-$$

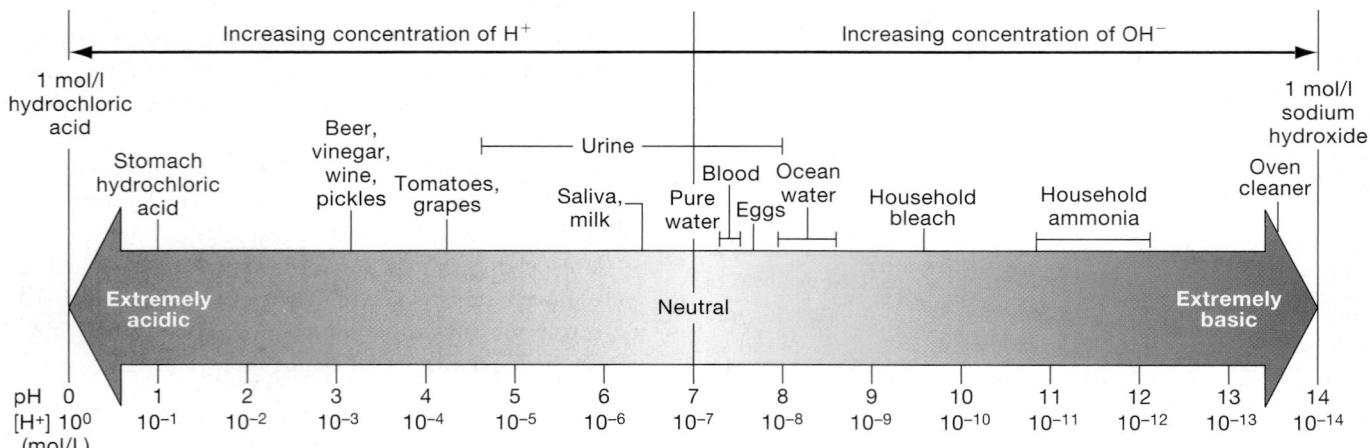

●**FIGURE 2–9**
pH and Hydrogen Ion Concentration. The scale is logarithmic; an increase or decrease of one unit corresponds to a tenfold change in H^+ concentration.

Strong bases have a variety of industrial and household uses. Drain openers (Drano™) and lye are two familiar examples.

Weak acids and *weak bases* fail to dissociate completely. At equilibrium, a significant number of molecules remain intact in the solution. For the same number of molecules in solution, weak acids and weak bases therefore have less of an impact on pH than do strong acids and strong bases. *Carbonic acid* (H_2CO_3) is a weak acid found in body fluids. In solution, carbonic acid reversibly dissociates into a hydrogen ion and a *bicarbonate ion*, HCO_3^-:

$$H_2CO_3 \rightleftharpoons H^+ + HCO_3^-$$

☐ SALTS

A **salt** is an ionic compound consisting of any cation except a hydrogen ion and any anion except a hydroxide ion. Because they are held together by ionic bonds, many salts dissociate completely in water, releasing cations and anions. For example, sodium chloride (table salt) dissociates immediately in water, releasing Na^+ and Cl^- ions. Sodium and chloride are the most abundant ions in body fluids. However, many other ions are present in lesser amounts as a result of the dissociation of other inorganic compounds. Ionic concentrations in the body are regulated by mechanisms we shall describe in Chapters 26 and 27.

The ionization of sodium chloride does not affect the local concentrations of hydrogen ions or hydroxide ions, so NaCl, like many salts, is a "neutral" solute. Through their interactions with water molecules, however, other salts may indirectly affect the concentrations of H^+ and OH^- ions. Thus, the dissociation of some salts makes a solution slightly acidic or slightly basic.

☐ BUFFERS AND pH CONTROL

Buffers are compounds that stabilize the pH of a solution by removing or replacing hydrogen ions. *Buffer systems* typically involve a weak acid and its related salt, which functions as a weak base. For example, the carbonic acid–bicarbonate buffer system (which will be detailed in Chapter 27) consists of carbonic acid and sodium bicarbonate, $NaHCO_3$, otherwise known as baking soda. Buffers and buffer systems in body fluids maintain the pH within normal limits. Figure 2–9● shows the pH of several body fluids. Antacids such as Alka-Seltzer® and Rolaids® use sodium bicarbonate to neutralize excess hydrochloric acid in the stomach. The effects of neutralization are most evident when you add a strong acid to a strong base. For example, by adding hydrochloric acid to sodium hydroxide, you neutralize both the strong acid and the strong base:

$$HCl + NaOH \rightarrow H_2O + NaCl$$

This reaction produces water and a salt—in this case, the neutral salt sodium chloride.

✓ Why does a solution of table salt conduct electricity, but a sugar solution does not?

✓ How does an antacid help decrease stomach discomfort?

Answers start on page Q-1

2–4 ORGANIC COMPOUNDS

Objective

■ Discuss the structures and functions of carbohydrates, lipids, proteins, nucleic acids, and high-energy compounds.

Organic compounds always contain the elements carbon and hydrogen, and generally oxygen as well. Many organic molecules are made up of long chains of carbon atoms linked by covalent bonds. The carbon atoms typically form additional covalent bonds with hydrogen or oxygen atoms and, less commonly, with nitrogen, phosphorus, sulfur, iron, or other elements.

Many organic molecules are soluble in water. Although the previous discussion focused on inorganic acids and bases, there are also important organic acids and bases. For example, *lactic acid* is an organic acid, generated by active muscle tissues, that must be neutralized by the buffers in body fluids.

In this discussion, we consider four major classes of organic compounds: *carbohydrates, lipids, proteins,* and *nucleic acids*. We also introduce *high-energy compounds*, which are insignificant in terms of their abundance in the body but vital to the survival of our cells. In addition, the human body contains small quantities of many other organic compounds whose structures and functions we shall consider in later chapters.

Although organic compounds are diverse, certain groupings of atoms occur again and again, even in very different types of molecules. These *functional groups* greatly influence the properties of any molecule they are part of. Table 2–4 details the functional groups you will encounter in this chapter.

☐ CARBOHYDRATES

A **carbohydrate** is an organic molecule that contains carbon, hydrogen, and oxygen in a ratio near 1:2:1. Familiar carbohydrates include the *sugars* and *starches* that make up roughly half of the typical U.S. diet. Our tissues can break down most carbohydrates, which typically account for less than 1 percent of the total body weight. Although they may have other functions, carbohydrates are most important as sources of energy that are catabolized rather than stored. We will focus on *monosaccharides, disaccharides,* and *polysaccharides.*

TABLE 2–4 IMPORTANT FUNCTIONAL GROUPS

Functional Group	Structural Formula*	Importance	Examples
Carboxyl group, —COOH	R...—C=O with OH	Acts as an acid, releasing H⁺ to become R-COO⁻	Fatty acids, amino acids
Amino group, —NH₂	R—N with H and H	Can accept or release H⁺, depending on pH; can form bonds with other molecules	Amino acids
Hydroxyl group, —OH	R—O—H	Strong bases dissociate to release hydroxide ions (OH⁻); may link molecules through dehydration synthesis	Carbohydrates, fatty acids, amino acids
Phosphate group, —PO₄	R—O—P—O⁻ with O and O⁻	May link other molecules to form larger structures; may store energy in high-energy bonds	Phospholipids, nucleic acids, high-energy compounds

* The term *R-group* is used to denote the rest of the molecule, whatever that might be. The R-group is also known as a *side chain*.

Monosaccharides

A *simple sugar*, or **monosaccharide** (mon-ō-SAK-uh-rīd; *mono-*, single + *sakcharon*, sugar), is a carbohydrate containing from three to seven carbon atoms. A monosaccharide can be called a *triose* (three-carbon), *tetrose* (four-carbon), *pentose* (five-carbon), *hexose* (six-carbon), or *heptose* (seven-carbon). The hexose **glucose** (GLOO-kōs), $C_6H_{12}O_6$, is the most important metabolic "fuel" in the body. The atoms in a glucose molecule may form a straight chain (Figure 2–10a●) or a ring (Figure 2–10b●). In the body, the ring form is more common. A three-dimensional model shows the arrangement of atoms in the ring most clearly (Figure 2–10c●).

The three-dimensional structure of an organic molecule is an important characteristic, because it usually determines the molecule's fate or function. Some molecules have the same molecular formula—in other words, the same types and numbers of atoms—but different structures. Such molecules are called **isomers**. The body usually treats different isomers as distinct molecules. For example, the monosaccharides glucose and fructose are isomers. *Fructose* is a hexose found in many fruits and in secretions of the male reproductive tract. Although its chemical formula, $C_6H_{12}O_6$, is the same as that of glucose, separate enzymes and reaction sequences control its breakdown and synthesis. Monosaccharides such as glucose and fructose dissolve readily in water and are rapidly distributed throughout the body by blood and other body fluids. **AM** The Pharmaceutical Use of Isomers

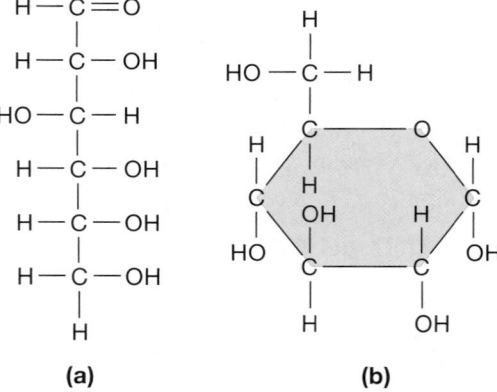

(a)   (b)

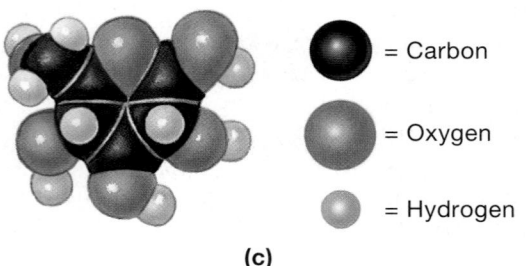

● = Carbon

● = Oxygen

● = Hydrogen

(c)

●FIGURE 2–10
Glucose. **(a)** The straight-chain structural formula. **(b)** The ring form that is most common in nature. **(c)** A three-dimensional model that shows the organization of the atoms in the ring.

Disaccharides and Polysaccharides

Carbohydrates other than simple sugars are complex molecules composed of monosaccharide building blocks. Two monosaccharides joined together form a **disaccharide** (dī-SAK-uh-rīd; *di-*, two). Disaccharides such as *sucrose* (table sugar) have a sweet taste and, like monosaccharides, are quite soluble in water. The formation of sucrose (Figure 2–11a●) involves dehydration synthesis, a process introduced earlier in the chapter. Dehydration synthesis, or condensation, links molecules together by the removal of a water molecule. The breakdown of sucrose into simple sugars is an example of hydrolysis, the functional opposite of dehydration synthesis (Figure 2–11b●).

Many foods contain disaccharides, but all carbohydrates except monosaccharides must be disassembled through hydrolysis before they can provide useful energy. Most popular junk foods, such as candies and sodas, abound in monosaccharides (commonly fructose) and disaccharides (generally sucrose). Some people cannot tolerate sugar for medical reasons; others avoid it in an effort to control their weight. (Excess sugars are converted to fat for long-term storage.) Many such people use *artificial sweeteners* in their foods and beverages. These compounds have a very sweet taste, but they either cannot be broken down in the body or are used in insignificant amounts. **AM** Artificial Sweeteners

Dehydration synthesis can continue adding monosaccharides or combining disaccharides to create increasingly complex carbohydrates. These large molecules are called **polysaccharides** (pol-ē-SAK-uh-rīdz; *poly-*, many). Polysaccharide chains can be straight or highly branched. *Cellulose*, a structural component of many plants, is a polysaccharide that our bodies cannot digest. Foods such as celery, which contains cellulose, water, and little else, contribute to the bulk of diges-

tive wastes but are useless as a source of energy. (In fact, you expend more energy when you chew a stalk of celery than you obtain by digesting it.)

Starches are large polysaccharides formed from glucose molecules. Most starches are manufactured by plants. Our digestive tract can break these molecules into monosaccharides. Starches such as those in potatoes and grains are a major dietary energy source.

The polysaccharide **glycogen** (GLĪ-kō-jen), or *animal starch*, has many side branches, all consisting of chains of glucose molecules (Figure 2–12●). Like most other starches, glycogen does not dissolve in water or other body fluids. Liver tissues and muscle tissues make and store glycogen. When these tissues have a high demand for glucose, glycogen molecules are broken down; when the demand is low, liver and muscle tissues absorb glucose from the bloodstream and rebuild glycogen reserves.

Despite their metabolic importance, carbohydrates account for less than 3 percent of our total body weight. Table 2–5 summarizes information about the carbohydrates.

☐ LIPIDS

Like carbohydrates, **lipids** (*lipos*, fat) contain carbon, hydrogen, and oxygen, and the carbon-to-hydrogen ratio is commonly 1:2. However, lipids contain much less oxygen than do carbohydrates with the same number of carbon atoms. The hydrogen-to-oxygen ratio is therefore very large; a representative lipid, such as lauric acid, has a formula of $C_{12}H_{24}O_2$. Lipids may also contain small quantities of other elements, such as phosphorus, nitrogen, or sulfur. Familiar lipids include *fats, oils,* and *waxes*. Most lipids are insoluble in water, but special transport mechanisms carry them in the circulating blood.

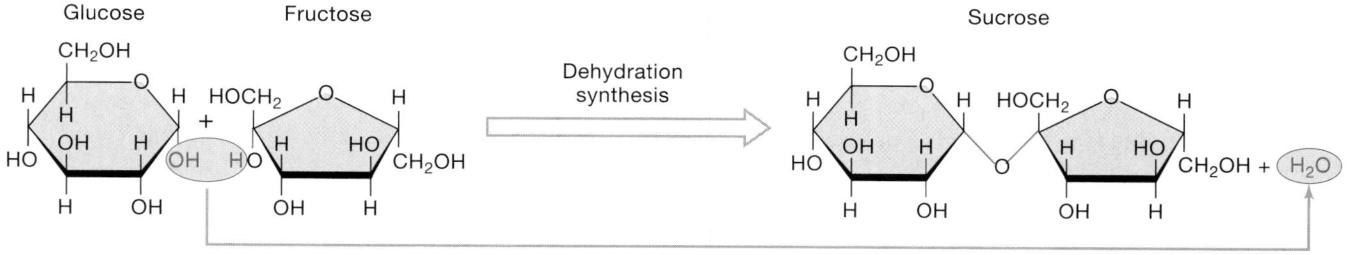

(a) During dehydration synthesis, two molecules are joined by the removal of a water molecule

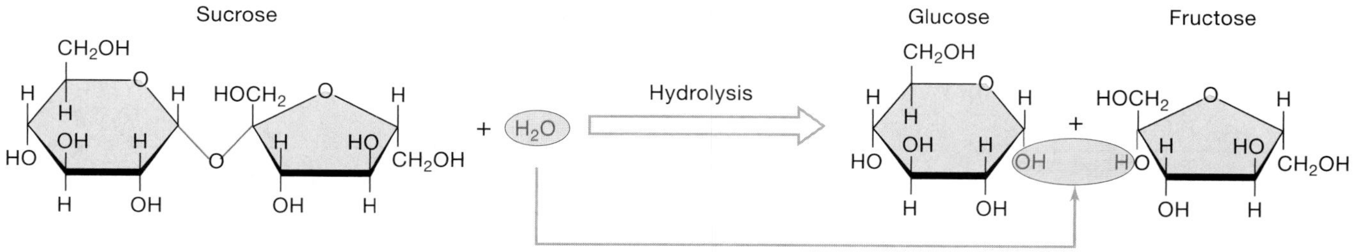

(b) Hydrolysis reverses the steps of dehydration synthesis; a complex molecule is broken down by the addition of a water molecule.

●**FIGURE 2–11**
The Formation and Breakdown of Complex Sugars. These reactions are performed by enzymes in the cell.

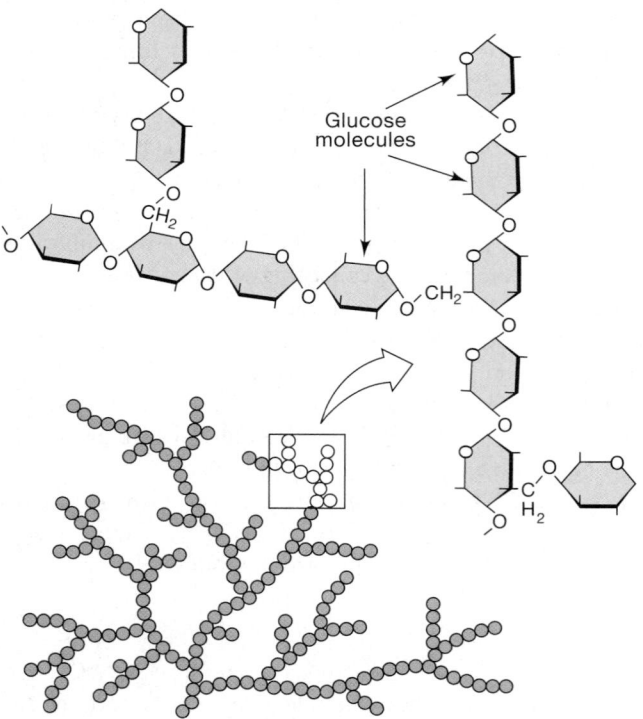

Glucose molecules

•FIGURE 2–12
The Structure of a Polysaccharide. Liver and muscle cells store glucose as the polysaccharide glycogen, a long, branching chain of glucose molecules. This figure uses a different method of representing a carbon ring structure: At each corner of the solid hexagon is a carbon atom. The position of an oxygen atom in each glucose ring is shown.

Lipids form essential structural components of all cells. In addition, lipid deposits are important as energy reserves. On average, lipids provide roughly twice as much energy as carbohydrates do, gram for gram, when broken down in the body. When the supply of lipids exceeds the demand for energy, the excess is stored in fat deposits. For this reason, there has been great interest in developing *fat substitutes* that provide less energy, but have the same taste and texture as lipids. **AM** Fat Substitutes

Lipids normally account for 12–18 percent of the total body weight of adult males and 18–24 percent of that of adult women. Many kinds of lipids exist in the body; major groups are presented in Table 2–6. We will consider five classes of lipids: *fatty acids, eicosanoids, glycerides, steroids,* and *phospholipids and glycolipids.*

Fatty Acids

Fatty acids are long carbon chains with hydrogen atoms attached. One end of the carbon chain always bears a *carboxylic* (kar-bok-SIL-ik) *acid group,* COOH (Table 2–4). The name *carboxyl* should help you remember that a carbon and a hydroxyl (–OH) group are the important structural features of fatty acids. The end opposite the carboxylic acid group is known as a hydrocarbon *tail* of the fatty acid. Figure 2–13a• shows a representative fatty acid, *lauric acid.*

When a fatty acid is in solution, only the carboxyl end associates with water molecules, because that is the only hydrophilic portion of the molecule. The hydrocarbon tail is hydrophobic, so fatty acids have a very limited solubility in water. In general, the longer the hydrocarbon tail, the lower the solubility of the molecule.

In a *saturated* fatty acid, each carbon atom in the hydrocarbon tail has four single covalent bonds. If some of the carbon-to-carbon bonds are double covalent bonds, the fatty acid is said to be *unsaturated.* These terms refer to the ratio of hydrogen atoms to carbon atoms in the fatty acid. Replacing a double bond between carbon atoms with a single bond allows the molecule to accept two more hydrogen atoms; hence, a double-bonded chain is unsaturated (Figure 2–13b•). A *monounsaturated* fatty acid has a single unsaturated bond in the hydrocarbon tail. A *polyunsaturated* fatty acid contains multiple unsaturated bonds.

Eicosanoids

Eicosanoids (ī-KŌ-sa-noydz) are lipids derived from *arachidonic* (ah-rak-i-DON-ik) *acid,* a fatty acid that must be absorbed in the diet because it cannot be synthesized by the body. The two major classes of eicosanoids are (1) *prostaglandins* and (2) *leukotrienes.* We consider only prostaglandins here, because virtually all tissues synthesize and respond to them. Leukotrienes are produced primarily by cells involved with coordinating the responses to injury or disease. We will consider leukotrienes in Chapters 18 and 22.

Prostaglandins (pros-tuh-GLAN-dinz) are short-chain fatty acids in which five of the carbon atoms are joined in a ring (Figure 2–14•). These compounds are released by cells to coordinate or direct local cellular activities. Almost every tissue in the body produces and responds to prostaglandins, which are ex-

TABLE 2–5 Carbohydrates in the Body			
Structure	Examples	Primary Function	Remarks
Monosaccharides (simple sugars)	Glucose, fructose	Energy source	Manufactured in the body and obtained from food; distributed in body fluids
Disaccharides	Sucrose, lactose, maltose	Energy source	Sucrose is table sugar, lactose is in milk, and maltose is malt sugar; all must be broken down to monosaccharides before absorption
Polysaccharides	Glycogen	Storage of glucose molecules	Glycogen is in animal cells; other starches and cellulose are in plant cells

TABLE 2–6 REPRESENTATIVE LIPIDS AND THEIR FUNCTIONS IN THE BODY

Lipid Type	Example(s)	Primary Function(s)	Remarks
Fatty acids	Lauric acid	Energy source	Absorbed from food or synthesized in cells; transported in the blood
Eicosanoids	Prostaglandins, leukotrienes	Chemical messengers coordinating local cellular activities	Prostaglandins are produced in most body tissues
Glycerides	Monoglycerides, diglycerides, triglycerides	Energy source, energy storage, insulation, and physical protection	Stored in fat deposits; must be broken down to fatty acids and glycerol before they can be used as an energy source
Steroids	Cholesterol	Structural component of cell membranes, hormones, digestive secretions in bile	All have the same carbon ring framework
Phospholipids, glycolipids	Lecithin (a phospholipid)	Structural components of cell membranes	Derived from fatty acids and nonlipid components

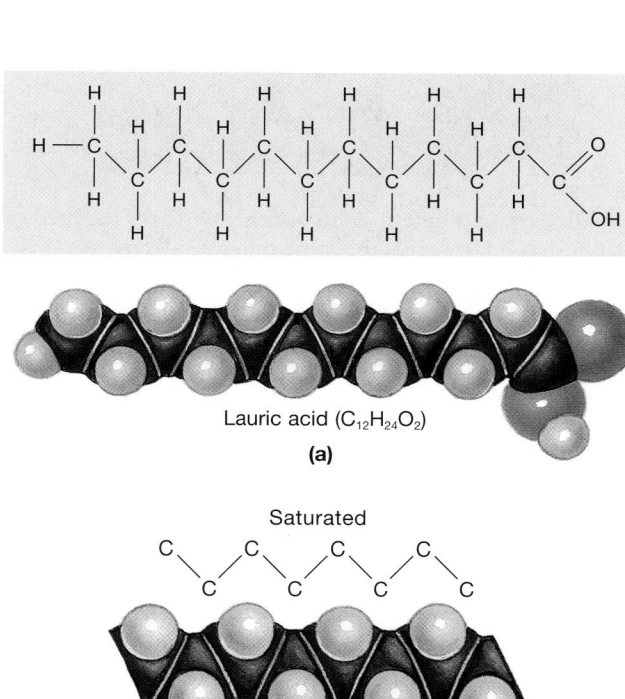

Lauric acid ($C_{12}H_{24}O_2$)

(a)

Saturated

Unsaturated

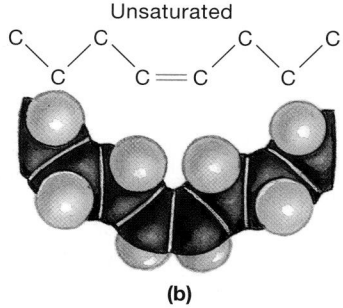

(b)

●**FIGURE 2–13**
Fatty Acids. **(a)** Lauric acid demonstrates two structural characteristics common to all fatty acids: a long chain of carbon atoms and a carboxylic acid group (–COOH) at one end. **(b)** A fatty acid is saturated or unsaturated. Unsaturated fatty acids have double covalent bonds; their presence causes a sharp bend in the molecule.

●**FIGURE 2–14**
Prostaglandins. Prostaglandins are unusual short-chain fatty acids.

tremely powerful and effective even in minute quantities. The effects of prostaglandins vary with their structure and the site of their release. Consider two examples:

1. Prostaglandins released by damaged tissues stimulate nerve endings and produce the sensation of pain (Chapter 15).
2. Prostaglandins released in the uterus help trigger the start of labor contractions (Chapter 29).

The body uses several types of chemical messengers. Those that are produced in one part of the body but have effects on distant parts, are called *hormones*. Hormones are distributed throughout the body in the bloodstream, whereas most prostaglandins affect only the area in which they are produced. As a result, prostaglandins are often called *local hormones*. The distinction is not a rigid one, however, as some prostaglandins also enter the bloodstream and affect other areas to some degree. We will discuss hormones and prostaglandins in Chapter 18.

Glycerides

Individual fatty acids cannot be strung together in a chain by dehydration synthesis, as monosaccharides can. But they can be attached to another compound, **glycerol** (GLI-se-rol), through a similar reaction. The result is a lipid known as a **glyceride** (GLI-se-rīd). Dehydration synthesis can produce a **monoglyceride** (mo-nō-GLI-se-rīd), consisting of glycerol plus one fatty acid.

Fatty Acids and Health

✚ Humans love fatty foods. The smooth, creamy texture of fatty substances, together with their appealing taste, makes fats a welcome part of our diet. Unfortunately, a diet containing large amounts of saturated fatty acids has been shown to increase the risk of heart disease and other circulatory problems. Saturated fats are found in popular foods like fatty meat and dairy products (including such favorites as butter, cheese, and ice cream).

Some unsaturated fats, by contrast, are thought to decrease the risk of heart disease. Vegetable oils contain a mixture of monounsaturated and polyunsaturated fatty acids. Current research indicates that monounsaturated fats may be more effective than polyunsaturated fats in lowering the risk of heart disease. According to current research, perhaps the healthiest choice is oleic acid, an 18-carbon monounsaturated fatty acid particularly abundant in olive and canola oils. Surprisingly, compounds called *trans* fatty acids, produced during the manufacturing of some margarine's and vegetable shortenings, appear to increase the risk of heart disease. Margarine's containing these substances may be no healthier for you than butter, so it is important to check the label before you buy.

Eskimos have lower rates of heart disease than do other populations, even though the Eskimo diet is high in fats and cholesterol. Interestingly, the fatty acids in the Eskimo diet have an unsaturated bond three carbon's before the last, or omega, carbon, a position known as "omega minus 3," or *omega-3*. Fish flesh and fish oils, a substantial portion of the

Eskimo diet, contain omega-3 fatty acids. Why does the presence of omega-3 fatty acids (or some other unidentified component of fish) in the diet reduce the risks of heart disease, rheumatoid arthritis, and other inflammatory diseases? The answer is not yet apparent, but as you can imagine, there is a great deal of interest in this area of research.

Subsequent reactions can yield a **diglyceride** (glycerol + two fatty acids) and then a **triglyceride** (glycerol + three fatty acids), as in Figure 2–15●. Hydrolysis breaks the glycerides into fatty acids and glycerol. Compare Figure 2–15● with Figure 2–11● to convince yourself that dehydration synthesis and hydrolysis operate the same way, whether the molecules involved are carbohydrates or lipids.

Triglycerides, also known as *triacylglycerols* or *neutral fats*, have three important functions:

1. *Energy Source.* Fatty deposits in specialized sites of the body represent a significant energy reserve. In times of need, the triglycerides are disassembled, yielding fatty acids that can be broken down to provide energy.
2. *Insulation.* Fat deposits under the skin serve as insulation, preventing heat loss to the environment. Heat loss across a layer of lipids is only about one-third that through other tissues.
3. *Protection.* A fat deposit around a delicate organ such as a kidney provides a cushion that protects against shocks or blows.

Triglycerides are stored in the body as lipid droplets within cells. The droplets absorb and accumulate lipid-soluble vitamins, drugs, or toxins that appear in body fluids. This accumulation has both positive and negative effects. For example, the body's lipid reserves retain both valuable lipid-soluble vitamins (A, D, E, K) and potentially dangerous lipid-soluble pesticides, such as DDT.

Steroids

Steroids are large lipid molecules that share a distinctive carbon framework. They differ in the carbon chains that are attached to the basic structure. Figure 2–16a● shows the structural formula of the steroid **cholesterol** (ko-LES-ter-ol; *chole-*, bile + *stereos*, solid). Cholesterol and related steroids are important for the following reasons:

- All animal cell membranes contain cholesterol. Cells need cholesterol to maintain their cell membranes, as well as for cell growth and division.
- Steroid hormones are involved in the regulation of sexual function. Examples include the sex hormones, such as *testosterone* and the *estrogens* (Figure 2–16b,c●).
- Steroid hormones are important in the regulation of tissue metabolism and mineral balance. Examples include the hormones of the adrenal cortex, called *corticosteroids*, and *calcitriol*, a hormone important in the regulation of calcium ion concentrations in the body.
- Steroid derivatives called *bile salts* are required for the normal processing of dietary fats. Bile salts, which are produced in the liver and secreted in bile, interact with lipids in the intestinal tract and facilitate the digestion and absorption of lipids.

Cholesterol is obtained from two sources: (1) by absorption from animal products in the diet and (2) by synthesis within the

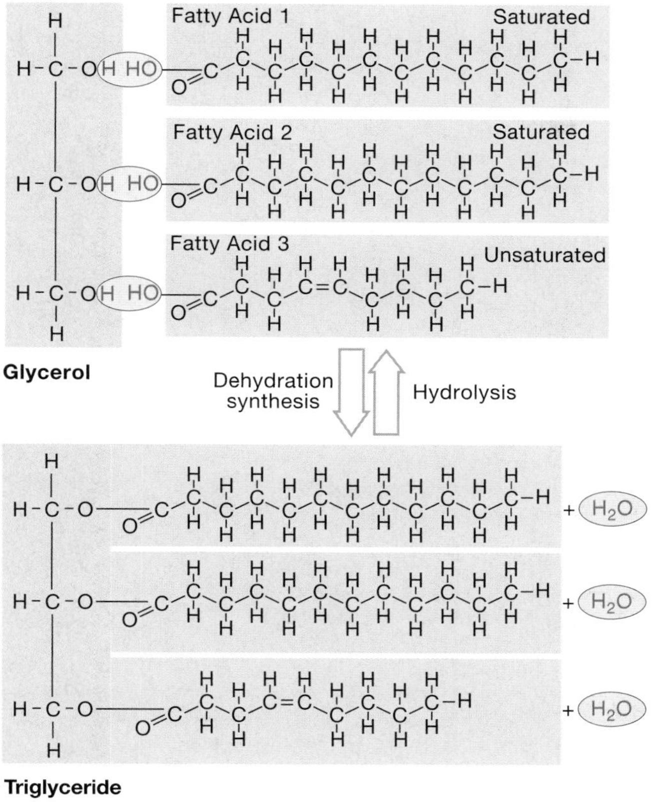

●FIGURE 2–15
Triglyceride Formation. The formation of a triglyceride involves the attachment of fatty acids to the carbon's of a glycerol molecule. In this example, a triglyceride is formed by the attachment of one unsaturated and two saturated fatty acids to a glycerol molecule.

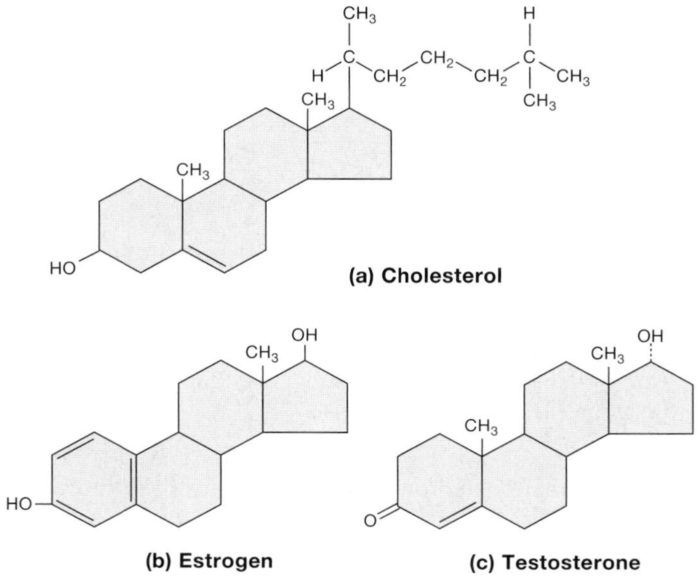

(a) Cholesterol

(b) Estrogen   **(c) Testosterone**

●FIGURE 2–16
Steroids. All steroids share a complex four-ring structure. Individual steroids differ in the side chains attached to the carbon rings.

body. Meat, cream, and egg yolks are especially rich dietary sources of cholesterol. A diet high in cholesterol can be harmful, because a strong link exists between high blood cholesterol levels and heart disease. Current nutritional advice suggests that you limit cholesterol intake to less than 300 mg per day. This amount represents a 40 percent reduction for the average adult in the United States. Unfortunately, because the body can synthesize cholesterol as well, blood cholesterol levels can be difficult to control by dietary restriction alone. We will examine the connection between blood cholesterol levels and heart disease more closely in later chapters.

Phospholipids and Glycolipids

Phospholipids (FOS-fō-lip-idz) and **glycolipids** (GLĪ-kō-lip-idz) are structurally related, and our cells can synthesize both types of lipids, primarily from fatty acids. In a *phospho*lipid, a *phosphate group* (PO_4^{3-}) links a diglyceride to a nonlipid group (Figure 2–17a●). In a *glyco*lipid, a carbohydrate is attached to a diglyceride (Figure 2–17b●). Note that placing -*lipid* last in these names indicates that the molecule consists primarily of lipid.

The long hydrocarbon tails of phospholipids and glycolipids are hydrophobic, but the opposite ends, the nonlipid *heads*, are hydrophilic. In water, large numbers of these molecules tend to form droplets, or *micelles* (mī-SELZ), with the hydrophilic portions on the outside (Figure 2–17c●). Most meals contain a mixture of lipids and other organic molecules, and micelles form as the food breaks down in your digestive tract. In addition to phospholipids and glycolipids, micelles may contain other insoluble lipids, such as steroids, glycerides, and long-chain fatty acids.

Cholesterol, phospholipids, and glycolipids are called *structural lipids*, because they help form and maintain intracellular structures called membranes. At the cellular level, *membranes* are sheets or layers composed primarily of hydrophobic lipids. A membrane surrounds each cell and separates the aqueous solution inside the cell from the aqueous solution in the extracellular environment. Because the two solutions are separated by a membrane, their compositions can be very different. A variety of internal membranes subdivide the interior of the cell into specialized compartments, each with a distinctive chemical nature and, as a result, a different function.

✓ A food contains organic molecules with the elements C, H, and O in a ratio of 1:2:1. What class of compounds do these molecules belong to, and what are their major functions in the body?

✓ When two monosaccharides undergo a dehydration synthesis reaction, which type of molecule is formed?

✓ Which kind of lipid would be found in a sample of fatty tissue taken from beneath the skin?

✓ Which lipids would you find in human cell membranes?

Answers start on page Q-1

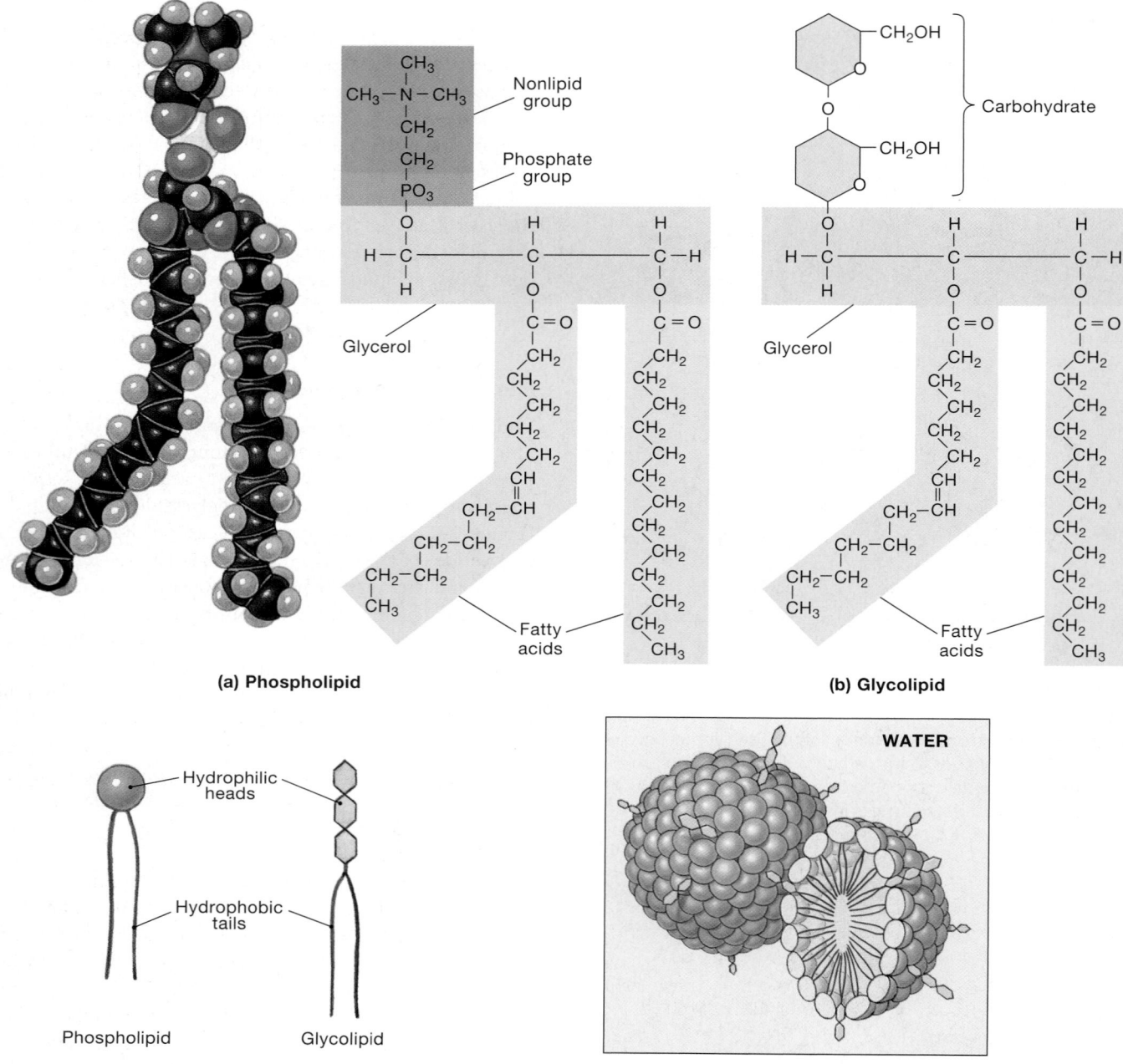

(a) **Phospholipid**

(b) **Glycolipid**

(c) **Micelle structure**

●**FIGURE 2–17**
Phospholipids and Glycolipids. **(a)** The phospholipid *lecithin*. In a phospholipid, a phosphate group links a nonlipid molecule to a diglyceride. **(b)** In a glycolipid, a carbohydrate is attached to a diglyceride. **(c)** In large numbers, phospholipids and glycolipids form micelles, with the hydrophilic heads facing the water molecules and the hydrophobic tails on the inside of each droplet.

☐ PROTEINS

Proteins are the most abundant organic components of the human body and in many ways the most important. The human body may contain as many as 2 million different proteins, and they account for about 20 percent of the total body weight. All proteins contain carbon, hydrogen, oxygen, and nitrogen; smaller quantities of sulfur may also be present.

Proteins perform a variety of essential functions, which can be classified into seven major categories:

1. *Support.* Structural proteins create a three-dimensional framework for the body, providing strength, organization, and support for cells, tissues, and organs.

2. *Movement.* Contractile proteins are responsible for muscular contraction; related proteins are responsible for the movement of individual cells.

3. *Transport.* Insoluble lipids, respiratory gases, special minerals such as iron, and several hormones cannot be transported in the blood, unless they are first bound to special *transport pro-*

teins. Other specialized proteins transport materials from one part of a cell to another.

4. *Buffering.* Proteins provide a considerable buffering action and thereby help prevent dangerous changes in pH in our cells and tissues.

5. *Metabolic Regulation. Enzymes* accelerate chemical reactions in cells. The sensitivity of enzymes to environmental factors is extremely important in controlling the pace and direction of metabolic operations.

6. *Coordination and Control.* Protein hormones can influence the metabolic activities of every cell in the body or affect the function of specific organs or organ systems.

7. *Defense.* The tough, waterproof proteins of the skin, hair, and nails protect the body from environmental hazards. Proteins called *antibodies,* components of the *immune response,* help protect us from disease. Special *clotting proteins* restrict bleeding after an injury to the cardiovascular system.

Structure of Proteins

Proteins consist of long chains of organic molecules called **amino acids**. Twenty types of amino acids occur in significant quantities in the body. A typical protein contains 1000 amino acids; the largest protein complexes have 100,000 or more. Each amino acid consists of a central carbon atom to which four groups are attached (Figure 2–18●):

- a hydrogen atom
- an *amino group* $(-NH_2)$
- a carboxylic acid group (–COOH)
- a variable group, known as an *R group* or *side chain*.

The name *amino acid* refers to the presence of the *amino* group and the carboxylic *acid* group, which all amino acids have in common. It is the different R groups that distinguish one amino acid from another, giving each its own chemical properties.

All amino acids are relatively small water-soluble molecules. The 20 amino acids commonly found in proteins are shown in Appendix III, enabling you to compare their R groups. In the normal pH range of body fluids, the carboxylic acid groups on many amino acids release hydrogen ions. When this occurs, the carboxylic acid group changes from –COOH to $-COO^-$, and the amino acids become negatively charged.

Figure 2–19● shows two representative amino acids: *glycine* and *alanine*. As the figure indicates, dehydration synthesis can link two amino acids. The process creates a covalent bond between the carboxylic acid group of one amino acid and the amino group of another. Such a bond is known as a **peptide bond**. Molecules consisting of amino acids held together by peptide bonds are called **peptides**. The molecule created in this example is called a *dipeptide,* because it contains two amino acids.

The chain can be lengthened by the addition of more amino acids. Attaching a third amino acid produces a *tripeptide;* next are *tetrapeptides, pentapeptides,* and so forth. Tripeptides and larger peptide chains are called **polypeptides**. Polypeptides containing

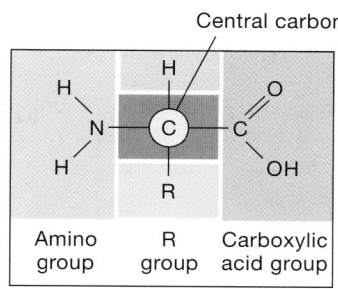

Central carbon

Amino group | R group | Carboxylic acid group

Structure of an amino acid

●FIGURE 2–18
Amino Acids.   Each amino acid consists of a central carbon atom to which four different groups are attached: a hydrogen atom, an amino group $(-NH_2)$, a carboxylic acid group (–COOH), and a variable group designated R.

more than 100 amino acids are usually called proteins. You may be familiar with the names of several important proteins, including *hemoglobin* in red blood cells and *keratin* in fingernails and hair. Because each protein contains amino acids that are negatively charged, the entire protein has a net negative charge. For that reason, proteins are often indicated by the abbreviation Pr^-.

Protein Shape

The characteristics of a particular protein are determined in part by the R groups on its component amino acids. But the properties of a protein are more than just the sum of the properties of

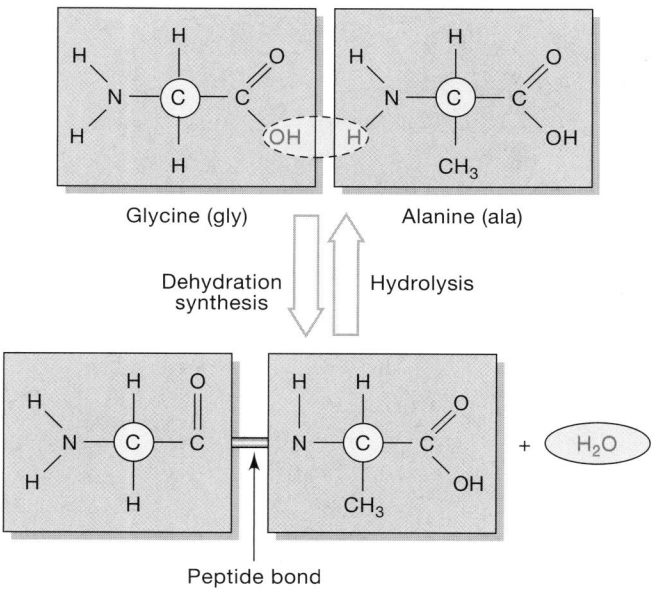

Glycine (gly)   Alanine (ala)

Dehydration synthesis   Hydrolysis

$+ \; H_2O$

Peptide bond

Peptide bond formation

●FIGURE 2–19
Peptide Bonds.   The amino acids glycine and alanine, linked to form a dipeptide. Peptides form as dehydration synthesis creates a peptide bond between the carboxylic acid group of one amino acid and the amino group of another.

its parts, for polypeptides can have highly complex shapes. Proteins have four levels of structural complexity:

1. **Primary structure** is the sequence of amino acids along the length of a single polypeptide. The primary structure of a short peptide chain is diagrammed in Figure 2–20a●.
2. **Secondary structure** results from bonds that develop between atoms at different parts of the polypeptide chain. Hydrogen

bonding, for example, may create a simple spiral, known as an *alpha-helix*, or a flat *pleated sheet* (Figure 2–20b●). Which one forms depends on the sequence of amino acids in the peptide chain and where hydrogen bonding occurs along the length of the peptide. The alpha-helix is the most common form, but a single polypeptide chain may have both helical and pleated sections.

3. **Tertiary structure** is the complex coiling and folding that gives the protein its final three-dimensional shape (Figure 2–20c●). Tertiary

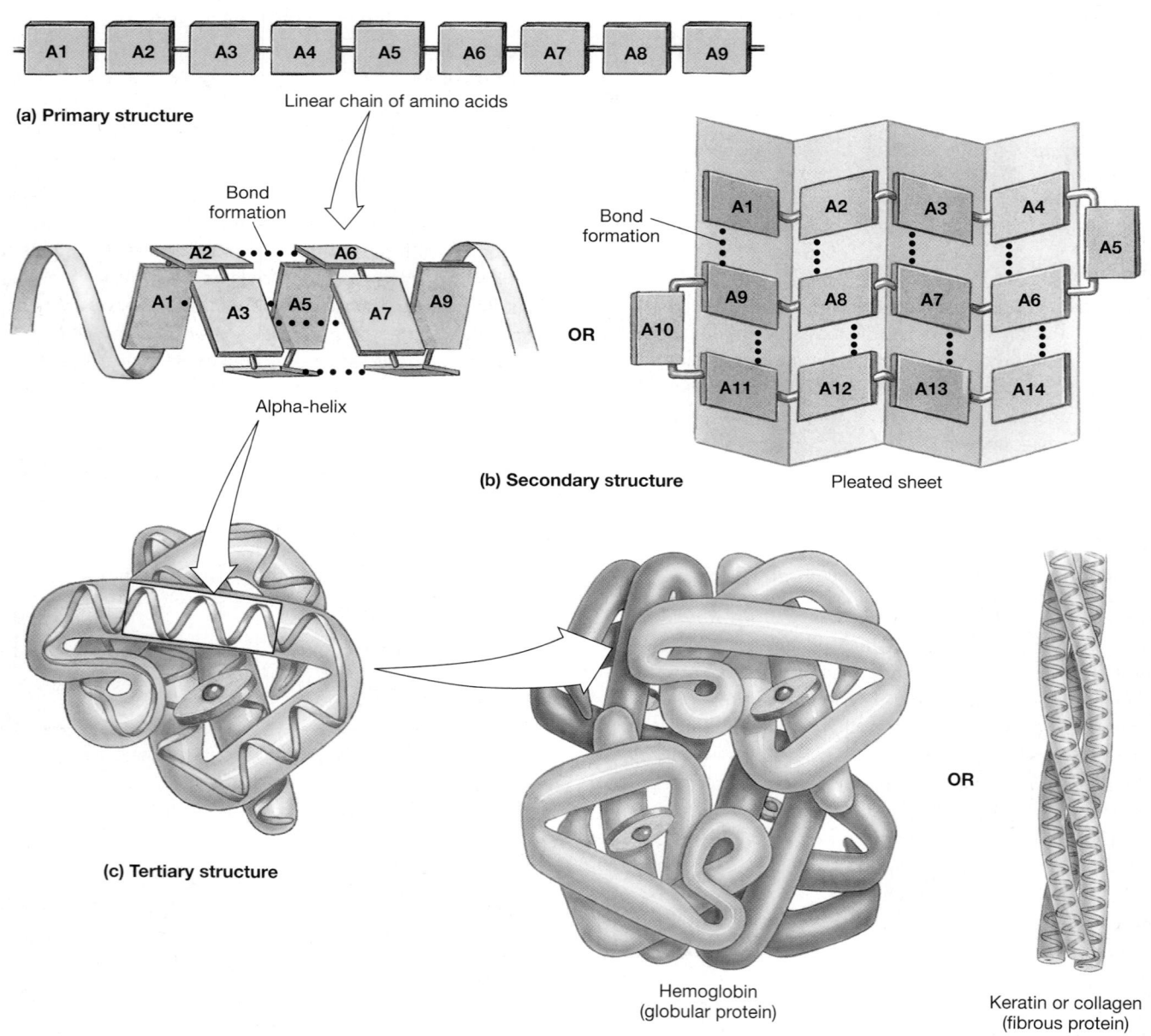

●**FIGURE 2–20**
Protein Structure. **(a)** The primary structure of a polypeptide is the sequence of amino acids (A1, A2, A3, and so on) along its length. **(b)** Secondary structure is primarily the result of hydrogen bonding along the length of the polypeptide chain. Such bonding often produces a simple spiral (an alpha-helix) or a flattened arrangement known as a pleated sheet. **(c)** Tertiary structure is the coiling and folding of a polypeptide. Within the cylindrical segments of this globular protein, the polypeptide chain is arranged in an alpha-helix. **(d)** Quaternary structure develops when separate polypeptide subunits interact to form a larger molecule. A single hemoglobin molecule contains four globular subunits. Hemoglobin transports oxygen in the blood; the oxygen binds reversibly to the heme units. In keratin and collagen, three fibrous subunits intertwine. Keratin is a tough, water-resistant protein in skin, hair, and nails. Collagen is the principal extracellular protein in most organs.

structure results primarily from interactions between the polypeptide chain and the surrounding water molecules, and to a lesser extent from interactions between the R groups of amino acids in different parts of the molecule. Most such interactions are relatively weak. One, however, is very strong: the *disulfide bond, a* covalent bond that may form between two molecules of the amino acid *cysteine,* each located at a different site along the chain. Disulfide bonds create permanent loops or coils in a polypeptide chain.

4. **Quaternary structure** is the interaction between individual polypeptide chains to form a protein complex (Figure 2–20d●). Each of the polypeptide subunits has its own secondary and tertiary structures. The protein *hemoglobin* contains four globular subunits. Hemoglobin is found within red blood cells, where it binds and transports oxygen. In *keratin* and *collagen,* three alpha-helical polypeptides are wound together like the strands of a rope. Keratin is the tough, water-resistant protein at the surface of the skin and in nails and hair. Collagen is the most abundant structural protein; collagen fibers form the framework that supports cells in most tissues.

FIBROUS AND GLOBULAR PROTEINS Proteins fall into two general structural classes on the basis of their overall shape and properties:

- **Fibrous proteins** take the form of extended sheets or strands. These shapes are usually the product of secondary structure (as in the case of proteins that exhibit the pleated-sheet configuration) or quaternary structure (as in the case of keratin and collagen). Fibrous proteins are tough, durable, and generally insoluble; in the body, they usually play structural roles.

- **Globular proteins** are compact, are generally rounded, and readily enter solution. The unique shape of each globular protein is the product of its tertiary structure. *Myoglobin,* a protein in muscle cells, is a globular protein, as is hemoglobin, the oxygen-carrying pigment in your red blood cells. Many enzymes, hormones, and other molecules that circulate in the bloodstream are globular proteins, as are the enzymes that control chemical reactions inside cells. These proteins can function only so long as they remain in solution.

SHAPE AND FUNCTION Proteins are extremely versatile and have a variety of functions. The shape of a protein determines its functional properties, and the ultimate determinant of shape is the sequence of amino acids. The 20 common amino acids can be linked in an astonishing number of combinations, creating proteins of enormously varied shape and function. Changing the identity of a single one of the 10,000 or more amino acids in a protein can significantly alter the protein's functional properties. For example, several cancers and *sickle cell anemia,* a blood disorder, result from single changes in the amino acid sequences of complex proteins.

The tertiary and quaternary shapes of complex proteins depend not only on their amino acid sequence, but also on the local environmental characteristics. Small changes in the ionic composition, temperature, or pH of their surroundings can thus affect the function of proteins. Protein shape can also be affected by hydrogen bonding to other molecules in solution. The significance of these factors is most striking when we consider the function of enzymes, for these proteins are essential to the metabolic operations that are underway in every one of our cells. Table 2–7 identifies the major types of proteins in the body.

TABLE 2–7 REPRESENTATIVE PROTEINS AND THEIR FUNCTIONS

Protein Type	Example(s)	Representative Location(s)	Function(s)
Structural proteins	Keratin	Skin surface, hair, nails	Provides strength and waterproofing
	Collagen	Dermis of skin, tendons	Provides strength
Contractile proteins	Actin, myosin	Muscle cells	Perform contraction and movement
Transport proteins	Albumin	Circulating blood	Transports fatty acids and steroid and thyroid hormones
	Transferrin	Circulating blood	Transports iron
	Apolipoproteins	Circulating blood	Transport glycerides
	Hemoglobin	Circulating blood	Transports oxygen in blood (in red blood cells)
Buffers	Intracellular and extracellular proteins	In cells and body fluids	Stabilize pH
Enzymes	Hydrolases	All cells	Catalyze hydrolysis of organic molecules
	Kinases	All cells	Attach phosphate groups to organic substrates
	Proteases	All cells; digestive secretions of stomach, pancreas	Break down proteins
	Carbohydrases	All cells; digestive secretions of salivary glands, pancreas	Break down carbohydrates
	Lipases	All cells; digestive secretions of pancreas	Break down lipids
Hormones	Insulin, glucagon	Circulating blood	Coordinate and/or control of metabolic activities
Antibodies (immunoglobulins)	Gamma globulins	Circulating blood	Attack foreign proteins and pathogens

Enzyme Function

Among the most important of all the body's proteins are the enzymes, first introduced earlier in this chapter. These molecules catalyze the reactions that sustain life: Almost everything that happens inside the human body does so because a specific enzyme makes it possible. **AM** Metabolic Anomalies

The reactants in enzymatic reactions are called **substrates**. As in other types of chemical reactions, the interactions among substrates yield specific products. Before an enzyme can function as a catalyst—to accelerate a chemical reaction without itself being permanently changed or consumed—the substrates must bind to a special region of the enzyme. This region, called the **active site**, is typically a groove or pocket into which one or more substrates nestle, like a key fitting into a lock. The physical fit is reinforced by weak electrical attractive forces, such as hydrogen bonding. The tertiary or quaternary structure of the enzyme molecule determines the shape of the active site. Although enzymes are proteins, any organic or inorganic compound that will bind to the active site can be a substrate.

Figure 2–21● diagrams enzyme function in a synthesis reaction. Substrates bind to the enzyme at its active site (step 1). It appears likely that substrate binding results in a temporary, reversible change in the shape of the protein and that this change furthers the reaction by placing physical stresses on the substrate molecules. The enzyme then promotes product formation (step 2). The completed product then detaches from the active site (step 3), and the enzyme is free to repeat the process. Enzymes work quickly, cycling rapidly between substrates and products. For example, an enzyme providing energy during a muscular contraction performs its reaction sequence 100 times per second.

The example in Figure 2–21● is an enzyme that catalyzes a synthesis reaction. Other enzymes may catalyze decomposition reactions or exchange reactions. Regardless of the reaction they catalyze, all enzymes share three basic characteristics:

1. *Specificity.* Each enzyme catalyzes only one type of reaction and can accommodate only one type of substrate molecule. This property is called **specificity**. The specificity of an enzyme is determined by the ability of a particular substrate to bind to the active site, which makes up a relatively small portion of the entire protein. Thus, differences in an enzymes structure that do not affect the active site or change the response of the enzyme to substrate binding do not affect the function of the enzyme. In fact, different tissues typically contain enzymes that differ slightly in structure, but catalyze the same reaction. Such enzyme variants are called **isozymes**.

2. *Saturation Limits.* The rate of an enzymatic reaction is directly proportional to the concentration of substrate molecules and enzymes. An enzyme molecule must encounter appropriate substrates before it can catalyze a reaction; the higher the substrate concentration, the more frequent encounters will be. When substrate concentrations are high enough that every enzyme molecule is cycling through its reaction sequence at top speed, further increases in substrate concentration will not affect the rate of reaction unless additional enzyme molecules are

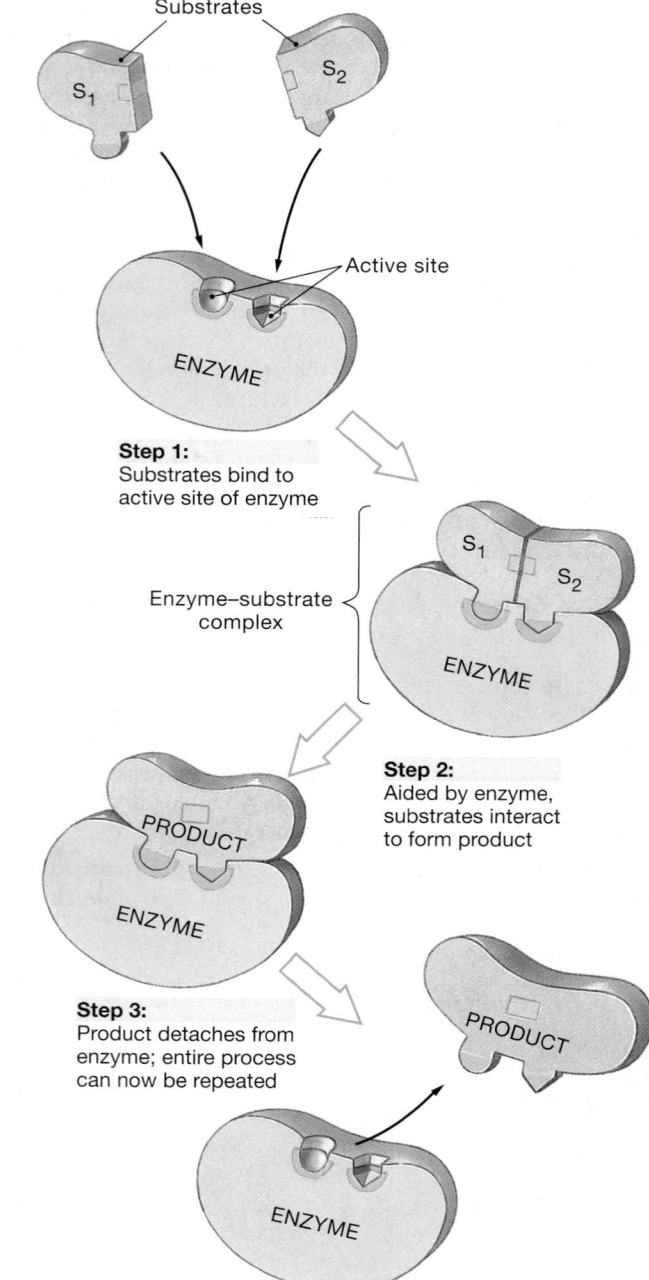

●**FIGURE 2–21**
A Simplified View of Enzyme Structure and Function.
Each enzyme contains a specific active site somewhere on its exposed surface.

provided. The substrate concentration required to have the maximum rate of reaction is called the *saturation limit.* An enzyme that has reached its saturation limit is said to be **saturated**. To increase the reaction rate further, the cell must increase the number of enzyme molecules available. This is one important way that cells promote specific reactions.

3. *Regulation.* A variety of factors can turn enzymes "on" or "off" and thereby control reaction rates inside the cell. We shall consider

only one example here: the presence or absence of *cofactors*. In fact, virtually anything that changes the tertiary or quaternary shape of an enzyme may turn it on or off. Each cell contains an assortment of enzymes, and any particular enzyme may be active under one set of conditions and inactive under another. Because the change is immediate, enzyme activation or inactivation is an important method of short-term control over reaction rates and pathways.

COFACTORS AND ENZYME FUNCTION A **cofactor** is an ion or a molecule that must bind to the enzyme before substrates can also bind. Without a cofactor, the enzyme is intact, but nonfunctional; with the cofactor, the enzyme can catalyze a specific reaction. Examples of cofactors include ions such as calcium (Ca^{2+}) and magnesium (Mg^{2+}), which bind at the enzymes active site. Cofactors may also bind at other sites, as long as they produce a change in the shape of the active site that makes substrate binding possible.

Coenzymes are nonprotein organic molecules that function as cofactors. Our bodies convert many vitamins into essential coenzymes. *Vitamins*, detailed in Chapter 25, are structurally related to lipids or carbohydrates, but have unique functional roles. The human body cannot synthesize most of the vitamins it needs, so you must obtain them from your diet.

TEMPERATURE AND PH Each enzyme works best at specific temperatures and pH values. As temperatures rise, protein shape changes and enzyme function deteriorates. Death occurs at very high body temperatures (above 43°C, or 110°F), because proteins undergo **denaturation**, a change in their tertiary or quaternary structure. Because denatured proteins are nonfunctional, the loss of structural proteins and enzymes soon causes irreparable damage to organs and organ systems.

Denaturation can be temporary or permanent. You see denaturation when you fry an egg. As the temperature rises, the proteins in the clear white denature. Eventually, the proteins become completely and irreversibly denatured, forming an insoluble white mass.

Enzymes are equally sensitive to changes in pH. *Pepsin*, an enzyme that breaks down proteins in the contents of your stomach, works best at a pH of 2.0 (strongly acidic). Your small intestine contains *trypsin*, another enzyme that attacks proteins. Trypsin works only in an alkaline environment, with an optimum pH of 7.7 (weakly basic).

Glycoproteins and Proteoglycans

Glycoproteins (GLĪ-kō-prō-tēnz) and **proteoglycans** (prō-tē-ō-GLĪ-kanz) are combinations of protein and carbohydrate molecules. Glyco*proteins* are large proteins with small carbohydrate groups attached. These molecules may function as enzymes, antibodies, hormones, or protein components of cell membranes. Glycoproteins in cell membranes play a major role in the identification of normal versus abnormal cells, as well as in the initiation and coordination of the immune response (Chapter 22). Glycoprotein secretions called *mucins* absorb water to form **mucus**. Mucus coats the surfaces of the respiratory and digestive tracts, providing lubrication. Proteo*glycans* are large polysaccha-ride molecules linked by polypeptide chains. The proteoglycans in tissue fluids give them a syrupy consistency.

✓ Proteins are chains of which small organic molecules?

✓ Which level of protein structure would be affected by an agent that breaks hydrogen bonds?

✓ Why does boiling a protein affect its structural and functional properties?

✓ How might a change in an enzymes active site affect its function?

Answers start on page Q-1

 Review protein folding and its relationship to enzyme structure and function by visiting the **Companion Website:** Chapter 2/Reviewing Concepts/Multiple Choice.

☐ NUCLEIC ACIDS

Nucleic (noo-KLĀ-ik) **acids** are large organic molecules composed of carbon, hydrogen, oxygen, nitrogen, and phosphorus. Nucleic acids store and process information at the molecular level, inside cells. The two classes of nucleic acid molecules are (1) **deoxyribonucleic** (dē-ok-sē-rī-bō-noo-KLĀ-ik) **acid**, or **DNA**, and (2) **ribonucleic** (rī-bō-noo-KLĀ-ik) **acid**, or **RNA**. As we shall see, these two classes of nucleic acids differ in composition, structure, and function.

The DNA in our cells determines our inherited characteristics, such as eye color, hair color, and blood type. DNA affects all aspects of body structure and function, because DNA molecules encode the information needed to build proteins. By directing the synthesis of structural proteins, DNA controls the shape and physical characteristics of our bodies. By controlling the manufacture of enzymes, DNA regulates not only protein synthesis, but all aspects of cellular metabolism, including the creation and destruction of lipids, carbohydrates, and other vital molecules.

Several forms of RNA cooperate to manufacture specific proteins by using the information provided by DNA. We will detail the functional relationships between DNA and RNA in Chapter 3.

Structure of Nucleic Acids

A nucleic acid consists of nucleotides linked by dehydration synthesis. Each **nucleotide** has three components: (1) a sugar, (2) a phosphate group, and (3) a **nitrogenous** (nitrogen-containing) **base**. The sugar is a *pentose* (five-carbon sugar)—either *ribose* (in RNA) or *deoxyribose* (in DNA). Each pentose is attached to a phosphate group and to a nitrogenous base. Five nitrogenous bases occur in nucleic acids: **adenine** (**A**), **guanine** (**G**), **cytosine** (**C**), **thymine** (**T**), and **uracil** (**U**) (Figure 2–22●). Adenine and guanine are double-ringed molecules called *purines;* the other three bases are single-ringed molecules called *pyrimidines*. Both RNA and DNA contain adenine, guanine, and cytosine. Uracil occurs only in RNA and thymine only in DNA.

In the formation of a nucleic acid, a nitrogenous base is attached to a pentose molecule. A nucleotide forms when a phosphate group

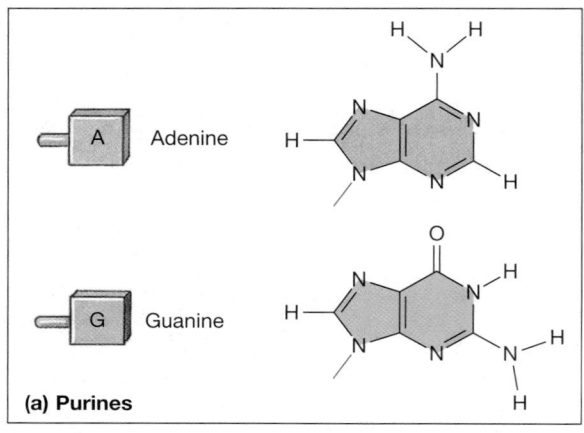

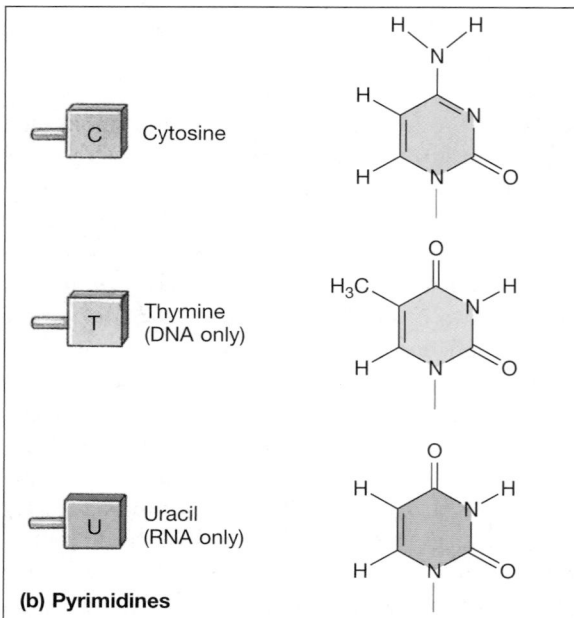

●**FIGURE 2–22**
Purines and Pyrimidines.   (a) Purines and (b) pyrimidines are the nitrogenous bases in nucleic acids.

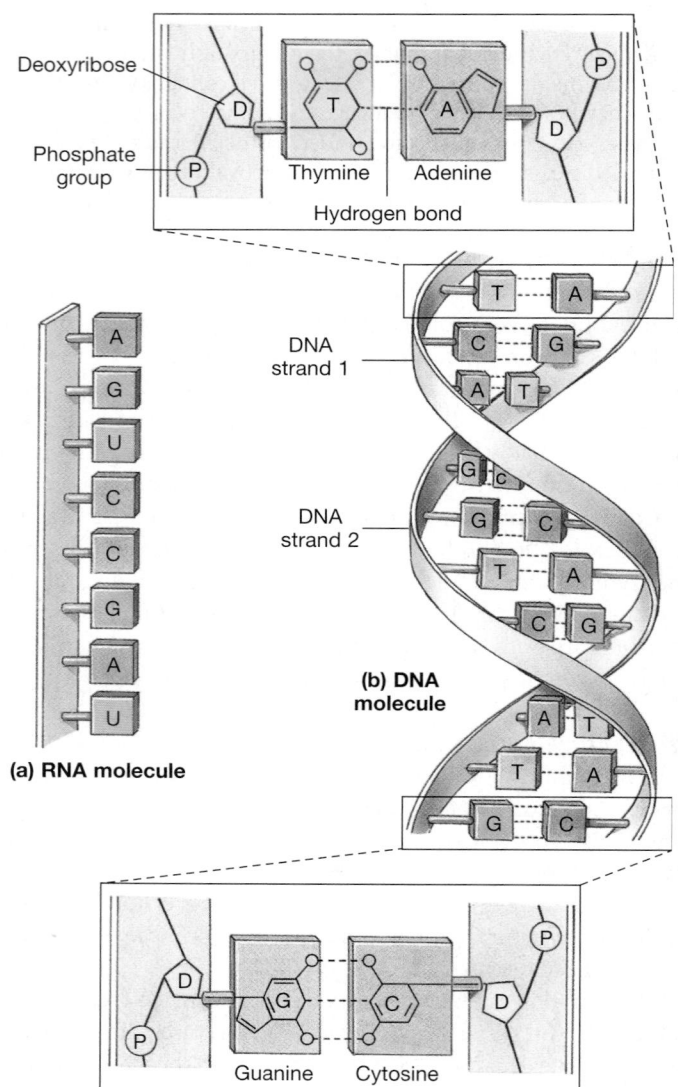

●**FIGURE 2–23**
Nucleic Acids: RNA and DNA.   Nucleic acids are long chains of nucleotides. Each molecule starts at the sugar of the first nucleotide and ends at the phosphate group of the last member of the chain. **(a)** An RNA molecule has a single nucleotide chain. Its shape is determined by the sequence of nucleotides and by the interactions among them. **(b)** A DNA molecule has a pair of nucleotide chains linked by hydrogen bonding between complementary base pairs.

binds to the sugar. Dehydration synthesis then attaches the phosphate group of one nucleotide to the carbohydrate of another. The "backbone" of a nucleic acid molecule is thus a linear sugar-to-phosphate-to-sugar sequence, with the nitrogenous bases projecting to one side.

The primary role of nucleic acids is the storage and transfer of information—specifically, information essential to the synthesis of proteins within our cells. Regardless of whether we are speaking of DNA or RNA, *it is the sequence of nitrogenous bases that carries the information.*

RNA and DNA

Important structural differences distinguish RNA from DNA. A molecule of RNA consists of a single chain of nucleotides (Figure 2–23a●). Its shape depends on the order of the nucleotides and the interactions among them. Our cells have three types of RNA:

(1) *messenger RNA* (*mRNA*), (2) *transfer RNA* (*tRNA*), and (3) *ribosomal RNA* (*rRNA*). These types have different shapes and functions, but all three are required for the synthesis of proteins, as you will see in Chapter 3.

A DNA molecule consists of a *pair* of nucleotide chains (Figure 2–23b●). Hydrogen bonding between opposing nitrogenous bases holds the two strands together. The shapes of the nitrogenous bases allow adenine to bond only to thymine and cytosine to bond only to guanine. As a result, the combinations adenine–thymine and cytosine–guanine are known as **complementary base pairs**, and the two nucleotide chains of the DNA molecule are known as

TABLE 2–8 A COMPARISON OF RNA WITH DNA

Characteristic	RNA	DNA
Sugar	Ribose	Deoxyribose
Nitrogenous bases	Adenine	Adenine
	Guanine	Guanine
	Cytosine	Cytosine
	Uracil	Thymine
Number of nucleotides in typical molecule	Varies from fewer than 100 to about 50,000	Always more than 45 million
Shape of molecule	Varies with hydrogen bonding along the length of the strand; three main types (mRNA, rRNA, tRNA)	Paired strands coiled in a double helix
Function	Performs protein synthesis as directed by DNA	Stores genetic information that controls protein synthesis

complementary strands. Through a sequence of events described in the next chapter, the cell uses one of the two complementary DNA strands to provide the information for the formation of a single strand of messenger RNA. The instructions needed to synthesize a specific protein are spelled out in the sequence of nitrogenous bases along the mRNA strand. Ribosomal RNA "reads" those instructions and, with the help of transfer RNA, assembles amino acids in the proper sequence to create the desired protein.

The two strands of DNA twist around one another in a double helix that resembles a spiral staircase. Each step of the staircase corresponds to one complementary base pair. Table 2–8 compares RNA with DNA.

☐ HIGH-ENERGY COMPOUNDS

To perform their vital functions, cells must use energy, obtained by breaking down organic substrates (catabolism). To be useful, that energy must be transferred from molecule to molecule or from one part of the cell to another.

The usual method of energy transfer involves the creation of *high-energy bonds* by enzymes within our cells. A high-energy bond is a covalent bond whose breakdown releases energy the cell can harness. In your cells, a high-energy bond generally connects a phosphate group (PO_4^{3-}) to an organic molecule. The resulting complex is called a **high-energy compound**.

The attachment of a phosphate group to another molecule is called **phosphorylation** (fos-for-i-LĀ-shun). This process does not necessarily produce high-energy bonds. The creation of a high-energy compound requires (1) a phosphate group, (2) enzymes capable of catalyzing the reactions involved, and (3) suitable organic substrates to which the phosphate can be added.

The most important such substrate is *adenosine*, a combination of adenine and ribose, with two phosphate groups attached. This compound, **adenosine diphosphate** (**ADP**), is created by the phosphorylation of the nucleotide *adenosine monophosphate* (*AMP*), a building block of nucleic acids. A significant energy input is required to convert AMP to ADP, and the second phosphate is attached by a high-energy bond. Even more energy is re-

quired to add a third phosphate and thereby create the high-energy compound **adenosine triphosphate** (**ATP**). Figure 2–24● shows the structure of ATP.

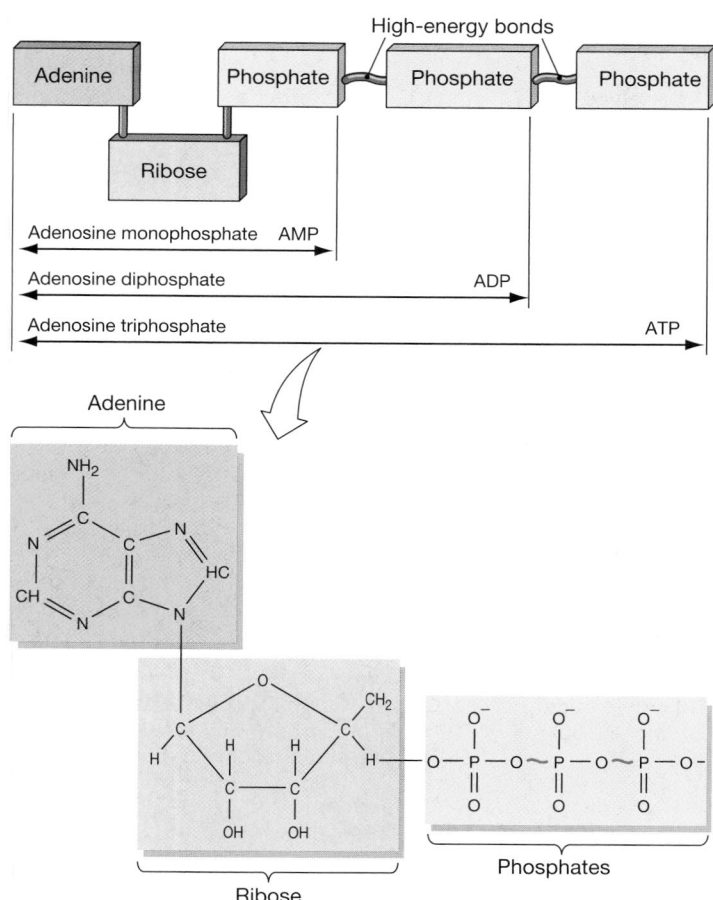

●**FIGURE 2–24**
The Structure of ATP. A molecule of ATP consists of adenosine (adenine plus ribose) to which three phosphate groups have been joined. Cells most often transfer energy by attaching a third phosphate group to ADP with a high-energy bond and then removing that phosphate group at another site, where the associated release of energy performs cellular work.

The conversion of ADP to ATP and the reversion of ATP to ADP are the most common methods of energy transfer in our cells:

$$ADP + phosphate\ group + energy \rightleftharpoons ATP + H_2O$$

The conversion of ATP to ADP requires an enzyme known as **adenosine triphosphatase**, or **ATPase**. Throughout life, our cells continuously generate ATP from ADP and use the energy provided by the ATP to perform vital functions, such as the synthesis of proteins or the contraction of muscles.

Although ATP is the most abundant high-energy compound, there are others—typically, other nucleotides that have undergone phosphorylation. For example, *guanosine triphosphate* (*GTP*) and *uridine triphosphate* (*UTP*) are nucleotide-based high-energy compounds that transfer energy in specific enzymatic reactions.

Table 2–9 summarizes information about the inorganic and organic compounds covered in this chapter.

SUMMARY TABLE 2–9 CLASSES OF INORGANIC AND ORGANIC COMPOUNDS			
Class	**Building Blocks**	**Sources**	**Functions**
INORGANIC			
Water (pp. 41–44)	Hydrogen and oxygen atoms	Absorbed as liquid water or generated by metabolism	Solvent; transport medium for dissolved materials and heat; cooling through evaporation; medium for chemical reactions; reactant in hydrolysis
Acids, bases, salts (pp. 44–45)	H^+, OH^-, various anions and cations	Obtained from the diet or generated by metabolism	Structural components; buffers; sources of ions
Dissolved gases (p. 41)	O, C, N, and other atoms	Atmosphere	O_2: required for cellular metabolism; CO_2: generated by cells as a waste product; NO: chemical messenger involved in cardiovascular, nervous, and lymphatic systems
ORGANIC			
Carbohydrates (pp. 45–47)	C, H, O, in some cases N; CHO in a 1:2:1 ratio	Obtained from the diet or manufactured in the body	Energy source; some structural role when attached to lipids or proteins; energy storage
Lipids (pp. 47–51)	C, H, O, in some cases N or P; CHO not in 1:2:1 ratio	Obtained from the diet or manufactured in the body	Energy source; energy storage; insulation; structural components; chemical messengers; protection
Proteins (pp. 52–57)	C, H, O, N, commonly S	20 common amino acids; roughly half can be manufactured in the body, others must be obtained from the diet	Catalysts for metabolic reactions; structural components; movement; transport; buffers; defense; control and coordination of activities
Nucleic acids (pp. 57–59)	C, H, O, N, and P; nucleotides composed of phosphates, sugars, and nitrogenous bases	Obtained from the diet or manufactured in the body	Storage and processing of genetic information
High-energy compounds (pp. 59–60)	Nucleotides joined to phosphates by high-energy bonds	Synthesized by all cells	Storage or transfer of energy

2-5 CHEMICALS AND CELLS

The human body is more than a collection of chemicals. Biochemical building blocks form functional units called **cells**. ∞ p. 6 Each cell behaves like a miniature organism, responding to internal and external stimuli. A phospholipid membrane separates the cell from its environment, and internal membranes create compartments with specific functions. Proteins form an internal supporting framework and, as enzymes, accelerate and control the chemical reactions that maintain homeostasis. Nucleic acids direct the synthesis of all cellular proteins, including the enzymes that enable the cell to synthesize a wide variety of other substances. Carbohydrates provide energy (transferred by high-energy compounds) to support vital activities, and they form part of specialized compounds such as proteoglycans and glycolipids.

Cells are dynamic structures that adapt to changes in their environment. Such adaptation may involve changes in the chemical organization of the cell—changes that are easily made because organic molecules other than DNA are temporary rather than permanent components of the cell. Their continuous removal and replacement are part of the process of **metabolic turnover**.

Most of the organic molecules in the cell are replaced at intervals ranging from hours to months. The average time between synthesis and recycling is known as the *turnover rate*. Table 2-10 lists the turnover rates of the organic components of representative cells.

TABLE 2-10	TURNOVER RATES	
Cell Type	Component	Average Recycling Time*
Liver	Total protein	5-6 days
	Enzymes	1 hour to several days, depending on the enzyme
	Glycogen	1-2 days
	Cholesterol	5-7 days
Muscle cell	Total protein	30 days
	Glycogen	12-24 hours
Neuron	Phospholipids	200 days
	Cholesterol	100+ days
Fat cell	Triglycerides	15-20 days

*Most values were obtained from studies on mammals other than humans.

✓ A large organic molecule is composed of the sugar ribose, nitrogenous bases, and phosphate groups. Which nucleic acid is this?

✓ What molecule is produced by the phosphorylation of ADP?

Answers start on page Q-1

Chapter Review

SELECTED CLINICAL TERMINOLOGY

Terms Discussed in This Chapter

cholesterol: A steroid, important in the structure of cellular membranes, that, in high concentrations, increases the risk of heart disease. (*p. 50*)

mole (mol), millimole (mmol): A quantity of an element or compound that has a weight in grams equal to its atomic or molecular weight, respectively; 1 mmol = 0.001 mol. (*p. 33*)

omega-3 fatty acids: Fatty acids, abundant in fish flesh and fish oils, that have a double bond three carbon's away from the end of the hydrocarbon chain. Their presence in the diet has been linked to reduced risks of heart disease and other conditions. (*p. 50*)

radioisotopes: Isotopes, with unstable nuclei, which spontaneously emit subatomic particles or radiation in measurable amounts. (*p. 33*)

AM Additional Terms Discussed in the *Applications Manual*

albinism
equivalent (Eq)
familial hypercholesterolemia
galactosemia
nuclear imaging
phenylketonuria
radiopharmaceuticals
tracer

STUDY OUTLINE

2-1 ATOMS, MOLECULES, AND BONDS   p. 31

1. Atoms are the smallest units of matter. They consist of **protons**, **neutrons**, and **electrons**. *(Figure 2–1)*

ATOMIC STRUCTURE   p. 32

2. The number of protons in an atom is its **atomic number**. Each **element** includes all the atoms that have the same number of protons and thus the same atomic number.
3. Within an atom, an **electron cloud** surrounds the nucleus. *(Figure 2–1; Table 2–1)*
4. The **mass number** of an atom is the total number of protons and neutrons in its nucleus. **Isotopes** are atoms of the same element whose nuclei contain different numbers of neutrons.
5. Electrons occupy a series of **energy levels**, commonly illustrated as **electron shells**. The electrons in the outermost energy level determine an atom's chemical properties. *(Figure 2–2)*

CHEMICAL BONDS   p. 34

6. Atoms can combine through chemical reactions that create **chemical bonds**. A **molecule** is any chemical structure consisting of atoms held together by covalent bonds. A **compound** is a chemical substance made up of atoms of two or more elements.
7. An **ionic bond** results from the attraction between **ions**, atoms that have gained or lost electrons. **Cations** are positively charged; **anions** are negatively charged. *(Figure 2–3)*
8. Atoms that share electrons to form a molecule are held together by **covalent bonds**. A sharing of one pair of electrons is a **single covalent bond**; a sharing of two pairs is a **double covalent bond**. A bond with equal sharing of electrons is a **nonpolar covalent bond**; a bond with unequal sharing of electrons is a **polar covalent bond**. *(Figures 2–4, 2–5)*
9. A **hydrogen bond** is a weak, but important, force that can affect the shapes and properties of molecules. *(Figure 2–6)*
10. Matter can exist as a *solid*, a *liquid*, or a *gas*, depending on the nature of the interactions among the component atoms or molecules.
11. The **molecular weight** of a molecule is the sum of the atomic weights of the component atoms.

FOCUS: Chemical Notation   p. 38

12. **Chemical notation** is the shorthand that allows us to describe chemical compounds and their reactions. *(Table 2–2)*

2-2 CHEMICAL REACTIONS   p. 39

1. A chemical reaction occurs when **reactants** combine to generate one or more **products**. Collectively, all the **chemical reactions** in the body constitute its **metabolism**. Through metabolism, cells capture, store, and use energy to maintain homeostasis and to support essential functions.

BASIC ENERGY CONCEPTS   p. 39

2. **Work** is the movement of an object or a change in its physical structure. **Energy** is the capacity to perform work.
3. **Kinetic energy** is the energy of motion. **Potential energy** is stored energy that results from the position or structure of an object. Conversions from potential to kinetic energy are not 100 percent efficient; every such energy conversion releases *heat*.

TYPES OF REACTIONS   p. 39

4. A chemical reaction is classified as a **decomposition**, **a synthesis**, or an **exchange reaction**.
5. Cells gain energy to power their functions by **catabolism**, the breakdown of complex molecules. Much of this energy supports **anabolism**, the synthesis of new molecules.

REVERSIBLE REACTIONS   p. 40

6. All chemical reactions are theoretically reversible. At **equilibrium**, the rates of two opposing reactions are in balance.

ENZYMES, ENERGY, AND CHEMICAL REACTIONS   p. 40

7. **Activation energy** is the amount of energy required to start a reaction. **Enzymes** are **catalysts**—substances that accelerate chemical reactions without themselves being permanently changed or used up. Enzymes promote chemical reactions by lowering the activation energy requirements. *(Figure 2–7)*
8. **Exergonic reactions** release energy; **endergonic reactions** require energy.

2-3 INORGANIC COMPOUNDS   p. 41

1. **Nutrients** are essential chemical substances normally obtained from the diet; **metabolites** are molecules synthesized or broken down by chemical reactions inside our bodies. Nutrients and metabolites can be categorized as inorganic or organic. Unlike organic compounds, **inorganic compounds** generally do not contain carbon and hydrogen atoms as their primary structural components.

WATER AND ITS PROPERTIES   p. 41

2. Water is the most important constituent of the body.
3. A **solution** is a uniform mixture of two or more substances. It consists of a medium, or **solvent**, in which atoms, ions, or molecules of another substance, or **solute**, are dispersed. In *aqueous solutions*, water is the solvent. *(Figure 2–8)*
4. Many inorganic compounds, called **electrolytes**, undergo *ionization*, or *dissociation*, in water to form ions *(Figure 2–8; Table 2–3)*. Compounds that interact readily with water molecules are called **hydrophilic**; those that do not interact with water molecules are called **hydrophobic**.
5. The **pH** of a solution indicates the concentration of hydrogen ions it contains. Solutions are classified as **neutral**, **acidic**, or **basic** (alkaline) on the basis of pH. *(Figure 2–9)*

INORGANIC ACIDS AND BASES   p. 44

6. An **acid** releases hydrogen ions; a **base** removes hydrogen ions from a solution. *Strong acids* and *strong bases* ionize completely, whereas *weak acids* and *weak bases* do not.

SALTS   p. 45

7. A **salt** is an electrolyte whose cation is not hydrogen (H^+) and whose anion is not hydroxide (OH^-).

BUFFERS AND PH CONTROL   p. 45

8. **Buffers** remove or replace hydrogen ions in solution. Buffers and *buffer systems* maintain the pH of body fluids within normal limits.

2–4 ORGANIC COMPOUNDS   p. 45

1. Carbon and hydrogen are the main constituents of **organic compounds**, which generally contain oxygen as well. Four major classes of organic compounds are *carbohydrates, lipids, proteins,* and *nucleic acids. High-energy compounds* are not abundant, but are vital to the survival of our cells. *(Table 2–4)*

CARBOHYDRATES   p. 45

2. **Carbohydrates** are most important as an energy source for metabolic processes. The three major types of carbohydrates are **monosaccharides** (*simple sugars*), **disaccharides**, and **polysaccharides**. Disaccharides and polysaccharides form from monosaccharides by **dehydration synthesis**. *(Figures 2-10–2-12; Table 2–5)*

LIPIDS   p. 47

3. **Lipids** are water-insoluble molecules that include *fats, oils,* and *waxes.* The five important classes of lipids are **fatty acids**, **eicosanoids**, **glycerides**, **steroids**, and **phospholipids** and **glycolipids**. *(Figures 2-13–2-17; Table 2–6)*

4. **Triglycerides** (*neutral fats*) consist of three fatty acid molecules attached by dehydration synthesis to a molecule of **glycerol**. **Diglycerides** consist of two fatty acids and glycerol. **Monoglycerides** consist of one fatty acid plus glycerol. *(Figure 2–15)*

5. Steroids (1) are involved in the structure of cell membranes, (2) include sex hormones and hormones regulating metabolic activities, and (3) are important in lipid digestion. *(Figure 2–16)*

PROTEINS   p. 52

6. **Proteins** perform a great variety of functions in the body. Six important types of proteins are *structural proteins, contractile proteins, transport proteins, enzymes, buffering proteins,* and *antibodies.*

7. Proteins are chains of **amino acids**. Each amino acid consists of an *amino group*, a *carboxylic acid group*, a hydrogen atom, and an *R group* (*side chain*) attached to a central carbon atom. A protein, or **polypeptide**, is a linear sequence of amino acids held together by **peptide bonds**. *(Figures 2–18, 2–19)*

8. The four levels of protein structure are **primary structure** (amino acid sequence), **secondary structure** (amino acid interactions, such as hydrogen bonds), **tertiary structure** (complex folding, *disulfide bonds*, and interaction with water molecules), and **quaternary structure** (formation of protein complexes from individual subunits). **Globular proteins**, such as *myoglobin*, are generally rounded and water-soluble. **Fibrous proteins**, such as *keratin* and *collagen*, are tough, durable, and generally insoluble. *(Figure 2–20)*

9. The reactants in an enzymatic reaction, called **substrates**, interact to yield a product by binding to the enzymes **active site**. **Cofactors** are ions or molecules that must bind to the enzyme before substrate binding can occur. **Coenzymes** are organic cofactors commonly derived from *vitamins*. *(Figure 2–21)*

10. The shape of a protein determines its functional characteristics. Each protein works best at an optimal combination of temperature and pH and will undergo temporary or permanent **denaturation** at temperatures or pH values outside the normal range.

   Protein folding: **Companion Website:** Chapter 2/Reviewing Concepts/Multiple Choice.

NUCLEIC ACIDS   p. 57

11. **Nucleic acids** store and process information at the molecular level. The two kinds of nucleic acids are **deoxyribonucleic acid** (**DNA**) and **ribonucleic acid** (**RNA**). *(Figures 2–22, 2–23; Table 2–8)*

12. Nucleic acids are chains of **nucleotides**. Each nucleotide contains a sugar, a phosphate group, and a **nitrogenous base**. The sugar is *ribose* in RNA and *deoxyribose* in DNA. DNA, which is a two-stranded double helix, contains the nitrogenous bases **adenine**, **guanine**, **cytosine**, and **thymine**. RNA, which consists of a single strand, contains **uracil** instead of thymine.

HIGH-ENERGY COMPOUNDS   p. 59

13. Cells store energy in the *high-energy bonds* of **high-energy compounds** for later use. The most important high-energy compound is **ATP** (**adenosine triphosphate**). Cells make ATP by adding a phosphate group to **ADP** (**adenosine diphosphate**) through **phosphorylation**. When ATP is broken down to ADP and phosphate, energy is released. The cell can use this energy to power essential activities. *(Figure 2–24; Summary Table 2–9)*

2–5 CHEMICALS AND CELLS   p. 61

1. Biochemical building blocks form functional units called **cells**.

2. The continuous removal and replacement of cellular organic molecules (other than DNA), a process called **metabolic turnover**, allows cells to change and to adapt to changes in their environment. *(Table 2–10)*

REVIEW QUESTIONS

 More assessment questions are available to you on the Companion Website. You will find Matching, Multiple Choice, True/False, and other quizzes to help further your understanding of the material covered in this chapter. To access the site, go to www.aw.com/martini.

LEVEL 1 Reviewing Facts and Terms

1. The lightest of an atom's main constituents
   (a) carries a negative charge
   (b) carries a positive charge
   (c) plays no part in the atom's chemical reactions
   (d) is found only in the nucleus

2. Isotopes of an element differ from each other in the number of
   (a) protons in the nucleus
   (b) neutrons in the nucleus
   (c) electrons in the outer shells
   (d) a, b, and c are all correct

3. The number and arrangement of electrons in an atom's outer energy level determines the atom's
   (a) atomic weight
   (b) atomic number
   (c) molecular weight
   (d) chemical properties

4. The bond between sodium and chlorine in the compound sodium chloride (NaCl) is
   (a) an ionic bond
   (b) a single covalent bond
   (c) a nonpolar covalent bond
   (d) a double covalent bond

Match each numbered item with the most closely related lettered item. Use letters for answers in the spaces provided.

_____ 5. atomic number   a. sum of atomic weights
_____ 6. mass number   b. insoluble in water
_____ 7. covalent bond   c. synthesis
_____ 8. ionic bond   d. catalyst
_____ 9. molecular weight   e. sharing of electrons
_____ 10. catabolism   f. A + B $\rightleftharpoons$ AB
_____ 11. anabolism   g. stabilize pH
_____ 12. exchange reaction   h. number of protons
_____ 13. reversible reaction i. decomposition
_____ 14. hydrophobic   j. carbohydrates, lipids, proteins
_____ 15. acid   k. loss or gain of electrons
_____ 16. enzyme   l. water, salts
_____ 17. buffer   m. H$^+$ donor
_____ 18. organic compounds   n. number of protons plus number of neutrons
_____ 19. inorganic compounds o. AB + CD $\rightarrow$ AD + CB

20. When atoms complete their outer electron shell by sharing electrons, they form
    (a) ionic bonds
    (b) covalent bonds
    (c) hydrogen bonds
    (d) anions and cations

21. All the chemical reactions that occur in the human body are collectively referred to as
    (a) anabolism   (b) catabolism
    (c) metabolism   (d) homeostasis

22. Which of the following equations illustrates a typical decomposition reaction?
    (a) A + B $\rightarrow$ AB
    (b) AB + CD $\rightarrow$ AD + CB
    (c) 2A$_2$ + B$_2$ $\rightarrow$ 2A$_2$B
    (d) AB $\rightarrow$ A + B

23. The speed, or rate, of a chemical reaction is influenced by
    (a) the presence of catalysts
    (b) the temperature
    (c) the concentration of the reactants
    (d) a, b, and c are all correct

24. A pH of 7.8 in the human body typifies a condition referred to as
    (a) acidosis   (b) alkalosis
    (c) dehydration   (d) homeostasis

25. A(n) _____ is a solute that dissociates to release hydrogen ions, and a(n) _____ is a solute that removes hydrogen ions from solution.
    (a) base, acid   (b) salt, base
    (c) acid, salt   (d) acid, base

26. Hydrophilic molecules readily associate with
    (a) lipid molecules
    (b) hydrophobic molecules
    (c) water molecules
    (d) a, b, and c are all correct

27. Carbohydrates, lipids, and proteins are classified as
    (a) organic molecules
    (b) inorganic molecules
    (c) acids
    (d) salts

28. Complementary base pairing in DNA includes the pairs
    (a) adenine–uracil and cytosine–guanine
    (b) adenine–thymine and cytosine–guanine
    (c) adenine–guanine and cytosine–thymine
    (d) guanine–uracil and cytosine–thymine

29. What are the three stable fundamental particles in atoms?

30. What four major classes of organic compounds are found in the body?

31. List three important functions of triglycerides (neutral fats) in the body.

32. List seven major functions performed by proteins.

33. (a) What three basic components make up a nucleotide of DNA?
    (b) What three basic components make up a nucleotide of RNA?

34. What three components are required to create the high-energy compound ATP?

LEVEL 2 Reviewing Concepts

35. If an isotope of oxygen has 8 protons, 10 neutrons, and 8 electrons, its mass number is
    (a) 26   (b) 16
    (c) 18   (d) 8

36. Of the following choices, the pH of the least acidic solution is
    (a) 6.0   (b) 4.5
    (c) 2.3   (d) 1.0

37. What is the difference between potential energy and kinetic energy?

38. Explain how enzymes function in chemical reactions.

39. What is a salt? How does a salt differ from an acid or a base?

40. Explain the differences among nonpolar covalent bonds, polar covalent bonds, and ionic bonds.

41. Explain the role of water molecules in polysaccharide formation.

42. Why does pure water have a neutral pH?

43. What role do buffer systems play in the human body?

44. A sample that contains an organic molecule has the following constituents: carbon, hydrogen, oxygen, nitrogen, and phosphorus. Is the molecule a carbohydrate, a lipid, a protein, or a nucleic acid?

LEVEL 3 Critical Thinking and Clinical Applications

45. Using the periodic table of the elements (Appendix II), determine the following information about calcium:

(a) number of protons   (b) number of electrons
(c) number of neutrons   (d) atomic number
(e) atomic weight   (f) number of electrons in each shell

46. The element sulfur has an atomic number of 16 and an atomic weight of 32. How many neutrons are in the nucleus of a sulfur atom? If sulfur forms covalent bonds with hydrogen, how many hydrogen atoms can bond to one sulfur atom?

47. An important buffer system in the human body involves carbon dioxide (CO_2) and bicarbonate ion (HCO_3^-) in the reaction

$$CO_2 + H_2O \rightleftharpoons H_2CO_3 \rightleftharpoons H^+ + HCO_3^-$$

If a person becomes excited and exhales large amounts of CO_2, how will the pH of the person's body be affected?

48. A student needs to prepare 1 liter of a solution of NaCl that contains 1.2 moles of ions per liter of solution. How many moles of NaCl should the student use to prepare the solution? What will that quantity weigh, in grams?

49. The concentration of salts in the body fluids averages 0.9%. Solutions of this concentration are often used in intravenous drips. How many grams of sodium chloride must you add to a liter of water to prepare a saline solution of 0.9% concentration?

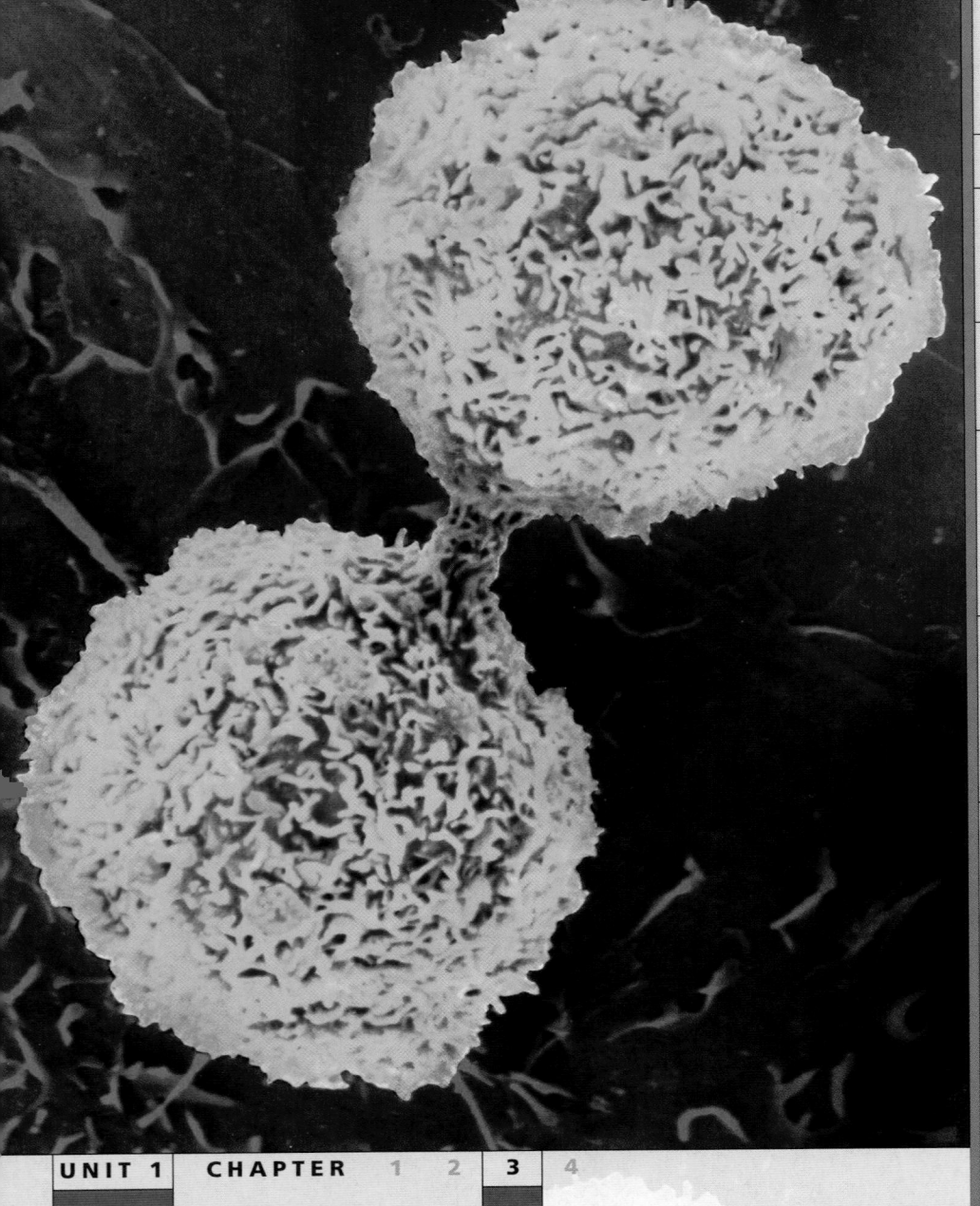

UNIT 1   CHAPTER   1   2   **3**   4

THE CELLULAR LEVEL OF ORGANIZATION

Many familiar structures, from pyramids to patchwork quilts, are made up of numerous small, similar components. The human body is built on the same principle; it is made up of several trillion tiny units called *cells*. Cells, however, differ from many other building blocks. For instance, they vary widely in size and appearance.

An individual brick can perform few (if any) of the functions of a building. But a cell can perform many of the functions of the body of which it is a part. Indeed, cells are the smallest entities that can perform all the basic life functions discussed in Chapter 1. Yet unlike a single-celled organism such as an amoeba, a cell in the human body must depend on its neighbors, and the actions of one cell can affect all others. Our cells have not only to survive, but also to work together in relative harmony, coordinating their activities. Each cell has a particular role to play, and in its own way each cell helps the body to maintain homeostasis.

3–1 AN INTRODUCTION TO CELLS

Objective

- List the main points of the cell theory.

Cells are very small indeed—a typical cell is only about 0.1 mm in diameter. As a result, no one could actually examine the structure of a cell until relatively effective microscopes were invented in the 17th century. In 1665, Robert Hooke inspected thin slices of cork and found that they consisted of millions of small, irregular units. Hooke described his observations in the publication *Micrographia* in the following terms:

> I could exceedingly plainly perceive it to be all perforated and porous, much like a Honey-comb, but that the pores of it were not regular. ... these pores, or cells, ... were indeed the first *microscopical* pores I ever saw, and perhaps, that were ever seen, for I had not met with any Writer or Person, that had made any mention of them before this.

Hooke used the term *cell* because the many small, bare spaces reminded him of the rooms, or cells, in a monastery or prison. Although Hooke saw only the outlines of the cells, and not the cells themselves, he stimulated considerable interest in the microscopic world and in the nature of cellular life. The research that he began over 175 years ago has, over time, produced the *cell theory* in its current form. The basic concepts of this theory can be summarized as follows:

1. Cells are the building blocks of all plants and animals.
2. All cells come from the division of preexisting cells.
3. Cells are the smallest units that perform all vital physiological functions.
4. Each cell maintains homeostasis at the cellular level.
5. Homeostasis at the level of the tissue, organ, organ system, and organism reflects the combined and coordinated actions of many cells.

Cells have a variety of forms and functions. Figure 3–1● gives examples of the range of sizes and shapes of cells in the human body. The relative proportions of the cells in the figure are correct, but all have been magnified roughly 500 times. Together, these and other types of cells create and maintain all anatomical structures and perform all vital physiological functions.

The human body contains trillions of cells, and all our activities—from running to thinking—result from the combined and coordinated responses of millions or even billions of cells. Yet each cell also functions as an individual entity, responding to a variety of environmental cues. As a result, anyone interested in understanding how the human body functions must first become familiar with basic concepts of cell biology. **AM** The Nature of Pathogens

Many insights into human physiology arose from studies of the functioning of individual cells. What we have learned over the last 50 years has given us a new understanding of cellular physiology and the mechanisms of homeostatic control. Today, the study of cellular structure and function, or **cytology**, is part of the broader discipline of **cell biology**, which incorporates aspects of biology, chemistry, and physics.

The two most common methods used to study cell and tissue structure are *light microscopy* and *electron microscopy*. A major difference between the two techniques is the level of resolution. Light microscopy permits us to see individual cells and the largest of their internal structures. Electron microscopy allows us to discern the smallest internal structures within cells. Within the pages that follow, you will find examples of each type of microscopy. **AM** Methods of Microanatomy

The human body contains two general classes of cells: sex cells and somatic cells. **Sex cells** (also called *germ cells* or *reproductive cells*) are either the *sperm* of males or the *oocytes* of females. The fusion of a sperm and an oocyte at fertilization is the first step in the creation of a new individual. **Somatic cells** (*soma*, body) include all the other cells in the human body. In this chapter, we focus on somatic cells; we shall discuss sex cells in Chapters 28 and 29, which deal with the reproductive system and development, respectively.

In the rest of this chapter, we describe the structure of a typical somatic cell, consider some of the ways in which cells interact with their environment, and discuss how somatic cells reproduce. It is important to keep in mind that the "typical" somatic cell is like the "average" person: Any description masks enormous

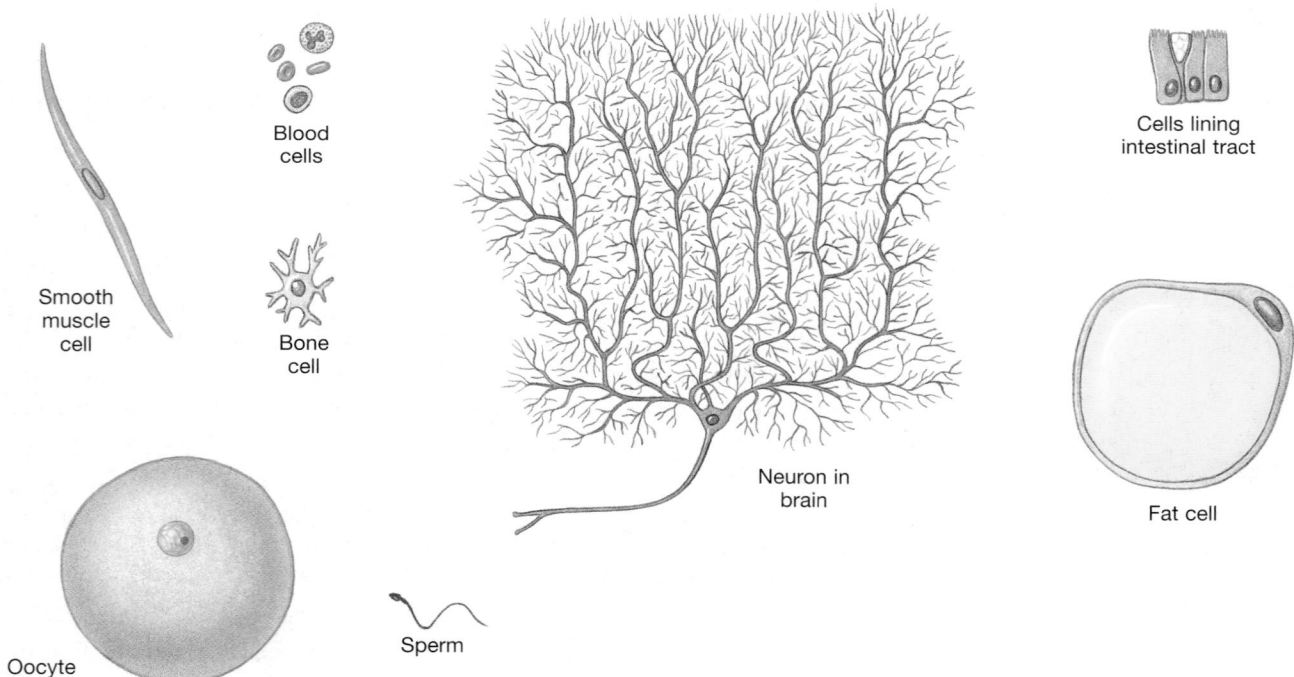

●**FIGURE 3–1**
The Diversity of Cells in the Human Body. The cells of the body, including these examples, have many different shapes and a variety of special functions. All the cells are shown with the dimensions they would have if magnified approximately 500 times.

individual variations. Our model cell will share features with most cells of the body, without being identical to any one. Figure 3–2● shows such a cell, and Table 3–1 summarizes the structures and functions of its parts.

Our representative cell is surrounded by a watery medium known as the **extracellular fluid**. The extracellular fluid in most tissues is called **interstitial** (in-ter-STISH-ul) **fluid** (*interstitium*, something standing between). A *cell membrane* separates the cell contents, or *cytoplasm*, from the extracellular fluid. The cytoplasm can itself be subdivided into (1) the *cytosol*, a liquid, and (2) intracellular structures collectively known as *organelles* (organ-ELZ; "little organs").

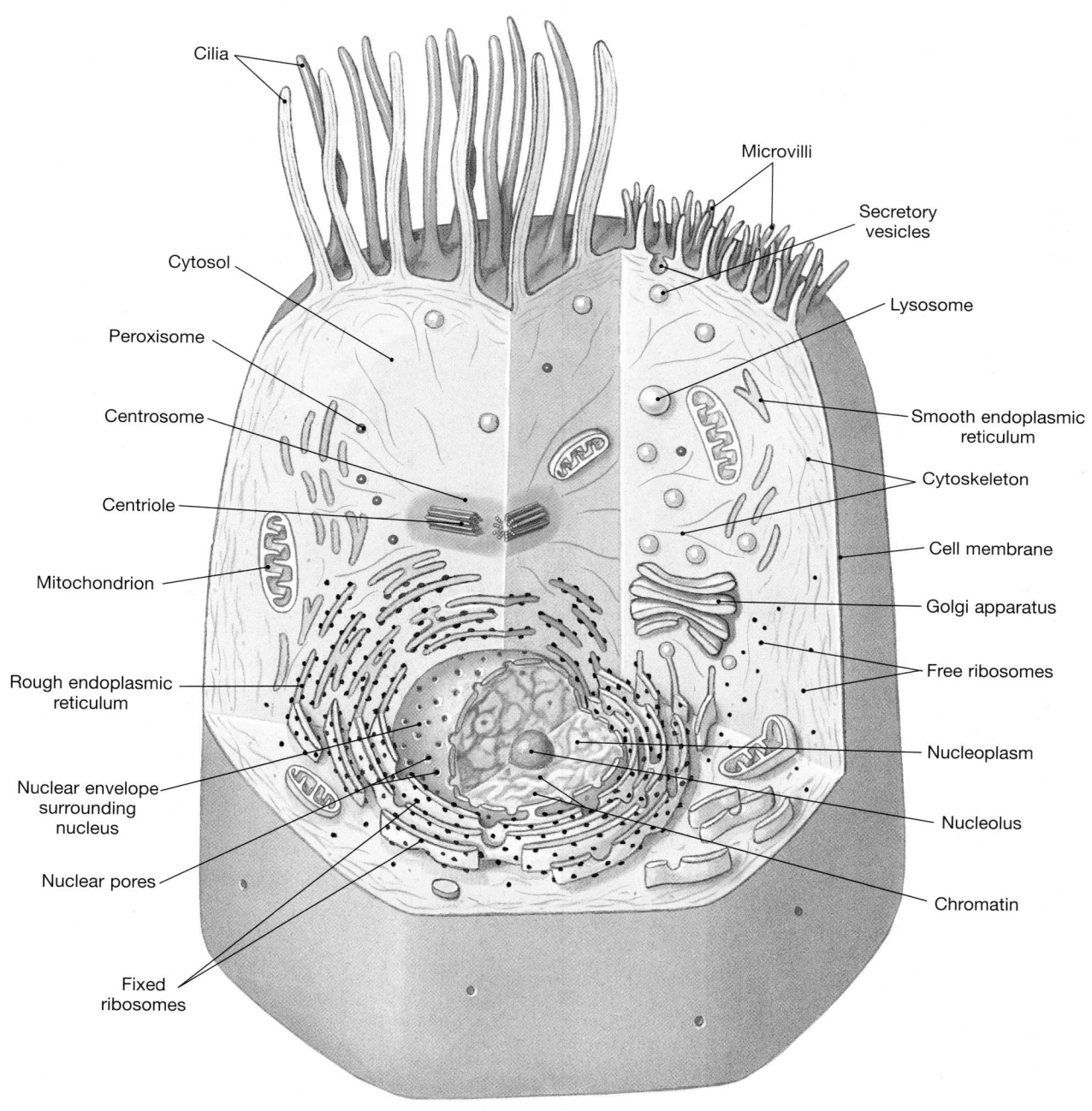

●**FIGURE 3–2**
The Anatomy of a Representative Cell.   See Table 3–1 for a summary of the functions associated with the various cell structures.

TABLE 3–1 THE ORGANELLES OF A REPRESENTATIVE CELL

Appearance	Structure	Composition	Function(s)
	CELL MEMBRANE	Lipid bilayer containing phospholipids, steroids, proteins, and carbohydrates	Isolation; protection; sensitivity; support; controls entry and exit of materials
	CYTOSOL	Fluid component of cytoplasm	Distributes materials by diffusion
	NONMEMBRANOUS ORGANELLES		
	Cytoskeleton Microtubule Microfilament	Proteins organized in fine filaments or slender tubes	Strength and support; movement of cellular structures and materials
	Microvilli	Membrane extensions containing microfilaments	Increase surface area to facilitate absorption of extracellular materials
	Centrosome Centriole	Cytoplasm containing two centrioles at right angles; each centriole is composed of 9 microtubule triplets in a 9 + 0 array	Essential for movement of chromosomes during cell division; organization of microtubules in cytoskeleton
	Cilia	Membrane extensions containing microtubule doublets in a 9 + 2 array	Movement of materials over cell surface
	Ribosomes	RNA + proteins; fixed ribosomes bound to rough endoplasmic reticulum, free ribosomes scattered in cytoplasm	Protein synthesis
	Proteasomes	Hollow cylinders of proteolytic enzymes with regulatory proteins at ends	Breakdown and recycling of damaged or abnormal intracellular proteins
	MEMBRANOUS ORGANELLES		
	Endoplasmic reticulum (ER)	Network of membranous channels extending throughout the cytoplasm	Synthesis of secretory products; intracellular storage and transport
	Rough ER (RER)	Has ribosomes bound to membranes	Modification and packaging of newly synthesized proteins
	Smooth ER (SER)	Lacks attached ribosomes	Lipid and carbohydrate synthesis
	Golgi apparatus	Stacks of flattened membranes (cisternae) containing chambers	Storage, alteration, and packaging of secretory products and lysosomal enzymes
	Lysosomes	Vesicles containing digestive enzymes	Intracellular removal of damaged organelles or pathogens
	Peroxisomes	Vesicles containing degradative enzymes	Catabolism of fats and other organic compounds; neutralization of toxic compounds generated in the process
	Mitochondria	Double membrane, with inner membrane folds (cristae) enclosing important metabolic enzymes	Produce 95% of the ATP required by the cell
	NUCLEUS Nuclear envelope	Nucleoplasm containing nucleotides, enzymes, nucleoproteins, and chromatin; surrounded by double membrane (nuclear envelope)	Control of metabolism; storage and processing of genetic information; control of protein synthesis
	Nucleolus	Dense region in nucleoplasm containing DNA and RNA	Site of rRNA synthesis and assembly of ribosomal subunits

3-2 THE CELL MEMBRANE

Objective

■ Describe the chief structural features of the cell membrane.

We will begin our look at the anatomy of cells by discussing the first structure you encounter when viewing cells through a microscope. The outer boundary of the cell is the **cell membrane**, also called the **plasma membrane** or *plasmalemma* (*lemma*, husk). Its general functions include the following:

- **Physical Isolation.** The cell membrane is a physical barrier that separates the inside of the cell from the surrounding extracellular fluid. Conditions inside and outside the cell are very different, and those differences must be maintained to preserve homeostasis. For example, the cell membrane keeps enzymes and structural proteins inside the cell.
- **Regulation of Exchange with the Environment.** The cell membrane controls the entry of ions and nutrients, such as glucose; the elimination of wastes; and the release of secretions.
- **Sensitivity.** The cell membrane is the first part of the cell affected by changes in the composition, concentration, or pH of

the extracellular fluid. It also contains a variety of receptors that allow the cell to recognize and respond to specific molecules in its environment. For instance, the cell membrane may receive chemical signals from other cells. The binding of just one molecule may trigger the activation or deactivation of enzymes that affect many cellular activities.

- **Structural Support.** Specialized connections between cell membranes or between membranes and extracellular materials give tissues a stable structure. For example, the cells at the surface of the skin are bound together, while those in the deepest layers are attached to extracellular protein fibers in underlying tissues.

The cell membrane is extremely thin and delicate, ranging from 6 to 10 nm in thickness. This membrane contains lipids, proteins, and carbohydrates.

☐ MEMBRANE LIPIDS

Figure 3-3● shows the structure of the cell membrane. Lipids form most of the surface area of the membrane, although they account for only about 42 percent of its weight. The cell membrane is called a **phospholipid bilayer**, because the phospholipid molecules in it form two layers. Recall from the previous chapter

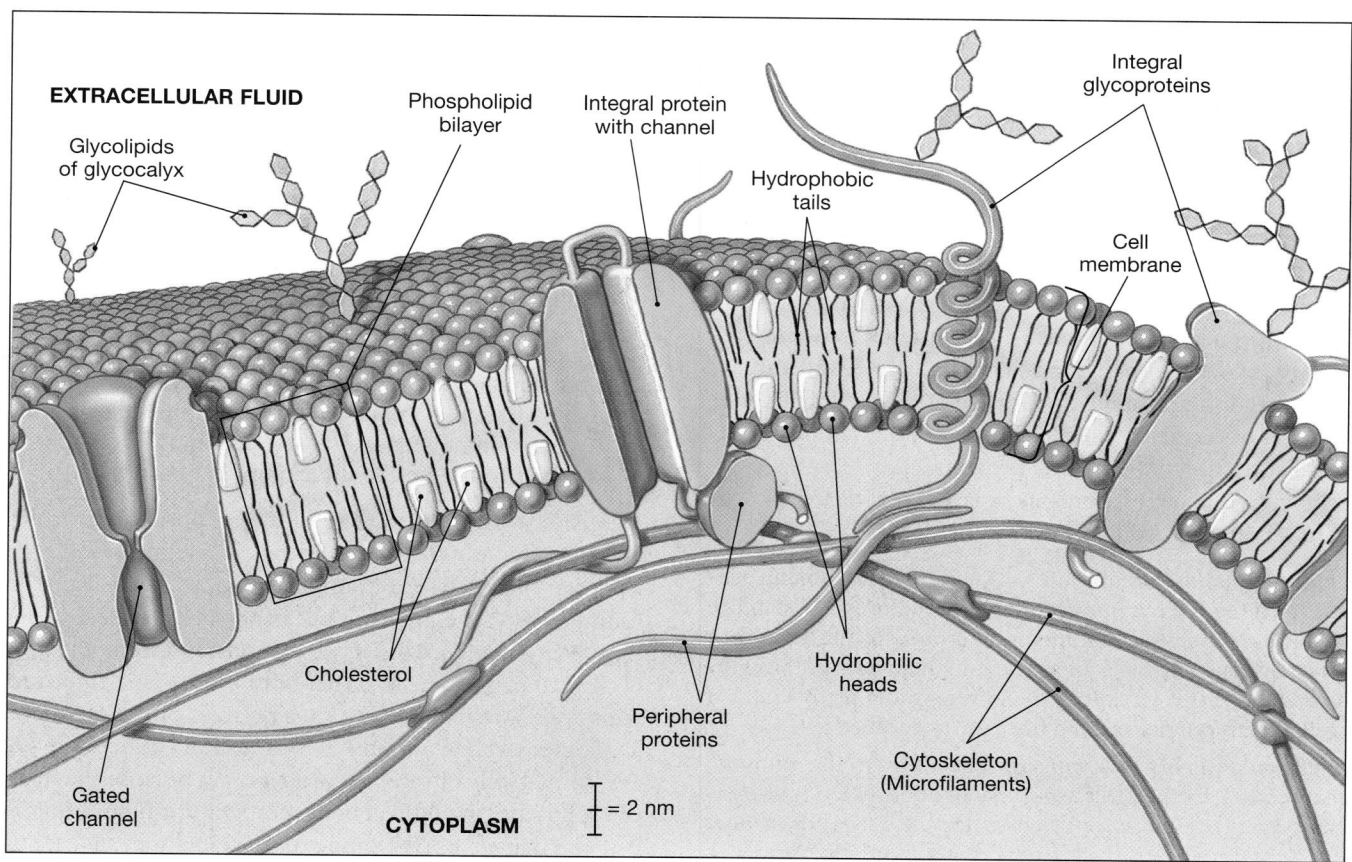

●**FIGURE 3-3**
The Cell Membrane

that a phospholipid has both a hydrophilic end (the phosphate portion) and a hydrophobic end (the lipid portion). ∞ p. 51 In each half of the bilayer, the phospholipids lie with their hydrophilic heads at the membrane surface and their hydrophobic tails on the inside. Thus, the hydrophilic heads of the two layers are in contact with the aqueous environments on either side of the membrane—the interstitial fluid on the outside and the cytosol on the inside—and the hydrophobic tails form the interior of the membrane. The lipid bilayer also contains cholesterol and small quantities of other lipids, but these have relatively little effect on the general properties of the cell membrane.

Notice the similarities in lipid organization between the cell membrane and a micelle (Figure 2–17c●, p. 52). Ions and water-soluble compounds cannot enter the interior of a micelle, because the lipid tails of the phospholipid molecules are hydrophobic and will not associate with water molecules. For the same reason, water and solutes cannot cross the lipid portion of the cell membrane. Thus, the hydrophobic compounds in the center of the membrane isolate the cytoplasm from the surrounding fluid environment. Such isolation is important because the composition of cytoplasm is very different from that of extracellular fluid, and the cell cannot survive if the differences are eliminated.

◻ MEMBRANE PROTEINS

Proteins, which are much denser than lipids, account for roughly 55 percent of the weight of a cell membrane. There are two general classes of membrane proteins. **Integral proteins** are part of the membrane structure and cannot be removed without damaging or destroying the membrane. Most integral proteins span the width of the membrane one or more times, and are therefore known as *transmembrane proteins*. **Peripheral proteins** are bound to the inner or outer surface of the membrane and are easily separated from it. In other words, they stick to the inner or outer surfaces of the membrane like Post-it® notes. Integral proteins greatly outnumber peripheral proteins.

Membrane proteins (Figure 3–4●) may have a variety of specialized functions, including the following:

1. *Anchoring Proteins.* Membrane proteins called **anchoring proteins** attach the cell membrane to other structures and stabilize its position. Inside the cell, membrane proteins are bound to the *cytoskeleton*, a network of supporting filaments in the cytoplasm. Outside the cell, other membrane proteins may attach the cell to extracellular protein fibers or to another cell.

2. *Recognition Proteins (Identifiers).* The cells of the immune system recognize other cells as normal or abnormal on the basis of the presence or absence of characteristic **recognition proteins**. Many important recognition proteins are glycoproteins. ∞ p. 57 (We will discuss one group, the *MHC proteins* involved in the immune response, in Chapter 22.)

3. *Enzymes.* Enzymes in cell membranes may be integral or peripheral proteins. These enzymes catalyze reactions in the extra-

cellular fluid or in the cytosol, depending on the location of the protein and its active site. For example, dipeptides are broken down into amino acids by enzymes on the exposed membranes of cells that line the intestinal tract.

4. *Receptor Proteins.* **Receptor proteins** in the cell membrane are sensitive to the presence of specific extracellular molecules called **ligands** (LĪ-gandz). A ligand can be anything from a small ion, like calcium (Ca^{2+}), to a relatively large and complex hormone. A receptor protein exposed to an appropriate ligand will bind to it, and that binding may trigger changes in the activity of the cell. For example, the binding of the hormone *insulin* to a specific membrane receptor protein is the key step that leads to an increase in the rate of glucose absorption by the cell. Cell membranes differ in the type and number of receptor proteins they contain, and these differences account for their differing sensitivities to hormones and other potential ligands.

5. *Carrier Proteins.* **Carrier proteins** bind solutes and transport them across the cell membrane. The transport process involves a change in the shape of the carrier protein. The shape changes when solute binding occurs, and the protein returns to its original shape when the solute is released. Carrier proteins may require ATP as an energy source. ∞ p. 59 For example, virtually all cells have carrier proteins that can bring glucose into the cytoplasm without expending ATP, but these cells must expend ATP to transport ions such as sodium and calcium across the cell membrane and out of the cytoplasm.

6. *Channels.* Some integral proteins contain a central pore, or **channel**, that forms a passageway completely across the cell membrane. The channel permits the movement of water and small solutes across the cell membrane. Ions do not dissolve in lipids, so they cannot cross the phospholipid bilayer. Thus, ions and other small water-soluble materials can cross the membrane only by passing through channels. Many channels are highly specific—that is, they permit the passage of only one particular ion. The movement of ions through channels is involved in a variety of physiological mechanisms. Although channels account for about 0.2 percent of the total surface area of the membrane, they are extremely important in physiological processes like nerve impulse transmission and muscle contraction, and you will learn more about them in Chapters 10 and 12.

Membranes are neither rigid nor uniform. At each location, the inner and outer surfaces of the cell membrane may differ in important respects. For example, some cytoplasmic enzymes are found only on the inner surface of the membrane, and some receptors are found exclusively on its outer surface. The properties of the cell membrane also vary with time. Some embedded proteins are always confined to specific areas of the cell membrane. These areas, called *rafts*, mark the location of anchoring proteins and some kinds of receptor proteins. Yet because the membrane phospholipids are fluid at body temperature, many other integral proteins drift across the surface of the membrane like ice cubes in a bowl of punch. In addition, the composition of the entire cell membrane can change over time, because large areas of the membrane surface are continually being removed and recycled in the process of metabolic turnover. ∞ p. 61

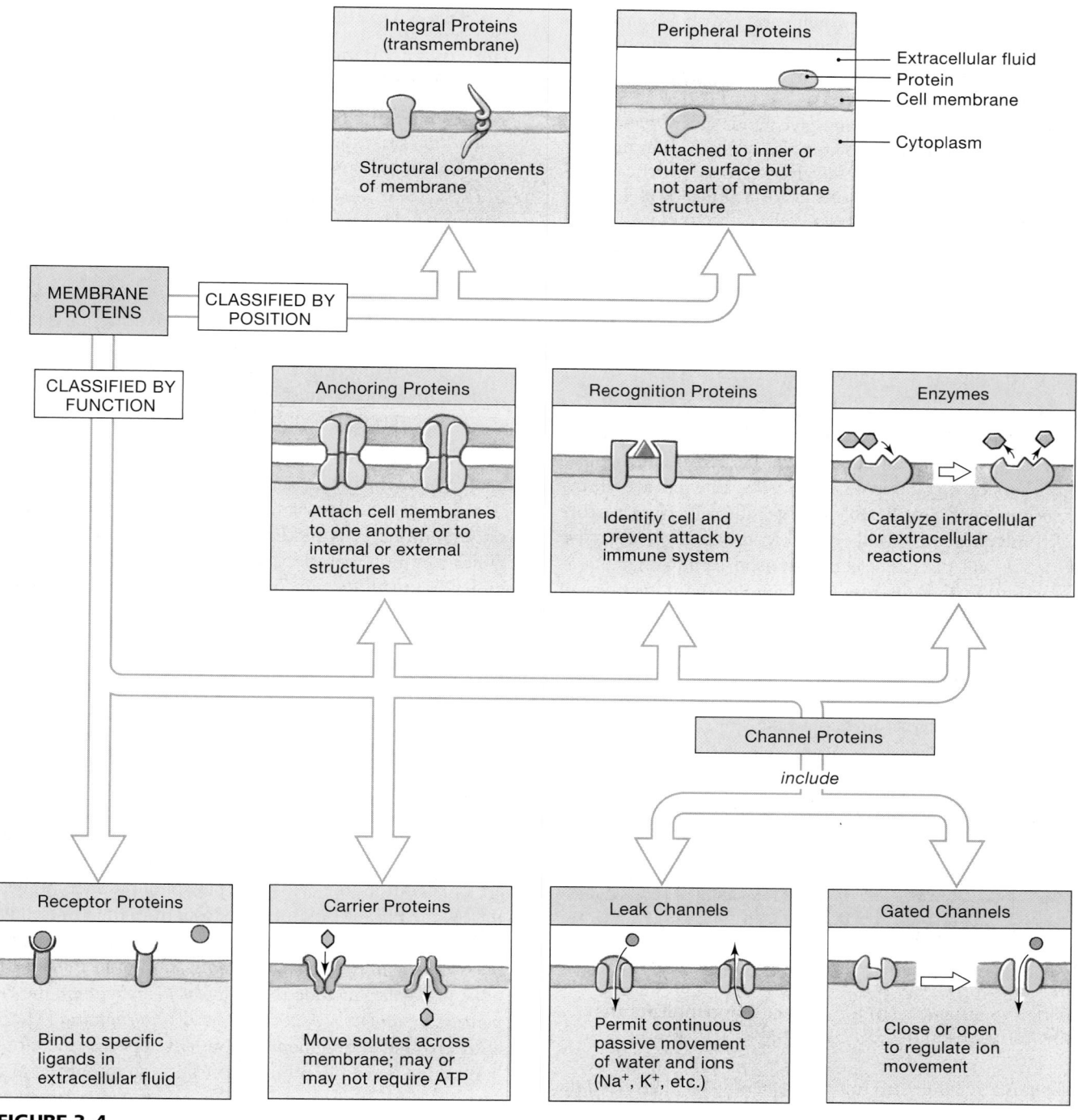

●**FIGURE 3–4**
Membrane Proteins

☐ MEMBRANE CARBOHYDRATES

Carbohydrates account for roughly 3 percent of the weight of a cell membrane. The carbohydrates in the cell membrane are components of complex molecules such as *proteoglycans, glycoproteins,* and *glycolipids.* ∞ p. 57 The carbohydrate portions of these large molecules extend beyond the outer surface of the membrane, forming a layer known as the **glycocalyx** (glī-kō-KĀ-liks; *calyx,* cup). The glycocalyx has a variety of important functions, including the following:

- **Lubrication and Protection.** The glycoproteins and glycolipids form a viscous layer that lubricates and protects the cell membrane.

- **Anchoring and Locomotion.** Because the components are sticky, the glycocalyx can help anchor the cell in place. It also participates in the locomotion of specialized cells.

- **Specificity in Binding.** Glycoproteins and glycolipids can function as receptors, binding specific extracellular compounds. Such binding can alter the properties of the cell surface and indirectly affect the cell's behavior.

- **Recognition.** Glycoproteins and glycolipids are recognized as normal or abnormal by cells involved with the immune response. The characteristics of the glycocalyx are genetically determined. For example, the presence or absence of membrane glycolipids on circulating red blood cells determines your blood type (A, B, AB, or O). The body's immune system can recognize its own membrane glycoproteins and glycolipids as "self" rather than as "foreign." This recognition system keeps your immune system from attacking your blood cells, while still enabling it to recognize and destroy foreign blood cells, should they appear in the bloodstream.

The cell membrane serves as a barrier between the cytosol and the extracellular fluid. In order for the cell to survive, dissolved substances and larger compounds must be permitted to move across this barrier. Metabolic wastes must be able to leave the cytosol, and nutrients must be able to enter the cell. The structure of the cell membrane is ideally suited to this need for selective transport. We will discuss selective transport and other membrane functions further, after we have completed our overview of cellular anatomy.

✓ Which component of the cell membrane is primarily responsible for the membrane's ability to form a physical barrier between the cell's internal and external environments?

✓ Which type of integral protein allows water and small ions to pass through the cell membrane?

Answers start on page Q-1

3–3 THE CYTOPLASM

Objective

- Describe the organelles of a typical cell, and indicate the specific functions of each.

Cytoplasm is a general term for the material located between the cell membrane and the membrane surrounding the nucleus. A colloid with a consistency that varies between that of thin maple syrup and almost-set gelatin, cytoplasm contains many more proteins than does extracellular fluid. ∞ p. 43 As an indication of the importance of proteins to the cell, 15 to 30 percent of the cell's weight can be attributed to proteins. The cytoplasm contains cytosol and organelles. **Cytosol,** or *intracellular fluid,* contains dissolved nutrients, ions, soluble and insoluble proteins, and waste products. The cell membrane separates the cytosol from the surrounding extracellular fluid. **Organelles** are structures suspended within the cytosol that perform specific functions within the cell.

THE CYTOSOL

The most important differences between cytosol and extracellular fluid are:

1. The concentration of potassium ions is higher in the cytosol than in the extracellular fluid. Conversely, the concentration of sodium ions is lower in the cytosol than in the extracellular fluid.

2. The cytosol contains a high concentration of suspended proteins. Many of the proteins are enzymes that regulate metabolic operations; others are associated with the various organelles. The consistency of the cytosol is determined in large part by these proteins. ∞ p. 56

3. The cytosol usually contains small quantities of carbohydrates and large reserves of amino acids and lipids. The carbohydrates are broken down to provide energy, and the amino acids are used to manufacture proteins. Lipids are used primarily as a source of energy when carbohydrates are unavailable.

The cytosol may also contain masses of insoluble materials known as **inclusions**. Among the most common inclusions are stored nutrients. For example, glycogen granules in liver or in skeletal muscle cells and lipid droplets in fat cells are inclusions. Other common inclusions are the pigment granules, such as the brown pigment *melanin* and the orange pigment *carotene.*

THE ORGANELLES

Organelles are the internal structures that perform most of the tasks needed to keep a cell alive and functioning normally. Each organelle has specific functions related to cell structure, growth, maintenance, and metabolism. Cellular organelles can be divided into two broad categories. **Nonmembranous organelles** are not completely enclosed by membranes, and all of their components are in direct contact with the cytosol. **Membranous organelles** are isolated from the cytosol by phospholipid membranes, just as the cell membrane isolates the cytosol from the extracellular fluid.

The cell's nonmembranous organelles include the *cytoskeleton, microvilli, centrioles, cilia, ribosomes,* and *proteasomes.* Membranous organelles include the *endoplasmic reticulum*, the *Golgi apparatus, lysosomes, peroxisomes,* and *mitochondria.* The *nucleus,* also surrounded by a membranous envelope and therefore, strictly speaking, a membranous organelle, is so important and has so many vital functions that we will consider it in a separate section.

The Cytoskeleton

The **cytoskeleton** functions as the cell's skeleton. It provides an internal protein framework that gives the cytoplasm strength and flexibility. The cytoskeleton of all cells includes *microfilaments, intermediate filaments,* and *microtubules.* Muscle cells contain *thick filaments* in addition to the other cytoskeletal elements. The filaments of the cytoskeleton form a dynamic network. The organizational details are as yet poorly understood, because the network is extremely delicate and thus hard to study in an intact state. Figure 3–5a● is based on our current knowledge of cytoskeletal structure.

We shall consider only a few of the many functions of the cytoskeleton in this section. In addition to the functions described here, the cytoskeleton plays a role in the metabolic organization of the cell by determining where in the cytoplasm key enzymatic reactions occur and where specific proteins are synthesized. For example, many intracellular enzymes, especially those involved with metabolism and energy production, and the ribosomes and RNA molecules responsible for the synthesis of proteins, are attached to the microfilaments and microtubules of the cytoskeleton. The varied metabolic functions of the cytoskeleton are now a subject of intensive research.

MICROFILAMENTS The most fragile of the cytoskeletal elements are the microfilaments. These protein strands are generally less than 6 nm in diameter. Typical microfilaments are composed of the protein **actin**. In most cells, the actin filaments are common in the periphery of the cell, but relatively rare in the region immediately surrounding the nucleus. In cells that form a layer or lining, such as the lining of the intestinal tract, actin filaments also form a layer, the *terminal web*, just inside the membrane at the exposed surface of the cell.

Microfilaments have three major functions:

1. Microfilaments anchor the cytoskeleton to integral proteins of the cell membrane. They provide additional mechanical strength to the cell and attach the cell membrane to the enclosed cytoplasm.

2. Microfilaments, interacting with other proteins, determine the consistency of the cytoplasm. The consistency can vary by region, depending on the nature of the interactions. For example, when microfilaments form a dense, flexible network, the cytoplasm has a gelatinous consistency; when they are widely dispersed, the cytoplasm is more fluid.

3. Actin can interact with the protein **myosin** to produce active movement of a portion of a cell or to change the shape of the entire cell.

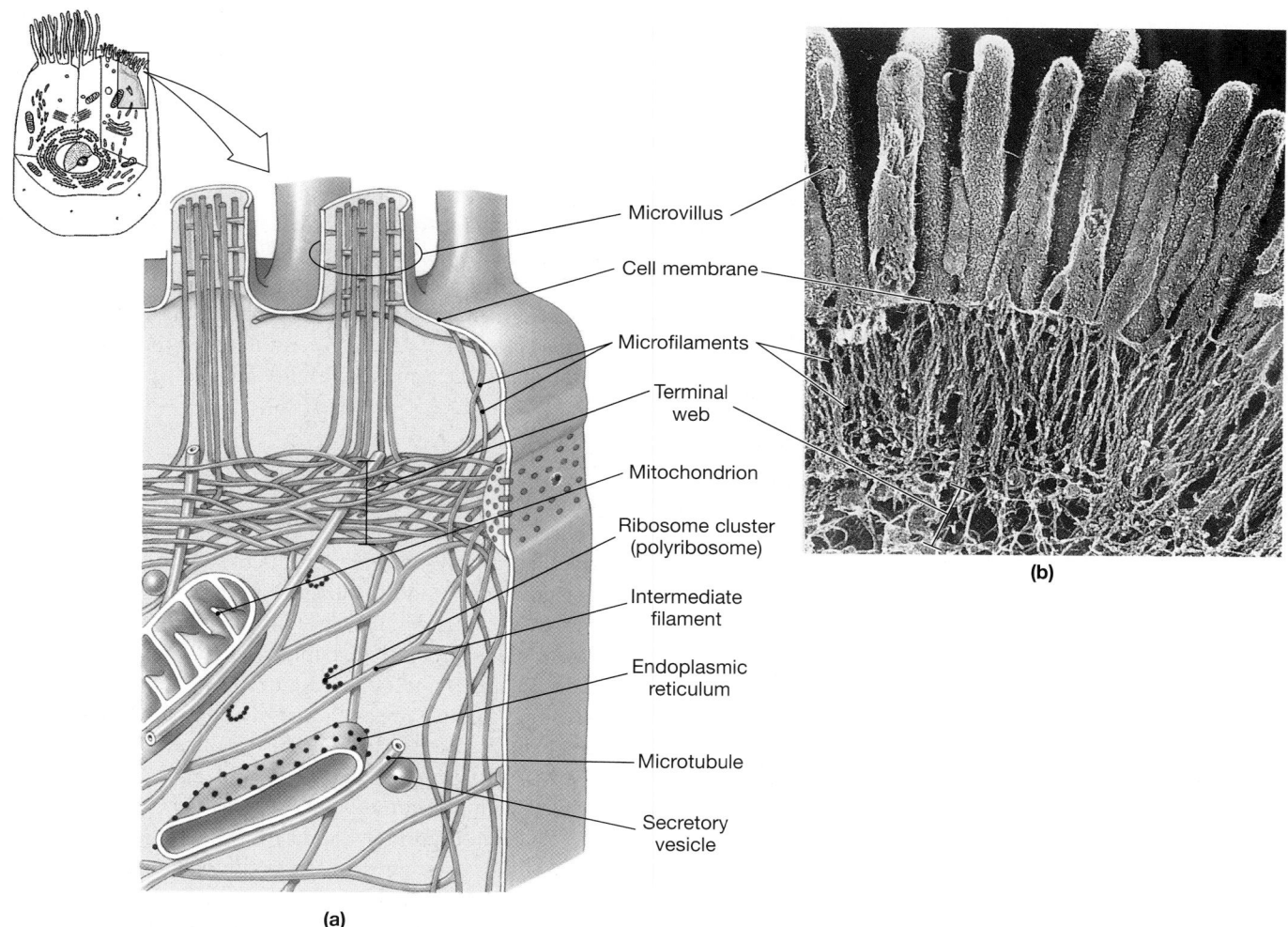

Microvillus

Cell membrane

Microfilaments

Terminal web

Mitochondrion

Ribosome cluster (polyribosome)

Intermediate filament

Endoplasmic reticulum

Microtubule

Secretory vesicle

(a)

(b)

●**FIGURE 3–5**
The Cytoskeleton. **(a)** The cytoskeleton provides strength and structural support for the cell and its organelles. Interactions between cytoskeletal components are also important in moving organelles and in changing the shape of the cell. **(b)** An SEM image of the microfilaments and microvilli of an intestinal cell.

INTERMEDIATE FILAMENTS The protein composition of **intermediate filaments** varies from one type of cell to another. These filaments, which range from 7 to 11 nm in diameter, are intermediate in size between microfilaments and thick filaments. Intermediate filaments (1) strengthen the cell and help maintain its shape, (2) stabilize the positions of organelles, and (3) stabilize the position of the cell with respect to surrounding cells through specialized attachment to the cell membrane. Intermediate filaments, which are insoluble, are the most durable of the cytoskeletal elements. Many cells contain specialized intermediate filaments with unique functions. For example, the collagen fibers in superficial layers of the skin are intermediate filaments that make these layers strong and able to resist stretching. ∞ pp. 54–55

MICROTUBULES All our cells contain **microtubules**, hollow tubes built from the globular protein **tubulin**. Microtubules are the largest components of the cytoskeleton, with diameters of about 25 nm. The microtubular array in a cell is centered near the nucleus, in a region known as the *centrosome* (see Figure 3–2●, p. 69.) From the centrosome, microtubules extend outward into the periphery of the cell. The number and distribution of microtubules present in the cell can change over time. Each microtubule forms by the aggregation of tubulin molecules, growing out from its origin at the centrosome. The entire structure persists for a time and then disassembles into individual tubulin molecules again. At any given moment, roughly half of the tubulin molecules in the cell are tied up in microtubules, and the rest are awaiting recycling.

Microtubules have the following functions:

1. Microtubules form the primary components of the cytoskeleton, giving the cell strength and rigidity and anchoring the position of major organelles.

2. The disassembly of microtubules provides a mechanism for changing the shape of the cell, perhaps assisting in cell movement.

3. Microtubules can serve as a kind of monorail system to move vesicles or other organelles within the cell. The movement is effected by proteins called *molecular motors*. These proteins, which bind to the structure being moved, gradually also bind to a microtubule and walk their way along it like a person climbing a rope hand over hand. The direction of movement depends on which of several known motor proteins is involved. For example, the molecular motors kinesin and dynein carry materials in opposite directions: If *kinesin* moves toward one end of a microtubule, *dynein* will move toward the other. Regardless of the direction of transport or the nature of the motor, the process requires ATP and is essential to normal cellular function.

4. During cell division, microtubules form the *spindle apparatus*, which distributes the duplicated chromosomes to opposite ends of the dividing cell. We shall consider this process in more detail in a later section.

5. Microtubules form structural components of organelles, such as *centrioles* and *cilia*.

THICK FILAMENTS **Thick filaments** are relatively massive bundles of myosin protein subunits that may reach 15 nm in diameter.

They appear only in muscle cells, where they interact with actin filaments to produce powerful contractions.

Microvilli

Many cells have small, finger-shaped projections of the cell membrane on their exposed surfaces (Figure 3–5b●). These projections, called **microvilli**, greatly increase the surface area of the cell exposed to the extracellular environment. Accordingly, they cover the surfaces of cells that are actively absorbing materials from the extracellular fluid, such as the cells lining the digestive tract. Microvilli have extensive connections with the cytoskeleton: A core of microfilaments stiffens each microvillus and anchors it to the cytoskeleton at the terminal web.

Centrioles

All animal cells capable of undergoing cell division contain a pair of **centrioles,** cylindrical structures composed of short microtubules (Figure 3–6a●). The microtubules form nine groups, three in each group. Each of the nine triplets is connected to its nearest neighbors on either side. Because there are no central microtubules, this organization is called a 9 + 0 *array*. (An axial structure with radial spokes leading toward the microtubular groups has also been observed, but its function is not known.)

During cell division, the centrioles form the spindle apparatus associated with the movement of DNA strands. Mature red blood cells, skeletal muscle cells, cardiac muscle cells, and typical neurons have no centrioles; as a result, these cells are incapable of dividing.

Centrioles are intimately associated with the cytoskeleton. The **centrosome**, the cytoplasm surrounding the centrioles, is the heart of the cytoskeletal system. Microtubules of the cytoskeleton generally begin at the centrosome and radiate through the cytoplasm.

Cilia

Cilia (singular, *cilium*) are relatively long, slender extensions of the cell membrane. They are found on cells lining the respiratory tract, on cells lining the reproductive tract, and at various other locations in the body. Cilia have an internal arrangement similar to that of centrioles. However, in cilia, there are nine *pairs* of microtubules (rather than triplets), and they surround a central pair (Figure 3–6b●)—an organization known as a 9 + 2 *array*. The microtubules are anchored to a compact **basal body** situated just beneath the cell surface. The organization of microtubules in the basal body resembles the 9 + 0 array of a centriole: nine triplets with no central pair.

Cilia are important because they can "beat" rhythmically to move fluids or secretions across the cell surface (Figure 3–6c●). The cilium is relatively stiff during the effective *power stroke* and flexible during the *return stroke*. The ciliated cells along your trachea beat their cilia in synchronized waves to move sticky mucus and trapped dust particles toward the throat and away from delicate respiratory surfaces. If the cilia are damaged or immobilized by heavy smoking or a metabolic problem, the cleansing action is

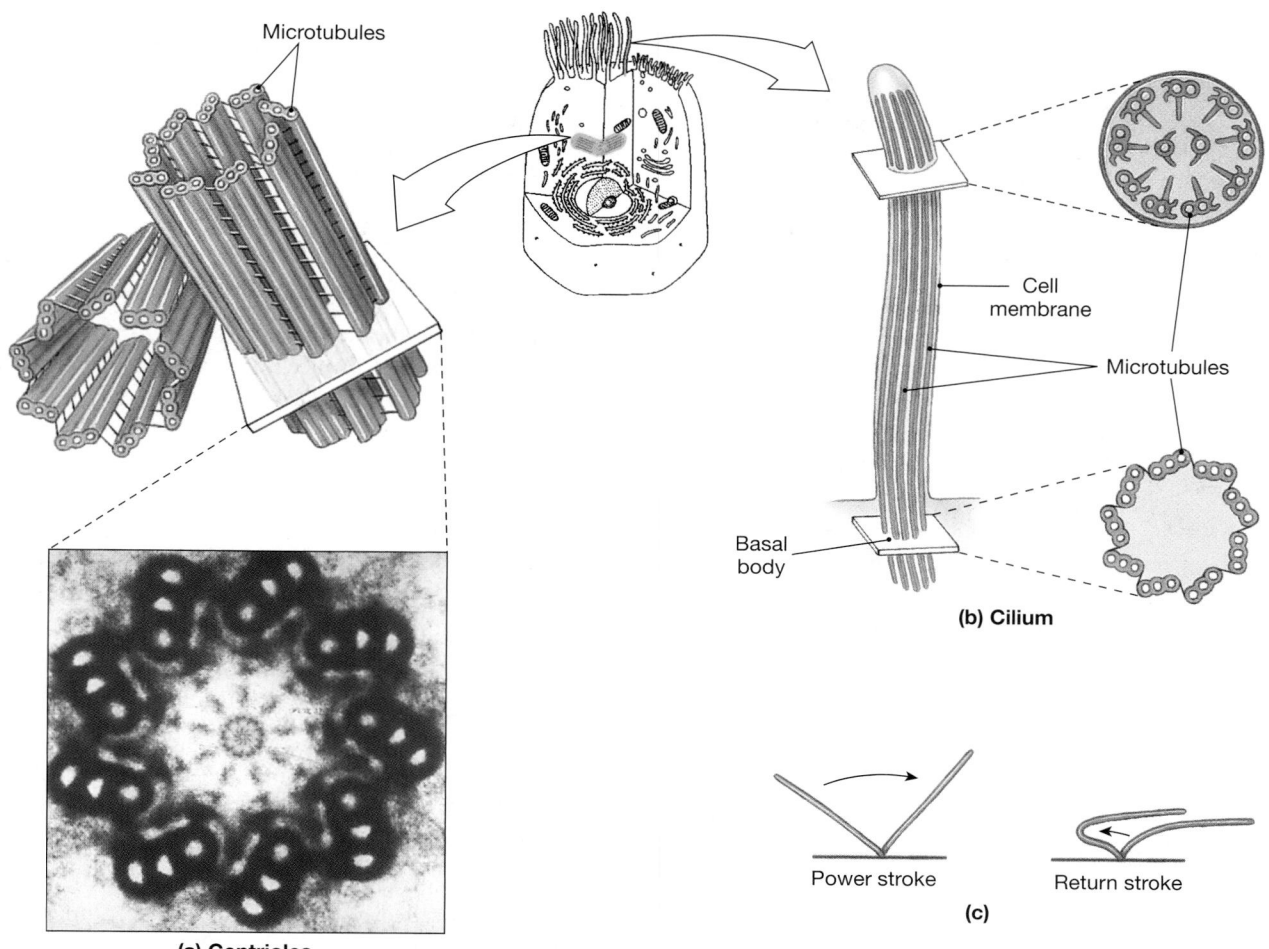

●**FIGURE 3–6**
Centrioles and Cilia. **(a)** A centriole consists of nine microtubule triplets (known as a 9 + 0 array). The centrosome contains a pair of centrioles oriented at right angles to one another. **(b)** A cilium contains nine pairs of microtubules surrounding a central pair (9 + 2 array). The basal body to which the cilium is anchored has a structure similar to that of a centriole. **(c)** A single cilium swings forward and then returns to its original position. During the power stroke, the cilium is relatively stiff; during the return stroke, it bends and moves back to its original position.

lost and the irritants will no longer be removed. As a result, chronic cough and respiratory infections develop. Ciliated cells are also responsible for moving oocytes along the uterine tubes, and wafting sperm from the testes into the male reproductive tract.

Ribosomes

Proteins are produced within cells, using information provided by the DNA of the nucleus. **Ribosomes** are the organelles responsible for protein synthesis. The number of ribosomes in a particular cell varies with the type of cell and its demand for new proteins. For example, liver cells, which manufacture blood proteins, contain far more ribosomes than do fat cells, which primarily synthesize lipids.

Individual ribosomes are not visible with the light microscope. In an electron micrograph, they appear as dense granules approximately 25 nm in diameter. Each ribosome consists of roughly 60 percent RNA and 40 percent protein.

A functional ribosome consists of two subunits that are normally separate and distinct. The subunits differ in size; one is called a **small ribosomal subunit** and the other a **large ribosomal subunit** (Figure 3–7●). These subunits contain special proteins and **ribosomal RNA (rRNA)**, one of the RNA types introduced in Chapter 2. ∞ p. 58 Before protein synthesis can begin, a small and a large ribosomal subunit must join together with a strand of *messenger RNA* (*mRNA*, another type of RNA).

Two major types of functional ribosomes are found in cells: (1) free ribosomes and (2) fixed ribosomes (Figure 3–2●, p. 69). **Free ribosomes** are scattered throughout the cytoplasm. The proteins they manufacture enter the cytosol. **Fixed ribosomes** are attached to the *endoplasmic reticulum* (*ER*), a membranous organelle. Proteins manufactured by fixed ribosomes enter the ER, where they are modified and packaged for secretion. We shall examine ribosomal structure and functions in later sections, when we deal with the endoplasmic reticulum and protein synthesis.

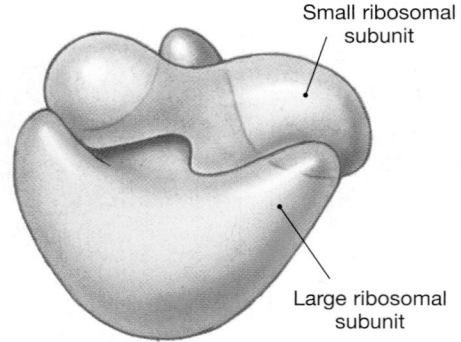

●**FIGURE 3–7**
Ribosomes. A diagrammatic view of the structure of an intact ribosome. The subunits are separate unless the ribosome is engaged in protein synthesis.

Proteasomes

Free ribosomes produce proteins within the cytoplasm; proteasomes remove them. **Proteasomes** are organelles that contain an assortment of protein-digesting enzymes, or *proteases*. ∞ p. 55 The enzymes aggregate to form a hollow cylinder, with the active sites facing the inner chamber. Special regulatory proteins block the ends of the cylinder and prevent the entry of normal intracellular proteins. Cytoplasmic enzymes attach chains of *ubiquitin*, a molecular "tag," to proteins destined for recycling. Tagged proteins are quickly transported into the proteasome. Once inside, they are rapidly disassembled into amino acids and small peptides, which can be released into the cytoplasm.

Proteasomes are responsible for removing and recycling damaged or denatured proteins, and for breaking down abnormal proteins, such as those produced within cells infected by viruses. They also play a key role in the immune response, as we shall see in Chapter 22.

✓ Cells lining the small intestine have numerous fingerlike projections on their free surface. What are these structures, and what is their function?

✓ What are the major differences between cytosol and extracellular fluid?

Answers start on page Q-1

The Endoplasmic Reticulum

The **endoplasmic reticulum** (en-dō-PLAZ-mik re-TIK-ū-lum), or **ER,** is a network of intracellular membranes connected to the *nuclear envelope*, which surrounds the nucleus. The name *endoplasmic reticulum* may seem odd, but is in fact very descriptive. *Endo* means "within," *plasm* refers to the cytoplasm, and a *reticulum* is a network. (Breaking unfamiliar words into their component parts in this way often makes it easier to understand and remember their meaning.) The ER has four major functions:

1. *Synthesis.* Specialized regions of the ER synthesize proteins, carbohydrates, and lipids.

2. *Storage.* The ER can store synthesized molecules or materials absorbed from the cytosol without affecting other cellular operations.

3. *Transport.* Materials can travel from place to place in the ER.

4. *Detoxification.* Drugs or toxins can be absorbed by the ER and neutralized by enzymes within it.

The ER (Figure 3–8●) forms hollow tubes, flattened sheets, and chambers called **cisternae** (sis-TUR-nē; singular, *cisterna*, a reservoir for water). Two types of ER exist: (1) *smooth endoplasmic reticulum* and (2) *rough endoplasmic reticulum*.

SMOOTH ENDOPLASMIC RETICULUM The term "smooth" refers to the fact that no ribosomes are associated with the **smooth endoplasmic reticulum (SER).** The SER has a variety of func-

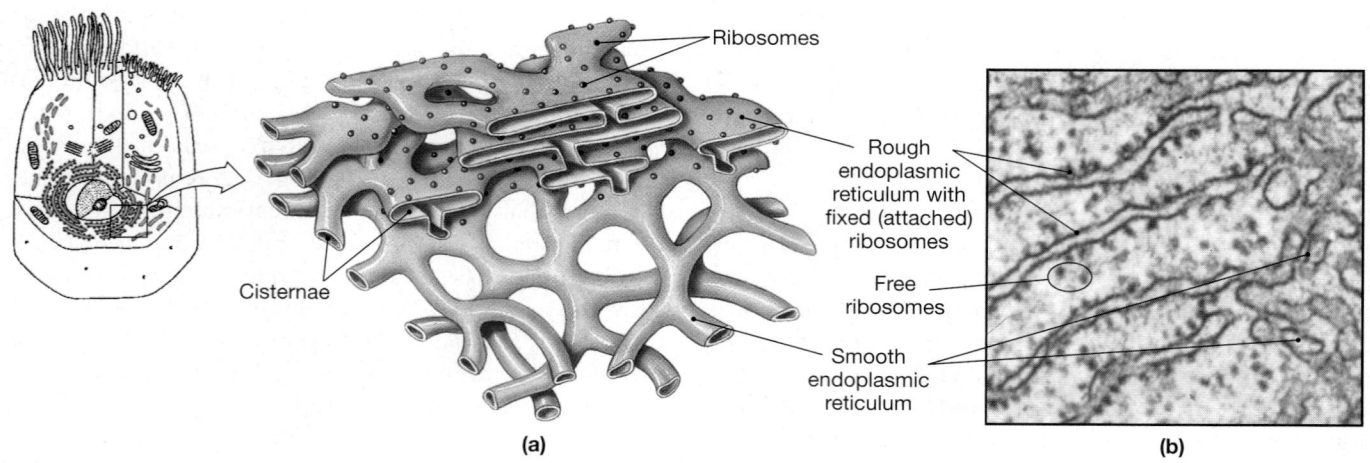

(a)   (b)

●**FIGURE 3–8**
The Endoplasmic Reticulum. **(a)** The three-dimensional relationships between the rough and smooth endoplasmic reticula. **(b)** Rough endoplasmic reticulum and free ribosomes in the cytoplasm of a cell. (TEM × 111,000)

tions, all associated with the synthesis of lipids and carbohydrates. For example, the SER is responsible for

- synthesizing the phospholipids and cholesterol needed for maintenance and growth of the cell membrane, ER, nuclear membrane, and Golgi apparatus in all cells.
- synthesizing steroid hormones, such as *androgens* and *estrogens* (the dominant sex hormones in males and in females, respectively) in the reproductive organs.
- synthesizing and storing glycerides, especially triacylglycerides, in liver cells and fat cells.
- synthesizing and storing glycogen in skeletal muscle and liver cells.

In skeletal muscle cells, neurons, and many other types of cell, the SER also adjusts the composition of the cytosol by absorbing and storing ions, such as Ca^{2+}, or larger molecules. In addition, the SER is responsible for the detoxification or inactivation of drugs by liver and kidney cells.

ROUGH ENDOPLASMIC RETICULUM The **rough endoplasmic reticulum (RER)** functions as a combination workshop and shipping depot. It is where many newly synthesized proteins undergo chemical modification and where they are packaged for export to their next destination, the *Golgi apparatus.*

The ribosomes on the outer surface of the rough endoplasmic reticulum are fixed ribosomes (Figure 3–8•). Their presence gives the RER a beaded, grainy, or rough appearance. Both free and fixed ribosomes synthesize proteins by using instructions provided by messenger RNA. The new polypeptide chains produced at fixed ribosomes are released into the cisternae of the RER. Inside the RER, each protein assumes its

secondary and tertiary structures. ∞ p. 54 Some of the proteins are enzymes that will function inside the endoplasmic reticulum. Other proteins are chemically modified by the attachment of carbohydrates, creating glycoproteins. Most of the proteins and glycoproteins produced by the RER are packaged into small membranous sacs that pinch off the tips of the cisternae. These **transport vesicles** subsequently deliver their contents to the Golgi apparatus.

THE RER AND SER IN SPECIALIZED CELLS The amount of endoplasmic reticulum and the proportion of RER to SER vary with the type of cell and its ongoing activities. For example, pancreatic cells that manufacture digestive enzymes contain an extensive RER, but the SER is relatively small. The situation is just the reverse in the cells that synthesize steroid hormones in the reproductive system.

The Golgi Apparatus

If a transport vesicle carries a newly synthesized protein or glycoprotein that is destined for export from the cell, it travels from the ER to an organelle that looks a bit like a stack of dinner plates. This organelle, the **Golgi** (GŌL-jē) **apparatus** (Figure 3–9•), typically consists of five or six flattened membranous discs called *cisternae.* There may be several of these organelles within a single cell, most often found near the nucleus.

The Golgi apparatus has three major functions:

1. It modifies and packages secretions, such as hormones or enzymes, for release through exocytosis.
2. It renews or modifies the cell membrane.
3. It packages special enzymes within vesicles for use in the cytosol.

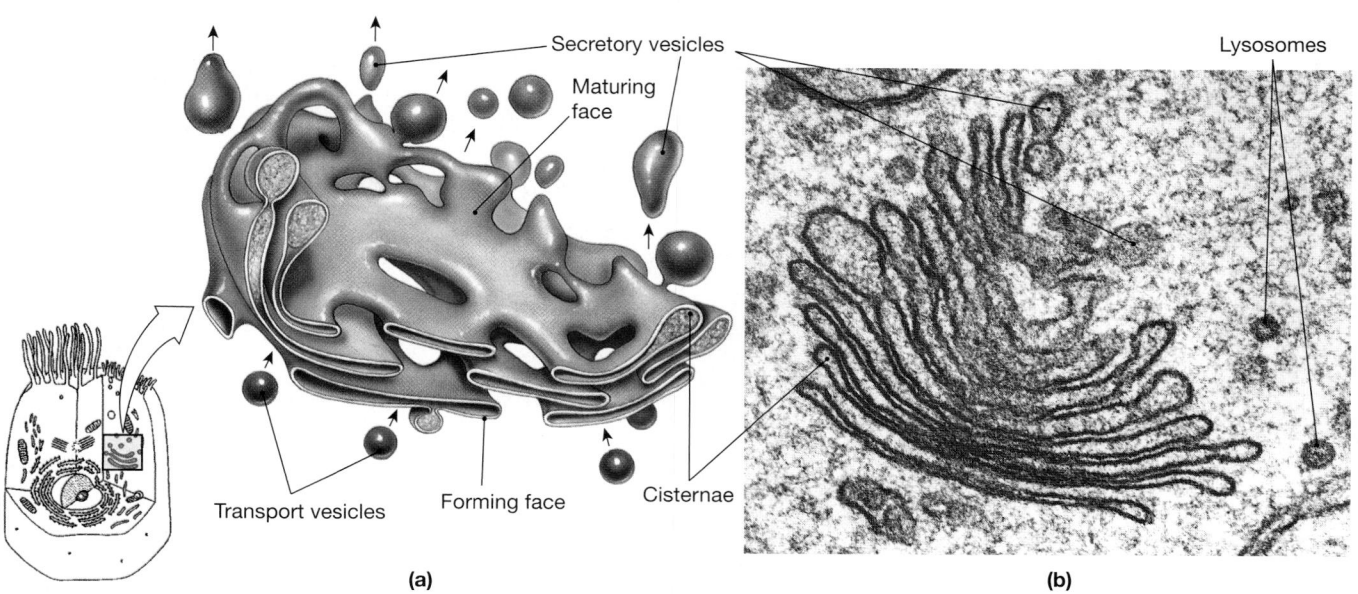

•FIGURE 3–9
The Golgi Apparatus.   **(a)** A three-dimensional view of the Golgi apparatus with a cut edge corresponding to **(b)**, a sectional view of the Golgi apparatus of an active secretory cell. (TEM × 57,660)

Figure 3–10a● diagrams the role of the Golgi apparatus in packaging secretions. Proteins and glycoproteins that reach the Golgi apparatus are synthesized in the RER. Transport vesicles then deliver these products to the Golgi apparatus. The vesicles generally arrive at a cisterna known as the *forming* (or *cis*) *face*. The transport vesicles then fuse with the Golgi membrane, emptying their contents into the cisterna. Inside the Golgi apparatus, enzymes modify the arriving proteins and glycoproteins. For example, the enzymes may change the carbohydrate structure of a glycoprotein, or they may attach a phosphate group, sugar, or fatty acid to a protein.

One of the unique things about this organelle is that compounds which enter the cisternae are constantly in motion, traveling up the stack from the ER toward the cell membrane. Small vesicles move material from one cisterna to the next. Ultimately, the product arrives at the *maturing* (or *trans*) *face*, which usually faces the cell surface. At the maturing face, three types of vesicles form that carry materials away from the Golgi apparatus:

1. *Secretory Vesicles.* **Secretory vesicles** are vesicles containing secretions that will be discharged from the cell. These vesicles fuse with the cell membrane and empty their contents into the extracellular environment (Figure 3–10b●). This process, known as *exocytosis*, is discussed in greater detail later in this chapter.

2. *Membrane Renewal Vesicles.* When vesicles produced at the Golgi apparatus fuse with the surface of the cell, they are adding new lipids and proteins to the cell membrane. At the same time, other areas of the cell membrane are being removed and recycled. The Golgi apparatus can thus change the properties of the cell membrane over time. For example, new glycoprotein receptors can be added, making the cell more sensitive to a particular stimulus. Alternatively, receptors can be removed and not replaced, making the cell less sensitive to specific ligands. Such changes can profoundly alter the sensitivity and functions of the cell.

3. *Lysosomes.* Vesicles called *lysosomes* that remain in the cytoplasm contain digestive enzymes. Their varied functions will be detailed next.

Lysosomes

Cells often need to break down and recycle large organic molecules and even complex structures like organelles. The breakdown process requires the use of powerful enzymes, and it often generates toxic chemicals that could damage or kill the cell. **Lysosomes** (LĪ-sō-sōmz; *lyso-*, dissolution + *soma*, body) are special vesicles that provide an isolated environment for potentially dangerous chemical reactions. These vesicles, produced at the Golgi apparatus, contain digestive enzymes. Lysosomes are small, often spherical bodies, and their contents look dense and dark in electron micrographs (Figure 3–9b●). *Primary lysosomes* contain inactive enzymes (Figure 3–11●). Enzyme activation occurs when the lysosome fuses with the membranes of transport vesicles or damaged organelles targeted for destruction, such as mitochondria or fragments of the ER. This fusion creates a *secondary lysosome*, which contains active enzymes. These enzymes then break down the lysosomal contents. Nutrients reenter the cytosol, and the remaining material is eliminated from the cell by exocytosis.

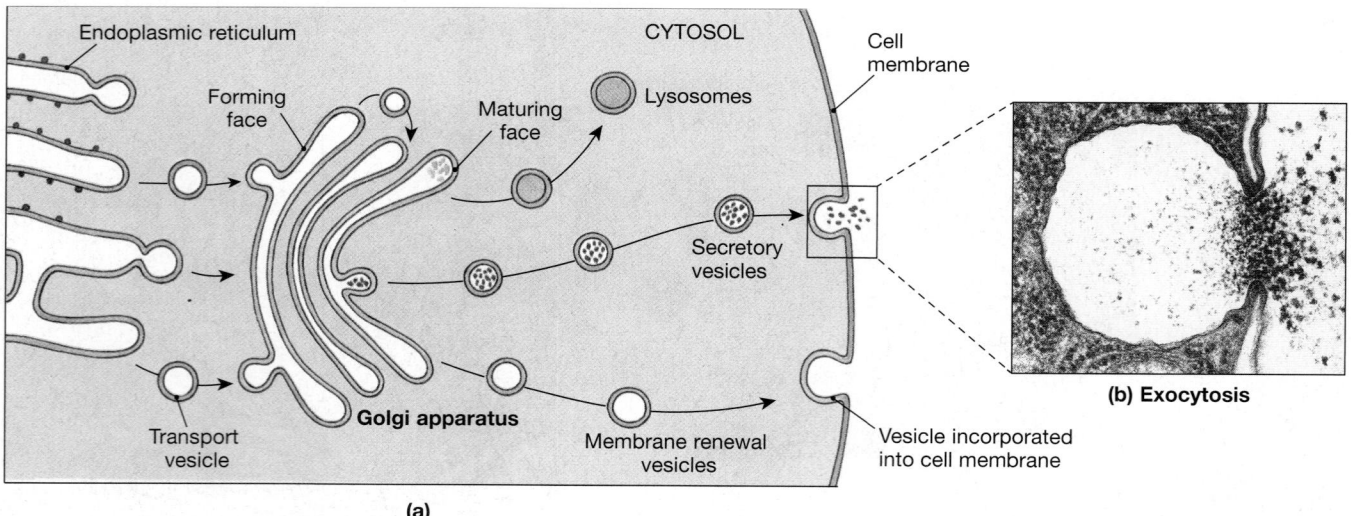

(a)

●FIGURE 3–10
Functions of the Golgi Apparatus. (a) Transport vesicles carry the secretory product from the endoplasmic reticulum to the Golgi apparatus (simplified to clarify the relationships between the membranes). Small vesicles move membrane and materials between the Golgi cisternae. At the maturing face, three functional categories of vesicles develop. Secretory vesicles carry the secretion from the Golgi to the cell surface, where exocytosis releases the contents into the extracellular fluid. Other vesicles add surface area and integral proteins to the cell membrane. Lysosomes, which remain in the cytoplasm, are vesicles filled with enzymes. (b) Exocytosis at the surface of a cell.

Lysosomes also function in the defense against disease. Cells may remove bacteria, as well as liquids and organic debris, from the extracellular fluid by enclosing them in a small portion of the cell membrane, which is then pinched off to form a vesicle. (This method of transporting substances across the cell membrane, called *endocytosis*, will be discussed later in this chapter.) Lysosomes then fuse with the vesicles. Lysosomal enzymes break down the contents of the lysosome and release usable substances, such as sugars or amino acids. In this way, the cell both protects itself against pathogenic organisms and obtains valuable nutrients.

Lysosomes perform essential cleanup and recycling functions inside the cell. For example, when muscle cells are inactive, lysosomes gradually break down their contractile proteins. (This mechanism accounts for the reduction in muscle mass seen among retired athletes.) The process is usually precisely controlled, but the regulatory mechanism fails in a damaged or dead cell. Lysosomes then disintegrate, releasing active enzymes into the cytosol. These enzymes rapidly destroy the proteins and organelles of the cell in a process called **autolysis** (aw-TAH-li-sis; *auto-*, self). We do not know how to control lysosomal activities or why the enclosed enzymes do not digest the lysosomal walls unless the cell is damaged.

Problems with lysosomal enzyme production cause more than 30 serious diseases affecting children. In these conditions, called *lysosomal storage diseases*, the lack of a specific lysosomal enzyme results in the buildup of waste products and debris normally removed and recycled by lysosomes. Affected individuals may die when vital cells, such as those of the heart, can no longer function. **AM** Lysosomal Storage Diseases

Peroxisomes

Peroxisomes are smaller than lysosomes and carry a different group of enzymes. In contrast to lysosomes, which are produced at the Golgi apparatus, new peroxisomes are produced by the growth and subdivison of existing peroxisomes. Their enzymes are produced at free ribosomes and transported from the cytosol into the peroxisomes by carrier proteins.

Peroxisomes absorb and break down fatty acids and other organic compounds. As they do so, peroxisomes generate hydrogen peroxide (H_2O_2), a potentially dangerous free radical. ∞ p. 36 Other enzymes within the peroxisome then break down the hydrogen peroxide to oxygen and water. Peroxisomes thus protect the cell from the potentially damaging effects of free radicals produced during catabolism. While these organelles are present in all cells, their numbers are highest in metabolically active cells, such as liver cells.

Membrane Flow

When the temperature changes markedly, you change your clothes. Similarly, when a cell's environment changes, it alters the structure and properties of its cell membrane. With the exception of mitochondria, all membranous organelles in the cell are either interconnected or in communication through the movement of vesicles. The RER and SER are continuous and are connected to the nuclear envelope. Transport vesicles connect the ER with the Golgi apparatus, and secretory vesicles link the Golgi apparatus with the cell membrane. Finally, vesicles forming at the exposed surface of the cell remove and recycle segments of the cell membrane. This continuous movement and exchange is called **membrane flow**. In

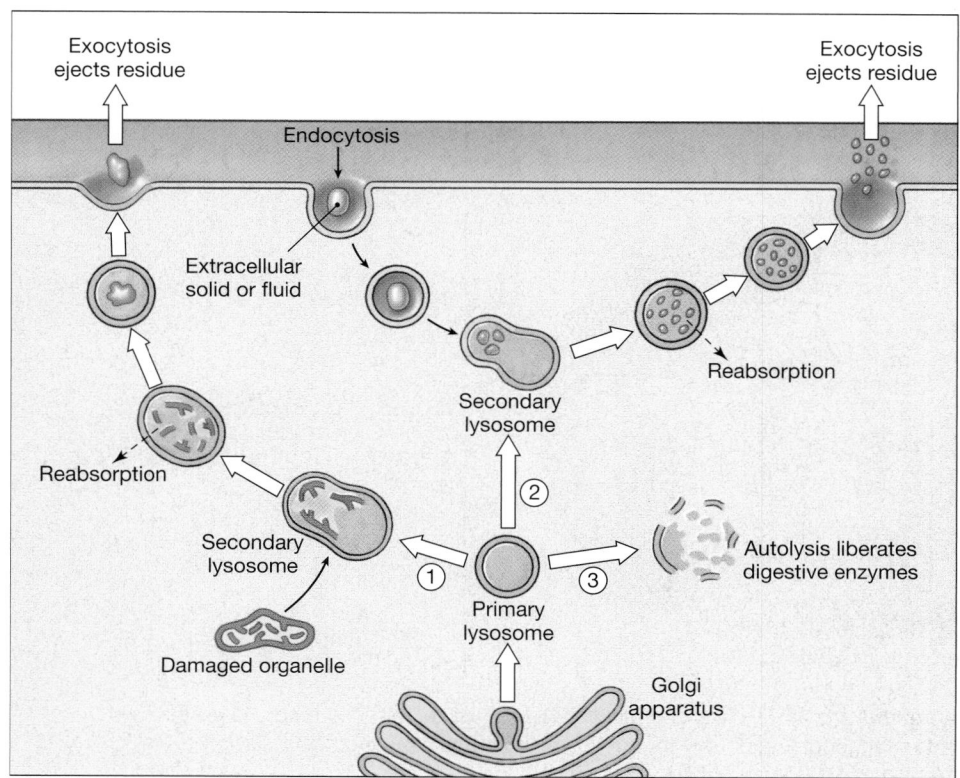

●**FIGURE 3–11**
Lysosome Functions. Primary lysosomes, formed at the Golgi apparatus, contain inactive enzymes. Activation may occur under any of three basic conditions: ① when the primary lysosome fuses with the membrane of another organelle, such as a mitochondrion, ② when the primary lysosome fuses with an endosome containing fluid or solid materials from outside the cell, or ③ in autolysis, when the lysosomal membrane breaks down following injury to, or death of, the cell.

an actively secreting cell, an area equal to the entire membrane surface may be replaced each hour.

Membrane flow is an example of the dynamic nature of cells. It provides a mechanism by means of which cells change the characteristics of their cell membranes—the lipids, receptors, channels, anchors, and enzymes—as they grow, mature, or respond to a specific environmental stimulus.

Mitochondria

Cells, like other living things, require energy to carry out the functions of life. The organelles responsible for energy production are the **mitochondria** (mī-tō-KON-drē-uh; singular, *mitochondrion; mitos,* thread + *chondrion,* granule). These are small structures that vary widely in shape, from long and slender to short and fat. The number of mitochondria in a particular cell varies with the cell's energy demands. Red blood cells lack mitochondria altogether, whereas these organelles may account for 20 percent of the volume of an active liver cell.

Mitochondria have an unusual double membrane (Figure 3–12a●). The outer membrane surrounds the organelle. The inner membrane contains numerous folds called **cristae**. Cristae increase the surface area exposed to the fluid contents, or **matrix**,

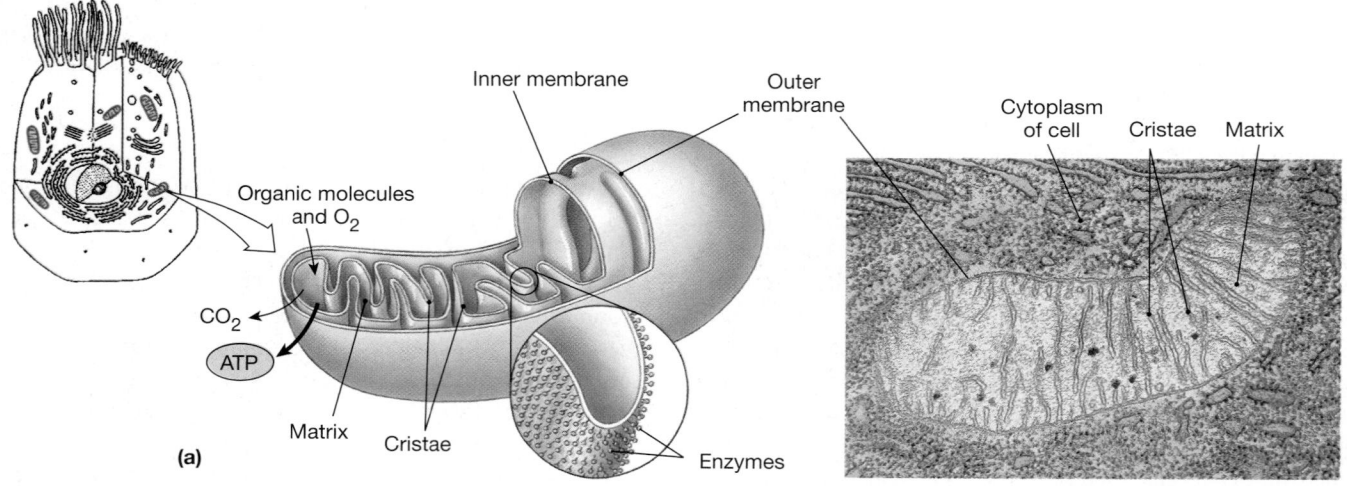

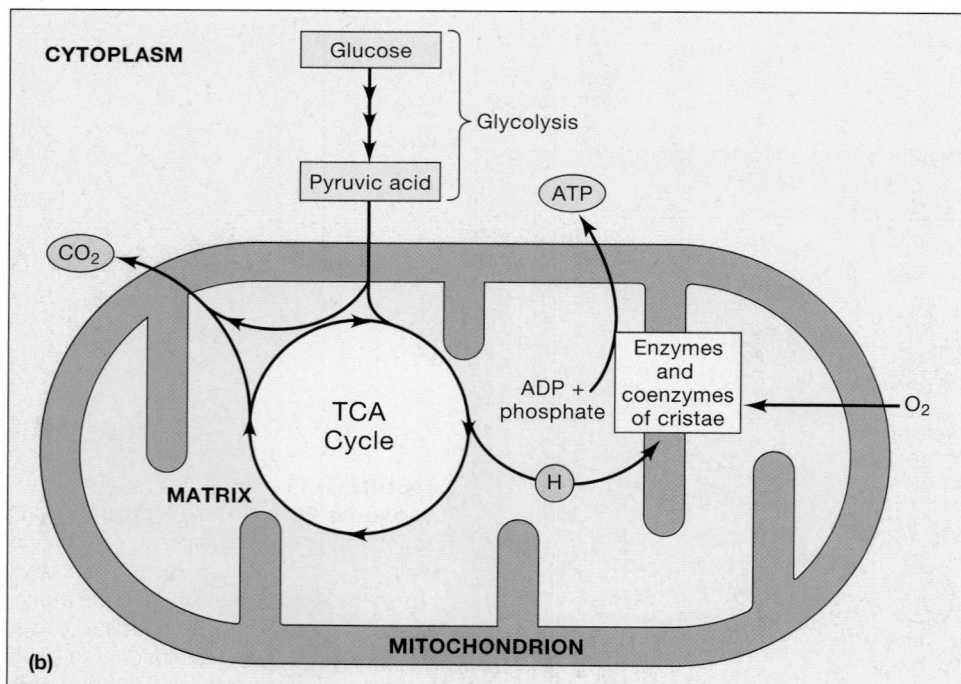

●**FIGURE 3–12**
Mitochondria. **(a)** The three-dimensional organization and a color enhanced TEM of a typical mitochondrion in section. (TEM × 46,332) **(b)** An overview of the role of mitochondria in energy production. Mitochondria absorb short carbon chains (such as pyruvic acid) and oxygen and generate carbon dioxide and ATP.

of the mitochondrion. Metabolic enzymes in the matrix catalyze the reactions that provide energy for cellular functions.

Most of the chemical reactions that release energy occur in the mitochondria, but most of the cellular activities that require energy occur in the surrounding cytoplasm. Cells must therefore store energy in a form that can be moved from place to place. Recall from Chapter 2 that cellular energy is stored and transferred in the form of *high-energy bonds*, such as those which attach a phosphate group (PO_4^{3-}) to adenosine diphosphate (ADP), creating the high-energy compound *adenosine triphosphate* (ATP). Cells can break the high-energy bond under controlled conditions, reconverting ATP to ADP and phosphate and thereby releasing energy for the cell's use.

MITOCHONDRIAL ENERGY PRODUCTION Most cells generate ATP and other high-energy compounds through the breakdown of carbohydrates, especially glucose. We shall examine the entire process in Chapter 25, but a few basic concepts now will help you follow discussions of muscle contraction, neuron function, and endocrine function in Chapters 10–18.

Although most of the ATP production occurs inside mitochondria, the first steps take place in the cytosol. In this reaction sequence, called **glycolysis**, each glucose molecule is broken down into two molecules of *pyruvic acid*. The pyruvic acid molecules are then absorbed by mitochondria (Figure 3–12b•).

In the mitochondrial matrix, a CO_2 molecule is removed from each absorbed pyruvic acid molecule; the remainder enters the **tricarboxylic acid cycle**, or **TCA cycle** (also known as the *Krebs cycle* and the *citric acid cycle*). The TCA cycle is an enzymatic pathway that systematically breaks down the absorbed pyruvic acid in the presence of oxygen. The remnants of pyruvic acid molecules contain carbon, oxygen, and hydrogen atoms. The carbon and oxygen atoms are released as carbon dioxide, which diffuses out of the cell. The hydrogen atoms are delivered to carrier proteins in the cristae. The electrons from the hydrogen atoms are then removed and passed along a chain of coenzymes. The energy released during these steps performs the enzymatic conversion of ADP to ATP. ∞ p. 60

Because mitochondrial activity requires oxygen, this method of ATP production is known as **aerobic metabolism** (*aero-*, air + *bios*, life), or *cellular respiration*. Aerobic metabolism in mitochondria produces about 95 percent of the ATP needed to keep a cell alive. (Enzymatic reactions in the cytosol produce the rest.)

Several inheritable disorders result from abnormal mitochondrial activity. While not totally self-sufficient, mitochondria do carry their own DNA and manufacture many of their own proteins under the direction of the genes on this DNA. The mitochondria involved in congenital diseases contain abnormal DNA, and the enzymes they produce reduce the efficiency of ATP production. Cells throughout the body may be affected, but symptoms involving muscle cells, neurons, and the receptor cells in the eye are most common, because these cells have especially high energy demands. **AM** Mitochondrial DNA, Disease, and Evolution

✓ Certain cells in the ovaries and testes contain large amounts of smooth endoplasmic reticulum (SER). Why?

✓ Microscopic examination of a cell's contents reveals many mitochondria. What does this observation imply about the cell's energy requirements?

Answers start on page Q-1

3–4 THE NUCLEUS

Objectives

▪ Explain the functions of the cell nucleus.

▪ Discuss the nature and importance of the genetic code.

▪ Summarize the process of protein synthesis.

The **nucleus** is usually the largest and most conspicuous structure in a cell; under a light microscope, it is often the only organelle visible. The nucleus serves as the control center for cellular operations. A single nucleus stores all the information needed to direct the synthesis of the more than 100,000 different proteins in the human body. The nucleus determines the structure of the cell and what functions it can perform by controlling which proteins are synthesized, under what circumstances, and in what amounts. A cell without a nucleus could be compared to a car without a driver, but the nucleus is even more important: A car can sit idle for years, but a cell without a nucleus will disintegrate within three or four months.

Most cells contain a single nucleus, but exceptions exist. For example, skeletal muscle cells have many nuclei, whereas mature red blood cells have none. Figure 3–13• details the structure of a typical nucleus. Surrounding the nucleus and separating it from the cytosol is a **nuclear envelope**, a double membrane with its two layers separated by a narrow **perinuclear space** (*peri-*, around). At several locations, the nuclear envelope is connected to the rough endoplasmic reticulum, as shown in Figure 3–2•.

To direct processes that take place in the cytosol, the nucleus must receive information about conditions and activities in other parts of the cell. Chemical communication between the nucleus and the cytosol occurs through **nuclear pores**. These pores, which cover about 10 percent of the surface of the nucleus, are large enough to permit the movement of ions and small molecules, but are too small for the free passage of proteins or DNA. Each nuclear pore contains regulatory proteins that govern the transport of specific proteins and RNA into or out of the nucleus.

☐ CONTENTS OF THE NUCLEUS

The fluid contents of the nucleus are called the *nucleoplasm*. The nucleoplasm contains the **nuclear matrix**, a network of fine filaments that provides structural support and may be involved in the regulation of genetic activity. The nucleoplasm also contains ions, enzymes, RNA and DNA nucleotides, small amounts of RNA, and DNA.

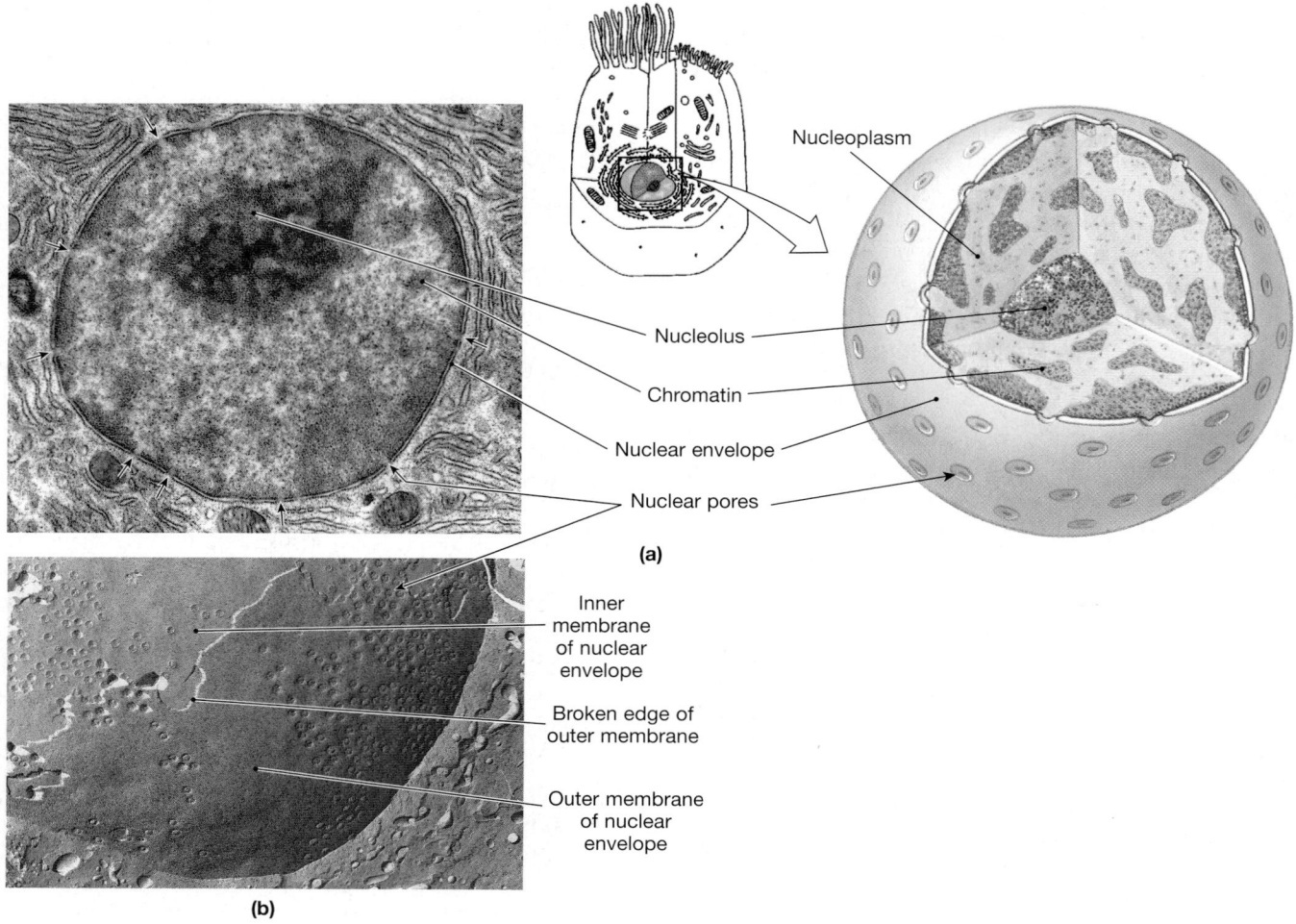

Nucleoplasm

Nucleolus

Chromatin

Nuclear envelope

Nuclear pores

(a)

Inner membrane of nuclear envelope

Broken edge of outer membrane

Outer membrane of nuclear envelope

(b)

•**FIGURE 3–13**
The Nucleus. **(a)** Important nuclear structures. (TEM × 4828) **(b)** This cell was frozen and then broken apart to make its internal structures visible. The technique, called *freeze fracture* or *freeze-etching*, provides a unique perspective on the internal organization of cells. The nuclear envelope and nuclear pores are visible. The fracturing process broke away part of the outer membrane of the nuclear envelope, and the cut edge of the nucleus can be seen. (SEM × 9240)

Most nuclei contain several dark-staining areas called **nucleoli** (noo-KLĒ-ō-lī; singular, *nucleolus*). Nucleoli are transient nuclear organelles that synthesize ribosomal RNA. They also assemble the ribosomal subunits, which reach the cytoplasm by carrier-mediated transport at the nuclear pores. Nucleoli are composed of RNA, enzymes, and proteins called **histones**. The nucleoli form around areas of DNA that contain the instructions for producing ribosomal proteins and RNA when those instructions are being carried out. Nucleoli are most prominent in cells that manufacture large amounts of proteins, such as liver and muscle cells, because those cells need large numbers of ribosomes.

It is the DNA in the nucleus that stores the instructions for protein synthesis, and this DNA is organized into structures called **chromosomes** (*chroma*, color). In humans, the nuclei of somatic cells contain 23 pairs of chromosomes. One member of

each pair is derived from the mother, and one from the father. Each chromosome contains DNA strands bound to histones. The DNA strands coil around the histones, allowing a great deal of DNA to be packaged in a small space. Interactions between the DNA and the histones help determine the information available to the cell at any moment.

The structure of a chromosome is illustrated in Figure 3–14•. At intervals, the DNA strands wind around the histones, forming a complex known as a **nucleosome**. The entire chain of nucleosomes may coil around other proteins. The degree of coiling determines whether the chromosome is long and thin or short and fat. In cells that are not dividing, the chromosomal material is loosely coiled within the nucleus, forming a tangle of fine filaments known as **chromatin**, which gives the nucleus a clumped, grainy appearance. Chromosomes don't become visible as individual units until just before cell division begins.

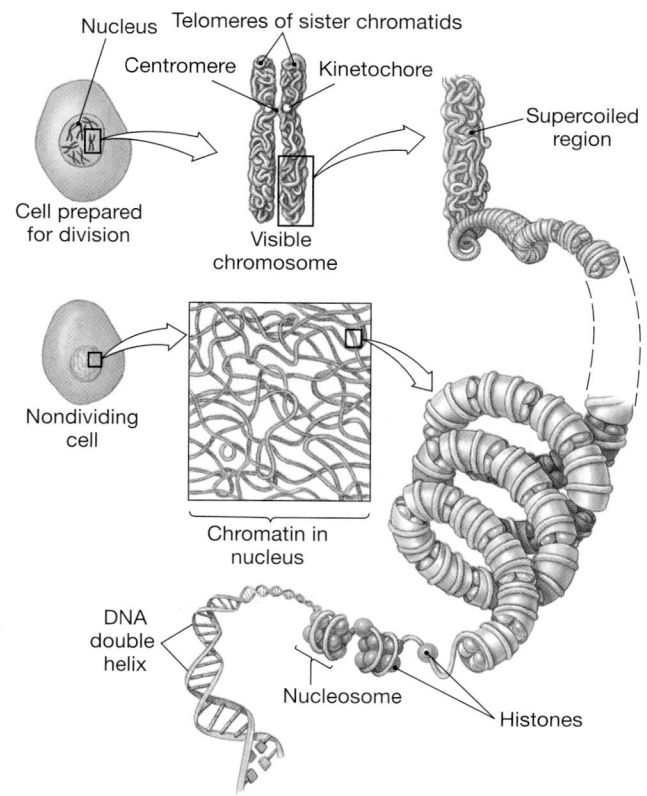

●**FIGURE 3–14**
Chromosome Structure. DNA strands are coiled around histones to form nucleosomes. Nucleosomes form coils that may be very tight or rather loose. In cells that are not dividing, the DNA is loosely coiled, forming a tangled network known as chromatin. When the coiling becomes tighter, as it does in preparation for cell division, the DNA becomes visible as distinct structures called chromosomes.

☐ INFORMATION STORAGE IN THE NUCLEUS

As we saw in Chapter 2, each protein molecule consists of a unique sequence of amino acids. ∞ p. 53 Any "recipe" for a protein, therefore, must specify the order of amino acids in the polypeptide chain. This information is stored in the chemical structure of the DNA strands in the nucleus. The chemical "language" used by the cell is known as the **genetic code**. An understanding of the genetic code has enabled us to determine how cells build proteins and how various structural and functional characteristics are inherited from generation to generation. Researchers are now experimenting with the manipulation of the genetic information in human cells. The techniques they develop may eventually revolutionize the treatment of some serious diseases.

To understand how the genetic code works, recall the basic structure of nucleic acids described in Chapter 2. ∞ p. 58 A single DNA molecule consists of a pair of DNA strands held together by hydrogen bonding between complementary nitrogenous bases. Information is stored in the sequence of nitrogenous bases along the length of the DNA strands. Those nitrogenous bases

are adenine (A), thymine (T), cytosine (C), and guanine (G). The genetic code is called a *triplet code*, because a sequence of three nitrogenous bases specifies the identity of a single amino acid. Thus, the information encoded in the sequence of nitrogenous bases must be read in groups of three. For example, the DNA triplet thymine–guanine–thymine (TGT) codes for the amino acid cysteine. More than one triplet may represent the same amino acid, however. For example, the DNA triplet thymine–guanine–cytosine (TGC) also codes for cysteine.

A **gene** is the functional unit of heredity; it contains all the DNA triplets needed to produce specific proteins. The number of triplets varies from gene to gene, depending on the size of the polypeptide represented. A relatively short polypeptide chain might require fewer than 100 triplets, whereas the instructions for building a large protein might involve 1000 or more triplets. Not all of the DNA molecule is devoted to carrying instructions for proteins; some segments contain instructions for the synthesis of transfer RNA or ribosomal RNA, and others have no apparent function.

⚕ Every nucleated somatic cell in the body carries a set of 46 chromosomes that are copies of the set formed at fertilization. Not all the DNA of these chromosomes codes for proteins, however, and a significant percentage of DNA segments have no known function. Some of the "useless" segments contain the same nucleotide sequence repeated over and over. The number of segments and the number of repetitions vary from individual to individual. The chance that any two individuals, other than identical twins, will have the same pattern of repeating DNA segments is less than one in 9 billion. The identification of individuals can therefore be made on the basis of a DNA pattern analysis, just as it can on the basis of a fingerprint. Skin scrapings, blood, semen, hair, or other tissues can be used as the DNA source. Information from *DNA fingerprinting* has been used to convict (and to acquit) persons accused of violent crimes, such as rape or murder. The science of molecular biology has thus become a viable addition to the crime-fighting arsenal.

☐ GENE ACTIVATION AND PROTEIN SYNTHESIS

Each DNA molecule contains thousands of genes and therefore holds the information needed to synthesize thousands of proteins. Normally, the genes are tightly coiled, and bound histones keep the genes inactive. Before a gene can affect a cell, the portion of the DNA molecule containing that gene must be uncoiled and the histones temporarily removed.

The factors controlling this process, called **gene activation**, are only partially understood. We know, however, that every gene contains segments responsible for regulating its own activity. In effect, these are triplets that say "do (or do not) read this message," "message starts here," or "message ends here." The "read me," "don't read me," and "start" signals form a special region of DNA called the *promoter*, or control segment, at the start of each gene. Each gene ends with a "stop" signal. Gene activation begins with the temporary disruption of the weak bonds between the nitrogenous bases of the two DNA strands and the removal of the histone that guards the promoter.

After the complementary strands have separated and the histone has been removed, the enzyme **RNA polymerase** binds to the promoter of the gene. This binding is the first step in the process of **transcription**, the production of RNA from a DNA template. The term *transcription* is appropriate, as it means "to copy out" or "rewrite." All three types of RNA are formed through the transcription of DNA, but we shall focus here on the transcription of mRNA, which carries the information needed to synthesize proteins. **Messenger RNA (mRNA)** is absolutely vital, because the DNA cannot leave the nucleus. Instead, its information is copied to messenger RNA, which *can* leave the nucleus and carry the information to the cytoplasm, where protein synthesis occurs.

The Transcription of mRNA

The two DNA strands in a gene are complementary. One of the two strands contains the triplets that specify the sequence of amino acids in the polypeptide. This is the **coding strand**. The other strand, called the **template strand**, contains complementary triplets that will be used as a template for mRNA production. The resulting mRNA will have a nucleotide sequence identical to that of the coding strand, but with uracil substituted for thymine. Figure 3–15● details transcription:

STEP 1: Once the DNA strands have separated and the promoter has been exposed, transcription can begin. The key event is the attachment of RNA polymerase to the template strand.

STEP 2: RNA polymerase promotes hydrogen bonding between the nitrogenous bases of the template strand and complementary nucleotides in the nucleoplasm. This enzyme begins at a "start" signal in the promoter region. It then strings nucleotides together by covalent bonding. The RNA polymerase interacts with only a small portion of the template strand at any one time as it travels along the DNA strand. The complementary strands separate in front of the enzyme as it moves one nucleotide at a time, and they reassociate behind it. The enzyme collects additional nucleotides and attaches them to the growing chain. The nucleotides involved are those characteristic of RNA, not of DNA; RNA polymerase can attach adenine, guanine, cytosine, or uracil, but never thymine. Thus, wherever an A occurs in the DNA strand, the polymerase will attach a U rather than a T to the growing mRNA strand. In this way, RNA polymerase assembles a complete strand of mRNA. The nucleotide sequence of the template strand determines the nucleotide sequence of the mRNA strand. Thus, each DNA triplet will correspond to a sequence of three nucleotide bases in the mRNA strand. Such a three-base mRNA sequence is called a **codon** (KŌ-don). Codons contain nitrogenous bases that are complementary to those of the triplets in the template strand. For example, if the DNA triplet is TCG, the corresponding mRNA codon

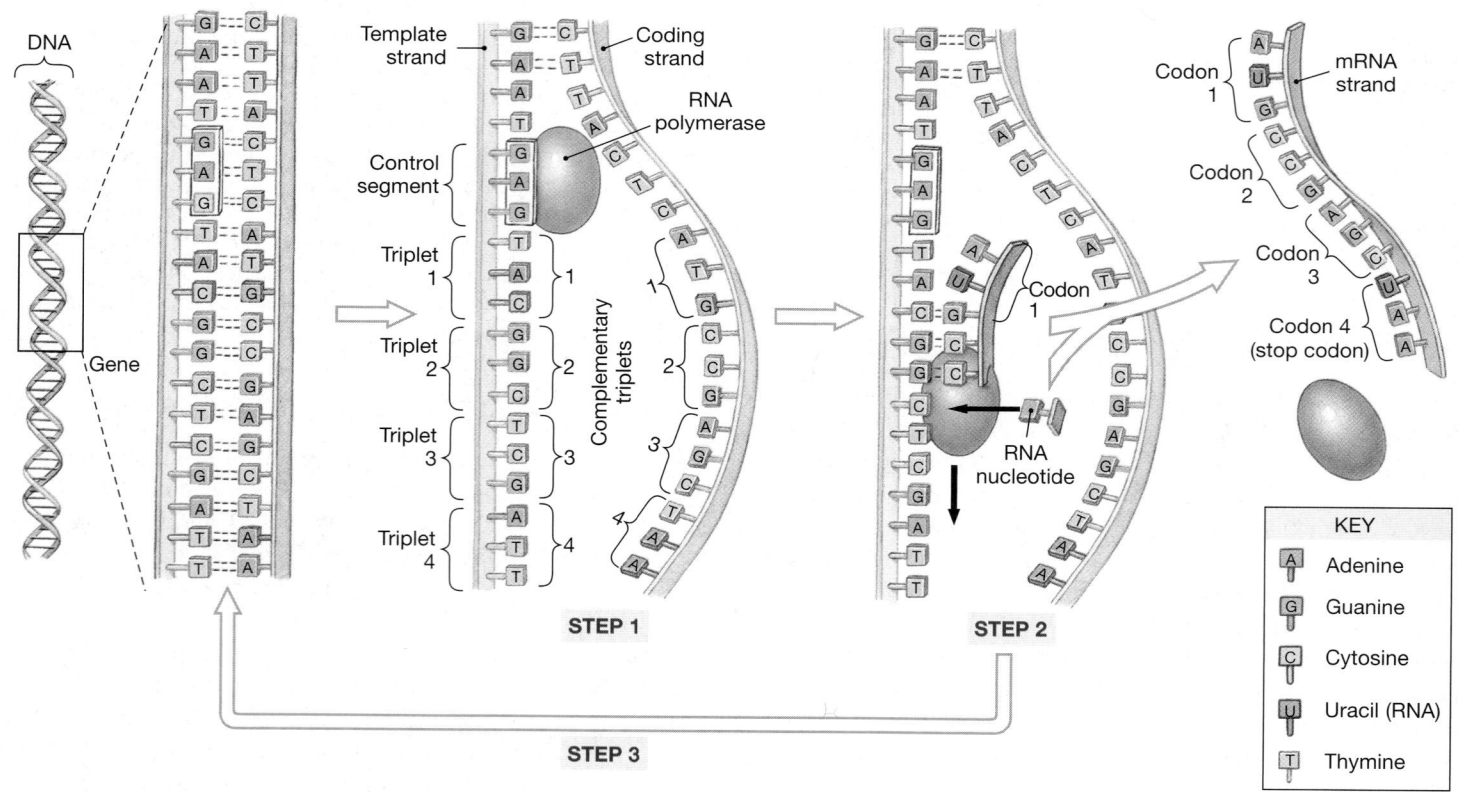

●**FIGURE 3–15**
mRNA Transcription. A small portion of a single DNA molecule, containing a single gene available for transcription. STEP 1: The two DNA strands separate, and RNA polymerase binds to the promoter of the gene. STEP 2: The RNA polymerase moves from one nucleotide to another along the length of the template strand. At each site, complementary RNA nucleotides form hydrogen bonds with the DNA nucleotides of the template strand. The RNA polymerase then strings the arriving nucleotides together into a strand of mRNA. STEP 3: On reaching the stop signal at the end of the gene, the RNA polymerase and the mRNA strand detach, and the two DNA strands reassociate.

will be AGC. This method of copying ensures that the mRNA exactly matches the coding strand of the gene.

STEP 3: At the "stop" signal, the enzyme and the mRNA strand detach from the DNA strand, and transcription ends. The complementary DNA strands now complete their reassociation as hydrogen bonding occurs between complementary base pairs.

Each gene includes a number of triplets that contain extraneous information, not needed to build a functional protein. As a result, the mRNA strand assembled during transcription, sometimes called immature mRNA or *pre-mRNA*, must be "edited" before it leaves the nucleus to direct protein synthesis. In this **RNA processing**, nonsense regions, called **introns**, are snipped out, and the remaining, coding segments, or **exons**, are spliced together. The process creates a much shorter, functional strand of mRNA that then enters the cytoplasm through one of the nuclear pores.

Intron removal is extremely important and tightly regulated. This is understandable because, if an error occurs in the editing, the resulting protein will be abnormal and the results potentially disastrous. But in addition, we now know that by changing the editing instructions and removing different introns, a single gene can produce mRNAs that code for several different proteins. How this variable editing is regulated remains a mystery.

Translation

Protein synthesis is the assembling of functional polypeptides in the cytoplasm. Figure 3–16● presents an overview of this process. Protein synthesis occurs through **translation**, the formation of a linear chain of amino acids, using the information provided by an mRNA strand. Again, the name is appropriate: To *translate* is to present the same information in a new language; in this case, a message written in the "language" of nucleic acids (the sequence of nitrogenous bases) is being translated by ribosomes into the "language" of proteins (the sequence of amino acids in a polypeptide chain). Each mRNA codon designates a particular amino acid to be incorporated into the polypeptide chain.

The amino acids are provided by **transfer RNA (tRNA)**, a relatively small and mobile type of RNA. Each tRNA molecule binds and delivers an amino acid of a specific type. More than 20 kinds of transfer RNA exist—at least one for each of the amino acids used in protein synthesis.

A tRNA molecule has a tail that binds an amino acid. Roughly midway along its length, the nucleotide chain of the tRNA forms a tight loop. It is this loop that can interact with an mRNA strand. The loop contains three nitrogenous bases that form an **anticodon**. During translation, the anticodon will bond complementarily with an appropriate mRNA codon. The base sequence of the anticodon indicates the type of amino acid carried by the tRNA. For example, a tRNA with the anticodon GGC always carries the amino acid *proline*, whereas a tRNA with the anticodon CGG carries *alanine*. Table 3–2 lists examples of several codons and anticodons that specify individual amino acids and summarizes the relationships among DNA, codons, and anticodons.

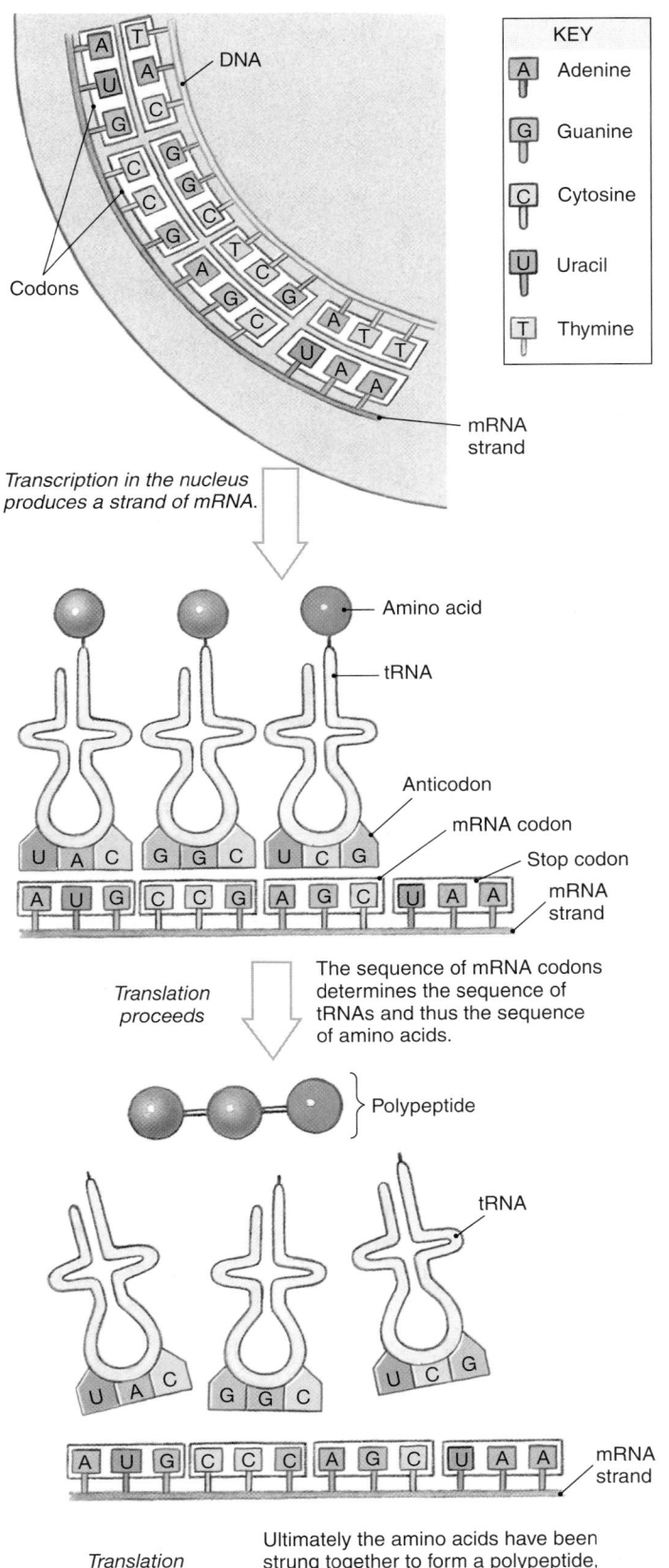

●**FIGURE 3–16**
An Overview of Protein Synthesis

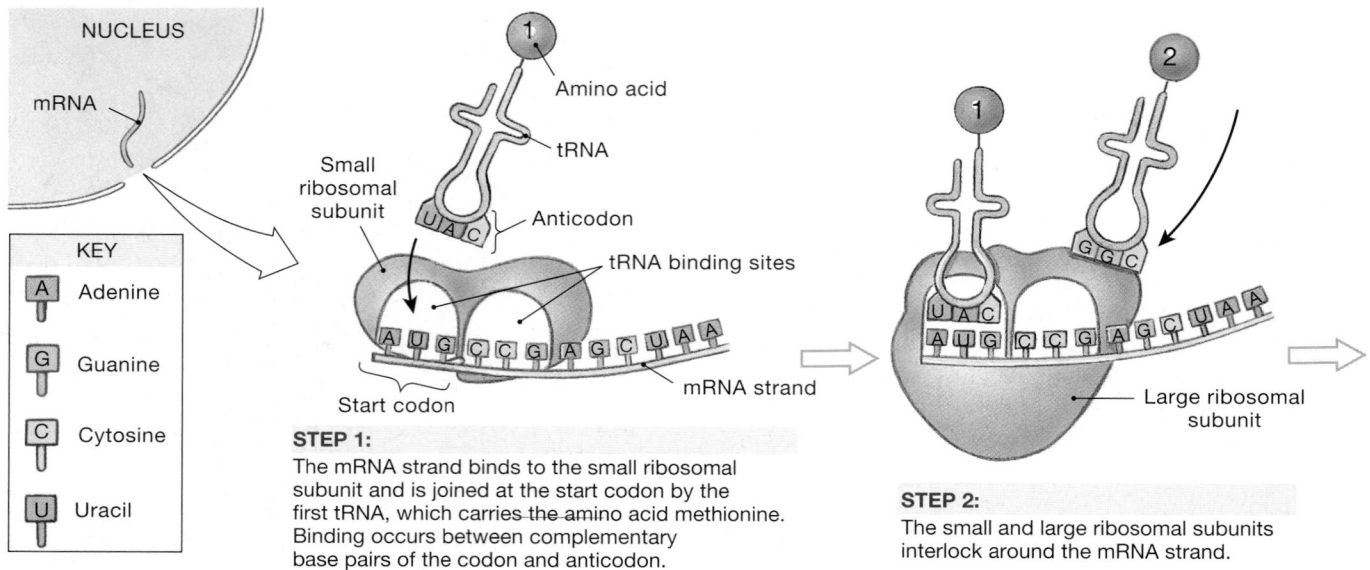

●FIGURE 3–17
The Process of Translation. For clarity, the components are not drawn to scale and their three-dimensional relationships have been simplified.

TABLE 3–2 EXAMPLES OF THE TRIPLET CODE

DNA Triplets

Coding Strand	Template Strand	mRNA Codon	tRNA Anticodon	Amino Acid
TTT	AAA	UUU	AAA	Phenylalanine
TTA	AAT	UUA	AAU	Leucine
TGT	ACA	UGU	ACA	Cysteine
GTT	CAA	GUU	CAA	Valine
ATG	TAC	AUG	UAC	Methionine
AGC	TCG	AGC	UCG	Serine
CCG	GGC	CCG	GGC	Proline
GCC	CGG	GCC	CGG	Alanine

The tRNA molecules thus provide the physical link between codons and amino acids. During translation, each codon along the mRNA strand will bind a complementary anticodon on a tRNA molecule. Thus, if the mRNA has the codons AUG–CCG–AGC, it will bind to tRNAs with anticodons UAC–GGC–UCG. The amino acid sequence of the polypeptide chain created is determined by the sequence of delivery by tRNAs, and that sequence depends on the arrangement of codons along the mRNA strand. In this case, the amino acid sequence in the resulting polypeptide would be methionine–proline–serine.

The translation process is illustrated in Figure 3–17●:

STEP 1: Translation begins as the mRNA strand binds to a small ribosomal subunit. The first codon, or *start codon*, of the mRNA strand always has the base sequence AUG. It binds a tRNA with the complementary anticodon sequence UAC. This tRNA, which carries the amino acid *methionine*, attaches to the first of two tRNA binding sites on the small ribosomal subunit. (The initial methionine will be removed from the finished protein.)

STEP 2: When this tRNA binding occurs, a large ribosomal subunit joins the complex to create a complete ribosome. The mRNA strand nestles in the gap between the small and the large ribosomal subunits.

STEP 3: A second tRNA now arrives at the second tRNA binding site of the ribosome, and its anticodon binds to the next codon of the mRNA strand.

STEP 4: Enzymes of the large ribosomal subunit then break the linkage between the tRNA and its amino acid. At the same time, the enzymes attach the amino acid to its neighbor by means of a peptide bond. The ribosome then moves one codon down the mRNA strand. The cycle is then repeated with the arrival of another molecule of tRNA. The tRNA already stripped of its amino acid drifts away. It will soon bind to another amino acid and be ready to participate in protein synthesis again.

STEP 5: The polypeptide chain continues to grow by the addition of amino acids until the ribosome reaches a "stop" signal, or *stop codon*, at the end of the mRNA strand. The ribosomal subunits now detach, leaving an intact strand of mRNA and a completed polypeptide.

Translation proceeds swiftly, producing a typical protein in about 20 seconds. The mRNA strand remains intact, and it can interact with other ribosomes to create additional copies of the same polypeptide chain. The process does not continue indefinitely, however, because after a few minutes to a few hours, mRNA strands are broken down and the nucleotides are recycled. However, large numbers of protein chains can be produced during that time. Although only two mRNA codons are "read" by a ribosome at any one time, the entire strand may contain thousands of codons. As a result, many ribosomes can bind to a single mRNA strand. At any moment, each ribosome will be reading a different part of the same message, but each

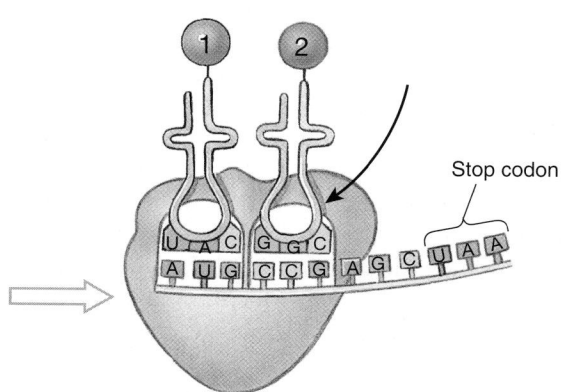

STEP 3:
A second tRNA arrives at the adjacent binding site of the ribosome. The anticodon of the second tRNA binds to the next mRNA codon.

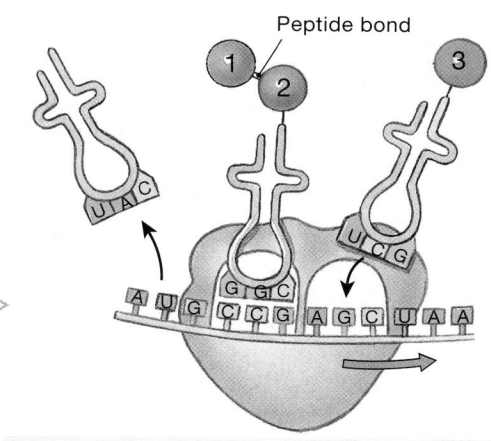

STEP 4:
The first amino acid is detached from its tRNA and is joined to the second amino acid by a peptide bond. The ribosome moves one codon farther along the mRNA strand; the first tRNA detaches as another tRNA arrives.

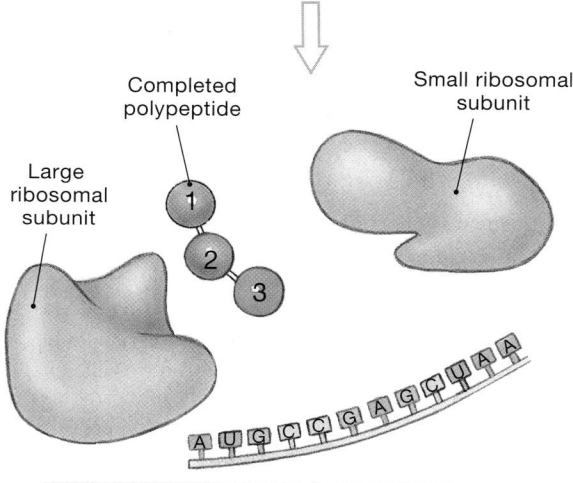

STEP 5:
The chain elongates until the stop codon is reached; the components then separate.

will end up constructing a copy of the same protein as the others. The arrangement is similar to a line of people at a buffet lunch; all the people will assemble the same meal, but each person is always a step behind the person ahead. A series of ribosomes attached to the same mRNA strand is called a *polyribosome*, or *polysome* (Figure 3–5a●).

Mutations are permanent changes in a cell's DNA that affect the nucleotide sequence of one or more genes. The simplest is a *point mutation*, a change in a single nucleotide that affects one codon. The triplet code has some flexibility, because several different codons can specify the same amino acid. But a point mutation that produces a codon that specifies a different amino acid will usually change the structure of the completed protein. A single change in the amino acid sequence of a structural protein or enzyme can prove fatal. Certain cancers and two potentially lethal blood disorders discussed in Chapter 19, *thalassemia* and *sickle-cell anemia*, result from variations in a single nucleotide.

More than 100 inherited disorders have been traced to abnormalities in enzyme or protein structure that reflect single changes in nucleotide sequence. More elaborate mutations, such as additions or deletions of nucleotides, can affect multiple codons in one gene or in several adjacent genes, or they can affect the structure of one or more chromosomes.

Because most mutations occur during DNA replication, they are most likely to involve cells that are undergoing cell division. A single cell, a group of cells, or an entire individual may be affected. This last prospect occurs when the changes are made early in development. For example, a mutation affecting the DNA of an individual's sex cells will be inherited by that individual's children. Our understanding of genetic structure is opening the possibility of diagnosing and correcting some of these problems. **AM** Genetic Engineering and **Gene Therapy**

☐ HOW THE NUCLEUS CONTROLS CELL STRUCTURE AND FUNCTION

As we noted at the start of this section, the DNA of the nucleus controls the cell by directing the synthesis of specific proteins. Through the control of protein synthesis, virtually every aspect of cell structure and function can be regulated. Two levels of control are involved:

1. The DNA of the nucleus has *direct* control over the synthesis of structural proteins, such as cytoskeletal components, membrane proteins (including receptors), and secretory products. By issuing appropriate instructions, in the form of mRNA strands, the nucleus can alter the internal structure of the cell, its sensitivity to substances in its environment, or its secretory functions to meet changing needs.

2. The DNA of the nucleus has *indirect* control over all other aspects of cellular metabolism, because it regulates the synthesis of enzymes. By ordering or stopping the production of appropriate enzymes, the nucleus can regulate all metabolic activities

and functions of the cell. For example, the nucleus can accelerate the rate of glycolysis by increasing the number of needed enzymes in the cytoplasm.

This brings us to a central question: How does the nucleus "know" what genes to activate? Although we don't have all the answers, we know that in many cases gene activation or deactivation is triggered by changes in the characteristics of the surrounding cytoplasm. These changes may affect the nucleus by altering the local environment, and this alone may be enough to turn specific genes on or off. Alternatively, messengers or hormones may enter the nucleus through the nuclear pores and bind to specific receptors or promotors along the DNA strands. Thus, there is continual chemical communication between the cytoplasm and the nucleus. That communication is relatively selective, thanks to the restrictive characteristics of the nuclear pores and the barrier posed by the nuclear envelope.

Of course, there is also continual communication between the cytoplasm and the extracellular fluid across the cell membrane, and what crosses the cell membrane today may alter gene activity tomorrow. In the next section, we shall examine how the cell membrane selectively regulates the passage of materials in and out of the cell.

✓ How does the nucleus control the activities of a cell?

✓ What process would be affected by the lack of the enzyme RNA polymerase?

Answers start on page Q-1

 The processes of transcription and translation can be viewed on the **Companion Website**: Chapter 3/Tutorials.

3–5 HOW THINGS GET INTO AND OUT OF CELLS

Objectives

- Specify the routes by which different ions and molecules can enter or leave a cell and the factors that may restrict such movement.
- Describe the various transport mechanisms that cells use to facilitate the absorption or removal of specific substances.
- Explain the origin and significance of the transmembrane potential.

The cell membrane is a barrier that isolates the cytoplasm from the extracellular fluid. Because the cell membrane is an effective barrier, conditions inside the cell can be considerably different from conditions outside the cell. However, the effectiveness of this barrier poses a difficulty for the cell. Cells are not self-sufficient. Each day they require nutrients that can be used to provide the energy they need to stay alive and function normally. In the course of their daily activities, they also generate waste

products that must be eliminated. Whereas your body has passageways and openings for nutrients, gases, and wastes, the cell is surrounded by a continuous, relatively uniform membrane. So how do materials—whether nutrients or waste products—get across the cell membrane without damaging it or reducing its effectiveness as a barrier? To answer this question, we must take a closer look at the structure and function of the cell membrane.

The **permeability** of the cell membrane is the property that determines precisely which substances can enter or leave the cytoplasm. A membrane through which nothing can pass is described as **impermeable**. A membrane through which any substance can pass without difficulty is **freely permeable**. The permeability of cell membranes lies somewhere between those extremes, so cell membranes are said to be **selectively permeable**.

A selectively permeable membrane permits the free passage of some materials and restricts the passage of others. The distinction may be based on size, electrical charge, molecular shape, lipid solubility, or other factors. Cells differ in their permeabilities, depending on what lipids and proteins are present in the cell membrane and how these components are arranged.

Passage across the membrane is either passive or active. *Passive processes* move ions or molecules across the cell membrane with no expenditure of energy by the cell. *Active processes* require that the cell expend energy, generally in the form of ATP.

Transport processes are also categorized by the mechanism involved. The three major categories are:

1. *Diffusion.* Diffusion, which results from the random motion and collisions of ions and molecules. Diffusion is a passive process.
2. *Carrier-mediated Transport.* Carrier-mediated transport, which requires the presence of specialized integral membrane proteins. Carrier-mediated transport can be passive or active, depending on the substance transported and the nature of the transport mechanism.
3. *Vesicular Transport.* Vesicular transport, which involves the movement of materials within small membranous sacs, or *vesicles*. Vesicular transport is always an active process.

□ DIFFUSION

Ions and molecules are constantly in motion, colliding and bouncing off one another and off obstacles in their paths. The movement is random: A molecule can bounce in any direction. One result of this continuous random motion is that, over time, the molecules in any given space will tend to become evenly distributed. This distribution process is called **diffusion**. As the molecules move around, there will be a net movement of material from areas of higher concentration to areas of lower concentration. The difference between the high and low concentrations is a **concentration gradient**. Diffusion tends to eliminate that gradient.

After the gradient has been eliminated, the molecular motion continues, but net movement no longer occurs in any particular direction. For convenience, we restrict use of the term *diffusion* to

●**FIGURE 3–18**
Diffusion.   Placing a colored sugar cube in a glass of water establishes a steep concentration gradient. As the cube dissolves, many sugar molecules are in one location and none are elsewhere. As diffusion occurs, the molecules spread through the solution. Eventually, diffusion eliminates the concentration gradient. The sugar cube has dissolved completely, and the molecules are distributed evenly. Molecular motion continues, but there is no directional movement.

the directional movement that eliminates concentration gradients—a process sometimes called *net diffusion*. Because diffusion tends to spread materials from a region of higher concentration to one of lower concentration, it is often described as proceeding "down a concentration gradient" or "downhill."

All of us have experienced the effects of diffusion, which occurs in air as well as in water. The scent of fresh flowers in a vase sweetens the air in a large room; a drop of ink spreads to color an entire glass of water. Each case begins with a very high concentration of molecules in a localized area. Consider a colored sugar cube dropped in water (Figure 3–18●). Placing the cube in a large volume of clear water establishes a steep concentration gradient for the ingredients as they dissolve: The sugar and dye concentration is high near the cube and negligible elsewhere. As diffusion proceeds, the sugar and dye molecules spread through the solution until they are distributed evenly.

Diffusion is important in body fluids because it tends to eliminate local concentration gradients. For example, every cell in the body generates carbon dioxide, and the intracellular concentration is relatively high. Carbon dioxide concentrations are lower in the surrounding interstitial fluid and lower still in the circulating blood. Because cell membranes are freely permeable to carbon dioxide, it can diffuse down its concentration gradient—traveling from the cell's interior into the interstitial fluid and from the interstitial fluid into the bloodstream, for eventual delivery to the lungs.

To be effective, the diffusion of nutrients, waste products, and dissolved gases must be able to keep pace with the demands of active cells. Important factors that influence diffusion rates include the following:

- **Distance.** Concentration gradients are eliminated quickly over short distances. The greater the distance, the longer is the time required. In the human body, diffusion distances are generally small. For example, few cells are farther than 125 μm from a blood vessel.

- **Molecule Size.** Ions and small organic molecules such as glucose diffuse more rapidly than do large proteins.

- **Temperature.** The higher the temperature, the faster is the diffusion rate. The human body maintains a temperature of about 37°C (98.6°F), and diffusion proceeds more rapidly at this temperature than at normal environmental temperatures.

- **Gradient Size.** The larger the concentration gradient, the faster diffusion proceeds. When cells become more active, the intracellular concentration of oxygen declines. This change increases the concentration gradient for oxygen between the inside of the cell (relatively low) and the interstitial fluid outside (relatively high). The rate of oxygen diffusion into the cell then increases.

- **Electrical Forces.** It is a basic physical principle that opposite electrical charges (+ and −) are attracted to each other; similar charges (+ and + *or* − and −) repel each other. The interior of the cell membrane has a net negative charge relative to the exterior surface, due in part to the high concentration of proteins in the cell (p. 53). This negative charge tends to pull positive ions from the extracellular fluid into the cell, while opposing the entry of negative ions. For example, interstitial fluid contains higher concentrations of sodium ions (Na^+) and chloride ions (Cl^-) than does cytosol. Diffusion of the positively charged sodium ions into the cell is therefore favored by both the concentration gradient, or *chemical gradient*, and the electrical gradient. In contrast, diffusion of the negatively charged chloride ions into the cell is favored by the chemical gradient, but opposed by the electrical gradient. For any ion, the net result of the chemical and electrical forces acting on it is called the *electrochemical gradient*.

Diffusion Across Cell Membranes

In extracellular fluids, water and dissolved solutes diffuse freely. A cell membrane, however, acts as a barrier that selectively restricts diffusion: Some substances pass through easily, while others cannot penetrate the membrane. An ion or a molecule can diffuse across a cell membrane only by (1) passing through a membrane channel or (2) crossing the lipid portion of the membrane (Figure 3–19●). If there is an electrochemical gradient for a particular ion or molecule, whether that substance moves across the membrane will depend on its size relative to the sizes of membrane channels and its lipid solubility.

Ions and most water-soluble compounds are not lipid soluble, so they must pass through a membrane channel to enter the cytoplasm. Membrane channels are very small—on average, about 0.8 nm in diameter. Water molecules can enter or exit freely, as can ions such as sodium, potassium, calcium, hydrogen, and chloride, but even a small organic molecule, such as glucose, is too big to fit through the channels.

Alcohol, fatty acids, and steroids can enter cells easily, because they can diffuse through the lipid portions of the membrane. Dissolved gases, such as oxygen and carbon dioxide, and lipid-soluble drugs also enter and leave our cells by diffusing through the phospholipid bilayer.

Many clinically important drugs affect cell membranes. In general, the potency of an anesthetic is directly correlated with its lipid solubility. Presumably, high lipid solubility accelerates the drug's entry into cells and enhances its ability to block ion channels or change other properties of cell membranes. The most important clinical result is a reduction in the sensitivity and responsiveness of neurons and muscle cells. Local anesthetics, such as *procaine* and *lidocaine*, affect nerve cells by blocking sodium channels in their cell membranes. This blockage reduces or eliminates the responsiveness of these cells to painful (or any other) stimuli. Lipid solubility is also involved in the action of general anesthetics, such as *chloroform* and *ether*. **AM** Drugs and the Cell Membrane

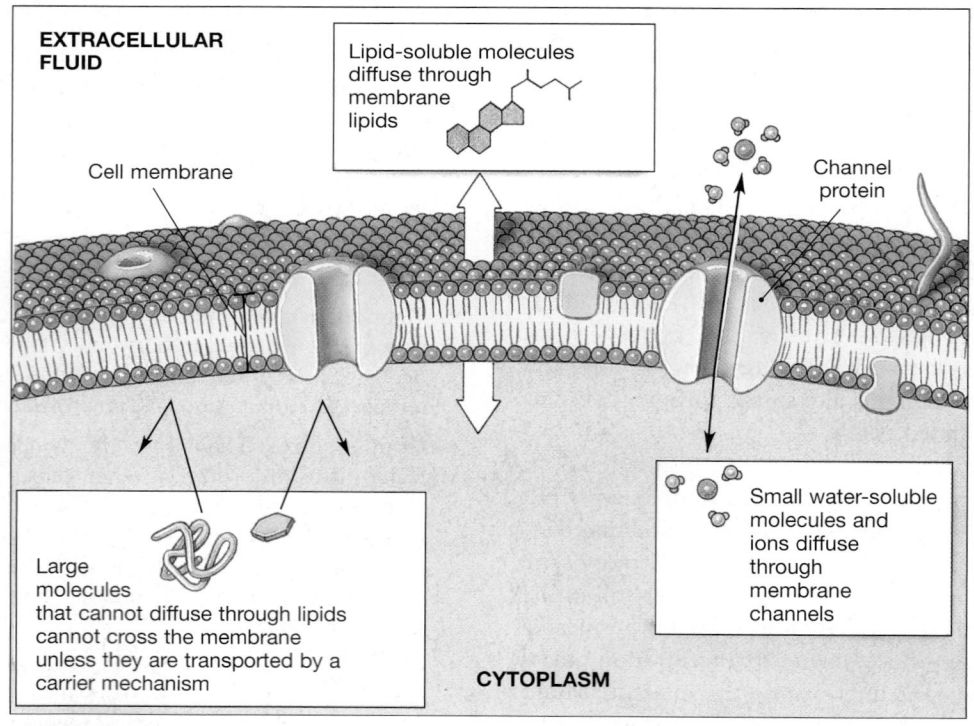

EXTRACELLULAR FLUID

Lipid-soluble molecules diffuse through membrane lipids

Cell membrane

Channel protein

Large molecules that cannot diffuse through lipids cannot cross the membrane unless they are transported by a carrier mechanism

Small water-soluble molecules and ions diffuse through membrane channels

CYTOPLASM

●**FIGURE 3–19**
Diffusion across the Cell Membrane

Osmosis: A Special Case of Diffusion

The net diffusion of water across a membrane is so important that it is given a special name: **osmosis** (oz-MŌ-sis; *osmos*, thrust). For convenience, we shall always use the term *osmosis* when we consider water movement and restrict the use of the term *diffusion* to the movement of solutes.

Intracellular and extracellular fluids are solutions that contain a variety of dissolved materials. Each solute diffuses as though it were the only material in solution. For example, the diffusion of sodium ions occurs only in response to the existence of a concentration gradient for sodium. A concentration gradient for another ion will have no effect on the rate or direction of sodium ion diffusion.

Some solutes diffuse into the cytoplasm, others diffuse out, and a few, such as proteins, are unable to diffuse across the cell membrane at all. Yet if we ignore the individual identities and simply count ions and molecules, we find that the *total* concentration of dissolved ions and molecules on either side of the cell membrane stays the same. This state of equilibrium persists because a typical cell membrane is freely permeable to water.

To understand the basis for such equilibrium, consider that whenever a solute concentration gradient exists, a concentration gradient for *water* exists also. Because dissolved solute molecules occupy space that would otherwise be taken up by water molecules, the higher the solute concentration, the lower the water concentration. As a result, *water molecules tend to diffuse across a membrane toward the solution containing the higher solute concentration*, because this movement is down the concentration gradient for water. Water movement will continue until water concentrations—and thus solute concentrations—are the same on either side of the membrane.

Remember these three characteristics of osmosis:

1. Osmosis is the diffusion of water molecules across a membrane.
2. Osmosis occurs across a selectively permeable membrane that is freely permeable to water, but not freely permeable to solutes.
3. In osmosis, water will flow across a membrane toward the solution that has the higher concentration of solutes, because that is where the concentration of water is lowest.

OSMOSIS AND OSMOTIC PRESSURE Figure 3–20● diagrams the process of osmosis. STEP 1 shows two solutions (A and B), with different solute concentrations, separated by a selectively permeable membrane. As osmosis occurs, water molecules cross the membrane until the solute concentrations in the two solutions are identical (STEP 2a). Thus, the volume of solution B increases while that of solution A decreases. The greater the initial difference in solute concentrations, the stronger is the osmotic flow. The **osmotic pressure** of a solution is an indication of the force with which pure water moves into that solution as a result of its solute concentration. We can measure a solution's osmotic pressure in several ways. For example, an opposing pressure can prevent the osmotic flow of water into the solution. Pushing against a fluid generates **hydrostatic pressure**. In STEP 2b, hydrostatic pressure opposes the osmotic pressure of solution B, so no net osmotic flow occurs.

Osmosis eliminates solute concentration differences much more quickly than we might predict on the basis of diffusion rates for other molecules. When water molecules cross a membrane, they move in groups held together by hydrogen bonding. So, whereas solute molecules usually diffuse through membrane channels one at a time, water molecules move together in large numbers. This phenomenon is called *bulk flow*.

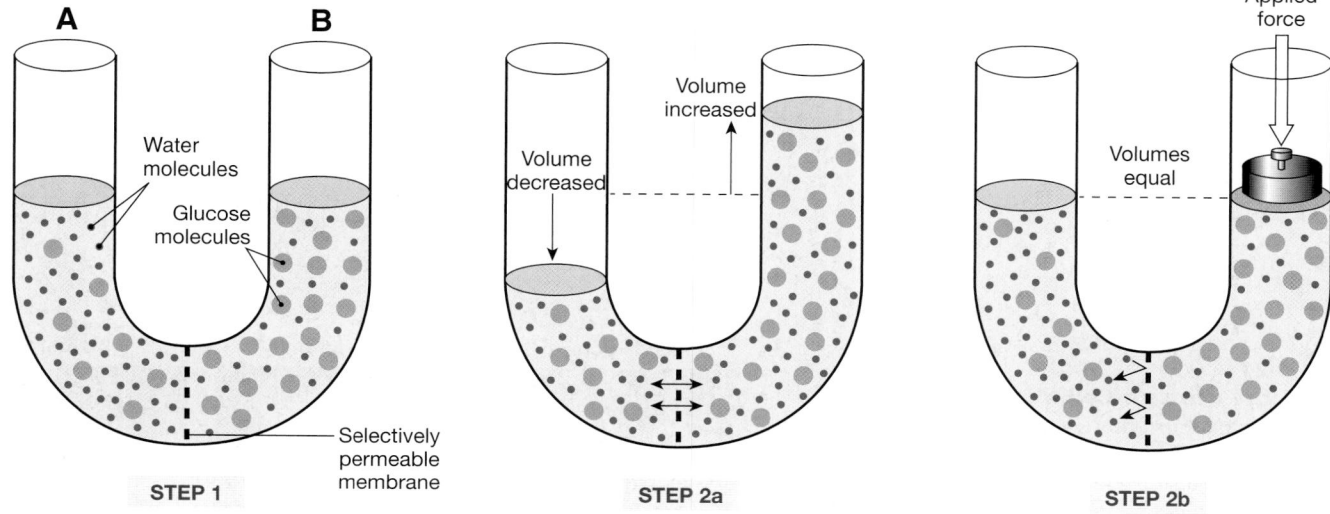

●FIGURE 3–20
Osmosis.   STEP 1: Two solutions containing different solute concentrations are separated by a selectively permeable membrane. Water molecules (small blue dots) begin to cross the membrane toward solution B, the solution with the higher concentration of solutes (larger pink circles). STEP 2a: At equilibrium, the solute concentrations on the two sides of the membrane are equal. The volume of solution B has increased at the expense of that of solution A. STEP 2b: Osmosis can be prevented by resisting the change in volume. The osmotic pressure of solution B is equal to the amount of hydrostatic pressure required to stop the osmotic flow.

OSMOLARITY AND TONICITY The total solute concentration in an aqueous solution is the solution's **osmolarity**, or **osmotic concentration**. The nature of the solutes, however, is often as important as the total osmolarity. Therefore, when we describe the effects of various osmotic solutions on cells, we usually use the term **tonicity** instead of osmolarity. If a solution does not cause an osmotic flow of water into or out of a cell, the solution is called **isotonic** (*iso*, same + *tonos*, tension).

Although often used interchangeably, the terms *osmolarity* and *tonicity* do not always mean the same thing. Osmolarity refers to the solute concentration of the solution, while tonicity is a description of how the solution affects the cell. To understand this distinction more clearly, consider a solution that has the same osmolarity as the intracellular fluid, but a higher concentration of one or more individual ions. If any of those ions can cross the cell membrane and diffuse into the cell, the osmolarity of the intracellular fluid will increase and that of the extracellular solution will decrease. Osmosis will then occur, moving water into the cell. If the process continues, the cell will gradually inflate like a water balloon. This stresses the cell membrane, which will eventually burst. In this case, the extracellular solution and the intracellular fluid were equal in osmolarity, but they were not isotonic.

Figure 3–21a● shows the appearance of a red blood cell in an isotonic solution. If a red blood cell is in a **hypotonic** solution, water will flow into the cell, causing it to swell up like a balloon (Figure 3–21b●). The cell may rupture, releasing its contents. This event is **hemolysis** (*hemo-*, blood + *lysis*, dissolution). A cell in a **hypertonic** solution will lose water by osmosis. As it does, the cell shrivels and dehydrates. The shrinking of red blood cells is called **crenation** (Figure 3–21c●).

It is often necessary to give patients large volumes of fluid to combat severe blood loss or dehydration. One fluid frequently administered is a 0.9 percent (0.9 g/dl) solution of sodium chloride (NaCl). This solution, which approximates the normal osmotic concentration of the extracellular fluids, is called *normal saline*. It is used because sodium and chloride are the most abundant ions in the extracellular fluid. Little net movement of either ion across cell membranes occurs; thus, normal saline is essentially isotonic with respect to body cells. An alternative treatment involves the use of an isotonic saline solution containing *dextran*, a carbohydrate that cannot cross cell membranes. The dextran molecules elevate the osmolarity of the blood, and as osmosis draws water into the blood vessels from the extracellular fluid, the blood volume increases further.

✓ How would a decrease in the concentration of oxygen in the lungs affect the diffusion of oxygen into the blood?

✓ Some pediatricians recommend the use of a 10 percent salt solution to relieve congestion for infants with stuffy noses. What effect would such a solution have on the cells lining the nasal cavity, and why?

Answers start on page Q-1

☐ CARRIER-MEDIATED TRANSPORT

In **carrier-mediated transport**, integral proteins bind specific ions or organic substrates and carry these substances across the cell membrane. All forms of carrier-mediated transport share the following characteristics with enzymes (p. 56):

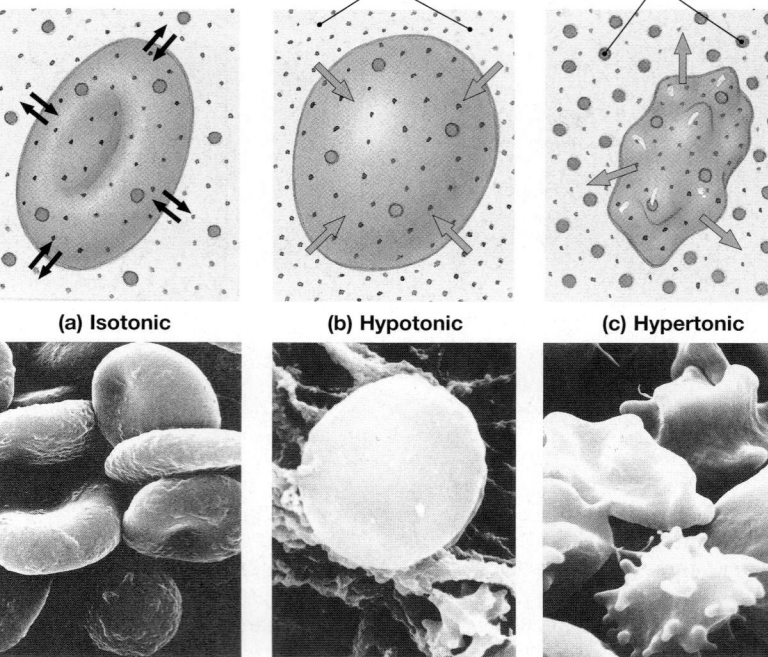

(a) Isotonic   **(b) Hypotonic**   **(c) Hypertonic**

●**FIGURE 3–21**
Osmotic Flow across a Cell Membrane. Black arrows indicate an equilibrium with no net water movement. Blue arrows indicate the direction of osmotic water movement. **(a)** Because these red blood cells are immersed in an isotonic saline solution, no osmotic flow occurs and the cells appear normal. **(b)** Immersion in a hypotonic saline solution results in the osmotic flow of water into the cells. The swelling may continue until the cell membrane ruptures, or lyses. **(c)** Exposure to a hypertonic solution results in the movement of water out of the cells. The red blood cells shrivel and become crenated. (SEM × 833)

1. *Specificity.* Each carrier protein in the cell membrane will bind and transport only certain substances. For example, the carrier protein that transports glucose will not transport other simple sugars.

2. *Saturation Limits.* The availability of substrate molecules and carrier proteins limits the rate of transport into or out of the cell, just as enzymatic reaction rates are limited by the availability of substrates and enzymes. When all the available carrier proteins are operating at maximum speed, the carriers are said to be *saturated*. The rate of transport cannot increase further, regardless of the size of the concentration gradient.

3. *Regulation.* Just as enzyme activity often depends on the presence of cofactors, the binding of other molecules, such as hormones, can affect the activity of carrier proteins. Hormones thus provide an important means of coordinating carrier protein activity throughout the body. The interplay between hormones and cell membranes will be examined when we deal with the endocrine system (Chapter 18) and metabolism (Chapter 25).

Many examples of carrier-mediated transport involve the movement of a single substrate molecule across the cell membrane. A few carrier mechanisms transport more than one substrate at a time. In **cotransport**, or *symport*, the carrier transports two substances in the same direction simultaneously, either into or out of the cell. In **countertransport**, or *antiport*, one substance moves into the cell and the other moves out.

We will consider two major examples of carrier-mediated transport here: (1) *facilitated diffusion* and (2) *active transport*.

Facilitated Diffusion

Many essential nutrients, such as glucose and amino acids, are insoluble in lipids and too large to fit through membrane channels. These substances can be passively transported across the membrane by carrier proteins in a process called **facilitated diffusion** (Figure 3–22●). The molecule to be transported must first bind to a **receptor site** on the protein. The shape of the protein then changes, moving the molecule to the inside of the cell membrane, where it is released into the cytoplasm.

As in the case of simple diffusion, no ATP is expended in facilitated diffusion: The molecules simply move from an area of higher concentration to one of lower concentration. However, facilitated diffusion differs from simple diffusion because once the carrier proteins are saturated, the rate of transport cannot increase, regardless of further changes in the concentration gradient.

All cells move glucose across their membranes through facilitated diffusion. However, several different carrier proteins are involved. In muscle cells, fat cells, and many other types of cell, the glucose transporter functions only when stimulated by the hormone *insulin*. Inadequate production of this hormone is one cause of *diabetes mellitus*, a metabolic disorder that we shall discuss in Chapter 18.

Active Transport

In **active transport**, a high-energy bond (in ATP or another high-energy compound) provides the energy needed to move ions or molecules across the membrane. Despite the energy cost, active transport offers one great advantage: It is not dependent on a concentration gradient. As a result, the cell can import or export specific substrates, *regardless of their intracellular or extracellular concentrations*.

All cells contain carrier proteins called **ion pumps**, which actively transport the cations sodium (Na^+), potassium (K^+), calcium (Ca^{2+}), and magnesium (Mg^{2+}) across their cell membranes. Specialized cells can transport additional ions, such as iodide (I^-), chloride (Cl^-), and iron (Fe^{2+}). Many of these carrier proteins move a specific cation or anion in one direction only, either into or out of the cell. In a few instances, one carrier protein will move more than one kind of ion at the same time. If one kind of ion moves in one direction and the other moves in the opposite direction, the carrier protein is called an **exchange pump**.

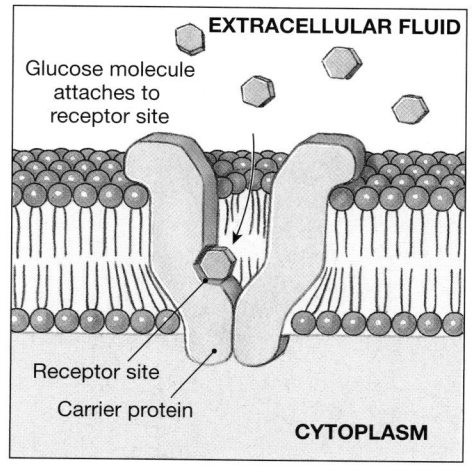

Glucose molecule attaches to receptor site

EXTRACELLULAR FLUID

Receptor site

Carrier protein

CYTOPLASM

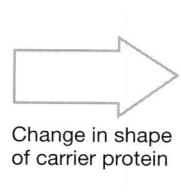

Change in shape of carrier protein

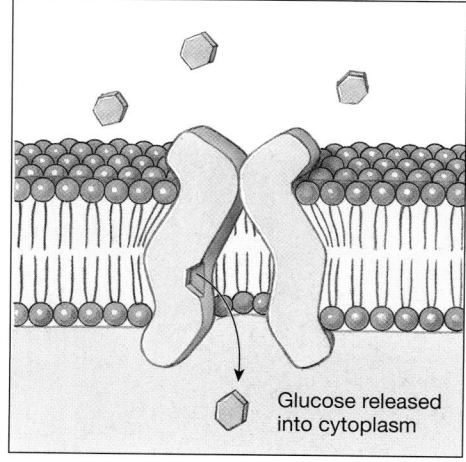

Glucose released into cytoplasm

●**FIGURE 3–22**
Facilitated Diffusion. In facilitated diffusion, an extracellular molecule, such as glucose, binds to a receptor site on a carrier protein. The binding alters the shape of the protein, which then releases the molecule to diffuse into the cytoplasm.

THE SODIUM–POTASSIUM EXCHANGE PUMP Sodium and potassium ions are the principal cations in body fluids. Sodium ion concentrations are high in the extracellular fluids, but low in the cytoplasm. The distribution of potassium in the body is just the opposite: low in the extracellular fluids and high in the cytoplasm. As a result, sodium ions slowly diffuse into the cell, and potassium ions diffuse out through leak channels. Homeostasis within the cell depends on the ejection of sodium ions and the recapture of lost potassium ions. This exchange is accomplished through the activity of a **sodium–potassium exchange pump**. The carrier protein involved in the process is called *sodium–potassium ATPase*.

The sodium–potassium exchange pump exchanges intracellular sodium for extracellular potassium (Figure 3–23●). On average, for each ATP molecule consumed, three sodium ions are ejected and two potassium ions are reclaimed by the cell. If ATP is readily available, the rate of transport depends on the concentration of sodium ions in the cytoplasm. When the concentration rises, the pump becomes more active. The energy demands are impressive: Sodium–potassium ATPase may use up to 40 percent of the ATP produced by a resting cell.

SECONDARY ACTIVE TRANSPORT In **secondary active transport**, the transport mechanism itself does not require energy, but the cell often needs to expend ATP at a later time to preserve homeostasis. As does facilitated transport, a secondary active transport mechanism moves a specific substrate down a concentration gradient. Unlike the proteins in facilitated transport, however, these carrier proteins can also move another substrate at the same time, without regard to its concentration gradient. In effect, the concentration gradient for one substance provides the driving force needed by the carrier protein, and the second substance gets a free ride.

The concentration gradient for sodium ions most often provides the driving force for cotransport mechanisms that move materials into the cell. For example, sodium-linked cotransport is important in the absorption of glucose and amino acids along

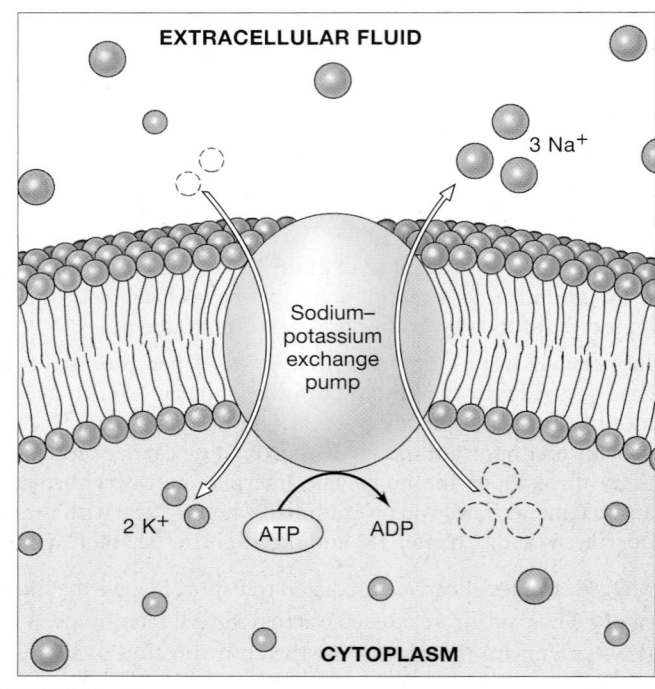

●**FIGURE 3–23**
The Sodium–Potassium Exchange Pump. The operation of the sodium–potassium exchange pump is an example of active transport. For each ATP converted to ADP, the protein called sodium–potassium ATPase carries three Na^+ out of the cell and two K^+ into the cell.

the intestinal tract. Although the transport activity proceeds without directly expending energy, the cell must expend ATP to pump the arriving sodium ions out of the cell by using the sodium–potassium exchange pump (Figure 3–24●). Sodium ions are also involved with many countertransport mechanisms. Sodium–calcium countertransport is responsible for keeping intracellular calcium ion concentrations very low.

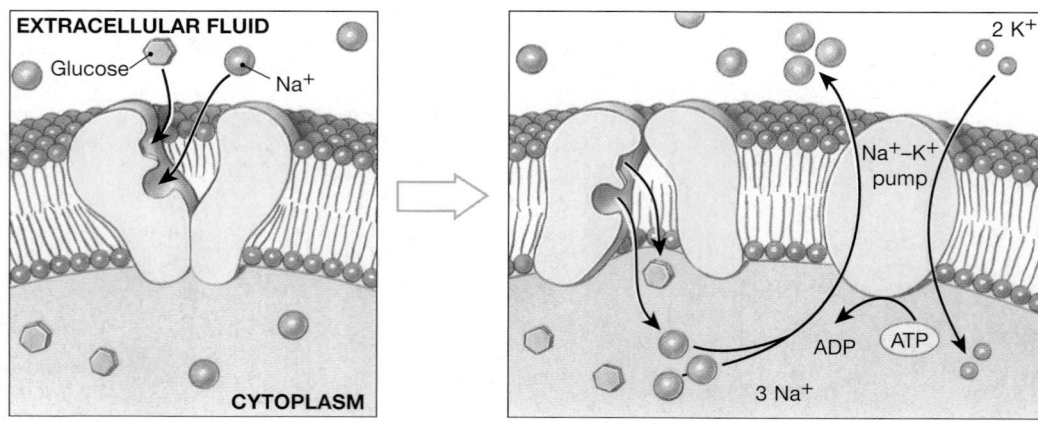

●**FIGURE 3–24**
Secondary Active Transport. In secondary active transport, glucose transport by this carrier protein will occur only after the carrier has bound a sodium ion. In three cycles, three glucose molecules and three sodium ions are transported into the cytoplasm. The cell now pumps the sodium ions across the cell membrane via the sodium–potassium exchange pump, at a cost of one ATP molecule.

☐ VESICULAR TRANSPORT

In **vesicular transport**, materials move into or out of the cell by means of **vesicles**, small membranous sacs that form at, or fuse with, the cell membrane. Because large volumes of fluid and solutes are transported in this way, this process is also known as *bulk transport*. The two major categories of vesicular transport are (1) *endocytosis* and (2) *exocytosis*.

Endocytosis

As we saw earlier in this chapter, extracellular materials can be packaged in vesicles at the cell surface and imported into the cell (p. 81). This process, called **endocytosis**, involves relatively large volumes of extracellular material and requires energy in the form of ATP. The three major types of endocytosis are (1) *receptor-mediated endocytosis*, (2) *pinocytosis*, and (3) *phagocytosis*. All three are active processes that require energy in the form of ATP.

Vesicles produced by receptor-mediated endocytosis or by pinocytosis are called *endosomes*; those produced by phagocytosis are called *phagosomes*. The contents of endosomes and phagosomes remain isolated within the vesicle. The movement of materials into the surrounding cytoplasm may involve active transport, simple or facilitated diffusion, or the destruction of the vesicle membrane.

RECEPTOR-MEDIATED ENDOCYTOSIS A selective process, **receptor-mediated endocytosis** involves the formation of small vesicles at the surface of the membrane. This process produces vesicles that contain a specific target molecule in high concentrations. Receptor-mediated endocytosis begins when materials in the extracellular fluid bind to receptors on the membrane surface (Figure 3–25●). Most receptor molecules are glycoproteins, and each binds to a specific ligand, or target, such as a transport protein or a hormone. Some receptors are distributed widely over the surface of the cell membrane; others are restricted to specific regions or in depressions on the cell surface.

Receptors bound to ligands cluster together. Once an area of the cell membrane has become covered with ligands, it forms grooves or pockets that move to one area of the cell and then pinch off to form an endosome. The endosomes produced in this way are called **coated vesicles**, because they are surrounded by a protein–fiber network that originally carpeted the inner membrane surface beneath the receptor–ligand clusters. This coating is essential to endosome formation and movement. Inside the cell, the coated vesicles fuse with primary lysosomes filled with digestive enzymes, creating secondary lysosomes, as described earlier in the chapter (p. 80). The lysosomal enzymes then free the ligands from their receptors, and the ligands enter the cytosol by diffusion or active transport. The vesicle membrane pinches off from the secondary lysosome and returns to the cell surface, where its receptors are ready to bind more ligands.

Many important substances, including cholesterol and iron ions (Fe^{2+}), are distributed through the body attached to special transport proteins. These proteins are too large to pass through membrane pores, but they can and do enter cells by receptor-mediated endocytosis.

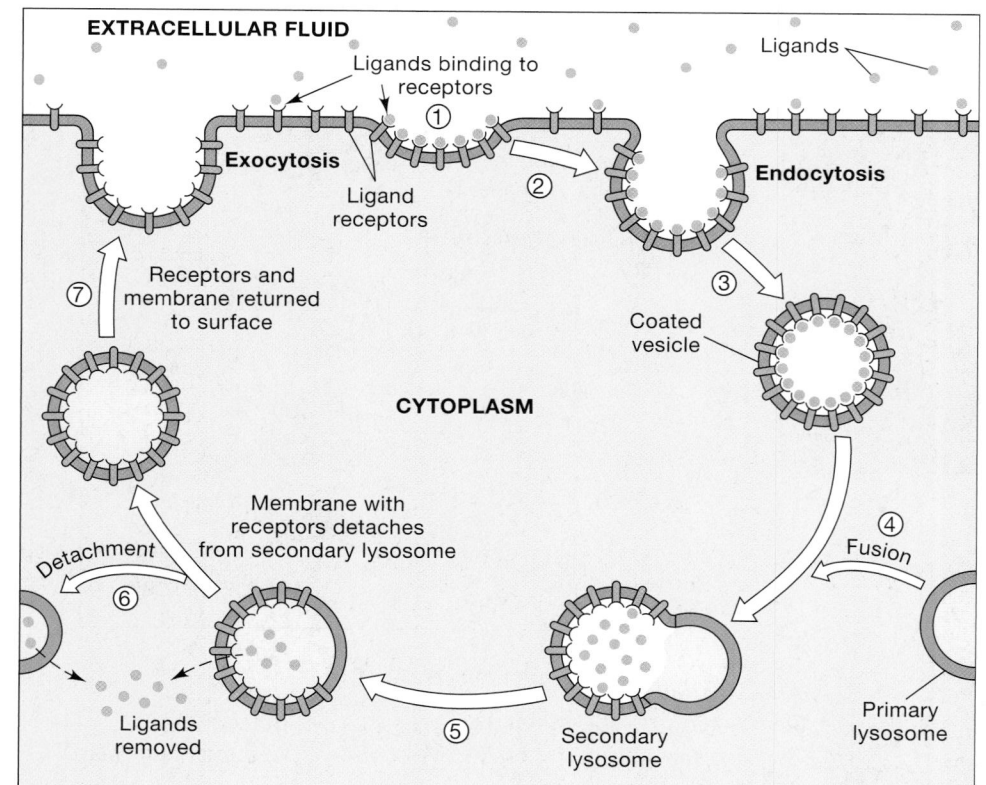

EXTRACELLULAR FLUID

Ligands binding to receptors ①

Ligands

Exocytosis ② Endocytosis

Ligand receptors

⑦ Receptors and membrane returned to surface

③

Coated vesicle

CYTOPLASM

Detachment ⑥

Membrane with receptors detaches from secondary lysosome

④ Fusion

Ligands removed

⑤ Secondary lysosome

Primary lysosome

●**FIGURE 3–25**
Receptor-Mediated Endocytosis. In this process, ① specific target molecules called ligands bind to receptors, generally glycoproteins, in the cell membrane. ② Membrane areas coated with ligands pinch off to form ③ vesicles that ④ fuse with primary lysosomes. ⑤ The ligands are freed from the receptors and, if necessary, broken down by enzymes before diffusing or being transported into the surrounding cytoplasm. ⑥ The membrane containing the receptor molecules separates from the membrane of the lysosome and ⑦ returns to the cell surface to bind additional ligands.

PINOCYTOSIS *Cell drinking,* or **pinocytosis** (pi-nō-sī-TŌ-sis), is the formation of endosomes filled with extracellular fluid. This process is not as selective as receptor-mediated endocytosis, because no receptor proteins are involved. The target appears to be the fluid contents in general rather than specific bound ligands. In pinocytosis, a deep groove or pocket forms in the cell membrane and then pinches off (Figure 3–26a●). The steps involved in the formation and fate of an endosome created by pinocytosis are similar to the steps in receptor-mediated endocytosis, except that ligand binding is not the trigger.

PHAGOCYTOSIS *Cell eating,* or **phagocytosis** (fa-gō-sī-TŌ-sis), produces phagosomes containing solid objects that may be as large as the cell itself. In this process, cytoplasmic extensions called **pseudopodia** (soo-dō-PŌ-dē-ah; *pseudo-,* false + *podon,* foot; singular *pseudopodium*) surround the object, and their membranes fuse to form a vesicle (Figure 3–26b●). The phagosome then fuses with many lysosomes, whereupon its contents are digested by lysosomal enzymes. Most cells display pinocytosis;

phagocytosis is performed only by specialized cells, such as the *macrophages* that protect tissues by engulfing bacteria, cell debris, and other abnormal materials.

Exocytosis

Exocytosis (ek-sō-sī-TŌ-sis), introduced in our discussion of the Golgi apparatus (p. 80), is the functional reverse of endocytosis. In exocytosis, a vesicle created inside the cell fuses with the cell membrane and discharges its contents into the extracellular environment (Figure 3–26b●). The ejected material may be secretory products, such as mucins or hormones, or waste products, such as those accumulating in endocytic vesicles. In a few specialized cells, endocytosis produces vesicles on one side of the cell that are discharged through exocytosis on the opposite side. This method of bulk transport is common in cells lining capillaries, which use a combination of pinocytosis and exocytosis to transfer fluid and solutes from the bloodstream into the surrounding tissues.

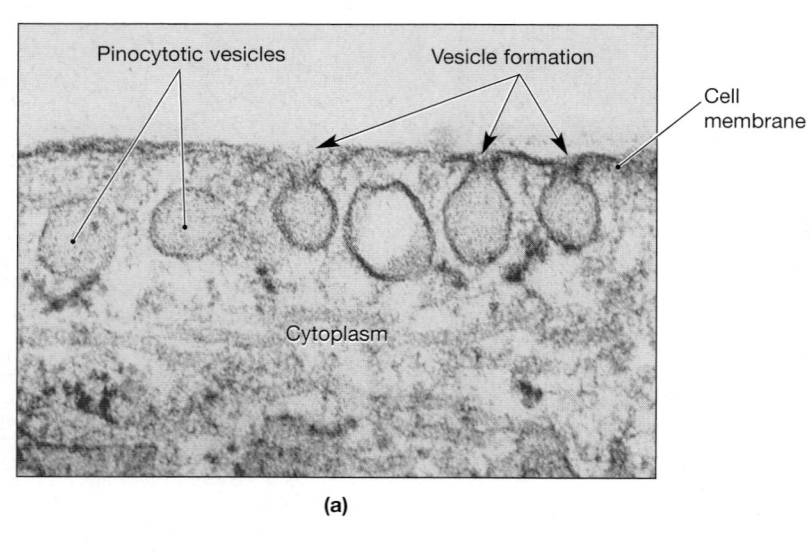

(a)

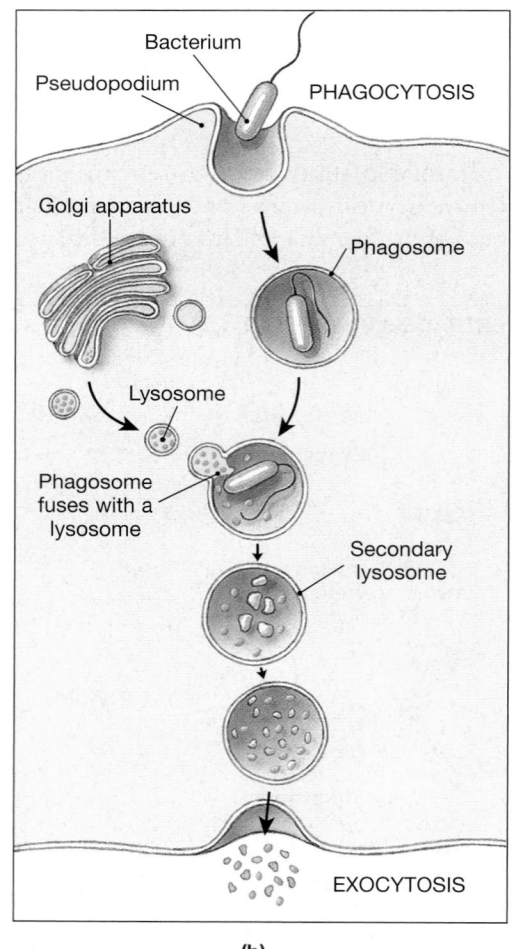

(b)

●**FIGURE 3–26**
Pinocytosis and Phagocytosis. **(a)** An electron micrograph showing pinocytosis at the surface of a cell. **(b)** Material brought into the cell by phagocytosis is enclosed in a phagosome and subsequently exposed to lysosomal enzymes. After nutrients are absorbed from the vesicle, the residue is discharged by exocytosis.

Many different mechanisms are moving materials into and out of the cell at any moment. Before proceeding further, review and compare the mechanisms summarized in Table 3–3.

☐ THE TRANSMEMBRANE POTENTIAL

As noted previously, the inside of the cell membrane has a slight negative charge with respect to the outside. The cause is a slight excess of positively charged ions outside the cell membrane and a slight excess of negatively charged ions (especially proteins) inside the cell membrane. This unequal charge distribution is created by differences in the permeability of the membrane to various ions, as well as by active transport mechanisms.

Although the positive and negative charges are attracted to each other and would normally rush together, they are kept apart by the phospholipid membrane. When positive and negative charges are held apart, a **potential difference** is said to exist between them. We refer to the potential difference across a cell membrane as the **transmembrane potential**.

The unit of measurement of potential difference is the *volt* (V). Most cars, for example, have 12-V batteries. The transmembrane potentials of cells are much smaller, typically in the vicinity of 0.07 V. Such a value is usually expressed as 70 mV, or 70 *millivolts* (thousandths of a volt). The transmembrane potential in an undisturbed cell is called the **resting potential**. Each type of cell has a characteristic resting potential between -10 mV (-0.01 V) and -100 mV (-0.1 V), with the minus sign signifying that the inside of the cell membrane contains an excess of negative charges compared with the outside. Examples include fat cells (-40 mV), thyroid cells (-50 mV), neurons (-70 mV), skeletal muscle cells (-85 mV), and cardiac muscle cells (-90 mV).

If the lipid barrier were removed, the positive and negative charges that it had separated would rush together and the potential difference would be eliminated. The cell membrane thus acts like a dam across a stream. Just as a dam resists the water pressure that builds up on the upstream side, a cell membrane resists electrochemical forces that would otherwise drive ions into or out of the cell. The water retained behind a dam and the ions

---

SUMMARY TABLE 3–3 MECHANISMS INVOLVED IN MOVEMENT ACROSS CELL MEMBRANES

Mechanism	Process	Factors Affecting Rate	Substances Involved (location)
Diffusion	Molecular movement of solutes; direction determined by relative concentrations	Size of gradient; size of molecules; charge; lipid solubility, temperature	Small inorganic ions; lipid-soluble materials (all cells)
Osmosis	Movement of water molecules toward solution containing relatively higher solute concentration; requires selectively permeable membrane	Concentration gradient; opposing osmotic or hydrostatic pressure	Water only (all cells)
Carrier-Mediated Transport			
Facilitated diffusion	Carrier proteins passively transport solutes across a membrane down a concentration gradient	Size of gradient, temperature and availability of carrier protein	Glucose and amino acids (all cells, but several different regulatory mechanisms exist)
Active transport	Carrier proteins actively transport solutes across a membrane, regardless of any concentration gradients	Availability of carrier, substrate, and ATP	Na^+, K^+, Ca^{2+}, Mg^{2+} (all cells); other solutes by specialized cells
Secondary active transport	Carrier proteins passively transport two solutes, with one (normally Na^+) moving down its concentration gradient; the cell must later expend ATP to eject the Na^+	Availability of carrier, substrates, and ATP	Glucose and amino acids (specialized cells)
Vesicular Transport			
Endocytosis	Creation of membranous vesicles containing fluid or solid material	Stimulus and mechanics incompletely understood; requires ATP	Fluids, nutrients (all cells); debris, pathogens (specialized cells)
Exocytosis	Fusion of vesicles containing fluids or solids (or both) within the cell membrane	Stimulus and mechanics incompletely understood; requires ATP	Fluids, debris (all cells)

held on either side of the cell membrane have *potential energy*—stored energy that can be released to do work. People have designed many ways to make use of the potential energy stored behind a dam—for example, turning a mill wheel or a turbine. Similarly, cells have ways of utilizing the potential energy stored in the transmembrane potential. For example, it is the transmembrane potential that makes possible the transmission of information in the nervous system, and thus our perceptions and thoughts. As we will see in later chapters, changes in the transmembrane potential also trigger the contraction of muscles and the secretion of glands.

✓ During digestion in the stomach, the concentration of hydrogen ions (H^+) rises to many times that in cells of the stomach. Which transport process could be responsible?

✓ If the cell membrane were freely permeable to sodium ions (Na^+), how would the transmembrane potential be affected?

✓ When they encounter bacteria, certain types of white blood cells engulf the bacteria and bring them into the cell. What is this process called?

Answers start on page Q-1

3–6 THE CELL LIFE CYCLE

Objectives

- Describe the stages of the cell life cycle.
- Describe the process of mitosis and explain its significance.
- Discuss the regulation of the cell life cycle and the relationship between cell division and cancer.
- Define differentiation and explain its importance.

The period between fertilization and physical maturity involves tremendous changes in organization and complexity. At fertilization, a single cell is all there is; at maturity, your body has roughly 75 trillion cells. This amazing transformation involves a form of cellular reproduction called **cell division**. The division of a single cell produces a pair of **daughter cells**, each half the size of the original. Before dividing, each of the daughter cells will grow to the size of the original cell.

Even when development is complete, cell division continues to be essential to survival. Cells are highly adaptable, but they can be damaged by physical wear and tear, toxic chemicals, temperature changes, and other environmental stresses. And cells, like individuals, are subject to aging. The life span of a cell varies from hours to decades, depending on the type of cell and the stresses involved. Many cells appear to be programmed to self-destruct after a certain period of time as a result of the activation of specific "suicide genes" in the nucleus. The genetically controlled death of cells is called **apoptosis** (ap-op-TŌ-sis or ap-ō-TŌ-sis; *ptosis*, a falling away). Several genes involved in the regulation of this process have

been identified. For example, a gene called *bcl-2* appears to prevent apoptosis and to keep a cell alive and functional. If something interferes with the function of this gene, the cell self-destructs.

Because a typical cell does not live nearly as long as a typical person, cell populations must be maintained over time by cell division. For cell division to be successful, the genetic material in the nucleus must be duplicated accurately, and one copy must be distributed to each daughter cell. The duplication of the cell's genetic material is called **DNA replication**, and nuclear division is called **mitosis** (mī-TŌ-sis). Mitosis occurs during the division of somatic cells. The production of sex cells involves a different process, **meiosis** (mī-Ō-sis), to be described in Chapter 28.

Figure 3–27● depicts the life cycle of a typical cell. That life cycle includes a fairly brief period of mitosis alternating with an *interphase* of variable duration.

◻ INTERPHASE

Most cells spend only a small part of their time actively engaged in cell division. Somatic cells spend the majority of their functional lives in a state known as **interphase**. During interphase, a cell performs all its normal functions and, if necessary, prepares for cell division. In a cell preparing to divide, interphase can be divided into the G_1, S, and G_2 *phases*. An interphase cell in the G_0 *phase* is not preparing for division, but is performing all other normal cell functions. Some mature cells, such as skeletal muscle cells and most neurons, remain in G_0 indefinitely and never divide. In contrast, *stem cells*, which divide repeatedly with very brief interphase periods, never enter G_0.

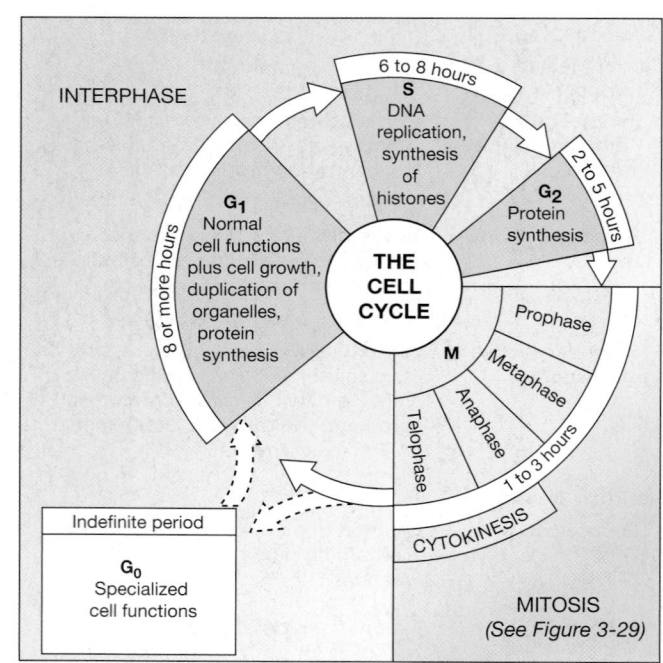

●**FIGURE 3–27**
The Cell Life Cycle

The G₁ Phase

A cell that is ready to divide first enters the **G₁ phase**. In this phase, the cell makes enough mitochondria, cytoskeletal elements, endoplasmic reticulum, ribosomes, Golgi membranes, and cytosol for two functional cells. Centriole replication begins in G₁ and commonly continues until G₂. In cells dividing at top speed, G₁ may last just 8–12 hours. Such cells pour all their energy into mitosis, and all other activities cease. If G₁ lasts for days, weeks, or months, preparation for mitosis occurs as the cells perform their normal functions.

The S Phase

When the activities of G₁ have been completed, the cell enters the **S phase**. Over the next 6–8 hours, the cell duplicates its chromosomes. This involves DNA replication and the synthesis of histones and other proteins in the nucleus. The goal of DNA replication, which occurs in cells preparing to undergo either mitosis or meiosis, is to copy the genetic information in the nucleus. The cell ends up with two identical sets of chromosomes. In mitosis, one set is given to each of the two daughter cells.

DNA REPLICATION Each DNA molecule consists of a pair of DNA strands joined by hydrogen bonding between complementary nitrogenous bases. ∞ p. 58 Figure 3–28● diagrams DNA replication. The process begins when enzymes called *helicases* unwind the strands and disrupt the weak bonds between the bases. As the strands unwind, molecules of **DNA polymerase** bind to the exposed nitrogenous bases. This enzyme (1) promotes bonding between the nitrogenous bases of the DNA strand and complementary DNA nucleotides dissolved in the nucleoplasm and (2) links the nucleotides by covalent bonds.

Many molecules of DNA polymerase work simultaneously along the DNA strands (Figure 3–28●). DNA polymerase can work in only one direction along a strand of DNA, but the two strands in a DNA molecule are oriented differently. As a result, the DNA polymerase on one strand works toward the site where the strands are unzipping, but those on the other strand work away from it. As the two original strands gradually separate, the DNA polymerase bound to one strand (the upper strand in the figure) adds nucleotides to make a single, continuous complementary copy of that strand. This copy grows toward the "zipper" from right to left, adding nucleotides in the sequence

$$9 \leftarrow 8 \leftarrow 7 \leftarrow 6 \leftarrow 5 \leftarrow 4 \leftarrow 3 \leftarrow 2 \leftarrow 1$$

The 1 is added first, then 2 to the left of 1, and so on.

DNA polymerase on the other original strand, however, can work only toward its free end, away from the unzipping site. For example, consider the lower strand in Figure 3–28●. The first DNA polymerase to bind to it must work from left to right, adding nucleotides in the sequence $1 \rightarrow 2 \rightarrow 3 \rightarrow 4 \rightarrow 5$. But as the original strands continue to unzip, additional nucleotides are continuously exposed. This molecule of DNA polymerase cannot go into reverse; it can only continue working from left to right. Thus, a second molecule of DNA polymerase must bind closer to the point of unzipping and assemble a complementary copy that grows in the sequence $6 \rightarrow 7 \rightarrow 8 \rightarrow 9$, until it bumps into the segment created by the first DNA polymerase. The two segments are then spliced together by enzymes called **ligases** (LĪ-gās-ez; *liga*, to tie).

Eventually, the unzipping completely separates the original strands. The copying ends, the last splicing is done, and two identical DNA molecules have formed.

The G₂ Phase

Once DNA replication has ended, there is a brief (2–5-hour) **G₂ phase** devoted to last-minute protein synthesis and to the completion of centriole replication. The cell then enters the **M phase**, and mitosis begins.

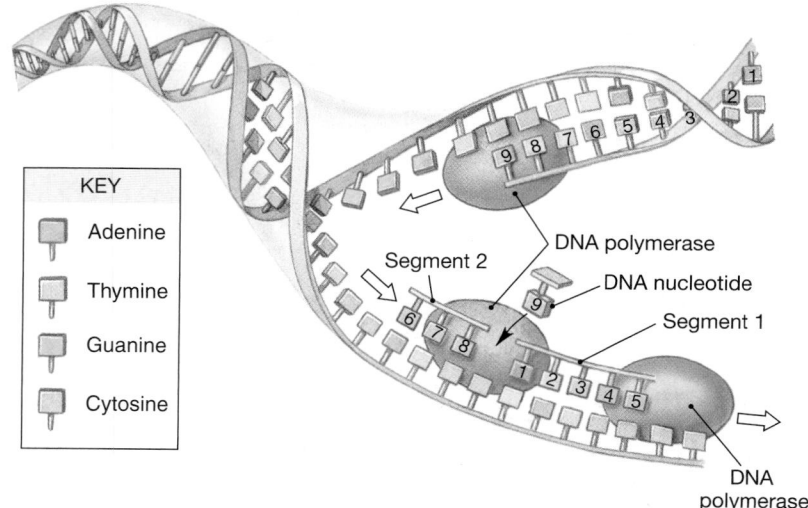

●**FIGURE 3–28**
DNA Replication. In DNA replication, the DNA strands unwind, and DNA polymerase begins attaching complementary DNA nucleotides along each strand. On one original strand, the complementary copy is produced as a continuous strand. Along the other original strand, the copy begins as a series of short segments spliced together by ligases. This process ultimately produces two identical copies of the original DNA molecule.

KEY
- Adenine
- Thymine
- Guanine
- Cytosine

Segment 2
DNA polymerase
DNA nucleotide
Segment 1
DNA polymerase

☐ MITOSIS

Mitosis separates the duplicated chromosomes of a cell into two identical nuclei. The term *mitosis* specifically refers to the division and duplication of the cell's nucleus; division of the cytoplasm to form two distinct new cells involves a separate, but related, process known as **cytokinesis** (sī-tō-ki-NĒ-sis; *cyto-*, cell + *kinesis*, motion). Figure 3–29a● depicts interphase, and Figure 3–29b–f● summarizes the four stages of mitosis: *prophase* (early and late), *metaphase, anaphase,* and *telophase.* Bear in mind that, although we describe mitosis in stages, it is really one smooth, continuous process.

Stage 1: Prophase

Prophase (PRŌ-fāz; *pro*, before) begins when the chromosomes coil so tightly that they become visible as individual structures under a light microscope. As a result of DNA replication during the S phase, two copies of each chromosome now exist. Each copy, called a **chromatid** (KRŌ-ma-tid), is physically connected to its duplicate copy at a single point, the **centromere** (SEN-trō-mēr). The centromere is surrounded by a protein complex known as the **kinetochore** (ki-NE-tō-kor).

As the chromosomes appear, the nucleoli disappear. The disappearance occurs in late prophase, often called *prometaphase.* Around this time, the two pairs of centrioles replicated during the G₁–G₂ period move toward opposite poles of the nucleus. An array of microtubules called **spindle fibers** extends between the centriole pairs. Smaller microtubules called *astral rays* radiate into the surrounding cytoplasm. Late in prophase, the nuclear

envelope disappears. The spindle fibers now form among the chromosomes, and the kinetochore of each chromatid becomes attached to a spindle fiber. Once that attachment occurs, the spindle fiber is called a *chromosomal microtubule.*

Stage 2: Metaphase

Metaphase (MET-a-fāz; *meta*, after) begins as the chromatids move to a narrow central zone called the **metaphase plate**. Metaphase ends when all the chromatids are aligned in the plane of the metaphase plate.

Stage 3: Anaphase

Anaphase (AN-a-fāz; *ana*, apart) begins when the centromere of each chromatid pair splits and the chromatids separate. The two **daughter chromosomes** are now pulled toward opposite ends of the cell along the chromosomal microtubules. This movement involves an interaction between the kinetochore and the microtubule. Anaphase ends when the daughter chromosomes arrive near the centrioles at opposite ends of the cell.

Stage 4: Telophase

During **telophase** (TĒL-ō-fāz; *telo*, end), each new cell prepares to return to the interphase state. The nuclear membranes form, the nuclei enlarge, and the chromosomes gradually uncoil. Once the chromosomes have relaxed and the fine filaments of chromatin become visible again, the nucleoli reappear and the nuclei resemble those of interphase cells. This stage marks the end of mitosis.

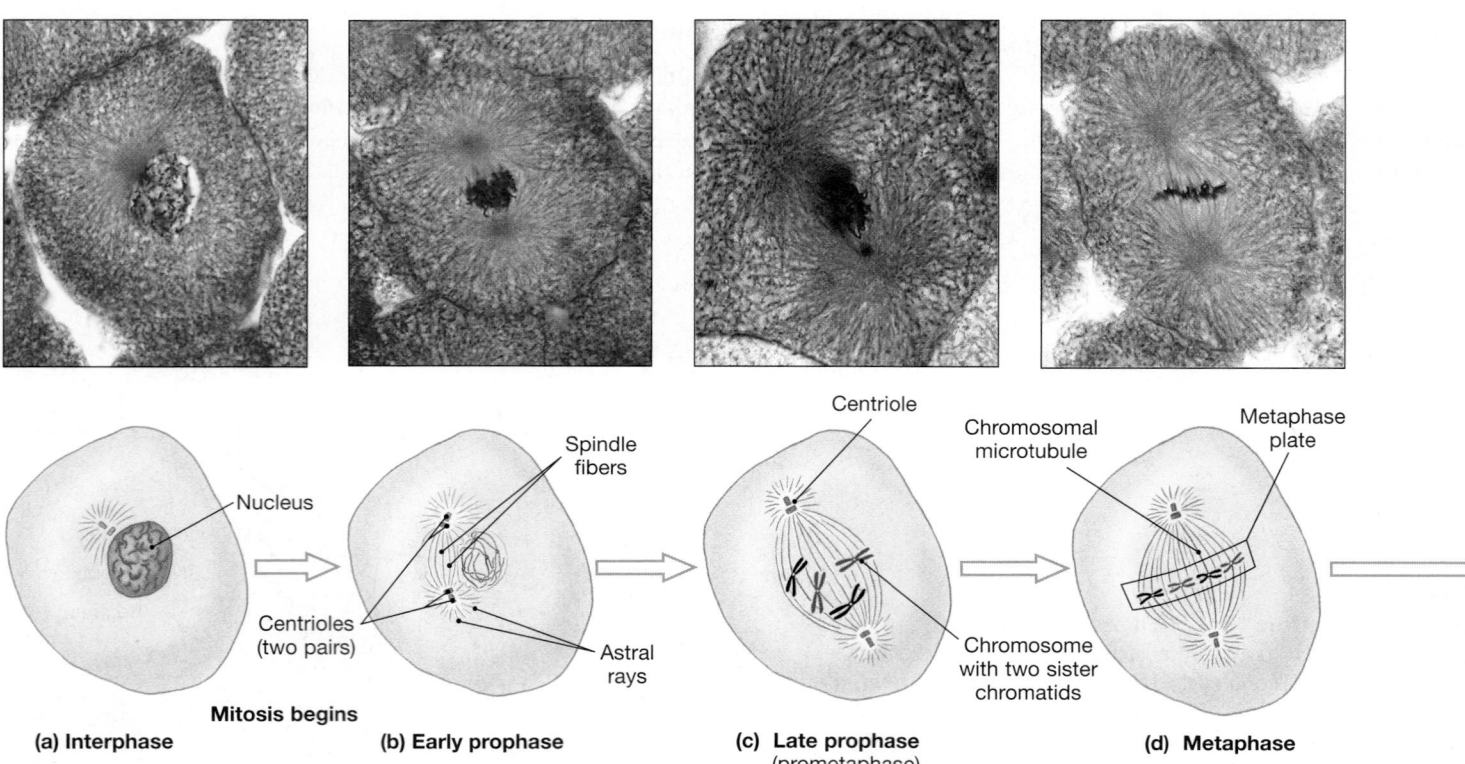

(a) Interphase

Mitosis begins

(b) Early prophase

(c) Late prophase (prometaphase)

(d) Metaphase

●**FIGURE 3–29**
Interphase, Mitosis, and Cytokinesis. Diagrammatic and microscopic views of representative cells undergoing cell division. (LM × 450)

☐ CYTOKINESIS

Cytokinesis is the cytoplasmic division of the daughter cells. This process usually begins in late anaphase. As the daughter chromosomes approach the ends of the spindle apparatus, the cytoplasm constricts along the plane of the metaphase plate, forming a *cleavage furrow*. Cytokinesis continues throughout telophase and is usually completed sometime after a nuclear membrane has re-formed around each daughter nucleus. The completion of cytokinesis marks the end of cell division, creating two separate and complete cells, each surrounded by its own cell membrane.

☐ THE MITOTIC RATE AND ENERGETICS

The preparations for cell division that occur between G_1 and the M phase are difficult to recognize in a light micrograph. However, the start of mitosis is easy to recognize, because the chromosomes become condensed and highly visible. The frequency of cell division can thus be estimated by the number of cells in mitosis at any time. As a result, we often use the term **mitotic rate** when we discuss rates of cell division. In general, the longer the life expectancy of a cell type, the slower the mitotic rate. Long-lived cells, such as muscle cells and neurons, either never divide or do so only under special circumstances. Other cells, such as those covering the surface of the skin or the lining of the digestive tract, are subject to attack by chemicals, pathogens, and abrasion. They survive for only days or even hours. Special cells called **stem cells** maintain these cell populations through repeated cycles of cell division.

Stem cells are relatively unspecialized, and their only function is the production of daughter cells. Each time a stem cell divides, one of its daughter cells develops functional specializations while the other prepares for further stem cell divisions. The rate of stem cell division can vary with the type of tissue and the demand for new cells. In heavily abraded skin, stem cells may divide more than once a day, but stem cells in adult connective tissues may remain inactive for years.

Dividing cells use an unusually large amount of energy. For example, they must synthesize new organic materials and move organelles and chromosomes within the cell. All these processes require ATP in substantial amounts. Cells that do not have adequate energy sources cannot divide. In a person who is starving, normal cell growth and maintenance grind to a halt. For this reason, prolonged starvation slows wound healing, lowers resistance to disease, thins the skin, and changes the lining of the digestive tract.

☐ REGULATION OF THE CELL LIFE CYCLE

In normal tissues, the rate of cell division balances the rate of cell loss or destruction. Mitotic rates are genetically controlled, and many different stimuli may be responsible for activating genes that promote cell division. Some of the stimuli are internal, and many cells set their own pace of mitosis and cell division. An important internal trigger is the level of **M-phase promoting factor** (**MPF**), also known as *maturation-promoting factor*. MPF is assembled from two parts: a cell division cycle protein called *Cdc2* and a second protein called *cyclin*. Cyclin levels climb as the cell life cycle proceeds. When levels are high enough, MPF appears in the cytoplasm and mitosis gets under way.

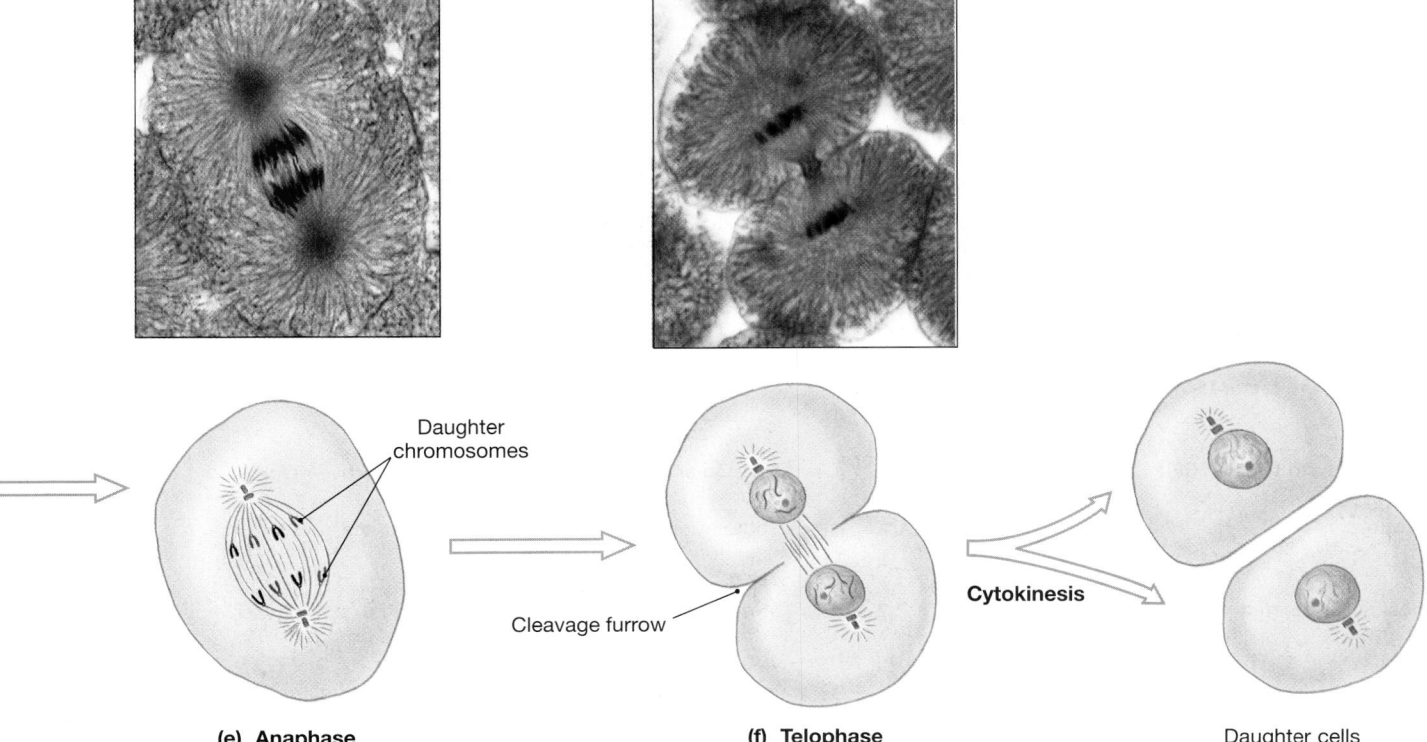

(e) Anaphase   Daughter chromosomes

(f) Telophase   Cleavage furrow   Cytokinesis   Daughter cells

Various extracellular compounds—generally, peptides—can stimulate the division of specific types of cells. These compounds include several hormones and a variety of **growth factors**. Table 3–4 lists some of the stimulatory compounds and their target tissues; we shall discuss these hormones and factors in later chapters. Each compound appears to exert its effects by binding to receptors on the cell membrane. Such binding initiates a series of biochemical events that initiate and promote cell division. The effects of these stimulatory factors may be opposed by a poorly understood class of peptides called *chalones* (KĀ-lōnz).

Many of the peptide growth factors bind to membrane receptors at the cell surface. The presence or absence of the appropriate binding protein determines whether a particular cell will respond to a particular growth factor. The mechanism inside the cell appears to be similar, however, regardless of the type of membrane receptor. Binding at the membrane surface triggers the activation and release of intermediaries known as *Ras proteins*. These proteins in turn activate intracellular enzymes and promote gene activation. Growth factors that do not use Ras proteins may use other intermediaries or may enter the cell and exert their effects directly on the nucleus.

Genetic mechanisms for inhibiting cell division have recently been identified. The genes involved are known as *repressor genes*. One gene, called *p53*, controls a protein that resides in the nucleus and activates genes that direct the production of growth-inhibiting factors inside the cell. Roughly half of all cancers are associated with abnormal forms of the *p53* gene.

There are indications that in humans, the *number* of cell divisions performed by a cell and its descendants is regulated at the chromosome level by structures called **telomeres**. Telomeres are terminal segments of DNA with associated proteins. These DNA-protein complexes bend and fold repeatedly to form caps at the ends of chromosomes. Telomeres have several functions, notably to attach chromosomes to the nuclear matrix and to protect the ends of the chromosomes from damage during mitosis. The telomeres themselves, however, are subject to wear and tear over the years. Each time a cell divides during adult life, some of the repeating segments break off, and the telomeres get shorter. When they get too short, repressor gene activity tells the cell to stop dividing. (See the Box on Telomeres, Aging, and Cancer.)

▢ CELL DIVISION AND CANCER

When the rates of cell division and growth exceed the rate of cell death, a tissue begins to enlarge. A **tumor**, or **neoplasm**, is a mass or swelling produced by abnormal cell growth and division. In a **benign tumor**, the cells usually remain within a connective-tissue capsule. Such a tumor seldom threatens an individual's life and can usually be surgically removed if its size or position disturbs tissue function.

Cells in a **malignant tumor** no longer respond to normal control mechanisms. These cells do not remain confined within a connective tissue capsule, but spread into surrounding tissues. The tumor of origin is called the *primary tumor* (or *primary neoplasm*), and the spreading process is called **invasion**. Malignant cells may also travel to distant tissues and organs and establish *secondary tumors*. This dispersion, called **metastasis** (me-TAS-ta-sis; *meta*, after + *stasis*, standing still), is very difficult to control.

TABLE 3–4 REPRESENTATIVE CHEMICAL FACTORS AFFECTING CELL DIVISION

Factor	Source(s)	Effect(s)	Target(s)
M-phase-promoting factor (maturation-promoting factor)	Forms within cytoplasm from Cdc2 and cyclin	Triggers start of mitosis	Regulatory mechanism active in all dividing cells
Growth hormone	Anterior lobe of the pituitary gland	Stimulation of growth, cell division, differentiation	All cells, especially in epithelia and connective tissues
Prolactin	Anterior lobe of the pituitary gland	Stimulation of cell growth, division, development	Gland and duct cells of mammary glands
Nerve growth factor (NGF)	Salivary glands; other sources suspected	Stimulation of nerve cell repair and development	Neurons and neuroglia
Epidermal growth factor (EGF)	Duodenal glands; other sources suspected	Stimulation of stem cell divisions and epithelial repairs	Epidermis of skin
Fibroblast growth factor (FGF)	Unknown	Division and differentiation of fibroblasts and related cells	Connective tissues
Erythropoietin	Kidneys (primary source)	Stimulation of stem cell divisions and maturation of red blood cells	Bone marrow
Thymosins and related compounds	Thymus	Stimulation of division and differentiation of lymphocytes (especially T cells)	Thymus and other lymphoid tissues and organs
Chalones	Many tissues	Inhibition of cell division	Cells in the immediate area

Telomerase, Aging, and Cancer

Each telomere contains a sequence of roughly 8000 nitrogenous bases, but they are multiple copies of the same base sequence, TTAGGG, repeated over and over again. Telomeres are not formed by DNA polymerase; instead, they are created by an enzyme called *telomerase*. Telomerase is functional early in life, but by adulthood it has become inactive. As a result, the telomere segments lost during each mitotic division are not replaced. Eventually, shortening of the telomere reaches a point at which the cell ceases to divide.

This mechanism is clearly a major factor in the aging process, since many of the signs of age result from the gradual loss of functional stem cell populations. Experiments are in progress to determine whether activating telomerase (or a suspected alternative repair enzyme) can forestall or reverse the effects of aging.

This would seem to be a very promising area of research. Activate telomerase, and forget aging—sounds good, doesn't it? Unfortunately, there's always a catch: In adults, telomerase activation is a key step in the development of cancer.

If for some reason a cell with short telomeres does *not* respond normally to repressor genes, it will continue to divide. The result is mechanical damage to the DNA strands, chromosomal abnormalities, and mutations. Interestingly, one of the first consequences of such damage is the abnormal activation of telomerase. Once this occurs, the abnormal cells can continue dividing indefinitely. Telomerase is active in at least 90 percent of all cancer cells. Research is therefore underway to find out how to turn off telomerase that has been improperly activated.

Cancer is an illness characterized by mutations that disrupt normal control mechanisms and produce potentially malignant cells. Cancer develops in the series of steps diagrammed in Figure 3–30●. Initially, the cancer cells are restricted to the primary tumor. In most cases, all the cells in the tumor are the daughter cells of a single malignant cell. Normal cells often become malignant when a mutation occurs in a gene involved with cell growth, differentiation, or division. The modified genes are called **oncogenes** (ON-kō-gēnz). **AM** Cancer: A Closer Look

Cancer cells gradually lose their resemblance to normal cells. They change size and shape, typically becoming abnormally large or small. At first, the growth of the primary tumor distorts the tissue, but the basic tissue organization remains intact. Metastasis begins with invasion as tumor cells "break out" of the primary tumor and invade the surrounding tissue. They may then enter the lymphatic system and accumulate in nearby lymph nodes. When metastasis involves the penetration of blood vessels, the cancer cells circulate throughout the body.

Responding to cues that are as yet unknown, cancer cells in the bloodstream ultimately escape out of the blood vessels to establish secondary tumors at other sites. These tumors are

extremely active metabolically, and their presence stimulates the growth of blood vessels into the area. The increased circulatory supply provides additional nutrients to the cancer cells and further accelerates tumor growth and metastasis.

As malignant tumors grow, organ function begins to deteriorate. The malignant cells may no longer perform their original functions, or they may perform normal functions in an unusual way. For example, endocrine cancer cells may produce normal hormones, but in excessively large amounts. Cancer cells do not use energy very efficiently. They grow and multiply at the expense of healthy tissues, competing for space and nutrients with normal cells. This competition accounts for the starved appearance of many patients in the late stages of cancer. Death may occur as a result of the compression of vital organs when nonfunctional cancer cells have killed or replaced the healthy cells in those organs, or when the cancer cells have starved normal tissues of essential nutrients.

The growth of blood vessels into the tumor is a vital step in the development and spread of the cancer. Without those vessels, the growth and metastasis of the cancer cells would be limited by the availability of oxygen and nutrients. A polypeptide called

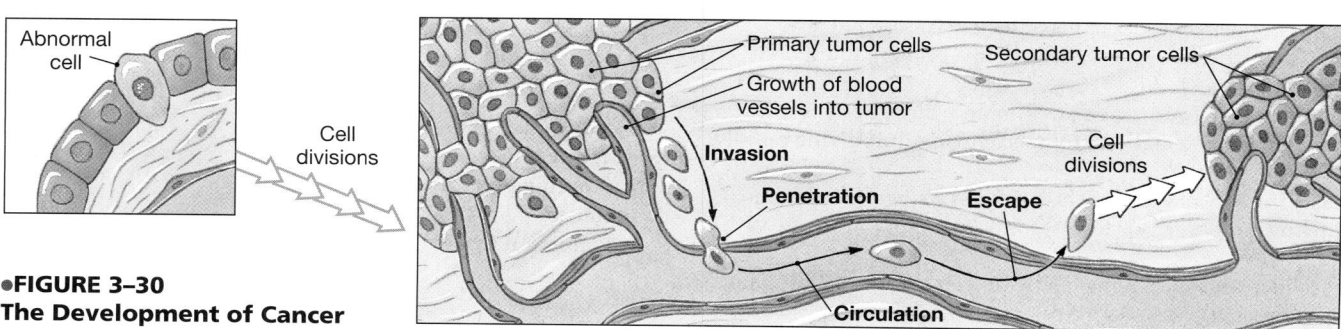

●**FIGURE 3–30**
The Development of Cancer

antiangiogenesis factor can prevent the growth of blood vessels and can slow the growth of cancers. This chemical, produced in normal human cartilage, can be extracted in large quantities from sharks, whose skeletons are entirely cartilaginous. Until molecular biological techniques were available to synthesize it, sharks were collected to obtain antiangiogenesis factor for use in experimental cancer therapies. We will return to the subject of cancer in later chapters that deal with specific systems.

□ CELL DIVERSITY AND DIFFERENTIATION

The liver cells, fat cells, and neurons of an individual all contain the same set of chromosomes and genes, but in each case a different set of genes has been turned *off*. In other words, liver cells differ from fat cells because liver cells have one set of genes accessible for transcription and fat cells another.

When a gene is functionally eliminated, the cell loses the ability to create a particular protein—and thus to perform any functions involving that protein. Each time another gene switches off, the cell's functional abilities become more restricted. This specialization process is called **differentiation**.

Fertilization produces a single cell with all its genetic potential intact. Repeated cell divisions follow, and differentiation begins as the number of cells increases. Differentiation produces specialized cells with limited capabilities. These cells form organized collections known as *tissues,* each with discrete functional roles. In Chapter 4, we will examine the structure and function of tissues and will consider the role of tissue interactions in the maintenance of homeostasis.

In most cases, the differentiation process is irreversible: Once genes are turned off, they won't be turned back on. However, some cells, such as stem cells, are relatively undifferentiated. These cells can differentiate into any of several different types of cell, depending on local conditions. For example, stem cells in many parts of the body can differentiate into fat cells when more nutrients are consumed than the body can use. Researchers are gradually discovering what chemical cues are responsible for controlling the differentiation of specific cell types. The ability to take a person's stem cells and create new heart muscle cells or neurons on demand may one day revolutionize the treatment of heart attacks and strokes.

✓ A cell is actively manufacturing enough organelles to serve two functional cells. This cell is probably in which phase of its life cycle?

✓ During DNA replication, a nucleotide is deleted from a sequence that normally codes for a polypeptide. What effect will this deletion have on the amino acid sequence of the polypeptide?

✓ What would happen if spindle fibers failed to form in a cell during mitosis?

Answers start on page Q-1

Chapter Review

SELECTED CLINICAL TERMINOLOGY

Terms Discussed in This Chapter

benign tumor: A mass or swelling in which the cells usually remain within a connective-tissue capsule; rarely life threatening. *(p. 104)*

cancer: An illness caused by mutations leading to the uncontrolled growth and replication of affected cells. *(p. 105)*

dextran: A carbohydrate that cannot cross cell membranes; commonly administered in solution to patients after blood loss or dehydration. *(p. 94)*

DNA fingerprinting: Identifying an individual on the basis of repeating nucleotide sequences in his or her DNA. *(p. 85)*

invasion: The spread of cancer cells from a primary tumor into surrounding tissues. *(p. 104)*

malignant tumor: A mass or swelling in which the cells no longer respond to normal control mechanisms, but divide rapidly. *(p. 104)*

metastasis (me-TAS-ta-sis): The spread of malignant cells into distant tissues and organs, where secondary tumors subsequently develop. *(p. 104)*

normal saline: A solution that approximates the normal osmotic concentration of extracellular fluids. *(p. 94)*

oncogene (ON-kō-jēn): A cancer-causing gene created by a somatic mutation in a normal gene involved with growth, differentiation, or cell division. *(p. 105)*

primary tumor *(primary neoplasm):* The site at which a cancer cell initially develops. *(p. 104)*

secondary tumor: A colony of cancerous cells formed by metastasis. *(p. 104)*

tumor (neoplasm): A mass or swelling produced by abnormal cell growth and division. *(p. 104)*

AM Additional Terms Discussed in the *Applications Manual*

carcinogen
genetic engineering
karyotyping
mutagen
recombinant DNA

STUDY OUTLINE

3-1 AN INTRODUCTION TO CELLS   p. 67

1. Contemporary *cell theory* incorporates several basic concepts: (1) **Cells** are the building blocks of all plants and animals; (2) cells are produced by the division of preexisting cells; (3) cells are the smallest units that perform all vital physiological functions; (4) each cell maintains homeostasis at the cellular level; and (5) homeostasis at the tissue, organ, organ system, and organism levels reflects the combined and coordinated actions of many cells. *(Figure 3–1)*

2. **Cytology**, the study of the structure and function of cells, is part of **cell biology**.

3. The human body contains two types of cells: **sex cells** (*sperm* and *oocytes*) and **somatic cells** (all other cells). *(Figure 3–2; Table 3–1)*

4. A typical cell is surrounded by **extracellular fluid**—specifically, the **interstitial fluid** of the tissue. The cell's outer boundary is the **cell membrane**, or **plasma membrane**.

3-2 THE CELL MEMBRANE   p. 71

1. The cell membrane's functions include physical isolation, regulation of exchange with the environment, sensitivity, and structural support.

3-2 MEMBRANE LIPIDS   p. 71

2. The cell membrane, which is a **phospholipid bilayer**, contains lipids, proteins, and carbohydrates. Phospholipids are the largest contributors to membrane structure. *(Figure 3–3)*

MEMBRANE PROTEINS   p. 72

3. **Integral proteins** are part of the membrane itself; **peripheral proteins** are attached to, but can separate from, the membrane.

4. Membrane proteins can act as anchors (**anchoring proteins**), identifiers (**recognition proteins**), enzymes, receptors (**receptor proteins**), carriers (**carrier proteins**), or **channels**. *(Figure 3–4)*

MEMBRANE CARBOHYDRATES   p. 73

5. The **glycocalyx** is formed by the carbohydrate portions of the *proteoglycans, glycoproteins,* and *glycolipids.* Functions include lubrication and protection, anchoring and locomotion, specificity in binding, and recognition.

3-3 THE CYTOPLASM   p. 74

1. The **cytoplasm** contains the fluid **cytosol** and the **organelles** that the cytosol surrounds.

THE CYTOSOL   p. 74

2. Cytosol differs from extracellular fluid in composition and in the presence of **inclusions**.

THE ORGANELLES   p. 74

3. **Nonmembranous organelles** are not enclosed by membranes, and they are always in contact with the cytosol. They include the *cytoskeleton, microvilli, centrioles, cilia, ribosomes,* and *proteasomes.* *(Table 3–1)*

4. **Membranous organelles** are surrounded by lipid membranes that isolate them from the cytosol. They include the *endoplasmic reticulum,* the *Golgi apparatus, lysosomes, peroxisomes,* and *mitochondria. (Table 3–1)*

5. The **cytoskeleton** gives the cytoplasm strength and flexibility. It has four components: **microfilaments** (typically made of **actin**), **intermediate filaments**, **microtubules** (made of **tubulin**), and **thick filaments** (made of **myosin**). *(Figure 3–5)*

6. **Microvilli** are small projections of the cell membrane that increase the surface area exposed to the extracellular environment. *(Figure 3–5)*

7. **Centrioles** direct the movement of chromosomes during cell division and organize the cytoskeleton. The **centrosome** is the cytoplasm surrounding the centrioles. *(Figure 3–6)*

8. **Cilia**, anchored by a **basal body**, beat rhythmically to move fluids or secretions across the cell surface. *(Figure 3–6)*

9. **Ribosomes**, responsible for manufacturing proteins, are composed of a **small** and a **large ribosomal subunit**, both of which contain **ribosomal RNA (rRNA)**. **Free ribosomes** are in the cytoplasm, and **fixed ribosomes** are attached to the endoplasmic reticulum. *(Figures 3–7, 3–2)*

10. **Proteasomes** remove and break down damaged or abnormal proteins that have been tagged with *ubiquitin.*

11. The **endoplasmic reticulum (ER)** is a network of intracellular membranes that function in synthesis, storage, transport, and detoxification. The ER forms hollow tubes, flattened sheets, and chambers called **cisternae**. **Rough endoplasmic reticulum (RER)** contains ribosomes on its outer surface and forms **transport vesicles**; **smooth endoplasmic reticulum (SER)** is involved in lipid synthesis. *(Figure 3–8)*

12. The **Golgi apparatus** forms **secretory vesicles** and new membrane components, as well as packaging *lysosomes*. Secretions are discharged from the cell by exocytosis. *(Figures 3–9, 3–10)*

13. **Lysosomes**, vesicles filled with digestive enzymes, are responsible for the **autolysis** of injured cells. *(Figures 3–9, 3–11)*

14. **Peroxisomes** carry enzymes that neutralize toxins.

15. **Membrane flow** refers to the continuous movement and recycling of the membrane between the ER, vesicles, the Golgi apparatus, and the cell membrane.

16. **Mitochondria** are responsible for ATP production through aerobic respiration. The **matrix**, or fluid contents of a mitochondrion, lie inside the **cristae**, or folds of an inner membrane. *(Figure 3–12)*

3-4 THE NUCLEUS   p. 83

1. The **nucleus** is the control center of cellular operations. It is surrounded by a **nuclear envelope** (a double membrane with a **perinuclear space**), through which it communicates with the cytosol by way of **nuclear pores**. *(Figures 3–2, 3–13)*

CONTENTS OF THE NUCLEUS   p. 83

2. The nucleus contains a supportive **nuclear matrix**; one or more **nucleoli** typically are present.

3. The nucleus controls the cell by directing the synthesis of specific proteins, using information stored in **chromosomes**, which consist of DNA bound to **histones**. In nondividing cells, chromosomes form a tangle of filaments called **chromatin**. *(Figure 3–14)*

INFORMATION STORAGE IN THE NUCLEUS   p. 85

4. The cell's information storage system, the **genetic code**, is called a *triplet code* because a sequence of three nitrogenous bases codes for a single amino acid. Each **gene** contains all the triplets needed to produce a specific polypeptide chain.

GENE ACTIVATION AND PROTEIN SYNTHESIS   p. 85

5. As **gene activation** begins, **RNA polymerase** must bind to the gene.

6. **Transcription** is the formation of RNA from a DNA template. After transcription, a strand of **messenger RNA (mRNA)** carries instructions from the nucleus to the cytoplasm. *(Figure 3–15)*

7. During **translation,** a functional polypeptide is constructed by using the information contained in the sequence of **codons** along an mRNA strand. The sequence of codons determines the sequence of amino acids in the polypeptide. (*Figure 3–16*)

8. By complementary base pairing of **anticodons** to mRNA codons, **transfer RNA (tRNA)** molecules bring amino acids to the ribosomal complex. (*Figure 3–17*)

Transcription and translation: **Companion Website:** Chapter 3/ Tutorials.

HOW THE NUCLEUS CONTROLS CELL STRUCTURE AND FUNCTION   p. 89

9. The DNA of the nucleus has both direct and indirect control over protein synthesis.

3–5   HOW THINGS GET INTO AND OUT OF CELLS   p. 90

1. **Permeability** is the ease with which substances can cross the cell membrane. Nothing can pass through an **impermeable** barrier; anything can pass through a **freely permeable** barrier. Cell membranes are **selectively permeable.**

DIFFUSION   p. 90

2. **Diffusion** is the net movement of material from an area where its concentration is higher to an area where its concentration is lower. Diffusion occurs until the **concentration gradient** is eliminated. (*Figures 3–18, 3–19*)

3. **Osmosis** is the diffusion of water across a membrane in response to differences in solute concentration. **Osmotic pressure** is the force of water movement into a solution as the result of solute concentration. **Hydrostatic pressure** can oppose osmotic pressure. Water molecules undergo *bulk flow,* movement in groups across a membrane. (*Figure 3–20*)

4. **Tonicity** describes the effects of osmotic solutions on cells. A solution that does not cause an osmotic flow is **isotonic.** A solution that causes water to flow into a cell is **hypotonic** and can lead to **hemolysis.** A solution that causes water to flow out of a cell is **hypertonic** and can lead to **crenation.** (*Figure 3–21*)

CARRIER-MEDIATED TRANSPORT   p. 94

5. Carrier-mediated trasport is a transport process that involves the binding and transporting of specific ions by integral proteins. Cotransport moves two substances in the same direction; countertransport moves them in opposite directions.

6. In **facilitated diffusion,** compounds are transported across a membrane after binding to a **receptor site** of a carrier protein. (*Figure 3–22*)

7. **Active transport** mechanisms consume ATP, but are independent of concentration gradients. Some **ion pumps** are **exchange pumps. Secondary active transport** may involve **cotransport** or **countertransport.** (*Figure 3–23*)

VESICULAR TRANSPORT   p. 97

8. In **vesicular transport,** material moves into or out of the cell in membranous **vesicles.** Movement into the cell is accomplished through **endocytosis,** an active process that can take three forms: **receptor-mediated endocytosis** (by means of **coated vesicles**), **pinocytosis,** or **phagocytosis** (using **pseudopodia**). The ejection of materials from the cytoplasm is accomplished by **exocytosis.** (*Figures 3–25, 3–26; Table 3–2, Summary Table 3–3*)

THE TRANSMEMBRANE POTENTIAL   p. 99

9. The potential difference between the two sides of a cell membrane is a **transmembrane potential.** The transmembrane potential in an undisturbed cell is the cell's **resting potential.**

3–6   THE CELL LIFE CYCLE   p. 100

1. **Cell division** is the reproduction of cells. **Apoptosis** is the genetically controlled death of cells. **Mitosis** is the nuclear division of somatic cells. Sex cells are produced by **meiosis.**

INTERPHASE   p. 100

2. Most somatic cells spend most of their time in **interphase,** which includes the G_1, **S** (**DNA replication**), and G_2 **phases.** (*Figures 3–27, 3–28*)

MITOSIS   p. 102

3. Mitosis proceeds in four stages: **prophase, metaphase, anaphase,** and **telophase.** (*Figure 3–29*)

CYTOKINESIS   p. 103

4. During **cytokinesis,** the cytoplasm is divided and cell division ends. (*Figure 3–29*)

THE MITOTIC RATE AND ENERGETICS   p. 103

5. In general, the longer the life expectancy of a cell type, the slower is the **mitotic rate. Stem cells** undergo frequent mitoses to replace other, more specialized cells.

REGULATION OF THE CELL LIFE CYCLE   p. 103

6. A variety of **growth factors** can stimulate cell division and growth. (*Summary Table 3–4*)

CELL DIVISION AND CANCER   p. 104

7. Produced by abnormal cell growth and division, a **tumor,** or **neoplasm,** can be **benign** or **malignant.** Malignant cells may spread locally (by **invasion**) or to distant tissues and organs (through **metastasis**). The resultant illness is called **cancer.** Malignancy is often caused by modified genes called **oncogenes.** (*Figure 3–30*)

CELL DIVERSITY AND DIFFERENTIATION   p. 106

8. **Differentiation,** a process of specialization, results from the inactivation of particular genes in different cells, producing populations of cells with limited capabilities. Specialized cells form organized collections called *tissues,* each of which has certain functional roles.

REVIEW QUESTIONS

More assessment questions are available to you on the Companion Website. You will find Matching, Multiple Choice, True/False, and other quizzes to help further your understanding of the material covered in this chapter. To access the site, go to www.aw.com/martini.

LEVEL 1 Reviewing Facts and Terms

1. All of the following membrane transport mechanisms are passive processes, except
   (a) diffusion   (b) facilitated diffusion
   (c) vesicular transport (d) osmosis

2. The cell membrane includes
   (a) integral proteins
   (b) glycolipids
   (c) phospholipids
   (d) all of the above

3. _____ ion concentrations are high in the extracellular fluids, and _____ ion concentrations are high in the cytoplasm.
   (a) Calcium, magnesium   (b) Chloride, sodium
   (c) Potassium, sodium   (d) Sodium, potassium

4. In a resting transmembrane potential, the inside of the cell is _____, and the cell exterior is _____:
   (a) slightly negative, slightly positive
   (b) slightly positive, slightly negative
   (c) slightly positive, neutral
   (d) slightly negative, neutral

5. The organelle responsible for a variety of functions centering around the synthesis of lipids and carbohydrates is
   (a) the Golgi apparatus
   (b) the rough endoplasmic reticulum
   (c) the smooth endoplasmic reticulum
   (d) mitochondria

6. The reaction sequence in which glucose is broken down into pyruvic acid is
   (a) aerobic metabolism   (b) the TCA cycle
   (c) mitochondrial   (d) glycolysis
       energy production

7. The construction of a functional polypeptide by using the information in an mRNA strand is
   (a) translation   (b) transcription
   (c) replication   (d) gene activation

8. Our somatic cell nuclei contain _____ pairs of chromosomes.
   (a) 8   (b) 16
   (c) 23   (d) 46

9. Transcription ends in the production of a(n)
   (a) DNA molecule   (b) protein
   (c) tRNA molecule   (d) mRNA molecule

10. The genetically controlled death of cells is called
    (a) differentiation   (b) replication
    (c) apoptosis   (d) metastasis

11. The interphase of the cell life cycle is divided into
    (a) prophase, metaphase, anaphase, and telophase
    (b) G_0, G_1, S, and G_2
    (c) mitosis and cytokinesis
    (d) a, b, and c are correct

12. List the five basic concepts that make up the modern-day cell theory.

13. What are four general functions of the cell membrane?

14. What are the primary functions of membrane proteins?

15. By what three major transport mechanisms do substances get into and out of cells?

16. List four important factors that influence diffusion rates.

17. Define osmosis.

18. List (a) the nonmembranous organelles and (b) the membranous organelles of a typical cell.

19. What are the four major functions of the endoplasmic reticulum?

20. List the four stages of mitosis in their order of occurrence.

LEVEL 2 Reviewing Concepts

21. Diffusion is important in body fluids because it tends to
    (a) increase local concentration gradients
    (b) eliminate local concentration gradients
    (c) move substances against concentration gradients
    (d) create concentration gradients

22. Osmotic pressure differs from hydrostatic pressure because the osmotic pressure of a solution is an indication of the force of water movement resulting from
    (a) the solute concentration
    (b) the volume of water
    (c) the permeability of the membrane
    (d) a, b, and c are correct

23. When a cell is placed in a(n) _____ solution, the cell will lose water through osmosis. This process results in the _____ of red blood cells.
    (a) hypotonic, crenation   (b) hypertonic, crenation
    (c) isotonic, hemolysis   (d) hypotonic, hemolysis

24. Suppose that a DNA segment has the following nucleotide sequence: CTC–ATA–CGA–TTC–AAG–TTA. Which nucleotide sequences would a complementary mRNA strand have?
    (a) GAG–UAU–GAU–AAC–UUG–AAU
    (b) GAG–TAT–GCT–AAG–TTC–AAT
    (c) GAG–UAU–GCU–AAG–UUC–AAU
    (d) GUG–UAU–GGA–UUG–AAC–GGU

25. How many amino acids are coded in the DNA segment in Review Question 24?
    (a) 18   (b) 9
    (c) 6   (d) 3

26. What general characteristics are important in carrier-mediated transport mechanisms?

27. (a) What are the similarities between facilitated diffusion and active transport?
    (b) What are the differences?

28. What role does the sodium–potassium exchange pump play in stabilizing the resting membrane potential?

29. How does the cytosol differ in composition from interstitial fluid?

30. Differentiate between transcription and translation.

31. List, in sequence, the phases of the interphase stage of the cell life cycle, and briefly describe what happens in each.

32. List the stages of mitosis, and briefly describe the events that occur in each.

33. (a) What is cytokinesis? (b) What is its role in the cell cycle?

LEVEL 3 Critical Thinking and Clinical Applications

34. The transport of a certain molecule exhibits the following characteristics: (1) The molecule moves down its concentration gradient; (2) at concentrations above a given level, the rate of transport does not increase; and (3) cellular energy is not required for transport to occur. Which transport process is at work?

35. Solutions A and B are separated by a selectively permeable barrier. Over time, the level of fluid on side A increases. Which solution initially had the higher concentration of solute?

36. In kidney dialysis, a person's blood is passed through a bath that contains several ions and molecules. The blood is separated from the dialysis fluid by a membrane that allows water, small ions, and small molecules to pass, but does not allow large proteins or blood cells to pass. What should the composition of dialysis fluid be for it to remove urea (a small molecule) without changing the blood volume (removing water from the blood)?

37. Which organelles are involved in membrane flow? Trace the route of a single integral membrane protein from formation to incorporation into the cell membrane.

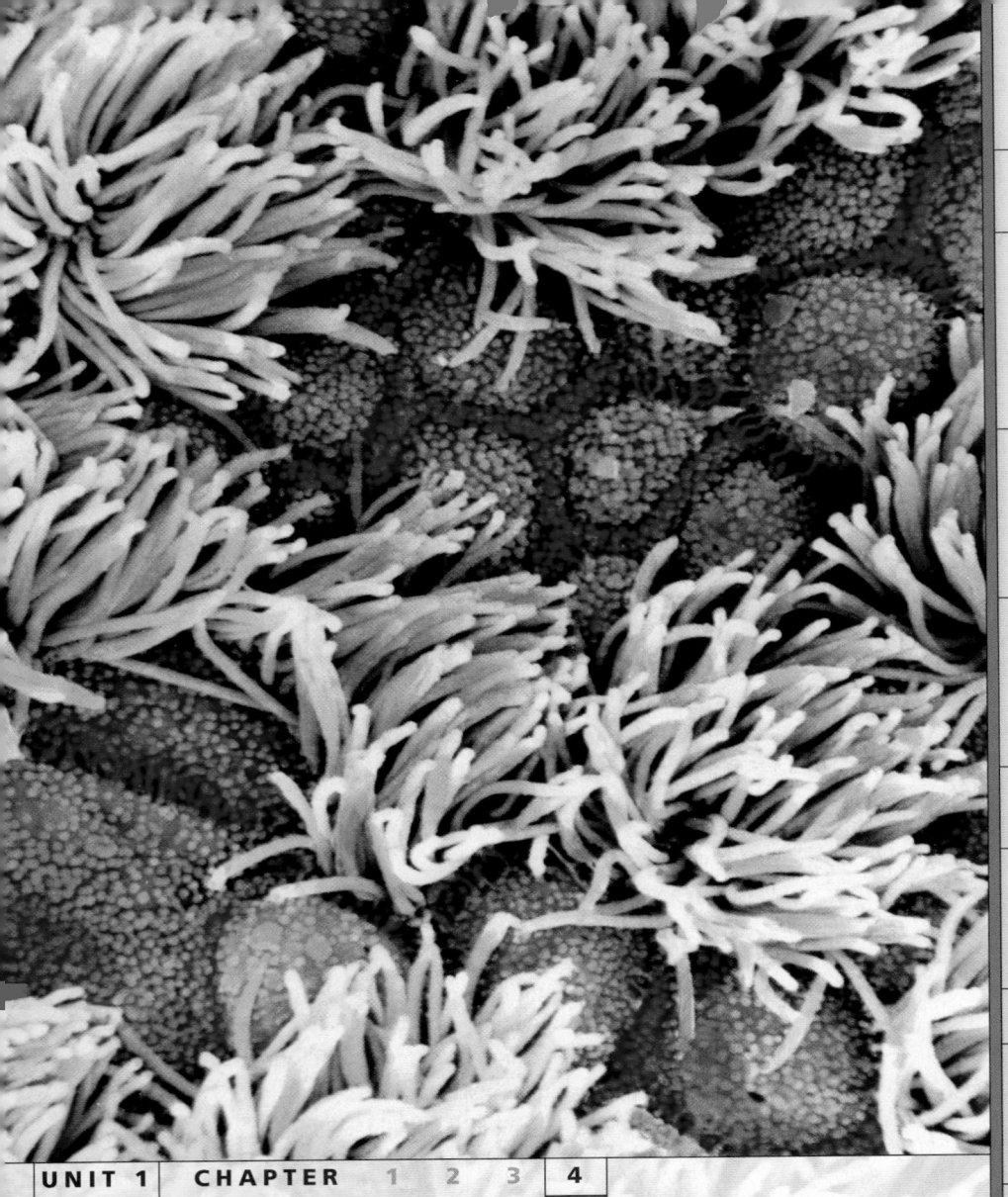

UNIT 1   CHAPTER   1   2   3   4

THE TISSUE LEVEL OF ORGANIZATION

There is a saying that you can't be all things to all people. By the same token, you can't be all people: You can't be an architect, an astronaut, a nuclear physicist, a ballet dancer, a jazz saxophonist, a cardiac surgeon, and a soccer star simultaneously. Cells are in the same predicament. As complex and seemingly independent as a single cell is, not one cell in your body is capable of performing all the functions you need in order to survive. Instead, through differentiation, cells become specialists that perform a limited number of functions very efficiently. ∞ p. 106 Although the human body contains trillions of cells, differentiation produces only about 200 types of cells. These varied cell types carry out all the vital functions that keep you alive.

4–1 TISSUES OF THE BODY: AN INTRODUCTION

Objective

- Identify the four major types of tissues in the body and describe their roles.

To work efficiently, several different types of cell must coordinate their efforts. The combination of different cell types creates **tissues**—collections of specialized cells and cell products that perform a relatively limited number of functions. The study of tissues is called **histology**. Histologists recognize four basic types of tissue:

1. *Epithelial tissue*, which covers exposed surfaces, lines internal passageways and chambers, and forms glands.
2. *Connective tissue*, which fills internal spaces, provides structural support for other tissues, transports materials within the body, and stores energy reserves.
3. *Muscle tissue*, which is specialized for contraction, includes the skeletal muscles of the body, as well as the muscle of the heart and the muscular lining of hollow organs.
4. *Neural tissue*, which carries information from one part of the body to another in the form of electrical impulses. **ATLAS** Embryology Summary 1: The Formation of Tissues

4–2 EPITHELIAL TISSUE

Objectives

■ Discuss the types and functions of epithelial cells.

■ Describe the relationship between form and function for each type of epithelium.

It is convenient to begin our discussion with **epithelial tissue**, because it includes the surface of your skin, a relatively familiar feature. Epithelial tissue includes *epithelia* and *glands*. **Epithelia** (e-pi-THĒ-lē-a; singular, *epithelium*) are layers of cells that cover internal or external surfaces. **Glands** are structures that produce fluid secretions; they are either attached to or derived from epithelia.

Epithelia cover every exposed surface of the body. Epithelia form the surface of the skin and line the digestive, respiratory, reproductive, and urinary tracts—in fact, they line all passageways that communicate with the outside world. The more delicate epithelia line internal cavities and passageways, such as the chest cavity, fluid-filled spaces in the brain, the inner surfaces of blood vessels, and the chambers of the heart.

Epithelia have several important characteristics:

- **Cellularity.** Epithelia are composed almost entirely of cells bound closely together by interconnections known as *cell junctions*. In other tissue types, the cells are often widely separated by extracellular materials.

- **Polarity.** An epithelium has an exposed surface, which faces the exterior of the body or some internal space and a base, which is attached to adjacent tissues. The two surfaces differ in membrane structure and function, as well as in the way that organelles and other cytoplasmic components are distributed between them. The term for this uneven distribution of membrane functions and organelles between the exposed and attached surfaces is **polarity**. Just as a magnet has a positive and a negative pole, each with different properties, epithelium has an exposed surface and a *basal lamina*, each with different structures and functions.

- **Attachment.** The base of an epithelium is bound to a thin **basal lamina**, or *basement membrane*. The basal lamina is a complex structure produced by the basal surface of the epithelium and the underlying connective tissue.

- **Avascularity.** Epithelia are **avascular** (ā-VAS-kū-lar; *a-*, without + *vas*, vessel); that is, they do not contain blood vessels. Epithelial cells must therefore obtain nutrients by diffusion or absorption across either the exposed or the attached epithelial surface.

- **Regeneration.** Epithelial cells that are damaged or lost at the exposed surface are continuously replaced through the divisions of stem cells in the epithelium. Although regeneration is a characteristic of other tissues as well, the rates of cell division and replacement are typically much higher in epithelia than in other tissues.

■ FUNCTIONS OF EPITHELIAL TISSUE

Epithelia perform four essential functions:

1. *Provide Physical Protection.* Epithelia protect exposed and internal surfaces from abrasion, dehydration, and destruction by chemical or biological agents.

2. *Control Permeability.* Any substance that enters or leaves your body has to cross an epithelium. Some epithelia are relatively impermeable; others are easily crossed by compounds as large as proteins. Many epithelia contain the molecular "machinery" needed for selective absorption or secretion. The epithelial barrier can be regulated and modified in response to stimuli. For example, hormones can affect the transport of ions and nutrients through epithelial cells. Even physical stress can alter the structure and properties of epithelia; think of calluses that form on your hands when you do rough physical work for a while.

3. *Provide Sensation.* Most epithelia are extremely sensitive to stimulation, because they have a large sensory nerve supply. These sensory nerves continually provide information about the external and internal environments. For example, the lightest touch of a mosquito will stimulate sensory neurons that tell you where to swat. A *neuroepithelium* is an epithelium that is specialized to perform a particular sensory function; neuroepithelia contain sensory cells that provide the sensations of smell, taste, sight, equilibrium, and hearing.

4. *Produce Specialized Secretions.* Epithelial cells that produce secretions are called *gland cells*. Individual gland cells are typically scattered among other cell types in an epithelium. In a **glandular epithelium**, most or all of the epithelial cells produce secretions, which are either discharged onto the surface of the epithelium to provide physical protection or released into the surrounding interstitial fluid and blood to act as chemical messengers.

■ SPECIALIZATIONS OF EPITHELIAL CELLS

Epithelial cells have several structural specializations that distinguish them from other body cells. For the epithelium as a whole to perform the functions listed in the previous section, individual epithelial cells may be specialized for (1) the movement of fluids over the epithelial surface, providing protection and lubrication; (2) the movement of fluids through the epithelium, to control permeability; or (3) the production of secretions that provide physical protection or act as chemical messengers. Specialized epithelial cells generally possess a strong polarity; one common type of epithelial polarity is shown in Figure 4–1●.

The cell is often divided into two functional regions: (1) the *apical surface*, where the cell is exposed to an internal or external environment; and (2) the *basolateral surface*, which includes both the base, where the cell attaches to underlying epithelial cells or deeper tissues, and the sides, where the cell contacts its neighbors.

Many epithelial cells that line internal passageways have microvilli on their exposed surfaces. ∞ p. 76 Just a few may be present, or microvilli may carpet the entire surface. Microvilli are especially abundant on epithelial surfaces where absorption and secretion take place, such as along portions of the digestive and

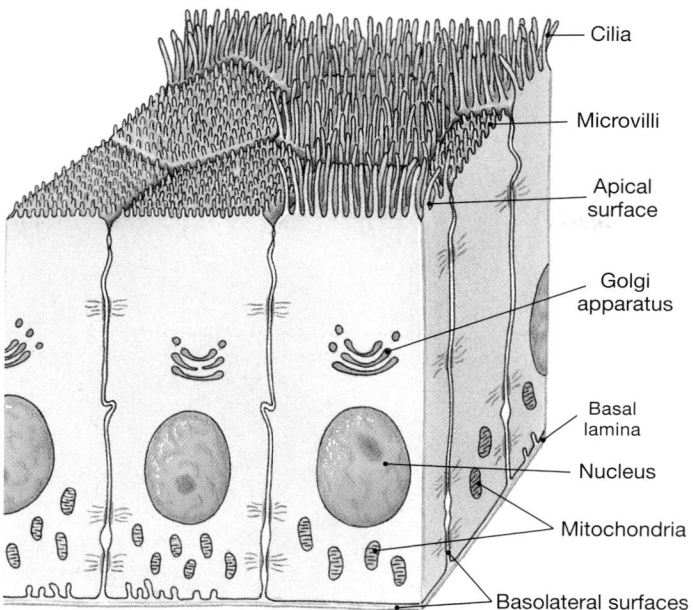

Cilia

Microvilli

Apical surface

Golgi apparatus

Basal lamina

Nucleus

Mitochondria

Basolateral surfaces

•**FIGURE 4–1**
The Polarity of Epithelial Cells. Many epithelial cells have an uneven distribution of organelles between the free surface (here, the top) and the basal lamina. Often, the free surface bears microvilli; sometimes it has cilia. In some epithelia, such as the lining of the kidney tubules, mitochondria are concentrated near the base of the cell, probably to provide energy for the cell's transport activities.

urinary tracts. The epithelial cells in these locations are transport specialists. A cell with microvilli has at least 20 times more surface area available for transporting materials than a cell without them.

The photo on p. 110 shows the surface of a **ciliated epithelium**. A typical ciliated cell contains about 250 cilia that beat in a coordinated fashion. As though on an escalator, substances are moved over the epithelial surface by the synchronized beating of the cilia. For example, the ciliated epithelium that lines the respiratory tract moves mucus up from the lungs and toward the throat. The sticky mucus traps inhaled particles, including dust, pollen, and pathogens; the ciliated epithelium carries the mucus and the trapped debris to the throat, where they can be swallowed. Injury to the cilia or to the epithelial cells, most commonly by abrasion or exposure to toxic compounds such as the nicotine in cigarette smoke, can stop ciliary movement and block the protective flow of mucus.

☐ MAINTAINING THE INTEGRITY OF EPITHELIA

An epithelium must form a complete cover or lining to be effective as a barrier. Three factors help maintain the physical integrity of an epithelium: (1) intercellular connections, (2) attachment to the basal lamina, and (3) epithelial maintenance and repair.

Intercellular Connections

Cells in an epithelium are firmly attached to one another, and the epithelium as a unit is attached to extracellular fibers of the basal lamina. Many cells in your body form permanent or temporary bonds with other cells or extracellular material. Epithelial cells, however, are specialists in intercellular connection (Figure 4–2a•).

Intercellular connections involve either extensive areas of opposing cell membranes or specialized attachment sites, discussed shortly. Large areas of opposing cell membranes are interconnected by transmembrane proteins called **cell adhesion molecules (CAMs)**, which bind to each other and to extracellular materials. For example, CAMs on the attached base of an epithelium help bind the cell to the underlying basal lamina. The membranes of adjacent cells may also be bonded by **intercellular cement**, a thin layer of proteoglycans that contain polysaccharide derivatives known as *glycosaminoglycans*, most notably **hyaluronan** (*hyaluronic acid*).

Cell junctions are specialized areas of the cell membrane that attach a cell to another cell or to extracellular materials. The three most common types of cell junctions are (1) tight junctions, (2) gap junctions, and (3) desmosomes.

At a **tight junction**, the lipid portions of the two cell membranes are tightly bound together by interlocking membrane proteins (Figure 4–2b•). This kind of attachment is so tight that these junctions prevent the passage of water and solutes between the cells. When the epithelium lines a tube, such as the intestinal tract, the apical surfaces of the epithelial cells are exposed to the space inside the tube, a passageway called the **lumen** (LOO-men). Tight junctions effectively isolate the contents of the lumen from the basolateral surfaces of the cell. For example, tight junctions near the apical surfaces of cells that line the digestive tract keep enzymes, acids, and wastes in the lumen from reaching the basolateral surfaces and digesting or otherwise damaging the underlying tissues and organs.

Some epithelial functions require rapid intercellular communication. At a **gap junction** (Figure 4–2c•), two cells are held together by interlocking membrane proteins called *connexons*. Because these are channel proteins, they leave a narrow passageway that lets small molecules and ions pass from cell to cell. Gap junctions are common among epithelial cells, where the movement of ions helps coordinate functions such as the beating of cilia. Gap junctions are also common in other tissues. For example, gap junctions in cardiac muscle tissue and smooth muscle tissue are essential to the coordination of muscle cell contractions.

Most epithelial cells are subject to mechanical stresses—stretching, bending, twisting, or compression—so they must have durable interconnections. At a **desmosome** (DEZ-mō-sōm; *desmos*, ligament + *soma*, body), CAMs and proteoglycans link the opposing cell membranes. Desmosomes are very strong and can resist stretching and twisting.

A typical desmosome is formed by two cells. Within each cell is a complex known as a *dense area*, which is connected to the cytoskeleton (Figure 4–2d•). It is this connection to the cytoskeleton that gives the desmosome—and the epithelium—its strength. For example, desmosomes are abundant between cells in the superficial layers of the skin. As a result, damaged skin cells are usually lost in sheets rather than as individual cells. (That is why your skin peels rather than comes off as a powder after a sunburn.)

There are several different types of desmosomes:

- *Belt desmosomes* form continuous bands that encircle cells and bind them to their neighbors. The bands are attached to the microfilaments of the terminal web. ⊂⊃ p. 75

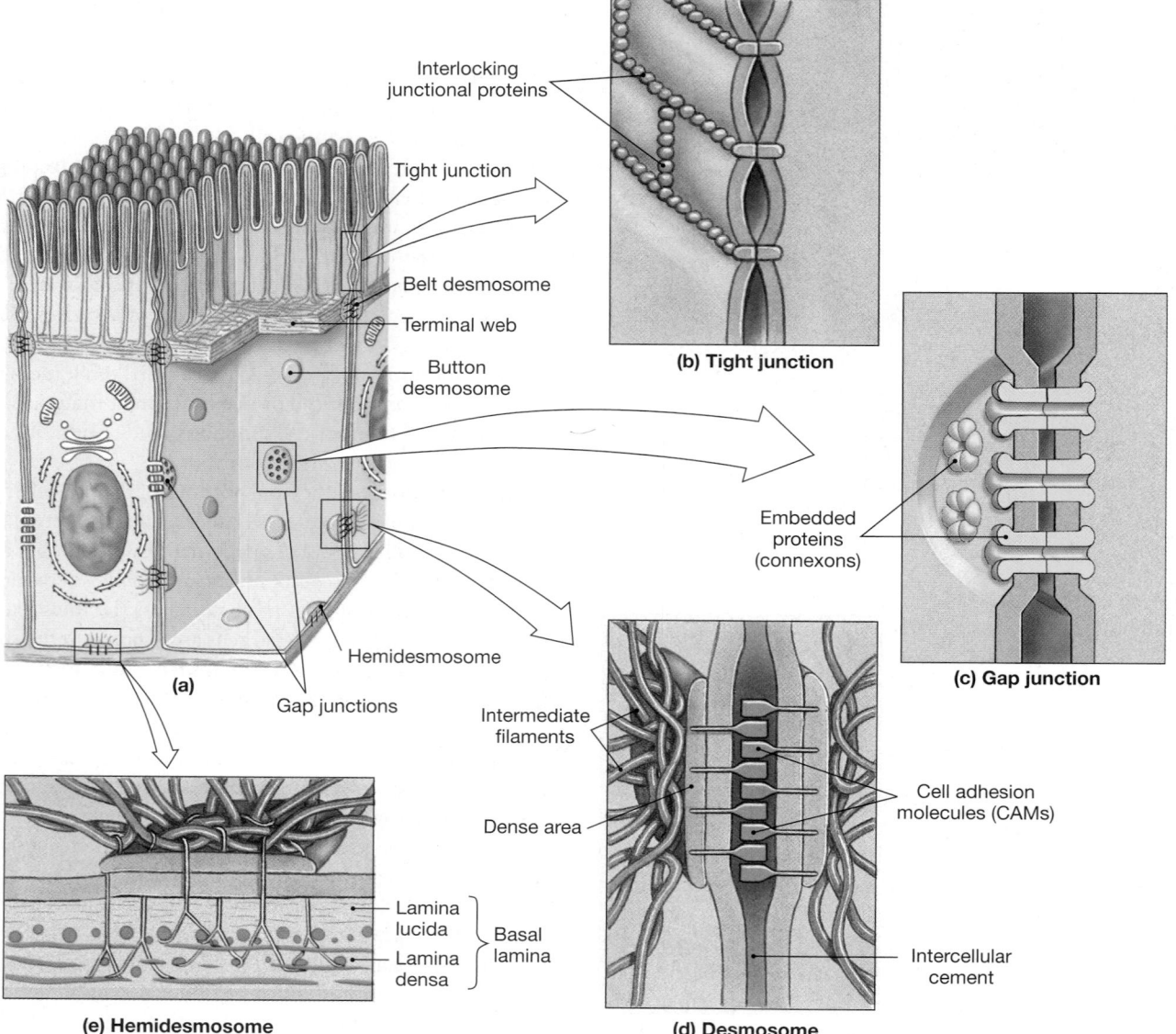

●FIGURE 4–2
Intercellular Connections. **(a)** A diagrammatic view of an epithelial cell, showing the major types of intercellular connections. **(b)** A tight junction is formed by the fusion of the outer layers of two cell membranes. Bands of tight junctions encircle the apical portion of many epithelial cells, preventing the diffusion of fluids and solutes between the cells. **(c)** Gap junctions permit the free diffusion of ions and small molecules between two cells. **(d)** A desmosome has a more organized network of intermediate filaments. Desmosomes attach one cell to another. A continuous belt of desmosomes lies deep to the tight junctions. This belt is tied to the microfilaments of the terminal web. **(e)** Hemidesmosomes attach a cell to extracellular structures, such as the protein fibers in the basal lamina.

- *Button desmosomes* are small discs connected to bands of intermediate fibers. The intermediate fibers function as cross-braces to stabilize the shape of the cell.
- *Hemidesmosomes* resemble half of a button desmosome. Rather than attaching one cell to another, a hemidesmosome attaches a cell to extracellular filaments in the basal lamina (Figure 4–2e●). This attachment helps stabilize the position of the epithelial cell and anchors it to underlying tissues.

Attachment to the Basal Lamina

Not only do epithelial cells hold onto one another, but they also remain firmly connected to the rest of the body. The basal surface of each epithelium is attached to a special two-part **basal lamina**. The layer closer to the epithelium, the *lamina lucida*

(LA-mi-nah LOO-si-dah; *lamina*, thin layer + *lucida*, clear), contains glycoproteins and a network of fine protein filaments (Figure 4–2e●). Secreted by the adjacent layer of epithelial cells, the lamina lucida acts as a barrier that restricts the movement of proteins and other large molecules from the underlying connective tissue into the epithelium.

The deeper layer of the basal lamina, the *lamina densa*, contains bundles of coarse protein fibers produced by connective tissue cells. The lamina densa gives the basement membrane its strength. Attachments between the fibers of the basal lamina and those of the reticular lamina hold the two layers together, and hemidesmosomes attach the epithelial cells to the composite basal lamina. The lamina densa also acts as a filter that determines what substances can diffuse between the adjacent tissues and the epithelium.

Epithelial Maintenance and Repair

Epithelial cells lead hard lives, for they are exposed to disruptive enzymes, toxic chemicals, pathogenic bacteria, and mechanical abrasion. Thus, an epithelium must continuously repair and renew itself. Consider the lining of the small intestine, where epithelial cells are exposed to a variety of enzymes, as well as to abrasion from partially digested food. In this extreme environment, an epithelial cell may last just a day or two before it is shed or destroyed. The only way the epithelium can maintain its structure over time is by the continual division of *stem cells*. ∞ p. 103 Most stem cells, also called **germinative cells**, are located near the basal lamina, in a relatively protected location. ATLAS Embryological Summary 2: The Development of Epithelia

✓ List five important characteristics of epithelial tissue.

✓ An epithelial surface bears many microvilli. What is the probable function of this epithelium?

✓ What is the functional significance of gap junctions?

Answers start on page Q-1

☐ CLASSIFICATION OF EPITHELIA

There are many different specialized types of epithelia. You can easily sort these into categories based on (1) the cell shape, and (2) the number of cell layers between the base and the exposed surface of the epithelium.

Three cell shapes are identified: *squamous, cuboidal,* and *columnar*. For classification purposes, one looks at the superficial cells in a section perpendicular to both the exposed surface and the basal lamina. In sectional view, squamous cells appear thin and flat, cuboidal cells look like little boxes, and columnar cells are tall and relatively slender rectangles.

Once you have determined whether the superficial cells are squamous, cuboidal, or columnar, you then look at the number of cell layers. There are only two options: *simple* or *stratified*.

Using the two criteria of cell shape (squamous, cuboidal, or columnar) and number of cell layers (simple or stratified), we can describe almost every epithelium in the body (Table 4–1).

If only one layer of cells covers the basal lamina, that layer is a **simple epithelium**. Simple epithelia are necessarily thin. All the cells have the same polarity, so the distance from the nucleus to the basal lamina does not change from one cell to the next. Because they are so thin, simple epithelia are also fragile. A single layer of cells cannot provide much mechanical protection, so simple epithelia are located only in protected areas inside the body. They line internal compartments and passageways, including the ventral body cavities, the heart chambers, and blood vessels.

Simple epithelia are also characteristic of regions in which secretion or absorption occurs, such as the lining of the intestines and the gas-exchange surfaces of the lungs. In these places, thinness is an advantage, for it reduces the time required for materials to cross the epithelial barrier.

TABLE 4–1 CLASSIFYING EPITHELIA

	Squamous	Cuboidal	Columnar
Simple	Simple squamous epithelium	Simple cuboidal epithelium	Simple columnar epithelium
Stratified	Stratified squamous epithelium	Stratified cuboidal epithelium	Stratified columnar epithelium

In a **stratified epithelium**, several layers of cells cover the basal lamina. Stratified epithelia are generally located in areas that need protection from mechanical or chemical stresses, such as the surface of the skin and the lining of the mouth.

Squamous Epithelia

The cells in a **squamous epithelium** (SKWĀ-mus; *squama*, plate or scale) are thin, flat, and somewhat irregular in shape, like pieces of a puzzle (Figure 4–3●). In sectional view, the disc-shaped nucleus occupies the thickest portion of each cell. From the surface, the cells look like fried eggs laid side by side.

A **simple squamous epithelium** is the body's most delicate type of epithelium (Figure 4–3a●). This type of epithelium is located in protected regions where absorption or diffusion takes place or where a slick, slippery surface reduces friction. Examples are the respiratory exchange surfaces *(alveoli)* of the lungs, the lining of the ventral body cavities, and the lining of the heart and blood vessels. A smooth lining is extremely important; for example, any irregularity in the lining of a blood vessel will result in the formation of a potentially dangerous blood clot.

Special names have been given to the simple squamous epithelia that line chambers and passageways that do not communicate with the outside world. The simple squamous epithelium that lines the ventral body cavities is a **mesothelium** (mez-ō-THĒ-lē-um; *mesos*, middle). The pleura, peritoneum, and pericardium each contain a superficial layer of mesothelium. The simple squamous epithelium lining the inner surface of the heart and all blood vessels is an **endothelium** (en-dō-THĒ-lē-um; *endo*, inside).

A **stratified squamous epithelium** (Figure 4–3b●) is generally located where mechanical stresses are severe. The cells form a series of layers, like a stack of plywood sheets or a ream of paper. The surface of the skin and the lining of the mouth, esophagus, and anus are areas where this type of epithelium protects against physical and chemical attacks. On exposed body surfaces, where mechanical stress and dehydration are potential problems, apical layers of epithelial cells are packed with filaments of the protein *keratin*. As a result, superficial layers are both tough and water resistant; the epithelium is said to be *keratinized*. A *nonkeratinized* stratified squamous epithelium resists abrasion, but will dry out and deteriorate unless kept moist. Nonkeratinized stratified squamous epithelia are situated in the oral cavity, pharynx, esophagus, anus, and vagina.

Cuboidal Epithelia

The cells of a **cuboidal epithelium** resemble hexagonal boxes. (In typical sectional views they appear square.) The spherical nuclei are near the center of each cell, and the distance between adjacent nuclei is roughly equal to the height of the epithelium. A **simple cuboidal epithelium** provides limited protection and occurs where secretion or absorption takes place. Such an epithelium lines portions of the kidney tubules (Figure 4–4a●). In the pancreas and salivary glands,

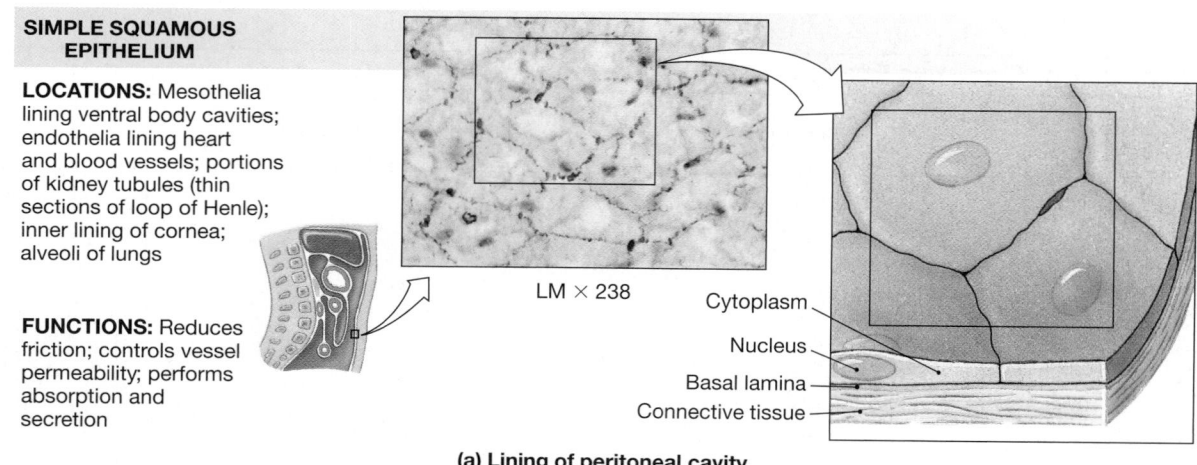

SIMPLE SQUAMOUS EPITHELIUM

LOCATIONS: Mesothelia lining ventral body cavities; endothelia lining heart and blood vessels; portions of kidney tubules (thin sections of loop of Henle); inner lining of cornea; alveoli of lungs

FUNCTIONS: Reduces friction; controls vessel permeability; performs absorption and secretion

LM × 238

Cytoplasm
Nucleus
Basal lamina
Connective tissue

(a) Lining of peritoneal cavity

●FIGURE 4–3
Squamous Epithelia.
(a) A superficial view of the simple squamous epithelium (mesothelium) that lines the peritoneal cavity. The three-dimensional drawing shows the epithelium in superficial and sectional views. **(b)** Sectional and diagrammatic views of the stratified squamous epithelium that covers the tongue.

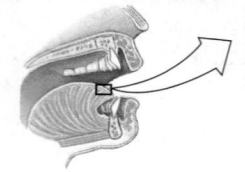
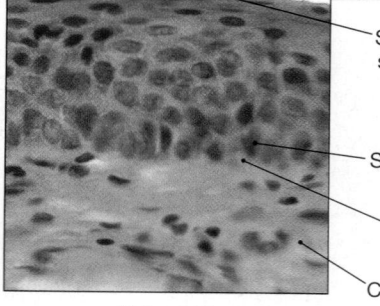

STRATIFIED SQUAMOUS EPITHELIUM

LOCATIONS: Surface of skin; lining of mouth, throat, esophagus, rectum, anus, and vagina

FUNCTIONS: Provides physical protection against abrasion, pathogens, and chemical attack

LM × 310

Squamous superficial cells
Stem cells
Basal lamina
Connective tissue

(b) Surface of tongue

SIMPLE CUBOIDAL EPITHELIUM

LOCATIONS: Glands; ducts; portions of kidney tubules; thyroid gland

FUNCTIONS: Limited protection, secretion, absorption

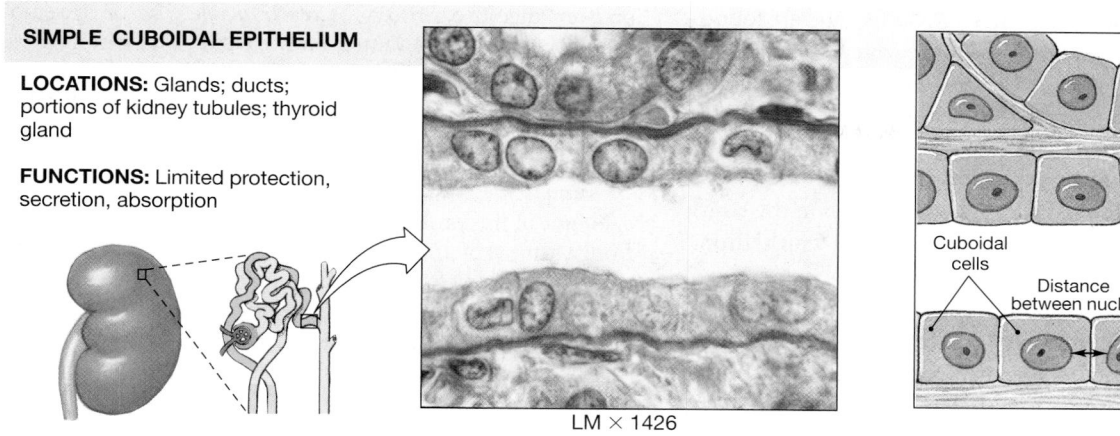

LM × 1426

Cuboidal cells · Height · Connective tissue · Distance between nuclei · Width

(a) Kidney tubule

STRATIFIED CUBOIDAL EPITHELIUM

LOCATIONS: Lining of some ducts (rare)

FUNCTIONS; Protection, secretion, absorption

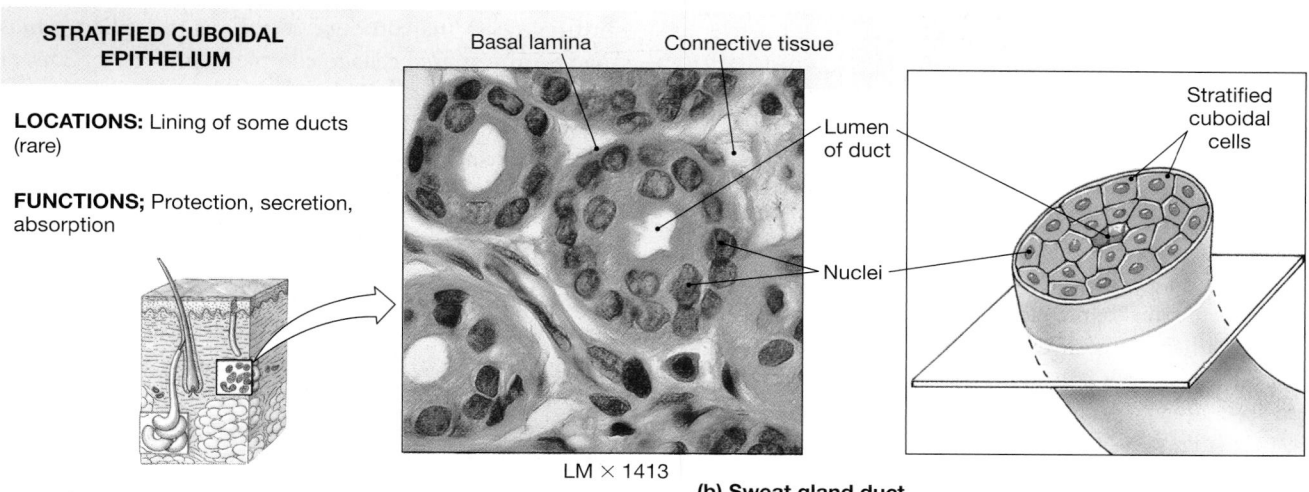

Basal lamina · Connective tissue · Lumen of duct · Nuclei · Stratified cuboidal cells

LM × 1413

(b) Sweat gland duct

TRANSITIONAL EPITHELIUM

LOCATIONS: Urinary bladder; renal pelvis; ureters
FUNCTIONS: Permits expansion and recoil after stretching

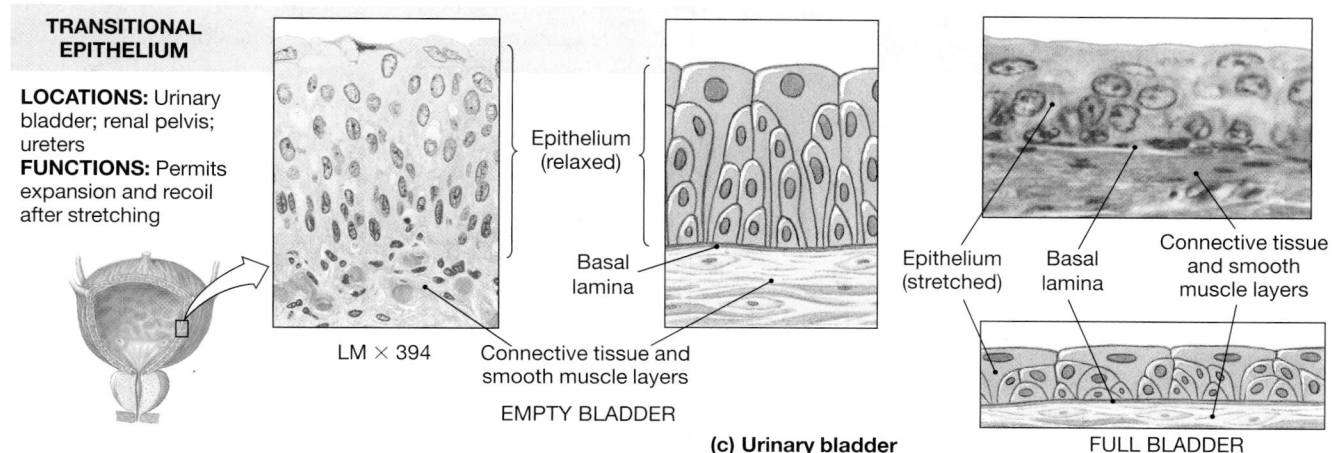

Epithelium (relaxed) · Basal lamina · Connective tissue and smooth muscle layers · Epithelium (stretched)

LM × 394

EMPTY BLADDER

(c) Urinary bladder

FULL BLADDER

●**FIGURE 4–4**
Cuboidal Epithelia. **(a)** A section through the simple cuboidal epithelial cells of a kidney tubule. **(b)** A sectional view of the stratified cuboidal epithelium that lines a sweat gland duct in the skin. **(c)** Left: The lining of the empty urinary bladder, showing a transitional epithelium in the relaxed state. Right: The lining of the full bladder, showing the effects of stretching on the appearance of cells in the epithelium.

simple cuboidal epithelia secrete enzymes and buffers and are found lining portions of the ducts that discharge those secretions. The thyroid gland contains chambers, called *thyroid follicles*, that are lined by a cuboidal secretory epithelium. Thyroid hormones are stored in the follicles and released as needed into the bloodstream.

Stratified cuboidal epithelia are relatively rare; they are located along the ducts of sweat glands (Figure 4–4b●) and in the larger ducts of the mammary glands. A **transitional epithelium** (Figure 4–4c●) is unusual because, unlike most epithelia, it tolerates repeated cycles of stretching and recoil without damage. It is called transitional because the appearance of the epithelium changes as the stretching occurs. A transitional epithelium is situated in regions of the urinary system, such as the urinary bladder, where large changes in volume occur. In an empty urinary bladder, the epithelium seems to have many layers, and the superficial cells are typically plump cuboidal cells. The multilayered appearance results from overcrowding. In the full urinary bladder, when the volume of urine has stretched the lining to its limits, the epithelium appears flattened, and more like a simple epithelium.

Columnar Epithelia

In a typical sectional view, **columnar epithelial cells** appear rectangular. In reality, the densely packed columnar cells are hexagonal, but they are taller and more slender than cells in a cuboidal epithelium (Figure 4–5●). The elongated nuclei are crowded into a narrow band close to the basal lamina. The height of the epithelium is several times the distance between adjacent nuclei. A **simple columnar epithelium** typically occurs where absorption or secretion is under way, such as in the small intestine (Figure 4–5a●). In the stomach and large intestine, the secretions of simple columnar epithelia protect against chemical stresses.

Portions of the respiratory tract contain a **pseudostratified columnar epithelium**, a columnar epithelium that includes several types of cells with varying shapes and functions. The distances between the cell nuclei and the exposed surface vary, so the epithelium appears to be layered, or stratified (Figure 4–5b●). It is not truly stratified, though, because all the epithelial cells contact the basal lamina. Pseudostratified columnar epithelial cells typically possess cilia. Epithelia of this type line most of the nasal cavity, the trachea (windpipe), the bronchi, and portions of the male reproductive tract.

Stratified columnar epithelia are relatively rare, providing protection along portions of the pharynx, epiglottis, anus, and urethra, as well as along a few large excretory ducts. The epithelium has either two layers (Figure 4–5c●) or multiple layers. In the latter case, only the superficial cells are columnar.

✚ *Exfoliative cytology* (eks-FŌ-lē-a-tiv; *ex*, from + *folium*, leaf) is the study of cells shed or collected from epithelial surfaces. The cells are examined for a variety of reasons—for example, to check for cellular changes that indicate cancer or to identify the pathogens involved in an infection. The cells are collected by sampling the fluids that cover the epithelia lining the res-

piratory, digestive, urinary, or reproductive tract or by removing fluid from one of the ventral body cavities. The sampling procedure is often called a *Pap test*, named after Dr. George Papanicolaou, who pioneered its use. The most familiar Pap test is that for cervical cancer, which involves the scraping of cells from the tip of the *cervix*, the portion of the uterus that projects into the vagina.

Amniocentesis is another important test based on exfoliative cytology. In this procedure, exfoliated epithelial cells are collected from a sample of *amniotic fluid*, the fluid that surrounds and protects a developing fetus. Examination of these cells can determine whether the fetus has a genetic abnormality, such as *Down syndrome*, that affects the number or structure of chromosomes.

☐ GLANDULAR EPITHELIA

Many epithelia contain gland cells that are specialized for secretion. Collections of epithelial cells (or structures derived from epithelial cells) that produce secretions are called *glands*. They range from scattered cells to complex glandular organs. Some of these glands, called **endocrine glands**, release their secretions into the interstitial fluid. Others, known as **exocrine glands**, release their secretions into passageways called **ducts** that open onto the epithelial surface.

Endocrine Glands

An endocrine gland produces *endocrine* (*endo-*, inside + *krinein*, to secrete) *secretions*, which are released directly into the surrounding interstitial fluid. These secretions, also called *hormones*, enter the bloodstream for distribution throughout the body. Hormones regulate or coordinate the activities of various tissues, organs, and organ systems. Examples of endocrine glands include the thyroid gland and the pituitary gland. Because their secretions are not released into ducts, endocrine glands are often called *ductless glands*.

Endocrine cells may be part of an epithelial surface, such as the lining of the digestive tract, or they may be found in separate organs, such as the pancreas, thyroid gland, thymus, and pituitary gland. We will consider endocrine cells, organs, and hormones further in Chapter 18.

Exocrine Glands

Exocrine glands produce *exocrine* (*exo-*, outside) *secretions* that are discharged onto an epithelial surface. Most exocrine secretions reach the surface through tubular ducts, which empty onto the skin surface or onto an epithelium lining an internal passageway that communicates with the exterior. Enzymes entering the digestive tract, perspiration on the skin, tears in the eyes, and milk produced by mammary glands are examples of exocrine secretions delivered to epithelial surfaces by ducts.

Exocrine glands exhibit several different methods of secretion; therefore, they are classified by their mode and type of secretion, as well as by the structure of the gland cells and associated ducts.

SIMPLE COLUMNAR EPITHELIUM

LOCATIONS: Lining of stomach, intestine, gallbladder, uterine tubes, and collecting ducts of kidneys

FUNCTIONS: Protection, secretion, absorption

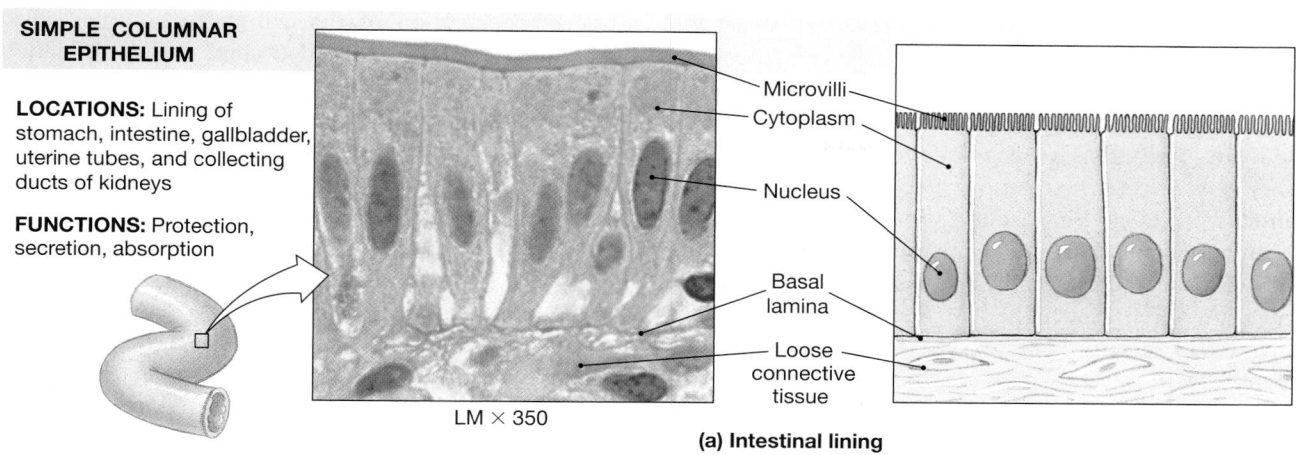

LM × 350

Microvilli
Cytoplasm
Nucleus
Basal lamina
Loose connective tissue

(a) Intestinal lining

PSEUDOSTRATIFIED CILIATED COLUMNAR EPITHELIUM

LOCATIONS: Lining of nasal cavity, trachea, and bronchi; portions of male reproductive tract

FUNCTIONS: Protection, secretion

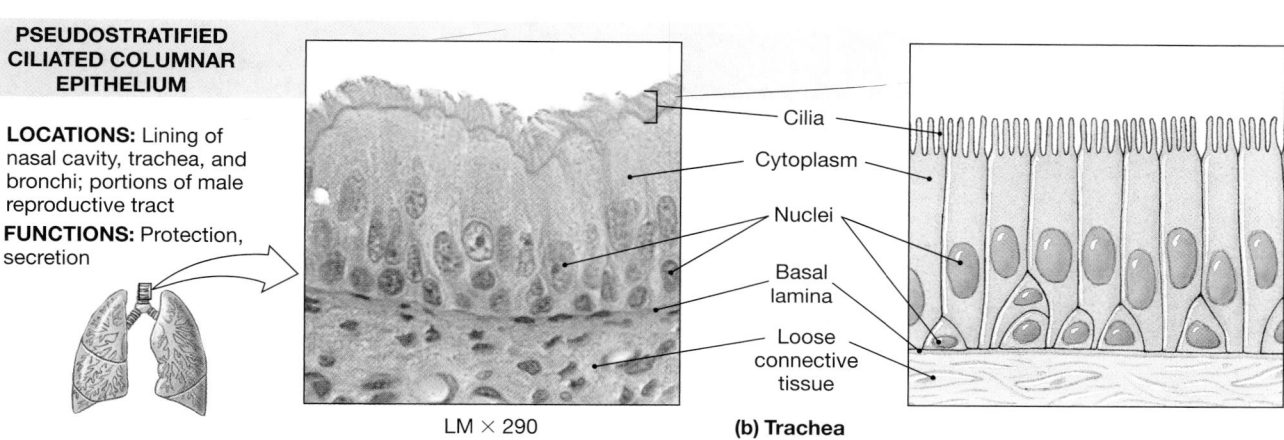

LM × 290

Cilia
Cytoplasm
Nuclei
Basal lamina
Loose connective tissue

(b) Trachea

STRATIFIED COLUMNAR EPITHELIUM

LOCATIONS: Small areas of the pharynx, epiglottis, anus, mammary gland, salivary gland ducts, and urethra

FUNCTION: Protection

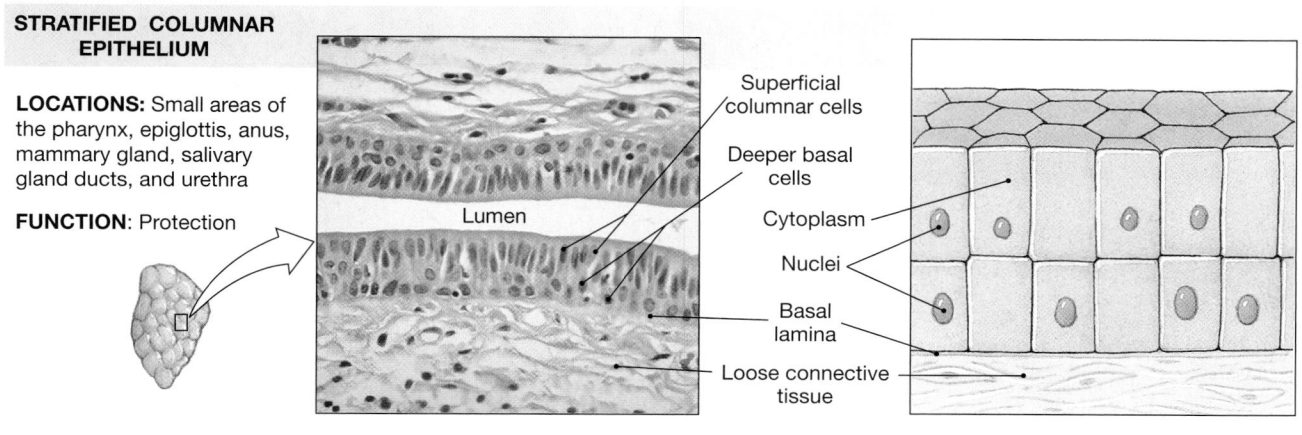

Lumen

Superficial columnar cells
Deeper basal cells
Cytoplasm
Nuclei
Basal lamina
Loose connective tissue

(c) Salivary gland duct

●**FIGURE 4–5**
Columnar Epithelia. Note the thickness of the epithelium and the location and orientation of the nuclei. **(a)** The simple columnar epithelium lining the small intestine. **(b)** The pseudostratified ciliated columnar epithelium of the respiratory tract. Note the uneven layering of the nuclei. **(c)** A stratified columnar epithelium occurs along some large ducts, such as this salivary gland duct.

MODES OF SECRETION A glandular epithelial cell releases its secretions by (1) merocrine secretion, (2) apocrine secretion, or (3) holocrine secretion. In **merocrine secretion** (MER-u-krin; *meros*, part), the product is released from secretory vesicles by exocytosis (Figure 4–6a●). This is the most common mode of secretion. One type of merocrine secretion, *mucin*, mixes with water to form **mucus**, an effective lubricant, a protective barrier, and a sticky trap for foreign particles and microorganisms. The mucous secretions of merocrine glands coat passageways in the digestive and respiratory tracts. In the skin, merocrine sweat glands produce the watery perspiration that helps cool you on a hot day.

Apocrine secretion (AP-ō-krin; *apo-*, off) involves the loss of cytoplasm, as well as the secretory product (Figure 4–6b●).

The apical portion of the cytoplasm becomes packed with secretory vesicles and is then shed. Milk production in the mammary glands involves a combination of merocrine and apocrine secretions.

Merocrine and apocrine secretions leave a cell relatively intact and able to continue secreting. **Holocrine secretion** (HOL-ō-krin; *holos*, entire), by contrast, destroys the gland cell. During holocrine secretion, the entire cell becomes packed with secretory products and then bursts (Figure 4–6c●), releasing the secretion, but killing the cell. Further secretion depends on the replacement of destroyed gland cells by the division of stem cells. Sebaceous glands, associated with hair follicles, produce an oily hair coating by means of holocrine secretion.

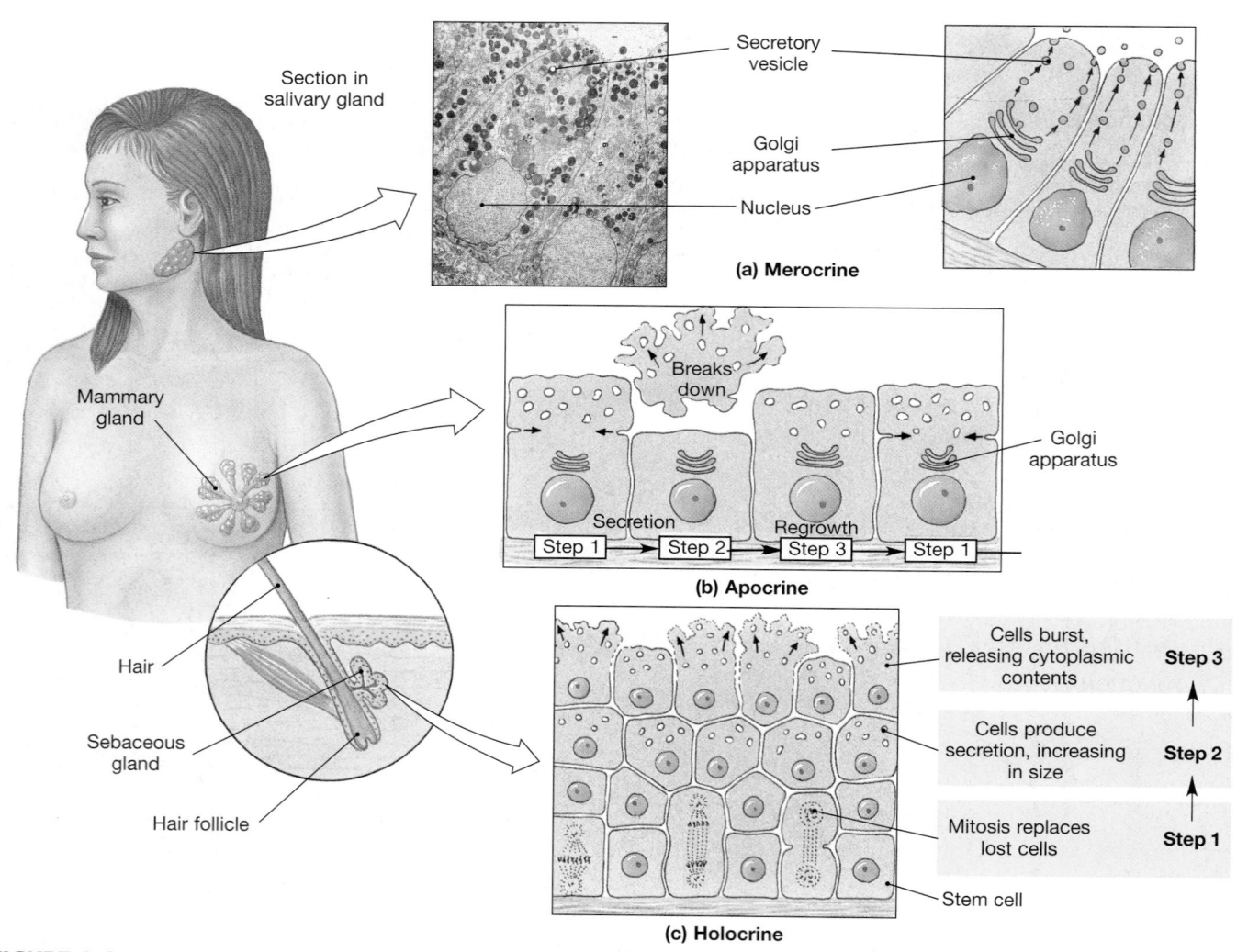

●**FIGURE 4–6**
Mechanisms of Glandular Secretion.   **(a)** In merocrine secretion, secretory vesicles are discharged at the apical surface of the gland cell by exocytosis. **(b)** Apocrine secretion involves the loss of apical cytoplasm. Inclusions, secretory vesicles, and other cytoplasmic components are shed in the process. The gland cell then undergoes growth and repair before it releases additional secretions. **(c)** Holocrine secretion occurs as superficial gland cells burst. Continued secretion involves the replacement of these cells through the mitotic division of underlying stem cells.

TYPES OF SECRETIONS Exocrine glands are also categorized by the types of secretion produced:

1. **Serous glands** secrete a watery solution that contains enzymes. The parotid salivary glands are serous glands.

2. **Mucous glands** secrete mucins that hydrate to form mucus. The sublingual salivary glands and the submucosal glands of the small intestine are mucous glands.

3. **Mixed exocrine glands** contain more than one type of gland cell and may produce two different exocrine secretions, one serous and the other mucous. The submandibular salivary glands are mixed exocrine glands.

GLAND STRUCTURE The final method of classifying exocrine glands is by structure. In epithelia that have independent, scattered gland cells, the individual secretory cells are called **unicellular glands. Multicellular glands** include glandular epithelia and aggregations of gland cells that produce exocrine or endocrine secretions.

The only **unicellular exocrine glands** in the body are **goblet cells,** which secrete mucins. Goblet cells are scattered among other epithelial cells. For example, both the pseudostratified ciliated columnar epithelium that lines the trachea and the columnar epithelium of the small and large intestines have an abundance of goblet cells.

The simplest **multicellular exocrine gland** is a *secretory sheet*, in which gland cells form an epithelium that releases secretions into an inner compartment. The continuous secretion of mucinsecreting cells that line the stomach, for instance, protects that organ from its own acids and enzymes. Most other multicellular exocrine glands are in pockets set back from the epithelial surface; their secretions travel through one or more ducts to the surface. Examples include the salivary glands, which produce mucins and digestive enzymes.

Three characteristics are used to describe the structure of multicellular exocrine glands (Figure 4–7●):

1. *The Shape of the Secretory Portion of the Gland.* Glands whose glandular cells form tubes are *tubular*. Those which form blind pockets are *alveolar* (al-VĒ-ō-lar; *alveolus*, sac) or *acinar* (A-si-nar; *acinus*, chamber). Glands whose secretory cells form both tubes and pockets are called *tubuloalveolar* or *tubuloacinar*.

2. *The Structure of the Duct.* A gland is *simple* if it has a single duct that does not divide on its way to the gland cells. The gland is *compound* if the duct divides one or more times on its way to the gland cells.

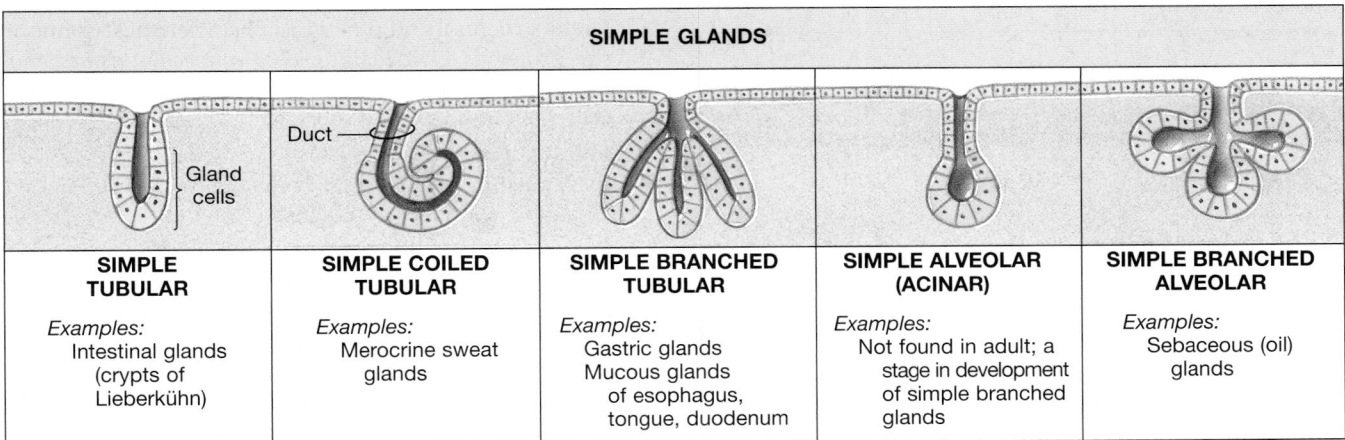

SIMPLE GLANDS				
SIMPLE TUBULAR	**SIMPLE COILED TUBULAR**	**SIMPLE BRANCHED TUBULAR**	**SIMPLE ALVEOLAR (ACINAR)**	**SIMPLE BRANCHED ALVEOLAR**
Examples: Intestinal glands (crypts of Lieberkühn)	*Examples:* Merocrine sweat glands	*Examples:* Gastric glands Mucous glands of esophagus, tongue, duodenum	*Examples:* Not found in adult; a stage in development of simple branched glands	*Examples:* Sebaceous (oil) glands

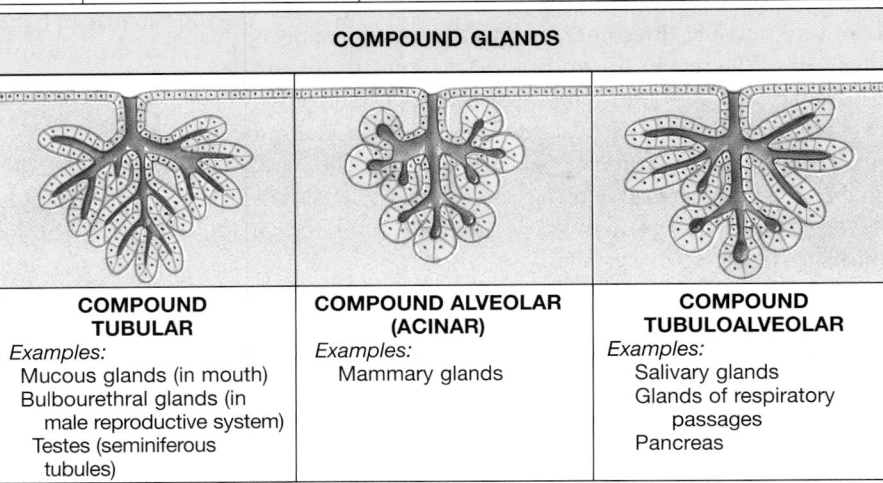

COMPOUND GLANDS		
COMPOUND TUBULAR	**COMPOUND ALVEOLAR (ACINAR)**	**COMPOUND TUBULOALVEOLAR**
Examples: Mucous glands (in mouth) Bulbourethral glands (in male reproductive system) Testes (seminiferous tubules)	*Examples:* Mammary glands	*Examples:* Salivary glands Glands of respiratory passages Pancreas

●**FIGURE 4–7**
A Structural Classification of Exocrine Glands

3. *The Relationship between the Ducts and the Glandular Areas.* A gland is *branched* if several secretory areas (tubular or acinar) share a duct. ("Branched" refers to the glandular areas and not to the duct.)

The vast majority of glands in the body produce either exocrine or endocrine secretions. However, a few complex organs, including the digestive tract and the pancreas, produce both kinds of secretions. We will consider the organization of these glands in Chapters 18 and 24.

✓ With a light microscope, you examine a tissue and see a simple squamous epithelium on the outer surface. Can this be a sample of the skin surface?

✓ Why do the pharynx, esophagus, anus, and vagina have the same epithelial organization?

✓ The secretory cells of sebaceous glands fill with secretions and then rupture, releasing their contents. Which type of secretion is this?

✓ A gland has no ducts to carry the glandular secretions, and the gland's secretions are released directly into the extracellular fluid. Which type of gland is this?

Answers start on page Q-1

 The classification system for epithelial tissue can be reviewed by visiting the **Companion Website**: Chapter 4/Reviewing Concepts/Labeling.

4–3 CONNECTIVE TISSUES

Objective

■ Compare the structures and functions of the various types of connective tissues.

It is impossible to discuss epithelial tissue without mentioning an associated type of tissue: **connective tissue**. Recall that the reticular layer of the basal lamina of all epithelial tissues is created by connective tissue; in essence, connective tissue connects the epithelium to the rest of the body. Other connective tissues include bone, fat, and blood, as well as tissues that provide structure, store energy reserves, and transport materials throughout the body. Connective tissues vary widely in appearance and function, but they all share three basic components: (1) specialized cells, (2) extracellular protein fibers, and (3) a fluid known as **ground substance**. The extracellular fibers and ground substance together constitute the **matrix**, which surrounds the cells. Whereas cells make up the bulk of epithelial tissue, the matrix typically accounts for most of the volume of connective tissues.

ATLAS Embryology Summary 3: The Origins of Connective Tissues

Connective tissues are situated throughout the body, but are never exposed to the outside environment. Many connective tissues are highly vascular (that is, they have many blood vessels) and contain sensory receptors that provide pain, pressure, temperature, and other sensations. Among the specific functions of connective tissues are the following:

• Establishing a structural framework for the body.

• Transporting fluids and dissolved materials.

• Protecting delicate organs.

• Supporting, surrounding, and interconnecting other types of tissue.

• Storing energy reserves, especially in the form of lipids.

• Defending the body from invading microorganisms.

☐ CLASSIFICATION OF CONNECTIVE TISSUES

Connective tissues are classified on the basis of their physical properties. The three general categories of connective tissue are connective tissue proper, fluid connective tissues, and supporting connective tissues (Figure 4–8●):

1. **Connective tissue proper** includes those connective tissues with many types of cells and extracellular fibers in a syrupy ground substance. This broad category of connective tissue encompasses tissues that differ in the number of cell types they contain and in the relative properties and proportions of fibers and ground substance. For example, both *adipose tissue* (fat) and *tendons* are connective tissue proper, but they have very different structural and functional characteristics. Connective tissue proper is divided into *loose connective tissues* and *dense connective tissues* on the basis of the relative proportions of cells, fibers, and ground substance.

2. **Fluid connective tissues** have distinctive populations of cells suspended in a watery matrix that contains dissolved proteins. Two types exist: *blood* and *lymph*.

3. **Supporting connective tissues** differ from connective tissue proper in having a less diverse cell population and a matrix containing much more densely packed fibers. Supporting connective tissues protect soft tissues and support the weight of part or all of the body. The two types of supporting connective tissues are *cartilage* and *bone*. The matrix of cartilage is a gel whose characteristics vary with the predominant type of fiber. The matrix of bone is said to be **calcified**, because it contains mineral deposits (primarily calcium salts) that provide rigidity.

☐ CONNECTIVE TISSUE PROPER

Connective tissue proper contains extracellular fibers, a viscous (syrupy) ground substance, and a varied cell population. Some of the cells, including *fibroblasts, adipocytes,* and *mesenchymal cells,* are involved with local maintenance, repair, and energy storage. These cells are permanent residents of the connective tissue. Other cells, including *macrophages, mast cells, lymphocytes, plasma cells,* and *microphages,* defend and repair damaged tissues. These cells are not permanent residents; they migrate through healthy connective tissues and aggregate at sites of tissue injury.

The number of cells and cell types in a tissue at any moment varies with local conditions. Figure 4–9● summarizes the cells and fibers of connective tissue proper.

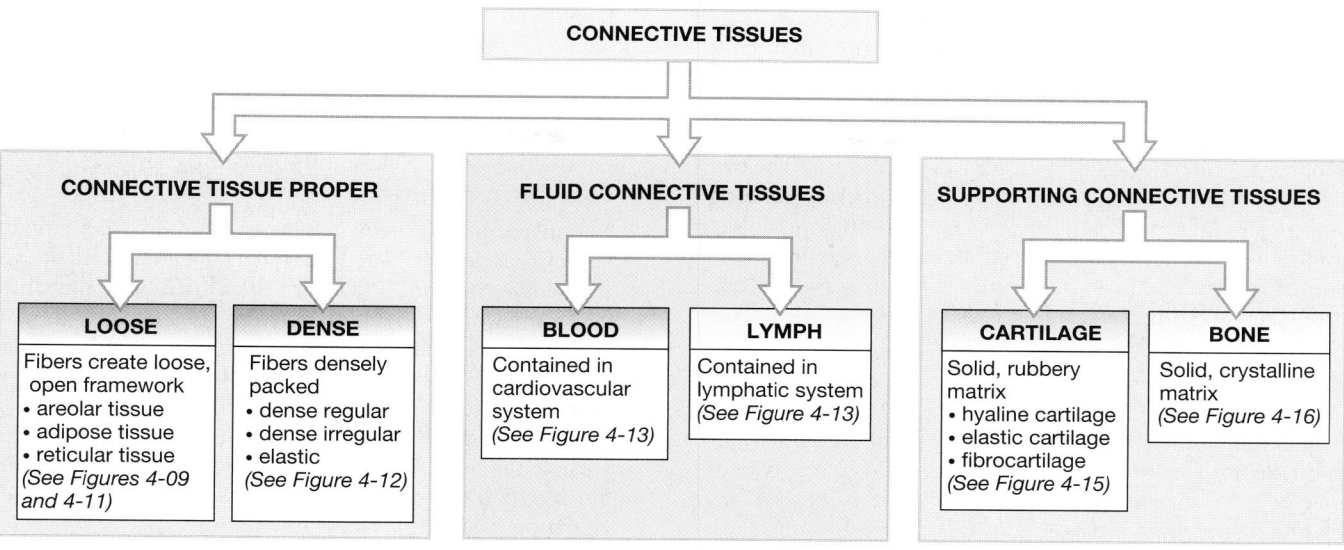

●**FIGURE 4–8**
A Classification of Connective Tissues

Components of Connective Tissue Proper

THE CELL POPULATION Fibroblasts (FĪ-brō-blasts) are the most abundant permanent residents of connective tissue proper and are the only cells that are *always* present in it. Fibroblasts secrete hyaluronan (a polysaccharide derivative) and proteins. In connective tissue proper, extracellular fluid, hyaluronan, and proteins interact to form the proteoglycans that make ground substance viscous. (You may recall hyaluronan as one of the ingredients in the intercellular cement that helps lock epithelial cells together.) Each fibroblast also secretes protein subunits that interact to form large extracellular fibers. ∞ p. 55

In addition to fibroblasts, connective tissues proper may contain several other types of cell:

- **Macrophages** (MAK-rō-fā-jez; *phagein*, to eat) are large, amoeboid cells scattered throughout the matrix. These scavengers

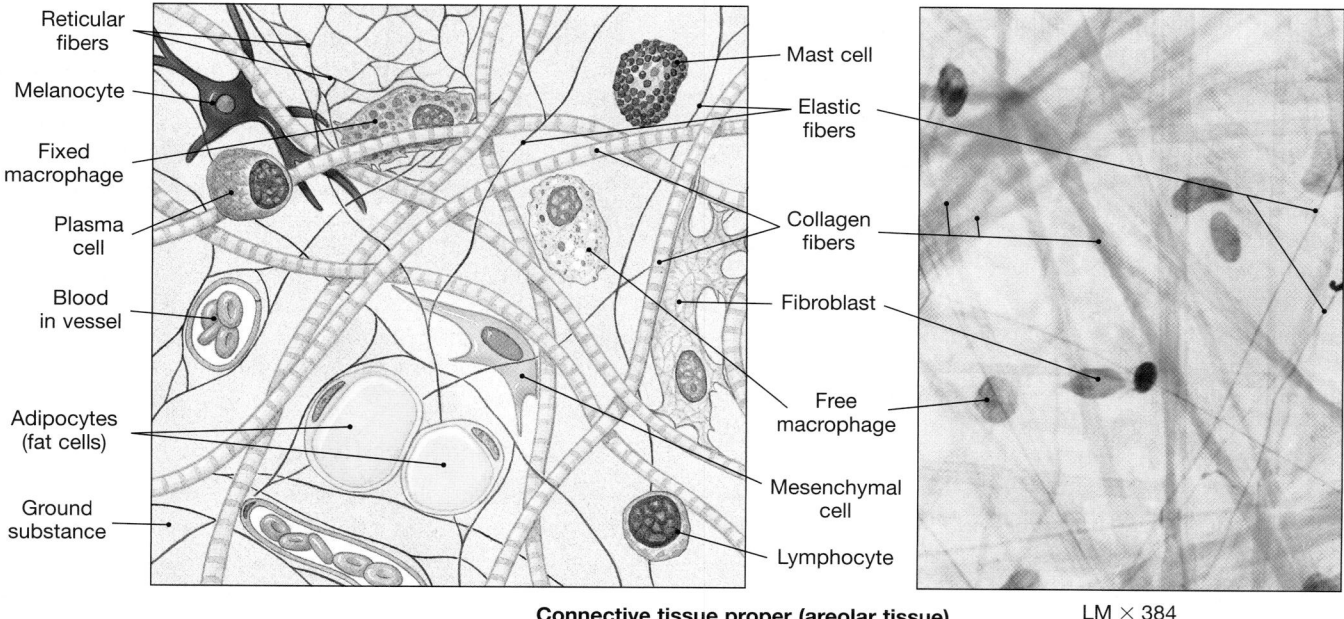

Connective tissue proper (areolar tissue)   LM × 384

●**FIGURE 4–9**
The Cells and Fibers of Connective Tissue Proper. Diagrammatic and histological views of the cell types and fibers of connective tissue proper. (Microphages, not shown, are common only in damaged or abnormal tissues.)

engulf pathogens or damaged cells that enter the tissue. (The name literally means "big eater.") Although not abundant, macrophages are important in mobilizing the body's defenses. When stimulated, they release chemicals that activate the immune system and attract large numbers of additional macrophages and other cells involved in tissue defense. The two classes of macrophage are *fixed macrophages*, which spend long periods in a tissue, and *free macrophages*, which migrate rapidly through the tissue. In effect, fixed macrophages provide a "frontline" defense that can be reinforced by the arrival of free macrophages and other specialized cells.

- **Adipocytes** (AD-i-pō-sīts) are also known as *adipose cells*, or fat cells. A typical adipocyte contains a single, enormous lipid droplet. The nucleus, other organelles, and cytoplasm are squeezed to one side, making a sectional view of the cell resemble a class ring. The number of fat cells varies from one type of connective tissue to another, from one region of the body to another, and among individuals.

- **Mesenchymal cells** are stem cells that are present in many connective tissues. These cells respond to local injury or infection by dividing to produce daughter cells that differentiate into fibroblasts, macrophages, or other connective tissue cells.

- **Melanocytes** (me-LAN-ō-sīts) synthesize and store the brown pigment **melanin** (ME-la-nin), which gives tissue a dark color. Melanocytes are common in the epithelium of the skin, where they play a major role in determining skin color. Melanocytes are also abundant in connective tissues of the eye and the dermis of the skin, although the number present differs by body region and among individuals.

- **Mast cells** are small, mobile connective tissue cells that are common near blood vessels. The cytoplasm of a mast cell is filled with granules containing **histamine** (HIS-tuh-mēn) and **heparin** (HEP-uh-rin). These chemicals, released after injury or infection, stimulate local inflammation. (If you have ever had a cold, you are familiar with the inflammatory effects of histamine; people often take antihistamines to reduce cold symptoms.) *Basophils*, blood cells that enter damaged tissues and enhance the inflammation process, also contain histamine and heparin.

- **Lymphocytes** (LIM-fō-sīts) migrate throughout the body, traveling through connective tissues and other tissues. Their numbers increase markedly wherever tissue damage occurs. Some lymphocytes may then develop into **plasma cells**, which produce *antibodies*—proteins involved in defending the body against disease.

- **Microphages** (*neutrophils* and *eosinophils*) are phagocytic blood cells that normally move through connective tissues in small numbers. When an infection or injury occurs, chemicals released by macrophages and mast cells attract numerous microphages to the site.

CONNECTIVE TISSUE FIBERS Three types of fibers occur in connective tissue: *collagen, reticular,* and *elastic*. Fibroblasts form all three by secreting protein subunits that interact in the matrix.

1. **Collagen fibers** are long, straight, and unbranched. They are the most common fibers in connective tissue proper. Each collagen fiber consists of a bundle of fibrous protein subunits, wound together like the strands of a rope. Like a rope, a collagen fiber is flexible, but it is stronger than steel when pulled from either end. *Tendons*, which connect skeletal muscles to bones, consist almost entirely of collagen fibers. Typical *ligaments* are similar to tendons, but they connect one bone to another. Tendons and ligaments can withstand tremendous forces. Uncontrolled muscle contractions or skeletal movements are more likely to break a bone than to snap a tendon or a ligament.

2. **Reticular fibers** (*reticulum*, network) contain the same protein subunits as do collagen fibers, but arranged differently. Thinner than collagen fibers, reticular fibers form a branching, interwoven framework that is tough yet flexible. Because they form a network rather than share a common alignment, reticular fibers resist forces applied from many directions. This interwoven network, called a *stroma*, stabilizes the relative positions of the functional cells, or **parenchyma** (pa-RENG-ki-ma) of organs such as the liver. Reticular fibers also stabilize the positions of an organ's blood vessels, nerves, and other structures, despite changing positions and the pull of gravity.

3. **Elastic fibers** contain the protein *elastin*. Elastic fibers are branched and wavy. After stretching, they will return to their original length. **Elastic ligaments** are dominated by elastic fibers. They are relatively rare, but have important functions, such as interconnecting vertebrae.

GROUND SUBSTANCE Ground substance fills the spaces between cells and surrounds connective tissue fibers (Figure 4–9●). In connective tissue proper, ground substance is clear, viscous, and colorless. Its typical consistency is like that of maple syrup, due to the presence of proteoglycans and glycoproteins. ∞ p. 57 Ground substance is dense enough that bacteria have trouble moving through it—imagine swimming in molasses. This density slows the spread of pathogens and makes them easier for phagocytes to catch.

Marfan's syndrome is an inherited condition caused by the production of an abnormal form of *fibrillin*, a glycoprotein that is important to the strength and elasticity of connective tissues. Because most organs contain connective tissues, the effects of this defect are widespread. The most visible sign of Marfan's syndrome involves the skeleton; most individuals with the condition are tall and have abnormally long limbs and fingers. The most serious consequences involve the cardiovascular system; roughly 90 percent of people with Marfan's syndrome have structural abnormalities in their cardiovascular system. The most dangerous possibility is that the weakened elastic connective tissues in the walls of major arteries, such as the aorta, may burst, causing a sudden, fatal loss of blood.

Embryonic Connective Tissues

Mesenchyme, or *embryonic connective tissue*, is the first connective tissue to appear in a developing embryo. Mesenchyme contains an abundance of star-shaped stem cells (mesenchymal cells separated by a matrix with very fine protein filaments; see Figure 4–10a●). Mesenchyme gives rise to all other connective tissues. **Mucous connective tissue** (Figure 4–10b●), or Wharton's jelly, is a loose connective tissue found in many parts of the embryo, including the umbilical cord.

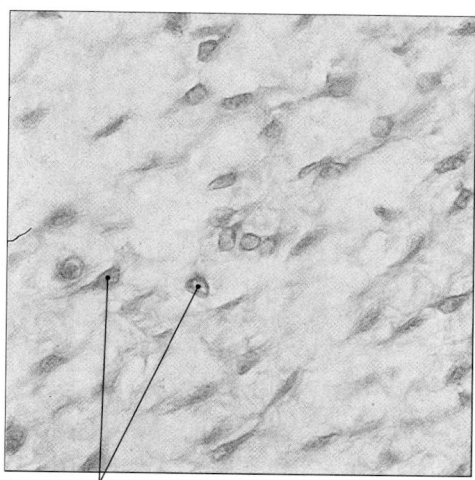

Mesenchymal cells
(a) Mesenchyme (LM × 136)

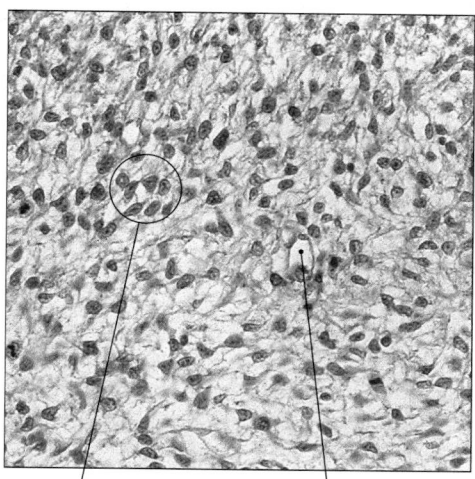

Mesenchymal cells   Blood vessel

(b) Mucous connective tissue (LM × 136)

●**FIGURE 4–10**
Connective Tissues in Embryos. **(a)** Mesenchyme, the first connective tissue to appear in an embryo. **(b)** Mucous connective tissue from the umbilical cord of a fetus. Mucous connective tissue in this location is also known as *Wharton's jelly.*

Adults have neither form of embryonic connective tissue. However, many adult connective tissues contain scattered mesenchymal stem cells that can assist in tissue repair after an injury.

Loose Connective Tissues

Loose connective tissues are the "packing materials" of the body. They fill spaces between organs, cushion and stabilize specialized cells in many organs, and support epithelia. These tissues surround and support blood vessels and nerves, store lipids, and provide a route for the diffusion of materials. Loose connective tissues include mucous connective tissue in embryos and *areolar tissue, adipose tissue,* and *reticular tissue* in adults.

AREOLAR TISSUE Areolar tissue (*areola*, little space) is the least specialized connective tissue in adults. It may contain all the cells and fibers of any connective tissue proper in a very loosely organized array (Figure 4–9●). Areolar tissue has an open framework; ground substance accounts for most of its volume. This syrupy fluid absorbs shocks. Because its fibers are loosely organized, areolar tissue can distort without damage. The presence of elastic fibers makes it resilient, so areolar tissue returns to its original shape after external pressure is relieved.

Areolar tissue forms a layer that separates the skin from deeper structures. In addition to providing padding, the elastic properties of this layer allow a considerable amount of independent movement. Thus, if you pinch the skin of your arm, you will not affect the underlying muscle. Conversely, contractions of the underlying muscle do not pull against your skin; as the muscle bulges, the areolar tissue stretches. Because this tissue has an extensive blood supply, the areolar tissue layer under the skin is a common injection site for drugs.

The capillaries (thin-walled blood vessels) in areolar tissue deliver oxygen and nutrients and remove carbon dioxide and waste products. They also carry wandering cells to and from the tissue. Epithelia commonly cover areolar tissue; fibroblasts maintain the reticular lamina of the basal lamina that separates the two kinds of tissue. The epithelial cells rely on the diffusion of oxygen and nutrients across that membrane from capillaries in the underlying connective tissue.

ADIPOSE TISSUE The distinction between areolar tissue and fat, or **adipose tissue**, is somewhat arbitrary. Adipocytes account for most of the volume of adipose tissue (Figure 4–11a●), but only a fraction of the volume of areolar tissue. Adipose tissue provides padding, absorbs shocks, acts as an insulator to slow heat loss through the skin, and serves as packing or filler around structures. Adipose tissue is common under the skin of the flanks, buttocks, and breasts. It fills the bony sockets behind the eyes, surrounds the kidneys, and is common beneath the mesothelial lining of the pericardial and abdominal cavities. Most of the adipose tissue in the body is called **white fat**, because it has a pale, yellow-white color.

In infants and young children, the adipose tissue between the shoulder blades, around the neck, and possibly elsewhere in the upper body is highly vascularized, and the individual adipocytes contain numerous mitochondria. Together, these characteristics give the tissue a deep, rich color from which the name **brown fat** is derived. When these cells are stimulated by the nervous system, lipid breakdown accelerates. The cells do not capture the energy that is released. Instead, it is absorbed by the surrounding tissues as heat. The heat warms the circulating blood, which distributes the heat throughout the body. In this way, an infant can increase metabolic heat generation by 100 percent very quickly. In adults, who have little if any brown fat, body temperature is elevated primarily by shivering, which involves rapidly oscillating contractions within large skeletal muscles. These contractions consume energy and generate heat that warms the body. Shivering is not an effective mechanism for infants, because their skeletal muscles are relatively small, weak, and poorly controlled.

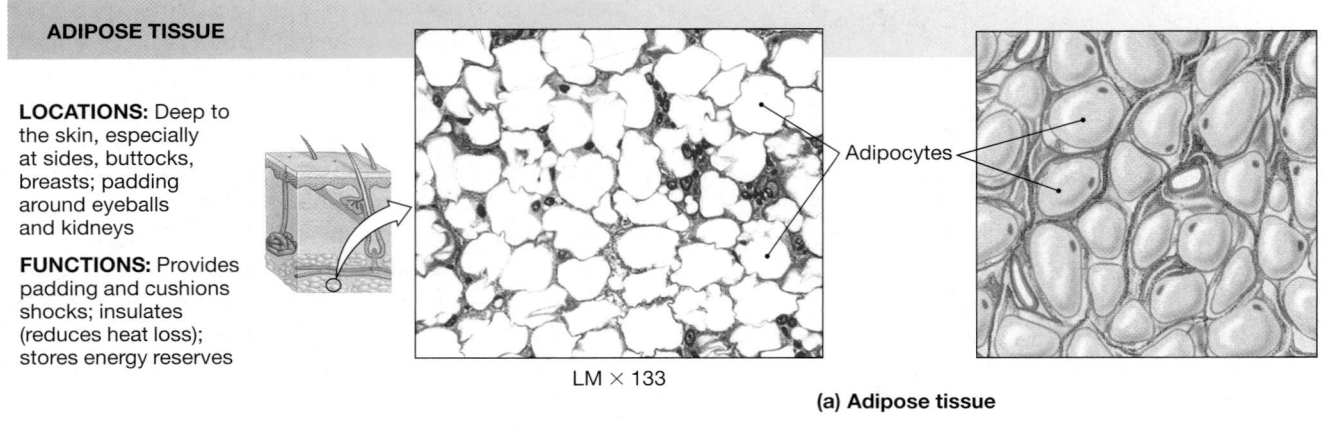

ADIPOSE TISSUE

LOCATIONS: Deep to the skin, especially at sides, buttocks, breasts; padding around eyeballs and kidneys

FUNCTIONS: Provides padding and cushions shocks; insulates (reduces heat loss); stores energy reserves

Adipocytes

LM × 133

(a) Adipose tissue

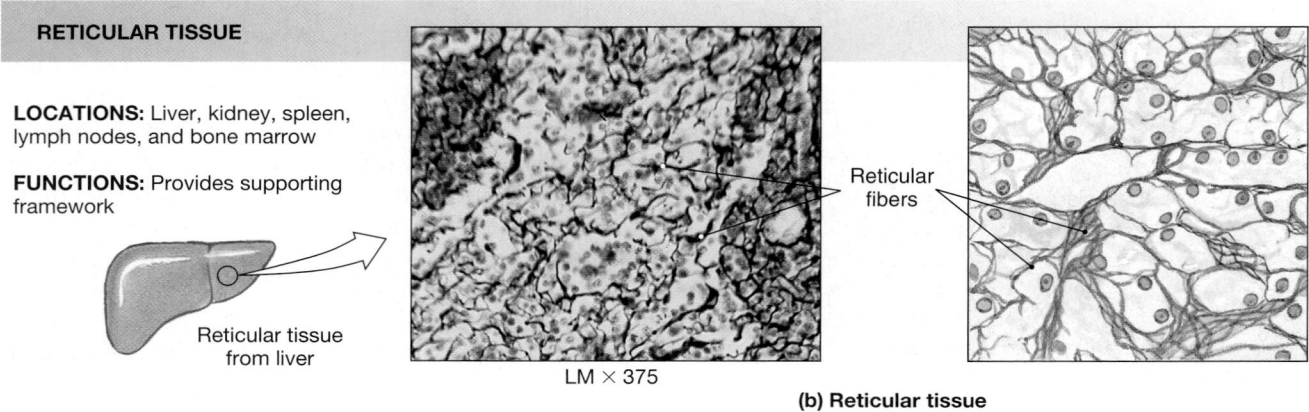

RETICULAR TISSUE

LOCATIONS: Liver, kidney, spleen, lymph nodes, and bone marrow

FUNCTIONS: Provides supporting framework

Reticular tissue from liver

Reticular fibers

LM × 375

(b) Reticular tissue

•**FIGURE 4–11**
Adipose and Reticular Tissues. **(a)** Adipose tissue is a loose connective tissue dominated by adipocytes. In standard histological preparations, the tissue looks empty because the lipids in the fat cells dissolve in the alcohol used in tissue processing. **(b)** Reticular tissue has an open framework of reticular fibers, which are usually very difficult to see because of the large numbers of cells around them.

Adipocytes are metabolically active cells; their lipids are constantly being broken down and replaced. When nutrients are scarce, adipocytes deflate like collapsing balloons. Often, such deflation occurs during a weight-loss program. Because the cells are not killed but merely reduced in size, the lost weight can easily be regained in the same areas of the body. In adults, adipocytes are incapable of dividing. The number of fat cells in peripheral tissues is established in the first few weeks of a newborn's life, perhaps in response to the amount of fats in the diet. However, that is not the end of the story, because loose connective tissues also contain mesenchymal cells. If circulating-lipid levels are chronically elevated, the mesenchymal cells will divide, giving rise to cells that differentiate into fat cells. As a result, areas of areolar tissue can become adipose tissue in times of nutritional plenty, even in adults. In the procedure known as *liposuction*, unwanted adipose tissue is surgically removed. Because adipose tissue can regenerate through the differentiation of mesenchymal cells, liposuction provides only a temporary solution to the problem of excess weight.

RETICULAR TISSUE As mentioned earlier, organs such as the spleen and liver contain **reticular tissue,** in which reticular fibers create a complex three-dimensional stroma (Figure 4–11b●). The stroma supports the parenchyma (functional cells) of these organs. This fibrous framework is also found in the lymph nodes and bone marrow. Fixed macrophages and fibroblasts are associated with the reticular fibers, but these cells are seldom visible, because the organs are dominated by specialized cells with other functions.

Dense Connective Tissues

Most of the volume of **dense connective tissues** is occupied by fibers. Dense connective tissues are often called **collagenous** (ko-LA-jin-us) **tissues**, because collagen fibers are the dominant type of fiber in them. The body has two types of dense connective tissues: (1) dense regular connective tissue and (2) dense irregular connective tissue.

In **dense regular connective tissue**, the collagen fibers are parallel to each other, packed tightly, and aligned with the forces applied to the tissue. **Tendons** are cords of dense regular connective tissue that attach skeletal muscles to bones (Figure 4–12a●). The collagen fibers run along the longitudinal axis of the tendon and transfer the pull of the contracting muscle to the bone. **Ligaments**

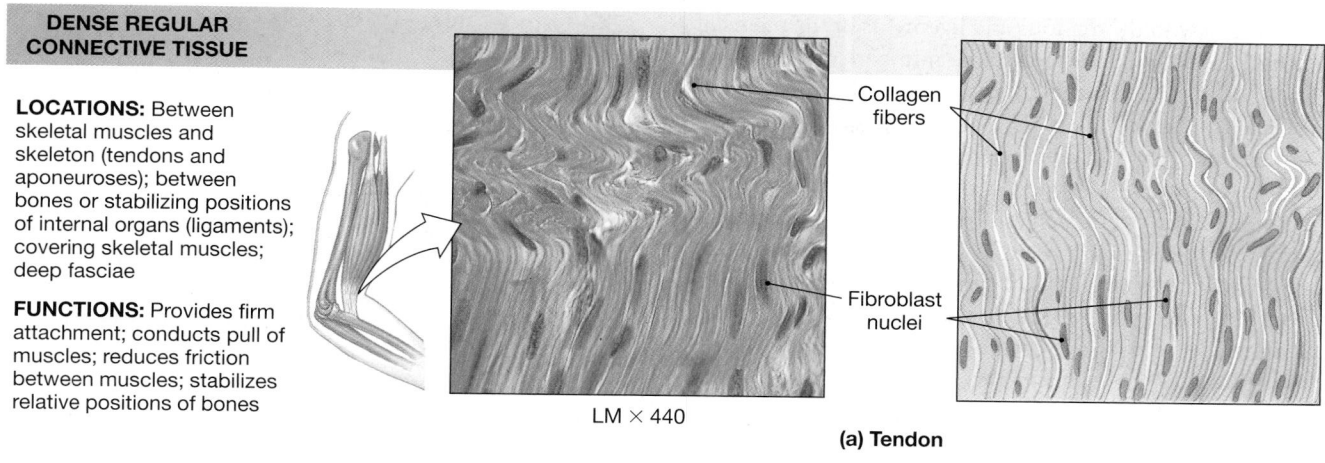

DENSE REGULAR CONNECTIVE TISSUE

LOCATIONS: Between skeletal muscles and skeleton (tendons and aponeuroses); between bones or stabilizing positions of internal organs (ligaments); covering skeletal muscles; deep fasciae

FUNCTIONS: Provides firm attachment; conducts pull of muscles; reduces friction between muscles; stabilizes relative positions of bones

LM × 440

(a) Tendon

Collagen fibers

Fibroblast nuclei

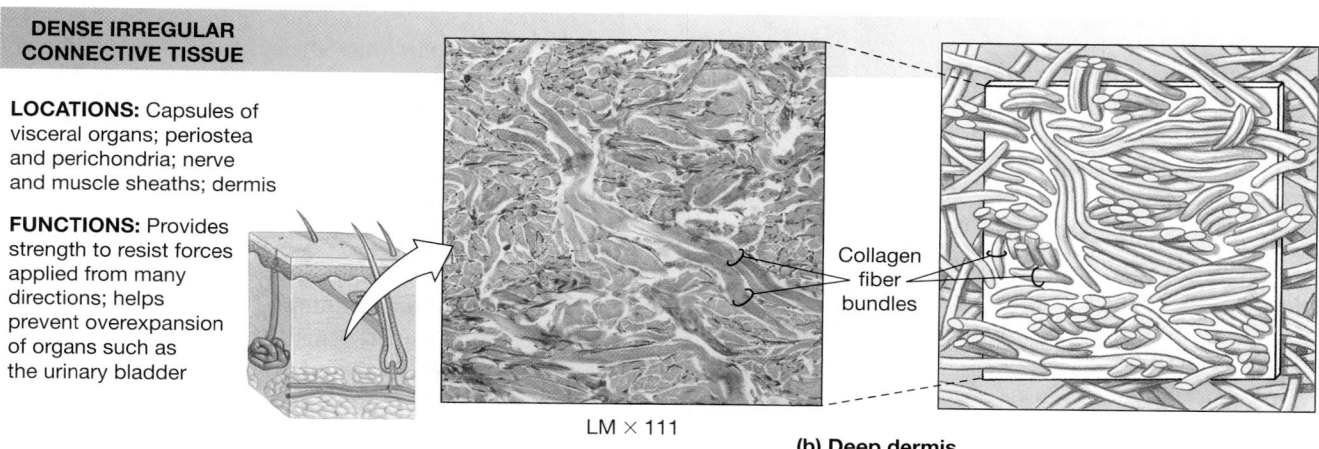

DENSE IRREGULAR CONNECTIVE TISSUE

LOCATIONS: Capsules of visceral organs; periostea and perichondria; nerve and muscle sheaths; dermis

FUNCTIONS: Provides strength to resist forces applied from many directions; helps prevent overexpansion of organs such as the urinary bladder

LM × 111

(b) Deep dermis

Collagen fiber bundles

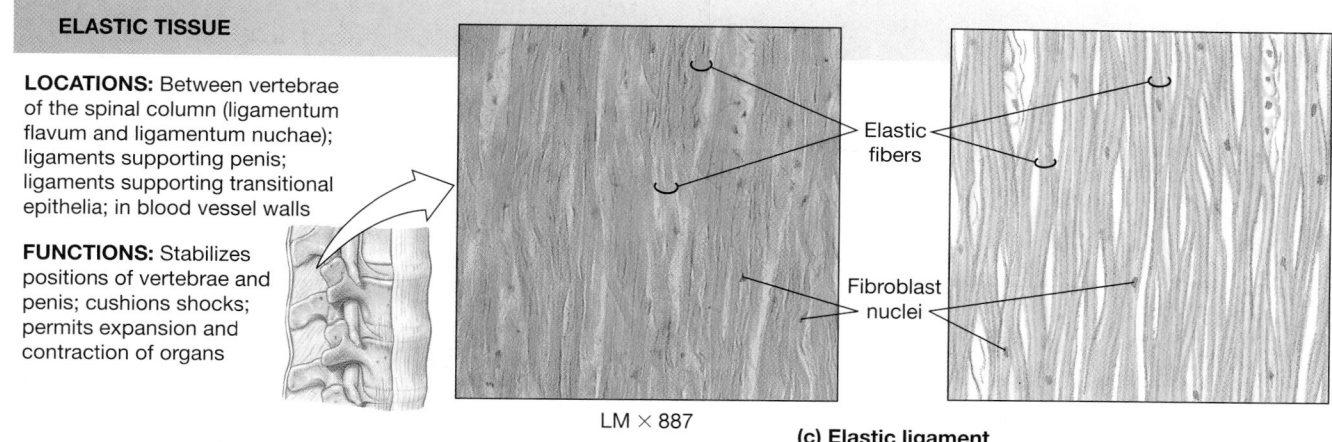

ELASTIC TISSUE

LOCATIONS: Between vertebrae of the spinal column (ligamentum flavum and ligamentum nuchae); ligaments supporting penis; ligaments supporting transitional epithelia; in blood vessel walls

FUNCTIONS: Stabilizes positions of vertebrae and penis; cushions shocks; permits expansion and contraction of organs

LM × 887

(c) Elastic ligament

Elastic fibers

Fibroblast nuclei

•FIGURE 4–12
Dense Connective Tissues. **(a)** The dense regular connective tissue in a tendon. Notice the densely packed, parallel bundles of collagen fibers. The fibroblast nuclei are flattened between the bundles. **(b)** The deep dermis of the skin contains a thick layer of dense irregular connective tissue. **(c)** An elastic ligament, an example of elastic tissue. Elastic ligaments extend between the vertebrae of the vertebral column. The bundles are fatter than those of a tendon or a ligament composed of collagen.

resemble tendons, but connect one bone to another or stabilize the positions of internal organs. An **aponeurosis** (AP-ō-noo-RŌ-sis; plural, *aponeuroses*) is a tendinous sheet that attaches a broad, flat muscle to another muscle or to several bones of the skeleton. It can also stabilize the positions of tendons and ligaments. Aponeuroses are associated with large muscles of the lower back and abdomen and with the tendons and ligaments of the palms of the hands and the soles of the feet. Large numbers of fibroblasts are scattered among the collagen fibers of tendons, ligaments, and aponeuroses.

In contrast, the fibers in **dense irregular connective tissue** form an interwoven meshwork in no consistent pattern (Figure 4–12b•). These tissues strengthen and support areas subjected to stresses from many directions. A layer of dense irregular connective tissue gives skin its strength. Cured leather (animal skin) is an excellent illustration of the interwoven nature of this tissue. Except at joints, dense irregular connective tissue forms a sheath around cartilages (the *perichondrium*) and bones (the *periosteum*). Dense irregular connective tissue also forms a thick fibrous layer called a **capsule**, which surrounds internal organs such as the liver, kidneys, and spleen and which encloses the cavities of joints.

Dense regular and dense irregular connective tissues contain variable amounts of elastic fibers. When elastic fibers outnumber collagen fibers, the tissue has a springy, resilient nature that allows it to tolerate cycles of extension and recoil. For instance, many elastic fibers are present in the connective tissue that supports transitional epithelia, in the walls of large blood vessels such as the aorta, and around the respiratory passageways.

Elastic tissue is a dense regular connective tissue dominated by elastic fibers. **Elastic ligaments**, which are almost completely dominated by elastic fibers, help stabilize the positions of the vertebrae of the spinal column (Figure 4–12c•).

✓ Lack of vitamin C in the diet interferes with the ability of fibroblasts to produce collagen. What effect might this interference have on connective tissue?

✓ Many allergy sufferers take antihistamines to relieve their allergy symptoms. Which type of cell produces the molecule that this medication blocks?

✓ Which type of connective tissue contains primarily triglycerides?

Answers start on page Q-1

☐ FLUID CONNECTIVE TISSUES

Blood and *lymph* are connective tissues with distinctive collections of cells. The fluid matrix that surrounds the cells also includes many types of suspended proteins that do not form insoluble fibers under normal conditions.

In **blood**, the watery matrix is called **plasma**. Plasma contains blood cells and fragments of cells, collectively known as *formed elements* (Figure 4–13•). There are three types of formed elements: (1) red blood cells, (2) white blood cells, and (3) platelets.

A single cell type, the **red blood cell**, or **erythrocyte** (e-RITH-rō-sīt; *erythros*, red), accounts for almost half the volume of blood and is the reason we associate the color red with blood. Red blood cells are responsible for the transport of oxygen and, to a lesser degree, of carbon dioxide in the blood.

Plasma also contains small numbers of **white blood cells**, or **leukocytes** (LOO-kō-sīts; *leuko*, white). White blood cells include the phagocytic microphages (*neutrophils* and *eosinophils*), *basophils*, *lymphocytes*, and *monocytes*. White blood cells are important components of the immune system, which protects the body from infection and disease.

The third type of formed element in blood consists not of whole cells, but of tiny membrane-enclosed packets of cytoplasm called **platelets**. These cell fragments, which contain enzymes and special proteins, function in the clotting response that seals breaks in the endothelial lining.

Recall from Chapter 3 that the human body contains a large volume of extracellular fluid. This fluid includes three major subdivisions: *plasma, interstitial fluid,* and *lymph*. Plasma is normally confined to the vessels of the circulatory system, and contractions of the heart keep it in motion. **Arteries** are vessels that carry blood away from the heart toward the fine, thin-walled **capillaries**. **Veins** are vessels that drain the capillaries and return blood to the heart, completing the circuit of blood. In tissues, filtration moves water and small solutes out of the capillaries and into the interstitial fluid, which bathes the body's cells. The major difference between plasma and interstitial fluid is that plasma contains numerous suspended proteins that are too large to pass into the interstitial fluid.

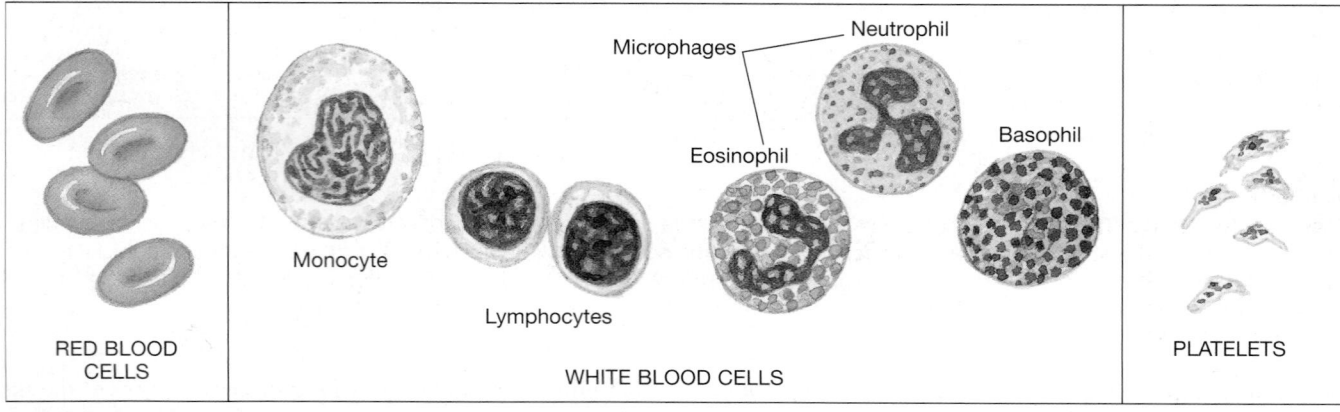

•FIGURE 4–13
Formed Elements of the Blood

Lymph forms as interstitial fluid enters **lymphatic vessels**, small passageways that return it to the cardiovascular system. As fluid passes along the lymphatic vessels, cells of the immune system monitor the composition of the lymph and respond to signs of injury or infection. The number of cells in lymph may vary, but ordinarily 99 percent of them are lymphocytes. The rest are primarily macrophages or microphages. This recirculation of fluid from the cardiovascular system, through the interstitial fluid, to the lymph, and then back to the cardiovascular system is a continuous process that is essential to homeostasis. It helps eliminate local differences in the levels of nutrients, wastes, or toxins, maintains blood volume, and alerts the immune system to infections that may be underway in peripheral tissues.

☐ SUPPORTING CONNECTIVE TISSUES

Cartilage and *bone* are called supporting connective tissues because they provide a strong framework that supports the rest of the body. In these connective tissues, the matrix contains numerous fibers and, in some cases, deposits of insoluble calcium salts.

Cartilage

The matrix of **cartilage** is a firm gel that contains polysaccharide derivatives called **chondroitin sulfates** (kon-DROY-tin;

chondros, cartilage). Chondroitin sulfates form complexes with proteins in the ground substance, producing proteoglycans. Cartilage cells, or **chondrocytes** (KON-drō-sīts), are the only cells in the cartilage matrix. They occupy small chambers known as **lacunae** (la-KOO-nē; *lacus*, pool). The physical properties of cartilage depend on the type and abundance of extracellular fibers, as well as on the proteoglycans of the matrix.

Unlike other connective tissues, cartilage is avascular, so all exchange of nutrients and waste products must occur by diffusion through the matrix. Blood vessels do not grow into cartilage because chondrocytes produce a chemical that discourages their formation. This chemical, named **antiangiogenesis factor** (*anti-*, against + *angeion*, vessel + *genno*, to produce), is now being tested as a potential anticancer agent. ∞ pp. 105–106

A cartilage is generally set apart from surrounding tissues by a fibrous **perichondrium** (pe-rē-KON-drē-um; *peri-*, around). The perichondrium contains two distinct layers: an outer, fibrous region of dense irregular connective tissue and an inner, cellular layer. The fibrous layer provides mechanical support and protection and attaches the cartilage to other structures. The cellular layer is important to the growth and maintenance of the cartilage.

CARTILAGE GROWTH Cartilage grows by two mechanisms: (1) *interstitial growth* and (2) *appositional growth* (Figure 4–14●):

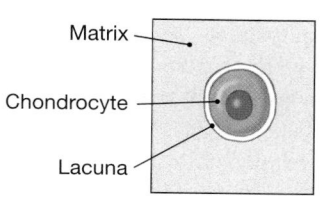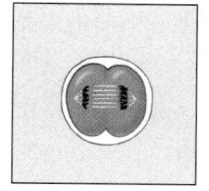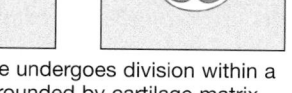

Matrix
Chondrocyte
Lacuna

Chondrocyte undergoes division within a lacuna surrounded by cartilage matrix.

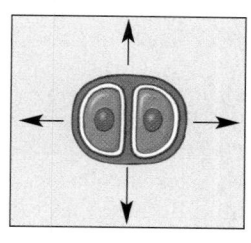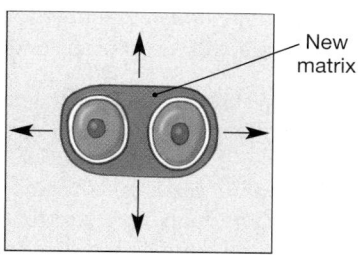

New matrix

As daughter cells secrete additional matrix, they move apart, expanding the cartilage from within.

(a) Interstitial growth

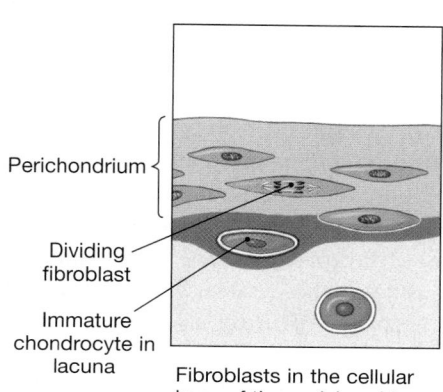

Perichondrium
Dividing fibroblast
Immature chondrocyte in lacuna

Fibroblasts in the cellular layer of the perichondrium differentiate into chondrocytes.

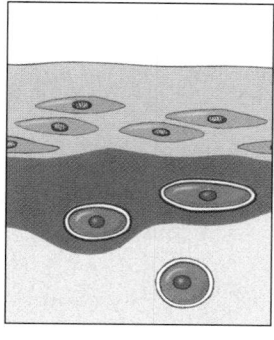

These chondrocytes secrete new matrix.

Fibroblasts
New matrix
Immature chondrocyte
Older matrix
Mature chondrocyte

As matrix enlarges, more fibroblasts are incorporated; they are replaced by divisions of cells in the perichondrium.

(b) Appositional growth

●**FIGURE 4–14**
The Formation and Growth of Cartilage. **(a)** Interstitial growth. The cartilage expands from within as chondrocytes in the matrix divide, grow, and produce new matrix. **(b)** Appositional growth. The cartilage grows at its external surface through the differentiation of fibroblasts into chondrocytes in the cellular layer of the perichondrium.

In **interstitial growth**, chondrocytes in the cartilage matrix undergo cell division, and the daughter cells produce additional matrix (Figure 4–14a●). This process enlarges the cartilage from within, like inflating a balloon by forcing more air into it.

In **appositional growth**, new layers of cartilage are added to the surface (Figure 4–14b●). In this process, cells of the inner layer of the perichondrium undergo repeated cycles of division. The innermost cells then differentiate into immature chondrocytes that begin producing cartilage matrix. As they become surrounded by and embedded in new matrix, they differentiate into mature chondrocytes. Appositional growth gradually increases the size of the cartilage by adding to its outer surface. Interstitial growth is most important during embryonic development. The process begins early in development and continues through adolescence.

Neither interstitial nor appositional growth occurs in the cartilages of normal adults. However, appositional growth may occur in unusual circumstances, such as after cartilage has been damaged or under excessive stimulation with *growth hormone* from the pituitary gland. Minor damage to cartilage can be repaired by appositional growth at the damaged surface. After more severe damage, the injured portion of the cartilage will be replaced by a dense fibrous patch.

TYPES OF CARTILAGE The body contains three major types of cartilage: hyaline cartilage, elastic cartilage, and fibrocartilage.

1. **Hyaline cartilage** (HĪ-uh-lin; *hyalos*, glass) is the most common type of cartilage. Except inside joint cavities, hyaline cartilage is covered by a dense perichondrium (Figure 4–15a●). The matrix of hyaline cartilage contains closely packed collagen fibers, making it tough but somewhat flexible. Because the fibers are not in large bundles and do not stain darkly, they are not always apparent in light microscopy (Figure 4–15b●). Examples in adults include the connections between the ribs and the sternum; the nasal cartilages and the supporting cartilages along the conducting passageways of the respiratory tract; and the *articular cartilages*, which cover opposing bone surfaces within many joints, such as the elbow and knee.

2. **Elastic cartilage** (Figure 4–15c●) contains numerous elastic fibers that make it extremely resilient and flexible. These cartilages usually have a yellowish color on gross dissection. Elastic cartilage forms the external flap (the *auricle*, or *pinna*) of the outer ear, the epiglottis, an airway to the middle ear cavity (the *auditory tube*), and small cartilages in the larynx (the *cuneiform cartilages*).

3. **Fibrocartilage** has little ground substance, and its matrix is dominated by densely interwoven collagen fibers (Figure 4–15d●), making this tissue extremely durable and tough. Fibrocartilaginous pads lie between the spinal vertebrae, between the pubic bones of the pelvis, and around or in a few joints and tendons. In these positions, fibrocartilage resists compression, absorbs shocks, and prevents damaging bone-to-bone contact. Cartilage heals poorly, and damaged fibro-

cartilage in joints such as the knee can interfere with normal movements.

Several complex joints, including the knee, contain both hyaline cartilage and fibrocartilage. The hyaline cartilage covers bony surfaces, and fibrocartilage pads in the joint prevent contact between bones during movement. Injuries to these joints can produce tears in the fibrocartilage pads, and the tears do not heal. Eventually, joint mobility is severely reduced. Surgery generally produces only a temporary or incomplete repair.

Recent advances in tissue culture have enabled researchers to grow fibrocartilage in the laboratory. Chondrocytes removed from the knees of injured dogs are cultured in an artificial framework of collagen fibers. Eventually, they produce masses of fibrocartilage that can be inserted into the damaged joints. Over time, the pads change shape and grow, restoring normal joint function. In the future, this technique may be used to treat severe joint injuries in humans.

Bone

Because we shall examine the detailed histology of **bone**, or **osseous** (OS-ē-us) **tissue** (*os*, bone), in Chapter 6, here we focus only on significant differences between cartilage and bone. The volume of ground substance in bone is very small. Roughly two-thirds of the matrix of bone consists of a mixture of calcium salts, primarily calcium phosphate, with lesser amounts of calcium carbonate. The rest of the matrix is dominated by collagen fibers. This combination gives bone truly remarkable properties. By themselves, calcium salts are hard, but rather brittle. Collagen fibers are stronger, but relatively flexible. In bone, the minerals are organized around the collagen fibers. The result is a strong, somewhat flexible combination that is highly resistant to shattering. In its overall properties, bone can compete with the best steel-reinforced concrete. In essence, the collagen fibers in bone act like the steel reinforcing rods, and the mineralized matrix acts like the concrete.

Figure 4–16● shows the general organization of osseous tissue. Lacunae in the matrix contain **osteocytes** (OS-tē-ō-sīts), or bone cells. The lacunae are typically organized around blood vessels that branch through the bony matrix. Although diffusion cannot occur through the hard matrix, osteocytes communicate with the blood vessels and with one another by means of slender cytoplasmic extensions. These extensions run through long, slender passageways in the matrix. Called **canaliculi** (kan-a-LIK-ū-lē; little canals), these passageways form a branching network for the exchange of materials between blood vessels and osteocytes.

Except in joint cavities, where they are covered by a layer of hyaline cartilage, bone surfaces are sheathed by a **periosteum** (pe-rē-OS-tē-um), a layer composed of fibrous (outer) and cellular (inner) layers. The periosteum assists in the attachment of a bone to surrounding tissues and to associated

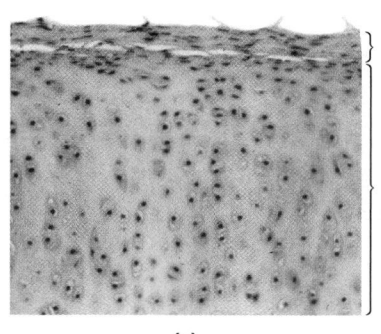

Perichondrium

Hyaline cartilage

(a)

●**FIGURE 4–15**
The Perichondrium and Types of Cartilage. **(a)** A perichondrium separates cartilage from other tissues. **(b)** Hyaline cartilage. Note the translucent matrix and the absence of prominent fibers. **(c)** Elastic cartilage. The closely packed elastic fibers are visible between the chondrocytes. **(d)** Fibrocartilage. The collagen fibers are extremely dense, and the chondrocytes are relatively far apart.

HYALINE CARTILAGE

LOCATIONS: Between tips of ribs and bones of sternum; covering bone surfaces at synovial joints; supporting larynx (voice box), trachea, and bronchi; forming part of nasal septum

FUNCTIONS: Provides stiff but somewhat flexible support; reduces friction between bony surfaces

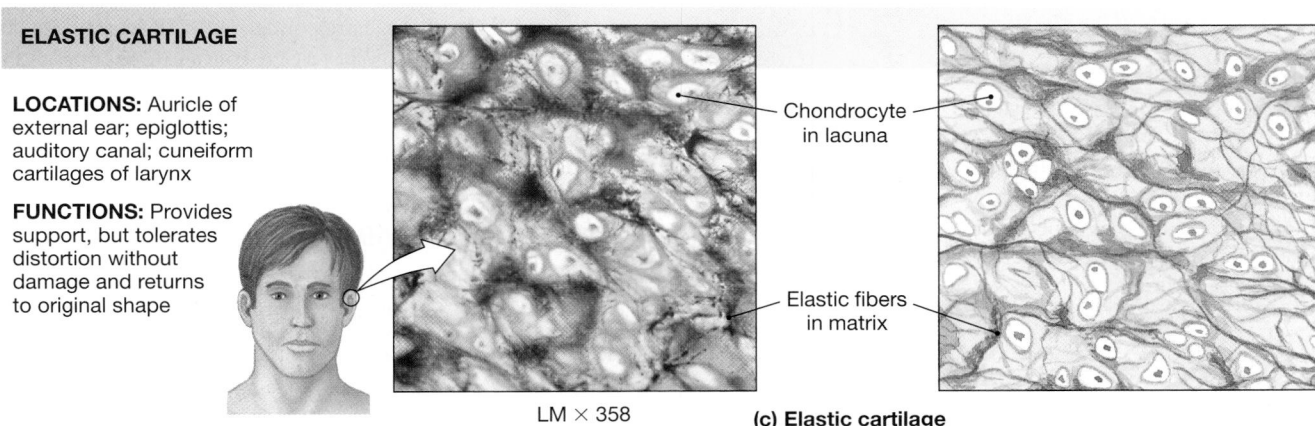

Chondrocytes in lacunae

Matrix

LM × 500

(b) Hyaline cartilage

ELASTIC CARTILAGE

LOCATIONS: Auricle of external ear; epiglottis; auditory canal; cuneiform cartilages of larynx

FUNCTIONS: Provides support, but tolerates distortion without damage and returns to original shape

Chondrocyte in lacuna

Elastic fibers in matrix

LM × 358

(c) Elastic cartilage

FIBROCARTILAGE

LOCATIONS: Pads within knee joint; between pubic bones of pelvis; intervertebral discs

FUNCTIONS: Resists compression; prevents bone-to-bone contact; limits relative movement

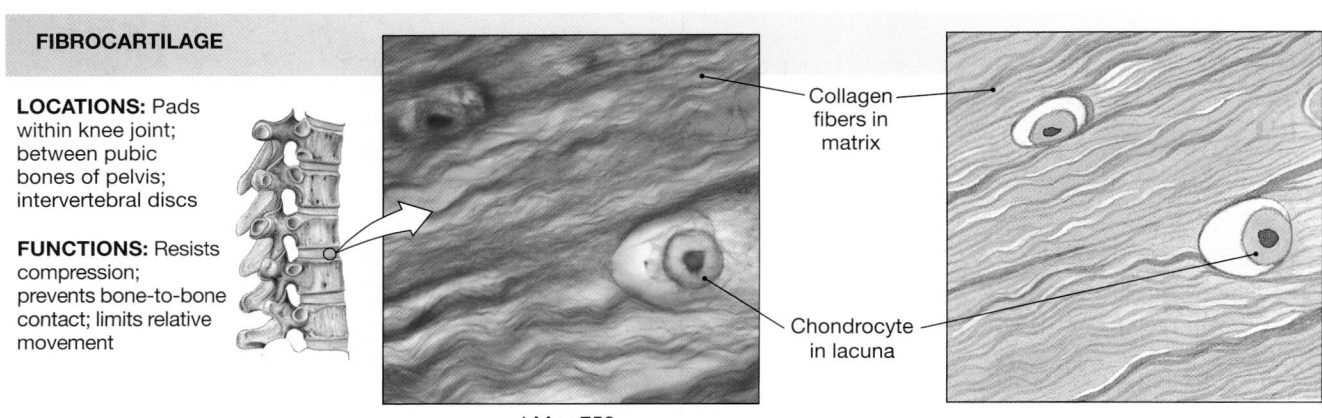

Collagen fibers in matrix

Chondrocyte in lacuna

LM × 750

(d) Fibrocartilage

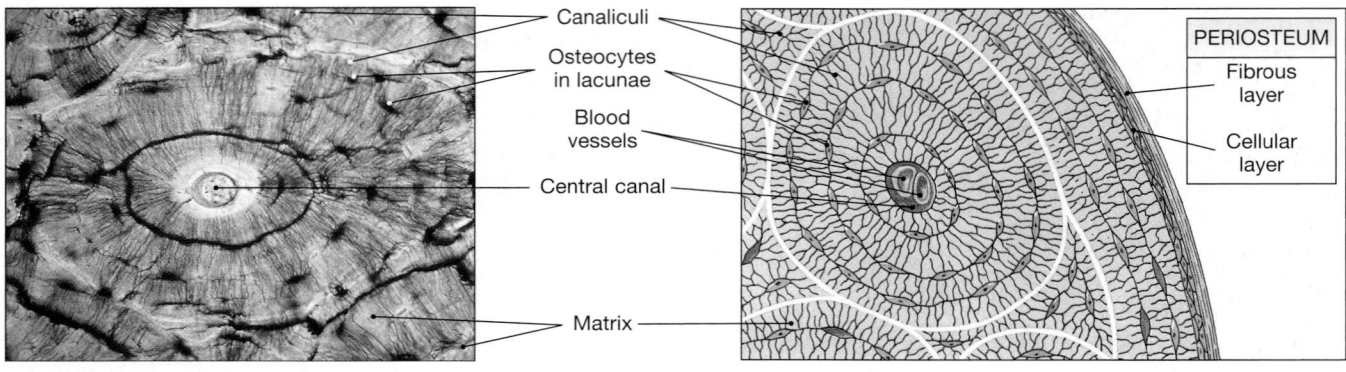

●FIGURE 4–16
Bone. The osteocytes in bone are generally organized in groups around a central space that contains blood vessels. Bone dust produced during grinding fills the lacunae and the central canal, making them appear dark. (LM × 362)

tendons and ligaments. The cellular layer functions in appositional bone growth and participates in repairs after an injury. Unlike cartilage, bone undergoes extensive remodeling throughout life, and complete repairs can be made even after severe damage has occurred. Bones also respond to the stresses placed on them, growing thicker and stronger with exercise and becoming thin and brittle with inactivity.

Table 4–2 summarizes the similarities and differences between cartilage and bone.

✓ Why does cartilage heal so slowly?

✓ If a person has a herniated intervertebral disc, which type of cartilage has been damaged?

✓ Which two types of connective tissue have a matrix that is fluid?

Answers start on page Q-1

 The classification system for connective tissue can be reviewed by visiting the **Companion Website**: Chapter 4/Reviewing Concepts/Labeling.

4–4 MEMBRANES

Objective

■ Explain how epithelial and connective tissues combine to form four types of membranes, and specify the functions of each.

A membrane is a physical barrier. There are many different types of anatomical membranes—you encountered cell membranes in Chapter 3, and you will find many other examples in later chapters. The membranes we are concerned with here all line or cover body surfaces. Each consists of an epithelium supported by connective tissue. Four such membranes occur in the body: (1) *mucous membranes*, (2) *serous membranes*, (3) the *cutaneous membrane*, and (4) *synovial membranes*.

□ MUCOUS MEMBRANES

Mucous membranes, or **mucosae** (mū-KŌ-suh), line passageways and chambers, including the digestive, respiratory, reproduc-

TABLE 4–2 A COMPARISON OF CARTILAGE AND BONE		
Characteristic	**Cartilage**	**Bone**
Structural Features		
Cells	Chondrocytes in lacunae	Osteocytes in lacunae
Ground substance	Chondroitin sulfate (in proteoglycan) and water	A small volume of liquid surrounding insoluble crystals of calcium salts (calcium phosphate and calcium carbonate)
Fibers	Collagen, elastic, and reticular fibers (proportions vary)	Collagen fibers predominate
Vascularity	None	Extensive
Covering	Perichondrium, two-part	Periosteum, two-part
Strength	Limited: bends easily, but hard to break	Strong: resists distortion until breaking point
Metabolic Features		
Oxygen demands	Low	High
Nutrient delivery	By diffusion through matrix	By diffusion through cytoplasm and fluid in canaliculi
Growth	Interstitial and appositional	Appositional only
Repair capabilities	Limited	Extensive

tive, and urinary tracts, that communicate with the exterior (Figure 4–17a●). The epithelial surfaces of these passageways must be kept moist to reduce friction and, in many cases, facilitate absorption or secretion. The epithelial surfaces are lubricated either by mucus, produced by goblet cells or multicellular glands, or by exposure to fluids such as urine or semen. The areolar tissue component of a mucous membrane is called the **lamina propria** (PRŌ-prē-uh). We shall consider the organization of specific mucous membranes in greater detail in later chapters.

Many mucous membranes are lined by simple epithelia that perform absorptive or secretory functions, such as the simple columnar epithelium of the digestive tract. Other types of epithelia may be involved, however. For example, a stratified squamous epithelium is part of the mucous membrane of the mouth, and the mucous membrane along most of the urinary tract has a transitional epithelium.

☐ SEROUS MEMBRANES

Serous membranes line the sealed, internal subdivisions of the ventral body cavity—cavities that are not open to the exterior. These membranes consist of a mesothelium supported by areolar tissue (Figure 4–17b●). As you may recall from Chapter 1, the three types of serous membranes are (1) the *pleura*, which lines the pleural cavities and covers the lungs; (2) the *peritoneum*, which lines the peritoneal cavity and covers the surfaces of the enclosed organs; and (3) the *pericardium*, which lines the pericardial cavity and covers the heart. ∞ p. 22 Serous membranes are very thin, but they are firmly attached to the body wall and to the organs they cover. When looking at an organ such as the heart or stomach, you are really seeing the tissues of the organ through a transparent serous membrane.

Each serous membrane can be divided into a *parietal portion*, which lines the inner surface of the cavity, and an opposing *visceral portion*, or **serosa**, which covers the outer surfaces of visceral organs. These organs often move or change shape as they perform their various functions, and the parietal and visceral surfaces of a serous membrane are in close contact at all times. Thus, the primary function of any serous membrane is to minimize friction between the opposing parietal and visceral surfaces. Friction is kept to a minimum because mesothelia are very thin and permeable; tissue fluids continuously diffuse onto the exposed surface, keeping it moist and slippery.

The fluid formed on the surfaces of a serous membrane is called a *transudate* (TRAN-sū-dāt; *trans*, across). In healthy individuals,

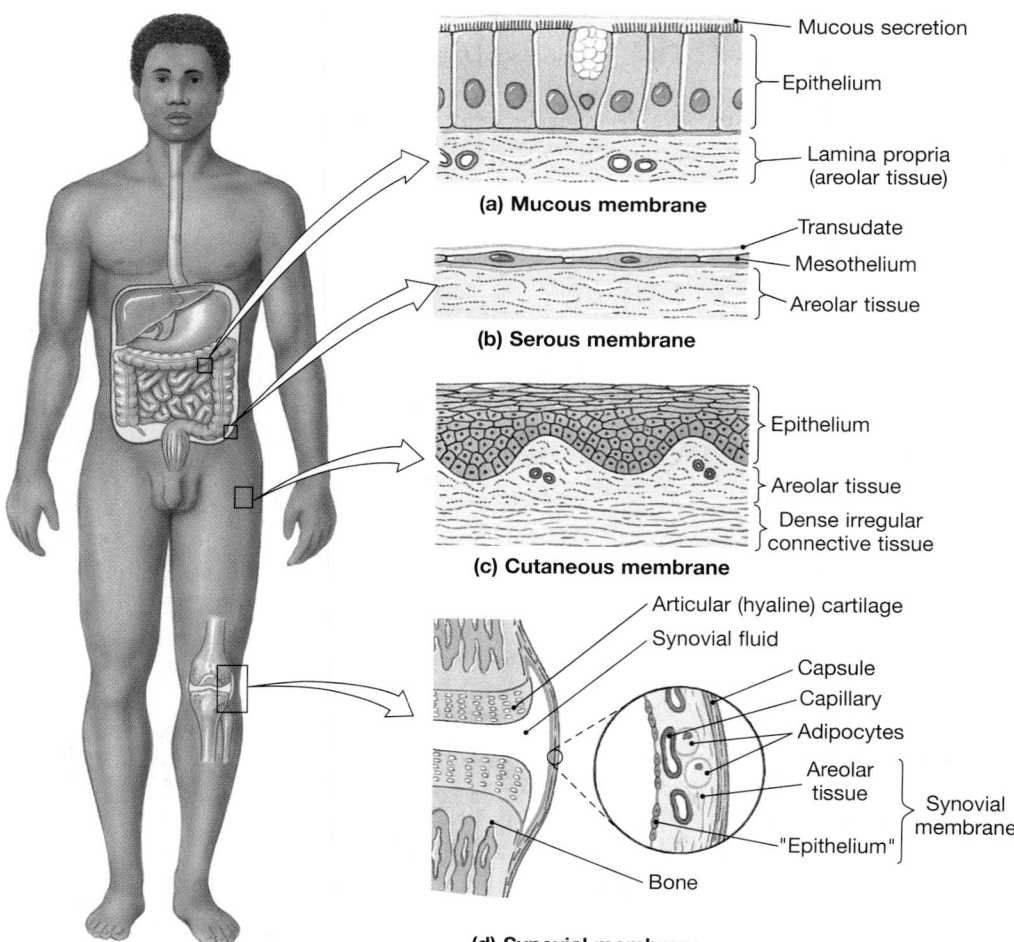

●**FIGURE 4–17**
Membranes. **(a)** Mucous membranes are coated with the secretions of mucous glands. These membranes line the digestive, respiratory, urinary, and reproductive tracts. **(b)** Serous membranes line the ventral body cavities (the peritoneal, pleural, and pericardial cavities). **(c)** The cutaneous membrane, or skin, covers the outer surface of the body. **(d)** Synovial membranes line joint cavities and produce the fluid within the joint.

(a) Mucous membrane
- Mucous secretion
- Epithelium
- Lamina propria (areolar tissue)

(b) Serous membrane
- Transudate
- Mesothelium
- Areolar tissue

(c) Cutaneous membrane
- Epithelium
- Areolar tissue
- Dense irregular connective tissue

(d) Synovial membrane
- Articular (hyaline) cartilage
- Synovial fluid
- Capsule
- Capillary
- Adipocytes
- Areolar tissue
- Synovial membrane
- "Epithelium"
- Bone

Problems with Serous Membranes

Several clinical conditions, including infection and chronic irritation, can cause the abnormal buildup of fluid in a ventral body cavity.

Pleuritis, or *pleurisy*, is an inflammation of the pleural cavities. At first the membranes become dry, and the opposing membranes may scratch against one another, producing a sound known as a *pleural rub*. In general, friction between opposing layers of serous membranes may promote the formation of adhesions—fibrous connections that lock the membranes together and eliminate the friction. Adhesions also severely restrict the movement of the affected organ or organs and may compress blood vessels or nerves. However, adhesions seldom form between the serous membranes of the pleural cavities. More commonly, continued inflammation and rubbing lead to a gradual increase in fluid production to levels well above normal. Fluid then accumulates in the pleural cavities, producing a condition known as *pleural effusion*. Pleural effusion is also caused by heart conditions that elevate the pressure in blood vessels of the lungs. As fluids build up in the pleural cavities, the lungs are compressed, making breathing difficult. The combination of severe pleural effusion and heart disease can be lethal.

Pericarditis is an inflammation of the pericardium. This condition typically leads to *pericardial effusion*, an abnormal accumulation of the fluid in the pericardial cavity. When sudden or severe, the fluid buildup can seriously reduce the efficiency of the heart and restrict blood flow through major vessels.

Peritonitis, an inflammation of the peritoneum, can follow infection of, or injury to, the peritoneal lining. Peritonitis is a potential complication of any surgical procedure in which the peritoneal cavity is opened. Liver disease, kidney disease, or heart failure can cause an increase in the rate of fluid movement through the peritoneal lining. *Ascites* (a-SĪ-tēz), the accumulation of peritoneal fluid, creates a characteristic abdominal swelling. The distortion of internal organs by the contained fluid can lead to symptoms such as heartburn, indigestion, and low-back pain.

the total volume of transudate is extremely small, just enough to prevent friction between the walls of the cavities and the surfaces of internal organs. But after an injury or in certain disease states, the volume of transudate may increase dramatically, complicating existing medical problems or producing new ones.

THE CUTANEOUS MEMBRANE

The **cutaneous membrane**, or skin, covers the surface of the body. It consists of a stratified squamous epithelium and a layer of areolar tissue reinforced by underlying dense connective tissue (Figure 4–17c●). In contrast to serous and mucous membranes, the cutaneous membrane is thick, relatively waterproof, and usually dry. We will examine the cutaneous membrane further in Chapter 5.

SYNOVIAL MEMBRANES

Adjacent bones often interact at joints, or *articulations*. At an articulation, the two articulating bones are very close together if not in contact. Joints that permit significant amounts of movement are complex structures. Such a joint is surrounded by a fibrous capsule, and the ends of the articulating bones lie within a *joint cavity* filled with **synovial** (sin-Ō-vē-ul) **fluid**. (Figure 4–17d●). The synovial fluid is produced by a **synovial membrane**, which lines the joint cavity. A synovial membrane consists of an extensive area of areolar tissue containing a matrix of interwoven collagen fibers, proteoglycans, and glycoproteins. An incomplete layer of macrophages and specialized fibroblasts separates the areolar tissue from the joint cavity. These cells regulate the composition of the synovial fluid. Although this lining is often called an epithelium, it differs from true epithelia in four respects: (1) It develops within a connective tissue, (2) no basal lamina is present, (3) gaps of up to 1 mm may separate adjacent cells, and (4) fluid and solutes are continuously exchanged between the synovial fluid and capillaries in the underlying connective tissue.

Even though the adjacent ends of the bones are covered by a smooth layer of articular cartilage, the surfaces must be lubricated to keep friction from damaging the opposing surfaces. The necessary lubrication is provided by the synovial fluid, which is similar in composition to the ground substance in loose connective tissues. Synovial fluid circulates from the areolar tissue into the joint cavity and percolates through the articular cartilages, providing oxygen and nutrients to the chondrocytes. Joint movement is important in stimulating the formation and circulation of synovial fluid: if a synovial joint is immobilized for long periods, the articular cartilages and the synovial membrane undergo degenerative changes. (We will discuss this problem further in Chapter 9.)

4–5 THE CONNECTIVE TISSUE FRAMEWORK OF THE BODY

Objective

- Describe how connective tissue establishes the framework of the body.

Connective tissues create the internal framework of the body. Layers of connective tissue connect the organs within the dorsal and ventral body cavities with the rest of the body. These layers (1) provide strength and stability, (2) maintain the relative positions of internal organs, and (3) provide a route for the distribution of blood vessels, lymphatic vessels, and nerves. **Fasciae** (FASH-ē-ē); singular, *fascia*) are connective tissue layers and wrappings that support and surround organs. We can divide the fasciae into three types of layers: the superficial fascia, the deep fascia, and the subserous fascia. The functional anatomy of these layers is illustrated in Figure 4–18●.

1. The **superficial fascia**, or **subcutaneous layer** (*sub*, below + *cutis*, skin) is also termed the *hypodermis* (*hypo*, below + *dermis*, skin). This layer of areolar tissue and fat separates the skin from underlying tissues and organs, provides insulation and padding, and lets the skin and underlying structures move independently.

2. The **deep fascia** consists of dense irregular connective tissue. The organization of the fibers resembles that of plywood: All the fibers in one layer run in the same direction, but the orientation of the fibers changes from layer to layer. This arrangement helps the tissue resist forces applied from many directions. The tough capsules that surround most organs, including the kidneys and the organs in the thoracic and peritoneal cavities, are bound to the deep fascia. The perichondrium around cartilages, the periosteum around bones and the ligaments that interconnect them, and the connective tissues of muscle, including tendons, are also connected to the deep fascia. The dense connective tissue components are interwoven. For example, the deep fascia around a muscle blends into the tendon, whose fibers intermingle with those of the periosteum. This arrangement creates a strong, fibrous network and ties structural elements together.

3. The **subserous fascia** is a layer of areolar tissue that lies between the deep fascia and the serous membranes that line body cavities. Because this layer separates the serous membranes from the deep fascia, movements of muscles or muscular organs do not severely distort the delicate lining.

✓ Which cavities in the body are lined by serous membranes?

✓ The lining of the nasal cavity is normally moist, contains numerous goblet cells, and rests on a layer of connective tissue called the lamina propria. Which type of membrane is this?

✓ A sheet of tissue has many layers of collagen fibers that run in different directions in successive layers. Which type of tissue is this?

Answers start on page Q-1

4–6 MUSCLE TISSUE

Objective

■ Describe the three types of muscle tissue and the special structural features of each type.

Epithelia cover surfaces and line passageways; connective tissues support weight and interconnect parts of the body. Together, these tissues provide a strong, interwoven framework within which the organs of the body can function. Several vital functions involve movement of one kind or another—movement of materials along the digestive tract, movement of blood around the cardiovascular system, or movement of the body from one place to another. Movement is produced by a **muscle tissue,**

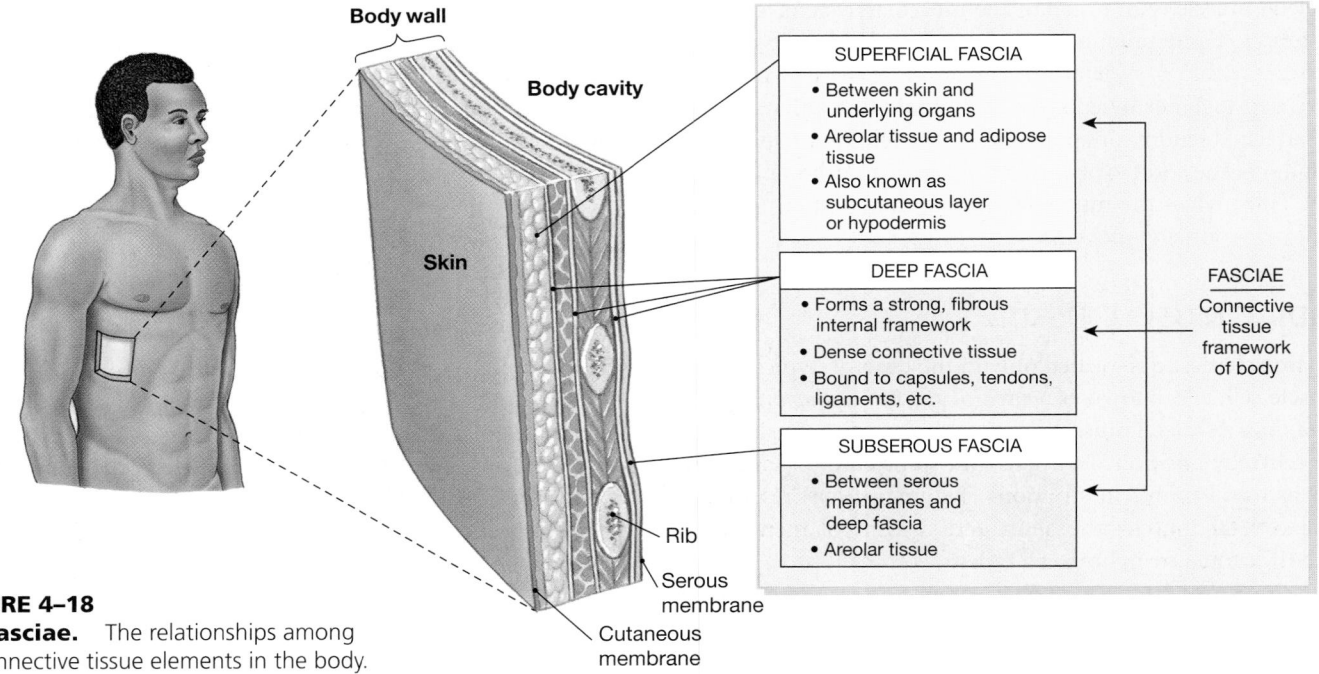

●**FIGURE 4–18**
The Fasciae. The relationships among the connective tissue elements in the body.

which is specialized for contraction. Muscle cells possess organelles and properties distinct from those of other cells. There are three types of muscle tissue: (1) *skeletal muscle*, which forms the large skeletal muscles responsible for gross body movements and locomotion; (2) *cardiac muscle*, found only in the heart and responsible for the circulation of blood; and (3) *smooth muscle*, found in the walls of visceral organs and a variety of other locations, where it provides elasticity, contractility, and support. The contraction mechanism is similar in all three types of muscle tissue, but the muscle cells differ in their internal organization. We shall examine only general characteristics at this point, because each type of muscle is described more fully in Chapter 10.

☐ SKELETAL MUSCLE TISSUE

Skeletal muscle tissue contains very large muscle cells—up to 0.3 meter (1 ft) or more in length. Because the individual muscle cells are relatively long and slender, they are usually called **muscle fibers**. Each muscle fiber is described as *multinucleate*, because, instead of having a single nucleus, it has several hundred distributed just inside the cell membrane (Figure 4–19a●). Skeletal muscle fibers are incapable of dividing, but new muscle fibers are produced through the divisions of **satellite cells**, stem cells that persist in adult skeletal muscle tissue. As a result, skeletal muscle tissue can at least partially repair itself after an injury.

As noted in Chapter 3, the cytoskeleton contains actin and myosin filaments. ∞ p. 75 In skeletal muscle fibers, however, these filaments are organized into repeating groups that give the cells a *striated*, or banded, appearance. The *striations*, or bands, are readily apparent in light micrographs. Skeletal muscle fibers do not usually contract unless stimulated by nerves, and the nervous system provides voluntary control over their activities. Thus, skeletal muscle is called **striated voluntary muscle**.

A skeletal muscle is an organ of the muscular system, and although muscle tissue predominates, it contains all four types of body tissue. Within a skeletal muscle, adjacent skeletal muscle fibers are tied together by collagen and elastic fibers that blend into the attached tendon or aponeurosis. The tendon or aponeurosis conducts the force of contraction, often to a bone of the skeleton. Thus, when the muscles contract, they pull on the attached bone, producing movement.

☐ CARDIAC MUSCLE TISSUE

Cardiac muscle tissue is located only in the heart. A typical cardiac muscle cell, also known as a **cardiocyte**, or *cardiac myocyte*, is smaller than a skeletal muscle cell. A typical cardiac muscle cell has one centrally positioned nucleus, but some cardiocytes have as many as five. Prominent striations (Figure 4–19b●) resemble those of skeletal muscle; the actin and myosin filaments are arranged the same way in both cell types.

Cardiac muscle cells form extensive connections with one another. The connections occur at specialized regions known as **intercalated discs**. At an intercalated disc, the membranes are locked together by desmosomes, intercellular cement, and gap junctions. As a result, cardiac muscle tissue consists of a branching network of interconnected muscle cells. The desmosomes and intercellular cement lock the cells together during a contraction. Ion movement through gap junctions helps coordinate the contractions of the cardiac muscle cells. Cardiac muscle tissue has a very limited ability to repair itself. Although some cardiac muscle cells do divide after an injury to the heart, the repairs are incomplete and some heart function is usually lost.

Cardiac muscle cells do not rely on nerve activity to start a contraction. Instead, specialized cardiac muscle cells called *pacemaker cells* establish a regular rate of contraction. Although the nervous system can alter the rate of pacemaker cell activity, it does not provide voluntary control over individual cardiac muscle cells. Therefore, cardiac muscle is called **striated involuntary muscle**.

☐ SMOOTH MUSCLE TISSUE

Smooth muscle tissue is located (1) in the walls of blood vessels, (2) around hollow organs such as the urinary bladder, and (3) in layers around the respiratory, circulatory, digestive, and reproductive tracts. A smooth muscle cell is a small, spindle-shaped cell with tapering ends and a single, oval nucleus (Figure 4–19c●). Smooth muscle cells can divide; hence, smooth muscle tissue can regenerate after an injury.

The actin and myosin filaments in smooth muscle cells are organized differently from those of skeletal and cardiac muscles. One result of this difference is that smooth muscle tissue has no striations. Smooth muscle cells may contract on their own, with gap junctions between adjacent cells coordinating the contractions of individual cells. The contraction of some smooth muscle tissue can be controlled by the nervous system, but contractile activity is not under voluntary control. Imagine the degree of effort that would be required to exert conscious control over the smooth muscles along the 8 m of digestive tract, not to mention the miles of blood vessels! Because the nervous system usually does not provide voluntary control over smooth muscle contractions, smooth muscle is known as **nonstriated involuntary muscle**.

4–7 NEURAL TISSUE

Objective

■ Discuss the basic structure and role of neural tissue.

Neural tissue, which is also known as *nervous tissue* or *nerve tissue*, is specialized for the conduction of electrical impulses from one region of the body to another. Ninety-eight percent of the neural tissue in the body is concentrated in the brain and spinal cord, which are the control centers of the nervous system.

Neural tissue contains two basic types of cells: (1) **neurons** (NOO-ronz; *neuro*, nerve) and (2) several kinds of supporting cells, collectively called **neuroglia** (noo-ROG-lē-uh or

SKELETAL MUSCLE TISSUE

Cells are long, cylindrical, striated, and multinucleate.

LOCATIONS: Combined with connective tissues and nervous tissue in skeletal muscles

FUNCTIONS: Moves or stabilizes the position of the skeleton; guards entrances and exits to the digestive, respiratory, and urinary tracts; generates heat; protects internal organs

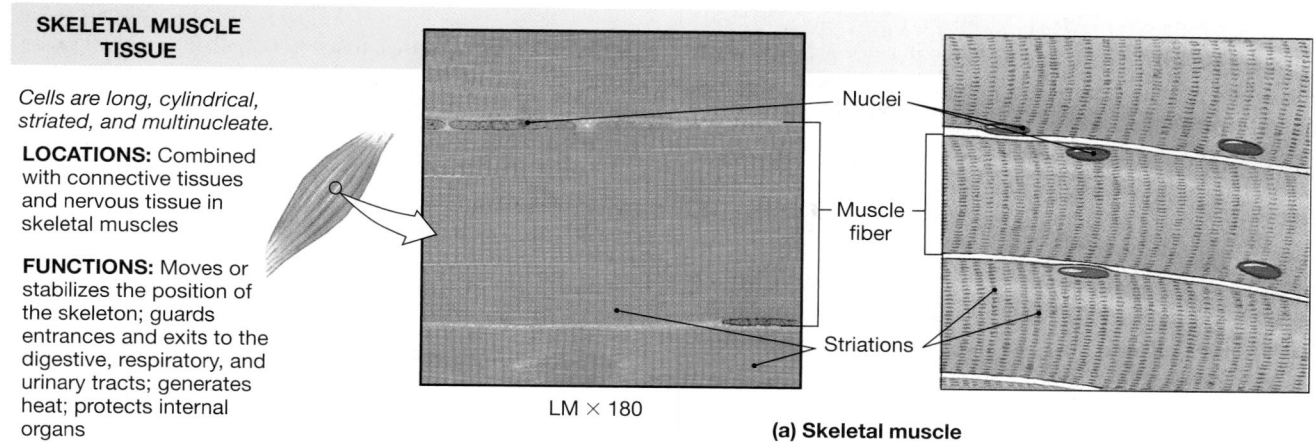

LM × 180

Nuclei

Muscle fiber

Striations

(a) Skeletal muscle

CARDIAC MUSCLE TISSUE

Cells are short, branched, and striated, usually with a single nucleus; cells are interconnected by intercalated discs.

LOCATION: Heart

FUNCTIONS: Circulates blood; maintains blood (hydrostatic) pressure

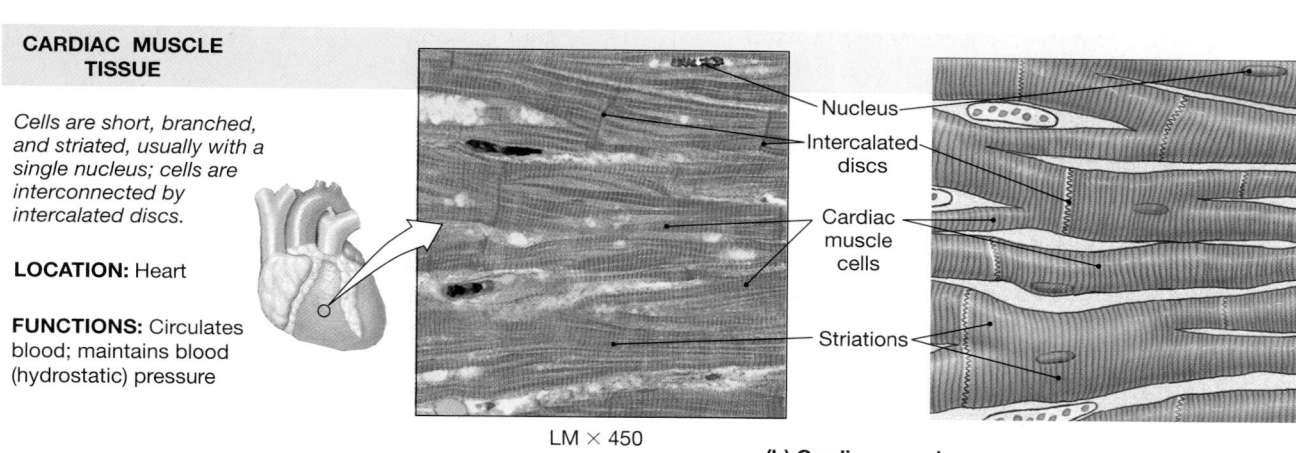

LM × 450

Nucleus

Intercalated discs

Cardiac muscle cells

Striations

(b) Cardiac muscle

SMOOTH MUSCLE TISSUE

Cells are short, spindle-shaped, and nonstriated, with a single, central nucleus

LOCATIONS: Encircles blood vessels; found in the walls of digestive, respiratory, urinary, and reproductive organs

FUNCTIONS: Moves food, urine, and reproductive tract secretions; controls diameter of respiratory passageways; regulates diameter of blood vessels

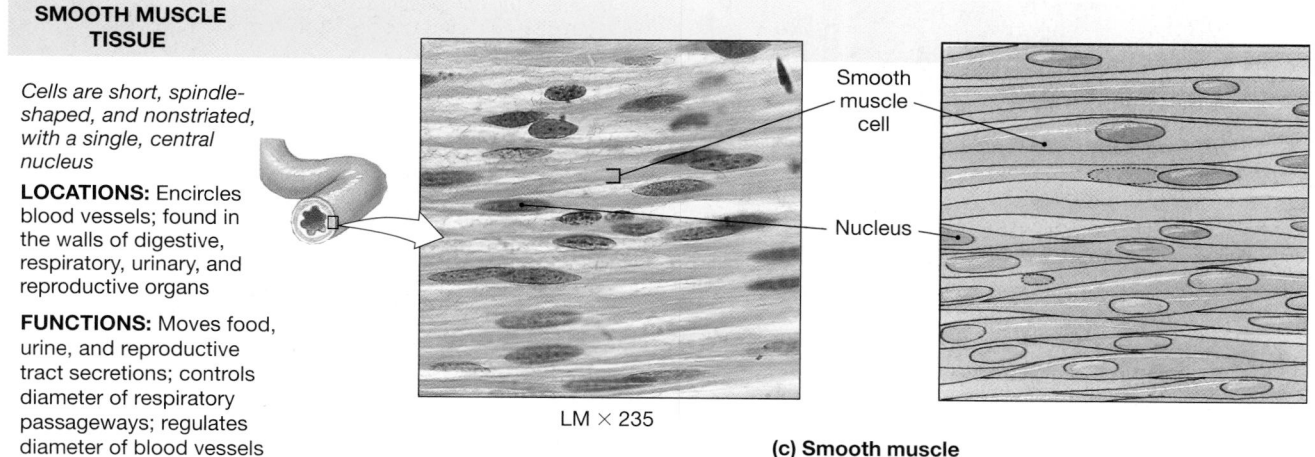

LM × 235

Smooth muscle cell

Nucleus

(c) Smooth muscle

●**FIGURE 4–19**
Muscle Tissue. **(a)** Skeletal muscle fibers are large, have multiple, peripherally located nuclei, and exhibit prominent striations (banding) and an unbranched arrangement. **(b)** Cardiac muscle cells differ from skeletal muscle fibers in three major ways: size (cardiac muscle cells are smaller), organization (cardiac muscle cells branch), and number and location of nuclei (a typical cardiac muscle cell has one centrally placed nucleus). Both skeletal and cardiac muscle cells contain actin and myosin filaments in an organized array that produces striations. **(c)** Smooth muscle cells are small and spindle shaped, with a central nucleus. They do not branch or have striations.

noo-rō-GLĒ-uh), or *glial cells* (*glia*, glue). Our conscious and unconscious thought processes reflect the communication among neurons in the brain. Such communication involves the propagation of electrical impulses, in the form of changes in the transmembrane potential. Information is conveyed both by the frequency and by the pattern of the impulses. Neuroglia support and repair neural tissue and supply nutrients to neurons.

The longest cells in your body are neurons, many of which are as much as a meter (39 in.) long. Most neurons cannot divide under normal circumstances, so they have a very limited ability to repair themselves after injury. A typical neuron has a large **cell body** with a large nucleus and a prominent nucleolus (Figure 4–20●). Extending from the cell body are many branching processes (projections or outgrowths) termed **dendrites** (DEN-drīts; *dendron*, a tree) and one **axon**. The dendrites receive information, typically from other neurons, and the axon carries that information to other cells. Because axons tend to be very long and slender, they are also called **nerve fibers**. In Chapter 12, we will further examine the properties of neural tissue.

✓ Which type of muscle tissue has small, tapering cells with single nuclei and no obvious striations?

✓ A tissue contains irregularly shaped cells with many fibrous projections, some several centimeters long. These are probably which type of cell?

✓ If skeletal muscle cells in adults are incapable of dividing, how is new skeletal muscle formed?

Answers start on page Q-1

4–8 TISSUE INJURIES AND AGING

Objective

■ Describe how injuries and aging affect the tissues of the body.

Tissues are not isolated; they combine to form organs with diverse functions. Therefore, any injury affects several types of tissue simultaneously. These tissues must respond in a coordinated way to preserve homeostasis.

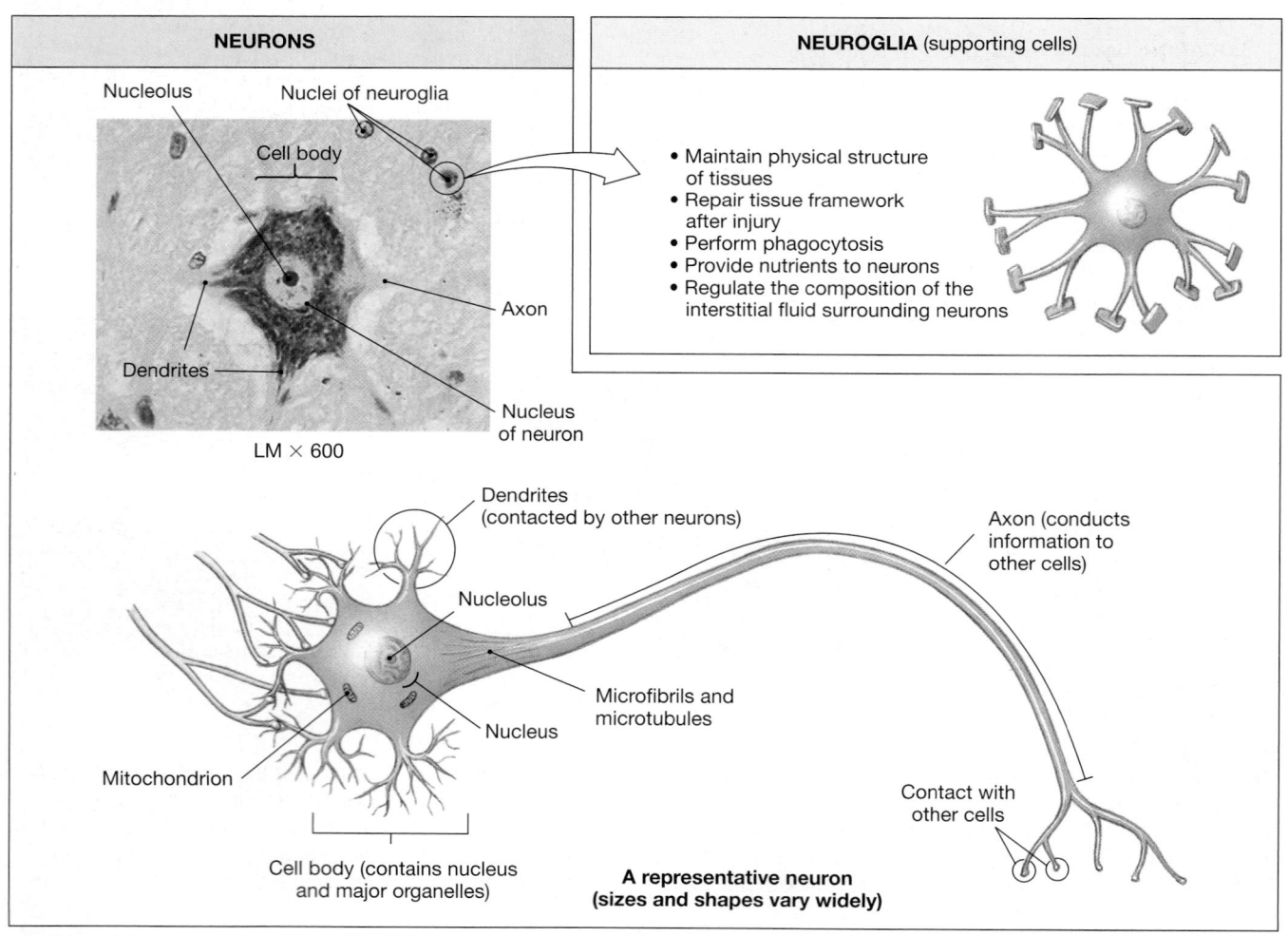

●**FIGURE 4–20**
Neural Tissue

☐ INFLAMMATION AND REGENERATION

The restoration of homeostasis after a tissue has been injured involves two related processes: inflammation and repair. First, immediately after the injury, the area is isolated while damaged cells, tissue components, and any dangerous microorganisms are cleaned up. This phase, which coordinates the activities of several types of tissue, is called **inflammation**, or the **inflammatory response**. It produces several familiar sensations, including swelling, redness, warmth, and pain. An inflammation resulting from the presence of pathogens, such as harmful bacteria, is called an **infection**.

Second, the damaged tissues are replaced or repaired to restore normal function. The repair process is called **regeneration**. Inflammation and regeneration are controlled at the tissue level. The two phases overlap; isolation establishes a framework that guides the cells responsible for reconstruction, and repairs are under way well before cleanup operations have ended.

We will now consider the basics of the repair process after an injury. At this time, we shall focus on the interaction among different tissues. Our example includes two connective tissues (areolar tissue and blood), an epithelium (the endothelia of blood vessels), a muscle tissue (smooth muscle in the vessel walls), and neural tissue (sensory nerve endings). In later chapters, especially Chapters 5 and 22, we will examine inflammation and regeneration in more detail.

First Phase: Inflammation

Many stimuli—including impact, abrasion, distortion, chemical irritation, infection by pathogenic organisms (such as bacteria or viruses), and extreme temperatures (hot or cold)—can produce inflammation. Each of these stimuli kills cells, damages fibers, or injures the tissue in some other way. Such changes alter the chemical composition of the interstitial fluid: Damaged cells release prostaglandins, proteins, and potassium ions, and the injury itself may have introduced foreign proteins or pathogens into the body.

Tissue conditions soon become even more abnormal. **Necrosis** (ne-KRŌ-sis), the tissue degeneration that occurs after cells have been hurt or destroyed, begins several hours after the original injury. The damage is caused by lysosomal enzymes. Through widespread autolysis, lysosomes release enzymes that first destroy the injured cells and then attack surrounding tissues. ∞ p. 80 The result may be an accumulation of debris, fluid, dead and dying cells, and necrotic tissue components collectively known as **pus**. An accumulation of pus in an enclosed tissue space is an **abscess**.

These tissue changes trigger the inflammatory response by stimulating mast cells—connective tissue cells introduced on page 124. Figure 4–21● follows the events set in motion by the activation of mast cells. Although in this example the injury has occurred in areolar tissue, the process would be basically the same after an injury to any connective tissue proper. Because all organs have connective tissues, inflammation can occur anywhere in the body.

When an injury occurs that damages fibers and cells, mast cells release a variety of chemicals. These chemicals, including histamine and prostaglandins, trigger changes in local circulation. In response, the smooth muscle tissue that surrounds local blood vessels relaxes, and the vessels *dilate*, or enlarge in diameter. This dilation increases blood flow through the tissue, turning the re-

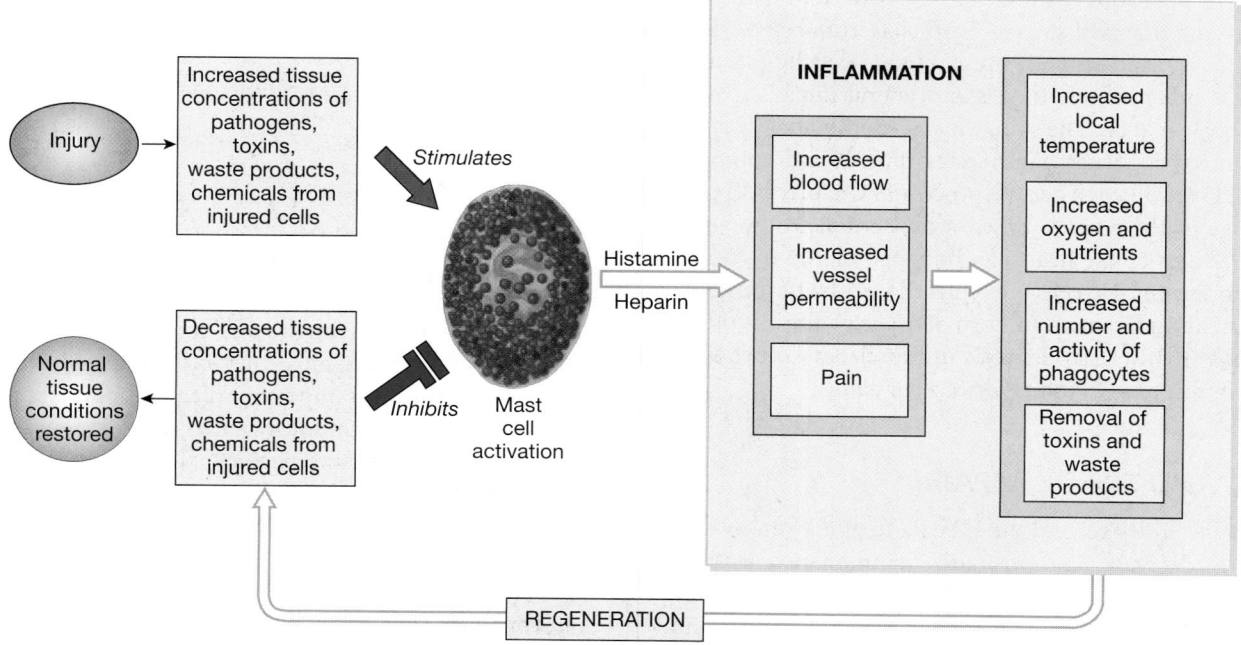

●**FIGURE 4–21**
An Introduction to Inflammation

gion red and making it warm to the touch. The combination of abnormal tissue conditions and chemicals released by mast cells stimulates sensory nerve endings that produce sensations of pain. At the same time, the chemicals released by mast cells make the endothelial cells of local capillaries more permeable. Plasma, including blood proteins, now diffuses into the injured tissue, so the area becomes swollen.

The increased blood flow accelerates the delivery of nutrients and oxygen and the removal of dissolved waste products and toxic chemicals. It also brings white blood cells to the region. These phagocytic cells migrate to the site of the injury and assist in defense and cleanup operations. Macrophages and microphages protect the tissue from infection and perform cleanup by engulfing both debris and bacteria. Over a period of hours to days, the cleanup process generally succeeds in eliminating the inflammatory stimulus.

Second Phase: Regeneration

By the time the inflammation phase is over, the situation is under control and no further damage will occur. However, many cells in the area either have already died or will soon die as a consequence of the original injury. As tissue conditions return to normal, fibroblasts move into the necrotic area, laying down a network of collagen fibers that stabilizes the injury site. This process produces a dense, collagenous framework known as *scar tissue* or *fibrous tissue*. Over time, scar tissue is usually "remodeled" and gradually assumes a more normal appearance. The cell population in the area gradually increases; some cells migrate to the site, and others are produced by the division of mesenchymal stem cells.

Each organ has a different ability to regenerate after injury—an ability that can be directly linked to the pattern of tissue organization in the injured organ. Epithelia, connective tissues (except cartilage), and smooth muscle tissue usually regenerate well, whereas other muscle tissues and neural tissue regenerate relatively poorly if at all. The skin, which is dominated by epithelia and connective tissues, regenerates rapidly and completely after injury. (We will consider the process in Chapter 5.) In contrast, damage to the heart is much more serious. Although the connective tissues of the heart can be repaired, the majority of damaged cardiac muscle cells are replaced only by fibrous tissue. The permanent replacement of normal tissue by fibrous tissue is called *fibrosis* (fī-BRŌ-sis). Fibrosis in muscle and other tissues may occur in response to injury, disease, or aging.

□ AGING AND TISSUE REPAIR

Tissues change with age, and the speed and effectiveness of tissue repairs decrease. Repair and maintenance activities throughout the body slow down; the rate of energy consumption in general declines. All these changes reflect various hor-

monal alterations occurring with age, often coupled with a reduction in physical activity and the adoption of a more sedentary lifestyle. These factors combine to alter the structure and chemical composition of many tissues. Epithelia get thinner and connective tissues more fragile. Individuals bruise easily and bones become brittle; joint pain and broken bones are common in the elderly. Because cardiac muscle cells and neurons cannot be replaced, cumulative damage can eventually cause major health problems, such as cardiovascular disease or a deterioration in mental functioning.

In later chapters, we will consider the effects of aging on specific organs and systems. Some of these effects are genetically programmed. For example, the chondrocytes of older individuals produce a slightly different form of proteoglycan than do those of younger people. This difference probably accounts for the thinner and less resilient cartilage of older people. In some cases, the tissue degeneration can be temporarily slowed or even reversed. The age-related reduction in bone strength, a condition called *osteoporosis*, typically results from a combination of inactivity, low dietary calcium levels, and a reduction in circulating sex hormones. A program of exercise, calcium supplements, and hormone replacement therapies can generally maintain healthy bone structure for many years.

□ AGING AND CANCER INCIDENCE

Cancer rates increase with age, and roughly 25 percent of all people in the United States develop cancer at some point in their lives. It has been estimated that 70–80 percent of cancer cases result from chemical exposure, environmental factors, or some combination of the two, and 40 percent of those cancers are caused by cigarette smoke. Each year in the United States, more than 500,000 individuals die of cancer, making it second only to heart disease as a cause of death. We discussed the development and growth of cancer in Chapter 3. ∞ p. 104
AM Cancer: A Closer Look

This chapter concludes the introductory portion of this text. In combination, the four basic tissue types described here form all of the organs and systems discussed in subsequent chapters. The *Systems Overview* that follows this chapter will help you make the transition from atoms, molecules, cells, and tissues to organ systems. One of the most important themes in this text is that organ systems interact continuously—they do not function in isolation. Thus, to understand specifics about one system, you need to know something about all of the others. The *Systems Overview* section provides a general orientation in more detail than was possible in Chapter 1. You will find this section useful as a reference throughout the rest of the text when you need some quick information about a particular system or region.

Tissue Structure and Disease

Physicians who specialize in the study of disease processes are called **pathologists** (pa-THOL-o-jists). Diagnosis, rather than treatment, is usually the main focus of their activities. In their analyses, pathologists integrate anatomical and histological observations to determine the nature and severity of a disease. Because disease processes affect the histological organization of tissues and organs, **biopsies**, or tissue samples, often play a key role in their diagnoses.

Figure 4–22● diagrams the histological changes induced by one relatively common irritating stimulus, cigarette smoke. The first abnormality to be observed is **dysplasia** (dis-PLĀ-zē-uh), a change in the normal shape, size, and organization of tissue cells. Dysplasia is generally a response to chronic irritation or inflammation, and the changes are reversible. The normal trachea (windpipe) and its branches are lined by a pseudostratified ciliated columnar epithelium. The cilia move a mucous layer that traps foreign particles and moistens incoming air. The drying and chemical effects of smoking first paralyze the cilia, halting the movement of mucus (Figure 4–22a●). As mucus builds up, the individual coughs to dislodge it (the well-known "smoker's cough").

Epithelia and connective tissues may undergo more radical changes in structure, caused by the division and differentiation of stem cells. **Metaplasia** (me-tuh-PLĀ-zē-uh) is a structural change that dramatically alters the character of the tissue. In our heavy-smoking example, over time the epithelial cells lose their cilia altogether.

As metaplasia progresses, the epithelial cells produced by stem cell divisions no longer differentiate into ciliated columnar cells. Instead, they form a stratified squamous epithelium that provides greater resistance to drying and chemical irritation (Figure 4–22b●). This epithelium protects the underlying tissues more effectively, but it eliminates the moisturization and cleaning properties of the epithelium. Cigarette smoke will now have an even greater effect on more delicate portions of the respiratory tract. Fortunately, metaplasia is reversible, and the epithelium gradually returns to normal if the individual quits smoking.

In **anaplasia** (a-nuh-PLĀ-zē-uh), tissue organization breaks down. Tissue cells change size and shape, typically becoming unusually large or small (Figure 4–22c●). Anaplasia occurs in smokers who develop one form of lung cancer; the cells divide more frequently, but not all divisions proceed in the normal way. Many of the tumor cells have abnormal chromosomes. Unlike dysplasia and metaplasia, anaplasia is irreversible.

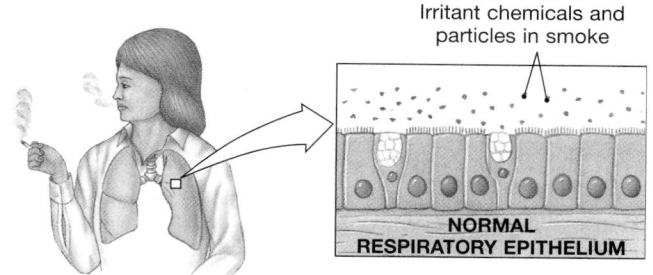

Irritant chemicals and particles in smoke

NORMAL RESPIRATORY EPITHELIUM

Reversible

(a) The cilia of respiratory epithelial cells are damaged and paralyzed by exposure to cigarette smoke. These changes cause the local buildup of mucus and reduce the effectiveness of the epithelium in protecting deeper, more delicate portions of the respiratory tract.

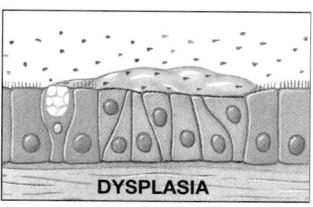
DYSPLASIA

Reversible

(b) In metaplasia, a tissue changes its structure. In this case the stressed respiratory surface converts to a stratified epithelium that protects underlying connective tissues but does nothing for other areas of the respiratory tract.

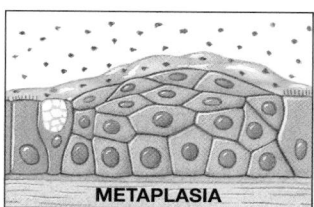
METAPLASIA

Irreversible

(c) In anaplasia, the tissue cells become tumor cells; anaplasia produces a cancerous tumor.

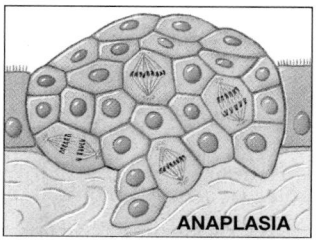
ANAPLASIA

●**FIGURE 4–22**
Changes in a Tissue under Stress

Chapter Review

SELECTED CLINICAL TERMINOLOGY

Terms Discussed in This Chapter

abscess: The accumulation of pus within an enclosed tissue space. *(p. 139)*

adhesions: Restrictive fibrous connections that can result from surgery, infection, or other injuries to serous membranes. *(p. 134)*

anaplasia (a-nuh-PLĀ-zē-uh): An irreversible change in the size and shape of tissue cells. *(p. 141)*

antiangiogenesis factor: A secretion, produced by chondrocytes, that inhibits the growth of blood vessels. *(p. 129)*

ascites (a-SĪ-tēz): The accumulation of fluid in the peritoneal cavity, usually caused by liver or kidney disease or heart failure. *(p. 134)*

dysplasia (dis-PLĀ-zē-uh): A change in the normal shape, size, and organization of tissue cells. *(p. 141)*

exfoliative cytology: The study of cells shed or collected from epithelial surfaces. *(p. 118)*

liposuction: A surgical procedure to remove unwanted adipose tissue by sucking it out through a tube. *(p. 126)*

metaplasia (me-tuh-PLĀ-zē-uh): A structural change that alters the character of a tissue. *(p. 141)*

necrosis: Tissue degeneration that occurs after cells have been injured or destroyed; a result of the release of lysosomal enzymes through autolysis. *(p. 139)*

pathologists (pa-THOL-o-jists): Physicians who specialize in the study of disease processes. *(p. 141)*

pericarditis: An inflammation of the pericardial lining that may lead to the accumulation of pericardial fluid (a *pericardial effusion*). *(p. 134)*

peritonitis: An inflammation of the peritoneum after infection or injury. *(p. 134)*

pleural effusion: The accumulation of fluid within the pleural cavities as a result of chronic infection or inflammation of the pleura. *(p. 134)*

pleuritis *(pleurisy)*: An inflammation of the pleural cavities. This condition may cause the production of a sound known as a *pleural rub. (p. 134)*

regeneration: The repairing of injured tissues that follows inflammation. *(p. 139)*

AM Additional Terms Discussed in the *Applications Manual*

chemotherapy
immunotherapy
oncologists
remission

STUDY OUTLINE

4–1 TISSUES OF THE BODY: AN INTRODUCTION   p. 111

1. **Tissues** are collections of specialized cells and cell products that are organized to perform a relatively limited number of functions. The four *tissue types* are *epithelial tissue, connective tissue, muscle tissue,* and *neural tissue.*
2. **Histology** is the study of tissues.

4–2 EPITHELIAL TISSUE   p. 112

1. **Epithelial tissue** includes epithelia and glands. An **epithelium** is an **avascular** layer of cells which forms a barrier that provides protection and regulates permeability. **Glands** are secretory structures derived from epithelia. Epithelial cells may show **polarity**, an uneven distribution of cytoplasmic components.

FUNCTIONS OF EPITHELIAL TISSUE   p. 112

2. Epithelia furnish physical protection, control permeability, provide sensation, and produce specialized secretions. Gland cells are epithelial cells that produce secretions. In **glandular epithelia**, most cells produce secretions.
3. A **basal lamina** attaches epithelia to underlying connective tissues.

SPECIALIZATIONS OF EPITHELIAL CELLS   p. 112

4. Epithelial cells are specialized to allow them to perform secretory or transport functions and to maintain the physical integrity of the epithelium. *(Figure 4–1)*
5. Many epithelial cells have microvilli.
6. The coordinated beating of the cilia on a **ciliated epithelium** moves materials across the epithelial surface.

MAINTAINING THE INTEGRITY OF EPITHELIA   p. 113

7. Cells can attach to other cells or to extracellular protein fibers by means of **cell adhesion molecules** (CAMs) or at specialized attachment sites called **cell junctions**. The three major types of cell junctions are **tight junctions**, **desmosomes**, and **gap junctions**. *(Figure 4–2)*
8. The inner surface of each epithelium is connected to a two-part basal lamina consisting of a *lamina lucida* and a *lamina densa*. Divisions by **germinative cells** continually replace the short-lived epithelial cells.

CLASSIFICATION OF EPITHELIA   p. 115

9. Epithelia are classified on the basis of the number of cell layers and the shape of the cells at the apical surface.
10. A **simple epithelium** has a single layer of cells covering the basal lamina; a **stratified epithelium** has several layers. The cells in a **squamous epithelium** are thin and flat. Cells in a **cuboidal epithelium** resemble little hexagonal boxes; those in a **columnar epithelium** are taller and more slender. *(Figures 4–3 to 4–5)*

GLANDULAR EPITHELIA   p. 118

11. **Exocrine glands** discharge secretions onto the body surface or into **ducts**, which communicate with the exterior. *Hormones*, the secretions of **endocrine glands**, are released by gland cells into the surrounding interstitial fluid.
12. A glandular epithelial cell may release its secretions by merocrine, apocrine, or holocrine modes. In **merocrine secretion**, the most common mode, the product is released through exocytosis. **Apocrine secretion** involves the loss of both secretory product and cytoplasm. Unlike the other two methods,

holocrine secretion destroys the cell, which becomes packed with secretions and bursts. *(Figure 4–6)*

13. In epithelia that contain scattered gland cells, individual secretory cells are called **unicellular glands. Multicellular glands** are organs which contain glandular epithelia that produce exocrine or endocrine secretions.

14. Exocrine glands can be classified on the basis of structure as **unicellular exocrine glands** (**goblet cells**) or as **multicellular exocrine glands.** Multicellular exocrine glands can be further classified according to structure. *(Figure 4–7)*

Epithelial tissue classification: **Companion Website:** Chapter 4/Reviewing Concepts/Labeling.

4–3 CONNECTIVE TISSUES   p. 122

1. **Connective tissues** are internal tissues with many important functions: establishing a structural framework; transporting fluids and dissolved materials; protecting delicate organs; supporting, surrounding, and interconnecting tissues; storing energy reserves; and defending the body from microorganisms. *(Figure 4–8)*

2. All connective tissues contain both specialized cells and a **matrix**, composed of extracellular protein fibers and a **ground substance.**

CLASSIFICATION OF CONNECTIVE TISSUES   p. 122

3. **Connective tissue proper** is connective tissue that contains varied cell populations and fiber types surrounded by a syrupy ground substance.

4. **Fluid connective tissues** have a distinctive population of cells suspended in a watery ground substance that contains dissolved proteins. The two types of fluid connective tissues are *blood* and *lymph.*

5. **Supporting connective tissues** have a less diverse cell population than connective tissue proper does and a dense ground substance with closely packed fibers. The two types of supporting connective tissues are *cartilage* and *bone.*

CONNECTIVE TISSUE PROPER   p. 122

6. Connective tissue proper contains fibers, a viscous ground substance, and a varied population of cells, including **fibroblasts, macrophages, adipocytes, mesenchymal cells, melanocytes, mast cells, lymphocytes,** and **microphages.**

7. The three types of fibers in connective tissue are **collagen fibers, reticular fibers,** and **elastic fibers.**

8. The first connective tissue to appear in an embryo is **mesenchyme,** or *embryonic connective tissue.*

9. Connective tissue proper is classified as **loose connective tissue** or **dense connective tissue.** Loose connective tissues are **mesenchyme** and **mucous connective tissues** in the embryo, **areolar tissue, adipose tissue,** including **white fat** and **brown fat,** and **reticular tissue.** Most of the volume in dense connective tissue consists of fibers. The two types of dense connective tissue are **dense regular connective tissue** and **dense irregular connective tissue** in the adult. *(Figures 4–9 to 4–12)*

FLUID CONNECTIVE TISSUES   p. 128

10. **Blood** and **lymph** are connective tissues that contain distinctive collections of cells in a fluid matrix.

11. Blood contains *formed elements*: **red blood cells** (*erythrocytes*), **white blood cells** (*leukocytes*), and **platelets.** The watery matrix of blood is called **plasma.** *(Figure 4–13)*

12. **Arteries** carry blood from the heart and toward **capillaries,** where water and small solutes move into the interstitial fluid of surrounding tissues. **Veins** return blood to the heart.

13. Lymph forms as interstitial fluid enters the **lymphatic vessels,** which return lymph to the cardiovascular system.

SUPPORTING CONNECTIVE TISSUES   p. 129

14. Cartilage and bone are called supporting connective tissues because they support the rest of the body.

15. **Cartilage** grows by two mechanisms: **interstitial growth** and **appositional growth.** *(Figure 4–14)*

16. The matrix of cartilage is a firm gel that contains **chondroitin sulfates** (used to form proteoglycans) and cells called **chondrocytes.** Chondrocytes occupy chambers called **lacunae.** A fibrous **perichondrium** separates cartilage from surrounding tissues. The three types of cartilage are **hyaline cartilage, elastic cartilage,** and **fibrocartilage.** *(Figure 4–15)*

17. Chondrocytes rely on diffusion through the avascular matrix to obtain nutrients.

18. **Bone,** or **osseous tissue,** has **osteocytes,** little ground substance, and a dense, mineralized matrix. Osteocytes are situated in lacunae. The matrix consists of calcium salts and collagen fibers, giving it unique properties. *(Figure 4–16; Table 4–1)*

19. Osteocytes depend on diffusion through **canaliculi** for nutrient intake.

20. Each bone is surrounded by a **periosteum** with fibrous and cellular layers.

Connective tissue classification: **Companion Website:** Chapter 4/Reviewing Concepts/Labeling.

4–4 MEMBRANES   p. 132

1. Membranes form a barrier or interface. Epithelia and connective tissues combine to form membranes that cover and protect other structures and tissues. The four types of membranes are *mucous, serous, cutaneous,* and *synovial. (Figure 4–17)*

MUCOUS MEMBRANES   p. 132

2. **Mucous membranes** line cavities that communicate with the exterior. They contain areolar tissue called the **lamina propria.** *(Figure 4–17)*

SEROUS MEMBRANES   p. 133

3. **Serous membranes** line the body's sealed internal cavities. They form a fluid called a *transudate. (Figure 4–17)*

THE CUTANEOUS MEMBRANE   p. 134

4. The **cutaneous membrane,** or skin, covers the body surface. *(Figure 4–17)*

SYNOVIAL MEMBRANES   p. 134

5. **Synovial membranes** form an incomplete lining within the cavities of synovial joints. *(Figure 4–17)*

4–5 THE CONNECTIVE TISSUE FRAMEWORK OF THE BODY   p. 134

1. Internal organs and systems are tied together by a network of connective tissue proper. This network consists of the **superficial fascia** (the **subcutaneous layer,** or **hypodermis,** separating the skin from underlying tissues and organs), the **deep fascia** (dense connective tissue), and the **subserous fascia** (the layer between the deep fascia and the serous membranes that line body cavities). *(Figure 4–18)*

4–6 MUSCLE TISSUE   p. 135

1. **Muscle tissue** is specialized for contraction. The three types of muscle tissue are *skeletal muscle, cardiac muscle,* and *smooth muscle. (Figure 4–19)*

SKELETAL MUSCLE TISSUE   p. 136

2. The cells of **skeletal muscle tissue** are **multinucleated**. Skeletal muscle, or **striated voluntary muscle**, produces new fibers by the division of **satellite cells**.

CARDIAC MUSCLE TISSUE   p. 136

3. **Cardiocytes**, the cells of **cardiac muscle tissue**, occur only in the heart. Cardiac muscle, or **striated involuntary muscle**, relies on *pacemaker cells* for regular contraction.

SMOOTH MUSCLE TISSUE   p. 136

4. **Smooth muscle tissue**, or **nonstriated involuntary muscle**, is not striated. Smooth muscle cells can divide and therefore regenerate after injury has occurred.

4–7   NEURAL TISSUE   p. 136

1. **Neural tissue** conducts electrical impulses, which convey information from one area of the body to another.
2. Cells in neural tissue are either neurons or neuroglia. **Neurons** transmit information as electrical impulses. Several kinds of **neuroglia** exist, but their basic functions include supporting neural tissue and helping supply nutrients to neurons. *(Figure 4–20)*

3. A typical neuron has a **cell body**, **dendrites**, and an **axon**, or **nerve fiber**. The axon carries information to other cells.

4–8   TISSUE INJURIES AND AGING   p. 138

INFLAMMATION AND REGENERATION   p. 139

1. Any injury affects several types of tissue simultaneously, and they respond in a coordinated manner. Homeostasis is restored by two processes: inflammation and regeneration.
2. **Inflammation**, or the **inflammatory response**, isolates the injured area while damaged cells, tissue components, and any dangerous microorganisms (which could cause **infection**) are cleaned up. **Regeneration** is the repair process that restores normal function. *(Figure 4–21)*

AGING AND TISSUE REPAIR   p. 140

3. Tissues change with age. Repair and maintenance become less efficient, and the structure and chemical composition of many tissues are altered.

AGING AND CANCER INCIDENCE   p. 140

4. The incidence of cancer increases with age, with roughly three-quarters of all cases caused by exposure to chemicals or by other environmental factors, such as cigarette smoke. *(Figure 4–22)*

REVIEW QUESTIONS

 More assessment questions are available to you on the Companion Website. You will find Matching, Multiple Choice, True/False, and other quizzes to help further your understanding of the material covered in this chapter. To access the site, go to www.aw.com/martini.

LEVEL 1 Reviewing Facts and Terms

Match each numbered item with the most closely related lettered item. Use letters for answers in the spaces provided.

_____ 1. histology
_____ 2. microvilli
_____ 3. gap junction
_____ 4. tight junction
_____ 5. ground substance
_____ 6. lamina lucida and lamina densa
_____ 7. germinative cells
_____ 8. mesothelium
_____ 9. endothelium
_____ 10. mucus
_____ 11. destroys gland cells
_____ 12. hormones
_____ 13. goblet cells
_____ 14. adipocytes
_____ 15. macrophages
_____ 16. mast cells
_____ 17. bone-to-bone attachment
_____ 18. muscle-to-bone attachment
_____ 19. skeletal muscle
_____ 20. cardiac muscle

a. hyaluronan
b. lines heart and blood vessels
c. repair and renewal
d. ligament
e. endocrine secretion
f. merocrine secretion
g. absorption and secretion
h. fat cells
i. holocrine secretion
j. study of tissues
k. unicellular exocrine glands
l. tendon
m. intercellular connection
n. histamine and heparin
o. interlocking of membrane proteins
p. lines ventral body cavities
q. intercalated discs
r. striated, voluntary
s. defense and repair
t. basal lamina

21. The four basic types of tissue in the body are
    (a) epithelial, connective, muscle, and neural
    (b) simple, cuboidal, squamous, and stratified
    (c) fibroblasts, adipocytes, melanocytes, and mesenchyme
    (d) lymphocytes, macrophages, microphages, and adipocytes

22. Which of the following structures would you expect to see on the apical surface of an epithelium?
    (a) thick filaments
    (b) mitochondria
    (c) cilia
    (d) a, b, and c are correct

23. A type of junction common in cardiac and smooth muscle tissues is the
    (a) desmosome   (b) basal lamina
    (c) tight junction   (d) gap junction

24. The most abundant connections between cells in the superficial layers of the skin are
    (a) connexons   (b) gap junctions
    (c) desmosomes   (d) tight junctions

25. _____ membranes have an epithelium that is stratified and supported by dense connective tissue.
    (a) Synovial   (b) Serous
    (c) Cutaneous   (d) Mucous

26. Mucous secretions that coat the passageways of the digestive and respiratory tracts result from _____ secretion.
    (a) apocrine   (b) merocrine
    (c) holocrine   (d) endocrine

27. Matrix is a characteristic of which type of tissue?
    (a) epithelial   (b) neural
    (c) muscle   (d) connective

28. The three basic types of fibers in connective tissue are
    (a) tendons, ligaments, and elastic ligaments
    (b) loose, dense, and irregular
    (c) cartilage, bone, and collagen
    (d) collagen, reticular, and elastic

29. Which of the following epithelia most easily permits diffusion?
    (a) stratified squamous   (b) simple squamous
    (c) transitional   (d) simple columnar

30. Two examples of connective tissue proper are
    (a) cartilage and bone
    (b) elastic tissue and bone
    (c) areolar tissue and adipose tissue
    (d) collagen and fibrin

31. The three major types of cartilage in the body are
    (a) collagen, reticular, and elastic
    (b) areolar, adipose, and reticular
    (c) hyaline, elastic, and fibrocartilage
    (d) tendons, reticular, and elastic

32. The primary function of serous membranes in the body is to
    (a) minimize friction between opposing surfaces
    (b) line cavities that communicate with the exterior
    (c) perform absorptive and secretory functions
    (d) cover the surface of the body

33. The type of cartilage growth characterized by adding new layers of cartilage to the surface is
    (a) interstitial growth   (b) appositional growth
    (c) intramembranous growth   (d) longitudinal growth

34. Intercalated discs and pacemaker cells are characteristic of
    (a) smooth muscle tissue   (b) cardiac muscle tissue
    (c) skeletal muscle tissue   (d) a, b, and c are correct

35. Axons, dendrites, and a cell body are characteristic of cells located in
    (a) neural tissue   (b) muscle tissue
    (c) connective tissue   (d) epithelial tissue

36. The repair process necessary to restore normal function in damaged tissues is
    (a) isolation   (b) regeneration
    (c) reconstruction   (d) a, b, and c are correct

37. What are the four major characteristics of epithelial tissue?

38. What are the four essential functions of epithelial tissue?

39. Differentiate between endocrine and exocrine glands.

40. What three cell shapes describe almost every epithelium in the body?

41. What three methods do various glandular epithelial cells use to release their secretions?

42. List three basic components of connective tissues.

43. What six basic functions do connective tissues perform in the body?

44. What are the three types of fibers in connective tissues?

45. Give four major examples of dense regular connective tissue.

46. What are the four kinds of membranes composed of epithelial and connective tissue that cover and protect other structures and tissues in the body?

47. What are the three types of muscle tissue in the body?

48. What two cell populations make up neural tissue? What is the function of each?

LEVEL 2 Reviewing Concepts

49. On surfaces of the body where mechanical stresses are severe, what would you expect to see as the dominant epithelium? Why?

50. Where would you expect to find a transitional epithelium?

51. Why is the presence of cilia on the respiratory surfaces important?

52. Why does holocrine secretion require continuous cell division?

53. What is the difference between an exocrine and an endocrine secretion?

54. A significant structural feature in the digestive system is the presence of tight junctions near the exposed surfaces of cells lining the digestive tract. Why are these junctions so important?

55. Why are dead skin cells generally shed in thick sheets rather than individually?

56. Describe the fluid connective tissues in the human body. Compare them with the supporting connective tissues. What are the main differences?

57. Why are infections always a serious threat after a severe burn or abrasion?

58. Compare the three types of muscle tissue. List three similarities and three differences among them.

59. After a weight-loss program, why is the lost weight often regained quickly in the same areas of the body?

60. Differentiate among the three major types of cartilage in the body.

61. Cartilage heals poorly and in many instances does not heal or recover at all after a severe injury. Why not?

62. What characteristics make the cutaneous membrane different from serous and mucous membranes?

63. What is a transudate? Where might one form, and what is its function?

64. Why is cardiac muscle tissue that has been damaged by injury or disease incapable of regeneration?

65. Where are you most likely to find elastic ligaments in the body, and why?

LEVEL 3 Critical Thinking and Clinical Applications

66. Assuming that you had the necessary materials to perform a detailed chemical analysis of body secretions, how could you determine whether a secretion was merocrine or apocrine?

67. After many years of chronic smoking, Mr. Butts is plagued by a hacking cough. Explain the causes of this cough.

68. A biology student accidentally loses the labels of two prepared slides she is studying. One is a slide of an intestine, the other of an esophagus. You volunteer to help her sort them out. How would you decide which slide is which?

69. You are asked to develop a two-step scheme that can be used to identify the three types of muscle tissue. What would the two steps be?

70. Using what you know about the tissues of the body, what tissues must be replaced in a total hip replacement.

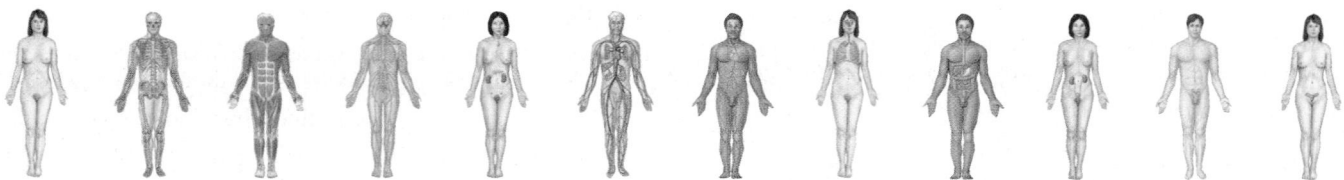

Systems Overview

Your perspective has gradually changed over the preceding chapters. Atoms can only be imagined or indirectly examined through experimental procedures. Cellular details often escape detection unless an electron microscope is used. Tissue structure, however, can be examined with a light microscope, and by this time you may already be able to identify some tissues with the unaided eye, based on your experiences in the laboratory. For example, once you have handled adipose tissue, with its lumpy, greasy texture, it would be difficult to mistake it for any other type of tissue.

Organs are combinations of tissues that perform complex functions. A great deal of information concerning organs and organ structure can be obtained by dissection and direct examination. In organ systems, several organs work together in a coordinated fashion. We can easily observe the functions of intact organ systems as they perform, direct, or moderate the activities of individual human beings. As a result, at the start of this course you probably knew much more about the major organs and systems than you did about cell and tissue structure.

Figure 1● presents four views of the composition of the human body, reflecting the changes in your perspective over the last four chapters. In Chapter 2 the body was treated as a collection of elements (Figure 1a●) that combine to form molecules. Cells, described in Chapter 3, are composed of organic molecules, inorganic molecules, ions, and water. Figure 1b● indicates the proportions of water and organic molecules in the body as a whole.

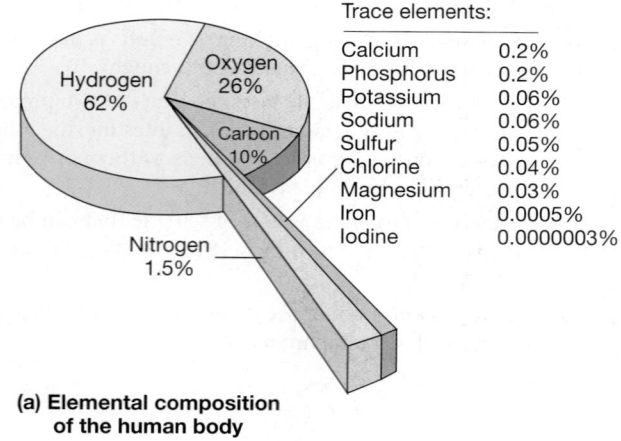

Trace elements:

Calcium	0.2%
Phosphorus	0.2%
Potassium	0.06%
Sodium	0.06%
Sulfur	0.05%
Chlorine	0.04%
Magnesium	0.03%
Iron	0.0005%
Iodine	0.0000003%

(a) Elemental composition of the human body

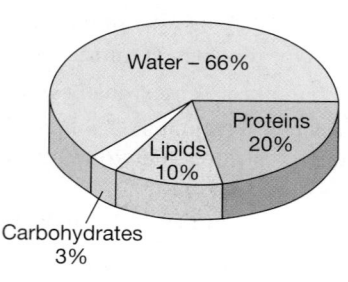

(b) Molecular composition of the human body

●**FIGURE 1**
Composition of the Human Body

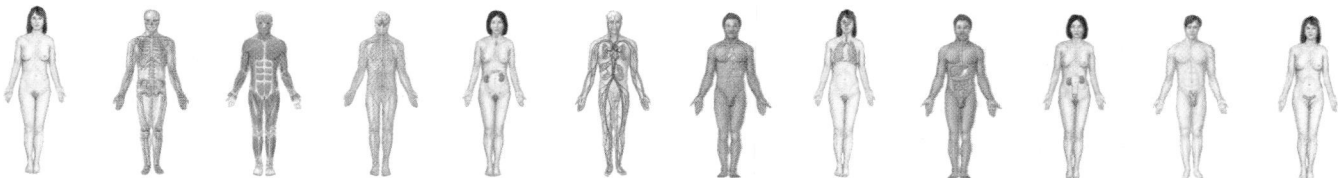

Chapter 4 described the association of roughly 200 types of cells in four body tissues (Figure 1c●). These tissues combine to form thousands of different organs. Some are quite large and distinctive; the liver, an organ of the digestive tract, weighs about 1.6 kg (3.5 lb), and some skeletal muscles can be even larger. Other organs are tiny and far more numerous; the skin contains roughly 3 million sweat glands that are barely large enough to see without a magnifying glass. Regardless of their size, all of these organs contain all four tissue types, although the proportions vary from organ to organ. For example, all four tissue types contribute extensively to the structure of the stomach, but most of the heart is composed of cardiac muscle tissue. Figure 1d● considers the human body at the organ system level, the focus of the rest of this textbook.

Despite their structural and functional differences, all organ systems share certain characteristics:

1. *Specialization for performance of a limited number of functions.* In other words, there is a division of labor among organ systems.

2. *Functional independence* when responding to local environmental stimuli.

3. *Dependence on other organ systems* for nutrient supply, oxygen, and waste removal.

4. *Integration of activity* through neural and hormonal mechanisms.

Figure 2●, on the pages that follow, summarizes the components, organization, and functions of the 11 organ systems in the human body. This information will provide a framework for later chapters dealing with specific systems. You should refer to them while you work through those chapters, whenever you need a reminder of the general functions or locations of specific organs.

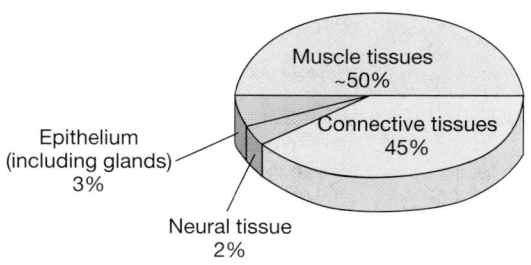

(c) Tissue composition of the human body

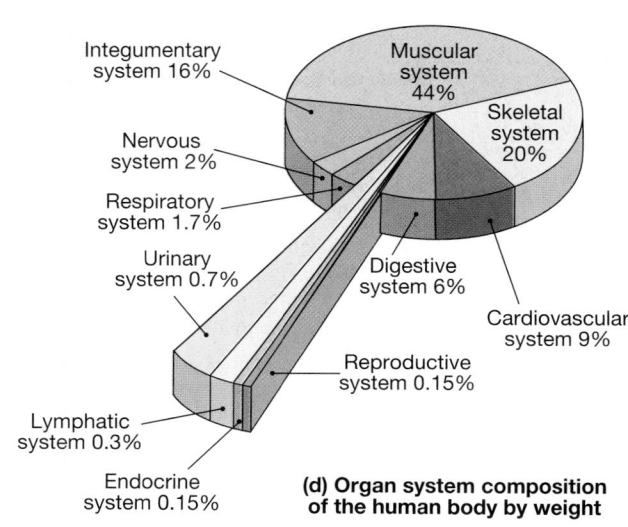

(d) Organ system composition of the human body by weight

●FIGURE 1
Composition of the Human Body (*continued*)

**•FIGURE 2
Organ System
Components and
Functions**

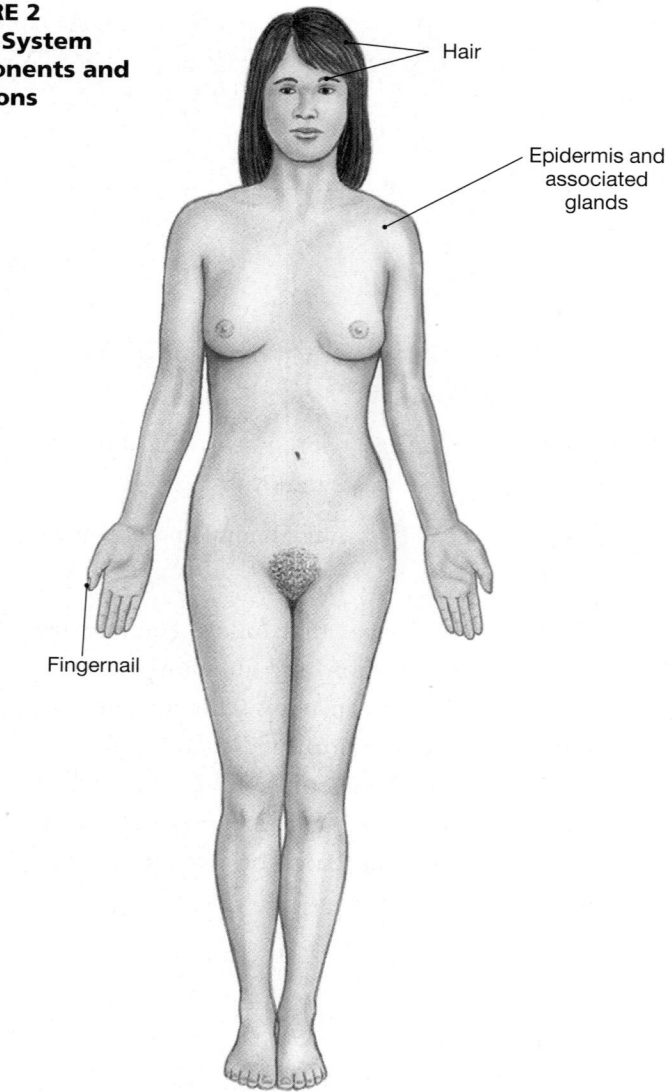

Hair

Epidermis and
associated
glands

Fingernail

(a) THE INTEGUMENTARY SYSTEM

Protects against environmental hazards;
helps control body temperature

Organ/Component	Primary Functions
Cutaneous Membrane	
Epidermis	Covers surface; protects deeper tissues
Dermis	Nourishes epidermis; provides strength; contains glands
Hair Follicles	Produce hair; innervation provides sensation
Hairs	Provide some protection for head
Sebaceous Glands	Secrete lipid coating that lubricates hair shaft and epidermis
Sweat Glands	Produce perspiration for evaporative cooling
Nails	Protect and stiffen distal tips of digits
Sensory Receptors	Provide sensations of touch, pressure, temperature, pain
Subcutaneous Layer	Stores lipids; attaches skin to deeper structures

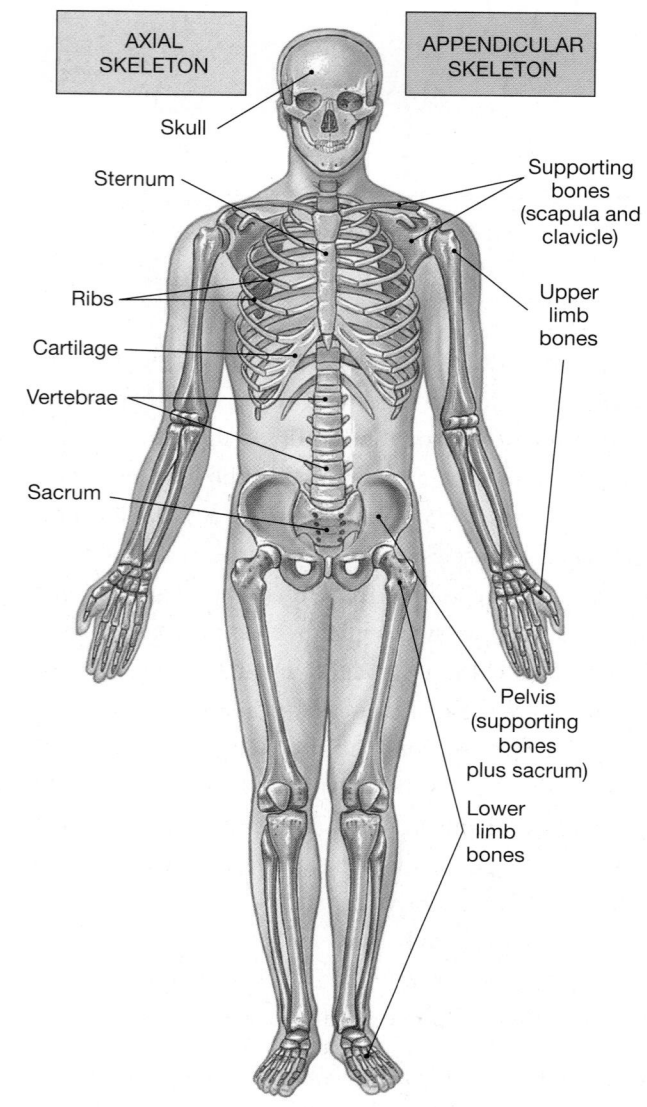

| AXIAL SKELETON | | APPENDICULAR SKELETON |

Skull

Sternum

Supporting bones (scapula and clavicle)

Ribs

Cartilage

Upper limb bones

Vertebrae

Sacrum

Pelvis (supporting bones plus sacrum)

Lower limb bones

(b) THE SKELETAL SYSTEM

Provides support; protects tissues; stores
minerals; forms blood

Organ/Component	Primary Functions
Bones, Cartilages, and Joints	Support, protect soft tissues; bones store minerals
Axial Skeleton (skull, vertebrae, ribs, sternum, sacrum, cartilages, and ligaments)	Protects brain, spinal cord, sense organs, and soft tissues of thoracic cavity; supports the body weight over the lower limbs
Appendicular Skeleton (supporting bones, cartilages, and ligaments of the limbs)	Provides internal support and positioning of the limbs; supports and moves axial skeleton
Bone Marrow	Acts as primary site of blood cell production (red blood cells, white blood cells); stores lipid reserves

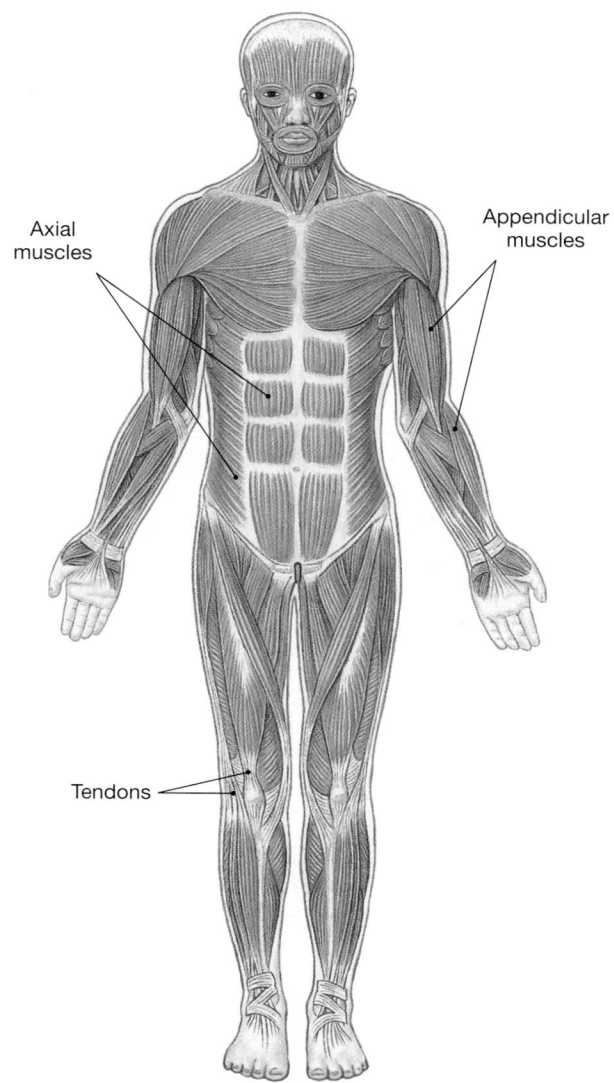

Axial muscles

Appendicular muscles

Tendons

(c) THE MUSCULAR SYSTEM

Produces movement and locomotion; provides support; generates heat

Organ/Component	Primary Functions
Skeletal Muscles (700)	Provide skeletal movement; control entrances and exits of digestive tract; produce heat; support skeletal position; protect soft tissues
Axial Muscles	Support and position axial skeleton
Appendicular Muscles	Support, move, and brace limbs
Tendons, Aponeuroses	Harness forces of contraction to perform specific tasks

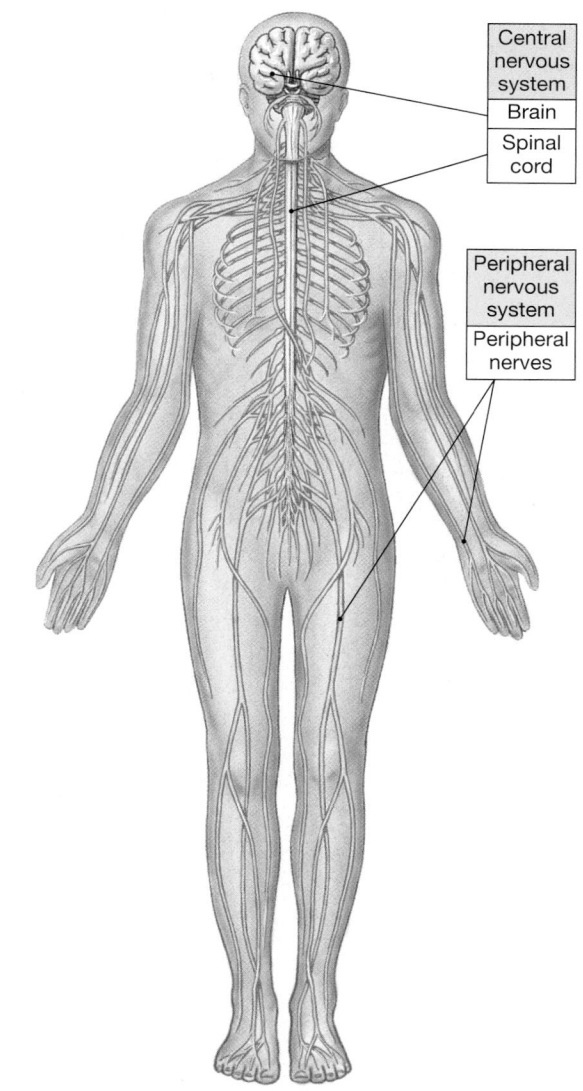

Central nervous system
Brain
Spinal cord

Peripheral nervous system
Peripheral nerves

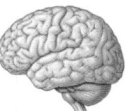

(d) THE NERVOUS SYSTEM

Directs immediate response to stimuli, usually by coordinating the activities of other organ systems

Organ/Component	Primary Functions
Central Nervous System (CNS)	Acts as control center for nervous system: processes information; provides short-term control over activities of other systems
Brain	Performs complex integrative functions; controls both voluntary and autonomic activities
Spinal Cord	Relays information to and from brain; performs less-complex integrative functions; directs many simple involuntary activities
Peripheral Nervous System (PNS)	Links CNS with other systems and with sense organs

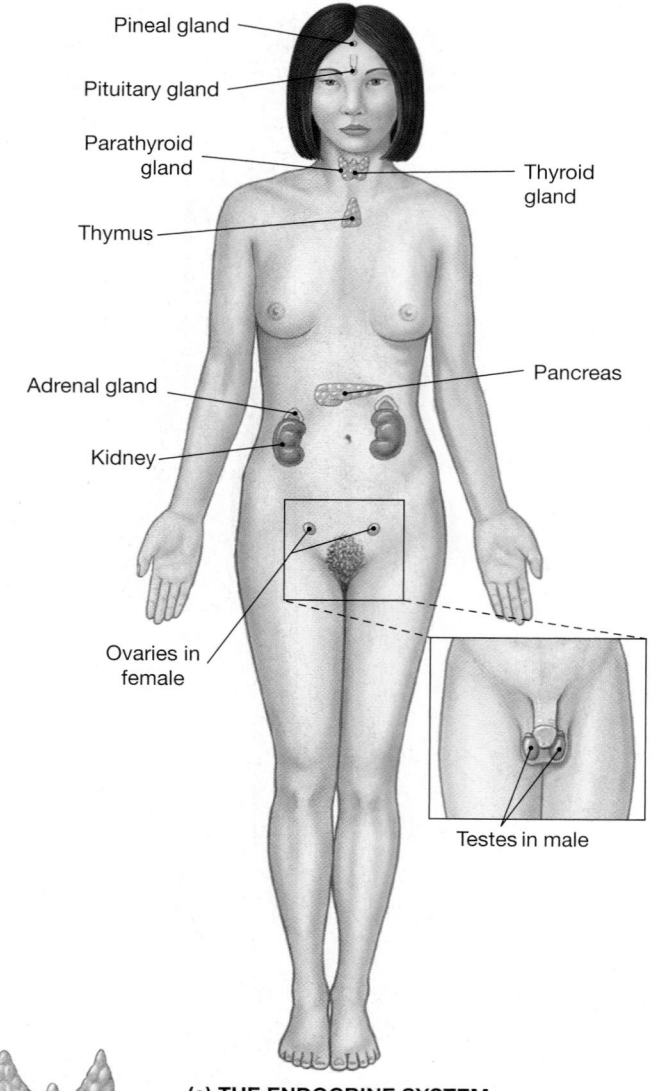

Pineal gland
Pituitary gland
Parathyroid gland
Thyroid gland
Thymus
Adrenal gland
Pancreas
Kidney
Ovaries in female

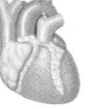

Testes in male

(e) THE ENDOCRINE SYSTEM

Directs long-term changes in other organ systems

Organ/Component	Primary Functions
Pineal Gland	May control timing of reproduction and set day-night rhythms
Pituitary Gland	Controls other endocrine glands; regulates growth and fluid balance
Thyroid Gland	Controls tissue metabolic rate; regulates calcium levels
Parathyroid Glands	Regulate calcium levels (with thyroid)
Thymus	Controls maturation of lymphocytes
Adrenal Glands	Adjust water balance, tissue metabolism, cardiovascular and respiratory activity
Kidneys	Control red blood cell production and assist in calcium regulation
Pancreas	Regulates blood glucose levels
Gonads	
Testes	Support male sexual characteristics and reproductive functions (see part k)
Ovaries	Support female sexual characteristics and reproductive functions (see part l)

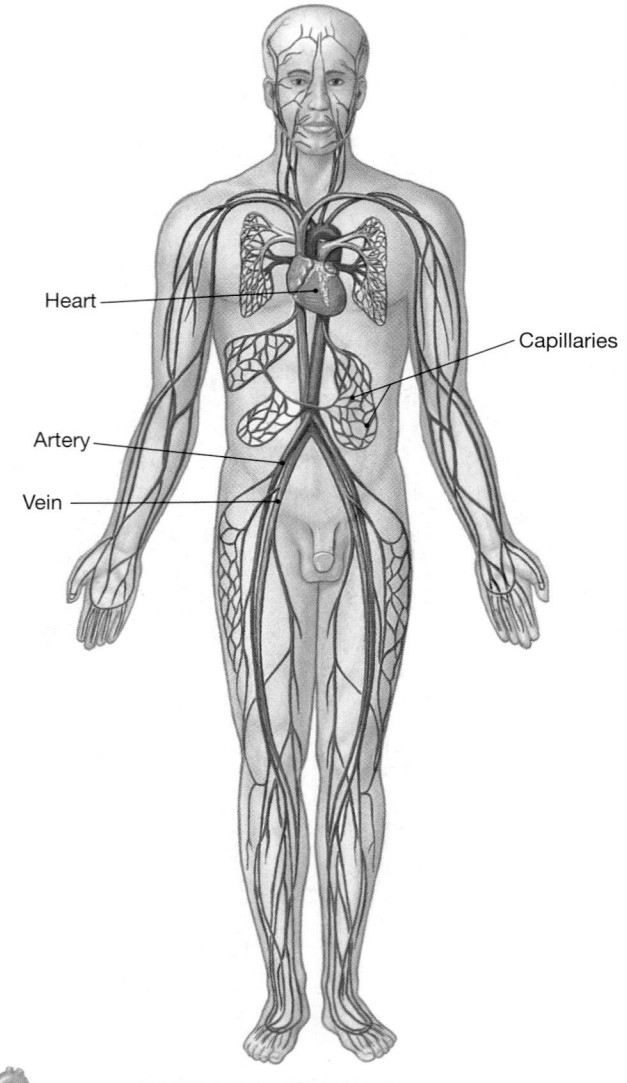

Heart
Capillaries
Artery
Vein

(f) THE CARDIOVASCULAR SYSTEM

Transports cells and dissolved materials, including nutrients, wastes, and gases

Organ/Component	Primary Functions
Heart	Propels blood; maintains blood pressure
Blood Vessels	Distribute blood around the body
Arteries	Carry blood from heart to capillaries
Capillaries	Permit diffusion between blood and interstitial fluids
Veins	Return blood from capillaries to the heart
Blood	Transports oxygen, carbon dioxide, and blood cells; delivers nutrients and hormones; removes waste products; assists in temperature regulation and defense against disease

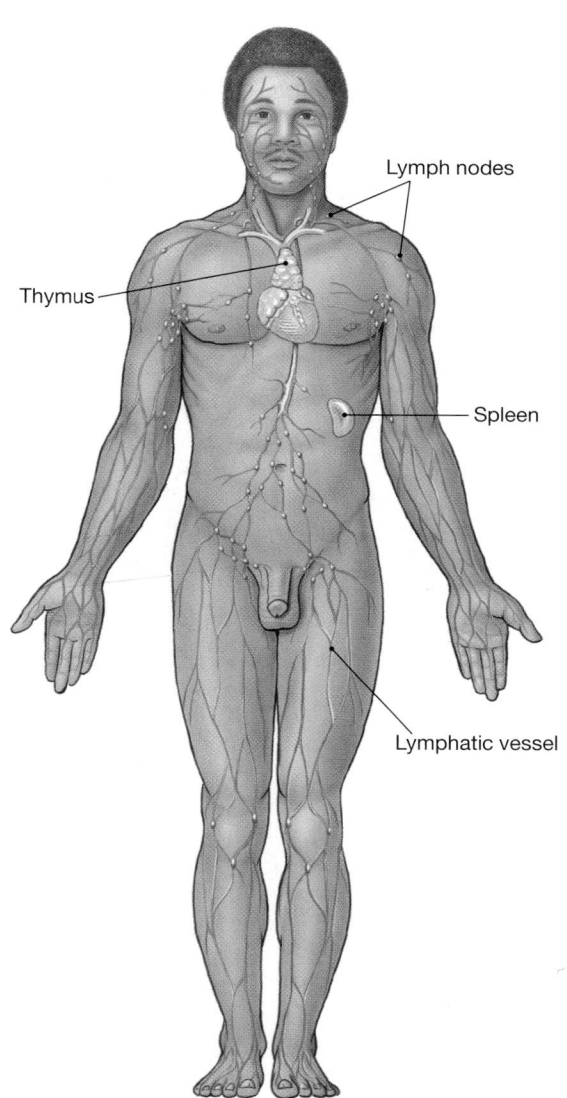

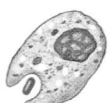

(g) THE LYMPHATIC SYSTEM

Defends against infection and disease; returns tissue fluid to the bloodstream

Organ/Component	Primary Functions
Lymphatic Vessels	Carry lymph (water and proteins) and lymphocytes from peripheral tissues to veins of the cardiovascular system
Lymph Nodes	Monitor the composition of lymph; engulf pathogens; stimulate immune response
Spleen	Monitors circulating blood; engulfs pathogens; stimulates immune response
Thymus	Controls development and maintenance of one class of lymphocytes (T cells)

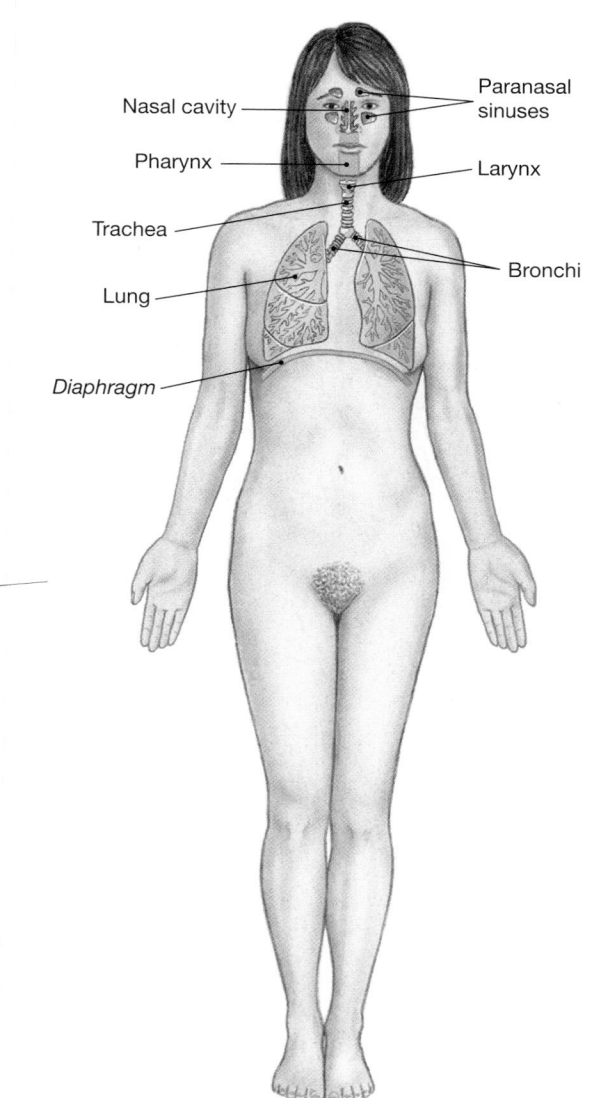

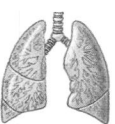

(h) THE RESPIRATORY SYSTEM

Delivers air to sites where gas exchange can occur between the air and circulating blood

Organ/Component	Primary Functions
Nasal Cavities, Paranasal Sinuses	Filter, warm, humidify air; detect smells
Pharynx	Conducts air to larynx; a chamber shared with the digestive tract *(see part i)*
Larynx	Protects opening to trachea and contains vocal cords
Trachea	Filters air, traps particles in mucus; cartilages keep airway open
Bronchi	(Same functions as trachea)
Lungs	Responsible for air movement through volume changes during movements of ribs and diaphragm; include airways and alveoli
Alveoli	Act as sites of gas exchange between air and blood

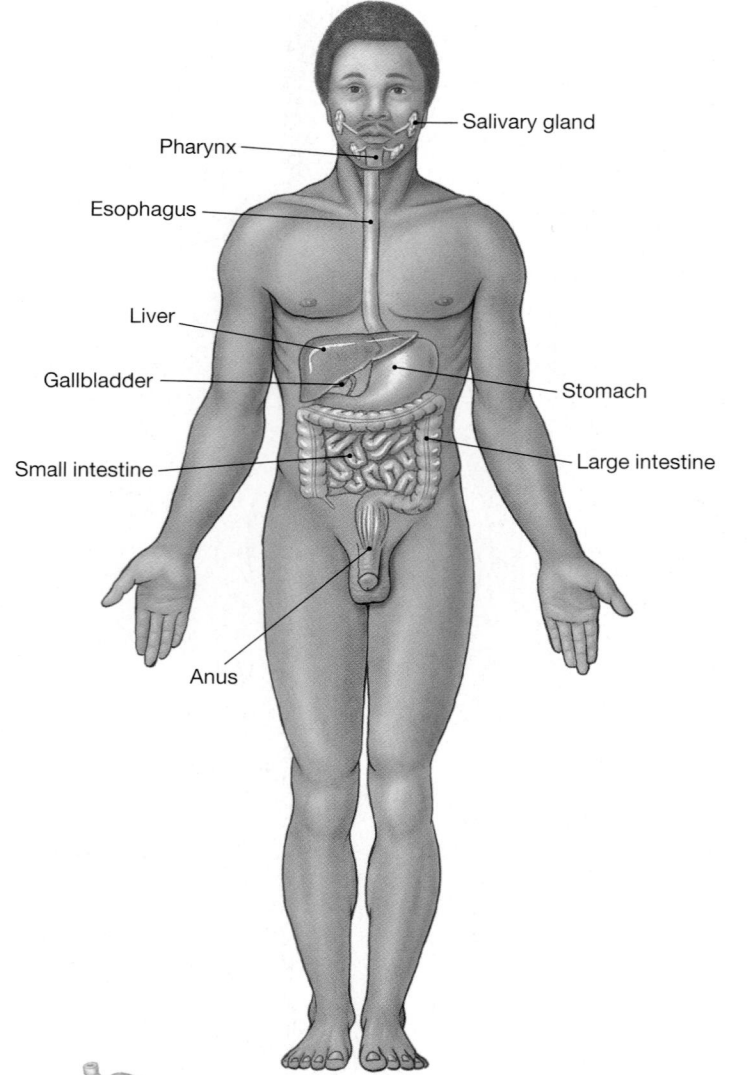

Pharynx

Salivary gland

Esophagus

Liver

Gallbladder

Stomach

Small intestine

Large intestine

Anus

(i) THE DIGESTIVE SYSTEM

Processes food; absorbs nutrients; eliminates waste products

Organ/Component	Primary Functions
Salivary Glands	Provide buffers and lubrication; produce enzymes that begin digestion
Pharynx	Conducts solid food and liquids to esophagus; chamber shared with respiratory tract (see part h)
Esophagus	Delivers food to stomach
Stomach	Secretes acids and enzymes
Small Intestine	Secretes digestive enzymes, buffers, and hormones; absorbs nutrients
Liver	Secretes bile; regulates nutrient composition of blood
Gallbladder	Stores bile for release into small intestine
Pancreas	Secretes digestive enzymes and buffers; contains endocrine cells (see part e)
Large Intestine	Removes water from fecal material; stores wastes

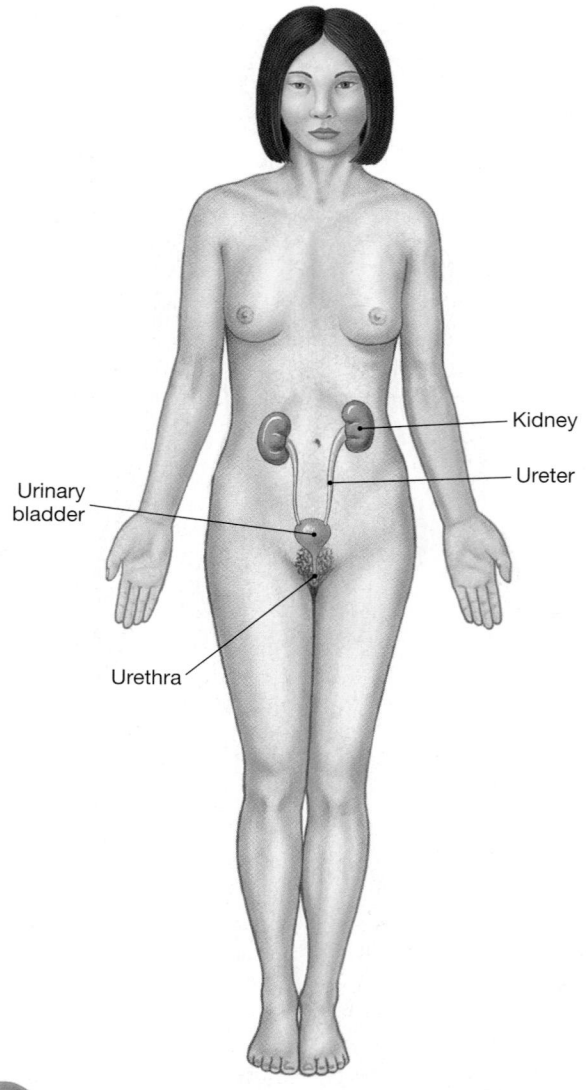

Kidney

Ureter

Urinary bladder

Urethra

(j) THE URINARY SYSTEM

Eliminates excess water, salts, and waste products

Organ/Component	Primary Functions
Kidneys	Form and concentrate urine; regulate blood pH and ion concentrations; perform endocrine functions (see part e)
Ureters	Conduct urine from kidneys to urinary bladder
Urinary Bladder	Stores urine for eventual elimination
Urethra	Conducts urine to exterior

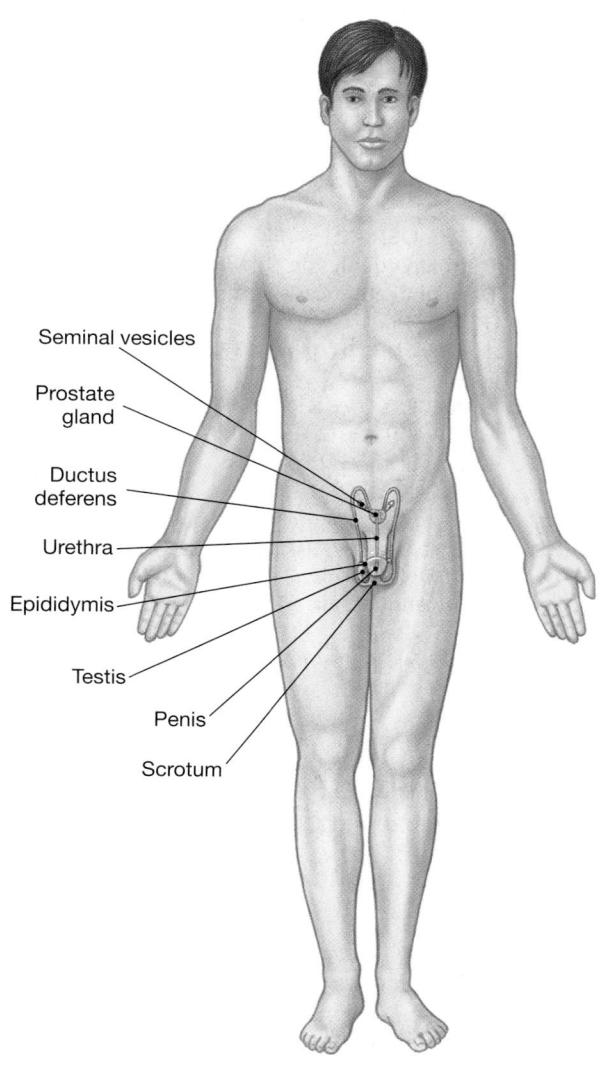

(k) THE MALE REPRODUCTIVE SYSTEM

Produces sex cells and hormones

Organ/Component	Primary Functions
Testes	Produce sperm and hormones (*see part e*)
Accessory Organs	
Epididymis	Acts as site of sperm maturation
Ductus Deferens (Sperm Duct)	Conducts sperm between epididymis and prostate gland
Seminal Vesicles	Secrete fluid that makes up much of the volume of semen
Prostate Gland	Secretes fluid and enzymes
Urethra	Conducts semen to exterior
External Genitalia	
Penis	Contains erectile tissue; deposits sperm in vagina of female; produces pleasurable sensations during sexual activities
Scrotum	Surrounds the testes and controls their temperature

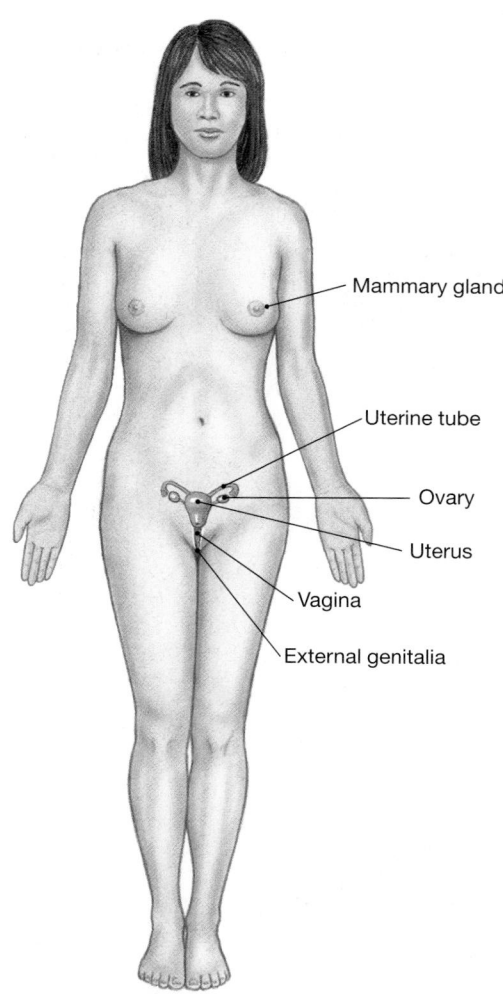

(l) THE FEMALE REPRODUCTIVE SYSTEM

Produces sex cells and hormones

Organ/Component	Primary Functions
Ovaries	Produce oocytes and hormones (*see part e*)
Uterine Tubes	Deliver oocyte or embryo to uterus; normal site of fertilization
Uterus	Site of embryonic development and exchange between maternal and embryonic bloodstreams
Vagina	Site of sperm deposition; acts as birth canal at delivery; provides passageway for fluids during menstruation
External Genitalia	
Clitoris	Contains erectile tissue; produces pleasurable sensations during sexual activities
Labia	Contain glands that lubricate entrance to vagina
Mammary Glands	Produce milk that nourishes newborn infant

UNIT 2   CHAPTER   5   6   7   8   9   10   11

CLINICAL NOTES

- Variations in Skin Pigmentation 161
- Dermatitis 163
- Bedsores 165
- Liposuction 166
- Hair Loss 169
- Seborrheic Dermatitis 170

CLINICAL DISCUSSIONS

- Skin Cancers, Melanomas, and SPF 162
- Drug Administration through the Skin 164

5

THE INTEGUMENTARY SYSTEM

Of all the body systems, the **integumentary system** is the only one we see every day. The components of this system include the *cutaneous membrane*, or skin, and the associated hairs, nails, and exocrine glands. Thus, when you look at yourself or another person, what you see is the integumentary system and the clothing used to cover it.

Because this part of you is so highly visible, you probably devote a lot of time to improving its appearance. Washing your face, brushing and trimming your hair, showering, and applying deodorant are activities that modify the appearance or properties of the integumentary system. And when something goes wrong with this system, the effects are immediately apparent to everyone. You will notice even a minor skin condition or blemish at once, whereas you may ignore more serious problems in other systems. This can actually help physicians, because changes in the color, flexibility, elasticity, or sensitivity of the skin can indicate problems with another system. **AM** Examination of the Skin

5–1 THE INTEGUMENTARY SYSTEM: AN OVERVIEW

Objectives

- List the components of the integumentary system, and describe their physical relationship to each other and to the subcutaneous layer.
- Specify the general functions of the integumentary system.

No other organ system is as accessible, large, varied in function, and underappreciated as the integumentary system. Often referred to simply as the **integument** (in-TEG-ū-ment), this system accounts for about 16 percent of your total body weight. Its surface, 1.5–2 square meters in area, is continuously abused, abraded, attacked by microorganisms, irradiated by sunlight, and exposed to environmental chemicals. The integumentary system is your first line of defense against an often hostile environment—the place where you and the outside world meet.

The integumentary system has two major components: the **cutaneous membrane**, or skin, and the accessory structures.

1. The cutaneous membrane itself has two components: the **epidermis** (*epi-*, above), or superficial epithelium, and the **dermis**, an underlying area of connective tissues.

2. The **accessory structures** include hair, nails, and multicellular exocrine glands. These structures are located primarily in the dermis and protrude through the epidermis to the skin surface.

The integument does not function in isolation. An extensive network of blood vessels branches through the dermis, and sensory receptors that monitor touch, pressure, temperature, and pain provide valuable information to the central nervous system about the state of the body. The general structure of the integument is shown in Figure 5–1●. Deep to the dermis, the loose connective tissue of the **subcutaneous layer**, also known as the superficial fascia or *hypodermis*, separates the integument from the deep fascia around other organs, such as muscles and bones. ∞ p. 135 Although the subcutaneous layer is often not considered a part of the integument, we will consider it in this chapter because its connective tissue fibers are interwoven with those of the dermis.

The general functions of the skin and subcutaneous layer include:

- *Protection* of underlying tissues and organs against shocks, abrasion, and chemical attack.
- *Excretion* of salts, water, and organic wastes by integumentary glands.
- *Maintenance* of normal body temperature through either insulation or evaporative cooling, as needed.

- *Synthesis* of *vitamin D_3*, a steroid that is subsequently converted to the hormone calcitriol, important to normal calcium metabolism.
- *Storage* of nutrients. Lipids are stored in adipocytes in the dermis and in adipose tissue in the subcutaneous layer.
- *Detection* of touch, pressure, pain, and temperature stimuli and the relaying of that information to the nervous system. (The cutaneous senses, which provide information about the external environment, will be considered further in Chapter 15.)

We will now consider the various components of the integument.

5–2 THE EPIDERMIS

Objectives

- Describe the main structural features of the epidermis, and explain their functional significance.
- Explain what accounts for individual and racial differences in skin, such as skin color.
- Discuss the effects of ultraviolet radiation on the skin and the role played by melanocytes.

The epidermis consists of a stratified squamous epithelium (Figures 5–1 and 5–2●). Recall from Chapter 4 that such an

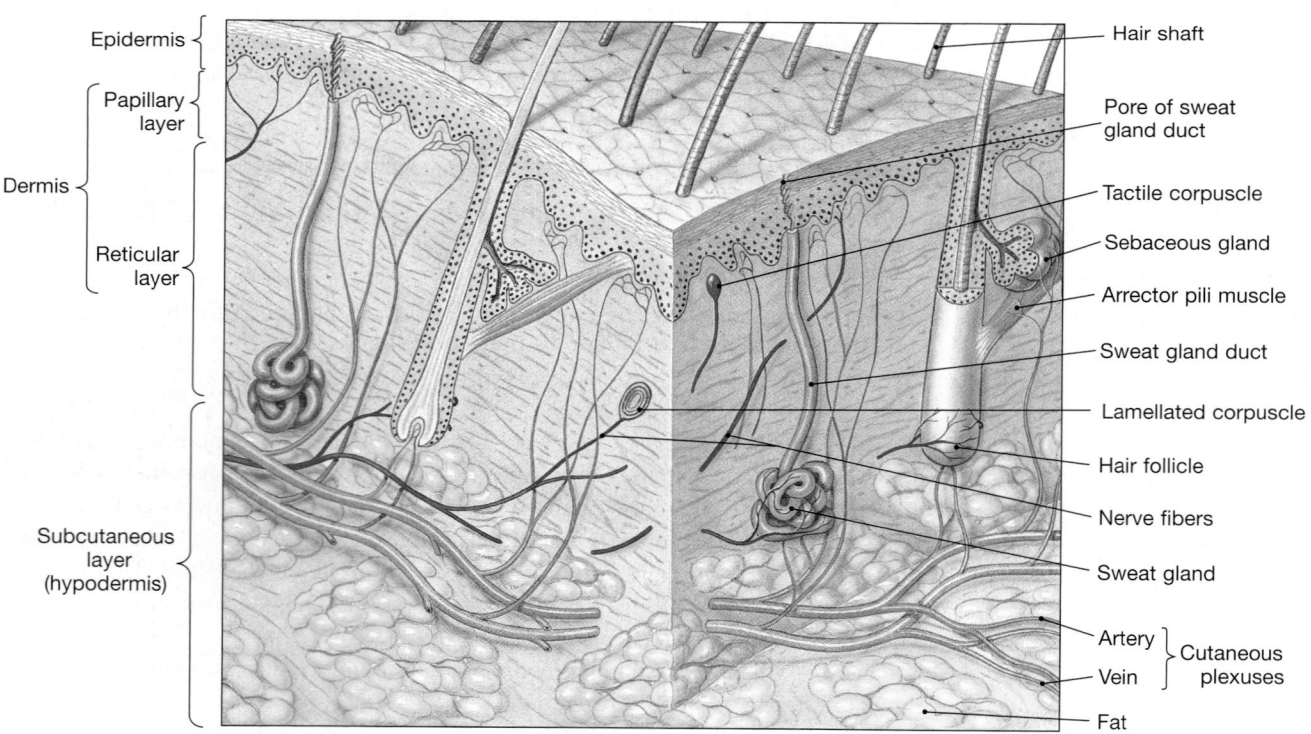

Epidermis

Papillary layer

Dermis

Reticular layer

Subcutaneous layer (hypodermis)

Hair shaft

Pore of sweat gland duct

Tactile corpuscle

Sebaceous gland

Arrector pili muscle

Sweat gland duct

Lamellated corpuscle

Hair follicle

Nerve fibers

Sweat gland

Artery
Vein
} Cutaneous plexuses

Fat

●**FIGURE 5–1**
The Components of the Integumentary System. The relationships among the major components of the integumentary system (with the exception of nails, shown in *Figure 5–13*).

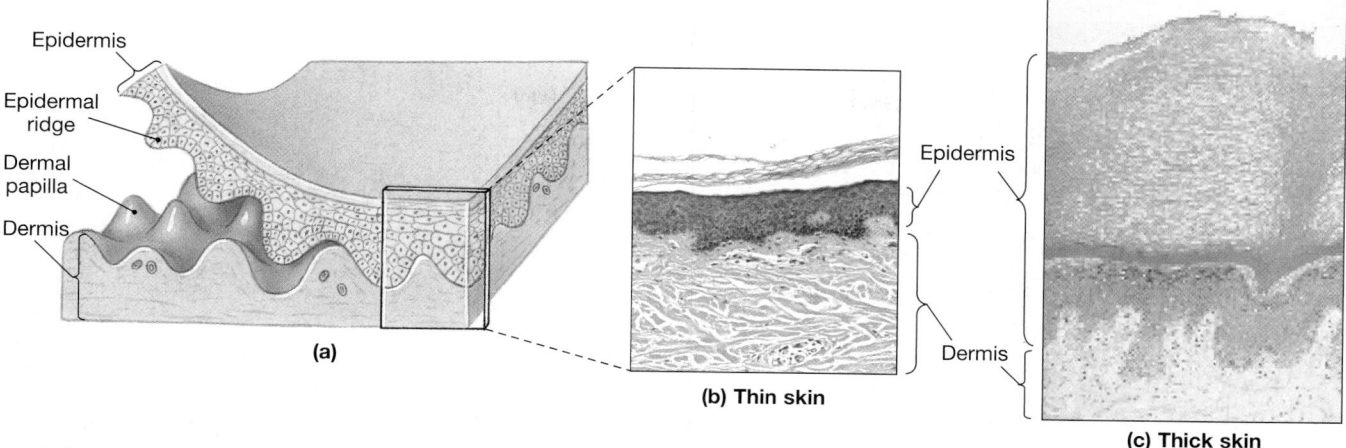

●**FIGURE 5–2**
Thin Skin and Thick Skin. **(a)** The basic organization of the epidermis. The proportions of the various layers vary with the location sampled. **(b)** Thin skin covers most of the exposed body surface. **(c)** Thick skin covers the surfaces of the palms and soles.

epithelium provides considerable mechanical protection and also helps keep microorganisms outside the body. ∞ p. 116 Like all other epithelia, the epidermis is avascular. Lacking blood vessels, the epidermal cells rely on the diffusion of nutrients and oxygen from capillaries within the dermis. As a result, the epidermal cells with the highest metabolic demands are found close to the basal lamina, where the diffusion distance is short. The superficial cells, far removed from the source of nutrients, are either inert or dead.

The body's most abundant epithelial cells, called **keratinocytes** (ker-A-tin-ō-sīts), form several layers. The precise boundaries between these layers can be difficult to see in a light micrograph. In **thick skin**, on the palms of the hands and soles of the feet, five layers can be distinguished, whereas only four layers are recognizable in the **thin skin** that covers the rest of the body. The terms *thick* and *thin* refer to the relative thickness of the epidermis, not to the integument as a whole.

Most of the body is covered by thin skin. In a sample of thin skin (Figure 5–2a,b●), the epidermis is a mere 0.08 mm thick. The epidermis in a sample of thick skin (Figure 5–2c●) can be as thick as 0.5 mm, or the thickness of a standard paper towel.

☐ LAYERS OF THE EPIDERMIS

Figure 5–3● shows the layers in a section of the epidermis of thick skin. The various layers have Latin names. The word *stratum* (plural, *strata*) means "layer"; the rest of the name refers to the function or appearance of the layer. In order, from the basal lamina toward the free surface, are the *stratum germinativum*, the *stratum spinosum*, the *stratum granulosum*, the *stratum lucidum*, and the *stratum corneum*.

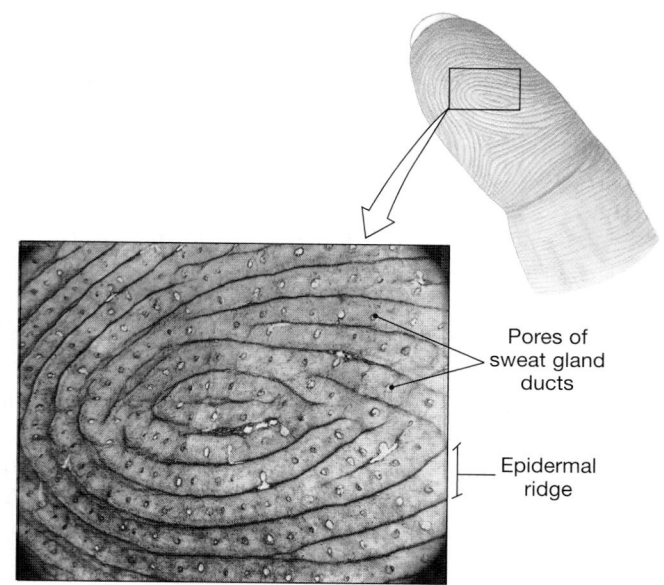

Pores of sweat gland ducts

Epidermal ridge

●**FIGURE 5–3**
The Epidermal Ridges of Thick Skin. Fingerprints reveal the pattern of epidermal ridges. This scanning electron micrograph shows the ridges on a fingertip. The pits are the openings of the ducts of merocrine sweat glands. [Reproduced from R. G. Kessel and R. H. Kardon, *Tissues and Organs: A Text-Atlas of Scanning Electron Microscopy*, W. H. Freeman & Co., 1979]

Stratum Germinativum

The innermost epidermal layer is the **stratum germinativum** (STRA-tum jer-mi-na-TĒ-vum), or *stratum basale* (Figure 5–4●). Hemidesmosomes attach the cells of this layer to the basal lamina

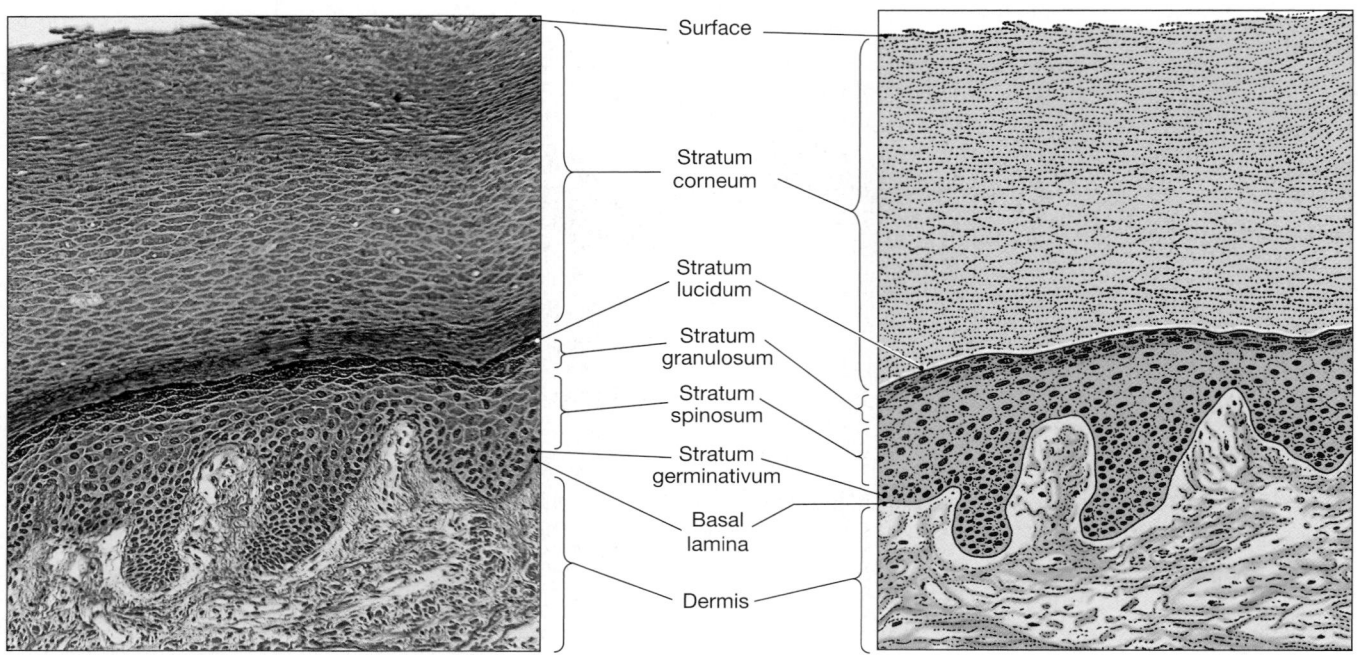

●**FIGURE 5–4**
The Structure of the Epidermis. A portion of the epidermis in thick skin, showing the major stratified layers of epidermal cells. Langerhans cells cannot be distinguished in standard histological preparations. (LM × 210)

that separates the epidermis from the areolar tissue of the adjacent dermis. ∞ p. 113   The stratum germinativum forms **epidermal ridges**, which extend into the dermis, increasing the area of contact between the two regions. Dermal projections called **dermal papillae** (singular, *papilla*; a nipple-shaped mound) extend between adjacent epidermal ridges (Figure 5–2a●). This interlocking arrangement increases the strength of the bond between the epidermis and dermis. The strength of the attachment is proportional to the surface area of the basal lamina: The more folds, the larger the surface area becomes.

The contours of the skin surface follow the ridge patterns, which vary from small conical pegs (in thin skin) to the complex whorls seen on the thick skin of the palms and soles. Ridges on the palms and soles increase the surface area of the skin and increase friction, ensuring a secure grip. Ridge shapes are genetically determined. Your epidermal ridges are unique to you and do not change during your lifetime. Fingerprints are ridge patterns on the tips of the fingers that can be used to identify individuals (Figure 5–3●). These patterns have been utilized in criminal investigations for more than a century.

Large **basal cells**, or *germinative cells*, dominate the stratum germinativum. Basal cells are stem cells whose divisions replace the more superficial keratinocytes that are lost or shed at the epithelial surface. Skin surfaces that lack hair also contain specialized epithelial cells known as **Merkel cells** scattered among the cells of the stratum germinativum. Merkel cells are sensitive to touch; when compressed, they release chemicals that stimulate sensory nerve

endings. (The skin contains many other kinds of sensory receptors, as we shall see in later sections.) The brown tones of skin result from the synthetic activities of pigment cells called *melanocytes*, ∞ p. 124 which are distributed throughout the stratum germinativum, with cell processes extending into more superficial layers.

Stratum Spinosum

Each time a stem cell divides, one of the daughter cells is pushed superficial to the stratum germinativum into the **stratum spinosum**, which means "spiny layer" (Figure 5–4●). The stratum spinosum consists of 8 to 10 layers of cells, with the keratinocytes bound together by desmosomes. ∞ p. 113 Standard histological procedures shrink the cytoplasm, but the cytoskeletal elements and desmosomes remain intact, so the cells look like miniature pincushions. Some of the cells entering this layer from the stratum germinativum continue to divide, further increasing the thickness of the epithelium. The stratum spinosum also contains **Langerhans cells**, participants in the immune response. These cells are responsible for stimulating a defense against (1) microorganisms that manage to penetrate the superficial layers of the epidermis and (2) superficial skin cancers.

Stratum Granulosum

The region superficial to the stratum spinosum is the **stratum granulosum**, or "grainy layer" (Figure 5–4●). The stratum granulosum consists of three to five layers of keratinocytes displaced from the stratum spinosum. By the time cells reach this layer, most have stopped dividing. They begin making large amounts of the proteins **keratin** (KER-a-tin; *keros*, horn) and **keratohyalin** (ker-a-tō-HĪ-a-lin). In humans, keratin, a tough, fibrous protein, is the basic structural component of hair and nails. ∞ p. 55 As keratin fibers develop, the cells grow thinner and flatter, their membranes thickening and becoming less permeable. Keratohyalin forms dense granules in the cytoplasm that promote dehydration of the cell as well as aggregation and cross-linking of the keratin fibers. The nuclei and other organelles then disintegrate, and the cells die. Further dehydration creates a tightly interlocked layer of cells that consist of keratin fibers surrounded by keratohyalin.

Stratum Lucidum

In the thick skin of the palms and soles, a glassy **stratum lucidum** (clear layer) covers the stratum granulosum (Figure 5–4•). The cells in the stratum lucidum are flattened, densely packed, and filled with keratin.

Stratum Corneum

At the exposed surface of both thick skin and thin skin is the **stratum corneum** (KOR-nē-um; *cornu*, horn) (Figure 5–4•). It normally contains 15 to 30 layers of keratinized cells. **Keratinization**, or *cornification*, is the formation of protective, superficial layers of cells filled with keratin. The process occurs on all exposed skin surfaces except the anterior surfaces of the eyes. The dead cells in each layer of the stratum corneum remain tightly interconnected by desmosomes. The connections are so secure that keratinized cells are generally shed in large groups or sheets rather than individually.

It takes 15 to 30 days for a cell to move from the stratum germinativum to the stratum corneum. The dead cells generally remain in the exposed stratum corneum layer for an additional two weeks before they are shed or washed away. This arrangement places the deeper portions of the epithelium and underlying tissues beneath a protective barrier of dead, durable, and expendable cells. Normally, the surface of the stratum corneum is relatively dry, so it is unsuitable for the growth of many microorganisms. Maintenance of this barrier involves coating the surface with lipid secretions from sebaceous glands. Exfoliation techniques scrub away the superficial layers of the stratum corneum, leaving a thinner layer of protective keratinized cells covering the deeper layers of the epidermis. In an individual with pale skin, the removal of superficial layers that are dry and pigmented leaves the skin looking pinker, softer, and more flexible. However, it also eliminates protective layers and leaves the skin more susceptible to damage from sunlight and dehydration.

The stratum corneum is water resistant, but not waterproof. Water from interstitial fluids slowly penetrates the surface, to be evaporated into the surrounding air. You lose roughly 500 ml (about 1 pt) of water in this way each day. The process is called **insensible perspiration**, to distinguish it from the **sensible perspiration** produced by active sweat glands. Damage to the epidermis can increase the rate of insensible perspiration. If the damage breaks connections between superficial and deeper layers of the epidermis, fluid will accumulate in pockets, or *blisters*, within the epidermis. (Blisters also form between the epidermis and dermis if that attachment is disrupted.) If the damage to the epidermis causes the stratum corneum to lose its effectiveness as a water barrier, the rate of insensible perspiration skyrockets and a potentially dangerous fluid loss occurs. This is a serious consequence of severe burns and a complication in the condition known as *xerosis* (excessively dry skin). AM Disorders of Keratin Production

When the skin is immersed in water, osmotic forces may move water into or out of the epithelium. ∞ p. 93 Sitting in a freshwater bath causes water to move into the epidermis, because fresh water is hypotonic (has fewer dissolved materials) compared with body fluids. The epithelial cells may swell to four times their normal volumes, a phenomenon particularly noticeable in the thickly keratinized areas of the palms and soles. Swimming in the ocean reverses the direction of osmotic flow, because the ocean is a hypertonic solution. As a result, water leaves the body, crossing the epidermis from the underlying tissues. The process is slow, but long-term exposure to seawater endangers survivors of a shipwreck by accelerating dehydration.

✓ Excessive shedding of cells from the outer layer of skin in the scalp causes dandruff. What is the name of that skin layer?

✓ A splinter pierces the palm of your hand and lodges in the third layer of the epidermis. Identify this layer.

✓ Why does swimming in fresh water for an extended period cause epidermal swelling?

✓ Some criminals sand the tips of their fingers so as not to leave recognizable fingerprints. Would this practice permanently remove fingerprints? Why or why not?

Answers start on page Q-1

Review the anatomy of the skin and follow the process of wound healing by visiting the **Companion Website**: Chapter 5/ Tutorials.

☐ SKIN COLOR

The color of your skin is a result of an interaction between (1) the epidermal pigmentation and (2) the dermal circulation.

Epidermal Pigmentation

The epidermis contains variable quantities of two pigments: carotene and melanin. **Carotene** (KAR-uh-tēn) is an orange-yellow pigment that normally accumulates in epidermal cells. It is most apparent in cells of the stratum corneum of light-skinned individuals, but it also accumulates in fatty tissues in the dermis. Carotene is found in a variety of orange vegetables, such as

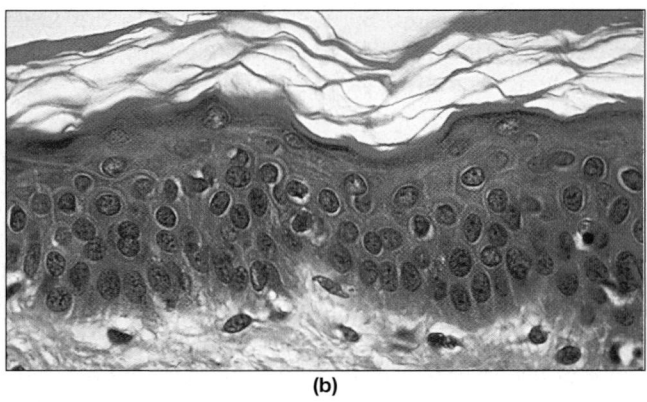

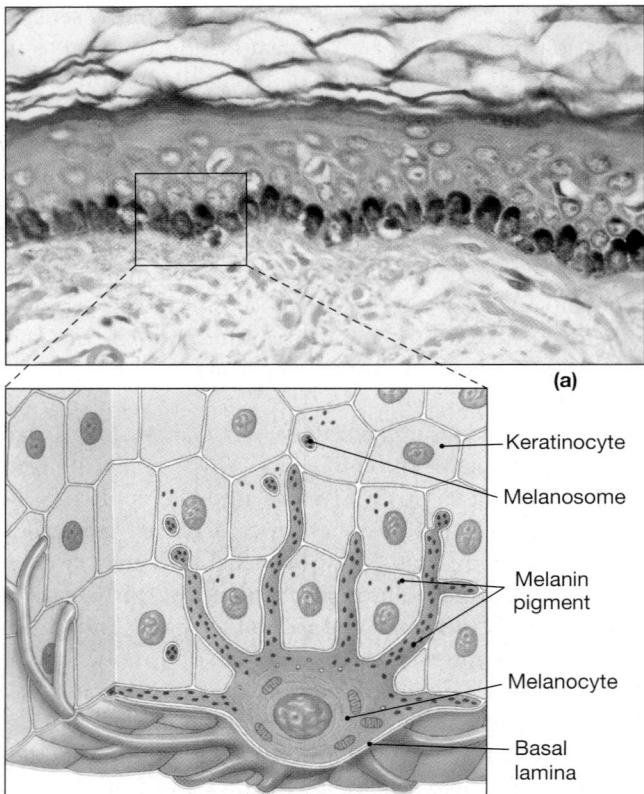

●**FIGURE 5–5**
Melanocytes. **(a)** The location and orientation of
melanocytes in the stratum germinativum of a dark-skinned
person. **(b)** Comparable micrograph of the skin of a pale-
skinned person.

carrots and squashes. The skin of a vegetarian with a special fondness for carrots can actually turn orange from an overabundance
of carotenes. The color change is very striking in a pale-skinned individual, but less obvious in persons with darker skin pigmentation. Carotene can be converted to vitamin A, which is required for
(1) the normal maintenance of epithelia and (2) the synthesis of
photoreceptor pigments in the eye. **Melanin** is a brown, yellow-
brown, or black pigment produced by melanocytes.

MELANOCYTES Melanocytes are located in the stratum germinativum, squeezed between or deep to the epithelial cells
(Figure 5–5●). Melanocytes manufacture the pigment melanin
from molecules of the amino acid *tyrosine*. The melanin is packaged in intracellular vesicles called *melanosomes*. These vesicles
travel within the processes of melanocytes and are transferred
intact to keratinocytes. The transfer of pigmentation colors the
keratinocyte temporarily, until the melanosomes are destroyed
by fusion with lysosomes. In individuals with pale skin, this
transfer occurs in the stratum germinativum and stratum spinosum, and the cells of more superfcal layers lose their pigmentation. In dark-skinned people, the melanosomes are larger and
the transfer may occur in the stratum granulosum as well; the
pigmentation is thus darker and more persistent.

Regardless of ethnic skin color variation, the ratio of
melanocytes to germinative cells ranges between 1:4 and 1:20,
depending on the region of the body. The skin covering most

areas of the body has about 1000 melanocytes per square millimeter. The cheeks and forehead, the nipples, and the genital region (the scrotum of males and the labia majora of females) have
higher concentrations (about 2000 per square millimeter). The
differences in skin color among individuals and even among
races do not reflect different numbers of melanocytes, but merely different levels of synthetic activity. Even the melanocytes of
albino individuals are distributed normally, although the cells are
incapable of producing melanin.

The melanin in keratinocytes protects your epidermis and
dermis from the harmful effects of sunlight, which contains significant amounts of **ultraviolet (UV) radiation**. A small amount
of UV radiation is beneficial, because it stimulates synthetic activity in the epidermis (a process discussed in a later section).
However, UV radiation can damage DNA, causing mutations
and promoting the development of cancer. In the keratinocytes,
melanosomes become concentrated in the region around the nucleus, where the melanin pigments provide some UV protection
for the DNA in those cells.

UV radiation can also produce the immediate effects of mild
or even serious burns, damaging both the epidermis and the dermis. Thus, the presence of a layer of pigment in the deep layers of
the epidermis helps protect both epidermal and dermal tissues.
However, although melanocytes respond to UV exposure by increasing their activity, the response is not rapid enough to prevent sunburn the first day you spend at the beach. Melanin

synthesis accelerates slowly, peaking about 10 days after the initial exposure. Individuals of any skin color can suffer sun damage to the integument, but dark-skinned individuals have greater initial protection against the effects of UV radiation.

Over time, cumulative damage to the integument by UV exposure can harm fibroblasts, causing impaired maintenance of the dermis. The resulting structural alterations lead to premature wrinkling. In addition, skin cancers can develop from chromosomal damage in germinative cells or melanocytes. One of the major consequences of the global depletion of the ozone layer in Earth's upper atmosphere is likely the sharp increase in the rate of skin cancers (such as *malignant melanoma*) that has been seen in recent years. Such an increase has been reported in Australia, which has already experienced a significant loss of ozone, as well as in the United States, Canada, and parts of Europe, which have experienced a more moderate ozone loss. For this reason, limiting UV exposure through a combination of protective clothing and sunscreens (or, better yet, sunblocks) is recommended during outdoor activities.

Dermal Circulation

Blood contains red blood cells filled with the pigment *hemoglobin*, which binds and transports oxygen in the bloodstream. When bound to oxygen, hemoglobin is bright red, giving blood vessels in the dermis a reddish tint that is most apparent in lightly pigmented individuals. If those vessels are dilated, as they are during inflammation, the red tones become much more pronounced.

When its blood supply is temporarily reduced, the skin becomes relatively pale; a Caucasian who is frightened may "turn white" as a result of a sudden drop in blood supply to the skin. During a sustained reduction in circulatory supply, the oxygen levels in the tissues decline. Hemoglobin then releases oxygen and turns a much darker red. Seen from the surface, the skin takes on a bluish coloration called **cyanosis** (sī-uh-NŌ-sis; *kyanos*, blue). In individuals of any skin color, cyanosis is most apparent in areas of thin skin, such as the lips or beneath the nails. It can occur in response to extreme cold or as a result of circulatory or respiratory disorders, such as heart failure or severe asthma. Parenting guides often tell parents to watch for "blue lips," because it is a sign that their young children are cold.

It is obvious that many different colors are present in the integument, even on a single individual. For example, you may notice *freckles*, or small areas of pigment, on the skin of pale-skinned individuals. These spots, which typically have an irregular border, represent the area serviced by melanocytes that are producing larger-than-average amounts of melanin. Freckles tend to be most abundant on surfaces such as the face, probably owing to its greater exposure to the sun. *Lentigos* are similar to freckles, but have regular borders and contain abnormal melanocytes. *Senile lentigos*, or *liver spots*, are variably pigmented areas that develop on exposed skin in older Caucasians.

Several diseases that have primary impacts on other systems may have secondary effects on skin color and pigmenta-

tion. Because the skin is easily observed, these color changes can be useful in diagnosis. For example,

- In *jaundice* (JAWN-dis), the liver is unable to excrete bile, so a yellowish pigment accumulates in body fluids. In advanced stages, the skin and whites of the eyes turn yellow.

- Some tumors affecting the pituitary gland result in the secretion of large amounts of *melanocyte-stimulating hormone (MSH)*. This hormone causes a darkening of the skin, as if the individual has an extremely deep bronze tan.

- In *Addison's disease*, the pituitary gland secretes large quantities of *adrenocorticotropic hormone (ACTH)*, a hormone that is structurally similar to MSH. The effect of ACTH on skin coloration is similar to that of MSH. John F. Kennedy suffered from Addison's disease during his term as president of the United States.

- In *vitiligo* (vi-ti-LĪ-gō), individuals lose their melanocytes. The condition develops in about 1 percent of the population, and its incidence increases among individuals with thyroid gland disorders, Addison's disease, or several other disorders. It is suspected that vitiligo develops when the immune defenses malfunction and antibodies attack normal melanocytes. The primary problem with vitiligo is cosmetic, especially for individuals with darkly pigmented skin. Pop-music star Michael Jackson is said to have vitiligo.

☐ THE EPIDERMIS AND VITAMIN D₃

Although strong sunlight can damage epithelial cells and deeper tissues, limited exposure to sunlight is beneficial. When exposed to ultraviolet radiation, epidermal cells in the stratum spinosum and stratum germinativum convert a cholesterol-related steroid into **vitamin D₃**, or **cholecalciferol** (kō-le-kal-SIF-e-rol). The liver then converts cholecalciferol into an intermediary product used by the kidneys to synthesize the hormone **calcitriol** (kal-si-TRĪ-ol). Calcitriol is essential for the normal absorption of calcium and phosphorus by the small intestine. An inadequate supply of calcitriol leads to impaired bone maintenance and growth.

If present in the diet, cholecalciferol can also be absorbed by the digestive tract. This fact accounts for the use of the term *vitamin*, even though the body can synthesize its own cholecalciferol. *Vitamin* usually refers to an essential organic nutrient that the body either cannot make or makes in insufficient amounts. Children who live in areas with overcast skies and whose diet lacks cholecalciferol can have abnormal bone development. This condition has largely been eliminated in the United States, because dairy companies add cholecalciferol, usually identified as "vitamin D," to the milk sold in grocery stores. In Chapter 6, we shall consider the hormonal control of bone growth in greater detail.

☐ EPIDERMAL GROWTH FACTOR

Epidermal growth factor (EGF) is one of the peptide growth factors introduced in Chapter 3. p. 104 Produced by the salivary glands and glands of the duodenum, EGF has

Skin Cancers, Melanomas, and SPF

Almost everyone has several benign tumors of the skin; freckles and moles are examples (p. 161). Less common, but equally harmless, are benign tumors affecting dermal circulation. *Skin cancers*, which are more dangerous, are the most common form of cancer.

An *actinic keratosis* is a scaly area on sun-damaged skin. It is an indication that sun damage is occurring, but it is not a sign of skin cancer. In contrast, **basal cell carcinoma** is a malignant cancer that originates in the stratum germinativum (Figure 5–6a●). This form is the most common skin cancer. Roughly two-thirds of these cancers appear in body areas subjected to chronic UV exposure. Genetic factors have been identified that predispose people to this condition. **Squamous cell carcinomas** are less common, but almost totally restricted to areas of sun-exposed skin. Metastasis seldom occurs in squamous cell carcinomas and virtually never in basal cell carcinomas, and most people survive these cancers. The usual treatment involves the surgical removal of the tumor, and 95 percent of patients survive for five years or longer after treatment. (This statistic, the 5-year survival rate, is a common method of reporting long-term prognoses.)

Compared with these common and seldom life-threatening cancers, **malignant melanomas** (mel-a-NŌ-maz) (Figure 5–6b●) are extremely dangerous. In this condition, cancerous melanocytes grow rapidly and metastasize through the lymphatic system. The outlook for long-term survival changes dramatically, depending on when the condition is diagnosed. If the cancer is localized, the 5-year survival rate is 99 percent; if it is widespread, the survival rate drops to 14 percent.

To detect melanoma at an early stage, you must examine your skin and you must know what to look for. The mnemonic ABCD makes it easy to remember the key points of detection:

- **A** is for *asymmetry*: Melanomas tend to be irregular in shape. Typically, they are raised; they may ooze or bleed.
- **B** is for *border*: The border of a melanoma is generally unclear, irregular, and in some cases notched.

- **C** is for *color*: A melanoma is generally mottled, with many colors (any combination of tan, brown, black, red, pink, white, and blue).
- **D** is for *diameter*: A growth more than about 5 mm (0.2 in.) in diameter, or roughly the area covered by the eraser on a pencil, is dangerous.

A new experimental treatment for melanoma uses genetic-engineering technology to manufacture antibodies that target the melanocyte-stimulating hormone (MSH) receptors on the surfaces of melanocytes. Melanocytes coated with these antibodies are recognized and attacked by cells of the immune system.

Fair-skinned individuals who live in the tropics are most susceptible to all forms of skin cancer, because their melanocytes are unable to shield them from UV radiation. Sun damage can be prevented by avoiding exposure to the sun during the middle of the day and by using a sunblock (not a tanning oil)—a practice that also delays the cosmetic problems of aging and wrinkling. *Everyone* who spends any time out in the sun should choose a broad-spectrum sunblock with a sun protection factor (SPF) of at least 15; blondes, redheads, and people with very pale skin are better off with an SPF of 20 to 30. (The risks are the same for those who spend time in a tanning salon or tanning bed.) The protection offered by these sunscreens is afforded by both organic molecules that absorb UV and inorganic pigments that absorb, scatter, and reflect UV. The higher the SPF factor, the more of these chemicals the product contains, and the fewer UV rays are able to penetrate to the skin's surface. Wearing a hat with a brim and panels to shield the neck and face provides added protection.

The use of sunscreens will be even more important as the ozone gas in the upper atmosphere is further destroyed by our industrial emissions. Ozone absorbs UV radiation before it reaches Earth's surface; in doing so, ozone assists the melanocytes in preventing skin cancer. Australia, the continent that is most affected by the depletion of ozone above the South Pole (the "ozone hole"), is already reporting an increased incidence of skin cancers.

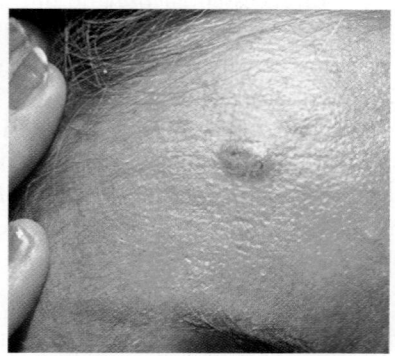

(a) Basal cell carcinoma

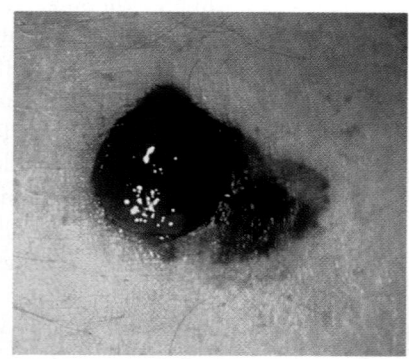

(b) Melanoma

●**FIGURE 5–6**
Skin Cancers.   **(a)** Basal cell carcinoma.
(b) Melanoma.

widespread effects on epithelia, especially the epidermis. Among the roles of EGF are:

- Promoting the divisions of germinative cells in the stratum germinativum and stratum spinosum.
- Accelerating the production of keratin in differentiating epidermal cells.
- Stimulating epidermal development and epidermal repair after injury.
- Stimulating synthetic activity and secretion by epithelial glands.

Epidermal growth factor has such a pronounced effect that it can be used to stimulate the growth and division of epidermal cells outside the body. In this technique, known as *tissue culture*, sheets of epidermal cells are produced for use in the treatment of extensive burns. Burned areas can be covered by epidermal sheets "grown" from a small sample of intact skin from another part of the burn victim's body. (We will consider this treatment later in the chapter when we discuss burns.)

✓ Why does exposure to sunlight or sunlamps darken skin?

✓ Why does the skin appear red when it is warm?

✓ In some cultures, women must be covered completely, except for their eyes, when they go outside. These women exhibit a high incidence of problems with their bones. Why?

Answers start on page Q-1

5–3 THE DERMIS

Objective

- Describe the structure and functions of the dermis.

The dermis lies between the epidermis and the subcutaneous layer (Figure 5–1●, p. 156). The dermis has two major components: (1) a superficial *papillary layer* and (2) a deeper *reticular layer*.

☐ DERMAL ORGANIZATION

The **papillary layer** consists of areolar tissue. This region contains the capillaries and sensory neurons that supply the surface of the skin. The papillary layer derives its name from the dermal papillae that project between the epidermal ridges (Figure 5–2a●, p. 157).

The **reticular layer**, deep to the papillary layer, consists of an interwoven meshwork of dense, irregular connective tissue. ∞ p. 128 Bundles of collagen fibers extend beyond the reticular layer to blend into those of the superficial papillary layer, so the boundary between the two layers is indistinct. Collagen fibers of the reticular layer also extend into the deeper subcutaneous layer. In addition to extracellular protein fibers, the dermis contains all the cells of connective tissue proper. ∞ pp. 123–124 Accessory organs of epidermal origin, such as hair follicles and sweat glands, extend into the dermis. In addition, the reticular and papillary layers of the dermis contain networks of blood vessels, lymph vessels, and nerve fibers (Figure 5–1●).

Because of the abundance of sensory receptors in the skin, regional infection or inflammation can be very painful. **Dermatitis** (der-muh-TĪ-tis) is an inflammation of the skin that primarily involves the papillary layer. The inflammation typically begins in a part of the skin exposed to infection or irritated by chemicals, radiation, or mechanical stimuli. Dermatitis may cause no discomfort, or it may produce an annoying itch, as in poison ivy. Other forms of the condition can be quite painful, and the inflammation can spread rapidly across the entire integument. **AM** Dermatitis

Wrinkles and Stretch Marks

Individual collagen fibers are very strong and resist stretching, but they are easily bent or twisted. Elastic fibers permit stretching and will then recoil to their original length. These two types of fiber are interwoven within the dermis. As a result, the dermis tolerates limited stretching. Your skin remains flexible and somewhat elastic, but the collagen fibers stop the distortion before tissue damage occurs. The water content of the skin also helps maintain its flexibility and resilience, properties collectively known as *skin turgor*. One of the symptoms of dehydration is the loss of skin turgor, exposed by pinching the skin on the back of the hand. A dehydrated dermis will remain peaked when pinched, whereas hydrated skin will flatten out. Aging, hormones, and the destructive effects of ultraviolet radiation permanently reduce the amount of elastin in the dermis, producing wrinkles and sagging skin. The extensive distortion of the dermis that occurs over the abdomen during pregnancy or after a substantial weight gain can exceed the elastic capabilities of the skin. The resulting damage to the dermis prevents it from recoiling to its original size after delivery or weight loss. The skin then wrinkles and creases, creating a network of **stretch marks**.

Tretinoin (Retin-A™*)* is a derivative of vitamin A that can be applied to the skin as a cream or gel. This drug was originally developed to treat acne, but it also increases blood flow to the dermis and stimulates dermal repair. As a result, the rate of wrinkle formation decreases, and existing wrinkles become smaller. The degree of improvement varies among individuals.

Lines of Cleavage

Most of the collagen and elastic fibers at any location are arranged in parallel bundles. Their orientation depends on the stress placed on the skin during normal movement: The bundles are aligned to resist applied forces. The resulting pattern of fiber bundles

Drug Administration through the Skin

The integument covers the body, making it the first barrier to substances entering the body. Occasionally, we compromise this barrier in order to administer therapeutic drugs transdermally. As the name implies, this method of introducing drugs requires the movement of the substance across the epidermal cell membranes. Recall that cell membranes are composed of a lipid bilayer, creating a barrier preventing water from rapidly entering or leaving the cell. However, lipid-soluble compounds will readily cross this barrier. If you dissolve a drug in an oil or some other lipid-soluble solvent, the drug can be carried across the cell membranes of the epidermis. The movement is slow, primarily due to the density of the stratum corneum, but once a drug reaches the underlying connective tissues, it will be absorbed into the circulation.

A useful technique for long-term drug administration is the placement of a sticky, drug-containing patch over an area of thin skin. To overcome the slow rate of diffusion, the patch must contain an extremely high concentration of the drug. This procedure is called *transdermal administration*. A single patch may work for several days, making daily pills unnecessary. Several drugs are now routinely administered transdermally:

- Transdermal scopolamine, a drug that affects the nervous system, can control the nausea associated with motion sickness.
- Transdermal nitroglycerin is used to improve blood flow within heart muscle to prevent a heart attack.
- Transdermal estrogens are administered to women to reduce symptoms of menopause.
- Transdermal medications and drugs are used to control high blood pressure.
- Transdermal nicotine is used to suppress the urge to smoke cigarettes. Nicotine is the addictive compound in tobacco. The nicotine from the patch depresses the craving for a cigarette; the transdermal dosage can gradually be reduced in small, controlled steps.

Dimethyl sulfoxide (DMSO) is a transdermal drug intended for the treatment of injuries to the muscles and joints of domesticated animals, such as horses or cows. It is a solvent that rapidly crosses the skin; drugs dissolved in DMSO are carried into the body along with it. Although DMSO has not been tested or approved for the treatment of humans in the United States, either for joint or muscle injuries or as a transdermal solvent, it can be prescribed in Canada and Europe. The long-term risks associated with DMSO are unknown. Reported short-term side effects include nausea, vomiting, cramps, and chills.

Drugs can also be administered by packaging them in *liposomes*, artificially produced lipid droplets. Liposomes containing DNA fragments have been used experimentally to introduce normal genes into abnormal human cells. For example, genes carried by liposomes have been used to alter the cell membranes of skin cancer cells so that the tumor cells will be attacked by the immune system. Another experimental procedure involves the creation of a transient change in skin permeability by administering a brief pulse of electricity. The electrical pulse temporarily changes the positions of the cells in the stratum corneum, creating channels that allow the drug to penetrate.

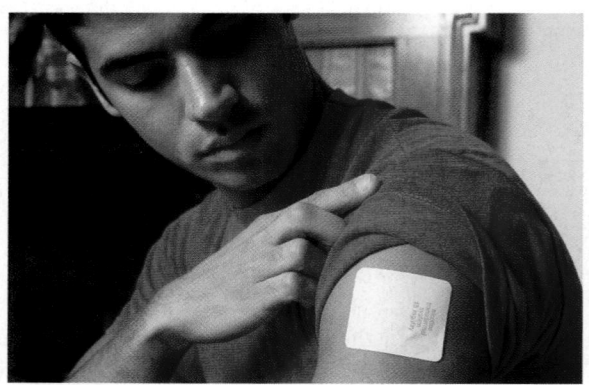

establishes the **lines of cleavage** of the skin (Figure 5–7●). Lines of cleavage are clinically significant, because a cut parallel to a cleavage line will usually remain closed, whereas a cut at right angles to a cleavage line will be pulled open as cut elastic fibers recoil. Surgeons choose their incision patterns accordingly, for a neatly closed incision made parallel to the lines of cleavage will heal faster and with less scarring than will an incision that crosses lines of cleavage.

☐ DERMAL CIRCULATION AND INNERVATION

The Dermal Blood Supply

Arteries supplying the skin form a network in the subcutaneous layer along its border with the reticular layer of the dermis. This network is called the *cutaneous plexus*. Tributaries of these arteries

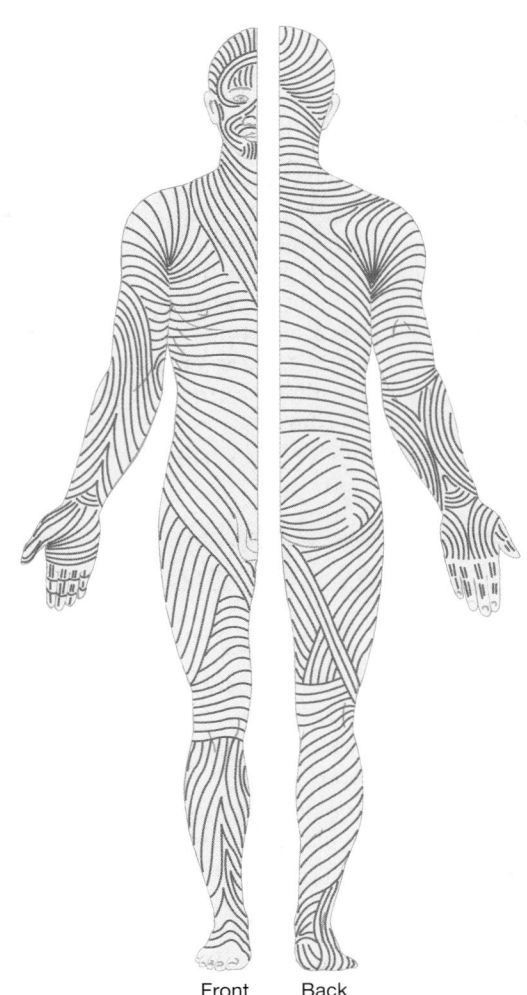

●**FIGURE 5–7**
Lines of Cleavage of the Skin.   Lines of cleavage follow the pattern of fiber bundles in the skin. They reflect the orientation of collagen fiber bundles in the dermis.

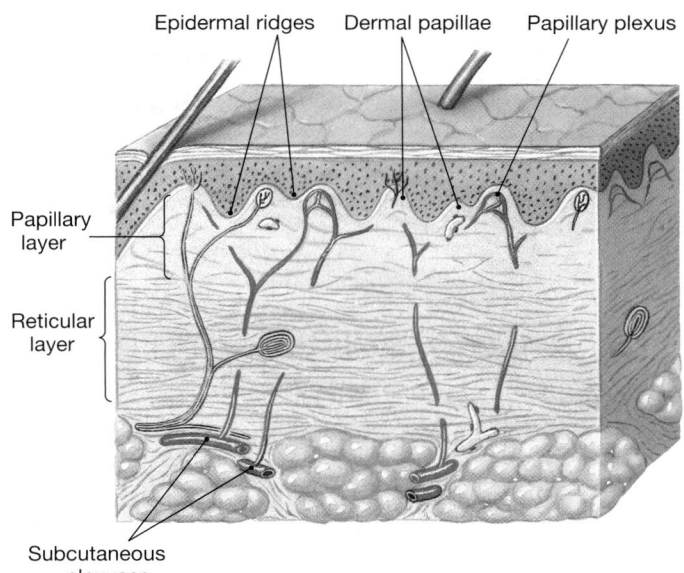

●**FIGURE 5–8**
Dermal Circulation

cial blood vessels. Such sores most commonly affect the skin near joints or projecting bones, where dermal blood vessels are pressed against deeper structures. The chronic lack of circulation kills epidermal cells, removing a barrier to bacterial infection. Eventually, dermal tissues deteriorate as well. (This type of tissue degeneration, or *necrosis*, occurs in any tissue deprived of adequate blood flow. ⬭ p. 139) Bedsores can be prevented or treated by frequently changing the position of the body, to vary the pressures applied to specific blood vessels.

The Innervation of the Skin

The integument is filled with sensory receptors. Anything that comes in contact with the skin, from the lightest touch of a mosquito to the weight of a loaded backpack, initiates a nerve impulse that reaches our conscious awareness. Nerve fibers in the skin control blood flow, adjust gland secretion rates, and monitor sensory receptors in the dermis and the deeper layers of the epidermis. We have already noted the presence of Merkel cells in the deeper layers of the epidermis. These cells are monitored by sensory terminals known as *tactile discs*, or *Merkel's discs*. The epidermis also contains the extensions of sensory neurons that provide sensations of pain and temperature. The dermis contains similar receptors, as well as other, more specialized receptors. Examples shown in Figure 5–1● include receptors sensitive to light touch—*tactile (Meissner's) corpuscles*, located in dermal papillae—and receptors sensitive to deep pressure and vibration—*lamellated (pacinian) corpuscles*, in the reticular layer.

This partial list of the receptors found in the skin should be enough to highlight the importance of the integument as a sensory structure. We will return to the topic in Chapter 15 and consider not only what receptors are present, but how they function.

supply the adipose tissues of the subcutaneous layer and the tissues of the integument. As small arteries travel toward the epidermis, branches supply the hair follicles, sweat glands, and other structures in the dermis. On reaching the papillary layer, the small arteries form another branching network, the *papillary plexus*, which provides arterial blood to capillary loops that follow the contours of the epidermis–dermis boundary. These capillaries empty into a network of small veins that form a venous plexus deep to the papillary plexus. This network is in turn connected to a larger venous plexus in the subcutaneous layer (Figure 5–8●). Trauma to the skin often results in a *contusion*, the rupture of dermal blood vessels. Blood leaks into the dermis, and the area develops a dark blue color responsible for the term "black and blue."

⚕ Problems with dermal circulation affect the epidermis and the dermis. An **ulcer** is a localized shedding of an epithelium. **Decubitis ulcers**, or *bedsores*, affect bedridden or mobile patients with circulatory restrictions, especially when splints, casts, or bedding continuously presses against superfi-

✓ Where are the capillaries and sensory neurons that supply the epidermis located?

✓ What accounts for the ability of the dermis to undergo repeated stretching?

Answers start on page Q-1

5–4 THE SUBCUTANEOUS LAYER

Objective

■ Describe the structure and functions of the subcutaneous layer.

The connective tissue fibers of the reticular layer are extensively interwoven with those of the subcutaneous layer, or **hypodermis**. The boundary between the two is generally indistinct (Figure 5–1●, p. 156). Although the subcutaneous layer is not a part of the integument, it is important in stabilizing the position of the skin in relation to underlying tissues, such as skeletal muscles or other organs, while permitting independent movement.

The subcutaneous layer consists of areolar and adipose tissues and is quite elastic. Only the superficial region contains large arteries and veins. The venous circulation of this region contains a substantial amount of blood, and much of this volume can be shifted to the general circulation by constricting the veins. For that reason, the skin is often described as a *blood reservoir*. The rest of the subcutaneous layer contains a limited number of capillaries and no vital organs. This last characteristic makes **subcutaneous injection**—by means of a **hypodermic needle**—a useful method of administering drugs.

Most infants and small children have extensive "baby fat," which helps them reduce heat loss. Subcutaneous fat also serves as a substantial energy reserve and a shock absorber for the rough-and-tumble activities of our early years. As we grow, the distribution of our subcutaneous fat changes. The greatest changes occur in response to circulating sex hormones. Beginning at puberty, men accumulate subcutaneous fat at the neck, on the arms, along the lower back, and over the buttocks. In contrast, women accumulate subcutaneous fat at the breasts, buttocks, hips, and thighs. In adults of either gender, the subcutaneous layer of the backs of the hands and the upper surfaces of the feet contain few fat cells, whereas distressing amounts of adipose tissue can accumulate in the abdominal region, producing a prominent "potbelly."

The accumulation of excessive amounts of adipose tissue increases the risks of diabetes, strokes, and other serious conditions. Dietary restrictions and increased activity levels are often successful in promoting weight loss and reducing these risks. However, there are surgical options, and one of them is **liposuction**, the removal of subcutaneous adipose tissue through a tube inserted deep to the skin. Adipose tissue tears relatively easily, and suction applied to the tube rips chunks of tissue from the body. After liposuction, the skin is loose fitting, and until it recoils, a tight-fitting surgical garment is usually worn. Liposuction is relatively common and is increasing in popularity. In 2000, an estimated 175,000 to 300,000 liposuction procedures were performed in the United States.

Although it might sound like an easy way to remove unwanted fat, in practice liposuction can be dangerous. There are risks from anesthesia, bleeding (adipose tissue is quite vascular), infection, and fluid loss. The death rate from liposuction procedures is 1 in 5000, very high for what is basically cosmetic surgery that provides only a temporary solution to a chronic problem. As noted in Chapter 4, unless there are changes in diet and lifestyle, the damaged adipose tissue will repair itself, and areas of areolar tissue will convert to adipose tissue. ∞ p. 126 Over time, the surgery will have to be repeated.

5–5 ACCESSORY STRUCTURES

Objectives

■ Explain the mechanisms that produce hair and that determine hair texture and color.

■ Discuss the various kinds of glands in the skin and the secretions of those glands.

■ Explain how the sweat glands of the integumentary system play a major role in regulating body temperature.

■ Describe the anatomical structure of nails and how they are formed.

The accessory structures of the integument include hair follicles, sebaceous glands, sweat glands, and nails. During embryological development, these structures originate from the epidermis, so they are also known as *epidermal derivatives*. Although located in the dermis, they project through the epidermis to the integumentary surface.
ATLAS Embryology Summary 5: The Development of the Integumentary System

☐ HAIR FOLLICLES AND HAIR

Hairs project above the surface of the skin almost everywhere, except over the sides and soles of the feet, the palms of the hands, the sides of the fingers and toes, the lips, and portions of the external genitalia. The human body has about 2.5 million hairs, and 75 percent of them are on the general body surface, not on the head. Hairs are nonliving structures produced in organs called **hair follicles**. Hair production is a complex process involving both the dermis and epidermis.

Hair Structure

Each hair is a long, cylindrical structure that extends outward, past the epidermal surface (Figure 5–9●). The **hair root**—the portion that anchors the hair into the skin—begins at the base of the hair, at the *hair bulb*, and extends distally to the point at which the internal organization of the hair is complete, about

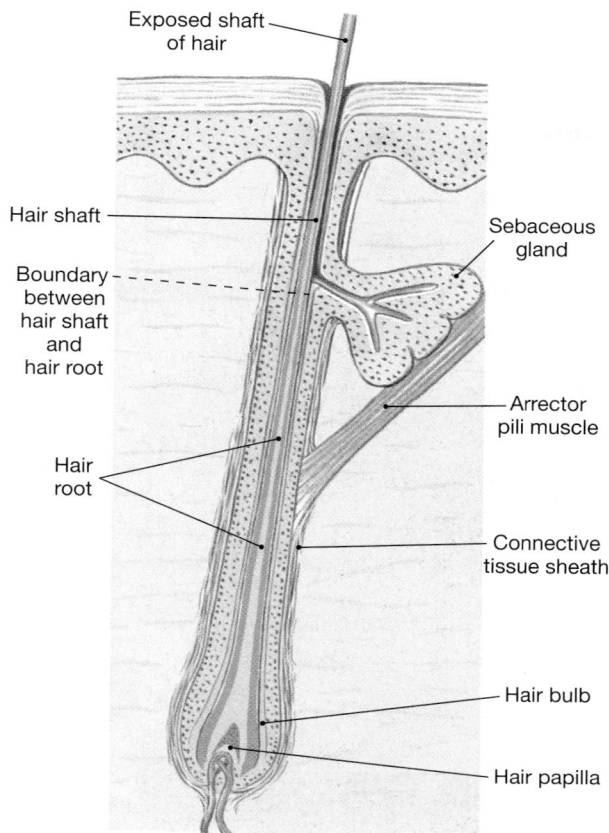

Exposed shaft of hair

Hair shaft

Boundary between hair shaft and hair root

Hair root

Sebaceous gland

Arrector pili muscle

Connective tissue sheath

Hair bulb

Hair papilla

●**FIGURE 5–9**
The Anatomy of a Single Hair

halfway to the skin surface. The **hair shaft**, part of which we see on the surface, extends from this halfway point to the exposed tip of the hair (Figures 5–9, 5–10●). The **cuticle** is the layer that forms the outer surface of the shaft. It consists of an overlapping layer of dead, keratinized cells. The cuticle covers the **cortex**, an intermediate layer. The cells in the cortex and cuticle contain thick layers of **hard keratin**, which give the hair its stiffness. The **medulla**, or core, of the hair contains flexible **soft keratin.**

Hair follicles, which produce hairs, extend deep into the dermis, typically projecting into the underlying subcutaneous layer. At the base of a hair follicle is a small **hair papilla**, a peg of connective tissue containing capillaries and nerves. The **hair bulb** consists of epithelial cells that surround the hair papilla (Figure 5–9●).

Types of Hairs

Hairs first appear after roughly three months of embryonic development. These hairs, collectively known as *lanugo* (la-NOO-gō), are extremely fine and unpigmented. Most lanugo hairs are shed before birth. They are replaced by one of two types of hairs in the adult integument: (1) vellus hairs or (2) terminal hairs. **Vellus hairs** are the fine "peach fuzz" hairs located over much of the body surface. **Terminal hairs** are heavy,

more deeply pigmented, and sometimes curly. The hairs on your head, including your eyebrows and eyelashes, are terminal hairs that are present throughout life. Hair follicles may alter the structure of the hairs in response to circulating hormones. For example, vellus hairs are present at the armpits, pubic area, and limbs until puberty. Thereafter, the follicles produce terminal hairs, in response to circulating sex hormones.

Hair Color

Variations in hair color reflect differences in structure and variations in the pigment produced by melanocytes at the hair papilla. Different forms of melanin give a dark brown, yellow-brown, or red coloration to the hair. These structural and biochemical characteristics are genetically determined, but hormonal and environmental factors influence the condition of your hair. As pigment production decreases with age, hair color lightens. White hair results from the combination of a lack of pigment and the presence of air bubbles in the medulla of the hair shaft. As the proportion of white hairs increases, the individual's overall hair color is described as gray. Because hair itself is dead and inert, any changes in its coloration are gradual. We are able to change our hair color by using harsh chemicals that disrupt the cuticle and permit dyes to enter and stain the cortex and medulla. These color treatments damage the hair, by disrupting the protective cuticle layer and dehydrating and weakening the hair shaft. As a result, the hair becomes thin and brittle. Cream rinses and oil treatments attempt to reduce the effects of this structural damage by rehydrating and recoating the shaft.

Functions of Hair

The hairs on your body have important functions. The roughly 500,000 hairs on your head protect your scalp from ultraviolet light, help cushion a (light) blow to the head, and insulate the skull. The hairs guarding the entrances to your nostrils and external ear canals help prevent the entry of foreign particles and insects, and your eyelashes perform a similar function for the surface of the eye. A **root hair plexus** of sensory nerves surrounds the base of each hair follicle (Figure 5–10a●). As a result, you can feel the movement of the shaft of even a single hair. This sensitivity provides an early-warning system that may help prevent injury. For example, you may be able to swat a mosquito before it reaches your skin.

Ribbons of smooth muscle called **arrector pili** (a-REK-tor PI-lē; plural, *arrectores pilorum*) muscles extend from the papillary layer of the dermis to the connective tissue sheath surrounding the hair follicle. When stimulated, the arrector pili pulls on the follicles and forces the hairs to stand erect. Contraction may be the result of emotional states, such as fear or rage, or a response to cold, producing "goose bumps." In a furry mammal, this action increases the thickness of the insulating coat, rather like putting on an extra sweater. Although humans do not receive any comparable insulating benefits, the reflex persists.

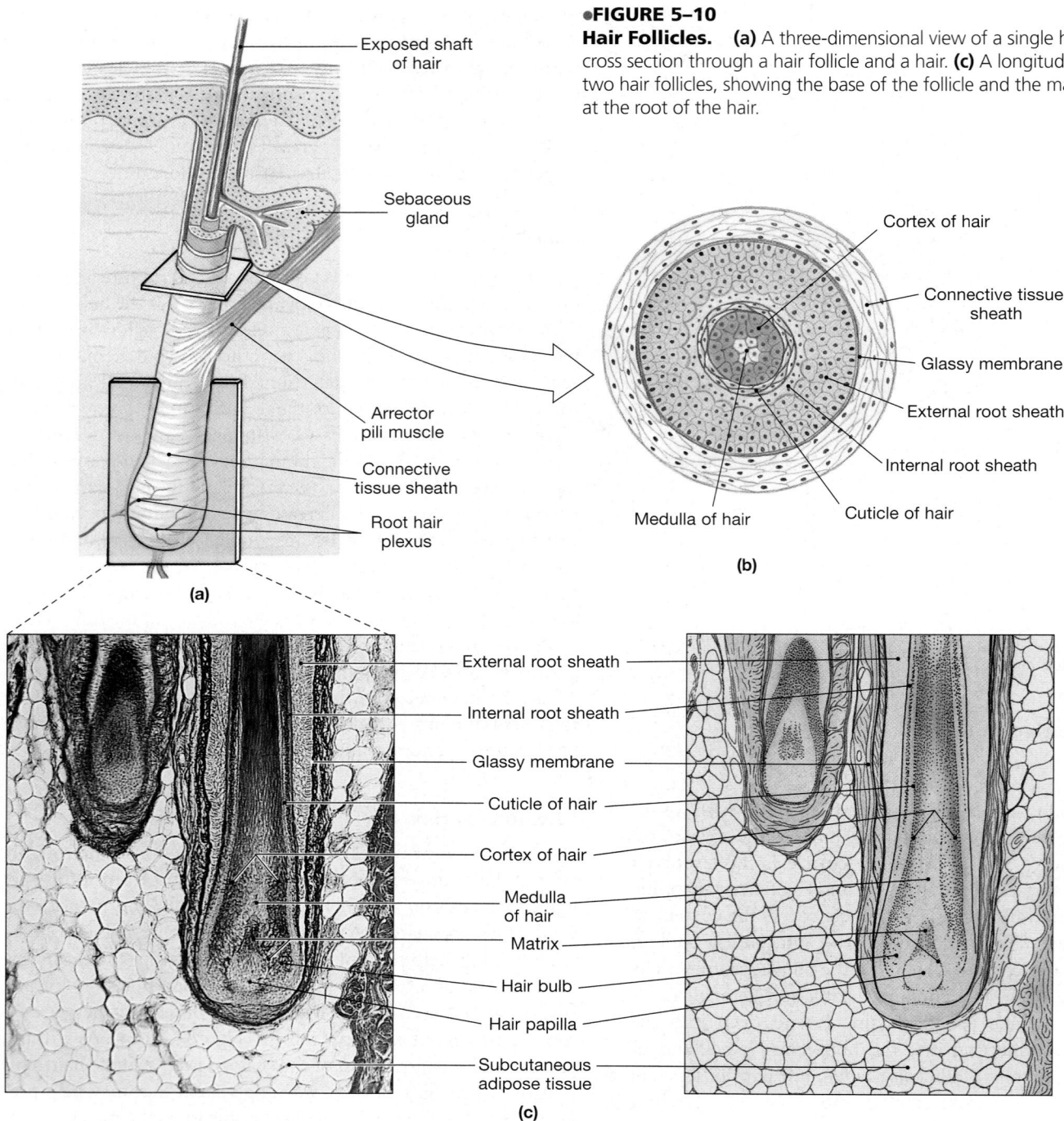

●**FIGURE 5–10**
Hair Follicles. **(a)** A three-dimensional view of a single hair follicle. **(b)** A cross section through a hair follicle and a hair. **(c)** A longitudinal section along two hair follicles, showing the base of the follicle and the matrix and papilla at the root of the hair.

Exposed shaft of hair

Sebaceous gland

Arrector pili muscle

Connective tissue sheath

Root hair plexus

(a)

Cortex of hair

Connective tissue sheath

Glassy membrane

External root sheath

Internal root sheath

Cuticle of hair

Medulla of hair

(b)

External root sheath

Internal root sheath

Glassy membrane

Cuticle of hair

Cortex of hair

Medulla of hair

Matrix

Hair bulb

Hair papilla

Subcutaneous adipose tissue

(c)

Hair Production and Follicle Structure

Hair production involves a specialization of keratinization. It begins at the deepest part of the hair follicle (Figure 5–10a●). The epithelial layer involved is the **hair matrix** (Figure 5–10c●). Basal cells near the center of the hair matrix divide, producing daughter cells that are gradually pushed toward the surface. Those cells closest to the center of the matrix form the medulla, cells closer to the edge of the developing hair form the cortex, and those at the outer margins form the cuticle. As cell divisions continue at the hair matrix, the daughter cells are pushed toward the surface of the skin, and

the hair gets longer. Keratinization is completed by the time these cells approach the surface. At this point, which corresponds to the start of the hair shaft, the cells of the medulla, cortex, and cuticle are dead, and the keratinization process is at an end.

The epithelial cells of the follicle walls are organized into several concentric layers (Figure 5–10b,c●). Outward from the hair cuticle, these layers include:

- The **internal root sheath**, which surrounds the hair root and the deeper portion of the shaft. The internal root sheath is produced by the cells at the periphery of the hair matrix. The cells of this sheath disintegrate quickly. This layer does not extend the entire length of the follicle.

- The **external root sheath**, which in longitudinal section extends from the skin surface to the hair matrix. Over most of that distance, it has all the cell layers found in the superficial epidermis. However, where the external root sheath joins the hair matrix, all the cells resemble those of the stratum germinativum.

- The **glassy membrane**, a thickened basal lamina wrapped in a dense connective tissue sheath. This membrane is in contact with the surrounding connective tissues of the dermis.

Growth and Replacement of Hair

Our hairs grow and are shed according to a **hair growth cycle**. A hair in the scalp grows for two to five years, at a rate of about 0.33 mm per day. Variations in the growth rate and in the duration of the hair growth cycle account for individual differences in the length of uncut hair.

While hair is growing, the cells of the hair root absorb nutrients and incorporate them into the hair structure. Clipping or collecting hair for analysis can be helpful in diagnosing several disorders. For example, hairs of individuals with lead poisoning or other heavy-metal poisoning contain high quantities of those metal ions. Hair samples can also be used for identification purposes through DNA fingerprinting. p. 85

As it grows, the root is firmly attached to the matrix of the follicle. At the end of the growth cycle, the follicle becomes inactive. The hair is now a **club hair**. The follicle gets smaller, and over time the connections between the hair matrix and the club hair root break down. When another cycle begins, the follicle produces a new hair; the old club hair is pushed to the surface and is shed.

If you are a healthy adult, you lose about 50 hairs from your head each day. Sustained losses of more than 100 hairs per day generally indicate that something is wrong.

Temporary increases in hair loss can result from drugs, dietary factors, radiation, an excess of vitamin A, high fever, stress, or hormonal factors related to pregnancy. In males, changes in the level of the sex hormones circulating in the blood can affect the scalp, causing a shift from terminal to vellus hair production, beginning at the temples and the crown of the head. This alteration is called **male pattern baldness**. Some cases of male pattern baldness respond to drug therapies, such as the topical application of *minoxidil* (*Rogaine*™). **AM** Baldness and Hirsuitism

✔ What happens when the arrector pili muscles contract?

✔ A person gets a burn on the forearm that destroys the epidermis and extensive areas of the deep dermis. When the injury heals, will hair grow again in the affected area?

Answers start on page Q-1

☐ GLANDS IN THE SKIN

The skin contains two types of exocrine glands: (1) *sebaceous glands* and (2) *sweat glands*. Sebaceous glands produce an oily lipid that coats hair shafts and the epidermis. Sweat glands produce a watery solution and perform other special functions.

Sebaceous (Oil) Glands

Sebaceous (se-BĀ-shus) **glands**, or *oil glands*, are holocrine glands that discharge a waxy, oily secretion into hair follicles (Figure 5–11). Sebaceous glands that communicate with a single follicle share a duct and thus are classified as simple branched alveolar glands. p. 121 The gland cells produce large quantities of lipids as they mature. The lipid product is released through holocrine secretion, a process that involves the rupture of the secretory cells. p. 120

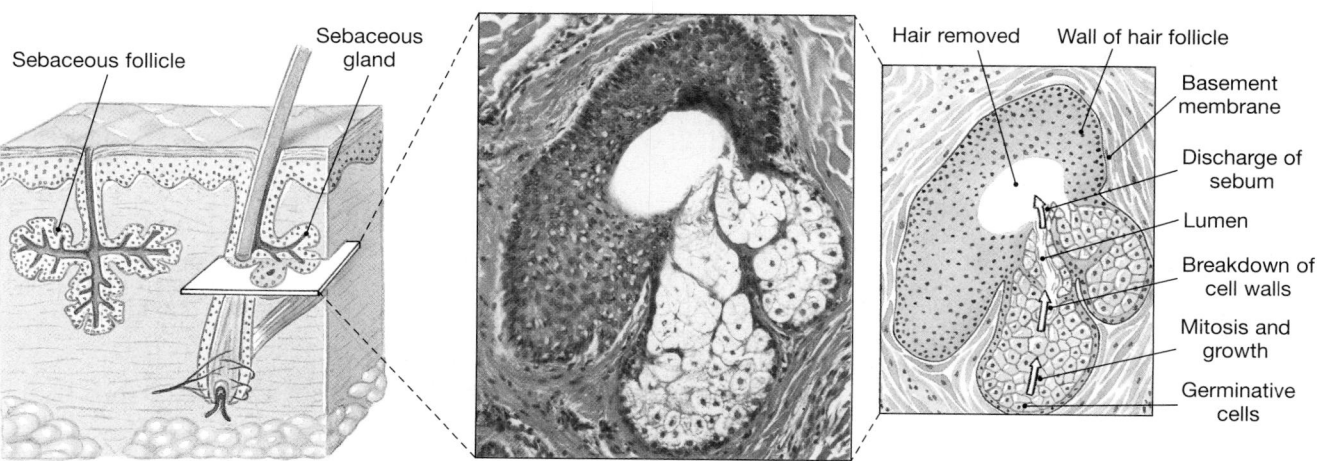

Sebaceous gland (LM × 150)

●**FIGURE 5–11**
Sebaceous Glands and Follicles. The structure of sebaceous glands and sebaceous follicles in the skin.

The lipids released from gland cells enter the lumen, or open passageway of the gland. The arrector pili muscles that erect the hair then contract, squeezing the sebaceous gland and forcing the lipids into the hair follicle and onto the surface of the skin. The secretion of lipids, called **sebum** (SĒ-bum), is a mixture of triacylglycerides, cholesterol, proteins, and electrolytes. Sebum inhibits the growth of bacteria, lubricates and protects the keratin of the hair shaft, and conditions the surrounding skin. Keratin is a tough protein, but dead, keratinized cells become dry and brittle once exposed to the environment. It is interesting to reflect on our custom of washing and shampooing to remove the oily secretions of sebaceous glands, only to add other lipids to the hair in the form of cream rinse and to the skin in the form of creams and lotions. **Sebaceous follicles** are large sebaceous glands that are not associated with hair follicles. Their ducts discharge sebum directly onto the epidermis. Sebaceous follicles are located on the face, back, chest, nipples, and external genitalia.

Surprisingly, sebaceous glands are very active during the last few months of fetal development. Their secretions, mixed with shed epidermal cells, form a protective superficial layer that coats the skin surface. This layer is called the *vernix caseosa*. Sebaceous gland activity all but ceases after birth, but increases again at puberty in response to rising levels of sex hormones.

Seborrheic dermatitis is an inflammation around abnormally active sebaceous glands. The affected area becomes red, and some epidermal scaling usually occurs. Sebaceous glands of the scalp are most often involved. In infants, mild cases are called *cradle cap*. This condition is one cause of dandruff in adults. Anxiety, stress, and food allergies can aggravate the problem. **AM** Folliculitis and Acne

Sweat Glands

The skin contains two types of sweat glands, or **sudoriferous glands:** (1) *apocrine sweat glands* and (2) *merocrine sweat glands* (Figure 5–12●). These names refer to the mechanism of secretion, introduced in Chapter 4. p. 120

APOCRINE SWEAT GLANDS In the armpits (axillae), around the nipples, and in the groin, **apocrine sweat glands** secrete their products into hair follicles (Figure 5–12a●). These coiled, tubular glands produce a sticky, cloudy, and potentially odorous secretion. The name *apocrine* was originally chosen because it was thought that the gland cells use an apocrine method of secretion. Although we now know that they rely on merocrine secretion, the name has not changed.

Apocrine sweat glands begin secreting at puberty. The sweat produced is a nutrient source for bacteria, which intensify its odor. Special **myoepithelial cells** (*myo-*, muscle) surround the secretory cells in these glands. Myoepithelial cells contract, squeezing the gland and thereby discharging the accumulated secretion into the hair follicles. The secretory activities of the gland cells and the contractions of myoepithelial cells are controlled by the nervous system and by circulating hormones.

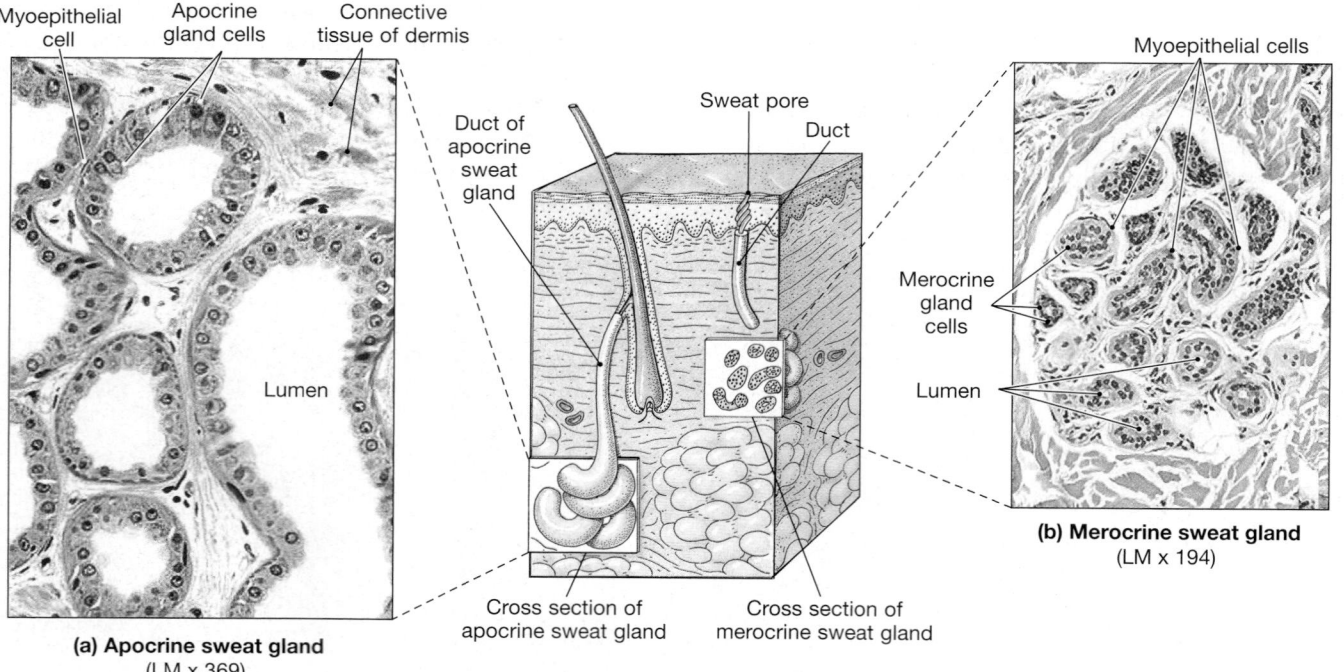

●FIGURE 5–12
Sweat Glands. **(a)** Apocrine sweat glands, located in the axillae, groin, and nipples, produce a thick, odorous fluid. **(b)** Merocrine sweat glands produce a watery fluid.

MEROCRINE (ECCRINE) SWEAT GLANDS Merocrine sweat glands, also known as **eccrine** (EK-rin) **sweat glands** (Figure 5–12•) are far more numerous and widely distributed than apocrine glands. The adult integument contains 2–5 million merocrine sweat glands, which are smaller than apocrine sweat glands and do not extend as far into the dermis. The palms and soles have the highest numbers, with the palm possessing an estimated 500 merocrine sweat glands per square centimeter (3000 per in.^{2}). These coiled, tubular glands discharge their secretions directly onto the surface of the skin.

As we learned earlier, the sweat produced by merocrine sweat glands is called sensible perspiration. Sweat is 99 percent water, but it also contains some electrolytes (chiefly sodium chloride), a number of organic nutrients, an antibiotic, and various waste products. It has a pH of 4–6.8, and the presence of sodium chloride gives sweat a salty taste. (See Appendix IV for a complete analysis of the composition of normal sweat.)

The functions of merocrine sweat gland activity include:

- **Cooling the Surface of the Skin to Reduce Body Temperature.** This is the primary function of sensible perspiration, and the degree of secretory activity is regulated by neural and hormonal mechanisms. When all the merocrine sweat glands are working at their maximum, the rate of perspiration can exceed a gallon per hour; dangerous fluid and electrolyte losses can occur. For this reason, athletes in endurance sports must pause at regular intervals to drink fluids.
- **Excreting Water and Electrolytes.** A number of ingested drugs are excreted as well.
- **Providing Protection from Environmental Hazards.** Sweat dilutes harmful chemicals and discourages the growth of microorganisms in two ways: (1) by flushing them from the surface or making it difficult for them to adhere to the epidermal surface and (2) through the action of *dermicidin*, a small peptide that has powerful antibiotic properties.

Other Integumentary Glands

Merocrine sweat glands are widely distributed across the body surface, and sebaceous glands are located wherever there are hair follicles. Apocrine sweat glands are located in relatively restricted areas. The skin also contains a variety of specialized glands that are restricted to specific locations. We will encounter many of them in later chapters, but we cite two important examples here:

1. The **mammary glands** of the breasts are anatomically related to apocrine sweat glands. A complex interaction between sex hormones and pituitary hormones controls their development and secretion. We shall discuss mammary gland structure and function in Chapter 28.
2. **Ceruminous** (se-ROO-mi-nus) **glands** are modified sweat glands in the passageway of the external ear. Their secretions combine with those of nearby sebaceous glands, forming a mixture called **cerumen**, or earwax. Together with tiny hairs along

the ear canal, earwax helps trap foreign particles, preventing them from reaching the eardrum.

Control of Glandular Secretions and the Homeostatic Role of the Integument

Sebaceous glands and apocrine sweat glands can be collectively turned on or off by the autonomic nervous system, but no regional control is possible. When one sebaceous or apocrine gland is activated, so are all the other glands of that type in the body. Merocrine sweat glands are much more precisely controlled, and the amount of secretion and the area of the body involved can vary independently. For example, when you are nervously awaiting an anatomy and physiology exam, only your palms may begin to sweat.

As we have seen, the primary function of sensible perspiration is to cool the surface of the skin and to reduce body temperature. This is a key component of *thermoregulation*, the process of maintaining temperature homeostasis, when the environmental temperature is high. When you sweat in the hot sun, all your merocrine glands are working together. The blood vessels beneath your epidermis are flushed with blood, and your skin reddens. The surface of your skin is warm and wet. As the moisture evaporates, your skin cools. If your body temperature then falls below normal, sensible perspiration ceases, blood flow to the skin is reduced, and the cool, dry surface releases little heat into the environment. We introduced the negative feedback mechanisms of thermoregulation in Chapter 1 pp. 11–12; we provide additional details in Chapter 25.

✔ What are the functions of sebaceous secretions?

✔ Deodorants are used to mask the effects of secretions from which type of skin gland?

✔ Which type of skin glands are most affected by the hormonal changes that occur during puberty?

Answers start on page Q-1

☐ NAILS

Nails form on the dorsal surfaces of the tips of the fingers and toes (Figure 5–13a•). The nails protect the exposed tips of the fingers and toes and help limit their distortion when they are subjected to mechanical stress—for example, when you run or grasp objects. The **nail body**, the visible portion of the nail, covers an area of epidermis called the **nail bed** (Figure 5–13b•). The nail body is recessed deep to the level of the surrounding epithelium and is bounded on either side by **lateral nail grooves** (depressions) and **lateral nail folds**. The **free edge** of the nail—the distal portion that continues past the nail bed—extends over the **hyponychium** (hī-pō-NIK-ē-um), an area of thickened stratum corneum.

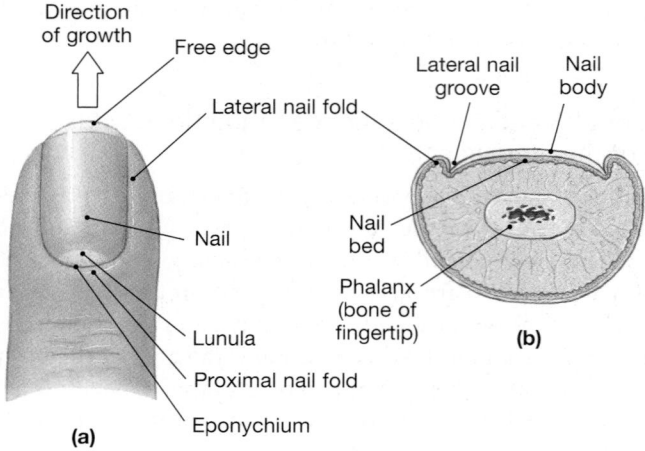

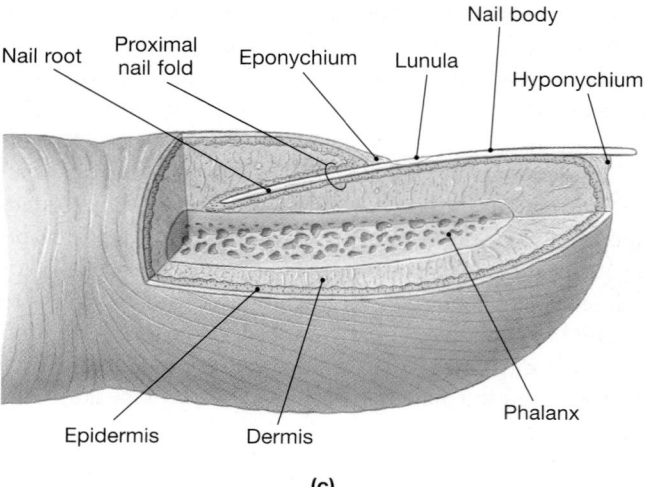

•FIGURE 5–13
The Structure of a Nail.　**(a)** A superficial view.　**(b)** A cross-sectional view. **(c)** A longitudinal section.

Nail production occurs at the **nail root**, an epidermal fold not visible from the surface (Figure 5–13c•). The deepest portion of the nail root lies very close to the bone of the fingertip. A portion of the stratum corneum of the nail root extends over the exposed nail, forming the **eponychium** (ep-ō-NIK-ē-um; *epi-*, over + *onyx*, nail), or **cuticle.** Underlying blood vessels give the nail its characteristic pink color. Near the root, these vessels may be obscured, leaving a pale crescent known as the **lunula** (LOO-nū-la; *luna*, moon) (Figure 5–13a,c•).

The body of the nail consists of dead, tightly compressed cells packed with keratin. The cells producing the nails can be affected by conditions that alter body metabolism, so changes in the shape, structure, or appearance of the nails can assist in diagnosis. For example, the nails may turn yellow in individuals who

have chronic respiratory disorders, thyroid gland disorders, or AIDS. Nails may become pitted and distorted as a result of *psoriasis* (a condition marked by rapid stem cell division in the stratum germinativum) and concave as a result of some blood disorders. **AM** Disorders of Keratin Production

5–6 LOCAL CONTROL OF INTEGUMENTARY FUNCTION

Objective

■ Explain how the skin responds to injury and repairs itself.

The integumentary system displays a significant degree of functional independence—it often responds directly and automatically to local influences without the involvement of the nervous or endocrine system. For example, when the skin is subjected to mechanical stresses, stem cells in the stratum germinativum divide more rapidly and the depth of the epithelium increases. That is why calluses form on your palms when you perform manual labor. A more dramatic display of local regulation can be seen after an injury to the skin.

☐ INJURY AND REPAIR

The skin can regenerate effectively even after considerable damage has occurred, because stem cells persist in both the epithelial and connective tissue components. Germinative cell divisions replace lost epidermal cells, and mesenchymal cell divisions replace lost dermal cells. The process can be slow. When large surface areas are involved, problems of infection and fluid loss complicate the situation. The relative speed and effectiveness of skin repair vary with the type of wound involved. A slender, straight cut, or *incision*, may heal relatively quickly compared with a deep scrape, or *abrasion*, which involves a much greater surface area to be repaired. **AM** Trauma to the Skin

Figure 5–14• illustrates the four stages in the regeneration of the skin after an injury. When damage extends through the epidermis and into the dermis, bleeding generally occurs (STEP 1). The blood clot, or **scab**, that forms at the surface temporarily restores the integrity of the epidermis and restricts the entry of additional microorganisms into the area (STEP 2). The bulk of the clot consists of an insoluble network of *fibrin*, a fibrous protein that forms from blood proteins during the clotting response. The clot's color reflects the presence of trapped red blood cells. Cells of the stratum germinativum undergo rapid divisions and begin to migrate along the sides of the wound in an attempt to replace the missing epidermal cells. Meanwhile, macrophages patrol the damaged area of the dermis, phagocytizing any debris and pathogens.

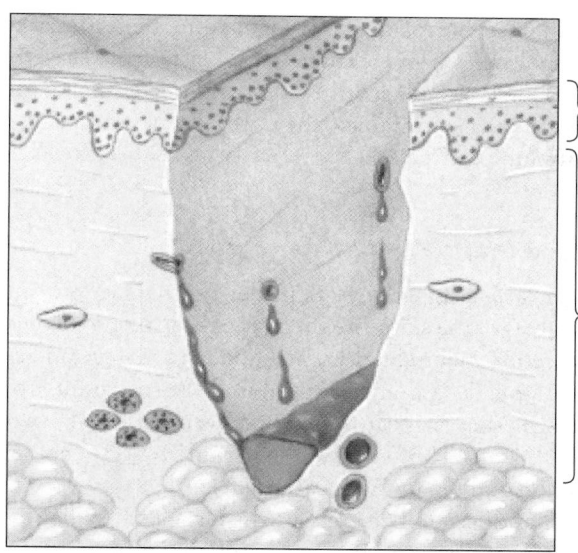

Epidermis

Dermis

STEP 1:
Bleeding occurs at the site of injury immediately after the injury, and mast cells in the region trigger an inflammatory response.

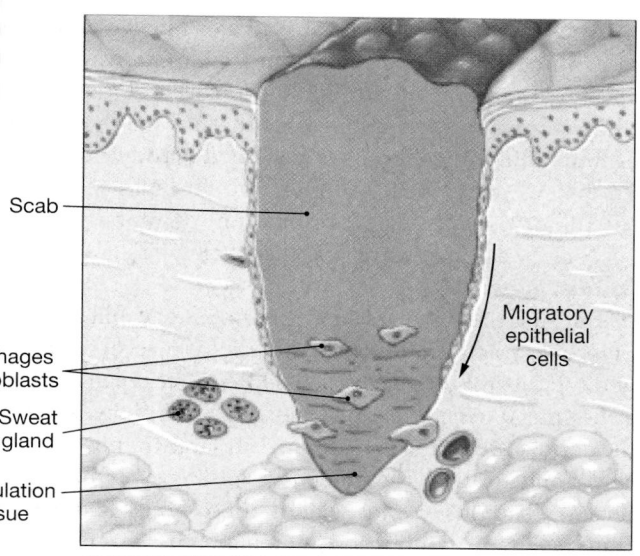

Scab

Macrophages and fibroblasts

Sweat gland

Granulation tissue

Migratory epithelial cells

STEP 2:
After several hours, a scab has formed and cells of the stratum germinativum are migrating along the edges of the wound. Phagocytic cells are removing debris, and more of these cells are arriving via the enhanced circulation in the area. Clotting around the edges of the affected area partially isolates the region.

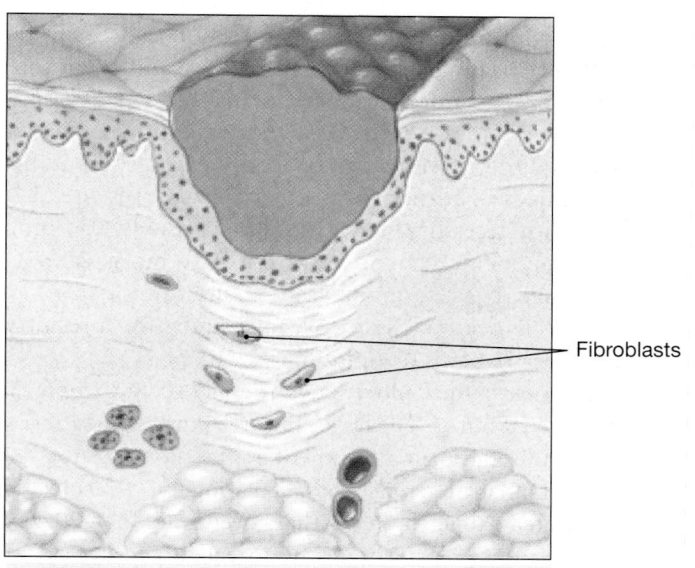

Fibroblasts

STEP 3:
One week after the injury, the scab has been undermined by epidermal cells migrating over the meshwork produced by fibroblast activity. Phagocytic activity around the site has almost ended, and the fibrin clot is disintegrating.

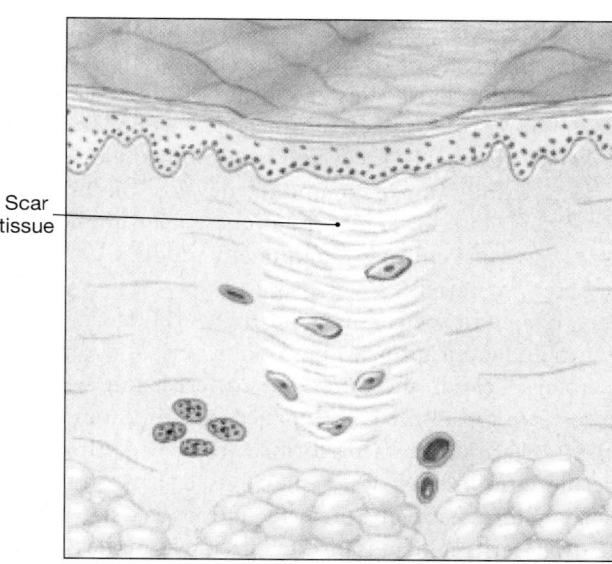

Scar tissue

STEP 4:
After several weeks, the scab has been shed, and the epidermis is complete. A shallow depression marks the injury site, but fibroblasts in the dermis continue to create scar tissue that will gradually elevate the overlying epidermis.

●**FIGURE 5–14**
Integumentary Repair

If the wound occupies an extensive area or involves a region covered by thin skin, dermal repairs must be underway before epithelial cells can cover the surface. Fibroblast and mesenchymal cell divisions produce mobile cells that invade the deeper areas of injury. Endothelial cells of damaged blood vessels also begin to divide, and capillaries follow the fibroblasts, providing a circulatory supply. The combination of blood clot, fibroblasts, and an extensive capillary network is called **granulation tissue**.

Over time, the clot dissolves and the number of capillaries declines. Fibroblast activity leads to the appearance of collagen fibers and typical ground substance (STEP 3). The repairs do not restore the integument to its original condition, however, because the dermis will contain an abnormally large number of collagen fibers and relatively few blood vessels. Severely damaged hair follicles, sebaceous or sweat glands, muscle cells, and nerves are seldom repaired, and they too are replaced by fibrous tissue. The formation of this rather inflexible, fibrous, noncellular **scar tissue** can be considered a practical limit to the healing process (STEP 4).

We do not know what regulates the extent of scar tissue formation, and the process is highly variable. For example, surgical procedures performed on a fetus do not leave scars, perhaps because damaged fetal tissues do not produce the same types of growth factors that adult tissues do. In some adults, most often those with dark skin, scar tissue formation may continue beyond the requirements of tissue repair. The result is a flattened mass of scar tissue that begins at the site of injury and grows into the surrounding dermis. This thickened area of scar tissue, called a **keloid** (KĒ-loyd), is covered by a shiny, smooth epidermal surface. Keloids most commonly develop on the upper back, shoulders, anterior chest, or earlobes. They are harmless, and some aboriginal cultures intentionally produce keloids as a form of body decoration.

In fact, people in societies around the world adorn the skin with culturally significant markings of one kind or another. Tattoos, piercings, keloids and other scar patterns, and even high-fashion makeup are all used to "enhance" the appearance of the integument. Scarification is performed by several African cultures, resulting in a series of complex, raised scars on the skin. Polynesian cultures have long preferred tattoos as a sign of status and beauty. A dark pigment is inserted deep within the dermis of the skin by tapping on a needle, shark tooth, or bit of bone. Because the pigment is inert, if infection does not occur (a potentially serious complication), the markings remain for the life of the individual, clearly visible through the overlying epidermis. American popular culture has recently rediscovered tattoos as a fashionable form of body adornment. The colored inks that are commonly used are less durable, and older tattoos can fade or lose their definition.

Tattoos can now be partially or completely removed. The removal process takes time (10 or more sessions may be required to remove a large tattoo), and scars often remain. To remove the tat-

too, an intense, narrow beam of light from a laser breaks down the ink molecules in the dermis. Each blast of the laser that destroys the ink also burns the surrounding dermal tissue. Although the burns are minor, they accumulate and result in the formation of localized scar tissue.

Burns and Grafts

Burns are special injuries, in that they can damage the integrity of large areas of the skin. The integumentary functions of protection, excretion, maintenance, detection, and storage and synthesis of vitamin D are all compromised. Burns result from the exposure of skin to heat, radiation, electrical shock, or strong chemical agents. The severity of the burn reflects the depth of penetration and the total area affected.

First- and second-degree burns are also called *partial-thickness burns*, because damage is restricted to the superficial layers of the skin. Only the surface of the epidermis is affected by a *first-degree burn*. In this type of burn, including most sunburns, the skin reddens and can be painful. The redness, a sign called **erythema** (er-i-THĒ-ma), results from inflammation of the sun-damaged tissues. In a *second-degree burn*, the entire epidermis and perhaps some of the dermis are damaged. Accessory structures such as hair follicles and glands are generally not affected, but blistering, pain, and swelling occur. If the blisters rupture at the surface, infection can easily develop. Healing typically takes one to two weeks, and some scar tissue may form. *Full-thickness burns*, or *third-degree burns*, destroy the epidermis and dermis, extending into subcutaneous tissues. Despite swelling, these burns are less painful than second-degree burns, because sensory nerves are destroyed along with accessory structures, blood vessels, and other dermal components. Extensive third-degree burns cannot repair themselves, because granulation tissue cannot form and epithelial cells are unable to cover the injury. As a result, the affected area remains open to infection.

Each year in the United States, roughly 10,000 people die from the effects of burns. The larger the area burned, the more significant are the effects on integumentary function. Figure 5–15● presents a standard reference for calculating the percentage of total surface area damaged. Burns that cover more than 20 percent of the skin surface represent serious threats to life, because they affect the following functions:

- **Fluid and Electrolyte Balance.** Even areas with partial-thickness burns lose their effectiveness as barriers to fluid and electrolyte losses. In full-thickness burns, the rate of fluid loss through the skin may reach five times the normal level.

- **Thermoregulation.** Increased fluid loss means increased evaporative cooling. More energy must be expended to keep the body temperature within acceptable limits.

- **Protection from Attack.** The epidermal surface, damp from uncontrolled fluid losses, encourages bacterial growth. If the

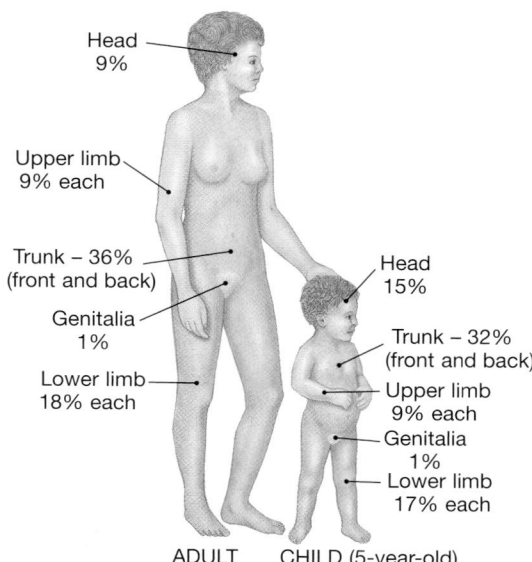

ADULT   CHILD (5-year-old)

●**FIGURE 5–15**
A Quick Method of Estimating the Percentage of Surface Area Affected by Burns. This method is called the *rule of nines*, because the surface area in adults is divided into multiples of nine. The rule must be modified for children, because their proportions are quite different.

skin is broken at a blister or at the site of a third-degree burn, infection is likely. Widespread bacterial infection, or **sepsis** (*septikos*, rotting), is the leading cause of death in burn victims.

Effective treatment of full-thickness burns focuses on the following four procedures:

1. Replacing lost fluids and electrolytes.
2. Providing sufficient nutrients to meet increased metabolic demands for thermoregulation and healing.
3. Preventing infection by cleaning and covering the burn while administering antibiotic drugs.
4. Assisting tissue repair.

Because large full-thickness burns cannot heal unaided, surgical procedures are necessary to encourage healing. In a **skin graft**, areas of intact skin are transplanted to cover the site of the burn. A *split-thickness graft* takes a shaving of the epidermis and superficial portions of the dermis. A *full-thickness graft* involves the epidermis and both layers of the dermis.

With fluid-replacement therapies, infection control methods, and grafting techniques, young patients with burns over 80 percent of the body have about a 50 percent chance of recovery. Recent advances in cell culturing may improve survival rates further. A small section of undamaged epidermis can be removed and

grown in the laboratory. Over time, germinative cell divisions produce large sheets of epidermal cells that can cover the burn area. From initial samples the size of a postage stamp, square meters of epidermis have been grown and transplanted onto body surfaces. Although questions remain about the strength and flexibility of the repairs, skin cultivation is a substantial advance in the treatment of serious burns. **AM** Synthetic Skin

5–7 AGING AND THE INTEGUMENTARY SYSTEM

Objective

■ Summarize the effects of the aging process on the skin.

Aging affects all the components of the integumentary system:

- The epidermis thins as germinative cell activity declines, making older people more prone to injury and skin infections.
- The number of Langerhans cells decreases to about 50 percent of levels seen at maturity (roughly, age 21). This decrease may reduce the sensitivity of the immune system and further encourage skin damage and infection.
- Vitamin D_3 production declines by about 75 percent. The result can be reduced calcium and phosphate absorption, eventually leading to muscle weakness and a reduction in bone strength.
- Melanocyte activity declines, and in Caucasians the skin becomes very pale. With less melanin in the skin, people become more sensitive to exposure to the sun and more likely to experience sunburn.
- Glandular activity declines. The skin becomes dry and often scaly, because sebum production is reduced. Merocrine sweat glands are also less active, and with impaired perspiration, older people cannot lose heat as fast as younger people can. Thus, the elderly are at greater risk of overheating in warm environments.
- The blood supply to the dermis is reduced at the same time that the sweat glands are becoming less active. Because blood flow decreases, the skin becomes cool, which in turn can stimulate thermoreceptors, making a person feel cold even in a warm room. However, with reduced circulation and sweat gland function, the elderly are less able than younger people to lose body heat. As a result, overexertion or exposure to high temperatures (such as those in a sauna or hot tub) can cause dangerously high body temperatures.
- Hair follicles stop functioning or produce thinner, finer hairs. With decreased melanocyte activity, these hairs are gray or white.
- The dermis thins, and the elastic fiber network decreases in size. The integument therefore becomes weaker and less resilient, and sagging and wrinkling occur. These effects are most pronounced in areas of the body that have been exposed to the sun.

- With changes in levels of sex hormones, secondary sexual characteristics in hair and body-fat distribution begin to fade. As a consequence, people aged 90–100 of both sexes and all races tend to look alike.

- Skin repairs proceed relatively slowly, and recurring infections may result. Skin repairs proceed most rapidly in young, healthy individuals. For example, barring infection, it might take three to four weeks to complete the repairs to a blister site in a young adult; the same repairs at age 65–75 could take six to eight weeks.

✓ What do you call the combination of fibrin clots, fibroblasts, and the extensive network of capillaries in healing tissue?

✓ Why can skin regenerate effectively even after considerable damage?

✓ Older individuals do not tolerate the summer heat as well as they did when they were young, and they are more prone to heat-related illness. What accounts for this change?

Answers start on page Q-1

Chapter Review

SELECTED CLINICAL TERMINOLOGY

Terms Discussed in This Chapter

basal cell carcinoma: A skin cancer caused by malignant stem cells within the stratum germinativum. (*p. 162*)

cyanosis (sī-uh-NŌ-sis): Bluish skin color as a result of reduced oxygenation of the blood in superficial vessels. (*p. 161*)

decubitis ulcers (*bedsores*): Ulcers that occur in areas subject to restricted circulation, especially common in bedridden persons. (*p. 165*)

dermatitis: An inflammation of the skin that primarily involves the papillary region of the dermis. (*p. 163*)

granulation tissue: A combination of fibrin, fibroblasts, and capillaries that forms during tissue repair after inflammation. (*p. 174*)

hypodermic needle: A needle used to administer drugs via subcutaneous injection. (*p. 166*)

keloid (KĒ-loyd): A thickened area of scar tissue covered by a shiny, smooth epidermal surface. (*p. 174*)

male pattern baldness: Hair loss in an adult male due to changes in levels of circulating sex hormones. (*p. 169*)

malignant melanoma (mel-uh-NŌ-muh): A skin cancer originating in malignant melanocytes. (*p. 161 and AM*)

scab: A blood clot that forms at the surface of a wound to the skin. (*p. 172*)

seborrheic dermatitis: An inflammation around abnormally active sebaceous glands. (*p. 170*)

sepsis: A dangerous, widespread bacterial infection; the leading cause of death in burn patients. (*p. 175*)

skin graft: The transplantation of a section of skin (partial thickness or full thickness) to cover an extensive injury, such as a third-degree burn. (*p. 175*)

squamous cell carcinoma: A skin cancer resulting from chronic exposure to excessive amounts of UV radiation in sunlight. (*p. 162*)

ulcer: A localized shedding of an epithelium. (*p. 165*)

AM Additional Terms Discussed in the *Applications Manual*

acne
contact dermatitis
diaper rash
eczema
erysipelas
folliculitis
furuncle
pruritis
psoriasis
skin signs
urticaria
xerosis

STUDY OUTLINE

5–1 THE INTEGUMENTARY SYSTEM: AN OVERVIEW p. 155

1. The **integument**, or **integumentary system**, consists of the **cutaneous membrane** or *skin* (which includes the **epidermis** and the **dermis**) and the **accessory structures**. Beneath the dermis lies the **subcutaneous layer**. (*Figure 5–1*)

2. Functions of the integument include protection, excretion, temperature maintenance, nutrient storage, vitamin D_3 synthesis, and sensory detection.

5–2 THE EPIDERMIS   p. 156

1. **Thin skin**, formed by four layers of **keratinocytes**, covers most of the body. Heavily abraded body surfaces may be covered by **thick skin**, formed by five layers. *(Figure 5–2)*
2. The epidermis provides mechanical protection, prevents fluid loss, and helps keep microorganisms out of the body.

LAYERS OF THE EPIDERMIS   p. 157

3. Cell divisions in the **stratum germinativum**, the innermost epidermal layer, replace more superficial cells. *(Figure 5–3)*
4. As epidermal cells age, they pass through the **stratum spinosum**, the **stratum granulosum**, the **stratum lucidum** (in thick skin), and the **stratum corneum**. In the process, they accumulate large amounts of **keratin**. Ultimately, the cells are shed or lost. *(Figure 5–4)*
5. **Epidermal ridges**, interlocked with **dermal papillae** of the underlying dermis, improve our gripping ability and increase the skin's sensitivity.
6. **Langherhans cells** in the stratum spinosum are part of the immune system. **Merkel cells** in the stratum germinativum provide sensory information about objects that touch the skin.

Wound healing process: **Companion Website**: Chapter 5/ Tutorials.

SKIN COLOR   p. 159

7. The color of the epidermis depends on two factors: blood supply and epidermal pigmentation.
8. The epidermis contains the pigments **carotene** and **melanin**. **Melanocytes**, which produce melanin, protect us from **ultraviolet (UV) radiation**. *(Figure 5–5)*
9. Interruptions of the dermal blood supply can lead to **cyanosis**.

THE EPIDERMIS AND VITAMIN D₃   p. 161

10. Epidermal cells synthesize **vitamin D$_3$**, or **cholecalciferol**, when exposed to the UV radiation in sunlight.

EPIDERMAL GROWTH FACTOR   p. 161

11. **Epidermal growth factor** (EGF) promotes growth, division, and repair of the epidermis and the secretion of epithelial glands.

5–3 THE DERMIS   p. 163

DERMAL ORGANIZATION   p. 163

1. The dermis consists of the superficial **papillary layer** and the deeper **reticular layer**. *(Figures 5–1, 5–2)*
2. The papillary layer of the dermis contains blood vessels, lymphatics, and sensory nerves that supply the epidermis. The reticular layer consists of a meshwork of collagen and elastic fibers oriented to resist tension in the skin.
3. Extensive distension of the dermis can cause **stretch marks**.
4. The pattern of collagen and elastic fiber bundles forms **lines of cleavage**. *(Figure 5–6)*

DERMAL CIRCULATION AND INNERVATION   p. 164

5. Arteries of the **cutaneous plexus** branch into the subcutaneous layer and the papillary dermis.
6. Integumentary sensory receptors detect both light touch and pressure.

5–4 THE SUBCUTANEOUS LAYER   p. 166

1. The subcutaneous layer, or **hypodermis**, stabilizes the skin's position against underlying organs and tissues. *(Figure 5–1)*

5–5 ACCESSORY STRUCTURES   p. 166

HAIR FOLLICLES AND HAIR   p. 166

1. **Hairs** originate in complex organs called **hair follicles**. Each hair has a **root** and a **shaft**. At the base of the root are a **hair papilla**, surrounded by a **hair bulb**, and a **root hair plexus** of sensory nerves. Hairs have a **medulla**, or core of soft keratin, surrounded by a **cortex** of hard keratin. The **cuticle** is a superficial layer of dead cells that protects the hair. *(Figures 5–9, 5–10)*
2. Our bodies have both **vellus hairs** ("peach fuzz") and heavy **terminal hairs**. A hair that has stopped growing is called a **club hair**.
3. The **arrector pili** muscles can erect the hairs. *(Figure 5–10)*
4. Our hairs grow and are shed according to the **hair growth cycle**. A typical hair grows for two to five years and is subsequently shed.

GLANDS IN THE SKIN   p. 169

5. A typical **sebaceous gland** discharges the waxy **sebum** into a **lumen** and, ultimately, into a hair follicle. **Sebaceous follicles** are large sebaceous glands that empty directly onto the surface of the epidermis. *(Figure 5–12)*
6. The two classes of sweat glands, or **sudoriferous glands**, are apocrine and merocrine sweat glands. **Apocrine sweat glands** produce an odorous secretion. The more numerous **merocrine**, or *eccrine*, **sweat glands** produce a watery secretion known as **sensible perspiration**. *(Figure 5–12)*
7. **Mammary glands** of the breasts are structurally similar to apocrine sweat glands. **Ceruminous glands** in the ear produce a waxy **cerumen**.

NAILS   p. 171

8. The **nail body** of a **nail** covers the **nail bed**. Nail production occurs at the **nail root**, which is overlain by the **cuticle**, or **eponychium**. The **free edge** of the nail extends over the **hyponychium**. *(Figure 5–13)*

5–6 LOCAL CONTROL OF INTEGUMENTARY FUNCTION   p. 172

INJURY AND REPAIR   p. 172

1. The skin can regenerate effectively even after considerable damage. The process includes the formation of a **scab**, **granulation tissue**, and **scar tissue**. *(Figure 5–14)*

5–7 AGING AND THE INTEGUMENTARY SYSTEM   p. 175

1. With aging, the integument thins, blood flow decreases, cellular activity decreases, and repairs occur more slowly.

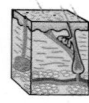

THE **INTEGUMENTARY SYSTEM** IN PERSPECTIVE

Organ/Component	Primary Functions
Cutaneous Membrane	
Epidermis	Covers surface; protects deeper tissues
Dermis	Nourishes epidermis; provides strength; contains glands
Hair Follicles	Produce hair; innervation provides sensation
Hairs	Provide some protection for head
Sebaceous Glands	Secrete lipid coating that lubricates hair shaft and epidermis
Sweat Glands	Produce perspiration for evaporative cooling
Nails	Protect and stiffen distal tips of digits
Sensory Receptors	Provide sensations of touch, pressure, temperature, pain
Subcutaneous Layer	Stores lipids; attaches skin to deeper structures

INTEGRATION WITH OTHER SYSTEMS

The integumentary system forms the external surface of the body and provides protection from dehydration, environmental chemicals, and impacts. The integument is separated and insulated from the rest of the body by the subcutaneous layer, but is interconnected with the rest of the body by an extensive circulatory network of blood and lymphatic vessels and is richly supplied with sensory nerve endings. As a result, although the protective mechanical functions of the skin can be discussed independently, its physiological activities are always tightly integrated with those of other systems.

The figure on the opposite page reviews the components and functions of the integumentary system and diagrams the major functional relationships between that system and other systems.

 ## CLINICAL PATTERNS

The skin is the most visible organ of the body. As a result, abnormalities are easily recognized. A bruise, for example, typically creates a swollen and discolored area where the walls of blood vessels have been damaged. Changes in skin color, skin tone, and the overall condition of the skin commonly accompany illness or disease. These changes can assist in the diagnosis of conditions involving other systems. For example, extensive bruising without any obvious cause may indicate a blood-clotting disorder; yellowish skin and mucous membranes may signify *jaundice*, which generally indicates some type of liver disorder. The general condition of the skin can also be significant. In addition to color changes, changes in the flexibility, elasticity, dryness, or sensitivity of the skin commonly follow the malfunctions of other organ systems.

The *Applications Manual*, discusses the diagnosis and treatment of major conditions affecting the integumentary system. For additional information and updates on clinical topics affecting this system, visit the Destinations page on the Companion Website.

 ## MEDIA CONNECTIONS

I. Objective: Explore the role of the integumentary system in thermoregulation.

Completion time = 12 minutes

Prepare a basic flow chart that indicates the role of the skin in thermoregulation, based on Figure 1–5 as well as what you have learned in this chapter. The skin has two methods of increasing the rate of heat exchange with the environment: by adjusting dermal blood flow and by sweat gland activity. Do you think that the two are linked? Do you always sweat when you are warm? How warm must your skin be before you will start sweating? Researchers are investigating the regulation of dermal blood flow while core temperature is within a range known as the "neutral zone." What is the neutral zone? Why is it significant? To answer these questions, go to the **Companion Website**, Chapter 5 Media Connections, and click on the keyword **THERMOREGULATION**. After reading the material presented on the neutral zone, modify your flow chart to incorporate the new information.

II. Objective: Develop an understanding of the integration of blood flow and temperature changes in the health of the dermis.

Completion time = 10 minutes

Decubitis ulcers, or bedsores, are formed by increased pressure on the skin, shutting off blood flow to the dermis and causing ischemia. Adding localized heat has been proposed as a way to prevent bedsores. How would this help? Write a short description of how the integument and underlying tissues respond to heat and pressure. Include a discussion of how heat might be helpful in treating bedsores. For help with this, visit the **Companion Website**, Chapter 5 Media Connections, and click on the keyword **ULCER**.

SKELETAL SYSTEM

- Provides structural support
- Synthesizes vitamin D_3, essential for calcium and phosphorus absorption (bone maintenance and growth)

MUSCULAR SYSTEM

- Contractions of skeletal muscles pull against skin of face, producing facial expressions important in communication
- Synthesizes vitamin D_3, essential for normal calcium absorption (calcium ions play an essential role in muscle contraction)

NERVOUS SYSTEM

- Controls blood flow and sweat gland activity for thermoregulation
- Stimulates contraction of arrector pili muscles to elevate hairs
- Receptors in dermis and deep epidermis provide sensations of touch, pressure, vibration, temperature, and pain

ENDOCRINE SYSTEM

- Sex hormones stimulate sebaceous gland activity
- Male and female sex hormones influence growth, distribution of subcutaneous fat, and apocrine sweat gland activity
- Adrenal hormones alter dermal blood flow and help mobilize lipids from adipocytes
- Synthesizes vitamin D_3, precursor of calcitriol

CARDIOVASCULAR SYSTEM

- Provides oxygen and nutrients; delivers hormones and cells of immune system
- Carries away carbon dioxide, waste products, and toxins
- Provides heat to maintain normal skin temperature
- Stimulation of mast cells produces localized changes in blood flow and capillary permeability

INTEGUMENTARY SYSTEM

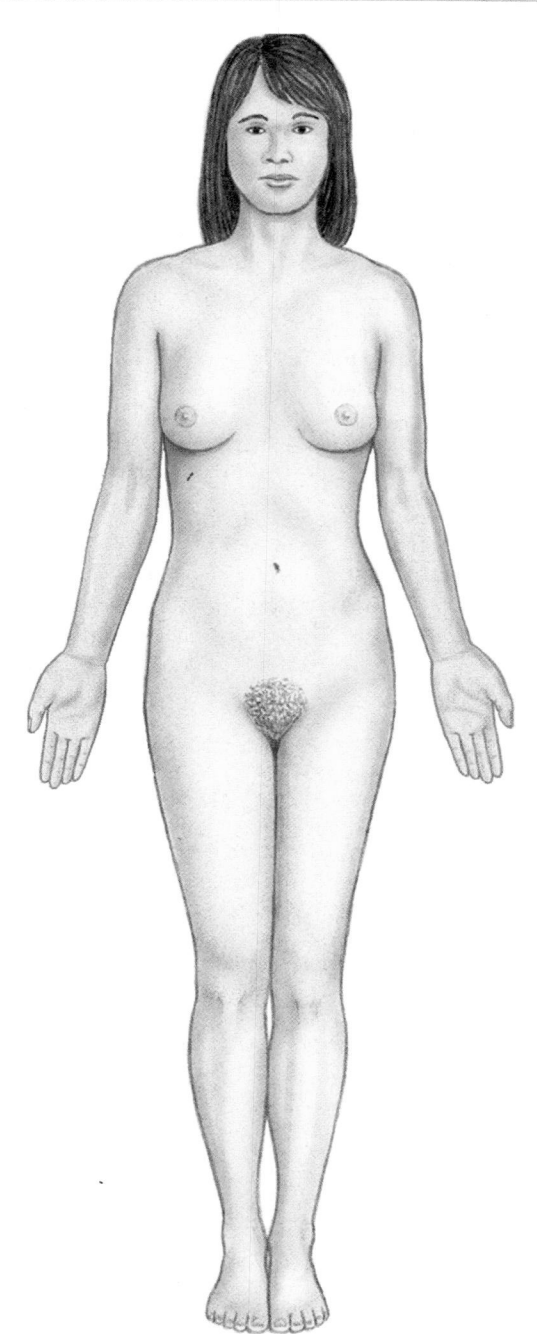

FOR ALL SYSTEMS
Provides mechanical protection against environmental hazards

LYMPHATIC SYSTEM

- Assists in defending the integument by providing additional macrophages and mobilizing lymphocytes
- Provides physical barriers that prevent entry of pathogens
- Langerhans cells and macrophages resist infection
- Mast cells trigger inflammation and initiate the immune response

RESPIRATORY SYSTEM

- Provides oxygen and eliminates carbon dioxide
- Hairs guard entrance to nasal cavity

DIGESTIVE SYSTEM

- Provides nutrients for all cells and lipids for storage by adipocytes
- Synthesizes vitamin D_3, needed for absorption of calcium and phosphorus

URINARY SYSTEM

- Excretes waste products
- Maintains normal pH and ion composition of body fluids
- Assists in excretion of water and solutes
- Keratinized epidermis limits fluid loss through skin

REPRODUCTIVE SYSTEM

- Covers external genitalia
- Provides sensations that stimulate sexual behaviors
- Mammary gland secretions provide nourishment for newborn infant
- Sex hormones affect hair distribution, adipose tissue distribution in subcutaneous layer, and mammary gland development

REVIEW QUESTIONS

More assessment questions are available to you on the Companion Website. You will find Matching, Multiple Choice, True/False, and other quizzes to help further your understanding of the material covered in this chapter. To access the site, go to www.aw.com/martini.

LEVEL 1 Reviewing Facts and Terms

1. The two major components of the integumentary system are
   (a) the cutaneous membrane and the accessory structures
   (b) the epidermis and the hypodermis
   (c) the hair and the nails
   (d) the dermis and the subcutaneous layer

2. Beginning at the basal lamina and traveling toward the free surface, the epidermis includes the following layers:
   (a) corneum, lucidum, granulosum, spinosum, germinativum
   (b) granulosum, lucidum, spinosum, germinativum, corneum
   (c) germinativum, spinosum, granulosum, lucidum, corneum
   (d) lucidum, granulosum, spinosum, germinativum, corneum

3. The protein that permits stretch and recoil of the skin is
   (a) collagen   (b) melanin
   (c) keratin   (d) elastin

4. The primary pigments contained in the epidermis are
   (a) carotene and xanthophyll (b) carotene and melanin
   (c) melanin and chlorophyll   (d) xanthophyll and melanin

5. The two major components of the dermis are the
   (a) superficial fascia and cutaneous membrane
   (b) epidermis and hypodermis
   (c) papillary layer and reticular layer
   (d) stratum germinativum and stratum corneum

6. The cutaneous plexus and papillary plexus consist of
   (a) a network of arteries providing the dermal blood supply
   (b) a network of nerves providing dermal sensations
   (c) specialized cells for cutaneous sensations
   (d) gland cells that release cutaneous secretions

7. The accessory structures of the integument include the
   (a) blood vessels, glands, muscles, and nerves
   (b) Merkel cells, lamellated corpuscles, and tactile corpuscles
   (c) hair, skin, and nails
   (d) hair follicles, nails, sebaceous glands, and sweat glands

8. The portion of the hair follicle where cell divisions occur is the
   (a) shaft   (b) matrix
   (c) root hair plexus   (d) cuticle

9. The two types of exocrine glands in the skin are
   (a) merocrine and sweat glands
   (b) sebaceous and sweat glands
   (c) apocrine and sweat glands
   (d) eccrine and sweat glands

10. Apocrine sweat glands can be controlled by
    (a) the autonomic nervous system
    (b) regional control mechanisms
    (c) the endocrine system
    (d) a and c are correct

11. The primary function of sensible perspiration is to
    (a) get rid of wastes
    (b) protect the skin from dryness
    (c) maintain electrolyte balance
    (d) reduce body temperature

12. The stratum corneum of the nail root, which extends over the exposed nail, is called the
    (a) hyponychium   (b) eponychium
    (c) cuticle   (d) cerumen

13. Muscle weakness and a reduction in bone strength in the elderly result from decreased
    (a) vitamin D_3 production (b) melanin production
    (c) sebum production   (d) dermal blood supply

14. Older persons are more sensitive to exposure to the sun and more likely to get sunburned than are younger persons because
    (a) melanocyte activity declines with age
    (b) vitamin D_3 production declines with age
    (c) glandular activity declines with age
    (d) skin thickness decreases with age

15. In which layer(s) of the epidermis does cell division occur?

16. What are the protein precursors of keratin, and in which layer(s) of the epidermis are these proteins produced?

17. What is the function of the arrector pili muscle?

18. What widespread effects does epidermal growth factor (EGF) have on the integument?

19. What two major layers constitute the dermis, and what components are in each layer?

20. Beginning at the hair cuticle, what three cell layers make up the walls of the hair follicle?

21. List the four stages in the regeneration of the skin after an injury.

LEVEL 2 Reviewing Concepts

22. How do insensible perspiration and sensible perspiration differ?

23. During the transdermal administration of drugs, why are fat-soluble drugs more desirable than drugs that are water soluble?

24. In our society, a tan body is associated with good health. However, medical research constantly warns about the dangers of excessive exposure to the sun. What are the benefits of a tan?

25. Why is it important for a surgeon to choose an incision pattern according to the lines of cleavage of the skin?

26. Why is regional infection or inflammation of the skin usually very painful?

27. Why is a subcutaneous injection with a hypodermic needle a useful method of administering drugs?

28. How are changes in the shape, structure, or appearance of the nails clinically significant?

29. Why is scab formation important in homeostasis?

30. How does the application of creams and oils mimic the natural functioning of the skin?

LEVEL 3 Critical Thinking and Clinical Applications

31. A new mother notices that her six-month-old child has a yellow-orange complexion. Fearful that the child may have jaundice, she takes him to her pediatrician. After examining the child, the pe-

diatrician declares him perfectly healthy and advises the mother to watch the child's diet. What could have been responsible for the change in skin color?

32. Vanessa's 80-year-old grandmother sets her thermostat at 26°C (80°F) and wears a sweater on balmy spring days. When asked why, the grandmother says she is cold. Can you give one possible cause for her feeling cold?

33. Exposure to optimum amounts of sunlight is necessary for proper bone maintenance and growth in children.
    (a) What does sunlight do to promote bone maintenance and growth?
    (b) If a child lives in an area where exposure to sunlight is rare because of pollution or overcast skies, what can be done to minimize impaired maintenance and growth of bone?

34. One of the factors to which lie detectors respond is an increase in skin conductivity due to the presence of moisture. Explain the physiological basis for the use of this indicator.

35. Many people change the natural appearance of their hair, either by coloring it or altering the degree of curl in it. Which layers of the hair do you suppose are affected by the chemicals added during these procedures? Why are the effects of the procedures not permanent?

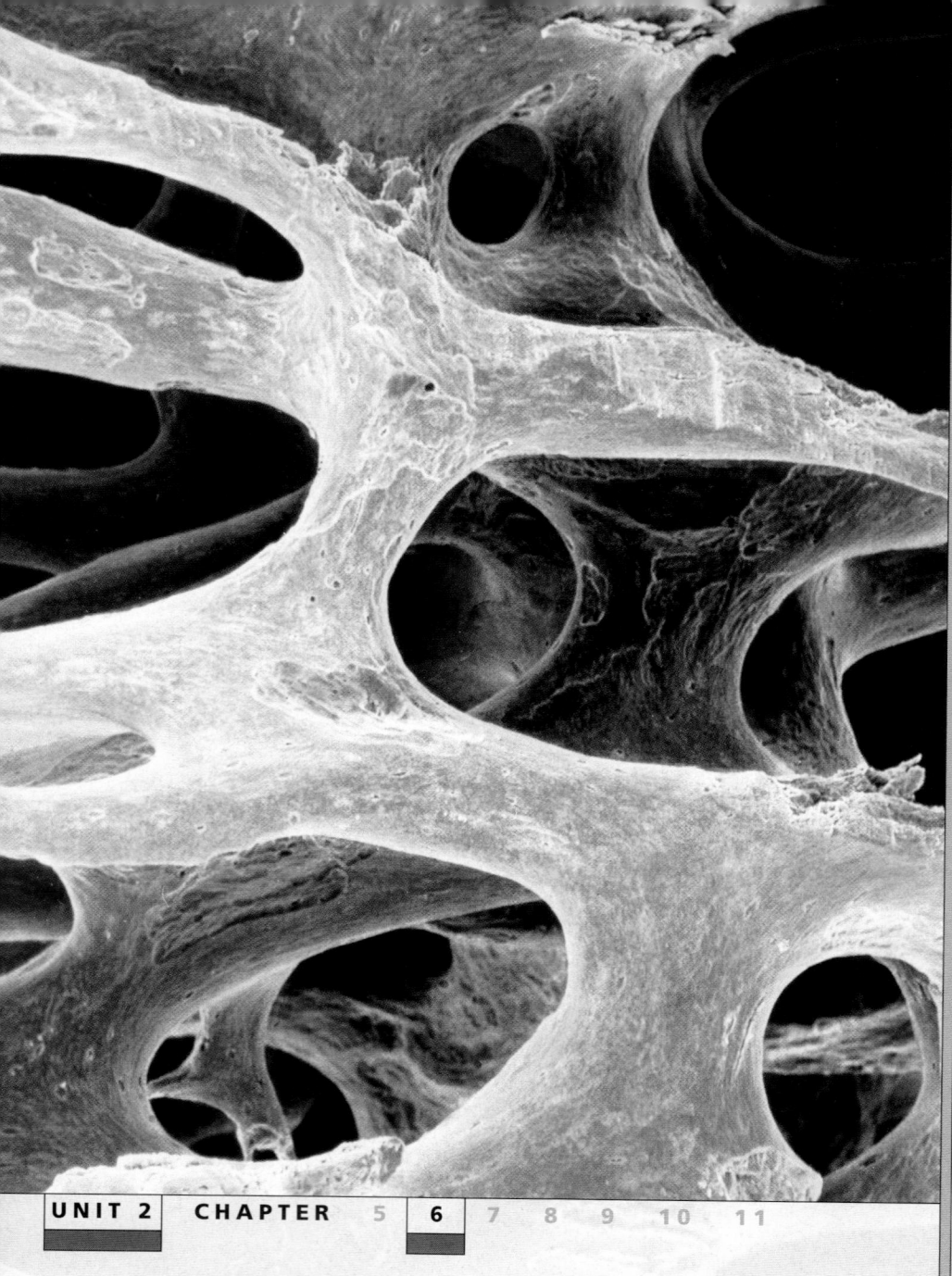

UNIT 2 CHAPTER 5 6 7 8 9 10 11

CLINICAL DISCUSSIONS

■ Abnormal Bone Growth and Development 198

OSSEOUS TISSUE AND SKELETAL STRUCTURE

The human form is distinctive and unique in many respects. A human has eyes at the front of a head roughly the shape of a bowling ball, balanced above the trunk on a relatively slim and delicate neck that lets the head rotate and bend to bring objects into clearer view. The body stands erect on two relatively long lower limbs, freeing the upper limbs for grasping or manipulating things in the environment. The extra height also gives the eyes a greater field of vision, but it makes balance more of a problem than it is for mammals standing on four legs. Head, trunk, and limbs, balance and manipulation—these features are totally dependent on the support provided by the *skeletal system.* **AM**
Examination of the Skeletal System

6–1 THE SKELETAL SYSTEM: AN INTRODUCTION

Objective

■ Describe the functions of the skeletal system.

The skeletal system includes the bones of the skeleton and the cartilages, ligaments, and other connective tissues that stabilize or connect the bones. Skeletal elements are more than just props, or racks from which muscles hang; they have a great variety of vital functions. In addition to supporting the weight of the body, bones work together with muscles to maintain body position and to produce controlled, precise movements. Without the skeleton to pull against, contracting muscle fibers could not make us sit, stand, walk, or run.

The skeletal system has five primary functions:

1. *Support.* The skeletal system provides structural support for the entire body. Individual bones or groups of bones provide a framework for the attachment of soft tissues and organs.

2. *Storage of Minerals and Lipids.* As we shall learn in Chapter 25, minerals are inorganic ions that contribute to the osmotic concentration of body fluids. They also participate in various physiological processes, and several minerals are important as enzyme cofactors. Calcium is the most abundant mineral in the human body. The calcium salts of bone are a valuable mineral reserve that maintains normal concentrations of calcium and phosphate ions in body fluids. In addition to acting as a mineral reserve, the bones of the skeleton store energy reserves as lipids in areas filled with *yellow marrow.*

3. *Blood Cell Production.* Red blood cells, white blood cells, and other blood elements are produced in *red marrow,* which fills the internal cavities of many bones. We shall describe the

role of bone marrow in blood cell formation when we deal with the cardiovascular and lymphatic systems (Chapters 19 and 22).

4. *Protection.* Many soft tissues and organs are surrounded by skeletal elements. The ribs protect the heart and lungs, the skull encloses the brain, the vertebrae shield the spinal cord, and the pelvis cradles delicate digestive and reproductive organs.

5. *Leverage.* Many bones function as levers that can change the magnitude and direction of the forces generated by skeletal muscles. The movements produced range from the dainty motion of a fingertip to changes in the position of the entire body.

Chapters 6–9 consider the structure and function of the skeletal system. We begin by describing bone, or osseous tissue, a supporting connective tissue introduced in Chapter 4. ∞ p. 130 You will find that all of the features and properties of the skeletal system ultimately depend on the unique and dynamic properties of bone. The bone specimens that you study in lab or that you are familiar with from skeletons of dead organisms are only the dry remains of this living tissue. They bear the same relationship to the bone in a living organism as a kiln-dried 2-by-4 does to a living oak.

6–2 A CLASSIFICATION OF BONES

Objective

- Classify bones according to their shapes and internal organization, and give examples of each type.

A description of bone may indicate its general shape or its internal organization. Before considering specific bones of the skeleton, you must be familiar with both classification methods.

□ BONE SHAPES

Every adult skeleton contains 206 major bones, which we can divide into six broad categories according to their individual shapes (Figure 6–1●):

1. **Long bones** are relatively long and slender (Figure 6–1a●). Long bones are located in the arm and forearm, thigh and leg, palms,

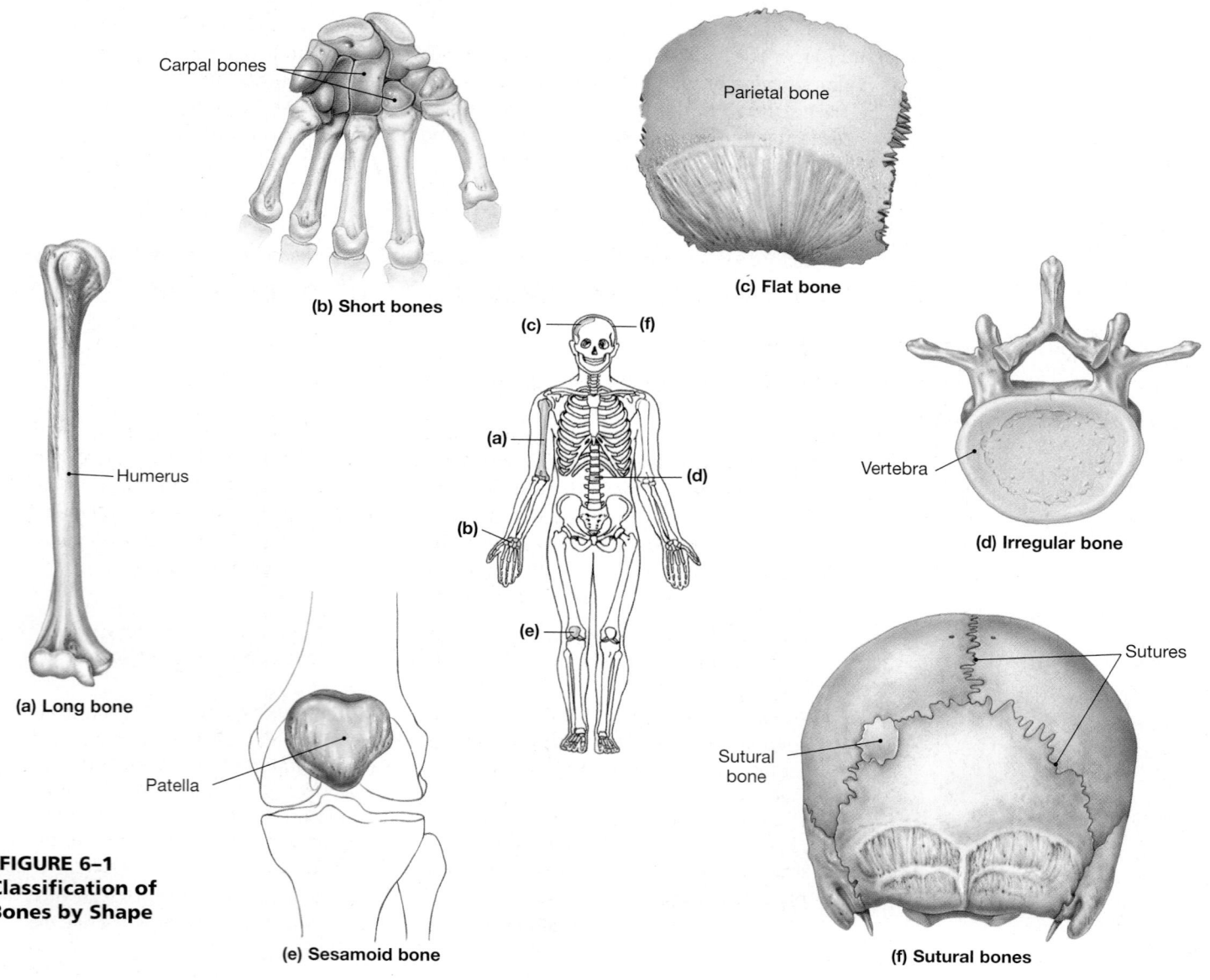

●FIGURE 6–1
Classification of Bones by Shape

(a) Long bone — Humerus

(b) Short bones — Carpal bones

(c) Flat bone — Parietal bone

(d) Irregular bone — Vertebra

(e) Sesamoid bone — Patella

(f) Sutural bones — Sutures, Sutural bone

soles, fingers, and toes. The femur, the long bone of the thigh, is the largest and heaviest bone in the body.

2. **Short bones** are small and boxy (Figure 6–1b●). Examples of short bones include the carpal bones (wrists) and tarsal bones (ankles).

3. **Flat bones** have thin, roughly parallel surfaces. Flat bones form the roof of the skull (Figure 6–1c●), the sternum, the ribs, and the scapula. They provide protection for underlying soft tissues and offer an extensive surface area for the attachment of skeletal muscles.

4. **Irregular bones** have complex shapes with short, flat, notched, or ridged surfaces (Figure 6–1d●). The spinal vertebrae, the bones of the pelvis, and several skull bones are irregular bones.

5. **Sesamoid bones** are generally small, flat, and shaped somewhat like a sesame seed (Figure 6–1e●). They develop inside tendons and are most commonly located near joints at the knees, the hands, and the feet. Everyone has sesamoid *patellae* (pa-TEL-ē singular, *patella*, a small shallow dish), or kneecaps, but individuals vary in

terms of the location and abundance of other sesamoid bones. This variation accounts for disparities in the total number of bones in the skeleton. (Sesamoid bones may form in at least 26 locations.)

6. **Sutural bones**, or *Wormian bones*, are small, flat, irregularly shaped bones between the flat bones of the skull (Figure 6–1f●). There are individual variations in the number, shape, and position of the sutural bones. Their borders are like pieces of a puzzle, and they range in size from a grain of sand to a quarter.

☐ BONE STRUCTURE

In considering individual bones, the internal structure of the bone is nearly as important as its general shape. Each bone in the skeleton contains two forms of osseous tissue: (1) compact bone and (2) spongy bone. **Compact bone**, or *dense bone*, is relatively solid, whereas **spongy bone**, or *cancellous* (KAN-sel-us) *bone*, forms an open network of struts and plates. Compact bone is always located on the surface of a bone, where it forms a sturdy protective layer. Spongy bone makes up the interior of a bone. The relationship between compact bone and spongy bone and their relative proportions vary with the shape of the bone. We shall consider two extremes: long bones and flat bones.

Figure 6–2● introduces the anatomy of the femur, a representative long bone with an extended tubular shaft, or **diaphysis** (dī-AF-i-sis). At each end is an expanded area known as the **epiphysis** (ē-PIF-i-sis). The diaphysis is connected to each epiphysis at a narrow zone known as the **metaphysis** (me-TAF-i-sis). The wall of the diaphysis consists of a layer of compact bone that surrounds a central space called the **marrow cavity**, or *medullary cavity* (*medulla*, innermost part) (Figure 6–2a●). The epiphyses consist largely of spongy bone with a thin covering, or **cortex**, of compact bone. The epiphyses of the femur form complex joints at the hip and knee. These joints have large joint cavities lined by synovial membranes and filled with synovial fluid. p. 134

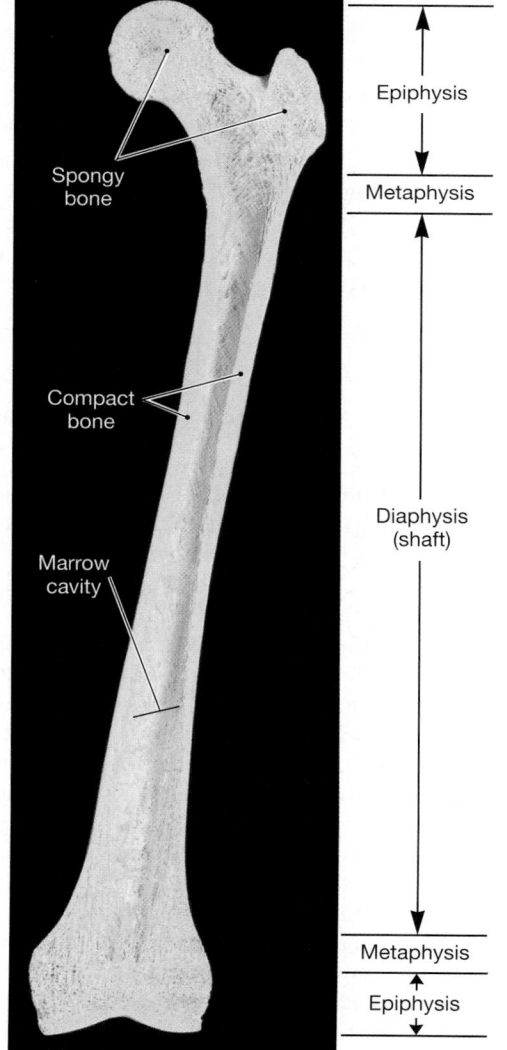

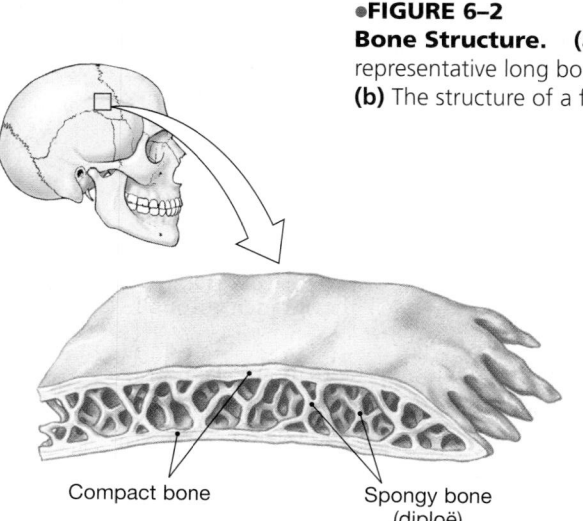

●FIGURE 6–2
Bone Structure. **(a)** The structure of a representative long bone in longitudinal section. **(b)** The structure of a flat bone.

(a)

(b)

The portion of an epiphysis within such a joint cavity is covered by a thin layer of hyaline cartilage called the *articular cartilage.*

The marrow cavity of the diaphysis and the spaces between the struts and plates of the epiphyses contain **bone marrow**, a loose connective tissue. **Yellow bone marrow** is dominated by fat cells. **Red bone marrow** consists largely of a mixture of mature and immature red blood cells, white blood cells, and the stem cells that produce them. Yellow bone marrow is an important energy reserve; areas of red bone marrow are important sites of blood cell formation.

Figure 6–2b● details the structure of a flat bone from the skull, such as one of the *parietal bones.* A flat bone resembles a spongy bone sandwich, with layers of compact bone covering a core of spongy bone. Although bone marrow is present within the spongy bone, there is no marrow cavity. The layer of spongy bone between the layers of compact bone is called the *diploë* (DIP-lō-ē).

Many people think of the skeleton as a rather dull collection of bony props. This is far from the truth. Our bones are complex, dynamic organs that constantly change to adapt to the demands we place on them. We will now consider the internal organization of a typical bone and will see how that seemingly inert structure is remodeled—that is, replaced with new bone—and repaired.

6–3 BONE HISTOLOGY

Objectives

- Identify the cell types in bone, and list their major functions.
- Compare the structures and functions of compact bone and spongy bone.

Osseous tissue is a supporting connective tissue. (You may wish to review the sections on dense connective tissues, cartilage, and bone in Chapter 4.) ∞ pp. 129–132 Like other connective tissues, osseous tissue contains specialized cells and a matrix consisting of extracellular protein fibers and a ground substance. The matrix of bone tissue is solid and sturdy, owing to the deposition of calcium salts around the protein fibers.

In Chapter 4, which introduced the organization of bone tissue, we discussed the following four characteristics of bone:

1. The matrix of bone is very dense and contains deposits of calcium salts.
2. The matrix contains bone cells, or *osteocytes*, within pockets called *lacunae.* (The spaces that chondrocytes occupy in cartilage are also called lacunae. ∞ p. 129) The lacunae of bone are typically organized around blood vessels that branch through the bony matrix.
3. *Canaliculi*, narrow passageways through the matrix, extend between the lacunae and nearby blood vessels, forming a branching network for the exchange of nutrients, waste products, and gases.
4. Except at joints, the outer surfaces of bones are covered by a *periosteum*, which consists of outer fibrous and inner cellular layers.

We now take a closer look at the organization of the matrix and cells of bone.

☐ THE MATRIX OF BONE

Calcium phosphate, $Ca_3(PO_4)_2$, accounts for almost two-thirds of the weight of bone. Calcium phosphate interacts with calcium hydroxide, $Ca(OH)_2$, to form crystals of **hydroxyapatite**, $Ca_{10}(PO_4)_6(OH)_2$. As they form, these crystals incorporate other calcium salts, such as calcium carbonate ($CaCO_3$), and ions such as sodium, magnesium, and fluoride. Roughly one-third of the weight of bone is contributed by collagen fibers. Cells account for only 2 percent of the mass of a typical bone.

Calcium phosphate crystals are very hard, but relatively inflexible and quite brittle. They can withstand compression, but are likely to shatter when exposed to bending, twisting, or sudden impacts. Collagen fibers, by contrast, are remarkably strong; when subjected to tension (pull), they are stronger than steel. Flexible as well as tough, they can easily tolerate twisting and bending, but offer little resistance to compression. When compressed, they simply bend out of the way.

In bone, collagen fibers provide an organic framework on which hydroxyapatite crystals can form. These crystals form small plates and rods that are locked into the collagen fibers. The result is a protein–crystal combination with properties intermediate between those of collagen and those of pure mineral crystals. The protein–crystal interactions allow bone to be strong, somewhat flexible, and highly resistant to shattering. In its overall properties, bone is on a par with the best steel-reinforced concrete. In fact, bone is far superior to concrete, because it can be remodeled easily and can repair itself after injury.

☐ CELLS IN BONE

Although osteocytes are most abundant, bone contains four types of cell. (Figure 6–3a●):

1. **Osteocytes** (*osteon*, bone) are mature bone cells that account for most of the cell population. Each osteocyte occupies a **lacuna**, a pocket sandwiched between layers of matrix (Figure 6–3b,c●). The layers are called **lamellae** (lah-MEL-lē; singular, lamella, a thin plate). Osteocytes cannot divide, and a lacuna never contains more than one osteocyte. Narrow passageways called **canaliculi** penetrate the lamellae, radiating through the matrix and connecting lacunae with one another and with sources of nutrients, such as the central canal.

   Canaliculi contain cytoplasmic extensions of osteocytes. Neighboring osteocytes are linked by gap junctions, which permit the exchange of ions and small molecules, including nutrients and hormones, between the cells. The interstitial fluid that surrounds the osteocytes and their extensions provides an additional route for the diffusion of nutrients and waste products.

   Osteocytes have two major functions: (1) They maintain and monitor the protein and mineral content of the surrounding matrix. As osteocytes release chemicals that dissolve the minerals in the adjacent matrix, those minerals enter the circulation.

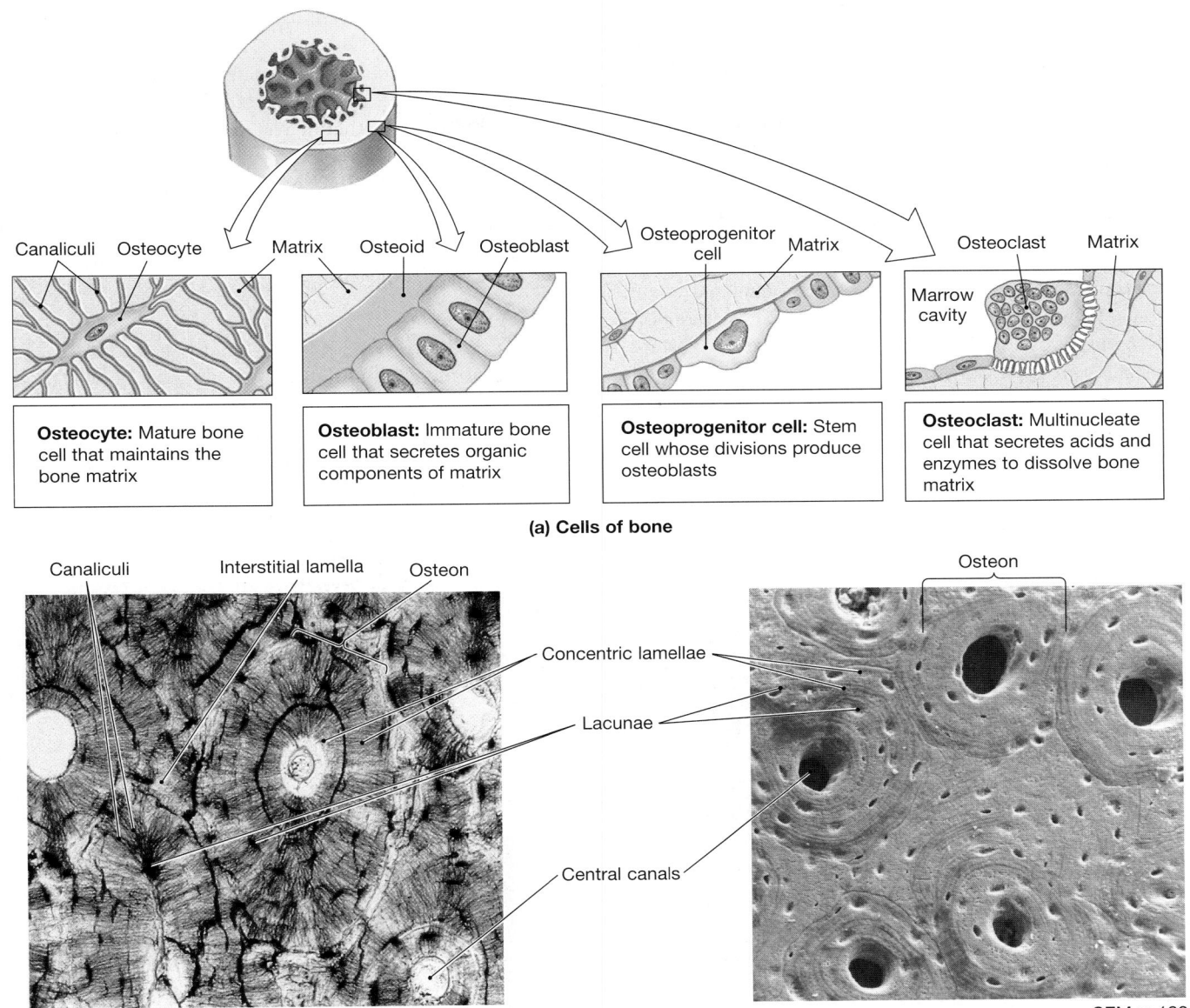

(a) Cells of bone

Canaliculi   Osteocyte   Matrix   Osteoid   Osteoblast

Osteocyte: Mature bone cell that maintains the bone matrix

Osteoblast: Immature bone cell that secretes organic components of matrix

Osteoprogenitor cell   Matrix

Osteoprogenitor cell: Stem cell whose divisions produce osteoblasts

Osteoclast   Matrix

Marrow cavity

Osteoclast: Multinucleate cell that secretes acids and enzymes to dissolve bone matrix

Canaliculi   Interstitial lamella   Osteon

Concentric lamellae

Lacunae

Central canals

(b) LM × 343

Osteon

(c) SEM × 182

●**FIGURE 6–3**
The Histology of Compact Bone. **(a)** Bone cells. **(b)** A thin section through compact bone. By this procedure, the intact matrix and central canals appear white, and the lacunae and canaliculi are shown in black. (LM × 343) **(c)** Several osteons in compact bone. (SEM × 182) Reproduced from R. G. Kessel and R. H. Kardon, *Tissues and Organs: A Text-Atlas of Scanning Electron Microscopy*, W. H. Freeman & Co., 1979.

Ultimately, the recycled minerals are replaced through the deposition of new hydroxyapatite crystals. (2) They can participate in the repair of damaged bone. If released from their lacunae, osteocytes can convert to a less specialized type of cell, such as an osteoblast or an osteoprogenitor cell.

2. **Osteoblasts** (OS-tē-ō-blasts; *blast*, precursor) produce new bone matrix in a process called **osteogenesis** (os-tē-ō-JEN-e-sis; *gennan*, to produce). Osteoblasts make and release the proteins and other organic components of the matrix. Before calcium salts deposit, this organic matrix is called **osteoid** (OS-tē-oyd). Osteoblasts also assist in elevating local concentrations of calcium phosphate and promoting the deposition of calcium

salts in the organic matrix. This process converts osteoid to bone. Osteocytes develop from osteoblasts that have become completely surrounded by bone matrix.

3. Bone contains small numbers of mesenchymal cells called **osteoprogenitor** (os-tē-ō-prō-JEN-i-tor) **cells** (*progenitor*, ancestor). These stem cells divide to produce daughter cells that differentiate into osteoblasts. Osteoprogenitor cells maintain populations of osteoblasts and are important in the repair of a *fracture* (a break or a crack in a bone). Osteoprogenitor cells are located in the inner, cellular layer of the periosteum, in an inner layer, or *endosteum*, that lines marrow cavities, and in the lining of vascular passageways in the matrix.

4. **Osteoclasts** (os-tē-ō-clasts; *clast*, to break) remove bone matrix. These are giant cells with 50 or more nuclei. Osteoclasts are not related to osteoprogenitor cells or their descendants. Instead, they are derived from the largest white blood cells, monocytes, which in other tissues become free macrophages. ⚬⚬ p. 123 Acids and proteolytic (protein-digesting) enzymes secreted by osteoclasts dissolve the matrix and release the stored minerals. This erosion process, called **osteolysis** (os-tē-OL-i-sis) or *resorption*, is important in the regulation of calcium and phosphate concentrations in body fluids.

In living bone, osteoclasts are constantly removing matrix and osteoblasts are always adding to it. The balance between the opposing activities of osteoblasts and osteoclasts is very important. When osteoclasts remove calcium salts faster than osteoblasts deposit them, bones weaken. When osteoblast activity predominates, bones become stronger and more massive. This opposition causes some interesting differences in skeletal components among individuals. Those who subject their bones to muscular stress through weight training or strenuous exercise develop not only stronger muscles, but also stronger bones. Alternatively, declining muscular activity due to immobility leads to a reduction in bone mass at sites of muscle attachment. We will investigate this phenomenon further in a later section of the chapter.

☐ COMPACT BONE AND SPONGY BONE

The composition of the matrix in compact bone is the same as that in spongy bone, but the two differ in their arrangement of osteocytes, canaliculi, and lamellae.

Compact Bone

The basic functional unit of mature compact bone is the **osteon** (OS-tē-on), or *Haversian system* (Figures 6–3c and 6–4a●). In an osteon, the osteocytes are arranged in concentric layers around a **central canal**, or *Haversian canal*. This canal contains one or more blood vessels (normally a capillary and a *venule*, a very

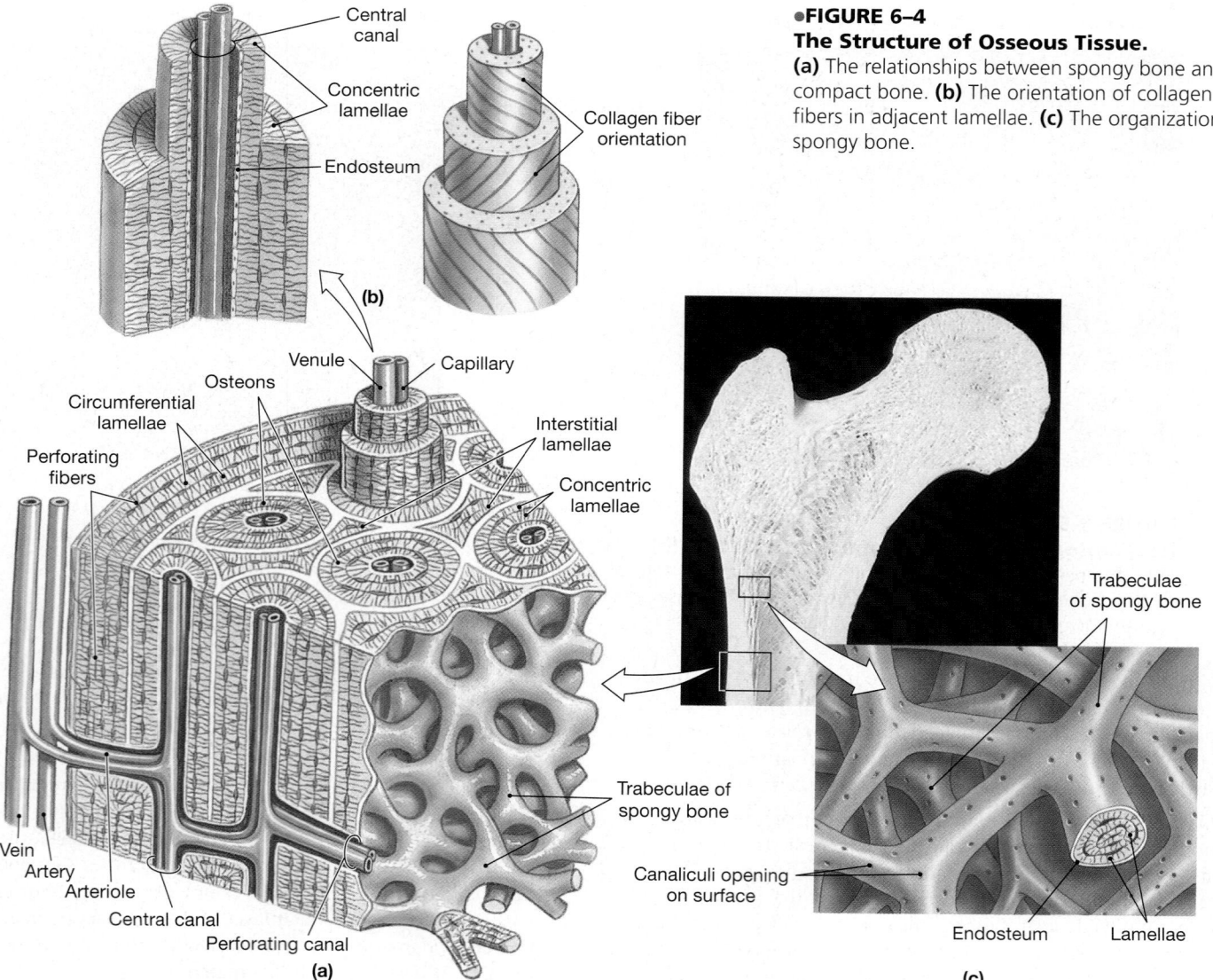

●**FIGURE 6–4**
The Structure of Osseous Tissue.
(a) The relationships between spongy bone and compact bone. **(b)** The orientation of collagen fibers in adjacent lamellae. **(c)** The organization of spongy bone.

small vein) that carry blood to and from the osteon. Central canals generally run parallel to the surface of the bone. Other passageways, known as **perforating canals**, or the *canals of Volkmann*, extend roughly perpendicular to the surface. Blood vessels in these canals supply blood to osteons deeper in the bone and to tissues of the marrow cavity.

The lamellae of each osteon form a series of nested cylinders around the central canal. In transverse section, these *concentric lamellae* create a targetlike pattern, with the central canal as the bull's-eye. Canaliculi radiating through the lamellae interconnect the lacunae of the osteon with one another and with the central canal. *Interstitial lamellae* fill in the spaces between the osteons in compact bone. These lamellae are remnants of osteons whose matrix components have been almost completely recycled by osteoclasts. *Circumferential lamellae* (*circum-*, around + *ferre*, to bear) are found at the outer and inner surfaces of the bone, where they are covered by the periosteum and endosteum, respectively (Figure 6–4a,b●). These lamellae are produced during the growth of the bone.

FUNCTION OF COMPACT BONE Compact bone is thickest where stresses arrive from a limited range of directions. All osteons in compact bone are aligned the same way, making such bones very strong when stressed along the axis of alignment. You might envision a single osteon as a drinking straw with very thick walls. When you attempt to push the ends of a straw together or to pull them apart, the straw is quite strong. But if you hold the ends and push from the side, the straw will break relatively easily.

The osteons in the diaphysis of a long bone are parallel to the long axis of the shaft. Thus, the shaft does not bend, even when extreme forces are applied to either end. (The femur can withstand 10–15 times the body's weight without breaking.) Yet a much smaller force applied to the side of the shaft can break the femur. The majority of breaks that occur in this bone are caused by a sudden sideways force, such as those applied during a football tackle or a hockey check.

Figure 6–5● shows the distribution of forces applied to the compact bone and spongy bone of the femur. The hip joint consists of the head of the femur and a corresponding socket on the lateral surface of the hip bone. The femoral head projects medially, and the body weight compresses the medial side of the diaphysis. Because the force is applied off center, the bone must resist the tendency to bend into a lateral bow. The other side of the shaft, which resists this bending, is placed under a stretching load, or *tension*. Notice that the central part of the bone is subjected neither to compression nor to tension. The powerful forces are restricted to the compact bone of the shaft, so the presence of the marrow cavity does not reduce the bone's strength.

Spongy Bone

In spongy bone, lamellae are not arranged in osteons. The matrix in spongy bone forms struts and plates called **trabeculae** (tra-BEK-ū-lē) (Figure 6–4a,c●). The thin trabeculae branch, creating an open network. There are no capillaries or venules in

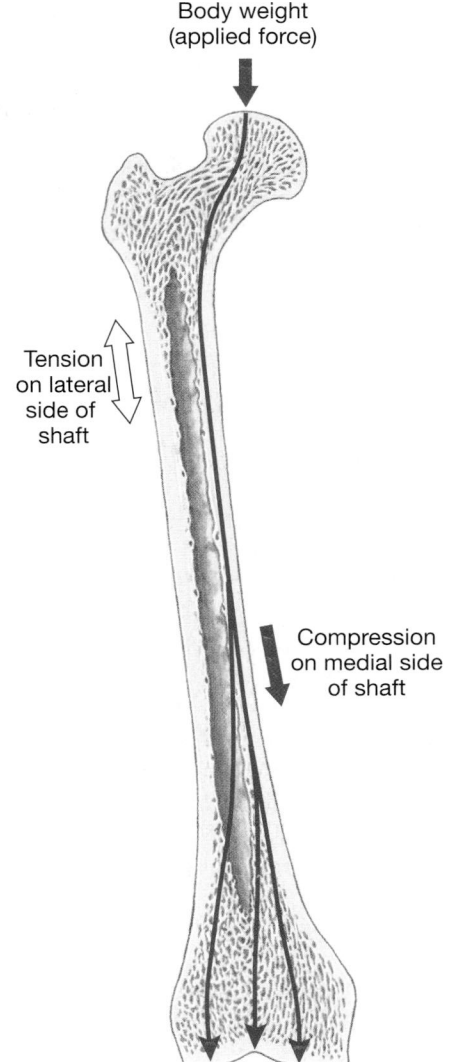

●FIGURE 6–5
The Distribution of Forces on a Long Bone. The femur, or thigh bone, has a diaphysis (shaft) with walls of compact bone and epiphyses filled with spongy bone. The body weight is transferred to the femur at the hip joint. Because the hip joint is off center relative to the axis of the shaft, the body weight is distributed along the bone such that the medial (inner) portion of the shaft is compressed and the lateral (outer) portion is stretched.

the matrix of spongy bone. Nutrients reach the osteocytes by diffusion along canaliculi that open onto the surfaces of trabeculae. Red marrow is found between the trabeculae of spongy bone, and blood vessels within this tissue deliver nutrients to the trabeculae and remove wastes generated by the osteocytes.

FUNCTION OF SPONGY BONE Spongy bone is located where bones are not heavily stressed or where stresses arrive from many directions. The trabeculae are oriented along stress lines, but with extensive cross-bracing. For example, at the proximal epiphysis of the femur, trabeculae transfer forces from the hip to the compact bone of the femoral shaft; at the distal epiphysis, trabeculae transfer weight from the shaft to the leg, across the knee joint.

In addition to being able to withstand stresses applied from many directions, spongy bone is much lighter than compact bone. Spongy bone reduces the weight of the skeleton and makes it easier for muscles to move the bones. Finally, the framework of trabeculae supports and protects the cells of the bone marrow. The red bone marrow within the spongy bone of the femoral epiphyses is a key site of blood cell formation.

☐ THE PERIOSTEUM AND ENDOSTEUM

Except within joint cavities, the superficial layer of compact bone that covers all bones is wrapped by a **periosteum**, a membrane with a fibrous outer layer and a cellular inner layer (Figure 6–6a●). The periosteum (1) isolates the bone from surrounding tissues, (2) provides a route for the circulatory and nervous supply, and (3) actively participates in bone growth and repair.

Near joints, the periosteum becomes continuous with the connective tissues that lock the bones together. At a synovial joint, the periosteum is continuous with the joint capsule. The fibers of the periosteum are also interwoven with those of the tendons attached to the bone. As the bone grows, these tendon fibers are cemented into the circumferential lamellae by osteoblasts from the cellular layer of the periosteum. Collagen fibers incorporated into bone tissue from tendons and ligaments, as well as from the superficial periosteum, are called *perforating fibers (Sharpey's fibers)*. This method of attachment bonds the tendons and ligaments into the general structure of the bone, providing a much stronger attachment than would otherwise be possible. An extremely powerful pull on a tendon or ligament will usually break a bone rather than snap the collagen fibers at the bone surface.

The **endosteum**, an incomplete cellular layer, lines the marrow cavity (Figure 6–6b●). This layer, active during bone growth, repair, and remodeling, covers the trabeculae of spongy bone and lines the inner surfaces of the central canals. The endosteum consists of a simple flattened layer of osteoprogenitor cells that covers the bone matrix, generally without any intervening connective tissue fibers. Where the cellular layer is not complete, the matrix is exposed. At these exposed sites, osteoclasts and osteoblasts can remove or deposit matrix components. The osteoclasts generally occur in shallow depressions *(Howship's lacunae)* that they have eroded into the matrix.

✓ How would the strength of a bone be affected if the ratio of collagen to hydroxyapatite increased?

✓ A sample of bone has concentric lamellae surrounding a central canal. Is the sample from the cortex or the marrow cavity of a long bone?

✓ If the activity of osteoclasts exceeds the activity of osteoblasts in a bone, how will the mass of the bone be affected?

Answers start on page Q–1

 Bone markings can be viewed on the **Anatomy CD-ROM**: Skeletal System/Axial Dissections.

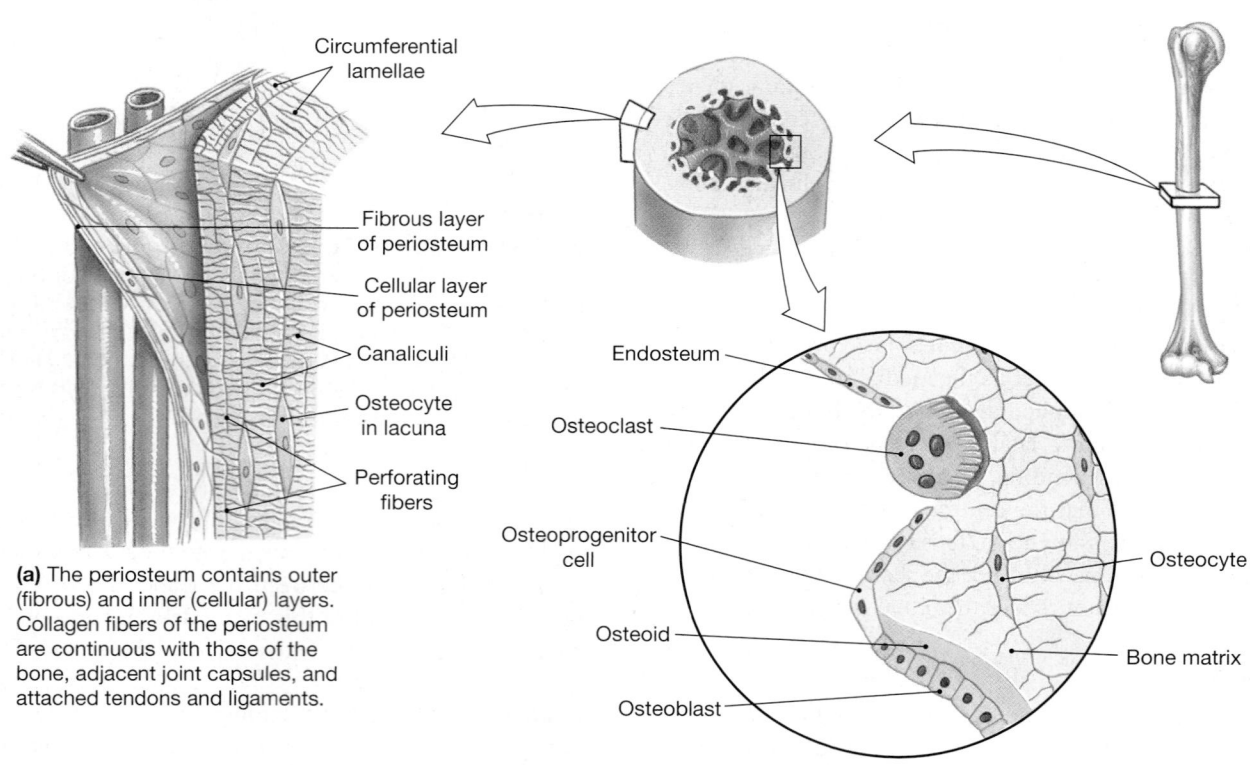

(a) The periosteum contains outer (fibrous) and inner (cellular) layers. Collagen fibers of the periosteum are continuous with those of the bone, adjacent joint capsules, and attached tendons and ligaments.

Circumferential lamellae
Fibrous layer of periosteum
Cellular layer of periosteum
Canaliculi
Osteocyte in lacuna
Perforating fibers

Endosteum
Osteoclast
Osteoprogenitor cell
Osteoid
Osteoblast
Osteocyte
Bone matrix

(b) The endosteum is an incomplete cellular layer containing epithelial cells, osteoblasts, osteoprogenitor cells, and osteoclasts.

●**FIGURE 6–6**
The Periosteum and Endosteum

6–4 BONE DEVELOPMENT AND GROWTH

Objectives

▪ Compare the mechanisms of intramembranous ossification and endochondral ossification.

▪ Discuss the timing of bone development and growth, and account for the differences in the internal structure of the bones of adults.

The growth of the skeleton determines the size and proportions of your body. The bony skeleton begins to form about six weeks after fertilization, when the embryo is approximately 12 mm (0.5 in.) long. (At this stage, the existing skeletal elements are cartilaginous.) During subsequent development, the bones undergo a tremendous increase in size. Bone growth continues through adolescence, and portions of the skeleton generally do not stop growing until roughly age 25. The entire process is carefully regulated, and a breakdown in regulation will ultimately affect all body systems. In this section, we consider the physical process of osteogenesis (bone formation) and bone growth. In the next section, we shall examine the maintenance and replacement of mineral reserves in the adult skeleton.

The process of replacing other tissues with bone is called **ossification**. The term refers specifically to the formation of bone. The process of **calcification**, the deposition of calcium salts, occurs during ossification, but it can also occur in other tissues. When calcification occurs in tissues other than bone, the result is a calcified tissue, such as calcified cartilage, that does not resemble bone. Two major forms of ossification exist: (1) intramembranous and (2) endochondral. In *intramembranous ossification*, bone develops directly from mesenchyme or fibrous connective tissue. In *endochondral ossification*, bone replaces existing cartilage.

☐ INTRAMEMBRANOUS OSSIFICATION

Intramembranous (in-tra-MEM-bra-nus) **ossification**, or *dermal ossification*, begins when osteoblasts differentiate within a mesenchymal or fibrous connective tissue. This type of ossification normally occurs in the deeper layers of the dermis. The bones that result are often called **dermal bones**. Examples of dermal bones are the flat bones of the skull, the mandible (lower jaw), and the clavicle (collarbone). In response to abnormal stresses, bone may form in other dermal areas or in connective tissues within tendons, around joints, in the kidneys, or in skeletal muscles. Dermal bones forming in abnormal locations are called *heterotopic bones* (hetero-, different + topos, place), or *ectopic bones* (ektos, outside). These bones can form in odd places, such as the testes or the whites of the eyes. **AM** Heterotopic Bone Formation

The steps in the process of intramembranous ossification (Figure 6–7●) can be summarized as follows:

STEP 1: Mesenchymal cells first cluster together and start to secrete the organic components of the matrix. The resulting osteoid then becomes mineralized through the crystallization of

●**FIGURE 6–7**
Intramembranous Ossification

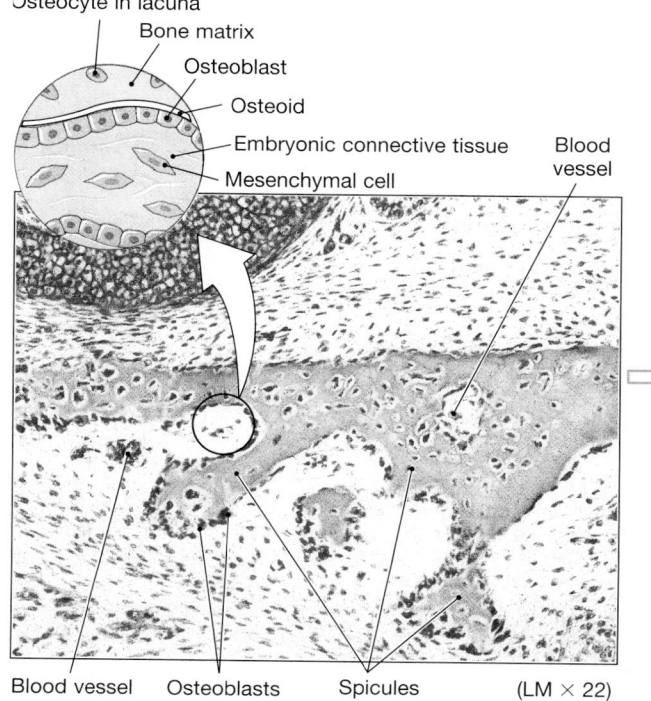

Osteocyte in lacuna
Bone matrix
Osteoblast
Osteoid
Embryonic connective tissue
Mesenchymal cell
Blood vessel

Blood vessel   Osteoblasts   Spicules   (LM × 22)

STEP 1:
Mesenchymal cells aggregate, differentiate, and begin the ossification process. The bone expands as a series of spicules that spread into surrounding tissues.

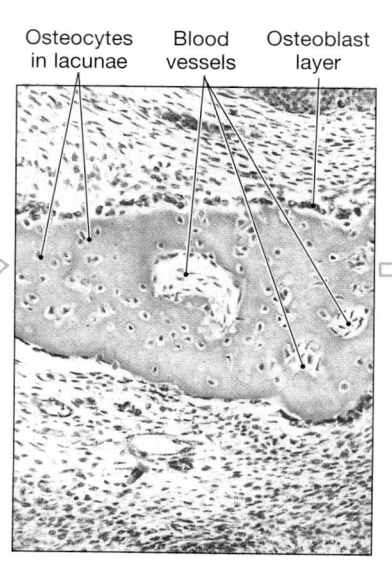

Osteocytes in lacunae   Blood vessels   Osteoblast layer

(LM × 23)

STEP 2:
As the spicules interconnect, they trap blood vessels within the bone.

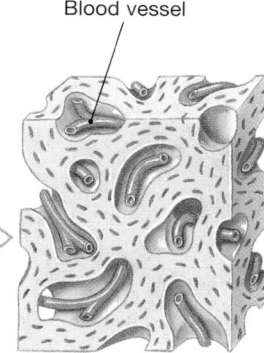

Blood vessel

STEP 3:
Over time, the bone assumes the structure of spongy bone. Areas of spongy bone may later be removed, creating marrow cavities. Through remodeling, spongy bone formed in this way can be converted to compact bone.

calcium salts. (The enzyme *alkaline phosphatase* plays a role in this process.) As calcification occurs, the mesenchymal cells differentiate into osteoblasts. The location in a tissue where ossification begins is called an **ossification center**. The developing bone grows outward from the ossification center in small struts called **spicules**. As ossification proceeds, it traps some osteoblasts inside bony pockets; these cells differentiate into osteocytes. Meanwhile, mesenchymal cell divisions continue to produce additional osteoblasts.

STEP 2: Bone growth is an active process, and osteoblasts require oxygen and a reliable supply of nutrients. Blood vessels begin to grow into the area. As spicules meet and fuse together, some of these blood vessels become trapped within the developing bone.

STEP 3: Initially, the intramembranous bone consists only of spongy bone. Subsequent remodeling around trapped blood vessels can produce osteons typical of compact bone. As the rate of growth slows, the connective tissue around the bone becomes organized into the fibrous layer of the periosteum. The osteoblasts closest to the bone surface become less active, but remain as the inner, cellular layer of the periosteum.

☐ ENDOCHONDRAL OSSIFICATION

During development, most bones originate as hyaline cartilages. Each cartilage is a miniature model of the bone which will occu-

py that particular position in the adult skeleton. These cartilage models are gradually converted to bone through the process of **endochondral** (en-dō-KON-drul) **ossification** (*endo-*, inside + *chondros*, cartilage). As an example, let us follow the steps in limb bone development. By the time an embryo is six weeks old, the proximal bone of the limb—either the humerus (arm) or femur (thigh)—is present, but is composed entirely of hyaline cartilage. This cartilage model continues to grow by expansion of the cartilage matrix (*interstitial growth*) and the production of new cartilage at the outer surface (*appositional growth*). ∞ p. 130 Steps in the growth and ossification of a limb bone are diagrammed in Figure 6–8a•:

STEP 1: As the cartilage enlarges, chondrocytes near the center of the shaft begin to increase greatly in size. As these cells enlarge, their lacunae expand and the matrix is reduced to a series of thin struts that soon begin to calcify. The enlarged chondrocytes are now deprived of nutrients, because diffusion cannot occur through calcified cartilage. These chondrocytes become surrounded by calcified cartilage, die, and disintegrate.

STEP 2: Blood vessels grow into the perichondrium surrounding the shaft of the cartilage. (We introduced the structure of the perichondrium and its role in cartilage formation in Chapter 4. ∞ p. 129 The cells of the inner layer of the perichondrium in this region then differentiate into osteoblasts and begin producing a

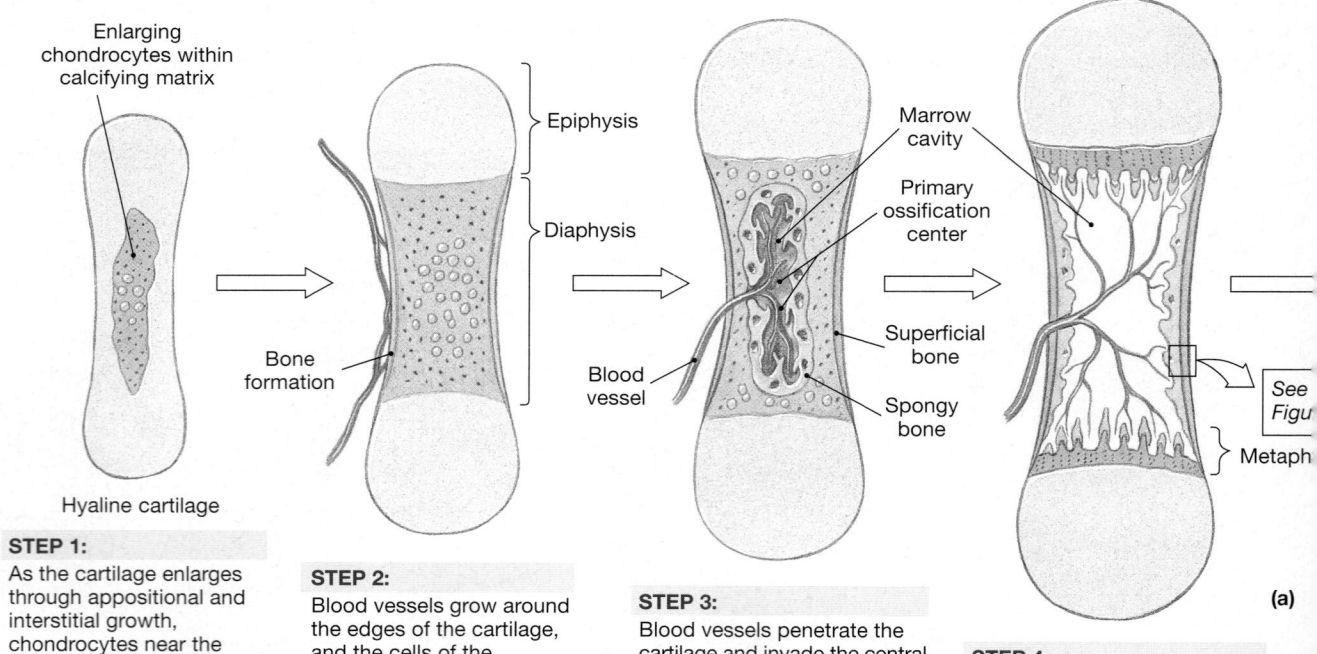

STEP 1:
As the cartilage enlarges through appositional and interstitial growth, chondrocytes near the center of the shaft increase greatly in size. The matrix is reduced to a series of small struts that soon begin to calcify. The enlarged chondrocytes then die and disintegrate, leaving cavities within the cartilage.

STEP 2:
Blood vessels grow around the edges of the cartilage, and the cells of the perichondrium convert to osteoblasts. The shaft of the cartilage then becomes ensheathed in a superficial layer of bone.

STEP 3:
Blood vessels penetrate the cartilage and invade the central region. Fibroblasts migrating with the blood vessels differentiate into osteoblasts and begin producing spongy bone at a primary center of ossification. Bone formation then spreads along the shaft toward both ends.

STEP 4:
Remodeling occurs as growth continues, creating a marrow cavity. The bone of the shaft becomes thicker, and the cartilage near each epiphysis is replaced by shafts of bone. Further growth involves increases in length (Steps 5 and 6) and diameter (Figure 6-10).

(a)

•**FIGURE 6–8**
Endochondral Ossification. **(a)** Steps in endochondral ossification.
(b) A light micrograph showing the interface between the degenerating cartilage and the advancing osteoblasts. **ATLAS** Plate 8.17

thin layer of bone around the shaft of the cartilage. The perichondrium is now technically a periosteum, because it covers bone rather than cartilage.

STEP 3: While these changes are underway, the blood supply to the periosteum increases, and capillaries and fibroblasts migrate into the heart of the cartilage, invading the spaces left by the disintegrating chondrocytes. The calcified cartilaginous matrix breaks down; the fibroblasts differentiate into osteoblasts that replace it with spongy bone. Bone development begins at this site, called the **primary center of ossification**, and spreads toward both ends of the cartilaginous model. While its diameter is small, the entire diaphysis is filled with spongy bone.

STEP 4: As the bone enlarges, osteoclasts appear and begin eroding the trabeculae in the center of the diaphysis, creating a marrow cavity. Further growth involves two distinct processes: (1) an increase in length, and (2) an enlargement in diameter by appositional growth. (We will consider appositional growth in a later section.)

STEP 5: The next major change occurs when the centers of the epiphyses begin to calcify. Capillaries and osteoblasts migrate into these areas, creating **secondary ossification centers**. The time of appearance of secondary ossification centers varies from one bone to another and from individual to individual. Secondary ossification centers may occur at birth in both ends of the humerus (arm), femur (thigh), and tibia (leg), but the ends of some other bones remain cartilaginous until early adulthood.

STEP 6: The epiphyses eventually become filled with spongy bone. A thin cap of the original cartilage model remains exposed to the joint cavity as the **articular cartilage**. This cartilage prevents damaging bone-to-bone contact within the joint. At the metaphysis, a relatively narrow cartilaginous region called the **epiphyseal cartilage**, or *epiphyseal plate*, now separates the epiphysis from the diaphysis. Figure 6–8b● shows the interface between the degenerating cartilage and the advancing osteoblasts.

As long as the epiphyseal cartilage continues to enlarge, the bone will continue to increase in length. On the shaft side, osteoblasts continuously invade the cartilage and replace it with bone. On the epiphyseal side, new cartilage is continuously added. The osteoblasts are therefore moving toward the epiphysis, which is being pushed away by the expansion of the epiphyseal cartilage. The situation is like a pair of joggers, one in front of the other. As long as they are running at the same speed, they can run for miles without colliding. The osteoblasts don't catch up to the epiphysis, but both the osteoblasts and the epiphysis "run away" from the primary ossification center. As they do, the bone grows longer and longer.

At puberty, the combination of rising levels of sex hormones, growth hormone, and thyroid hormones stimulates bone growth dramatically. Osteoblasts now begin producing bone at a rate faster than that of epiphyseal cartilage expansion. As a result, the

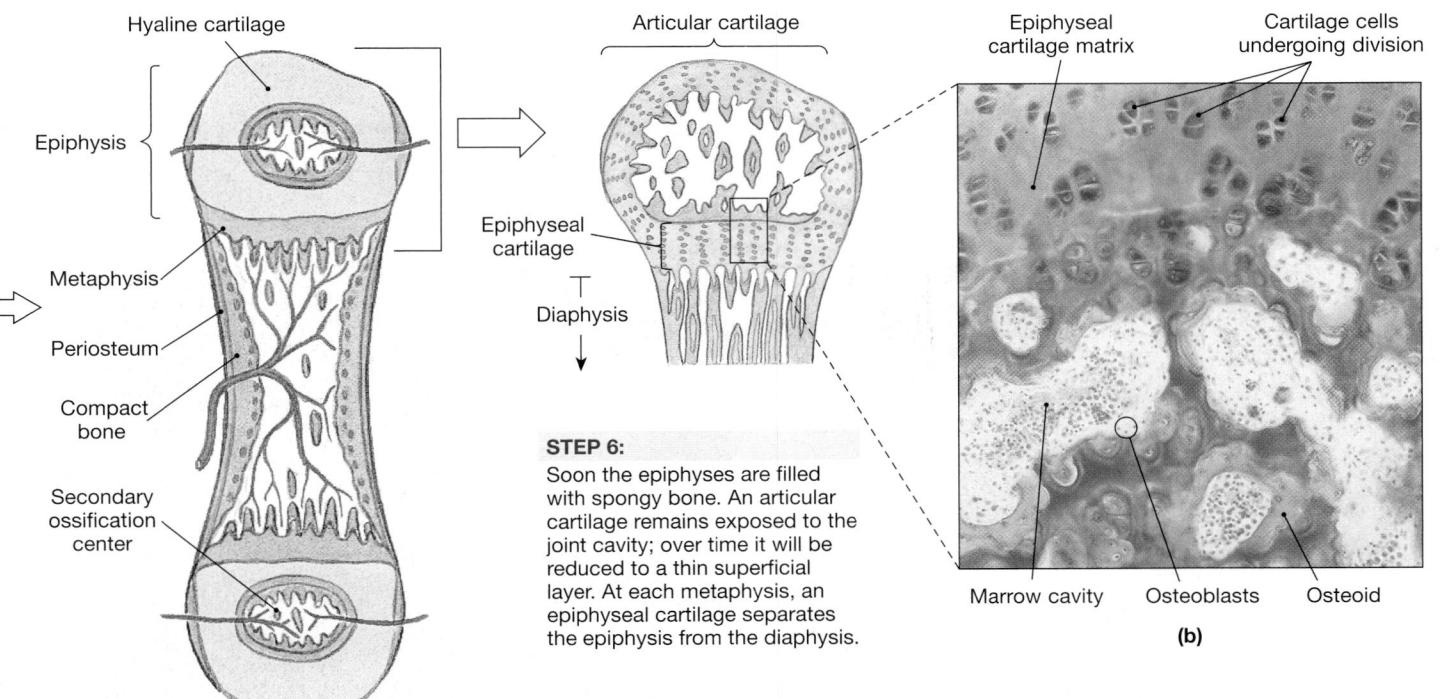

STEP 5:
Capillaries and osteoblasts migrate into the epiphyses, creating secondary ossification centers.

STEP 6:
Soon the epiphyses are filled with spongy bone. An articular cartilage remains exposed to the joint cavity; over time it will be reduced to a thin superficial layer. At each metaphysis, an epiphyseal cartilage separates the epiphysis from the diaphysis.

(b)

epiphyseal cartilage gets narrower and narrower, until it ultimately disappears. The timing of this event can be monitored by comparing the width of the epiphyseal cartilages in successive X rays. In adults, the former location of this cartilage is often detectable in X rays as a distinct **epiphyseal line**, which remains after epiphyseal growth has ended (Figure 6–9●). The completion of epiphyseal growth is called *closure*.

Appositional Growth

A superficial layer of bone forms early in endochondral ossification (Figure 6–8●, step 2). Thereafter, the developing bone increases in diameter through appositional growth at the outer surface (Figure 6–10a●). In this process, cells of the inner layer of the periosteum differentiate into osteoblasts and contribute to the growth of the bone matrix. Eventually, they become surrounded by matrix and differentiate into osteocytes. Over much of the surface, appositional growth adds a series of layers that form circumferential lamellae. In time, the deeper lamellae are recycled and replaced with osteons typical of compact bone. However, blood vessels and collagen fibers of the periosteum can become enclosed within the matrix. Osteons then form around the smaller vessels, as indicated in Figure 6–10a●.

While bone matrix is being added to the outer surface of the growing bone, osteoclasts are removing bone matrix at the inner surface, albeit at a slower rate. As a result, the marrow cavity gradually enlarges as the bone gets larger in diameter (Figure 6–10b●).

✓ During intramembranous ossification, which type(s) of tissue is (are) replaced by bone?

✓ In endochondral ossification, what is the original source of osteoblasts?

✓ How could X rays of the femur be used to determine whether a person has reached full height?

Answers start on page Q-1

 Review the steps in intramembranous and endochondral ossification by visiting the **Companion Website**: Chapter 6/ Tutorials.

☐ THE BLOOD AND NERVE SUPPLIES

Osseous tissue is highly vascular, and the bones of the skeleton have an extensive blood supply. In a typical bone such as the humerus, three major sets of blood vessels develop (Figure 6–11●):

1. *The Nutrient Artery and Vein.* The blood vessels that supply the diaphysis form by invading the cartilage model as endochondral ossification begins. Most bones have only one *nutrient artery* and one *nutrient vein*, but a few bones, including the femur, have more than one of each. The vessels enter the bone through one or more round passageways called *nutrient foramina* in the diaphysis. Branches of these large vessels form smaller perforating canals and extend along the length of the shaft into the osteons of the surrounding cortex.

2. *Metaphyseal Vessels.* The *metaphyseal vessels* supply blood to the inner (diaphyseal) surface of each epiphyseal cartilage, where that cartilage is being replaced by bone.

3. *Periosteal Vessels.* Blood vessels from the periosteum are incorporated into the developing bone surface as described previously (Figure 6–10a●). These *periosteal vessels* provide blood to the superficial osteons of the shaft. During endochondral bone formation, branches of periosteal vessels also enter the epiphyses, providing blood to the secondary ossification centers.

Following the closure of the epiphyses, all three sets of vessels become extensively interconnected.

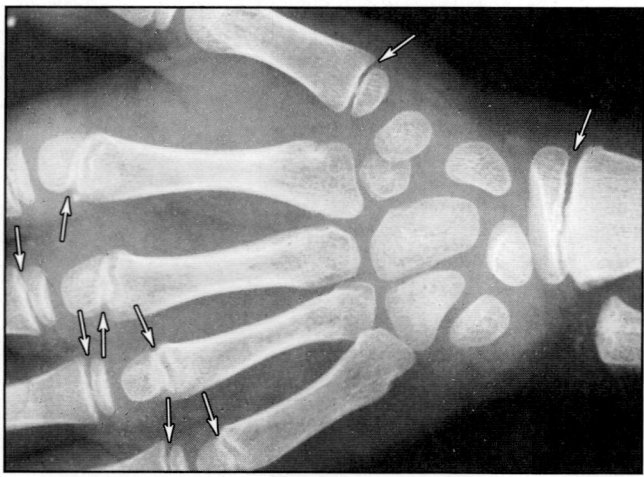

(a) Epiphyseal cartilages

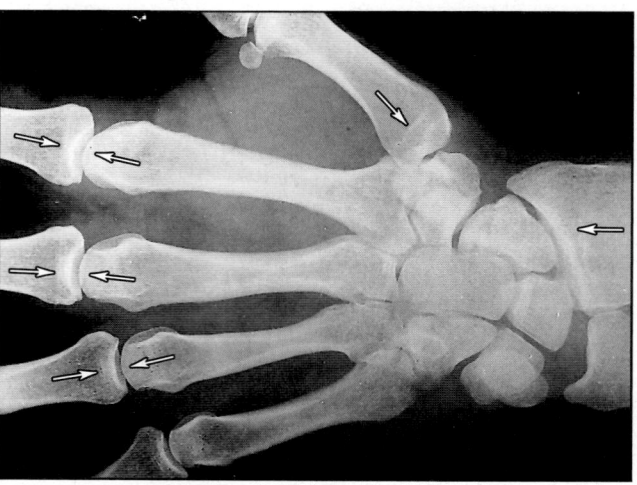

(b) Epiphyseal lines

●**FIGURE 6–9**
Bone Growth at an Epiphyseal Cartilage. (a) An X ray of growing epiphyseal cartilages (arrows). (b) Epiphyseal lines in an adult (arrows).

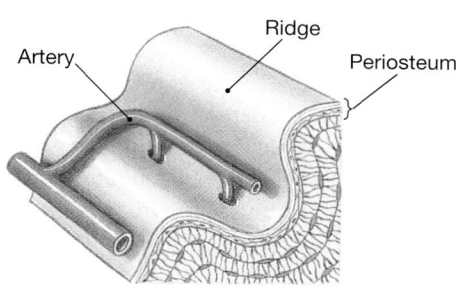

STEP 1:
Bone formation at the surface of the bone produces ridges that parallel a blood vessel.

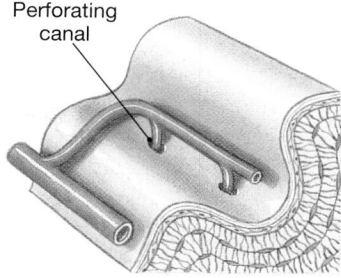

STEP 2:
The ridges enlarge and create a deep pocket.

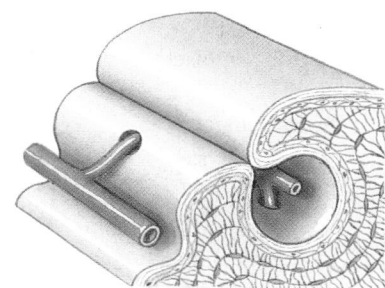

STEP 3:
The ridges meet and fuse, trapping the vessel inside the bone.

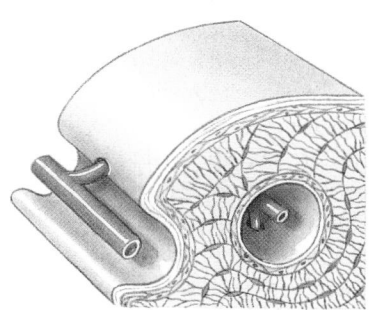

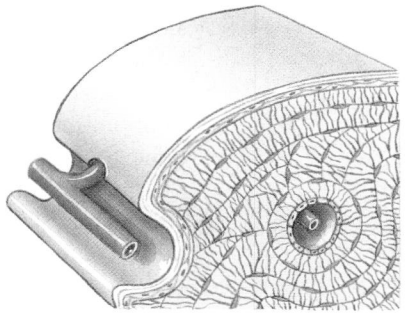

STEPS 4–6:
Bone deposition then proceeds inward toward the vessel, creating a typical osteon. Meanwhile, additional circumferential lamellae are deposited and the bone continues to increase in diameter. As it does, additional blood vessels will be enclosed.

(a) Steps in appositional bone growth

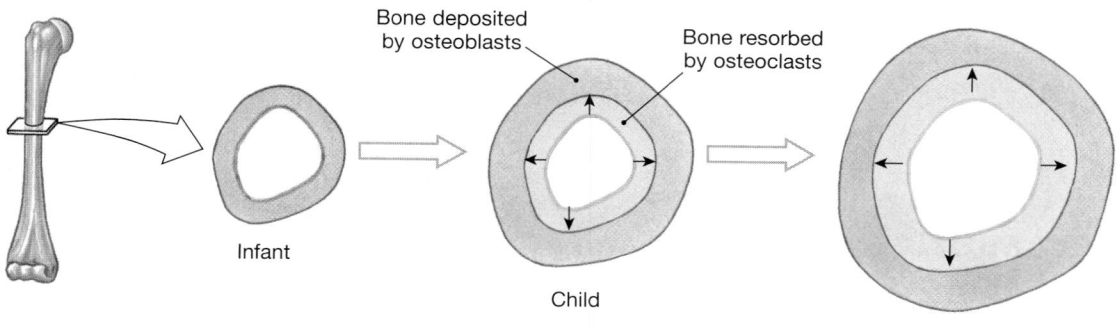

(b) Appositional growth and remodeling

●**FIGURE 6–10**
Appositional Bone Growth. **(a)** Three-dimensional diagrams illustrating the mechanism responsible for increasing the diameter of a growing bone. **(b)** A bone grows in diameter as new bone is added to the outer surface. At the same time, osteoclasts resorb bone on the inside, enlarging the marrow cavity.

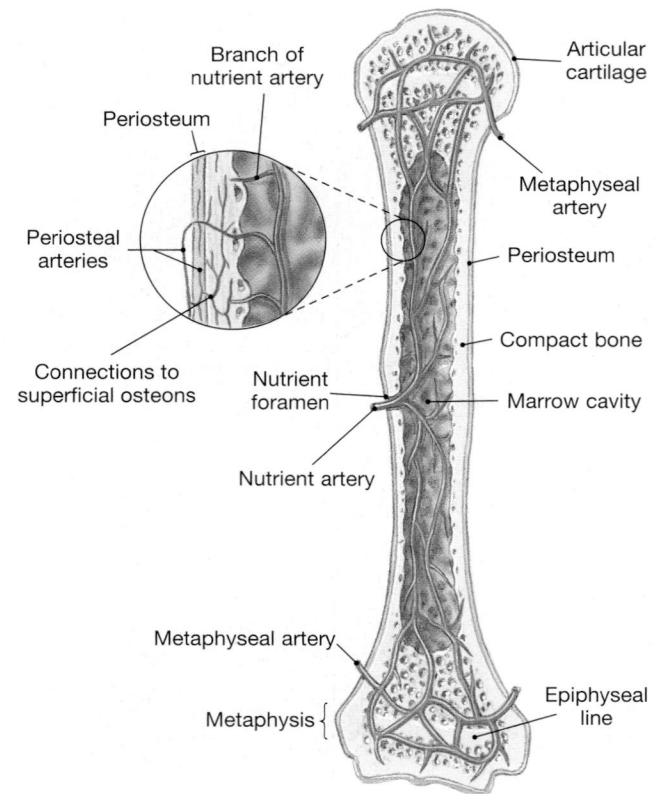

●**FIGURE 6–11**
The Circulatory Supply to a Mature Bone

The periosteum contains an extensive network of lymphatic vessels and sensory nerves. The lymphatics collect lymph from branches that enter the bone and reach individual osteons via the perforating canals. The sensory nerves penetrate the cortex with the nutrient artery to innervate the endosteum, marrow cavity, and epiphyses. Because of the rich sensory innervation, injuries to bones are usually very painful.

6–5 THE DYNAMIC NATURE OF BONE

Objectives

■ Describe the remodeling and homeostatic mechanisms of the skeletal system.

■ Discuss the effects of nutrition, hormones, exercise, and aging on bone development and on the skeletal system.

■ Describe the types of fractures, and explain how they heal.

The organic and mineral components of the bone matrix are continuously being recycled and renewed through the process of **remodeling**. Bone remodeling goes on throughout life, as part of normal bone maintenance. Remodeling can replace the matrix, but leave the bone as a whole unchanged, or it may change the shape, internal architecture, or mineral content of the bone.

Through this remodeling process, older mineral deposits are removed from bone and released into the circulation at the same time that circulating minerals are being absorbed and deposited.

Bone remodeling involves an interplay among the activities of osteocytes, osteoblasts, and osteoclasts. In adults, osteocytes are continuously removing and replacing the surrounding calcium salts. Osteoclasts and osteoblasts also remain active, even after the epiphyseal cartilages have closed. Normally, their activities are balanced: As quickly as osteoblasts form one osteon, osteoclasts remove another by osteolysis. The turnover rate of bone is quite high. In young adults, almost one-fifth of the adult skeleton is recycled and replaced each year. Not every part of every bone is affected; the rate of turnover differs regionally and even locally. For example, the spongy bone in the head of the femur may be replaced two or three times each year, whereas the compact bone along the shaft remains largely unchanged.

Because of their biochemical similarity to calcium, heavy-metal ions such as strontium or cobalt, or more exotic forms like uranium or plutonium, can be incorporated into the matrix of bone. Osteoblasts do not differentiate between these heavy metal ions and calcium. Therefore, when they are present in the bloodstream, these heavy metal ions will be deposited into the matrix of bone. Some of these ions are potentially dangerous, and the continual turnover of bone matrix can have detrimental health effects as ions that are absorbed and accumulated are released into the circulation over a period of years. This was one of the major complications in the aftermath of the Chernobyl nuclear reactor incident in 1986. Radioactive compounds released in the meltdown of the reactor were deposited into the bones of exposed individuals. Over time, the radiation released by their own bones resulted in cases of leukemia and other potentially fatal cancers.

☐ EFFECTS OF EXERCISE ON BONE

The turnover and recycling of minerals give each bone the ability to adapt to new stresses. The sensitivity of osteoblasts to electrical events has been theorized as the mechanism that controls the internal organization and structure of bone. Whenever a bone is stressed, the mineral crystals generate minute electrical fields. Osteoblasts are apparently attracted to these electrical fields and, once in the area, begin to produce bone. This finding has led to the successful use of small electrical fields in stimulating the repair of severe fractures.

Because bones are adaptable, their shapes reflect the forces applied to them. For example, bumps and ridges on the surface of a bone mark the sites where tendons are attached. If muscles become more powerful, the corresponding bumps and ridges enlarge to withstand the increased forces. Heavily stressed bones become thicker and stronger, whereas bones that are not subjected to ordinary stresses become thin and brittle. Regular exercise is therefore important as a stimulus that maintains normal bone structure. Champion weight lifters have massive bones with thick, prominent ridges where muscles attach. In nonathletes (especially couch potatoes), moderate amounts of physical activity and weight-bearing activities are essential to stimulate normal bone maintenance and to maintain adequate bone strength.

Degenerative changes in the skeleton occur after relatively brief periods of inactivity. For example, you may use a crutch to take weight off an injured leg while you wear a cast. After a few weeks, your unstressed bones will lose up to a third of their mass. The bones rebuild just as quickly when you resume normal weight loading. However, the removal of calcium salts can be a serious health hazard both for astronauts remaining in a weightless environment and for bedridden or paralyzed patients who spend months or years without stressing their skeleton.

☐ HORMONAL AND NUTRITIONAL EFFECTS ON BONE

Normal bone growth and maintenance depend on a combination of nutritional and hormonal factors:

- Normal bone growth and maintenance cannot occur without a constant dietary source of calcium and phosphate salts. Lesser amounts of other minerals, such as magnesium, fluoride, iron, and manganese, are also required.
- The hormone *calcitriol*, synthesized in the kidneys, is essential for normal calcium and phosphate ion absorption in the digestive tract. Calcitriol synthesis is dependent on the availability of a related steroid, *cholecalciferol* (vitamin D_3), which may be synthesized in the skin or absorbed from the diet. p. 161
- Adequate levels of vitamin C must be present in the diet. This vitamin, which is required by certain key enzymatic reactions in collagen synthesis, also stimulates osteoblast differentiation. One of the signs of vitamin C deficiency, a condition called *scurvy*, is a loss of bone mass and strength.
- Three other vitamins have significant effects on bone structure. Vitamin A, which stimulates osteoblast activity, is particularly important for normal bone growth in children. Vitamins K and B_{12} are required for the synthesis of proteins in normal bone.

- *Growth hormone*, produced by the pituitary gland, and *thyroxine*, from the thyroid gland, stimulate bone growth. Growth hormone stimulates protein synthesis and cell growth throughout the body. Thyroxine stimulates cell metabolism and increases the rate of osteoblast activity. In proper balance, these hormones maintain normal activity at the epiphyseal cartilages until roughly the time of puberty.
- At puberty, rising levels of sex hormones (*estrogens* in females and *androgens* in males) stimulate osteoblasts to produce bone faster than the rate at which epiphyseal cartilage expands. Over time, the epiphyseal cartilages narrow and eventually close. The timing of epiphyseal closure differs from bone to bone and from individual to individual. The toes may complete ossification by age 11, but parts of the pelvis or the wrist may continue to enlarge until roughly age 25. Differences in male and female sex hormones account for gender variation and for related variations in body size and proportions. Because estrogens cause faster epiphyseal closure than do androgens, women are generally shorter than men at maturity.

Two other hormones—*calcitonin* (kal-si-TŌ-nin), from the thyroid gland, and *parathyroid hormone*, from the parathyroid gland—are important in the homeostatic control of calcium and phosphate levels in body fluids. We consider the interactions of these hormones in the next section. The major hormones affecting the growth and maintenance of the skeletal system are summarized in Table 6–1.

The skeletal system is unique in that it remains after life, providing clues to the gender, lifestyle, and environmental conditions experienced by the individual. Not only do the bones reflect the physical stresses placed on the body, but they also provide clues as to the person's health and diet. Hormonal deficiencies produce characteristic symptoms that can be detected by forensic scientists and physical anthropologists. Their work relies heavily on the appearance, strength, and

TABLE 6–1 HORMONES INVOLVED IN THE REGULATION OF BONE GROWTH AND MAINTENANCE

Hormone	Primary Source	Effects on Skeletal System
Calcitriol	Kidneys	Promotes calcium and phosphate ion absorption along the digestive tract
Growth hormone	Pituitary gland	Stimulates osteoblast activity and the synthesis of bone matrix
Thyroxine	Thyroid gland (follicle cells)	With growth hormone, stimulates osteoblast activity and synthesis of bone matrix
Sex hormones	Ovaries (estrogens) / Testes (androgens)	Stimulate osteoblast activity and synthesis of bone matrix
Parathyroid hormone	Parathyroid glands	Stimulates osteoclast (and osteoblast) activity; elevates calcium ion concentrations in body fluids
Calcitonin	Thyroid gland (C cells)	Inhibits osteoclast activity; promotes calcium loss at kidneys; reduces calcium ion concentrations in body fluids

Abnormal Bone Growth and Development

A variety of endocrine or metabolic problems can result in characteristic skeletal changes. In **pituitary growth failure** (*pituitary dwarfism*), inadequate production of growth hormone leads to reduced epiphyseal cartilage activity and abnormally short bones. This condition is becoming increasingly rare in the United States, because children can be treated with human growth hormone.

Gigantism results from an overproduction of growth hormone before puberty. (The world record for height is 272 cm, or 8 ft, 11 in., reached by Robert Wadlow, of Alton, Illinois, who died at age 22 in 1940. Wadlow weighed 216 kg, or 475 lb.) If growth hormone levels rise abnormally after epiphyseal cartilages close, the skeleton does not grow larger, but cartilage growth and alterations in soft-tissue structure lead to changes in bone density

and in physical features, such as the contours of the face. These physical changes occur in the disorder called **acromegaly**.

Several inherited metabolic conditions that affect many systems influence the growth and development of the skeletal system. These conditions produce characteristic variations in body proportions. For example, many individuals with **Marfan's syndrome** are very tall and have long, slender limbs, due to excessive cartilage formation at the epiphyseal cartilages. ∞ p. 124 Although this is an obvious physical distinction, the characteristic body proportions are not in themselves dangerous. However, the underlying mutation, which affects the structure of connective tissue, commonly causes life-threatening problems for the cardiovascular system. **AM** Congenital Disorders of the Skeleton

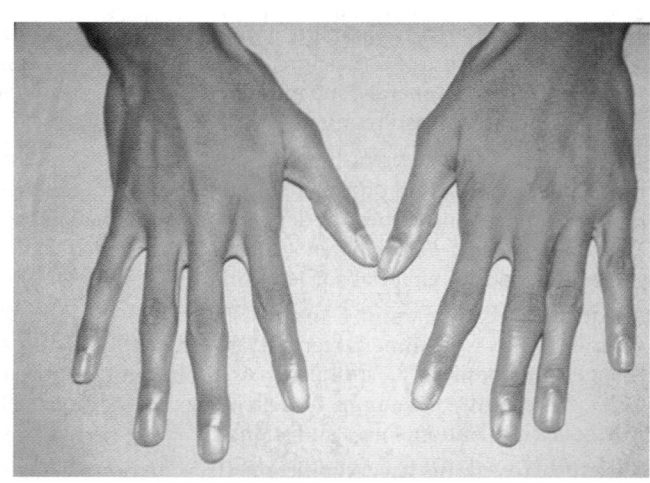

composition of bone. Combining the physical clues provided by the skeleton with modern molecular techniques, such as DNA fingerprinting, can provide a wealth of information.

✓ Why would you expect the arm bones of a weight lifter to be thicker and heavier than those of a jogger?

✓ A child who enters puberty several years later than the average age is generally taller than average as an adult. Why?

✓ A seven-year-old child has a pituitary tumor involving the cells that secrete growth hormone (GH), resulting in increased levels of GH. How will this condition affect the child's growth?

Answers start on page Q-1

☐ THE SKELETON AS A CALCIUM RESERVE

The bones of the skeleton are important mineral reservoirs (Figure 6–12●). For the moment, we shall focus on the homeostatic regula-

tion of the calcium ion concentration in body fluids; we shall consider other minerals in later chapters. Calcium is the most abundant mineral in the human body. A typical human body contains 1–2 kg (2.2–4.4 lb) of calcium, with roughly 99 percent of it deposited in the skeleton.

Calcium ions play a role in a variety of physiological processes, so the body must tightly control calcium ion concentrations in order to prevent damage to essential physiological systems. Even small variations from the normal concentration will affect cellular operations. Larger changes can cause a clinical crisis. Calcium ions are particularly important to the membranes and to the intracellular activities of neurons and muscle cells, especially cardiac muscle cells. If the calcium concentration of body fluids increases by 30 percent, neurons and muscle cells become relatively unresponsive. If calcium levels decrease by 35 percent, neurons become so excitable that convulsions can occur. A 50 percent reduction in calcium concentration generally causes death. Calcium ion concentration is so closely regulated, however, that daily fluctuations of more than 10 percent are highly unusual.

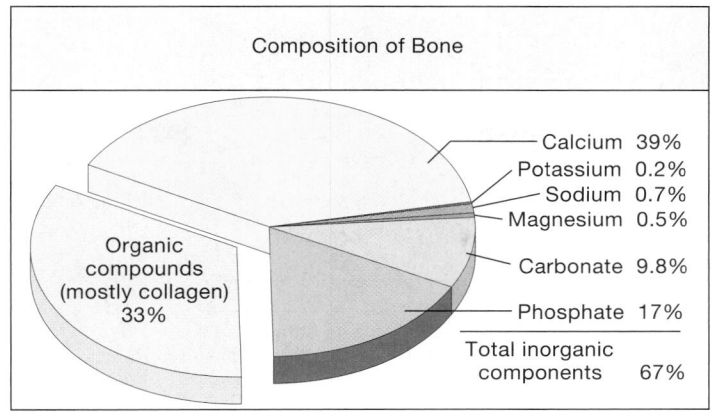

●FIGURE 6–12
A Chemical Analysis of Bone

Hormones and Calcium Balance

Calcium ion homeostasis is maintained by a negative feedback system involving a pair of hormones with opposing effects. These hormones, parathyroid hormone and calcitonin, coordinate the storage, absorption, and excretion of calcium ions. Three target sites are involved: (1) the bones (storage), (2) the digestive tract (absorption), and (3) the kidneys (excretion). Figure 6–13a● indicates factors that elevate calcium levels in the blood; Figure 6–13b● indicates factors that depress calcium levels.

When calcium ion concentrations in the blood fall below normal, cells of the **parathyroid glands**, embedded in the thyroid gland in the neck, release **parathyroid hormone** (**PTH**) into the bloodstream. Parathyroid hormone has three major effects, all of which increase blood calcium levels:

1. *Stimulating osteoclast activity* and enhancing the recycling of minerals by osteocytes. (PTH also stimulates osteoblast activity, but to a lesser degree.)

2. *Increasing the rate of intestinal absorption of calcium ions* by enhancing the action of calcitriol. Under normal circumstances, calcitriol is always present, and parathyroid hormone controls its impact on the intestinal epithelium.

3. *Decreasing the rate of excretion of calcium ions at the kidneys.*

Under these conditions, more calcium ions enter body fluids, and losses are restricted. The calcium ion concentration increases to normal levels, and homeostasis is restored.

If the calcium ion concentration of the blood instead rises above normal, special cells (*parafollicular cells*, or *C cells*) in the thyroid gland secrete **calcitonin**. This hormone has two major functions, which together act to decrease calcium ion concentrations in body fluids:

1. *Inhibiting osteoclast activity.*

2. *Increasing the rate of excretion of calcium ions at the kidneys.*

Under these conditions, less calcium *enters* body fluids because osteoclasts leave the mineral matrix alone. More calcium *leaves* body fluids because osteoblasts continue to produce new bone matrix while calcium ion excretion at the kidneys accelerates.

The net result is a decline in the calcium ion concentration of body fluids, restoring homeostasis.

By providing a calcium reserve, the skeleton plays the primary role in the homeostatic maintenance of normal calcium ion concentrations of body fluids. This function can have a direct effect on the shape and strength of the bones in the skeleton. When large numbers of calcium ions are mobilized in body fluids, the bones become weaker; when calcium salts are deposited, the bones become denser and stronger.

Because the bone matrix contains protein fibers as well as mineral deposits, changes in mineral content do not necessarily affect the shape of the bone. In *osteomalacia* (os-tē-ō-ma-LĀ-shē-uh; *malakia*, softness), the bones appear normal, although they are weak and flexible owing to poor mineralization. *Rickets*, a form of osteomalacia affecting children, generally results from a vitamin D₃ deficiency caused by inadequate exposure to sunlight and an inadequate dietary supply of the vitamin. The bones of children with rickets are so poorly mineralized that they become very flexible. Affected individuals develop a bowlegged appearance as the thigh bones and leg bones bend under the weight of the body. Homogenized milk is fortified with vitamin D in the United States specifically to prevent rickets.

✓ Why does a child who has rickets have difficulty walking?

✓ What effect would increased PTH secretion have on blood calcium levels?

✓ How does calcitonin help lower the calcium ion concentration of blood?

Answers start on page Q-1

☐ FRACTURE REPAIR

Despite its mineral strength, bone can crack or even break if subjected to extreme loads, sudden impacts, or stresses from unusual directions. The damage produced constitutes a **fracture**. (See "FOCUS: A Classification of Fractures," on page 203.) Most

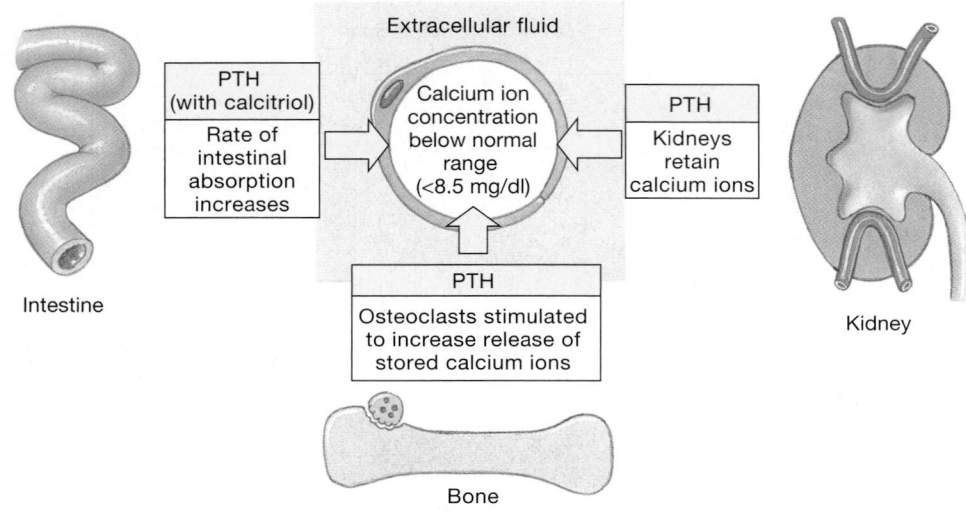

(a) Factors that increase blood calcium levels

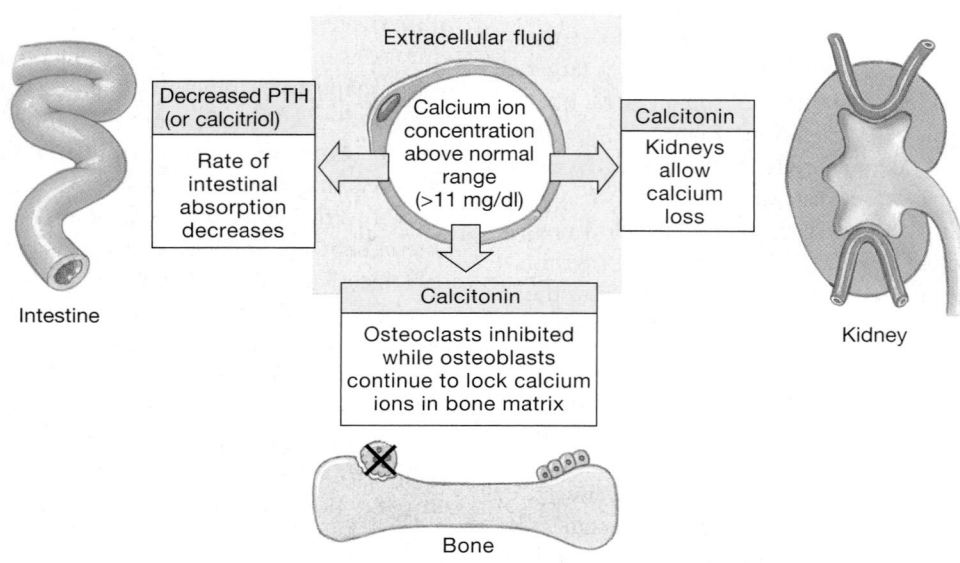

(b) Factors that decrease blood calcium levels

•FIGURE 6–13
Factors That Alter the Concentration of Calcium Ions in Body Fluids

fractures heal even after severe damage, provided that the blood supply and the cellular components of the endosteum and periosteum survive. Steps in the repair process are illustrated in Figure 6–14•:

STEP 1: In even a small fracture, many blood vessels are broken and extensive bleeding occurs. A large blood clot, or **fracture hematoma**, soon closes off the injured vessels and leaves a fibrous meshwork in the damaged area. The disruption of circulation kills osteocytes around the fracture, broadening the area affected. Dead bone soon extends along the shaft in either direction from the break.

STEP 2: In adults, the cells of the periosteum and endosteum are relatively inactive. When a fracture occurs, the cells of the intact endosteum and periosteum undergo rapid mitoses, and the daughter cells migrate into the fracture zone. An **external callus** (*callum*, hard skin), or enlarged collar of cartilage and bone, forms and encircles the bone at the level of the fracture. An extensive **internal callus** organizes within the marrow cavity and between the broken ends of the shaft. At the center of the external callus, cells differentiate into chondrocytes and produce blocks of cartilage. At the edges of each callus, the cells differentiate into osteoblasts and begin creating a bridge between the bone fragments

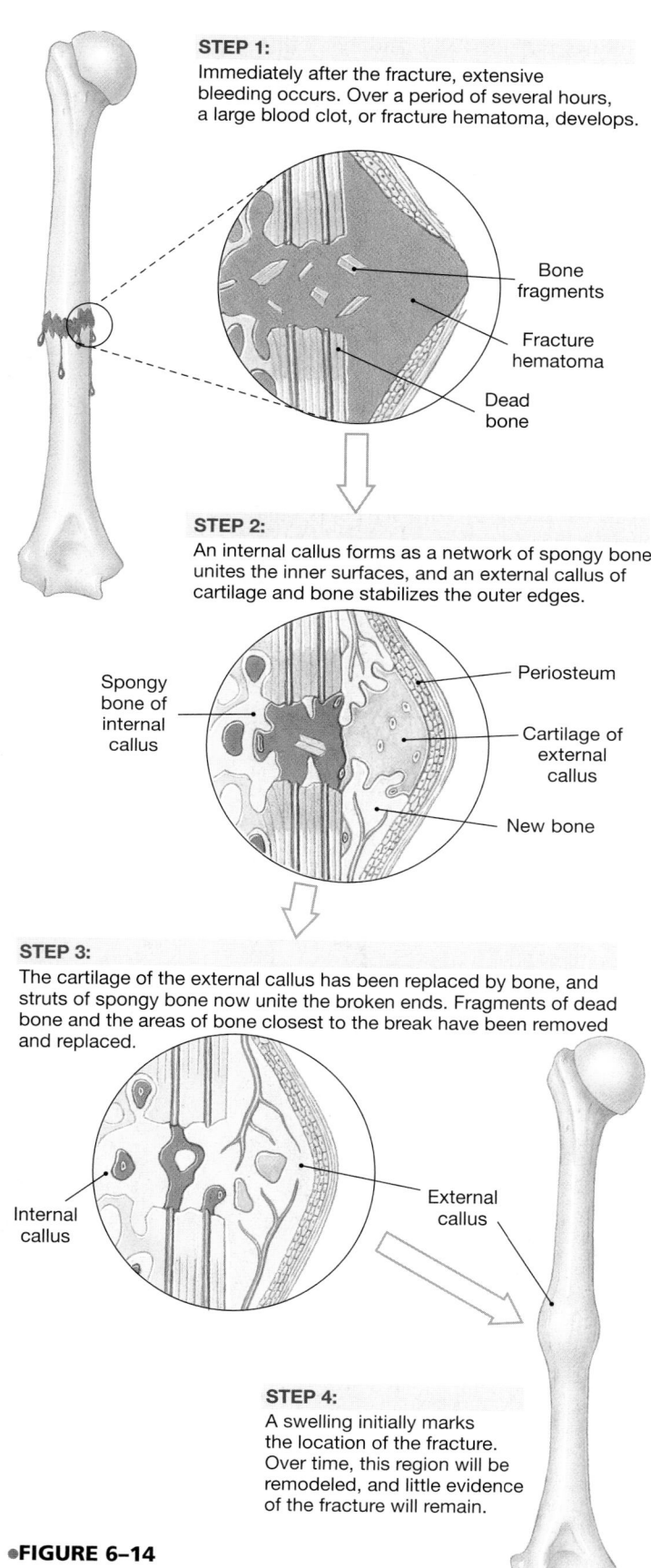

STEP 1:
Immediately after the fracture, extensive bleeding occurs. Over a period of several hours, a large blood clot, or fracture hematoma, develops.

Bone fragments

Fracture hematoma

Dead bone

STEP 2:
An internal callus forms as a network of spongy bone unites the inner surfaces, and an external callus of cartilage and bone stabilizes the outer edges.

Spongy bone of internal callus

Periosteum

Cartilage of external callus

New bone

STEP 3:
The cartilage of the external callus has been replaced by bone, and struts of spongy bone now unite the broken ends. Fragments of dead bone and the areas of bone closest to the break have been removed and replaced.

Internal callus

External callus

STEP 4:
A swelling initially marks the location of the fracture. Over time, this region will be remodeled, and little evidence of the fracture will remain.

•FIGURE 6–14
Steps in the Repair of a Fracture

on either side of the fracture. At this point, the broken ends have been temporarily stabilized.

STEP 3: As the repair continues, osteoblasts replace the central cartilage of the external callus with spongy bone. When this conversion is complete, the external and internal calluses form an extensive and continuous brace at the fracture site. Struts of spongy bone now unite the broken ends. The surrounding area is gradually reshaped as fragments of dead bone are removed and replaced. The ends of the fracture are now held firmly in place and can withstand normal stresses from muscle contractions. If the fracture required external support in the form of a cast, that support can be removed at this stage.

STEP 4: Osteoclasts and osteoblasts continue to remodel the region of the fracture for a period ranging from four months to well over a year. When the remodeling is complete, the bone of the calluses is gone and only living compact bone remains. The repair may be "good as new," and it may leave no indications that a fracture ever occurred, or the bone may be slightly thicker and stronger than normal at the fracture site. Under comparable stresses, a second fracture will generally occur at a different site.

Comparable events occur after more complex breaks or even after the transplantation of a bone fragment. Each year in the United States, roughly 200,000 people receive *bone grafts*—transplants of bone—to stimulate bone repair. The bone fragments are commonly taken from another part of the body, such as the hip. However, because the inserted bone is ultimately destroyed and replaced, dead and sterilized bone fragments from other donors or even from other species of animals can be used to establish a framework for the repair process. Steel plates, rods, and screws are sometimes used in severe fractures to help hold the ends in place as calluses form. In situations in which the fracture is so severe that normal repair processes cannot occur, even with bone grafts, surgeons may rely on synthetic bone or the use of strong electrical fields to stimulate osteoblast activity.

AM Stimulation of Bone Growth and Repair

6–6 BONE MARKINGS (SURFACE FEATURES)

Objective

■ Identify the major types of bone markings, and explain their functional significance.

Each bone in the body has characteristic external and internal features. Elevations or projections form where tendons and ligaments attach and where adjacent bones articulate (that is, form joints). Depressions, grooves, and tunnels in bone indicate sites where blood vessels or nerves lie alongside or penetrate the bone. Detailed examination of these **bone markings**, or *surface features*, can yield an abundance of anatomical information. For example, anthropologists, criminologists, and pathologists can often determine the size, age, gender, and general appearance of an individual on the basis of incomplete skeletal remains.

We will ignore minor variations of individual bones to focus on prominent features that identify the bone (Table 6–2). These markings provide fixed landmarks that can help us determine the position of the soft-tissue components of other systems. Specific anatomical terms are used to describe the various elevations and depressions.

TABLE 6–2 AN INTRODUCTION TO SKELETAL TERMINOLOGY

General Description	Anatomical Term	Definition
Elevations and projections (general)	Process	Any projection or bump
	Ramus	An extension of a bone making an angle with the rest of the structure
Processes formed where tendons or ligaments attach	Trochanter	A large, rough projection
	Tuberosity	A smaller, rough projection
	Tubercle	A small, rounded projection
	Crest	A prominent ridge
	Line	A low ridge
	Spine	A pointed process
Processes formed for articulation with adjacent bones	Head	The expanded articular end of an epiphysis, separated from the shaft by the neck
	Neck	A narrow connection between the epiphysis and the diaphysis
	Condyle	A smooth, rounded articular process
	Trochlea	A smooth, grooved articular process shaped like a pulley
	Facet	A small, flat articular surface
Depressions	Fossa	A shallow depression
	Sulcus	A narrow groove
Openings	Foramen	A rounded passageway for blood vessels or nerves
	Canal	A passageway through the substance of a bone
	Fissure	An elongate cleft
	Sinus or antrum	A chamber within a bone, normally filled with air

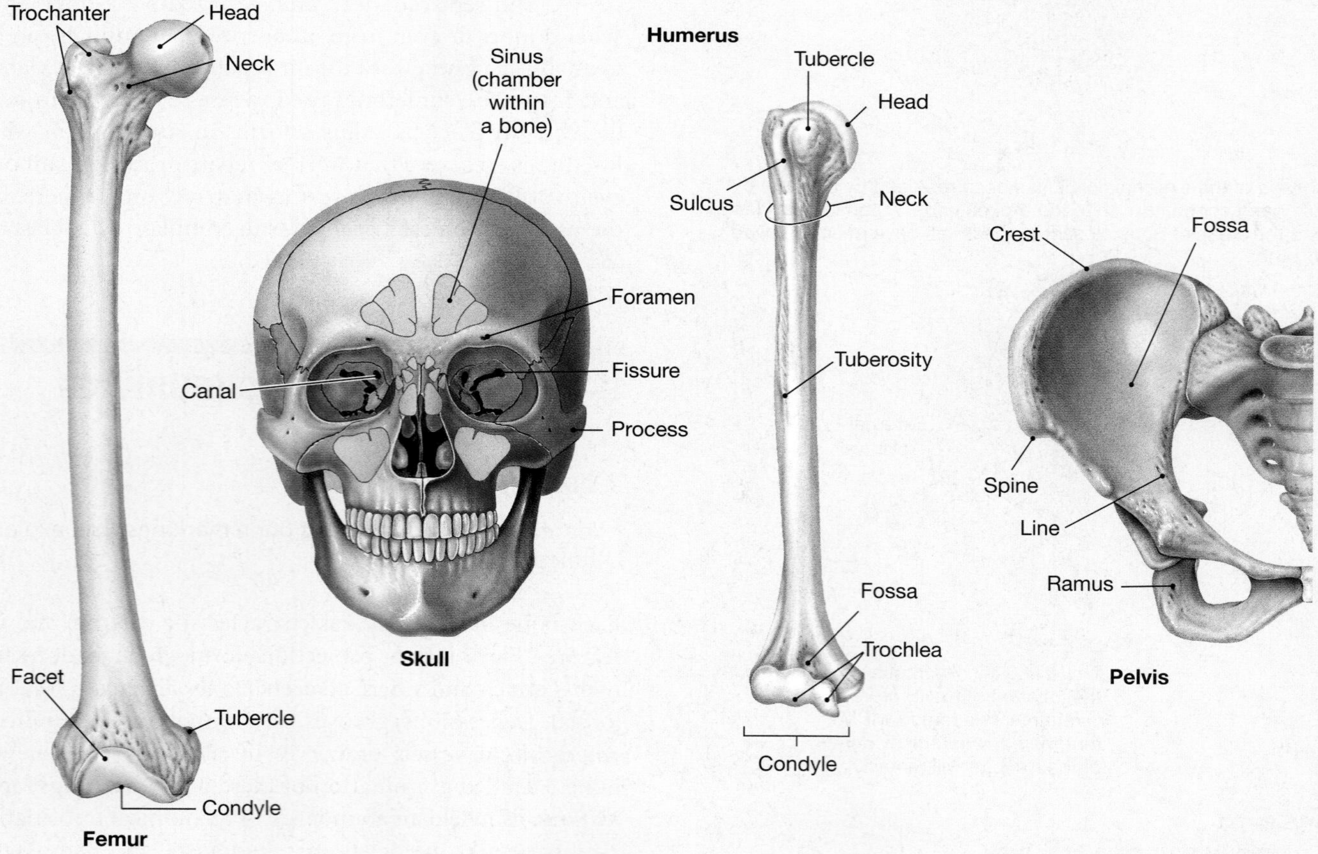

Femur

Skull

Humerus

Pelvis

A Classification of Fractures | focus

Fractures are classified according to their external appearance, their location, and the nature of the crack or break in the bone. Important types of fracture are indicated here by representative X rays. **Closed**, or *simple*, fractures are completely internal. They do not involve a break in the skin and are usually relatively simple to treat, as the ends of the bone remain close together. **Open**, or *compound*, fractures project through the skin. They are more dangerous than closed fractures, due to the possibility of infection or uncontrolled bleeding. Many fractures fall into more than one category. For example, a *Colles' fracture* is a transverse fracture that, depending on the injury, may also be an open or closed *comminuted fracture*. Comminuted fractures are characterized by a shattering of the affected bone. These fractures are common in industries in which heavy equipment is used, such as farming and construction. Representative examples of the common types of fractures are shown in Figure 6–15●. Identifying the fracture in these images takes some practice. Look for the dark line that interrupts the homogenous appearance of the bone of the shaft.

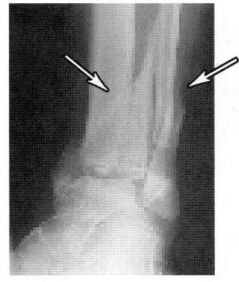

A **Pott's fracture** occurs at the ankle and affects both bones of the leg.

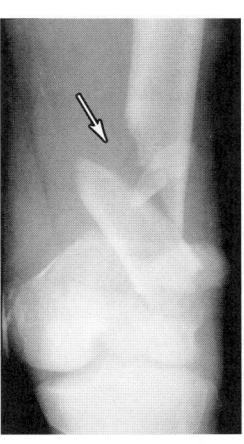

Comminuted fractures, such as this fracture of the femur, shatter the affected area into a multitude of bony fragments.

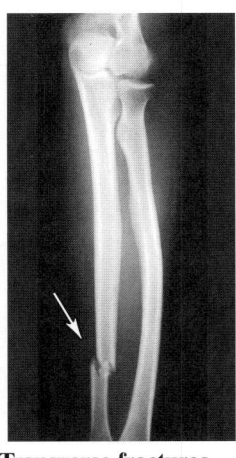

Transverse fractures, such as this fracture of the ulna, break a shaft bone across its long axis.

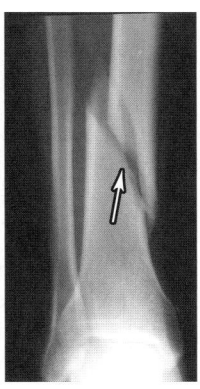

Spiral fractures, such as this fracture of the tibia, are produced by twisting stresses that spread along the length of the bone.

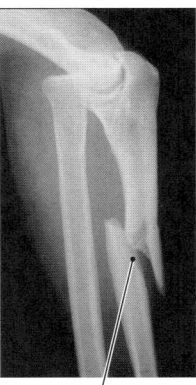

Displaced fractures produce new and abnormal bone arrangements; **nondisplaced fractures** retain the normal alignment of the bones or fragments.

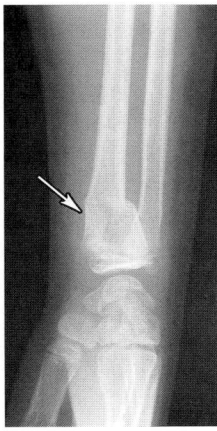

A **Colles' fracture**, a break in the distal portion of the radius, is typically the result of reaching out to cushion a fall.

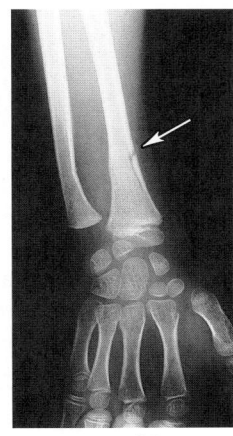

In a **greenstick fracture**, such as this fracture of the radius, only one side of the shaft is broken, and the other is bent. This type of fracture generally occurs in children, whose long bones have yet to ossify fully.

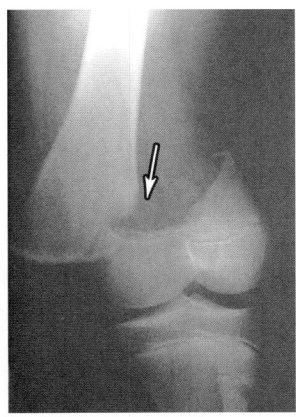

Epiphyseal fractures, such as this fracture of the femur, tend to occur where the bone matrix is undergoing calcification and chondrocytes are dying. A clean transverse fracture along this line generally heals well. Unless carefully treated, fractures between the epiphysis and the epiphyseal cartilage can permanently stop growth at this site.

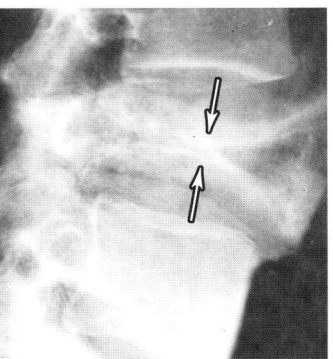

Compression fractures occur in vertebrae subjected to extreme stresses, such as those produced by the forces that arise when you land on your seat in a fall.

●**FIGURE 6–15**
Major Types of Fractures

6–7 AGING AND THE SKELETAL SYSTEM

Objective

■ Summarize the effects of the aging process on the skeletal system.

The bones of the skeleton become thinner and weaker as a normal part of the aging process. Inadequate ossification is called **osteopenia** (os-tē-ō-PĒ-nē-uh; *penia*, lacking), and all of us become slightly osteopenic as we age. This reduction in bone mass begins between the ages of 30 and 40. Over that period, osteoblast activity begins to decline, while osteoclast activity continues at previous levels. Once the reduction begins, women lose roughly 8 percent of their skeletal mass every decade, whereas the skeletons of men deteriorate at about 3 percent per decade. Not all parts of the skeleton are equally affected. Epiphyses, vertebrae, and the jaws lose more than their share, resulting in fragile limbs, a reduction in height, and the loss of teeth.

When the reduction in bone mass is sufficient to compromise normal function, the condition is known as **osteoporosis** (os-tē-ō-por-Ō-sis; *porosus*, porous). The fragile bones that result are likely to break when exposed to stresses that younger individuals could easily tolerate. For example, a hip fracture can occur when a woman in her nineties simply tries to stand. Any fractures that do occur lead to a loss of independence and an immobility that further weakens the skeleton. The extent of the loss of spongy bone mass due to osteoporosis is shown in Figure 6–16●; the reduction in compact bone mass is equally severe.

Sex hormones are important in maintaining normal rates of bone deposition. Over age 45, an estimated 29 percent of women and 18 percent of men have osteoporosis. The condition accelerates after menopause, owing to a decline in circulating estrogens. Because men continue to produce androgens until relatively late in life, severe osteoporosis is less common in males below age 60 than in females in that same age group.

Osteoporosis can also develop as a secondary effect of many cancers. Cancers of the bone marrow, breast, or other tissues release a chemical known as **osteoclast-activating factor**. This compound increases both the number and activity of osteoclasts and produces severe osteoporosis. **AM** Osteoporosis and Age-Related Skeletal Abnormalities

---

✔ At which point in fracture repair would you find an external callus?

✔ Why is osteoporosis more common in older women than in older men?

Answers start on page Q-1

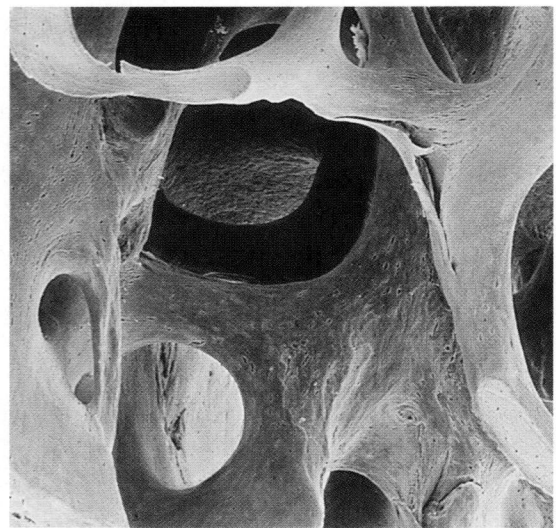

(a) Normal spongy bone (SEM × 25)

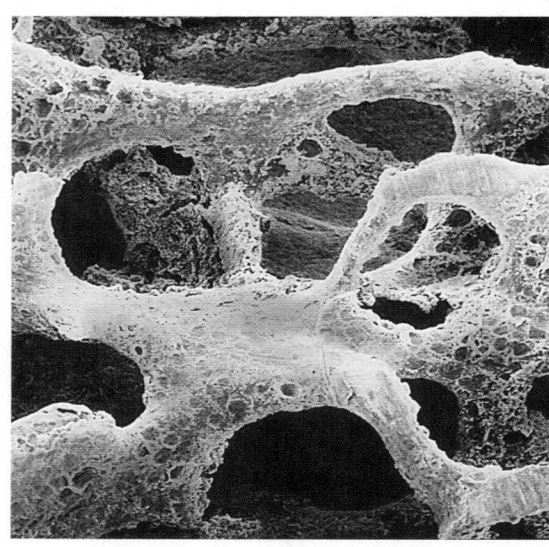

(b) Spongy bone in osteoporosis (SEM × 21)

●**FIGURE 6–16**
The Effects of Osteoporosis.   **(a)** Normal spongy bone from the epiphysis of a young adult. **(b)** Spongy bone from a person with osteoporosis.

Chapter Review

SELECTED CLINICAL TERMINOLOGY

Terms Discussed in This Chapter

acromegaly: A condition caused by excess secretion of growth hormone after puberty. Skeletal abnormalities develop, affecting the cartilages and various small bones. *(p. 198)*

external callus: A toughened layer of connective tissue that encircles and stabilizes a bone at a fracture site. *(p. 200)*

fracture: A crack or break in a bone. *(p. 199)*

fracture hematoma: A large blood clot that closes off the injured vessels around a fracture and leaves a fibrous meshwork in the damaged area of bone; the first step in fracture repair. *(p. 200)*

gigantism: A condition resulting from an overproduction of growth hormone before puberty. *(p. 198)*

internal callus: A bridgework of bone trabeculae that unites the broken ends of a bone on the marrow side of a fracture. *(p. 200)*

Marfan's syndrome: An inherited condition linked to defective production of *fibrillin*, a connective tissue glycoprotein. Extreme height and long, slender limbs are the most obvious physical indications of Marfan's syndrome; cardiovascular problems are the most dangerous aspects of the condition. *(p. 198)*

osteoclast-activating factor: A compound, released by cancers of the bone marrow, breast, or other tissues, that produces a severe osteoporosis. *(p. 204)*

osteomalacia (os-tē-ō-ma-LĀ-sheē-uh): A softening of bone due to a decrease in its mineral content. *(p. 199)*

osteopenia (os-tē-ō-PĒ-nē-uh): Inadequate ossification, leading to thinner, weaker bones. *(p. 204)*

osteoporosis (os-tē-ō-por-Ō-sis): A reduction in bone mass to a degree that compromises normal function. *(p. 204)*

pituitary growth failure (*pituitary dwarfism*): A disorder caused by inadequate production of growth hormone. *(p. 198)*

rickets: A childhood disorder that reduces the amount of calcium salts in the skeleton; typically characterized by a bowlegged appearance, because the leg bones bend under the body's weight. *(p. 199)*

scurvy: A condition involving weak, brittle bones as a result of a vitamin C deficiency. *(p. 197)*

AM Additional Terms Discussed in the *Applications Manual*

achondroplasia
hyperostosis
osteogenesis imperfecta
osteomyelitis
osteopetrosis
Paget's disease, or **osteitis deformans**

STUDY OUTLINE

6–1 THE SKELETAL SYSTEM: AN INTRODUCTION p. 183

1. The skeletal system includes the bones of the skeleton and the cartilages, ligaments, and other connective tissues that stabilize or connect them. The functions of the skeletal system include support, storage of minerals and lipids, blood cell production, protection, and leverage.

6–2 A CLASSIFICATION OF BONES p. 184

BONE SHAPES p. 184

1. Bones may be categorized as **long bones, short bones, flat bones, irregular bones, sesamoid bones,** and **sutural bones** (*Wormian bones*). (*Figure 6–1*)

BONE STRUCTURE p. 185

2. The two types of bone are **compact** (*dense*) **bone** and **spongy** (*cancellous*) **bone.**
3. A representative long bone has a **diaphysis, epiphyses, metaphyses, articular cartilages,** and a **marrow cavity.** (*Figure 6–2*)
4. The marrow cavity and spaces within spongy bone contain either **yellow bone marrow** (for lipid storage) or **red bone marrow** (for blood cell formation).

6–3 BONE HISTOLOGY p. 186

1. **Osseous tissue** is a supporting connective tissue with a solid matrix. The minerals are deposited in **lamellae** and ensheathed by a **periosteum.**

THE MATRIX OF BONE p. 186

2. Bone matrix consists largely of crystals of **hydroxyapatite.**

CELLS IN BONE p. 186

3. **Osteocytes,** located in **lacunae,** are mature bone cells. Adjacent osteocytes are interconnected by **canaliculi. Osteoblasts** synthesize the bony matrix by **osteogenesis. Osteoclasts** dissolve the bony matrix through **osteolysis. Osteoprogenitor cells** differentiate into osteoblasts. (*Figure 6–3*)

COMPACT BONE AND SPONGY BONE p. 188

4. The basic functional unit of compact bone is the **osteon,** containing osteocytes arranged around a **central canal. Perforating canals** extend perpendicularly to the bone surface. (*Figure 6–4*)
5. Spongy bone contains **trabeculae,** typically in an open network. (*Figure 6–4*)
6. Compact bone is located where stresses come from a limited range of directions, such as along the diaphysis of some bones. Spongy bone is located where stresses are few or come from many directions, such as at the epiphyses of some bones. (*Figure 6–5*)

THE PERIOSTEUM AND ENDOSTEUM p. 190

7. A bone is covered by a **periosteum** and lined with an **endosteum.** (*Figure 6–6*)

Bone markings: **Anatomy CD-ROM:** Skeletal System/Axial Dissections

6-4 BONE DEVELOPMENT AND GROWTH   p. 191

1. **Ossification** is the process of converting other tissues to bone. **Calcification** is the process of depositing calcium salts within a tissue.

INTRAMEMBRANOUS OSSIFICATION   p. 191

2. **Intramembranous ossification** begins when osteoblasts differentiate within connective tissue. The process produces dermal bones. Such ossification begins at an **ossification center.** *(Figure 6–7)*

ENDOCHONDRAL OSSIFICATION   p. 192

3. **Endochondral ossification** begins with a cartilage model that is gradually replaced by bone at the metaphysis. In this way, bone length increases. *(Figure 6–8)*
4. The timing of *closure* of the **epiphyseal cartilage** differs among bones and among individuals. *(Figure 6–9)*
5. Bone diameter increases through *appositional growth*. *(Figure 6–10)*

Intramembranous and endochondral ossification: **Companion Website**: Chapter 6/Tutorials.

THE BLOOD AND NERVE SUPPLIES   p. 194

6. Three major sets of blood vessels provide an extensive supply of blood to bone. *(Figure 6–11)*

6-5 THE DYNAMIC NATURE OF BONE   p. 196

1. The organic and mineral components of bone are continuously recycled and renewed through **remodeling.**

EFFECTS OF EXERCISE ON BONE   p. 196

2. The shapes and thicknesses of bones reflect the stresses applied to them.

HORMONAL AND NUTRITIONAL EFFECTS ON BONE   p. 197

3. Normal osteogenesis requires a reliable source of minerals, vitamins, and hormones.
4. *Growth hormone* and *thyroxine* stimulate bone growth. Calcitonin and parathyroid hormone control blood calcium levels. *(Table 6–1)*

THE SKELETON AS A CALCIUM RESERVE   p. 198

5. Calcium is the most common mineral in the human body, with roughly 99 percent of it located in the skeleton. *(Figure 6–12)*
6. Interactions among the bones, intestinal tract, and kidneys affect the calcium ion concentration. *(Figure 6–13)*
7. Two hormones, **calcitonin** and **parathyroid hormone** (**PTH**), regulate calcium ion homeostasis. Calcitonin leads to a decline in the calcium concentration, whereas parathyroid hormone increases the calcium concentration. *(Figure 6–14)*

FRACTURE REPAIR   p. 199

8. A **fracture** is a crack or a break in a bone. The repair of a fracture involves the formation of a **fracture hematoma,** an **external callus,** and an **internal callus.** *(Figure 6–14)*

6-6 BONE MARKINGS (SURFACE FEATURES)   p. 201

1. Each bone has characteristic **bone markings,** including elevations, projections, depressions, grooves, and tunnels. *(Table 6–2)*

6-7 AGING AND THE SKELETAL SYSTEM   p. 204

1. The effects of aging on the skeleton include **osteopenia** and **osteoporosis.** *(Figure 6–16)*

REVIEW QUESTIONS

More assessment questions are available to you on the Companion Website. You will find Matching, Multiple Choice, True/False, and other quizzes to help further your understanding of the material covered in this chapter. To access the site, go to www.aw.com/martini.

LEVEL 1 Reviewing Facts and Terms

1. Hematopoiesis occurs in the bones of the skeleton in areas of
   (a) yellow marrow
   (b) red marrow
   (c) the matrix of bone tissue
   (d) the ground substance

2. Two-thirds of the weight of bone is accounted for by
   (a) crystals of calcium phosphate
   (b) collagen fibers
   (c) osteocytes
   (d) calcium carbonate

3. The membrane found wrapping the bones, except at the joint cavity, is the
   (a) periosteum
   (b) endosteum
   (c) perforating fibers
   (d) a, b, and c are correct

4. The basic functional units of mature compact bone are called
   (a) lacunae
   (b) osteocytes
   (c) osteons
   (d) canaliculi

5. Unlike compact bone, spongy bone contains concentric lamellae that form struts or plates called
   (a) canaliculi
   (b) canals of Volkmann
   (c) osteons
   (d) trabeculae

6. The vitamins essential for normal adult bone maintenance and repair are
   (a) A and E
   (b) C and D
   (c) B and E
   (d) B complex and K

7. The hormones that coordinate the storage, absorption, and excretion of calcium ions are
   (a) growth hormone and thyroxine
   (b) calcitonin and parathyroid hormone
   (c) calcitriol and cholecalciferol
   (d) estrogens and androgens

8. The deposition of calcium salts in tissues other than bone is referred to as
   (a) endochondral ossification
   (b) intramembranous ossification
   (c) calcification
   (d) osteogenesis

9. The primary reason that osteoporosis accelerates after menopause in women is
   (a) reduced levels of circulating estrogens
   (b) reduced levels of vitamin C
   (c) diminished osteoclast activity
   (d) increased osteoblast activity

10. A child with rickets would have
    (a) oversized facial bones
    (b) long limbs
    (c) weak, brittle bones
    (d) bowlegs

11. What are the five primary functions of the skeletal system?

12. List the four distinctive cell populations of osseous tissue.

13. What is the functional difference between an osteoblast and an osteoclast?

14. What are the primary parts of a typical long bone?

15. What is the primary difference between intramembranous ossification and endochondral ossification?

16. List the organic and inorganic components of bone matrix.

17. (a) What nutritional factors are essential for normal bone growth and maintenance?

    (b) What hormonal factors are necessary for normal bone growth and maintenance?

18. Which three organs or tissues interact to assist in the regulation of calcium ion concentration in body fluids?

19. What major effects of parathyroid hormone oppose those of calcitonin?

20. What are the major functions of the hormone calcitonin?

LEVEL 2 Reviewing Concepts

21. If spongy bone has no osteons, how do nutrients reach the osteocytes?

22. Why are stresses or impacts to the side of the shaft in a long bone more dangerous than stress applied to the long axis of the shaft?

23. Why do extended periods of inactivity cause degenerative changes in the skeleton?

24. What are the functional relationships between the skeleton, on the one hand, and the digestive and urinary systems, on the other?

25. During the growth of a long bone, how is the epiphysis forced farther from the shaft?

26. Why would a physician concerned about the growth patterns of a young child request an X ray of the hand?

27. How might damage to the thyroid gland influence calcium regulation in the body?

28. Why does a second fracture in the same bone tend to occur at a site different from that of the first fracture?

29. What type of bone growth occurs as bones gain mass in response to weight training?

30. How might bone markings be useful in identifying the remains of a criminal who has been shot and killed?

31. What purpose do elevations or projections serve on bones?

LEVEL 3 Critical Thinking and Clinical Applications

32. While playing on her swing set, 10-year-old Sally falls and breaks her right leg. At the emergency room, the doctor tells her parents that the proximal end of the tibia where the epiphysis meets the diaphysis is fractured. The fracture is properly set and eventually heals. During a routine physical when she is 18, Sally learns that her right leg is 2 cm shorter than her left, probably because of her accident. What might account for this difference?

33. Sherry is a pregnant teenager. Her diet before she was pregnant consisted mostly of junk food, and that hasn't changed since she became pregnant. Approximately 8–10 weeks into her pregnancy, she falls and breaks her arm. She doesn't understand why the bone broke, because it wasn't a hard fall. Test results determine that a significant amount of bone demineralization is occurring. Explain what is happening to Sherry.

34. Would you expect to see changes in blood levels of the hormones calcitonin and PTH as a result of vitamin D_3 deficiency? Explain.

35. Why might a person in kidney failure exhibit symptoms similar to those of osteoporosis?

36. In physical anthropology, cultural conclusions can be drawn from a thorough examination of the skeletons of ancient peoples. What sorts of clues might bones provide as to their owner's lifestyle?

UNIT 2   CHAPTER   5 6 **7** 8 9 10 11

CLINICAL NOTES
- TMJ Syndrome 224
- Craniostenosis 228
- Spina Bifida 231
- Whiplash 231
- Thoracentesis 239

CLINICAL DISCUSSIONS
- Sinus Problems and Septal Defects 226

7

THE AXIAL SKELETON

Bone, a tissue found only in vertebrate animals, first appeared more than 500 million years ago in primitive fishes. One hundred million years later, it had become very common in many groups of fishes. This was dermal bone that provided a mineral reserve and armor plating to protect against predators. The internal skeletons of these fishes were composed of cartilage. In the evolutionary branch that led toward humans, bones replaced most of the cartilages.

As we saw in Chapter 6, bones are remarkable structures. They are as strong as reinforced concrete, but are considerably lighter. ∞ p. 186 Better yet, bones are remodeled and reshaped to meet metabolic demands or to adapt to changing activity patterns. The basic features of the human skeleton have been shaped by evolution, but the detailed characteristics of each bone reflect the stresses placed on it. As a result, the skeleton changes in the course of a lifetime. Examples discussed in Chapter 6 included the proportional changes at puberty and the gradual osteopenia of aging. In this chapter, we will consider additional examples, such as the changes that occurred in the shape of your vertebral column as you made the transition from crawling to walking.

7-1 DIVISIONS OF THE SKELETON

Objective

■ Identify the bones of the axial skeleton and specify their functions.

The skeletal system typically includes 206 separate bones and a number of associated cartilages and ligaments. Learning the names, positions, and markings of bones is essential to understanding the other systems of the body. In many instances, structures located near or associated with a skeletal component are named after the bone. Examples include the femoral artery, which runs along the femur (the bone of the thigh), and the frontal lobe of the brain, which is covered by the frontal bone of the skull. Our descriptions of specific bones of the skeleton will make use of anatomical terminology introduced in earlier chapters, so you may wish to review the terms summarized in Tables 1–3, 1–4, and 6–1 before proceeding. ∞ pp. 18, 20, and 197

The skeletal system is divided into axial and appendicular divisions. The **axial skeleton** (Figure 7–1•) forms the longitudinal axis of the body. The axial skeleton has 80 bones, roughly 40 percent of the bones in the human body. The axial components are:

- The *skull* (8 *cranial bones* and 14 *facial bones*).
- Bones associated with the skull (6 *auditory ossicles* and the *hyoid bone*).
- The *thoracic cage* (the *sternum* and 24 *ribs*).
- The *vertebral column* (24 *vertebrae*, the *sacrum*, and the *coccyx*).

The axial skeleton provides a framework that supports and protects organs in the dorsal and ventral body cavities. It also provides an extensive surface area for the attachment of muscles that (1) adjust the positions of the head, neck, and trunk; (2) perform respiratory movements; and (3) stabilize or position parts of the appendicular skeleton. The joints of the axial skeleton permit limited movement, but they are very strong and heavily reinforced with ligaments.

The **appendicular skeleton** consists of 126 bones. This division includes the bones of the limbs and of the pectoral and pelvic girdles that attach the limbs to the trunk. We will examine the appendicular skeleton in Chapter 8; the current chapter describes the functional anatomy of the axial skeleton. We begin with the skull.

7–2 THE SKULL

Objectives

- Identify the bones of the cranium and face, and explain the significance of the markings on the individual bones.
- Describe the structure of the nasal complex and the functions of the individual bones.
- Explain the functions of the paranasal sinuses.
- Describe key structural differences among the skulls of infants, children, and adults.

The bones of the **skull** protect the brain and guard the entrances to the digestive and respiratory systems. The skull contains 22 bones: 8 form the **cranium**, or *braincase*, and 14 are associated with the face (Figure 7–2•). Seven additional bones are associated with the skull: 6 auditory ossicles are situated within the *temporal bones* of

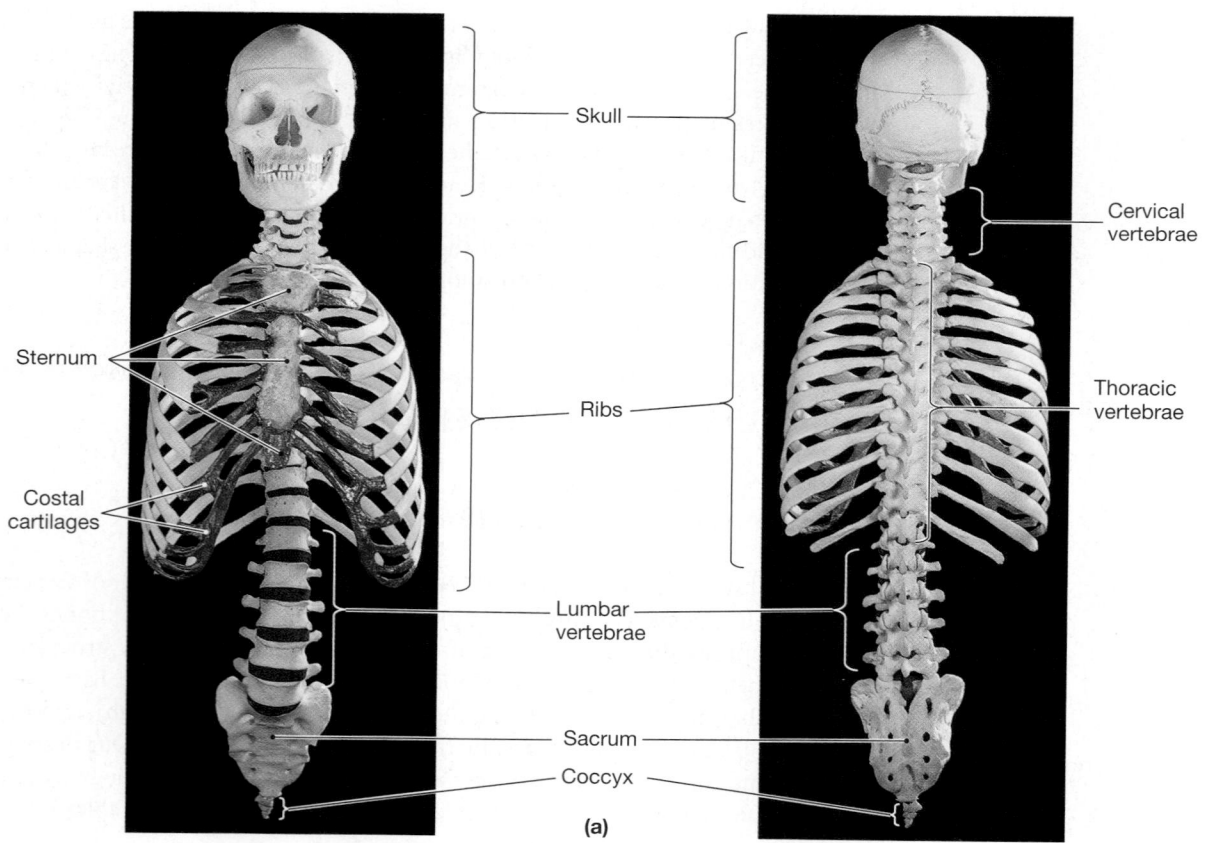

Skull

Cervical vertebrae

Sternum

Ribs

Thoracic vertebrae

Costal cartilages

Lumbar vertebrae

Sacrum

Coccyx

(a)

•**FIGURE 7–1**
The Axial Skeleton. **(a)** Anterior and posterior views. The bones associated with the skull are not visible. **(b)** An anterior view of the entire skeleton, with the axial components highlighted. The numbers in the boxes indicate the number of bones in the adult skeleton. **ATLAS** Plate 1

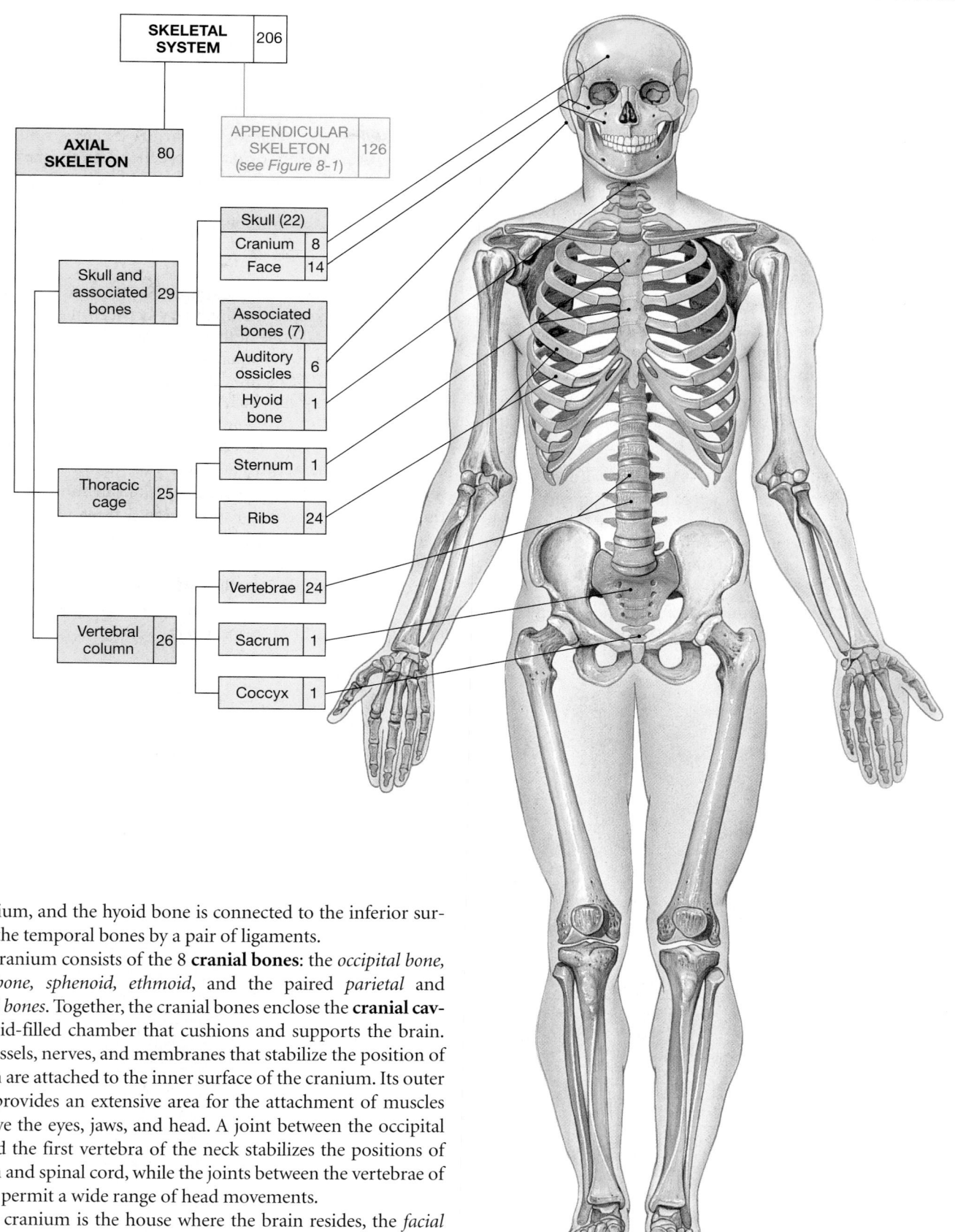

SKELETAL SYSTEM	206

AXIAL SKELETON	80

APPENDICULAR SKELETON (see Figure 8-1)	126

Skull and associated bones	29

Skull (22)	
Cranium	8
Face	14

Associated bones (7)	
Auditory ossicles	6
Hyoid bone	1

Thoracic cage	25

Sternum	1
Ribs	24

Vertebral column	26

Vertebrae	24
Sacrum	1
Coccyx	1

(b)

the cranium, and the hyoid bone is connected to the inferior surfaces of the temporal bones by a pair of ligaments.

The cranium consists of the 8 **cranial bones**: the *occipital bone, frontal bone, sphenoid, ethmoid,* and the paired *parietal* and *temporal bones.* Together, the cranial bones enclose the **cranial cavity**, a fluid-filled chamber that cushions and supports the brain. Blood vessels, nerves, and membranes that stabilize the position of the brain are attached to the inner surface of the cranium. Its outer surface provides an extensive area for the attachment of muscles that move the eyes, jaws, and head. A joint between the occipital bone and the first vertebra of the neck stabilizes the positions of the brain and spinal cord, while the joints between the vertebrae of the neck permit a wide range of head movements.

If the cranium is the house where the brain resides, the *facial complex* is the front porch. **Facial bones** protect and support the entrances to the digestive and respiratory tracts. The superficial facial bones (the paired *maxillary, lacrimal, nasal,* and *zygomatic* *bones* and the single *mandible*) (Figure 7–2●) provide areas for the attachment of muscles that control facial expressions and assist in manipulating food. The deeper facial bones (the *palatine bones, inferior nasal conchae,* and *vomer*) help separate the oral and nasal cavities, increase the surface area of the nasal cavities, or help form the **nasal septum** (*septum,* wall), which subdivides the nasal cavity.

Several bones of the skull contain air-filled chambers called **sinuses**. Sinuses have two major functions: (1) The presence of a sinus makes a bone much lighter than it would otherwise be, and (2) the mucous membrane lining the sinuses produces mucus that moistens and cleans the air in and adjacent to the sinus. We shall consider the sinuses as we discuss specific bones.

Joints, or *articulations,* form where two bones interconnect. Except where the mandible contacts the cranium, the connections between the skull bones of adults are immovable joints called **sutures**. At a suture, bones are tied firmly together with dense fibrous connective tissue. Each suture of the skull has a name, but at this point you need to know only four major sutures:

1. *Lambdoid Suture.* The **lambdoid** (LAM-doyd) **suture** (Greek *lambda,* L + *eidos,* shape) arches across the posterior surface of the skull (Figure 7–3a●). This suture separates the occipital bone from the two parietal bones. One or more **sutural bones** (*Wormian bones*) may be present along the lambdoid suture. ∞ p. 185

2. *Coronal Suture.* The **coronal suture** attaches the frontal bone to the parietal bones of either side (Figure 7–3b●). The occipital,

parietal, and frontal bones form the **calvaria** (kal-VAR-ē-uh), or skullcap. A cut through the body that parallels the coronal suture produces a *coronal section,* or *frontal section* (Figure 1–10●). ∞ p. 19

3. *Sagittal Suture.* The **sagittal suture** extends from the lambdoid suture to the coronal suture, between the parietal bones (Figure 7–3b●). A cut through the body that parallels the sagittal suture produces a parasagittal section. If the slice passes along the midline of the suture, it is a midsagittal section, dividing the body into equal left and right halves. ∞ p. 19

4. *Squamous Sutures.* A **squamous** (SKWĀ-mus) **suture** on each side of the skull is the boundary between the temporal bone and the parietal bone of that side. Figure 7–3a● shows the intersection between the squamous sutures and the lambdoid suture. Figure 7–3c● shows the path of the squamous suture on the right side of the skull.

"FOCUS: The Individual Bones of the Skull," on pages 216–223, considers each skull bone. Refer to the figures there and to Figures 7–3 and 7–4● (the adult skull in superficial and sectional views, respectively) to develop a three-dimensional perspective of the relationships among the bones. The ridges and *foramina* (fō-RAM-in-uh; passageways; singular, *foramen*) detailed here mark the attachment sites of muscles discussed in Chapter 11, or the passage of nerves and blood vessels discussed in Chapters 14 and 21, respectively.

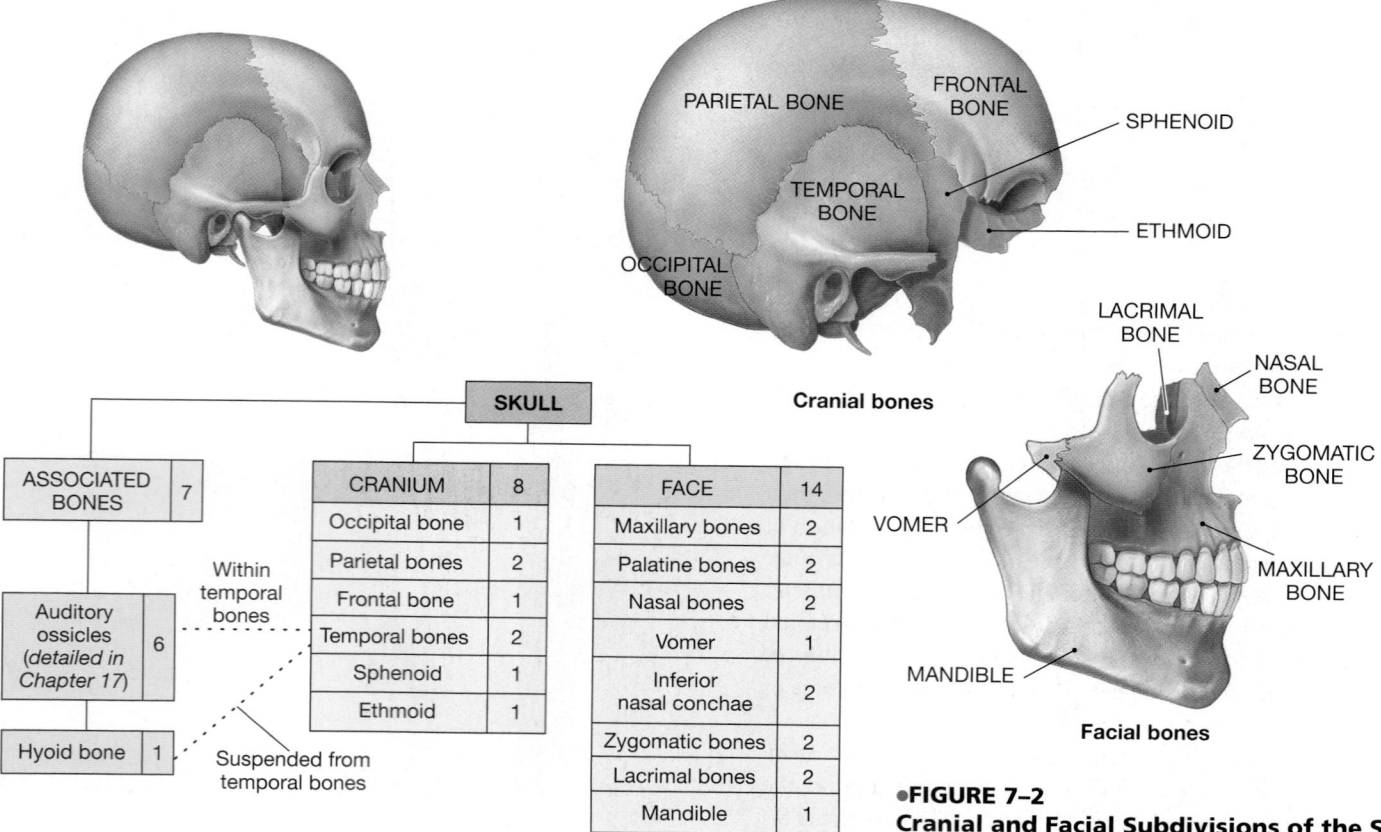

ASSOCIATED BONES	7
Auditory ossicles (*detailed in Chapter 17*)	6
Hyoid bone	1

Within temporal bones

Suspended from temporal bones

SKULL	

CRANIUM	8
Occipital bone	1
Parietal bones	2
Frontal bone	1
Temporal bones	2
Sphenoid	1
Ethmoid	1

FACE	14
Maxillary bones	2
Palatine bones	2
Nasal bones	2
Vomer	1
Inferior nasal conchae	2
Zygomatic bones	2
Lacrimal bones	2
Mandible	1

Cranial bones

Facial bones

●**FIGURE 7–2**
Cranial and Facial Subdivisions of the Skull.
The seven associated bones are not illustrated.

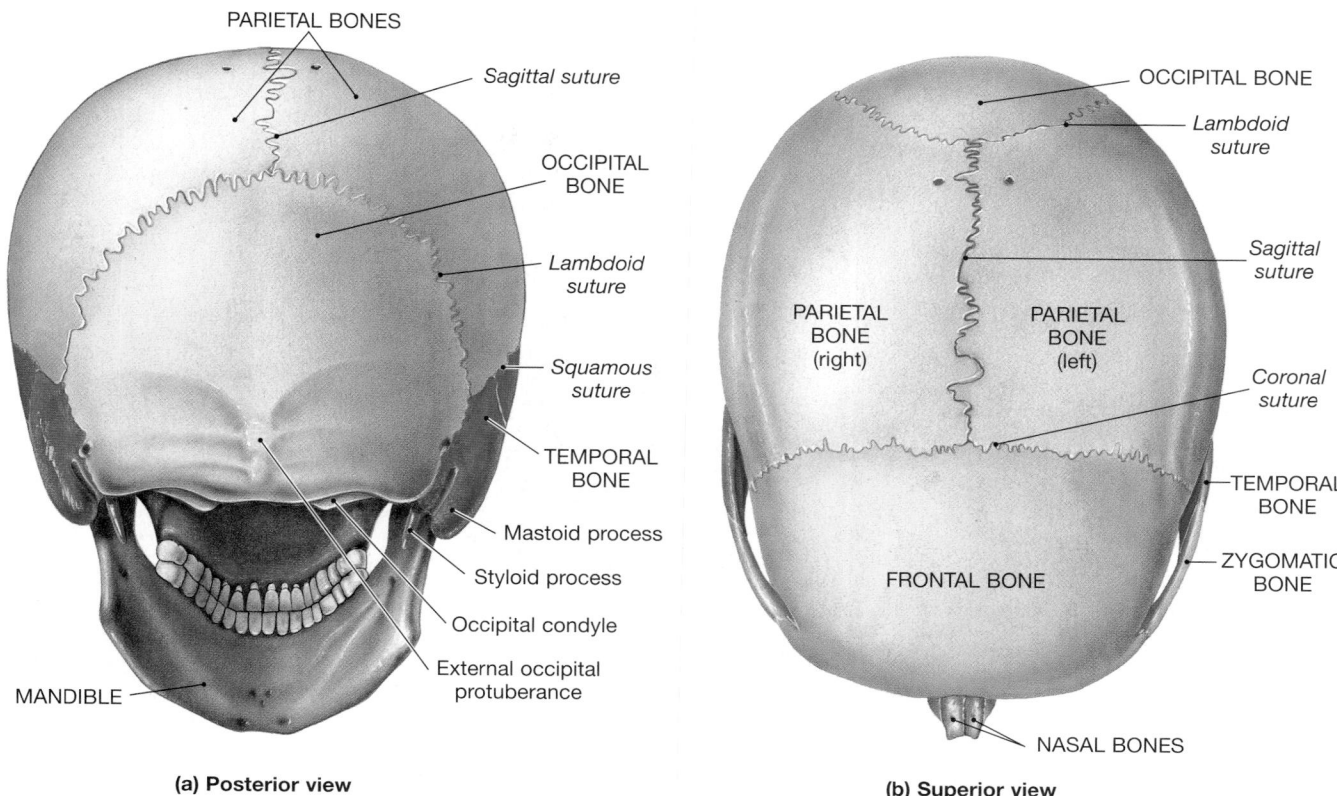

PARIETAL BONES

Sagittal suture

OCCIPITAL BONE

Lambdoid suture

Squamous suture

TEMPORAL BONE

Mastoid process

Styloid process

Occipital condyle

External occipital protuberance

MANDIBLE

(a) Posterior view

OCCIPITAL BONE

Lambdoid suture

Sagittal suture

PARIETAL BONE (right)

PARIETAL BONE (left)

Coronal suture

TEMPORAL BONE

ZYGOMATIC BONE

FRONTAL BONE

NASAL BONES

(b) Superior view

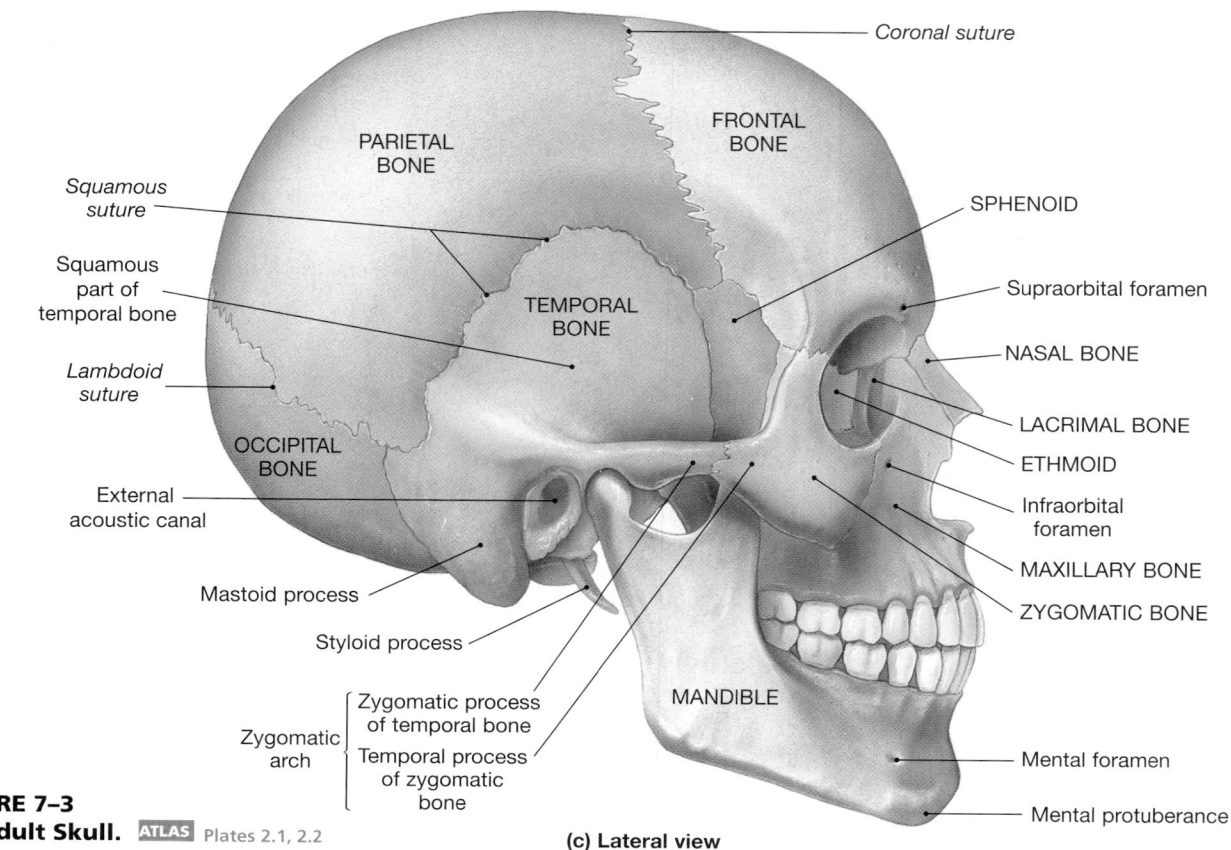

Coronal suture

PARIETAL BONE

FRONTAL BONE

Squamous suture

Squamous part of temporal bone

Lambdoid suture

TEMPORAL BONE

OCCIPITAL BONE

External acoustic canal

Mastoid process

Styloid process

Zygomatic arch

Zygomatic process of temporal bone

Temporal process of zygomatic bone

SPHENOID

Supraorbital foramen

NASAL BONE

LACRIMAL BONE

ETHMOID

Infraorbital foramen

MAXILLARY BONE

ZYGOMATIC BONE

MANDIBLE

Mental foramen

Mental protuberance

(c) Lateral view

•FIGURE 7–3
The Adult Skull. ATLAS Plates 2.1, 2.2

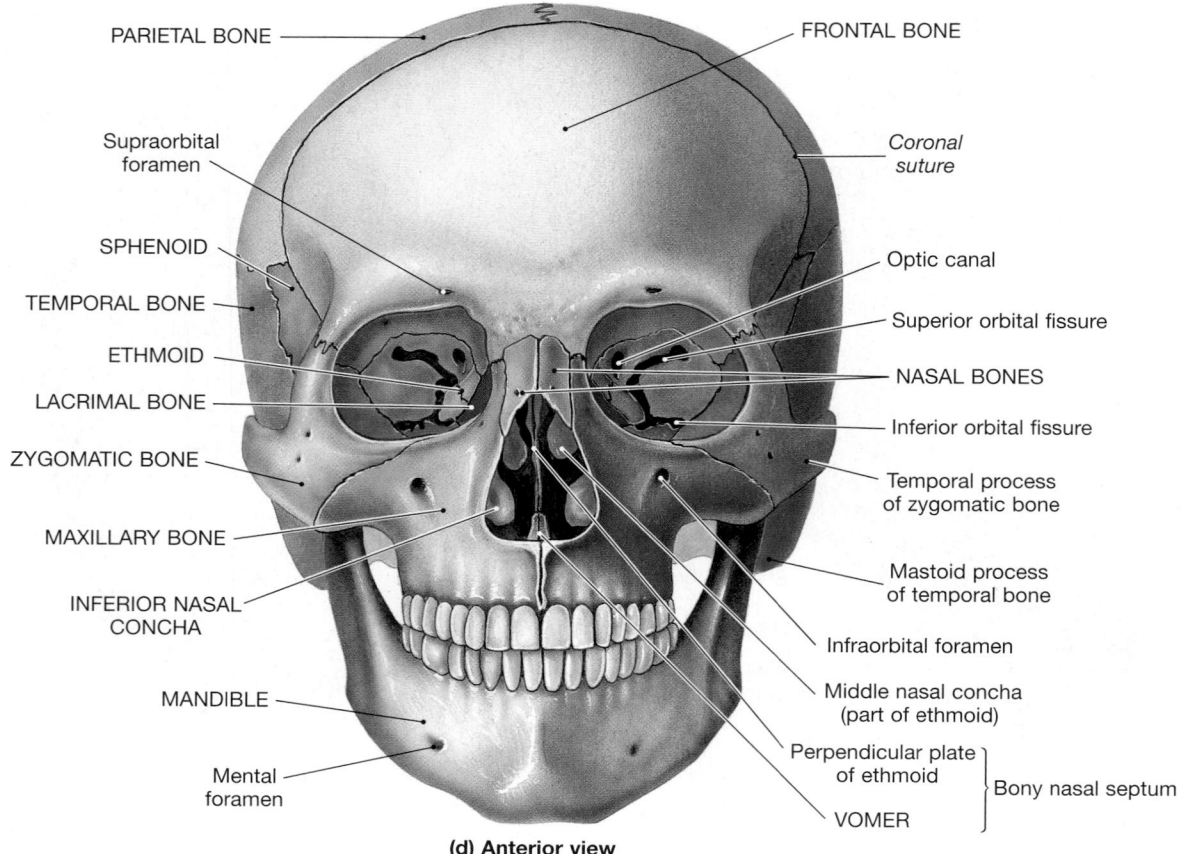

PARIETAL BONE

FRONTAL BONE

Supraorbital foramen

Coronal suture

SPHENOID

Optic canal

TEMPORAL BONE

Superior orbital fissure

ETHMOID

NASAL BONES

LACRIMAL BONE

Inferior orbital fissure

ZYGOMATIC BONE

Temporal process of zygomatic bone

MAXILLARY BONE

Mastoid process of temporal bone

INFERIOR NASAL CONCHA

Infraorbital foramen

MANDIBLE

Middle nasal concha (part of ethmoid)

Mental foramen

Perpendicular plate of ethmoid } Bony nasal septum

VOMER

(d) Anterior view

FIGURE 7–3
The Adult Skull *(continued)*.

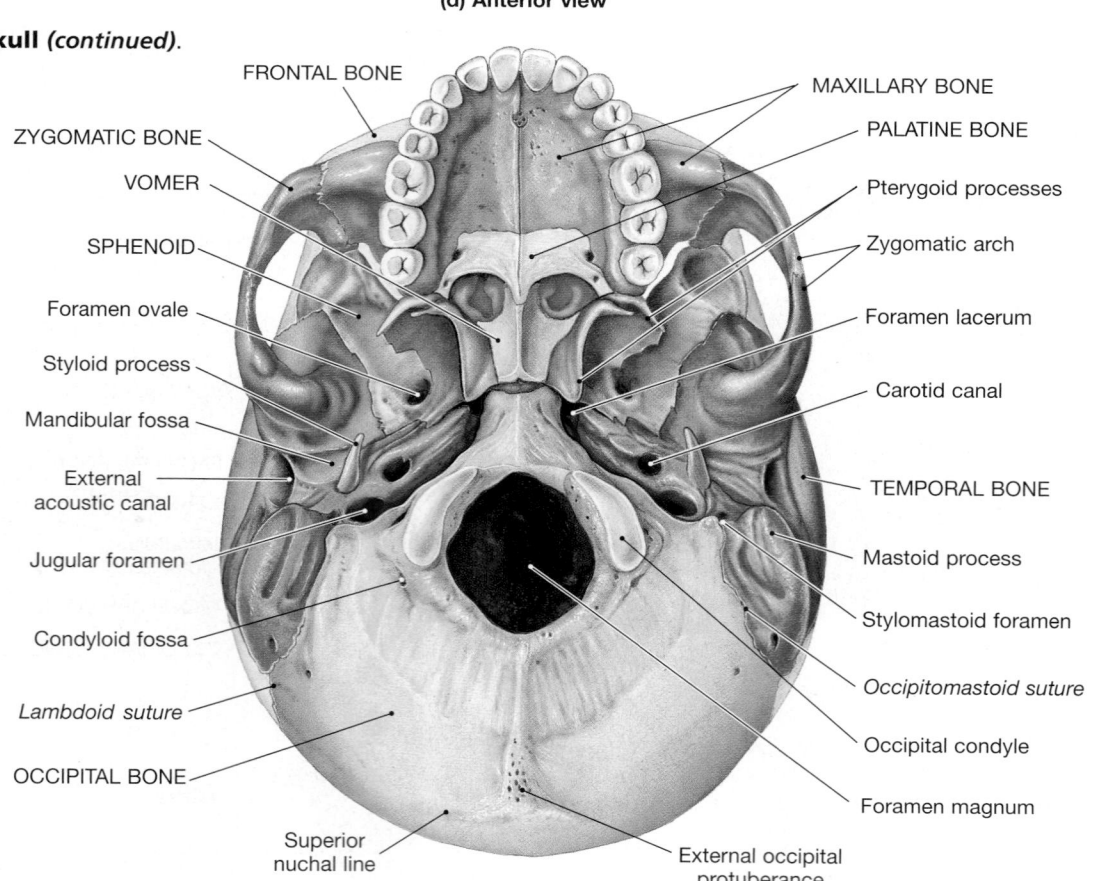

FRONTAL BONE

MAXILLARY BONE

ZYGOMATIC BONE

PALATINE BONE

VOMER

Pterygoid processes

SPHENOID

Zygomatic arch

Foramen ovale

Foramen lacerum

Styloid process

Carotid canal

Mandibular fossa

External acoustic canal

TEMPORAL BONE

Jugular foramen

Mastoid process

Condyloid fossa

Stylomastoid foramen

Lambdoid suture

Occipitomastoid suture

OCCIPITAL BONE

Occipital condyle

Foramen magnum

Superior nuchal line

External occipital protuberance

(e) Inferior view

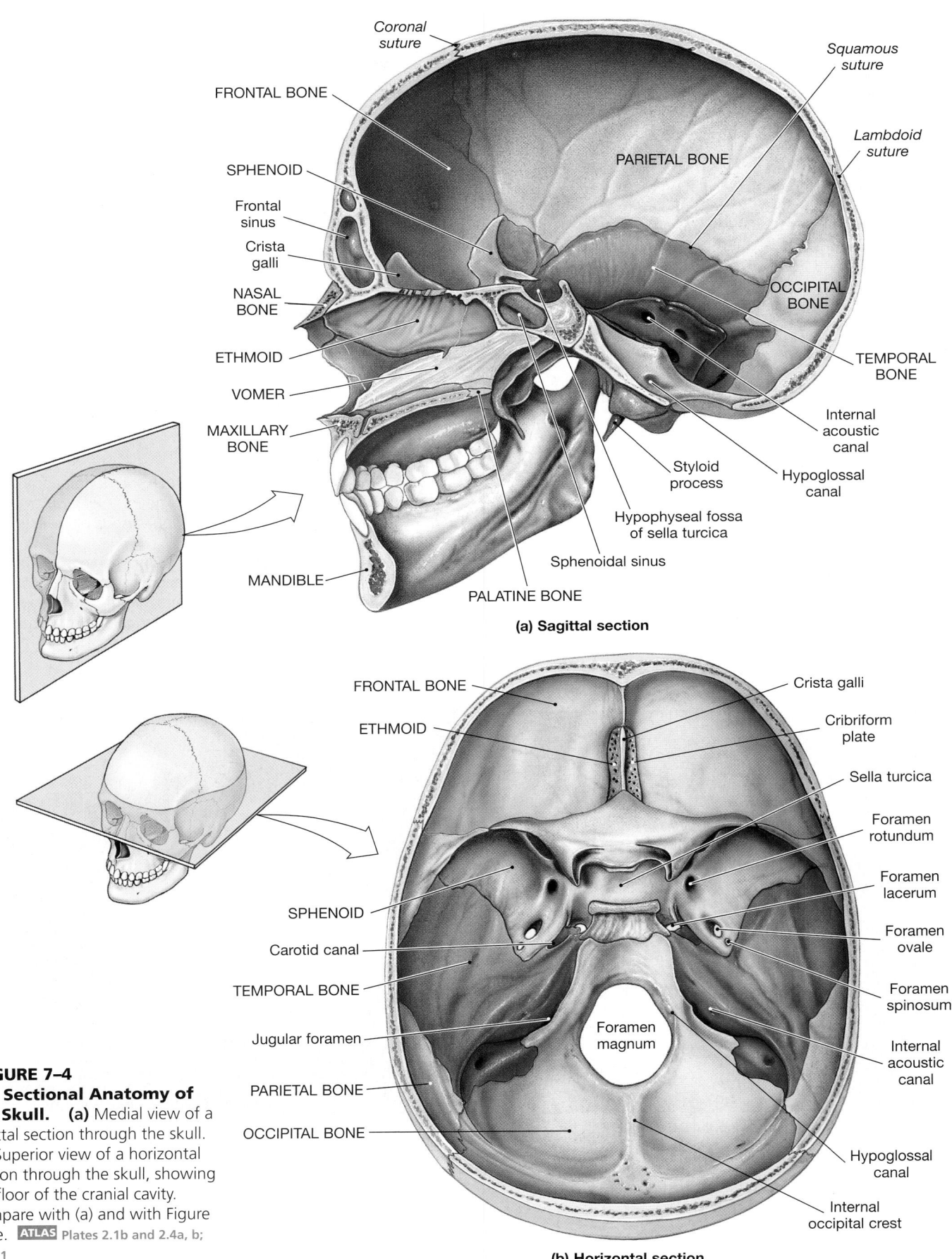

Coronal suture

FRONTAL BONE

SPHENOID

Frontal sinus

Crista galli

NASAL BONE

ETHMOID

VOMER

MAXILLARY BONE

MANDIBLE

PALATINE BONE

Sphenoidal sinus

Hypophyseal fossa of sella turcica

Styloid process

Squamous suture

PARIETAL BONE

Lambdoid suture

OCCIPITAL BONE

TEMPORAL BONE

Internal acoustic canal

Hypoglossal canal

(a) Sagittal section

FRONTAL BONE

ETHMOID

SPHENOID

Carotid canal

TEMPORAL BONE

Jugular foramen

PARIETAL BONE

OCCIPITAL BONE

Crista galli

Cribriform plate

Sella turcica

Foramen rotundum

Foramen lacerum

Foramen ovale

Foramen spinosum

Internal acoustic canal

Hypoglossal canal

Internal occipital crest

Foramen magnum

•**FIGURE 7–4**
The Sectional Anatomy of the Skull. **(a)** Medial view of a sagittal section through the skull. **(b)** Superior view of a horizontal section through the skull, showing the floor of the cranial cavity. Compare with (a) and with Figure 7–3e. **ATLAS** Plates 2.1b and 2.4a, b; Scan 1

(b) Horizontal section

focus | The Individual Bones of the Skull

Each of these bones can be explored further, using the related images in the Atlas and on the Anatomy CD. It is helpful to use the CD to manipulate the images of each bone while studying its structure. Foramina and fissures are present for the passage of vessels and nerves. The vessels are detailed in Chapter 21; the nerves are shown in the Focus box on cranial nerves in Chapter 14.

Cranial Bones

The Occipital Bone (Figure 7–5a●)

General Functions: The **occipital bone** forms much of the posterior and inferior surfaces of the cranium.

Articulations: The occipital bone articulates with the parietal bones, the temporal bones, the sphenoid, and the first cervical vertebra (the atlas) (Figures 7–3a–c,e and 7–4●).

Regions/Landmarks: The **external occipital protuberance** is a small bump at the midline on the inferior surface.

The **occipital crest**, which begins at the external occipital protuberance, marks the attachment of a ligament that helps stabilize the vertebrae of the neck.

The **occipital condyles** are the site of articulation between the skull and the first vertebra of the neck.

The *inferior* and *superior nuchal* (NOO-kul) *lines* are ridges that intersect the occipital crest. They mark the attachment sites of muscles and ligaments that stabilize the articulation at the occipital condyles and balance the weight of the head over the vertebrae of the neck.

The concave internal surface of the occipital bone (Figure 7–4a●) closely follows the contours of the brain. The grooves follow the paths of major blood vessels, and the ridges mark the attachment sites of membranes that stabilize the position of the brain.

Foramina: The **foramen magnum** connects the cranial cavity with the spinal cavity, which is enclosed by the vertebral column (Figure 7–4b●). This foramen surrounds the connection between the brain and spinal cord.

The **jugular foramen** lies between the occipital bone and the temporal bone (Figure 7–3e●). The *internal jugular vein* passes through this foramen, carrying venous blood from the brain.

The **hypoglossal canals** (Figure 7–3e●) begin at the lateral base of each occipital condyle and end on the inner surface of the occipital bone near the foramen magnum. The *hypoglossal nerves*, cranial nerves that control the tongue muscles, pass through these canals.

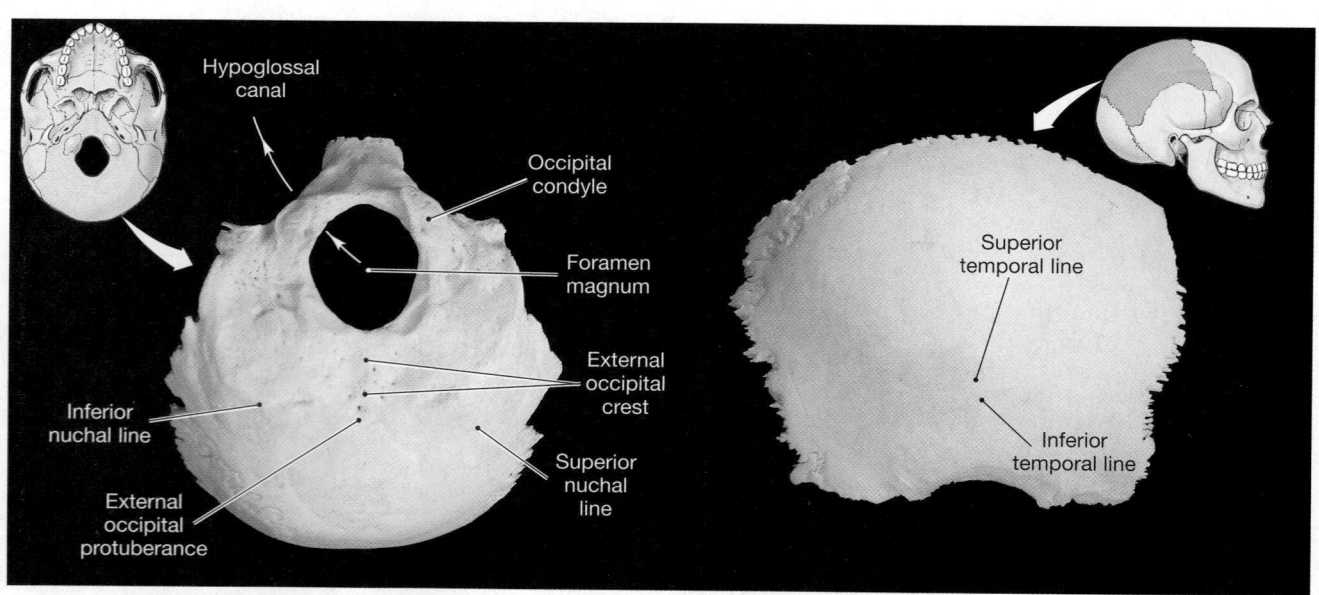

(a) Occipital bone, inferior view

(b) Right parietal bone, lateral view

●**FIGURE 7–5**
The Occipital and Parietal Bones

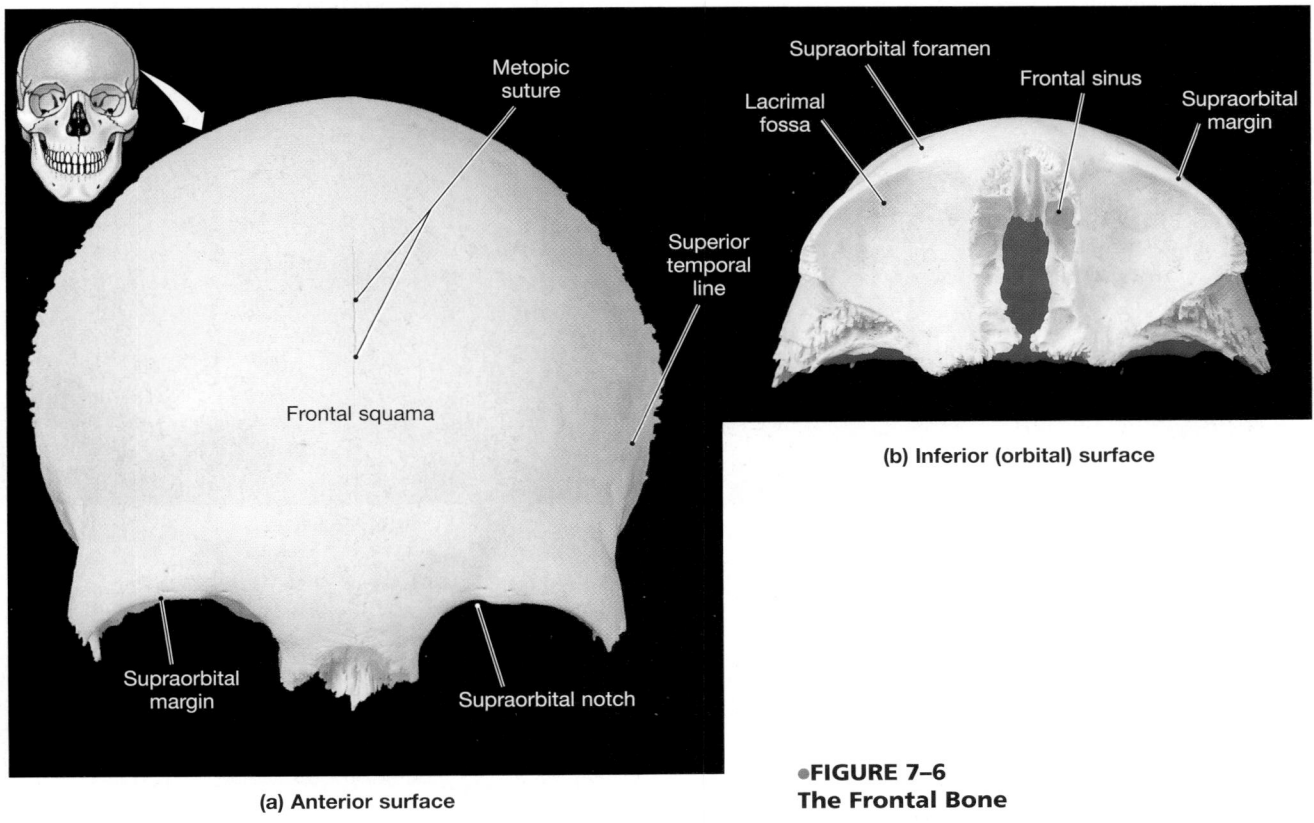

Metopic suture

Frontal squama

Superior temporal line

Supraorbital margin

Supraorbital notch

(a) Anterior surface

Supraorbital foramen

Lacrimal fossa

Frontal sinus

Supraorbital margin

(b) Inferior (orbital) surface

●**FIGURE 7–6**
The Frontal Bone

The Parietal Bones (Figure 7–5b●)

General Functions: The **parietal bones** form part of the superior and lateral surfaces of the cranium.

Articulations: The parietal bones articulate with one another and with the occipital, temporal, frontal, and sphenoid bones (Figures 7–3a–d and 7–4●).

Regions/Landmarks: The *superior* and *inferior temporal lines* are low ridges that mark the attachment sites of the *temporalis muscle*, a large muscle that closes the mouth.

Grooves on the inner surface of the parietal bones mark the paths of cranial blood vessels (Figure 7–4a●).

The Frontal Bone (Figure 7–6a,b●)

General Functions: The **frontal bone** forms the anterior portion of the cranium and the roof of the *orbits* (eye sockets). Mucous secretions of the *frontal sinuses* within this bone help flush the surfaces of the nasal cavities.

Articulations: The frontal bone articulates with the parietal, sphenoid, ethmoid, nasal, lacrimal, maxillary, and zygomatic bones (Figures 7–3b–e and 7–4●).

Regions/Landmarks: The **frontal squama,** or forehead, forms the anterior, superior portion of the cranium and provides surface area for the attachment of facial muscles.

The *superior temporal line* is continuous with the superior temporal line of the parietal bone.

The **supraorbital margin** is a thickening of the frontal bone that helps protect the eye.

The **lacrimal fossa** on the anterior and medial surface of the orbit is a shallow depression that marks the location of the *lacrimal* (tear) *gland*, which lubricates the surface of the eye.

The **frontal sinuses** are extremely variable in size and time of appearance. They generally appear after age 6, but some people never develop them. We will describe the frontal sinuses and other sinuses of the cranium and face in a later section.

Foramina: The **supraorbital foramen** provides passage for blood vessels that supply the eyebrow, eyelids, and frontal sinuses. In some cases, this foramen is incomplete; the vessels then cross the orbital rim within a **supraorbital notch**.

Remarks: During development, the bones of the cranium form by the fusion of separate centers of ossification. At birth, the fusions have not been completed: Two frontal bones articulate along the *metopic suture*. Although the suture generally disappears by age 8 as the bones fuse, the adult skull commonly retains traces of the suture line. This suture, or what remains of it, runs down the center of the frontal squama.

FOCUS *continues* →

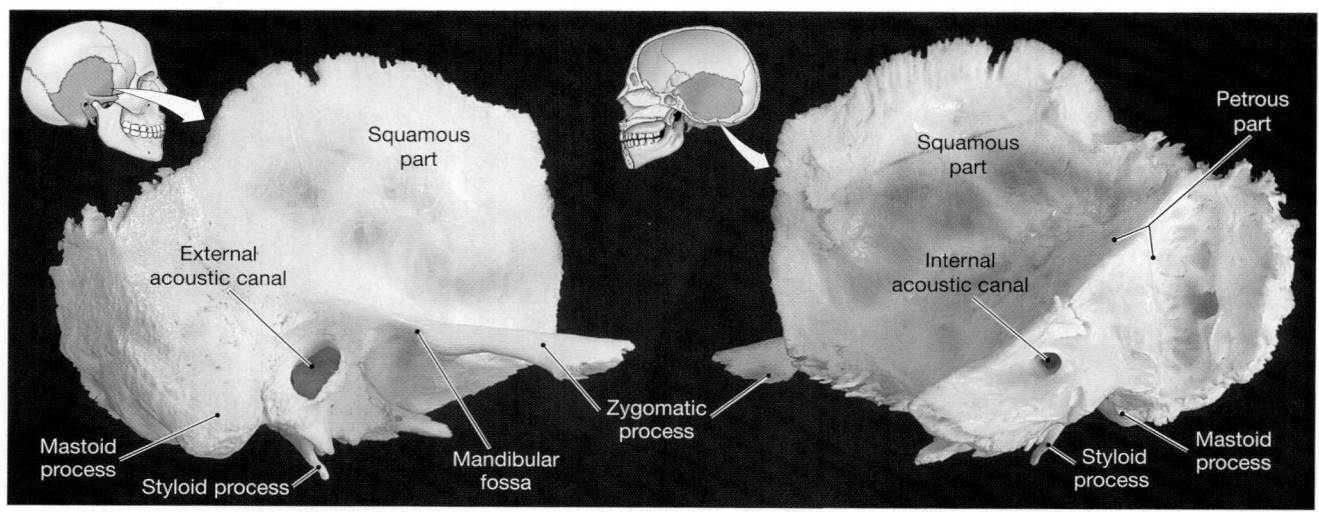

(a) Lateral view

(b) Medial view

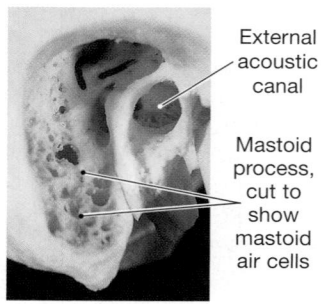

(c) Mastoid air cells

•**FIGURE 7–7**
The Temporal Bones. **(a,b)** The right temporal bone. **(c)** A cutaway view of the mastoid air cells

The Temporal Bones (Figure 7–7a,b•)

General Functions: The **temporal bones** (1) form part of both the lateral walls of the cranium and the *zygomatic arches,* (2) form the only articulations with the mandible, (3) surround and protect the sense organs of the inner ear, and (4) are attachment sites for muscles that close the jaws and move the head.

Articulations: The temporal bones articulate with the zygomatic, sphenoid, parietal, and occipital bones of the cranium and with the mandible (Figures 7–3 and 7–4•).

Regions/Landmarks: The **squamous part,** or *squama,* of the temporal bone is the convex, irregular surface that borders the squamous suture.

The **zygomatic process,** inferior to the squamous portion, articulates with the *temporal process* of the zygomatic bone. Together, these processes form the **zygomatic arch,** or cheekbone.

The **mandibular fossa** on the inferior surface marks the site of articulation with the mandible.

The **mastoid process** is an attachment site for muscles that rotate or extend the head. *Mastoid air cells* within the mastoid process are connected to the middle ear cavity (Figure 7–7c•). If

pathogens invade the mastoid air cells, *mastoiditis* develops. Symptoms include severe earaches, fever, and swelling behind the ear.

The **styloid** (STĪ-loyd; *stylos,* pillar) **process,** near the base of the mastoid process, is attached to ligaments that support the hyoid bone and to the tendons of several muscles associated with the hyoid bone, the tongue, and the pharynx.

The **petrous part** of the temporal bone, located on its internal surface, encloses the structures of the *inner ear*—sense organs that provide information about hearing and balance. The **auditory ossicles** are located in the *tympanic cavity,* or *middle ear,* a cavity within the petrous part. These tiny bones—three on each side—transfer sound vibrations from the delicate *tympanic membrane,* or eardrum, to the inner ear. (We shall discuss each bone in Chapter 17.)

Foramina (Figure 7–3e•): The jugular foramen, between the temporal and occipital bones, provides passage for the internal jugular vein.

The **carotid canal** provides passage for the internal carotid artery, a major artery to the brain. As it leaves the carotid canal, the internal carotid artery passes through the anterior portion of the foramen lacerum.

The **foramen lacerum** (LA-se-rum; *lacerare,* to tear) is a jagged slit extending between the sphenoid and the petrous portion of the temporal bone and containing hyaline cartilage and small arteries that supply the inner surface of the cranium. The *auditory tube,* an airway that connects the pharynx to the tympanic cavity, passes through the posterior portion of the foramen lacerum.

The **external acoustic canal,** or *external acoustic meatus,* on the lateral surface ends at the tympanic membrane (which disintegrates during the preparation of a dried skull).[1]

The **stylomastoid foramen** lies posterior to the base of the styloid process. The *facial nerve* passes through this foramen to control the facial muscles.

The **internal acoustic canal,**[1] or *internal acoustic meatus,* begins on the medial surface of the petrous part of the temporal bone. It carries blood vessels and nerves to the inner ear and conveys the facial nerve to the stylomastoid foramen.

---

[1]The names for these passageways vary widely; the terms *auditory* and *acoustic* are used interchangeably, as are *canal* and *meatus.*

The Sphenoid (Figure 7–8a,b•)

General Functions: The **sphenoid**, or *sphenoidal bone*, forms part of the floor of the cranium, unites the cranial and facial bones, and acts as a cross brace that strengthens the sides of the skull. Mucous secretions of the *sphenoidal sinuses* within this bone help clean the surfaces of the nasal cavities.

Articulations: The sphenoid articulates with the ethmoid and the frontal, occipital, parietal, and temporal bones of the cranium and the palatine bones, zygomatic bones, maxillary bones, and vomer of the face (Figures 7–3c–e and 7–4•).

Regions/Landmarks: The shape of the sphenoid has been compared to a bat with its wings extended. Although this bone is relatively large, much of it is hidden by more superficial bones.

The **body** forms the central axis of the sphenoid.

The **sella turcica** (TUR-si-kuh), or Turkish saddle, is a bony, saddle-shaped enclosure on the superior surface of the body. The **hypophyseal** (hī-pō-FIZ-ē-ul) **fossa** is the depression within the sella turcica. The *pituitary gland* occupies this fossa.

The **sphenoidal sinuses** are on either side of the body, inferior to the sella turcica.

The **lesser wings** extend horizontally anterior to the sella turcica.

The **greater wings** extend laterally from the body and form part of the cranial floor. A sharp *sphenoidal spine* lies at the posterior, lateral corner of each greater wing. Anteriorly, each greater wing contributes to the posterior wall of the orbit.

The **pterygoid** (TER-i-goyd; *pterygion*, wing) **processes** are vertical projections that originate on either side of the body. Each pterygoid process forms a pair of *pterygoid plates*, which are attachment sites for muscles that move the lower jaw and soft palate.

Foramina: The **optic canals** permit passage of the optic nerves from the eyes to the brain.

A **superior orbital fissure, foramen rotundum, foramen ovale** (ō-VAH-lē), and **foramen spinosum** penetrate each greater wing. These passages carry blood vessels and nerves to the orbit, face, and jaws, respectively.

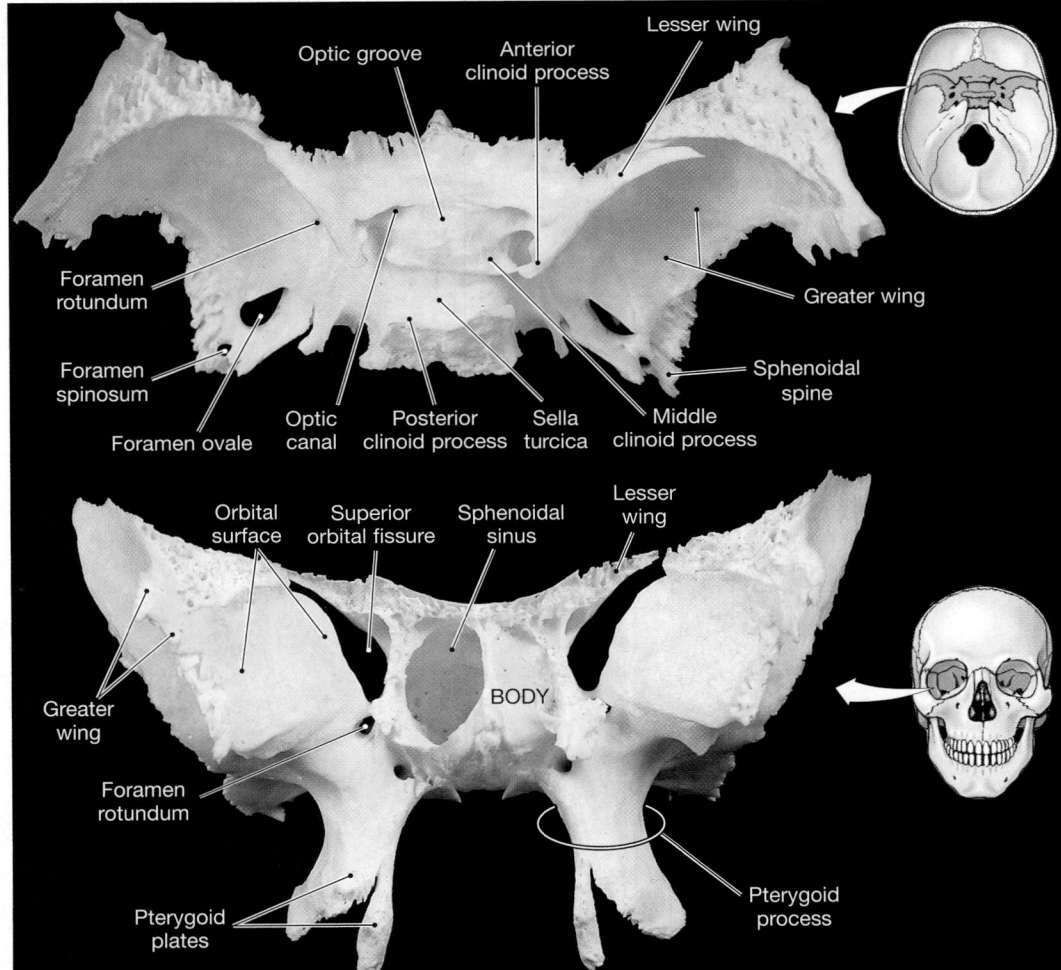

(a) Superior surface

Optic groove — Anterior clinoid process — Lesser wing

Foramen rotundum

Foramen spinosum

Foramen ovale — Optic canal — Posterior clinoid process — Sella turcica — Middle clinoid process

Greater wing

Sphenoidal spine

(b) Anterior surface

Orbital surface — Superior orbital fissure — Sphenoidal sinus — Lesser wing

Greater wing

Foramen rotundum

BODY

Pterygoid process

Pterygoid plates

•**FIGURE 7–8**
The Sphenoid

FOCUS *continues* →

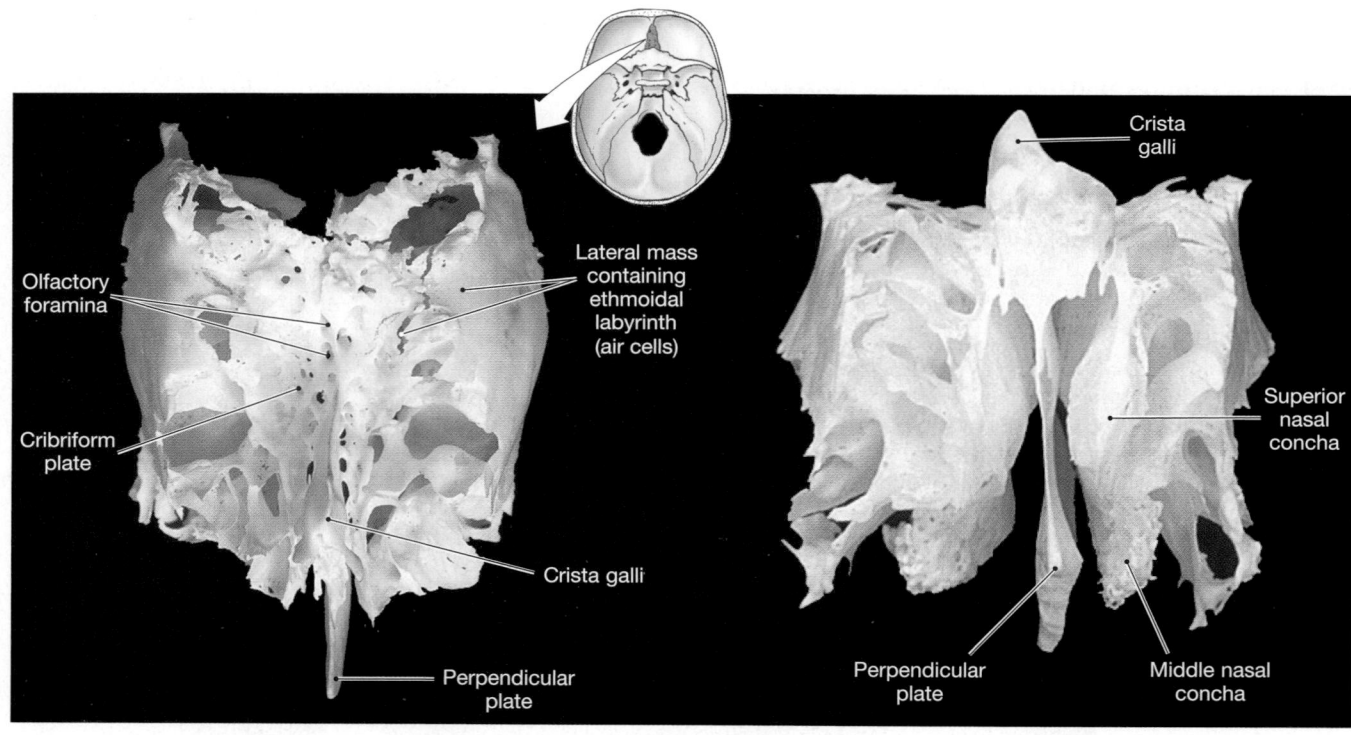

(a) Superior surface

(b) Posterior surface

●FIGURE 7–9
The Ethmoid

The Ethmoid (Figure 7–9a,b●)

General Functions: The **ethmoid,** or *ethmoidal bone,* forms the anteromedial floor of the cranium, the roof of the nasal cavity, and part of the nasal septum and medial orbital wall. Mucous secretions from a network of sinuses, or *ethmoidal air cells,* within this bone flush the surfaces of the nasal cavities.

Articulations: The ethmoid articulates with the frontal bone and sphenoid of the cranium and with the nasal, lacrimal, palatine, and maxillary bones and the inferior nasal conchae and vomer of the face (Figures 7–3c,d and 7–4●).

Regions/Landmarks: The ethmoid has three parts: (1) the cribriform plate, (2) the paired lateral masses, and (3) the perpendicular plate.

The **cribriform plate** (*cribrum,* sieve) forms the anteromedial floor of the cranium and the roof of the nasal cavity. The **crista galli** (*crista,* crest + *gallus,* chicken; cock's comb) is a bony ridge that projects superior to the cribriform plate. The *falx cerebri,* a membrane that stabilizes the position of the brain, attaches to this ridge.

The **lateral masses** contain the **ethmoidal labyrinth,** which consists of the interconnected **ethmoidal air cells** that open into the nasal cavity on each side. The **superior nasal conchae** (KONG-kē; singular, *concha,* a snail shell) and the **middle nasal conchae** are delicate projections of the lateral masses.

The **perpendicular plate** forms part of the nasal septum, along with the vomer and a piece of hyaline cartilage.

Foramina: The **olfactory foramina** in the cribriform plate permit passage of the olfactory nerves, which provide the sense of smell.

Remarks: *Olfactory* (smell) *receptors* are located in the epithelium that covers the inferior surfaces of the cribriform plate, the medial surfaces of the superior nasal conchae, and the superior portion of the perpendicular plate.

The nasal conchae break up the airflow in the nasal cavity, creating swirls, turbulence, and eddies that have three major functions: (1) The swirling throws any particles in the air against the sticky mucus that covers the walls of the nasal cavity; (2) the turbulence slows air movement, providing time for warming, humidification, and dust removal before the air reaches more delicate portions of the respiratory tract; and (3) the eddies direct air toward the superior portion of the nasal cavity, adjacent to the cribriform plate, where the olfactory receptors are located.

Facial Bones
The Maxillary Bones (Figure 7–10a,b●)

General Functions: The **maxillary bones,** or *maxillae,* support the upper teeth and form the inferior orbital rim, the lateral margins of the external nares, the upper jaw, and most of the hard palate. The *maxillary sinuses* in these bones produce mucous secretions that flush the inferior surfaces of the nasal cavities. The maxillary bones are the largest facial bones, and the maxillary sinuses are the largest sinuses.

Articulations: The maxillary bones articulate with the frontal bones and ethmoid, with one another, and with all the other facial bones except the mandible (Figures 7–3c–e and 7–4a●).

Regions/Landmarks: The **orbital rim** protects the eye and other structures in the orbit. The *anterior nasal spine* is found at the anterior portion of the maxillary bone, at its articulation with the maxillary bone of the other side. It is an attachment point for the cartilaginous anterior portion of the nasal septum.

The **alveolar processes** that border the mouth support the upper teeth.

The **palatine processes** form most of the **hard palate**, or bony roof of the mouth. One type of *cleft palate*, a developmental disorder, results when the maxillary bones fail to meet along the midline of the hard palate. **ATLAS** Embryology Summary 6: The Development of the Skull

The **maxillary sinuses** lighten the portion of the maxillary bone superior to the teeth.

The **nasolacrimal canal**, formed by the maxillary and lacrimal bones, protects the *lacrimal sac* and the *nasolacrimal duct*, which carries tears from the orbit to the nasal cavity.

Foramina: The **infraorbital foramen** marks the path of a major sensory nerve that reaches the brain by way of the foramen rotundum of the sphenoid.

The **inferior orbital fissure** (Figure 7–3d●), which lies between the maxillary bone and the sphenoid, permits passage of cranial nerves and blood vessels.

The Palatine Bones (Figure 7–10b,c●)

General Functions: The **palatine bones** form the posterior portion of the hard palate and contribute to the floor of each orbit.

Articulations: The palatine bones articulate with one another, with the maxillary bones, with the sphenoid and ethmoid, with the inferior nasal conchae, and with the vomer (Figures 7–3e and 7–4a●).

Regions/Landmarks: The palatine bones are L shaped. The **horizontal plate** forms the posterior part of the hard palate; the **perpendicular plate** extends from the horizontal plate to the **orbital process**, which forms part of the floor of the orbit. This process contains a small sinus that normally opens into the sphenoidal sinus.

Foramina: Small blood vessels and nerves supplying the roof of the mouth penetrate the lateral portion of the horizontal plate.

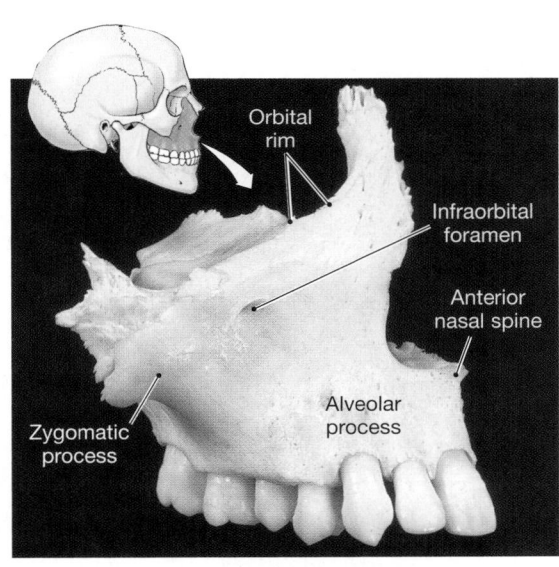

(a) Lateral surface

●**FIGURE 7–10**
The Maxillary and Palatine Bones. **(a)** An anterolateral view of the right maxillary bone. **(b)** Superior view of a horizontal section; note the size and orientation of the maxillary sinus. **(c)** An anterior view of the two palatine bones. **ATLAS** Plate 2.4e–g; Scan 1c,d

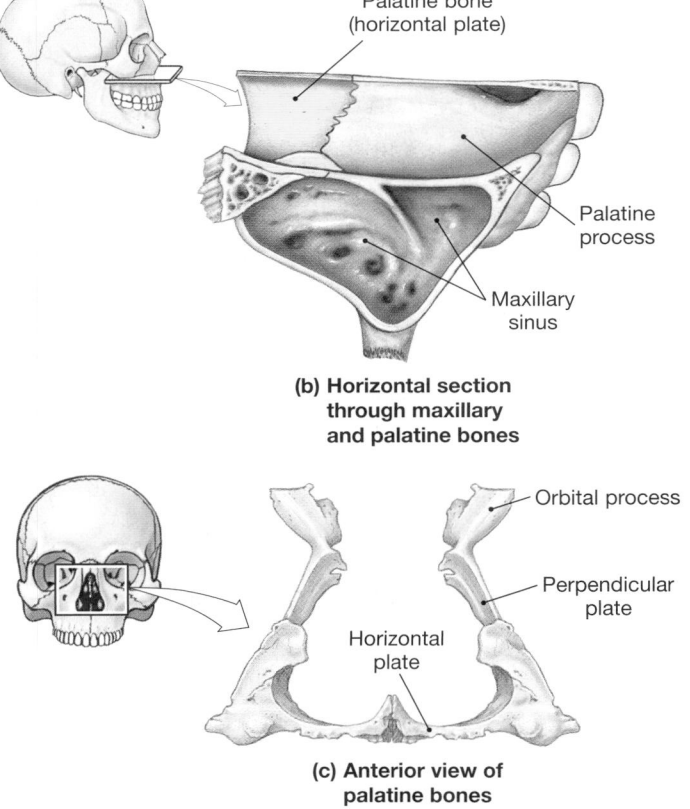

(b) Horizontal section through maxillary and palatine bones

(c) Anterior view of palatine bones

FOCUS *continues* →

The Nasal Bones (Figure 7–11●)

General Functions: The **nasal bones** support the superior portion of the bridge of the nose. They are connected to cartilages that support the distal portions of the nose. These flexible cartilages, and associated soft tissues, extend to the superior border of the **external nares** (NA-rēz; singular, *naris*), the entrances to the nasal cavity.

Articulations: The paired nasal bones articulate with one another, with the ethmoid, and with the frontal and maxillary bones (Figures 7–3b–d and 7–4a●).

The Vomer (Figure 7–11●)

General Functions: The **vomer** forms the inferior portion of the bony nasal septum.

Articulations: The vomer articulates with the sphenoid, ethmoid, palatine bones, and maxillary bones, and with the cartilaginous part of the nasal septum, which extends into the fleshy part of the nose (Figures 7–3d,e and 7–4a●).

The Inferior Nasal Conchae (Figure 7–11●)

General Functions: The **inferior nasal conchae** create turbulence in air passing through the nasal cavity, and increase the epithelial surface area to promote warming and humidification of inhaled air.

Articulations: The inferior nasal conchae articulate with the ethmoid, maxillary, palatine, and lacrimal bones (Figure 7–3d●).

The Zygomatic Bones (Figure 7–11●)

General Functions: The **zygomatic bones** contribute to the rim and lateral wall of the orbit and form part of the zygomatic arch.

Articulations: The zygomatic bones articulate with the sphenoid, and the frontal, temporal, and maxillary bones (Figure 7–3b–e●).

Regions/Landmarks: The **temporal process** curves posteriorly to meet the zygomatic process of the temporal bone.

Foramina: The **zygomaticofacial foramen** on the anterior surface of each zygomatic bone carries a sensory nerve that innervates the cheek.

The Lacrimal Bones (Figure 7–11●)

General Functions: The **lacrimal bones** form part of the medial wall of the orbit.

Articulations: The smallest facial bones, the lacrimal bones articulate with the frontal and maxillary bones, and with the ethmoid (Figure 7–3c,d●).

Regions/Landmarks: The **lacrimal sulcus**, a groove along the anterior, lateral surface of the lacrimal bone, marks the location of the lacrimal sac. The lacrimal sulcus leads to the nasolacrimal canal, which begins at the orbit and opens into the nasal cavity. As noted earlier, this canal is formed by the lacrimal bone and the maxillary bone.

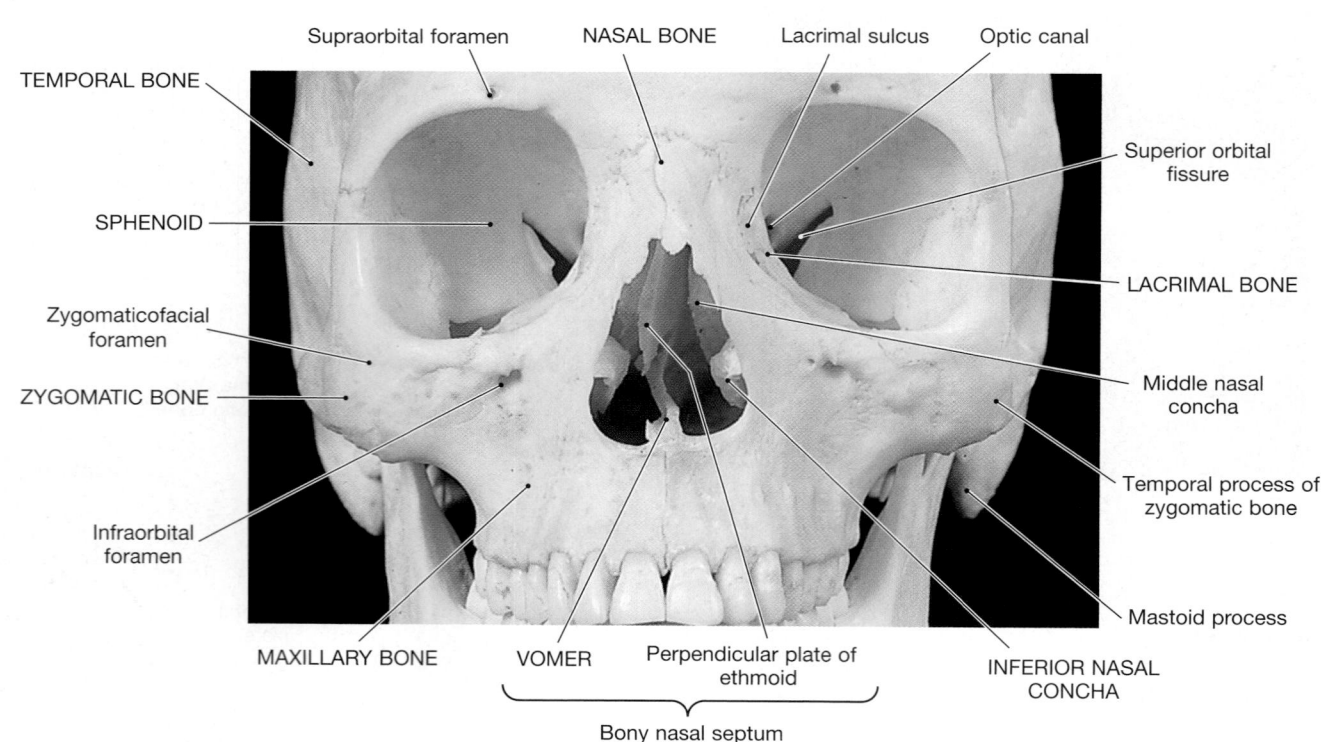

●FIGURE 7–11
The Smaller Bones of the Face

The Mandible (Figure 7–12a,b•)

General Functions: The **mandible** forms the lower jaw.

Articulations: The mandible articulates with the mandibular fossae of the temporal bones (Figures 7–3c,e and 7–7a•).

Regions/Landmarks: The **body** of the mandible is the horizontal portion of that bone.
   The **alveolar processes** support the lower teeth.
   The *mental protuberance*, or chin, is the attachment site for several facial muscles.
   A prominent depression on the medial surface marks the position of the *submandibular salivary gland*.
   The *mylohyoid line* marks the insertion of the *mylohyoid muscle*, which supports the floor of the mouth.

The **ramus** of the mandible is the ascending part that begins at the *mandibular angle* on either side. On each ramus,

1. The **condylar process** articulates with the temporal bone at the *temporomandibular joint*.
2. The **coronoid** (ko-RŌ-noyd) **process** is the insertion point for the *temporalis muscle*, a powerful muscle that closes the jaws.
3. The **mandibular notch** is the depression that separates the condylar and coronoid processes.

Foramina: The **mental foramina** (*mentalis*, chin) are openings for nerves that carry sensory information from the lips and chin to the brain.
   The **mandibular foramen** is the entrance to the *mandibular canal*, a passageway for blood vessels and nerves that service the lower teeth. Before they work on the lower teeth, dentists typically anesthetize the sensory nerve that enters this canal.

The Hyoid Bone (Figure 7–12c•)

General Functions: The **hyoid bone** supports the larynx and is the attachment site for muscles of the larynx, pharynx, and tongue.

Articulations: *Stylohyoid ligaments* connect the *lesser horns* to the styloid processes of the temporal bones.

Regions/Processes: The **body** of the hyoid is an attachment site for muscles of the larynx, tongue, and pharynx.
   The **greater horns**, or *greater cornua*, help support the larynx and are attached to muscles that move the tongue.
   The **lesser horns**, or *lesser cornua*, are attached to the stylohyoid ligaments; from these ligaments, the hyoid and larynx hang beneath the skull like a child's swing from the limb of a tree.

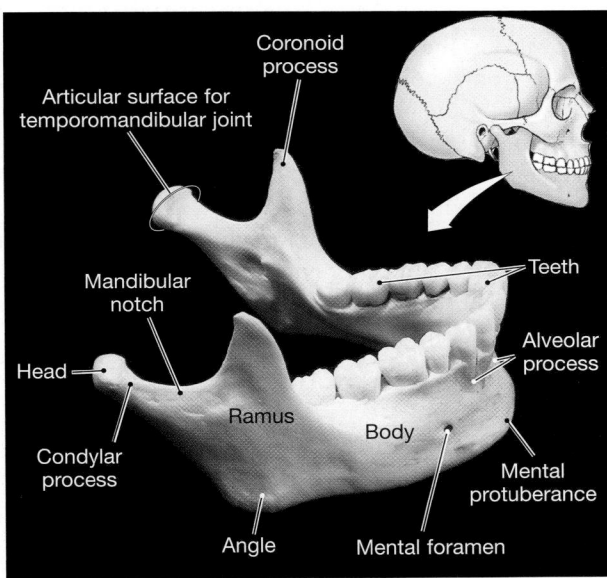

(a) Lateral view

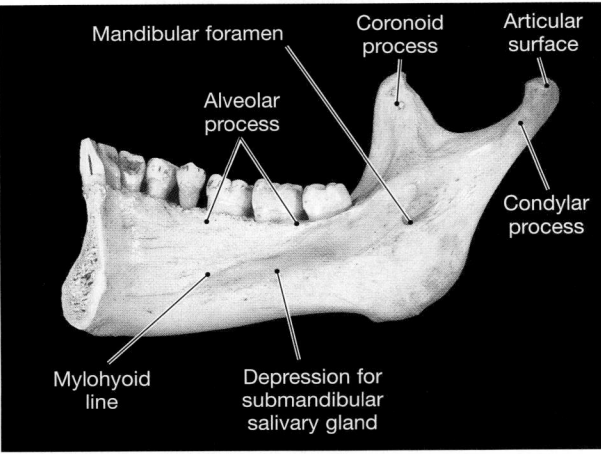

(b) Medial view

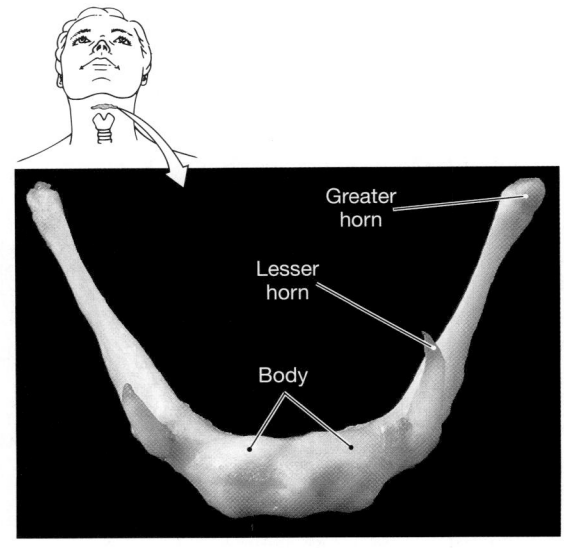

(c) Anterior-superior view

•**FIGURE 7–12**
The Mandible and Hyoid Bone. **(a)** A lateral and slightly superior view of the mandible. **(b)** A medial view of the right mandible. **(c)** An anterior view of the hyoid bone.

⚕ The *temporomandibular joint* (*TMJ*), between each temporal bone and the mandible, is quite mobile, allowing your jaw to move while you chew or talk. The disadvantage of such mobility is that your jaw can easily be dislocated by forceful forward or lateral displacement. The connective tissue sheath, or *capsule*, that surrounds the joint is relatively loose, and the opposing bone surfaces are separated by a fibrocartilage pad. In **TMJ syndrome**, or *myofacial pain syndrome*, the mandible is pulled slightly out of alignment, generally by spasms in one of the jaw muscles. The individual experiences facial pain that radiates around the ear on the affected side and an inability to open the mouth fully. TMJ syndrome is a repeating cycle of muscle spasm → misalignment → pain → muscle spasm. It has been linked to involuntary behaviors, such as grinding of the teeth during sleep, and to emotional stress. Treatment focuses on breaking the cycle of muscle spasm and pain and, when necessary, providing emotional support. The application of heat to the affected joint, coupled with the use of anti-inflammatory drugs, local anesthetics, or both, may help. If teeth grinding is suspected, special mouth guards may be worn at night.

✓ In which bone is the foramen magnum located?

✓ Tomás suffers a blow to the skull that fractures the right superior lateral surface of his cranium. Which bone is fractured?

✓ Which bone contains the depression called the sella turcica? What is located in this depression?

Answers start on page Q-1

☐ SUMMARY: THE FORAMINA AND FISSURES OF THE SKULL

Table 7–1 summarizes information about the foramina and fissures introduced thus far. This reference source will be especially important to you in later chapters when you deal with the nervous and cardiovascular systems.

☐ THE ORBITAL AND NASAL COMPLEXES

The facial bones not only protect and support the openings of the digestive and respiratory systems, but also protect the delicate sense organs responsible for vision and smell. Together, certain cranial bones and facial bones form the *orbital complex*, which surrounds each eye, and the *nasal complex*, which surrounds the nasal cavities.

The **orbits** are the bony recesses that contain the eyes. Each orbit is formed by the seven bones of the **orbital complex** (Figure 7–13●). The frontal bone forms the roof, and the maxillary bone provides most of the orbital floor. The orbital rim and the first portion of the medial wall are formed by the maxillary bone, the lacrimal bone, and the lateral mass of the ethmoid. The lateral mass articulates with the sphenoid and a small process of the palatine bone. Several prominent foramina and fissures penetrate the sphenoid or lie between it and the maxillary bone. Laterally, the sphenoid and maxillary bone articulate with the zygomatic bone, which forms the lateral wall and rim of the orbit.

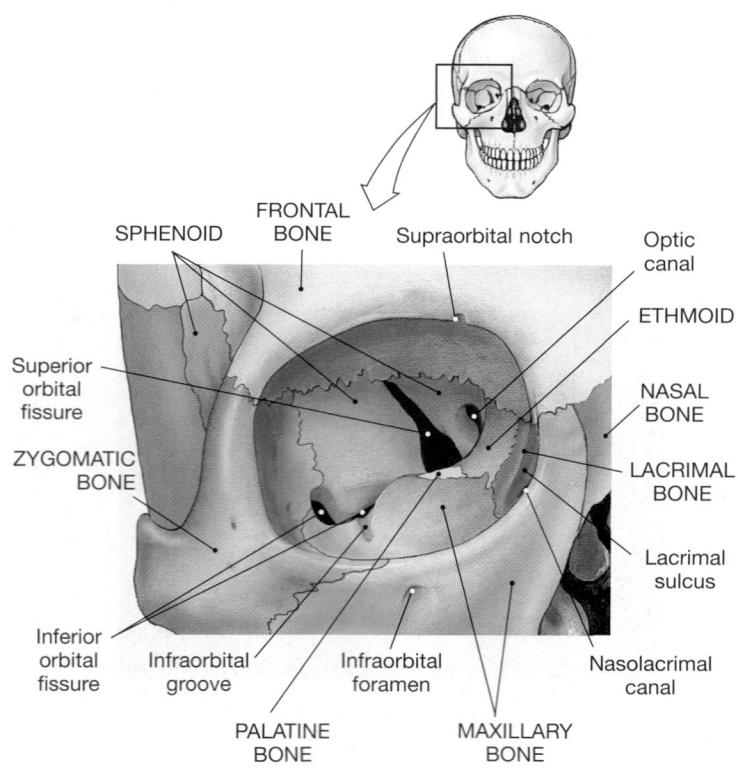
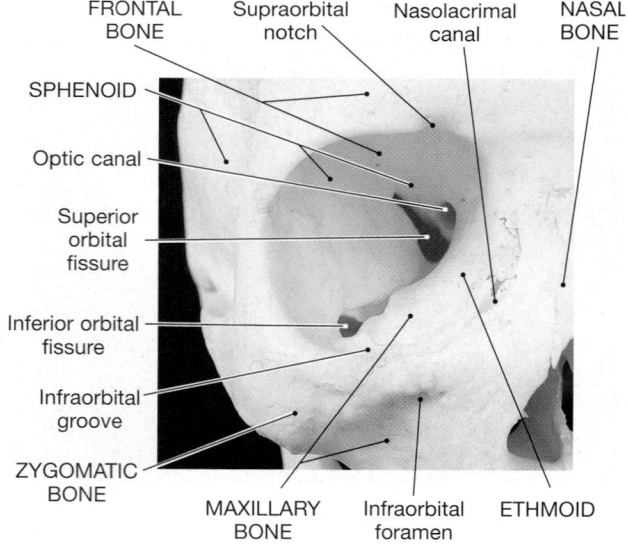

●**FIGURE 7–13**
The Orbital Complex. The right orbital region. **ATLAS** Plate 2.3

SUMMARY TABLE 7–1 A KEY TO THE FORAMINA AND FISSURES OF THE SKULL

Bone	Foramen/Fissure	Major Structures Using Passageway	
		Neural Tissue*	Vessels and Other Structures
OCCIPITAL BONE	Foramen magnum	Medulla oblongata (most caudal portion of brain) and accessory nerve (XI), which provides motor control over several neck and back muscles	Vertebral arteries to brain; supporting membranes around central nervous system
	Hypoglossal canal	Hypoglossal nerve (XII) provides motor control to muscles of the tongue	
With temporal bone	Jugular foramen	Glossopharyngeal nerve (IX), vagus nerve (X), accessory nerve (XI). Nerve IX provides taste sensation; X is important for visceral functions; XI innervates important muscles of the back and neck	Internal jugular vein; important vein returning blood from brain to heart
FRONTAL BONE	Supraorbital foramen (or notch)	Supraorbital nerve, sensory branch of ophthalmic nerve, innervating the eyebrow, eyelid, and frontal sinus	Supraorbital artery delivers blood to same region
LACRIMAL BONE	Lacrimal sulcus, nasolacrimal canal (with maxillary bone)		Lacrimal sac and tear duct; drains into nasal cavity
TEMPORAL BONE	Stylomastoid foramen	Facial nerve (VII) provides motor control of facial muscles	
	Carotid canal		Internal carotid artery; supplying blood to brain
	External acoustic canal		Air in canal conducts sound to eardrum
	Internal acoustic canal	Vestibulocochlear nerve (VIII) from sense organs for hearing and balance. Facial nerve (VII) enters here, exits at stylomastoid foramen	Internal acoustic artery to inner ear
SPHENOID	Optic canal	Optic nerve (II) brings information from the eye to the brain	Ophthalmic artery brings blood into orbit
	Superior orbital fissure	Oculomotor nerve (III), trochlear nerve (IV), ophthalmic branch of trigeminal nerve (V), abducens nerve (VI). Ophthalmic nerve provides sensory information about eye and orbit; other nerves control muscles that move the eye	Ophthalmic vein returns blood from orbit
	Foramen rotundum	Maxillary branch of trigeminal nerve (V) provides sensation from the face	
	Foramen ovale	Mandibular branch of trigeminal nerve (V) controls the muscles that move the lower jaw and provides sensory information from that area	
	Foramen spinosum		Vessels to membranes around central nervous system
With temporal and occipital bones	Foramen lacerum		Internal carotid artery after leaving carotid canal; auditory tube; small vessels; hyaline cartilage
With maxillary bone	Inferior orbital fissure	Maxillary branch of trigeminal nerve (V). *See Foramen rotundum*	
ETHMOID	Olfactory foramina	Olfactory nerve (I) provides sense of smell	
MAXILLARY BONE	Infraorbital foramen	Infraorbital nerve, maxillary branch of trigeminal nerve (V) from the inferior orbital fissure to face	Infraorbital artery with same distribution
MANDIBLE	Mental foramen	Mental nerve, sensory branch of the mandibular nerve, provides sensation from the chin and lips	Mental vessels to chin and lips
	Mandibular foramen	Inferior alveolar nerve, sensory branch of mandibular nerve, provides sensation from the gums, teeth	Inferior alveolar vessels supply same region
ZYGOMATIC BONE	Zygomaticofacial foramen	Zygomaticofacial nerve, sensory branch of mandibular nerve to cheek	

* Twelve pairs of cranial nerves exist, numbered I–XII. Their functions and distribution are detailed in Chapter 14, pp. 493–503.

The **nasal complex** includes the bones that enclose the nasal cavities and the **paranasal sinuses**, air-filled chambers connected to the nasal cavities. The frontal bone, sphenoid, and ethmoid form the superior wall of the nasal cavities. The lateral walls are formed by the maxillary and lacrimal bones, the ethmoid (the superior and middle nasal conchae), and the inferior nasal conchae (Figure 7–14●). Much of the anterior margin of the nasal cavity is formed by the soft tissues of the nose, but the bridge of the nose is supported by the maxillary and nasal bones.

Paranasal Sinuses

The sphenoid, ethmoid, and the frontal, palatine, and maxillary bones contain the paranasal sinuses. Figure 7–14a● shows the location of the frontal and sphenoidal sinuses. Ethmoidal air cells and maxillary sinuses are shown in Figure 7–14b●. (The tiny palatine sinuses, not shown, generally open into the sphenoidal sinuses.) The paranasal sinuses lighten the skull bones and pro-

vide an extensive area of mucous epithelium. The mucous secretions are released into the nasal cavities. The ciliated epithelium passes the mucus back toward the throat, where it is eventually swallowed. Incoming air is humidified and warmed as it flows across this thick carpet of mucus. Foreign particulate matter, such as dust or microorganisms, becomes trapped in the sticky mucus and is then swallowed. This mechanism helps protect the more delicate portions of the respiratory tract.

☐ THE SKULLS OF INFANTS AND CHILDREN

Many different centers of ossification are involved in the formation of the skull. As development proceeds, the centers fuse, producing a smaller number of composite bones. For example, the sphenoid begins as 14 separate ossification centers. At birth, fusion has not been completed: There are two frontal bones, four occipital bones, and several sphenoid and temporal elements.

Sinus Problems and Septal Defects

When an irritant is introduced into the nasal passages, our bodies work to remove the irritant. Large particles or strong chemical agents can cause us to sneeze, expelling a sizable amount of air quickly and forcefully. With luck, this sweeps the offending particles or chemicals out with the air. Smaller particles or milder irritants trigger the production of large amounts of mucus by the epithelium of the paranasal sinuses. The mucus stream flushes the nasal surfaces clean, often removing irritants such as pepper, pollen grains, or dust. A sinus infection, however, is another matter entirely. A viral or bacterial infection produces an inflammation of the mucous membrane of the nasal cavity. As swelling occurs, the communicating passageways narrow. Drainage of mucus slows, congestion increases, and the individual experiences headaches and a feeling of pressure within the facial bones. This condition of sinus inflammation and congestion is called **sinusitis**. The maxillary sinuses are commonly involved. Because gravity does little to assist mucus drainage from these sinuses, the effectiveness of the flushing action is reduced and pressure on the sinus walls typically increases.

The relief of pain associated with sinusitis is the basis of a large over-the-counter (OTC) drug market in the United States. Every major pharmaceutical company has at least one product designed specifically to relieve sinus pressure. The active ingredients in these preparations are compounds that dry the epithelial linings, reduce pain, and restrict further swelling. Usually included are the antihistamine *chlorpheniramine maleate* and the nasal decongestant *pseudoephedrine HCl* or *phenylephrine HCl*. (Another formerly common decongestant, phenylpropanolamine HCl, was recently eliminated from cold and sinus remedies at the request of the FDA because it was linked to an increased risk of hemorrhagic stroke in young women.) A relative newcomer to the market is an inhalant

spray containing a steroid compound that reduces swelling (*oxymetazoline HCl*). It is interesting to note that, with the exception of this inhalant, the ingredients and dosages differ very little from one product to the next. Marketing and packaging play a major role in determining which OTC remedy dominates the market at any given time.

Temporary sinus problems may accompany allergies or the exposure of the mucous epithelium to chemical irritants or invading microorganisms. Chronic sinusitis may occur as the result of a **deviated** (nasal) **septum**. In this condition, the nasal septum has a bend in it, generally at the junction between the bony and cartilaginous regions. Septal deviation often blocks the drainage of one or more sinuses, producing chronic cycles of infection and inflammation. A deviated septum can result from developmental abnormalities or from injuries to the nose. Many boxers suffer from a deviated septum, as their sport subjects them to numerous blows to the soft tissues of the nose. (The condition is fairly common among ice hockey players as well.) It can usually be corrected or improved by surgery.

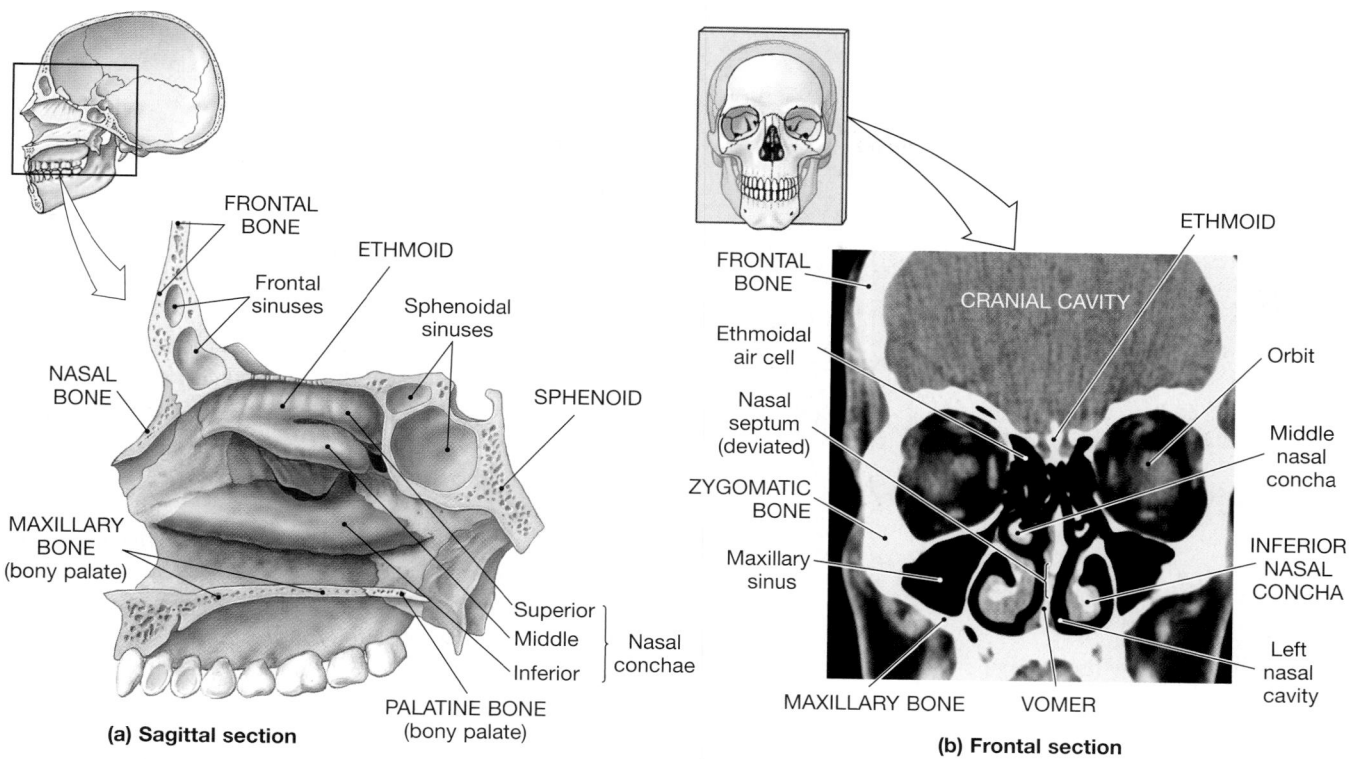

●FIGURE 7–14

The Nasal Complex. **(a)** A sagittal section through the skull, with the nasal septum removed to show major features of the wall of the right nasal cavity. The sphenoidal sinuses are visible. **(b)** An MRI scan showing a frontal section through the ethmoidal air cells and maxillary sinuses, part of the paranasal sinuses. **ATLAS** Scans 1c–e, 2a, 3a

The skull organizes around the developing brain. As the time of birth approaches, the brain enlarges rapidly. Although the bones of the skull are also growing, they fail to keep pace. At birth, the cranial bones are connected by areas of fibrous connective tissue. The connections are quite flexible, and the skull can be distorted without damage. Such distortion normally occurs during delivery and eases the passage of the infant through the birth canal. The largest fibrous areas between the cranial bones are known as **fontanels** (fon-tuh-NELZ; sometimes spelled *fontanelles*) (Figure 7–15●):

- The *anterior fontanel* is the largest fontanel. It lies at the intersection of the frontal, sagittal, and coronal sutures in the anterior portion of the skull.

- The *occipital fontanel* is at the junction between the lambdoid and sagittal sutures.

- The *sphenoidal fontanels* are at the junctions between the squamous sutures and the coronal suture.

- The *mastoid fontanels* are at the junctions between the squamous sutures and the lambdoid suture.

The anterior fontanel is often referred to as the "soft spot" on newborns, and is often the only fontanel easily seen by new parents. Because it is composed of fibrous connective tissue and covers a major blood vessel, the anterior fontanel pulses as the heart beats. In warmer climates, this fontanel is sometimes used

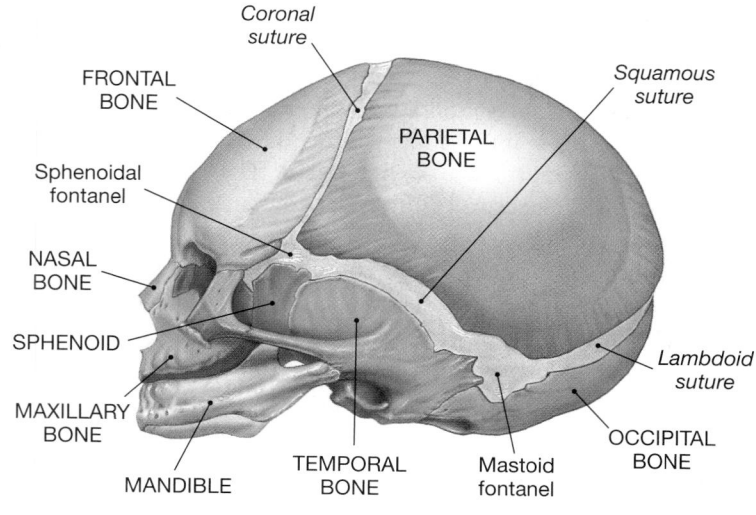

(a) Lateral view

●FIGURE 7–15

The Skull of an Infant. **(a)** A lateral view. The skull of an infant contains more individual bones than that of an adult. Many of the bones eventually fuse; thus, the adult skull has fewer bones. The flat bones of the skull are separated by areas of fibrous connective tissue, allowing for cranial expansion and the distortion of the skull during birth. The large fibrous areas are called fontanels. By about age four, these areas will disappear.

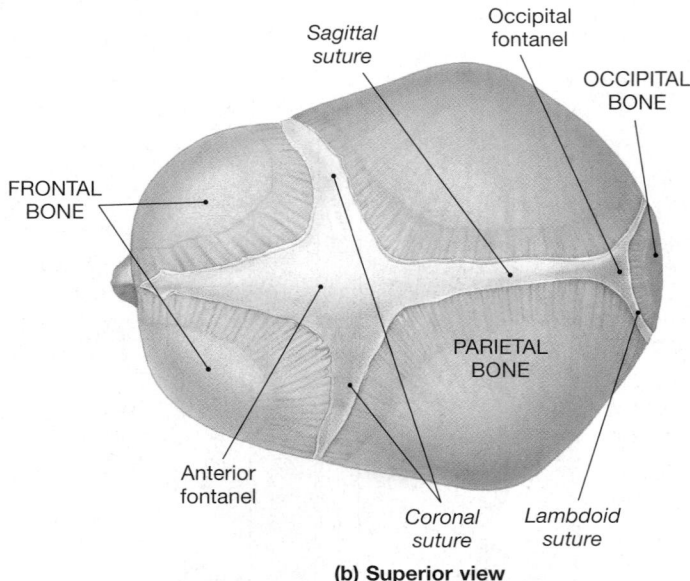

Sagittal suture

Occipital fontanel

OCCIPITAL BONE

FRONTAL BONE

PARIETAL BONE

Anterior fontanel

Coronal suture

Lambdoid suture

(b) Superior view

FIGURE 7–15 (continued). **The Skull of an Infant.** **(b)** A superior view

to determine whether an infant is dehydrated, as the surface becomes indented when blood volume is low.

The occipital, sphenoidal, and mastoid fontanels disappear within a month or two after birth. The anterior fontanel generally persists until the child is nearly two years old. Even after the fontanels disappear, the bones of the skull remain separated by fibrous connections.

The skulls of infants and adults differ in terms of the shape and structure of cranial elements. This difference accounts for variations in proportions as well as in size. The most significant growth in the skull occurs before age five, because at that time the brain stops growing and the cranial sutures develop. As a result, the cranium of a young child, compared with the skull as a whole, is relatively larger than that of an adult. The growth of the cranium is generally coordinated with the expansion of the brain. If one or more sutures form before the brain stops growing, the skull will be abnormal in shape, size, or both.

✚ Unusual distortions of the skull result from *craniostenosis* (krā-nē-ō-sten-Ō-sis; *stenosis*, narrowing), the premature closure of one or more fontanels. As the brain continues to enlarge, the rest of the skull distorts to accommodate it. A long and narrow head is produced by early closure of the sagittal suture, whereas a very broad skull results if the coronal suture forms prematurely. Early closure of all cranial sutures restricts the development of the brain, and surgery must be performed to prevent brain damage. If brain enlargement stops due to genetic or developmental abnormalities, however, skull growth ceases as well. This condition, which results in an undersized head, is called *microcephaly* (mī-krō-SEF-uh-lē).

7–3 THE VERTEBRAL COLUMN

Objectives

- Identify and describe the curvatures of the spinal column and their functions.
- Identify the vertebral regions, and describe the distinctive structural and functional characteristics of each vertebral group.

The rest of the axial skeleton consists of the vertebral column, ribs, and sternum. The adult **vertebral column**, or *spine*, consists of 26 bones: the **vertebrae** (24), the **sacrum**, and the **coccyx** (KOK-siks), or tailbone. The vertebrae provide a column of support, bearing the weight of the head, neck, and trunk and ultimately transferring the weight to the appendicular skeleton of the lower limbs. The vertebrae also protect the spinal cord and help maintain an upright body position, as in sitting or standing.

The vertebral column is divided into cervical, thoracic, lumbar, sacral, and coccygeal regions (Figure 7–16●). Seven **cervical vertebrae** constitute the neck and extend inferiorly to the trunk. Twelve **thoracic vertebrae** form the superior portion of the back; each articulates with one or more pairs of ribs. Five **lumbar vertebrae** form the inferior portion of the back; the fifth articulates with the sacrum, which in turn articulates with the coccyx. The cervical, thoracic, and lumbar regions consist of individual vertebrae. During development, the sacrum originates as a group of five vertebrae, and the coccyx begins as three to five very small vertebrae. In general, the vertebrae of the sacrum are completely fused by age 25–30. Ossification of the distal coccygeal vertebrae is not complete before puberty, and thereafter fusion occurs at a variable pace. The total length of the vertebral column of an adult averages 71 cm (28 in.). **ATLAS** EMBRYOLOGY SUMMARY 7: The Development of the Vertebral Column

▢ SPINAL CURVATURE

The vertebral column is not straight and rigid. A lateral view shows four **spinal curves** (Figure 7–16●): the (1) **cervical curve**, (2) **thoracic curve**, (3) **lumbar curve**, and (4) **sacral curve**.

If you have ever watched an infant, you may have noted that its body axis forms a loose comma or a C, with the back curving posteriorly. The C shape results from the thoracic and sacral curves. These are called **primary curves**, because they appear late in fetal development, or **accommodation curves**, because they accommodate the thoracic and abdominopelvic viscera. The primary curves are present in the vertebral column at birth. The lumbar and cervical curves, known as **secondary curves**, do not appear until several months after birth. These curves are also called **compensation curves**, because they help shift the weight of the trunk over the lower limbs. The cervical curve develops as the infant learns to balance the head upright. The lumbar curve develops with the ability to stand. Both compensations become

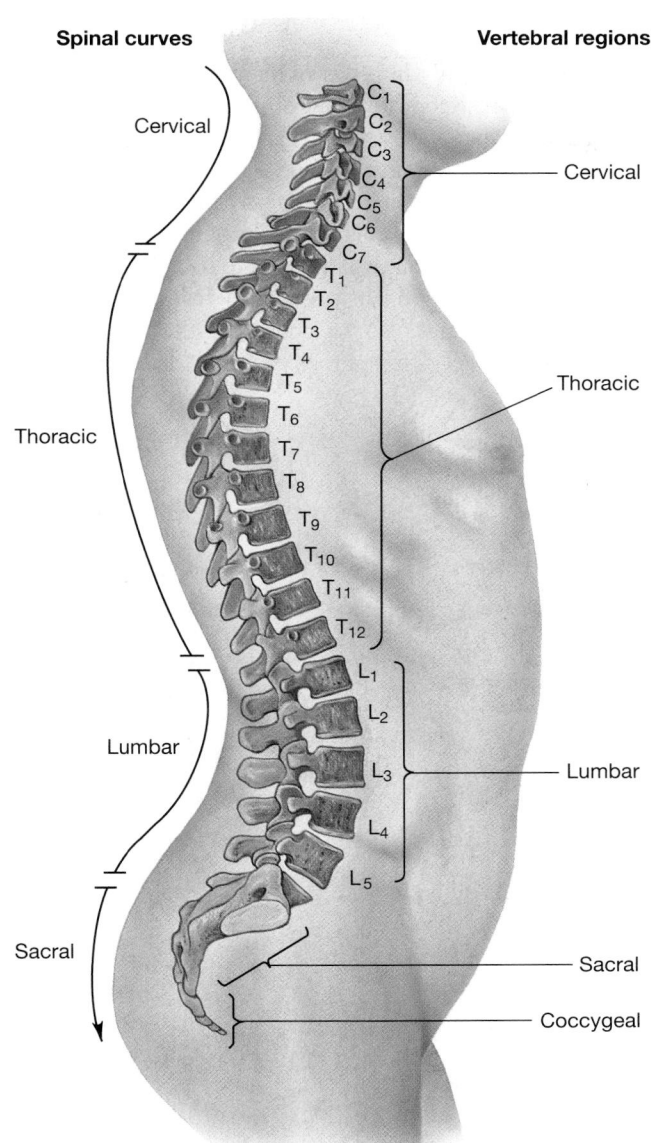

Spinal curves

Cervical

Thoracic

Lumbar

Sacral

Vertebral regions

C₁
C₂
C₃
C₄
C₅
C₆
C₇

T₁
T₂
T₃
T₄
T₅
T₆
T₇
T₈
T₉
T₁₀
T₁₁
T₁₂

L₁
L₂
L₃
L₄
L₅

Cervical

Thoracic

Lumbar

Sacral

Coccygeal

●**FIGURE 7–16**
The Vertebral Column. The major divisions of the adult vertebral column; notice the four spinal curves. **ATLAS** Plate 4.4 and Scan 3c, d

accentuated as the toddler learns to walk and run. All four curves are fully developed by age 10.

Several abnormal distortions of spinal curvature may appear during childhood and adolescence. *Kyphosis* (kī-FŌ-sis) is an exaggerated thoracic curvature, *lordosis* (lor-DŌ-sis) an exaggerated lumbar curvature, and *scoliosis* (skō-lē-Ō-sis) an abnormal lateral curvature. Figure 7–17● demonstrates these abnormal curvatures. **AM** Kyphosis, Lordosis, and Scoliosis

When you stand, the weight of your body must be transmitted through the vertebral column to the hips and ultimately to the lower limbs. Yet most of your body weight lies anterior to the vertebral column. The various curves bring that weight in line with the body axis. Consider what you do automatically when standing with a heavy object hugged to your chest. You avoid toppling forward by exaggerating the lumbar curve and moving the weight back toward the body axis. This posture can lead to discomfort at the base of the spinal column. For example, many women in the last three months of pregnancy develop chronic back pain from the changes in lumbar curvature that must adjust for the increasing weight of the fetus. In many parts of the world, people often balance heavy objects on their head (as in the photo on page 208). This practice increases the load on the vertebral column, but the spinal curves are not affected because the weight is aligned with the axis of the spine.

●**FIGURE 7–17**
Abnormal Curvatures of the Spine.

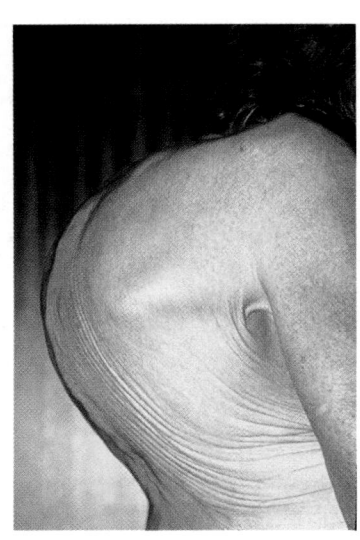

(a) Kyphosis

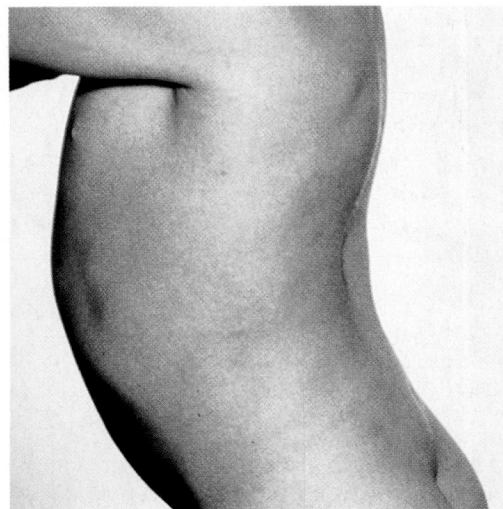

(b) Lordosis

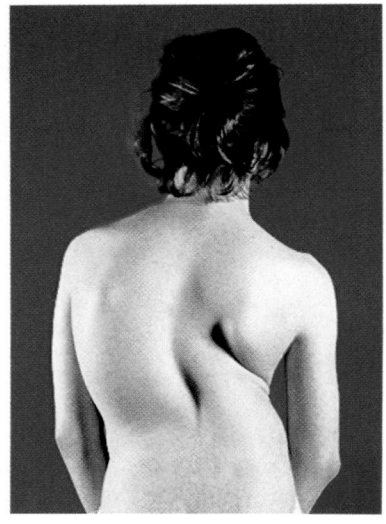

(c) Scoliosis

☐ VERTEBRAL ANATOMY

Each vertebra consists of three basic parts: (1) a *vertebral body*, (2) a *vertebral arch*, and (3) *articular processes* (Figure 7–18●).

The Vertebral Body

The **vertebral body**, or *centrum* (plural, *centra*), is the part of a vertebra that transfers weight along the axis of the vertebral column. The bodies of adjacent vertebrae are interconnected by ligaments, but are separated by pads of fibrocartilage, the **intervertebral discs**.

The Vertebral Arch

The **vertebral arch** forms the posterior margin of each **vertebral foramen**. Together, the vertebral foramina of successive vertebrae form the **vertebral canal**, which encloses the spinal cord. The vertebral arch has walls, called **pedicles** (PE-di-kulz), and a roof, formed by flat layers called **laminae** (LA-mi-nē; singular, *lamina*, a thin plate). The pedicles arise along the posterior and lateral margins of the body. The laminae on either side extend dorsally and medially to complete the roof.

A **spinous process** projects posteriorly from the point where the vertebral laminae fuse to complete the vertebral arch. You can see—and feel—the spinous processes through the skin of

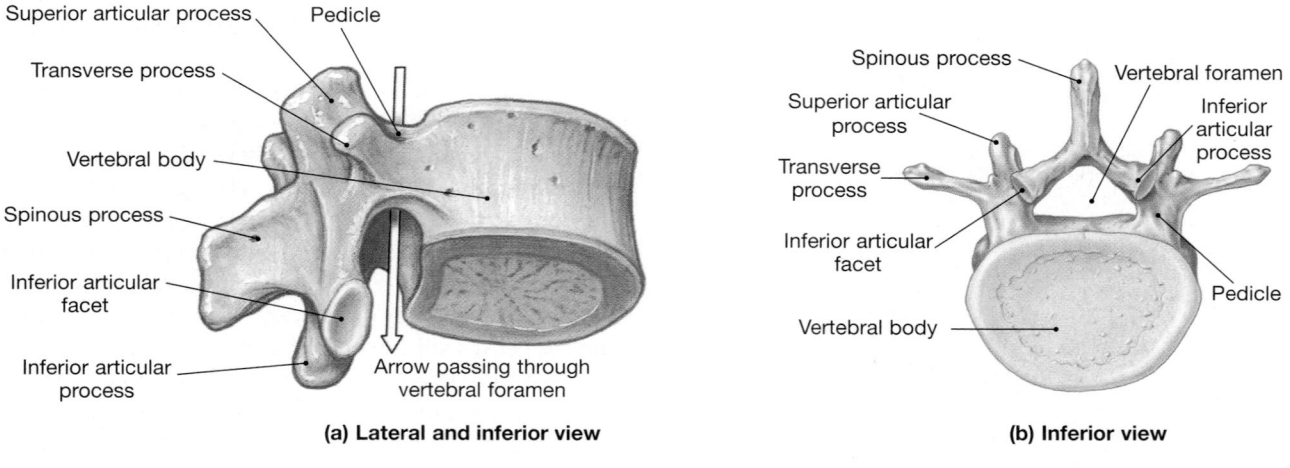

(a) Lateral and inferior view

(b) Inferior view

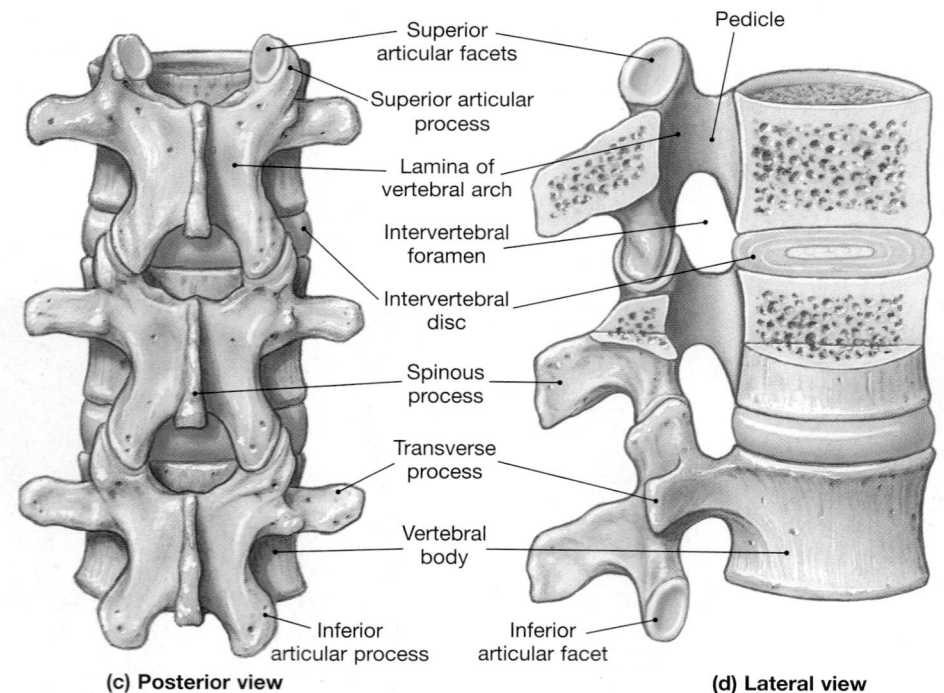

(c) Posterior view

(d) Lateral view

●**FIGURE 7–18**
Vertebral Anatomy. The anatomy of a typical vertebra and the arrangement of articulations between vertebrae. **(a)** A lateral and slightly inferior view of a vertebra. **(b)** An inferior view of a vertebra. **(c)** A posterior view of three articulated vertebrae. **(d)** A lateral and sectional view of three articulated vertebrae.

the back when the spine is flexed. **Transverse processes** project laterally or dorsolaterally on both sides from the point where the laminae join the pedicles. These processes are sites of muscle attachment, and they may also articulate with the ribs.

During the third week of embryonic development, the vertebral arches form around the developing spinal cord. In the condition called *spina bifida* (SPĪ-nuh BI-fi-duh; *bifidus*, cut into two parts), a portion of the spinal cord develops abnormally, and as a result, the adjacent vertebral arches do not form. Because the vertebral arch is incomplete, the membranes, or *meninges*, that line the dorsal body cavity bulge outward. This is the most common developmental abnormality of the nervous system, occurring at a rate of up to 4 cases per thousand births.

The region affected and the severity of the condition vary widely. Mild cases involving the sacral and lumbar regions may pass unnoticed, because neural function is not compromised significantly and "baby fat" may mask the fact that some of the spinous processes are missing. When spina bifida is detected, surgical repairs can close the gap in the vertebral wall. Severe cases, involving the entire spinal column and skull, reflect major problems with the formation of the spinal cord and brain. These neural problems usually kill the fetus before delivery; infants born with such developmental defects seldom survive more than a few hours or days.

The Articular Processes

Like the transverse processes, the **articular processes** arise at the junction between the pedicles and the laminae. A **superior** and an **inferior articular process** lie on each side of the vertebra. The superior articular processes articulate with the inferior articular processes of a more superior vertebra (or the occipital condyles, in the case of the first cervical vertebra). The inferior articular processes articulate with the superior articular processes of a more inferior vertebra (or the sacrum, in the case of the last lumbar vertebra).

Vertebral Articulation

The inferior articular processes of one vertebra articulate with the superior articular processes of the next vertebra. Each articular process has a smooth concave surface called an **articular facet**. The superior processes have articular facets on their dorsal surfaces, whereas the inferior processes articulate along their ventral surfaces.

Adjacent vertebral bodies are separated by intervertebral discs, and gaps separate the pedicles of successive vertebrae. These gaps, called **intervertebral foramina**, permit the passage of nerves running to or from the enclosed spinal cord.

✓ Why does the vertebral column of an adult have fewer vertebrae than that of a newborn?

✓ What is the importance of the secondary curves of the spine?

✓ When you run your finger along a person's spine, what part of the vertebrae are you feeling just beneath the skin?

Answers start on page Q-1

 Review the anatomy of individual vertebrae, as well as the anatomy of the rest of the axial skeleton by visiting the **Anatomy CD-ROM**: Skeletal System/Axial Dissections.

□ VERTEBRAL REGIONS

When referring to a specific vertebra, we use a capital letter to indicate the vertebral region: C, T, L, S, and Co indicate the cervical, thoracic, lumbar, sacral, and coccygeal regions, respectively. In addition, we use a subscript number to indicate the relative position of the vertebra within that region, with 1 indicating the vertebra closest to the skull. For example, C_3 is the third cervical vertebra, with C_1 in contact with the skull; and L_4 is the fourth lumbar vertebra, with L_1 in contact with the last thoracic vertebra (Figure 7–16, p. 229●). We will use this shorthand throughout the text.

Although each vertebra bears characteristic markings and articulations, we will focus on the general characteristics of each region and on how regional variations determine the vertebral group's function.

Cervical Vertebrae

Most mammals—whether giraffes, whales, or mice—have the same number of cervical vertebrae as humans. These seven cervical vertebrae (Figure 7–19a●) are the smallest in the vertebral column. They extend from the occipital bone of the skull to the thorax. The body of a cervical vertebra is small compared with the size of the vertebral foramen (Figure 7–19b●). At this level, the spinal cord still contains most of the axons that connect the brain to the rest of the body. As you proceed caudally along the vertebral canal, you will note that the diameter of the spinal cord decreases, and so does the diameter of the vertebral arch. However, cervical vertebrae support only the weight of the head, so the vertebral body can be relatively small and light. As you continue toward the sacrum, you find that the loading increases and the vertebral bodies gradually enlarge.

In a typical cervical vertebra, the superior surface of the body is concave from side to side, and it slopes, with the anterior edge inferior to the posterior edge (Figure 7–19c●). Vertebra C_1 has no spinous process. The spinous processes of the other cervical vertebrae are relatively stumpy, generally shorter than the diameter of the vertebral foramen. In the case of vertebrae C_2–C_6, the tip of each process bears a prominent notch (Figure 7–19b●). A notched spinous process is said to be **bifid**.

Laterally, the transverse processes are fused to the **costal processes**, which originate near the ventrolateral portion of the vertebral body. The costal and transverse processes encircle prominent, round **transverse foramina**. These passageways protect the *vertebral arteries* and *vertebral veins*, important blood vessels that service the brain.

The preceding description would be adequate to identify all but the first two cervical vertebrae. When cervical vertebrae C_3–C_7 articulate, their interlocking bodies permit more flexibility than do those of other regions. The first two cervical vertebrae are unique, and the seventh is modified. Table 7–2 summarizes the features of cervical vertebrae.

Compared with the cervical vertebrae, your head is relatively massive. It sits atop the cervical vertebrae like a soup bowl on the tip of a finger. With this arrangement, small muscles

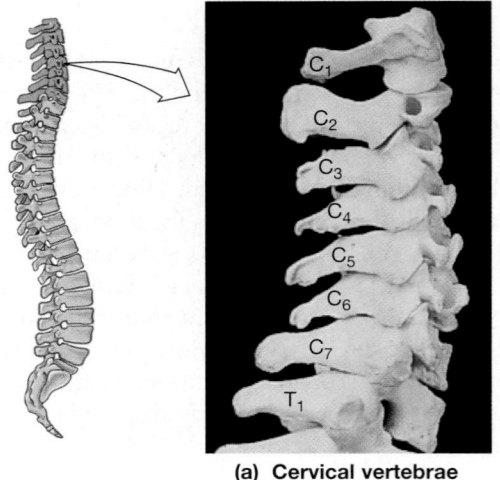

(a) Cervical vertebrae

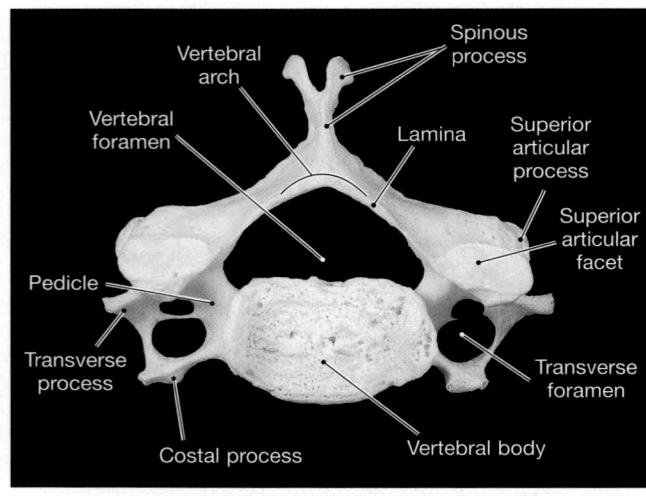

(b) Typical cervical vertebra (superior view)

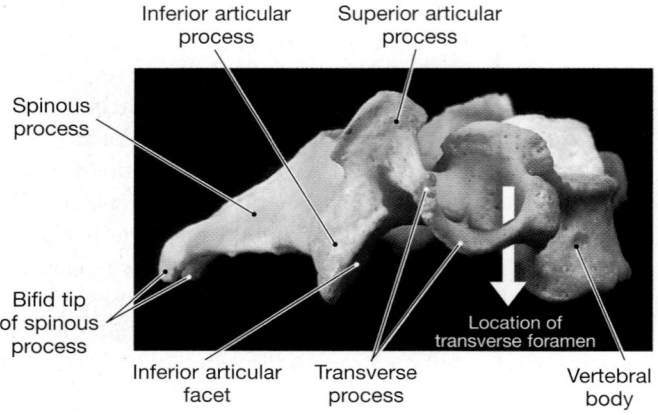

(c) Typical cervical vertebra (lateral view)

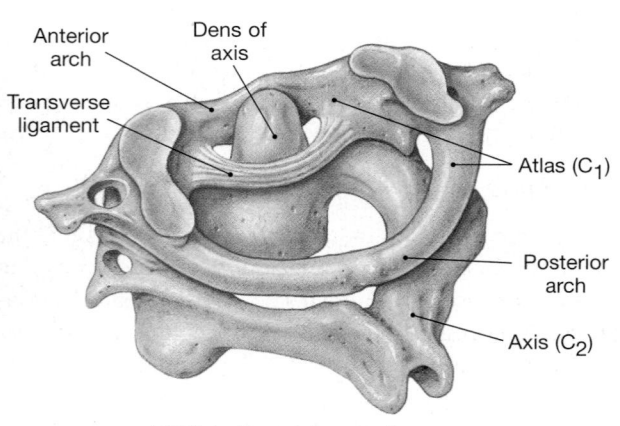

(d) The atlas–axis complex

●FIGURE 7–19
The Cervical Vertebrae.   **(a)** A lateral view of the cervical vertebrae, C_1–C_7. **(b)** A superior view of a representative cervical vertebra showing characteristics of C_3–C_6. Notice the typical features listed in Table 7–2. **(c)** A lateral view of the same vertebra. **(d)** The atlas (C_1) and axis (C_2). **ATLAS** Plate 4.5a–e, Scan 3

can produce significant effects by tipping the balance one way or another. But if you change position suddenly, as in a fall or during rapid acceleration (a jet takeoff) or deceleration (a car crash), the balancing muscles are not strong enough to stabilize the head. A dangerous partial or complete dislocation of the cervical vertebrae can result, with injury to muscles and ligaments and potential injury to the spinal cord. The term **whiplash** is used to describe such an injury, because the movement of the head resembles the cracking of a whip.

THE ATLAS (C_1) The **atlas**, cervical vertebra C_1, holds up the head, articulating with the occipital condyles of the skull (Figure 7–19d●). This vertebra is named after Atlas, who, according to Greek myth, holds the world on his shoulders. The articulation between the occipital condyles and the atlas is a joint that permits you to nod (such as when you indicate "yes"). The atlas can easi-

ly be distinguished from other vertebrae by (1) the lack of a body and spinous process and (2) the presence of a large, round vertebral foramen bounded by **anterior** and **posterior arches**.

The atlas articulates with the second cervical vertebra, the *axis*. This articulation permits rotation (as when you shake your head to indicate "no").

THE AXIS (C_2) During development, the body of the atlas fuses to the body of the second cervical vertebra, called the **axis** (C_2) (Figure 7–19d●). This fusion creates the prominent **dens** (DENZ; *dens*, tooth), or *odontoid* (ō-DON-toyd; *odontos*, tooth) *process*, of the axis. A transverse ligament binds the dens to the inner surface of the atlas, forming a pivot for rotation of the atlas and skull. Important muscles controlling the position of the head and neck attach to the especially robust spinous process of the axis.

TABLE 7–2 REGIONAL DIFFERENCES IN VERTEBRAL STRUCTURE AND FUNCTION

Feature	Type (Number)		
	Cervical Vertebrae (7)	Thoracic Vertebrae (12)	Lumbar Vertebrae (5)
Location	Neck	Chest	Inferior portion of back
Body	Small, oval, curved faces	Medium, heart-shaped, flat faces; facets for rib articulations	Massive, oval, flat faces
Vertebral foramen	Large	Smaller	Smallest
Spinous process	Long; split tip; points inferiorly	Long, slender; not split; points inferiorly	Blunt, broad, points posteriorly
Transverse process	Has transverse foramen	All but two (T_{11}, T_{12}) have facets for rib articulations	Short; no articular facets or transverse foramina
Functions	Support skull, stabilize relative positions of brain and spinal cord, and allow controlled head movement	Support weight of head, neck, upper limbs, and chest; articulate with ribs to allow changes in volume of thoracic cage	Support weight of head, neck, upper limbs, and trunk
Typical appearance (superior view)			

In children, the fusion between the dens and axis is incomplete. Impacts or even severe shaking can cause dislocation of the dens and severe damage to the spinal cord. In adults, a blow to the base of the skull can be equally dangerous, because a dislocation of the axis–atlas joint can force the dens into the base of the brain, with fatal results.

THE VERTEBRA PROMINENS (C_7) The transition from one vertebral region to another is not abrupt, and the last vertebra of one region generally resembles the first vertebra of the next. The **vertebra prominens**, or seventh cervical vertebra (C_7), has a long, slender spinous process that ends in a broad tubercle that you can feel through the skin at the base of the neck. This vertebra is the interface between the cervical curve, which arches anteriorly, and the thoracic curve, which arches posteriorly. (See Figures 7–16 and 7–20a●.) The transverse processes of C_7 are large, providing additional surface area for muscle attachment. The **ligamentum nuchae** (li-guh-MEN-tum NOO-kē; *nucha*, nape), a large elastic ligament, begins at the vertebra prominens and extends to an in-

sertion along the occipital crest of the skull. Along the way, it attaches to the spinous processes of the other cervical vertebrae. When your head is upright, this ligament acts like the string on a bow, maintaining the cervical curvature without muscular effort. If you have bent your neck forward, the elasticity in the ligamentum nuchae helps return your head to an upright position.

Thoracic Vertebrae

There are 12 thoracic vertebrae (Figure 7–20●). A typical thoracic vertebra has a distinctive heart-shaped body that is more massive than that of a cervical vertebra. The vertebral foramen is relatively smaller, and the long, slender spinous process projects posteriorly and inferiorly. The spinous processes of T_{10}, T_{11}, and T_{12} increasingly resemble those of the lumbar series as the transition between the thoracic and lumbar curves approaches. Because the inferior thoracic and lumbar vertebrae carry so much weight, the transition between the thoracic and lumbar curves is difficult to stabilize. As a result, compression fractures or compression–dislocation fractures incurred after a hard fall tend to involve the last thoracic and first two lumbar vertebrae.

Each thoracic vertebra articulates with ribs along the dorsolateral surfaces of the body. The **costal facets** on the vertebral bodies articulate with the heads of the ribs. The location and structure of the articulations vary somewhat from vertebra to vertebra (Figure 7–20a●). Vertebrae T_1–T_8 each articulate with two pairs of ribs, so their vertebral bodies have two costal facets (*superior* and *inferior*) on each side. Vertebrae T_9–T_{11} have a single costal facet on each side, and each vertebra articulates with a single pair of ribs.

The transverse processes of vertebrae T_1–T_{10} are relatively thick and contain **transverse costal facets** for rib articulation (Figure 7–20b,c●). Thus, rib pairs 1 through 10 contact their ver-

tebrae at two points: a costal facet and a transverse costal facet. Table 7–2, p. 233, summarizes the features of thoracic vertebrae.

Lumbar Vertebrae

The five lumbar vertebrae are the largest vertebrae. The body of a typical lumbar vertebra (Figure 7–21●) is thicker than that of a thoracic vertebra, and the superior and inferior surfaces are oval rather than heart shaped. Other noteworthy features are that (1) lumbar vertebrae do not have costal facets; (2) the slender transverse processes, which lack transverse costal facets, project dorsolaterally; (3) the vertebral foramen is triangular; (4) the stumpy spinous processes project dorsally; (5) the superior articular processes face medially ("up and

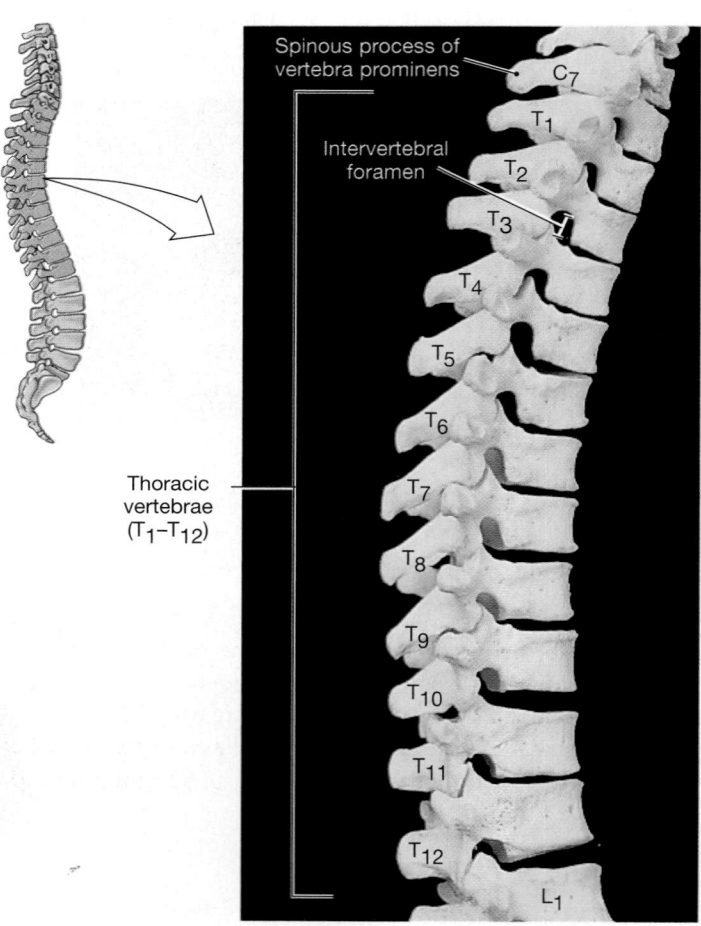

(a) Thoracic vertebrae, lateral view

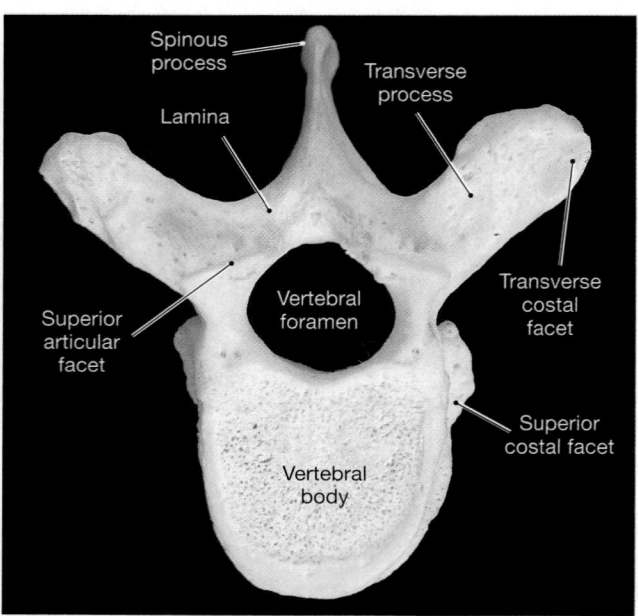

(b) Thoracic vertebra, superior view

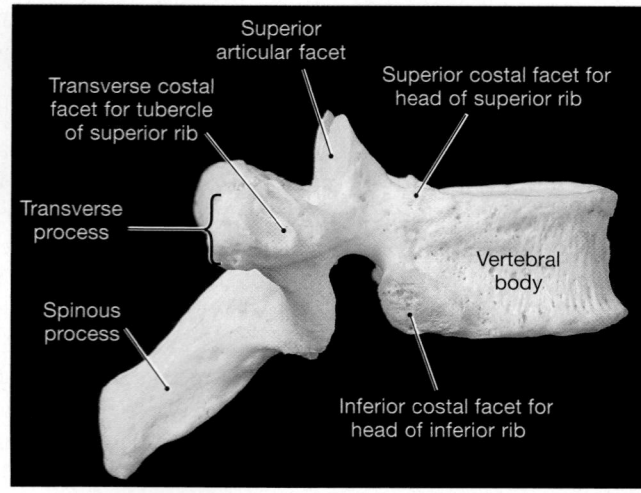

(c) Thoracic vertebra, lateral view

●**FIGURE 7–20**
The Thoracic Vertebrae. (a) A lateral view of the thoracic region of the vertebral column. The vertebra prominens (C_7) resembles T_1, but lacks facets for rib articulation. Vertebra T_{12} resembles the first lumbar vertebra (L_1), but has a facet for rib articulation. (b) Thoracic vertebra, superior view. (c) Thoracic vertebra, lateral view. Notice the characteristic features listed in Table 7–2.

in"); and (6) the inferior articular processes face laterally ("down and out").

The lumbar vertebrae bear the most weight. Their massive spinous processes provide surface area for the attachment of lower back muscles that reinforce or adjust the lumbar curve. Table 7–2, p. 233, summarizes the characteristics of lumbar vertebrae.

The Sacrum

The sacrum consists of the fused components of five sacral vertebrae. These vertebrae begin fusing shortly after puberty and, in general, are completely fused at age 25–30. The sacrum protects the reproductive, digestive, and urinary organs and, via paired articulations, attaches the axial skeleton to the pelvic girdle of the appendicular skeleton (Figure 7–1b●, p. 211).

The broad posterior surface of the sacrum (Figure 7–22a●) provides an extensive area for the attachment of muscles, especially those that move the thigh. The superior articular processes of the first sacral vertebra articulate with the last lumbar vertebra. The **sacral canal** is a passageway that begins between these articular processes and extends the length of the sacrum. Nerves

and membranes that line the vertebral canal in the spinal cord continue into the sacral canal.

The **median sacral crest** is a ridge formed by the fused spinous processes of the sacral vertebrae. The laminae of the fifth sacral vertebra fail to contact one another at the midline; they form the **sacral cornua** (KORN-ū-uh; singular, *cornu*; (*cornua*, horns). These ridges form the margins of the **sacral hiatus** (hī-Ā-tus), the opening at the inferior end of the sacral canal. This opening is covered by connective tissues. Four pairs of **sacral foramina** open on either side of the median sacral crest. The intervertebral foramina of the fused sacral vertebrae open into these passageways. The **lateral sacral crest** on each side is a ridge that represents the fused transverse processes of the sacral vertebrae. The sacral crests provide surface area for the attachment of muscles of the lower back and hip.

The sacrum is curved, with a convex posterior surface (Figure 7–22b●). The degree of curvature is more pronounced in males than in females. The **auricular surface** is a thickened, flattened area lateral to the superior portion of the lateral sacral crest. The auricular surface is the site of articulation with the pelvic girdle (the *sacroiliac joint*). The **sacral tuberosity** is a roughened area

●**FIGURE 7–21**
The Lumbar Vertebrae. **(a)** A lateral view of the lumbar vertebrae and sacrum. **(b)** A lateral view of a typical lumbar vertebra. **(c)** A superior view of the same vertebra.

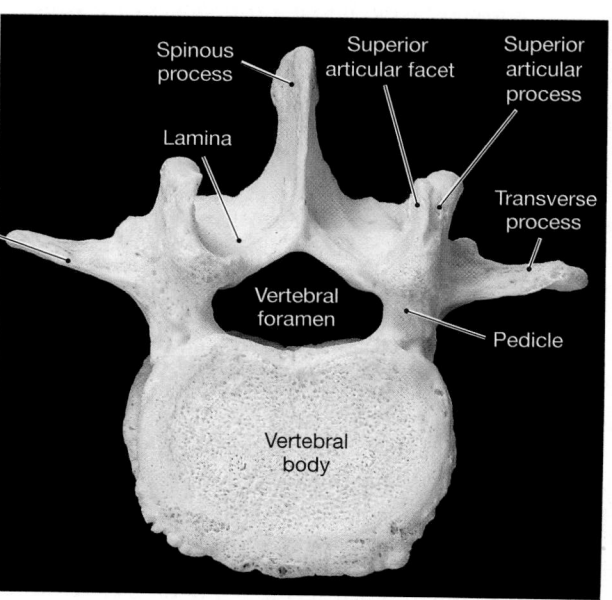

L1
L2
L3
L4
L5

Sacrum
Coccyx

(a)

Superior articular process
Spinous process
Inferior articular facet
Inferior articular process
Pedicle
Vertebral body
Transverse process

(b) Lateral view

Spinous process
Lamina
Superior articular facet
Superior articular process
Transverse process
Vertebral foramen
Pedicle
Vertebral body

(c) Superior view

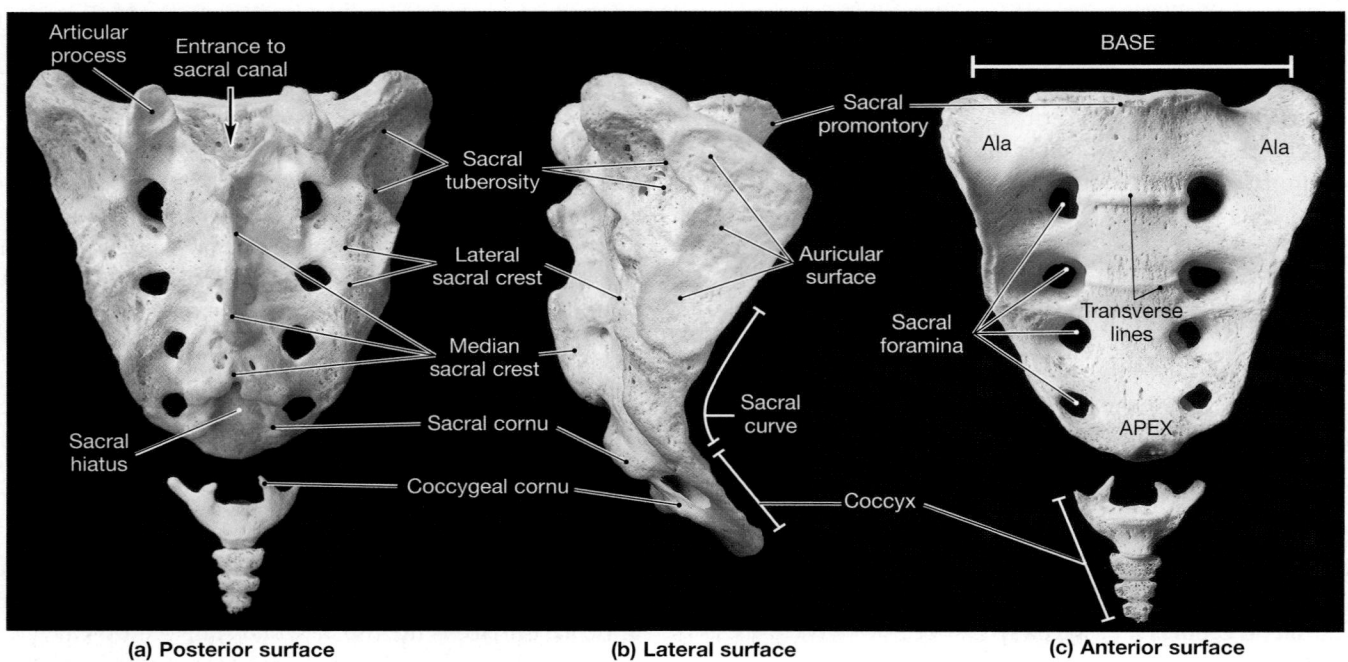

•FIGURE 7–22
The Sacrum and Coccyx. **(a)** A posterior view. **(b)** A lateral view from the right side. **(c)** An anterior view.

between the lateral sacral crest and the auricular surface. It marks the attachment site of ligaments that stabilize the sacroiliac joint.

The subdivisions of the sacrum are most clearly seen in anterior view (Figure 7–22c•). The narrow, inferior portion is the sacral **apex**, whereas the broad superior surface forms the **base**. The **sacral promontory**, a prominent bulge at the anterior tip of the base, is an important landmark in females during pelvic examinations and during labor and delivery. Prominent *transverse lines* mark the former boundaries of individual vertebrae that fuse during the formation of the sacrum. At the base of the sacrum, a broad sacral **ala**, or *wing*, extends on either side. The anterior and superior surfaces of each ala provide an extensive area for muscle attachment. At the apex, a flattened area marks the site of articulation with the coccyx.

The Coccyx

The small coccyx consists of three to five (typically, four) coccygeal vertebrae that have generally begun fusing by age 26 (Figure 7–22•). The coccyx provides an attachment site for a number of ligaments and for a muscle that constricts the anal opening. The first two coccygeal vertebrae have transverse processes and unfused vertebral arches. The prominent laminae of the first coccygeal vertebrae are known as the **coccygeal cornua**. These laminae curve to meet the cornua of the sacrum. The coccygeal vertebrae do not fuse completely until late in adulthood. In very old people, the coccyx may fuse with the sacrum.

✓ Joe suffered a hairline fracture at the base of the dens. Which bone is fractured, and where is it located?

✓ Examining a human vertebra, you notice that, in addition to the large foramen for the spinal cord, two smaller foramina are on either side of the bone in the region of the transverse processes. From which region of the vertebral column is this vertebra?

✓ Why are the bodies of the lumbar vertebrae so large?

Answers start on page Q-1

View cervical, thoracic, and lumbar vertebrae on the **Anatomy CD-ROM**: Skeletal System/Axial Dissections.

7–4 THE THORACIC CAGE

Objective

■ Explain the significance of the articulations between the thoracic vertebrae and ribs and between the ribs and sternum.

The skeleton of the chest, or **thoracic cage**, provides bony support to the walls of the thoracic cavity. It consists of the thoracic vertebrae, the ribs, and the sternum (breastbone) (Figure 7–23•). The ribs and the sternum form the *rib cage*, whose movements are important in respiration. The thoracic cage as a whole serves two functions:

1. It protects the heart, lungs, thymus, and other structures in the thoracic cavity.

2. It serves as an attachment point for muscles involved in (1) respiration, (2) maintenance of the position of the vertebral column, and (3) movements of the pectoral girdle and upper limbs.

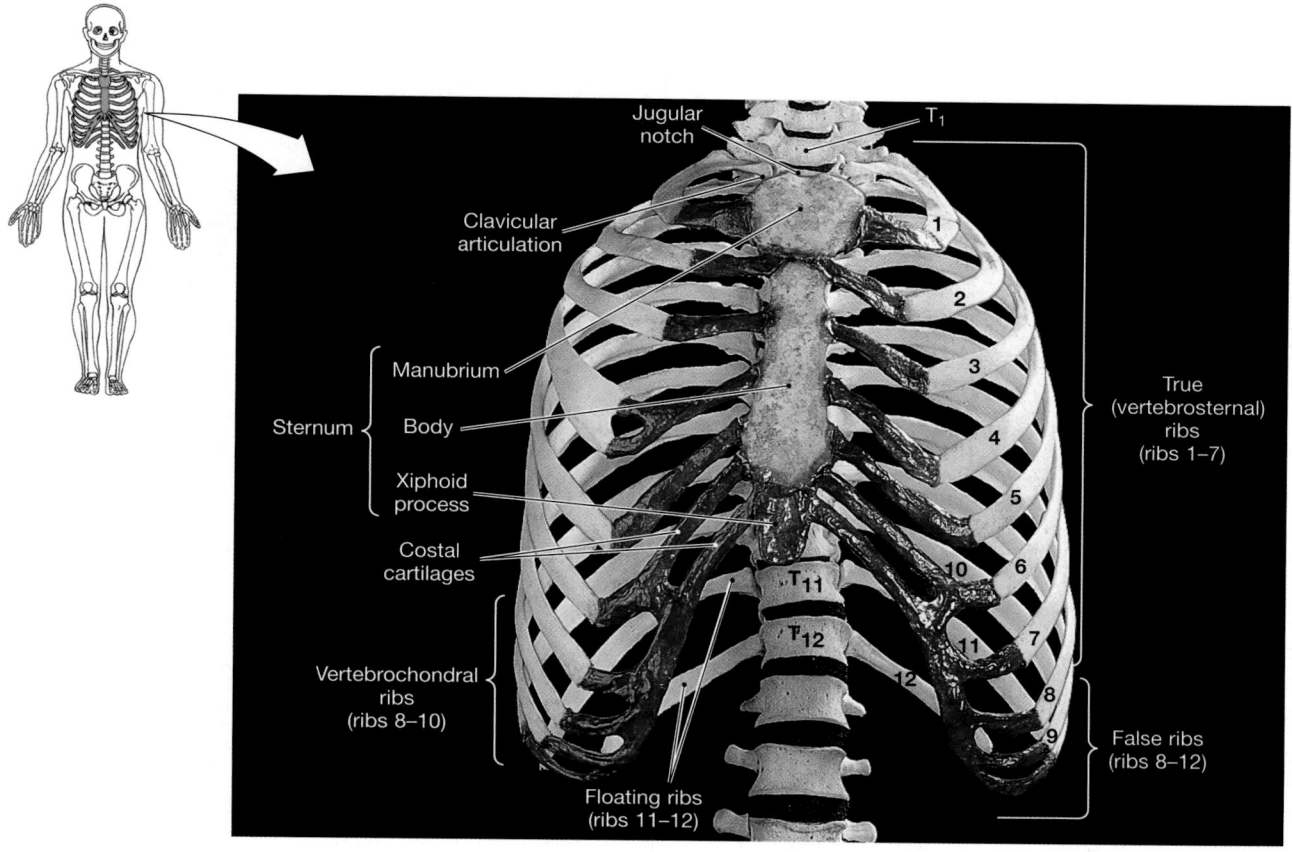

Jugular notch
T₁
Clavicular articulation
1
2
3
True (vertebrosternal) ribs (ribs 1–7)
Manubrium
Sternum
Body
4
5
Xiphoid process
Costal cartilages
10
T₁₁
6
T₁₂
11
7
Vertebrochondral ribs (ribs 8–10)
12
8
9
False ribs (ribs 8–12)
Floating ribs (ribs 11–12)

(a) Anterior view

●**FIGURE 7–23**
The Thoracic Cage. **(a)** An anterior view of the thoracic cage and sternum. **(b)** A posterior view of the thoracic cage.

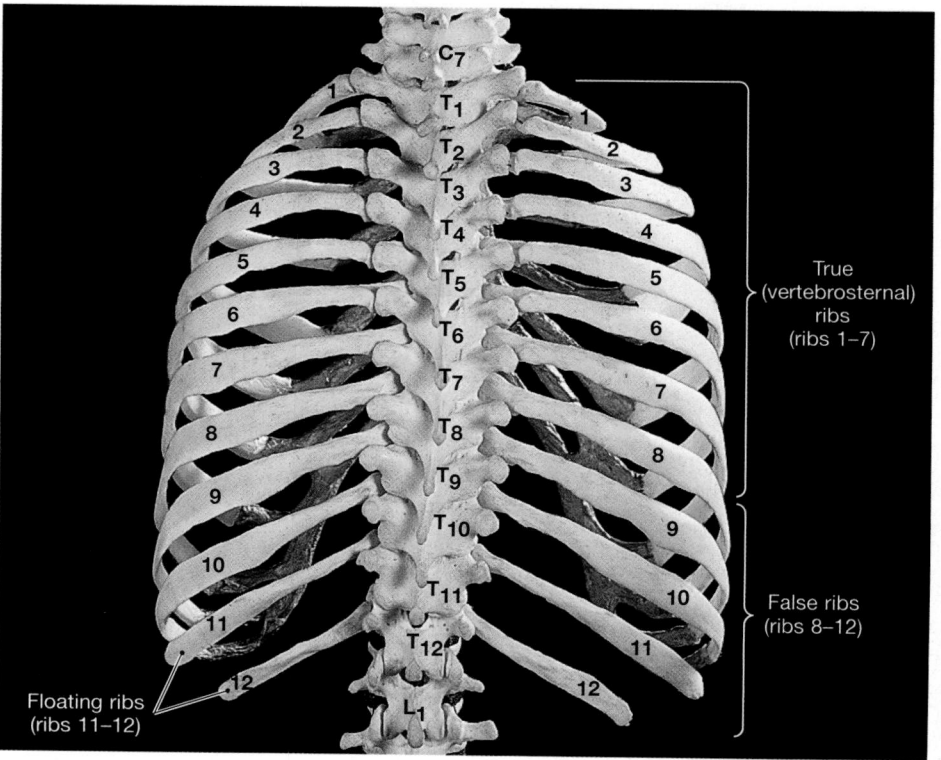

C₇
1
T₁
1
2
T₂
2
3
T₃
3
4
T₄
4
5
T₅
5
True (vertebrosternal) ribs (ribs 1–7)
6
T₆
6
7
T₇
7
8
T₈
8
9
T₉
9
T₁₀
10
T₁₁
10
11
T₁₂
11
False ribs (ribs 8–12)
12
L₁
12
Floating ribs (ribs 11–12)

(b) Posterior view

THE RIBS

Ribs, or *costae*, are elongate, curved, flattened bones that originate on or between the thoracic vertebrae and end in the wall of the thoracic cavity. Each of us, regardless of gender, has 12 pairs of ribs (Figure 7–23●). The first seven pairs are called **true ribs**, or *vertebrosternal ribs*. They reach the anterior body wall and are connected to the sternum by separate cartilaginous extensions, the **costal cartilages**. Beginning with the first rib, the vertebrosternal ribs gradually increase in length and in radius of curvature.

Ribs 8–12 are called **false ribs**, because they do not attach directly to the sternum. The costal cartilages of ribs 8–10, the *vertebrochondral ribs*, fuse together and merge with the cartilages of rib pair 7 before they reach the sternum (Figure 7–23a●). The last two pairs of ribs (11 and 12) are called *floating ribs*, because they have no connection with the sternum, or *vertebral ribs*, because they are attached only to the vertebrae (Figure 7–23a,b●).

Figure 7–24a● shows the superior surface of a typical rib. The *vertebral end* of the rib articulates with the vertebral column at the **head**, or *capitulum* (ka-PIT-ū-lum). A ridge divides the articular surface of the head into superior and inferior articular facets (Figure 7–24b●). From the head, a short **neck** leads to the **tubercle**, or *tuberculum* (too-BER-kū-lum), a small elevation that projects dorsally. The inferior portion of the tubercle contains an articular facet that contacts the transverse process of the thoracic vertebra. Ribs 1 and 10 originate at costal facets on vertebrae T_1 and T_{10}, respectively, and their tubercular facets articulate with the transverse

costal facets on those vertebrae. The heads of ribs 2–9 articulate with costal facets on two adjacent vertebrae; their tubercular facets articulate with the transverse costal facets of the inferior member of the vertebral pair. Ribs 11 and 12, which originate at T_{11} and T_{12}, do not have tubercular facets and do not contact the transverse processes of T_{11} or T_{12}. The difference in rib orientation can be seen by comparing Figure 7–20a, p. 234, with Figure 7–23b●.

The bend, or *angle*, of the rib is the site where the tubular **body**, or *shaft*, begins curving toward the sternum. The internal rib surface is concave, and a prominent *costal groove* along its inferior border marks the path of nerves and blood vessels. The superficial surface is convex and provides an attachment site for muscles of the pectoral girdle and trunk. The *intercostal muscles*, which move the ribs, are attached to the superior and inferior surfaces.

With their complex musculature, dual articulations at the vertebrae, and flexible connection to the sternum, the ribs are quite mobile. Note how the ribs curve away from the vertebral column to angle inferiorly (Figure 7–23●). A typical rib acts as if it were the handle on a bucket, lying just below the horizontal plane. Pushing the handle down forces it inward; pulling it up swings it outward (Figure 7–24c●). In addition, because of the curvature of the ribs, the same movements change the position of the sternum. Depressing the ribs pulls the sternum inward, whereas elevation moves it outward. As a result, movements of the ribs affect both the width and the depth of the thoracic cage, increasing or decreasing its volume accordingly.

●**FIGURE 7–24**
The Ribs. **(a)** Details of rib structure and the articulations between the ribs and thoracic vertebrae. **(b)** A posterior view of the head of a representative rib from the right side (ribs 2–9). **(c)** The effect of rib movement on the thoracic cavity, similar to the movement of a bucket handle.

Transverse costal facet

Tubercle of rib

Angle

Neck

Vertebral end

Demifacet

Head (capitulum)

(a) Superior view

Head

Neck

Tubercle

Attachment to costal cartilage (sternal end)

Articular facets

Body

Costal groove

Angle

(b) Posterior view

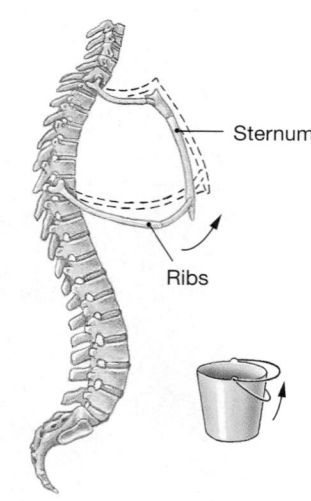

Sternum

Ribs

(c)

The ribs can bend and move to cushion shocks and absorb blows, but severe or sudden impacts can cause painful rib fractures. Because the ribs are tightly bound in connective tissues, a cracked rib can heal without a cast or splint. But compound fractures of the ribs can send bone splinters or fragments into the thoracic cavity, with potential damage to internal organs.

Surgery on the heart, lungs, or other organs in the thorax typically involves entering the thoracic cavity. The mobility of the ribs and the cartilaginous connections with the sternum allow the ribs to be temporarily moved out of the way. "Rib spreaders" are used to push the ribs apart in much the same way that a jack lifts a car off the ground for a tire change. If more extensive access is required, the cartilages of the sternum can be cut and the entire sternum folded out of the way. Once replaced, scar tissue reunites the cartilages, and the ribs heal fairly rapidly.

After thoracic surgery, *chest tubes* may be inserted through the wall of the thoracic cavity to permit drainage of fluids. To install a chest tube or obtain a sample of pleural fluid, a physician must penetrate the wall of the thorax. This process, called *thoracentesis* (tho-ra-sen-TĒ-sis) or *thoracocentesis* (tho-ra-kō-sen-TĒ-sis; *kentesis*, perforating), involves the penetration of the thoracic wall along the superior border of one of the ribs. Penetration at this location avoids damage to vessels and nerves within the costal groove.

☐ THE STERNUM

The adult **sternum**, or breastbone, is a flat bone that forms in the anterior midline of the thoracic wall (Figure 7–23a●). The sternum has three components:

1. The broad, triangular **manubrium** (ma-NOO-brē-um) articulates with the *clavicles* (collarbones) and the cartilages of the first pair of ribs. The manubrium is the widest and most superi-

or portion of the sternum. Only the first pair of ribs is attached by cartilage to this portion of the sternum. The **jugular notch**, located between the clavicular articulations, is a shallow indentation on the superior surface of the manubrium.

2. The tongue-shaped **body** attaches to the inferior surface of the manubrium and extends inferiorly along the midline. Individual costal cartilages from rib pairs 2–7 are attached to this portion of the sternum.

3. The **xiphoid** (ZĪ-foyd) **process**, the smallest part of the sternum, is attached to the inferior surface of the body. The muscular *diaphragm* and *rectus abdominis muscles* attach to the xiphoid process.

Ossification of the sternum begins at 6 to 10 centers, and fusion is not completed until at least age 25. Before that age, the sternal body consists of four separate bones. In adults, their boundaries appear as a series of transverse lines crossing the sternum. The xiphoid process is generally the last sternal component to ossify and fuse. Its connection to the sternal body can be broken by impact or by strong pressure, creating a spear of bone that can severely damage the liver. Cardiopulmonary resuscitation (CPR) training strongly emphasizes proper hand positioning to reduce the chances of breaking ribs or the xiphoid process.

✓ How could you distinguish between true ribs and false ribs?

✓ Improper administration of cardiopulmonary resuscitation (CPR) can result in a fracture of which bone(s)?

✓ What are the main differences between vertebrosternal and vertebrochondral ribs?

Answers start on page Q-1

Chapter Review

SELECTED CLINICAL TERMINOLOGY

Terms Discussed in This Chapter

chest tube: A drain installed after thoracic surgery to permit removal of blood and pleural fluid. (*p. 239*)
craniostenosis (krā-nē-ō-sten-Ō-sis): The premature closure of one or more fontanels, which can lead to unusual distortions of the skull. (*p. 228*)
deviated (nasal) septum: A bent nasal septum that slows or prevents sinus drainage. (*p. 226*)
kyphosis (kī-FŌ-sis): An abnormal exaggeration of the thoracic curve that produces a humpbacked appearance. (*p. 229 and AM*)
lordosis (lor-DŌ-sis): An abnormal lumbar curve that gives a swaybacked appearance. (*p. 229 and AM*)
microcephaly (mī-krō-SEF-uh-lē): An undersized head resulting from genetic or developmental abnormalities. (*p. 228*)

scoliosis (skō-lē-Ō-sis): An abnormal lateral curvature of the spine. (*p. 229 and AM*)
sinusitis: Inflammation and congestion of the sinuses. (*p. 226*)
spina bifida (SPĪ-nuh BI-fi-duh): A condition resulting from the failure of the vertebral laminae to unite during development; commonly associated with developmental abnormalities of the brain and spinal cord. (*p. 231 and AM*)
thoracentesis (tho-ra-sen-TĒ-sis), or thoracocentesis (tho-ra-kō-sen-TĒ-sis): The penetration of the thoracic wall along the superior border of one of the ribs. (*p. 239*)
TMJ syndrome: A painful condition resulting from a misalignment of the mandible at the temporomandibular joint. (*p. 224*)
whiplash: An injury resulting from a sudden change in the body position that can injure the cervical vertebrae. (*p. 232*)

STUDY OUTLINE

7–1 DIVISIONS OF THE SKELETON   p. 209

1. The skeletal system consists of the axial skeleton and the appendicular skeleton. The **axial skeleton** can be divided into the *skull*, the *auditory ossicles* and *hyoid bone*, the *vertebral column*, and the *thoracic cage*. (*Figure 7–1*)
2. The **appendicular skeleton** includes the pectoral and pelvic girdles, which support the upper and lower limbs.

7–2 THE SKULL   p. 210

1. The **skull** consists of the **cranium** and the bones of the face. The cranium, composed of **cranial bones**, encloses the **cranial cavity**, a division of the dorsal body cavity. The **facial bones** protect and support the entrances to the digestive and respiratory tracts. (*Figure 7–2*)
2. Prominent superficial landmarks on the skull include the **lambdoid**, **coronal**, **sagittal**, and **squamous sutures**. (*Figure 7–3*)

Focus: The Individual Bones of the Skull   p. 216

CRANIAL BONES   p. 216

3. The cranial bones are the **occipital bone**, the two **parietal bones**, the **frontal bone**, the two **temporal bones**, the **sphenoid**, and the **ethmoid**. (*Figures 7–3 to 7–9*)
4. The occipital bone surrounds the **foramen magnum**. (*Figures 7–3 to 7–5*)
5. The frontal bone contains the **frontal sinuses**. (*Figures 7–4, 7–6*)
6. The **auditory ossicles** are located in a cavity within the temporal bone. (*Figure 7–7*)

FACIAL BONES   p. 220

7. The bones of the face are the **maxillary bones**, the **palatine bones**, the **nasal bones**, the **vomer**, the **inferior nasal conchae**, the **zygomatic bones**, the **lacrimal bones**, the **mandible**, and the **hyoid bone**. (*Figures 7–3, 7–4, 7–10 to 7–12*)
8. The left and right maxillary bones, or *maxillae*, are the largest facial bones; they form the upper jaw and most of the **hard palate**. (*Figures 7–3, 7–4, 7–10*)
9. The palatine bones are small L-shaped bones that form the posterior portions of the hard palate and contribute to the floor of the orbital cavities. (*Figures 7–3, 7–4, 7–10*)
10. The paired nasal bones extend to the superior border of the **external nares**. (*Figures 7–3, 7–4, 7–11*)
11. The vomer forms the inferior portion of the **nasal septum**. (*Figures 7–3, 7–4, 7–11*)
12. The **temporal process** of the zygomatic bone articulates with the **zygomatic process** of the temporal bone to form the **zygomatic arch**. (*Figures 7–3, 7–7, 7–11*)
13. The paired lacrimal bones, the smallest bones of the face, are situated medially in each **orbit**. (*Figures 7–3, 7–11*)
14. The mandible is the bone of the lower jaw. (*Figures 7–3, 7–4, 7–12*)
15. The **hyoid bone**, suspended by *stylohyoid ligaments*, supports the larynx. (*Figure 7–12*)

SUMMARY: THE FORAMINA AND FISSURES OF THE SKULL   p. 224

16. Features of the adult skull are summarized in *Table 7–1*.

THE ORBITAL AND NASAL COMPLEXES   p. 224

17. Seven bones form each **orbital complex**. (*Figure 7–13*)
18. The **nasal complex** includes the bones that enclose the nasal cavities and the **paranasal sinuses**, hollow airways that connect with the nasal passages. (*Figure 7–14*)

THE SKULLS OF INFANTS AND CHILDREN   p. 226

19. Fibrous connective-tissue **fontanels** permit the skulls of infants and children to continue growing. (*Figure 7–15*)

7–3 THE VERTEBRAL COLUMN   p. 228

1. The **vertebral column** consists of the vertebrae, sacrum, and coccyx. We have 7 **cervical vertebrae** (the first articulates with the skull), 12 **thoracic vertebrae** (which articulate with ribs), and 5 **lumbar vertebrae** (the last articulates with the sacrum). The **sacrum** and **coccyx** consist of fused vertebrae. (*Figure 7–16*)

SPINAL CURVATURE   p. 228

2. The spinal column has four **spinal curves**. The **thoracic** and **sacral curves** are called **primary**, or **accommodation**, **curves**; the **lumbar** and **cervical curves** are known as **secondary**, or **compensation**, **curves**. (*Figure 7–16*)

VERTEBRAL ANATOMY   p. 230

3. A typical vertebra has a **body** and a **vertebral arch**, and articulates with adjacent vertebrae at the **superior** and **inferior articular processes**. (*Figure 7–17*)
4. Adjacent vertebrae are separated by **intervertebral discs**. Spaces between successive **pedicles** form the **intervertebral foramina**. (*Figure 7–17*)

🔘 Anatomy of vertebrae and axial skeleton: **Anatomy CD-ROM**: Skeletal System/Axial Dissections.

VERTEBRAL REGIONS   p. 231

5. Cervical vertebrae are distinguished by the shape of the body, the relative size of the vertebral foramen, the presence of **costal processes** with **transverse foramina**, and notched **spinous processes**. These vertebrae include the **atlas**, **axis**, and **vertebra prominens**. (*Figure 7–18; Table 7–2*)
6. Thoracic vertebrae have a distinctive heart-shaped body; long, slender spinous processes; and articulations for the ribs. (*Figures 7–19, 7–22; Table 7–2*)
7. The lumbar vertebrae are the most massive and least mobile of the vertebrae; they are subjected to the greatest strains. (*Figure 7–20; Table 7–2*)
8. The sacrum protects reproductive, digestive, and urinary organs and articulates with the pelvic girdle and with the fused elements of the coccyx. (*Figure 7–21*)

🔘 Cervical, thoracic, and lumbar vertebrae: **Anatomy CD-ROM**: Skeletal System/Axial Dissections.

7–4 THE THORACIC CAGE   p. 236

1. The skeleton of the **thoracic cage** consists of the thoracic vertebrae, the ribs, and the sternum. The **ribs** and **sternum** form the *rib cage*. (*Figure 7–22*)

THE RIBS   p. 238

2. Ribs 1–7 are **true**, or *vertebrosternal*, **ribs**. Ribs 8–12 are called **false ribs**; they include the *vertebrochondral ribs* (ribs 8–10) and two pairs of *floating (vertebral) ribs* (ribs 11–12). A typical rib has a **head**, or *capitulum*; a **neck**; a **tubercle**, or *tuberculum*; an *angle*; and a **body**, or *shaft*. A *costal groove* marks the path of nerves and blood vessels. (*Figures 7–22, 7–23*)

THE STERNUM   p. 239

3. The sternum consists of the **manubrium**, **body**, and **xiphoid process**. (*Figure 7–22*)

REVIEW QUESTIONS

 More assessment questions are available to you on the Companion Website. You will find Matching, Multiple Choice, True/False, and other quizzes to help further your understanding of the material covered in this chapter. To access the site, go to www.aw.com/martini.

LEVEL 1 Reviewing Facts and Terms

Match each numbered item with the most closely related lettered item. Use letters for answers in the spaces provided.

_____ 1. foramina	**a.** skullcap
_____ 2. sinuses	**b.** tear ducts
_____ 3. sutures	**c.** atlas
_____ 4. calvaria	**d.** lumbar and cervical
_____ 5. auditory ossicles	**e.** air-filled chambers
_____ 6. hypophyseal fossa	**f.** axis
_____ 7. lacrimal bones	**g.** ear bones
_____ 8. accommodation curve	**h.** ribs
_____ 9. compensation curve	**i.** passageways
_____ 10. costae	**j.** sella turcica
_____ 11. C_1	**k.** thoracic and sacral
_____ 12. C_2	**l.** immovable joints

13. Which list contains *only* facial bones?

   (a) mandible, maxillary, nasal, zygomatic
   (b) frontal, occipital, zygomatic, parietal
   (c) occipital, sphenoid, temporal, lacrimal
   (d) frontal, parietal, occipital, sphenoid

14. The unpaired facial bones include the

   (a) lacrimal and nasal   (b) vomer and mandible
   (c) maxillary and mandible (d) zygomatic and palatine

15. The boundaries between skull bones are immovable joints called

   (a) foramina   (b) fontanels
   (c) lacunae   (d) sutures

16. The joint between the frontal and parietal bones is correctly called the _____ suture.

   (a) parietal   (b) lambdoid
   (c) squamous   (d) coronal

17. Of the following bones, which is unpaired?

   (a) vomer   (b) maxillary
   (c) palatine   (d) nasal

18. The styloid process, zygomatic process, and auditory ossicles are associated with the

   (a) parietal bone   (b) occipital bone
   (c) sphenoid   (d) temporal bone

19. The bone that contains the cribriform plate is the

   (a) ethmoid   (b) occipital
   (c) parietal   (d) sphenoid

20. The membranous areas between the cranial bones of the fetal skull are

   (a) fontanels   (b) sutures
   (c) Wormian bones   (d) foramina

21. The part(s) of the vertebra that transfer(s) weight along the axis of the vertebral column is (are) the

   (a) vertebral arch (b) lamina
   (c) pedicles   (d) body

22. Which five bones make up the facial complex?

23. What seven bones constitute the orbital complex?

24. What is the primary function of the vomer?

25. Which bones contain the paranasal sinuses?

LEVEL 2 Reviewing Concepts

26. The atlas (C_1) can be distinguished from the other vertebrae by

   (a) the presence of anterior and posterior vertebral arches
   (b) the lack of a body
   (c) the presence of superior facets and inferior articular facets
   (d) a, b, and c are correct

27. What is the relationship between the temporal bone and the ear?

28. What is the relationship between the ethmoid and the nasal cavity?

29. What purpose do the fontanels serve during birth?

30. Unlike that of other animals, the skull of humans is balanced above the vertebral column, allowing an upright posture. What anatomical adaptations are necessary in humans to maintain the balance of the skull?

31. The structural features and skeletal components of the sternum make it a part of the axial skeleton that is important in a variety of clinical situations. If you were teaching this information to prospective nursing students, what clinical applications would you cite?

32. Why are ruptured intervertebral discs more common in lumbar vertebrae and dislocations and fractures more common in cervical vertebrae?

33. Why are the upper seven pairs of ribs called *true* ribs and the lower five pairs called *false* ribs?

34. Describe how ribs function in breathing.

35. Why is it important to keep your back straight when you lift a heavy object?

LEVEL 3 Critical Thinking and Clinical Applications

36. Tess is diagnosed with a disease that affects the membranes surrounding the brain. The physician tells Tess' family that the disease is caused by an airborne virus. Explain how this virus could have entered the cranium.

37. Joe is 40 years old and 30 pounds overweight. Like many middle-aged men, Joe carries most of this extra weight in his abdomen and jokes with his friends about his "beer gut." During an annual physical, Joe's physician advises him that his spine is developing an abnormal curvature. Why is the curvature of Joe's spine changing, and what is this condition called?

38. While working at an excavation, an archaeologist finds several small skull bones. She examines the frontal, parietal, and occipital bones and concludes that the skulls are those of children not yet one year old. How can she tell their ages from an examination of their bones?

| UNIT 2 | CHAPTER | 5 | 6 | 7 | **8** | 9 | 10 | 11 |

CLINICAL NOTES

THE APPENDICULAR SKELETON

In the previous chapter, we discussed the 80 bones of the axial skeleton. Appended to these bones are the remaining 60 percent of the bones that make up the skeletal system. The **appendicular skeleton** includes the bones of the limbs and the supporting elements, or *girdles*, that connect them to the trunk (Figure 8–1●). To appreciate the role that the appendicular skeleton plays in your life, try making a mental list of all the things you have done with your arms or legs in the past day. Standing, walking, writing, turning pages, eating, dressing, shaking hands, waving—the list quickly becomes unwieldy. Your axial skeleton protects and supports internal organs and participates in vital functions, such as respiration. But it is your appendicular skeleton that gives you control over your environment, changes your position in space, and provides mobility.

The appendicular skeleton is dominated by the long bones that support the limbs. ∞ p. 184 Each long bone shares common features with other long bones. For example, one epiphysis is usually called the *head*, the diaphysis is called the *shaft*, and the head and shaft are normally separated by a *neck*. For simplicity, when an illustration shows a single bone, we shall use those terms without including the name of the bone. Thus, a photo of the humerus will be labeled *head* rather than with the complete name, *head of the humerus* or *humeral head*. When more than one bone is shown, we shall give the complete name to avoid confusion. The descriptions in this chapter emphasize surface features that have functional importance, such as the attachment sites for skeletal muscles and the paths of major nerves and blood vessels, or that provide landmarks which define areas and locate structures of the body.

8–1 THE PECTORAL GIRDLE AND UPPER LIMBS

Objectives

- Identify the bones that form the pectoral girdle, their functions, and their superficial features.
- Identify the bones of the upper limbs, their functions, and their superficial features.

Each arm articulates (that is, forms a joint) with the trunk at the **pectoral girdle**, or *shoulder girdle*. The pectoral girdle consists of two S-shaped *clavicles* (KLAV-i-kulz; collarbones) and two broad, flat *scapulae* (SKAP-ū-lē; singular, *scapula*, SKAP-ū-luh; or shoulder blade). The medial, anterior end of the clavicle articulates with the manubrium of the sternum. ∞ p. 239

These articulations are the *only* direct connections between the pectoral girdle and the axial skeleton. Skeletal muscles support and position the scapulae, which have no direct bony or ligamentous connections to the thoracic cage. As a result, the shoulders are extremely mobile, but not very strong.

☐ THE PECTORAL GIRDLE

Movements of the clavicles and scapulae position the shoulder joints and provide a base for arm movement. Once the shoulder joints are in position, muscles that originate on the pectoral girdle

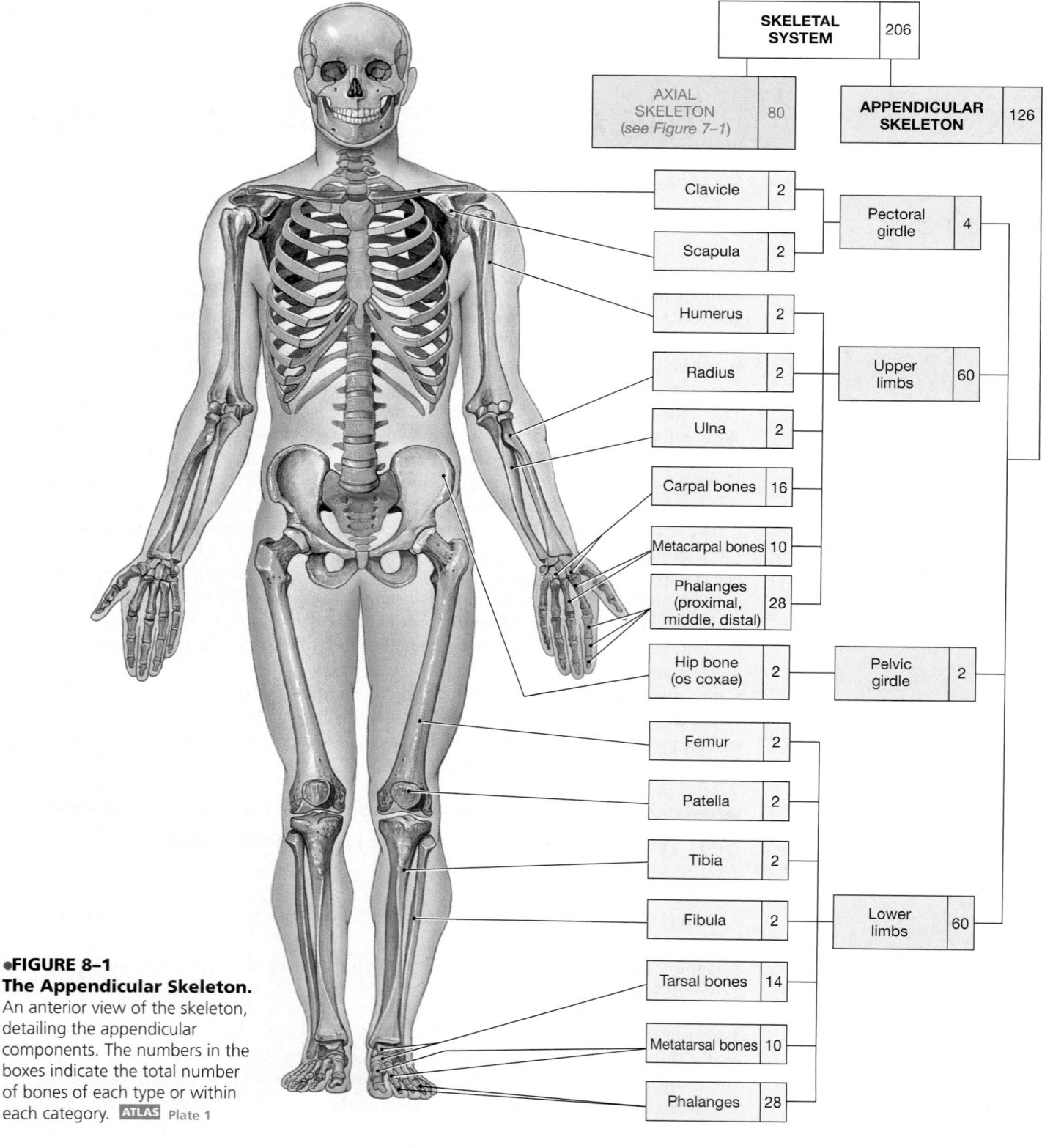

●FIGURE 8–1
The Appendicular Skeleton.
An anterior view of the skeleton, detailing the appendicular components. The numbers in the boxes indicate the total number of bones of each type or within each category. **ATLAS** Plate 1

help move the upper limbs. The surfaces of the scapulae and clavicles are extremely important as sites for muscle attachment. Where they attach, major muscles leave marks in the form of bony ridges and flanges. Other bone markings, such as sulci or foramina, indicate the positions of nerves or the passage of blood vessels that control the muscles and nourish the muscles and bones.

The Clavicles

The **clavicles** are S-shaped bones that originate at the superior, lateral border of the manubrium of the sternum, lateral to the jugular notch (Figure 8–2●). From the roughly pyramidal **sternal end**, each clavicle curves laterally and posteriorly for roughly half its length. It then forms a smooth posterior curve to articulate with a process of the scapula, the *acromion* (a-KRŌ-mē-on). The flat, **acromial end** of the clavicle is broader than the sternal end (Figure 8–2b,c●).

The smooth, superior surface of the clavicle lies just beneath the skin. The rough, inferior surface of the acromial end is marked by prominent lines and tubercles identified in Figure 8–2c●. These surface features are attachment sites for muscles and ligaments of the shoulder. The combination of the direction of curvature and the differences between superior and inferior surfaces make it relatively easy to tell a left clavicle from a right clavicle.

You can explore the interaction between scapulae and clavicles. With your fingers in your jugular notch, locate the clavicle on either side. When you move your shoulders, you can feel the clavicles change their positions. Because the clavicles are so close to the skin, you can trace one laterally until it articulates with the scapula.

The clavicles are relatively small and fragile, and therefore fractures of the clavicle are fairly common. For example, you can fracture a clavicle as the result of a simple fall if you land on your hand with your arm outstretched. Fortunately, in view of the clavicle's vulnerability, most clavicular fractures heal rapidly without a cast.

The Scapulae

The anterior surface of the **body** of each **scapula** forms a broad triangle (Figure 8–3a●). The three sides of the triangle are the **superior border**; the **medial border**, or *vertebral border*; and the **lateral border**, or *axillary border* (*axilla*, armpit). Muscles that position the scapula attach along these edges. The corners of the triangle are called the *superior angle*, the *inferior angle*, and the *lateral angle*. The lateral angle, or *head* of the scapula, forms a broad process that supports the cup-shaped **glenoid cavity**. At the glenoid cavity, the scapula articulates with the *humerus*, the proximal bone of the upper limb. This articulation is the shoulder joint, also known as the *glenohumeral joint*. The anterior surface of the body of the scapula is relatively smooth and concave. The depression in the anterior surface is called the **subscapular fossa**.

Two large scapular processes extend beyond the margin of the glenoid cavity (Figure 8–3b●) superior to the head of the humerus. The smaller, anterior projection is the **coracoid** (KOR-uh-koyd) **process**. The **acromion** is the larger, posterior process. If you run your fingers along the superior surface of the shoulder joint, you will feel this process. The acromion articulates with the clavicle at the *acromioclavicular joint*. Both the acromion and the coracoid process are attached to ligaments and tendons associated with the shoulder joint.

The acromion is continuous with the **scapular spine** (Figure 8–3c●), a ridge crosses that posterior surface of the scapular body before ending at the medial border. The scapular spine divides the convex posterior surface of the body into two regions. The area

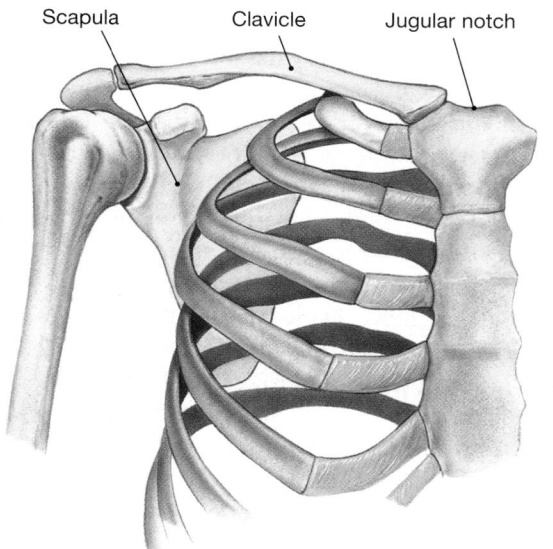

(a) Anterior view

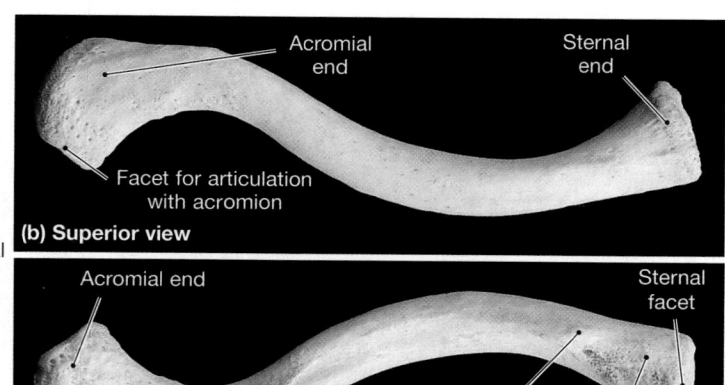

●**FIGURE 8–2**
The Clavicle. **(a)** The position of the clavicle, anterior view. **(b)** Superior and **(c)** inferior views of the right clavicle. Stabilizing ligaments attach to the *conoid tubercle* and the *costal tuberosity*. ATLAS Plate 5.4

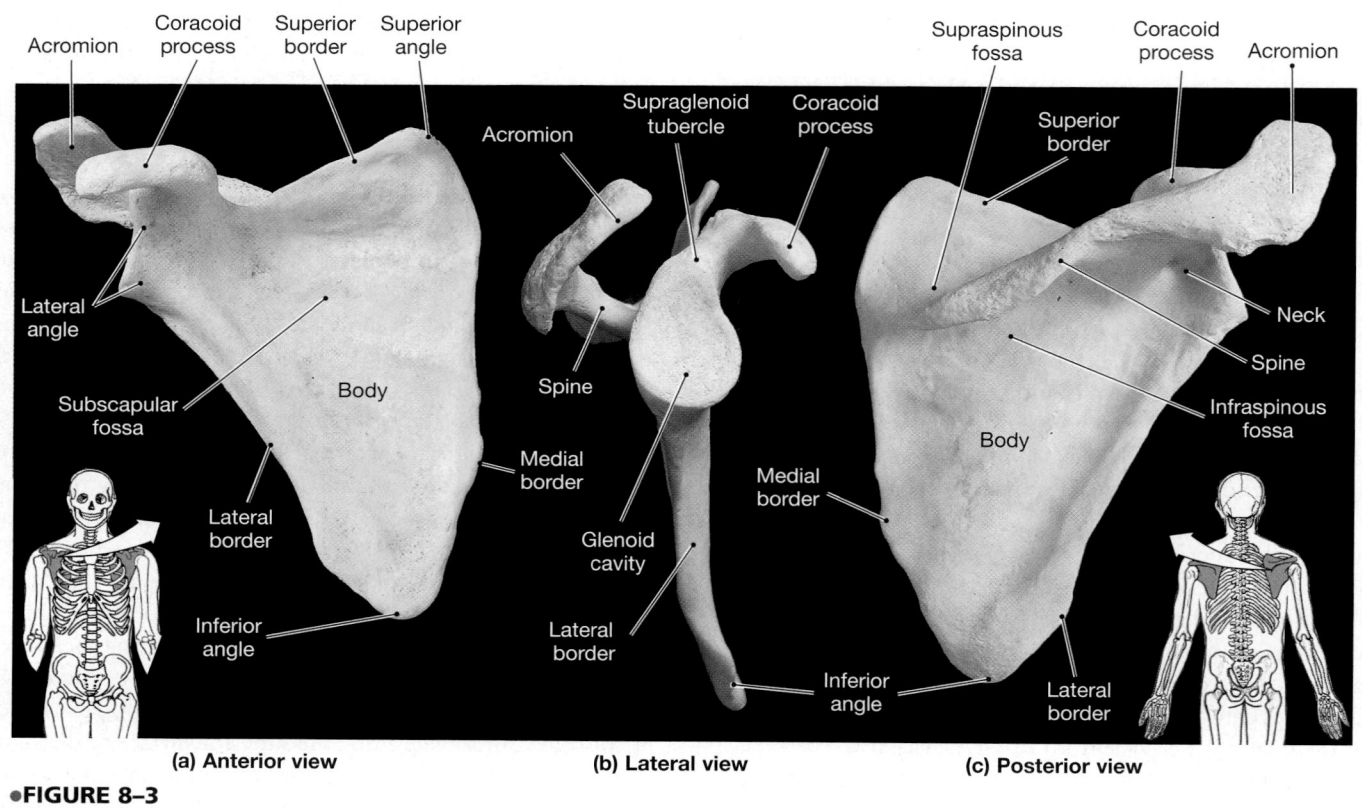

(a) **Anterior view**   (b) **Lateral view**   (c) **Posterior view**

●**FIGURE 8–3**
The Scapula.   (a) Anterior, (b) lateral, and (c) posterior views of the right scapula.   **ATLAS** Plate 5.5

superior to this spine constitutes the **supraspinous fossa** (*supra*, above); the region inferior to the spine is the **infraspinous fossa** (*infra*, beneath). The entire posterior surface is marked by small ridges and lines where smaller muscles attach to the scapula.

✓ Why would a broken clavicle affect the mobility of the scapula?

✓ Which bone articulates with the scapula at the glenoid cavity?

Answers start on page Q-1

 To clarify the interaction between the pectoral girdle and the proximal bone of the arm, visit the **Anatomy CD-ROM:** Skeletal System/Appendicular Dissections/Clavicle & Scapula and Humerus.

☐ THE UPPER LIMBS

The skeleton of the upper limbs consists of the bones of the arms, forearms, wrists, and hands. Notice that in anatomical descriptions, the term *arm* refers only to the proximal portion of the upper limb (from shoulder to elbow), not to the entire limb. The anatomical usage and common usage are thus quite different. We shall examine the bones of the right upper limb. The arm, or *brachium*, contains one bone, the **humerus**, which extends from the scapula to the elbow. At the proximal end of the humerus, the round **head** articulates with the scapula (Figure 8–4●). At its distal

end, the humerus articulates with the *radius* and the *ulna*, the bones of the forearm (or *antebrachium*).

The Humerus

The prominent **greater tubercle** is a rounded projection on the lateral surface of the epiphysis, near the margin of the humeral head (Figure 8–4●). The greater tubercle establishes the lateral contour of the shoulder. You can verify its position by feeling for a bump situated a few centimeters from the tip of the acromion. The **lesser tubercle** is a smaller projection that lies on the anterior, medial surface of the epiphysis, separated from the greater tubercle by the **intertubercular groove**, or *intertubercular sulcus*. Both tubercles are important sites for muscle attachment; a large tendon runs along the groove. Lying between the tubercles and the articular surface of the head, the **anatomical neck** marks the extent of the joint capsule. The narrower **surgical neck** corresponds to the metaphysis of the growing bone. The name reflects the fact that fractures typically occur at this site.

The proximal shaft of the humerus is round in section. The **deltoid tuberosity** is a large, rough elevation on the lateral surface of the shaft, approximately halfway along its length. It is named after the *deltoid muscle*, which attaches to it.

On the posterior surface, the deltoid tuberosity ends at the **radial groove** (Figure 8–4b●). This depression marks the path of the *radial nerve*, a large nerve that provides both sensory in-

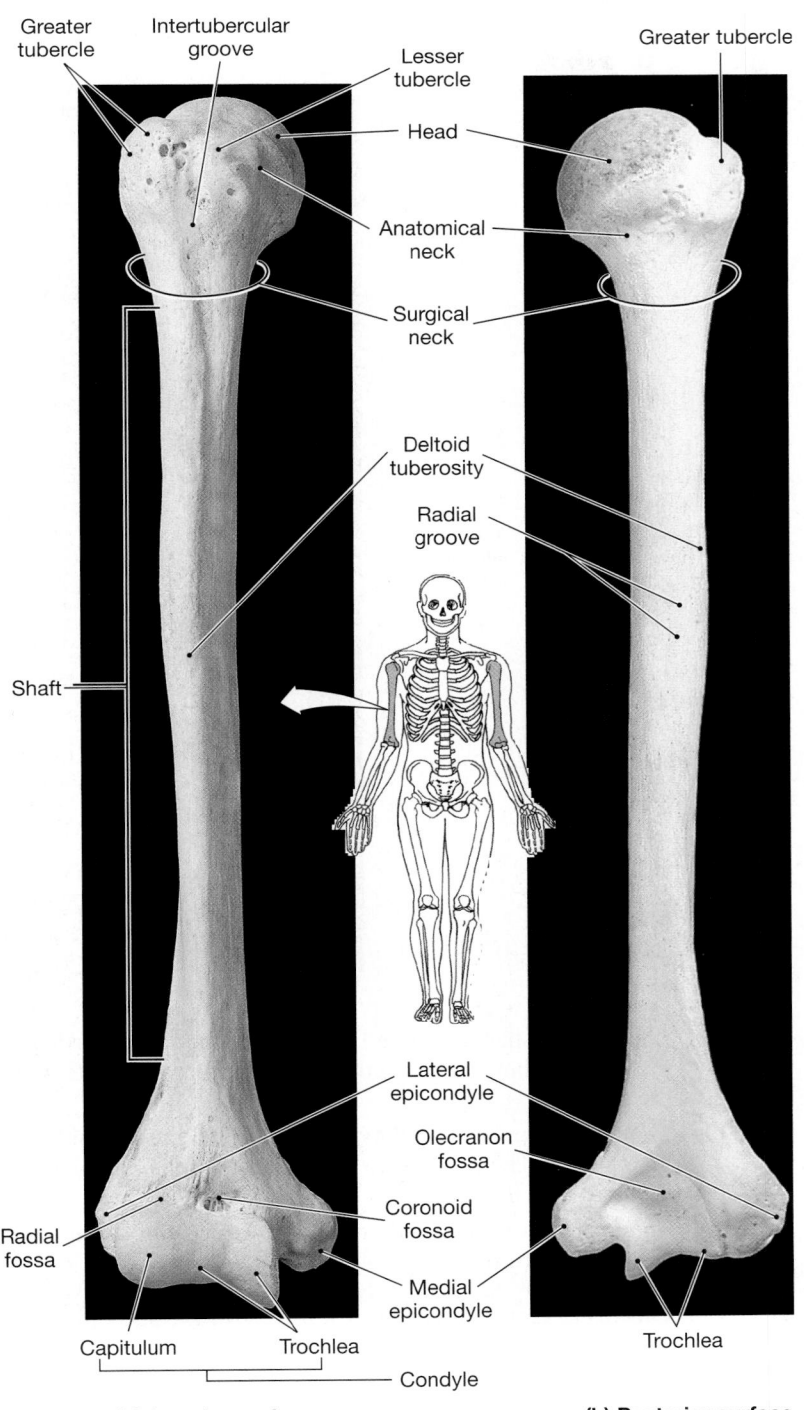

Greater tubercle

Intertubercular groove

Lesser tubercle

Head

Anatomical neck

Surgical neck

Greater tubercle

Deltoid tuberosity

Radial groove

Shaft

Lateral epicondyle

Olecranon fossa

Coronoid fossa

Medial epicondyle

Radial fossa

Capitulum

Trochlea

Condyle

Trochlea

(a) Anterior surface

(b) Posterior surface

•FIGURE 8–4
The Humerus. **(a)** The anterior and **(b)** posterior surfaces of the right humerus. **ATLAS** Plates 6.2c and 6.4c,d

formation from the posterior surface of the limb and motor control over the large muscles that straighten the elbow. Distal to the radial groove, the posterior surface of the humerus is relatively flat. Near the distal articulation with the bones of the forearm, the shaft expands to either side at the **medial** and

lateral epicondyles. *Epicondyles* are processes that develop proximal to an articulation and provide additional surface area for muscle attachment. The *ulnar nerve* crosses the posterior surface of the medial epicondyle. A blow at the humeral side of the elbow joint can strike this nerve and produce a temporary numbness and paralysis of muscles on the anterior surface of the forearm. Because of the odd sensation, this area is sometimes called the *funny bone*.

At the **condyle**, the humerus articulates with the bones of the forearm—the radius and the ulna. The condyle is divided into two articular regions: the trochlea and the capitulum (Figure 8–4•). The **trochlea** (*trochlea*, a pulley), the spool-shaped medial portion of the condyle, extends from the base of the **coronoid** (*corona*, crown) **fossa** on the anterior surface to the **olecranon** (ō-LEK-ruh-non) **fossa** on the posterior surface (Figure 8–4b•). These depressions accept projections from the ulnar surface as the elbow approaches the limits of its range of motion. The rounded **capitulum** forms the lateral surface of the condyle. A shallow **radial fossa** superior to the capitulum accommodates a portion of the radial head as the forearm approaches the humerus. The prominent lateral head and the differences between the medial and lateral condyles make it relatively easy to tell a left humerus from a right humerus.

The Ulna

The *ulna* and *radius* are parallel bones that support the forearm. In the anatomical position, the **ulna** lies medial to the radius. The **olecranon**, the superior end of the ulna, is the point of the elbow (Figure 8–5a•). On the anterior surface of the proximal epiphysis (Figure 8–5b•), the **trochlear notch** of the ulna articulates with the trochlea of the humerus at the elbow joint. (The fact that this notch forms a "U" in lateral view may help you to remember the name *ulna*.)

The olecranon forms the superior lip of the trochlear notch, and the **coronoid process** forms its inferior lip. At the limit of *extension*, with the forearm and arm forming a straight line, the olecranon swings into the olecranon fossa on the posterior surface of the humerus. At the limit of *flexion*, with the arm and forearm forming a V, the coronoid process projects into the coronoid fossa on the anterior humeral surface. Lateral to the coronoid process, a smooth **radial notch** accommodates the head of the radius at the *proximal radioulnar joint*.

Viewed in cross section, the shaft of the ulna is roughly triangular. The **interosseous membrane**, a fibrous sheet, connects the lateral margin of the ulna to the radius. Near the wrist, the shaft of the ulna narrows before ending at a disc-shaped **ulnar head**, or *head of the ulna*. The posterior, lateral surface of the ulnar head bears a short **styloid process** (*styloid*, long and pointed). A

●**FIGURE 8–5**
The Radius and Ulna. The right radius and ulna in **(a)** posterior and **(b)** anterior views.
ATLAS Plates 6.2c and 6.4g

triangular *articular disc* attaches to the styloid process; this cartilage separates the ulnar head from the bones of the wrist. The lateral surface of the ulnar head articulates with the distal end of the radius to form the *distal radioulnar joint*.

The Radius

The **radius** is the lateral bone of the forearm. The disc-shaped **radial head**, or *head of the radius*, articulates with the capitulum of the humerus. During flexion, the radial head swings into the radial fossa of the humerus. A narrow neck extends from the radial head to the **radial tuberosity**, which marks the attachment site of the *biceps brachii muscle*, a large muscle on the anterior surface of the arm. The shaft of the radius curves along its length. It also enlarges, and the distal portion of the radius is considerably larger than the distal portion of the ulna. The **ulnar notch** on the medial surface of the distal end of the radius marks the site of articulation with the head of the ulna. The distal end of the radius articulates with the bones of the wrist. The **styloid process** on the lateral surface of the radius helps stabilize this joint. If you are looking at an isolated radius or ulna, you can quickly identify whether it is left or right by finding the radial notch (ulna) or ulnar notch (radius) and remembering that the radius lies lateral to the ulna.

The Carpal Bones

The *carpus*, or wrist, contains eight **carpal bones**. These bones form two rows, one with four **proximal carpal bones** and the other with four **distal carpal bones**.

The proximal carpal bones are the scaphoid bone, lunate bone, triquetrum, and pisiform bone (Figure 8–6a●).

- The **scaphoid bone** is the proximal carpal bone on the lateral border of the wrist; it is the carpal bone closest to the styloid process of the radius.

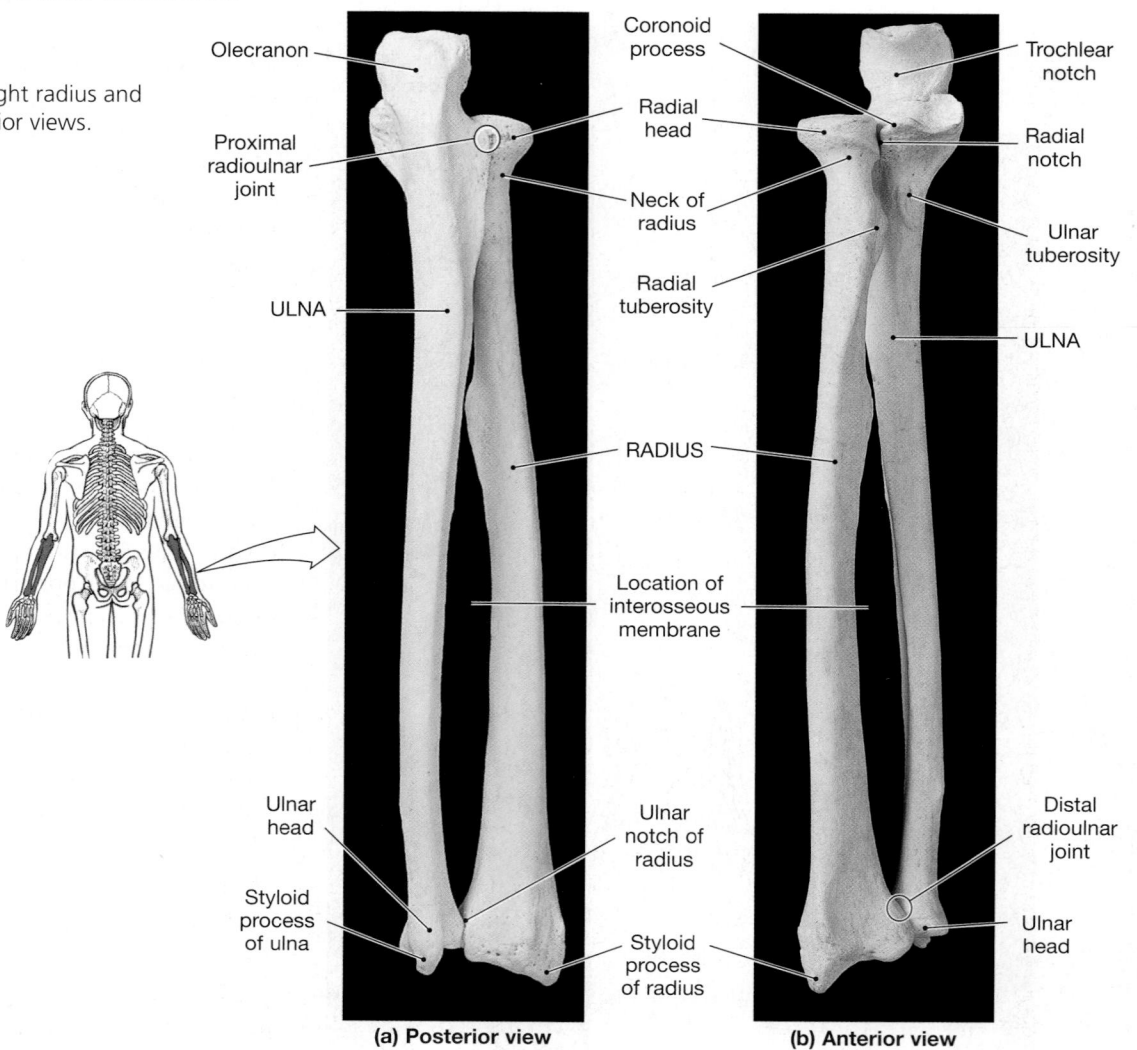

(a) Posterior view   **(b) Anterior view**

Labels (a) Posterior view: Olecranon, Proximal radioulnar joint, ULNA, Ulnar head, Styloid process of ulna, Coronoid process, Radial head, Neck of radius, Radial tuberosity, RADIUS, Location of interosseous membrane, Ulnar notch of radius, Styloid process of radius

Labels (b) Anterior view: Trochlear notch, Radial notch, Ulnar tuberosity, ULNA, Distal radioulnar joint, Ulnar head

- The comma-shaped **lunate** (*luna*, moon) **bone** lies medial to the scaphoid bone and, like the scaphoid bone, articulates with the radius.
- The **triquetrum** is a small pyramid-shaped bone medial to the lunate bone. The triquetrum articulates with the articular disc that separates the ulnar head from the wrist.
- The small, pea-shaped **pisiform** (PIS-i-form) **bone** sits anterior to the triquetrum.

The distal carpal bones are the trapezium, trapezoid bone, capitate bone, and hamate bone (Figure 8–6b●).

- The **trapezium** is the lateral bone of the distal row; its proximal surface articulates with the scaphoid bone.
- The wedge-shaped **trapezoid bone** lies medial to the trapezium. Like the trapezium, it has a proximal articulation with the scaphoid bone.
- The **capitate bone**, the largest carpal bone, sits between the trapezoid bone and the hamate bone.
- The **hamate** (*hamatum*, hooked) **bone** is the medial distal carpal bone.

It may help you to identify the eight carpal bones if you remember the sentence "Sam Likes To Push The Toy Car Hard." In

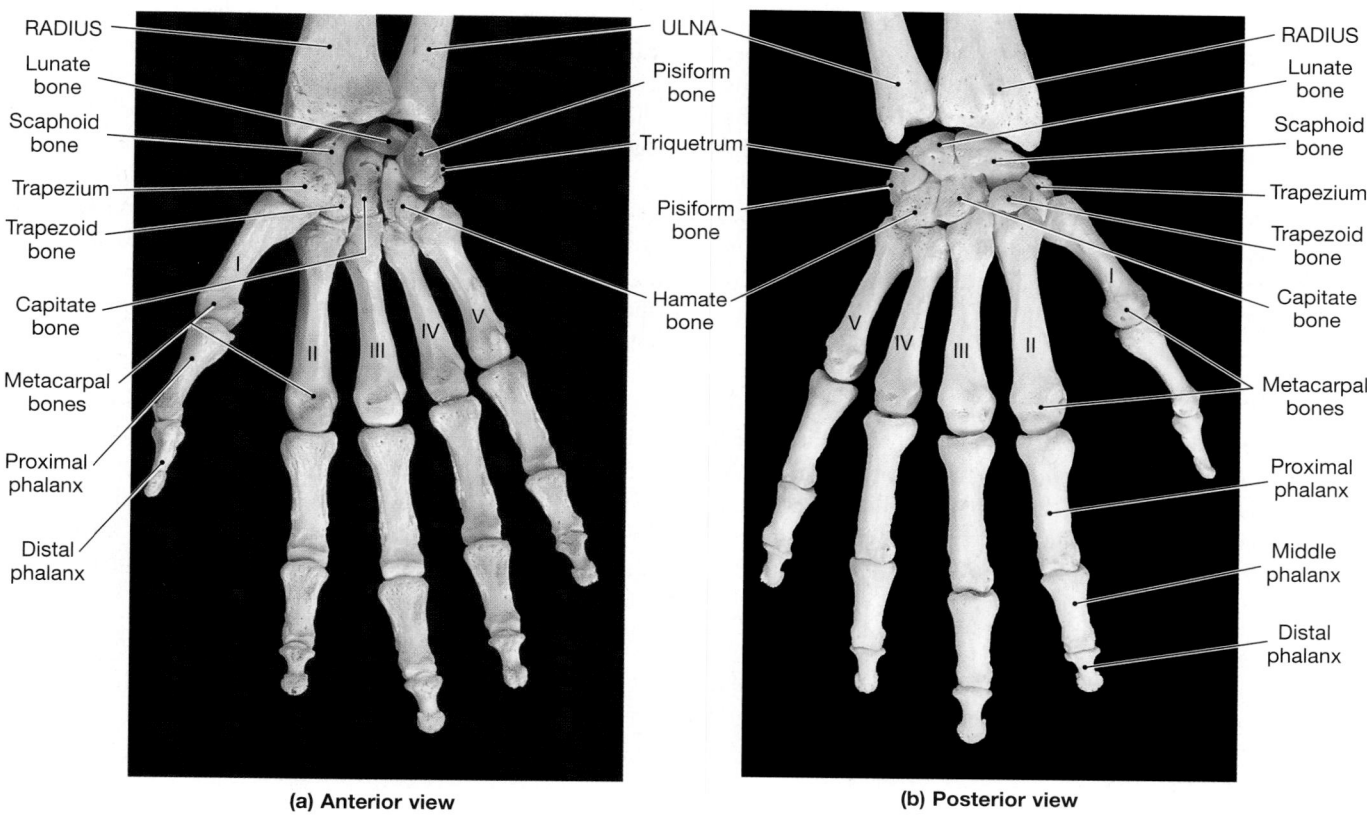

(a) Anterior view

(b) Posterior view

●**FIGURE 8–6**
Bones of the Wrist and Hand. (a) Anterior and (b) posterior views of the right hand. [ATLAS] Plate 6.5

lateral-to-medial order, the first four words stand for the proximal carpal bones (scaphoid, lunate, triquetrum, pisiform) and the last four for the distal carpal bones (trapezium, trapezoid, capitate, hamate).

The carpal bones articulate with one another at joints that permit limited sliding and twisting. Ligaments interconnect the carpal bones and help stabilize the wrist joint. The tendons of muscles that flex the fingers pass across the anterior surface of the wrist, sandwiched between the intercarpal ligaments and a broad, superficial transverse ligament called the *flexor retinaculum.* Inflammation of the connective tissues between the flexor retinaculum and the carpal bones can compress the tendons and adjacent sensory and motor nerves, producing pain and a loss of wrist mobility. This condition, called *carpal tunnel syndrome,* is discussed further in Chapter 11.

The Hand

Five **metacarpal** (met-uh-KAR-pul; *metacarpus,* hand) **bones** articulate with the distal carpal bones and support the hand (Figure 8–6●). Roman numerals I–V are used to identify the metacarpal bones, beginning with the lateral metacarpal bone that articulates with the trapezium. Hence, metacarpal I articulates with the proximal bone of the thumb.

Distally, the metacarpal bones articulate with the proximal finger bones. Each hand has 14 finger bones, or **phalanges** (fa-LAN-jēz; singular, *phalanx*). The first finger, known as the **pollex** (POL-eks), or thumb, has two phalanges (*proximal* and *distal*). Each of the other fingers has three phalanges (*proximal, middle,* and *distal*).

✓ The rounded projections on either side of the elbow are parts of which bone?

✓ Which bone of the forearm is lateral in the anatomical position?

✓ Bill accidentally fractures his first distal phalanx with a hammer. Which finger is broken?

Answers start on page Q-1

8–2 THE PELVIC GIRDLE AND LOWER LIMBS

Objectives

■ Identify the bones that form the pelvic girdle, their functions, and their superficial features.

■ Identify the bones of the lower limbs, their functions, and their superficial features.

■ Discuss structural and functional differences between the pelvis of females and that of males.

Because they must withstand the stresses involved in weight bearing and locomotion, the bones of the **pelvic girdle** are more massive than those of the pectoral girdle. For similar reasons, the bones of the lower limbs are more massive than those of the upper limbs. The pelvic girdle consists of the two hipbones. The *pelvis* is a composite structure that includes the hipbones of the appendicular skeleton and the sacrum and coccyx of the axial skeleton. ∞ pp. 235–236 Each lower limb consists of the *femur* (thigh), the *patella* (kneecap), the *tibia* and *fibula* (leg), and the bones of the ankle (*tarsal bones*) and foot.

□ THE PELVIC GIRDLE

The pelvic girdle consists of the paired hipbones, which are called the **ossa coxae**, or *innominate bones*. Each hipbone, or **os coxae** (singular), forms by the fusion of three bones: an **ilium** (IL-ē-um; plural, *ilia*), an **ischium** (IS-kē-um; plural, *ischia*), and a **pubis** (PŪ-bis) (Figure 8–7●). The ilia have a sturdy articulation with the

auricular surfaces of the sacrum, attaching the pelvic girdle to the axial skeleton. ∞ p. 235 Anteriorly, the medial surfaces of the hipbones are interconnected by a pad of fibrocartilage at a joint called the *pubic symphysis*. On the lateral surface of each hipbone, the **acetabulum** (a-se-TAB-ū-lum; *acetabulum*, vinegar cup), a socket with a curved surface, articulates with the head of the femur (Figure 8–7a●). A ridge of bone forms the lateral and superior margins of the acetabulum, which has a diameter of about 5 cm (2 in.). The anterior and inferior portion of the ridge is incomplete; the gap is called the **acetabular notch**. The smooth, C-shaped articular surface of the acetabulum is the **lunate surface**.

The ilium, ischium, and pubis meet inside the acetabulum, as though it were a pie sliced into three pieces. Superior to the acetabulum, the ilium forms a broad, curved surface that provides an extensive area for the attachment of muscles, tendons, and ligaments (Figure 8–7a●). Landmarks along the margin of the ilium include the *iliac spines*, which mark the attachment sites of important muscles and ligaments; the *gluteal lines*, which mark the attachment of

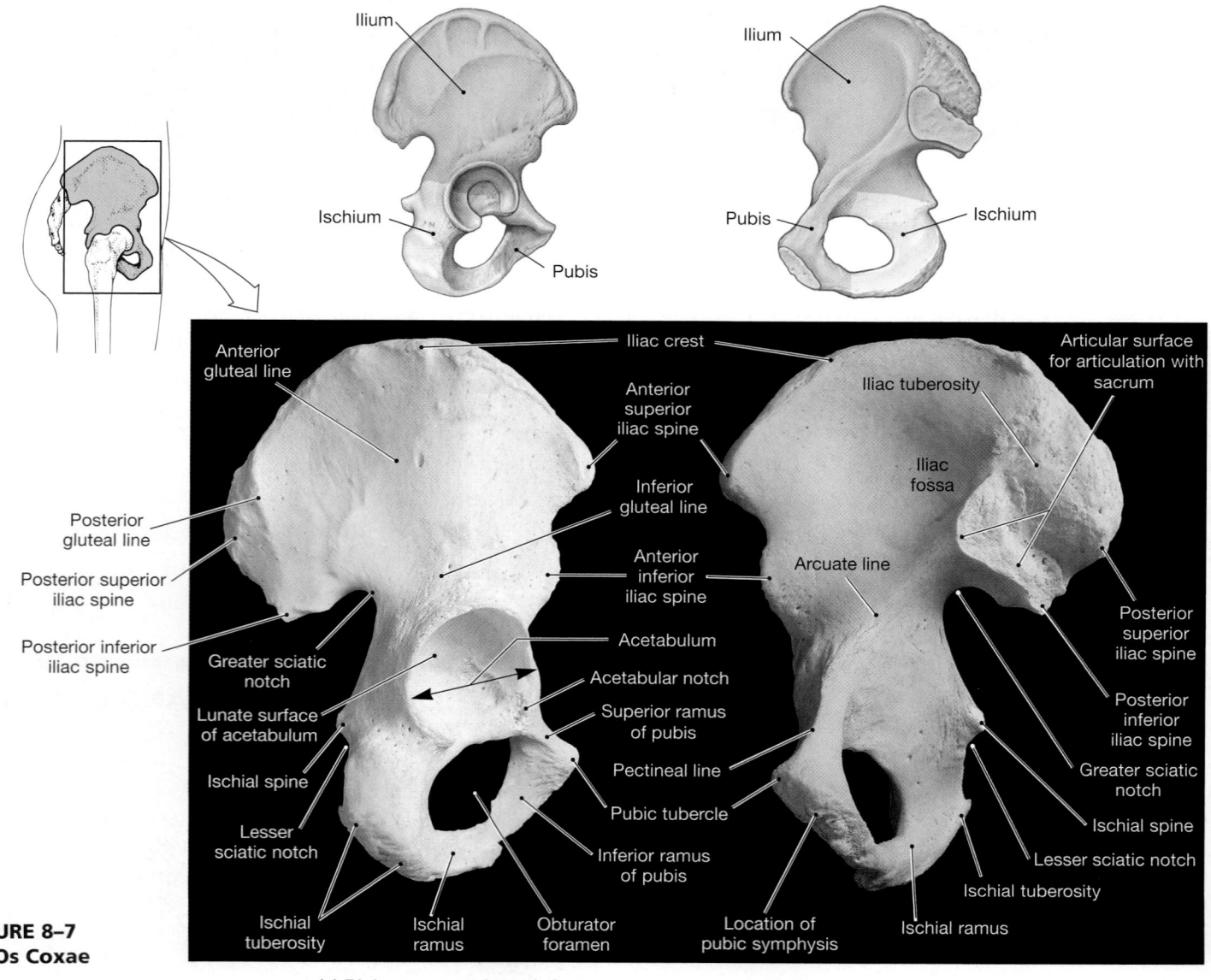

●**FIGURE 8–7**
The Os Coxae

(a) **Right os coxae, lateral view**

(b) **Right os coxae, medial view**

large hip muscles; and the **greater sciatic** (sī-A-tik) **notch**, through which a major nerve (the *sciatic nerve*) reaches the lower limb.

Near the edge of the acetabulum, the ilium fuses with the ischium. Posterior to the acetabulum, the prominent **ischial spine** projects superior to the *lesser sciatic notch*, through which blood vessels, nerves, and a small muscle pass. The **ischial tuberosity**, a roughened projection, is located at the posterior and lateral edge of the ischium. When you are seated, the ischial tuberosities bear your body's weight.

The narrow **ischial ramus** (branch) continues until it meets the **inferior ramus** of the pubis. The inferior pubic ramus extends between the ischial ramus and the *pubic tubercle*, a small, elevated area anterior and lateral to the pubic symphysis. There the inferior pubic ramus meets the **superior ramus** of the pubis, which originates near the acetabulum. The anterior, superior surface of the superior ramus bears the *pectineal line*, a ridge that ends at the pubic tubercle. The pubic and ischial rami encircle the **obturator** (OB-tū-rā-tor) **foramen**, a space that is closed by a sheet of collagen fibers whose inner and outer surfaces provide a firm base for the attachment of muscles of the hip.

The broadest part of the ilium extends between the **arcuate line**, which is continuous with the pectineal line, and the **iliac crest**. These prominent ridges mark the attachments of ligaments and muscles (Figure 8–7b●). The area between the arcuate line and the iliac crest forms a shallow depression known as the **iliac fossa**.

The concave surface of the iliac fossa helps support the abdominal organs and provides additional area for muscle attachment.

In medial view, the anterior and medial surface of the pubis contains a roughened area that marks the site of articulation with the pubis of the opposite side (Figure 8–7b●). At this articulation, the pubic symphysis, the two pubic bones are attached to a median fibrocartilage pad. Posteriorly, the **auricular surface** of the ilium articulates with the auricular surface of the sacrum at the *sacroiliac joint*. ∞ p. 235 Ligaments arising at the **iliac tuberosity**, a roughened area superior to the auricular surface, stabilize this joint.

The Pelvis

Figure 8–8● shows anterior and posterior views of the **pelvis**, which consists of the two ossa coxae, the sacrum, and the coccyx. An extensive network of ligaments connects the lateral borders of the sacrum with the iliac crest, the ischial tuberosity, the ischial spine, and the arcuate line. Other ligaments tie the ilia to the posterior lumbar vertebrae. These interconnections increase the stability of the pelvis.

The pelvis may be divided into the **false** (*greater*) **pelvis** and the **true** (*lesser*) **pelvis** (Figure 8–9a,b●). The true pelvis encloses the *pelvic cavity*, a subdivision of the abdominopelvic cavity. ∞ p. 20 The superior limit of the true pelvis is a line that extends from either side of the base of the sacrum, along the arcuate line and pectineal line to the pubic symphysis. The bony edge of the true pelvis is called the **pelvic brim**, or *linea terminalis*, and the enclosed space is the **pelvic inlet**. The false pelvis consists of the expanded, bladelike portions of each ilium superior to the pelvic brim.

The **pelvic outlet** is the opening bounded by the coccyx, the ischial tuberosities, and the inferior border of the pubic symphysis

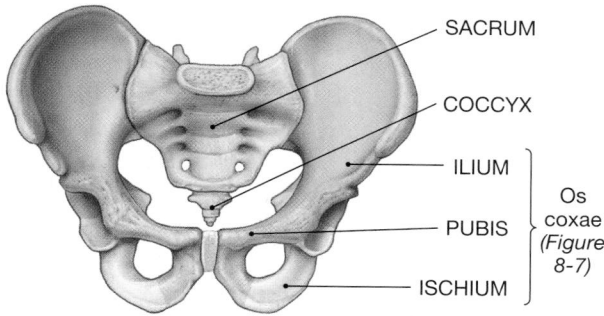

Label	
SACRUM	
COCCYX	
ILIUM	Os coxae (Figure 8-7)
PUBIS	
ISCHIUM	

●**FIGURE 8–8**
The Pelvis. The pelvis of an adult male. *(See Figure 7–22, p. 236, for a detailed view of the sacrum and coccyx.)*

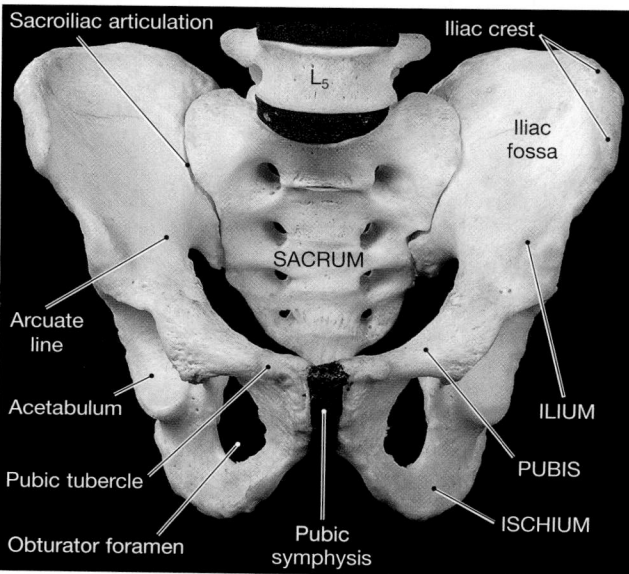

(a) Anterior view

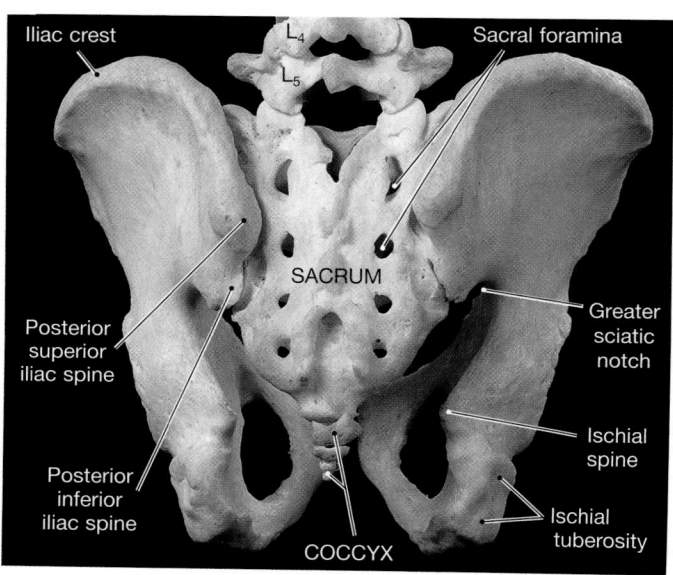

(b) Posterior view

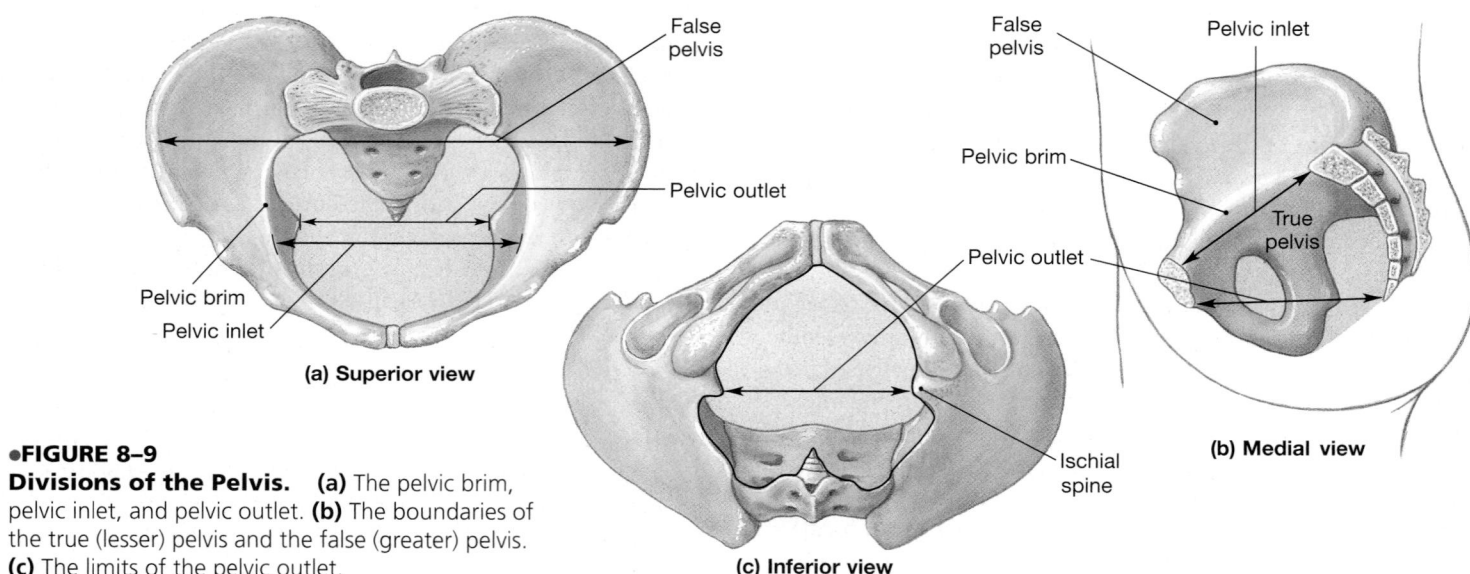

●FIGURE 8–9
Divisions of the Pelvis. **(a)** The pelvic brim, pelvic inlet, and pelvic outlet. **(b)** The boundaries of the true (lesser) pelvis and the false (greater) pelvis. **(c)** The limits of the pelvic outlet.

(Figure 8–9b,c●). The region bounded by the inferior edges of the pelvis is called the *perineum* (per-i-NĒ-um). Perineal muscles form the floor of the pelvic cavity and support the organs in the true pelvis.

The shape of the pelvis of a female is somewhat different from that of a male (Figure 8–10●). Some of the differences are the result of variations in body size and muscle mass. For example, in females, the pelvis is generally smoother, is lighter, and has less prominent markings. Females have other skeletal adaptations for childbearing, including:

- An enlarged pelvic outlet.
- Less curvature on the sacrum and coccyx, which, in males, arc into the pelvic outlet.

- A wider, more circular pelvic inlet.
- A relatively broad, low pelvis.
- Ilia that project farther laterally, but do not extend as far superior to the sacrum.
- A broader *pubic angle* (the inferior angle between the pubic bones), greater than 100°.

These adaptations are related to the support of the weight of the developing fetus and uterus and the passage of the newborn through the pelvic outlet during delivery. In addition, the hormone *relaxin*, produced during pregnancy, loosens the pubic symphysis, allowing relative movement between the hipbones that can further increase the size of the pelvic inlet and outlet.

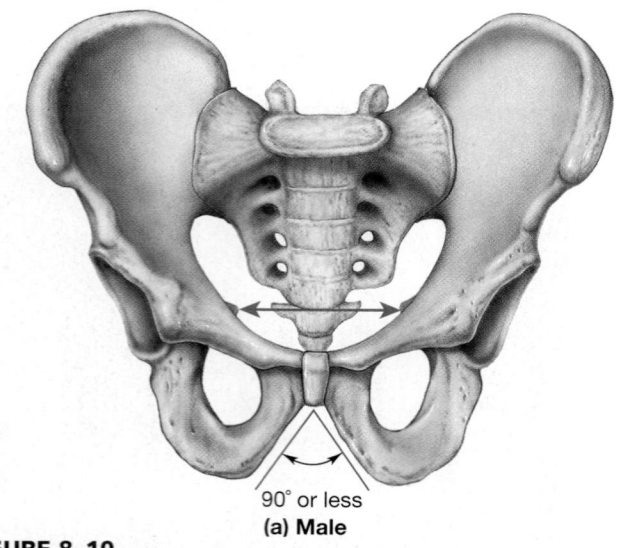

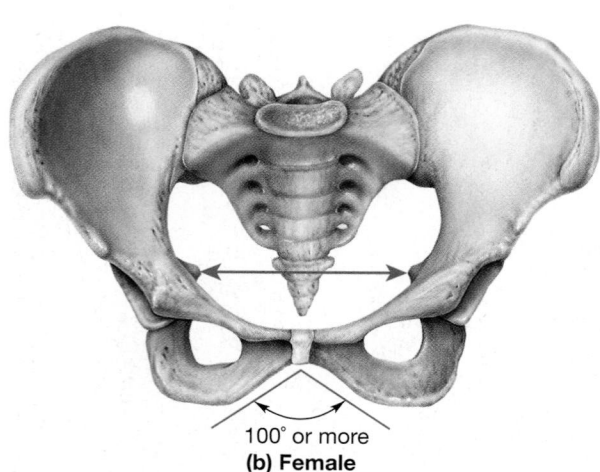

●FIGURE 8–10
Anatomical Differences in the Pelvis of a Male and a Female. Representative pelvises of a male **(a)** and a female **(b)** in anterior view. Notice the much sharper pubic angle (indicated by the black arrows) and the smaller pelvic outlet (red arrows) in the pelvis of a male as compared with that of a female.

✓ Which three bones make up the os coxae?

✓ How is the pelvis of females adapted for childbearing?

✓ When you are seated, which part of the pelvis bears your body's weight?

Answers start on page Q-1

The relationship between the pelvic girdle and the lower limb can be viewed on the **Anatomy CD-ROM**: Skeletal System/ Appendicular Dissections/Pelvis.

☐ THE LOWER LIMBS

The skeleton of each lower limb consists of a femur, a patella, a tibia and fibula, and the bones of the ankle and foot. Once again, anatomical terminology differs from common usage. In anatom-

ical terms, *leg* refers only to the distal portion of the limb, not to the entire lower limb. Thus, we will use *thigh* and *leg*, whereas people unfamiliar with proper terminology might use *upper leg* and *lower leg*, respectively.

The functional anatomy of the lower limbs is different from that of the upper limbs, primarily because the lower limbs transfer the body weight to the ground. We now examine the bones of the right lower limb.

The Femur

The **femur** is the longest and heaviest bone in the body (Figure 8–11●). It articulates with the os coxae at the hip joint and with the tibia of the leg at the knee joint. The rounded epiphysis, or **femoral head**, articulates with the pelvis at the acetabulum. A ligament attaches the acetabulum to the femur at the **fovea capitis**,

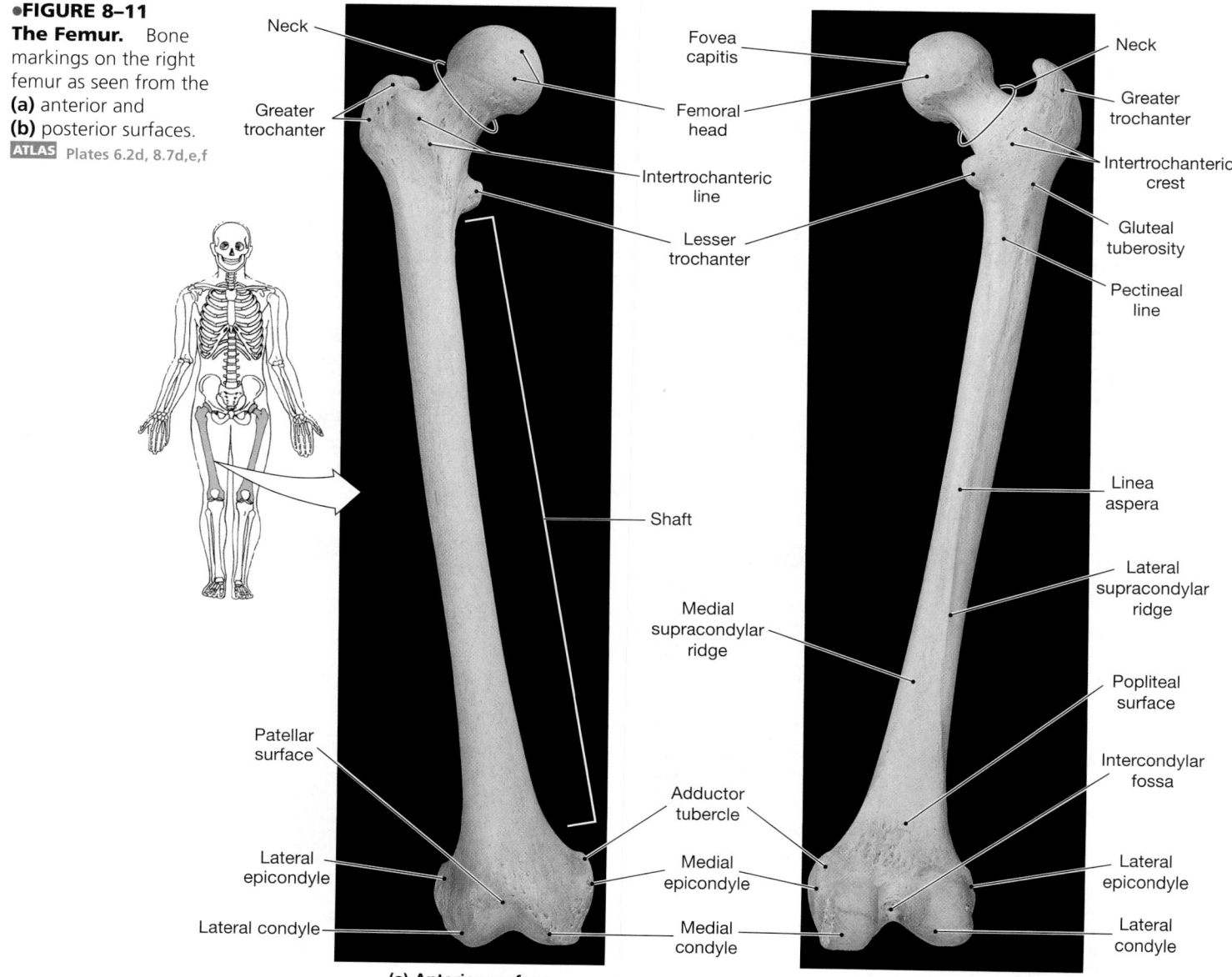

●FIGURE 8–11
The Femur. Bone markings on the right femur as seen from the **(a)** anterior and **(b)** posterior surfaces.
ATLAS Plates 6.2d, 8.7d,e,f

(a) Anterior surface

(b) Posterior surface

a small pit in the center of the femoral head. The **neck** of the femur joins the **shaft** at an angle of about 125°. The **greater** and **lesser trochanters** are large, rough projections that extend laterally from the junction of the neck and shaft. Both trochanters develop where large tendons attach to the femur. On the anterior surface of the femur, the raised **intertrochanteric** (in-ter-trō-kan-TER-ik) **line** marks the edge of the articular capsule. This line continues around to the posterior surface as the **intertrochanteric crest**.

The **linea aspera** (*aspera*, rough), a prominent elevation, runs along the center of the posterior surface of the femur, marking the attachment site of powerful hip muscles (Figure 8–11b●). As it approaches the knee joint, the linea aspera divides into a pair of ridges that continue to the **medial** and **lateral epicondyles**. These smoothly rounded projections form superior to the **medial** and **lateral condyles**, which participate in the knee joint. The two condyles are separated by a deep **intercondylar fossa**.

The medial and lateral condyles extend across the inferior surface of the femur, but the intercondylar fossa does not reach the anterior surface (Figure 8–11a●). The anterior and inferior surfaces of the two condyles are separated by the **patellar surface**, a smooth articular surface over which the patella glides.

The Patella

The **patella** is a large sesamoid bone that forms within the tendon of the *quadriceps femoris*, a group of muscles that extend (straighten) the knee. The patella has a rough, convex anterior surface and a broad **base** (Figure 8–12a●). The roughened surface reflects the attachment of the quadriceps tendon (anterior and superior surfaces) and the *patellar ligament* (anterior and inferior surfaces). The patellar ligament connects the **apex** of the patella to the tibia. The posterior surface (Figure 8–12b●) presents two concave facets for articulation with the medial and lat-

eral condyles of the femur. The patellae are cartilaginous at birth, but start to ossify after the individual begins walking, when thigh and leg movements become more powerful. Ossification usually begins at age 2 or 3 and ends at roughly the time of puberty.

Normally, the patella glides across the patellar surface. Its direction of movement is superior–inferior (up and down) but not medial–lateral (side to side). *Runner's knee*, or *patellofemoral stress syndrome*, develops from improper tracking of the patella across the patellar surface of the femur. In this syndrome, the patella is forced outside of its normal track, so that it shifts laterally; the movement is often associated with increased compression forces or with lateral muscles in the quadriceps group overpowering the medial muscles. Running on hard surfaces or slanted surfaces, such as the intertidal area of a beach or the shoulder of a road, and inadequate arch support are often responsible. The misalignment puts lateral pressure on the knee, resulting in swelling and tenderness after exercise.

The Tibia

The **tibia** (TI-bē-uh), or shinbone, is the large medial bone of the leg (Figure 8–13a●). The medial and lateral condyles of the femur articulate with the **medial** and **lateral tibial condyles** at the proximal end of the tibia. The **intercondylar eminence** is a ridge that separates the condyles (Figure 8–13b●). The anterior surface of the tibia near the condyles bears a prominent, rough **tibial tuberosity**, which you can feel through the skin. This tuberosity marks the attachment of the patellar ligament.

The **anterior margin** is a ridge that begins at the tibial tuberosity and extends distally along the anterior tibial surface. You can also easily feel the anterior margin of the tibia through the skin. As it approaches the ankle joint, the tibia broadens, and the medial border ends in the **medial malleolus** (ma-LĒ-o-lus; *malleolus*, hammer), a large process familiar to you as the medial

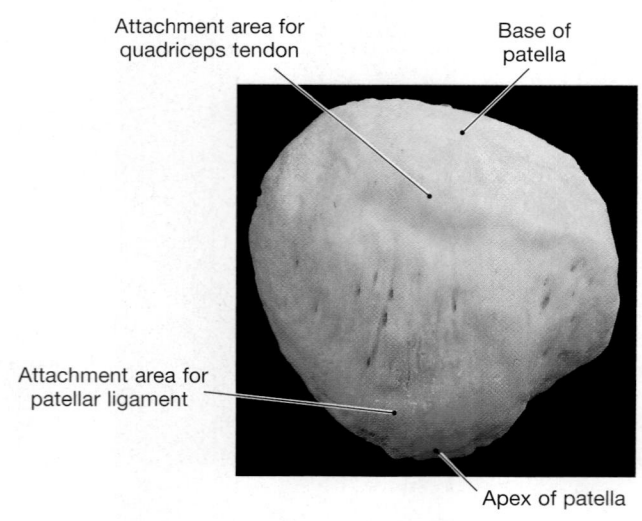

Attachment area for quadriceps tendon
Base of patella
Attachment area for patellar ligament
Apex of patella

(a) Anterior view

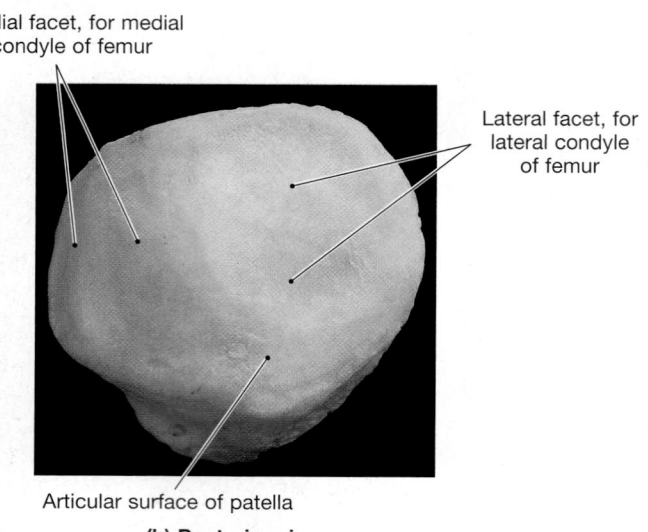

Medial facet, for medial condyle of femur
Lateral facet, for lateral condyle of femur
Articular surface of patella

(b) Posterior view

●**FIGURE 8–12**
The Right Patella

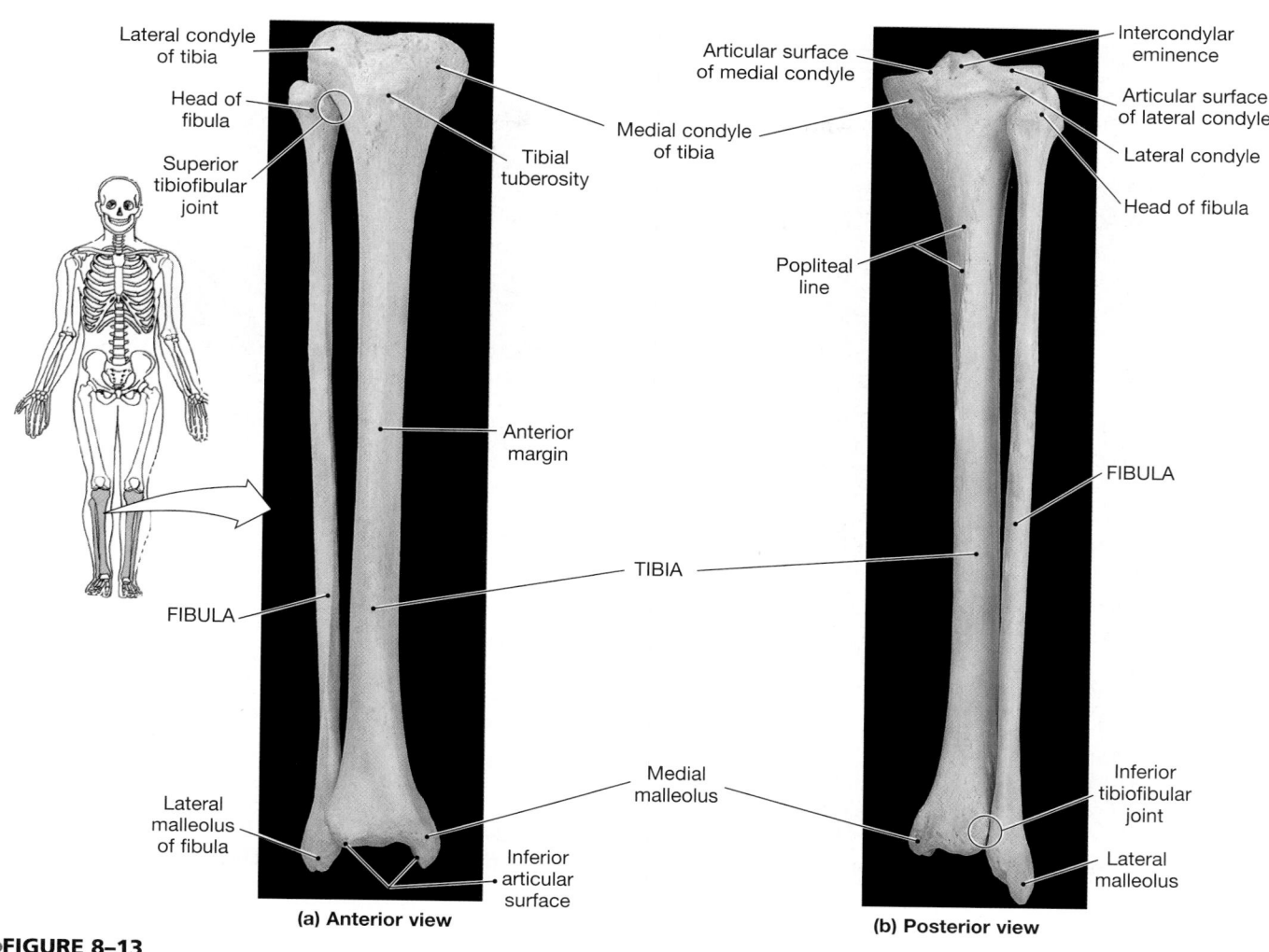

●**FIGURE 8–13**
The Tibia and Fibula. **(a)** Anterior and **(b)** posterior views of the right tibia and fibula. ATLAS Plates 6.2, 8.10

projection at the ankle. The inferior surface of the tibia articulates with the proximal bone of the ankle; the medial malleolus provides medial support for this joint.

The Fibula

The slender **fibula** (FIB-ū-luh) parallels the lateral border of the tibia (Figure 8–13a,b●). The head of the fibula articulates with the tibia. The articular facet is located on the anterior, inferior surface of the lateral tibial condyle. The medial border of the thin shaft is bound to the tibia by the **interosseous membrane**, which extends to the lateral margin of the tibia. This membrane helps stabilize the positions of the two bones and provides additional surface area for muscle attachment.

As its relatively small diameter suggests, the fibula does not help transfer weight to the ankle and foot. In fact, it does not even articulate with the femur. However, the fibula is important as a site for the attachment of muscles that move the foot and toes. In addition, the distal tip of the fibula extends lateral to the ankle joint. This fibular process, the **lateral malleolus**, provides

lateral stability to the ankle. However, forceful movement of the foot outward and backward can dislocate the ankle, breaking both the lateral malleolus of the fibula and the medial malleolus of the tibia. This injury is called a *Pott's fracture*. ⊂⊃ p. 203

The Tarsal Bones

The ankle, or *tarsus*, consists of seven **tarsal bones** (Figure 8–14●). The large **talus** transmits the weight of the body from the tibia toward the toes. The articulation between the talus and the tibia occurs across the superior and medial surfaces of the **trochlea**, a pulley-shaped articular process. The lateral surface of the trochlea articulates with the lateral malleolus of the fibula.

The **calcaneus** (kal-KĀ-nē-us), or heel bone, is the largest of the tarsal bones. When you stand normally, most of your weight is transmitted from the tibia, to the talus, to the calcaneus and then to the ground. The posterior portion of the calcaneus is a rough, knob-shaped projection. This is the attachment site for the *calcaneal tendon* (*Achilles tendon* or *calcanean tendon*), which arises at strong calf muscles. If you are standing, these muscles can lift the

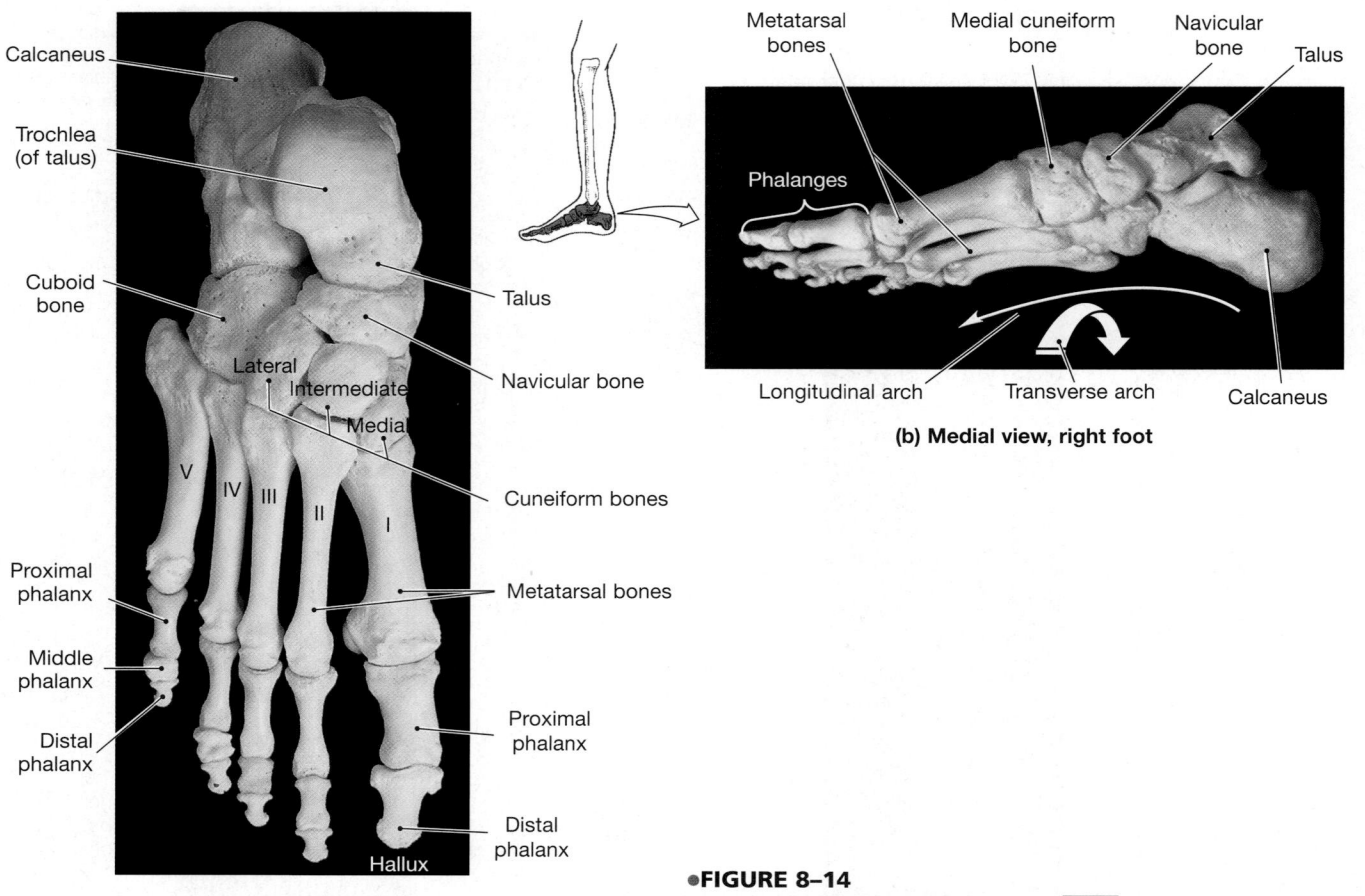

(a) Superior view, right foot

(b) Medial view, right foot

•FIGURE 8–14
Bones of the Ankle and Foot. ATLAS Plates 8.14b,d, 8.15; Scan 8

heel off the ground so that you stand on tiptoes. The superior and anterior surfaces of the calcaneus bear smooth facets for articulation with other tarsal bones.

The **cuboid bone** articulates with the anterior surface of the calcaneus. The **navicular bone** is anterior to the talus, on the medial side of the ankle. It articulates with the talus and with the three *cuneiform bones*. These are wedge-shaped bones arranged in a row, with articulations between them. They are named according to their position: **medial cuneiform**, **intermediate cuneiform**, and **lateral cuneiform**. Proximally, the cuneiform bones articulate with the anterior surface of the navicular bone. The lateral cuneiform bone also articulates with the medial surface of the cuboid bone. The distal surfaces of the cuboid bone and the cuneiform bones articulate with the metatarsal bones of the foot. To remember the names of the tarsal bones in the order presented, try the memory aid "Tom Can Control Not Much In Life."

The Foot

The **metatarsal bones** are five long bones that form the distal portion of the foot, or *metatarsus* (Figure 8–14•). The metatarsal bones are identified by Roman numerals I–V, proceeding from

medial to lateral across the sole. Proximally, metatarsal bones I–III articulate with the three cuneiform bones, and metatarsal bones IV and V articulate with the cuboid bone. Distally, each metatarsal bone articulates with a different proximal phalanx. The **phalanges**, or toe bones (Figure 8–14•), have the same anatomical organization as the fingers. The toes contain 14 phalanges. The **hallux**, or great toe, has two phalanges (*proximal* and *distal*), and the other four toes have three phalanges apiece (*proximal*, *middle*, and *distal*).

Running, while beneficial to overall health, places the foot bones under more stress than does walking. Stress fractures of the foot usually involve one of the metatarsal bones. These fractures are caused either by improper placement of the foot while running or by poor arch support. In a fitness regime that includes street running, it is essential to provide proper support for the bones of the foot. An entire running-shoe market has arisen around the amateur runner's need for good arch support.

ARCHES OF THE FOOT Weight transfer occurs along the **longitudinal arch** of the foot (Figure 8–14b•). Ligaments and tendons maintain this arch by tying the calcaneus to the distal portions of the metatarsal bones. However, the lateral, or *calcaneal*, portion of the longitudinal arch has much less curvature than the medial, *talar* portion, in part because the talar portion has considerably more elasticity. As a result, the medial plantar surface of the foot remains elevated, and the muscles, nerves, and blood vessels that supply the inferior surface are not squeezed between the metatarsal bones and the ground. In the condition known as *flatfeet*, normal arches are lost ("fall") or never form. Individuals with this condition cannot walk long distances without discomfort; hence, they are not allowed to enlist in the U.S. Army.

The elasticity of the talar portion of the longitudinal arch absorbs the shocks from sudden changes in weight loading. For example, the stresses that running or ballet dancing places on the toes are cushioned by the elasticity of this portion of the arch. The degree of curvature changes from the medial to the lateral borders of the foot, so a **transverse arch** also exists.

When you stand normally, your body weight is distributed evenly between the calcaneus and the distal ends of the metatarsal bones. The amount of weight transferred forward depends on the position of the foot and the placement of one's body weight. During *flexion* at the ankle, a movement also called *dorsiflexion*, all your body weight rests on the calcaneus; an example would be when you "dig in your heels." During *extension* at the ankle, also known as *plantar flexion*, the talus and calcaneus transfer your weight to the metatarsal bones and phalanges through the more anterior tarsal bones; this occurs when you stand on tiptoe.

The arches of the foot are usually present at birth. Sometimes, however, they fail to develop properly. In **congenital talipes equinovarus** (*clubfoot*), abnormal muscle development distorts growing bones and joints. One or both feet may be involved, and the condition can be mild, moderate, or severe. In most cases, the tibia, ankle, and foot are affected; the longitudinal arch is exaggerated, and the feet are turned medially and inverted. If both feet are involved, the soles face one another. This condition, which affects 2 in 1000 births, is roughly twice as common in boys as girls. Prompt treatment with casts or other supports in infancy helps alleviate the problem, and fewer than half the cases require surgery. Kristi Yamaguchi, an Olympic gold medalist in figure skating, was born with clubfeet. **AM** Problems with the Ankle and Foot

---

✓ The fibula neither participates in the knee joint nor bears weight. When it is fractured, however, walking becomes difficult. Why?

✓ While jumping off the back steps at his house, 10-year-old Joey lands on his right heel and breaks his foot. Which foot bone is most likely broken?

✓ Which foot bone transmits the weight of the body from the tibia toward the toes?

Answers start on page Q-1

8–3 INDIVIDUAL VARIATION IN THE SKELETAL SYSTEM

Objectives

■ Explain how study of the skeleton can reveal significant information about an individual.

■ Summarize the skeletal differences between males and females.

■ Describe briefly how the aging process affects the skeletal system.

A comprehensive study of a human skeleton can reveal important information about the individual. We can estimate a person's muscular development and muscle mass from the appearance of various ridges and from the general bone mass. Details such as the condition of the teeth or the presence of healed fractures give an indication of the individual's medical history. Two important details, gender and age, can be determined or closely estimated on the basis of measurements indicated in Tables 8–1 and 8–2. **ATLAS** EMBRYOLOGY SUMMARY 8: The Development of the Appendicular Skeleton

Table 8–1 identifies characteristic differences between the skeletons of males and females, but not every skeleton shows every feature in classic detail. Many differences, including markings on the skull, cranial capacity, and general skeletal features, reflect differences in average body size, muscle mass, and muscular strength. The general changes in the skeletal system that take place with age are summarized in Table 8–2. Note that these changes begin at age 3 months and continue throughout life. For example, the epiphyseal cartilages begin to fuse at about age 3, and degenerative changes in the normal skeletal system, such as a reduction in mineral content in the bony matrix, typically do not begin until age 30–45. The timing of epiphyseal closure is a key factor determining adult body size. Young people are often involved in sports that demand physical strength at a time when their long bones are still growing. Their conditioning programs should avoid very heavy weight training, due to the risk of crushing the epiphyseal cartilages and thus shortening the stature.

An understanding of individual variation and of the normal timing of skeletal development is important in clinical diagnosis and treatment. Several professions focus on specific aspects of skeletal form and function. Each specialty has a different perspective with its own techniques, traditions, and biases. For example, a person with back pain may consult an *orthopedist*, an *osteopath*, or a *chiropractor*.

TABLE 8–1 GENDER DIFFERENCES IN THE HUMAN SKELETON

Region and Feature	Male (compared with female)	Female (compared with male)
SKULL		
General appearance	Heavier, rougher	Lighter, smoother
Forehead	More sloping	More vertical
Sinuses	Larger	Smaller
Cranium	About 10% larger (average)	About 10% smaller
Mandible	Larger, more robust	Smaller, lighter
Teeth	Larger	Smaller
PELVIS		
General appearance	Narrower, more robust, rougher	Broader, lighter, smoother
Pelvic inlet	Heart shaped	Oval to round
Iliac fossa	Deeper	Shallower
Ilium	More vertical; Extends farther superior to sacral articulation	Less vertical; less extension superior to sacroiliac joint
Angle inferior to public symphysis	Under 90°	100° or more (see Figure 8–10, p. 252)
Acetabulum	Directed laterally	Faces slightly anteriorly as well as laterally
Obturator foramen	Oval	Triangular
Ischial spine	Points medially	Points posteriorly
Sacrum	Long, narrow triangle with pronounced sacral curvature	Broad, short triangle with less curvature
Coccyx	Points anteriorly	Points inferiorly
OTHER SKELETAL ELEMENTS		
Bone weight	Heavier	Lighter
Bone markings	More prominent	Less prominent

TABLE 8–2 AGE-RELATED CHANGES IN THE SKELETON

Region and Feature	Event(s)	Age (Years)
GENERAL SKELETON		
Bony matrix	Reduction in mineral content; increased risk of osteoporosis	Begins at ages 30–45; values differ for males versus females between ages 45 and 65; similar reductions occur in both genders after age 65
Markings	Reduction in size, roughness	Gradual reduction with increasing age and decreasing muscular strength and mass
SKULL		
Fontanels	Closure	Completed by age 2
Metopic suture	Fusion	2–8
Occipital bone	Fusion of ossification centers	1–4
Styloid process	Fusion with temporal bone	12–16
Hyoid bone	Complete ossification and fusion	25–30
Teeth	Loss of "baby teeth"; appearance of secondary dentition; eruption of permanent molars	Detailed in Chapter 24 (digestive system)
Mandible	Loss of teeth; reduction in bone mass; change in angle at mandibular notch	Accelerates in later years (60+)
VERTEBRAE		
Curvature	Development of major curves	3 months–10 years
Intervertebral discs	Reduction in size, percentage contribution to height	Accelerates in later years (60+)
LONG BONES		
Epiphyseal cartilages	Fusion	Begins about age 3; ranges vary, but general analysis permits determination of approximate age
PECTORAL AND PELVIC GIRDLES		
Epiphyses	Fusion	Relatively narrow ranges (14–16, 16–18, 22–25) that increase accuracy of age estimates

Chapter Review

SELECTED CLINICAL TERMINOLOGY

Terms Discussed in This Chapter

carpal tunnel syndrome: An inflammation of the tissues at the anterior wrist, causing compression of adjacent tendons and nerves. Symptoms are pain and a loss of wrist mobility. *(p. 249)*
congenital talipes equinovarus, or *clubfoot*:   A congenital deformity affecting one or both feet. It develops secondary to abnormalities in muscular development. *(p. 257)*
flatfeet: The loss or absence of a longitudinal arch. *(p. 257)*

AM Additional Terms Discussed in the *Applications Manual*

dancer's fracture

STUDY OUTLINE

1. The **appendicular skeleton** includes the bones of the upper and lower limbs and the pectoral and pelvic girdles, which connect the limbs to the trunk. *(Figure 8–1)*

8–1 THE PECTORAL GIRDLE AND UPPER LIMBS   p. 243

1. Each upper limb articulates with the trunk via the **pectoral girdle**, or *shoulder girdle*, which consists of the **scapulae** and **clavicles**.

THE PECTORAL GIRDLE   p. 244

2. On each side, a clavicle and scapula position the shoulder joint, help move the upper limb, and provide a base for muscle attachment. *(Figures 8–2, 8–3)*
3. Both the **coracoid process** and the **acromion** of the scapula are attached to ligaments and tendons of the shoulder joint. *(Figure 8–3)*

Pectoral girdle interaction with proximal bone of the arm: **Anatomy CD-ROM:** Skeletal System/Appendicular Dissections/Clavicle & Scapula and Humerus.

THE UPPER LIMBS   p. 246

4. The scapula articulates with the **humerus** at the shoulder *(glenohumeral)* joint. The **greater** and **lesser tubercles** are important sites of muscle attachment. *(Figure 8–4)*
5. The humerus articulates with the **radius** and **ulna**, the bones of the forearm, at the elbow joint. *(Figure 8–5)*
6. The **carpal bones** of the wrist, or **carpus**, form two rows. The distal row articulates with the five **metacarpal bones.** Four of the fingers contain three **phalanges**; the **pollex** (thumb) has only two phalanges. *(Figure 8–6)*

8–2 THE PELVIC GIRDLE AND LOWER LIMBS   p. 249

1. The bones of the **pelvic girdle** are more massive than those of the pectoral girdle.

THE PELVIC GIRDLE   p. 250

2. The pelvic girdle consists of two **ossa coxae.** Each os coxae forms through the fusion of an **ilium**, an **ischium**, and a **pubis.** *(Figure 8–7)*

3. The ilium is the largest hipbone. Inside the **acetabulum**, the ilium is fused to the ischium (posteriorly) and the pubis (anteriorly). The **pubic symphysis** limits movement between the pubic bones of the left and right hipbones. *(Figures 8–7, 8–8)*
4. The **pelvis** consists of the hipbones, the sacrum, and the coccyx. It is subdivided into the **false** (*greater*) **pelvis** and the **true** (*lesser*) **pelvis.** *(Figures 8–8 to 8–10)*

Pelvic girdle and relationship to the proximal leg: **Anatomy CD-ROM:** Skeletal System/Appendicular Dissections/Pelvis.

THE LOWER LIMBS   p. 253

5. The **femur** is the longest bone in the body. It articulates with the **tibia** at the knee joint. *(Figures 8–11, 8–13)*
6. The **patella** is a large sesamoid bone. *(Figure 8–12)*
7. The **fibula** parallels the tibia laterally. *(Figure 8–13)*
8. The **tarsus**, or ankle, has seven **tarsal bones.** *(Figure 8–14)*
9. The basic organizational pattern of the **metatarsal bones** and **phalanges** of the foot resembles that of the hand. All the toes have three phalanges, except for the **hallux**, which has two. *(Figure 8–14)*
10. When a person stands normally, most of the body weight is transferred to the **calcaneus**, and the rest is passed on to the five metatarsal bones. Weight transfer occurs along the **longitudinal arch**; there is also a **transverse arch.** *(Figure 8–14)*

8–3 INDIVIDUAL VARIATION IN THE SKELETAL SYSTEM   p. 257

1. Studying a human skeleton can reveal important information, such as the person's race, medical history, weight, gender, body size, muscle mass, and age. *(Tables 8–1, 8–2)*
2. Age-related changes and events take place in the skeletal system. These changes begin at about age 1 and continue throughout life. *(Table 8–2)*

REVIEW QUESTIONS

More assessment questions are available to you on the Companion Website. You will find Matching, Multiple Choice, True/False, and other quizzes to help further your understanding of the material covered in this chapter. To access the site, go to www.aw.com/martini.

LEVEL 1 Reviewing Facts and Terms

1. The only direct connection between the pectoral girdle and the axial skeleton is where the
   (a) clavicle articulates with the humerus
   (b) clavicle articulates with the manubrium of the sternum
   (c) os coxae articulates with the femur
   (d) vertebral column articulates with the skull

2. The presence of tubercles on bones indicates the positions of
   (a) tendons and ligaments
   (b) muscle attachment
   (c) ridges and flanges
   (d) a, b, and c are correct

3. At the glenoid cavity, the scapula articulates with the proximal end of the
   (a) humerus   (b) radius
   (c) ulna   (d) femur

4. In anatomical position, the ulna lies
   (a) medial to the radius
   (b) lateral to the radius
   (c) inferior to the radius
   (d) superior to the radius

5. The proximal carpal bones include the
   (a) trapezium, trapezoid, capitate, and hamate
   (b) scaphoid, capitate, lunate, and hamate
   (c) trapezium, triquetrum, trapezoid, and pisiform
   (d) scaphoid, lunate, triquetrum, and pisiform

6. The point of the elbow is actually the _____ of the ulna.
   (a) styloid process   (b) olecranon
   (c) coronoid process   (d) trochlear notch

7. The bones of the hand articulate distally with the
   (a) carpal bones   (b) ulna and radius
   (c) metacarpal bones   (d) phalanges

8. Each os coxae of the pelvic girdle consists of the following three fused bones:
   (a) ulna, radius, and humerus
   (b) ilium, ischium, and pubis
   (c) femur, tibia, and fibula
   (d) hamate, capitate, and trapezium

9. The large foramen between the pubic and ischial rami is the
   (a) foramen magnum   (b) suborbital foramen
   (c) acetabulum   (d) obturator foramen

10. Which of the following is an adaptation for childbearing?
   (a) inferior angle of 100° or more between the pubic bones
   (b) a relatively broad, low pelvis
   (c) less curvature of the sacrum and coccyx
   (d) a, b, and c are correct

11. The epiphysis of the femur articulates with the pelvis at the
   (a) pubic symphysis   (b) acetabulum
   (c) sciatic notch   (d) obturator foramen

12. The large medial bone of the leg is the
   (a) tibia   (b) femur
   (c) fibula   (d) humerus

13. The tarsal bone that accepts weight and distributes it to the heel or toes is the
   (a) cuneiform   (b) calcaneus
   (c) talus   (d) navicular

14. The calcaneus is the attachment site of the
   (a) calcaneal tendon   (b) muscles of the calf
   (c) talus   (d) a, b, and c are correct

15. Name the skeletal components of the pectoral girdle.

16. Define the appendicular skeleton.

17. Which bones make up the arm and forearm?

18. Which two movements are associated with the proximal radioulnar articulation?

19. What anatomical structures are responsible for the condition known as carpal tunnel syndrome?

20. Which bones constitute the lower limb?

21. Name the components of each os coxae.

22. Which seven bones make up the ankle (tarsus)?

23. Distinguish between the pollex and the hallux.

LEVEL 2 Reviewing Concepts

24. Why are injuries of the clavicle common?

25. What is the difference in skeletal structure between the pelvic girdle and the pelvis?

26. How do anatomists distinguish between the false (greater) pelvis and the true (lesser) pelvis?

27. Why is the tibia, but not the fibula, involved in the transfer of weight to the ankle and foot?

28. Describe how the arches of the foot assist in weight distribution.

29. Why would an instructor teaching self-defense advise a student to strike an assailant's clavicle in an attack?

30. Why is it necessary for the bones of the pelvic girdle to be more massive than the bones of the pectoral girdle?

LEVEL 3 Critical Thinking and Clinical Applications

31. To settle a bet, you need to measure the length of your lower limb (femur and tibia). What landmarks would you use to make the measurement?

32. While Fred, a fireman, is fighting a fire in a building, part of the ceiling collapses, and a beam strikes him on his left shoulder. He is rescued, but has a great deal of pain in his shoulder. He cannot move his arm properly, especially in the anterior direction. His clavicle is not broken, and his humerus is intact. What is the probable nature of Fred's injury?

33. Cindy is anxiously awaiting the birth of her first child. As she gets closer to term, she has an ultrasound scan. After seeing the results, her physician makes some calculations and informs Cindy that she will probably have to have a cesarean section. What clues might tell the physician that Cindy can't deliver the baby by natural childbirth?

34. Your son is just learning to ride a two-wheeled bike. In his first solo attempt, he hits the curb and falls forward over the handlebars. You watch him hit the sidewalk hands first, followed by the rest of his body. What skeletal elements might be injured in this fall? Trace the pathway of stress on his bones, and indicate the most likely area for fracture.

UNIT 2   CHAPTER   5 6 7 8 **9** 10 11

CLINICAL NOTES
■ Bursitis 267

CLINICAL DISCUSSIONS
■ Problems with Intervertebral Discs 276
■ Knee Injuries 280

ARTICULATIONS

In the last two chapters, you have become familiar with the individual bones of the skeleton. These bones provide strength, support, and protection for softer tissues of the body. However, your daily life demands more of the skeleton—it also has to facilitate and adapt to body movements. Think of your activities in a typical day: You breathe, talk, walk, sit, stand, and change positions innumerable times. In each case, your skeleton is directly involved. Because the bones of the skeleton are rigid and relatively inflexible, movements can occur only at **articulations**, or joints, where two bones interconnect. The characteristic structure of a joint determines the type of movement that may occur. Each joint reflects a compromise between the need for strength and the need for mobility.

This chapter compares the relationships between articular form and function. We will use several examples that range from the relatively immobile but very strong (the intervertebral articulations) to the highly mobile but relatively weak (the shoulder).

9–1 A CLASSIFICATION OF JOINTS

Objectives

- Contrast the major categories of joints, and explain the relationship between structure and function for each category.
- Describe the basic structure of a synovial joint, identifying possible accessory structures and their functions.

Two classification methods are used to categorize joints. The first—the one we will use in this chapter, considers the amount of movement permitted, a property known as the *range of motion* (Table 9–1). This functional category is further subdivided on the basis of the anatomical structure of the joint or the range of motion permitted:

1. An *immovable joint* is a **synarthrosis** (sin-ar-THRŌ-sis; *syn*, together + *arthros*, joint). A synarthrosis can be *fibrous* or *cartilaginous*, depending on the nature of the connection. Over time, the two bones may fuse.

2. A *slightly movable joint* is an **amphiarthrosis** (am-fē-ar-THRŌ-sis; *amphi*, on both sides). An amphiarthrosis is *fibrous* or *cartilaginous*, depending on the nature of the connection between the opposing bones.

3. A *freely movable joint* is a **diarthrosis** (dī-ar-THRŌ-sis; *dia*, through), or *synovial joint*. Diarthroses are subdivided according to the amount or range of movement permitted.

TABLE 9–1 A FUNCTIONAL CLASSIFICATION OF ARTICULATIONS

Functional Category	Structural Category	Description	Example
Synarthrosis (no movement)	**Fibrous**		
	Suture	Fibrous connections plus interlocking projections	Between the bones of the skull
	Gomphosis	Fibrous connections plus insertion in alveolar process	Between the teeth and jaws
	Cartilaginous		
	Synchondrosis	Interposition of cartilage plate	Epiphyseal cartilages
	Bony fusion		
	Synostosis	Conversion of other articular form to solid mass of bone	Portions of the skull, epiphyseal lines
Amphiarthrosis (little movement)	**Fibrous**		
	Syndesmosis	Ligamentous connection	Between the tibia and fibula
	Cartilaginous		
	Symphysis	Connection by a fibrocartilage pad	Between right and left pubic bones of pelvis; between adjacent vertebral bodies along vertebral column
Diarthrosis (free movement)	**Synovial**	Complex joint bounded by joint capsule and containing synovial fluid	Numerous; subdivided by range of movement (see Figure 9–6)
	Monaxial	Permits movement in one plane	Elbow, ankle
	Biaxial	Permits movement in two planes	Ribs, wrist
	Triaxial	Permits movement in all three planes	Shoulder, hip

The second classification scheme relies solely on the anatomical organization of the joint, without regard to the degree of movement permitted. In this framework, joints are classified as *bony, fibrous, cartilaginous,* or *synovial* (Table 9–2). The two classifications are loosely correlated. Many anatomical patterns are seen among immovable or slightly movable joints, but there is only one type of freely movable joint, and all synovial joints are diarthroses. We will use the functional classification rather than the anatomical one because we are interested primarily in how joints work.

☐ SYNARTHROSES (IMMOVABLE JOINTS)

At a synarthrosis, the bony edges are quite close together and may even interlock. These extremely strong joints are located where movement between the bones must be prevented. There are four major types of synarthrotic joints:

1. *Sutures.* A **suture** (*sutura,* a sewing together) is a synarthrotic joint located only between the bones of the skull. The edges of the bones are interlocked and bound together at the suture by dense connective tissue.

2. *Gomphoses.* A **gomphosis** (gom-FŌ-sis; *gomphosis,* a bolting together) is a synarthrosis that binds the teeth to bony sockets in the maxillary bone and mandible. The fibrous connection between a tooth and its socket is a *periodontal* (pe-rē-ō-DON-tal) *ligament* (*peri,* around + *odontos,* tooth).

3. *Synchondroses.* A **synchondrosis** (sin-kon-DRŌ-sis; *syn,* together + *chondros,* cartilage) is a rigid, cartilaginous bridge between two articulating bones. The epiphyseal cartilage is a synchondrosis that connects the diaphysis of a long bone with an epiphysis, even though the two bones are part of the same skeletal element. ⚭ p. 193 Another example is the cartilaginous connection between the ends of the first pair of vertebrosternal ribs and the sternum.

4. *Synostoses.* A **synostosis** (sin-os-TŌ-sis) is a totally rigid, immovable joint created when two separate bones fuse and the boundary between them disappears. The *metopic suture* of the frontal bone and the epiphyseal lines of mature bones are synostoses. ⚭ pp. 217, 194

☐ AMPHIARTHROSES (SLIGHTLY MOVABLE JOINTS)

An amphiarthrosis is another compromise between mobility and strength. It permits more movement than a synarthrosis, but is much stronger than a freely movable joint. The articulating bones are connected by collagen fibers or cartilage. There are two major types of amphiarthrotic joints:

1. At a **syndesmosis** (sin-dez-MŌ-sis; *desmos,* a band or ligament), bones are connected by a ligament. One example is the distal articulation between the tibia and fibula (Figure 8–13●). ⚭ p. 255

2. At a **symphysis,** or *symphyseal joint,* the articulating bones are separated by a wedge or pad of fibrocartilage. The articulation between the bodies of vertebrae (at the *intervertebral disc*) and the connection between the two pubic bones (the *pubic symphysis*) are examples of this type of joint.

TABLE 9–2 A Structural Classification of Articulations

Structure	Type	Functional Category	Example
Bony fusion	**Synostosis** *(illustrated)*	Synarthrosis	Frontal bone / Metopic suture (fusion)
Fibrous joint	**Suture** *(illustrated)* **Gomphosis** **Syndesmosis**	Synarthrosis Synarthrosis Amphiarthrosis	Skull Lambdoid suture
Cartilaginous joint	**Synchondrosis** **Symphysis** *(illustrated)*	Synarthrosis Amphiarthrosis	Pelvis Pubic symphysis
Synovial joint	**Monaxial** **Biaxial** **Triaxial** *(illustrated)*	Diarthroses	Synovial joint

□ DIARTHROSES (FREELY MOVABLE JOINTS)

Diarthroses, or **synovial** (si-NŌ-vē-ul) **joints**, permit a wider range of motion than do other types of joints. A synovial joint is surrounded by a fibrous **articular capsule**, and a *synovial membrane* lines the walls of the articular cavity. This membrane does not cover the articulating surfaces within the joint. Recall that a synovial membrane consists of areolar tissue covered by an incomplete epithelial layer. The synovial fluid that fills the joint cavity originates in the areolar tissue of the synovial membrane. ∞ p. 134 Figure 9–1● introduces the structure of synovial joints, which are typically located at the ends of long bones, such as those of the upper and lower limbs.

Articular Cartilages

Under normal conditions, the bony surfaces at a synovial joint cannot contact one another, because the articulating surfaces are covered by special **articular cartilages**. Articular cartilages resemble hyaline cartilages elsewhere in the body. However, articular cartilages have no perichondrium (the fibrous sheath described in Chapter 4), and the matrix contains more water than that of other cartilages. ∞ p. 129

The surfaces of the articular cartilages are slick and smooth. This feature alone can reduce friction during movement at the joint. However, even when pressure is applied across a joint, the smooth articular cartilages do not touch one another, because they are separated by a thin film of synovial fluid in the joint cavity (Figure 9–1a●). This fluid acts as a lubricant, minimizing friction.

Normal synovial joint function cannot continue if the articular cartilages are damaged. When such damage occurs, the matrix may begin to break down. The exposed surface will then change from a slick, smooth, gliding surface to a rough feltwork of bristly collagen fibers. This feltwork drastically increases friction at the joint.

Synovial Fluid

Synovial fluid resembles interstitial fluid, but contains a high concentration of proteoglycans secreted by fibroblasts of the synovial membrane. A thick, viscous solution with the consistency of heavy molasses, the synovial fluid within a joint has three primary functions:

1. *Lubrication.* The articular cartilages are like sponges filled with synovial fluid. When part of an articular cartilage is compressed, some of the synovial fluid is squeezed out of the cartilage and into the space between the opposing surfaces. This thin layer of fluid markedly reduces friction between moving surfaces, just as a thin film of water reduces friction between a car's tires and a highway. When the compression stops, synovial fluid is sucked back into the articular cartilages.

2. *Nutrient Distribution.* The total quantity of synovial fluid in a joint is normally less than 3 ml, even in a large joint such as the knee. This small volume of fluid must circulate continuously to provide nutrients and a waste-disposal route for the chondrocytes

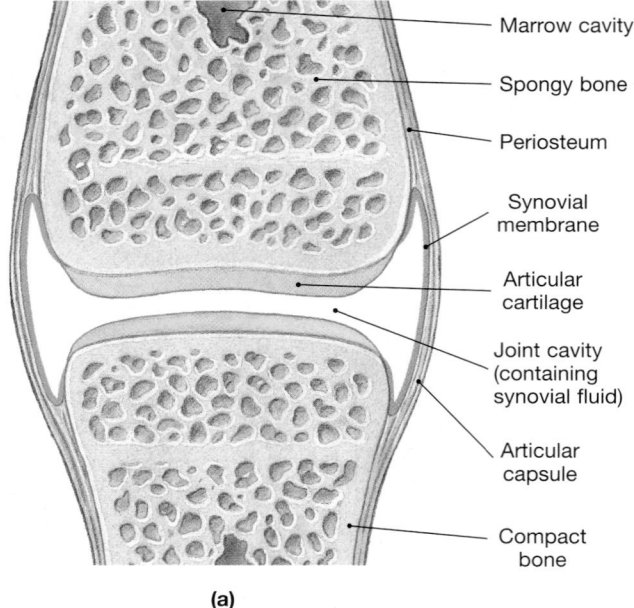

(a)

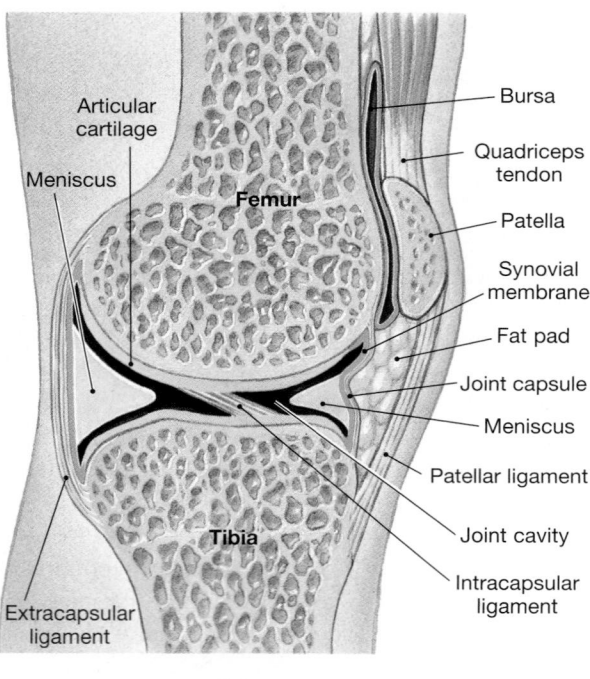

(b)

●**FIGURE 9–1**
The Structure of a Synovial Joint. (a) A diagrammatic view of a simple articulation. (b) A simplified sectional view of the knee joint.

of the articular cartilages. The synovial fluid circulates whenever the joint moves, and the compression and reexpansion of the articular cartilages pump synovial fluid into and out of the cartilage matrix. As the synovial fluid flows through the areolar tissue of the synovial membrane, waste products are absorbed and additional nutrients are obtained by diffusion across capillary walls.

3. *Shock Absorption.* Synovial fluid cushions shocks in joints that are subjected to compression. For example, your hip, knee, and

ankle joints are compressed as you walk and are more severely compressed when you jog or run. When the pressure across a joint suddenly increases, the synovial fluid lessens the shock by distributing it evenly across the articular surfaces as well as outward to the articular capsule.

Accessory Structures

Synovial joints may have a variety of accessory structures, including pads of cartilage or fat, ligaments, tendons, and bursae (Figure 9–1b●).

CARTILAGES AND FAT PADS In several joints, including the knee (Figure 9–1b●), accessory structures such as menisci and fat pads may lie between the opposing articular surfaces. A **meniscus** (men-IS-kus; a crescent; plural, *menisci*) is a pad of fibrocartilage situated between opposing bones within a synovial joint. Menisci, or *articular discs*, may subdivide a synovial cavity, channel the flow of synovial fluid, or allow for variations in the shapes of the articular surfaces.

   Fat pads are localized masses of adipose tissue covered by a layer of synovial membrane. They are commonly superficial to the joint capsule (Figure 9–1b●). Fat pads protect the articular cartilages and act as packing material for the joint. When the bones move, the fat pads fill in the spaces created as the joint cavity changes shape.

LIGAMENTS The capsule that surrounds the entire joint is continuous with the periostea of the articulating bones. **Accessory ligaments** support, strengthen, and reinforce synovial joints. *Intrinsic ligaments*, or *capsular ligaments*, are localized thickenings of the joint capsule. *Extrinsic ligaments* are separate from the joint capsule. These ligaments may be located either outside or inside the joint capsule, and are called *extracapsular* or *intracapsular* ligaments, respectively.

   Ligaments are very strong. In a **sprain**, a ligament is stretched to the point at which some of the collagen fibers are torn, but the ligament as a whole survives and the joint is not damaged. With excessive force, one of the attached bones usually breaks before the ligament tears. In general, a broken bone heals much more quickly and effectively than does a torn ligament.

TENDONS Although not part of the articulation itself, tendons passing across or around a joint may limit the joint's range of motion and provide mechanical support for it. For example, tendons associated with the muscles of the arm provide much of the bracing for the shoulder joint.

BURSAE Bursae (BUR-sē; singular, *bursa*, a pouch) are small, fluid-filled pockets in connective tissue. They contain synovial fluid and are lined by a synovial membrane. Bursae may be connected to the joint cavity or may be separate from it. They form where a tendon or ligament rubs against other tissues. Located around most synovial joints, such as the shoulder joint, bursae reduce friction and act as shock absorbers. *Synovial tendon sheaths* are tubular bursae that surround tendons where they cross bony surfaces. Bursae may also appear deep to the skin, covering a bone or lying within other connective tissues exposed to friction or pressure. Bursae that develop in abnormal locations, or because of abnormal stresses, are called *adventitious bursae.*

When bursae become inflamed, causing pain in the affected area whenever the tendon or ligament moves, the condition is called **bursitis**. Inflammation can result from the friction due to repetitive motion, pressure over the joint, irritation by chemical stimuli, infection, or trauma. Bursitis associated with repetitive motion typically occurs at the shoulder; musicians, golfers, baseball pitchers, and tennis players may develop bursitis there. The most common pressure-related bursitis is a **bunion**. Bunions form over the base of the great toe as a result of friction and distortion of the first metatarsophalangeal joint by tight shoes, especially narrow shoes with pointed toes.

We have special names for bursitis at other locations, indicating the occupations most often associated with them. In "housemaid's knee," which accompanies prolonged kneeling, the affected bursa lies between the patella and the skin. The condition of "student's elbow" is a form of bursitis that can result from propping your head up with your arm on a desk while you read your anatomy and physiology textbook.

Factors That Stabilize Joints

A joint cannot be both highly mobile and very strong. The greater the range of motion at a joint, the weaker it becomes. A synarthrosis, the strongest type of joint, permits no movement, whereas a diarthrosis, such as the shoulder, is far weaker, but permits a broad range of movement. Any mobile diarthrosis will be damaged by movement beyond its normal range of motion. Several factors are responsible for limiting the range of motion, stabilizing the joint, and reducing the chance of injury:

- The collagen fibers of the joint capsule and any accessory, extracapsular, or intracapsular ligaments.
- The shapes of the articulating surfaces and menisci, which may prevent movement in specific directions.
- The presence of other bones, skeletal muscles, or fat pads around the joint.
- Tension in tendons attached to the articulating bones. When a skeletal muscle contracts and pulls on a tendon, movement in a specific direction may be either encouraged or opposed.

The pattern of stabilizing structures varies among joints. For example, the hip joint is stabilized by the shapes of the bones (the head of the femur projects into the acetabulum), a heavy capsule, intracapsular and extracapsular ligaments, tendons, and massive muscles. It is therefore very strong and stable. In contrast, the elbow, another stable joint, gains its stability primarily from the interlocking of the articulating bones, with additional support from the capsule and associated ligaments. In general, the more stable the joint, the more restricted is its range of motion. The shoulder joint, the most mobile synovial joint, relies only on the surrounding ligaments, muscles, and tendons for stability. It is thus fairly weak.

When reinforcing structures cannot protect a joint from extreme stresses, a **dislocation**, or **luxation** (luks-Ā-shun), results. In a dislocation, the articulating surfaces are forced out of position. The displacement can damage the articular cartilages, tear ligaments, or distort the joint capsule. Although the *inside* of a joint has no pain receptors, nerves that monitor the capsule, ligaments, and tendons are quite sensitive, so dislocations are very painful. The damage accompanying a partial dislocation, or **subluxation** (sub-luks-Ā-shun), is less severe. People who are "double jointed" have joints that are weakly stabilized. Although their joints permit a greater range of motion than do those of other individuals, they are more likely to suffer partial or complete dislocations.

✓ What common characteristics do typical synarthrotic and amphiarthrotic joints share?

✓ In a newborn infant, the large bones of the skull are joined by fibrous connective tissue. Which type of joints are these? The bones later grow, interlock, and form immovable joints. Which type of joints are these?

✓ Why would improper circulation of synovial fluid lead to the degeneration of articular cartilages in the affected joint?

Answers start on page Q-1

Review the structure of synovial joints by visiting the **Companion Website:** Chapter 9/Reviewing Concepts/Labeling.

9–2 ARTICULAR FORM AND FUNCTION

Objectives

- Describe the dynamic movements of the skeleton.
- List the types of synovial joints, and discuss how the characteristic motions of each type are related to its anatomical structure.

To *understand* human movement, you must be aware of the relationship between structure and function at each articulation. To *describe* human movement, you need a frame of reference that permits accurate and precise communication. We can classify the synovial joints according to their anatomical and functional properties. To demonstrate the basis for that classification, we will use a simple model to describe the movements that occur at a typical synovial joint.

☐ DESCRIBING DYNAMIC MOTION

Take a pencil (or a pen) as your model, and stand it upright on the surface of a desk or table (Figure 9–2a). The pencil represents a bone, and the desk is an articular surface. A little imagination and a lot of twisting, pushing, and pulling will demonstrate

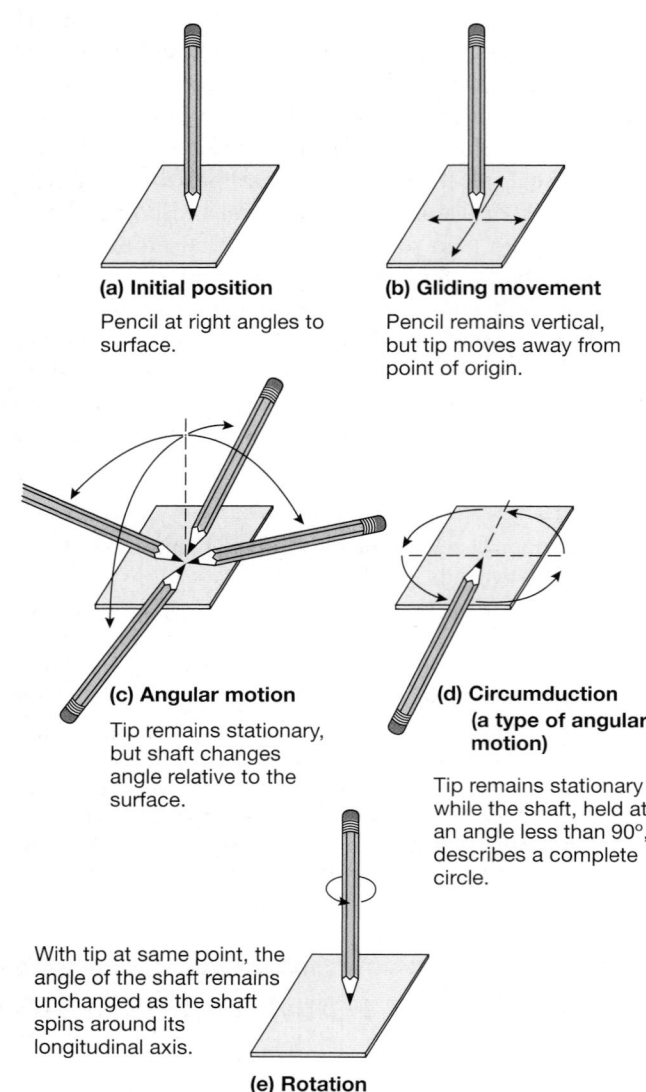

(a) Initial position
Pencil at right angles to surface.

(b) Gliding movement
Pencil remains vertical, but tip moves away from point of origin.

(c) Angular motion
Tip remains stationary, but shaft changes angle relative to the surface.

(d) Circumduction (a type of angular motion)
Tip remains stationary while the shaft, held at an angle less than 90°, describes a complete circle.

(e) Rotation
With tip at same point, the angle of the shaft remains unchanged as the shaft spins around its longitudinal axis.

•FIGURE 9–2
A Simple Model of Articular Motion

that there are only three ways to move the model. Considering them one at a time will provide a frame of reference for us to analyze complex movements:

Possible Movement 1: *The point can move.* If you hold the pencil upright, without securing the point, you can push the pencil across the surface. This kind of motion, *gliding* (Figure 9–2b●), is an example of **linear motion**. You could slide the point forward or backward, from side to side, or diagonally. However you move the pencil, the motion can be described by using two lines of reference. One line represents forward–backward motion, the other left–right movement. For example, a simple movement along one axis could be described as "forward 1 cm" or "left 2 cm." A diagonal movement could be described with both axes, as in "backward 1 cm and to the right 2.5 cm."

Possible Movement 2: *The shaft can change its angle with the surface.* With the tip held in position, you can move the free (eraser) end of the pencil forward and backward, from side to side,

or at some intermediate angle. These movements, which change the angle between the shaft and the articular surface, are examples of **angular motion** (Figure 9–2c●). We can describe such motion by the angle the shaft makes with the surface.

Any angular movement can be described with reference to the same two axes (forward–backward, left–right) and the angular change (in degrees). In one instance, however, a special term is used to describe a complex angular movement. Grasp the free end of the pencil and move it in any direction until the shaft is no longer vertical. Now swing that end through a complete circle (Figure 9–2d●). This movement, which corresponds to the path of your arm when you draw a large circle on a chalkboard, is very difficult to describe. Anatomists avoid the problem by using a special term, **circumduction** (sir-kum-DUK-shun; *circum*, around), for this type of angular motion.

Possible Movement 3: *The shaft can rotate.* If you prevent movement of the base of the pencil and keep the shaft vertical, you can spin the shaft around its longitudinal axis. This movement is called **rotation** (Figure 9–2e●). Several articulations permit partial rotation, but none can rotate freely. Such a movement would hopelessly tangle the blood vessels, nerves, and muscles that cross the joint.

An articulation that permits movement along only one axis is called **monaxial** (mon-AKS-ē-ul). In the pencil model, if an articulation permits only angular movement in the forward–backward plane or prevents any movement other than rotation around its longitudinal axis, it is monaxial. If movement can occur along two axes, the articulation is **biaxial** (bī-AKS-ē-ul). If the pencil could undergo angular motion in the forward–backward *and* left–right planes, but not rotation, it would be biaxial. The most mobile joints permit a combination of angular movement and rotation. These joints are said to be **triaxial** (trī-AKS-ē-ul).

Joints that permit gliding allow only small amounts of movement. These joints may be called *nonaxial,* because they permit only small sliding movements, or *multiaxial,* because sliding may occur in any direction.

☐ TYPES OF MOVEMENTS

In descriptions of motion at synovial joints, phrases such as "bend the leg" or "raise the arm" are not sufficiently precise. Anatomists use descriptive terms that have specific meanings. We will consider these motions with reference to the basic categories of movement discussed previously: gliding, angular motion, and rotation.

Linear Motion (Gliding)

In **gliding**, two opposing surfaces slide past one another, as in possible movement 1. Gliding occurs between the surfaces of articulating carpal bones, between tarsal bones, and between the clavicles and the sternum. The movement can occur in almost any direction, but the amount of movement is slight, and rotation is generally prevented by the capsule and associated ligaments.

Angular Motion

Examples of angular motion include *flexion, extension, abduction, adduction,* and *circumduction.* Descriptions of these movements are based on reference to an individual in the anatomical position.
∞ p. 15

FLEXION AND EXTENSION Flexion (FLEK-shun) is movement in the anterior–posterior plane that reduces the angle between the articulating elements. **Extension** occurs in the same plane, but it increases the angle between articulating elements (Figure 9–3a●). These terms are usually applied to the movements of the long bones of the limbs, but they are also used to describe movements of the axial skeleton. For example, when you bring your head toward your chest, you flex the intervertebral joints of the neck. When you bend down to touch your toes, you flex the intervertebral joints of the spine. Extension reverses these movements, returning you to the anatomical position. When a person is in the anatomical position, all of the major joints of the axial and appendicular skeletons other than the ankle are at full extension. (Special terms used to describe movements of the ankle joint are introduced shortly.)

Flexion of the shoulder joint or hip joint moves the limbs anteriorly, whereas extension moves them posteriorly. Flexion of the wrist joint moves the hand anteriorly, and extension moves it posteriorly. In each of these examples, extension can be continued past the anatomical position. Extension past the anatomical position is called **hyperextension** (Figure 9–3a●). When you hyperextend your neck, you can gaze at the ceiling. Hyperextension of many joints, such as the elbow or the knee, is prevented by ligaments, bony processes, or soft tissues.

ABDUCTION AND ADDUCTION Abduction (*ab,* from) is movement *away from the longitudinal axis of the body* in the frontal plane (Figure 9–3b●). For example, swinging the upper limb to the side is abduction of the limb. Moving it back to the anatomical position constitutes **adduction** (*ad,* to). Adduction of the wrist moves the heel of the hand and fingers *toward* the body, whereas abduction moves them farther away. Spreading the fingers or toes apart abducts them, because they move *away* from a central digit (Figure 9–3c●). Bringing them together constitutes adduction. (Fingers move toward or away from the middle finger; toes move toward or away from the second toe.) Abduction and adduction always refer to movements of the appendicular skeleton, not to those of the axial skeleton.

CIRCUMDUCTION We introduced a special type of angular motion, circumduction, in our model. Moving your arm in a loop is circumduction (Figure 9–3d●), as when you draw a large circle on a chalkboard. Your hand moves in a circle, but your arm does not rotate.

Rotation

Rotational movements are also described with reference to a figure in the anatomical position. Rotation of the head may involve **left rotation** or **right rotation** (Figure 9–4a,b●). Limb rotation may be described by reference to the longitudinal axis of the trunk. During **medial rotation**, also known as *internal rotation* or *inward rotation,* the anterior surface of a limb turns toward the long axis of the trunk. The reverse movement is called **lateral rotation**, *external rotation,* or *outward rotation* (Figure 9–4b●).

The proximal articulation between the radius and the ulna permits rotation of the radial head. As the shaft of the radius rotates, the distal epiphysis of the radius rolls across the anterior surface of the ulna. This movement, called **pronation** (prō-NĀ-shun), turns the wrist and hand from palm facing front to palm facing back (Figure 9–4c●). The opposing movement, in which the palm is turned anteriorly, is **supination** (soo-pi-NĀ-shun). The forearm is supinated in the anatomical position. This view makes it easier to follow the path of the blood vessels, nerves, and tendons, which rotate with the radius during pronation.

Special Movements

Several special terms apply to specific articulations or unusual types of movement (Figure 9–5●):

- **Inversion** (*in,* into + *vertere,* to turn) is a twisting motion of the foot that turns the sole inward, elevating the medial edge of the sole. The opposite movement is called **eversion** (ē-VER-zhun; *e,* out).

- **Dorsiflexion** is flexion at the ankle joint and elevation of the sole, as when you dig in your heel. **Plantar flexion** (*planta,* sole), the opposite movement, extends the ankle joint and elevates the heel, as when you stand on tiptoe. However, it is also acceptable (and simpler) to use "flexion and extension at the ankle," rather than "dorsiflexion and plantar flexion."

- **Opposition** is movement of the thumb toward the surface of the palm or the pads of other fingers. Opposition permits you to grasp and hold objects between your thumb and palm. It involves movement at the first carpometacarpal and metacarpophalangeal joints. Flexion at the 5th metacarpophalangeal joint can assist this movement.

- **Protraction** entails moving a part of the body anteriorly in the horizontal plane. **Retraction** is the reverse movement. You protract your jaw when you grasp your upper lip with your lower teeth, and you protract your clavicles when you cross your arms.

- **Elevation** and **depression** occur when a structure moves in a superior or an inferior direction, respectively. You depress your mandible when you open your mouth; you elevate your mandible as you close your mouth. Another familiar elevation occurs when you shrug your shoulders.

- **Lateral flexion** occurs when your vertebral column bends to the side. This movement is most pronounced in the cervical and thoracic regions.

□ A STRUCTURAL CLASSIFICATION OF SYNOVIAL JOINTS

Synovial joints are described as *gliding, hinge, pivot, ellipsoidal, saddle,* or *ball-and-socket joints* on the basis of the shapes of the articulating surfaces. Each type of joint permits a different type

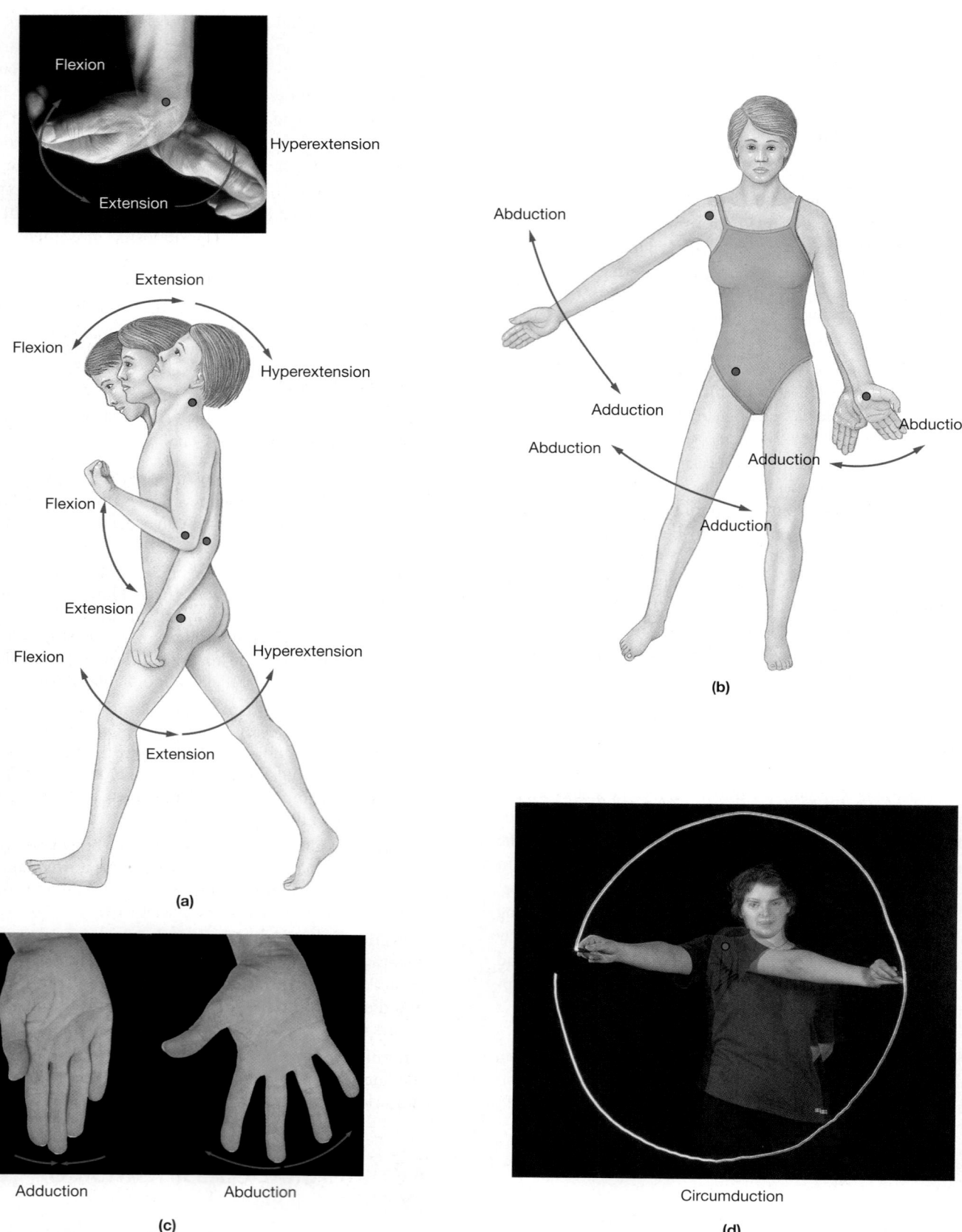

Flexion

Hyperextension

Extension

Extension

Flexion

Hyperextension

Flexion

Extension

Flexion

Hyperextension

Extension

(a)

Abduction

Adduction

Abduction

Adduction

Abduction

Adduction

(b)

Adduction   Abduction

(c)

Circumduction

(d)

•FIGURE 9–3
Angular Movements. The red dots indicate the locations of the joints involved in the movement illustrated.

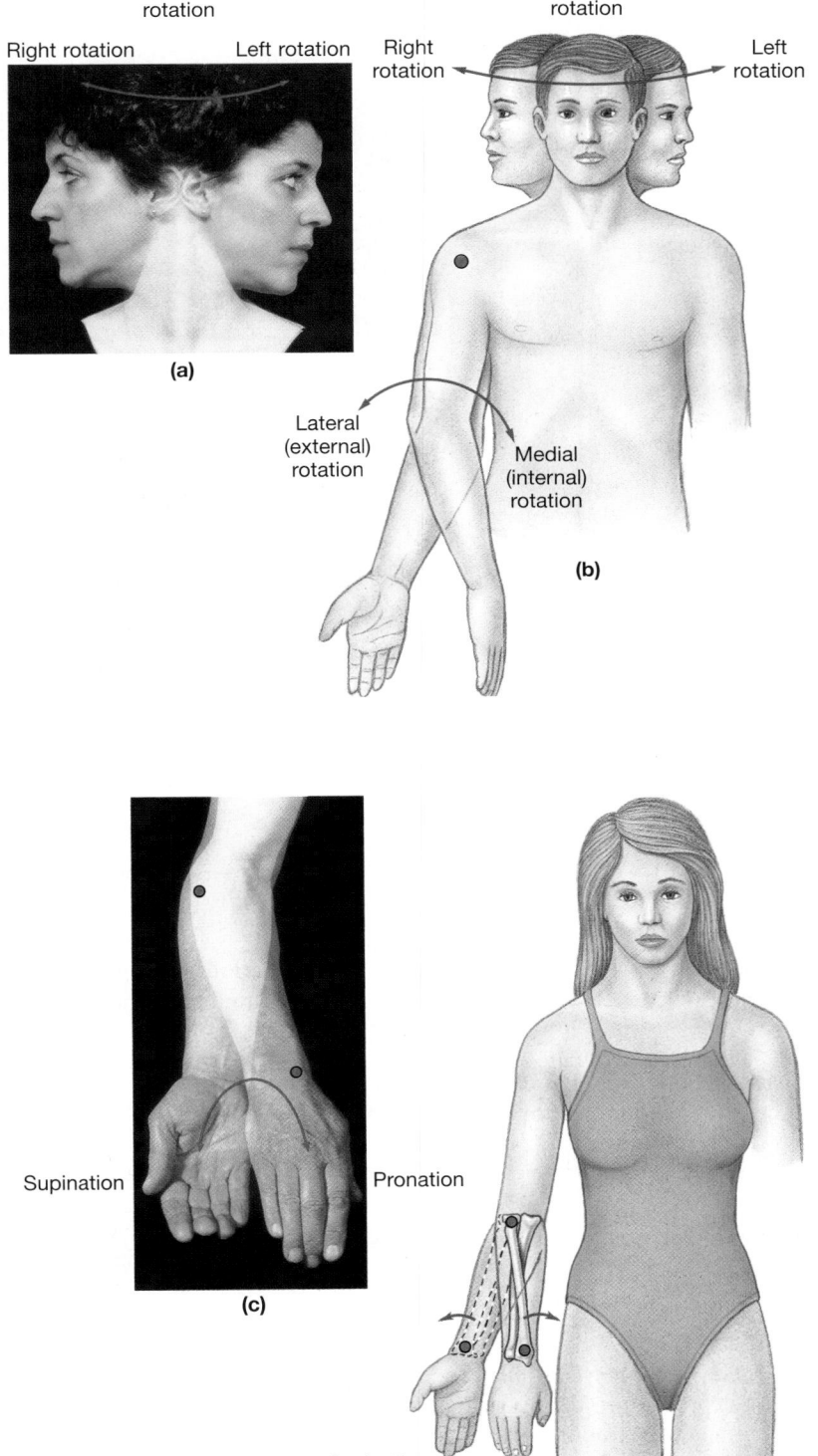

•FIGURE 9–4
Rotational Movements

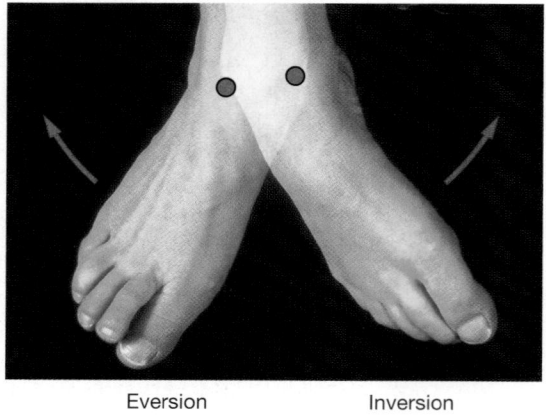

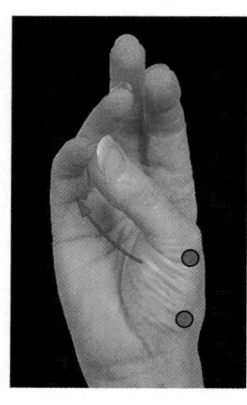

Eversion   Inversion

Opposition

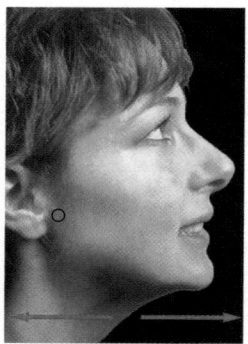

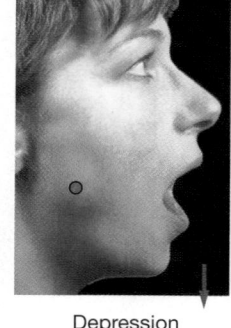

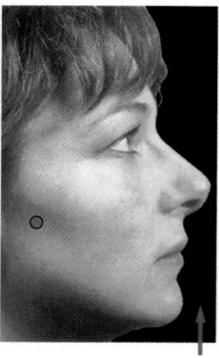

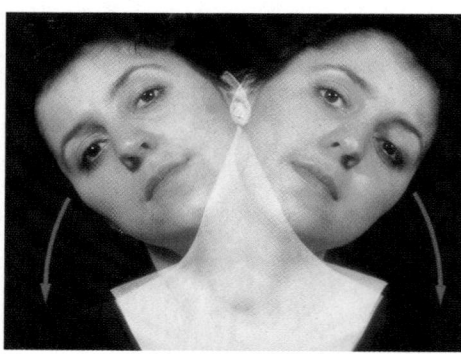

Retraction   Protraction

Depression   Elevation

Lateral flexion

●**FIGURE 9–5**
Special Movements

and range of motion. Figure 9–6● lists the structural categories and the types of movement each permits.

- **Gliding joints** (Figure 9–6a●), also called *planar joints*, have flattened or slightly curved faces. The relatively flat articular surfaces slide across one another, but the amount of movement is very slight. Although rotation is theoretically possible at such a joint, ligaments usually prevent or restrict such movement.

- **Hinge joints** (Figure 9–6b●) permit angular motion in a single plane, like the opening and closing of a door.

- **Pivot joints** (Figure 9–6c●) also are monaxial, but they permit only rotation.

- In an **ellipsoidal joint** (Figure 9–6d●), or *condyloid joint*, an oval articular face nestles within a depression in the opposing surface. With such an arrangement, angular motion occurs in two planes: along or across the length of the oval.

- **Saddle joints** (Figure 9–6e●), or *sellaris joints*, have saddle-shaped articular faces. Each face is concave on one axis and convex on the other, and the opposing faces nest together. This arrangement permits angular motion, including circumduction, but prevents rotation.

- In a **ball-and-socket joint** (Figure 9–6f●), the round head of one bone rests within a cup-shaped depression in another. All

combinations of angular and rotational movements, including circumduction and rotation, can be performed at ball-and-socket joints.

✓ When you do jumping jacks, which lower limb movements are necessary?

✓ Which movements are associated with hinge joints?

Answers start on page Q-1

Review the movements associated with synovial joints by visiting the **Companion Website:** Chapter 9/Reviewing Concepts/Multiple Choice.

9–3 REPRESENTATIVE ARTICULATIONS

Objectives

- Describe the articulations between the vertebrae of the vertebral column.

- Describe the structure and function of the shoulder, elbow, hip, and knee joints.

- Explain the relationship between joint strength and mobility, using specific examples.

Types of Synovial Joints		Movement	Examples
Gliding joint Clavicle Manubrium		Slight nonaxial or multiaxial	• Acromioclavicular and claviculosternal joints • Intercarpal and intertarsal joints • Vertebrocostal joints • Sacroiliac joints
Hinge joint Humerus Ulna		Monaxial	• Elbow joint • Knee joint • Ankle joint • Interphalangeal joint
Pivot joint Atlas Axis		Monaxial (rotation)	• Atlas/axis • Proximal radioulnar joint
Ellipsoidal joint Scaphoid bone Ulna Radius		Biaxial	• Radiocarpal joint • Metacarpophalangeal joints 2–5 • Metatarsophalangeal joints
Saddle joint III II Metacarpal bone of thumb Trapezium		Biaxial	• First carpometacarpal joint
Ball-and-socket joint Humerus Scapula		Triaxial	• Shoulder joint • Hip joint

SUMMARY TABLE 9–3 ARTICULATIONS OF THE AXIAL SKELETON

Element	Joint	Type of Articulation	Movements
SKULL			
Cranial and facial bones of skull	Various	Synarthroses (suture or synostosis)	None
Maxillary bone/teeth and mandible/teeth	Alveolar	Synarthrosis (gomphosis)	None
Temporal bone/mandible	Temporomandibular	Combined gliding joint and hinge diarthrosis	Elevation, depression, and lateral gliding
VERTEBRAL COLUMN			
Occipital bone/atlas	Atlanto-occipital	Ellipsoidal diarthrosis	Flexion/extension
Atlas/axis	Atlanto-axial	Pivot diarthrosis	Rotation
Other vertebral elements	Intervertebral (*between vertebral bodies*)	Amphiarthrosis (symphysis)	Slight movement
	Intervertebral (*between articular processes*)	Gliding diarthrosis	Slight rotation and flexion/extension
L_5/sacrum	Between L_5 body and sacral body	Amphiarthrosis (symphysis)	Slight movement
	Between inferior articular processes of L_5 and articular processes of sacrum	Gliding diarthrosis	Slight flexion/extension
Sacrum/os coxae	Sacroiliac	Gliding diarthrosis	Slight movement
Sacrum/coccyx	Sacrococcygeal	Gliding diarthrosis (*may become fused*)	Slight movement
Coccygeal bones		Synarthrosis (synostosis)	No movement
THORACIC CAGE			
Bodies of T_1–T_{12} and heads of ribs	Costovertebral	Gliding diarthrosis	Slight movement
Transverse processes of T_1–T_{10}	Costovertebral	Gliding diarthrosis	Slight movement
Ribs and costal cartilages		Synarthrosis (synchondrosis)	No movement
Sternum and first costal cartilage	Sternocostal (1st)	Synarthrosis (synchondrosis)	No movement
Sternum and costal cartilages 2–7	Sternocostal (2nd–7th)	Gliding diarthrosis*	Slight movement

*Commonly converts to synchondrosis in elderly individuals.

In this section, we consider articulations that demonstrate important functional principles. We consider only representative articulations here, but Tables 9–3 and 9–4 (pp. 274 and 282) summarize data on most articulations in the body.

☐ INTERVERTEBRAL ARTICULATIONS

The articulations between the superior and inferior articular processes of adjacent vertebrae are gliding joints that permit small movements associated with flexion and rotation of the vertebral column. Little gliding occurs between adjacent ver-

tebral bodies. From axis to sacrum, the vertebrae are separated and cushioned by pads of fibrocartilage called **intervertebral discs** (Figure 9–7●). Thus, the bodies of vertebrae form symphyseal joints. Intervertebral discs and symphyseal joints are not found in the sacrum or coccyx, where vertebrae have fused, or between the first and second cervical vertebrae. The first cervical vertebra has no vertebral body and no intervertebral disc; the only articulation between the first two cervical vertebrae is a pivot joint that permits much more rotation than do the symphyseal joints between other cervical vertebrae.

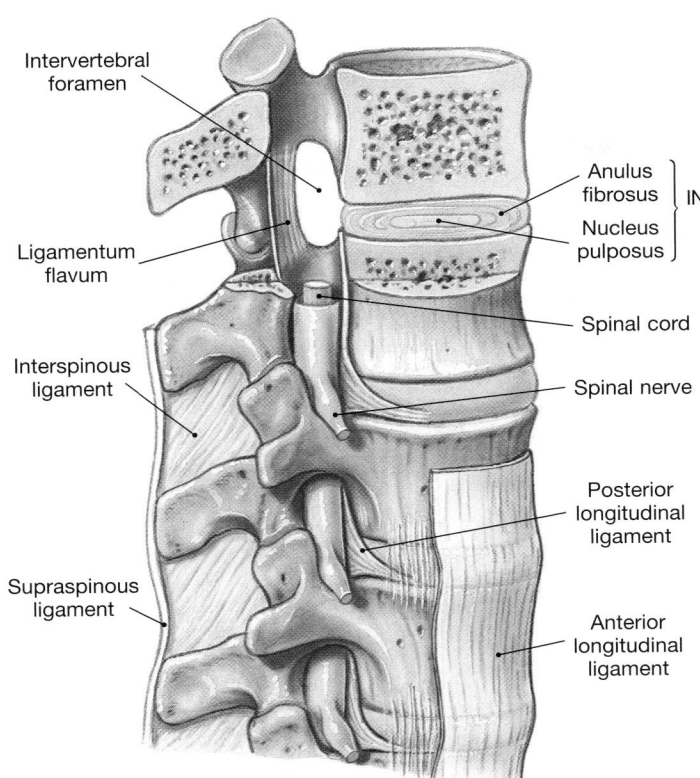

Intervertebral foramen

Ligamentum flavum

Interspinous ligament

Supraspinous ligament

Anulus fibrosus
Nucleus pulposus
INTERVERTEBRAL DISC

Spinal cord

Spinal nerve

Posterior longitudinal ligament

Anterior longitudinal ligament

•FIGURE 9–7
Intervertebral Articulations. ATLAS Scans 3b,c,d

The Intervertebral Discs

Each intervertebral disc has a tough outer layer of fibrocartilage, the **anulus fibrosus** (AN-ū-lus fī-BRŌ-sus). The collagen fibers of this layer attach the disc to the bodies of adjacent vertebrae. The anulus fibrosus surrounds the **nucleus pulposus** (pul-PŌ-sus), a soft, elastic, and gelatinous core (Figure 9–7a•). The nucleus pulposus gives the disc resiliency and lets it absorb shocks.

Movement of the vertebral column compresses the nucleus pulposus and displaces it in the opposite direction. This displacement permits smooth gliding movements by each vertebra, while maintaining their alignment. The discs make a significant contribution to an individual's height: They account for roughly one-quarter the length of the vertebral column superior to the sacrum. As we grow older, the water content of the nucleus pulposus in each disc decreases. The discs gradually become less effective as a cushion, and the chances of vertebral injury increase. Loss of water by the discs also causes shortening of the vertebral column, accounting for the characteristic decrease in height with advancing age.

Intervertebral Ligaments

Numerous ligaments are attached to the bodies and processes of all vertebrae, binding them together and stabilizing the vertebral col-

umn (Figure 9–7a,b•). Ligaments interconnecting adjacent vertebrae include:

- The *anterior longitudinal ligament*, which connects the anterior surfaces of adjacent vertebral bodies.
- The *posterior longitudinal ligament*, which parallels the anterior longitudinal ligament and connects the posterior surfaces of adjacent vertebral bodies.
- The *ligamentum flavum* (plural, *ligamenta flava*), which connects the laminae of adjacent vertebrae.
- The *interspinous ligament*, which connects the spinous processes of adjacent vertebrae.
- The *supraspinous ligament*, which interconnects the tips of the spinous processes from C_7 to the sacrum. The *ligamentum nuchae*, which extends from vertebra C_7 to the base of the skull, is continuous with the supraspinous ligament. ∞ p. 233

Vertebral Movements

The following movements can occur across the intervertebral joints of the vertebral column: (1) flexion, or bending anteriorly; (2) extension, or bending posteriorly; (3) lateral flexion, or bending laterally; and (4) rotation. Table 9–3 summarizes information about other articulations of the axial skeleton.

✓ Which regions of the vertebral column do not contain intervertebral discs? Why is the absence of discs significant?

✓ Which vertebral movements are involved in (a) bending forward, (b) bending to the side, and (c) moving the head to signify "no"?

Answers start on page Q-1

☐ THE SHOULDER JOINT

The shoulder joint, or *glenohumeral joint*, permits the greatest range of motion of any joint. Because it is also the most frequently dislocated joint, it provides an excellent demonstration of the principle that stability must be sacrificed to obtain mobility.

This joint is a ball-and-socket diarthrosis formed by the articulation of the head of the humerus with the glenoid cavity of the scapula (Figure 9–9a•). The surface of the glenoid cavity is covered by a fibrocartilaginous **glenoid labrum** (*labrum*, lip or edge),

Problems with Intervertebral Discs

An intervertebral disc compressed beyond its normal limits can be temporarily or permanently damaged. If the posterior longitudinal ligaments are weakened, as often occurs with advancing age, the compressed nucleus pulposus may distort the anulus fibrosus, forcing it partway into the vertebral canal. This condition is called a **slipped disc** (Figure 9–8a●), although the disc does not actually slip. A disc problem can occur at any age as a result of an accidental injury, such as a hard fall or a "whiplash" injury to the neck. However, with advanced age, the supporting ligaments may become so weak that the problem could occur without warning or apparent cause.

If the nucleus pulposus breaks through the anulus fibrosus, it too may protrude into the vertebral canal. This condition is called a **herniated disc** (Figure 9–8b●). When a disc herniates, sensory nerves are distorted and the protruding mass can also compress the nerves passing through the adjacent intervertebral foramen. The result is severe backache, abnormal posture, a burning or tingling sensation in the lower back and lower limbs, and, in some cases, a partial loss of control over skeletal muscles innervated by the compressed nerves. **AM** Diagnosing and Treating Disc Problems

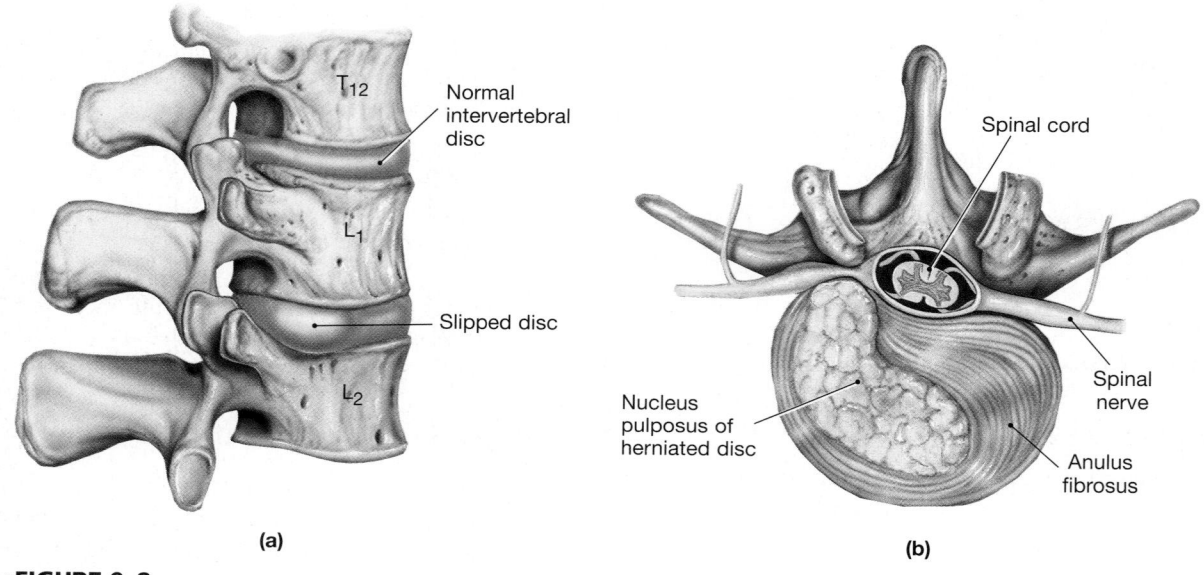

●**FIGURE 9–8**
Damage to the Intervertebral Discs.
(a) A lateral view of the lumbar region of the spinal column, showing a distorted intervertebral disc (a "slipped" disc).
(b) A sectional view through a herniated disc, showing the release of the nucleus pulposus and its effect on the spinal cord and adjacent nerves.

which extends past the bony rim and deepens the socket (Figure 9–9b●). The relatively loose articular capsule extends from the scapula, proximal to the glenoid labrum, to the anatomical neck of the humerus. Somewhat oversized, the articular capsule permits an extensive range of motion. The bones of the pectoral girdle provide some stability to the superior surface, because the acromion and coracoid process project laterally superior to the head of the humerus. However, most of the stability at this joint is provided by the surrounding skeletal muscles, with help from their associated tendons and various ligaments. Bursae reduce friction between the tendons and other tissues at the joint.

The major ligaments that help stabilize the shoulder joint are the *glenohumeral, coracohumeral, coracoacromial, coracoclavicular,* and *acromioclavicular ligaments.* The acromioclavicular ligament

reinforces the capsule of the acromioclavicular joint and supports the superior surface of the shoulder. A **shoulder separation** is a relatively common injury involving partial or complete dislocation of the acromioclavicular joint. This injury can result from a blow to the superior surface of the shoulder. The acromion is forcibly depressed, but the clavicle is held back by strong muscles.

The muscles that move the humerus do more to stabilize the shoulder joint than do all the ligaments and capsular fibers combined. Muscles originating on the trunk, pectoral girdle, and humerus cover the anterior, superior, and posterior surfaces of the capsule. The tendons of the *supraspinatus, infraspinatus, subscapularis,* and *teres minor muscles* reinforce the joint capsule and limit range of movement. These muscles, known as the muscles of the *rotator cuff,* are the primary mechanism for supporting the shoulder

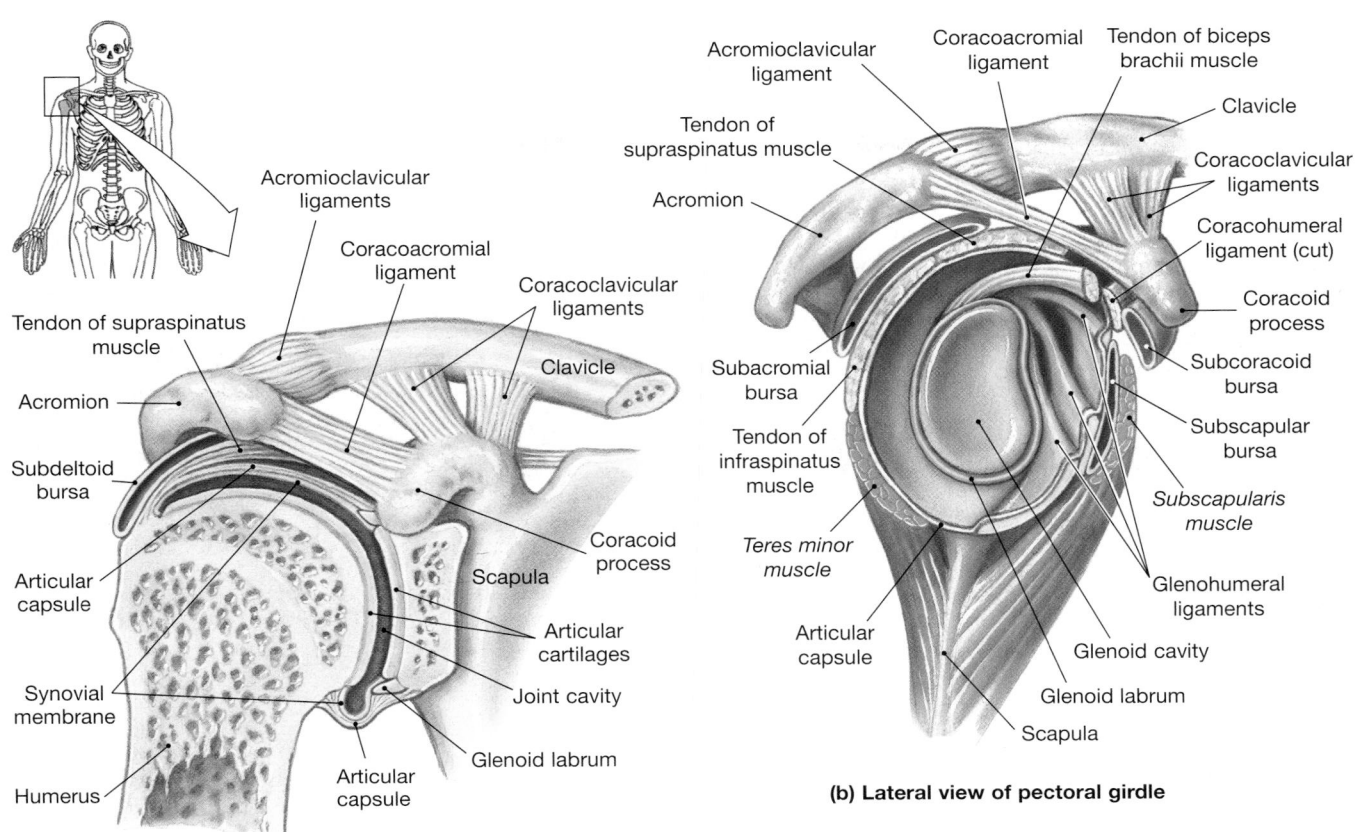

(a) Anterior view, frontal section

(b) Lateral view of pectoral girdle

●**FIGURE 9–9**
The Shoulder Joint. **(a)** A sectional view showing major structural features. **(b)** A lateral view of the shoulder joint with the humerus removed. **ATLAS** Plate 6.1d

joint and limiting its range of movement. Damage to the rotator cuff typically occurs when individuals are engaged in sports that place severe strains on the shoulder. White-water kayakers, baseball pitchers, and quarterbacks are all at high risk for rotator cuff injuries.

The anterior, superior, and posterior surfaces of the shoulder joint are reinforced by ligaments, muscles, and tendons, but the inferior capsule is poorly reinforced. As a result, a dislocation caused by an impact or a violent muscle contraction is most likely to occur at this site. Such a dislocation can tear the inferior capsular wall and the glenoid labrum. The healing process typically leaves a weakness that increases the chances for future dislocations.

As at other joints, bursae at the shoulder reduce friction where large muscles and tendons pass across the joint capsule. The shoulder has a relatively large number of important bursae, such as the *subacromial bursa*, the *subcoracoid bursa*, the *subdeltoid bursa*, and the *subscapular bursa* (Figure 9–9●). A tendon of the biceps brachii muscle runs through the shoulder joint. ∞ p. 248 As it passes through the articular capsule, it is surrounded by a tubular bursa that is continuous with the joint cavity. Inflammation of any of these extracapsular bursae can restrict motion and produce the painful symptoms of bursitis (p. 267).

☐ **THE ELBOW JOINT**

The elbow joint is a complex hinge joint that involves the humerus, radius, and ulna (Figure 9–10●). The largest and strongest articulation at the elbow is the *humeroulnar joint*, where the trochlea of the humerus articulates with the trochlear notch of the ulna. This joint works like a door hinge, with physical limitations imposed on the range of motion. In the case of the elbow, the shape of the trochlear notch of the ulna determines the plane of movement, and the combination of the notch and the olecranon limits the degree of extension permitted. ∞ p. 247 At the smaller *humeroradial joint*, the capitulum of the humerus articulates with the head of the radius.

Muscles that extend the elbow attach to the rough surface of the olecranon. These muscles are primarily under the control of the *radial nerve*, which passes along the radial groove of the humerus. ∞ p. 246 The large *biceps brachii muscle* covers the anterior surface of the arm. Its tendon is attached to the radius at the radial tuberosity. Contraction of this muscle produces supination of the forearm and flexion at the elbow.

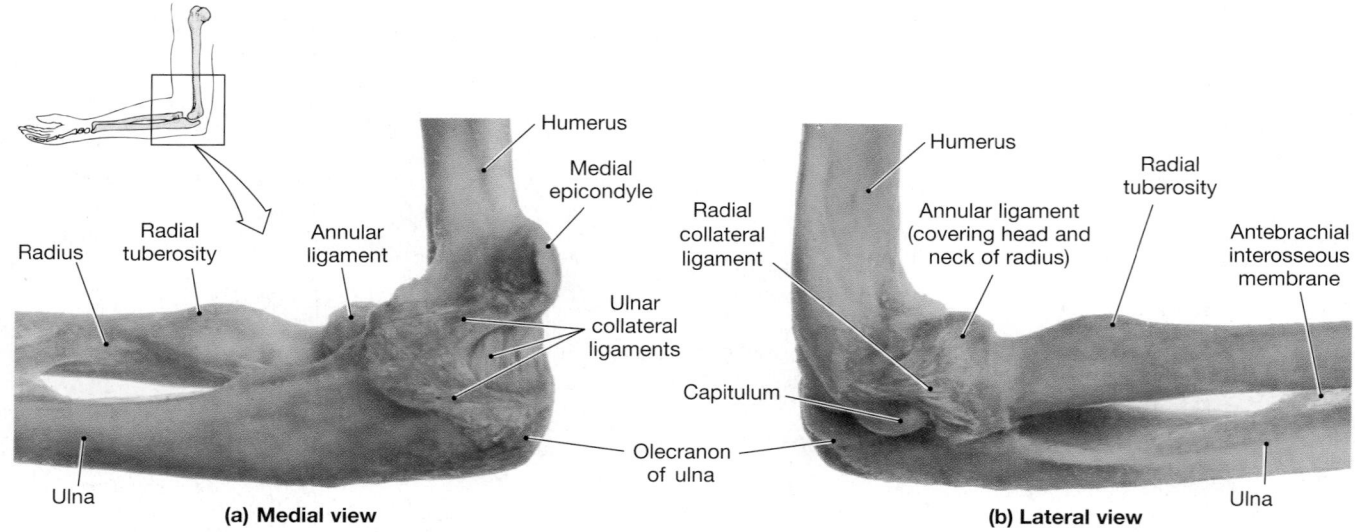

●**FIGURE 9–10**
The Elbow Joint. The right elbow joint. **(a)** A medial view, showing ligaments that stabilize the joint. **(b)** A lateral view.
ATLAS Plates 6.4e-i; Scans 6a,b

The elbow joint is extremely stable because (1) the bony surfaces of the humerus and ulna interlock, (2) the articular capsule is very thick, and (3) the capsule is reinforced by strong ligaments. The *radial collateral ligament* stabilizes the lateral surface of the joint. It extends between the lateral epicondyle of the humerus and the *annular ligament*, which binds the head of the radius to the ulna. The medial surface of the joint is stabilized by the *ulnar collateral ligament*, which extends from the medial epicondyle of the humerus anteriorly to the coronoid processes of the ulna and posteriorly to the olecranon (Figure 9–10●).

Despite the strength of the capsule and ligaments, the elbow joint can be damaged by severe impacts or unusual stresses. For example, if you fall on your hand with a partially flexed elbow, contractions of muscles that extend the elbow may break the ulna at the center of the trochlear notch. Less violent stresses can produce dislocations or other injuries to the elbow, especially if epiphyseal growth has not been completed. For example, parents in a hurry may drag a toddler along behind them, exerting an upward, twisting pull on the elbow joint that can result in a partial dislocation known as *nursemaid's elbow*.

Table 9–4 summarizes the characteristics of the joints of the upper limb.

✓ Which tissues or structures provide most of the stability for the shoulder joint?

✓ Would a tennis player or a jogger be more likely to develop inflammation of the subscapular bursa? Why?

✓ A football player received a blow to the upper surface of his shoulder, causing a shoulder separation. What does this mean?

✓ Terry suffers an injury to his forearm and elbow. After the injury, he notices an unusually large degree of motion between the radius and the ulna at the elbow. Which ligament did Terry most likely damage?

Answers start on page Q-1

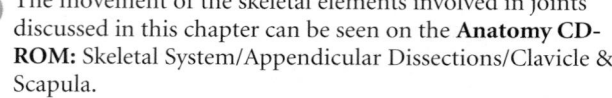 The movement of the skeletal elements involved in joints discussed in this chapter can be seen on the **Anatomy CD-ROM:** Skeletal System/Appendicular Dissections/Clavicle & Scapula.

☐ THE HIP JOINT

The hip joint, or *coxal joint*, is a sturdy ball-and-socket diarthrosis that permits flexion and extension, adduction and abduction, circumduction, and rotation. Figure 9–11● introduces the structure of the hip joint. The acetabulum, a deep fossa, accommodates the head of the femur. ∞ p. 253 Within the acetabulum, a fibrocartilage pad extends like a horseshoe to either side of the acetabular notch (Figure 9–11a●). The *acetabular labrum*, a projecting rim of fibrocartilage, increases the depth of the joint cavity.

The articular capsule of the hip joint is extremely dense and strong. It extends from the lateral and inferior surfaces of the pelvic girdle to the intertrochanteric line and intertrochanteric crest of the femur, enclosing both the head and neck of the femur. ∞ p. 254 This arrangement helps keep the femoral head from moving too far from the acetabulum.

Four broad ligaments reinforce the articular capsule (Figure 9–11●). Three of them—the *iliofemoral, pubofemoral,* and *ischiofemoral ligaments*—are regional thickenings of the capsule. The *transverse acetabular ligament* crosses the acetabu-

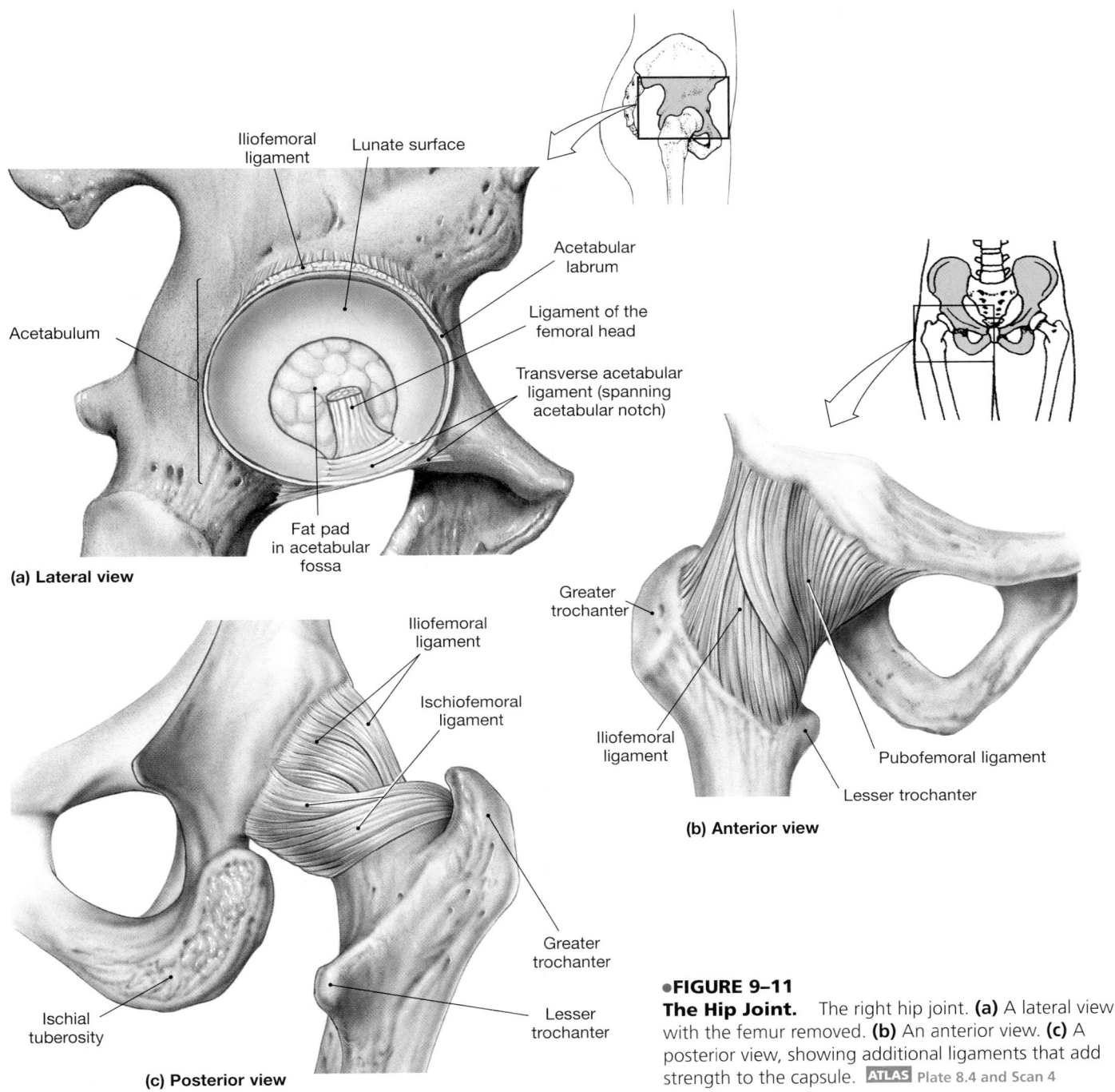

(a) Lateral view

Iliofemoral ligament

Lunate surface

Acetabulum

Acetabular labrum

Ligament of the femoral head

Transverse acetabular ligament (spanning acetabular notch)

Fat pad in acetabular fossa

(b) Anterior view

Greater trochanter

Iliofemoral ligament

Pubofemoral ligament

Lesser trochanter

(c) Posterior view

Iliofemoral ligament

Ischiofemoral ligament

Ischial tuberosity

Greater trochanter

Lesser trochanter

●**FIGURE 9–11**
The Hip Joint. The right hip joint. **(a)** A lateral view with the femur removed. **(b)** An anterior view. **(c)** A posterior view, showing additional ligaments that add strength to the capsule. **ATLAS** Plate 8.4 and Scan 4

lar notch, filling in the gap in the inferior border of the acetabulum. A fifth ligament, the *ligament of the femoral head*, or *ligamentum teres* (*teres*, long and round), originates along the transverse acetabular ligament (Figure 9–11a●) and attaches to the fovea capitis, a small pit at the center of the femoral head. ∞ p. 253 This ligament tenses only when the hip is flexed and the thigh is undergoing lateral rotation. Much more important stabilization is provided by the bulk of the surrounding muscles, aided by ligaments and capsular fibers.

The combination of an almost complete bony socket, a strong articular capsule, supporting ligaments, and muscular padding makes the hip joint an extremely stable joint. The head of the femur is well supported, but the ball-and-socket joint is not directly aligned with the weight distribution along the shaft. Stress must be transferred at an angle from the joint, along the relatively thin femoral neck to the length of the femur. ∞ p. 254 Fractures of the femoral neck or between the greater and lesser trochanters of the femur are more common than hip dislocations. As we noted in Chapter 6, however, hip fractures are relatively common in elderly individuals with severe osteoporosis. p. 204

☐ THE KNEE JOINT

The hip joint passes weight to the femur, and the knee joint transfers the weight from the femur to the tibia. The shoulder is mobile; the hip, stable; and the knee,…? If you had to choose one word, it would probably be "complicated." Although the knee functions as a hinge joint, the articulation is far more complex than that of the elbow or even the ankle. The rounded condyles of the femur roll across the superior surface of the tibia, so the points of contact are constantly changing. The joint permits flexion, extension, and very limited rotation.

Structurally, the knee resembles three separate joints: two between the femur and tibia (medial condyle to medial condyle and lateral condyle to lateral condyle) and one between the patella and the patellar surface of the femur (Figure 9–12●).

The Articular Capsule and Joint Cavity

The articular capsule at the knee joint is thin and in some areas incomplete, but it is strengthened by various ligaments and the tendons of associated muscles. A pair of fibrocartilage pads, the **medial** and **lateral menisci**, lie between the femoral and tibial surfaces (Figures 9–1b, p. 266, and 9–12b,c●). The menisci (1) act as cushions, (2) conform to the shape of the articulating surfaces as the femur changes position, and (3) provide lateral stability to the joint. Prominent fat pads cushion the margins of the joint and assist the many bursae in reducing friction between the patella and other tissues.

Supporting Ligaments

A complete dislocation of the knee is very rare, largely because seven major ligaments stabilize the knee joint:

1. The tendon from the muscles responsible for extending the knee passes over the anterior surface of the joint. The patella is embedded in this tendon, and the *patellar ligament* continues to its attachment on the anterior surface of the tibia. The patellar ligament and two ligamentous bands known as the *patellar retinaculae* support the anterior surface of the knee joint (Figure 9–12a●).

2, 3. Two *popliteal ligaments* extend between the femur and the heads of the tibia and fibula (Figure 9–12b●). These ligaments reinforce the knee joint's posterior surface.

4, 5. Inside the joint capsule, the *anterior cruciate ligament* (ACL) and *posterior cruciate ligament* (PCL) attach the intercondylar area of the tibia to the condyles of the femur. *Anterior* and *posterior* refer to the sites of origin of these ligaments on the tibia. They cross one another as they proceed to their destinations on the femur (Figure 9–12b,c●). (The term *cruciate* is derived from the Latin word

Knee Injuries

Athletes place tremendous stresses on their knees. Ordinarily, the medial and lateral menisci move as the position of the femur changes. Placing a lot of weight on the knee while it is partially flexed can trap a meniscus between the tibia and femur, resulting in a break or tear in the cartilage. In the most common injury, the lateral surface of the leg is driven medially, tearing the medial meniscus. In addition to being quite painful, the torn cartilage may restrict movement at the joint. It can also lead to chronic problems and the development of a "trick knee"—a knee that feels unstable. Sometimes the meniscus can be heard and felt popping in and out of position when the knee is extended.

Less common knee injuries involve tearing one or more stabilizing ligaments or damaging the patella. Torn ligaments can be difficult to correct surgically, and healing is slow. The patella can be injured in a number of ways. If the leg is immobilized (as it might be in a football pileup) while you try to extend the knee, the muscles are powerful enough to pull the patella apart. Impacts to the anterior surface of the knee can also shatter the patella. Treatment of a fractured patella is difficult and time con-

suming. The fragments must be surgically removed and the tendons and ligaments repaired. The joint must then be immobilized. Total knee replacements are rarely performed on young people, but they are becoming increasingly common among elderly patients with severe arthritis. **AM** Arthroscopic Surgery and the Diagnosis of Joint Injuries

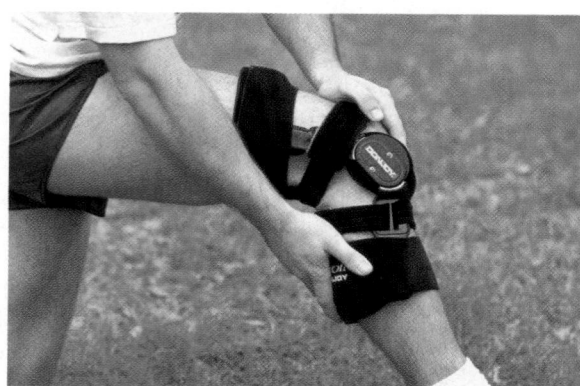

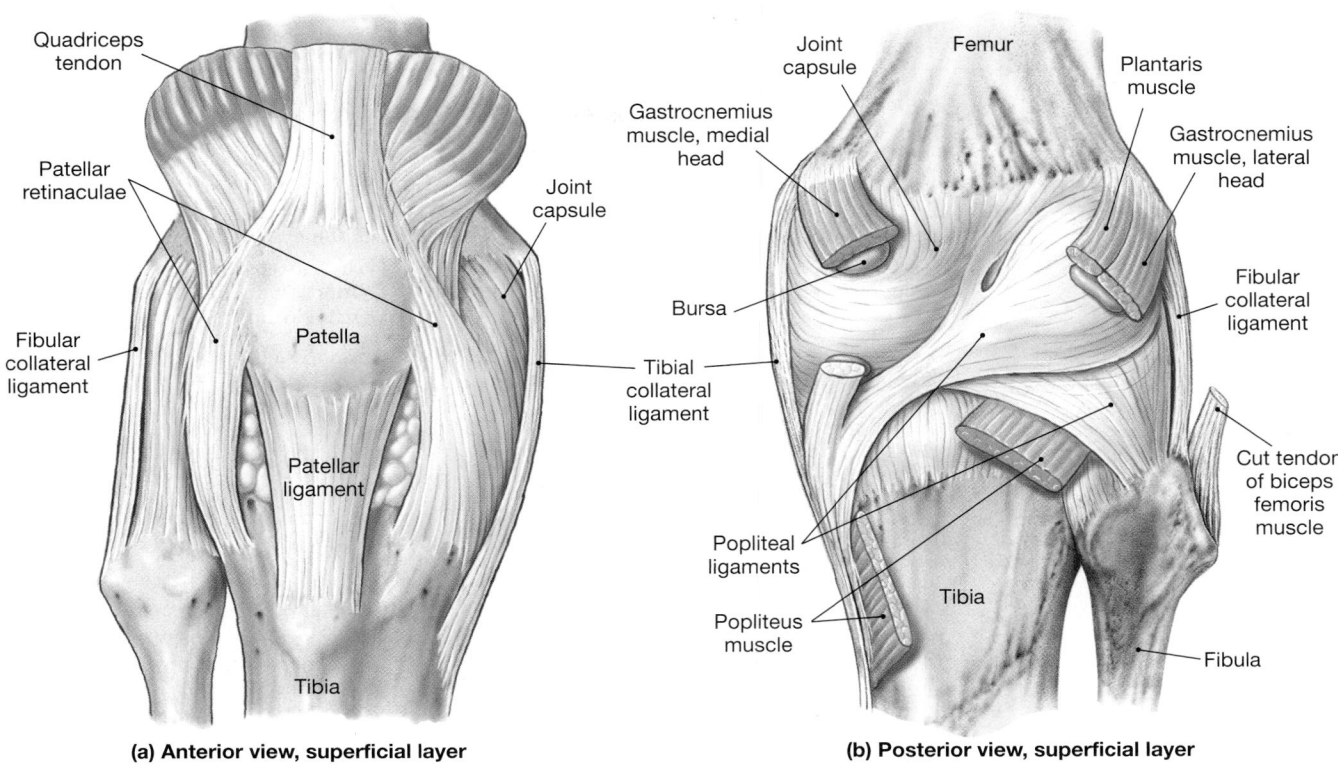

(a) Anterior view, superficial layer

(b) Posterior view, superficial layer

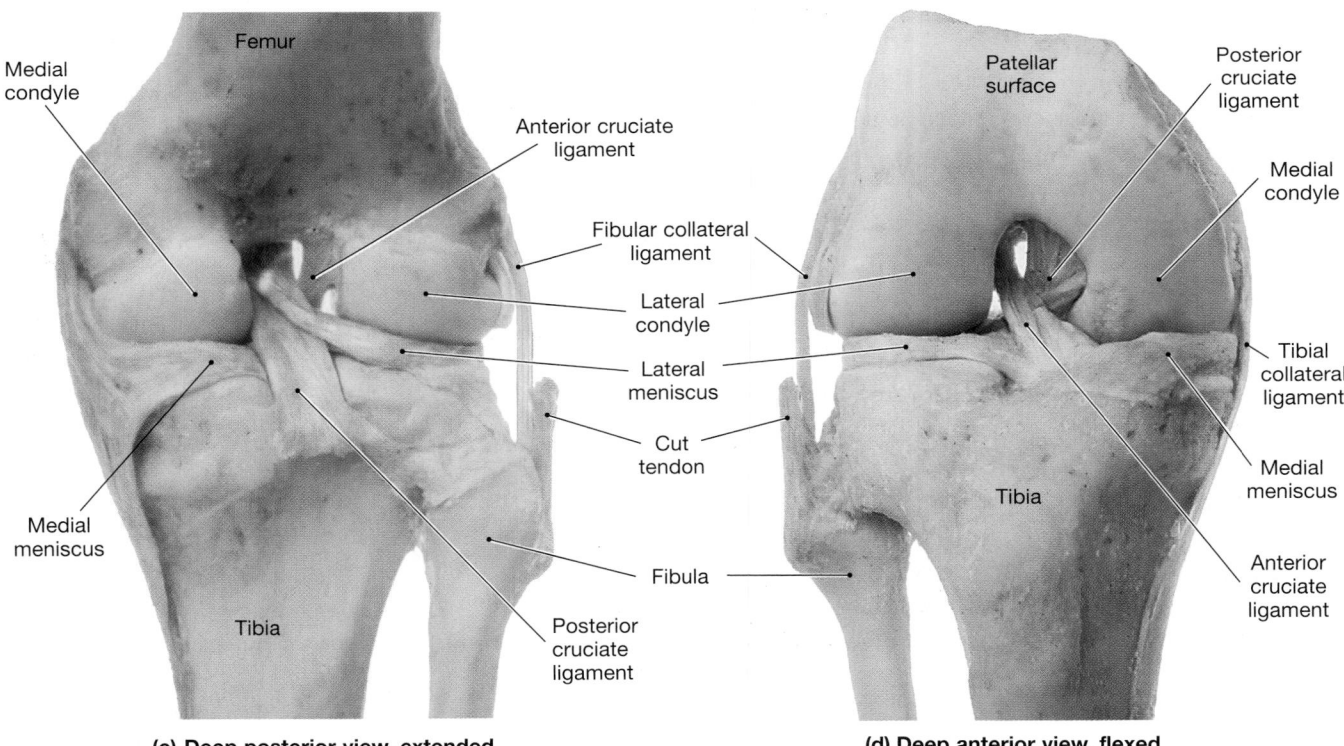

(c) Deep posterior view, extended

(d) Deep anterior view, flexed

●**FIGURE 9–12**
The Knee Joint.   The right knee. Superficial anterior **(a)** and posterior **(b)** views of the extended knee joint. **(c)** A deep posterior view, at full extension. **(d)** An anterior view, at full flexion. **ATLAS** Plates 8.9–8.11; Scans 7a-f

SUMMARY TABLE 9–4 ARTICULATIONS OF THE APPENDICULAR SKELETON

	Element	Joint	Type of Articulation	Movements
ARTICULATIONS OF THE PECTORAL GIRDLE AND UPPER LIMB				
	Sternum/clavicle	Sternoclavicular	Gliding diarthrosis*	Protraction/retraction, elevation/depression, slight rotation
	Scapula/clavicle	Acromioclavicular	Gliding diarthrosis	Slight movement
	Scapula/humerus	Shoulder, or glenohumeral	Ball-and-socket diarthrosis	Flexion/extension, adduction/abduction, circumduction, rotation
	Humerus/ulna and humerus/radius	Elbow (humeroulnar and humeroradial)	Hinge diarthrosis	Flexion/extension
	Radius/ulna	Proximal radioulnar	Pivot diarthrosis	Rotation
		Distal radioulnar	Pivot diarthrosis	Pronation/supination
	Radius/carpal bones	Radiocarpal	Ellipsoidal diarthrosis	Flexion/extension, adduction/abduction, circumduction
	Carpal bone to carpal bone	Intercarpal	Gliding diarthrosis	Slight movement
	Carpal bone to metacarpal bone (I)	Carpometacarpal of thumb	Saddle diarthrosis	Flexion/extension, adduction/abduction, circumduction, opposition
	Carpal bone to metacarpal bone (II–V)	Carpometacarpal	Gliding diarthrosis	Slight flexion/extension, adduction/abduction
	Metacarpal bone to phalanx	Metacarpophalangeal	Ellipsoidal diarthrosis	Flexion/extension, adduction/abduction, circumduction
	Phalanx/phalanx	Interphalangeal	Hinge diarthrosis	Flexion/extension
ARTICULATIONS OF THE PELVIC GIRDLE AND LOWER LIMB				
	Sacrum/ilium of os coxae	Sacroiliac	Gliding diarthrosis	Slight movement
	Os coxae/os coxae	Pubic symphysis	Amphiarthrosis (symphysis)	None†
	Os coxae/femur	Hip	Ball-and-socket diarthrosis	Flexion/extension, adduction/abduction, circumduction, rotation
	Femur/tibia	Knee	Complex, functions as hinge	Flexion/extension, limited rotation
	Tibia/fibula	Tibiofibular (proximal)	Gliding diarthrosis	Slight movement
		Tibiofibular (distal)	Gliding diarthrosis and amphiarthrotic syndesmosis	Slight movement
	Tibia and fibula with talus	Ankle, or talocrural	Hinge diarthrosis	Flexion/extension (dorsiflexion/plantar flexion)
	Tarsal bone to tarsal bone	Intertarsal	Gliding diarthrosis	Slight movement
	Tarsal bone to metatarsal bone	Tarsometatarsal	Gliding diarthrosis	Slight movement
	Metatarsal bone to phalanx	Metatarsophalangeal	Ellipsoidal diarthrosis	Flexion/extension, adduction/abduction
	Phalanx/phalanx	Interphalangeal	Hinge diarthrosis	Flexion/extension

*A "double gliding joint," with two joint cavities separated by an articular cartilage.
†During pregnancy, hormones weaken the symphysis and permit movement important to childbirth; see Chapter 29.

crucialis, meaning a cross.) The ACL and the PCL limit the anterior and posterior movement of the femur and maintain the alignment of the femoral and tibial condyles.

6, 7. The *tibial collateral ligament* reinforces the medial surface of the knee joint, and the *fibular collateral ligament* reinforces the lateral surface (Figure 9–12●). These ligaments tighten only at full extension, the position in which they stabilize the joint.

At full extension, a slight lateral rotation of the tibia tightens the anterior cruciate ligament and jams the lateral meniscus between the tibia and femur. The knee joint is essentially locked in the extended position. With the joint locked, a person can stand for prolonged periods without using (and tiring) the muscles that extend the knee. Unlocking the knee joint requires muscular contractions that medially rotate the tibia or laterally rotate the femur. If the locked knee is struck from the side, the lateral meniscus can tear and the supporting ligaments can be seriously damaged.

The knee is structurally complex and is subjected to severe stresses in the course of normal activities. Painful knee injuries are all too familiar to both amateur and professional athletes. Treatment is often costly and prolonged, and repairs seldom make the joint "good as new."

Table 9–4 summarizes information about the articulations of the pelvic girdle and lower limb.

✓ Where would you find the following ligaments: iliofemoral ligament, pubofemoral ligament, and ischiofemoral ligament?

✓ What symptoms would you expect to see in an individual who has damaged the menisci of the knee joint?

✓ Why is "clergyman's knee" (a type of bursitis) common among carpet layers and roofers?

Answers start on page Q-1

 The movement of the skeletal elements involved in joints discussed in this chapter can be seen on the **Anatomy CD-ROM:** Skeletal System/Appendicular Dissections/Hip.

9–4 AGING AND ARTICULATIONS

Objective

■ Describe the effects of aging on articulations, and discuss the most common clinical problems that develop as a result.

Joints are subjected to heavy wear and tear throughout our lifetimes, and problems with joint function are relatively common, especially in older individuals. **Rheumatism** (ROO-muh-tizm) is a general term that indicates pain and stiffness affecting the skeletal system, the muscular system, or both. Several major forms of rheumatism exist. **Arthritis** (ar-THRĪ-tis) encompasses all the rheumatic diseases that affect synovial joints. Arthritis al-

ways involves damage to the articular cartilages, but the specific cause can vary. For example, arthritis can result from bacterial or viral infection, injury to the joint, metabolic problems, or severe physical stresses.

Osteoarthritis (os-tē-ō-ar-THRĪ-tis), also known as *degenerative arthritis* or *degenerative joint disease* (*DJD*), generally affects individuals age 60 or older. Osteoarthritis can result from cumulative wear and tear at the joint surfaces or from genetic factors affecting collagen formation. In the U.S. population, 25 percent of women and 15 percent of men over age 60 show signs of this disease.

Rheumatoid arthritis is an inflammatory condition that affects roughly 2.5 percent of the adult population. At least some cases occur when the immune response mistakenly attacks the joint tissues. Such a condition, in which the body attacks its own tissues, is called an *autoimmune disease*. Allergies, bacteria, viruses, and genetic factors have all been proposed as contributing to or triggering the destructive inflammation.

In *gouty arthritis*, crystals form within the synovial fluid of joints. The crystals accumulate over time and eventually begin to interfere with normal movement. This form of arthritis is named after the metabolic disorder known as *gout*, discussed further in Chapter 25. In gout, the crystals are derived from uric acid (a metabolic waste product), and the joint most often affected is the metatarsal–phalangeal joint of the great toe. Gout is relatively rare, but other forms of gouty arthritis are much more common—some degree of calcium salt deposition occurs in 30–60% of those over age 85. The cause is unknown, but the condition appears to be linked to age-related changes in the articular cartilages.

Regular exercise, physical therapy, and drugs that reduce inflammation, such as aspirin, can often slow the progress of osteoarthritis. Surgical procedures can realign or redesign the affected joint. In extreme cases involving the hip, knee, elbow, or shoulder, the defective joint can be replaced by an artificial one.

AM Rheumatism, Arthritis, and Synovial Function

Degenerative changes comparable to those seen in arthritis may result from joint immobilization. When motion ceases, so does the circulation of synovial fluid, and the cartilages begin to suffer. **Continuous passive motion (CPM)** of any injured joint appears to encourage the repair process by improving the circulation of synovial fluid. The movement is often performed by a physical therapist or a machine during the recovery process.

With age, bone mass decreases and bones become weaker, so the risk of fractures increases. p. 204 If osteoporosis develops, the bones may weaken to the point at which fractures occur in response to stresses that could easily be tolerated by normal bones. Hip fractures are among the most dangerous fractures seen in elderly persons, with or without osteoporosis. These fractures, most often involving individuals over age 60, may be accompanied by hip dislocation or by pelvic fractures.

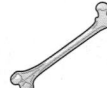

THE **SKELETAL SYSTEM** IN PERSPECTIVE

Organ/Component	Primary Functions
Bones, Cartilages, and Joints	Support, protect soft tissues; bones store minerals
Axial Skeleton (skull, vertebrae, ribs, sternum, sacrum, cartilages, and ligaments)	Protects brain, spinal cord, sense organs, and soft tissues of thoracic cavity; supports the body weight over the lower limbs
Appendicular Skeleton (supporting bones, cartilages, and ligaments of the limbs)	Provides internal support and positioning of the limbs; supports and moves axial skeleton
Bone Marrow	Acts as primary site of blood cell production (red blood cells, white blood cells); stores lipid reserves

INTEGRATION WITH OTHER SYSTEMS

Although the bones you study in the lab may seem to be rigid and permanent structures, the living skeleton is dynamic and undergoes continuous remodeling. The balance between osteoblast and osteoclast activity is delicate and subject to change at a moment's notice. When osteoblast activity predominates, the bones thicken and strengthen; when osteoclast activity predominates, the bones get thinner and weaker. The balance between bone formation and bone recycling varies with (1) the age of the individual, (2) the physical stresses applied to the bone, (3) circulating hormone levels, (4) rates of calcium and phosphorus absorption and excretion, and (5) genetic or environmental factors. Most of these variables involve some interaction between the skeletal system and other systems.

In fact, the skeletal system is intimately associated with other systems. For instance, the bones of the skeleton are attached to the muscular system, extensively connected to the cardiovascular and lymphatic systems, and largely under the physiological control of the endocrine system. The digestive and urinary systems also play important roles in providing the calcium and phosphate minerals needed for bone growth. In return, the skeleton represents a reserve of calcium, phosphate, and other minerals that can compensate for changes in the dietary supply of those ions. The figure on the opposite page reviews the components and functions of the skeletal system and diagrams the major functional relationships between that system and other systems.

CLINICAL PATTERNS

Because the skeletal system is dependent on other systems, skeletal system disorders can reflect problems originating within the skeletal system itself (such as bone tumors or inherited conditions affecting bone formation) or secondary problems that reflect changes in other systems. *Rickets*, a condition characterized by inadequate bone mineralization, was discussed on p. 199. Rickets is an example of a skeletal problem that develops when other systems—especially the integumentary system and the digestive system—are unable to function normally.

The *Applications Manual* considers the diagnosis and treatment of major conditions affecting the skeletal system. For additional information and updates on clinical topics affecting this system, visit the Destinations page on the Companion Website.

MEDIA CONNECTIONS

I. Objective: Investigate factors that affect bone density and the rate of bone turnover.

Completion time = 10 minutes

Many systems of the body are involved in maintaining overall skeletal health. Despite this, bone density begins to decline in adults in their mid-thirties. What factors are involved in this change? List the organ systems that are responsible for maintaining the skeletal system, and briefly describe each interaction. This list will be easier to create if you work through a case study involving bone density. Go to the **Companion Website**, Chapter 9 Media Connections and click on the keyword **Osteoporosis**.

II. Objective: Explore the hormonal and cardiovascular requirements of developing bone.

Completion time = 10 minutes

Developing new bone is not common in adult skeletons, making it difficult to study. Fibrodysplasia ossificans progressiva (*FOP*) is a genetic disorder of connective tissue that causes disfigurement and loss of function through heterotopic ossification. Studying the progression of this disease has uncovered interactions between the skeletal, endocrine, and cardiovascular systems. Go to the **Companion Website**, Chapter 9 Media Connections and click on the keyword **FOP**. Prepare a short essay outlining the hormonal and cardiovascular controls on bone growth using the information provided on this website.

INTEGUMENTARY SYSTEM

- Synthesizes vitamin D$_3$, essential for calcium and phosphorus absorption (bone maintenance and growth)

- Provides structural support

MUSCULAR SYSTEM

- Stabilizes bone positions
- Tension in tendons stimulates bone growth and maintenance

- Provides calcium needed for normal muscle contraction
- Bones act as levers to produce body movements

NERVOUS SYSTEM

- Regulates bone position by controlling muscle contractions

- Provides calcium for neural function
- Protects brain, spinal cord
- Receptors at joints provide information about body position

ENDOCRINE SYSTEM

- Skeletal growth regulated by growth hormone, thyroid hormones, and sex hormones
- Calcium mobilization regulated by parathyroid hormone and calcitonin

- Protects endocrine organs, especially in brain, chest, and pelvic cavity

CARDIOVASCULAR SYSTEM

- Provides oxygen, nutrients, hormones, blood cells
- Removes waste products and carbon dioxide

- Provides calcium needed for cardiac muscle contraction
- Blood cells produced in bone marrow
- Axial skeleton protects heart and great vessels

SKELETAL SYSTEM

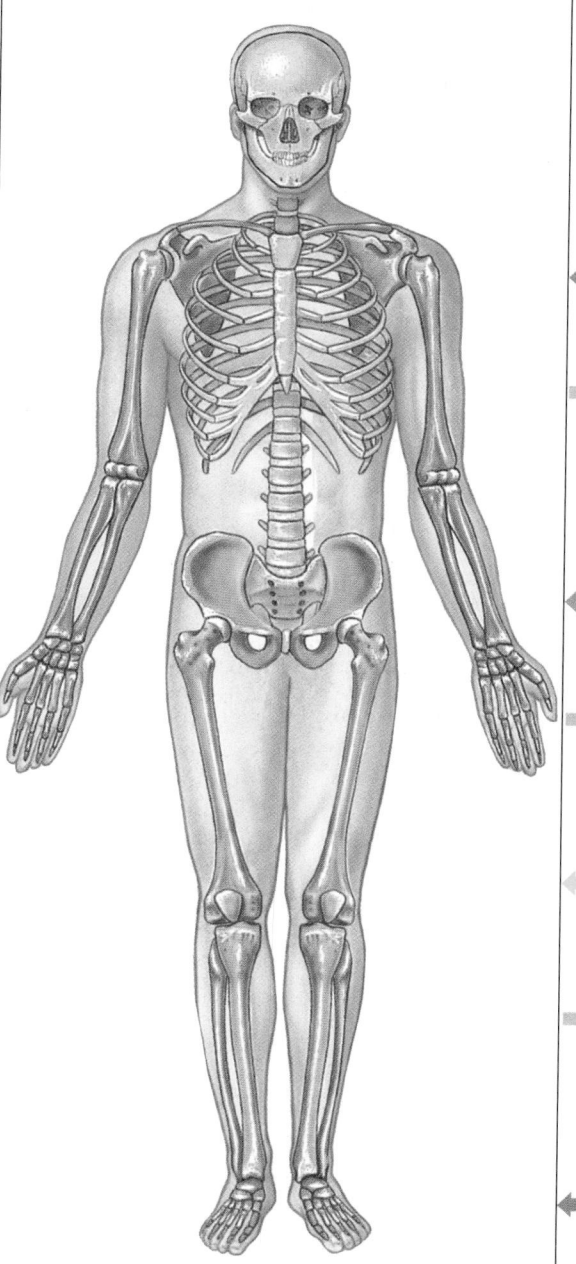

FOR ALL SYSTEMS
Provides mechanical support; stores energy reserves; stores calcium and phosphate reserves

LYMPHATIC SYSTEM

- Lymphocytes assist in the defense and repair of bone following injuries

- Lymphocytes and other cells of the immune response are produced and stored in bone marrow

RESPIRATORY SYSTEM

- Provides oxygen and eliminates carbon dioxide

- Movements of ribs important in breathing
- Axial skeleton surrounds and protects lungs

DIGESTIVE SYSTEM

- Provides nutrients, calcium, and phosphate

- Ribs protect portions of liver, stomach, and intestines

URINARY SYSTEM

- Conserves calcium and phosphate needed for bone growth
- Disposes of waste products

- Axial skeleton provides some protection for kidneys and ureters
- Pelvis protects urinary bladder and proximal urethra

REPRODUCTIVE SYSTEM

- Sex hormones stimulate growth and maintenance of bones
- Surge of sex hormones at puberty causes acceleration of growth and closure of epiphyseal cartilages

- Pelvis protects reproductive organs of female, protects portion of ductus deferens and accessory glands in male

Healing of hip fractures proceeds very slowly, and the powerful muscles that surround the hip joint often prevent proper alignment of the bone fragments. Fractures at the greater or lesser trochanter generally heal well if the joint can be stabilized; steel frames, pins, screws, or some combination of these devices may be needed to preserve alignment and to permit healing to proceed normally.

Although severe hip fractures are most common among those over age 60, in recent years the frequency of hip fractures has increased dramatically among young, healthy professional athletes. **AM** Hip Fractures, Aging, and Professional Athletes

9–5 BONES AND MUSCLES

The skeletal and muscular systems are structurally and functionally interdependent. Their interactions are so extensive that they are often considered to be parts of a single *musculoskeletal system.*

The two systems have direct physical connections. Many of the anatomical landmarks identified in this chapter are attachment sites of skeletal muscles, and the connective tissues that surround individual muscle fibers are continuous with those that form the organic framework of an attached bone. Muscles and bones are also physiologically linked through their mutual dependence on stable calcium ion concentrations in body fluids. With most of the body's calcium tied up in the skeleton, bone abnormalities can have a direct effect on the muscle.

The bones, in turn, are directly affected by muscular activity. When a muscle enlarges due to regular exercise, the increased forces exerted on the skeleton will make bones become stronger and more massive. The reverse holds true as well, for both muscle and bones decrease in size and strength after periods of inactivity.

In the next two chapters, we shall examine the structure and function of the muscular system and discuss how muscular contractions perform specific movements.

Chapter Review

SELECTED CLINICAL TERMINOLOGY

Terms Discussed in This Chapter

arthritis (ar-THRĪ-tis): A group of rheumatic diseases that affect synovial joints. Arthritis always involves damage to the articular cartilages, but the specific cause can vary. The diseases of arthritis are usually classified as either *degenerative* or *inflammatory.* (*p. 283* and *AM)*

bunion: The most common pressure-related bursitis, involving a tender nodule formed around bursae over the base of the great toe. (*p. 267)*

bursitis: An inflammation of a bursa, causing pain whenever the associated tendon or ligament moves. (*p. 267)*

continuous passive motion (CPM): A therapeutic procedure involving the passive movement of an injured joint to stimulate the circulation of synovial fluid. The goal is to prevent degeneration of the articular cartilages. (*p. 283* and *AM)*

herniated disc: A condition caused by intervertebral compression severe enough to rupture the anulus fibrosus and release the nucleus pulposus, which may protrude beyond the intervertebral space. (*p. 276)*

luxation (luks-Ā-shun): A dislocation; a condition in which the articulating surfaces are forced out of position. (*p. 267)*

osteoarthritis (os-tē-ō-ar-THRĪ-tis) (*degenerative arthritis* or *degenerative joint disease, DJD*): An arthritic condition resulting from (1) cumulative wear and tear on joint surfaces or (2) a genetic predisposition. In the U.S. population, 25 percent of women and 15 percent of men over age 60 show signs of this disease. (*p. 283* and *AM)*

rheumatism (ROO-muh-tizm): A general term that indicates pain and stiffness affecting the skeletal system, the muscular system, or both. (*p. 283* and *AM)*

rheumatoid arthritis: An inflammatory arthritis that affects roughly 2.5 percent of the adult U.S. population. The cause is uncertain, although allergies, bacteria, viruses, and genetic factors have all been proposed. The primary symptom is *synovitis*—swelling and inflammation of the synovial membrane. (*p. 283* and *AM)*

shoulder separation: The partial or complete dislocation of the acromioclavicular joint. (*p. 276)*

slipped disc: A common name for a condition caused by the distortion of an intervertebral disc. The distortion applies pressure to spinal nerves, causing pain and limiting range of motion. (*p. 276)*

sprain: A condition in which a ligament is stretched to the point at which some of the collagen fibers are torn. The ligament remains functional, and the structure of the joint is not affected. (*p. 266)*

subluxation (sub-luks-Ā-shun): A partial dislocation; the displacement of articulating surfaces sufficient to cause discomfort, but resulting in less physical damage to the joint than that which occurs during a dislocation. (*p. 267)*

AM Additional Terms Discussed in the *Applications Manual*

ankylosis
arthroscope
arthroscopic surgery
bony crepitus
laminectomy
meniscectomy
monoarthritic
polyarthritic

STUDY OUTLINE

9-1 A CLASSIFICATION OF JOINTS p. 263

1. **Articulations** (joints) exist wherever two bones interconnect.
2. *Immovable joints* are **synarthroses**; *slightly movable joints* are **amphiarthroses**; and joints that are *freely movable* are called **diarthroses** or **synovial joints.** (*Table 9–1*)
3. Alternatively, joints are classified structurally, as *bony, fibrous, cartilaginous,* or *synovial.* (*Table 9–2*)

SYNARTHROSES (IMMOVABLE JOINTS) p. 264

4. The four major types of synarthroses are a **suture** (skull bones bound together by dense connective tissue), a **gomphosis** (teeth bound to bony sockets by *periodontal ligaments*), a **synchondrosis** (two bones joined by a rigid cartilaginous bridge), and a **synostosis** (two bones completely fused).

AMPHIARTHROSES (SLIGHTLY MOVABLE JOINTS) p. 264

5. The two major types of amphiarthroses are a **syndesmosis** (bones connected by a ligament) and a **symphysis** (bones separated by fibrocartilage).

DIARTHROSES (FREELY MOVABLE JOINTS) p. 265

6. The bony surfaces at diarthroses are enclosed within an **articular capsule**, covered by **articular cartilages**, and lubricated by **synovial fluid.**
7. Other synovial structures include **menisci**, or *articular discs*; **fat pads**; **accessory ligaments**; and **bursae.** (*Figure 9–1*)
8. A **dislocation** occurs when articulating surfaces are forced out of position.

Structure of synovial joints: **Companion Website:** Chapter 9/ Reviewing Concepts/Labeling.

9-2 ARTICULAR FORM AND FUNCTION p. 267

DESCRIBING DYNAMIC MOTION p. 267

1. Possible movements are **linear motion**, **angular motion**, and **rotation.** (*Figure 9–2*)
2. Joints are called **monaxial**, **biaxial**, or **triaxial**, depending on the planes of movement they allow.

TYPES OF MOVEMENTS p. 268

3. In **gliding**, two surfaces slide past one another.
4. Important terms that describe angular motion are **flexion**, **extension**, **hyperextension**, **circumduction**, **abduction**, and **adduction.** (*Figure 9–3*)
5. Rotational movement can be **left** or **right**, **medial** (*internal*) or **lateral** (*external*), or, in the bones of the forearm, **pronation** or **supination.** (*Figure 9–4*)
6. Movements of the foot include **inversion** and **eversion.** The ankle undergoes flexion and extension, also known as **dorsiflexion** and **plantar flexion**, respectively.
7. **Opposition** is the thumb movement that enables us to grasp objects. (*Figure 9–5*)
8. **Protraction** involves moving something anteriorly; **retraction** involves moving it posteriorly. **Depression** and **elevation** occur when we move a structure inferiorly and superiorly, respectively.

Lateral flexion occurs when the vertebral column bends to one side. (*Figure 9–5*)

A STRUCTURAL CLASSIFICATION OF SYNOVIAL JOINTS p. 269

9. **Gliding joints** permit limited movement, generally in a single plane. (*Figure 9–6*)
10. **Hinge joints** are monaxial joints that permit only angular movement in one plane. (*Figure 9–6*)
11. **Pivot joints** are monaxial joints that permit only rotation. (*Figure 9–6*)
12. **Ellipsoidal joints** are biaxial joints with an oval articular face that nestles within a depression in the opposing surface. (*Figure 9–6*)
13. **Saddle joints** are biaxial joints with articular faces that are concave on one axis and convex on the other. (*Figure 9–6*)
14. **Ball-and-socket joints** are triaxial joints that permit rotation as well as other movements. (*Figure 9–6*)

Movement of synovial joints: **Companion Website:** Chapter 9/ Reviewing Concepts/Multiple Choice.

9-3 REPRESENTATIVE ARTICULATIONS p. 272

INTERVERTEBRAL ARTICULATIONS p. 274

1. The articular processes of vertebrae form gliding joints with those of adjacent vertebrae. The bodies form symphyseal joints that are separated and cushioned by **intervertebral discs**, which contain an inner **nucleus pulposus** and an outer **anulus fibrosus.** Several ligaments stabilize the vertebral column. (*Figures 9–7, 9–8; Summary Table 9–3*)

THE SHOULDER JOINT p. 275

2. The shoulder joint, or *glenohumeral joint*, is formed by the glenoid cavity and the head of the humerus. This articulation permits the greatest range of motion of any joint. It is a ball-and-socket diarthrosis with various stabilizing ligaments. Strength and stability are sacrificed in favor of mobility. (*Figure 9–9; Summary Table 9–4*)

THE ELBOW JOINT p. 277

3. The elbow joint permits only flexion–extension. It is a hinge diarthrosis whose capsule is reinforced by strong ligaments. (*Figure 9–10; Summary Table 9–4*)

Clavicle & Scapula and related joint movement: **Anatomy CD-ROM:** Skeletal System/Appendicular Dissections/Clavicle & Scapula.

THE HIP JOINT p. 278

4. The hip joint is a ball-and-socket diarthrosis formed by the union of the acetabulum with the head of the femur. The joint permits flexion–extension, adduction–abduction, circumduction, and rotation; it is stabilized by numerous ligaments. (*Figure 9–11; Summary Table 9–4*)

THE KNEE JOINT p. 281

5. The knee joint is a hinge joint formed by the union of the condyles of the femur with the superior condylar surfaces of the tibia. The joint permits flexion–extension and limited rotation, and it has various supporting ligaments. *(Figure 9–12; Summary Table 9–4)*
Hip and related joint movement: **Anatomy CD-ROM:** Skeletal System/Appendicular Dissections/Hip.

9–4 AGING AND ARTICULATIONS p. 283

1. Problems with joint function are relatively common, especially in older individuals. **Rheumatism** is a general term for pain and

stiffness affecting the skeletal system, the muscular system, or both; several major forms exist. **Arthritis** encompasses all the rheumatic diseases that affect synovial joints. Both conditions become increasingly common with age.

9–5 BONES AND MUSCLES p. 284

1. The interactions of the skeletal and muscular systems are so extensive that we often refer to a single *musculoskeletal system.*

REVIEW QUESTIONS

More assessment questions are available to you on the Companion Website. You will find Matching, Multiple Choice, True/False, and other quizzes to help further your understanding of the material covered in this chapter. To access the site, go to www.aw.com/martini.

LEVEL 1 Reviewing Facts and Terms

1. A synarthrosis located between the bones of the skull is a
   (a) symphysis   (b) syndesmosis
   (c) synchondrosis   (d) suture

2. The articulation between adjacent vertebral bodies is a
   (a) syndesmosis   (b) symphysis
   (c) synchondrosis   (d) synostosis

3. The anterior articulation between the two pubic bones is a
   (a) synchondrosis   (b) synostosis
   (c) symphysis   (d) synarthrosis

4. Joints typically located at the end of long bones are
   (a) synarthroses   (b) amphiarthroses
   (c) diarthroses   (d) symphyses

5. The function of the articular cartilage is
   (a) to reduce friction
   (b) to prevent bony surfaces from contacting one another
   (c) to provide lubrication
   (d) a and b are correct

6. The structures responsible for channeling the flow of synovial fluid are
   (a) menisci   (b) bursae
   (c) carpal tunnels   (d) articular cartilages

7. The structures that limit the range of motion of a joint and provide mechanical support across or around the joint are
   (a) bursae   (b) tendons
   (c) menisci   (d) a, b, and c are correct

8. A partial dislocation of an articulating surface is a
   (a) circumduction   (b) hyperextension
   (c) subluxation   (d) supination

9. Abduction and adduction always refer to movements of the
   (a) axial skeleton
   (b) appendicular skeleton
   (c) skull
   (d) vertebral column

10. Rotation of the forearm that makes the palm face posteriorly is
    (a) supination   (b) pronation
    (c) proliferation   (d) projection

11. A saddle joint permits _____ movement, but prevents _____ movement.
    (a) rotational, gliding   (b) angular, linear
    (c) linear, rotational   (d) angular, rotational

12. Standing on tiptoe is an example of _____ at the ankle.
    (a) elevation   (b) flexion
    (c) extension   (d) retraction

13. Examples of monaxial joints, which permit angular movement in a single plane, are the
    (a) intercarpal and intertarsal joints
    (b) shoulder and hip joints
    (c) elbow and knee joints
    (d) a, b, and c are correct

14. Joints that connect the fingers and toes with the metacarpal bones and metatarsal bones, respectively, are
    (a) ellipsoidal joints   (b) saddle joints
    (c) pivot joints   (d) hinge joints

15. Movements that occur at the shoulder and the hip represent the actions that occur at a _____ joint.
    (a) hinge   (b) ball-and-socket
    (c) pivot   (d) gliding

16. The anulus fibrosus and nucleus pulposus are structures associated with the
    (a) intervertebral discs   (b) knee and elbow
    (c) shoulder and hip   (d) carpal and tarsal bones

17. Subacromial, subcoracoid, and subscapular bursae reduce friction in the _____ joint.
    (a) hip   (b) knee
    (c) elbow   (d) shoulder

18. List the types of movement that may be permitted at a synovial joint.

LEVEL 2 Reviewing Concepts

19. The hip is an extremely stable joint because it has
    (a) a complete bony socket
    (b) a strong articular capsule
    (c) supporting ligaments
    (d) a, b, and c are correct

20. Complete dislocation of the knee is an extremely rare event because
    (a) the knee is protected by the patella
    (b) the femur articulates with the tibia at the knee
    (c) the knee contains seven major ligaments
    (d) the knee contains fat pads to absorb shocks

21. How does a meniscus (articular disc) function in a joint?

22. The greater the range of motion at a joint, the weaker the joint becomes. Why?

23. How do articular cartilages differ from other cartilages in the body?

24. Differentiate between a slipped disc and a herniated disc.

25. How would you explain to your grandmother the characteristic decrease in height with advancing age?

26. When the triceps brachii muscle contracts, what movements does it produce?

27. List the four different types of synarthroses and give an example of each.

LEVEL 3 Critical Thinking and Clinical Applications

28. While playing tennis, Dave "overturns" his ankle. He experiences swelling and pain. After being examined, he is told that he has no torn ligaments and that the structure of the ankle is not affected. On the basis of the symptoms and the examination results, what happened to Dave's ankle?

29. During a basketball game, Bob injured his right knee when he jumped to retrieve the ball and then landed off balance on his right leg. Since then, he has pain and limited mobility of his right knee joint. What type of injury did Bob sustain?

30. A high school student comes to the emergency room complaining of persistent pain beneath her right shoulder blade. In talking with her, you discover that she has been spending many hours trying to improve her pitching skills for her school's softball team. What is likely causing the pain?

UNIT 2 CHAPTER 5 6 7 8 9 10 11

CLINICAL NOTES
- Interference with Neural Control Mechanisms 306
- Rigor Mortis 306
- Muscle Atrophy 322

CLINICAL DISCUSSIONS
- The Muscular Dystrophies 300
- Tetanus 311
- Delayed-Onset Muscle Soreness 323

10

MUSCLE TISSUE

Think for a moment what life would be like without muscle tissue. Imagine being unable to sit, stand, walk, speak, or grasp objects. Imagine, too, how your internal functions would be affected. Blood would not circulate, because you would have no heartbeat to propel it through the vessels. You would be unable to breathe, speak, or eat, and food could not move along your digestive tract. In fact, there would be practically no movement along any of your internal passageways.

This is not to say that all life depends on muscle tissue. Some large organisms get by very nicely without it; we call them plants. But life as *we* live it would be impossible, because virtually all of our dynamic interactions with the environment involve muscle tissue.

10–1 SKELETAL MUSCLE TISSUE AND THE MUSCULAR SYSTEM

Objective

■ Specify the functions of skeletal muscle tissue.

Muscle tissue, one of the four primary types of tissue, consists chiefly of muscle cells that are highly specialized for contraction. Three types of muscle tissue exist: (1) *skeletal muscle*, (2) *cardiac muscle*, and (3) *smooth muscle*. ∞ p. 136 Without these muscle tissues, nothing in the body would move, and no body movement could occur. Skeletal muscle tissue moves the body by pulling on bones of the skeleton, making it possible for us to walk, dance, bite an apple, or play the ukulele. Cardiac muscle tissue pushes blood through the circulatory system. Smooth muscle tissue pushes fluids and solids along the digestive tract and regulates the diameters of small arteries, among other functions.

This chapter deals primarily with the structure and function of skeletal muscle tissue, in preparation for our discussion of the muscular system (Chapter 11). This chapter will also provide an overview of the differences among skeletal, cardiac, and smooth muscle tissues.

Skeletal muscle tissue forms **skeletal muscles**, organs that also contain connective tissues, nerves, and blood vessels. Each cell in skeletal muscle tissue is a single muscle *fiber*. Skeletal muscles are directly or indirectly attached to the bones of the skeleton. Our skeletal muscles perform the following five functions:

1. *Produce Skeletal Movement.* Skeletal muscle contractions pull on tendons and move the bones of the skeleton. The effects range from simple motions such as extending the

arm or breathing to the highly coordinated movements of swimming, skiing, or typing.

2. *Maintain Posture and Body Position.* Tension in our skeletal muscles maintains body posture—for example, holding your head in position when you read a book or balancing your body weight above your feet when you walk. Without constant muscular activity, we could not sit upright or stand.

3. *Support Soft Tissues.* The abdominal wall and the floor of the pelvic cavity consist of layers of skeletal muscle. These muscles support the weight of visceral organs and shield internal tissues from injury.

4. *Guard Entrances and Exits.* The openings of the digestive and urinary tracts are encircled by skeletal muscles. These muscles provide voluntary control over swallowing, defecation, and urination.

5. *Maintain Body Temperature.* Muscle contractions require energy; whenever energy is used in the body, some of it is converted to heat. The heat released by working muscles keeps our body temperature in the range required for normal functioning.

We will begin our discussion with the gross anatomy of a typical skeletal muscle. We will then consider, at the microscopic level, the structural features that make contractions possible.

10–2 ANATOMY OF SKELETAL MUSCLE

Objectives

- Describe the organization of muscle at the tissue level.
- Explain the unique characteristics of skeletal muscle fibers.
- Identify the structural components of a sarcomere.

Figure 10–1● illustrates the appearance and organization of a representative skeletal muscle. As we noted, a skeletal muscle contains connective tissues, blood vessels, and nerves, as well as skeletal muscle tissue.

☐ ORGANIZATION OF CONNECTIVE TISSUES

Three layers of connective tissue are part of each muscle: (1) an epimysium, (2) a perimysium, and (3) an endomysium. These layers and the relationships among them are diagrammed in Figure 10–1●.

The entire muscle is surrounded by the **epimysium** (ep-i-MĪZ-ē-um; *epi-*, on + *mys*, muscle), a dense layer of collagen fibers. The epimysium separates the muscle from surrounding tissues and organs. It is connected to the deep fascia, a dense connective tissue layer (See Figure 4–18●). ∞ p. 135

The connective tissue fibers of the **perimysium** (per-i-MĪZ-ē-um; *peri-*, around) divide the skeletal muscle into a series of compartments, each containing a bundle of muscle fibers called a **fascicle** (FAS-si-kl; *fasciculus*, a bundle). In addition to possessing collagen and elastic fibers, the perimysium contains blood

vessels and nerves that maintain blood flow and innervate the fascicles. Each fascicle receives branches of these blood vessels and nerves.

Within a fascicle, the delicate connective tissue of the **endomysium** (en-dō-MĪZ-ē-um; *endo-*, inside) surrounds the individual skeletal muscle cells, or *muscle fibers*, and interconnects adjacent muscle fibers. This connective tissue layer contains (1) capillary networks that supply blood to the muscle fibers; (2) **satellite cells**, embryonic stem cells that function in the repair of damaged muscle tissue; and (3) nerve fibers that control the muscle. All of these are in direct contact with the individual muscle fibers. ∞ p. 136 **AM** Disruption of Normal Muscle Organization

The collagen fibers of the endomysium and perimysium are interwoven and blend into one another. At each end of the muscle, the collagen fibers of the epimysium, perimysium, and endomysium come together to form a bundle known as a **tendon** or a broad sheet called an **aponeurosis**. Tendons and aponeuroses usually attach skeletal muscles to bones. Where they contact the bone, the collagen fibers extend into the bone matrix, providing a firm attachment. As a result, any contraction of the muscle will exert a pull on the attached bone (or bones).

☐ BLOOD VESSELS AND NERVES

The connective tissues of the epimysium and perimysium contain the blood vessels and nerves that supply the muscle fibers. Muscle contraction requires tremendous quantities of energy. An extensive vascular network delivers the necessary oxygen and nutrients and carries away the metabolic wastes generated by active skeletal muscles. The blood vessels and the nerve supply generally enter the muscle together and follow the same branching pattern through the perimysium. Within the endomysium, arterioles supply blood to a capillary network that surrounds each individual muscle fiber.

Skeletal muscles contract only under stimulation from the central nervous system. Axons, or *nerve fibers*, penetrate the epimysium, branch through the perimysium, and enter the endomysium to innervate individual muscle fibers. Skeletal muscles are often called voluntary muscles, because we have voluntary control over their contractions. Many skeletal muscles may also be controlled at a subconscious level. For example, skeletal muscles involved with breathing, such as the *diaphragm*, usually work outside our conscious awareness.

Next, we will examine the microscopic structure of a typical skeletal muscle fiber and will relate that microstructure to the physiology of the contraction process.

☐ SKELETAL MUSCLE FIBERS

Skeletal muscle fibers are quite different from the "typical" cells we described in Chapter 3. One obvious difference is size: Skeletal muscle fibers are enormous. A muscle fiber from a thigh muscle could have a diameter of 100 μm and a length equal to that of the

•**FIGURE 10–1**
The Organization of Skeletal Muscles. A skeletal muscle consists of fascicles (bundles of muscle fibers) enclosed by the epimysium. The bundles are separated by connective tissue fibers of the perimysium, and within each bundle the muscle fibers are surrounded by the endomysium. Each muscle fiber has many superficial nuclei, as well as mitochondria and other organelles (detailed in *Figure 10–3*).

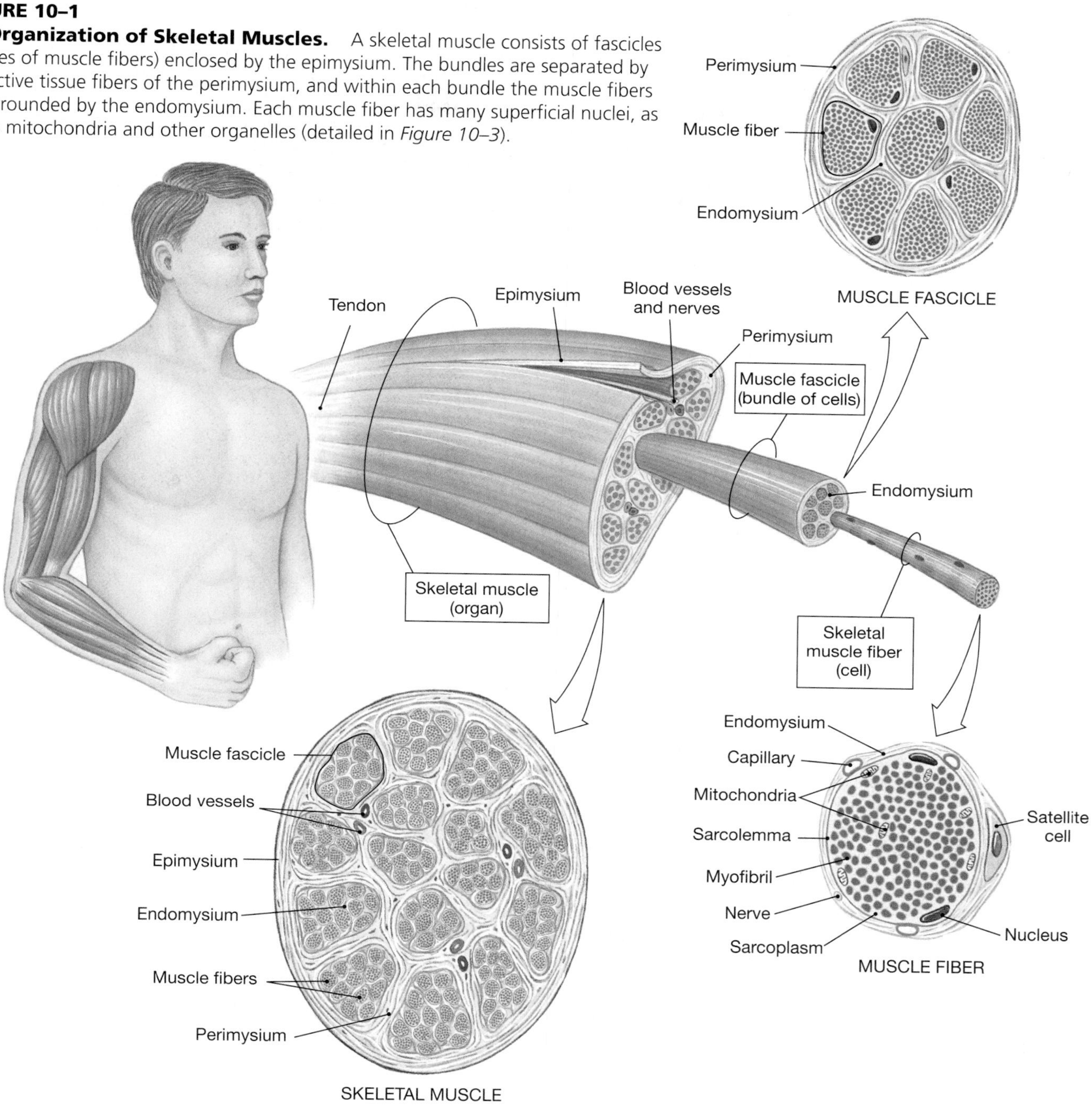

MUSCLE FASCICLE

- Perimysium
- Muscle fiber
- Endomysium

- Tendon
- Epimysium
- Blood vessels and nerves
- Perimysium
- Muscle fascicle (bundle of cells)
- Endomysium
- Skeletal muscle (organ)
- Skeletal muscle fiber (cell)

MUSCLE FIBER

- Endomysium
- Capillary
- Mitochondria
- Sarcolemma
- Myofibril
- Nerve
- Sarcoplasm
- Satellite cell
- Nucleus

SKELETAL MUSCLE

- Muscle fascicle
- Blood vessels
- Epimysium
- Endomysium
- Muscle fibers
- Perimysium

entire muscle (up to 30 cm, or 12 in.). A second obvious difference is that skeletal muscle fibers are *multinucleate*: Each skeletal muscle fiber contains hundreds of nuclei just internal to the cell membrane (Figure 10–2•). The genes in these nuclei direct the production of enzymes and structural proteins required for normal muscle contraction, and the presence of multiple copies of these genes speeds up the process. This feature is important because metabolic turnover tends to be very rapid in skeletal muscle fibers. ∞ p. 61

The distinctive features of size and multiple nuclei are related. During development, groups of embryonic cells called **myoblasts** fuse, forming individual skeletal muscle fibers (Figure

10–2a•). Each nucleus in a skeletal muscle fiber reflects the contribution of a single myoblast. Some myoblasts, however, do not fuse with developing muscle fibers. These unfused cells remain in adult skeletal muscle tissue as the satellite cells seen in Figures 10–1 and 10–2a,b•. After an injury, satellite cells may enlarge, divide, and fuse with damaged muscle fibers, thereby assisting in the regeneration of the tissue.

The Sarcolemma and Transverse Tubules

The **sarcolemma** (sar-kō-LEM-uh; *sarkos*, flesh + *lemma*, husk), or cell membrane of a muscle fiber, surrounds the

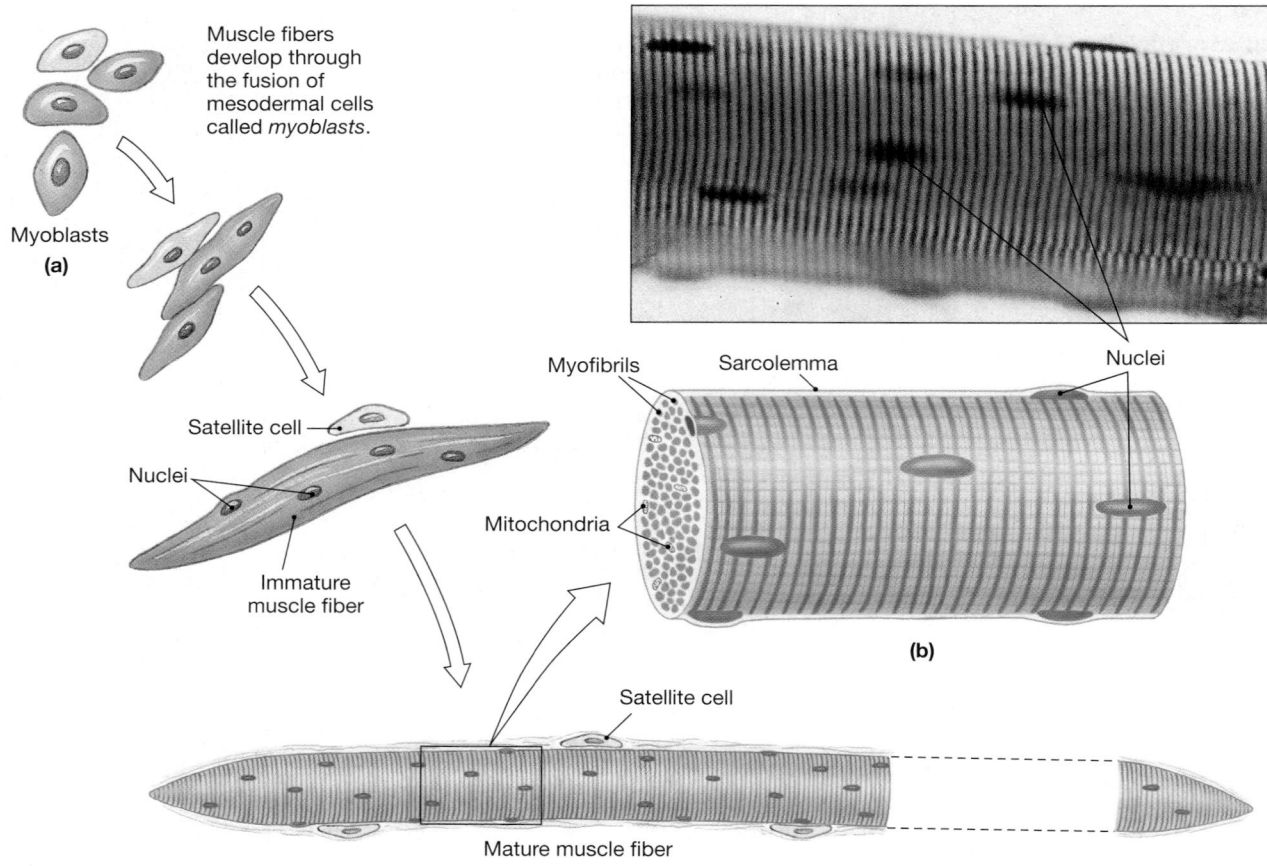

Muscle fibers develop through the fusion of mesodermal cells called *myoblasts*.

Myoblasts
(a)

Satellite cell

Nuclei

Immature
muscle fiber

Myofibrils

Sarcolemma

Nuclei

Mitochondria

(b)

Satellite cell

Mature muscle fiber

●**FIGURE 10–2**
The Formation and Nuclei of a Skeletal Muscle. **(a)** The formation of muscle fibers by the fusion of myoblasts. Notice the multiple nuclei. **(b)** A micrograph and diagrammatic view of one muscle fiber. (LM × 612)

sarcoplasm (SAR-kō-plazm), or cytoplasm of the muscle fiber (Figure 10–3a●). Like other cell membranes, the sarcolemma has a characteristic transmembrane potential due to the unequal distribution of positive and negative charges across the membrane. ∞ p. 99 In a skeletal muscle fiber, a sudden change in the transmembrane potential is the first step that leads to a contraction.

A skeletal muscle fiber is very large, but all regions of the cell must contract simultaneously. Thus, the signal to contract must be distributed quickly throughout the interior of the cell. This signal is conducted through the transverse tubules. **Transverse tubules**, or **T tubules**, are narrow tubes that are continuous with the sarcolemma and extend into the sarcoplasm at right angles to the cell surface (Figure 10–3b●). Filled with extracellular fluid, T tubules form passageways through the muscle fiber, like a series of tunnels through a mountain. The T tubules have the same general properties as the sarcolemma. Electrical impulses conducted by the sarcolemma travel along the T tubules. These impulses, or *action potentials*, are the trigger for muscle fiber contraction.

Myofibrils

Inside the muscle fiber, branches of the transverse tubules encircle cylindrical structures called **myofibrils** (Figure 10–3●). A

myofibril is 1–2 μm in diameter and as long as the entire cell. Each skeletal muscle fiber contains hundreds to thousands of myofibrils.

Myofibrils consist of bundles of **myofilaments**, protein filaments composed primarily of actin and myosin. The actin forms the bulk of **thin filaments**, and the myosin forms **thick filaments**. We introduced both types of filaments in Chapter 3. ∞ p. 75 Myofibrils, which can actively shorten, are responsible for skeletal muscle fiber contraction. At each end of the skeletal muscle fiber, the myofibrils are anchored to the inner surface of the sarcolemma. In turn, the outer surface of the sarcolemma is attached to collagen fibers of the tendon or aponeurosis of the skeletal muscle. As a result, when the myofibrils contract, the entire cell shortens. In doing so, it pulls on the tendon. Scattered among and around the myofibrils are mitochondria and granules of glycogen, the storage form of glucose. Glucose breakdown through glycolysis and mitochondrial activity provides the ATP needed to power muscular contractions. ∞ pp. 59–60

The Sarcoplasmic Reticulum

Wherever a transverse tubule encircles a myofibril, the tubule is tightly bound to the membranes of the sarcoplasmic reticu-

●**FIGURE 10–3**
The Structure of a Skeletal Muscle Fiber.
The internal organization of a muscle fiber.

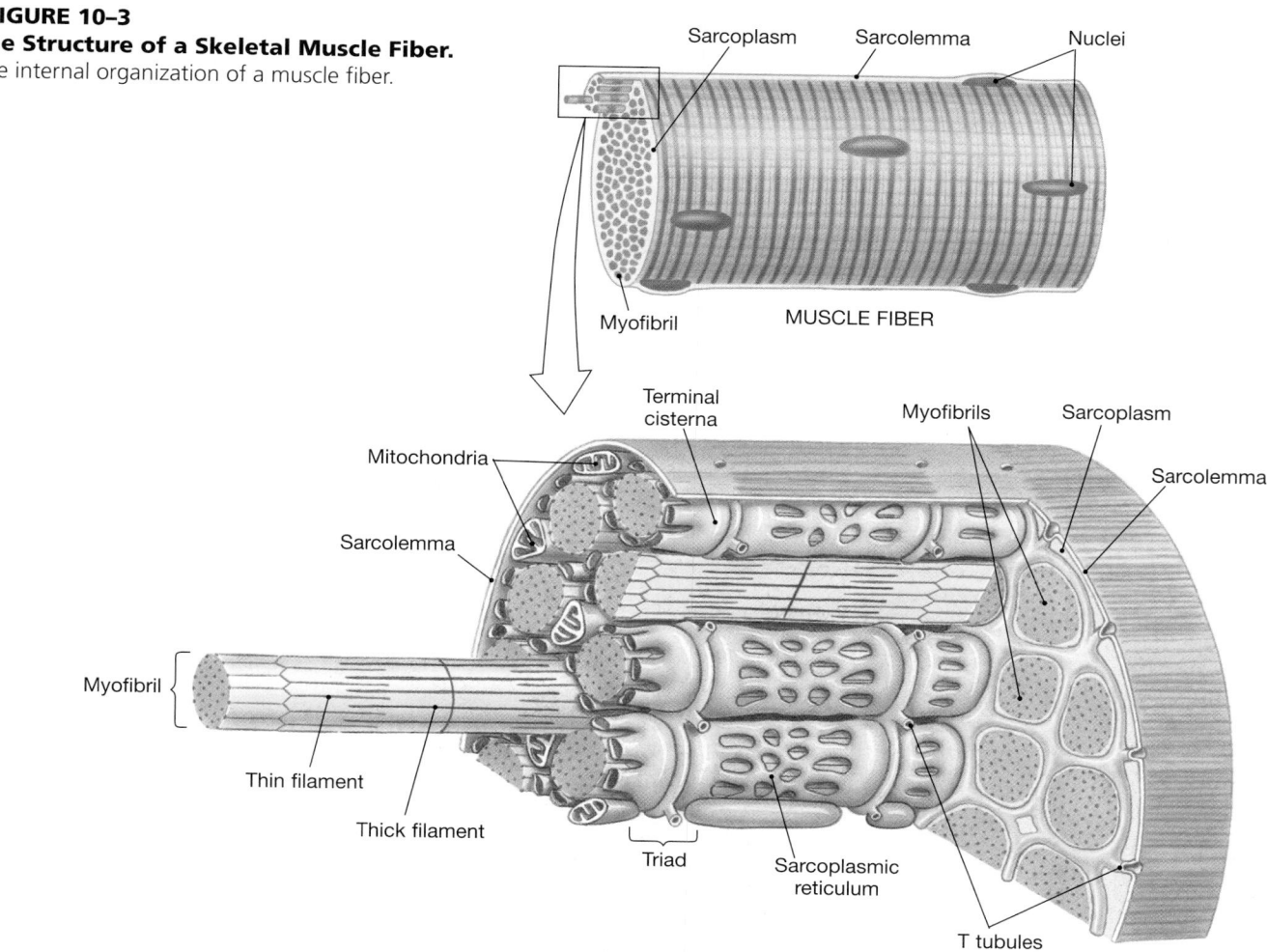

lum. The **sarcoplasmic reticulum (SR)** is a membrane complex similar to the smooth endoplasmic reticulum of other cells. In skeletal muscle fibers, the SR forms a tubular network around each individual myofibril (Figure 10–3b●). On either side of a T tubule, the tubules of the SR enlarge, fuse, and form expanded chambers called **terminal cisternae.** The combination of a pair of terminal cisternae plus a transverse tubule is known as a **triad.** Although the membranes of the triad are tightly bound together, their fluid contents are separate and distinct.

In Chapter 3, we noted the existence of special ion pumps that keep the intracellular concentration of calcium ions (Ca^{2+}) very low. ∞ p. 95 Most cells pump the calcium ions across their cell membranes and into the extracellular fluid. Although skeletal muscle fibers do pump Ca^{2+} out of the cell in this way, they also remove calcium ions from the sarcoplasm by actively transporting them into the terminal cisternae of the sarcoplasmic reticulum. The sarcoplasm of a resting skeletal muscle fiber contains very low concentrations of Ca^{2+}, around 10^{-7} mmol/l. The free Ca^{2+} concentration levels inside the terminal cisternae may be as much as 1000 times higher. In addition, cisternae contain the protein *calsequestrin*, which reversibly binds Ca^{2+}. Including both the free calcium and the bound calcium, the total concentration of Ca^{2+} inside cisternae can be 40,000 times that of the surrounding sarcoplasm.

A muscle contraction begins when stored calcium ions are released into the sarcoplasm. These ions then diffuse into individual contractile units called *sarcomeres.*

Sarcomeres

As we have seen, myofibrils are bundles of thin and thick filaments. These myofilaments are organized into repeating functional units called **sarcomeres** (SAR-kō-mērz; *sarkos*, flesh + *meros*, part).

A myofibril consists of approximately 10,000 sarcomeres, end to end. Each sarcomere has a resting length of 1.6–2.6 μm. Sarcomeres are the smallest functional units of the muscle fiber. Interactions between the thick and thin filaments of sarcomeres are responsible for muscle contraction. A sarcomere contains (1) thick filaments, (2) thin filaments, (3) proteins that stabilize the positions of the thick and thin filaments, and (4) proteins that regulate the interactions between thick and thin filaments.

Differences in the size, density, and distribution of thick filaments and thin filaments account for the banded appearance of

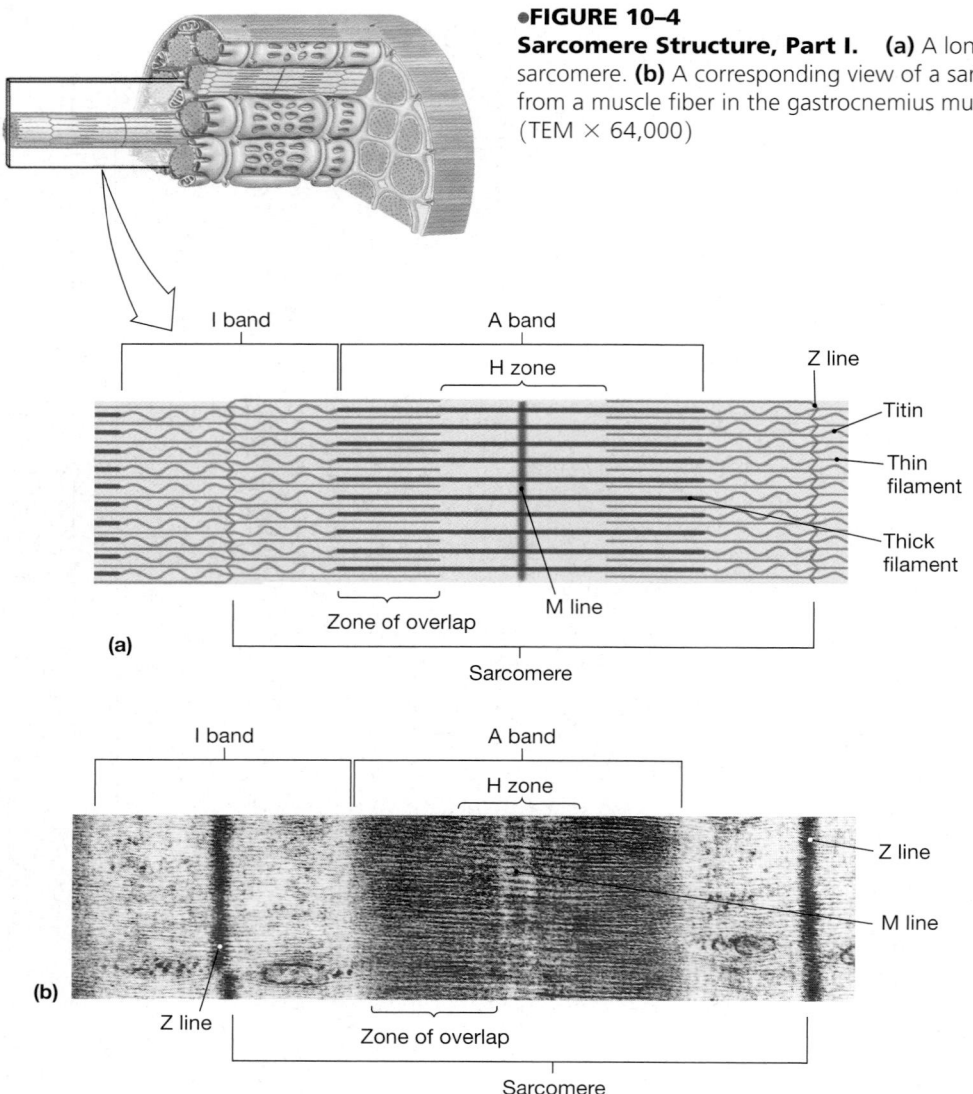

●**FIGURE 10–4**
Sarcomere Structure, Part I. **(a)** A longitudinal section of a sarcomere. **(b)** A corresponding view of a sarcomere in a myofibril from a muscle fiber in the gastrocnemius muscle of the calf. (TEM × 64,000)

each myofibril (Figure 10–4●). There are dark bands (**A bands**) and light bands (**I bands**). The names of these bands are derived from *anisotropic* and *isotropic*, which refer to the appearance of the bands when they are viewed under polarized light. You may find it helpful to remember that A bands are d**A**rk and that I bands are l**I**ght in a typical light micrograph.

THE A BAND The thick filaments are located at the center of a sarcomere, in the A band. The length of the A band is equal to the length of a typical thick filament. The A band, which also includes portions of thin filaments, contains the following three subdivisions (Figure 10–4●):

1. *The M Line.* The central portion of each thick filament is connected to its neighbors by proteins of the **M line**. These dark-staining proteins help stabilize the positions of the thick filaments.

2. *The H Zone.* In a resting sarcomere, the **H zone**, or *H band*, is a lighter region on either side of the M line. The H zone contains thick filaments, but no thin filaments.

3. *The Zone of Overlap.* In the **zone of overlap**, thin filaments are situated between the thick filaments. In this region, each thin filament is surrounded by three thick filaments, and each thick filament is surrounded by six thin filaments.

The cross sectional views in Figure 10–5● should help you visualize these features of the three-dimensional structure of the sarcomere.

THE I BAND Each I band, which contains thin filaments, but not thick filaments, extends from the A band of one sarcomere to the A band of the next sarcomere. **Z lines** mark the boundary between adjacent sarcomeres. The Z lines consist of proteins called *actinins*, which interconnect thin filaments of adjacent sarcomeres. From the Z lines at either end of the sarcomere, thin filaments extend toward the M line and into the zone of overlap. Strands of the protein **titin** extend from the tips of the thick filaments to attachment sites at the Z line (Figures 10–4a and 10–5●). Titin helps keep the thick and thin filaments in proper alignment; it also helps the muscle fiber

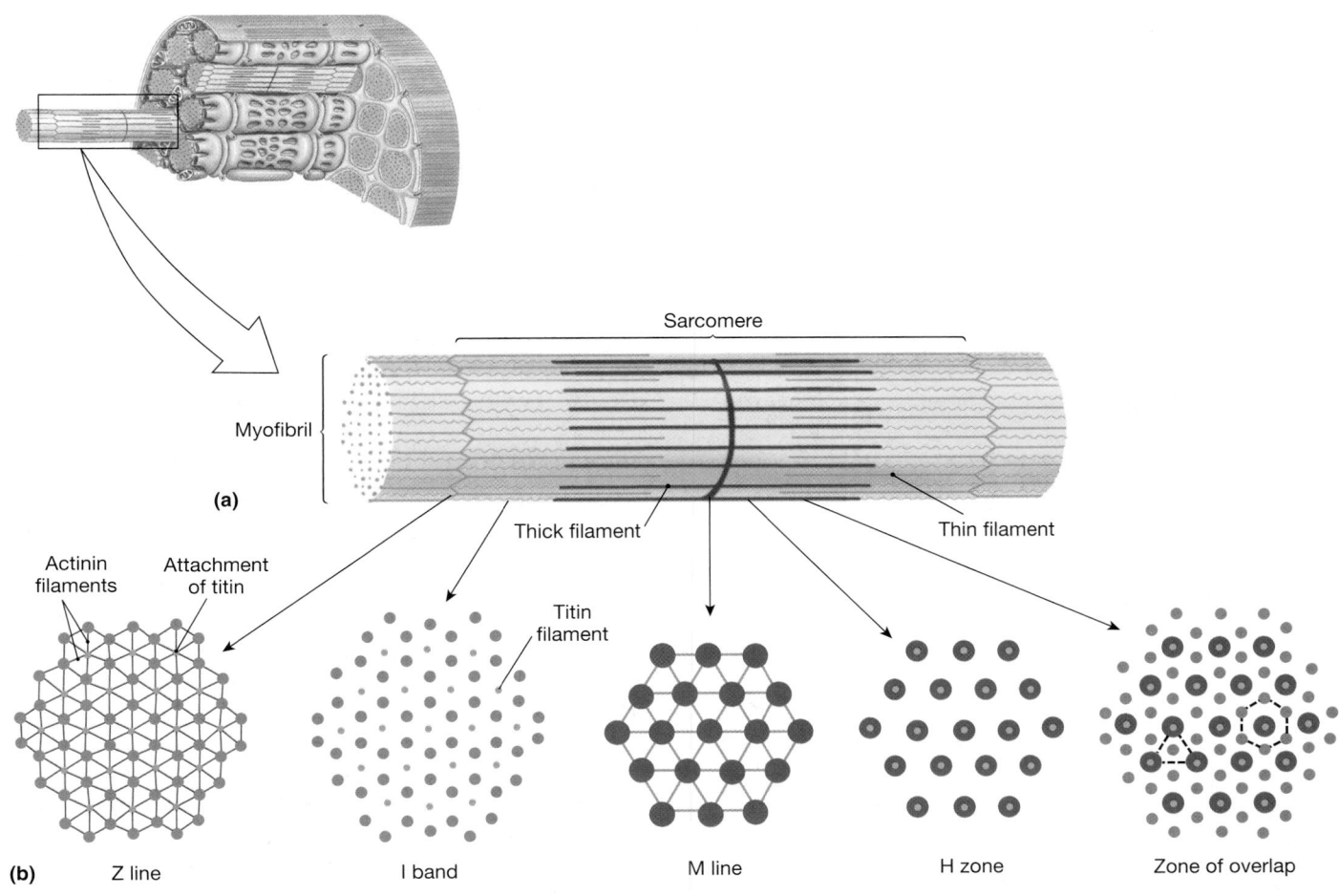

●FIGURE 10–5
Sarcomere Structure, Part II.   **(a)** A superficial view of a sarcomere. **(b)** Cross-sectional views of different portions of a sarcomere. Dashed lines show the relationships between thick and thin filaments in the zone of overlap.

resist extreme stretching that would otherwise disrupt the contraction mechanism.

Two transverse tubules encircle each sarcomere, and triads are located on each side of the M line at the zone of overlap. As a result, calcium ions released by the SR enter the regions where thick and thin filaments can interact.

Each Z line is surrounded by a meshwork of intermediate filaments that interconnect adjacent myofibrils. The myofibrils closest to the sarcolemma are bound to attachment sites on the inside of the membrane. Because the Z lines of all the myofibrils are aligned in this way, the muscle fiber as a whole has a banded appearance (Figure 10–2b●). These bands, or *striations*, are visible with the light microscope, so skeletal muscle tissue is also known as striated muscle. ∞ p. 136

Figure 10–6● reviews the levels of organization we have discussed so far. We now consider the molecular structure of the myofilaments responsible for muscle contraction.

THIN FILAMENTS A typical thin filament is 5–6 nm in diameter and 1 μm in length (Figure 10–7a●). A single thin filament contains four proteins: F actin, nebulin, tropomyosin, and troponin:

1. **F actin** is a twisted strand composed of two rows of 300–400 individual globular molecules of **G actin** (Figure 10–7b●). A long strand of **nebulin** spirals along the F actin strand in the cleft between the rows of G actin molecules. Nebulin holds the F actin strand together; as thin filaments develop, the length of the nebulin molecule probably determines the length of the F actin strand. Each molecule of G actin contains an **active site** that can bind to myosin much as a substrate molecule binds to the active site of an enzyme. ∞ p. 56 Under resting conditions, myosin binding is prevented by the *troponin–tropomyosin complex.*

2. Strands of **tropomyosin** (trō-pō-MĪ-ō-sin; *trope*, turning) cover the active sites and prevent actin–myosin interaction (Figure 10–7b●). A tropomyosin molecule is a double stranded protein that covers seven active sites. It is bound to one molecule of troponin midway along its length.

3. A **troponin** (TRŌ-pō-nin) molecule consists of three globular subunits. One subunit binds to tropomyosin, locking them together as a troponin–tropomyosin complex; a second subunit

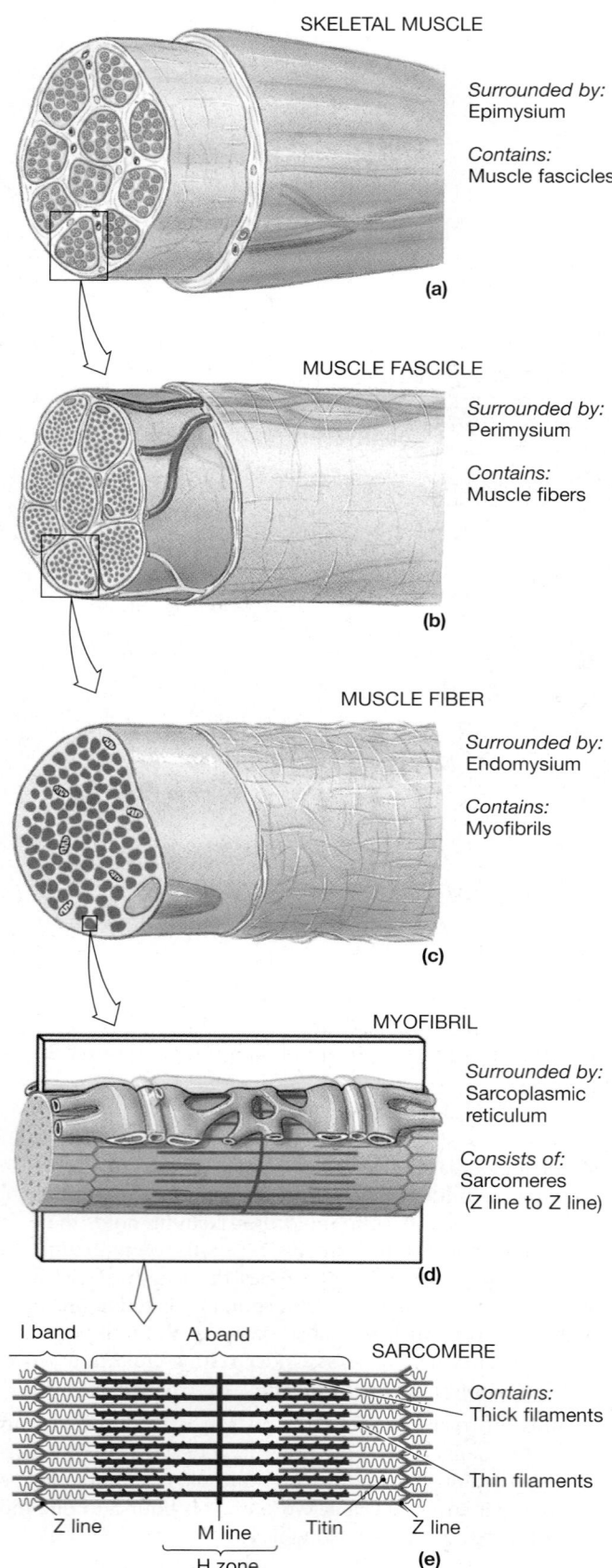

SKELETAL MUSCLE

Surrounded by:
Epimysium

Contains:
Muscle fascicles

(a)

MUSCLE FASCICLE

Surrounded by:
Perimysium

Contains:
Muscle fibers

(b)

MUSCLE FIBER

Surrounded by:
Endomysium

Contains:
Myofibrils

(c)

MYOFIBRIL

Surrounded by:
Sarcoplasmic
reticulum

Consists of:
Sarcomeres
(Z line to Z line)

(d)

I band   A band   SARCOMERE

Contains:
Thick filaments

Thin filaments

Z line   M line   Titin   Z line
H zone

(e)

●**FIGURE 10–6**
Levels of Functional Organization in a Skeletal Muscle Fiber

binds to one G actin, holding the troponin–tropomyosin complex in position; the third subunit has a receptor that binds a calcium ion. In a resting muscle, intracellular Ca^{2+} concentrations are very low, and that binding site is empty.

A contraction cannot occur unless the position of the troponin–tropomyosin complex changes, exposing the active sites on F actin. The necessary change in position takes place when calcium ions bind to receptors on the troponin molecules.

At either end of the sarcomere, the thin filaments are attached to the Z line (Figure 10–7a●). Although it is called a *line* because it looks like a dark line on the surface of the myofibril, the Z line in sectional view is more like an open meshwork (Figure 10–5b●). For this reason, the Z line is often called the *Z disc*.

THICK FILAMENTS Thick filaments are 10–12 nm in diameter and 1.6 μm long (Figure 10–7c●). A thick filament contains roughly 500 myosin molecules, each made up of a pair of myosin subunits twisted around one another (Figure 10–7d●). The long **tail** is bound to other myosin molecules in the thick filament. The free **head**, which projects outward toward the nearest thin filament, has two globular protein subunits. When the myosin heads interact with thin filaments during a contraction, they are known as **cross-bridges**. The connection between the head and the tail functions as a hinge that lets the head pivot at its base. When pivoting occurs, the head swings toward or away from the M line. As we will see in a later section, this pivoting is the key step in muscle contraction.

All the myosin molecules are arranged with their tails pointing toward the M line (Figure 10–7c●). The H zone includes a central region where there are no myosin heads. Elsewhere, the myosin heads are arranged in a spiral, each facing one of the surrounding thin filaments.

Each thick filament has a core of titin. On either side of the M line, a strand of titin extends the length of the thick filament and then continues across the I band to the Z line on that side. The portion of the titin strand exposed within the I band is highly elastic and will recoil after stretching. In the normal resting sarcomere, the titin strands are completely relaxed; they become tense only when some external force stretches the sarcomere.

☐ THE SLIDING FILAMENT THEORY

When a skeletal muscle fiber contracts, (1) the H zones and I bands get smaller, (2) the zones of overlap get larger, (3) the Z lines move closer together, and (4) the width of the A band remains constant. These observations make sense only if the thin filaments are sliding toward the center of each sarcomere, alongside the thick filaments (Figure 10–8a●). The contraction ends with the disappearance of the I bands, at which point the Z lines are in contact with the ends of the thick filaments.

During a contraction, sliding occurs in every sarcomere along the myofibril (Figure 10–8b●). As a result, the myofibril gets shorter. Because myofibrils are attached to the sarcolemma at

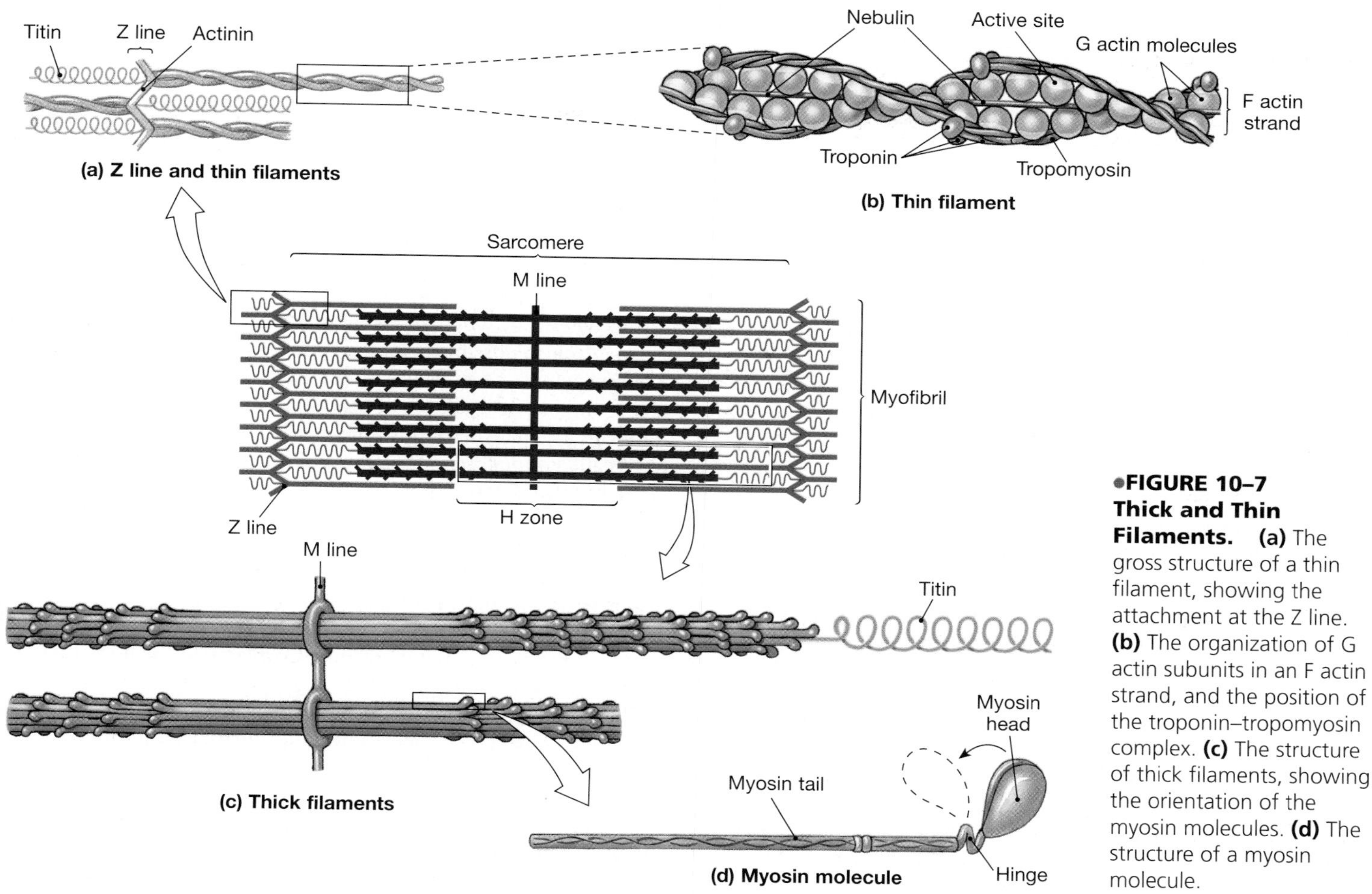

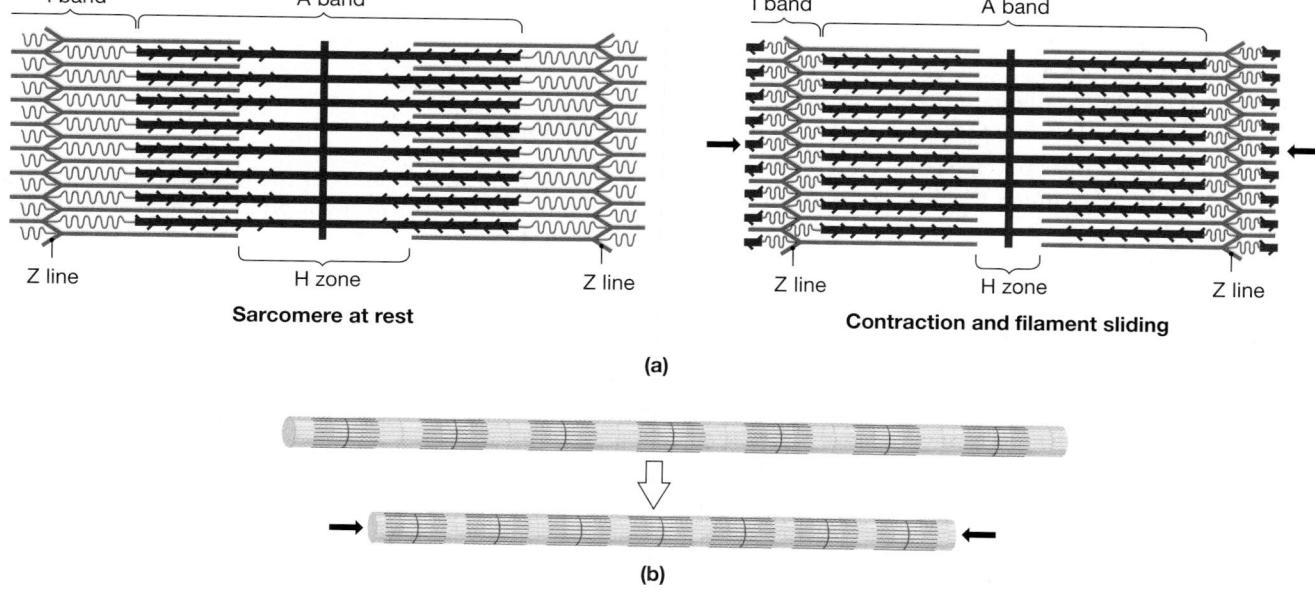

Titin   Z line   Actinin

(a) Z line and thin filaments

Nebulin   Active site   G actin molecules

F actin strand

Troponin   Tropomyosin

(b) Thin filament

Sarcomere

M line

Myofibril

Z line

H zone

M line   Titin

Myosin head

Myosin tail

(c) Thick filaments

Hinge

(d) Myosin molecule

•FIGURE 10–7
Thick and Thin Filaments. **(a)** The gross structure of a thin filament, showing the attachment at the Z line. **(b)** The organization of G actin subunits in an F actin strand, and the position of the troponin–tropomyosin complex. **(c)** The structure of thick filaments, showing the orientation of the myosin molecules. **(d)** The structure of a myosin molecule.

•FIGURE 10–8
Changes in the Appearance of a Sarcomere during the Contraction of a Skeletal Muscle Fiber. **(a)** During a contraction, the A band stays the same width, but the Z lines move closer together and the I band gets smaller. **(b)** When the ends of a myofibril are free to move, the sarcomeres shorten simultaneously and the ends of the myofibril are pulled toward its center.

I band   A band

Z line   H zone   Z line

Sarcomere at rest

I band   A band

Z line   H zone   Z line

Contraction and filament sliding

(a)

(b)

The Muscular Dystrophies

The **muscular dystrophies** (DIS-trō-fēz) are inherited diseases that produce progressive muscle weakness and deterioration. One of the most common and best understood conditions is **Duchenne's muscular dystrophy** (**DMD**). This form of muscular dystrophy appears in childhood, commonly between the ages of 3 and 7. The condition affects males only; the incidence is roughly 30 per 100,000 male births. A progressive muscular weakness develops, and the child typically requires a wheelchair by age 12. Most individuals die before age 20, due to respiratory paralysis or cardiac problems. Skeletal muscles are primarily affected, although for some reason the facial muscles continue to function normally. In later stages of the disease, the facial muscles and cardiac muscle tissue may also become involved.

The skeletal muscle fibers in a person with DMD are structurally different from those of other individuals. Abnormal membrane permeability, cholesterol content, rates of protein synthesis, and enzyme composition have been reported. Individuals with DMD also lack the protein *dystrophin*, found in normal muscle fibers. Dystrophin is a large protein that attaches thin filaments of the sarcomeres to an anchoring protein on the sarcolemma. Dystrophin provides mechanical strength to the muscle fiber and connects the myofibrils to the sarcolemma; it is also suspected to play a role in the regulation of calcium ion channels in the sarcolemma. In children with DMD, calcium channels remain open for an extended period, and calcium levels rise in the sarcoplasm to the point at which key proteins denature. The muscle fiber then degenerates. Researchers have recently identified and cloned the gene for dystrophin production; that gene is located on the X chromosome. Rats with DMD have been cured by the insertion of this gene into their muscle fibers, a technique that may eventually be used to treat humans.

The inheritance of DMD is sex linked: Women carrying the defective genes are unaffected, but each of their male children will have a 50 percent chance of developing DMD. Now that the specific location of the gene has been identified, it is possible to determine whether a woman is carrying the defective gene. It is also possible to use an innovative prenatal test to determine whether a fetus has inherited the condition. In this procedure, a small sample of fluid is collected from the membranous sac that surrounds the fetus. The fluid contains fetal cells called *amniocytes*, which are collected and cultivated in the laboratory. Researchers then insert a gene called *MyoD*, which triggers the differentiation of the amniocytes into skeletal muscle fibers. These cells can then be tested not only for the signs of muscular dystrophy, but also for indications of other inherited muscular disorders.

Attempts have been made to give children with DMD injections that contain donated normal myoblasts, in the hope that these myoblasts will fuse with developing muscle fibers and will provide normal dystrophin-production genes. Results have been inconclusive. Animal studies indicate that symptoms of DMD develop when fewer than 20 percent of the muscle fibers contain normal dystrophin. In one recent study that used muscle biopsies, 1–10 percent of the muscle fibers of treated children contained normal dystrophin. Efforts to improve the delivery method and increase the treatment's effects are under way. **AM** Myotonic Dystrophy

---

each Z line and at either end of the muscle fiber, when myofibrils get shorter, so does the muscle fiber.

The explanation for the physical changes that occur during contraction is called the **sliding filament theory**. The sliding filament theory accurately describes what happens to the sarcomere during a contraction, but it does not explain the mechanism involved. To do so, we must examine the contraction process in greater detail.

✓ How would severing the tendon attached to a muscle affect the muscle's ability to move a body part?

✓ Why does skeletal muscle appear striated when viewed through a microscope?

✓ Where would you expect the greatest concentration of Ca^{2+} in resting skeletal muscle to be?

Answers start on page Q-1

The gross anatomy of skeletal muscle can be viewed on the **IP CD-ROM**: Muscular System/Anatomy Review.

10–3 THE CONTRACTION OF SKELETAL MUSCLE

Objectives

- Identify the components of the neuromuscular junction, and summarize the events involved in the neural control of skeletal muscles.
- Explain the key steps involved in the contraction of a skeletal muscle fiber.

Most of the rest of this chapter is devoted to how muscles contract and how those contractions are harnessed to do what you want them to do. Before we can proceed, you have to understand some basic physical principles that apply to muscle cells. Muscle cells are specialized to contract. The individual cells in muscle tissue are tied together, primarily by collagen fibers. When the muscle cells contract, they pull on those fibers the way a group of people might pull on a rope. The pull, called *tension*, is an active force: Energy must be expended to produce it. Tension is applied *to* some object, whether a rope, a rubber band, or a book on a tabletop.

Tension applied to an object tends to pull the object toward the source of the tension. However, before movement can occur, the applied tension must overcome the object's *resistance*, a passive force that opposes movement. The amount of resistance can depend on the weight of the object, its shape, friction, and other factors. When the applied tension exceeds the resistance, the object moves. In contrast, *compression*, or a push applied to an object, tends to force the object away from the source of the compression. Again, no movement can occur until the applied compression exceeds the resistance of the object. Muscle cells can use energy to shorten and generate tension, through interactions between thick and thin filaments, but not to lengthen and generate compression.

With that background, we can investigate the mechanics of muscle contraction in some detail. Figure 10–9● provides an overview of the "big picture" we will be examining.

- Normal skeletal muscle contraction occurs only when skeletal muscle fibers are activated by neurons whose cell bodies are in the central nervous system. A neuron can activate a muscle fiber through stimulation of its sarcolemma. What follows is called *excitation–contraction coupling*.
- The first step in excitation–contraction coupling is the release of calcium ions from the cisternae of the sarcoplasmic reticulum.
- The calcium ions then trigger interactions between thick filaments and thin filaments, resulting in muscle fiber contraction.
- These filament interactions produce tension and consume energy.

☐ THE CONTROL OF SKELETAL MUSCLE ACTIVITY

Skeletal muscle fibers contract only under the control of the nervous system. Communication between the nervous system and a skeletal muscle fiber occurs at a specialized intercellular connection known as a **neuromuscular junction (NMJ)**, or *myoneural junction*. One such junction is shown in Figure 10–10a●.

Each skeletal muscle fiber is controlled by a neuron at a single neuromuscular junction midway along the fiber's length. Figure 10–10b● summarizes key features of this structure. A single axon branches within the perimysium to form a number of fine branches. Each branch ends at an expanded **synaptic terminal**. The cytoplasm of the synaptic terminal contains mitochondria and vesicles filled with molecules of **acetylcholine** (as-ē-til-KŌ-lēn), or **ACh**. Acetylcholine is a *neurotransmitter*, a chemical released by a neuron to change the permeability or other properties of another cell membrane. In this case, the release of ACh from the synaptic terminal can alter the permeability of the sarcolemma and trigger the contraction of the muscle fiber.

The **synaptic cleft**, a narrow space, separates the synaptic terminal of the neuron from the opposing sarcolemmal surface. This surface, which contains membrane receptors that bind

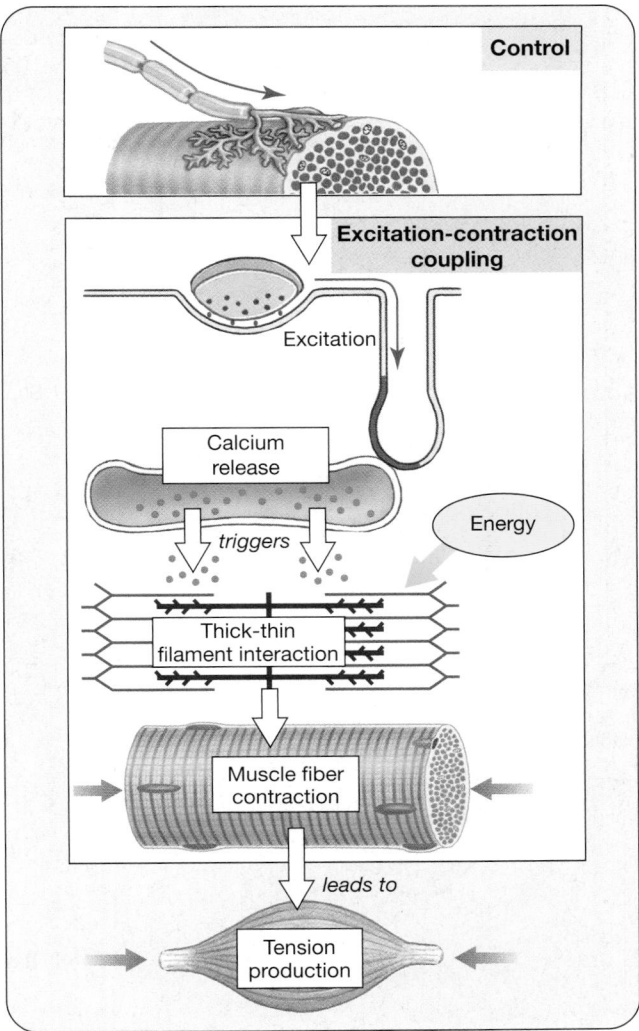

●FIGURE 10–9
An Overview of the Process of Skeletal Muscle Contraction. The major factors are indicated here as a series of interrelated steps and processes. Each factor will be described further in a related section of the text. A simplified version of this figure will appear in later figures as a Navigator icon; its presence indicates that we are taking another step in the discussion.

ACh, is known as the **motor end plate**. The motor end plate has deep creases called *junctional folds*, which increase its surface area and thus the number of available ACh receptors. The synaptic cleft and the sarcolemma also contain molecules of the enzyme **acetylcholinesterase** (**AChE**, or *cholinesterase*), which breaks down ACh.

When a neuron stimulates a muscle fiber, the process occurs in a series of steps (Figure 10–10c●):

STEP 1: *The Arrival of an Action Potential.* The stimulus for ACh release is the arrival of an electrical impulse, or **action potential**, at the synaptic terminal. An action potential is a sudden change in the transmembrane potential that travels along the length of the axon.

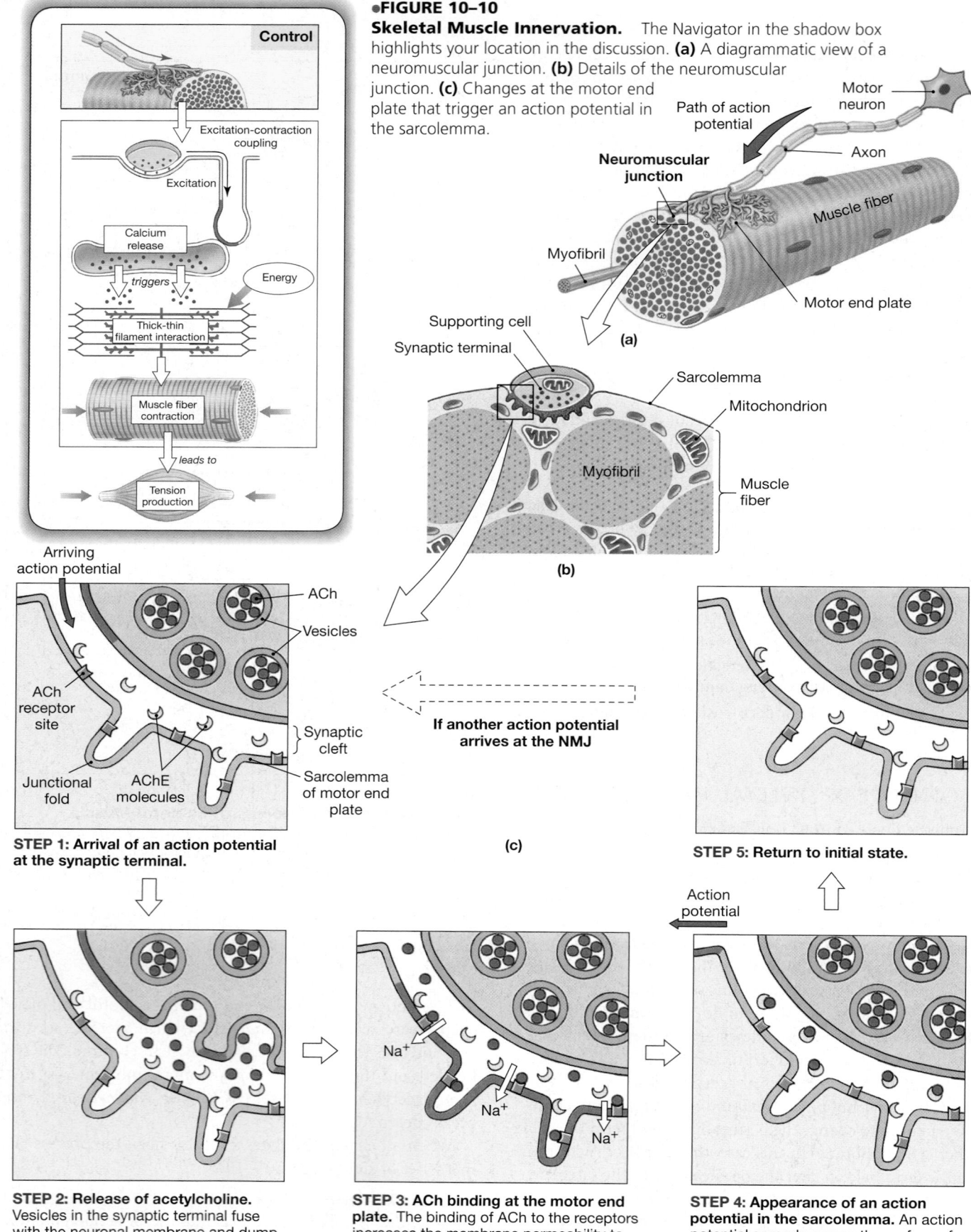

Control

Excitation-contraction coupling

Excitation

Calcium release

triggers

Energy

Thick-thin filament interaction

Muscle fiber contraction

leads to

Tension production

●FIGURE 10–10
Skeletal Muscle Innervation.   The Navigator in the shadow box highlights your location in the discussion. **(a)** A diagrammatic view of a neuromuscular junction. **(b)** Details of the neuromuscular junction. **(c)** Changes at the motor end plate that trigger an action potential in the sarcolemma.

Motor neuron

Path of action potential

Axon

Neuromuscular junction

Muscle fiber

Myofibril

Motor end plate

(a)

Supporting cell

Synaptic terminal

Sarcolemma

Mitochondrion

Myofibril

Muscle fiber

(b)

Arriving action potential

ACh

Vesicles

ACh receptor site

Synaptic cleft

Junctional fold

AChE molecules

Sarcolemma of motor end plate

STEP 1: Arrival of an action potential at the synaptic terminal.

If another action potential arrives at the NMJ

(c)

STEP 5: Return to initial state.

Action potential

STEP 2: Release of acetylcholine.
Vesicles in the synaptic terminal fuse with the neuronal membrane and dump their contents into the synaptic cleft.

Na⁺
Na⁺
Na⁺

STEP 3: ACh binding at the motor end plate. The binding of ACh to the receptors increases the membrane permeability to sodium ions. Sodium ions then rush into the cell.

STEP 4: Appearance of an action potential in the sarcolemma. An action potential spreads across the surface of the sarcolemma. While this occurs, AChE removes the ACh.

STEP 2: *The Release of ACh.* When the action potential reaches the synaptic terminal, permeability changes in the membrane trigger the exocytosis of ACh into the synaptic cleft. This is accomplished when vesicles in the synaptic terminal fuse with the cell membrane of the neuron.

STEP 3: *ACh Binding at the Motor End Plate.* ACh molecules diffuse across the synaptic cleft and bind to ACh receptors on the surface of the sarcolemma at the motor end plate. ACh binding changes the permeability of the motor end plate to sodium ions. Recall from Chapter 3 that the extracellular fluid contains a high concentration of sodium ions, whereas sodium ion concentrations inside the cell are very low. ∞ p. 96 When the membrane permeability to sodium increases, sodium ions rush into the sarcoplasm. This influx continues until AChE removes the ACh from the receptors.

STEP 4: *Appearance of an Action Potential in the Sarcolemma.* The sudden inrush of sodium ions results in the generation of an action potential in the sarcolemma. This electrical impulse originates at the edges of the motor end plate, sweeps across the entire membrane surface, and travels along each T tubule. The arrival of an action potential at the synaptic terminal thus leads to the appearance of an action potential in the sarcolemma.

STEP 5: *Return to Initial State.* Even before the action potential has spread across the entire sarcolemma, the ACh has been broken down by AChE. Some of the breakdown products will be absorbed by the synaptic terminal and used to resynthesize ACh for subsequent release. This sequence of events can now be repeated should another action potential arrive at the synaptic terminal.

☐ EXCITATION–CONTRACTION COUPLING

The link between the generation of an action potential in the sarcolemma and the start of a muscle contraction is called **excitation–contraction coupling**. This coupling occurs at the triads. On reaching a triad, an action potential triggers the release of Ca^{2+} from the cisternae of the sarcoplasmic reticulum. The change in the permeability of the SR to Ca^{2+} is temporary, lasting only about 0.03 second. Yet within a millisecond, the Ca^{2+} concentration in and around the sarcomere reaches 100 times resting levels. Because the terminal cisternae are situated at the zones of overlap, where the thick and thin filaments interact, the effect of calcium release on the sarcomere is almost instantaneous.

Troponin is the lock that keeps the active sites inaccessible. Calcium is the key to that lock. Recall from Figure 10–7 that troponin binds to actin and to tropomyosin, and the tropomyosin molecules cover the active sites and prevent interactions between thick filaments and thin filaments. Each troponin molecule also has a binding site for calcium; this site is empty when the muscle fiber is at rest. Figure 10–11● illustrates what happens when calcium ions bind to troponin. Calcium binding changes the shape of the troponin molecule and weakens the bond between troponin and actin. The troponin molecule then changes position, swinging the tropomyosin strand away from the active sites. At this point, the **contraction cycle** begins.

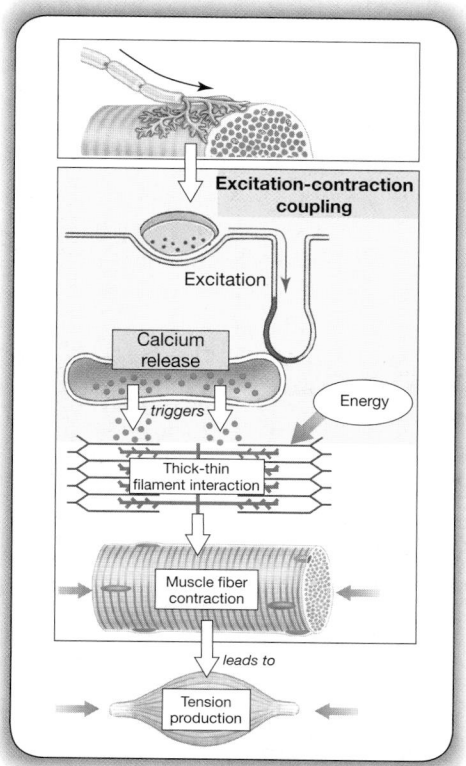

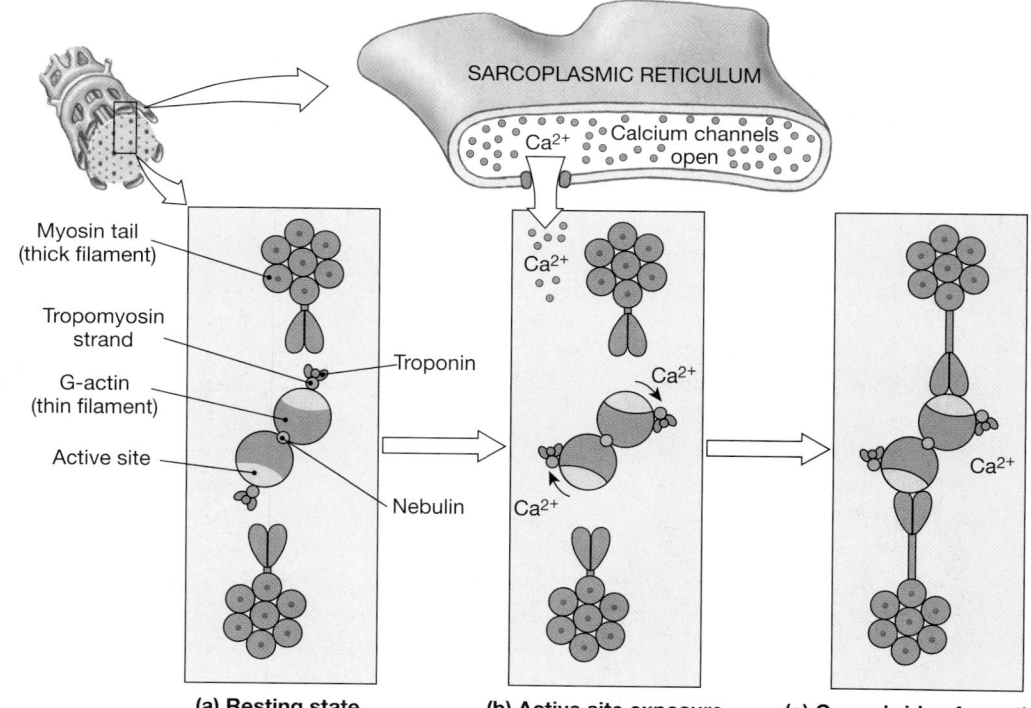

(a) Resting state   (b) Active site exposure   (c) Cross-bridge formation

●**FIGURE 10–11**
The Exposure of Active Sites.   **(a)** In a resting sarcomere, the tropomyosin strands cover the active sites on the thin filaments, preventing cross-bridge formation. **(b)** When calcium ions enter the sarcomere, they bind to troponin, which rotates and swings the tropomyosin away from the active sites. **(c)** Cross-bridge formation then occurs, and the contraction cycle begins.

The Contraction Cycle

Figure 10–12● details the molecular events that occur during the contraction cycle. In the resting sarcomere, each myosin head is already "energized"—charged with the energy that will be used to power a contraction. The myosin head functions as ATPase, an enzyme that can break down ATP. ∞ p. 60 At the start of the contraction cycle, each myosin head has already split a molecule of ATP and stored the energy released in the process. The breakdown products, ADP and phosphate (often represented as P), remain bound to the myosin head.

●**FIGURE 10–12**
The Contraction Cycle

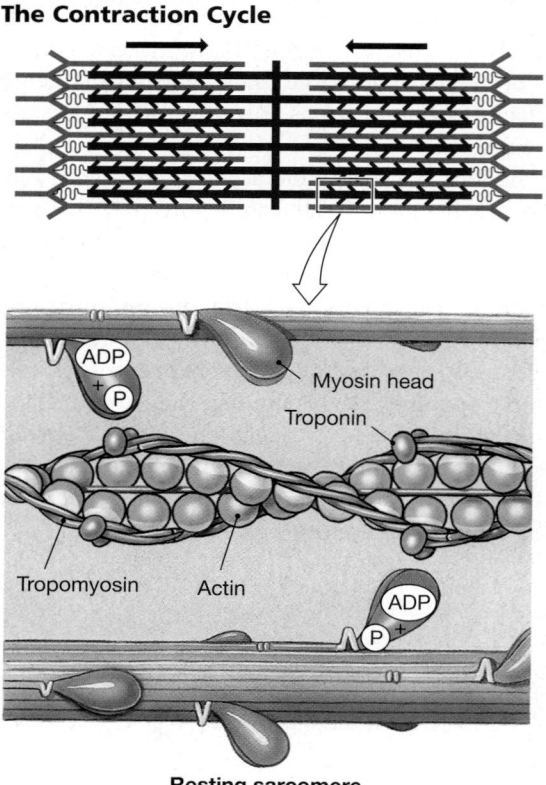

Resting sarcomere

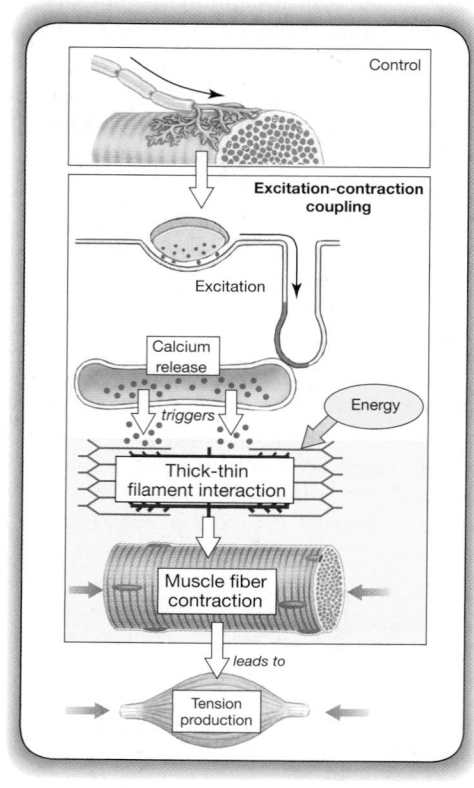

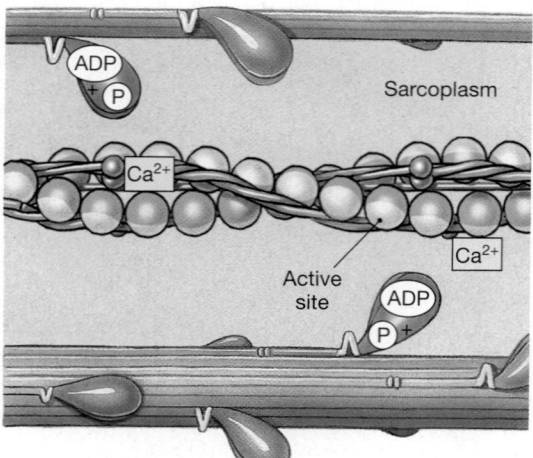

STEP 1: Active-site exposure

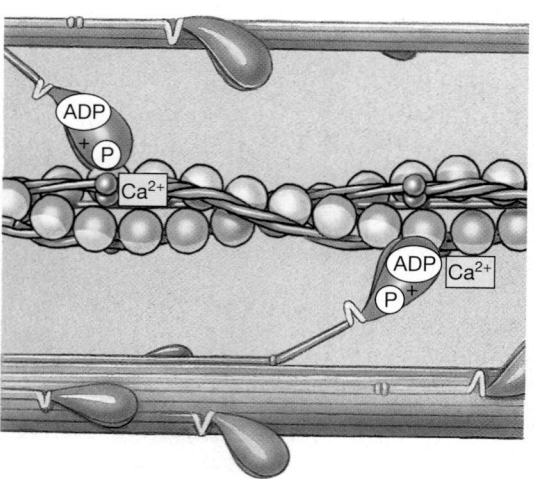

STEP 2: Cross-bridge attachment

The contraction cycle involves five interlocking steps (Figure 10–12●):

STEP 1: *Exposure of Active Sites.* The calcium ions entering the sarcoplasm bind to troponin. This binding weakens the bond between the troponin–tropomyosin complex and actin. The troponin molecule then changes position, pulling the tropomyosin molecule away from the active sites and allowing interaction with the energized myosin heads.

STEP 2: *Formation of Cross-Bridges.* When the active sites are exposed, the energized myosin heads bind to them, forming cross-bridges.

STEP 3: *Pivoting.* In the resting sarcomere, each myosin head points away from the M line. In this position, the myosin head is "cocked" like the spring in a mousetrap. Cocking the myosin head requires energy, which is obtained by breaking down ATP into ADP and a phosphate group. In the cocked position, both the ADP and the phosphate are still bound to the myosin head. After cross-bridge formation, the stored energy is released as the myosin head pivots toward the M line. This action is called the *power stroke*; when it occurs, the ADP and phosphate group are released.

STEP 4: *Detachment of Cross-Bridges.* When an ATP binds to the myosin head, the link between the active site on the actin molecule and the myosin head is broken. The active site is now exposed and able to form another cross-bridge.

STEP 5: *Reactivation of Myosin.* Myosin reactivation occurs when the free myosin head splits the ATP into ADP and a phosphate group. The energy released in this process is used to recock the myosin head. The entire cycle can now be repeated. If calcium ion concentrations remain elevated and ATP reserves are sufficient, each myosin head will repeat the cycle about five times per second. Each power stroke shortens the sarcomere by about 1 percent, so each second the sarcomere can shorten by roughly 5 percent. Because all the sarcomeres contract together, the entire muscle shortens at the same rate.

To appreciate the overall effect, imagine that you are pulling on a large rope. You are the myosin head, and the rope is a thin filament. You reach forward, grab the rope with both hands, and pull it toward you. This action corresponds to cross-bridge attachment and pivoting. You now release the rope, reach forward, and grab it once again. By repeating the cycle over and over, you can gradually pull in the rope.

Now consider several people lined up, all pulling on the same rope, as in a tug-of-war. Each person reaches forward, grabs the rope, pulls it, releases it, and then grabs it again to repeat the cycle. The individual actions are not coordinated: At any one moment, some people are grabbing, some are pulling, and others are letting go. The amount of tension produced is a function of how many people are pulling at any given instant. A comparable situation applies to tension in a muscle fiber, where the myosin heads along a thick filament work together to pull a thin filament toward the center of the sarcomere.

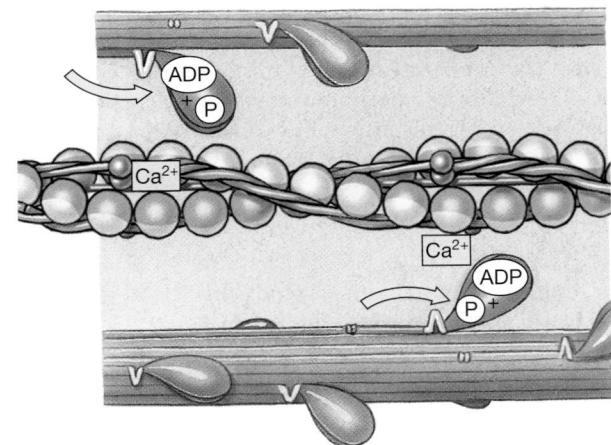

STEP 5: Myosin reactivation

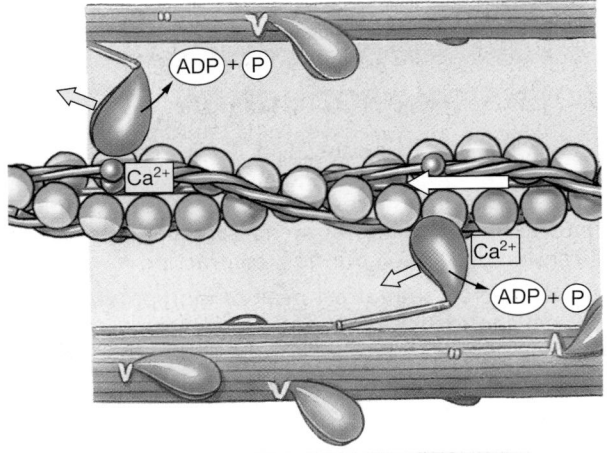

STEP 3: Pivoting of myosin head

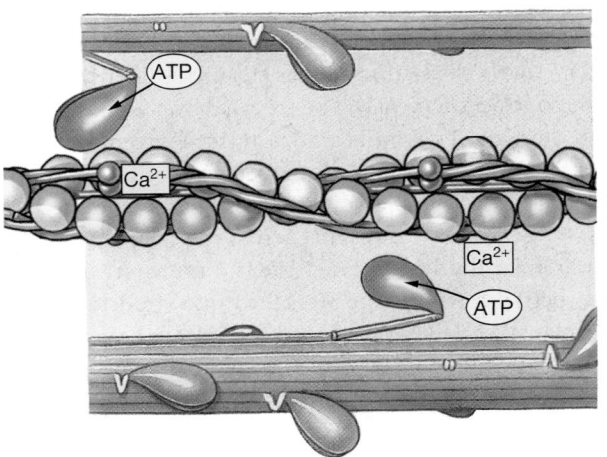

STEP 4: Cross-bridge detachment

Each myofibril consists of a string of sarcomeres, and in a contraction all of the thin filaments are pulled toward the centers of the sarcomeres. You might wonder how this can happen, since the thin filaments on one side of the Z line are being pulled in one direction and the thin filaments on the other side are being pulled in the opposite direction. In fact, if you securely anchor the two ends of the myofibril so that they cannot move, thick and thin filament interactions generate tension but sliding cannot occur. If you release both ends, the myofibril will shorten as both ends move toward the center of the myofibril, as illustrated in Figure 10–8b●.

Anything that interferes with neural function or with excitation–contraction coupling will cause muscular paralysis. Two examples are worth noting:

1. **Botulism** results from the consumption of contaminated canned or smoked foods that contain a toxin. The toxin, produced by bacteria, prevents the release of ACh at the synaptic terminals, leading to a potentially fatal muscular paralysis.

2. The progressive muscular paralysis of **myasthenia gravis** results from the loss of ACh receptors at the junctional folds. The primary cause is a misguided attack on the ACh receptors by the immune system. Genetic factors play a role in predisposing individuals to this condition.
AM Problems with the Control of Muscle Activity

☐ RELAXATION

The duration of a contraction depends on (1) the duration of stimulation at the neuromuscular junction, (2) the presence of free calcium ions in the sarcoplasm, and (3) the availability of ATP. A single stimulus has only a brief effect on a muscle fiber because the ACh that is released after a single action potential arrives at the synaptic terminal does not remain intact for long. Whether it is bound to the sarcolemma or is free in the synaptic cleft, the released ACh is rapidly broken down and inactivated by AChE. Inside the muscle fiber, the permeability changes in the SR are also very brief. A contraction will therefore continue only if additional action potentials arrive at the synaptic terminal in rapid succession. When they do, the continual release of ACh into the synaptic cleft produces a series of action potentials in the sarcolemma that keeps Ca^{2+} levels elevated in the sarcoplasm. Under these conditions, the contraction cycle will be repeated over and over.

If just one action potential arrives at the neuromuscular junction, Ca^{2+} concentrations in the sarcoplasm will quickly return to normal resting levels. Two mechanisms are involved in this process: (1) active Ca^{2+} transport across the cell membrane into the extracellular fluid and (2) active Ca^{2+} transport into the SR. Of the two, transport into the SR is the more important. Virtually as soon as the calcium ions have been released, the SR returns to its normal permeability and begins actively absorbing Ca^{2+} from the surrounding sarcoplasm. As Ca^{2+} concentrations in the sarcoplasm fall, (1) calcium ions detach from troponin, (2) troponin returns to its original position, and (3) the active sites are re-covered by tropomyosin. The contraction has ended.

Once the contraction has ended, the sarcomere does not automatically return to its original length. Sarcomeres shorten actively, but there is no active mechanism for reversing the process. External forces must act on the contracted muscle fiber to stretch the myofibrils and sarcomeres to their original dimensions. We will describe those forces in a later section.

When death occurs, circulation ceases and the skeletal muscles are deprived of nutrients and oxygen. Within a few hours, the skeletal muscle fibers have run out of ATP and the sarcoplasmic reticulum becomes unable to pump Ca^{2+} out of the sarcoplasm. Calcium ions diffusing into the sarcoplasm from the extracellular fluid or leaking out of the sarcoplasmic reticulum then trigger a sustained contraction. Without ATP, the cross-bridges cannot detach from the active sites. Skeletal muscles throughout the body become locked in the contracted position. Because all the skeletal muscles are involved, the individual becomes "stiff as a board." This physical state, **rigor mortis**, lasts until the lysosomal enzymes released by autolysis break down the myofilaments 15–25 hours later. The timing is dependent on environmental factors, such as temperature. Forensic pathologists can estimate the time of death on the basis of the degree of rigor mortis and the characteristics of the local environment.

Before you proceed, review the entire sequence of events from neural activation through excitation–contraction coupling to the completion of a contraction. Table 10–1 provides a summary of the contraction process, from ACh release to the end of the contraction.

✓ How would a drug that interferes with cross-bridge formation affect muscle contraction?

✓ What would you expect to happen to a resting skeletal muscle if the sarcolemma suddenly became very permeable to Ca^{2+}?

✓ Predict what would happen to a muscle if the motor end plate failed to produce acetylcholinesterase.

Answers start on page Q-1

 The sliding filament theory can be reviewed on the **IP CD-ROM**: Muscular System/Sliding Filament Theory.

10–4 TENSION PRODUCTION

Objectives

■ Describe the mechanism responsible for tension production in a muscle fiber, and discuss the factors that determine the peak tension developed during a contraction.

■ Discuss the factors that affect peak tension production during the contraction of an entire skeletal muscle, and explain the significance of the motor unit in this process.

■ Compare the different types of muscle contractions.

When sarcomeres shorten in a contraction, they shorten the muscle fiber. This shortening exerts tension on the connective

SUMMARY TABLE 10–1 STEPS INVOLVED IN SKELETAL MUSCLE CONTRACTION

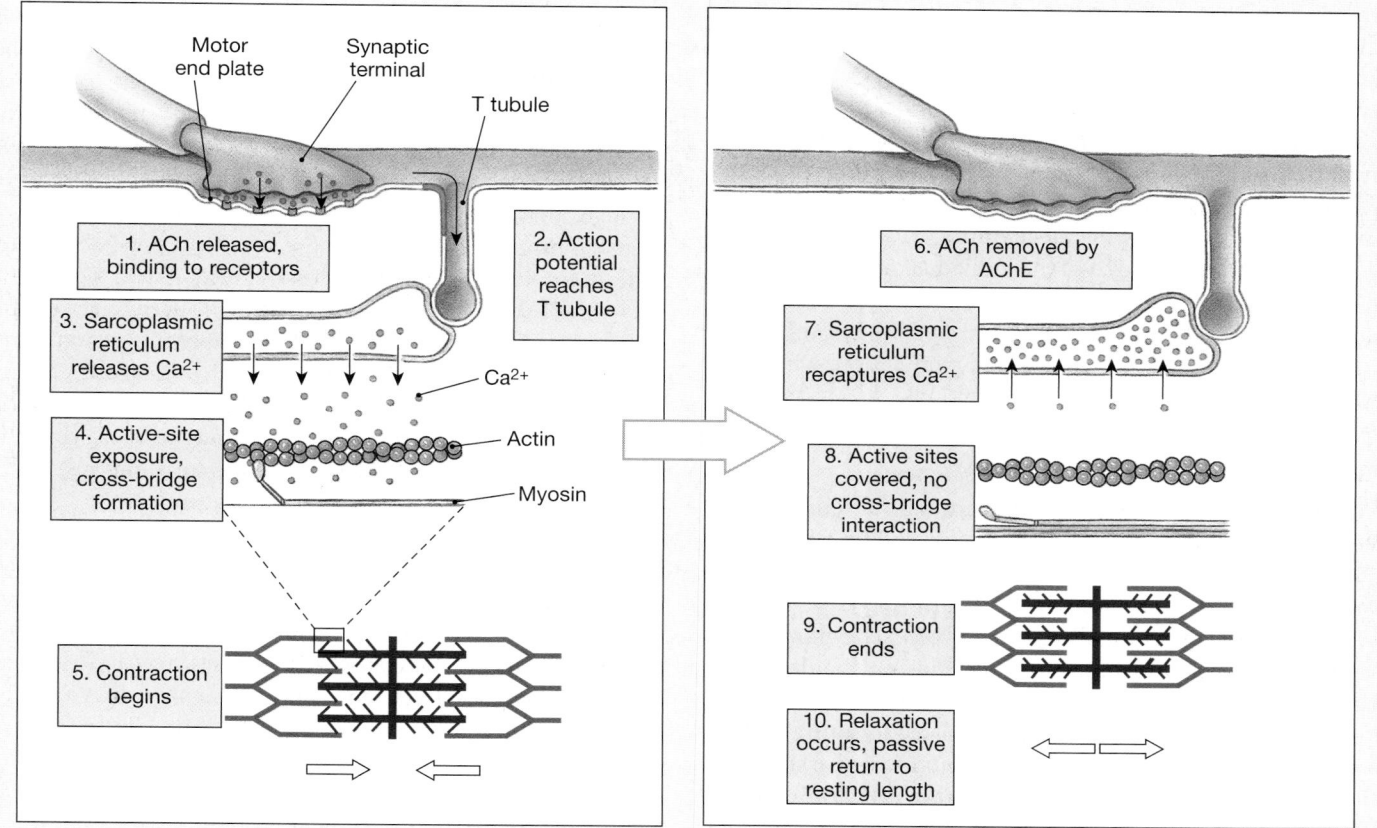

STEPS THAT INITIATE A CONTRACTION:

1. At the neuromuscular junction (NMJ), ACh released by the synaptic terminal binds to receptors on the sarcolemma.

2. The resulting change in the transmembrane potential of the muscle fiber leads to the production of an action potential that spreads across the entire surface of the muscle fiber and along the T tubules.

3. The sarcoplasmic reticulum (SR) releases stored calcium ions, increasing the calcium concentration of the sarcoplasm in and around the sarcomeres.

4. Calcium ions bind to troponin, producing a change in the orientation of the troponin–tropomyosin complex that exposes active sites on the thin (actin) filaments. Cross-bridges form when myosin heads bind to active sites on F actin.

5. The contraction begins as repeated cycles of cross-bridge binding, pivoting, and detachment occur, powered by the hydrolysis of ATP. These events produce filament sliding, and the muscle fiber shortens.

STEPS THAT END A CONTRACTION:

6. Action potential generation ceases as ACh is broken down by acetylcholinesterase (AChE).

7. The SR reabsorbs calcium ions, and the concentration of calcium ions in the sarcoplasm declines.

8. When calcium ion concentrations approach normal resting levels, the troponin–tropomyosin complex returns to its normal position. This change re-covers the active sites and prevents further cross-bridge interaction.

9. Without cross-bridge interactions, further sliding cannot take place, and the contraction ends.

10. Muscle relaxation occurs, and the muscle returns passively to its resting length.

tissue fibers attached to the muscle fiber. The tension produced by an individual muscle fiber can vary, and in the next section we will consider the specific factors involved. In a subsequent section, we shall see that the tension produced by an entire skeletal

muscle can vary even more widely, because not only can tension production vary among the individual muscle fibers, but the total number of stimulated muscle fibers can change from moment to moment.

☐ TENSION PRODUCTION BY MUSCLE FIBERS

The amount of tension produced by an individual muscle fiber ultimately depends on the number of pivoting cross-bridges. There is no mechanism to regulate the amount of tension produced in that contraction by changing the number of contracting sarcomeres. When calcium ions are released, they are released from all triads in the muscle fiber. Thus, a muscle fiber is either "on" (producing tension) or "off" (relaxed). This feature of muscle mechanics is known as the **all-or-none principle**. However, tension production at the level of the individual muscle fiber *does* vary, depending on (1) the resting length at the time of stimulation, which determines the degree of overlap between thick and thin filaments, and (2) the frequency of stimulation, which affects the internal concentration of calcium ions.

Length–Tension Relationships

When many people pull on a rope, the amount of tension produced is proportional to the number of people pulling. Similarly, in a skeletal muscle fiber, the amount of tension generated during a contraction depends on the number of pivoting cross-bridges in all the sarcomeres along all the myofibrils. The number of cross-bridges that can form, in turn, depends on the degree of overlap between thick filaments and thin filaments within these sarcomeres. When the muscle fiber is stimulated to contract, only myosin heads in the zones of overlap can bind to active sites and produce tension. The tension produced by the entire muscle fiber can thus be related to the structure of individual sarcomeres.

A sarcomere works most efficiently within an optimal range of lengths (Figure 10–13a●). When the resting sarcomere length is within this range, the maximum number of cross-bridges can form and the tension produced is highest. If the resting sarcomere length falls outside of the range—if the sarcomere is compressed and shortened, or stretched and lengthened—it cannot produce as much tension when stimulated. This is because the amount of tension produced is largely determined by the number of cross-bridges that form. An increase in sarcomere length reduces the tension produced by reducing the size of the zone of overlap and the number of potential cross-bridge interactions (Figure 10–13b●). When the zone of overlap is reduced to zero, thin and thick filaments cannot interact at all. Under these conditions, the muscle fiber cannot produce any tension and a contraction cannot occur. Such extreme stretching of a muscle fiber is normally prevented by the titin filaments in the muscle fiber, which tie the thick filaments to the Z lines, and by the surrounding connective tissues, which limit the degree of muscle stretch.

A decrease in the resting sarcomere length reduces efficiency because the stimulated sarcomere cannot shorten very much before the thin filaments extend across the center of the sarcomere and collide with or overlap the thin filaments of the opposite side (Figure 10–13c●). This disrupts the precise three-dimensional relationship between thick and thin filaments (shown in Figure 10–5●) and interferes with the binding of myosin heads to active sites. Because the number of cross bridges is reduced, tension declines in the stimulated muscle fiber. Tension production falls to zero when the resting sarcomere is as short as it can be (Figure

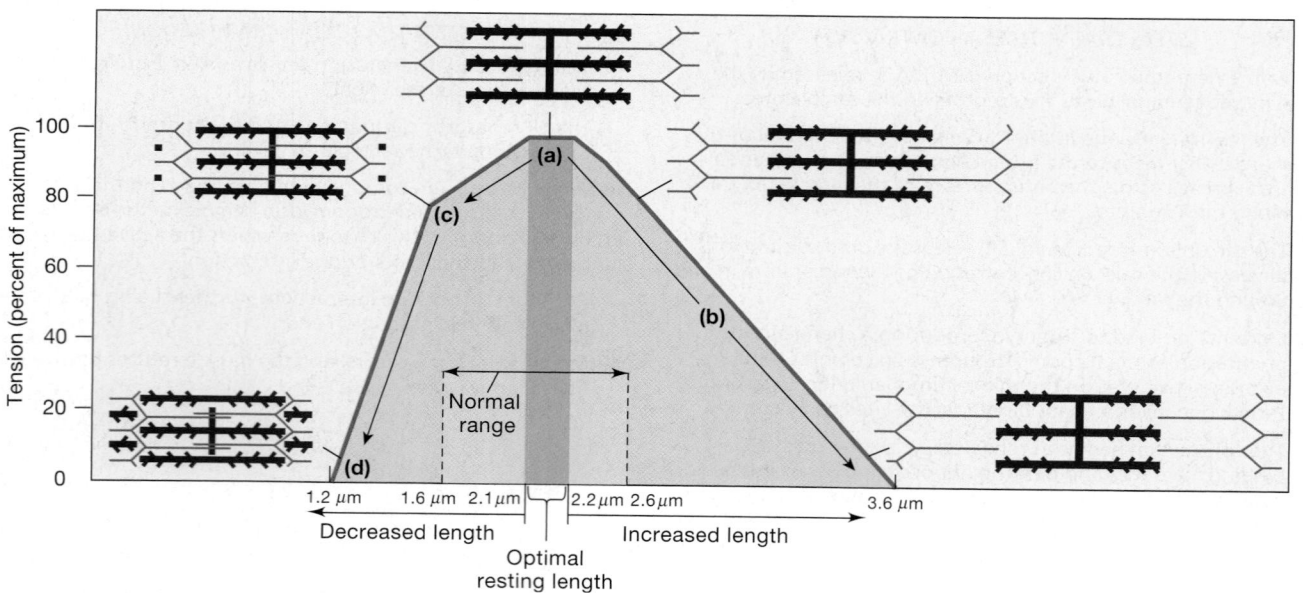

●FIGURE 10–13
The Effect of Sarcomere Length on Tension. **(a)** Maximum tension is produced when the zone of overlap is large but the thin filaments do not extend across the sarcomere's center. **(b)** If the sarcomeres are stretched too far, the zone of overlap is reduced or disappears, and cross-bridge interactions are reduced or cannot occur. **(c)** At short resting lengths, thin filaments extending across the center of the sarcomere interfere with the normal orientation of thick and thin filaments, reducing tension production. **(d)** When the thick filaments contact the Z lines, the sarcomere cannot shorten—the myosin heads cannot pivot and tension cannot be produced. The light purple area represents the normal range of resting sarcomere lengths.

10-13d•). At this point, the thick filaments are jammed against the Z lines and the sarcomere cannot shorten further. Although cross-bridge binding can still occur, the myosin heads cannot pivot and generate tension, because the thin filaments cannot move.

In summary, skeletal muscle fibers contract most forcefully when stimulated over a narrow range of resting lengths. The normal range of sarcomere lengths in the body (Figure 10-13•) is 75 to 130 percent of the optimal length. The arrangement of skeletal muscles, connective tissues, and bones normally prevents extreme compression or excessive stretching. For example, straightening your elbow stretches your *biceps brachii muscle*, but the bones and ligaments of the elbow end this movement before the muscle fibers stretch too far. During an activity such as walking, in which muscles contract and relax cyclically, muscle fibers are stretched to a length very close to "ideal" before they are stimulated to contract. When muscles must contract over a larger range of resting lengths, they often "team up" to improve efficiency. (We shall discuss the mechanical principles involved in Chapter 11.)

The Frequency of Stimulation

A single stimulation produces a single contraction, or *twitch*, that may last 7–100 milliseconds, depending on the muscle stimulated. Although they can be produced by electrical stimulation in a laboratory and can generate heat when you are shivering, muscle twitches are too brief to be part of any normal activity. The duration of a contraction can be extended by repeated stimulation, and a muscle fiber undergoing such a sustained contraction produces more tension than it does in a single twitch. To understand why, we need to take a closer look at tension production during a twitch contraction and then follow the changes that occur as the rate of stimulation increases. This is a subject with real importance, as all of your consciously and subconsciously directed muscular activities—standing, walking, running, reaching, and so forth—involve sustained muscular contractions rather than twitches.

TWITCH CONTRACTIONS A **twitch** is a single stimulus–contraction–relaxation sequence in a muscle fiber. Twitches vary in duration, depending on the type of muscle, its location, internal and external environmental conditions, temperature, and other factors. Twitches in an eye muscle fiber can be as brief as 7.5 msec, but a twitch in a muscle fiber from the *soleus*, a small calf muscle, lasts about 100 msec. Figure 10-14a• is a **myogram**, or graph of twitch tension development in muscle fibers from various skeletal muscles.

Figure 10-14b• details the phases of a 40-msec twitch in a muscle fiber from the *gastrocnemius muscle*, a prominent calf muscle. A single twitch can be divided into a *latent period*, a *contraction phase*, and a *relaxation phase*:

1. The **latent period** begins at stimulation and typically lasts about 2 msec. Over this period, the action potential sweeps across the sarcolemma and the sarcoplasmic reticulum releases calcium ions. The muscle fiber does not produce tension during the latent period, because the contraction cycle has yet to begin.

2. In the **contraction phase**, tension rises to a peak. As the tension rises, calcium ions are binding to troponin, active sites on thin filaments are being exposed, and cross-bridge interactions are occurring. The contraction phase ends roughly 15 msec after stimulation.

3. The **relaxation phase** then continues for about another 25 msec. During this period, calcium levels are falling, active sites are being covered by tropomyosin, and the number of active cross-bridges is declining. As a result, tension falls to resting levels.

TREPPE If a skeletal muscle is stimulated a second time immediately after the relaxation phase has ended, the contraction that occurs will develop a slightly higher maximum tension than did the contraction after the first stimulation. The increase in peak tension indicated in Figure 10-15a• will continue over the first 30–50 stimulations. Thereafter, the amount of tension produced will remain constant. Because the tension rises in stages, like the steps in a staircase, this phenomenon is called **treppe** (TREP-e), a German word meaning "stairs." The rise is thought to result from a gradual increase in the concentration of calcium ions in the sarcoplasm, in part because the ion pumps in the sarcoplasmic reticulum are unable to recapture the ions in the time between stimulations.

WAVE SUMMATION AND INCOMPLETE TETANUS If a second stimulus arrives before the relaxation phase has ended, a second, more powerful contraction occurs. The addition of one twitch to another in this way constitutes the **summation of twitches**, or **wave summation** (Figure 10-15b•). The duration of a single twitch determines the maximum time available to produce wave summation. For example, if a twitch lasts 20 msec (1/50 sec), subsequent stimuli must be separated by less than 20 msec—a stimulation rate of more than 50 stimuli per second. Rather than refer to the stimulation rate, we usually use the *frequency*, which is a number per unit time. In this instance, a stimulus frequency of greater than 50 per second produces wave summation, whereas a stimulus frequency below 50 per second will produce individual twitches and treppe.

If the stimulation continues and the muscle is never allowed to relax completely, tension will rise until it reaches a peak value roughly four times the maximum produced by treppe (Figure 10-15c•). A muscle producing peak tension during rapid cycles of contraction and relaxation is in **incomplete tetanus** (*tetanos*, convulsive tension).

COMPLETE TETANUS Complete tetanus is obtained by increasing the stimulation rate until the relaxation phase is eliminated (Figure 10-15d•). During complete tetanus, action potentials arrive so rapidly that the sarcoplasmic reticulum does not have time to reclaim the calcium ions. The high Ca^{2+} concentration in the cytoplasm prolongs the contraction state, making it continuous. Virtually all normal muscular contractions involve complete tetanus of the participating muscle fibers.

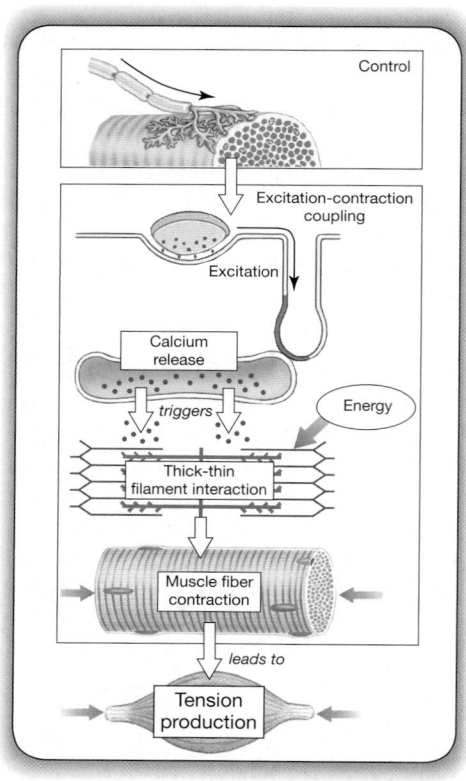

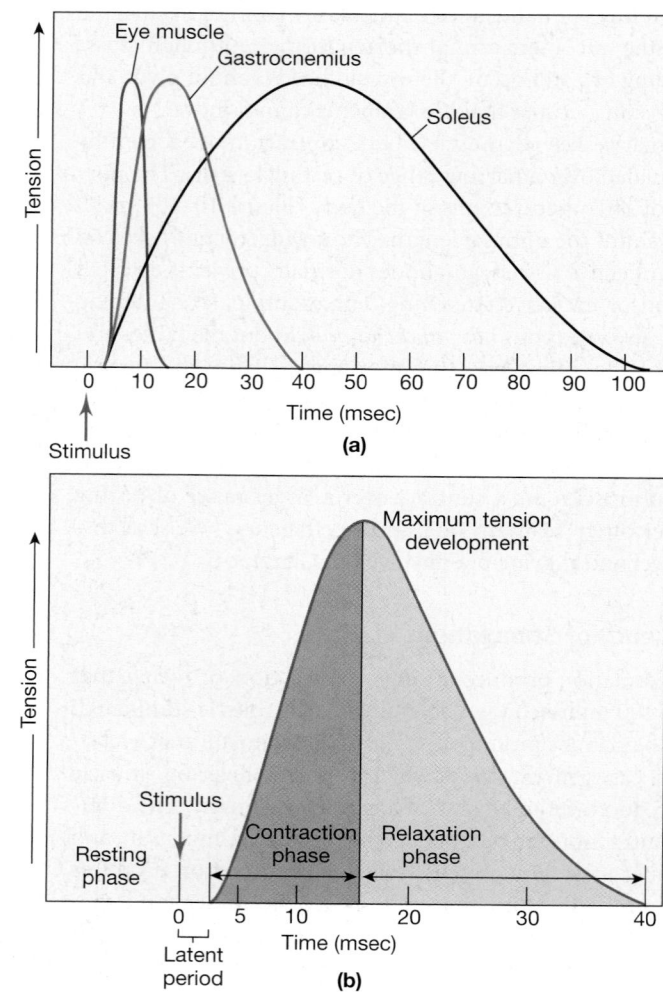

•FIGURE 10–14
The Twitch and the Development of Tension. (a) A myogram showing differences in tension over time for a twitch contraction in different skeletal muscles. (b) The details of tension over time for a single twitch contraction in the gastrocnemius muscle. Notice the presence of a latent period, which corresponds to the time needed for the conduction of an action potential and the subsequent release of calcium ions by the sarcoplasmic reticulum.

☐ TENSION PRODUCTION BY SKELETAL MUSCLES

Now that you are familiar with the basic mechanisms of muscle contraction at the level of the individual muscle fiber, we can begin to examine the performance of skeletal muscles—the organs of the muscular system. In this section, we will consider the coordinated contractions of an entire population of skeletal muscle fibers. The amount of tension produced in the skeletal muscle *as a whole* is determined by (1) the *internal* and *external* *tension* produced by the stimulated muscle fibers and (2) the total number of muscle fibers stimulated.

Internal and External Tension

We have already discussed the relationship between resting sarcomere length and tension production in individual muscle fibers. When we consider an entire skeletal muscle, we see that the

situation is complicated by the fact that the muscle fibers are not directly connected to the structures they pull against. The extracellular fibers of the endomysium, perimysium, epimysium, and tendons are flexible and somewhat elastic. When a skeletal muscle contracts, the myofibrils in the muscle fibers generate *internal tension*. This internal tension is applied to the extracellular fibers. The tension in the extracellular fibers is called *external tension*.

As the external tension rises, the extracellular fibers stretch; for this reason, they are called *series elastic elements*. The series elastic elements behave like fat rubber bands. They stretch easily at first, but as they elongate, they become stiffer and more effective at transferring tension to the resistance. As a result, external tension does not climb as quickly as internal tension when a contraction occurs.

To understand this relationship, attach a rubber band to one of the rings in a three-ring notebook. Put a finger through the

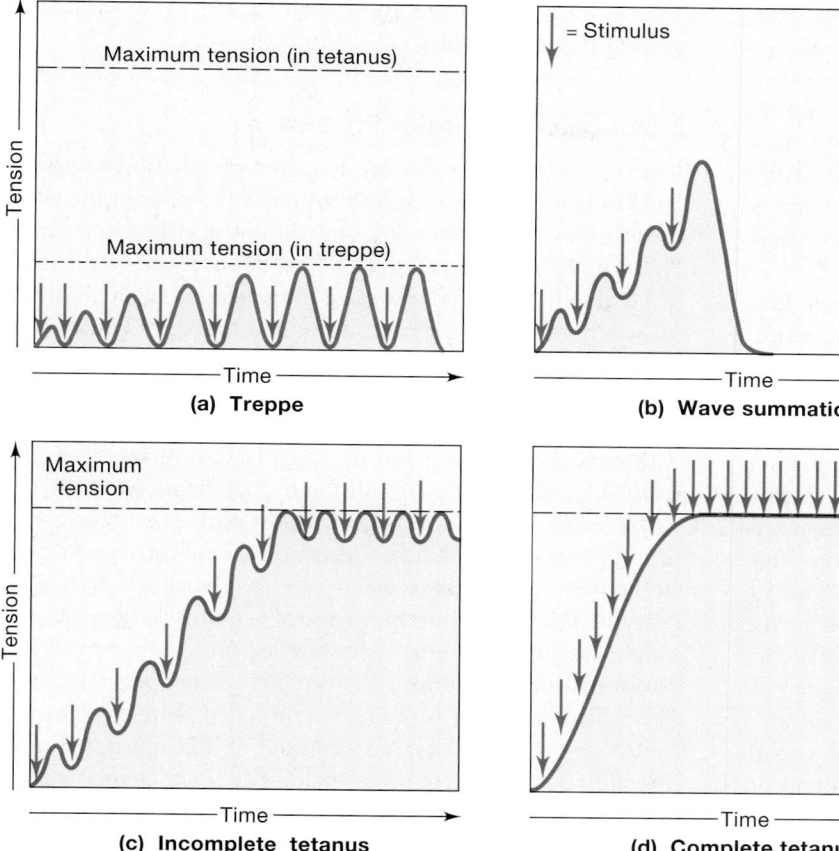

(a) **Treppe**

(b) **Wave summation**

(c) **Incomplete tetanus**

(d) **Complete tetanus**

●**FIGURE 10–15**
Effects of Repeated Stimulations.
(a) Treppe is an increase in peak tension with each successive stimulus delivered shortly after the completion of the relaxation phase of the preceding twitch. **(b)** Wave summation occurs when successive stimuli arrive before the relaxation phase (the downturn of the curve) has been completed. **(c)** Incomplete tetanus occurs if the rate of stimulation increases further. Tension production will rise to a peak, and the periods of relaxation will be very brief. **(d)** During complete tetanus, the frequency of stimulation is so high that the relaxation phase is eliminated; tension plateaus at maximal levels.

Tetanus

Children are often told to be careful of rusty nails. Parents should worry most not about the rust or the nail, but about an infection caused by the very common bacterium *Clostridium tetani*, which can cause **tetanus**. Although they have the same name, the disease tetanus has no relation to the normal muscle response to neural stimulation. The *Clostridium* bacterium occurs in soil and virtually everywhere else in the environment, but it thrives in tissues that contain low amounts of oxygen. For this reason, a deep puncture wound, such as that from a nail, carries a much greater risk of producing tetanus than does a shallow, open cut that bleeds freely.

When active in body tissues, these bacteria release a powerful toxin that affects the central nervous system. Motor neurons, which control skeletal muscles throughout the body, are particularly sensitive to it. The toxin suppresses the mechanism that regulates motor neuron activity. The result is a sustained, powerful contraction of skeletal muscles throughout the body.

The incubation period (the time from exposure to the development of symptoms) is generally less than 2 weeks. The most common early complaints are headache, muscle stiffness, and difficulty in swallowing. Because it soon becomes difficult to open the mouth, the disease is also called *lockjaw*. Widespread muscle spasms typically develop within 2 or 3 days of the initial symptoms and continue for a week before subsiding. After 2–4 weeks, patients who survive recover with no aftereffects.

Severe tetanus has a 40–60 percent mortality rate; that is, for every 100 people who develop severe tetanus, 40 to 60 die. Fortunately, immunization is effective in preventing the disease. Approximately 500,000 cases of tetanus occur worldwide each year, but only about 100 of them occur in the United States, thanks to an effective immunization program. ("Tetanus shots" are recommended, with booster shots every 10 years.) Severe symptoms in unimmunized patients can be prevented by early administration of an antitoxin, in most cases *human tetanus immune globulin*. Such treatment does not reduce symptoms that have already appeared, however.

rubber band and use it to pull the notebook across a table. Your finger represents the muscle fibers; the rubber band, the attached tendon; and the notebook, a bone of the skeleton. When you first apply tension, the rubber band stretches and becomes stiffer. Over this period, external tension rises. The notebook starts to move when the rubber band becomes sufficiently taut—that is, when the tension in the rubber band (the external tension) overcomes friction and the weight of the notebook (the resistance).

If you now relax your hand, the rubber band will pull your finger toward the notebook. The same thing happens in a muscle: When the contraction ends, the series elastic elements recoil and pull on the muscle. This recoil helps return the muscle to its original resting length.

A myogram performed in the laboratory generally measures the tension in a tendon and so is reporting external tension rather than internal tension. Figure 10–16a● compares the internal and external tensions during a single, brief contraction that represents a twitch in the stimulated muscle fibers. A single twitch is so brief in duration that there isn't enough time for the external tension to rise as high as the internal tension. Twitches are therefore ineffective in terms of performing useful work.

Notice, however, that the external tension remains elevated until the relaxation phase has ended. If a second twitch occurs before the external tension returns to zero, the external tension will peak at a higher level, because more of the internal tension will be conveyed to the series elastic elements. Think of pushing a child on a swing: You push once to start the swing moving; if you push a second time, the child swings higher because the energy of the second push is added to the energy remaining from the first. This mechanism of elevated external tension is now thought to be the primary basis of wave summation: Each successive contraction begins before the external tension has fallen to resting levels, so the external contraction continues to rise until it reaches peak levels. During a tetanic contraction, there is sufficient time for internal and external tensions to equalize, so the peak tension measured in the myogram represents the peak internal tension (Figure 10–16b●).

Motor Units and Tension Production

During a normal contraction, activated muscle fibers are stimulated to complete tetanus. As a result, the tension in the skeletal muscle as a whole rises smoothly, following the curve seen in Figure 10–15d●. The amount of tension produced by the muscle as a whole is the sum of the tensions generated by the individual muscle fibers, since they are all pulling together. Thus, by controlling the number of activated muscle fibers, you can control the amount of tension produced by the skeletal muscle.

A typical skeletal muscle contains thousands of muscle fibers. Although some motor neurons control a few muscle fibers, most control hundreds of them. All the muscle fibers controlled by a single motor neuron constitute a **motor unit**. The size of a motor unit is an indication of how fine the control of movement can be. In the muscles of the eye, where precise control is extremely important, a motor neuron may control 4–6 muscle fibers. We have much less precise control over our leg muscles, where a single motor neuron may control 1000–2000 muscle fibers. The muscle fibers of each motor unit are intermingled with those of other motor units (Figure 10–17a●). Because of this intermingling, the direction of pull exerted on the tendon does not change when the number of activated motor units changes.

When you decide to perform a specific arm movement, specific groups of motor neurons in the spinal cord are stimulated. The contraction begins with the activation of the smallest motor units in the stimulated muscle. These motor units generally contain muscle fibers that contract relatively slowly. Over time, larger motor units containing faster and more powerful muscle fibers are activated, and tension production rises steeply. The smooth, but steady, increase in muscular tension produced by increasing the number of active motor units is called **recruitment**, or **multiple motor unit summation**.

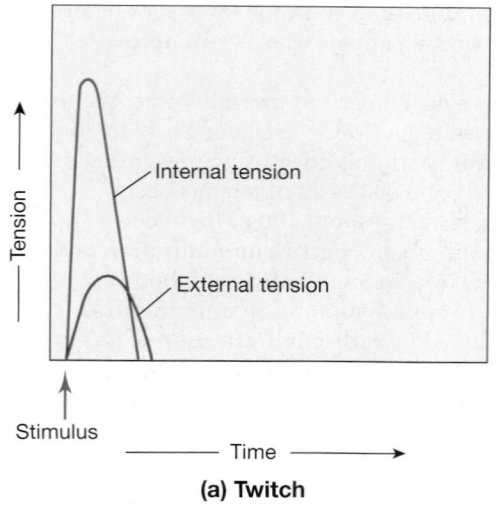

(a) Twitch

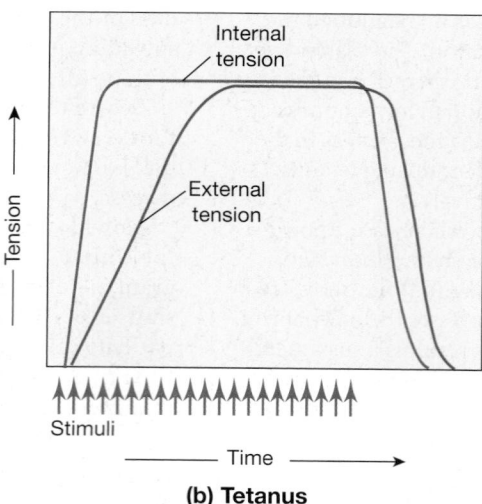

(b) Tetanus

●**FIGURE 10–16**
Internal and External Tension.
Internal tension rises as the muscle fiber contracts. External tension rises more slowly as the series elastic elements are stretched. **(a)** During a single-twitch contraction, external tension cannot rise as high as internal tension before the relaxation phase begins. **(b)** During a tetanic contraction, external tension soon plateaus at a level roughly equivalent to internal tension. External tension remains elevated for the duration of the contraction.

Peak tension production occurs when all motor units in the muscle contract in a state of complete tetanus. Such powerful contractions do not last long, however, because the individual muscle fibers soon use up their available energy reserves. During a sustained tetanic contraction, motor units are activated on a rotating basis, so some of them are resting and recovering while others are actively contracting. This "relay team" approach, called *asynchronous motor unit summation*, lets each motor unit recover somewhat before it is stimulated again (Figure 10–17b●). As a result, when your muscles contract for sustained periods, they produce slightly less than maximal tension.

Muscle Tone

In any skeletal muscle, some motor units are always active, even when the entire muscle is not contracting. Their contractions do not produce enough tension to cause movement, but they do tense and firm the muscle. This resting tension in a skeletal muscle is called **muscle tone**. A muscle with little muscle tone appears limp and flaccid, whereas one with moderate muscle tone is firm and solid. The identity of the stimulated motor units changes constantly, so a constant tension in the attached tendon is maintained, but individual muscle fibers can relax.

Resting muscle tone stabilizes the positions of bones and joints. For example, in muscles involved with balance and posture, enough motor units are stimulated to produce the tension needed to maintain body position. Muscle tone also helps prevent sudden, uncontrolled changes in the positions of bones and joints. In addition to bracing the skeleton, the elastic nature of muscles and tendons lets skeletal muscles act as shock absorbers that cushion the impact of a sudden bump or shock. Heightened muscle tone accelerates the recruitment process during a voluntary contraction, because some of the motor units are already stimulated. Strong muscle tone also makes skeletal muscles appear firm and well defined, even at rest.

Activated muscle fibers use energy, so the greater the muscle tone, the higher the "resting" rate of metabolism. Increasing this rate is one of the significant effects of exercise in a weight-loss program. In such a program, you lose weight by reducing energy intake in food while maximizing your energy use each day. Although exercise consumes a lot of energy very quickly, the period of activity is usually quite brief. In contrast, elevated muscle tone increases resting energy consumption by a small amount, but the effects last indefinitely.

Isotonic and Isometric Contractions

We can classify muscle contractions as *isotonic* or *isometric* on the basis of their pattern of tension production.

ISOTONIC CONTRACTIONS In an **isotonic contraction** (*iso-*, equal + *tonos*, tension), tension rises and the skeletal muscle's length changes. Lifting an object off a desk, walking, and running involve isotonic contractions.

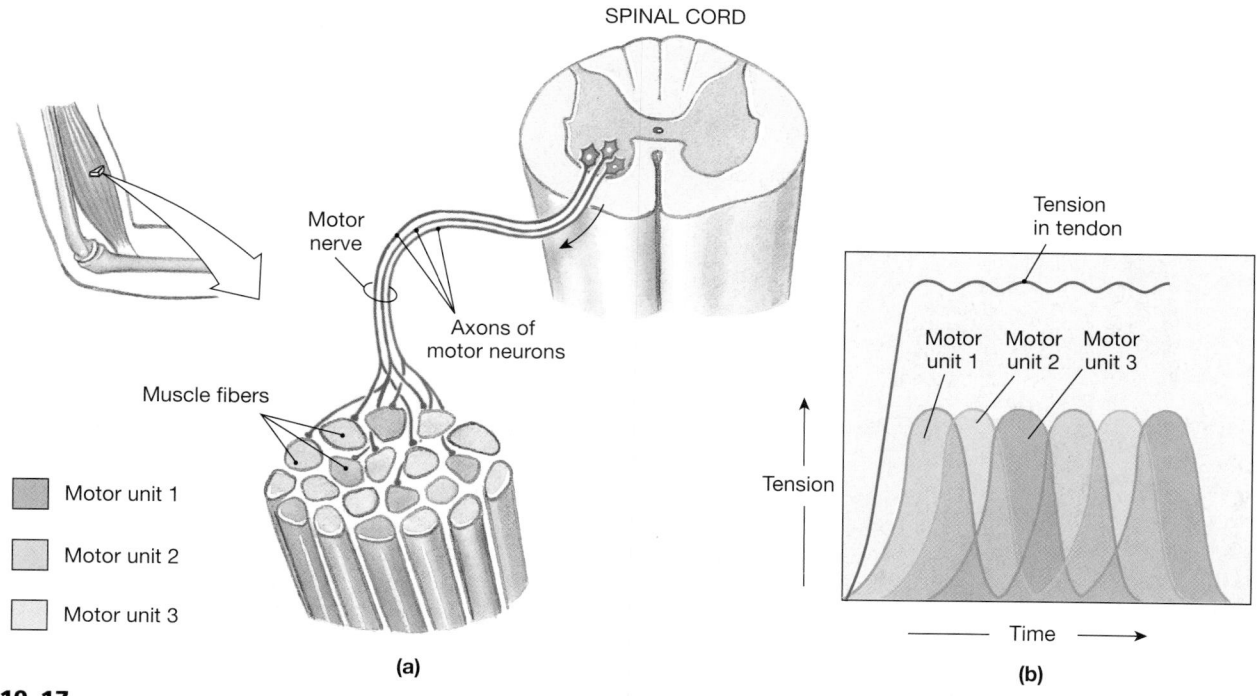

●FIGURE 10–17
The Arrangement of Motor Units in a Skeletal Muscle. **(a)** Muscle fibers of different motor units are intermingled, so the forces applied to the tendon remain roughly balanced regardless of which muscle groups are stimulated. **(b)** The tension applied to the tendon remains relatively constant, even though individual motor units cycle between contraction and relaxation.

Two types of isotonic contractions exist: (1) concentric and (2) eccentric. In a **concentric contraction**, the muscle tension exceeds the resistance and the muscle shortens. Consider the experiment summarized in Figure 10–18●. A skeletal muscle 1 cm² in cross-sectional area can produce roughly 4 kg of tension in complete tetanus. If we hang a 2-kg weight from that muscle and stimulate it, the muscle will shorten (Figure 10–18a●). Before the muscle can shorten, the cross-bridges must produce enough tension to overcome the resistance—in this case, the 2-kg weight. During this initial period, internal tension in the muscle fibers rises until the external tension in the tendon exceeds the amount of resistance. As the muscle shortens, the internal and external tensions in the skeletal muscle remain constant at a value that just exceeds the resistance (Figure 10–18b●). The term *isotonic* originated from this type of experiment.

In the body, however, the situation is more complicated. For example, muscles are not always positioned directly above the resistance, and they are attached to bones rather than to static weights. Changes in the relative positions of the muscle and the articulating bones, the effects of gravity, and other mechanical and physical factors interact to increase or decrease the amount of resistance the muscle must overcome as a movement proceeds. Nevertheless, at any time during a concentric contraction, the tension produced exceeds that resistance.

The speed of shortening varies with the difference between the amount of tension produced and the amount of resistance. If all the muscle units are stimulated and the resistance is relatively small, the muscle will shorten very quickly. In contrast, if the muscle barely produces enough tension to overcome the resistance, it will shorten very slowly.

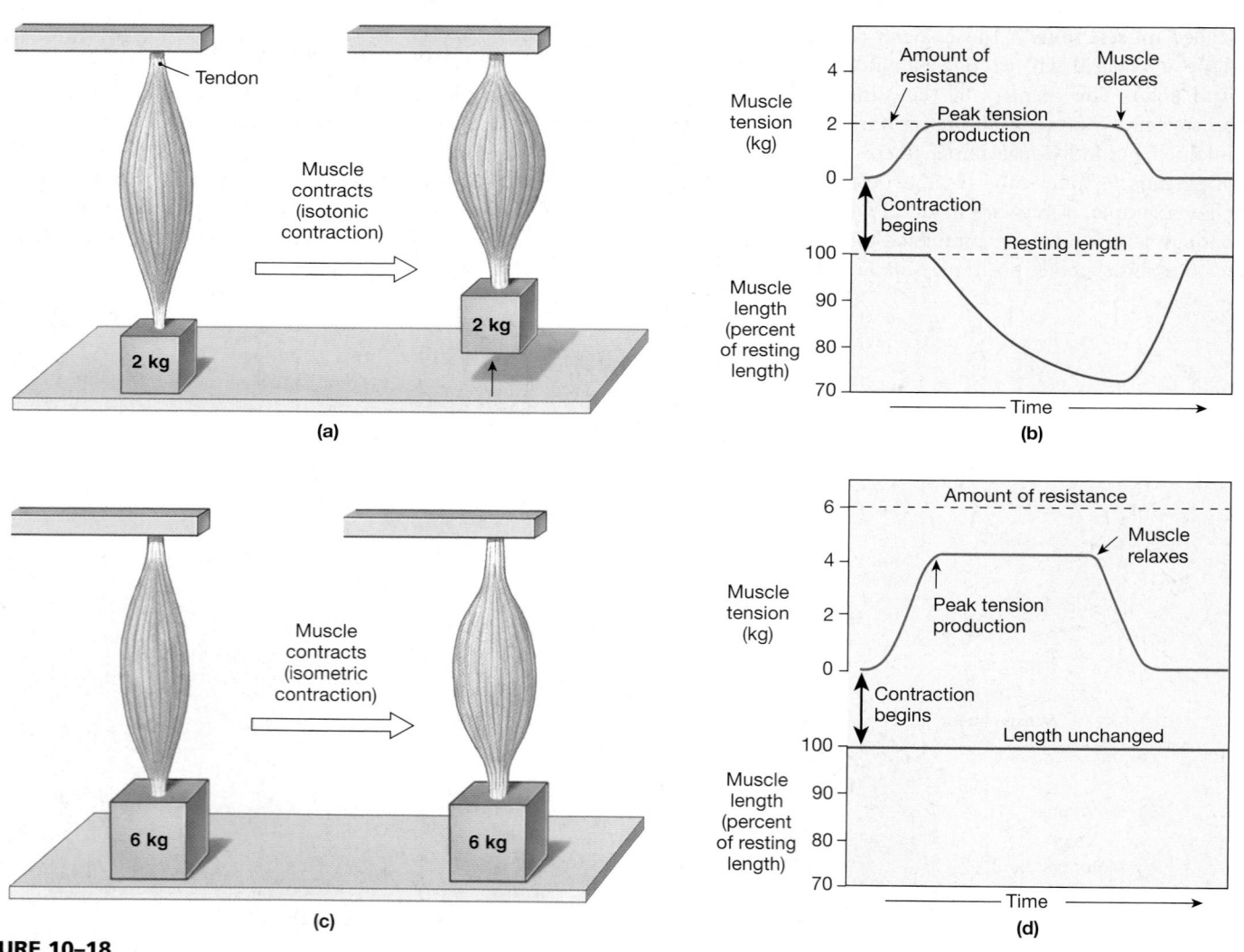

●**FIGURE 10–18**

Isotonic and Isometric Contractions. **(a, b)** This muscle is attached to a weight less than its peak tension capabilities. On stimulation, it develops enough tension to lift the weight. Tension remains constant for the duration of the contraction, although the length of the muscle changes. This is an example of isotonic contraction. **(c, d)** The same muscle is attached to a weight that exceeds its peak tension capabilities. On stimulation, tension will rise to a peak, but the muscle as a whole cannot shorten. This is an isometric contraction.

In an **eccentric contraction**, the peak tension developed is less than the resistance, and the muscle elongates owing to the contraction of another muscle or the pull of gravity. Think of a tug-of-war team trying to stop a moving car. Although everyone pulls as hard as they can, the rope slips through their fingers. The speed of elongation depends on the difference between the amount of tension developed by the active muscle fibers and the amount of resistance. In our analogy, the team might slow down a small car, but would have little effect on a large truck.

Eccentric contractions are very common, and they are an important part of a variety of movements. In these movements, you exert precise control over the amount of tension produced. By varying the tension in an eccentric contraction, you can control the rate of elongation, just as you can vary the tension in a concentric contraction. For example, precisely controlled eccentric contractions occur each time you walk down stairs or settle into a chair. During physical training, people commonly perform cycles of concentric and eccentric contractions, as when you hold a weight in your hand and flex and extend your elbow.

ISOMETRIC CONTRACTIONS In an **isometric contraction** (*metric*, measure), the muscle as a whole does not change length, and the tension produced never exceeds the resistance. Figure 10–18c● shows what happens if we attach a weight heavier than 4 kg to the experimental muscle and then stimulate the muscle. Although cross-bridges form and tension rises to peak values, the muscle cannot overcome the resistance of the weight and so cannot shorten (Figure 10–18d●). Examples of isometric contractions include holding a heavy weight above the ground, pushing against a locked door, and trying to pick up a car. These are rather unusual movements. However, many of the reflexive muscle contractions that keep your body upright when you stand or sit involve the isometric contractions of muscles that oppose the force of gravity.

You may have noticed that when you perform an isometric contraction, the contracting muscle bulges (but not as much as it does during an isotonic contraction). In an isometric contraction, although the muscle *as a whole* does not shorten, the individual muscle fibers shorten until the tendons are taut and the external tension equals the internal tension generated by the muscle fibers. The muscle fibers cannot shorten further, because the external tension does not exceed the resistance.

Normal daily activities therefore involve a combination of isotonic and isometric muscular contractions. As you sit and read this text, isometric contractions of postural muscles stabilize your vertebrae and maintain your upright position. When you turn a page, the movements of your arm, forearm, hand, and fingers are produced by a combination of concentric and eccentric isotonic contractions.

Resistance and Speed of Contraction

You can lift a light object more rapidly than you can lift a heavy one because resistance and the speed of contraction are inversely related. If the resistance is less than the tension produced, an isotonic, concentric contraction will occur; the muscle will shorten. The heavier the resistance, the longer it takes for the movement to begin, because muscle tension, which increases gradually, must exceed the resistance before shortening can occur (Figure 10–19●). The contraction itself proceeds more slowly. At the cellular level, the speed of cross-bridge pivoting is reduced as the load increases.

For each muscle, an optimal combination of tension and speed exists for any given resistance. If you have ever ridden a 10-speed bicycle, you are probably already aware of this fact. When you are cruising along comfortably, your thigh and leg muscles are working at an optimal combination of speed and tension. When you come to a hill, the resistance increases. Your muscles must now develop more tension, and they move more slowly; they are no longer working at optimal efficiency. You then shift to a lower gear. The load on your muscles decreases, the speed increases, and the muscles are once again working efficiently.

Muscle Relaxation and the Return to Resting Length

As we noted earlier, there is no active mechanism for muscle fiber elongation. The sarcomeres in a muscle fiber can shorten and develop tension, but the power stroke cannot be reversed to push the Z lines farther apart. After a contraction, a muscle fiber returns to its original length through a combination of elastic forces, opposing muscle contractions, and gravity.

ELASTIC FORCES When the contraction ends, some of the energy initially "spent" in stretching the series elastic elements is recovered as they recoil. The recoil of the series elastic elements gradually helps return the muscle fiber to its original resting length.

OPPOSING MUSCLE CONTRACTIONS The contraction of opposing muscles can return a muscle to its resting length more quickly than

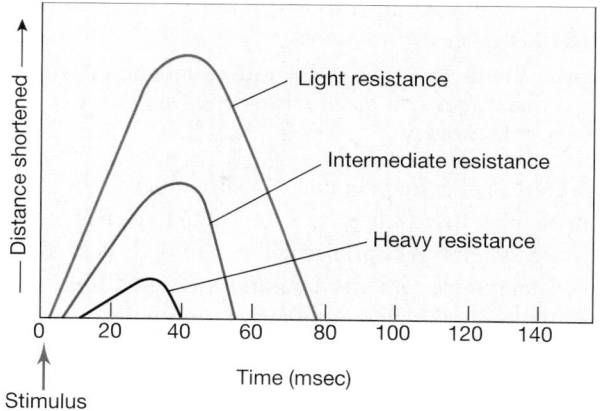

●**FIGURE 10–19**
Resistance and Speed of Contraction. The heavier the resistance on a muscle, the longer it will take for the muscle to begin to shorten and the less the muscle will shorten.

elastic forces can. Consider the muscles of the arm that flex or extend the elbow. Contraction of the *biceps brachii muscle* on the anterior part of the arm flexes the elbow; contraction of the *triceps brachii muscle* on the posterior part of the arm extends the elbow. When the biceps brachii muscle contracts, the triceps brachii muscle is stretched. When the biceps brachii muscle relaxes, contraction of the triceps brachii muscle extends the elbow and stretches the muscle fibers of the biceps brachii muscle to their original length.

GRAVITY Gravity may assist opposing muscle groups in quickly returning a muscle to its resting length after a contraction. For example, imagine the biceps brachii muscle fully contracted with the elbow pointed at the ground. When the muscle relaxes, gravity will pull the forearm down and stretch the muscle. Although gravity can provide assistance in stretching muscles, some active muscle tension is needed to control the rate of movement and to prevent damage to the joint. In the previous example, eccentric contraction of the biceps brachii muscle can control the movement.

✓ Why is it difficult to contract a muscle that has been overstretched?

✓ During treppe, why does tension in a muscle gradually increase even though the strength and frequency of the stimulus are constant?

✓ Can a skeletal muscle contract without shortening? Explain.

Answers on start on page Q-1

 Skeletal muscle contraction can be reviewed on the **IP CD-ROM**: Muscular System/Contraction of Motor Units and Contraction of Whole Muscle.

10–5 ENERGY USE AND MUSCULAR ACTIVITY

Objectives

■ Describe the mechanisms by which muscle fibers obtain the energy to power contractions.

■ Describe the factors that contribute to muscle fatigue, and discuss the stages and mechanisms involved in the muscle's subsequent recovery.

A single muscle fiber may contain 15 billion thick filaments. When that muscle fiber is actively contracting, each thick filament breaks down roughly 2500 ATP molecules per second. Because even a small skeletal muscle contains thousands of muscle fibers, the ATP demands of a contracting skeletal muscle are enormous. In practical terms, the demand for ATP in a contracting muscle fiber is so high that it would be impossible to have all the necessary energy available as ATP before the contraction begins. Instead, a resting muscle fiber contains only enough ATP and other high-energy compounds to sustain a contraction until additional ATP can be generated. Throughout the rest of the contraction, the muscle fiber will generate ATP at roughly the same rate as it is used.

□ ATP AND CP RESERVES

The primary function of ATP is the transfer of energy from one location to another rather than the long-term storage of energy. At rest, a skeletal muscle fiber produces more ATP than it needs. Under these conditions, ATP transfers energy to creatine. *Creatine* is a small molecule that muscle cells assemble from fragments of amino acids. The energy transfer creates another high-energy compound, **creatine phosphate (CP)**, or *phosphorylcreatine*:

$$ATP + creatine \rightarrow ADP + creatine\ phosphate$$

During a contraction, each myosin head breaks down ATP, producing ADP and a phosphate group. The energy stored in creatine phosphate is then used to "recharge" ADP, converting it back to ATP through the reverse reaction:

$$ADP + creatine\ phosphate \rightarrow ATP + creatine$$

The enzyme that facilitates this reaction is **creatine phosphokinase** (**CPK** or **CK**). When muscle cells are damaged, CPK leaks across the cell membranes and into the bloodstream. Thus, a high blood concentration of CPK usually indicates serious muscle damage.

The energy reserves of a representative muscle fiber are indicated in Table 10–2. A resting skeletal muscle fiber contains about six times as much creatine phosphate as ATP, but when a muscle fiber is undergoing a sustained contraction, these energy reserves are exhausted in only about 15 seconds. The muscle fiber must then rely on other mechanisms to convert ADP to ATP.

□ ATP GENERATION

We learned in Chapter 3 that most cells in the body generate ATP through aerobic metabolism in mitochondria and through glycolysis in the cytoplasm. ⊂⊃ p. 83

Aerobic Metabolism

Aerobic metabolism normally provides 95 percent of the ATP demands of a resting cell. In this process, mitochondria absorb oxygen, ADP, phosphate ions, and organic substrates from the surrounding cytoplasm. The substrates then enter the *TCA* (*tricarboxylic acid*) *cycle* (also known as the *citric acid cycle* or the *Krebs cycle*), an enzymatic pathway that breaks down organic molecules. The carbon atoms are released as carbon dioxide. The hydrogen atoms are shuttled to respiratory enzymes in the inner mitochondrial membrane, where their electrons are removed. After a series of intermediate steps, the protons and electrons are combined with oxygen to form water. Along the way, large amounts of energy are released and used to make ATP. The entire process is very efficient: For each molecule of pyruvic acid "fed" to the TCA cycle, the cell will gain 17 ATP molecules.

Resting skeletal muscle fibers rely almost exclusively on the aerobic metabolism of fatty acids to generate ATP. These fatty

TABLE 10–2 SOURCES OF ENERGY STORED IN A TYPICAL MUSCLE FIBER

Energy Stored As	Utilized through	Initial Quantity	Number of Twitches Supported by Each Energy Source Alone	Duration of Isometric Tetanic Contraction Supported by Each Energy Source Alone
ATP	ATP → ADP + P	3 mmol	10	2 sec
CP	ADP + CP → ATP + C	20 mmol	70	15 sec
Glycogen	Glycolysis (anaerobic) Aerobic metabolism	100 mmol	670 12,000	130 sec 2400 sec (40 min)

acids are absorbed from the circulation. When the muscle starts contracting, the mitochondria begin breaking down molecules of pyruvic acid instead of fatty acids. The pyruvic acid is provided by the enzymatic pathway of glycolysis, which breaks down glucose in the cytoplasm. The glucose can come either from the surrounding interstitial fluid or through the breakdown of glycogen reserves within the sarcoplasm. Because a typical skeletal muscle fiber contains large amounts of glycogen, the shift from fatty acid metabolism to glucose metabolism makes it possible for the cell to continue contracting for an extended period, even without an external source of nutrients.

Glycolysis

Glycolysis is the breakdown of glucose to pyruvic acid in the cytoplasm of a cell. It is called an **anaerobic** process, because it does not require oxygen. Glycolysis provides a net gain of 2 ATP molecules and generates 2 pyruvic acid molecules from each glucose molecule. The ATP produced by glycolysis is therefore only a small fraction of that produced by aerobic metabolism, in which the breakdown of the 2 pyruvic acid molecules in mitochondria would generate 34 ATP molecules. Thus, when energy demands are relatively low and oxygen is readily available, glycolysis is important only because it provides the substrates for aerobic metabolism. Yet, because it can proceed in the absence of oxygen, glycolysis becomes an important source of energy when energy demands are at a maximum and the availability of oxygen limits the rate of mitochondrial ATP production.

The glucose broken down under these conditions is obtained primarily from the reserves of glycogen in the sarcoplasm. Glycogen, diagrammed in Figure 2–12●, is a polysaccharide chain of glucose molecules. ∞ p. 47 Typical skeletal muscle fibers contain large glycogen reserves, which may account for 1.5 percent of the total muscle weight. When the muscle fiber begins to run short of ATP and CP, enzymes split the glycogen molecules apart, releasing glucose, which can be used to generate more ATP. When energy demands are low and oxygen is abundant, glycolysis provides substrates for anaerobic metabolism and aerobic metabolism provides the ATP needed for contraction. However, during peak periods of muscular activity, energy demands are extremely high and oxygen supplies are very limited. Under these conditions, glycolysis provides most of the ATP needed to sustain muscular contraction.

□ ENERGY USE AND THE LEVEL OF MUSCLE ACTIVITY

As the level of muscular activity increases, the pattern of energy production and use changes:

- In a resting skeletal muscle (Figure 10–20a●), the demand for ATP is low. More than enough oxygen is available for the mitochondria to meet that demand, and they produce a surplus of ATP. The extra ATP is used to build up reserves of CP and glycogen. Resting muscle fibers absorb fatty acids and glucose that are delivered by the bloodstream. The fatty acids are broken down in the mitochondria, and the ATP that is generated is used to convert creatine to creatine phosphate and glucose to glycogen.

- At moderate levels of activity (Figure 10–20b●), the demand for ATP increases. This demand is met by the mitochondria. As the rate of mitochondrial ATP production rises, so does the rate of oxygen consumption. Oxygen availability is not a limiting factor, because oxygen can diffuse into the muscle fiber fast enough to meet mitochondrial needs. But all the ATP produced is needed by the muscle fiber, and no surplus is available. The skeletal muscle now relies primarily on the aerobic metabolism of pyruvic acid to generate ATP. The pyruvic acid is provided by glycolysis, which breaks down glucose molecules obtained from glycogen in the muscle fiber. If glycogen reserves are low, the muscle fiber can also break down other substrates, such as lipids or amino acids. As long as the demand for ATP can be met by mitochondrial activity, the ATP provided by glycolysis makes a relatively minor contribution to the total energy budget of the muscle fiber.

- At peak levels of activity, ATP demands are enormous and mitochondrial ATP production rises to a maximum. This maximum rate is determined by the availability of oxygen, and oxygen cannot diffuse into the muscle fiber fast enough to enable the mitochondria to produce the required ATP. At peak levels of exertion, mitochondrial activity can provide only about one-third of the ATP needed. The remainder is produced through glycolysis (Figure 10–20c●).

When glycolysis produces pyruvic acid faster than it can be utilized by the mitochondria, pyruvic acid levels rise in the sarcoplasm. Under these conditions, pyruvic acid is converted to **lactic acid**, a related three-carbon molecule.

The anaerobic process of glycolysis enables the cell to generate additional ATP when the mitochondria are unable to meet the

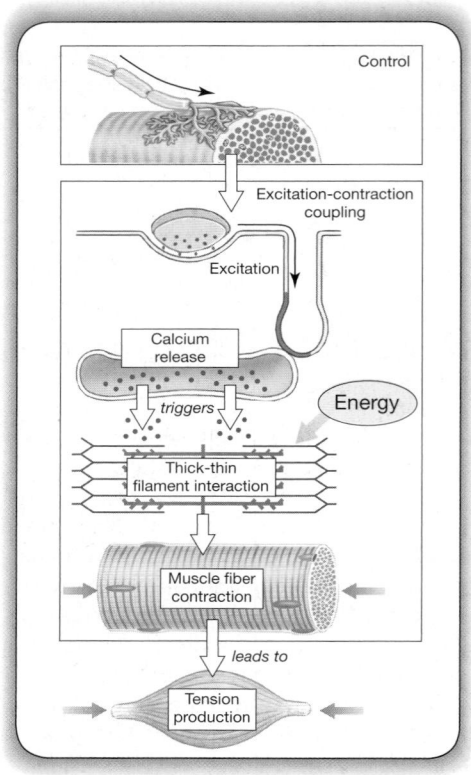

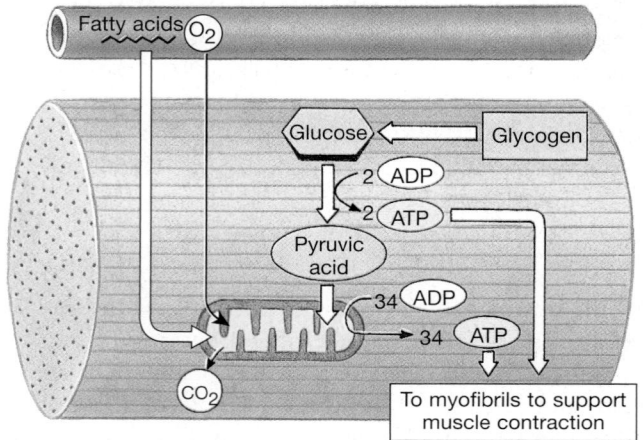

(a) **Resting muscle: Fatty acids are catabolized; the ATP produced is used to build energy reserves of ATP, CP, and glycogen.**

(b) **Moderate activity: Glucose and fatty acids are catabolized; the ATP produced is used to power contraction.**

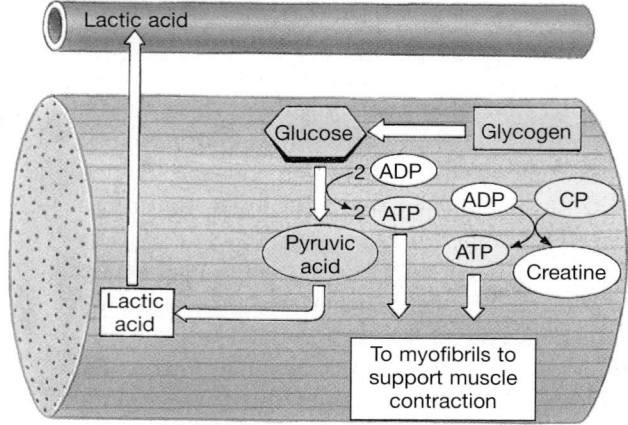

(c) **Peak activity: Most ATP is produced through glycolysis, with lactic acid as a by-product. Mitochondrial activity (not shown) now provides only about one-third of the ATP consumed.**

●FIGURE 10–20
Muscle Metabolism. **(a)** A resting muscle breaks down fatty acids by aerobic metabolism to make ATP. Surplus ATP is used to build reserves of creatine phosphate (CP) and glycogen. **(b)** At modest activity levels, mitochondria can meet ATP demands through the aerobic metabolism of fatty acids and glucose. **(c)** At peak activity levels, mitochondria cannot get enough oxygen to meet ATP demands. Most of the ATP is provided by glycolysis, leading to the production of lactic acid.

current energy demands. However, anaerobic energy production has its drawbacks:

• Lactic acid is an organic acid that dissociates in body fluids into a hydrogen ion and a negatively charged *lactate ion*. The production of lactic acid can therefore lower the intracellular pH.

Buffers in the sarcoplasm can resist pH shifts, but these defenses are limited. Eventually, changes in pH will alter the functional characteristics of key enzymes. The muscle fiber then cannot continue to contract.

• Glycolysis is a relatively inefficient way to generate ATP. Under anaerobic conditions, each glucose molecule generates 2 pyruvic acid molecules, which are converted to lactic acid. In return, the cell gains 2 ATP molecules through glycolysis. Had those 2 pyruvic acid molecules been catabolized aerobically in a mitochondrion, the cell would have gained 34 additional ATP.

☐ MUSCLE FATIGUE

An active skeletal muscle is said to be **fatigued** when it can no longer continue to perform at the required level of activity. Many factors are involved in promoting muscle fatigue. For example, muscle fatigue has been correlated with (1) depletion of metabolic reserves within the muscle fibers, (2) a decline in pH within the muscle fibers and the muscle as a whole, altering enzyme activities, and (3) a sense of weariness and a reduction in the desire to continue the activity, due to the effects of low blood pH and pain on the brain. After prolonged exertion, additional muscular factors may be involved, such as physical damage to the sarcolemma and sarcoplasmic reticulum. Muscle fatigue is cumulative—the effects become more pronounced as more neurons and muscle fibers are affected. The result is a gradual reduction in the capabilities and performance of the entire skeletal muscle.

If the muscle fiber is contracting at moderate levels and ATP demands can be met through aerobic metabolism, fatigue will not occur until glycogen, lipid, and amino acid reserves are depleted. This type of fatigue affects the muscles of long-distance athletes, such as marathon runners, after hours of exertion.

When a muscle produces a sudden, intense burst of activity at peak levels, most of the ATP is provided by glycolysis. After just seconds to minutes, the rising lactic acid levels lower the tissue pH, and the muscle can no longer function normally. Athletes who run sprints, such as the 100-meter dash, get this type of muscle fatigue. We will return to the topics of fatigue, athletic training, and metabolic activity later in the chapter.

Normal muscle function requires (1) substantial intracellular energy reserves, (2) a normal circulatory supply, (3) normal blood oxygen levels, and (4) a blood pH within normal limits. Anything that interferes with one or more of those factors will promote premature muscle fatigue. For example, reduced blood flow from tight clothing, heart problems, or a loss of blood slows the delivery of oxygen and nutrients, accelerates the buildup of lactic acid, and promotes muscle fatigue.

☐ THE RECOVERY PERIOD

When a muscle fiber contracts, the conditions in the sarcoplasm are changed. Energy reserves are consumed, heat is released, and, if the contraction was at peak levels, lactic acid is generated. In the **recovery period**, the conditions in muscle fibers are returned to normal, preexertion levels. It may take several hours for muscle fibers to recover from a period of moderate activity. After sustained activity at higher levels, complete recovery can take a week.

Lactic Acid Removal and Recycling

Glycolysis enables a skeletal muscle to continue contracting even when mitochondrial activity is limited by the availability of oxygen. As we have seen, however, lactic acid production is not an ideal way to generate ATP. It squanders the glucose reserves of the muscle fibers, and it is potentially dangerous because lactic acid can alter the pH of the blood and tissues.

During the recovery period, when oxygen is available in abundance, lactic acid can be recycled by conversion back to pyruvic acid. The pyruvic acid can then be used either by mitochondria to generate ATP or as a substrate for enzyme pathways that synthesize glucose and rebuild glycogen reserves.

During a period of exertion, lactic acid diffuses out of the muscle fibers and into the bloodstream. The process continues after the exertion has ended, because intracellular lactic acid concentrations are still relatively high. The liver absorbs the lactic acid and converts it to pyruvic acid. Roughly 30 percent of these pyruvic acid molecules are broken down in the TCA cycle, providing the ATP needed to convert the other pyruvic acid molecules to glucose. (We shall cover these processes more fully in Chapter 25.) The glucose molecules are then released into the circulation, where they are absorbed by skeletal muscle fibers and used to rebuild their glycogen reserves. This shuffling of lactic acid to the liver and glucose back to muscle cells is called the **Cori cycle**.

The Oxygen Debt

During the recovery period, the body's oxygen demand remains elevated above normal resting levels. The more ATP required, the more oxygen will be needed. The amount of oxygen required to restore normal, preexertion conditions is called the **oxygen debt**, or *excess postexercise oxygen consumption (EPOC)*.

Skeletal muscle fibers, which must restore ATP, creatine phosphate, and glycogen concentrations to their former levels, and liver cells, which generate the ATP needed to convert excess lactic acid to glucose, are responsible for most of the additional oxygen consumption. However, several other tissues also increase their rate of oxygen consumption and ATP generation during the recovery period. For example, sweat glands increase their secretory activity until normal body temperature is restored. While the oxygen debt is being repaid, the breathing rate and depth are increased. As a result, you continue to breathe heavily long after you stop exercising.

Heat Loss

Muscular activity generates substantial amounts of heat. When a catabolic process occurs, such as the breakdown of glycogen or the reactions of glycolysis, the muscle fiber captures only a portion of the released energy. ∞ p. 39 The rest is released as heat. A resting muscle fiber relying on aerobic metabolism captures about 42 percent of the energy released in catabolism. The other 58 percent warms the sarcoplasm, interstitial fluid, and circulating blood. Active skeletal muscles release roughly 85 percent of the heat needed to maintain normal body temperature.

When muscles become active, their consumption of energy skyrockets. As anaerobic energy production becomes the primary method of ATP generation, muscle fibers become less efficient at capturing energy. At peak levels of exertion, only about 30 percent of the released energy is captured as ATP; the remaining 70 percent warms the muscle and surrounding tissues. Body temperature soon climbs, and heat loss at the skin accelerates through mechanisms introduced in Chapters 1 and 5. ∞ **pp. 12, 171**

◻ HORMONES AND MUSCLE METABOLISM

Metabolic activities in skeletal muscle fibers are adjusted by hormones of the endocrine system. *Growth hormone* from the pituitary gland and *testosterone* (the primary sex hormone in males) stimulate the synthesis of contractile proteins and the enlargement of skeletal muscles. *Thyroid hormones* elevate the rate of energy consumption by resting and active skeletal muscles. During a sudden crisis, hormones of the adrenal gland, notably *epinephrine* (adrenaline), stimulate muscle metabolism and increase both the duration of stimulation and the force of contraction. (We shall further examine the effects of hormones on muscle and other tissues in Chapter 18.)

10–6 MUSCLE PERFORMANCE

Objectives

- Relate the types of muscle fibers to muscle performance.
- Distinguish between aerobic and anaerobic endurance, and explain their implications for muscular performance.

Muscle performance can be considered in terms of **power**, the maximum amount of tension produced by a particular muscle or muscle group, and **endurance**, the amount of time during which the individual can perform a particular activity. Two major factors determine the performance capabilities of any skeletal muscle: (1) the types of muscle fibers in the muscle and (2) physical conditioning or training.

◻ TYPES OF SKELETAL MUSCLE FIBERS

The human body has three major types of skeletal muscle fibers: *fast fibers*, *slow fibers*, and *intermediate fibers*.

Fast Fibers

Most of the skeletal muscle fibers in the body are called **fast fibers**, because they can contract in 0.01 sec or less after stimulation. Fast fibers are large in diameter and contain densely packed myofibrils, large glycogen reserves, and relatively few mitochondria. The tension produced by a muscle fiber is directly proportional to the number of sarcomeres, so muscles dominated by fast fibers produce powerful contractions. However, fast fibers fatigue rapidly because their contractions use ATP in massive amounts, so prolonged activity is supported primarily by anaerobic metabolism. Several other names are used to refer to these muscle fibers, including *white muscle fibers*, *fast-twitch glycolytic fibers*, and *Type II-A fibers*.

Slow Fibers

Slow fibers are only about half the diameter of fast fibers and take three times as long to contract after stimulation. These fibers are specialized to enable them to continue contracting for extended periods, long after a fast muscle would have become fatigued. The most important specializations improve mitochondrial performance. For example, one of the main characteristics of slow muscle tissue is that it contains a more extensive network of capillaries than is typical of fast muscle tissue and so has a dramatically higher oxygen supply.

Slow fibers also contain the red pigment **myoglobin** (MĪ-ō-glō-bin). This globular protein is structurally related to hemoglobin, the oxygen-carrying pigment in blood. Both myoglobin and hemoglobin are red pigments that reversibly bind oxygen molecules. Although other muscle fiber types contain small amounts of myoglobin, it is most abundant in slow fibers. As a result, resting slow fibers contain substantial oxygen reserves that can be mobilized during a contraction. Because slow fibers have both an extensive capillary supply and a high concentration of myoglobin, skeletal muscles dominated by slow fibers are dark red. They are also known as *red muscle fibers*, *slow-twitch oxidative fibers*, and *Type I fibers*.

With oxygen reserves and a more efficient blood supply, the mitochondria of slow fibers can contribute more ATP during contraction. Thus, slow fibers are less dependent on anaerobic metabolism than are fast fibers. Some of the mitochondrial energy production involves the breakdown of stored lipids rather than glycogen, so glycogen reserves of slow fibers are smaller than those of fast fibers. Slow fibers also contain more mitochondria than do fast fibers. Figure 10–21● compares the appearance of fast and slow fibers.

Intermediate Fibers

The properties of **intermediate fibers** are intermediate between those of fast fibers and slow fibers. In appearance, intermediate fibers most closely resemble fast fibers, for they contain little myoglobin and are relatively pale. They have a more extensive capillary network around them, however, and are more resistant to fatigue than are fast fibers. Intermediate fibers are also known as *fast-twitch oxidative fibers* and *Type II-B fibers*.

The three types of muscle fibers are compared in Table 10–3. In muscles that contain a mixture of fast and intermediate fibers, the proportion can change with physical conditioning. For example, if a muscle is used repeatedly for endurance events, some of the fast fibers will develop the appearance and functional capabilities of intermediate fibers. The muscle as a whole will thus become more resistant to fatigue.

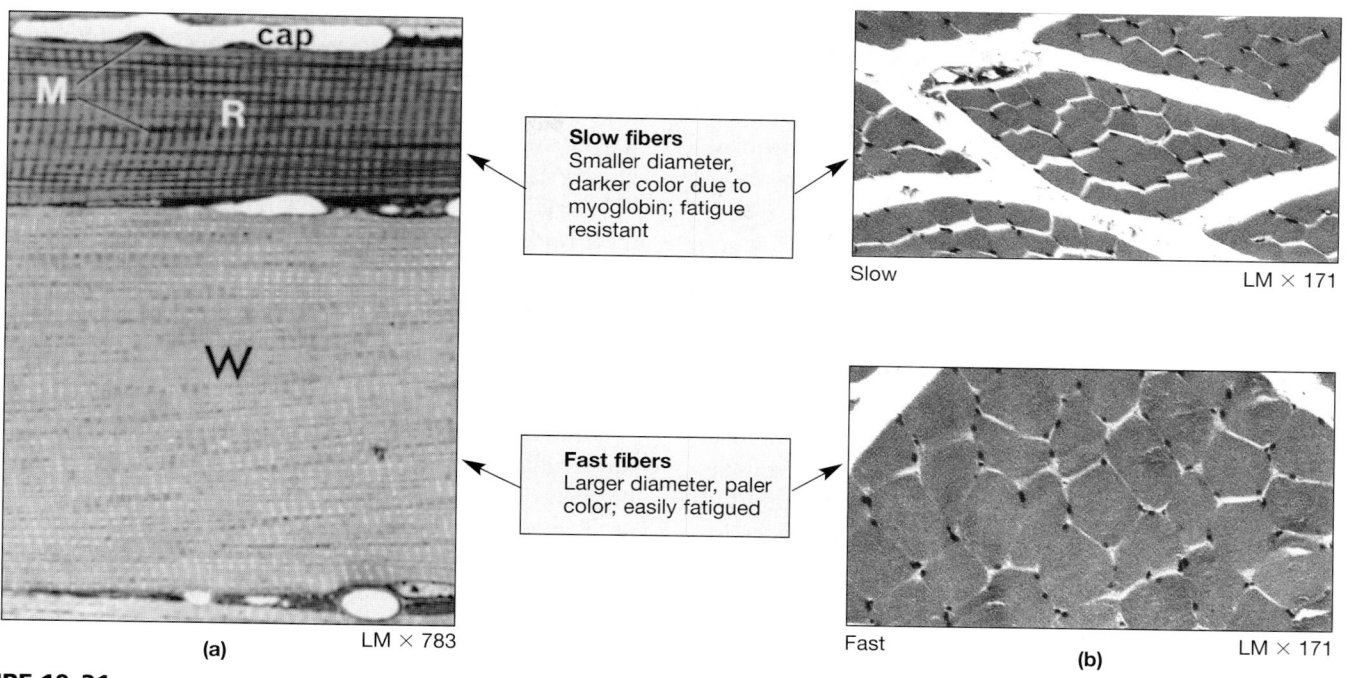

●FIGURE 10–21
Fast versus Slow Fibers. **(a)** The slender, slow fiber (R for red) has more mitochondria (M) and a more extensive capillary supply (cap) than does the fast fiber (W for white). **(b)** Notice the difference in the sizes of slow fibers, above, and of fast fibers, below.

☐ MUSCLE PERFORMANCE AND THE DISTRIBUTION OF MUSCLE FIBERS

The percentages of fast, intermediate, and slow fibers in a skeletal muscle can be quite variable. Muscles dominated by fast fibers appear pale and are often called **white muscles**. Chicken breasts contain "white meat" because chickens use their wings only for brief intervals, as when fleeing from a predator, and the power for flight comes from the anaerobic process of glycolysis in the fast

fibers of their breast muscles. As we learned earlier, the extensive blood vessels and myoglobin in slow fibers give these fibers a reddish color; muscles dominated by slow fibers are therefore known as **red muscles**. Chickens walk around all day, and the movements are powered by aerobic metabolism in the slow fibers of the "dark meat" of their legs.

Most human muscles contain a mixture of fiber types and so appear pink. However, there are no slow fibers in muscles of the eye or

TABLE 10–3 PROPERTIES OF SKELETAL MUSCLE FIBER TYPES

Property	Slow	Intermediate	Fast
Cross-sectional diameter	Small	Intermediate	Large
Tension	Low	Intermediate	High
Contraction speed	Slow	Fast	Fast
Fatigue resistance	High	Intermediate	Low
Color	Red	Pink	White
Myoglobin content	High	Low	Low
Capillary supply	Dense	Intermediate	Scarce
Mitochondria	Many	Intermediate	Few
Glycolytic enzyme concentration in sarcoplasm	Low	High	High
Substrates used for ATP generation during contraction	Lipids, carbohydrates, amino acids (aerobic)	Primarily carbohydrates (anaerobic)	Carbohydrates (anaerobic)
Alternative names	Type I, S (slow), red, SO (slow oxidizing), slow-twitch oxidative	Type II-B, FR (fast resistant), fast-twitch oxidative	Type II-A, FF (fast fatigue), white, fast-twitch glycolytic

hand, where swift, but brief, contractions are required. Many back and calf muscles are dominated by slow fibers; these muscles contract almost continuously to maintain an upright posture. The percentage of fast versus slow fibers in each muscle is genetically determined. As we noted earlier, the proportion of intermediate fibers to fast fibers can increase as a result of athletic training.

▢ MUSCLE HYPERTROPHY

As a result of repeated, exhaustive stimulation, muscle fibers develop more mitochondria, a higher concentration of glycolytic enzymes, and larger glycogen reserves. Such muscle fibers have more myofibrils than do muscles that are less used, and each myofibril contains more thick and thin filaments. The net effect is **hypertrophy**, or an enlargement of the stimulated muscle. The number of muscle fibers does not change significantly, but the muscle as a whole enlarges because each muscle fiber increases in diameter.

Hypertrophy occurs in muscles that have been repeatedly stimulated to produce near-maximal tension. The intracellular changes that occur increase the amount of tension produced when these muscles contract. A champion weight lifter or bodybuilder is an excellent example of hypertrophied muscular development.

A skeletal muscle that is not regularly stimulated by a motor neuron loses muscle tone and mass. The muscle becomes flaccid, and the muscle fibers become smaller and weaker. This reduction in muscle size, tone, and power is called **atrophy**. Individuals paralyzed by spinal injuries or other damage to the nervous system will gradually lose muscle tone and size in the areas affected. Even a temporary reduction in muscle use can lead to muscular atrophy; you can easily observe this effect by comparing "before and after" limb muscles in someone who has worn a cast. Muscle atrophy is initially reversible, but dying muscle fibers are not replaced. In extreme atrophy, the functional losses are permanent. That is why physical therapy is crucial for people who are temporarily unable to move normally.

Because skeletal muscles depend on their motor neurons for stimulation, disorders that affect the nervous system can indirectly affect the muscular system. In polio, a virus attacks motor neurons in the spinal cord and brain, producing muscular paralysis and atrophy. **AM** Problems with the Control of Muscle Activity

▢ PHYSICAL CONDITIONING

Physical conditioning and training schedules enable athletes to improve both power and endurance. In practice, the training schedule varies, depending on whether the activity is supported primarily by aerobic or anaerobic energy production.

Anaerobic endurance is the length of time muscular contraction can continue to be supported by glycolysis and by the existing energy reserves of ATP and CP. Anaerobic endurance is limited by (1) the amount of ATP and CP on hand, (2) the amount of glycogen available for breakdown, and (3) the ability of the muscle to tolerate the lactic acid generated during the anaerobic period. Typically, the onset of muscle fatigue occurs within 2 minutes of the start of maximal activity.

Activities that require above-average levels of anaerobic endurance include a 50-meter dash or swim, a pole vault, and a weight-lifting competition. These activities involve the contractions of fast fibers. The energy for the first 10–20 seconds of activity comes from the ATP and CP reserves of the cytoplasm. As these reserves dwindle, glycogen breakdown and glycolysis provide additional energy. Athletes training to improve anaerobic endurance perform frequent, brief, intensive workouts that stimulate muscle hypertrophy.

Aerobic endurance is the length of time a muscle can continue to contract while supported by mitochondrial activities. Aerobic endurance is determined primarily by the availability of substrates for aerobic respiration, which the muscle fibers can obtain by breaking down carbohydrates, lipids, or amino acids. Initially, many of the nutrients catabolized by the muscle fiber are obtained from reserves in the sarcoplasm. Prolonged aerobic activity, however, must be supported by nutrients provided by the circulating blood.

During exercise, blood vessels in the skeletal muscles dilate, increasing blood flow and thus bringing oxygen and nutrients to the active muscle tissue. Warm-up periods are therefore important not only in that they take advantage of treppe, the increase in tension production noted on page 309, but also because they stimulate circulation in the muscles before the serious workout begins. Because glucose is a preferred energy source, aerobic athletes such as marathon runners typically "load" or "bulk up" on carbohydrates for the last three days before an event. They may also consume glucose-rich "sports drinks" during a competition. (We shall consider the risks and benefits of these practices in Chapter 25.)

Training to improve aerobic endurance generally involves sustained low levels of muscular activity. Examples include jogging, distance swimming, and other exercises that do not require peak tension production. Improvements in aerobic endurance result from altering the characteristics of muscle fibers and improving the performance of the cardiovascular system:

1. *Altering the Characteristics of Muscle Fibers.* The composition of fast and slow fibers in each muscle is genetically determined, and individual differences are significant. These variations affect aerobic endurance, because a person with more slow fibers in a particular muscle will be better able to perform under aerobic conditions than will a person with fewer. However, skeletal muscle cells respond to changes in the pattern of neural stimulation. Fast fibers trained for aerobic competition develop the characteristics of intermediate fibers, and this change improves aerobic endurance.

2. *Improving Cardiovascular Performance.* Cardiovascular activity affects muscular performance by delivering oxygen and nutrients to active muscles. Physical training alters cardiovascular function by accelerating blood flow, thus improving oxygen and nutrient availability. (We shall examine factors involved in improving cardiovascular performance in Chapter 21.)

Aerobic activities do not promote muscle hypertrophy. Many athletes train using a combination of aerobic and anaerobic exer-

Delayed-Onset Muscle Soreness

You have probably experienced muscle soreness the day after a period of physical exertion. Considerable controversy exists over the source and significance of this pain, which is known as *delayed-onset muscle soreness* (DOMS) and has several interesting characteristics:

- DOMS is distinct from the soreness you experience immediately after you stop exercising. The initial short-term soreness is probably related to the biochemical events associated with muscle fatigue.

- DOMS generally begins several hours after the exercise period ends and may last 3 or 4 days.

- The amount of DOMS is highest when the activity involves eccentric contractions. Activities dominated by concentric or isometric contractions produce less soreness.

- Levels of CPK and myoglobin are elevated in the blood, indicating damage to muscle cell membranes. The nature of the activity (eccentric, concentric, or isometric) has no effect on these levels, nor can the levels be used to predict the degree of soreness experienced.

Three mechanisms have been proposed to explain DOMS:

1. Small tears may exist in the muscle tissue, leaving muscle fibers with damaged membranes. The sarcolemma of each damaged muscle fiber permits the loss of enzymes, myoglobin, and other chemicals that may stimulate pain receptors in the region.

2. The pain may result from muscle spasms in the affected skeletal muscles. In some studies, stretching the muscle involved can reduce the degree of soreness.

3. The pain may result from tears in the connective tissue framework and tendons of the skeletal muscle.

Some evidence supports each of these mechanisms, but it is unlikely that any one tells the entire story. For example, muscle fiber damage is certainly supported by biochemical findings, but if that were the only factor, the type of activity would be correlated with the level of pain experienced.

---

cises so that their muscles will enlarge and both anaerobic and aerobic endurance will improve. These athletes alternate an aerobic activity, such as swimming, with sprinting or weight lifting. The combination is known as *interval training* or *cross-training*. Interval training is particularly useful for persons engaged in racquet sports, such as tennis or squash, which are dominated by aerobic activities, but are punctuated by brief periods of anaerobic effort. **AM** Power, Endurance, and Energy Reserves

✓ Why would a sprinter experience muscle fatigue before a marathon runner would?

✓ Which activity would be more likely to create an oxygen debt: swimming laps or lifting weights?

✓ Which type of muscle fibers would you expect to predominate in the large leg muscles of someone who excels at endurance activities, such as cycling or long-distance running?

Answers start on page Q-1

 Review muscle fatigue on the **IP CD-ROM**: Muscular System/ Muscle Metabolism.

10–7 CARDIAC MUSCLE TISSUE

Objective

- Identify the structural and functional differences between skeletal muscle and cardiac muscle cells.

We introduced **cardiac muscle tissue** in Chapter 4 and briefly compared its properties with those of other types of muscle. **Cardiac muscle cells**, also called *cardiocytes* or *cardiac myocytes*, are found only in the heart. Table 10–4 compares skeletal, cardiac, and smooth muscle tissues in greater detail.

Like skeletal muscle fibers, cardiac muscle cells contain organized myofibrils, and the presence of many aligned sarcomeres gives the cells a striated appearance. However, cardiac muscle cells are much smaller than skeletal muscle fibers, and significant structural and functional differences exist between the two.

TABLE 10–4 A COMPARISON OF SKELETAL, CARDIAC, AND SMOOTH MUSCLE TISSUES

Property	Skeletal Muscle	Cardiac Muscle	Smooth Muscle
Fiber dimensions (diameter × length)	100 μm × up to 30 cm	10–20 μm × 50–100 μm	5–10 μm × 30–200 μm
Nuclei	Multiple, near sarcolemma	Generally single, centrally located	Single, centrally located
Filament organization	In sarcomeres along myofibrils	In sarcomeres along myofibrils	Scattered throughout sarcoplasm
SR	Terminal cisternae in triads at zones of overlap	SR tubules contact T tubules at Z lines	Dispersed throughout sarcoplasm, no T tubules
Control mechanism	Neural, at single neuromuscular junction	Automaticity (pacemaker cells)	Automaticity (pacesetter cells), neural or hormonal control
Ca^{2+} source	Release from SR	Extracellular fluid and release from SR	Extracellular fluid and release from SR
Contraction	Rapid onset; may be tetanized; rapid fatigue	Slower onset; cannot be tetanized; resistant to fatigue	Slow onset; may be tetanized; resistant to fatigue
Energy source	Aerobic metabolism at moderate levels of activity; glycolysis (anaerobic) during peak activity	Aerobic metabolism, usually lipid or carbohydrate substrates	Primarily aerobic metabolism

☐ STRUCTURAL CHARACTERISTICS OF CARDIAC MUSCLE TISSUE

Important structural differences between skeletal muscle fibers and cardiac muscle cells include the following:

- Cardiac muscle cells are relatively small, averaging 10–20 μm in diameter and 50–100 μm in length.

- A typical cardiac muscle cell (Figure 10–22a,b●) has a single, centrally placed nucleus, although a few may have two or more nuclei.

- The T tubules in a cardiac muscle cell are short and broad, and there are no triads. The T tubules encircle the sarcomeres at the Z lines rather than at the zone of overlap.

- The SR of a cardiac muscle cell lacks terminal cisternae, and its tubules contact the cell membrane as well as the T tubules (Figure 10–22c●). As in skeletal muscle fibers, the appearance of an action potential triggers the release of calcium from the SR and the contraction of sarcomeres; it also increases the permeability of the sarcolemma to extracellular calcium ions.

- Cardiac muscle cells are almost totally dependent on aerobic metabolism to obtain the energy they need to continue contracting. The sarcoplasm of a cardiac muscle cell thus contains large numbers of mitochondria and abundant reserves of myoglobin (to store oxygen). Energy reserves are maintained in the form of glycogen and lipid inclusions.

- Each cardiac muscle cell contacts several others at specialized sites known as **intercalated** (in-TER-ka-lā-ted) **discs**. ∞ p. 136 Intercalated discs play a vital role in the function of cardiac muscle, as we shall learn next.

Intercalated Discs

At an intercalated disc (Figure 10–22a,b●), the cell membranes of two adjacent cardiac muscle cells are extensively intertwined and bound together by gap junctions and desmosomes. ∞ p. 113 These connections help stabilize the relative positions of adjacent cells and maintain the three-dimensional structure of the tissue. The gap junctions allow ions and small molecules to move from one cell to another. This arrangement creates a direct electrical connection between the two muscle cells. An action potential can travel across an intercalated disc, moving quickly from one cardiac muscle cell to another.

Myofibrils in the two interlocking muscle cells are firmly anchored to the membrane at the intercalated disc. Because their myofibrils are essentially locked together, the two muscle cells can "pull together" with maximum efficiency. Because the cardiac muscle cells are mechanically, chemically, and electrically connected to one another, the entire tissue resembles a single, enormous muscle cell. For this reason, cardiac muscle has been called a *functional syncytium* (sin-SISH-um; a fused mass of cells).

☐ FUNCTIONAL CHARACTERISTICS OF CARDIAC MUSCLE TISSUE

In Chapter 20, we shall examine cardiac muscle physiology in detail; here, we will briefly summarize four major functional specialties of cardiac muscle:

1. Cardiac muscle tissue contracts without neural stimulation. This property is called **automaticity**. The timing of contractions is normally determined by specialized cardiac muscle cells called **pacemaker cells**.

2. Innervation by the nervous system can alter the pace established by the pacemaker cells and adjust the amount of tension produced during a contraction.

3. Cardiac muscle cell contractions last roughly 10 times longer than do those of skeletal muscle fibers.

4. The properties of cardiac muscle cell membranes differ from those of skeletal muscle fiber membranes. As a result, individual twitches cannot undergo wave summation, and cardiac muscle tissue cannot produce tetanic contractions. This difference is

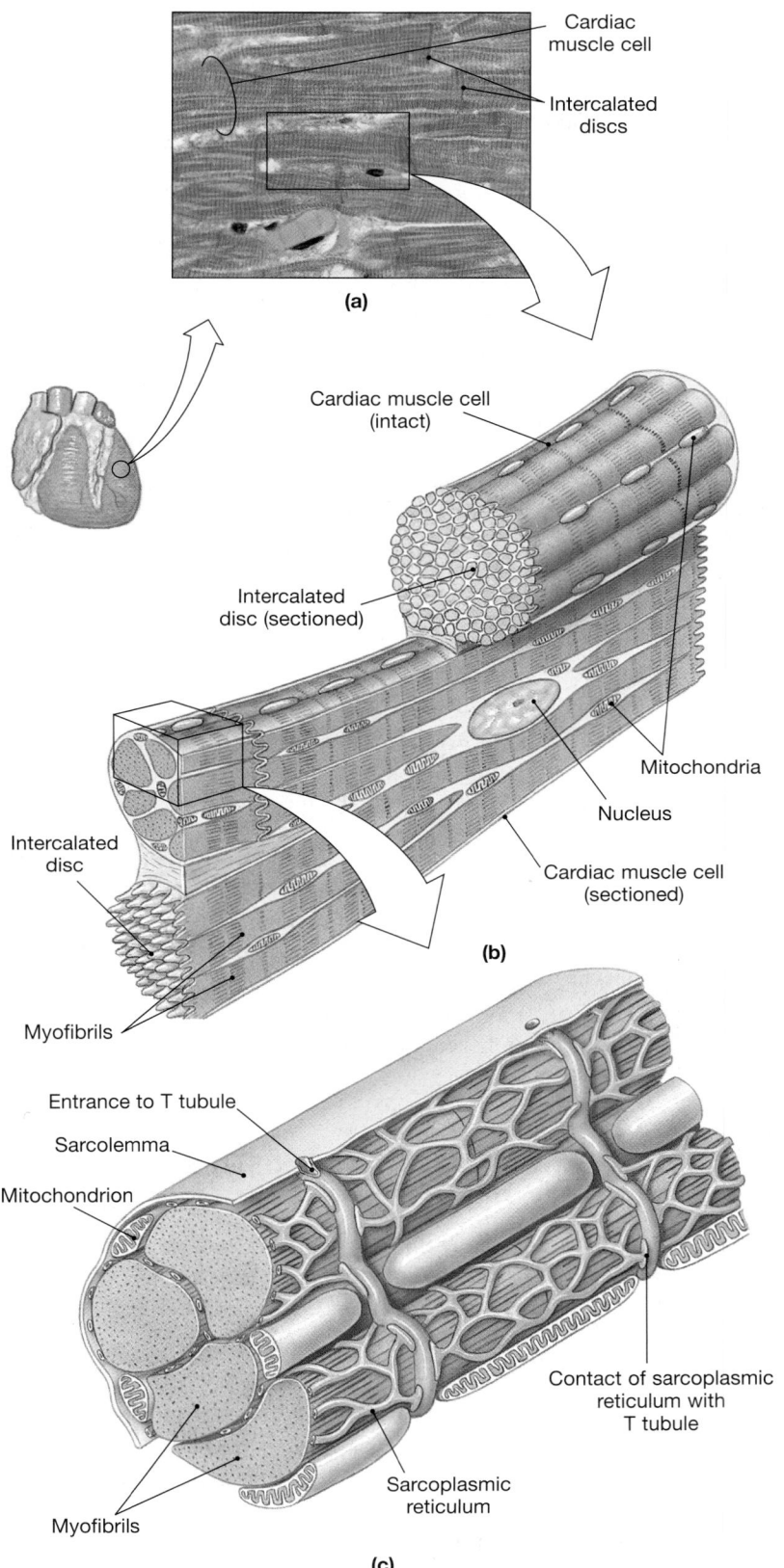

important, because a heart in a sustained tetanic contraction could not pump blood.

10–8 SMOOTH MUSCLE TISSUE

Objectives

- Identify the structural and functional differences between skeletal muscle and smooth muscle cells.
- Discuss the role that smooth muscle tissue plays in systems throughout the body.

Smooth muscle tissue is present in almost every organ, forming sheets, bundles, or sheaths around other tissues. Smooth muscles around blood vessels regulate blood flow through vital organs. In the digestive and urinary systems, rings of smooth muscle, called *sphincters*, regulate the movement of materials along internal passageways. Smooth muscles in bundles, layers, or sheets play a variety of other roles:

- **Integumentary System.** Smooth muscles around blood vessels regulate the flow of blood to the superficial dermis; smooth muscles of the arrector pili elevate hairs. ∞ p. 167

- **Cardiovascular System.** Smooth muscles encircling blood vessels control the distribution of blood and help regulate the blood pressure.

- **Respiratory System.** Smooth muscle contraction or relaxation alters the diameters of the respiratory passageways and changes the resistance to airflow.

- **Digestive System.** Extensive layers of smooth muscle in the walls of the digestive tract play an essential role in moving materials along the tract. Smooth muscle in the walls of the gallbladder contract to eject bile into the digestive tract.

- **Urinary System.** Smooth muscle tissue in the walls of small blood vessels alters the rate of filtration at the kidneys. Layers of smooth muscle in the walls of the ureters transport urine to the urinary bladder; the contraction of the smooth muscle in the wall of the urinary bladder forces urine out of the body.

- **Reproductive System.** Layers of smooth muscle help move sperm along the reproductive tract in males and cause the ejection of glandular secretions from the accessory glands into the reproductive tract. In females, layers of smooth muscle help move oocytes (and perhaps sperm) along the reproductive tract, and contraction of the smooth muscle in the walls of the uterus expels the fetus at delivery.

Figure 10–23a● shows typical smooth muscle tissue as seen by light microscopy. Smooth muscle tissue differs from both skeletal and cardiac muscle tissues in structure and function (Table 10–4).

●**FIGURE 10–22**
Cardiac Muscle Tissue. **(a)** A light micrograph of a cardiac muscle tissue. Notice the striations and the intercalated discs. **(b, c)** The structure of a cardiac muscle cell; compare with Figure 10–3.

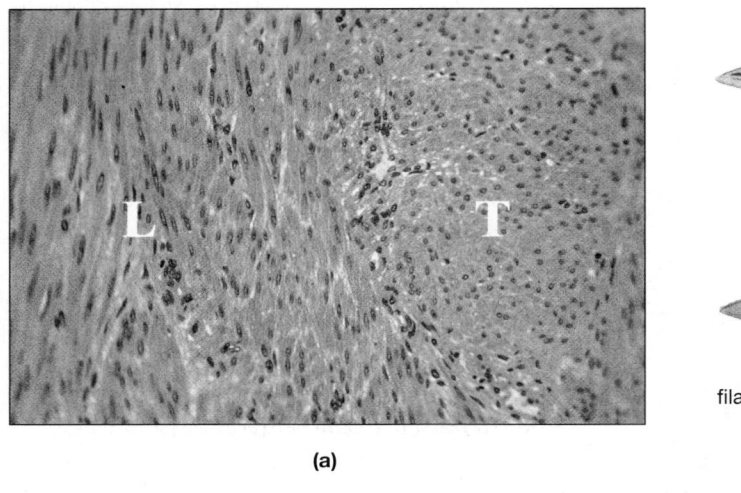

(a)

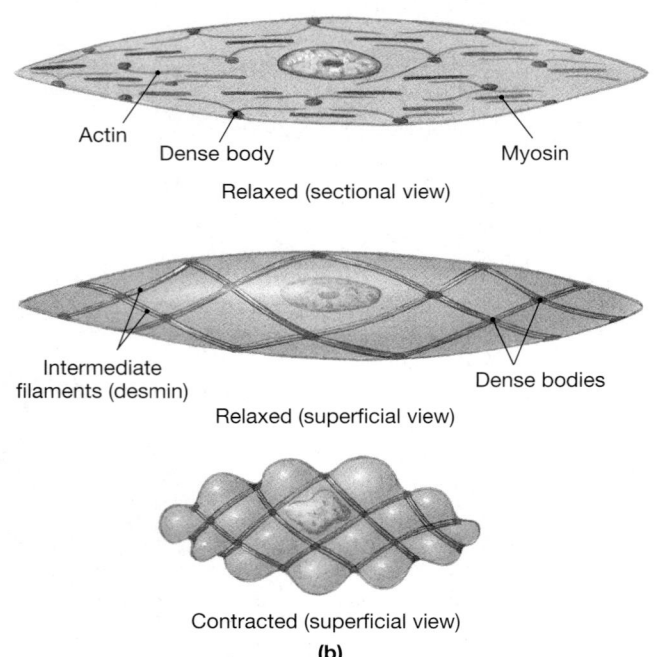

Actin   Dense body   Myosin

Relaxed (sectional view)

Intermediate
filaments (desmin)   Dense bodies

Relaxed (superficial view)

Contracted (superficial view)

(b)

●FIGURE 10–23
Smooth Muscle Tissue. **(a)** Many visceral organs contain several layers of smooth muscle tissue oriented in different directions. Here, a single sectional view shows a smooth muscle cell in both longitudinal (L) and transverse (T) sections. **(b)** A single relaxed smooth muscle cell is spindle shaped and has no striations. Note the changes in cell shape as contraction occurs.

☐ STRUCTURAL CHARACTERISTICS OF SMOOTH MUSCLE TISSUE

Actin and myosin are present in all three types of muscle tissue. In skeletal and cardiac muscle cells, these proteins are organized in sarcomeres, with thin and thick filaments. The internal organization of a smooth muscle cell is very different:

• Smooth muscle cells are relatively long and slender, ranging from 5 to 10 μm in diameter and from 30 to 200 μm in length.

• Each cell is spindle shaped and has a single, centrally located nucleus.

• A smooth muscle fiber has no T tubules, and the sarcoplasmic reticulum forms a loose network throughout the sarcoplasm. Smooth muscle cells also lack myofibrils and sarcomeres. As a result, this tissue also has no striations and is called **nonstriated** muscle.

• Thick filaments are scattered throughout the sarcoplasm of a smooth muscle cell. The myosin proteins are organized differently than in skeletal or cardiac muscle cells, and smooth muscle cells have more cross-bridges per thick filament.

• The thin filaments in a smooth muscle cell are attached to **dense bodies**, structures distributed throughout the sarcoplasm in a network of intermediate filaments composed of the protein *desmin* (Figure 10–23b●). Some of the dense bodies are firmly attached to the sarcolemma. The dense bodies and intermediate filaments anchor the thin filaments such that, when sliding occurs between thin and thick filaments, the cell shortens. Dense bodies are not arranged in straight lines, so when a contraction occurs, the muscle cell twists like a corkscrew.

• Adjacent smooth muscle cells are bound together at dense bodies, transmitting the contractile forces from cell to cell throughout the tissue.

• Although smooth muscle cells are surrounded by connective tissue, the collagen fibers never unite to form tendons or aponeuroses, as they do in skeletal muscles.

☐ FUNCTIONAL CHARACTERISTICS OF SMOOTH MUSCLE TISSUE

Smooth muscle tissue differs from other muscle tissue in its (1) excitation–contraction coupling, (2) length–tension relationships, (3) control of contractions, and (4) smooth muscle tone.

Excitation–Contraction Coupling

The trigger for smooth muscle contraction is the appearance of free calcium ions in the cytoplasm. On stimulation, a blast of calcium ions enters the cell from the extracellular fluid, and additional calcium ions are released by the sarcoplasmic reticulum. The net result is a rise in calcium ion concentrations throughout the cell. Once in the sarcoplasm, the calcium ions interact with **calmodulin**, a calcium-binding protein. Calmodulin then activates the enzyme **myosin light chain kinase**, which breaks down ATP and initiates the contraction. This situation is quite different from that in skeletal and cardiac muscles, in which the trigger for contraction is the binding of calcium ions to troponin.

Length–Tension Relationships

Because the thick and thin filaments are scattered and are not organized into sarcomeres, tension development and resting length in smooth muscle are not directly related. A stretched smooth muscle soon adapts to its new length and retains the ability to contract on demand. This ability to function over a wide range of lengths is called **plasticity**. Smooth muscle can contract over a range of lengths four times greater than that of skeletal muscle. Plasticity is especially important in digestive organs that undergo great changes in volume, such as the stomach. Despite the lack of sarcomere organization, smooth muscle contractions can be just as powerful as those of skeletal muscles. Like skeletal muscle fibers, smooth muscle cells often undergo sustained tetanic contractions.

Control of Contractions

Many smooth muscle cells are not innervated by motor neurons, and the neurons that do innervate smooth muscles are not under voluntary control. The nature of the connection with the nervous system provides a means of categorizing smooth muscle cells as either multiunit or visceral. **Multiunit smooth muscle cells** are innervated in motor units comparable to those of skeletal muscles, but each smooth muscle cell may be connected to several motor neurons rather than to just one. In contrast, many **visceral smooth muscle cells** lack a direct contact with any motor neuron.

Multiunit smooth muscle cells resemble skeletal muscle fibers and cardiac muscle cells, in that neural activity produces an action potential that is propagated over the sarcolemma. However, the contractions of these smooth muscle cells are more leisurely than are those of skeletal or cardiac muscle cells. Multiunit smooth muscle cells are located in the iris of the eye, where they regulate the diameter of the pupil; along portions of the male reproductive tract; within the walls of large arteries; and in the arrector pili muscles of the skin. Multiunit smooth muscle cells do not typically occur in the digestive tract.

Visceral smooth muscle cells are arranged in sheets or layers. Within each layer, adjacent muscle cells are connected by gap junctions. Because they are connected in this way, whenever one muscle cell contracts, the electrical impulse that triggered the contraction can travel to adjacent smooth muscle cells. The contraction therefore spreads in a wave that soon involves every smooth muscle cell in the layer. The initial stimulus may be the activation of a motor neuron that contacts one of the muscle cells in the region. But smooth muscle cells will also contract or relax in response to chemicals, hormones, local concentrations of oxygen or carbon dioxide, or physical factors such as extreme stretching or irritation.

Many visceral smooth muscle networks show rhythmic cycles of activity in the absence of neural stimulation. These cycles are characteristic of the smooth muscle cells in the wall of the digestive tract, where **pacesetter cells** undergo spontaneous depolarization and trigger the contraction of entire muscular sheets. Visceral smooth muscle cells are located in the walls of the digestive tract, the gallbladder, the urinary bladder, and many other internal organs.

Smooth Muscle Tone

Both multiunit and visceral smooth muscle tissues show a normal background level of activity, or smooth muscle tone. The regulatory mechanisms just detailed stimulate contraction and increase muscle tone. Neural, hormonal, or chemical factors can also stimulate smooth muscle relaxation, producing a decrease in muscle tone. For example, smooth muscle cells at the entrances to capillaries regulate the amount of blood flow into each vessel. If the tissue becomes starved for oxygen, the smooth muscle cells relax, whereupon blood flow increases, delivering additional oxygen. As conditions return to normal, the smooth muscle regains its normal muscle tone.

✓ What feature of cardiac muscle tissue allows the heart to act as a functional syncytium?

✓ Why are cardiac and smooth muscle contractions more affected by changes in extracellular Ca^{2+} than are skeletal muscle contractions?

✓ Smooth muscle can contract over a wider range of resting lengths than skeletal muscle can. Why?

Answers start on page Q-1

Chapter Review

SELECTED CLINICAL TERMINOLOGY

Terms Discussed in This Chapter

botulism: A severe, potentially fatal paralysis of skeletal muscles, resulting from the consumption of a bacterial toxin. *(p. 306 and AM)*

Duchenne's muscular dystrophy (DMD):   One of the most common and best understood of the muscular dystrophies. *(p. 300 and AM)*

muscular dystrophies (DIS-trō-fēz):   A varied collection of congenital diseases that produce progressive muscle weakness and deterioration. *(p. 300 and AM)*

myasthenia gravis (mī-as-THĒ-nē-uh GRA-vis):   A general muscular weakness resulting from a reduction in the number of ACh receptors on the motor end plate. *(p. 306 and AM)*

polio: A disease resulting from the destruction of motor neurons by a certain virus and characterized by the paralysis and atrophy of motor units. *(p. 322 and AM)*

rigor mortis: A state following death during which muscles are locked in the contracted position, making the body extremely stiff. *(p. 306)*

tetanus: A disease caused by a bacterial toxin that leads to the production of sustained, powerful contractions of skeletal muscles. *(p. 311)*

AM Additional Terms Discussed in the *Applications Manual*

cholinesterase inhibitor
chronic fatigue syndrome
fibromyalgia syndrome
trichinosis

STUDY OUTLINE

10–1 SKELETAL MUSCLE TISSUE AND THE MUSCULAR SYSTEM p. 291

1. The three types of muscle tissue are skeletal muscle, cardiac muscle, and smooth muscle.
2. **Skeletal muscles** attach to bones directly or indirectly. Their functions are to (1) produce skeletal movement, (2) maintain posture and body position, (3) support soft tissues, (4) guard entrances and exits, and (5) maintain body temperature.

10–2 ANATOMY OF SKELETAL MUSCLE p. 292

ORGANIZATION OF CONNECTIVE TISSUES p. 292

1. Each muscle cell or fiber is surrounded by an **endomysium**. Bundles of muscle fibers are sheathed by a **perimysium**, and the entire muscle is covered by an **epimysium**. At the ends of the muscle are **tendons** or **aponeuroses** that attach the muscle to bones. *(Figure 10–1)*

BLOOD VESSELS AND NERVES p. 292

2. The epimysium and perimysium contain the blood vessels and nerves that supply the muscle fibers.

SKELETAL MUSCLE FIBERS p. 292

3. A skeletal muscle fiber has a **sarcolemma**, or cell membrane; **sarcoplasm** (cytoplasm); and **sarcoplasmic reticulum** (**SR**), similar to the smooth endoplasmic reticulum of other cells. **Transverse** (**T**) **tubules** and **myofibrils** aid in contraction. Filaments in a myofibril are organized into repeating functional units called **sarcomeres**. *(Figures 10–3 to 10–6)*
4. **Myofilaments** called **thin filaments** and **thick filaments** form myofibrils. *(Figures 10–2 to 10–6)*
5. Thin filaments consist of **F actin**, **nebulin**, **tropomyosin**, and **troponin**. Tropomyosin molecules cover **active sites** on the **G actin** subunits that form the F actin strand. Troponin binds to G actin and tropomyosin and holds the tropomyosin in position. *(Figure 10–7)*
6. Thick filaments consist of a bundle of myosin molecules around a **titin** core. Each myosin molecule has an elongate **tail** and a globular **head**, which forms **cross-bridges** during contraction. In a resting muscle cell, the interactions between the active sites on G actin and myosin heads are prevented by tropomyosin. *(Figure 10–7)*

THE SLIDING-FILAMENT THEORY p. 298

7. The **sliding-filament theory** explains how the relationship between thick and thin filaments changes as the muscle contracts. *(Figure 10–8)*

Gross anatomy of skeletal muscle: **IP CD-ROM**: Muscular System/Anatomy Review.

10–3 THE CONTRACTION OF SKELETAL MUSCLE p. 300

1. When muscle cells contract, they create **tension**. *(Figure 10–19)*

THE CONTROL OF SKELETAL MUSCLE ACTIVITY p. 301

2. The activity of a muscle fiber is controlled by a neuron at a **neuromuscular** *(myoneural)* **junction** (**NMJ**). *(Figure 10–10)*
3. When an **action potential** arrives at the **synaptic terminal**, **acetylcholine** (**ACh**) is released into the **synaptic cleft**. The binding of ACh to receptors on the opposing *junctional folds* leads to the generation of an action potential in the sarcolemma. *(Figure 10–10)*

EXCITATION–CONTRACTION COUPLING p. 303

4. **Excitation–contraction coupling** occurs as the passage of an action potential along a T tubule triggers the release of Ca^{2+} from the cisternae of the SR at triads. *(Figure 10–11)*
5. Release of Ca^{2+} initiates a **contraction cycle** of attachment, pivoting, detachment, and return. The calcium ions bind to troponin, which changes position and moves tropomyosin away from the active sites of actin. Cross-bridges of myosin heads then bind to actin. Next, each myosin head pivots at its base, pulling the actin filament toward the center of the sarcomere. *(Figure 10–12)*

RELAXATION p. 306

6. Acetylcholinesterase (AChE) breaks down ACh and limits the duration of muscle stimulation. *(Summary Table 10–1)*

Sliding filament theory: **IP CD-ROM**: Muscular System/Sliding Filament Theory.

10–4 TENSION PRODUCTION p. 306

TENSION PRODUCTION BY MUSCLE FIBERS p. 308

1. Muscle fibers function via the **all-or-none principle**. The amount of tension produced by a muscle fiber depends on the number of cross-bridges formed.
2. Skeletal muscle fibers can contract most forcefully when stimulated over a narrow range of resting lengths. *(Figure 10–13)*
3. A **twitch** is a cycle of contraction and relaxation produced by a single stimulus. *(Figure 10–12)*
4. Repeated stimulation after the relaxation phase has been completed produces **treppe**. *(Figure 10–14)*
5. Repeated stimulation before the relaxation phase ends may produce **summation of twitches** (**wave summation**), in which one twitch is added to another; **incomplete tetanus**, in which tension peaks because the muscle is never allowed to relax completely; or **complete tetanus**, in which the relaxation phase is eliminated. *(Figure 10–15)*

TENSION PRODUCTION BY SKELETAL MUSCLES p. 310

6. *Internal tension* is generated inside contracting skeletal muscle fibers; *external tension* is generated in the extracellular fibers. *(Figure 10–16)*

7. The number and size of a muscle's **motor units** determine how precisely controlled its movements are. *(Figure 10–17)*

8. Resting **muscle tone** stabilizes bones and joints.

9. Normal activities generally include both **isotonic contractions** (in which the tension in a muscle rises and the length of the muscle changes) and **isometric contractions** (in which tension rises, but the length of the muscle remains constant). *(Figure 10–18)*

10. Resistance and speed of contraction are inversely related. *(Figure 10–19)*

11. The return to resting length after a contraction may involve elastic elements, the contraction of opposing muscle groups, and gravity.

Skeletal muscle contraction: **IP CD-ROM: Muscular System/ Contraction of Motor Units and Contraction of Whole Muscle.**

10–5 ENERGY USE AND MUSCULAR ACTIVITY   p. 316

1. Muscle contractions require large amounts of energy. *(Table 10-2)*

ATP AND CP RESERVES   p. 316

2. **Creatine phosphate** (**CP**) can release stored energy to convert ADP to ATP. *(Table 10–2)*

ATP GENERATION   p. 316

3. At rest or at moderate levels of activity, **aerobic metabolism** can provide most of the ATP needed to support muscle contractions. *(Figure 10–20)*

4. At peak levels of activity, the cell relies heavily on the **anaerobic** process of **glycolysis** to generate ATP, because the mitochondria cannot obtain enough oxygen to meet the existing ATP demands. *(Figure 10–20)*

ENERGY USE AND THE LEVEL OF MUSCULAR ACTIVITY   p. 317

5. As muscular activity changes, the pattern of energy production and use changes.

MUSCLE FATIGUE   p. 319

6. A fatigued muscle can no longer contract, because of changes in pH due to the buildup of lactic acid, the exhaustion of energy resources, or other factors.

THE RECOVERY PERIOD   p. 319

7. The **recovery period** begins immediately after a period of muscle activity and continues until conditions inside the muscle have returned to preexertion levels. The **oxygen debt**, or *excess postexercise oxygen consumption* (*EPOC*), created during exercise is the amount of oxygen required during the recovery period to restore the muscle to its normal condition.

HORMONES AND MUSCLE METABOLISM   p. 320

8. Circulating hormones may alter metabolic activities in skeletal muscle fibers.

10–6 MUSCLE PERFORMANCE   p. 320

TYPES OF SKELETAL MUSCLE FIBERS   p. 320

1. The three types of skeletal muscle fibers are **fast fibers**, **slow fibers**, and **intermediate fibers**. *(Table 10–3; Figure 10–21)*

2. Fast fibers, which are large in diameter, contain densely packed myofibrils, large glycogen reserves, and relatively few mitochondria. They produce rapid and powerful contractions of relatively brief duration. *(Figure 10–21)*

3. Slow fibers are about half the diameter of fast fibers and take three times as long to contract after stimulation. Specializations such as abundant mitochondria, an extensive capillary supply, and high concentrations of **myoglobin** enable slow fibers to continue contracting for extended periods. *(Figure 10–21)*

4. Intermediate fibers are very similar to fast fibers, but have a greater resistance to fatigue.

MUSCLE PERFORMANCE AND THE DISTRIBUTION OF MUSCLE FIBERS   p. 321

5. Muscles dominated by fast fibers appear pale and are called **white muscles**.

6. Muscles dominated by slow fibers are rich in **myoglobin** and appear as **red muscles**.

MUSCLE HYPERTROPHY   p. 322

7. Training to develop anaerobic endurance can lead to **hypertrophy** (enlargement) of the stimulated muscles.

PHYSICAL CONDITIONING   p. 322

8. **Anaerobic endurance** is the time over which muscular contractions can be sustained by glycolysis and reserves of ATP and CP.

9. **Aerobic endurance** is the time over which a muscle can continue to contract while supported by mitochondrial activities.

Muscle fatigue: **IP CD-ROM: Muscular System/Muscle Metabolism.**

10–7 CARDIAC MUSCLE TISSUE   p. 323

STRUCTURAL CHARACTERISTICS OF CARDIAC MUSCLE TISSUE   p. 324

1. **Cardiac muscle tissue** is located only in the heart. **Cardiac muscle cells** are small, have one centrally located nucleus, have short, broad T-tubules, and are dependent on aerobic metabolism. **Intercalated discs** are found where cell membranes connect. *(Figure 10–22; Table 10–4)*

FUNCTIONAL CHARACTERISTICS OF CARDIAC MUSCLE TISSUE   p. 324

2. Cardiac muscle cells contract without neural stimulation (**automaticity**), and their contractions last longer than those of skeletal muscle.

3. Because cardiac muscle twitches do not exhibit wave summation, cardiac muscle tissue cannot produce tetanic contractions.

10–8 SMOOTH MUSCLE TISSUE   p. 325

STRUCTURAL CHARACTERISTICS OF SMOOTH MUSCLE TISSUE   p. 326

1. **Smooth muscle tissue** is nonstriated, involuntary muscle tissue.

2. Smooth muscle cells lack sarcomeres and the resulting striations. The thin filaments are anchored to **dense bodies**. *(Figure 10–23; Table 10–4)*

FUNCTIONAL CHARACTERISTICS OF SMOOTH MUSCLE TISSUE   p. 326

3. Smooth muscle contracts when calcium ions interact with **calmodulin**, which activates **myosin light chain kinase**.

4. Smooth muscle functions over a wide range of lengths (**plasticity**).

5. **Multiunit smooth muscle cells** are innervated by more than one motor neuron. **Visceral smooth muscle cells** are not always innervated by motor neurons. Neurons that innervate smooth muscle cells are not under voluntary control.

REVIEW QUESTIONS

 More assessment questions are available to you on the Companion Website. You will find Matching, Multiple Choice, True/False, and other quizzes to help further your understanding of the material covered in this chapter. To access the site, go to www.aw.com/martini.

LEVEL 1 Reviewing Facts and Terms

1. The connective tissue coverings of a skeletal muscle, listed from superficial to deep, are
   (a) endomysium, perimysium, and epimysium
   (b) endomysium, epimysium, and perimysium
   (c) epimysium, endomysium, and perimysium
   (d) epimysium, perimysium, and endomysium

2. The contraction of a muscle exerts a pull on a bone because
   (a) muscles are attached to bones by ligaments
   (b) muscles are directly attached to bones
   (c) muscles are attached to bones by tendons
   (d) a, b, and c are correct

3. The detachment of the myosin cross-bridges is directly triggered by
   (a) the repolarization of T tubules
   (b) the attachment of ATP to myosin heads
   (c) the hydrolysis of ATP
   (d) calcium ions

4. A muscle producing peak tension during rapid cycles of contraction and relaxation is said to be in
   (a) incomplete tetanus   (b) treppe
   (c) complete tetanus   (d) a twitch

5. In the sarcomere, the I band is composed of
   (a) thick and thin filaments   (b) titin
   (c) thick filaments only   (d) thin filaments and titin

6. The type of contraction in which the tension rises, but the resistance does not change, is
   (a) a wave summation   (b) a twitch
   (c) an isotonic contraction   (d) an isometric contraction

7. Large-diameter, densely packed myofibrils, large glycogen reserves, and few mitochondria are characteristics of
   (a) slow fibers   (b) intermediate fibers
   (c) fast fibers   (d) red muscles

8. An action potential can travel quickly from one cardiac muscle cell to another because of the presence of
   (a) gap junctions   (b) tight junctions
   (c) intercalated discs   (d) a and c are correct

9. List the three types of muscle tissue in the body.

10. What are the five functions of skeletal muscle?

11. What three layers of connective tissue are part of each muscle? What functional role does each layer play?

12. What is a motor unit?

13. What structural feature of a skeletal muscle fiber is responsible for conducting action potentials into the interior of the cell?

14. Under resting conditions, what two proteins prevent interactions between cross-bridges and active sites?

15. What five interlocking steps are involved in the contraction process?

16. What two factors affect the amount of tension produced when a skeletal muscle contracts?

17. What neurotransmitter is responsible for muscle contraction? What enzyme breaks down this neurotransmitter?

18. What forms of energy reserves do resting skeletal muscle fibers contain?

19. What two mechanisms are used to generate ATP from glucose in muscle cells?

20. Define hypertrophy.

21. What is the calcium-binding protein in smooth muscle tissue?

LEVEL 2 Reviewing Concepts

22. An activity that would require anaerobic endurance is
   (a) a 50-meter dash
   (b) a pole vault
   (c) a weight-lifting competition
   (d) a, b, and c are correct

23. Areas of the body where you would not expect to find slow fibers include the
   (a) back and calf muscles   (b) eye and hand
   (c) chest and abdomen   (d) a, b, and c are correct

24. When contraction occurs,
   (a) the H and I bands get smaller
   (b) the Z lines move closer together
   (c) the width of the A band remains constant
   (d) a, b, and c are correct

25. The amount of tension produced during a contraction is *not* affected by
   (a) the amount of ATP available to the muscle
   (b) the degree of overlap between thick and thin filaments
   (c) the interaction of troponin and tropomyosin
   (d) the number of cross-bridge interactions in a muscle fiber

26. Describe the basic sequence of events that occurs at a neuromuscular junction.

27. Describe the relationship between lactic acid and fatigue.

28. Describe the graphic events seen on a myogram as tension is developed in a stimulated calf muscle fiber during a twitch.

29. What three processes are involved in repaying the oxygen debt during a muscle's recovery period?

30. Why is cardiac muscle an example of a functional syncytium?

31. How does cardiac muscle tissue contract without neural stimulation?

32. Atracurium is a drug that blocks the binding of ACh to receptors. Give an example of a site where such binding normally occurs, and predict the physical effect of this drug.

33. Describe what happens to muscles that are not "used" on a regular basis. What can be done to offset this effect?

34. The time of a murder victim's death may be estimated by the flexibility of the body. Explain why.

35. Many visceral smooth muscle cells lack motor neuron innervation. How are their contractions coordinated and controlled?

36. A motor unit in a skeletal muscle contains 1500 muscle fibers. Would this muscle be involved in fine, delicate movements or powerful, gross movements? Explain.

LEVEL 3 Critical Thinking and Clinical Applications

37. Many potent insecticides contain toxins, called organophosphates, that interfere with the action of the enzyme acetylcholinesterase. Ivan is using an insecticide containing organophosphates and is very careless. He does not use gloves or a dust mask and absorbs some of the chemical through his skin. He inhales a large amount as well. What symptoms would you expect to observe in Ivan as a result of the organophosphate poisoning?

38. A rare (hypothetical) genetic disease causes the body to produce antibodies that compete with acetylcholine for receptors on the motor end plate. Patients with this disease exhibit varying degrees of muscle weakness and flaccid paralysis in the affected muscles. If you could administer a drug that inhibits acetylcholinesterase or a drug that inhibits acetylcholine, which would you use to alleviate these symptoms?

39. Thirty minutes after Mary has completed a 10-km race, she begins to notice soreness and stiffness in her leg muscles. She wonders whether she may have damaged the muscles during the race. She visits her physician, who orders a blood test. How could the physician tell from a blood test whether muscle damage had occurred?

UNIT 2   **CHAPTER**   5   6   7   8   9   10   **11**

11

THE MUSCULAR SYSTEM

The **muscular system** includes all the skeletal muscles that can be controlled voluntarily. Most of the muscle tissue in the body is part of this system, and approximately 700 skeletal muscles have been identified. Some are attached to bony processes, others to broad sheets of connective tissue, but all are directly or indirectly associated with the skeletal system. Many professions, from dancing to coaching to physical therapy, rely on an understanding of the functional anatomy of skeletal muscles. Learning the placement and function of all 700 skeletal muscles is beyond the scope of most career needs; therefore, we shall focus on a relatively small but representative set of muscles, about 20 percent of the total. To simplify memorization, we have organized these muscles into anatomical and functional groups.

The shape or appearance of each muscle provides clues to its primary function. Muscles involved with locomotion and posture work across joints, producing skeletal movement. Those which support soft tissue form slings or sheets between relatively stable bony elements, whereas those which guard an entrance or exit completely encircle the opening.

At the level of the individual skeletal muscle, two factors interact to determine the effects of its contraction: (1) the anatomical arrangement of the muscle fibers and (2) the way the muscle attaches to the bones of the skeletal system. We can understand the performance of muscles in the body in terms of basic mechanical laws. In this chapter, we examine the functional anatomy of the muscular system.

11-1 MUSCLE ORGANIZATION AND FUNCTION

Objectives

- Describe the arrangement of fascicles in the various types of muscles, and explain the resulting functional differences.
- Describe the classes of levers and how they make muscles more efficient.

Although most skeletal muscle fibers contract at comparable rates and shorten to the same degree, variations in microscopic and macroscopic organization can dramatically affect the power, range, and speed of movement produced when a muscle contracts.

☐ ORGANIZATION OF SKELETAL MUSCLE FIBERS

Muscle fibers in a skeletal muscle form bundles called *fascicles*. ∞ p. 292 The muscle fibers in a single fascicle are parallel, but the organization of fascicles in the skeletal muscle can vary, as can the relationship between the fascicles and the associated tendon. Four patterns of fascicle organization form *parallel muscles, convergent muscles, pennate muscles*, and *circular muscles*.

Parallel Muscles

In a **parallel muscle**, the fascicles are parallel to the long axis of the muscle. Most of the skeletal muscles in the body are parallel muscles. Some are flat bands with broad attachments (*aponeuroses*) at each end; others are plump and cylindrical with tendons at one or both ends. In the latter case, the muscle is spin-

dle-shaped (Figure 11–1a●), with a central **body**, also known as the *belly*, or *gaster* (GAS-ter; stomach). The *biceps brachii muscle* of the arm is a parallel muscle with a central body. When a parallel muscle contracts, it gets shorter, and larger in diameter. You can see the bulge of the contracting biceps brachii on the anterior surface of your arm when you flex your elbow.

A skeletal muscle cell can contract until it has shortened by roughly 30 percent. Because the fibers in a parallel muscle are parallel to the long axis of the muscle, when the fibers contract together, the entire muscle shortens by the same amount. If the muscle is 10 cm long and one end is held in place, the other end will move 3 cm when the muscle contracts. The tension developed during this contraction depends on the total number of myofibrils the muscle contains. Because the myofibrils are distributed evenly through the sarcoplasm of each cell, we can use the cross-sectional area of the resting muscle to estimate the ten-

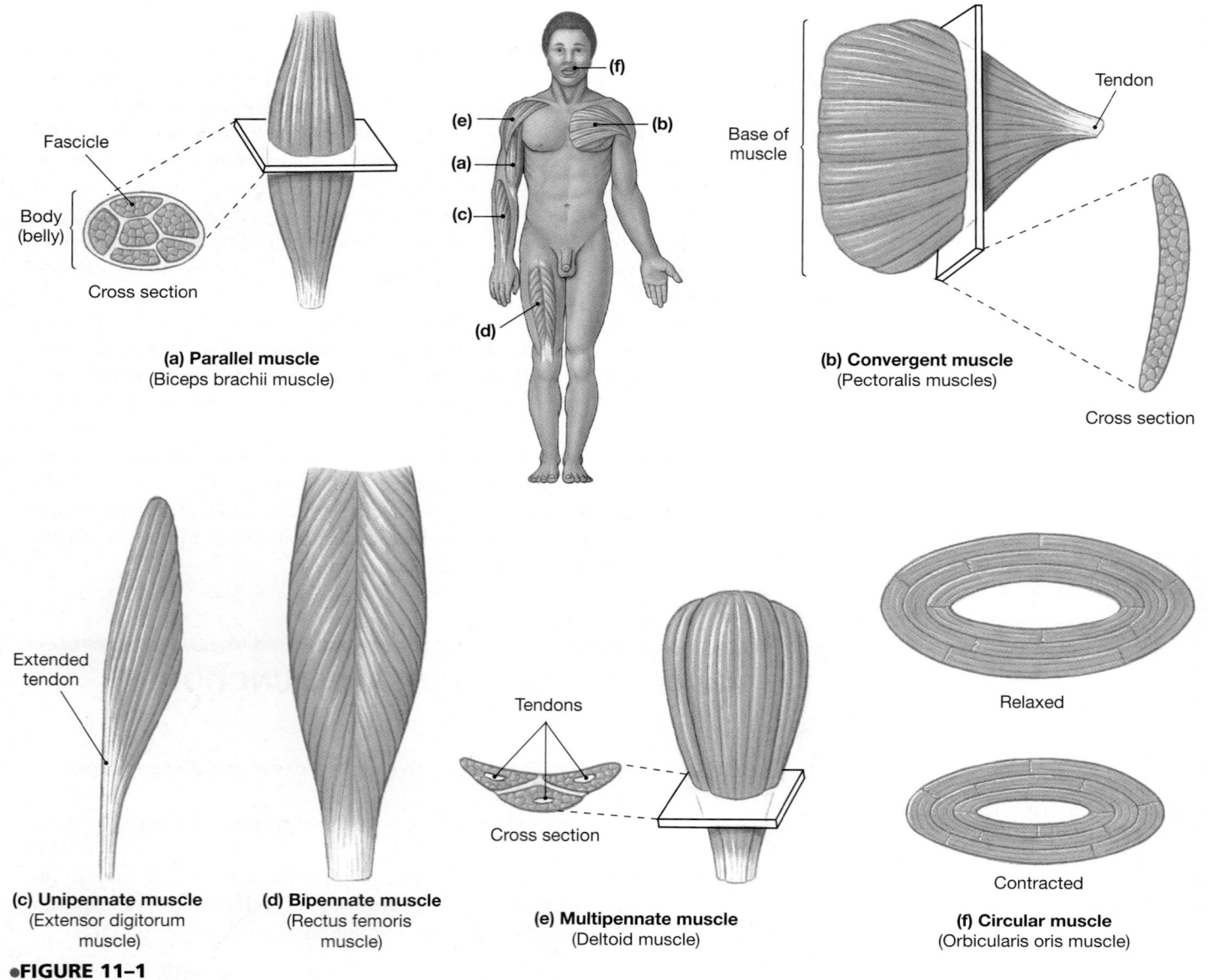

Fascicle

Body (belly)

Cross section

(a) Parallel muscle
(Biceps brachii muscle)

(f)
(e)
(a)
(b)
(c)
(d)

Tendon

Base of muscle

(b) Convergent muscle
(Pectoralis muscles)

Cross section

Extended tendon

(c) Unipennate muscle
(Extensor digitorum muscle)

(d) Bipennate muscle
(Rectus femoris muscle)

Tendons

Cross section

(e) Multipennate muscle
(Deltoid muscle)

Relaxed

Contracted

(f) Circular muscle
(Orbicularis oris muscle)

●**FIGURE 11–1**
Different Arrangements of Skeletal Muscle Fibers

sion. A parallel muscle 6.25 cm² (1 in.²) in cross-sectional area can develop approximately 23 kg (50 lb) of tension.

Convergent Muscles

In a **convergent muscle**, the muscle fibers are spread over a broad area, but all the fasicles converge at one common attachment site. They may pull on a tendon, an aponeurosis, or a slender band of collagen fibers known as a **raphe** (RĀ-fē; seam). The muscle fibers typically spread out, like a fan or a broad triangle, with a tendon at the apex. The prominent chest muscles of the *pectoralis group* have this shape (Figure 11–1b●). A convergent muscle is versatile, because the stimulation of only one portion of the muscle can change the direction of pull. However, when the entire muscle contracts, the muscle fibers do not pull as hard on the attachment site as would a parallel muscle of the same size. This is because convergent muscle fibers pull in different directions, rather than all pulling in the same direction.

Pennate Muscles

In a **pennate muscle** (*penna*, feather), the fascicles form a common angle with the tendon. Because the muscle cells pull at an angle, contracting pennate muscles do not move their tendons as far as parallel muscles do. But a pennate muscle contains more muscle fibers than does a parallel muscle of the same size, and so produces more tension. (As we saw in Chapter 10, tension is proportional to the number of contracting sarcomeres—the more muscle fibers, the more myofibrils and sarcomeres.) 🔗 pp. 310–312

If all the muscle fibers are on the same side of the tendon, the pennate muscle is *unipennate*. The *extensor digitorum muscle*, a forearm muscle that extends the finger joints, is unipennate (Figure 11–1c●). More commonly, a pennate muscle has fibers on both sides of the tendon. Such a muscle is called *bipennate*. The *rectus femoris muscle*, a prominent muscle that extends the knee, is bipennate (Figure 11–1d●). If the tendon branches within a pennate muscle, the muscle is said to be *multipennate*. The triangular *deltoid muscle* of the shoulder is multipennate (Figure 11–1e●).

Circular Muscles

In a **circular muscle**, or **sphincter** (SFINK-ter), the fibers are concentrically arranged around an opening or a recess. When the muscle contracts, the diameter of the opening decreases. Circular muscles guard entrances and exits of internal passageways such as the digestive and urinary tracts. An example is the *orbicularis oris muscle* of the mouth (Figure 11–1f●).

☐ LEVERS

Skeletal muscles do not work in isolation. When a muscle is attached to the skeleton, the nature and site of the connection will determine the force, speed, and range of the movement pro-

duced. These characteristics are interdependent, and the relationships can explain a great deal about the general organization of the muscular and skeletal systems.

The force, speed, or direction of movement produced by contraction of a muscle can be modified by attaching the muscle to a lever. A **lever** is a rigid structure—such as a board, a crowbar, or a bone—that moves on a fixed point called the **fulcrum**. A child's teeter-totter, or seesaw, provides a familiar example of lever action. In the body, each bone is a lever and each joint is a fulcrum. Levers can change (1) the direction of an applied force, (2) the distance and speed of movement produced by an applied force, and (3) the effective strength of an applied force. You are familiar with the advantages of lever systems if you have ever used a shovel to roll over a rock. Working across the fulcrum—where the blade of the shovel contacts the ground—the leverage provided by the long handle makes it easy to move a rock too heavy to lift.

Classes of Levers

Three classes of levers are found in the human body (Figure 11–2●). The seesaw is a **first-class lever**. The body has few first-class levers. One, involved with extension of the neck, is shown in Figure 11–2a●. In such a lever, the fulcrum lies between the applied force (AF) and the resistance (R).

In a **second-class lever** (Figure 11–2b●), the resistance is located between the applied force and the fulcrum. A familiar example is a loaded wheelbarrow. The weight of the load is the resistance, and the upward lift on the handle is the applied force. Because in this arrangement the force is always farther from the fulcrum than the resistance is, a small force can balance a larger weight. That is, the effective force is increased. Notice, however, that when a force moves the handle, the load moves more slowly and covers a shorter distance. The body has few second-class levers. In performing ankle extension (plantar flexion), the calf muscles use a second-class lever (Figure 11–2b●).

Third-class levers are the most common levers in the body. In this lever system, a force is applied between the resistance and the fulcrum (Figure 11–2c●). An example is a ladder that you raise to lean against a building. The fulcrum is the base of the ladder, in contact with the ground. Force is applied where you grasp the ladder, and the resistance is the weight of the ladder between your hands and the free end. The effect is the reverse of that for a second-class lever: Speed and distance traveled are increased at the expense of effective force. In the example shown (the biceps brachii muscle, which flexes the elbow), the resistance is six times farther from the fulcrum than is the applied force. The effective force is reduced to the same degree. The muscle must generate 180 kg of tension at its attachment to the forearm to support 30 kg held in the hand. However, the distance traveled and the speed of movement are increased by that same 6:1 ratio: The load will travel 45 cm when the point of attachment moves 7.5 cm.

Although not every muscle operates as part of a lever system, the presence of levers provides speed and versatility far in excess

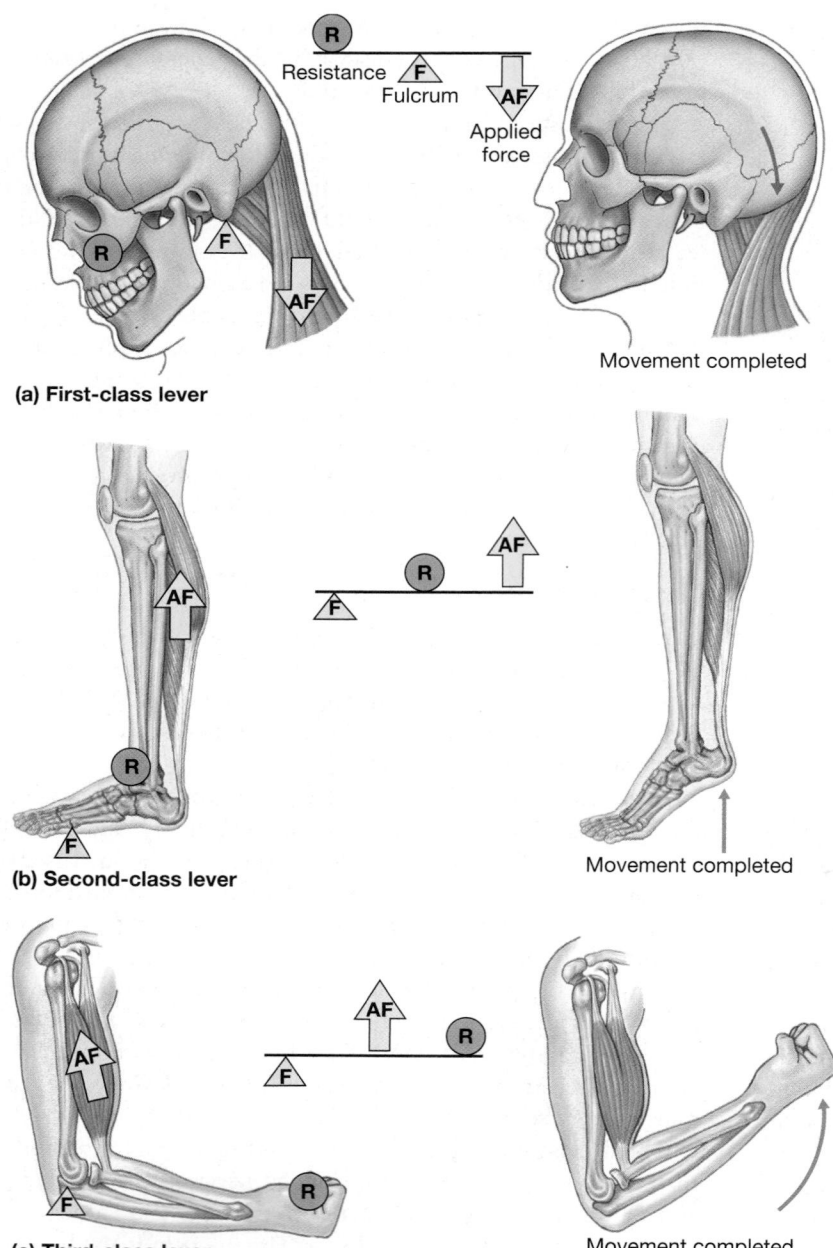

(a) First-class lever

Resistance
Fulcrum
Applied force

Movement completed

(b) Second-class lever

Movement completed

(c) Third-class lever

Movement completed

●**FIGURE 11–2**
The Three Classes of Levers. **(a)** In a first-class lever, the applied force and the resistance are on opposite sides of the fulcrum. **(b)** In a second-class lever, the resistance lies between the applied force and the fulcrum. **(c)** In a third-class lever, the force is applied between the resistance and the fulcrum.

of what we would predict on the basis of muscle physiology alone. Skeletal muscle fibers resemble one another closely, and their abilities to contract and generate tension are quite similar. Consider a skeletal muscle that can contract in 500 msec and shorten 1 cm while it exerts a 10-kg pull. Without using a lever, this muscle would be performing efficiently only when moving a 10-kg weight a distance of 1 cm. By using a lever, however, the same muscle operating at the same efficiency could move 20 kg a distance of 0.5 cm, 5 kg a distance of 2 cm, or 1 kg a distance of 10 cm.

✓ Why does a pennate muscle generate more tension than does a parallel muscle of the same size?

✓ Which type of muscle would you expect to be guarding the opening between the stomach and the small intestine?

✓ The joint between the occipital bone of the skull and the first cervical vertebra (atlas) is part of which type of lever system?

Answers start on p. Q-1

11–2 MUSCLE TERMINOLOGY

Objectives

■ Predict the actions of a muscle on the basis of the relative positions of its origin and insertion.

■ Explain how muscles interact to produce or oppose movements.

■ Explain how the name of a muscle can help identify its location, appearance, or function.

This chapter focuses on the functional anatomy of skeletal muscles and muscle groups. You must learn a number of new terms, and this section attempts to give you assistance in understanding them. It may help you to create a vocabulary list from the terms introduced. Once you are familiar with the basic terminology, the names and actions of skeletal muscles are easily understood.

☐ ORIGINS AND INSERTIONS

In Chapter 10 we noted that when both ends of a muscle are free to move, the ends move toward the center of the muscle during a contraction. ∞ p. 298 In the body, however, the ends of a muscle are always attached to other structures that may limit their movement. In most cases one end is fixed in position, and during a contraction the free end moves toward the fixed end. The place where the fixed end attaches to a bone, cartilage, or connective tissue is called the **origin** of the muscle. The site where the movable end attaches to another structure is called the **insertion** of the muscle. The origin is typically proximal to the insertion. When a muscle contracts, it produces a specific **action**, or movement. Actions are described using the terms introduced in Chapter 9 (flexion, extension, adduction, and so forth).

As an example, consider the *gastrocnemius muscle* (in the calf), which extends from the distal portion of the femur to the calcaneus. As Figure 11–2b● shows, when the gastrocnemius muscle contracts, it pulls the calcaneus toward the knee. As a result, we say that the gastrocnemius muscle has its origin at the femur and its insertion at the calcaneus; its action can be described as "extension at the ankle" or "plantar flexion."

The decision as to which end is the origin and which is the insertion is usually based on movement from the anatomical position. Part of the fun of studying the muscular system is that you can actually do the movements and think about the muscles involved. As a result, laboratory discussions of the muscular system tend to resemble disorganized aerobics classes.

When the origins and insertions cannot be determined easily on the basis of common movement or position, other rules are used. If a muscle extends between a broad aponeurosis and a narrow tendon, the aponeurosis is the origin and the tendon is the insertion. If several tendons are at one end and just one is at the other, the muscle has multiple origins and a single insertion. These simple rules cannot cover every situation. Knowing which end is the origin and which is the insertion is ultimately less important than knowing where the two ends attach and what the muscle accomplishes when it contracts.

Most muscles originate at a bone, but some originate at a connective-tissue sheath or band. Examples of these sheaths or bands include *intermuscular septa* (components of the deep fascia that may separate adjacent skeletal muscles), the interosseous membranes of the forearm or leg, and the fibrous sheet that spans the obturator foramen of the pelvis.

□ ACTIONS

Almost all skeletal muscles either originate or insert on the skeleton. When a muscle moves a portion of the skeleton, that movement may involve flexion, extension, adduction, abduction, protraction, retraction, elevation, depression, rotation, circumduction, pronation, supination, inversion, eversion, lateral flexion, or opposition. (Before proceeding, you may want to review the discussions of planes of motion and Figures 9–2 to 9–5.) ⊂⊃ pp. 268–272

Actions can be described by one of two methods. The first, used by most undergraduate textbooks and references such as *Gray's Anatomy*, describes actions in terms of the bone affected. Thus, a muscle such as the biceps brachii muscle is said to perform "flexion of the forearm." The second method, of increasing use among specialists such as kinesiologists, identifies the joint involved. With this method, the action of the biceps brachii muscle would be "flexion at (or of) the elbow." Both methods are valid, and each has its advantages. In general, we will use the latter method.

When complex movements occur, muscles commonly work in groups rather than individually. Their cooperation improves the efficiency of a particular movement. For example, large muscles of the limbs produce flexion or extension over an extended range of motion. Although these muscles cannot develop much tension at full extension, they are generally paired with one or more smaller muscles that provide assistance until the larger muscle can perform at maximum efficiency. At the start of the movement, the smaller muscle is producing maximum tension, while the larger muscle is producing minimum tension. The impor-

tance of this smaller "assistant" decreases as the movement proceeds and the efficiency of the primary muscle increases.

On the basis of size and range of motion, muscles are described as follows:

- An **agonist**, or **prime mover**, is a muscle whose contraction is chiefly responsible for producing a particular movement. The biceps brachii muscle is an agonist that produces flexion at the elbow.

- An **antagonist** is a muscle whose action opposes that of a particular agonist. The *triceps brachii muscle* is an agonist that extends the elbow. It is therefore an antagonist of the biceps brachii muscle, and the biceps brachii is an antagonist of the triceps brachii. Agonists and antagonists are functional opposites; if one produces flexion, the other will produce extension. When an agonist contracts to produce a particular movement, the corresponding antagonist will be stretched, but it will usually not relax completely. Instead, it will contract eccentrically, with the tension adjusted to control the speed of the movement and ensure its smoothness. ⊂⊃ p. 315 You may find it easiest to learn about muscles in agonist–antagonist pairs (flexors–extensors, abductors–adductors) that act at a specific joint. This method highlights the functions of the muscles involved, and it can help organize the information into a logical framework. The tables in this chapter are arranged to facilitate such an approach.

- When a **synergist** (*syn-*, together + *ergon*, work) contracts, it helps a larger agonist work efficiently. Synergists may provide additional pull near the insertion or may stabilize the point of origin. Their importance in assisting a particular movement may change as the movement progresses. In many cases, they are most useful at the start, when the agonist is stretched and unable to develop maximum tension. For example, the *latissimus dorsi muscle* is a large trunk muscle that extends, adducts, and medially rotates the arm at the shoulder joint. A much smaller muscle, the *teres* (TER-ēz) *major muscle*, assists in starting such movements when the shoulder joint is at full flexion. Synergists may also assist an agonist by preventing movement at another joint and thereby stabilizing the origin of the agonist. Such synergists are called **fixators**.

□ NAMES OF SKELETAL MUSCLES

Except for the *platysma* and the *diaphragm*, the complete names of all skeletal muscles include the term *muscle*. Although the full name, such as the biceps brachii muscle, will usually appear in the text, for simplicity only the descriptive name (biceps brachii) will be used in figures and tables.

You need not learn every one of the approximately 700 muscles in the human body, but you will have to become familiar with the most important ones. Fortunately, anatomists assigned names to the muscles that provide clues to their identification. If you can learn to recognize the clues, you will find it easier to remember the names and identify the muscles. The name of a muscle may include information about its fascicle organization, location, relative position, structure, size, shape, origin and insertion, or action. When faced with a new muscle name, it is helpful to first identify the descriptive portions of the name. Once these are identified and defined, the name of the muscle will be easier to remember.

Fascicle Organization

A muscle name may refer to the orientation of the muscle fibers within a particular skeletal muscle. **Rectus** means "straight," and rectus muscles are parallel muscles whose fibers generally run along the long axis of the body. Because we have several rectus muscles, the name typically includes a second term that refers to a precise region of the body. For example, the *rectus abdominis muscle* is located on the abdomen, and the *rectus femoris muscle* on the thigh. Other directional indicators include **transversus** and **oblique**, for muscles whose fibers run across or at an oblique angle to the longitudinal axis of the body, respectively.

Location

Table 11–1 includes a useful summary of terms that designate specific regions of the body. The terms are common as modifiers that help identify individual muscles, as in the case of the rectus muscles. In a few cases, the muscle is such a prominent feature of the region that the regional name alone will identify it. Examples include the *temporalis muscle* of the head and the *brachialis* (brā-kē-A-lis) *muscle* of the arm.

Relative Position

Muscles visible at the body surface are often called **externus** or **superficialis**, whereas deeper muscles are termed **internus** or **profundus**. Superficial muscles that position or stabilize an

TABLE 11–1 MUSCLE TERMINOLOGY

Terms Indicating Direction Relative to Axes of the Body	Terms Indicating Specific Regions of the Body*	Terms Indicating Structural Characteristics of the Muscle	Terms Indicating Actions
Anterior (front)	Abdominis (abdomen)	**Origin**	**General**
Externus (superficial)	Anconeus (elbow)	Biceps (two heads)	Abductor
Extrinsic (outside)	Auricularis (auricle of ear)	Triceps (three heads)	Adductor
Inferioris (inferior)	Brachialis (brachium)	Quadriceps (four heads)	Depressor
Internus (deep, internal)	Capitis (head)		Extensor
Intrinsic (inside)	Carpi (wrist)	**Shape**	Flexor
Lateralis (lateral)	Cervicis (neck)	Deltoid (triangle)	Levator
Medialis/medius (medial, middle)	Cleido-/-clavius (clavicle)	Orbicularis (circle)	Pronator
Oblique	Coccygeus (coccyx)	Pectinate (comblike)	Rotator
Posterior (back)	Costalis (ribs)	Piriformis (pear-shaped)	Supinator
Profundus (deep)	Cutaneous (skin)	Platys- (flat)	Tensor
Rectus (straight, parallel)	Femoris (femur)	Pyramidal (pyramid)	
Superficialis (superficial)	Genio- (chin)	Rhomboid	**Specific**
Superioris (superior)	Glosso-/-glossal (tongue)	Serratus (serrated)	Buccinator (trumpeter)
Transversus (transverse)	Hallucis (great toe)	Splenius (bandage)	Risorius (laugher)
	Ilio- (ilium)	Teres (long and round)	Sartorius (like a tailor)
	Inguinal (groin)	Trapezius (trapezoid)	
	Lumborum (lumbar region)		
	Nasalis (nose)	**Other Striking Features**	
	Nuchal (back of neck)	Alba (white)	
	Oculo- (eye)	Brevis (short)	
	Oris (mouth)	Gracilis (slender)	
	Palpebrae (eyelid)	Lata (wide)	
	Pollicis (thumb)	Latissimus (widest)	
	Popliteus (posterior to knee)	Longissimus (longest)	
	Psoas (loin)	Longus (long)	
	Radialis (radius)	Magnus (large)	
	Scapularis (scapula)	Major (larger)	
	Temporalis (temples)	Maximus (largest)	
	Thoracis (thoracic region)	Minimus (smallest)	
	Tibialis (tibia)	Minor (smaller)	
	Ulnaris (ulna)	-tendinosus (tendinous)	
	Uro- (urinary)	Vastus (great)	

*For other regional terms, refer to Figure 1–7, p. 16, which deals with anatomical landmarks.

organ are called **extrinsic**; muscles located entirely within an organ are **intrinsic**.

Structure, Size, and Shape

Some muscles are named after distinctive structural features. The biceps brachii muscle, for example, has two tendons of origin (*bi-*, two + *caput*, head); the triceps brachii muscle has three; and the *quadriceps group*, four. Shape is sometimes an important clue to the name of a muscle. For example, the *trapezius* (tra-PĒ-zē-us), *deltoid, rhomboid* (rom-BOYD), and *orbicularis* (or-bik-ū-LAR-is) muscles look like a trapezoid, a triangle, a rhomboid, and a circle, respectively. Long muscles are called **longus** (long) or **longissimus** (longest), and **teres** muscles are both long and round. Short muscles are called **brevis**. Large ones are called **magnus** (big), **major** (bigger), or **maximus** (biggest); small ones are called **minor** (smaller) or **minimus** (smallest).

Origin and Insertion

Many names tell you the specific origin and insertion of each muscle. In such cases, the first part of the name indicates the origin, the second part the insertion. The *genioglossus muscle*, for example, originates at the chin (*geneion*) and inserts in the tongue (*glossus*). The names may be long and difficult to pronounce, but Table 11–1 and the anatomical terms introduced in Chapter 1 can help you identify and remember them. ⧂ p. 18

Action

Many muscles are named *flexor, extensor, retractor, abductor,* and so on. These are such common actions that the names almost always include other clues as to the appearance or location of the muscle. For example, the *extensor carpi radialis longus muscle* is a long muscle along the radial (lateral) border of the forearm. When it contracts, its primary function is extension at the carpus (wrist).

A few muscles are named after the specific movements associated with special occupations or habits. The *sartorius* (sar-TO-rē-us) *muscle* is active when you cross your legs. Before sewing machines were invented, a tailor would sit on the floor cross-legged, and the name of this muscle was derived from *sartor*, the Latin word for "tailor." On the face, the *buccinator* (BUK-si-nā-tor) *muscle* compresses the cheeks—for example, when you purse your lips and blow forcefully. *Buccinator* translates as "trumpet player." Another facial muscle, the *risorius* (ri-SO-rē-us) *muscle*, was supposedly named after the mood expressed. The Latin term *risor*, however, means "laughter"; a more appropriate description for the effect would be "grimace."

☐ AXIAL AND APPENDICULAR MUSCLES

The separation of the skeletal system into axial and appendicular divisions provides a useful guideline for subdividing the muscular system as well:

1. The **axial musculature** arises on the axial skeleton. It positions the head and spinal column and also moves the rib cage, assisting in the movements that make breathing possible. It does not play a role in movement or support of either the pectoral or pelvic girdle or the limbs. This category encompasses roughly 60 percent of the skeletal muscles in the body.

2. The **appendicular musculature** stabilizes or moves components of the appendicular skeleton and includes the remaining 40 percent of all skeletal muscles.

Figure 11–3● provides an overview of the major axial and appendicular muscles of the human body. These are the superficial muscles, which tend to be relatively large. The superficial muscles cover deeper, smaller muscles that cannot be seen unless the overlying muscles are either removed or *reflected*—that is, cut and pulled out of the way. Later figures that show deep muscles in specific regions will indicate whether superficial muscles have been removed or reflected for the sake of clarity.

Paying attention to patterns of origin, insertion, and action, we shall now examine representatives of both muscular divisions. The discussion assumes that you already understand skeletal anatomy. As you examine the figures in this chapter, you will find that some bony and cartilaginous landmarks are labeled for orientation purposes. These labels are shown in italics, to differentiate these landmarks from the muscles that are the primary focus of each figure. Should you need further review of skeletal anatomy, figure captions in this chapter indicate the relevant figure numbers in Chapters 7, 8, or 9.

The tables that follow also contain information about the innervation of the individual muscles. **Innervation** is the distribution of nerves to a region or organ; the tables indicate the nerves that control each muscle. Many of the muscles of the head and neck are innervated by cranial nerves, which originate at the brain and pass through the foramina of the skull. Alternatively, spinal nerves are connected to the spinal cord and pass through the intervertebral foramina. For example, spinal nerve L_1 passes between vertebrae L_1 and L_2. Spinal nerves may form a complex network after exiting the spinal cord; one branch of this network may contain axons from several spinal nerves. Thus, many tables identify the spinal nerves involved as well as the names of the peripheral nerves.

✓ The *gracilis muscle* is attached to the anterior surface of the tibia at one end and to the pubis and ischium of the pelvis at the other. When the muscle contracts, flexion occurs at the hip. Which attachment point is the muscle's origin?

✓ Muscle A abducts the humerus, and muscle B adducts the humerus. What is the relationship between these two muscles?

✓ What does the name *flexor carpi radialis longus* tell you about this muscle?

Answers start on p. Q-1

 Muscle organization and lever function can be reviewed on the **Companion Website**: Chapter 11/Tutorials.

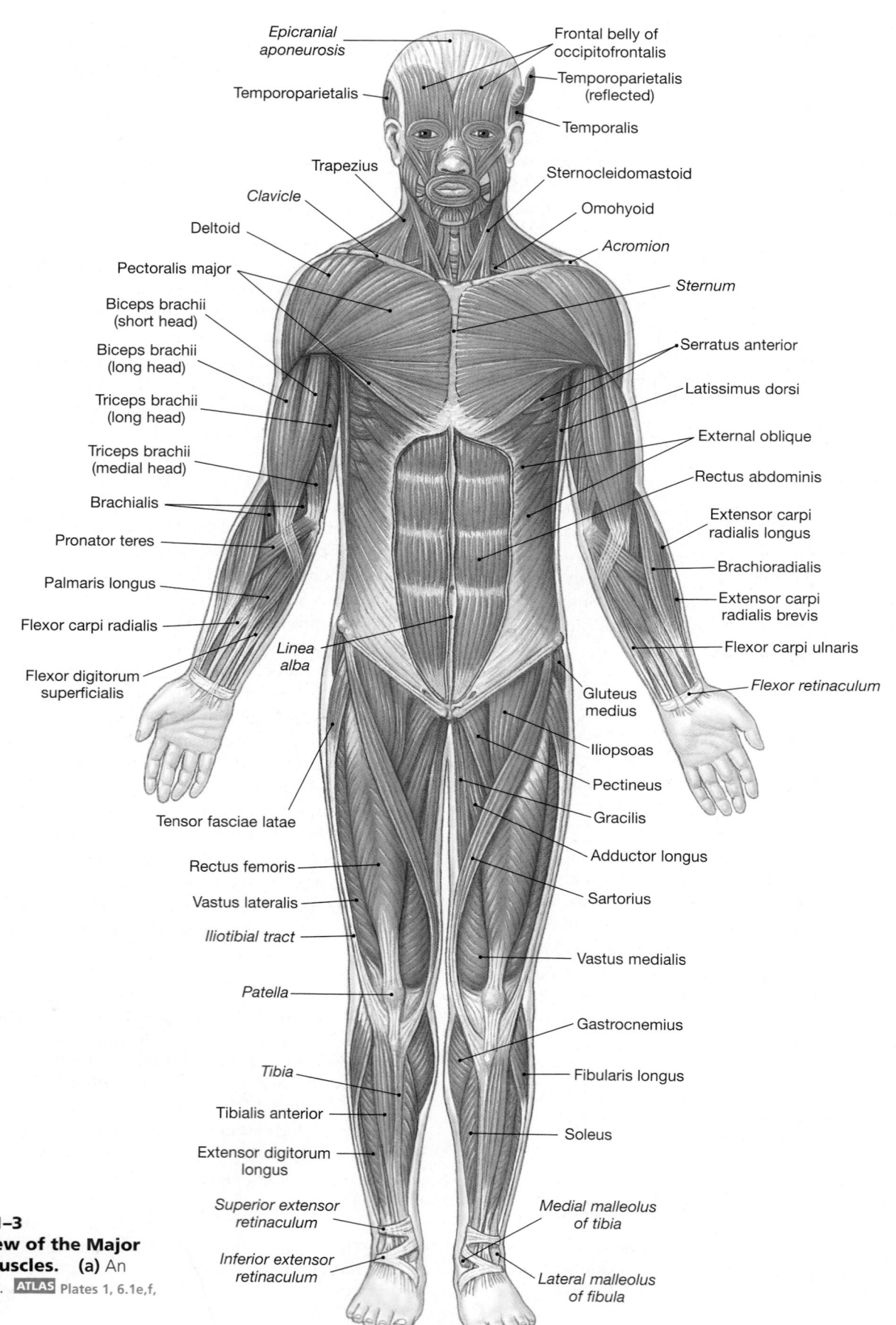

Epicranial aponeurosis
Frontal belly of occipitofrontalis
Temporoparietalis
Temporoparietalis (reflected)
Temporalis
Trapezius
Sternocleidomastoid
Clavicle
Omohyoid
Deltoid
Acromion
Pectoralis major
Sternum
Biceps brachii (short head)
Serratus anterior
Biceps brachii (long head)
Latissimus dorsi
Triceps brachii (long head)
External oblique
Triceps brachii (medial head)
Rectus abdominis
Brachialis
Extensor carpi radialis longus
Pronator teres
Brachioradialis
Palmaris longus
Extensor carpi radialis brevis
Flexor carpi radialis
Flexor carpi ulnaris
Linea alba
Flexor retinaculum
Flexor digitorum superficialis
Gluteus medius
Iliopsoas
Pectineus
Gracilis
Tensor fasciae latae
Adductor longus
Rectus femoris
Sartorius
Vastus lateralis
Iliotibial tract
Vastus medialis
Patella
Gastrocnemius
Tibia
Fibularis longus
Tibialis anterior
Soleus
Extensor digitorum longus
Superior extensor retinaculum
Medial malleolus of tibia
Inferior extensor retinaculum
Lateral malleolus of fibula

●**FIGURE 11–3**
An Overview of the Major Skeletal Muscles. **(a)** An anterior view. **ATLAS** Plates 1, 6.1e,f, 7.1–7.3

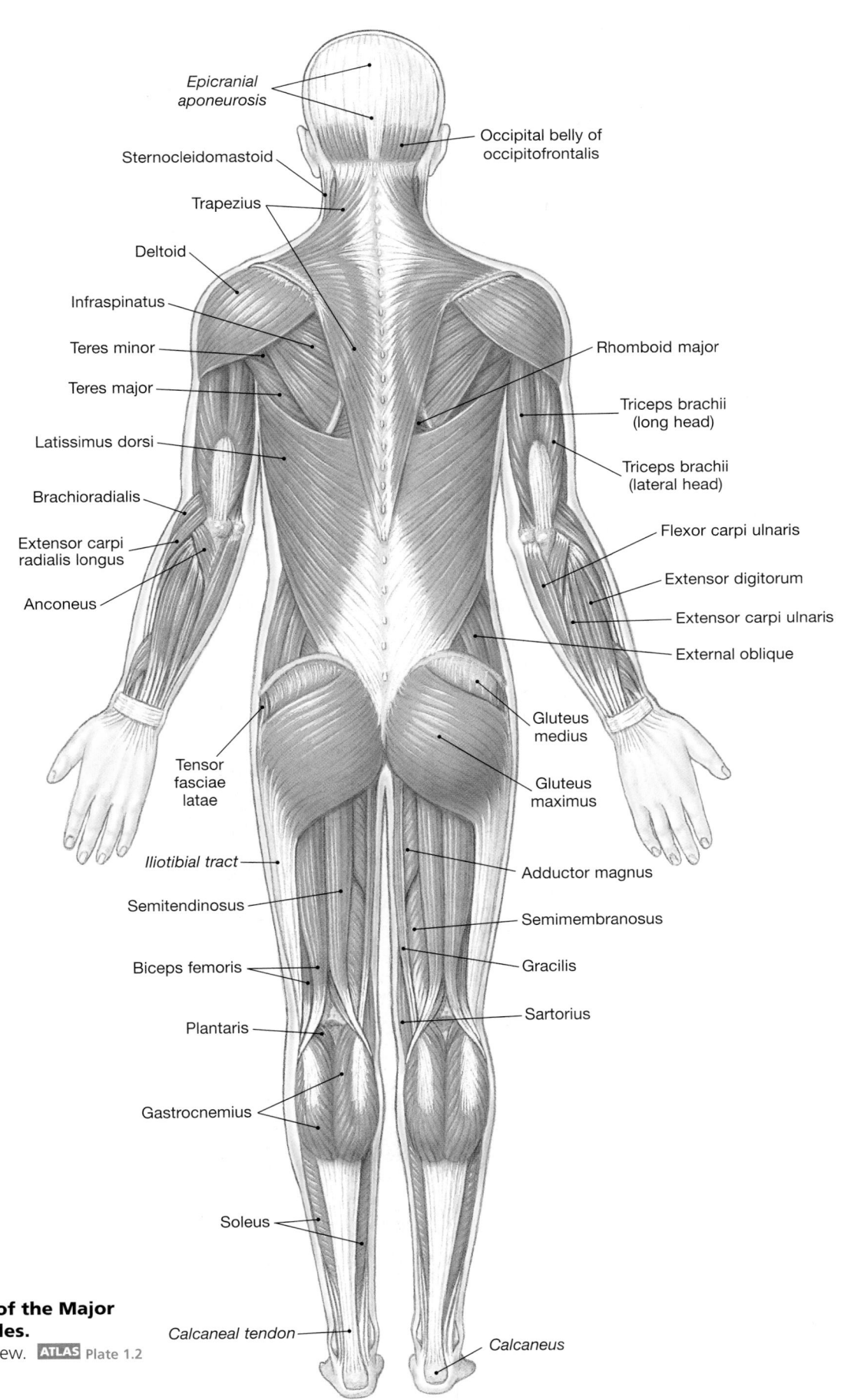

Epicranial aponeurosis

Occipital belly of occipitofrontalis

Sternocleidomastoid

Trapezius

Deltoid

Infraspinatus

Teres minor

Teres major

Latissimus dorsi

Brachioradialis

Extensor carpi radialis longus

Anconeus

Rhomboid major

Triceps brachii (long head)

Triceps brachii (lateral head)

Flexor carpi ulnaris

Extensor digitorum

Extensor carpi ulnaris

External oblique

Tensor fasciae latae

Gluteus medius

Gluteus maximus

Iliotibial tract

Semitendinosus

Biceps femoris

Plantaris

Gastrocnemius

Adductor magnus

Semimembranosus

Gracilis

Sartorius

Soleus

Calcaneal tendon

Calcaneus

●**FIGURE 11–3**
**An Overview of the Major
Skeletal Muscles.**

(b) A posterior view. **ATLAS** Plate 1.2

11–3 THE AXIAL MUSCLES

Objectives

■ Identify the principal axial muscles of the body, together with their origins, insertions, actions, and innervation.

The axial muscles fall into logical groups on the basis of location, function, or both. The groups do not always have distinct anatomical boundaries. For example, a function such as extension of the vertebral column involves muscles along its entire length and movement at each of the intervertebral joints. We shall discuss the axial muscles in four groups:

1. *The Muscles of the Head and Neck.* This group includes muscles that move the face, tongue, and larynx. They are therefore responsible for verbal and nonverbal communication—laughing, talking, frowning, smiling, whistling, and so on. You also use these muscles when you eat—especially in sucking and chewing—and even when you look for something to eat, by controlling your eye movements. The group does not include muscles of the neck that are involved with movements of the vertebral column.

2. *The Muscles of the Vertebral Column.* This group includes numerous flexors, extensors, and rotators of the vertebral column.

3. *The Oblique and Rectus Muscles.* This group forms the muscular walls of the thoracic and abdominopelvic cavities between the first thoracic vertebra and the pelvis. In the thoracic area these muscles are partitioned by the ribs, but over the abdominal surface they form broad muscular sheets. The neck also has oblique and rectus muscles. Although they do not form a complete muscular wall, they share a common developmental origin with the oblique and rectus muscles of the trunk. **ATLAS** EMBRYOLOGY SUMMARY 9: The Development of the Muscular System

4. *The Muscles of the Pelvic Floor.* These muscles extend between the sacrum and pelvic girdle. The group forms the *perineum*, a muscular sheet that closes the pelvic outlet.

☐ MUSCLES OF THE HEAD AND NECK

We can divide the muscles of the head and neck into several functional groups. The *muscles of facial expression*, the *muscles of mastication* (chewing), the *muscles of the tongue*, and the *muscles of the pharynx* originate on the skull or hyoid bone.

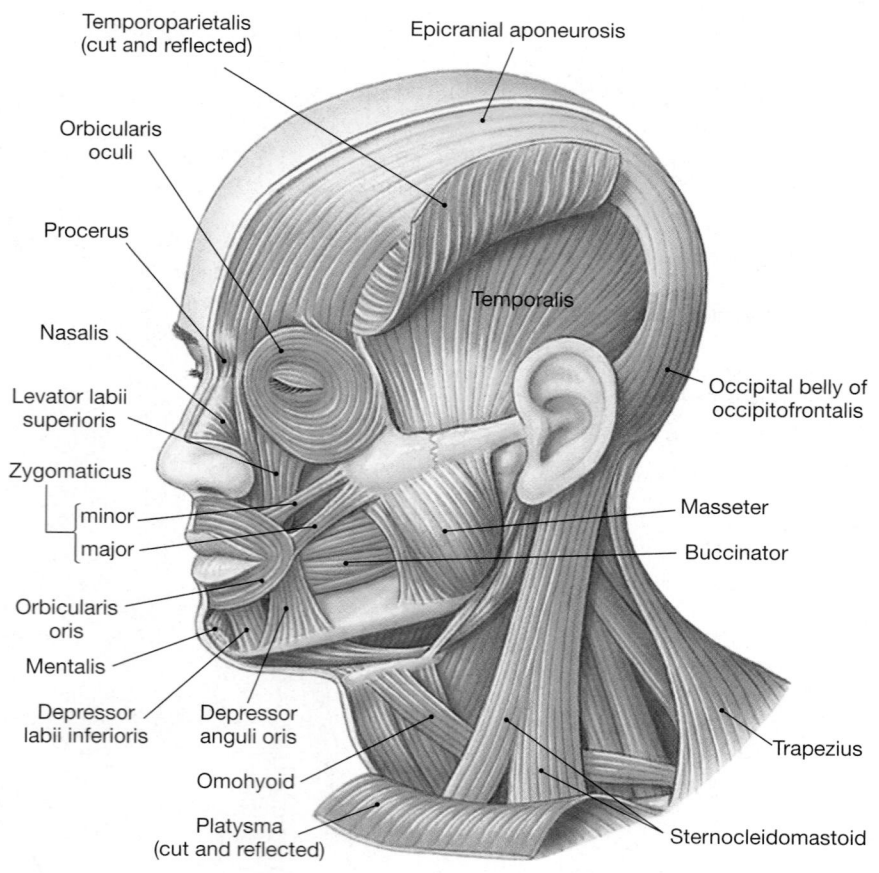

Temporoparietalis
(cut and reflected)

Epicranial aponeurosis

Orbicularis
oculi

Procerus

Nasalis

Temporalis

Levator labii
superioris

Zygomaticus

Occipital belly of
occipitofrontalis

minor

major

Masseter

Orbicularis
oris

Buccinator

Mentalis

Depressor
labii inferioris

Depressor
anguli oris

Omohyoid

Trapezius

Platysma
(cut and reflected)

Sternocleidomastoid

●FIGURE 11–4
Muscles of Facial Expression.
See also Figure 7–3, pp. 213–214.
(a) An anterolateral view. **(b)** An anterior view. **ATLAS** Plates 5.1, 5.2

(a)

Muscles involved with sight and hearing also are based on the skull. Here, we will consider the *extra-ocular muscles*—those associated with movements of the eye. We shall discuss the intrinsic eye muscles, which control the diameter of the pupil and the shape of the lens, and the tiny skeletal muscles associated with the auditory ossicles, in Chapter 17. In the neck, the *extrinsic muscles of the larynx* adjust the position of the hyoid bone and larynx. We shall examine the intrinsic laryngeal muscles, including those of the vocal cords, in Chapter 23 (the respiratory system).

Muscles of Facial Expression

The muscles of facial expression originate on the surface of the skull (Figure 11–4● and Table 11–2, p. 344). At their insertions, the fibers of the epimysium are woven into those of the superficial fascia and the dermis of the skin: When they contract, the skin moves. These muscles are innervated by the seventh cranial nerve, the *facial nerve.*

The largest group of facial muscles is associated with the mouth. The **orbicularis oris** muscle constricts the opening, and other muscles move the lips or the corners of the mouth. The **buccinator** muscle has two functions related to eating (in addition to its importance to musicians). During chewing, it cooperates with the masticatory muscles by moving food back across the teeth from the space inside the cheeks. In infants, the buccinator provides suction for suckling at the breast.

Smaller groups of muscles control movements of the eyebrows and eyelids, the scalp, the nose, and the external ear. The **epicranium** (ep-i-KRĀ-nē-um; *epi-*, on + *kranion*, skull), or scalp, contains the **temporoparietalis** muscle and the **occipitofrontalis** muscle, which has a *frontal belly* and an *occipital belly*. The two bellies are separated by the **epicranial aponeurosis**, a thick, collagenous sheet. The **platysma** (pla-TIZ-muh; *platys*, flat) covers the anterior surface of the neck, extending from the base of the neck to the periosteum of the mandible and the fascia at the corner of the mouth. One of the effects of aging is the loss of muscle tone in the platysma, resulting in a looseness of the skin of the anterior throat.

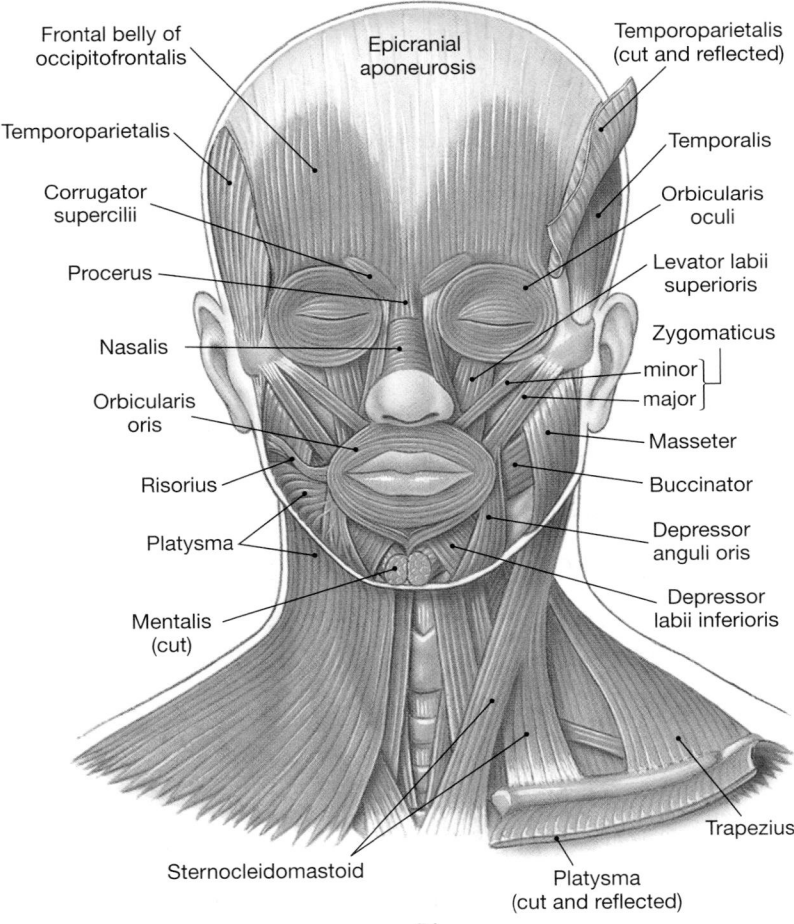

(b)

TABLE 11–2 MUSCLES OF FACIAL EXPRESSION (FIGURE 11–4)

Region/Muscle	Origin	Insertion	Action	Innervation
MOUTH				
Buccinator	Alveolar processes of maxillary bone and mandible	Blends into fibers of orbicularis oris	Compresses cheeks	Facial nerve (VII)
Depressor labii inferioris	Mandible between the anterior midline and the mental foramen	Skin of lower lip	Depresses lower lip	As above
Levator labii superioris	Inferior margin of orbit, superior to the infraorbital foramen	Orbicularis oris	Elevates upper lip	As above
Mentalis	Incisive fossa of mandible	Skin of chin	Elevates and protrudes lower lip	As above
Orbicularis oris	Maxillary bone and mandible	Lips	Compresses, purses lips	As above
Risorius	Fascia surrounding parotid salivary gland	Angle of mouth	Draws corner of mouth to the side	As above
Depressor anguli oris	Anterolateral surface of mandibular body	Skin at angle of mouth	Depresses corner of mouth	As above
Zygomaticus major	Zygomatic bone near zygomaticomaxillary suture	Angle of mouth	Retracts and elevates corner of mouth	As above
Zygomaticus minor	Zygomatic bone posterior to zygomaticotemporal suture	Upper lip	Retracts and elevates upper lip	As above
EYE				
Corrugator supercilii	Orbital rim of frontal bone near nasal suture	Eyebrow	Pulls skin inferiorly and anteriorly; wrinkles brow	As above
Levator palpebrae superioris (see Figure 11–5)	Tendinous band around optic foramen	Upper eyelid	Elevates upper eyelid	Oculomotor nerve (III)*
Orbicularis oculi	Medial margin of orbit	Skin around eyelids	Closes eye	Facial nerve (VII)
NOSE				
Procerus	Nasal bones and lateral nasal cartilages	Aponeurosis at bridge of nose and skin of forehead	Moves nose, changes position and shape of nostrils	As above
Nasalis	Maxillary bone and alar cartilage of nose	Bridge of nose	Compresses bridge, depresses tip of nose; elevates corners of nostrils	As above
EAR				
Temporoparietalis	Fascia around external ear	Epicranial aponeurosis	Tenses scalp, moves auricle of ear	As above
SCALP (EPICRANIUM)				
Occipitofrontalis				
Frontal belly	Epicranial aponeurosis bridge of nose	Skin of eyebrow and wrinkles forehead	Raises eyebrows,	As above
Occipital belly	Superior nuchal line	Epicranial aponeurosis	Tenses and retracts scalp	As above
NECK				
Platysma	Superior thorax between cartilage of 2nd rib and acromion of scapula	Mandible and skin of cheek	Tenses skin of neck; depresses mandible	As above

*This muscle originates in association with the extrinsic eye muscles, so its innervation is unusual.

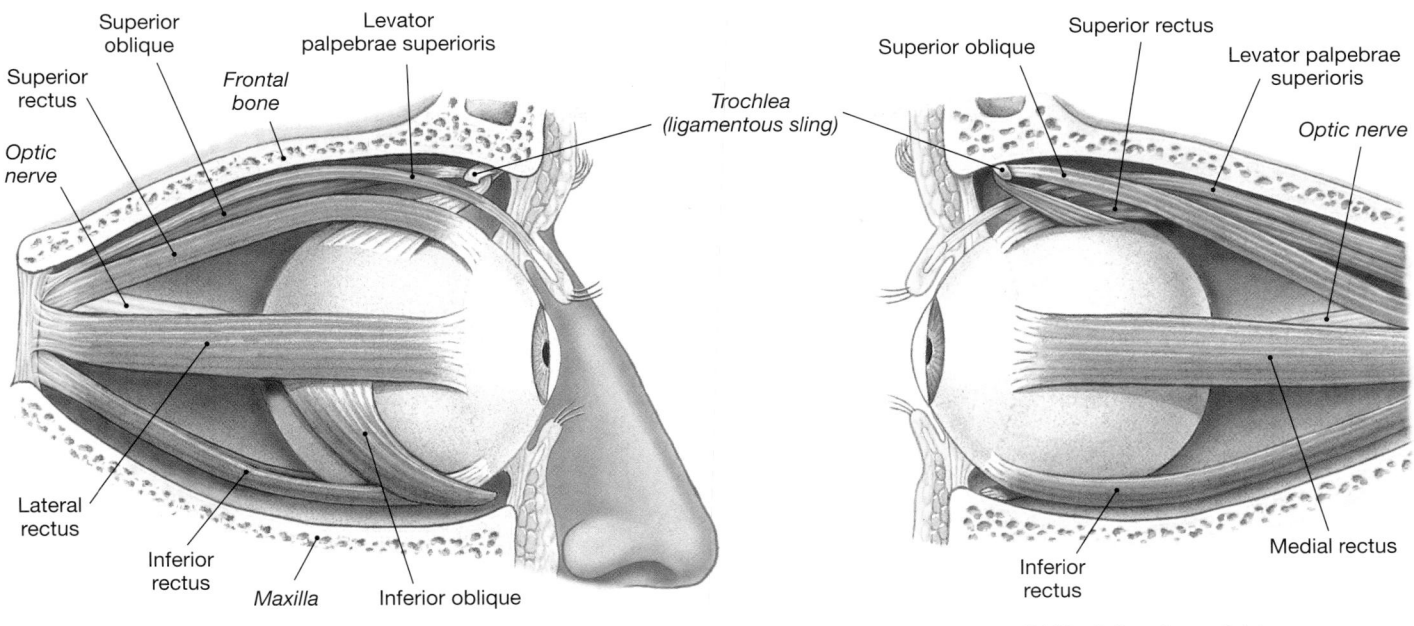

●**FIGURE 11–5**
Extrinsic Eye Muscles. *See also Figure 7–13, p. 224.*
ATLAS Plates 3.2, 3.3

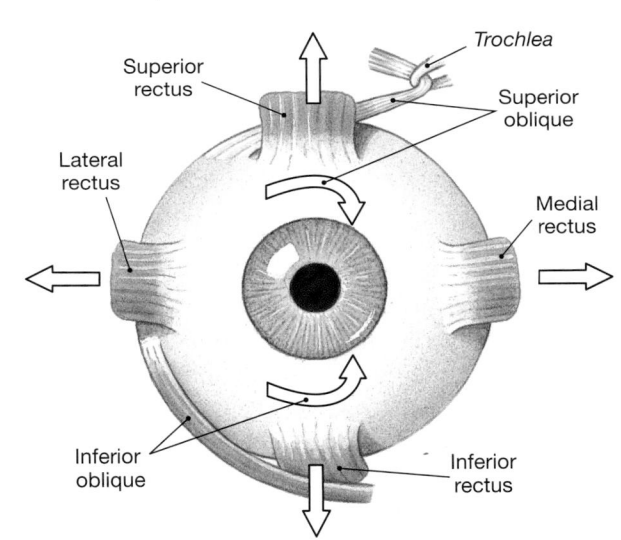

Extrinsic Eye Muscles

Six extrinsic eye muscles, also known as the **extra-ocular muscles** or *oculomotor muscles*, originating on the surface of the orbit control the position of each eye. These muscles, shown in Figure 11–5● and detailed in Table 11–3, are the **inferior rectus**, **medial rectus**, **superior rectus**, **lateral rectus**, **inferior oblique**, and **superior oblique** muscles. The extrinsic eye muscles are innervated by the third (*oculomotor*), fourth (*trochlear*), and sixth (*abducens*) cranial nerves.

TABLE 11–3 EXTRINSIC EYE MUSCLES (FIGURE 11–5)

Muscle	Origin	Insertion	Action	Innervation
Inferior rectus	Sphenoid around optic canal	Inferior, medial surface of eyeball	Eye looks down	Oculomotor nerve (III)
Medial rectus	As above	Medial surface of eyeball	Eye looks medially	As above
Superior rectus	As above	Superior surface of eyeball	Eye looks up	As above
Lateral rectus	As above	Lateral surface of eyeball	Eye looks to the side	Abducens nerve (VI)
Inferior oblique	Maxillary bone at anterior portion of orbit	Inferior, lateral surface of eyeball	Eye rolls, looks up and to the side	Oculomotor nerve (III)
Superior oblique	Sphenoid around optic canal	Superior, lateral surface of eyeball	Eye rolls, looks down and to the side	Trochlear nerve (IV)

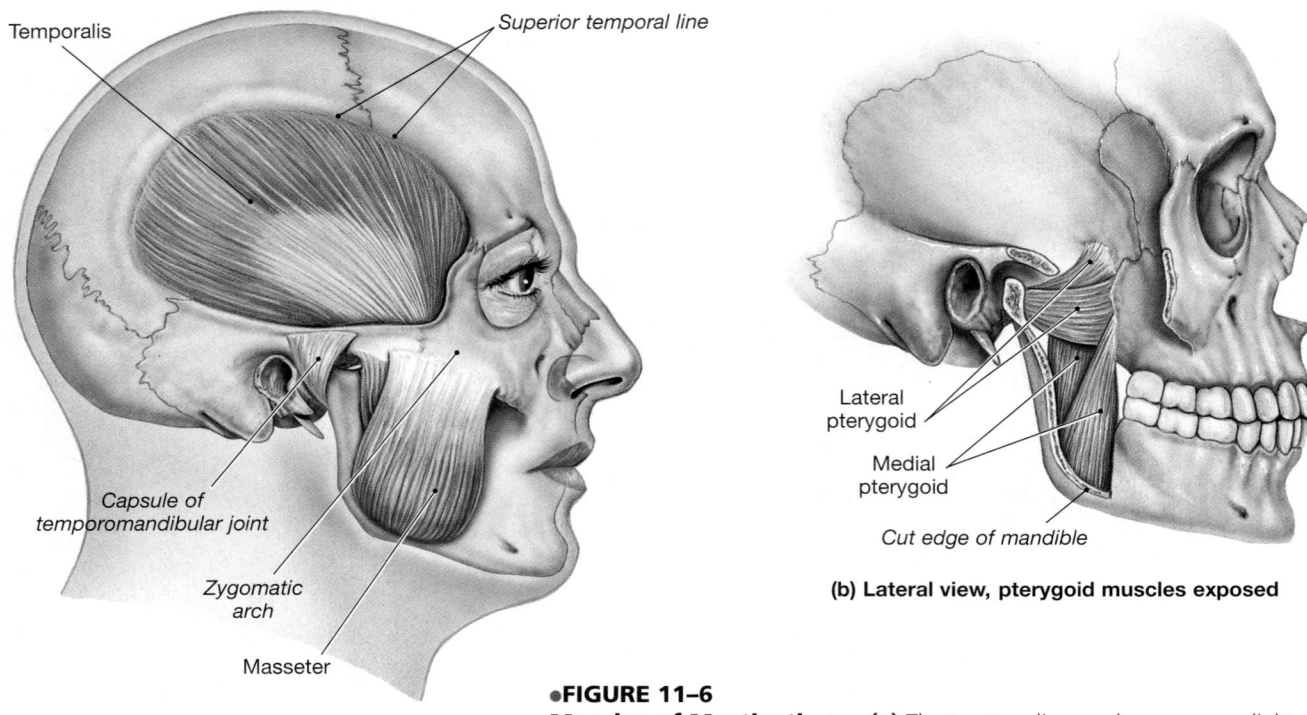

Temporalis

Superior temporal line

Capsule of
temporomandibular joint

Zygomatic
arch

Masseter

(a) Lateral view

Lateral
pterygoid

Medial
pterygoid

Cut edge of mandible

(b) Lateral view, pterygoid muscles exposed

●**FIGURE 11–6**
Muscles of Mastication. **(a)** The temporalis muscle passes medial to the zygomatic arch to insert on the coronoid process of the mandible. The masseter inserts on the angle and lateral surface of the mandible. **(b)** The location and orientation of the pterygoid muscles can be seen after the overlying muscles, along with a portion of the mandible, are removed. *See also Figures 7–3 and 7–12, pp. 213–214, 223.* **ATLAS** Plate 5.2

Muscles of Mastication

The muscles of mastication (Figure 11–6● and Table 11–4) move the mandible at the temporomandibular joint. The large **masseter** muscle is the strongest jaw muscle. The **temporalis** muscle assists in elevation of the mandible. Gritting your teeth while resting your hand on the side of your face below and then above the zygomatic arch will allow you to feel these muscles in action. The **pterygoid** muscles, used in various combinations, can elevate, depress, or protract the mandible or slide it from side to side, a movement called *lateral excursion*. These

movements are important in making efficient use of your teeth while you chew foods of various consistencies. The muscles of mastication are innervated by the fifth cranial nerve, the *trigeminal nerve.*

Muscles of the Tongue

The muscles of the tongue have names ending in *glossus,* the Greek word for "tongue." Once you can recall the structures referred to by *palato-, stylo-, genio-,* and *hyo-,* the names will be self explanatory. The **palatoglossus** muscle originates at

TABLE 11–4 MUSCLES OF MASTICATION (FIGURE 11–6)

Muscle	Origin	Insertion	Action	Innervation
Masseter	Zygomatic arch	Lateral surface of mandibular ramus	Elevates mandible and closes the jaws	Trigeminal nerve (V), mandibular branch
Temporalis	Along temporal lines of skull	Coronoid process of mandible	Elevates mandible	As above
Pterygoids (medial and lateral)	Lateral pterygoid plate	Medial surface of mandibular ramus	*Medial*: Elevates the mandible and closes the jaws, or performs lateral excursion	As above
			Lateral: Opens jaws, protrudes mandible, or performs lateral excursion	As above

What's New in Anatomy?

Like most people, you might assume that every anatomical structure in the human body was described centuries ago and that nothing is "new" in the field of anatomy. Many people were surprised in 1996, however, when anatomical researchers at the University of Maryland documented the existence of a previously unknown skeletal muscle. This "new" muscle, named the *sphenomandibularis muscle*, extends from the lateral surface of the sphenoid to the mandible. The sphenomandibularis muscle assists the muscles of mastication.

The research began with a computer analysis of the Virtual Human database, a digitized photographic atlas of cross-sectional anatomy (which also provided the basis of the animations on the Anatomy CD). That initial work was then supported by careful cadaver dissections. Although the discovery of the sphenomandibularis muscle remains controversial (it may have been described previously as a portion of the temporalis muscle), it is a good example of how modern technologies are providing new perspectives on the human body.

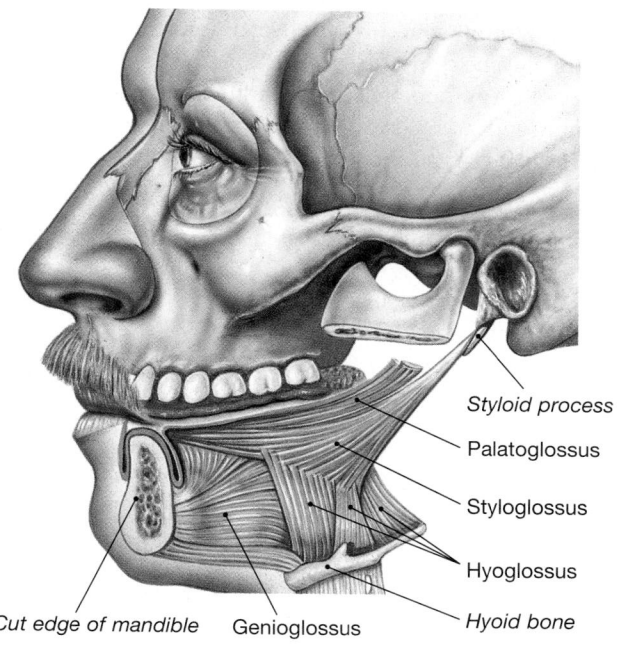

Styloid process
Palatoglossus
Styloglossus
Hyoglossus
Hyoid bone

Cut edge of mandible   Genioglossus

●**FIGURE 11–7**
Muscles of the Tongue. *See also Figure 7–3, pp. 213–214.*

the palate, the **styloglossus** muscle at the styloid process of the temporal bone, the **genioglossus** muscle at the chin, and the **hyoglossus** muscle at the hyoid bone (Figure 11–7●). These muscles, used in various combinations, move the tongue in the delicate and complex patterns necessary for speech, and manipulate food within the mouth in preparation for swallowing. Most are innervated by the *hypoglossal nerve* (XII), a cranial nerve whose name indicates its function and location (Table 11–5).

Muscles of the Pharynx

The muscles of the pharynx (Figure 11–8● and Table 11–6) are responsible for initiating the swallowing process. The **pharyngeal constrictor** muscles (*superior, middle,* and *inferior*) move materials into the esophagus. The **laryngeal elevator** muscles elevate the larynx. The two **palatal muscles**—the *tensor veli palatini* and the *levator veli palatini*—elevate the soft palate and adjacent portions of the pharyngeal wall and also pull open the entrance to the auditory tube. As a result, swallowing repeatedly can open the entrance to the auditory tube and help you adjust to pressure changes when you fly or dive.

TABLE 11–5 MUSCLES OF THE TONGUE (FIGURE 11–7)

Muscle	Origin	Insertion	Action	Innervation
Genioglossus	Medial surface of mandible around chin	Body of tongue, hyoid bone	Depresses and protracts tongue	Hypoglossal nerve (XII)
Hyoglossus	Body and greater horn of hyoid bone	Side of tongue	Depresses and retracts tongue	As above
Palatoglossus	Anterior surface of soft palate	As above	Elevates tongue, depresses soft palate	Cranial branch of accessory nerve (XI)
Styloglossus	Styloid process of temporal bone	Along the side to tip and base of tongue	Retracts tongue, elevates side	Hypoglossal nerve (XII)

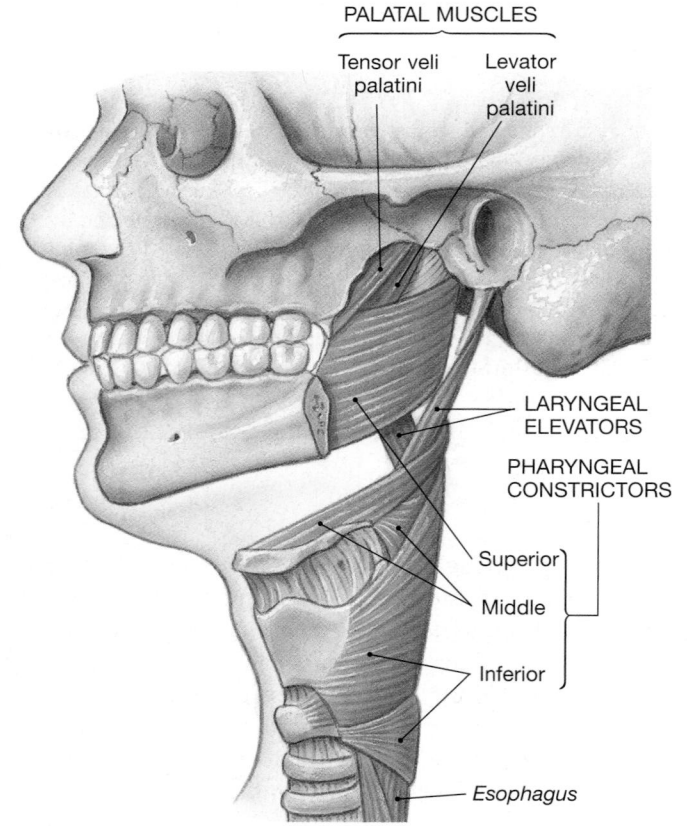

PALATAL MUSCLES

Tensor veli palatini   Levator veli palatini

LARYNGEAL ELEVATORS

PHARYNGEAL CONSTRICTORS

Superior

Middle

Inferior

Esophagus

●**FIGURE 11–8**
Muscles of the Pharynx. A lateral view. *See also Figure 7–4.*

Anterior Muscles of the Neck

The anterior muscles of the neck include (1) muscles that control the position of the larynx, (2) muscles that depress the mandible

and tense the floor of the mouth, and (3) muscles that provide a stable foundation for muscles of the tongue and pharynx (Figure 11–9● and Table 11–7). The **digastric** (dī-GAS-trik) muscle has two bellies, as the name implies (*di-*, two + *gaster*, stomach). One belly extends from the chin to the hyoid bone; the other continues from the hyoid bone to the mastoid portion of the temporal bone. Depending on which belly contracts and whether fixator muscles are stabilizing the position of the hyoid bone, the digastric muscle can open the mouth by depressing the mandible, or it can elevate the larynx by raising the hyoid bone. The digastric muscle overlies the broad, flat **mylohyoid** muscle, which provides a muscular floor to the mouth, aided by the deeper **geniohyoid** muscles that extend between the hyoid bone and the chin. The **stylohyoid** muscle forms a muscular connection between the hyoid bone and the styloid process of the skull. The **sternocleidomastoid** (ster-nō-klī-dō-MAS-toyd) muscle extends from the clavicle and the sternum to the mastoid region of the skull (Figures 11–4, pp. 342–343, and 11–9●). The **omohyoid** (ō-mō-HĪ-oyd) muscle attaches to the scapula, the clavicle and first rib, and the hyoid bone. The other members of this group are straplike muscles that extend between the sternum and larynx (*sternothyroid*) or hyoid bone (*sternohyoid*), and between the larynx and hyoid bone (*thyrohyoid*).

✓ If you were contracting and relaxing your masseter muscle, what would you probably be doing?

✓ Which facial muscle would you expect to be well developed in a trumpet player?

✓ Why can swallowing help alleviate the pressure sensations at the eardrum when you are in an airplane that is changing altitude?

Answers start on p. Q-1

The muscles of the head and neck can be reviewed on the **Anatomy CD-ROM:** Muscular System/Axial Dissections/Head & Face.

TABLE 11–6 MUSCLES OF THE PHARYNX (FIGURE 11–8)

Muscle	Origin	Insertion	Action	Innervation
PHARYNGEAL CONSTRICTORS				
Superior constrictor	Pterygoid process of sphenoid, medial surfaces of mandible	Median raphe attached to occipital bone	Constrict pharynx to propel bolus into esophagus	Branches of pharyngeal plexus (X)
Middle constrictor	Horns of hyoid bone	Median raphe	As above	As above
Inferior constrictor	Cricoid and thyroid cartilages of larynx	As above	As above	As above
LARYNGEAL ELEVATORS *	Ranges from soft palate, to cartilage around inferior portion of auditory tube, to styloid process of temporal bone	Thyroid cartilage	Elevate larynx	Branches of pharyngeal plexus (IX and X)
PALATAL MUSCLES				
Levator veli palatini	Petrous part of temporal bone; tissues around the auditory tube	Soft palate	Elevates soft palate	Branches of pharyngeal plexus (X)
Tensor veli palatini	Sphenoidal spine; tissues around the auditory tube	As above	As above	V

*The palatopharyngeus, salpingopharyngeus, and stylopharyngeus, assisted by the thyrohyoid, geniohyoid, stylohyoid, and hyoglossus muscles, discussed in Tables 11–5 and 11–7.

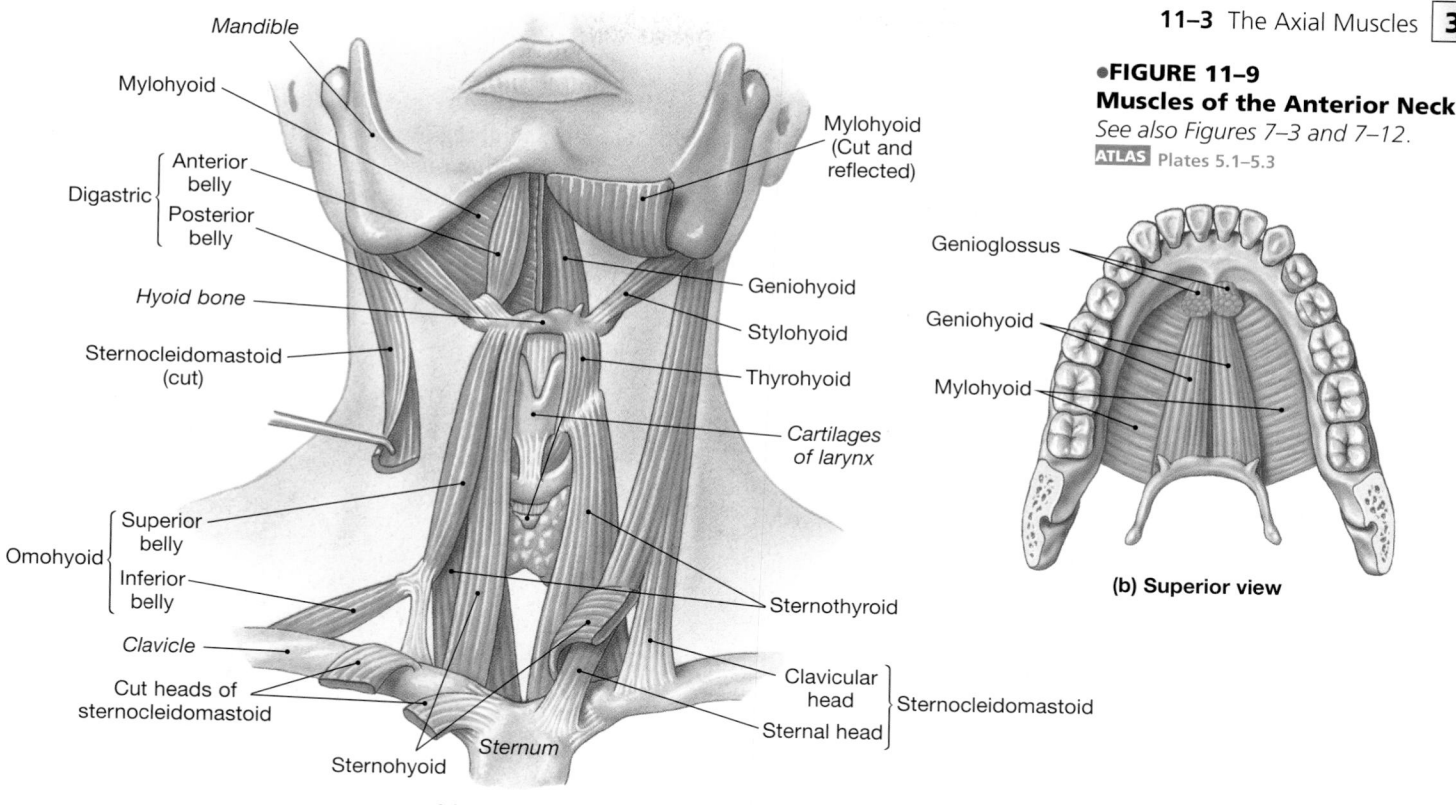

●FIGURE 11–9
Muscles of the Anterior Neck.
See also Figures 7–3 and 7–12.
ATLAS Plates 5.1–5.3

(a) Anterior view

(b) Superior view

TABLE 11–7   ANTERIOR MUSCLES OF THE NECK (FIGURE 11–9)

Muscle	Origin	Insertion	Action	Innervation
Digastric	Two bellies: *anterior* from inferior surface of mandible at chin; *posterior* from mastoid region of temporal bone	Hyoid bone	Depresses mandible or elevates larynx	*Anterior belly*: Trigeminal nerve (V), mandibular branch *Posterior belly*: Facial nerve (VII)
Geniohyoid	Medial surface of mandible at chin	Hyoid bone	As above and pulls hyoid bone anteriorly	Cervical nerve C_1 via hypoglossal nerve (XII)
Mylohyoid	Mylohyoid line of mandible	Median connective tissue band (raphe) that runs to hyoid bone	Elevates floor of mouth and hyoid bone or depresses mandible	Trigeminal nerve (V), mandibular branch
Omohyoid (superior and inferior bellies united at central tendon anchored to clavicle and first rib)	Superior border of scapula near scapular notch	Hyoid bone	Depresses hyoid bone and larynx	Cervical spinal nerves C_2–C_3
Sternohyoid	Clavicle and manubrium	Hyoid bone	As above	Cervical spinal nerves C_1–C_3
Sternothyroid	Dorsal surface of manubrium and first costal cartilage	Thyroid cartilage of larynx	As above	As above
Stylohyoid	Styloid process of temporal bone	Hyoid bone	Elevates larynx	Facial nerve (VII)
Thyrohyoid	Thyroid cartilage of larynx	Hyoid bone	Elevates thyroid, depresses hyoid bone	Cervical spinal nerves C_1–C_2 via hypoglossal nerve (XII)
Sternocleido-mastoid	Two bellies: *clavicular head* attaches to sternal end of clavicle; *sternal head* attaches to manubrium	Mastoid region of skull and lateral portion of superior nuchal line	Together, they flex the neck; alone, one side bends head toward shoulder and turns face to opposite side	Accessory nerve (XI) and cervical spinal nerves (C_2–C_3) of cervical plexus

☐ MUSCLES OF THE VERTEBRAL COLUMN

The muscles of the vertebral column are covered by more superficial back muscles, such as the trapezius and latissimus dorsi muscles (Figure 11–3b●, p. 341). The **erector spinae**, or *spinal extensors*, include superficial and deep layers. The superficial layer can be divided into **spinalis**, **longissimus**, and **ilio-**

costalis divisions (Figure 11–10● and Table 11–8). In the inferior lumbar and sacral regions, the boundary between the longissimus and iliocostalis muscles is indistinct. When contracting together, the erector spinae extend the vertebral column. When the muscles on only one side contract, the result is lateral flexion of the vertebral column.

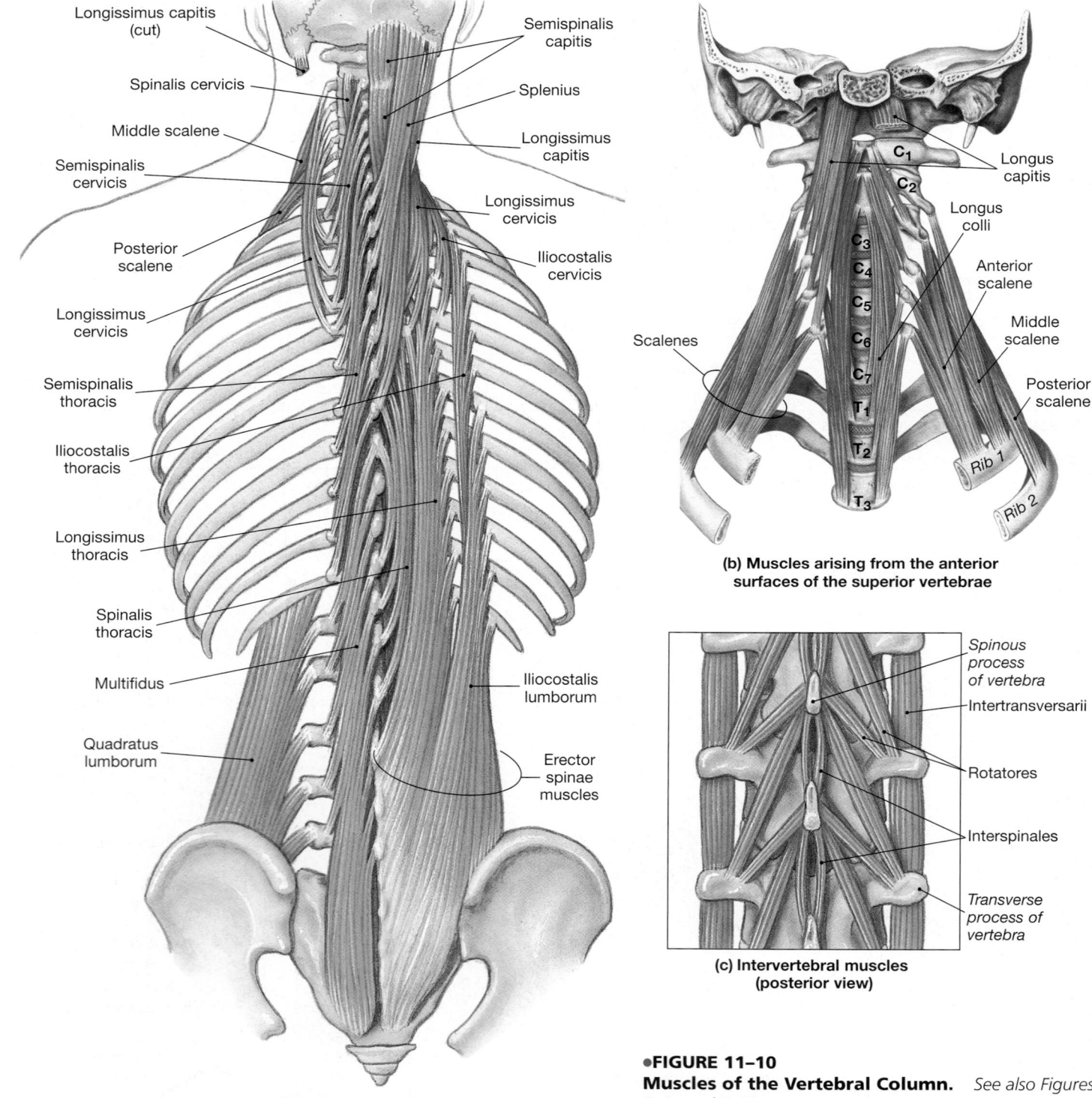

(a) Posterior view of the superficial muscle layer

(b) Muscles arising from the anterior surfaces of the superior vertebrae

(c) Intervertebral muscles (posterior view)

●**FIGURE 11–10**
Muscles of the Vertebral Column. *See also Figures 7–1a and 7–22.*

TABLE 11-8 MUSCLES OF THE VERTEBRAL COLUMN (FIGURE 11–10)

Group and Muscle(s)	Origin	Insertion	Action	Innervation
SUPERFICIAL LAYER				
Splenius (Splenius capitis, splenius cervicis)	Spinous processes and ligaments connecting inferior cervical and superior thoracic vertebrae	Mastoid process, occipital bone of skull, and superior cervical vertebrae	Together, the two sides extend neck; alone, each rotates and laterally flexes neck to that side	Cervical spinal nerves
Erector spinae				
Spinalis group				
Spinalis cervicis	Inferior portion of ligamentum nuchae and spinous process of C$_7$	Spinous process of axis	Extends neck	As above
Spinalis thoracis	Spinous processes of inferior thoracic and superior lumbar vertebrae	Spinous processes of superior thoracic vertebrae	Extends vertebral column	Thoracic and lumbar spinal nerves
Longissimus group				
Longissimus capitis	Transverse processes of inferior cervical and superior thoracic vertebrae	Mastoid process of temporal bone	Together, the two sides extend head; alone, each rotates and laterally flexes neck to that side	Cervical and thoracic spinal nerves
Longissimus cervicis	Transverse processes of superior thoracic vertebrae	Transverse processes of middle and superior cervical vertebrae	As above	As above
Longissimus thoracis	Broad aponeurosis and transverse processes of inferior thoracic and superior lumbar vertebrae; joins iliocostalis	Transverse processes of superior vertebrae and inferior surfaces of ribs	Extends vertebral column; alone, each produces lateral flexion to that side	Thoracic and lumbar spinal nerves
Iliocostalis group				
Iliocostalis cervicis	Superior borders of vertebrosternal ribs near the angles	Transverse processes of middle and inferior cervical vertebrae	Extends or laterally flexes neck, elevates ribs	Cervical and superior thoracic spinal nerves
Iliocostalis thoracis	Superior borders of inferior seven ribs medial to the angles	Upper ribs and transverse process of last cervical vertebra	Stabilizes thoracic vertebrae in extension	Thoracic spinal nerves
Iliocostalis lumborum	Iliac crest, sacral crests, and spinous processes	Inferior surfaces of inferior seven ribs near their angles	Extends vertebral column, depresses ribs	Inferior thoracic and lumbar spinal nerves
DEEP LAYER (TRANSVERSOSPINALIS)				
Semispinalis group				
Semispinalis capitis	Articular processes of inferior cervical and traverse processes of superior thoracic vertebrae	Occipital bone, between nuchal lines	Together, the two sides extend head; alone, each extends and laterally flexes neck	Cervical spinal nerves
Semispinalis cervicis	Transverse processes of T$_1$–T$_5$ or T$_6$	Spinous processes of C$_2$–C$_5$	Extends vertebral column and rotates toward opposite side	As above
Semispinalis thoracis	Transverse processes of T$_6$–T$_{10}$	Spinous processes of C$_5$–T$_4$	As above	Thoracic spinal nerves
Multifidus	Sacrum and transverse processes of each vertebra	Spinous processes of the third or fourth more superior vertebrae	As above	Cervical, thoracic, and lumbar spinal nerves
Rotatores	Transverse processes of each vertebra	Spinous processes of adjacent, more superior vertebra	As above	As above
Interspinales	Spinous processes of each vertebra	Spinous processes of more superior vertebra	Extends vertebral column	As above
Intertransversarii	Transverse processes of each vertebra	Transverse process of more superior vertebra	Laterally flexes the vertebral column	As above
SPINAL FLEXORS				
Longus capitis	Transverse processes of cervical vertebrae	Base of the occipital bone	Together, the two sides flex the neck; alone, each rotates head to that side	Cervical spinal nerves
Longus colli	Anterior surfaces of cervical and superior thoracic vertebrae	Transverse processes of superior cervical vertebrae	Flexes or rotates neck; limits hyperextension	As above
Quadratus lumborum	Iliac crest and iliolumbar ligament	Last rib and transverse processes of lumbar vertebrae	Together, they depress ribs; alone, each side laterally flexes vertebral column	Thoracic and lumbar spinal nerves

Deep to the spinalis muscles, smaller muscles interconnect and stabilize the vertebrae. These muscles include the **semispinalis** muscles and the **multifidus**, **interspinales**, **intertransversarii**, and **rotatores** muscles (Figure 11–10a●). In various combinations, they produce slight extension or rotation of the vertebral column. They are also important in making delicate adjustments in the positions of individual vertebrae, and they stabilize adjacent vertebrae. If injured, these muscles can start a cycle of pain–muscle stimulation– contraction–pain. Pressure on adjacent spinal nerves results, leading to sensory losses as well as limiting mobility. Many of the warm-up and stretching exercises recommended before athletic events are intended to prepare these small but very important muscles for their supporting role.

The muscles of the vertebral column include many dorsal extensors, but few ventral flexors. The vertebral column does not need a massive series of flexor muscles, because (1) many of the large trunk muscles flex the vertebral column when they contract, and (2) most of the body weight lies anterior to the vertebral column, so gravity tends to flex the spine. However, a few spinal flexors are associated with the anterior surface of the vertebral column. In the neck, the **longus capitis** and the **longus colli** muscles rotate or flex the neck, depending on whether the muscles of one or both sides are contracting (Figure 11–11b●). In the lumbar region, the large **quadratus lumborum** muscles flex the vertebral column and depress the ribs (Figure 11–11a●).

☐ OBLIQUE AND RECTUS MUSCLES

The oblique and rectus muscles lie within the body wall, between the spinous processes of vertebrae and the ventral midline (Figures 11–3, pp. 340–341, and 11–11● and Table 11–9, p. 353).

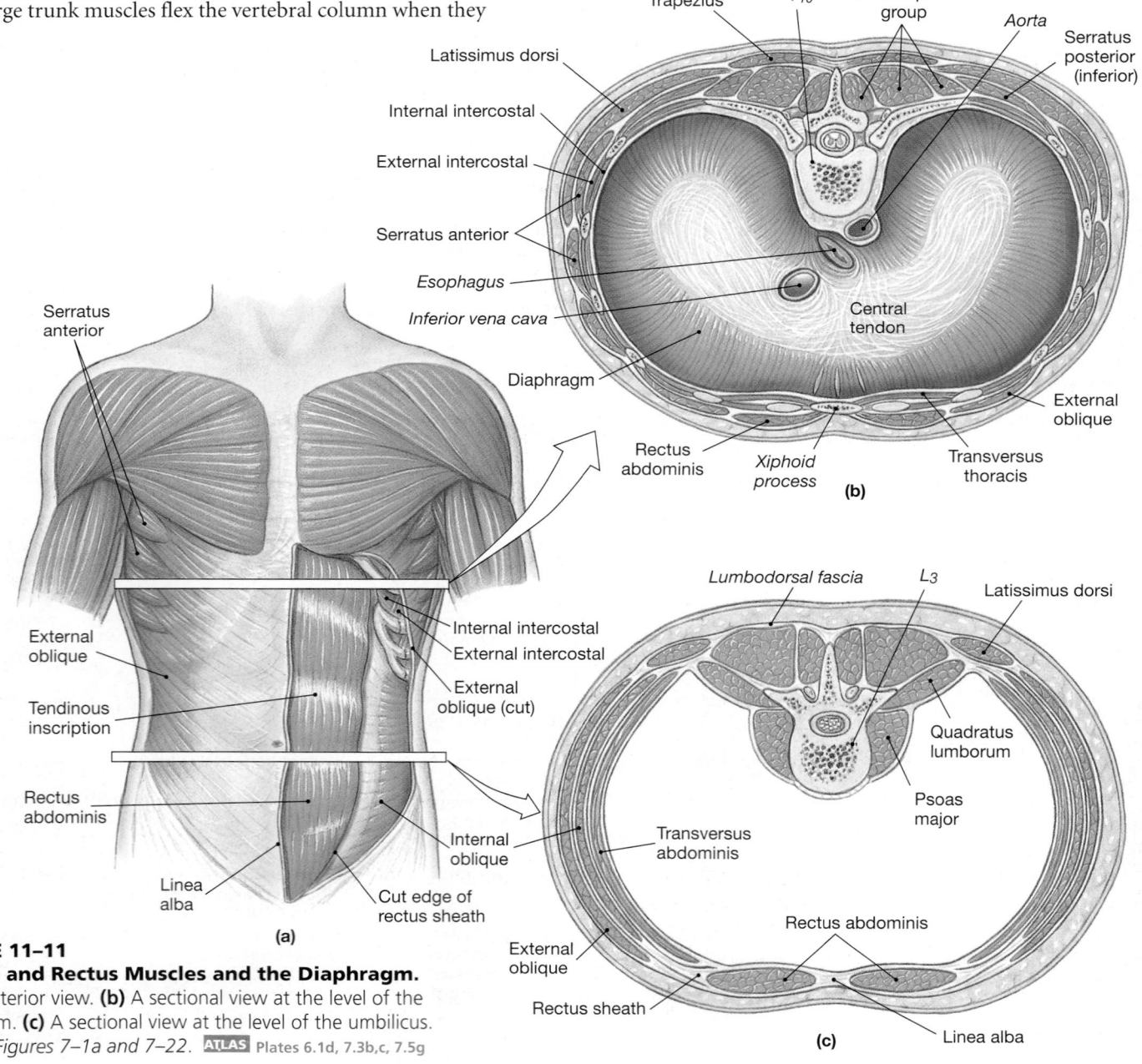

●**FIGURE 11–11**
Oblique and Rectus Muscles and the Diaphragm.
(a) An anterior view. (b) A sectional view at the level of the diaphragm. (c) A sectional view at the level of the umbilicus. *See also Figures 7–1a and 7–22.* **ATLAS** Plates 6.1d, 7.3b,c, 7.5g

TABLE 11–9 OBLIQUE AND RECTUS MUSCLES[*] (FIGURE 11–11)

Group and Muscle(s)	Origin	Insertion	Action	Innervation
OBLIQUE GROUP				
Cervical region				
Scalenes (anterior, middle, and posterior)	Transverse and costal processes of cervical vertebrae	Superior surfaces of first two ribs	Elevate ribs or flex neck	Cervical spinal nerves
Thoracic region				
External intercostals	Inferior border of each rib	Superior border of more inferior rib	Elevate ribs	Intercostal nerves (branches of thoracic spinal nerves)
Internal intercostals	Superior border of each rib	Inferior border of the preceding rib	Depress ribs	As above
Transversus thoracis	Posterior surface of sternum	Cartilages of ribs	As above	As above
Serratus posterior				
superior (see Figure 11-13a)	Spinous processes of C_7–T_3 and ligamentum nuchae	Superior borders of ribs 2–5 near angles	Elevates ribs, enlarges thoracic cavity	Thoracic nerves (T_1–T_4)
inferior	Aponeurosis from spinous processes of T_{10}–L_3	Inferior borders of ribs 8–12	Pulls ribs inferiorly; also pulls outward, opposing diaphragm	Thoracic nerves (T_9–T_{12})
Abdominal region				
External oblique	External and inferior borders of ribs 5–12	Linea alba and iliac crest	Compresses abdomen, depresses ribs, flexes or bends spine	Intercostal, iliohypogastric, and ilioinguinal nerves
Internal oblique	Lumbodorsal fascia and iliac crest	Inferior ribs, xiphoid process, and linea alba	As above	As above
Transversus abdominis	Cartilages of ribs 6–12, iliac crest, and lumbodorsal fascia	Linea alba and pubis	Compresses abdomen	As above
RECTUS GROUP				
Cervical region	See muscles in Table 11–6			
Thoracic region				
Diaphragm	Xiphoid process, cartilages of ribs 4–10, and anterior surfaces of lumbar vertebrae	Central tendinous sheet	Contraction expands thoracic cavity, compresses abdominopelvic cavity	Phrenic nerves (C_3–C_5)
Abdominal region				
Rectus abdominis	Superior surface of pubis around symphysis	Inferior surfaces of costal cartilages (ribs 5–7) and xiphoid process	Depresses ribs, flexes vertebral column	Intercostal nerves (T_7–T_{12})

[*]Where appropriate, spinal nerves involved are given in parentheses.

The oblique muscles compress underlying structures or rotate the vertebral column, depending on whether one or both sides contract. The rectus muscles are important flexors of the vertebral column, acting in opposition to the erector spinae. The oblique and rectus muscles share embryological origins; we can divide them into cervical, thoracic, and abdominal groups.

The oblique group includes the **scalene** muscles of the neck (Figure 11–10b●) and the **intercostal** and **transversus** muscles of the thorax (Figure 11–11a,b●). The scalene muscles (*anterior, middle,* and *posterior*) elevate the first two ribs and assist in flexion of the neck. In the thorax, the oblique muscles extend between the ribs, with the **external intercostal** muscles covering the **internal intercostal** muscles. Both groups of intercostal muscles aid in respira-

tory movements of the ribs. A small **transversus thoracis** muscle crosses the inner surface of the rib cage and is separated from the pleural cavity by the parietal pleura, a *serous membrane.* ∞ p. 133 The sternum occupies the place where we might otherwise expect thoracic rectus muscles to be.

The same basic pattern of musculature extends unbroken across the abdominopelvic surface (Figure 11–11a,c●). Here, the muscles are called the **external oblique, internal oblique, transversus abdominis,** and **rectus abdominis** muscles. The rectus abdominis muscle inserts at the xiphoid process and originates near the pubic symphysis. This muscle is longitudinally divided by the **linea alba** (white line), a median collagenous partition (Figure 11–3a●, p. 340). The rectus abdominis muscle is separated into segments by transverse bands of collagen fibers called **tendinous inscriptions**. Each segment contains muscle fibers that extend longitudinally, originating and inserting on the tendinous inscriptions. Due to the bulging of enlarged muscle fibers between the tendinous inscriptions, body builders often refer to the rectus abdominis as the "six-pack."

The Diaphragm

The term *diaphragm* refers to any muscular sheet that forms a wall. When used without a modifier, however, **diaphragm**, or *diaphragmatic muscle*, specifies the muscular partition that separates the abdominopelvic and thoracic cavities (Figure 11–11b●). We include this muscle here because it develops in association with the other muscles of the chest wall. The diaphragm is a major muscle of respiration.

☐ MUSCLES OF THE PELVIC FLOOR

The muscles of the pelvic floor extend from the sacrum and coccyx to the ischium and pubis. These muscles (1) support the organs of the pelvic cavity, (2) flex the sacrum and coccyx, and (3) control the movement of materials through the urethra and anus (Figure 11–12● and Table 11–10).

The boundaries of the **perineum**, the muscular sheet that forms the pelvic floor, are established by the inferior margins of the pelvis. If you draw a line between the ischial tuberosities, you will divide the perineum into two triangles: an anterior **urogenital triangle** and a posterior **anal triangle**. The superficial muscles of the urogenital triangle are the muscles of the external genitalia. They cover deeper muscles that strengthen the pelvic floor and encircle the urethra. These muscles constitute

TABLE 11–10 MUSCLES OF THE PELVIC FLOOR (FIGURE 11–12)

Group and Muscle(s)	Origin	Insertion	Action	Innervation
UROGENITAL TRIANGLE				
Superficial muscles				
Bulbospongiosus:				
Males	Collagen sheath at base of penis; fibers cross over urethra	Median raphe and central tendon of perineum	Compresses base and stiffens penis; ejects urine or semen	Pudendal nerve, perineal branch (S_2–S_4)
Females	Collagen sheath at base of clitoris; fibers run on either side of urethral and vaginal opening	Central tendon of perineum	Compresses and stiffens clitoris; narrows vaginal opening	As above
Ischiocavernosus	Ischial ramus and tuberosity	Pubic symphysis anterior to base of penis or clitoris	Compresses and stiffens penis or clitoris	As above
Superficial transverse perineal	Ischial ramus	Central tendon of perineum	Stabilizes central tendon of perineum	As above
Deep muscles: **Urogenital diaphragm**				
Deep transverse perineal	Ischial ramus	Median raphe of urogenital diaphragm	As above	As above
External urethral sphincter:				
Males	Ischial and pubic rami	To median raphe at base of penis; inner fibers encircle urethra	Closes urethra; compresses prostate and bulbourethral glands	As above
Females	Ischial and pubic rami	To median raphe; inner fibers encircle urethra	Closes urethra; compresses vagina and greater vestibular glands	As above
ANAL TRIANGLE				
Pelvic diaphragm				
Coccygeus	Ischial spine	Lateral, inferior borders of sacrum and coccyx	Flexes coccygeal joints; tenses and supports pelvic floor	Inferior sacral nerves (S_4–S_5)
Levator ani: Iliococcygeus	Ischial spine, pubis	Coccyx and median raphe	Tenses floor of pelvis; flexes coccygeal joints; elevates and retracts anus	Pudendal nerve (S_2–S_4)
Pubococcygeus	Inner margins of pubis	As above	As above	As above
External anal sphincter	Via tendon from coccyx	Encircles anal opening	Closes anal opening	Pudendal nerve, hemorrhoidal branch (S_2–S_4)

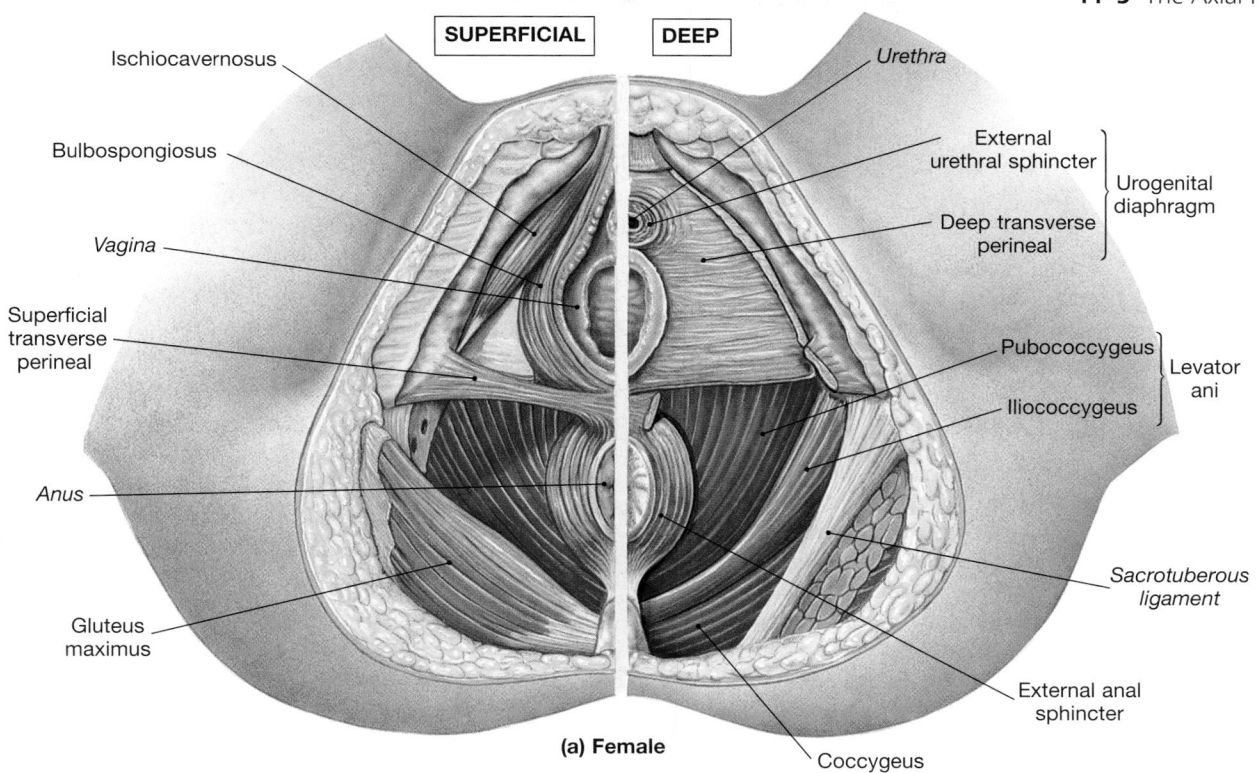

SUPERFICIAL	DEEP

Ischiocavernosus

Bulbospongiosus

Vagina

Superficial transverse perineal

Anus

Gluteus maximus

Urethra

External urethral sphincter

Deep transverse perineal

} Urogenital diaphragm

Pubococcygeus

Iliococcygeus

} Levator ani

Sacrotuberous ligament

External anal sphincter

Coccygeus

(a) Female

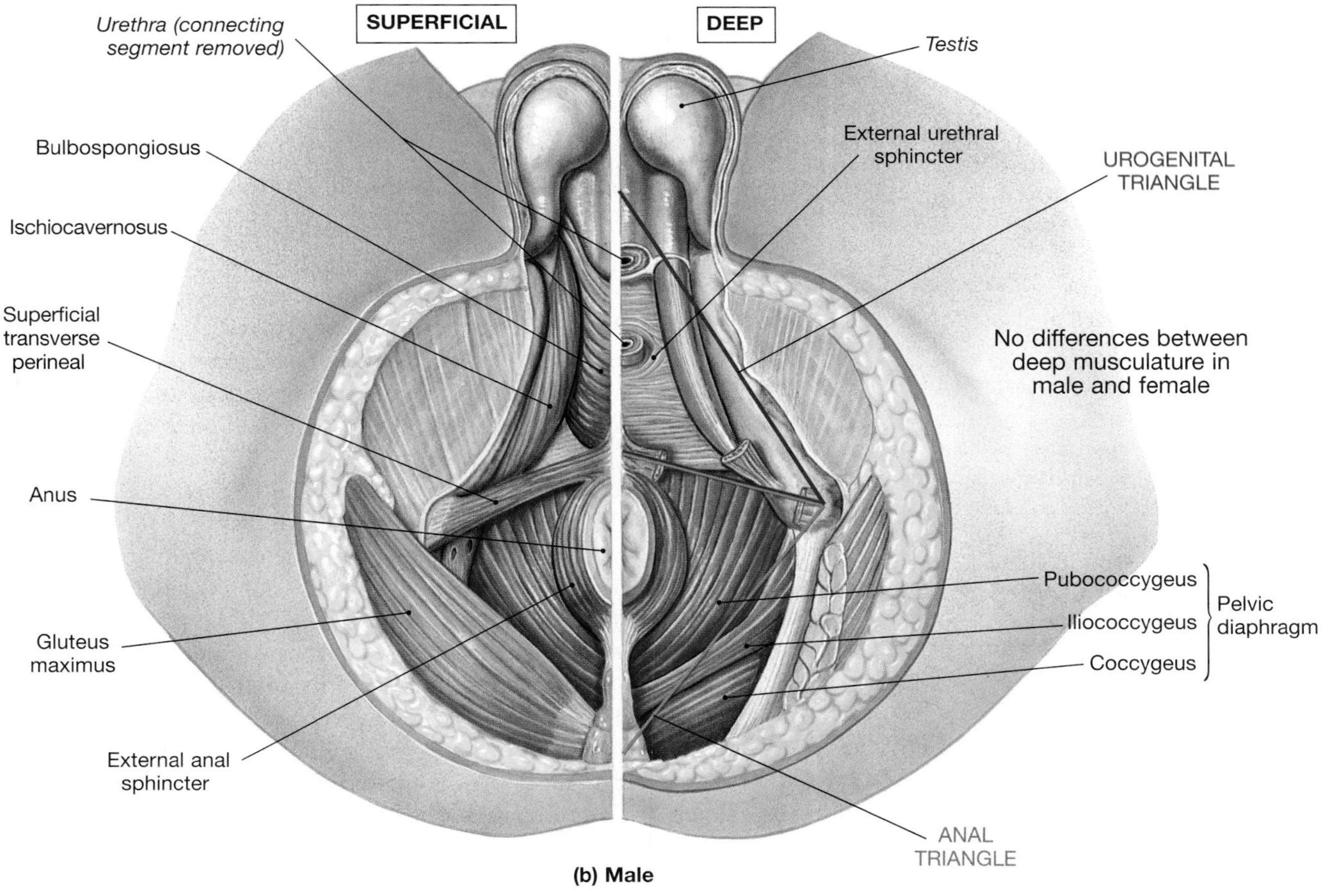

SUPERFICIAL	DEEP

Urethra (connecting segment removed)

Bulbospongiosus

Ischiocavernosus

Superficial transverse perineal

Anus

Gluteus maximus

External anal sphincter

Testis

External urethral sphincter

UROGENITAL TRIANGLE

No differences between deep musculature in male and female

Pubococcygeus

Iliococcygeus

Coccygeus

} Pelvic diaphragm

ANAL TRIANGLE

(b) Male

●**FIGURE 11–12**
Muscles of the Pelvic Floor. *See also Figures 7–1a, 8–8, and 8–9.*

the **urogenital diaphragm**, a deep muscular layer that extends between the pubic bones.

An even more extensive muscular sheet, the **pelvic diaphragm**, forms the muscular foundation of the anal triangle. This layer, covered by the urogenital diaphragm, extends as far as the pubic symphysis.

The urogenital and pelvic diaphragms do not completely close the pelvic outlet, for the urethra, vagina, and anus pass through them to open on the external surface. Muscular sphincters surround their openings and permit voluntary control of urination and defecation. Muscles, nerves, and blood vessels also pass through the pelvic outlet as they travel to or from the lower limbs.

When your abdominal muscles contract forcefully, pressures in your abdominopelvic cavity can skyrocket, and those pressures are applied to internal organs. If you exhale at the same time, the pressure is relieved, because your diaphragm can move upward as your lungs collapse. But during vigorous isometric exercises or when you lift a weight while holding your breath, pressure in the abdominopelvic cavity can rise high enough to cause a variety of problems, among them a hernia.

A **hernia** develops whenever an organ protrudes through an abnormal opening. The most common hernias are inguinal and diaphragmatic hernias. *Inguinal hernias* typically occur in males, at the *inguinal canal*, the site where blood vessels, nerves, and reproductive ducts pass through the abdominal wall to reach the testes. Elevated abdominal pressure can force open the inguinal canal and push a portion of the intestine into the pocket created. *Diaphragmatic hernias* develop when visceral organs, such as a portion of the stomach, are forced into the left pleural cavity. If herniated structures become trapped or twisted, surgery may be required to prevent serious complications, such as intestinal blockage or tissue degeneration due to the interruption of blood flow. **AM** Hernias

---

✓ Damage to the external intercostal muscles would interfere with what important process?

✓ If someone hit you in your rectus abdominis muscle, how would your body position change?

✓ After spending an afternoon carrying heavy boxes from his basement to his attic, Joe complains that the muscles in his back hurt. Which muscle(s) is (are) most likely sore?

Answers start on p. Q-1

The muscles of the vertebral column and the pelvic floor can be reviewed on the **Anatomy CD-ROM**: Muscular System/Axial Dissections/Thorax & Abdomen.

11-4 THE APPENDICULAR MUSCLES

Objectives

- Identify the principal appendicular muscles of the body, together with their origins, insertions, actions, and innervation.

- Compare the major muscle groups of the upper and lower limbs, and relate their differences to their functional roles.

The appendicular musculature positions and stabilizes the pectoral and pelvic girdles and moves the upper and lower limbs. There are two major groups of appendicular muscles: (1) *the muscles of the shoulders and upper limbs* and (2) *the muscles of the pelvic girdle and lower limbs*. The functions and required ranges of motion are very different from one group to another. In addition to increasing the mobility of the arms, the muscular connections between the pectoral girdle and the axial skeleton must act as shock absorbers. For example, while you jog, you can still perform delicate hand movements, because the muscular connections between the axial and appendicular skeletons smooth out the bounces in your stride. In contrast, the pelvic girdle has evolved to transfer weight from the axial to the appendicular skeleton. A muscular connection would reduce the efficiency of the transfer, and the emphasis is on strength rather than versatility. Figure 11–13• provides an introduction to the organization of the appendicular muscles.

◻ MUSCLES OF THE SHOULDERS AND UPPER LIMBS

Muscles associated with the shoulders and upper limbs can be divided into four groups: (1) *muscles that position the pectoral girdle*, (2) *muscles that move the arm*, (3) *muscles that move the forearm and hand*, and (4) *muscles that move the hand and fingers*.

Muscles That Position the Pectoral Girdle

The large, superficial **trapezius** muscles cover the back and portions of the neck, reaching to the base of the skull. These muscles originate along the midline of the neck and back and insert on the clavicles and the scapular spines (Figures 11–13 and 11–14a•, p. 359). The trapezius muscles are innervated by more than one nerve (Table 11–11), and specific regions can be made to contract independently. As a result, their actions are quite varied.

Removing the trapezius muscle reveals the **rhomboid** and **levator scapulae** muscles (Figure 11–14a•). These muscles are attached to the dorsal surfaces of the cervical and thoracic vertebrae. They insert along the vertebral border of each scapula, between the superior and inferior angles. Contraction of a rhomboid muscle adducts (retracts) the scapula on that side. The levator scapulae muscle, as its name implies, elevates the scapula.

On the chest, the **serratus anterior** muscle originates along the anterior surfaces of several ribs (Figures 11–3, pp. 340–341, and 11–14a,b•, p. 359). This fan-shaped muscle inserts along the anterior margin of the vertebral border of the scapula. When the serratus anterior muscle contracts, it abducts (protracts) the scapula and swings the shoulder anteriorly.

Two other deep chest muscles arise along the ventral surfaces of the ribs on either side. The **subclavius** (sub-KLĀ-vē-us; *sub-*, below + *clavius*, clavicle) muscle inserts on the inferior border of

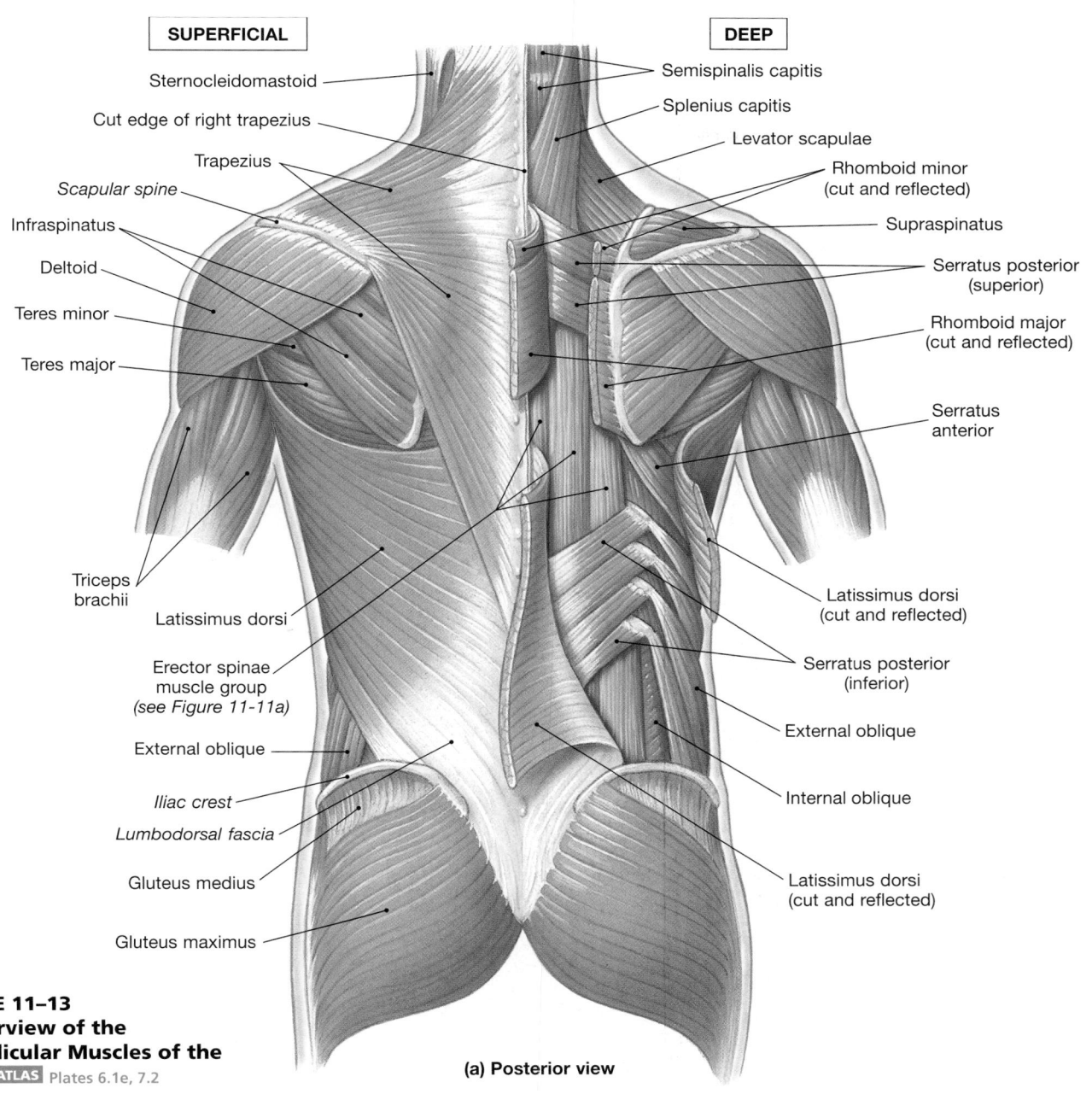

SUPERFICIAL

DEEP

Sternocleidomastoid

Cut edge of right trapezius

Trapezius

Scapular spine

Infraspinatus

Deltoid

Teres minor

Teres major

Triceps brachii

Latissimus dorsi

Erector spinae muscle group
(see Figure 11-11a)

External oblique

Iliac crest

Lumbodorsal fascia

Gluteus medius

Gluteus maximus

Semispinalis capitis

Splenius capitis

Levator scapulae

Rhomboid minor
(cut and reflected)

Supraspinatus

Serratus posterior
(superior)

Rhomboid major
(cut and reflected)

Serratus anterior

Latissimus dorsi
(cut and reflected)

Serratus posterior
(inferior)

External oblique

Internal oblique

Latissimus dorsi
(cut and reflected)

•FIGURE 11–13
An Overview of the Appendicular Muscles of the Trunk. **ATLAS** Plates 6.1e, 7.2

(a) Posterior view

TABLE 11–11 MUSCLES THAT POSITION THE PECTORAL GIRDLE (FIGURES 11–13, 11–14)

Muscle	Origin	Insertion	Action	Innervation
Levator scapulae	Transverse processes of first four cervical vertebrae	Vertebral border of scapula near superior angle	Elevates scapula	Cervical nerves C_3–C_4 and dorsal scapular nerve (C_5)
Pectoralis minor	Anterior-superior surfaces of ribs 3–5	Coracoid process of scapula	Depresses and protracts shoulder; rotates scapula so glenoid cavity moves inferiorly (downward rotation); elevates ribs if scapula is stationary	Medial pectoral nerve (C_8, T_1)
Rhomboid major	Spinous processes of superior thoracic vertebrae	Vertebral border of scapula from spine to inferior angle	Adducts scapula and performs downward rotation	Dorsal scapular nerve (C_5)
Rhomboid minor	Spinous processes of vertebrae C_7–T_1	Vertebral border of scapula near spine	As above	As above

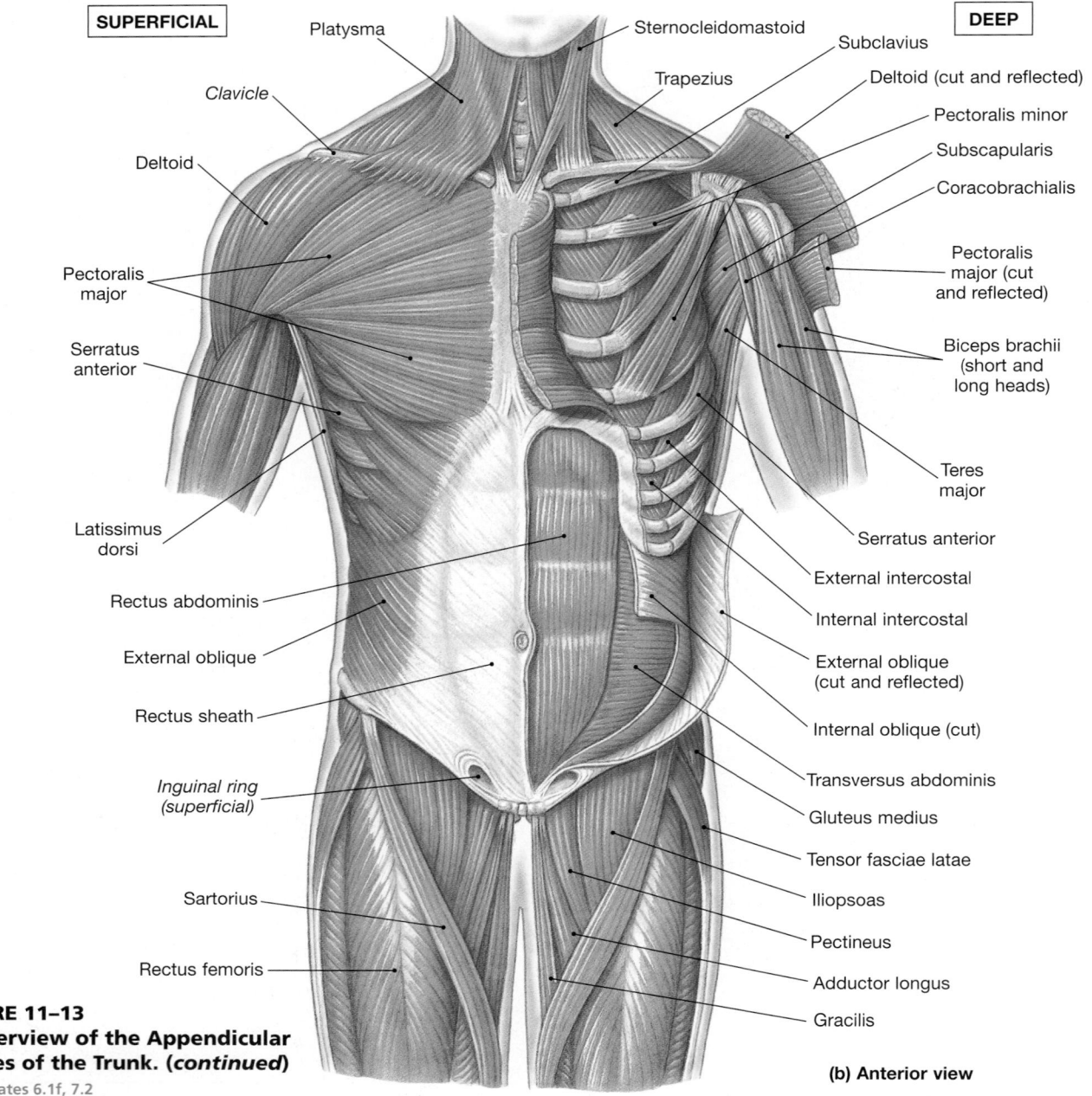

SUPERFICIAL | **DEEP**

Platysma
Sternocleidomastoid
Subclavius
Clavicle
Trapezius
Deltoid (cut and reflected)
Deltoid
Pectoralis minor
Subscapularis
Coracobrachialis
Pectoralis major
Pectoralis major (cut and reflected)
Serratus anterior
Biceps brachii (short and long heads)
Latissimus dorsi
Teres major
Rectus abdominis
Serratus anterior
External oblique
External intercostal
Rectus sheath
Internal intercostal
External oblique (cut and reflected)
Inguinal ring (superficial)
Internal oblique (cut)
Transversus abdominis
Gluteus medius
Tensor fasciae latae
Sartorius
Iliopsoas
Pectineus
Rectus femoris
Adductor longus
Gracilis

•FIGURE 11–13
An Overview of the Appendicular Muscles of the Trunk. (continued)
ATLAS Plates 6.1f, 7.2

(b) Anterior view

TABLE 11–11 MUSCLES THAT POSITION THE PECTORAL GIRDLE (CONTINUED)				
Muscle	**Origin**	**Insertion**	**Action**	**Innervation**
Serratus anterior	Anterior and superior margins of ribs 1–8 or 1–9	Anterior surface of vertebral border of scapula	Protracts shoulder; rotates scapula so glenoid cavity moves superiorly (upward rotation)	Long thoracic nerve (C_5–C_7)
Subclavius	First rib	Clavicle (inferior border)	Depresses and protracts shoulder	Nerve to subclavius (C_5–C_6)
Trapezius	Occipital bone, ligamentum nuchae, and spinous processes of thoracic vertebrae	Clavicle and scapula (acromion and scapular spine)	Depends on active region and state of other muscles; may (1) elevate, retract, depress, or rotate scapula upward (2) elevate clavicle, or (3) extend neck	Accessory nerve (XI) and cervical spinal nerves (C_3–C_4)

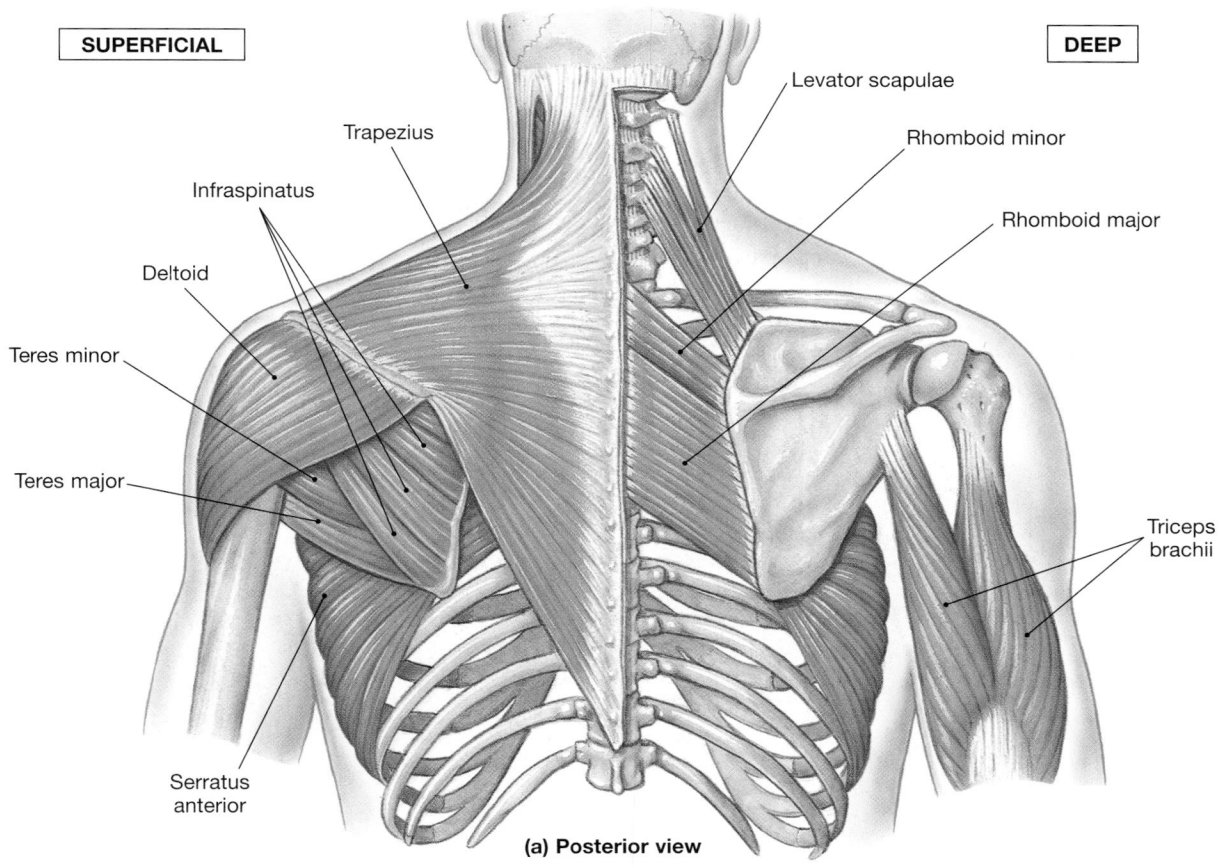

SUPERFICIAL

DEEP

Trapezius

Levator scapulae

Infraspinatus

Rhomboid minor

Deltoid

Rhomboid major

Teres minor

Teres major

Triceps
brachii

Serratus
anterior

(a) Posterior view

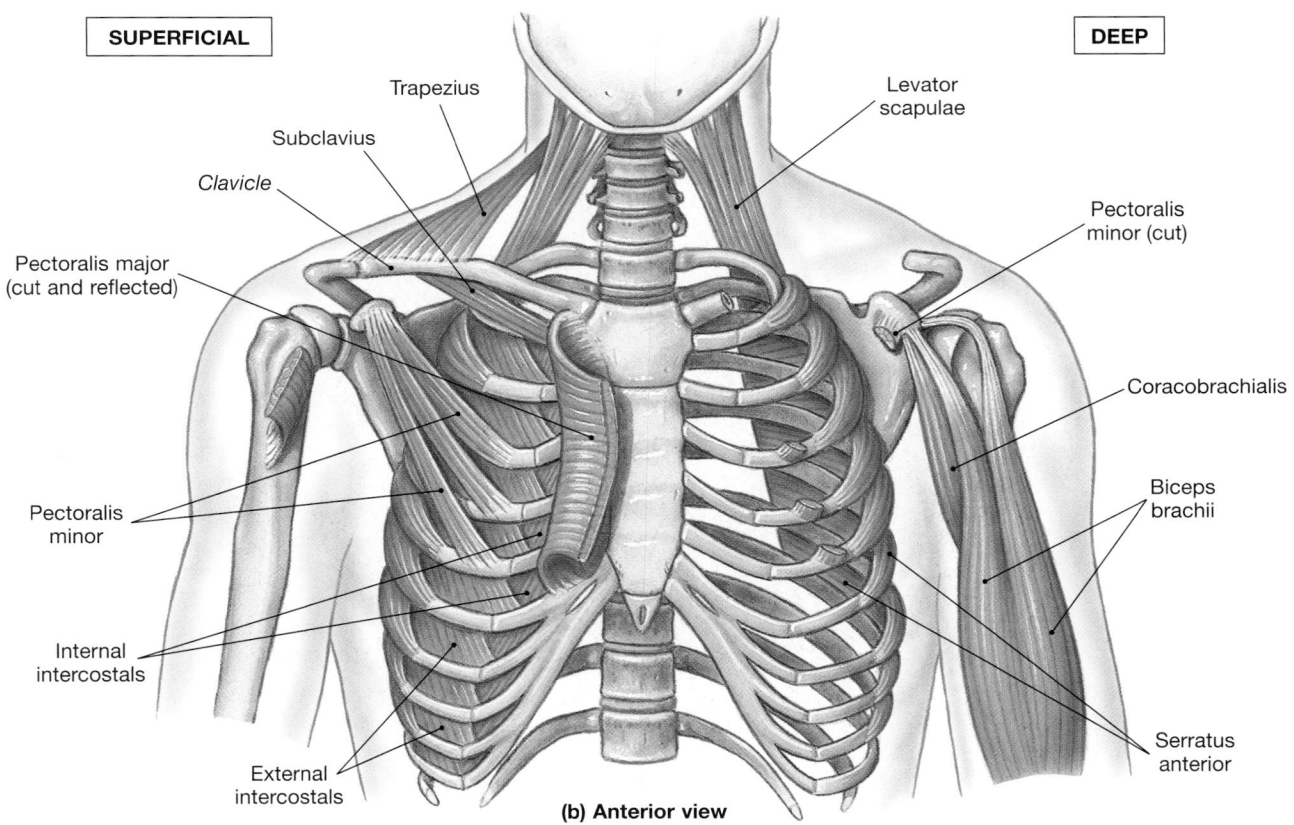

SUPERFICIAL

DEEP

Trapezius

Levator
scapulae

Subclavius

Clavicle

Pectoralis
minor (cut)

Pectoralis major
(cut and reflected)

Coracobrachialis

Pectoralis
minor

Biceps
brachii

Internal
intercostals

Serratus
anterior

External
intercostals

(b) Anterior view

●**FIGURE 11–14**
Muscles That Position the Pectoral Girdle. *See also Figures 8–2, 8–3, 8–4, and 9–9.* [ATLAS] Plates 6.1e,f, 7.1,
7.2, 7.3a,b

the clavicle (Figure 11–14b●). When it contracts, it depresses and protracts the scapular end of the clavicle. Because ligaments connect this end to the shoulder joint and scapula, those structures move as well. The **pectoralis** (pek-to-RA-lis) **minor** muscle attaches to the coracoid process of the scapula. The contraction of this muscle generally complements that of the subclavius muscle.

Muscles That Move the Arm

The muscles that move the arm (Figures 11–13 to 11–15●) are easiest to remember when they are grouped by their actions at the shoulder joint (Table 11–12). The **deltoid** muscle is the major abductor, but the **supraspinatus** (soo-pra-spī-NA-tus)

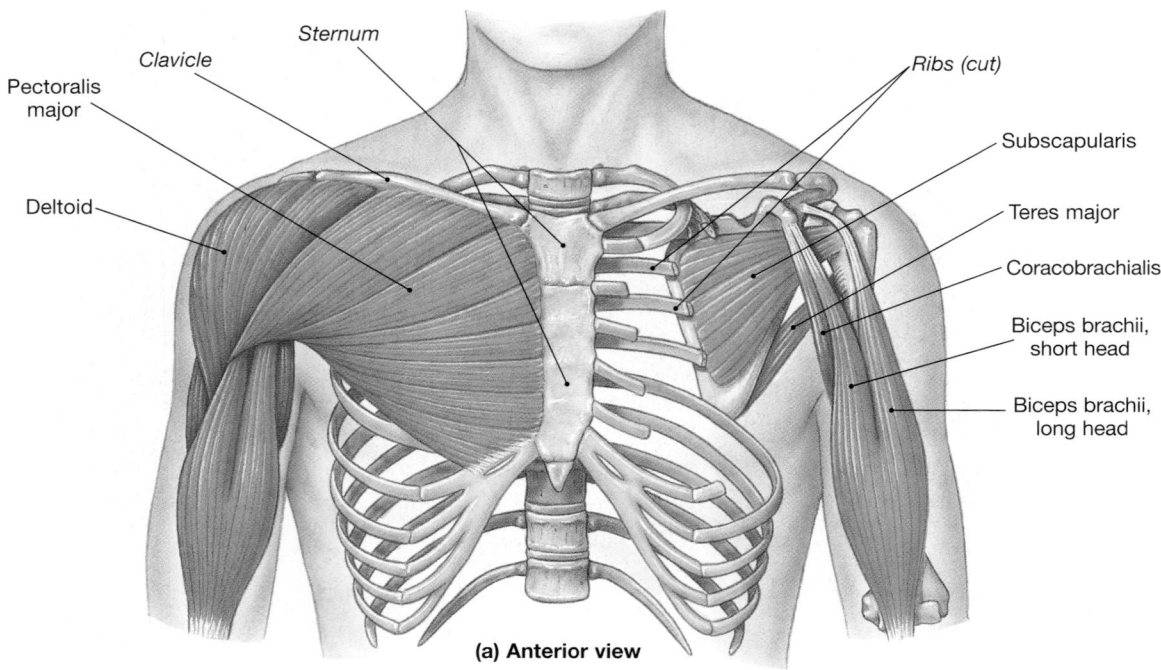

(a) Anterior view

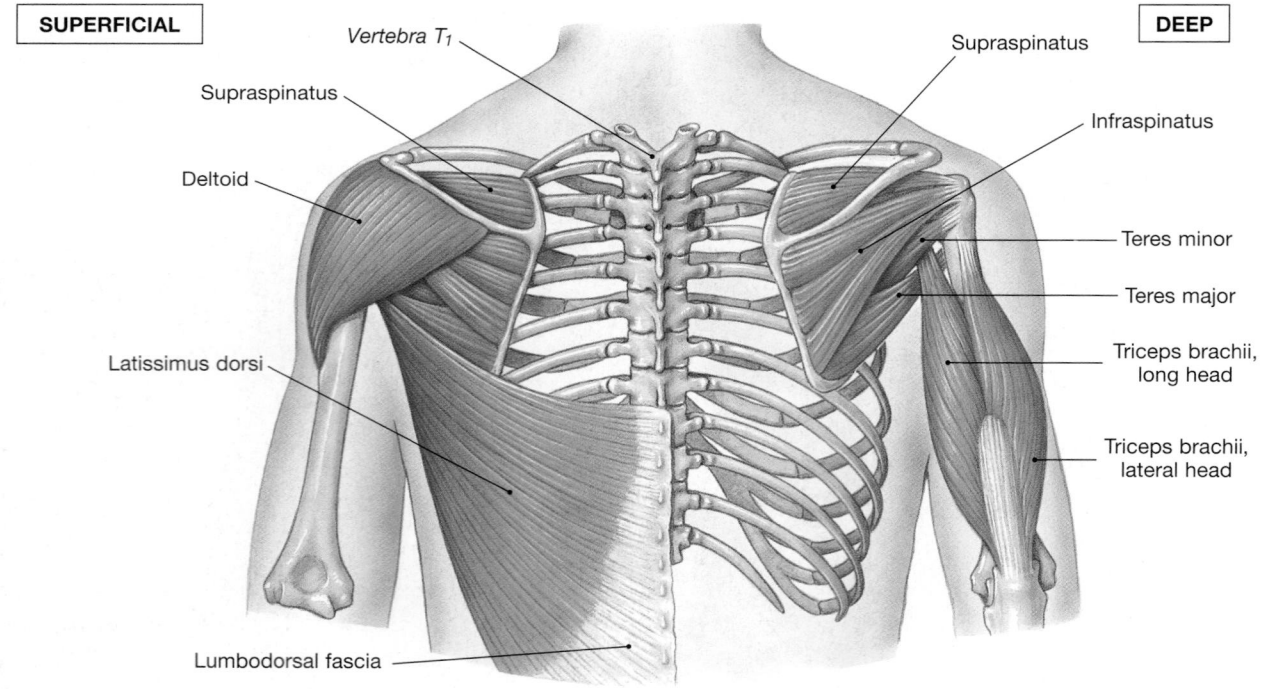

(b) Posterior view

●**FIGURE 11–15**
Muscles That Move the Arm. *See also Figures 7–22, 8–3, and 9–9.* **ATLAS** Plates 6.1e,f, 7.1, 7.2, 7.3a,b

TABLE 11–12 MUSCLES THAT MOVE THE ARM (FIGURES 11–13 TO 11–15)

Muscle	Origin	Insertion	Action	Innervation
Coracobrachialis	Coracoid process	Medial margin of shaft of humerus	Adduction and flexion at shoulder	Musculocutaneous nerve (C_5–C_7)
Deltoid	Clavicle and scapula (acromion and adjacent scapular spine)	Deltoid tuberosity of humerus	*Whole muscle*: abduction at shoulder; *anterior part*: flexion and medial rotation; *posterior part*: extension and lateral rotation	Axillary nerve (C_5–C_6)
Supraspinatus	Supraspinous fossa of scapula	Greater tubercle of humerus	Abduction at the shoulder	Suprascapular nerve (C_5)
Infraspinatus	Infraspinous fossa of scapula	Greater tubercle of humerus	Lateral rotation at shoulder	Suprascapular nerve (C_5–C_6)
Subscapularis	Subscapular fossa of scapula	Lesser tubercle of humerus	Medial rotation at shoulder	Subscapular nerves (C_5–C_6)
Teres major	Inferior angle of scapula	Passes medially to reach the medial lip of intertubercular groove of humerus	Extension, adduction, and medial rotation at shoulder	Lower subscapular nerve (C_5–C_6)
Teres minor	Lateral border of scapula	Passes laterally to reach the greater tubercle of humerus	Lateral rotation at shoulder	Axillary nerve (C_5)
Triceps brachii (long head)	See Table 11–13			
Latissimus dorsi	Spinous processes of inferior thoracic and all lumbar vertebrae, ribs 8–12, and lumbodorsal fascia	Floor of intertubercular groove of the humerus	Extension, adduction, and medial rotation at shoulder	Thoracodorsal nerve (C_6–C_8)
Pectoralis major	Cartilages of ribs 2–6, body of sternum, and inferior, medial portion of clavicle	Crest of greater tubercle and lateral lip of intertubercular groove of humerus	Flexion, adduction, and medial rotation at shoulder	Pectoral nerves (C_5–T_1)

muscle assists at the start of this movement. The **subscapularis** and **teres major** muscles produce medial rotation at the shoulder, whereas the **infraspinatus** and the **teres minor** muscles produce lateral rotation. All these muscles originate on the scapula. The small **coracobrachialis** (KOR-uh-kō-brā-kē-A-lis) muscle is the only muscle attached to the scapula that produces flexion and adduction at the shoulder.

The **pectoralis major** muscle extends between the anterior portion of the chest and the crest of the greater tubercle of the humerus. The **latissimus dorsi** (la-TIS-i-mus DOR-sē) muscle extends between the thoracic vertebrae at the posterior midline and the intertubercular groove of the humerus (Figure 11–15b●). The pectoralis major muscle produces flexion at the shoulder joint, and the latissimus dorsi muscle produces extension. These two muscles can also work together to produce adduction and medial rotation of the humerus at the shoulder.

Collectively, the supraspinatus, infraspinatus, subscapularis, and teres minor muscles and their associated tendons form the **rotator cuff**. Sports that involve throwing a ball, such as baseball or football, place considerable strain on the rotator cuff, and rotator cuff injuries are relatively common.

AM Sports Injuries

Muscles That Move the Forearm and Hand

Although most of the muscles that insert on the forearm and hand originate on the humerus, the biceps brachii and triceps brachii muscles are noteworthy exceptions. The **biceps brachii** muscle and the *long head* of the **triceps brachii** muscle originate on the scapula and insert on the bones of the forearm (Figure 11–16●). The triceps brachii muscle inserts on the olecranon. Contraction of the triceps brachii muscle extends the elbow, as when you do push-ups. The biceps brachii muscle inserts on the radial tuberosity, a roughened area on the anterior surface of the radius. ∞ p. 248 Contraction of the biceps brachii muscle flexes the elbow and supinates the forearm. With the forearm pronated (palm facing back), the biceps brachii muscle cannot function effectively. As a result, you are strongest when you flex your elbow with a supinated forearm; the biceps brachii muscle then makes a prominent bulge.

The biceps brachii muscle plays an important role in the stabilization of the shoulder joint. The short head originates on the coracoid process and provides support to the posterior surface of the capsule. The long head originates at the supraglenoid tubercle, inside the shoulder joint. ∞ p. 246 After crossing the head of the humerus, it passes along the intertubercular groove. In this position, the tendon helps to hold the head of the humerus within the glenoid cavity while arm movements are under way.

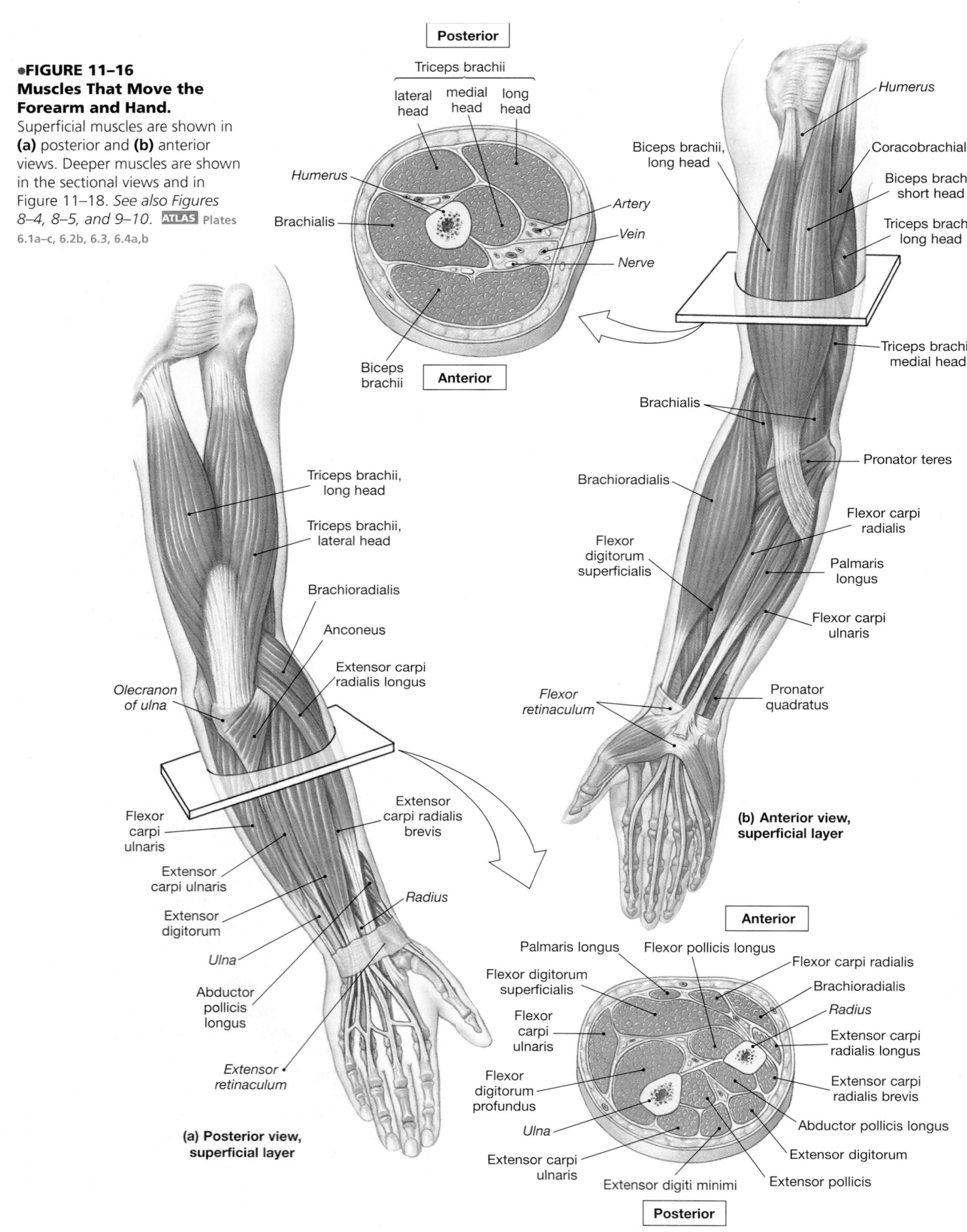

●FIGURE 11–16
Muscles That Move the Forearm and Hand.
Superficial muscles are shown in **(a)** posterior and **(b)** anterior views. Deeper muscles are shown in the sectional views and in Figure 11–18. *See also Figures 8–4, 8–5, and 9–10.* **ATLAS** Plates 6.1a–c, 6.2b, 6.3, 6.4a,b

Posterior

Triceps brachii

lateral head   medial head   long head

Humerus

Brachialis

Artery

Vein

Nerve

Biceps brachii

Anterior

Humerus

Biceps brachii, long head

Coracobrachialis

Biceps brachii, short head

Triceps brachii, long head

Triceps brachii, medial head

Brachialis

Pronator teres

Brachioradialis

Flexor carpi radialis

Flexor digitorum superficialis

Palmaris longus

Flexor carpi ulnaris

Flexor retinaculum

Pronator quadratus

(b) Anterior view, superficial layer

Triceps brachii, long head

Triceps brachii, lateral head

Brachioradialis

Anconeus

Extensor carpi radialis longus

Olecranon of ulna

Flexor carpi ulnaris

Extensor carpi ulnaris

Extensor digitorum

Ulna

Abductor pollicis longus

Extensor retinaculum

Extensor carpi radialis brevis

Radius

(a) Posterior view, superficial layer

Anterior

Palmaris longus

Flexor pollicis longus

Flexor carpi radialis

Flexor digitorum superficialis

Brachioradialis

Flexor carpi ulnaris

Radius

Extensor carpi radialis longus

Flexor digitorum profundus

Extensor carpi radialis brevis

Ulna

Abductor pollicis longus

Extensor carpi ulnaris

Extensor digiti minimi

Extensor pollicis

Extensor digitorum

Posterior

More muscles are shown in Figure 11–16● and listed in Table 11–13. The **brachialis** and **brachioradialis** (brā-kē-ō-rā-dē-A-lis) muscles flex the elbow and are opposed by the **anconeus** muscle and the triceps brachii muscle, respectively.

The **flexor carpi ulnaris**, **flexor carpi radialis**, and **palmaris longus** muscles are superficial muscles that work together to produce flexion of the wrist. The flexor carpi radialis muscle flexes and *ab*ducts, and the flexor carpi ulnaris muscle flexes and

TABLE 11–13 MUSCLES THAT MOVE THE FOREARM AND HAND (FIGURE 11–16)

Muscle	Origin	Insertion	Action	Innervation
ACTION AT THE ELBOW				
Flexors				
Biceps brachii	*Short head* from the coracoid process; *long head* from the supraglenoid tubercle (both on the scapula)	Tuberosity of radius	Flexion at elbow and shoulder; supination	Musculocutaneous nerve (C_5–C_6)
Brachialis	Anterior, distal surface of humerus	Tuberosity of ulna	Flexion at elbow	As above and radial nerve (C_7–C_8)
Brachioradialis	Ridge superior to the lateral epicondyle of humerus	Lateral aspect of styloid process of radius	As above	Radial nerve (C_5–C_6)
Extensors				
Anconeus	Posterior, inferior surface of lateral epicondyle of humerus	Lateral margin of olecranon on ulna	Extension at elbow	Radial nerve (C_7–C_8)
Triceps brachii lateral head	Superior, lateral margin of humerus	Olecranon of ulna	As above	Radial nerve (C_6–C_8)
long head	Infraglenoid tubercle of scapula	As above	As above, plus extension and adduction at the shoulder	As above
medial head	Posterior surface of humerus inferior to radial groove	As above	Extension at elbow	As above
PRONATORS/SUPINATORS				
Pronator quadratus	Anterior and medial surfaces of distal portion of ulna	Anterolateral surface of distal portion of radius	Pronation	Median nerve (C_8–T_1)
Pronator teres	Medial epicondyle of humerus and coronoid process of ulna	Mid-lateral surface of radius	As above	Median nerve (C_6–C_7)
Supinator	Lateral epicondyle of humerus, annular ligament, and ridge near radial notch of ulna	Anterolateral surface of radius distal to the radial tuberosity	Supination	Deep radial nerve (C_6–C_8)
ACTION AT THE HAND				
Flexors				
Flexor carpi radialis	Medial epicondyle of humerus	Bases of second and third metacarpal bones	Flexion and abduction at wrist	Median nerve (C_6–C_7)
Flexor carpi ulnaris	Medial epicondyle of humerus; adjacent medial surface of olecranon and anteromedial portion of ulna	Pisiform bone, hamate bone, and base of fifth metacarpal bone	Flexion and adduction at wrist	Ulnar nerve (C_8–T_1)
Palmaris longus	Medial epicondyle of humerus	Palmar aponeurosis and flexor retinaculum	Flexion at wrist	Median nerve (C_6–C_7)
Extensors				
Extensor carpi radialis longus	Lateral supracondylar ridge of humerus	Base of second metacarpal bone	Extension and abduction at wrist	Radial nerve (C_6–C_7)
Extensor carpi radialis brevis	Lateral epicondyle of humerus	Base of third metacarpal bone	As above	As above
Extensor carpi ulnaris	Lateral epicondyle of humerus; adjacent dorsal surface of ulna	Base of fifth metacarpal bone	Extension and adduction at wrist	Deep radial nerve (C_6–C_8)

●FIGURE 11–17
Muscles That Move the Hand and Fingers.
Middle and deep muscle layers of the right forearm; for superficial muscles, see Figure 11–17. *See also Figure 8–5.*

*ad*ducts. *Pitcher's arm* is an inflammation at the origins of the flexor carpi muscles at the medial epicondyle of the humerus. This condition results from forcibly flexing the wrist just before releasing a baseball.

The **extensor carpi radialis** muscles and the **extensor carpi ulnaris** muscle have a similar relationship to that between the flexor carpi muscles. That is, the extensor carpi radialis muscles produce extension and *ab*duction, whereas the extensor carpi ulnaris muscle produces extension and *ad*duction.

The **pronator teres** and **supinator** muscles originate on both the humerus and ulna. These muscles rotate the radius without either flexing or extending the elbow. The **pronator quadratus** muscle originates on the ulna and assists the pronator teres muscle in opposing the actions of the supinator or biceps brachii muscles. The muscles involved in pronation and supination are shown in Figure 11–17●. During pronation, the tendon of the biceps brachii muscle rotates with the radius. As a result, this muscle cannot assist in flexion of the elbow when the forearm is pronated.

As you study the muscles listed in Table 11–13, notice that, in general, the extensor muscles lie along the posterior and lateral surfaces of the arm, whereas the flexors are on the anterior and medial surfaces.

Muscles That Move the Hand and Fingers

Several superficial and deep muscles of the forearm flex and extend the finger joints (Figure 11–17● and Table 11–14). These muscles stop before reaching the wrist, and only their tendons cross the articulation. The muscles are relatively large, and keeping them clear of the joints ensures maximum mobility at both the wrist and hand. The tendons that cross the dorsal and ventral surfaces of the wrist pass through **synovial tendon sheaths**, elongated bursae that reduce friction. ∞ p. 267

The muscles of the forearm provide strength and crude control of the hand and fingers. These muscles are known as the *extrinsic muscles of the hand*. Fine control of the hand involves small *intrinsic muscles*, which originate

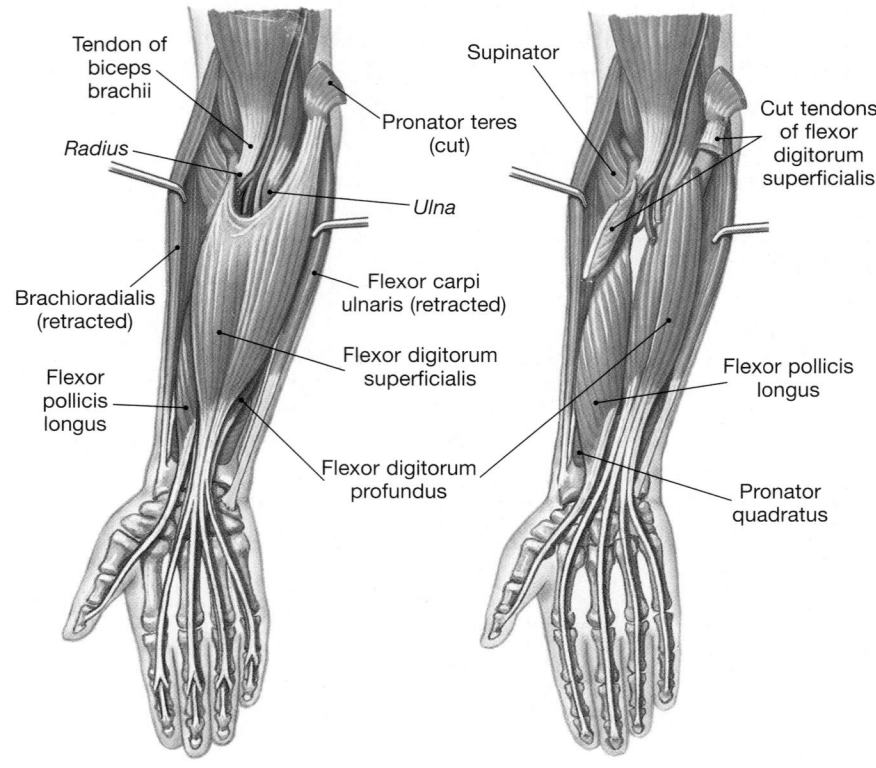

(a) Anterior view, middle layer　　**(b) Anterior view, deepest layer**

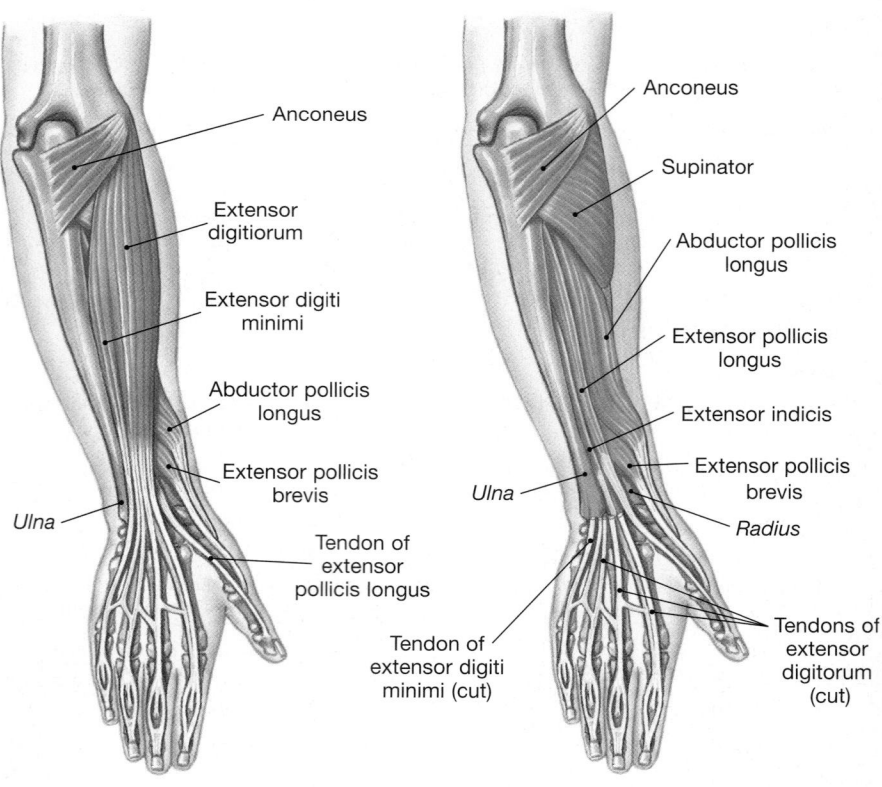

(c) Posterior view, middle layer　　**(d) Posterior view, deepest layer**

TABLE 11–14 MUSCLES THAT MOVE THE HAND AND FINGERS (FIGURE 11–17)

Muscle	Origin	Insertion	Action	Innervation
Abductor pollicis longus	Proximal dorsal surfaces of ulna and radius	Lateral margin of first metacarpal bone	Abduction at joints of thumb and wrist	Deep radial nerve (C_6–C_7)
Extensor digitorum	Lateral epicondyle of humerus	Posterior surfaces of the phalanges, fingers 2–5	Extension at finger joints and wrist	Deep radial nerve (C_6–C_8)
Extensor pollicis brevis	Shaft of radius distal to origin of adductor pollicis longus	Base of proximal phalanx of thumb	Extension at joints of thumb; abduction at wrist	Deep radial nerve (C_6–C_7)
Extensor pollicis longus	Posterior and lateral surfaces of ulna and interosseous membrane	Base of distal phalanx of thumb	As above	Deep radial nerve (C_6–C_8)
Extensor indicis	Posterior surface of ulna and interosseous membrane	Posterior surface of phalanges of index finger (2), with tendon of extensor digitorum	Extension and adduction at joints of index finger	As above
Extensor digiti minimi	Via extensor tendon to lateral epicondyle of humerus and from intermuscular septa	Posterior surface of proximal phalanx of little finger (5)	Extension at joints of little finger	As above
Flexor digitorum superficialis	Medical epicondyle of humerus; adjacent anterior surfaces of ulna and radius	Midlateral surfaces of middle phalanges of fingers 2–5	Flexion at proximal interphalangeal, metacarpophalangeal, and wrist joints	Median nerve (C_7–T_1)
Flexor digitorum profundus	Medial and posterior surfaces of ulna, medial surface of coronoid process, and interosseus membrane	Bases of distal phalanges of fingers 2–5	Flexion at distal interphalangeal joints and, to a lesser degree, proximal interphalangeal joints and wrist	Palmar interosseous nerve, from median nerve, and ulnar nerve (C_8–T_1)
Flexor pollicis longus	Anterior shaft of radius, interosseous membrane	Base of distal phalanx of thumb	Flexion at joints of thumb	Median nerve (C_8–T_1)

on the carpal and metacarpal bones. No muscles originate on the phalanges, and only tendons extend across the distal joints of the fingers. The intrinsic muscles of the hand are detailed in Figure 11–18● and Table 11–15 (p. 367).

The fascia of the forearm thickens on the posterior surface of the wrist, forming the **extensor retinaculum** (ret-i-NAK-ū-lum), a wide band of connective tissue. The extensor retinaculum holds the tendons of the extensor muscles in place. On the anterior surface, the fascia also thickens to form another wide band of connective tissue, the **flexor retinaculum**, which stabilizes the tendons of the flexor muscles. Inflammation of the retinacula and synovial tendon sheaths can restrict movement and irritate the distal portions of the *median nerve*, a mixed (sensory and motor) nerve that innervates the hand. This condition, known as *carpal tunnel syndrome*, causes chronic pain.

Working at a computer all day may seem innocuous; however, it does carry some health risks. The continual fine movements of the fingers when typing can cause inflammation of the flexor tendon sheaths. *Tenosynovitis* is the term used to describe inflammation of a tendon sheath. **Carpal tunnel syndrome** results from tenosynovitis of the synovial tendon sheath surrounding the flexor tendons of the palm. This common condition often strikes persons engaged in repetitive hand movements, such as typing, working at a computer keyboard, or

playing the piano. The inflammation leads to compression of the median nerve. Symptoms include pain (especially on palmar flexion), a tingling sensation or numbness on the palm, and weakness in the *abductor pollicis brevis*. Treatment involves the administration of anti-inflammatory drugs such as aspirin, the injection of anti-inflammatory agents such as *glucocorticoids* (steroid hormones produced by the adrenal cortex), and the use of a splint to prevent wrist flexion and to stabilize the region.

Carpal tunnel syndrome is an example of a *cumulative trauma disorder*, or *overuse syndrome*. These disorders are caused by repetitive movements of the arms, hands, and fingers. Such musculoskeletal problems now account for over 50 percent of all work-related injuries in the United States.

✔ Which muscle are you using when you shrug your shoulders?

✔ Baseball pitchers sometimes suffer from rotator cuff injuries. Which muscles are involved in this type of injury?

✔ Which two movements would injury to the flexor carpi ulnaris muscle impair?

Answers start on p. Q-1

 The muscles of the shoulders and upper limbs can be reviewed on the **Anatomy CD-ROM**: Muscular System/Appendicular Dissections/Shoulder & Upper Arm.

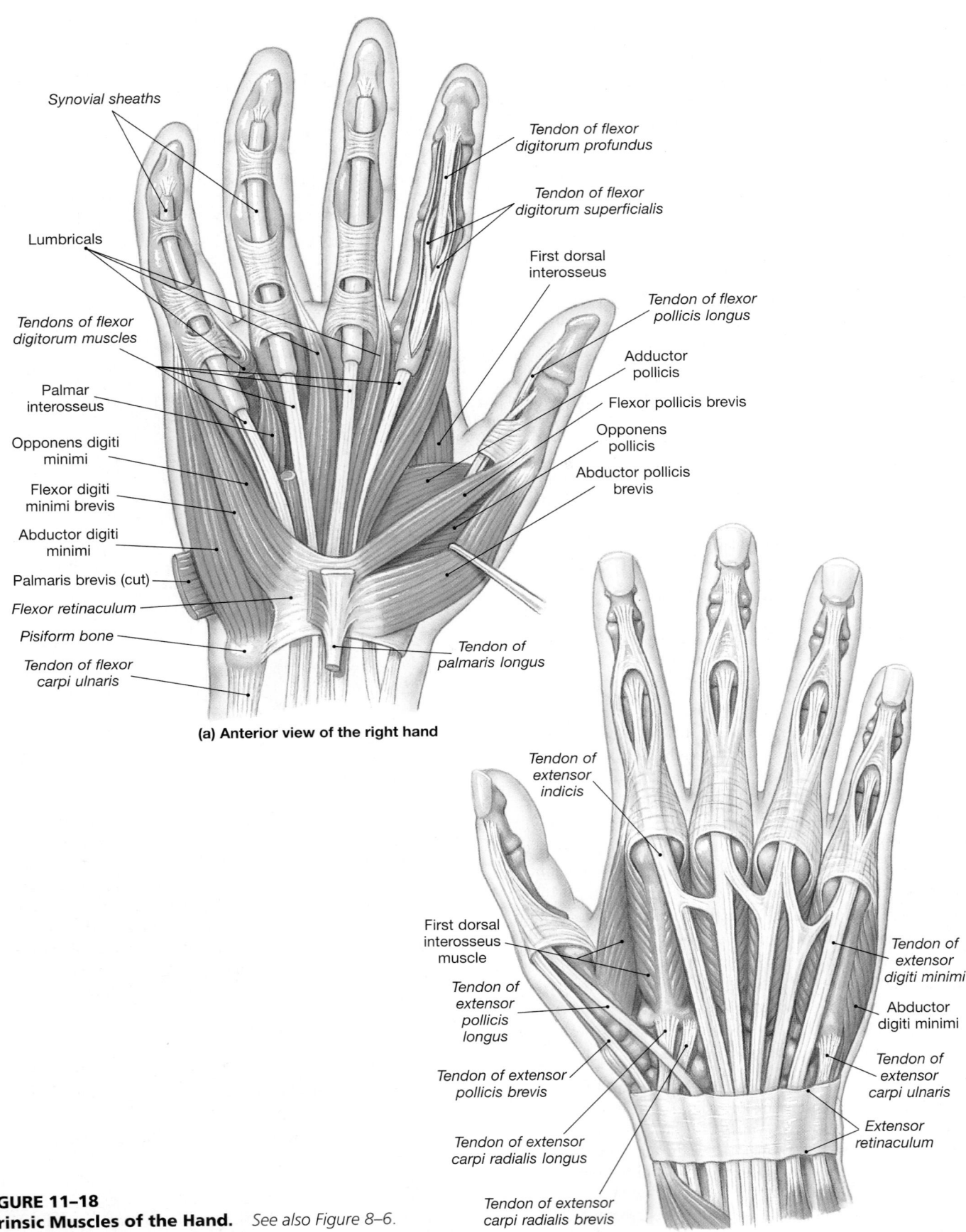

Synovial sheaths

Lumbricals

Tendons of flexor
digitorum muscles

Palmar
interosseus

Opponens digiti
minimi

Flexor digiti
minimi brevis

Abductor digiti
minimi

Palmaris brevis (cut)

Flexor retinaculum

Pisiform bone

*Tendon of flexor
carpi ulnaris*

Tendon of flexor
digitorum profundus

Tendon of flexor
digitorum superficialis

First dorsal
interosseus

*Tendon of flexor
pollicis longus*

Adductor
pollicis

Flexor pollicis brevis

Opponens
pollicis

Abductor pollicis
brevis

*Tendon of
palmaris longus*

(a) Anterior view of the right hand

Tendon of
extensor
indicis

First dorsal
interosseus
muscle

*Tendon of
extensor
pollicis
longus*

*Tendon of extensor
pollicis brevis*

*Tendon of extensor
carpi radialis longus*

*Tendon of extensor
carpi radialis brevis*

*Tendon of
extensor
digiti minimi*

Abductor
digiti minimi

*Tendon of
extensor
carpi ulnaris*

*Extensor
retinaculum*

(b) Posterior view of the right hand

●**FIGURE 11–18**
Intrinsic Muscles of the Hand. *See also Figure 8–6.*
ATLAS Plates 6.6a,c

TABLE 11–15 INTRINSIC MUSCLES OF THE HAND (FIGURE 11–18)

Muscle	Origin	Insertion	Action	Innervation
Adductor pollicis	Metacarpal and carpal bones	Proximal phalanx of thumb	Adduction of thumb	Ulnar nerve, deep branch (C_8–T_1)
Opponens pollicis	Trapezium and flexor retinaculum	First metacarpal bone	Opposition of thumb	Median nerve (C_6–C_7)
Palmaris brevis	Palmar aponeurosis	Skin of medial border of hand	Moves skin on medial border toward midline of palm	Ulnar nerve, superficial branch (C_8)
Abductor digiti minimi	Pisiform bone	Proximal phalanx of little finger	Abduction of little finger and flexion at its metacarpophalangeal joint	Ulnar nerve, deep branch (C_8–T_1)
Abductor pollicis brevis	Transverse carpal ligament, scaphoid bone, and trapezium	Radial side of base of proximal phalanx of thumb	Abduction of thumb	Median nerve (C_6–C_7)
Flexor pollicis brevis	Flexor retinaculum, trapezium, capitate bone, and ulnar side of first metacarpal bone	Radial and ulnar sides of proximal phalanx of thumb	Flexion and adduction of thumb	Branches of median and ulnar nerves
Flexor digiti minimi brevis	Hamate bone	Proximal phalanx of little finger	Flexion at joints of little finger	Ulnar nerve, deep branch (C_8–T_1)
Opponens digiti minimi	As above	Fifth metacarpal bone	Opposition of fifth metacarpal bone	As above
Lumbrical (4)	Tendons of flexor digitorum profundus	Tendons of extensor digitorum to digits 2–5	Flexion at metacarpophalangeal joints 2–5; extension at proximal and distal interphalangeal joints, digits 2–5	No. 1 and no. 2 by median nerve; no. 3 and no. 4 by ulnar nerve, deep branch
Dorsal interosseus (4)	Each originates from opposing faces of two metacarpal bones (I and II, II and III, III and IV, IV and V)	Bases of proximal phalanges of fingers 2–4	Abduction at metacarpophalangeal joints of fingers 2 and 4; flexion at metacarpophalangeal joints; extension at interphalangeal joints	Ulnar nerve, deep branch (C_8–T_1)
Palmar interosseus* (3–4)	Sides of metacarpal bones II, IV, and V	Bases of proximal phalanges of fingers 2, 4, and 5	Adduction at metacarpophalangeal joints of fingers 2, 4, and 5; flexion at metacarpophalangeal joints; extension at interphalangeal joints	As above

*The deep, medial portion of the flexor pollicis brevis originating on the first metacarpal bone is sometimes called the *first palmar interosseus muscle*; it inserts on the ulnar side of the phalanx and is innervated by the ulnar nerve.

◻ MUSCLES OF THE PELVIS AND LOWER LIMBS

The pelvic girdle is tightly bound to the axial skeleton, permitting little relative movement. In our discussion of the axial musculature, we therefore encountered few muscles that can influence the position of the pelvis. The muscles that position the lower limbs can be divided into three functional groups: (1) *muscles that move the thigh*, (2) *muscles that move the leg*, and (3) *muscles that move the foot and toes*.

Muscles That Move the Thigh

Table 11–16 lists the muscles that move the thigh. **Gluteal muscles** cover the lateral surfaces of the ilia (Figures 11–13, pp. 357, 358, and 11–19a,b,c●). The **gluteus maximus** muscle is the largest and most posterior of the gluteal muscles. Its origin includes parts of the ilium; the sacrum, coccyx, and associated ligaments; and the lumbodorsal fascia (Figure 11–13●). Acting alone, this massive muscle produces extension and lateral rotation at the hip joint. The gluteus maximus shares an insertion with the **tensor fasciae latae** (FASH-ē-ē LĀ-tā) muscle, which originates on the iliac crest and the anterior superior iliac spine. Together, these muscles pull on the **iliotibial** (il-ē-ō-TIB-ē-ul) **tract**, a band of collagen fibers that extends along the lateral surface of the thigh and inserts on the tibia. This tract provides a lateral brace for the knee that becomes particularly important when you balance on one foot.

The **gluteus medius** and **gluteus minimus** muscles (Figure 11–19b,c●) originate anterior to the origin of the gluteus maximus muscle and insert on the greater trochanter of the

TABLE 11–16 MUSCLES THAT MOVE THE THIGH (FIGURE 11–19)

Group and Muscle(s)	Origin	Insertion	Action	Innervation
GLUTEAL GROUP				
Gluteus maximus	Iliac crest, posterior gluteal line, and lateral surface of ilium; sacrum, coccyx, and lumbodorsal fascia	Iliotibial tract and gluteal tuberosity of femur	Extension and lateral rotation at hip	Inferior gluteal nerve (L_5–S_2)
Gluteus medius	Anterior iliac crest of ilium, lateral surface between posterior and anterior gluteal lines	Greater trochanter of femur	Abduction and medial rotation at hip	Superior gluteal nerve (L_4–S_1)
Gluteus minimus	Lateral surface of ilium between inferior and anterior gluteal lines	As above	As above	As above
Tensor fasciae latae	Iliac crest and lateral surface of anterior superior iliac spine	Iliotibial tract	Flexion and medial rotation at hip; tenses fascia lata, which laterally supports the knee	As above
LATERAL ROTATOR GROUP				
Obturators (externus and internus)	Lateral and medial margins of obturator foramen	Trochanteric fossa of femur (externus); medial surface of greater trochanter (internus)	Lateral rotation at hip	Obturator nerve (externus: L_3–L_4) and special nerve from sacral plexus (internus: L_5–S_2)
Piriformis	Anterolateral surface of sacrum	Greater trochanter of femur	Lateral rotation and abduction at hip	Branches of sacral nerves (S_1–S_2)
Gemelli (superior and inferior)	Ischial spine and tuberosity	Medial surface of greater trochanter with tendon of obturator internus	Lateral rotation at hip	Nerves to obturator internus and quadratus femoris
Quadratus femoris	Lateral border of ischial tuberosity	Intertrochanteric crest of femur	As above	Special nerve from sacral plexus (L_4–S_1)
ADDUCTOR GROUP				
Adductor brevis	Inferior ramus of pubis	Linea aspera of femur	Adduction, flexion, and medial rotation at hip	Obturator nerve (L_3–L_4)
Adductor longus	Inferior ramus of pubis anterior to adductor brevis	As above	As above	As above
Adductor magnus	Inferior ramus of pubis posterior to adductor brevis and ischial tuberosity	Linea aspera and adductor tubercle of femur	Adduction at hip; superior part produces flexion and medial rotation; inferior part produces extension and lateral rotation	Obturator and sciatic nerves
Pectineus	Superior ramus of pubis	Pectineal line inferior to lesser trochanter of femur	Flexion, medial rotation, and adduction at hip	Femoral nerve (L_2–L_4)
Gracilis	Inferior ramus of pubis	Medial surface of tibia inferior to medial condyle	Flexion at knee; adduction and medial rotation at hip	Obturator nerve (L_3–L_4)
ILIOPSOAS GROUP				
Iliacus	Iliac fossa of ilium	Femur distal to lesser trochanter; tendon fused with that of psoas major	Flexion at hip	Femoral nerve (L_2–L_3)
Psoas major	Anterior surfaces and transverse processes of vertebrae T_{12}–L_5	Lesser trochanter in company with iliacus	Flexion at hip or lumbar intervertebral joints	Branches of the lumbar plexus (L_2–L_3)

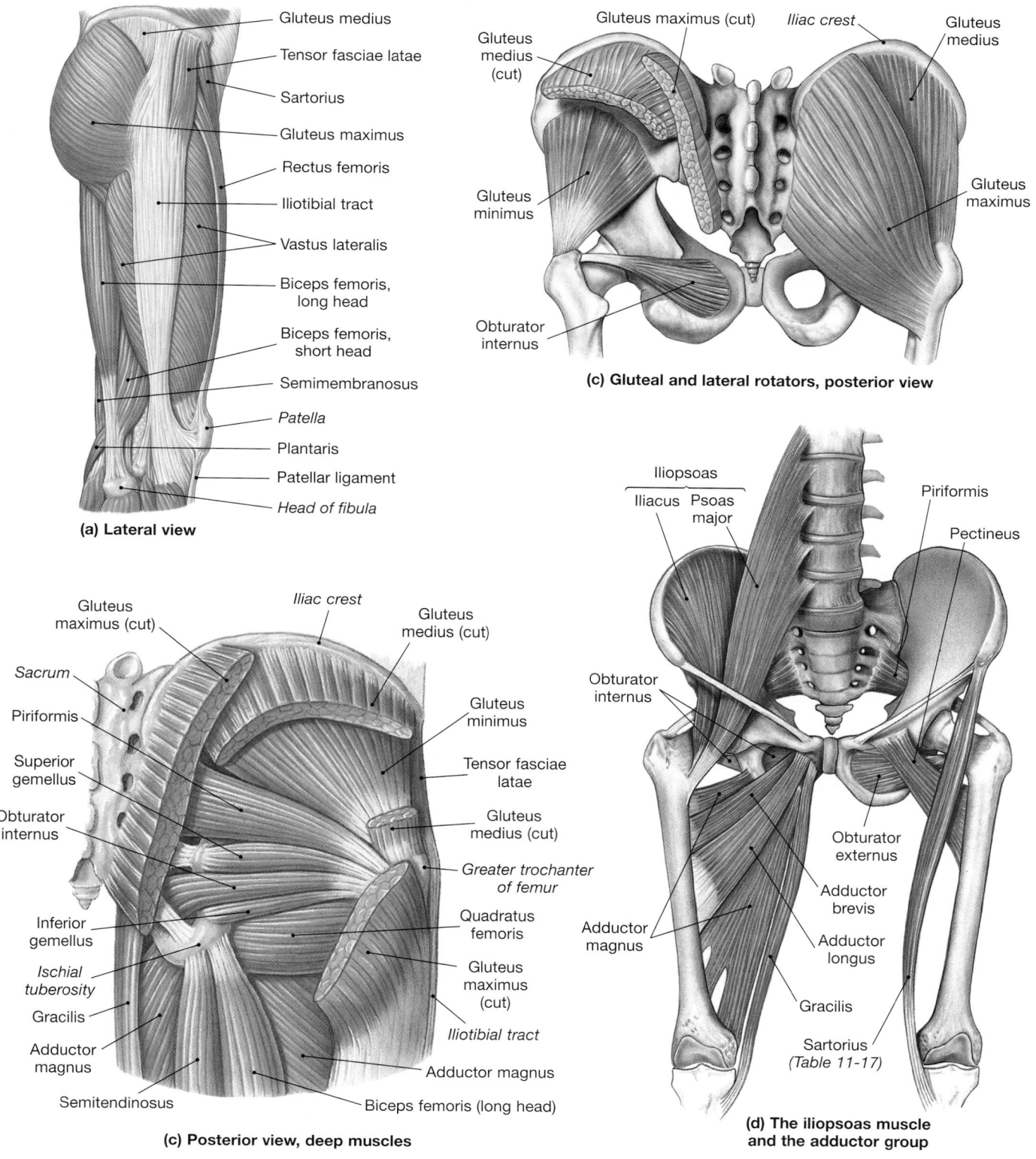

Gluteus medius
Tensor fasciae latae
Sartorius
Gluteus maximus
Rectus femoris
Iliotibial tract
Vastus lateralis
Biceps femoris, long head
Biceps femoris, short head
Semimembranosus
Patella
Plantaris
Patellar ligament
Head of fibula

(a) Lateral view

Gluteus medius (cut)
Gluteus maximus (cut)
Iliac crest
Gluteus medius
Gluteus minimus
Gluteus maximus
Obturator internus

(c) Gluteal and lateral rotators, posterior view

Gluteus maximus (cut)
Iliac crest
Gluteus medius (cut)
Sacrum
Piriformis
Superior gemellus
Obturator internus
Inferior gemellus
Ischial tuberosity
Gracilis
Adductor magnus
Semitendinosus
Gluteus minimus
Tensor fasciae latae
Gluteus medius (cut)
Greater trochanter of femur
Quadratus femoris
Gluteus maximus (cut)
Iliotibial tract
Adductor magnus
Biceps femoris (long head)

(c) Posterior view, deep muscles

Iliopsoas
Iliacus Psoas major
Piriformis
Pectineus
Obturator internus
Obturator externus
Adductor brevis
Adductor magnus
Adductor longus
Gracilis
Sartorius *(Table 11-17)*

(d) The iliopsoas muscle and the adductor group

●**FIGURE 11–19**
Muscles That Move the Thigh. *See also Figures 8–7, 8–8, 8–11, and 9–11.* **ATLAS** Plates 8.1, 8.2, 8.4, 8.5b, 8.6; Scan 4a

femur. The anterior gluteal line on the lateral surface of the ilium marks the boundary between these muscles.

The **lateral rotators** originate at or inferior to the horizontal axis of the acetabulum. There are six lateral rotator muscles in all, of which the **piriformis** (pir-i-FOR-mis) muscle and the **obturator** muscles are dominant (Figure 11–19c,d●).

The **adductors** (Figure 11–19c,d●) originate inferior to the horizontal axis of the acetabulum. This muscle group includes the **adductor magnus**, **adductor brevis**, **adductor longus**, **pectineus** (pek-TIN-ē-us), and **gracilis** (GRAS-i-lis) muscles. All but the adductor magnus originate both anterior and inferior to the joint, so they perform hip flexion as well as adduction. The adductor magnus muscle can produce either adduction and flexion or adduction and extension, depending on the region stimulated. The adductor magnus muscle can also produce medial or lateral rotation at the hip. The other muscles, which insert on low ridges along the posterior surface of the femur, produce medial rotation. When an athlete suffers a *pulled groin*, the problem is a *strain*—a muscle tear or break— in one of these adductor muscles.

The internal surface of the pelvis is dominated by a pair of muscles. The large **psoas** (SŌ-us) **major** muscle originates alongside the inferior thoracic and lumbar vertebrae, and its insertion lies on the lesser trochanter of the femur. Before reaching this insertion, its tendon merges with that of the **iliacus** (il-Ē-uh-kus) muscle, which nestles within the iliac fossa. These two powerful hip flexors are often referred to collectively as the **iliopsoas** (il-ē-ō-SŌ-us) muscle.

Muscles That Move the Leg

As in the upper limb, muscle distribution in the lower limb exhibits a pattern (Figure 11–20● and Table 11–17). Extensor muscles are located along the anterior and lateral surfaces of the leg, and flexors lie along the posterior and medial surfaces. Although the flexors and adductors originate on the pelvic girdle, most extensors originate on the femoral surface.

The *flexors of the knee* include the **biceps femoris**, **semimembranosus** (sem-ē-mem-bra-NŌ-sus), **semitendinosus** (sem-ē-tendi-NŌ-sus), and **sartorius** muscles (Figure 11–20●). These muscles originate along the edges of the pelvis and insert on the tibia and fibula, and their contractions produce flexion at the knee. The sartorius muscle is the only knee flexor that originates superior to the acetabulum, and its insertion lies along the medial surface of the tibia.

TABLE 11–17 MUSCLES THAT MOVE THE LEG (FIGURE 11–20)

Muscle	Origin	Insertion	Action	Innervation
FLEXORS OF THE KNEE				
Biceps femoris	Ischial tuberosity and linea aspera of femur	Head of fibula, lateral condyle of tibia	Flexion at knee; extension and lateral rotation at hip	Sciatic nerve; tibial portion (S_1–S_3; to long head) and common fibular branch (L_5–S_2; to short head)
Semimembranosus	Ischial tuberosity	Posterior surface of medial condyle of tibia	Flexion at knee; extension and medial rotation at hip	Sciatic nerve (tibial portion; L_5–S_2)
Semitendinosus	As above	Proximal, medial surface of tibia near insertion of gracilis	As above	As above
Sartorius	Anterior superior iliac spine	Medial surface of tibia near tibial tuberosity	Flexion at knee; flexion and lateral rotation at hip	Femoral nerve (L_2–L_3)
Popliteus	Lateral condyle of femur	Posterior surface of proximal tibial shaft	Medial rotation of tibia (or lateral rotation of femur); flexion at knee	Tibial nerve (L_4–S_1)
EXTENSORS OF THE KNEE				
Rectus femoris	Anterior inferior iliac spine and superior acetabular rim of ilium	Tibial tuberosity via patellar ligament	Extension at knee; flexion at hip	Femoral nerve (L_2–L_4)
Vastus intermedius	Anterolateral surface of femur and linea aspera (distal half)	As above	Extension at knee	As above
Vastus lateralis	Anterior and inferior to greater trochanter of femur and along linea aspera (proximal half)	As above	As above	As above
Vastus medialis	Entire length of linea aspera of femur	As above	As above	As above

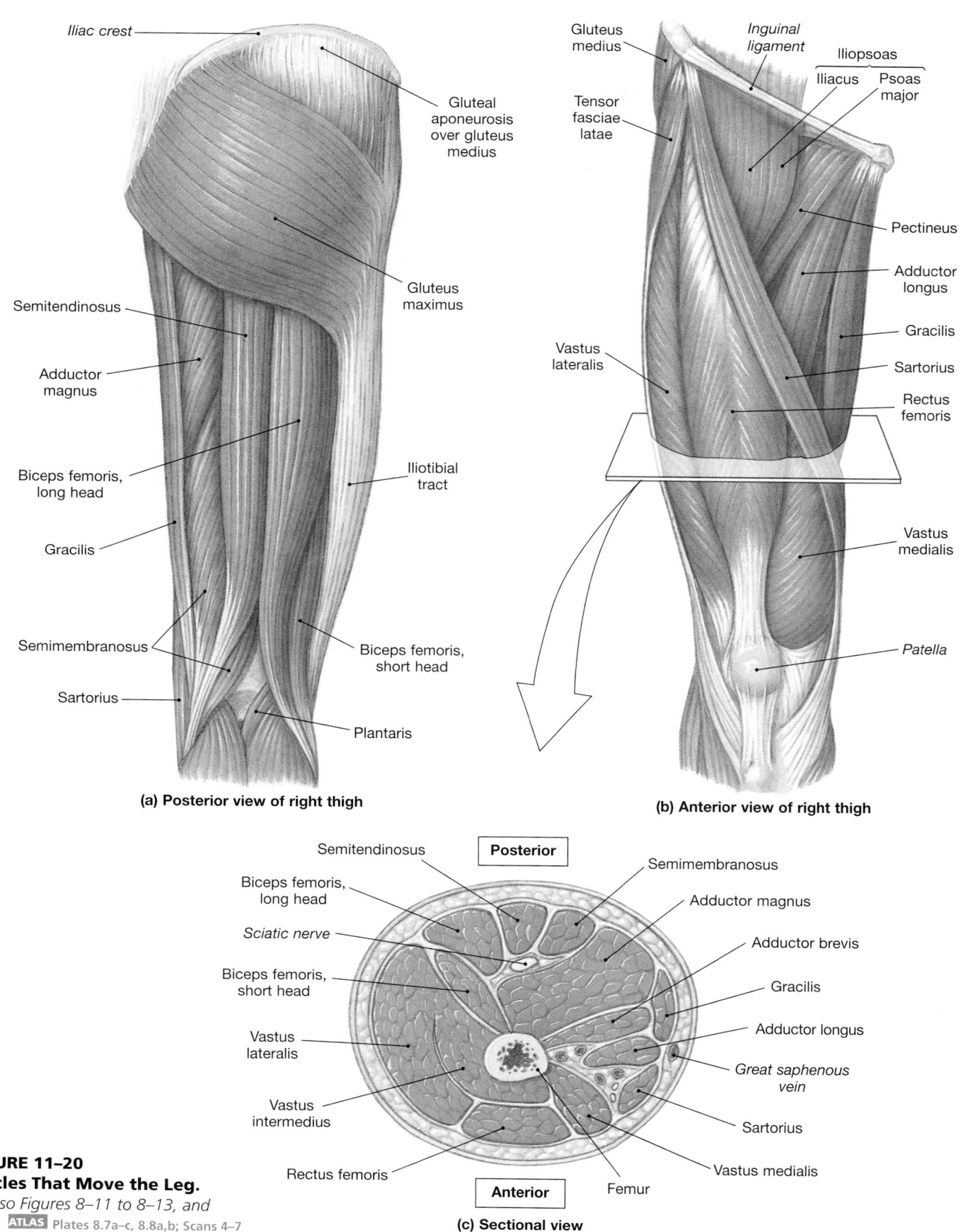

(a) Posterior view of right thigh

Iliac crest

Gluteal aponeurosis over gluteus medius

Semitendinosus

Adductor magnus

Biceps femoris, long head

Gracilis

Semimembranosus

Sartorius

Gluteus maximus

Iliotibial tract

Biceps femoris, short head

Plantaris

(b) Anterior view of right thigh

Gluteus medius

Inguinal ligament

Iliopsoas

Iliacus

Psoas major

Tensor fasciae latae

Vastus lateralis

Pectineus

Adductor longus

Gracilis

Sartorius

Rectus femoris

Vastus medialis

Patella

(c) Sectional view

Posterior

Semitendinosus

Biceps femoris, long head

Sciatic nerve

Biceps femoris, short head

Vastus lateralis

Vastus intermedius

Rectus femoris

Semimembranosus

Adductor magnus

Adductor brevis

Gracilis

Adductor longus

Great saphenous vein

Sartorius

Vastus medialis

Femur

Anterior

●FIGURE 11–20
Muscles That Move the Leg.
See also Figures 8–11 to 8–13, and 9–12. **ATLAS** Plates 8.7a–c, 8.8a,b; Scans 4–7

When the sartorius contracts, it produces flexion at the knee and lateral rotation at the hip—for example, when you cross your legs.

Because the biceps femoris, semimembranosus, and semitendinosus muscles originate on the pelvic surface inferior and posterior to the acetabulum, their contractions produce not only flexion at the knee, but also extension at the hip. These three muscles are often called the **hamstrings**. A *pulled hamstring* is a relatively common sports injury caused by a strain affecting one of the hamstring muscles.

The knee joint can be locked at full extension by a slight lateral rotation of the tibia. ∞ p. 283 The small **popliteus** (pop-LI-tē-us) muscle originates on the femur near the lateral condyle and inserts on the posterior tibial shaft. When flexion is initiated, this muscle contracts to produce a slight medial rotation of the tibia that unlocks the knee joint.

Collectively, the four *knee extensors* make up the **quadriceps femoris**: the three **vastus** muscles, which originate along the shaft of the femur, and the **rectus femoris**. Together, the vastus muscles cradle the rectus femoris muscle the way a bun surrounds a hot dog (Figure 11–20c●). All four muscles insert on the patella via the quadriceps tendon. The force of their contraction is relayed to the tibial tuberosity by way of the patellar ligament. The rectus femoris muscle originates on the anterior inferior iliac spine and the superior acetabular rim—so in addition to extending the knee, it assists in flexion of the hip.

Muscles That Move the Foot and Toes

The extrinsic muscles that move the foot and toes are shown in Figure 11–21● and listed in Table 11–18). Most of the muscles that move the ankle produce the plantar flexion involved with walking

---

TABLE 11–18 EXTRINSIC MUSCLES THAT MOVE THE FOOT AND TOES (FIGURE 11–21)

Muscle	Origin	Insertion	Action	Innervation
ACTION AT THE ANKLE				
Flexors (Dorsiflexors)				
Tibialis anterior	Lateral condyle and proximal shaft of tibia	Base of first metatarsal bone and medial cuneiform bone	Flexion (dorsiflexion) at ankle; inversion of foot	Deep fibular nerve $(L_4–S_1)$
Extensors (Plantar flexors)				
Gastrocnemius	Femoral condyles	Calcaneus via calcaneal tendon	Extension (plantar flexion) at ankle; inversion of foot; flexion at knee	Tibial nerve $(S_1–S_2)$
Fibularis brevis	Midlateral margin of fibula	Base of fifth metatarsal bone	Eversion of foot and extension (plantar flexion) at ankle	Superficial fibular nerve $(L_4–S_1)$
Fibularis longus	Lateral condyle of tibia, head and proximal shaft of fibula	Base of first metatarsal bone and medial cuneiform bone	Eversion of foot and extension (plantar flexion) at ankle; supports longitudinal arch	As above
Plantaris	Lateral supracondylar ridge	Posterior portion of calcaneus	Extension (plantar flexion) at ankle; flexion at knee	Tibial nerve $(L_4–S_1)$
Soleus	Head and proximal shaft of fibula and adjacent posteromedial shaft of tibia	Calcaneus via calcaneal tendon (with gastrocnemius)	Extension (plantar flexion) at ankle	Sciatic nerve, tibial branch $(S_1–S_2)$
Tibialis posterior	Interosseous membrane and adjacent shafts of tibia and fibula	Tarsal and metatarsal bones	Adduction and inversion of foot; extension (plantar flexion) at ankle	As above
ACTION AT THE TOES				
Digital flexors				
Flexor digitorum longus	Posteromedial surface of tibia	Inferior surfaces of distal phalanges, toes 2–5	Flexion at joints of toes 2–5	Sciatic nerve, tibial branch $(L_5–S_1)$
Flexor hallucis longus	Posterior surface of fibula	Inferior surface, distal phalanx of great toe	Flexion at joints of great toe	As above
Digital extensors				
Extensor digitorum longus	Lateral condyle of tibia, anterior surface of fibula	Superior surfaces of phalanges, toes 2–5	Extension at joints of toes 2–5	Deep fibular nerve $(L_4–S_1)$
Extensor hallucis longus	Anterior surface of fibula	Superior surface, distal phalanx of great toe	Extension at joints of great toe	As above

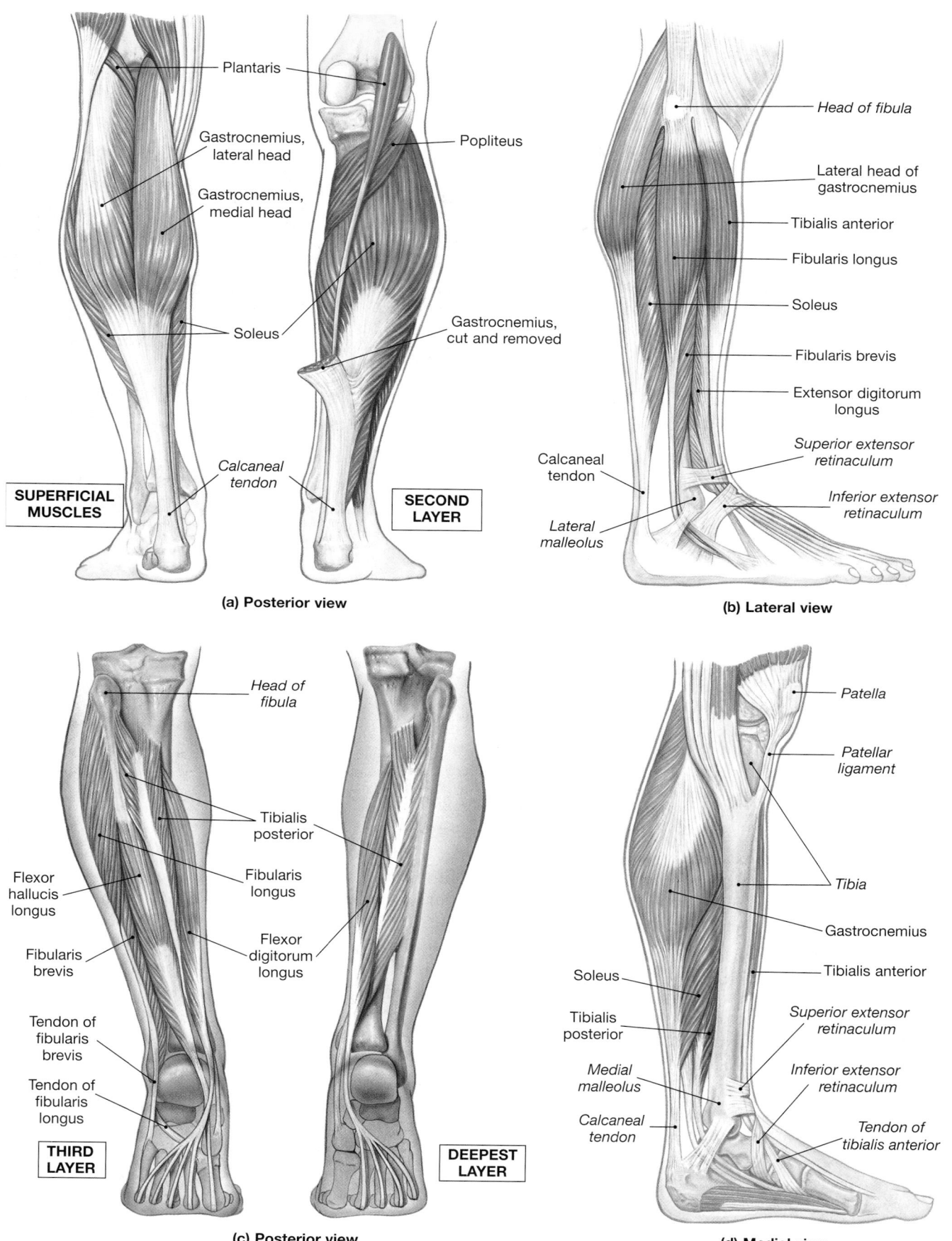

Plantaris

Gastrocnemius, lateral head

Gastrocnemius, medial head

Soleus

SUPERFICIAL MUSCLES

Calcaneal tendon

Popliteus

Gastrocnemius, cut and removed

SECOND LAYER

(a) Posterior view

Head of fibula

Lateral head of gastrocnemius

Tibialis anterior

Fibularis longus

Soleus

Fibularis brevis

Extensor digitorum longus

Superior extensor retinaculum

Calcaneal tendon

Inferior extensor retinaculum

Lateral malleolus

(b) Lateral view

Head of fibula

Tibialis posterior

Fibularis longus

Flexor hallucis longus

Flexor digitorum longus

Fibularis brevis

Tendon of fibularis brevis

Tendon of fibularis longus

THIRD LAYER

DEEPEST LAYER

(c) Posterior view

Patella

Patellar ligament

Tibia

Gastrocnemius

Tibialis anterior

Soleus

Superior extensor retinaculum

Tibialis posterior

Inferior extensor retinaculum

Medial malleolus

Calcaneal tendon

Tendon of tibialis anterior

(d) Medial view

●**FIGURE 11–21**

Extrinsic Muscles That Move the Foot and Toes. *See also Figures 8–13 and 8–14.* ATLAS Plates 8.12, 8.13

Intramuscular Injections

Drugs are commonly injected into tissues rather than directly into the circulation. This method enables the physician to introduce a large amount of a drug at one time, yet have it enter the circulation gradually. In an **intramuscular (IM) injection**, the drug is introduced into the mass of a large skeletal muscle. Uptake is generally faster and accompanied by less tissue irritation than when drugs are administered *intradermally* or *subcutaneously* (injected into the dermis or subcutaneous layer, respectively). 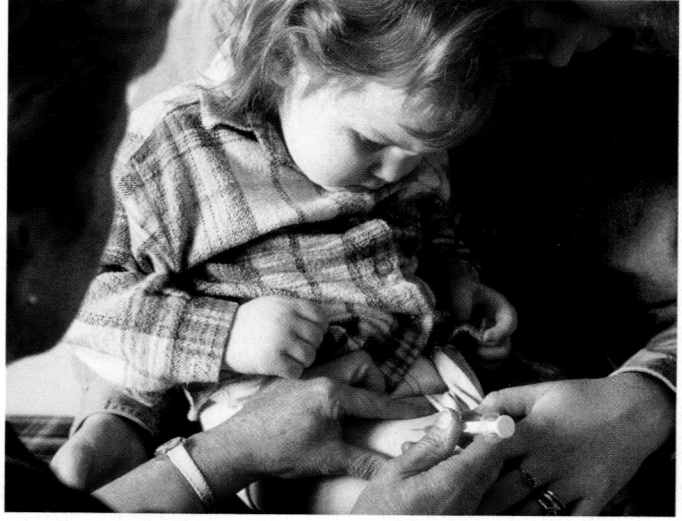 pp. 163, 166 Up to 5 ml of fluid may be injected at one time, and multiple injections are possible.

The most common complications involve accidental injection into a blood vessel or piercing of a nerve. The sudden entry of massive quantities of a drug into the bloodstream can have fatal consequences, and damage to a nerve can cause motor paralysis or sensory loss. As a result, the site of injection must be selected with care. Bulky muscles that contain few large vessels or nerves make ideal targets. The gluteus medius muscle or the posterior, lateral, superior part of the gluteus maximus muscle is commonly selected. The deltoid muscle of the arm, about 2.5 cm (1 in.) distal to the acromion, is another popular site. Probably the most satisfactory from a technical point of view is the vastus lateralis muscle of the thigh; an injection into this thick muscle will not encounter vessels or nerves. This is the preferred injection site in infants and young children, whose gluteal and deltoid muscles are relatively small. The site is also used in elderly patients or others with atrophied gluteal and deltoid muscles.

and running movements. The **gastrocnemius** (gas-trok-NĒ-mē-us; *gaster*, stomach + *kneme*, knee) muscle of the calf is an important plantar flexor, but the slow muscle fibers of the underlying **soleus** (SŌ-lē-us) muscle are more powerful. These muscles are best seen in posterior and lateral views (Figure 11–21a,b●). The gastrocnemius muscle arises from two heads located on the medial and lateral epicondyles of the femur just proximal to the knee. The *fabella*, a sesamoid bone, is generally present within the lateral head of the gastrocnemius muscle. The gastrocnemius and soleus muscles share a common tendon, the **calcaneal tendon**, commonly known as the *Achilles tendon* or *calcanean tendon*.

The term "Achilles tendon" comes from Greek mythology. Achilles was an almost invincible warrior who had but one spot where he could be wounded: the calcaneal tendon. His mother had dipped him in the River Styx as an infant to make him invulnerable, but she held him by the ankles and forgot to dip the heel of his foot. This oversight proved fatal for Achilles, who was killed in battle by an arrow through the tendon that now bears his name.

Deep to the gastrocnemius and soleus muscles lie a pair of **fibularis** muscles, or *peroneus* muscles (Figure 11–21b,c●). The fibularis muscles produce eversion and extension (plantar flexion) at the ankle. Inversion is caused by the contraction of the **tibialis** (tib-ē-A-lis) muscles. The large **tibialis anterior** muscle (Figure 11–21b,d●) flexes the ankle and opposes the gastrocnemius muscle.

Important digital muscles originate on the surface of the tibia, the fibula, or both (Figure 11–21b,c,d●). Large synovial tendon sheaths surround the tendons of the tibialis anterior, **extensor digitorum longus**, and **extensor hallucis longus** muscles, where they cross the ankle joint. The positions of these sheaths are stabilized by superior and inferior **extensor retinacula** (Figure 11–21b,d●).

Intrinsic muscles of the foot originate on the tarsal and metatarsal bones (Figure 11–22● and Table 11–19). Their contractions move the toes and contribute to the maintenance of the longitudinal arch of the foot. 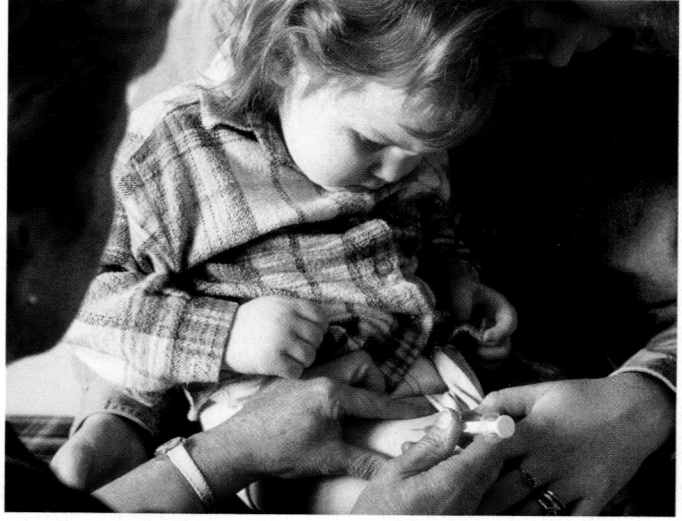 p. 256

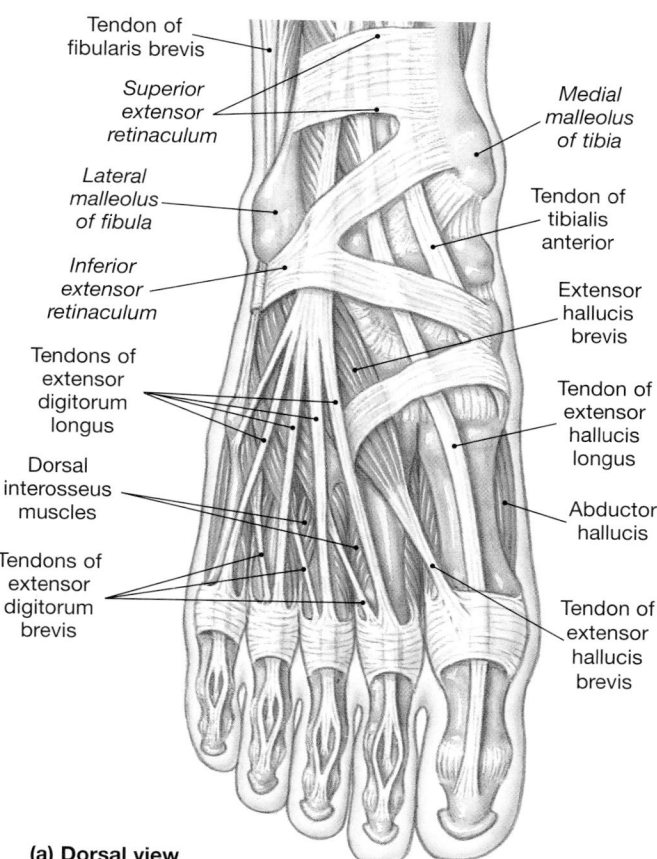

Tendon of fibularis brevis

Superior extensor retinaculum

Lateral malleolus of fibula

Inferior extensor retinaculum

Tendons of extensor digitorum longus

Dorsal interosseus muscles

Tendons of extensor digitorum brevis

Medial malleolus of tibia

Tendon of tibialis anterior

Extensor hallucis brevis

Tendon of extensor hallucis longus

Abductor hallucis

Tendon of extensor hallucis brevis

(a) Dorsal view

●**FIGURE 11–22**
Intrinsic Muscles of the Foot. *See also Figure*
8–14. **ATLAS** Plates 8.14, 8.15a, 8.16; Scan 8a–c

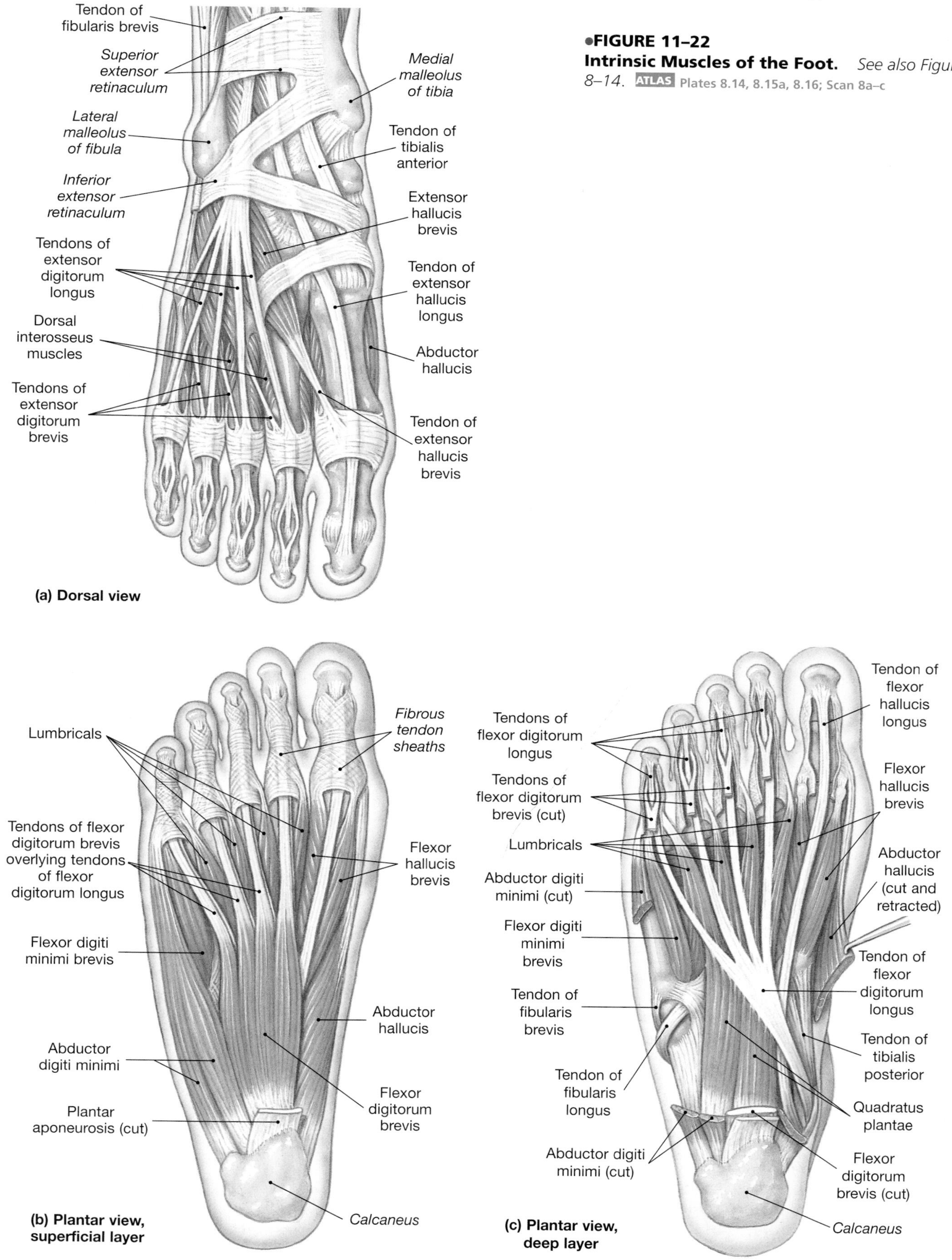

Lumbricals

Tendons of flexor digitorum brevis overlying tendons of flexor digitorum longus

Flexor digiti minimi brevis

Abductor digiti minimi

Plantar aponeurosis (cut)

Fibrous tendon sheaths

Flexor hallucis brevis

Abductor hallucis

Flexor digitorum brevis

Calcaneus

(b) Plantar view, superficial layer

Tendons of flexor digitorum longus

Tendons of flexor digitorum brevis (cut)

Lumbricals

Abductor digiti minimi (cut)

Flexor digiti minimi brevis

Tendon of fibularis brevis

Tendon of fibularis longus

Abductor digiti minimi (cut)

Tendon of flexor hallucis longus

Flexor hallucis brevis

Abductor hallucis (cut and retracted)

Tendon of flexor digitorum longus

Tendon of tibialis posterior

Quadratus plantae

Flexor digitorum brevis (cut)

Calcaneus

(c) Plantar view, deep layer

TABLE 11–19 INTRINSIC MUSCLES OF THE FOOT (FIGURE 11–22)

Muscle	Origin	Insertion	Action	Innervation
Extensor digitorum brevis	Calcaneus (superior and lateral surfaces)	Dorsal surfaces of toes 1–4	Extension at metatarso-phalangeal joints of toes 1–4	Deep fibular nerve (L_5–S_1)
Abductor hallucis	Calcaneus (tuberosity on inferior surface)	Medial side of proximal phalanx of great toe	Abduction at metatarsophalangeal joint of great toe	Medial plantar nerve (L_4–L_5)
Flexor digitorum brevis	As above	Sides of middle phalanges, toes 2–5	Flexion at proximal interphalangeal joints of toes 2–5	As above
Abductor digiti minimi	As above	Lateral side of proximal phalanx, toe 5	Abduction at metatarsophalangeal joint of toe 5	Lateral plantar nerve (L_4–L_5)
Quadratus plantae	Calcaneus (medial, inferior surfaces)	Tendon of flexor digitorum longus	Flexion at joints of toes 2–5	As above
Lumbrical (4)	Tendons of flexor digitorum longus	Insertions of extensor digitorum longus	Flexion at metatarsophalangeal joints; extension at proximal interphalangeal joints of toes 2–5	Medial plantar nerve (1), lateral plantar nerve (2–4)
Flexor hallucis brevis	Cuboid and lateral cuneiform bones	Proximal phalanx of great toe	Flexion at metatarsophalangeal joint of great toe	Medial plantar nerve (L_4–L_5)
Adductor hallucis	Bases of metatarsal bones II–IV and plantar ligaments	As above	Adduction at metatarsophalangeal joint of great toe	Lateral plantar nerve (S_1–S_2)
Flexor digiti minimi brevis	Base of metatarsal bone V	Lateral side of proximal phalanx of toe 5	Flexion at metatarsophalangeal joint of toe 5	As above
Dorsal interosseus (4)	Sides of metatarsal bones	Medial and lateral sides of toe 2; lateral sides of toes 3 and 4	Abduction at metatarsophalangeal joints of toes 3 and 4	As above
Plantar interosseus (3)	Bases and medial sides of metatarsal bones	Medial sides of toes 3–5	Adduction at metatarso-phalangeal joints of toes 3–5	As above

Outside of mythology, damage to the calcaneal tendon isn't a fatal problem. But although it is among the largest, strongest tendons in the body, its rupture is relatively common. The muscles are very strong, and they have to apply a lot of force to the calcaneus to lift the body weight; the second-class lever system provides them with little mechanical advantage. The amount of force required increases if one is trying to accelerate or decelerate quickly. As a result, sprinters can rupture the calcaneal tendon as they push from the starting blocks on takeoff. It is also not uncommon for a stumble or fall in the elderly to snap this tendon as it absorbs the shock of a sudden impact. Rupture of the calcaneal tendon produces immediate pain and a swollen, sustained contraction of the calf muscles. Because the calcaneal tendon forms the posterior margin of the distal leg, its rupture produces a prominent and unmistakable gap in the profile of the limb. The individual finds it easy to perform ankle flexion, but finds ankle exten-sion very difficult, except with the aid of gravity. Surgery may be necessary to reposition and reconnect the broken ends of the tendon to promote healing.

✓ Which leg movement would be impaired by injury to the obturator muscle?

✓ You often hear of athletes who suffer a pulled hamstring. To what does this phrase refer?

✓ How would you expect a torn calcaneal tendon to affect movement of the foot? What muscles are the antagonists of the muscles that pull on the calcaneal tendon?

Answers start on p. Q-1

The muscles of the pelvis and lower limbs can be reviewed on the **Anatomy CD-ROM**: Muscular System/Appendicular Dissections/Pelvis & Abdomen, /Thigh, and /Lower Leg.

Compartment Syndrome

In the limbs, the interconnections between the superficial fascia, the deep fascia of the muscles, and the periostea of the appendicular skeleton are quite substantial. The muscles within a limb are in effect isolated in **compartments** formed by dense collagenous sheets (Figure 11–23●). Blood vessels and nerves traveling to specific muscles within the limb enter and branch within the appropriate compartments.

When a crushing injury, severe contusion, or muscle strain occurs, the blood vessels in one or more compartments may be damaged. The compartments become swollen with blood and fluid leaked from damaged vessels. The connective-tissue partitions are very strong; the accumulated fluid cannot escape, so pressure rises within the affected compartments. Eventually, the pressures can become so high that they compress regional blood vessels and eliminate the circulatory supply to the muscles and nerves of the compartment. This compression produces **ischemia** (is-KĒ-mē-uh), or "blood starvation," known as **compartment syndrome**.

Slicing into the compartment along its longitudinal axis or implanting a drain are emergency measures used to relieve the pressure. If such steps are not taken, the contents of the compartment will suffer severe damage. Nerves in the affected compartment will be destroyed after 2–4 hours of ischemia, although they can regenerate to some degree if the circulation is restored. After 6 hours or more, the muscle tissue will also be destroyed, and no regeneration can occur. The muscles will be replaced by scar tissue, and shortening of the connective tissue fibers may result in *contracture*, a permanent contraction of an entire muscle following the atrophy of individual muscle fibers.

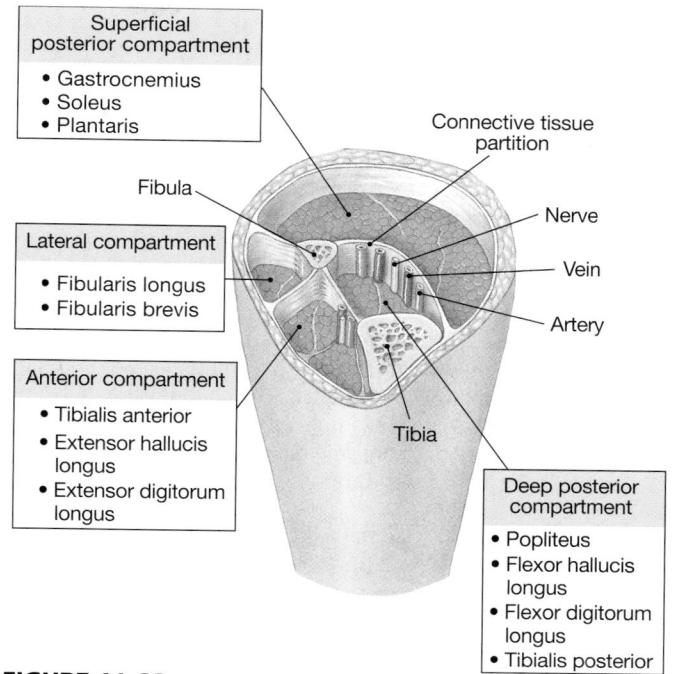

Superficial posterior compartment
- Gastrocnemius
- Soleus
- Plantaris

Connective tissue partition
Fibula
Nerve
Vein
Artery

Lateral compartment
- Fibularis longus
- Fibularis brevis

Anterior compartment
- Tibialis anterior
- Extensor hallucis longus
- Extensor digitorum longus

Tibia

Deep posterior compartment
- Popliteus
- Flexor hallucis longus
- Flexor digitorum longus
- Tibialis posterior

●**FIGURE 11–23**
Musculoskeletal Compartments.   A section through the leg, with the muscles partially removed. A section through the thigh or arm would show a comparable arrangement of dense connective-tissue partitions. The anterior and lateral compartments of the leg contain muscles that flex (dorsiflex) the ankle and extend the toes, and the posterior compartments contain the muscles that extend (plantar flex) the ankle and flex the toes.

11–5 AGING AND THE MUSCULAR SYSTEM

As the body ages, the size and power of all muscle tissues decrease. The effects on the muscular system can be summarized as follows:

- **Skeletal Muscle Fibers Become Smaller in Diameter.** This reduction in size reflects primarily a decrease in the number of myofibrils. In addition, the muscle fibers contain smaller ATP, CP, and glycogen reserves and less myoglobin. The overall effect is a reduction in skeletal muscle size, strength, and endurance, combined with a tendency to fatigue rapidly. Because cardiovascular performance also decreases with age, blood flow to active muscles does not increase with exercise as rapidly as it does in younger people. These factors interact to produce decreases of 30–50 percent in anaerobic and aerobic performance by age 65.

- **Skeletal Muscles Become Less Elastic.** Aging skeletal muscles develop increasing amounts of fibrous connective tissue, a process called **fibrosis**. Fibrosis makes the muscle less flexible, and the collagen fibers can restrict movement and circulation.

- **Tolerance for Exercise Decreases.** A lower tolerance for exercise results in part from the tendency toward rapid fatigue and in part from the reduction in thermoregulatory ability described in Chapter 5. ⊂⊃ p. 175 Individuals over age 65 cannot eliminate the heat their muscles generate during contraction as effectively as younger people can and thus are subject to overheating.

- **The Ability to Recover from Muscular Injuries Decreases.** The number of satellite cells steadily decreases with age, and the amount of fibrous tissue increases. As a result, when an injury occurs, repair capabilities are limited. Scar tissue formation is the usual result.

The *rate* of decline in muscular performance is the same in all individuals, regardless of their exercise patterns or lifestyle. Therefore, to be in good shape late in life, you must be in *very* good shape early in life. Regular exercise helps control body weight, strengthens bones, and generally improves the quality of life at all ages. Extremely demanding exercise is not as important as regular exercise. In fact, extreme exercise in the elderly can damage tendons, bones, and joints. Although exercise has obvious effects on the quality of life, there is no clear evidence that it prolongs life expectancy.

THE **MUSCULAR SYSTEM** IN PERSPECTIVE

Organ/Component	Primary Functions
Skeletal Muscles (700)	Provide skeletal movement; control entrances and exits of digestive tract; produce heat; support skeletal position; protect soft tissues
Axial muscles	Support and position axial skeleton
Appendicular muscles	Support, move, and brace limbs
Tendons, Aponeuroses	Harness forces of contraction to perform specific tasks

INTEGRATION WITH OTHER SYSTEMS

To operate at maximum efficiency, the muscular system must be supported by many other systems. The changes that occur during exercise provide a good example of such interaction. As we noted earlier, active muscles consume oxygen and generate carbon dioxide and heat. Responses of other systems include:

- **Cardiovascular System.** Blood vessels in the active muscles and in the skin dilate, and the heart rate increases. These ad-

justments accelerate oxygen delivery and carbon dioxide removal at the muscle and bring heat to the skin for radiation into the environment.

- **Respiratory System.** The respiratory rate and the depth of respiration increase. Air moves into and out of the lungs more quickly, keeping pace with the increased rate of blood flow through the lungs.

- **Integumentary System.** Blood vessels dilate, and sweat gland secretion increases. This combination helps promote evaporation at the surface of the skin and removes the excess heat generated by muscular activity.

- **Nervous and Endocrine Systems.** The responses of other systems are directed by controlling the heart rate, respiratory rate, and sweat gland activity.

Even when the body is at rest, the muscular system has extensive interactions with other systems. The figure on the opposite page summarizes the range of interactions between the muscular system and other vital systems.

CLINICAL PATTERNS

Skeletal muscles contract only under the command of the nervous system. For this reason, clinical observation of muscular activity provides not only direct information about the muscular system, but also indirect information about the nervous system. In general, muscular system disorders are divided into *primary disorders*, which result from problems with the muscular system itself, and *secondary disorders*, which appear as the result of problems originating in other systems. Examples of primary disorders include

muscle infections, muscle trauma, and inherited disorders such as the muscular dystrophies (p. 300). Secondary disorders include a wide range of nervous system disorders that affect the coordination or control of muscle contraction, nutritional or metabolic problems that affect the energy supply available to the muscles, and cardiovascular disorders that restrict or reduce blood flow to skeletal muscles. Examples of these categories are presented and discussed in the *Applications Manual*.

MEDIA CONNECTIONS

I. Objective: Explore the functional dependence of the muscular system on other body systems.

Completion time = 15 minutes

The muscular system depends on other systems of the body in many ways. To investigate this dependence, visit the **InterActive Physiology CD-ROM**: Muscular System/Muscle Metabolism. Complete screens 3 through 8, noting places where the information presented relates to what you have learned in this text. Where have you encountered ATP, and what do you know about it? Where is the energy for muscle contraction generated? (Review Chapter 3.) Begin a table defining the functions of other systems essential to skeletal muscle contraction. Continue with screens 10 through 15, 22, and 23 of the tutorial, and think about what compounds muscles need to generate the ATP that powers their contractions. What waste products are created, and how do the cells remove them? Extend your table, using information from the tutorial as well as the text, to include other systems that interact with contracting muscle.

II. Objective: Investigate the interaction between the muscular and skeletal systems under normal conditions and during space flight.

Completion time = 10 minutes

As skeletal muscles contract, they apply stresses to the skeletal system. These stresses can cause homeostatic adjustments in the structure of the stressed bones. Briefly explain the basics of this interaction, using Section 6–5 as a guide. During prolonged space flight, under zero gravity, there are functional and structural changes in both the skeletal and muscular systems. Given your knowledge of the interrelationships between these systems, predict what kinds of changes would occur when muscles and bones are weightless. Check your predictions by accessing the musculoskeletal experimental data from the Space Shuttle/Mir log. To do this, visit the **Companion Website**, Chapter 11 Media Connections, and click on the keyword **NASA**.

INTEGUMENTARY SYSTEM

- Removes excess body heat
- Synthesizes vitamin D_3 for calcium and phosphate absorption
- Protects underlying muscles

- Skeletal muscles pulling on skin of face produce facial expressions

SKELETAL SYSTEM

- Maintains normal calcium and phosphate levels in body fluids
- Supports skeletal muscles
- Provides sites of attachment

- Provides movement and support
- Stresses exerted by tendons maintain bone mass
- Stabilizes bones and joints

NERVOUS SYSTEM

- Controls skeletal muscle contractions
- Adjusts activities of respiratory and cardiovascular systems during periods of muscular activity

---

- Muscle spindles monitor body position
- Facial muscles express emotion
- Intrinsic laryngeal muscles permit speech

ENDOCRINE SYSTEM

- Hormones adjust muscle metabolism and growth
- Parathyroid hormone and calcitonin regulate calcium and phosphate ion concentrations

---

- Skeletal muscles provide protection for some endocrine organs

CARDIOVASCULAR SYSTEM

- Delivers oxygen and nutrients
- Removes carbon dioxide, lactic acid, and heat

- Skeletal muscle contractions assist in moving blood through veins
- Protects deep blood vessels

MUSCULAR SYSTEM

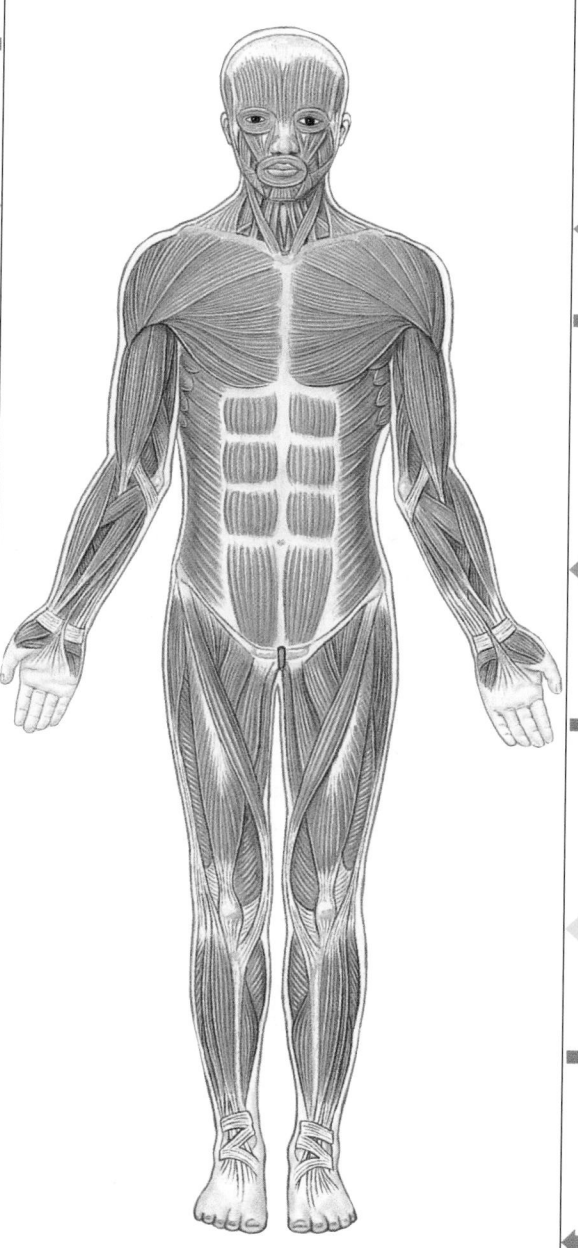

FOR ALL SYSTEMS
Generates heat that maintains normal body temperature

LYMPHATIC SYSTEM

- Defends skeletal muscles against infection and assists in tissue repairs after injury

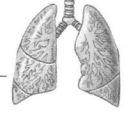

- Protects superficial lymph nodes and the lymphatic vessels in the abdominopelvic cavity

RESPIRATORY SYSTEM

- Provides oxygen and eliminates carbon dioxide

- Muscles generate carbon dioxide
- Muscles control entrances to respiratory tract, fill and empty lungs, control airflow through larynx, and produce sounds

DIGESTIVE SYSTEM

- Provides nutrients
- Liver regulates blood glucose and fatty acid levels and removes lactic acid from circulation

---

- Protects and supports soft tissues in abdominal cavity
- Controls entrances to and exits from digestive tract

URINARY SYSTEM

- Removes waste products of protein metabolism
- Assists in regulation of calcium and phosphate concentrations

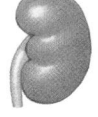

- External sphincter controls urination by constricting urethra

REPRODUCTIVE SYSTEM

- Reproductive hormones accelerate skeletal muscle growth

- Contractions of skeletal muscles eject semen from male reproductive tract
- Muscle contractions during sex act produce pleasurable sensations

Chapter Review

SELECTED CLINICAL TERMINOLOGY

Terms Discussed in This Chapter

carpal tunnel syndrome: An inflammation of the sheath surrounding the flexor tendons of the palm that leads to nerve compression and pain. *(p. 365)*

compartment syndrome: Ischemia resulting from accumulated blood and fluid trapped within a musculoskeletal compartment. *(p. 377)*

diaphragmatic hernia *(hiatal hernia):* A hernia that occurs when abdominal organs are forced into the thoracic cavity. *(p. 356)*

fibrosis: A process in which muscle tissue is replaced by fibrous connective tissue, making muscles weaker and less flexible. *(p. 377)*

hernia: A condition wherein an organ or a body part protrudes through an abnormal opening. *(p. 356)*

inguinal hernia: A condition in which the inguinal canal enlarges and abdominal contents are forced into it. *(p. 356)*

intramuscular (IM) injection: The administration of a drug by injecting it into the mass of a large skeletal muscle. *(p. 374)*

ischemia (is-KĒ-mē-uh): A condition of "blood starvation" resulting from the compression of regional blood vessels. *(p. 377)*

rotator cuff: The muscles that surround the shoulder joint; a common site of sports injuries. *(p. 361)*

AM Additional Terms Discussed in the *Applications Manual*

bone bruise

bursitis

muscle cramps

sprains

strains

stress fractures

tendinitis

STUDY OUTLINE

11–1 MUSCLE ORGANIZATION AND FUNCTION   p. 333

1. The **muscular system** includes all the skeletal muscle tissue that can be controlled voluntarily.

ORGANIZATION OF SKELETAL MUSCLE FIBERS   p. 334

2. A muscle can be classified as a **parallel muscle, convergent muscle, pennate muscle,** or **circular muscle (sphincter)** according to the arrangement of fibers and fascicles in it. A pennate muscle may be *unipennate, bipennate,* or *multipennate. (Figure 11–1)*

LEVERS   p. 335

3. A **lever** is a rigid structure that moves on a fixed point called the **fulcrum.** Levers can change the direction and effective strength of an applied force and the distance and speed of the movement it produces.

4. Levers are classified as **first-class, second-class,** or **third-class levers.** Third-class levers are the most common type in the body. *(Figure 11–2)*

11–2 MUSCLE TERMINOLOGY   p. 336

ORIGINS AND INSERTIONS   p. 336

1. Each muscle can be identified by its **origin, insertion,** and **action.**

ACTIONS   p. 337

2. According to function, a muscle can be classified as an **agonist,** or **prime mover;** an **antagonist;** a **synergist;** or a **fixator.**

NAMES OF SKELETAL MUSCLES   p. 337

3. The names of muscles commonly provide clues to their fascicle organization, location, relative position, structure, size, shape, origin and insertion, or action. *(Table 11–1)*

AXIAL AND APPENDICULAR MUSCLES   p. 339

4. The **axial musculature** arises on the axial skeleton; it positions the head and spinal column and moves the rib cage. The **appendicular musculature** stabilizes or moves components of the appendicular skeleton. *(Figure 11–3)*

5. **Innervation** refers to the identity of the nerve that controls a given muscle.

Muscle organization: **Companion Website:** Chapter 11/Tutorials.

11–3 THE AXIAL MUSCLES   p. 342

1. The axial muscles fall into logical groups on the basis of location, function, or both.

MUSCLES OF THE HEAD AND NECK   p. 342

2. The muscles of facial expression are the **orbicularis oris, buccinator,** and **occipitofrontalis** muscles and the **platysma.** *(Figure 11–4; Table 11–2)*

3. Six **extrinsic eye muscles** (**oculomotor muscles** or *extra-ocular muscles*) control eye movements: the **inferior** and **superior rectus** muscles, the **lateral** and **medial rectus** muscles, and the **inferior** and **superior oblique** muscles. *(Figure 11–5; Table 11–3)*

4. The muscles of mastication (chewing) are the **masseter, temporalis,** and **pterygoid** muscles. *(Figure 11–6; Table 11–4)*

5. The muscles of the tongue are necessary for speech and swallowing and assist in mastication. They are the **genioglossus, hyoglossus, palatoglossus,** and **styloglossus** muscles. *(Figure 11–7; Table 11–5)*

6. The muscles of the pharynx constrict the pharyngeal walls (**pharyngeal constrictors**), elevate the larynx (**laryngeal elevators**), or raise the soft palate (**palatal muscles**). *(Figure 11–8; Table 11–6)*

7. The muscles of the neck control the position of the larynx, depress the mandible, and provide a foundation for the muscles of the tongue and pharynx. The neck muscles include the **digastric** and **sternocleidomastoid** muscles and seven muscles that originate or insert on the hyoid bone. *(Figure 11–9; Table 11–7)*

Head and neck muscles: **Anatomy CD-ROM:** Muscular System/Axial Dissections/Head & Face.

MUSCLES OF THE VERTEBRAL COLUMN   p. 350

8. The superficial muscles of the spine can be classified into the **spinalis**, **longissimus**, and **iliocostalis** divisions. *(Figure 11–10; Table 11–8)*

9. Other muscles of the spine include the **longus capitis** and **longus colli** muscles of the neck and the **quadratus lumborum** muscle of the lumbar region. *(Figure 11–10; Table 11–8)*

OBLIQUE AND RECTUS MUSCLES   p. 352

10. The oblique muscles include the **scalene** muscles and the **intercostal** and **transversus** muscles. The **external** and **internal intercostal** muscles are important in respiratory movements of the ribs. Also important to respiration is the **diaphragm**. *(Figures 11–10, 11–11; Table 11–9)*

MUSCLES OF THE PELVIC FLOOR   p. 354

11. The **perineum** can be divided into an anterior **urogenital triangle** and a posterior **anal triangle**. The pelvic floor consists of the **urogenital diaphragm** and the **pelvic diaphragm**. *(Figure 11–12; Table 11–10)*

Vertebral column and pelvic floor muscles: **Anatomy CD-ROM:** Muscular System/Axial Dissections/Thorax & Abdomen.

11–4 THE APPENDICULAR MUSCLES   p. 356

MUSCLES OF THE SHOULDERS AND UPPER LIMBS   p. 356

1. The **trapezius** muscle affects the positions of the shoulder girdle, head, and neck. Other muscles inserting on the scapula include the **rhomboid**, **levator scapulae**, **serratus anterior**, **subclavius**, and **pectoralis minor** muscles. *(Figures 11–13, 11–14; Table 11–11)*

2. The **deltoid** and the **supraspinatus** muscles are important abductors. The **subscapularis** and **teres major** muscles produce medial rotation at the shoulder; the **infraspinatus** and **teres minor** muscles perform lateral rotation; and the **coracobrachialis** muscle produces flexion and adduction at the shoulder joint. *(Figures 11–13 to 11–15; Table 11–12)*

3. The **pectoralis major** muscle flexes the shoulder joint, and the **latissimus dorsi** muscle extends it. *(Figures 11–13 to 11–15; Table 11–12)*

4. The actions of the **biceps brachii** muscle and the **triceps brachii** muscle (long head) affect the elbow joint. The **brachialis** and **brachioradialis** muscles flex the elbow, opposed by the **anconeus** muscle. The **flexor carpi ulnaris**, **flexor carpi radialis**,

and **palmaris longus** muscles cooperate to flex the wrist. They are opposed by the **extensor carpi radialis** muscles and the **extensor carpi ulnaris** muscle. The **pronator teres** and **pronator quadratus** muscles pronate the forearm and are opposed by the **supinator** muscle. *(Figures 11–15 to 11–18; Tables 11–13 to 11–15)*

Shoulders and upper limb muscles: **Anatomy CD-ROM:** Muscular System/Appendicular Dissections/Shoulder & Upper Arm.

MUSCLES OF THE PELVIS AND LOWER LIMBS   p. 367

5. **Gluteal muscles** cover the lateral surfaces of the ilia. The largest is the **gluteus maximus** muscle, which shares an insertion with the **tensor fasciae latae**. Together, these muscles pull on the **iliotibial tract**. *(Figures 11–13, 11–19; Table 11–16)*

6. The **piriformis** muscle and the **obturator** muscles are the most important **lateral rotators**. The **adductors** can produce a variety of movements. *(Figure 11–19; Table 11–16)*

7. The **psoas major** and **iliacus** muscles merge to form the **iliopsoas** muscle, a powerful flexor of the hip. *(Figures 11–19, 11–20; Table 11–16)*

8. The flexors of the knee include the **biceps femoris**, **semimembranosus**, and **semitendinosus** muscles (the three *hamstrings*) and the **sartorius** muscle. The **popliteus** muscle unlocks the knee joint. *(Figures 11–20, 11–21; Table 11–17)*

9. Collectively, the knee extensors are known as the **quadriceps femoris**. This group consists of the three **vastus** muscles and the **rectus femoris** muscle. *(Figure 11–20; Table 11–17)*

10. The **gastrocnemius** and **soleus** muscles produce plantar flexion. A pair of **fibularis** muscles produces eversion as well as extension (plantar flexion) at the ankle. *(Figure 11–21; Table 11–18)*

11. Smaller muscles of the calf and shin position the foot and move the toes. Precise control of the phalanges is provided by muscles originating at the tarsal and metatarsal bones. *(Figure 11–22; Table 11–19)*

Pelvis and lower limb muscles: **Anatomy CD-ROM:** Muscular System/Appendicular Dissections/Pelvis & Abdomen, /Thigh, and /Lower Leg

11–5 AGING AND THE MUSCULAR SYSTEM   p. 377

1. With aging, the size and power of all muscle tissues decrease. Skeletal muscles undergo fibrosis, the tolerance for exercise decreases, and repair of injuries slows.

REVIEW QUESTIONS

 More assessment questions are available to you on the Companion Website. You will find Matching, Multiple Choice, True/False, and other quizzes to help further your understanding of the material covered in this chapter. To access the site, go to www.aw.com/martini.

LEVEL 1 Reviewing Facts and Terms

1. Muscle fibers in a single fascicle are arranged
   - **(a)** randomly
   - **(b)** in parallel
   - **(c)** in concentric rings
   - **(d)** in sequence

2. Most muscles in the body are _____ muscles.
   - **(a)** convergent
   - **(b)** pennate
   - **(c)** circular
   - **(d)** parallel

3. Skeletal muscle fibers produce maximum tension over a relatively narrow range of
   - **(a)** complex movements
   - **(b)** extensions
   - **(c)** flexions
   - **(d)** sarcomere lengths

4. The bones, which serve as levers in the body, change
   - **(a)** the direction of an applied force
   - **(b)** the distance and speed of movement produced by a force
   - **(c)** the effective strength of a force
   - **(d)** a, b, and c are correct

5. A general rule about the difference between an origin and an insertion is
   (a) the origin moves while the insertion remains stationary
   (b) each muscle begins and ends at the origin
   (c) the origin remains stationary while the insertion moves
   (d) each muscle begins and ends at the insertion

6. The descriptor used for a muscle that helps larger prime movers work efficiently is
   (a) antagonist   (b) fixator
   (c) agonist   (d) synergist

7. The muscles of facial expression are innervated by cranial nerve
   (a) VII   (b) V
   (c) IV   (d) VI

8. The strongest masticatory muscle is the _____ muscle.
   (a) pterygoid   (b) masseter
   (c) temporalis   (d) mandible

9. The muscles of the tongue are innervated by the
   (a) hypoglossal nerve (XII)
   (b) trochlear nerve (IV)
   (c) abducens nerve (VI)
   (d) a, b, and c are correct

10. The muscle that rotates the eye medially is the _____ muscle.
    (a) superior oblique   (b) inferior rectus
    (c) medial rectus   (d) lateral rectus

11. Important flexors of the vertebral column that act in opposition to the erector spinae are the _____ muscles.
    (a) rectus   (b) longus capitis
    (c) longus colli   (d) scalene

12. The linea alba is a median collagenous partition that longitudinally divides the _____ muscle.
    (a) external oblique   (b) rectus abdominis
    (c) external intercostal (d) rectus femoris

13. The muscular partition that separates the abdominopelvic and thoracic cavities is the
    (a) masseter   (b) perineum
    (c) diaphragm   (d) transversus abdominis muscle

14. The major extensor of the elbow is the _____ muscle.
    (a) triceps brachii   (b) biceps brachii
    (c) deltoid   (d) subscapularis

15. The muscles that rotate the radius without producing either flexion or extension of the elbow are the _____ muscles.
    (a) brachialis and brachioradialis
    (b) pronator teres and supinator
    (c) biceps brachii and triceps brachii
    (d) a, b, and c are correct

16. The powerful flexors of the hip are the _____ muscles.
    (a) piriformis   (b) obturators
    (c) pectineus   (d) iliopsoas

17. Knee extensors known as the quadriceps consist of the _____.
    (a) three vastus muscles and the rectus femoris muscle
    (b) biceps femoris, gracilis, and sartorius muscles
    (c) popliteus, iliopsoas, and gracilis muscles
    (d) gastrocnemius, tibialis, and peroneus muscles

18. What two factors interact to determine the effects of a muscle contraction?

19. List the four fascicle organizations that produce the different patterns of skeletal muscles.

20. What is an aponeurosis? Give two examples.

21. Which four muscle groups make up the axial musculature?

22. Which axial muscles are used in mastication?

23. What three functions are accomplished by the muscles of the pelvic floor?

24. Which four muscles are associated with the rotator cuff?

25. What three functional groups make up the muscles of the lower limbs?

LEVEL 2 Reviewing Concepts

26. Of the following examples, the one that illustrates the action of a second-class lever is
    (a) knee extension
    (b) ankle extension (plantar flexion)
    (c) flexion at the elbow
    (d) a, b, and c are correct

27. An example of an antagonist for the prime mover that produces flexion at the elbow is the _____ muscle.
    (a) brachioradialis   (b) triceps brachii
    (c) brachialis   (d) biceps brachii

28. The removal of the trapezius muscle exposes the
    (a) serratus anterior and subclavian muscles
    (b) infraspinatus and teres minor muscles
    (c) deltoid and supraspinatus muscles
    (d) rhomboid and levator scapulae muscles

29. How do first-, second-, and third-class levers differ?

30. What three actions are used to identify muscle groups? Give a brief description of each action.

31. What is the functional difference between the axial musculature and the appendicular musculature?

32. The muscles of the vertebral column include many dorsal extensors, but few ventral flexors. Why?

33. Which nerves innervate the muscles of respiration?

34. Why does a convergent muscle exhibit more versatility when contracting than does a parallel muscle?

35. Why can a pennate muscle generate more tension than can a parallel muscle of the same size?

36. Why is it difficult to lift a heavy object when the elbow is at full extension?

37. If cranial nerve V (the trigeminal nerve) is damaged or severed, what function in the body will be affected?

38. Which types of movements are affected when the hamstrings are injured?

39. Which three muscular sites are most desirable for intramuscular injections? Why?

LEVEL 3 Critical Thinking and Clinical Applications

40. Mary sees Jill coming toward her and immediately contracts her frontalis and procerus muscles. She also contracts her levator labii on the right side. Is Mary glad to see Jill? How can you tell?

41. Jeff is interested in toning his abdomen, especially the "love handles" on his flanks. What muscles would you tell Jeff to exercise

to accomplish his goal? What movements would best exercise these muscles?

42. Shelly gives her son an ice-cream cone. The boy grasps the cone with his right hand, opens his mouth, and begins to lick at the ice cream. Which muscles does he use to perform these actions?

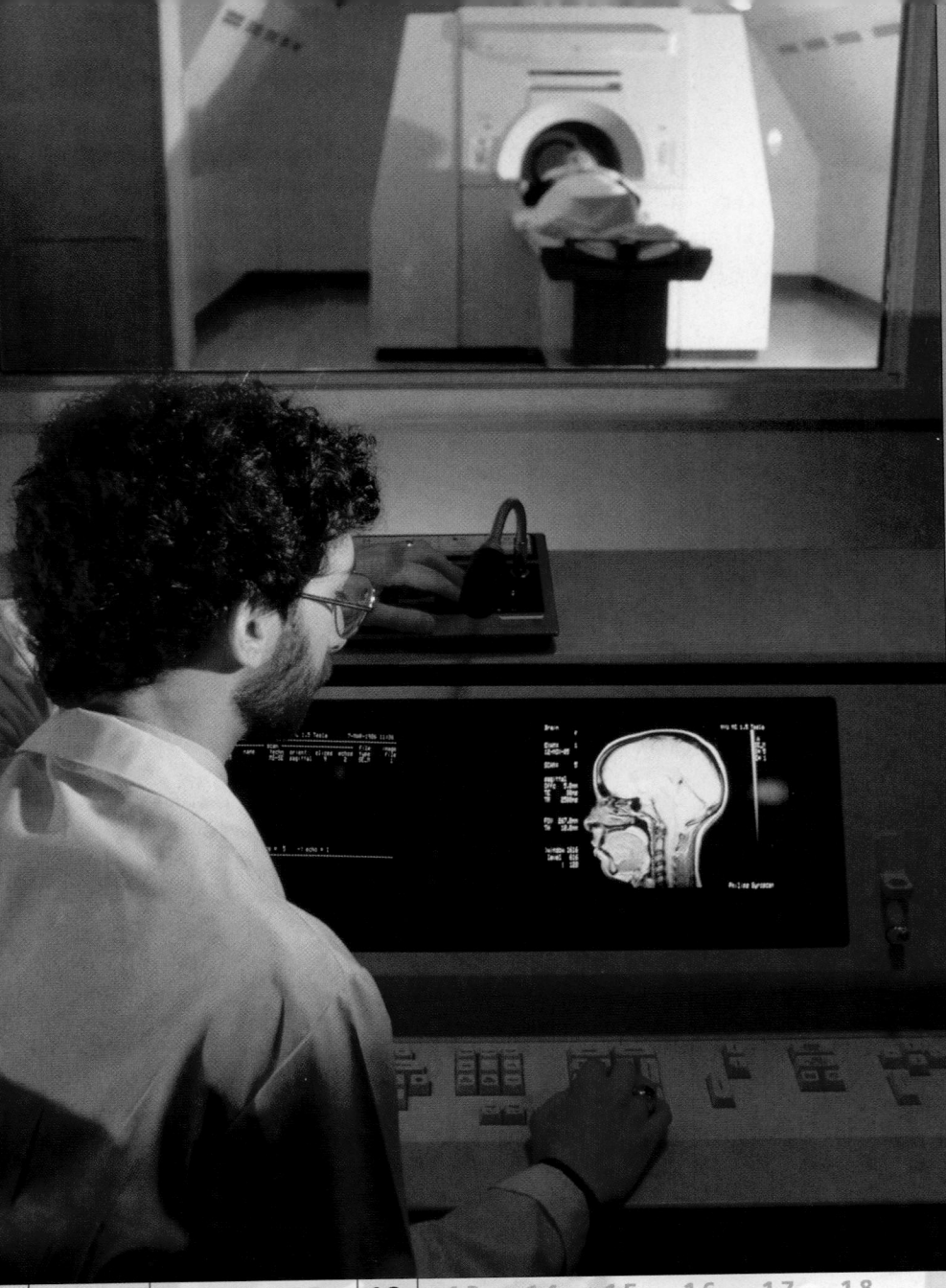

UNIT 3   **CHAPTER** **12**   13   14   15   16   17   18

CLINICAL NOTES
- Rabies 390
- CNS Tumors and Neuroglia 395
- Demyelinating Diseases 395

CLINICAL DISCUSSIONS
- Drugs and Synaptic Function 415
- Factors that Disrupt Neural Function 423

NEURAL TISSUE

In the next seven chapters, our attention will shift to mechanisms that coordinate the activities of the body's organ systems. Each moment the world changes a little bit, and you must recognize and respond to many of those changes. In many cases you respond immediately, yet even the simplest response requires some level of cooperation among systems. For example, when it gets hot, your body starts to sweat; blood vessels in the skin dilate, and sweat glands increase their rates of secretion. The nervous system detects the rise in body temperature and controls the response, all at the subconscious level. Yet if these automatic responses fail to prevent a further increase in body temperature, you will suddenly become consciously aware that you feel hot. You will be acutely uncomfortable, and you will probably look around for a shady spot (and perhaps a cold drink). This series of responses requires the integration of sensory information and the coordination of a hierarchy of motor responses—simple, automatic responses directed at the subconscious level, followed by complex, variable responses that are under conscious control. All of these functions are performed by the *nervous system*.

The nervous system accounts for a mere 3 percent of the total body weight, but it is by far the most complex organ system. It is vital not only to life but also to our awareness and appreciation of life. This chapter examines the structure of neural tissue and introduces important functional principles that govern neural activities. In subsequent chapters, we will consider increasing levels of structural and functional complexity.

The nervous system is not the only system involved with the control of body functions and the coordination of activities involving multiple systems. The *endocrine system* shares many characteristics with the nervous system, and the two systems usually act in a complementary fashion. The nervous system provides swift, but generally brief, responses to stimuli through temporary modifications in the activities of other organ systems. For example, a muscular contraction begins within a few milliseconds of neural stimulation, and the muscle fibers relax almost as quickly when the stimulation ends. The endocrine system, in contrast, adjusts the metabolic operations of other systems, usually to produce longer-term physiological and physical changes. For example, the endocrine system coordinates growth and maturation, sexual development, pregnancy, and responses to chronic environmental stresses. It may also act over comparatively short periods, as during a "fight or flight" emergency, but even then the endocrine response lags far behind that of the nervous system.

Even though endocrine responses develop more slowly than those of the nervous system, they last much longer. This is in part due to the fact that the endocrine system sends its commands in the form of chemical messengers, or *hormones*, that must be distributed by the bloodstream. That mechanism takes a lot more time, and although a single hormone can have multiple targets throughout the body, its effects aren't nearly as precise and finely tuned as those produced by the release of a neurotransmitter at a synapse.

12–1 AN OVERVIEW OF THE NERVOUS SYSTEM

Objective

■ List the two major anatomical divisions of the nervous system, and describe the characteristics of each division.

The nervous system includes all the **neural tissue** in the body. ∞ p. 136 The basic functional units of the nervous system are individual cells called **neurons**. Supporting cells, or **neuroglia** (noo-RŌG-lē-uh or noo-rō-GLĒ-uh; *glia*, glue), separate and protect the neurons, provide a supportive framework for neural tissue, act as phagocytes, and help regulate the composition of the interstitial fluid. Neuroglia, also called *glial cells*, far outnumber neurons and account for roughly half the volume of the nervous system.

Neural tissue, with supporting blood vessels and connective tissues, forms the organs of the nervous system: the brain; the spinal cord; the receptors in complex sense organs, such as the eye and ear; and the *nerves* that interconnect these organs and link the nervous system with other systems. In the *Systems Overview* (pp. 146–153), we introduced the two major anatomical divisions of the nervous system: (1) the central nervous system and (2) the peripheral nervous system.

The **central nervous system** (**CNS**) consists of the spinal cord and brain. These are complex organs that include not only neural tissue, but also blood vessels and the various connective tissues that provide physical protection and support. The CNS is responsible for integrating, processing, and coordinating sensory data and motor commands. Sensory data convey information about conditions inside or outside the body. Motor commands control or adjust the activities of peripheral organs, such as skeletal muscles. When you stumble, the CNS integrates information regarding your balance and the position of your limbs and then coordinates your recovery by sending motor commands to appropriate skeletal muscles—all in a split second and without conscious effort. The CNS—specifically, the brain—is also the seat of higher functions, such as intelligence, memory, learning, and emotion.

The **peripheral nervous system** (**PNS**) includes all the neural tissue outside the CNS. The PNS delivers sensory information to the CNS and carries motor commands to peripheral tissues and systems. Bundles of axons, or *nerve fibers*, carry sensory information and motor commands in the PNS. Such bundles, with associated blood vessels and connective tissues, are called *peripheral nerves*, or simply **nerves**. Nerves connected to the brain are called **cranial nerves**; those attached to the spinal cord are called **spinal nerves**.

Figure 12–1● diagrams the functional divisions of the nervous system. The PNS is divided into afferent and efferent divisions. The **afferent division** (*ad*, to + *ferre*, to carry) of the PNS brings sensory information to the CNS from receptors in peripheral tissues and organs. **Receptors** are sensory structures that either detect changes in the internal environment or respond to the presence of specific stimuli. These structures range from the dendrites (slender cytoplasmic extensions) of single cells to complex organs. Receptors may be neurons or specialized cells of other tissues, such as the Merkel cells of the epidermis. ∞ p. 158

The **efferent division** (*effero*, to bring out) of the PNS carries motor commands from the CNS to muscles and glands. These target organs, which respond by *doing* something, are called **effectors**. The efferent division has both somatic and autonomic components.

- The **somatic nervous system** (**SNS**) controls skeletal muscle contractions. *Voluntary* contractions are under conscious control. For example, you exert conscious control over your arm as you raise a full glass of water to your lips. *Involuntary* contractions may be simple, automatic responses or complex movements, but they are controlled at the subconscious level, outside your conscious awareness. For instance, if you accidentally place your hand on a hot stove, you will withdraw it immediately, usually before you even notice any pain. This type of automatic response is called a **reflex**.

- The **autonomic nervous system** (**ANS**), or *visceral motor system*, provides automatic regulation of smooth muscle, cardiac muscle, and glandular secretions at the subconscious level. The ANS includes a *sympathetic division* and a *parasympathetic division*, which commonly have antagonistic effects. For example, activity of the sympathetic division accelerates the heart rate, whereas parasympathetic activity slows the heart rate. **AM** The Neurological Examination

12–2 NEURONS

Objectives

■ Sketch and label the structure of a typical neuron, and describe the functions of each component.
■ Classify neurons on the basis of their structure and function.

Before we consider how neurons relay information, we must take a closer look at the structure of an individual neuron. In this section, we shall consider (1) the structure of a "model" neuron and (2) the structural and functional classifications of neurons.

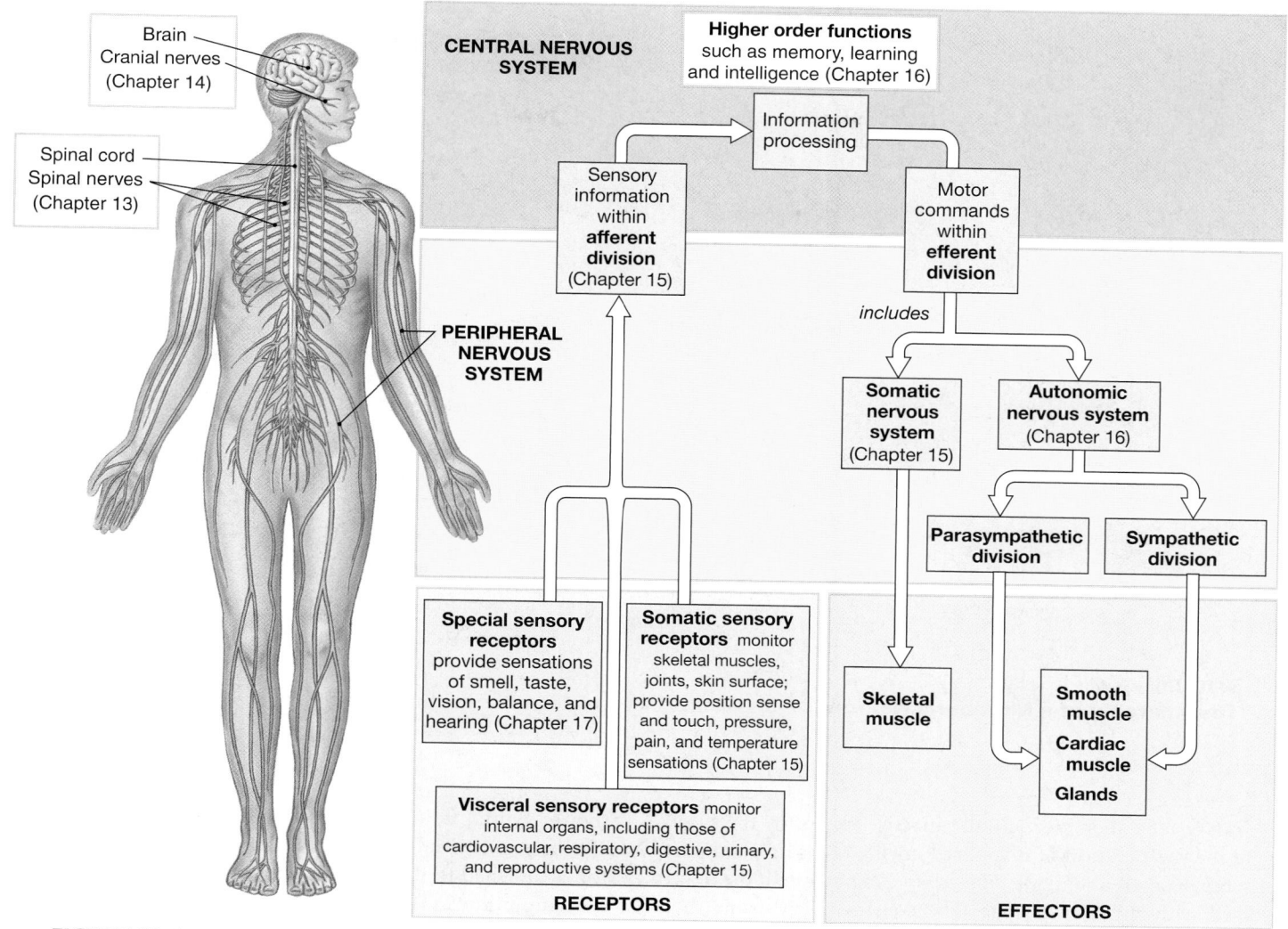

Brain
Cranial nerves
(Chapter 14)

Spinal cord
Spinal nerves
(Chapter 13)

CENTRAL NERVOUS SYSTEM

Higher order functions such as memory, learning and intelligence (Chapter 16)

Information processing

PERIPHERAL NERVOUS SYSTEM

Sensory information within **afferent division** (Chapter 15)

Motor commands within **efferent division**

includes

Somatic nervous system (Chapter 15)

Autonomic nervous system (Chapter 16)

Parasympathetic division

Sympathetic division

Special sensory receptors provide sensations of smell, taste, vision, balance, and hearing (Chapter 17)

Somatic sensory receptors monitor skeletal muscles, joints, skin surface; provide position sense and touch, pressure, pain, and temperature sensations (Chapter 15)

Skeletal muscle

Smooth muscle

Cardiac muscle

Glands

Visceral sensory receptors monitor internal organs, including those of cardiovascular, respiratory, digestive, urinary, and reproductive systems (Chapter 15)

RECEPTORS

EFFECTORS

•**FIGURE 12–1**
A Functional Overview of the Nervous System

☐ THE STRUCTURE OF NEURONS

Figure 12–2● shows the structure of a representative neuron. Neurons have a variety of shapes. The one shown is a *multipolar neuron,* the most common type of neuron in the central nervous system. Each multipolar neuron has a large *cell body,* that is connected to a single, elongate *axon* and several short, branched *dendrites.*

The Cell Body

The **cell body,** or *soma* (plural, *somata*), contains a relatively large, round nucleus with a prominent nucleolus (Figure 12–2●). The cytoplasm surrounding the nucleus constitutes the **perikaryon** (per-i-KAR-ē-on; *karyon,* nucleus). The cytoskeleton of the perikaryon contains **neurofilaments** and **neurotubules,** which are similar to the microfilaments and microtubules of other types of cells. Bundles of neurofilaments, called **neurofibrils,** extend into the dendrites and axon, providing internal support for these slender processes.

The perikaryon contains organelles that provide energy and synthesize organic materials, especially the chemical neurotransmitters that are important in cell-to-cell communication. ∞ p. 301 The numerous mitochondria, free and fixed ribosomes, and membranes of the rough endoplasmic reticulum (RER) give the perikaryon a coarse, grainy appearance. Mitochondria generate ATP to meet the high energy demands of an active neuron. The ribosomes and RER synthesize proteins. Some areas of the perikaryon contain clusters of RER and free ribosomes. These regions, which stain darkly, are called *Nissl bodies,* because they were first described by the German microscopist Franz Nissl. Nissl bodies account for the gray color of areas containing neuron cell bodies—the *gray matter* seen in gross dissection.

Most neurons lack centrioles, important organelles involved in the organization of the cytoskeleton and the movement of chromosomes during mitosis. ∞ p. 76 As a result, typical CNS neurons cannot divide; thus, they cannot be replaced if lost to injury or disease. Although neural stem cells persist in the adult

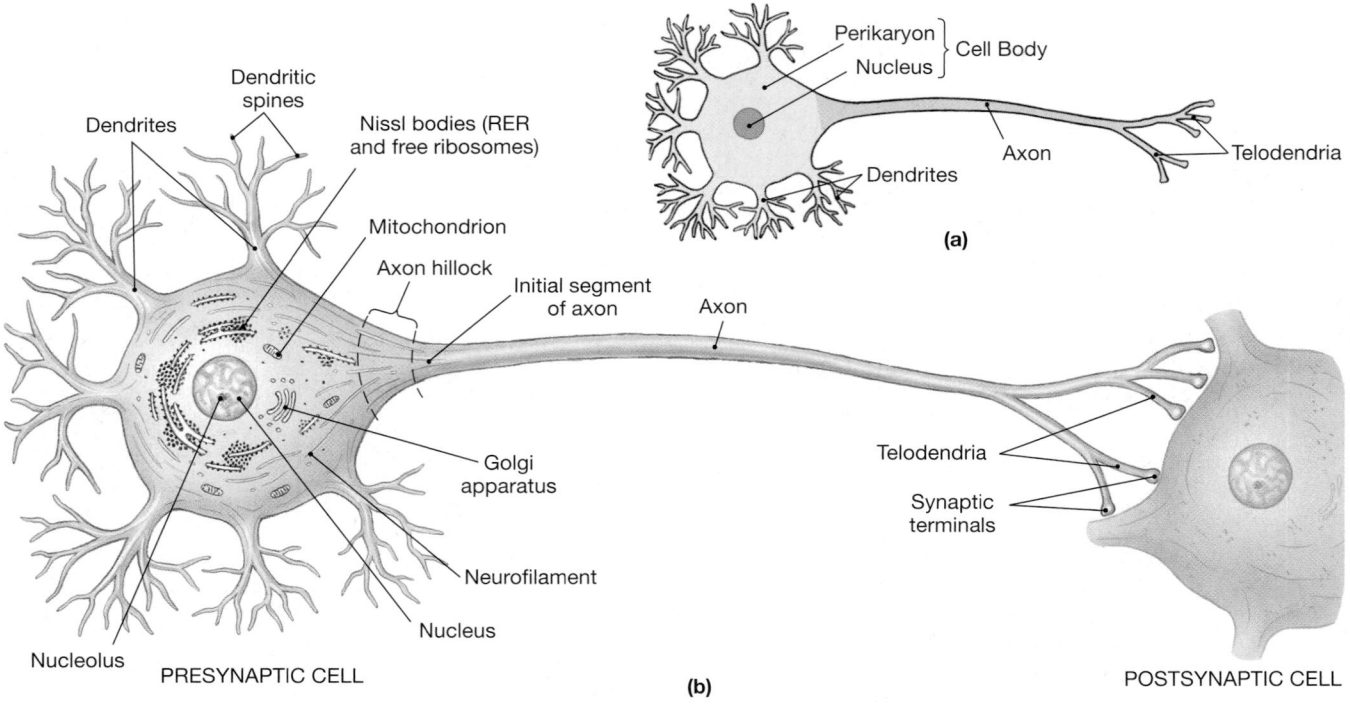

•FIGURE 12–2
The Anatomy of a Multipolar Neuron. A diagrammatic view of a neuron, showing major organelles.

nervous system, they are typically inactive except in the nose, where the regeneration of olfactory (smell) receptors maintains our sense of smell, and in the *hippocampus*, a portion of the brain involved with memory storage. The control mechanisms that trigger neural stem cell activity are now being investigated, with the goal of preventing or reversing neuron loss due to trauma, disease, or aging.

The Dendrites

A variable number of slender, sensitive processes known as **dendrites** extend out from the cell body (Figure 12–2•). Typical dendrites are highly branched, and each branch bears fine processes called *dendritic spines*. In the CNS, a neuron receives information from other neurons primarily at the dendritic spines, which represent 80–90 percent of the neuron's total surface area.

The Axon

An **axon** is a long cytoplasmic process capable of propagating an electrical impulse known as an *action potential*. ∞ p. 301 The **axoplasm** (AK-sō-plazm), or cytoplasm of the axon, contains neurofibrils, neurotubules, small vesicles, lysosomes, mitochondria, and various enzymes. The axoplasm is surrounded by the **axolemma** (*lemma*, husk), a specialized portion of the cell membrane. In the CNS, the axolemma may be exposed to

the interstitial fluid or covered by the processes of neuroglia. The base, or **initial segment**, of the axon in a multipolar neuron is attached to the cell body at a thickened region known as the **axon hillock** (Figure 12–2•).

An axon may branch along its length, producing side branches collectively known as **collaterals**. Collaterals enable a single neuron to communicate with several other cells. The main axon trunk and any collaterals end in a series of fine extensions, or **telodendria** (te-lō-DEN-drē-uh; *telo-*, end + *dendron*, tree) (Figure 12–2•). The telodendria of an axon end at **synaptic terminals**.

The Synapse

Each synaptic terminal is part of a **synapse**, a specialized site where the neuron communicates with another cell. Two cells meet at every synapse: (1) the *presynaptic cell*, which has the synaptic terminal and sends a message, and (2) the *postsynaptic cell*, which receives the message (Figure 12–3•). The communication between cells at a synapse most commonly involves the release of chemicals called **neurotransmitters** by the synaptic terminal. These chemicals, released by the presynaptic cell, affect the activity of the postsynaptic cell. As we saw in Chapter 10, this release is triggered by electrical events, such as the arrival of an action potential. Neurotransmitters are typically packaged in *synaptic vesicles*.

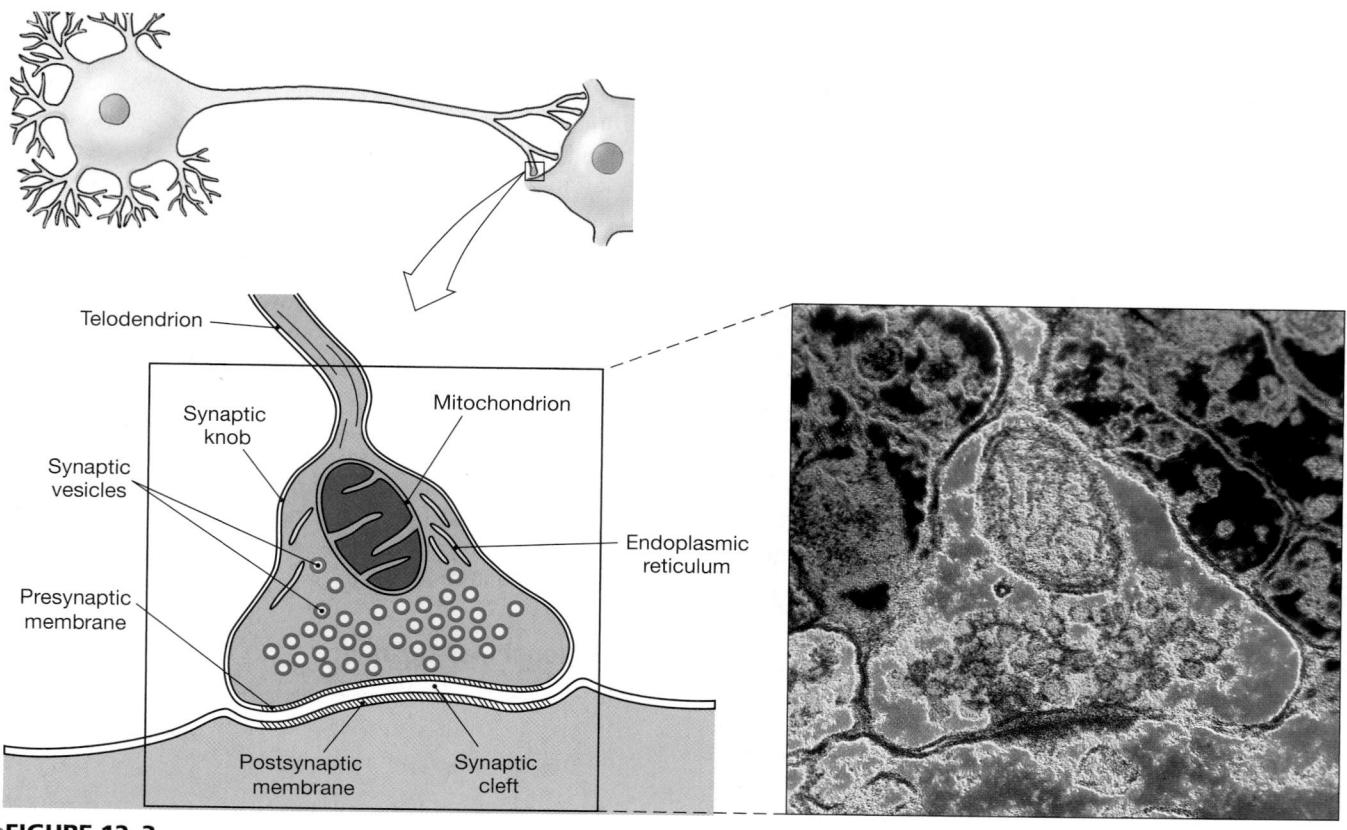

●**FIGURE 12–3**
The Structure of a Typical Synapse. A diagrammatic view of a synapse and a micrograph of a typical synapse between two neurons. (TEM, color enhanced, × 222,000)

Labels in figure:
Telodendrion
Synaptic knob
Synaptic vesicles
Presynaptic membrane
Mitochondrion
Endoplasmic reticulum
Postsynaptic membrane
Synaptic cleft

The presynaptic cell is usually a neuron. (Specialized receptor cells may form synaptic connections with the dendrites of neurons, a process that will be described in Chapter 15.) The postsynaptic cell can be either a neuron or another type of cell. When one neuron communicates with another, the synapse may occur on a dendrite, on the cell body, or along the length of the axon of the receiving cell. A synapse between a neuron and a muscle cell is called a **neuromuscular junction**. ◯⊃ p. 301 At a **neuroglandular junction**, a neuron controls or regulates the activity of a secretory (gland) cell. Neurons also innervate a variety of other cell types, such as adipocytes (fat cells). We will consider the nature of that innervation in later chapters.

The structure of the synaptic terminal varies with the type of postsynaptic cell. A relatively simple, round **synaptic knob** occurs where the postsynaptic cell is another neuron.* At a synapse, a narrow synaptic cleft separates the **presynaptic membrane**, where neurotransmitters are released, from the **postsynaptic membrane**, which bears receptors for neurotransmitters. The synaptic terminal at a neuromuscular junction is much more complex. We shall deal primarily with the

structure of synaptic knobs in this chapter, leaving the details of other types of synaptic terminals to later chapters.

AXOPLASMIC TRANSPORT Each synaptic knob contains mitochondria, portions of the endoplasmic reticulum, and thousands of vesicles filled with neurotransmitter molecules. Breakdown products of neurotransmitters released at the synapse are reabsorbed and reassembled at the synaptic knob, which also receives a continuous supply of neurotransmitters synthesized in the cell body, along with enzymes and lysosomes. These materials travel the length of the axon along neurotubules, pushed along by "molecular motors," called *kinesins*, that run on ATP. The movement of materials between the cell body and synaptic knobs is called **axoplasmic transport**. Some materials travel slowly, at rates of a few millimeters per day. This transport mechanism is known as the "slow stream." Vesicles containing neurotransmitters move much more rapidly, traveling in the "fast stream" at 5–10 mm per hour.

Axoplasmic transport occurs in both directions. The flow of materials from the cell body to the synaptic knob is *anterograde* (AN-ter-ō-grād) *flow*. At the same time, other substances are being transported toward the cell body in *retrograde* (RET-rō-grād) *flow* (*retro*, backward). If debris or unusual chemicals appear in the synaptic knob, retrograde flow soon delivers them to the cell body. The arriving materials may then alter the activity of the cell by turning appropriate genes on or off.

---

*The term *synaptic knob* is widely recognized and will be used throughout this text. However, the same structures are also called terminal buttons, terminal boutons, end bulbs, or neuropods.

Rabies is perhaps the most dramatic example of a clinical condition directly related to retrograde flow. A bite from a rabid animal injects the rabies virus into peripheral tissues, where virus particles quickly enter synaptic knobs and peripheral axons. Retrograde flow then carries the virus into the CNS, with fatal results. Many toxins, including heavy metals, some pathogenic bacteria, and other viruses, also bypass CNS defenses by exploiting axoplasmic transport. **AM** Axoplasmic Transport and Disease

☐ THE CLASSIFICATION OF NEURONS

The neurons in the nervous system are quite variable in form. They can be grouped by structure or by function.

Structural Classification

Neurons are classified as anaxonic, bipolar, unipolar, or multipolar on the basis of the relationship of the dendrites to the cell body and the axon:

1. **Anaxonic** (an-ak-SON-ik) **neurons** are small, and have no anatomical features that distinguish dendrites from axons; all the cell processes look alike (Figure 12–4a●). Anaxonic neurons are located in the brain and in special sense organs. Their functions are poorly understood.

2. **Bipolar neurons** have two distinct processes—one dendritic process that branches extensively at its distal tip and one axon, with the cell body between them (Figure 12–4b●). Bipolar neurons are rare, but occur in special sense organs, where they relay informa-

tion about sight, smell, or hearing from receptor cells to other neurons. Bipolar neurons are smaller than unipolar or multipolar neurons; the largest measure less than 30 mm from end to end.

3. In a **unipolar neuron**, or *pseudounipolar neuron*, the dendrites and axon are continuous—basically, fused—and the cell body lies off to one side (Figure 12–4c●). In such a neuron, the initial segment lies where the dendrites converge. The rest of the process, which carries action potentials, is usually considered to be an axon. Most sensory neurons of the peripheral nervous system are unipolar. Their axons may extend a meter or more, ending at synapses in the central nervous system. The longest are those carrying sensations from the tips of the toes to the spinal cord.

4. **Multipolar neurons** have two or more dendrites and a single axon. Multipolar neurons (Figure 12–4d●) are the most common type of neuron in the CNS. For example, all the motor neurons that control skeletal muscles are multipolar neurons. The axons of multipolar neurons can be as long as those of unipolar neurons; the longest carry motor commands to small muscles that move the toes.

Functional Classification

Alternatively, neurons are categorized by function as (1) sensory neurons, (2) motor neurons, and (3) interneurons. Their relationships are diagrammed in Figure 12–5●.

SENSORY NEURONS Sensory neurons, or *afferent neurons*, form the afferent division of the PNS. They deliver information from sensory receptors to the CNS. The cell bodies of sensory neurons are located in peripheral *sensory ganglia*. (A *ganglion* is a collection of

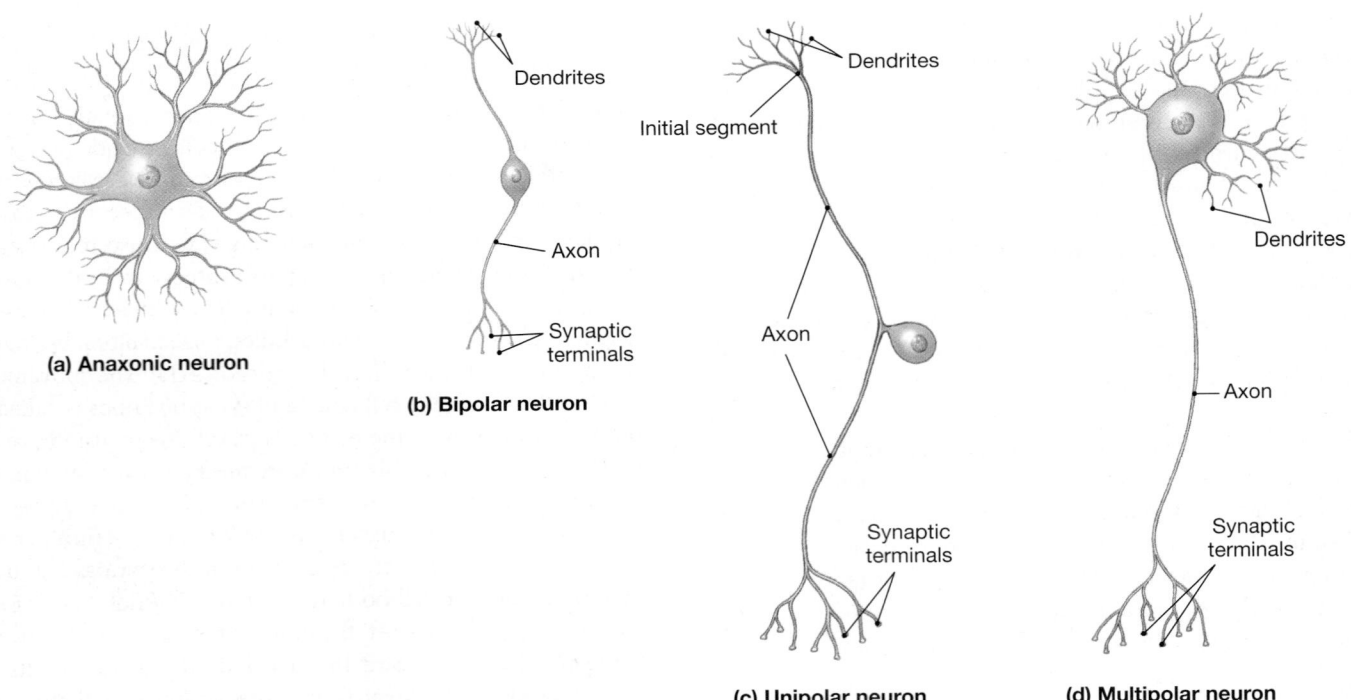

(a) Anaxonic neuron

(b) Bipolar neuron
Dendrites
Axon
Synaptic terminals

(c) Unipolar neuron
Dendrites
Initial segment
Axon
Synaptic terminals

(d) Multipolar neuron
Dendrites
Dendrites
Axon
Synaptic terminals

●**FIGURE 12–4**
A Structural Classification of Neurons.   The neurons are not drawn to scale; typical anaxonic neurons and bipolar neurons are many times smaller than typical unipolar or multipolar neurons.

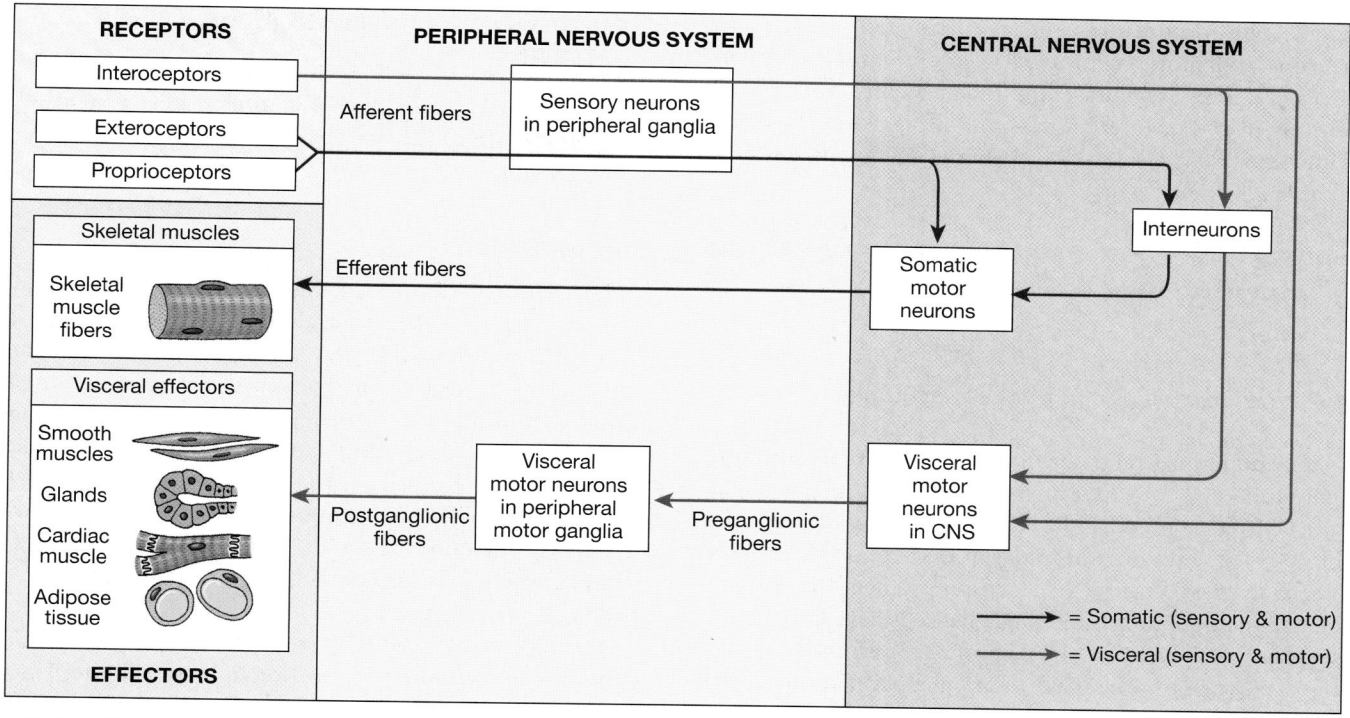

●**FIGURE 12–5**
A Functional Classification of Neurons. The boxes for sensory neurons, motor neurons, and interneurons indicate the locations of the cell bodies.

neuron cell bodies in the PNS.) Sensory neurons are unipolar neurons with processes, known as **afferent fibers**, that extend between a sensory receptor and the spinal cord or brain. Sensory neurons collect information concerning the external or internal environment. The human body has about 10 million such neurons. **Somatic sensory neurons** monitor the outside world and our position within it. **Visceral sensory neurons** monitor internal conditions and the status of other organ systems.

Sensory receptors are either the processes of specialized sensory neurons or cells monitored by sensory neurons. Receptors are broadly categorized as follows:

1. **Exteroceptors** (*extero-*, outside) provide information about the external environment in the form of touch, temperature, or pressure sensations and the more complex senses of sight, smell, and hearing.

2. **Proprioceptors** (prō-prē-ō-SEP-torz) monitor the position and movement of skeletal muscles and joints.

3. **Interoceptors** (*intero-*, inside) monitor the digestive, respiratory, cardiovascular, urinary, and reproductive systems and provide sensations of taste, deep pressure, and pain.

MOTOR NEURONS Motor neurons, or *efferent neurons*, form the efferent division of the PNS. These neurons carry instructions from the CNS to peripheral effectors in a peripheral tissue, organ, or organ system. Your body has about half a million motor neurons. Axons traveling away from the CNS are called **efferent fibers**. As we learned earlier, the two major efferent systems are

the somatic nervous system (SNS) and the autonomic (visceral) nervous system (ANS). The somatic nervous system includes all the **somatic motor neurons** that innervate skeletal muscles. You have conscious control over the activity of the SNS. The cell body of a somatic motor neuron lies in the CNS, and its axon extends into the periphery to innervate skeletal muscle fibers at neuromuscular junctions.

You do not have conscious control over the activities of the ANS. **Visceral motor neurons** innervate all peripheral effectors other than skeletal muscles. Thus, the ANS innervates smooth muscle, cardiac muscle, glands, and adipose tissue throughout the body. The axons of visceral motor neurons in the CNS innervate a second set of visceral motor neurons in peripheral *autonomic ganglia*. The neurons whose cell bodies are located in those ganglia innervate and control peripheral effectors.

To get from the CNS to a visceral effector, such as a smooth muscle cell, the signal must travel along one axon, be relayed across a synapse, and then travel along a second axon to its final destination. The axons extending from the CNS to an autonomic ganglion are called *preganglionic fibers*. Axons connecting the ganglion cells with the peripheral effectors are known as *postganglionic fibers*.

INTERNEURONS The 20 billion or so **interneurons**, or *association neurons*, outnumber all other types of neurons combined. Most are located within the brain and spinal cord; some are in autonomic ganglia. Interneurons are responsible for the distribution of sensory information and the coordination of motor activity.

For example, one or more interneurons are situated between sensory neurons and motor neurons. The more complex the response to a given stimulus, the greater the number of interneurons involved. Interneurons are also involved with all higher functions, such as memory, planning, and learning.

12–3 NEUROGLIA

Objective

■ Describe the locations and functions of neuroglia.

The organization of neural tissue in the CNS differs significantly from that in the PNS, owing primarily to their distinctive populations of neuroglia. The CNS has a greater variety of neuroglia than the PNS has. Although histological descriptions have been available for the past century, the technical problems involved in isolating and manipulating individual glial cells have limited our understanding of their functions. Figure 12–6● summarizes information known about the major neuroglia populations in the CNS and PNS.

☐ NEUROGLIA OF THE CENTRAL NERVOUS SYSTEM

The central nervous system has four types of neuroglia: (1) *ependymal cells*, (2) *astrocytes*, (3) *oligodendrocytes*, and (4) *microglia*.

Ependymal Cells

A fluid-filled central passageway extends along the longitudinal axis of the spinal cord and brain. The thickness of the wall and the diameter of this passageway vary from one region to another. The narrow passageway in the spinal cord is called the *central canal*. In the brain, the passageway forms enlarged chambers called *ventricles*. The central canal and ventricles are lined by a cellular layer of epithelial cells called the **ependyma** (e-PEN-di-muh) and are filled with **cerebrospinal fluid (CSF)**. This fluid, which also surrounds the brain and spinal cord, provides a protective cushion and transports dissolved gases, nutrients, wastes, and other materials.

The ependyma consists of **ependymal cells**, which are cuboidal to columnar in sectional view (Figure 12–7a,b●). During embryonic development and early childhood, the free

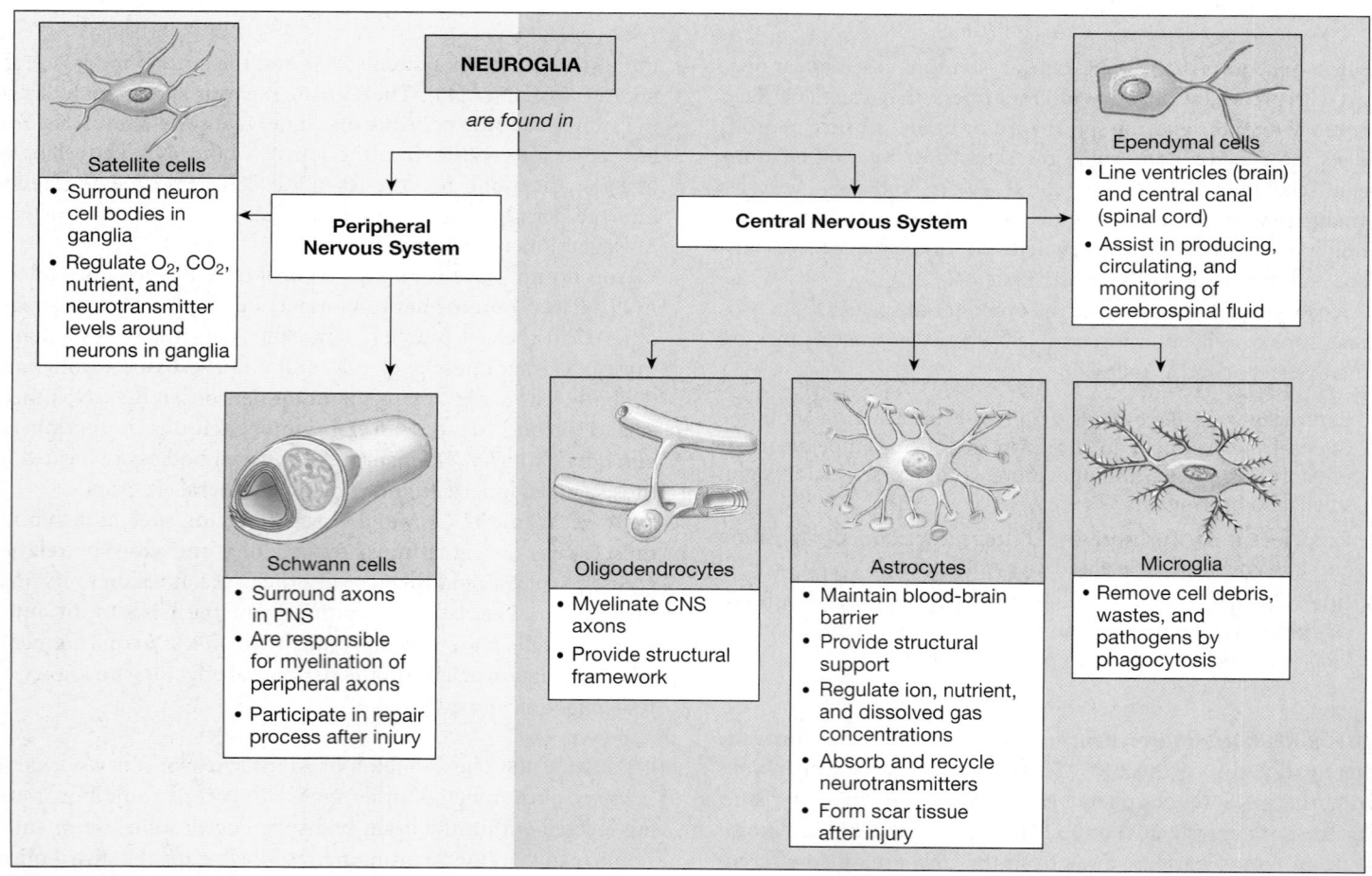

●**FIGURE 12–6**
An Introduction to Neuroglia

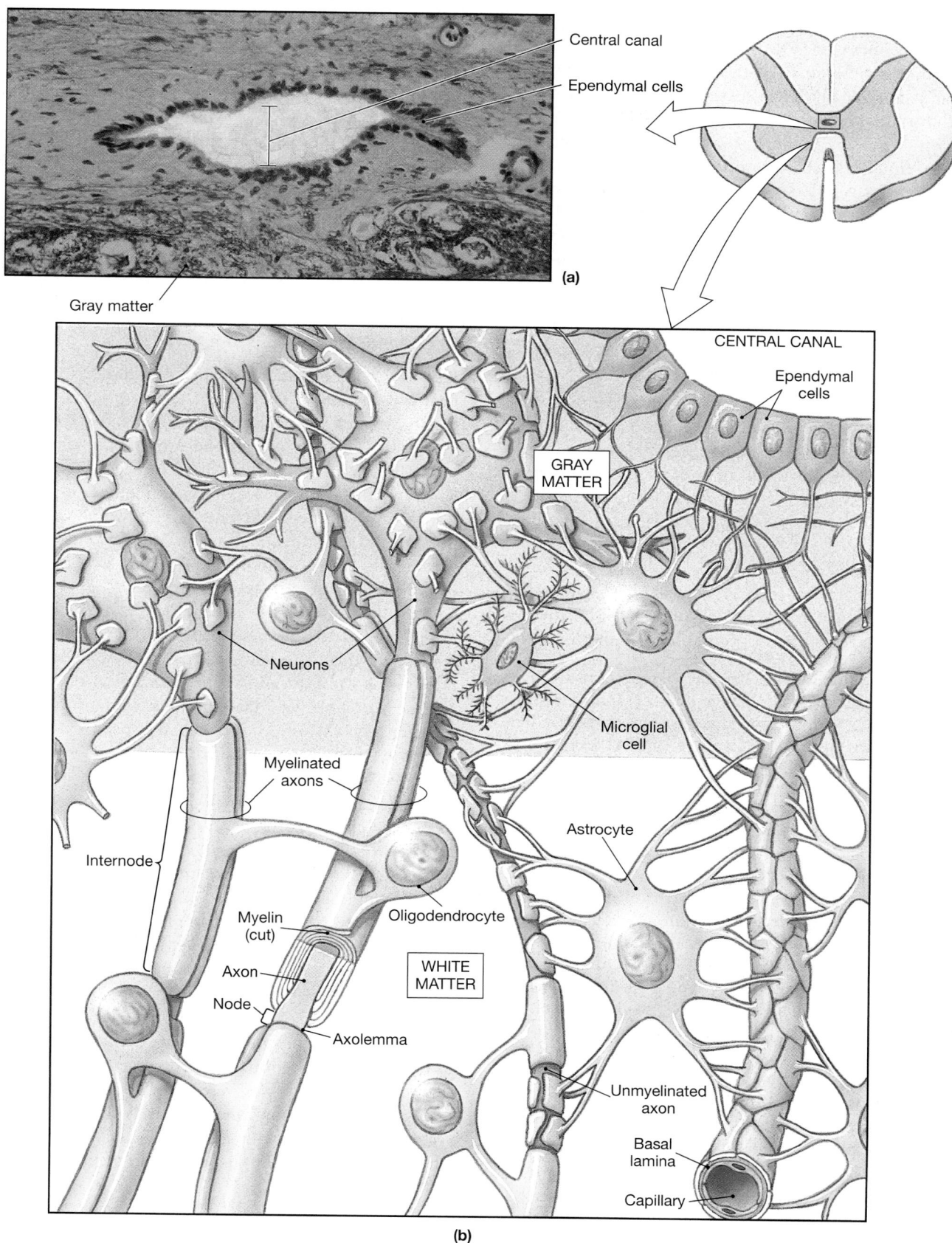

Central canal

Ependymal cells

Gray matter

(a)

CENTRAL CANAL

Ependymal cells

GRAY MATTER

Neurons

Microglial cell

Astrocyte

Myelinated axons

Internode

Myelin (cut)

Oligodendrocyte

Axon

WHITE MATTER

Node

Axolemma

Unmyelinated axon

Basal lamina

Capillary

(b)

●FIGURE 12–7
Neuroglia in the CNS. **(a)** Light micrograph showing the ependymal lining of the central canal of the spinal cord. (LM × 236) **(b)** A diagrammatic view of neural tissue in the CNS, showing relationships between neuroglia and neurons.

surfaces of ependymal cells are covered with cilia. The cilia persist in adults only within the ventricles of the brain, where they assist in the circulation of the CSF. In other areas, the ependymal cells typically have scattered microvilli. In a few parts of the brain, specialized ependymal cells participate in the secretion of the CSF. Other regions of the ependyma may have sensory functions, such as monitoring the composition of the CSF.

Unlike typical epithelial cells, ependymal cells have slender processes that branch extensively and make direct contact with neuroglia in the surrounding neural tissue. The functions of these connections are not known. During early embryonic development, stem cells line the central canal and ventricles; the divisions of these stem cells give rise to neurons and all CNS neuroglia other than microglia. **ATLAS** Embryology Summary 10: An Introduction to the Development of the Nervous System

Evidence indicates that the ependyma in adults contains stem cells that can divide to produce additional neurons. The specific regulatory mechanisms are now being investigated. If we could control this process, the treatment of strokes and several other debilitating nervous system disorders would be revolutionized.

Astrocytes

Astrocytes (AS-trō-sīts; *astro-*, star + *cyte*, cell) are the largest and most numerous neuroglia in the CNS (Figure 12–7b●). Astrocytes have a variety of functions, many of them poorly understood:

- **Maintaining the Blood–Brain Barrier.** Compounds dissolved in the circulating blood do not have free access to the interstitial fluid of the CNS. Neural tissue must be physically and biochemically isolated from the general circulation, because hormones or other chemicals in the blood can alter neuron function. The endothelial cells lining CNS capillaries control the chemical exchange between the blood and interstitial fluid. These cells create a **blood–brain barrier**, which isolates the CNS from the general circulation.

  The slender cytoplasmic extensions of astrocytes end in expanded "feet," processes that wrap around capillaries. Astrocytic processes form a complete blanket around the capillaries, interrupted only where other neuroglia come in contact with the capillary walls. Chemicals secreted by astrocytes are somehow responsible for maintaining the special permeability characteristics of endothelial cells. (We shall discuss the blood–brain barrier further in Chapter 14.)

- **Creating a Three-Dimensional Framework for the CNS.** Astrocytes are packed with microfilaments that extend across the breadth of the cell and its processes. This extensive cytoskeletal reinforcement assists astrocytes in providing a structural framework for the neurons of the brain and spinal cord.

- **Repairing Damaged Neural Tissue.** In the CNS, damaged neural tissue seldom regains normal function. However, astrocytes moving into the site of injury can make structural repairs that stabilize the tissue and prevent further injury. We shall detail neural damage and subsequent repair in a later section.

- **Guiding Neuron Development.** Astrocytes in the embryonic brain appear to be involved in directing both the growth and interconnection of developing neurons.

- **Controlling the Interstitial Environment.** Although much remains to be learned about astrocyte physiology, evidence indicates that astrocytes adjust the composition of interstitial fluid by several means: (1) regulating the concentration of sodium ions, potassium ions, and carbon dioxide; (2) providing a rapid-transit system for the transport of nutrients, ions, and dissolved gases between capillaries and neurons; (3) controlling the volume of blood flow through the capillaries; (4) absorbing and recycling some neurotransmitters; and (5) releasing chemicals that enhance or suppress communication across synaptic terminals.

Oligodendrocytes

Like astrocytes, **oligodendrocytes** (o-li-gō-DEN-drō-sīts; *oligo-*, few) possess slender cytoplasmic extensions, but the cell bodies of oligodendrocytes are smaller, and have fewer processes, than astrocytes (Figure 12–7b●). The processes of oligodendrocytes generally are in contact with the exposed surfaces of neurons. The functions of processes ending at the neuron cell body have yet to be determined. Much more is known about the processes that end on the surfaces of axons. Many axons in the CNS are completely sheathed in these processes, which insulate them from contact with the extracellular fluid.

Near the tip of each process, the axolemma expands to form an enormous membranous pad, and the cytoplasm there becomes very thin. This flattened "pancake" somehow gets wound around the axon, forming concentric layers of cell membrane (Figure 12–7b●). The membranous wrapping of insulation, called **myelin** (MĪ-e-lin), increases the speed at which an action potential travels along the axon.

Many oligodendrocytes cooperate in the formation of a **myelin sheath** along the length of an axon. The axon is then said to be **myelinated**. Each oligodendrocyte myelinates segments of several axons. The relatively large areas of the axon that are thus wrapped in myelin are called **internodes** (*inter*, between). Internodes typically range from 1–2 mm in length. Small gaps of a few micrometers separate adjacent internodes. These gaps are called **nodes**, or *nodes of Ranvier* (rahn-vē-Ā). When an axon branches, the branches originate at nodes.

In dissection, myelinated axons appear glossy white, primarily because of the lipids that are present. Regions dominated by myelinated axons constitute the **white matter** of the CNS. Not all axons in the CNS are myelinated, however. **Unmyelinated** axons may not be completely covered by the processes of neuroglia. Such axons are common where relatively short axons and collaterals form synapses with densely packed neuron cell bodies. These areas have a dusky gray color, and they constitute the **gray matter** of the CNS.

In sum, oligodendrocytes play a role in structural organization by tying clusters of axons together, and these neuroglia also improve the functional performance of neurons by wrapping axons within a myelin sheath.

Microglia

The least numerous and smallest neuroglia in the CNS are **microglia** (mī-KRŌG-lē-uh). Their slender cytoplasmic processes have many fine branches (Figure 12–7b●). These cells are capable of migrating through neural tissue. Microglia appear early in embryonic development, originating from mesodermal stem cells related to those stem cells which produce monocytes and macrophages. ∞ pp. 123, 128 Microglia migrate into the CNS as the nervous system forms. Thereafter, they remain isolated in neural tissue, where they act as a wandering police force and janitorial service, engulfing cellular debris, waste products, and pathogens.

Tumors of the brain, spinal cord, and associated membranes result in approximately 90,000 deaths in the United States each year. Tumors that originate in the central nervous system are called *primary CNS tumors*, to distinguish them from *secondary CNS tumors*, which arise from the metastasis of cancer cells that originate elsewhere. Roughly 75 percent of CNS tumors are primary tumors. In adults, primary CNS tumors result from the divisions of abnormal neuroglia rather than from the divisions of abnormal neurons, because typical neurons in adults cannot divide. However, through the divisions of stem cells, neurons increase in number until children reach age 4. As a result, primary CNS tumors involving abnormal neurons can occur in young children. Symptoms of CNS tumors vary with the location affected. Treatment may involve surgery, radiation, chemotherapy, or a combination of these procedures.

□ NEUROGLIA OF THE PERIPHERAL NERVOUS SYSTEM

The cell bodies of neurons in the PNS are clustered in masses called **ganglia** (singular, *ganglion*). Neuronal cell bodies and most axons in the PNS are completely insulated from their surroundings by the processes of neuroglia. The two types of neuroglia involved are called *satellite cells* and *Schwann cells*.

Satellite cells, or *amphicytes* (AM-fi-sīts), surround the neuron cell bodies in ganglia. They regulate the environment around the neurons, much as astrocytes do in the CNS. **Schwann cells**, or *neurilemmal cells* (*neurilemmocytes*), form a sheath around peripheral axons. Wherever a Schwann cell covers an axon, the outer surface of the Schwann cell is called the **neurilemma** (noo-ri-LEM-uh). Most axons in the PNS, myelinated or unmyelinated, are shielded from contact with interstitial fluids by Schwann cells.

Whereas an oligodendrocyte in the CNS may myelinate portions of several adjacent axons (Figure 12–7●), a Schwann cell can myelinate only one segment of a single axon (Figure 12–8a●). However, a Schwann cell can *enclose* segments of several unmyelinated axons (Figure 12–8b●). A series of Schwann cells is required to enclose an axon along its entire length.

Demyelination is the progressive destruction of myelin sheaths, both in the CNS and in the PNS. The result is a loss of sensation and motor control that leaves affected regions numb and paralyzed. Many unrelated conditions that result in the destruction of myelin can cause symptoms of demyelination. Several important demyelinating disorders are *heavy-metal poisoning*, *diphtheria*, *multiple sclerosis (MS)*, and *Guillain–Barré syndrome*. **AM** Demyelination Disorders

□ NEURAL RESPONSES TO INJURIES

A neuron responds to injury in a very limited, stereotyped fashion. In the cell body, the Nissl bodies disperse and the nucleus moves away from its centralized location as the cell increases its rate of protein synthesis. If the neuron recovers its functional abilities, it will return to a normal appearance. The key to recovery seems to be the events under way in the axon. Consider the response of a neuron to mechanical stresses, such as the pressure applied during a crushing injury. The pressure produces a local decrease in blood flow and oxygen availability, and the affected membrane becomes unexcitable. If the pressure is released after an hour or two, the neuron will recover within a few weeks. More severe or prolonged pressure will produce effects similar to those caused by cutting off the distal portion of the axon.

In the PNS, the Schwann cells participate in the repair of damaged nerves. In the process known as **Wallerian degeneration** (Figure 12–9●), the axon distal to the injury site degenerates, and macrophages migrate into the area to phagocytize the debris. The Schwann cells do not degenerate. Instead, they proliferate and form a solid cellular cord that follows the path of the original axon. As the neuron recovers, its axon grows into the site of injury, and the Schwann cells wrap around the axon.

If it continues to grow into the periphery alongside the appropriate cord of Schwann cells, the axon may eventually reestablish its normal synaptic contacts. If it stops growing or wanders off in some new direction, normal function will not return. The growing axon is most likely to arrive at its appropriate destination if the cut edges of the original nerve bundle remain in contact.

Limited regeneration can occur in the CNS, but the situation is more complicated because (1) many more axons are likely to be involved, (2) astrocytes produce scar tissue that can prevent axon growth across the damaged area, and (3) astrocytes release chemicals that block the regrowth of axons.

✓ What would damage to the afferent division of the PNS affect?

✓ A tissue sample shows unipolar neurons. Are these more likely to be sensory neurons or motor neurons?

✓ Which type of neuroglia would occur in larger than normal numbers in the brain tissue of a person with a CNS infection?

Answers start on page Q-1

 Review neural anatomy on the **IP CD-ROM**: Nervous System I/Anatomy Review.

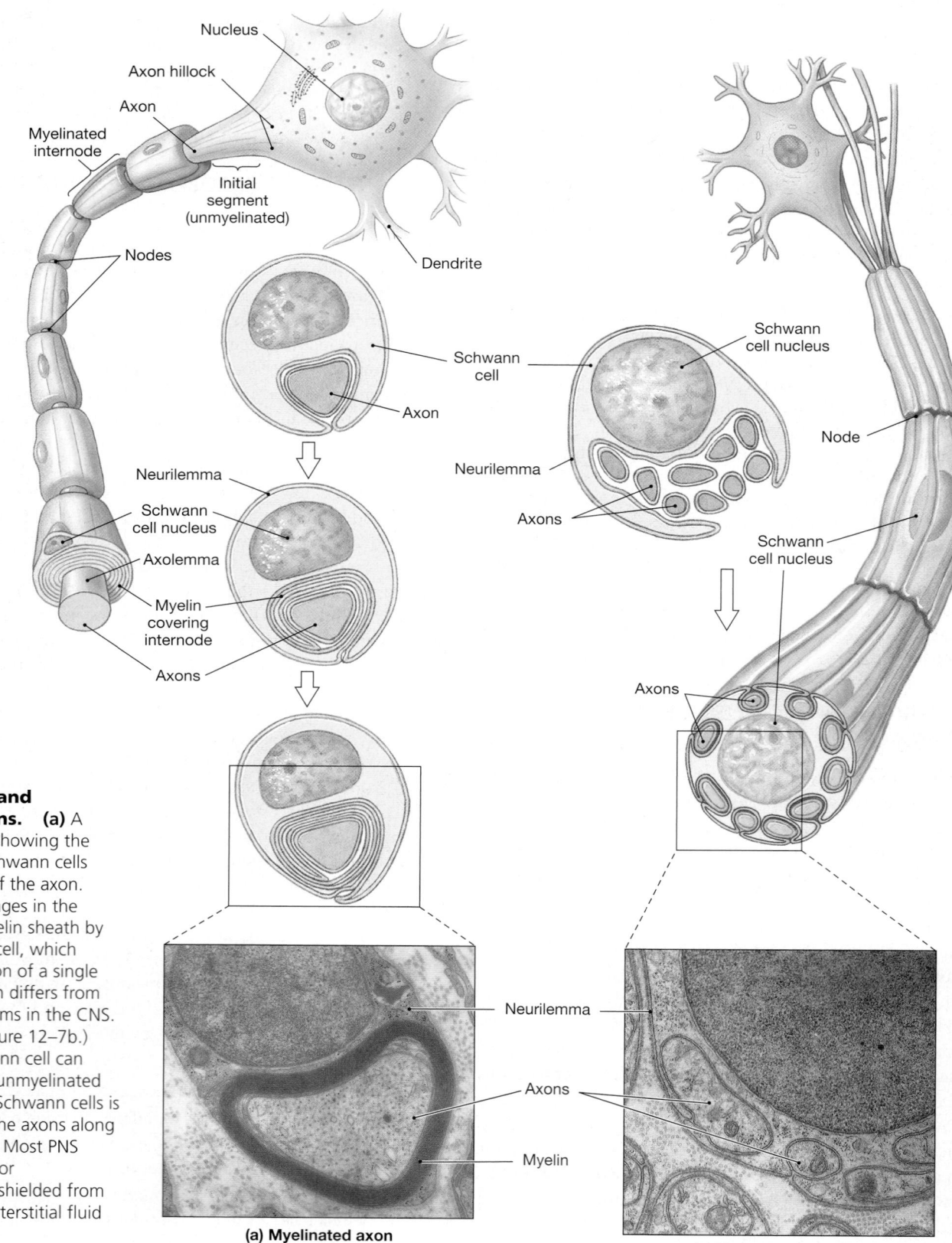

(Compare with Figure 12–7b.)

●FIGURE 12–8
Schwann Cells and Peripheral Axons. (a) A myelinated axon, showing the organization of Schwann cells along the length of the axon. Also shown are stages in the formation of a myelin sheath by a single Schwann cell, which myelinates a portion of a single axon. This situation differs from the way myelin forms in the CNS. (Compare with Figure 12–7b.) (b) A single Schwann cell can enfold a group of unmyelinated axons. A series of Schwann cells is required to cover the axons along their entire length. Most PNS axons, myelinated or unmyelinated, are shielded from contact with the interstitial fluid by Schwann cells.

(a) Myelinated axon

(b) Unmyelinated axon

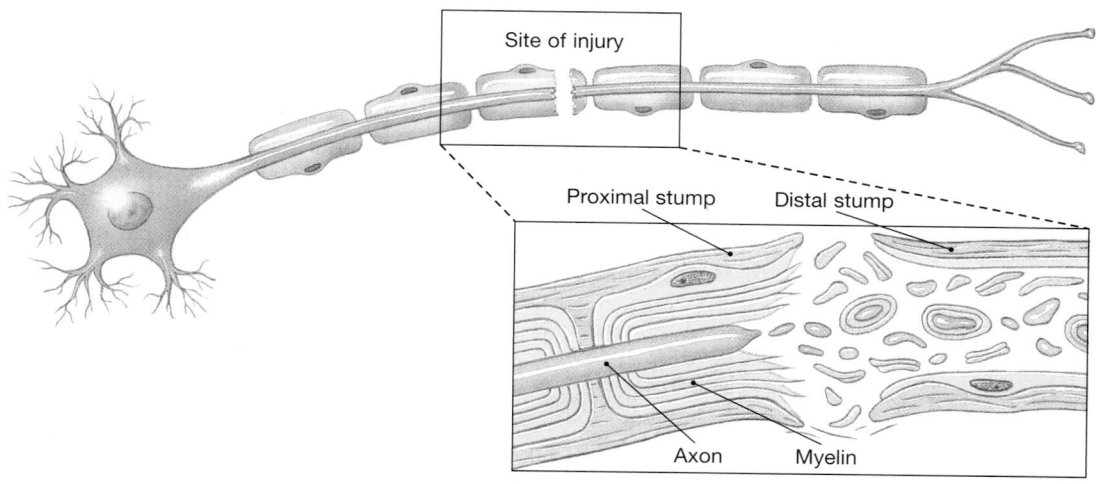

Site of injury

Proximal stump   Distal stump

Axon   Myelin

STEP 1:
Fragmentation of axon and myelin occurs in distal stump.

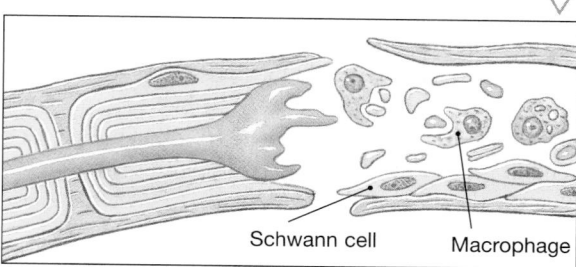

Schwann cell   Macrophage

STEP 2:
Schwann cells form cord, grow into cut, and unite stumps. Macrophages engulf degenerating axon and myelin.

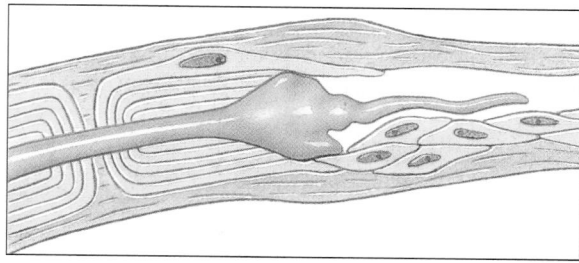

STEP 3:
Axon sends buds into network of Schwann cells and then starts growing along cord of Schwann cells.

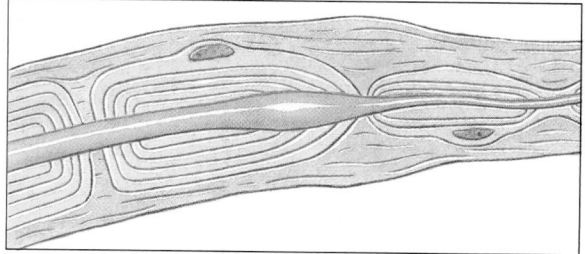

●**FIGURE 12–9**
Peripheral Nerve Regeneration after Injury

STEP 4:
Axon continues to grow into distal stump and is enfolded by Schwann cells.

12–4 NEUROPHYSIOLOGY: IONS AND ELECTRICAL SIGNALS

Objectives

- Explain how the resting potential is created and maintained.
- Describe the events involved in the generation and propagation of an action potential.
- Discuss the factors that affect the speed with which action potentials are propagated.

In Chapter 3, we introduced the concepts of the *transmembrane potential* and the *resting potential*, two characteristic features of all cells. p. 99 In this section, we shall focus on the membrane properties of neurons; many of the principles discussed also apply to other types of cells.

Figure 12–10• introduces the important membrane processes we shall be examining.

- The *resting potential* is the transmembrane potential of a resting cell. All neural activities begin with a change in the resting potential of a neuron.
- A typical stimulus produces a temporary, localized change in the resting potential. The effect, which decreases with distance from the stimulus, is called a *graded potential*.
- If the graded potential is sufficiently large, it produces an *action potential* in the membrane of the axon. An action potential is an electrical impulse that is propagated across the surface of the membrane and that does not diminish as it moves away from its source. This impulse travels along the axon to one or more synapses.
- *Synaptic activity* then produces graded potentials in the cell membrane of the postsynaptic cell. The process typically involves

the release of neurotransmitters, such as ACh, that bind to receptors on the postsynaptic cell membrane. The mechanism is comparable to that of the neuromuscular junction, described in Chapter 10. p. 301

- The response of the postsynaptic cell ultimately depends on what the stimulated receptors do and what other stimuli are influencing the cell at the same time. The integration of stimuli at the level of the individual cell is the simplest form of *information processing* in the nervous system.

When you understand each of the foregoing processes, you will know how neurons process information and communicate with one another and with peripheral effectors.

◻ THE TRANSMEMBRANE POTENTIAL

Chapter 3 introduced three important concepts regarding the transmembrane potential:

1. The intracellular fluid (cytosol) and extracellular fluid differ markedly in ionic composition. The extracellular fluid (ECF) contains high concentrations of sodium ions (Na^+) and chloride ions (Cl^-), whereas the cytosol contains high concentrations of potassium ions (K^+) and negatively charged proteins.
2. If the cell membrane were freely permeable, diffusion would continue until all the ions were evenly distributed across the membrane, at equilibrium. But an even distribution does not occur, because cells have selectively permeable membranes. p. 90 Ions cannot freely cross the lipid portions of the cell membrane; they can enter or leave the cell only through membrane channels. Many kinds of membrane channels exist, each with its own properties. At the resting potential, or transmembrane potential of an undisturbed cell, ion movement occurs through *leak channels*—membrane channels that are always

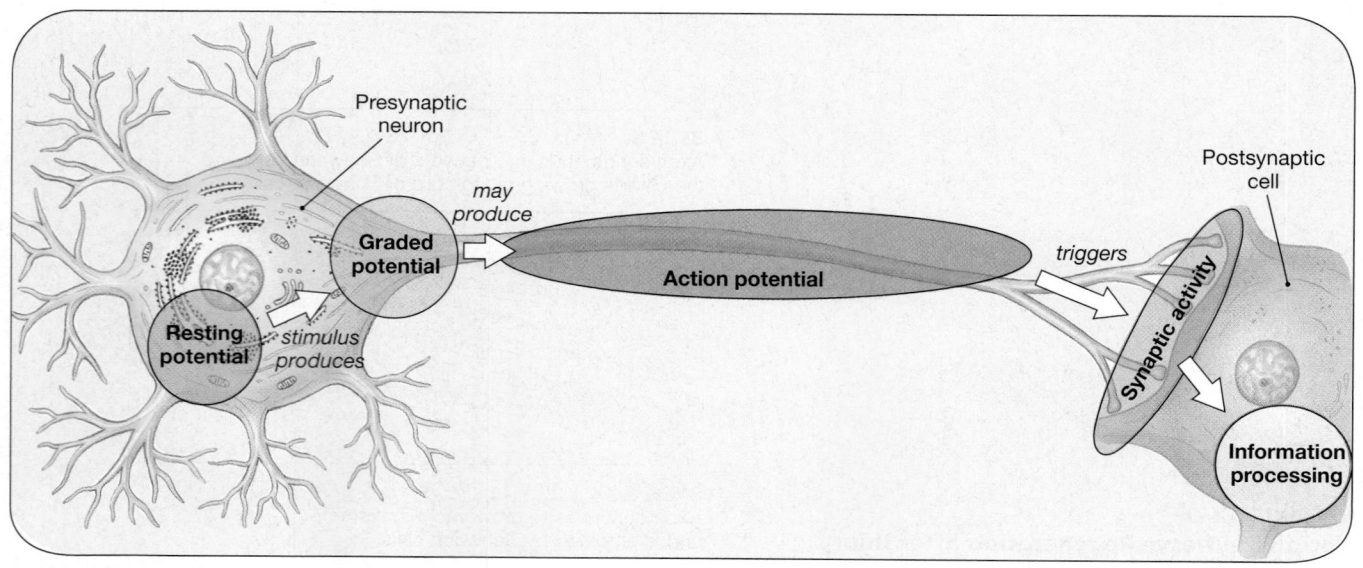

•**FIGURE 12–10**
An Overview of Neural Activities. The important membrane processes are shown in order of their presentation in the text. This figure will be repeated, in simplified form, as a Navigator icon in key figures as we change topics.

open. ∞ p. 72 There are also active transport mechanisms that move specific ions into or out of the cell.

3. The cell's passive and active mechanisms do not ensure an equal distribution of charges across its membrane. This is because membrane permeability varies by ion. For example, negatively charged proteins inside the cell cannot cross the membrane, and it is easier for K^+ to diffuse out of the cell through a potassium channel than it is for Na^+ to enter the cell through a sodium channel. As a result, the inner surface has an excess of negative charges with respect to the outer surface.

Both passive and active forces act across the cell membrane to determine the transmembrane potential at any moment. Figure 12–11● provides a brief overview of the state of the membrane at the normal resting potential.

Passive Forces

The passive forces acting across the membrane are both chemical and electrical in nature.

CHEMICAL GRADIENTS Because the intracellular concentration of potassium ions is relatively high, potassium ions tend to move *out of* the cell through open potassium channels. The movement is driven by a concentration gradient, or *chemical gradient*. Similarly, a chemical gradient for sodium ions tends to drive those ions *into* the cell.

ELECTRICAL GRADIENTS Because the cell membrane is much more permeable to potassium than to sodium, potassium ions leave the cytoplasm more rapidly than sodium ions enter. As a result,

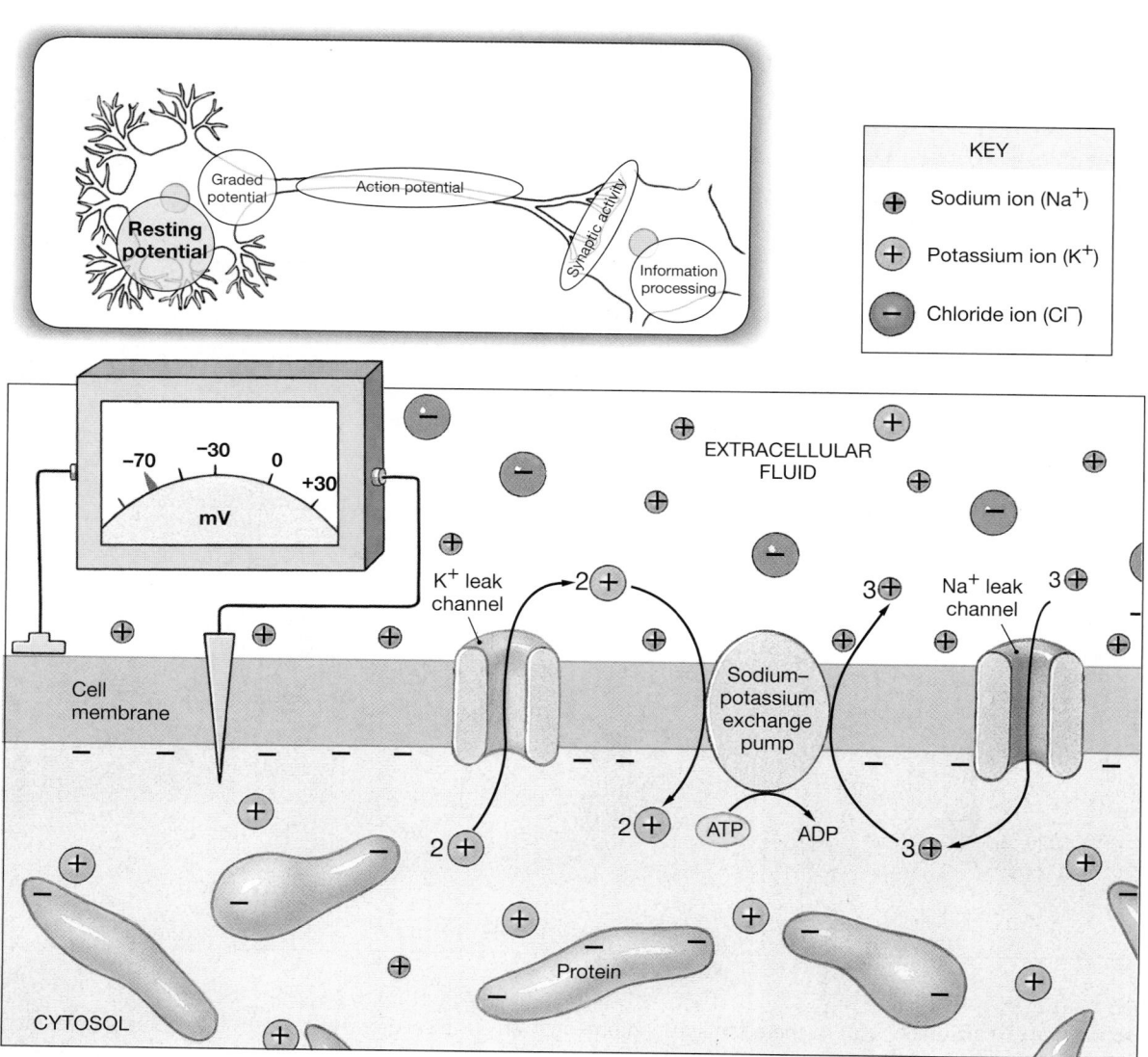

●**FIGURE 12–11**
An Introduction to the Resting Potential. The resting potential is the transmembrane potential of an undisturbed cell. The phospholipid bilayer of the cell membrane is shown as a simple blue band. The Navigator icon in the shadow box highlights the resting potential to indicate "You are here!"

the cytosol along the interior of the membrane exhibits a net loss of positive charges, leaving an excess of negatively charged proteins. At the same time, the extracellular fluid near the outer surface of the cell membrane displays a net gain of positive charges. The positive and negative charges are separated by the cell membrane, which restricts the free movement of ions. Whenever positive and negative ions are held apart, a *potential difference* arises.

The size of a potential difference is measured in volts (V) or millivolts (mV; thousandths of a volt). The resting potential varies widely, depending on the type of cell, but averages about 0.07 V for many cells, including most neurons. We shall use this value in our discussion, usually expressing it as -70 mV (Figure 12–11●). The minus sign signifies that the inner surface of the cell membrane is negatively charged with respect to the exterior.

Positive and negative charges attract one another. If nothing separates them, oppositely charged ions will move together and eliminate the potential difference between them. A movement of charges to eliminate a potential difference is called a **current**. If a barrier, such as a cell membrane, separates the oppositely charged ions, the amount of current depends on how easily the

ions can cross the membrane. The **resistance** of the membrane is a measure of how much the membrane restricts ion movement. If the resistance is high, the current is very small, because few ions can cross the membrane. If the resistance is low, the current is very large, because ions flood across the membrane. The resistance of a cell membrane can be changed by the opening or closing of ion channels. The ensuing changes result in currents that bring ions into or out of the cytoplasm.

THE ELECTROCHEMICAL GRADIENT Electrical gradients can reinforce or oppose the chemical gradient for each ion. The **electrochemical gradient** for a specific ion is the sum of the chemical and electrical forces acting on that ion across the cell membrane. The electrochemical gradients for K^+ and Na^+ are the primary factors affecting the resting potential of most cells, including neurons. We shall look at the forces acting on each ion independently.

The intracellular concentration of potassium ions is relatively high, whereas the extracellular concentration is very low (Figure 12–12a●). Therefore, the chemical gradient for potassium ions

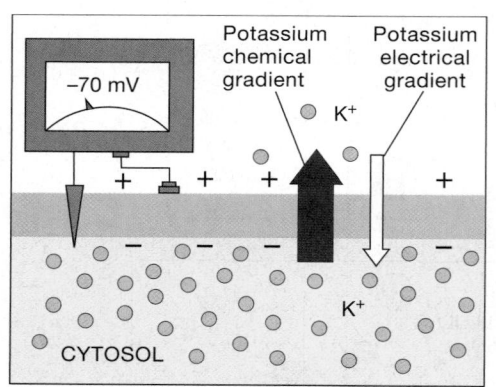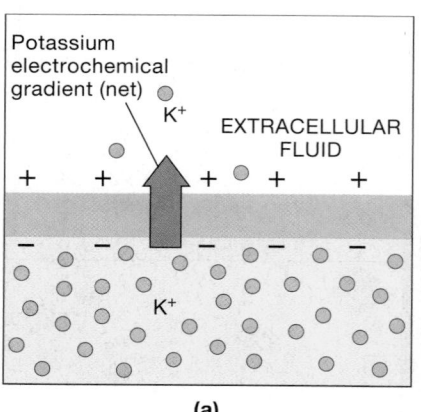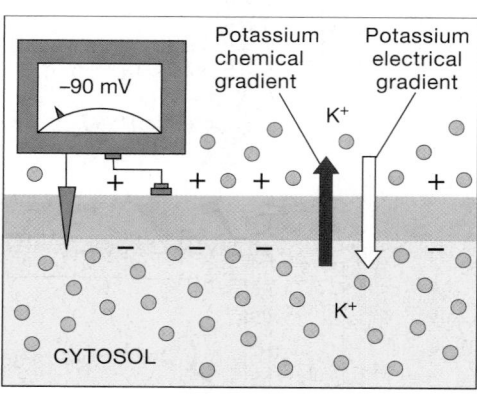

(a)

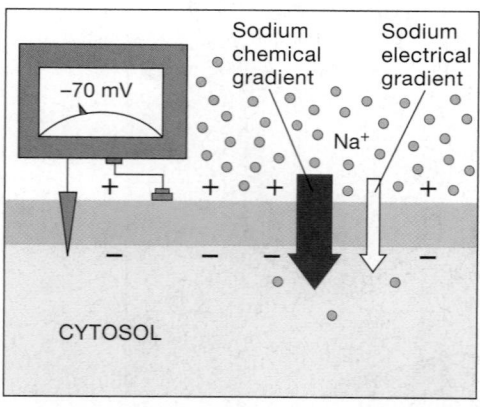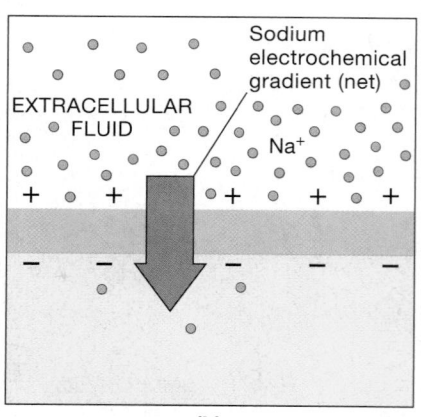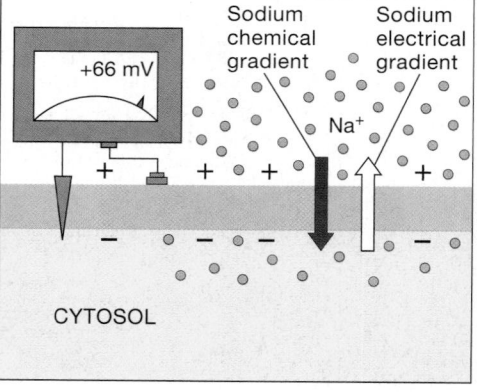

(b)

●**FIGURE 12–11**

Electrochemical Gradients. **(a)** At the normal resting potential, an electrical gradient opposes the chemical gradient for potassium ions. The net electrochemical gradient tends to force potassium ions out of the cell. If the cell membrane were freely permeable to potassium ions, the outflow of K^+ would continue until the equilibrium potential was reached. **(b)** At the normal resting potential, chemical and electrical gradients combine to drive sodium ions into the cell. If the cell membrane were freely permeable to sodium ions, the influx of Na^+ would continue until the equilibrium potential was reached. The chemical and electrical gradients would then be equal and opposite in direction, and no net movement of Na^+ would occur across the membrane.

tends to drive them out of the cell. However, the electrical gradient opposes this movement, because the K^+ are attracted to the negative charges on the inside of the cell membrane and repelled by the positive charges on the outside of the cell membrane. The chemical gradient is strong enough to overpower the electrical gradient, but this weakens the force driving K^+ out of the cell.

If the cell membrane were freely permeable to K^+, but impermeable to other positively charged ions, potassium ions would continue to leave the cell until the electrical gradient (opposing the exit of K^+ from the cell) was as strong as the chemical gradient (driving K^+ out of the cell). The transmembrane potential at which there is no net movement of a particular ion across the cell membrane is called the *equilibrium potential* for that ion. For potassium ions, this equilibrium would occur at a transmembrane potential of about -90 mV. The resting membrane potential is typically -70 mV, a value very close to the equilibrium potential for K^+; the difference is due primarily to the continuous leakage of Na^+ into the cell.

The sodium ion concentration in the extracellular fluid is relatively high, whereas that inside the cell is extremely low (Figure 12–12b●). This means that there is a strong chemical gradient driving Na^+ into the cell. In addition, the extracellular sodium ions are attracted by the excess of negative charges on the inner surface of the cell membrane. This means that there are electrical forces as well as chemical forces acting to drive Na^+ into the cell.

If the cell membrane were freely permeable to Na^+, these ions would continue to enter the cell until the interior of the cell membrane contained enough excess positive charges to reverse the electrical gradient. In other words, ion movement would continue until the interior developed such a strongly positive charge that repulsion between the positive charges would prevent any further net movement of Na^+ into the cell. The equilibrium potential for Na^+ is approximately $+66$ mV. The resting potential is nowhere near that value, because the resting membrane permeability to Na^+ is very low, and ion pumps in the cell membrane are able to eject sodium ions as fast as they cross the membrane.

An electrochemical gradient is a form of *potential energy*. ∞ p. 39 Potential energy is stored energy—the energy of position, as in a stretched spring, a charged battery, or water behind a dam. Without a cell membrane, diffusion would eliminate all electrochemical gradients. In effect, the cell membrane acts like a dam across a river. Without the dam, water would simply respond to gravity and flow downstream, gradually losing energy. With the dam in place, even a small opening will release water under tremendous pressure. Similarly, any stimulus that increases the permeability of the cell membrane to sodium or potassium ions will produce sudden and dramatic ion movement. For example, a stimulus that opens sodium ion channels will trigger an immediate rush of Na^+ into the cell. The nature of the stimulus does not determine the amount of ion movement; if the stimulus opens the door, the electrochemical gradient will do the rest.

Active Forces: The Sodium–Potassium Exchange Pump

We can compare a cell to a leaky fishing boat with a load of fish on its deck. The hull represents the cell membrane; the fish, K^+; and the ocean water, Na^+. As the boat rumbles and rolls, water comes in through the cracks and fish fall overboard. To stay afloat and keep the catch, the fisher has to work hard, bailing out the water and recapturing the lost fish.

Similarly, at the normal resting potential, the cell must expend energy, in the form of ATP, to bail out sodium ions that leak in and to recapture potassium ions that leak out. The ATP is used by an ion pump, introduced in Chapter 3, that depends on the carrier protein *sodium–potassium ATPase.* ∞ p. 96 This pump exchanges 3 intracellular sodium ions for 2 extracellular potassium ions. At the normal resting potential, the pump's primary significance is that it ejects sodium ions as quickly as they enter the cell. Thus, the activity of the exchange pump exactly balances the passive forces of diffusion, and the resting potential remains stable.

Table 12–1 provides a summary of what we have learned about the resting potential.

☐ CHANGES IN THE TRANSMEMBRANE POTENTIAL

The resting potential is the transmembrane potential of an "undisturbed" cell. Yet cells are dynamic structures that continually modify their activities either in response to external stimuli or to perform specific functions. The transmembrane potential is equally dynamic, rising or falling in response to temporary changes in membrane permeability. Those changes result from the opening or closing of specific membrane channels.

SUMMARY TABLE 12–1 THE RESTING POTENTIAL

- Because the cell membrane is highly permeable to potassium ions, the resting potential is fairly close to -90 mV, the equilibrium potential for K^+.

- Although the electrochemical gradient for sodium ions is very large, the membrane's permeability to these ions is very low. Consequently, Na^+ has only a small effect on the normal resting potential, making it just slightly less negative than it would otherwise be.

- The sodium–potassium exchange pump ejects 3 Na^+ ions for every 2 K^+ ions that it brings into the cell. It thus serves to stabilize the resting potential when the ratio of Na^+ entry to K^+ loss through passive channels is 3:2.

- At the normal resting potential, these passive and active mechanisms are in balance. The resting potential varies widely with the type of cell. A typical neuron has a resting potential of approximately -70 mV.

Membrane Channels

Membrane channels control the movement of ions across the cell membrane. Our discussion will focus on the permeability of the membrane to sodium and potassium ions, which are the primary determinants of the transmembrane potential of many cell types, including neurons. Sodium and potassium ion channels are either passive or active. **Passive channels**, or **leak channels**, are always open. However, their permeability can vary from moment to moment as the proteins that make up the channel change shape in response to local conditions. Sodium and potassium leak channels are important in establishing the normal resting potential of the cell, as we noted earlier (Figure 12–11●).

Cell membranes also contain **active channels**, often called **gated channels**. These channels open or close in response to specific stimuli. Each gated channel can be in one of three states: (1) closed, but capable of opening, (2) open (**activated**), or (3) closed and incapable of opening (**inactivated**).

Three classes of gated channels exist: chemically regulated channels, voltage-regulated channels, and mechanically regulated channels:

1. **Chemically regulated channels** open or close when they bind specific chemicals (Figure 12–13a●). The receptors that bind acetylcholine (ACh) at the neuromuscular junction are chemically regulated channels. ∞ p. 301 Chemically regulated channels

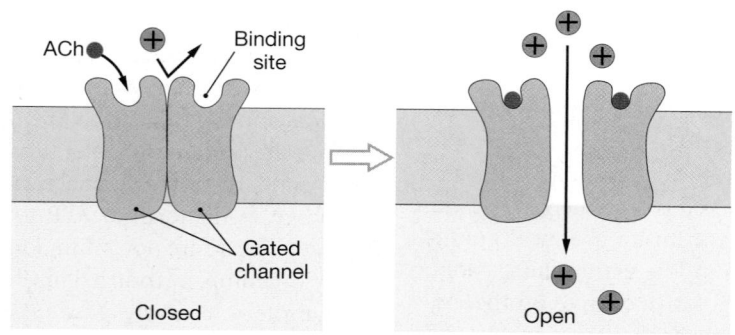

(a) Chemically regulated channel

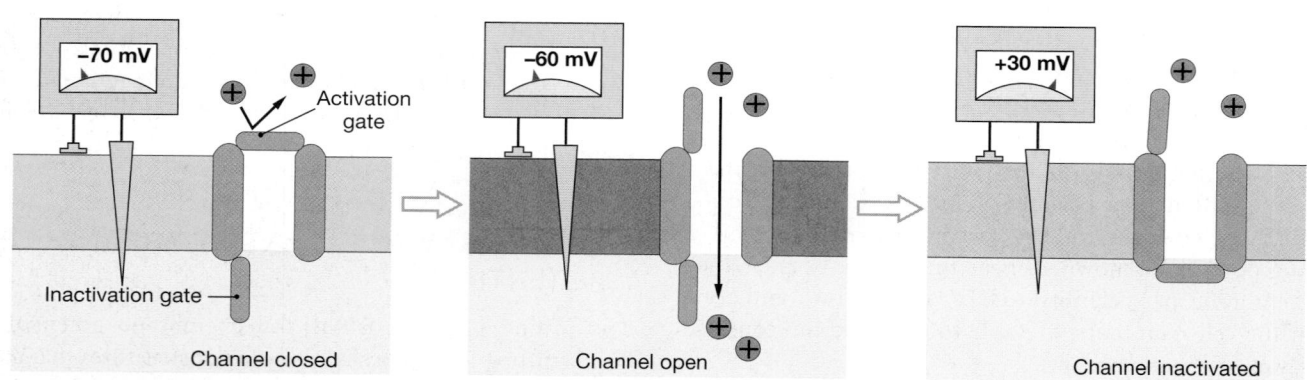

(b) Voltage-regulated channel

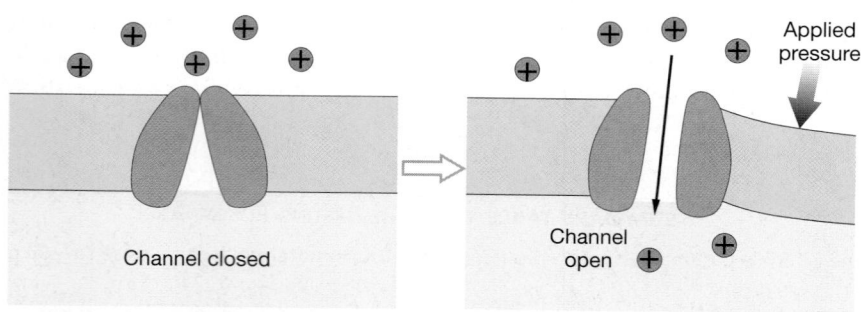

(c) Mechanically regulated channel

●**FIGURE 12–13**
Gated Channels. We use Na^+ channels here, but comparable gated channels regulate the movements of other cations and anions. **(a)** A chemically regulated Na^+ channel that opens in response to the presence of ACh at a binding site. **(b)** A voltage-regulated Na^+ channel that responds to changes in the transmembrane potential. At the normal resting potential, the channel is closed. At a membrane potential of −60 mV, the channel opens. At +30 mV, it is inactivated. Later we shall consider factors that alter the membrane potential. **(c)** Mechanically regulated channels open in response to distortion of the membrane.

are most abundant on the dendrites and cell body of a neuron, the areas where most synaptic communication occurs.

2. **Voltage-regulated channels** are characteristic of areas of **excitable membrane**, a membrane capable of generating and conducting an action potential. Examples of excitable membranes are the axons of unipolar and multipolar neurons and the sarcolemma (including T tubules) of skeletal muscle fibers and cardiac muscle cells. ∞ pp. 294, 324 Voltage-regulated channels open or close in response to changes in the transmembrane potential. The most important voltage-regulated channels, for our purposes, are voltage-regulated sodium channels, potassium channels, and calcium channels. The sodium channels have two gates that function independently: an *activation gate* that opens on stimulation, letting sodium ions into the cell, and an *inactivation gate* that closes to stop the entry of sodium ions (Figure 12–13b●).

3. **Mechanically regulated channels** open or close in response to physical distortion of the membrane surface (Figure 12–13c●). Such channels are important in sensory receptors that respond to touch, pressure, or vibration. We will discuss these receptors in more detail in Chapter 15.

At the resting potential, most gated channels are closed. The opening of gated channels alters the rate of ion movement across the cell membrane and thus changes the transmembrane potential. The distribution of membrane channels can vary from one region of the cell membrane to another, affecting how and where a cell responds to specific stimuli. For example, chemically regulated sodium channels are widespread on the surfaces of a neuron, but voltage-regulated sodium channels are restricted to the axon, its branches, and the synaptic terminals, while mechanically regulated channels are typically located only on the dendrites of sensory neurons. The functional implications of these differences in distribution will become apparent in later sections.

□ GRADED POTENTIALS

Graded potentials, or *local potentials*, are changes in the transmembrane potential that cannot spread far from the area surrounding the site of stimulation. Table 12–2 summarizes information about graded potentials.

Any stimulus that opens a gated channel will produce a graded potential. Figure 12–14 shows what happens when a membrane is exposed to a chemical that opens chemically regulated sodium channels. (For clarity, only gated channels are shown; leak channels are present, but they are not involved in the production of graded potentials.) Sodium ions enter the cell and are attracted to the negative charges along the inner surface of the membrane. The arrival and spreading out of additional positive charges shifts the transmembrane potential toward 0 mv (STEP 1). Any shift from the resting potential toward 0 mV is called a **depolarization**, a term that applies to changes in potential from −70 mV to smaller negative values (−65 mV, −45 mV, −10 mV), as well as to membrane potentials above 0 mV (+10 mV, +30 mV).

At the resting potential, sodium ions are drawn to the outer surface of the cell membrane, attracted by the excess of negative ions on the inside of the membrane. As the cell membrane depolarizes, sodium ions are released from its outer surface. These ions, accompanied by other extracellular sodium ions, then move toward the open channels, replacing ions that have already entered the cell. This movement of positive charges parallel to the inner and outer surfaces of a membrane is called a **local current** (STEP 2).

The degree of depolarization decreases with distance, because the cytosol offers considerable resistance to ion movement and because some of the sodium ions entering the cell then move back across the membrane through sodium leak channels. At some distance from the entry point, the effects on the transmembrane potential are undetectable (STEP 2). The maximum change in the transmembrane potential is proportional to the size of the stimulus, because that determines the number of open sodium channels. The more open channels, the more sodium ions enter the cell, the greater the membrane area affected, and the greater the degree of depolarization.

When a chemical stimulus is removed and normal membrane permeability is restored, the transmembrane potential soon returns to resting levels. The process of restoring the normal resting potential after depolarization is called **repolarization** (Figure 12–15a●). Repolarization typically involves a combination of ion movement through membrane channels and the activities of ion pumps, especially the sodium–potassium exchange pump.

Opening a gated potassium channel would have the opposite effect on the transmembrane potential as opening a gated sodium channel: The rate of potassium outflow would increase, and the interior of the cell would lose positive ions. The loss of positive ions produces **hyperpolarization**, an increase in the negativity of the resting potential from −70 mV to perhaps −80 mV or more (Figure 12–15b●). Again, a local current would distribute the effect to adjacent portions of the cell membrane, and the effect would decrease with distance from the open channel or channels.

SUMMARY TABLE 12–2 GRADED POTENTIALS

Graded potentials, whether depolarizing or hyperpolarizing, share four basic characteristics:

1. The transmembrane potential is most affected at the site of stimulation, and the effect decreases with distance.

2. The effect spreads passively, owing to local currents.

3. The graded change in membrane potential may involve either depolarization or hyperpolarization. The nature of the change is determined by the properties of the membrane channels involved. For example, in a resting membrane, the opening of sodium channels will cause depolarization, whereas the opening of potassium channels will cause hyperpolarization.

4. The stronger the stimulus, the greater is the change in the transmembrane potential and the larger is the area affected.

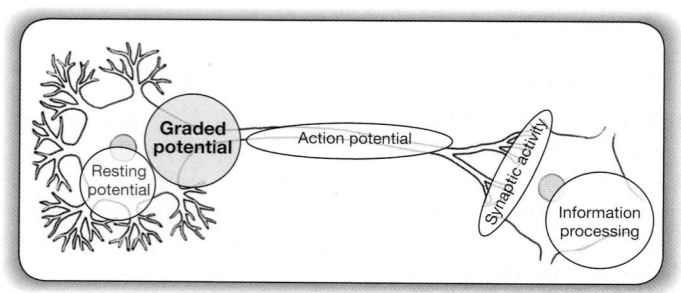

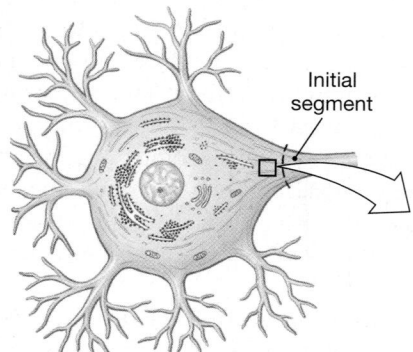

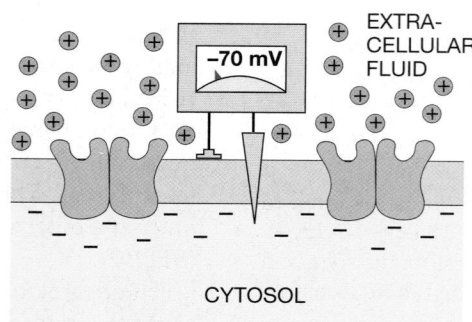

Resting membrane with closed chemically
regulated sodium ion channels

STEP 1:
Membrane exposed to chemical that opens the sodium ion channels

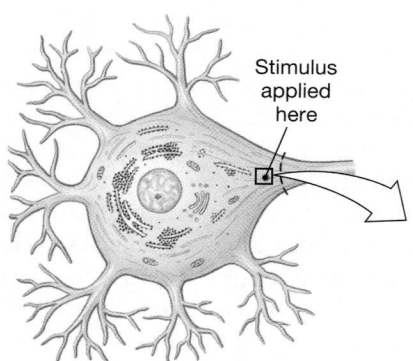

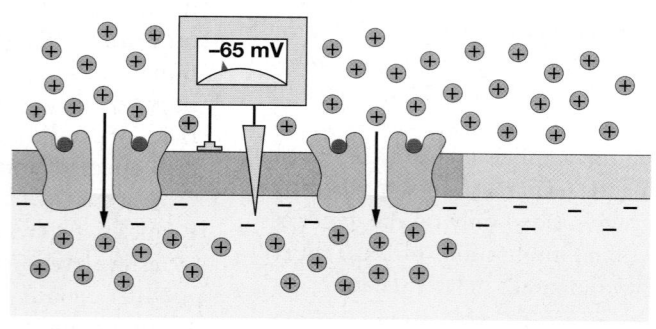

STEP 2:
Spread of sodium ions inside cell membrane produces a local current that depolarizes
adjacent portions of the cell membrane

●**FIGURE 12–14**
Graded Potentials. The depolarization radiates in all directions away from the source of stimulation. For clarity, only gated channels are shown; leak channels are present, but are not responsible for the production of graded potentials. Color changes in the phospholipid bilayer indicate that the resting potential has been disturbed and that the transmembrane potential is no longer −70 mV. Notice that the Navigator Icon now highlights the graded potential.

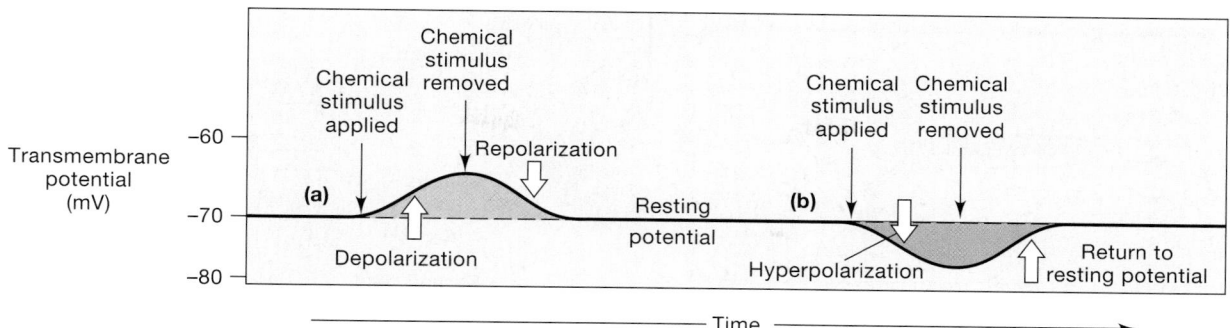

●**FIGURE 12–15**
Depolarization and Hyperpolarization. **(a)** Depolarization and repolarization in response to the application and removal of a stimulus that opens chemically regulated sodium channels. **(b)** Hyperpolarization in response to the application of a stimulus that opens chemically regulated potassium channels. When the stimulus is removed, the membrane potential returns to the resting level.

The Distribution and Importance of Graded Potentials

Graded potentials occur in the membranes of many types of cells, such as epithelial cells, gland cells, adipocytes, and a variety of sensory receptors. They are often the trigger for specific cell functions; for example, a graded potential at the surface of a gland cell may trigger the exocytosis of secretory vesicles. Similarly, it is the graded depolarization of the motor end plate by ACh that triggers an action potential in adjacent portions of the sarcolemma. The motor end plate supports graded potentials, whereas the rest of the sarcolemma consists of excitable membrane.

This brings us to an interesting observation: Each neuron receives information in the form of graded potentials on its dendrites and cell body, and releases neurotransmitter in response to graded potentials at synaptic terminals which may be a meter away from the cell body. Yet even the largest graded potentials affect an area perhaps only 1 mm in diameter. In a typical sensory or motor neuron, the axon is so long that graded potentials on the dendrites and cell body can have no direct effect on the synaptic terminals. Instead, an action potential links the two. In a representative neuron, an action potential first appears in the initial segment of the axon. It develops in response to a graded potential that spreads to the initial segment from the adjacent axon hillock of the cell body (Figure 12–2●, p. 388). After it forms, the action potential is propagated along the length of the axon and its telodendria. The arrival of an action potential at a synaptic terminal causes a graded potential in the presynaptic membrane, and this can trigger the release of neurotransmitters. .

☐ ACTION POTENTIALS

Action potentials are propagated changes in the transmembrane potential that, once initiated, affect an entire excitable membrane. The first step in the generation of an action potential is the opening of voltage-regulated sodium ion channels at one site, usually the initial segment of the axon. The movement of sodium ions into the cell depolarizes adjacent sites, triggering the opening of additional voltage-regulated channels. The result is a chain reaction that spreads across the surface of the membrane like a line of falling dominoes. In this way, the impulse is propagated along the length of the axon, ultimately reaching the synaptic terminals.

The All-or-None Principle

The stimulus that initiates an action potential is a depolarization large enough to open voltage-regulated sodium channels. That opening occurs at a transmembrane potential known as the **threshold**. Threshold for an axon is typically between −60 mV and −55 mV, corresponding to a depolarization of 10–15 mV. A stimulus that shifts the resting membrane potential from −70 mV to −62 mV will not produce an action potential, only a graded depolarization. When such a stimulus is removed, the transmembrane potential returns to the resting level. The depolarization of the initial segment of the axon is caused by local currents resulting from the graded depolarization of the axon hillock.

The initial depolarization acts like pressure on the trigger of a gun. If a slight pressure is applied, the gun will not fire. It will fire only when a certain minimum pressure is applied to the trigger. Once the pressure on the trigger reaches this threshold, the firing pin drops and the gun discharges. At that point, it no longer matters whether the pressure was applied gradually or suddenly or whether it was caused by the precise movement of just one finger or by the clenching of the entire hand. The speed and range of the bullet that leaves the gun do not change, regardless of the forces that were applied to the trigger.

In the case of an axon or another area of excitable membrane, a graded depolarization is analogous to the pressure on the trigger, and the action potential is like the firing of the gun. All stimuli that bring the membrane to threshold generate identical action potentials. In other words, the properties of the action potential are independent of the relative strength of the depolarizing stimulus, as long as that stimulus exceeds the threshold. This concept is called the **all-or-none principle**, because a given stimulus either triggers a typical action potential or does not produce one at all. The all-or-none principle applies to all excitable membranes.

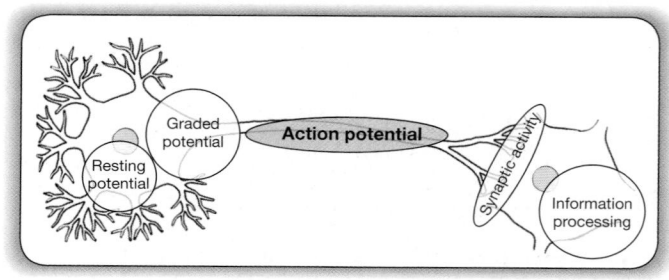

•FIGURE 12–16
The Generation of an Action Potential. For clarity, only gated channels are shown.

We will now take a closer look at the mechanisms whereby action potentials are generated and propagated. The two are closely related, both in terms of time and space: an action potential must be generated at one site before it can be propagated away from that site.

Generation of Action Potentials

Figure 12–16• diagrams the steps involved in the generation of an action potential from the resting state. At the normal resting potential, the activation gates of the voltage-regulated sodium channels are closed. The steps are as follows:

STEP 1: *Depolarization to Threshold.* Before an action potential can begin, an area of excitable membrane must be depolarized to its threshold by local currents.

STEP 2: *The Activation of Sodium Channels and Rapid Depolarization.* At threshold, the sodium activation gates open and the cell membrane becomes much more permeable to Na^+. Driven by the large electrochemical gradient, sodium ions rush into the cytoplasm, and rapid depolarization occurs at the site. In less than a millisecond, the inner membrane surface has changed; it now contains more positive ions than negative ones, and the transmembrane potential has changed from -60 mV to positive values closer to the equilibrium potential for sodium ions.

Notice that the first two steps in the generation of an action potential are an example of positive feedback: A small depolarization triggers a larger depolarization.

STEP 3: *The Inactivation of Sodium Channels and the Activation of Potassium Channels.* As the transmembrane potential approaches $+30$ mV, the inactivation gates of the voltage-regulated sodium channels begin closing. This step is known as **sodium channel inactivation**. While it is under way, voltage-regulated potassium channels are opening. At a transmembrane potential of $+30$ mV, the cytosol along the interior of the membrane contains an excess of positive charges. Thus, in contrast to the situation in the resting membrane (p. 401), both the electrical *and* chemical gradients favor the movement of K^+ out of the cell. The sudden loss of positive charges then shifts the transmembrane potential back toward resting levels, and repolarization begins.

STEP 4: *The Return to Normal Permeability.* The voltage-regulated sodium channels remain inactivated until the membrane has repolarized to near threshold levels. At this time, they regain their normal status: closed, but capable of opening. The voltage-regulated potassium

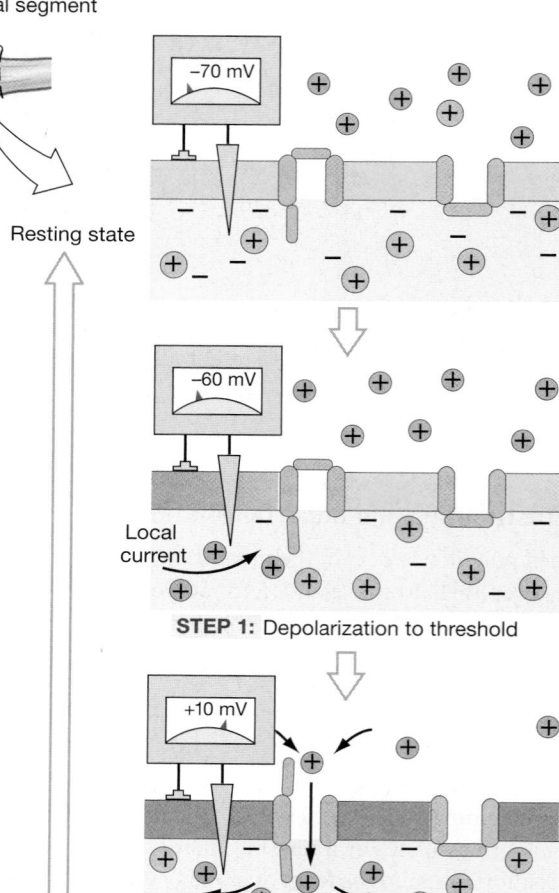

Resting state

STEP 1: Depolarization to threshold

Local current

STEP 2: Activation of sodium channels and rapid depolarization

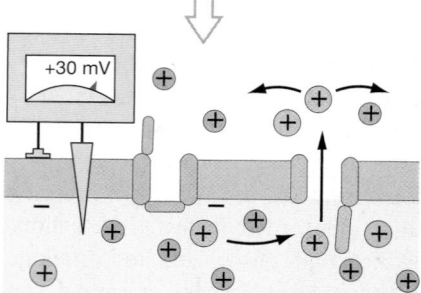

STEP 3: Inactivation of sodium channels and activation of potassium channels

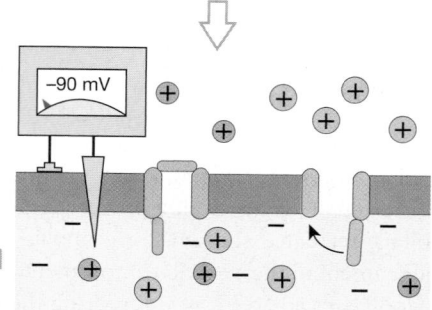

STEP 4: The return to normal permeability

channels begin closing as the membrane reaches the normal resting potential (about −70 mV), but the process takes at least a millisecond. Over that period, potassium ions continue to move out of the cell at a faster rate than when they are at rest, producing a brief hyperpolarization that brings the transmembrane potential very close to the equilibrium potential for potassium (−90 mV). As the voltage-regulated potassium channels close, the transmembrane potential returns to normal resting levels. The membrane is now in a prestimulation condition, and the action potential is over.

THE REFRACTORY PERIOD From the time an action potential begins until the normal resting potential has stabilized, the membrane will not respond normally to additional depolarizing stimuli. This period is known as the **refractory period** of the membrane. From the moment the voltage-regulated sodium channels open at threshold until sodium channel inactivation ends, the membrane cannot respond to further stimulation, because all the voltage-regulated sodium channels either are already open or are inactivated. This portion of the refractory period, the **absolute refractory period**, lasts 0.4–1.0 msec; the smaller the axon diameter, the longer the duration. The **relative refractory period** begins when the sodium channels regain their normal resting condition and continues until the transmembrane potential stabilizes at resting levels. Another action potential can occur over this period if the membrane is sufficiently depolarized. That depolarization, however, requires a larger-than-normal stimulus, because (1) the local current must deliver enough Na$^+$ to counteract the loss of positively charged K$^+$ through voltage-regulated K$^+$ channels and

(2) the membrane is hyperpolarized to some degree through most of the relative refractory period.

THE ROLE OF THE SODIUM–POTASSIUM EXCHANGE PUMP In an action potential, depolarization results from the influx of Na$^+$, and repolarization involves the loss of K$^+$. Over time, the sodium–potassium exchange pump returns intracellular and extracellular ion concentrations to prestimulation levels. Compared with the total number of ions inside and outside the cell, however, the number involved in a single action potential is insignificant. Tens of thousands of action potentials can occur before intracellular ion concentrations change enough to disrupt the entire mechanism. Thus, the exchange pump is not essential to any single action potential.

However, a maximally stimulated neuron can generate action potentials at a rate of 1000 per second. Under these circumstances, the exchange pump is needed if ion concentrations are to remain within acceptable limits over a prolonged period. The sodium–potassium exchange pump requires energy in the form of ATP. Each time the pump exchanges two extracellular potassium ions for three intracellular sodium ions, one molecule of ATP must be broken down to ADP. The transmembrane protein of the exchange pump is called *sodium–potassium ATPase*, because it provides the energy to pump ions by splitting a phosphate group from a molecule of ATP, forming ADP. If the cell runs out of ATP, or if sodium–potassium ATPase is inactivated by a metabolic poison, a neuron will soon lose its ability to function.

Table 12–3 summarizes the generation of an action potential.

SUMMARY TABLE 12–3 GENERATION OF ACTION POTENTIALS

STEP 1: **Depolarization to Threshold**
- A graded depolarization brings an area of excitable membrane to threshold (−60 mV).

STEP 2: **Activation of Sodium Channels and Rapid Depolarization**
- The voltage-regulated sodium channels open (sodium channel activation).
- Sodium ions, driven by electrical attraction and the chemical gradient, flood into the cell.
- The transmembrane potential goes from −60 mV, the threshold level, toward +30 mV.

STEP 3: **Inactivation of Sodium Channels and Activation of Potassium Channels**
- The voltage-regulated sodium channels close (sodium channel inactivation occurs) at +30 mV.
- The voltage-regulated potassium channels are now open, and potassium ions diffuse out of the cell.
- Repolarization begins.

STEP 4: **Return to Normal Permeability**
- The voltage-regulated sodium channels regain their normal properties in 0.4–1.0 msec. The membrane is now capable of generating another action potential if a larger than normal stimulus is provided.
- The voltage-regulated potassium channels begin closing at −70 mV. Because they do not all close at the same time, potassium loss continues, and a temporary hyperpolarization to approximately −90 mV occurs.
- At the end of the relative refractory period, all voltage-regulated channels have closed, and the membrane is back to its resting state.

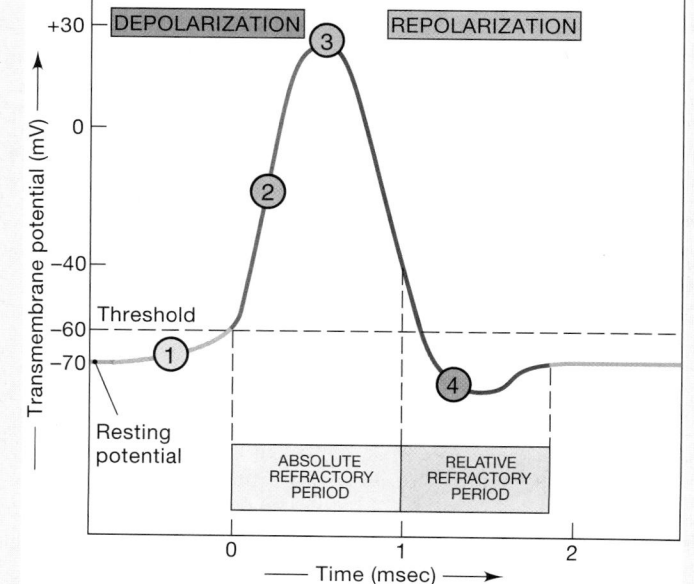

Propagation of Action Potentials

The sequence of events described earlier occurs in a relatively small portion of the total membrane surface. But we have already noted that, unlike graded potentials, which diminish rapidly with distance, action potentials spread to affect the entire excitable membrane. To understand the basic principle involved, imagine that you are standing by the doors of a movie theater at the start of a long line. Everyone is waiting for the doors to open. The manager steps outside and says to you, "Let everyone know that we're opening in 15 minutes." How would you spread the news?

If you treated the line as an inexcitable membrane, you would shout "The doors open in 15 minutes!" as loudly as you could. The closest people in the line would hear the news very clearly, but those farther away might not hear the entire message, and those at the end of the line might not hear you at all. If, on the other hand, you treated the crowd as an excitable membrane, you would give the message to another person in line, with instructions to pass it on. In that way, the message would travel along the line undiminished, until everyone had heard the news. Such a message "moves" as each person repeats it to someone else. Distance is not a factor; the line can contain 50 people or 5000.

The situation just described is comparable to the way an action potential spreads across an excitable membrane. An action potential (message) is relayed from one location to another in a series of steps. At each step, the message is repeated. Because the same events take place over and over, the term **propagation** is preferable to the term *conduction*, which suggests a flow of charge similar to that which takes place in a conductor such as a copper wire. (In fact, axons are relatively poor conductors of electricity.)

CONTINUOUS PROPAGATION The basic mechanism by which an action potential is propagated along an unmyelinated axon is shown in Figure 12–17●. For convenience, we shall consider the membrane as a series of adjacent segments. The action potential begins at the initial segment. For a brief moment at the peak of the action potential, the transmembrane potential becomes positive rather than negative (STEP 1). A local current then develops as sodium ions begin moving in the cytosol and the extracellular fluid (STEP 2). The local current spreads in all directions, depolarizing adjacent portions of the membrane. The axon hillock cannot respond with an action potential. (Like the rest of the cell body, it lacks voltage-gated channels.) But when the initial segment of the axon is depolarized to threshold, an action potential develops there. The process then continues in a chain reaction (STEPS 3, 4). Eventually, the most distant portions of the cell membrane will be affected. As in our "movie line" model, the message is being relayed from one location to another. At each step along the way, the message is retold, so distance has no effect on the process. The action potential reaching the synaptic knob is identical to the one generated at the initial segment, and the net effect is the same as if a single action potential had traveled across the surface of the membrane. This form of action potential propagation is known as **continuous propagation**.

Each time a local current develops, the action potential moves forward, not backward, because the previous segment of the axon is still in the absolute refractory period. As a result, an action potential always proceeds away from the site of generation

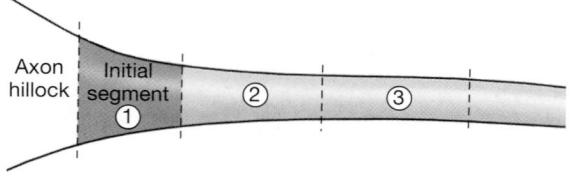

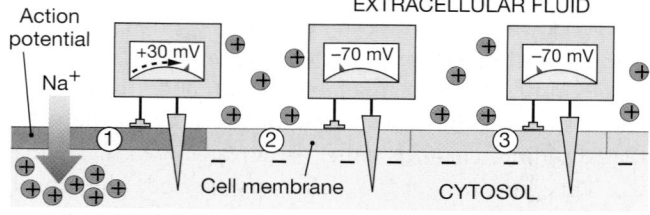

STEP 1:
As an action potential develops in the initial segment, the transmembrane potential depolarizes to +30 mV.

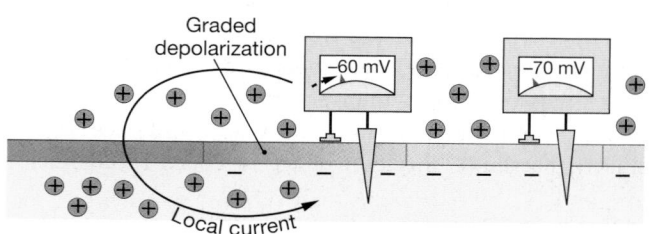

STEP 2:
A local current depolarizes the adjacent portion of the membrane to threshold.

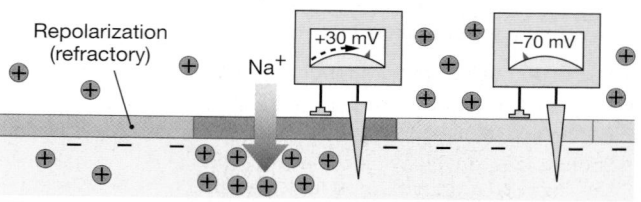

STEP 3:
An action potential develops at this location, and the initial segment enters the refractory period.

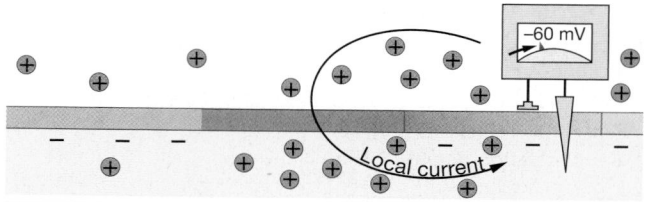

STEP 4:
A local current depolarizes the adjacent portion of the membrane to threshold, and the cycle is repeated.

●**FIGURE 12–17**
Propagation of an Action Potential along an Unmyelinated Axon. The axon can be viewed as a series of adjacent segments.

and cannot reverse direction. For a second action potential to occur at the same site, a second stimulus must be applied.

In continuous propagation, an action potential appears to move across the surface of the membrane in a series of tiny steps. Even though the events at any one location take only about a millisecond, the sequence must be repeated at each step along the way. Continuous propagation along unmyelinated axons occurs at a speed of about 1 meter per second (approximately 2 mph).

SALTATORY PROPAGATION In a myelinated axon, the axolemma is wrapped in a myelin sheath that is complete, except at the nodes. Continuous propagation cannot occur along a myelinated axon, because myelin increases resistance to the flow of ions across the membrane. Ions can readily cross the cell membrane just at the nodes. As a result, only the nodes can respond to a depolarizing stimulus.

When an action potential appears at the initial segment of a myelinated axon, the local current skips the internodes and depolarizes the closest node to threshold (Figure 12–18●). Because the nodes may be 1–2 mm apart in a large myelinated axon, the action potential "jumps" from node to node rather than moving along the axon in a series of tiny steps. This process is called **saltatory propagation** (*saltare*, leaping). Imagine relaying a message along a line of people spaced 5 meters apart. Each person shouts the message to the next person in line; by the time the message has been repeated four times, it has moved 20 meters. In our model of continuous propagation, in which people were closely packed, the message would have moved only a few meters by the time it had been repeated four times. Saltatory propagation in the CNS and PNS carries nerve impulses along an axon much more rapidly than does continuous propagation. It also uses proportionately less energy, because less surface area is involved and fewer sodium ions need to be pumped out of the cytoplasm.

Table 12–4 reviews the key differences between graded potentials and action potentials.

Axon Diameter and Propagation Speed

Myelin greatly increases the propagation speed of action potentials. The diameter of the axon also affects the propagation speed, although the effects are less dramatic. Axon diameter is important because in order to depolarize adjacent portions of the cell membrane, ions must move through the cytoplasm. Cytoplasm offers resistance to ion movement, although much less resistance than the cell membrane. In this instance, an axon behaves like an electrical cable: The larger the diameter, the lower the resistance. (That is why motors with large current demands, such as the starter on a car, an electric stove, or a big air conditioner, use such thick wires.)

Axons are classified into three groups according to the relationships among the diameter, myelination, and propagation speed:

1. **Type A fibers** are the largest axons, with diameters ranging from 4 to 20 μm. These are myelinated axons that carry action potentials at speeds of up to 140 meters per second, or over 300 mph.

2. **Type B fibers** are smaller myelinated axons, with diameters of 2–4 μm. Their propagation speeds average around 18 meters per second, or roughly 40 mph.

3. **Type C fibers** are unmyelinated and less than 2 μm in diameter. These axons propagate action potentials at the leisurely pace of 1 meter per second, or a mere 2 mph.

We can understand the relative importance of myelin by noting that in going from Type C to Type A fibers, the diameter increases tenfold but the propagation speed increases by 140 times.

Type A fibers carry to the CNS sensory information about position, balance, and delicate touch and pressure sensations from the skin surface. The motor neurons that control skeletal muscles also send their commands over large, myelinated Type A axons. Type B fibers and Type C fibers carry information to the CNS. They deliver temperature, pain, and general touch and pressure

TABLE 12–4 A Comparison of Graded Potentials and Action Potentials

Graded Potentials	Action Potentials
Depolarizing or hyperpolarizing	Always depolarizing
No threshold value	Must depolarize to threshold before action potential begins
Amount of depolarization or hyperpolarization depends on intensity of stimulus	All-or-none phenomenon; all stimuli that exceed threshold will produce identical action potentials
Passive spread from site of stimulation	Action potential at one site depolarizes adjacent sites to threshold
Effect on membrane potential decreases with distance from stimulation site	Propagated across entire membrane surface without decrease in strength
No refractory period	Refractory period
Occur in most cell membranes	Occur only in excitable membranes of specialized cells such as neurons and muscle cells

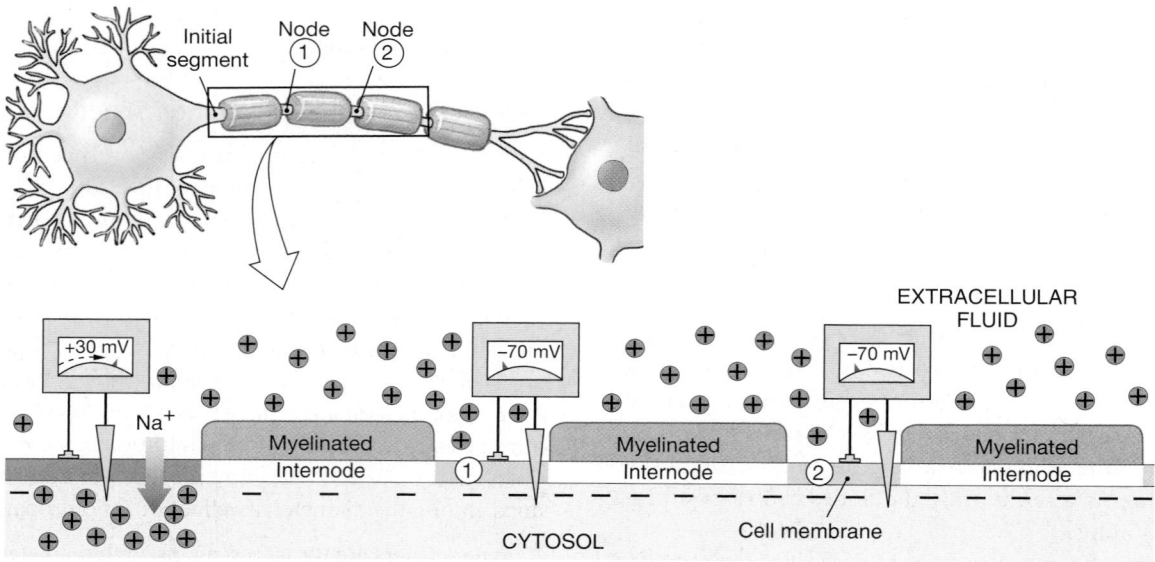

STEP 1: Action potential at initial segment

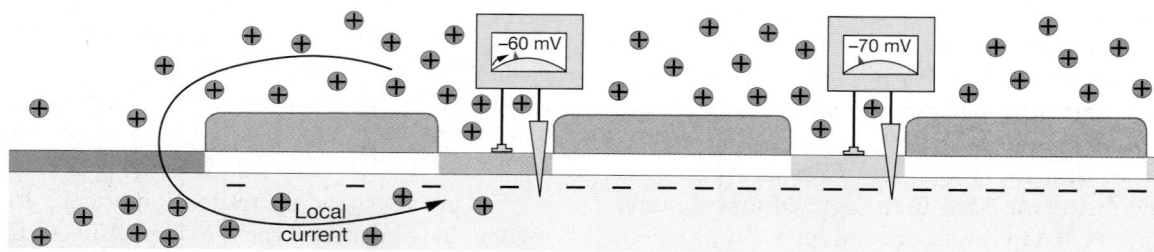

STEP 2: Depolarization to threshold at node 1

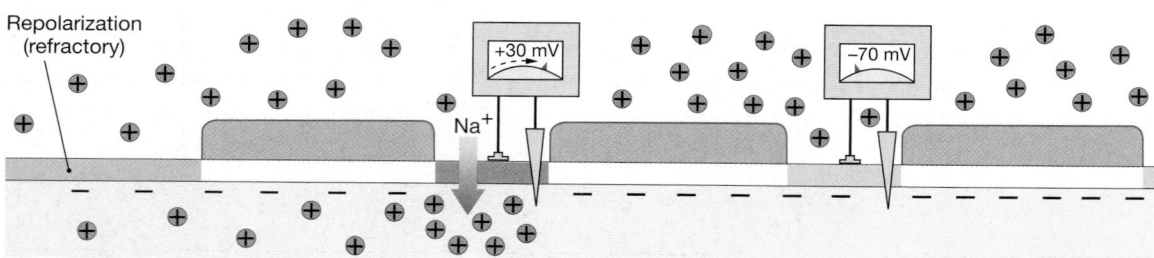

STEP 3: Action potential at node 1

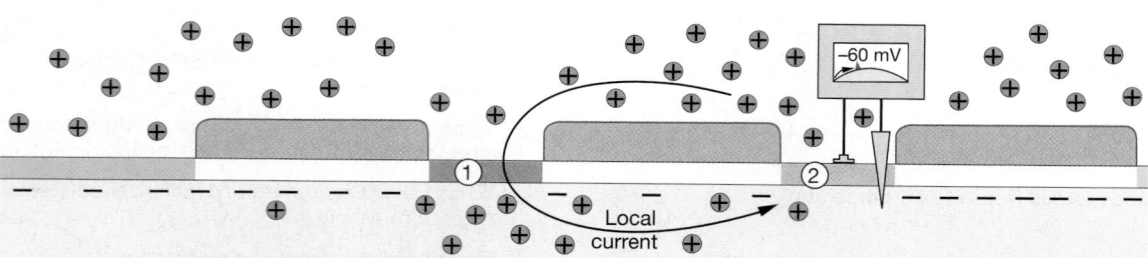

STEP 4: Depolarization to threshold at node 2

●**FIGURE 12–18**
Saltatory Propagation along a Myelinated Axon. This process will continue along the entire length of the axon.

sensations and carry instructions to smooth muscle, cardiac muscle, glands, and other peripheral effectors.

When we need to tell a friend urgent news or receive an immediate response, we usually pick up the telephone. For general correspondence, we often send a first-class letter or an e-mail message, both of which produce fast, but not immediate, responses. If we have to distribute an enormous volume of information to a huge number of people and are in no particular rush, bulk mail offers efficiency at a considerable savings. Instead of representing a compromise between time and money, information transfer in the nervous system reflects a compromise between time and space. Axons carry the information, and the larger the axon, the faster the rate of transmission. But if all sensory information were carried by Type A fibers, your peripheral nerves would be the size of garden hoses and your spinal cord would have the diameter of a garbage can. Instead, only around one-third of all axons carrying sensory information are myelinated, and most sensory information arrives over slender Type C fibers.

Myelination improves coordination and control by decreasing the time between the reception of a sensation and the initiation of an appropriate response. Myelination begins relatively late in development, and the myelination of sensory and motor fibers is not completed until early adolescence. In growing children, the pace of myelination and the pathways involved can be quite variable, contributing to the observed range of physical capabilities in a given age group.

Action Potentials in Muscle Tissues

Although neurons, skeletal muscle fibers, multiunit smooth muscle cells, and cardiac muscle cells have excitable membranes, their action potentials differ in several important ways:

- **The Potential Difference Is Larger in Muscle Cells Than in Axons.** The resting potential of a skeletal, cardiac, or multiunit smooth muscle cell is -85 mV to -90 mV, close to the potassium equilibrium potential. Threshold values in these cells are comparable to those of axons.

- **Action Potentials Last Longer in Muscle Cells Than in Axons.** An action potential in an axon may last 0.5 msec, versus 7.5 msec in a skeletal muscle fiber, 50 msec in a multiunit smooth muscle cell, and 250–300 msec in a cardiac muscle cell from one of the ventricles.

- **Despite the Large Diameters of Muscle Cells, Action Potentials Are Propagated Relatively Slowly.** Action potentials of muscle cells travel by continuous propagation at speeds of 3–5 meters per second (6–10 mph).

Visceral smooth muscle cells do not have typical excitable membranes, and most have neither stable resting potentials nor action potentials comparable to those of other types of muscle cells. When visceral smooth muscle cells are stimulated to contract, changes in the transmembrane potential reflect primarily changes in the membrane permeability to calcium ions.

✓ How would a chemical that blocks the sodium channels in neuron cell membranes affect a neuron's ability to depolarize?

✓ What effect would decreasing the concentration of extracellular potassium ions have on the transmembrane potential of a neuron?

✓ One axon propagates action potentials at 50 meters per second; another carries them at 1 meter per second. Which axon is myelinated?

Answers start on page Q-1

Action potential generation and transmission can be reviewed on the **IP CD-ROM**: Nervous System I/The Action Potential.

12–5 SYNAPTIC ACTIVITY

Objectives

- Describe the structure of a synapse, and explain the mechanism involved in synaptic activity.
- Describe the major types of neurotransmitters and neuromodulators, and discuss their effects on postsynaptic membranes.

In the nervous system, messages move from one location to another in the form of action potentials along axons. These electrical events are also known as **nerve impulses**. To be effective, a message must not only be propagated along an axon but also transferred in some way to another cell. At a synapse between two neurons, the impulse passes from the **presynaptic neuron** to the **postsynaptic neuron**. A synapse may also involve other types of postsynaptic cells. For example, the neuromuscular junction is a synapse where the postsynaptic cell is a skeletal muscle fiber.

☐ GENERAL PROPERTIES OF SYNAPSES

At a synapse, a change in the transmembrane potential of the synaptic terminal affects the activity of another cell. A synapse may be *electrical*, with direct physical contact between the cells, or *chemical*, involving a neurotransmitter.

Electrical Synapses

Electrical synapses are located in the CNS and PNS, but they are extremely rare. They are present in some areas of the brain, including the *vestibular nuclei*, in the eye, and in at least one pair of PNS ganglia (the *ciliary ganglia*). At an electrical synapse, the presynaptic and postsynaptic membranes are locked together at gap junctions (Figure 4–2•, p. 114). The lipid portions of the opposing membranes, separated by only 2 nm, are held in position by binding between integral membrane proteins called *connexons*. These proteins contain pores that permit the passage of ions between the cells. Because the two cells are linked in this way, changes in the transmembrane potential of one cell will produce local currents that affect the other cell as if the two shared a common membrane. As a result, an electrical synapse propagates action potentials quickly and efficiently from one cell to the next.

Chemical Synapses

The situation at a **chemical synapse** is far more dynamic than that at an electrical synapse, because the cells are not directly coupled. For example, an action potential that reaches an electrical synapse will *always* be propagated to the next cell. But at a chemical synapse, an arriving action potential *may or may not* release enough neurotransmitter to bring the postsynaptic neuron to threshold. In addition, other factors may intervene and make the postsynaptic cell more or less sensitive to arriving stimuli. In essence, the postsynaptic cell at a chemical synapse is not a slave to the presynaptic neuron; its activity can be adjusted, or "tuned," by a variety of factors.

Chemical synapses are by far the most abundant type of synapse. Most synapses between neurons and all communications between neurons and other types of cells involve chemical synapses. Normally, communication across a chemical synapse can occur in only one direction: from the presynaptic membrane to the postsynaptic membrane.

Although acetylcholine is the neurotransmitter that has received the most attention, there are other important chemical transmitters. Neurotransmitters are often classified as excitatory or inhibitory on the basis of their effects on postsynaptic membranes. **Excitatory neurotransmitters** cause depolarization and promote the generation of action potentials, whereas **inhibitory neurotransmitters** cause hyperpolarization and suppress the generation of action potentials.

This classification is useful, but not always precise. For example, acetylcholine typically produces a depolarization in the postsynaptic membrane, but the acetylcholine released at neuromuscular junctions in the heart has an inhibitory effect, producing a transient hyperpolarization of the postsynaptic membrane. This situation highlights an important aspect of neurotransmitter function: *The effect of a neurotransmitter on the postsynaptic membrane depends on the properties of the receptor, not on the nature of the neurotransmitter.*

We shall begin our discussion of chemical synapses with a closer look at a synapse that releases the neurotransmitter **acetylcholine (ACh)**. We shall then introduce other important neurotransmitters that will be encountered in later chapters.

■ CHOLINERGIC SYNAPSES

Synapses that release ACh are known as **cholinergic synapses**. The neuromuscular junction is an example of a cholinergic synapse. ∞ p. 301 ACh, the most widespread (and best-studied) neurotransmitter, is released (1) at all neuromuscular junctions involving skeletal muscle fibers, (2) at many synapses in the CNS, (3) at all neuron-to-neuron synapses in the PNS, and (4) at all neuromuscular and neuroglandular junctions of the parasympathetic division of the ANS.

At a cholinergic synapse between two neurons, the presynaptic and postsynaptic membranes are separated by a synaptic cleft that averages 20 nm (0.02 μm) in width. Most of the ACh in the synaptic knob is packaged in synaptic vesicles, each containing several thousand molecules of the neurotransmitter. A single synaptic knob may contain a million such vesicles.

Events at a Cholinergic Synapse

Figure 12–19● diagrams the events that occur at a cholinergic synapse after an action potential arrives at a synaptic knob. For convenience, we assume that this synapse is adjacent to the initial segment of the axon.

STEP 1: *An Action Potential Arrives and Depolarizes the Synaptic Knob* (Figure 12–19a●). The normal stimulus for neurotransmitter release is the depolarization of the synaptic knob by the arrival of an action potential.

STEP 2: *Extracellular Calcium Ions Enter the Synaptic Knob, Triggering the Exocytosis of ACh* (Figure 12–19b●). The depolarization of the synaptic knob opens voltage-regulated calcium channels. In the brief period during which these channels remain open, calcium ions rush into the knob. Their arrival triggers exocytosis and the release of ACh into the synaptic cleft. The ACh is released in packets of roughly 3000 molecules, the average number of ACh molecules in a single vesicle. The release of ACh stops very soon, because the calcium ions that triggered exocytosis are rapidly removed from the cytoplasm by active transport mechanisms. They are either pumped out of the cell or transfered into mitochondria, vesicles, or the endoplasmic reticulum.

STEP 3: *ACh Binds to Receptors and Depolarizes the Postsynaptic Membrane* (Figure 12–19c●). The ACh released through exocytosis diffuses across the synaptic cleft toward receptors on the postsynaptic membrane. These receptors are chemically regulated ion channels. The primary response is an increased permeability to Na^+, producing a depolarization that lasts about 20 msec.*

This depolarization is a graded potential: The greater the amount of ACh released at the presynaptic membrane, the larger the depolarization. If the depolarization brings the excitable membrane surrounding the synapse to threshold, an action potential will appear in the postsynaptic neuron.

STEP 4: *ACh Is Removed by AChE* (Figure 12–19d●). The effects on the postsynaptic membrane are temporary, because the synaptic cleft and the postsynaptic membrane contain *acetylcholinesterase* (*AChE*, or *cholinesterase*). Roughly half of the ACh released at the presynaptic membrane is broken down before it reaches receptors on the postsynaptic membrane. ACh molecules that succeed in binding to receptor sites are generally broken down within 20 msec of their arrival.

---

*These channels also let potassium ions out of the cell, but because sodium ions are driven by a much stronger electrochemical gradient, the net effect is a slight depolarization of the postsynaptic membrane.

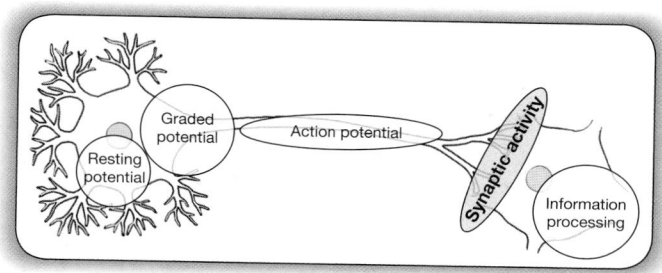

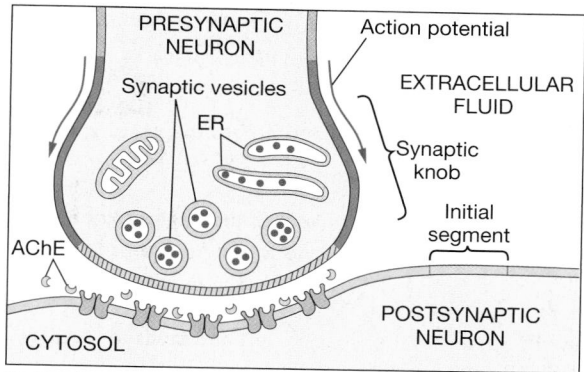

STEP 1: An action potential arrives and depolarizes the synaptic knob.

The enzyme AChE breaks down molecules of ACh into **acetate** and **choline**. The choline is actively absorbed by the synaptic knob and is used to synthesize more ACh, using acetate provided by *coenzyme A (CoA)*. (Recall from Chapter 2 that coenzymes derived from vitamins are required in many enzymatic reactions. ∞ p. 57) Acetate diffusing away from the synapse can be absorbed and metabolized by the postsynaptic cell or by other cells and tissues.

Table 12–5 summarizes the events that occur at a cholinergic synapse.

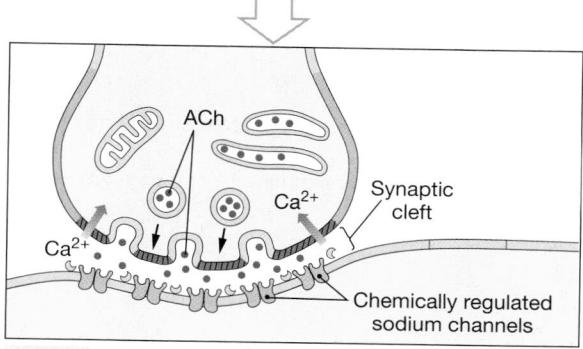

STEP 2: Extracellular Ca^{2+} enters the synaptic cleft triggering the exocytosis of ACh.

Synaptic Delay

A 0.2–0.5-msec **synaptic delay** occurs between the arrival of the action potential at the synaptic knob and the effect on the postsynaptic membrane. Most of that delay reflects the time involved in calcium influx and neurotransmitter release, not in the neurotransmitter's diffusion—the synaptic cleft is narrow, and neurotransmitters can diffuse across it in very little time.

Although a delay of 0.5 msec is not very long, in that time an action potential may travel more than 7 cm (about 3 in.) along a myelinated axon. When information is being passed along a chain of interneurons in the CNS, the cumulative synaptic delay may exceed the propagation time along the axons. Reflexes, which provide rapid and automatic responses to stimuli, involve a small number of synapses. The fewer synapses involved, the shorter the total synaptic delay and the faster the response. The fastest reflexes have just one synapse, with a sensory neuron directly controlling a motor neuron. The muscle spindle reflexes, discussed in Chapter 13, are an important example.

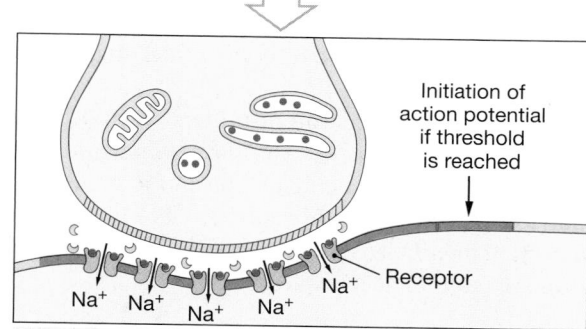

STEP 3: ACh binds to receptors and depolarizes the postsynaptic membrane.

Synaptic Fatigue

Because ACh molecules are recycled, the synaptic knob is not totally dependent on the ACh synthesized in the cell body and delivered by axoplasmic transport. But under intensive stimulation, resynthesis and transport mechanisms may be unable to keep pace with the demand for neurotransmitter. **Synaptic fatigue** then occurs, and the synapse remains inactive until ACh has been replenished.

☐ OTHER NEUROTRANSMITTERS

It has become clear that the nervous system relies on a complex form of chemical communication. Where it was once thought that neurons responded to a single neurotransmitter, we now

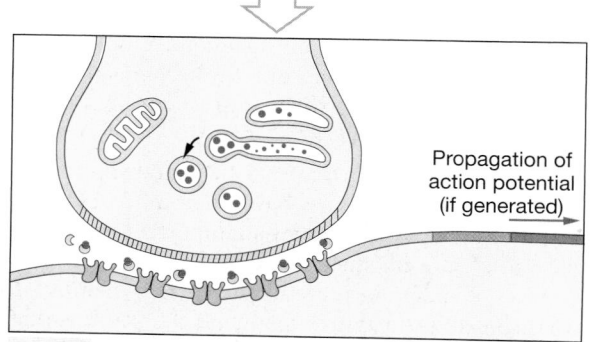

STEP 4: ACh is removed by AChE (acetylcholinesterase).

●**FIGURE 12–19**
The Function of a Cholinergic Synapse

SUMMARY TABLE 12–5 SYNAPTIC ACTIVITY

THE SEQUENCE OF EVENTS AT A TYPICAL CHOLINERGIC SYNAPSE:

STEP 1:

• An arriving action potential depolarizes the synaptic knob.

STEP 2:

• Calcium ions enter the cytoplasm of the synaptic knob.
• ACh is released through exocytosis of neurotransmitter vesicles.

STEP 3:

• ACh diffuses across the synaptic cleft and binds to receptors on the postsynaptic membrane.
• Chemically regulated sodium channels on the postsynaptic surface are activated, producing a graded depolarization.
• ACh release ceases because calcium ions are removed from the cytoplasm of the synaptic knob.

STEP 4:

• The depolarization ends as ACh is broken down into acetate and choline by AChE.
• The synaptic knob reabsorbs choline from the synaptic cleft and uses it to resynthesize ACh.

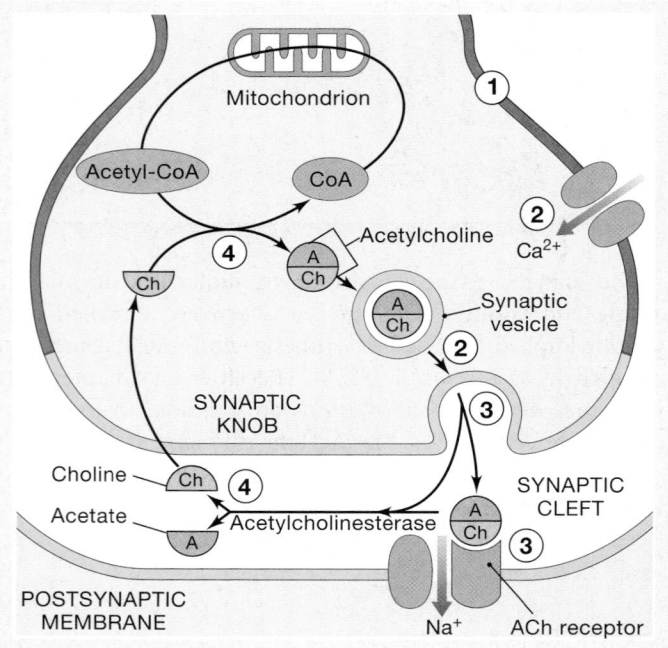

realize that each neuron is continuously exposed to a variety of neurotransmitters. Some, such as norepinephrine (NE), usually have excitatory effects; others, such as gamma amino butyric acid (GABA), usually have inhibitory effects. Yet in all cases, the observed effects depend on the receptor rather than the structure of the neurotransmitter.

Major categories of neurotransmitters include *biogenic amines, amino acids, neuropeptides, dissolved gases,* and a variety of other compounds. General information about these categories is summarized in Table 12–6. We shall consider only a few of the most important neurotransmitters here; in later chapters, we shall encounter additional examples.

• **Norepinephrine** (nor-ep-i-NEF-rin), or **NE**, is a neurotransmitter that is widely distributed in the brain and in portions of the ANS. Norepinephrine is also called *noradrenaline,* and synapses that release NE are known as **adrenergic synapses**. Norepinephrine typically has an excitatory, depolarizing effect on the postsynaptic membrane, but the mechanism is quite distinct from that of ACh. We shall consider specifics in Chapter 16.

• **Dopamine** (DŌ-puh-mēn), a CNS neurotransmitter released in many areas of the brain, may have either inhibitory or excitatory effects. Inhibitory effects play an important role in our precise control of movements. For example, dopamine release in one portion of the brain prevents the overstimulation of neurons that control skeletal muscle tone. If the neurons that produce dopamine are damaged or destroyed, the result can be the characteristic rigidity and stiffness of *Parkinson's disease,* a condition we shall describe in Chapter 14. At other sites, dopamine release has excitatory effects. Cocaine inhibits the removal of

dopamine from synapses in specific areas of the brain. The resulting rise in dopamine concentrations at these synapses is responsible for the "high" experienced by cocaine users.

• **Serotonin** (ser-o-TŌ-nin) is another important CNS neurotransmitter. Inadequate serotonin production can have widespread effects on a person's attention and emotional states and may be responsible for many cases of severe chronic depression. *Fluoxetine (Prozac™), Paxil™, Zoloft™,* and related antidepressant drugs inhibit the reabsorption of serotonin by synaptic knobs (hence their classification as *selective* serotonin *reuptake* inhibitors, or *SSRI's*). This inhibition leads to increased serotonin concentrations at synapses; over time, the increase may relieve the symptoms of depression. Interactions among serotonin, norepinephrine, and other neurotransmitters are thought to be involved in the regulation of sleep and wake cycles.

• **Gamma aminobutyric** (a-MĒ-nō-bū-TĒR-ik) **acid**, or **GABA**, generally has an inhibitory effect. Although roughly 20% of the synapses in the brain release GABA, its functions remain incompletely understood. In the CNS, GABA release appears to reduce anxiety, and some antianxiety drugs work by enhancing this effect.

The functions of many neurotransmitters are not well understood. In a clear demonstration of the principle "the more you look, the more you see," at least 50 neurotransmitters have been identified, including certain amino acids, peptides, polypeptides, prostaglandins, and ATP. In addition, two gases, nitric oxide and carbon monoxide, are now known to be important neurotransmitters. Nitric oxide (NO) is generated by synaptic terminals that innervate smooth muscle in the walls of blood vessels in the PNS

Drugs and Synaptic Function

Many drugs interfere with key steps in synaptic activity. These drugs (1) interfere with neurotransmitter synthesis, (2) alter the rate of neurotransmitter release, (3) prevent neurotransmitter inactivation, or (4) prevent neurotransmitter binding to receptors. Our discussion here is limited to clinically important compounds that act at cholinergic synapses (Figure 12–20•).

Some drugs affect the synaptic terminals. *Botulinus toxin* is responsible for the primary symptom of *botulism*, a widespread paralysis of skeletal muscles. ∞ p. 306 *Botulinus* toxin blocks the release of ACh at the presynaptic membrane of cholinergic neurons. Curiously, the same substance that causes this potentially fatal illness is now widely used by plastic surgeons and dermatologists for cosmetic purposes. Minute quantities of the toxin (available under the name "Botox") are injected under the skin to paralyze small facial muscles that cause wrinkles. (The treatment must be repeated every few months.) The venom of the black widow spider has the opposite effect, causing a massive release of ACh that produces intense muscular cramps and spasms.

Other drugs primarily affect the postsynaptic membrane. **Anticholinesterase drugs**, or *cholinesterase inhibitors*, block the breakdown of ACh by acetylcholinesterase. The result is an exaggerated and prolonged stimulation of the postsynaptic membrane. At the neuromuscular junctions, this abnormal stimulation produces an extended and extreme state of contraction. Military nerve gases block cholinesterase activity for weeks, although few persons exposed are likely to live long enough to regain normal synaptic function. Most animals utilize ACh as a neurotransmitter, and anticholinesterase (AChE) drugs, such as *malathion*, are in widespread use in insect-control projects.

Drugs such as **atropine** and **d-tubocurarine** prevent ACh from binding to the postsynaptic receptors. They work on different types of ACh receptors. Atropine can be administered to counteract the effects of AChE poisoning on ACh receptors in smooth and cardiac muscles; d-tubocurarine is a derivative of *curare*, a plant extract used by certain South American tribes to paralyze their prey. Curare and related compounds induce paralysis by preventing the stimulation of the neuromuscular junction by ACh. Other compounds, including **nicotine**, an active ingredient in cigarette smoke, bind to the receptor sites and stimulate the postsynaptic membrane. No enzymes exist that remove these compounds, and their effects are relatively prolonged. **AM** Neuroactive Drugs

Figure 12–20• also indicates the sites of action of other chemicals mentioned in this chapter. Several toxins found in

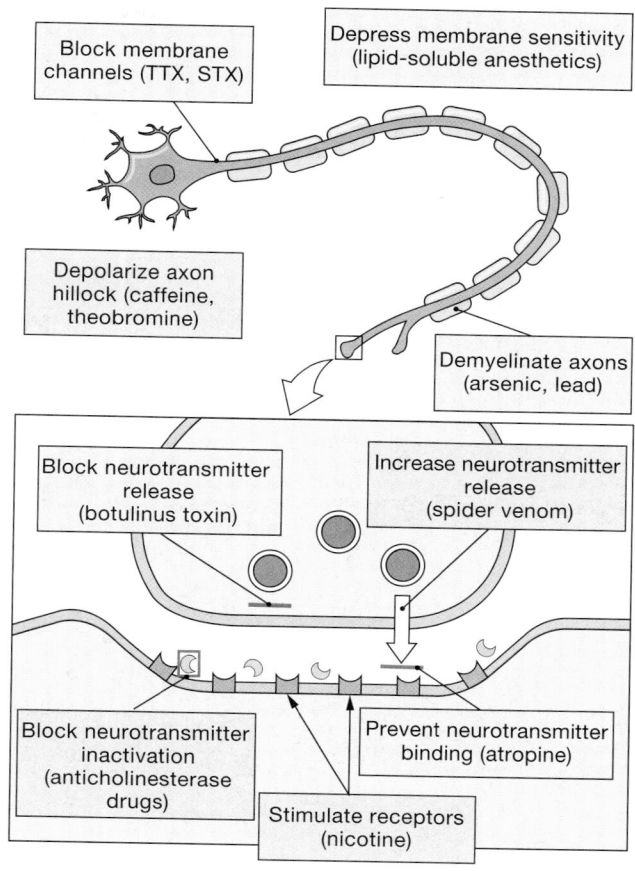

•**FIGURE 12–20**
The Mechanism of Drug Action at a Cholinergic Synapse. Factors that facilitate neural function and make neurons more excitable are shown in violet. Factors that inhibit or depress neural function are in blue.

seafood, including *tetrodotoxin* (TTX), *saxitoxin* (STX), and *ciguatoxin* (CTX), block sodium ion channels. At low doses, these toxins can produce abnormal sensations and interfere with muscle control. Higher doses cause death, generally through the paralysis of respiratory muscles. Neurophysiology research laboratories have long used TTX to study the electrical properties of neurons. After blocking sodium ion channels, researchers can monitor subtle changes in the transmembrane potential that occur at or above the threshold—changes that would otherwise be masked by an influx of sodium ions. **AM** Neurotoxins in Seafood

and at synapses in several regions of the brain. Carbon monoxide (CO), best known as a component of automobile exhaust, is also generated by specialized synaptic knobs in the brain, where it functions as a neurotransmitter. Our knowledge of the signifi-

cance of these compounds and the mechanisms involved in their synthesis and release remains incomplete. Table 12–6 lists major neurotransmitters of the brain and spinal cord and their primary effects (if known).

TABLE 12-6 REPRESENTATIVE NEUROTRANSMITTERS AND NEUROMODULATORS

Class and Neurotransmitter	Chemical Structure	Mechanism of Action	Location(s)	Comments
Acetylcholine		Primarily direct, through binding to chemically regulated channels	CNS: Synapses throughout brain and spinal cord. PNS: Neuromuscular junctions; preganglionic synapses of ANS; neuroglandular junctions of parasympathetic division and (rarely) sympathetic division of ANS; amacrine cells of retina	Widespread in CNS and PNS; best known and most studied of the neurotransmitters
BIOGENIC AMINES Norepinephrine		Indirect: G proteins and second messengers	CNS: Cerebral cortex, hypothalamus, brain stem, cerebellum, spinal cord. PNS: Most neuromuscular and neuroglandular junctions of sympathetic division of ANS	Involved in attention and consciousness, control of body temperature, and regulation of pituitary gland secretion
Epinephrine		Indirect: G proteins and second messengers	CNS: Thalamus, hypothalamus, midbrain, spinal cord	Uncertain functions
Dopamine		Indirect: G proteins and second messengers	CNS: Hypothalamus, midbrain, limbic system, cerebral cortex, retina	Regulation of subconscious motor function; receptor abnormalities have been linked to development of schizophrenia
Serotonin		Primarily indirect: G proteins and second messengers	CNS: Hypothalamus, limbic system, cerebellum, spinal cord, retina	Important in emotional states, moods, and body temperature; several illicit hallucinogenic drugs, such as *ecstasy*, target serotonin receptors
Histamine		Indirect: G proteins and second messengers	CNS: Neurons in hypothalamus, with axons projecting throughout the brain	Receptors are primarily on presynaptic membranes; functions in sexual arousal, pain threshold, pituitary hormone secretion, thirst, and blood pressure control
AMINO ACIDS *Excitatory:* Glutamate		Indirect: G proteins and second messengers. Direct: opens calcium channels on pre- and postsynaptic membranes	CNS: Cerebral cortex and brain stem	Important in memory and learning; most important excitatory neurotransmitter in the brain
Aspartate		Direct or indirect (G proteins), depending on type of receptor	CNS: Cerebral cortex, retina, and spinal cord	Used by pyramidal cells that provide voluntary motor control over skeletal muscles
Inhibitory: Gamma aminobutyric acid (GABA)		Direct or indirect (G proteins), depending on type of receptor	CNS: Cerebral cortex, cerebellum, interneurons throughout brain and spinal cord	Direct effects: open Cl⁻ channels; indirect effects: open K⁺ channels and block entry of Ca^{2+}
Glycine		Direct: Opens Cl⁻ channels	CNS: Interneurons in brain stem, spinal cord, and retina	Produces postsynaptic inhibition; the poison *strychnine* produces fatal convulsions by blocking glycine receptors

TABLE 12–6 CONTINUED

Class and Neurotransmitter	Chemical Structure	Mechanism of Action	Location(s)	Comments
NEUROPEPTIDES Substance P		Indirect: G proteins and second messengers	CNS: Synapses of pain receptors within spinal cord, hypothalamus, and other areas of the brain PNS: Enteric nervous system (network of neurons along the digestive tract)	Important in pain pathway, regulation of pituitary gland function, and control of digestive tract reflexes
Neuropeptide Y -		As above PNS: sympathetic neurons	CNS: hypothalamus	Stimulates appetite and food intake
Opioids Endorphins		Indirect: G proteins and second messengers	CNS: Thalamus, hypothalamus, brain stem, retina	Pain control; emotional and behavioral effects poorly understood
Enkephalins	*36-amino-acid peptide*	As above	CNS: Basal nuclei, hypothalamus, midbrain, pons, medulla oblongata, spinal cord	As above
Endomorphin	*31-amino-acid peptide*	As above	CNS: Thalamus, hypothalamus, basal nuclei	As above
Dynorphin		As above	CNS: hypothalamus, midbrain, medulla oblongata	As above
PURINES ATP, GTP	*(see Figure 2–24)*	Direct or indirect (G proteins), depending on type of receptor	CNS: Spinal cord PNS: Autonomic ganglia	
Adenosine	*(see Figure 2–24)*	Indirect: G proteins and second messengers	CNS: Cerebral cortex, hippocampus, cerebellum	Produces drowsiness; stimulatory effect of caffeine is due to inhibition of adenosine activity
HORMONES ADH, oxytocin, insulin, glucagon, secretin, CCK, GIP, VIP, inhibins, ANP, BNP, and many others		Typically indirect: G proteins and second messengers	CNS: Brain (widespread)	Numerous, complex, and incompletely understood
GASES Carbon monoxide (CO)	$C = O$	Indirect: By diffusion to enzymes activating second messengers	CNS: Brain PNS: Some neuromuscular and neuroglandular junctions	Localization and function poorly understood
Nitric oxide (NO)	$N = O$	As above	CNS: Brain, especially at blood vessels PNS: Some sympathetic neuromuscular and neuroglandular junctions	
LIPIDS Anandamide		Indirect: G proteins and second messengers	CNS: cerebral cortex, hippocampus, cerebellum	Euphoria, drowsiness; receptors are targeted by the active ingredient in marijuana

☐ NEUROMODULATORS

Although it is convenient to discuss each synapse as if it were releasing only one chemical, synaptic knobs may release a mixture of active compounds, either through diffusion across the membrane or through exocytosis, in the company of neurotransmitter molecules. These compounds may have a variety of functions. Those which alter the rate of neurotransmitter release by the presynaptic neuron or change the postsynaptic cell's response to neurotransmitters are called **neuromodulators** (noo-rō-MOD-ū-lā-torz). These substances are typically **neuropeptides**, small peptide chains synthesized and released by the synaptic knob. Most neuromodulators act by binding to receptors in the presynaptic or postsynaptic membranes and activating cytoplasmic enzymes.

Neuromodulators called **opioids** (Ō-pē-oydz) have effects similar to those of the drugs *opium* and *morphine*, because they bind to the same group of postsynaptic receptors. Four classes of opioids are identified in the CNS: (1) the **endorphins** (en-DOR-finz), (2) the **enkephalins** (en-KEF-a-linz), (3) the **endomorphins**, and (4) the **dynorphins** (dī-NOR-finz). The primary function of opioids is probably the relief of pain—they inhibit the release of the neurotransmitter *substance P* at synapses that relay pain sensations. Dynorphins have far more powerful analgesic (pain-relieving) effects than morphine or the other opioids.

In general, neuromodulators (1) have long-term effects that are relatively slow to appear, (2) trigger responses that involve a number of steps and intermediary compounds, (3) may affect the presynaptic membrane, the postsynaptic membrane, or both, and (4) can be released alone or in the company of a neurotransmitter. In practice, it can be very difficult to distinguish neurotransmitters from neuromodulators on either biochemical or functional grounds: A neuropeptide may function in one site as a neuromodulator and in another as a neurotransmitter. For this reason, Table 12–6 does not distinguish between neurotransmitters and neuromodulators.

☐ HOW NEUROTRANSMITTERS AND NEUROMODULATORS WORK

Functionally, neurotransmitters and neuromodulators fall into one of three groups: (1) *Compounds that have a direct effect on membrane potential,* (2) *compounds that have an indirect effect on membrane potential,* or (3) *lipid-soluble gases that exert their effects inside of the cell.*

Compounds that have direct effects on membrane potential exert those effects by opening or closing gated ion channels (Figure 12–21a●). Examples include ACh and the amino acids *glycine* and *aspartate*. These neurotransmitters alter ion movement across the membrane; they are therefore said to have *ionotropic effects.* A few neurotransmitters, notably glutamate, GABA, NE, and serotonin, have both direct and indirect effects, because these compounds target two different classes of receptors. The direct effects are ionotropic; the indirect effects, which involve changes in the metabolic activity of the postsynaptic cell, are called *metabotropic.*

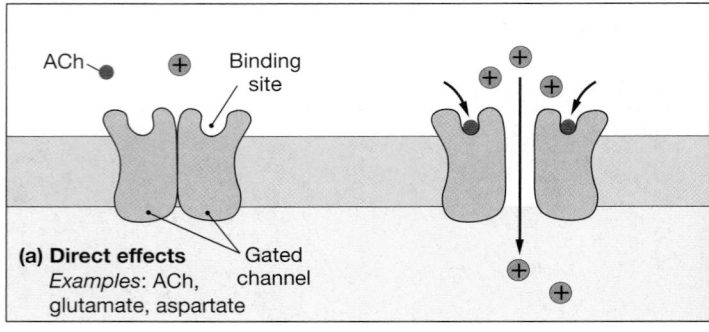

(a) Direct effects
Examples: ACh, glutamate, aspartate

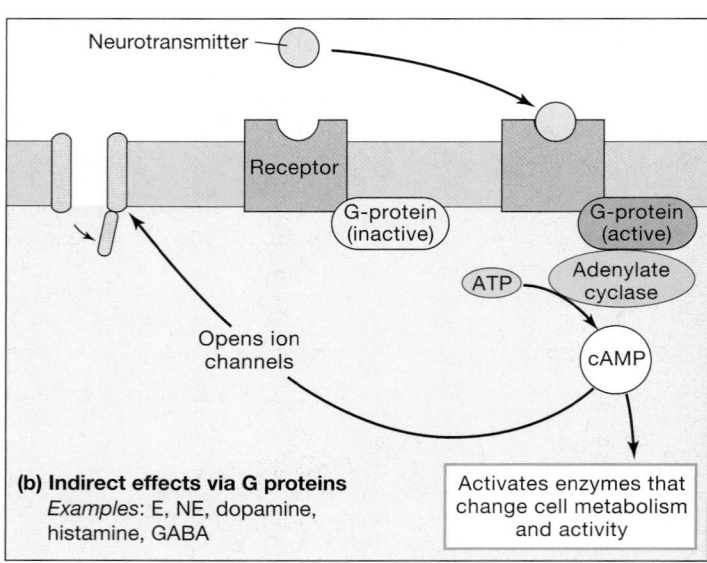

(b) Indirect effects via G proteins
Examples: E, NE, dopamine, histamine, GABA

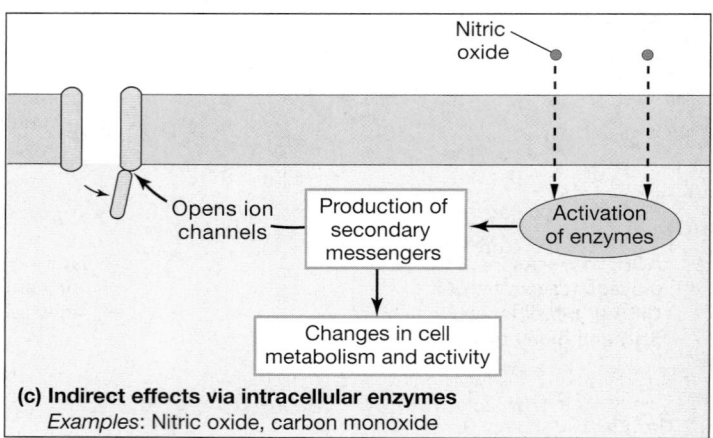

(c) Indirect effects via intracellular enzymes
Examples: Nitric oxide, carbon monoxide

●**FIGURE 12–21**
Neurotransmitter Functions. **(a)** Effects on membrane channels. **(b)** Effects mediated by G proteins. **(c)** Effects mediated by intracellular enzymes.

Compounds that have an indirect effect on membrane potential work through intermediaries known as *second messengers*. The neurotransmitter represents a *first messenger*, because it delivers the message to receptors on the cell membrane or within the cell. Second messengers are ions or molecules that are produced or released inside the cell when a first messenger binds to one of these receptors.

Many neurotransmitters, including epinephrine, norepinephrine, dopamine, serotonin, histamine, and GABA, as well as many neuromodulators, bind to receptors in the cell membrane. In these instances, the link between the first messenger and the second messenger involves a **G protein**, an enzyme complex coupled to a membrane receptor. The name *G protein* refers to the fact that these proteins bind GTP, a high-energy compound introduced in Chapter 2. ∞ p. 60 There are several types of G proteins, but each type includes an enzyme that is "turned on" when an extracellular compound binds to the associated receptor at the cell surface.

Figure 12–21b• shows one possible result of this binding: the activation of the enzyme **adenylate cyclase**, also known as *adenylyl cyclase*. This enzyme converts ATP, the energy currency of the cell, to *cyclic-AMP*, a ring-shaped form of the compound AMP that was introduced in Chapter 2. ∞ p. 59 The conversion takes place at the inner surface of the cell membrane. Cyclic-AMP (cAMP) is a second messenger that may open membrane channels, activate intracellular enzymes, or do both, depending on the nature of the postsynaptic cell. This is only an overview of the function of one type of G protein; we shall examine several types of G proteins more closely in later chapters.

Two lipid-soluble gases, nitric oxide (NO) and carbon monoxide (CO), are now known to be important neurotransmitters in specific regions of the brain. Because they can diffuse through lipid membranes, these gases can enter the cell and bind to enzymes on the inner surface of the membrane or elsewhere in the cytoplasm (Figure 12–21c•). These enzymes then promote the appearance of second messengers that can affect cellular activity.

✓ What effect would blocking voltage-regulated calcium channels at a cholinergic synapse have on synaptic communication?

✓ One pathway in the central nervous system consists of three neurons, another of five neurons. If the neurons in the two pathways are identical, which pathway will transmit impulses more rapidly?

Answers start on page Q-1

Synaptic activity can be reviewed on the **IP CD-ROM**: Nervous System II/Synaptic Potentials and Cellular Integration, and Synaptic Transmission.

12–6 INFORMATION PROCESSING

Objective

- Discuss the interactions that make possible the processing of information in neural tissue.

At each synapse, an action potential arriving at a synaptic knob triggers chemical or electrical events that affect another cell. A single neuron may receive information across thousands of synapses. Some of the neurotransmitters arriving at the postsynaptic cell at any moment may be excitatory, whereas others may be inhibitory. The net effects on the transmembrane potential of the cell body—

specifically, in the area of the axon hillock—will determine how the postsynaptic neuron responds from moment to moment. If a depolarization occurs at the axon hillock, it will spread to affect the transmembrane potential at the initial segment. If threshold is reached at the initial segment, an action potential will be generated and propagated along the axon.

The transmembrane potential at the axon hillock and initial segment therefore represents an integration of all the excitatory and inhibitory stimuli affecting the neuron at that moment. Excitatory and inhibitory stimuli are integrated through interactions between *postsynaptic potentials*. This interaction is the simplest level of **information processing** in the nervous system.

☐ POSTSYNAPTIC POTENTIALS

Postsynaptic potentials are graded potentials that develop in the postsynaptic membrane in response to a neurotransmitter. Two major types of postsynaptic potentials develop at neuron-to-neuron synapses: (1) excitatory postsynaptic potentials and (2) inhibitory postsynaptic potentials.

An **excitatory postsynaptic potential**, or **EPSP**, is a graded depolarization caused by the arrival of a neurotransmitter at the postsynaptic membrane. An EPSP results from the opening of chemically regulated membrane channels that lead to depolarization of the cell membrane. For example, the graded depolarization produced by the binding of ACh is an EPSP. Because it is a graded potential, an EPSP affects only the area immediately surrounding the synapse, as in Figure 12–13•, page 402.

We have already noted that not all neurotransmitters have an excitatory (depolarizing) effect. An **inhibitory postsynaptic potential**, or **IPSP**, is a graded hyperpolarization of the postsynaptic membrane. For example, an IPSP may result from the opening of chemically regulated potassium channels. While the hyperpolarization continues, the neuron is said to be **inhibited**, because a larger-than-usual depolarizing stimulus must be provided to bring the membrane potential to threshold. A stimulus sufficient to shift the transmembrane potential by 10 mV (from −70 mV to −60 mV) would normally produce an action potential. If the transmembrane potential were reset at −85 mV by an IPSP, however, the same stimulus would depolarize it to only −75 mV, which is below threshold.

Summation

An individual EPSP has a small effect on the transmembrane potential; a typical EPSP produces a depolarization of about 0.5 mV at the postsynaptic membrane. Before an action potential will appear in the initial segment, local currents must depolarize that region by at least 10 mV. A single EPSP will therefore not result in an action potential, even if the synapse is on the axon hillock. But individual EPSPs combine through the process of **summation**, which integrates the effects of graded potentials that affect one portion of the cell membrane. The graded potentials may be EPSPs, IPSPs, or both. We will consider EPSPs in our example.

Two forms of summation exist: (1) temporal summation and (2) spatial summation.

Temporal summation (*tempus*, time) is the addition of stimuli occurring in rapid succession. Temporal summation occurs at a *single synapse* that is active *repeatedly*. Consider the fact that you cannot fill a bathtub with a single bucket of water, but you can fill it if you keep using the bucket over and over. The water in each bucketful corresponds to the sodium ions that enter the cytoplasm during an EPSP. Now suppose that a second EPSP arrives before the effects of the first have disappeared (Figure 12–22a●). A typical EPSP lasts about 20 msec, but under maximum stimulation one action potential can reach the synaptic knob each millisecond. Every time an action potential arrives, a group of vesicles discharges ACh into the synaptic cleft. Every time more ACh molecules arrive at the postsynaptic membrane, more chemically regulated channels open and the degree of depolarization increases. In this way, a series of small steps can eventually bring the initial segment to its threshold.

Spatial summation occurs when simultaneous stimuli at different locations have a cumulative effect on the transmembrane potential. Spatial summation involves *multiple synapses* that are active *simultaneously*. In terms of our bucket analogy, you could fill a bathtub immediately if 50 friends emptied their buckets into it at the same time.

The activity of one synapse produces a graded potential with localized effects. If more than one synapse is active at the same time, all will "pour" sodium ions across the postsynaptic membrane. At each active synapse, the sodium ions that produce the EPSP spread out along the inner surface of the membrane and mingle with those entering at other synapses. As a result, the effects on the initial segment are cumulative (Figure 12–22b●). The degree of depolarization depends on how many synapses are active at any moment and on their distance from the initial segment. As in temporal summation, an action potential appears when the transmembrane potential at the initial segment reaches threshold.

FACILITATION Spatial or temporal summation of EPSPs may not depolarize the initial segment to threshold, but every step closer to threshold makes it easier for the *next* stimulus to trigger an action potential. A neuron that is brought closer to threshold is said to be **facilitated**. The larger the degree of facilitation, the smaller is the additional stimulus needed to trigger an action potential. In a highly facilitated neuron, even a small depolarizing stimulus produces an action potential.

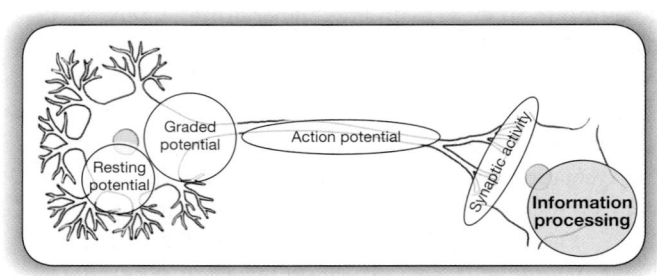

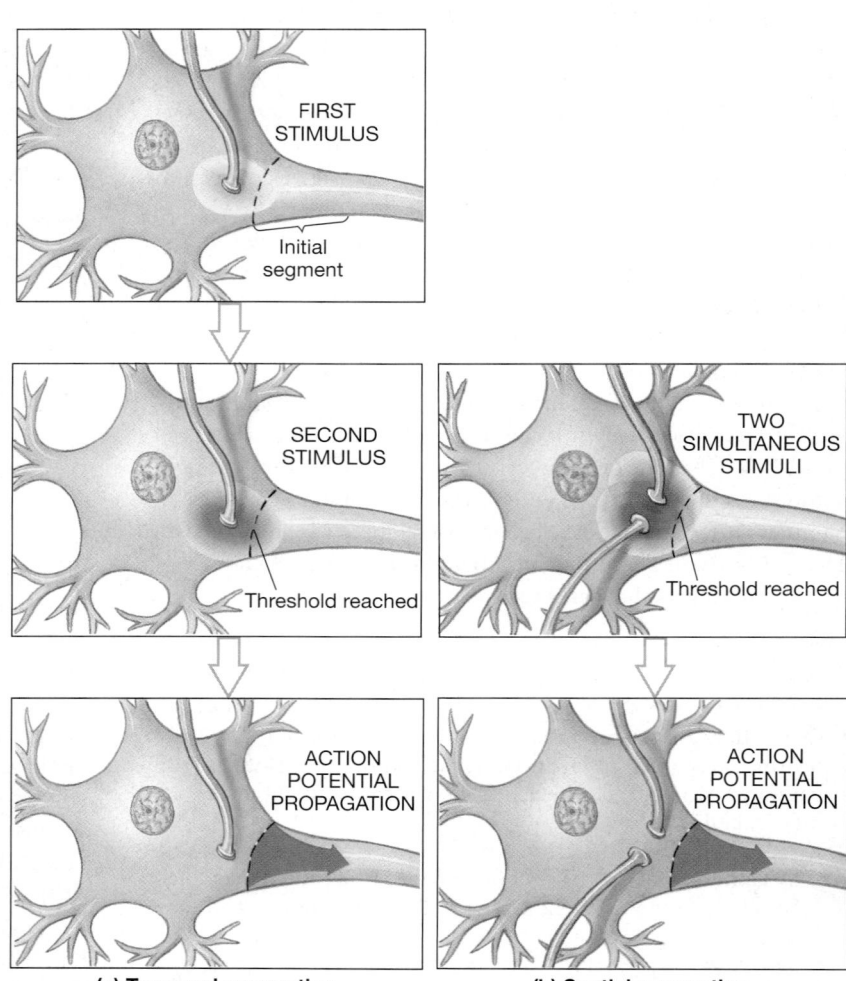

(a) Temporal summation   **(b) Spatial summation**

●**FIGURE 12–22**
Temporal and Spatial Summation.   **(a)** Temporal summation occurs on a membrane that receives two depolarizing stimuli from the same source in rapid succession. The effects of the second stimulus are added to those of the first. **(b)** Spatial summation occurs when sources of stimulation arrive simultaneously, but at different locations. Local currents spread the depolarizing effects, and areas of overlap experience the combined effects.

Facilitation also results from the exposure of a neuron to certain drugs in the extracellular fluid. For example, the nicotine in cigarettes stimulates postsynaptic ACh receptors, producing prolonged EPSPs that facilitate CNS neurons. The active ingredients of coffee and colas (caffeine), cocoa (theobromine), and tea (theophylline) also cause facilitation, but they act in a different way. These compounds lower the threshold at the initial segment, so that a smaller-than-usual depolarization will cause an action potential. Once an action potential reaches the synaptic knob, these compounds also increase the amount of ACh released. The nervous system becomes more excitable, and you feel more alert. "Coffee makes you jumpy" is not an exaggeration: After several cups, you may jump at a sudden noise because your reflexes are primed to respond to the slightest stimulus. Since nicotine and caffeine stimulate the nervous system in different ways, the combination of cigarettes and coffee has a dramatic effect on the CNS.

SUMMATION OF EPSPS AND IPSPS Like EPSPs, IPSPs summate spatially and temporally. EPSPs and IPSPs reflect the activation of different types of chemically regulated channels, producing opposing effects on the transmembrane potential. The antagonism between IPSPs and EPSPs is an important mechanism for cellular information processing. In terms of our bucket analogy, EPSPs put water into the bathtub, and IPSPs take water out. If more buckets add water than remove water, the water level in the tub will rise. If more buckets remove water, the level will fall. If a bucket of water is removed every time another bucket is dumped in, the level will remain stable. Comparable interactions between EPSPs and IPSPs (Figure 12–23●) determine the transmembrane potential at the boundary between the axon hillock and the initial segment.

Neuromodulators, hormones, or both can change the postsynaptic membrane's sensitivity to excitatory or inhibitory neurotransmitters. By shifting the balance between EPSPs and IPSPs, these compounds promote facilitation or inhibition of CNS and PNS neurons.

CHLORIDE CHANNELS AND EPSP–IPSP SUMMATION Many chemicals that inhibit EPSP formation open chemically regulated chloride channels rather than potassium channels. The opening of chemically regulated chloride channels in a resting membrane is difficult to detect, because the equilibrium potential for chloride ions is -70 mV, the same as the typical resting potential. However, with the chloride channels open, Cl^- is free to cross the membrane if the transmembrane potential changes.

Consider what happens if the cell receives a depolarizing stimulus that would, under other circumstances, shift the transmembrane potential from -70 mV to -65 mV. As soon as the potential climbs above -70 mV, chloride ions start rushing into the cell. The situation now is like a bathtub with a large hole just above the normal water level; to raise the water level, you must add much more water—enough to overcome the water loss through the hole. In a similar way, once the chloride ion channels are open, a much larger-than-normal stimulus is needed to depolarize the membrane to threshold.

☐ PRESYNAPTIC INHIBITION AND FACILITATION

Inhibitory or excitatory responses may occur at an *axoaxonal synapse*, a synapse between the axons of two neurons. An axoaxonal synapse that occurs on the synaptic knob can modify the rate of neurotransmitter release at the presynaptic membrane. In one form of **presynaptic inhibition**, the release of GABA inhibits the opening of voltage-regulated calcium channels in the synaptic knob (Figure 12–24a●). This inhibition reduces the amount of neurotransmitter released when an action potential arrives at the synaptic knob and thus limits the effects on the postsynaptic membrane.

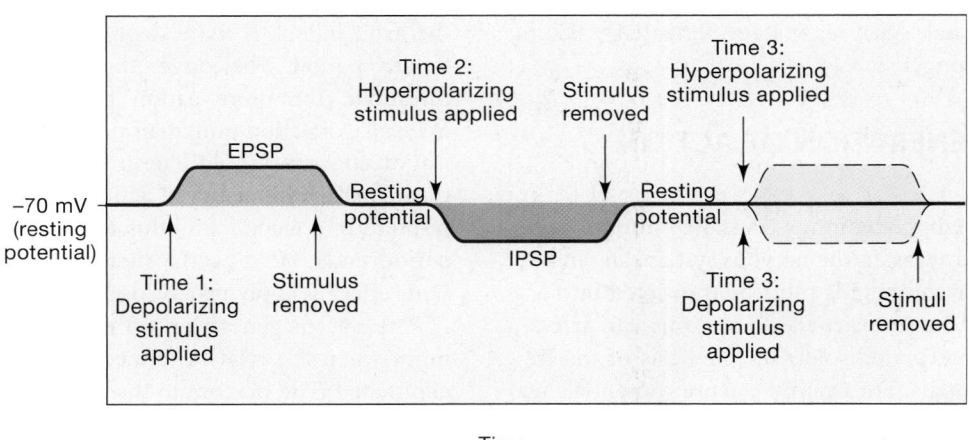

●**FIGURE 12–23**
EPSP–IPSP Interactions. At time 1, a small depolarizing stimulus produces an EPSP. At time 2, a small hyperpolarizing stimulus produces an IPSP of comparable magnitude. If the two stimuli are applied simultaneously, as they are at time 3, summation occurs. Because the two are equal in size but have opposite effects, the membrane potential remains at the resting level. If the EPSP were larger, a net depolarization would result; if the IPSP were larger, a net hyperpolarization would result instead.

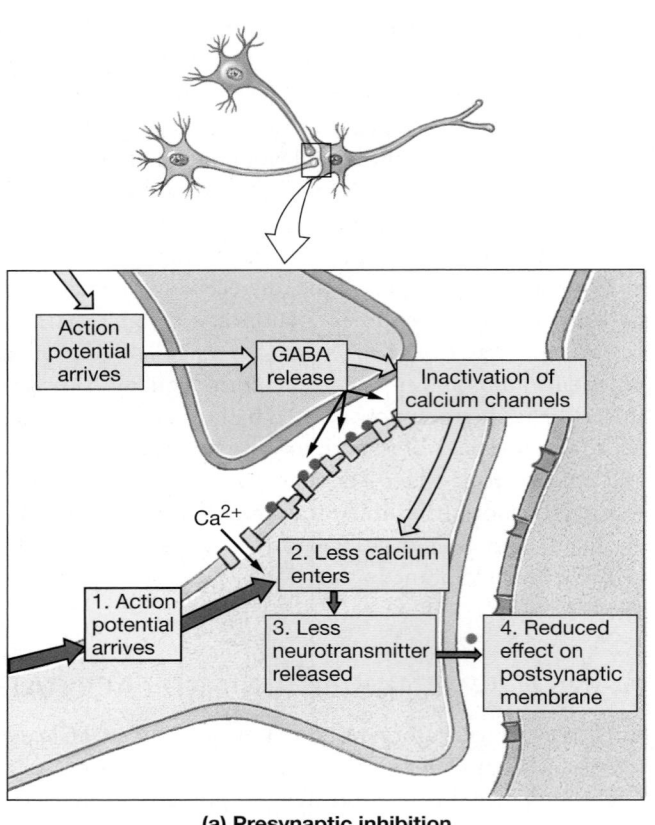

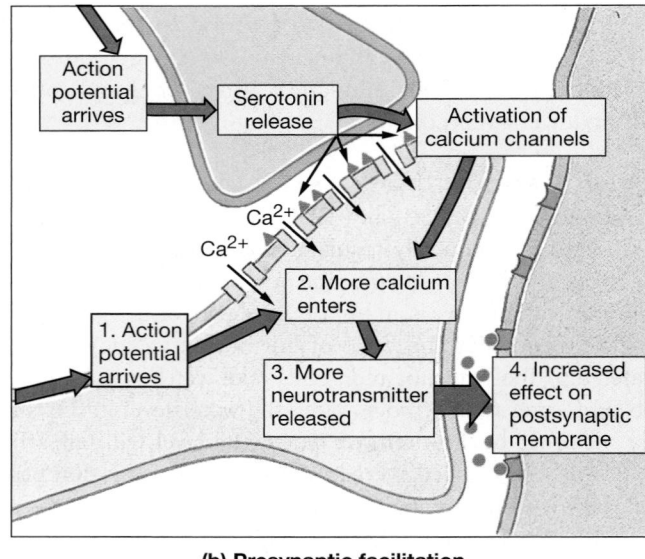

(a) Presynaptic inhibition

(b) Presynaptic facilitation

•FIGURE 12–24
Presynaptic Inhibition and Facilitation. (a) Steps in presynaptic inhibition. (b) Steps in presynaptic facilitation.

In **presynaptic facilitation**, activity at an axoaxonal synapse increases the amount of neurotransmitter released when an action potential arrives at the synaptic knob (Figure 12–24b●). This increase enhances and prolongs the neurotransmitter's effects on the postsynaptic membrane. The neurotransmitter *serotonin* is involved in presynaptic facilitation. In the presence of serotonin released at an axoaxonal synapse, voltage-regulated calcium channels remain open longer.

◻ THE RATE OF GENERATION OF ACTION POTENTIALS

In a computer, complex information is translated into a simple binary code of ones and zeros. In the nervous system, all sensory information and motor commands must be translated into action potentials that can be propagated along axons. On arrival, the message is often interpreted solely on the basis of the frequency of action potentials. For example, action potentials arriving at a neuromuscular junction at the rate of 1 per second may produce a series of isolated twitches in the associated skeletal muscle fiber, whereas at the rate of 100 per second they will cause a sustained tetanic contraction. Similarly, a few action potentials per second along a sensory fiber may be perceived as a featherlight touch, whereas hundreds of action potentials per second along that same axon may be perceived as unbearable

pressure. In this section, we shall examine factors that vary the rate of generation of action potentials. We shall consider the functional significance of these changes in later chapters.

If a graded potential briefly depolarizes the axon hillock such that the initial segment reaches its threshold, an action potential will be propagated along the axon. Now consider what happens if the axon hillock *remains* depolarized past threshold for an extended period. The longer the initial segment remains above threshold, the more action potentials it will produce. The *frequency* of action potentials depends on the degree of depolarization above threshold: The greater the degree of depolarization, the higher the frequency of action potentials. The membrane can respond to a second stimulus as soon as the absolute refractory period ends. Holding the membrane above threshold has the same effect as applying a second, larger than normal stimulus.

The rate of generation of action potentials reaches a maximum when the relative refractory period has been completely eliminated. The maximum theoretical frequency is therefore established by the duration of the absolute refractory period. The absolute refractory period is shortest in large-diameter axons, in which the *theoretical* maximum frequency of action potentials is 2500 per second. However, the highest frequencies recorded from axons in the body range between 500 and 1000 per second.

Table 12–7 summarizes the basic principles of information processing.

SUMMARY TABLE 12–7 INFORMATION PROCESSING

- The neurotransmitters released at a synapse may have excitatory or inhibitory effects. The effect on the axon's initial segment reflects a summation of the stimuli that arrive at any moment. The frequency of generation of action potentials is an indication of the degree of sustained depolarization at the axon hillock.
- Neuromodulators can alter the rate of neurotransmitter release or the response of a postsynaptic neuron to specific neurotransmitters.
- Neurons may be facilitated or inhibited by extracellular chemicals other than neurotransmitters or neuromodulators.
- The effect of a presynaptic neuron's activation on a postsynaptic neuron at a synapse may be altered by other neurons, through presynaptic inhibition or presynaptic facilitation.

Factors That Disrupt Neural Function

Two key factors can disrupt normal CNS and PNS functions: (1) changes in the extracellular environment and (2) an inability to meet the metabolic demands of active neurons.

Environmental Factors

Environmental factors, such as changes in pH, ionic composition, or temperature, can alter the resting membrane potential or disrupt the metabolic operations that support the generation of action potentials:

- *Changes in hydrogen ion concentration (pH) can have dramatic effects on neural activity.* The normal extracellular pH averages 7.35–7.45. If the pH rises, neurons are facilitated: At a pH near 7.8, they begin to generate action potentials spontaneously, producing severe convulsions. If the pH drops, neurons are inhibited: At a pH around 7.0, the rate of generation of action potentials is so low that the nervous system shuts down and the individual becomes completely unresponsive.

- *Changes in the ionic composition of the extracellular fluids have a direct effect on neural function.* Fluctuations in Na^+ or K^+ concentrations, such as those caused by dehydration or kidney disease, may facilitate or inhibit neural activity by depolarizing or hyperpolarizing the cell membrane. For example, an elevated extracellular K^+ concentration, a condition called **hyperkalemia** (hī-per-ka-LĒ-mē-uh; *kalium*, potassium + *haima*, blood) reduces the chemical gradient for K^+ across the cell membrane. The rate of potassium loss decreases as a result, and the retention of positive charges gradually depolarizes the membrane.

- *Hyperkalemia has damaging effects on all excitable membranes.* Slight hyperkalemia causes facilitation, but extreme hyperkalemia interferes with repolarization and suppresses the generation of action potentials. The result is a general paralysis of skeletal muscles and death by cardiac arrest. Abnormally high or low extracellular Ca^{2+} concentrations affect synaptic function by reducing or enhancing calcium entry into the synaptic knob, thereby modifying the amount of neurotransmitter released. There are also direct effects on the excitability of membranes.

- *Changes in body temperature have a direct effect on the activity of neurons.* This is one reason the regulation of body temperature is so important. If your body temperature rises, your neurons become more excitable; an individual with a high fever may experience hallucinations or convulsions. If your body temperature falls, your neurons become inhibited; an individual whose body temperature has fallen outside normal limits will be lethargic and confused and may lose consciousness.

Metabolic Processes

The brain accounts for just 2 percent of your body weight, but it accounts for 18 percent of your resting energy consumption. Active neurons need ATP to support (1) the synthesis, release, and recycling of neurotransmitter molecules, (2) the movement of materials to and from the cell body by axoplasmic transport, (3) the maintenance of the normal resting potential, and (4) the recovery from action potentials through the activity of the sodium–potassium exchange pump. Each time an action potential occurs, sodium ions enter and potassium ions leave the cell; over time, ATP must be expended to maintain normal cytoplasmic ion concentrations. When impulses are generated at high frequencies, the energy demands are enormous.

Neurons normally derive ATP solely through aerobic mechanisms. Because their cytoplasm does not contain glycogen reserves, neurons are totally dependent on a continuous and reliable supply of both oxygen and glucose from the blood. In cases of severe malnutrition, neural function deteriorates as the body's energy reserves are exhausted.

Neural function is also impaired if the local circulation is restricted or, worse yet, shut off. If the circulation to a region is interrupted for just a few seconds, neurons in that region will be injured. The longer the interruption, the more severe the injury will be. In a **stroke**, the blood supply to the brain is interrupted by a circulatory blockage or by vascular rupture. The degree of functional impairment after a stroke is determined by (1) the location and size of the region deprived of circulation and (2) the duration of the circulatory interruption. In subsequent chapters, we will consider the origins and treatment of strokes in greater detail.

Several inherited abnormalities in neural function are caused by metabolic problems in neurons. For example, *Tay–Sachs disease* is a genetic abnormality involving the metabolism of *gangliosides*, glycolipids that are important components of neuron cell membranes. **AM** Tay–Sachs Disease

Chapter Review

SELECTED CLINICAL TERMINOLOGY

Terms Discussed in This Chapter

anticholinesterase drug: A drug that blocks the breakdown of ACh by AChE. *(p. 415)*

atropine: A drug that prevents ACh from binding to the postsynaptic membrane of cardiac muscle and smooth muscle cells. *(p. 415)*

demyelination: The destruction of the myelin sheaths around axons in the CNS and PNS. *(p. 395 and AM)*

d-tubocurarine: A drug, derived from curare, that prevents ACh from binding to the postsynaptic membrane of skeletal muscle fibers. *(p. 415)*

endorphins (en-DOR-finz): Neuropeptides produced in the brain and spinal cord that appear to relieve pain and to affect mood. *(p. 418)*

hyperkalemia (hī-per-ka-LĒ-mē-uh): An abnormal physiological state resulting from a high extracellular concentration of potassium. *(p. 423)*

neurotoxin: A compound that disrupts normal nervous system function by interfering with the generation or propagation of action potentials. Examples include *tetrodotoxin (TTX)*, *saxitoxin (STX)*, and *ciguatoxin (CTX)*. *(p. 415 and AM)*

nicotine: A compound, found in tobacco, that binds to specific ACh receptor sites and stimulates the postsynaptic membrane. *(p. 415)*

rabies: A fatal disease caused by a virus that reaches the CNS by retrograde flow along peripheral axons. *(p. 390 and AM)*

Tay–Sachs disease: A genetic abnormality involving the metabolism of gangliosides, important components of neuron cell membranes. The result is a gradual deterioration of neurons due to the buildup of metabolic by-products and the release of lysosomal enzymes. *(p. 423 and AM)*

AM Additional Terms Discussed in the *Applications Manual*

diphtheria
Guillain–Barré syndrome
multiple sclerosis (MS)

STUDY OUTLINE

12–1 AN OVERVIEW OF THE NERVOUS SYSTEM   p. 386

1. The nervous and endocrine systems coordinate the activities of other organ systems. The *nervous system* provides swift, but brief, responses to stimuli; the *endocrine system* adjusts metabolic operations and directs long-term changes.
2. The nervous system includes all the **neural tissue** in the body. The basic functional unit is the **neuron**. The anatomical divisions of the nervous system are the **central nervous system (CNS)** (the brain and spinal cord) and the **peripheral nervous system (PNS)** (all the neural tissue outside the CNS). Bundles of *axons (nerve fibers)* in the PNS are called **nerves**.
3. Functionally, the PNS can be broken down into an **afferent division**, which brings sensory information from **receptors** to the CNS, and an **efferent division**, which carries motor commands to muscles and glands called **effectors**. *(Figure 12–1)*
4. The efferent division of the PNS includes the **somatic nervous system (SNS)** (which controls skeletal muscle contractions) and the **autonomic nervous system (ANS)** (which regulates smooth muscle, cardiac muscle, and glandular activity). *(Figure 12–1)*

12–2 NEURONS   p. 386

THE STRUCTURE OF NEURONS   p. 387

1. The **perikaryon** of a multipolar neuron contains organelles, including **neurofilaments**, **neurotubules**, and **neurofibrils**. The **axon hillock** connects the **initial segment** of the **axon** to the **cell body**, or **soma**. The **axoplasm** contains numerous organelles. *(Figure 12–2)*
2. **Collaterals** may branch from an axon, with **telodendria** branching from the axon's tip.
3. A **synapse** is a site of intercellular communication. A **synaptic knob** is the most common type of synaptic terminal. **Neurotransmitters** are released from the synaptic knob of the **presynaptic neuron** and affect the postsynaptic cell, which may be a neuron or another type of cell. *(Figures 12–2, 12–3)*

THE CLASSIFICATION OF NEURONS   p. 390

4. Neurons are described histologically as **anaxonic**, **unipolar**, **bipolar**, or **multipolar**. *(Figure 12–4)*
5. The three functional categories of neurons are sensory neurons, motor neurons, and interneurons. *(Figure 12–5)*
6. **Sensory neurons** form the afferent division of the PNS. They deliver information received from **exteroceptors**, **proprioceptors**, or **interoceptors** to the CNS. *(Figure 12–5)*
7. **Motor neurons** form the efferent division of the PNS. These neurons stimulate or modify the activity of a peripheral tissue, organ, or organ system. *(Figure 12–5)*
8. **Interneurons** (*association neurons*) are always located in the CNS and may be situated between sensory and motor neurons. They distribute sensory inputs and coordinate motor outputs. *(Figure 12–5)*

12–3 NEUROGLIA   p. 392

NEUROGLIA OF THE CENTRAL NERVOUS SYSTEM   p. 392

1. The four types of **neuroglia**, or *glial cells*, in the CNS are (1) **ependymal cells**, with functions related to the **cerebrospinal fluid (CSF)**, (2) **astrocytes**, the largest and most numerous neuroglias, (3) **oligodendrocytes**, which are responsible for the **myelination** of CNS axons, and (4) **microglia**, or phagocytic cells. *(Figures 12–6, 12–7)*

NEUROGLIA OF THE PERIPHERAL NERVOUS SYSTEM   p. 395

2. Neuron cell bodies in the PNS are clustered into **ganglia**. *(Figure 12–8)*
3. **Satellite cells**, or *amphicytes*, surround neuron cell bodies within ganglia. **Schwann cells** ensheath axons in the PNS. A single Schwann cell may myelinate one segment of an axon or enfold segments of several unmyelinated axons. *(Figure 12–8)*

NEURAL RESPONSES TO INJURIES p. 395

4. In the PNS, functional repair may follow **Wallerian degeneration**. In the CNS, many factors complicate the repair process and reduce the chances of functional recovery. *(Figure 12–9)*

  Neural anatomy: **IP CD-ROM:** Nervous System I/Anatomy Review.

12–4 NEUROPHYSIOLOGY: IONS AND ELECTRICAL SIGNALS   p. 398

THE TRANSMEMBRANE POTENTIAL p. 398

1. The **electrochemical gradient** is the sum of all chemical and electrical forces acting across the cell membrane. *(Figures 12–10, 12–11, 12–12)*

2. The **sodium–potassium exchange pump** stabilizes the resting potential at approximately -70 mV. *(Summary Table 12–1)*

CHANGES IN THE TRANSMEMBRANE POTENTIAL p. 401

3. The cell membrane contains **passive (leak) channels**, which are always open, and **active (gated) channels**, which open or close in response to specific stimuli. *(Figure 12–11)*

4. The three types of gated channels are **chemically regulated channels**, **voltage-regulated channels**, and **mechanically regulated channels**. *(Figure 12–13)*

GRADED POTENTIALS p. 403

5. A localized **depolarization** or **hyperpolarization** is a **graded potential** (a change in potential that decreases with distance). *(Figures 12–14, 12–15; Summary Table 12–2)*

ACTION POTENTIALS p. 405

6. An **action potential** appears when a region of excitable membrane depolarizes to its **threshold**. The steps involved are, in order, membrane depolarization and the activation of sodium channels, sodium channel inactivation, potassium channel activation, and the return to normal permeability. *(Figure 12–15; Summary Table 12–3; Table 12–4)*

7. The generation of an action potential follows the **all-or-none principle**. The **refractory period** lasts from the time an action potential begins until the normal resting potential has returned. *(Table 12–4)*

8. In **continuous propagation**, an action potential spreads across the entire excitable membrane surface in a series of small steps. *(Figure 12–17)*

9. In **saltatory propagation**, an action potential appears to leap from node to node, skipping the intervening membrane surface. Saltatory propagation carries nerve impulses many times more rapidly than does continuous propagation. *(Figure 12–18; Table 12–4)*

10. Axons are classified as **Type A fibers**, **Type B fibers**, or **Type C fibers** on the basis of their diameter, myelination, and propagation speed.

11. Compared with action potentials in neural tissue, those in muscle tissue have (1) higher resting potentials, (2) longer lasting action potentials, and (3) slower propagation of action potentials.

  Action potential: **IP CD-ROM:** Nervous System I/The Action Potential.

12-5 SYNAPTIC ACTIVITY   p. 411

1. An action potential traveling along an axon is a **nerve impulse**. At a synapse between two neurons, information passes from the **presynaptic neuron** to the **postsynaptic neuron** or other type of postsynaptic cell.

GENERAL PROPERTIES OF SYNAPSES p. 411

2. A synapse is either electrical (with direct physical contact between cells) or chemical (involving a neurotransmitter).

3. **Electrical synapses** occur in the CNS and PNS, but they are rare. At an electrical synapse, the presynaptic and postsynaptic cell membranes are bound by interlocking membrane proteins at a gap junction. Pores within these proteins permit the passage of local currents, and the two neurons act as if they share a common cell membrane.

4. **Chemical synapses** are more common than electrical synapses. **Excitatory neurotransmitters** cause depolarization and promote the generation of action potentials, whereas **inhibitory neurotransmitters** cause hyperpolarization and suppress the generation of action potentials.

5. The effect of a neurotransmitter on the postsynaptic membrane depends on the properties of the receptor, not on the nature of the neurotransmitter.

CHOLINERGIC SYNAPSES p. 412

6. **Cholinergic synapses** release the neurotransmitter **acetylcholine (ACh)**. Communication moves from the presynaptic neuron to the postsynaptic neuron across a synaptic cleft. A **synaptic delay** occurs because calcium influx and the release of the neurotransmitter takes an appreciable length of time. *(Figure 12–19)*

7. The **choline** released during the breakdown of ACh in the synaptic cleft is reabsorbed and recycled by the synaptic knob. If stores of ACh are exhausted, **synaptic fatigue** can occur. *(Summary Table 12–5)*

OTHER NEUROTRANSMITTERS p. 413

8. **Adrenergic synapses** release **norepinephrine (NE)**, also called *noradrenaline*. Other important neurotransmitters include **dopamine**, **serotonin**, and **gamma aminobutyric acid (GABA)**. *(Table 12–6)*

NEUROMODULATORS p. 418

9. **Neuromodulators** influence the postsynaptic cell's response to neurotransmitters.

HOW NEUROTRANSMITTERS AND NEUROMODULATORS WORK p. 419

10. Neurotransmitters can have a direct effect or an indirect effect on membrane potential, or they can exert their effects via lipid-soluble gases that diffuse across the cell membrane. *(Figure 12–21)*

  Synaptic activity: **IP CD-ROM:** Nervous System II/Synaptic Potentials and Cellular Integration, and Synaptic Transmission.

12–6 INFORMATION PROCESSING   p. 419

1. Excitatory and inhibitory stimuli are integrated through interactions between **postsynaptic potentials**. This interaction is the simplest level of **information processing** in the nervous system.

POSTSYNAPTIC POTENTIALS p. 419

2. A depolarization caused by a neurotransmitter is an **excitatory postsynaptic potential (EPSP)**. Individual EPSPs can combine through **summation**, which can be **temporal** (occurring at a single synapse when a second EPSP arrives before the effects of the first have disappeared) or **spatial** (resulting from the cumulative effects of multiple synapses at various locations). *(Figure 12–22)*

3. Hyperpolarization of the postsynaptic membrane is an **inhibitory postsynaptic potential (IPSP)**.

4. The most important determinants of neural activity are EPSP–IPSP interactions. *(Figure 12–23)*

PRESYNAPTIC INHIBITION AND FACILITATION   p. 421

5. In **presynaptic inhibition**, GABA release at an *axoaxonal synapse* inhibits the opening of voltage-regulated calcium channels in the synaptic knob. This inhibition reduces the amount of neurotransmitter released when an action potential arrives at the synaptic knob. (*Figure 12–24*)

6. In **presynaptic facilitation**, activity at an axoaxonal synapse increases the amount of neurotransmitter released when an action potential arrives at the synaptic knob. This increase enhances and prolongs the effects of the neurotransmitter on the postsynaptic membrane. (*Figure 12–24*)

THE RATE OF GENERATION OF ACTION POTENTIALS   p. 422

7. The neurotransmitters released at a synapse have excitatory or inhibitory effects. The effect on the initial segment reflects an integration of the stimuli arriving at any moment. The frequency of generation of action potentials depends on the degree of depolarization above threshold at the axon hillock. (*Table 12–7*)

8. Neuromodulators can alter the rate of release of neurotransmitters or the response of a postsynaptic neuron to specific neurotransmitters. Neurons may be facilitated or inhibited by extracellular chemicals other than neurotransmitters or neuromodulators. (*Summary Table 12–7*)

9. The effect of a presynaptic neuron's activation on a postsynaptic neuron at a synapse may be altered by other neurons. (*Table 12–7*)

10. The greater the degree of sustained depolarization at the axon hillock, the higher the frequency of generation of action potentials. At a frequency of about 1000 per second, the relative refractory period has been eliminated and further depolarization will have no effect. (*Summary Table 12–7*)

REVIEW QUESTIONS

More assessment questions are available to you on the Companion Website. You will find Matching, Multiple Choice, True/False, and other quizzes to help further your understanding of the material covered in this chapter. To access the site, go to www.aw.com/martini.

LEVEL 1 Reviewing Facts and Terms

1. Regulation by the nervous system provides
   (a) relatively slow, but long-lasting, responses to stimuli
   (b) swift, long-lasting responses to stimuli
   (c) swift, but brief, responses to stimuli
   (d) relatively slow, short-lived responses to stimuli

2. The afferent division of the PNS
   (a) brings sensory information to the CNS
   (b) carries motor commands to muscles and glands
   (c) processes and integrates sensory data
   (d) is the seat of higher functions in the body

3. The part of the nervous system that controls voluntary contractions of skeletal muscles is the
   (a) somatic nervous system
   (b) autonomic nervous system
   (c) visceral motor system
   (d) sympathetic division of the ANS

4. Smooth muscle, cardiac muscle, and glands are among the targets of the
   (a) somatic nervous system   (c) afferent division of the PNS
   (b) sensory neurons   (d) autonomic nervous system

5. In the CNS, a neuron typically receives information from other neurons at its
   (a) axon   (b) Nissl bodies
   (c) dendrites   (d) nucleus

6. Neuroglia responsible for maintaining the blood–brain barrier are the
   (a) microglia   (b) ependymal cells
   (c) astrocytes   (d) oligodendrocytes

7. Phagocytic cells in neural tissue of the CNS are
   (a) astrocytes   (b) ependymal cells
   (c) oligodendrocytes   (d) microglia

8. Substances transported from an axon terminal to the cell body of the same neuron are delivered by
   (a) axoplasmic transport   (b) synaptic conduction
   (c) retrograde flow   (d) active transport

9. All the motor neurons that control skeletal muscles are
   (a) multipolar neurons
   (b) myelinated bipolar neurons
   (c) unipolar, unmyelinated sensory neurons
   (d) anaxonic neurons

10. The neural cells responsible for the analysis of sensory inputs and coordination of motor outputs are
    (a) neuroglia   (b) interneurons
    (c) sensory neurons   (d) motor neurons

11. Depolarization of a neuron cell membrane will shift the membrane potential toward
    (a) 0 mV   (b) −70 mV
    (c) −90 mV   (d) a, b, and c are correct

12. The primary determinant of the resting membrane potential is
    (a) the membrane permeability to sodium
    (b) the membrane permeability to potassium
    (c) intracellular negatively charged proteins
    (d) negatively charged chloride ions in the ECF

13. Receptors that bind acetylcholine at the postsynaptic membrane are
    (a) chemically regulated channels
    (b) voltage-regulated channels
    (c) passive channels
    (d) mechanically regulated channels

14. Gated channels that open or close in response to a change in the transmembrane potential are
    (a) mechanically regulated channels
    (b) voltage-regulated channels
    (c) chemically regulated channels
    (d) a, b, and c are correct

15. Action potentials vary in strength, depending on the intensity of the initial stimulus:
    (a) true   (b) false

16. Changes in the transmembrane potential that are not restricted to the area surrounding the site of stimulation are
    (a) action potentials   (b) graded potentials
    (c) inhibitory potentials   (d) hyperpolarizing potentials

17. Neuromodulators are compounds that influence the
    (a) postsynaptic cell's response to a neurotransmitter
    (b) synaptic vesicles in the synaptic knob
    (c) release of calcium ions into the axoplasm
    (d) a, b, and c are correct

18. A transient hyperpolarization of the postsynaptic membrane is
    (a) a refractory period   (b) an EPSP
    (c) an IPSP   (d) an action potential

19. A neuron that is brought closer to threshold is considered to be
    (a) facilitated   (b) inhibited
    (c) fatigued   (d) enabled

20. What are the major components of (a) the central nervous system? (b) the peripheral nervous system?

21. What two major cell populations are found in the nervous system? What is the primary function of each type of cell?

22. Which two types of neuroglia insulate neuron cell bodies and axons in the PNS from their surroundings?

23. What three *functional* groups of neurons are found in the nervous system? What is the function of each type of neuron?

LEVEL 2 Reviewing Concepts

24. If the resting membrane potential is −70 mV and the threshold is −55 mV, a membrane potential of −60 mV will
    (a) produce an action potential
    (b) make it easier to produce an action potential
    (c) make it harder to produce an action potential
    (d) hyperpolarize the membrane

25. A graded potential
    (a) decreases with distance from the point of stimulation
    (b) spreads passively because of local currents
    (c) may involve either depolarization or hyperpolarization
    (d) a, b, and c are correct

26. For an action potential to begin, an area of excitable membrane must
    (a) have its voltage-regulated gates inactivated
    (b) be hyperpolarized
    (c) be depolarized to threshold level
    (d) not be in a relative refractory period

27. During an absolute refractory period, the membrane
    (a) continues to hyperpolarize
    (b) cannot respond to further stimulation
    (c) can respond to a larger-than-normal depolarizing stimulus
    (d) will respond to summated stimulation

28. A neuron exhibiting facilitation requires a _____ additional stimulus to trigger an action potential.
    (a) smaller depolarizing   (b) larger depolarizing
    (c) smaller hyperpolarizing   (d) larger hyperpolarizing

29. The loss of positive ions from the interior of a neuron produces
    (a) depolarization   (b) threshold
    (c) hyperpolarization   (d) an action potential

30. The continuous propagation of an action potential cannot occur in
    (a) myelinated axons   (b) unmyelinated axons
    (c) Type A fibers   (d) Type B fibers

31. Why can't most neurons in the CNS be replaced when they are lost to injury or disease?

32. What purpose do collaterals serve in the nervous system?

33. What is the difference between axoplasmic transport and retrograde flow?

34. How does a neuron become hyperpolarized?

35. What is the *functional* difference among voltage-regulated, chemically regulated, and mechanically regulated channels?

36. What four basic characteristics are associated with graded potentials?

37. State the all-or-none principle of action potentials.

38. Describe the steps involved in the generation of an action potential.

39. What is meant by saltatory propagation? How does it differ from continuous propagation?

40. What is the relationship between axon diameter and propagation speed?

41. What are the functional differences among type A, B, and C fibers?

42. How does an action potential in a skeletal muscle fiber differ from that in a neuron?

43. Why is an electrical synapse a more efficient carrier of nerve impulses from cell to cell than is a chemical synapse?

44. Describe the steps that take place at a typical cholinergic synapse.

45. How does the action of a neurotransmitter differ from that of a neuromodulator?

46. What is the difference between temporal summation and spatial summation?

47. Which functions of neurons necessitate the support of energy from ATP?

48. When a runner experiences "runner's high," why is the suppression of pain common?

49. Multiple sclerosis (MS) is a demyelination disorder. How does this condition produce muscular paralysis and sensory losses?

LEVEL 3 Critical Thinking and Clinical Applications

50. If neurons in the CNS lack centrioles and are unable to divide, how can a person develop brain cancer?

51. Harry has a kidney condition that causes changes in his body's electrolyte level (concentration of ions in the extracellular fluid). As a result, he is exhibiting tachycardia, an abnormally fast heart rate. Which ion is involved, and how does a change in its concentration cause Harry's symptoms?

52. Twenty neurons synapse with a single receptor neuron. Fifteen of the 20 neurons release neurotransmitters that produce EPSPs at the postsynaptic membrane, and the other 5 release neurotransmitters that produce IPSPs. Each time one of the neurons is stimulated, it releases enough neurotransmitter to produce a 2-mV change in potential at the postsynaptic membrane. If the threshold of the postsynaptic neuron is 10 mV, how many of the excitatory neurons must be stimulated to produce an action potential in the receptor neuron if all 5 inhibitory neurons are stimulated? (Assume that spatial summation occurs.)

53. The myelination of peripheral neurons occurs rapidly throughout the first year of life. How can this process explain the increased abilities of infants during their first year?

54. A drug that blocks ATPase is introduced into an experimental neuron preparation. The neuron is then repeatedly stimulated, and recordings are made of the response. What effect would you expect to observe?

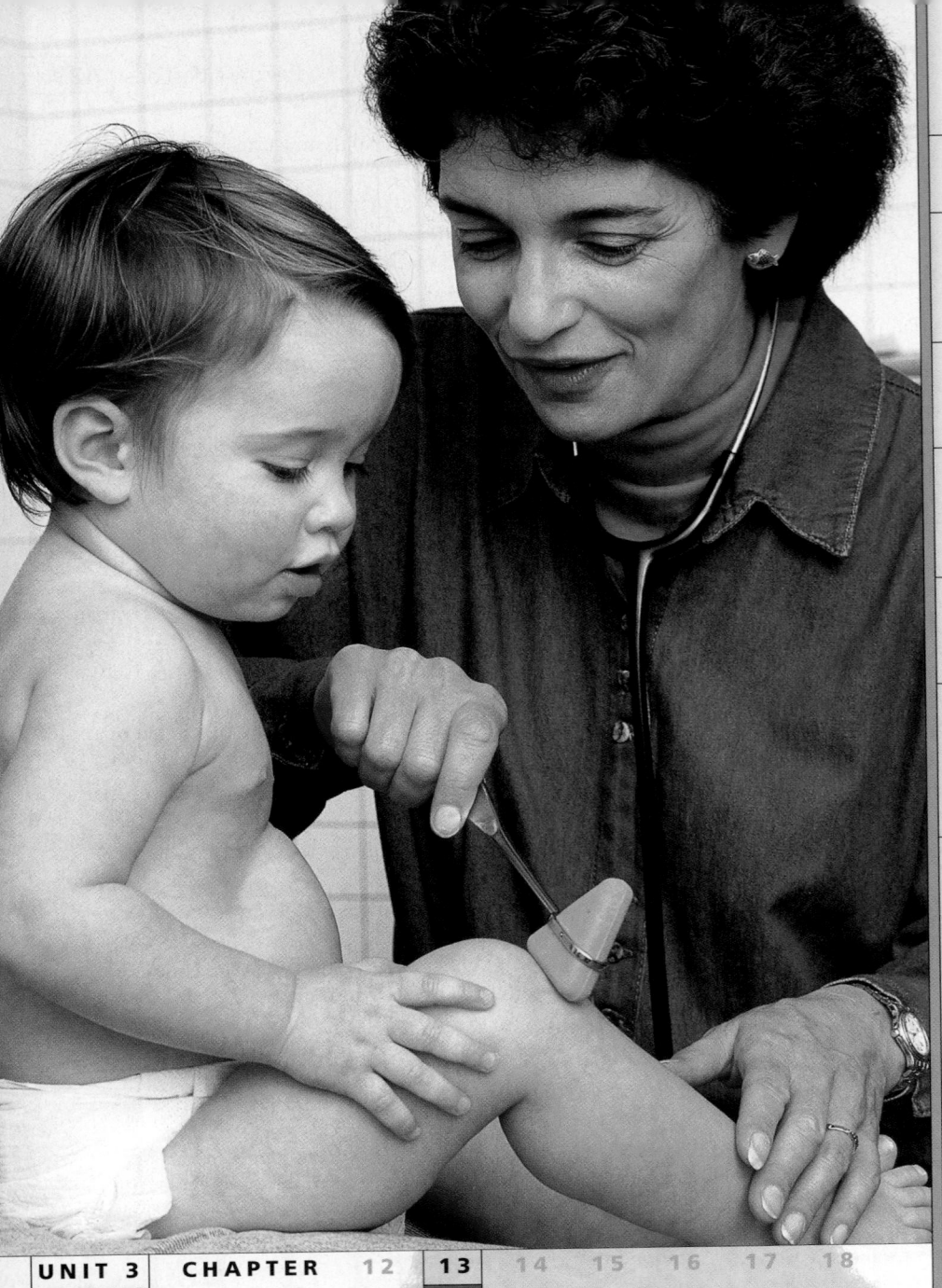

UNIT 3   CHAPTER   12   **13**   14   15   16   17   18

CLINICAL NOTES
- Meningitis 434
- Epidural Anesthesia 434
- Peripheral Neuropathies 440

CLINICAL DISCUSSIONS
- Spinal Taps and Myelography 435
- Damage to Spinal Tracts 438
- Reflexes and Diagnostic Testing 456

13

THE SPINAL CORD AND SPINAL NERVES

Consider the many different types of sensory information your nervous system receives each moment. You are probably sitting down while reading this, and you are focusing on the visual information provided by your eyes. Yet at the same time you are aware of background sounds, how comfortable (or uncomfortable) your chair might be, the temperature of the room, and many other sensations that are noted and "tuned out" at the subconscious level. All of these arriving sensations have to be routed appropriately, interpreted consciously or subconsciously, and then either ignored or acted on. Even when you are at rest in an empty room, your nervous system is receiving a flood of sensory information. Most of it may be routine and relatively unimportant—but there's always the chance for a surprise that requires an immediate response.

How does the central nervous system discriminate between important and unimportant information? How does it select the appropriate response to one particular situation and stimulus? We shall begin to sort out the answers to these questions in this chapter. Given the number of sensory neurons involved (somewhere around 10 million), every arriving sensation cannot be distributed to every neuron in the CNS. Instead, arriving sensations are routed according to the nature of the stimulus, so that, for example, information from receptors involved in hearing arrives at centers devoted specifically to processing auditory sensations.

Organization is usually the key to success in any complex environment, be it a large corporation or your own body. Because a large corporation has both a system to distribute mail on specific topics and executive assistants who decide whether the issue can be ignored or easily dealt with, only the most complex and important problems make it to the desk of the president. The nervous system works in much the same way: It has input pathways that route sensations and processing centers that prioritize and distribute the information. There are also several levels that issue motor responses. Just as in a corporation, the president—your conscious mind—gets involved only in a fraction of the day-to-day activities. The other decisions are handled at lower levels that operate outside of your conscious awareness. This is a very efficient system, and it works only because it is so highly organized. We will therefore begin with an overview of the organization of the nervous system.

13–1 GENERAL ORGANIZATION OF THE NERVOUS SYSTEM

Objective

- Describe the basic structural and organizational characteristics of the nervous system.

The central nervous system (CNS) consists of the brain and spinal cord. Most of the neural tissue in the body is found in the CNS. The rest forms the peripheral nervous system (PNS). Neurons, axons, and glial cells are not randomly scattered in the CNS and PNS. Instead, they form masses or bundles with distinct anatomical boundaries, and they are identified by specific terms. We shall use these terms in all later chapters, so we offer a brief overview here (Figure 13–1●).

In the peripheral nervous system,

- Neuron cell bodies are located in *ganglia*.
- Axons are bundled together in *nerves*, with *spinal nerves* connected to the spinal cord and *cranial nerves* connected to the brain.

In the central nervous system,

- A collection of neuron cell bodies with a common function is called a **center**. A center with a discrete anatomical boundary is a **nucleus**. Portions of the brain surface are covered by a thick layer of gray matter called the **neural cortex**. The term *higher centers* refers to the most complex integration centers, nuclei, and cortical areas of the brain.
- The white matter of the CNS contains bundles of axons that share common origins, destinations, and functions. These bundles are called **tracts**. Tracts in the spinal cord form larger groups called **columns**.
- The centers and tracts that link the brain with the rest of the body are called pathways. For example, **sensory pathways** distribute information from peripheral receptors to processing centers in the brain, and **motor pathways** begin at CNS centers concerned with motor control and end at the effectors they control.

Because the nervous system has so many components and does so much, our discussion will span the next four chapters. If our primary interest were the anatomy of this system, we would probably start with an examination of the central nervous system (brain and spinal cord) and then consider the peripheral nervous

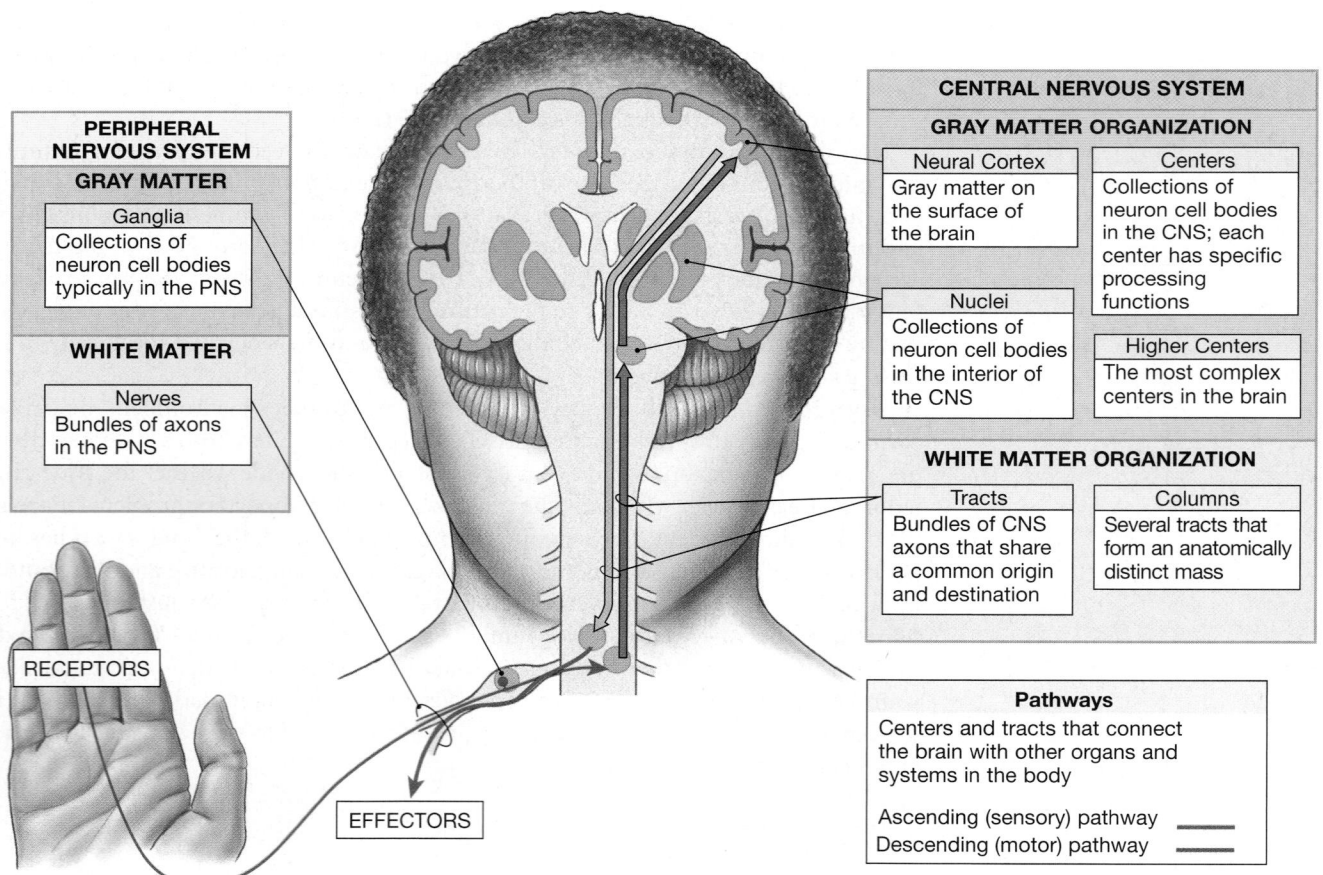

●FIGURE 13–1
An Introduction to the Anatomical Organization of the Nervous System

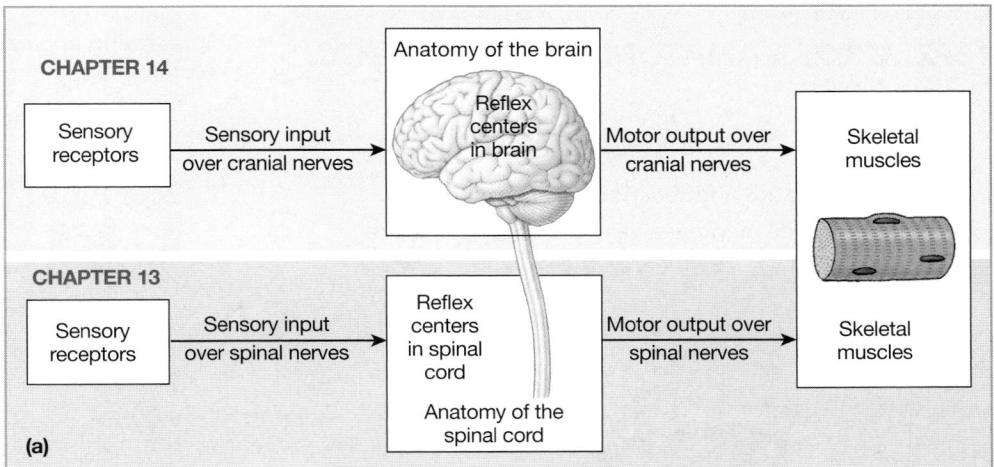

●**FIGURE 13–2**
An Overview of Chapters 13–16.
(a) Basic neuroanatomy and simple reflexes. **(b)** Neural pathways and integrative functions.

system (cranial nerves and spinal nerves). But our primary interest is how the nervous system *functions*, so we shall consider the system from that perspective. The basic approach has been diagrammed in Figure 13–2●.

In the chapters that follow, we shall be looking at increasing levels of structural and functional complexity. Chapter 12 provided the foundation by considering the function of individual neurons. In the current chapter, we consider the spinal cord and spinal nerves and the basic wiring of relatively simple *spinal reflexes*—rapid, automatic responses triggered by specific stimuli. Spinal reflexes are controlled in the spinal cord; whether they involve a single spinal segment or multiple segments, they can function without any input from the brain. For example, a reflex controlled in the spinal cord makes you drop a scalding-hot frying pan after you have accidentally grabbed it. By the time the information reaches your brain and you become consciously aware of the pain, the pan has already been released. Although there are much more complex spinal reflexes, this functional pattern still applies; a reflex provides a quick, automatic response under a specific set of conditions.

Your spinal cord is structurally and functionally integrated with your brain. Chapter 14 provides an overview of the major components and functions of the brain and cranial nerves. It also discusses the *cranial reflexes*, relatively localized reflex responses comparable in organization and complexity to those of the spinal cord.

Chapters 15 and 16 consider the nervous system as an integrated unit. Chapter 15 deals with the interplay between centers in the brain and spinal cord that occurs in the processing of sensory information. It then examines the conscious and subconscious control of skeletal muscle activity provided by the *somatic nervous system* (SNS).

Chapter 16 continues with a discussion of the control of visceral functions by the autonomic nervous system (ANS). The ANS, which has processing centers in the brain, spinal cord, and peripheral nervous system, is responsible for the control of visceral effectors, such as peripheral smooth muscles, cardiac mus-

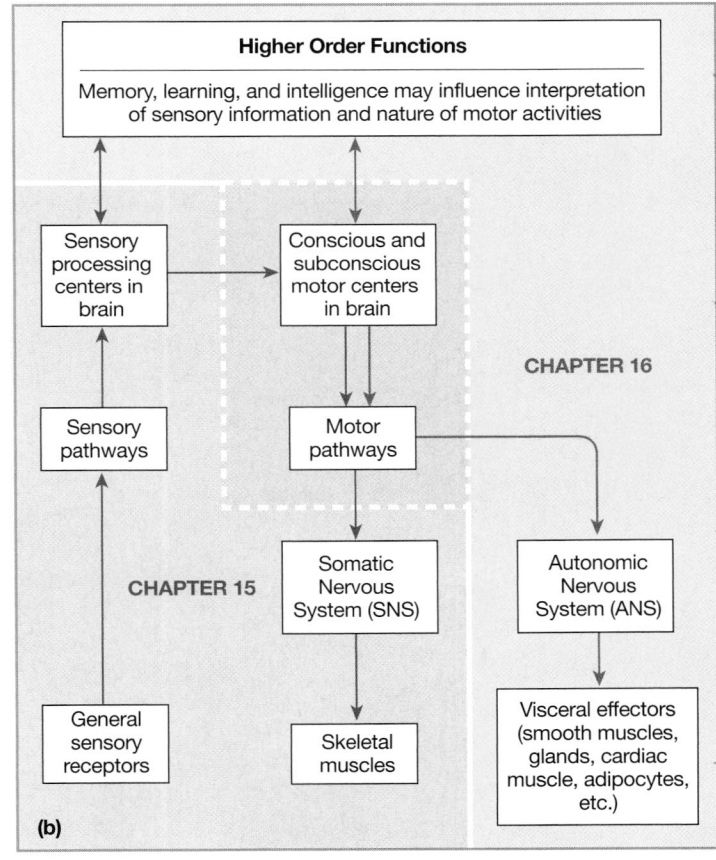

cle, and glands. We then conclude this section of the book by examining what are often called *higher order functions*: memory, learning, consciousness, and personality. These fascinating topics are difficult to investigate, but they can affect activity along the sensory and motor pathways and alter our perception of those activities.

With these basic principles, definitions, and strategies in mind, we can begin our examination of the levels of functional organization in the nervous system.

13–2 GROSS ANATOMY OF THE SPINAL CORD

Objectives

- Discuss the structure and functions of the spinal cord.
- Describe the three meningeal layers that surround the central nervous system.
- Explain the roles of white matter and gray matter in processing and relaying sensory information and motor commands.

The adult spinal cord (Figure 13–3a●) measures approximately 45 cm (18 in.) in length and has a maximum width of roughly 14 mm (0.55 in.). Note that the cord itself is not as long as the

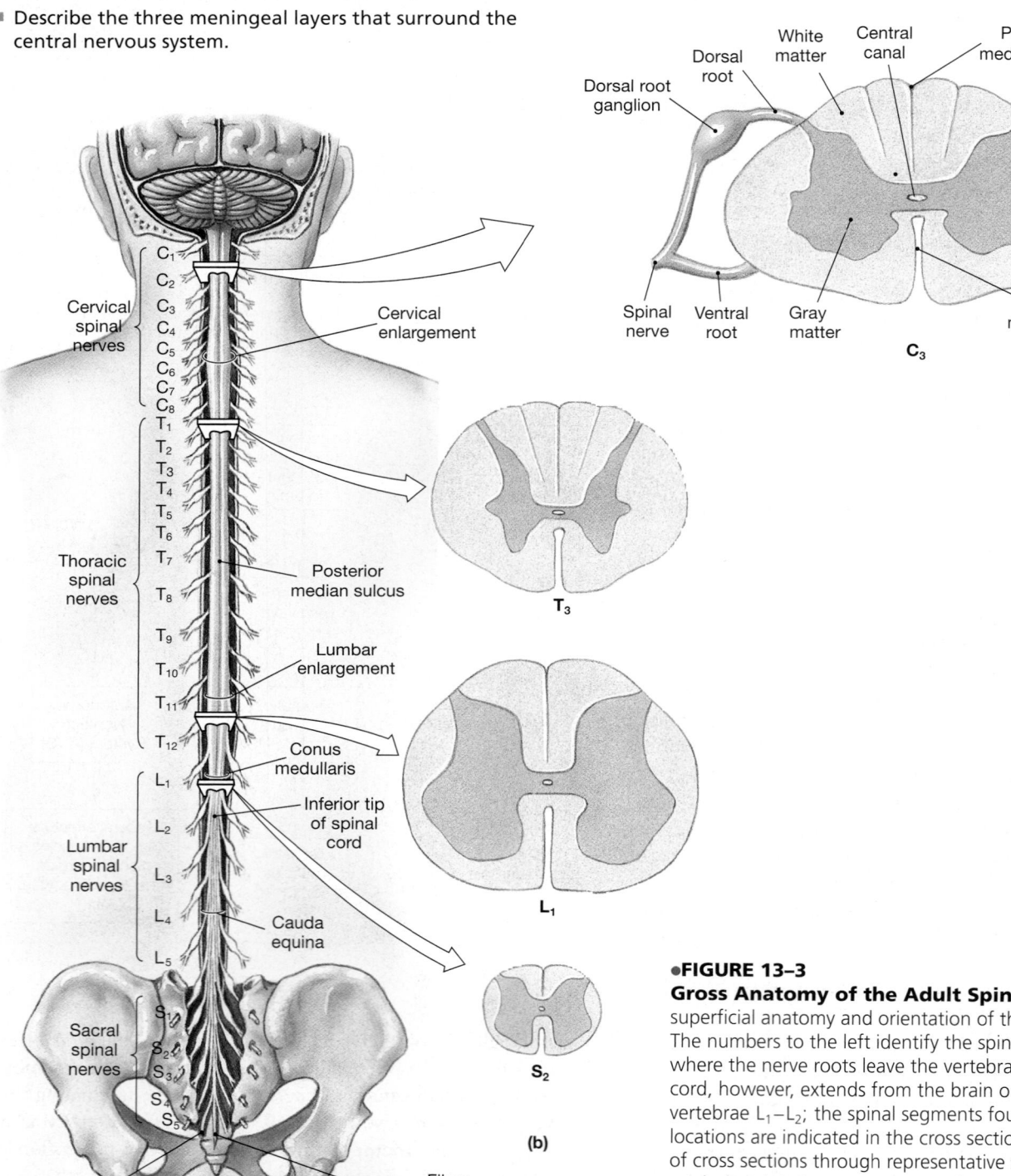

●FIGURE 13–3

Gross Anatomy of the Adult Spinal Cord. (a) The superficial anatomy and orientation of the adult spinal cord. The numbers to the left identify the spinal nerves and indicate where the nerve roots leave the vertebral canal. The spinal cord, however, extends from the brain only to the level of vertebrae L_1–L_2; the spinal segments found at representative locations are indicated in the cross sections. (b) Inferior views of cross sections through representative segments of the spinal cord, showing the arrangement of gray matter and white matter. **ATLAS** Plates 4.1–4.3; Scan 3c,d

vertebral column—instead, the adult spinal cord ends between vertebrae L₁ and L₂. The posterior (dorsal) surface of the spinal cord bears a shallow longitudinal groove, the **posterior median sulcus**. The **anterior median fissure** is a deeper groove along the anterior (ventral) surface.

The amount of gray matter is greatest in segments of the spinal cord that deal with the sensory and motor control of the limbs. These areas are expanded, forming the **enlargements** of the spinal cord. The **cervical enlargement** supplies nerves to the shoulder girdles and upper limbs; the **lumbar enlargement** provides innervation to structures of the pelvis and lower limbs. Inferior to the lumbar enlargement, the spinal cord becomes tapered and conical; this region is the **conus medullaris**. The **filum terminale** ("terminal thread"), a slender strand of fibrous tissue, extends from the inferior tip of the conus medullaris. It continues along the length of the vertebral canal as far as the second sacral vertebra, where it provides longitudinal support to the spinal cord as a component of the *coccygeal ligament*.

Figure 13–3b● provides a series of sectional views that demonstrate the variations in the relative mass of gray matter and white matter in the cervical, thoracic, lumbar, and sacral regions of the spinal cord. The entire spinal cord can be divided into 31 segments on the basis of the origins of the spinal nerves. Each segment is identified by a letter and number designation, the same method used to identify vertebrae. For example, C₃, the segment in the uppermost section of Figure 13–3b●, is the third cervical segment.

Every spinal segment is associated with a pair of **dorsal root ganglia** (Figure 13–3b●), situated near the spinal cord. These ganglia contain the cell bodies of sensory neurons. The axons of the neurons form the **dorsal roots**, which bring sensory information into the spinal cord. A pair of **ventral roots** contains the axons of motor neurons that extend into the periphery to control somatic and visceral effectors. On both sides, the dorsal and ventral roots of each segment pass between the vertebral canal and the periphery at the *intervertebral foramen* between successive vertebrae. The dorsal root ganglion lies between the pedicles of the adjacent vertebrae. (You may wish to review the description of vertebral anatomy in Chapter 7. ∞ pp. 230-231)

Distal to each dorsal root ganglion, the sensory and motor roots are bound together into a single **spinal nerve**. Spinal nerves are classified as **mixed nerves**—that is, they contain both afferent (sensory) and efferent (motor) fibers. There are 31 pairs of spinal nerves, each identified by its association with adjacent vertebrae. For example, we may speak of "cervical spinal nerves" or even "cervical nerves" when we make a general reference to spinal nerves of the neck. When we indicate specific spinal nerves, it is customary to give them a regional number, as indicated in Figure 13–3●. Each spinal nerve inferior to the first thoracic vertebra takes its name from the vertebra immediately preceding it. Thus, spinal nerve T₁ emerges immediately inferior to vertebra T₁, spinal nerve T₂ follows vertebra T₂, and so forth.

The arrangement differs in the cervical region, because the first pair of spinal nerves, C₁, passes between the skull and the first cervical vertebra. For this reason, each cervical nerve takes its name from the vertebra immediately *following* it. In other words, cervical nerve C₂ *precedes* vertebra C₂, and the same system is used for the rest of the cervical series. The transition from one numbering system to another occurs between the last cervical vertebra and first thoracic vertebra. The spinal nerve found at this location has been designated C₈. Therefore, although there are only seven cervical vertebrae, there are *eight* cervical nerves.

The spinal cord continues to enlarge and elongate until an individual is approximately four years old. Up to that time, enlargement of the spinal cord keeps pace with the growth of the vertebral column. Throughout this period, the ventral and dorsal roots are very short, and they enter the intervertebral foramina immediately adjacent to their spinal segment. After age four, the vertebral column continues to elongate, but the spinal cord does not. This vertebral growth moves the intervertebral foramina, and thus the spinal nerves, farther and farther from their original positions relative to the spinal cord. As a result, the dorsal and ventral roots gradually elongate, and the correspondence between the spinal segment and the vertebral segment is lost. For example, in adults, the sacral segments of the spinal cord are at the level of vertebrae L₁–L₂.

Because the adult spinal cord extends only to the level of the first or second lumbar vertebra, the dorsal and ventral roots of spinal segments L₂ to S₅ extend inferiorly, past the inferior tip of the conus medullaris. When seen in gross dissection, the filum terminale and the long ventral and dorsal roots resemble a horse's tail. As a result, early anatomists called this complex the **cauda equina** (KAW-duh ek-WĪ-nuh; *cauda*, tail + *equus*, horse).

☐ SPINAL MENINGES

The vertebral column and its surrounding ligaments, tendons, and muscles isolate the spinal cord from the external environment. The delicate neural tissues must also be protected from damaging contacts with the surrounding bony walls of the vertebral canal. The **spinal meninges** (men-IN-jēz; singular, *meninx*, membrane), a series of specialized membranes surrounding the spinal cord, provide the necessary physical stability and shock absorption. Blood vessels branching within these layers deliver oxygen and nutrients to the spinal cord.

The relationships among the spinal meninges are shown in Figure 13–4a●. The spinal meninges consist of three layers: (1) the *dura mater*, (2) the *arachnoid*, and (3) the *pia mater*. At the foramen magnum of the skull, the spinal meninges are continuous with the **cranial meninges**, which surround the brain. (We shall discuss the cranial meninges, which have the same three layers, in Chapter 14.)

Bacterial or viral infection can cause **meningitis**, or inflammation of the meningeal membranes. Meningitis is dangerous because it can disrupt the normal circulatory and cerebrospinal fluid supplies, damaging or killing neurons and neuroglia in the affected areas. Although an initial diagnosis may specify the meninges of the spinal cord (*spinal meningitis*) or brain (*cerebral meningitis*), in later stages the entire meningeal system is usually affected. **AM** Meningitis

The Dura Mater

The tough, fibrous **dura mater** (DOO-ruh MĀ-ter; *dura*, hard + *mater*, mother) is the layer that forms the outermost covering of the spinal cord (Figure 13–4a●). This layer contains dense collagen fibers that are oriented along the longitudinal axis of the cord. Between the dura mater and the walls of the vertebral canal lies the **epidural space**, a region that contains loose connective tissue, blood vessels, and a protective padding of adipose tissue (Figure 13–4b●).

The spinal dura mater does not have extensive, firm connections to the surrounding vertebrae. Longitudinal stability is provided by localized attachment sites at either end of the vertebral canal. Cranially, the outer layer of the spinal dura mater fuses with the periosteum of the occipital bone around the margins of the foramen magnum. There, the spinal dura mater becomes continuous with the cranial dura mater. Within the sacral canal, the spinal dura mater tapers from a sheath to a dense cord of collagen fibers that blends with components of the filum terminale to form the **coccygeal ligament**.

The coccygeal ligament continues along the sacral canal, ultimately blending into the periosteum of the coccyx. Lateral support for the spinal dura mater is provided by loose connective tissue and adipose tissue within the epidural space. In addition, this dura mater extends between adjacent vertebrae at each intervertebral foramen, fusing with the connective tissues that surround the spinal nerves.

Anesthetics are often injected into the epidural space. Introduced in this way, a drug should affect only the spinal nerves in the immediate area of the injection. The result is an **epidural block**—a temporary sensory loss or a sensory and motor paralysis, depending on the anesthetic selected. Epidural blocks in the inferior lumbar or sacral regions may be used to control pain during childbirth. **AM** Spinal Anesthesia

The Arachnoid

In most anatomical and histological preparations, a narrow **subdural space** separates the dura mater from deeper meningeal layers. It is likely, however, that in life no such space exists and that the inner surface of the dura mater is in contact with the outer surface of the **arachnoid** (a-RAK-noyd; *arachne*, spider), the middle meningeal layer (Figure 13–4b●). The inner surface of the dura mater and the outer surface of the arachnoid are covered by simple squamous epithelia. The arachnoid includes this epithelium, called the *arachnoid membrane*, and the *arachnoid trabeculae*, a delicate network of collagen and elastic fibers that extends between the arachnoid membrane and the outer surface of the pia

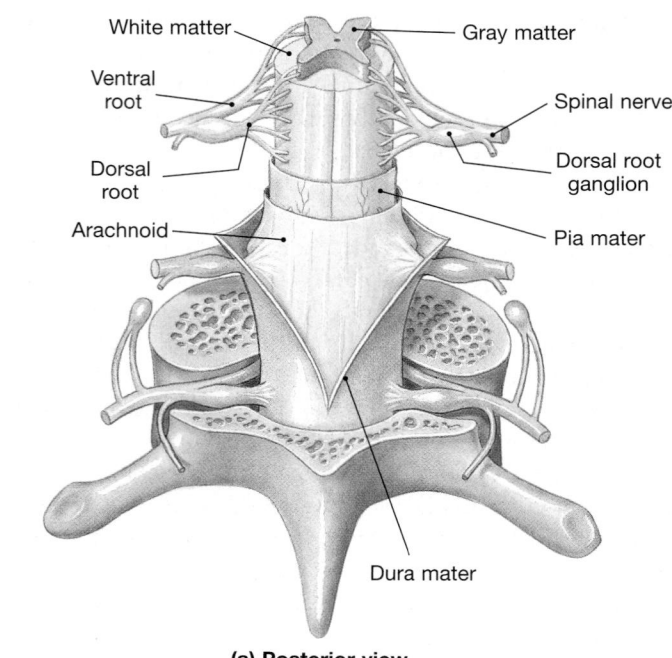

White matter — Gray matter
Ventral root
Dorsal root — Spinal nerve
Arachnoid — Dorsal root ganglion
— Pia mater
Dura mater

(a) Posterior view

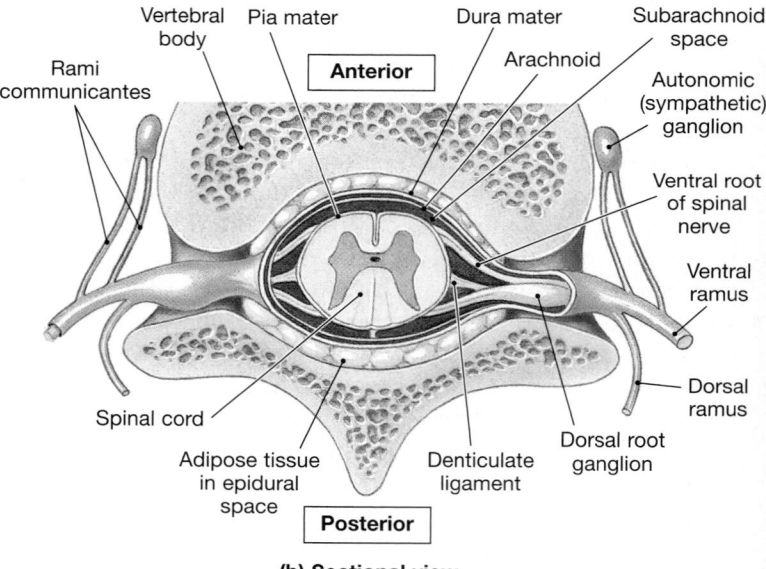

Vertebral body — Pia mater — Dura mater — Subarachnoid space
Rami communicantes — **Anterior** — Arachnoid — Autonomic (sympathetic) ganglion
Ventral root of spinal nerve
Ventral ramus
Spinal cord — Dorsal ramus
Adipose tissue in epidural space — Denticulate ligament — Dorsal root ganglion
Posterior

(b) Sectional view

●**FIGURE 13–4**
The Spinal Cord and Spinal Meninges. **(a)** A posterior view of the spinal cord, showing the meningeal layers, superficial landmarks, and distribution of gray matter and white matter. **(b)** A sectional view through the spinal cord and meninges, showing the peripheral distribution of spinal nerves.

mater. The intervening region is called the **subarachnoid space**. It is filled with **cerebrospinal fluid** (CSF), which acts as a shock absorber and a diffusion medium for dissolved gases, nutrients, chemical messengers, and waste products.

The spinal arachnoid extends inferiorly as far as the filum terminale, and the dorsal and ventral roots of the cauda equina lie within the fluid-filled subarachnoid space. In adults, the

Spinal Taps and Myelography

Tissue samples, or **biopsies**, are taken from many organs to assist in diagnosis. For example, when a liver or skin disorder is suspected, small plugs of tissue are removed and examined for signs of cell damage or abnormal cell growth, or are used to identify the microorganism causing an infection. Unlike many other tissues, however, neural tissue consists largely of cells rather than extracellular fluids or fibers. Samples of neural tissue are seldom removed for analysis, because the body usually does not replace extracted or damaged neurons. Instead, small volumes of cerebrospinal fluid are collected and analyzed. Cerebrospinal fluid is intimately associated with the neural tissue of the CNS. Accordingly, when pathogens, cell debris, or metabolic wastes are present in the CNS they can be detected in the cerebrospinal fluid.

The withdrawal of CSF, a procedure called a *spinal tap*, must be done with care to avoid injuring the spinal cord. The adult spinal cord extends inferiorly only as far as vertebra L_1 or L_2. Between vertebra L_2 and the sacrum, the spinal meninges remain intact, but they enclose only the relatively sturdy components of the cauda equina and a significant quantity of CSF. When the vertebral column is flexed, a needle can be inserted between the inferior lumbar vertebrae and into the subarachnoid space with minimal risk to the cauda equina. Risks are low primarily because (1) the volume of CSF is much greater than the volume of neural tissue, and (2) the needle is inserted with a plug inside it, and it will usually push aside, rather than penetrate, any nerve roots encountered. (The primary purpose of the plug, called a *stylet*, is to prevent the introduction of superficial tissues into the CSF; the stylet is withdrawn after insertion.) In this procedure, known as a **lumbar puncture**, 3–9 ml of fluid are taken from the subarachnoid space between vertebrae L_3 and L_4 (Figure 13–5a●). The fluid is not withdrawn by suction, which could pull nerve roots against the tip of the needle and injure them. Instead, the fluid drips out under its own pressure, and the process resembles the collection of maple syrup from a maple tree. Spinal taps are performed when CNS infection is suspected or when severe back pain, headaches, disc problems, and some types of strokes are diagnosed.

Myelography is the introduction of radiopaque dyes into the CSF of the subarachnoid space. Because the dyes are opaque to X rays, the CSF appears white on an X-ray image (Figure 13–5b●). Any tumors, inflammation, or adhesions that distort or divert CSF circulation appear in silhouette. Spinal taps and myelography allow diagnoses such as severe infection, inflammation, or leukemia (cancer of the white blood cells). Treatment may then involve injecting antibiotics, steroids, or anticancer drugs into the subarachnoid space.

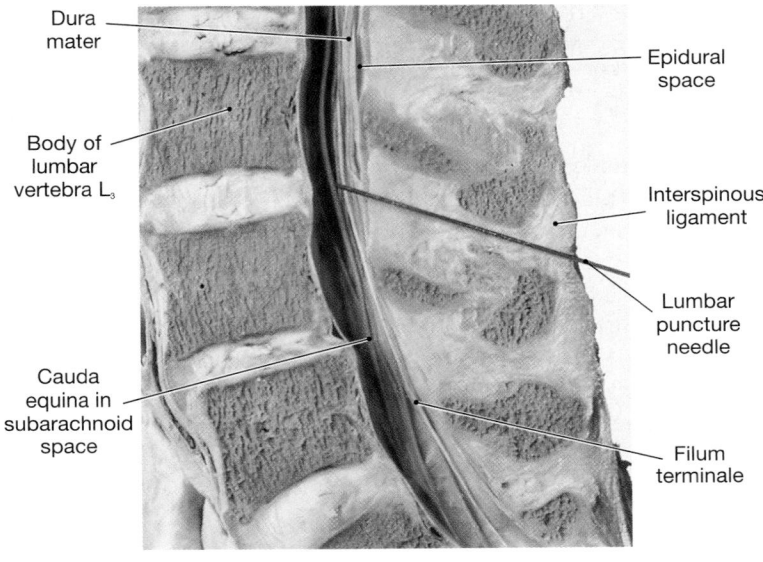

Dura mater
Epidural space
Body of lumbar vertebra L_3
Interspinous ligament
Lumbar puncture needle
Cauda equina in subarachnoid space
Filum terminale

(a)

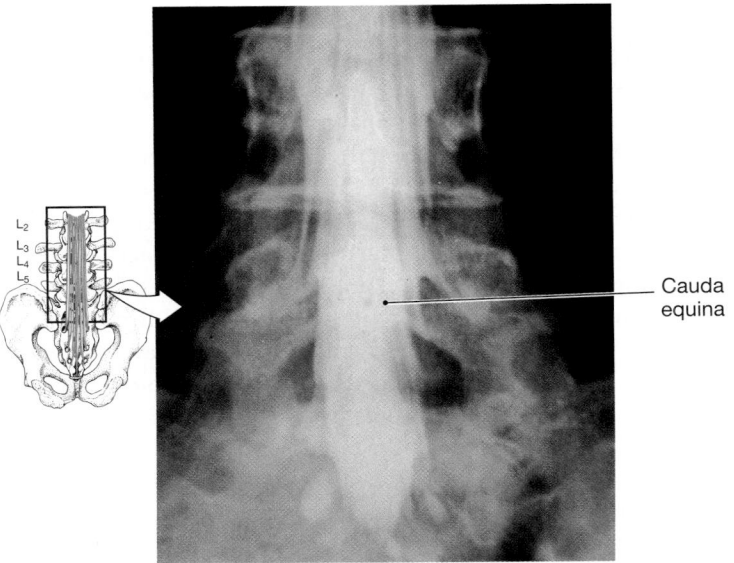

L_2
L_3
L_4
L_5

Cauda equina

(b)

●**FIGURE 13–5**
Spinal Taps and Myelography.   **(a)** The lumbar puncture needle is in the subarachnoid space, between the nerves of the cauda equina. The needle has been inserted between the third and fourth lumbar spinous processes, pointing at a superior angle. **(b)** A myelogram—an X-ray image of the spinal cord after a radiopaque dye has been introduced into the CSF—of the inferior lumbar region

withdrawal of cerebrospinal fluid, a procedure known as a **spinal tap**, involves the insertion of a needle into the subarachnoid space in the inferior lumbar region. This placement avoids damage to the spinal cord. Spinal taps are performed when CNS infection is suspected or to diagnose severe back pain, headaches, disc problems, and some types of strokes.

The Pia Mater

The subarachnoid space bridges the gap between the arachnoid epithelium and the innermost meningeal layer, the **pia mater** (*pia*, delicate + *mater*, mother). The pia mater consists of a meshwork of elastic and collagen fibers that is firmly bound to the underlying neural tissue (Figure 13–4●). These connective-tissue fibers are extensively interwoven with those that span the subarachnoid space, firmly binding the arachnoid to the pia mater. The blood vessels servicing the spinal cord run along the surface of the spinal pia mater, within the subarachnoid space.

Along the length of the spinal cord, paired **denticulate ligaments** extend from the pia mater through the arachnoid to the dura mater (Figures 13–4b and 13–6●). Denticulate ligaments, which originate along either side of the spinal cord, prevent lateral (side-to-side) movement. The dural connections at the foramen magnum and the coccygeal ligament prevent longitudinal (superior–inferior) movement.

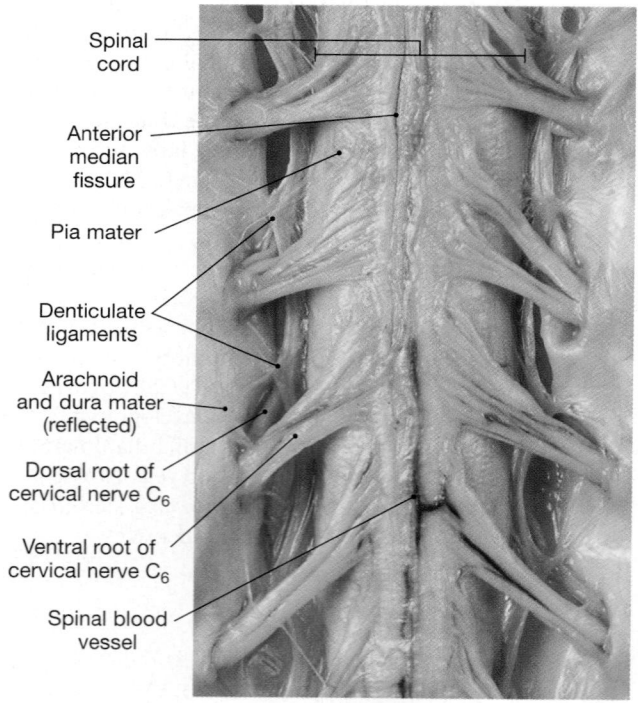

Spinal cord

Anterior median fissure

Pia mater

Denticulate ligaments

Arachnoid and dura mater (reflected)

Dorsal root of cervical nerve C$_6$

Ventral root of cervical nerve C$_6$

Spinal blood vessel

Anterior view

●**FIGURE 13–6**
The Cervical Spinal Cord. An anterior view of the spinal cord and spinal nerve roots in the vertebral canal. The dura mater and arachnoid membrane have been cut and reflected; notice the blood vessels that run in the subarachnoid space, bound to the outer surface of the delicate pia mater.

The spinal meninges accompany the dorsal and ventral roots as these roots pass through the intervertebral foramina. As the sectional views of Figure 13–4b● indicate, the meningeal membranes are continuous with the connective tissues that surround the spinal nerves and their peripheral branches.

✓ Damage to which root of a spinal nerve would interfere with motor function?

✓ Where is the cerebrospinal fluid that surrounds the spinal cord located?

Answers start on page Q-1

 View the gross anatomy of the spinal cord on the **Anatomy CD-ROM:** Nervous System/Spinal Cord Dissections. Pay special attention to the connection of the cord to the brain and the nerve plexuses.

■ SECTIONAL ANATOMY OF THE SPINAL CORD

To understand the functional organization of the spinal cord, you must become familiar with its sectional organization. Together, the anterior median fissure and the posterior median sulcus mark the division between the left and right sides of the spinal cord. The superficial white matter contains large numbers of myelinated and unmyelinated axons. The gray matter, dominated by the cell bodies of neurons, neuroglia, and unmyelinated axons, surrounds the narrow **central canal** and forms an H or butterfly shape (Figure 13–7●). The projections of gray matter toward the outer surface of the spinal cord are called **horns**.

Organization of Gray Matter

The cell bodies of neurons in the gray matter of the spinal cord are organized into functional groups called *nuclei*. **Sensory nuclei** receive and relay sensory information from peripheral receptors. **Motor nuclei** issue motor commands to peripheral effectors. Although sensory and motor nuclei appear rather small in transverse section, they may extend for a considerable distance along the length of the spinal cord.

A frontal section along the length of the central canal of the spinal cord separates the sensory (posterior, or dorsal) nuclei from the motor (anterior, or ventral) nuclei. The **posterior gray horns** contain somatic and visceral sensory nuclei, whereas the **anterior gray horns** contain somatic motor nuclei. The **lateral gray horns**, located only in the thoracic and lumbar segments, contain visceral motor nuclei. The **gray commissures** (*commissura*, a joining together) posterior to and anterior to the central canal contain axons that cross from one side of the cord to the other before they reach a destination in the gray matter.

Figure 13–7a● shows the relationship between the function of a particular nucleus (sensory or motor) and its relative position in the gray matter of the spinal cord. The nuclei within each gray horn are also organized. For example, the anterior gray horns of the cervical enlargement contain nuclei whose motor neurons control the muscles of the upper limbs. On each side of the spinal cord, in

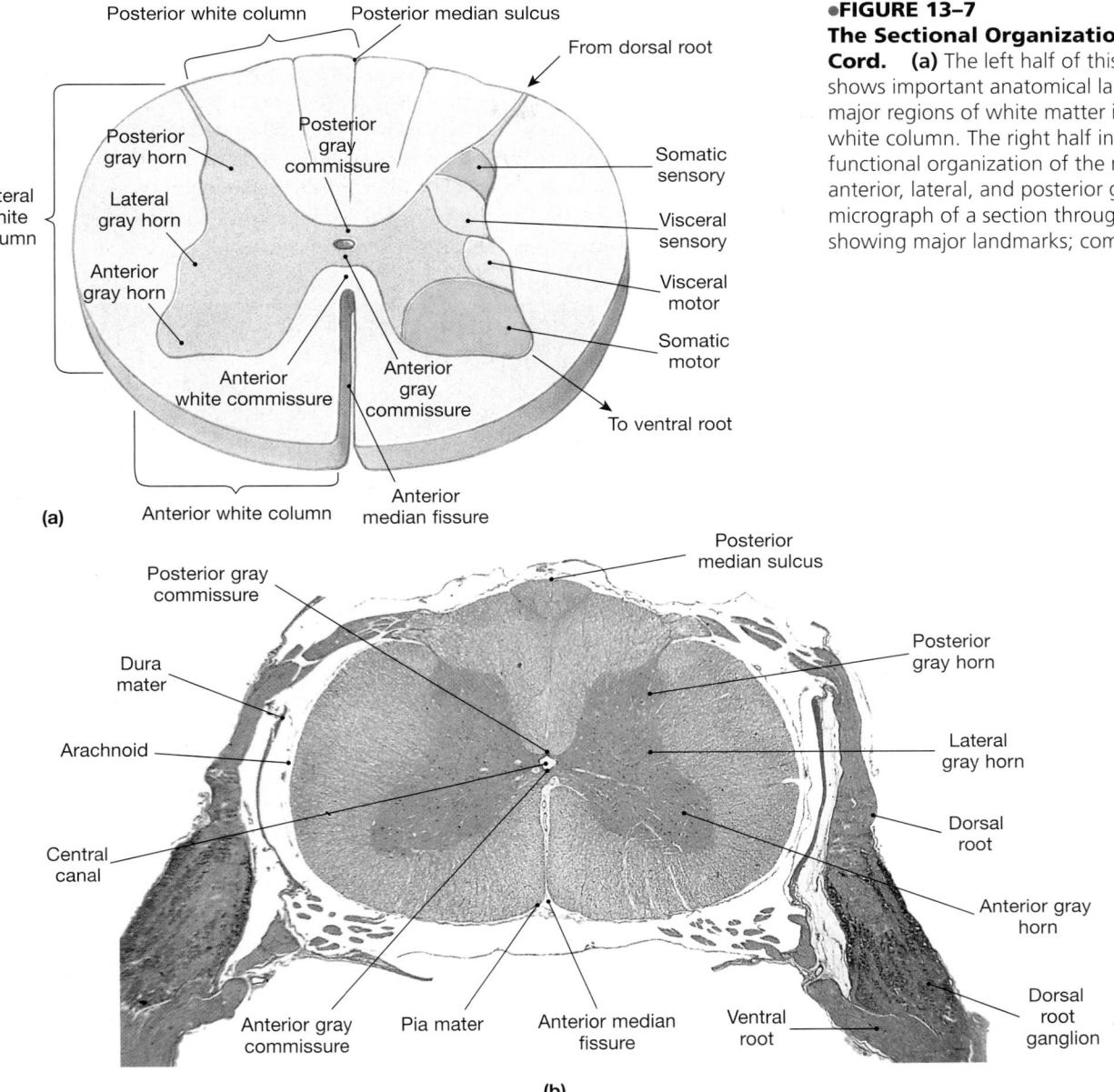

(a)

(b)

medial to lateral sequence, are somatic motor nuclei that control (1) muscles that position the pectoral girdle, (2) muscles that move the arm, (3) muscles that move the forearm and hand, and (4) muscles that move the hand and fingers. Within each of these regions, the motor neurons that control flexor muscles are grouped separately from those that control extensor muscles. Because the spinal cord is so highly organized, we can predict the muscles that will be affected by damage to a specific area of gray matter.

Organization of White Matter

The white matter on each side of the spinal cord can be divided into three regions called **columns**, or *funiculi* (Figure 13–7a●). The **posterior white columns** lie between the posterior gray horns and the posterior median sulcus. The **anterior white**

columns lie between the anterior gray horns and the anterior median fissure. The anterior white columns are interconnected by the **anterior white commissure**, a region where axons cross from one side of the spinal cord to the other. The white matter between the anterior and posterior columns on each side makes up the **lateral white column**.

Each column contains tracts whose axons share functional and structural characteristics. A **tract**, or *fasciculus* (fa-SIK-ū-lus; bundle), is a bundle of axons in the CNS that are relatively uniform with respect to diameter, myelination, and conduction speed. All the axons within a tract relay the same type of information (sensory or motor) in the same direction. Short tracts carry sensory or motor signals between segments of the spinal cord, and longer tracts connect the spinal cord with the brain. **Ascending tracts** carry sensory information toward the brain, and **descending tracts** convey motor commands to the spinal cord. We shall examine specific tracts and their functions in Chapters 15 and 16.

✓ A person with polio has lost the use of his leg muscles. In which area of his spinal cord would you expect the virus-infected motor neurons to be?

✓ Which portion of the spinal cord would be affected by a disease that damages myelin sheaths?

Answers start on page Q-1

Damage to Spinal Tracts

Multiple sclerosis (MS), a disorder introduced in Chapter 12, produces muscular paralysis and sensory losses through demyelination. ∞ p. 395 The initial symptoms appear as the result of myelin degeneration within the white matter of the lateral and posterior columns of the spinal cord or along tracts within the brain. During subsequent attacks, the effects may become more widespread. The cumulative sensory and motor losses can eventually lead to a generalized muscular paralysis. **AM** Multiple Sclerosis

The viral disease *polio* causes paralysis due to the destruction of somatic motor neurons. This disorder, introduced in Chapter 10, has been almost eliminated in the Western Hemisphere by an aggressive immunization program. Immunization continues because polio still occurs in other areas of the world. The disease could be brought into the United States at any time, leading to an epidemic among unimmunized children.

Paralysis can also result from physical damage to the spinal cord due to a severe auto crash or other accident. The damaged tracts seldom undergo even partial repair. Extensive damage to the spinal cord at or superior to the fifth cervical vertebra eliminates sensation and motor control of the upper and lower limbs. The extensive paralysis produced is called *quadriplegia. Paraplegia*, the loss of motor control of the lower limbs, may follow damage to the thoracic spinal cord. **AM** Spinal Cord Injuries and Experimental Treatments

Less severe injuries affecting the spinal cord or cauda equina produce symptoms of sensory loss or motor paralysis that reflect the specific nuclei, tracts, or spinal nerves involved. We shall explain one example, the loss of peripheral sensation along the distribution of a spinal nerve, in a later section.

13-3 SPINAL NERVES

Objectives

- Describe the major components of a spinal nerve.
- Relate the distribution pattern of spinal nerves to the regions they innervate.

Every segment of the spinal cord is attached to a pair of spinal nerves. A series of connective tissue layers surrounds each spinal nerve and continues along all its peripheral branches. These layers, best seen in sectional view (Figure 13–8●), are comparable to those associated with skeletal muscles (∞ p. 293). The **epineurium**, or outermost layer, consists of a dense network of collagen fibers. The fibers of the **perineurium**, the middle layer, extend inward from the epineurium. These connective tissue partitions divide the nerve into a series of compartments that

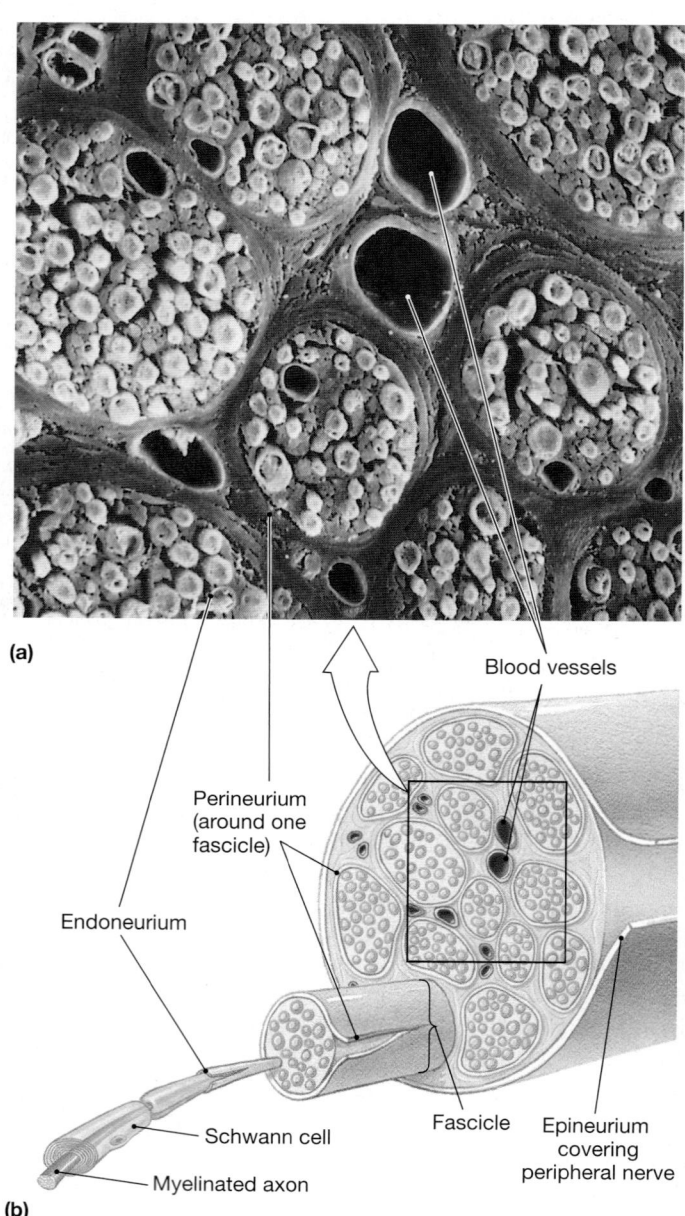

(a)

Blood vessels

Perineurium (around one fascicle)

Endoneurium

Schwann cell

Myelinated axon

Fascicle

Epineurium covering peripheral nerve

(b)

●**FIGURE 13–8**
A Peripheral Nerve. **(a)** Electron micrograph and **(b)** diagrammatic views of a typical peripheral nerve and its connective tissue wrappings, the perineurium, endoneurium, and epineurium. (SEM x 340) Reproduced from R. G. Kessel and R. H. Kardon, *Tissues and Organs: A Text-Atlas of Scanning Electron Microscopy*, W. H. Freeman & Co., 1979.

contain bundles of axons, or *fascicles*. Delicate connective tissue fibers of the **endoneurium**, the innermost layer, extend from the perineurium and surround individual axons.

Arteries and veins penetrate the epineurium and branch within the perineurium. Capillaries leaving the perineurium branch in the endoneurium and supply the axons and Schwann cells of the nerve and the fibroblasts of the connective tissues.

If a peripheral axon is damaged, but not displaced, normal function may eventually return as the cut stump grows across the site of injury, away from the cell body and along its former path. ∞ p. 395 Repairs made after an entire peripheral *nerve* has been damaged are generally incomplete, primarily because of problems with axon alignment and regrowth. Various technologically sophisticated procedures designed to improve nerve regeneration and repair are currently under evaluation. An entire family of *nerve growth factors* has been discovered in recent years. Their use alone or in combination with other therapies may ultimately revolutionize the treatment of damaged neural tissue inside and outside the CNS. **AM** Damage and Repair of Peripheral Nerves

▢ PERIPHERAL DISTRIBUTION OF SPINAL NERVES

Figure 13–9● shows the distribution, or pathway, of a typical spinal nerve that originates from the thoracic or superior lumbar segments of the spinal cord. The spinal nerve forms just lateral to the intervertebral foramen, where the dorsal and ventral roots unite. We shall now follow the nerve's distribution in the periphery; we consider the branches of a spinal nerve from the thoracic region.

The ventral root of each spinal nerve contains the axons of somatic motor and visceral motor neurons (Figure 13–9a●). Distally, the first branch from the spinal nerve carries visceral motor fibers to a nearby *sympathetic ganglion*, part of the *sympathetic division* of the autonomic nervous system. (Among its other functions, the sympathetic division is responsible for elevating the metabolic rate and for increasing alertness.) Because preganglionic axons are myelinated, this branch has a light color and is therefore known as the **white ramus** ("branch"). Postganglionic fibers that

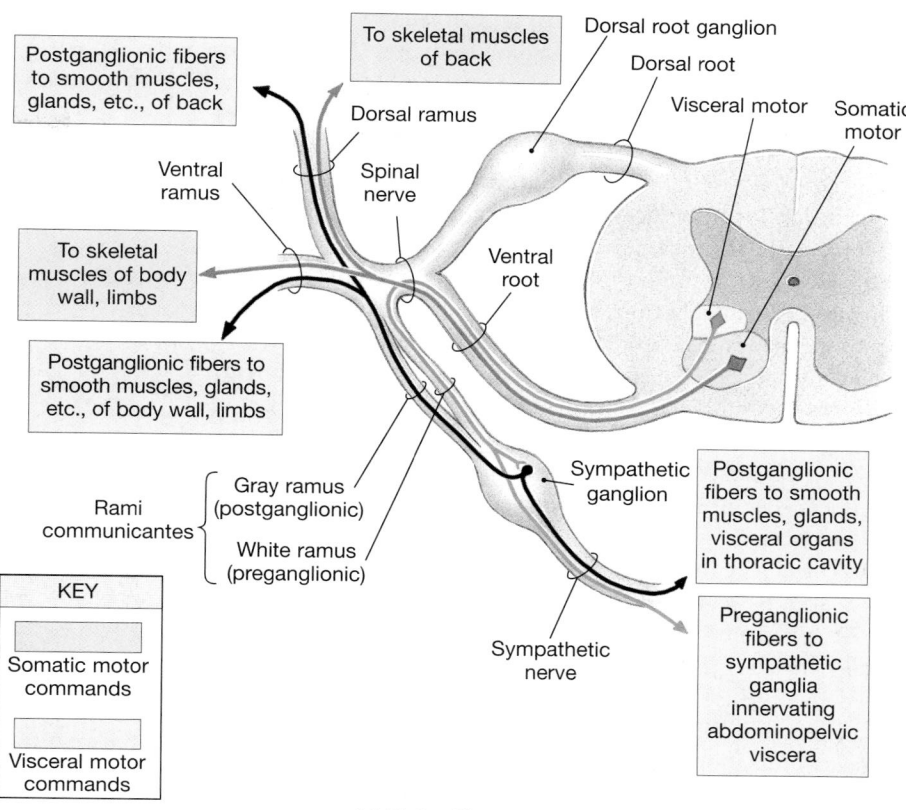

(a) Motor fibers

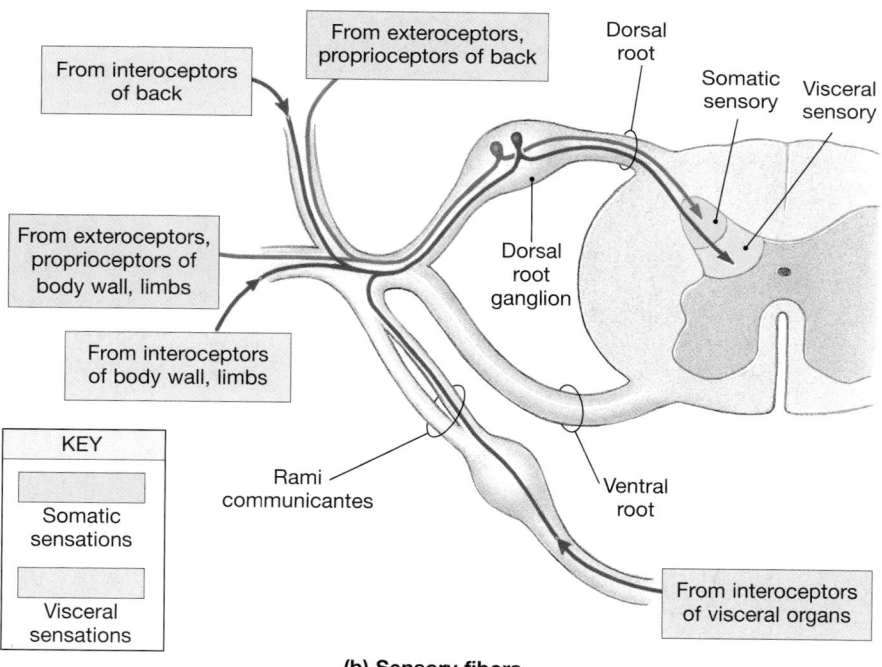

(b) Sensory fibers

●**FIGURE 13–9**
Peripheral Distribution of Spinal Nerves. (a) The distribution of motor fibers in the major branches of a representative thoracic or superior lumbar spinal nerve. (Although the gray ramus is normally proximal to the white ramus, this diagrammatic view makes it easier to follow the relationships between preganglionic and postganglionic fibers.) (b) A comparable view of the distribution of sensory fibers.

innervate smooth muscles, glands, and organs in the thoracic cavity extend directly from the ganglion to their respective effector organs. These axons form a series of **sympathetic nerves**.

Postganglionic fibers innervating glands and smooth muscles in the body wall or limbs return from the ganglion to rejoin the spinal nerve. These fibers, which are unmyelinated and have a darker color, form the **gray ramus**. The gray ramus is typically proximal to the white ramus; together, they are known as the *rami communicantes* (RĀ-mī ko-mū-ni-KAN-tēz), or "communicating branches." The **dorsal ramus** of each spinal nerve contains somatic motor and visceral motor fibers that innervate the skin and skeletal muscles of the back. The axons in the relatively large **ventral ramus** supply the ventrolateral body surface, structures in the body wall, and the limbs.

The dorsal, ventral, and white rami also contain sensory (afferent) fibers (Figure 13–9b•). Somatic sensory information arrives over the dorsal and ventral rami; visceral sensory information reaches the dorsal root through the dorsal, ventral, and white rami.

The specific region of the skin surface monitored by a single pair of spinal nerves is known as a **dermatome**. Each pair of spinal nerves services its own dermatome (Figure 13–10•), although the precise boundaries of adjacent dermatomes overlap to some degree. Dermatomes are clinically important, because damage or infection of a spinal nerve or dorsal root ganglion will produce a characteristic loss of sensation in the skin.

Peripheral *nerve palsies*, or **peripheral neuropathies**, are regional losses of sensory and motor function as a result of nerve trauma or compression. You have experienced a mild, temporary palsy if your arm or leg has ever "fallen asleep" after you leaned or sat in an uncomfortable position. The location of the affected dermatomes provides clues to the location of injuries along the spinal cord, but the information is not precise. More exact conclusions can be drawn if there is loss of motor control, on the basis of the origin and distribution of the peripheral nerves originating at nerve plexuses. For example, in the condition *shingles*, a virus infects dorsal root ganglia, causing a painful rash whose distribution corresponds to that of the affected sensory nerves. **AM** Peripheral Neuropathies; Shingles and Hansen's Disease **ATLAS** Embryology Summary 11: The Development of the Spinal Cord and Spinal Nerves

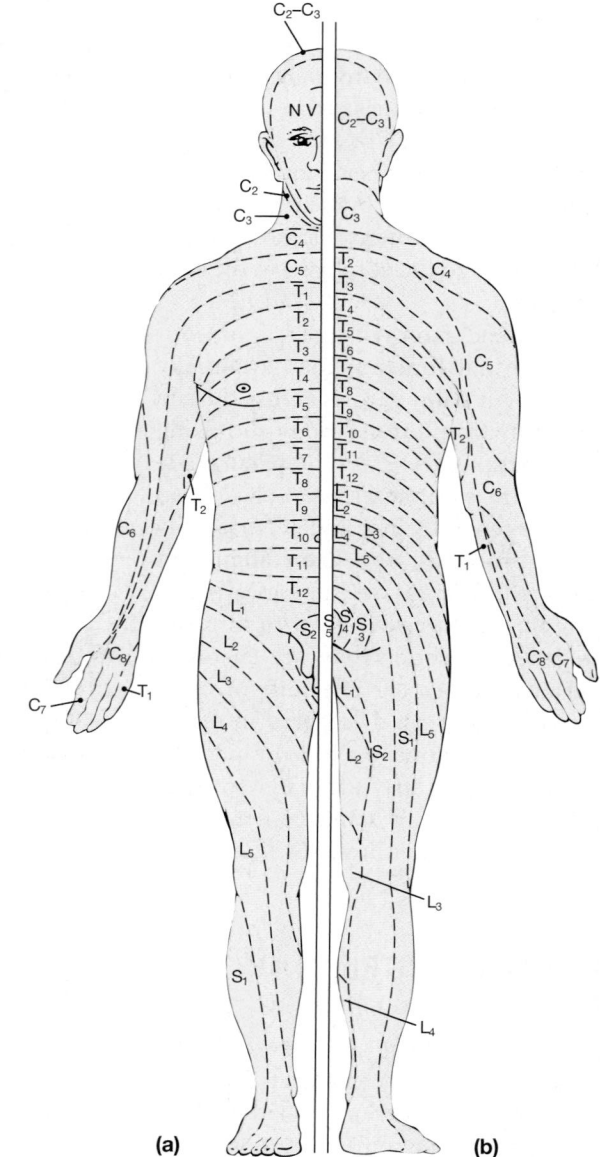

•**FIGURE 13–10**
Dermatomes. **(a)** Anterior and **(b)** posterior distributions of dermatomes on the surface of the skin.

▫ NERVE PLEXUSES

The simple distribution pattern of dorsal and ventral rami in Figure 13–9• applies to spinal nerves T_2–T_{12}. But in segments controlling the skeletal musculature of the neck, upper limbs, or lower limbs, the situation is more complicated. During development, small skeletal muscles innervated by different ventral rami typically fuse to form larger muscles with compound origins. The anatomical distinctions between the component muscles may disappear, but separate ventral rami continue to provide sensory innervation and motor control to each part of the compound muscle. As they converge, the ventral rami of adjacent spinal nerves blend their fibers, producing a series of compound nerve trunks. Such a complex interwoven network of nerves is called a **nerve plexus** (PLEK-sus; *plexus*, braid). The ventral rami form four major plexuses: (1) the *cervical plexus*, (2) the *brachial plexus*, (3) the *lumbar plexus*, and (4) the *sacral plexus* (Figure 13–11•). Because they form from the fusion of ventral rami, the nerves arising at these plexuses contain sensory as well as motor fibers (Figure 13–9b•).

In Chapter 11, we introduced the peripheral nerves that control the major axial and appendicular muscles. You may find it helpful to refer to the tables in that chapter as we proceed to review the innervation of the skeletal muscle groups. ∞ pp. 342–376

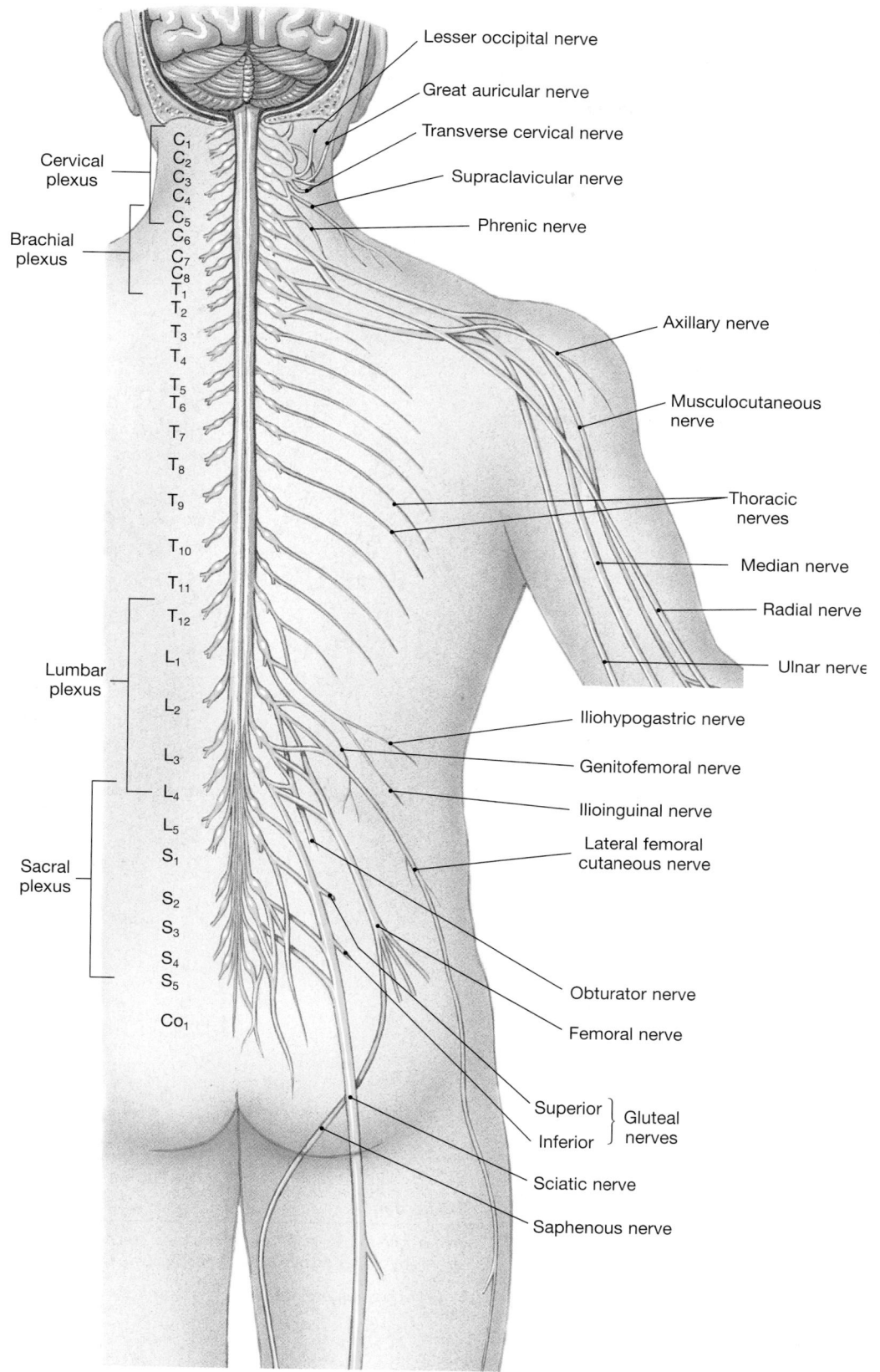

Lesser occipital nerve

Great auricular nerve

Transverse cervical nerve

Supraclavicular nerve

Phrenic nerve

Cervical plexus

Brachial plexus

C_1
C_2
C_3
C_4
C_5
C_6
C_7
C_8
T_1
T_2
T_3
T_4
T_5
T_6
T_7
T_8
T_9
T_{10}
T_{11}
T_{12}
L_1
L_2
L_3
L_4
L_5
S_1
S_2
S_3
S_4
S_5
Co_1

Lumbar plexus

Sacral plexus

Axillary nerve

Musculocutaneous nerve

Thoracic nerves

Median nerve

Radial nerve

Ulnar nerve

Iliohypogastric nerve

Genitofemoral nerve

Ilioinguinal nerve

Lateral femoral cutaneous nerve

Obturator nerve

Femoral nerve

Superior ⎱ Gluteal
Inferior ⎰ nerves

Sciatic nerve

Saphenous nerve

●FIGURE 13–11
Peripheral Nerves and Nerve Plexuses. ATLAS Plate 4.1

The Cervical Plexus

The **cervical plexus** consists of the ventral rami of spinal nerves C_1–C_5 (Figures 13–11, 13–12•; Table 13–1). The branches of the cervical plexus innervate the muscles of the neck and extend into the thoracic cavity, where they control the diaphragmatic muscles. The **phrenic nerve**, the major nerve of the cervical plexus, provides the entire nerve supply to the diaphragm, a key respiratory muscle. Other branches of this nerve plexus are distributed to the skin of the neck and the superior part of the chest.

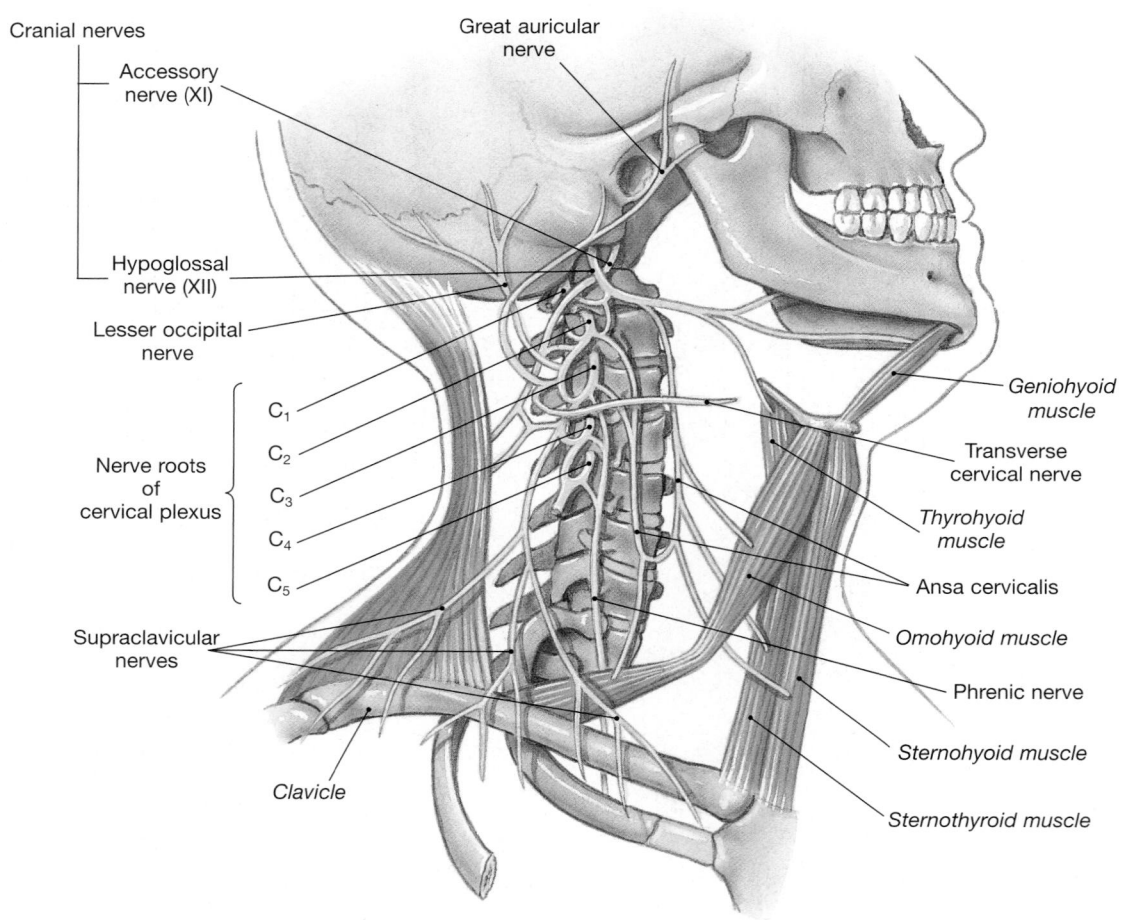

•**FIGURE 13–12**
The Cervical Plexus. **ATLAS** Plates 5.2, 5.3

TABLE 13–1 THE CERVICAL PLEXUS		
Nerve(s)	Spinal Segments	Distribution
Ansa cervicalis (superior and inferior branches)	C_1–C_4	Five of the extrinsic laryngeal muscles: sternothyroid, sternohyoid, omohyoid, geniohyoid, and thyrohyoid muscles (via XII)
Lesser occipital, transverse cervical, supraclavicular, and great auricular nerves	C_2–C_3	Skin of upper chest, shoulder, neck, and ear
Phrenic nerve	C_3–C_5	Diaphragm
Cervical nerves	C_1–C_5	Levator scapulae, scalene, sternocleidomastoid, and trapezius muscles (with XI)

The Brachial Plexus

The **brachial plexus** innervates the pectoral girdle and upper limb, with contributions from the ventral rami of spinal nerves $C_5–T_1$ (Figures 13–11 and 13–13•; Table 13–2). The nerves that form this plexus originate from trunks and cords. **Trunks** are large bundles of axons contributed by several spinal nerves. **Cords** are smaller branches that originate at trunks. Both trunks and cords are named according to their location relative to the axillary artery, a large artery supplying the upper limb. Hence we have *superior, middle,* and *inferior trunks* and *lateral, medial,* and *posterior cords*. The lateral cord forms the **musculocutaneous nerve** exclusively and, together with the medial cord, contributes to the **median nerve**. The **ulnar nerve** is the other major nerve of the medial cord. The posterior cord gives rise to the **axillary nerve** and the **radial nerve**. Table 13–2 provides further information about these and other major nerves of the brachial plexus.

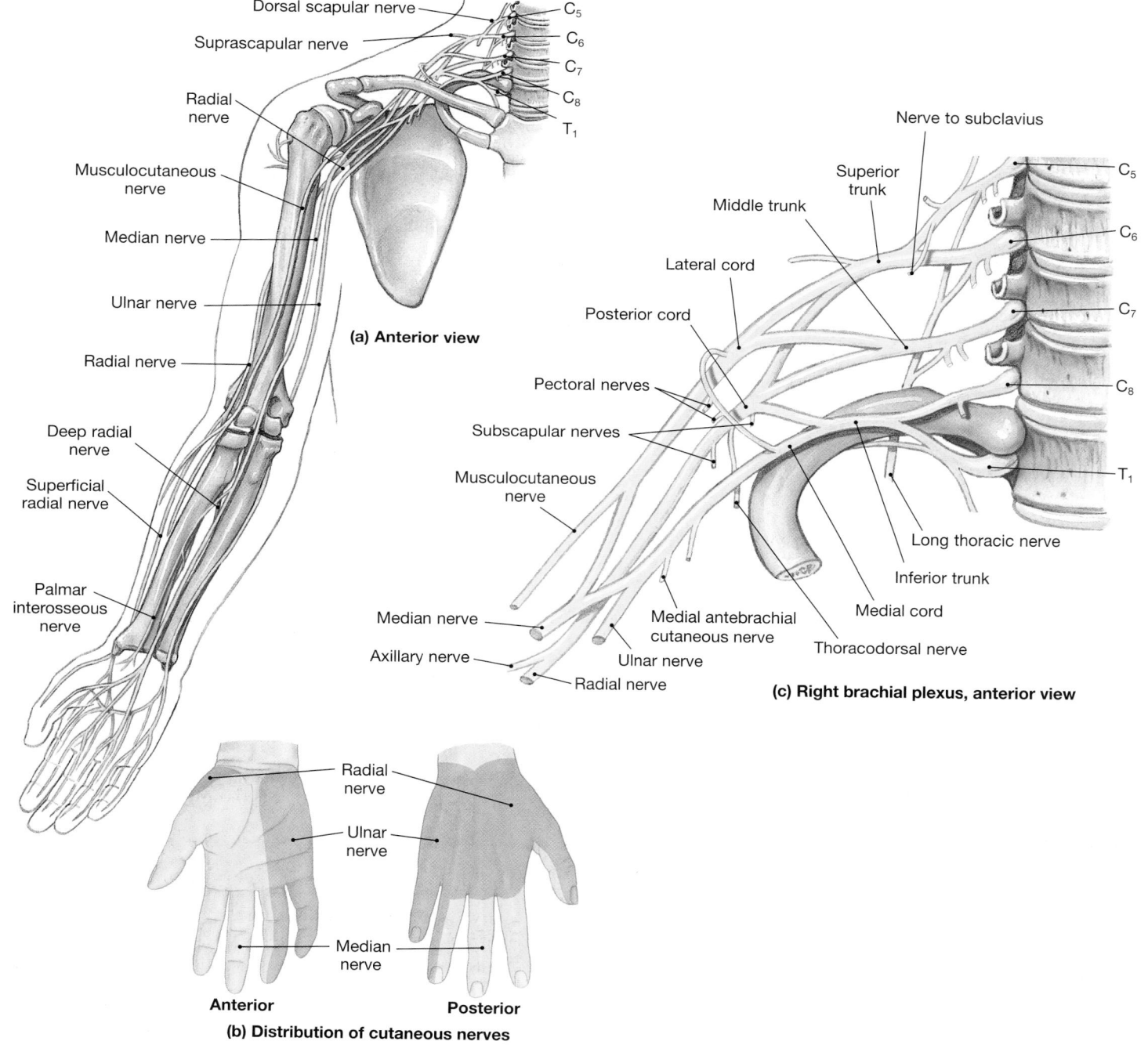

(a) Anterior view

(b) Distribution of cutaneous nerves

(c) Right brachial plexus, anterior view

•**FIGURE 13–13**
The Brachial Plexus. (a) Major nerves originating at the right brachial plexus. (b) The right brachial plexus. **ATLAS** Plates 6.1a-c, 6.2a,b

TABLE 13–2 THE BRACHIAL PLEXUS

Nerve(s)	Spinal Segments	Distribution
Nerve to subclavius	$C_4–C_6$	Subclavius muscle
Dorsal scapular nerve	C_5	Rhomboid and levator scapulae muscles
Long thoracic nerve	$C_5–C_7$	Serratus anterior muscle
Suprascapular nerve	C_5, C_6	Supraspinatus and infraspinatus muscles; sensory from shoulder joint and scapula
Pectoral nerves (medial and lateral)	$C_5–T_1$	Pectoralis muscles
Subscapular nerves	C_5, C_6	Subscapularis and teres major muscles
Thoracodorsal nerve	$C_6–C_8$	Latissimus dorsi muscle
Axillary nerve	C_5, C_6	Deltoid and teres minor muscles; sensory from the skin of the shoulder through the *lateral brachial cutaneous nerve*
Medial antebrachial cutaneous nerve	C_8, T_1	Sensory from skin over anterior, medial surface of arm and forearm
Radial nerve	$C_5–T_1$	Many extensor muscles on the arm and forearm (triceps brachii, extensor carpi radialis, and extensor carpi ulnaris muscles) and the brachioradialis muscle, supinator muscle, digital extensor muscles and abductor pollicis muscle; sensory from skin over the posterolateral surface of the limb through the *posterior brachial cutaneous nerve* (arm), *posterior antebrachial cutaneous nerve* (forearm) and *superficial radial nerve* (radial half of hand)
Musculocutaneous nerve	$C_5–C_7$	Flexor muscles on the arm (biceps brachii, brachialis, and coracobrachialis muscles); sensory from skin over lateral surface of the forearm through the *lateral antebrachial cutaneous nerve*
Median nerve	$C_6–T_1$	Flexor muscles on the forearm (flexor carpi radialis and palmaris longus muscles); pronator quadratus and pronator teres muscles; digital flexors (through the *palmar interosseous nerve*); sensory from skin over anterolateral surface of the hand
Ulnar nerve	C_8, T_1	Flexor carpi ulnaris muscle, flexor digitorum profundus muscle, adductor pollicis muscle, and small digital muscles; sensory from skin over medial surface of the hand through the *palmar cutaneous nerve* and the *superficial volar nerve*

The Lumbar and Sacral Plexuses

The **lumbar plexus** and the **sacral plexus** arise from the lumbar and sacral segments of the spinal cord, respectively. The nerves arising at these plexuses innervate the pelvic girdle and lower limbs (Figures 13–11 and 13–14●). The individual nerves that form the lumbar and sacral plexuses are listed in Table 13–3.

The lumbar plexus contains axons from the ventral rami of spinal nerves $T_{12}–L_4$. The major nerves of this plexus are the **genitofemoral nerve**, the **lateral femoral cutaneous nerve**, and the **femoral nerve**. The sacral plexus contains axons from the ventral rami of spinal nerves $L_4–S_4$. Two major nerves arise at this plexus: the **sciatic nerve** and the **pudendal nerve**. The sciatic nerve passes posterior to the femur, deep to the long head of the biceps femoris muscle. As it approaches the knee, the sciatic nerve divides into two branches: the **fibular nerve** (or *peroneal nerve*) and the **tibial nerve**. The *sural nerve*, formed by branches of the fibular nerve, is a sensory nerve innervating the lateral portion of the foot. A section of this nerve is often removed for use in nerve grafts.

In discussions of motor performance, a distinction is usually made between the conscious ability to control motor function—something that requires communication and feedback between the brain and spinal cord—and automatic motor responses coordinated entirely within the spinal cord. These automatic responses, called reflexes, are relatively stereotyped motor responses to specific stimuli. Although simple in structure, reflexes are like Lego™ blocks—they can be combined to build relatively elaborate motor patterns. The rest of this chapter looks at how sensory neurons, interneurons, and motor neurons interconnect, and how these interconnections produce both simple and complex reflexes.

✓ An anesthetic blocks the function of the dorsal rami of the cervical spinal nerves. Which areas of the body will be affected?

✓ Injury to which of the nerve plexuses would interfere with the ability to breathe?

✓ Compression of which nerve produces the sensation that your leg has "fallen asleep"?

Answers start on page Q-1

 To review the functions of the spinal cord and related plexus, visit the **Companion Website:** Chapter 13/Reviewing Facts & Terms/Multiple Choice.

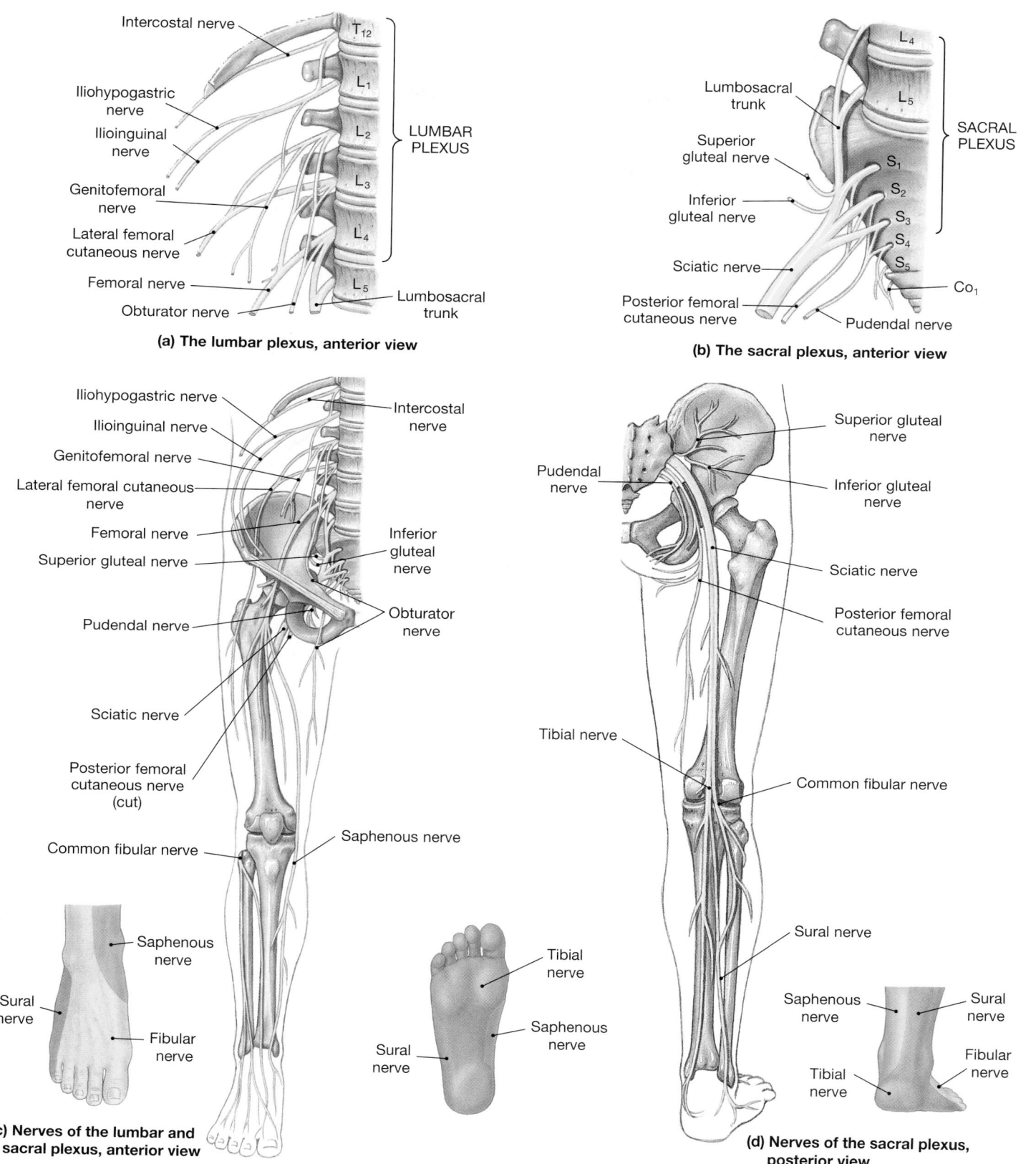

(a) The lumbar plexus, anterior view

(b) The sacral plexus, anterior view

(c) Nerves of the lumbar and sacral plexus, anterior view

(d) Nerves of the sacral plexus, posterior view

•FIGURE 13–14
The Lumbar and Sacral Plexuses. (a) The right lumbar plexus. (b) The right sacral plexus. (c) The major branches of the right lumbar plexus. (d) The major branches of the right sacral plexus. ATLAS Plates 8.7c, 8.8a, 8.13a

TABLE 13–3 THE LUMBAR AND SACRAL PLEXUSES

Nerve(s)	Spinal Segment(s)	Distribution
LUMBAR PLEXUS		
Iliohypogastric nerve	T_{12}, L_1	Abdominal muscles (external and internal oblique muscles, transversus abdominis muscle); skin over inferior abdomen and buttocks
Ilioinguinal nerve	L_1	Abdominal muscles (with iliohypogastric nerve); skin over superior, medial thigh and portions of external genitalia
Genitofemoral nerve	L_1, L_2	Skin over anteromedial surface of thigh and portions of external genitalia
Lateral femoral cutaneous nerve	L_2, L_3	Skin over anterior, lateral, and posterior surfaces of thigh
Femoral nerve	L_2–L_4	Anterior muscles of thigh (sartorius muscle and quadriceps group); flexors and adductors of hip (pectineus and iliopsoas muscles); skin over anteromedial surface of thigh, medial surface of leg and foot
Obturator nerve	L_2–L_4	Adductors of hip (adductors magnus, brevis, and longus muscles); gracilis muscle; skin over medial surface of thigh
Saphenous nerve	L_2–L_4	Skin over medial surface of leg
SACRAL PLEXUS		
Gluteal nerves: Superior	L_4–S_2	Abductors of hip (gluteus minimus, gluteus medius, and tensor fasciae latae muscles)
Inferior		Extensor of hip (gluteus maximus muscle)
Posterior femoral cutaneous nerve	S_1–S_3	Skin of perineum and posterior surfaces of thigh and leg
Sciatic nerve:	L_4–S_3	Two of the hamstrings (semimembranosus and semitendinosus muscles); adductor magnus muscle (with obturator nerve)
Tibial nerve		Flexors of knee and extensors (plantar flexors) of ankle (popliteus, gastrocnemius, soleus, and tibialis posterior muscles and the long head of the biceps femoris muscle); flexors of toes; skin over posterior surface of leg, plantar surface of foot
Fibular nerve		Biceps femoris muscle (short head); fibularis muscles (brevis and longus) and tibialis anterior muscle; extensors of toes; skin over anterior surface of leg and dorsal surface of foot; skin over lateral portion of foot (through the *sural nerve*)
Pudendal nerve	S_2–S_4	Muscles of perineum, including urogenital diaphragm and external anal and urethral sphincter muscles; skin of external genitalia and related skeletal muscles (bulbospongiosus and ischiocavernosus muscles)

13–4 PRINCIPLES OF FUNCTIONAL ORGANIZATION

Objectives

- Discuss the significance of neuronal pools, and describe the major patterns of interaction among neurons within and among these pools.
- Describe the steps in a neural reflex.
- Classify the types of reflexes, and explain the functions of each.

The human body has about 10 million sensory neurons, one-half million motor neurons, and 20 *billion* interneurons. The sensory neurons deliver information to the CNS, the motor neurons distribute commands to peripheral effectors, such as skeletal muscles, and the interneurons interpret, plan, and coordinate the incoming and outgoing signals.

☐ NEURONAL POOLS

The billions of interneurons of the CNS are organized into a much smaller number of **neuronal pools**—functional groups of interconnected neurons. A neuronal pool may be diffuse, involving neurons in several regions of the brain, or localized, with neurons restricted to one specific location in the brain or spinal cord. Estimates of the actual number of neuronal pools range between a few hundred and a few thousand. Each has a limited number of input sources and output destinations, and each may contain both excitatory and inhibitory neurons. The output of the entire neuronal pool may stimulate or depress activity in other parts of the brain or spinal cord, affecting the interpretation of sensory information or the coordination of motor commands.

The pattern of interaction among neurons provides clues to the functional characteristics of a neuronal pool. It is customary to refer to such a "wiring diagram" as a *neural circuit*, just as we refer to electrical circuits in the wiring of a house. We can distinguish five circuit patterns:

1. **Divergence** is the spread of information from one neuron to several neurons (Figure 13–15a●) or from one pool to multiple pools. Divergence permits the broad distribution of a specific input. Considerable divergence occurs when sensory neurons bring information into the CNS, for the information is distributed to neuronal pools throughout the spinal cord and brain. For example, visual information arriving from the eyes reaches your conscious awareness at the same time it is distributed to areas of the brain that control posture and balance at the subconscious level.

2. In **convergence**, several neurons synapse on the same postsynaptic neuron (Figure 13–15b●). Several patterns of activity in the presynaptic neurons can therefore have the same effect on the postsynaptic neuron. Through convergence, the same motor neurons can be subject to both conscious and subconscious control. For example, the movements of your diaphragm and ribs are now being controlled by your brain at the subconscious level. But the same motor neurons can also be controlled consciously, as when you take a deep breath and hold it. Two neuronal pools are involved, both synapsing on the same motor neurons.

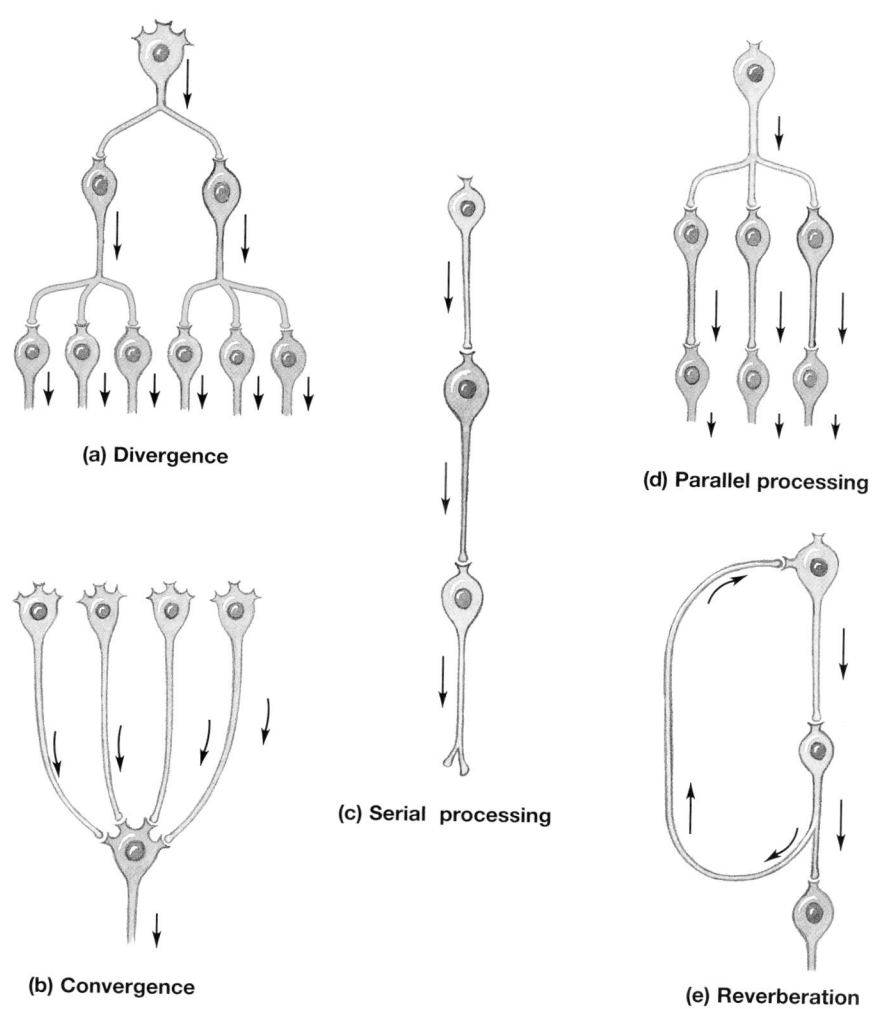

(a) Divergence

(b) Convergence

(c) Serial processing

(d) Parallel processing

(e) Reverberation

●**FIGURE 13–15**

The Organization of Neuronal Pools. **(a)** Divergence, a mechanism for spreading stimulation to multiple neurons or neuronal pools in the CNS. **(b)** Convergence, a mechanism providing input to a single neuron from multiple sources. **(c)** Serial processing, in which neurons or pools work sequentially. **(d)** Parallel processing, in which neurons or pools process information simultaneously. **(e)** Reverberation, a positive feedback mechanism.

3. Information may be relayed in a stepwise fashion, from one neuron to another or from one neuronal pool to the next. This pattern, called **serial processing** (Figure 13–15c●), occurs as sensory information is relayed from one part of the brain to another. For example, pain sensations en route to your conscious awareness make two stops along the way, at neuronal pools along the pain pathway.

4. **Parallel processing** occurs when several neurons or neuronal pools process the same information at one time (Figure 13–15d●). Divergence must take place before parallel processing can occur. Thanks to parallel processing, many responses can occur simultaneously. For example, when you step on a sharp object, sensory neurons that distribute the information to a number of neuronal pools are stimulated. As a result of parallel processing, you might withdraw your foot, shift your weight, move your arms, feel the pain, and shout "Ouch!" all at the same time.

5. Some neural circuits utilize positive feedback to produce **reverberation**. In this arrangement, collateral branches of axons somewhere along the sequence extend back toward the source of an impulse and further stimulate the presynaptic neurons. Once a reverberating circuit has been activated, it will continue to function until synaptic fatigue or inhibitory stimuli break the cycle. Reverberation can occur within a single neuronal pool, or it may involve a series of interconnected pools. A simple example of reverberation is shown in Figure 13–15e●. Reverberation is like a positive feedback loop involving neurons: Once the circuit is activated, it continues to stimulate itself. Highly complicated examples of reverberation among neuronal pools in the brain may help maintain consciousness, muscular coordination, and normal breathing.

The functions of the nervous system depend on the interactions among the neurons organized in neuronal pools. The most complex neural processing steps occur in the spinal cord and brain.

The simplest, which occur within the PNS and the spinal cord, control basic *reflexes*—stereotyped responses to stimuli. Reflexes are a bit like Legos™: Individually, they are quite simple, but they can be combined in a great variety of ways to create very complex responses. Reflexes are thus the basic building blocks of neural function, as you will see in the next section.

☐ AN INTRODUCTION TO REFLEXES

Conditions inside or outside the body can change rapidly and unexpectedly. **Reflexes** are rapid, automatic responses to specific stimuli. Reflexes preserve homeostasis by making rapid adjustments in the function of organs or organ systems. The response shows little variability: Each time a particular reflex is activated, it usually produces the same motor response. In Chapter 1, we introduced the basic functional components involved in all types of homeostatic regulation: a *receptor*, an *integration center*, and an *effector*. ∞ p. 11 In the current chapter, we consider *neural reflexes*, in which sensory fibers deliver information from peripheral receptors to the CNS and motor fibers carry motor commands to peripheral effectors. In Chapter 18, we shall examine *endocrine reflexes*, in which the commands to peripheral tissues and organs are delivered by hormones in the bloodstream.

The Reflex Arc

The "wiring" of a single reflex is called a **reflex arc**. A reflex arc begins at a receptor and ends at a peripheral effector, such as a muscle fiber or a gland cell. Figure 13–16● diagrams the five steps in a neural reflex:

STEP 1: *The Arrival of a Stimulus and Activation of a Receptor.* A *receptor* is either a specialized cell or the dendrites of a sensory

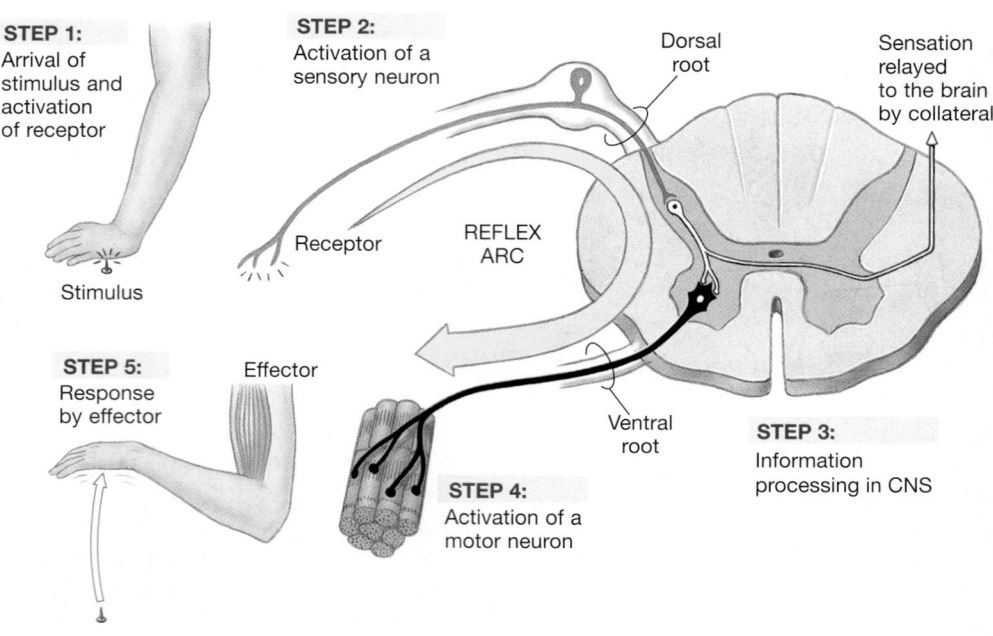

●**FIGURE 13–16**
Components of a Reflex Arc

neuron. Receptors are sensitive to physical or chemical changes in the body and to changes in the external environment. We introduced general categories of sensory receptors in Chapter 12. ∞ p. 391 When you lean on a tack, for example, pain receptors in the palm of your hand are activated. These receptors, the dendrites of sensory neurons, respond to stimuli that cause or accompany tissue damage. (We shall discuss the link between receptor stimulation and sensory neuron activation further in Chapter 15.)

STEP 2: *The Activation of a Sensory Neuron.* The stimulation of pain receptors leads to the formation and propagation of action potentials along the axons of sensory neurons. This information reaches your spinal cord by way of a dorsal root. In our example, steps 1 and 2 involve the same cell. However, the two steps may involve different cells. For example, reflexes triggered by loud sounds begin when receptor cells in the inner ear release neurotransmitters that stimulate sensory neurons.

STEP 3: *Information Processing.* Information processing begins when excitatory neurotransmitter molecules, released by the synaptic knob of a sensory neuron, arrive at the postsynaptic membrane of an interneuron. The neurotransmitter then produces an excitatory postsynaptic potential (EPSP), which is integrated with other stimuli arriving at the interneuron at that moment. ∞ p. 419 An interneuron is not always present in a reflex arc; in the simplest reflexes, the sensory neuron innervates a motor neuron directly. In that case, the motor neuron performs the information processing as summation occurs at the axon hillock. On the other hand, complex reflexes involve several interneurons, some releasing excitatory neurotransmitters (*excitatory interneurons*) and others inhibitory neurotransmitters (*inhibitory interneurons*).

STEP 4: *The Activation of a Motor Neuron.* The axons of the stimulated motor neurons carry action potentials into the periphery—in this example, through the ventral root of a spinal nerve.

STEP 5: *The Response of a Peripheral Effector.* The release of neurotransmitters by the motor neurons at synaptic knobs then leads to a response by a peripheral effector—in this case, a skeletal muscle whose contraction pulls your hand away from the tack.

A reflex response generally removes or opposes the original stimulus; in this case, the contracting muscle pulls your hand away from the painful stimulus. This reflex arc is therefore an example of *negative feedback.* ∞ p. 11 By opposing potentially harmful changes in the internal or external environment, reflexes play an important role in homeostatic maintenance. The immediate reflex response is typically not the only response to a stimulus. In the example given, you might wince, say "Ouch," and shake your hand. These other responses, which are directed by your brain, involve multiple synapses and take longer to organize and coordinate.

Classification of Reflexes

Reflexes are classified on the basis of (1) their development, (2) the site of information processing, (3) the nature of the resulting motor response, or (4) the complexity of the neural circuit involved. These categories are not mutually exclusive—they represent different ways of describing a single reflex (Figure 13–17●).

DEVELOPMENT OF REFLEXES **Innate reflexes** result from the connections that form between neurons during development. Such reflexes generally appear in a predictable sequence, from the simplest reflex responses (withdrawal from pain) to more complex motor patterns (chewing, suckling, or tracking objects with the eyes). The neural connections responsible for the basic motor patterns of an innate reflex are genetically or developmentally programmed. Examples include the reflexive removal of your hand from a hot stovetop and blinking when your eyelashes are touched.

More complex, learned motor patterns are called **acquired reflexes.** An experienced driver steps on the brake when trouble appears ahead; a professional skier must make equally quick adjustments in body position while racing. These motor responses are rapid and automatic, but they were learned rather than

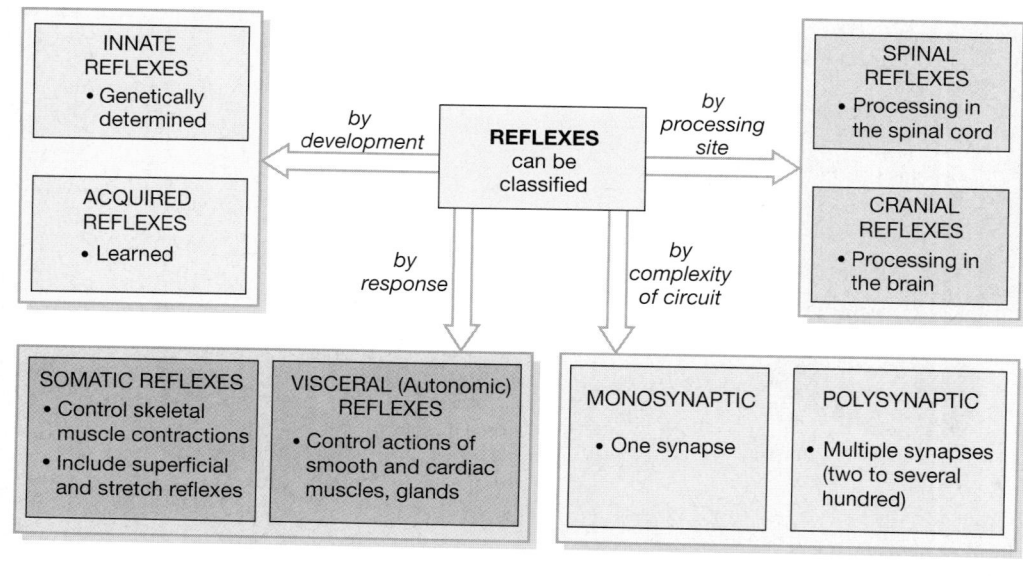

●FIGURE 13–17
Methods of Classifying Reflexes

preestablished. Such reflexes are enhanced by repetition. The distinction between innate and acquired reflexes is not absolute: Some people can learn motor patterns more quickly than others, and the differences probably have a genetic basis.

Most reflexes, whether innate or acquired, can be modified over time or suppressed through conscious effort. For example, while walking a tightrope over the Grand Canyon, you might ignore a bee sting on your hand, although under other circumstances you would probably withdraw your hand immediately, shouting and thrashing as well.

PROCESSING SITES In a **spinal reflex**, the important interconnections and processing events occur in the spinal cord. We shall discuss these reflexes in detail in the next section. Reflexes processed in the brain are called **cranial reflexes**. For example, the reflex movements in response to a sudden loud noise or bright light are cranial reflexes directed by nuclei in the brain. We will encounter other cranial reflexes in Chapters 14, 16, and 17.

NATURE OF THE RESPONSE **Somatic reflexes** provide a mechanism for the involuntary control of the muscular system. *Superficial reflexes* are triggered by stimuli at the skin or mucous membranes. *Stretch reflexes* are triggered by the sudden elongation of a tendon; a familiar example is the patellar, or "knee-jerk," reflex that is usually tested during physical exams. These reflexes are also known as *deep tendon reflexes*, or *myotactic reflexes*. **Visceral reflexes**, or *autonomic reflexes*, control the activities of other systems. We will consider somatic reflexes in detail in this chapter and visceral reflexes in Chapter 16.

The movements directed by somatic reflexes are not delicate or precise. You might therefore wonder why they exist at all, because we have voluntary control over the same muscles. In fact, somatic reflexes are absolutely vital, primarily because they are *immediate*. Making decisions and coordinating voluntary responses take time, and in an emergency—you slip while descending a flight of stairs or lean your hand against a knife edge—any delay increases the likelihood of a severe injury. Thus, somatic reflexes provide a rapid response that, if necessary, can be modified later by voluntary motor commands.

COMPLEXITY OF THE CIRCUIT In the simplest reflex arc, a sensory neuron synapses directly on a motor neuron, which serves as the processing center. Such a reflex is a **monosynaptic reflex**. Transmission across a chemical synapse always involves a synaptic delay, but with only one synapse, the delay between the stimulus and the response is minimized. Circuit 1 in Figure 13–18a● depicts a typical monosynaptic reflex. Stretch reflexes are the most common monosynaptic reflexes.

Most other types of reflexes have at least one interneuron between the sensory neuron and the motor neuron, as does circuit 2 of Figure 13–18b●. An example of this type of reflex is the withdrawal reflex diagrammed in Figure 13–16●. Such **polysynaptic reflexes** have a longer delay between stimulus and response. The length of the delay is proportional to the number of synapses involved. Polysynaptic reflexes can produce far more complicated responses than monosynaptic reflexes, because the interneurons can control several muscle groups.

●**FIGURE 13–18**
Neural Organization and Simple Reflexes. **(a)** A monosynaptic reflex (circuit 1) involves a sensory neuron and a central motor neuron. In this example, stimulation of the receptor will lead to a reflexive contraction in a skeletal muscle. **(b)** A polysynaptic reflex (circuit 2) involves a sensory neuron, interneurons, and motor neurons. In this example, the stimulation of the receptor leads to the coordinated contraction of two skeletal muscles.

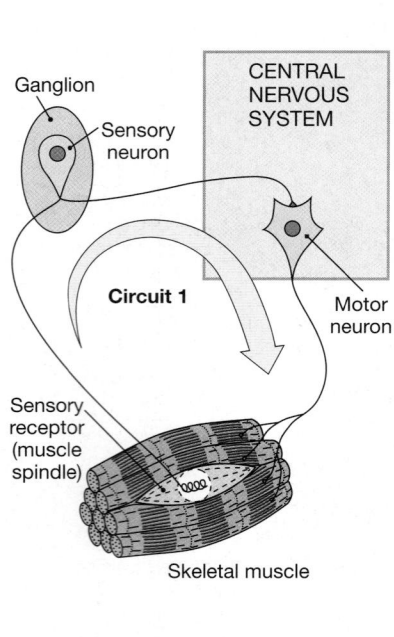

(a) Monosynaptic reflex

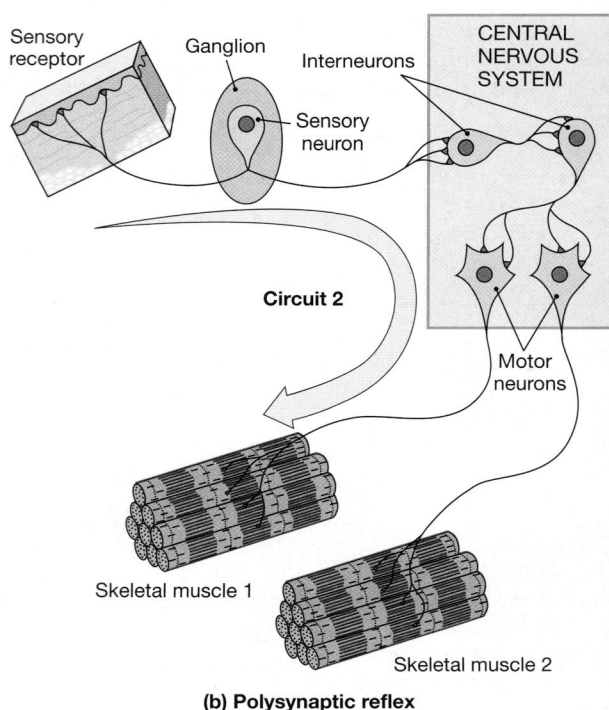

(b) Polysynaptic reflex

13–5 SPINAL REFLEXES

Objectives

- Distinguish among the types of motor responses produced by various reflexes.
- Explain how reflexes interact to produce complex behaviors.

Spinal reflexes range in complexity from simple monosynaptic reflexes involving a single segment of the spinal cord to polysynaptic reflexes that involve many segments. In the most complicated spinal reflexes, called **intersegmental reflex arcs**, many segments interact to produce a coordinated, highly variable motor response.

☐ MONOSYNAPTIC REFLEXES

In a monosynaptic reflex, there is little delay between sensory input and motor output. These reflexes control the most-rapid, stereotyped motor responses of the nervous system to specific stimuli.

The Stretch Reflex

The best-known monosynaptic reflex is the **stretch reflex** (Figure 13–19●), which provides automatic regulation of skeletal muscle length. The stimulus (increasing muscle length) activates a sensory neuron, which triggers an immediate motor response (contraction of the stretched muscle) that counteracts the stimulus. Action potentials traveling toward and away from the spinal cord are conducted along large, myelinated Type A fibers. ∞ p. 409 The entire reflex is completed within 20–40 msec.

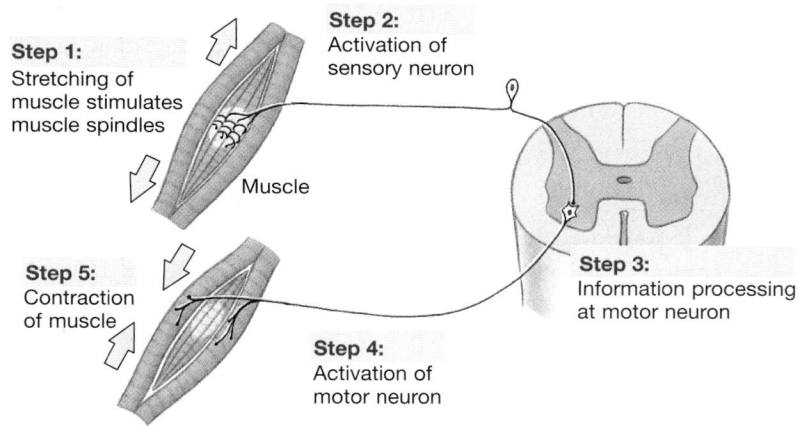

Step 1: Stretching of muscle stimulates muscle spindles

Muscle

Step 2: Activation of sensory neuron

Step 3: Information processing at motor neuron

Step 4: Activation of motor neuron

Step 5: Contraction of muscle

●**FIGURE 13–19**
Components of the Stretch Reflex. A diagram of the steps in a stretch reflex.

The **patellar reflex**, or *knee-jerk reflex* (Figure 13–20●), is a familiar example of a stretch reflex. This reflex is triggered by stretching the quadriceps muscles, which extend the knee. You can demonstrate the patellar reflex by tapping on the patellar tendon of a flexed knee when the thigh and leg muscles are relaxed. A tap on the patellar ligament stretches specialized sensory structures, called *muscle spindles*, in the quadriceps group. (The sensory mechanism will be described in the next section.) The stretching of muscle spindles produces a sudden burst of activity in the sensory neurons that monitor them. This in turn leads to stimulation of motor neurons that control the motor units in the quadriceps group. The result is a rapid increase in the muscle tone of the quadriceps. The rise is usually so sudden

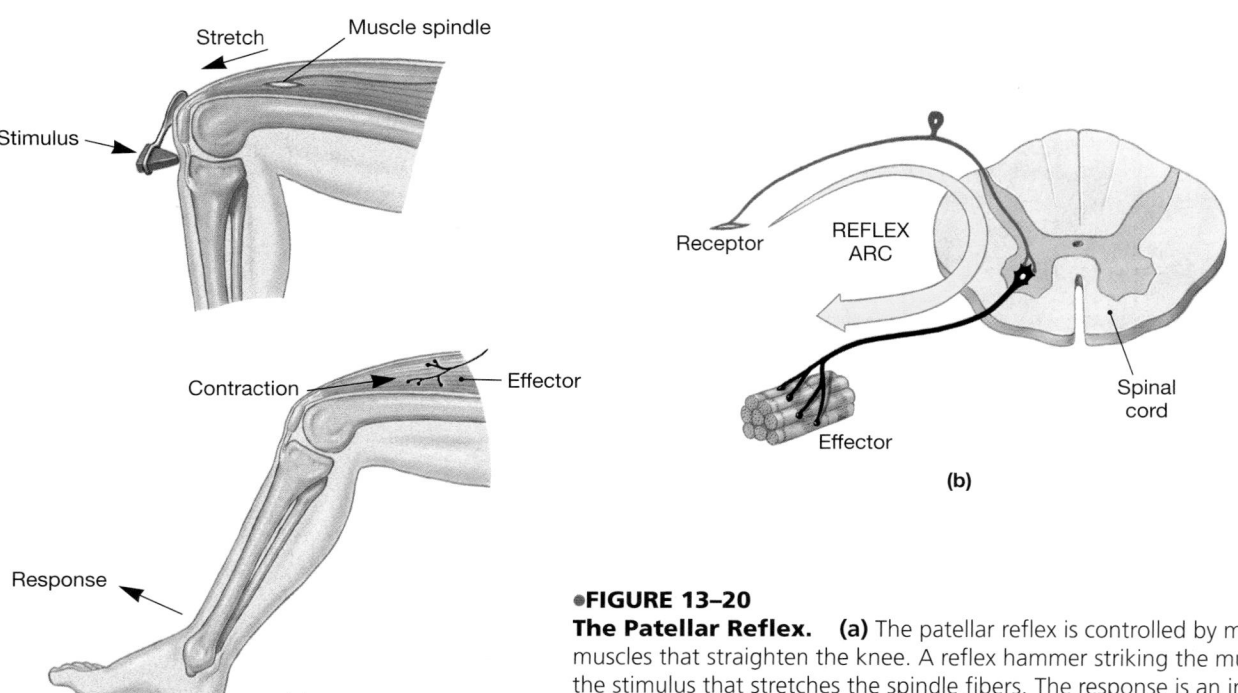

Stretch

Muscle spindle

Stimulus

Contraction

Effector

Response

(a)

Receptor

REFLEX ARC

Spinal cord

Effector

(b)

●**FIGURE 13–20**
The Patellar Reflex. **(a)** The patellar reflex is controlled by muscle spindles in the muscles that straighten the knee. A reflex hammer striking the muscle tendon provides the stimulus that stretches the spindle fibers. The response is an immediate increase in muscle tone and a reflexive kick. **(b)** The basic wiring of a monosynaptic reflex arc.

that it produces a noticeable kick. As the muscles contract, the muscle spindles return to their resting length. The rate of action potential generation in the sensory neurons then declines, causing a drop in muscle tone, and the leg then swings back.

Muscle Spindles

The sensory receptors involved in the stretch reflex are **muscle spindles**. Each consists of a bundle of small, specialized skeletal muscle fibers called **intrafusal muscle fibers** (Figure 13–21a●). The muscle spindle is surrounded by larger **extrafusal muscle fibers**, which are responsible for the resting muscle tone and, at greater levels of stimulation, for the contraction of the entire muscle.

Each intrafusal fiber is innervated by both sensory and motor neurons. The dendrites of the sensory neuron spiral around the central portion of the intrafusal fiber. Axons from spinal motor neurons form neuromuscular junctions on either end of this fiber. Motor neurons innervating intrafusal fibers are called **gamma (γ) motor neurons**; their axons are called **gamma efferents**. An intrafusal fiber has one set of myofibrils at each end. Instead of extending the length of the muscle fiber, as in extrafusal fibers, these myofibrils run from the end of the intrafusal fiber only to the sarcolemma in a central region that is closely monitored by the sensory neuron. The gamma efferents give the CNS the ability to adjust the sensitivity of the muscle spindle. Before seeing how this is accomplished, we shall consider the normal functioning of this sensory receptor and its effects on the surrounding extrafusal fibers.

The sensory neuron is always active, conducting impulses to the CNS. The axon enters the CNS in a dorsal root and synapses on motor neurons in the anterior gray horn of the spinal cord. Collaterals distribute the information to the brain, providing information about the state of the muscle spindle. Stretching the central portion of the intrafusal fiber distorts the dendrites and stimulates the sensory neuron, increasing the frequency of action potential generation. Compressing the central portion inhibits the sensory neuron, decreasing the frequency of action potential generation.

The axon of the sensory neuron synapses on CNS motor neurons that control the extrafusal muscle fibers of the same muscle. An increase in stimulation of the sensory neuron, caused by stretching of the intrafusal fiber, will increase stimulation to the motor neuron controlling the surrounding extrafusal fibers, so muscle tone increases. A decrease in the stimulation of the sensory neuron, due to compression of the intrafusal fiber, will lead to a decrease in the stimulation of the motor neuron controlling the surrounding extrafusal fibers, so muscle tone decreases.

When a skeletal muscle is stretched, its muscle spindles elongate and its muscle tone increases (Figure 13–21b●). This increase provides automatic resistance that reduces the chance of muscle damage due to overstretching. The patellar reflex and similar reflexes serve this function. When a skeletal muscle is compressed, its muscle spindles are also compressed, and its muscle tone de-

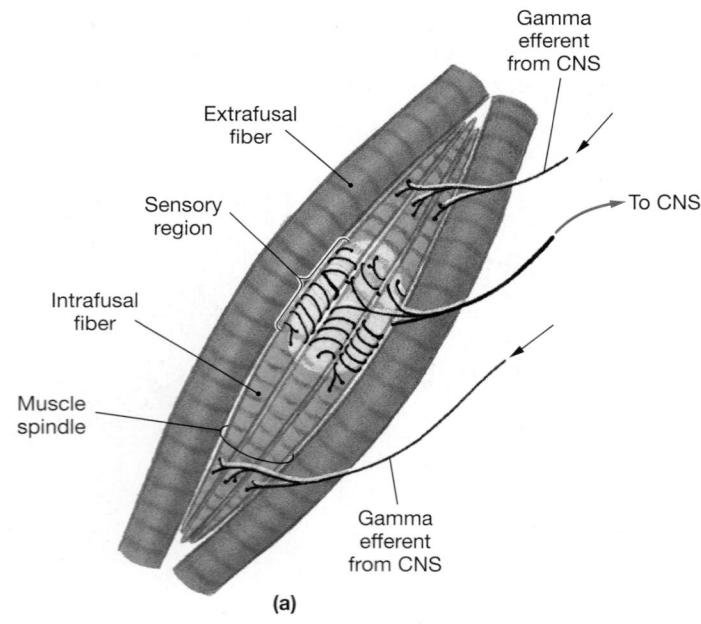

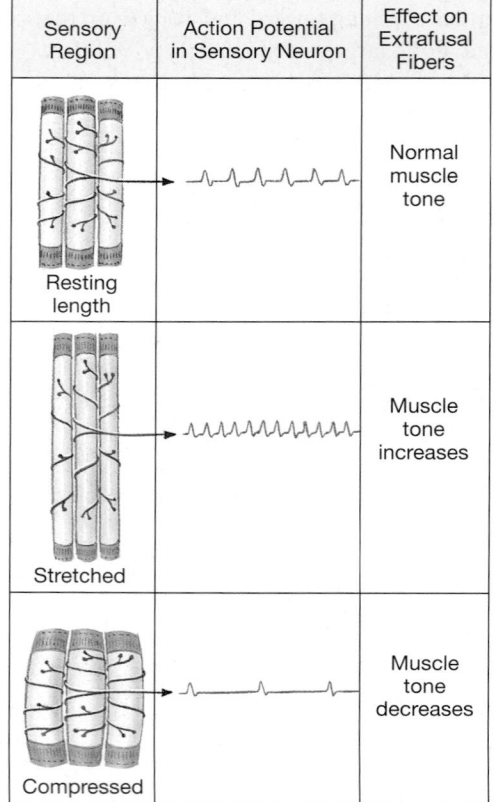

Sensory Region	Action Potential in Sensory Neuron	Effect on Extrafusal Fibers
Resting length		Normal muscle tone
Stretched		Muscle tone increases
Compressed		Muscle tone decreases

(b)

●**FIGURE 13–21**
Intrafusal Fibers. **(a)** The structure of a muscle spindle. **(b)** The effect of changes in muscle spindle length on muscle tone in the surrounding extrafusal fibers.

creases (Figure 13–21b●). This decrease reduces resistance to the movement under way. For example, if your elbow is flexed and you let gravity extend it, the triceps brachii muscle, which is compressed by this movement, relaxes.

Many stretch reflexes are **postural reflexes**—reflexes that help us to maintain a normal upright posture. For example, standing involves a cooperative effort on the part of many muscle groups. Some of these muscles work in opposition to one another, exerting forces that keep the body's weight balanced over the feet. If the body leans forward, stretch receptors in the calf muscles are stimulated. Those muscles then respond by contracting, thereby returning the body to an upright position. If the muscles overcompensate and the body begins to lean back, the calf muscles relax. But then stretch receptors in muscles of the shins and thighs are stimulated, and the problem is corrected immediately.

Postural muscles generally have a firm muscle tone and extremely sensitive stretch receptors. As a result, very fine adjustments are continually being made, and you are not aware of the cycles of contraction and relaxation that occur. Stretch reflexes are only one type of postural reflex; there are many complex polysynaptic postural reflexes.

Now that you understand the basic stretch reflex, we can return to the role of the gamma efferents, which let the CNS adjust the sensitivity of muscle spindles. Gamma efferents play a vital role whenever voluntary contractions change the length of a muscle. For example, suppose you start to flex your right elbow by contracting the biceps brachii muscle. From the previous description of the stretch reflex, you would expect that as the biceps brachii muscle contracts, the extrafusal fibers would shorten and the muscle spindles would be compressed. This would be a problem, since the compression would produce a reflexive decrease in muscle tone that would make it difficult for you to complete the movement. However, no such compression actually occurs. Instead, impulses arriving over gamma efferents cause the contraction of myofibrils in the intrafusal fibers as the biceps brachii muscle shortens. The myofibrils pull on the sarcolemma in the central portion of the intrafusal fiber—the region monitored by the sensory neuron—until that membrane is stretched to its normal resting length. As a result, the muscle spindles remain sensitive to any externally imposed changes in muscle length. For instance, if someone drops a ball into your palm when your elbow is partially flexed, the muscle spindles will automatically adjust the muscle tone to compensate for the increased load.

✓ What is the minimum number of neurons in a reflex arc?

✓ One of the first somatic reflexes to develop is the suckling reflex. Which type of reflex is this?

✓ How would the stimulation of the muscle spindles involved in the patellar (knee-jerk) reflex by gamma motor neurons affect the speed of the reflex?

Answers start on page Q-1

☐ POLYSYNAPTIC REFLEXES

Polysynaptic reflexes can produce far more complicated responses than can monosynaptic reflexes. One reason is that the interneurons involved can control several muscle groups. Moreover, these interneurons may produce either excitatory or inhibitory postsynaptic potentials (EPSPs or IPSPs) at CNS motor nuclei, so the response can involve the stimulation of some muscles and the inhibition of others.

The Tendon Reflex

The stretch reflex regulates the length of a skeletal muscle. The **tendon reflex** monitors the external tension produced during a muscular contraction and prevents tearing or breaking of the tendons. The sensory receptors for this reflex have not been identified, but they are distinct from both muscle spindles and proprioceptors in tendons. The receptors are stimulated when the collagen fibers are stretched to a dangerous degree. In the spinal cord, these neurons stimulate inhibitory interneurons that innervate the motor neurons controlling the skeletal muscle. The greater the tension in the tendon, the greater the inhibitory effect on the motor neurons. As a result, skeletal muscles generally cannot develop enough tension to break their tendons.

Withdrawal Reflexes

Withdrawal reflexes move affected parts of the body away from a source of stimulation. The strongest withdrawal reflexes are triggered by painful stimuli, but these reflexes are sometimes initiated by the stimulation of touch receptors or pressure receptors.

The **flexor reflex**, a representative withdrawal reflex, affects the muscles of a limb (Figure 13–22a●). Recall from Chapters 9 and 11 that flexion is a reduction in the angle between two articulating bones and that the contractions of flexor muscles perform this movement. ∞ pp. 269, 339 When you step on a tack, a dramatic flexor reflex is produced in the affected leg. When the pain receptors in your foot are stimulated, the sensory neurons activate interneurons in the spinal cord that stimulate motor neurons in the anterior gray horns. The result is a contraction of flexor muscles that yanks your foot off the ground.

When a specific muscle contracts, opposing muscles must relax to permit the movement. For example, the flexor muscles that bend the knee are opposed by extensor muscles that straighten it out. A potential conflict exists: In theory, the contraction of a flexor muscle should trigger a stretch reflex in the extensors that would cause them to contract, opposing the movement. Interneurons in the spinal cord prevent such competition through **reciprocal inhibition**. When one set of motor neurons is stimulated, those neurons that control antagonistic muscles are inhibited. The term *reciprocal* refers to the fact that the system works both ways: When the flexors contract, the extensors relax; when the extensors contract, the flexors relax.

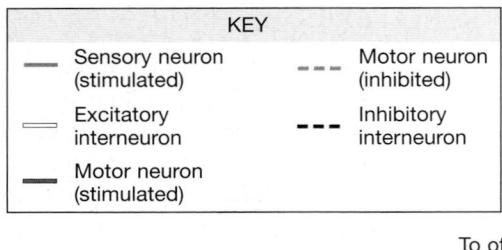

KEY

——— Sensory neuron (stimulated)	- - - Motor neuron (inhibited)
▭ Excitatory interneuron	- - - Inhibitory interneuron
▬ Motor neuron (stimulated)	

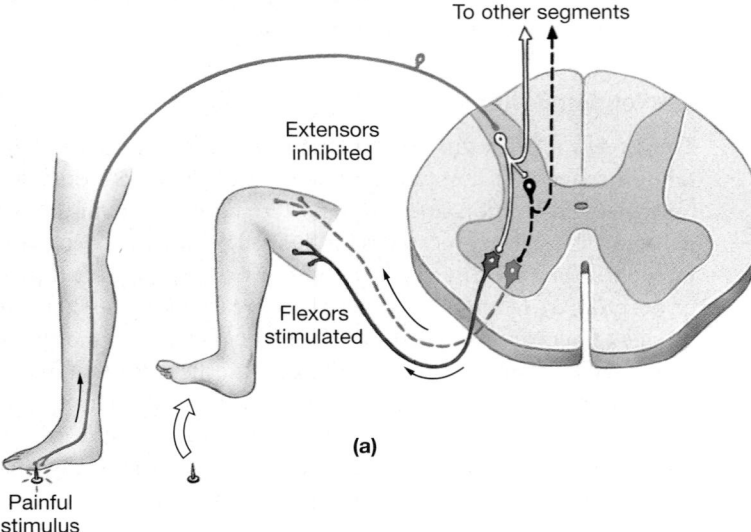

To other segments

Extensors
inhibited

Flexors
stimulated

(a)

Painful
stimulus

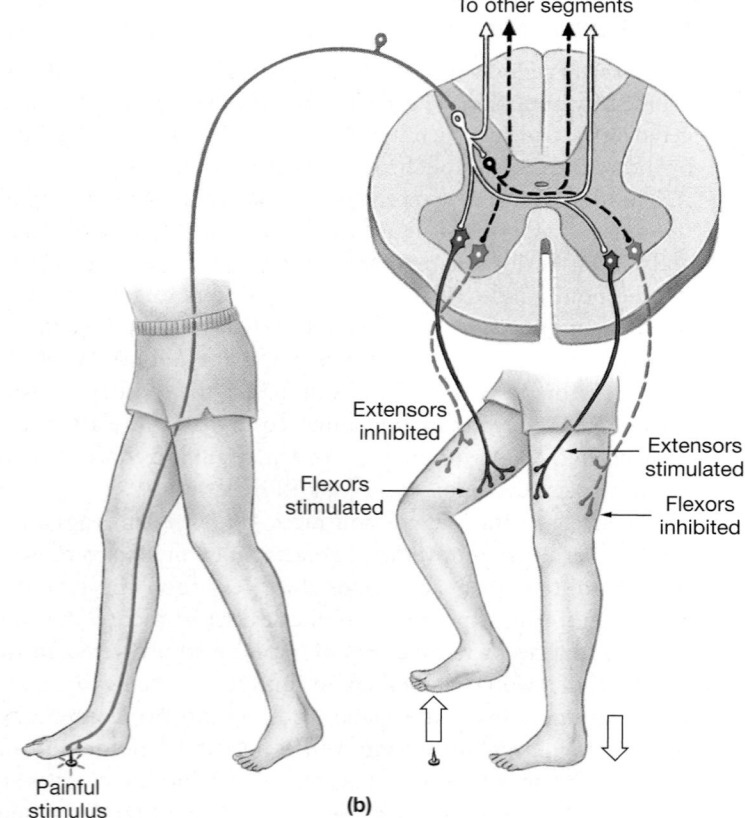

To other segments

Extensors
inhibited

Flexors
stimulated

Extensors
stimulated

Flexors
inhibited

Painful
stimulus

(b)

●**FIGURE 13–22**
The Flexor and Crossed Extensor Reflexes. **(a)** The flexor reflex, an example of a withdrawal reflex. **(b)** The crossed extensor reflex.

Withdrawal reflexes are much more complex than any monosynaptic reflex. They also show tremendous versatility, because the sensory neurons activate many pools of interneurons. If the stimuli are strong, interneurons will carry excitatory and inhibitory impulses up and down the spinal cord, affecting motor neurons in many segments. The end result is always the same: a coordinated movement away from the stimulus. But the distribution of the effects and the strength and character of the motor responses depend on the intensity and location of the stimulus. When you step on something sharp, mild discomfort might provoke a brief contraction in muscles of your ankle and foot. More powerful stimuli would produce coordinated muscular contractions affecting the positions of your ankle, foot, and leg. Severe pain would also stimulate contractions of your shoulder, trunk, and arm muscles. These contractions could persist for several seconds, owing to the activation of reverberating circuits. In contrast, monosynaptic reflexes are relatively invariable and brief; the patellar reflex is completed in roughly 20 msec.

Crossed Extensor Reflexes

The stretch, tendon, and withdrawal reflexes are *ipsilateral reflex arcs* (*ipsi*, same + *lateral*, side): The sensory stimulus and the motor response occur on the same side of the body. The **crossed extensor reflex** (Figure 13–22b●) is called a *contralateral reflex arc* (*contra*, opposite), because the motor response occurs on the side opposite the stimulus.

The crossed extensor reflex complements the flexor reflex, and the two occur simultaneously. When you step on a tack, while the flexor reflex pulls the affected foot away from the ground, the crossed extensor reflex straightens the other leg to support your body weight. In the crossed extensor reflex, the axons of interneurons responding to the pain cross to the other side of the spinal cord and stimulate motor neurons that control the extensor muscles of the uninjured leg. Your opposite leg straightens to support the shifting weight. Reverberating circuits use positive feedback to ensure that the movement lasts long enough to be effective, despite the absence of motor commands from higher centers of the brain.

Polysynaptic reflexes range in complexity from a simple tendon reflex to the complex and variable reflexes associated with standing, walking, and running. Yet all polysynaptic reflexes share the same five basic characteristics:

1. *They Involve Pools of Interneurons.* Processing occurs in pools of interneurons before motor neurons are activated. The result may be excitation or inhibition; the tendon reflex produces inhibition of motor neurons, whereas the flexor and crossed extensor reflexes direct specific muscle contractions.

2. *They Are Intersegmental in Distribution.* The interneuron pools extend across spinal segments and may activate muscle groups in many parts of the body.

3. *They Involve Reciprocal Inhibition.* Reciprocal inhibition coordinates muscular contractions and reduces resistance to movement. In the flexor and crossed extensor reflexes, the contraction of one muscle group is associated with the inhibition of opposing muscles.

4. *They Have Reverberating Circuits, Which Prolong the Reflexive Motor Response.* Positive feedback between interneurons that innervate motor neurons and the processing pool maintains the stimulation even after the initial stimulus has faded.

5. *Several Reflexes May Cooperate to Produce a Coordinated, Controlled Response.* As a reflex movement gets under way, antagonistic reflexes are inhibited. For example, during the stretch reflex, antagonistic muscles are inhibited; in the tendon reflex, antagonistic muscles are stimulated. In complex polysynaptic reflexes, commands may be distributed along the length of the spinal cord, producing a well-coordinated response.

13–6 INTEGRATION AND CONTROL OF SPINAL REFLEXES

Objective

- Explain how higher centers control and modify reflex responses.

Reflex motor behaviors occur automatically, without instructions from higher centers. However, higher centers can have a profound effect on the performance of a reflex. Processing centers in the brain can facilitate or inhibit reflex motor patterns based in the spinal cord. Descending tracts originating in the brain synapse on interneurons and motor neurons throughout the spinal cord. These synapses are continuously active, producing EPSPs or IPSPs at the postsynaptic membrane.

☐ VOLUNTARY MOVEMENTS AND REFLEX MOTOR PATTERNS

Spinal reflexes produce consistent, stereotyped motor patterns that are triggered by specific external stimuli. However, the same motor patterns can also be activated as needed by centers in the brain. By making use of these preexisting patterns, relatively few descending fibers can control complex motor functions. For example, the motor patterns for walking, running, and jumping are directed primarily by neuronal pools in the spinal cord. The descending pathways from the brain provide appropriate facilita-

tion, inhibition, or "fine-tuning" of the established patterns. This is a very efficient system that is similar to a "Macro" in word processing: A single command triggers a complex, predetermined sequence of events.

When complicated voluntary movements are under way, the neurons involved with spinal reflexes assist with muscular coordination and control. For instance, the descending tracts that stimulate motor neurons controlling the biceps brachii muscle, brachialis muscle, and other elbow flexors also stimulate (1) the gamma efferents to the intrafusal fibers in these muscles and (2) inhibitory interneurons to relax antagonistic muscle groups, such as the triceps brachii muscle. The gamma efferents regulate the sensitivity of the stretch reflex, and the interneurons are the same as those activated in the withdrawal and crossed extensor reflexes. As muscle contraction proceeds, tendon reflexes keep any tension that is produced within tolerable limits.

Motor control therefore involves a series of interacting levels. At the lowest level are monosynaptic reflexes that are rapid, but stereotyped and relatively inflexible. At the highest level are centers in the brain that can modulate or build on reflexive motor patterns.

☐ REINFORCEMENT AND INHIBITION

A single EPSP may not depolarize the postsynaptic neuron sufficiently to generate an action potential, but it does make that neuron more sensitive to other excitatory stimuli. This process of *facilitation* was introduced in Chapter 12. Alternatively, an IPSP will make the neuron less responsive to excitatory stimulation, through the process of *inhibition.* ∞ p. 419 By stimulating excitatory or inhibitory interneurons within the brain stem or spinal cord, higher centers can adjust the sensitivity of reflexes by creating EPSPs or IPSPs at the motor neurons involved in reflex responses.

When many of the excitatory synapses are chronically active, the postsynaptic neuron can enter a state of generalized facilitation. This facilitation of reflexes can result in **reinforcement**, an enhancement of spinal reflexes. For example, a voluntary effort to pull apart clasped hands elevates the general state of facilitation along the spinal cord, reinforcing all spinal reflexes. In testing reflexes, if a clinical stimulus fails to elicit a particular reflex response, there can be many reasons for the failure: The person may be consciously suppressing the response, the nerves involved may be damaged, or there may be underlying problems inside the CNS. The clinician may then ask the patient to perform an action designed to provide reinforcement. Reinforced reflexes are usually too strong to suppress consciously; if the reflex still fails to appear, the likelihood of nerve or CNS damage is increased, and more sophisticated tests, such as nerve conduction studies or scans may be ordered.

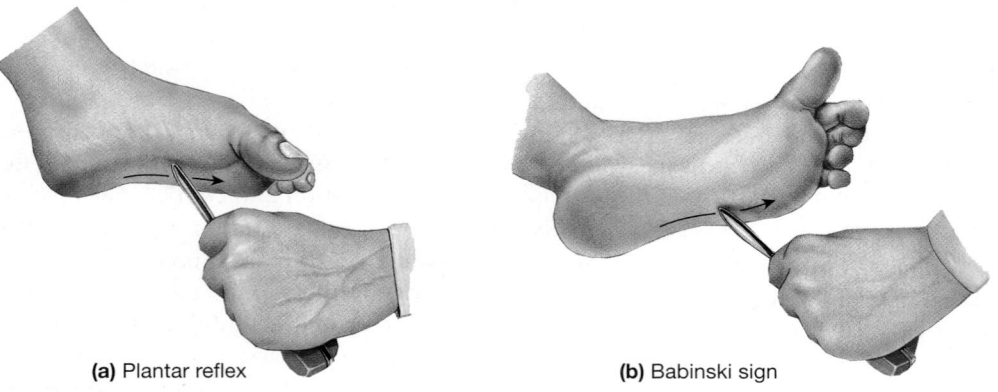

(a) Plantar reflex (b) Babinski sign

●**FIGURE 13–23**
The Babinski Reflexes. **(a)** The plantar reflex (negative Babinski reflex), a curling of the toes, is seen in healthy adults. **(b)** The Babinski sign (positive Babinski reflex) occurs in the absence of descending inhibition. It is normal in infants, but pathological in adults.

Other descending fibers have an inhibitory effect on spinal reflexes. In adults, stroking the foot on the side of the sole produces a curling of the toes, called a **plantar reflex**, or *negative Babinski reflex*, after about a 1-second delay (Figure 13–23a●). Stroking an infant's foot on the side of the sole produces a fanning of the toes known as the **Babinski sign**, or *positive Babinski reflex*. This response disappears as descending motor pathways develop. If either the higher centers or the descending tracts are damaged, the Babinski sign will reappear in an adult (Figure 13–23b●). As a result, this reflex is often tested if CNS injury is suspected.

✓ A weight lifter is straining to lift a 200-kg barbell. Shortly after he lifts it to chest height, his muscles appear to relax and he drops the barbell. Which reflex has occurred?

✓ During a withdrawal reflex, what happens to the limb on the side opposite the stimulus? What is this response called?

✓ After injuring her back, Tina exhibits a positive Babinski reflex. What does this imply about Tina's injury?

Answers start on page Q-1

Test your understanding of reflexes by visiting the **Companion Website:** Chapter 13/Reviewing Concepts/ Labeling.

Reflexes and Diagnostic Testing

A neurological examination evaluates the sensory, motor, behavioral, and cognitive functions of the nervous system. The *Applications Manual* discusses the techniques involved; they range from asking questions to monitoring brain function through sophisticated scanning procedures.

Many somatic reflexes can be assessed through careful observation and the use of simple tools. The procedures are easy to perform, and the results can provide valuable information about the location of damage to the spinal cord or spinal nerves. By testing a series of spinal and cranial reflexes, a physician can assess the function of sensory pathways and motor centers throughout the spinal cord and brain. Neurologists test many reflexes; only a few are so generally useful that physicians make them part of a standard physical examination. Representative examples are shown in Figure 13–24●

and Table 13–4 (pp. 457-458) lists somatic reflexes that can be used in this way.

The *abdominal reflex* (Figure 13–24a●) is a superficial reflex that is normally present in adults. In this reflex, a light stroking of the skin produces a reflexive twitch in the abdominal muscles that moves the navel toward the stimulus. The reflex is facilitated by descending commands; it disappears after descending tracts have been damaged. The *patellar reflex* (Figure 13–20a●), *biceps reflex* (Figure 13–24b●), *triceps reflex* (Figure 13–24c●), and *ankle-jerk reflex* (Figure 13–24d●) are stretch reflexes controlled by specific segments of the spinal cord. Testing these reflexes provides information about the corresponding spinal segments. For example, a normal patellar reflex indicates that spinal nerves and spinal segments L_2-L_4 are undamaged.

AM Abnormal Reflex Activity

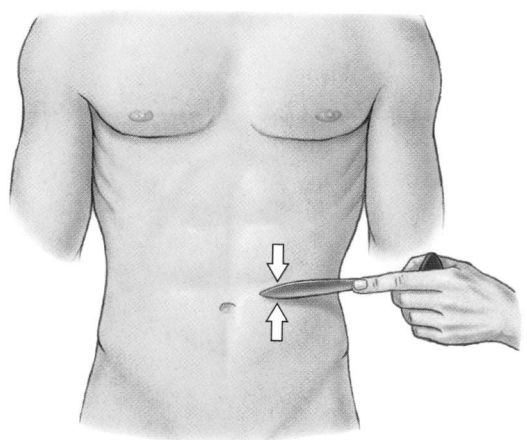

(a) Abdominal reflex

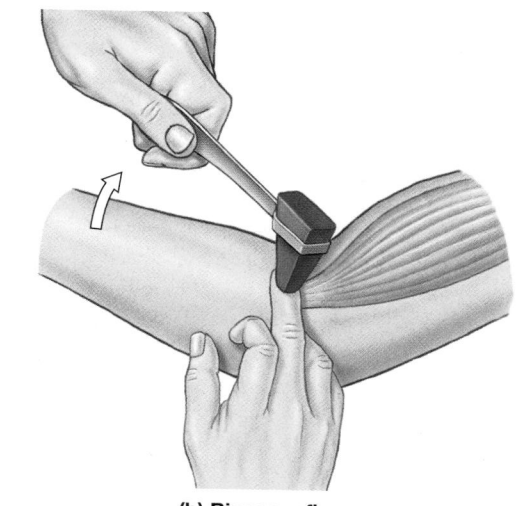

(b) Biceps reflex

•**FIGURE 13–24**
Somatic Reflexes. **(a)** The abdominal reflex, an example of a superficial reflex. The biceps reflex **(b)**, triceps reflex **(c)**, and ankle jerk reflex **(d)** are examples of stretch reflexes.

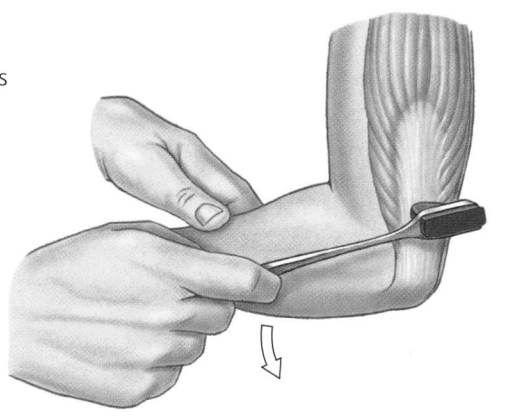

(c) Triceps reflex

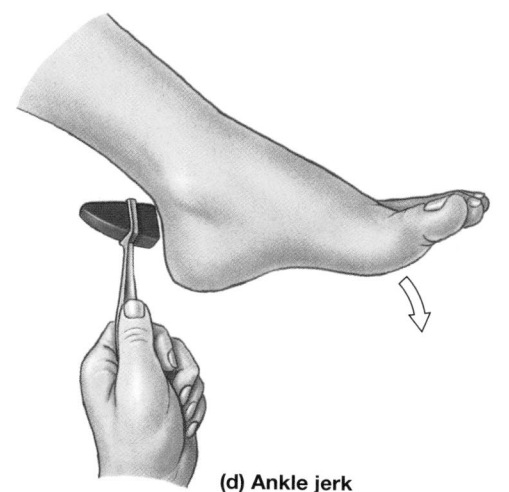

(d) Ankle jerk

TABLE 13–4 REFLEXES USED IN DIAGNOSTIC TESTING

Reflex	Stimulus	Afferent Nerve(s)	Spinal Segment	Efferent Nerve(s)	Normal Response
SUPERFICIAL REFLEXES					
Abdominal reflex (*Figure 13–24a*)	Light stroking of skin of abdomen	T_7–T_{12}, depending on region stroked	T_7–T_{12}, at level of arrival	Same as afferent	Contraction of abdominal muscles that pulls navel toward the stimulus
Cremasteric reflex	Stroking of skin of upper thigh	Femoral nerve	L_1	Genitofemoral nerve	Contraction of cremaster, elevation of scrotum
Plantar reflex (*Figure 13–23*)	Longitudinal stroking of sole of foot	Tibial nerve	S_1, S_2	Tibial nerve	Flexion at toe joints
Anal reflex	Stroking of region around the anus	Pudendal nerve	S_4, S_5	Pudendal nerve	Constriction of external anal sphincter

TABLE 13–4 (continued) REFLEXES USED IN DIAGNOSTIC TESTING

Reflex	Stimulus	Afferent Nerve(s)	Spinal Segment	Efferent Nerve(s)	Normal Response
STRETCH REFLEXES					
Biceps reflex (*Figure 13–24b*)	Tap to tendon of biceps brachii muscle near its insertion	Musculocutaneous nerve	C_5, C_6	Musculocutaneous nerve	Flexion at elbow
Triceps reflex (*Figure 13–24c*)	Tap to tendon of triceps brachii muscle near its insertion	Radial nerve	C_6, C_7	Radial nerve	Extension at elbow
Brachioradialis reflex	Tap to forearm near styloid process of the radius	Radial nerve	C_5, C_6	Radial nerve	Flexion at elbow, supination, and flexion at finger joints
Patellar reflex (*Figure 13–20a*)	Tap to patellar tendon	Femoral nerve	L_2–L_4	Femoral nerve	Extension at knee
Ankle-jerk reflex (*Figure 13–24d*)	Tap to calcaneal tendon	Tibial nerve	S_1, S_2	Tibial nerve	Extension (plantar flexion) at ankle

Chapter Review

SELECTED CLINICAL TERMINOLOGY

Terms Discussed in This Chapter

Babinski sign (*positive Babinski reflex*): A spinal reflex in infants, consisting of a fanning of the toes and produced by stroking the foot on the side of the sole; in adults, an abnormal plantar reflex that indicates CNS injury. (*p. 458*)

epidural block: The injection of anesthetic into the epidural space to eliminate sensory and motor innervation via spinal nerves in the area of injection. (*p. 434* and *AM*)

lumbar puncture: A spinal tap performed between adjacent lumbar vertebrae inferior to the conus medullaris. (*p. 435*)

meningitis: An inflammation of the meninges involving the spinal cord (*spinal meningitis*) or the brain (*cerebral meningitis*); generally caused by bacterial or viral pathogens. (*p. 434* and *AM*)

myelography: A diagnostic procedure in which a radiopaque dye is introduced into the cerebrospinal fluid to obtain an X ray of the spinal cord and cauda equina. (*p. 435*)

nerve growth factor: A peptide factor that promotes the growth and maintenance of neurons. Other factors that are important to neuron growth and repair include BDNF, NT-3, NT-4, and GAP-43. (*p. 439* and *AM*)

nerve palsies (peripheral neuropathies): Regional losses of sensory and motor function as a result of nerve trauma or compression. Common palsies include *radial nerve palsy, ulnar palsy, sciatica*, and *fibular palsy*. (*p. 440* and *AM*)

paraplegia: Paralysis involving a loss of motor control of the lower, but not the upper, limbs. (*p.438*)

patellar reflex (knee-jerk reflex): A stretch reflex resulting from the stimulation of stretch receptors in the quadriceps muscles. (*p. 451*)

plantar reflex (*negative Babinski reflex*): A spinal reflex in adults, consisting of a curling of the toes and produced by stroking the foot on the side of the sole. (*p. 455*)

quadriplegia: Paralysis involving the loss of sensation and motor control of the upper and lower limbs. (*p. 438* and *AM*)

shingles: A condition caused by the infection of neurons in dorsal root ganglia by the virus *Herpes varicella-zoster*. The primary symptom is a painful rash along the sensory distribution of the affected spinal nerves. (*p. 440* and *AM*)

spinal tap: A procedure in which cerebrospinal fluid is removed from the subarachnoid space through a needle, generally inserted between the lumbar vertebrae. (*p. 435*)

AM Additional Terms Discussed in the *Applications Manual*

abdominal reflex
areflexia
clonus
Hansen's disease
hyperreflexia
hyporeflexia
mass reflex
paresthesia
sciatica
spinal shock

STUDY OUTLINE

13–1 INTRODUCTION ORGANIZATION OF THE NERVOUS SYSTEM   p. 430

1. The CNS consists of the brain and spinal cord, with the remainder of the nervous tissue forming the PNS. *(Figure 13–2)*
2. In the CNS white matter, axons are arranged in **tracts** and **columns**.

13–2 GROSS ANATOMY OF THE SPINAL CORD   p. 432

1. The adult spinal cord includes localized **enlargements**, which provide innervation to the limbs. The spinal cord has 31 segments, each associated with a pair of **dorsal roots** and a pair of **ventral roots**. *(Figure 13–4)*
2. The **filum terminale** (a strand of fibrous tissue), which originates at the **conus medullaris**, ultimately becomes part of the *coccygeal ligament*. *(Figure 13–4)*
3. **Spinal nerves** are **mixed nerves**: They contain both afferent (sensory) and efferent (motor) fibers.

SPINAL MENINGES   p. 433

4. The **spinal meninges** provide physical stability and shock absorption for neural tissues of the spinal cord; the **cranial meninges** surround the brain. *(Figure 13–5)*
5. The **dura mater** covers the spinal cord; inferiorly, it tapers into the **coccygeal ligament**. The **epidural space** separates the dura mater from the walls of the vertebral canal. *(Figure 13–5)*
6. Interior to the inner surface of the dura mater are the **subdural space,** the **arachnoid** (the second meningeal layer), and the **subarachnoid space**. The subarachnoid space contains **cerebrospinal fluid (CSF)**, which acts as a shock absorber and a diffusion medium for dissolved gases, nutrients, chemical messengers, and waste products. *(Figures 13–5, 13–7)*
7. The **pia mater**, a meshwork of elastic and collagen fibers, is the innermost meningeal layer. **Denticulate ligaments** extend from the pia mater to the dura mater. *(Figures 13–5, 13–6)*

SECTIONAL ANATOMY OF THE SPINAL CORD   p. 436

8. The white matter of the spinal cord contains myelinated and unmyelinated axons, whereas the gray matter contains cell bodies of neurons and neuroglia and unmyelinated axons. The projections of gray matter toward the outer surface of the cord are called **horns**. *(Figure 13–5)*
9. The **posterior gray horns** contain somatic and visceral sensory nuclei; nuclei in the **anterior gray horns** deal with somatic motor control. The **lateral gray horns** contain visceral motor neurons. The **gray commissures** contain axons that cross from one side of the spinal cord to the other. *(Figure 13–5)*
10. The white matter can be divided into six **columns** *(funiculi)*, each of which contains **tracts** *(fasciculi)*. **Ascending tracts** relay information from the spinal cord to the brain, and **descending tracts** carry information from the brain to the spinal cord. *(Figure 13–8)*

   Gross anatomy of the spinal cord: **Anatomy CD-ROM:** Nervous System/Spinal Cord Dissections.

13–3 SPINAL NERVES   p. 438

1. There are 31 pairs of spinal nerves. Each has an **epineurium** (outermost layer), a **perineurium**, and an **endoneurium** (innermost layer). *(Figure 13–9)*

PERIPHERAL DISTRIBUTION OF SPINAL NERVES   p. 439

2. A typical spinal nerve has a **white ramus** (containing myelinated axons), a **gray ramus** (containing unmyelinated fibers that innervate glands and smooth muscles in the body wall or limbs), a **dorsal ramus** (providing sensory and motor innervation to the skin and muscles of the back), and a **ventral ramus** (supplying the ventrolateral body surface, structures in the body wall, and the limbs). Each pair of nerves monitors a region of the body surface called a **dermatome**. *(Figures 13–10, 13–11)*

NERVE PLEXUSES   p. 440

3. A complex, interwoven network of nerves is a **nerve plexus**. The four large plexuses are the **cervical plexus**, the **brachial plexus**, the **lumbar plexus**, and the **sacral plexus**. *(Figures 13–12 to 13–15; Tables 13–1 to 13–3)*

13–4 PRINCIPLES OF FUNCTIONAL ORGANIZATION   p. 446

1. The body has sensory neurons, which deliver information to the CNS; motor neurons, which distribute commands to peripheral effectors; and interneurons, which interpret information and coordinate responses.

NEURONAL POOLS   p. 446

2. A functional group of interconnected neurons is a **neuronal pool**.
3. The neural circuit patterns are **divergence**, **convergence**, **serial processing, parallel processing**, and **reverberation**. *(Figure 13–15)*

AN INTRODUCTION TO REFLEXES   p. 448

4. **Reflexes** are rapid, automatic responses to stimuli. A *neural reflex* involves sensory fibers delivering information to the CNS and motor fibers carrying commands to the effectors via the PNS.
5. A **reflex arc** is the neural "wiring" of a single reflex. *(Figure 13–16)*
6. The five steps involved in a neural reflex are (1) the arrival of a stimulus and activation of a receptor, (2) the activation of a sensory neuron, (3) information processing, (4) the activation of a motor neuron, and (5) a response by an effector. *(Figure 13–16)*
7. Reflexes are classified according to (1) their development, (2) the site of information processing, (3) the nature of the resulting motor response, and (4) the complexity of the neural circuit. *(Figure 13–17)*
8. **Innate reflexes** result from the connections that form between neurons during development. **Acquired reflexes** are learned and typically are more complex.
9. Reflexes processed in the brain are **cranial reflexes**. In a **spinal reflex**, the important interconnections and processing events occur in the spinal cord.
10. **Somatic reflexes** control skeletal muscles; **visceral reflexes** *(autonomic reflexes)* control the activities of other systems.
11. In a **monosynaptic reflex**—the simplest reflex arc—a sensory neuron synapses directly on a motor neuron, which acts as the processing center. In a **polysynaptic reflex**, which has at least one interneuron between the sensory afferent and the motor efferent, there is a longer delay between stimulus and response. *(Figure 13–18)*

13-5 SPINAL REFLEXES   p. 451

1. Spinal reflexes range from simple monosynaptic reflexes to more complex polysynaptic and **intersegmental reflexes**, in which many segments interact to produce a coordinated motor response.

MONOSYNAPTIC REFLEXES   p. 451

2. The **stretch reflex** (such as the **patellar**, or **knee-jerk**, **reflex**) is a monosynaptic reflex that automatically regulates skeletal muscle length and muscle tone. The sensory receptors involved are **muscle spindles**. (*Figures 13–19 to 13–21*)

3. A **postural reflex** maintains one's normal upright posture.

Spinal cord and related plexuses: **Companion Website:** Chapter 13/Reviewing Facts & Terms/Multiple Choice.

POLYSYNAPTIC REFLEXES   p. 453

4. Polysynaptic reflexes can produce more complicated responses than can monosynaptic reflexes. Examples include the **tendon reflex** (which monitors the tension produced during muscular contractions and prevents damage to tendons) and **withdrawal reflexes** (which move affected portions of the body away from a source of stimulation). The **flexor reflex** is a withdrawal reflex affecting the muscles of a limb. The **crossed extensor reflex** complements withdrawal reflexes. (*Figure 13–22*)

5. All polysynaptic reflexes (1) involve pools of interneurons, (2) are intersegmental in distribution, (3) involve reciprocal inhibition, and (4) have reverberating circuits, which prolong the reflexive motor response. (5) Several reflexes may cooperate to produce a coordinated response.

13-6 INTEGRATION AND CONTROL OF SPINAL REFLEXES   p. 455

1. The brain can facilitate or inhibit reflex motor patterns based in the spinal cord.

VOLUNTARY MOVEMENTS AND REFLEX MOTOR PATTERNS   p. 455

2. Motor control involves a series of interacting levels. Monosynaptic reflexes form the lowest level; at the highest level are the centers in the brain that can modulate or build on reflexive motor patterns.

REINFORCEMENT AND INHIBITION   p. 455

3. Facilitation can produce an enhancement of spinal reflexes known as **reinforcement**. Spinal reflexes may also be inhibited, as when the **plantar reflex** in adults replaces the **Babinski sign** in infants. (*Figure 13–23*)

Reflexes: **Companion Website:** Chapter 13/Reviewing Concepts/Labeling.

REVIEW QUESTIONS

More assessment questions are available to you on the Companion Website. You will find Matching, Multiple Choice, True/False, and other quizzes to help further your understanding of the material covered in this chapter. To access the site, go to www.aw.com/martini.

LEVEL 1 Reviewing Facts and Terms

1. The expanded area of the spinal cord that supplies nerves to the pectoral girdle and arms is the
   (a) conus medullaris
   (b) filum terminale
   (c) lumbar enlargement
   (d) cervical enlargement

2. The ventral roots of each spinal segment
   (a) bring sensory information into the spinal cord
   (b) control peripheral effectors
   (c) contain the axons of somatic motor and visceral motor neurons
   (d) both b and c are correct

3. Spinal nerves are called mixed nerves because they
   (a) contain sensory and motor fibers
   (b) exit at intervertebral foramina
   (c) are associated with a pair of dorsal root ganglia
   (d) are associated with dorsal and ventral roots

4. The _____ extend from the pia mater to the dura mater, preventing lateral movement of the spinal cord.
   (a) dorsal roots
   (b) ventral roots
   (c) denticulate ligaments
   (d) coccygeal ligaments

5. The tough, fibrous outermost covering of the spinal cord is the
   (a) arachnoid
   (b) pia mater
   (c) dura mater
   (d) epidural block

6. The white matter of the spinal cord is dominated by
   (a) unmyelinated axons
   (b) cell bodies of neurons, neuroglia, and unmyelinated axons
   (c) Schwann cells and satellite cells
   (d) myelinated axons

7. The outermost layer of connective tissue that surrounds each spinal nerve is the
   (a) perineurium
   (b) epineurium
   (c) endoneurium
   (d) epimysium

8. A sensory region monitored by the dorsal rami of a single spinal segment is
   (a) a ganglion
   (b) a fascicle
   (c) a dermatome
   (d) a ramus

9. The major nerve of the cervical plexus that innervates the diaphragm is the
   (a) median nerve
   (b) axillary nerve
   (c) phrenic nerve
   (d) fibular nerve

10. The genitofemoral, femoral, and lateral femoral cutaneous nerves are major nerves of the
    (a) lumbar plexus
    (b) sacral plexus
    (c) brachial plexus
    (d) cervical plexus

11. The synapsing of several neurons on the same postsynaptic neuron is called
    (a) serial processing
    (b) reverberation
    (c) divergence
    (d) convergence

12. The reflexes that control the most-rapid, stereotyped motor responses of the nervous system to stimuli are
    (a) monosynaptic reflexes
    (b) polysynaptic reflexes
    (c) tendon reflexes
    (d) extensor reflexes

13. An example of a stretch reflex triggered by passive muscle movement is the
    (a) tendon reflex
    (b) patellar reflex
    (c) flexor reflex
    (d) ipsilateral reflex

14. The contraction of flexor muscles and the relaxation of extensor muscles illustrates the principle of
    (a) reverberating circuitry   (b) generalized facilitation
    (c) reciprocal inhibition   (d) reinforcement

15. Reflex arcs in which the sensory stimulus and the motor response occur on the same side of the body are
    (a) contralateral   (b) paraesthetic
    (c) ipsilateral   (d) monosynaptic

16. All polysynaptic reflexes
    (a) have reverberating circuits that prolong the reflexive motor response
    (b) involve pools of interneurons
    (c) involve reciprocal inhibition
    (d) a, b, and c are correct

17. Proceeding inward from the outermost layer, number the following in the correct sequence:
    (a) _____ walls of vertebral   (b) _____ pia mater
    canal   (d) _____ arachnoid membrane
    (c) _____ dura mater   (f) _____ subarachnoid space
    (e) _____ subdural space   (h) _____ spinal cord
    (g) _____ epidural space

18. Beginning with the outermost layer, list the connective-tissue layers that surround each spinal nerve.

LEVEL 2 Reviewing Concepts

19. Explain the anatomical significance of the fact that spinal cord growth ceases at age four.

20. List, in sequence, the five steps involved in a neural reflex.

21. Polysynaptic reflexes can produce far more complicated responses than can monosynaptic reflexes because
    (a) the response time is quicker
    (b) the response is initiated by highly sensitive receptors
    (c) motor neurons carry impulses at a faster rate than do sensory neurons
    (d) the interneurons involved can control several muscle groups

22. Why do cervical nerves outnumber cervical vertebrae?

23. If the anterior gray horns of the spinal cord were damaged, what type of control would be affected?

24. List all areas of the CNS where cerebrospinal fluid (CSF) is located. What are the functions of CSF?

25. What five characteristics are common to all polysynaptic reflexes?

26. Predict the effects on the body of a spinal cord transection at C_7. How would these effects differ from those of a spinal cord transection at T_{10}?

27. Suppose that you feel something brushing against the following parts of your body:
    (a) abdomen   (b) left forearm
    (c) upper back   (d) right ankle
    For each case, indicate which ramus would be involved.

28. Which plexus has the primary responsibility for innervating the following muscles?
    (a) diaphragm   (b) pelvic diaphragm
    (c) deltoid   (d) gastrocnemius

29. Why is response time in a monosynaptic reflex much faster than response time in a polysynaptic reflex?

30. What would happen if the dorsal root of a spinal nerve were damaged or transected?

31. How does the stimulation of a sensory neuron that innervates an extrafusal muscle fiber affect muscle tone?

32. Why is it important that a spinal tap be done between the third and fourth lumbar vertebrae?

LEVEL 3 Critical Thinking and Clinical Applications

33. Mary complains that when she wakes up in the morning, her thumb and forefinger are always "asleep." She mentions this condition to her physician, who asks Mary whether she sleeps with her wrists flexed. She replies that she does. The physician tells Mary that sleeping in that position may compress a portion of one of her peripheral nerves, producing her symptoms. Which nerve is involved?

34. The improper use of crutches can produce a condition known as "crutch paralysis," characterized by a lack of response by the extensor muscles of the arm, and a condition known as "wrist drop," consisting of an inability to extend the fingers and wrist. Which nerve is involved?

35. During childbirth, some women are given a local anesthetic that temporarily deadens sensory neurons in the region of the genitals. Which nerve does this anesthetic affect?

36. While playing football, Ramón is tackled hard, and as he tries to get up, he finds that he cannot flex his left thigh or extend his left knee. Which nerve innervating the lower limb may be damaged, and how would the damage affect sensory perception in Ramón's left leg?

UNIT 3   CHAPTER   12   13   **14**   15   16   17   18

CLINICAL NOTES
- Problems with the Circulation of CSF 471
- Cerebrovascular Disease 472
- Penetrating the Blood–Brain and Blood–CSF Barriers 472
- Parkinson's Disease 487

CLINICAL DISCUSSIONS
- Epidural and Subdural Hemorrhages 469
- Hydrocephalus 471
- Aphasia and Dyslexia 490
- Disconnection Syndrome 492

14

THE BRAIN AND CRANIAL NERVES

The brain is probably the most fascinating organ in the body, yet we know relatively little about its structural and functional complexities. What we do know is that all our dreams, passions, plans, and memories result from brain activity. The brain contains tens of billions of neurons organized into hundreds of neuronal pools; it has a complex three-dimensional structure and performs a bewildering array of functions. If an individual's heart, liver, lung, or kidney stops working, heroic measures, including organ transplantation, can be taken to preserve the person's life. But if the brain permanently stops working, the person is classified as "brain dead," and for all intents and purposes ceases to exist.

The brain is far more complex than the spinal cord, and it can respond to stimuli with greater versatility. That versatility results from (1) the tremendous number of neurons and neuronal pools in the brain and (2) the complexity of the interconnections among these neurons and neuronal pools. Because the interconnections are complex and the pools are large, the response can be varied to meet changing circumstances. But this versatility has a price: A response cannot be immediate, precise, and adaptable all at the same time. Adaptability requires multiple processing steps, and every synapse adds to the delay between stimulus and response. One of the major functions of spinal reflexes is to provide an *immediate* response that can be fine-tuned or elaborated on by more versatile—but slower—processing centers in the brain. For example, the flexor reflex will pull your hand from a hot stove before you become consciously aware that something is wrong. ∞ p. 453 In effect, this spinal reflex prevents further injury and gives the brain time to decide what to do next.

In this chapter, we consider the brain and the cranial nerves and their primary functional roles. The chapter organization, which parallels that of Chapter 13, will make it easy for you to compare the functional organization of the brain with that of the spinal cord.

14-1 AN INTRODUCTION TO THE ORGANIZATION OF THE BRAIN

Objectives

- Name the major regions of the brain, and describe their functions.
- Name the three primary brain vesicles, and indicate which adult structures they give rise to.
- Name the ventricles of the brain, and describe their locations and the connections between them.

The adult human brain contains almost 98 percent of the body's neural tissue. A "typical" brain weighs 1.4 kg (3 lb) and has a volume of 1200 cc (71 in.^{3}). Brain size varies considerably among individuals. The brains of males are, on average, about 10 percent larger than those of females, owing to differences in average body size. No correlation exists between brain size and intelligence. Individuals with the smallest brains (750 cc) and the largest brains (2100 cc) are functionally normal.

☐ A PREVIEW OF MAJOR REGIONS AND LANDMARKS

The adult brain is dominated in size by the cerebrum (Figure 14–1●). Viewed from the anterior and superior surfaces, the **cerebrum** (SER-e-brum, or se-RĒ-brum) of the adult brain can be divided into large, paired **cerebral hemispheres**. The surfaces of the cerebral hemispheres are highly folded and covered by **neural cortex** (*cortex*, rind or bark), a superficial layer of gray matter. This *cerebral cortex* forms a series of elevated ridges, or **gyri** (JĪ-rī; singular, *gyrus*) that serve to increase its surface area. The gyri are separated by shallow depressions called **sulci** (SUL-sī) or by deeper grooves called **fissures**. The cerebrum is the seat of most higher mental functions. Conscious thoughts, sensations, intellect, memory, and complex movements all originate in the cerebrum.

The hemispheres of the **cerebellum** (ser-e-BEL-um) are partially hidden by the cerebral hemispheres, but this is the second-largest part of the brain. Like the cerebrum, the cerebellum is covered by a layer of cortical gray matter, the *cerebellar cortex*. The cerebellum adjusts ongoing movements on the basis of comparisons between arriving sensations and sensations experienced previously, allowing you to perform the same movements over and over.

The other major regions of the brain can best be examined after the cerebral hemispheres have been removed (Figure 14–2● p. 466). The walls of the **diencephalon** (dī-en-SEF-a-lon; *dia*, through) are composed of the **left thalamus** and **right thalamus** (THAL-a-mus; plural, *thalami*). Each thalamus contains relay and processing centers for sensory information. The **hypothalamus** (*hypo-*, below), or floor of the diencephalon, contains centers involved with emotions, autonomic function, and hormone production. The *infundibulum*, a narrow stalk, connects the hypothalamus to the **pituitary gland**, a component of the endocrine system. As we shall see in Chapter 18, the pituitary gland is the primary link between the nervous and endocrine systems.

The diencephalon is a structural and functional link between the cerebral hemispheres and the components of the brain stem. The **brain stem** contains a variety of important processing centers and nuclei that relay information headed to or from the cerebrum or cerebellum. It includes the *mesencephalon, pons,* and *medulla oblongata.**

---

*Some sources consider the brain stem to include the diencephalon. We will use the more restrictive definition here.

- The **mesencephalon** (*mesos*, middle), or midbrain, contains nuclei that process visual and auditory information and control reflexes triggered by these stimuli. For example, your immediate, reflexive responses to a loud, unexpected noise (eye movements and head turning) are directed by nuclei in the midbrain. This region also contains centers that help maintain consciousness.

- The term *pons* is Latin for "bridge"; the **pons** of the brain connects the cerebellum to the brain stem. In addition to tracts and relay centers, the pons also contains nuclei involved with somatic and visceral motor control.

- The spinal cord connects to the brain at the **medulla oblongata**. Near the pons, the posterior wall of the medulla oblongata is thin and membranous. The inferior portion of the medulla oblongata resembles the spinal cord in that it has a narrow central canal. The medulla oblongata relays sensory information to the thalamus and to centers in other portions of the brain stem. The medulla oblongata also contains major centers that regulate autonomic function, such as heart rate, blood pressure, and digestion.

The boundaries and general functions of the diencephalon and brain stem are listed in Figure 14–2●. When we consider the individual components of the brain, we will begin at the inferior portion of the medulla oblongata. This region has the simplest organization found anywhere in the brain, and in many respects it resembles the spinal cord. We will then ascend to regions of increasing structural and functional complexity until we reach the cerebral cortex, whose functions and capabilities are as yet poorly understood.

☐ EMBRYOLOGY OF THE BRAIN

To understand the internal organization of the adult brain, we must consider its embryological origins. The central nervous system (CNS) begins as a hollow tube known as the *neural tube*. This tube has a fluid-filled internal cavity, the *neurocoel*. In the cephalic portion of the neural tube, three areas enlarge rapidly through expansion of the neurocoel. This enlargement creates three prominent divisions called **primary brain vesicles**. The primary brain vesicles are named for their relative positions: the **prosencephalon** (prōz-en-SEF-a-lon; *proso*, forward + *enkephalos*, brain), or "forebrain"; the mesencephalon, or "midbrain"; and the **rhombencephalon** (rom-ben-SEF-a-lon), or "hindbrain."

The fate of the three primary divisions of the brain is summarized in Table 14–1. The prosencephalon and rhombencephalon are subdivided further, forming **secondary brain vesicles**. The prosencephalon forms the **telencephalon** (tel-en-SEF-a-lon; *telos*, end) and the diencephalon. The telencephalon will ultimately form the cerebrum of the adult brain. The walls of the mesencephalon thicken, and the neurocoel becomes a relatively narrow passageway, comparable to the central canal of the spinal cord. The portion of the rhombencephalon adjacent to the mesencephalon forms the **metencephalon** (met-en-SEF-a-lon; *meta*, after). The dorsal portion of the metencephalon will become the cerebellum, and the ventral portion will develop into the pons.

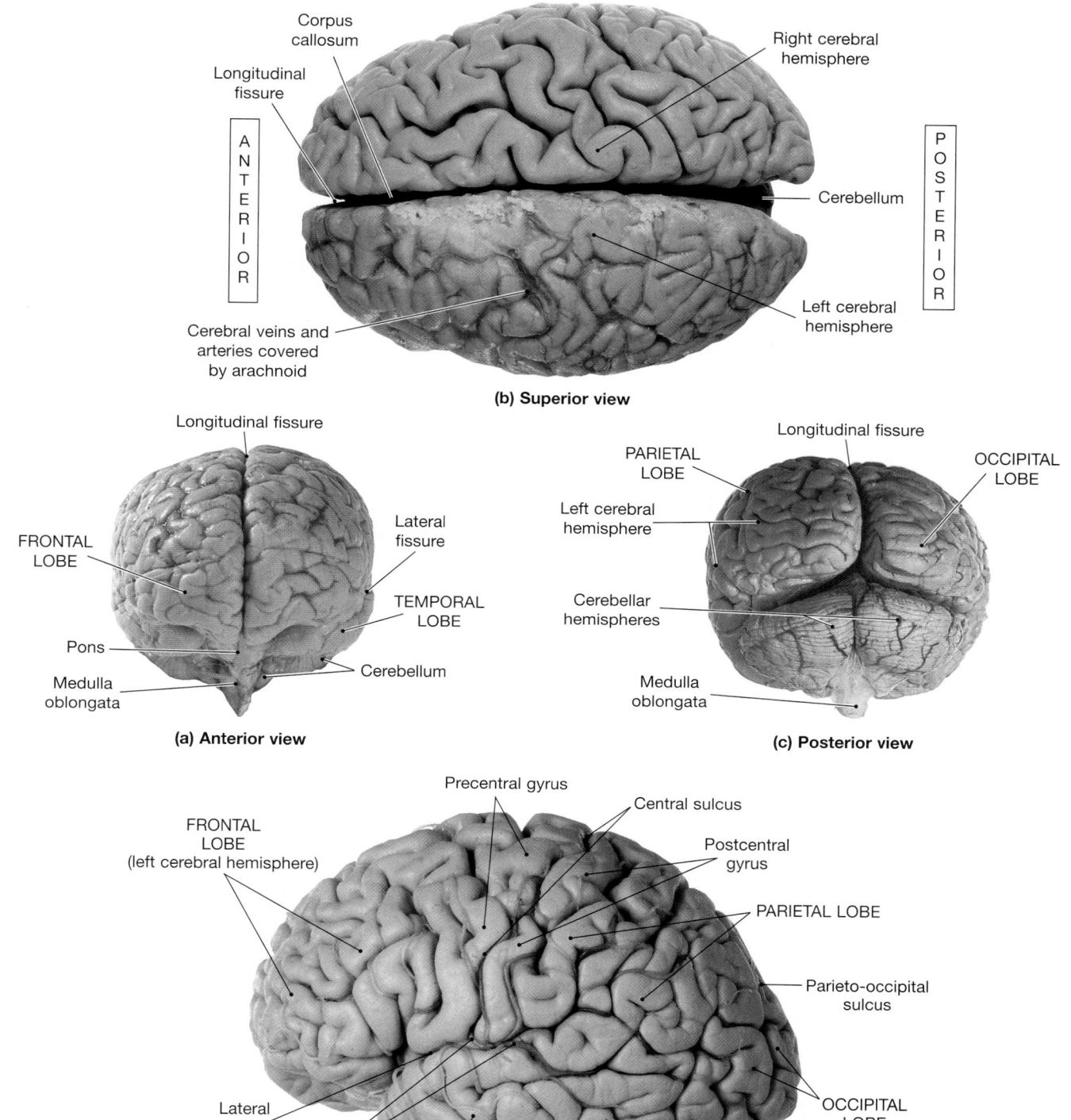

(b) **Superior view**

(a) **Anterior view**

(c) **Posterior view**

(d) **Lateral view**

●**FIGURE 14–1**
The Adult Brain. **(a)** An anterior view of the brain.
(b) A superior view, which is dominated by the paired cerebral
hemispheres. **(c)** A posterior view. **(d)** A lateral view.

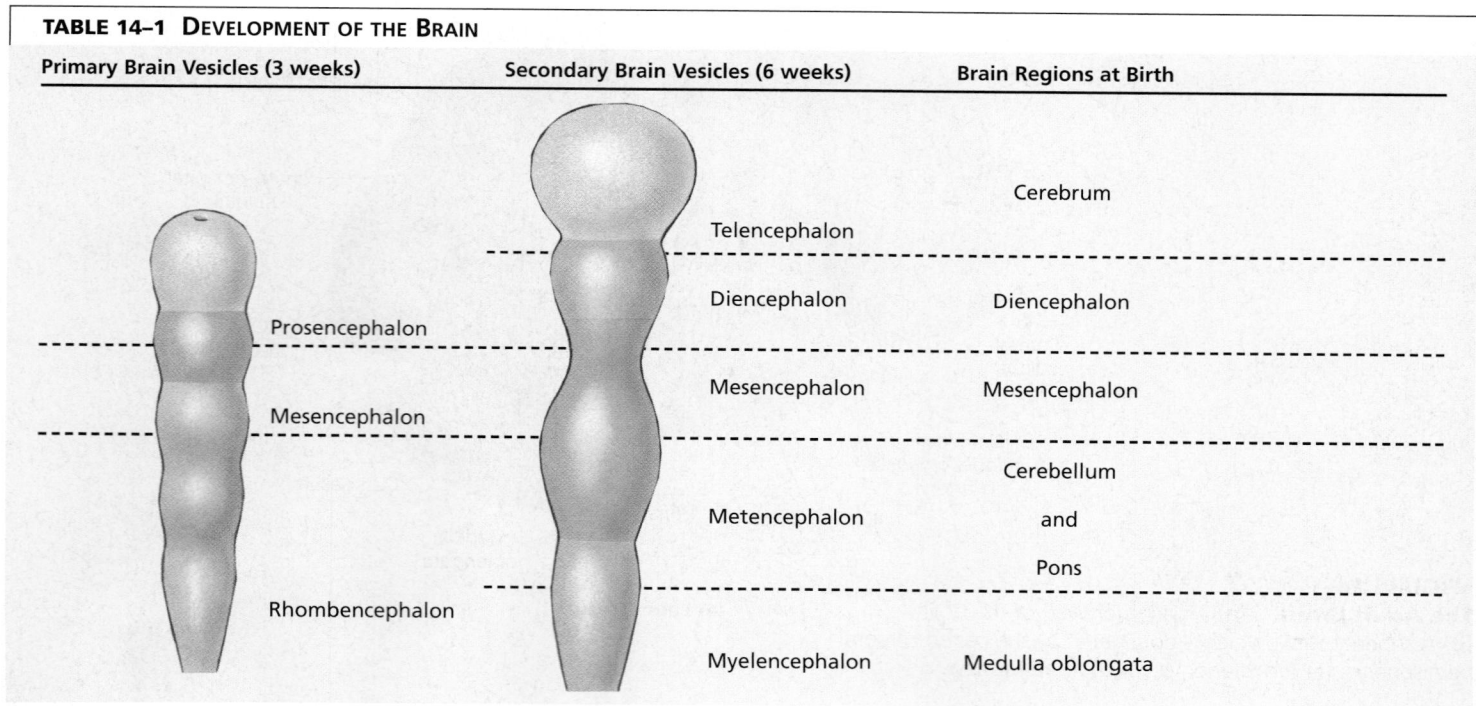

CEREBRUM
- Conscious thought processes, intellectual functions
- Memory storage and processing
- Conscious and subconscious regulation of skeletal muscle contractions

● FIGURE 14–2
An Introduction to Brain Functions

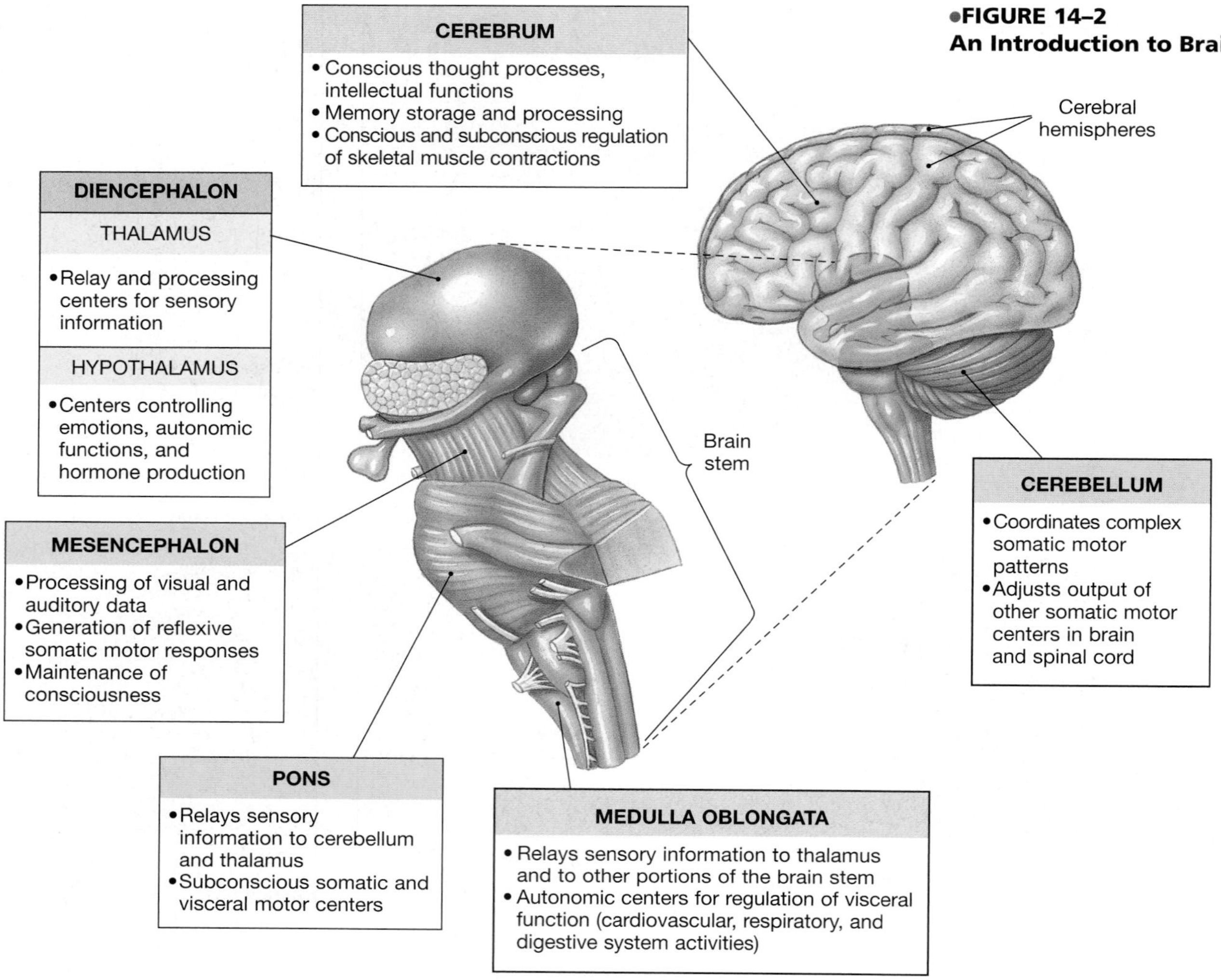

Cerebral hemispheres

DIENCEPHALON

THALAMUS

- Relay and processing centers for sensory information

HYPOTHALAMUS

- Centers controlling emotions, autonomic functions, and hormone production

MESENCEPHALON
- Processing of visual and auditory data
- Generation of reflexive somatic motor responses
- Maintenance of consciousness

Brain stem

CEREBELLUM
- Coordinates complex somatic motor patterns
- Adjusts output of other somatic motor centers in brain and spinal cord

PONS
- Relays sensory information to cerebellum and thalamus
- Subconscious somatic and visceral motor centers

MEDULLA OBLONGATA
- Relays sensory information to thalamus and to other portions of the brain stem
- Autonomic centers for regulation of visceral function (cardiovascular, respiratory, and digestive system activities)

TABLE 14–1 DEVELOPMENT OF THE BRAIN

Primary Brain Vesicles (3 weeks)	Secondary Brain Vesicles (6 weeks)	Brain Regions at Birth
Prosencephalon	Telencephalon	Cerebrum
	Diencephalon	Diencephalon
Mesencephalon	Mesencephalon	Mesencephalon
Rhombencephalon	Metencephalon	Cerebellum and Pons
	Myelencephalon	Medulla oblongata

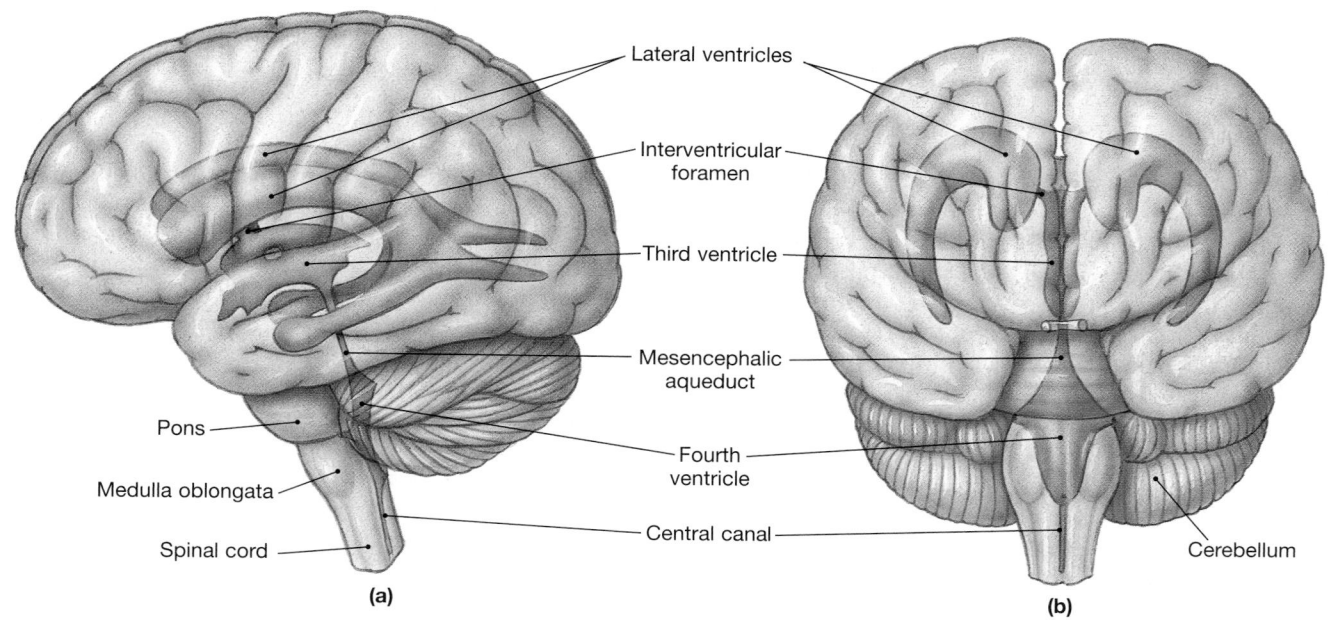

(a)   **(b)**

●**FIGURE 14–3**
Ventricles of the Brain.   The orientation and extent of the ventricles as they would appear if the brain were transparent.
(a) A lateral view. **(b)** An anterior view.   **ATLAS** Plate 2.6; Scans 1, 2

The portion of the rhombencephalon closer to the spinal cord forms the **myelencephalon** (mī-el-en-SEF-a-lon; *myelon*, spinal cord), which will become the medulla oblongata.   **ATLAS** Embryology Summary 12. The Development of the Brain and Cranial Nerves

☐ VENTRICLES OF THE BRAIN

During development, the neurocoel within the cerebral hemispheres, diencephalon, metencephalon, and medulla oblongata expands to form chambers called **ventricles** (VEN-tri-kls). The ventricles are lined by cells of the *ependyma*. ∞ pp. 392-393

Each cerebral hemisphere contains a large **lateral ventricle**. The **septum pellucidum**, a thin medial partition, separates the two lateral ventricles. Because there are two lateral ventricles (first and second), the ventricle in the diencephalon is called the **third ventricle**. Although the two lateral ventricles are not directly connected, each communicates with the third ventricle of the diencephalon through an **interventricular foramen** (*foramen of Monro*) (Figure 14–3●).

The mesencephalon has a slender canal known as the **mesencephalic aqueduct** (or *the aqueduct of the midbrain, aqueduct of Sylvius*, or *cerebral aqueduct*). This passageway connects the third ventricle with the **fourth ventricle**. The superior portion of the fourth ventricle lies between the posterior surface of the pons and the anterior surface of the cerebellum. The fourth ventricle extends into the superior portion of the medulla oblongata. This ventricle then narrows and becomes continuous with the central canal of the spinal cord.

The ventricles are filled with cerebrospinal fluid (CSF). The CSF continuously circulates from the ventricles and central canal into the *subarachnoid space* of the surrounding cranial meninges. The CSF passes between the interior and exterior of the CNS through foramina in the roof of the fourth ventricle.

✔ What brain regions make up the brain stem?
✔ Which embryological structures are destined to form the cerebellum and pons?

Answers start on page Q-1

 Review the orientation of the ventricles on the **Anatomy CD-ROM**: Nervous System/Brain Dissections.

14–2 PROTECTION AND SUPPORT OF THE BRAIN

Objectives

■ Explain how the brain is protected and supported.
■ Discuss the formation, circulation, and functions of the cerebrospinal fluid.

The delicate tissues of the brain are protected from mechanical forces by (1) the bones of the cranium, (2) the *cranial meninges*, and (3) cerebrospinal fluid. In addition, the neural tissue of the brain is biochemically isolated from the general circulation by the *blood–brain barrier*. Refer to Figures 7–3 and 7–4● (pp. 213–215) for a review of the bones of the cranium. We will discuss the other protective factors here.

▢ THE CRANIAL MENINGES

The layers that make up the cranial meninges—the cranial dura mater, arachnoid, and pia mater—are continuous with those of the spinal meninges. ∞ p. 433 However, the cranial meninges have distinctive anatomical and functional characteristics:

- The cranial *dura mater* consists of outer and inner fibrous layers. The outer layer is fused to the periosteum of the cranial bones. As a result, there is no epidural space superficial to the dura mater, as there is along the spinal cord. The outer, or *endosteal*, and inner, or *meningeal*, layers of the cranial dura mater are typically separated by a slender gap that contains tissue fluids and blood vessels, including several large venous sinuses. The veins of the brain open into these sinuses, which deliver the venous blood to the *internal jugular veins* of the neck.

- The cranial *arachnoid* consists of the arachnoid membrane, an epithelial layer, and the cells and fibers of the arachnoid trabeculae that cross the subarachnoid space to the pia mater. The arachnoid membrane covers the brain, providing a smooth surface that does not follow the brain's underlying folds. This membrane is in contact with the inner epithelial layer of the dura mater. The subarachnoid space extends between the arachnoid membrane and the pia mater.

- The *pia mater* sticks to the surface of the brain, anchored by the processes of astrocytes. It extends into every fold, and accompanies the branches of cerebral blood vessels as they penetrate the surface of the brain to reach internal structures.

Functions of the Cranial Meninges

The brain is cradled within the cranium. The overall shape of the brain roughly corresponds to that of the cranial cavity (Figure 14–4a●). The massive cranial bones provide mechanical protection, but they also pose a threat. The brain is like a person driving a car. If the car hits a tree, the car protects the driver from contact with the tree, but serious injury will occur unless a seat belt or airbag protects the driver from contact with the car. The cranial meninges protect the brain effectively. Tough, fibrous extensions of the dura mater, called *dural folds*, act like safety belts that hold the brain in position. The cerebrospinal fluid in the subarachnoid space acts like an airbag that cushions against sudden jolts and shocks.

Cranial trauma is a head injury resulting from impact with another object. Each year in the United States, roughly 8 million cases of cranial trauma occur, but only 1 case in 8 results in serious brain damage. The percentage is relatively low because the cranial meninges and CSF are so effective in protecting the brain. **AM** Cranial Trauma

Dural Folds

In several locations, the inner layer of the dura mater extends into the cranial cavity, forming a sheet that dips inward and then returns. These **dural folds** provide additional stabilization and support to the brain. **Dural sinuses** are large collecting veins located between the two layers of a dural fold. The three largest dural folds are called the falx cerebri, the tentorium cerebelli, and the falx cerebelli (Figure 14–4b●):

1. The **falx cerebri** (FALKS ser-Ē-brē; *falx*, curving or sickle-shaped) is a fold of dura mater that projects between the cerebral hemispheres in the longitudinal fissure. Its inferior portions attach anteriorly to the crista galli and posteriorly to the *internal occipital crest*, a ridge along the inner surface of the occipital bone. The **superior sagittal sinus** and the **inferior sagittal sinus**, two large venous sinuses, lie within this dural fold. The posterior margin of the falx cerebri intersects the tentorium cerebelli.

2. The **tentorium cerebelli** (ten-TOR-ē-um ser-e-BEL-ē; *tentorium*, a covering) separates and protects the cerebellar hemispheres from those of the cerebrum. It extends across the cranium at right angles to the falx cerebri. The **transverse sinus** lies within the tentorium cerebelli.

3. The **falx cerebelli** divides the two cerebellar hemispheres along the midsagittal line inferior to the tentorium cerebelli.

▢ CEREBROSPINAL FLUID

Cerebrospinal fluid (CSF) completely surrounds and bathes the exposed surfaces of the CNS. The CSF has several important functions, including the following:

- **Cushioning Delicate Neural Structures.**

- **Supporting the Brain.** In essence, the brain is suspended inside the cranium and floats in the CSF. A human brain weighs about 1400 g in air, but only about 50 g when supported by CSF.

- **Transporting Nutrients, Chemical Messengers, and Waste Products.** Except at the choroid plexus, where CSF is produced, the ependymal lining is freely permeable and the CSF is in constant chemical communication with the interstitial fluid of the CNS.

Because free exchange occurs between the interstitial fluid and the CSF, changes in CNS function can produce changes in the composition of the CSF. As we noted in Chapter 13, a *spinal tap* can provide useful clinical information about CNS injury, infection, or disease. ∞ p. 435

The Formation of CSF

The **choroid plexus** (*choroid*, vascular coat; *plexus*, network) consists of a combination of specialized ependymal cells and permeable capillaries for the production of cerebrospinal fluid. Two extensive folds of the choroid plexus originate in the roof of the third ventricle and extend through the interventricular foramina. These folds cover the floors of the lateral ventricles (Figure 14–5a●). In the inferior brain stem, a region of the choroid plexus in the roof of the fourth ventricle projects between the cerebellum and the pons.

Specialized ependymal cells, interconnected by tight junctions, surround the capillaries of the choroid plexus. The ependymal cells secrete CSF into the ventricles; they also remove waste products from the CSF and adjust its composition over

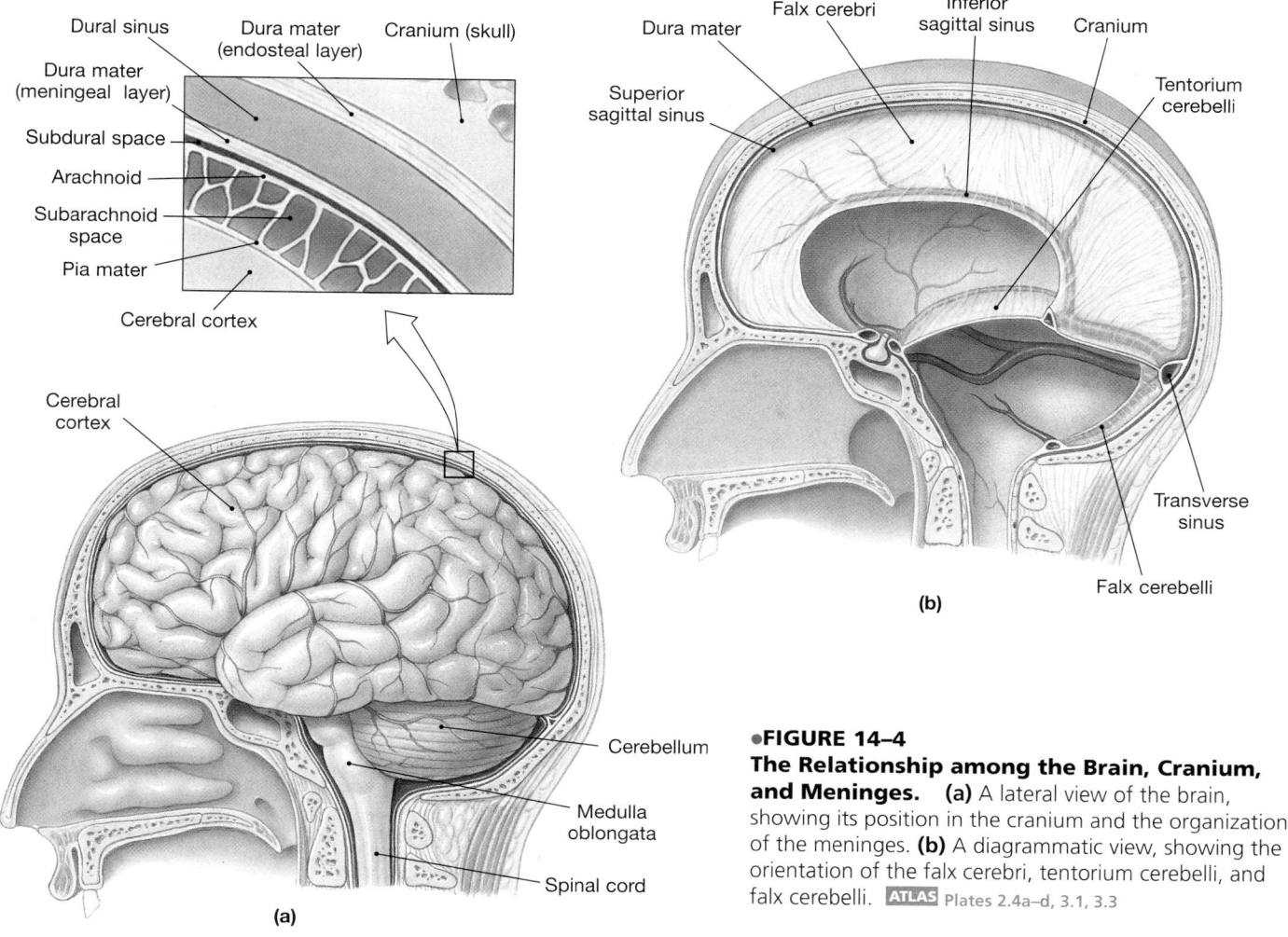

Dural sinus
Dura mater (endosteal layer)
Cranium (skull)
Dura mater (meningeal layer)
Subdural space
Arachnoid
Subarachnoid space
Pia mater
Cerebral cortex

Cerebral cortex

Cerebellum

Medulla oblongata

Spinal cord

(a)

Dura mater
Falx cerebri
Inferior sagittal sinus
Cranium
Superior sagittal sinus
Tentorium cerebelli
Transverse sinus
Falx cerebelli

(b)

•FIGURE 14–4
The Relationship among the Brain, Cranium, and Meninges. **(a)** A lateral view of the brain, showing its position in the cranium and the organization of the meninges. **(b)** A diagrammatic view, showing the orientation of the falx cerebri, tentorium cerebelli, and falx cerebelli. **ATLAS** Plates 2.4a–d, 3.1, 3.3

Epidural and Subdural Hemorrhages

A severe head injury may damage meningeal vessels and cause bleeding into the cranial cavity. The most serious cases involve an arterial break, because arterial blood pressure is relatively high. If blood is forced between the dura mater and the cranium, the condition is known as an **epidural hemorrhage**. The elevated fluid pressure then distorts the underlying tissues of the brain. The individual loses consciousness for a period lasting from minutes to hours after the injury, and death follows in untreated cases. An epidural hemorrhage involving a damaged vein does not produce massive symptoms immediately, and the individual may not have neurological problems for hours, days, or even weeks after the original injury. As a result, the problem may not be noticed until the nervous tissue has been severely damaged by distortion, compression, and secondary hemorrhaging. Epidural hemor-

rhages are rare, occurring in fewer than 1 percent of head injuries. However, the mortality rate is 100 percent in untreated cases and over 50 percent even after the blood pool has been removed and the damaged vessels have been closed.

The term **subdural hemorrhage** may be misleading, because in many cases blood enters the meningeal layer of the dura mater, flowing deep to the epithelium that contacts the arachnoid membrane rather than between the dura mater and the arachnoid. Subdural hemorrhages are roughly twice as common as epidural hemorrhages. The most common source of blood is a small vein or one of the dural sinuses. Because the venous blood pressure in a subdural hemorrhage is lower than that in an arterial epidural hemorrhage, the distortion produced is gradual and the effects on brain function can be quite variable and difficult to diagnose.

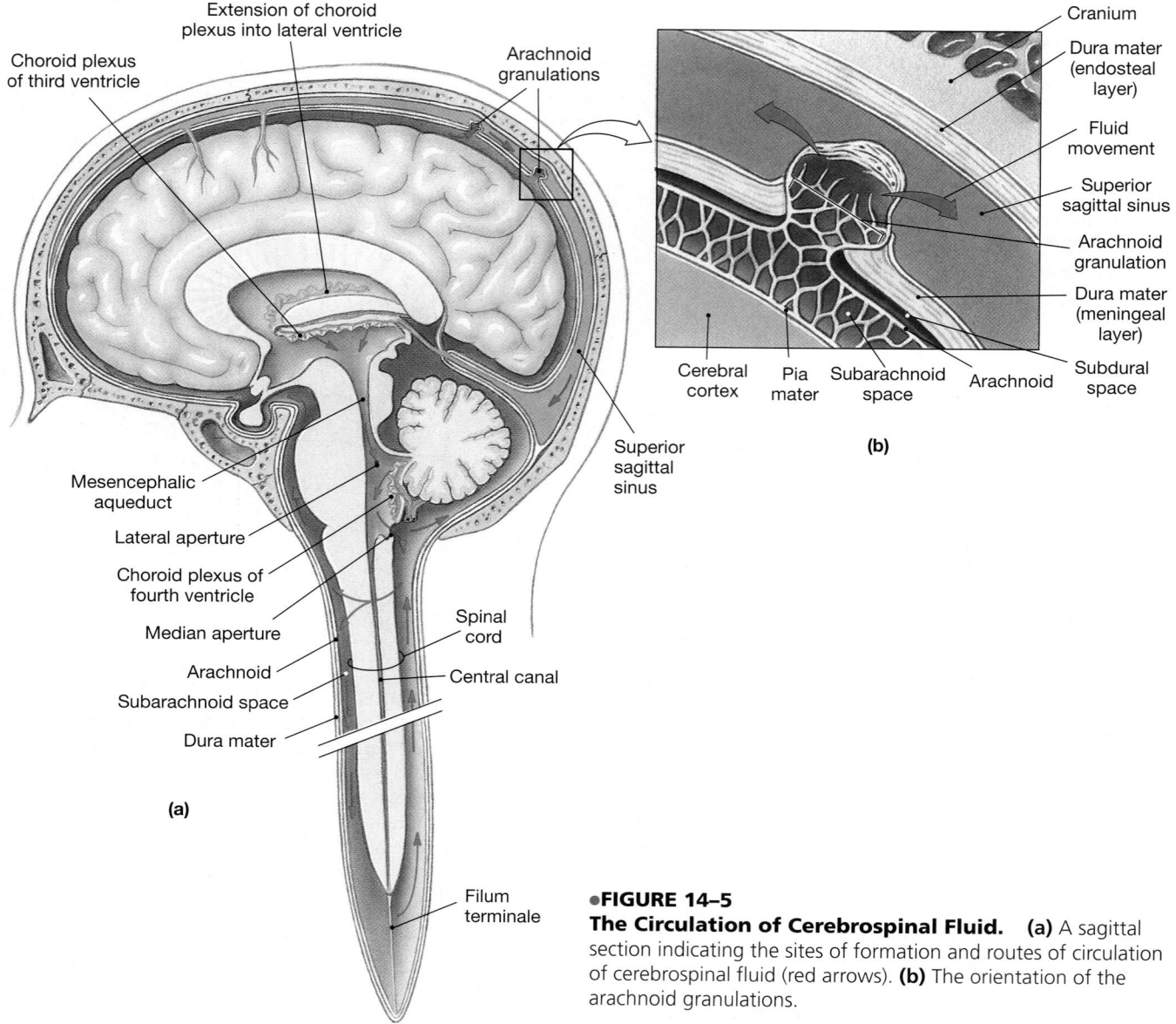

Choroid plexus
of third ventricle

Extension of choroid
plexus into lateral ventricle

Arachnoid
granulations

Mesencephalic
aqueduct

Lateral aperture

Choroid plexus of
fourth ventricle

Median aperture

Arachnoid

Subarachnoid space

Dura mater

Spinal
cord

Central canal

Superior
sagittal
sinus

Filum
terminale

(a)

Cranium

Dura mater
(endosteal
layer)

Fluid
movement

Superior
sagittal sinus

Arachnoid
granulation

Dura mater
(meningeal
layer)

Subdural
space

Cerebral
cortex

Pia
mater

Subarachnoid
space

Arachnoid

(b)

●FIGURE 14–5
The Circulation of Cerebrospinal Fluid. **(a)** A sagittal
section indicating the sites of formation and routes of circulation
of cerebrospinal fluid (red arrows). **(b)** The orientation of the
arachnoid granulations.

time. The differences in composition between CSF and blood
plasma (blood with the cellular elements removed) are quite pronounced. For example, the blood contains high concentrations of
soluble proteins, but the CSF does not. The concentrations of individual ions and the levels of amino acids, lipids, and waste
products are also different.

Circulation of CSF

The choroid plexus produces CSF at a rate of about 500 ml/day.
The total volume of CSF at any moment is approximately 150 ml;
thus, the entire volume of CSF is replaced roughly every eight
hours. Despite this rapid turnover, the composition of CSF is
closely regulated and the rate of removal normally keeps pace
with the rate of production.

The CSF circulates from the choroid plexus through the ventricles and the central canal of the spinal cord (Figure 14–5a●). As
the CSF circulates, diffusion between it and the interstitial fluid
of the CNS is unrestricted between and across the ependymal
cells. The CSF reaches the subarachnoid space through two
lateral apertures and a single **median aperture**, openings in the
roof of the fourth ventricle. Cerebrospinal fluid then flows
through the subarachnoid space surrounding the brain, spinal
cord, and cauda equina.

Fingerlike extensions of the arachnoid membrane, called the
arachnoid villi, penetrate the meningeal layer of the dura mater
and extend into the superior sagittal sinus. In adults, clusters of
villi form large **arachnoid granulations** (Figure 14–5b●). Cerebrospinal fluid is absorbed into the venous circulation at the
arachnoid granulations.

Hydrocephalus

The adult brain is surrounded by the relatively inflexible bones of the cranium. The enclosed cranial cavity contains two fluids—blood and cerebrospinal fluid—and the soft tissues of the brain. Because the total volume cannot change, when the volume of blood or CSF increases, the volume of the brain must decrease. For example, if a dural or subarachnoid blood vessel ruptures, the fluid volume increases as blood collects in the cranial cavity. The rising intracranial pressure compresses the brain, leading to neural dysfunction that can end in unconsciousness and, ultimately, death.

Any change in the rate of CSF production is normally matched by an increase in the rate of CSF removal at the arachnoid granulations. If this equilibrium is disturbed, clinical problems appear as the intracranial pressure changes. The volume of CSF will increase if the rate of formation accelerates or the rate of removal decreases. In either event, the increased fluid volume compresses and distorts the brain. Increased rates of formation can accompany head injuries, but the most common problems arise from masses, such as tumors or abscesses, scarring from meningeal infection, or from developmental abnormalities. These conditions restrict the normal circulation and reabsorption of CSF. Because CSF production continues, the ventricles gradually expand, distorting the surrounding neural tissues and causing brain function to deteriorate.

Infants are especially sensitive to changes in intracranial pressure, because the arachnoid granulations do not appear until roughly three years of age. (Until then, CSF is reabsorbed into small vessels in the subarachnoid space and beneath the ependyma lining the ventricles.) As in adults, if intracranial pressure becomes abnormally high, the ventricles will expand. But in infants, the cranial sutures have yet to fuse, so the skull can enlarge to accommodate the extra fluid volume. The result is a condition called **hydrocephalus**, or "water on the brain" (Figure 14–6●). Infant hydrocephalus typically results from

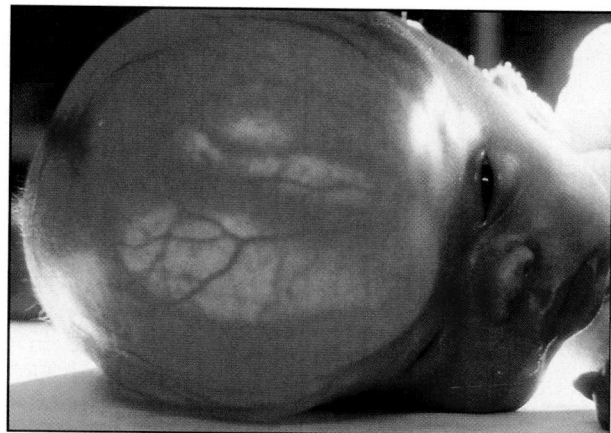

●**FIGURE 14–6**
Hydrocephalus.   This infant has hydrocephalus, a condition generally caused by impaired circulation and impaired removal of cerebrospinal fluid. The buildup of CSF leads to the distortion of the brain and enlargement of the cranium.

some interference with normal CSF circulation, such as a blockage of the mesencephalic aqueduct or a constriction of the connection between the subarachnoid spaces of the cranial and spinal meninges. Untreated infants commonly suffer some degree of mental retardation.

Successful treatment generally involves the installation of a *shunt*, a bypass that either avoids the blockage site or drains the excess CSF. The goal is reduction of the intracranial pressure. The shunt may be removed if (1) further growth of the brain eliminates the blockage or (2) the intracranial pressure decreases after the arachnoid granulations develop when the child reaches three years of age.

If the normal circulation or reabsorption of CSF is interrupted, a variety of clinical problems may appear. For example, a problem with the reabsorption of CSF in infancy is responsible for symptoms of *hydrocephalus,* or "water on the brain." Infants with this condition have enormously expanded skulls due to the presence of an abnormally large volume of CSF. (See the clinical discussion above, on "Hydrocephalus.") In adults, a failure of reabsorption or a blockage of CSF circulation can distort and damage the brain.

☐ THE BLOOD SUPPLY TO THE BRAIN

As we noted in Chapter 12, neurons have a high demand for energy, but they have neither energy reserves, in the form of

carbohydrates or lipids, nor oxygen reserves, in the form of myoglobin. ⚭ p. 423 Your brain, with billions of neurons, is an extremely active organ with a continuous demand for nutrients and oxygen. These demands are met by an extensive circulatory supply. Arterial blood reaches the brain through the *internal carotid arteries* and the *vertebral arteries*. Most of the venous blood from the brain leaves the cranium in the *internal jugular veins*, which drain the dural sinuses. A head injury that damages cerebral blood vessels may cause bleeding into the dura mater, either near the dural epithelium or between the outer layer of the dura mater and the bones of the skull. These are serious conditions, because the blood entering these spaces compresses and distorts the relatively soft tissues of the brain.

Cerebrovascular diseases are cardiovascular disorders that interfere with the normal blood supply to the brain. The particular distribution of the vessel involved determines the symptoms, and the degree of oxygen or nutrient starvation determines their severity. A **cerebrovascular accident** (**CVA**), or *stroke*, occurs when the blood supply to a portion of the brain is shut off. Affected neurons begin to die in a matter of minutes.

The Blood–Brain Barrier

Neural tissue in the CNS is isolated from the general circulation by the **blood–brain barrier**. This barrier exists because the endothelial cells that line the capillaries of the CNS are extensively interconnected by tight junctions. These junctions prevent the diffusion of materials between adjacent endothelial cells. In general, only lipid-soluble compounds (including carbon dioxide; oxygen; ammonia; lipids, such as steroids or prostaglandins; and small alcohols) can diffuse across the membranes of endothelial cells into the interstitial fluid of the brain and spinal cord. Water and ions must pass through channels in the apical and basal cell membranes. Larger, water-soluble compounds can cross the capillary walls only by active or passive transport.

The restricted permeability characteristics of the endothelial lining of brain capillaries are in some way dependent on chemicals secreted by astrocytes—cells that are in close contact with CNS capillaries. ∞ p. 392 The outer surfaces of the endothelial cells are covered by the processes of astrocytes. Because the astrocytes release chemicals that control the permeabilities of the endothelium to various substances, these cells play a key supporting role in the blood–brain barrier. If the astrocytes are damaged or stop stimulating the endothelial cells, the blood–brain barrier disappears.

The choroid plexus is not part of the neural tissue of the brain, so no astrocytes are in contact with the endothelial cells there. As a result, capillaries in the choroid plexus are highly permeable. Substances do not have free access to the CNS, however, because a **blood–CSF barrier** is created by specialized ependymal cells. These cells, interconnected by tight junctions, surround the capillaries of the choroid plexus.

Transport across the blood–brain and blood–CSF barriers is selective and directional. Even the passage of small ions, such as sodium, hydrogen, potassium, or chloride, is controlled. As a result, the pH and concentrations of sodium, potassium, calcium, and magnesium ions in the blood and CSF are different. Some organic compounds are readily transported, and others cross only in minute amounts. Neurons have a constant need for glucose. This need must be met regardless of the relative concentrations in the blood and interstitial fluid. Even when circulating glucose levels are low, endothelial cells continue to transport glucose from the blood to the interstitial fluid of the brain. In contrast, only trace amounts of circulating norepinephrine, epinephrine, dopamine, or serotonin pass into the interstitial fluid or CSF of the brain. This limitation is important, because these compounds are neurotransmitters—their entry from the bloodstream (where concentrations can be relatively high) could result in the uncontrolled stimulation of neurons throughout the brain.

The blood–brain barrier remains intact throughout the CNS, with four noteworthy exceptions:

1. In portions of the hypothalamus, the capillary endothelium is extremely permeable. This permeability exposes hypothalamic nuclei to circulating hormones and permits the diffusion of hypothalamic hormones into the circulation.

2. Capillaries in the posterior lobe of the pituitary gland, which is continuous with the floor of the hypothalamus, are highly permeable. At this site, the hormones *ADH* and *oxytocin*, produced by hypothalamic neurons, are released into the circulation.

3. Capillaries in the *pineal gland* are also very permeable. The pineal gland, an endocrine structure, is located in the posterior, superior surface of the diencephalon. The capillary permeability allows pineal secretions into the general circulation.

4. Capillaries at the choroid plexus are extremely permeable. Although the capillary characteristics of the blood–brain barrier are lost there, the transport activities of specialized ependymal cells in the choroid plexus maintain the blood–CSF barrier.

Physicians must sometimes get specific compounds into the interstitial fluid of the brain to fight CNS infections or to treat other neural disorders. To do this, they must understand the limitations of the blood–brain and blood–CSF barriers. For example, *Parkinson's disease*, discussed on page 487, results from inadequate release of the neurotransmitter dopamine within the basal nuclei. Although dopamine will not cross the blood–brain or blood–CSF barrier, a related compound, *L-dopa*, passes readily and is converted to dopamine inside the brain. In treatments for meningitis or other CNS infections, the antibiotics *penicillin* and *tetracycline* are excluded from the brain, but *sulfisoxazole* and *sulfadiazine* enter the CNS very rapidly.

When the CNS is damaged, the blood–brain barrier breaks down. As a result, the locations of injury sites, infections, and tumors can be determined by injecting into the bloodstream tracers that normally cannot enter the CNS tissues. One common method is to use the plasma protein *albumin* labeled with radioactive iodine. Any radioactivity in the brain is easily detected in a brain scan. Several antibiotics that do not normally enter the CSF may do so during CNS infections such as meningitis. These antibiotics which include the penicillins, aminoglycosides, and cephalosporins, are used to treat CNS infections.

✓ What would happen if an interventricular foramen became blocked?

✓ How would decreased diffusion across the arachnoid granulations affect the volume of cerebrospinal fluid in the ventricles?

✓ Many water-soluble molecules that are relatively abundant in the blood occur in small amounts or not at all in the extracellular fluid of the brain. Why?

Answers start on page Q-1

 Review the structures associated with CSF formation and flow on the **Companion Website:** Chapter 14/Reviewing Concepts/Labeling.

14–3 THE MEDULLA OBLONGATA

Objectives

- Describe the anatomical differences between the medulla oblongata and the spinal cord.
- List the main components of the medulla oblongata and specify their functions.

The medulla oblongata is continuous with the spinal cord. Figure 14–7● shows the position of the medulla oblongata in relation to the other components of the brain stem and the diencephalon. This figure includes the origins of 11 of the 12 pairs of cranial nerves. For the moment, we shall identify the individual cranial nerves by Roman numerals rather than by name. (The full names and functions of these nerves will be introduced in a later section.)

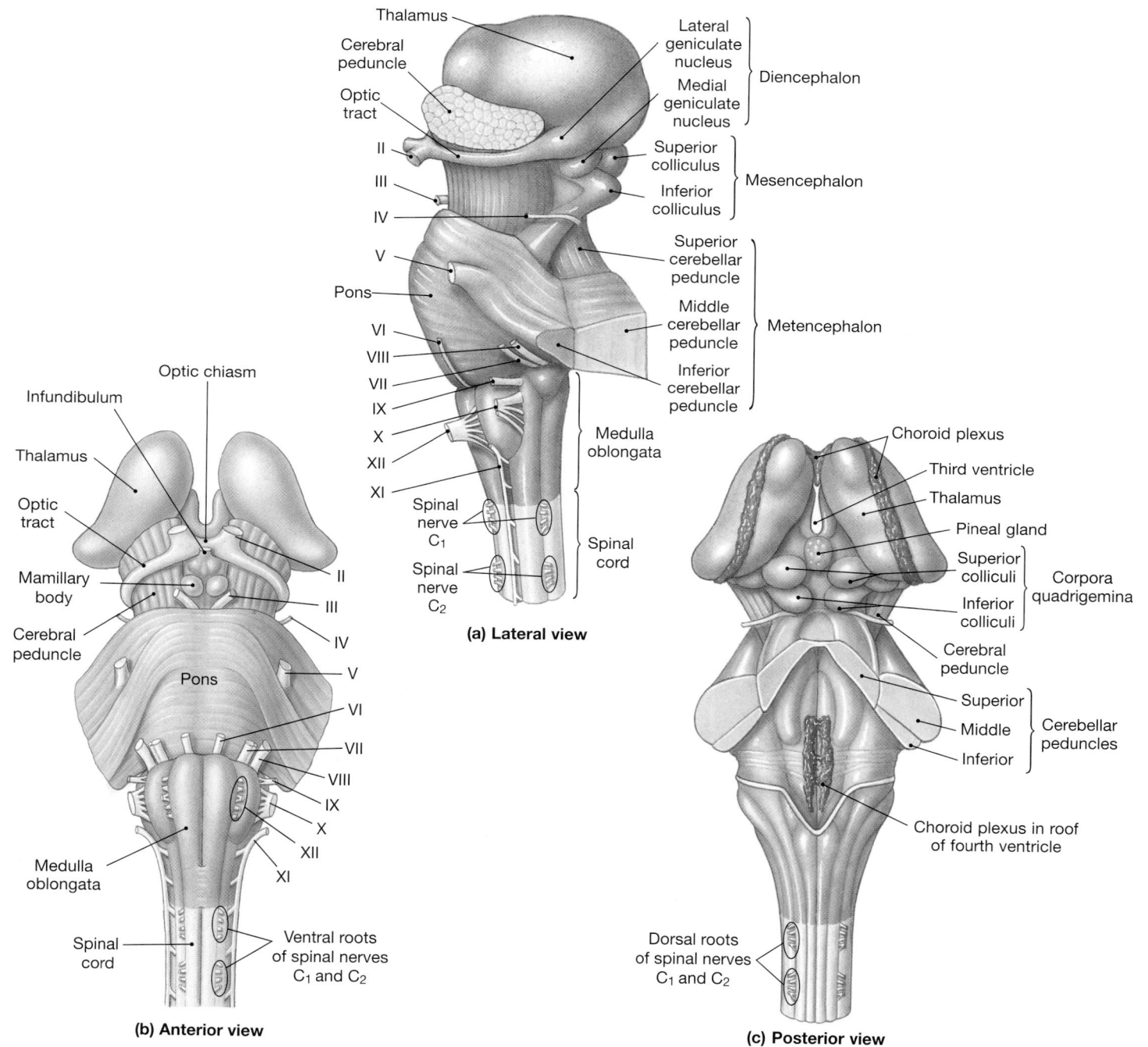

(a) Lateral view

(b) Anterior view

(c) Posterior view

●FIGURE 14–7
The Diencephalon and Brain Stem. (a) A lateral view, as seen from the left side. (b) A anterior view. (c) A posterior view.

In sectional view, the inferior portion of the medulla oblongata resembles the spinal cord, with a small central canal. However, the gray matter and white matter organization is more complex. As one ascends the medulla oblongata, the central canal opens into the fourth ventricle, and the similarity to the spinal cord all but disappears (Figure 14–8●).

The medulla oblongata is a very busy place—all communication between the brain and spinal cord involves tracts that ascend or descend through the medulla oblongata. In addition, the medulla oblongata is a center for the coordination of relatively complex autonomic reflexes and the control of visceral functions. Figure 14–8c● and Table 14–2 summarize its major components.

The medulla oblongata includes three groups of nuclei that we shall encounter in later chapters:

1. *Autonomic Nuclei Controlling Visceral Activities.* The **reticular formation** is a loosely organized mass of gray matter that contains embedded nuclei. It extends from the medulla oblongata to the mesencephalon. The portion of the reticular formation in the medulla oblongata contains nuclei and centers responsible for the regulation of vital autonomic functions. These **reflex centers** receive inputs from cranial nerves, the cerebral cortex, and the brain stem. Their output controls or adjusts the activities of one or more peripheral systems. There are two major groups of reflex centers. The **cardiovascular centers** adjust the

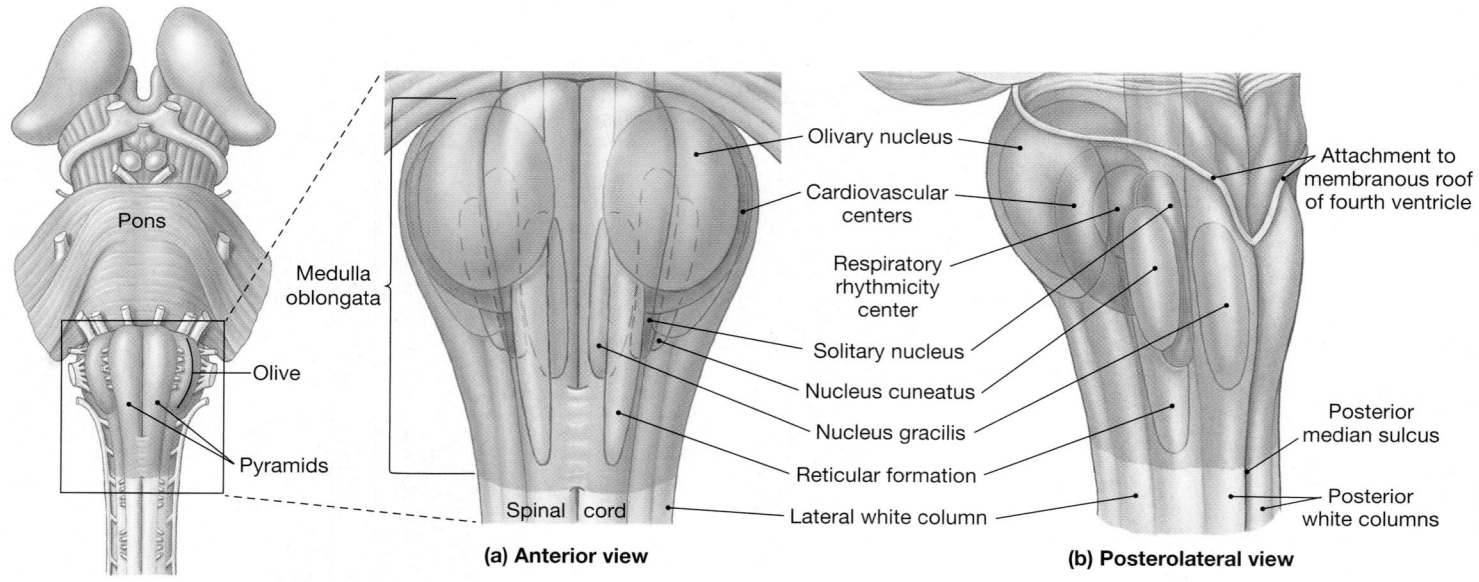

(a) Anterior view

(b) Posterolateral view

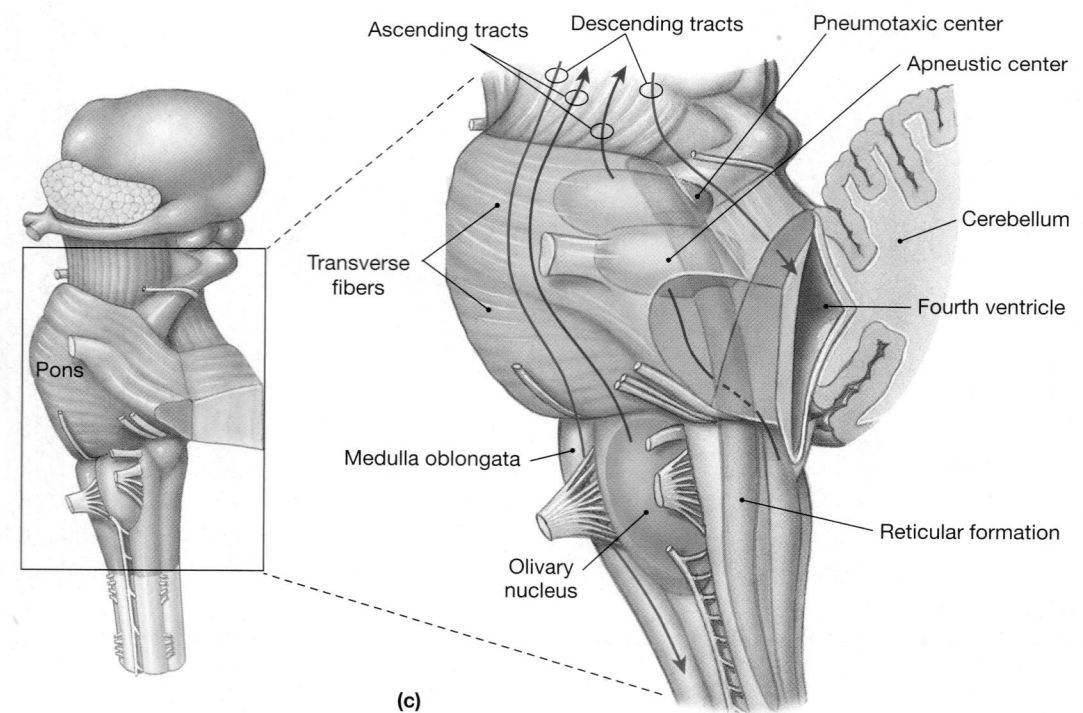

(c)

●**FIGURE 14–8**
The Medulla Oblongata and Pons. **(a, b)** Nuclei and tracts in the medulla oblongata. **(c)** Nuclei and tracts in the pons.

TABLE 14–2 COMPONENTS AND FUNCTIONS OF THE MEDULLA OBLONGATA AND PONS

Region/Subdivision	Component(s)	Function(s)
MEDULLA OBLONGATA		
Gray matter	Nucleus gracilis Nucleus cuneatus	Relay somatic sensory information to the thalamus
	Olivary nuclei	Relay information from the red nucleus, other nuclei of the mesencephalon, and the cerebral cortex to the cerebellum
	Reflex centers	
	Cardiac centers	Regulate heart rate and force of contraction
	Vasomotor centers	Regulate distribution of blood flow
	Respiratory rhythmicity centers	Set the pace of respiratory movements
	Other nuclei/centers	Contain sensory and motor nuclei of cranial nerves VIII (in part), IX, X, XI (in part) and XII; relay ascending sensory information from the spinal cord or higher centers
White matter	Ascending and descending tracts	Link the brain with the spinal cord
PONS		
Gray matter	Nuclei associated with cranial nerves V, VI, VII, and VIII (in part)	Relay sensory information and issue somatic motor commands
	Apneustic and pneumotaxic centers	Adjust activities of the respiratory rhythmicity centers in the medulla oblongata
	Relay centers	Relay sensory and motor information to the cerebellum
White matter	Ascending tracts	Carry sensory information from the nucleus cuneatus and nucleus gracilis to the thalamus
	Descending tracts	Carry motor commands from higher centers to motor nuclei of cranial or spinal nerves

heart rate, the strength of cardiac contractions, and the flow of blood through peripheral tissues. (In terms of function, the cardiovascular centers are subdivided into **cardiac** (*kardia*, heart) and **vasomotor** (*vas*, canal) **centers**, but their anatomical boundaries are difficult to determine.) The **respiratory rhythmicity centers** set the basic pace for respiratory movements. Their activity is regulated by inputs from the apneustic and pneumotaxic centers of the pons.

2. *Sensory and Motor Nuclei of Cranial Nerves.* The medulla oblongata contains sensory and motor nuclei associated with five of the cranial nerves (VIII, IX, X, XI, and XII). These cranial nerves provide motor commands to muscles of the pharynx, neck, and back as well as to the visceral organs of the thoracic and peritoneal cavities. Cranial nerve VIII carries sensory information from receptors in the inner ear to the vestibular and cochlear nuclei, which extend from the pons into the medulla oblongata.

3. *Relay Stations Along Sensory and Motor Pathways.* The **nucleus gracilis** and the **nucleus cuneatus** pass somatic sensory information to the thalamus. Tracts leaving these nuclei cross to the opposite side of the brain before reaching their destinations. This crossing over is called *decussation* (dē-kuh-SĀ-shun; *decussatio*, crossing over). The **solitary nucleus** on either side receives visceral sensory information that reaches the CNS from the spinal nerves and cranial nerves. This information is integrated and forwarded to other autonomic centers in the medulla oblongata and

elsewhere. The **olivary nuclei** relay information to the cerebellar cortex about somatic motor commands as they are issued by motor centers at higher levels. The bulk of the olivary nuclei create the **olives**, prominent olive-shaped bulges along the ventrolateral surface of the medulla oblongata.

14–4 THE PONS

Objective

■ List the main components of the pons and specify their functions.

The pons links the cerebellum with the mesencephalon, diencephalon, cerebrum, and spinal cord. Important features and regions of the pons are indicated in Figures 14–7 and 14–8● and Table 14–2. The pons contains four groups of components:

1. *Sensory and Motor Nuclei of Cranial Nerves.* These cranial nerves (V, VI, VII, and VIII) innervate the jaw muscles, the anterior surface of the face, one of the extraocular muscles (the lateral rectus), and the sense organs of the inner ear (the *vestibular* and *cochlear nuclei*).

2. *Nuclei Involved with the Control of Respiration.* On each side of the pons, the reticular formation in this region contains two respiratory centers: the *apneustic center* and the *pneumotaxic center*. These centers modify the activity of the *respiratory rhythmicity center* in the medulla oblongata.

3. *Nuclei and Tracts that Process and Relay Information Heading To or From the Cerebellum.* The pons links the cerebellum with the brain stem, cerebrum, and spinal cord.

4. *Ascending, Descending, and Transverse Tracts.* Longitudinal tracts interconnect other portions of the CNS. The middle cerebellar peduncles are connected to the **transverse fibers**, which cross the anterior surface of the pons. These fibers are axons that link nuclei of the pons (*pontine nuclei*) with the cerebellar hemisphere of the opposite side.

14–5 THE CEREBELLUM

Objective

■ List the main components of the cerebellum and specify their functions.

The cerebellum (Figure 14–9● and Table 14–3) is an automatic processing center. It has two primary functions:

1. *Adjusting the Postural Muscles of the Body.* The cerebellum coordinates rapid, automatic adjustments that maintain balance and equilibrium. These alterations in muscle tone and position are made by modifying the activities of motor centers in the brain stem.

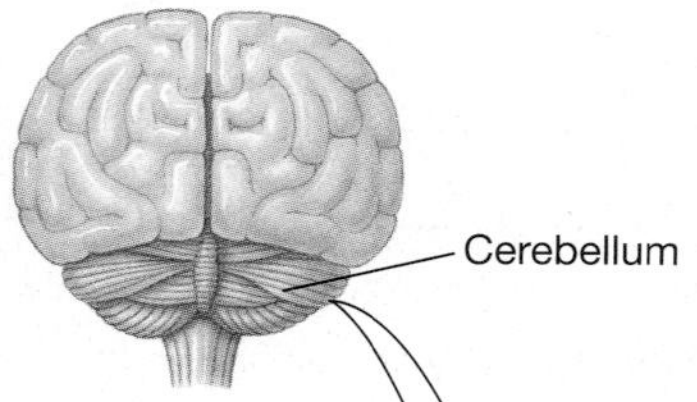

●**FIGURE 14–9**
The Cerebellum. **(a)** The posterior, superior surface of the cerebellum, showing major anatomical landmarks and regions. **(b)** A sectional view of the cerebellum, showing the arrangement of gray matter and white matter.

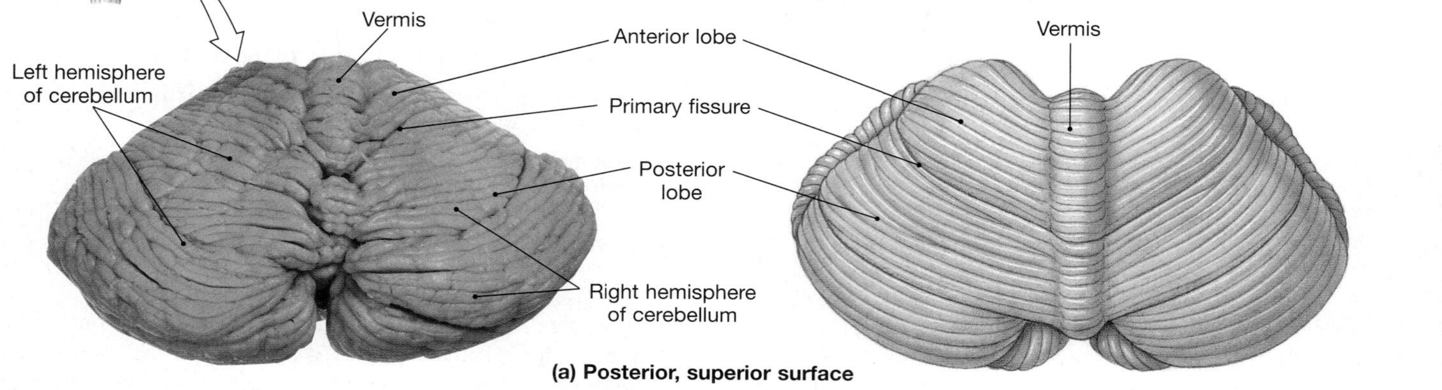

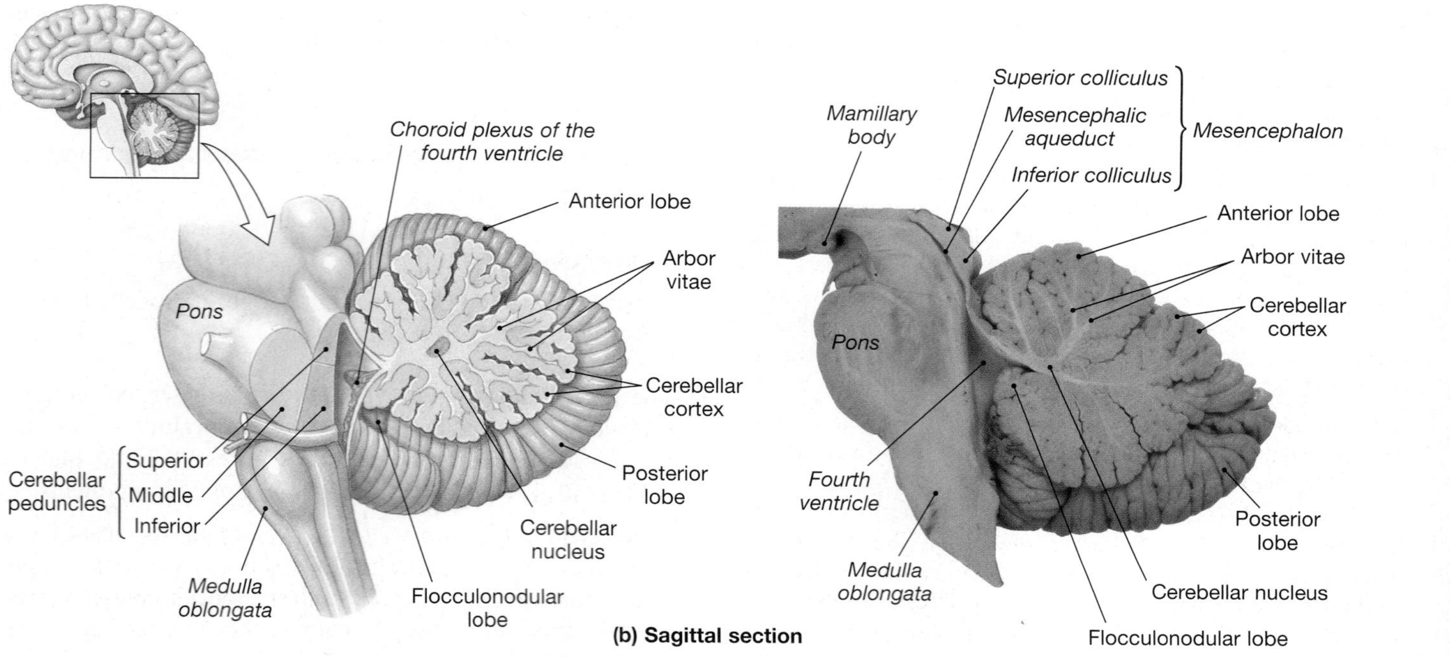

Cerebellum

Vermis

Left hemisphere of cerebellum

Anterior lobe

Primary fissure

Posterior lobe

Right hemisphere of cerebellum

Vermis

(a) Posterior, superior surface

Choroid plexus of the fourth ventricle

Anterior lobe

Arbor vitae

Pons

Cerebellar cortex

Cerebellar peduncles {Superior, Middle, Inferior}

Posterior lobe

Medulla oblongata

Flocculonodular lobe

Cerebellar nucleus

Superior colliculus

Mesencephalic aqueduct } *Mesencephalon*

Inferior colliculus

Mamillary body

Anterior lobe

Arbor vitae

Cerebellar cortex

Pons

Fourth ventricle

Medulla oblongata

Cerebellar nucleus

Posterior lobe

Flocculonodular lobe

(b) Sagittal section

TABLE 14-3 COMPONENTS OF THE CEREBELLUM

Subdivision	Region/Nuclei	Function(s)
Gray matter	Cerebellar cortex	Involuntary coordination and control of ongoing body movements
	Cerebellar nuclei	As above
White matter	Arbor vitae	Connects cerebellar cortex and nuclei with cerebellar peduncles
	Cerebellar peduncles	
	Superior	Link the cerebellum with mesencephalon, diencephalon, and cerebrum
	Middle	Contain transverse fibers and carry communications between the cerebellum and pons
	Inferior	Link the cerebellum with the medulla oblongata and spinal cord
	Transverse fibers	Interconnect pontine nuclei with the cerebellar hemisphere on the opposite side

2. *Programming and Fine-Tuning Movements Controlled at the Conscious and Subconscious Levels.* The cerebellum refines learned movement patterns. This function is performed indirectly by regulating activity along motor pathways at the cerebral cortex, basal nuclei, and motor centers in the brain stem. The cerebellum compares the motor commands with proprioceptive information (position sense) and performs any adjustments needed to make the movement smooth.

The cerebellum has a complex, highly convoluted surface composed of neural cortex. The **folia** (FŌ-lē-uh; leaves), or folds of the cerebellum surface, are less prominent than the folds in the surfaces of the cerebral hemispheres. The **anterior** and **posterior lobes** are separated by the **primary fissure** (Figure 14–9a●). Along the midline, a narrow band of cortex known as the **vermis** (VER-mis; worm) separates the **cerebellar hemispheres**. The slender **flocculonodular** (flok-ū-lō-NOD-ū-lar) **lobe** lies between the roof of the fourth ventricle and the cerebellar hemispheres and vermis (Figure 14–9b●).

Like the cerebrum, the cerebellum has a superficial layer of neural cortex. The cerebellar cortex contains huge, highly branched **Purkinje** (pur-KIN-jē) **cells.** The extensive dendrites of each Purkinje cell receive input from up to 200,000 synapses. Internally, the white matter of the cerebellum forms a branching array that in sectional view resembles a tree. Anatomists call it the **arbor vitae,** or "tree of life."

The cerebellum receives proprioceptive information from the spinal cord and monitors all proprioceptive, visual, tactile, balance, and auditory sensations received by the brain. Most axons that carry sensory information do not synapse in the cerebellar nuclei but pass through the deeper layers of the cerebellum on their way to the Purkinje cells of the cerebellar cortex. Information about the motor commands issued at the conscious and subconscious levels reaches the Purkinje cells indirectly, after being relayed by nuclei in the pons or by the **cerebellar nuclei** embedded within the arbor vitae.

Tracts that link the cerebellum with the brain stem, cerebrum, and spinal cord leave the cerebellar hemispheres as the superior, middle, and inferior cerebellar peduncles. The **superior cerebel-** lar peduncles link the cerebellum with nuclei in the midbrain, diencephalon, and cerebrum. The **middle cerebellar peduncles** are connected to a broad band of fibers that cross the ventral surface of the pons at right angles to the axis of the brain stem. The middle cerebellar peduncles also connect the cerebellar hemispheres with sensory and motor nuclei in the pons. The **inferior cerebellar peduncles** permit communication between the cerebellum and nuclei in the medulla oblongata and carry ascending and descending cerebellar tracts from the spinal cord.

The cerebellum can be permanently damaged by trauma or stroke, or temporarily affected by drugs such as alcohol. The result is **ataxia** (a-TAK-sē-uh; *ataxia*, lack of order), a disturbance in balance. In severe ataxia, the individual cannot sit or stand without assistance. **AM** Cerebellar Dysfunction

---

✓ Which part of the brain has a worm (vermis) and a tree (arbor vitae)?

✓ The medulla oblongata is one of the smallest sections of the brain, yet damage there can cause death, whereas similar damage in the cerebrum might go unnoticed. Why?

✓ If the respiratory center of the pons were damaged, what respiratory controls might be lost?

Answers start on page Q-1

14-6 THE MESENCEPHALON

Objective

■ List the main components of the mesencephalon and specify their functions.

The external anatomy of the mesencephalon, or midbrain, is shown in Figure 14–7●, and the major nuclei are listed in Table 14–4 and shown in Figure 14–10●. The **tectum,** or roof of the mesencephalon, contains two pairs of sensory nuclei known collectively as the **corpora quadrigemina** (KOR-po-ra quad-ri-JEM-i-nuh). These nuclei, the superior and inferior colliculi, process visual and

TABLE 14–4 COMPONENTS AND FUNCTIONS OF THE MESENCEPHALON

Subdivision	Region/Nuclei	Functions
GRAY MATTER		
Tectum (roof)	Superior colliculi	Integrate visual information with other sensory inputs; initiate reflex responses to visual stimuli
	Inferior colliculi	Relay auditory information to medial geniculate nuclei; initiate reflex responses to auditory stimuli
Walls and floor	Red nuclei	Subconscious control of upper limb position and background muscle tone
	Substantia nigra	Regulates activity in the basal nuclei
	Reticular formation (headquarters)	Automatic processing of incoming sensations and outgoing motor commands; can initiate involuntary motor responses to stimuli; helps maintain consciousness (RAS)
	Other nuclei/centers	Nuclei associated with two cranial nerves (III, IV)
WHITE MATTER		
	Cerebral peduncles	Connect primary motor cortex with motor neurons in brain and spinal cord; carry ascending sensory information to thalamus

auditory sensations. Each **superior colliculus** (kol-IK-ū-lus; *colliculus*, a small hill) receives visual inputs from the lateral geniculate nucleus of the thalamus on that side. The **inferior colliculus** receives auditory data from nuclei in the medulla oblongata and pons. Some of this information may be forwarded to the medial geniculate on the same side. The superior colliculi control the reflex movements of the eyes, head, and neck in response to visual stimuli, such as a bright light. The inferior colliculi control reflex move-

ments of the head, neck, and trunk in response to auditory stimuli, such as a loud noise.

On each side, the mesencephalon contains a red nucleus and the substantia nigra. The **red nucleus** contains numerous blood vessels, which give it a rich red color. This nucleus, which receives information from the cerebrum and cerebellum, issues subconscious motor commands that affect upper limb position and background muscle tone. The **substantia nigra** (NĪ-gruh; *nigra*, black)

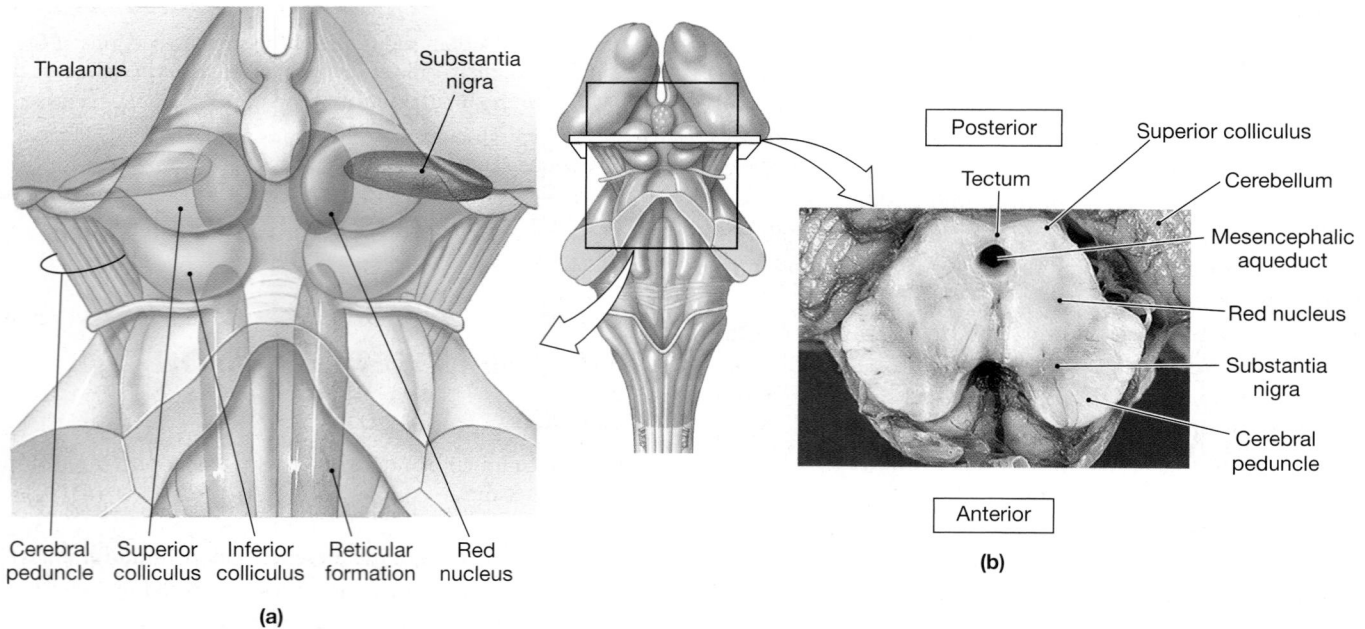

●FIGURE 14–10
The Mesencephalon. (a) A posterior view. Superficial structures are labeled on the left; underlying nuclei are colored and labeled on the right. (b) A superior view of a section at the level of the mesencephalon. **ATLAS** Plates 2.5, 3.1, 3.3; Scan 1b

is a nucleus that lies lateral to the red nucleus (Figure 14–10a●). The gray matter in this region contains darkly pigmented cells, giving it a black color.

The nerve fiber bundles on the ventrolateral surfaces of the mesencephalon (Figures 14–7, p. 473, and 14–10b●) are the **cerebral peduncles** (*peduncles*, little feet). They contain (1) descending fibers that go to the cerebellum by way of the pons and (2) descending fibers that carry voluntary motor commands issued by the cerebral hemispheres.

The mesencephalon also contains the headquarters of the **reticular activating system** (**RAS**), a specialized component of the reticular formation. Stimulation of the mesencephalic portion of the RAS makes you more alert and attentive. We will consider the role of the RAS in the maintenance of consciousness in Chapter 16.

14–7 THE DIENCEPHALON

Objective

- List the main components of the diencephalon and specify their functions.

The diencephalon plays a vital role in integrating conscious and unconscious sensory information and motor commands. It consists of the epithalamus, thalamus, and hypothalamus. Figures 14–7, 14–11, and 14–12● show the position of the diencephalon and its relationship to landmarks on the brain stem.

The *epithalamus* is the roof of the diencephalon superior to the third ventricle. The anterior portion of the epithalamus contains an extensive area of choroid plexus that extends through the interventricular foramina into the lateral ventricles. The posterior portion of the epithalamus contains the **pineal gland** (Figure 14–7b●), an endocrine structure that secretes the hormone **melatonin**. Melatonin is important in the regulation of day–night cycles and also in the regulation of reproductive functions. (We will describe the role of melatonin in Chapter 18.)

Most of the neural tissue in the diencephalon is concentrated in the two *thalami*, which form the lateral walls, and the *hypothalamus*, which forms the floor. Ascending sensory information from the spinal cord and cranial nerves (other than the olfactory tract) synapses in a nucleus in the left or right thalamus before reaching the cerebral cortex and our conscious awareness. The hypothalamus contains centers involved with emotions and visceral processes that affect the cerebrum as well as other components of the brain stem. It also controls a variety of autonomic functions and forms the link between the nervous and endocrine systems.

☐ THE THALAMUS

The thalamus is the final relay point for ascending sensory information that will be projected to the primary sensory cortex. It acts as a filter, passing on only a small portion of the arriving sen-

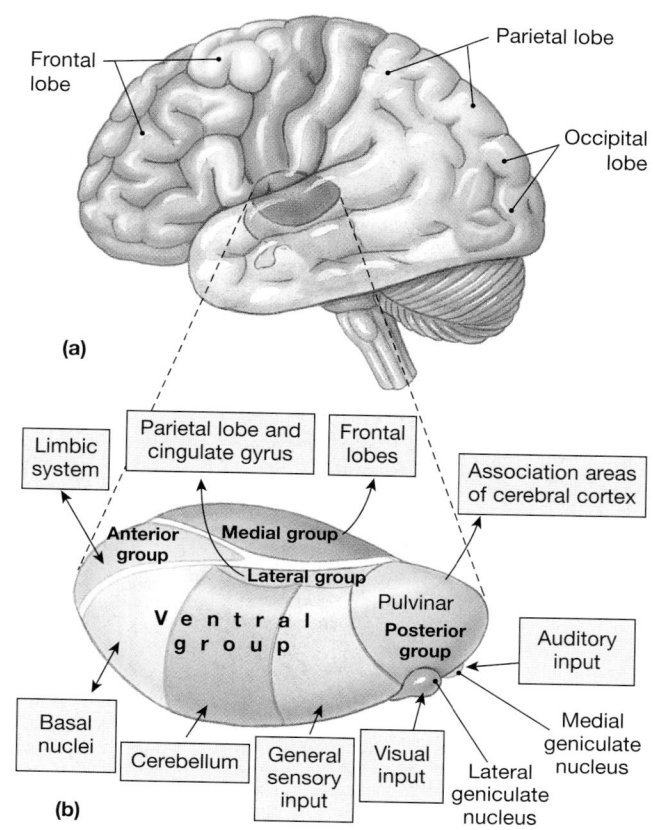

●**FIGURE 14–11**
The Thalamus. **(a)** A lateral view of the brain, color coded to indicate the regions that receive input from the thalamic nuclei. **(b)** An enlarged view of the thalamic nuclei of the left side.

sory information. The thalamus also coordinates the activities of the basal nuclei and the cerebral cortex by relaying information between them.

The left thalamus and right thalamus are separated by the third ventricle. Each thalamus consists of a rounded mass of *thalamic nuclei* (Figure 14–11●). Viewed in midsagittal section (Figure 14–12a●), each thalamus extends from the anterior commissure to the inferior base of the pineal gland. A projection of gray matter called an **intermediate mass** extends into the ventricle from the thalamus on either side. In roughly 70 percent of the human population, the two thalami fuse in the midline, although no fibers cross the midline.

Functions of Thalamic Nuclei

The thalamic nuclei deal primarily with the relay of sensory information to the basal nuclei and cerebral cortex. The five major groups of thalamic nuclei, listed in Table 14–5 and shown in Figure 14–11●, are the anterior, medial, ventral, posterior, and lateral groups:

1. The *anterior group* includes the **anterior nuclei**, which are part of the *limbic system*. This system, which is involved with emotion and motivation, is discussed in a later section.

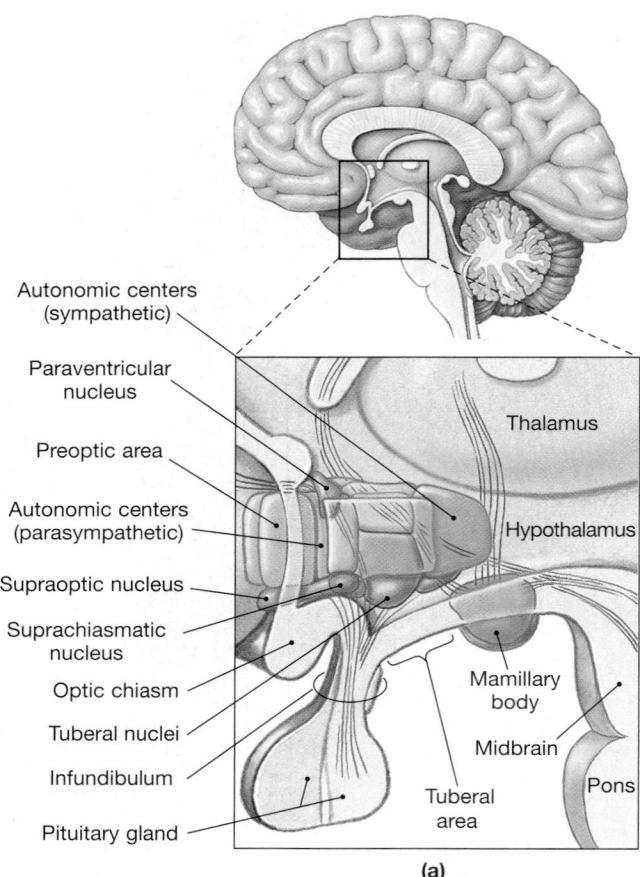

●**FIGURE 14–12**
The Hypothalamus in Sagittal Section. **(a)** A diagrammatic view of the hypothalamus, showing the locations of major nuclei and centers. **(b)** The hypothalamus and adjacent portions of the brain.

Autonomic centers (sympathetic)

Paraventricular nucleus

Preoptic area

Autonomic centers (parasympathetic)

Supraoptic nucleus

Suprachiasmatic nucleus

Optic chiasm

Tuberal nuclei

Infundibulum

Pituitary gland

Thalamus

Hypothalamus

Mamillary body

Midbrain

Pons

Tuberal area

(a)

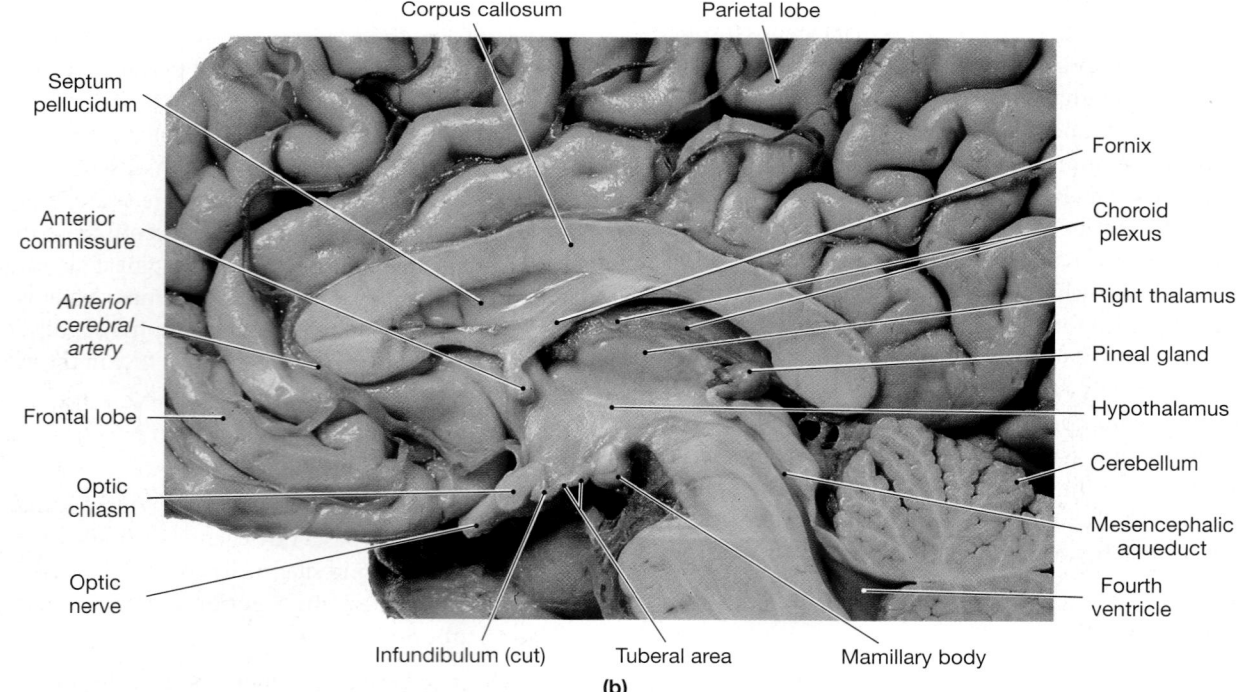

Corpus callosum

Parietal lobe

Septum pellucidum

Anterior commissure

Anterior cerebral artery

Frontal lobe

Optic chiasm

Optic nerve

Infundibulum (cut)

Tuberal area

Mamillary body

Fornix

Choroid plexus

Right thalamus

Pineal gland

Hypothalamus

Cerebellum

Mesencephalic aqueduct

Fourth ventricle

(b)

TABLE 14–5 THE THALAMUS

Group/Nuclei	Function(s)
ANTERIOR GROUP	Part of the limbic system
MEDIAL GROUP	Integrates sensory information for projection to the frontal lobes
VENTRAL GROUP	Projects sensory information to the primary sensory cortex; relays information from cerebellum and basal nuclei to motor area of cerebral cortex
POSTERIOR GROUP	
Pulvinar	Integrates sensory information for projection to association areas of cerebral cortex
Lateral geniculate nuclei	Project visual information to the visual cortex
Medial geniculate nuclei	Project auditory information to the auditory cortex
LATERAL GROUP	Integrates sensory information and influences emotional states

2. The nuclei of the *medial group* provide a conscious awareness of emotional states by connecting emotional centers in the hypothalamus with the *frontal lobes* of the cerebral hemispheres. The medial group also receives and relays sensory information from other portions of the thalamus.

3. The nuclei of the *ventral group* relay information from the *basal nuclei* of the cerebrum and the cerebellum to somatic motor areas of the cerebral cortex. Ventral group nuclei also relay sensory information about touch, pressure, pain, temperature, and proprioception (position) to the sensory areas of the cerebral cortex.

4. The *posterior group* includes the pulvinar and the geniculate nuclei. The **pulvinar** integrates sensory information for projection to the cerebral cortex. The **lateral geniculate** (je-NIK-ū-lāt) **nucleus** (*genicula*, little knee) of each thalamus receives visual information over the *optic tract*, which originates at the eyes. The output of the lateral geniculate nucleus goes to the *occipital lobes* of the cerebral hemispheres and to the mesencephalon. The **medial geniculate nucleus** relays auditory information to the appropriate area of the cerebral cortex from specialized receptors of the inner ear.

5. The nuclei of the *lateral group* form feedback loops with the limbic system, described in a later section, and the *parietal lobes* of the cerebral hemispheres. The lateral group affects emotional states and the integration of sensory information.

☐ THE HYPOTHALAMUS

The hypothalamus extends from the area superior to the *optic chiasm*, a crossover where the optic tracts from the eyes arrive at the brain, to the posterior margins of the **mamillary bodies** (*mamilla*, little breast; Figure 14–12a●). The mamillary bodies process sensory information, including olfactory sensations that provide the sense of smell. They also contain motor nuclei that control reflex movements associated with eating, such as chewing, licking, and swallowing.

Immediately posterior to the optic chiasm, a narrow stalk called the **infundibulum** (in-fun-DIB-ū-lum; *infundibulum*, funnel) extends inferiorly, connecting the floor of the hypothalamus to the pituitary gland (Figure 14–12b●).

The floor of the hypothalamus between the infundibulum and the mamillary bodies is the **tuberal area** (*tuber*, swelling). The tuberal area contains nuclei that are involved with the control of pituitary gland function.

Functions of the Hypothalamus

The hypothalamus contains important control and integrative centers, in addition to those associated with the limbic system. These centers are shown in Figure 14–12a● and their functions are summarized in Table 14–6. Hypothalamic centers may be stimulated by (1) sensory information from the cerebrum, brain stem, and spinal cord; (2) changes in the compositions of the CSF and interstitial fluid; or (3) chemical stimuli in the circulating blood that move rapidly across highly permeable capillaries to enter the hypothalamus (where there is no blood–brain barrier).

The hypothalamus performs the following functions:

1. *The Subconscious Control of Skeletal Muscle Contractions.* The hypothalamus directs somatic motor patterns associated with rage, pleasure, pain, and sexual arousal by stimulating centers in other portions of the brain. For example, the changes in facial expression that accompany rage and the basic movements associated with sexual activity are controlled by hypothalamic centers.

TABLE 14–6 COMPONENTS AND FUNCTIONS OF THE HYPOTHALAMUS

Region/Nucleus	Function
Supraoptic nucleus	Secretes ADH, restricting water loss at the kidneys
Paraventricular nucleus	Secretes oxytocin
Preoptic areas	Regulate body temperature
Suprachiasmatic nucleus	Coordinates day–night cycles of activity
Tuberal nuclei	Release hormones that control endocrine cells of the anterior pituitary gland
Autonomic centers	Control medullary nuclei that regulate heart rate and blood pressure
Mamillary bodies	Control feeding reflexes (licking, swallowing, etc.)

2. *The Control of Autonomic Function.* The hypothalamus adjusts and coordinates the activities of autonomic centers in the pons and medulla oblongata that regulate heart rate, blood pressure, respiration, and digestive functions.

3. *The Coordination of Activities of the Nervous and Endocrine Systems.* The hypothalamus coordinates neural and endocrine activities by inhibiting or stimulating endocrine cells in the pituitary gland through the production of *regulatory hormones*. These hormones are produced at the tuberal area and are released into local capillaries for transport to the anterior lobe of the pituitary gland.

4. *The Secretion of Hormones.* The hypothalamus secretes *antidiuretic hormone (ADH)*, which is produced by the **supraoptic nucleus** and restricts water loss at the kidneys, and *oxytocin*, which is produced by the **paraventricular nucleus** and stimulates smooth muscle contractions in the uterus and mammary glands of females and the prostate gland of males. These hormones are transported along axons that pass through the infundibulum to the posterior lobe of the pituitary gland. There the hormones are released into the blood for distribution throughout the body.

5. *The Production of Emotions and Behavioral Drives.* Specific hypothalamic centers produce sensations that lead to conscious or subconscious changes in behavior. For example, stimulation of the **feeding center** produces the sensation of hunger, and stimulation of the **thirst center** produces the sensation of thirst. These unfocused "impressions" originating in the hypothalamus are called **drives**. The conscious sensations are only part of the hypothalamic response. For instance, the thirst center also orders the release of ADH by neurons in the supraoptic nucleus.

6. *Coordination Between Voluntary and Autonomic Functions.* When you think about a dangerous or stressful situation, your heart rate and respiratory rate go up and your body prepares for an emergency. These autonomic adjustments are made by the hypothalamus.

7. *The Regulation of Body Temperature.* The **preoptic area** of the hypothalamus coordinates the activities of other CNS centers and regulates other physiological systems to maintain normal body temperature. If body temperature falls, the preoptic area sends instructions to the *vasomotor center*, an autonomic center in the medulla oblongata that controls blood flow by regulating the diameter of peripheral blood vessels. In response, the vasomotor center decreases the blood supply to the skin, reducing the rate of heat loss.

8. *The Control of Circadian Rhythms.* The **suprachiasmatic nucleus** coordinates daily cycles of activity that are linked to the day–night cycle. This nucleus receives input from the retina of the eye, and its output adjusts the activities of other hypothalamic nuclei, the pineal gland, and the reticular formation.

14–8 THE LIMBIC SYSTEM

Objective

- Identify the main components of the limbic system and specify their locations and functions.

The **limbic system** (*limbus*, border) includes nuclei and tracts along the border between the cerebrum and diencephalon. This system is a functional grouping rather than an anatomical one. Functions of the limbic system include (1) establishing emotional states; (2) linking the conscious, intellectual functions of the cerebral cortex with the unconscious and autonomic functions of the brain stem; and (3) facilitating memory storage and retrieval. Whereas the sensory cortex, motor cortex, and association areas of the cerebral cortex enable you to perform complex tasks, it is largely the limbic system that makes you *want* to do them. For this reason, the limbic system is also known as the *motivational system*.

Figure 14–13● focuses on major components of the limbic system. The **amygdaloid** (ah-MIG-da-loyd; *amygdale*, almond) **body** appears to act as an interface between the limbic system, the cerebrum, and various sensory systems. It plays a role in the regulation of heart rate, in the control of the "fight or flight" response, and in linking emotions with specific memories. The **limbic lobe** of the cerebral hemisphere consists of the superficial folds, or *gyri*, and underlying structures adjacent to the diencephalon. The gyri curve along the *corpus callosum*, a fiber tract that links the two cerebral hemispheres, and continue onto the medial surface of the cerebrum lateral to the diencephalon (Figure 14–13a●). There are three gyri in the limbic lobe. The **cingulate** (SIN-gū-lāt) **gyrus** (*cingulum*, girdle or belt) sits superior to the corpus callosum. The **dentate gyrus** and the **parahippocampal** (pa-ra-hip-ō-KAM-pal) **gyrus** form the posterior and inferior portions of the limbic lobe. These gyri conceal the **hippocampus**, a nucleus inferior to the floor of the lateral ventricle. To early anatomists, this structure resembled a sea horse (*hippocampus*); it is important in learning, especially in the storage and retrieval of new long-term memories.

The **fornix** (FOR-niks) is a tract of white matter that connects the hippocampus with the hypothalamus (Figures 14–10c,d and 14–13b●). From the hippocampus, the fornix curves medially, meeting its counterpart from the opposing hemisphere. The fornix proceeds anteriorly, inferior to the corpus callosum, before curving toward the hypothalamus. Many fibers of the fornix end in the mamillary bodies of the hypothalamus.

Several other nuclei in the wall (thalamus) and floor (hypothalamus) of the diencephalon are components of the limbic system. The *anterior nucleus* of the thalamus relays information from the mamillary body (of the hypothalamus) to the cingulate gyrus on that side. The boundaries between the hypothalamic nuclei of the limbic system are often poorly defined, but experimental stimulation has outlined a number of important hypothalamic centers responsible for the emotions of rage, fear, pain, sexual arousal, and pleasure. The stimulation of specific regions of the hypothalamus can also produce heightened alertness and a generalized excitement or generalized lethargy and sleep. These responses are caused by the stimulation or inhibition of the reticular formation. Although the reticular formation extends the length of the brain stem, its headquarters lies in the mesencephalon.

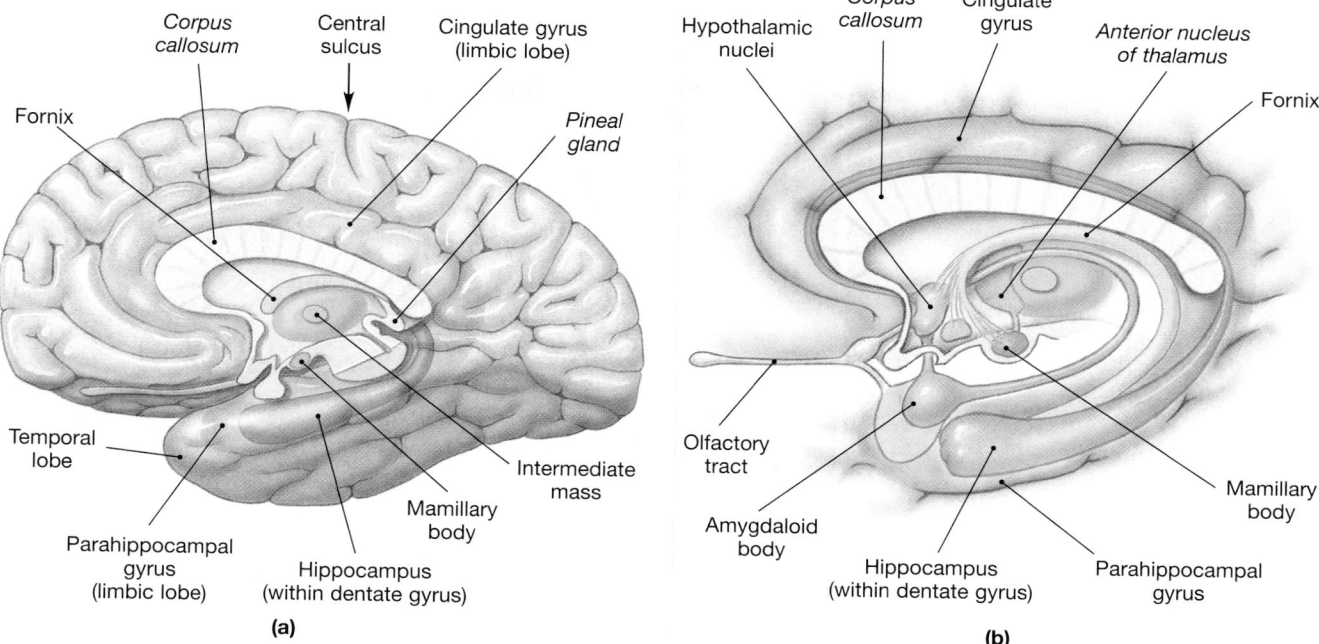

●FIGURE 14–13
The Limbic System. **(a)** A diagrammatic sagittal section through the cerebrum, showing the cortical areas associated with the limbic system. The parahippocampal gyrus is shown as though transparent to make deeper limbic components visible. **(b)** A three-dimensional reconstruction of the limbic system, showing the relationships among the major components.

Table 14–7 summarizes the organization and functions of the limbic system. Figure 14–14●, which shows the brain in midsagittal and frontal sections, includes many of the limbic structures, as well as other components of the cerebrum, diencephalon, and brain stem that were introduced in earlier sections.

TABLE 14–7 THE LIMBIC SYSTEM

FUNCTIONS

Processing of memories; creation of emotional states, drives, and associated behaviors

CEREBRAL COMPONENTS

Cortical areas: limbic lobe (cingulate gyrus, dentate gyrus, and parahippocampal gyrus)

Nuclei: hippocampus, amygdaloid body

Tracts: fornix

DIENCEPHALIC COMPONENTS

Thalamus: anterior nuclear group

Hypothalamus: centers concerned with emotions, appetites (thirst, hunger), and related behaviors *(see Table 14–6)*

OTHER COMPONENTS

Reticular formation: network of interconnected nuclei throughout brain stem

---

✓ Reflex movements of the eyes, head and neck are controlled by which area(s) of the mesencephalon?

✓ Damage to the lateral geniculate nuclei of the thalamus would interfere with the functions of which senses?

✓ Which area of the diencephalon would be stimulated by changes in body temperature?

✓ Damage to the amygdaloid body would interfere with regulation of what division of the autonomic nervous system?

Answers start on page Q-1

14–9 THE CEREBRUM

Objectives

■ Identify the major anatomical subdivisions of the cerebrum.

■ Locate the motor, sensory, and association areas of the cerebral cortex, and discuss their functions.

■ Discuss the origin and significance of the major categories of brain waves seen in an electroencephalogram.

The cerebrum is the largest region of the brain. Conscious thoughts and all intellectual functions originate in the cerebral

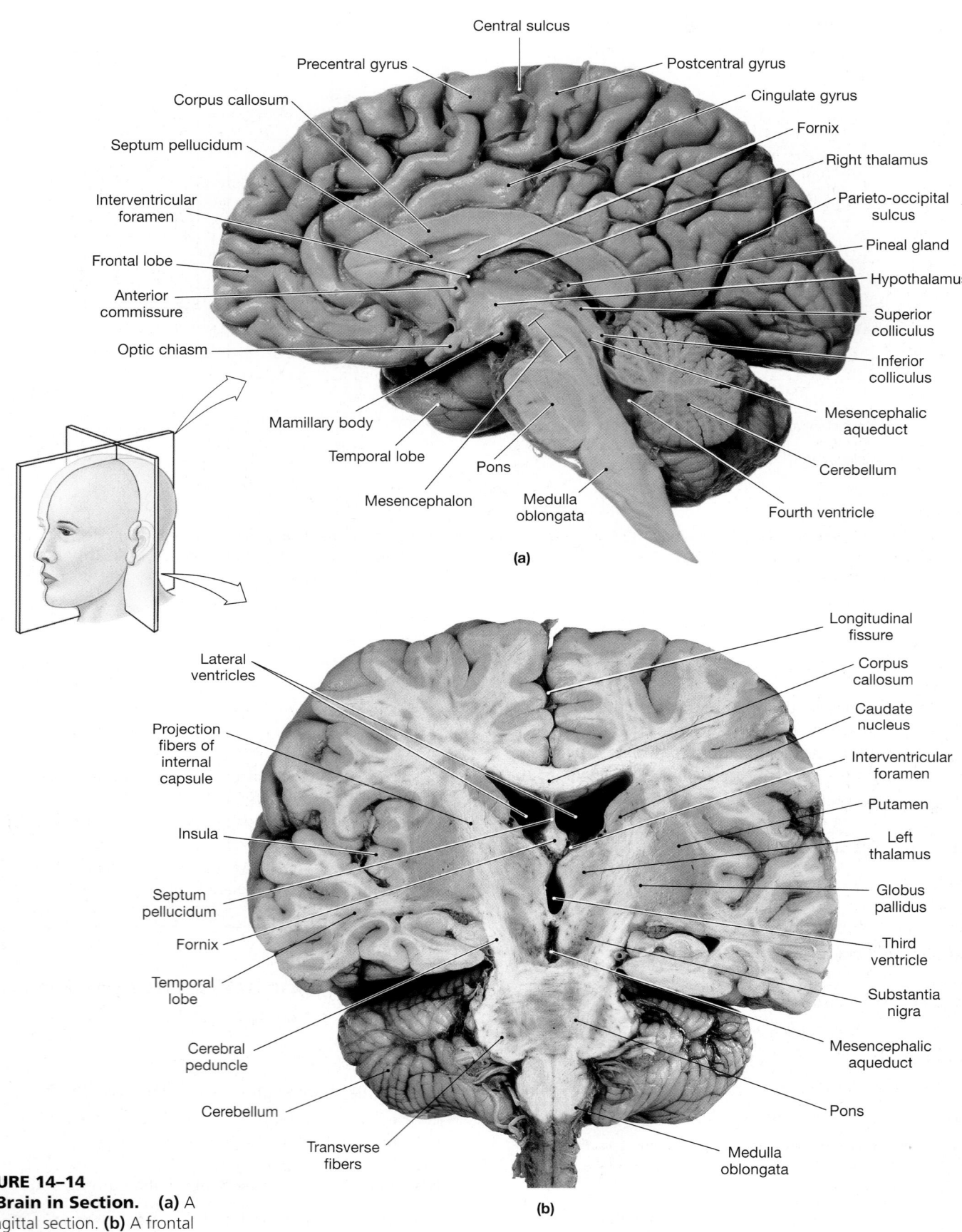

(a)

(b)

●**FIGURE 14–14**
The Brain in Section. **(a)** A
midsagittal section. **(b)** A frontal
section. **ATLAS** Plate 3.1; Scans 1d,e, 2a–d

hemispheres. Much of the cerebrum is involved in the processing of somatic sensory and motor information.

Figure 14–1b,d●, p. 465, provides a perspective on the cerebrum and its relationships with other regions of the brain. Gray matter in the cerebrum is located in the *cerebral cortex* and in deeper *basal nuclei*. The white matter of the cerebrum lies deep to the neural cortex and around the basal nuclei.

☐ THE CEREBRAL CORTEX

As we have previously seen, a blanket of neural cortex covers the paired cerebral hemispheres, which dominate the superior and lateral surfaces of the cerebrum (Figure 14–1●). The gyri increase the surface area of the cerebral hemispheres, and thus the number of cortical neurons they contain; the total surface area of the cerebral hemispheres is roughly equivalent to 2200 cm^2 (2.5 ft^2) of flat surface. The entire brain has enlarged over the course of human evolution, but the cerebral hemispheres have enlarged at a much faster rate than has the rest of the brain, because complex analytical and integrative functions require large numbers of neurons. Since the neurons involved are in the superficial layer of cortex, it is there that the expansion has been most pronounced. The only solution available, other than an enlargement of the entire skull, was for the cortical layer to fold like a crumpled piece of paper.

The two cerebral hemispheres are almost completely separated by a deep **longitudinal fissure** (Figure 14–1a●). They remain connected by a thick band of white matter called the *corpus callosum*. Each cerebral hemisphere can be divided into *lobes*, or regions, named after the overlying bones of the skull. Your brain has a unique pattern of sulci and gyri, as individual as a fingerprint, but the boundaries between lobes are reliable landmarks. On each hemisphere, the **central sulcus**, a deep groove, divides the anterior **frontal lobe** from the more posterior **parietal lobe**. The roughly horizontal **lateral sulcus** separates the frontal lobe from the **temporal lobe**. Pushing the temporal lobe to the side exposes the **insula** (IN-sū-luh; *insula*, island), an "island" of cortex that is otherwise invisible. The more posterior **parieto-occipital sulcus** separates the parietal lobe from the **occipital lobe**.

Each lobe contains functional regions whose boundaries are less clearly defined. Some of these regions deal with sensory information and others with motor commands. Keep in mind three points about the cerebral lobes:

1. *Each Cerebral Hemisphere Receives Sensory Information From, and Sends Motor Commands To, the Opposite Side of the Body.* For example, the motor areas of the left cerebral hemisphere control muscles on the right side, and the right cerebral hemisphere controls muscles on the left side. This crossing over has no known functional significance.

2. *The Two Hemispheres Have Different Functions, Even Though They Look Almost Identical.* We will discuss these differences in a later section.

3. *The Assignment of a Specific Function to a Specific Region of the Cerebral Cortex is Imprecise.* Because the boundaries are indis-

tinct and have considerable overlap, one region may have several functions. Some aspects of cortical function, such as consciousness, cannot easily be assigned to any single region. However, we know that normal individuals use all portions of the brain.

☐ THE WHITE MATTER OF THE CEREBRUM

The interior of the cerebrum consists primarily of white matter. The axons can be roughly classified as association fibers, commissural fibers, and projection fibers (Figure 14–15●).

- **Association fibers** interconnect areas of neural cortex within a single cerebral hemisphere. Shorter association fibers are called **arcuate** (AR-kū-āt) **fibers**, because they curve in an arc to pass from one gyrus to another. Longer association fibers are organized into discrete bundles, or *fasciculi*. The **longitudinal fasciculi** connect the frontal lobe to the other lobes of the same hemisphere.

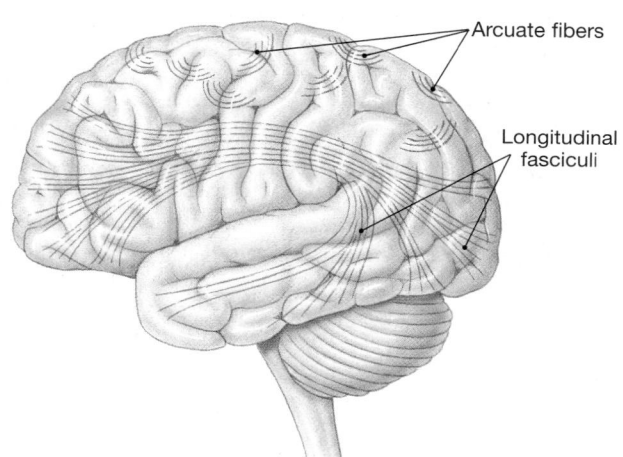

(a) Lateral view

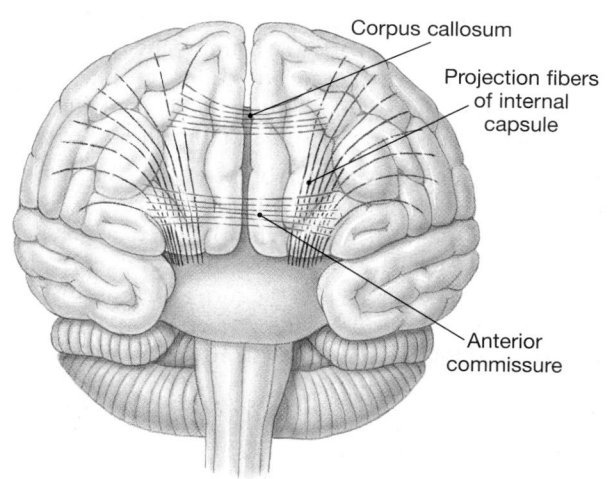

(b) Anterior view

●**FIGURE 14–15**
The White Matter of the Cerebrum *(See also Figure 14–14b.)*

●**FIGURE 14–16**
The Basal Nuclei. **(a)** The relative positions of the basal nuclei in the intact brain. **(b)** Frontal sections. **(c)** A horizontal section; compare this with **(d)**, the view in dissection.

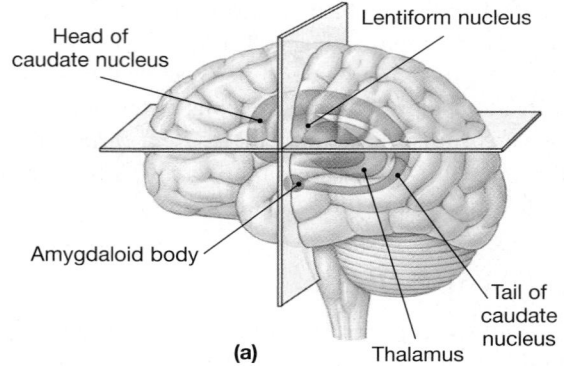

(a)

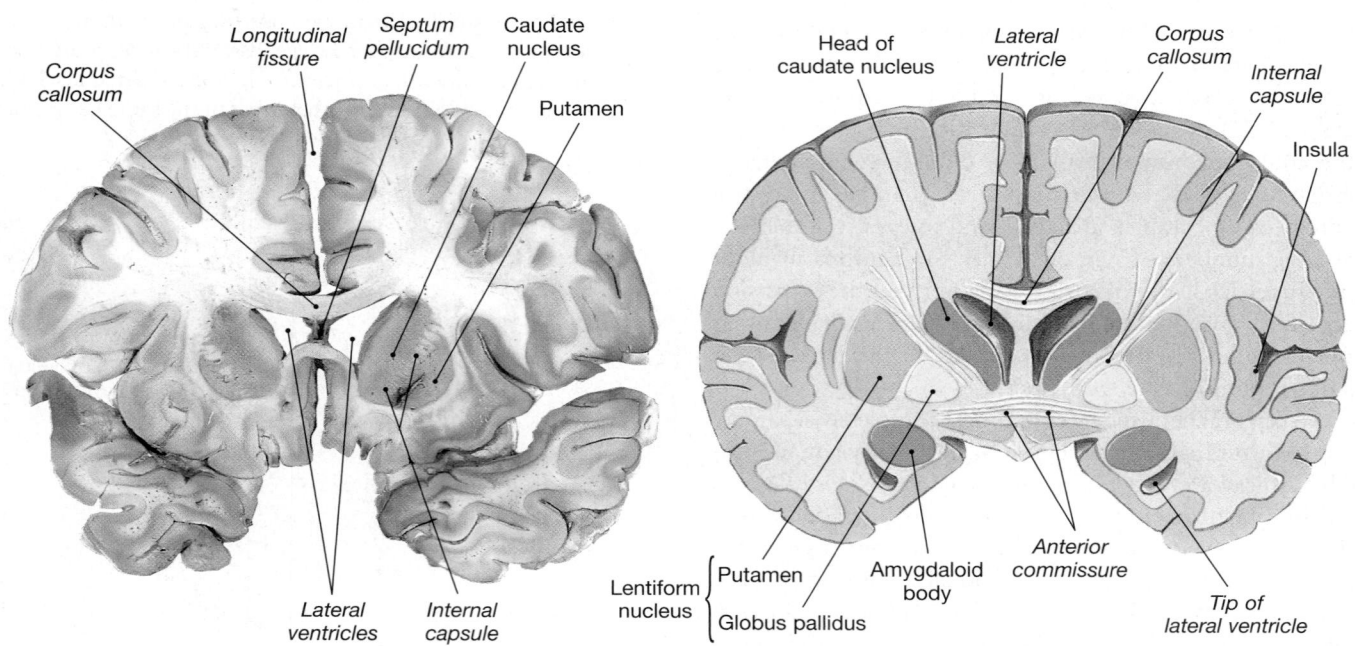

(b) Frontal section

- **Commissural** (kom-MIS-ū-rul) **fibers** (*commissura*, crossing over) interconnect and permit communication between the cerebral hemispheres. Bands of commissural fibers linking the hemispheres include the **corpus callosum** and the **anterior commissure**. The corpus callosum alone contains more than 200 million axons, carrying some 4 billion impulses per second!

- **Projection fibers** link the cerebral cortex to the diencephalon, brain stem, cerebellum, and spinal cord. All projection fibers must pass through the diencephalon, where axons heading to sensory areas of the cerebral cortex pass among the axons descending from motor areas of the cortex. In gross dissection, the ascending fibers and descending fibers look alike. The entire collection of fibers is known as the **internal capsule**.

▢ **THE BASAL NUCLEI**

While your cerebral cortex is consciously directing a complex movement or solving some intellectual puzzle, other centers of your cerebrum, diencephalon, and brain stem are processing sensory information and issuing motor commands outside your conscious awareness. Many of these activities, which occur at the subconscious level, are directed by the *basal nuclei*, or *cerebral nuclei*.

The **basal nuclei** are masses of gray matter that lie within each hemisphere deep to the floor of the lateral ventricle (Figure 14–16●). They are embedded in the white matter of the cerebrum, and the radiating projection fibers and commissural fibers travel around or between these nuclei. Historically, the basal nuclei have been considered part of a larger functional group known as the *basal ganglia*. This group included the basal nuclei of the cerebrum and the associated motor nuclei in the diencephalon and mesencephalon. Although we will consider the functional interactions among these components in Chapter 15, we will avoid the term 'basal ganglia' because ganglia are otherwise restricted to the PNS.

The **caudate nucleus** has a massive head and a slender, curving tail that follows the curve of the lateral ventricle. The head of the caudate nucleus lies anterior to the **lentiform** (lens-shaped)

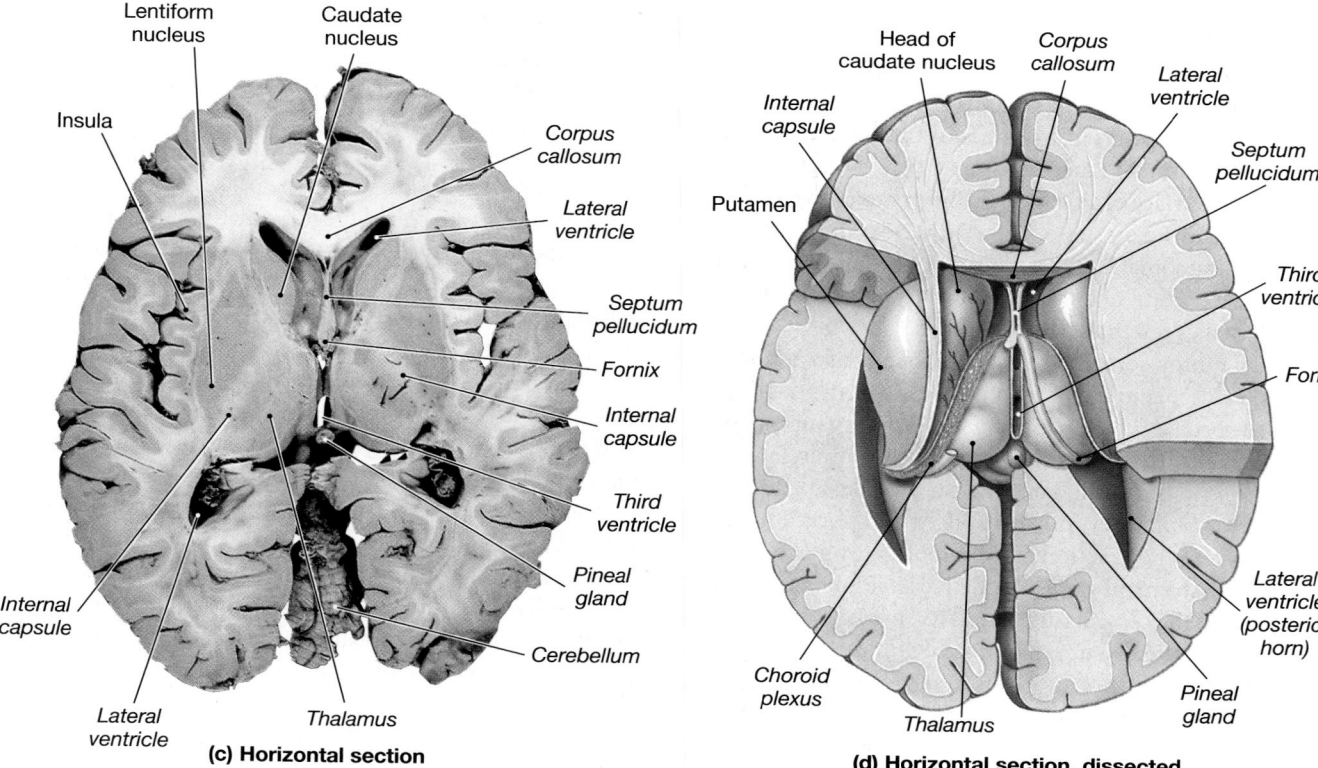

(c) Horizontal section

Lentiform nucleus
Caudate nucleus
Insula
Corpus callosum
Lateral ventricle
Septum pellucidum
Fornix
Internal capsule
Third ventricle
Pineal gland
Cerebellum
Internal capsule
Lateral ventricle
Thalamus

(d) Horizontal section, dissected

Head of caudate nucleus
Corpus callosum
Internal capsule
Lateral ventricle
Putamen
Septum pellucidum
Third ventricle
Fornix
Choroid plexus
Lateral ventricle (posterior horn)
Pineal gland
Thalamus

nucleus. The lentiform nucleus consists of a medial **globus pallidus** (GLŌ-bus PAL-i-dus; pale globe) and a lateral **putamen** (pū-TĀ-men). The term *corpus striatum* (striated body) has been used to refer to the caudate and lentiform nuclei, or to the caudate and putamen. The name refers to the striated (striped) appearance of the internal capsule as it passes among these nuclei. The amygdaloid nucleus, part of the limbic system, lies anterior to the tail of the caudate nucleus and inferior to the lentiform nucleus.

Functions of the Basal Nuclei

The basal nuclei are involved with the subconscious control of skeletal muscle tone and the coordination of learned movement patterns. Under normal conditions, these nuclei do not initiate particular movements. But once a movement is under way, the basal nuclei provide the general pattern and rhythm, especially for movements of the trunk and proximal limb muscles.

Information arrives at the caudate nucleus and putamen from sensory, motor, and integrative areas of the cerebral cortex. Processing occurs in these nuclei and in the adjacent globus pallidus. Most of the output of the basal nuclei leaves the globus pallidus and synapses in the thalamus. Nuclei in the thalamus then project the information to appropriate areas of the cerebral cortex. The basal nuclei alter the motor commands issued by the cerebral cortex through this feedback loop. For example:

- When you walk, your basal nuclei control the cycles of arm and thigh movements that occur between the time you decide to "start" walking and the time you give the "stop" order.

- As you begin a voluntary movement, your basal nuclei control and adjust muscle tone, particularly in the appendicular muscles, to set your body position. When you decide to pick up a pencil, you consciously reach and grasp with your forearm, wrist, and hand while the basal nuclei operate at the subconscious level to position your shoulder and stabilize your arm.

Neurons in the substantia nigra of the mesencephalon inhibit the activity of the basal nuclei by releasing the neurotransmitter *dopamine*. ∞ p. 414 If the substantia nigra is damaged or the neurons secrete less dopamine, basal nuclei become more active. The result is a gradual, generalized increase in muscle tone and the appearance of symptoms characteristic of **Parkinson's disease**. Persons with Parkinson's disease have difficulty starting voluntary movements, because opposing muscle groups do not relax; they must be overpowered. Once a movement is under way, every aspect must be voluntarily controlled through intense effort and concentration. **AM** The Basal Nuclei and Parkinson's Disease

✓ What name is given to axons carrying information between the brain and spinal cord, and what brain regions do they pass through?

✓ What symptoms would you expect to observe in an individual who has damage to the basal nuclei?

Answers start on page Q-1

The entire anatomy of the brain, from medulla through cerebrum can be reviewed on the **Anatomy CD-ROM:** Nervous System/Brain Dissections.

☐ MOTOR AND SENSORY AREAS OF THE CORTEX

The major motor and sensory regions of the cerebral cortex are listed in Table 14–8 and shown in Figure 14–17●. The central sulcus separates the motor and sensory areas of the cortex. The **precentral gyrus** of the frontal lobe forms the anterior border of the central sulcus. The surface of this gyrus is the **primary motor cortex**. Neurons of the primary motor cortex direct voluntary movements by controlling somatic motor neurons in the brain stem and spinal cord. These cortical neurons are called **pyramidal cells**, because their cell bodies resemble little pyramids.

The primary motor cortex is like the keyboard of a piano. If you strike a specific piano key, you produce a specific sound; if you stimulate a specific motor neuron in the primary motor cortex, you generate a contraction in a specific skeletal muscle.

Like the gauges in the dashboard of a car, the sensory areas of the cerebral cortex report key information. At each location, sensory information is reported in the pattern of neuron activity in the cortex. The **postcentral gyrus** of the parietal lobe forms the posterior border of the central sulcus, and its surface contains the **primary sensory cortex**. Neurons in this region receive somatic sensory information from receptors for touch, pressure, pain, vibration, taste, or temperature. We are consciously aware of these sensations only when nuclei in the thalamus relay the information to the primary sensory cortex.

Sensations of sight, sound, smell, and taste arrive at other portions of the cerebral cortex (Figure 14–17a●). The **visual cortex** of the occipital lobe receives visual information, and the **auditory cortex** and **olfactory cortex** of the temporal lobe receive information about hearing and smell, respectively. The **gustatory cortex**, which receives information from taste receptors of the tongue and pharynx, lies in the anterior portion of the insula and adjacent portions of the frontal lobe.

TABLE 14–8 THE CEREBRAL CORTEX

Lobe/Region	Function
FRONTAL LOBE	
Primary motor cortex	Voluntary control of skeletal muscles
PARIETAL LOBE	
Primary sensory cortex	Conscious perception of touch, pressure, vibration, pain, temperature, and taste
OCCIPITAL LOBE	
Visual cortex	Conscious perception of visual stimuli
TEMPORAL LOBE	
Auditory cortex and olfactory cortex	Conscious perception of auditory and olfactory stimuli
ALL LOBES	
Association areas	Integration and processing of sensory data; processing and initiation of motor activities

Association Areas

The sensory and motor regions of the cortex are connected to nearby **association areas**, regions of the cortex that interpret incoming data or coordinate a motor response (Figure 14–17a●). Like the information provided by the gauges in a car, the arriving information must be noticed and interpreted before it can be acted on. *Sensory association areas* are cortical regions that monitor and interpret the information that arrives at the sensory areas of the cortex. Examples include the somatic sensory association area, visual association area, and auditory association area.

The **somatic sensory association area** monitors activity in the primary sensory cortex. It is the somatic sensory association area that allows you to recognize a touch as light as the arrival of a mosquito on your arm (and gives you a chance to swat the mosquito before it bites).

The special senses of smell, sight, and hearing involve separate areas of the sensory cortex, and each has its own association area. These areas monitor and interpret arriving sensations. For example, the **visual association area** monitors the patterns of activity in the visual cortex and interprets the results. You see the symbols c, a, and r when the stimulation of receptors in your eyes leads to the stimulation of neurons in your visual cortex. Your visual association area recognizes that these are letters and that c + a + r = car. An individual with a damaged visual association area could scan the lines of a printed page and see rows of symbols that are clear, but would perceive no meaning from the symbols. Similarly, the **auditory association area** monitors sensory activity in the auditory cortex; word recognition occurs in this association area.

The **somatic motor association area**, or **premotor cortex**, is responsible for the coordination of learned movements. The primary motor cortex does nothing on its own, any more than a piano keyboard can play itself. The neurons in the primary motor cortex must be stimulated by neurons in other parts of the cerebrum. When you perform a voluntary movement, the instructions are relayed to the primary motor cortex by the premotor cortex. With repetition, the proper pattern of stimulation becomes stored in your premotor cortex. You can then perform the movement smoothly and easily by triggering the *pattern* rather than by controlling the individual neurons. This principle applies to any learned movement, from something as simple as picking up a glass to something as complex as playing the piano. One area of the premotor cortex, the *frontal eye field*, controls learned eye movements, such as when you scan these lines of type. Someone with damage to the frontal eye field can understand written letters and words but cannot read, because his or her eyes cannot follow the lines on a printed page.

Integrative Centers

Integrative centers are areas that receive information from many association areas and direct extremely complex motor activities. These centers also perform complicated analytical functions. For example, the *prefrontal cortex* of the frontal lobe (Figure 14–17b●) integrates information from sensory association areas and performs abstract intellectual functions, such as predicting the consequences of possible responses.

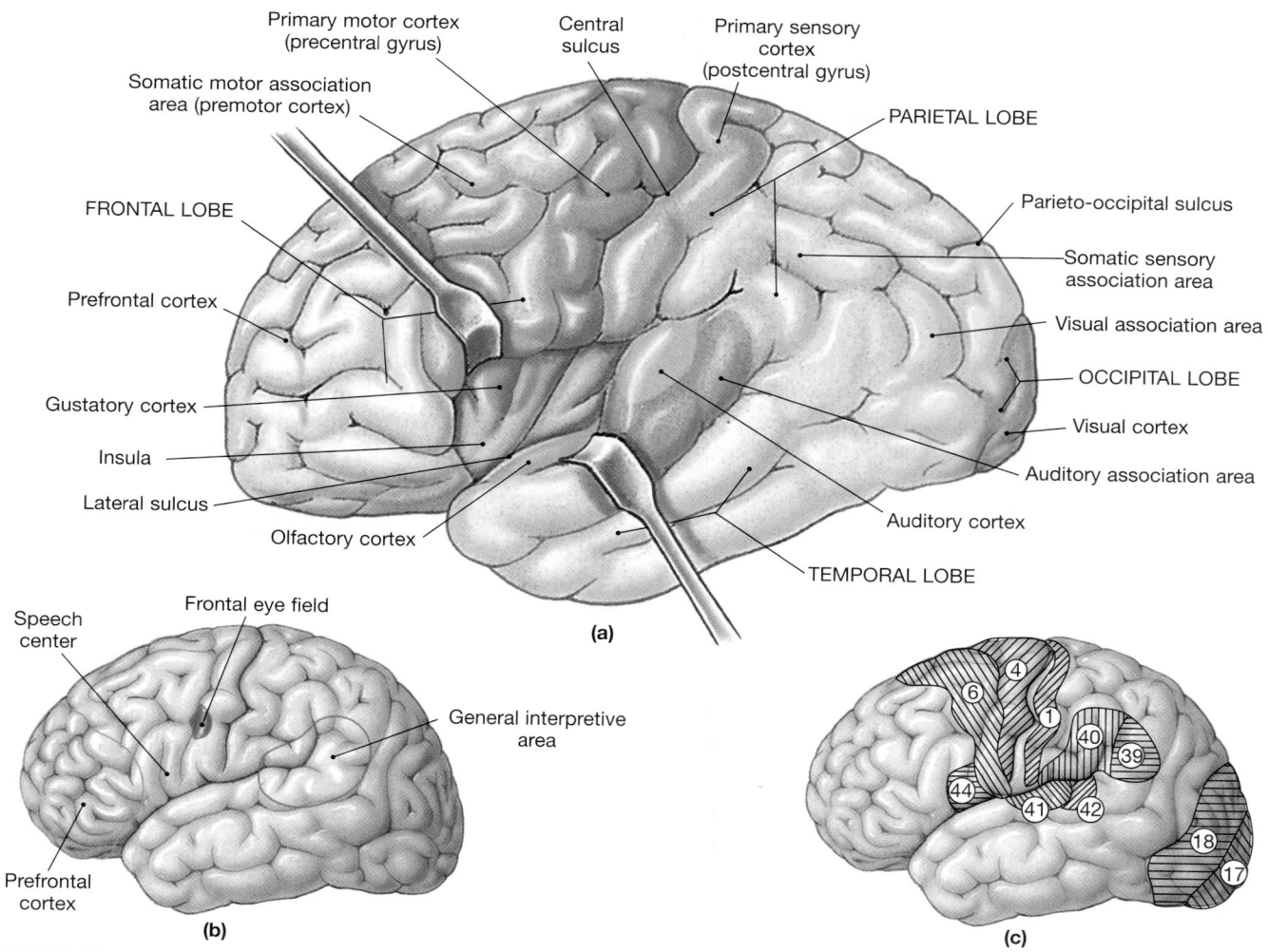

•FIGURE 14–17
The Cerebral Hemispheres.   **(a)** Major anatomical landmarks on the surface of the left cerebral hemisphere. The lateral sulcus has been pulled apart to expose the insula. **(b)** The left hemisphere generally contains the general interpretive area and the speech center. The prefrontal cortex of each hemisphere is involved with conscious intellectual functions. **(c)** Regions of the cerebral cortex as determined by histological analysis. Several of the 47 regions described by Brodmann are shown for comparison with the results of functional mapping.

Integrative centers are located in the lobes and cortical areas of both cerebral hemispheres. Integrative centers concerned with the performance of complex processes, such as speech, writing, mathematical computation, or understanding spatial relationships, are restricted to either the left or the right hemisphere. These centers include the *general interpretive area* and the *speech center*. The corresponding regions on the opposite hemisphere are also active, but their functions are less well defined.

THE GENERAL INTERPRETIVE AREA The **general interpretive area**, or *Wernicke's area* (Figure 14–17b•), also called the *gnostic area*, receives information from all the sensory association areas. This analytical center is present in only one hemisphere (typically the left). This region plays an essential role in your personality by integrating sensory information and coordinating access to complex visual and auditory memories. Damage to the general interpretive area affects the ability to interpret what is seen or

heard, even though the words are understood as individual entities. For example, if your general interpretive area were damaged, you might still understand the meaning of the spoken words *sit* and *here*, because word recognition occurs in the auditory association areas. But you would be totally bewildered by the request *sit here*. Damage to another portion of the general interpretive area might leave you able to see a chair clearly, and to know that you recognize it, but you would be unable to name it because the connection to your visual association area has been disrupted.

THE SPEECH CENTER Some of the neurons in the general interpretive area innervate the **speech center**, also called *Broca's area* or the *motor speech area* (Figure 14–17b•). This center lies along the edge of the premotor cortex in the same hemisphere as the general interpretive area (usually the left). The speech center regulates the patterns of breathing and vocalization needed for normal speech. This regulation involves coordinating the activities of the respiratory muscles, the laryngeal and pharyngeal muscles, and the muscles of the tongue, cheeks, lips, and jaws. A person with a damaged speech center can make sounds but not words.

The motor commands issued by the speech center are adjusted by feedback from the auditory association area, also called the *receptive speech area*. Damage to the related sensory areas can cause a variety of speech-related problems. Some affected individuals have difficulty speaking although they know exactly which words to use; others talk constantly but use all the wrong words.

THE PREFRONTAL CORTEX The **prefrontal cortex** of the frontal lobe coordinates information relayed from the association areas of the entire cortex. In doing so, it performs such abstract intellectual functions as predicting the consequences of events or actions. Damage to the prefrontal cortex leads to difficulties in estimating temporal relationships between events. Questions such as "How long ago did this happen?" or "What happened first?" become difficult to answer.

The prefrontal cortex has extensive connections with other cortical areas and with other portions of the brain. Feelings of frustration, tension, and anxiety are generated at the prefrontal cortex as it interprets ongoing events and makes predictions about future situations or consequences. If the connections between the prefrontal cortex and other brain regions are severed, the frustrations, tensions, and anxieties are removed. During the middle of the twentieth century, this rather drastic procedure, called **prefrontal lobotomy**, was used to "cure" a variety of mental illnesses, especially those associated with violent or antisocial behavior. After a lobotomy, the patient would no longer be concerned about what had previously been a major problem, whether psychological (hallucinations) or physical (severe pain). However, the individual was often equally unconcerned about tact, decorum, and toilet training. Drugs have been developed to target specific pathways and regions of the CNS, so lobotomies are no longer used to change behavior.

BRODMANN'S AREAS Early in the 20th century, several attempts were made to describe and classify regional differences in the histological organization of the cerebral cortex. It was hoped that the patterns of cellular organization could be correlated to specific functions. By 1919, at least 200 patterns had been described, but most of the classification schemes have since been abandoned. However, the cortical map prepared by Korbinian Brodmann in 1909 has proved useful to neuroanatomists. Brodmann, a German neurologist, described 47 patterns of cellular organization in the cerebral cortex. Several of these *Brodmann's areas* are shown in Figure 14–17c●. Some correspond to known functional areas. For example, Brodmann's area 44 corresponds to the speech center and area 41 to the auditory cortex; area 4 follows the contours of the primary motor cortex. In other cases, the correspondence is less precise. For instance, Brodmann's area 42 forms only a small portion of the auditory association area.

Hemispheric Lateralization

Each of the two cerebral hemispheres is responsible for specific functions that are not ordinarily performed by the opposite hemisphere. This regional specialization has been called *hemispheric lateralization*. Figure 14–18● indicates the major functional differences between the hemispheres. In most people, the left hemisphere contains the general interpretive and speech centers and is responsible for language-based skills. For example, reading, writing, and speaking are dependent on processing done in the left cerebral hemisphere. In addition, the premotor cortex involved

---

Aphasia and Dyslexia

Aphasia (*a-*, without + *phasia*, speech) is a disorder affecting the ability to speak or read. **Global aphasia** results from extensive damage to the general interpretive area or to the associated sensory tracts. Affected individuals are unable to speak, to read, or to understand the speech of others. Global aphasia often accompanies a severe stroke or tumor that affects a large area of cortex, including the speech and language areas. Recovery is possible when the condition results from edema or hemorrhage, but the process often takes months or even years.
AM Aphasia

Dyslexia (*lexis*, diction) is a disorder affecting the comprehension and use of words. **Developmental dyslexia** affects children; estimates indicate that up to 15 percent of children in the United States have some degree of dyslexia. For unknown reasons, the problem is much more common among left-handed children than right-handed children. Children with dyslexia have difficulty reading and writing, although their other intellectual functions

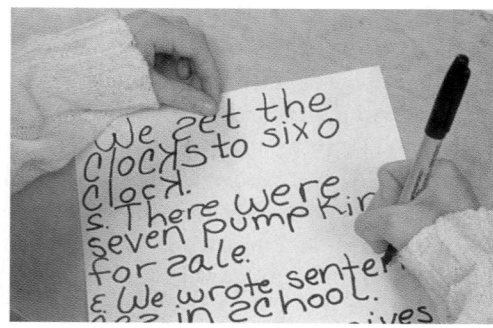

may be normal or above normal. Their writing looks uneven and disorganized; letters are typically written in the wrong order (*dig* becomes *gid*) or reversed (*E* becomes Ǝ). Recent evidence suggests that at least some forms of dyslexia result from problems in processing, sorting, and integrating visual or auditory information.

with the control of hand movements is larger on the left side for right-handed individuals than for left-handed ones. The left hemisphere is also important in performing analytical tasks, such as mathematical calculations and logical decision making. For these reasons, the left cerebral hemisphere has been called the *dominant hemisphere*, or the *categorical hemisphere*.

The right cerebral hemisphere analyzes sensory information and relates the body to the sensory environment. Interpretive centers in this hemisphere permit you to identify familiar objects by touch, smell, sight, taste, or feel. For example, the right hemisphere plays a dominant role in recognizing faces and in understanding three-dimensional relationships. It is also important in analyzing the emotional context of a conversation—for instance, determining whether the phrase "get lost" was intended as a threat or a question. Individuals with a damaged right hemisphere may be unable to add emotional inflections to their own words.

Left-handed persons represent 9 percent of the human population; in most cases, although the primary motor cortex of the right hemisphere controls motor function for the dominant hand, the centers involved with speech and analytical function are in the left hemisphere. Interestingly, an unusually high per-

centage of musicians and artists are left handed. The complex motor activities performed by these individuals are directed by the primary motor cortex and association areas on the right cerebral hemisphere, near the association areas involved with spatial visualization and emotions.

Monitoring Brain Activity: The Electroencephalogram

The primary sensory cortex and the primary motor cortex have been mapped by direct stimulation in patients undergoing brain surgery. The functions of other regions of the cerebrum can be revealed by the behavioral changes that follow localized injuries or strokes, and the activities of specific regions can be examined by a PET scan or sequential MRI scans.

The electrical activity of the brain is commonly monitored to assess brain activity. Neural function depends on electrical events within the cell membrane of neurons. The brain contains billions of neurons, and their activity generates an electrical field that can be measured by placing electrodes on the brain or on the outer surface of the skull. The electrical activity changes constantly, as nuclei and cortical areas are stimulated or quiet down. A printed report of the electrical activity of the brain is called an **electroencephalogram** (**EEG**). The electrical patterns observed are called *brain waves*.

Typical brain waves are shown in Figure 14–19a–d●. **Alpha waves** occur in the brains of healthy, awake adults who are resting with their eyes closed. Alpha waves disappear during sleep, but they also vanish when the individual begins to concentrate on some specific task. During attention to stimuli or tasks, alpha waves are replaced by higher frequency **beta waves**. Beta waves are typical of individuals who are either concentrating on a task, under stress, or in a state of psychological tension. **Theta waves** may appear transiently during sleep in normal adults but are most often observed in children and in intensely frustrated adults. The presence of theta waves under other circumstances may indicate the presence of a brain disorder, such as a tumor. **Delta waves** are very-large-amplitude, low-frequency waves. They are normally seen during deep sleep in individuals of all ages. Delta waves are also seen in the brains of infants (in whom cortical development is still incomplete) and in the awake adult when a tumor, vascular blockage, or inflammation has damaged portions of the brain.

Electrical activity in the two hemispheres is generally synchronized by a "pacemaker" mechanism that appears to involve the thalamus. Asynchrony between the hemispheres can therefore be used to detect localized damage or other cerebral abnormalities. For example, a tumor or injury affecting one hemisphere typically changes the pattern in that hemisphere, and the patterns of the two hemispheres are no longer aligned. A

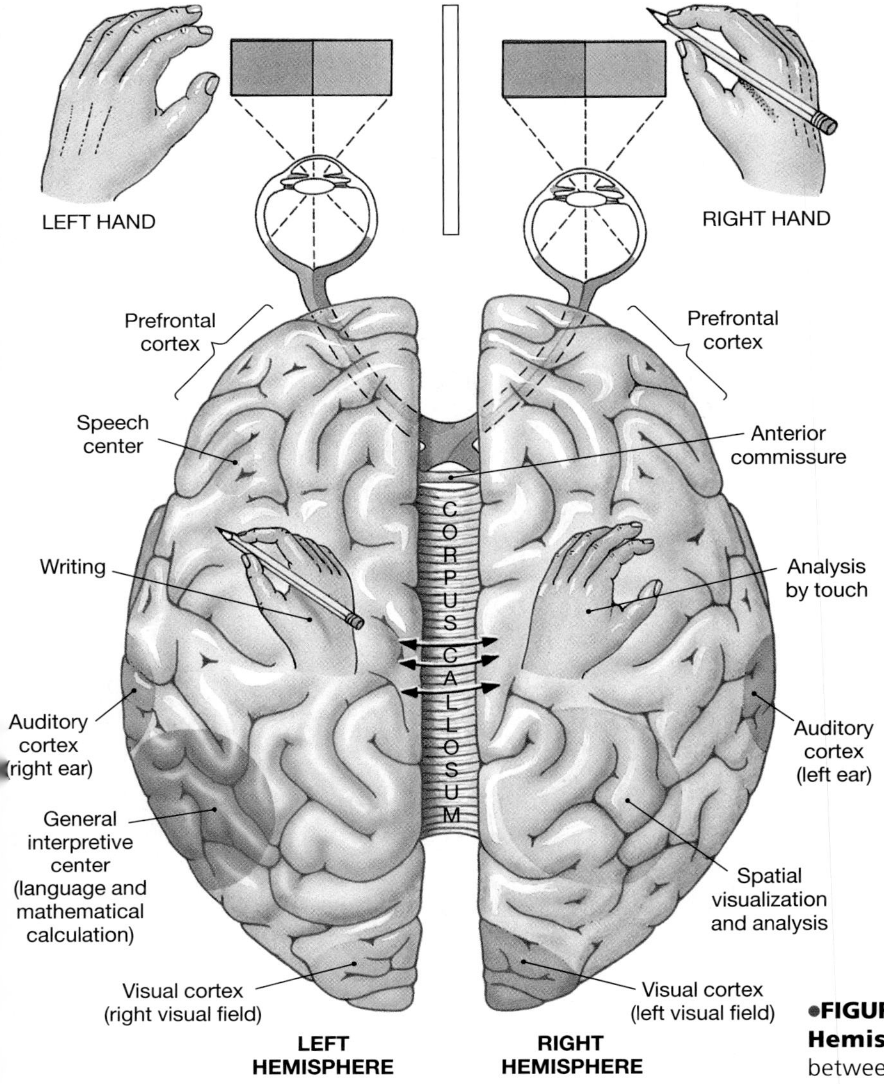

●**FIGURE 14–18**
Hemispheric Lateralization. Functional differences between the left and right cerebral hemispheres.

Disconnection Syndrome

The functional differences between the hemispheres become apparent if the corpus callosum is cut, a procedure sometimes performed to treat epileptic seizures that cannot be controlled by other methods. This surgery produces symptoms of **disconnection syndrome**. In this condition, the two hemispheres function independently, each "unaware" of stimuli or motor commands that involve its counterpart.

Individuals with this syndrome exhibit some rather interesting changes in their mental abilities. For example, objects touched by the left hand can be recognized but not verbally identified, because the sensory information arrives at the right hemisphere but the speech center is on the left. The object can be verbally identified if felt with the right hand, but the person cannot say whether it is the same object previously touched with the left hand. Sensory information from the left side of the body arrives at the right hemisphere and cannot reach the general interpretive area. Thus, conscious decisions are made without regard to sensations from the left side.

Two years after a surgical sectioning of the corpus callosum, the most striking behavioral abnormalities have disappeared and the person may test normally. In addition, individuals born without a functional corpus callosum do not have sensory, motor, or intellectual problems. In some way, the CNS adapts to these situations, probably by increasing the amount of information transferred across the anterior commissure.

---

seizure is a temporary cerebral disorder accompanied by abnormal movements, unusual sensations, inappropriate behavior, or some combination of these symptoms. Clinical conditions characterized by seizures are known as seizure disorders, or *epilepsies*. Seizures of all kinds are accompanied by a marked change in the pattern of electrical activity monitored in an electroencephalogram. The change begins in one portion of the cerebral cortex but may subsequently spread across the entire cortical surface, like a wave on the surface of a pond.

The nature of the symptoms produced depends on the region of the cortex involved. If a seizure affects the primary motor cortex, movements will occur; if it affects the auditory cortex, the individual will hear strange sounds. **AM** Seizures and Epilepsies

✓ Shelly suffers a head injury that damages her primary motor cortex. Where is this area located?

✓ Which senses would be affected by damage to the temporal lobes of the cerebrum?

✓ After suffering a stroke, Jake is unable to speak. He can understand what is said to him, and he can understand written messages, but he cannot express himself verbally. Which part of his brain has been affected by the stroke?

✓ Paul is having a difficult time remembering facts and recalling long-term memories. Which part of his cerebrum is probably involved?

Answers start on page Q-1

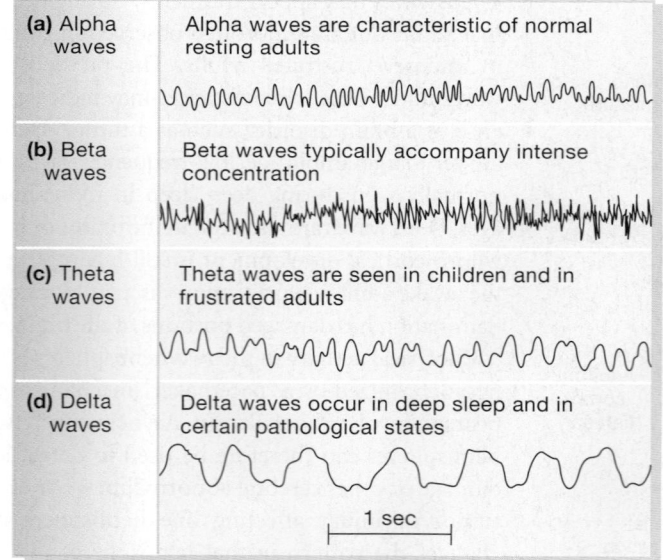

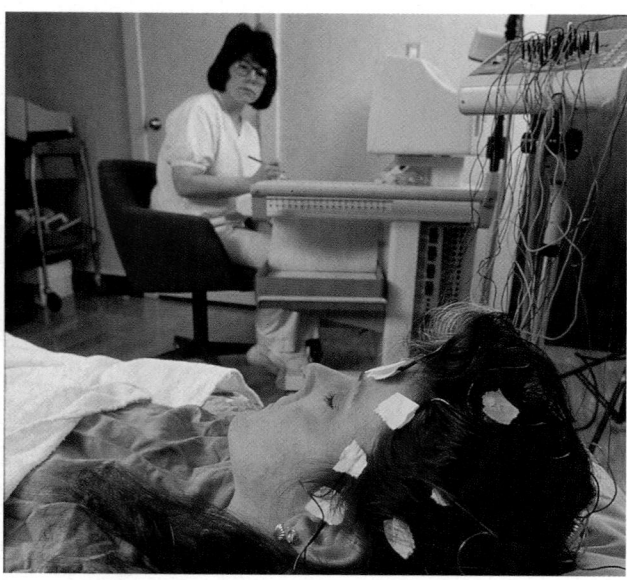

(e)

●**FIGURE 14–19**
Brain Waves. **(a)** Alpha waves are characteristic of healthy resting adults. **(b)** Beta waves typically accompany intense concentration. **(c)** Theta waves are seen in children and in frustrated adults. **(d)** Delta waves occur during deep sleep and in certain pathological states. (These four waves are not drawn to the same scale.) **(e)** A patient wired for EEG monitoring.

Cranial Nerves | focus:

Cranial nerves are PNS components that connect directly to the brain rather than to the spinal cord. The twelve pairs of cranial nerves are located on the ventrolateral surface of the brain stem (Figure 14–20●), each with a name related to its appearance or function.

The number assigned to a cranial nerve roughly corresponds to the nerve's position along the longitudinal axis of the brain, beginning at the cerebrum. Roman numerals are usually used. (You may sometimes encounter these numerals preceded by the prefix N or CN.)

Each cranial nerve attaches to the brain near the associated sensory or motor nuclei. The sensory nuclei act as switching centers, with the postsynaptic neurons relaying the information to other nuclei or to processing centers in the cerebral or cerebellar cortex. In a similar fashion, the motor nuclei receive convergent inputs from higher centers or from other nuclei along the brain stem.

In this section, we classify cranial nerves as primarily sensory, special sensory, motor, or mixed (sensory and motor). In this classification, sensory nerves carry somatic sensory information, including touch, pressure, vibration, temperature, or pain. Special sensory nerves carry the sensations of smell, sight, hearing, or balance. Motor nerves are dominated by the axons of somatic motor neurons; mixed nerves have a mixture of sensory and motor fibers. This is a useful method of classification, but it is based on the primary function, and a cranial nerve can have important secondary functions. Three examples are worth noting:

1. The olfactory receptors, the visual receptors, and the receptors of the inner ear are innervated by cranial nerves that are dedicated almost entirely to carrying special sensory information. The sensation of taste, considered to be one of the special senses, is carried by axons that form only a small part of large cranial nerves that have other primary functions.

2. As elsewhere in the PNS, a nerve containing tens of thousands of motor fibers that lead to a skeletal muscle will also contain sensory fibers from muscle spindles and tendon organs in that muscle. We assume that these sensory fibers are present but ignore them in the classification of the nerve.

3. Regardless of their other functions, several cranial nerves (III, VII, IX, and X) distribute autonomic fibers to peripheral ganglia, just as spinal nerves deliver them to ganglia along the spinal cord. We shall note the presence of small numbers of autonomic fibers (and shall discuss them further in Chapter 16) but ignore them in the classification of the nerve.

●FIGURE 14–20
Origins of the Cranial Nerves. **(a)** An inferior view of the brain. **(b)** A diagrammatic view, showing the attachment of the 12 pairs of cranial nerves.**(c)** Superior view of the cranial fossae after removal of the brain. Several cranial nerves are visible; compare with parts (a) and (b).

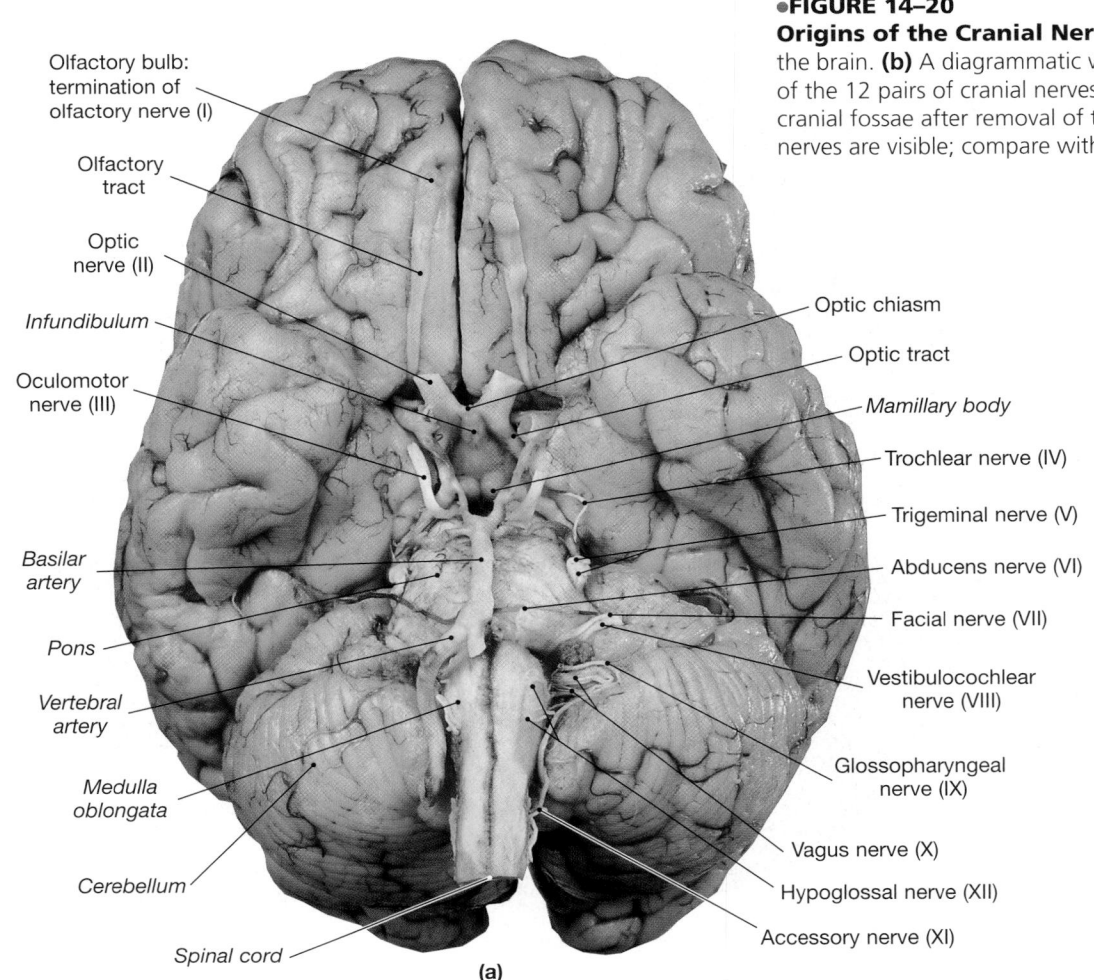

Olfactory bulb: termination of olfactory nerve (I)
Olfactory tract
Optic nerve (II)
Infundibulum
Oculomotor nerve (III)
Basilar artery
Pons
Vertebral artery
Medulla oblongata
Cerebellum
Spinal cord

Optic chiasm
Optic tract
Mamillary body
Trochlear nerve (IV)
Trigeminal nerve (V)
Abducens nerve (VI)
Facial nerve (VII)
Vestibulocochlear nerve (VIII)
Glossopharyngeal nerve (IX)
Vagus nerve (X)
Hypoglossal nerve (XII)
Accessory nerve (XI)

(a)

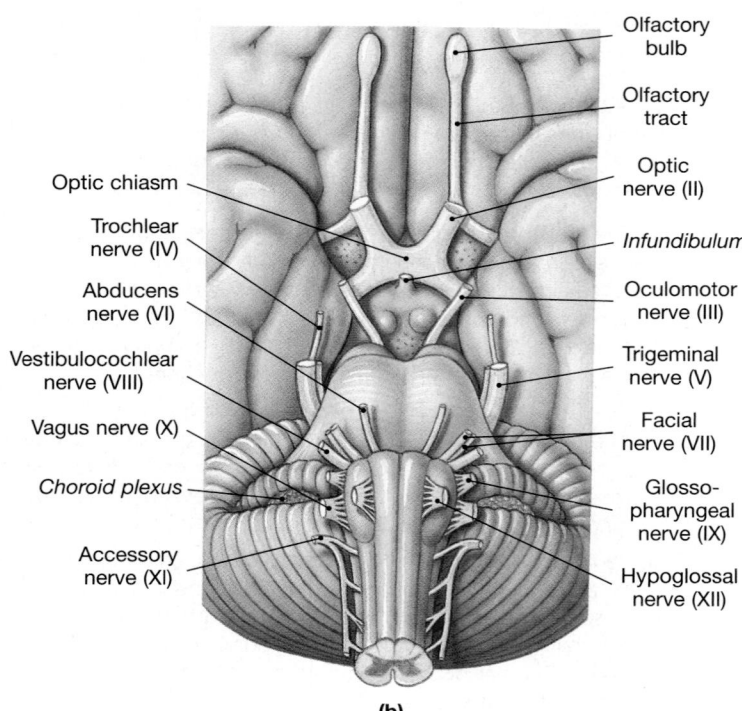

(b)

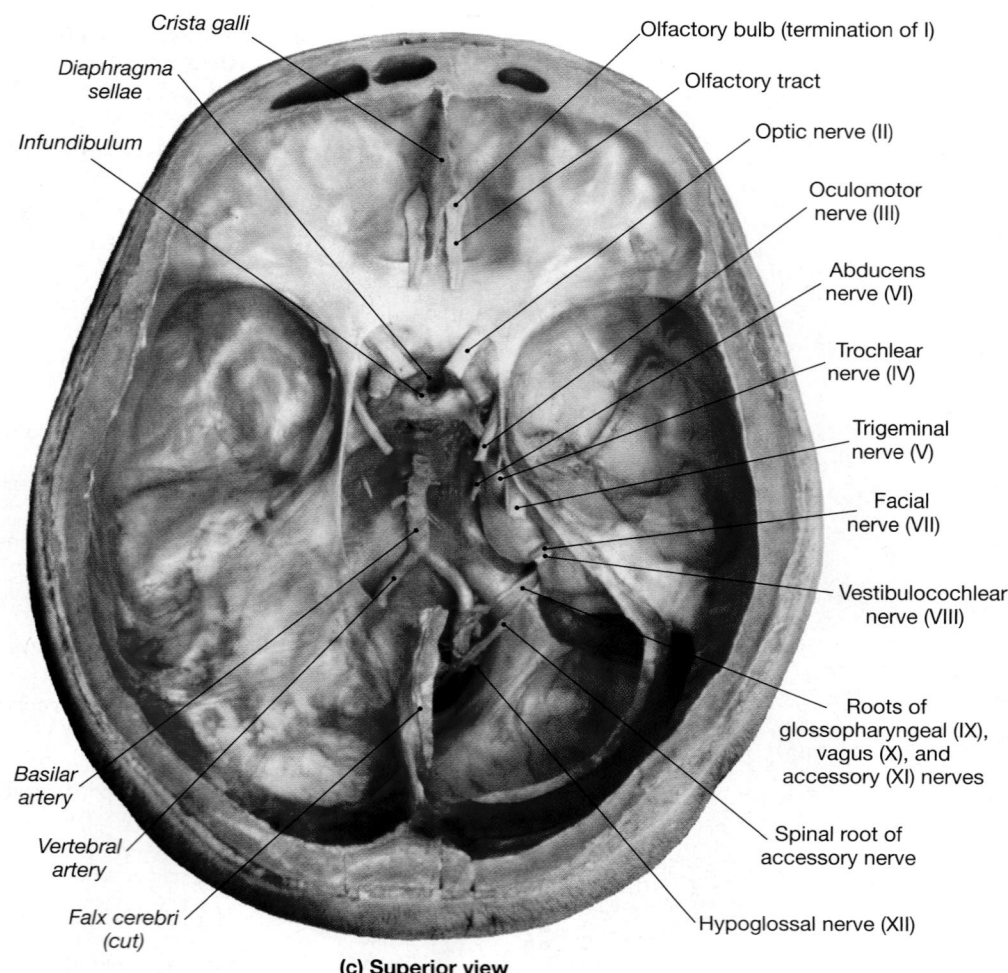

(c) Superior view

●**FIGURE 14–20**
Origins of the Cranial Nerves (*continued*).

The Olfactory Nerves (I)

Primary function: Special sensory (smell)
Origin: Receptors of olfactory epithelium
Pass through: Olfactory foramina in cribriform plate of ethmoid
∞ pp. 215, 220
Destination: Olfactory bulbs

The first pair of cranial nerves (Figure 14–21●) carries special sensory information responsible for the sense of smell. The olfactory receptors are specialized neurons in the epithelium covering the roof of the nasal cavity, the superior nasal conchae, and the superior parts of the nasal septum. Axons from these sensory neurons collect to form 20 or more bundles that penetrate the cribriform plate of the ethmoid bone. These bundles are components of the **olfactory nerves** (I). Almost at once these bundles enter the **olfactory bulbs**, neural masses on either side of the crista galli. The olfactory afferents synapse within the olfactory bulbs. The axons of the postsynaptic neurons proceed to the cerebrum along the slender **olfactory tracts** (Figures 14–20 and 14–21●).

Because the olfactory tracts look like typical peripheral nerves, anatomists about one hundred years ago misidentified these tracts as the first cranial nerve. Later studies demonstrated that the olfactory tracts and bulbs are part of the cerebrum, but by then the numbering system was already firmly established. Anatomists were left with a forest of tiny olfactory nerve bundles lumped together as nerve I.

The olfactory nerves are the only cranial nerves attached directly to the cerebrum. The rest originate or terminate within nuclei of the diencephalon or brain stem, and the ascending sensory information synapses in the thalamus before reaching the cerebrum.

The Optic Nerves (II)

Primary function: Special sensory (vision)
Origin: Retina of eye
Pass through: Optic canals of sphenoid ∞ p. 219
Destination: Diencephalon via the optic chiasm

The **optic nerves** (II) carry visual information from special sensory ganglia in the eyes. These nerves (Figure 14–22●) contain about 1 million sensory nerve fibers. The optic nerves pass through the optic canals of the sphenoid. Then they converge at the ventral, anterior margin of the diencephalon, at the **optic chiasm** (*chiasma*, a crossing). At the optic chiasm, fibers from the nasal half of each retina cross over to the opposite side of the brain.

The reorganized axons continue toward the lateral geniculate nuclei of the thalamus as the **optic tracts** (Figures 14–20 and 14–22●). After synapsing in the lateral geniculates, projection fibers deliver the information to the visual cortex of the occipital lobe. With this arrangement, each cerebral hemisphere receives visual information from the lateral half of the retina of the eye on that side and from the medial half of the retina of the eye of the opposite side. Relatively few axons in the optic tracts bypass the lateral geniculate nuclei and synapse in the superior colliculus of the midbrain. We will consider that pathway in Chapter 17.

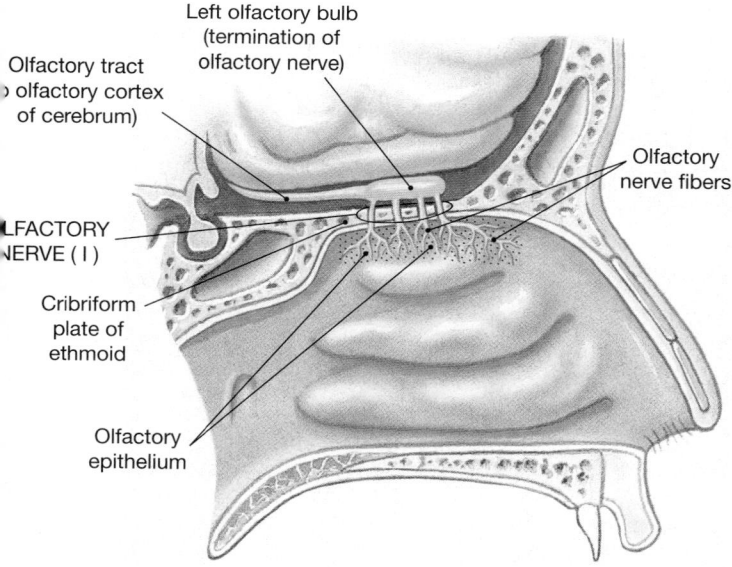

●**FIGURE 14–21**
The Olfactory Nerve

Left olfactory bulb (termination of olfactory nerve)
Olfactory tract (to olfactory cortex of cerebrum)
OLFACTORY NERVE (I)
Cribriform plate of ethmoid
Olfactory epithelium
Olfactory nerve fibers

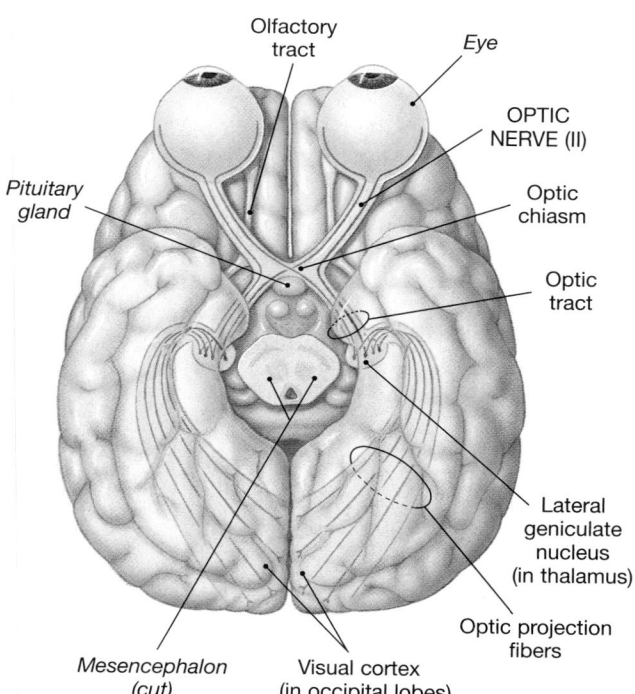

●**FIGURE 14–22**
The Optic Nerve

Olfactory tract
Eye
OPTIC NERVE (II)
Pituitary gland
Optic chiasm
Optic tract
Lateral geniculate nucleus (in thalamus)
Optic projection fibers
Mesencephalon (cut)
Visual cortex (in occipital lobes)

FOCUS *continues* →

The Oculomotor Nerves (III)

Primary function: Motor (eye movements)

Origin: Mesencephalon

Pass through: Superior orbital fissures of sphenoid ∞ **pp. 214, 219, 224**

Destination: *Somatic motor*: superior, inferior, and medial rectus muscles; inferior oblique muscle; levator palpebrae superioris muscle. *Visceral motor*: intrinsic eye muscles

The mesencephalon contains the motor nuclei controlling the third and fourth cranial nerves. Each **oculomotor nerve** (III) innervates four of the six extra-ocular muscles that move the eye and the levator palpebrae superioris muscle, which raises the upper eyelid (Figure 14–23•). On each side of the brain, nerve III emerges from the ventral surface of the mesencephalon and penetrates the posterior wall of the orbit at the superior orbital fissure. Individuals with damage to this nerve often complain of pain over the eye and double vision, because the movements of the left and right eyes cannot be coordinated properly.

The oculomotor nerve also delivers preganglionic autonomic fibers to neurons of the **ciliary ganglion**. The neurons of the ciliary ganglion control intrinsic eye muscles. These muscles change the diameter of the pupil, adjusting the amount of light entering the eye, and change the shape of the lens to focus images on the retina.

The Trochlear Nerves (IV)

Primary function: Motor (eye movements)

Origin: Mesencephalon

Pass through: Superior orbital fissures of sphenoid ∞ **pp. 214, 219, 224**

Destination: Superior oblique muscle

A **trochlear** (TRŌK-lē-ar; *trochlea*, a pulley) **nerve** (IV), the smallest cranial nerve, innervates the superior oblique muscle of each eye (Figure 14–23•). The trochlea is a pulley-shaped, ligamentous sling. Each superior oblique muscle passes through a trochlea on its way to its insertion on the surface of the eye. An individual with damage to nerve IV or to its nucleus will have difficulty looking down and to the side.

The Abducens Nerves (VI)

Primary function: Motor (eye movements)

Origin: Pons

Pass through: Superior orbital fissures of sphenoid ∞ **p. 214, 219, 224**

Destination: Lateral rectus muscle

The **abducens** (ab-DŪ-senz) **nerves** (VI) innervate the lateral rectus muscles, the sixth pair of extra-ocular muscles. This muscle makes the eye look to the side; in essence, the *abducens* causes *abduction* of the eye. Each abducens nerve emerges from the inferior surface of the brain stem at the border between the pons and the medulla oblongata (Figure 14–23•). Along with the oculomotor and trochlear nerves from that side, it reaches the orbit through the superior orbital fissure.

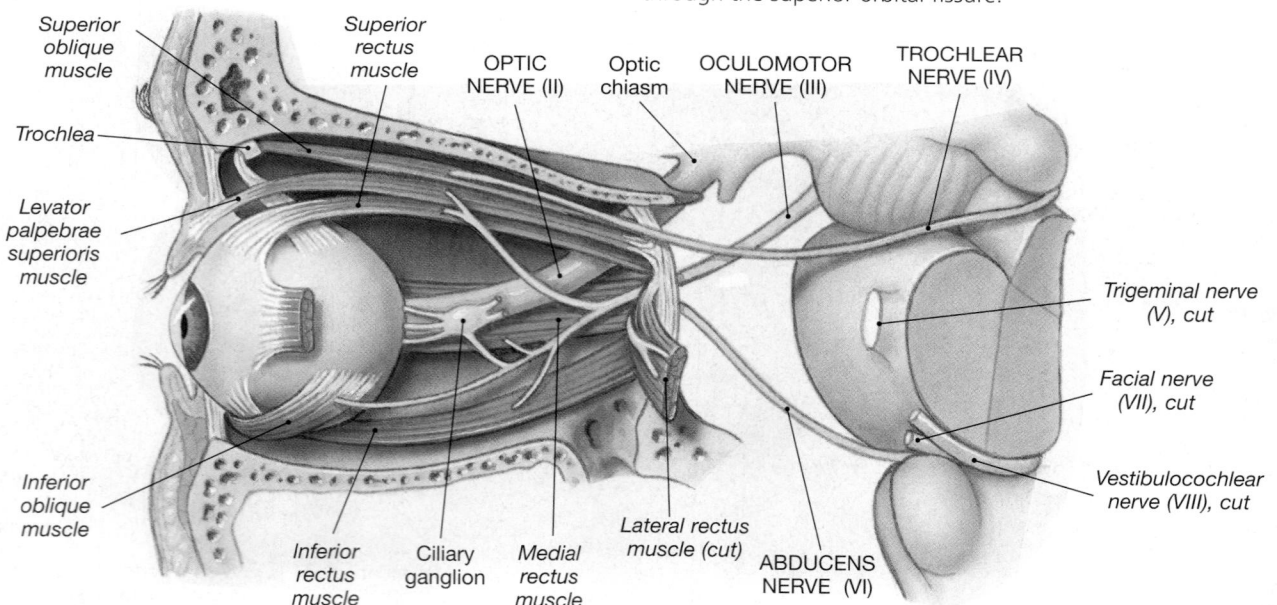

Superior oblique muscle

Superior rectus muscle

OPTIC NERVE (II)

Optic chiasm

OCULOMOTOR NERVE (III)

TROCHLEAR NERVE (IV)

Trochlea

Levator palpebrae superioris muscle

Trigeminal nerve (V), cut

Facial nerve (VII), cut

Vestibulocochlear nerve (VIII), cut

Inferior oblique muscle

Inferior rectus muscle

Ciliary ganglion

Medial rectus muscle

Lateral rectus muscle (cut)

ABDUCENS NERVE (VI)

•**FIGURE 14–23**
Cranial Nerves Controlling the Extra-ocular Muscles. **ATLAS** Plate 3.2

The Trigeminal Nerves (V)

Primary function: Mixed (sensory and motor) to face

Origin: *Ophthalmic branch* (sensory): orbital structures, nasal cavity, skin of forehead, upper eyelid, eyebrow, nose (part). *Maxillary branch* (sensory): lower eyelid, upper lip, gums, and teeth; cheek; nose, palate, and pharynx (part). *Mandibular branch* (mixed): sensory from lower gums, teeth, and lips; palate and tongue (part); motor from motor nuclei of pons

Pass through (on each side): Ophthalmic branch through superior orbital fissure, maxillary branch through foramen rotundum, mandibular branch through foramen ovale ⚭ pp. 214, 215, 219, 224

Destination: Ophthalmic, maxillary and mandibular branches to sensory nuclei in pons; mandibular branch also innervates muscles of mastication ⚭ p. 346

The pons contains the nuclei associated with three cranial nerves (V, VI, and VII) and contributes to a fourth (VIII). The **trigeminal** (trī-JEM-i-nal) **nerves** (V), the largest cranial nerves, are mixed nerves. Each provides both somatic sensory information from the head and face and motor control over the muscles of mastication. Sensory (dorsal) and motor (ventral) roots originate on the lateral surface of the pons (Figure 14–24●). The sensory branch is larger, and the enormous **semilunar ganglion** contains the cell bodies of the sensory neurons. As the name implies, the trigeminal has three major branches; the relatively small motor root contributes to only one of the three. **Tic douloureux** (doo-loo-ROO; *douloureux,* painful) is a painful condition affecting the area innervated by the maxillary and mandibular branches of the trigeminal nerve. Sufferers complain of debilitating pain triggered by contact with the lip, tongue, or gums. The cause of the condition is unknown. **AM** Tic Douloureux

The trigeminal nerve branches are associated with the *ciliary, sphenopalatine, submandibular,* and *otic ganglia.* These are autonomic (parasympathetic) ganglia whose neurons innervate structures of the face. However, although its nerve fibers may pass around or through these ganglia, the trigeminal nerve does not contain visceral motor fibers. We discussed the ciliary ganglion on page 496 and will describe the other ganglia next, with the branches of the *facial nerves* (VII) and the *glossopharyngeal nerves* (IX).

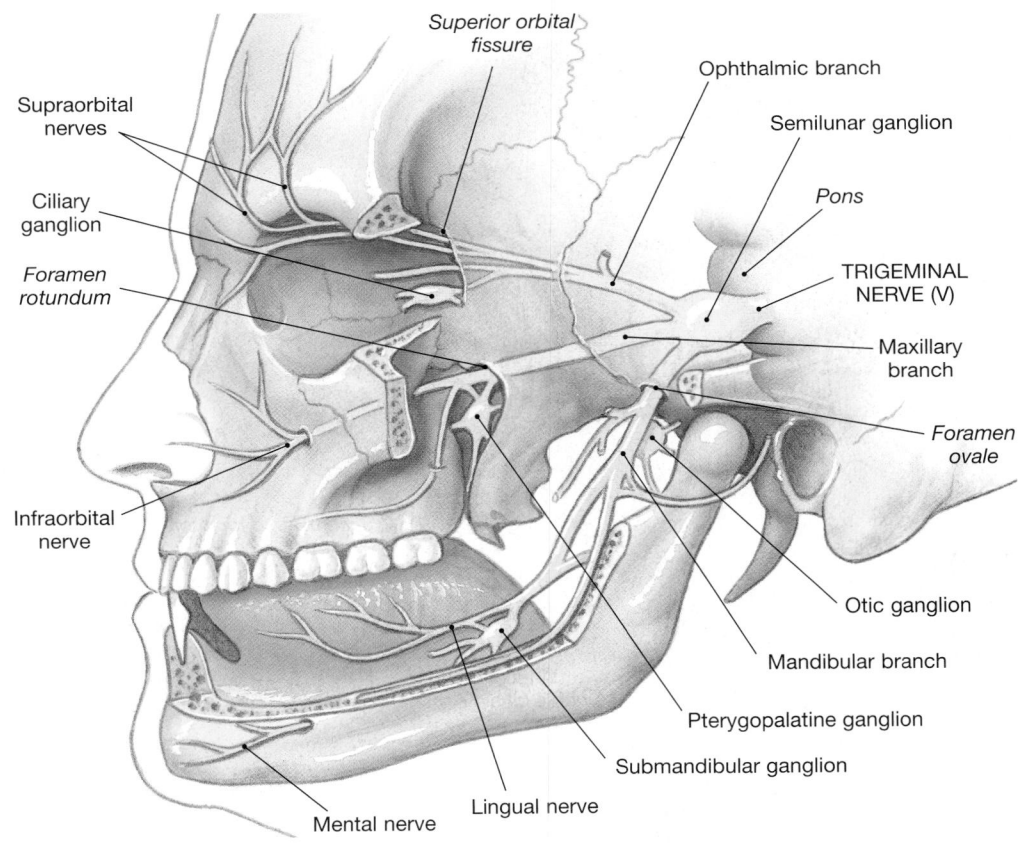

●**FIGURE 14–24**
The Trigeminal Nerve

FOCUS *continues* →

The Facial Nerves (VII)

Primary function: Mixed (sensory and motor) to face

Origin: *Sensory*: taste receptors on anterior two thirds of tongue. *Motor*: motor nuclei of pons

Pass through: Internal acoustic canals to the *facial canals*, which end at the stylomastoid foramina ∞ **pp. 214, 215, 218**

Destination: *Sensory*: sensory nuclei of pons. *Somatic motor*: muscles of facial expression. ∞ **pp. 342-343** *Visceral motor*: lacrimal (tear) gland and nasal mucous glands by way of the pterygopalatine ganglion; submandibular and sublingual salivary glands by way of the submandibular ganglion

The **facial nerves** (VII) are mixed nerves. The cell bodies of the sensory neurons are located in the **geniculate ganglia**, and the motor nuclei are in the pons. On each side, the sensory and motor roots emerge from the pons and enter the internal acoustic canal of the temporal bone. Each facial nerve then passes through the facial canal to reach the face by way of the stylomastoid foramen. The nerve then splits to form the temporal, zygomatic, buccal, mandibular, and cervical branches (Figure 14–25●).

The sensory neurons monitor proprioceptors in the facial muscles, provide deep pressure sensations over the face, and receive taste information from receptors along the anterior two-thirds of the tongue. Somatic motor fibers control the superficial muscles of the scalp and face and deep muscles near the ear.

The facial nerves carry preganglionic autonomic fibers to the sphenopalatine and submandibular ganglia. Postganglionic fibers from the **pterygopalatine ganglia** innervate the lacrimal glands and small glands of the nasal cavity and pharynx. The **submandibular ganglia** innervate the *submandibular* and *sublingual* (*sub-*, under, + *lingua*, tongue) *salivary glands*.

Bell's palsy is a cranial nerve disorder that results from an inflammation of a facial nerve. The condition is probably due to a viral infection. Symptoms include paralysis of facial muscles on the affected side and loss of taste sensations from the anterior two-thirds of the tongue. The condition is usually painless and in most cases goes away after a few weeks or months.

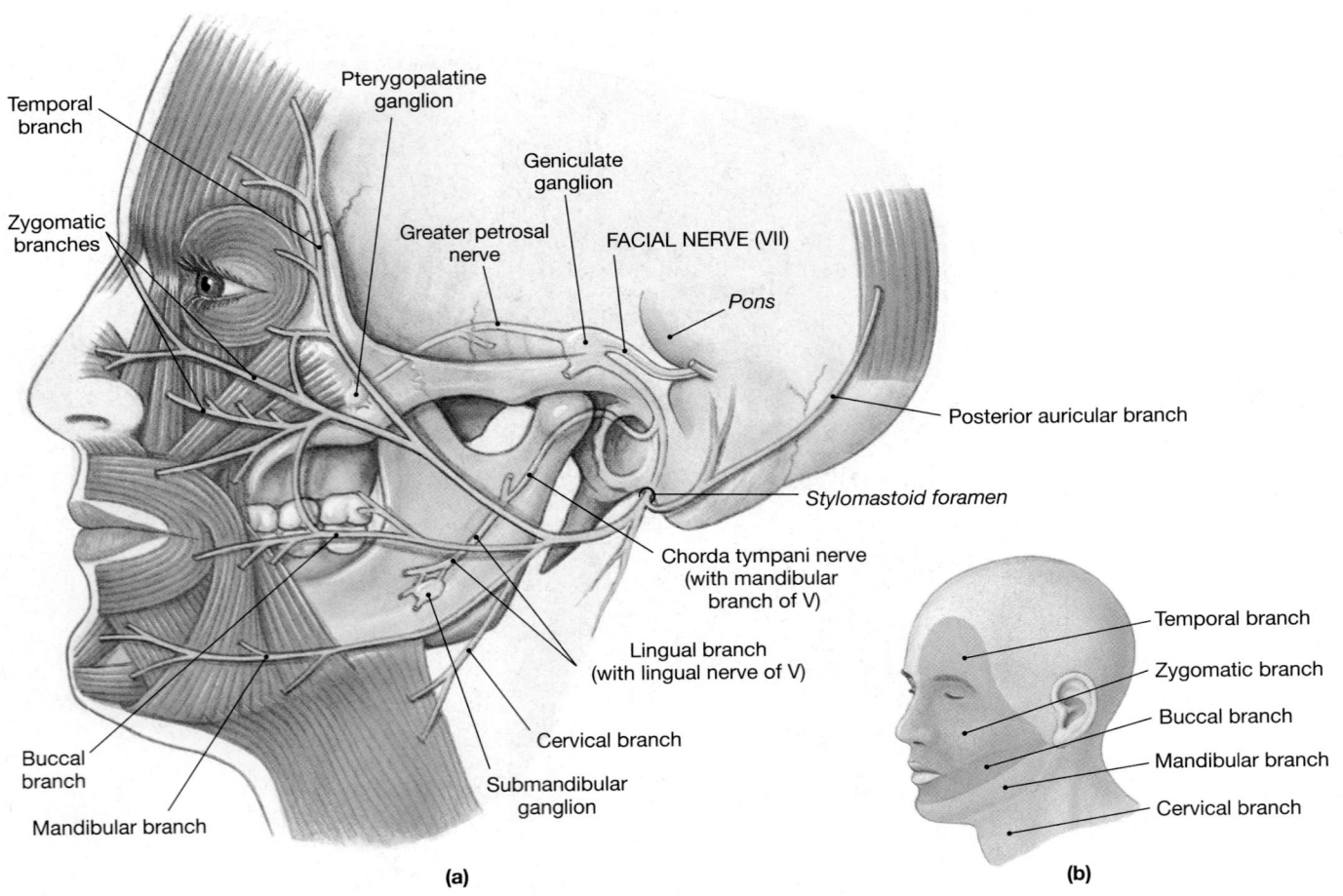

(a)

(b)

●**FIGURE 14–25**
The Facial Nerve. **(a)** The origin and branches of the facial nerve. **(b)** The superficial distribution of the five major branches of the facial nerve.

The Vestibulocochlear Nerves (VIII)

Primary function: Special sensory: balance and equilibrium (vestibular branch) and hearing (cochlear branch)

Origin: Monitor receptors of the inner ear (vestibule and cochlea)

Pass through: Internal acoustic canals of temporal bones ⚭ pp. 215, 218

Destination: Vestibular and cochlear nuclei of pons and medulla oblongata

The **vestibulocochlear nerves** (VIII) are also known as the *acoustic nerves*, the *auditory nerves*, and the *stato-acoustic nerves*. We will use *vestibulocochlear*, because this term indicates the names of the two major branches: the vestibular branch and the cochlear branch. Each vestibulocochlear nerve lies posterior to the origin of the facial nerve, straddling the boundary between the pons and the medulla oblongata (Figure 14–26●). This nerve reaches the sensory receptors of the inner ear by entering the internal acoustic canal in company with the facial nerve. Each vestibulocochlear nerve has two distinct bundles of sensory fibers. The **vestibular branch** (*vestibulum*, cavity) originates at the receptors of the *vestibule*, the portion of the inner ear concerned with balance sensations. The sensory neurons are located in an adjacent sensory ganglion, and their axons target the **vestibular nuclei** of the pons and medulla oblongata. These afferents convey information about the orientation and movement of the head. The **cochlear branch** (*cochlea*, snail shell) monitors the receptors in the *cochlea*, the portion of the inner ear that provides the sense of hearing. The cell bodies of the sensory neurons are located within a peripheral ganglion (the *spiral ganglion*), and their axons synapse within the **cochlear nuclei** of the pons and medulla oblongata. Axons leaving the vestibular and cochlear nuclei relay the sensory information to other centers or initiate reflexive motor responses. We will discuss balance and the sense of hearing in Chapter 17.

The Glossopharyngeal Nerves (IX)

Primary function: Mixed (sensory and motor) to head and neck

Origin: *Sensory*: posterior one-third of the tongue, part of the pharynx and palate, carotid arteries of the neck. *Motor*: motor nuclei of medulla oblongata

Pass through: Jugular foramina between the occipital bone and the temporal bones ⚭ pp. 214, 215

Destination: *Sensory*: sensory nuclei of medulla oblongata. *Somatic motor*: pharyngeal muscles involved in swallowing. *Visceral motor*: parotid salivary gland by way of the otic ganglion

The medulla oblongata contains the sensory and motor nuclei of cranial nerves IX, X, XI, and XII, in addition to the vestibular nucleus of nerve VIII. The **glossopharyngeal** (glos-ō-fah-RIN-jē-al; *glossum*, tongue) **nerves** (IX) innervate the tongue and pharynx. Each glossopharyngeal nerve penetrates the cranium within the jugular foramen, with nerves X and XI.

The glossopharyngeal nerves are mixed nerves, but sensory fibers are most abundant. The sensory neurons on each side are in the **superior** (*jugular*) **ganglion** and **inferior** (*petrosal*) **ganglion** (Figure 14–27●). The sensory fibers carry general sensory information from the lining of the pharynx and the soft palate to a nucleus in the medulla oblongata. These nerves also provide taste sensations from the posterior third of the tongue and have special receptors that monitor the blood pressure and dissolved gas concentrations in major blood vessels.

The somatic motor fibers control the pharyngeal muscles involved in swallowing. Visceral motor fibers synapse in the **otic ganglion**, and postganglionic fibers innervate the parotid salivary gland of the cheek.

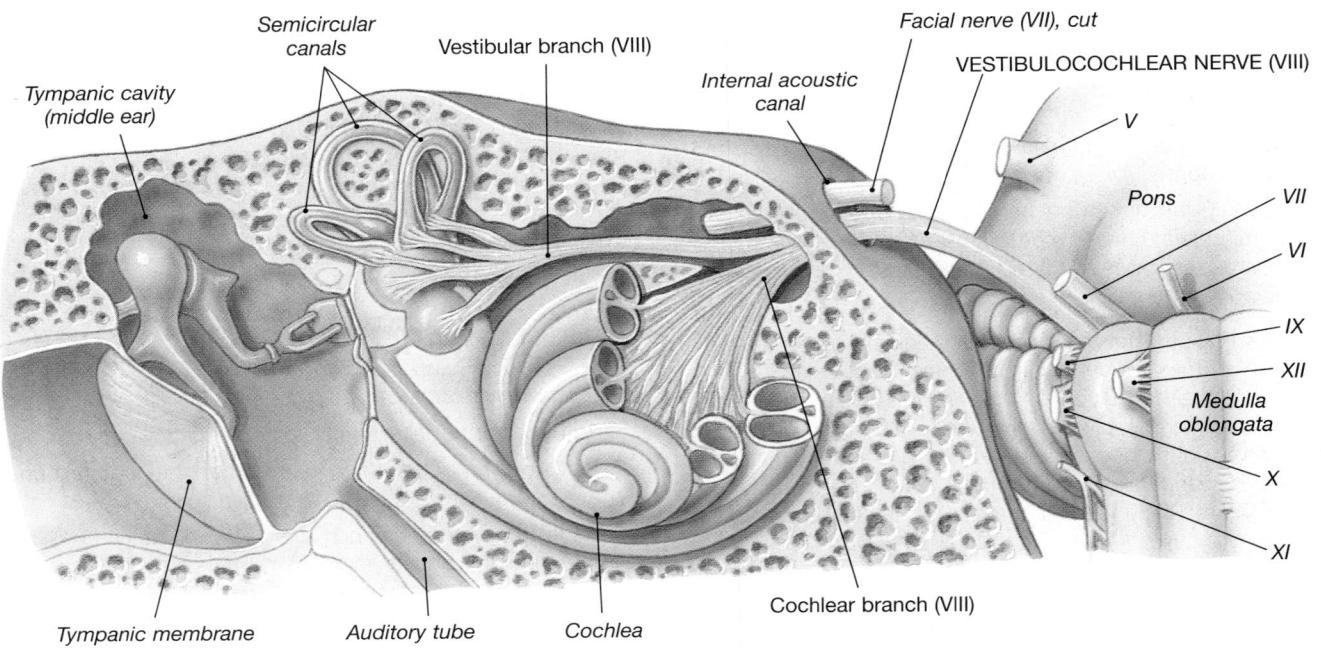

Semicircular canals
Vestibular branch (VIII)
Facial nerve (VII), cut
VESTIBULOCOCHLEAR NERVE (VIII)
Tympanic cavity (middle ear)
Internal acoustic canal
V
Pons
VII
VI
IX
XII
Medulla oblongata
X
XI
Tympanic membrane
Auditory tube
Cochlea
Cochlear branch (VIII)

●**FIGURE 14–26**
The Vestibulocochlear Nerve

FOCUS *continues* →

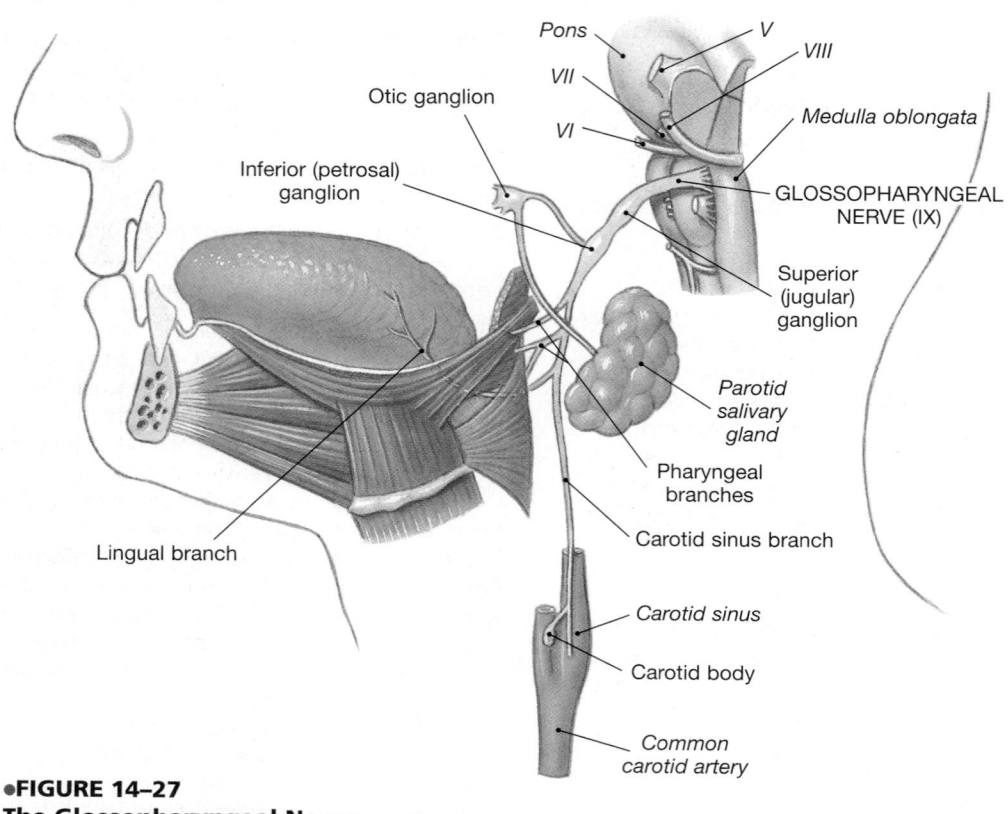

●FIGURE 14–27
The Glossopharyngeal Nerve

The Vagus Nerves (X)

Primary function: Mixed (sensory and motor), widely distributed in the thorax and abdomen

Origin: *Sensory*: pharynx (part), auricle and external auditory canal, diaphragm, and visceral organs in thoracic and abdominopelvic cavities. *Motor*: motor nuclei in medulla oblongata

Pass through: Jugular foramina between the occipital bone and the temporal bones ∞ **pp. 214, 215**

Destination: *Sensory*: sensory nuclei and autonomic centers of medulla oblongata. *Visceral motor*: muscles of the palate, pharynx, digestive, respiratory, and cardiovascular systems in the thoracic and abdominal cavities

The **vagus** (VĀ-gus) **nerves** (X) arise immediately posterior to the attachment of the glossopharyngeal nerves. Many small rootlets contribute to their formation, and developmental studies indicate that these nerves probably represent the fusion of several smaller cranial nerves during our evolutionary history. As the name suggests (*vagus*, wanderer), the vagus nerves branch and radiate extensively. Figure 14–28● shows only the general pattern of distribution.

Sensory neurons are located in the superior **jugular ganglion** and the inferior **nodose** (NŌ-dōs) **ganglion** (*node*, knot). Each vagus nerve provides somatic sensory information about the external acoustic canal, a portion of the ear, and the diaphragm, and special sensory information from pharyngeal taste receptors. But most of

the vagal afferents carry visceral sensory information from receptors along the esophagus, respiratory tract, and abdominal viscera as distant as the last portions of the large intestine. This visceral sensory information is vital to the autonomic control of visceral function.

The motor components of the vagus are equally diverse. Each vagus nerve carries preganglionic autonomic (parasympathetic) fibers that affect the heart and control smooth muscles and glands within the areas monitored by its sensory fibers, including the stomach, intestines, and gallbladder. Difficulty in swallowing is one of the most common signs of damage to either nerve IX or X, because damage to either one prevents the coordination of the swallowing reflex.

The Accessory Nerves (XI)

Primary function: Motor to muscles of the neck and upper back

Origin: Motor nuclei of spinal cord and medulla oblongata

Pass through: Jugular foramina between the occipital bone and the temporal bones ∞ **pp. 214, 215**

Destination: Internal branch innervates voluntary muscles of palate, pharynx, and larynx; external branch controls sternocleidomastoid and trapezius muscles

The **accessory nerves** (XI) are also known as the *spinal accessory nerves* or the *spinoaccessory nerves*. Unlike other cranial nerves, each accessory nerve has some motor fibers that originate in the lateral

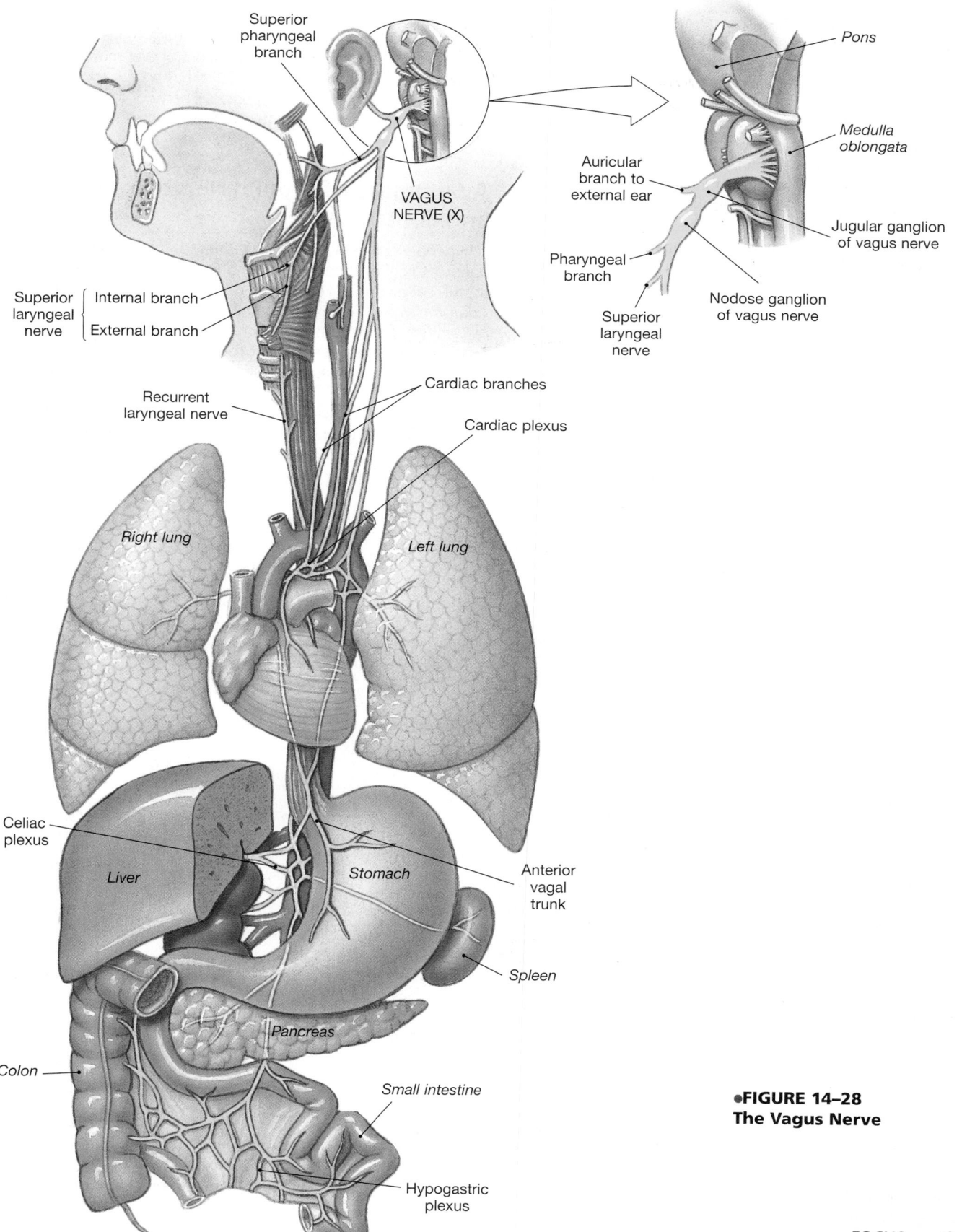

Superior
pharyngeal
branch

Pons

Medulla
oblongata

Auricular
branch to
external ear

Jugular ganglion
of vagus nerve

Pharyngeal
branch

Nodose ganglion
of vagus nerve

Superior
laryngeal
nerve

VAGUS
NERVE (X)

Superior
laryngeal
nerve

Internal branch

External branch

Recurrent
laryngeal nerve

Cardiac branches

Cardiac plexus

Right lung

Left lung

Celiac
plexus

Liver

Stomach

Anterior
vagal
trunk

Spleen

Pancreas

Colon

Small intestine

Hypogastric
plexus

●**FIGURE 14–28**
The Vagus Nerve

FOCUS *continues* →

part of the anterior gray horns of the first five cervical segments of the spinal cord (Figure 14–29●). These somatic motor fibers form the **spinal root** of nerve XI. They enter the cranium through the foramen magnum. They then join the motor fibers of the **cranial root**, which originates at a nucleus in the medulla oblongata. The composite nerve leaves the cranium through the jugular foramen and divides into two branches.

The **internal branch** of nerve XI joins the vagus nerve and innervates the voluntary swallowing muscles of the soft palate and pharynx and the intrinsic muscles that control the vocal cords. The **external branch** of nerve XI controls the sternocleidomastoid and trapezius muscles of the neck and back. ∞ pp. 348, 356 The motor fibers of this branch originate in the lateral gray part of the anterior horns of cervical spinal nerves C_1 to C_5.

The Hypoglossal Nerves (XII)

Primary function: Motor (tongue movements)

Origin: Motor nuclei of medulla oblongata

Pass through: Hypoglossal canals of occipital bone ∞ pp. 214, 215, 216

Destination: Muscles of the tongue ∞ pp. 346-347

Each **hypoglossal** (hī-pō-GLOS-al) **nerve** (XII) leaves the cranium through the hypoglossal canal. The nerve then curves to reach the skeletal muscles of the tongue (Figure 14–29●). This cranial nerve provides voluntary motor control over movements of the tongue. Its condition is checked by having you stick out your tongue. Damage to one hypoglossal nerve or to its associated nuclei causes the tongue to veer toward the affected side.

A Summary of Cranial Nerve Branches and Functions

Few people can easily recall the names, numbers, and functions of the cranial nerves. Mnemonic devices may prove useful. For example, you may find "*Oh, Once One Takes The Anatomy Final, Very Good Vacations Are Heavenly*" easy to remember; the first letter of each word corresponds to the first letter of a cranial nerve, proceeding from nerves I to XII. Table 14–9 summarizes the basic distribution and function of each cranial nerve.

Cranial nerves are clinically important, in part because they can provide clues to underlying CNS problems. As a result, a number of standardized tests for cranial nerve function are used. **AM** Cranial Nerve Tests

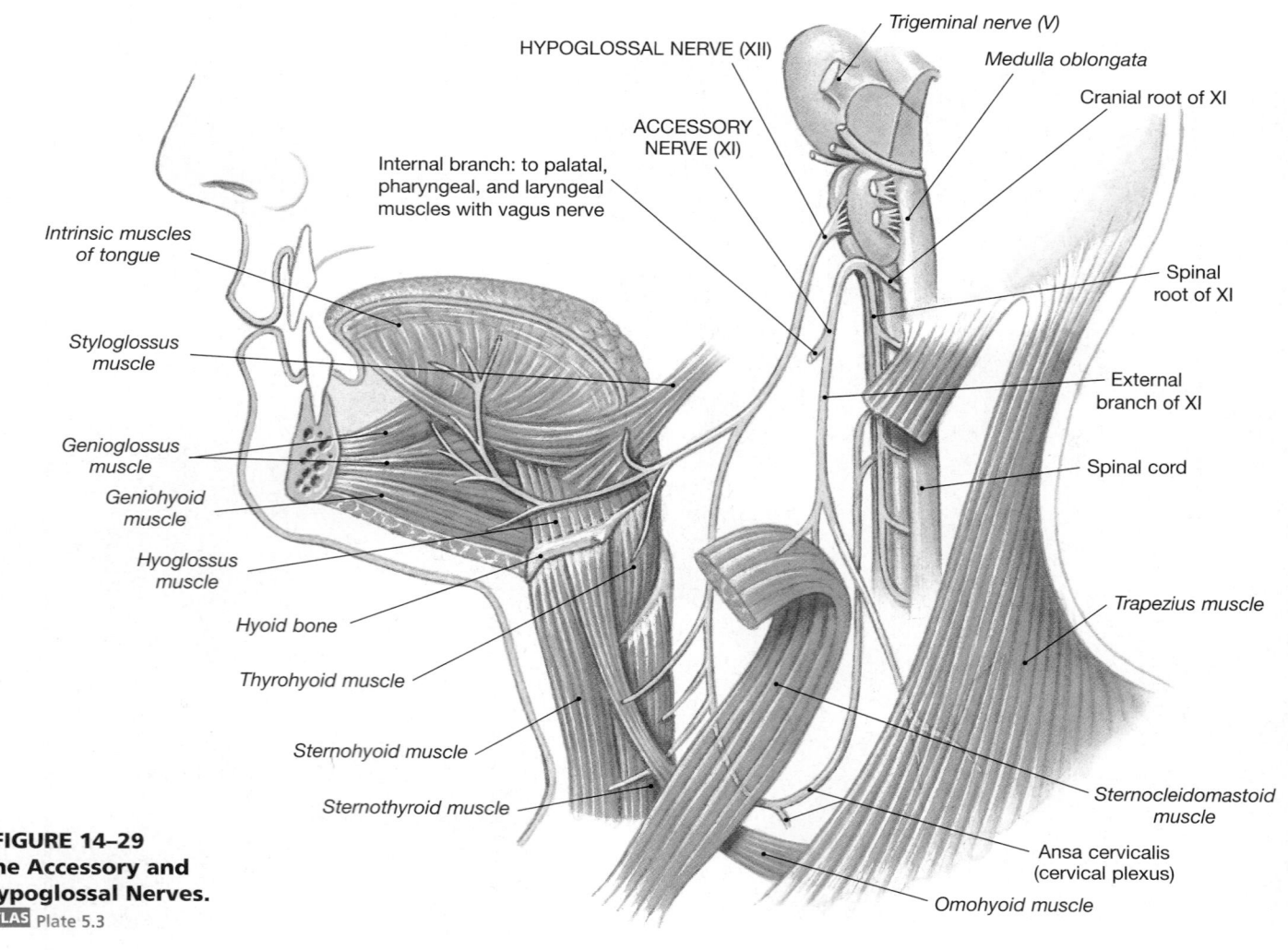

HYPOGLOSSAL NERVE (XII)

Trigeminal nerve (V)

Medulla oblongata

Cranial root of XI

ACCESSORY NERVE (XI)

Internal branch: to palatal, pharyngeal, and laryngeal muscles with vagus nerve

Intrinsic muscles of tongue

Spinal root of XI

Styloglossus muscle

External branch of XI

Genioglossus muscle

Spinal cord

Geniohyoid muscle

Hyoglossus muscle

Trapezius muscle

Hyoid bone

Thyrohyoid muscle

Sternohyoid muscle

Sternothyroid muscle

Sternocleidomastoid muscle

●**FIGURE 14–29**
The Accessory and Hypoglossal Nerves.
ATLAS Plate 5.3

Ansa cervicalis (cervical plexus)

Omohyoid muscle

SUMMARY TABLE 14–9 The Cranial Nerve Branches and Functions

Cranial Nerve (Number)	Sensory Ganglion	Branch	Primary Function	Foramen	Innervation
Olfactory (I)			Special sensory	Olfactory foramina of ethmoid	Olfactory epithelium
Optic (II)			Special sensory	Optic canal	Retina of eye
Oculomotor (III)			Motor	Superior orbital fissure	Inferior, medial, superior rectus, inferior oblique and levator palpebrae superioris muscles; intrinsic eye muscles
Trochlear (IV)			Motor	Superior orbital fissure	Superior oblique muscle
Trigeminal (V)	Semilunar		Mixed	Superior orbital fissure	Areas associated with the jaws
		Ophthalmic	Sensory	Superior orbital fissure	Orbital structures, nasal cavity, skin of forehead, upper eyelid, eyebrows, nose (part)
		Maxillary		Foramen rotundum	Lower eyelid; superior lip, gums, and teeth; cheek, nose (part), palate, and pharynx (part)
		Mandibular		Foramen ovale	*Sensory*: inferior gums, teeth, lips, palate (part), and tongue (part) *Motor*: muscles of mastication
Abducens (VI)			Motor	Superior orbital fissure	Lateral rectus muscle
Facial (VII)	Geniculate		Mixed	Internal acoustic canal to facial canal; exits at stylomastoid foramen	*Sensory*: taste receptors on anterior 2/3 of tongue *Motor*: muscles of facial expression, lacrimal gland, submandibular gland, sublingual salivary glands
Vestibulocochlear (Acoustic) (VIII)		Cochlear	Special sensory	Internal acoustic canal	Cochlea (receptors for hearing)
		Vestibular			Vestibule (receptors for motion and balance)
Glossopharyngeal (IX)	Superior (jugular) and inferior (petrosal)		Mixed	Jugular foramen	*Sensory*: posterior 1/3 of tongue; pharynx and palate (part); receptors for blood pressure, pH, oxygen, and carbon dioxide concentrations *Motor*: pharyngeal muscles and parotid salivary gland
Vagus (X)	Jugular and nodose		Mixed	Jugular foramen	*Sensory*: pharynx; auricle and external acoustic canal; diaphragm; visceral organs in thoracic and abdominopelvic cavities *Motor*: palatal and pharyngeal muscles and visceral organs in thoracic and abdominopelvic cavities
Accessory (XI)		Internal	Motor	Jugular foramen	Skeletal muscles of palate, pharynx, and larynx (with vagus nerve)
		External	Motor	Jugular foramen	Sternocleidomastoid and trapezius muscles
Hypoglossal (XII)			Motor	Hypoglossal canal	Tongue musculature

14–10 CRANIAL REFLEXES

Objective

■ Describe representative examples of cranial reflexes that produce somatic responses or visceral responses to specific stimuli.

Cranial reflexes are reflex arcs that involve the sensory and motor fibers of cranial nerves. Numerous examples of cranial reflexes will be encountered in later chapters, and this section will simply provide an overview and general introduction.

Table 14–10 lists representative examples of cranial reflexes and their functions. These reflexes are clinically important because they provide a quick and easy method for observing the condition of cranial nerves and specific nuclei and tracts in the brain. The somatic reflexes mediated by the cranial nerves are seldom more complex than the somatic reflexes of the spinal cord that were discussed in Chapter 13. ∞ p. 451 Table 14–10 includes four representative somatic reflexes: the corneal reflex, the tympanic reflex, the auditory reflexes, and the vestibulo-ocular reflexes. These cranial reflexes are often used to check for damage to the cranial nerves or the associated processing centers in the brain.

The brain stem contains many reflex centers that control visceral motor activity. The motor output of these reflexes is distributed by the autonomic nervous system. As you will see in Chapter 16, the cranial nerves carry most of the commands issued by the parasympathetic division of the ANS, whereas spinal nerves T_1-L_2 carry the sympathetic commands. Many of the centers that coordinate autonomic reflexes are located in the medulla oblongata. These centers can direct very complex visceral motor responses that are essential to the control of respiratory, digestive, and cardiovascular functions.

TABLE 14–10 CRANIAL REFLEXES

Reflex	Stimulus	Afferents	Central Synapse	Efferents	Response
SOMATIC REFLEXES					
Corneal reflex	Contact with corneal surface	V (trigeminal)	Motor nucleus for VII (facial)	VII (facial)	Blinking of eyelids
Tympanic reflex	Loud noise	VIII (vestibulocochlear)	Inferior colliculus	VII	Reduced movement of auditory ossicles
Auditory reflexes	Loud noise	VIII	Motor nuclei of brain stem and spinal cord	III, IV, VI, VII, X, and cervical nerves	Eye and/or head movements triggered by sudden sounds
Vestibulo-ocular reflexes	Rotation of head	VIII	Motor nuclei controlling eye muscles	III, IV, VI	Opposite movement of eyes to stabilize field of vision
VISCERAL REFLEXES					
Direct light reflex	Light striking photoreceptors	II (optic)	Superior colliculus	III (oculomotor)	Constriction of ipsilateral pupil
Consensual light reflex	Light striking photoreceptors	II	Superior colliculus	III	Constriction of contralateral pupil

Chapter Review

SELECTED CLINICAL TERMINOLOGY

Terms Discussed in This Chapter

aphasia: A disorder that impairs the ability to speak or read. (p. 490)

ataxia: A disturbance of balance that, in severe cases, leaves the individual unable to stand without assistance. (p. 477)

Bell's palsy: A condition resulting from an inflammation of the facial nerve. (p. 498)

cerebrovascular accident (CVA), or *stroke*: A condition in which the blood supply to a portion of the brain is blocked off. (p. 472)

cranial trauma: A head injury resulting from violent contact with another object. (p. 468 and *AM*)

disconnection syndrome: A condition in which the two cerebral hemispheres function independently. (p. 492)

dyslexia: A disorder affecting the comprehension and use of words. (p. 490)

electroencephalogram (EEG): A printed record of the brain's electrical activity over time. (p. 491)

epidural hemorrhage: A condition involving bleeding between the dura mater and the cranium, generally resulting from cranial trauma. (p. 469)

hydrocephalus, or "water on the brain": A condition resulting from interference with the normal circulation and/or reabsorption of cerebrospinal fluid. (p. 471)

Parkinson's disease *(paralysis agitans):* A condition characterized by a pronounced increase in muscle tone, resulting from the excitation of neurons in the basal nuclei. (p. 487 and *AM*)

seizure: A temporary disorder of cerebral function, accompanied by abnormal movements, unusual sensations, and/or inappropriate behavior. *(p. 492 and AM)*

subdural hemorrhage: A condition in which blood accumulates under the dural epithelium in contact with the arachnoid membrane. *(p. 469)*

tic douloureux (doo-loo-ROO): A disorder of the maxillary and mandibular branches of nerve V characterized by almost totally debilitating pain triggered by contact with the lip, tongue, or gums. *(p. 497 and AM)*

AM Additional Terms Discussed in the *Applications Manual*

decerebrate rigidity
dysmetria
epilepsy
intention tremor
spasticity
tremor

STUDY OUTLINE

14–1 AN INTRODUCTION TO THE ORGANIZATION OF THE BRAIN   p. 463

A PREVIEW OF MAJOR REGIONS AND LANDMARKS   p. 464

1. The six regions in the adult brain are the cerebrum, diencephalon, mesencephalon ("midbrain"), pons, cerebellum, and medulla oblongata. *(Figures 14–1, 14–2; Table 14–1)*
2. The brain contains extensive areas of **neural cortex**, a layer of gray matter on the surfaces of the cerebrum and cerebellum.

EMBRYOLOGY OF THE BRAIN   p. 464

3. The brain forms from three swellings at the superior tip of the developing *neural tube*: the **prosencephalon**, the mesencephalon, and the **rhombencephalon**. The prosencephalon ("forebrain") forms the **telencephalon** (which becomes the cerebrum) and diencephalon; the rhombencephalon ("hindbrain") forms the **metencephalon** (cerebellum and pons) and **myelencephalon** (medulla oblongata). *(Table 14–1)*

VENTRICLES OF THE BRAIN   p. 467

4. The central passageway of the brain expands to form chambers called **ventricles** that contain cerebrospinal fluid. *(Figure 14–3)*

14–2 PROTECTION AND SUPPORT OF THE BRAIN   p. 467

THE CRANIAL MENINGES   p. 468

1. The cranial meninges (the *dura mater, arachnoid,* and *pia mater*) are continuous with those of the spinal cord.
2. Folds of dura mater, including the **falx cerebri, tentorium cerebelli,** and **falx cerebelli,** stabilize the position of the brain. *(Figure 14–4)*

CEREBROSPINAL FLUID   p. 468

3. Cerebrospinal fluid (CSF) (1) cushions delicate neural structures, (2) supports the brain, and (3) transports nutrients, chemical messengers, and waste products.
4. Cerebrospinal fluid is produced at the **choroid plexus**, reaches the subarachnoid space through the **lateral** and **median apertures** and diffuses across the **arachnoid granulations** into the **superior sagittal sinus.** *(Figure 14–5)*

THE BLOOD SUPPLY TO THE BRAIN   p. 471

5. The **blood–brain barrier** isolates neural tissue from the general circulation.
6. The blood–brain barrier is incomplete in parts of the hypothalamus, the pituitary gland, the pineal gland, and the choroid plexus.

CSF formation and flow: Companion Website: Chapter 14/Reviewing Concepts/Labeling.

14–3 THE MEDULLA OBLONGATA   p. 473

1. The medulla oblongata connects the brain and spinal cord. It contains relay stations such as the **olivary nuclei,** and **reflex centers,** including the **cardiovascular** and **respiratory rhythmicity centers.** The **reticular formation** begins in the medulla oblongata and extends into more superior portions of the brain stem. *(Figures 14–7, 14–8; Table 14–2)*

14–4 THE PONS   p. 475

1. The pons contains (1) sensory and motor nuclei for four cranial nerves; (2) nuclei that help control respiration; (3) nuclei and tracts linking the cerebellum with the brain stem, cerebrum, and spinal cord; and (4) ascending, descending and transverse tracts. *(Figure 14–8; Table 14–2)*

14–5 THE CEREBELLUM   p. 476

1. The cerebellum adjusts postural muscles and programs and tunes ongoing movements. The **cerebellar hemispheres** consist of the **anterior** and **posterior lobes,** the **vermis,** and the **flocculonodular lobe.** *(Figure 14–9; Table 14–3)*
2. The **superior, middle,** and **inferior cerebellar peduncles** link the cerebellum with the brain stem, diencephalon, cerebrum, and spinal cord and interconnect the two cerebral hemispheres.

14–6 THE MESENCEPHALON   p. 477

1. The **tectum** (roof of the mesencephalon) contains the **corpora quadrigemina** (**superior colliculi** and **inferior colliculi**). The mesencephalon contains the **red nucleus,** the **substantia nigra,** the **cerebral peduncles,** and the headquarters of the **RAS.** *(Figure 14–10; Table 14–4)*

14–7 THE DIENCEPHALON   p. 479

1. The diencephalon is composed of the epithalamus, the hypothalamus, and the thalamus. *(Figures 14–10 to 14–13)*

THE THALAMUS   p. 479

2. The thalamus is the final relay point for ascending sensory information and coordinates the activities of the cerebral cortex and basal nuclei. *(Figures 14–11, 14–12; Table 14–5)*

THE HYPOTHALAMUS   p. 481

3. The hypothalamus can (1) control somatic motor activities at the subconscious level, (2) control autonomic function, (3) coordinate activities of the nervous and endocrine systems, (4) secrete hormones, (5) produce emotions and behavioral **drives,** (6) coordinate voluntary and autonomic functions, (7) regulate body temperature, and (8) coordinate circadian cycles of activity. *(Figure 14–12; Table 14–6)*

14–8 THE LIMBIC SYSTEM p. 482

1. The **limbic system**, or *motivational system*, includes the **amygdaloid body**, **cingulate gyrus**, **denticulate gyrus**, **parahippocampal gyrus**, **hippocampus**, and **fornix**. The functions of the limbic system involve emotional states and related behavioral drives (*Figures 14–13, 14–14; Table 14–7*)

14–9 THE CEREBRUM p. 483

THE CEREBRAL CORTEX p. 485

1. The cortical surface contains **gyri** (elevated ridges) separated by **sulci** (shallow depressions) or **fissures** (deeper grooves). The **longitudinal fissure** separates the two **cerebral hemispheres**. The **central sulcus** separates the **frontal** and **parietal lobes**. Other sulci form the boundaries of the **temporal** and **occipital lobes**. (*Figure 14–1*)

THE WHITE MATTER OF THE CEREBRUM p. 485

2. The white matter of the cerebrum contains **association fibers**, **commissural fibers**, and **projection fibers**. (*Figure 14–15*)

THE BASAL NUCLEI p. 486

3. The **basal nuclei** include the **caudate nucleus**, **globus pallidus**, and **putamen**; they control muscle tone and coordinate learned movement patterns and other somatic motor activities. (*Figure 14–16*)

Anatomy of the brain: Anatomy CD-ROM: Nervous System/Brain Dissections.

MOTOR AND SENSORY AREAS OF THE CORTEX p. 488

4. The **primary motor cortex** of the **precentral gyrus** directs voluntary movements. The **primary sensory cortex** of the **postcentral gyrus** receives somatic sensory information from touch, pressure, pain, taste, and temperature receptors. (*Figure 14–17; Table 14–8*)

5. **Association areas**, such as the **somatic sensory association area**, **visual association area**, and **somatic motor association area** (**premotor cortex**), control our ability to understand sensory information and coordinate a motor response. (*Figure 14–17*)

6. The **general interpretive area** receives information from all the sensory association areas. It is present in only one hemisphere—generally the left. (*Figure 14–17*)

7. The **speech center** regulates the patterns of breathing and vocalization needed for normal speech. (*Figure 14–17*)

8. The **prefrontal cortex** coordinates information from the secondary and special association areas of the entire cortex and performs abstract intellectual functions. (*Figure 14–17*)

9. The left hemisphere typically contains the general interpretive and speech centers and is responsible for language-based skills. The right hemisphere is typically responsible for spatial relationships and analyses. (*Figure 14–18*)

10. Brain activity is measured using an **electroencephalogram**. **Alpha waves** appear in healthy resting adults; **beta waves** occur when adults are concentrating; **theta waves** appear in children; and **delta waves** are normal during sleep. (*Figure 14–19*)

Focus: Cranial Nerves p. 493

11. We have 12 pairs of cranial nerves. Each nerve attaches to the ventrolateral surface of the brainstem near the associated sensory or motor nuclei. (*Figure 14–20*)

12. The **olfactory nerves** (I) carry sensory information responsible for the sense of smell. The olfactory afferents synapse within the **olfactory bulbs**. (*Figures 14–20, 14–21*)

13. The **optic nerves** (II) carry visual information from special sensory receptors in the eyes. (*Figures 14–20, 14–22*)

14. The **oculomotor nerves** (III) are the primary source of innervation for four of the extraocular muscles. (*Figure 14–23*)

15. The **trochlear nerves** (IV), the smallest cranial nerves, innervate the superior oblique muscles of the eyes. (*Figure 14–23*)

16. The **trigeminal nerves** (V), the largest cranial nerves, are mixed nerves with *ophthalmic*, *maxillary*, and *mandibular branches*. (*Figure 14–24*)

17. The **abducens nerves** (VI) innervate the lateral rectus muscles. (*Figure 14–23*)

18. The **facial nerves** (VII) are mixed nerves that control muscles of the scalp and face. They provide pressure sensations over the face and receive taste information from the tongue. (*Figure 14–25*)

19. The **vestibulocochlear nerves** (VIII) contain the **vestibular branch**, which monitors sensations of balance, position, and movement, and the **cochlear branch**, which monitors hearing receptors. (*Figure 14–26*)

20. The **glossopharyngeal nerves** (IX) are mixed nerves that innervate the tongue and pharynx and control the action of swallowing. (*Figure 14–27*)

21. The **vagus nerves** (X) are mixed nerves that are vital to the autonomic control of visceral function. (*Figure 14–28*)

22. The **accessory nerves** (XI) have **internal branches**, which innervate voluntary swallowing muscles of the soft palate and pharynx, and **external branches**, which control muscles associated with the pectoral girdle. (*Figure 14–29*)

23. The **hypoglossal nerves** (XII) provide voluntary motor control over tongue movements. (*Figure 14–29*)

24. The branches and functions of the cranial nerves are summarized in *Table 14–9*.

14–10 CRANIAL REFLEXES p. 504

1. Cranial reflexes involve sensory and motor fibers of cranial nerves (*Table 14–10*).

REVIEW QUESTIONS

More assessment questions are available to you on the Companion Website. You will find Matching, Multiple Choice, True/False, and other quizzes to help further your understanding of the material covered in this chapter. To access the site, go to www.aw.com/martini.

LEVEL 1 Reviewing Facts and Terms

1. The regulation of autonomic function, such as heart rate and blood pressure, originates in the
   (a) cerebrum (b) cerebellum
   (c) diencephalon (d) medulla oblongata

2. The smooth surface that covers the brain but does not follow the underlying neural convolutions or sulci is the
   (a) neural cortex (b) dura mater
   (c) pia mater (d) arachnoid membrane

3. The meningeal layer that adheres to the surface contour of the brain, extending into every fold and curve, is the
   (a) pia mater   (b) dura mater
   (c) arachnoid membrane (d) neural cortex

4. The dural fold that divides the two cerebellar hemispheres is the
   (a) transverse sinus   (b) falx cerebri
   (c) tentorium cerebelli (d) falx cerebelli

5. Cerebrospinal fluid is produced and secreted in the
   (a) hypothalamus   (b) choroid plexus
   (c) medulla oblongata (d) crista galli

6. The primary purpose of the blood–brain barrier is to
   (a) provide the brain with oxygenated blood
   (b) drain venous blood via the internal jugular veins
   (c) isolate neural tissue in the CNS from the general circulation
   (d) a, b, and c are correct

7. The centers in the pons that modify the activity of the respiratory rhythmicity centers in the medulla oblongata are the
   (a) apneustic and pneumotaxic centers
   (b) inferior and superior peduncles
   (c) cardiac and vasomotor centers
   (d) nucleus gracilis and nucleus cuneatus

8. The two hormones secreted by the hypothalamus are
   (a) epinephrine and norepinephrine
   (b) antidiuretic hormone and oxytocin
   (c) melatonin and serotonin
   (d) FSH and ATP

9. The final relay point for ascending sensory information that will be projected to the primary sensory cortex is the
   (a) hypothalamus   (b) thalamus
   (c) spinal cord   (d) medulla oblongata

10. The establishment of emotional states is a function of the
    (a) limbic system   (b) tectum
    (c) mamillary bodies (d) thalamus

11. Coordination of learned movement patterns at the subconscious level is performed by
    (a) the cerebellum
    (b) the substantia nigra
    (c) association fibers
    (d) the hypothalamus

12. The two cerebral hemispheres are functionally different, even though anatomically they appear the same.
    (a) true   (b) false

13. What are the three important functions of the CSF?

14. Which three areas in the brain are not isolated from the general circulation by the blood–brain barrier?

15. Using the mnemonic device "Oh, Once One Takes The Anatomy Final, Very Good Vacations Are Heavenly," list the 12 pairs of cranial nerves and their functions.

LEVEL 2 Reviewing Concepts

16. Why can the brain respond to stimuli with greater versatility than the spinal cord?

17. Briefly summarize the overall function of the cerebellum.

18. Damage to the corpora quadrigemina would interfere with the
    (a) control of autonomic function
    (b) regulation of body temperature
    (c) processing of visual and auditory sensations
    (d) conscious control of skeletal muscles

19. If symptoms characteristic of Parkinson's disease appear, which part of the mesencephalon is inhibited from secreting a neurotransmitter? Which neurotransmitter is it?

20. What role does the hypothalamus play in the body?

21. Stimulation of which part of the brain would produce sensations of hunger and thirst?

22. Which structure in the brain would your A & P instructor be referring to when talking about a nucleus that resembles a sea horse and that appears to be important in the storage and retrieval of long-term memories? In which functional system of the brain is it located?

23. What are the principal functional differences between the right and left hemispheres of the cerebrum?

24. If the corpus callosum is cut,
    (a) cross-referencing of sensory information is inhibited
    (b) the individual talks constantly but uses the wrong words
    (c) objects touched with the left hand cannot be recognized but can be verbally identified
    (d) objects touched with the right hand cannot be verbally identified

25. A person with a damaged visual association area may be
    (a) unable to scan the lines of a page or see rows of clear symbols
    (b) declared legally blind
    (c) unable to see letters but able to identify words and their meanings
    (d) able to see letters quite clearly but unable to recognize or interpret them

26. What kinds of problems are associated with the presence of lesions in Wernicke's area and Broca's area?

LEVEL 3 Critical Thinking and Clinical Applications

27. Smelling salts can sometimes help restore consciousness after a person has fainted. The active ingredient of smelling salts is ammonia, and it acts by irritating the lining of the nasal cavity. Propose a mechanism by which smelling salts would raise a person from the unconscious state to the conscious state.

28. A police officer has just stopped Bill on suspicion of driving while intoxicated. The officer asks Bill to walk the yellow line on the road and then to place the tip of his index finger on the tip of his nose. How would these activities indicate Bill's level of sobriety? Which part of the brain is being tested by these activities?

29. While having some dental work performed, Tyler is given an injection of local anesthetic in his lower jaw. His dentist tells him not to eat until the anesthetic wears off, not because of his teeth but because of his tongue. Why is the dentist giving Tyler this advice?

30. Doris develops a clot that blocks the right branch of the middle cerebral artery, a blood vessel that serves the anterior portion of the right cerebral hemisphere. What symptoms would you expect to observe as a result of this blockage?

31. Infants have little to no control of the movements of their head. One of the consequences of this is that they are susceptible to shaken-baby syndrome. Shaken-baby syndrome is caused by vigorous shaking of an infant or young child by the arms, legs, chest, or shoulders. Forceful shaking can cause brain damage leading to mental retardation, speech and learning disabilities, paralysis, seizures, hearing loss, and even death. What areas of the brain might be affected in this syndrome, resulting in each of these symptoms?

UNIT 3   **CHAPTER**   12   13   14   **15**   16   17   18

NEURAL INTEGRATION I: SENSORY PATHWAYS AND THE SOMATIC NERVOUS SYSTEM

It is said that "big cities never sleep." In New York City or Los Angeles at 3:00 A.M., shops are open, deliveries are made, people are on the street, and traffic moves briskly. The central nervous system (CNS) is much more complex than any city and far busier. Information continuously flows among the brain, spinal cord, and peripheral nerves. At any moment, millions of sensory neurons are delivering information to processing centers in the CNS, and millions of motor neurons are controlling or adjusting the activities of peripheral effectors. This process continues 24 hours a day, whether we are awake or asleep. We may sleep soundly, but our nervous system does not; many brain stem centers are active throughout our lives, performing vital autonomic functions. This chapter examines the flow of sensory information to the CNS and the output of somatic motor commands. The next chapter will consider the visceral motor pathways and the higher-order functions that involve all parts of the brain and tie the sensory, somatic, and visceral systems together.

15–1 AN OVERVIEW OF SENSORY PATHWAYS AND THE SOMATIC NERVOUS SYSTEM

Objective

- Specify the components of the afferent and efferent divisions of the nervous system, and explain what is meant by the somatic nervous system.

Figure 15–1● provides an overview of the topics we will cover in this chapter. Specialized cells called *sensory receptors* monitor specific conditions in the body or the external environment. When stimulated, a receptor passes information to the CNS in the form of action potentials along the axon of a sensory neuron. The afferent division of the nervous system includes the receptors, sensory neurons and *sensory pathways*—the nerves, nuclei, and tracts that deliver somatic and visceral sensory information to their final destinations inside the CNS. Somatic and visceral sensory information often travel along the same pathway. Somatic sensory information is distributed to either the primary sensory cortex of the cerebral hemispheres or appropriate areas of the cerebellar hemispheres. Visceral sensory information is distributed primarily to reflex centers in the brain stem and diencephalon.

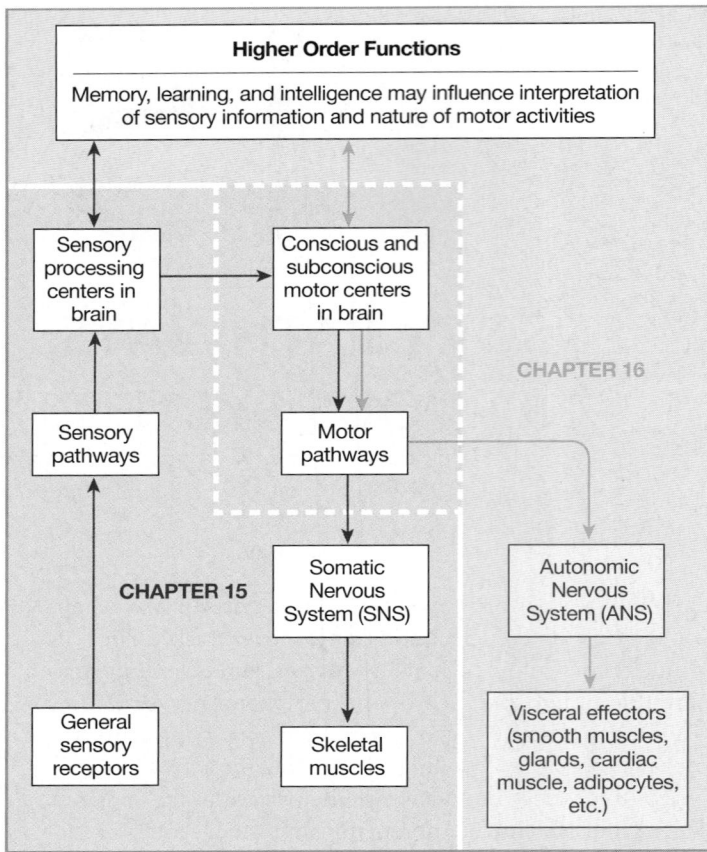

•FIGURE 15–1
An Overview of Neural Integration

The efferent division of the nervous system includes the nuclei, motor tracts, and motor neurons that control peripheral effectors. This chapter considers the somatic motor portion of the efferent division. Somatic motor commands—whether they arise at the conscious or subconscious levels—travel along *somatic motor pathways,* which consist of motor nuclei, tracts, and nerves. The motor neurons and pathways that control skeletal muscles form the somatic nervous system (SNS).

Chapter 16 begins with a discussion of the visceral motor portion of the efferent division. All visceral motor commands are carried into the PNS by the autonomic nervous system (ANS). Both somatic and visceral motor commands may be issued in response to arriving sensory information, but these commands may be modified on the basis of planning, memories, and learning—the so-called *higher order functions* of the brain that we shall consider at the close of Chapter 16.

15–2 SENSORY RECEPTORS AND THEIR CLASSIFICATION

Objectives

- Explain why receptors respond to specific stimuli and how the organization of a receptor affects its sensitivity.

- Identify the receptors for the general senses, and describe how they function.

Sensory receptors are specialized cells or cell processes that provide your central nervous system with information about conditions inside or outside the body. The term **general senses** is used to describe our sensitivity to temperature, pain, touch, pressure, vibration, and proprioception. General sensory receptors are distributed throughout the body, and they are relatively simple in structure. Some of the information they send to the CNS reaches the primary sensory cortex and our conscious awareness. As we noted in Chapter 12, sensory information is interpreted on the basis of the frequency of arriving action potentials. ∞ p. 422 For example, when pressure sensations are arriving, the harder the pressure, the higher the frequency of action potentials. The arriving information is called a **sensation**. The conscious awareness of a sensation is called a **perception**.

The **special senses** are **olfaction** (smell), **vision** (sight), **gustation** (taste), **equilibrium** (balance), and **hearing**. These sensations are provided by receptors that are structurally more complex than those of the general senses. Special sensory receptors are located in **sense organs** such as the eye or ear, where the receptors are protected by surrounding tissues. The information these receptors provide is distributed to specific areas of the cerebral cortex (the auditory cortex, the visual cortex, and so forth) and to centers throughout the brain stem. We shall consider the special senses further in Chapter 17. **AM** Analyzing Sensory Disorders

☐ SENSORY RECEPTORS

Sensory receptors represent the interface between the nervous system and the internal and external environments. A sensory receptor detects an arriving stimulus and translates it into an action potential that can be conducted to the CNS. This translation process is called *transduction.* If transduction does not occur, as far as you are concerned, the stimulus doesn't exist. For example, bees can see ultraviolet light that you can't see, and dogs can respond to sounds that you can't hear. In each case the stimuli are there—but your receptors cannot detect them. **AM** Transduction: A Closer Look

The Detection of Stimuli

Each receptor has a characteristic sensitivity. For example, a touch receptor is very sensitive to pressure but relatively insensitive to chemical stimuli, whereas a taste receptor is sensitive to dissolved chemicals but insensitive to pressure. This feature is called *receptor specificity.*

Specificity may result from the structure of the receptor cell, or from the presence of accessory cells or structures that shield the receptor cell from other stimuli. The simplest receptors are the dendrites of sensory neurons. The branching tips of these dendrites, called **free nerve endings**, are not protected by accessory structures. Free nerve endings extend through a tissue the way grass roots extend into the soil. They can be stimulated by many different stimuli and therefore exhibit little receptor specificity. For example, free nerve endings that respond to tissue damage by providing pain sensations may be stimulated by chemical stimulation, pressure, temperature changes, or trauma. Complex receptors, such as the eye's visual receptors, are protected by accessory cells and connective tissue layers. These cells are seldom exposed to any stimulus other than light and so provide very specific information.

The area monitored by a single receptor cell is its *receptive field* (Figure 15–2●). Whenever a sufficiently strong stimulus arrives in the receptive field, the CNS receives the information "stimulus arriving at receptor X." The larger the receptive field, the poorer your ability to localize a stimulus. For example, a touch receptor on the general body surface may have a receptive field 7 cm (2.5 in.) in diameter. As a result, you can describe a light touch there as affecting only a general area, not an exact spot. On the tongue or fingertips, where the receptive fields are less than a millimeter in diameter, you can be very precise about the location of a stimulus.

An arriving stimulus can take many forms. It may be a physical force (such as pressure), a dissolved chemical, a sound, or a light. Regardless of the nature of the stimulus, however, sensory information must be sent to the CNS in the form of action potentials, which are electrical events.

As we learned earlier, transduction is the translation of an arriving stimulus into an action potential by a sensory receptor. Transduction begins when a stimulus changes the transmembrane potential of the receptor cell. This change, called a *receptor potential*, is either a graded depolarization or a hyperpolarization. The stronger the stimulus, the larger the receptor potential.

The typical receptors for the general senses are the dendrites of sensory neurons, and the sensory neuron is the receptor cell. Any receptor potential that depolarizes the cell membrane will bring the membrane closer to threshold. A receptor potential large enough to produce an action potential is called a *generator potential*.

Sensations of taste, hearing, equilibrium, and vision are provided by specialized receptor cells that communicate with sensory neurons across chemical synapses. The receptor cells

develop graded receptor potentials in response to stimulation, and the change in membrane potential alters the rate of neurotransmitter release at the synapse. The result is a depolarization or hyperpolarization of the sensory neuron. If sufficient depolarization occurs, an action potential appears in the sensory neuron. In this case, the receptor potential and the generator potential occur in different cells: The receptor potential develops in the receptor cell, and the generator potential appears later, in the sensory neuron.

Whenever a generator potential appears, action potentials develop in the axon of a sensory neuron. For reasons discussed in Chapter 12, the greater the degree of sustained depolarization produced by the generator potential, the higher the frequency of action potentials in the afferent fiber. p. 422 The arriving information is then processed and interpreted by the CNS at the conscious and subconscious levels.

The Interpretation of Sensory Information

Sensory information that arrives at the CNS is routed according to the location and nature of the stimulus. In previous chapters, we emphasized the fact that axons in the CNS are organized in bundles with specific origins and destinations. Along sensory pathways, a series of neurons relays information from one point (the receptor) to another (a neuron at a specific site in the cerebral cortex). For example, sensations of touch, pressure, pain, and temperature arrive at the primary sensory cortex; visual, auditory, gustatory, and olfactory sensations reach the visual, auditory, gustatory, and olfactory regions of the cortex, respectively.

The link between receptor and cortical neuron is called a *labeled line*. Each labeled line consists of axons carrying information about one modality, or type of stimulus (touch, pressure, light, sound, and so forth). A labeled line begins at receptors in a specific part of the body. The CNS interprets the modality entirely on the basis of the labeled line over which it arrives. As a result, you cannot tell the difference between a true sensation and a false one generated somewhere along the line. For example, when you rub your eyes, you commonly see flashes of light. Although the stimulus is mechanical rather than visual, any activity along the optic nerve is projected to the visual cortex and experienced as a visual perception.

The identity of the active labeled line indicates the type of stimulus. Where it arrives within the sensory cortex determines its perceived location. For example, if activity in a labeled line that carries touch sensations stimulates the facial region of your primary sensory cortex, you perceive a touch on the face. *All other characteristics* of the stimulus—its strength, duration, and variation—are conveyed by the frequency and pattern of action potentials. The translation of complex sensory information into meaningful patterns of action potentials is called *sensory coding*.

Some sensory neurons, called **tonic receptors**, are always active. The frequency with which these receptors generate action potentials indicates the background level of stimulation. When the stimulus increases or decreases, the rate of action potential generation changes

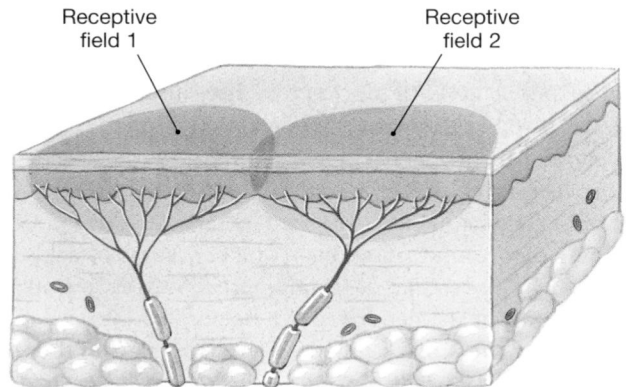

●**FIGURE 15–2**
Receptors and Receptive Fields. Each receptor monitors a specific area known as the receptive field.

accordingly. Other receptors are normally inactive, but become active for a short time whenever a change occurs in the conditions they are monitoring. These receptors, called **phasic receptors**, provide information about the intensity and rate of change of a stimulus. Receptors that combine phasic and tonic coding can convey extremely complicated sensory information.

Adaptation

Adaptation is a reduction in sensitivity in the presence of a constant stimulus. You seldom notice the rumble of the tires when you ride in a car, or the background noise of the air conditioner, because your nervous system quickly adapts to stimuli that are painless and constant. *Peripheral adaptation* occurs when the level of receptor activity changes. The receptor responds strongly at first, but thereafter its activity gradually declines, in part because the size of the generator potential gradually decreases. This response is characteristic of phasic receptors, which are hence also called **fast-adapting receptors**. Temperature receptors (*thermoreceptors*) are phasic receptors; you seldom notice room temperature unless it changes suddenly. Tonic receptors show little peripheral adaptation and so are called **slow-adapting receptors**. Pain receptors (*nociceptors*) are slow-adapting receptors, which is one reason why pain sensations remind you of an injury long after the initial damage has occurred.

Adaptation also occurs inside the CNS along the sensory pathways. For example, a few seconds after you have been exposed to a new smell, conscious awareness of the stimulus virtually disappears, although the sensory neurons are still quite active. This process is known as *central adaptation*. Central adaptation generally involves the inhibition of nuclei along a sensory pathway.

Peripheral adaptation reduces the amount of information that reaches the CNS. Central adaptation at the subconscious level further restricts the amount of detail that arrives at the cerebral cortex. Most of the incoming sensory information is processed in centers along the spinal cord or brain stem at the subconscious level. Although this processing can produce reflexive motor responses, we are seldom consciously aware of either the stimuli or the responses.

The output from higher centers can increase receptor sensitivity or facilitate transmission along a sensory pathway. The reticular activating system in the mesencephalon helps focus our attention and thus heightens or reduces our awareness of arriving sensations. ∞ p. 479 This adjustment of sensitivity can occur under conscious or subconscious direction. When you "listen carefully," your sensitivity and awareness of auditory stimuli increase. Output from higher centers can also inhibit transmission along a sensory pathway. This inhibition occurs when you enter a noisy factory or walk along a crowded city street, as you automatically tune out the high level of background noise.

In this discussion, we have introduced basic concepts of receptor function and sensory processing. We next consider how those concepts apply to the general senses.

☐ THE GENERAL SENSES

Receptors for the general senses are scattered throughout the body and are relatively simple in structure. The simple classification scheme introduced in Chapter 12 divides them into exteroceptors, proprioceptors, and interoceptors. ∞ p. 391 *Exteroceptors* provide information about the external environment; *proprioceptors* report the positions of skeletal muscles and joints; and *interoceptors* monitor visceral organs and functions.

A more detailed classification system divides the general sensory receptors into four types by the nature of the stimulus that excites them: *nociceptors* (pain), *thermoreceptors* (temperature), *mechanoreceptors* (physical distortion), and *chemoreceptors* (chemical concentration). Each class of receptors has distinct structural and functional characteristics. The difference between a somatic receptor and a visceral receptor is its location, not its structure. A pain receptor in the gut looks and acts like a pain receptor in the skin, but the two sensations are delivered to separate locations in the CNS. However, proprioception is a purely somatic sensation—there are no proprioceptors in the visceral organs of the thoracic and abdominopelvic cavities. Your mental map of your body doesn't include these organs; you cannot tell, for example, where your spleen, appendix, or pancreas is at the moment. The visceral organs also have fewer pain, temperature, and touch receptors than one finds elsewhere in the body, and the sensory information you receive is poorly localized because the receptive fields are very large and may be widely separated.

Although general sensations are widely distributed in the CNS, most of the processing occurs in centers along the sensory pathways in the spinal cord or brain stem. Only about 1 percent of the information provided by afferent fibers reaches the cerebral cortex and our conscious awareness. For example, we usually do not feel the clothes we wear or hear the hum of the engine in our car.

Nociceptors

Pain receptors, or **nociceptors**, are especially common in the superficial portions of the skin, in joint capsules, within the periostea of bones, and around the walls of blood vessels. Other deep tissues and most visceral organs have few nociceptors. Pain receptors are free nerve endings with large receptive fields (Figure 15–2●). As a result, it is often difficult to determine the exact source of a painful sensation.

Three populations of nociceptors have been identified: (1) those sensitive to extremes of temperature, (2) those sensitive to mechanical damage, and (3) those sensitive to dissolved chemicals, such as chemicals released by injured cells. Very strong stimuli, however, will excite all three receptor types. For that reason, people describing very painful sensations—whether caused by acids, heat, or a deep cut—use similar descriptive terms, such as "burning."

Stimulation of the dendrites of a nociceptor causes depolarization. When the initial segment of the axon reaches threshold, an action potential heads toward the CNS. The sensory neurons use the amino acid *glutamate* or the neuropeptide *Substance P* as neurotransmitters.

Two types of axons—Type A and Type C fibers—carry painful sensations. ∞ p. 409 Myelinated Type A fibers carry sensations of **fast pain**, or *prickling pain*. An injection or a deep cut produces this type of pain. These sensations very quickly reach the CNS, where they often trigger somatic reflexes. They are also relayed to the primary sensory cortex and so receive conscious attention. In most cases, the arriving information permits the stimulus to be localized to an area several inches in diameter.

Slower, Type C fibers carry sensations of **slow pain**, or *burning and aching pain*. These sensations cause a generalized activation of the reticular formation and thalamus. The individual becomes aware of the pain but has only a general idea of the area affected.

Pain receptors are tonic receptors. Significant peripheral adaptation does not occur, and the receptors continue to respond as long as the painful stimulus remains. Painful sensations cease only after tissue damage has ended. However, central adaptation may reduce the *perception* of the pain while the pain receptors are still

Acute and Chronic Pain

Pain management poses a number of problems for clinicians. Painful sensations can result from tissue damage or sensory nerve irritation; it can originate where it is perceived, be referred from another location, or represent a false signal generated along the sensory pathway. The treatment differs in each case, and an accurate diagnosis is an essential first step. **Acute pain** is the result of tissue injury; the cause is apparent, and local treatment of the injury is typically effective in relieving the pain. The most effective solution is to stop the damage, end the stimulation, and suppress the painful sensations at the injury site. Pain sensations are suppressed when topical or locally injected anesthetics inactivate nociceptors in the immediate area. Analgesic drugs can also be administered. They work in many different ways; we will consider only a few examples here.

Tissue injury results in damage to cell membranes. A fatty acid called arachidonic acid escapes from injured membranes. In interstitial fluid, an enzyme called *cyclo-oxygenase* converts arachidonic acid molecules to prostaglandins; it is these prostaglandins that stimulate nociceptors in the area. Aspirin, ibuprofen, and related analgesics reduce inflammation and suppress pain by blocking the action of cyclo-oxygenase.

Chronic pain is more difficult to categorize and treat. It includes (1) pain from an injury that persists after tissue structure has been repaired; (2) pain from a chronic disease, such as cancer; and (3) pain without an apparent cause. Chronic pain in part reflects permanent facilitation of the pain pathways and the creation of a reverberating "pain memory." Complex psychological and physiological components are also involved. For example, many chronic pain patients develop a tolerance for pain medications, and insomnia and depression are common complaints. Chronic pain can be helped by antidepressants, which affect neurotransmitter levels. Counseling may help the person focus attention outward rather than inward; the outward focus can lessen the perceived level of pain and reduce the amount of pain medication required. Curiously, developing a second, acute source of pain, such as a herpes-zoster infection, can reduce the perception of preexisting chronic pain.

In some cases, chronic pain and severe acute pain can be suppressed by inhibition of the central pain pathway. Analgesics related to morphine reduce pain by mimicking the action of

endorphins. The perception of pain may be altered, although the pain remains. For example, patients on morphine report being aware of painful sensations, but they are not distressed by them. Surgical steps can be taken to control severe pain; for instance, (1) the sensory innervation of an area can be destroyed by an electrical current, (2) the dorsal roots carrying the painful sensations can be cut (a *rhizotomy*), (3) the ascending tracts in the spinal cord can be severed (a *tractotomy*), or (4) thalamic or limbic centers can be stimulated or destroyed. These options, listed in order of increasing degree of effect, surgical complexity, and associated risk, are used only when other methods of pain control have failed to provide relief.

In the Chinese technique of *acupuncture* to control pain, fine needles are inserted at specific locations and are either heated or twirled by the therapist. Several theories have been proposed to account for the positive effects, but none is widely accepted or proven. It has been suggested that the pain relief may result from the release of endorphins, but how acupuncture stimulates endorphin release is not known; the acupuncture points do not correspond to the distribution of any of the major peripheral nerves.

Many other aspects of pain generation and control remain a mystery. Up to 30 percent of patients who receive a nonfunctional medication subsequently experience a significant reduction in pain. It has been suggested that this *placebo effect* results from endorphin release triggered by the expectation of pain relief. Although the medication has no direct effect, the indirect effect can be quite significant and complicates the evaluation of analgesic medications.

stimulated. This effect involves the inhibition of centers in the thalamus, reticular formation, lower brain stem, and spinal cord.

An understanding of the origins of pain sensations and an ability to control or reduce pain levels have always been among the most important aspects of medical treatment. After all, it is usually pain that induces someone to seek treatment; conditions that are not painful are typically ignored or tolerated. Although we often use the term *pain pathways*, it is becoming clear that pain distribution and perception are extremely complex—more so than had previously been imagined. **AM** Pain Mechanisms, Pathways, and Control: A Closer Look

The sensory neurons that bring pain sensations into the CNS release *glutamate* or *Substance P* as neurotransmitters. These neurotransmitters produce facilitation of neurons along the pain pathways. As a result, the level of pain experienced (especially chronic pain) can be out of proportion to the amount of painful stimuli or the apparent tissue damage. This effect may be one reason why people differ so widely in their perception of the pain associated with childbirth, headaches, or back pain. This facilitation is also presumed to play a role in phantom limb pain; the sensory neurons may be inactive, but the hyperexcitable interneurons may continue to generate pain sensations.

The level of pain felt by an individual can be reduced by the release of endorphins and enkephalins within the CNS. As you may recall from Chapter 12, endorphins and enkephalins are neuromodulators whose release inhibits activity along pain pathways in the brain. ∞ p. 418 These compounds, structurally similar to morphine, are found in the limbic system, hypothalamus, and reticular formation. The pain centers in these areas also use Substance P as a neurotransmitter. Endorphins bind to the presynaptic membrane and prevent the release of Substance P, thereby reducing the conscious perception of pain, although the painful stimulus remains.

Thermoreceptors

Temperature receptors, or **thermoreceptors**, are free nerve endings located in the dermis of the skin, in skeletal muscles, in the liver, and in the hypothalamus. Cold receptors are three or four times more numerous than warm receptors. No structural differences between warm and cold thermoreceptors have been identified.

Temperature sensations are conducted along the same pathways that carry pain sensations. They are sent to the reticular formation, the thalamus, and (to a lesser extent) the primary sensory cortex. Thermoreceptors are phasic receptors: They are very active when the temperature is changing, but they quickly adapt to a stable temperature. When you enter an air-conditioned classroom on a hot summer day or a warm lecture hall on a brisk fall evening, the temperature change seems extreme at first, but you quickly become comfortable as adaptation occurs.

Mechanoreceptors

Mechanoreceptors are sensitive to stimuli that distort their cell membranes. These membranes contain *mechanically regulated*

ion channels whose gates open or close in response to stretching, compression, twisting, or other distortions of the membrane. There are three classes of mechanoreceptors:

1. **Tactile receptors**, which provide sensations of touch, pressure, and vibration. These sensations are closely related. Touch sensations provide information about shape or texture, whereas pressure sensations indicate the degree of mechanical distortion. Vibration sensations indicate a pulsing or oscillating pressure. The receptors involved may be specialized in some way. For example, rapidly adapting tactile receptors are best suited for detecting vibration. But your interpretation of a sensation as touch rather than pressure is typically a matter of the degree of stimulation and not of differences in the type of receptor stimulated.

2. **Baroreceptors** (bar-ō-rē-SEP-torz; *baro-*, pressure), which detect pressure changes in the walls of blood vessels and in portions of the digestive, reproductive, and urinary tracts.

3. **Proprioceptors,** which monitor the positions of joints and muscles. These are the most structurally and functionally complex of the general sensory receptors.

TACTILE RECEPTORS Fine touch and pressure receptors provide detailed information about a source of stimulation, including its exact location, shape, size, texture, and movement. These receptors are extremely sensitive and have relatively narrow receptive fields. **Crude touch and pressure receptors** provide poor localization and, because they have relatively large receptive fields, give little additional information about the stimulus.

Tactile receptors range in complexity from free nerve endings to specialized sensory complexes with accessory cells and supporting structures. Figure 15–3● shows six types of tactile receptors in the skin:

1. Free nerve endings sensitive to touch and pressure are situated between epidermal cells (Figure 15–3a●). There appear to be no structural differences between these receptors and the free nerve endings that provide temperature or pain sensations. These are the only sensory receptors on the corneal surface of the eye, but in other portions of the body surface, more specialized tactile receptors are probably more important. Free nerve endings that provide touch sensations are tonic receptors with small receptive fields.

2. Wherever hairs are located, the nerve endings of the **root hair plexus** monitor distortions and movements across the body surface (Figure 15–3b●). When a hair is displaced, the movement of the follicle distorts the sensory dendrites and produces action potentials. These receptors adapt rapidly, so they are best at detecting initial contact and subsequent movements. For example, you generally feel your clothing only when you move or when you consciously focus on tactile sensations from the skin.

3. **Tactile discs**, or *Merkel's* (MER-kelz) *discs*, are fine touch and pressure receptors (Figure 15–3c●). They are extremely sensitive tonic receptors, with very small receptive fields. The dendritic processes of a single myelinated afferent fiber make close contact with unusually large epithelial cells in the stratum ger-

minativum of the skin; we described these *Merkel cells* in Chapter 5. ∞ p. 158

4. **Tactile corpuscles**, or *Meissner's* (MĪS-nerz) *corpuscles*, perceive sensations of fine touch and pressure and low-frequency vibration. They adapt to stimulation within a second after contact. Tactile corpuscles are fairly large structures, measuring roughly 100 μm in length and 50 μm in width. These receptors are most abundant in the eyelids, lips, fingertips, nipples, and external genitalia. The dendrites are highly coiled and interwoven, and they are surrounded by modified Schwann cells. A fibrous capsule surrounds the entire complex and anchors it within the dermis (Figure 15–3d●).

5. **Lamellated** (LAM-e-lā-ted; *lamella*, a little thin plate) **corpuscles,** or *pacinian* (pa-SIN-ē-an) *corpuscles*, are sensitive to deep pressure. Because they are fast-adapting receptors, they are most sensitive to pulsing or high-frequency vibrating stimuli. A single dendrite lies within a series of concentric layers of collagen fibers

and supporting cells (specialized fibroblasts) (Figure 15–3e●). The entire corpuscle may reach 4 mm in length and 1 mm in diameter. The concentric layers, separated by interstitial fluid, shield the dendrite from virtually every source of stimulation other than direct pressure. Lamellated corpuscles adapt quickly because distortion of the capsule soon relieves pressure on the sensory process. Somatic sensory information is provided by lamellated corpuscles located throughout the dermis, notably in the fingers, mammary glands, and external genitalia, in the superficial and deep fasciae, and in joint capsules. Visceral sensory information is provided by lamellated corpuscles in mesenteries, in the pancreas, and in the walls of the urethra and urinary bladder.

6. **Ruffini** (roo-FĒ-nē) **corpuscles** are also sensitive to pressure and distortion of the skin, but they are located in the reticular (deep) dermis. These receptors are tonic and show little if any adaptation. The capsule surrounds a core of collagen fibers that are continuous with those of the surrounding dermis

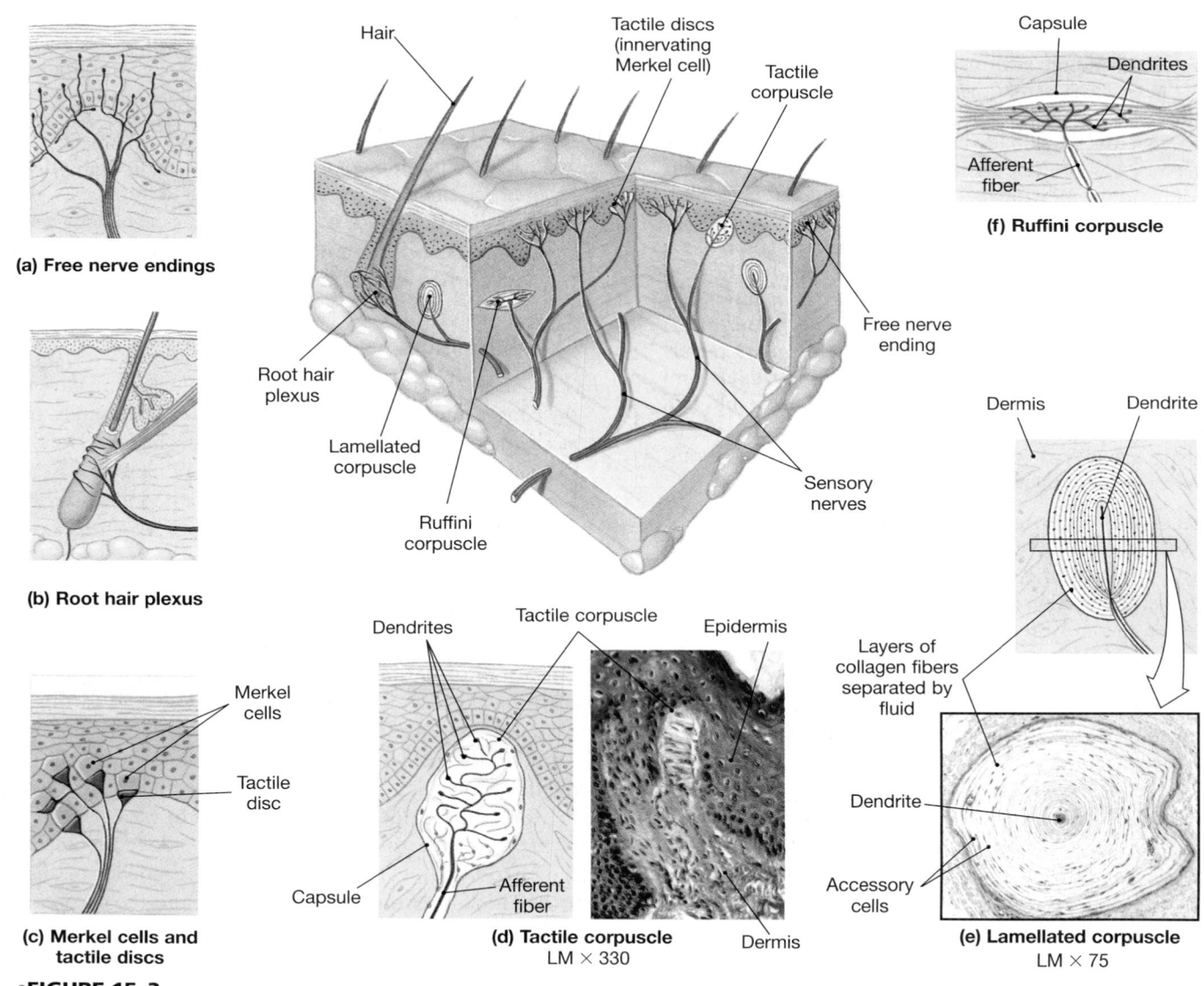

(a) Free nerve endings

(b) Root hair plexus

(c) Merkel cells and tactile discs

(d) Tactile corpuscle
LM × 330

(e) Lamellated corpuscle
LM × 75

(f) Ruffini corpuscle

●**FIGURE 15–3**
Tactile Receptors in the Skin

(Figure 15–3f●). In the capsule, a network of dendrites is intertwined with the collagen fibers. Any tension or distortion of the dermis tugs or twists the capsular fibers, stretching or compressing the attached dendrites and altering the activity in the myelinated afferent fiber.

Our sensitivity to tactile sensations may be altered by infection, disease, or damage to sensory neurons or pathways. As a result, mapping tactile responses can sometimes aid clinical assessment. Sensory losses with clear regional boundaries indicate trauma to spinal nerves. For example, sensory loss along the boundary of a dermatome can help identify the affected spinal nerve or nerves. ∞ p. 440 Regional sensitivity to light touch can be checked by gentle contact with a fingertip or a slender wisp of cotton. Vibration receptors are tested by applying the base of a tuning fork to the skin. We discuss more detailed procedures, such as the *two-point discrimination test*, in the *Applications Manual*. **AM** Assessment of Tactile Sensitivities

Tickle and itch sensations are closely related to the sensations of touch and pain. The receptors involved are free nerve endings, and the information is carried by unmyelinated Type C fibers. Tickle sensations, which are usually (but not always) described as pleasurable, are produced by a light touch that moves across the skin. Psychological factors are involved in the interpretation of tickle sensations, and tickle sensitivity differs greatly among individuals. Little is known about the labeled line involved, but the sensations travel over the spinothalamic pathway. Itching is probably produced by the stimulation of the same receptors. Specific "itch spots" can be mapped in the skin, the inner surfaces of the eyelids, and the mucous membrane of the nose. Itch sensations are absent from other mucous membranes and from deep tissues and viscera. Itching is extremely unpleasant, even more unpleasant than pain. Individuals with extreme itching will scratch even when pain is the result. Itch receptors can be stimulated by the injection of histamine or proteolytic enzymes into the epidermis and superficial dermis. The precise receptor mechanism is unknown.

BARORECEPTORS Baroreceptors monitor changes in pressure. A baroreceptor consists of free nerve endings that branch within the elastic tissues in the wall of a distensible organ, such as a blood vessel or a portion of the respiratory, digestive, or urinary tract. When the pressure changes, the elastic walls of the tract recoil or expand. This movement distorts the dendritic branches and alters the rate of action potential generation. Baroreceptors respond immediately to a change in pressure, but they adapt rapidly, and the output along the afferent fibers gradually returns to normal.

Figure 15–4● provides an overview of the distribution of baroreceptors in the body. Baroreceptors monitor blood pressure in the walls of major vessels, including the carotid artery (at the *carotid sinus*) and the aorta (at the *aortic sinus*). The information plays a major role in regulating cardiac function and adjusting blood flow to vital tissues. Baroreceptors in the lungs monitor the degree of lung expansion. This information is relayed to the respiratory rhythmicity centers, which set the pace of respiration. Comparable stretch receptors in the digestive and urinary tracts

trigger a variety of visceral reflexes, including those of urination and defecation. We shall describe those baroreceptor reflexes in chapters that deal with specific physiological systems.

PROPRIOCEPTORS Proprioceptors monitor the position of joints, the tension in tendons and ligaments, and the state of muscular contraction. There are three major groups of proprioceptors:

1. *Muscle Spindles.* Muscle spindles monitor skeletal muscle length and trigger stretch reflexes. ∞ p. 452
2. *Golgi Tendon Organs.* **Golgi tendon organs** are similar in function to Ruffini corpuscles but are located at the junction between a skeletal muscle and its tendon. In a Golgi tendon organ, dendrites branch repeatedly and wind around the densely packed collagen fibers of the tendon. These receptors are stimulated by tension in the tendon; they thus monitor the external tension developed during muscle contraction.
3. *Receptors in Joint Capsules.* Joint capsules are richly innervated by free nerve endings that detect pressure, tension, and movement at the joint. Your sense of body position results from the integration of information from these receptors with the information provided by muscle spindles, Golgi tendon organs, and the receptors of the inner ear.

Proprioceptors do not adapt to constant stimulation, and each receptor continuously sends information to the CNS. A relatively small proportion of the arriving proprioceptive information reaches your conscious awareness; most proprioceptive information is processed at subconscious levels.

Chemoreceptors

Specialized chemoreceptive neurons can detect small changes in the concentration of specific chemicals or compounds. In general, **chemoreceptors** respond only to water-soluble and lipid-soluble substances that are dissolved in the surrounding fluid. These receptors exhibit peripheral adaptation over a period of seconds, and central adaptation may also occur. Figure 15–5● indicates the locations and functions of chemoreceptors.

The chemoreceptors included in the general senses do not send information to the primary sensory cortex, so we are not consciously aware of the sensations they provide. The arriving sensory information is routed to brain stem centers that deal with the autonomic control of respiratory and cardiovascular functions. Neurons in the respiratory centers of the brain respond to the concentration of hydrogen ions (pH) and carbon dioxide molecules in the cerebrospinal fluid. Chemoreceptive neurons are also located in the **carotid bodies**, near the origin of the internal carotid arteries on each side of the neck, and in the **aortic bodies**, between the major branches of the aortic arch. These receptors monitor the pH and the carbon dioxide and oxygen concentrations of arterial blood. The afferent fibers leaving the carotid or aortic bodies reach the respiratory centers by traveling within cranial nerves IX (glossopharyngeal) and X (vagus).

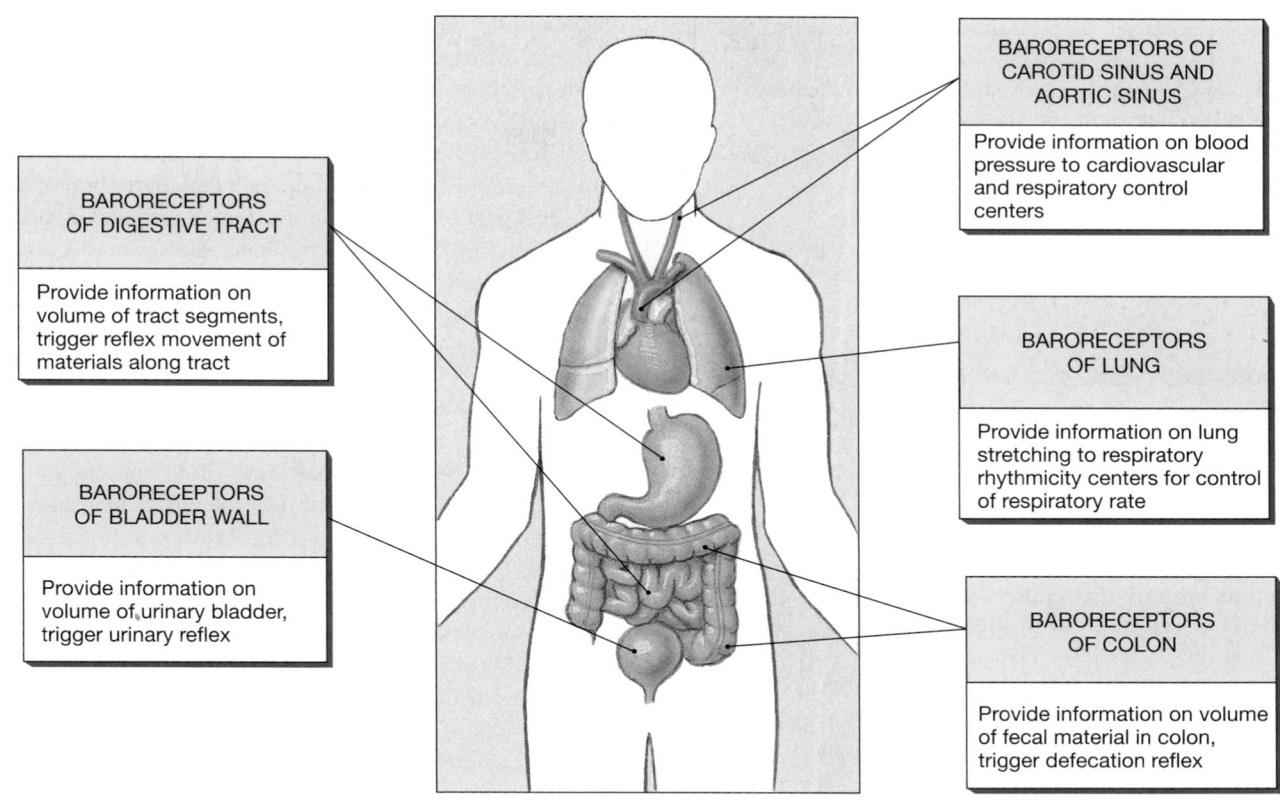

•FIGURE 15–4

Baroreceptors and the Regulation of Visceral Function. Baroreceptors provide information essential to the regulation of visceral (autonomic) activities, including cardiovascular function, urination, defecation, and respiration.

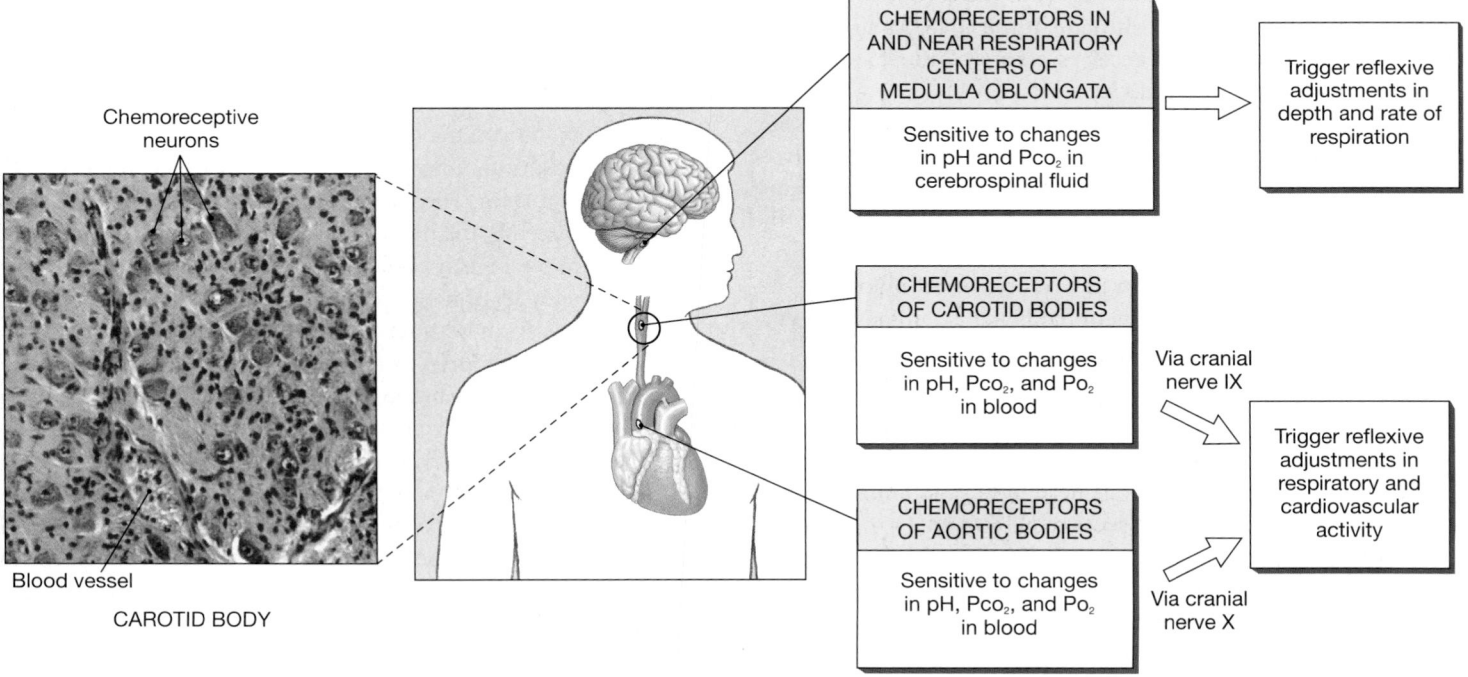

•FIGURE 15–5

Chemoreceptors. Chemoreceptors are located inside the CNS, on the ventrolateral surfaces of the medulla oblongata, and in the aortic and carotid bodies. These receptors are involved in the autonomic regulation of cardiovascular and respiratory function. The micrograph shows the histological appearance of the chemoreceptive neurons in the carotid body. (LM × 1150)

✓ Receptor A has a circular receptive field with a diameter of 2.5 cm. Receptor B has a circular receptive field 7.0 cm in diameter. Which receptor will provide more precise sensory information?

✓ When the nociceptors in your hand are stimulated, what sensation do you perceive?

✓ What would happen to you if the information from proprioceptors in your legs were blocked from reaching the CNS?

Answers start on page Q-1

15–3 THE ORGANIZATION OF SENSORY PATHWAYS

Objectives

- Identify the major sensory pathways.
- Explain how we can distinguish among sensations that originate in different areas of the body.

A sensory neuron that delivers sensations to the CNS is often called a **first-order neuron**. The cell body of a first-order general sensory neuron is located in a dorsal root ganglion or cranial nerve ganglion. In the CNS, the axon of that sensory neuron synapses on an interneuron known as a **second-order neuron**, which may be located in the spinal cord or brain stem. If the sensation is to reach our conscious awareness, the second-order neuron will synapse on a **third-order neuron** in the thalamus. Somewhere along its length, the axon of the second-order neuron crosses over to the opposite side of the CNS. As a result, the right side of the thalamus receives sensory information from the left side of the body, and vice versa.

The axons of the third-order neurons ascend without crossing over and synapse on neurons of the primary sensory cortex of the cerebral hemisphere. As a result, the right cerebral hemisphere receives sensory information from the left side of the body, and the left cerebral hemisphere receives sensations from the right side. The reason for this crossover is unknown. Although it has no apparent functional benefits, crossover occurs along sensory and motor pathways in all vertebrates.

◻ SOMATIC SENSORY PATHWAYS

Somatic sensory pathways carry sensory information from the skin and musculature of the body wall, head, neck, and limbs. We shall consider three major somatic sensory pathways: (1) the *posterior column pathway*, (2) the *anterolateral pathway*, and (3) the *spinocerebellar pathway*. These pathways utilize pairs of spinal tracts, symmetrically arranged on opposite sides of the spinal cord. All the axons within a tract share a common origin and destination.

Figure 15–6● indicates the relative positions of the spinal tracts involved. It may help you to know at the outset that tract names often give clues to their function. For example, if the name of a tract begins with *spino-*, the tract must *start* in the spinal cord and *end* in the brain. It must therefore be an ascending tract that carries

sensory information. The rest of the name indicates the tract's destination. Thus, a *spinothalamic tract* begins in the spinal cord and carries sensory information to the thalamus of the brain.

If, on the other hand, the name of a tract ends in -*spinal*, the tract *ends* in the spinal cord and *starts* in a higher center of the brain. It must therefore be a descending tract that carries motor commands. The first part of the name indicates the nucleus or cortical area of the brain where the tract originates. For example, a *corticospinal tract* carries motor commands from the cerebral cortex to the spinal cord. Such tracts will be considered later in the chapter.

The Posterior Column Pathway

The **posterior column pathway** carries sensations of highly localized ("fine") touch, pressure, vibration, and proprioception (Figure 15–7a●). This pathway, also known as the *dorsal column/medial lemniscus pathway*, begins at a peripheral receptor and ends at the primary sensory cortex of the cerebral hemispheres. The spinal tracts involved are the left and right **fasciculus gracilis** (*gracilis*, delicate) and the left and right **fasciculus cuneatus** (*cuneus*, wedge-shaped). On each side of the posterior median sulcus, the fasciculus gracilis is medial to the fasciculus cuneatus.

The axons of the first-order neurons reach the CNS within the dorsal roots of spinal nerves and the sensory roots of cranial nerves. The axons ascending within the posterior column are organized according to the region innervated. Axons carrying sensations from the inferior half of the body ascend within the fasciculus gracilis and synapse in the nucleus gracilis of the medulla oblongata. Axons carrying sensations from the superior half of the trunk, upper limbs, and neck ascend in the fasciculus cuneatus and synapse in the nucleus cuneatus. ∞ p. 475

Axons of the second-order neurons of the nucleus gracilis and nucleus cuneatus ascend to the thalamus. As they ascend, these axons cross over to the opposite side of the brain stem. The crossing of an axon from the left side to the right side, or vice versa, is called **decussation**. Once on the opposite side of the brain, the axons enter a tract called the **medial lemniscus** (*lemniskos*, ribbon). As it ascends, the medial lemniscus runs alongside a smaller tract that carries sensory information from the face, relayed from the sensory nuclei of the trigeminal nerve (V).

The axons in these tracts synapse on third-order neurons in one of the ventral nuclei of the thalamus. ∞ p. 481 These nuclei sort the arriving information according to (1) the nature of the stimulus and (2) the region of the body involved. Processing in the thalamus determines whether we perceive a given sensation as fine touch rather than as pressure or vibration.

Our ability to localize the sensation—to determine precisely where on the body a specific stimulus originated—depends on the projection of information from the thalamus to the primary sensory cortex. Sensory information from the toes arrives at one end of the primary sensory cortex, and information from the head arrives at the other. When neurons in one portion of your primary sensory cortex are stimulated, you become consciously aware of sensations originating at a specific location. If your primary sensory cortex were damaged or the projection fibers were cut, you could detect a light touch but would be unable to determine its source.

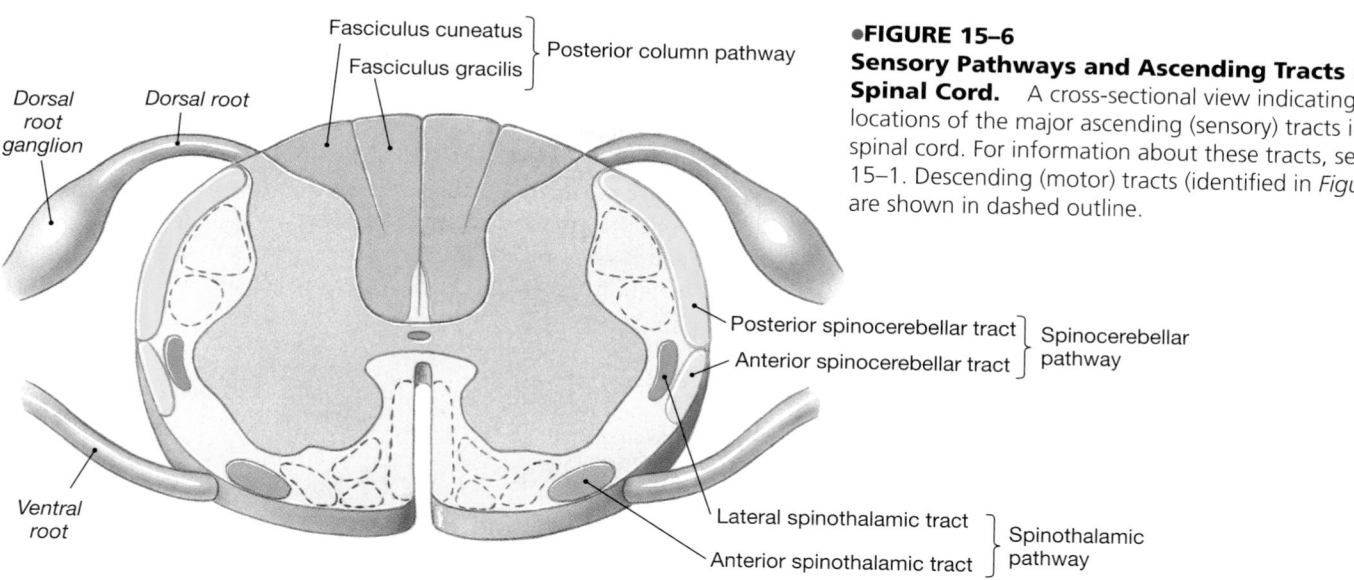

•FIGURE 15–6
Sensory Pathways and Ascending Tracts in the Spinal Cord. A cross-sectional view indicating the locations of the major ascending (sensory) tracts in the spinal cord. For information about these tracts, see Table 15–1. Descending (motor) tracts (identified in *Figure 15–10*) are shown in dashed outline.

Abnormal Sensations

The perception that an arriving stimulus is painful rather than cold, hot, or vibrating depends on which second-order and third-order neurons are stimulated. The ability to localize that stimulus to a specific location in the body depends on the stimulation of an appropriate area of the primary sensory cortex. Any abnormality along the pathway can result in inappropriate sensations or inaccurate localization of the source. Consider these examples:

- An individual can experience painful sensations that are not real. For example, a person may continue to experience pain in an amputated limb. This *phantom limb pain* is caused by activity in the sensory neurons or interneurons along the spinothalamic pathway. The neurons involved were once part of the labeled line that monitored conditions in the intact limb.

- An individual can feel pain in an uninjured part of the body when the pain actually originates at another location. For example, strong visceral pain sensations arriving at a segment of the spinal cord can stimulate interneurons that are part of the spinothalamic pathway. Activity in these interneurons leads to the stimulation of the primary sensory cortex, so the individual feels pain in a specific part of the body surface. This phenomenon is called **referred pain**. Two familiar examples are (1) the pain of a heart attack, which is frequently felt in the left arm, and (2) the pain of appendicitis, which is generally felt first in the area around the navel and then in the right lower quadrant. These and additional examples are shown in Figure 15–7•.

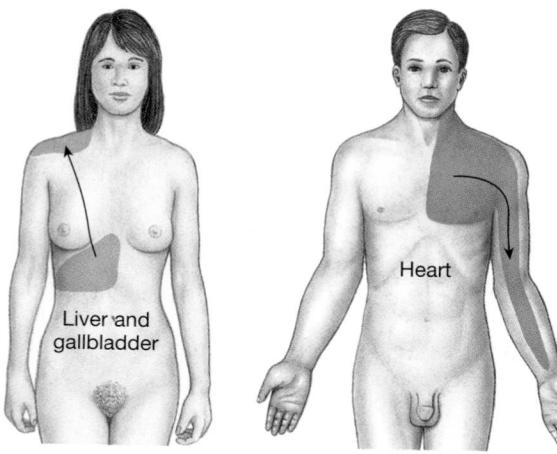

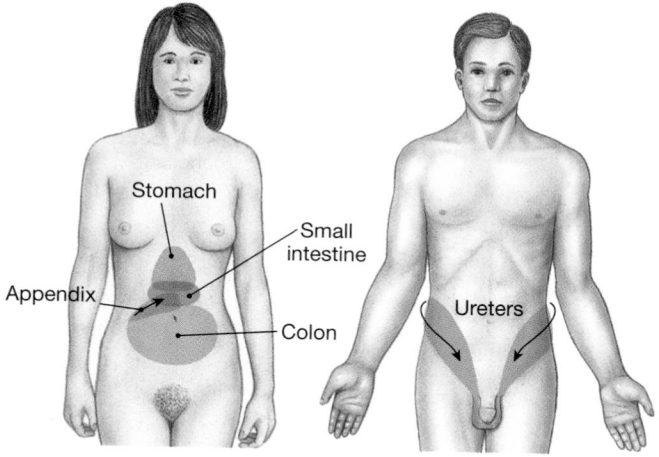

•FIGURE 15–7
Referred Pain. Pain sensations from visceral organs are often perceived as involving specific regions of the body surface innervated by the same spinal segments. Each region of perceived pain is labeled according to the organ at which the pain originates.

The same sensations are reported whether the cortical neurons are activated by axons ascending from the thalamus or by direct electrical stimulation. Researchers have electrically stimulated the primary sensory cortex in awake individuals during brain surgery and asked the subjects where they thought the stimulus originated. The results were used to create a functional map of the primary sensory cortex. This map, shown in Figure 15–8●, is called a **sensory homunculus** ("little man").

The proportions of the sensory homunculus are very different from those of any individual. For example, the face is huge and distorted, with enormous lips and tongue, whereas the back is relatively tiny. These distortions occur because the area of sensory cortex devoted to a particular body region is proportional not to the region's absolute size but to the *number of sensory receptors* it contains. In other words, many more cortical neurons are required to process sensory information arriving from the tongue, which has tens of thousands of taste and touch receptors, than to analyze sensations originating on the back, where touch receptors are few and far between.

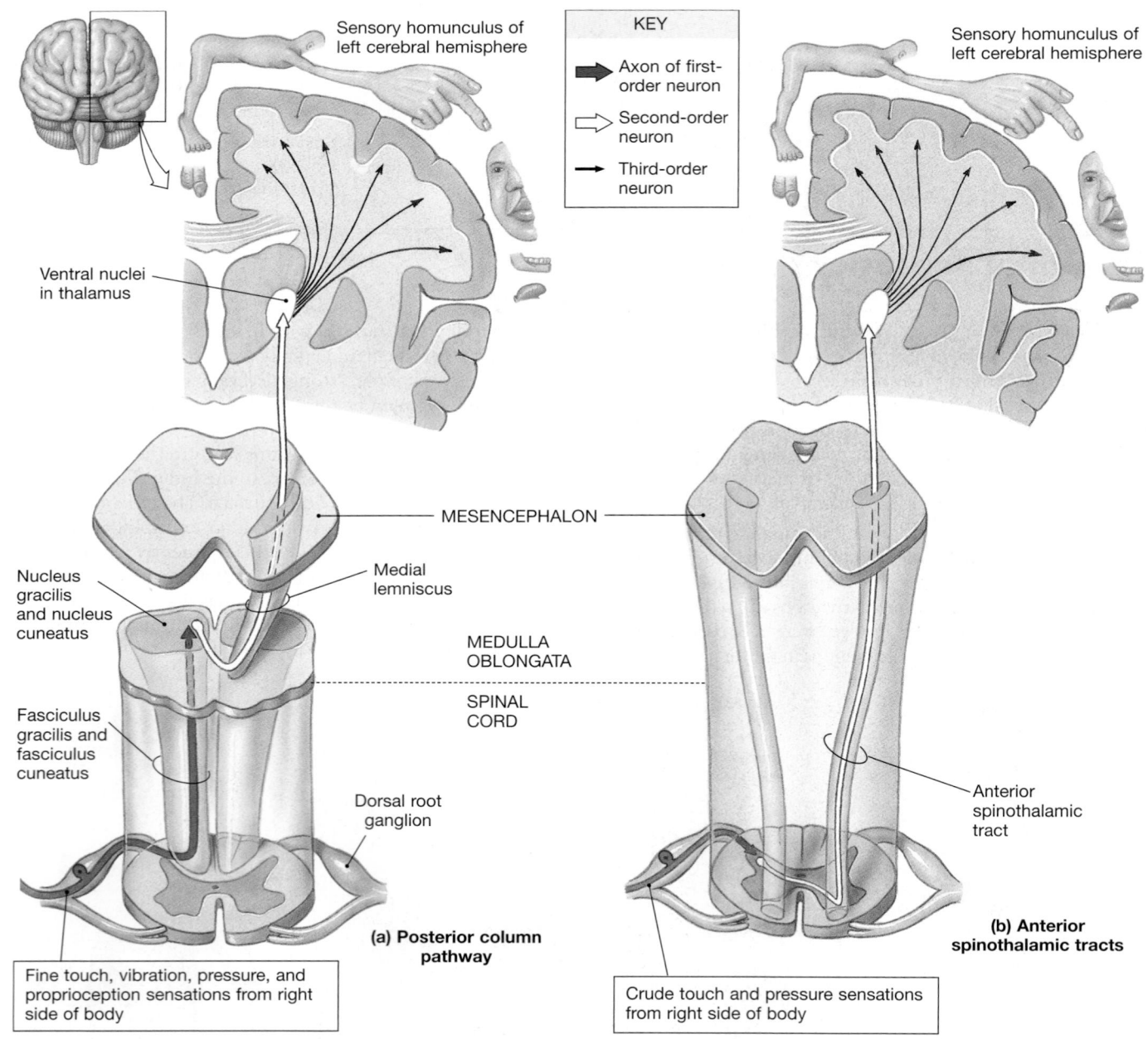

KEY

→ Axon of first-order neuron

⇨ Second-order neuron

→ Third-order neuron

Sensory homunculus of left cerebral hemisphere

Ventral nuclei in thalamus

Nucleus gracilis and nucleus cuneatus

Medial lemniscus

Fasciculus gracilis and fasciculus cuneatus

Dorsal root ganglion

MESENCEPHALON

MEDULLA OBLONGATA

SPINAL CORD

Anterior spinothalamic tract

(a) Posterior column pathway

Fine touch, vibration, pressure, and proprioception sensations from right side of body

(b) Anterior spinothalamic tracts

Crude touch and pressure sensations from right side of body

●**FIGURE 15–8**

The Posterior Column Pathway and the Spinothalamic Tracts. For clarity, only the pathways for sensations originating on the right side of the body are shown. **(a)** The posterior column pathway delivers fine touch, vibration, and proprioception information to the primary sensory cortex on the opposite side of the body. **(b)** The anterior spinothalamic tracts carry sensations of crude touch and pressure to the primary sensory cortex on the opposite side of the body. **(c)** The lateral spinothalamic tracts carry sensations of pain and temperature to the primary sensory cortex on the opposite side of the body.

The Anterolateral Pathway

The **anterolateral pathway** provides conscious sensations of poorly localized ("crude") touch, pressure, pain, and temperature. In this pathway, the axons of first-order sensory neurons enter the spinal cord and synapse on second-order neurons within the posterior gray horns. The axons of these interneurons cross to the opposite side of the spinal cord before ascending. This pathway includes relatively small tracts that deliver sensations to reflex centers in the brain stem as well as larger tracts that carry sensations destined for the cerebral cortex.

Sensations bound for the cerebral cortex ascend within the anterior or lateral spinothalamic tracts. The **anterior spinothalamic tracts** carry crude touch and pressure sensations (Figure 15–8b●), whereas the **lateral spinothalamic tracts** carry pain and tempera-

ture sensations (Figure 15–8c●). These tracts end at third-order neurons in the ventral nucleus group of the thalamus. After the sensations have been sorted and processed, they are relayed to the primary sensory cortex.

The Spinocerebellar Pathway

The cerebellum receives proprioceptive information about the position of skeletal muscles, tendons, and joints from the **spinocerebellar pathway** (Figure 15–9●). This information does not reach our conscious awareness. The axons of first-order sensory neurons synapse on interneurons in the dorsal gray horns of the spinal cord. The axons of these second-order neurons ascend in one of the spinocerebellar tracts.

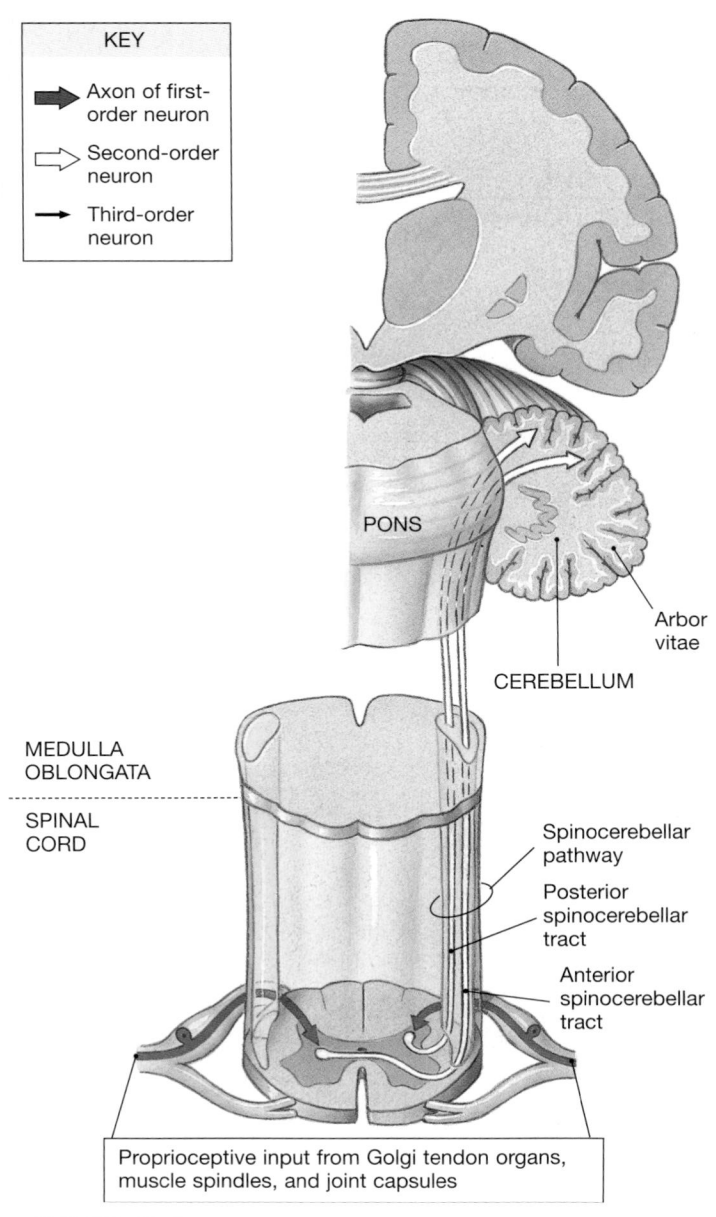

(c) Lateral spinothalamic tracts

Pain and temperature sensations from right side of body

●**FIGURE 15–9**
The Spinocerebellar Pathway

Proprioceptive input from Golgi tendon organs, muscle spindles, and joint capsules

- The **posterior spinocerebellar tracts** contain axons that do not cross over to the opposite side of the spinal cord. These axons reach the cerebellar cortex via the inferior cerebellar peduncle of that side.

- The **anterior spinocerebellar tracts** are dominated by axons that have crossed over to the opposite side of the spinal cord, although they do contain a significant number of uncrossed axons as well. The sensations carried by the anterior spinocerebellar tracts reach the cerebellar cortex via the superior cerebellar peduncle. Interestingly, many of the axons that cross over and ascend to the cerebellum then cross over again within the cerebellum, synapsing on the same side as the original stimulus. The functional significance of this "double cross" is unknown.

The information carried by the spinocerebellar pathway ultimately arrives at the *Purkinje cells* of the cerebellar cortex. ∞ p. 477

Proprioceptive information from each part of the body is relayed to a specific portion of the cerebellar cortex. We shall consider the integration of proprioceptive information and the role of the cerebellum in somatic motor control in a later section.

Table 15–1 reviews the somatic sensory pathways discussed in this section.

□ VISCERAL SENSORY PATHWAYS

Visceral sensory information is collected by interoceptors monitoring visceral tissues and organs, primarily within the thoracic and abdominopelvic cavities. These interoceptors include nociceptors, thermoreceptors, tactile receptors, baroreceptors, and chemoreceptors, although none of them are as numerous as they are in somatic tissues. The axons of the first-order neurons usual-

TABLE 15–1 PRINCIPAL ASCENDING (SENSORY) PATHWAYS

Pathway/Tract	Sensation(s)	Location of Neuron Cell Bodies		
		First-Order	Second-Order	Third-Order
POSTERIOR COLUMN PATHWAY				
Fasciculus gracilis	Proprioception and fine touch, pressure, and vibration from inferior half of body	Dorsal root ganglia of inferior half of the body; axons enter CNS in dorsal roots and join fasciculus gracilis	Nucleus gracilis of medulla oblongata; axons cross over before entering medial lemniscus	Ventral nuclei of thalmus
Fasciculus cuneatus	Proprioception and fine touch, and ventral pressure, and vibration from superior half of body	Dorsal root ganglia of superior half of body; axons enter CNS in dorsal roots and join fasciculus cuneatus	Nucleus cuneatus of medulla oblongata; axons cross over before entering medial lemniscus	As above
SPINOTHALAMIC PATHWAY				
Lateral spinothalamic tracts	Pain and temperature	Dorsal root ganglia; axons enter CNS in dorsal roots	Interneurons in posterior gray horn; axons enter lateral spinothalamic tract on opposite side	Ventral nuclei of thalamus
Anterior spinothalamic tracts	Crude touch and pressure	As above	Interneurons in posterior gray horn; axons enter anterior spinothalamic tract on opposite side	As above
SPINOCEREBELLAR PATHWAY				
Posterior spinocerebellar tracts	Proprioception	Dorsal root ganglia; axons enter CNS in dorsal roots	Interneurons in posterior gray horn; axons enter posterior spinothalamic tract on same side	Not present
Anterior spinocerebellar tracts	Proprioception	As above	Interneurons in same spinal section; axons enter anterior spinocerebellar tract on the same or opposite side	Not present

ly travel in company with autonomic motor fibers innervating the same visceral structures (Figure 13–9●, p. 439).

Cranial nerves V, VII, IX, and X carry visceral sensory information from the mouth, palate, pharynx, larynx, trachea, esophagus, and associated vessels and glands. pp. 497-502 This information is delivered to the **solitary nucleus**, a large nucleus in the medulla oblongata. The solitary nucleus is a major processing and sorting center for visceral sensory information; it has extensive connections with the various cardiovascular and respiratory centers as well as with the reticular formation.

The dorsal roots of spinal nerves $T_1–L_2$ carry visceral sensory information provided by receptors in organs located between the diaphragm and the pelvic cavity. The dorsal roots of spinal nerves $S_2–S_4$ carry visceral sensory information from organs in the inferior portion of the pelvic cavity, including the last portion of the large intestine, the urethra and base of the urinary bladder, and the prostate gland (males) or the cervix of the uterus and adjacent portions of the vagina (females).

The first-order neurons deliver the visceral sensory information to interneurons whose axons ascend within the anterolateral pathway. Most of the sensory information is delivered to the solitary nucleus, and because it never reaches the primary sensory cortex we remain unaware of these sensations.

✓ As a result of pressure on her spinal cord, Jill cannot feel touch or pressure on her lower limbs. Which spinal tract is being compressed?

✓ Which spinal tract carries action potentials generated by nociceptors?

✓ Which cerebral hemisphere receives impulses conducted by the right fasciculus gracilis?

Answers start on page Q-1

The somatic sensory pathways can be reviewed on the **Companion Website**: Chapter 15/Reviewing Concepts/Labeling.

Final Destination	Site of Cross-Over
Primary sensory cortex on side opposite stimulus	Axons of second-order neurons before entering the medial lemniscus
As above	As above
Primary sensory cortex on side opposite stimulus	Axons of second-order neurons at level of entry
As above	As above
Cerebellar cortex on side of stimulus	None
Cerebellar cortex on side opposite (and side of) stimulus	Axons of most second-order neurons cross over before entering tract; many re-cross at cerebellum

15–4 THE SOMATIC NERVOUS SYSTEM

Objectives

- Describe the components, processes, and functions of the somatic motor pathways.
- Describe the levels of information processing involved in motor control.

Motor commands issued by the CNS are distributed by the somatic nervous system (SNS) and the autonomic nervous system (ANS). The somatic nervous system, also called the *somatic motor system*, controls the contractions of skeletal muscles. The output of the SNS is under voluntary control. The autonomic nervous system, or *visceral motor system*, controls visceral effectors, such as smooth muscle, cardiac muscle, and glands. In Chapter 16, we will examine the organization of the ANS; our interest here is the structure of the SNS. Throughout this discussion we will use the terms *motor neuron* and *motor control* to refer specifically to somatic motor neurons and pathways that control skeletal muscles.

Somatic motor pathways always involve at least two motor neurons: an **upper motor neuron**, whose cell body lies in a CNS processing center, and a **lower motor neuron**, whose cell body lies in a nucleus of the brain stem or spinal cord. The upper motor neuron synapses on the lower motor neuron, which in turn innervates a single motor unit in a skeletal muscle. Activity in the upper motor neuron may facilitate or inhibit the lower motor neuron. Activation of the lower motor neuron triggers a contraction in the innervated muscle. Only the axon of the lower motor neuron extends outside the CNS. Destruction of or damage to a lower motor neuron eliminates voluntary and reflex control over the innervated motor unit.

Conscious and subconscious motor commands control skeletal muscles by traveling over three integrated motor pathways: the *corticospinal pathway*, the *medial pathway*, and the *lateral pathway*. Figure 15–10● indicates the positions of the associated motor tracts in the spinal cord. Activity within these motor pathways is monitored and adjusted by the basal nuclei and cerebellum. Their output stimulates or inhibits the activity of either (1) motor nuclei or (2) the primary motor cortex.

☐ THE CORTICOSPINAL PATHWAY

The **corticospinal pathway**, sometimes called the *pyramidal system* (Figure 15–11●) provides voluntary control over skeletal muscles. This system begins at the *pyramidal cells* of the primary motor cortex. ∞ p. 488 The axons of these upper motor neurons descend into the brain stem and spinal cord to synapse on lower motor neurons that control skeletal muscles. In general, the corticospinal pathway is direct: The upper motor neurons synapse directly on the lower motor neurons. However, the corticospinal pathway also works indirectly, as it innervates centers of the medial and lateral pathways.

The corticospinal pathway contains three pairs of descending tracts: (1) the *corticobulbar tracts*, (2) the *lateral corticospinal tracts*, and (3) the *anterior corticospinal tracts*. These tracts enter the white matter of the internal capsule, descend into the brain stem, and emerge on either side of the mesencephalon as the *cerebral peduncles*.

The Corticobulbar Tracts

Axons in the **corticobulbar** (kor-ti-kō-BUL-bar) **tracts** (*bulbar*, brain stem) synapse on lower motor neurons in the motor nuclei of cranial nerves III, IV, V, VI, VII, IX, XI, and XII. The corticobulbar tracts provide conscious control over skeletal muscles that move the eye, jaw, and face and some muscles of the neck and pharynx. The corticobulbar tracts also innervate the motor centers of the medial and lateral pathways.

The Corticospinal Tracts

Axons in the **corticospinal tracts** synapse on lower motor neurons in the anterior gray horns of the spinal cord. As they descend, the corticospinal tracts are visible along the ventral surface of the medulla oblongata as a pair of thick bands, the **pyramids**. Along the length of the pyramids, roughly 85 percent of the axons cross the midline (decussate) to enter the descending **lateral corticospinal tracts** on the opposite side of the spinal cord. The other 15 percent continue uncrossed along the spinal cord as the **anterior corticospinal tracts**. At the spinal segment it targets, an axon in the anterior corticospinal tract crosses over to the opposite side of the spinal cord in the anterior white commissure before synapsing on lower motor neurons in the anterior gray horns.

The Motor Homunculus

The activity of pyramidal cells in a specific portion of the primary motor cortex will result in the contraction of specific peripheral muscles. The identities of the stimulated muscles depend on the region of motor cortex that is active. As in the primary sensory cortex, the primary motor cortex corresponds point by point with specific regions of the body. The cortical areas have been mapped out in diagrammatic form, creating a **motor homunculus**. Figure 15–11● shows the motor homunculus of the left cerebral hemisphere and the corticospinal pathway controlling skeletal muscles on the right side of the body.

The proportions of the motor homunculus are quite different from those of the actual body, because the motor area devoted to

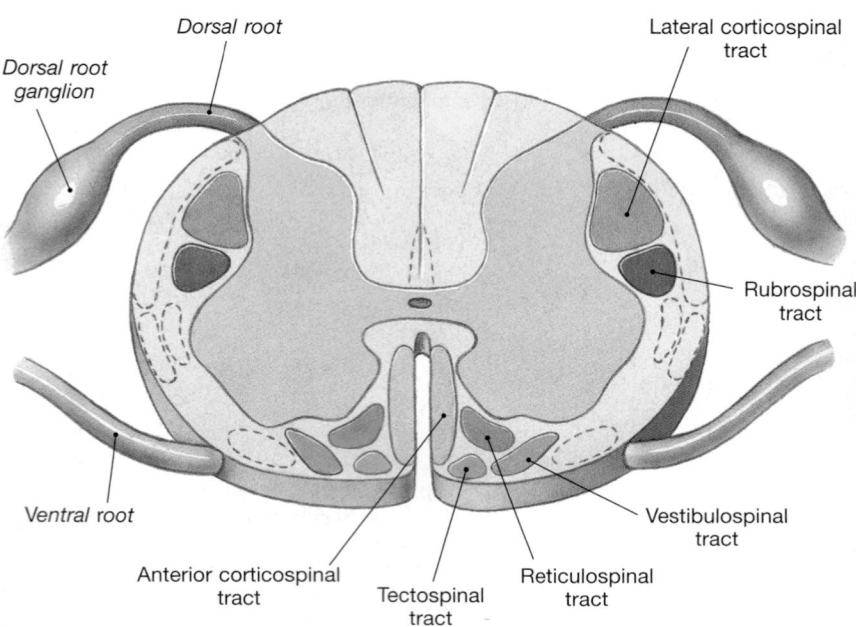

●**FIGURE 15–10**
Descending (Motor) Tracts in the Spinal Cord.
A cross-sectional view indicating the locations of the major descending (motor) tracts that contain the axons of upper motor neurons. The origins and destinations of these tracts are listed in Table 15–2. Sensory tracts (shown in *Figure 15–2*) appear in dashed outline.

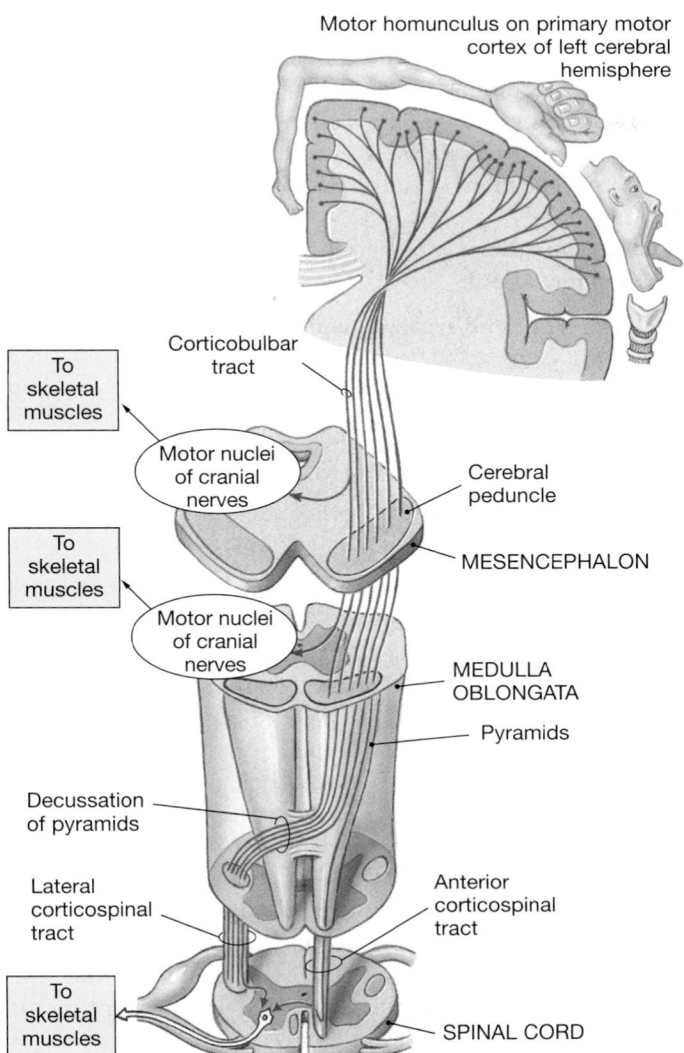

Motor homunculus on primary motor cortex of left cerebral hemisphere

Corticobulbar tract

To skeletal muscles

Motor nuclei of cranial nerves

To skeletal muscles

Motor nuclei of cranial nerves

Cerebral peduncle

MESENCEPHALON

MEDULLA OBLONGATA

Pyramids

Decussation of pyramids

Lateral corticospinal tract

Anterior corticospinal tract

To skeletal muscles

SPINAL CORD

●**FIGURE 15–11**
The Corticospinal Pathway.   The corticospinal pathway originates at the primary motor cortex. The corticobulbar tracts end at the motor nuclei of cranial nerves on the opposite side of the brain. Most fibers in this pathway cross over in the medulla and enter the lateral corticospinal tracts; the rest descend in the anterior corticospinal tracts and cross over after reaching target segments in the spinal cord.

a specific region of the cortex is proportional to the number of motor units involved in the region's control rather than to its actual size. As a result, the homunculus provides an indication of the degree of fine motor control available. For example, the hands, face, and tongue, all of which are capable of varied and complex movements, appear very large, whereas the trunk is relatively small. These proportions are similar to those of the sensory homunculus (Figure 15–8●, p. 520). The sensory and motor homunculi differ in other respects because some highly sensitive regions, such as the sole of the foot, contain few motor units, and some areas with an abundance of motor units, such as the eye muscles, are not particularly sensitive.

The term **cerebral palsy** refers to a number of disorders that affect voluntary motor performance; they appear during infancy or childhood and persist throughout the life of the affected individual. The cause may be trauma associated with premature or unusually stressful birth, maternal exposure to drugs, including alcohol, or a genetic defect that causes the improper development of motor pathways. Problems with labor and delivery may result from the compression or interruption of placental circulation or oxygen supplies. If the oxygen concentration in fetal blood declines significantly for as little as 5–10 minutes, CNS function can be permanently impaired. The cerebral cortex, cerebellum, basal nuclei, hippocampus, and thalamus are likely targets, producing abnormalities in motor skills, posture and balance, memory, speech, and learning abilities.

☐ THE MEDIAL AND LATERAL PATHWAYS

Several centers in the cerebrum, diencephalon, and brain stem may issue somatic motor commands as a result of processing performed at a subconscious level. These centers and their associated tracts were long known as the *extrapyramidal system (EPS)*, because it was thought that they operated independently of, and in parallel with, the *pyramidal system* (corticospinal pathway). This classification scheme is both inaccurate and misleading, because motor control is integrated at all levels through extensive feedback loops and interconnections. It is more appropriate to group these nuclei and tracts in terms of their primary functions: The components of the **medial pathway** help control gross movements of the trunk and proximal limb muscles, whereas those of the **lateral pathway** help control the distal limb muscles that perform more precise movements.

The medial and lateral pathways can modify or direct skeletal muscle contractions by stimulating, facilitating, or inhibiting lower motor neurons. It is important to note that the axons of upper motor neurons in the medial and lateral pathways synapse on the same lower motor neurons innervated by the corticospinal pathway. This means that the various motor pathways interact not only within the brain, through interconnections between the primary motor cortex and motor centers in the brain stem, but also through excitatory or inhibitory interactions at the level of the lower motor neuron.

The Medial Pathway

The medial pathway is primarily concerned with the control of muscle tone and gross movements of the neck, trunk, and proximal limb muscles. The upper motor neurons of the medial pathway are located in the *vestibular nuclei*, the *superior* and *inferior colliculi*, and the *reticular formation*.

The vestibular nuclei receive information, over the vestibulocochlear nerve (VIII), from receptors in the inner ear that monitor the position and movement of the head. These nuclei respond to changes in the orientation of the head, sending motor commands that alter the muscle tone, extension, and position of the neck, eyes, head, and limbs. The primary goal is to maintain posture and balance. The descending fibers in the spinal cord constitute the **vestibulospinal tracts**.

The superior and inferior colliculi are located in the *tectum*, or roof of the mesencephalon (Figure 14–10●, p. 478). The colliculi receive visual (superior) and auditory (inferior) sensations. Axons of upper motor neurons in the colliculi descend in the **tectospinal tracts**. These axons cross to the opposite side immediately, before descending to synapse on lower motor neurons in the brain stem or spinal cord. Axons in the tectospinal tracts direct reflexive changes in the position of the head, neck, and upper limbs in response to bright lights, sudden movements, or loud noises.

The reticular formation is a loosely organized network of neurons that extends throughout the brain stem. ∞ p. 474 The reticular formation receives input from almost every ascending and descending pathway. It also has extensive interconnections with the cerebrum, the cerebellum, and brain stem nuclei. Axons of upper motor neurons in the reticular formation descend into the **reticulospinal tracts** without crossing to the opposite side. The effects of reticular formation stimulation are determined by the region stimulated. For example, the stimulation of upper motor neurons in one portion of the reticular formation produces eye movements, whereas the stimulation of another portion activates respiratory muscles.

The Lateral Pathway

The lateral pathway is primarily concerned with the control of muscle tone and the more precise movements of the distal parts of the limbs. The upper motor neurons of the lateral pathway lie within the red nuclei of the mesencephalon. ∞ p. 478 Axons of upper motor neurons in the red nuclei cross to the opposite side of the brain and descend into the spinal cord in the **rubrospinal tracts** (*ruber*, red). In humans, the rubrospinal tracts are small and extend only to the cervical spinal cord. There they provide motor control over distal muscles of the upper limbs; normally, their role is insignificant as compared with that of the lateral corticospinal tracts. However, the rubrospinal tracts can be important in maintaining motor control and muscle tone in the upper limbs if the lateral corticospinal tracts are damaged.

Table 15–2 reviews the major motor tracts we discussed in this section.

☐ THE BASAL NUCLEI AND CEREBELLUM

The basal nuclei and cerebellum are responsible for coordination and feedback control over muscle contractions, whether those contractions are consciously or subconsciously directed.

The Basal Nuclei

The basal nuclei provide the background patterns of movement involved in voluntary motor activities. For example, they may control muscles that determine the background position of the trunk or limbs, or they may direct rhythmic cycles of movement, as in walking or running. These nuclei do not exert direct control over lower motor neurons. Instead, they adjust the activities of upper motor neurons in the various motor pathways based on input from all portions of the cerebral cortex, as well as from the substantia nigra.

The basal nuclei use two major pathways to adjust or establish patterns of movement:

1. One group of axons synapses on thalamic neurons, which then send their axons to the premotor cortex, the motor association area that directs activities of the primary motor cortex. This arrangement creates a feedback loop that changes the sensitivity of the pyramidal cells and alters the pattern of instructions carried by the corticospinal tracts.

2. A second group of axons synapses in the reticular formation, altering the excitatory or inhibitory output of the reticulospinal tracts.

Two distinct populations of neurons exist: one that stimulates neurons by releasing acetylcholine (ACh) and another that inhibits neurons by the release of gamma aminobutyric acid (GABA). Under normal conditions, the excitatory interneurons are kept inactive and the tracts leaving the basal nuclei have an inhibitory effect on upper motor neurons. In *Parkinson's disease* (Chapter 14), the excitatory neurons become more active, leading to problems with the voluntary control of movement. ∞ p. 487

If the primary motor cortex is damaged, the individual loses the ability to exert fine control over skeletal muscles. However, some voluntary movements can still be controlled by the basal nuclei. In effect, the medial and lateral pathways function as they usually do, but the corticospinal pathway cannot fine-tune the movements. For example, after damage to the primary motor cortex, the basal nuclei can still receive information about planned movements from the prefrontal cortex and can perform preparatory movements of the trunk and limbs. But because the corticospinal pathway is inoperative, precise movements of the forearms, wrists, and hands cannot occur. An individual in this condition can stand, maintain balance, and even walk, but all movements are hesitant, awkward, and poorly controlled.

The Cerebellum

The cerebellum monitors proprioceptive (position) sensations, visual information from the eyes, and vestibular (balance) sensations from the inner ear as movements are under way. Axons relaying proprioceptive information reach the cerebellar cortex in the spinocerebellar tracts. Visual information is relayed from the superior colliculi, and balance information is relayed from the vestibular nuclei. The output of the cerebellum affects upper motor neuron activity in the corticospinal, medial, and lateral pathways.

All motor pathways send information to the cerebellum when motor commands are issued. As the movement proceeds, the cerebellum monitors proprioceptive and vestibular information and compares the arriving sensations with those experienced

TABLE 15–2 PRINCIPAL DESCENDING (MOTOR) PATHWAYS

Tract	Location of Upper Motor Neurons	Destination	Site of Cross-Over	Action
CORTICOSPINAL PATHWAY				
Corticobulbar tracts	Primary motor cortex (cerebral hemisphere)	Lower motor neurons of cranial nerve nuclei in brain stem	Brain stem	Conscious motor control of skeletal muscles
Lateral corticospinal tracts	As above	Lower motor neurons of anterior gray horns of spinal cord	Pyramids of medulla oblongata	As above
Anterior corticospinal tracts	As above	As above	Level of lower motor neuron	As above
MEDIAL PATHWAY				
Vestibulospinal tracts	Vestibular nuclei (at border of pons and medulla oblongata)	As above	None (uncrossed)	Subconscious regulation of balance and muscle tone
Tectospinal tracts	Tectum (mesencephalon: superior and inferior colliculi)	Lower motor neurons of anterior gray horns (cervical spinal cord only)	Brain stem (mesencephalon)	Subconscious regulation of eye, head, neck, and upper limb position in response to visual and auditory stimuli
Reticulospinal tracts	Reticular formation (network of nuclei in brain stem)	Lower motor neurons of anterior gray horns of spinal cord	None (uncrossed)	Subconscious regulation of reflex activity
LATERAL PATHWAY				
Rubrospinal tracts	Red nuclei of mesencephalon	As above	Brain stem (mesencephalon)	Subconscious regulation of upper limb muscle tone and movement

during previous movements. It then adjusts the activities of the upper motor neurons involved. In general, any voluntary movement begins with the activation of far more motor units than are required—or even desirable. The cerebellum provides the necessary inhibition, reducing the number of motor commands to an efficient minimum. As the movement proceeds, the pattern and degree of inhibition change, producing the desired result.

The patterns of cerebellar activity are learned by trial and error, over many repetitions. Many of the basic patterns are established early in life; examples include the fine balancing adjustments you make while standing and walking. The ability to fine-tune a complex pattern of movement improves with practice, until the movements become fluid and automatic. Consider the relaxed, smooth movements of acrobats, golfers, and sushi chefs. These people move without thinking about the details of their movements. This ability is important, because when you concentrate on voluntary control, the rhythm and pattern of the movement usually fall apart as your primary motor cortex starts overriding the commands of the basal nuclei and cerebellum.

Amyotrophic lateral sclerosis (ALS), formerly known as *Lou Gehrig's disease*, is a progressive, degenerative disorder that affects motor neurons in the spinal cord, brain stem, and cerebral hemispheres. The degeneration affects both upper and lower motor neurons. Because a motor neuron and its dependent muscle fibers are so intimately related, the destruction of CNS neurons causes atrophy of the associated skeletal muscles. The specificity of this disease is remarkable: Motor neurons are destroyed, but sensory neurons, sensory pathways, and intellectual functions remain unaffected. Roughly 90 percent of ALS cases have no known cause. About 10 percent of cases can be linked to genetic factors, but the nature of the genetic abnormality is uncertain. Most people with ALS do not develop symptoms until after age 50, although inherited forms may affect individuals in their early teens. Affected individuals typically die within five years after symptoms appear. There is no effective treatment, only supportive care. Noted physicist Stephen Hawking has this condition. **AM** Amyotrophic Lateral Sclerosis

---

✓ For what anatomical reason does the left side of the brain control motor function on the right side of the body?

✓ An injury involving the superior portion of the motor cortex would affect which region of the body?

✓ What effect would increased stimulation of the motor neurons of the red nucleus have on muscle tone?

Answers start on p. Q-1

 The motor pathways of the SNS can be investigated on the **Companion Website**: Chapter 15/Reviewing Facts & Terms/ Multiple Choice.

☐ LEVELS OF PROCESSING AND MOTOR CONTROL

All sensory and motor pathways involve a series of synapses, one after the other. Along the way, the information is distributed to processing centers operating at the subconscious level. Think about what happens when you stumble—you often recover your balance even as you become aware that a problem exists. Long before your cerebral cortex could assess the situation, evaluate possible responses (shift weight *here*, move leg *there*, etc.), and issue appropriate motor commands, monosynaptic and poly-synaptic reflexes, perhaps adjusted by the brain stem and cerebellum, successfully prevented a fall. This is a general pattern; spinal and cranial reflexes provide rapid, involuntary, preprogrammed responses that preserve homeostasis over the short term. Voluntary responses are more complex, but they require more time to prepare and execute.

Figure 15–12● reviews the primary sites of somatic motor control. Cranial and spinal reflexes control the most basic motor activities. Integrative centers in the brain perform more elaborate processing, and as we move from the medulla oblongata to the cerebral cortex, the motor patterns become increasingly complex and variable. The most complex and variable motor activities are directed by the primary motor cortex of the cerebral hemispheres.

During development, the spinal and cranial reflexes are the first to appear. More complex reflexes and motor patterns develop as CNS neurons multiply, enlarge, and interconnect. The process proceeds relatively slowly, as billions of neurons establish trillions of synaptic connections. At birth, neither the cerebral nor the cerebellar cortex is fully functional. The behavior of newborn infants is directed primarily by centers in the diencephalon and brain stem.

Among the anatomical factors that contribute to the maturation of the CNS and the refinement of motor skills are the following:

- The number of neurons in the cerebral cortex continues to increase until at least age one.
- The brain as a whole grows in size and complexity until at least age four.
- The myelination of CNS axons continues at least until age one–two, and peripheral myelination may continue through puberty. Myelination reduces the delay between stimulus and response and thereby improves coordination and motor control. ∞ p. 409

As these events occur, cortical neurons continue to establish new synaptic interconnections that will have a long-term effect on mental capabilities.

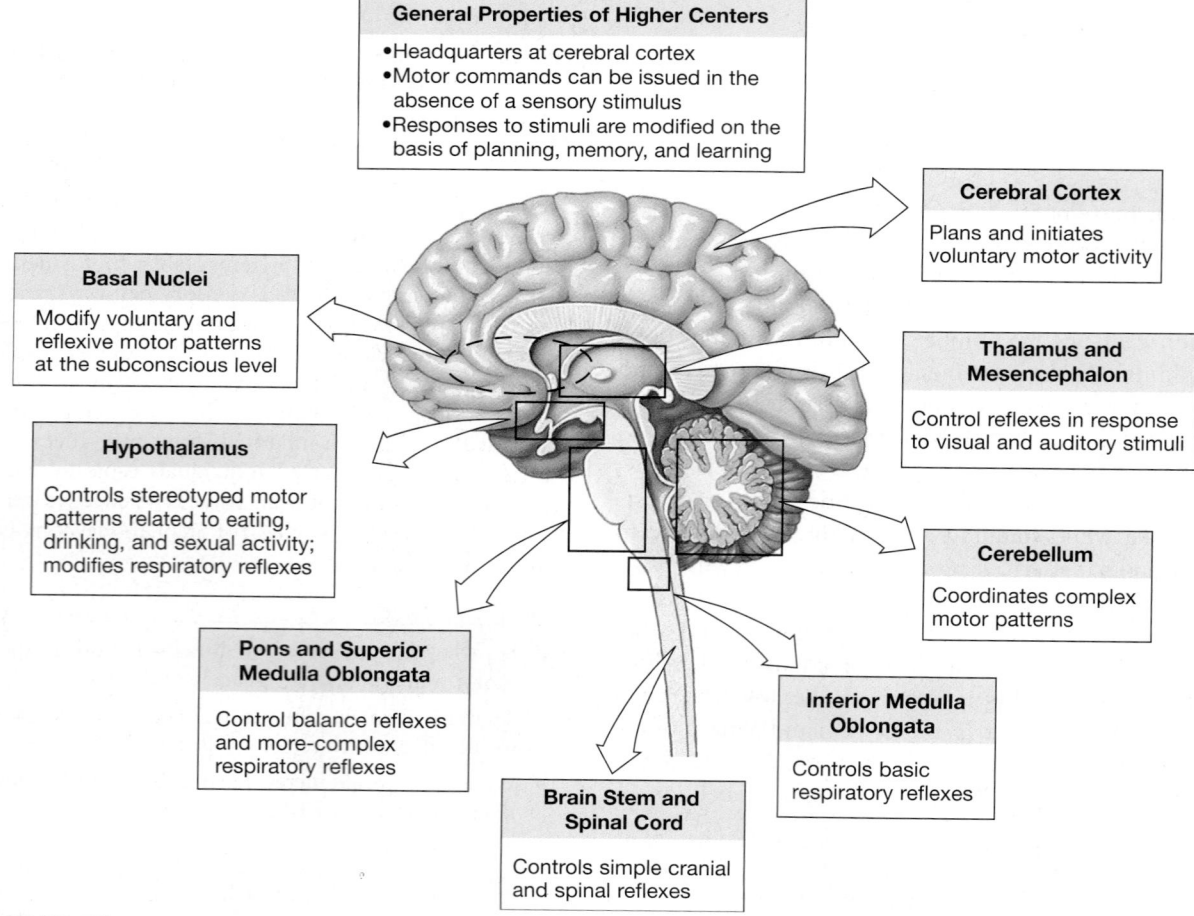

●**FIGURE 15–12**
Centers of Somatic Motor Control. Somatic motor control involves many regions of the brain. As we ascend toward the cerebral cortex, the motor patterns generated become increasingly complex and variable.

✚ Although it may sound strange, physicians generally take a newborn infant into a dark room and shine a light against the skull. They are checking for **anencephaly** (an-en-SEF-uh-lē), a rare condition in which the brain fails to develop at levels above the mesencephalon or lower diencephalon.

In most such cases, the cranium also fails to develop, and diagnosis is easy, but in some cases, a normal skull forms. In such instances, the cranium is empty and translucent enough to transmit light. Unless the condition is discovered right away, the parents may take the infant home, unaware of the problem.

All the normal behavior patterns expected of a newborn are present, including suckling, stretching, yawning, crying, kicking, sticking fingers in the mouth, and tracking movements with the eyes. However, death will occur naturally within days or months.

This tragic condition provides a striking demonstration of the role of the brain stem in controlling complex motor patterns. During normal development, these patterns become incorporated into variable and versatile behaviors as control centers and analytical centers appear in the cerebral cortex.

Chapter Review

SELECTED CLINICAL TERMINOLOGY

Terms Discussed in This Chapter

amyotrophic lateral sclerosis (ALS): A progressive, degenerative disorder affecting motor neurons of the spinal cord, brain stem, and cerebral hemispheres. *(p. 527 and AM)*
anencephaly (an-en-SEF-a-lē): A rare condition in which the brain fails to develop at levels above the mesencephalon or inferior part of the diencephalon. *(p. 529)*
cerebral palsy: A disorder that affects voluntary motor performance and arises in infancy or childhood as a result of prenatal trauma, drug exposure, or a congenital defect. *(p. 525)*

AM Additional Terms Discussed in the *Applications Manual*

anesthesia
hypesthesia
paresthesia

STUDY OUTLINE

15–1 AN OVERVIEW OF SENSORY PATHWAYS AND THE SOMATIC NERVOUS SYSTEM   p. 509

1. The brain, spinal cord, and peripheral nerves continuously communicate with each other and with the internal and external environments. Information arrives via sensory receptors and ascends within the **afferent division**, while motor commands descend and are distributed by the **efferent division**. *(Figure 15–1)*

15–2 SENSORY RECEPTORS AND THEIR CLASSIFICATION   p. 510

1. A **sensory receptor** is a specialized cell or cell process that monitors specific conditions within the body or in the external environment. Arriving information is called a **sensation**; conscious awareness of a sensation is a **perception**.
2. The **general senses** are pain, temperature, physical distortion, and chemical detection. Receptors for these senses are distributed throughout the body. **Special senses**, located in specific **sense organs**, are structurally more complex.

SENSORY RECEPTORS   p. 510

3. Each receptor cell monitors a specific receptive field. **Transduction** begins when a large enough stimulus changes the *receptor potential* reaching *generator potential*. *(Figure 15–2)*
4. **Tonic receptors** are always active. **Phasic receptors** provide information about the intensity and rate of change of a stimulus. **Adaptation** is a reduction in sensitivity in the presence of a constant stimulus. Tonic receptors are **slow-acting receptors**, while phasic receptors are **fast-acting receptors**.

THE GENERAL SENSES   p. 512

5. Three types of **nociceptor** are found in the body. They provide information on pain as related to extremes of temperature, mechanical damage, and dissolved chemicals. Mylenated Type A fibers carry **fast pain.** Slower Type C fibers carry **slow pain.** *(Figure 15–2)*
6. **Thermoreceptors** are found in the dermis. **Mechanoreceptors** are sensitive to distortion of their membranes, and include **tactile receptors**, **baroreceptors**, and **proprioceptors**. There are six types of tactile receptor in the skin, and three groups of proprioceptors. Chemoreceptors include **carotid bodies** and **aortic bodies**. *(Figures 15–3, 15–4, 15–5)*

15–3 THE ORGANIZATION OF SENSORY PATHWAYS   p. 518

1. Sensory neurons that deliver sensation to the CNS are referred to as **first-order neurons.** These synapse on **second-order neurons** in the brain stem or spinal cord. The next neuron in this chain is a **third-order neuron,** found in the thalamus.

SOMATIC SENSORY PATHWAYS   p. 518

2. Three major somatic sensory pathways carry sensory information from the skin and musculature of the body wall, head, neck, and limbs: the posterior column pathway, the anterolateral pathway, and the spinocerebellar pathway. *(Figure 15–6)*
3. The **posterior column pathway** carries fine touch, pressure, and proprioceptive sensations. The axons ascend within the **fasciculus gracilis** and **fasciculus cuneatus** and relay information to the thalamus via the **medial lemniscus.** Before the axons enter the medial lemniscus, they cross over to the opposite side

of the brain stem. This crossing over is called **decussation**. *(Figure 15–8; Table 15–1)*

4. The **anterolateral pathway** carries poorly localized sensations of touch, pressure, pain, and temperature. The axons involved decussate in the spinal cord and ascend within the **anterior** and **lateral spinothalamic tracts** to the ventral nuclei of the thalamus. *(Figure 15–7; Table 15–1)*

5. The **spinocerebellar pathway**, including the **posterior** and **anterior spinocerebellar tracts**, carries sensations to the cerebellum concerning the position of muscles, tendons, and joints. *(Figure 15–9; Table 15–1)*

VISCERAL SENSORY PATHWAYS   p. 522

6. Visceral sensory pathways carry information collected by interoceptors. Sensory information from cranial nerves V, VII, IX, and X is delivered to the **solitary nucleus** in the medulla oblongata. Dorsal roots of spinal nerves T_1–L_2 carry visceral sensory information from organs between the diaphragm and the pelvic cavity. Dorsal roots of spinal nerves S_2–S_4 carry sensory information below this area.

 Somatic sensory pathways: **Companion Website**: Chapter 15/Reviewing Facts & Terms/Labeling.

15–4 THE SOMATIC NERVOUS SYSTEM   p. 523

1. Somatic motor pathways always involve an **upper motor neuron** (whose cell body lies in a CNS processing center) and a **lower motor neuron** (whose cell body is located in a motor nucleus of the brain stem or spinal cord). *(Figure 15–10)*

THE CORTICOSPINAL PATHWAY   p. 524

2. The neurons of the primary motor cortex are *pyramidal cells*. The **corticospinal pathway** provides voluntary skeletal muscle control. The **corticobulbar tracts** terminate at the cranial nerve nuclei; the **corticospinal tracts** synapse on motor neurons in the anterior gray horns of the spinal cord. The corticospinal tracts are visible along the medulla as a pair of thick bands, the **pyramids**, where most of the axons decussate to enter the descending **lateral corticospinal tracts**. Those that do not cross over enter the **anterior corticospinal tracts**. The corticospinal

pathway provides a rapid, direct mechanism for controlling skeletal muscles. *(Figure 15–11; Table 15–2)*

THE MEDIAL AND LATERAL PATHWAYS   p. 525

3. The **medial** and **lateral pathways** include several other centers that issue motor commands as a result of processing performed at a subconscious level. *(Table 15–2)*

4. The medial pathway primarily controls gross movements of the trunk and proximal limbs. It includes the vestibulospinal, tectospinal, and reticulospinal tracts. The **vestibulospinal tracts** carry information related to maintaining balance and posture. Commands carried by the **tectospinal tracts** change the position of the head, neck, and upper limbs in response to bright lights, sudden movements, or loud noises. Motor commands carried by the **reticulospinal tracts** vary according to the region stimulated. *(Table 15–2)*

5. The lateral pathway consists of the **rubrospinal tracts**, which primarily control muscle tone and movements of the distal muscles of the upper limbs. *(Table 15–2)*

THE BASAL NUCLEI AND CEREBELLUM   p. 526

6. The basal nuclei adjust the motor commands issued in other processing centers and provide background patterns of movement involved in voluntary motor activities.

7. The cerebellum monitors proprioceptive sensations, visual information, and vestibular sensations. The integrative activities performed by neurons in the cortex and nuclei of the cerebellum are essential for the precise control of movements.

 Motor pathways of the SNS: **Companion Website**: Chapter 15/Reviewing Facts & Terms/Multiple Choice.

LEVELS OF PROCESSING AND MOTOR CONTROL   p. 528

8. Spinal and cranial reflexes provide rapid, involuntary, preprogrammed responses that preserve homeostasis. Voluntary responses are more complex, and require more time to prepare and execute. *(Figure 15–12)*

9. During development, the spinal and cranial reflexes are first to appear. Complex reflexes develop over years, as the CNS matures and the brain grows in size and complexity.

REVIEW QUESTIONS

 More assessment questions are available to you on the Companion Website. You will find Matching, Multiple Choice, True/False, and other quizzes to help further your understanding of the material covered in this chapter. To access the site, go to www.aw.com/martini.

LEVEL 1 Reviewing Facts and Terms

1. A sensory receptor characterized as a free nerve ending, using the amino acid glutamate and the neuropeptide *Substance P*, would most likely be a
   (a) chemoreceptor   (b) mechanoreceptor
   (c) thermoreceptor   (d) nociceptor

2. The ability to localize a specific stimulus depends on the organized distribution of sensory information to the
   (a) spinothalamic pathway   (b) posterior column tract
   (c) spinocerebellar pathway   (d) primary sensory cortex

3. The corticospinal tract
   (a) carries motor commands from the cerebral cortex to the spinal cord
   (b) carries sensory information from the spinal cord to the brain
   (c) starts in the spinal cord and ends in the brain
   (d) *a, b,* and *c* are correct

4. Destruction of or damage to a lower motor neuron in the somatic nervous system produces
   (a) the inability to localize a stimulus
   (b) a subconscious response to stimulation
   (c) paralysis of the innervated motor unit
   (d) a stimulation of the innervated muscle

5. What three steps are necessary for transduction to occur?

6. What are the four types of sensory receptors for the general senses? What is the nature of the stimuli that excite each type?

7. Identify six types of tactile receptors located in the skin and describe their sensitivities.

8. What three types of mechanoreceptors respond to stretching, compression, twisting, or other distortions of the cell membrane?

9. What are the three major somatic sensory pathways, and what is the function of each pathway?

10. Which three pairs of descending tracts make up the corticospinal pathway?

11. Which three motor pathways make up the medial pathway?

12. What are the two primary functional roles of the cerebellum?

13. What three anatomical factors contribute to the maturation of the CNS and the refinement of motor skills?

LEVEL 2 Reviewing Concepts

14. Describe the relationship among first-, second-, and third-order neurons.

15. Damage to the medial lemniscus, a component of the mesencephalon, leads to
    (a) a loss of the ability to pinpoint the location of sensations of touch, pressure, vibrations, and position
    (b) a loss of the ability to generally experience sensations of touch, pressure, vibrations, and position
    (c) a loss of cerebellar motor control and posture
    (d) an immediate loss of all proprioceptive information

16. Differentiate between a tonic receptor and a phasic receptor.

17. How is the autonomic nervous system *functionally* different from the somatic nervous system?

18. What is a motor homunculus? How does it differ from a sensory homunculus?

19. What effect does injury to the primary motor cortex have on peripheral muscles?

20. By which structures and in which part of the brain is the level of muscle tone in the body's skeletal muscles controlled? How is this control exerted?

21. Explain the phenomenon of *referred pain* in terms of labeled lines and the organization of sensory pathways.

LEVEL 3 Critical Thinking and Clinical Applications

22. Kelly is having difficulty controlling her eye movements and has lost some control of her facial muscles. After an examination and testing, Kelly's physician tells her that her cranial nerves are perfectly normal but that a small tumor is putting pressure on certain fiber tracts in her brain. This pressure is the cause of Kelly's symptoms. Where is the tumor most likely located?

23. While playing lacrosse, Frank took a blow to the back of his neck. This caused swelling of the posterior spinal cord at C_7, T_1. What tracts lie in this area? What symptoms might you expect Frank to experience while the swelling persists?

24. Researchers studying an illegal drug find that it causes individuals to experience tactile stimulations that aren't actually present. Perceptions of touch are exaggerated to the point of pain, and individuals taking the drug report increased irritability of the skin. Which neurotransmitter might this drug be mimicking, and which general sensory receptors are involved?

UNIT 3   CHAPTER   12   13   14   15   **16**   17   18

CLINICAL NOTES

- Nerve Damage and Sympathetic
  Deficits   538
- Amnesia   554

CLINICAL DISCUSSIONS

- Modifying Autonomic Nervous
  System Function   552
- Alzheimer's Disease   557

NEURAL INTEGRATION II: THE AUTONOMIC NERVOUS SYSTEM AND HIGHER ORDER FUNCTIONS

Your conscious thoughts, plans, and actions represent a tiny fraction of the activities of the nervous system. In practical terms, your conscious thoughts and the somatic nervous system (SNS), which operates under conscious control, seldom have a direct effect on your long-term survival. Of course, the somatic nervous system can be important in moving you out of the way of a speeding bus or pulling your hand from a hot stove—but it was your conscious movements that put you in jeopardy in the first place. If all consciousness were eliminated, vital physiological processes would continue virtually unchanged; a night's sleep is not a life-threatening event. Longer, deeper states of unconsciousness are not necessarily more dangerous, as long as nourishment and other basic care is provided. People who have suffered severe brain injuries can survive in a coma for decades.

Survival under these conditions is possible because routine homeostatic adjustments in physiological systems are made by the **autonomic nervous system (ANS)**. It is the ANS that coordinates cardiovascular, respiratory, digestive, urinary, and reproductive functions. In doing so, the ANS adjusts internal water, electrolyte, nutrient, and dissolved gas concentrations in body fluids—and it does so without instructions or interference from the conscious mind.

Figure 16–1● provides an overview of the material covered in this chapter. We'll begin by completing our discussion of the efferent division of the nervous system with an examination of the ANS. Our understanding of this system has had a profound effect on the practice of medicine. For example, in 1960, the five-year survival rate for patients surviving their first heart attack was very low, primarily because it was difficult and sometimes impossible to control high blood pressure. Forty years later, many survivors of heart attacks lead normal lives. This dramatic change occurred as we learned to manipulate the ANS with specific drugs and clinical procedures.

The interpretation of arriving sensory information and the commands issued by the SNS and ANS can be influenced or modified in response to conscious planning, memories, and learning—the so-called higher order functions of the brain. We will conclude this chapter with an examination of these functions. We will also consider the effects of aging on this system and the interactions between the nervous system and other systems.

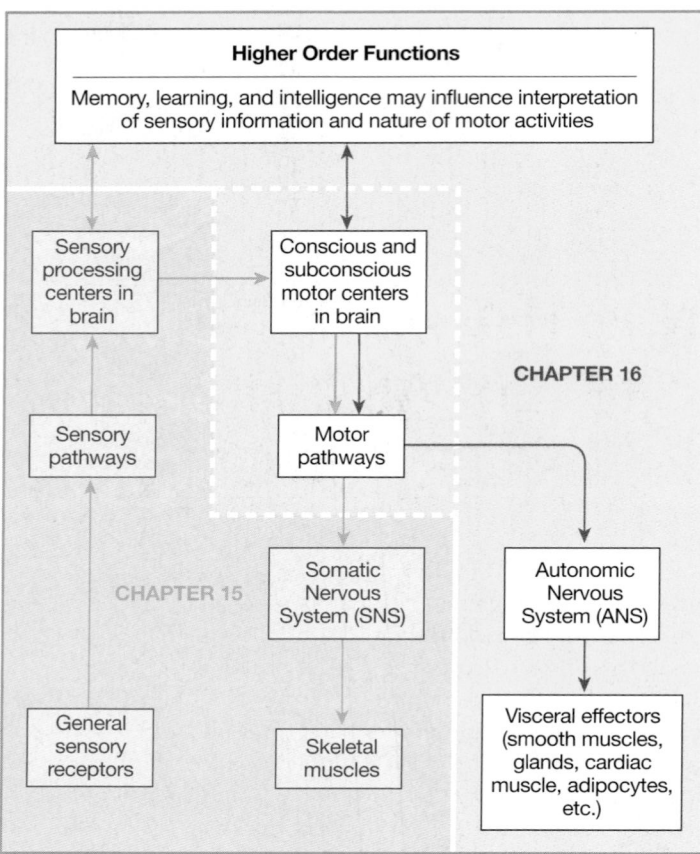

•FIGURE 16–1
An Overview of Neural Integration. This figure illustrates the relationships between Chapters 15 and 16 and indicates the major topics considered in this chapter.

16–1 AN OVERVIEW OF THE ANS

Objectives

- Compare the organization of the autonomic nervous system with that of the somatic nervous system.
- List the divisions of the ANS, and give the functions of each.

It will be useful to compare the organization of the ANS with that of the SNS, which controls our skeletal muscles. We will focus on (1) the neural interactions that direct motor output and (2) the subdivisions of the ANS, based on structural and functional patterns of peripheral innervation.

Figure 16–2• compares the organization of the somatic and autonomic nervous systems. Both are efferent divisions that carry motor commands; the SNS controls skeletal muscles and the ANS controls visceral effectors. The primary structural difference between the two is that, in the SNS, motor neurons of the central nervous system exert direct control over skeletal muscles (Figure 16–2a•). In the ANS, by contrast, motor neurons of the central nervous system synapse on visceral motor neurons in au-

tonomic ganglia, and it is these ganglionic neurons that control visceral effectors (Figure 16–2b•).

The visceral motor neurons in the CNS are known as **preganglionic neurons**, and the axons of these neurons are called **preganglionic fibers**. Preganglionic fibers leave the CNS and synapse on **ganglionic neurons**—visceral motor neurons in peripheral ganglia. These ganglia, which contain hundreds to thousands of ganglionic neurons, are called **autonomic ganglia**. The ganglionic neurons innervate visceral effectors such as cardiac muscle, smooth muscles, glands, and adipose tissues. The axons of ganglionic neurons are called **postganglionic fibers**, because they begin at the autonomic ganglia and extend to the peripheral target organs.

Somatic or visceral sensory information can trigger visceral reflexes, and the motor commands of those reflexes are distributed by the ANS. Sometimes those motor commands control the activities of target organs. For example, in cold weather, it is the ANS that stimulates the arrector pili muscles and gives you "goosebumps." ∞ p. 167 In other cases, the motor commands may alter some ongoing activity. A sudden, loud noise can startle you and make you jump, but thanks to the ANS, that sound can also increase your heart rate dramatically and temporarily stop all digestive gland secretion. These changes in visceral activity occur in response to the release of neurotransmitters by postganglionic fibers. As we noted in Chapter 12, whether a specific neurotransmitter produces a stimulation or an inhibition of activity depends on the response of the membrane receptors, and we will consider the major types of receptors later in the chapter.

We will begin with the anatomy and physiology of the ANS. We will then consider the nature of *visceral reflexes*, polysynaptic reflexes that regulate visceral function. Finally, we will discuss the higher centers involved with the coordination of visceral functions and the integration of ANS activity with the other activities of the nervous system.

☐ DIVISIONS OF THE ANS

The ANS contains two subdivisions: the *sympathetic division* and the *parasympathetic division*. Most often, the two divisions have opposing effects; if the sympathetic division causes excitation, the parasympathetic causes inhibition. However, this is not always the case, because (1) the two divisions may work independently, with some structures innervated by only one division or the other, and (2) the two divisions may work together, each controlling one stage of a complex process. In general, the parasympathetic division predominates under resting conditions, and the sympathetic division "kicks in" only during periods of exertion, stress, or emergency.

The ANS also includes the *enteric nervous system (ENS)*, an extensive network of neurons and nerve networks located in the walls of the digestive tract. Although the activities of the enteric nervous system are influenced by the sympathetic and parasympathetic divisions, many complex visceral reflexes are initiated and coordinated locally, without instructions from the

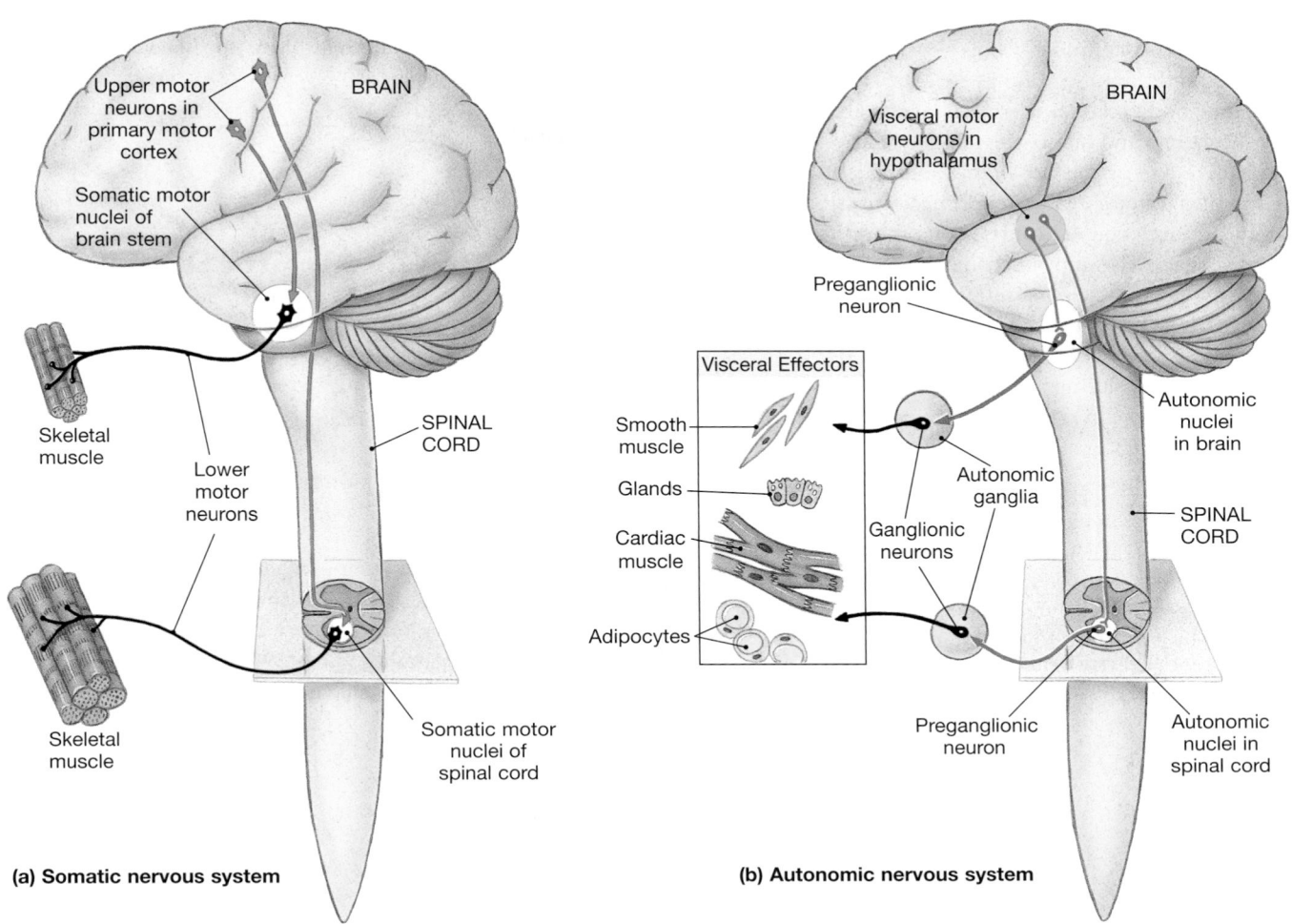

●**FIGURE 16–2**
The Organization of the Somatic and Autonomic Nervous Systems

CNS. Altogether, this system has roughly 100 million neurons—at least as many as the spinal cord—and all of the neurotransmitters found in the brain. In this chapter, we focus on the sympathetic and parasympathetic divisions that integrate and coordinate visceral functions throughout the body. We will consider the activities of the enteric nervous system when we discuss visceral reflexes later in this chapter and when we examine the control of digestive functions in Chapter 24.

The Sympathetic Division

In the **sympathetic division**, or *thoracolumbar* (tho-ra-kō-LUM-bar) *division*, preganglionic fibers from the thoracic and superior lumbar segments of the spinal cord synapse in ganglia near the spinal cord. In this division of the ANS, therefore, the preganglionic fibers are short and the postganglionic fibers are long.

The sympathetic division prepares the body for heightened levels of somatic activity. When fully activated, this division produces what is known as the "fight or flight" response, which readies the body for a crisis that may require sudden, intense physical activity. An increase in sympathetic activity generally stimulates tissue metabolism and increases alertness. Imagine walking down a long,

dark alley and hearing strange noises in the darkness ahead. Your body responds immediately, and you become more alert and aware of your surroundings. Your metabolic rate rises quickly, to as much as twice the resting level. Your digestive and urinary activities are suspended temporarily, and blood flow to your skeletal muscles increases. You begin breathing more quickly and more deeply. Both your heart rate and blood pressure increase, circulating your blood more rapidly. You feel warm and begin to perspire. The general pattern can be summarized as follows: (1) heightened mental alertness, (2) increased metabolic rate, (3) reduced digestive and urinary functions, (4) activation of energy reserves, (5) increased respiratory rate and dilation of respiratory passageways, (6) increased heart rate and blood pressure, and (7) activation of sweat glands.

The Parasympathetic Division

In the **parasympathetic division**, or *craniosacral* (krā-nē-ō-SĀ-krul) *division*, preganglionic fibers originate in the brain stem and the sacral segments of the spinal cord, and they synapse in ganglia very close to (or within) the target organs. Thus, in this division of the ANS, the preganglionic fibers are long and the postganglionic fibers are short. The parasympathetic division stimulates visceral activity; for example, it is responsible for the state of "rest and re-pose" that follows a big dinner. General parasympathetic activation conserves energy and promotes sedentary activities, such as digestion. Your body relaxes, energy demands are minimal, and both your heart rate and blood pressure are relatively low. Meanwhile, your digestive organs are highly stimulated. Your salivary glands and other secretory glands are active; your stomach is contracting; and smooth muscle contractions move materials along your digestive tract. This movement promotes defecation; at the same time, smooth muscle contractions along your urinary tract promote urination. The overall pattern is as follows: (1) decreased metabolic rate, (2) decreased heart rate and blood pressure, (3) increased secretion by salivary and digestive glands, (4) increased motility and blood flow in the digestive tract, and (5) stimulation of urination and defecation.

✓ How many motor neurons are required to conduct an action potential from the spinal cord to smooth muscles in the wall of the intestine?

✓ While out for a brisk walk, Julie is suddenly confronted by an angry dog. Which division of the autonomic nervous system is responsible for the physiological changes that occur in Julie as she turns and runs?

✓ On the basis of anatomy, how could you distinguish the sympathetic division from the parasympathetic division of the autonomic nervous system?

Answers start on page Q-1

16–2 THE SYMPATHETIC DIVISION

Objectives

■ Describe the structures and functions of the sympathetic division of the autonomic nervous system.

■ Describe the mechanisms of neurotransmitter release in the sympathetic division.

■ Describe the effects of sympathetic neurotransmitters on target organs and tissues.

Figure 16–3● summarizes the overall organization of the sympathetic division of the ANS. This division consists of preganglionic neurons that are located between segments T_1 and L_2 of the spinal cord, and ganglionic neurons that are located in ganglia near the vertebral column. The cell bodies of the preganglionic

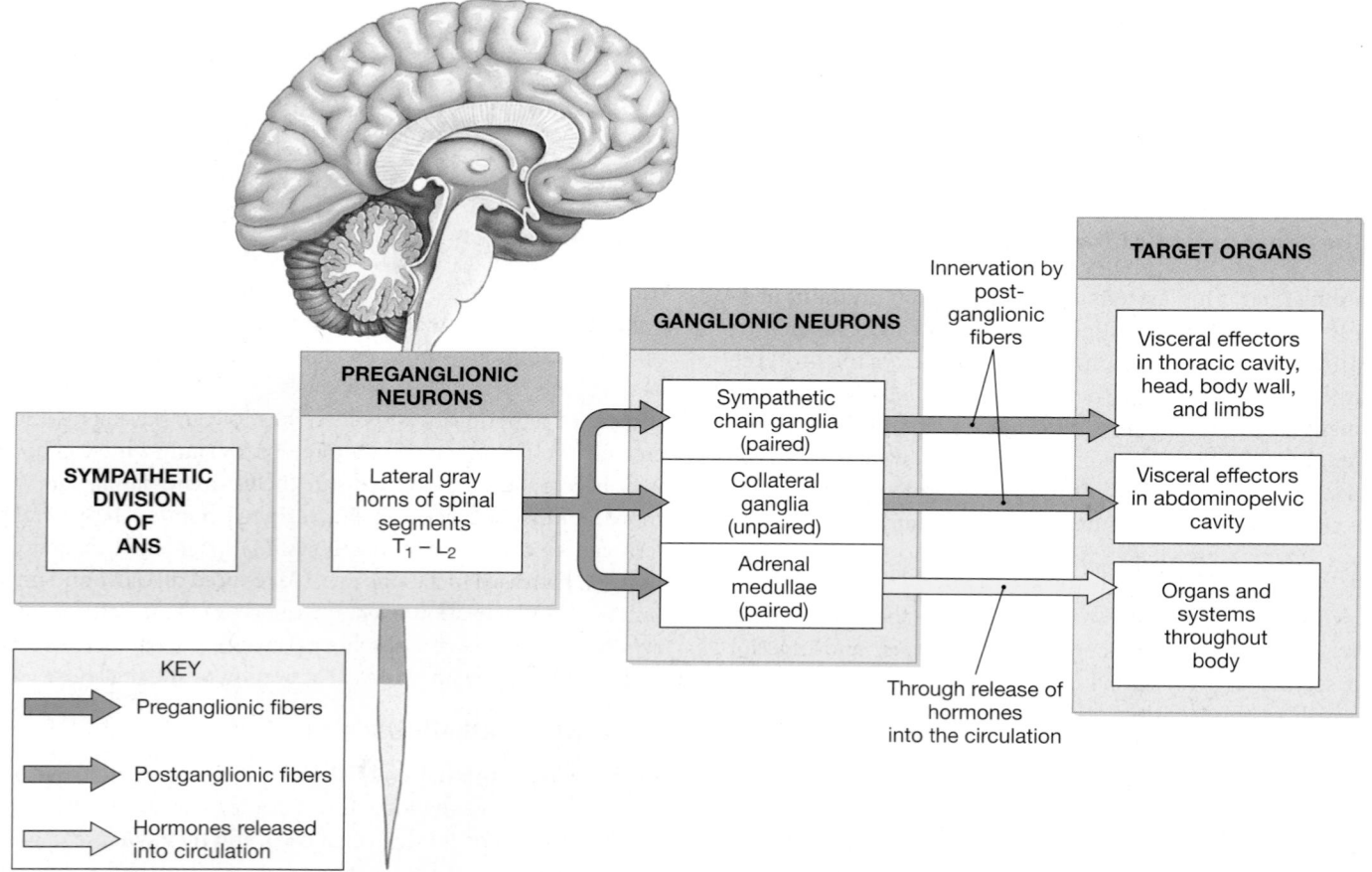

●**FIGURE 16–3**
The Organization of the Sympathetic Division of the ANS

neurons are situated in the lateral gray horns, and their axons enter the ventral roots of these segments. The ganglionic neurons occur in three locations:

1. *Sympathetic Chain Ganglia.* **Sympathetic chain ganglia**, also called *paravertebral ganglia* or *lateral ganglia*, lie on both sides of the vertebral column. Neurons in these ganglia control effectors in the body wall, inside the thoracic cavity, and in the head and limbs (Figure 16–4a●).

2. *Collateral Ganglia.* **Collateral ganglia**, also known as *prevertebral ganglia*, are anterior to the vertebral bodies (Figure 16–4b●). Collateral ganglia contain ganglionic neurons that innervate tissues and organs in the abdominopelvic cavity.

3. *The Adrenal Medullae.* The center of each adrenal gland, an area known as the *adrenal medulla*, is a modified sympathetic ganglion (Figure 16–4c●). The ganglionic neurons of the adrenal medullae have very short axons; when stimulated, they release their neurotransmitters into the bloodstream. This change in the release site—from a synapse to a capillary—allows the neurotransmitters to function as hormones that affect target cells throughout the body.

In the sympathetic division, the preganglionic fibers are relatively short, because the ganglia are located relatively near the spinal cord. In contrast, postganglionic fibers are relatively long, except at the adrenal medullae.

☐ ORGANIZATION AND ANATOMY OF THE SYMPATHETIC DIVISION

The ventral roots of spinal segments T_1 to L_2 contain sympathetic preganglionic fibers. The basic pattern of sympathetic innervation in these regions was described in Figure 13–7a●, p. 437. After passing through the intervertebral foramen, each ventral root gives rise to a myelinated *white ramus*, which carries myelinated preganglionic fibers into a nearby sympathetic chain ganglion. These fibers may synapse within the sympathetic chain ganglia, at one of the collateral ganglia, or in the adrenal medullae (Figure 16–4●). Extensive divergence occurs, with one preganglionic fiber synapsing on two dozen or more ganglionic neurons. Preganglionic fibers running between the sympathetic chain ganglia interconnect them, making the chain resemble a string of pearls. Each ganglion in the sympathetic chain innervates a particular body segment or group of segments.

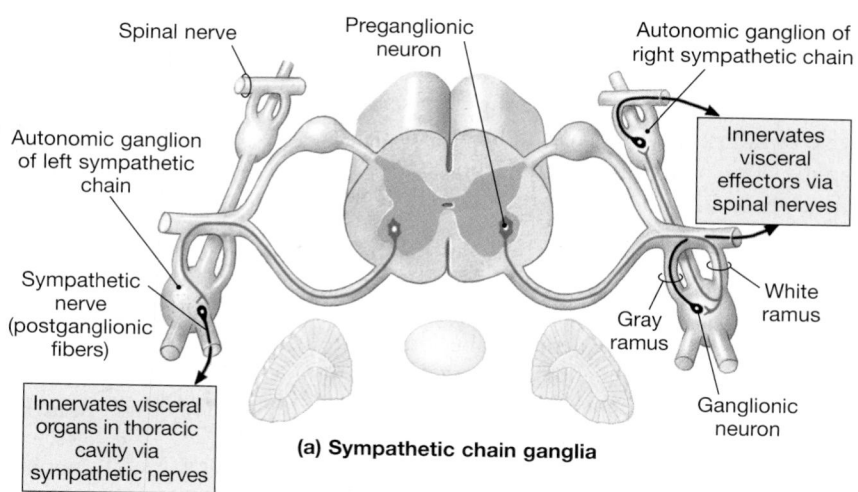

(a) Sympathetic chain ganglia

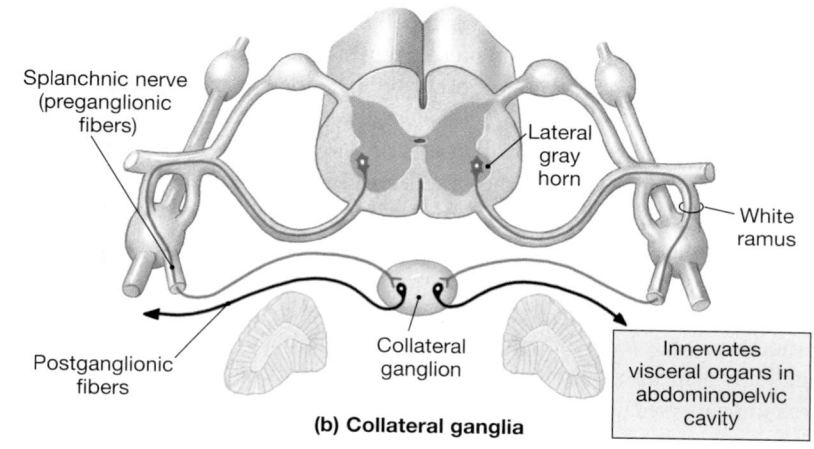

(b) Collateral ganglia

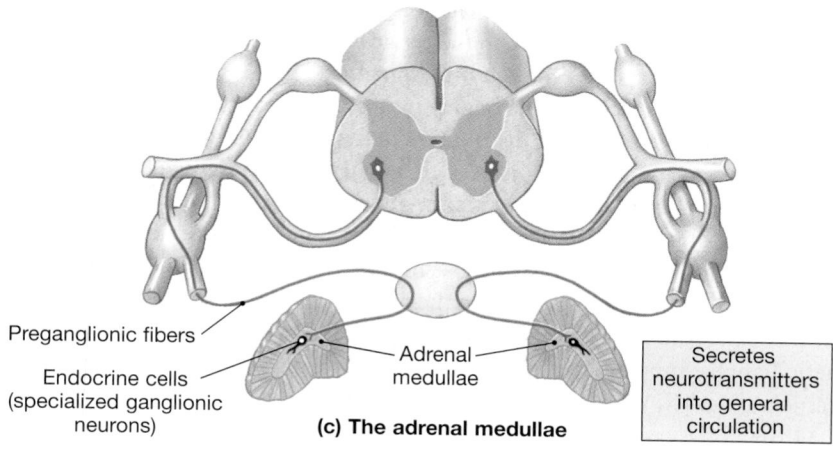

(c) The adrenal medullae

●**FIGURE 16–4**
Sympathetic Pathways. Superior views of sections through the thoracic spinal cord, showing the three major patterns of distribution for preganglionic and postganglionic fibers.

KEY
Preganglionic neurons = —
Ganglionic neurons = —

Sympathetic Chain Ganglia

If a preganglionic fiber carries motor commands that target structures in the body wall or thoracic cavity, or in the head, neck, or limbs, it will synapse in one or more sympathetic chain ganglia. The paths of the unmyelinated postganglionic fibers differ, depending on whether their targets lie in the body wall or within the thoracic cavity.

- Postganglionic fibers that control visceral effectors in the body wall, head, neck, or limbs enter the *gray ramus* and return to the spinal nerve for subsequent distribution (Figure 16–4a•, right). These postganglionic fibers innervate effectors such as the sweat glands of the skin and the smooth muscles in superficial blood vessels.
- Postganglionic fibers innervating structures in the thoracic cavity, such as the heart and lungs, form bundles known as **sympathetic nerves** (Figure 16–4a•, left).

Although Figure 16–4a• shows sympathetic nerves on the left side and spinal nerve distribution on the right for the sake of clarity, in reality *both* innervation patterns occur on *each* side of the body.

Figure 16–5• provides a more detailed view of the structure of the ganglion chain and the sympathetic division as a whole. The left side of the image shows the distribution to the skin and to skeletal muscles and other tissues of the body wall, whereas the right side depicts the innervation of visceral organs.

Each sympathetic chain contains 3 cervical, 10–12 thoracic, 4–5 lumbar, and 4–5 sacral ganglia, plus 1 coccygeal ganglion. (The numbers vary due to the occasional fusion of adjacent ganglia.) Preganglionic neurons are limited to spinal cord segments T_1–L_2, and these spinal nerves have both white rami (myelinated preganglionic fibers) and gray rami (unmyelinated postganglionic fibers). The neurons in the cervical, inferior lumbar, and sacral sympathetic chain ganglia are innervated by preganglionic fibers that run along the axis of the chain. In turn, these chain ganglia provide postganglionic fibers, through gray rami, to the cervical, lumbar, and sacral spinal nerves. As a result, although only spinal nerves T_1–L_2 have white rami, every spinal nerve has a gray ramus that carries sympathetic postganglionic fibers for distribution in the body wall.

About 8 percent of the axons in each spinal nerve are sympathetic postganglionic fibers. As a result, the spinal nerves, which provide somatic motor innervation to skeletal muscles of the body wall and limbs, also distribute sympathetic postganglionic fibers (Figures 16–4a and 16–5•). In the head and neck, postganglionic sympathetic fibers leaving the superior cervical sympathetic ganglia supply the regions and structures innervated by cranial nerves III, VII, IX, and X. ⟣ pp. 496, 498, 499, 500

In sum,

- The cervical, inferior lumbar, and sacral chain ganglia receive preganglionic innervation by preganglionic fibers from spinal segments T_1–L_2, and every spinal nerve receives a gray ramus from a ganglion of the sympathetic chain.
- Only the thoracic and superior lumbar ganglia (T_1–L_2) receive preganglionic fibers from white rami.

- Every spinal nerve receives a gray ramus from a ganglion of the sympathetic chain.

⚕ If the ventral roots of thoracic spinal nerves are damaged, there will be no sympathetic motor function on the affected side of the head, neck, or trunk. Yet damage to the ventral roots of cervical spinal nerves will produce voluntary muscle paralysis on the affected side but will leave sympathetic function intact, because the preganglionic fibers innervating the cervical ganglia originate in the white rami of thoracic segments, which are undamaged.

In contrast, damage to the cervical ganglia or thoracic segments can eliminate sympathetic innervation to the face, although sensation and muscle control remain unaffected. The affected side of the face becomes flushed, although sweating does not occur, and the pupil of the eye constricts. This combination of symptoms is known as *Horner's syndrome*. **AM** Hypersensitivity and Sympathetic Function

Collateral Ganglia

The abdominopelvic viscera receive sympathetic innervation by way of sympathetic preganglionic fibers that pass through the sympathetic chain without synapsing. They synapse in separate collateral ganglia (Figures 16–3 and 16–4•). Preganglionic fibers that innervate the collateral ganglia form the **splanchnic** (SPLANK–nik) **nerves**, which lie in the dorsal wall of the abdominal cavity. Although they originate as paired ganglia (left and right), the two usually fuse together, and in the adult collateral ganglia are typically single rather than paired.

Postganglionic fibers leaving the collateral ganglia extend throughout the abdominopelvic cavity, innervating a variety of visceral tissues and organs. The general pattern is (1) a reduction of blood flow and energy use by visceral organs that are not important to short-term survival, such as the digestive tract, and (2) the release of stored energy reserves.

The splanchnic nerves innervate three collateral ganglia (Figure 16–5•). Preganglionic fibers from the seven inferior thoracic segments end at the **celiac** (SĒ-lē-ak) **ganglion** or the **superior mesenteric ganglion**. These ganglia are embedded in an extensive network of autonomic nerves. Preganglionic fibers from the lumbar segments form splanchnic nerves that end at the **inferior mesenteric ganglion**. All three ganglia are named for their association with adjacent arteries:

- The celiac ganglion is named after the *celiac trunk*, a major artery supplying the stomach, spleen, and liver. The celiac ganglion most commonly consists of a pair of interconnected masses of gray matter situated at the base of that artery. The celiac ganglion may also form a single mass or many small, interwoven masses. Postganglionic fibers from this ganglion innervate the stomach, liver, gallbladder, pancreas, and spleen.
- The superior mesenteric ganglion is located near the base of the *superior mesenteric artery*, which provides blood to the stomach, small intestine, and pancreas. Postganglionic fibers

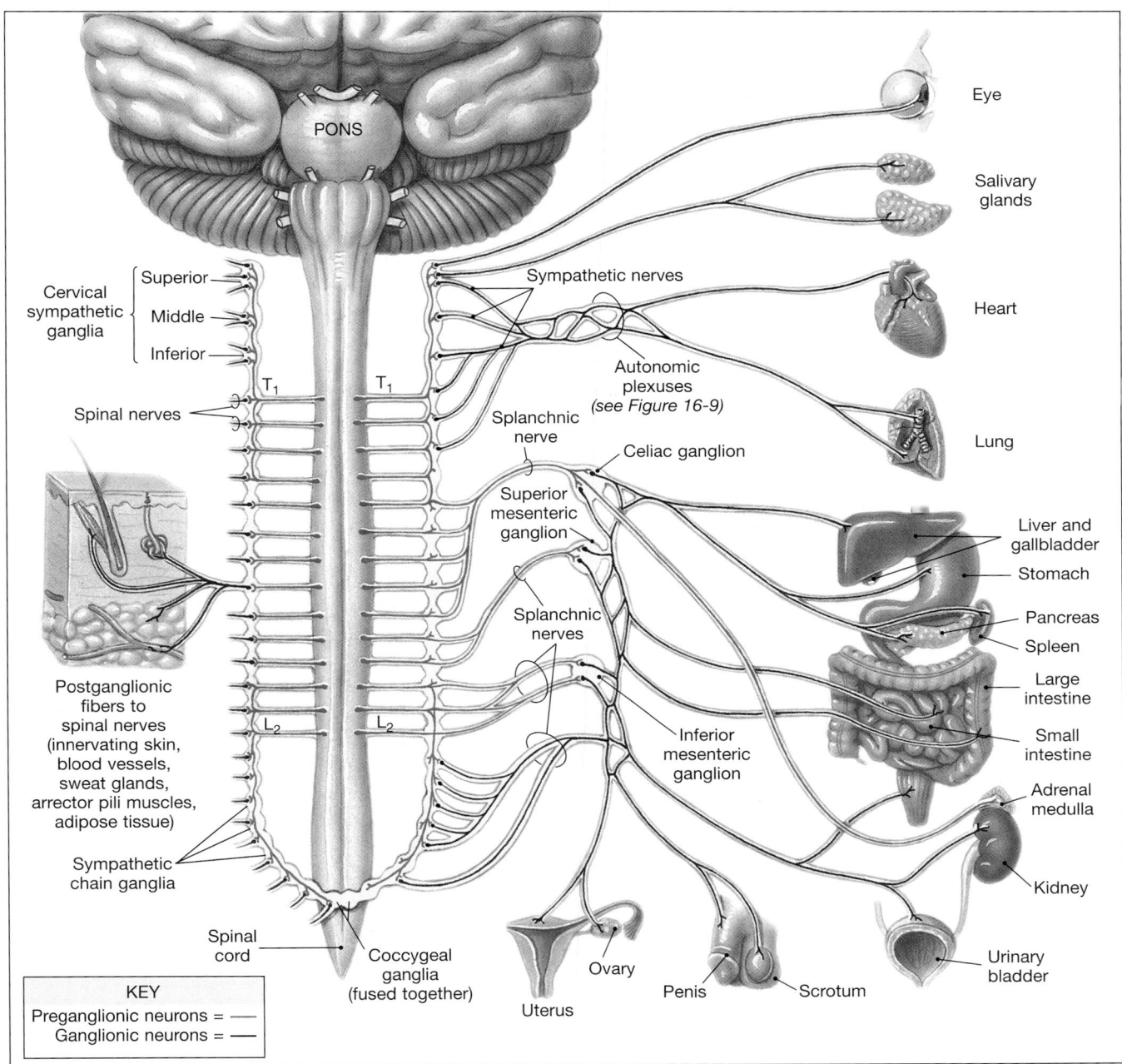

●**FIGURE 16–5**
The Distribution of Sympathetic Innervation.   The distribution of sympathetic fibers is the same on both sides of the body. For clarity, the innervation of somatic structures is shown on the left, and the innervation of visceral structures on the right.

leaving the superior mesenteric ganglion innervate the small intestine and the initial segments of the large intestine.

- The inferior mesenteric ganglion is located near the base of the *inferior mesenteric artery*, which supplies the large intestine and other organs in the inferior portion of the abdominopelvic cavity. Postganglionic fibers from this ganglion provide sympathetic innervation to the terminal portions of the large intestine, the kidney and urinary bladder, and the sex organs.

The Adrenal Medullae

Preganglionic fibers entering an adrenal gland proceed to its center, a region called the **adrenal medulla** (Figures 16–4c and 16–5●). The adrenal medulla is a modified sympathetic ganglion where preganglionic fibers synapse on *neuroendocrine cells,* specialized neurons that secrete hormones into the bloodstream. The neuroendocrine cells of the adrenal medullae secrete the neurotransmitters *epinephrine (E)* and *norepinephrine (NE).* Epinephrine, or *adrenaline,* accounts for 75–80 percent of the secretory output; the rest is NE.

The bloodstream then carries the neurotransmitters throughout the body, causing changes in the metabolic activities of many different cells. These effects resemble those produced by the stimulation of sympathetic postganglionic fibers. They differ, however, in two respects: (1) Cells not innervated by sympathetic postganglionic fibers are affected as well; and (2) the effects last much longer than those produced by direct sympathetic innervation, because the hormones continue to diffuse out of the bloodstream for an extended period.

☐ SYMPATHETIC ACTIVATION

The sympathetic division can change the activities of tissues and organs by releasing NE at peripheral synapses and by distributing E and NE throughout the body in the bloodstream. The visceral motor fibers that target specific effectors, such as smooth muscle fibers in blood vessels of the skin, can be activated in reflexes that do not involve other visceral effectors. In a crisis, however, the entire division responds. This event, called **sympathetic activation**, is controlled by sympathetic centers in the hypothalamus. The effects are not limited to peripheral tissues; sympathetic activation also alters CNS activity.

When sympathetic activation occurs, an individual experiences the following changes:

- Increased alertness via stimulation of the reticular activating system, causing the individual to feel "on edge."

- A feeling of energy and euphoria, often associated with a disregard for danger and a temporary insensitivity to painful stimuli.

- Increased activity in the cardiovascular and respiratory centers of the pons and medulla oblongata, leading to elevations in blood pressure, heart rate, breathing rate, and depth of respiration.

- A general elevation in muscle tone through stimulation of the medial and lateral pathways, so the person *looks* tense and may begin to shiver.

- The mobilization of energy reserves, through the accelerated breakdown of glycogen in muscle and liver cells and the release of lipids by adipose tissues.

These changes, plus the peripheral changes already noted, complete the preparations necessary for the individual to cope with a stressful situation.

☐ NEUROTRANSMITTERS AND SYMPATHETIC FUNCTION

We have examined the distribution of sympathetic impulses and the general effects of sympathetic activation. We now consider the cellular basis of these effects on peripheral organs.

Neurotransmitter Release

The stimulation of sympathetic preganglionic neurons leads to the release of ACh at synapses with ganglionic neurons. Synapses that use ACh as a transmitter are called *cholinergic*. ∞ p. 412 The effect on the ganglionic neurons is always excitatory.

The stimulation of these ganglionic neurons leads to the release of neurotransmitters at specific target organs. The synaptic terminals are typically different from neuromuscular junctions of the somatic nervous system. Instead of forming synaptic knobs, telodendria form a branching network. Each branch resembles a string of pearls. Each "pearl," a swollen segment called a **varicosity**, is packed with neurotransmitter vesicles. Chains of varicosities pass along or near the surfaces of the effector cells (Figure 16–6●). There are no specialized postsynaptic membranes, but membrane receptors are scattered across the surfaces of the target cells.

Most sympathetic ganglionic neurons release NE at their varicosities. Neurons that use NE as a neurotransmitter are called *adrenergic*. ∞ p. 414 The sympathetic division also contains a small, but significant, number of ganglionic neurons that release ACh rather than NE. Varicosities releasing ACh are located in the body wall, the skin, the brain, and skeletal muscles.

The NE released by varicosities affects its targets until it is reabsorbed or inactivated by enzymes. From 50 to 80 percent of the NE is reabsorbed by varicosities and either reused or broken down by the enzyme *monoamine oxidase (MAO)*. The rest of the NE diffuses out of the area or is broken down by the enzyme *catechol-O-methyltransferase (COMT)* in surrounding tissues.

In general, the effects of NE on the postsynaptic membrane persist for a few seconds, significantly longer than the 20-msec duration of ACh effects. (As usual, the responses of the target cells vary with the nature of the receptor on the postsynaptic membrane.) The effects of NE or E released by the adrenal medullae last even longer, because (1) the bloodstream does not contain MAO or COMT, and (2) most tissues contain relatively low concentrations of those enzymes. After the adrenal medullae are stimulated, tissue concentrations of NE and E throughout the body may remain elevated for as long as 30 seconds, and the effects may persist for several minutes.

Sympathetic Stimulation and the Release of NE and E

The effects of sympathetic stimulation result primarily from interactions with membrane receptors sensitive to NE and E. There are two classes of sympathetic receptors: *alpha receptors* and *beta receptors*. In general, norepinephrine stimulates alpha receptors more than it does beta receptors, whereas epinephrine stimulates both classes of receptors. Thus, localized sympathetic activity, involving the release of NE at varicosities, primarily affects alpha receptors located near the active varicosities. By contrast, generalized sympathetic activation and the release of E by the adrenal medulla affect alpha and beta receptors throughout the body.

Alpha receptors and beta receptors are *G proteins*. As we saw in Chapter 12, the effects of stimulating such a receptor depend on the production of *second messengers*, intracellular intermediaries with varied functions. ∞ pp. 418-419

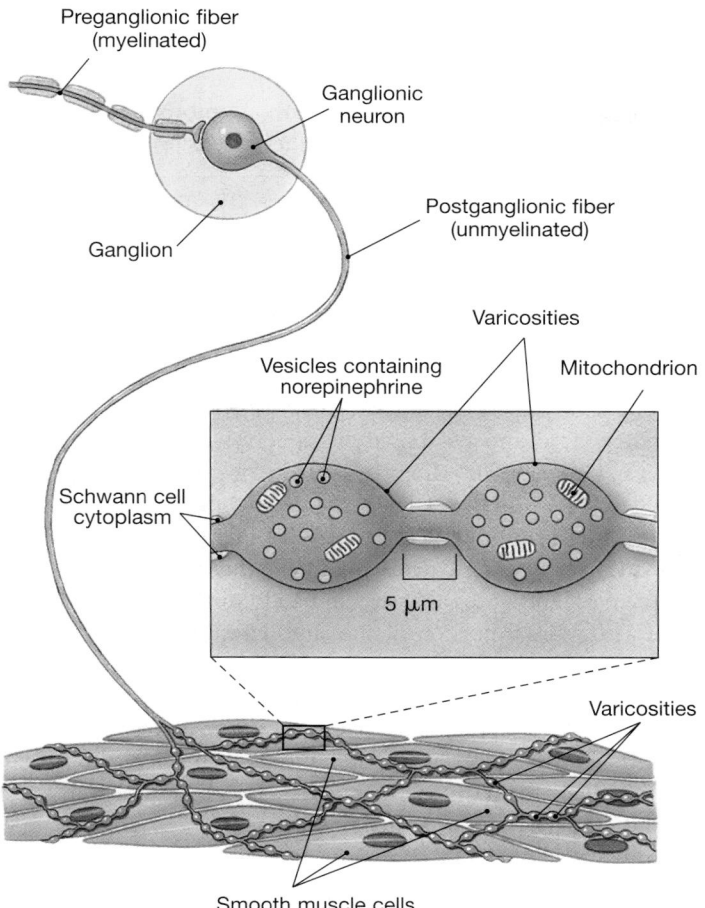

•FIGURE 16–6
Sympathetic Varicosities

The stimulation of **alpha (α) receptors** activates enzymes on the inside of the cell membrane. There are two types of alpha receptors: alpha-1 (α_1) and alpha-2 (α_2).

- The function of the more common type of alpha receptor, α_1, is the release of intracellular calcium ions from reserves in the endoplasmic reticulum. This action generally has an excitatory effect on the target cell. For example, the stimulation of α_1 receptors on the surfaces of smooth muscle cells is responsible for the constriction of peripheral blood vessels and the closure of sphincters along the digestive tract.

- Stimulation of α_2 receptors results in a lowering of cyclic-AMP (cAMP) levels in the cytoplasm. Cyclic-AMP is an important second messenger that can activate or inactivate key enzymes. ∞ p. 419 This reduction generally has an inhibitory effect on the cell. The presence of α_2 receptors in the parasympathetic division helps coordinate sympathetic and parasympathetic activities. When the sympathetic division is active, the NE released binds to parasympathetic neuromuscular and neuroglandular junctions and inhibits their activity.

Beta (β) receptors are located on the membranes of cells in many organs, including skeletal muscles, the lungs, the heart, and the liver. The stimulation of beta receptors triggers changes in the metabolic activity of the target cell. These changes occur indirectly, as the beta receptor is a G protein whose stimulation results in an increase in intracellular cAMP levels. There are two major types of beta receptors: beta-1 (β_1) and beta-2 (β_2).

- The stimulation of β_1 receptors leads to an increase in metabolic activity. For example, the stimulation of β_1 receptors on skeletal muscles accelerates the metabolic activities of the muscles. The stimulation of β_1 receptors in the heart causes an increase in heart rate and in the force of contraction.

- The stimulation of β_2 receptors causes inhibition, triggering a relaxation of smooth muscles along the respiratory tract. As a result, the diameter of the respiratory passageways increases, making breathing easier. This response accounts for the effectiveness of the inhalers used to treat asthma.

- A third type of beta receptor, beta-3 (β_3) is found in adipose tissue. Stimulation of β_3 receptors leads to *lipolysis*, the breakdown of triglycerides stored within adipocytes. The fatty acids generated through lipolysis are released into the circulation for use by other tissues.

Sympathetic Stimulation and the Release of ACh and NO

Although the vast majority of sympathetic postganglionic fibers are adrenergic (releasing NE), a few are cholinergic (releasing ACh). These postganglionic fibers innervate sweat glands of the skin and the blood vessels to skeletal muscles and the brain. The activation of these sympathetic fibers stimulates sweat gland secretion and dilates the blood vessels.

It may seem strange that sympathetic terminals release ACh, which is the neurotransmitter used by the parasympathetic nervous system. However, neither the body wall nor skeletal muscles are innervated by the parasympathetic division, and NE and ACh differ significantly in function. For example, the effects of ACh release last for a much shorter period than do those of NE release; ACh stimulates sweat gland secretion much more than does NE; and ACh causes the dilation of most small peripheral arteries (*vasodilation*), whereas NE causes their constriction (*vasoconstriction*). The distribution of cholinergic fibers within the sympathetic division provides a method of stimulating sweat gland secretion and selectively enhancing blood flow to skeletal muscles, while the adrenergic terminals reduce the blood flow to other tissues in the body wall.

The sympathetic division also includes *nitroxidergic* synapses, which release *nitric oxide (NO)* as a neurotransmitter. Such synapses occur where neurons innervate smooth muscles in the walls of blood vessels in many regions, notably in skeletal muscles and the brain. The activity of these synapses promotes immediate vasodilation and thereby increased blood flow through the region.

☐ SUMMARY: THE SYMPATHETIC DIVISION

To summarize our discussion of the sympathetic division:

1. The sympathetic division of the ANS includes two sets of sympathetic chain ganglia, one on each side of the vertebral column; three collateral ganglia anterior to the vertebral column; and two adrenal medullae.

2. The preganglionic fibers are short, because the ganglia are close to the spinal cord. The postganglionic fibers are longer and extend a considerable distance before reaching their target organs. (In the case of the adrenal medullae, very short axons end at capillaries that carry their secretions to the bloodstream.)

3. The sympathetic division shows extensive divergence, and a single preganglionic fiber may innervate two dozen or more ganglionic neurons in different ganglia. As a result, a single sympathetic motor neuron in the CNS can control a variety of visceral effectors and can produce a complex and coordinated response.

4. All preganglionic neurons release ACh at their synapses with ganglionic neurons. Most postganglionic fibers release NE, but a few release ACh or NO.

5. The effector response depends on the second messengers activated when NE or E binds to alpha receptors or beta receptors.

✓ Where do the nerves that synapse in collateral ganglia originate?

✓ How would a drug that stimulates acetylcholine receptors affect the sympathetic nervous system?

✓ An individual with high blood pressure is given a medication that blocks beta receptors. How could this medication help correct that person's condition?

Answers start on page Q-1

 Reinforce your understanding of the structure and functions of the sympathetic division of the ANS by visiting the **Companion Website:** Chapter 16/Reviewing Facts & Terms/Multiple Choice.

16–3 THE PARASYMPATHETIC DIVISION

Objectives

- Describe the structures and functions of the parasympathetic division of the autonomic nervous system.
- Describe the mechanisms of neurotransmitter release in the parasympathetic division.
- Describe the effects of parasympathetic neurotransmitters on target organs and tissues.

The parasympathetic division of the ANS (Figure 16–7●) consists of:

1. *Preganglionic Neurons in the Brain Stem and in Sacral Segments of the Spinal Cord.* The mesencephalon, pons, and medulla oblongata contain autonomic nuclei associated with cranial nerves III, VII, IX, and X. In sacral segments of the spinal cord, the autonomic nuclei lie in the lateral gray horns of spinal segments S_2–S_4.

2. *Ganglionic Neurons in Peripheral Ganglia within or Adjacent to the Target Organs.* Preganglionic fibers of the parasympathetic division do not diverge as extensively as do those of the sympathetic division. A typical preganglionic fiber synapses on six to eight ganglionic neurons. These neurons may be situated in a **terminal ganglion**, located near the target organ, or in an **intramural ganglia** (*murus*, wall), which is embedded in the tissues of the target organ.

In contrast to the pattern in the sympathetic division, all these ganglionic neurons are located in the same ganglion, and their postganglionic fibers influence the same target organ. Thus, the effects of parasympathetic stimulation are more specific and localized than are those of the sympathetic division.

☐ ORGANIZATION AND ANATOMY OF THE PARASYMPATHETIC DIVISION

Parasympathetic preganglionic fibers leave the brain as components of cranial nerves III (oculomotor), VII (facial), IX (glossopharyngeal), and X (vagus) (Figure 16–8●). These fibers carry the cranial parasympathetic output. Parasympathetic fibers in the oculomotor, facial, and glossopharyngeal nerves control visceral structures in the head. These fibers synapse in the *ciliary, pterygopalatine, submandibular,* and *otic ganglia.* ∞ pp. 496, 498, 500 Short postganglionic fibers continue to their peripheral targets. The vagus nerve provides preganglionic parasympathetic innervation to structures in the neck and in the thoracic and abdominopelvic cavity as distant as the distal portion of the large intestine. The vagus nerve alone provides roughly 75 percent of all parasympathetic outflow.

The preganglionic fibers in the sacral segments of the spinal cord carry the sacral parasympathetic output. These fibers do not join the ventral roots of the spinal nerves. Instead, the preganglionic fibers form distinct **pelvic nerves**, which innervate intramural ganglia in the walls of the kidney, urinary bladder, terminal portions of the large intestine, and sex organs.

☐ PARASYMPATHETIC ACTIVATION

The major effects produced by the parasympathetic division include the following:

- Constriction of the pupils, to restrict the amount of light that enters the eyes, and focusing the lenses of the eyes on nearby objects.
- Secretion by digestive glands, including salivary glands, gastric glands, duodenal glands, intestinal glands, the pancreas, and the liver.
- The secretion of hormones that promote the absorption and utilization of nutrients by peripheral cells.
- Changes in blood flow and glandular activity associated with sexual arousal.
- An increase in smooth muscle activity along the digestive tract.
- The stimulation and coordination of defecation.
- Contraction of the urinary bladder during urination.
- Constriction of the respiratory passageways.
- A reduction in heart rate and in the force of contraction.
- Sexual arousal and the stimulation of sexual glands in both genders.

These functions center on relaxation, food processing, and energy absorption. The parasympathetic division has been called the *anabolic system,* because its stimulation leads to a general increase in the nutrient content of the blood. (*Anabolic* comes from

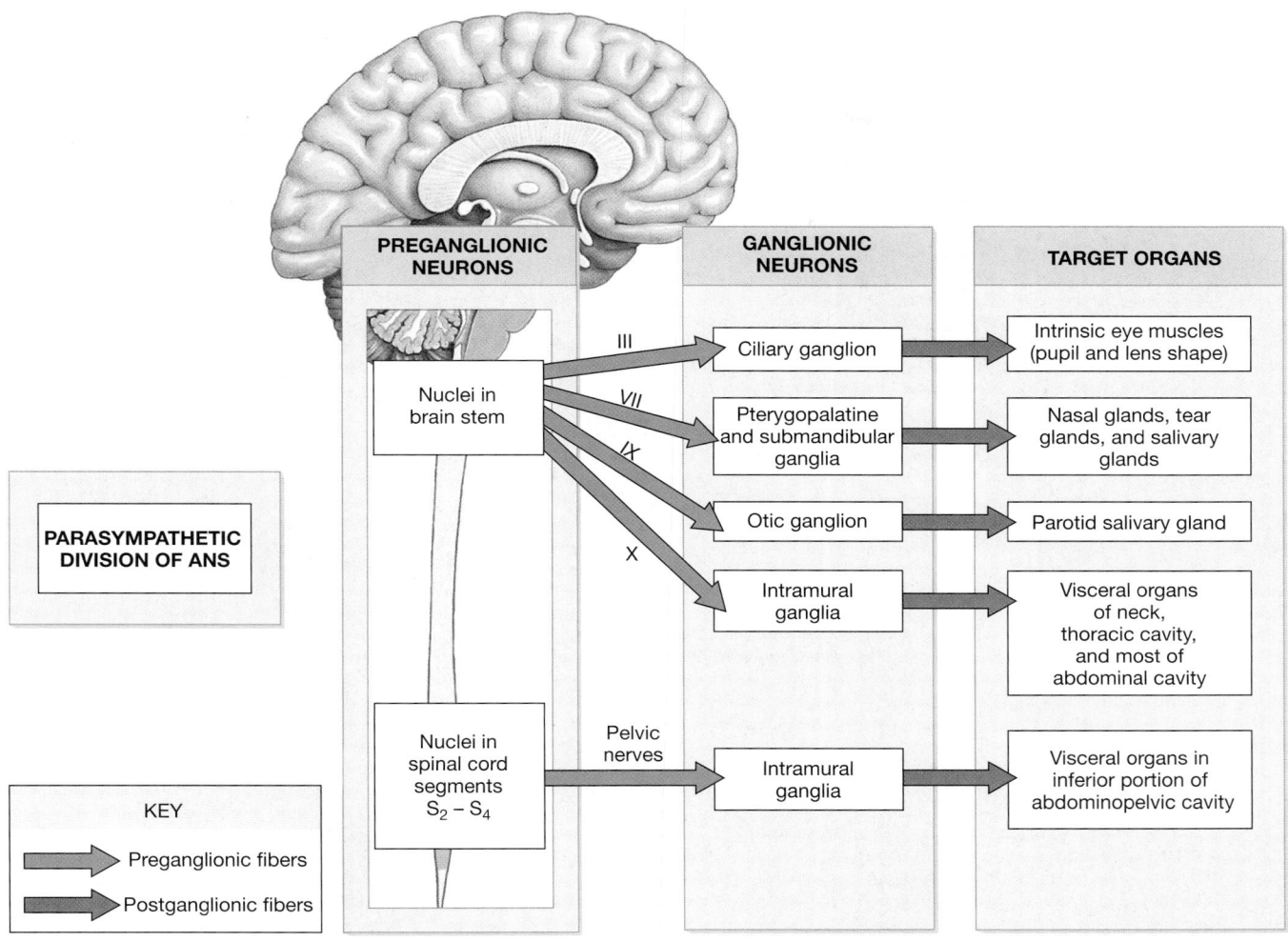

•**FIGURE 16–7**
The Organization of the Parasympathetic Division of the ANS

the Greek word *anabole,* which means "a rising up.") Cells throughout the body respond to this increase by absorbing nutrients and using them to support growth, cell division, and the creation of energy reserves in the form of lipids or glycogen.

☐ NEUROTRANSMITTERS AND PARASYMPATHETIC FUNCTION

All parasympathetic neurons release ACh as a neurotransmitter. The effects on the postsynaptic cell can vary widely, however, due to variations in the type of receptor or the nature of the second messenger involved.

Neurotransmitter Release

The neuromuscular and neuroglandular junctions of the parasympathetic division are small and have narrow synaptic clefts. The effects of stimulation are short lived, because most of the ACh released is inactivated by *acetylcholinesterase (AChE)* at the synapse. Any ACh

diffusing into the surrounding tissues will be inactivated by the enzyme *tissue cholinesterase,* also called *pseudocholinesterase.* As a result, the effects of parasympathetic stimulation are quite localized, and they last a few seconds at most.

Membrane Receptors and Responses

Although all the synapses (neuron to neuron) and neuromuscular or neuroglandular junctions (neuron to effector) of the parasympathetic division use the same transmitter, ACh, two types of ACh receptors occur on the postsynaptic membranes:

1. **Nicotinic** (nik-ō-TIN-ik) **receptors** are located on the surfaces of ganglion cells of both the parasympathetic and sympathetic divisions, as well as at neuromuscular junctions of the somatic nervous system. Exposure to ACh always causes excitation of the ganglionic neuron or muscle fiber by the opening of chemically gated channels in the postsynaptic membrane.

2. **Muscarinic** (mus-kar-IN-ik) **receptors** are located at cholinergic neuromuscular or neuroglandular junctions in the

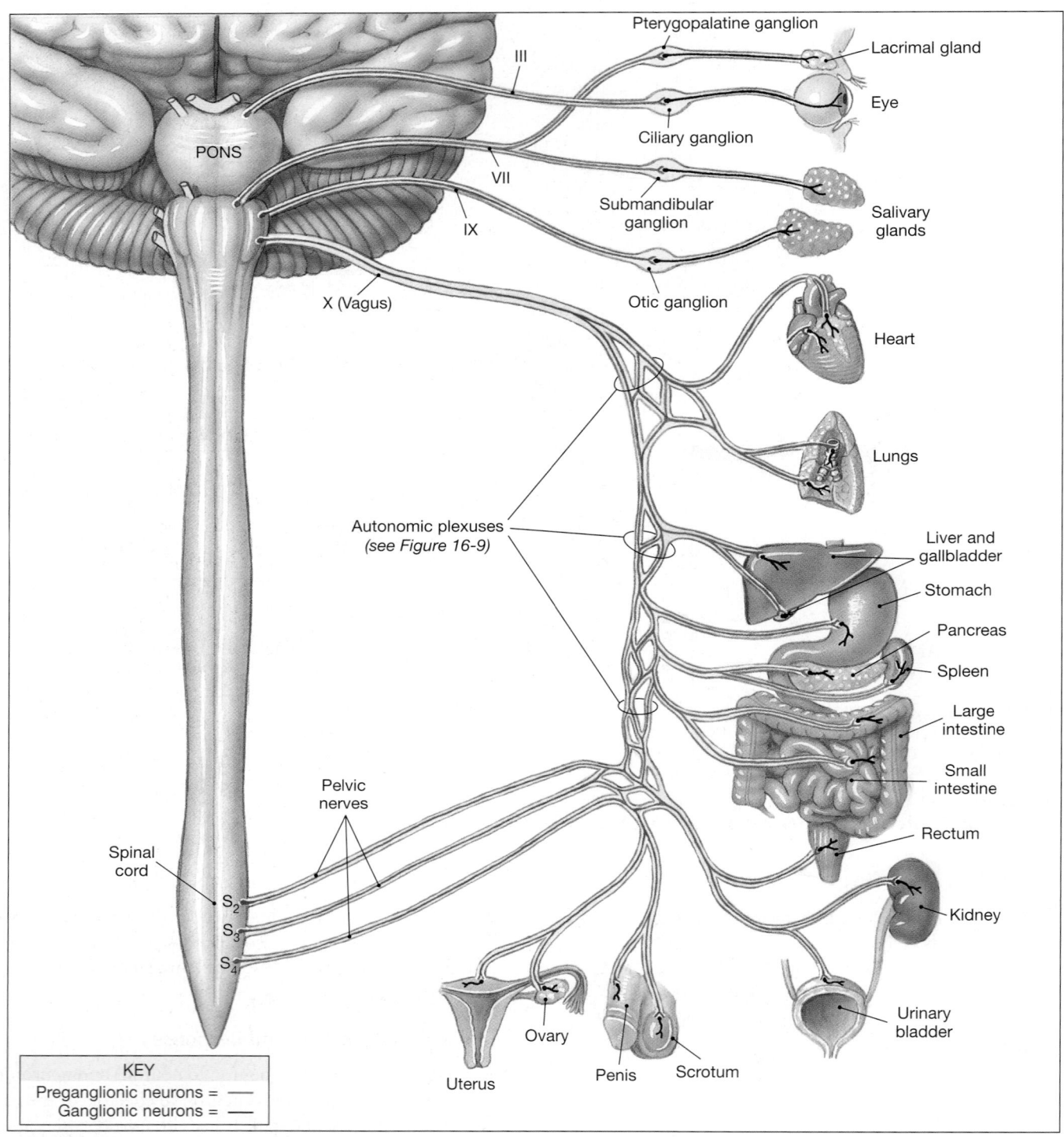

●**FIGURE 16–8**
The Distribution of Parasympathetic Innervation

parasympathetic division, as well as at the few cholinergic junctions in the sympathetic division. Muscarinic receptors are G proteins, and their stimulation produces longer-lasting effects than does the stimulation of nicotinic receptors. The response, which reflects the activation or inactivation of specific enzymes, can be excitatory or inhibitory.

The names *nicotinic* and *muscarinic* originated with researchers who found that dangerous environmental toxins bind to these receptor sites. Nicotinic receptors bind *nicotine*, a powerful toxin that can be obtained from a variety of sources, including tobacco leaves. Muscarinic receptors are stimulated by *muscarine*, a toxin produced by some poisonous mushrooms.

These compounds have discrete actions, targeting either the autonomic ganglia and skeletal neuromuscular junctions (nicotine) or the parasympathetic neuromuscular or neuroglandular junctions (muscarine). They produce dangerously exaggerated, uncontrolled responses that parallel those produced by normal receptor stimulation. For example, nicotine poisoning occurs if as little as 50 mg of the compound is ingested or absorbed through the skin. The symptoms reflect widespread autonomic activation: vomiting, diarrhea, high blood pressure, rapid heart rate, sweating, and profuse salivation. Because the neuromuscular junctions of the somatic nervous system are stimulated, convulsions occur. In severe cases, the stimulation of nicotinic receptors inside the CNS can lead to coma and death within minutes. The symptoms of muscarine poisoning are almost entirely restricted to the parasympathetic division: salivation, nausea, vomiting, diarrhea, constriction of respiratory passages, low blood pressure, and an abnormally slow heart rate.

Table 16–1 summarizes details about the adrenergic and cholinergic receptors of the ANS.

☐ SUMMARY: THE PARASYMPATHETIC DIVISION

In summary,

- The parasympathetic division includes visceral motor nuclei associated with cranial nerves III, VII, IX, and X and with sacral segments $S_2 - S_4$.

- Ganglionic neurons are located within or next to their target organs.

- The parasympathetic division innervates areas serviced by the cranial nerves and organs in the thoracic and abdominopelvic cavities.

- All parasympathetic neurons are cholinergic. Ganglionic neurons have nicotinic receptors, which are excited by ACh. Muscarinic receptors at neuromuscular or neuroglandular junctions produce either excitation or inhibition, depending on the nature of the enzymes activated when ACh binds to the receptor.

- The effects of parasympathetic stimulation are generally brief and restricted to specific organs and sites.

✓ Which nerve is responsible for the parasympathetic innervation of the lungs, heart, stomach, liver, pancreas, and parts of the small and large intestines?

✓ How would the stimulation of muscarinic receptors in cardiac muscle affect the heart?

✓ Why is the parasympathetic division sometimes referred to as the anabolic system?

Answers start on page Q-1

 Compare the structure of the parasympathetic division with that of the sympathetic division by visiting the **Companion Website:** Chapter 16/Critical Thinking/Essay.

16–4 INTERACTIONS BETWEEN THE SYMPATHETIC AND PARASYMPATHETIC DIVISIONS

Objectives

- Discuss the relationship between the two divisions of the autonomic nervous system and the significance of dual innervation.

- Explain the importance of autonomic tone.

SUMMARY TABLE 16–1 Adrenergic and Cholinergic Receptors of the ANS			
Receptor	**Location(s)**	**Response(s)**	**Mechanism**
ADRENERGIC			
α_1	Widespread, found in most tissues	Excitation, stimulation of metabolism	Activation of enzymes; release of intracellular Ca^{2+}
α_2	Sympathetic neuromuscular or neuroglandular junctions	Inhibition of effector cell	Reduction of cAMP concentrations
	Parasympathetic neuromuscular or neuroglandular junctions	Inhibition of neurotransmitter release	Reduction of cAMP concentrations
β_1	Heart, kidneys, liver, adipose tissue*	Stimulation, increased energy consumption	Enzyme activation
β_2	Smooth muscle in vessels of heart and skeletal muscle; smooth muscle layers in intestines, lungs, bronchi	Inhibition, relaxation	Enzyme activation
CHOLINERGIC			
Nicotinic	All autonomic synapses between preganglionic and ganglionic neurons; neuromuscular junctions of SNS	Stimulation, excitation; muscular contraction	Opening of chemically regulated Na^+ channels
Muscarinic	All parasympathetic and cholinergic sympathetic neuromuscular or neuroglandular junctions	Variable	Enzyme activation causing changes in membrane permeability to K^+

*Adipocytes also contain an additional receptor type, β_3, not found in other tissues. Stimulation of β_3 receptors causes lipolysis.

The sympathetic division has widespread impact, reaching visceral organs and tissues throughout the body. The parasympathetic division innervates only visceral structures that are serviced by the cranial nerves or that lie within the abdominopelvic cavity. Although some organs are innervated by just one division, most vital organs receive **dual innervation**, receiving instructions from both the sympathetic and the parasympathetic divisions. Where dual innervation exists, the two divisions commonly have opposing effects. Dual innervation with opposing effects is most evident in the digestive tract, heart, and lungs. At other sites, the responses may be separate or complementary. Secretory control of the salivary glands and the sexual functions of the male reproductive tract are examples.

ANATOMY OF DUAL INNERVATION

Parasympathetic postganglionic fibers from the ciliary, pterygopalatine, submandibular, and otic ganglia of the head accompany the cranial nerves to their peripheral destinations. The sympathetic innervation reaches the same structures by traveling directly from the superior cervical ganglia of the sympathetic chain.

In the thoracic and abdominopelvic cavities, the sympathetic postganglionic fibers mingle with parasympathetic preganglionic fibers at a series of nerve networks collectively called *autonomic plexuses*: the cardiac plexus, the pulmonary plexus, the esophageal plexus, the celiac plexus, the inferior mesenteric plexus, and the hypogastric plexus (Figure 16–9●). Nerves leaving these networks travel with the blood vessels and lymphatic vessels that supply visceral organs.

Autonomic fibers entering the thoracic cavity intersect at the **cardiac plexus** and the **pulmonary plexus**. These plexuses contain sympathetic and parasympathetic fibers bound for the heart and lungs, respectively, as well as the parasympathetic ganglia whose output affects those organs. The **esophageal plexus** contains descending branches of the vagus nerve and splanchnic nerves leaving the sympathetic chain on either side.

Parasympathetic preganglionic fibers of the vagus nerve enter the abdominopelvic cavity with the esophagus. There the fibers enter the **celiac plexus**, also known as the *solar plexus*. The celiac plexus and associated smaller plexuses, such as the **inferior mesenteric plexus**, innervate viscera within the abdominal cavity. The **hypogastric plexus** contains the parasympathetic outflow of the pelvic nerves, sympathetic postganglionic fibers from the inferior mesenteric ganglion, and splanchnic nerves from the sacral sympathetic chain. This plexus innervates the digestive, urinary, and reproductive organs of the pelvic cavity.

A COMPARISON OF THE DIVISIONS OF THE ANS

Figure 16–10● and Table 16–2 compare key structural features of the sympathetic and parasympathetic divisions of the ANS. The distinctions have physiological and functional correlates. Table 16–3 provides a functional comparison of the two divisions, taking into account the effects of sympathetic or parasympathetic activity on specific organs and systems.

AUTONOMIC TONE

Even in the absence of stimuli, autonomic motor neurons show a resting level of spontaneous activity. The background level of activation determines an individual's **autonomic tone**. Autonomic tone is an important aspect of ANS function, just as muscle tone is a key aspect of SNS function. If a nerve is absolutely inactive under normal conditions, then all it can do is increase its activity on demand. But if the nerve maintains a background level of activity, it can increase or decrease its activity, providing a range of control options.

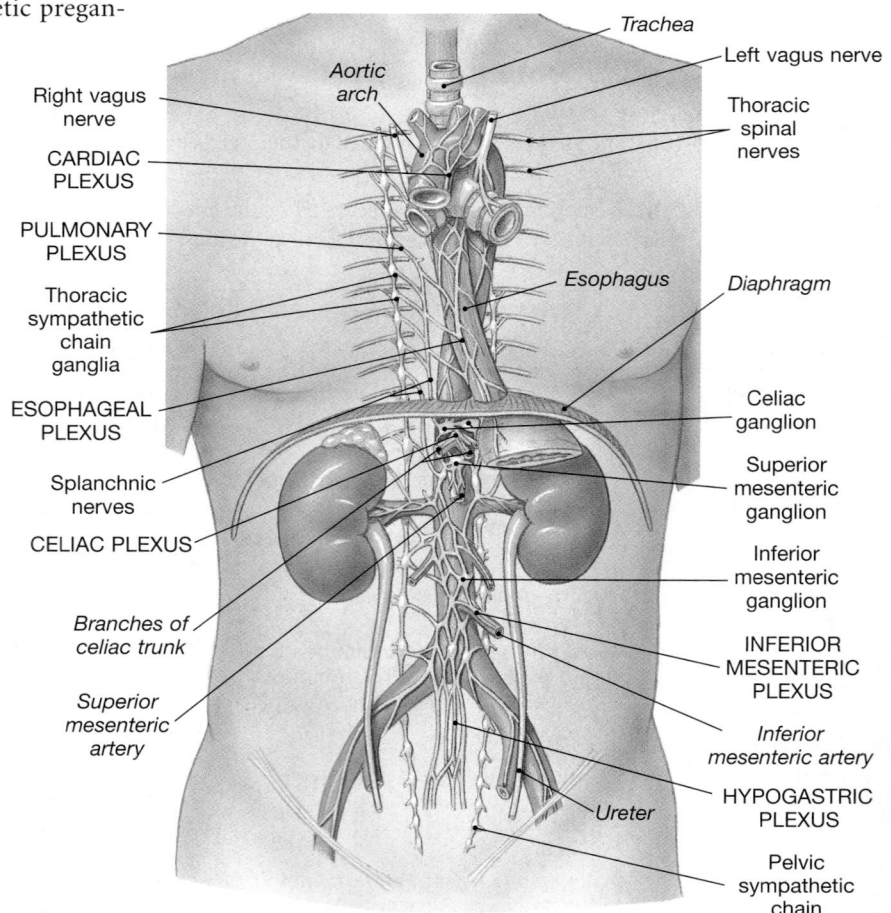

●**FIGURE 16–9**
The Autonomic Plexuses

Autonomic tone is significant where dual innervation occurs and the two ANS divisions have opposing effects (Table 16–3). It is even more important in situations in which dual innervation does not occur. To demonstrate how autonomic tone affects ANS function, we will consider one example of each situation.

The heart is an organ that receives dual innervation. Recall that the heart consists of cardiac muscle tissue and that its contractions are triggered by specialized pacemaker cells. ∞ p. 324 The two autonomic divisions have opposing effects on heart function. Acetylcholine released by postganglionic fibers of the parasympathetic division causes a reduction in heart rate, whereas NE released by varicosities of the sympathetic division accelerates the heartrate. Because autonomic tone exists, small amounts of both of these neurotransmitters are released continuously. However, parasympathetic innervation dominates under resting conditions. If you remove all autonomic innervation, the heart rate goes up; if you remove the parasympathetic innervation but leave the sympathetic innervation intact, the heart rate goes up a lot. The heart rate can be controlled very precisely to meet the demands of active tissues by making small adjustments in the balance between parasympathetic stimulation and sympathetic stimulation. In a crisis, the stimulation of the sympathetic innervation and the inhibition of the parasympathetic innervation accelerate the heart rate to the maximum extent possible.

Several organs are innervated by one division only. For example, most structures in the body wall, such as blood vessels, sweat glands, adipose tissue, and skeletal muscles, receive only sympathetic innervation.

The sympathetic control of blood vessel diameter demonstrates how autonomic tone allows fine adjustment of peripheral activities when the target organ is not innervated by both ANS divisions. It is very important that blood flow to specific organs is controlled to meet the tissue demands for oxygen and nutrients. When a blood vessel dilates, blood flow through it increases; when it constricts, blood flow is reduced. Sympathetic

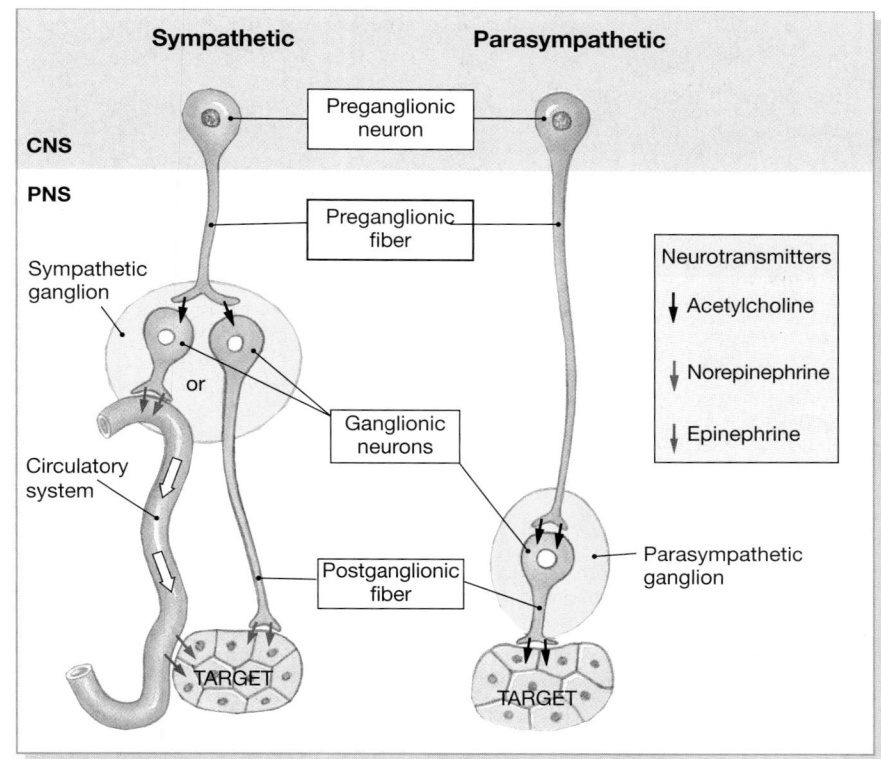

•**FIGURE 16–10**
Summary: The Anatomical Differences between the Sympathetic and Parasympathetic Divisions

SUMMARY TABLE 16–2	A STRUCTURAL COMPARISON OF THE SYMPATHETIC AND PARASYMPATHETIC DIVISIONS OF THE ANS	
Characteristic	**Sympathetic Division**	**Parasympathetic Division**
Location of CNS visceral motor neurons	Lateral gray horns of spinal segments T_1-L_2	Brain stem and spinal segments S_2-S_4
Location of PNS ganglia	Near vertebral column	Typically intramural
Preganglionic fibers		
Length	Relatively short	Relatively long
Neurotransmitter released	Acetylcholine	Acetylcholine
Postganglionic fibers		
Length	Relatively long	Relatively short
Neurotransmitter released	Normally NE; sometimes ACh	Acetylcholine
Neuromuscular or neuroglandular junction	Varicosities and enlarged terminal knobs that release transmitter near target cells	Junctions that release transmitter to special receptor surface
Degree of divergence from CNS to ganglion cells	Approximately 1:32	Approximately 1:6
General function(s)	Stimulates metabolism; increases alertness; prepares for emergency ("fight or flight")	Promotes relaxation, nutrient uptake, energy storage ("rest and repose")

SUMMARY TABLE 16–3 A FUNCTIONAL COMPARISON OF THE SYMPATHETIC AND PARASYMPATHETIC DIVISIONS OF THE ANS

Structure	Sympathetic Effects (receptor or synapse type)	Parasympathetic Effects (all muscarinic receptors)
EYE	Dilation of pupil (α_1); accommodation for distance vision (β_2)	Constriction of pupil; accommodation for close vision
Lacrimal glands	None (not innervated)	Secretion
SKIN		
Sweat glands	Increased secretion, palms and soles (α_1); generalized increase in secretion (cholinergic)	None (not innervated)
Arrector pili muscles	Contraction; erection of hairs (α_1)	As above
CARDIOVASCULAR SYSTEM		
Blood vessels		As above
To skin	Dilation (cholinergic); constriction (α_1)	
To skeletal muscles	Dilation (β_2 and cholinergic; nitroxidergic)	
To heart	Dilation (β_2); constriction (α_1, α_2)	
To lungs	Dilation (β_2); constriction (α_1)	
To digestive viscera	Constriction (α_1); dilation (β_2)	
To kidneys	Constriction, decreased urine production (α_1, α_2); dilation, increased urine production (β_1, β_2)	
To brain	Dilation (cholinergic and nitroxidergic)	
Veins	Constriction (α_1, β_2)	
Heart	Increased heart rate, force of contraction, and blood pressure (α_1, β_1)	Decreased heart rate, force of contraction, and blood pressure
ENDOCRINE SYSTEM		
Adrenal gland	Secretion of epinephrine, norepinephrine by adrenal medulla	None (not innervated)
Posterior lobe of pituitary gland	Secretion of ADH (β_1)	As above
Pancreas	Decreased insulin secretion (α_2)	Increased insulin secretion
Pineal gland	Increased melatonin secretion $(\beta)^*$	
RESPIRATORY SYSTEM		
Airways	Increased airway diameter (β_2)	Decreased airway diameter
Secretory glands	Mucous Secretion (α_1)	None
DIGESTIVE SYSTEM		
Salivary glands	Production of viscous secretion (α_1, β_1) containing mucins and enzymes	Production of copious, watery secretion
Sphincters	Constriction (α_1)	Dilation
General level of activity	Decreased (α_2, β_2)	Increased
Secretory glands	Inhibition (α_2)	Stimulation
Liver	Glycogen breakdown, glucose synthesis and release (α_1, β_2)	Glycogen synthesis
Pancreas	Decreased exocrine secretion (α_1)	Increased exocrine secretion
SKELETAL MUSCLES	Increased force of contraction, glycogen breakdown (β_2)	None (not innervated)
	Facilitation of ACh release at neuromuscular junction (α_2)	As above
ADIPOSE TISSUE	Lipolysis, fatty acid release $(\alpha_1, \beta_1, \beta_3)$	
URINARY SYSTEM		
Kidneys	Secretion of renin (β_1)	Uncertain effects on urine production
Urinary bladder	Constriction of internal sphincter; relaxation of urinary bladder (α_1, β_2)	Tensing of urinary bladder, relaxation of internal sphincter to eliminate urine
MALE REPRODUCTIVE SYSTEM	Increased glandular secretion and ejaculation (α_1)	Erection
FEMALE REPRODUCTIVE SYSTEM	Increased glandular secretion; contraction of pregnant uterus (α_1)	Variable (depending on hormones present)
	Relaxation of nonpregnant uterus (β_2)	As above

*The type of beta receptor has not yet been determined.

postganglionic fibers that release NE innervate the smooth muscle cells in the walls of peripheral vessels. The background sympathetic tone keeps these muscles partially contracted, so the vessels are ordinarily at roughly half their maximum diameter. When increased blood flow is needed, the rate of NE release decreases and sympathetic cholinergic fibers are stimulated. As a result, the smooth muscle cells relax, the vessels dilate, and blood flow quickly increases. By adjusting sympathetic tone and the activity of cholinergic fibers, the sympathetic division can exert precise control of vessel diameter over its entire range.

16–5 INTEGRATION AND CONTROL OF AUTONOMIC FUNCTIONS

Objectives

- Describe the hierarchy of interacting levels of control in the autonomic nervous system.
- Define a visceral reflex and explain the significance of such reflexes.

Figure 15–12●, p. 528, diagrammed the relationships among centers involved in somatic motor control. The lowest level of regulatory control consists of the lower motor neurons involved in cranial and spinal reflex arcs. The highest level consists of the pyramidal motor neurons of the primary motor cortex, operating with the feedback from the cerebellum and cerebral nuclei.

The ANS is also organized into a series of interacting levels. At the bottom are visceral motor neurons in the lower brain stem and spinal cord that are involved in cranial and spinal visceral reflexes. **Visceral reflexes** provide automatic motor responses that can be modified, facilitated, or inhibited by higher centers, especially those of the hypothalamus.

For example, when a light is shined in one of your eyes, a visceral reflex constricts the pupils of *both* eyes (the *consensual light reflex*). The visceral motor commands are distributed by parasympathetic fibers. In darkness, your pupils dilate; this *pupillary reflex* is directed by sympathetic postganglionic fibers. However, the motor nuclei directing pupillary constriction or dilation are also controlled by hypothalamic centers concerned with emotional states. When you are queasy or nauseated, your pupils constrict; when you are sexually aroused, your pupils dilate.

☐ VISCERAL REFLEXES

Each **visceral reflex arc** consists of a receptor, a sensory neuron, a processing center (one or more interneurons), and two visceral motor neurons (Figure 16–11●). All visceral reflexes are polysynaptic; they are either long reflexes or short reflexes. **Long reflexes** are the autonomic equivalents of the polysynaptic reflexes introduced in Chapter 13. ⬭ p. 453 Visceral sensory neurons deliver information to the CNS along the dorsal roots of spinal nerves, within the sensory branches of cranial nerves, and within the autonomic nerves that innervate visceral effectors. The processing steps involve interneurons within the CNS, and the ANS carries the motor commands to the appropriate visceral effectors.

Short reflexes bypass the CNS entirely; they involve sensory neurons and interneurons whose cell bodies are located within autonomic ganglia. The interneurons synapse on ganglionic neurons, and the motor commands are then distributed by postganglionic fibers. Short reflexes control very simple motor responses with localized effects. In general, short reflexes may control patterns of activity in one small part of a target organ, whereas long reflexes coordinate the activities of the entire organ.

In most organs, long reflexes are most important in regulating visceral activities, but this is not the case with the digestive tract and its associated glands. In these areas, short reflexes provide most of the control and coordination required for normal functioning. The neurons involved form the *enteric nervous system*, introduced on p. 534. The ganglia in the walls of the digestive tract contain the cell bodies of visceral sensory neurons, interneurons, and visceral motor neurons, and their axons form extensive nerve nets. Although parasympathetic innervation of the visceral motor neurons can stimulate and coordinate various digestive activities, the enteric nervous system is quite capable of controlling digestive functions independent of the central nervous system. We will consider the functions of the enteric nervous system further in Chapter 24.

As we examine other body systems in later chapters, we will encounter many examples of autonomic reflexes involved in

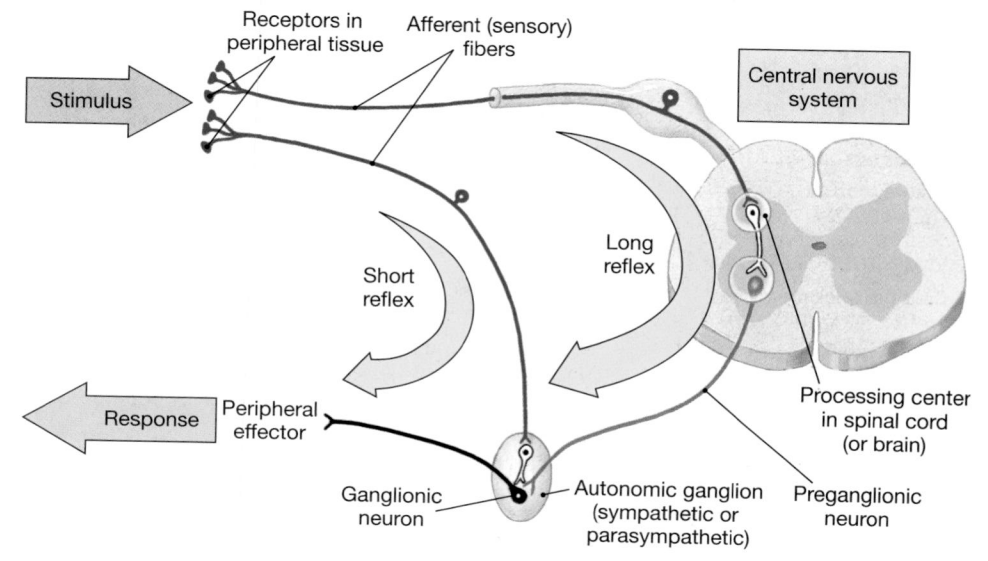

●**FIGURE 16–11**
Visceral Reflexes. Visceral reflexes have the same basic components as somatic reflexes, but all visceral reflexes are polysynaptic.

respiration, cardiovascular function, and other visceral activities. Some of the most important are previewed in Table 16–4. Notice that the parasympathetic division participates in reflexes that affect individual organs and systems. This specialization reflects the relatively specific and restricted pattern of innervation. In contrast, fewer sympathetic reflexes exist. The sympathetic division is typically activated as a whole, in part because it has such a high degree of divergence and in part because the release of hormones by the adrenal medullae produces widespread peripheral effects.

▢ HIGHER LEVELS OF AUTONOMIC CONTROL

The levels of activity in the sympathetic and parasympathetic divisions of the ANS are controlled by centers in the brain stem that deal with specific visceral functions. Figure 16–12● diagrams the levels of autonomic control. As in the SNS, in the ANS simple reflexes based in the spinal cord provide relatively rapid and automatic responses to stimuli. More complex sympathetic and parasympathetic reflexes are coordinated by processing centers in the medulla oblongata. In addition to the cardiovascular and respiratory centers, the medulla oblongata contains centers and nuclei involved with salivation, swallowing, digestive secretions, peristalsis, and urinary function. These centers are in turn subject to regulation by the hypothalamus. ∞ pp. 474, 482

The term *autonomic* was originally applied because the regulatory centers involved with the control of visceral function were thought to operate independent of other CNS activities. This view has been drastically revised in light of subsequent research. Because the hypothalamus interacts with all other portions of the brain, activity in the limbic system, thalamus, or cerebral cortex can have dramatic effects on autonomic function. For example, when you become angry, your heart rate accelerates, your blood pressure rises, and your respiratory rate increases; when you remember your last big dinner, your stomach "growls" and your mouth waters. **AM** Biofeedback

▢ THE INTEGRATION OF SNS AND ANS ACTIVITIES

Figure 16–13● and Table 16–5 indicate how the activities of the somatic nervous system, discussed in Chapter 15, and those of the autonomic nervous system are integrated. Although we have considered somatic and visceral portions of the nervous system separately, the two have many parallels, both in terms of organization and function. Integration occurs at the level of the brain stem, and both systems are under the control of higher centers.

TABLE 16–4 REPRESENTATIVE VISCERAL REFLEXES

Reflex	Stimulus	Response	Comments
PARASYMPATHETIC REFLEXES			
Gastric and intestinal reflexes (Chapter 24)	Pressure and physical contact	Smooth muscle contractions that propel food materials and mix with secretions	Via vagus nerve
Defecation (Chapter 24)	Distention of rectum	Relaxation of internal anal sphincter	Requires voluntary relaxation of external anal sphincter
Urination (Chapter 26)	Distention of urinary bladder	Contraction of walls of urinary bladder; relaxation of internal urethral sphincter	Requires voluntary relaxation of external urethral sphincter
Direct light and consensual light reflexes (Chapter 14)	Bright light shining in eye(s)	Constriction of pupils of both eyes	
Swallowing reflex (Chapter 24)	Movement of food and liquids into pharynx	Smooth muscle and skeletal muscle contractions	Coordinated by medullary swallowing center
Coughing reflex (Chapter 23)	Irritation of respiratory tract	Sudden explosive ejection of air	Coordinated by medullary coughing center
Baroreceptor reflex (Chapters 17, 20, 21)	Sudden rise in carotid blood pressure	Reduction in heart rate and force of contraction	Coordinated in cardiac center of medulla oblongata
Sexual arousal (Chapter 28)	Erotic stimuli (visual or tactile)	Increased glandular secretions, sensitivity	
SYMPATHETIC REFLEXES			
Cardioacceleratory reflex (Chapter 21)	Sudden decline in blood pressure in carotid artery	Increase in heart rate and force of contraction	Coordinated in cardiac center of medulla oblongata
Vasomotor reflexes (Chapter 21)	Changes in blood pressure in major arteries	Changes in diameter of peripheral vessels	Coordinated in vasomotor center in medulla oblongata
Pupillary reflex (Chapter 17)	Low light level reaching visual receptors	Dilation of pupil	
Ejaculation (in males) (Chapter 28)	Erotic stimuli (tactile)	Skeletal muscle contractions ejecting semen	

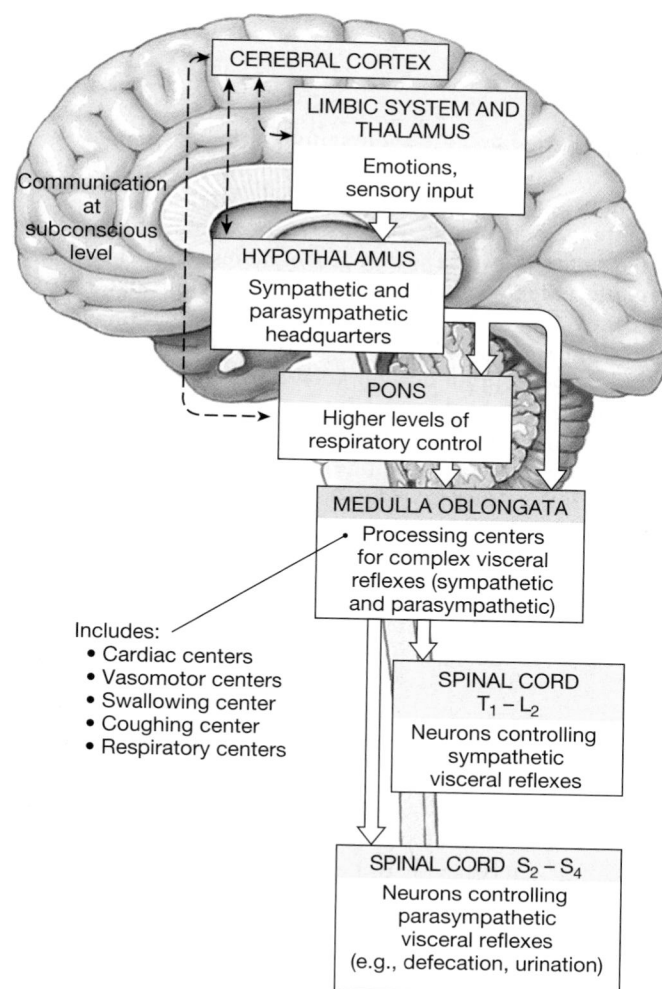

●**FIGURE 16–12**
Levels of Autonomic Control

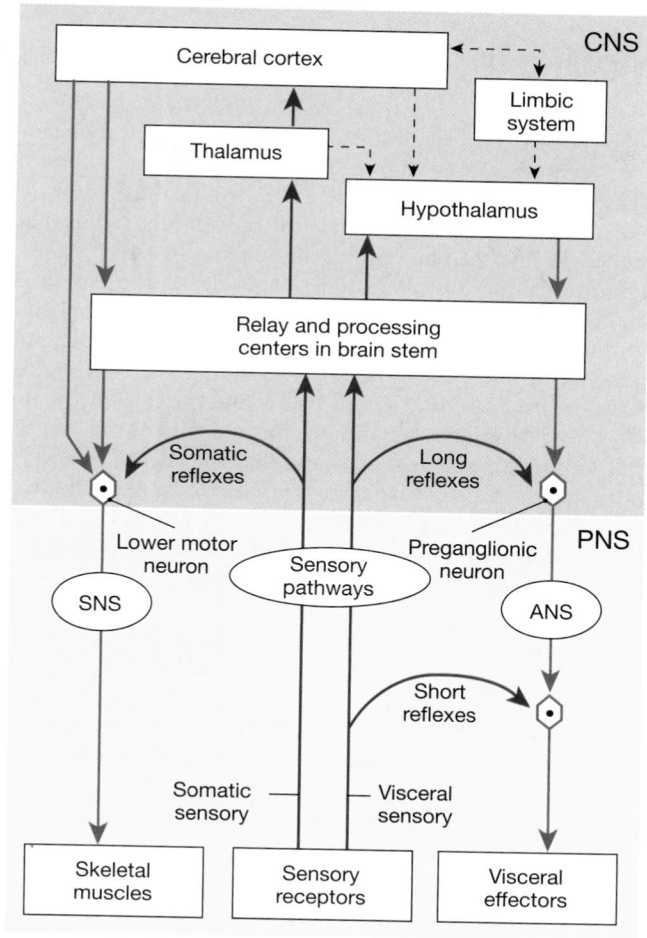

●**FIGURE 16–13**
A Comparison of Somatic and Autonomic Function.
The somatic and autonomic nervous systems are organized in parallel and are integrated at the level of the brain stem. Blue arrows indicate ascending sensory information; red arrows, descending motor commands; and dashed lines, communication and feedback.

SUMMARY TABLE 16–5 A COMPARISON OF THE ANS AND SNS		
Characteristic	**ANS**	**SNS**
Innervation	Visceral effectors, including cardiac muscle, smooth muscle, glands, fat cells	Skeletal muscles
Activation	In response to sensory stimuli or from commands of higher centers	In response to sensory stimuli or from commands of higher centers
Relay and processing centers	Brain stem	Brain stem and thalamus
Headquarters	Hypothalamus	Cerebral cortex
Feedback received from	Limbic system and thalamus	Cerebellum and basal nuclei
Control method	Adjustment of activity in brain stem processing centers that innervate preganglionic neurons	Direct (corticospinal) and indirect (medial and lateral) pathways that innervate lower motor neurons
Reflexes	Polysynaptic (short and long)	Monosynaptic and polysynaptic

Modifying Autonomic Nervous System Function

Drugs may be administered to counteract or reduce symptoms caused or aggravated by autonomic activities. These drugs are called **mimetic** if they mimic the activity of one of the normal autonomic neurotransmitters. Drugs that reduce the effects of autonomic stimulation by keeping the neurotransmitter from affecting the postsynaptic membranes are known as **blocking agents**. Mimetic drugs commonly have advantages over natural neurotransmitters. For example, norepinephrine and epinephrine must be administered into the bloodstream by injection or infusion, and the side effects are short lived. In contrast, **sympathomimetic drugs** may survive oral administration, produce longer-lasting effects, or have more specific actions than does E or NE. Examples include drugs applied topically to reduce hemorrhaging, by spray to reduce nasal congestion, by inhalation to dilate the respiratory passageways, or in drops to dilate the pupils.

Sympathetic blocking agents bind to the receptor sites and prevent a normal response to the presence of neurotransmitters or sympathomimetic drugs. **Alpha-blockers** eliminate the peripheral vasoconstriction (a reduction of the diameter of blood vessels) that accompanies sympathetic stimulation. **Beta-blockers** are effective and clinically useful for treating chronic high blood pressure and other forms of cardiovascular disease.

Parasympathomimetic drugs may be used to increase the activity along the digestive tract and to encourage defecation and urination. **Parasympathetic blocking agents** target the muscarinic receptors at neuromuscular or neuroglandular junctions. These drugs have diverse effects, but they are often used to control the diarrhea and cramps associated with various forms of food poisoning. **AM** Pharmacology and the Autonomic Nervous System

---

✓ What effect would the loss of sympathetic tone have on blood flow to a tissue?

✓ What physiological changes would you expect in a patient who is about to undergo a root canal and is quite anxious about the procedure?

✓ Harry has a brain tumor that is interfering with the function of his hypothalamus. Would you expect this tumor to interfere with autonomic function? Why or why not?

Answers start on page Q-1

16–6 HIGHER ORDER FUNCTIONS

Objectives

- Explain how memories are created, stored, and recalled.
- Distinguish between the levels of consciousness and unconsciousness, and identify the characteristics of brain activity associated with the different levels of sleep.

Higher order functions share the following three characteristics:

1. The cerebral cortex is required for their performance, and they involve complex interactions among areas of the cortex and between the cerebral cortex and other areas of the brain.

2. They involve both conscious and unconscious information processing.

3. They are not part of the programmed "wiring" of the brain; therefore, the functions are subject to modification and adjustment over time.

In Chapter 14, we considered functional areas of the cerebral cortex and the regional specializations of the left and right cere-

bral hemispheres. pp. 488–491 In this section, we consider the mechanisms of memory and learning and describe the neural interactions responsible for consciousness, sleep, and arousal. In the next section, we shall provide an overview of brain chemistry and its effects on behavior and personality.

☐ MEMORY

What was the topic of the last sentence you read? What do your parents look like? What is your Social Security number? When did Columbus reach the New World? What does a red traffic light mean? What does a hot dog taste like? Answering these questions involves accessing *memories*, stored bits of information gathered through experience. **Fact memories** are specific bits of information, such as the color of a stop sign or the smell of a perfume. **Skill memories** are learned motor behaviors. For example, you can probably remember how to light a match or open a screw-top jar. With repetition, skill memories become incorporated at the unconscious level. Examples include the complex motor patterns involved in skiing, playing the violin, and similar activities. Skill memories related to programmed behaviors, such as eating, are stored in appropriate portions of the brain stem. Complex skill memories involve the integration of motor patterns in the cerebral nuclei, cerebral cortex, and cerebellum.

Two classes of memories are identified. **Short-term memories**, or *primary memories*, do not last long, but while they persist, the information can be recalled immediately. Primary memories contain small bits of information, such as a person's name or a telephone number. Repeating a phone number or other bit of information reinforces the original short-term memory and helps ensure its conversion to a long-term memory. **Long-term memories** last much longer, in some cases for an entire lifetime. The conversion from short-term to long-term memory is called **memory consolidation**. There are two types of long-term memory: (1) *Secondary memories* are long-term memories that fade with time and may require considerable effort to recall. (2) *Tertiary memories* are long-term memories that seem to be part of consciousness, such as your name or the contours of your own body. Proposed relationships among these memory classes are diagrammed in Figure 16–14●.

Brain Regions Involved in Memory Consolidation and Access

The amygdaloid body and the hippocampus, two components of the limbic system (Figure 14–13●, p. 483), are essential to memory consolidation. Damage to the hippocampus leads to an inability to convert short-term memories to new long-term memories, although existing long-term memories remain intact and accessible. Tracts leading from the amygdaloid body to the hypothalamus may link memories to specific emotions.

The **nucleus basalis**, a cerebral nucleus near the diencephalon, plays an uncertain role in memory storage and retrieval. Tracts connect this nucleus with the hippocampus, amygdaloid body, and all areas of the cerebral cortex. Damage to this nucleus is associated with changes in emotional states, memory, and intellectual function (as we will see in the discussion of Alzheimer's disease later in this chapter).

Most long-term memories are stored in the cerebral cortex. Conscious motor and sensory memories are referred to the appropriate association areas. For example, visual memories are stored in the visual association area, and memories of voluntary motor activity are stored in the premotor cortex. Special portions of the occipital and temporal lobes are crucial to the memories of faces, voices, and words. In at least some cases, a specific memory probably depends on the activity of a single neuron. For example, in one portion of the temporal lobe an individual neuron responds to the sound of one word and ignores others. A specific neuron may also be activated by the proper combination of sensory stimuli associated with a particular individual, such as your grandmother. As a result, these neurons are called "grandmother cells."

Information on one subject is parceled out to many different regions of the brain. Your memories of cows are stored in the visual association area (what a cow looks like, how the letters *c-o-w* mean "cow"), the auditory association area (the "moo" sound and how the word *cow* sounds), the speech center (how to say the word *cow*), and the frontal lobes (how big cows are, what they eat). Related information, such as how you feel about cows and what milk tastes like, is stored in other locations. If one of those storage areas is damaged, your memory will be incomplete in some way. How these memories are accessed and assembled on demand remains a mystery.

Cellular Mechanisms of Memory Formation and Storage

Memory consolidation at the cellular level involves anatomical and physiological changes in neurons and synapses. For legal, ethical, and logistical reasons, it is not possible to perform much research on these mechanisms with human subjects. Research on other animals, commonly those with relatively simple nervous systems, has indicated that the following mechanisms may be involved:

- **Increased Neurotransmitter Release.** A synapse that is frequently active increases the amount of neurotransmitter that it stores, and it releases more on each stimulation. The more neurotransmitter released, the greater the effect on the postsynaptic neuron.

- **Facilitation at Synapses.** When a neural circuit is repeatedly activated, the synaptic terminals begin releasing neurotransmitter in small quantities continuously. The neurotransmitter binds to receptors on the postsynaptic membrane, producing a graded depolarization that brings the membrane closer to threshold. The facilitation that results affects all neurons in the circuit.

- **The Formation of Additional Synaptic Connections.** Evidence indicates that when one neuron repeatedly communicates with another, the axon tip branches and forms additional synapses on the postsynaptic neuron. As a result, the presynaptic neuron will have a greater effect on the transmembrane potential of the postsynaptic neuron.

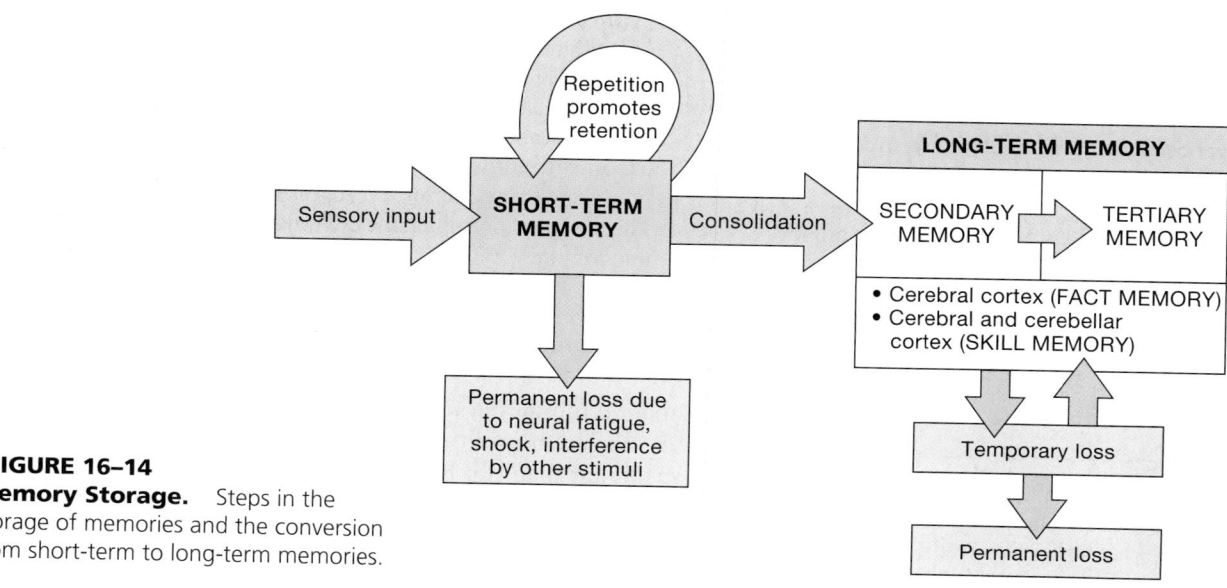

●**FIGURE 16–14**
Memory Storage. Steps in the storage of memories and the conversion from short-term to long-term memories.

These processes create anatomical changes that facilitate communication along a specific neural circuit. This facilitated communication is thought to be the basic method of memory storage. A single circuit that corresponds to a single memory has been called a **memory engram**. This definition is based on function rather than structure; we know too little about the organization and storage of memories to be able to describe the neural circuits involved.

Efficient conversion of a short-term memory into a memory engram takes time, usually at least an hour. Whether that conversion will occur depends on several factors, including the nature, intensity, and frequency of the original stimulus. Very strong, repeated, or exceedingly pleasant or unpleasant events are most likely to be converted to long-term memories. Drugs that stimulate the CNS, such as caffeine and nicotine, may enhance memory consolidation through facilitation; we discussed the membrane effects of those drugs in Chapter 12. ⊙⊙ pp. 415, 420-1

The hippocampus plays a key role in the consolidation of memories. The mechanism, which remains unknown, is linked to the presence of *NMDA* (N-methyl D-aspartate) *receptors*, which are chemically gated calcium channels. When activated by the neurotransmitter *glycine*, the gates open and calcium enters the cell. Blocking NMDA receptors in the hippocampus prevents long-term memory formation.

Amnesia is the loss of memory as a result of disease or trauma. The type of memory loss depends on the specific regions of the brain affected. For example, damage to the auditory association areas can make it difficult to remember sounds. Amnesia occurs suddenly or progressively, and recovery is complete, partial, or nonexistent, depending on the nature of the problem. In **retrograde amnesia** (*retro-*, behind), the individual loses memories of past events. Some degree of retrograde amnesia commonly follows a head injury; after a car wreck or a fall, many victims are unable to remember the moments preceding the accident. In **anterograde amnesia** (*antero-*, ahead), an individual may be unable to store additional memories, but earlier memories are intact and accessible. The problem appears to involve an inability to generate long-term memories. At least two drugs—*diazepam* (*Valium*®) and *Halcion*®—have been known to cause brief periods of anterograde amnesia. Brain injuries can cause more prolonged memory problems. A person with permanent anterograde amnesia lives in surroundings that are always new. Magazines can be read, chuckled over, and reread a few minutes later with equal pleasure, as if they had never been seen before. Clinicians must introduce themselves at every meeting, even if they have been treating the patient for years.

Post-traumatic amnesia (PTA) commonly develops after a head injury. The duration of the amnesia varies with the severity of the injury. This condition combines the characteristics of retrograde and anterograde amnesias; the individual can neither remember the past nor consolidate memories of the present. **AM** Amnesia

---

✓ List three characteristics of higher order functions.

✓ After suffering a head injury in an automobile accident, David has difficulty comprehending what he hears or reads. This symptom might indicate damage to which portion of his brain?

✓ As you recall facts while you take your A & P test, which type of memory are you using?

Answers start on page Q-1

☐ CONSCIOUSNESS

A conscious individual is alert and attentive; an unconscious individual is not. The difference is obvious, but there are many gradations of both the conscious and unconscious states. For example, a healthy conscious person can be almost asleep, wide awake, or high-strung and jumpy; a healthy sleeping person can be dozing lightly or so deeply asleep that he or she cannot easily be awakened. Healthy individuals cycle between the alert, conscious state and the asleep state each day. The degree of wakefulness at any moment is an indication of the level of ongoing CNS activity. When CNS function becomes abnormal or depressed, the state of wakefulness can be affected. For example, in a state of *coma*, the individual is unconscious and cannot be awakened, even by strong stimuli. As a result, clinicians are quick to note any change in the responsiveness of comatose patients. **AM** Altered States

Sleep

Conscious implies a state of awareness of and attention to external events and stimuli. *Unconscious* can imply a number of conditions, ranging from the deep, unresponsive state induced by anesthesia before major surgery to the light, drifting "nod" that occasionally plagues students who are reading anatomy and physiology textbooks. You are considered to be asleep when you are unconscious but can still be awakened by normal sensory stimuli.

Two general levels of sleep are recognized, each typified by characteristic patterns of brain wave activity (Figure 16–15●):

1. In **deep sleep**, also called *slow wave* or *non-REM* (*NREM*) *sleep*, your entire body relaxes and activity at the cerebral cortex is at a minimum. Heart rate, blood pressure, respiratory rate, and energy utilization decline by up to 30 percent.

2. During **rapid eye movement (REM) sleep**, active dreaming occurs, accompanied by changes in your blood pressure and respiratory rate. Although the EEG shows traces resembling those of the awake state, you become even less receptive to outside stimuli than in deep sleep, and muscle tone decreases markedly. Intense inhibition of somatic motor neurons probably prevents you from physically producing the responses you envision during the dream sequence. The neurons controlling the eye muscles escape this inhibitory influence, and your eyes move rapidly as the dream events unfold.

Periods of REM and deep sleep alternate throughout the night, beginning with a period of deep sleep that lasts about an hour and a half. Rapid eye movement periods initially average about 5 minutes in length, but they gradually increase to about 20 minutes over an eight-hour night. Each night we probably spend less than two hours dreaming, but variation among individuals is significant. For example, children devote more time to REM sleep than do adults, and extremely tired individuals have very short and infrequent REM periods.

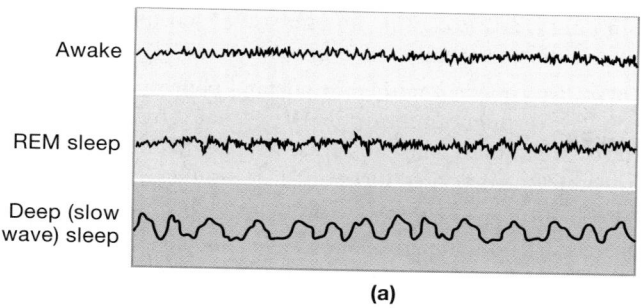

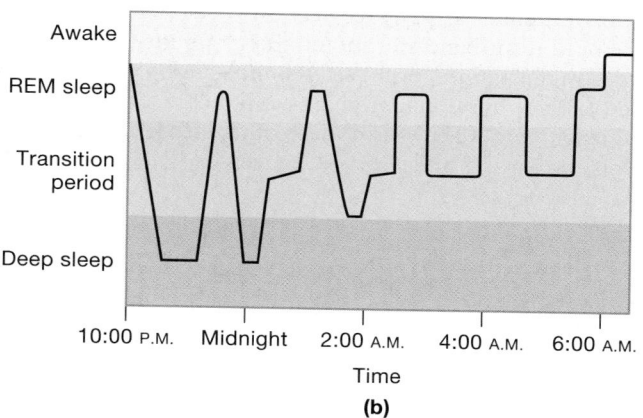

•**FIGURE 16–15**
Levels of Sleep. **(a)** EEG from the awake, REM, and deep (slow wave) sleep states. The EEG pattern during REM sleep resembles the alpha waves typical of awake adults. **(b)** Typical pattern of oscillation between sleep stages of a healthy young adult during a single night's sleep.

Sleep produces only minor changes in the physiological activities of other organs and systems, and none of these changes appear to be essential to normal function. The significance of sleep must lie in its impact on the CNS, but the physiological or biochemical basis remains to be determined. We do know that protein synthesis in neurons increases during sleep. Extended periods without sleep will lead to a variety of disturbances in mental function. Roughly 25 percent of the U.S. population experiences some form of a *sleep disorder*. Examples of such disorders include abnormal patterns or duration of REM sleep or unusual behaviors performed while sleeping, such as sleepwalking. In some cases, these problems begin to affect the individual's conscious activities. Slowed reaction times, irritability, and behavioral changes may result. **AM** Sleep Disorders

Arousal

Arousal, or awakening from sleep, appears to be one of the functions of the reticular formation. The reticular formation is especially well suited for providing "watchdog" services, because it has extensive interconnections with the sensory, motor, and integrative nuclei and pathways all along the brain stem.

THE RETICULAR ACTIVATING SYSTEM Your state of consciousness is determined by complex interactions between your brain stem and

cerebral cortex. One of the most important brain stem components is a diffuse network in the reticular formation known as the **reticular activating system (RAS)**. This network extends from the medulla oblongata to the mesencephalon (Figure 16–16•). The output of the RAS projects to thalamic nuclei that influence large areas of the cerebral cortex. When the RAS is inactive, so is the cerebral cortex; stimulation of the RAS produces a widespread activation of the cerebral cortex.

The mesencephalic portion of the RAS appears to be the center of the system. Stimulation of this area produces the most pronounced and long-lasting effects on the cerebral cortex. Stimulating other portions of the RAS seems to have an effect only to the degree that it changes the activity of the mesencephalic region. The greater the stimulation to the mesencephalic region of the RAS, the more alert and attentive the individual will be to incoming sensory information. The thalamic nuclei associated with the RAS may also play an important role in focusing attention on specific mental processes.

Sleep may be ended by any stimulus sufficient to activate the reticular formation and RAS. Arousal occurs rapidly, but the effects of a single stimulation of the RAS last less than a minute. Thereafter, consciousness can be maintained by positive feedback, because activity in the cerebral cortex, cerebral nuclei, and sensory and motor pathways will continue to stimulate the RAS.

After many hours of activity, the reticular formation becomes less responsive to stimulation. The individual becomes less alert and more lethargic. The precise mechanism remains unknown, but neural fatigue probably plays a relatively minor role in the reduction of RAS activity. Evidence suggests that the regulation of

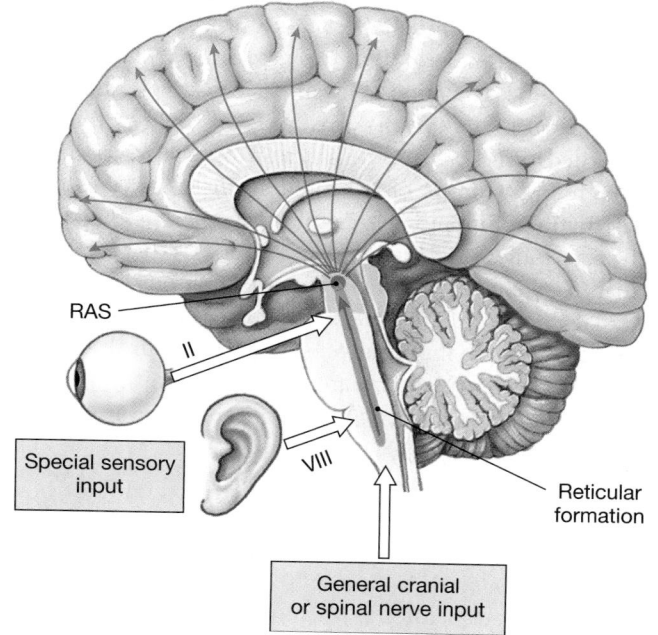

•**FIGURE 16–16**
The Reticular Activating System. The mesencephalic headquarters of the reticular formation receives collateral inputs from a variety of sensory pathways. Stimulation of this region produces arousal and heightened states of attentiveness.

awake–asleep cycles involves an interplay between brain stem nuclei that use different neurotransmitters. One group of nuclei stimulates the RAS with norepinephrine and maintains the awake, alert state. The other group, which depresses RAS activity with serotonin, promotes deep sleep. These "dueling" nuclei are located in the brain stem.

16–7 BRAIN CHEMISTRY AND BEHAVIOR

Objective

- Describe some of the ways in which the interactions of neurotransmitters influence brain function.

We discussed the general distribution of neurotransmitters in the central nervous system in Chapter 12. ∞ pp. 412-417 Of the roughly 50 known neurotransmitters, the most important and widespread are acetylcholine (ACh), norepinephrine (NE), glutamate, aspartate, glycine, gamma aminobutyric acid (GABA), dopamine, serotonin, histamine, Substance P, and the various opioids (see Table 12–6, p. 416). The distribution of neurons that release each of these compounds varies from region to region in the brain. However, tracts originating at each nucleus distribute these neurotransmitters throughout the CNS.

☐ NEUROTRANSMITTERS AND BRAIN FUNCTION

Changes in the normal balance between two or more neurotransmitters can also profoundly affect brain function. For example, an interplay between serotonin and norepinephrine appears to be involved in the regulation of awake–asleep cycles. Another example deals with *Huntington's disease*. The primary problem in this inherited disease is the destruction of ACh-secreting and GABA-secreting neurons in the cerebral nuclei. The reason for this destruction is unknown. Symptoms appear as the cerebral nuclei and frontal lobes slowly degenerate. An individual with Huntington's disease has difficulty controlling movements, and intellectual abilities gradually decline. **AM** Huntington's Disease

In many cases, the importance of a specific neurotransmitter has been revealed during the search for a mechanism for the effects of administered drugs. Three examples demonstrate patterns that are emerging:

1. An extensive network of tracts delivers serotonin to nuclei and higher centers throughout the brain. Compounds that enhance the effects of serotonin produce hallucinations; for example, *lysergic acid diethylamide (LSD)* is a powerful hallucinogenic drug that activates serotonin receptors in the brain stem, hypothalamus, limbic system, and spinal cord. Compounds that inhibit serotonin production or block its action cause severe depression and anxiety. The most effective antidepressive drug now in use, *fluoxetine (Prozac®)*, slows the removal of serotonin at synapses, causing an increase in serotonin concentrations at the postsynaptic membrane.

2. Norepinephrine is another important neurotransmitter with pathways throughout the brain. Drugs that stimulate NE release cause exhilaration, and those that depress NE release cause depression. One inherited form of depression has been linked to a defective enzyme involved in NE synthesis.

3. Disturbances in dopamine transmission have been linked to several neurological disorders. Inadequate dopamine production causes the motor problems of Parkinson's disease. Excessive production of dopamine may be associated with **schizophrenia**, a psychological disorder marked by pronounced disturbances of mood, thought patterns, and behavior. Amphetamines, or "speed," stimulate dopamine secretion and, in large doses, can produce symptoms resembling those of schizophrenia. (They also affect other neurotransmitter systems, producing a variety of changes in CNS function.) **AM** Pharmacology and Drug Abuse

☐ PERSONALITY AND OTHER MYSTERIES

The basic pathways and components of the CNS have been traced, and memory acquisition, learning, and associated mental processes can now be manipulated and explored in animal experiments. But the truly remarkable human characteristics of self-awareness and personality remain phenomena without discrete anatomical foundations. That is not to say that an anatomical basis does not exist, just that self-awareness and personality are more characteristics of the brain as an integrated system than functions of any specific nucleus or region. As a result, the origins of human consciousness and personality may remain a mystery for the immediate future. But our knowledge of the anatomy and biochemistry of specific tracts, nuclei, and regions will continue to provide useful clinical information.

In recent years, significant advances have been made through the correlation of anatomical or physiological deficits with observed behavioral disorders. In some cases, such as the use of L-dopa to treat Parkinson's disease, this procedure has led to new and effective forms of treatment.

16–8 AGING AND THE NERVOUS SYSTEM

Objective

- Summarize the effects of aging on the nervous system.

The aging process affects all body systems, and the nervous system is no exception. Anatomical and physiological changes begin shortly after maturity (probably by age 30) and accumulate over time. Although an estimated 85 percent of people above age 65 lead relatively normal lives, they exhibit noticeable changes in mental performance and in CNS function. Common age-related anatomical changes in the nervous system include the following:

Alzheimer's Disease

Alzheimer's disease is a progressive disorder characterized by the loss of higher order cerebral functions. It is the most common cause of **senile dementia**, or *senility*. Symptoms may appear at 50–60 years of age or later, although the disease occasionally affects younger individuals. Alzheimer's disease has widespread impact. An estimated 2 million people in the United States, including roughly 15 percent of those over age 65, and nearly half of those over age 85 have some form of the condition, and it causes approximately 100,000 deaths each year.

The link is uncertain, but the areas containing plaques and neurofibrillary tangles are the same regions involved with memory, emotions, and intellectual function. It remains to be determined whether these deposits cause Alzheimer's disease or are secondary signs of ongoing metabolic alterations with an environmental, hereditary, or infectious basis.

Genetic factors certainly play a major role. The late-onset form of Alzheimer's disease has been traced to a gene on chromosome 19 that codes for proteins involved in cholesterol transport. Less than 5 percent of people with Alzheimer's disease have the early-onset form; these individuals develop the condition before age 50. The early-onset form of Alzheimer's disease has been linked to genes on chromosomes 21 and 14. Interestingly, the majority of individuals with *Down syndrome* develop Alzheimer's disease relatively early in life. (Down syndrome results from an extra copy of chromosome 21; we will discuss this condition further in Chapter 29.) There is no cure for Alzheimer's disease, but a few medications and supplements slow its progress in many patients. Diagnosis involves excluding metabolic and anatomical conditions that can mimic dementia, a detailed history and physical, and evaluation of mental functioning. **AM** Alzheimer's Disease

- **A Reduction in Brain Size and Weight.** This reduction results primarily from a decrease in the volume of the cerebral cortex. The brains of elderly individuals have narrower gyri and wider sulci than do those of young persons, and the subarachnoid space is larger.

- **A Reduction in the Number of Neurons.** Brain shrinkage has been linked to a loss of cortical neurons, although evidence exists that neuronal loss does not occur (at least to the same degree) in brain stem nuclei.

- **A Decrease in Blood Flow to the Brain.** With age, fatty deposits gradually accumulate in the walls of blood vessels. Just as a kink in a garden hose or a clog in a drain reduces water flow, these deposits reduce the rate of blood flow through arteries. (This process, called *arteriosclerosis*, affects arteries throughout the body; we will discuss it further in Chapter 21.) The reduction in blood flow may not cause an acute cerebral crisis, but it does increase the chances that a stroke will occur.

- **Changes in the Synaptic Organization of the Brain.** In many areas, the number of dendritic branches, spines, and interconnections appears to decrease. Synaptic connections are lost, and the rate of neurotransmitter production declines.

- **Intracellular and Extracellular Changes in CNS Neurons.** Many neurons in the brain accumulate abnormal intracellular deposits, including lipofuscin and neurofibrillary tangles. **Lipofuscin** is a granular pigment that has no known function. **Neurofibrillary tangles** are masses of neurofibrils that form dense mats inside the cell body and axon. **Plaques** are extracellular accumulations of fibrillar proteins, surrounded by abnormal dendrites and axons. Both plaques and tangles contain deposits of several peptides, primarily two forms of **amyloid β (Aβ)** protein, and appear in brain regions such as the hippocampus, specifically associated with memory processing. The significance of these histological abnormalities is unknown.

Evidence indicates that they appear in all aging brains, but when present in excess, they seem to be associated with clinical abnormalities.

These anatomical changes are linked to functional changes. In general, neural processing becomes less efficient with age. Memory consolidation typically becomes more difficult, and secondary memories, especially those of the recent past, become harder to access. The sensory systems of the elderly—notably, hearing, balance, vision, smell, and taste—become less acute. Lights must be brighter, sounds louder, and smells stronger before they are perceived. Reaction times are slowed, and reflexes—even some withdrawal reflexes—weaken or disappear. The precision of motor control decreases, and it takes longer to perform a given motor pattern than it did 20 years earlier.

For roughly 85 percent of the elderly population, these changes do not interfere with their abilities to function in society. But for as yet unknown reasons, some elderly individuals become incapacitated by progressive CNS changes. By far the most common such incapacitating condition is *Alzheimer's disease.*

✓ You are asleep. What would happen to you if your reticular activating system were suddenly stimulated?

✓ What would be the effect of a drug that substantially increases the amount of serotonin released in the brain?

✓ One of the problems associated with aging is difficulty in recalling things or even a total loss of memory. What are some possible reasons for these changes?

Answers start on page Q-1

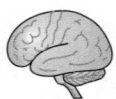

The **NERVOUS SYSTEM** in Perspective

Organ/Component	Primary Functions
Central Nervous System (CNS)	Acts as control center for nervous system: processes information; provides short-term control over activities of other systems
Brain	Performs complex integrative functions; controls both voluntary and autonomic activities
Spinal Cord	Relays information to and from brain; performs less-complex integrative functions; directs many simple involuntary activities
Peripheral Nervous System (PNS)	Links CNS with other systems and with sense organs

and performing the most complex integrative functions in the body. As part of this process, the nervous system monitors all other systems and issues commands that adjust their activities. However, the significance and impact of these commands varies greatly from one system to another. The normal functions of the muscular system simply cannot be performed without instructions from the nervous system. Yet the cardiovascular system is relatively independent—the nervous system merely coordinates and adjusts cardiovascular activities to meet the circulatory demands of other systems. In the final analysis the nervous system is like the conductor of an orchestra, keeping the rhythm and balancing the performances of each section to produce a symphony, instead of simply a very loud noise.

The figure on the opposite page diagrams the relationships between the nervous system and other physiological systems. We shall explore many of these relationships in greater detail in subsequent chapters.

INTEGRATION WITH OTHER SYSTEMS

Every moment of your life, billions of neurons in your nervous system are exchanging information across trillions of synapses

CLINICAL PATTERNS

Neural tissue is extremely delicate, and the characteristics of the extracellular environment must be kept within narrow homeostatic limits. When homeostatic regulatory mechanisms break down under the stress of genetic or environmental factors, infection, or trauma, symptoms of neurological disorders appear.

Literally hundreds of disorders affect the nervous system. These disorders can be roughly categorized into the following groups:

- *Infections*, which include diseases such as rabies (p. 390) and polio (p. 438)
- *Congenital disorders*, such as spina bifida (p. 231) and hydrocephalus (p. 471)
- *Degenerative disorders*, such as Parkinson's disease (p. 487) and Alzheimer's disease (p. 557)

- *Tumors* of neural origin (p. 395)
- *Trauma*, such as spinal cord injuries and concussions
- *Toxins*, such as heavy metals and the neurotoxins found in certain seafoods (p. 415)
- *Secondary disorders*, which are problems resulting from dysfunction in other systems; examples include strokes and several demyelination disorders (p. 395)

A *neurological examination* attempts to trace the source of the problem through evaluation of the sensory, motor, behavioral, and cognitive functions of the nervous system. Information on neurological exams, diagnostic tests, and a discussion of representative disorders in each of these classes can be found in the *Applications Manual*.

MEDIA CONNECTIONS

I. Objective: To explore the widespread importance of ion channels in regulating physiological functions.

Completion time = 15 minutes

As you have seen, ion channels determine many of the characteristics of neurons. Review the function of these channels by visiting the **InterActive Physiology CD-ROM:** Nervous System I/Ion Channels. After viewing screens 1 through 9, prepare a chart listing the various types of ion channels described and the functions of each. Compare this information with that presented in Section 12–4 of your text, and expand your chart as needed. Where else have you seen such channels used to regulate function? Add a section to your chart indicating in which systems ion channels appear and what roles they play. For help with this activity, review Chapter 10 of your text, and scan the appropriate sections of the **InterActive Physiology CD-ROM** (Muscular System/The Neuromuscular Junction and Sliding Filament Theory). As you proceed through the systems, add information to the last column of your chart as new regulatory uses for ion channels are explained.

II. Objective: To investigate the interactions of the nervous and muscular systems through a study of poliomyelitis.

Completion time = 10 minutes

The poliomyelitis virus is a small enterovirus with an affinity for motor neurons. Type 1 polio causes lesions, necrosis, and hemorrhage in the anterior horns of the spinal cord. Visit the **Companion Website**, Chapter 16 Media Connections. Click on the keyword **poliomyelitis** for a brief description of the virus and a photomicrograph of an infected spinal cord. Using this information along with Figures 13–7, 13–9, 15–10, and 15–11, suggest how polio virus infection might inhibit the normal functioning of the spinal cord. Which pathways do you think would most likely be affected? Post-polio syndrome is a progressive weakening of muscle occurring in patients 10 to 40 years after recovery from polio. Click on the keyword **Post-polio** to read what is known of this syndrome to date. What are the possible causes of post-polio myelitis?

INTEGUMENTARY SYSTEM

- Provides sensations of touch, pressure, pain, vibration, and temperature
- Hair provides some protection and insulation for skull and brain
- Protects peripheral nerves

- Controls contraction of arrector pili muscles and secretion of sweat glands

SKELETAL SYSTEM

- Provides calcium for neural function
- Protects brain and spinal cord

- Controls skeletal muscle contractions that produce bone thickening and maintenance and determine bone position

MUSCULAR SYSTEM

- Facial muscles express emotional state
- Intrinsic laryngeal muscles permit communication
- Muscle spindles provide proprioceptive sensations

- Controls skeletal muscle contractions
- Coordinates respiratory and cardiovascular activities

ENDOCRINE SYSTEM

- Many hormones affect CNS neural metabolism
- Reproductive hormones and thyroid hormone influence CNS development and behavior

---
- Controls pituitary gland and many other endocrine organs
- Secretes ADH and oxytocin

CARDIOVASCULAR SYSTEM

- Endothelial cells of capillaries maintain blood-brain barrier when stimulated by astrocytes
- Blood vessels (with ependymal cells) produce CSF

- Modifies heart rate and blood pressure
- Astrocytes stimulate maintenance of blood-brain barrier

NERVOUS SYSTEM

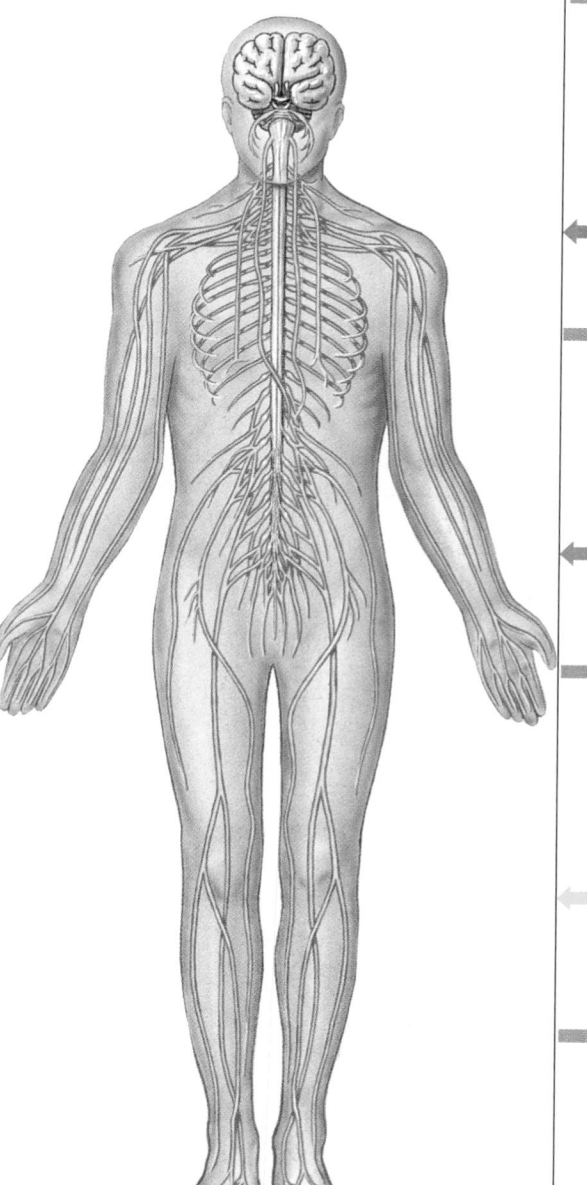

FOR ALL SYSTEMS

Monitors pressure, pain, and temperature; adjusts tissue blood flow patterns

LYMPHATIC SYSTEM

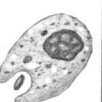

- Defends against infection and assists in tissue repairs

- Release of neurotransmitters and hormones affect sensitivity of immune response

RESPIRATORY SYSTEM

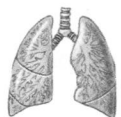

- Provides oxygen and eliminates carbon dioxide

- Controls pace and depth of respiration

DIGESTIVE SYSTEM

- Provides nutrients for energy production and neurotransmitter synthesis

---
- Regulates digestive tract movement and secretion

URINARY SYSTEM

- Eliminates metabolic wastes
- Regulates body fluid pH and electrolyte concentrations

- Adjusts renal blood pressure
- Controls urination

REPRODUCTIVE SYSTEM

- Sex hormones affect CNS development and sexual behaviors

- Controls sexual behaviors and sexual function

Chapter Review

SELECTED CLINICAL TERMINOLOGY

Terms Discussed in this Chapter

alpha-blockers: Drugs that eliminate the peripheral vasoconstriction that accompanies sympathetic stimulation. *(p. 552 and AM)*

Alzheimer's disease: A progressive disorder marked by the loss of higher-order cerebral functions. *(p. 558 and AM)*

amnesia: A temporary or permanent loss of memory. *(p. 554 and AM)*

anterograde amnesia: Loss of the ability to store new memories. *(p. 554)*

beta-blockers: Drugs that decrease the heart rate and force of contraction, lowering peripheral blood pressure. *(p. 552)*

Huntington's disease: An inherited disease marked by a progressive deterioration of mental abilities and by motor disturbances. *(p. 556 and AM)*

parasympathetic blocking agents: Drugs that target the muscarinic receptors at neuromuscular or neuroglandular junctions. *(p. 552 and AM)*

parasympathomimetic drugs: Drugs that mimic parasympathetic stimulation and increase the activity along the digestive tract. *(p. 552 and AM)*

post-traumatic amnesia (PTA): Loss of both past memories and the ability to consolidate new memories as a result of a head injury. *(p. 554)*

retrograde amnesia: Loss of memory of past events. *(p. 554)*

schizophrenia: A psychological disorder marked by pronounced disturbances of mood, thought patterns, and behavior. *(p. 556)*

senile dementia, or *senility:* A progressive loss of memory, spatial orientation, language, and personality as a consequence of aging. *(p. 557)*

sympathetic blocking agents: Drugs that bind to receptor sites, preventing a normal response to neurotransmitters or sympathomimetic drugs. *(p. 552 and AM)*

sympathomimetic drugs: Drugs that mimic the effects of sympathetic stimulation. *(p. 552 and AM)*

AM Additional Terms Discussed in the *Applications Manual*

delirium, or *acute confusional state* **(ACS)**
dementia
hypersomnia
insomnia
parasomnias
sleep apnea

STUDY OUTLINE

1. The **autonomic nervous system (ANS)** coordinates cardiovascular, respiratory, digestive, urinary, and reproductive functions. *(Figure 16–1)*

16–1 AN OVERVIEW OF THE ANS   p. 534

1. **Preganglionic neurons** in the CNS send axons to synapse on **ganglionic neurons** in **autonomic ganglia** outside the CNS. *(Figure 16–2)*

DIVISIONS OF THE ANS   p. 534

2. Preganglionic fibers from the thoracic and lumbar segments form the **sympathetic division**, or *thoracolumbar division* ("fight or flight" system), of the ANS. Preganglionic fibers leaving the brain and sacral segments form the **parasympathetic division**, or *craniosacral division* ("rest and repose" system).

16–2 THE SYMPATHETIC DIVISION   p. 536

1. The sympathetic division consists of preganglionic neurons between segments T_1 and L_2, ganglionic neurons in ganglia near the vertebral column, and specialized neurons in the adrenal glands. *(Figure 16–3)*

2. The two types of sympathetic ganglia are **sympathetic chain ganglia** *(paravertebral ganglia)* and **collateral ganglia** *(prevertebral ganglia)*. *(Figure 16–4)*

ORGANIZATION AND ANATOMY OF THE SYMPATHETIC DIVISION   p. 537

3. In spinal segments T_1–L_2, ventral roots give rise to the myelinated white ramus which, in turn, leads to the sympathetic chain ganglia *(Figures 16–4, 16–5)*.

4. **Postganglionic fibers** targeting structures in the body wall and limbs rejoin the spinal nerves and reach their destinations by way of the dorsal and ventral rami. *(Figures 16–4, 16–5)*

5. Postganglionic fibers targeting structures in the thoracic cavity form **sympathetic nerves**, which go directly to their visceral destination. Preganglionic fibers run between the sympathetic chain ganglia and interconnect them. *(Figures 16–4, 16–5)*

6. The abdominopelvic viscera receive sympathetic innervation via preganglionic fibers that synapse within collateral ganglia. The preganglionic fibers that innervate the collateral ganglia form the **splanchnic nerves**. *(Figures 16–4, 16–5)*

7. The **celiac ganglion** innervates the stomach, liver, gallbladder, pancreas, and spleen; the **superior mesenteric ganglion** innervates the small intestine and initial segments of the large intestine; and the **inferior mesenteric ganglion** innervates the kidney, urinary bladder, sex organs, and terminal portions of the large intestine. *(Figures 16–5, 16–9)*

8. Preganglionic fibers entering an adrenal gland synapse within the **adrenal medulla**. *(Figures 16–4, 16–5)*

SYMPATHETIC ACTIVATION   p. 540

9. In a crisis, the entire sympathetic division responds—an event called **sympathetic activation**. Its effects include increased alertness, a feeling of energy and euphoria, increased cardiovascular and respiratory activities, a general elevation in muscle tone, and a mobilization of energy reserves.

NEUROTRANSMITTERS AND SYMPATHETIC FUNCTION   p. 540

10. The stimulation of the sympathetic division has two distinctive results: the release of either ACh or *norepinephrine (NE)* at

specific locations, and the secretion of *epinephrine (E)* and NE into the general circulation.

11. Sympathetic ganglionic neurons end in telodendria studded with **varicosities** containing neurotransmitters. *(Figure 16–6)*

12. The two types of sympathetic receptors are **alpha receptors** and **beta receptors**.

13. Most postganglionic fibers are *adrenergic*; a few are *cholinergic* or *nitroxidergic*.

SUMMARY: THE SYMPATHETIC DIVISION   p. 541

14. The sympathetic division includes two sympathetic chain ganglia, three collateral ganglia and two adrenal medullae. *(Figure 16–10; Summary Tables 16–2, 16–3)*

Sympathetic division of the ANS: **Companion Website:** Chapter 16/Reviewing Facts & Terms/Multiple Choice.

16–3 THE PARASYMPATHETIC DIVISION   p. 542

1. The parasympathetic division includes preganglionic neurons in the brain stem and sacral segments of the spinal cord and ganglionic neurons in peripheral ganglia located within **(intramural)** or next to **(terminal)** target organs. *(Figure 16–7)*

ORGANIZATION AND ANATOMY OF THE PARASYMPATHETIC DIVISION   p. 542

2. Preganglionic fibers leave the brain as cranial nerves III, VI, IX, and X. Those leaving the sacral segments form **pelvic nerves**. *(Figure 16–8)*

PARASYMPATHETIC ACTIVATION   p. 542

3. The effects produced by the parasympathetic division center on relaxation, food processing, and energy absorption.

NEUROTRANSMITTERS AND PARASYMPATHETIC FUNCTION   p. 543

4. All parasympathetic preganglionic and postganglionic fibers release ACh. The effects are short lived, because ACh is inactivated by *acetylcholinesterase (AChE)* and by *tissue cholinesterase*.

5. Postsynaptic membranes have two types of ACh receptors. The stimulation of **muscarinic receptors** produces a longer-lasting effect than does the stimulation of **nicotinic receptors**. *(Summary Table 16–1)*

SUMMARY: THE PARASYMPATHETIC DIVISION   p. 545

6. The parasympathetic division innervates areas serviced by cranial nerves and organs in the thoracic and abdominopelvic cavities. *(Figure 16–10; Summary Tables 16–2, 16–3)*

Parasympathetic and sympathetic division comparison: **Companion Website:** Chapter 16/Critical Thinking/Essay.

16–4 INTERACTIONS BETWEEN THE SYMPATHETIC AND PARASYMPATHETIC DIVISIONS   p. 545

1. The sympathetic division has widespread influence on visceral and somatic structures.

2. The parasympathetic division innervates only visceral structures serviced by cranial nerves or lying within the abdominopelvic cavity. Organs with **dual innervation** receive input from both divisions.

ANATOMY OF DUAL INNERVATION   p. 546

3. In body cavities, the parasympathetic and sympathetic nerves intermingle to form a series of characteristic *autonomic plexuses* (nerve networks): the **cardiac**, **pulmonary**, **esophageal**, **celiac**, **inferior mesenteric**, and **hypogastric plexuses**. *(Figure 16–9)*

A COMPARISON OF THE DIVISIONS OF THE ANS   p. 546

4. Important physiological and functional differences exist between the sympathetic and parasympathetic divisions. *(Figure 16–10; Summary Tables 16–2, 16–3)*

AUTONOMIC TONE   p. 546

5. Even when stimuli are absent, autonomic motor neurons show a resting level of activation, the **autonomic tone**.

16–5 INTEGRATION AND CONTROL OF AUTONOMIC FUNCTIONS   p. 549

VISCERAL REFLEXES   p. 549

1. **Visceral reflex arcs** are the simplest function of the ANS, and can be either **long reflexes** (with interneurons) or **short reflexes** (bypassing the CNS). *(Figure 16–11)*

2. Parasympathetic reflexes govern respiration, cardiovascular functions and other visceral activities. *(Table 16–4)*

HIGHER LEVELS OF AUTONOMIC CONTROL   p. 550

3. Levels of activity in the sympathetic and parasympathetic divisions of the ANS are controlled by centers in the brain stem that deal with specific visceral functions. *(Figure 16–12)*

THE INTEGRATION OF SNS AND ANS ACTIVITIES   p. 550

4. The SNS and ANS are organized in parallel. Integration occurs at the level of the brain stem and higher centers. *(Figure 16–13; Summary Table 16–5)*

16–6 HIGHER ORDER FUNCTIONS   p. 552

1. Higher order functions (1) are performed by the cerebral cortex and involve complex interactions among areas of the cerebral cortex and between the cortex and other areas of the brain, (2) involve conscious and unconscious information processing, and (3) are subject to modification and adjustment over time.

MEMORY   p. 552

2. Memories can be classified as **short term** or **long term**.

3. The conversion from short-term to long-term memory is **memory consolidation**. *(Figure 16–14)*

4. **Amnesia** is the loss of memory as a result of disease or trauma.

CONSCIOUSNESS   p. 554

5. In **deep sleep** (*slow wave* or *non-REM sleep*), the body relaxes and cerebral cortex activity is low. In **rapid eye movement (REM) sleep**, active dreaming occurs. *(Figure 16–16)*

6. The **reticular activating system (RAS)**, a network in the reticular formation, is most important to arousal and the maintenance of consciousness. *(Figure 16–16)*

16–7 BRAIN CHEMISTRY AND BEHAVIOR   p. 556

NEUROTRANSMITTERS AND BRAIN FUNCTION   p. 556

1. Changes in the normal balance between two or more neurotransmitters can profoundly alter brain function.

PERSONALITY AND OTHER MYSTERIES   p. 556

2. Self-awareness and personality appear to be characteristic of the brain as an integrated system rather than functions of any specific component.

16–8 AGING AND THE NERVOUS SYSTEM   p. 556

1. Age-related changes in the nervous system include a reduction in brain size and weight, a reduction in the number of neurons, a decrease in blood flow to the brain, changes in the synaptic organization of the brain, and intracellular and extracellular changes in CNS neurons.

REVIEW QUESTIONS

More assessment questions are available to you on the Companion Website. You will find Matching, Multiple Choice, True/False, and other quizzes to help further your understanding of the material covered in this chapter. To access the site, go to www.aw.com/martini.

LEVEL 1 Reviewing Facts and Terms

1. All preganglionic autonomic fibers release _____ at their synaptic terminals, and the effects are always _____.
   (a) norepinephrine; inhibitory
   (b) norepinephrine; excitatory
   (c) acetylcholine; excitatory
   (d) acetylcholine; inhibitory

2. Ganglionic neurons that innervate tissues and organs in the abdominopelvic cavity are located in
   (a) sympathetic chain ganglia   (b) collateral ganglia
   (c) paravertebral ganglia   (d) lateral ganglia

3. The effect of the neurotransmitter on the target cell depends on the nature of the
   (a) neurotransmitter
   (b) receptor on the presynaptic membrane
   (c) receptor on the postsynaptic membrane
   (d) a, b, and c are correct

4. Approximately 75 percent of parasympathetic outflow is provided by the
   (a) vagus nerve   (b) sciatic nerve
   (c) glossopharyngeal nerves (d) pelvic nerves

5. The neurotransmitter at all synapses and neuromuscular or neuroglandular junctions in the parasympathetic division of the ANS is
   (a) epinephrine
   (b) norepinephrine
   (c) cyclic-AMP
   (d) acetylcholine

6. A progressive disorder characterized by the loss of higher order cerebral functions is
   (a) Parkinson's disease
   (b) parasomnia
   (c) Huntington's disease
   (d) Alzheimer's disease

7. How does the emergence of sympathetic fibers from the spinal cord differ from the emergence of parasympathetic fibers?

8. Starting in the spinal cord, trace an impulse through the sympathetic division of the ANS until it reaches a target organ in the abdominopelvic region.

9. Which three collateral ganglia serve as origins for ganglionic neurons that innervate organs or tissues in the abdominopelvic region?

10. What two distinctive results are produced by the stimulation of sympathetic ganglionic neurons?

11. What is the difference between a cholinergic synapse and an adrenergic synapse?

12. Which four pairs of cranial nerves are associated with the cranial segment of the parasympathetic division of the ANS?

13. Which four ganglia serve as origins for postganglionic fibers that deal with the control of visceral structures in the head?

14. Which six plexuses in the thoracic and abdominopelvic cavities innervate visceral organs, and what are the effects of sympathetic versus parasympathetic stimulation?

15. What are the components of a visceral reflex arc?

16. Which control centers in the brain stem influence the levels of activity in the sympathetic and parasympathetic divisions of the ANS?

17. What three characteristics are shared by higher order functions?

18. As a result of animal studies, what cellular mechanisms are thought to be involved in memory formation and storage?

19. What physiological activities distinguish non-REM sleep from REM sleep?

20. What anatomical and functional changes in the brain are linked to alterations that occur with aging?

LEVEL 2 Reviewing Concepts

21. Autonomic tone in autonomic motor neurons exists because ANS neurons
   (a) from both divisions commonly innervate the same organ
   (b) rarely innervate the same organs as somatic motor neurons do
   (c) are inactive unless stimulated by higher centers
   (d) are always active to some degree

22. Dual innervation refers to situations in which
   (a) vital organs receive instructions from both sympathetic and parasympathetic fibers
   (b) the atria and ventricles of the heart receive autonomic stimulation from the same nerves
   (c) sympathetic and parasympathetic fibers have similar effects
   (d) a, b, and c are correct

23. Damage to the hippocampus, a component of the limbic system, leads to
   (a) a loss of emotion due to forgetfulness
   (b) a loss of consciousness
   (c) a loss of long-term memory
   (d) an immediate loss of short-term memory

24. Why does sympathetic function remain intact even when the ventral roots of the cervical spinal nerves are damaged?

25. Why is the adrenal medulla considered to be a modified sympathetic ganglion?

26. How does the result of alpha-receptor stimulation differ from that of beta-receptor stimulation?

27. Why are the effects of parasympathetic stimulation quite localized and of short duration?

28. Why is autonomic tone a significant part of ANS function?

29. You are home alone at night when you hear what sounds like breaking glass. What physiological effects would this experience probably produce, and what would be their cause?

30. One patient suffers a cerebrovascular accident (CVA) in the left hemisphere; another suffers a CVA in the right hemisphere. What functions could be affected in each patient?

LEVEL 3 Critical Thinking and Clinical Applications

31. In some very severe cases of stomach ulcers, the branches of the vagus nerve (X) that lead to the stomach are surgically severed. How would this procedure help control the ulcers?

32. Mr. Martin is suffering from a condition known as ventricular tachycardia, in which his heart beats too quickly. Would an alpha-blocker or a beta-blocker help alleviate his problem? Why?

33. Phil is stung on his cheek by a wasp. Because Phil is allergic to wasp venom, his throat begins to swell and his respiratory passages constrict. Would acetylcholine or epinephrine be more helpful in relieving his symptoms? Why?

34. While studying the activity of smooth muscle in blood vessels, Shelly discovers that, when applied to muscle membrane, a molecule chemically similar to a neurotransmitter triggers an increase in intracellular calcium ions. Which neurotransmitter is the molecule mimicking, and to which receptors is it binding?

35. Researchers studying an illegal drug find that it causes individuals to see strange shapes and to hear sounds that aren't actually present. Perceptions of color are exaggerated, and individuals taking the drug report increased sexual appetites. Which neurotransmitter is this drug mimicking, and which part or parts of the brain are involved?

UNIT 3   CHAPTER   12   13   14   15   16   17   18

17

THE SPECIAL SENSES

Our knowledge of the world around us is limited to those characteristics that stimulate our sensory receptors. Although we may not realize it, our picture of the environment is incomplete. Colors we cannot distinguish guide insects to flowers; sounds we cannot hear and smells we cannot detect provide dolphins, dogs, and cats with important information about their surroundings.

What we *do* perceive varies considerably with the state of our nervous systems. For example, during sympathetic activation, we experience a heightened awareness of sensory information and hear sounds that would normally escape our notice. Yet, when concentrating on a difficult problem, we may remain unaware of relatively loud noises. Finally, our perception of any stimulus reflects activity in the cerebral cortex, and that activity can be inappropriate. In cases of phantom limb pain, for example, a person feels pain in a missing limb, and during an epileptic seizure, an individual may experience sights, sounds, or smells that have no physical basis.

Chapter 15 introduced basic principles of receptor function and sensory processing, in our discussion of the general senses and sensory pathways. We now turn our attention to the five *special senses:* olfaction, gustation, vision, equilibrium, and hearing. Although the sense organs involved are structurally more complex than those of the general senses, the same basic principles of receptor function apply. **ATLAS** Embryology Summary 13: The Development of Special Sense Organs

17–1 OLFACTION

Objectives

- Describe the sensory organs of smell, and trace the olfactory pathways to their destinations in the brain.

- Explain what is meant by olfactory discrimination, and briefly describe the physiology involved.

The sense of smell, more precisely called *olfaction*, is provided by paired **olfactory organs** (Figure 17–1●). These organs are located in the nasal cavity on either side of the nasal septum. The olfactory organs are made up of two layers: the olfactory epithelium and the lamina propria. The **olfactory epithelium** contains the **olfactory receptors**, supporting cells, and **basal cells** (*stem cells*). It covers the inferior surface of the cribriform plate, the superior portion of the perpendicular plate, and the superior nasal conchae of the ethmoid. ∞ p. 220 The underlying lamina propria consists of areolar tissue, numerous blood vessels, and nerves. This layer also contains **olfactory glands**, or *Bowman's glands*, whose secretions absorb water and form a thick, pigmented mucus. ∞ p. 133

When you draw air in through your nose, the air swirls and eddies within the nasal cavity, and this turbulence brings airborne compounds to your olfactory organs. A normal, relaxed inhalation carries a small sample (about 2 percent) of the inhaled air to the olfactory organs. Sniffing repeatedly increases the flow of air across the olfactory epithelium, intensifying the stimulation of the olfactory receptors. However, those receptors can be stimulated only by water-soluble and lipid-soluble materials that can diffuse into the overlying mucus.

◻ OLFACTORY RECEPTORS

Olfactory receptors are highly modified neurons. The exposed tip of each receptor cell forms a prominent knob that projects beyond the epithelial surface (Figure 17–1b●). The knob provides a base for up to 20 cilia that extend into the surrounding mucus. These cilia lie parallel to the epithelial surface, exposing their considerable surface area to dissolved compounds.

Olfactory reception occurs on the surfaces of the olfactory cilia as dissolved chemicals interact with receptors, called *odorant-binding proteins*, on the membrane surface. *Odorants* are chemicals that stimulate olfactory receptors. In general, odorants are small organic

molecules; the strongest smells are associated with molecules of high solubility both in water and in lipids. The receptors involved are G proteins; binding of an odorant to its receptor leads to the activation of adenylate cyclase, the enzyme that converts ATP to cyclic-AMP (cAMP). ∞ p. 419 The cAMP then opens sodium channels in the membrane, resulting in a localized depolarization. If sufficient depolarization occurs, an action potential is triggered in the axon and the information is relayed to the CNS.

Between 10 and 20 million olfactory receptors are packed into an area of roughly 5 cm^2 (0.8 in.^{2}). If we take into account the exposed ciliary surfaces, the actual sensory area probably approaches that of the entire body surface. Nevertheless, our olfactory sensitivities cannot compare with those of other vertebrates such as dogs, cats, or fishes. A German shepherd sniffing for smuggled drugs or explosives has an olfactory receptor surface 72 times greater than that of the nearby customs inspector!

◻ OLFACTORY PATHWAYS

The olfactory system is very sensitive. As few as four molecules of an odorous substance can activate an olfactory receptor. However, the activation of an afferent fiber does not guarantee a conscious awareness of the stimulus. Considerable convergence occurs along the olfactory pathway, and inhibition at the intervening synapses can prevent the sensations from reaching the *olfactory cortex* of the cerebral hemispheres. ∞ p. 488 The olfac-

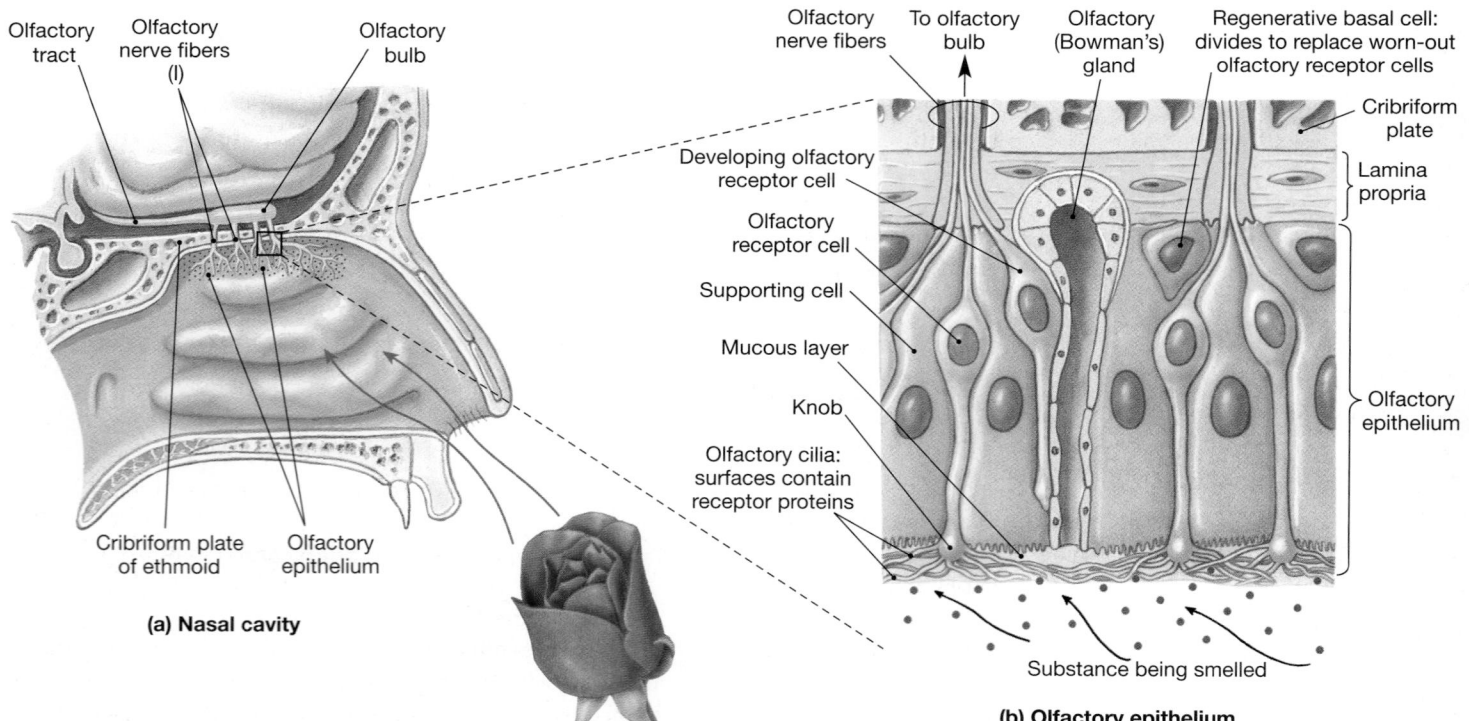

Olfactory tract
Olfactory nerve fibers (I)
Olfactory bulb
Cribriform plate of ethmoid
Olfactory epithelium

(a) Nasal cavity

Olfactory nerve fibers
To olfactory bulb
Olfactory (Bowman's) gland
Regenerative basal cell: divides to replace worn-out olfactory receptor cells
Cribriform plate
Lamina propria
Developing olfactory receptor cell
Olfactory receptor cell
Supporting cell
Mucous layer
Knob
Olfactory cilia: surfaces contain receptor proteins
Olfactory epithelium
Substance being smelled

(b) Olfactory epithelium

●**FIGURE 17–1**
The Olfactory Organs. **(a)** The olfactory organ on the left side of the nasal septum. **(b)** An olfactory receptor is a modified neuron with multiple cilia extending from its free surface.

tory receptors themselves adapt very little to a persistent stimulus. Rather, it is central adaptation which ensures that you quickly lose awareness of a new smell but retain sensitivity to others.

Axons leaving the olfactory epithelium collect into 20 or more bundles that penetrate the cribriform plate of the ethmoid bone to reach the *olfactory bulbs* of the cerebrum (Figure 17–1●), where the first synapse occurs. Efferent fibers from nuclei elsewhere in the brain also innervate neurons of the olfactory bulbs. This arrangement provides a mechanism for central adaptation or facilitation of olfactory sensitivity. Axons leaving the olfactory bulb travel along the olfactory tract to reach the olfactory cortex, the hypothalamus, and portions of the limbic system.

Olfactory stimulation is the only type of sensory information that reaches the cerebral cortex without first synapsing in the thalamus. The extensive limbic and hypothalamic connections help explain the profound emotional and behavioral responses that can be produced by certain smells. The perfume industry, which understands the practical implications of these connections, expends considerable effort to develop odors that trigger sexual responses.

☐ OLFACTORY DISCRIMINATION

The olfactory system can make subtle distinctions among 2000–4000 chemical stimuli. No apparent structural differences exist among the olfactory cells, but the epithelium as a whole contains receptor populations with distinct sensitivities. At least 50 "primary smells" are known, and it is almost impossible to describe these sensory impressions effectively. It appears likely that the CNS interprets each smell on the basis of the overall pattern of receptor activity.

Although the human olfactory organs can discriminate among many smells, acuity varies widely, depending on the nature of the odorant. Many odorants are detected in amazingly small concentrations. Beta-mercaptan, for example, is an odorant commonly added to natural gas, propane, and butane, which are otherwise odorless. Because we can smell beta-mercaptan in extremely low concentrations (a few parts per billion), its addition enables us to detect a gas leak almost at once and take steps to prevent an explosion.

Aging and Olfactory Sensitivity

The olfactory receptor population shows considerable turnover, with new receptor cells being produced by the division and differentiation of basal cells in the epithelium. This turnover is one of the few examples of neuronal replacement in adult humans. Despite the process, the total number of receptors declines with age, and the remaining receptors become less sensitive. As a result, elderly individuals have difficulty detecting odors in low concentrations. This decline in the number of receptors accounts for Grandmother's tendency to use excessive quantities of perfume and explains why Grandfather's aftershave seems so overdone: They must apply more to be able to smell it.

17–2 GUSTATION

Objectives

■ Describe the sensory organs of taste, and trace the gustatory pathways to their destinations in the brain.

■ Explain what is meant by gustatory discrimination, and briefly describe the physiology involved.

Gustation, or taste, provides information about the foods and liquids we consume. **Taste receptors**, or *gustatory* (GUS-tator-ē) *receptors*, are distributed over the superior surface of the tongue and adjacent portions of the pharynx and larynx. By the time we reach adulthood, the taste receptors on the pharynx and larynx have decreased in importance and abundance. The most important taste receptors are on the tongue, although a few remain on the epiglottis and adjacent areas of the pharynx. Taste receptors and specialized epithelial cells form sensory structures called **taste buds**. An adult has about 3000 taste buds.

The superior surface of the tongue bears epithelial projections called *lingual papillae* (pa-PIL-lē; *papilla*, a nipple-shaped mound). The human tongue bears three types of lingual papillae: (1) **filiform** (*filum*, thread) **papillae**, (2) **fungiform** (*fungus*, mushroom) **papillae**, and (3) **circumvallate** (sir-kum-VAL-āt) **papillae** (*circum-*, around + *vallum*, wall). The distribution of these lingual papillae varies by region (Figure 17–2a●). Filiform papillae provide friction that helps the tongue to move objects around in the mouth, but do not contain taste buds. Each small fungiform papilla contains about five taste buds; each large circumvallate papilla contains as many as 100 taste buds. The circumvallate papillae form a V near the posterior margin of the tongue.

☐ TASTE RECEPTORS

Taste buds are recessed into the surrounding epithelium, isolated from the relatively unprocessed contents of the mouth. Each taste bud (Figure 17–2b,c●) contains about 40 slender, spindle-shaped cells. There are at least four different types of cells within a taste bud. **Basal cells** appear to be stem cells. These cells divide to produce daughter cells that mature in stages corresponding to the other types of cell. The cells of the last stage are called **gustatory cells**. Each gustatory cell extends slender microvilli, sometimes called *taste hairs*, into the surrounding fluids through the **taste pore**, a narrow opening.

Despite this relatively protected position, it's still a hard life: A typical gustatory cell survives for only about 10 days before it is replaced. Although everyone agrees that gustatory cells are taste receptors, it remains uncertain as to whether the cells at earlier stages of development provide taste information as well. (Cells at all three stages are innervated by sensory neurons.)

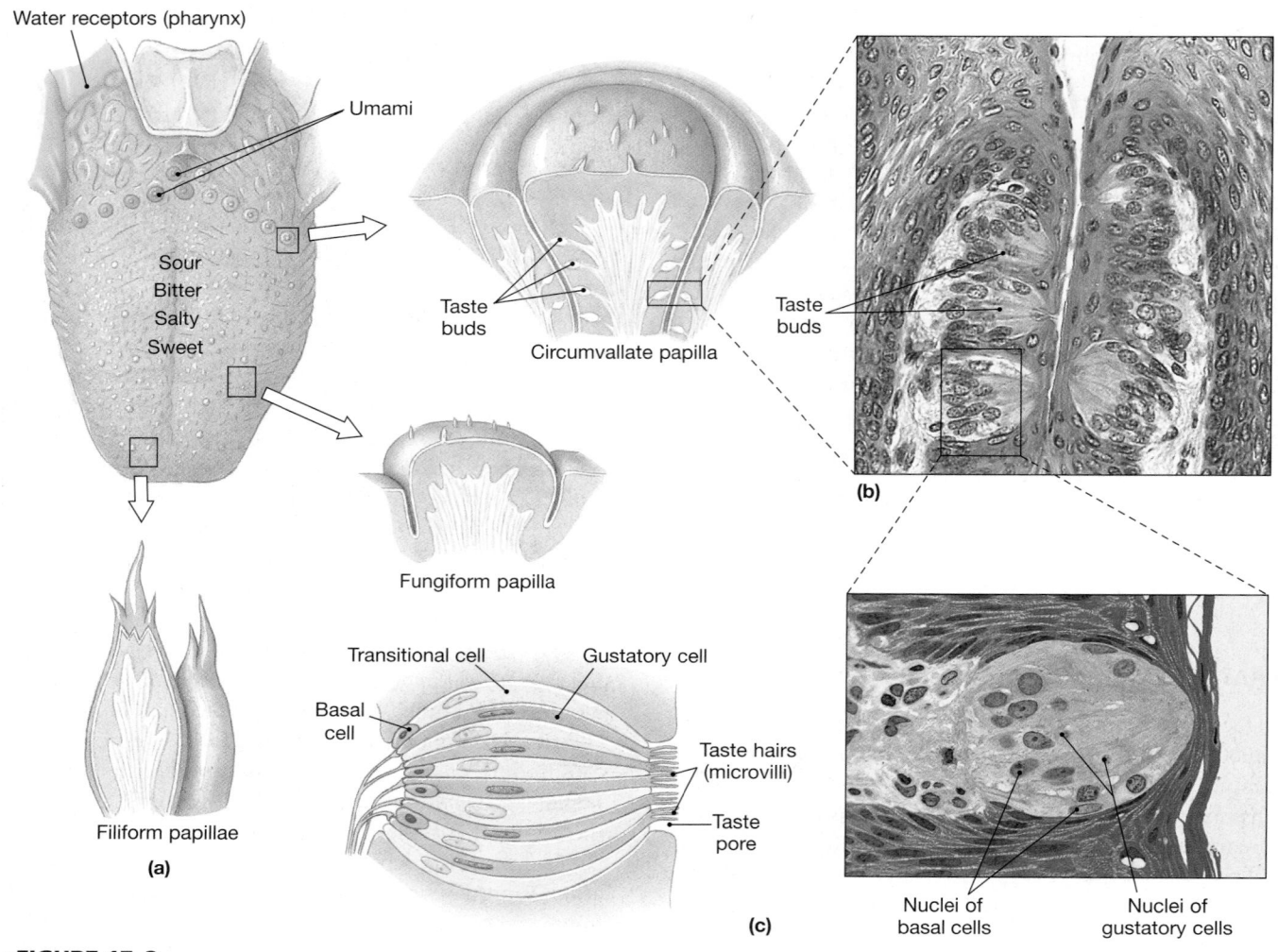

●**FIGURE 17–2**
Gustatory Reception. (a) Taste receptors are located in taste buds, which form pockets in the epithelium of fungiform or circumvallate papillae. (b) Taste buds in a circumvallate papilla. (c) A taste bud, showing receptor (gustatory) cells and supporting cells. The diagrammatic view shows details of the taste pore that are not visible in the micrograph.

◻ GUSTATORY PATHWAYS

Taste buds are monitored by cranial nerves VII (facial), IX (glossopharyngeal), and X (vagus). The facial nerve monitors all the taste buds located on the anterior two-thirds of the tongue, from the tip to the line of circumvallate papillae. The circumvallate papillae and the posterior one-third of the tongue are innervated by the glossopharyngeal nerve. The vagus nerve innervates taste buds scattered on the surface of the epiglottis. The sensory afferents carried by these cranial nerves synapse in the solitary nucleus of the medulla oblongata, and the axons of the postsynaptic neurons enter the medial lemniscus. There, the neurons join axons that carry somatic sensory information on touch, pressure, and proprioception. After another synapse in the thalamus, the information is projected to the appropriate portions of the primary sensory cortex.

A conscious perception of taste is produced as the information received from the taste buds is correlated with other sensory data. Information about the texture of food, along with taste-related sensations such as "peppery" or "burning hot," is provided by sensory afferents in the trigeminal nerve (V). In addition, the level of stimulation from the olfactory receptors plays an overwhelming role in taste perception. Thus, you are several thousand times more sensitive to "tastes" when your olfactory organs are fully functional. By contrast, when you have a cold and your nose is stuffed up, airborne molecules cannot reach your olfactory receptors, so meals taste dull and unappealing. This reduction in taste perception occurs even though the taste buds may be responding normally.

◻ GUSTATORY DISCRIMINATION

You are probably already familiar with the four **primary taste sensations**: sweet, salty, sour, and bitter. There is some behavioral evidence for differences in sensitivity to tastes along the axis of the tongue, with greatest sensitivity to salty–sweet anteriorly and sour–bitter posteriorly. However, there are no differences in the structure of the taste buds, and taste buds in all portions of the tongue provide all four primary taste sensations.

At least two additional tastes have been discovered in humans:

- **Umami.** Umami (oo-MAH-mē) is a pleasant taste that is characteristic of beef broth, and chicken broth, and parmesan cheese. This taste is produced by receptors sensitive to the presence of amino acids, especially glutamate, small peptides, and nucleotides. The distribution of these receptors is not known in detail, but they are present in taste buds of the circumvallate papillae.

- **Water.** Most people say that water has no flavor. However, research on humans and other vertebrates has demonstrated the presence of **water receptors**, especially in the pharynx. The sensory output of these receptors is processed in the hypothalamus and affects several systems that deal with water balance and the regulation of blood volume. For example, a mouthful of distilled water held for 20 minutes will inhibit ADH secretion and promote water loss at the kidneys.

The mechanism behind gustatory reception resembles that of olfaction. Dissolved chemicals contacting the taste hairs bind to receptor proteins of the gustatory cell. The different tastes involve different receptor mechanisms. Salt receptors and sour receptors are chemically gated ion channels whose stimulation produces depolarization of the cell. Receptors responding to stimuli that produce sweet, bitter, and umami sensations are G proteins called **gustducins**—protein complexes that use second messengers to produce their effects. The end result of taste receptor stimulation is the release of neurotransmitters by the receptor cell. The dendrites of the sensory afferents are tightly wrapped by folds of the receptor cell membrane, and neurotransmitter release leads to the generation of action potentials in the afferent fiber. Taste receptors adapt slowly, but central adaptation quickly reduces your sensitivity to a new taste.

The threshold for receptor stimulation varies for each of the primary taste sensations, and the taste receptors respond more readily to unpleasant than to pleasant stimuli. For example, we are almost a thousand times more sensitive to acids, which taste sour, than to either sweet or salty chemicals, and we are a hundred times more sensitive to bitter compounds than to acids. This sensitivity has survival value, because acids can damage the mucous membranes of the mouth and pharynx, and many potent biological toxins produce an extremely bitter taste.

Taste sensitivity differs significantly among individuals. Many conditions related to taste sensitivity are inherited. The best-known example involves sensitivity to the compound *phenylthiourea*, also known as *phenylthiocarbamide*, or **PTC**. Roughly 70 percent of Caucasians can taste this substance; the other 30 percent are unable to detect it.

Aging and Gustatory Sensitivity

Our tasting abilities change with age. We begin life with more than 10,000 taste buds, but the number begins declining dramatically by age 50. The sensory loss becomes especially significant because, as we have already noted, aging individuals also experience a decline in the number of olfactory receptors. As a result, many elderly people find that their food tastes bland and unappetizing, whereas children tend to find the same foods too spicy.

✓ When you first enter the A & P lab for dissection, you are very aware of the odor of preservatives. By the end of the lab period, the smell doesn't seem to be nearly as strong. Why?

✓ If you completely dry the surface of your tongue and then place salt or sugar crystals on it, you can't taste them. Why not?

✓ Your grandfather can't understand why foods he used to enjoy just don't taste the same anymore. Explain it to him.

Answers start on page Q-1

17–3 VISION

Objectives

- Identify the accessory structures of the eye, and explain their functions.
- Describe the internal structures of the eye, and explain their functions.
- Explain how we are able to distinguish colors and perceive depth.
- Explain how light stimulates the production of nerve impulses, and trace the visual pathways to their destinations in the brain.

We rely more on vision than on any other special sense. Our visual receptors are contained in the eyes, elaborate structures that enable us not only to detect light, but also to create detailed visual images. We shall begin our discussion of these fascinating organs by considering the *accessory structures* of the eye, which provide protection, lubrication, and support.

ACCESSORY STRUCTURES OF THE EYE

The **accessory structures** of the eye include the eyelids, the superficial epithelium of the eye, and the structures associated with the production, secretion, and removal of tears. Figure 17–3a● shows the superficial anatomy of the eye and its accessory structures.

Eyelids

The eyelids, or **palpebrae** (pal-PĒ-brē), are a continuation of the skin. The eyelids act like windshield wipers: Their continual blinking movements keep the surface of the eye lubricated and free from dust and debris. The eyelids can also close firmly to protect the delicate surface of the eye. The **palpebral fissure** is the gap that separates the free margins of the upper and lower eyelids. The two eyelids are connected, however, at the **medial canthus** (KAN-thus) and the **lateral canthus** (Figure 17–3●). The **eyelashes**, along the margins of the eyelids, are very robust hairs that help prevent foreign matter (including insects) from reaching the surface of the eye.

The eyelashes are associated with unusually large sebaceous glands. Along the inner margin of the lid, modified sebaceous glands called **tarsal glands**, or *Meibomian* (mī-BŌ-mē-an) *glands*, secrete a lipid-rich product that helps keep the eyelids from sticking together. At the medial canthus, the **lacrimal caruncle** (KAR-ung-kul), a mass of soft tissue, contains glands producing the thick secretions that contribute to the gritty deposits that sometimes appear after a good night's sleep. These various glands are subject to occasional invasion and infection by bacteria. A *chalazion* (kah-LĀ-zē-on; small lump), or cyst, generally results from the infection of a tarsal gland. An infection in a

sebaceous gland of one of the eyelashes, a tarsal gland, or one of the many sweat glands that open to the surface between the follicles produces a painful localized swelling known as a *sty*.

The skin covering the visible surface of the eyelid is very thin. Deep to the skin lie the muscle fibers of the *orbicularis oculi* and *levator palpebrae superioris muscles*. ∞ p. 344 These skeletal muscles are responsible for closing the eyelids and raising the upper eyelid, respectively.

The epithelium covering the inner surfaces of the eyelids and the outer surface of the eye is called the **conjunctiva** (kon-junk-TĪ-vuh). It is a mucous membrane covered by a specialized stratified squamous epithelium. The **palpebral conjunctiva** covers the inner surface of the eyelids, and the **ocular conjunctiva**, or *bulbar conjunctiva*, covers the anterior surface of the eye (Figure 17–3b●). The ocular conjunctiva extends to the edges of the **cornea** (KOR-nē-uh), a transparent part of the outer fibrous layer of the eye. The cornea is covered by a very delicate squamous *corneal epithelium*, five to seven cells thick, that is continuous with the ocular conjunctiva. A constant supply of fluid washes over the surface of the eyeball, keeping the ocular conjunctiva and cornea moist and clean. Goblet cells in the epithelium assist the accessory glands in providing a superficial lubricant that prevents friction and drying of the opposing conjunctival surfaces.

Conjunctivitis, or pinkeye, results from damage to, and irritation of, the conjunctival surface. The most obvious symptom, redness, is due to the dilation of blood vessels deep to the conjunctival epithelium. This condition may be caused by pathogenic infection or by physical or chemical irritation of the conjunctival surface.

●FIGURE 17–3
External Features and Accessory Structures of the Eye. **(a)** Gross and superficial anatomies of the accessory structures. **(b)** The organization of the lacrimal apparatus.

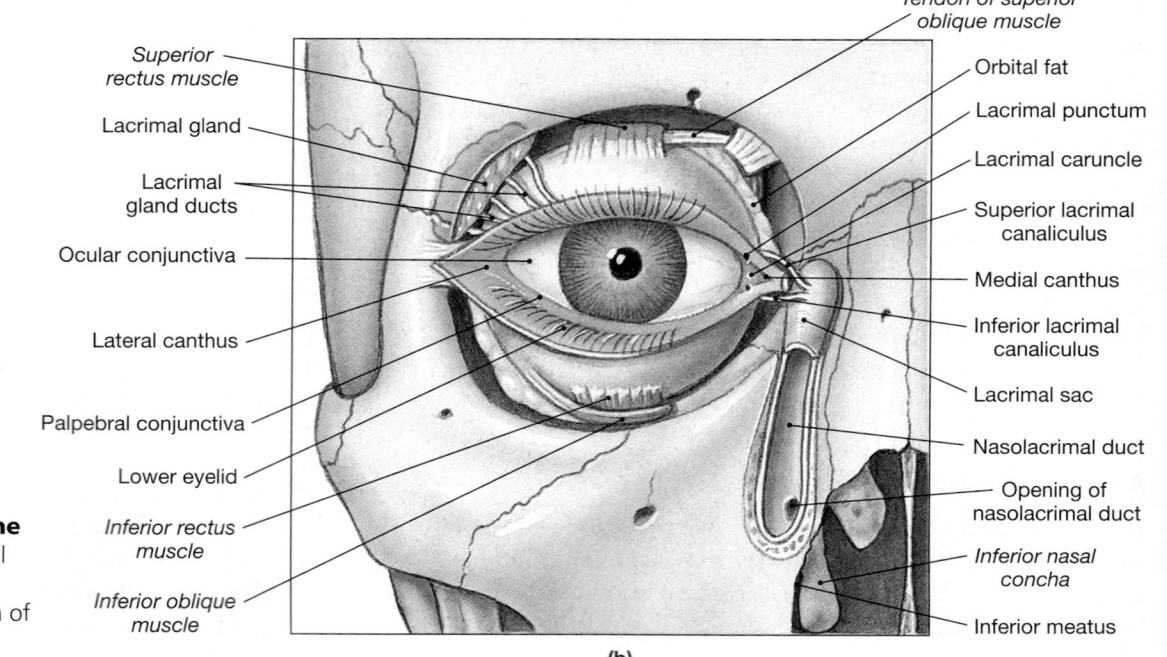

The Lacrimal Apparatus

A constant flow of tears keeps conjunctival surfaces moist and clean. Tears reduce friction, remove debris, prevent bacterial infection, and provide nutrients and oxygen to portions of the conjunctival epithelium. The **lacrimal apparatus** produces, distributes, and removes tears. The lacrimal apparatus of each eye consists of (1) a *lacrimal gland* with associated ducts, (2) *lacrimal canaliculi*, (3) a *lacrimal sac*, and (4) a *nasolacrimal duct*.

The pocket created where the palpebral conjunctiva becomes continuous with the ocular conjunctiva is known as the **fornix** (Figure 17–4a●). The lateral portion of the superior fornix receives 10–12 ducts from the **lacrimal gland**, or tear gland (Figure 17–3b●). This gland is about the size and shape of an almond, measuring roughly 12–20 mm (0.5–0.75 in.). It nestles within a depression in the frontal bone, just inside the orbit and superior and lateral to the eyeball. p. 217 The lacrimal gland normally provides the key ingredients and most of the volume of the tears that bathe the conjunctival surfaces. The nutrient and oxygen demands of the corneal cells are supplied by diffusion from the lacrimal secretions, which are watery and slightly alkaline. They contain the enzyme **lysozyme**, which attacks bacteria, and antibodies that attack pathogens before they enter the body.

The lacrimal gland produces about 1 ml of tears each day. Once the lacrimal secretions have reached the ocular surface, they mix with the products of accessory glands and the oily secretions of the tarsal glands. The result is a superficial "oil slick" that assists in lubrication and slows evaporation.

Blinking sweeps the tears across the ocular surface, and they accumulate at the medial canthus in an area known as the *lacrimal lake* (*lacus lacrimalis*), or "lake of tears." The lacrimal lake covers the **lacrimal caruncle**, which bulges anteriorly. The **lacrimal puncta** (singular, *punctum*), two small pores, drain the lacrimal lake. They empty into the **lacrimal canaliculi**, small canals that in turn lead to the **lacrimal sac** (Figure 17–3b●),

which nestles within the lacrimal sulcus of the orbit. p. 222 From the inferior portion of the lacrimal sac, the **nasolacrimal duct** passes through the *nasolacrimal canal*, formed by the lacrimal bone and the maxillary bone. The nasolacrimal duct delivers tears to the nasal cavity on that side. The duct empties into the *inferior meatus*, a narrow passageway inferior and lateral to the inferior nasal concha. When tears are produced in large quantities, as when a person cries, the situation resembles a major flood. Tears rushing into the nasal cavity produce a runny nose, and if the lacrimal puncta can't provide enough drainage, the lacrimal lake overflows and tears stream across the face.

□ THE EYE

The eyes are extremely sophisticated visual instruments—more versatile and adaptable than the most expensive cameras, yet light, compact, and durable. Each eye is a slightly irregular spheroid with an average diameter of 24 mm (almost 1 in.), a little smaller than a ping-pong ball, and a weight of about 8 g (0.28 oz). Within the orbit, the eyeball shares space with the extrinsic eye muscles, the lacrimal gland, and the cranial nerves and blood vessels that supply the eye and adjacent portions of the orbit and face. **Orbital fat** cushions and insulates the eye (Figures 17–3b and 17–4c●).

The wall of the eye contains three distinct layers, or *tunics* (Figure 17–4b●): (1) an outer *fibrous tunic*, (2) an intermediate *vascular tunic*, and (3) an inner *neural tunic*. The visual receptors, or *photoreceptors*, are located in the neural tunic. The eyeball is hollow; its interior can be divided into two cavities (Figure 17–4c●). The large **posterior cavity** is also called the *vitreous chamber*, because it contains the gelatinous *vitreous body*. The smaller **anterior cavity** is subdivided into two *chambers*, anterior and posterior. The shape of the eye is stabilized in part by the vitreous body and the clear *aqueous humor*, which fills the anterior cavity.

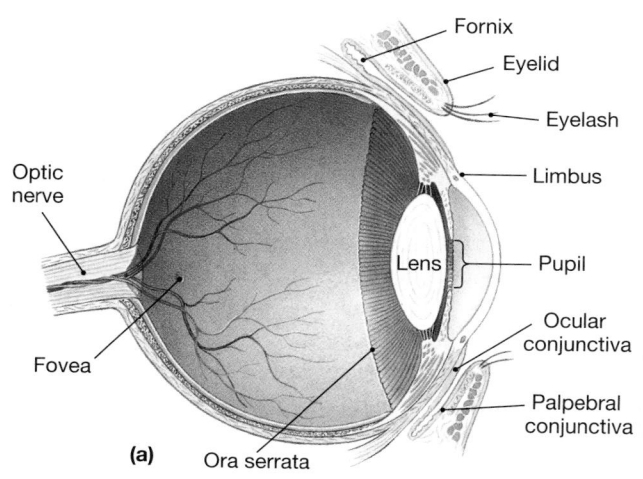
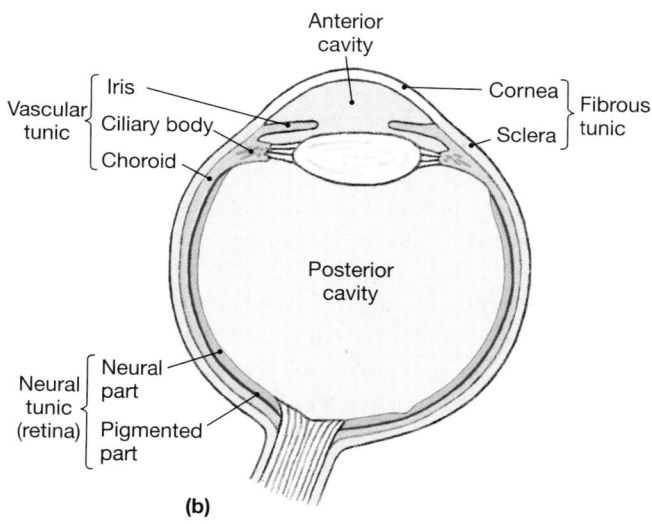

●**FIGURE 17–4**
The Sectional Anatomy of the Eye. **(a)** A sagittal section through the left eye, showing the position of the fornix. **(b)** A sagittal section, showing the three layers, or tunics, of the eye.

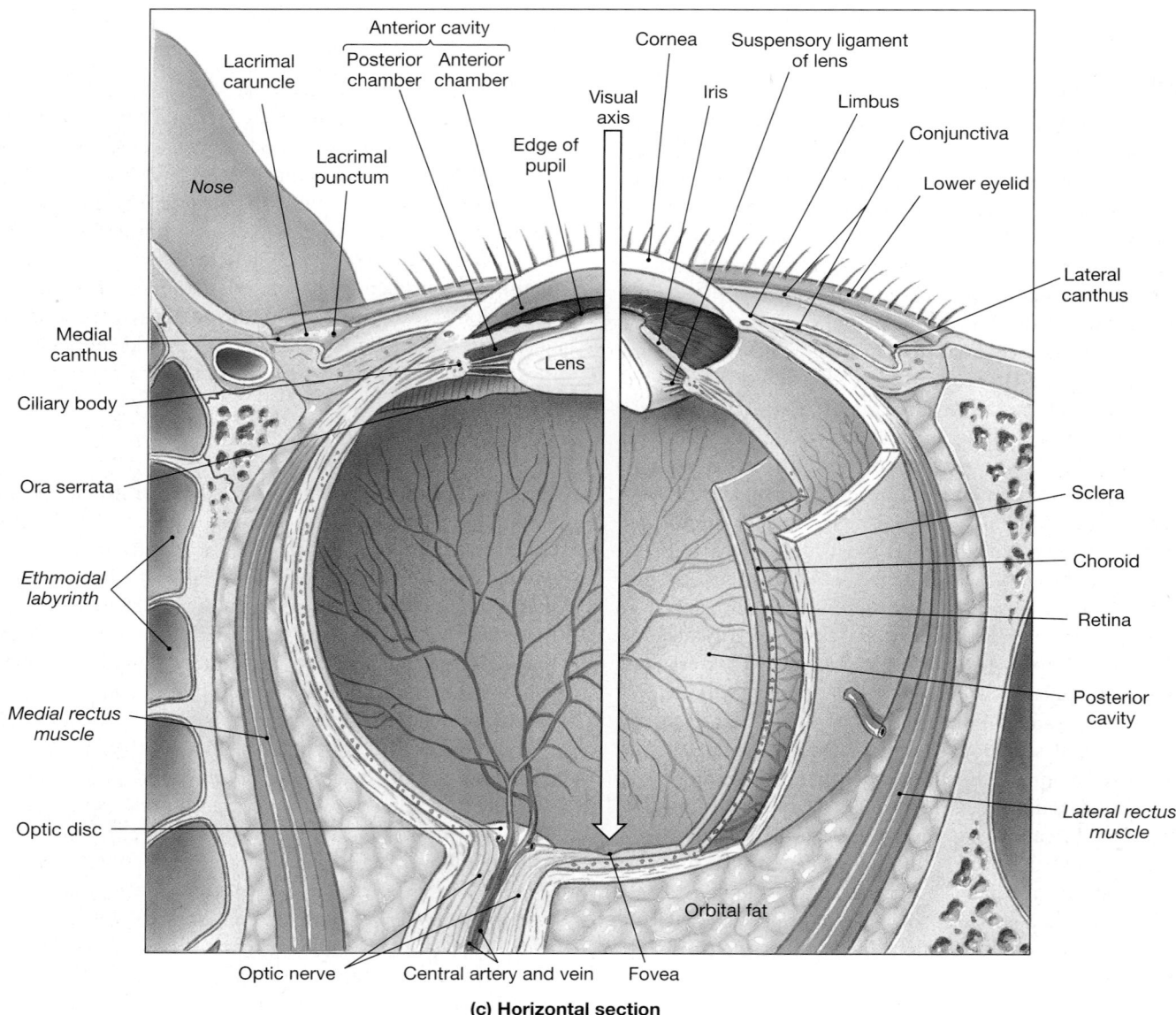

(c) Horizontal section

●**FIGURE 17–4**
The Sectional Anatomy of the Eye. (*continued*) (c) A horizontal section of the right eye. **ATLAS** Plates 3.2, 3.3

The Fibrous Tunic

The **fibrous tunic**, the outermost layer of the eye, consists of the *sclera* (SKLER-uh) and the *cornea*. The fibrous tunic (1) provides mechanical support and some degree of physical protection, (2) serves as an attachment site for the extrinsic eye muscles, and (3) contains structures that assist in the focusing process.

Most of the ocular surface is covered by the **sclera** (Figure 17–4b,c●), or "white of the eye," which consists of a dense fibrous connective tissue containing both collagen and elastic fibers. This layer is thickest over the posterior surface of the eye, near the exit of the optic nerve, and thinnest over the anterior surface. The six extrinsic eye muscles insert on the sclera, blending their collagen fibers with those of the fibrous tunic. ⚯ p. 345

The surface of the sclera contains small blood vessels and nerves that penetrate the sclera to reach internal structures. The network of small vessels interior to the ocular conjunctiva gener-

ally does not carry enough blood to lend an obvious color to the sclera, but on close inspection, the vessels are visible as red lines against the white background of collagen fibers.

The transparent cornea is structurally continuous with the sclera; the **limbus** (Figures 17–3a, 17–4a,c●) is the border between the two. Deep to the delicate corneal epithelium, the cornea consists primarily of a dense matrix containing multiple layers of collagen fibers, organized so as not to interfere with the passage of light. The cornea has no blood vessels; the superficial epithelial cells must obtain oxygen and nutrients from the tears that flow across their free surfaces. The cornea also has numerous free nerve endings, and it is the most sensitive portion of the eye.

⚕ Corneal damage may cause blindness even though the functional components of the eye—including the photoreceptors—are perfectly normal. The cornea has a very restricted ability to repair itself, so corneal injuries must be treated immediately to prevent serious vision losses. Restoring vision after corneal scarring generally requires the replacement of the cornea through a corneal transplant. Corneal replacement is probably the most common form of transplant surgery. Such transplants can be performed between unrelated individuals, because there are no blood vessels to carry white blood cells, which attack foreign tissues, into the area. Corneal grafts are obtained from the eyes of donors who have died from illness or accident. For best results, the tissues must be removed within five hours after the donor's death.

The Vascular Tunic (Uvea)

The **vascular tunic**, or **uvea**, contains numerous blood vessels, lymphatic vessels, and the intrinsic eye muscles (Figure 17–4b,c•). The functions of this middle layer include (1) providing a route for blood vessels and lymphatics that supply tissues of the eye; (2) regulating the amount of light that enters the eye; (3) secreting and reabsorbing the *aqueous humor* that circulates within the eye; and (4) controlling the shape of the *lens*, an essential part of the focusing process. The vascular tunic includes the *iris*, the *ciliary body*, and the *choroid*.

THE IRIS The **iris**, which we can see through the transparent corneal surface, contains blood vessels, pigment cells, and two layers of smooth muscle fibers (Figure 17–5•). When these muscles contract, they change the diameter of the **pupil**, or central opening of the iris. One group of smooth muscle fibers, the **pupillary constrictor muscles**, forms a series of concentric circles around the pupil. When these sphincter muscles contract, the diameter of the pupil decreases. A second group of smooth muscles, the **pupillary dilator muscles**, extends radially away from the edge of the pupil. Contraction of these muscles enlarges the pupil. Both muscle groups are controlled by the autonomic nervous system. For example, parasympathetic activation in re-

sponse to bright light causes the pupils to constrict (the *consensual light reflex*), and sympathetic activation in response to dim light causes the pupils to dilate . ⊂⊃ pp. 504, 550

The body of the iris consists of a highly vascular, pigmented, loose connective tissue. The anterior surface has no epithelial covering; instead, it has an incomplete layer of fibroblasts and melanocytes. Melanocytes are also scattered within the body of the iris. The posterior surface is covered by a pigmented epithelium that is part of the neural tunic and contains melanin granules. Your eye color is determined by the density and distribution of melanocytes on the anterior surface and interior of the iris, as well as by the density of the pigmented epithelium. When the connective tissue of the iris contains few melanocytes, light passes through it and bounces off the pigmented epithelium. The eye then appears blue. Individuals with green, brown, or black eyes have increasing numbers of melanocytes in the body and surface of the iris. The eyes of human albinos appear a very pale gray or blue gray.

THE CILIARY BODY At its periphery, the iris attaches to the anterior portion of the **ciliary body**, a thickened region that begins deep to the junction between the cornea and the sclera. The ciliary body extends posteriorly to the level of the **ora serrata** (Ō-ra ser-RA-tuh; serrated mouth), the serrated anterior edge of the thick, inner portion of the neural tunic (Figure 17–4a,c•). The bulk of the ciliary body consists of the **ciliary muscle**, a smooth muscular ring that projects into the interior of the eye. The epithelium of this muscle is thrown into numerous folds called **ciliary processes**. The **suspensory ligaments** of the lens attach to the tips of these processes. The connective-tissue fibers of these ligaments hold the lens posterior to the iris and centered on the pupil. As a result, any light passing through the pupil and headed for the photoreceptors will also pass through the lens.

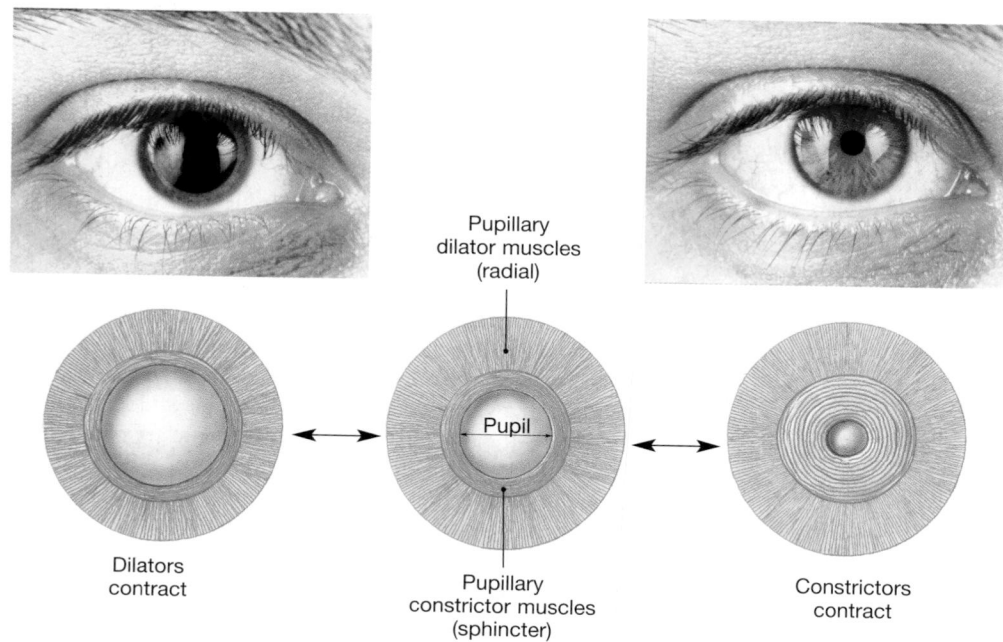

Pupillary dilator muscles (radial)

Pupil

Dilators contract

Pupillary constrictor muscles (sphincter)

Constrictors contract

•FIGURE 17–5
The Pupillary Muscles

THE CHOROID The **choroid** is a vascular layer that separates the fibrous and neural tunics posterior to the ora serrata (Figure 17–4c●). Covered by the sclera and attached to the outermost layer of the retina, the choroid contains an extensive capillary network that delivers oxygen and nutrients to the retina. The choroid also contains melanocytes, which are especially numerous near the sclera.

The Neural Tunic

The **neural tunic**, or **retina**, is the innermost layer of the eye. It consists of a thin, outer layer, called the *pigmented part*, and a thick inner layer, called the *neural part*. The pigmented part of the retina absorbs light that passes through the neural part, preventing light from bouncing back through the neural part and producing visual "echos." The pigment cells also have important biochemical interactions with the retina's light receptors, which are located in the neural part of the retina. In addition to light receptors, the neural part of the retina contains supporting cells and neurons that perform preliminary processing and integration of visual information, as well as blood vessels supplying the tissues that line the posterior cavity.

The two layers of the retina are normally very close together, but not tightly interconnected. The pigmented part of the retina continues over the ciliary body and iris; the neural part extends anteriorly only as far as the ora serrata. The neural part of the retina thus forms a cup that establishes the posterior and lateral boundaries of the posterior cavity (Figure 17–4b,c●).

A *retinopathy* is a disease of the retina. **Diabetic retinopathy** develops in many individuals with *diabetes mellitus*, an endocrine disorder that interferes primarily with glucose metabolism. Many systems are affected by diabetes, but serious cardiovascular problems are particularly common. Diabetic retinopathy which develops over a period of years, results from the degeneration, rupture, and excessive new growth of blood vessels of the retina. All of these processes can lead to detachment of the retina from the choroid. Visual acuity is gradually lost through damage to photoreceptors (which are deprived of oxygen and nutrients), leakage of blood into the posterior chamber, and the overgrowth of blood vessels. Laser therapy can seal leaking vessels and block new vessel growth. The posterior chamber can be drained and the cloudy fluid replaced by a suitably clear substitute. However, these are only temporary fixes that must be periodically repeated, since they fail to correct the underlying metabolic problems.

ORGANIZATION OF THE RETINA In sectional view, the retina is seen to contain several layers of cells (Figure 17–6a●). The outermost layer, closest to the pigmented part of the retina, contains the **photoreceptors**, or cells that detect light.

The eye has two types of photoreceptors: rods and cones. **Rods** do not discriminate among colors of light. Highly sensitive to light, they enable us to see in dimly lit rooms, at twilight, and in pale moonlight. **Cones** provide us with color vision. Three types of cones are present, and their stimulation in various combinations provides the perception of different colors. Cones give

us sharper, clearer images than rods do, but cones require more intense light. If you sit outside at sunset, you will probably be able to tell when your visual system shifts from cone-based vision (clear images in full color) to rod-based vision (relatively grainy images in black and white).

Rods and cones are not evenly distributed across the outer surface of the retina. Approximately 125 million rods form a broad band around the periphery of the retina. Roughly 6 million cones span the retina's posterior surface. Most of these cones are concentrated in the area where a visual image arrives after it passes through the cornea and lens. This region, which is known as the **macula lutea** (LOO-tē-uh; yellow spot), contains no rods. The highest concentration of cones occurs at the center of the macula lutea, an area called the **fovea** (FŌ-vē-uh; shallow depression), or *fovea centralis* (Figure 17–6c●). The fovea is the site of sharpest vision: When you look directly at an object, its image falls on this portion of the retina. A line drawn from the center of that object through the center of the lens to the fovea establishes the **visual axis** of the eye (Figure 17–4c●).

You are probably already aware of the visual consequences of this distribution. When you look directly at an object, you are focusing its image on your fovea, the center of color vision. You see a very good image as long as there is enough light to stimulate the cones. In very dim light, cones cannot function. For example, when you try to stare at a dim star, you are unable to see it. But if you look a little to one side rather than directly at the star, you will see it quite clearly. Shifting your gaze moves the image of the star from the fovea, where it does not provide enough light to stimulate the cones, to the periphery, where it can affect the more sensitive rods.

Rods and cones synapse with roughly 6 million neurons called **bipolar cells** (Figure 17–6a●), which in turn synapse within the layer of neurons called **ganglion cells** adjacent to the posterior cavity. A network of **horizontal cells** extends across the outer portion of the retina at the level of the synapses between photoreceptors and bipolar cells. A comparable layer of **amacrine** (AM-a-krin) **cells** occurs where bipolar cells synapse with ganglion cells. Horizontal and amacrine cells can facilitate or inhibit communication between photoreceptors and ganglion cells, thereby altering the sensitivity of the retina. The effect is comparable to adjusting the contrast on a television set. These cells play an important role in the eye's adjustment to dim or brightly lit environments.

Photoreceptors are entirely dependent on the diffusion of oxygen and nutrients from blood vessels in the choroid. In a **detached retina**, the neural part of the retina becomes separated from the pigmented part. This condition can result from a variety of factors, including a sudden blow to the eye. Unless the two parts of the neural tunic are reattached, the photoreceptors will degenerate and vision will be lost. The reattachment is generally performed by "welding" the two layers together by means of laser beams focused through the cornea. These beams heat the layers, thereby fusing them together at several points around the retina. However, the procedure destroys the photoreceptors and other cells at the "welds," producing permanent blind spots.

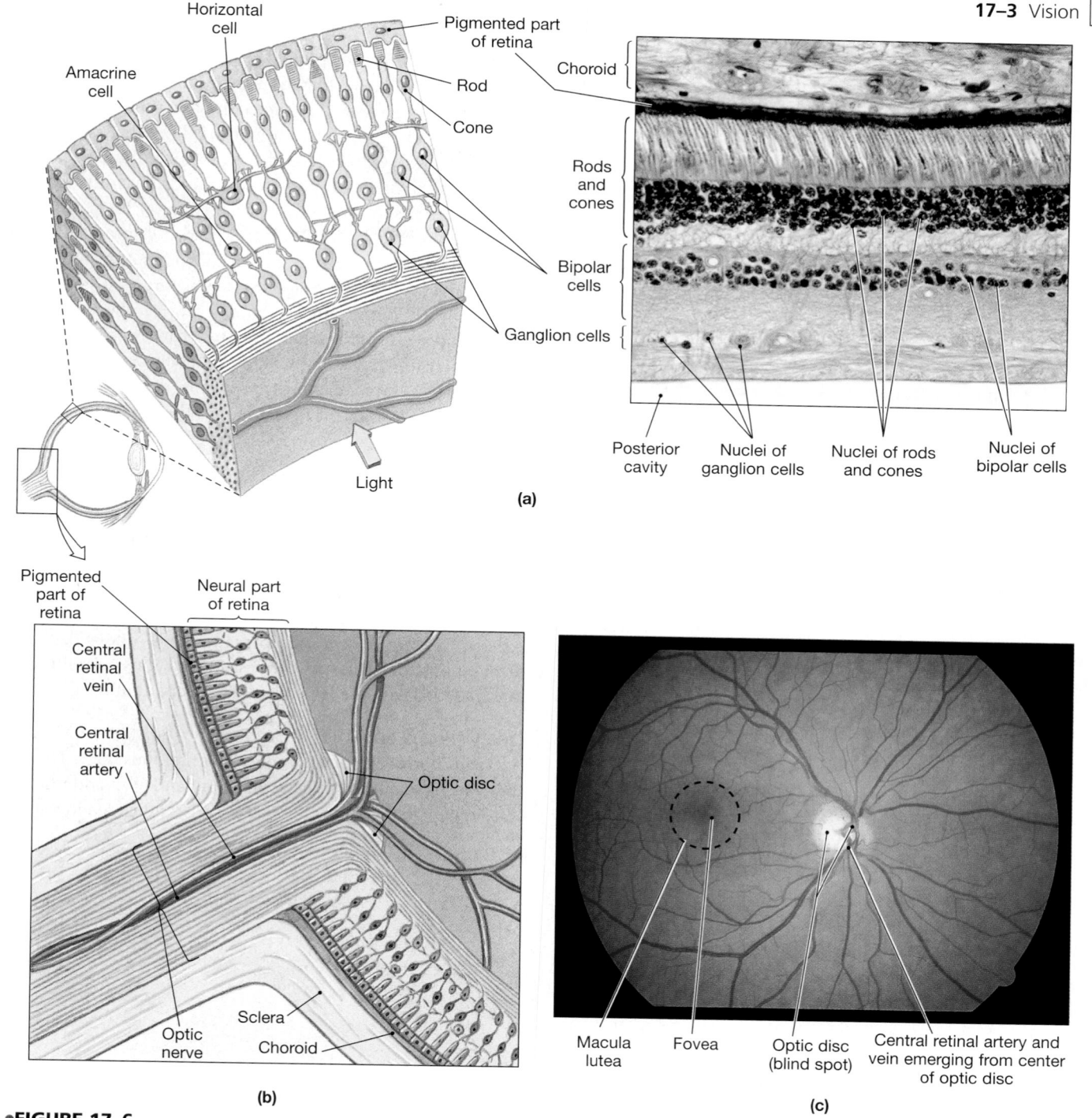

(a)

(b)

(c)

●**FIGURE 17–6**
The Organization of the Retina. (a) The cellular organization of the retina. The photoreceptors are closest to the choroid, rather than near the posterior cavity (vitreous chamber). (b) The optic disc in diagrammatic horizontal section. (c) A photograph of the retina as seen through the pupil.

THE OPTIC DISC Axons from an estimated 1 million ganglion cells converge on the **optic disc**, a circular region just medial to the fovea. The optic disc is the origin of the optic nerve (N II). From this point, the axons turn, penetrate the wall of the eye, and proceed toward the diencephalon (Figure 17–6b●). The *central retinal artery* and *central retinal vein*, which supply the retina, pass through the center of the optic nerve and emerge on the surface of the optic disc (Figure 17–6b,c●). The optic disc has no pho-

toreceptors or other structures that are typical of the rest of the retina. Because light striking this area goes unnoticed, the optic disc is commonly called the **blind spot**. You do not notice a blank spot in your field of vision, primarily because involuntary eye movements keep the visual image moving and allow your brain to fill in the missing information. However, a simple experiment, shown in Figure 17–7●, will demonstrate the presence and location of the blind spot in your field of vision.

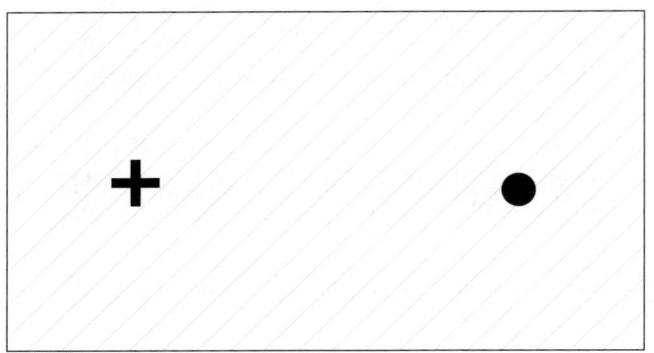

●FIGURE 17–7
The Optic Disc.   Close your left eye and stare at the cross with your right eye, keeping the cross in the center of your field of vision. Begin with the page a few inches away from your eye, and gradually increase the distance. The dot will disappear when its image falls on the blind spot, at your optic disc. To check the blind spot in your left eye, close your right eye and repeat the sequence while you stare at the dot.

The Chambers of the Eye

As we learned earlier, the ciliary body and lens divide the interior of the eye into a large posterior cavity, or vitreous chamber, and a smaller anterior cavity. The anterior cavity is subdivided into the **anterior chamber**, which extends from the cornea to the iris, and a **posterior chamber**, between the iris and the ciliary body and lens. The anterior and posterior chambers are filled with the fluid *aqueous humor*. The posterior cavity is filled with a gelatinous substance known as the *vitreous body*, or *vitreous humor*.

AQUEOUS HUMOR **Aqueous humor** is a fluid that circulates within the anterior cavity, passing from the posterior to the anterior

chamber through the pupil (Figure 17–8●). This fluid, which forms through active secretion by epithelial cells of the ciliary body's ciliary processes, is secreted into the posterior chamber at a rate of $1-2 \mu$l per minute. The epithelial cells regulate its composition, which resembles that of cerebrospinal fluid. Because aqueous humor circulates, it provides an important route for nutrient and waste transport, in addition to forming a fluid cushion.

The eye is filled with fluid, and fluid pressure in the aqueous humor helps retain the eye's shape. Fluid pressure also stabilizes the position of the retina, pressing the neural part against the pigmented part. In effect, the aqueous humor acts like the air inside a balloon. The eye's **intraocular pressure** can be measured in the anterior chamber, where the fluid pushes against the inner surface of the cornea. Intraocular pressure is most often checked by bouncing a tiny blast of air off the surface of the eye and measuring the deflection produced. Normal intraocular pressure ranges from 12 to 21 mm Hg.

Aqueous humor returns to the circulation from the anterior chamber. After filtering through a network of connective tissues located near the base of the iris, aqueous humor enters the **canal of Schlemm,** or *scleral venous sinus*, a passageway that extends completely around the eye at the level of the limbus. Collecting channels deliver the aqueous humor from this canal to veins in the sclera. The rate of removal normally keeps pace with the rate of generation at the ciliary processes, and aqueous humor is removed and recycled within a few hours of its formation.

THE VITREOUS BODY The posterior cavity of the eye contains the **vitreous body** (or *vitreous humor*), a gelatinous mass. The vitreous body helps stabilize the shape of the eye and gives additional physical support to the retina. Specialized cells embedded in the vitreous

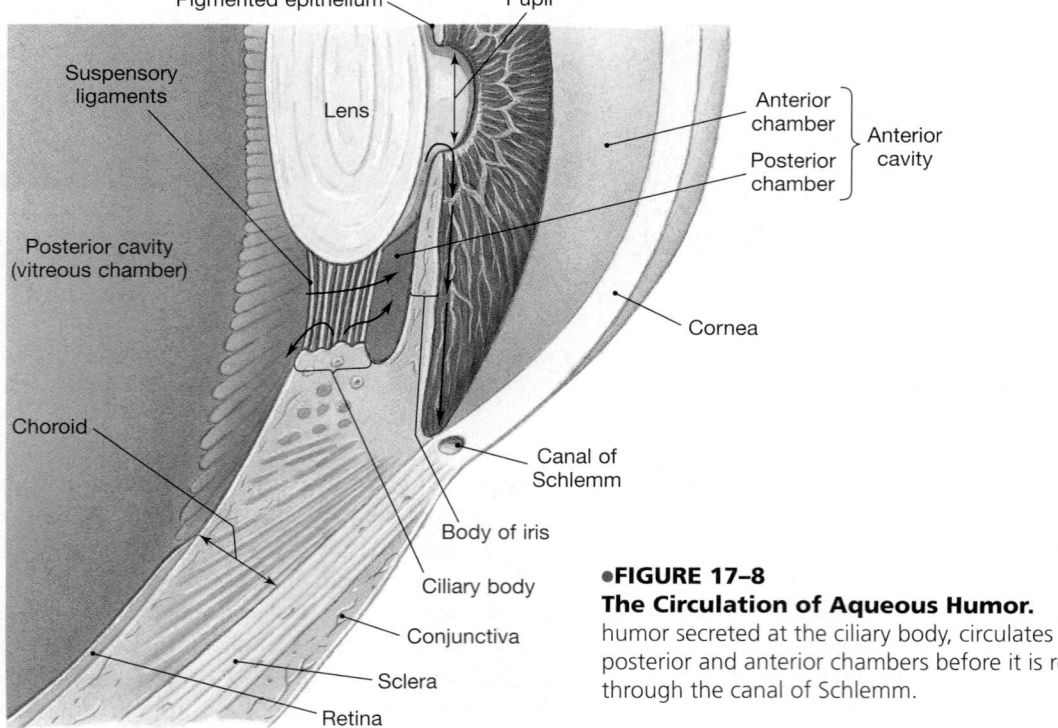

●FIGURE 17–8
The Circulation of Aqueous Humor.   Aqueous humor secreted at the ciliary body, circulates through the posterior and anterior chambers before it is reabsorbed through the canal of Schlemm.

Glaucoma

✝ If aqueous humor cannot enter the canal of Schlemm, **glaucoma** results. Although drainage is impaired, the production of aqueous humor continues, so the intraocular pressure begins to rise. Because the sclera is a fibrous coat, it cannot expand significantly. The increasing pressure therefore begins to distort soft tissues within the eye. The tough sclera cannot enlarge like an inflating balloon, but it does have one weak point-the optic disc, where the optic nerve penetrates the wall of the eye. Because the optic nerve must penetrate all three tunics, it is not wrapped in connective tissue. When intraocular pressures have risen to roughly twice normal levels, the distortion of the nerve fibers begins to block the propagation of action potentials, and vision begins to deteriorate. If this condition is not corrected, blindness eventually results.

Glaucoma affects roughly 2 percent of the population over age 35. In most cases, the primary factors responsible cannot be determined. Because glaucoma is a relatively common condition, with over 2 million cases in the United States alone, most eye exams include a test of intraocular pressure. Glaucoma may be treated by the application of drugs that constrict the pupil and tense the edge of the iris, making the surface more permeable to aqueous humor. Surgical correction in-

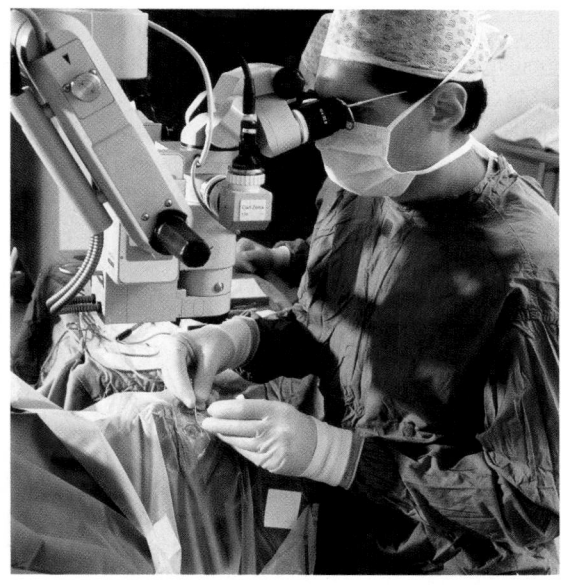

volves perforating the wall of the anterior chamber to encourage drainage. This procedure is now performed by laser surgery on an outpatient basis.

---

body produce the collagen fibers and proteoglycans that account for the gelatinous consistency of this mass. Unlike the aqueous humor, the vitreous body is formed during development of the eye and is not replaced. Aqueous humor produced in the posterior chamber freely diffuses through the vitreous body and across the surface of the retina.

The Lens

The **lens** lies posterior to the cornea, held in place by the suspensory ligaments that originate on the ciliary body of the choroid (Figures 17–4b, p. 571, and 17–8●). The primary function of the lens is to focus the visual image on the photoreceptors. The lens does so by changing its shape.

The lens consists of concentric layers of cells that are precisely organized. A dense fibrous capsule covers the entire lens. Many of the capsular fibers are elastic. Unless an outside force is applied, they will contract and make the lens spherical. Around the edges of the lens, the capsular fibers intermingle with those of the suspensory ligaments. The cells in the interior of the lens are called **lens fibers**. These highly specialized cells have lost their nuclei and other organelles. They are slender and elongate and are filled with transparent proteins called **crystallins**, which are responsible for both the clarity and the focusing power of the lens. Crystallins are extremely stable proteins that remain intact and functional for a lifetime without the need for replacement.

✝ The transparency of the lens depends on a precise combination of structural and biochemical characteristics. When

that balance becomes disturbed, the lens loses its transparency; the abnormal lens is known as a **cataract**. Cataracts can result from injuries, radiation, or reaction to drugs, but **senile cataracts**, a natural consequence of aging, are the most common form.

Over time, the lens turns yellowish and eventually begins to lose its transparency. As the lens becomes "cloudy," the individual needs brighter and brighter light for reading, and visual clarity begins to fade. If the lens becomes completely opaque, the person will be functionally blind, even though the photoreceptors are normal. Surgical procedures involve removal of the lens, either intact or in pieces, after it has been shattered with high-frequency sound. The missing lens is replaced by an artificial substitute, and vision is then fine-tuned with glasses or contact lenses.

REFRACTION The retina has about 130 million photoreceptors, each monitoring a specific location. A visual image results from the processing of information from all the receptors. The eye is often compared to a camera. To provide useful information, the lens of the eye, like a camera lens, must focus the arriving image. To say that an image is "in focus" means that the rays of light arriving from an object strike the sensitive surface of the retina (or film) in precise order so as to form a miniature image of the object. If the rays are not perfectly focused, the image is blurry. Focusing normally occurs in two steps, as light passes through first the cornea and then the lens.

Light is **refracted**, or bent, when it passes from one medium to a medium with a different density. You can demonstrate this effect by sticking a pencil into a glass of water. Because refraction occurs as

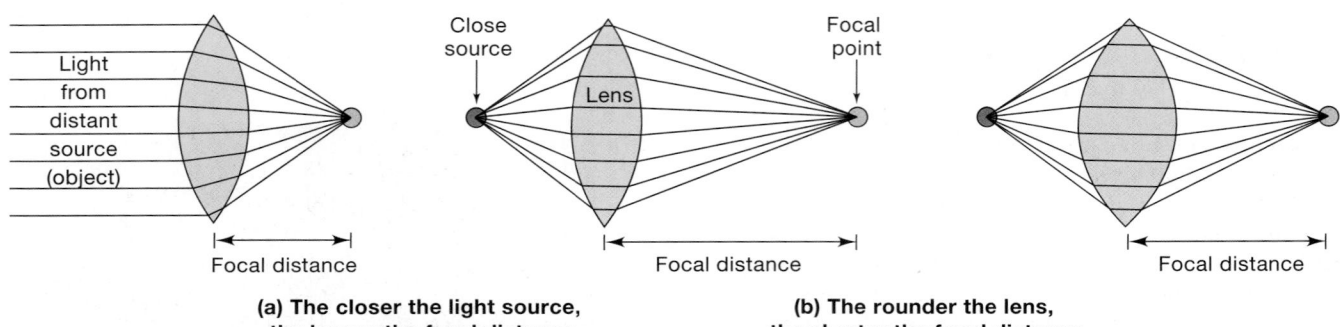

**(a) The closer the light source,
the longer the focal distance**

**(b) The rounder the lens,
the shorter the focal distance**

•**FIGURE 17–9**
Image Formation.   Light rays from a source are refracted when they reach the lens of the eye. The rays are then focused onto a single focal point. **(a)** The focal distance increases as the object nears the lens. **(b)** A rounder lens has a shorter focal distance than a flatter lens does.

the light passes into the air from the much denser water, the shaft of the pencil appears to bend sharply at the air–water interface.

In the human eye, the greatest amount of refraction occurs when light passes from the air into the corneal tissues, which have a density close to that of water. When you open your eyes under water, you cannot see clearly, because the air–water interface has been eliminated; light is passing from one watery medium to another.

Additional refraction takes place when the light passes from the aqueous humor into the relatively dense lens. The lens provides the extra refraction you need to focus the light rays from an object toward a **focal point**—a specific point of intersection on the retina. The distance between the center of the lens and its focal point is the **focal distance** of the lens. Whether in the eye or in a camera, the focal distance is determined by the following two factors:

1. *The Distance from the Object to the Lens.* The focal distance increases as an object moves closer to the lens (Figure 17–9a•).
2. *The Shape of the Lens.* The rounder the lens, the more refraction occurs, so a very round lens has a shorter focal distance than a flatter one (Figure 17–9b•).

ACCOMMODATION Using a camera, we can focus an image by moving the lens toward or away from the film. This method of focusing cannot work in our eyes, because the distance from the lens to the macula lutea cannot change. We focus images on the retina by changing the shape of the lens to keep the focal length constant, a process called **accommodation**. During accommodation, the lens becomes rounder to focus the image of a nearby object on the retina; the lens flattens when we focus on a distant object.

The lens is held in place by the suspensory ligaments that originate at the ciliary body. Smooth muscle fibers in the ciliary body act like sphincter muscles. When the ciliary muscle contracts, the ciliary body moves toward the lens, thereby reducing the tension in the suspensory ligaments. The elastic capsule then pulls the lens into a more spherical shape that increases the refractive power of the lens, enabling it to bring light from nearby objects into focus on the retina (Figure 17–10a•). When the ciliary muscle relaxes, the suspensory ligaments pull at the circumference of the lens, making the lens flatter (Figure 17–10b•).

The greatest amount of refraction is required to view objects that are very close to the lens. The inner limit of clear vision, known as the *near point of vision*, is determined by the degree of elasticity in the lens. Children can usually focus on something 7–9 cm from the eye, but over time the lens tends to become stiffer and less responsive. A young adult can usually focus on objects 15–20 cm away. As aging proceeds, this distance gradually increases; the near point at age 60 is typically about 83 cm.

If light passing through the cornea and lens fails to refract properly, the visual image will be distorted. In the condition called **astigmatism**, the degree of curvature in the cornea or lens varies from one axis to another. For example, the cornea may be more strongly curved vertically than horizontally. Astigmatism can be corrected by glasses or special contact lenses. Minor astigmatism is very common; the image distortion may be so minimal that people are unaware of the condition.

IMAGE REVERSAL Thus far, we have considered light that originates at a single point, either near or far from the viewer. An

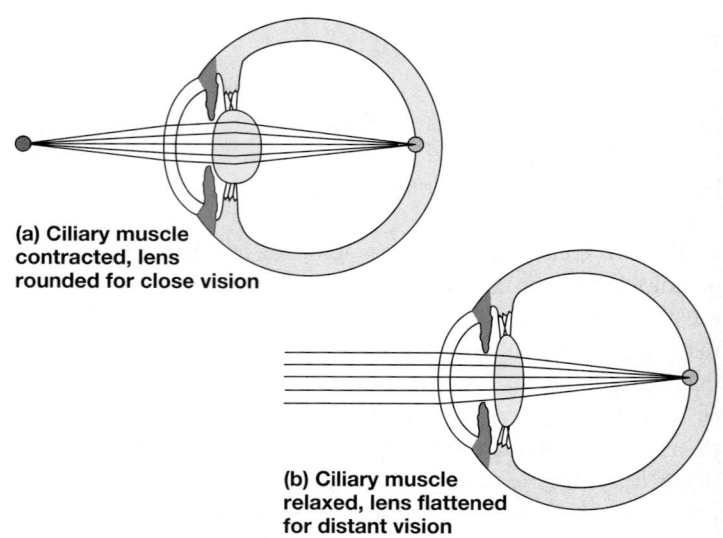

**(a) Ciliary muscle
contracted, lens
rounded for close vision**

**(b) Ciliary muscle
relaxed, lens flattened
for distant vision**

•**FIGURE 17–10**
Accommodation.   For the eye to form a sharp image, the focal distance must equal the distance between the center of the lens and the retina. **(a)** When the ciliary muscle contracts, the ligaments allow the lens to round. **(b)** When the ciliary muscle relaxes, the suspensory ligaments pull against the margins of the lens and flatten it.

Accommodation Problems

In the healthy eye, when the ciliary muscle is relaxed and the lens is flattened, a distant image will be focused on the retina's surface (Figure 17–11a●). This condition is called **emmetropia** (*emmetro-*, proper), or normal vision.

Figure 17–11b,c● diagrams two common problems with the accommodation mechanism. If the eyeball is too deep or the resting curvature of the lens is too great, the image of a distant object will form in front of the retina (Figure 17–11b●). The individual will see distant objects as blurry and out of focus. Vision at close range will be normal, because the lens will be able to round as needed to focus the image on the retina. Such individuals are said to be *nearsighted*. Their condition is more formally termed **myopia** (*myein*, to shut + *ops*, eye). Myopia can be treated by placing a *diverging* lens in front of the eye (Figure 17–11d●). This shape, typical of the lenses used in prescription corrective glasses, spreads the light rays apart as if the object were closer to the viewer.

If the eyeball is too shallow or the lens is too flat, **hyperopia** results (Figure 17–11c●). The ciliary muscle must contract to focus even a distant object on the retina, and at close range the lens cannot provide enough refraction to focus an image on the retina. Individuals with this problem are said to be *farsighted*, because they can see distant objects most clearly. Older individuals become farsighted as their lenses lose elasticity; this form of hyperopia is called **presbyo-**pia (*presbys*, old man). Hyperopia can be corrected by placing a *converging* lens in front of the eye. This lens provides the additional refraction needed to bring nearby objects into focus (Figure 17–11e●).

Variable success at correcting myopia and hyperopia has been achieved by surgically reshaping the cornea to alter its refractive powers. In **radial keratotomy**, several corneal incisions flatten the surface. Some scarring and localized corneal weakness may occur, and only two-thirds of patients are satisfied with the results. Many ophthalmological surgeons have expressed concerns that the scarring which develops after radial keratotomy creates weak points in the cornea that may increase the chances of a dangerous corneal rupture.

A newer procedure is **photorefractive keratectomy (PRK)**, in which a computer-guided laser shapes the cornea to exact specifications. Tissue is removed only to a depth of $10-20\ \mu m$—no more than about 10 percent of the cornea's thickness. The entire procedure can be done in less than a minute. Each year, an estimated 100,000 people undergo PRK therapy in the United States. Corneal scarring is rare, and hundreds of thousands of people have had this procedure. However, many still need reading glasses, and both immediate and long-term visual problems can occur. Advances in laser design and improvements in accuracy may make the procedure more popular during the next decade.

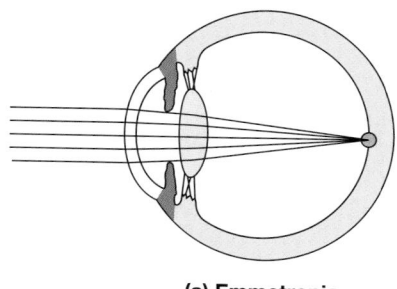

(a) Emmetropia

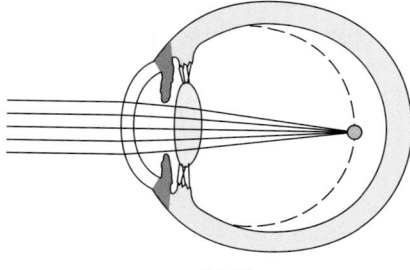

(b) Myopia

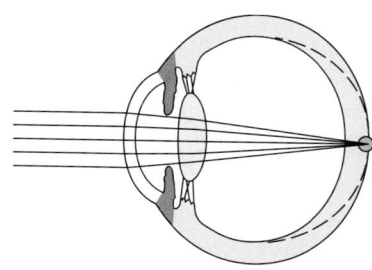

(c) Hyperopia

●**FIGURE 17–11**
Visual Abnormalities. **(a)** In normal vision, or emmetropia, the lens focuses the visual image on the retina. Common problems with the accommodation mechanism involve **(b)** myopia, an inability to lengthen the focal distance enough to focus the image of a distant object on the retina, and **(c)** hyperopia, an inability to shorten the focal distance adequately for nearby objects. These conditions can be corrected by placing appropriately shaped lenses in front of the eyes. **(d)** A diverging lens is used to correct myopia, and **(e)** a converging lens is used to correct hyperopia.

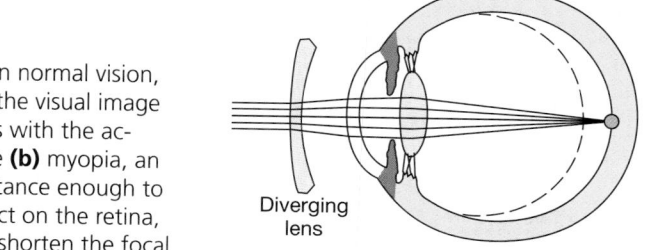

Diverging lens

(d) Myopia (corrected)

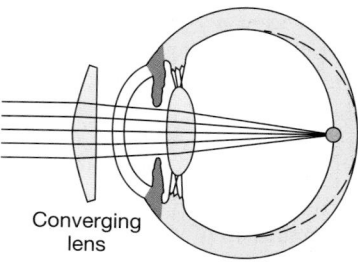

Converging lens

(e) Hyperopia (corrected)

object in view, however, is a complex light source, that must be treated as a large number of individual points. Light from each point is focused on the retina as indicated in Figure 17–12a,b•. The result is the creation of a miniature image of the original, but the image arrives upside down and backwards.

To understand why the image is reversed in this fashion, consider Figure 17–12c•, a sagittal section through an eye that is looking at a telephone pole. The image of the top of the pole lands at the bottom of the retina, and the image of the bottom hits the top of the retina. Now consider Figure 17–12d•, a horizontal section through an eye that is looking at a picket fence. The image of the left edge of the fence falls on the right side of the retina, and the image of the right edge falls on the left side of the retina. The brain compensates for this image reversal, and we are not consciously aware of any difference between the orientation of the image on the retina and that of the object.

VISUAL ACUITY Clarity of vision, or **visual acuity**, is rated against a "normal" standard. A person whose vision is rated 20/20 can see details at a distance of 20 feet as clearly as would an individual with normal vision. Vision noted as 20/15 is better than average, because at 20 feet the person is able to see details that would be clear to a normal eye only at a distance of 15 feet. Conversely, a person with 20/30 vision must be 20 feet from an object to discern details that a person with normal vision could make out at a distance of 30 feet.

When visual acuity falls below 20/200, even with the help of glasses or contact lenses, the individual is considered to be legally blind. There are probably fewer than 400,000 legally blind people in the United States; more than half are over 65 years of age. The term *blindness* implies a total absence of vision due to damage to the eyes or to the optic pathways. Common causes of blindness include diabetes mellitus, cataracts, glaucoma, corneal scarring, detachment of the retina, accidental injuries, and hereditary factors that are as yet poorly understood.

Abnormal blind spots, or **scotomas** (skō-TŌ-muhz), may appear in the field of vision at positions other than at the optic disc. Scotomas are permanent abnormalities that are fixed in position. They may result from a compression of the optic nerve, damage to photoreceptors, or central damage along the visual pathway. *Floaters*, small spots that drift across the field of vision, are generally temporary phenomena that result from blood cells or cellular debris in the vitreous body. They can be detected by staring at a blank wall or a white sheet of paper.

✓ Which layer of the eye would be affected first by the inadequate production of tears?

✓ When the lens of your eye is very round, are you looking at an object that is close to you or far from you?

✓ As Renee enters a dark room, most of the available light becomes focused on the fovea of her eye. Will she be able to see very clearly?

✓ How would a blockage of the canal of Schlemm affect your vision?

Answers start on page Q-1

Review the structures of the eye and the pathway of light by visiting the **Companion Website**: Chapter 17/Tutorials.

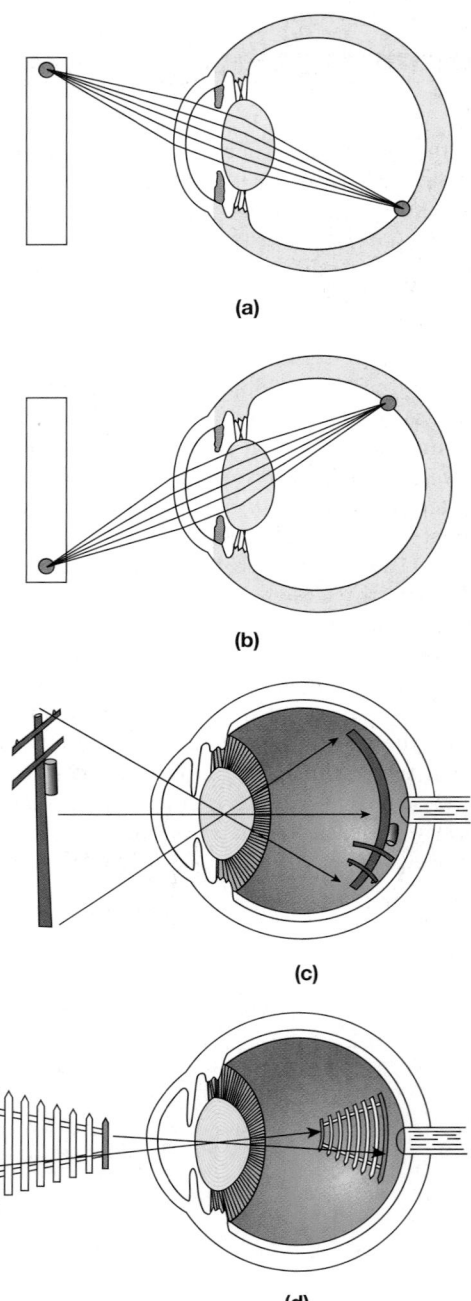

•FIGURE 17–12
Image Formation. **(a,b)** Light from each portion of an object is focused on a different part of the retina. The resulting image arrives **(c,d)** upside down and backwards.

◻ VISUAL PHYSIOLOGY

The rods and cones of the retina are called *photoreceptors* because they detect *photons*, basic units of visible light. A form of *radiant energy*, light energy is radiated in waves that have a characteristic *wavelength* (distance between wave peaks).

Our eyes are sensitive to wavelengths of 700–400 nm, the spectrum of visible light. This spectrum, seen in a rainbow, can be remembered by the acronym ROY G. BIV (*Red, Orange, Yellow,*

Green, Blue, Indigo, Violet). Photons of red light carry the least energy, and those from the violet portion of the spectrum contain the most. Rods provide the central nervous system with information about the presence or absence of photons, without regard to their wavelength. Cones provide information about the wavelength of arriving photons, giving us a perception of color.

Anatomy of Rods and Cones

Figure 17–13a● compares the structures of rods and cones. The elongated **outer segment** of a photoreceptor contains hundreds to thousands of flattened membranous plates, or **discs**. The names *rod* and *cone* refer to the outer segment's shape. In a rod, each disc is an independent entity, and the outer segment forms

an elongated cylinder. In a cone, the discs are infoldings of the cell membrane and the outer segment tapers to a blunt tip.

A narrow connecting stalk attaches the outer segment to the **inner segment**, a region that contains all the usual cellular organelles. The inner segment makes synaptic contact with other cells, and it is here that neurotransmitters are released.

VISUAL PIGMENTS The discs of the outer segment in both rods and cones contain special organic compounds called **visual pigments**. The absorption of photons by visual pigments is the first key step in the process of *photoreception*—the detection of light. Visual pigments are derivatives of the compound **rhodopsin** (rō-DOP-sin), or *visual purple*, the visual pigment found in rods. Rhodopsin consists of a protein, **opsin**, bound to the pig-

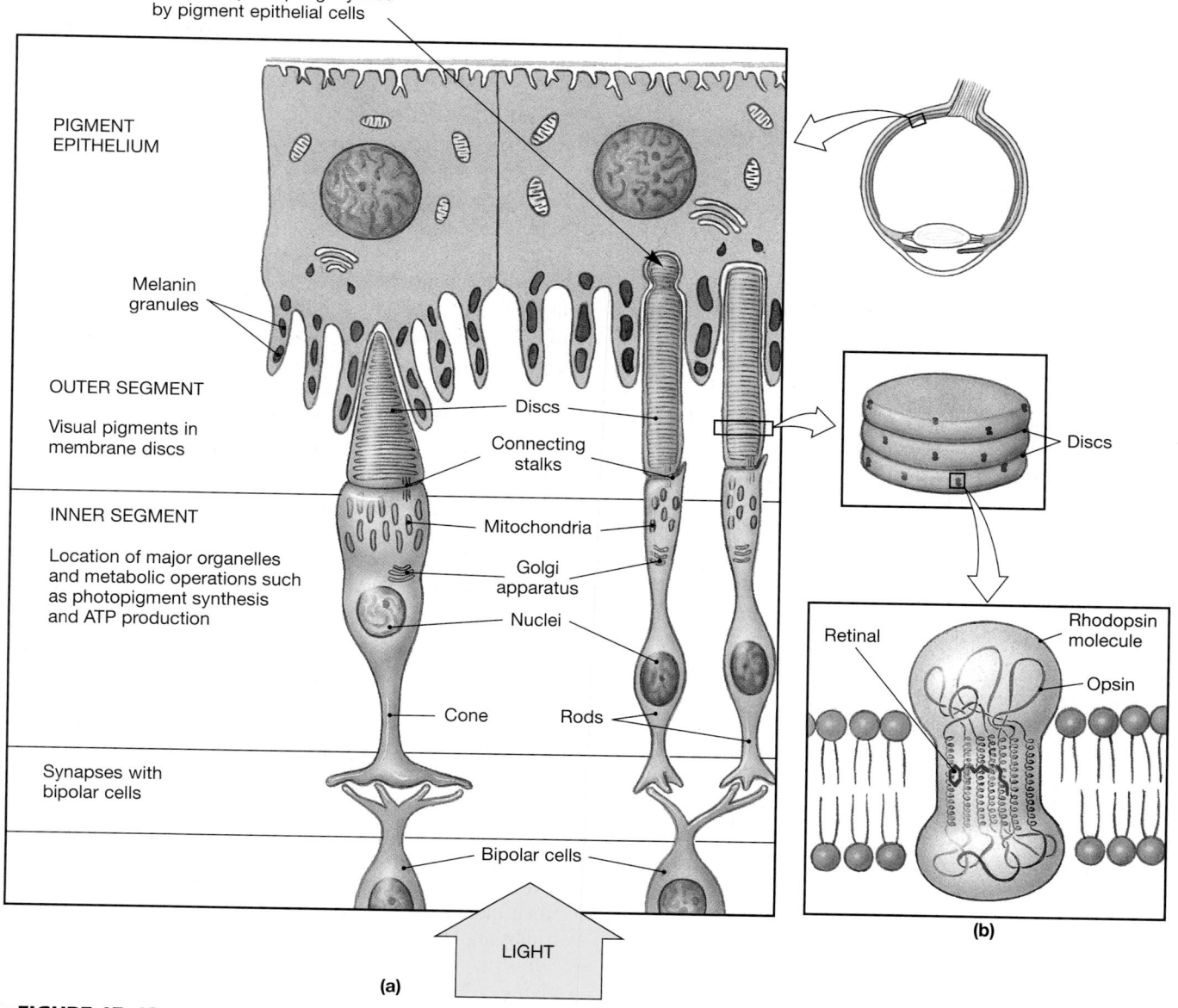

●**FIGURE 17–13**
Rods and Cones. **(a)** The structures of rods and cones. Notice the shapes of their outer segments. **(b)** The structure of a rhodopsin molecule.

ment **retinal** (RET-i-nal), or *retinene*, which is synthesized from **vitamin A**. One form of opsin is characteristic of all rods. Figure 17–13b● diagrams the structure of a rhodopsin molecule.

Cones contain the same retinal pigment that rods do, but in cones retinal is attached to other forms of opsin. The type of opsin present determines the wavelength of light that can be absorbed by retinal. Differential stimulation of these cone populations is the basis of color vision.

New discs containing visual pigment are continuously assembled at the base of the outer segment. A completed disc then moves toward the tip of the segment. After about 10 days, the disc will be shed in a small droplet of cytoplasm. Droplets with shed discs are absorbed by the pigment cells, which break down the membrane's components and reconvert the retinal to vitamin A. The vitamin A is then stored within the pigment cells for subsequent transfer to the photoreceptors.

The term **retinitis pigmentosa** (RP) refers to a collection of inherited retinopathies. Together, they are the most common inherited visual pathology, affecting approximately 1 individual in 3000. The visual receptors gradually deteriorate, and blindness eventually results. The mutations that are responsible change the structure of the photoreceptors—specifically, the visual pigments of the membrane discs. It is not known how the altered pigments lead to the destruction of photoreceptors.

Photoreception

The cell membrane in the outer segment of the photoreceptor contains chemically regulated sodium ion channels. In darkness, these gated channels are kept open in the presence of *cyclic-GMP* (*cyclic guanosine monophosphate*, or *cGMP*), a derivative of the high-energy compound *guanosine triphosphate* (GTP). Because the channels are open, the transmembrane potential is approximately -40 mV, rather than the -70 mV typical of resting neurons. At the -40-mV transmembrane potential, the photoreceptor is continuously releasing neurotransmitters (in this case, glutamate) across synapses at the inner segment. The inner segment also continuously pumps sodium ions out of the cytoplasm. The movement of sodium ions into the outer segment, on to the inner segment, and out of the cell is known as the *dark current* (Figure 17–14a●).

Rhodopsin-based photoreception begins when a photon strikes the retinal portion of a rhodopsin molecule. The bound retinal molecule has two possible configurations: the **11-*cis*** form and the **11-*trans*** form (Figure 17–14b●). Normally, the molecule is in the 11-*cis* form; on absorbing light, it changes to the more linear 11-*trans* form. This change in shape triggers a chain of enzymatic steps (Figure 17–14c●):

STEP 1: *Opsin Is Activated.* Rhodopsin is a receptor protein embedded in the membrane of the disc. Opsin is an enzyme that is inactive in the dark. When a photon of light strikes the rhodopsin, the pigment retinal in that molecule switches from the 11-*cis* to the 11-*trans* form, thereby activating the opsin portion of rhodopsin.

STEP 2: *Opsin Activates Transducin, Which in Turn Activates Phosphodiesterase.* **Transducin** is a G protein—a membrane-bound enzyme complex. ∞ p. 419 In this case, transducin is activated by opsin, and transducin in turn activates **phosphodiesterase** (**PDE**), which breaks down cGMP.

STEP 3: *Cyclic-GMP Levels Decline, and Gated Sodium Ion Channels Close.* The removal of cGMP from the gated sodium channels results in their inactivation. The rate of Na^+ entry into the cytoplasm then decreases. The arrival of a photon thus reduces the dark current.

STEP 4: *The Rate of Neurotransmitter Release Declines.* Because active transport continues to remove Na^+ from the cytoplasm, when the sodium channels close, the transmembrane potential drops toward -70 mV (Figure 17–14d●). As the membrane hyperpolarizes, the rate of neurotransmitter release decreases, indicating to the adjacent bipolar cell that the photoreceptor has absorbed a photon.

RECOVERY AFTER STIMULATION After absorbing a photon, retinal does not spontaneously convert back to the 11-*cis* form. Instead, the entire rhodopsin molecule must be broken down and reassembled. Shortly after the change in shape occurs, the rhodopsin molecule begins to break down into retinal and opsin, a process known as **bleaching** (Figure 17–15●). Before it can recombine with opsin, the retinal must be enzymatically converted to the 11-*cis* form. This conversion requires energy in the form of ATP (adenosine triphosphate), and it takes time.

Bleaching contributes to the lingering visual impression you have after you see a flashbulb go off. Following intense exposure to light, a photoreceptor cannot respond to further stimulation until its rhodopsin molecules have been regenerated. As a result, a "ghost" image remains on the retina. Bleaching is seldom noticeable under ordinary circumstances, because the eyes are constantly making small, involuntary changes in position that move the image across the retina's surface.

While the rhodopsin molecule is being reassembled, membrane permeability is returning to normal. Opsin is inactivated when bleaching occurs, and the breakdown of cGMP halts as a result. As other enzymes generate cGMP in the cytoplasm, the chemically gated sodium channels reopen.

As noted, the visual pigments of the photoreceptors are synthesized from vitamin A. The body contains vitamin A reserves sufficient for several months, and a significant amount is stored in the cells of the pigmented part of the retina. If dietary sources are inadequate, these reserves are gradually exhausted and the amount of visual pigment in the photoreceptors begins to decline. Daylight vision is affected, but in daytime the light is usually bright enough to stimulate any visual pigments that remain within the densely packed cone population of the fovea. As a result, the problem first becomes apparent at night, when the dim light proves insufficient to activate the rods. This condition, known as **night blindness**, can be treated by the administration of vitamin A. The body can convert the carotene pigments in many vegetables to vitamin A. Carrots are a particularly good source of carotene—hence the old adage that carrots are good for your eyes.

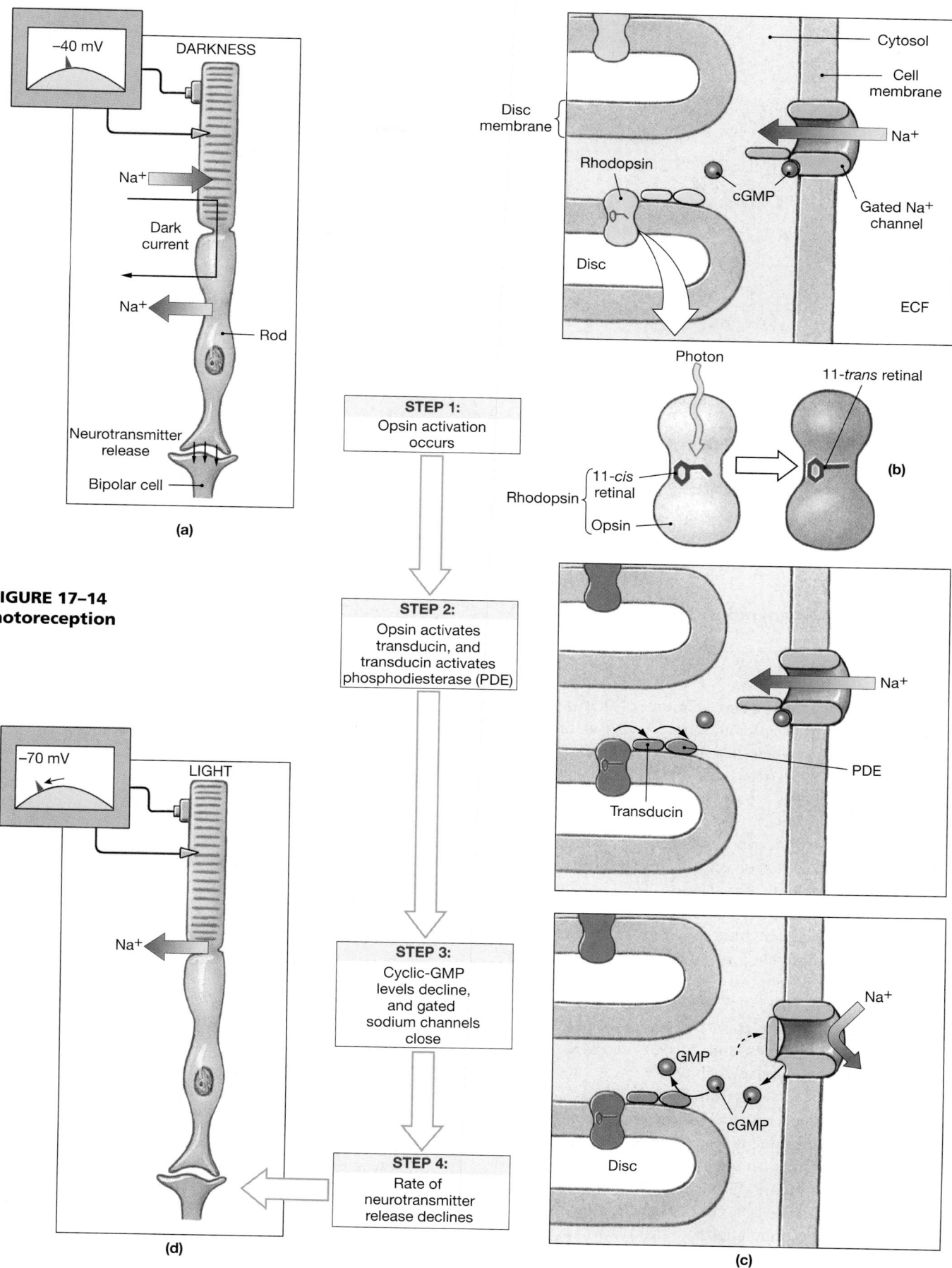

**•FIGURE 17–14
Photoreception**

(a)

DARKNESS
−40 mV
Na⁺
Dark
current
Na⁺
Rod
Neurotransmitter
release
Bipolar cell

STEP 1:
Opsin activation
occurs

STEP 2:
Opsin activates
transducin, and
transducin activates
phosphodiesterase (PDE)

STEP 3:
Cyclic-GMP
levels decline,
and gated
sodium channels
close

STEP 4:
Rate of
neurotransmitter
release declines

LIGHT
−70 mV
Na⁺

(d)

Cytosol
Cell
membrane
Disc
membrane
Rhodopsin
cGMP
Na⁺
Gated Na⁺
channel
Disc
ECF

Photon
11-*trans* retinal
11-*cis*
retinal
Rhodopsin
Opsin
(b)

PDE
Transducin
Na⁺

GMP
Na⁺
cGMP
Disc

(c)

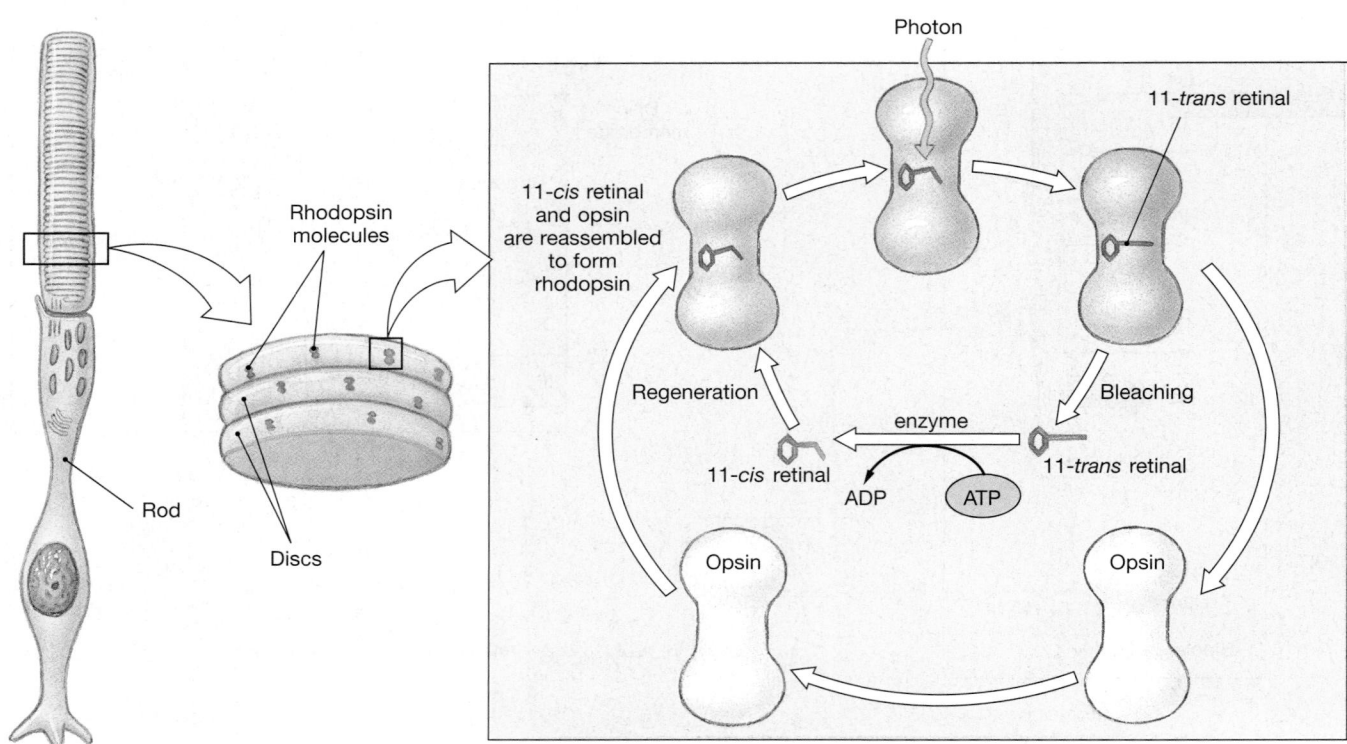

●FIGURE 17–15
Bleaching and Regeneration of Visual Pigments

Color Vision

An ordinary lightbulb or the sun emits photons of all wavelengths. These photons stimulate both rods and cones. When all three types of cones are stimulated, or when rods alone are stimulated, you see a "white" light. Your eyes also detect photons that reach your retina after they bounce off objects around you. If photons of all colors bounce off an object, the object will appear white to you; if all the photons are absorbed by the object (so that none reach the retina), the object will appear black. An object will appear to have a particular color if it reflects (or transmits) photons from one portion of the visible spectrum and absorbs the rest.

The three types of cones are **blue cones**, **green cones**, and **red cones**. Each type has a different form of opsin and a sensitivity to a different range of wavelengths. Their stimulation in various combinations is the basis for color vision. In an individual with normal vision, the cone population consists of 16 percent blue cones, 10 percent green cones, and 74 percent red cones. Although their sensitivities overlap, each type is most sensitive to a specific portion of the visual spectrum (Figure 17–16●).

Color discrimination occurs through the integration of information arriving from all three types of cones. For example, the perception of yellow results from a combination of inputs from green cones (which are highly stimulated), red cones (which are less strongly stimulated), and blue cones (which are relatively unaffected). If all three cone populations are stimulated, we perceive the color as white. Because we also perceive white if rods, rather

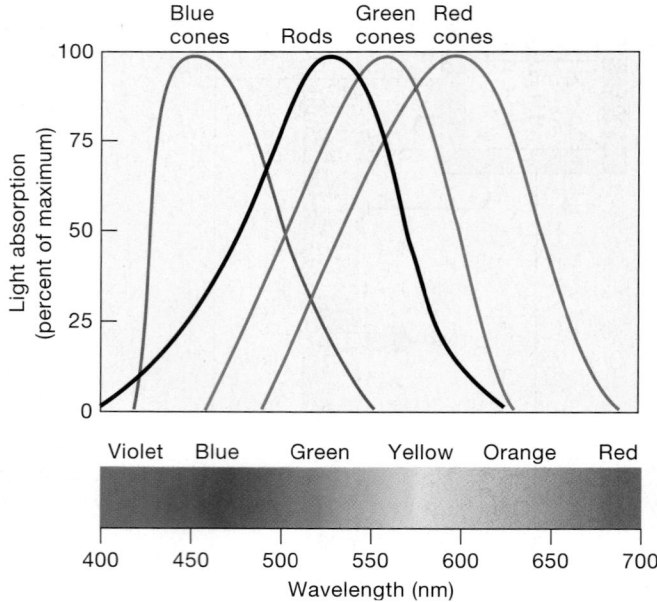

●FIGURE 17–16
Cones and Color Vision. A graph comparing the absorptive characteristics of blue, green, and red cones with those of typical rods. Notice that the sensitivities of the rods overlap those of the cones and that the three types of cones have overlapping sensitivity curves.

than cones, are stimulated, everything appears to be black and white when we enter dimly lit surroundings or walk by starlight.

Light and Dark Adaptation

The sensitivity of your visual system varies with the intensity of illumination. After 30 minutes or more in the dark, almost all visual pigments will be fully receptive to stimulation. This is the **dark-adapted state**. When dark adapted, the visual system is extremely sensitive. For example, a single rod will hyperpolarize in response to a single photon of light. Even more remarkable, if as few as seven rods absorb photons at one time, you will see a flash of light.

When the lights come on, at first they seem almost unbearably bright, but over the next few minutes your sensitivity decreases as bleaching occurs. Eventually, the rate of breakdown of visual pigments is balanced by the rate at which they re-form. This condition is the **light-adapted state**. If you moved from the depths of a cave to the full light of midday, your receptor sensitivity would decrease by about 25,000 times.

A variety of central responses further adjust light sensitivity. Constriction of the pupil, via the *pupillary constrictor reflex*, reduces the amount of light entering your eye to one-thirtieth the maximum dark-adapted level. Dilating the pupil fully can produce a thirtyfold increase in the amount of light entering the eye, and facilitating some of the synapses along the visual pathway

can perhaps triple its sensitivity. Hence, the entire system may increase its efficiency by a factor of more than 1 million.

□ THE VISUAL PATHWAY

The visual pathway begins at the photoreceptors and ends at the visual cortex of the cerebral hemispheres. In other sensory pathways we have examined, at most one synapse lies between a receptor and a sensory neuron that delivers information to the CNS. In the visual pathway, the message must cross two synapses (photoreceptor to bipolar cell and bipolar cell to ganglion cell) before it heads toward the brain. The extra synapse increases the synaptic delay, but it provides an opportunity for the processing and integration of visual information before it leaves the retina.

Processing by the Retina

Each photoreceptor in the retina monitors a specific receptive field. The retina contains about 130 million photoreceptors, 6 million bipolar cells, and 1 million ganglion cells. Thus, a considerable amount of convergence occurs at the start of the visual pathway. The degree of convergence differs between rods and cones. Regardless of the amount of convergence, each ganglion cell monitors a specific portion of the field of vision.

As many as a thousand rods may pass information via their bipolar cells to a single ganglion cell. The ganglion cells that mon-

Color Blindness

Persons who are unable to distinguish certain colors have a form of **color blindness**. The standard tests for color vision involve picking numbers or letters out of a complex and colorful picture (Figure 17–17●). Color blindness occurs when one or more classes of cones are nonfunctional. The cones may be absent, or they may be present, but unable to manufacture the necessary visual pigments. In the most common type of color blindness (red–green color blindness), the red cones are missing, so the individual cannot distinguish red light from green light.

Inherited color blindness involving one or two cone pigments is not unusual. The genes for the red and green cone pigments are located on the X chromosome. As we shall see in Chapter 29, women have two copies of this chromosome *(XX)*, whereas men have one X chromosome paired with a Y chromosome *(XY)*. As a result, men are much more likely to be color blind than are women, who must have the same abnormal cone pigment genes on both X chromosomes to exhibit color blindness. Ten percent of all men show some color blindness, whereas the incidence among women is only about 0.67 percent. Total color blindness is extremely rare; only 1 person in 300,000 fails to manufacture any cone pigments.

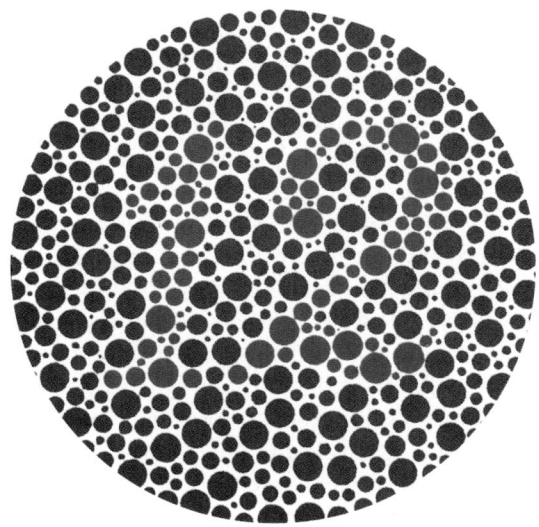

●**FIGURE 17–17**
A Standard Test for Color Vision. If you lack one or more populations of cones, you will be unable to distinguish the patterned image (the number 12).

itor rods, called **M cells** (*magnocells; magnus,* great), are relatively large. They provide information about the general form of an object, motion, and shadows in dim lighting. Because so much convergence occurs, the activation of an M cell indicates that light has arrived in a general area rather than at a specific location.

The loss of specificity due to convergence is partially overcome by the fact that the activity of ganglion cells varies according to the pattern of activity in their sensory field, which is generally circular. Typically, a ganglion cell responds differently to stimuli that arrive in the center of its receptive field than to stimuli that arrive at the edges (Figure 17–18●). Some ganglion cells (**on-center neurons**) are excited by light arriving in the center of their sensory field and are inhibited when light strikes the edges of their receptive field. Others (**off-center neurons**) are inhibited by light in the central zone, but are stimulated by illumination at the edges. On-center and off-center neurons provide information about which portion of their sensory field is illuminated.

Cones typically show very little convergence; in the fovea, the ratio of cones to ganglion cells is 1:1. The ganglion cells that monitor cones, called **P cells** (*parvo cells; parvus,* small), are smaller and more numerous than M cells. P cells are active in bright light, and they provide information about edges, fine detail, and color. Because little convergence occurs, the activation of a P cell means that light has arrived at one specific location. As a result, cones provide more precise information about a visual image than do rods. In videographic terms, images formed by rods have a coarse, grainy, pixelated appearance that blurs details; by contrast, images produced by cones are fine grained, of high density, sharp, and clear.

Central Processing of Visual Information

Axons from the entire population of ganglion cells converge on the optic disc, penetrate the wall of the eye, and proceed toward the diencephalon as the optic nerve (II). The two optic nerves, one from each eye, reach the diencephalon at the optic chiasm (Figure 17–19●). From that point, approximately half the fibers proceed toward the lateral geniculate nucleus of the same side of the brain, whereas the other half cross over to reach the lateral geniculate nucleus of the opposite side. ∞ p. 481 From each lateral geniculate nucleus, visual information travels to the occipital cortex of the cerebral hemisphere on that side. The bundle of projection fibers linking the lateral geniculates with the visual cortex is known as the **optic radiation.** Collaterals from the fibers synapsing in the lateral geniculate continue to subconscious processing centers in the diencephalon and brain stem. For example, the pupillary reflexes and reflexes that control eye movement are triggered by collaterals carrying information to the superior colliculi.

THE FIELD OF VISION The perception of a visual image reflects the integration of information that arrives at the visual cortex of the occipital lobes. Each eye receives a slightly different visual image, because (1) the foveas are 5–7.5 cm apart, and (2) the nose and eye socket block the view of the opposite side. **Depth perception,** an interpretation of the three-dimensional rela-

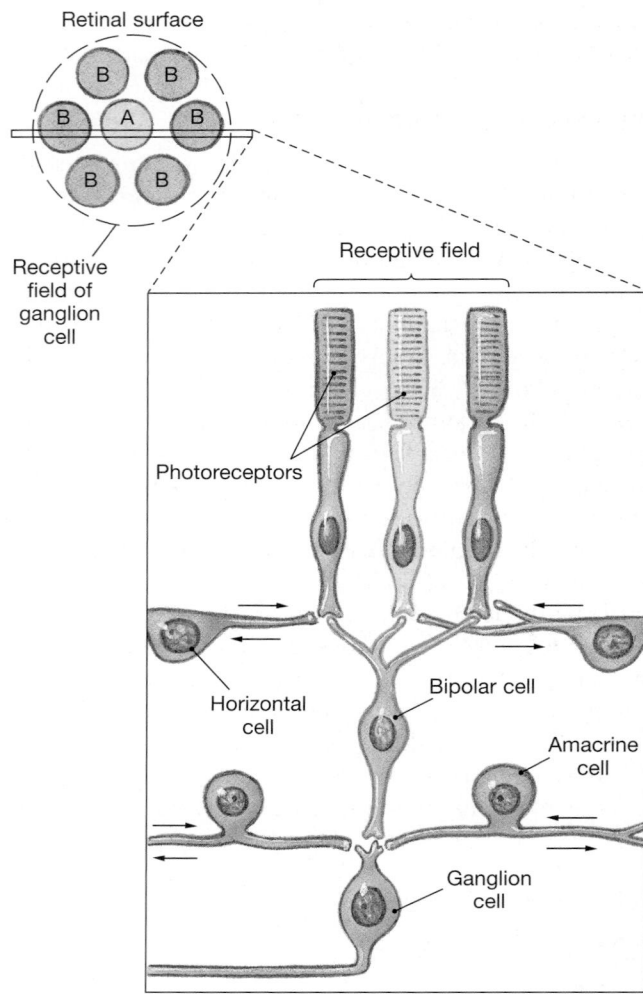

●FIGURE 17–18
Convergence and Ganglion Cell Function.
Photoreceptors are organized in groups within the field of vision. Each ganglion cell monitors a well-defined portion of that field. Some ganglion cells (on-center neurons) respond strongly to light arriving at the center of their receptive field (receptor A). Others (off-center neurons) respond most strongly to illumination of the edges of their receptive field (receptors B).

tionships among objects in view, is obtained by comparing the relative positions of objects within the images received by the left and right eyes.

When you look straight ahead, the visual images from your left and right eyes overlap (Figure 17–19●). The image received by the fovea of each eye lies in the center of the region of overlap. A vertical line drawn through the center of this region marks the division of visual information at the optic chiasm. Visual information from the left half of the combined field of vision reaches the visual cortex of your right occipital lobe; visual information from the right half of the combined field of vision arrives at the visual cortex of your left occipital lobe.

The cerebral hemispheres thus contain a map of the entire field of vision. As in the case of the primary sensory cortex, the map

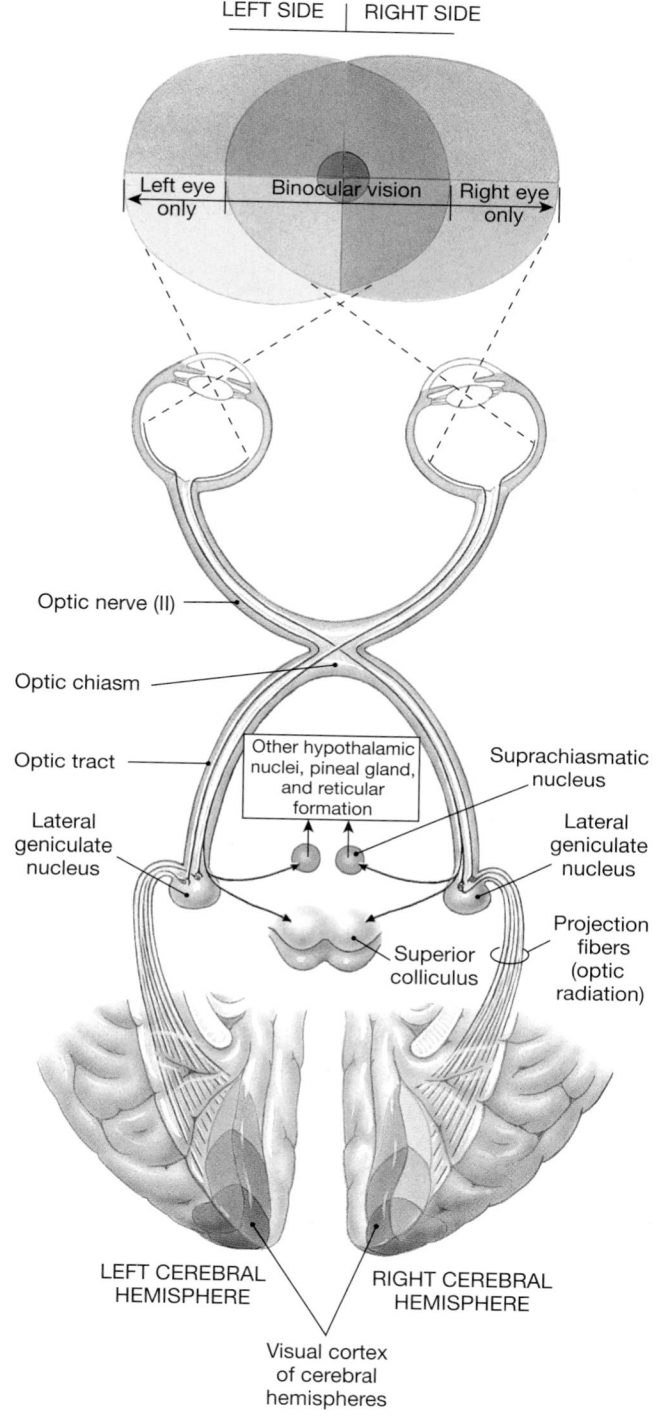

LEFT SIDE | RIGHT SIDE

Left eye only ← Binocular vision → Right eye only

Optic nerve (II)

Optic chiasm

Optic tract

Other hypothalamic nuclei, pineal gland, and reticular formation

Suprachiasmatic nucleus

Lateral geniculate nucleus

Lateral geniculate nucleus

Superior colliculus

Projection fibers (optic radiation)

LEFT CEREBRAL HEMISPHERE

RIGHT CEREBRAL HEMISPHERE

Visual cortex of cerebral hemispheres

●FIGURE 17–19
The Visual Pathways. A partial crossover of nerve fibers occurs at the optic chiasm. As a result, each hemisphere receives visual information from the medial half of the field of vision of the eye on that side and from the lateral half of the field of vision of the eye on the opposite side. Visual association areas integrate this information to develop a composite picture of the entire field of vision.

does not faithfully duplicate the relative areas within the sensory field. For example, the area assigned to the macula lutea and fovea covers about 35 times the surface it would cover if the map were

proportionally accurate. The map is also upside down and backward, duplicating the orientation of the visual image at the retina.

THE BRAIN STEM AND VISUAL PROCESSING Many centers in the brain stem receive visual information, either from the lateral geniculate nuclei or through collaterals from the optic tracts. Collaterals that bypass the lateral geniculates synapse in the superior colliculi or in the hypothalamus. The superior colliculi of the mesencephalon issue motor commands that control unconscious eye, head, or neck movements in response to visual stimuli. Visual inputs to the suprachiasmatic nucleus of the hypothalamus affect the function of other brain stem nuclei. ∞ p. 482 The suprachiasmatic nucleus and the *pineal gland* of the epithalamus establish a daily pattern of visceral activity that is tied to the day–night cycle. This **circadian rhythm** (*circa*, about + *dies*, day) affects your metabolic rate, endocrine function, blood pressure, digestive activities, awake–asleep cycle, and other physiological and behavioral processes.

✓ If you had been born without cones in your eyes, would you still be able to see? Explain.

✓ How could a diet deficient in vitamin A affect vision?

✓ What effect would a decrease in phosphodiesterase activity in photoreceptor cells have on vision?

Answers start on page Q-1

17–4 EQUILIBRIUM AND HEARING

Objectives

■ Describe the structures of the external and middle ears, and explain how they function.

■ Describe the parts of the inner ear and their roles in equilibrium and hearing.

■ Trace the pathways for the sensations of equilibrium and hearing to their respective destinations in the brain.

The special senses of equilibrium and hearing are provided by the *inner ear*, a receptor complex located in the petrous part of the temporal bone of the skull. *Equilibrium* sensations inform us of the position of the head in space by monitoring gravity, linear acceleration, and rotation. *Hearing* enables us to detect and interpret sound waves. The basic receptor mechanism for both senses is the same. The receptors, called *hair cells*, are simple mechanoreceptors. The complex structure of the inner ear and the different arrangement of accessory structures account for the abilities of the hair cells to respond to different stimuli and thus to provide the input for both senses.

☐ ANATOMY OF THE EAR

The ear is divided into three anatomical regions: the external ear, the middle ear, and the inner ear (Figure 17–20●). The *external*

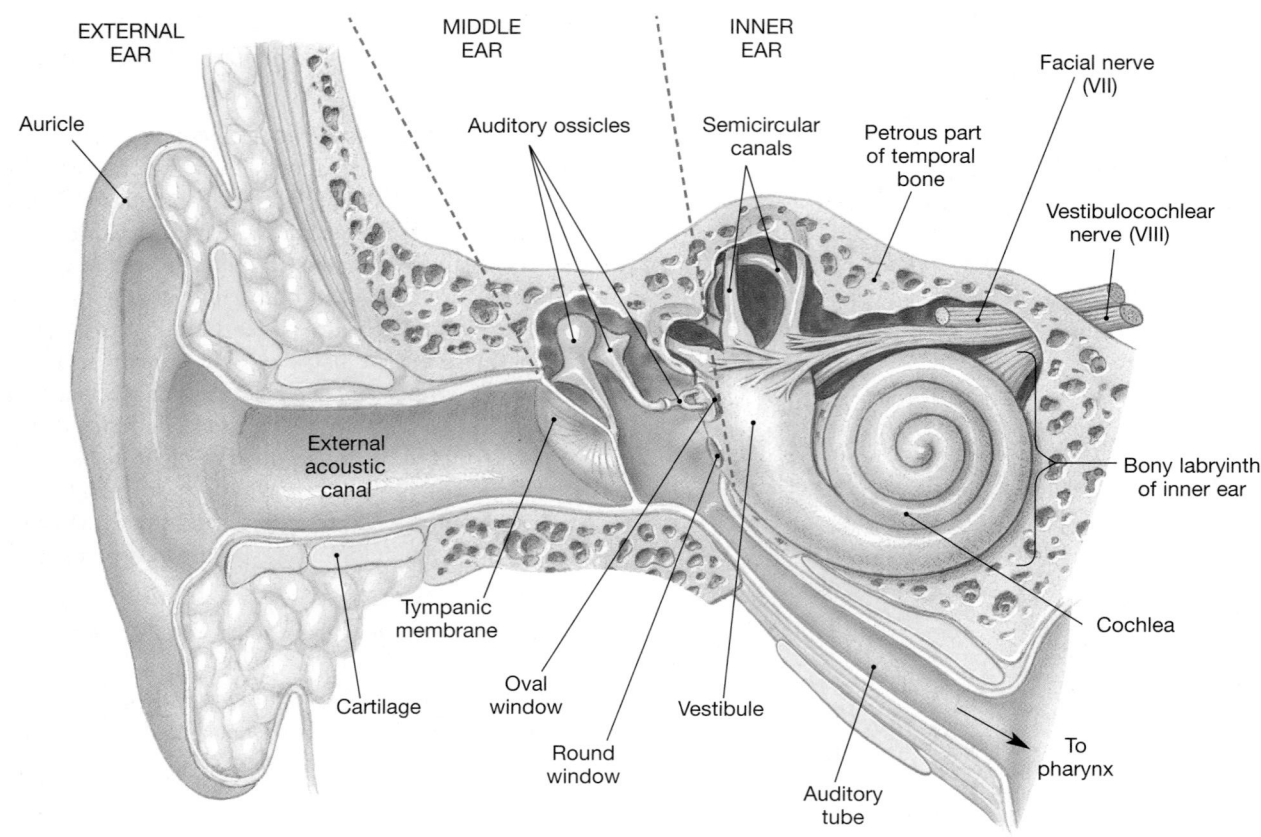

EXTERNAL EAR · MIDDLE EAR · INNER EAR

Auricle

Auditory ossicles

Semicircular canals

Petrous part of temporal bone

Facial nerve (VII)

Vestibulocochlear nerve (VIII)

Bony labryinth of inner ear

Cochlea

External acoustic canal

Tympanic membrane

Cartilage

Oval window

Round window

Vestibule

Auditory tube

To pharynx

●**FIGURE 17–20**
The Anatomy of the Ear.   The boundaries separating the three regions of the ear (external, middle, and inner) are roughly marked by the dashed lines.

ear—the visible portion of the ear—collects and directs sound waves toward the *middle ear*, a chamber located within the petrous portion of the temporal bone. Structures of the middle ear collect sound waves and transmit them to an appropriate portion of the *inner ear*, which contains the sensory organs for hearing and equilibrium.

The External Ear

The **external ear** includes the fleshy and cartilaginous **auricle**, or *pinna*, which surrounds the **external acoustic canal**, or *ear canal*. The auricle protects the opening of the canal and provides directional sensitivity to the ear. Sounds coming from behind the head are blocked by the auricle; sounds coming from the side or front are collected and channeled into the external acoustic canal. (When you "cup" your ear with your hand to hear a faint sound more clearly, you are exaggerating this effect.) The external acoustic canal is a passageway that ends at the **tympanic membrane**, also called the *tympanum* or *eardrum*. The tympanic membrane is a thin, semitransparent sheet that separates the external ear from the middle ear (Figure 17–20●).

The tympanic membrane is very delicate. The auricle and the narrow external acoustic canal provide some protection from accidental injury. In addition, **ceruminous glands**—integumentary glands along the external acoustic canal—secrete a waxy

material that helps deny access to foreign objects or small insects, as do many small, outwardly projecting hairs. These hairs also provide increased tactile sensitivity through their root hair plexuses. The waxy secretion of the ceruminous glands, called **cerumen**, also slows the growth of microorganisms in the external acoustic canal and reduces the chances of infection.

The Middle Ear

The **middle ear**, or **tympanic cavity**, is filled with air. Separated from the external acoustic canal by the tympanic membrane, the middle ear communicates both with the *nasopharynx* (the superior portion of the pharynx), through the **auditory tube**, and with the mastoid air cells, through a number of small connections (Figures 17–20 and 17–21a●). The auditory tube is also called the *pharyngotympanic tube* or the *Eustachian tube*. About 4 cm long, it consists of two portions. The portion near the connection to the middle ear is relatively narrow and is supported by elastic cartilage. The portion near the opening into the nasopharynx is relatively broad and funnel shaped. The auditory tube permits the equalization of pressures inside and outside the tympanic membrane. Unfortunately, the auditory tube can also allow microorganisms to travel from the nasopharynx into the mid-

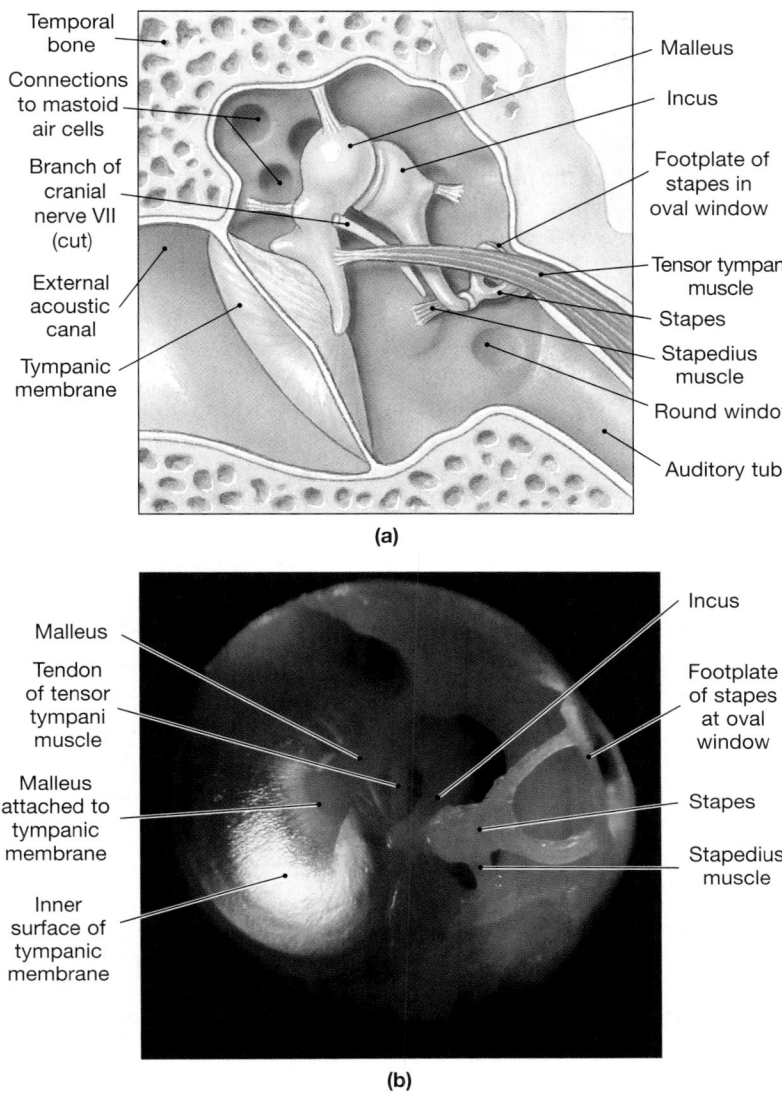

Temporal bone

Connections to mastoid air cells

Branch of cranial nerve VII (cut)

External acoustic canal

Tympanic membrane

Malleus

Incus

Footplate of stapes in oval window

Tensor tympani muscle

Stapes

Stapedius muscle

Round window

Auditory tube

(a)

Malleus

Tendon of tensor tympani muscle

Malleus attached to tympanic membrane

Inner surface of tympanic membrane

Incus

Footplate of stapes at oval window

Stapes

Stapedius muscle

(b)

●**FIGURE 17–21**
The Middle Ear. **(a)** The structures of the middle ear. **(b)** The tympanic membrane and auditory ossicles.

dle ear. Invasion by microorganisms can lead to an unpleasant middle ear infection known as *otitis media*. **AM** Otitis Media and Mastoiditis

THE AUDITORY OSSICLES The middle ear contains three tiny ear bones, collectively called **auditory ossicles**. The ear bones connect the tympanic membrane with one of the receptor complexes of the inner ear (Figures 17–20 and 17–21●). The three auditory ossicles are the malleus, the incus, and the stapes. The **malleus** (*malleus*, hammer) attaches at three points to the interior surface of the tympanic membrane. The **incus** (*incus*, anvil)—the middle ossicle—attaches the malleus to the **stapes** (*stapes*, stirrup)—the inner ossicle. The edges of the base of the stapes are bound to the edges of the *oval window*, an opening in the bone that surrounds the inner ear. The articulations between the auditory ossicles are the smallest synovial joints in the body. Each has a tiny capsule and supporting extracapsular ligaments.

Vibration of the tympanic membrane converts arriving sound waves into mechanical movements. The auditory ossicles act as levers that conduct those vibrations to the inner ear. The ossicles are connected in such a way that an in–out movement of the tympanic membrane produces a rocking motion of the stapes. The ossicles thus function as a lever system that collects the force applied to the tympanic membrane and focuses it on the oval window. Because the tympanic membrane is 22 times larger and heavier than the oval window, considerable amplification occurs, and we can hear very faint sounds. But that degree of amplification can be a problem when we are exposed to very loud noises. In the middle ear, two small muscles protect the tympanic membrane and ossicles from violent movements under very noisy conditions:

1. The **tensor tympani** (TEN-sor tim-PAN-ē) **muscle** is a short ribbon of muscle whose origin is the petrous portion of the temporal bone and the auditory tube and whose insertion is on the "handle" of the malleus. When the tensor tympani contracts,

the malleus is pulled medially, stiffening the tympanic membrane. This increased stiffness reduces the amount of movement possible. The tensor tympani muscle is innervated by motor fibers of the mandibular branch of the trigeminal nerve (V).

2. The **stapedius** (sta-PĒ-dē-us) **muscle**, innervated by the facial nerve (VII), originates from the posterior wall of the middle ear and inserts on the stapes. Contraction of the stapedius pulls the stapes, reducing movement of the stapes at the oval window.

The Inner Ear

The senses of equilibrium and hearing are provided by the receptors of the **inner ear** (Figures 17–20 and 17–22a●). The receptors lie within a collection of fluid-filled tubes and chambers known as the **membranous labyrinth** (*labyrinthos*, network of canals). The membranous labyrinth contains **endolymph** (EN-dō-limf), a fluid with electrolyte concentrations different from those of typical body fluids.

The **bony labyrinth** is a shell of dense bone that surrounds and protects the membranous labyrinth. The inner contours of the bony labyrinth closely follow the contours of the membranous labyrinth, and the outer walls are fused with the surrounding temporal bone. Between the bony and membranous labyrinths flows the **perilymph** (PER-i-limf), a liquid whose properties closely resemble those of cerebrospinal fluid. (See Appendix IV for a chemical analysis of perilymph, endolymph, and other body fluids.)

The bony labyrinth can be subdivided into the *vestibule*, three *semicircular canals*, and the *cochlea* (Figure 17–22a●). The **vestibule** (VES-ti-būl) consists of a pair of membranous sacs: the **saccule** (SAK-ūl) and the **utricle** (Ū-tri-kul), or *sacculus* and *utriculus*. Receptors in the saccule and utricle provide sensations of gravity and linear acceleration.

The **semicircular canals** enclose slender **semicircular ducts**. Receptors in the semicircular ducts are stimulated by rotation of the head. The combination of vestibule and semicircular canals is called the **vestibular complex**. The fluid-filled chambers within the vestibule are broadly continuous with those of the semicircular canals.

The **cochlea** (KOK-lē-uh; *cochlea*, a snail shell) is a spiral-shaped, bony chamber that contains the **cochlear duct** of the membranous labyrinth. Receptors within the cochlear duct provide the sense of hearing. The duct is sandwiched between a pair of perilymph-filled chambers. The entire complex makes turns around a central bony hub. In sectional view, the spiral arrangement resembles a snail shell.

The walls of the bony labyrinth consist of dense bone everywhere, except at two small areas near the base of the cochlear spiral (Figure 17–20●). The **round window** is a thin, membranous partition that separates the perilymph of the cochlear chambers from the air spaces of the middle ear. Collagen fibers connect the bony margins of the opening known as the **oval window** to the base of the stapes. When a sound vibrates the tympanic membrane, the movements are conducted to the

stapes by the other auditory ossicles. Movement of the stapes produces a rocking motion at the oval window, generating fluid waves that stimulate receptors in the cochlear duct, and we hear the sound.

Receptor Function in the Inner Ear

The sensory receptors of the inner ear are called **hair cells** (Figure 17–22b●). These receptors are surrounded by supporting cells and are monitored by sensory afferent fibers. The free surface of each hair cell supports 80–100 long **stereocilia**, which resemble very long microvilli. Each hair cell in the vestibule also contains a **kinocilium**, a single large cilium. Hair cells do not actively move their kinocilia or stereocilia. However, when an external force pushes against these processes, the distortion of the cell membrane alters the rate at which the hair cell releases chemical transmitters.

Hair cells provide information about the direction and strength of mechanical stimuli. The stimuli involved, however, are quite varied: gravity or acceleration in the vestibule, rotation in the semicircular canals, and sound in the cochlea. The sensitivities of the hair cells differ, because each of these regions has different accessory structures that determine which stimulus will provide the force to deflect the kinocilia and stereocilia. The importance of these accessory structures will become apparent when we consider the way sensations of equilibrium and hearing are produced.

☐ EQUILIBRIUM

As we just learned, equilibrium sensations are provided by receptors of the vestibular complex. The semicircular ducts convey information about rotational movements of the head. For example, when you turn your head to the left, receptors stimulated in the semicircular ducts tell you how rapid the movement is and in which direction. The saccule and the utricle convey information about your position with respect to gravity. If you stand with your head tilted to one side, these receptors will report the angle involved and whether your head tilts forward or backward. The saccule and the utricle are also stimulated by sudden acceleration. When your car accelerates from a stop, the saccular and utricular receptors give you the impression of increasing speed.

The Semicircular Ducts

Sensory receptors in the semicircular ducts respond to rotational movements of the head. These hair cells are active during a movement, but are quiet when the body is motionless. The **anterior**, **posterior**, and **lateral semicircular ducts** are continuous with the utricle (Figure 17–23a●). Each semicircular duct contains an **ampulla**, an expanded region that contains the hair cells. Hair cells attached to the wall of the ampulla form a raised structure

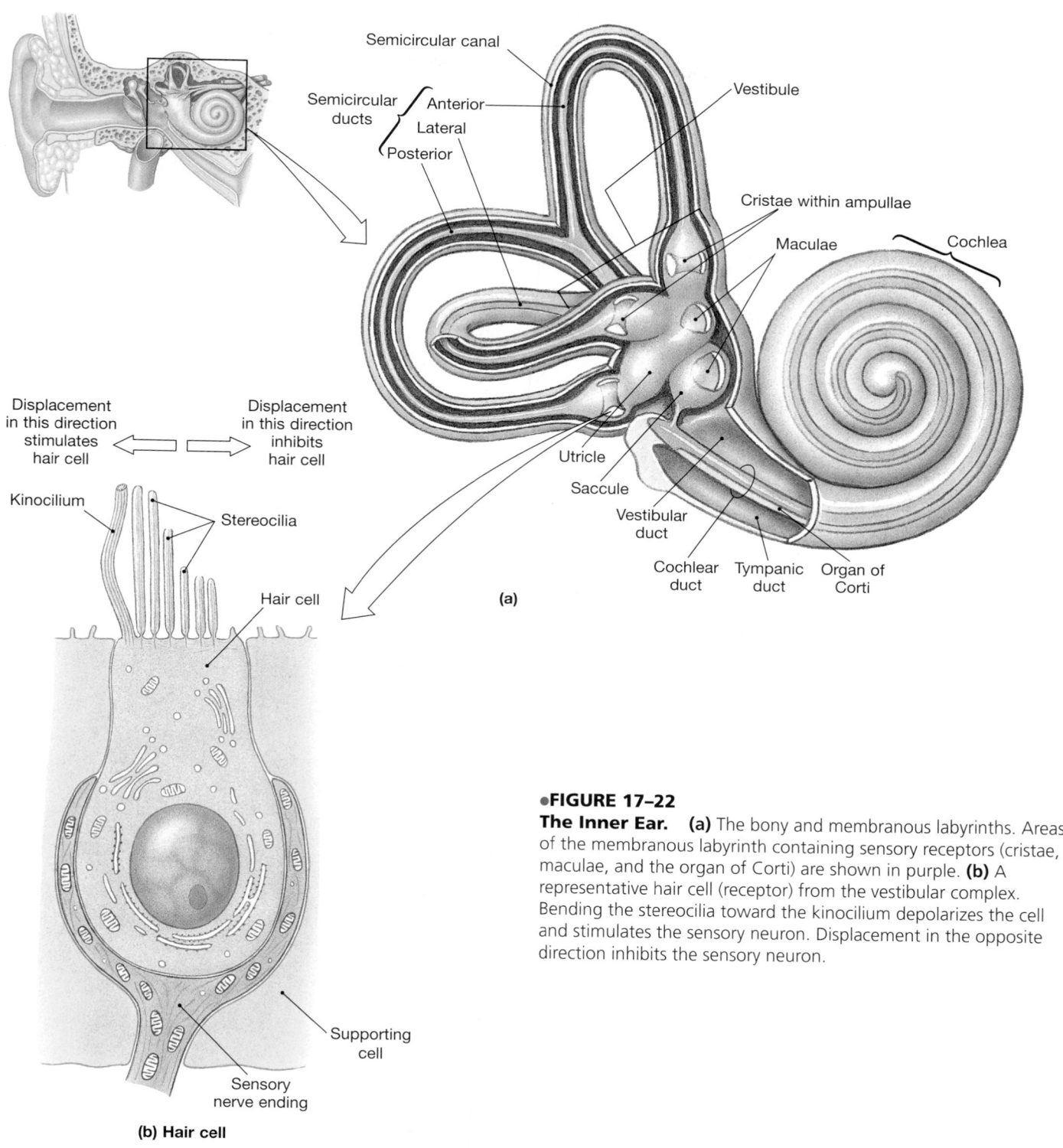

(a)

Semicircular canal

Semicircular ducts
- Anterior
- Lateral
- Posterior

Vestibule

Cristae within ampullae

Maculae

Cochlea

Utricle

Saccule

Vestibular duct

Cochlear duct

Tympanic duct

Organ of Corti

Displacement in this direction stimulates hair cell

Displacement in this direction inhibits hair cell

Kinocilium

Stereocilia

Hair cell

Supporting cell

Sensory nerve ending

(b) Hair cell

●**FIGURE 17–22**
The Inner Ear. **(a)** The bony and membranous labyrinths. Areas of the membranous labyrinth containing sensory receptors (cristae, maculae, and the organ of Corti) are shown in purple. **(b)** A representative hair cell (receptor) from the vestibular complex. Bending the stereocilia toward the kinocilium depolarizes the cell and stimulates the sensory neuron. Displacement in the opposite direction inhibits the sensory neuron.

known as a **crista** (Figure 17–23b●). The kinocilia and stereocilia of the hair cells are embedded in the **cupula** (KŪ-pū-luh), a gelatinous structure. Because the cupula has a density very close to that of the surrounding endolymph, it essentially floats above the receptor surface, nearly filling the ampulla. When your head rotates in the plane of the duct, the movement of endolymph

along the length of the semicircular duct pushes the cupula and distorts the receptor processes (Figure 17–23c●). Movement of fluid in one direction stimulates the hair cells, and movement in the opposite direction inhibits them. When the endolymph stops moving, the elastic nature of the cupula makes it bounce back to its normal position.

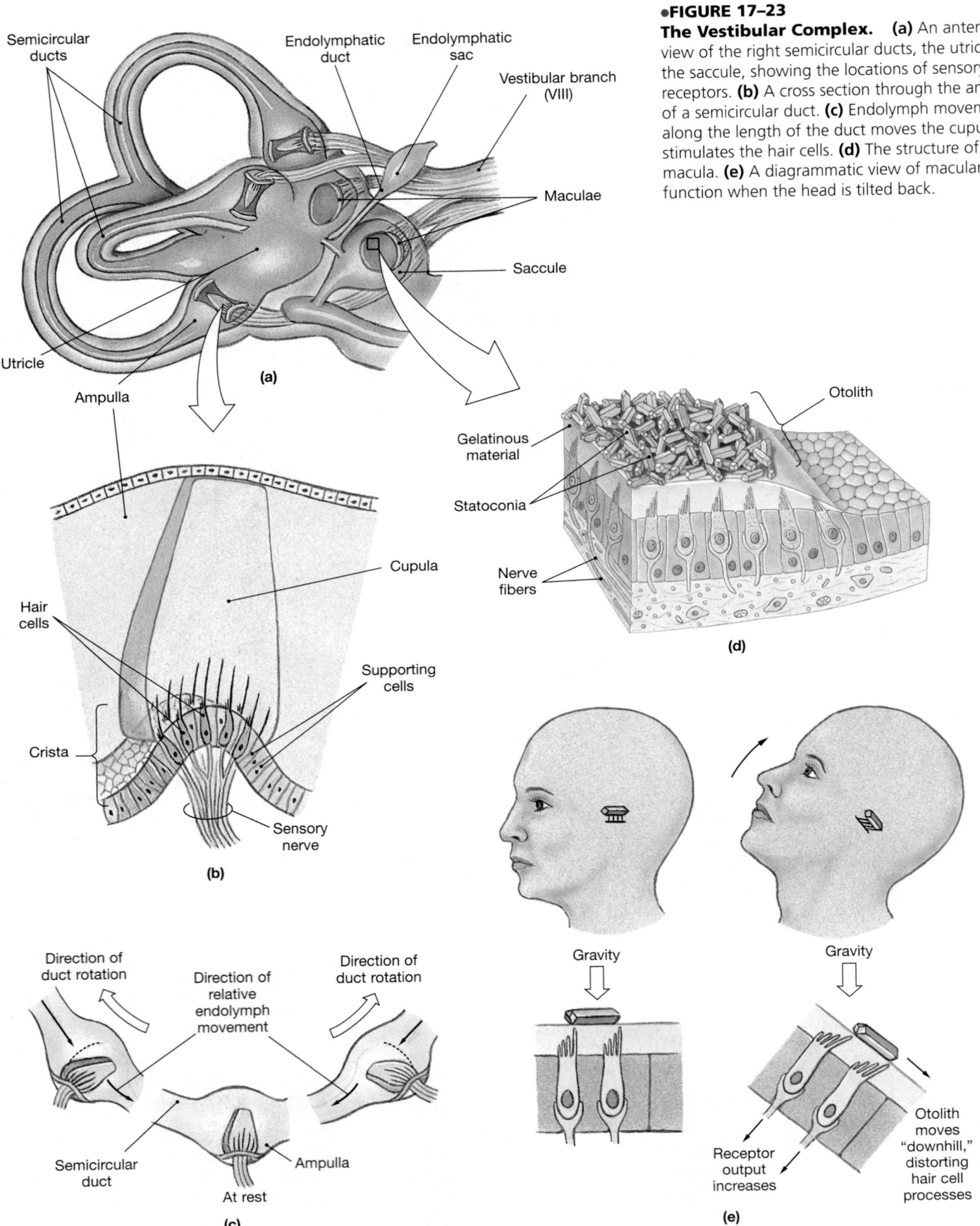

Semicircular ducts

Endolymphatic duct

Endolymphatic sac

Vestibular branch (VIII)

Maculae

Saccule

Utricle

Ampulla

(a)

●FIGURE 17–23
The Vestibular Complex. **(a)** An anterior view of the right semicircular ducts, the utricle, and the saccule, showing the locations of sensory receptors. **(b)** A cross section through the ampulla of a semicircular duct. **(c)** Endolymph movement along the length of the duct moves the cupula and stimulates the hair cells. **(d)** The structure of a macula. **(e)** A diagrammatic view of macular function when the head is tilted back.

Cupula

Hair cells

Supporting cells

Crista

Sensory nerve

(b)

Otolith

Gelatinous material

Statoconia

Nerve fibers

(d)

Direction of duct rotation

Direction of relative endolymph movement

Direction of duct rotation

Semicircular duct

Ampulla

At rest

(c)

Gravity

Gravity

Receptor output increases

Otolith moves "downhill," distorting hair cell processes

(e)

Even the most complex movement can be analyzed in terms of motion in three rotational planes. Each semicircular duct responds to one of these rotational movements. A horizontal rotation, as in shaking your head "no," stimulates the hair cells of the lateral semicircular duct. Nodding "yes" excites the anterior duct, and tilting your head from side to side activates the receptors in the posterior duct.

The Utricle and Saccule

The utricle and saccule provide equilibrium sensations, whether the body is moving or is stationary. The two chambers are connected by a slender passageway that is continuous with the narrow endolymphatic duct. The **endolymphatic duct** ends in a blind pouch called the **endolymphatic sac** (Figure 17–23a●). This sac projects through the dura mater that lines the temporal bone and into the subdural space. Portions of the cochlear duct secrete endolymph continuously, and at the endolymphatic sac excess fluids return to the general circulation.

The hair cells of the utricle and saccule are clustered in oval structures called **maculae** (MAK-ū-lē; *macula*, spot) (Figure 17–23a●). As they are in ampullae, the hair cell processes are embedded in a gelatinous mass. However, the surface of this gelatinous material contains densely packed calcium carbonate crystals known as **statoconia** (*statos*, standing + *conia*, dust). The complex as a whole (gelatinous matrix and statoconia) is called an **otolith** ("ear stone").

The macula of the saccule is diagrammed in Figure 17–23d●. When your head is in the normal, upright position, the statoconia sit atop the macula. Their weight presses on the macular surface, pushing the hair cell processes down rather than to one side or another. When your head is tilted, the pull of gravity on the statoconia shifts them to the side, thereby distorting the hair cell processes. The change in receptor activity tells the CNS that your head is no longer level (Figure 17–23e●).

A similar mechanism accounts for your perception of linear acceleration when you are in a car that speeds up suddenly. The statoconia lag behind, and the effect on the hair cells is comparable to tilting your head back. Under normal circumstances, your nervous system distinguishes between the sensations of tilting and linear acceleration by integrating vestibular sensations with visual information. Many amusement park rides confuse your sense of equilibrium by combining rapid rotation with changes in position and acceleration while providing restricted or erroneous visual information.

Pathways for Equilibrium Sensations

Hair cells of the vestibule and semicircular ducts are monitored by sensory neurons located in adjacent **vestibular ganglia**. Sensory fibers from these ganglia form the **vestibular branch** of the vestibulocochlear nerve (VIII). ∞ p. 499 These fibers innervate neurons within the pair of **vestibular nuclei** at the boundary between the pons and the medulla oblongata. The two vestibular nuclei have the following four functions:

1. Integrating sensory information about balance and equilibrium that arrives from both sides of the head.

2. Relaying information from the vestibular complex to the cerebellum.

3. Relaying information from the vestibular complex to the cerebral cortex, providing a conscious sense of head position and movement.

4. Sending commands to motor nuclei in the brain stem and spinal cord.

The reflexive motor commands that are issued by the vestibular nuclei are distributed to the motor nuclei for cranial nerves involved with eye, head, and neck movements (III, IV, VI, and XI). Instructions descending in the *vestibulospinal tracts* of the

Motion Sickness

The exceedingly unpleasant symptoms of **motion sickness** include headache, sweating, flushing of the face, nausea, vomiting, and various changes in mental perspective. (Sufferers may go from a state of giddy excitement to almost suicidal despair in a matter of moments.) It has been suggested that the condition results when central processing stations, such as the tectum of the mesencephalon, receive conflicting sensory information. Why and how these conflicting reports result in nausea, vomiting, and other symptoms are not known. Sitting below decks on a moving boat or reading in a car or airplane tends to provide the necessary conditions. Your eyes (which are tracking lines on a page) report that your position in space is not changing, but your semicircular ducts report that your

body is lurching and turning. To counter this effect, seasick sailors watch the horizon rather than their immediate surroundings, so that their eyes will provide visual confirmation of the movements detected by their inner ears. It is not known why some individuals are almost immune to motion sickness, whereas others find travel by boat or plane almost impossible.

Drugs commonly administered to prevent motion sickness include dimenhydrinate (*Dramamine*), scopolamine, and promethazine. These compounds appear to depress activity at the vestibular nuclei. Sedatives, such as prochlorperazine (*Compazine*), may also be effective. Scopolamine can be administered across the skin surface by using an adhesive patch (*Transderm Scop*™).

spinal cord adjust peripheral muscle tone and complement the reflexive movements of the head or neck. ∞ p. 526 These pathways are indicated in Figure 17–24●.

The automatic movements of the eye that occur in response to sensations of motion are directed by the *superior colliculus* of the mesencephalon. ∞ p. 478 These movements attempt to keep your gaze focused on a specific point in space, despite changes in body position and orientation. If your body is turning or spinning rapidly, your eyes will fix on one point for a moment and then jump ahead to another in a series of short, jerky movements.

This type of eye movement can occur even when the body is stationary if either the brain stem or the inner ear is damaged. An individual with this condition, which is called **nystagmus**, has trouble controlling his or her eye movements. Physicians commonly check for nystagmus by asking the patient to watch a small penlight as it is moved across the field of vision. **AM** Vertigo and Ménière's Disease

☐ HEARING

The receptors of the cochlear duct provide a sense of hearing that enables us to detect the quietest whisper, yet remain functional, in a crowded, noisy room. The receptors responsible for auditory sensations are hair cells similar to those of the vestibular complex. However, their placement within the cochlear duct and the organization of the surrounding accessory structures shield them from stimuli other than sound.

In conveying vibrations from the tympanic membrane to the oval window, the auditory ossicles convert pressure fluctuations in air to much greater pressure fluctuations in the perilymph of the cochlea. These fluctuations stimulate hair cells along the cochlear spiral. The *frequency* of the perceived sound is determined by *which part* of the cochlear duct is stimulated. The *intensity* (volume) of the perceived sound is determined by *how many* of the hair cells at that location are stimulated. We shall now consider the mechanics of this remarkably elegant process.

The Cochlear Duct

In sectional view (Figures 17–25a,b and 17–26a,b●), the cochlear duct, or *scala media*, lies between a pair of perilymphatic chambers: the **vestibular duct** (*scala vestibuli*) and the **tympanic duct** (*scala tympani*). The outer surfaces of these ducts are encased by the bony labyrinth everywhere, except at the oval window (the base of the vestibular duct) and the round window (the base of the tympanic duct). Because the vestibular and tympanic ducts are interconnected at the tip of the cochlear spiral, they really form one long and continuous perilymphatic chamber that begins at the oval window, extends through the vestibular duct, around the top of the cochlea, and along the tympanic duct, and ends at the round window.

The cochlear duct is an elongated tubelike structure suspended between the vestibular duct and the tympanic duct. The hair cells of the cochlear duct are located in a structure called

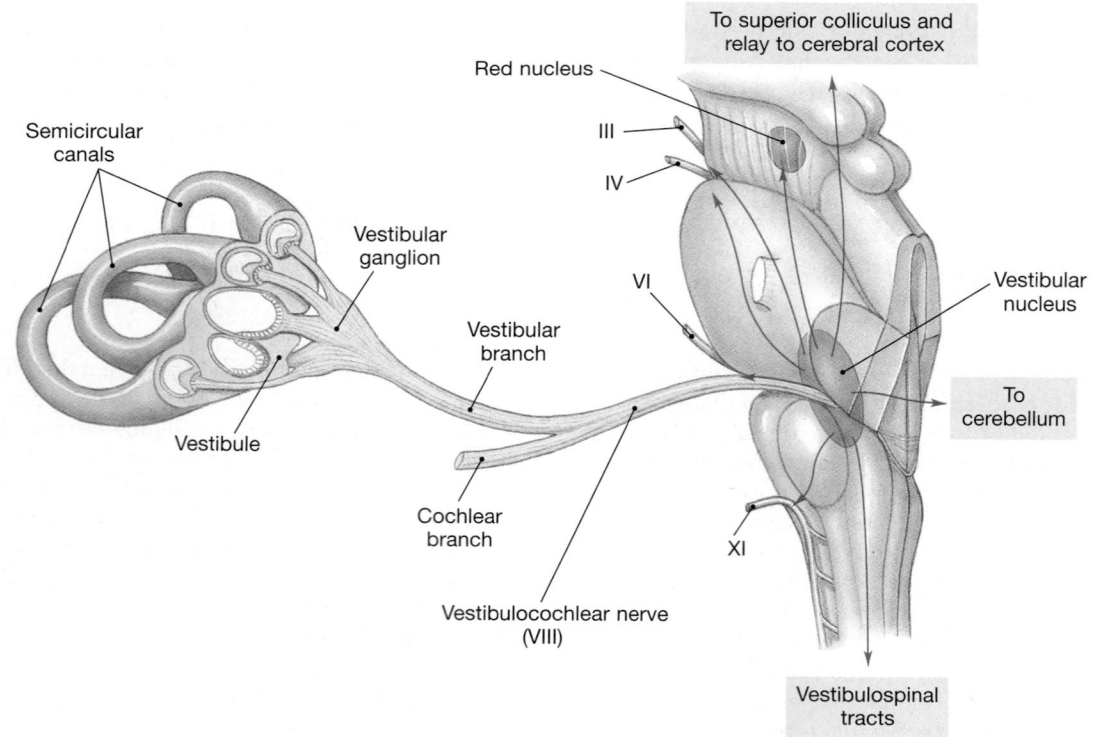

●FIGURE 17–24
Pathways for Equilibrium Sensations

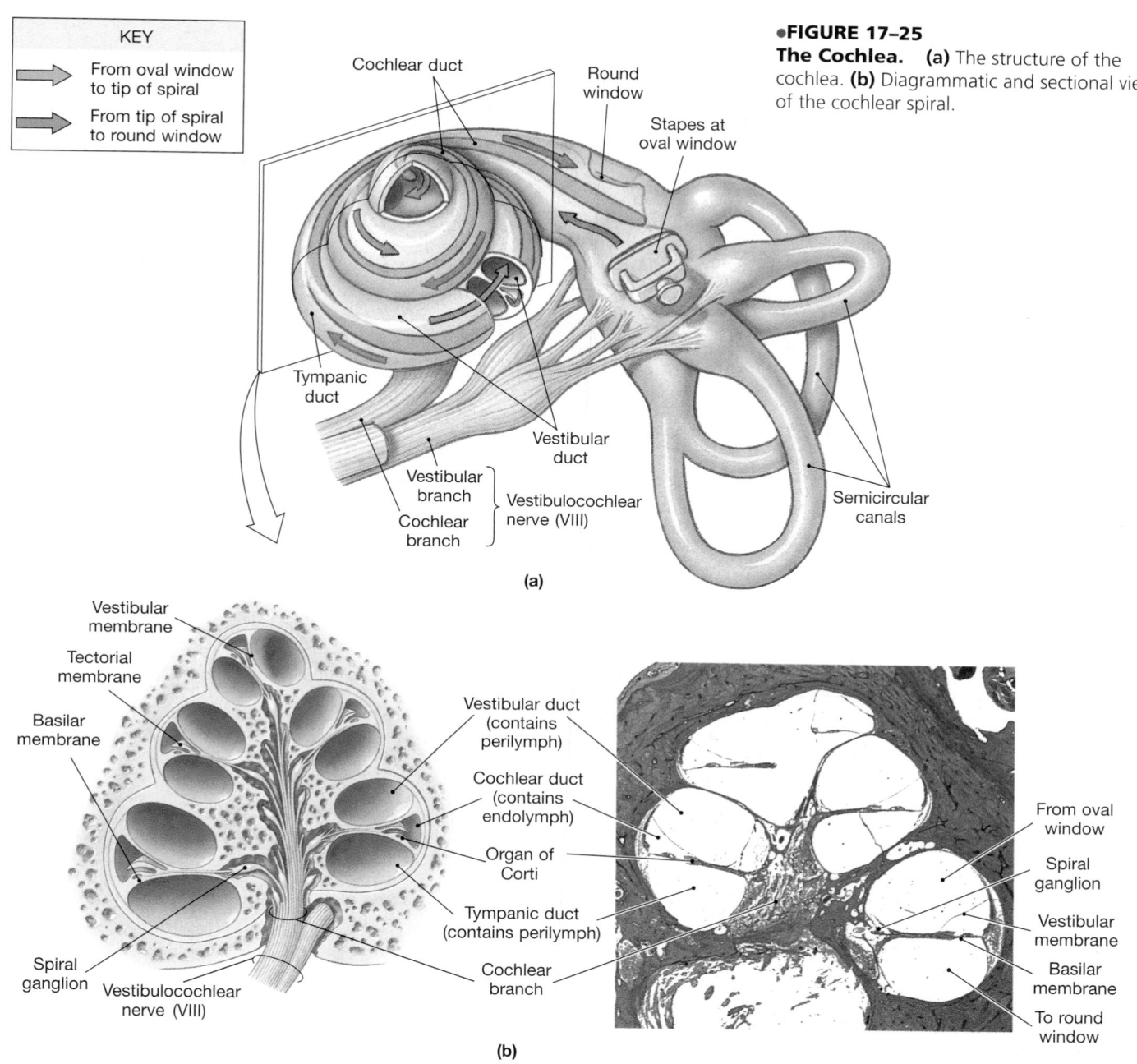

KEY

From oval window
to tip of spiral

From tip of spiral
to round window

Cochlear duct

Round
window

Stapes at
oval window

Tympanic
duct

Vestibular
duct

Vestibular
branch

Cochlear
branch

Vestibulocochlear
nerve (VIII)

Semicircular
canals

(a)

●**FIGURE 17–25**
The Cochlea. **(a)** The structure of the
cochlea. **(b)** Diagrammatic and sectional views
of the cochlear spiral.

Vestibular
membrane

Tectorial
membrane

Basilar
membrane

Spiral
ganglion

Vestibulocochlear
nerve (VIII)

Vestibular duct
(contains
perilymph)

Cochlear duct
(contains
endolymph)

Organ of
Corti

Tympanic duct
(contains perilymph)

Cochlear
branch

From oval
window

Spiral
ganglion

Vestibular
membrane

Basilar
membrane

To round
window

(b)

the **organ of Corti** (Figures 17–25b and 17–26a,b●). This sensory structure sits on the **basilar membrane**, a membrane that separates the cochlear duct from the tympanic duct. The hair cells are arranged in a series of longitudinal rows. They lack kinocilia, and their stereocilia are in contact with the overlying **tectorial** (tek-TOR-ē-al) **membrane** (*tectum*, roof). This membrane is firmly attached to the inner wall of the cochlear duct. When a portion of the basilar membrane bounces up and down, the stereocilia of the hair cells are pressed aga inst the tectorial membrane and become distorted. The basilar membrane moves in response to pressure fluctuations within the perilymph. These pressure changes are triggered by sound waves arriving at the tympanic membrane. To understand this process, we must consider the basic properties of sound.

An Introduction to Sound

Hearing is the detection of sound, which consists of waves of pressure conducted through a medium such as air or water. In air, each *pressure wave* consists of a region where the air molecules are crowded together and an adjacent zone where they are farther apart (Figure 17–27a●). These waves are sine waves—that is, S-shaped curves that repeat in a regular pattern—and travel through the air at about 1235 km/h (768 mph).

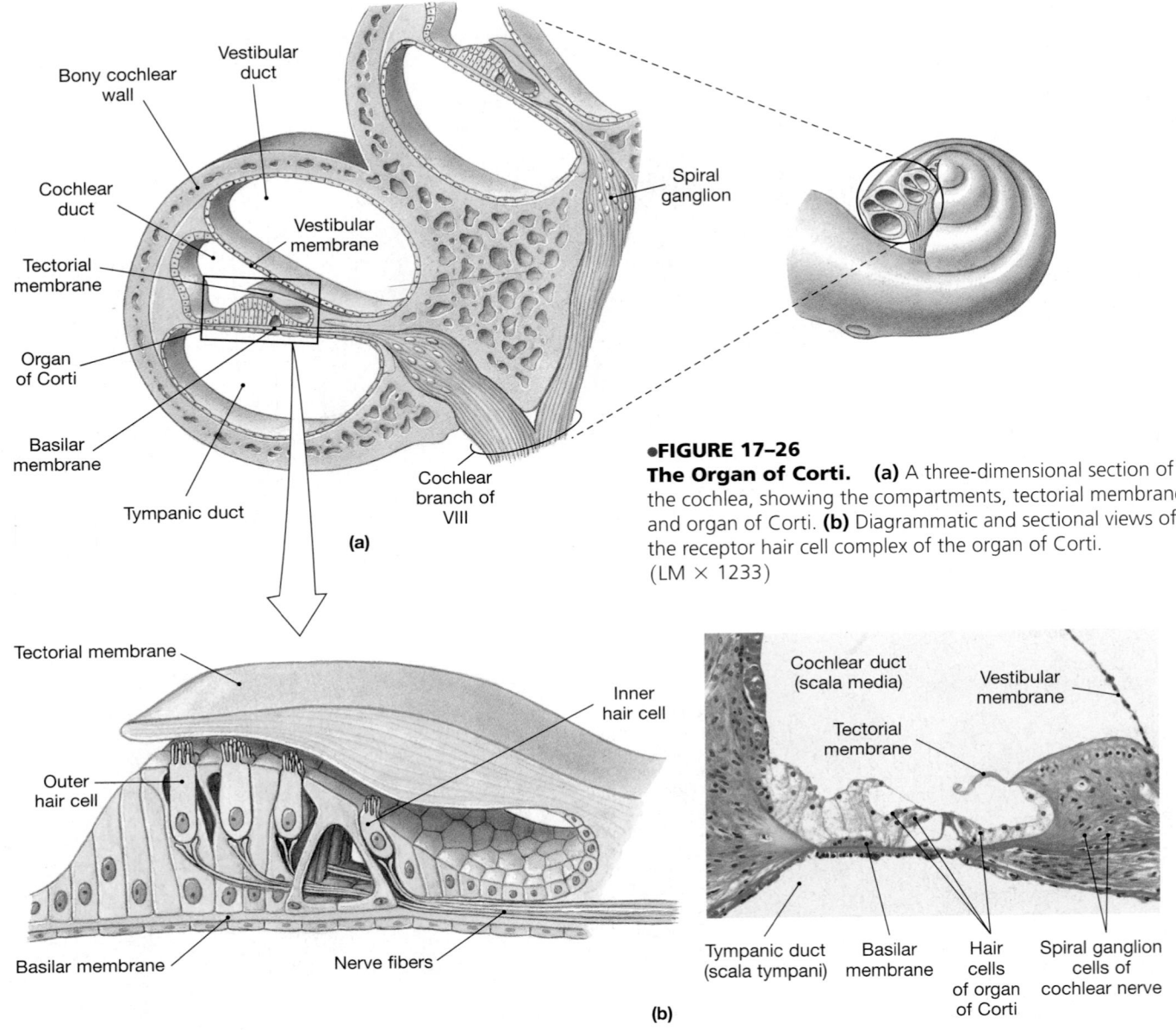

●FIGURE 17–26
The Organ of Corti. **(a)** A three-dimensional section of the cochlea, showing the compartments, tectorial membrane, and organ of Corti. **(b)** Diagrammatic and sectional views of the receptor hair cell complex of the organ of Corti. (LM × 1233)

The *wavelength* of sound is the distance between two adjacent wave crests (peaks) or, equivalently, the distance between two adjacent wave troughs (Figure 17–27b●). Wavelength is inversely related to **frequency**—the number of waves that pass a fixed reference point in a given time. Physicists use the term **cycles** rather than *waves*. Hence, the frequency of a sound is measured in terms of the number of cycles per second (cps), a unit called **hertz (Hz)**. What we perceive as the **pitch** of a sound is our sensory response to its frequency. A *high-frequency* sound (high pitch, short wavelength) might have a frequency of 15,000 Hz or more; a very *low-frequency* sound (low pitch, long wavelength) could have a frequency of 100 Hz or less.

It takes energy to produce sound waves. When you strike a tuning fork, it vibrates and pushes against the surrounding air, producing sound waves whose frequency depends on the instrument's frequency of vibration. The harder you strike the tuning fork, the more energy you provide and the louder the sound. The loudness increases because the sound waves carry more energy away with them. The energy content, or *power*, of a sound determines its **intensity**, or volume. Intensity is reported in **decibels** (DES-i-belz). Table 17–1 indicates the decibel levels of familiar sounds.

When sound waves strike an object, their energy is a physical pressure. You may have seen windows move in a room in which a stereo is blasting. The more flexible the object, the more easily it will respond to sound pressure. Even soft stereo music will vibrate a sheet of paper held in front of the speaker. Given the right combination of frequencies and intensities

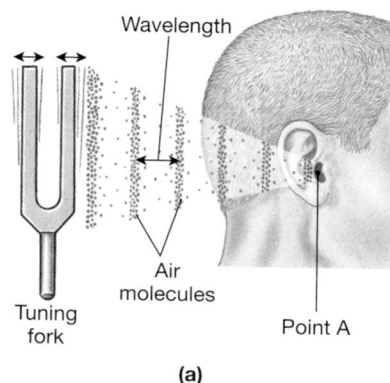

(a)

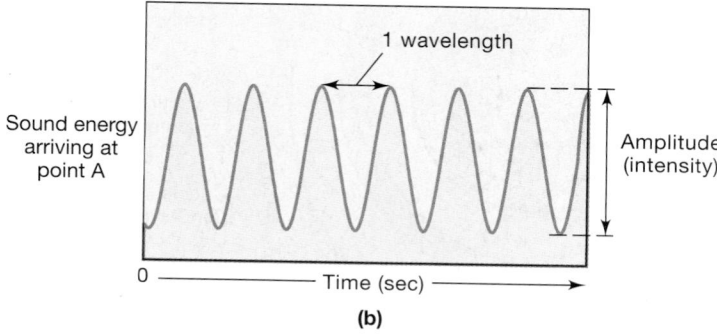

(b)

•**FIGURE 17-27**
Production of Sound. **(a)** Sound waves generated by a tuning fork travel through the air as pressure waves. The frequency of a sound wave is the number of waves that pass a fixed reference point in a given time. Frequencies are reported in cycles per second (cps), or hertz (Hz). **(b)** A graph of sound energy arriving at the tympanic membrane. The amount of energy in each wave determines the intensity of the sound and the amplitude of each wave. The distance between wave peaks is the wavelength. The number of waves arriving each second is the frequency, which we perceive as pitch.

TABLE 17-1 INTENSITIES OF VARIOUS SOUNDS

Typical Decibel Level	Example	Dangerous Time Exposure
0	Lowest audible sound	
30	Quiet library; soft whisper	
40	Quiet office; living room; bedroom away from traffic	
50	Light traffic at a distance; refrigerator; gentle breeze	
60	Air conditioner at 20 feet; conversation; sewing machine in operation	
70	Busy traffic; noisy restaurant	Some damage if continuous
80	Subway; heavy city traffic; alarm clock at 2 feet; factory noise	More than 8 hours
90	Truck traffic; noisy home appliances; shop tools; gas lawn mower	Less than 8 hours
100	Chain saw; boiler shop; pneumatic drill	2 hours
120	"Heavy metal" rock concert; sandblasting; thunderclap nearby	Immediate danger
140	Gunshot; jet plane	Immediate danger
160	Rocket launching pad	Hearing loss inevitable

of sound, an object will begin to vibrate at the same frequency as the sound, a phenomenon called *resonance*. The greater the intensity of a sound, the greater is the amount of vibration produced. For you to be able to hear any sound, your thin, flexible tympanic membrane must vibrate in resonance with the sound waves.

Probably more than 6 million people in the United States alone have at least a partial hearing deficit. **Conductive deafness** results from conditions in the outer or middle ear that block the normal transfer of vibrations from the tympanic membrane to the oval window. An external acoustic canal plugged with accumulated wax or trapped water can cause a temporary hearing loss. Scarring or perforation of the tympanic membrane and immobilization of one or more of the auditory ossicles are more serious causes of conductive deafness.

In **nerve deafness**, the problem lies within the cochlea or somewhere along the auditory pathway. The vibrations reach the oval window and enter the perilymph, but either the receptors cannot respond or their response cannot reach its central destinations. For example,

• Very loud (high-intensity) sounds can produce nerve deafness by breaking stereocilia off the surfaces of the hair cells. (The reflex contraction of the tensor tympani and stapedius muscles in response to a dangerously loud noise occurs in less than 0.1 second, but this may not be fast enough to prevent damage.)

• Drugs such as the aminoglycoside antibiotics (*neomycin* and *gentamicin*) can diffuse into the endolymph and kill hair cells. Because hair cells and sensory nerves can also be damaged by bacterial infection, the potential side effects must be balanced against the severity of infection.

Many treatment options are available for conductive deafness; treatment options for nerve deafness are relatively limited. Because many of these problems become progressively worse, early diagnosis improves the chances of successful treatment.

AM Testing and Treating Hearing Deficits

SUMMARY TABLE 17–2 STEPS IN THE PRODUCTION OF AUDITORY SENSATIONS

STEP ①: Sound waves arrive at the tympanic membrane.

STEP ②: Movement of the tympanic membrane causes displacement of the auditory ossicles.

STEP ③: Movement of the stapes at the oval window establishes pressure waves in the perilymph of the vestibular duct.

STEP ④: The pressure waves distort the basilar membrane on their way to the round window of the tympanic duct.

STEP ⑤: Vibration of the basilar membrane causes vibration of hair cells against the tectorial membrane.

STEP ⑥: Information about the region and intensity of stimulation is relayed to the CNS over the cochlear branch of cranial nerve VIII.

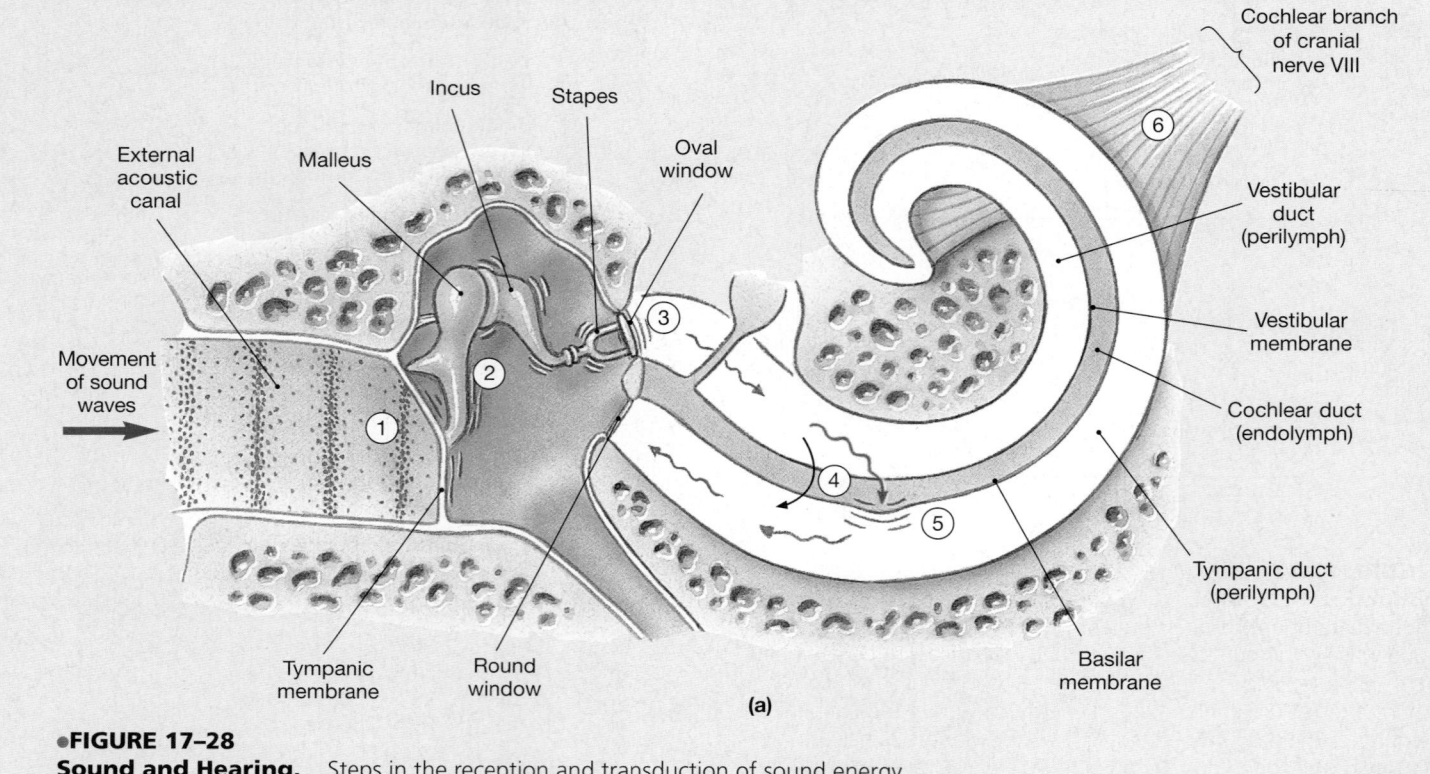

●**FIGURE 17–28**
Sound and Hearing. Steps in the reception and transduction of sound energy.

The Hearing Process

The process of hearing can be divided into six basic steps, summarized in Table 17–2 and diagrammed in Figure 17–28a●:

STEP 1: *Sound Waves Arrive at the Tympanic Membrane.* Sound waves enter the external acoustic canal and travel toward the tympanic membrane. The orientation of the canal provides some directional sensitivity. Sound waves approaching a particular side of the head have direct access to the tympanic membrane on that side, whereas sounds arriving from another direction must bend around corners or pass through the auricle or other body tissues.

STEP 2: *Movement of the Tympanic Membrane Causes Displacement of the Auditory Ossicles.* The tympanic membrane provides a surface for the collection of sound, and it vibrates in resonance to sound waves with frequencies between approximately 20 and 20,000 Hz (in a young child). When the tympanic membrane vibrates, so do the malleus and, through their articulations, the incus and stapes. In this way, the sound is amplified.

STEP 3: *Movement of the Stapes at the Oval Window Establishes Pressure Waves in the Perilymph of the Vestibular Duct.* Liquids

are incompressible: If you push down on one part of a water bed, the bed bulges somewhere else. Because the rest of the cochlea is sheathed in bone, pressure applied at the oval window can be relieved only at the round window. Although the stapes actually has a rocking movement, the in-out component is easiest to visualize and describe. Basically, when the stapes moves inward, the round window bulges outward. As the stapes moves in and out, vibrating at the frequency of the sound arriving at the tympanic membrane, it creates pressure waves within the perilymph.

STEP 4: *The Pressure Waves Distort the Basilar Membrane On Their Way to the Round Window of the Tympanic Duct.* The pressure waves established by the movement of the stapes travel through the perilymph of the vestibular and tympanic ducts to reach the round window. In doing so, the waves distort the basilar membrane. The location of maximum distortion varies with the frequency of the sound, owing to regional differences in the width and flexibility of the basilar membrane along its length. High-frequency sounds, which have a very short wavelength, vibrate the basilar membrane near the oval window. The lower the frequency of the sound, the longer is the wavelength, and the farther from the oval window the area of maximum distortion will be (Figure

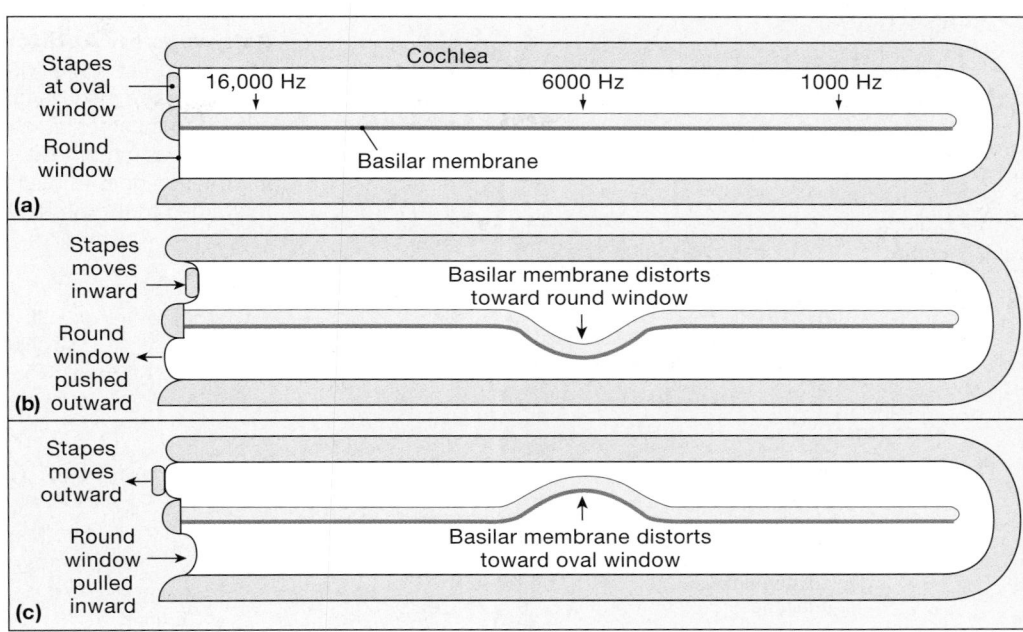

●FIGURE 17–29
Sound and Hearing. **(a)** The flexibility of the basilar membrane changes along its length, so pressure waves of different frequencies affect different parts of the membrane. **(b, c)** The effects of a vibration of the stapes at a frequency of 6000 Hz. When the stapes moves inward, as in part (b), the basilar membrane distorts toward the round window, which bulges into the middle-ear cavity. When the stapes moves outward, as in part (c), the basilar membrane rebounds and distorts toward the oval window.

17–29a–c●). Thus, information about frequency is translated into information about *position*.

The *amount* of movement at a given location depends on the amount of force applied by the stapes, which in turn is a function of the sound intensity. The louder the sound, the more the basilar membrane moves.

STEP 5: *Vibration of the Basilar Membrane Causes Vibration of Hair Cells Against the Tectorial Membrane.* Vibration of the affected region of the basilar membrane moves hair cells against the tectorial membrane. This movement leads to the displacement of the stereocilia, which in turn opens ion channels in their cell membranes. The resulting inrush of ions depolarizes the hair cells, leading to the release of neurotransmitters and thus to the stimulation of sensory neurons.

The hair cells of the organ of Corti are arranged in several rows. A very soft sound may stimulate only a few hair cells in a portion of one row. As the intensity of a sound increases, not only do these hair cells become more active, but additional hair cells—at first in the same row and then in adjacent rows—are stimulated as well. The number of hair cells responding in a given region of the organ of Corti thus provides information on the intensity of the sound.

STEP 6: *Information About the Region and Intensity of Stimulation Is Relayed to the CNS Over the Cochlear Branch of the Vestibulocochlear Nerve (VIII).* The cell bodies of the bipolar sensory neurons that monitor the cochlear hair cells are located at the center of the bony cochlea, in the **spiral ganglion** (Figure 17–26a●). From there, the information is carried by cranial nerve VIII to the cochlear nuclei of the medulla oblongata for subsequent distribution to other centers in the brain.

Auditory Pathways

Stimulation of hair cell activates sensory neurons whose cell bodies are in the adjacent spiral ganglion. The afferent fibers of those neurons form the **cochlear branch** of the vestibulocochlear nerve (VIII) (Figure 17–30●). These axons enter the medulla oblongata, where they synapse at the **cochlear nucleus.** From there, the information crosses to the opposite side of the brain and ascends to the inferior colliculus of the mesencephalon. This processing center coordinates a number of responses to acoustic stimuli, including auditory reflexes that involve skeletal muscles of the head, face, and trunk. These reflexes automatically change the position of your head in response to a sudden loud noise; you usually turn your head and your eyes toward the source of the sound.

Before reaching the cerebral cortex and your conscious awareness, ascending auditory sensations synapse in the medial geniculate nucleus of the thalamus. Projection fibers then deliver the information to the auditory cortex of the temporal lobe. Information travels to the cortex over labeled lines: High-frequency sounds activate one portion of the cortex, low-frequency sounds another. In effect, the auditory cortex contains a map of the organ of Corti. Thus, information about *frequency*, translated into information about *position* on the basilar membrane, is projected in that form onto the auditory cortex, where it is interpreted to produce your subjective sensation of pitch.

An individual whose auditory cortex is damaged will respond to sounds and have normal acoustic reflexes, but will find it difficult or impossible to interpret the sounds and recognize a pattern in them. Damage to the adjacent association area leaves the ability to detect the tones and patterns, but produces an inability to comprehend their meaning.

Auditory Sensitivity

Our hearing abilities are remarkable, but it is difficult to assess the absolute sensitivity of the system. The range from the softest audi-

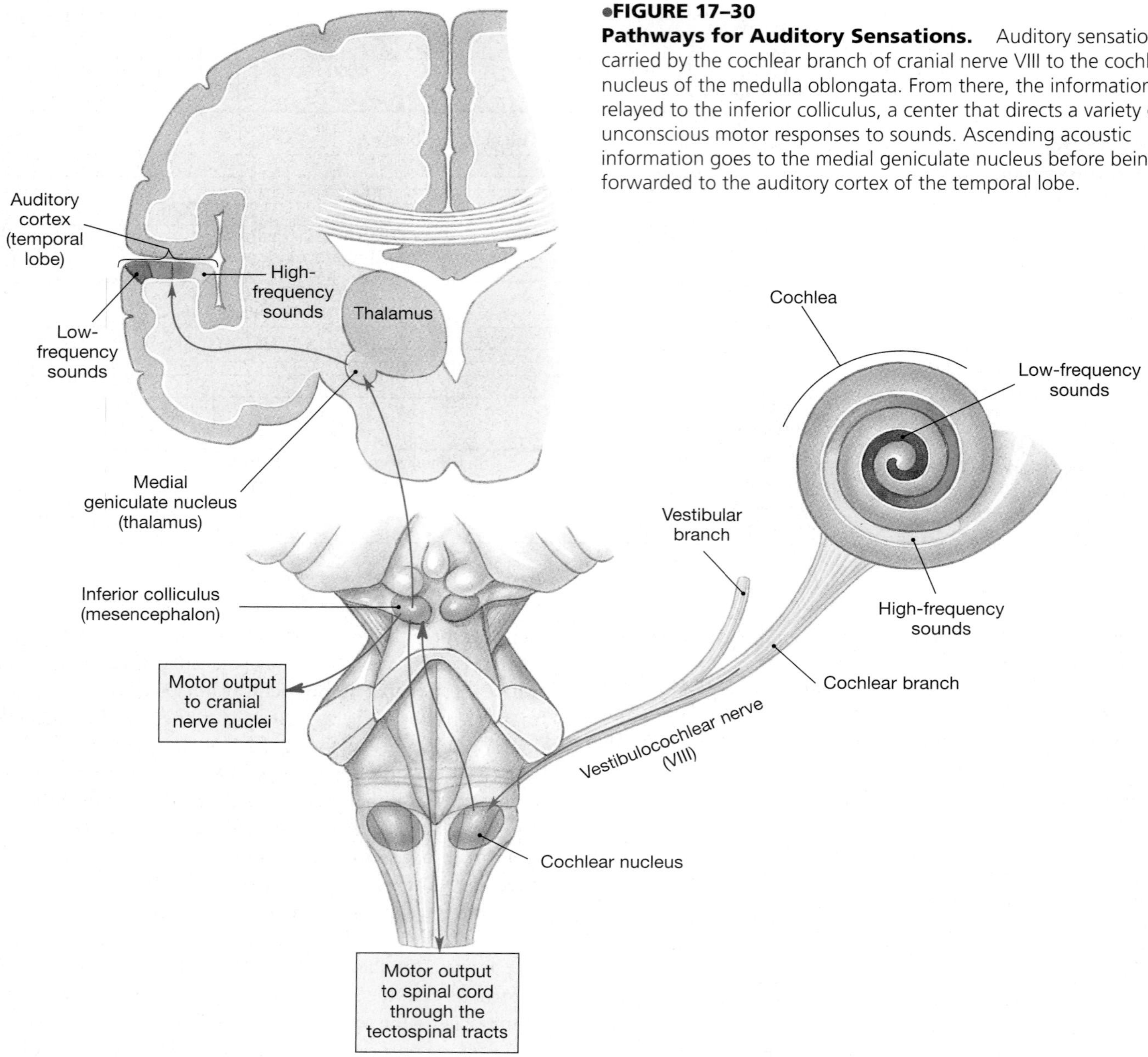

●FIGURE 17–30
Pathways for Auditory Sensations. Auditory sensations are carried by the cochlear branch of cranial nerve VIII to the cochlear nucleus of the medulla oblongata. From there, the information is relayed to the inferior colliculus, a center that directs a variety of unconscious motor responses to sounds. Ascending acoustic information goes to the medial geniculate nucleus before being forwarded to the auditory cortex of the temporal lobe.

Auditory cortex (temporal lobe)

High-frequency sounds

Thalamus

Low-frequency sounds

Medial geniculate nucleus (thalamus)

Inferior colliculus (mesencephalon)

Motor output to cranial nerve nuclei

Cochlea

Low-frequency sounds

Vestibular branch

High-frequency sounds

Cochlear branch

Vestibulocochlear nerve (VIII)

Cochlear nucleus

Motor output to spinal cord through the tectospinal tracts

ble sound to the loudest tolerable blast represents a trillionfold increase in power. The receptor mechanism is so sensitive that, if we were to remove the stapes, we could, in theory, hear air molecules bouncing off the oval window. We never use the full potential of this system, because body movements and our internal organs produce squeaks, groans, thumps, and other sounds that are tuned out by central and peripheral adaptation. When other environmental noises fade away, the level of adaptation drops and the system becomes increasingly sensitive. For example, when you relax in a quiet room, your heartbeat seems to get louder and louder as the auditory system adjusts to the level of background noise.

Young children have the greatest hearing range: They can detect sounds ranging from a 20-Hz buzz to a 20,000-Hz whine. With age, damage due to loud noises or other injuries accumulates. The tym-

panic membrane gets less flexible, the articulations between the ossicles stiffen, and the round window may begin to ossify. As a result, older individuals show some degree of hearing loss.

✓ If the round window were not able to bulge out with increased pressure in the perilymph, how would the perception of sound be affected?

✓ How would the loss of stereocilia from hair cells of the organ of Corti affect hearing?

✓ Why does a blockage of the auditory tube produce an earache?

Answers start on page Q-1

 The process of hearing is covered in detail on the **Companion Website**: Chapter 17/Tutorials.

Chapter Review

SELECTED CLINICAL TERMINOLOGY

Terms Discussed in This Chapter

astigmatism: Reduction in visual acuity due to a curvature irregularity in the cornea or lens. *(p. 578)*

cataract: An abnormal lens that has lost its transparency. *(p. 577)*

color blindness: A condition in which a person is unable to distinguish certain colors. *(p. 585)*

conductive deafness: Deafness resulting from conditions in the outer or middle ear that block the transfer of vibrations from the tympanic membrane to the oval window. *(p. 597 and AM)*

detached retina: Delamination of a portion of the neural retina, which separates the photoreceptor layer from the pigment layer. If untreated, blindness can result in the affected area. *(p. 574)*

diabetic retinopathy: Deterioration of the retinal photoreceptor layer due to vascular damage and the overgrowth and rupture of blood vessels on the retinal surface. *(p. 574)*

glaucoma: A condition characterized by increased intraocular pressure due to the impaired reabsorption of aqueous humor; can result in blindness. *(p. 577)*

hyperopia, or *farsightedness:* A condition in which nearby objects are blurry, but distant objects are clear. *(p. 579)*

motion sickness: A condition resulting from conflicting visual and equilibrium sensory stimuli. Symptoms can include headache, sweating, nausea, vomiting, and changes in mental state. *(p. 593)*

myopia, or *nearsightedness:* A condition in which vision at close range is normal, but distant objects appear blurry. *(p. 579)*

nerve deafness: Deafness resulting from problems within the cochlea or along the auditory pathway. *(p. 597 and AM)*

night blindness: Loss of visual acuity under dim light conditions due to inadequate visual pigment production, usually as a result of vitamin A deficiency. *(p. 582)*

nystagmus: Abnormal eye movements that may appear after the brain stem or inner ear is damaged. *(p. 594)*

otitis media: Infection and tissue inflammation within the middle ear cavity. *(p. 589 and AM)*

presbyopia: A type of hyperopia that develops with age as lenses become less elastic. *(p. 579)*

retinitis pigmentosa: A group of inherited retinopathies characterized by the progressive deterioration of photoreceptors, eventually resulting in blindness. *(p. 582)*

scotomas: Abnormal blind spots that are fixed in position. *(p. 580)*

AM Additional Terms Discussed in the *Applications Manual*

mastoiditis
Ménière's disease
vertigo

STUDY OUTLINE

17–1 OLFACTION   p. 565

OLFACTORY RECEPTORS   p. 566

1. The **olfactory organs** contain the **olfactory epithelium** with **olfactory receptors,** supporting cells, and **basal (stem) cells.** The surfaces of the olfactory organs are coated with the secretions of the **olfactory glands.** *(Figure 17–1)*
2. The olfactory receptors are modified neurons.
3. Olfactory reception involves detecting dissolved chemicals as they interact with odorant-binding proteins.

OLFACTORY PATHWAYS   p. 566

4. In olfaction, the arriving information reaches the arriving information centers without first synapsing in the thalamus. *(Figure 17–1)*

OLFACTORY DISCRIMINATION   p. 567

5. The olfactory system can distinguish thousands of chemical stimuli. The CNS interprets smells by the pattern of receptor activity.
6. The olfactory receptor population shows considerable turnover. The number of olfactory receptors declines with age.

17–2 GUSTATION   p. 567

1. **Taste** (gustatory) **receptors** are clustered in **taste buds.**
2. Taste buds are associated with epithelial projections on the dorsal surface of the tongue (**lingual papillae**). *(Figure 17–2)*

TASTE RECEPTORS   p. 567

3. Each taste bud contains **basal cells,** which appear to be stem cells, and **gustatory cells,** which extend *taste hairs* through a narrow **taste pore.** *(Figure 17–2)*

GUSTATORY PATHWAYS   p. 568

4. The taste buds are monitored by cranial nerves that synapse within the solitary nucleus of the medulla oblongata and then on to the thalamus and the primary sensory cortex.

GUSTATORY DISCRIMINATION   p. 568

5. The **primary taste sensations** are sweet, salt, sour, and bitter. Receptors also exist for **umami** and **water.**
6. Taste sensitivity exhibits significant individual differences, some of which are inherited.
7. The number of taste buds declines with age.

17–3 VISION   p. 569

ACCESSORY STRUCTURES OF THE EYE   p. 569

1. The **accessory structures** of the eye include the **eyelids** (**palpebrae**), separated by the **palpebral fissure,** the **eyelashes,** and the **tarsal glands.** *(Figures 17–3, 17–4)*
2. An epithelium called the **conjunctiva** covers most of the exposed surface of the eye. The **cornea** is transparent. *(Figures 17–3, 17–4)*
3. The secretions of the **lacrimal gland** contain **lysozyme.** Tears collect in the **lacrimal lake** and reach the inferior meatus of the nose after they pass through the **lacrimal puncta,** the **lacrimal canaliculi,** the **lacrimal sac,** and the **nasolacrimal duct.** *(Figures 17–3, 17–4)*

THE EYE   p. 571

4. The eye has three layers: an outer **fibrous tunic,** a middle **vascular tunic,** and an inner **neural tunic.** *(Figures 17–3, 17–4)*

5. The fibrous tunic consists of the **sclera,** the cornea, and the **limbus.** *(Figure 17–4)*

6. The vascular tunic, or **uvea,** includes the **iris,** the **ciliary body,** and the **choroid.** The iris contains muscle fibers that change the diameter of the **pupil.** The ciliary body contains the **ciliary muscle** and the **ciliary processes,** which attach to the **suspensory ligaments** of the *lens. (Figures 17–4, 17–5)*

7. The neural tunic, or **retina,** consists of an outer *pigmented part* and an inner *neural part*; the latter contains visual receptors and associated neurons. *(Figures 17–4, 17–6)*

8. The retina contains two types of **photoreceptors: rods** and **cones.**

9. Cones are densely clustered in the **fovea,** at the center of the **macula lutea.** *(Figure 17–6)*

10. The direct line to the CNS proceeds from the **photoreceptors** to **bipolar cells,** then to **ganglion cells,** and, finally, to the brain via the optic nerve. The axons of ganglion cells converge at the **optic disc,** or **blind spot. Horizontal cells** and **amacrine cells** modify the signals passed among other components of the retina. *(Figure 17–6)*

11. The ciliary body and lens divide the interior of the eye into a large **posterior cavity,** or *vitreous chamber,* and a smaller **anterior cavity.** The anterior cavity is subdivided into the **anterior chamber,** which extends from the cornea to the iris, and a **posterior chamber,** between the iris and the ciliary body and lens. *(Figure 17–8)*

12. The fluid **aqueous humor** circulates within the eye and reenters the circulation after diffusing through the walls of the anterior chamber and into the **canal of Schlemm.** *(Figure 17–8)*

13. The **lens** lies posterior to the cornea and forms the anterior boundary of the posterior cavity. This cavity contains the **vitreous body,** a gelatinous mass that helps stabilize the shape of the eye and support the retina. *(Figure 17–8)*

14. The lens focuses a visual image on the photoreceptors. A lens that has lost its transparency is a **cataract.**

15. Light is **refracted** (bent) when it passes through the cornea and lens. During **accommodation,** the shape of the lens changes to focus an image on the retina. "Normal" **visual acuity** is rated 20/20. *(Figures 17–9 to 17–12)*

Structures of the eye: **Companion Website:** Chapter 17/Tutorials.

VISUAL PHYSIOLOGY   p. 580

16. The two types of photoreceptors are **rods,** which respond to almost any photon, regardless of its energy content, and **cones,** which have characteristic ranges of sensitivity. *(Figure 17–13)*

17. Each photoreceptor contains an **outer segment** with membranous **discs.** A narrow stalk connects the outer segment to the **inner segment.** Light absorption occurs in the **visual pigments,** which are derivatives of **rhodopsin** (opsin plus the pigment retinal, which is synthesized from vitamin A). *(Figures 17–13 to 17–15)*

18. Color sensitivity depends on the integration of information from **red, green,** and **blue** cones. **Color blindness** is the inability to detect certain colors. *(Figures 17–16, 17–17)*

19. In the **dark-adapted state,** most visual pigments are fully receptive to stimulation. In the **light-adapted state,** the pupil constricts and **bleaching** of the visual pigments occurs.

THE VISUAL PATHWAY   p. 585

20. The ganglion cells that monitor rods, called **M cells** *(magnocells),* are relatively large. The ganglion cells that monitor cones, called **P cells** *(parvo cells),* are smaller and more numerous. *(Figure 17–18)*

21. Visual data from the left half of the combined field of vision arrive at the visual cortex of the right occipital lobe; data from the right half of the combined field of vision arrive at the visual cortex of the left occipital lobe. *(Figure 17–19)*

22. **Depth perception** is obtained by comparing relative positions of objects between the left- and right-eye images *(Figure 17–19)*

23. Visual inputs to the suprachiasmatic nucleus of the hypothalamus affect the function of other brain stem nuclei. This nucleus establishes a visceral **circadian rhythm,** which is tied to the day–night cycle and affects other metabolic processes.

17–4 EQUILIBRIUM AND HEARING   p. 587

1. The senses of equilibrium and hearing are provided by the receptors of the inner ear.

ANATOMY OF THE EAR   p. 587

2. The ear is divided into the **external ear,** the **middle ear,** and the **inner ear.** *(Figure 17–20)*

3. The external ear includes the **auricle,** or *pinna,* which surrounds the entrance to the **external acoustic canal,** which ends at the **tympanic membrane** *(eardrum). (Figure 17–20)*

4. The middle ear communicates with the nasopharynx via the **auditory** *(pharyngotympanic)* **tube.** The middle ear encloses and protects the **auditory ossicles.** *(Figures 17–20, 17–21)*

5. The **membranous labyrinth** (the chambers and tubes) of the inner ear contains the fluid **endolymph.** The **bony labyrinth** surrounds and protects the membranous labyrinth and can be subdivided into the **vestibule,** the **semicircular canals,** and the **cochlea.** *(Figures 17–20, 17–22)*

6. The vestibule of the inner ear encloses the **saccule** and **utricle.** The semicircular canals contain the **semicircular ducts.** The cochlea contains the **cochlear duct,** an elongated portion of the membranous labyrinth. *(Figure 17–22)*

7. The **round window** separates the perilymph from the air spaces of the middle ear. The **oval window** is connected to the base of the stapes. *(Figure 17–20)*

8. The basic receptors of the inner ear are **hair cells,** which provide information about the direction and strength of mechanical stimuli. *(Figure 17–22)*

EQUILIBRIUM   p. 590

9. The **anterior, posterior,** and **lateral semicircular ducts** are continuous with the utricle. Each duct contains an **ampulla** with a gelatinous **cupula** and associated sensory receptors. *(Figure 17–23)*

10. The saccule and utricle are connected by a passageway that is continuous with the **endolymphatic duct,** which terminates in the **endolymphatic sac.** In the saccule and utricle, hair cells cluster within **maculae,** where their cilia contact the **otolith** (densely packed mineral crystals, called **statoconia,** in a matrix). *(Figure 17–23)*

11. The vestibular receptors activate sensory neurons of the **vestibular ganglia.** The axons form the **vestibular branch** of the vestibulocochlear nerve (VIII), synapsing within the **vestibular nuclei.** *(Figure 17–24)*

HEARING   p. 594

12. The cochlear duct lies between the **vestibular duct** and the **tympanic duct.** The hair cells of the cochlear duct lie within the **organ of Corti.** *(Figures 17–25, 17–26)*

13. The energy content of a sound determines its **intensity**, measured in **decibels**. Sound waves travel toward the tympanic membrane, which vibrates; the auditory ossicles conduct these vibrations to the inner ear. Movement at the **oval window** applies pressure to the **perilymph** of the **vestibular duct**. *(Table 17–1, Figures 17–27, 17–28 Summary Table 17–2)*

14. Pressure waves distort the **basilar membrane** and push the hair cells of the **organ of Corti** against the **tectorial membrane**. The **tensor tympani** and **stapedius muscles** contract to reduce the amount of motion when very loud sounds arrive. *(Figures 17–28, 17–29)*

15. The sensory neurons are located in the **spiral ganglion** of the cochlea. The afferent fibers of these neurons form the **cochlear branch** of the vestibulocochlear nerve (VIII), synapsing at the **cochlear nucleus**. *(Figure 17–30)*

Process of hearing: **Companion Website:** Chapter 17/Tutorials.

REVIEW QUESTIONS

 More assessment questions are available to you on the Companion Website. You will find Matching, Multiple Choice, True/False, and other quizzes to help further your understanding of the material covered in this chapter. To access the site, go to www.aw.com/martini.

LEVEL 1 Reviewing Facts and Terms

1. A reduction in sensitivity in the presence of a constant stimulus is
   (a) transduction   (b) sensory coding
   (c) line labeling   (d) adaptation

2. The anterior, transparent part of the fibrous tunic is the
   (a) cornea   (b) iris
   (c) sclera   (d) retina

3. The thick, gellike fluid that helps support the structure of the eyeball is the
   (a) vitreous humor   (b) aqueous humor
   (c) ora serrata   (d) perilymph

4. Pupillary muscle groups are controlled by the ANS. Parasympathetic activation causes pupillary _____, and sympathetic activation causes _____.
   (a) dilation, constriction   (b) dilation, dilation
   (c) constriction, dilation   (d) constriction, constriction

5. The retina is the
   (a) vascular tunic   (b) fibrous tunic
   (c) neural tunic   (d) a, b, and c are correct

6. At sunset, your visual system adapts to
   (a) fovea vision   (b) rod-based vision
   (c) macular vision   (d) cone-based vision

7. A better-than-average visual acuity rating is
   (a) 20/20   (b) 20/30
   (c) 15/20   (d) 20/15

8. The malleus, incus, and stapes are the tiny bones located in the
   (a) outer ear   (b) middle ear
   (c) inner ear   (d) membranous labyrinth

9. Receptors in the saccule and utricle provide sensations of
   (a) angular acceleration   (b) hearing
   (c) vibration   (d) gravity and linear acceleration

10. The organ of Corti is located in the _____ of the inner ear.
    (a) utricle   (b) bony labyrinth
    (c) vestibule   (d) cochlea

11. Auditory information about the frequency and intensity of stimulation is relayed to the CNS over the cochlear branch of cranial nerve
    (a) IV   (b) VI
    (c) VIII   (d) X

12. What are the three types of papillae on the human tongue?

13. (a) What structures make up the fibrous tunic of the eye?
    (b) What are the functions of the fibrous tunic?

14. What structures make up the vascular tunic of the eye?

15. What are the three auditory ossicles in the middle ear, and what are their functions?

LEVEL 2 Reviewing Concepts

16. Trace the olfactory pathway from the time an odor reaches the olfactory epithelium until it reaches its final destination in the brain.

17. Why are olfactory sensations long lasting and an important part of our memories and emotions?

18. What is the usual result if a sebaceous gland of an eyelash or a tarsal gland becomes infected?

19. Jane makes an appointment with the optometrist for a vision test. Her test result is reported as 20/15. What does this mean? Is a rating of 20/20 better or worse?

20. Trace the pathway of a nerve impulse from the optic nerve to the visual cortex.

21. What six basic steps are involved in the process of hearing?

LEVEL 3 Critical Thinking and Clinical Applications

22. You are at a park watching some deer 35 feet away from you. A friend taps you on the shoulder to ask a question. As you turn to look at your friend, who is standing just 2 feet away, what changes would your eyes undergo?

23. Sally's driver's license indicates that she must wear glasses when she drives. She does not need glasses to read or to see objects that are close. What type of lenses are in Sally's glasses?

24. After attending a Fourth of July fireworks extravaganza, Millie finds it difficult to hear normal conversation and her ears keep "ringing." What is causing her hearing problems?

25. For a few seconds after you ride the express elevator from the 20th floor to the ground floor, you still feel as if you are descending, even though you have come to a stop. Why?

26. Juan has a disorder involving the saccule and the utricle. He is asked to stand with his feet together and arms extended forward. As long as he keeps his eyes open, he exhibits very little movement. But when he closes his eyes, his body begins to sway a great deal, and his arms tend to drift in the direction of the impaired vestibular receptors. Why does this occur?

UNIT 3   CHAPTER 12 13 14 15 16 17 **18**

CLINICAL NOTES
- Disorders of ADH Production 618
- Disorders of Parathyroid
  Function 626
- Disorders of Aldosterone Secretion 629
- Abnormal Activity in the Zona
  Reticularis 630

CLINICAL DISCUSSIONS
- Growth Hormone Abnormalities 621
- Thyroid Gland Disorders 624
- Abnormal Glucocorticoid
  Production 629
- Diabetes Mellitus 633
- Endocrine Disorders 638
- Endocrinology and Athletic
  Performance 642

18

THE ENDOCRINE SYSTEM

To preserve homeostasis, cellular activities must be coordinated throughout the body. Neurons monitor or control specific cells or groups of cells. However, the number innervated is only a small fraction of the total number of cells in the body, and the commands issued are very specific and of relatively brief duration. Many life processes are not short lived. Reaching adult stature takes decades. Maintenance of reproductive capabilities requires continual control for at least 30 years in the typical female, and even longer in the male. There is no way that the nervous system can regulate long-term processes like growth, development, or reproduction, which involve or affect metabolic activities in virtually every cell and tissue. This type of regulation is provided by the endocrine system, which uses chemical messengers to relay information and instructions between cells. To understand how these messages are generated and interpreted, we need to take a closer look at how cells communicate with one another.

18–1 INTERCELLULAR COMMUNICATION

Objective

- Compare the endocrine and nervous systems.

Mechanisms of intercellular communication are listed in Table 18–1. In a few specialized cases, cellular activities are coordinated by the exchange of ions and molecules from one cell to the next across gap junctions. This **direct communication** occurs between two cells of the same type, and the cells must be in extensive physical contact. The two cells communicate so closely that they function as a single entity. For example, gap junctions (1) coordinate ciliary movement among epithelial cells, (2) coordinate the contractions of cardiac muscle cells, and (3) facilitate the propagation of action potentials from one neuron to the next at electrical synapses.

Direct communication is highly specialized and relatively rare. Most of the communication between cells involves the release and receipt of chemical messages. Each cell continuously "talks" to its neighbors by releasing chemicals into the extracellular fluid. These chemicals tell cells what their neighbors are doing at any moment. The result is the coordination of tissue function at the local level. The use of chemical messengers to transfer information from cell to cell within a single tissue is called **paracrine communication**. The chemicals involved are called paracrine factors, also known as *cytokines*, or *local hormones*. Examples of paracrine factors include the prostaglandins, introduced in Chapter 2, and the various growth factors discussed in Chapter 3. ∞ pp. 48, 104

605

TABLE 18–1 MECHANISMS OF INTERCELLULAR COMMUNICATION

Mechanism	Transmission	Chemical Mediators	Distribution of Effects
Direct communication	Through gap junctions	Ions, small solutes, lipid-soluble materials	Usually limited to adjacent cells of the same type that are interconnected by connexons
Paracrine communication	Through extracellular fluid	Paracrine factors (cytokines)	Primarily limited to local area, where concentrations are relatively high. Target cells must have appropriate receptors
Endocrine communication	Through the circulatory system	Hormones	Target cells are primarily in other tissues and organs and must have appropriate receptors
Synaptic communication	Across synaptic clefts	Neurotransmitters	Limited to very specific area. Target cells must have appropriate receptors

Paracrine factors enter the bloodstream, but the concentrations are usually so low that distant cells and tissues are not affected. However, some paracrine factors, including several of the prostaglandins and related chemicals, have primary effects in their tissues of origin and secondary effects in other tissues and organs. When secondary effects occur, the paracrine factors are also acting as hormones. **Hormones** are chemical messengers that are released in one tissue and transported in the bloodstream to reach specific cells in other tissues. Whereas most cells release paracrine factors, typical hormones are produced only by specialized cells. Nevertheless, the difference between paracrine factors and hormones is one of degree rather than an absolute distinction. Paracrine factors can diffuse out of their tissue of origin and have widespread effects, and hormones can affect their tissues of origin as well as distant targets.

In intercellular communication, hormones are like letters and the circulatory system is the postal service. A hormone released into the bloodstream will be distributed throughout the body. Each hormone has **target cells**, specific cells that respond to its presence. These cells possess the receptors needed to bind and "read" the hormonal message. Although every cell in the body is exposed to the mixture of hormones in circulation at any moment, each individual cell will respond to only a few of the hormones present. The other hormones are treated like junk mail and ignored, because the cell lacks the receptors to read the messages they contain. The use of hormones to coordinate cellular activities in tissues in distant portions of the body is called **endocrine communication**.

Hormones alter cellular operations by changing the types, quantities, or activities of important enzymes and structural proteins. In other words, a hormone may

- stimulate the synthesis of an enzyme or a structural protein not already present in the cytoplasm by activating appropriate genes in the cell nucleus;

- increase or decrease the rate of synthesis of a particular enzyme or other protein by changing the rate of transcription or translation; or
- turn an existing enzyme "on" or "off" by changing its shape or structure.

Through one or more of these mechanisms, a hormone can modify the physical structure or biochemical properties of its target cells. Because the target cells can be anywhere in the body, a single hormone can alter the metabolic activities of multiple tissues and organs simultaneously. These effects may be slow to appear, but they typically persist for days. Consequently, hormones are effective in coordinating cell, tissue, and organ activities on a sustained, long-term basis. For example, circulating hormones keep body water content and levels of electrolytes and organic nutrients within normal limits 24 hours a day throughout our entire lives.

While the effects of a single hormone persist, a cell may receive additional instructions from other hormones. The result will be a further modification of cellular operations. Gradual changes in the quantities and identities of circulating hormones can produce complex changes in physical structure and physiological capabilities. Examples include the processes of embryological and fetal development, growth, and puberty. Hormonal regulation is thus quite suitable for directing gradual, coordinated processes, but it is totally unable to handle situations requiring split-second responses. That kind of crisis management is the job of the nervous system.

The nervous system also relies primarily on chemical communication, but it does not use the bloodstream to deliver messages. Instead, neurons release a neurotransmitter at a synapse very close to the target cells that bear the appropriate receptors. The command to release the neurotransmitter rapidly travels from one location to another in the form of action potentials propagated along axons. The nervous system thus acts like a telephone company, carrying high-speed "messages" from one location in the body to another and delivering them to a specific destination. The effects of neural stimulation are generally short lived, and they tend to be restricted to specific target cells—primarily because the neurotransmitter is rapidly broken down or recycled. This form of **synaptic communication** is ideal for crisis management: If you are in danger of being hit by a speeding bus, the nervous system can coordinate and direct your leap to safety. Once the crisis is over and the neural circuit quiets down, things soon return to normal.

Table 18–1 provides a summary of the ways our cells and tissues communicate with one another. Viewed from a general perspective, the differences between the nervous and endocrine systems seem relatively clear. In fact, these broad organizational and functional distinctions are the basis for treating them as two separate systems. Yet, when we consider them in detail, we see that the two systems are organized along parallel lines:

- Both systems rely on the release of chemicals that bind to specific receptors on their target cells.

- The two systems share many chemical messengers; for example, norepinephrine and epinephrine are called *hormones* when released into the bloodstream, but *neurotransmitters* when released across synapses.

- Both systems are regulated primarily by negative feedback control mechanisms.

- The two systems share a common goal: to preserve homeostasis by coordinating and regulating the activities of other cells, tissues, organs, and systems.

In this chapter, we introduce the components and functions of the endocrine system and explore the interactions between the nervous and endocrine systems. In later chapters, we will consider specific endocrine organs, hormones, and functions in detail.

The **endocrine system** includes all the endocrine cells and tissues of the body which produce hormones or paracrine factors that have effects beyond their tissues of origin. As we noted in Chapter 4, *endocrine cells* are glandular secretory cells that release their secretions into the extracellular fluid. This characteristic distinguishes them from *exocrine cells*, which secrete their products onto epithelial surfaces, generally by way of ducts. ∞ p. 118 The chemicals released by endocrine cells may affect only adjacent cells, as in the case of most paracrine factors, or they may affect cells throughout the body.

The components of the endocrine system are introduced in Figure 18–1●, which also lists the major hormones produced in each endocrine tissue and organ. Some of these organs, such as the pituitary gland, have endocrine secretion as a primary function; others, such as the pancreas, have many other functions in

18–2 AN OVERVIEW OF THE ENDOCRINE SYSTEM

Objectives

- Compare the cellular components of the endocrine system with those of other tissues and systems.

- Compare the major chemical classes of hormones.

- Explain the general mechanisms of hormonal action.

- Describe how endocrine organs are controlled.

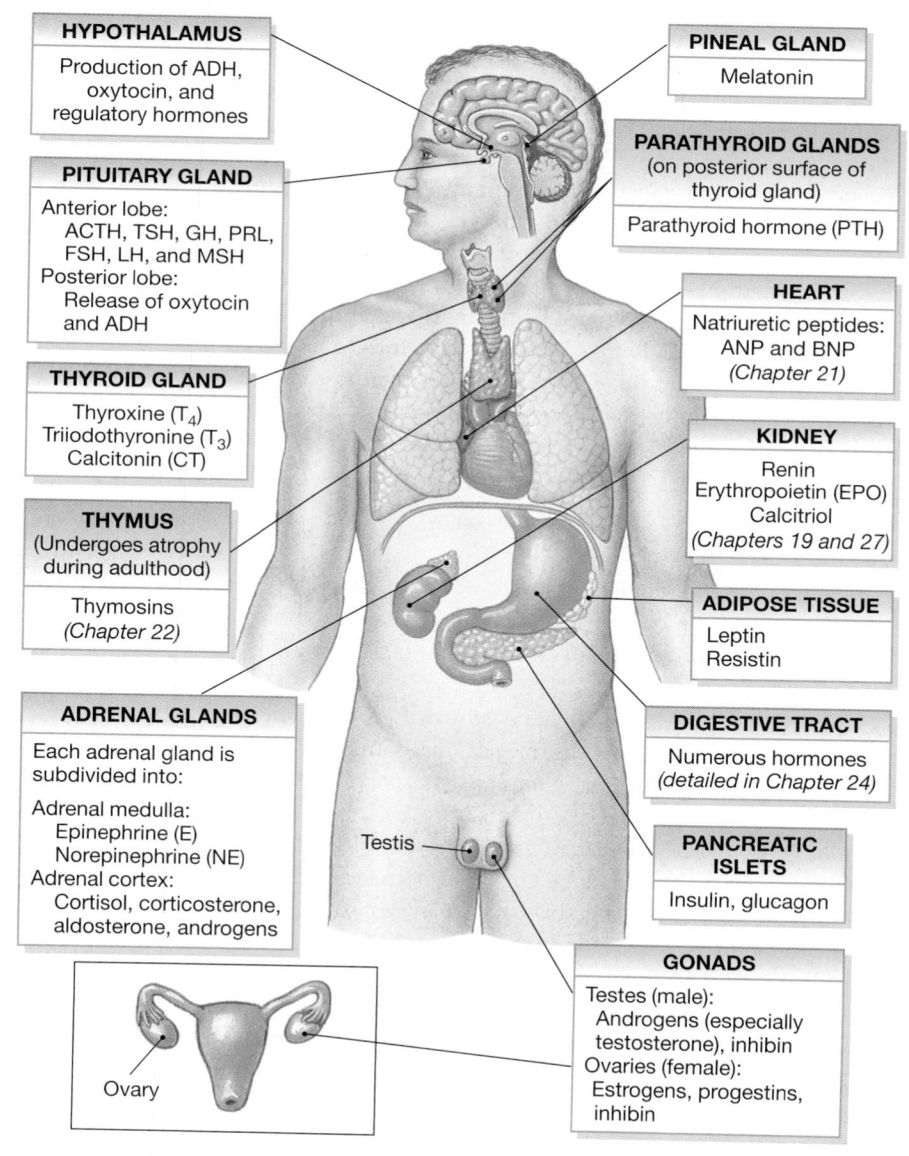

HYPOTHALAMUS
Production of ADH, oxytocin, and regulatory hormones

PITUITARY GLAND
Anterior lobe:
ACTH, TSH, GH, PRL, FSH, LH, and MSH
Posterior lobe:
Release of oxytocin and ADH

THYROID GLAND
Thyroxine (T_4)
Triiodothyronine (T_3)
Calcitonin (CT)

THYMUS
(Undergoes atrophy during adulthood)
Thymosins
(Chapter 22)

ADRENAL GLANDS
Each adrenal gland is subdivided into:
Adrenal medulla:
Epinephrine (E)
Norepinephrine (NE)
Adrenal cortex:
Cortisol, corticosterone, aldosterone, androgens

PINEAL GLAND
Melatonin

PARATHYROID GLANDS
(on posterior surface of thyroid gland)
Parathyroid hormone (PTH)

HEART
Natriuretic peptides:
ANP and BNP
(Chapter 21)

KIDNEY
Renin
Erythropoietin (EPO)
Calcitriol
(Chapters 19 and 27)

ADIPOSE TISSUE
Leptin
Resistin

DIGESTIVE TRACT
Numerous hormones
(detailed in Chapter 24)

PANCREATIC ISLETS
Insulin, glucagon

Testis

GONADS
Testes (male):
Androgens (especially testosterone), inhibin
Ovaries (female):
Estrogens, progestins, inhibin

Ovary

●**FIGURE 18–1**
The Endocrine System

addition to endocrine secretion. We consider the structure and functions of these endocrine organs in detail in chapters dealing with other systems. Examples include the hypothalamus (Chapter 14), the adrenal medullae (Chapter 16), the heart (Chapter 20), the thymus (Chapter 22), the pancreas and digestive tract (Chapter 24), the kidneys (Chapter 26), the reproductive organs (Chapter 28), and the placenta (Chapter 29).

☐ HORMONE STRUCTURE

Hormones can be divided into three groups on the basis of their chemical structure: (1) *amino acid derivatives*, (2) *peptide hormones*, and (3) *lipid derivatives*.

Amino Acid Derivatives

Amino acid derivatives are relatively small molecules that are structurally related to amino acids, the building blocks of proteins. ∞ p. 53 (The amino acids are shown in Appendix III.) This group of hormones, sometimes known as the *biogenic amines*, includes epinephrine, norepinephrine, dopamine, the thyroid hormones, and melatonin. The compounds epinephrine (E), norepinephrine (NE), and dopamine, which have similar structures, are sometimes called *catecholamines* (kat-e-KŌ-la-mēnz). As we saw in Chapter 16, both E and NE are secreted by the adrenal medullae during sympathetic activation. ∞ p. 540 Thyroid hormones are produced by the thyroid gland, and melatonin (mel-a-TŌ-nin) is produced by the pineal gland. Catecholamines and thyroid hormones are synthesized from molecules of the amino acid tyrosine (TĪ-rō-sēn). *Melatonin* is manufactured from molecules of the amino acid *tryptophan* (TRIP-tō-fan).

Peptide Hormones

Peptide hormones are chains of amino acids. In general, peptide hormones are produced as prohormones—inactive molecules that are converted to active hormones either before or after they are secreted.

Peptide hormones can be divided into two groups. One large and diverse group includes hormones that range from short polypeptide chains, such as *antidiuretic hormone* (ADH) and *oxytocin* (9 amino acids apiece), to small proteins, such as *growth hormone* (GH; 191 amino acids) and *prolactin* (PRL; 198 amino acids). This group includes all the hormones secreted by the hypothalamus, heart, thymus, digestive tract, pancreas, and posterior lobe of the pituitary gland, as well as most of the hormones secreted by the anterior lobe of the pituitary gland.

The second group of peptide hormones consists of glycoproteins. ∞ p. 57 These proteins are more than 200 amino acids long and have carbohydrate side chains. The glycoproteins include *thyroid-stimulating hormone (TSH), luteinizing hormone (LH)*, and *follicle-stimulating hormone (FSH)* from the anterior lobe of the pituitary gland, as well as several hormones produced in other organs.

Lipid Derivatives

There are two classes of *lipid derivatives*: (1) *steroid hormones*, derived from cholesterol, and (2) *eicosanoids*, derived from *arachidonic* (a-rak-i-DON-ik) *acid*, a 20-carbon fatty acid.

STEROID HORMONES Steroid hormones are lipids structurally similar to cholesterol (Figure 2–16a•, p. 51). Steroid hormones are released by male and female reproductive organs (*androgens* by the testes, *estrogens* and *progestins* by the ovaries), the adrenal glands (*corticosteroids*), and the kidneys (*calcitriol*). The individual hormones differ in the side chains attached to the basic ring structure.

In the blood, steroid hormones are bound to specific transport proteins in the plasma. For this reason, they remain in circulation longer than do secreted peptide hormones. The liver gradually absorbs these steroids and converts them to a soluble form that can be excreted in the bile or urine.

EICOSANOIDS Eicosanoids (ī-KŌ-sa-noydz) are small molecules with a five-carbon ring at one end. These compounds are important paracrine factors that coordinate cellular activities and affect enzymatic processes (such as blood clotting) that occur in extracellular fluids. Some of the eicosanoids also have secondary roles as hormones. Examples of important eicosanoids include the following:

- **Leukotrienes** (loo-kō-TRĪ-ēns) are released by activated white blood cells, or *leukocytes*. Leukotrienes are important in coordinating tissue responses to injury or disease.
- **Prostaglandins** are produced in most tissues of the body. Within each tissue, the prostaglandins released are involved primarily in coordinating local cellular activities. In some tissues, prostaglandins are converted to **thromboxanes** (throm-BOX-ānz) and **prostacyclins** (pros-ta-SĪ-klinz), which also have strong paracrine effects.

Our focus in this chapter is on circulating hormones whose primary functions are the coordination of activities in many tissues and organs. We shall consider eicosanoids in chapters that deal with individual tissues and organs, including Chapters 19 (the blood), 22 (the lymphatic system), and 28 (the reproductive system). Examples of each group of hormones discussed here are included in Figure 18–2•.

☐ SECRETION AND DISTRIBUTION OF HORMONES

Hormone release typically occurs where capillaries are abundant, and the hormones quickly enter the bloodstream for distribution throughout the body. Within the blood, hormones may circulate freely, or they may be bound to special carrier proteins. A freely circulating hormone remains functional for less than one hour, and sometimes for as little as two minutes. It is inactivated when (1) it diffuses out of the bloodstream and binds to receptors in

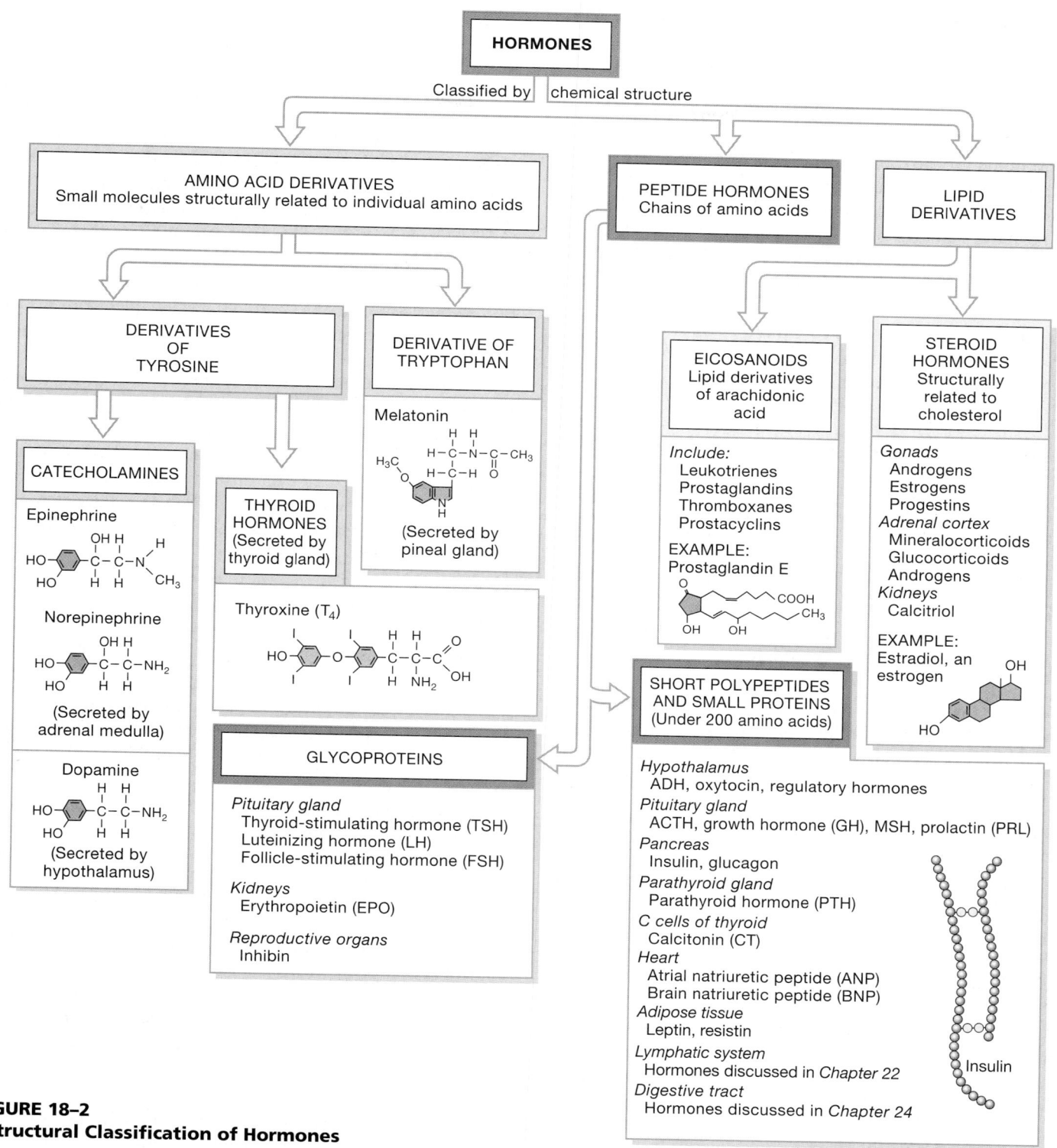

●**FIGURE 18–2**
A Structural Classification of Hormones

target tissues, (2) it is absorbed and broken down by cells of the liver or kidneys, or (3) it is broken down by enzymes in the plasma or interstitial fluids.

Thyroid hormones and steroid hormones remain in circulation much longer, because when these hormones enter the bloodstream, almost all of them become attached to special transport proteins. Less than 1 percent of these hormones circulate freely. An equilibrium state exists between the free hormones and the bound hormones: As the free hormones are removed and inactivated, they are replaced by the release of bound hormones. Thus, the bloodstream contains a substantial reserve (several weeks' supply) of these hormones at any time.

■ MECHANISMS OF HORMONE ACTION

To affect a target cell, a hormone must first interact with an appropriate receptor. Each cell has the receptors needed to respond to several different hormones, but cells in different tissues have different combinations of receptors. This arrangement accounts for the differential effects of hormones on specific tissues. For every cell, the presence or absence of a specific receptor determines the cell's hormonal sensitivities. If a cell has a receptor that will bind a particular hormone, that cell will respond to the hormone's presence. If a cell lacks the proper receptor for that hormone, the hormone will have no effect on that cell.

Hormone receptors are located either (1) on the cell membrane or (2) inside the cell. Using a few specific examples, we shall now introduce the basic mechanisms involved.

Hormones and the Cell Membrane

The receptors for catecholamines (E, NE, and dopamine), peptide hormones, and eicosanoids are in the cell membranes of their respective target cells. Because catecholamines and peptide hormones are not lipid soluble, they are unable to penetrate a cell membrane. Instead, these hormones bind to receptor proteins at the outer surface of the cell membrane. Eicosanoids, which *are* lipid soluble, diffuse across the membrane to reach receptor proteins on the inner surface of the membrane.

FIRST AND SECOND MESSENGERS Any hormone that binds to receptors in the cell membrane does not have direct effects on the target cell. For example, such a hormone does not begin building a protein or catalyzing a specific reaction. Instead, the hormone acts as a **first messenger**, a substance that causes the appearance of a **second messenger** in the cytoplasm. The second messenger functions inside the cell as an enzyme activator, inhibitor, or cofactor, with the net result being a change in the rates of various metabolic reactions. The most important second messengers are (1) *cyclic-AMP (cAMP)*, a derivative of ATP; (2) *cyclic-GMP (cGMP)*, a derivative of GTP, another high-energy compound; and (3) calcium ions.

The binding of a small number of hormone molecules to membrane receptors may lead to the appearance of thousands of second messengers in a cell. This process, which magnifies the effect of a hormone on the target cell, is called *amplification*. Moreover, the arrival of a single hormone may promote the release of more than one type of second messenger, or the production of a linked sequence of messengers known as a *receptor cascade*. Through such mechanisms, the hormone can alter many aspects of cell function at the same time.

The presence or absence of a hormone can also affect the nature and number of hormone receptor proteins in the cell membrane. **Down-regulation** is a process in which the presence of a hormone triggers a decrease in the number of hormone receptors. In down-regulation, when levels of a particular hormone are high, cells become less sensitive to it. Conversely, **up-regulation** is a process in which the absence of a hormone triggers an increase in the number of hormone receptors. In up-regulation, when levels of a particular hormone are low, cells become *more* sensitive to it.

The link between the first messenger and the second messenger generally involves a **G protein**, an enzyme complex coupled to a membrane receptor. The name *G protein* refers to the fact that these proteins bind GTP. ∞ p. 60 There are several types of G proteins, but in each type the G protein is activated when a hormone binds to its receptor at the membrane surface. Figure 18–3● diagrams important mechanisms that may be set in motion by the activation of a G protein. Each sequence affects levels of a second messenger in the cytoplasm.

G PROTEINS AND CAMP Many G proteins, when activated, exert their effects by changing the concentration of the second messenger *cyclic-AMP* (cAMP) within the cell. In most cases, the result is an increase in cAMP levels, and this accelerates metabolic activity within the cell. In a few instances, however, the result is a decrease in cAMP levels and an inhibition of cellular activity.

The steps involved in increasing cAMP levels are diagrammed in Figure 18–3● (left panel):

- The activated G protein activates the enzyme **adenylate cyclase**, also called *adenylyl cyclase*.
- Adenylate cyclase converts ATP to the ring-shaped molecule *cyclic-AMP*.
- Cyclic-AMP then functions as a second messenger, typically by activating a *kinase*. A kinase is an enzyme that performs *phosphorylation*, the attachment of a phosphate group (PO_4^{3-}) to another molecule.
- Generally, the kinases activated by cyclic-AMP phosphorylate proteins. The effect on the target cell depends on the nature of the proteins affected. The phosphorylation of membrane proteins, for example, can open ion channels. In the cytoplasm, many important enzymes can be activated only by phosphorylation.

The hormones calcitonin, parathyroid hormone, ADH, ACTH, epinephrine, FSH, LH, TSH, and glucagon all produce their effects by this mechanism. The increase is usually short lived, because the cytoplasm contains another enzyme, **phosphodiesterase** (PDE), which inactivates cyclic-AMP by converting it to AMP (adenosine monophosphate).

Figure 18–3● (center panel) outlines one example of how activation of a G protein can lower the concentration of cAMP within the cell. In this case, the activated G protein stimulates PDE activity and inhibits adenylate cyclase activity. Levels of cAMP then decline, because cAMP breakdown accelerates while cAMP synthesis is prevented. The decline has an inhibitory effect on the cell, because without phosphorylation, key enzymes remain inactive. This is the mechanism responsible for the inhibitory effects that follow the stimulation of α_2 receptors, as discussed in Chapter 16. ∞ pp. 540-541

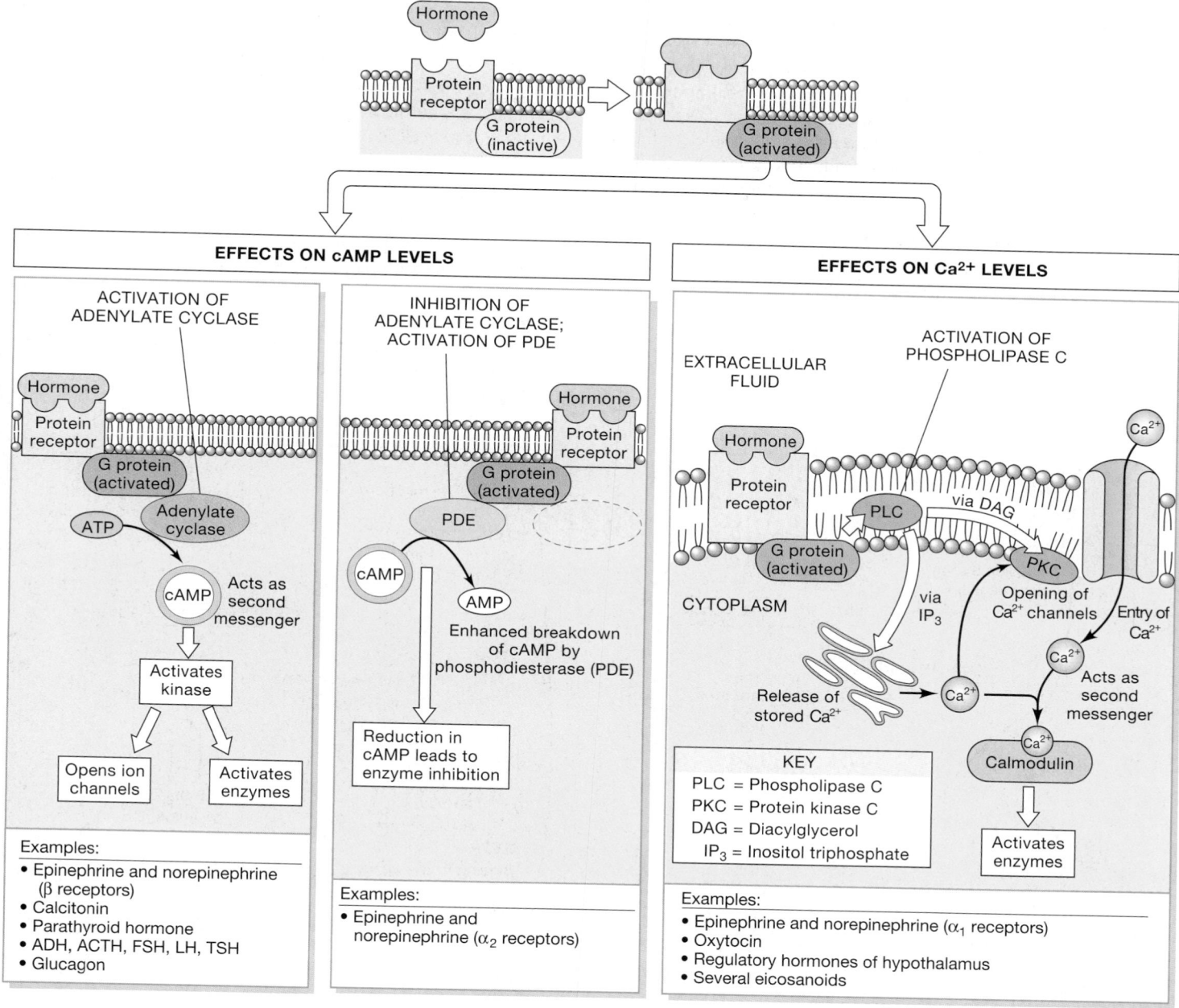

G Proteins and Hormone Activity. Peptide hormones, catecholamines, and eicosanoids bind to membrane receptors and activate G proteins. Three possible results are shown, each affecting the concentration of second messengers in the cytoplasm. G protein activation may lead to (1) activation of adenylate cyclase and subsequent formation of cyclic-AMP, (2) opening of calcium ion channels and calcium release into the cytoplasm, or (3) inhibition of cellular activities by reduction of second messenger concentrations.

G PROTEINS AND *CALCIUM IONS* An activated G protein can trigger the opening of calcium ion channels in the membrane or the release of calcium ions from intracellular stores. The steps involved are diagrammed in Figure 18–3● (right panel). The G protein first activates the enzyme *phospholipase C (PLC)*. This enzyme triggers a receptor cascade that begins with the production of **diacylglycerol (DAG)** and **inositol triphosphate (IP$_3$)** from membrane phospholipids. The cascade then proceeds as follows:

- Inositol triphosphate diffuses into the cytoplasm and triggers the release of Ca^{2+} from intracellular reserves, such as those in the smooth endoplasmic reticulum of many cells.

- The combination of diacylglycerol and intracellular calcium ions activates another membrane protein: **kinase C (PKC)**. The activation of PKC leads to the phosphorylation of calcium channel proteins, a process that opens the channels and permits the entry of extracellular Ca^{2+}. This sets up a positive feedback loop that rapidly elevates intracellular calcium ion concentrations.

- The calcium ions themselves serve as messengers, generally in combination with an intracellular protein called **calmodulin**. Once it has bound calcium ions, calmodulin can activate specific cytoplasmic enzymes. This chain of events is responsible for the stimulatory effects that follow the activation of α_1 receptors by epinephrine or norepinephrine. ∞ p. 541 Calmodulin activation is also involved in the responses to oxytocin and to several regulatory hormones secreted by the hypothalamus.

Hormones and Intracellular Receptors

Steroid hormones diffuse across the lipid part of the cell membrane and bind to receptors in the cytoplasm or nucleus. The hormone–receptor complexes then activate or deactivate specific genes (Figure 18–4a●). By this mechanism, steroid hormones can alter the rate of DNA transcription in the nucleus, and thus change the pattern of protein synthesis. Alterations in the synthesis of enzymes or structural proteins will directly affect both the metabolic activity and the structure of the target cell. For example, the sex hormone *testosterone* stimulates the production of enzymes and structural proteins in skeletal muscle fibers, causing an increase in muscle size and strength.

Thyroid hormones cross the cell membrane by diffusion or by a carrier mechanism. Once in the cytosol, these hormones bind to receptors within the nucleus and on mitochondria (Figure 18–4b●). The hormone–receptor complexes in the nucleus activate specific genes or change the rate of transcription. The change in rate affects the metabolic activities of the cell by increasing or decreasing the concentration of specific enzymes. Thyroid hormones bound to mitochondria increase the mitochondrial rates of ATP production.

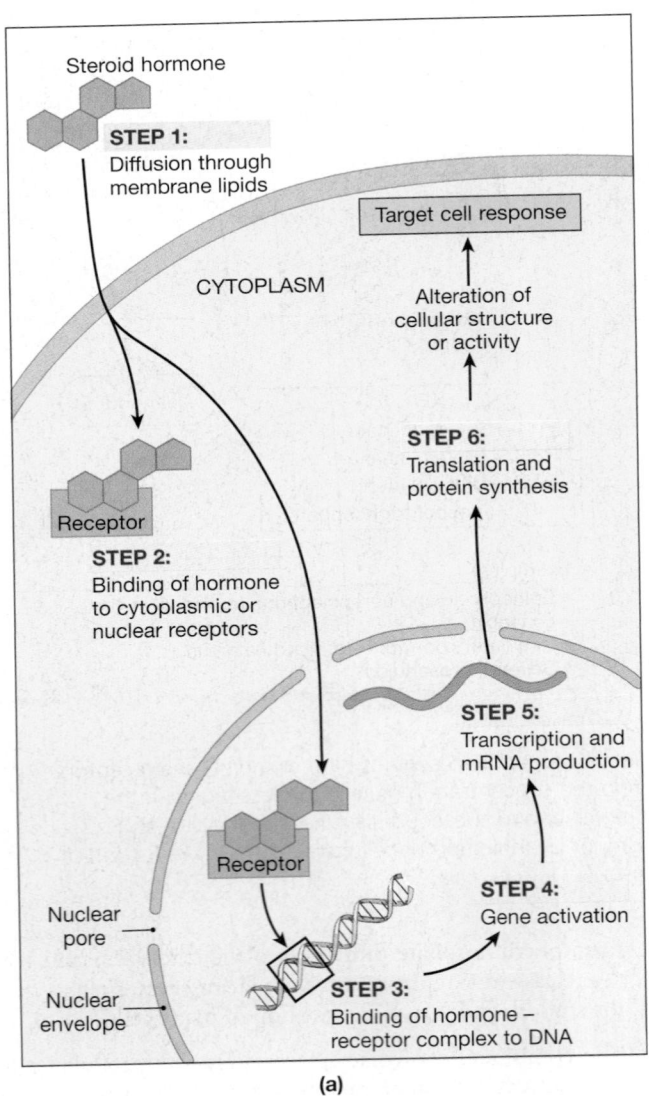

(a)

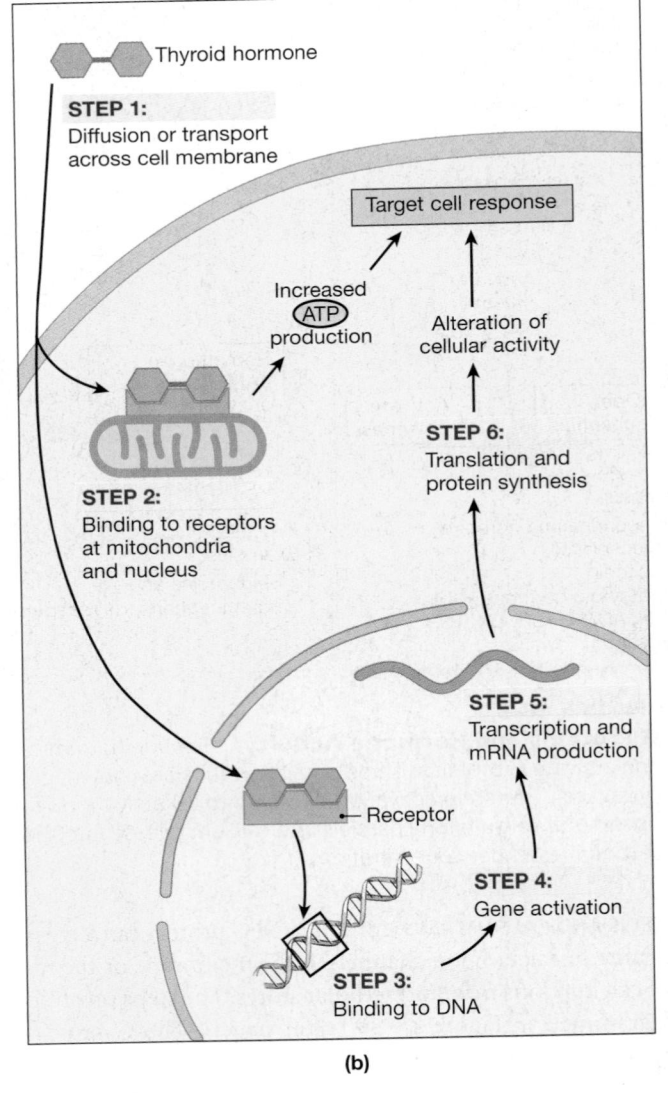

(b)

●**FIGURE 18–4**
Hormone Effects on Gene Activity. **(a)** Steroid hormones diffuse through the membrane lipids and bind to receptors in the cytoplasm or nucleus. The complex then binds to DNA in the nucleus and activates specific genes. **(b)** Thyroid hormones enter the cytoplasm and bind to receptors in the nucleus to activate specific genes. They also bind to receptors on mitochondria and accelerate ATP production.

☐ CONTROL OF ENDOCRINE ACTIVITY

As we noted earlier, the functional organization of the nervous system parallels that of the endocrine system in many ways. In Chapter 13, we considered the basic operation of neural reflex arcs, the simplest organizational units in the nervous system. ⟳ p. 448 The most direct arrangement was a monosynaptic reflex, such as the stretch reflex. Polysynaptic reflexes provide more complex and variable responses to stimuli, and higher centers, which integrate multiple inputs, can facilitate or inhibit these reflexes as needed.

Endocrine Reflexes

Endocrine reflexes are the functional counterparts of neural reflexes. They can be triggered by (1) *humoral stimuli* (changes in the composition of the extracellular fluid), (2) *hormonal stimuli* (the arrival or removal of a specific hormone), or (3) *neural stimuli* (the arrival of neurotransmitter at neuroglandular junctions). In most cases, endocrine reflexes are controlled by negative feedback mechanisms: A stimulus triggers the production of a hormone whose direct or indirect effects reduce the intensity of the stimulus.

Endocrine cells in a simple endocrine reflex involve only one hormone. The endocrine cells involved respond directly to changes in the composition of the extracellular fluid. The secreted hormone adjusts the activities of target cells and restores homeostasis. Simple endocrine reflexes control hormone secretion by the heart, pancreas, parathyroid gland, and digestive tract.

More complex endocrine reflexes involve one or more intermediary steps and two or more hormones. The hypothalamus, the highest level of endocrine control, integrates the activities of the nervous and endocrine systems. This integration involves three mechanisms, summarized in Figure 18–5●:

1. The hypothalamus secretes **regulatory hormones**, special hormones that control endocrine cells in the pituitary gland. (By convention, the use of *hormone* in the name indicates that the substance's identity is known; if the identity is not known, the term *factor* is used instead.) The hypothalamic regulatory hormones control the secretory activities of endocrine cells in the anterior lobe of the pituitary gland. The hormones released by the anterior lobe, in turn, control the activities of endocrine cells in the thyroid, adrenal cortex, and reproductive organs.

2. The hypothalamus itself acts as an endocrine organ. Hypothalamic neurons synthesize hormones, transport them along axons within the infundibulum, and release them into the circulation at the posterior lobe of the pituitary gland. We introduced two of these hormones, ADH and oxytocin, in Chapter 14. ⟳ p. 482

3. The hypothalamus contains autonomic centers that exert direct neural control over the endocrine cells of the adrenal medullae. When the sympathetic division is activated, the adrenal medullae release hormones into the bloodstream.

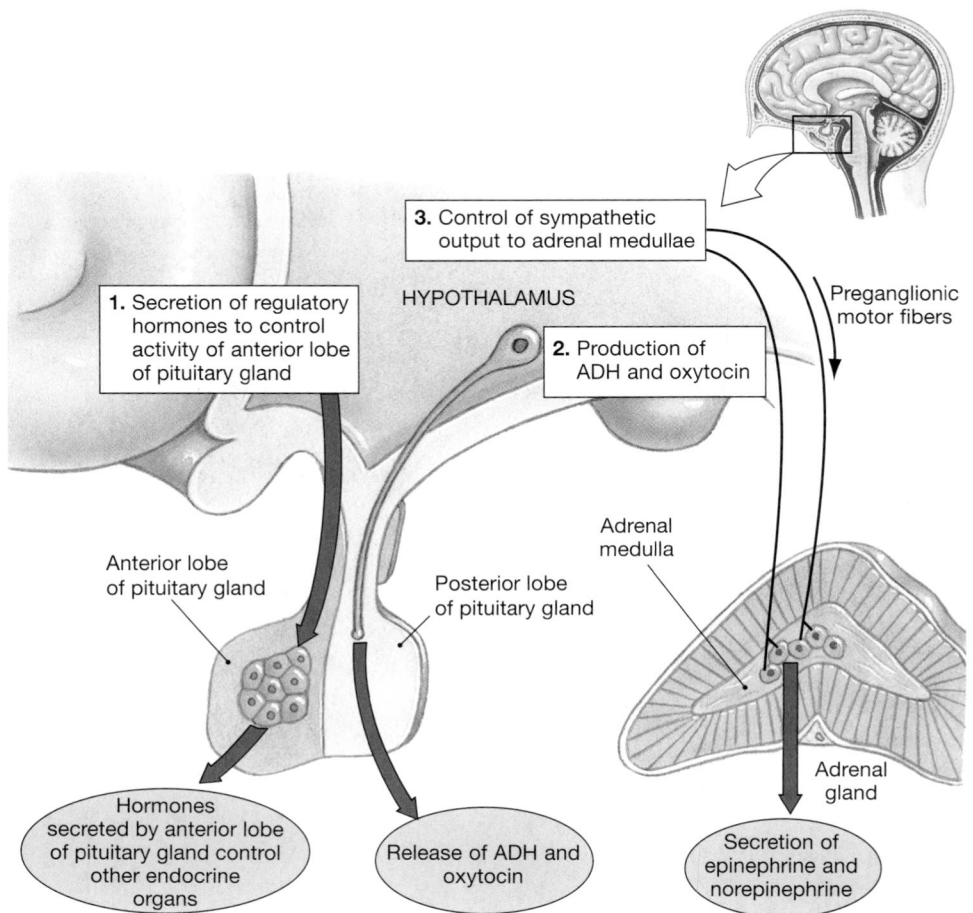

3. Control of sympathetic output to adrenal medullae

1. Secretion of regulatory hormones to control activity of anterior lobe of pituitary gland

HYPOTHALAMUS

2. Production of ADH and oxytocin

Preganglionic motor fibers

Anterior lobe of pituitary gland

Posterior lobe of pituitary gland

Adrenal medulla

Adrenal gland

Hormones secreted by anterior lobe of pituitary gland control other endocrine organs

Release of ADH and oxytocin

Secretion of epinephrine and norepinephrine

●**FIGURE 18–5**
Three Methods of Hypothalamic Control over Endocrine Function

The hypothalamus secretes regulatory hormones and ADH in response to changes in the composition of the circulating blood. The secretion of oxytocin, E, and NE involves both neural and hormonal mechanisms. For example, the adrenal medullae secrete E and NE in response to action potentials rather than to circulating hormones. Such pathways are called *neuroendocrine reflexes,* because they include both neural and endocrine components. We shall consider these reflex patterns in more detail as we deal with specific endocrine tissues and organs.

In Chapter 15, we noted that receptors provide complex information by varying the frequency and pattern of action potentials in a sensory neuron. In the endocrine system, complex commands are issued by changing the amount of hormone secreted and the pattern of hormone release. In a simple endocrine reflex, hormones are released continuously, but the rate of secretion rises and falls in response to humoral stimuli. For example, when blood glucose levels climb, the pancreas increases its secretion of *insulin,* a hormone that stimulates glucose absorption and utilization. As insulin levels rise, glucose levels decline; in turn, the stimulation of the insulin-secreting cells is reduced. As glucose levels return to normal, the rate of insulin secretion reaches resting levels. (We discussed the same pattern in Chapter 1 when we considered the negative feedback control of body temperature. ∞ p. 12)

In this example, the responses of the target cells change over time, because the impact of insulin is proportional to its concentration. However, the relationship between hormone concentration and target cell response is not always predictable. For instance, a hormone can have one effect at low concentrations and additional effects—or even different effects—at high concentrations. (We shall consider specific examples later in the chapter.)

Several hormones of the hypothalamus and pituitary gland are released in sudden bursts called *pulses,* rather than continuously. When hormones arrive in pulses, target cells may vary their response with the frequency of the pulses. For example, the target cell response to one pulse every three hours can differ from the response when pulses arrive every 30 minutes. The most complicated hormonal instructions issued by the hypothalamus involve changes in the frequency of pulses *and* in the amount secreted in each pulse.

✓ How could you distinguish between a neural response and an endocrine response on the basis of response time and duration?

✓ How would the presence of a molecule that blocks adenylate cyclase affect the activity of a hormone that produces its cellular effects by way of the second messenger cAMP?

✓ What primary factor determines each cell's hormonal sensitivities?

Answers start on page Q-1

Hormone mechanisms of action can be reviewed on the **Companion Website**: Chapter 18/Tutorials.

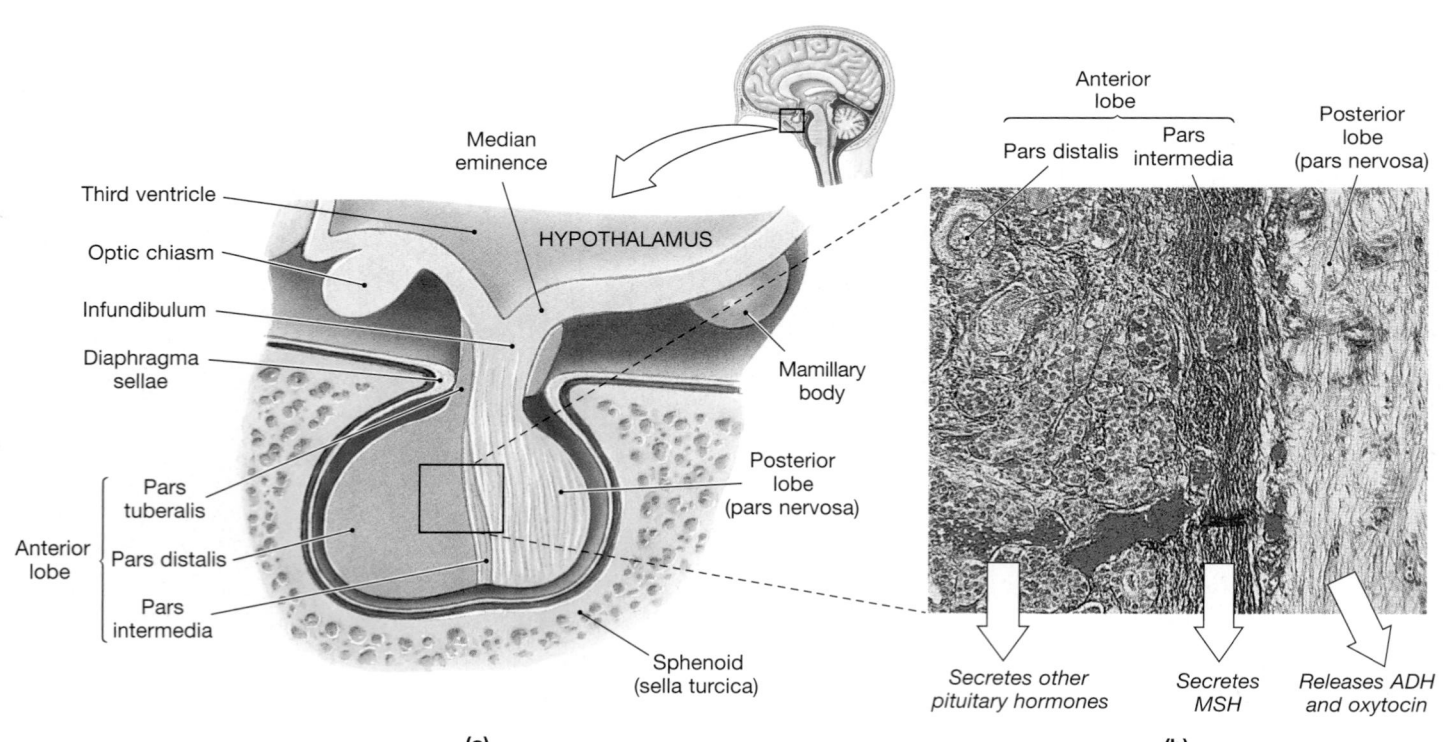

(a)

(b)

●FIGURE 18–6
The Anatomy and Orientation of the Pituitary Gland

18–3 THE PITUITARY GLAND

Objectives

■ Describe the location and structure of the pituitary gland, and explain its structural and functional relationships with the hypothalamus.

■ Identify the hormones produced by the anterior and posterior lobes of the pituitary, and specify the functions of those hormones.

■ Discuss the results of abnormal levels of pituitary hormone production.

Figure 18–6● shows the anatomical organization of the pituitary gland, or **hypophysis** (hī-POF-i-sis). This small, oval gland lies nestled within the *sella turcica*, a depression in the sphenoid bone (Figure 7–8●, p. 219). The pituitary gland hangs inferior to the hypothalamus, connected by the slender, funnel-shaped structure called the **infundibulum** (in-fun-DIB-ū-lum; funnel). The base of the infundibulum lies between the optic chiasm and the mamillary bodies. The pituitary gland is cradled by the sella turcica and held in position by the *diaphragma sellae*, a dural sheet that encircles the infundibulum. The diaphragma sellae locks the pituitary gland in position and isolates it from the cranial cavity.

The pituitary gland can be divided into posterior and anterior lobes on the basis of function and developmental anatomy. Nine important peptide hormones are released by the pituitary gland—seven by the anterior lobe and two by the posterior lobe. All nine hormones bind to membrane receptors, and all nine use cAMP as a second messenger. **ATLAS** Embryology Summary 14: The Development of the Endocrine System

☐ THE ANTERIOR LOBE

The **anterior lobe** of the pituitary gland, or **adenohypophysis** (ad-ē-nō-hī-POF-i-sis), contains a variety of endocrine cells. The anterior lobe can be subdivided into three regions: (1) the **pars distalis** (dis-TAL-is; distal part), which is the largest and most anterior portion of the pituitary gland; (2) an extension called the **pars tuberalis**, which wraps around the adjacent portion of the infundibulum; and (3) the slender **pars intermedia**, which forms a narrow band bordering the posterior lobe (Figure 18–6●). An extensive capillary network radiates through these regions, giving every endocrine cell immediate access to the circulatory system.

The Hypophyseal Portal System

By secreting specific regulatory hormones, the hypothalamus controls the production of hormones in the anterior lobe. At the median eminence (Figure 14–15●, p. 485), a swelling near the attachment of the infundibulum, hypothalamic neurons release regulatory factors into the surrounding interstitial fluids. Their secretions enter the bloodstream quite easily, because the en-

dothelial cells lining the capillaries in this region are unusually permeable. These **fenestrated** (FEN-es-trā-ted) **capillaries** (*fenestra*, window) allow relatively large molecules to enter or leave the circulatory system. The capillary networks in the median eminence are supplied by the *superior hypophyseal artery*.

Before leaving the hypothalamus, the capillary networks unite to form a series of larger vessels that spiral around the infundibulum to reach the anterior lobe of the pituitary gland. Once within the anterior lobe, these vessels form a second capillary network that branches among the endocrine cells (Figure 18–7●).

This vascular arrangement is unusual: A typical artery conducts blood from the heart to a capillary network, and a typical vein carries blood from a capillary network back to the heart. The vessels between the median eminence and the anterior lobe, however, carry blood from one capillary network to another. Blood vessels that link two capillary networks are called **portal vessels**; in this case, they have the histological structure of veins.

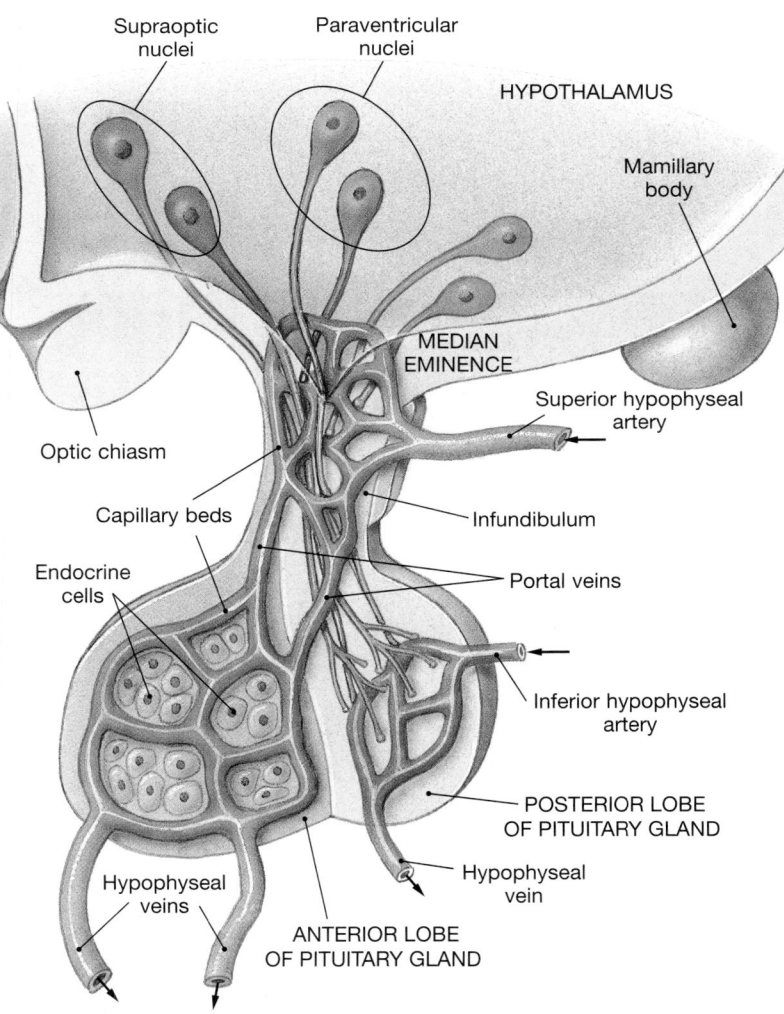

●**FIGURE 18–7**
The Hypophyseal Portal System

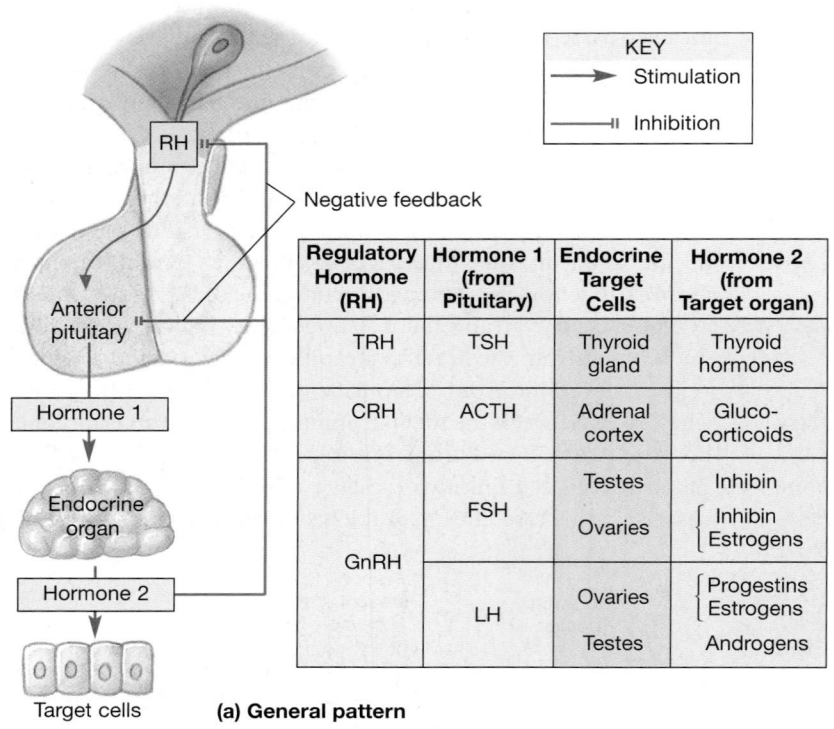

KEY
→ Stimulation
—⊣ Inhibition

Negative feedback

Regulatory Hormone (RH)	Hormone 1 (from Pituitary)	Endocrine Target Cells	Hormone 2 (from Target organ)
TRH	TSH	Thyroid gland	Thyroid hormones
CRH	ACTH	Adrenal cortex	Gluco-corticoids
GnRH	FSH	Testes	Inhibin
		Ovaries	{ Inhibin / Estrogens
	LH	Ovaries	{ Progestins / Estrogens
		Testes	Androgens

(a) General pattern

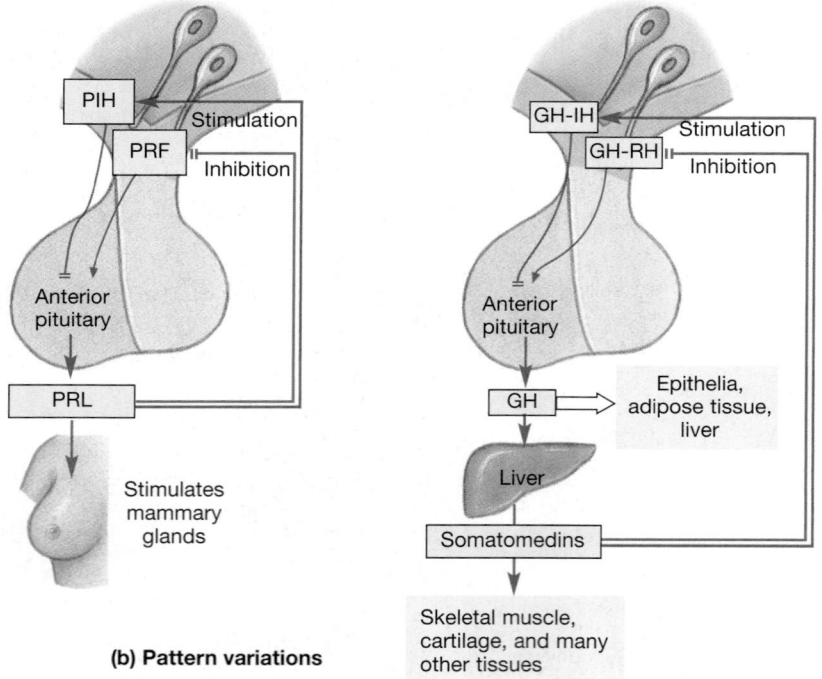

(b) Pattern variations

The entire complex is a **portal system**. Portal systems are named after their destinations; hence, this particular network is known as the **hypophyseal** (hī-pō-FI-sē-al) **portal system**.

Portal systems provide an efficient means of chemical communication by ensuring that all the blood entering the portal

●FIGURE 18–8
Feedback Control of Endocrine Secretion.
(a) A typical pattern of regulation when multiple endocrine organs are involved. The hypothalamus produces a releasing hormone (RH) to stimulate hormone production by other glands, and control is via negative feedback. **(b)** Variations on the theme outlined in part (a). Left: The regulation of prolactin (PRL) production by the anterior lobe. In this case, the hypothalamus produces both a releasing factor (PRF) and an inhibiting hormone (PIH); when one is stimulated, the other is inhibited. Right: The regulation of growth hormone (GH) production by the anterior lobe; when GH–RH release is inhibited, GH–IH release is stimulated.

vessels will reach the intended target cells before it returns to the general circulation. The communication is strictly one way, however, because any chemicals released by the cells "downstream" must do a complete tour of the circulatory system before they reach the capillaries at the start of the portal system.

Hypothalamic Control of the Anterior Lobe

Two classes of hypothalamic regulatory hormones exist: (1) releasing hormones and (2) inhibiting hormones. A **releasing hormone** (**RH**) stimulates the synthesis and secretion of one or more hormones at the anterior lobe, whereas an **inhibiting hormone** (**IH**) prevents the synthesis and secretion of hormones from the anterior lobe. An endocrine cell in the anterior lobe may be controlled by releasing hormones, inhibiting hormones, or some combination of the two. The regulatory hormones released at the hypothalamus are transported directly to the anterior lobe by the hypophyseal portal system.

The rate at which the hypothalamus secretes regulatory hormones is controlled by negative feedback. The primary regulatory patterns are diagrammed in Figure 18–8●; we shall refer to them as we examine specific pituitary hormones.

Hormones of the Anterior Lobe

We will discuss seven hormones whose functions and control mechanisms are reasonably well understood: *thyroid-stimulating hormone, adrenocorticotropic hormone, follicle-stimulating hormone, luteinizing hormone, prolactin, growth hormone,* and *melanocyte-stimulating hormone.* Of the six hormones produced by the pars distalis, four regulate the production of hormones by other endocrine glands. The names of these hormones indicate their activities, but many of the phrases are so long that abbreviations are often used instead.

The hormones of the anterior lobe are also called *tropic hormones* (*tropé*, a turning), because they "turn on" endocrine glands or support the functions of other organs. (Some sources call them *trophic hormones* [*trophé*, nourishment] instead.)

THYROID-STIMULATING HORMONE Thyroid-stimulating hormone (**TSH**), or *thyrotropin*, targets the thyroid gland and triggers the release of thyroid hormones. TSH is released in response to *thyrotropin-releasing hormone* (*TRH*) from the hypothalamus. As circulating concentrations of thyroid hormones rise, the rates of TRH and TSH production decline (Figure 18–8a●).

ADRENOCORTICOTROPIC HORMONE Adrenocorticotropic hormone (**ACTH**), also known as *corticotropin*, stimulates the release of steroid hormones by the *adrenal cortex*, the outer portion of the adrenal gland. ACTH specifically targets cells that produce *glucocorticoids* (gloo-kō-KOR-ti-koydz), hormones that affect glucose metabolism. ACTH release occurs under the stimulation of **corticotropin-releasing hormone** (**CRH**) from the hypothalamus. As glucocorticoid levels increase, the rates of CRH release and ACTH release decline (Figure 18–8a●).

THE GONADOTROPINS Follicle-stimulating hormone and luteinizing hormone are called **gonadotropins** (gō-nad-ō-TRŌ-pinz), because they regulate the activities of the *gonads*. (These organs—the testes and ovaries in males and females, respectively—produce reproductive cells as well as hormones.) The production of gonadotropins occurs under stimulation by gonadotropin-releasing hormone (GnRH) from the hypothalamus. An abnormally low production of gonadotropins produces **hypogonadism**. Children with this condition will not undergo sexual maturation, and adults with hypogonadism cannot produce functional sperm or oocytes.

- **Follicle-stimulating hormone** (**FSH**), or *follitropin*, promotes follicle development in females and, in combination with luteinizing hormone, stimulates the secretion of *estrogens* (ES-trō-jenz) by ovarian cells. *Estradiol* is the most important estrogen. In males, FSH stimulates *sustentacular cells*, specialized cells in the tubules where sperm differentiate. In response, the sustentacular cells promote the physical maturation of developing sperm. FSH production is inhibited by *inhibin*, a peptide hormone released by cells in the testes and ovaries (Figure 18–8a●). (The role of inhibin in suppressing the release of GnRH as well as FSH is under debate.)

- **Luteinizing** (LOO-tē-in-ī-zing) hormone (**LH**), or *lutropin*, induces *ovulation*, the production of reproductive cells in females. It also promotes the secretion, by the ovaries, of estrogens and the *progestins* (such as *progesterone*), which prepare the body for possible pregnancy. In males, this gonadotropin is sometimes called *interstitial cell–stimulating hormone (ICSH)*, because it stimulates the production of sex hormones by the *interstitial cells* of the testes. These sex hormones are called **androgens** (AN-drō-jenz; *andros*, man), the most important of which is *testosterone*. LH production, like FSH production, is stimulated

by GnRH from the hypothalamus. GnRH production is inhibited by estrogens, progestins, and androgens (Figure 18–8a●).

PROLACTIN Prolactin (*pro-*, before + *lac*, milk) (**PRL**), or *mammotropin*, works with other hormones to stimulate mammary gland development. In pregnancy and during the nursing period that follows delivery, PRL also stimulates milk production by the mammary glands. The functions of PRL in males are poorly understood, but evidence indicates that PRL helps regulate androgen production by making interstitial cells more sensitive to LH.

Prolactin production is inhibited by **prolactin-inhibiting hormone** (**PIH**)—the neurotransmitter dopamine. The hypothalamus also secretes a prolactin-releasing hormone, but the identity of this *prolactin-releasing factor* (*PRF*) is a mystery. Circulating PRL stimulates PIH release and inhibits the secretion of PRF (Figure 18–8b●).

Although PRL exerts the dominant effect on the glandular cells, normal development of the mammary glands is regulated by the interaction of several hormones. Prolactin, estrogens, progesterone, glucocorticoids, pancreatic hormones, and hormones produced by the placenta cooperate in preparing the mammary glands for secretion, and milk ejection occurs only in response to oxytocin release at the posterior lobe of the pituitary gland. We shall describe the functional development of the mammary glands in Chapter 28.

GROWTH HORMONE Growth hormone (**GH**), or **somatotropin** (*soma*, body), stimulates cell growth and replication by accelerating the rate of protein synthesis. Although virtually every tissue responds to some degree, skeletal muscle cells and chondrocytes (cartilage cells) are particularly sensitive to GH levels.

The stimulation of growth by GH involves two mechanisms. The primary mechanism, which is indirect, is best understood. Liver cells respond to the presence of GH by synthesizing and releasing **somatomedins**, or **insulinlike growth factors** (**IGFs**), which are peptide hormones that bind to receptor sites on a variety of cell membranes (Figure 18–8b●). In skeletal muscle fibers, cartilage cells, and other target cells, somatomedins increase the rate of uptake of amino acids and their incorporation into new proteins. These effects develop almost immediately after GH is released; they are particularly important after a meal, when the blood contains high concentrations of glucose and amino acids. In functional terms, cells can now obtain ATP easily through the aerobic metabolism of glucose, and amino acids are readily available for protein synthesis. Under these conditions, GH, acting through the somatomedins, stimulates protein synthesis and cell growth.

The direct actions of GH are more selective and tend not to appear until after blood glucose and amino acid concentrations have returned to normal levels:

- In epithelia and connective tissues, GH stimulates stem cell divisions and the differentiation of daughter cells. The subsequent growth of these daughter cells will be stimulated by somatomedins.

- In adipose tissue, GH stimulates the breakdown of stored triglycerides by adipocytes (fat cells), which then release fatty acids into the blood. As circulating fatty acid levels rise, many tissues stop breaking down glucose and start breaking down fatty acids to generate ATP. This process is termed a **glucose-sparing effect.**

- In the liver, GH stimulates the breakdown of glycogen reserves by liver cells, which then release glucose into the bloodstream. Because most other tissues are now metabolizing fatty acids rather than glucose, blood glucose concentrations begin to climb, perhaps to levels significantly higher than normal. The elevation of blood glucose levels by GH has been called a **diabetogenic effect,** because *diabetes mellitus,* an endocrine disorder we shall consider later in the chapter, is characterized by abnormally high blood glucose concentrations.

The production of GH is regulated by **growth hormone–releasing hormone** (**GH–RH,** or *somatocrinin*) and **growth hormone–inhibiting hormone** (**GH–IH,** or *somatostatin*) from the hypothalamus. Somatomedins stimulate GH–IH and inhibit GH–RH (Figure 18–8b●).

MELANOCYTE-STIMULATING HORMONE The pars intermedia may secrete two forms of **melanocyte-stimulating hormone (MSH),** or *melanotropin.* As the name indicates, MSH stimulates the melanocytes of the skin, increasing their production of melanin, a brown, black, or yellow-brown pigment. ∞ p. 160 The release of MSH is inhibited by dopamine.

Melanocyte stimulating hormone is important in the control of skin pigmentation in fishes, amphibians, reptiles, and many mammals other than primates. The pars intermedia in adult humans is virtually nonfunctional, and the circulating blood usually does not contain MSH. However, MSH is secreted by the human pars intermedia (1) during fetal development, (2) in very young children, (3) in pregnant women, and (4) in some diseases. The functional significance of MSH secretion under these circumstances is not known. The administration of a synthetic form of MSH causes the skin to darken, so MSH has been suggested as a means of obtaining a "sunless tan."

☐ THE POSTERIOR LOBE

The **posterior lobe** of the pituitary gland is also called the **neurohypophysis** (noo-rō-hī-POF-i-sis), or *pars nervosa* (nervous part), because it contains the axons of hypothalamic neurons. Neurons of the **supraoptic** and **paraventricular nuclei** manufacture antidiuretic hormone (ADH) and oxytocin, respectively. These products move along axons in the infundibulum to the basement membranes of capillaries in the posterior lobe by means of axoplasmic transport. ∞ p. 389

Antidiuretic Hormone

Antidiuretic hormone (ADH), also known as *arginine vasopressin* (*AVP*), is released in response to a variety of stimuli, most notably a rise in the electrolyte concentration in the blood or a fall in blood volume or blood pressure. A rise in the electrolyte concentration stimulates the secretory neurons directly. Because they respond to a change in the osmotic concentration of body fluids, these neurons are called *osmoreceptors.*

The primary function of ADH is to decrease the amount of water lost at the kidneys. With losses minimized, any water absorbed from the digestive tract will be retained, reducing the concentration of electrolytes in the extracellular fluid. In high concentrations, ADH also causes *vasoconstriction,* a constriction of peripheral blood vessels that helps elevate blood pressure. ADH release is inhibited by alcohol, which explains the increased fluid excretion that follows the consumption of alcoholic beverages.

Diabetes occurs in several forms, all characterized by excessive urine production (polyuria). Although diabetes can be caused by physical damage to the kidneys, most forms are the result of endocrine abnormalities. The two most prevalent forms are diabetes mellitus and diabetes insipidus. Diabetes mellitus is described on page 633. **Diabetes insipidus** generally develops because the posterior lobe of the pituitary gland no longer releases adequate amounts of ADH. Water conservation at the kidneys is impaired, and excessive amounts of water are lost in the urine. As a result, the individual is constantly thirsty, but the fluids consumed are not retained by the body.

Mild cases of diabetes insipidus may not require treatment if fluid and electrolyte intake keep pace with urinary losses. In severe cases, the fluid losses can reach 10 liters per day, and dehydration and electrolyte imbalances are fatal without treatment.

Oxytocin

In women, **oxytocin** (*oxy-,* quick + *tokos,* childbirth), or **OT,** stimulates smooth muscle tissue in the wall of the uterus, promoting labor and delivery. After delivery, oxytocin stimulates the contraction of myoepithelial cells around the secretory alveoli and the ducts of the mammary glands, promoting the ejection of milk.

Until the last stages of pregnancy, the uterine smooth muscles are relatively insensitive to oxytocin, but sensitivity becomes more pronounced as the time of delivery approaches. The trigger for normal labor and delivery is probably a sudden rise in oxytocin levels at the uterus. There is good evidence, however, that the oxytocin released at the posterior lobe plays only a supporting role and that most of the oxytocin involved is secreted by the uterus and fetus.

Oxytocin secretion and milk ejection are part of a neuroendocrine reflex. The stimulus is an infant suckling at the breast, and sensory nerves innervating the nipples relay the information to the hypothalamus. Oxytocin is then released into the circulation at the posterior lobe, and the myoepithelial cells respond by squeezing milk from the secretory alveoli into large collecting ducts. This *milk let-down reflex* can be modified by any factor that affects the hypothalamus. For example, anxiety, stress, and other factors can prevent the flow of milk, even when the mammary glands are fully functional. In contrast, nursing mothers can become conditioned to associate a baby's crying with suckling. These women may begin milk let-down as soon as they hear a baby cry.

Although the functions of oxytocin in sexual activity remain uncertain, it is known that circulating concentrations of oxytocin in both genders rise during sexual arousal and peak at orgasm. Evidence indicates that in men oxytocin stimulates smooth muscle contractions in the walls of the sperm duct (*ductus deferens*) and prostate gland. These actions may be important in *emission*—the ejection of secretions of the prostate gland, sperm, and the secretions of other glands into the male reproductive tract before ejaculation. Studies suggest that the oxytocin released during intercourse in females may stimulate smooth muscle contractions in the uterus and vagina that promote the transport of sperm toward the uterine tubes.

SUMMARY: The Hormones of the Pituitary Gland

Figure 18–9● and Table 18–2 summarize important information about the hormonal products of the pituitary gland. Review these carefully before considering the structure and function of other endocrine organs.

✓ If a person were dehydrated, how would the level of ADH released by the posterior lobe change?

✓ A blood sample shows elevated levels of somatomedins. Which pituitary hormone would you expect to be elevated as well?

✓ What effect would elevated circulating levels of cortisol, a steroid hormone from the adrenal cortex, have on the pituitary secretion of ACTH?

Answers start on page Q-1

 Reinforce your understanding of the hormones and glands of the endocrine system by visiting the **Companion Website**/Chapter 18/Tutorials.

●FIGURE 18–9
Pituitary Hormones and Their Targets

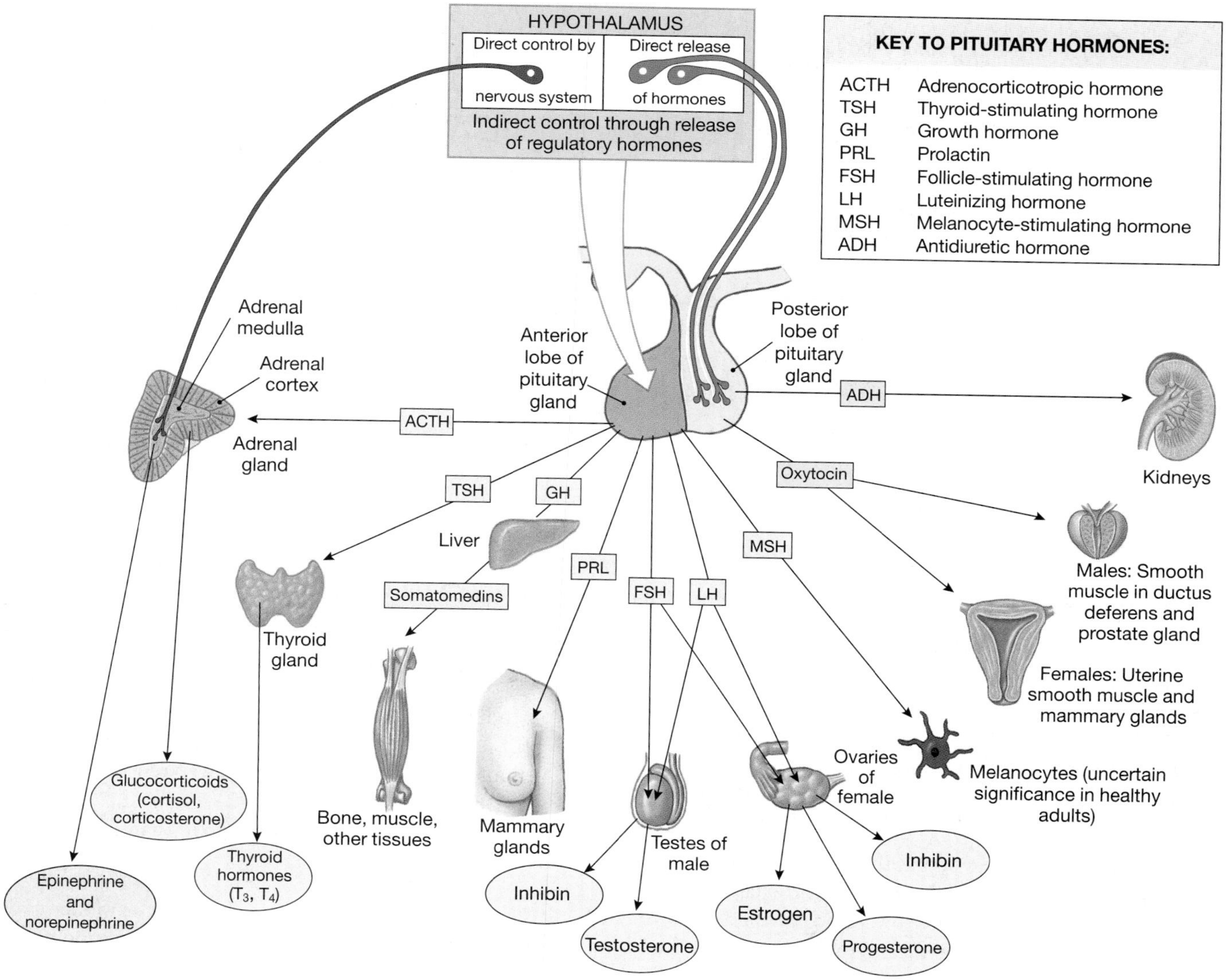

	SUMMARY TABLE 18–2 THE PITUITARY HORMONES			
Region/Area	Hormone(s)	Target(s)	Hormonal Effect(s)	Hypothalamic Regulatory Hormone
ANTERIOR LOBE (ADENOHYPOPHYSIS)				
Pars distalis	Thyroid-stimulating hormone (TSH)	Thyroid gland	Secretion of thyroid hormones (T_3, T_4)	Thyrotropin-releasing hormone (TRH)
	Adrenocorticotropic hormone (ACTH)	Adrenal cortex (zona fasciculata)	Secretion of glucocorticoids (cortisol, corticosterone)	Corticotropin-releasing hormone (CRH)
	Gonadotropins:			
	Follicle-stimulating hormone (FSH)	Follicle cells of ovaries Sustentacular cells of testes	Secretion of estrogen, follicle development Stimulation of sperm maturation	Gonadotropin-releasing hormone (GnRH) As above
	Luteinizing hormone (LH)	Follicle cells of ovaries Interstitial cells of testes	Ovulation, formation of corpus luteum, secretion of progesterone Secretion of testosterone	As above As above
	Prolactin (PRL)	Mammary glands	Production of milk	Prolactin-releasing factor (PRF) Prolactin-inhibiting hormone (PIH)
	Growth hormone (GH)	All cells	Growth, protein synthesis, lipid mobilization and catabolism	Growth-hormone-releasing hormone (GH–RH) Growth hormone-inhibiting hormone (GH–IH)
Pars intermedia (not active in normal adults)	Melanocyte-stimulating hormone (MSH)	Melanocytes	Increased melanin synthesis in epidermis	Melanocyte-stimulating hormone–inhibiting hormone (MSH–IH)
POSTERIOR LOBE (NEUROHYPOPHYSIS OR PARS NERVOSA)				
	Antidiuretic hormone (ADH)	Kidneys	Reabsorption of water, elevation of blood volume and pressure	None: Transported along axons from supraoptic nucleus to posterior lobe of the pituitary gland
	Oxytocin (OT)	Uterus, mammary glands (females) Ductus deferens and prostate gland (males)	Labor contractions, milk ejection Contractions of ductus deferens and prostate gland	None: Transported along axons from paraventricular nucleus to posterior lobe of the pituitary gland

18-4 THE THYROID GLAND

Objectives

- Describe the location and structure of the thyroid gland.
- Identify the hormones produced by the thyroid gland, specify the functions of those hormones, and discuss the results of abnormal levels of thyroid hormones.

The **thyroid gland** curves across the anterior surface of the trachea just inferior to the *thyroid* ("shield-shaped") *cartilage*, which forms most of the anterior surface of the larynx (Figure 18–11a●). The two **lobes** of the thyroid gland are united by a slender connection, the **isthmus** (IS-mus). You can easily feel the gland with your fingers. When something goes wrong with it, the thyroid gland typically becomes visible as it swells and distorts the surface of the neck. The size of the gland is quite variable, depending on heredity and environmental and nutritional factors, but its average weight is about 34 g (1.2 oz). An extensive blood supply gives the thyroid gland a deep red color.

☐ THYROID FOLLICLES AND THYROID HORMONES

The thyroid gland contains large numbers of **thyroid follicles**, spheres lined by a simple cuboidal epithelium

Growth Hormone Abnormalities

Growth hormone stimulates muscular and skeletal development. If it is administered before the epiphyseal cartilages have closed, it will cause an increase in height, weight, and muscle mass. In extreme cases, *gigantism* can result. ∞ p. 198 In **acromegaly** (*akron*, extremity + *megale*, great), an excessive amount of GH is released after puberty, when most of the epiphyseal cartilages have already closed. Cartilages and small bones respond to the hormone, however, resulting in abnormal growth of the hands, feet, lower jaw, skull, and clavicle. Figure 18–10● shows a typical acromegalic individual.

Children who are unable to produce adequate concentrations of GH have *pituitary growth failure*, a condition we introduced in Chapter 6. ∞ p. 198 The steady growth and maturation that typically precede and accompany puberty do not occur in these individuals, who have short stature, slow epiphyseal growth, and larger-than-normal adipose tissue reserves.

Normal growth patterns can be restored by the administration of GH. Before the advent of gene splicing and recombinant DNA techniques, GH had to be carefully extracted and purified from the pituitary glands of cadavers at considerable expense and risk of infectious disease. Now genetically manipulated bacteria are used to produce human GH in commercial quantities.

The current availability of purified human growth hormone has led to its use under medically questionable circumstances. For example, it is now being praised as an "antiaging" miracle cure. Although GH supplements do slow or even reverse the

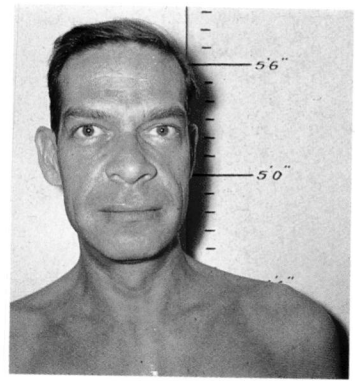

●FIGURE 18–10
Acromegaly. Acromegaly results from the overproduction of growth hormone after the epiphyseal cartilages have closed. Bone shapes change, and cartilaginous areas of the skeleton enlarge. Notice the broad facial features and the enlarged lower jaw.

losses in bone and muscle mass that accompany aging, little is known about adverse side effects that may accompany long-term use of the hormone in mature adults. GH is also being sought by some parents of short but otherwise healthy children. These parents view short stature as a handicap that merits treatment by a physician. Whether we are considering GH treatment of adults or children, it is important to remember that GH and the somatomedins affect many different tissues and have widespread metabolic impacts. For example, children exposed to GH may grow faster, but their body fat content declines drastically and sexual maturation is delayed. The decline is associated with metabolic changes in many organs. The range and significance of these metabolic side effects are now the subject of long-term studies.

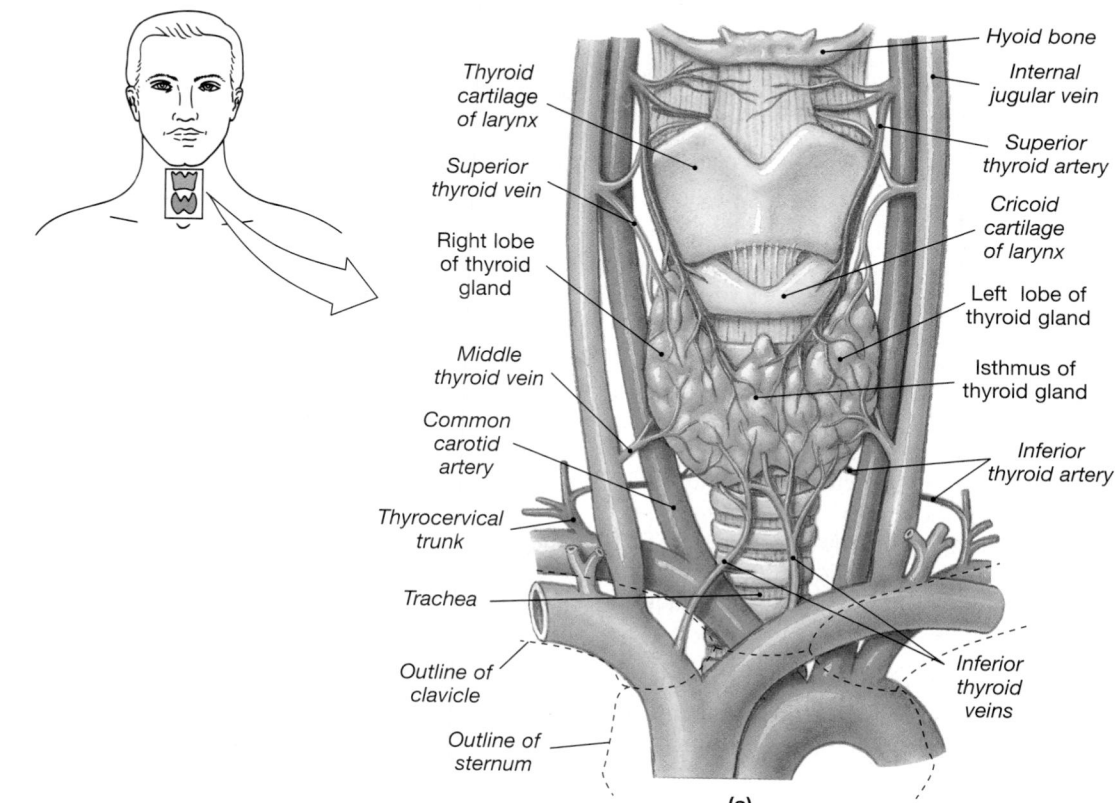

●FIGURE 18–11
The Thyroid Gland. **(a)** The location, anatomy, and blood supply of the thyroid gland.

- Thyroid cartilage of larynx
- Superior thyroid vein
- Right lobe of thyroid gland
- Middle thyroid vein
- Common carotid artery
- Thyrocervical trunk
- Trachea
- Outline of clavicle
- Outline of sternum

- Hyoid bone
- Internal jugular vein
- Superior thyroid artery
- Cricoid cartilage of larynx
- Left lobe of thyroid gland
- Isthmus of thyroid gland
- Inferior thyroid artery
- Inferior thyroid veins

(a)

●**FIGURE 18–11**
The Thyroid Gland.
(**continued**)
(**b**) A diagrammatic
view of a section
through the wall of the
thyroid gland.
(**c**) Histological details,
showing thyroid
follicles. (LM × 211)

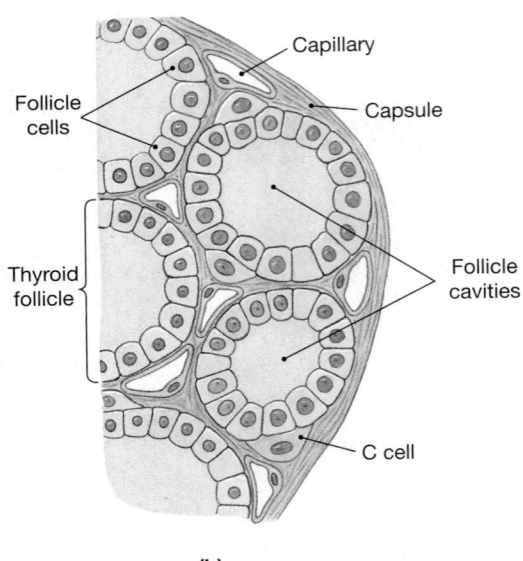

(b)

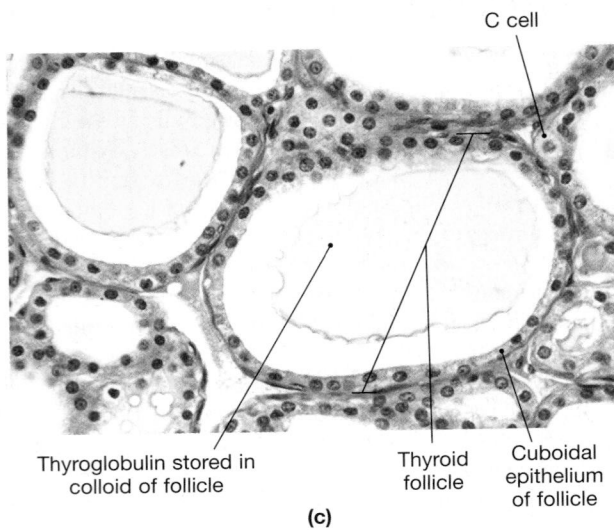

Thyroglobulin stored in colloid of follicle

Thyroid follicle

Cuboidal epithelium of follicle

(c)

(Figure 18–11b,c●). The follicle cells surround a **follicle cavity** that holds a viscous *colloid*, a fluid containing large quantities of suspended proteins. A network of capillaries surrounds each follicle, delivering nutrients and regulatory hormones to the glandular cells and accepting their secretory products and metabolic wastes.

Follicle cells synthesize a globular protein called **thyroglobulin** (thī-rō-GLOB-ū-lin) and secrete it into the colloid of the thyroid follicles (Figures 18–11c and 18–12a●). Each thyroglobulin molecule contains the amino acid *tyrosine*, the building block of thyroid hormones. The formation of thyroid hormones involves three basic steps:

1. Iodide ions are absorbed from the diet at the digestive tract and are delivered to the thyroid gland by the bloodstream. Carrier proteins in the basal membrane of the follicle cells transport iodide ions (I^-) into the cytoplasm. Normally, the follicle cells maintain intracellular concentrations of iodide that are many times higher than those in the extracellular fluid.

2. The iodide ions diffuse to the apical surface of each follicle cell, where they are converted to an activated form of iodide (I^+) by the enzyme *thyroid peroxidase*. This reaction sequence also attaches one or two iodide ions to the tyrosine molecules of thyroglobulin.

3. Tyrosine molecules to which iodide ions have been attached are paired, forming molecules of thyroid hormones that remain incorporated into thyroglobulin. The pairing process is probably performed by thyroid peroxidase. The hormone **thyroxine** (thī-ROKS-ēn), also known as *tetraiodothyronine*, or T_4, contains four iodide ions. **Triiodothyronine**, or T_3, is a related molecule containing three iodide ions. Eventually, each molecule of thyroglobulin contains four to eight molecules of T_3 or T_4 hormones or both.

The major factor controlling the rate of thyroid hormone release is the concentration of TSH in the circulating blood (Figure 18–12b●). TSH stimulates iodide transport into the follicle cells and stimulates the production of thyroglobulin and thyroid per-

oxidase. TSH also stimulates the release of thyroid hormones. Under the influence of TSH, the following steps occur:

1. Follicle cells remove thyroglobulin from the follicles by endocytosis.

2. Lysosomal enzymes break the thyroglobulin down, and the amino acids and thyroid hormones enter the cytoplasm. The amino acids are then recycled and used to synthesize thyroglobulin.

3. The released molecules of T_3 and T_4 diffuse across the basement membrane and enter the bloodstream. About 90 percent of all thyroid secretions is T_4; T_3 is secreted in comparatively small amounts.

4. Roughly 75 percent of the T_4 molecules and 70 percent of the T_3 molecules entering the bloodstream become attached to transport proteins called **thyroid-binding globulins** (**TBGs**). Most of the rest of the T_4 and T_3 in the circulation is attached to **transthyretin**, also known as *thyroid-binding prealbumin* (*TBPA*), or to *albumin*, one of the plasma proteins. Only the relatively small quantities of thyroid hormones that remain unbound—roughly 0.3 percent of the circulating T_3 and 0.03 percent of the circulating T_4—are free to diffuse into peripheral tissues.

An equilibrium exists between the bound and unbound thyroid hormones. At any moment, free thyroid hormones are being bound to carriers at the same rate at which bound hormones are being released. When unbound thyroid hormones diffuse out of the bloodstream and into other tissues, the equilibrium is disturbed. The carrier proteins then release additional thyroid hormones until a new equilibrium is reached. The bound thyroid hormones represent a substantial reserve: The bloodstream normally contains more than a week's supply of thyroid hormones.

TSH plays a key role in both the synthesis and the release of thyroid hormones. In the absence of TSH, the thyroid follicles become inactive, and neither synthesis nor secretion occurs. TSH binds to membrane receptors and, by stimulating adenylate cyclase, activates key enzymes involved in thyroid hormone production (Figure 18–3●, p. 611).

●**FIGURE 18–12**
The Thyroid Follicles. **(a)** The synthesis, storage, and secretion of thyroid hormones. For a detailed explanation of the numbered events, see the text. **(b)** The regulation of thyroid secretion.

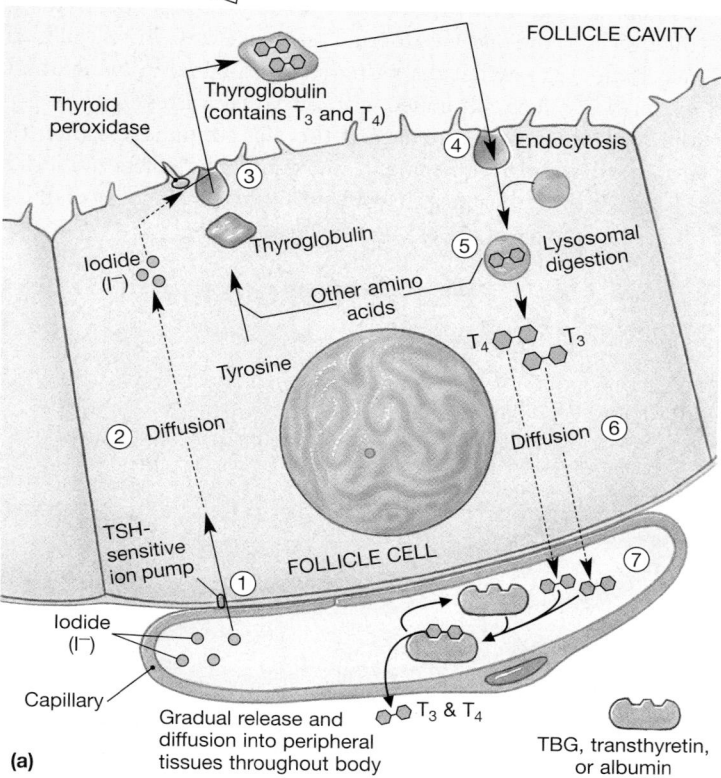

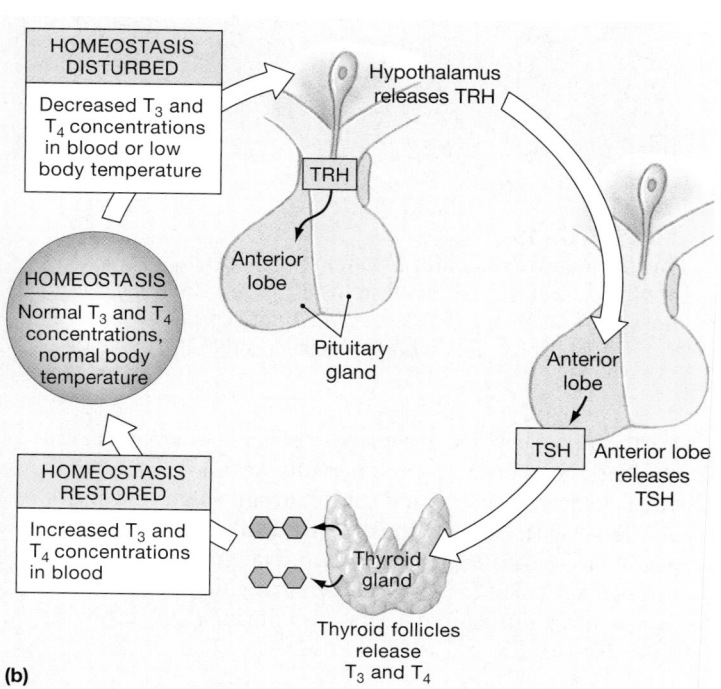

☐ FUNCTIONS OF THYROID HORMONES

Thyroid hormones readily cross cell membranes, and they affect almost every cell in the body. Inside a cell, they bind to (1) receptors in the cytoplasm, (2) receptors on the surfaces of mitochondria, and (3) receptors in the nucleus.

- Thyroid hormones bound to cytoplasmic receptors are essentially held in storage. If intracellular levels of thyroid hormones decline, the bound thyroid hormones are released into the cytoplasm.
- The thyroid hormones binding to mitochondria increase the rates of mitochondrial ATP production.
- The binding to receptors in the nucleus activates genes that control the synthesis of enzymes involved with energy transformation and utilization. One specific effect of binding to nuclear receptors is the accelerated production of sodium–potassium ATPase, the membrane protein responsible for the ejection of intracellular sodium and the recovery of extracellular potassium. As we noted in Chapter 3, this exchange pump consumes large amounts of ATP. ∞ p. 96

Thyroid hormones also activate genes that code for the synthesis of enzymes involved in glycolysis and ATP production. This effect, coupled with the direct effect of thyroid hormones on mitochondria, increases the metabolic rate of the cell. Because the cell consumes more energy and because energy use is measured in *calories*, the effect is called the **calorigenic effect** of thyroid hormones. When the metabolic rate increases, more heat is generated. In young children, TSH production increases in cold weather; the calorigenic effect may help them adapt to cold climates. (This response does not occur in adults.) In growing children, thyroid hormones are also essential to normal development of the skeletal, muscular, and nervous systems.

T_3 Versus T_4

The thyroid gland produces large amounts of T_4, but T_3 is primarily responsible for the observed effects of thyroid hormones: a strong, immediate, and short-lived increase in the rate of cellular metabolism. Peripheral tissues have two sources of T_3:

1. *Release by the Thyroid Gland.* At any moment, T_3 from the thyroid gland accounts for only 10–15 percent of the T_3 in peripheral tissues.
2. *The Conversion of T_4 to T_3.* Enzymes in the liver, kidneys, and other tissues can convert T_4 to T_3. Roughly 85–90 percent of the T_3 that reaches the target cells is produced by the conversion of T_4 within peripheral tissues. Table 18–3 summarizes the effects of thyroid hormones on major organs and systems.

Iodine and Thyroid Hormones

Iodine in the diet is absorbed at the digestive tract as I^-. The follicle cells in the thyroid gland absorb 120–150 μg of I^- each day, the minimum dietary amount needed to maintain normal thyroid function. The iodide ions are actively transported into the

TABLE 18–3	EFFECTS OF THYROID HORMONES ON PERIPHERAL TISSUES

1. Elevate rates of oxygen consumption and energy consumption; in children, may cause a rise in body temperature
2. Increase heart rate and force of contraction; generally cause a rise in blood pressure
3. Increase sensitivity to sympathetic stimulation
4. Maintain normal sensitivity of respiratory centers to changes in oxygen and carbon dioxide concentrations
5. Stimulate the formation of red blood cells and thereby enhance oxygen delivery
6. Stimulate the activity of other endocrine tissues
7. Accelerate the turnover of minerals in bone

thyroid follicle cells, so the concentration of I⁻ inside thyroid follicle cells is generally about 30 times higher than that in the blood plasma. If plasma I⁻ levels rise, so do levels inside the follicle cells.

The thyroid follicles contain most of the iodide reserves in the body. The active transport mechanism for iodide is stimulated by TSH. The resulting increase in the rate of iodide movement into the cytoplasm accelerates the formation of thyroid hormones.

The typical diet in the United States provides approximately 500 μg of iodide per day, roughly three times the minimum daily requirement. Much of the excess is due to the addition of I⁻ to the table salt sold in grocery stores as "iodized salt." Thus, iodide deficiency is seldom responsible for limiting the rate of thyroid hormone production. (This is not necessarily the case in other countries.) Excess I⁻ is removed from the blood at the kidneys, and each day a small amount of I⁻ (about 20 μg) is excreted by the liver into the *bile*, an exocrine product stored in the gallbladder. Iodide excreted at the kidneys is eliminated in urine; the I⁻ excreted in bile is eliminated with intestinal wastes. The losses in the bile, which continue even if the diet contains less than the minimum iodide requirement, can gradually deplete the iodide reserves in the thyroid. Thyroid hormone production then declines, regardless of the circulating levels of TSH.

☐ THE C CELLS OF THE THYROID GLAND: CALCITONIN

A second population of endocrine cells lies sandwiched between the cuboidal follicle cells and their basement membrane. These cells, which are larger than those of the follicular epithelium and do

Thyroid Gland Disorders

Normal production of thyroid hormones establishes the background rates of cellular metabolism. These hormones exert their primary effects on metabolically active tissues and organs, including skeletal muscles, the liver, and the kidneys. The inadequate production of thyroid hormones is called **hypothyroidism**. In an infant, hypothyroidism produces *cretinism*, a condition marked by inadequate skeletal and nervous development and a metabolic rate as much as 40 percent below normal levels (Figure 18–13a●). The condition affects approximately 1 birth out of every 5000. Cretinism developing later in childhood will retard growth and mental development and delay puberty.

Adults with hypothyroidism are lethargic and unable to tolerate cold temperatures. The symptoms of adult hypothyroidism, collectively known as *myxedema* (miks-e-DĒ-muh), include subcutaneous swelling, dry skin, hair loss, low body temperature, muscular weakness, and slowed reflexes. Hypothyroidism may also be associated with the enlargement of the thyroid gland, producing a distinctive swelling called a goiter (Figure 18–13b●). Hypothyroidism, myxedema, and goiter as the result of inadequate dietary iodide are very rare in the United States, in part due to the addition of iodine to table salt, but these conditions can be relatively common in poorer countries, especially landlocked ones (seafood is a good source of iodine).

Hyperthyroidism, or *thyrotoxicosis*, occurs when thyroid hormones are produced in excessive quantities. The metabolic rate climbs, and the skin becomes flushed and moist with perspiration. Blood pressure and heart rate increase, and the

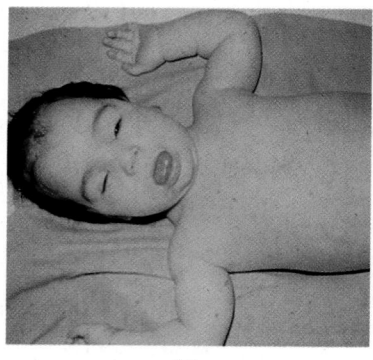 

(a)   (b)

●FIGURE 18–13
Thyroid Disorders.   (a) Cretinism, or congenital hypothyroidism, results from thyroid hormone insufficiency in infancy. (b) An enlarged thyroid gland, or goiter, can be associated with thyroid hyposecretion due to iodine insufficiency in adults.

heartbeat may become irregular as circulatory demands escalate. The effects on the central nervous system make the individual restless, excitable, and subject to shifts in mood and emotional states. Despite the drive for increased activity, the person has limited energy reserves and fatigues easily. *Graves' disease* is a form of hyperthyroidism that afflicted President George W. H. Bush and Barbara Bush during their stay in the White House. **AM** Thyroid Gland Disorders

not stain as clearly, are the **C** (**clear**) **cells**, or *parafollicular cells* (Figure 18–11b,c●). C cells produce the hormone **calcitonin** (**CT**), which aids in the regulation of Ca^{2+} concentrations in body fluids. We introduced the functions of this hormone in Chapter 6. ∞ p. 199 The net effect of calcitonin release is a drop in the Ca^{2+} concentration in body fluids, accomplished by (1) the inhibition of osteoclasts, which slows the rate of Ca^{2+} release from bone, and (2) the stimulation and then falloff of Ca^{2+} excretion at the kidneys.

The control of calcitonin secretion is an example of direct endocrine regulation: Neither the hypothalamus nor the pituitary gland is involved. The C cells respond directly to elevations in the Ca^{2+} concentration of blood. When those concentrations rise, calcitonin secretion increases. The Ca^{2+} concentrations then drop, eliminating the stimulus and "turning off" the C cells.

Calcitonin is probably most important during childhood, when it stimulates bone growth and mineral deposition in the skeleton. It also appears to be important in reducing the loss of bone mass (1) during prolonged starvation and (2) in the late stages of pregnancy, when the maternal skeleton competes with the developing fetus for calcium ions absorbed by the digestive tract. The role of calcitonin in the healthy nonpregnant adult is uncertain.

In several chapters, we have considered the importance of Ca^{2+} in controlling muscle cell and neuron activities. Calcium ion concentrations also affect the sodium permeabilities of excitable membranes. At high Ca^{2+} concentrations, sodium permeability decreases and membranes become less responsive. Such

problems are relatively rare. Problems caused by lower than normal Ca^{2+} concentrations are equally dangerous and are much more common. When calcium ion concentrations decline, sodium permeabilities increase and cells become extremely excitable. If calcium levels fall too far, convulsions or muscular spasms can result. Maintenance of adequate calcium levels involves the *parathyroid glands* and *parathyroid hormone*.

18–5 THE PARATHYROID GLANDS

Objective

■ Describe the location of the parathyroid glands, the functions of the hormone they produce, and the effects of abnormal levels of parathyroid hormone production.

There are normally two pairs of **parathyroid glands** embedded in the posterior surfaces of the thyroid gland (Figure 18–14a●). The cells of the adjacent glands are separated by the dense capsular fibers of the thyroid. Altogether, the four parathyroid glands weigh a mere 1.6 g (0.06 oz). The histological appearance of a parathyroid gland is shown in Figure 18–14b,c●. The parathyroid glands have at least two cell populations: The **chief cells** produce parathyroid hormone; the functions of the other cells, called *oxyphils*, are unknown.

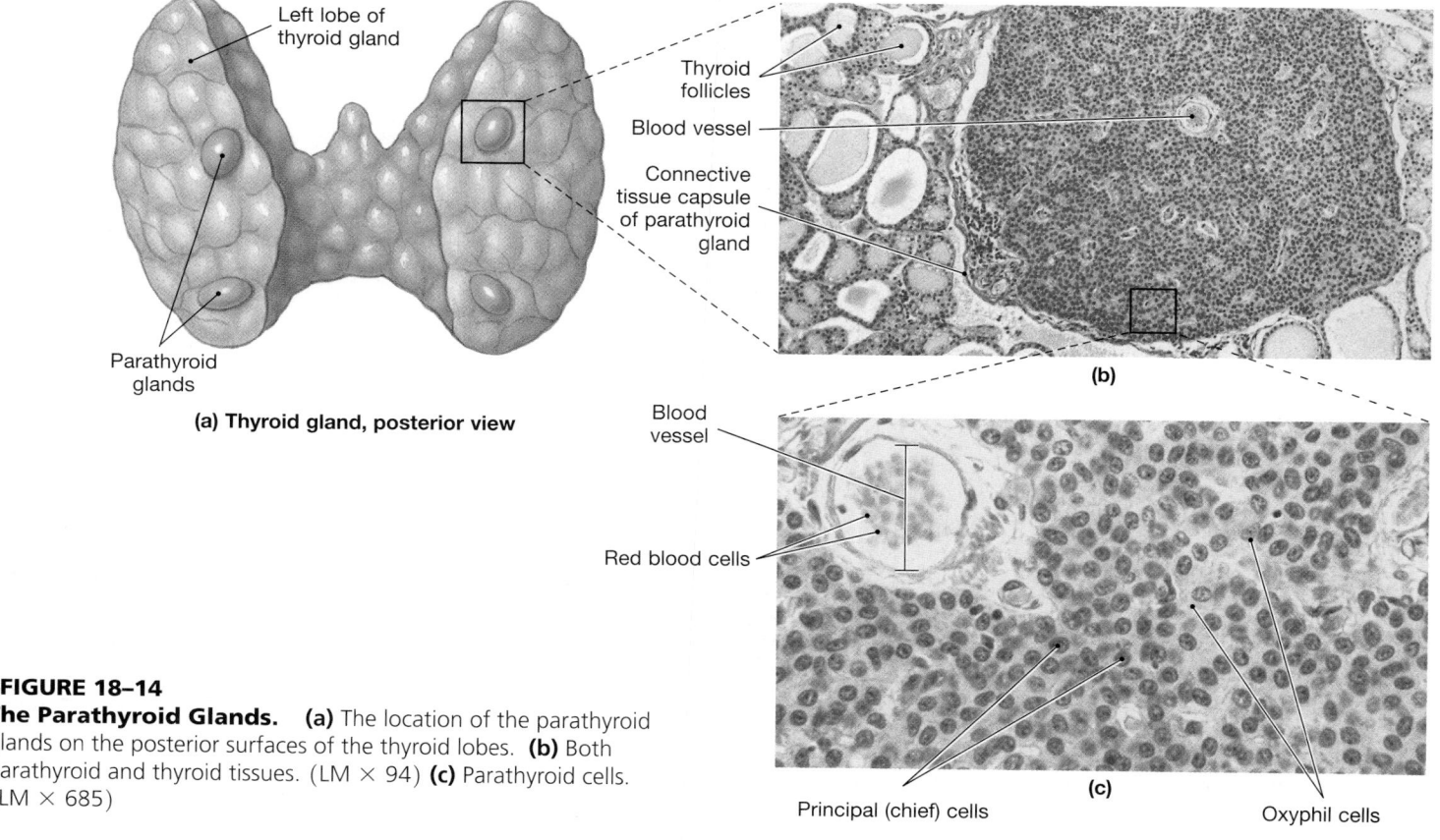

●**FIGURE 18–14**
The Parathyroid Glands. **(a)** The location of the parathyroid glands on the posterior surfaces of the thyroid lobes. **(b)** Both parathyroid and thyroid tissues. (LM × 94) **(c)** Parathyroid cells. (LM × 685)

Like the C cells of the thyroid gland, the chief cells monitor the circulating concentration of calcium ions. When the Ca^{2+} concentration of the blood falls below normal, the chief cells secrete **parathyroid hormone** (**PTH**), or *parathormone*. The net result of PTH secretion is an increase in Ca^{2+} concentration in body fluids. Parathyroid hormone has four major effects:

1. It stimulates osteoclasts, accelerating mineral turnover and the release of Ca^{2+} from bone.

2. It inhibits osteoblasts, reducing the rate of calcium deposition in bone.

3. It enhances the reabsorption of Ca^{2+} at the kidneys, reducing urinary losses.

4. It stimulates the formation and secretion of *calcitriol* at the kidneys. In general, the effects of calcitriol complement or enhance those of PTH, but one major effect of calcitriol is the enhancement of Ca^{2+} and PO_4^{3-} absorption by the digestive tract. ∞ p. 197

Figure 18–15● compares the functions of calcitonin and PTH. It is likely that PTH, aided by calcitriol, is the primary regulator of circulating calcium ion concentrations in healthy adults. Information about the hormones of the thyroid gland and parathyroid glands is summarized in Table 18–4.

When the parathyroid gland secretes inadequate or excessive amounts of parathyroid hormone, Ca^{2+} concentrations move outside normal homeostatic limits. Inadequate parathyroid hormone production, a condition called **hypoparathyroidism**, leads to low Ca^{2+} concentrations in body fluids. The most obvious symptoms involve neural and muscle tissues: The nervous system becomes more excitable, and the affected individual may experience *hypocalcemic tetany*, a dangerous condition characterized by prolonged muscle spasms that initially involve the limbs and face.

In **hyperparathyroidism**, Ca^{2+} concentrations become abnormally high. Bones grow thin and brittle, skeletal muscles become weak, CNS function is depressed, and nausea and vomiting occur. In severe cases, the patient may become comatose. **AM** Disorders of Parathyroid Function

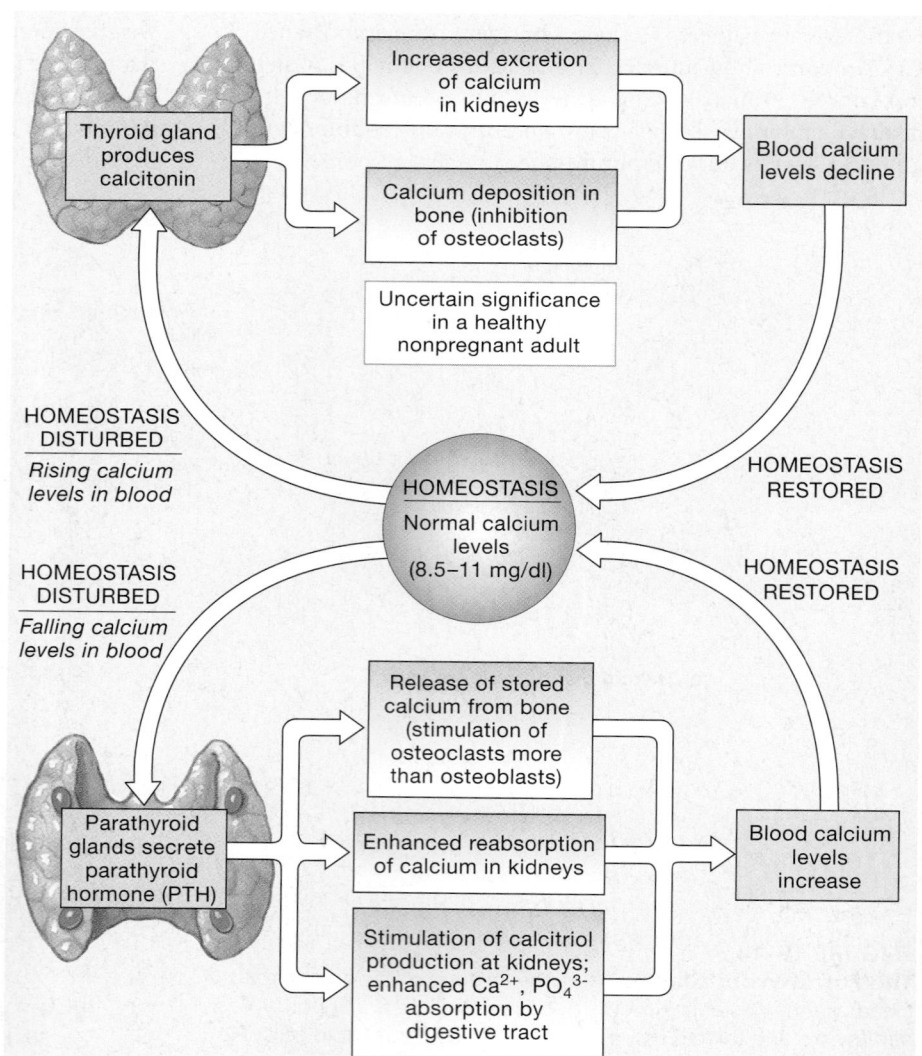

●**FIGURE 18–15**
The Homeostatic Regulation of Calcium Ion Concentrations

TABLE 18–4 HORMONES OF THE THYROID GLAND AND PARATHYROID GLANDS

Gland/Cells	Hormone(s)	Targets	Hormonal Effects	Regulatory Control
THYROID GLAND				
Follicular epithelium	Thyroxine (T_4), triiodothyronine (T_3)	Most cells	Increases energy utilization, oxygen consumption, growth, and development (see Table 18–3)	Stimulated by TSH from anterior lobe of the pituitary gland (see Figure 18–8a)
C cells	Calcitonin (CT)	Bone, kidneys	Decreases Ca^{2+} concentrations in body fluids (see Figure 18–13)	Stimulated by elevated blood Ca^{2+} levels; actions opposed by PTH
PARATHYROID GLANDS				
Chief cells	Parathyroid hormone (PTH)	Bone, kidneys	Increases Ca^{2+} concentrations in body fluids (see Figure 18–13)	Stimulated by low blood Ca^{2+} levels; PTH effects enhanced by calcitriol and opposed by calcitonin

18–6 THE ADRENAL GLANDS

Objectives

- Describe the location, structure, and general functions of the adrenal glands.
- Identify the hormones produced by the adrenal cortex and medulla, and specify the functions of those hormones.
- Discuss the results of abnormal levels of adrenal hormone production.

A yellow, pyramid-shaped **adrenal gland**, or *suprarenal* (soo-pra-RĒ-nal) *gland* (*supra-*, above + *renes*, kidneys), sits on the superior border of each kidney (Figure 18–16a●). Each adrenal gland lies at roughly the level of the 12th rib and is firmly attached to the superior portion of each kidney by a dense fibrous capsule. The adrenal gland on each side nestles between the kidney, the diaphragm, and the major arteries and veins that run along the dorsal wall of the abdominopelvic cavity. The adrenal glands project into the peritoneal cavity, and their anterior surfaces are covered by a layer of parietal peritoneum. Like other endocrine glands, the adrenal glands are highly vascularized.

A typical adrenal gland weighs about 7.5 g (0.19 oz), but its size can vary greatly as secretory demands change. The adrenal gland is divided into two parts: a superficial **adrenal cortex** and an inner **adrenal medulla** (Figure 18–16b●).

☐ THE ADRENAL CORTEX

The yellowish color of the adrenal cortex is due to the presence of stored lipids, especially cholesterol and various fatty acids. The adrenal cortex produces more than two dozen steroid hormones, collectively called **adrenocortical steroids**, or simply **corticosteroids**. In the bloodstream, these hormones are bound to transport proteins called *transcortins*.

Corticosteroids are vital: If the adrenal glands are destroyed or removed, the individual will die unless corticosteroids are administered. Corticosteroids, like other steroid hormones, exert their effects by determining which genes are transcribed in the nuclei of their target cells and at what rates. The resulting changes in the nature and concentration of enzymes in the cytoplasm affect cellular metabolism.

Deep to the adrenal capsule are three distinct regions, or zones, in the adrenal cortex: (1) an outer *zona glomerulosa*; (2) a middle *zona fasciculata*; and (3) an inner *zona reticularis* (Figure 18–16c●). Each zone synthesizes specific steroid hormones, as listed in Table 18–5.

The Zona Glomerulosa

The **zona glomerulosa** (glō-mer-ū-LŌ-suh), the outer region of the adrenal cortex, produces **mineralocorticoids (MCs)**, steroid hormones that affect the electrolyte composition of body fluids. **Aldosterone** is the principal mineralocorticoid produced by the adrenal cortex.

The zona glomerulosa accounts for about 15 percent of the volume of the adrenal cortex (Figure 18–16c●). A *glomerulus* is a little ball; as the term *zona glomerulosa* implies, the endocrine cells in this region form small, dense knots or clusters. This zone extends from the capsule to the radiating cords of the deeper zona fasciculata.

ALDOSTERONE Aldosterone secretion stimulates the conservation of sodium ions and the elimination of potassium ions. This hormone targets cells that regulate the ionic composition of excreted fluids. It causes the retention of sodium ions at the kidneys, sweat glands, salivary glands, and pancreas, preventing Na^+ loss in urine, sweat, saliva, and digestive secretions. The retention of Na^+ is accompanied by a loss of K^+. As a secondary effect, the reabsorption of Na^+ enhances the osmotic reabsorption of water at the kidneys, sweat glands, salivary glands, and pancreas. The effect at the kidneys is most dramatic when normal levels of ADH are present. In addition, aldosterone increases the sensitivity of salt receptors in the taste buds of the tongue. As a result, interest in (and consumption of) salty food increases.

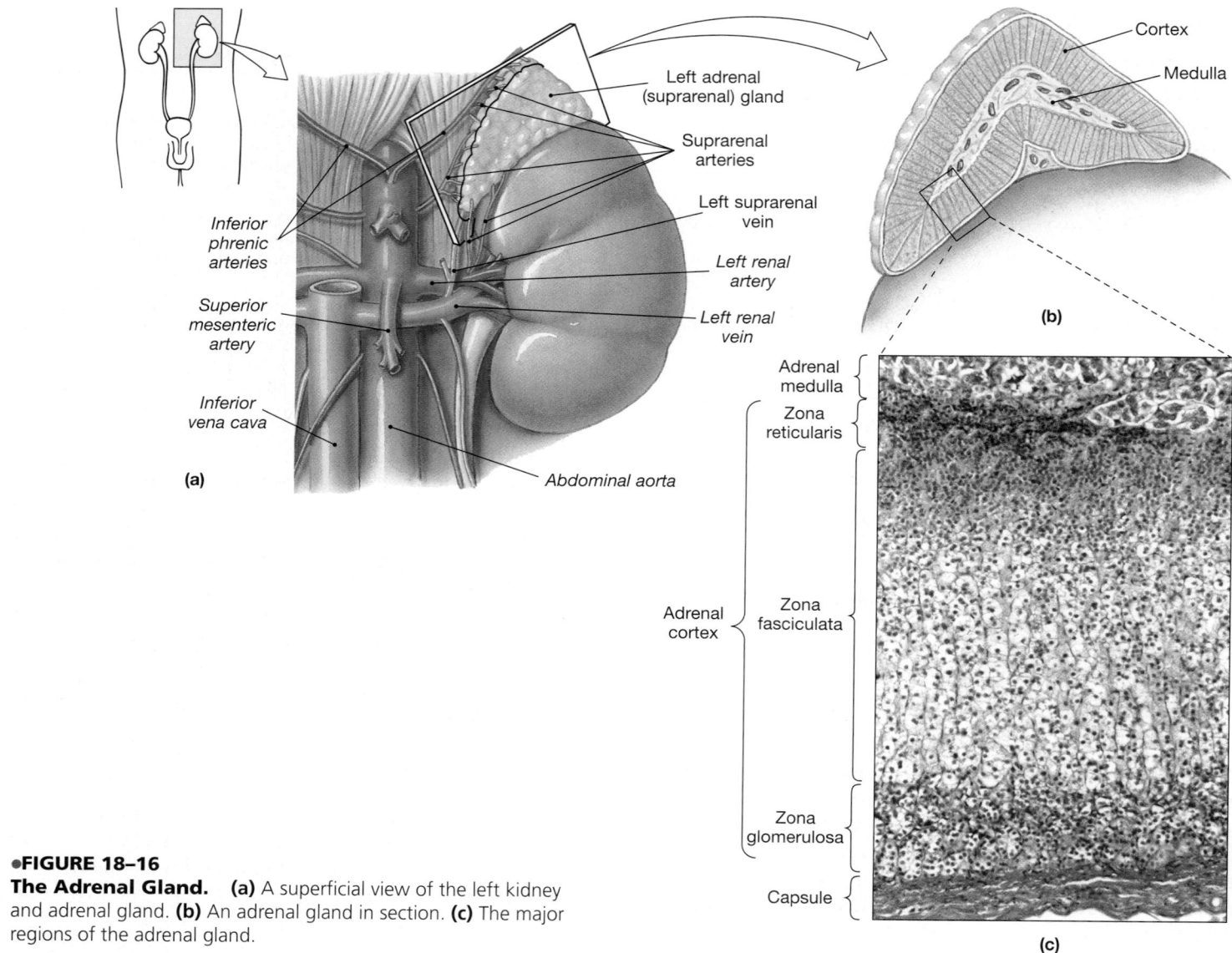

●**FIGURE 18–16**
The Adrenal Gland. (a) A superficial view of the left kidney
and adrenal gland. (b) An adrenal gland in section. (c) The major
regions of the adrenal gland.

TABLE 18–5   THE ADRENAL HORMONES

Region/Zone	Hormone(s)	Targets	Hormonal Effects	Regulatory Control
CORTEX				
Zona glomerulosa	Mineralocorticoids (MCs), primarily aldosterone	Kidneys	Increase renal reabsorption of Na^+ and water (especially in the presence of ADH) and accelerate urinary loss of K^+	Stimulated by angiotensin II; inhibited by natriuretic peptides (ANP and BNP)
Zona fasciculata	Glucocorticoids (GCs): cortisol (hydrocortisone), corticosterone	Most cells	Release amino acids from skeletal muscles, and lipids from adipose tissues; promote liver formation of glucose and glycogen; promote peripheral utilization of lipids; anti-inflammatory effects	Stimulated by ACTH from anterior lobe of pituitary gland (*see Figure 18–8a*)
Zona reticularis	Androgens		Uncertain significance under normal conditions	Stimulated by ACTH; significance uncertain
MEDULLA	Epinephrine, norepinephrine	Most cells	Increases cardiac activity, blood pressure, glycogen breakdown, blood glucose levels; releases lipids by adipose tissue	Stimulated during sympathetic activation by sympathetic preganglionic fibers (*see Chapter 16*)

Aldosterone secretion occurs in response to a drop in blood Na^+ content, blood volume, or blood pressure or to a rise in blood K^+ concentration. Changes in Na^+ or K^+ concentration have a direct effect on the zona glomerulosa, stimulating the release of aldosterone. Aldosterone release also occurs in response to *angiotensin II*. We will discuss this hormone, part of the *renin–angiotensin system*, later in this chapter.

In **hypoaldosteronism**, the zona glomerulosa fails to produce enough aldosterone, generally because the kidneys are not releasing adequate amounts of the enzyme *renin*. Affected individuals lose excessive amounts of water and Na^+ at the kidneys, leading to low blood volume and low blood pressure. Changes in electrolyte concentrations affect transmembrane potentials, eventually disrupting neural and muscular tissues throughout the body.

Hypersecretion of aldosterone results in the condition called **aldosteronism**. Under continued aldosterone stimulation, the kidneys retain sodium ions very effectively, but potassium ions are lost in large quantities. A crisis eventually ensues when low extracellular K^+ concentrations disrupt normal cardiac, neural, and kidney cell functions. **AM** Disorders of the Adrenal Cortex

The Zona Fasciculata

The **zona fasciculata** (fa-sik-ū-LA-tuh; *fasciculus*, little bundle) produces steroid hormones collectively known as **glucocorticoids** (**GCs**), due to their effects on glucose metabolism. This zone, which begins at the inner border of the zona glomerulosa and extends toward the adrenal medulla, contributes about 78 percent of the cortical volume. The endocrine cells are larger and contain more lipids than those of the zona glomerulosa, and the lipid droplets give the cytoplasm a pale, foamy appearance. The cells of the zona fasciculata form individual cords composed of stacks of cells. Adjacent cords are separated by flattened blood vessels with fenestrated walls.

THE GLUCOCORTICOIDS When stimulated by ACTH from the anterior lobe of the pituitary, the zona fasciculata secretes primarily **cortisol** (KOR-ti-sol), also called *hydrocortisone*, along with smaller amounts of the related steroid **corticosterone** (kor-ti-KOS-te-rōn). The liver converts some of the circulating cortisol to **cortisone**, another active glucocorticoid. Glucocorticoid secretion is regulated by negative feedback: The glucocorticoids released have an inhibitory effect on the production of corticotropin-releasing hormone (CRH) in the hypothalamus and of ACTH in the anterior lobe. This relationship was diagrammed in Figure 18–8a, p. 616.

EFFECTS OF GLUCOCORTICOIDS Glucocorticoids accelerate the rates of glucose synthesis and glycogen formation, especially in the liver. Adipose tissue responds by releasing fatty acids into the blood, and other tissues begin to break down fatty acids and proteins instead of glucose. This process is another example of a glucose-sparing effect (p. 618).

Glucocorticoids also show **anti-inflammatory** effects; that is, they inhibit the activities of white blood cells and other components

Abnormal Glucocorticoid Production

Addison's disease is caused by inadequate glucocorticoid production (see Figure 18–17a). The usual cause is the destruction of the zona fasciculata as the result of an *autoimmune response* (in which the body's defenses mistakenly attack healthy tissues) or an infection by the bacteria responsible for *tuberculosis*. Affected individuals become weak and lose weight; they cannot effectively use their lipid reserves to generate ATP, and blood glucose concentrations fall sharply within hours after a meal. In most cases the zona glomerulosa is affected as well, so symptoms similar to those of hypoaldosteronism may also appear.

Cushing's disease results from an overproduction of glucocorticoids, generally owing to the hypersecretion of ACTH. The symptoms resemble those of a protracted and exaggerated response to stress. Glucose metabolism is suppressed, lipid reserves are mobilized, and peripheral proteins are broken down. Lipid reserves are redistributed, and the distribution of body fat changes. Individuals develop a "moonfaced" appearance, due to the deposition of lipids in the subcutaneous tissues of the face (see Figure 18–17b). Skeletal muscles become weak as contractile proteins break down. **AM** Disorders of the Adrenal Cortex

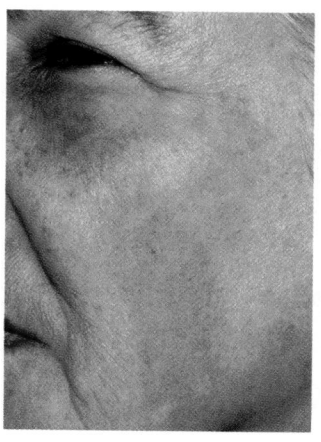 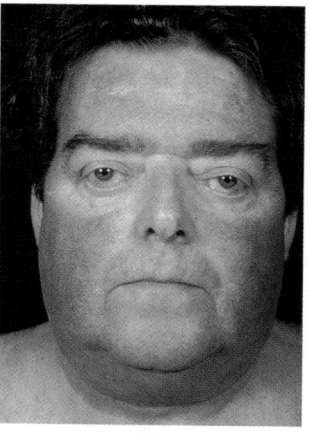

(a)   (b)

•FIGURE 18–17
Adrenal Abnormalities. (a) Addison's disease is caused by hyposecretion of corticosteroids, especially glucocorticoids. Pigment changes result from stimulation of melanocytes by ACTH, which is structurally similar to MSH. (b) Cushing's disease is caused by hypersecretion of glucocorticoids. Lipid reserves are mobilized and adipose tissue accumulates in the cheeks and at the base of the neck.

of the immune system. "Steroid creams" are commonly used to control irritating allergic rashes, such as those produced by poison ivy, and injections of glucocorticoids may be used to control more severe allergic reactions. Glucocorticoids slow the migration of phagocytic cells into an injury site and cause phagocytic cells already in the area to become less active. In addition, mast cells exposed to these steroids are less likely to release histamine and other chemicals that promote inflammation. ∞ p. 139 As a result, swelling and further irritation are dramatically reduced. On the negative side, the rate of wound healing decreases, and the weakening of the region's defenses makes it an easy target for infectious organisms. For that reason, topical steroids are used to treat superficial rashes, but are never applied to open wounds.

The Zona Reticularis

The **zona reticularis** (re-tik-ū-LAR-is; *reticulum*, network) forms a narrow band bordering each adrenal medulla (Figure 18–16c●). This zone accounts for only about 7 percent of the total volume of the adrenal cortex. The endocrine cells of the zona reticularis form a folded, branching network, and fenestrated blood vessels wind between the cells.

The zona reticularis normally produces small quantities of androgens, the sex hormones produced in large quantities by the testes in males. Once in the bloodstream, some of the androgens released by the zona reticularis are converted to estrogens, the dominant sex hormone in females. When secreted in normal amounts, neither the androgens nor the estrogens affect sexual characteristics, and the significance of adrenal sex hormone production remains uncertain. ACTH stimulates the zona reticularis to a slight degree, but the effects are generally insignificant.

The zona reticularis ordinarily produces a negligible amount of androgens. If a tumor forms there, androgen secretion may increase dramatically, producing symptoms of **adrenogenital syndrome.** In women, this condition leads to the gradual development of male secondary sex characteristics, including body and facial hair patterns, adipose tissue distribution, and muscle development. Tumors affecting the zona reticularis of males can result in the production of large quantities of estrogens. Affected males develop enlarged breasts and, in some cases, other female secondary sex characteristics. This array of symptoms is called **gynecomastia** (*gynaikos*, woman + *mastos*, breast).

□ THE ADRENAL MEDULLA

The boundary between the adrenal cortex and the adrenal medulla is irregular, and the supporting connective tissues and blood vessels are extensively interconnected. The adrenal medulla is reddish brown, owing in part to the many blood vessels in the area, and contains large, rounded cells—similar to those in sympathetic ganglia that are innervated by preganglionic sympathetic fibers. The secretory activities of the adren-

al medullae are controlled by the sympathetic division of the autonomic nervous system. ∞ p. 539

The adrenal medulla contains two populations of secretory cells: One produces epinephrine (adrenaline), the other norepinephrine (noradrenaline). Evidence suggests that the two types of cells are distributed in different areas of the medulla and that their secretory activities can be independently controlled. The secretions are packaged in vesicles that form dense clusters just inside cell membranes. The hormones in these vesicles are continuously released at low levels by exocytosis. Sympathetic stimulation dramatically accelerates the rate of exocytosis and hormone release.

Epinephrine and Norepinephrine

Epinephrine makes up 75–80 percent of the secretions from the adrenal medullae, the rest being norepinephrine. We described the peripheral effects of these hormones, which result from interaction with alpha and beta receptors on cell membranes, in Chapter 16. ∞ pp. 540-541 Stimulation of α_1 and β_1 receptors, the most common types, accelerates the utilization of cellular energy and the mobilization of energy reserves.

When the adrenal medulla is activated,

- Epinephrine and norepinephrine trigger a mobilization of glycogen reserves in skeletal muscles and accelerate the breakdown of glucose to provide ATP. This combination increases both muscular strength and endurance.

- In adipose tissue, stored fats are broken down into fatty acids, which are released into the bloodstream for use by other tissues for ATP production.

- In the liver, glycogen molecules are broken down. The resulting glucose molecules are released into the bloodstream, primarily for use by neural tissues, which cannot shift to fatty acid metabolism.

- In the heart, the stimulation of β_1 receptors triggers an increase in the rate and force of cardiac muscle contraction.

The metabolic changes that follow the release of catecholamines such as E and NE are at their peak 30 seconds after adrenal stimulation, and they linger for several minutes thereafter. As a result, the effects produced by the stimulation of the adrenal medullae outlast the other signs of sympathetic activation. **AM** Disorders of the Adrenal Medulla

---

✓ What symptoms would you expect to see in an individual whose diet lacks iodine?

✓ When a person's thyroid gland is removed, signs of decreased thyroid hormone concentration do not appear until about one week later. Why?

✓ The removal of the parathyroid glands would result in a decrease in the blood concentration of which important mineral?

✓ What effect would elevated cortisol levels have on the level of glucose in the blood?

Answers start on page Q-1

18–7 THE PINEAL GLAND

Objective

- Describe the location of the pineal gland and the functions of the hormone that it produces.

The **pineal gland**, part of the epithalamus, lies in the posterior portion of the roof of the third ventricle. ∞ p. 479 The pineal gland contains neurons, neuroglia, and special secretory cells called **pinealocytes** (pi-NĒ-al-ō-sīts). These cells synthesize the hormone melatonin from molecules of the neurotransmitter *serotonin*. Collaterals from the visual pathways enter the pineal gland and affect the rate of melatonin production, which is lowest during daylight hours and highest at night.

Several functions, including the following, have been suggested for melatonin in humans:

- **Inhibiting Reproductive Functions.** In some other mammals, melatonin slows the maturation of sperm, oocytes, and reproductive organs by reducing the rate of GnRH secretion. The significance of this effect in humans remains uncertain, but circumstantial evidence suggests that melatonin may play a role in the timing of human sexual maturation. For example, melatonin levels in the blood decline at puberty, and pineal tumors that eliminate melatonin production cause premature puberty in young children.

- **Protecting Against Damage by Free Radicals.** Melatonin is a very effective *antioxidant* that may protect CNS neurons from free radicals, such as nitric oxide (NO) or hydrogen peroxide (H_2O_2), that may be generated in active neural tissue.

- **Setting Circadian Rhythms.** Because pineal activity is cyclical, the pineal gland may also be involved with the maintenance of basic *circadian rhythms*—daily changes in physiological processes that follow a regular pattern. ∞ p. 482 Increased melatonin secretion in darkness has been suggested as a primary cause of *seasonal affective disorder* (*SAD*). This condition, characterized by changes in mood, eating habits, and sleeping patterns, can develop during the winter in people who live at high latitudes, where sunshine is scarce or lacking. **AM** Light and Behavior

18–8 THE PANCREAS

Objectives

- Describe the location and structure of the pancreas.
- Identify the hormones produced by the pancreas and specify the functions of those hormones.
- Discuss the results of abnormal levels of pancreatic hormone production.

The **pancreas** lies within the abdominopelvic cavity in the J-shaped loop between the stomach and the small intestine. It is a slender, pale organ with a nodular (lumpy) consistency (Figure 18–18a●). The adult pancreas is 20–25 cm (8–10 in.) long and weighs about 80 g (2.8 oz). We shall consider its anatomy further in Chapter 24, because it is primarily an exocrine organ that makes digestive enzymes.

The **exocrine pancreas**, roughly 99 percent of the pancreatic volume, consists of clusters of gland cells, called *pancreatic acini*, and their attached ducts. Together, the gland and duct cells secrete large quantities of an alkaline, enzyme-rich fluid that reaches the lumen of the digestive tract by traveling along a network of secretory ducts.

The **endocrine pancreas** consists of small groups of cells scattered among the exocrine cells. The endocrine clusters are known as **pancreatic islets**, or the *islets of Langerhans* (LAN-ger-hanz)

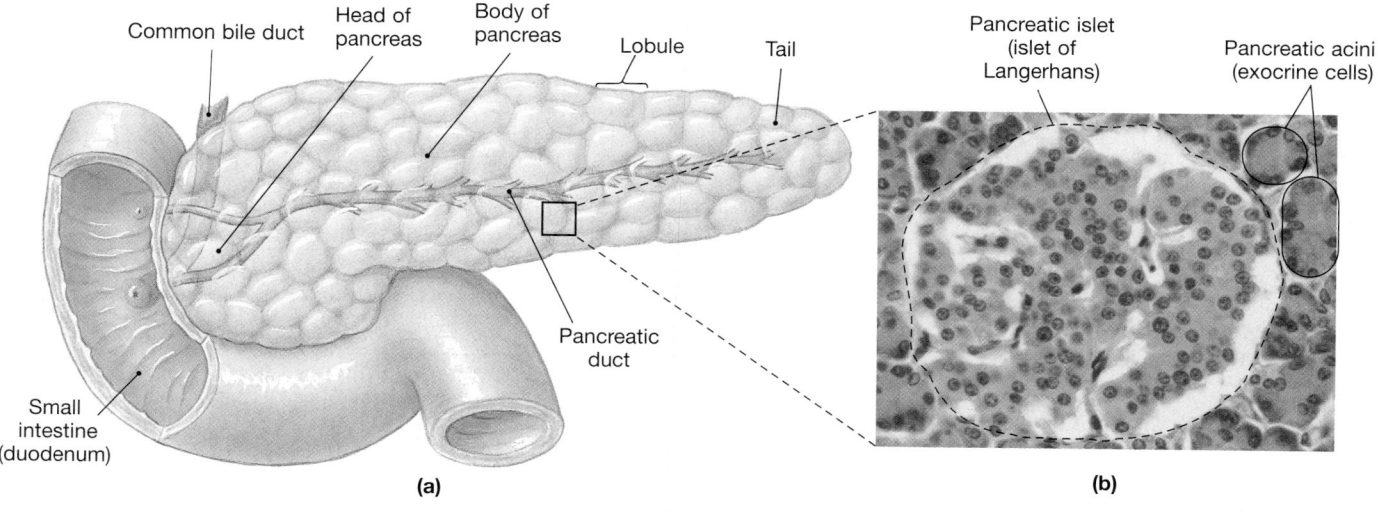

●FIGURE 18–18
The Endocrine Pancreas.   **(a)** The gross anatomy of the pancreas. **(b)** A pancreatic islet surrounded by exocrine cells.

(Figure 18–18b●). Pancreatic islets account for only about 1 percent of all cells in the pancreas. Nevertheless, a typical pancreas contains roughly 2 million pancreatic islets.

☐ THE PANCREATIC ISLETS

The pancreatic islets are surrounded by an extensive, fenestrated capillary network that carries its hormones into the bloodstream. Each islet contains four types of cells:

1. **Alpha cells** produce the hormone glucagon (GLOO-ka-gon). Glucagon raises blood glucose levels by increasing the rates of glycogen breakdown and glucose release by the liver.

2. **Beta cells** produce the hormone insulin (IN-suh-lin). Insulin lowers blood glucose levels by increasing the rate of glucose uptake and utilization by most body cells and increasing glycogen synthesis in skeletal muscles and the liver. Beta cells also secrete *amylin*, a recently discovered peptide hormone whose role is uncertain.

3. **Delta cells** produce a peptide hormone identical to growth hormone–inhibiting hormone (GH–IH), a hypothalamic regulatory hormone. GH–IH suppresses the release of glucagon and insulin by other islet cells and slows the rates of food absorption and enzyme secretion along the digestive tract.

4. **F cells** produce the hormone **pancreatic polypeptide** (**PP**). PP inhibits gallbladder contractions and regulates the production of some pancreatic enzymes, and may also help control the rate of nutrient absorption by the digestive tract.

We shall focus on insulin and glucagon, the hormones responsible for the regulation of blood glucose concentrations. These hormones interact to control blood glucose levels (Figure 18–19●). When blood glucose levels rise, beta cells secrete insulin, which then stimulates the transport of glucose across cell membranes. When blood glucose levels decline, alpha cells secrete glucagon, which stimulates glucose release by the liver.

☐ INSULIN

Insulin is a peptide hormone released by beta cells when glucose levels exceed normal levels (70–110 mg/dl). Secretion of this hormone is also stimulated by elevated levels of some amino acids, including arginine and leucine.

Insulin exerts its effects on cellular metabolism in a series of steps that begins when insulin binds to receptor proteins on the cell membrane. Binding leads to the activation of the receptor, which functions as a kinase, attaching phosphate groups to intracellular enzymes. The phosphorylation of enzymes then produces primary and secondary effects in the cell, the biochemical details of which remain unresolved.

One of the most important of these effects is the enhancement of glucose absorption and utilization. Insulin receptors are present in most cell membranes; such cells are called *insulin*

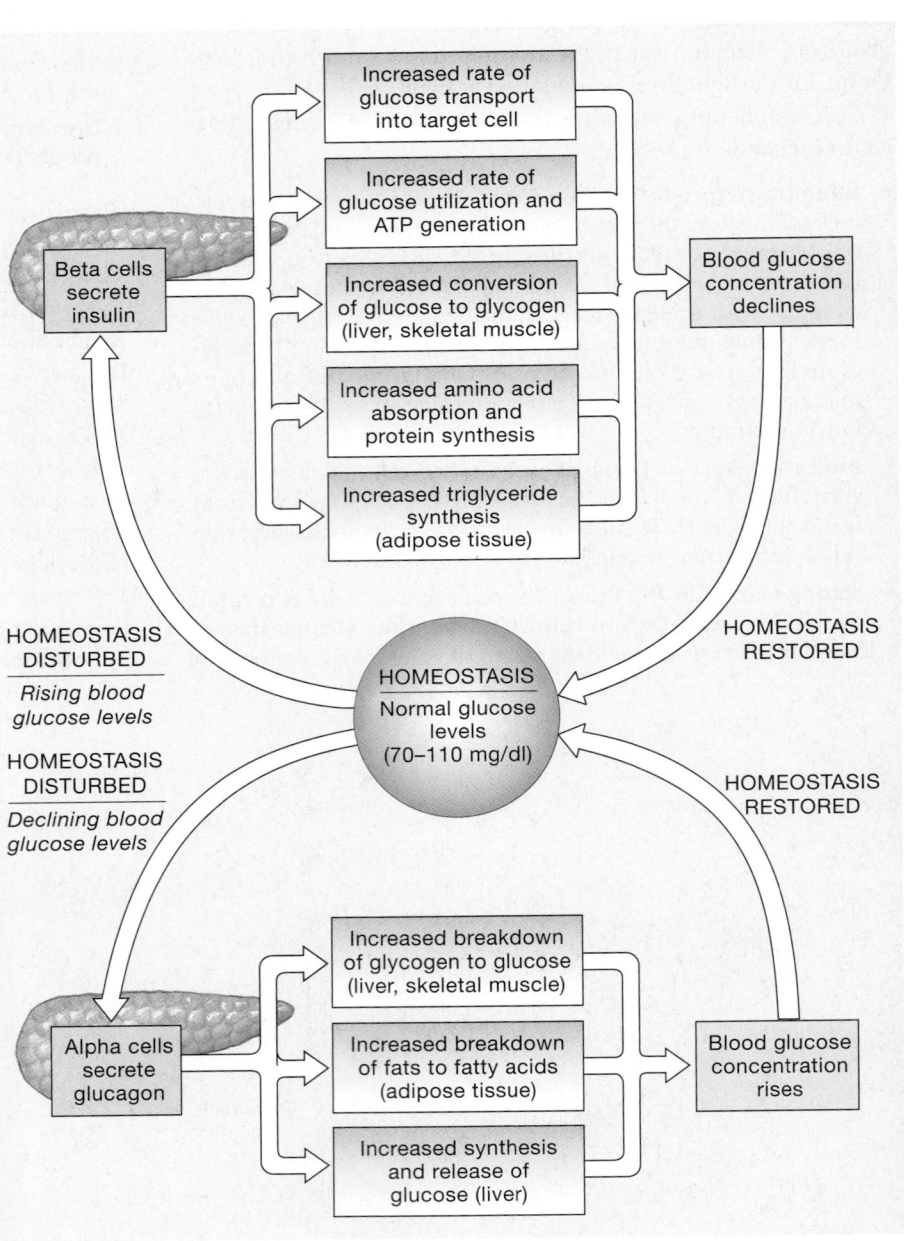

●**FIGURE 18–19**
The Regulation of Blood Glucose Concentrations

dependent. However, cells in the brain and kidneys, cells in the lining of the digestive tract, and red blood cells lack insulin receptors. These cells are called *insulin independent*, because they can absorb and utilize glucose without insulin stimulation.

The effects of insulin on its target cells include the following:

- **The Acceleration of Glucose Uptake (All Target Cells).** This effect results from an increase in the number of glucose transport proteins in the cell membrane. These proteins transport glucose into the cell by facilitated diffusion, a movement that follows the concentration gradient for glucose and for which ATP is not required.

- **The Acceleration of Glucose Utilization (All Target Cells) and Enhanced ATP Production.** This effect occurs for two reasons: (1) The rate of glucose use is proportional to its availability—when more glucose enters the cell, more is used. (2) Second messengers activate a key enzyme involved in the initial steps of glycolysis.

- **The Stimulation of Glycogen Formation (Skeletal Muscles and Liver Cells).** When excess glucose enters these cells, it is stored in the form of glycogen.

- **The Stimulation of Amino Acid Absorption and Protein Synthesis.**

- **The Stimulation of Triglyceride Formation in Adipose Tissues.** Insulin stimulates the absorption of fatty acids and glycerol by adipocytes, which store these components as triglycerides. Adipocytes also increase their absorption of glucose; excess glucose is used in the synthesis of additional triglycerides.

In sum, insulin is secreted when glucose is abundant; the hormone stimulates glucose utilization to support growth and the establishment of carbohydrate (glycogen) and lipid (triglyceride) reserves. The accelerated use of glucose soon brings circulating glucose levels within normal limits.

Diabetes Mellitus

Whether glucose is absorbed across the digestive tract or manufactured and released by the liver, very little glucose leaves the body intact once it has entered the bloodstream. Glucose does not get filtered out of the blood at the kidneys; instead, virtually all of it is reabsorbed, so urinary glucose losses are negligible.

Diabetes mellitus (MEL-i-tus; *mellitum*, honey) is characterized by glucose concentrations that are high enough to overwhelm the reabsorption capabilities of the kidneys. (The presence of abnormally high glucose levels in the blood in general is called **hyperglycemia** [hī-per-glī-SĒ-mē-ah].) Glucose appears in the urine (**glycosuria**; glī-kō-SOO-rē-a), and urine production generally becomes excessive (*polyuria*).

Diabetes mellitus can be caused by genetic abnormalities, and some of the genes responsible have been identified. Mutations that result in inadequate insulin production, the synthesis of abnormal insulin molecules, or the production of defective receptor proteins will produce comparable symptoms. In this setting, obesity accelerates the onset and severity of the disease. Diabetes mellitus can also result from other pathological conditions, injuries, immune disorders, or hormonal imbalances. There are two major types of diabetes mellitus: *insulin-dependent (Type 1) diabetes* and *non-insulin-dependent (Type 2) diabetes*. Type 1 diabetes can be controlled with varying success through the administration of insulin by injection or infusion by an insulin pump (photo). Dietary restrictions are most effective in treating Type 2 diabetes; for details, see the *Applications Manual*. **AM** Diabetes Mellitus

Probably because glucose levels cannot be stabilized adequately, even with treatment, persons with diabetes mellitus commonly develop chronic medical problems. These problems arise because the tissues involved are experiencing an energy crisis—in essence, most of the tissues are responding as they would during chronic starvation, breaking down lipids and even proteins because they are unable to absorb glucose from their surroundings. The most common examples of diabetes-related medical problems include the following:

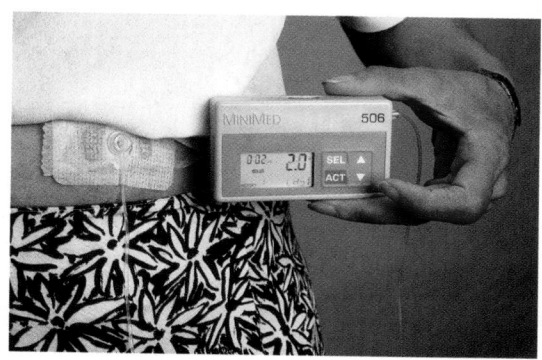

- The proliferation of capillaries and hemorrhaging at the retina may cause partial or complete blindness. This condition is called **diabetic retinopathy**.

- Changes occur in the clarity of the lens of the eye, producing cataracts.

- Small hemorrhages and inflammation at the kidneys cause degenerative changes that can lead to kidney failure. This condition, called **diabetic nephropathy**, is the primary cause of kidney failure. Treatment with drugs that improve blood flow to the kidneys can slow the progression to kidney failure.

- A variety of neural problems appear, including *peripheral neuropathies* and abnormal autonomic function. p. 440 These disorders, collectively termed **diabetic neuropathy**, are probably related to disturbances in the blood supply to neural tissues.

- Degenerative changes in cardiac circulation can lead to early heart attacks. For a given age group, heart attacks are three to five times more likely in diabetic individuals than in nondiabetic persons.

- Other changes in the vascular system can disrupt normal blood flow to the distal portions of the limbs. For example, a reduction in blood flow to the feet can lead to tissue death, ulceration, infection, and loss of toes or a major portion of one or both feet.

☐ GLUCAGON

When glucose concentrations fall below normal, alpha cells release glucagon and energy reserves are mobilized. When glucagon binds to a receptor in the cell membrane, the hormone activates adenylate cyclase. As we have seen, cAMP acts as a second messenger that activates cytoplasmic enzymes (p. 610). The primary effects of glucagon are as follows:

- **Stimulating the Breakdown of Glycogen in Skeletal Muscle and Liver Cells.** The glucose molecules released will either be metabolized for energy (skeletal muscle fibers) or released into the bloodstream (liver cells).
- **Stimulating the Breakdown of Triglycerides in Adipose Tissues.** The adipocytes then release the fatty acids into the bloodstream for use by other tissues.
- **Stimulating the Production of Glucose in the Liver.** Liver cells absorb amino acids from the bloodstream, convert them to glucose, and release the glucose into the circulation. This process of glucose synthesis in the liver is called *gluconeogenesis* (gloo-kō-nē-ō-JEN-e-sis).

The results are a reduction in glucose use and the release of more glucose into the bloodstream. Blood glucose concentrations soon rise toward normal levels.

Pancreatic alpha cells and beta cells monitor blood glucose concentrations, and the secretion of glucagon and insulin occur without endocrine or nervous instructions. Yet, because the alpha cells and beta cells are highly sensitive to changes in blood glucose levels, any hormone that affects blood glucose concentrations will indirectly affect the production of both insulin and glucagon. Insulin production

is also influenced by autonomic activity: Parasympathetic stimulation enhances insulin release, and sympathetic stimulation inhibits it.

Information about insulin, glucagon, and other pancreatic hormones is summarized in Table 18–6.

18–9 THE ENDOCRINE TISSUES OF OTHER SYSTEMS

Objective

- Describe the functions of the hormones produced by the kidneys, heart, thymus, testes, ovaries, and adipose tissue.

As we noted earlier, many organs that are part of other body systems have secondary endocrine functions. Examples are the intestines (digestive system), the kidneys (urinary system), the heart (cardiovascular system), the thymus (lymphatic system), the pancreas (digestive system), and the *gonads*—the testes in males and the ovaries in females (reproductive system).

Over the last decade, several new hormones from these endocrine tissues have been identified. In many cases, their structures and modes of action remain to be determined, and they have not been described in this chapter. However, in one instance, a significant new hormone was traced to an unexpected site of origin and led to the realization that the body's adipose tissue represents an important endocrine organ. Although all of the details have yet to be worked out, we shall consider the endocrine functions of adipose tissue in this section as well. Table 18–7 provides an overview of some of the hormones these organs produce.

TABLE 18–6 **HORMONES OF THE PANCREAS**

Structure/Cells	Hormone	Primary Targets	Hormonal Effects	Regulatory Control
PANCREATIC ISLETS				
Alpha cells	Glucagon	Liver, adipose tissues	Mobilizes lipid reserves; promotes glucose synthesis and glycogen breakdown in liver; elevates blood glucose concentrations	Stimulated by low blood glucose concentrations; inhibited by GH–IH from delta cells
Beta cells	Insulin	Most cells	Facilitates uptake of glucose by target cells; stimulates formation and storage of lipids and glycogen	Stimulated by high blood glucose concentrations, parasympathetic stimulation, and high levels of some amino acids; inhibited by GH–IH from delta cells and by sympathetic activation
Delta cells	GH–IH	Other islet cells, digestive epithelium	Inhibits insulin and glucagon secretion; slows rates of nutrient absorption and enzyme secretion along digestive tract	Stimulated by protein-rich meal; mechanism uncertain
F cells	Pancreatic polypeptide (PP)	Digestive organs	Inhibits gallbladder contraction; regulates production of pancreatic enzymes; influences rate of nutrient absorption by digestive tract	Stimulated by protein-rich meal and by parasympathetic stimulation

TABLE 18–7 REPRESENTATIVE HORMONES PRODUCED BY ORGANS OF OTHER SYSTEMS

Organ	Hormone(s)	Primary Target(s)	Hormonal Effects
Intestines	Many (secretin, gastrin, cholecystokinin, etc.)	Other regions and organs of the digestive system	Coordinate digestive activities
Kidneys	Erythropoietin (EPO)	Red bone marrow	Stimulates red blood cell production
	Calcitriol	Intestinal lining	Stimulates calcium and phosphate absorption; stimulates Ca^{2+} release from bone; inhibits PTH secretion
Heart	Natriuretic peptides (ANP and BNP)	Kidney, hypothalamus, adrenal gland	Increase water and salt loss at kidneys; decrease thirst; suppress secretion of ADH and aldosterone
Thymus	Thymosins (many)	Lymphocytes and other cells of the immune response	Coordinate and regulate immune response
Gonads	See Table 18–8		
Adipose tissues	Leptin	Hypothalamus	Suppression of appetite; permissive effects on GnRH and gonadotropin synthesis
	Resistin	Cells throughout the body	Suppression of insulin response

☐ THE INTESTINES

The intestines, which process and absorb nutrients, release a variety of hormones that coordinate the activities of the digestive system. Although the pace of digestive activities can be affected by the autonomic nervous system, most digestive processes are controlled locally. These hormones will be described in Chapter 24.

☐ THE KIDNEYS

The kidneys release the steroid hormone *calcitriol*, the peptide hormone *erythropoietin*, and the enzyme *renin*. Calcitriol is important for calcium ion homeostasis; erythropoietin and renin are involved in the regulation of blood volume and blood pressure.

Calcitriol

Calcitriol is a steroid hormone secreted by the kidneys in response to the presence of parathyroid hormone (PTH). *Cholecalciferol* (vitamin D_3) is a related steroid that is synthesized in the skin or absorbed from the diet (Figure 18-20a●). Cholecalciferol is converted to calcitriol, although not directly. The term *vitamin D* applies to the entire group of related steroids, including calcitriol, cholecalciferol, and various intermediate products.

The best-known function of calcitriol is the stimulation of calcium and phosphate ion absorption along the digestive tract. The effects of PTH on Ca^{2+} absorption result primarily from stimulation of calcitriol release. Calcitriol's other effects on calcium metabolism include (1) stimulating the formation and differentiation of osteoprogenitor cells and osteoclasts, (2) stimulating bone resorption by osteoclasts, (3) stimulating Ca^{2+} reabsorption at the kidneys, and (4) suppressing PTH production. Evidence indicates that calcitriol affects lymphocytes and keratinocytes in the skin; these effects have nothing to do with regulating calcium levels.

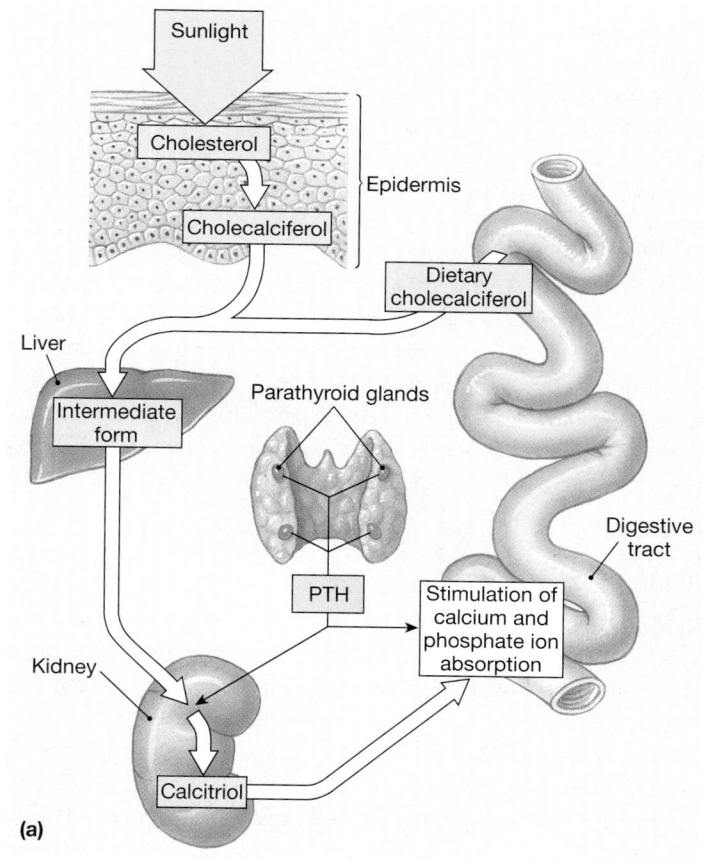

●**FIGURE 18–20**
Endocrine Functions of the Kidneys (a) The production of calcitriol. (b) The release of renin and EPO, and an overview of the renin-angiotensin system.

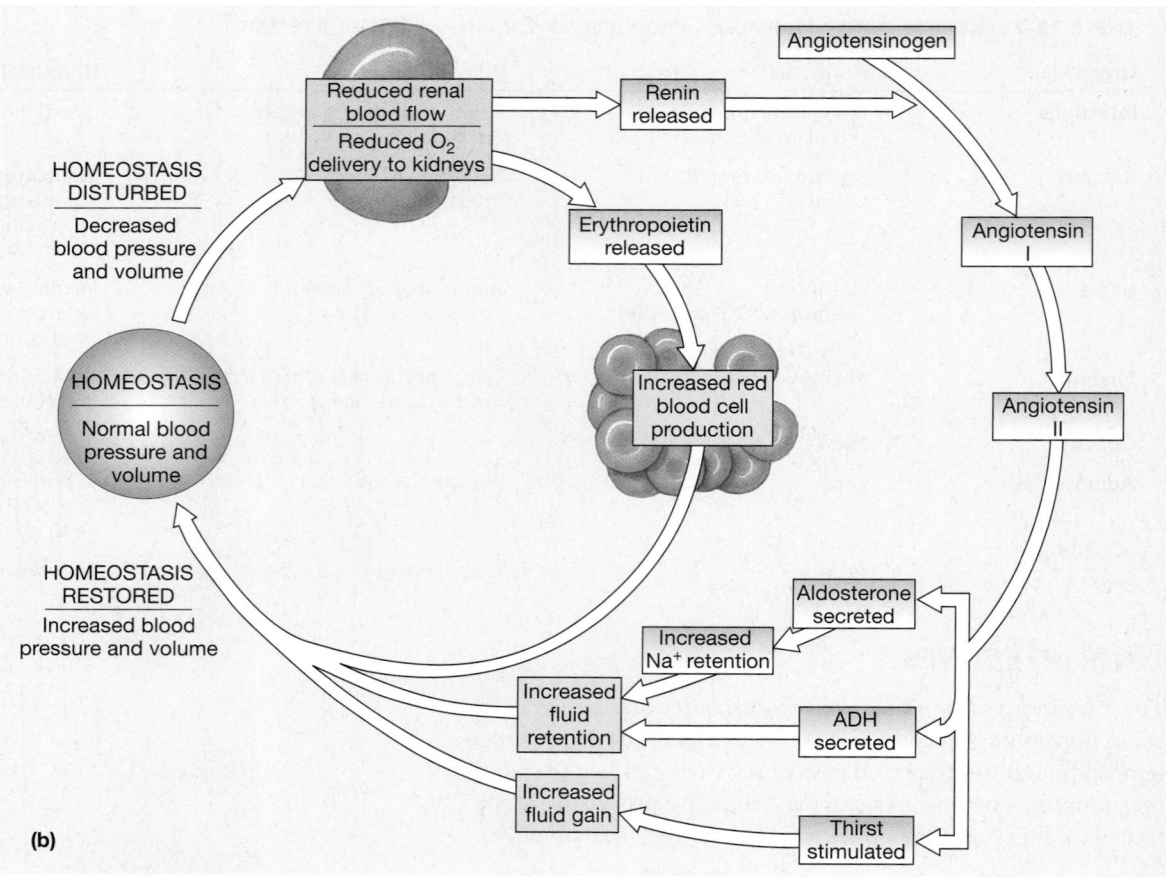

●FIGURE 18–20 (continued)

Erythropoietin

Erythropoietin (e-rith-rō-poy-ē-tin; *erythros*, red + *poiesis*, making), or **EPO**, is a peptide hormone released by the kidneys in response to low oxygen levels in kidney tissues. EPO stimulates the production of red blood cells by bone marrow. The increase in the number of red blood cells elevates blood volume. Because these cells transport oxygen, the increase in their number improves oxygen delivery to peripheral tissues. We will consider EPO again in Chapter 19.

Renin

Renin is released by specialized kidney cells in response to (1) sympathetic stimulation or (2) a decline in renal blood flow. Once in the bloodstream, renin functions as an enzyme that starts an enzymatic chain reaction known as the *renin–angiotensin system* (Figure 18-20b●). First, renin converts **angiotensinogen**, a plasma protein produced by the liver, to angiotensin I. In the capillaries of the lungs, **angiotensin I** is then modified to the hormone **angiotensin II**, which stimulates the secretion of aldosterone by the adrenal cortex and ADH at the posterior lobe of the pituitary gland. The combination of aldosterone and ADH restricts salt and water losses at the kidneys. Angiotensin II also stimulates thirst and elevates blood pressure.

Because renin plays such a key role in the renin–angiotensin system, many physiological and endocrinological references consider renin to be a hormone. We will take a closer look at the renin–angiotensin system in our discussion of the control of blood pressure and blood volume in Chapter 21.

☐ THE HEART

The endocrine cells in the heart are cardiac muscle cells in the walls of the *atria* (chambers that receive blood from the veins) and the *ventricles* (which pump blood to the rest of the body). If the blood volume becomes too great, these cells are stretched excessively, to the point where they begin to secrete **natriuretic peptides** (nā-trē-ū-RET-ik; *natrium*, sodium + *ouresis*, making water). In general, the effects of natriuretic peptides oppose those of angiotensin II: Natriuretic peptides promote the loss of Na^+ and water at the kidneys and inhibit renin release and the secretion of ADH and aldosterone. They also suppress thirst and prevent angiotensin II and norepinephrine from elevating blood pressure. The net result is a reduction in both blood volume and blood pressure, thereby reducing the stretching of the cardiac muscle cells in the heart walls. There are two different natriuretic peptides: *ANP* (atrial natriuretic peptide) and *BNP* (brain natriuretic peptide); we will discuss them further when we consider the control of blood pressure and volume in Chapters 21 and 26.

□ THE THYMUS

The **thymus** is located in the mediastinum, generally just posterior to the sternum. The thymus produces several hormones that are important to the development and maintenance of immune defenses. **Thymosin** (THĪ-mō-sin) is the name originally given to an extract from the thymus that promotes the development and maturation of *lymphocytes*, the white blood cells responsible for immunity. The thymic extract actually contains a blend of several complementary hormones; the term *thymosins* is now sometimes used to refer to all thymic hormones. We will consider the histological organization of the thymus and the functions of the thymosins in Chapter 22.

□ THE GONADS

In males, the **interstitial cells** of the testes produce the male hormones known as androgens. The most important of these androgens is **testosterone** (tes-TOS-ter-ōn). This hormone promotes the production of functional sperm, maintains the secretory glands of the male reproductive tract, stimulates growth, and determines male secondary sex characteristics, such as the distribution of facial hair and body fat. Testosterone also affects metabolic operations throughout the body, stimulating protein synthesis and muscle growth and producing aggressive behavioral responses. During embryonic development, the production of testosterone affects the development of CNS structures, including hypothalamic nuclei that will later influence sexual behaviors.

Sustentacular cells in the testes support the differentiation and physical maturation of sperm. Under FSH stimulation, these cells secrete the hormone **inhibin**, which inhibits the secretion of FSH at the anterior lobe and perhaps suppresses GnRH release at the hypothalamus.

Immature female reproductive cells, called *oocytes*, are produced in specialized structures called **follicles** that develop under FSH stimulation. The cells of these follicles, which form a layer around each developing oocyte, produce estrogens when stimulated by FSH and LH. **Estrogens** are steroid hormones that support the maturation of the oocytes and stimulate the growth of the lining of the uterus. The principal estrogen is **estradiol**. Circulating FSH stimulates the secretion of inhibin by follicle cells, and inhibin suppresses FSH release through a feedback mechanism comparable to that in males.

A surge in LH secretion during the ovarian cycle triggers ovulation, and LH then causes the remaining follicle cells to reorganize into a **corpus luteum** (LOO-tē-um; "yellow body"). These cells release a mixture of estrogens and hormones called **progestins**. **Progesterone** (prō-JES-ter-ōn), the principal progestin, has several important functions, including the following:

- It prepares the uterus for the arrival of a developing embryo, in case *fertilization* (the fusion of oocyte and sperm) occurs.
- It accelerates the movement of the oocyte or embryo to the uterus.
- It causes an enlargement of the mammary glands, working in combination with other hormones, such as estradiol, growth hormone, and prolactin.

During pregnancy, additional hormones produced by the placenta and uterus interact with those produced by the ovaries and the pituitary gland to promote normal fetal development and delivery. We shall consider the endocrinological aspects of pregnancy in Chapter 29.

Information about the reproductive hormones of the testes and ovaries is summarized in Table 18–8.

□ ADIPOSE TISSUES

Adipose tissue is a type of loose connective tissue introduced in Chapter 4. ∞ p. 125 Adipose tissue is known to produce two peptide hormones: *leptin* and *resistin*. **Leptin**, secreted by adipose tissues throughout the body, has several functions, the best

Structure/Cells	Hormone(s)	Primary Targets	Hormonal Effects	Regulatory Control
TESTES				
Interstitial cells	Androgens	Most cells	Support functional maturation of sperm, protein synthesis in skeletal muscles, male secondary sex characteristics, and associated behaviors	Stimulated by LH from anterior lobe of pituitary gland (see *Figure 18–8a*)
Sustentacular cells	Inhibin	Anterior lobe of pituitary gland	Inhibits secretion of FSH	Stimulated by FSH from anterior lobe of pituitary gland (see *Figure 18–8a*)
OVARIES				
Follicular cells	Estrogens	Most cells	Support follicle maturation, female secondary sex characteristics, and associated behaviors	Stimulated by FSH and LH from anterior lobe of pituitary gland (see *Figure 18–8a*)
	Inhibin	Anterior lobe of pituitary gland	Inhibits secretion of FSH	Stimulated by FSH from anterior lobe of pituitary gland (see *Figure 18–8a*)
Corpus luteum	Progestins	Uterus, mammary glands	Prepare uterus for implantation; prepare mammary glands for secretory activity	Stimulated by LH from anterior lobe of pituitary gland (see *Figure 18–8a*)

TABLE 18–8 HORMONES OF THE REPRODUCTIVE SYSTEM

Endocrine Disorders

Endocrine disorders fall into two basic categories: They reflect either abnormal hormone production or abnormal cellular sensitivity to hormones. The symptoms highlight the significance of normally "silent" hormonal contributions.

The characteristics of these disorders are summarized in Table 18–9. **AM** A Classification of Endocrine Disorders; Symptoms and Diagnosis of Endocrine Disorders

TABLE 18–9   CLINICAL IMPLICATIONS OF ENDOCRINE MALFUNCTIONS

Hormone	Underproduction Syndrome	Principal Symptoms	Overproduction Syndrome	Principal Symptoms
Growth hormone (GH)	Pituitary growth failure (pp. 198, 621)	Retarded growth, abnormal fat distribution, low blood glucose hours after a meal	Gigantism, acromegaly (pp. 198, 621 and AM)	Excessive growth
Antidiuretic hormone (ADH) or arginine vasopressin (AVP)	Diabetes insipidus (p. 618)	Polyuria, dehydration, thirst	SIADH (syndrome of inappropriate ADH secretion) (AM)	Increased body weight and water content
Thyroxine (T$_4$), triiodothyronine (T$_3$)	Myxedema, cretinism (p. 624 and AM)	Low metabolic rate; low body temperature; impaired physical and mental development	Hyperthyroidism, Graves' disease, (p. 624 and AM)	High metabolic rate and body temperature
Parathyroid hormone (PTH)	Hypoparathyroidism (p. 626 and AM)	Muscular weakness, neurological problems, formation of dense bones, tetany due to low blood Ca^{2+} concentrations	Hyperparathyroidism (p. 626 and AM)	Neurological, mental, muscular problems due to high blood Ca^{2+} concentrations; weak and brittle bones
Insulin	Diabetes mellitus (Type 1) (p. 633 and AM)	High blood glucose, impaired glucose utilization, dependence on lipids for energy; glycosuria	Excess insulin production or administration	Low blood glucose levels, possibly causing coma
Mineralocorticoids (MCs)	Hypoaldosteronism (p. 629 and AM)	Polyuria, low blood volume, high blood K$^+$, low blood Na$^+$ concentrations	Aldosteronism (p. 629 and AM)	Increased body weight due to Na$^+$ and water retention; low blood K$^+$ concentration
Glucocorticoids (GCs)	Addison's disease (p. 629 and AM)	Inability to tolerate stress, mobilize energy reserves, or maintain normal blood glucose concentrations	Cushing's disease (p. 629 and AM)	Excessive breakdown of tissue proteins and lipid reserves; impaired glucose metabolism
Epinephrine (E), norepinephrine (NE)	None identified		Pheochromocytoma (AM)	High metabolic rate, body temperature, and heart rate; elevated blood glucose levels
Estrogens (females)	Hypogonadism (p. 617)	Sterility, lack of secondary sex characteristics	Adrenogenital syndrome (p. 630)	Overproduction of androgens by zona reticularis of adrenal cortex leads to masculinization
			Precocious puberty (p. 643)	Premature sexual maturation and related behavioral changes
Androgens (males)	Hypogonadism (p. 617)	Sterility, lack of secondary sex characteristics	Adrenogenital syndrome (gynecomastia) (p. 630)	Abnormal production of estrogen, sometimes due to adrenal or interstitial cell tumors; leads to breast enlargement
			Precocious puberty (p. 643)	Premature sexual maturation and related behavioral changes

known being the feedback control of appetite. When you eat, adipose tissues absorb glucose and lipids and synthesize triglycerides for storage. At the same time, they release leptin into the bloodstream. Leptin binds to neurons in the hypothalamus that deal with emotion and appetite control. The result is a sense of satiation and the suppression of appetite.

Leptin was first discovered in a strain of obese mice that had a defective leptin gene. The administration of leptin to one of these overweight mice quickly turned it into a slim, athletic animal. The initial hope that leptin could be used to treat human obesity was soon dashed, however. Most obese people appear to have defective leptin receptors (or leptin pathways) in the appetite centers of the CNS. Their circulating leptin levels are already several times higher than in individuals of normal body weight, so the administration of additional leptin would have no effect. Researchers are now working on the structure of the receptor protein and the biochemistry of the pathway triggered by leptin binding.

Leptin also has a permissive effect on GnRH and gonadotropin synthesis. This effect explains (1) why thin girls commonly enter puberty relatively late, (2) why an increase in body fat content can improve fertility, and (3) why women stop menstruating when their body fat content becomes very low.

It is now known that adipose tissue also produces a second hormone, tentatively called *resistin*. **Resistin** reduces insulin sensitivity throughout the body; it has been proposed that this is the "missing link" between obesity and Type 2 diabetes. (See the Clinical Discussion on p. 633.) Experimental evidence from obese mice supports this linkage, but the data from human research remain inconclusive.

✓ Why does a person with Type 1 or Type 2 diabetes urinate frequently and have a pronounced thirst?

✓ What effect would increased levels of glucagon have on the amount of glycogen stored in the liver?

✓ Increased amounts of light would inhibit the production of which hormone?

Answers start on page Q-1

18–10 PATTERNS OF HORMONAL INTERACTION

Objectives

■ Explain how hormones interact to produce coordinated physiological responses.

■ Identify the hormones that are especially important to normal growth, and discuss their roles.

■ Define the general adaptation syndrome, and compare homeostatic responses with stress responses.

■ Describe the effects of hormones on behavior.

Although hormones are usually studied individually, the extracellular fluids contain a mixture of hormones whose concentrations change daily or even hourly. As a result, cells never respond to only one hormone; instead, they respond to multiple hormones simultaneously. When a cell receives instructions from two hormones at the same time, four outcomes are possible:

1. The two hormones may have **antagonistic**, or opposing, effects, as in the case of PTH and calcitonin or insulin and glucagon. The net result will depend on the balance between the two hormones. In general, when antagonistic hormones are present, the observed effects will be weaker than those produced by either hormone acting unopposed.

2. The two hormones may have additive effects, so that the net result is greater than the effect that each would produce acting alone. In some cases, the net result is greater than the *sum* of the hormones' individual effects. This phenomenon is called a **synergistic** (sin-er-JIS-tik; *synairesis*, a drawing together) effect. An example is the glucose-sparing action of GH and glucocorticoids.

3. One hormone can have a **permissive** effect on another. In such cases, the first hormone is needed for the second to produce its effect. For example, epinephrine does not change energy consumption unless thyroid hormones are also present in normal concentrations.

4. Finally, hormones may produce different, but complementary, results in specific tissues and organs. These **integrative** effects are important in coordinating the activities of diverse physiological systems. The differing effects of calcitriol and parathyroid hormone on tissues involved in calcium metabolism are illustrative.

In this section, we shall present three examples of the ways hormones interact to produce complex, well-coordinated results. The patterns introduced will provide the background needed to understand the more detailed discussions found in chapters that deal with cardiovascular function, metabolism, excretion, and reproduction.

☐ HORMONES AND GROWTH

Normal growth requires the cooperation of several endocrine organs. Six hormones—GH, thyroid hormones, insulin, PTH, calcitriol, and reproductive hormones—are especially important, although many others have secondary effects on growth. The circulating concentrations of these hormones are regulated independently. Every time the hormonal mixture changes, metabolic operations are modified to some degree. The modifications vary in duration and intensity, producing unique individual growth patterns:

1. *Growth Hormone (GH).* Growth hormone assists in the maintenance of normal blood glucose concentrations and in the mobilization of lipid reserves stored in adipose tissues. It is not the primary hormone involved, however, and an adult with a GH

deficiency but normal levels of thyroxine (T_4), insulin, and glucocorticoids will have no physiological problems. The effects of GH on protein synthesis and cellular growth are most apparent in children, in whom GH supports muscular and skeletal development. Undersecretion or oversecretion of GH can lead to pituitary growth failure or gigantism (p. 621).

2. *Thyroid Hormones.* Normal growth also requires appropriate levels of thyroid hormones. If these hormones are absent for the first year after birth, the nervous system will fail to develop normally, and mental retardation and other signs of cretinism will result. If T_4 concentrations decline later in life but before puberty, normal skeletal development will not continue.

3. *Insulin.* Growing cells need adequate supplies of energy and nutrients. Without insulin, the passage of glucose and amino acids across cell membranes will be drastically reduced or eliminated.

4. *Parathyroid Hormone (PTH) and Calcitriol.* Parathyroid hormone and calcitriol promote the absorption of calcium salts for subsequent deposition in bone. Without adequate levels of both hormones, bones can still enlarge, but will be poorly mineralized, weak, and flexible. For example, in *rickets*, a condition typically caused by inadequate calcitriol production in growing children, the limb bones are so weak that they bend under the body's weight. ∞ p. 199

5. *Reproductive Hormones.* The activity of osteoblasts in key locations and the growth of specific cell populations are affected by the presence or absence of reproductive hormones (androgens in males, estrogens in females). These sex hormones stimulate cell growth and differentiation in their target tissues. The targets differ for androgens and estrogens, and the differential growth induced by each accounts for gender-related differences in skeletal proportions and secondary sex characteristics.

☐ HORMONES AND STRESS

Any condition—physical or emotional—that threatens homeostasis is a form of **stress**. Many stresses are opposed by specific homeostatic adjustments. For example, a decline in body temperature will lead to shivering or changes in the pattern of blood flow in an attempt to restore normal body temperature.

In addition, the body has a *general* response to stress that can occur while other, more specific, responses are under way. Exposure to a wide variety of stress-causing factors will produce the same general pattern of hormonal and physiological adjustments. These responses are part of the **general adaptation syndrome (GAS)**, also known as the **stress response**. The GAS, first described by Hans Selye in 1936, can be divided into three phases: (1) the *alarm phase*, (2) the *resistance phase*, and (3) the *exhaustion phase* (Figure 18–21●).

The Alarm Phase

During the **alarm phase**, an immediate response to the stress occurs. This response is directed by the sympathetic division of the

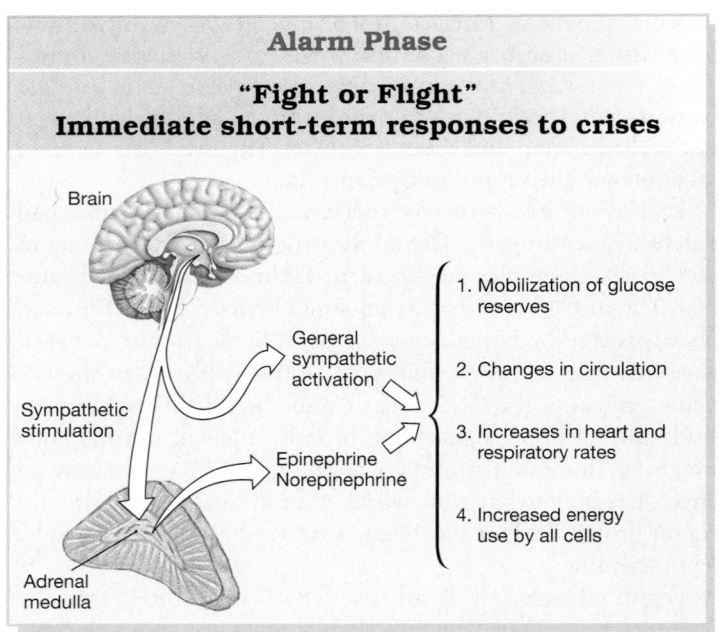

Alarm Phase

"Fight or Flight"
Immediate short-term responses to crises

Brain

General sympathetic activation

Sympathetic stimulation

Epinephrine
Norepinephrine

Adrenal medulla

1. Mobilization of glucose reserves

2. Changes in circulation

3. Increases in heart and respiratory rates

4. Increased energy use by all cells

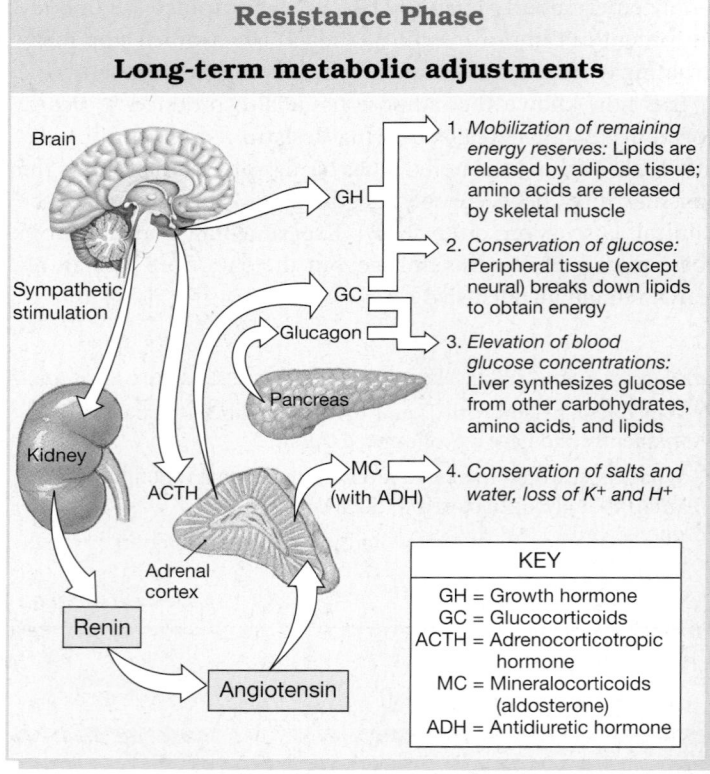

Resistance Phase

Long-term metabolic adjustments

Brain

Sympathetic stimulation

GH

GC

Glucagon

Pancreas

Kidney

ACTH

Adrenal cortex

Renin

Angiotensin

MC
(with ADH)

1. *Mobilization of remaining energy reserves:* Lipids are released by adipose tissue; amino acids are released by skeletal muscle

2. *Conservation of glucose:* Peripheral tissue (except neural) breaks down lipids to obtain energy

3. *Elevation of blood glucose concentrations:* Liver synthesizes glucose from other carbohydrates, amino acids, and lipids

4. *Conservation of salts and water, loss of K^+ and H^+*

KEY
GH = Growth hormone
GC = Glucocorticoids
ACTH = Adrenocorticotropic hormone
MC = Mineralocorticoids (aldosterone)
ADH = Antidiuretic hormone

Exhaustion Phase

Collapse of vital systems

Causes may include:
- Exhaustion of lipid reserves
- Inability to produce glucocorticoids
- Failure of electrolyte balance
- Cumulative structural or functional damage to vital organs

●**FIGURE 18–21**
The General Adaptation Syndrome

autonomic nervous system. In the alarm phase, (1) energy reserves are mobilized, mainly in the form of glucose, and (2) the body prepares to deal with the stress-causing factor by "fight or flight" responses. ⚬⚬ p. 535

Epinephrine is the dominant hormone of the alarm phase, and its secretion accompanies a generalized sympathetic activation. The characteristics of the alarm phase include:

- Increased mental alertness.
- Increased energy consumption by skeletal muscles and many other tissues.
- The mobilization of energy reserves (glycogen and lipids).
- Increased blood flow to skeletal muscles.
- Decreased blood flow to the skin, kidneys, and digestive organs.
- A drastic reduction in digestion and urine production.
- Increased sweat gland secretion.
- An increase in blood pressure, heart rate, and respiratory rate.

Although the effects of epinephrine are most apparent during the alarm phase, other hormones play supporting roles. For example, the reduction of water losses by circulatory changes, ADH production, and aldosterone secretion can be very important if the stress involves a loss of blood.

The Resistance Phase

The temporary adjustments of the alarm phase are often sufficient to remove or overcome a stress. But some stresses, such as starvation, acute illness, or severe anxiety, can persist for hours, days, or even weeks. If a stress lasts longer than a few hours, the individual will enter the **resistance phase** of the GAS.

Glucocorticoids are the dominant hormones of the resistance phase. Epinephrine, GH, and thyroid hormones are also involved. Energy demands in the resistance phase remain higher than normal, owing to the combined effects of these hormones.

Neural tissue has a high demand for energy, and neurons must have a reliable supply of glucose. If blood glucose concentrations fall too far, neural function will deteriorate. Glycogen reserves are adequate to maintain normal glucose concentrations during the alarm phase, but are nearly exhausted after several hours. The endocrine secretions of the resistance phase are coordinated to achieve four integrated results:

1. *The Mobilization of Lipid and Protein Reserves.* The hypothalamus produces GH–RH and CRH, stimulating the release of GH and, by means of ACTH, the secretion of glucocorticoids. Adipose tissues respond to GH and glucocorticoids by releasing stored fatty acids. Skeletal muscles respond to glucocorticoids by breaking down proteins and releasing amino acids into the bloodstream.

2. *The Conservation of Glucose for Neural Tissues.* Glucocorticoids and GH from the anterior lobe of the pituitary gland stimulate lipid metabolism in most tissues. These glucose-sparing effects maintain normal blood glucose levels even after long periods of starvation. Neural tissues do not alter their metabolic activities, however, and they continue to use glucose as an energy source.

3. *The Elevation and Stabilization of Blood Glucose Concentrations.* As blood glucose levels decline, glucagon and glucocorticoids stimulate the liver to manufacture glucose from other carbohydrates, from glycerol by way of triglycerides, and from amino acids provided by skeletal muscles. The glucose molecules are then released into the bloodstream, and blood glucose concentrations return to normal levels.

4. *The Conservation of Salts and Water and the Loss of K^+ and H^+.* Blood volume is conserved through the actions of ADH and aldosterone. As Na^+ is conserved, K^+ and H^+ are lost.

The body's lipid reserves are sufficient to maintain the resistance phase for a period of weeks or even months. (These reserves account for the ability to endure lengthy periods of starvation.) But the resistance phase cannot be sustained indefinitely. If starvation is the primary stress, the resistance phase ends as lipid reserves are exhausted and structural proteins become the primary energy source. If another factor is the cause, the resistance phase will end due to complications brought about by hormonal side effects. For example:

- Although the metabolic effects of glucocorticoids are essential to normal resistance, their anti-inflammatory action slows wound healing and increases the body's susceptibility to infection.
- The continued conservation of fluids under the influence of ADH and aldosterone stresses the cardiovascular system by producing elevated blood volumes and higher-than normal blood pressures.
- The adrenal cortex may become unable to continue producing glucocorticoids, quickly eliminating the ability to maintain blood glucose concentrations at acceptable levels.

Poor nutrition, emotional or physical trauma, chronic illness, or damage to key organs such as the heart, liver, or kidneys will hasten the end of the resistance phase.

The Exhaustion Phase

When the resistance phase ends, homeostatic regulation breaks down and the **exhaustion phase** begins. Unless corrective actions are taken almost immediately, the failure of one or more organ systems will prove fatal.

Mineral imbalances contribute to the existing problems with major systems. The production of aldosterone throughout the resistance phase results in a conservation of Na^+ at the expense of K^+. As the body's K^+ content declines, a variety of cells—notably neurons and muscle fibers—begin to malfunction. Although a single cause, such as heart failure, may be listed as the cause of death, the underlying problem is the inability to sustain the endocrine and metabolic adjustments of the resistance phase. Figure 18–21● summarizes the three phases of the GAS.

☐ HORMONES AND BEHAVIOR

As we have seen, many endocrine functions are regulated by the hypothalamus, and hypothalamic neurons monitor the levels of many circulating hormones. Other portions of the CNS are also quite sensitive to hormonal stimulation.

Endocrinology and Athletic Performance

Medical research involving humans is generally subject to tight ethical constraints and meticulous scientific scrutiny. Yet a clandestine, unscientific, and potentially quite dangerous program of "experimentation" with hormones is being pursued by athletes in many countries. The use of hormones to improve athletic performance is banned by the International Olympic Committee, the U.S. Olympic Committee, the National Collegiate Athletic Association, and the National Football League, and it is condemned by the American Medical Association and the American College of Sports Medicine. A significant number of amateur and professional athletes, however, persist in this dangerous practice. Synthetic forms of testosterone are used most often, but athletes may use any combination of testosterone, GH, EPO, and a variety of synthetic hormones.

ANDROGEN ABUSE The use of *anabolic steroids*, or androgens, has become popular with many amateur and professional athletes. The goal of steroid use is to increase one's muscle mass, endurance, and "competitive spirit." It has been suggested that as many as 30 percent of college and professional athletes and 10–20 percent of male high school athletes are using anabolic steroids (with or without GH) to improve their performance. Among U.S. bodybuilders, the proportion who use steroids may be as high as 80 percent. The use of steroids such as *androstenedione*, which the body can convert to testosterone, was highlighted during 1998–1999 when baseball slugger Mark McGwire admitted using it to improve his performance. (Performance-enhancing drugs are still legal in professional baseball, although they are banned in some other professional sports.)

One supposed justification for this practice has been the unfounded opinion that compounds manufactured in the body not only are safe, but are good for you. In reality, the administration of natural or synthetic androgens in abnormal amounts carries unacceptable health risks. Androgens affect many tissues in a variety of ways. Known complications include (1) the premature closure of epiphyseal cartilages, (2) various liver dysfunctions (including jaundice and liver tumors), (3) prostate gland enlargement and urinary tract obstruction, and (4) testicular atrophy and infertility. A link to heart attacks, impaired cardiac function, and strokes has also been suggested.

Moreover, the normal regulation of androgen production involves a feedback mechanism comparable to that described for adrenal steroids earlier in this chapter. GnRH stimulates the production of LH, and LH stimulates the secretion of testosterone and other androgens by the interstitial cells of the testes. Circulating androgens, in turn, inhibit the production of both GnRH and LH. Thus, when synthetic androgens are administered in high doses, they can suppress the normal production of testosterone and depress the manufacture of GnRH by the hypothalamus. *This suppression of GnRH release can be permanent.*

The use of androgenic "bulking agents" by female bodybuilders not only may add muscle mass, but can alter muscular proportions and secondary sex characteristics as well. For example, women taking steroids can develop irregular menstrual periods and changes in body hair distribution (including baldness).

Finally, androgen abuse can cause a generalized depression of the immune system.

EPO ABUSE Because it is now synthesized by recombinant DNA techniques, EPO is readily available. Endurance athletes, such as cyclists or marathon runners, may use it to boost the number of oxygen-carrying red blood cells in the bloodstream. Although this effect improves the oxygen content of blood, it also makes blood denser; thus, the heart must work harder to push it around the circulatory system. This can result in death due to heart failure or stroke in young and otherwise healthy individuals. Several competitors were disqualified from the 2002 Winter Olympics when their blood tests revealed abnormally high levels of EPO.

GHB AND CLENBUTEROL Androgens and EPO are hormones with reasonably well-understood effects. Because drug testing is now widespread in amateur and professional sports, athletes interested in "getting an edge" are experimenting with drugs not easily detected by standard tests. The long-term and short-term effects of these drugs are difficult to predict. Two examples are the recent use of *GHB* and *clenbuterol* by amateur athletes.

Gamma-hydroxybutyrate (GHB) was tested for use as an anesthetic in the 1960s. It was rejected, in part because it was linked to petit mal and grand mal seizures. In 1990, the drug appeared in health-food stores, where it was sold as an anabolic agent and diet aid. It has also been used as a "date rape" drug. According to the FDA, GHB and related compounds, sold or distributed under the names Renewtrient, Revivarant, Blue Nitro, Firewater, and Serenity, have recently been responsible for 145 serious illnesses and at least eight deaths. Symptoms include reduced heart rate, lowered body temperature, confusion, hallucinations, seizures, and coma at doses from 0.25 teaspoon to 4 tablespoons.

Clenbuterol, sometimes used to treat asthma, mimics epinephrine and stimulates β_2 receptors. This effect increases the diameter of the respiratory passageways and accelerates blood flow through active skeletal muscles. Although the drug is also rumored to have anabolic properties comparable to those of androgens, no evidence supports that rumor. Clenbuterol abuse is reportedly widespread, although exact numbers are difficult to obtain. Heavy usage can cause severe headaches, tremors, insomnia, and dangerous abnormal heartbeats.

The clearest demonstrations of the effects of specific hormones involve individuals whose endocrine glands are oversecreting or undersecreting. But even normal changes in circulating hormone levels can cause behavioral changes. In *precocious* (premature) *puberty*, sex hormones are produced at an inappropriate time, perhaps as early as age five or six. An affected child not only begins to develop adult secondary sex characteristics, but also undergoes significant behavioral changes. The "nice little kid" disappears, and the child becomes aggressive and assertive. These behavioral changes represent the effects of sex hormones on CNS function. Thus, behaviors that in normal teenagers are usually attributed to environmental stimuli, such as peer pressure, have a physiological basis as well. In adults, changes in the mixture of hormones reaching the CNS can have significant effects on intellectual capabilities, memory, learning, and emotional states.

Blood and tissue concentrations of many other hormones, including TSH, thyroid hormones, ADH, PTH, prolactin, and glucocorticoids, remain unchanged with advancing age. Although circulating hormone levels may remain within normal limits, some endocrine tissues become less responsive to stimulation. For example, in elderly individuals, smaller amounts of GH and insulin are secreted after a carbohydrate-rich meal. The reduction in levels of GH and other tropic hormones affects tissues throughout the body; these hormonal effects are associated with the reductions in bone density and muscle mass noted in earlier chapters.

Finally, age-related changes in other tissues affect their abilities to respond to hormonal stimulation. As a result, peripheral tissues may become less responsive to some hormones. This loss of sensitivity has been documented in the case of glucocorticoids and ADH.

18–11 AGING AND HORMONE PRODUCTION

The endocrine system shows relatively few functional changes with age. The most dramatic exception is the decline in the concentration of reproductive hormones. We noted the effects of these hormonal changes on the skeletal system in Chapter 6 (∞ p. 204); we will continue the discussion in Chapter 29.

✓ Insulin lowers the level of glucose in the blood, and glucagon causes glucose levels to rise. What is this type of hormonal interaction called?

✓ The lack of which hormones would inhibit skeletal formation?

✓ Why do levels of GH–RH and CRH rise during the resistance phase of the general adaptation syndrome?

Answers start on page Q-1

Chapter Review

SELECTED CLINICAL TERMINOLOGY

Terms Discussed in This Chapter

Addison's disease: A condition caused by the hyposecretion of glucocorticoids and mineralocorticoids; characterized by an inability to mobilize energy reserves and maintain normal blood glucose levels. *(p. 629 and AM)*

cretinism *(congenital hypothyroidism):* A condition caused by hypothyroidism in infancy; marked by inadequate skeletal and nervous development and a metabolic rate as much as 40 percent below normal levels. *(p. 624)*

Cushing's disease: A condition caused by the hypersecretion of glucocorticoids; characterized by the excessive breakdown and relocation of lipid reserves and proteins. *(p. 629 and AM)*

diabetes insipidus: A disorder that develops when the posterior lobe of the pituitary gland no longer releases adequate amounts of ADH or when the kidneys cannot respond to ADH. *(p. 618)*

diabetes mellitus: (mel-Ī-tus) A disorder that damages many organ systems; characterized by glucose concentrations high enough to overwhelm the kidneys' reabsorption capabilities. *(p. 633 and AM)*

diabetic retinopathy, nephropathy, neuropathy: Disorders of the retina, kidneys, and peripheral nerves, respectively, related to diabetes mellitus; the conditions most often afflict middle-aged or older diabetics. *(p. 633)*

general adaptation syndrome (GAS): The pattern of hormonal and physiological adjustments with which the body responds to all forms of stress. *(p. 640)*

glycosuria: The presence of glucose in the urine. *(p. 633)*

goiter: An abnormal enlargement of the thyroid gland. *(p. 624 and AM)*

hyperglycemia: Abnormally high blood glucose levels. *(p. 633)*

hypocalcemic tetany: Muscle spasms affecting the face and upper extremities; caused by low Ca^{2+} concentrations in body fluids. *(p. 626 and AM)*

insulin-dependent diabetes mellitus (IDDM), or *Type 1 diabetes*, or *juvenile-onset diabetes*: A type of diabetes mellitus; the primary cause is inadequate insulin production by the beta cells of the pancreatic islets. *(p. 633 and AM)*

myxedema: Symptoms of hyposecretion of thyroid hormones, including subcutaneous swelling, hair loss, dry skin, low body temperature, muscle weakness, and slowed reflexes. *(p. 624 and AM)*

non-insulin-dependent diabetes mellitus (NIDDM), or *Type 2 diabetes*, or *maturity-onset diabetes*: A type of diabetes mellitus in which insulin levels are normal or elevated, but peripheral tissues no longer respond normally. *(p. 633 and AM)*

polyuria: The production of excessive amounts of urine; a symptom of diabetes. *(p. 618)*

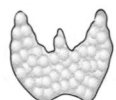

The **ENDOCRINE SYSTEM** in Perspective

Organ/Component	Primary Functions
Pineal Gland	May control timing of reproduction and set day-night rhythms
Pituitary Gland	Controls other endocrine glands; regulates growth and fluid balance
Thyroid Gland	Controls tissue metabolic rate; regulates calcium levels
Parathyroid Glands	Regulate calcium levels (with thyroid)
Thymus	Controls maturation of lymphocytes
Adrenal Glands	Adjust water balance, tissue metabolism, cardiovascular and respiratory activity
Kidneys	Control red blood cell production and assist in calcium regulation
Pancreas	Regulates blood glucose levels
Gonads	
Testes	Support male sexual characteristics and reproductive functions *(see Chapter 28)*
Ovaries	Support female sexual characteristics and reproductive functions *(see Chapter 28)*

INTEGRATION WITH OTHER SYSTEMS

The endocrine system provides long-term regulation and adjustment of homeostatic mechanisms that affect many body functions. For example, the endocrine system regulates fluid and electrolyte balance, cell and tissue metabolism, growth and development, and reproductive functions. It also assists the nervous system in responding to stressful stimuli through the General Adaptation Syndrome.

The relationships between the endocrine system and the other systems of the body are summarized in the figure on the opposite page. The most extensive interactions are with the nervous system, as you might expect. Although the hypothalamus modifies the activities of many endocrine organs via the anterior lobe of the pituitary gland, there are so many complex feedback loops involved that it is often hard to decide whether the endocrine system or the nervous system is really in charge. For example, many hormones also serve as neurotransmitters in the brain, spinal cord, and/or enteric nervous system. As a result, circulating hormones that cross the blood-brain barrier can have direct and widespread effects on neural—and neuroendocrine—activity.

CLINICAL PATTERNS

Homeostatic regulation of circulating hormone levels primarily involves negative feedback control mechanisms. The feedback loop involves an interplay between the endocrine organ and its target tissues, and endocrine disorders can result from abnormalities in the endocrine gland, the endocrine or neural regulatory mechanisms, or the target tissues. The net result may be overproduction (*hypersecretion*) or underproduction (*hyposecretion*) of hormones.

Primary disorders result from problems within the endocrine organ. The underlying cause may be a metabolic factor; hypothyroidism due to a lack of dietary iodine is an example. An endocrine organ may also malfunction due to physical damage that destroys cells or disrupts the normal blood supply. Congenital problems may

also affect the regulation, production, or release of hormones by endocrine cells.

Secondary disorders result from problems in other organs or target tissues. Such disorders often involve the hypothalamus or pituitary gland. For example, if the hypothalamus produces inadequate levels of TRH, the anterior lobe secretes minimal amounts of TSH, and the individual will show signs of hypothyroidism.

Abnormalities in target cells can affect their sensitivity or responsiveness to a particular hormone. For example, Type 2 diabetes is a condition that results from a reduction in the target cell's sensitivity to insulin. The origin and diagnosis of endocrine disorders is discussed further in the *Applications Manual*.

MEDIA CONNECTIONS

I. Objective: To relate hormonal control of blood sugar to the functioning of other systems.

Completion time = 15 minutes

Diabetes seems, on the surface, to be a relatively simple disorder in which insulin from the pancreas does not effectively control blood sugar levels. However, an increased concentration of glucose in the blood alters the osmotic pressure and viscosity of the blood, resulting in far-reaching effects on all other systems of the body. Think back on the systems covered thus far. How might an imbalance in blood glucose affect the nervous, muscular, and skeletal systems? Prepare a concept map, placing diabetes in the center and outlining the relationship of this pathology to those three systems. To complete this activity, go to the **Companion Website**, Chapter 18 Media Connections, and click on the keyword **diabetes**.

II. Objective: To understand the relationship between the chemical environment and hormone function.

Completion time = 15 minutes

Many industrial chemicals, as well as compounds produced naturally by certain plant species, can interfere with the normal operation of the endocrine system. The best-known examples are compounds that affect estrogen binding and physiological function. Some of these compounds compete with estrogen for receptor sites and mimic the physiological effects of estrogen. These substances are collectively known as *environmental estrogens*. Others, which block the action of estrogen, are called *anti-estrogens*. Environmental estrogens and anti-estrogens are known to interfere with the reproductive and general health of wildlife. Can they affect humans in the same way? Prepare a flow chart of the effects of estrogen on both sexes, using the information in Chapter 18. What might be the effects of environmental estrogens and anti-estrogens on males? On females? For recent research findings on this topic, visit the **Companion Website**, Chapter 18 Media Connections and click on the keyword **estrogen**.

INTEGUMENTARY SYSTEM

- Protects superficial endocrine organs
- Epidermis synthesizes cholecalciferol

- Sex hormones stimulate sebaceous gland activity, influence hair growth, and apocrine sweat gland activity
- PRL stimulates development of mammary glands
- Adrenal hormones alter dermal blood flow, stimulate release of lipids from adipocytes
- MSH stimulates melanocyte activity

SKELETAL SYSTEM

- Protects endocrine organs, especially in brain, chest, and pelvic cavity

---

- Skeletal growth regulated by several hormones
- Calcium mobilization regulated by parathyroid hormone and calcitonin
- Sex hormones speed growth and closure of epiphyseal cartilages at puberty and help maintain bone mass in adults

MUSCULAR SYSTEM

- Skeletal muscles protect some endocrine organs

- Hormones adjust muscle metabolism, energy production, and growth
- Regulate calcium and phosphate levels in body fluids
- Speed skeletal muscle growth

NERVOUS SYSTEM

- Hypothalamic hormones directly control pituitary and indirectly control secretions of other endocrine organs
- Controls adrenal medullae
- Secretes ADH and oxytocin

- Several hormones affect neural metabolism
- Hormones help regulate fluid and electrolyte balance
- Reproductive hormones influence CNS development and behaviors

CARDIOVASCULAR SYSTEM

- Circulatory system distributes hormones throughout the body
- Heart secretes ANP, BNP

- Erythropoietin regulates production of RBCs
- Several hormones elevate blood pressure
- Epinephrine elevates heart rate and contractile force

ENDOCRINE SYSTEM

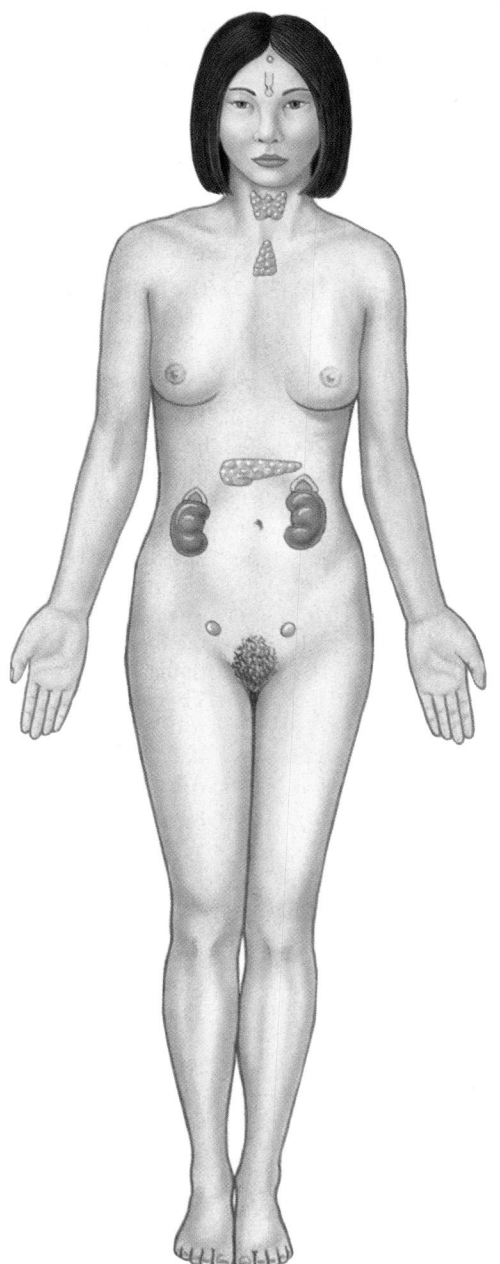

FOR ALL SYSTEMS
Adjusts metabolic rates and substrate utilization; regulates growth and development

LYMPHATIC SYSTEM

- Lymphocytes provide defense against infection and, with other WBCs, assist in repair after injury

---

- Glucocorticoids have anti-inflammatory effects
- Thymosins stimulate development of lymphocytes
- Many hormones affect immune function

RESPIRATORY SYSTEM

- Provides oxygen and eliminates carbon dioxide generated by endocrine cells
- Converting enzyme in lung capillaries converts angiotensin I to angiotensin II

- Epinephrine and norepinephrine stimulate respiratory activity and dilate respiratory passageways

DIGESTIVE SYSTEM

- Provides nutrients and substrates to endocrine cells
- Endocrine cells of pancreas secrete insulin and glucagon
- Liver produces and releases angiotensinogen

- E and NE stimulate constriction of sphincters and depress activity along digestive tract
- Hormones coordinate secretory activities along tract

URINARY SYSTEM

- Kidney cells (1) release renin and erythropoietin when local blood pressure declines and (2) produce calcitriol

---

- Aldosterone, ADH, ANP, and BNP adjust rates of fluid and electrolyte reabsorption in kidneys

REPRODUCTIVE SYSTEM

- Steroid sex hormones and inhibin suppress secretory activities in hypothalamus and pituitary gland

---

- Hypothalamic factors and pituitary hormones regulate sexual development and function
- Oxytocin stimulates smooth muscle contractions in uterus, prostate, and mammary glands

seasonal affective disorder (**SAD**): A condition characterized by depression, lethargy, an inability to concentrate, and altered sleep and eating habits; linked to enhanced melatonin levels in individuals exposed to only short periods of daylight. (*p. 631 and AM*)

thyrotoxicosis: A condition caused by the oversecretion of thyroid hormones (hyperthyroidism). Symptoms include increases in metabolic rate, blood pressure, and heart rate; excitability and emotional instability; and lowered energy reserves. (*p. 624 and AM*)

AM **Additional Terms Discussed in the *Applications Manual***

exophthalmos

hypoglycemia

ketoacidosis

ketone bodies

thyrotoxic crisis

STUDY OUTLINE

18–1 INTERCELLULAR COMMUNICATION   p. 605

1. In general, the nervous system performs short-term "crisis management," whereas the endocrine system regulates longer term, ongoing metabolic processes.
2. **Endocrine communication** is carried out by *endocrine cells* releasing chemicals called hormones, which alter the metabolic activities of many tissues and organs simultaneously. (*Table 18–1*) Hormones exert their effects by modifying the activities of **target cells.**
3. **Paracrine communication** involves the use of chemical messengers to transfer information from cell to cell within a single tissue.

18–2 AN OVERVIEW OF THE ENDOCRINE SYSTEM   p. 607

1. The endocrine system includes all the cells and endocrine tissues of the body that produce hormones or paracrine factors. (*Figure 18–1*)

HORMONE STRUCTURE   p. 608

2. Hormones can be divided into three groups: *amino acid derivatives*, structurally similar to amino acids; *peptide hormones*, chains of amino acids; and *lipid derivatives*, including **steroid hormones** and **eicosanoids.** (*Figure 18–2*)

SECRETION AND DISTRIBUTION OF HORMONES   p. 608

3. Hormones may circulate freely or may be bound to transport proteins. Free hormones are rapidly removed from the bloodstream.

MECHANISMS OF HORMONE ACTION   p. 610

4. Receptors for *catecholamines*, peptide hormones, and eicosanoids are in the cell membranes of target cells. Thyroid and steroid hormones cross the cell membrane and bind to receptors in the cytoplasm or nucleus, activating or inactivating specific genes. (*Figures 18–3, 18–4*)

CONTROL OF ENDOCRINE ACTIVITY   p. 613

5. **Endocrine reflexes** are the functional counterparts of neural reflexes. (*Figure 18–5*)
6. The hypothalamus regulates the activities of the nervous and endocrine systems by (1) secreting **regulatory hormones,** which control the activities of endocrine cells in the anterior lobe of the pituitary gland, (2) acting as an endocrine organ by releasing hormones into the bloodstream at the posterior lobe of the pituitary gland, and (3) exerting direct neural control over the endocrine cells of the adrenal medullae. (*Figure 18–5*)

🔲 Hormone mechanisms: **Companion Website:** Chapter 18/ Tutorials.

18–3 THE PITUITARY GLAND   p. 615

1. The **pituitary gland,** or **hypophysis,** releases nine important peptide hormones; all bind to membrane receptors and use cyclic-AMP as a second messenger. (*Figures 18–6, 18–9; Table 18–2*)

THE ANTERIOR LOBE   p. 615

2. The **anterior lobe** (**adenohypophysis**) of the pituitary gland can be subdivided into the **pars distalis,** the **pars intermedia,** and the **pars tuberalis.** (*Figure 18–6*)
3. At the median eminence of the hypothalamus, neurons release regulatory factors (either **releasing hormones RH** or **inhibiting hormones IH**) into the surrounding interstitial fluids through **fenestrated capillaries.** (*Figure 18–7*)
4. The **hypophyseal portal system** ensures that all the blood entering the portal vessels will reach the intended target cells before it returns to the general circulation. (*Figure 18–7*)
5. **Thyroid-stimulating hormone** (**TSH**) triggers the release of thyroid hormones. *Thyrotropin-releasing hormone (TRH)* promotes the secretion of TSH. (*Figure 18–8*)
6. **Adrenocorticotropic hormone** (**ACTH**) stimulates the release of *glucocorticoids* by the adrenal gland. Corticotropin-releasing hormone (CRH) causes the secretion of ACTH. (*Figure 18–8*)
7. **Follicle-stimulating hormone** (**FSH**) stimulates follicle development and estrogen secretion in females and sperm production in males. **Luteinizing hormone** (**LH**) causes *ovulation* and *progestin* production in females and androgen production in males. **Gonadotropin-releasing hormone** (**GnRH**) promotes the secretion of both FSH and LH. (*Figure 18–8*)
8. **Prolactin** (**PRL**), together with other hormones, stimulates both the development of the mammary glands and milk production. (*Figure 18–8*)
9. **Growth hormone** (**GH,** or **somatotropin**) stimulates cell growth and replication through the release of **somatomedins** or **IGFs** from liver cells. The production of GH is regulated by **growth hormone–releasing hormone** (**GH–RH**) and **growth hormone–inhibiting hormone** (**GH–IH**). (*Figure 18–8*)
10. **Melanocyte-stimulating hormone** (**MSH**) may be secreted by the pars intermedia during fetal development, early childhood, pregnancy, or certain diseases. This hormone stimulates melanocytes to produce melanin.

THE POSTERIOR LOBE   p. 618

11. The **posterior lobe** (**neurohypophysis**) of the pituitary gland contains the axons of hypothalamic neurons. Neurons of the **supraoptic** and **paraventricular nuclei** manufacture **antidiuretic hormone** (**ADH**) and **oxytocin,** respectively. ADH decreases the

amount of water lost at the kidneys and, in higher concentrations, elevates blood pressure. In women, oxytocin stimulates contractile cells in the mammary glands and has a stimulatory effect on smooth muscles in the uterus. *(Figure 18–9; Summary Table 18–2)*

 Hormones and glands of the endocrine system: **Companion Website**: Chapter 18/Tutorials.

18–4 THE THYROID GLAND   p. 620

1. The thyroid gland lies near the *thyroid cartilage* of the larynx and consists of two **lobes** connected by a narrow **isthmus.** *(Figure 18–11)*

THYROID FOLLICLES AND THYROID HORMONES   p. 620

2. The thyroid gland contains numerous **thyroid follicles.** Thyroid follicles release several hormones, including **thyroxine** (T_4) and **triiodothyronine** (T_3). *(Figures 18–11, 18–12; Table 18–4)*

3. Most of the thyroid hormones entering the bloodstream are attached to special **thyroid-binding globulins** (**TBGs**); the rest are attached to **transthyretin** or albumin. *(Figure 18–12)*

FUNCTIONS OF THYROID HORMONES   p. 623

4. Thyroid hormones are held in storage, bound to mitochondria (thereby increasing ATP production), or bound to receptors activating genes that control energy utilization. They also exert a **calorigenic effect.** *(Table 18–3)*

THE C CELLS OF THE THYROID GLAND: CALCITONIN   p. 624

5. The **C cells** of the thyroid follicles produce **calcitonin** (**CT**), which helps regulate Ca^{2+} concentrations in body fluids, especially during childhood and pregnancy. *(Figures 18–11, 18–13; Table 18–4)*

18–5 THE PARATHYROID GLANDS   p. 625

1. Four **parathyroid glands** are embedded in the posterior surface of the thyroid gland. The **chief cells** produce **parathyroid hormone** (**PTH**) in response to lower-than-normal concentrations of Ca^{2+}. The parathyroid glands, aided by *calcitriol*, are the primary regulators of Ca^{2+} levels in healthy adults. *(Figures 18–14, 18–15; Table 18–4)*

18–6 THE ADRENAL GLANDS   p. 627

1. One **adrenal** (*suprarenal*) **gland** lies along the superior border of each kidney. The gland is subdivided into the superficial **adrenal cortex** and the inner **adrenal medulla.** *(Figure 18–16)*

THE ADRENAL CORTEX   p. 627

2. The adrenal cortex manufactures steroid hormones called **adrenocortical steroids** (**corticosteroids**). The cortex can be subdivided into three areas: (1) the **zona glomerulosa**, which releases **mineralocorticoids** (**MCs**) principally **aldosterone**; (2) the **zona fasciculata**, which produces **glucocorticoids** (**GCs**), notably **cortisol** and **corticosterone**; and (3) the **zona reticularis**, which produces androgens of uncertain significance. *(Figures 18–16, 18–17; Table 18–5)*

THE ADRENAL MEDULLA   p. 630

3. The adrenal medulla produces epinephrine (75–80 percent of medullary secretion) and norepinephrine (20–25 percent). *(Figure 18–16; Table 18–5)*

18–7 THE PINEAL GLAND   p. 631

1. The **pineal gland** contains **pinealocytes**, which synthesize **melatonin.** Suggested functions include inhibiting reproductive function, protecting against damage by free radicals, and setting circadian rhythms.

18–8 THE PANCREAS   p. 631

THE PANCREATIC ISLETS   p. 632

1. The pancreas contains both exocrine and endocrine cells. Cells of the endocrine pancreas form clusters called **pancreatic islets** (*islets of Langerhans*). These islets contain **alpha cells** (which secrete the hormone **glucagon**), **beta cells** (which secrete **insulin**), **delta cells** (which secrete **GH–IH**), and **F cells** (which secrete **pancreatic polypeptide**). *(Figure 18–18; Table 18–6)*

INSULIN AND GLUCAGON   p. 632

2. Insulin lowers blood glucose by increasing the rate of glucose uptake and utilization; glucagon raises blood glucose by increasing the rates of glycogen breakdown and glucose manufacture in the liver. *(Figure 18–19, Table 18–6)*

18-9 THE ENDOCRINE TISSUES OF OTHER SYSTEMS   p. 634

THE INTESTINES   p. 635

1. The intestines produce hormones important to the coordination of digestive activities. *(Table 18–7)*

THE KIDNEYS   p. 635

2. Endocrine cells in the kidneys produce the hormones *calcitriol* and *erythropoietin* and the enzyme *renin*. *(Table 18–7)*

3. **Calcitriol** stimulates calcium (Ca^{2+}) and phosphate (PO_4^{3-}) ion absorption along the digestive tract. *(Figure 18-20)*

4. **Erythropoietin** (**EPO**) stimulates red blood cell production by the bone marrow. *(Figure 18-20)*

5. **Renin** converts **angiotensinogen** to **angiotensin I.** In the capillaries of the lungs, the latter compound is converted to **angiotensin II**, the hormone that (1) stimulates the adrenal production of aldosterone, (2) stimulates the pituitary release of ADH, (3) promotes thirst, and (4) elevates blood pressure. *(Figure 18–20)*

THE HEART   p. 636

6. Specialized muscle cells in the heart produce **natriuretic peptides** (*ANP* and *BNP*) when the blood volume becomes excessive. In general, their actions oppose those of angiotensin II. *(Table 18–7)*

THE THYMUS   p. 637

7. The thymus produces several hormones, collectively known as **thymosins**, which play a role in developing and maintaining normal immune defenses. *(Table 18–7)*

THE GONADS   p. 637

8. The **interstitial cells** of the testes produce androgens. **Testosterone** is the most important sex hormone in males. *(Table 18–8)*

9. In females, *oocytes* develop in **follicles**; follicle cells produce **estrogens**, especially **estradiol.** After ovulation, the remaining follicle cells reorganize into a **corpus luteum.** Those cells release a mixture of estrogens and **progestins**, especially **progesterone.** *(Table 18–8)*

ADIPOSE TISSUES   p. 637

10. Adipose tissues secrete **leptin** (a feedback control for appetite) and **resistin** (which reduces insulin sensitivity).

18-10 PATTERNS OF HORMONAL INTERACTION   p. 639

1. The hormones of the endocrine system often interact, producing (1) **antagonistic** (opposing) effects; (2) **synergistic** (additive) effects; (3) **permissive** effects, in which one hormone is necessary for another to produce its effect; or (4) **integrative** effects, in which hormones produce different, but complementary, results.

HORMONES AND GROWTH   p. 639

2. Normal growth requires the cooperation of several endocrine organs. Six hormones are especially important: GH, thyroid hormones, insulin, PTH, calcitriol, and reproductive hormones.

HORMONES AND STRESS   p. 640

3. Any condition that threatens homeostasis is a **stress**. Our bodies respond to a variety of stress-causing factors by the **general adaptation syndrome** (**GAS**), or **stress response**.

4. The GAS can be divided into three phases: (1) the **alarm phase** (an immediate, "fight or flight" response, under the direction of the sympathetic division of the ANS); (2) the **resistance phase**, dominated by glucocorticoids; and (3) the **exhaustion phase**, the eventual breakdown of homeostatic regulation and failure of one or more organ systems. (*Figure 18–21*)

HORMONES AND BEHAVIOR   p. 641

5. Many hormones affect the CNS; changes in the normal mixture of hormones can significantly alter intellectual capabilities, memory, learning, and emotional states.

18-11 AGING AND HORMONE PRODUCTION   p. 643

1. The endocrine system shows few functional changes with advanced age. The chief change is a decline in the concentration of reproductive hormones.

REVIEW QUESTIONS

 More assessment questions are available to you on the Companion Website. You will find Matching, Multiple Choice, True/False, and other quizzes to help further your understanding of the material covered in this chapter. To access the site, go to www.aw.com/martini.

LEVEL 1 Reviewing Facts and Terms

1. The use of a chemical messenger to transfer information from cell to cell within a single tissue is referred to as _____ communication.
   - (a) direct
   - (b) paracrine
   - (c) hormonal
   - (d) endocrine

2. Cyclic-AMP functions as a second messenger to
   - (a) build proteins and catalyze specific reactions
   - (b) activate adenylate cyclase
   - (c) open ion channels and activate key enzymes in the cytoplasm
   - (d) bind the hormone–receptor complex to DNA segments

3. Adrenocorticotropic hormone (ACTH) stimulates the release of
   - (a) thyroid hormones by the hypothalamus
   - (b) gonadotropins by the adrenal glands
   - (c) growth hormones by the hypothalamus
   - (d) steroid hormones by the adrenal glands

4. FSH production in males supports
   - (a) the maturation of sperm by stimulating sustentacular cells
   - (b) the development of muscles and strength
   - (c) the production of male sex hormones
   - (d) an increased desire for sexual activity

5. The hormone that induces ovulation in women and promotes the secretion of progesterone by the ovaries is
   - (a) interstitial cell–stimulating hormone
   - (b) estradiol
   - (c) luteinizing hormone
   - (d) prolactin

6. The two hormones released by the posterior lobe are
   - (a) GH and gonadotropin
   - (b) estrogen and progesterone
   - (c) GH and prolactin
   - (d) ADH and oxytocin

7. The primary function of ADH is to
   - (a) increase the amount of water lost at the kidneys
   - (b) decrease the amount of water lost at the kidneys
   - (c) dilate peripheral blood vessels to decrease blood pressure
   - (d) increase absorption along the digestive tract

8. The element required for normal thyroid function is
   - (a) magnesium
   - (b) calcium
   - (c) potassium
   - (d) iodine

9. Reduced fluid losses in urine due to the retention of Na^+ and water is a result of the action of
   - (a) antidiuretic hormone
   - (b) calcitonin
   - (c) aldosterone
   - (d) cortisone

10. The adrenal medulla produces the hormones
    - (a) cortisol and cortisone
    - (b) epinephrine and norepinephrine
    - (c) corticosterone and testosterone
    - (d) androgens and progesterone

11. The endocrine portion of the pancreas is the
    - (a) pancreatic islets
    - (b) pancreatic acini
    - (c) pancreatic duct
    - (d) entire pancreas

12. What three higher level mechanisms are involved in integrating the activities of the nervous and endocrine systems?

13. Which seven hormones are released by the anterior lobe of the pituitary gland?

14. What six hormones primarily affect growth?

15. What five primary effects result from the action of thyroid hormones?
16. What effects do calcitonin and parathyroid hormone have on blood calcium levels?
17. What three zones make up the adrenal cortex, and what kinds of hormones are produced by each zone?
18. Which two hormones are released by the kidneys, and what is the importance of each hormone?
19. What are the four opposing effects of atrial natriuretic peptide and angiotensin II?
20. What four cell populations make up the endocrine pancreas? Which hormone does each type of cell produce?

LEVEL 2 Reviewing Concepts

21. What is the primary difference in the way the nervous and endocrine systems communicate with their target cells?
22. How can a hormone modify the activities of its target cells?
23. What possible results occur when a cell receives instructions from two hormones at the same time?
24. What is an endocrine reflex? Compare endocrine and neural reflexes.
25. How would blocking the activity of phosphodiesterase affect a cell that responds to hormonal stimulation by the cAMP second-messenger system?
26. How does control of the adrenal medulla differ from control of the adrenal cortex?
27. Describe the effects of activation of the adrenal medulla on alpha and beta receptors.

28. Describe the four patterns of hormonal interaction.
29. (a) Describe the three phases of the general adaptation syndrome (GAS) that constitute the body's response to stress.
    (b) Which endocrine secretions play dominant roles in the alarm and resistance phases?

LEVEL 3 Critical Thinking and Clinical Applications

30. Roger has been extremely thirsty; he drinks numerous glasses of water every day and urinates a great deal. Name two disorders that could produce these symptoms. What test could a clinician perform to determine which disorder Roger has?
31. Julie is pregnant, but is not receiving prenatal care. She has a poor diet consisting mostly of fast food. She drinks no milk, preferring colas instead. How would this situation affect Julie's level of parathyroid hormone?
32. Sherry tells her physician that she has been restless and irritable lately. She has a hard time sleeping and complains of diarrhea and weight loss. During the examination, her physician notices a higher-than-normal heart rate and a fine tremor in her outstretched fingers. What tests could the physician suggest to make a positive diagnosis of Sherry's condition?
33. Patients receiving steroid hormone frequently retain large amounts of water. Why?
34. Pamela and her teammates are considering testosterone supplements to enhance their competitive skills. What natural effects of this hormone are they hoping to gain? What additional side effects might these women expect should they begin an anabolic steroid regime?

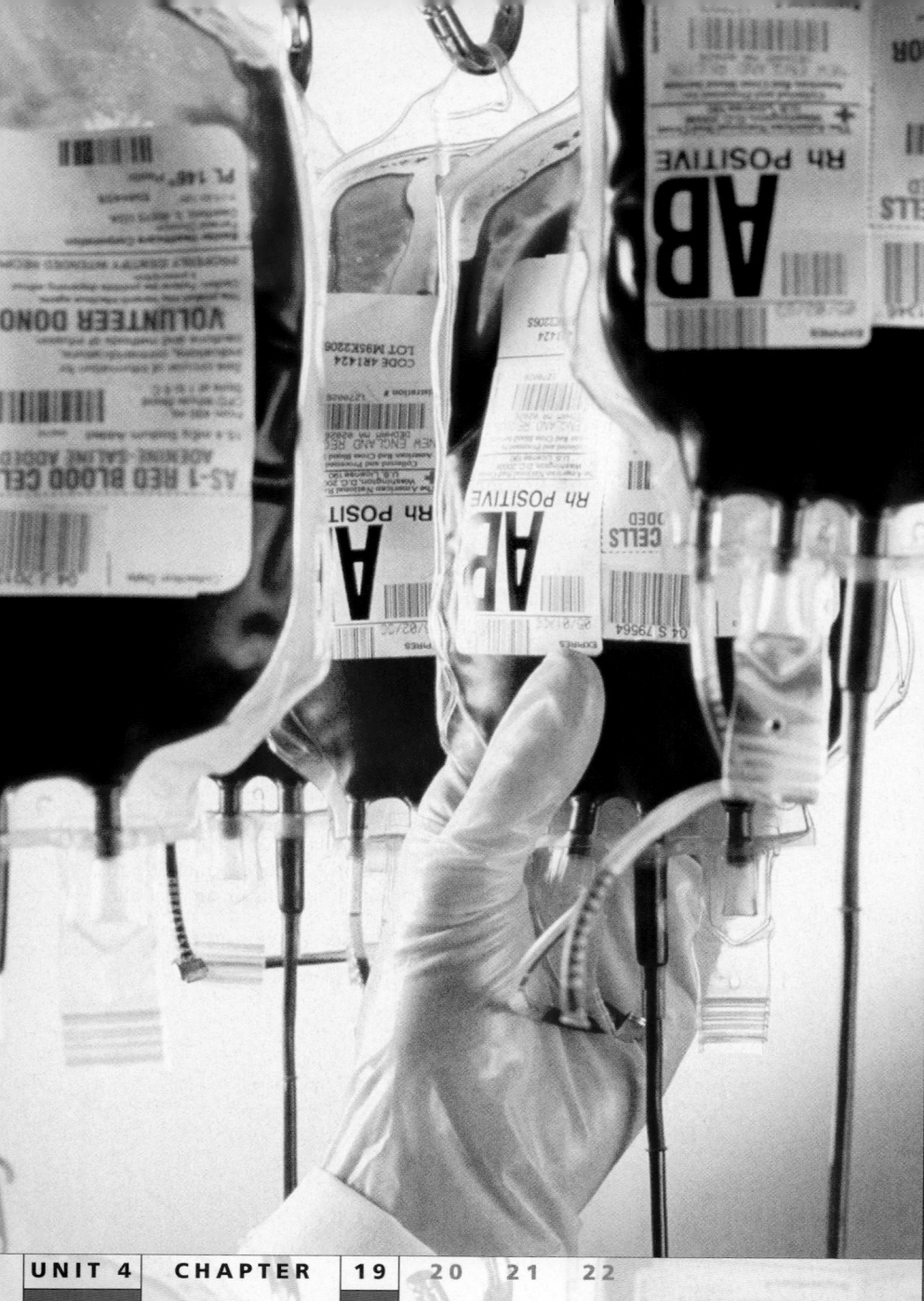

UNIT 4   CHAPTER   19   20   21   22

19

BLOOD

Out of sight, out of mind. That saying certainly applies to blood. As long as we don't see it, we don't think much about it. But looking at vials or bags of it in a hospital—or, worse, watching even a drop of our own oozing from a cut or scrape—makes most of us feel uneasy; some people faint at the sight. This anxiety probably comes from being reminded of how dependent we are on our relatively small supply of the precious fluid. If you have three half-gallon milk containers in your refrigerator at home, you probably have more milk than you do blood. You can easily spare a drop of blood or donate a pint, but if you suddenly lose more than 15 to 20 percent of your blood supply, you will need more in a hurry. Fortunately, blood transfusions are now fairly routine; but, as we will see in this chapter, not just anyone's blood will do. We'll also explore the many functions of this vital substance.

19–1 THE CARDIOVASCULAR SYSTEM: AN INTRODUCTION

Objective

■ List the components of the cardiovascular system and explain the major functions of this system.

Blood is the fluid component of the **cardiovascular system**, which also includes a pump (the heart) that circulates the fluid and a series of conducting pipes (the blood vessels) that carry it throughout the body. In the last chapter, we mentioned this system as the transport medium for hormones of the endocrine system, but that is only one of its many vital roles.

Small embryos don't need cardiovascular systems, because diffusion across their exposed surfaces provides adequate oxygen and removes waste products as rapidly as they are generated. By the time a human embryo reaches a few millimeters in length, however, developing tissues are consuming oxygen and nutrients faster than they can be provided by simple diffusion. At that stage, the cardiovascular system must begin functioning to provide a rapid-transport system for oxygen, nutrients, and waste products. It is the first system to become fully operational: The heart begins beating at the end of the third week of embryonic life, when many other systems have barely begun their development.

When the heart starts beating, blood begins circulating. The embryo can now more efficiently use the nutrients obtained from the maternal bloodstream, and its size doubles in the next week. In adults, circulating blood provides nutrients, oxygen, chemical instructions, and a mechanism for waste removal to each of the roughly 75 trillion cells in the human body. The blood also transports specialized cells that defend peripheral tissues from infection and disease. These services are essential—so much so that a region deprived of circulation dies in a matter of minutes.

In the next three chapters, we shall examine the individual components of the cardiovascular system. This chapter deals with the nature of the circulating blood, Chapter 20 covers the structure and function of the heart, and Chapter 21 examines the organization of blood vessels and the integrated functioning of the cardiovascular system.

The cardiovascular system is extensively integrated with the *lymphatic system*. The two systems are interconnected and interdependent. Fluid leaves the bloodstream, enters the tissues, and returns to the blood within the vessels of the lymphatic system. For this reason, the cardiovascular and lymphatic systems are often said to be components of a single *circulatory system*. The bloodstream carries cells, antibodies, and cytokines of the lymphatic system throughout the body, and the cardiovascular system assists the lymphatic system in defending the body against invasion by pathogens. In Chapter 22, we will describe the components of the lymphatic system and will consider the array of defenses that protect the body from internal and external hazards.

19–2 FUNCTIONS AND COMPOSITION OF BLOOD

Objectives

- Describe the important components and major functions of blood.
- Identify locations on the body used for blood collection, and list the basic physical characteristics of the blood samples drawn from these locations.

In this chapter, we examine the structure and functions of **blood**, a specialized fluid connective tissue that contains cells suspended in a fluid matrix. As you may recall, Chapter 4 contained an introduction to the components and properties of this connective tissue. ⚏ p. 128

☐ THE FUNCTIONS OF BLOOD

The functions of blood include the following:

- **The Transportation of Dissolved Gases, Nutrients, Hormones, and Metabolic Wastes.** Blood carries oxygen from the lungs to peripheral tissues and carbon dioxide from those tissues to the lungs. Blood distributes nutrients absorbed at the digestive tract

or released from storage in adipose tissue or in the liver. It carries hormones from endocrine glands toward their target cells, and it absorbs and carries the wastes produced by tissue cells to the kidneys for excretion.

- **The Regulation of the pH and Ion Composition of Interstitial Fluids.** Diffusion between interstitial fluids and blood eliminates local deficiencies or excesses of ions such as calcium or potassium. Blood also absorbs and neutralizes the acids generated by active tissues, such as the lactic acid produced by skeletal muscles.

- **The Restriction of Fluid Losses at Injury Sites.** Blood contains enzymes and other substances that respond to breaks in the vessel walls by initiating the process of *clotting*. A blood clot acts as a temporary patch and prevents further blood loss.

- **Defense Against Toxins and Pathogens.** Blood transports *white blood cells*, specialized cells that migrate into peripheral tissues to fight infections or remove debris. Blood also delivers *antibodies*, special proteins that attack invading organisms or foreign compounds.

- **The Stabilization of Body Temperature.** Blood absorbs the heat generated by active skeletal muscles and redistributes it to other tissues. If the body temperature is already high, that heat will be lost across the surface of the skin. If the body temperature is too low, the warm blood is directed to the brain and to other temperature-sensitive organs.

☐ THE COMPOSITION OF BLOOD

Blood has a characteristic and unique composition (Figure 19–1●). It is a fluid connective tissue with a matrix called **plasma** (PLAZ-muh). Plasma proteins are in solution rather than forming insoluble fibers like those in other connective tissues, such as loose connective tissue or cartilage. Because these proteins are in solution, plasma is slightly denser than water. Plasma is similar to interstitial fluid, although it contains a unique assortment of suspended proteins. There is a continuous exchange of fluid from the tissues into the blood and vice versa, driven by a combination of hydrostatic pressure, concentration gradients, and osmosis. These relationships will be considered further in Chapter 21.

Formed elements are blood cells and cell fragments that are suspended in plasma. Three types of formed elements exist: (1) red blood cells, (2) white blood cells, and (3) platelets. **Red blood cells** (**RBCs**), or **erythrocytes** (e-RITH-rō-sīts; *erythros*, red + *cyte*, cell), are the most abundant blood cells. These specialized cells are essential for the transport of oxygen in the blood. The less numerous **white blood cells** (**WBCs**), or **leukocytes** (LOO-kō-sīts; *leukos*, white), are cells involved with the body's defense mechanisms. There are five classes of leukocyte, each with slightly different functions. **Platelets** are small, membrane-bound cell fragments that contain enzymes and other substances important to the process of clotting.

Formed elements are produced through the process of **hemopoiesis** (hēm-ō-poy-E-sis), or *hematopoiesis*. **Hemocytoblasts**, or *pluripotent stem cells*, divide to produce *myeloid stem cells* and

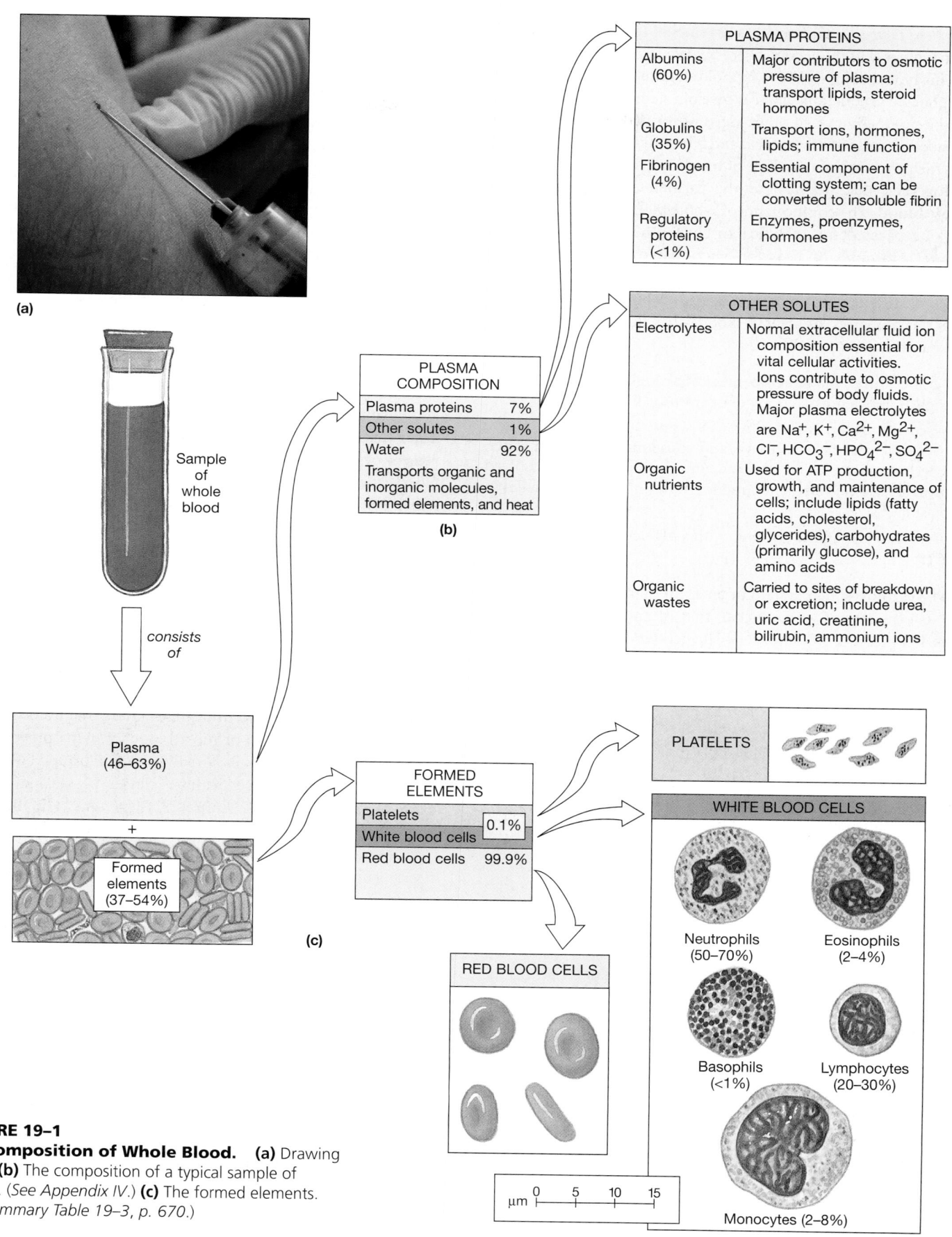

PLASMA PROTEINS	
Albumins (60%)	Major contributors to osmotic pressure of plasma; transport lipids, steroid hormones
Globulins (35%)	Transport ions, hormones, lipids; immune function
Fibrinogen (4%)	Essential component of clotting system; can be converted to insoluble fibrin
Regulatory proteins (<1%)	Enzymes, proenzymes, hormones

OTHER SOLUTES	
Electrolytes	Normal extracellular fluid ion composition essential for vital cellular activities. Ions contribute to osmotic pressure of body fluids. Major plasma electrolytes are Na^+, K^+, Ca^{2+}, Mg^{2+}, Cl^-, HCO_3^-, HPO_4^{2-}, SO_4^{2-}
Organic nutrients	Used for ATP production, growth, and maintenance of cells; include lipids (fatty acids, cholesterol, glycerides), carbohydrates (primarily glucose), and amino acids
Organic wastes	Carried to sites of breakdown or excretion; include urea, uric acid, creatinine, bilirubin, ammonium ions

Sample of whole blood

PLASMA COMPOSITION	
Plasma proteins	7%
Other solutes	1%
Water	92%
Transports organic and inorganic molecules, formed elements, and heat	

(b)

consists of

Plasma (46–63%)

+

Formed elements (37–54%)

FORMED ELEMENTS	
Platelets	0.1%
White blood cells	
Red blood cells	99.9%

(c)

PLATELETS

WHITE BLOOD CELLS

Neutrophils (50–70%)

Eosinophils (2–4%)

Basophils (<1%)

Lymphocytes (20–30%)

Monocytes (2–8%)

RED BLOOD CELLS

μm 0 5 10 15

●FIGURE 19–1
The Composition of Whole Blood. **(a)** Drawing blood. **(b)** The composition of a typical sample of plasma. (*See Appendix IV.*) **(c)** The formed elements. (*See Summary Table 19–3, p. 670.*)

lymphoid stem cells. These stem cells remain capable of division, but the daughter cells they generate are involved with the production of specific types of blood cells. Lymphoid stem cells are responsible for lymphocyte production, whereas myeloid stem cells are responsible for the production of all other kinds of formed elements. We will consider the fates of the myeloid and lymphoid stem cells as we discuss the formation of each type of formed element.

Together, the plasma and the formed elements constitute **whole blood**. The components of whole blood can be **fractionated**, or separated, for analytical or clinical purposes. We will encounter examples of uses for fractionated blood later in the chapter.

Whole blood from any source—venous blood, blood from peripheral capillaries, or arterial blood—has the same basic physical characteristics:

- **Temperature.** Blood temperature is roughly 38°C (100.4°F), slightly above normal body temperature.

- **Viscosity.** Blood is five times as viscous as water—that is, five times stickier, more cohesive, and resistant to flow than water. The high viscosity results from interactions among dissolved proteins, formed elements, and surrounding water molecules in plasma.

- **pH.** Blood is slightly alkaline, with a pH between 7.35 and 7.45, averaging 7.4.

The cardiovascular system of an adult man contains 5–6 liters (5.3–6.4 quarts) of whole blood; that of an adult woman contains 4–5 liters (4.2–5.3 quarts). The gender differences in blood volume primarily reflect differences in average body size. Blood volume in liters can be estimated for an individual of either gender by calculating 7 percent of the body weight in kilograms. For example, a 75-kg individual would have a blood volume of approximately 5.25 liters (5.4 quarts).

Fresh whole blood is generally collected from a superficial vein, such as the *median cubital vein* on the anterior surface of the elbow (Figure 19–1a●). The procedure is called **venipuncture** (VĒN-i-punk-chur; *vena*, vein + *punctura*, a piercing). It is a common sampling technique because (1) superficial veins are easy to locate, (2) the walls of veins are thinner than those of comparably sized arteries, and (3) blood pressure in the venous system is relatively low, so the puncture wound seals quickly. The most common clinical procedures examine venous blood.

Blood from peripheral capillaries can be obtained by puncturing the tip of a finger, the lobe of an ear, or (in infants) the great toe or heel. A small drop of capillary blood can be used to prepare a *blood smear*, a thin film of blood on a microscope slide. The blood smear is then stained with special dyes to show each type of formed element. Capillary blood can also be used to monitor levels of glucose, cholesterol, and hemoglobin, as well as to check the clotting system.

An **arterial puncture**, or "arterial stick," can be used for checking the efficiency of gas exchange at the lungs. Samples are generally drawn from the *radial artery* at the wrist or the *brachial artery* at the elbow.

19–3 PLASMA

Objective

- Specify the composition and functions of plasma.

The composition of whole blood is summarized in Figure 19–1b●. Plasma makes up 46–63 percent of the volume of whole blood, with water accounting for 92 percent of the plasma volume. Together, plasma and interstitial fluid account for most of the volume of extracellular fluid (ECF) in the body.

In many respects, the composition of plasma resembles that of interstitial fluid. The concentrations of the major plasma ions, for example, are similar to those of interstitial fluid and differ markedly from those inside cells. This similarity is understandable, as water, ions, and small solutes are continuously exchanged between plasma and interstitial fluids across the walls of capillaries. Normally, the capillaries deliver more liquid and solutes to a tissue than they remove. The excess flows through the tissue, into vessels of the lymphatic system, and eventually back to the bloodstream. The primary differences between plasma and interstitial fluid involve (1) the concentrations of dissolved proteins, because plasma proteins cannot cross capillary walls, and (2) the levels of respiratory gases (oxygen and carbon dioxide), due to the respiratory activities of tissue cells.

◻ PLASMA PROTEINS

Plasma contains significant quantities of dissolved proteins. On average, each 100 ml of plasma contains 7.6 g (0.3 oz) of protein, almost five times the concentration in interstitial fluid. The large size and globular shapes of most blood proteins prevent them from crossing capillary walls, and they remain trapped within the circulatory system. Three primary classes of plasma proteins exist: (1) *albumins* (al-BŪ-minz), (2) *globulins* (GLOB-ū-linz), and (3) *fibrinogen* (fī-BRIN-ō-jen). These three classes make up over 99% of the plasma proteins. The remainder consists of circulating enzymes, hormones, and prohormones.

Albumins

Albumins constitute roughly 60 percent of the plasma proteins. As the most abundant plasma proteins, they are major contributors to the osmotic pressure of plasma. Albumins are also important in the transport of fatty acids, thyroid hormones, some steroid hormones, and other substances.

Globulins

Globulins account for approximately 35 percent of the proteins in plasma. Examples of important plasma globulins include antibodies and transport globulins. **Antibodies**, also called **immunoglobulins** (i-mū-nō-GLOB-ū-linz), attack foreign proteins and pathogens. We will describe several classes of immunoglobulins in

Chapter 22. **Transport globulins** bind small ions, hormones, or compounds that might otherwise be lost at the kidneys or that have very low solubility in water. Important examples of transport globulins include the following:

- *Thyroid-binding globulin* and *transthyretin*, which transport thyroid hormones; *transcortin*, which transports ACTH; and *transcalciferin*, which transports calcitriol.
- *Metalloproteins*, which transport metal ions. *Transferrin*, for example, is a metalloprotein that transports iron (Fe^{2+}).
- *Apolipoproteins* (Ā-pō-lī-pō-PRŌ-tēnz), which carry triglycerides and other lipids in blood. When bound to lipids, an apolipoprotein becomes a **lipoprotein** (LĪ-pō-prō-tēn).
- *Steroid-binding proteins*, which transport steroid hormones in blood. For example, *testosterone-binding globulin* (*TeBG*) binds and transports testosterone.

Fibrinogen

The third type of plasma protein, **fibrinogen**, functions in clotting. Fibrinogen normally accounts for roughly 4 percent of plasma proteins. Under certain conditions, fibrinogen molecules interact, forming large, insoluble strands of **fibrin** (FĪ-brin). These fibers provide the basic framework for a blood clot. If steps are not taken to prevent clotting, the conversion of fibrinogen to fibrin will occur in a sample of plasma. This conversion removes the clotting proteins, leaving a fluid known as **serum**. The clotting process also removes calcium ions and other materials from solution, so plasma and serum differ in several significant ways. (See Appendix IV.) Thus, the results of a blood test generally indicate whether the sample source is plasma (P) or serum (S).

Other Plasma Proteins

The remaining 1 percent of plasma proteins is composed of specialized proteins whose levels vary widely. Peptide hormones, including insulin, prolactin (PRL), and the glycoproteins thyroid-stimulating hormone (TSH), follicle-stimulating hormone (FSH), and luteinizing hormone (LH), are normally present in circulating blood. ∞ Chapter 18 Their plasma concentrations rise and fall from day to day or even from hour to hour.

Origins of the Plasma Proteins

The liver synthesizes and releases more than 90 percent of the plasma proteins, including all albumins and fibrinogen, most globulins, and various prohormones. Antibodies are produced by *plasma cells*. Plasma cells are derived from *lymphocytes*, the primary cells of the lymphatic system. Peptide hormones are produced in a variety of endocrine organs.

Because the liver is the primary source of plasma proteins, liver disorders can alter the composition and functional properties of blood. For example, some forms of liver disease can lead to uncontrolled bleeding due to the inadequate synthesis of fibrinogen and other proteins involved in clotting.

Plasma expanders can be used to increase the blood volume temporarily, over a period of hours, while preliminary lab work is under way to determine a person's blood type. Isotonic electrolyte solutions such as normal saline can be used as a plasma expander, but its effects are short lived due to diffusion into interstitial fluid and cytosol. This fluid loss is slowed by the addition of solutes that cannot freely diffuse across cell membranes. One example is *Ringer's solution*, isotonic saline containing lactate ions. The effects of Ringer's solution fade gradually as the liver, skeletal muscles, and other tissues absorb and metabolize the lactate ions. Another option is the administration of isotonic saline solution containing purified human albumin. However, the plasma expanders in clinical use often contain large carbohydrate molecules, rather than proteins, to maintain proper osmotic concentration. (We noted the emergency use of the carbohydrate dextran in sodium chloride solutions in Chapter 3. ∞ p. 94) Although these carbohydrates are not metabolized, they are gradually removed from the bloodstream by phagocytes, and the blood volume slowly declines. Plasma expanders are easily stored, and their sterile preparation avoids viral or bacterial contamination, which can be a problem with donated plasma. Note that, although they provide a temporary solution to low blood volume, plasma expanders do not increase the amount of oxygen carried by the blood; that function is performed by red blood cells.

✓ Why is venipuncture a common technique for obtaining a blood sample?

✓ What would be the effects of a decrease in the amount of plasma proteins?

✓ Which plasma protein would you expect to be elevated during a viral infection?

Answers start on page Q-1

19–4 RED BLOOD CELLS

Objectives

- List the characteristics and functions of red blood cells.
- Describe the structure of hemoglobin, and indicate its functions.
- Describe the recycling system for aged or damaged red blood cells.
- Define erythropoiesis, identify the stages involved in red blood cell maturation, and describe the homeostatic regulation of red blood cell production.
- Explain the importance of blood typing and the basis for ABO and Rh incompatibilities.

The most abundant blood cells are the red blood cells (RBCs), which account for 99.9 percent of the formed elements. These cells give whole blood its deep red color. Red blood cells contain the red pigment *hemoglobin*, (HĒ-mō-glō-bin), which binds and transports oxygen and carbon dioxide.

☐ ABUNDANCE OF RBCs

A standard blood test reports the number of RBCs per microliter (μl) of whole blood as the *red blood cell count*. In adult males, 1 microliter, or 1 *cubic millimeter* (mm^3), of whole blood contains 4.5–6.3 million RBCs; in adult females, 1 microliter contains 4.2–5.5 million. A single drop of whole blood contains approximately 260 million RBCs, and the blood of an average adult has 25 trillion RBCs. The number of RBCs thus accounts for roughly one-third of all cells in the human body.

The **hematocrit** (he-MAT-ō-krit) is the percentage of whole blood occupied by cellular elements. The normal hematocrit in adult males averages 46 (range: 40–54); the average for adult females is 42 (range: 37–47). The gender difference in hematocrit primarily reflects the fact that androgens (male hormones) stimulate red blood cell production, whereas estrogens (female hormones) do not.

The hematocrit is determined by centrifuging a blood sample so that all the formed elements come out of suspension. Whole blood contains roughly 1000 red blood cells for each white blood cell. After centrifugation, the white blood cells and platelets form a very thin *buffy coat* above a thick layer of RBCs (Figure 19–1●). Because the hematocrit value is due almost entirely to the volume of RBCs, hematocrit is commonly reported as the *volume of packed red cells* (*VPRC*), or simply the *packed cell volume* (*PCV*).

Many conditions can affect the hematocrit. For example, the hematocrit increases in cases of dehydration, owing to a reduction in plasma volume, or after *erythropoietin* (*EPO*) stimulation. ∞ p. 636 The hematocrit can decrease as a result of internal bleeding or problems with RBC formation. As a result, the hematocrit alone does not provide specific diagnostic information. Yet a change in hematocrit is an indication that other, more specific tests are needed. (We will consider some of those tests later in the chapter.) **AM** Polycythemia

☐ STRUCTURE OF RBCs

Red blood cells are among the most specialized cells of the body. A red blood cell is very different from the "typical cell" we discussed in Chapter 3. Each RBC is a biconcave disc with a thin central region and a thicker outer margin (Figure 19–2●). An average RBC has a diameter of 7.8 μm and a maximum thickness of 2.6 μm, although the center narrows to about 0.8 μm. (Normal ranges of measurement are given in Figure 19–2d●.)

This unusual shape has three important effects on RBC function:

1. *It Gives Each RBC a Large Ratio of Surface Area to Volume.* The RBC carries oxygen bound to intracellular proteins. That oxygen must be absorbed or released quickly as the RBC passes through the capillaries of the lungs or peripheral tissues. The greater the surface area per unit volume, the faster is the exchange between the cell's interior and the surrounding plasma. The total surface area of the RBCs in the blood of a typical adult is about 3800 square meters, roughly 2000 times the total surface area of the body.

2. *It Enables RBCs to Form Stacks, Like Dinner Plates, That Smooth the Flow Through Narrow Blood Vessels.* These stacks form and dissociate repeatedly without affecting the cells involved. An entire stack can pass along a blood vessel only slightly larger than the diameter of a single RBC, whereas individual cells would bump the walls, bang together, and form logjams that could restrict or prevent blood flow.

3. *It Enables RBCs to Bend and Flex When Entering Small Capillaries and Branches.* Red blood cells are very flexible. By changing shape, individual RBCs can squeeze through capillaries as narrow as 4 μm.

During their differentiation, the RBCs of humans and other mammals lose most of their organelles, including nuclei; the cells retain only the cytoskeleton. (The RBCs of vertebrates other than mammals have nuclei.) Because they lack nuclei and ribosomes, circulating mammalian RBCs cannot divide or synthesize structural proteins or enzymes. As a result, the RBCs cannot perform repairs, and their life span is relatively short—normally less than 120 days. With few organelles and no ability to perform protein synthesis, their energy demands are low. In the absence of mitochondria, they obtain the energy they need through the anaerobic metabolism of glucose absorbed from the surrounding plasma. The absence of mitochondria ensures that absorbed oxygen will be carried to peripheral tissues, not "stolen" by mitochondria in the cell.

☐ HEMOGLOBIN

In effect, a developing red blood cell loses any organelle not directly associated with the cell's primary function: the transport of respiratory gases. Molecules of **hemoglobin** (**Hb**) account for over 95 percent of the intracellular proteins. The hemoglobin content of whole blood is reported in grams of Hb per 100 ml of whole blood (g/dl). Normal ranges are 14–18 g/dl in males and 12–16 g/dl in females. Hemoglobin is responsible for the cell's ability to transport oxygen and carbon dioxide.

Hemoglobin Structure

Each Hb molecule has a complex quaternary shape. The Hb molecule has two *alpha* (α) *chains* and two *beta* (β) *chains* of polypeptides. Each individual chain is a globular protein subunit that resembles the myoglobin in skeletal and cardiac muscle cells. Like myoglobin, each Hb chain contains a single molecule of **heme**, a pigment complex (Figure 19–3●). Each heme unit holds an iron ion in such a way that the iron can interact with an oxygen molecule, forming **oxyhemoglobin**, HbO_2. Blood containing RBCs filled with oxyhemoglobin is bright red. The iron–oxygen interaction is very weak; the two can easily be separated without damaging the heme unit or the oxygen molecule. The binding of an oxygen molecule to the iron in a heme unit is therefore completely reversible. A hemoglobin molecule in which the iron has separated from the oxygen molecule is called *deoxyhemoglobin*.

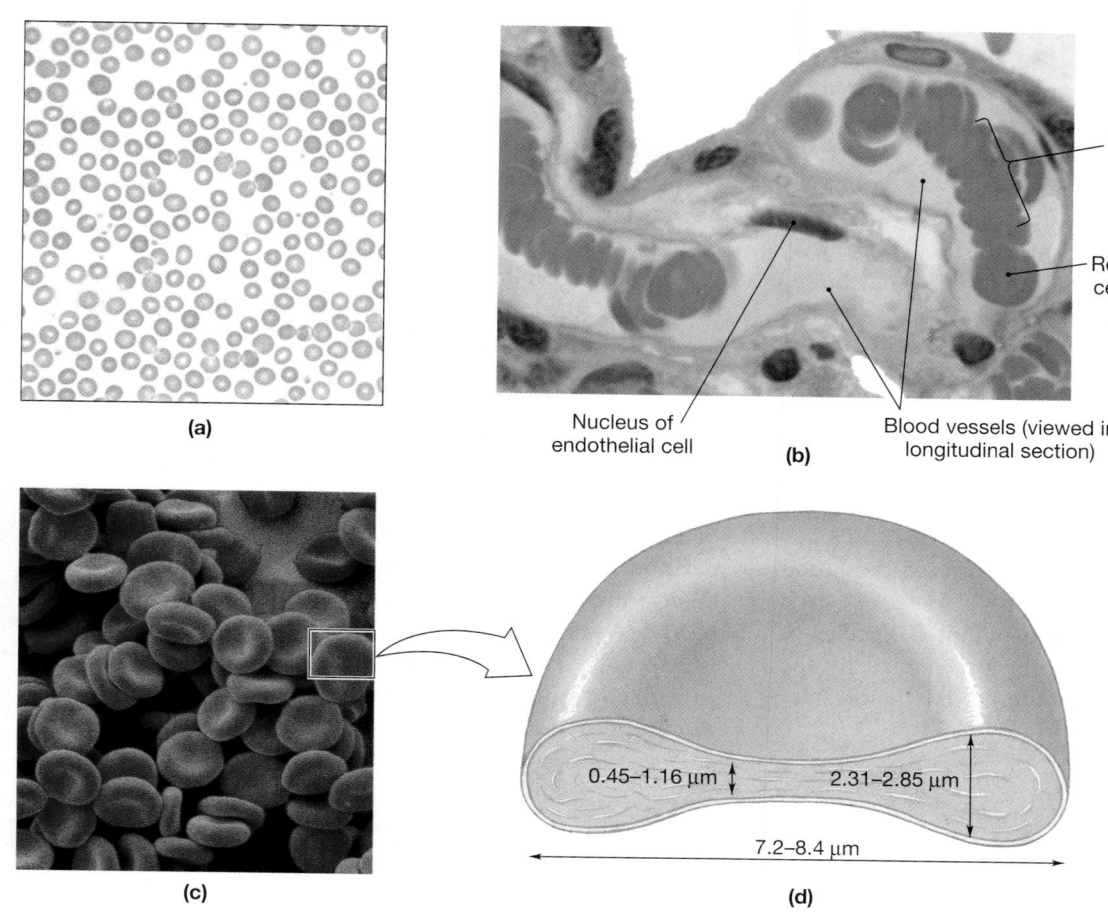

(a)

Stacked
RBCs

Red blood
cell (RBC)

Nucleus of
endothelial cell

Blood vessels (viewed in
longitudinal section)

(b)

(c)

0.45–1.16 μm

2.31–2.85 μm

7.2–8.4 μm

(d)

●FIGURE 19–2
The Anatomy of Red Blood Cells. **(a)** When viewed in a standard blood smear, red blood cells appear as two-dimensional objects, because they are flattened against the surface of the slide. **(b)** When traveling through relatively narrow capillaries, RBCs may stack like dinner plates. **(c)** The three-dimensional structure of red blood cells. **(d)** A sectional view of a mature red blood cell, showing the normal ranges for its dimensions.

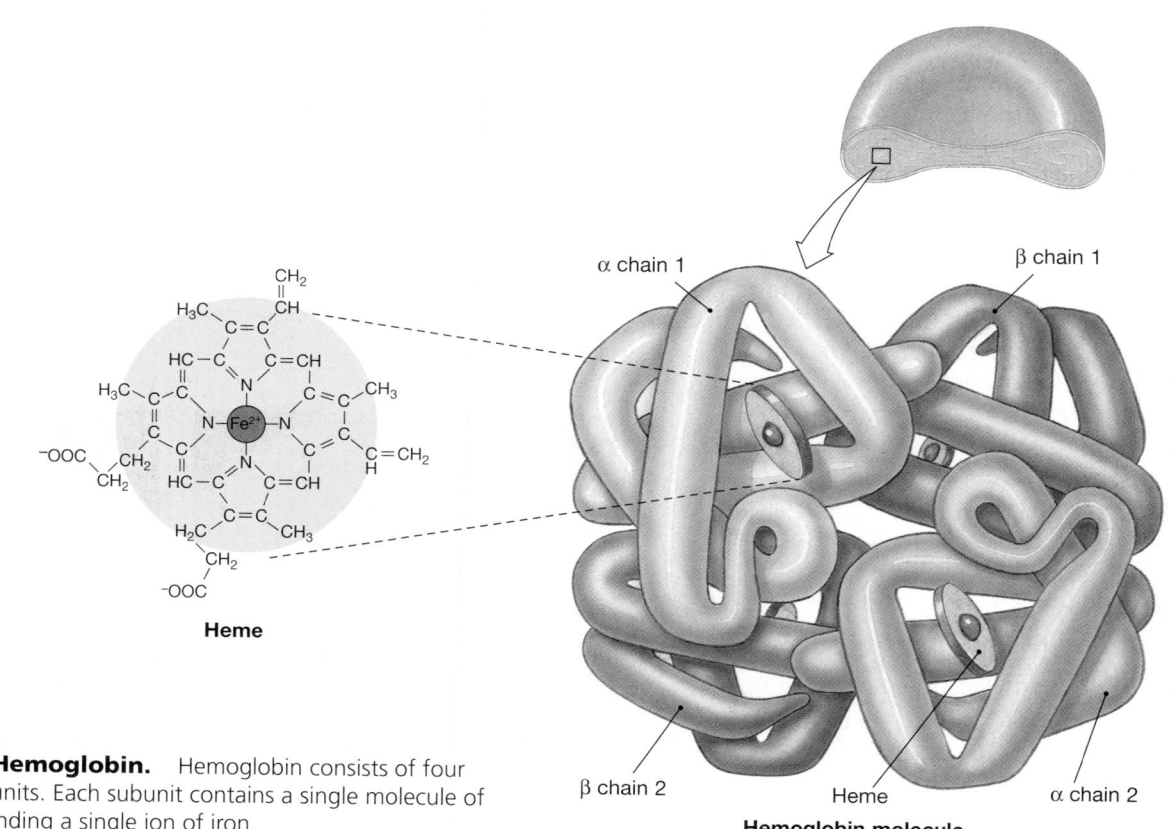

α chain 1

β chain 1

β chain 2

Heme

α chain 2

Hemoglobin molecule

CH_2
CH
H_3C
C=C
HC—C
N
C=CH
H_3C
C
C
CH_3
N
Fe²⁺
N
^-OOC
CH_2
C
C
CH_2
HC—C
N
C=CH
CH_2
CH
H_2C
C
C
CH_3
CH_2
^-OOC

Heme

●FIGURE 19–3
The Structure of Hemoglobin. Hemoglobin consists of four globular protein subunits. Each subunit contains a single molecule of heme—a ring surrounding a single ion of iron.

Blood containing RBCs filled with deoxyhemoglobin is dark red—almost burgundy.

The RBCs of an embryo or a fetus contain a different form of hemoglobin, known as *fetal hemoglobin*, which binds oxygen more readily than does the hemoglobin of adults. For this reason, the developing fetus can "steal" oxygen from the maternal bloodstream at the placenta. The conversion from fetal hemoglobin to the adult form begins shortly before birth and continues over the next year. The production of fetal hemoglobin can be stimulated in adults by the administration of drugs such as *hydroxyurea* or *butyrate*. This is one method of treatment for conditions, such as *sickle-cell anemia* or *thalassemia*, that result from the production of abnormal forms of adult hemoglobin.

Hemoglobin Function

There are about 280 million Hb molecules in each red blood cell. Because a Hb molecule contains four heme units, each RBC can potentially carry more than a billion molecules of oxygen. Roughly 98.5 percent of the oxygen carried by the blood travels through the bloodstream bound to Hb molecules inside RBCs.

The amount of oxygen bound to hemoglobin depends primarily on the oxygen content of the plasma. When plasma oxygen levels are low, hemoglobin releases oxygen. Under these conditions, typical of peripheral capillaries, plasma carbon dioxide levels are elevated. The alpha and beta chains of hemoglobin then bind carbon dioxide, forming **carbaminohemoglobin**. In the capillaries of the lungs, plasma oxygen levels are high and carbon dioxide levels are low. Upon reaching these capillaries, RBCs absorb oxygen (which is then bound to hemoglobin) and release carbon dioxide. We will revisit these processes in Chapter 23.

Normal activity levels can be sustained only when tissue oxygen levels are kept within normal limits. If the hematocrit is low or the Hb content of the RBCs is reduced, the condition of **anemia** exists. Anemia interferes with oxygen delivery to peripheral tissues. Every system is affected as organ function deteriorates owing to oxygen starvation. Anemic individuals become weak, lethargic, and often confused, because the brain is affected as well. Anemia occurs in many forms; we shall consider specific examples both in this chapter and in the *Applications Manual*.

Abnormal Hemoglobin

Several inherited disorders are characterized by the production of abnormal hemoglobin. Two of the best known are thalassemia and sickle-cell anemia (SCA).

The various forms of **thalassemia** (thal-ah-SĒ-mē-uh) result from an inability to produce adequate amounts of alpha or beta chains of hemoglobin. As a result, the rate of RBC production is slowed, and the mature RBCs are fragile and short lived. The scarcity of healthy RBCs reduces the oxygen-carrying capacity of the blood and leads to problems with the development and growth of systems throughout the body. Individuals with severe thalassemia must periodically undergo *transfusions*—the administration of blood components—to keep adequate numbers of RBCs in the bloodstream.

AM Thalassemia; Transfusions and Synthetic Blood

Sickle-cell anemia results from a mutation affecting the amino acid sequence of the beta chains of the Hb molecule. When blood contains abundant oxygen, the Hb molecules and the RBCs that carry them appear normal (Figure 19–4a●). But when the defective hemoglobin gives up enough of its bound oxygen, the adjacent Hb molecules interact and the cells become stiff and curved (Figure 19–4b●). This "sickling" makes the RBCs fragile and easily damaged. Moreover, an RBC that has folded to squeeze into a narrow capillary delivers its oxygen to the surrounding tissue, but the cell can become stuck as sickling occurs. A circulatory blockage results, and nearby tissues become starved for oxygen. **AM** Sickle-Cell Anemia

●**FIGURE 19–4**
"Sickling" in Red Blood Cells. **(a)** When fully oxygenated, the cells of an individual with the sickling trait appear relatively normal. **(b)** At lower oxygen concentrations, the RBCs change shape, becoming more rigid and sharply curved.

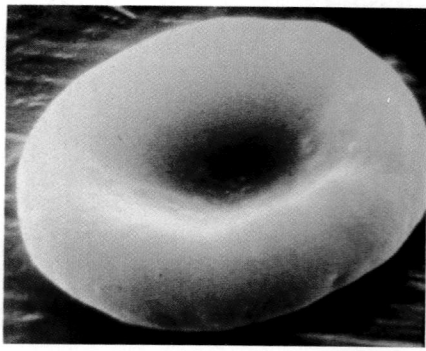
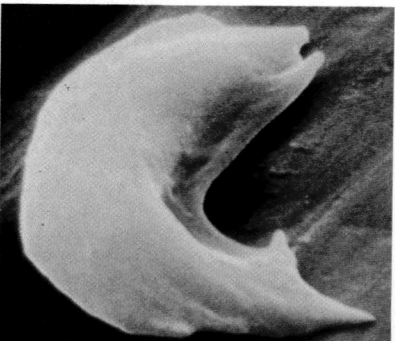

(a) Normal RBC   (b) Sickled RBC

☐ RBC LIFE SPAN AND CIRCULATION

A red blood cell is exposed to severe mechanical stresses. A single round-trip from the heart, through the peripheral tissues, and back to the heart usually takes less than a minute. In that time, an RBC gets pumped out of the heart and forced along vessels, where it bounces off the walls and collides with other RBCs. It forms stacks, contorts and squeezes through tiny capillaries, and then joins its comrades in a headlong rush back to the heart for another round.

With all this wear and tear and no repair mechanisms, a typical RBC has a relatively short life span. After it travels about 700 miles in 120 days, either its cell membrane ruptures or some other damage is detected by phagocyes, which engulf the RBC. The continuous elimination of RBCs usually goes unnoticed, because new ones enter the bloodstream at a comparable rate. About 1 percent of the circulating RBCs are replaced each day, and in the process approximately 3 million new RBCs enter the bloodstream *each second*!

Hemoglobin Conservation and Recycling

Phagocytic cells of the liver, spleen, and bone marrow monitor the condition of circulating RBCs, generally recognizing and engulfing them before they **hemolyze**, or rupture. These phagocytes also detect and remove Hb molecules and cell fragments from the relatively small number of RBCs that hemolyze in the bloodstream (about 10 percent of the total recycled each day).

If the Hb released by hemolysis is not phagocytized, its components will not be recycled. Hemoglobin remains intact only under the conditions inside RBCs. When hemolysis occurs, the Hb breaks down and the alpha and beta chains are filtered by the kidneys and eliminated in urine. When abnormally large numbers of RBCs break down in the bloodstream, urine may turn red or brown. This condition is called **hemoglobinuria**. **Hematuria** (hē-ma-TOO-rē-uh), the presence of intact RBCs in urine, occurs only after kidney damage or damage to vessels along the urinary tract.

Once an RBC has been engulfed and broken down by a phagocytic cell, each component of the Hb molecule has a different fate (Figure 19–5●). The globular proteins are disassembled into their component amino acids, which are then either metabolized by the cell or released into the bloodstream for use by other cells. Each heme unit is stripped of its iron and converted to **biliverdin** (bili-VER-din), an organic compound with a green color. (Bad bruises commonly develop a green tint due to biliverdin formation in the blood-filled tissues.) Biliverdin is then converted to **bilirubin** (bil-i-ROO-bin), an orange-yellow pigment, and released into the bloodstream. There, the bilirubin binds to albumin and is transported to the liver for excretion in bile.

If the bile ducts are blocked or the liver cannot absorb or excrete bilirubin, circulating levels of the compound climb rapidly. Bilirubin then diffuses into peripheral tissues, giving them a yellow color that is most apparent in the skin and over the sclera of the eyes. This combination of signs (yellow skin and eyes) is called **jaundice** (JAWN-dis). **AM** Bilirubin Tests and Jaundice

In the large intestine, bacteria convert bilirubin to related pigments called *urobilinogens* and *stercobilinogens*. Some of the urobilinogens are absorbed into the bloodstream and are subsequently excreted into urine. On exposure to oxygen, some of the urobilinogens and stercobilinogens are converted to **urobilins** and **stercobilins**. Urine is usually yellow because it contains urobilins; feces are yellow-brown or brown owing to the presence of urobilins and stercobilins in varying proportions.

Iron

Large quantities of free iron are toxic to cells, so iron in the body is generally bound to transport or storage proteins. Iron extracted from the heme molecules may be bound and stored in a phagocytic cell or released into the bloodstream, where it binds to **transferrin** (trans-FER-in), a plasma protein. Red blood cells developing in the bone marrow absorb the amino acids and transferrins from the bloodstream and use them to synthesize new Hb molecules. Excess transferrins are removed in the liver and spleen, and the iron is stored in two special protein–iron complexes: **ferritin** (FER-i-tin) and **hemosiderin** (hē-mō-SID-e-rin).

This recycling system is remarkably efficient. Although roughly 26 mg of iron is incorporated into Hb molecules each day, a dietary supply of 1–2 mg can keep pace with the incidental losses that occur at the kidneys and digestive tract.

Any impairment in iron uptake or metabolism can cause serious clinical problems, because RBC formation will be affected. *Iron-deficiency anemia*, which results from a lack of iron in the diet or from problems with iron absorption, is one example. Too much iron can also cause problems, owing to excessive buildup in secondary storage sites, such as the liver and cardiac muscle tissue. Excessive iron deposition in cardiac muscle cells has been linked to heart disease. **AM** Iron Deficiences and Excesses

☐ RBC PRODUCTION

Embryonic blood cells appear in the bloodstream during the third week of development. These cells divide repeatedly, rapidly increasing in number. The vessels of the embryonic *yolk sac* are the primary site of blood formation for the first eight weeks of development. As other organ systems appear, some of the embryonic blood cells move out of the bloodstream and into the liver, spleen, thymus, and bone marrow. These embryonic cells differentiate into stem cells that produce blood cells by their divisions.

The liver and spleen are the primary sites of hemopoiesis from the second to fifth months of development, but as the skeleton enlarges, the bone marrow becomes increasingly important. In adults, red bone marrow is the only site of red blood cell production as well as the primary site of white blood cell formation.

Red blood cell formation, or **erythropoiesis** (e-rith-rō-poy-Ē-sis), occurs only in *red bone marrow*, or **myeloid** (MĪ-e-loyd) **tissue** (*myelos*, marrow). This tissue is located in

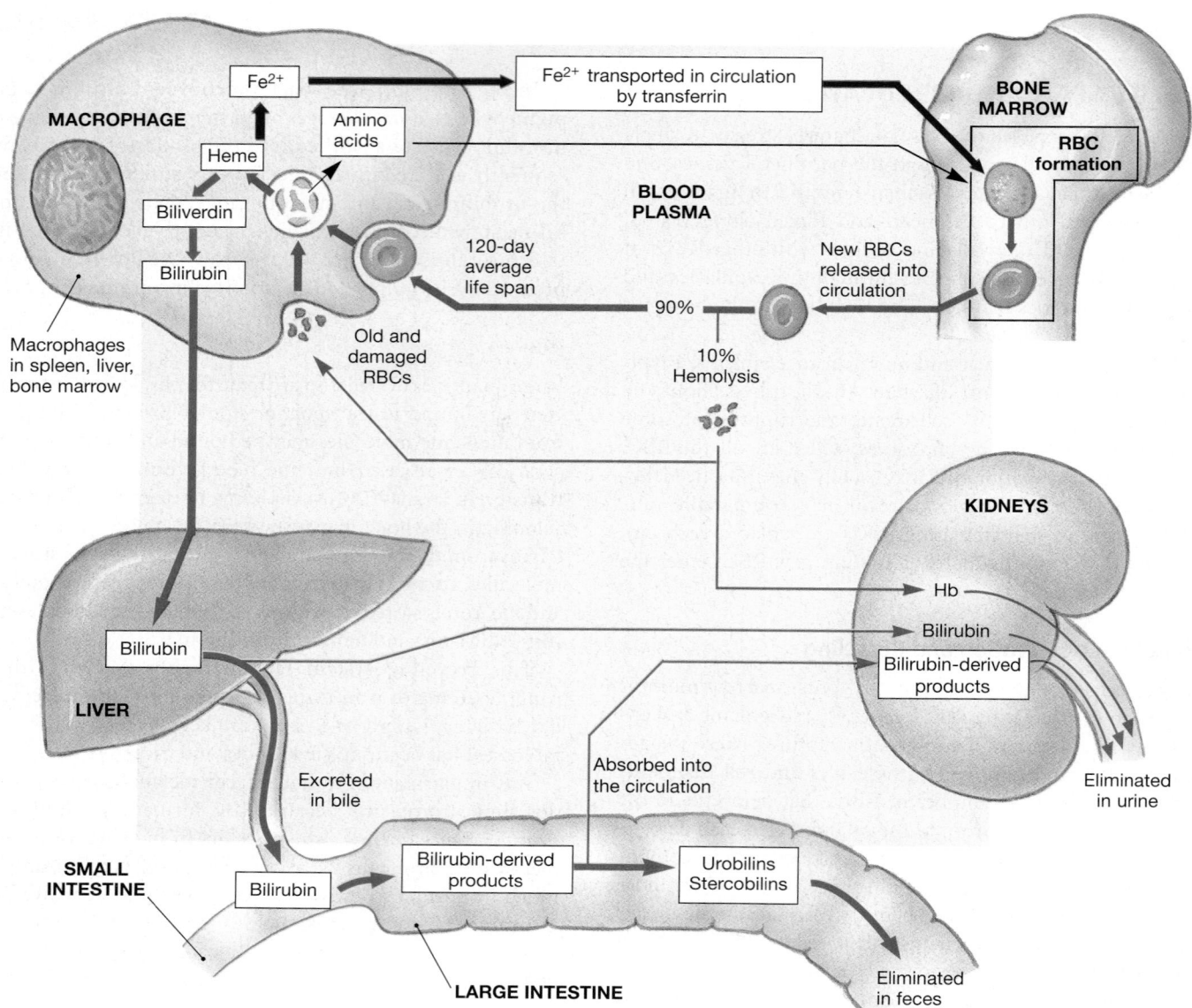

●FIGURE 19–5
Red Blood Cell Turnover.   The normal pathways for recycling amino acids and iron from aging or damaged RBCs, broken down by macrophages. The amino acids are absorbed, especially by developing cells in bone marrow. The iron is stored in many sites. The rings of the heme units are converted to bilirubin, absorbed by the liver, and excreted in bile or urine; some of the breakdown products produced in the large intestine are recirculated.

portions of the vertebrae, sternum, ribs, skull, scapulae, pelvis, and proximal limb bones. Other marrow areas contain a fatty tissue known as *yellow bone marrow*. ∞ p. 186 Under extreme stimulation, such as severe and sustained blood loss, areas of yellow marrow can convert to red marrow, increasing the rate of RBC formation.

Stages in RBC Maturation

During its maturation, a red blood cell passes through a series of stages. **Hematologists** (hē-ma-TOL-o-jists), specialists in blood formation and function, have given specific names to key stages. Divisions of hemocytoblasts in bone marrow produce (1) **myeloid stem cells**, which in turn divide to produce red blood cells and several classes of white blood cells, and (2)

lymphoid stem cells, which divide to produce the various classes of lymphocytes. Cells destined to become RBCs first differentiate into **proerythroblasts** and then proceed through various **erythroblast** stages (Figure 19–6●). Erythroblasts, which actively synthesize hemoglobin, are categorized on the basis of total size, the amount of hemoglobin present, and the size and appearance of the nucleus.

After roughly four days of differentiation, the erythroblast, now called a *normoblast*, sheds its nucleus and becomes a **reticulocyte** (re-TIK-ū-lō-sīt), which contains 80 percent of the Hb of a mature RBC. Hb synthesis then continues for two to three more days. During this period, while the cells are synthesizing hemoglobin and other proteins, their cytoplasm still contains RNA which can be seen under the microscope with certain stains. After two days in the bone marrow, reticulocytes enter the bloodstream. At this time,

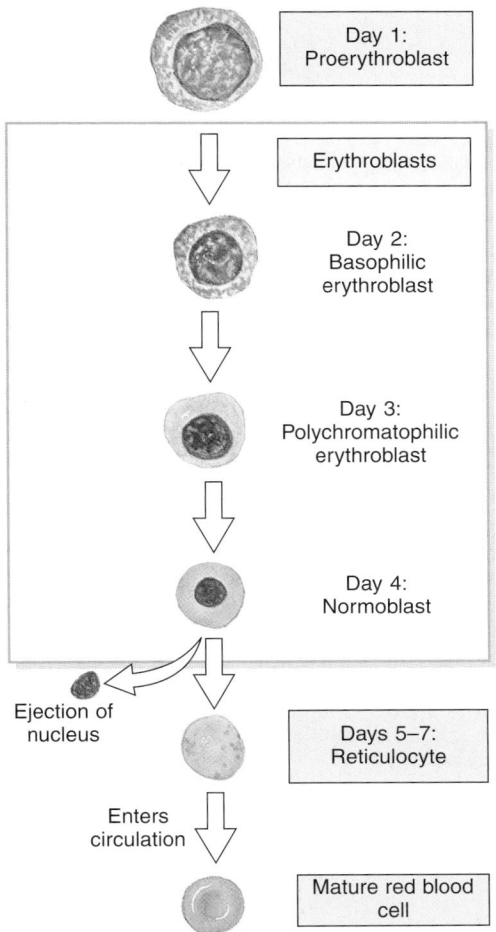

Day 1:
Proerythroblast

Erythroblasts

Day 2:
Basophilic
erythroblast

Day 3:
Polychromatophilic
erythroblast

Day 4:
Normoblast

Ejection of
nucleus

Days 5–7:
Reticulocyte

Enters
circulation

Mature red blood
cell

●FIGURE 19–6
Stages of RBC Maturation.   Red blood cells are produced in the red bone marrow. The color density indicates the abundance of hemoglobin in the cytoplasm. Note the reduction in the size of the cell and the size of the nucleus before a reticulocyte is formed.

reticulocytes normally account for about 0.8 percent of the RBC population in the blood and can still be detected by staining. After 24 hours in circulation, the reticulocytes complete their maturation and become indistinguishable from other mature RBCs.

Regulation of Erythropoiesis

For erythropoiesis to proceed normally, the red bone marrow must receive adequate supplies of amino acids, iron, and vitamins (including B_{12}, B_6, and folic acid) required for protein synthesis. For example, we obtain **vitamin B_{12}** from dairy products and meat, and its absorption requires the presence of *intrinsic factor* produced in the stomach. If vitamin B_{12} is not obtained from the diet, normal stem cell divisions cannot occur and *pernicious anemia* results. Thus, pernicious anemia is caused by a vitamin B_{12} deficiency, by a problem with the production of intrinsic factor, or by a problem with the absorption of vitamin B_{12} bound to intrinsic factor.

Erythropoiesis is stimulated directly by the peptide hormone erythropoietin (∞ p. 636) and indirectly by several hormones, including thyroxine, androgens, and growth hormone. As we noted earlier, estrogens do not stimulate erythropoiesis, a fact

that accounts for the differences between male and female hematocrit values.

Erythropoietin, also called **EPO** or **erythropoiesis-stimulating hormone**, is a glycoprotein that appears in the plasma when peripheral tissues, especially the kidneys, are exposed to low oxygen concentrations. The state of low tissue oxygen levels is called **hypoxia** (hī-POKS-ē-uh; *hypo-*, below + *ox-*, presence of oxygen). Erythropoietin is released (1) during anemia, (2) when blood flow to the kidneys declines, (3) when the oxygen content of air in the lungs declines, owing to disease or to high altitudes, and (4) when the respiratory surfaces of the lungs are damaged. Once in the bloodstream, EPO travels to areas of red bone marrow, where it stimulates stem cells and developing RBCs.

Erythropoietin has two major effects: (1) It stimulates increased cell division rates in erythroblasts and in the stem cells that produce erythroblasts, and (2) it speeds up the maturation of RBCs, mainly by accelerating the rate of Hb synthesis. Under maximum EPO stimulation, bone marrow can increase the rate of RBC formation tenfold, to about 30 million per second.

The ability to increase the rate of blood formation quickly and dramatically is important to a person recovering from a severe blood loss. But if EPO is administered to a healthy individual, as in the case of the cyclists and Olympic competitors mentioned in Chapter 18, the hematocrit may rise to 65 or more. ∞ p. 642 Such an increase can place an intolerable strain on the heart. Comparable problems can occur after **blood doping**, a practice in which athletes attempt to elevate their hematocrits by reinfusing packed RBCs that were removed and stored at an earlier date (Figure 19–7●). The goal is to improve oxygen delivery to muscles, thereby enhancing performance. The strategy can be dangerous, however, because it elevates blood viscosity and increases the workload on the heart. **AM** Erythrocytosis and Blood Doping

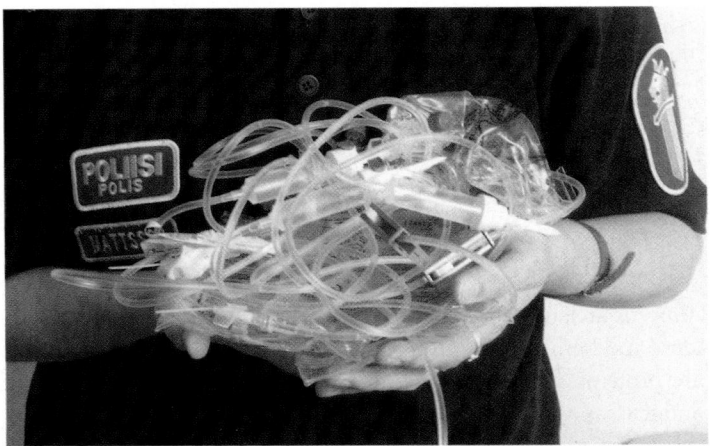

●FIGURE 19–7
Swifter, Higher, Stronger.   A Helsinki police officer displays syringes and empty infusion bags of banned blood-thinning agents that officials suspect may have been used by the Finnish cross-country ski team. Blood-thinning agents are often employed to counteract the increase in blood viscosity caused by blood doping or the use of EPO.

Blood Tests and RBCs

Blood tests provide information about the general health of an individual, usually with a minimum of trouble and expense. Several common blood tests focus on red blood cells, the most abundant formed elements. These *RBC tests* assess the number, size, shape, and maturity of circulating RBCs, providing an indication of the erythro-poietic activities under way. The tests can also be useful in detecting problems, such as internal bleeding, that may not produce other obvious symptoms. Table 19–1 lists examples of important blood tests and related terms. (See the *Applications Manual* for sample calculations.) **AM** Blood Tests and RBCs

TABLE 19–1 RBC TESTS AND RELATED TERMINOLOGY

Test	Determines	Terms Associated with Abnormal Values	
		Elevated	**Depressed**
Hematocrit (Hct)	Percentage of formed elements in whole blood Normal = 37–54	Polycythemia (may reflect erythrocytosis or leukocytosis)	Anemia
Reticulocyte count (Retic.)	Circulating percentage of reticulocytes Normal = 0.8%	Reticulocytosis	
Hemoglobin concentration (Hb)	Concentration of hemoglobin in blood Normal = 12–18 g/dl		Anemia
RBC count	Number of RBCs per μl of whole blood Normal = 4.2–6.3 million/μl	Erythrocytosis/polycythemia	Anemia
Mean corpuscular volume (MCV)	Average volume of single RBC Normal = 82–101 μm^3 (normocytic)	Macrocytic	Microcytic
Mean corpuscular hemoglobin concentration (MCHC)	Average amount of Hb in one RBC Normal = 27–34 pg/μl (normochromic)	Hyperchromic	Hypochromic

---

✓ How would the hematocrit change after an individual suffered a hemorrhage?

✓ Dave develops a blockage in his renal arteries that restricts blood flow to the kidneys. Will his hematocrit change?

✓ How would the level of bilirubin in the blood be affected by a disease that causes damage to the liver?

Answers start on page Q-1

 The information on red blood cells can be reviewed on the **Companion Website:** Chapter 19/Reviewing Facts & Terms/ Multiple Choice.

☐ BLOOD TYPES

Antigens are substances that can trigger an *immune response*, a defense mechanism that protects you from infection. Most antigens are proteins, although some other types of organic molecules are antigens as well. Your cell membranes contain **surface antigens**, substances that your immune defenses recognize as "normal." In other words, your immune system ignores these substances rather than attacking them as "foreign." Your **blood type** is a classification determined by the presence or absence of specific surface antigens in the RBC cell membranes. The surface antigens involved are integral membrane glycoproteins or glycolipids whose characteristics are ge-netically determined. The surface antigens of RBCs are often called **agglutinogens** (a-gloo-TIN-ō-jenz). Red blood cells have at least 50 kinds of surface antigens. Three of particular importance—**A**, **B**, and **Rh** (or **D**)—have been designated as surface antigens.

The RBCs of an individual have (1) either A or B surface antigens, (2) both A and B surface antigens, or (3) neither A nor B surface antigens (Figure 19–8a●). **Type A** blood has surface antigen A only, **Type B** has surface antigen B only, **Type AB** has both A and B, and **Type O** has neither A nor B. These blood types are not evenly distributed throughout the population. The average values for the U.S. population are as follows: Type O, 46 percent; Type A, 40 percent; Type B, 10 percent; and Type AB, 4 percent (Table 19–2).

The term **Rh positive** (Rh$^+$) indicates the presence of the Rh surface antigen, sometimes called the *Rh factor*. The absence of this antigen is indicated as **Rh negative** (Rh$^-$). When the complete blood type is recorded, the term *Rh* is usually omitted and the data are reported as O negative (O$^-$), A positive (A$^+$), and so on. As in the distribution of A and B surface antigens, Rh type differs by race and by region (Table 19–2).

You are probably aware that your blood type must be checked before you can give or receive blood. Your immune system ignores the surface antigens on your own RBCs. However, your plasma contains antibodies, sometimes called *agglutinins* (a-GLOO-ti-ninz), that will attack the antigens on "foreign"

TABLE 19–2 DIFFERENCES IN BLOOD GROUP DISTRIBUTION

Population	O	A	B	AB	RH⁺
		Percentage with Each Blood Type			
U.S. (AVERAGE)	46	40	10	4	85
African-American	49	27	20	4	95
Caucasian	45	40	11	4	85
Chinese-American	42	27	25	6	100
Filipino-American	44	22	29	6	100
Hawaiian	46	46	5	3	100
Japanese-American	31	39	21	10	100
Korean-American	32	28	30	10	100
NATIVE NORTH AMERICAN	79	16	4	<1	100
NATIVE SOUTH AMERICAN	100	0	0	0	100
AUSTRALIAN ABORIGINE	44	56	0	0	100

RBCs. When these antibodies attack, the foreign cells **agglutinate**, or clump together; the process is called **agglutination**. The plasma of a Type A, Type B, or Type O individual always contains either anti-A or anti-B antibodies, or both, even if the person has never been exposed to RBCs that carry foreign surface antigens. If you have Type A blood, your plasma contains anti-B antibodies, which will attack Type B surface antigens. If you have Type B blood, your plasma contains anti-A antibodies. The RBCs of an individual with Type O blood have neither A nor B surface antigens, and that person's plasma contains both anti-A and anti-B antibodies. A Type AB individual has RBCs with both A and B surface antigens, and the plasma does not contain anti-A or anti-B antibodies.

In contrast to the situation with surface antigens A and B, the plasma of an Rh-negative individual does not necessarily contain anti-Rh antibodies. These antibodies are present only if the individual has been **sensitized** by previous exposure to Rh-positive RBCs. Such exposure can occur accidentally during a transfusion, but it can also accompany a seemingly normal pregnancy involv-

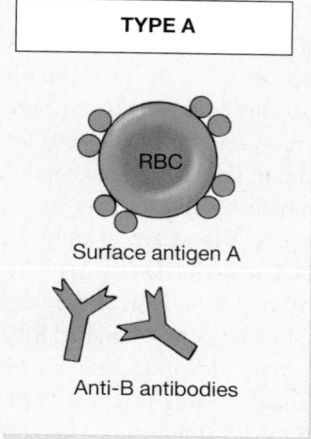 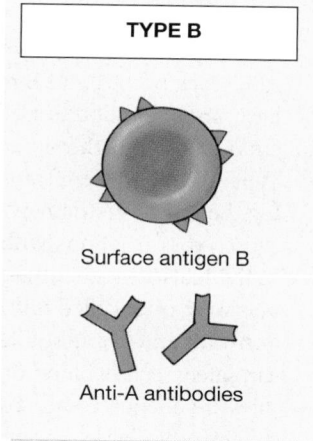 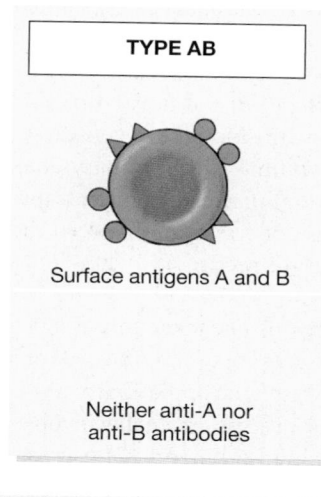 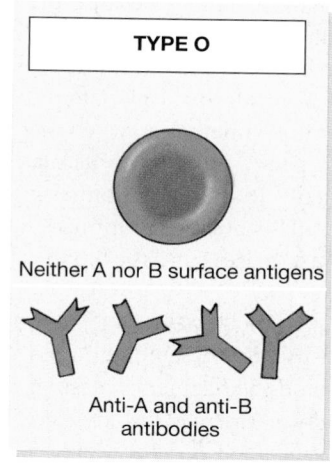

(a)

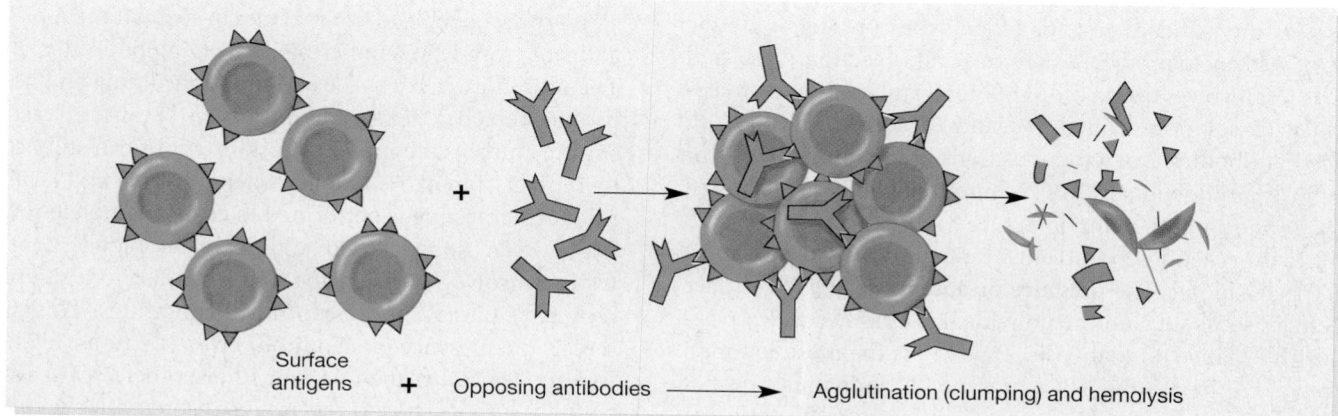

(b)

●FIGURE 19–8
Blood Typing and Cross-Reactions. The blood type depends on the presence of surface antigens (agglutinogens) on RBC surfaces. **(a)** The plasma contains antibodies (agglutinins) that will react with foreign surface antigens. The relative frequencies of each blood type in the U.S. population are indicated in Table 19–2. **(b)** In a cross-reaction, antibodies that encounter their target antigens lead to agglutination and hemolysis of the affected RBCs.

ing an Rh-negative mother and an Rh-positive fetus. (See the box, "Hemolytic Disease of the Newborn," on pages 665–666.)

Cross-Reactions

When an antibody meets its specific surface antigen, the RBCs agglutinate and may also hemolyze. This reaction is called a **cross-reaction** (Figure 19–8b and 19–9●). For instance, an anti-A antibody that encounters A surface antigens will cause the RBCs bearing the surface antigens to clump or even break up. Clumps and fragments of RBCs under attack form drifting masses that can plug small blood vessels in the kidneys, lungs, heart, or brain, damaging or destroying dependent tissues. Such cross-reactions, or *transfusion reactions*, can be prevented by ensuring that the blood types of the donor and the recipient are **compatible**—that is, that the donor's blood cells will not cross-react with the plasma of the recipient.

In practice, the surface antibodies on the donor's cells are more important in determining compatibility than are the antibodies in the donor's plasma. Unless large volumes of whole blood or plasma are transferred, cross-reactions between the donor's plasma and the recipient's blood cells will fail to produce significant agglutination. This is because the donated plasma will be diluted quickly through mixing with the relatively large plasma volume of the recipient. (One unit of whole blood, 500 ml, contains roughly 275 ml of plasma, only about 10 percent of the normal plasma volume.) Nonetheless, when increasing the blood's oxygen-carrying capacity rather than its plasma volume is the primary goal, packed RBCs, with a minimal amount of plasma, are often transfused. This practice minimizes the risk of a reaction between the donated plasma and the blood cells of the recipient.

TESTING FOR COMPATIBILITY Extra care must be taken to avoid potentially life-threatening cross-reactions between the donor's cells and the recipient's plasma. As a result, a compatibility test is usually performed in advance. This process normally involves two steps: (1) a determination of blood type and (2) a cross-match test. At least 50 surface antigens have been identified on RBCs, but the standard test for blood type considers only the three most likely to produce dangerous cross-reactions: A, B, and Rh. The test involves taking drops of blood and mixing them separately with solutions containing anti-A, anti-B, and anti-Rh (anti-D) antibodies. Any cross-reactions are then recorded. For example, if an individual's RBCs clump together when exposed to anti-A and to anti-B antibodies, the individual has Type AB blood. If no reactions occur after exposure, that person must have Type O blood. The presence or absence of the Rh surface antigen is also noted, and the individual is classified as Rh positive or Rh negative on that basis. Type O^+ is the most common blood type. Although the RBCs of Type O^+ individuals do not have surface antigens A or B, these cells do have the Rh antigen.

Standard blood-typing of both donor and recipient can be completed in a matter of minutes. However, in an emergency, there may not be time for preliminary testing. For example, a person with a severe gunshot wound may require 5 *liters* or more of blood before the damage can be repaired. Under these circumstances, Type O

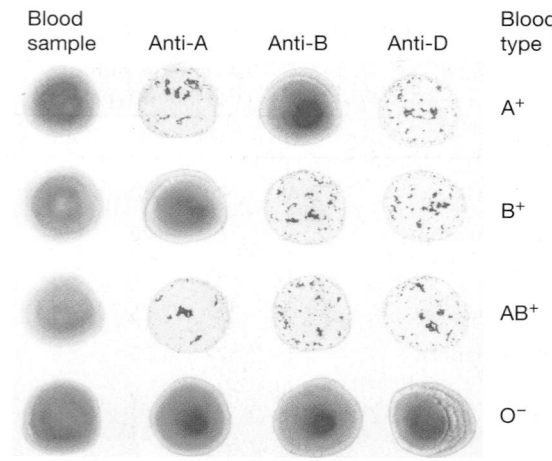

●**FIGURE 19–9**
Blood Type Testing. Test results for blood samples from four individuals. Drops are taken from the sample at the left and mixed with solutions containing antibodies to the surface antigens A, B, AB, and D (Rh). Clumping occurs when the sample contains the corresponding surface antigen(s). The individuals' blood types are shown at right.

blood (preferably O^-) will be administered. Because the donated RBCs lack both A and B surface antigens, the recipient's blood can have anti-A antibodies, anti-B antibodies, or both and still not cross-react with the donor's blood. (More than 100,000 units of Type O^- blood were transfused under emergency conditions during the war in Vietnam, with no deaths due to cross-reactions.) Because cross-reactions with Type O blood are so unlikely, Type O individuals are sometimes called *universal donors*. Type AB individuals were once called *universal recipients*, because they do not have anti-A or anti-B antibodies that would attack donated RBCs, and so can safely recieve blood of any type. However, now that blood supplies are adequate and cross-match testing is regularly performed, the term has largely been dropped. If the recipient's blood type is known to be AB, Type AB blood will be administered.

It is now possible to use enzymes to strip off the A or B surface antigens from RBCs and create Type O blood in the laboratory. The procedure is expensive and time consuming and has limited use in emergency treatment. In addition, because at least 48 other surface antigens are present, cross-reactions can still occur, even to Type O^- blood. As a result, whenever time and facilities permit, further testing is performed to ensure complete compatibility between donor blood and recipient blood. **Cross-match testing** involves exposing the donor's RBCs to a sample of the recipient's plasma under controlled conditions. This procedure reveals the presence of significant cross-reactions involving surface antigens other than A, B, or Rh. Another way to avoid compatibility problems is to replace lost blood with synthetic blood substitutes, which do not contain surface antigens that can trigger a cross-reaction. **AM** Transfusions and Synthetic Blood

Because blood groups are inherited, blood tests are also used as paternity tests and in crime detection. The blood collected cannot prove that a particular individual *is* a certain child's father or *is*

Hemolytic Disease of the Newborn

Genes controlling the presence or absence of any surface antigen in the membrane of a red blood cell are provided by both parents, so a child can have a blood type different from that of either parent. During pregnancy, when fetal and maternal circulatory systems are closely inter-

twined, the mother's antibodies may cross the placenta, attacking and destroying fetal RBCs. The resulting condition is called **hemolytic disease of the newborn** (**HDN**).

This disease has many forms, some so mild as to remain undetected. Those involving the Rh surface antigen are quite

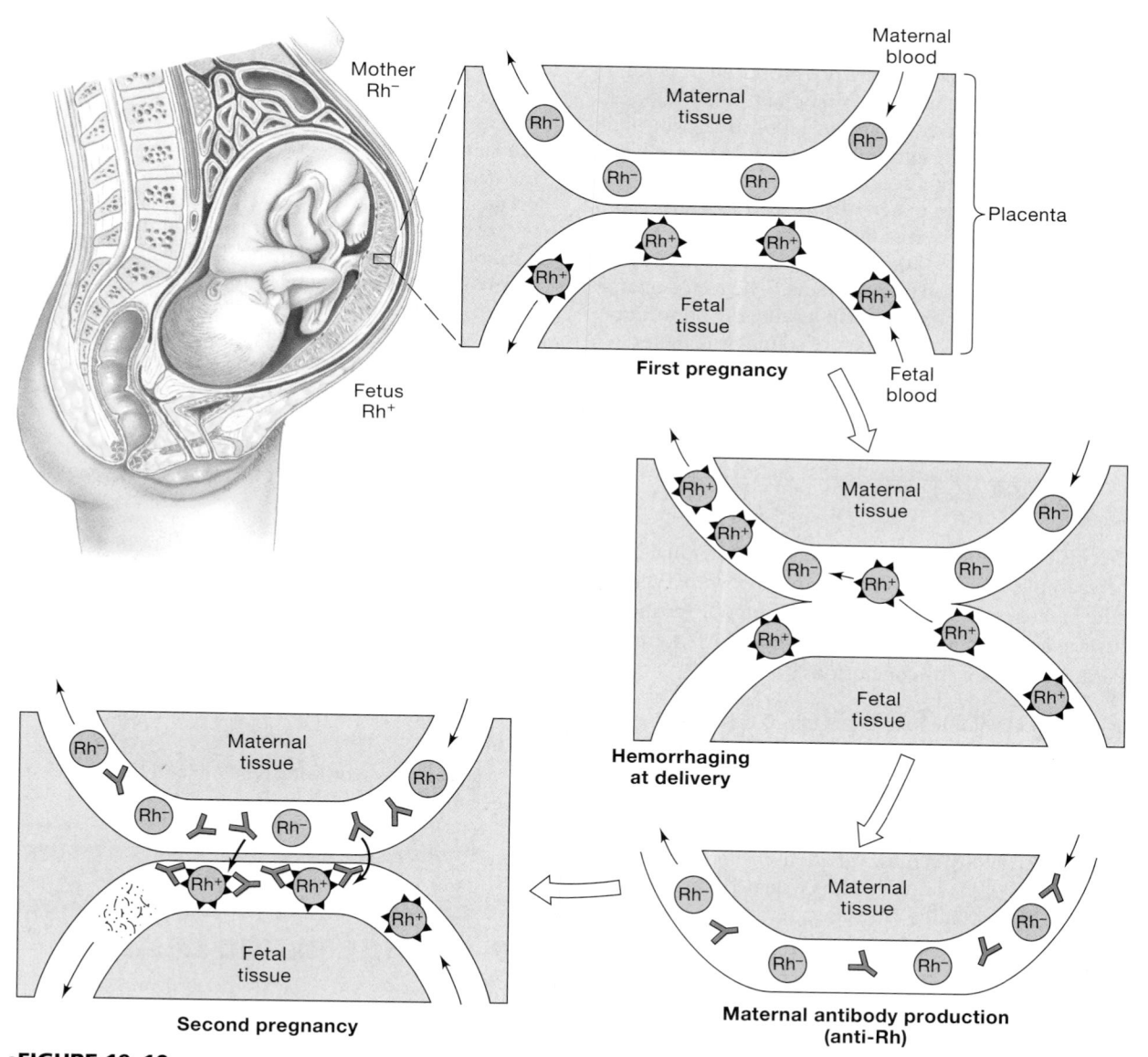

●**FIGURE 19–10**
Rh Factors and Pregnancy. When an Rh-negative woman has her first Rh-positive child, fetal and maternal blood mix at delivery when the placenta breaks down. The appearance of Rh-positive blood cells in the maternal bloodstream sensitizes the mother, stimulating the production of anti-Rh antibodies. If another pregnancy occurs with an Rh-positive fetus, maternal anti-Rh antibodies can cross the placenta and attack fetal blood cells, producing hemolytic disease of the newborn (HDN).

dangerous, because the anti-Rh antibodies are able to cross the placenta and enter the fetal bloodstream, whereas the anti-A and anti-B antibodies are not. An Rh-positive mother (who lacks anti-Rh antibodies) can carry an Rh-negative fetus without difficulty. Because maternal antibodies can cross the placenta, problems may appear when an Rh-negative woman carries an Rh-positive fetus. Sensitization generally occurs at delivery, when bleeding takes place at the placenta and uterus. This event, which mixes fetal and material blood, can stimulate the mother's immune system to produce anti-Rh antibodies. Within six months of delivery, roughly 20 percent of Rh-negative mothers who carried Rh-positive children have become sensitized.

Because the anti-Rh antibodies are not produced in significant amounts until after delivery, the first infant is not affected. (Some fetal RBCs cross into the maternal bloodstream during pregnancy, but generally not in numbers sufficient to stimulate antibody production.) But if a second pregnancy occurs involving an Rh-positive fetus, maternal anti-Rh antibodies produced after the first delivery cross the placenta and enter the fetal bloodstream (Figure 19–10●). These antibodies destroy fetal RBCs and produce a dangerous anemia. The fetal demand for blood cells increases, and they begin leaving the bone marrow and entering the bloodstream before completing their development. Because these immature RBCs are erythroblasts, HDN is also known as **erythroblastosis fetalis** (e-rith-rō-blas-TŌ-sis fē-TAL-is).

Without treatment, the fetus will probably die before delivery or shortly thereafter. A newborn with severe HDN is anemic, and the high concentration of circulating bilirubin produces jaundice. Because the maternal antibodies will remain active for one to two months after delivery, the entire blood volume of the infant may have to be replaced. Replacing the blood removes most of the maternal anti-Rh antibodies, as well as the affected RBCs, reducing complications and the chance of the infant's dying.

If the fetus is in danger of not surviving to full term, delivery may be induced after seven to eight months of development. In a severe case affecting a fetus at an earlier stage, one or more transfusions can be given while the fetus continues to develop in the uterus.

The maternal production of anti-Rh antibodies can be prevented by administering such antibodies (available under the name *RhoGam*) to the mother in the last three months of pregnancy and during and after delivery. (Anti-Rh antibodies are also given after a miscarriage or an abortion.) These antibodies will destroy any fetal RBCs that cross the placenta before they can stimulate an immune response in the mother. Sensitization therefore does not occur, so no anti-Rh antibodies are produced. This relatively simple procedure has almost entirely prevented HDN mortality caused by Rh incompatibilities.

guilty of a specific crime, but it can prove that the individual is *not* involved. For example, it is impossible for an adult with Type AB blood to be the parent of an infant with Type O blood. Testing for additional surface antigens, other than the standard ABO groups, can increase the accuracy of the conclusions.

Changes in blood volume have direct effects on other body systems. Clinicians use the terms **hypovolemic** (hī-pō-vō-LĒ-mik), **normovolemic** (nor-mō-vō-LĒ-mik), and **hypervolemic** (hī-per-vō-LĒ-mik) to refer to low, normal, and excessive blood volumes, respectively. The hypovolemic and hypervolemic conditions are potentially dangerous, because variations in blood volume affect other components of the cardiovascular system. For example, an abnormally large blood volume can place a severe stress on the heart, which must keep the extra fluid circulating through the lungs and throughout the body.

Hypovolemia is clinically much more common than hypervolemia. Short-term therapies to treat hypovolemia may include the use of a plasma expander (p. 655) or the infusion of blood, a procedure known as a **transfusion**. In a transfusion, blood components are provided to restore blood volume or to remedy a deficiency in blood composition. Whole blood, packed red blood cells, plasma, platelets, extracted proteins, or clotting factors, may be administered.

Two major complications from transfusions are (1) cross-reactions between plasma antibodies and donated red blood cells and (2) infection from bacterial or viral pathogens in the donated blood. Improved cross-match testing and screening for infected blood have helped reduce the frequency of transfusion-related problems. **AM** Transfusions and Synthetic Blood

✓ What are surface antigens on RBCs?

✓ Which blood type(s) can be transfused into a person with Type O blood?

✓ Why can't a person with Type A blood safely receive blood from a person with Type B blood?

Answers start on page Q-1

19–5 WHITE BLOOD CELLS

Objective

■ Categorize the various white blood cells on the basis of their structures and functions, and discuss the factors that regulate the production of each class.

Unlike red blood cells, white blood cells have nuclei and other organelles, but they lack hemoglobin. White blood cells (WBCs), or leukocytes, help defend the body against invasion by pathogens, and they remove toxins, wastes, and abnormal or damaged cells. Several types of WBCs can be differentiated from one another (and from RBCs) in a blood smear with the aid of either of two standard stains used in blood work: *Wright's stain* or *Geimsa stain*. Traditionally, WBCs have been divided into two groups on the basis of their appearance after such staining: (1) *granular leukocytes*, or *granulocytes* (with abundant stained granules)—the *neutrophils*, *eosinophils*, and *basophils*; and (2) *agranular leukocytes*, or *agranulocytes* (with few, if any, stained granules)—the *monocytes* and *lymphocytes*. This categorization is convenient but somewhat misleading, because the granules in "granular leukocytes" are secretory vesicles and lysosomes, and the "agranular leukocytes" contain vesicles and lysosomes as well; they are just smaller and difficult to see with the light microscope.

A typical microliter of blood contains 6000 to 9000 WBCs, compared with 4.2 to 6.3 million RBCs. Most of the WBCs in the body at any moment are in connective tissues proper or in organs of the lymphatic system. Circulating WBCs thus represent only a small fraction of the total WBC population.

☐ WBC CIRCULATION AND MOVEMENT

Unlike RBCs, WBCs circulate for only a short portion of their life span. White blood cells migrate through the loose and dense connective tissues of the body, using the bloodstream primarily to travel from one organ to another and for rapid transportation to areas of invasion or injury. As they travel along the miles of capillaries, WBCs can detect the chemical signs of damage to surrounding tissues. When problems are detected, these cells leave the bloodstream and enter the damaged area.

Circulating WBCs have four characteristics:

1. *They Are Capable of Amoeboid Movement. Amoeboid movement* is a gliding motion accomplished by the flow of cytoplasm into a slender cellular process extended in front of the cell. (The movement is so named because it is similar to that of an *amoeba*, a type of protozoan.) The mechanism is not fully understood, but it involves the continuous rearrangement of bonds between actin filaments in the cytoskeleton, and it requires calcium ions and ATP. This mobility allows WBCs to move along the walls of blood vessels and, when outside the bloodstream, through surrounding tissues.

2. *They Can Migrate Out of the Bloodstream.* When white blood cells in the bloodstream become activated, they contact and adhere to the vessel walls in a process called *margination*. After further interaction with the endothelial cells, the activated WBCs squeeze between adjacent endothelial cells and enter the surrounding tissue. This process is called *emigration*, or *diapedesis*.

3. *They Are Attracted to Specific Chemical Stimuli.* This characteristic, called **positive chemotaxis** (kē-mō-TAK-sis), guides WBCs to invading pathogens, damaged tissues, and other active WBCs.

4. *Neutrophils, Eosinophils, and Monocytes Are Capable of Phagocytosis.* These cells may engulf pathogens, cell debris, or other materials. Neutrophils and eosinophils are sometimes called *microphages*, to distinguish them from the larger macrophages in connective tissues. Macrophages are monocytes that have moved out of the bloodstream and have become actively phagocytic. ∞ pp. 123–4, 128

☐ TYPES OF WBCs

Neutrophils, eosinophils, basophils, and monocytes contribute to the body's *nonspecific defenses*. Such immune defenses are activated by a variety of stimuli, but they do not discriminate between one type of threat and another. Lymphocytes, in contrast, are responsible for *specific defenses:* the mounting of a counterattack against particular invading pathogens or foreign proteins on an individual basis. We will discuss the interactions among WBCs and the relationships between specific and nonspecific defenses in Chapter 22.

Neutrophils

Fifty to seventy percent of the circulating WBCs are **neutrophils** (NOO-trō-filz). This name reflects the fact that the granules of these WBCs are chemically neutral and thus are difficult to stain with either acidic or basic dyes. A mature neutrophil has a very dense, segmented nucleus that forms two to five lobes resembling beads on a string (Figure 19–11a●). This structure has given neutrophils another name: **polymorphonuclear** (pol-ē-mor-fō-NOO-klē-ar) **leukocytes** (*poly*, many + *morphe*, form), or *PMNs*. "Polymorphs," or "polys," as they are often called, are roughly 12 μm in diameter. Their cytoplasm is packed with pale granules containing lysosomal enzymes and bactericidal (bacteria-killing) compounds.

Neutrophils are highly mobile, and consequently are generally the first of the WBCs to arrive at the site of an injury. They are very active cells that specialize in attacking and digesting bacteria that have been "marked" with antibodies or with *complement proteins*—plasma proteins involved in tissue defenses. (We will discuss the complement system in Chapter 22.)

Upon encountering a bacterium, the neutrophil quickly engulfs it, and the metabolic rate of the neutrophil increases dramatically. This *respiratory burst* accompanies the production of highly reactive, destructive chemical agents, including *hydrogen peroxide* (H_2O_2) and *superoxide anions* (O_2^-), which can kill bacteria.

Meanwhile, the vesicle containing the engulfed pathogen fuses with lysosomes that contain digestive enzymes and small peptides called **defensins**. This process, which reduces the number of granules in the cytoplasm, is called **degranulation**. Defensins kill a variety of pathogens, including bacteria, fungi, and some viruses, by combining to form large channels in their cell membranes. The digestive enzymes then break down the bacterial remains. While actively engaged in attacking bacteria, a neutrophil releases prostaglandins and leukotrienes. ∞ p. 608 The prostaglandins increase capillary permeability in the affected region, thereby contributing to local inflammation and restricting the spread of injury and infection. Leukotrienes are hormones of the immune system that attract other phagocytes and help coordinate the immune response.

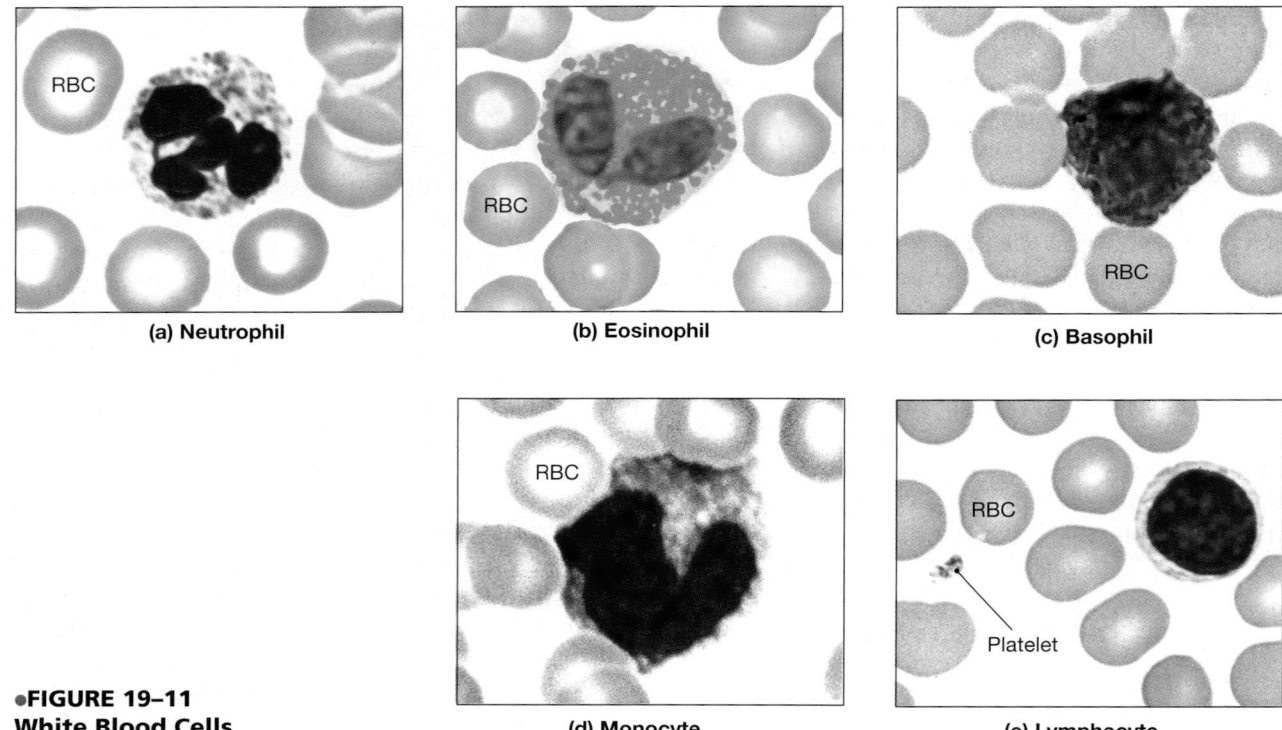

(a) Neutrophil

(b) Eosinophil

(c) Basophil

●FIGURE 19–11
White Blood Cells

(d) Monocyte

(e) Lymphocyte

Most neutrophils have a short life span, surviving in the bloodstream for only about 10 hours. When actively engulfing debris or pathogens, they may last 30 minutes or less. A neutrophil dies after engulfing one to two dozen bacteria, but its breakdown releases additional chemicals that attract other neutrophils to the site. A mixture of dead neutrophils, cellular debris, and other waste products form the *pus* associated with infected wounds.

Eosinophils

Eosinophils (ē-ō-SIN-ō-filz) were so named because their granules stain darkly with *eosin*, a red dye. The granules also stain with other acid dyes, so the name **acidophils** (a-SID-ō-filz) applies as well. Eosinophils, which generally represent 2–4 percent of the circulating WBCs, are similar in size to neutrophils. However, the combination of deep red granules and a bilobed (two-lobed) nucleus makes an eosinophil easy to identify (Figure 19–11b●).

Eosinophils attack objects that are coated with antibodies. They are phagocytic cells and will engulf antibody-marked bacteria, protozoa, or cellular debris. However, their primary mode of attack is the exocytosis of toxic compounds, including nitric oxide and cytotoxic enzymes, onto the surface of their targets. Eosinophils defend against large multicellular parasites, such as flukes or parasitic worms; these WBCs increase in number dramatically during a parasitic infection. **AM** The Nature of Pathogens

Because they are sensitive to circulating *allergens* (materials that trigger allergies), eosinophils increase in number during allergic reactions as well. Eosinophils are also attracted to sites of injury, where they release enzymes that reduce the degree of inflammation produced by mast cells and neutrophils, thus controlling the spread of inflammation to adjacent tissues.

Basophils

Basophils (BĀ-sō-filz) have numerous granules that stain darkly with basic dyes. In a standard blood smear, the inclusions are deep purple or blue (Figure 19–11c●). Measuring 8–10 μm in diameter, basophils are smaller than neutrophils or eosinophils. They are also relatively rare, accounting for less than 1 percent of the circulating WBC population.

Basophils migrate to injury sites and cross the capillary endothelium to accumulate in the damaged tissues, where they discharge their granules into the interstitial fluids. The granules contain *histamine*, which dilates blood vessels, and *heparin*, a compound that prevents blood from clotting. Stimulated basophils release these chemicals into the interstitial fluids, and their arrival enhances the local inflammation initiated by mast cells. ∞ pp. 124, 139 Although the same compounds are released by mast cells in damaged connective tissues, mast cells and basophils are distinct populations with separate origins. Other chemicals released by stimulated basophils attract eosinophils and other basophils to the area.

Monocytes

Monocytes (MON-ō-sīts) in blood are spherical cells that may exceed 15 μm in diameter, nearly twice the diameter of a typical red blood cell. When flattened in a blood smear, they look even larger, so monocytes are relatively easy to identify

(Figure 19–11d●). The nucleus is large and tends to be oval or kidney bean–shaped rather than lobed. Monocytes normally account for 2–8 percent of the circulating WBCs.

An individual monocyte uses the bloodstream as a highway, remaining in circulation for only about 24 hours before entering peripheral tissues to become a tissue macrophage. Macrophages are aggressive phagocytes, often attempting to engulf items as large as or larger than themselves. While phagocytically active, they release chemicals that attract and stimulate neutrophils, monocytes, and other phagocytic cells. Active macrophages also secrete substances that lure fibroblasts into the region. The fibroblasts then begin producing scar tissue, which will wall off the injured area.

Lymphocytes

Typical **lymphocytes** (LIM-fō-sīts) are slightly larger than RBCs and lack abundant, deeply stained granules. In fact, when you see a lymphocyte in a blood smear, you generally see just a thin halo of cytoplasm around a relatively large, round nucleus (Figure 19–11e●).

Lymphocytes account for 20–30 percent of the WBC population of blood. Lymphocytes continuously migrate from the bloodstream, through peripheral tissues, and back to the bloodstream. Circulating lymphocytes represent only a minute fraction of all lymphocytes, for at any moment most of your body's lymphocytes are in other connective tissues and in organs of the lymphatic system.

The circulating blood contains three functional classes of lymphocytes, which cannot be distinguished with a light microscope:

1. *T Cells.* **T cells** are responsible for *cell-mediated immunity*, a defense mechanism against invading foreign cells and tissues, and for the coordination of the immune response. T cells either enter peripheral tissues and attack foreign cells directly or control the activities of other lymphocytes.

2. *B Cells.* **B cells** are responsible for *humoral immunity*, a defense mechanism that involves the production and distribution of antibodies, which in turn attack foreign antigens throughout the body. Activated B cells differentiate into **plasma cells**, which are specialized to synthesize and secrete antibodies. Whereas the T cells responsible for cellular immunity must migrate to their targets, the antibodies produced by plasma cells in one location can destroy antigens almost anywhere in the body.

3. *NK Cells.* **Natural killer (NK) cells** are responsible for *immune surveillance*—the detection and subsequent destruction of abnormal tissue cells. These cells, sometimes known as *large granular lymphocytes*, are important in preventing cancer.

◻ THE DIFFERENTIAL COUNT AND CHANGES IN WBC PROFILES

A variety of disorders, including pathogenic infection, inflammation, and allergic reactions, cause characteristic changes in circulating populations of WBCs. By examining a stained blood smear, we can obtain a **differential count** of the WBC popula-

tion. The values reported indicate the number of each type of cell in a sample of 100 WBCs.

The normal range for each type of WBC is indicated in Table 19–3. The term **leukopenia** (loo-kō-PĒ-ne-uh; *penia*, poverty) indicates inadequate numbers of WBCs. **Leukocytosis** (loo-kō-sī-TO-sis) refers to excessive numbers of WBCs. A modest leukocytosis is normal during an infection. Extreme leukocytosis ($100,000/\mu l$ or more) generally indicates the presence of some form of **leukemia** (loo-KĒ-mē-uh). There are many types of leukemias. Treatment helps in some cases; unless treated, all are fatal. The endings *-penia* and *-osis* can also indicate low or high numbers of specific types of WBCs. For example, *lymphopenia* means too few lymphocytes, and *lymphocytosis* means too many. **AM** The Leukemias

◻ WBC PRODUCTION

Stem cells responsible for the production of WBCs originate in the bone marrow, with the divisions of hemocytoblasts. As we noted earlier, hemocytoblast divisions produce myeloid stem cells and lymphoid stem cells. Myeloid stem cell division creates **progenitor cells**, which give rise to all the formed elements except lymphocytes. One type of progenitor cell produces daughter cells that mature into RBCs; a second type produces cells that manufacture platelets. Neutrophils, eosinophils, basophils, and monocytes develop from daughter cells produced by a third type of progenitor cell.

All WBCs except monocytes complete their development in the bone marrow. Monocytes begin their differentiation in the bone marrow, enter the bloodstream, and complete development when they become free macrophages in peripheral tissues. Each of the other types of cell goes through a characteristic series of maturational stages, proceeding from *blast cells* to *myelocytes* to *band cells* before becoming mature WBCs. For example, a cell differentiating into a neutrophil goes from a myeloblast to a *neutrophilic myelocyte* and then becomes a *neutrophilic band cell*. Some band cells enter the bloodstream before completing their maturation; normally, 3–5 percent of all circulating WBCs are band cells. Figure 19–12● summarizes the relationships among the various WBC populations and compares the formation of WBCs and RBCs.

Many of the lymphoid stem cells responsible for the production of lymphocytes migrate from the bone marrow to peripheral **lymphoid tissues**, including the thymus, spleen, and lymph nodes. As a result, lymphocytes are produced in these organs as well as in the bone marrow. The process of lymphocyte production is called **lymphopoiesis**.

Regulation of WBC Production

Factors that regulate lymphocyte maturation are as yet incompletely understood. Prior to maturity, hormones produced by the thymus promote the differentiation and maintenance of T cell populations. The importance of the thymus in adulthood,

SUMMARY TABLE 19–3 FORMED ELEMENTS OF THE BLOOD

Cell	Abundance (average number per μl)	Appearance in a Stained Blood Smear	Functions	Remarks
RED BLOOD CELLS	5.2 million (range: 4.4–6.0 million)	Flattened, circular cell; no nucleus, mitochondria, or ribosomes; red	Transport oxygen from lungs to tissues and carbon dioxide from tissues to lungs	Remain in bloodstream; 120-day life expectancy; amino acids and iron recycled; produced in bone marrow
WHITE BLOOD CELLS	7000 (range: 6000–9000)			
Neutrophils	4150 (range: 1800–7300) Differential count: 50–70%	Round cell; nucleus lobed and may resemble a string of beads; cytoplasm contains large, pale inclusions	Phagocytic: Engulf pathogens or debris in tissues, release cytotoxic enzymes and chemicals	Move into tissues after several hours; may survive minutes to days, depending on tissue activity; produced in bone marrow
Eosinophils	165 (range: 0–700) Differential count: 2–4%	Round cell; nucleus generally in two lobes; cytoplasm contains large granules that generally stain bright red	Phagocytic: Engulf antibody-labeled materials, release cytotoxic enzymes, reduce inflammation	Move into tissues after several hours; survive minutes to days, depending on tissue activity; produced in bone marrow
Basophils	44 (range: 0–150) Differential count: <1%	Round cell; nucleus generally cannot be seen through dense, blue-stained granules in cytoplasm	Enter damaged tissues and release histamine and other chemicals that promote inflammation	Survival time unknown; assist mast cells of tissues in producing inflammation; produced in bone marrow
Monocytes	456 (range: 200–950) Differential count: 2–8%	Very large cell; kidney bean–shaped nucleus; abundant pale cytoplasm	Enter tissues to become macrophages; engulf pathogens or debris	Move into tissues after 1–2 days; survive for months or longer; produced primarily in bone marrow
Lymphocytes	2185 (range: 1500–4000) Differential count: 20–30%	Generally round cell, slightly larger than RBC; round nucleus; very little cytoplasm	Cells of lymphatic system, providing defense against specific pathogens or toxins	Survive for months to decades; circulate from blood to tissues and back; produced in bone marrow and lymphoid tissues
PLATELETS	350,000 (range: 150,000–500,000)	Round to spindle-shaped cytoplasmic fragment; contain enzymes and proenzymes; no nucleus	Hemostasis: Clump together and stick to vessel wall (platelet phase); activate intrinsic pathway of coagulation phase	Remain in bloodstream or in vascular organs; remain intact for 7–12 days; produced by megakaryocytes in bone marrow

especially with respect to aging, remains controversial. In adults, the production of B and T lymphocytes is regulated primarily by exposure to antigens (foreign proteins, cells, or toxins). When antigens appear, lymphocyte production escalates. We shall describe the control mechanisms in Chapter 22.

Several hormones are involved in the regulation of other WBC populations. Figure 19–12● diagrams the targets of these hormones, called **colony-stimulating factors** (**CSFs**), and the

target of erythropoietin (EPO), discussed on page 661. Four CSFs have been identified, each stimulating the formation of WBCs or both WBCs and RBCs. The abbreviation for each factor indicates its target:

1. **M-CSF** stimulates the production of monocytes.
2. **G-CSF** stimulates the production of granulocytes (neutrophils, eosinophils, and basophils).

●**FIGURE 19–12**
The Origins and Differentiation of Formed Elements.
Hemocytoblast divisions give rise to myeloid stem cells or lymphoid stem cells. Lymphoid stem cells produce the various lymphocytes. Myeloid stem cells produce progenitor cells that divide to produce the other classes of formed elements. The targets of EPO and the four colony-stimulating factors (CSFs) are indicated.

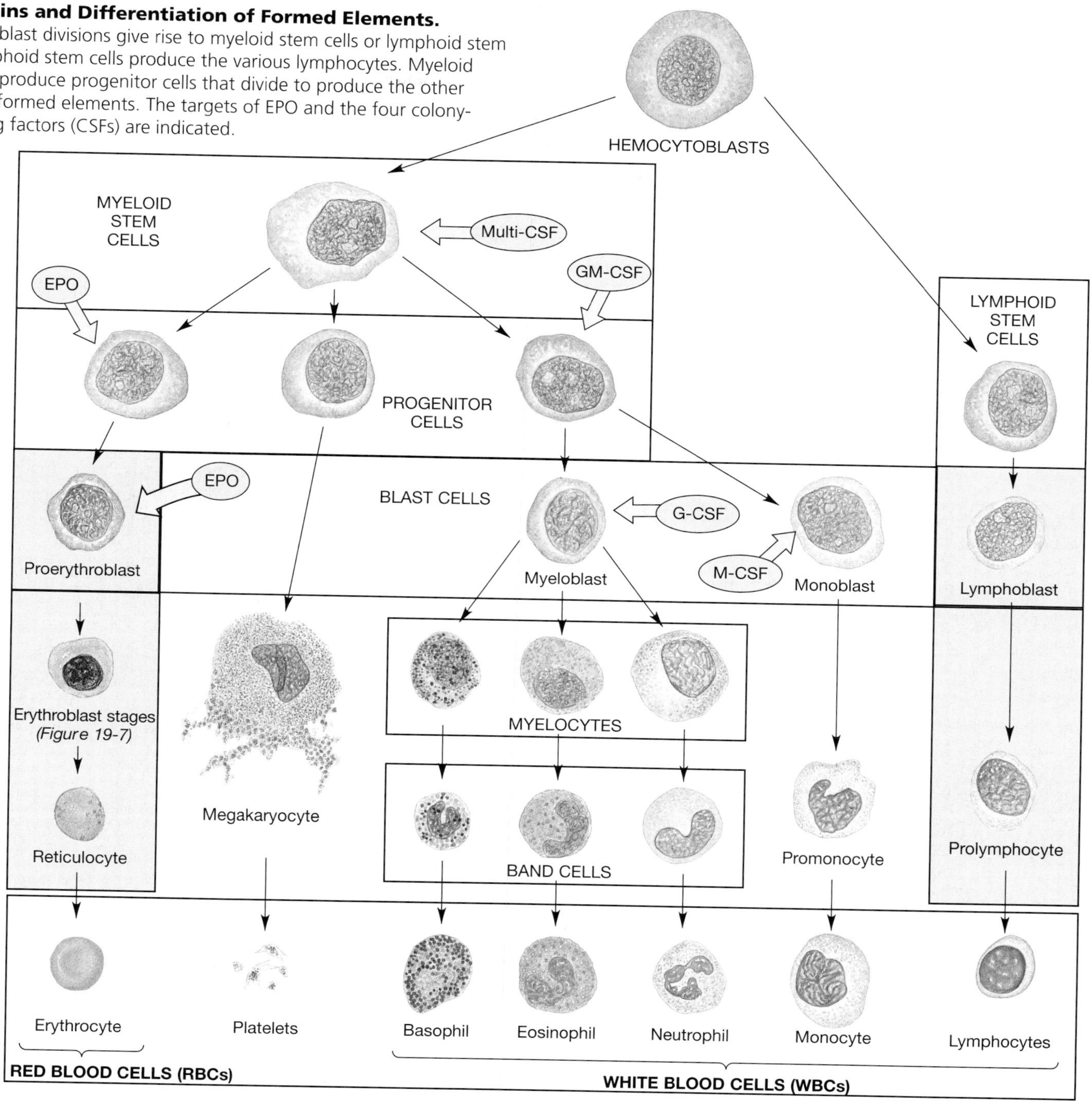

3. **GM-CSF** stimulates the production of both granulocytes and monocytes.

4. **Multi-CSF** accelerates the production of granulocytes, monocytes, platelets, and RBCs.

Chemical communication between lymphocytes and other WBCs assists in the coordination of the immune response. For example, active macrophages release chemicals that make lymphocytes more sensitive to antigens and that accelerate the devel-opment of specific immunity. In turn, active lymphocytes release multi-CSF and GM-CSF, reinforcing nonspecific defenses. Immune system hormones are currently being studied intensively because of their potential clinical importance. The molecular structures of many of the stimulating factors have been deter-mined, and several can be produced by genetic engineering. The U.S. Food and Drug Administration approved the administra-tion of synthesized forms of EPO, G-CSF, and GM-CSF to stimu-late the production of specific blood cell lines. For instance, a genetically engineered form of G-CSF, sold under the name *filgrastim* (*Neupogen*), is used to stimulate the production of neutrophils in patients undergoing cancer chemotherapy.

✓ Which type of white blood cell would you find in the greatest numbers in an infected cut?

✓ Which type of cell would you find in elevated numbers in a person who is producing large amounts of circulating antibodies to combat a virus?

✓ How do basophils respond during inflammation?

Answers start on page Q-1

19–6 PLATELETS

Objective

■ Describe the structure, function, and production of platelets.

Platelets (PLĀT-lets) are flattened discs, round when viewed from above and spindle shaped when seen in section or in a blood smear (Figure 19–11e●). They average about 4 μm in diameter and are roughly 1 μm thick. Platelets in nonmammalian vertebrates are nucleated cells called **thrombocytes** (THROM-bō-sīts; *thrombos*, clot). Because in humans they are cell fragments rather than individual cells, the term *platelet* is preferred when referring to our blood. Platelets are a major participant in a vascular *clotting system* that also includes plasma proteins and the cells and tissues of the blood vessels.

Platelets are continuously replaced. Each platelet circulates for 9–12 days before being removed by phagocytes, mainly in the spleen. Each microliter of circulating blood contains 150,000–500,000 platelets; 350,000/μl is the average concentration. Roughly one-third of the platelets in the body at any moment are held in the spleen and other vascular organs, rather than in the bloodstream. These reserves are mobilized during a circulatory crisis, such as severe bleeding.

An abnormally low platelet count (80,000/μl or less) is known as **thrombocytopenia** (throm-bō-sī-tō-PĒ-nē-uh). Thrombocytopenia generally indicates excessive platelet destruction or inadequate platelet production. Symptoms include bleeding along the digestive tract, within the skin, and occasionally inside the CNS. In **thrombocytosis** (throm-bō-sī-TŌ-sis), platelet counts can exceed 1,000,000/μl. Thrombocytosis generally results from accelerated platelet formation in response to infection, inflammation, or cancer.

□ PLATELET FUNCTIONS

The functions of platelets include:

• **The Transport of Chemicals Important to the Clotting Process.** By releasing enzymes and other factors at the appropriate times, platelets help initiate and control the clotting process.

• **The Formation of a Temporary Patch in the Walls of Damaged Blood Vessels.** Platelets clump together at an injury site, forming a *platelet plug*, which can slow the rate of blood loss while clotting occurs.

• **Active Contraction After Clot Formation Has Occurred.** Platelets contain filaments of actin and myosin. After a blood clot has formed, the contraction of platelet filaments shrinks the clot and reduces the size of the break in the vessel wall.

□ PLATELET PRODUCTION

Platelet production, or **thrombocytopoiesis**, occurs in the bone marrow. Normal bone marrow contains a number of **megakaryocytes** (meg-a-KAR-ē-ō-sīts; *mega-*, big + *karyon*, nucleus + *-cyte*, cell), enormous cells (up to 160 μm in diameter) with large nuclei. During their development and growth, megakaryocytes manufacture structural proteins, enzymes, and membranes. They then begin shedding cytoplasm in small membrane-enclosed packets. These packets are the platelets that enter the bloodstream. A mature megakaryocyte gradually loses all of its cytoplasm, producing about 4000 platelets before the nucleus is engulfed by phagocytes and broken down for recycling.

The rate of megakaryocyte activity and platelet formation is stimulated by (1) *thrombopoietin* (TPO), or *thrombocyte-stimulating factor*, a peptide hormone produced in the kidneys and perhaps other sites as well, which accelerates platelet formation and stimulates the production of megakaryocytes; (2) *interleukin-6* (IL-6), a hormone of the immune system, which stimulates platelet formation; and (3) multi CSF, which stimulates platelet production by promoting the formation and growth of megakaryocytes.

19–7 HEMOSTASIS

Objective

■ Discuss mechanisms that control blood loss after an injury, and describe the reaction sequences responsible for blood clotting.

The process of **hemostasis** (*haima*, blood + *stasis*, halt), the cessation of bleeding, prevents the loss of blood through the walls of damaged vessels. At the same time, it establishes a framework for tissue repairs. Hemostasis consists of three phases: (1) the *vascular phase*, (2) the *platelet phase*, and (3) the *coagulation phase*. In reality, however, the entire process is more like a chain reaction than separate, identifiable phases. Each step modifies the events already under way. As a result, it is easier to say when a particular step begins than when it ends.

□ THE VASCULAR PHASE

Cutting the wall of a blood vessel triggers a contraction in the smooth muscle fibers of the vessel wall (Figure 19–13●). This

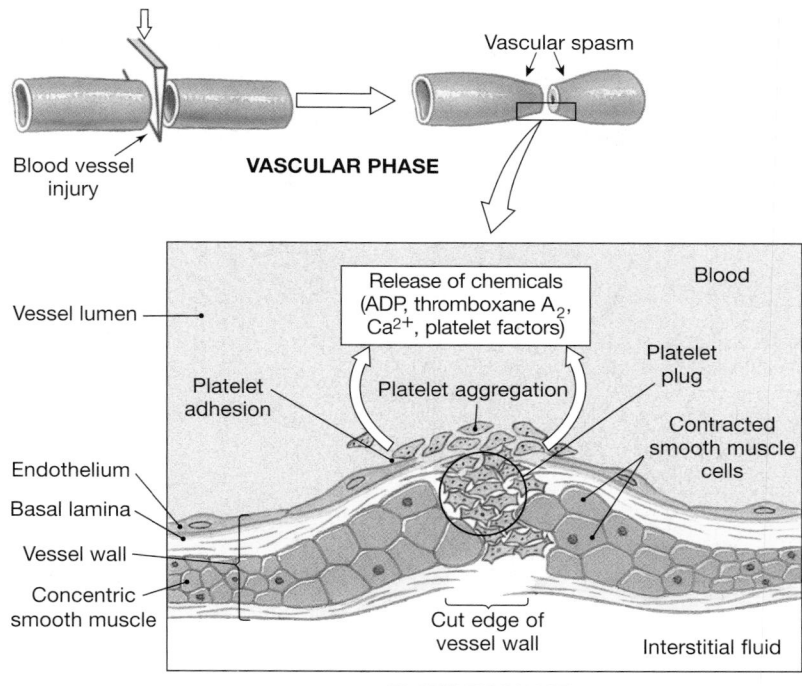

•**FIGURE 19–13**
The Vascular and Platelet Phases of Hemostasis

local contraction of the vessel is a **vascular spasm**, which decreases the diameter of the vessel at the site of injury. Such a constriction can slow or even stop the loss of blood through the wall of a small vessel. The vascular spasm lasts about 30 minutes, a period called the **vascular phase** of hemostasis.

During the vascular phase, changes occur in the endothelium of the vessel at the injury site:

• **The Endothelial Cells Contract and Expose the Underlying Basal Lamina to the Bloodstream.**

• **The Endothelial Cells Begin Releasing Chemical Factors and Local Hormones.** We shall discuss several of these factors, including *ADP*, *tissue factor*, and *prostacyclin*, in later sections. Endothelial cells also release **endothelins**, peptide hormones that (1) stimulate smooth muscle contraction and promote vascular spasms and (2) stimulate the division of endothelial cells, smooth muscle cells, and fibroblasts to accelerate the repair process.

• **The Endothelial Cell Membranes Become "Sticky."** In small capillaries, endothelial cells on opposite sides of the vessel may stick together and close off the passageway.

□ THE PLATELET PHASE

Platelets now begin to attach to sticky endothelial surfaces, to the basal lamina, and to exposed collagen fibers. This attachment marks the start of the **platelet phase** of hemostasis (Figure 19–13•). The attachment of platelets to exposed surfaces is called **platelet adhesion**. As more and more platelets arrive, they begin

sticking to one another as well. This process, called **platelet aggregation**, forms a **platelet plug** that may close the break in the vessel wall if the damage is not severe or the vessel is relatively small. Platelet aggregation begins within 15 seconds after an injury occurs.

As they arrive at the injury site, platelets become activated. The first sign of activation is that they change shape, becoming more spherical and developing cytoplasmic processes that extend toward adjacent platelets. At this time, the platelets begin releasing a wide variety of compounds, including (1) *adenosine diphosphate* (*ADP*), which stimulates platelet aggregation and secretion; (2) *thromboxane A_2* and *serotonin*, which stimulate vascular spasms; (3) *clotting factors*, proteins that play a role in blood clotting; (4) *platelet-derived growth factor* (*PDGF*), a peptide that promotes vessel repair; and (5) calcium ions, which are required for platelet aggregation and by several steps in the clotting process.

The platelet phase proceeds rapidly, because the ADP, thromboxane, and calcium ions that each arriving platelet releases stimulate further aggregation. This positive feedback loop ultimately produces a platelet plug that will be reinforced as clotting occurs. However, platelet aggregation must be controlled and restricted to the site of injury. Several key factors limit the growth of the platelet plug: (1) **prostacyclin**, a prostaglandin that inhibits platelet aggregation and is released by endothelial cells; (2) inhibitory compounds released by white blood cells entering the area; (3) circulating plasma enzymes that break down ADP near the plug; (4) compounds that, when abundant, inhibit plug formation (for example, serotonin at high concentrations will block the action of ADP); and (5) the development of a blood clot, which reinforces the platelet plug, but separates it from the general circulation.

□ THE COAGULATION PHASE

The vascular and platelet phases begin within a few seconds after the injury. The **coagulation** (cō-ag-ū-LĀ-shun) **phase** does not start until 30 seconds or more after the vessel has been damaged. **Coagulation**, or *blood clotting*, involves a complex sequence of steps leading to the conversion of circulating fibrinogen into the insoluble protein fibrin. As the fibrin network grows, it covers the surface of the platelet plug. Passing blood cells and additional platelets are trapped in the fibrous tangle, forming a **blood clot**, which effectively seals off the damaged portion of the vessel. Figure 19–14• shows the formation and structure of a blood clot.

Clotting Factors

Normal blood clotting cannot occur unless the plasma contains the necessary **clotting factors**, or **procoagulants**. Important clotting factors include Ca^{2+} and 11 different proteins. Many of the proteins are **proenzymes**, which, when converted to active enzymes, direct essential reactions in the clotting response.

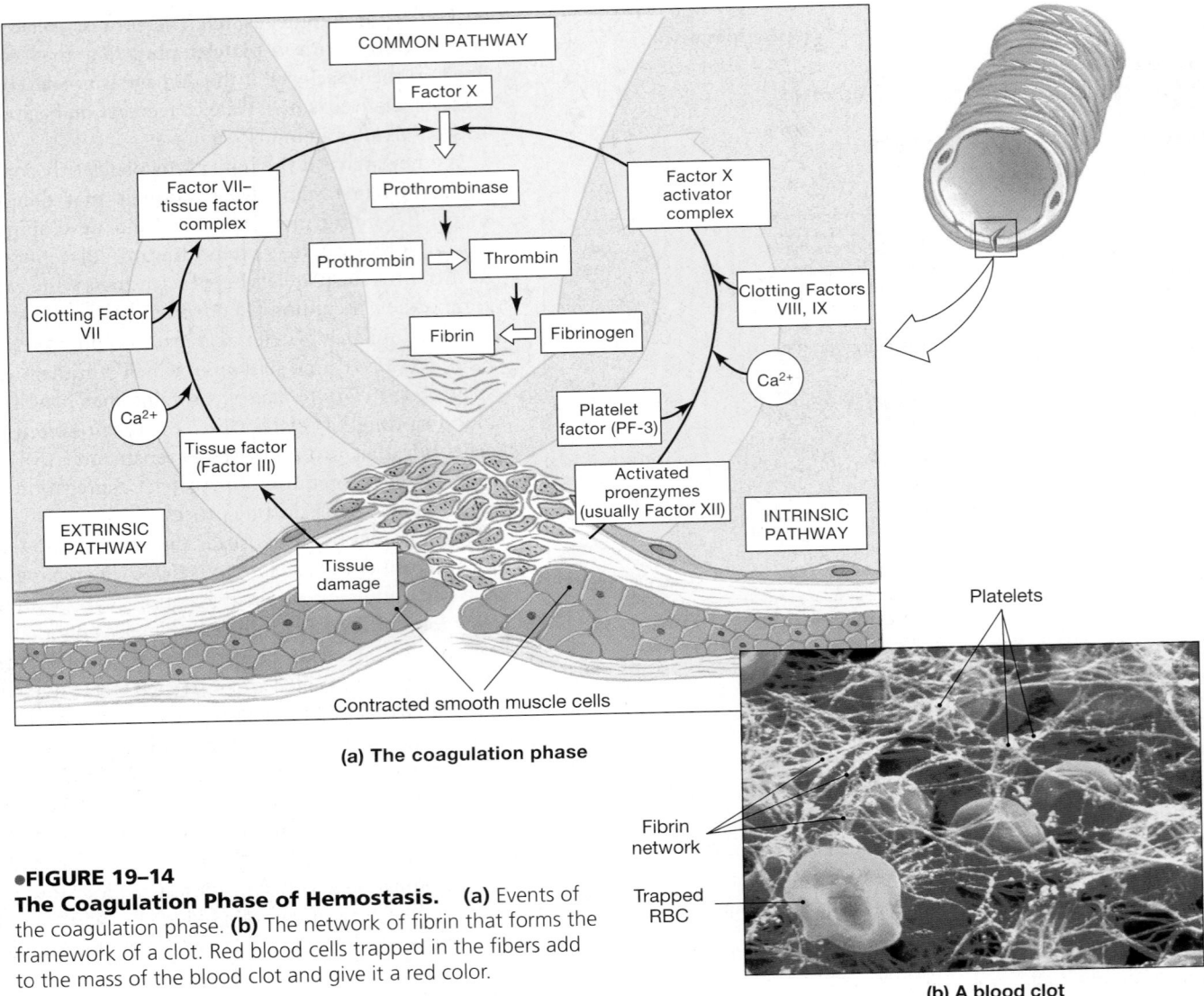

(a) The coagulation phase

Platelets

Fibrin network

Trapped RBC

(b) A blood clot

●FIGURE 19–14
The Coagulation Phase of Hemostasis. (a) Events of the coagulation phase. (b) The network of fibrin that forms the framework of a clot. Red blood cells trapped in the fibers add to the mass of the blood clot and give it a red color.

For reference, specific clotting factors are identified in Table 19–4 (p. 675). Many are identified by Roman numerals; Ca^{2+}, for example, is also known as clotting Factor IV. All but three of the clotting factors (Factors III, IV, and VIII) are synthesized and released by the liver, and all but two (Factors III and VIII) are always present in the bloodstream. Activated platelets release five clotting factors (Factors III, IV, V, VIII, and XIII) during the platelet phase. During the coagulation phase, enzymes and proenzymes interact. The activation of one proenzyme commonly creates an enzyme that activates a second proenzyme, and so on in a chain reaction, or *cascade*.

Figure 19–14a● surveys the cascades involved in the *extrinsic*, *intrinsic*, and *common pathways*. The extrinsic pathway begins outside the bloodstream, in the vessel wall; the intrinsic pathway begins inside the bloodstream, with the activation of a circulating proenzyme. These two pathways converge at the common pathway. **AM** The Clotting System: A Closer Look

The Extrinsic Pathway

The **extrinsic pathway** begins with the release of **Factor III**, also known as **tissue factor** (**TF**), by damaged endothelial cells or peripheral tissues. The greater the damage, the more tissue factor is released and the faster clotting occurs. Tissue factor then combines with Ca^{2+} and another clotting factor (Factor VII) to form an enzyme complex capable of activating Factor X, the first step in the common pathway.

The Intrinsic Pathway

The **intrinsic pathway** begins with the activation of proenzymes (usually Factor XII) exposed to collagen fibers at the injury site (or a glass surface on a slide or collection tube). This pathway proceeds with the assistance of **PF-3**, a platelet factor released by aggregating platelets. Platelets also release a variety of other factors that accelerate the reactions of the intrinsic pathway. After a

TABLE 19-4 CLOTTING FACTORS

Factor	Structure	Name	Source	Concentration in Plasma (μg/ml)	Pathway
I	Protein	Fibrinogen	Liver	2500–3500	Common
II	Protein	Prothrombin	Liver, requires vitamin K	100	Common
III	Lipoprotein	Tissue factor (TF)	Damaged tissue, activated platelets	0	Extrinsic
IV	Ion	Calcium ions	Bone, diet, platelets	100	Entire process
V	Protein	Proaccelerin	Liver, platelets	10	Extrinsic and intrinsic
VI	(No longer used)				
VII	Protein	Proconvertin	Liver, requires vitamin K	0.5	Extrinsic
VIII	Protein	Antihemophilic factor (AHF)	Platelets, endothelial cells	15	Intrinsic
IX	Protein	Plasma thromboplastin factor	Liver, requires vitamin K	3	Intrinsic
X	Protein	Stuart–Power factor	Liver, requires vitamin K	10	Extrinsic and intrinsic
XI	Protein	Plasma thromboplastin antecedent (PTA)	Liver	<5	Intrinsic
XII	Protein	Hageman factor	Liver	<5	Intrinsic; also activates plasmin
XIII	Protein	Fibrin-stabilizing factor (FSF)	Liver, platelets	20	Stabilizes fibrin, slows fibrinolysis

series of linked reactions, activated Factors VIII and IX combine to form an enzyme complex capable of activating Factor X.

The Common Pathway

The **common pathway** begins when enzymes from either the extrinsic or intrinsic pathway activate Factor X, forming the enzyme **prothrombinase**. Prothrombinase converts the proenzyme prothrombin into the enzyme **thrombin** (THROM-bin). Thrombin then completes the clotting process by converting fibrinogen, a plasma protein, to insoluble strands of fibrin.

Interactions among the Pathways

When a blood vessel is damaged, both the extrinsic and the intrinsic pathways respond. The extrinsic pathway is shorter and faster than the intrinsic pathway, and it is usually the first to initiate clotting. In essence, the extrinsic pathway produces a small amount of thrombin very quickly. This quick patch is reinforced by the intrinsic pathway, which produces more thrombin, but somewhat later.

The time required to complete clot formation varies with the site and the nature of the injury. In tests of the clotting system, blood held in fine glass tubes normally clots in 8–18 minutes (the *coagulation time*), and a small puncture wound typically stops bleeding in 1–4 minutes (the *bleeding time*).

Feedback Control of Blood Clotting

Thrombin generated in the common pathway stimulates blood clotting by (1) stimulating the formation of tissue factor and (2) stimulating the release of PF-3 by platelets. Thus, the activity of the common pathway stimulates both the intrinsic and extrinsic pathways. This positive feedback loop accelerates the clotting process, and speed can be very important in reducing blood loss after a severe injury.

Blood clotting is restricted by factors that either deactivate or remove clotting factors and other stimulatory agents from the blood. Examples include the following:

- Normal plasma contains several **anticoagulants**—enzymes that inhibit clotting. One, **antithrombin-III**, inhibits several clotting factors, including thrombin.

- **Heparin**, a compound released by basophils and mast cells, is a cofactor that accelerates the activation of antithrombin-III. Heparin is used clinically to impede or prevent clotting.

- **Thrombomodulin** is released by endothelial cells. This protein binds to thrombin and converts it to an enzyme that activates protein C. **Protein C** is a plasma protein that inactivates several clotting factors and stimulates the formation of *plasmin*, an enzyme that gradually breaks down fibrin strands.

- Prostacyclin released during the platelet phase inhibits platelet aggregation and opposes the stimulatory action of thrombin, ADP, and other factors.

- Other plasma proteins with anticoagulant properties include *alpha-2-macroglobulin*, which inhibits thrombin, and C_1 *inactivator*, which inhibits several clotting factors involved in the intrinsic pathway.

Abnormal Hemostasis

The clotting process involves a complex chain of events, and a disorder that affects any individual clotting factor can disrupt the entire process. As a result, many clinical conditions involve the clotting system, and it is often necessary to control or manipulate the clotting response. **AM** Testing the Clotting System

Excessive or Abnormal Blood Clotting

If the clotting response is inadequately controlled, blood clots will begin to form in the bloodstream rather than at the site of an injury. These blood clots do not stick to the wall of the vessel, but continue to drift around until either plasmin digests them or they become stuck in a small blood vessel. A drifting blood clot is a type of **embolus** (EM-bō-lus; *embolos*, plug), an abnormal mass within the bloodstream. An embolus that becomes stuck in a blood vessel blocks circulation to the area downstream, killing the affected tissues. The blockage is called an **embolism**, and the tissue damage caused by the circulatory interruption is an **infarct**. Infarcts at the brain are known as *strokes*; infarcts at the heart are called *myocardial infarctions*, or *heart attacks*.

An embolus in the arterial system can get stuck in capillaries in the brain, causing a stroke. An embolus in the venous system will probably become lodged in one of the capillaries of the lungs, causing a *pulmonary embolism*.

A **thrombus** (*thrombos*, clot), or blood clot attached to a vessel wall, begins to form when platelets stick to the wall of an intact blood vessel. Often, the platelets are attracted to areas called *plaques*, where endothelial and smooth muscle cells contain large quantities of lipids. (The mechanism of plaque formation will be discussed in Chapter 21.) The thrombus gradually enlarges, projecting into the lumen of the vessel and reducing its diameter. Eventually, the vessel may be completely blocked, or a large chunk of the clot may break off, creating an equally dangerous embolus. **AM** Disseminated Intravascular Coagulation

To treat these circulatory blockages, clinicians may attempt surgery to remove the obstruction or may use enzymes to attack blood clots and prevent further clot formation. Important anticoagulant drugs include:

- Heparin, which activates antithrombin-III.

- **Coumadin** (COO-ma-din), or *warfarin* (WAR-fa-rin), and **dicumarol** (dī-KOO-ma-rol), which depress the synthesis of several clotting factors by blocking the action of vitamin K.

- Recombinant DNA–synthesized **tissue plasminogen activator (t-PA)**, which stimulates plasmin formation.

- **Streptokinase** (strep-tō-KĪ-nās) and **urokinase** (ū-rō-KĪ-nās), enzymes that convert plasminogen to plasmin.

- Aspirin, which inactivates platelet enzymes involved with the production of thromboxanes and prostaglandins and inhibits the production of prostacyclin by endothelial cells. Daily ingestion of small quantities of aspirin reduces the sensitivity of the clotting process. This method has been proven to be effective in preventing heart attacks in people with significant heart disease.

It is sometimes necessary to control clotting within a blood sample to avoid changes in plasma composition. Blood samples can be stabilized temporarily by the addition of heparin or *EDTA* (*ethylenediaminetetroacetic acid*). EDTA removes Ca^{2+} from plasma, effectively preventing clotting. In units of whole blood held for extended periods in a blood bank, *citratephosphate dextrose* (CPD) is typically added. Like EDTA, CPD ties up plasma Ca^{2+}.

Inadequate Blood Clotting

Hemophilia (hē-mō-FĒL-ē-uh) is one of many inherited disorders characterized by the inadequate production of clotting factors. The condition affects about 1 in 10,000 people, 80–90 percent of whom are males. In hemophilia, the production of a single clotting factor (most commonly, Factor VIII) is inadequate; the severity of the condition depends on the degree of underproduction. In severe cases, extensive bleeding accompanies the slightest mechanical stress; hemorrhages occur spontaneously at joints and around muscles.

In many cases, transfusions of clotting factors can reduce or control the symptoms of hemophilia, but plasma samples from many individuals must be pooled (combined) to obtain adequate amounts of clotting factors. This procedure makes the treatment very expensive and increases the risk of blood-borne infections such as hepatitis or AIDS. Gene-splicing techniques have been used to manufacture clotting Factor VIII, an essential component of the intrinsic clotting pathway. As methods are developed to synthesize other clotting factors, treatment of the various forms of hemophilia will become safer and cheaper.

The condition known as **von Willebrand disease** is the most common inherited coagulation disorder. The *von Willebrand factor* (vWF) is a plasma protein that binds and stabilizes Factor VIII. The symptoms and severity of the bleeding vary widely. Many individuals with mild forms of von Willebrand disease remain unaware of any bleeding problems until they have an accident or undergo surgery. Treatment consists of the administration of pooled Factor VIII, rather than synthetic forms, because pooled Factor VIII contains normal vWF. In some forms of this disease in which normal vWF is produced, but plasma levels are abnormally low, bleeding can be controlled by the use of nasal sprays containing a synthetic form of antidiuretic hormone (ADH). The absorbed ADH appears to stimulate the release of vWF from endothelial cells.

Calcium Ions, Vitamin K, and Blood Clotting

Calcium ions and **vitamin K** affect almost every aspect of the clotting process. All three pathways (intrinsic, extrinsic, and common) require Ca^{2+}. Any disorder that lowers plasma Ca^{2+} concentrations will impair blood clotting.

Adequate amounts of vitamin K must be present for the liver to be able to synthesize four of the clotting factors, including prothrombin. Vitamin K is a fat-soluble vitamin, present in green vegetables, grain, and organ meats, that is absorbed with dietary lipids. It is obtained from the diet, and roughly half the daily requirement is manufactured by bacteria in the large intestine. A diet inadequate in fats or in vitamin K, or a disorder that affects fat digestion and absorption (such as problems with bile production), will lead to a vitamin K deficiency. This condition will cause the eventual breakdown of the common pathway due to a lack of clotting factors and, ultimately, deactivation of the entire clotting system.

Clot Retraction

Once the fibrin meshwork has appeared, platelets and red blood cells stick to the fibrin strands. The platelets then contract, and the entire clot begins to undergo **clot retraction**, or *syneresis* (sin-ER-ē-sis; "a drawing together"). Clot retraction, which occurs over a period of 30–60 minutes, (1) pulls the torn edges of the vessel closer together, reducing residual bleeding and stabilizing the injury site, and (2) reduces the size of the damaged area, making it easier for fibroblasts, smooth muscle cells, and endothelial cells to complete repairs.

☐ FIBRINOLYSIS

As the repairs proceed, the clot gradually dissolves. This process, called **fibrinolysis** (fī-bri-NOL-i-sis), begins with the activation of the proenzyme **plasminogen** by two enzymes: thrombin, produced by the common pathway, and **tissue plasminogen activator** (t-PA), released by damaged tissues at the site of injury. The activation of plasminogen produces the enzyme **plasmin** (PLAZ-min), which begins digesting the fibrin strands and eroding the foundation of the clot.

✓ A sample of bone marrow has unusually few megakaryocytes. What body process would you expect to be impaired as a result?

✓ Vitamin K is fat-soluble, and some dietary fat is required for its absorption. How could a diet of fruit juice and water have an effect on blood clotting?

✓ Unless chemically treated, blood will coagulate in a test tube. The process begins when Factor XII becomes activated. Which clotting pathway is involved in this process?

Answers start on page Q-1

To perform its vital functions, blood must be kept in motion. On average, an RBC completes two trips around the cardiovascular system each minute. The circulation of blood begins in the third week of embryonic development and continues throughout life. If the blood supply is cut off, dependent tissues may die in a matter of minutes. In Chapter 20, we shall examine the structure and function of the heart—the pump that maintains this vital blood flow.

Chapter Review

SELECTED CLINICAL TERMINOLOGY

Terms Discussed in This Chapter

anemia (a-NĒ-mē-uh): A condition in which the oxygen-carrying capacity of blood is reduced, owing to low hematocrit or low blood hemoglobin concentrations. (*p. 658 and AM*)

embolism: A condition in which a drifting blood clot becomes stuck in a blood vessel, blocking circulation to the area downstream. (*p. 676*)

hematocrit (hē-MAT-ō-krit): The value that indicates the percentage of whole blood occupied by cellular elements. (*p. 656*)

hematuria (hē-ma-TOO-rē-uh): The presence of red blood cells in urine. (*p. 659*)

hemoglobinuria: The presence of hemoglobin in urine. (*p. 659*)

hemolytic disease of the newborn (HDN): A condition in which fetal red blood cells have been destroyed by maternal antibodies. (*p. 665*)

hemophilia (hē-mō-FĒL-ē-uh): Inherited disorders characterized by the inadequate production of clotting factors. (*p. 676*)

hypervolemic (hī-per-vō-LĒ-mik): Having an excessive blood volume. (*p. 666*)

hypovolemic (hī-pō-vō-LĒ-mik): Having a low blood volume. (*p. 666*)

hypoxia (hī-POKS-ē-uh): Low tissue oxygen levels. (*p. 661 and AM*)

jaundice (JAWN-dis): A condition characterized by yellow skin and eyes, caused by abnormally high levels of plasma bilirubin; examples include *hemolytic jaundice* and *obstructive jaundice*. (*p. 659 and AM*)

leukemia (loo-KĒ-mē-uh): A condition characterized by extremely elevated levels of circulating white blood cells; includes both *myeloid* and *lymphoid* forms. (*p. 669 and AM*)

leukocytosis (loo-kō-sī-TŌ-sis): Excessive numbers of white blood cells in the bloodstream. (*p. 669*)

leukopenia (loo-kō-PĒ-nē-uh): Inadequate numbers of white blood cells in the bloodstream. (*p. 669*)

normochromic: Having red blood cells that contain normal amounts of hemoglobin. *(p. 662 and AM)*

normocytic: Having cells of normal size. *(p. 662 and AM)*

normovolemic (nor-mō-vō-LĒ-mik): Having a normal blood volume. *(p. 666)*

plaque: An abnormal area within a blood vessel where large quantities of lipids accumulate. *(p. 676)*

RBC tests: These tests include a *reticulocyte count, hematocrit, hemoglobin concentration, RBC count, mean corpuscular volume,* and *mean corpuscular hemoglobin concentration. (p. 662 and AM)*

sickle-cell anemia: An anemia resulting from the production of an abnormal form of hemoglobin; causes red blood cells to become sickle shaped at low oxygen levels. *(p. 658 and AM)*

thalassemia: A disorder resulting from the production of an abnormal form of hemoglobin. *(p. 658 and AM)*

thrombus: A blood clot attached to the lumenal surface of a blood vessel. *(p. 676)*

transfusion: A procedure in which blood components are given to someone whose blood volume has been reduced or whose blood is defective. *(p. 666 and AM)*

venipuncture (VĒN-i-punk-chur): The puncturing of a vein for any purpose, including the withdrawal of blood or the administration of medication. *(p. 654)*

AM Additional Terms Discussed in the *Applications Manual*

autologous marrow transplant
clotting system tests
erythrocytosis
heterologous marrow transplant
packed red cells
polycythemia
thrombocytopenic purpura

STUDY OUTLINE

19–1 THE CARDIOVASCULAR SYSTEM: AN INTRODUCTION   p. 651

1. The **cardiovascular system** provides a mechanism for the rapid transport of nutrients, waste products, respiratory gases, and cells within the body.

19–2 FUNCTIONS AND COMPOSITION OF BLOOD   p. 652

THE FUNCTIONS OF BLOOD   p. 652

1. **Blood** is a specialized fluid connective tissue. Its functions include (1) transporting dissolved gases, nutrients, hormones, and metabolic wastes; (2) regulating the pH and ion composition of interstitial fluids; (3) restricting fluid losses at injury sites; (4) defending the body against toxins and pathogens; and (5) regulating body temperature by absorbing and redistributing heat.

THE COMPOSITION OF BLOOD   p. 652

2. Blood contains **plasma** and **formed elements—red blood cells** (RBCs), **white blood cells** (WBCs), and **platelets.** The plasma and formed elements constitute **whole blood,** which can be **fractionated** for analytical or clinical purposes. *(Figure 19–1; Table 19–3)*

3. **Hemopoiesis** is the process of blood cell formation. Circulating stem cells called **hemocytoblasts** divide to form all types of blood cells.

4. Whole blood from any region of the body has roughly the same temperature, viscosity, and pH.

19–3 PLASMA   p. 654

1. Plasma accounts for 46–63 percent of the volume of blood; roughly 92 percent of plasma is water. *(Figure 19–1)*

2. Plasma has a higher dissolved oxygen concentration and more dissolved proteins than interstitial fluid has.

PLASMA PROTEINS   p. 654

3. The three primary classes of plasma proteins are *albumins, globulins,* and *fibrinogen.*

4. **Albumins** constitute about 60 percent of plasma proteins. **Globulins** constitute roughly 35 percent of plasma proteins; they include **antibodies** (**immunoglobulins**), which attack foreign proteins and pathogens, and **transport globulins,** which bind ions, hormones, and other compounds. **Fibrinogen** molecules are converted to **fibrin** in the clotting process. The removal of fibrinogen from plasma leaves a fluid called **serum.**

5. The liver synthesizes and releases more than 90 percent of the plasma proteins.

19–4 RED BLOOD CELLS   p. 655

ABUNDANCE OF RBCs   p. 656

1. Red blood cells (**erythrocytes**) account for slightly less than half the blood volume and 99.9 percent of the formed elements. The **hematocrit** value indicates the percentage of whole blood occupied by formed elements and is commonly reported as the **volume of packed red cells** (VPRC). *(Figure 19–1; Table 19–1)*

STRUCTURE OF RBCs   p. 656

2. Each RBC is a biconcave disc, providing a large surface-to-volume ratio. This shape allows the RBCs to stack, bend, and flex. *(Figure 19–1)*

3. Red blood cells lack organelles, retaining only the cytoskeleton. They typically degenerate after about 120 days in the bloodstream.

HEMOGLOBIN   p. 656

4. Molecules of **hemoglobin** (**Hb**) account for over 95 percent of the proteins in RBCs. Hemoglobin is a globular protein formed from two pairs of polypeptide subunits. Each subunit contains a single molecule of **heme** and can reversibly bind an oxygen molecule. Damaged or dead RBCs are recycled by phagocytes. *(Figures 19–3, 19–4)*

RBC LIFE SPAN AND CIRCULATION p. 659

5. RBCs are continuously damaged and replaced at a rate of approximately 3 million new RBCs entering the bloodstream per second. They are replaced before they **hemolyze**.

6. The components of hemoglobin are individually recycled, with the heme being stripped of its iron and converted to **biliverdin**, which is converted to **bilirubin**. If bile ducts are blocked, bilirubin builds up in skin and eyes, resulting in **jaundice**. *(Figure 19–5)*

7. Iron is also recycled by being stored in phagocytic cells or transported through the bloodstream, bound to **transferrin**.

RBC PRODUCTION p. 659

8. **Erythropoiesis**, the formation of red blood cells, occurs only in **red bone marrow** (myeloid tissue). The process speeds up under stimulation by **erythropoiesis-stimulating hormone (erythropoietin, EPO)**. Stages in RBC development include **erythroblasts** and **reticulocytes**. *(Figure 19–6)*

Red blood cells: **Companion Website**: Chapter 19/Reviewing Facts & Terms/Multiple Choice.

BLOOD TYPES p. 662

9. **Blood type** is determined by the presence or absence of specific **surface antigens** (*agglutinogens*) in the RBC cell membranes: antigens A, B, and **Rh** (**D**). Antibodies (*agglutinins*) in the plasma will react with RBCs bearing different surface antigens. When an antibody meets its specific surface antigen, the resulting reaction is a **cross-reaction**. *(Figures 19–8 to 19–10; Table 19–2)*

19–5 WHITE BLOOD CELLS p. 666

1. White blood cells (**leukocytes**) have nuclei and other organelles. They defend the body against pathogens and remove toxins, wastes, and abnormal or damaged cells.

WBC CIRCULATION AND MOVEMENT p. 667

2. White blood cells are capable of amoeboid movement, *margination*, and **positive chemotaxis**. Some WBCs are also capable of *phagocytosis*.

TYPES OF WBCs p. 667

3. *Granular leukocytes* (*granulocytes*) are subdivided into **neutrophils** (PMNs), **eosinophils**, and **basophils**. Fifty to 70 percent of circulating WBCs are neutrophils, which are highly mobile phagocytes. The much less common eosinophils are phagocytes attracted to foreign compounds that have reacted with circulating antibodies. The relatively rare basophils migrate to damaged tissues and release *histamine* and *heparin*, aiding the inflammation response. *(Figure 19–11)*

4. *Agranular leukocytes* (*agranulocytes*) are subdivided into **monocytes** and **lymphocytes**. Monocytes that migrate into peripheral tissues become tissue macrophages. Lymphocytes, the primary cells of the lymphatic system, include **T cells** (which enter peripheral tissues and attack foreign cells directly or affect the activities of other lymphocytes), **B cells** (which produce antibodies), and **natural killer** (NK) **cells** (which destroy abnormal tissue cells). *(Figure 19–11; Summary Table 19–3)*

THE DIFFERENTIAL COUNT AND CHANGES IN WBC PROFILES p. 669

5. A **differential count** of the WBC population can indicate a variety of disorders. **Leukemia** is indicated by extreme **leukocytosis**—that is, excessive numbers of WBCs. *(Summary Table 19–3)*

WBC PRODUCTION p. 669

6. Granulocytes and monocytes are produced by stem cells in the bone marrow that divide to create **progenitor cells**. Stem cells also originate in the bone marrow, but many migrate to peripheral **lymphoid tissues**. *(Figure 19–12)*

7. Factors that regulate lymphocyte maturation are not completely understood. Several **colony-stimulating factors** (CSFs) are involved in regulating other WBC populations and in coordinating RBC and WBC production. *(Figure 19–12)*

19–6 PLATELETS p. 672

1. Platelets are flattened discs that appear round from above and spindle shaped in section. They circulate for 9–12 days before being removed by phagocytes. *(Figure 19–11)*

PLATELET FUNCTIONS p. 672

2. The functions of platelets include (1) transporting chemicals important to the clotting process, (2) forming a temporary patch in the walls of damaged blood vessels, and (3) contracting after a clot has formed, in order to reduce the size of the break in the vessel wall.

PLATELET PRODUCTION p. 672

3. During **thrombocytopoiesis**, **megakaryocytes** in the bone marrow release packets of cytoplasm (platelets) into the circulating blood. The rate of platelet formation is stimulated by thrombopoietin, thrombocyte-stimulating factor, interleukin-6, and multi-CSF.

19–7 HEMOSTASIS p. 672

1. **Hemostasis** prevents the loss of blood through the walls of damaged vessels. It consists of three phases: the *vascular phase*, the *platelet phase*, and the *coagulation phase*.

THE VASCULAR PHASE p. 672

2. The **vascular phase** is a period of local blood vessel constriction, or **vascular spasm**, at the injury site. *(Figure 19–13)*

THE PLATELET PHASE p. 673

3. The **platelet phase** follows as platelets are activated, aggregate at the site, and adhere to the damaged surfaces. *(Figure 19–13)*

THE COAGULATION PHASE p. 673

4. The **coagulation phase** occurs as factors released by platelets and endothelial cells interact with **clotting factors** (through either the **extrinsic pathway**, the **intrinsic pathway**, or the **common pathway**) to form a **blood clot**. In this reaction sequence, suspended fibrinogen is converted to large, insoluble fibers of fibrin. *(Figure 19–14; Table 19–4)*

5. During **clot retraction**, platelets contract and pull the torn edges of the damaged vessel closer together.

FIBRINOLYSIS p. 677

6. During **fibrinolysis**, the clot gradually dissolves through the action of **plasmin**, the activated form of circulating **plasminogen**.

7. Clotting can be prevented by drugs that depress the clotting response or dissolve existing clots. Important **anticoagulant** drugs include heparin, **coumadin**, **dicumarol**, **t-PA**, **streptokinase**, **urokinase**, and aspirin.

REVIEW QUESTIONS

More assessment questions are available to you on the Companion Website. You will find Matching, Multiple Choice, True/False, and other quizzes to help further your understanding of the material covered in this chapter. To access the site, go to www.aw.com/martini.

LEVEL 1 Reviewing Facts and Terms

1. The formed elements of the blood include
   (a) plasma, fibrin, and serum
   (b) albumins, globulins, and fibrinogen
   (c) WBCs, RBCs, and platelets
   (d) a, b, and c are correct

2. Blood temperature is approximately _____, and blood pH averages _____.
   (a) 36°, 7.0   (b) 39°, 7.8
   (c) 38°C, 7.4   (d) 37°C, 7.0

3. Plasma contributes approximately _____ percent of the volume of whole blood, and water accounts for _____ percent of the plasma volume.
   (a) 55, 92   (b) 25, 55
   (c) 92, 55   (d) 35, 72

4. A hematocrit provides information on
   (a) blood type   (b) clotting factors
   (c) packed cell volume (d) plasma composition

5. In adults, the only site of red blood cell production, and the primary site of white blood cell formation, is the
   (a) liver   (b) spleen
   (c) thymus   (d) red bone marrow

6. The most numerous WBCs in a differential count of a healthy individual are
   (a) neutrophils   (b) basophils
   (c) lymphocytes   (d) monocytes

7. The differential count of a person who has an allergy would indicate a high number of
   (a) neutrophils   (b) eosinophils
   (c) basophils   (d) monocytes

8. Stem cells responsible for lymphopoiesis are located in the
   (a) thymus and spleen (b) lymph nodes
   (c) red bone marrow   (d) a, b, and c are correct

9. The first step in hemostasis is
   (a) clotting   (b) the platelet phase
   (c) fibrinolysis   (d) vascular spasm

10. _____ and _____ affect almost every aspect of the clotting process
    (a) calcium and vitamin K   (b) calcium and vitamin B_{12}
    (c) sodium and vitamin K   (d) sodium and vitamin B_{12}

11. What five major functions are performed by blood?

12. What three primary classes of plasma proteins are in the blood? What is the major function of each?

13. Which type of antibodies does plasma contain for each of the following blood types?
    (a) Type A   (b) Type B
    (c) Type AB   (d) Type O

14. What four characteristics of WBCs are important to their response to tissue invasion or injury?

15. Which kinds of WBCs contribute to the body's nonspecific defenses?

16. Which three classes of lymphocytes are the primary cells of the lymphatic system? What are the functions of each class?

17. Which kinds of WBCs are produced by each of the four colony-stimulating factors (CSFs)?

18. What are the three functions of platelets during the clotting process?

19. What four conditions cause the release of erythropoietin?

20. What five controls prevent clots from growing indefinitely?

21. What contribution from the intrinsic and the extrinsic pathways is necessary for the common pathway to begin?

22. Distinguish between an embolus and a thrombus.

LEVEL 2 Reviewing Concepts

23. Dehydration would cause
    (a) an increase in the hematocrit
    (b) a decrease in the hematocrit
    (c) no effect on the hematocrit
    (d) an increase in plasma volume

24. Erythropoietin directly stimulates RBC formation by
    (a) increasing rates of mitotic divisions in erythroblasts
    (b) speeding up the maturation of red blood cells
    (c) accelerating the rate of hemoglobin synthesis
    (d) a, b, and c are correct

25. A person with Type A blood has
    (a) antigen A in the plasma
    (b) anti-B antibodies in the plasma
    (c) anti-A antibodies on the red blood cells
    (d) antigen B on the red blood cell membranes

26. Hemolytic disease of the newborn may result if
    (a) the mother is Rh positive and the father is Rh negative
    (b) both the father and the mother are Rh negative
    (c) both the father and the mother are Rh positive
    (d) an Rh negative mother carries an Rh positive fetus

27. How do red blood cells differ from WBCs in both form and function?

28. How does blood defend against toxins and pathogens in the body?

29. What is the role of blood in the stabilization and maintenance of body temperature?

30. Describe the structure of hemoglobin. How does the structure relate to its function?

31. You are a respiratory therapist, and you need blood to check the efficiency of gas exchange in a patient's lungs. Which type of "stick" should you use, and from which blood vessels might you draw the sample?

32. Linda was given RhoGam during and after the delivery of her baby. What does the administration of RhoGam imply, and what effect does it have on Linda?

33. Why is aspirin sometimes prescribed for the prevention of vascular problems?

LEVEL 3 Critical Thinking and Clinical Applications

34. A test for prothrombin time is used to determine deficiencies in the extrinsic clotting pathway and is prolonged if any of the factors are deficient. A test for activated partial thromboplastin time is used in a similar fashion to detect deficiencies in the intrinsic clotting pathway. Which factor would be deficient if a person had a prolonged prothrombin time, but a normal partial thromboplastin time?

35. Which of the formed elements would increase after the donation of a pint of blood?

36. Mary has an enlarged spleen and her platelet count dropped to $50,000/\mu l$. What signs and symptoms will she exhibit?

37. Why do people with advanced kidney disease commonly become anemic?

38. After Randy was diagnosed with stomach cancer, nearly all of his stomach had to be removed. Postoperative treatment included regular injections of vitamin B_{12}. Why was this vitamin prescribed, and why was injection specified?

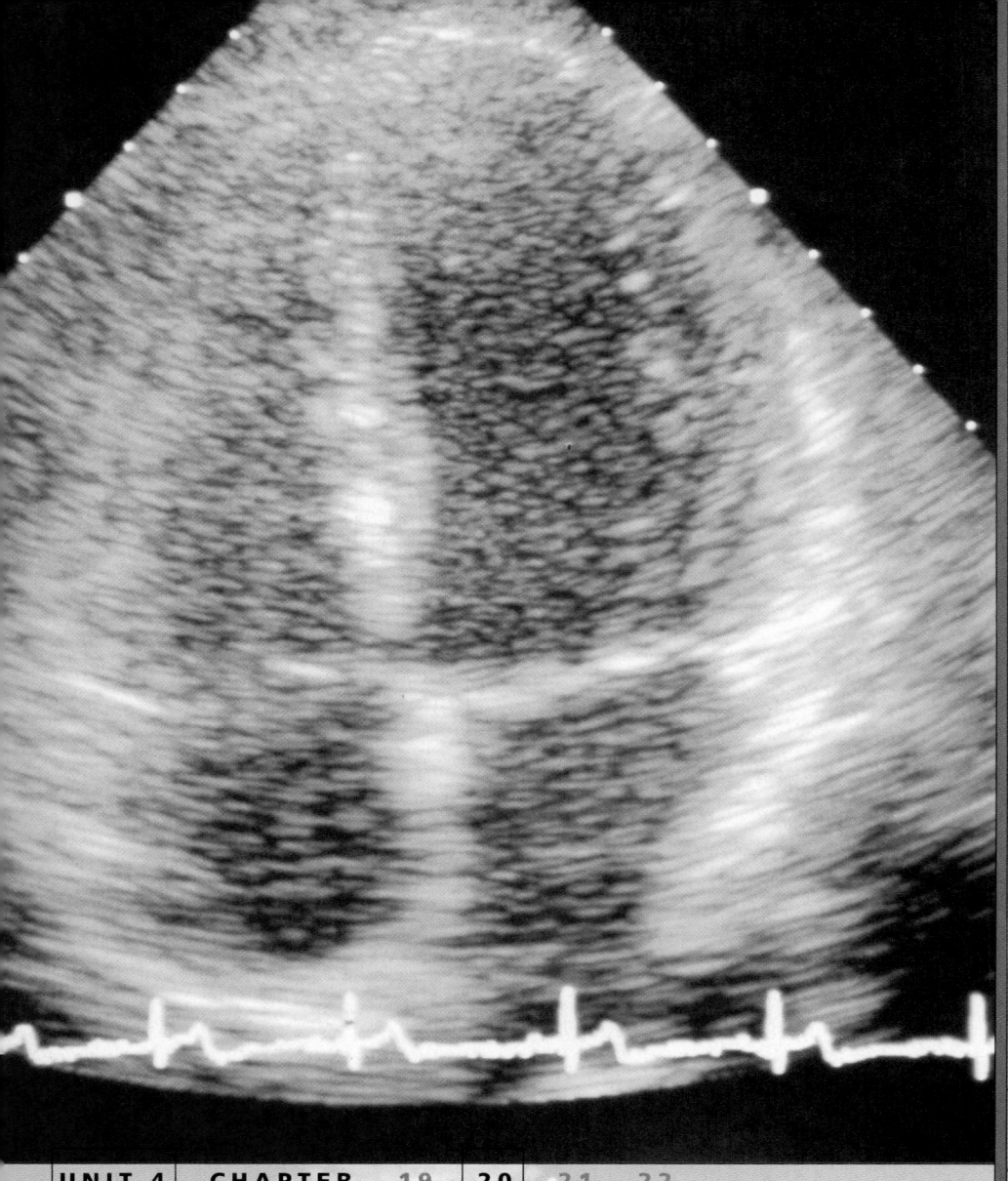

UNIT 4   CHAPTER   19   20   21   22

CLINICAL NOTES

- Pericarditis 685
- Valvular Heart Disease 692
- Abnormal Pacemaker Function 699
- Conduction Deficits 700
- Mitral Valve Prolapse 708
- Drugs and Contractility 714

CLINICAL DISCUSSIONS

- Coronary Artery Disease 695
- Heart Attacks 704
- Abnormal Conditions Affecting
  Cardiac Output 716

Every cell relies on the surrounding interstitial fluid for oxygen, nutrients, and waste disposal. The composition of the interstitial fluid in tissues throughout the body is kept stable through continuous exchange between the peripheral tissues and the bloodstream. Yet blood can help maintain homeostasis only as long as it stays in motion. If blood remains stationary, its oxygen and nutrient supplies are quickly exhausted, its capacity to absorb wastes is soon saturated, and neither hormones nor white blood cells can reach their intended targets. Thus, all the functions of the cardiovascular system ultimately depend on the **heart**.

Unlike most other muscles, the heart never rests. This extraordinary organ beats approximately 100,000 times each day, pumping roughly 8000 liters of blood—enough to fill forty 55-gallon drums, or 8800 quart-sized milk cartons. Try transferring a gallon of water by using a squeeze pump, and you'll appreciate just how hard the heart has to work to keep you alive. We begin this chapter by examining the structural features that enable the heart to perform so reliably. We will then consider the physiological mechanisms that regulate cardiac activity to meet the body's constantly changing demands.

20-1 THE ORGANIZATION OF THE CARDIOVASCULAR SYSTEM

Objective

■ Describe the organization of the cardiovascular system and of the heart.

Blood flows through a network of blood vessels that extend between the heart and peripheral tissues. Those blood vessels can be subdivided into a **pulmonary circuit**, which carries blood to and from the gas exchange surfaces of the lungs, and a **systemic circuit**, which transports blood to and from the rest of the body (Figure 20–1●). Each circuit begins and ends at the heart, and blood travels through these circuits in sequence. For example, blood returning to the heart from the systemic circuit must complete the pulmonary circuit before reentering the systemic circuit.

Blood is carried away from the heart by **arteries**, or *efferent vessels*, and returns to the heart by way of **veins**, or *afferent vessels*. Small, thin-walled vessels called **capillaries** interconnect the smallest arteries and the smallest veins. Capillaries are called **exchange vessels**, because their thin walls permit the exchange of nutrients, dissolved gases, and waste products between the blood and surrounding tissues.

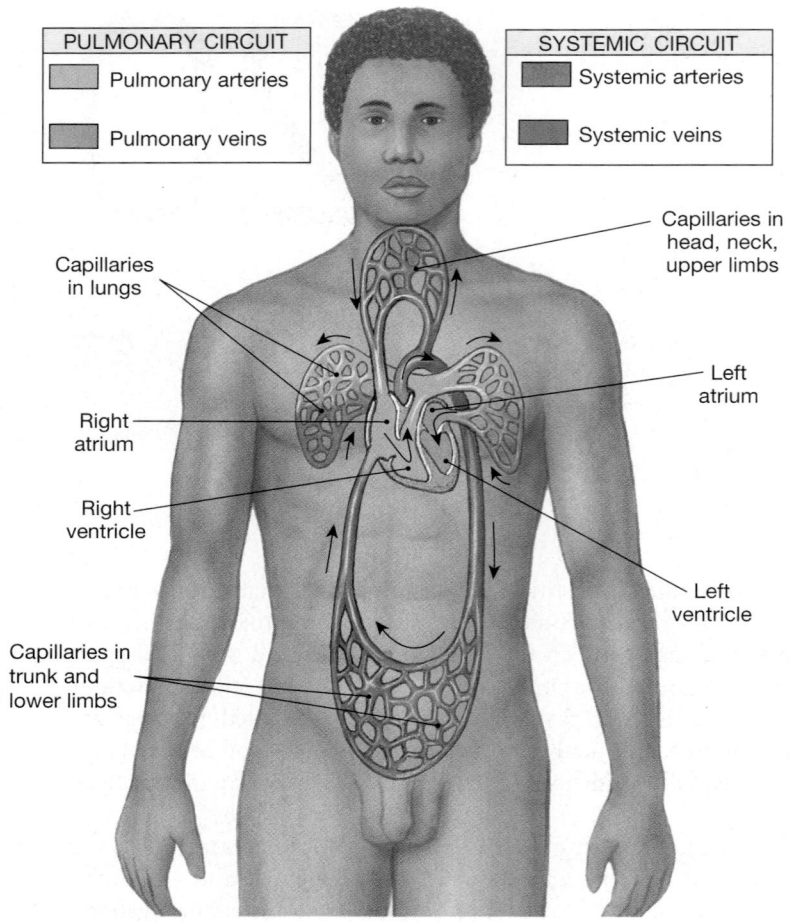

PULMONARY CIRCUIT
Pulmonary arteries
Pulmonary veins

SYSTEMIC CIRCUIT
Systemic arteries
Systemic veins

Capillaries in head, neck, upper limbs

Capillaries in lungs

Left atrium

Right atrium

Right ventricle

Left ventricle

Capillaries in trunk and lower limbs

●**FIGURE 20–1**
An Overview of the Cardiovascular System. Driven by the pumping of the heart, blood flows through separate pulmonary and systemic circuits. Each circuit begins and ends at the heart and contains arteries, capillaries, and veins.

Despite its impressive workload, the heart is a small organ, roughly the size of a clenched fist. The heart contains four muscular chambers, two associated with each circuit. The **right atrium** (Ā-trē-um; chamber; plural, *atria*) receives blood from the systemic circuit and passes it to the **right ventricle** (VEN-tri-kl; little belly), which pumps blood into the pulmonary circuit. The **left atrium** collects blood from the pulmonary circuit and empties it into the **left ventricle**, which then contracts, ejecting blood into the systemic circuit. When the heart beats, the atria contract, and then the ventricles contract. The two ventricles contract at the same time and eject equal volumes of blood into the pulmonary and systemic circuits.

20–2 ANATOMY OF THE HEART

Objectives

- Describe the location and general features of the heart.
- Describe the structure of the pericardium, and explain its functions.

- Trace the flow of blood through the heart, identifying the major blood vessels, chambers, and heart valves.
- Identify the layers of the heart wall.
- Describe the vascular supply to the heart.

The heart is located near the anterior chest wall, directly posterior to the sternum (Figure 20–2a●). The great veins and arteries are connected to the superior end of the heart at the attached **base**. The base sits posterior to the sternum at the level of the third costal cartilage, centered about 1.2 cm (0.5 in.) to the left side. The inferior, pointed tip of the heart is the free **apex** (Ā-peks). A typical adult heart measures approximately 12.5 cm (5 in.) from the base to the apex, which reaches the fifth intercostal space approximately 7.5 cm (3 in.) to the left of the midline. A midsagittal section through the trunk does not divide the heart into two equal halves, because (1) the center of the base lies slightly to the left of the midline, (2) a line drawn between the center of the base and the apex points further to the left, and (3) the entire heart is rotated to the left around this line, so that the right atrium and right ventricle dominate an anterior view of the heart.

The heart, surrounded by the **pericardial** (per-i-KAR-dē-al) **cavity**, sits in the anterior portion of the mediastinum. The **mediastinum**, the region between the two pleural cavities, also contains the thymus, esophagus, and trachea. ∞ p. 22 Figure 20–2b● is a sectional view that illustrates the position of the heart relative to other structures in the mediastinum.

☐ THE PERICARDIUM

The lining of the pericardial cavity is called the **pericardium**. To visualize the relationship between the heart and the pericardial cavity, imagine pushing your fist toward the center of a large balloon (Figure 20–2c●). The balloon represents the pericardium, and your fist is the heart. Your wrist, where the balloon folds back on itself, corresponds to the *base* of the heart, to which the *great vessels*, the largest veins and arteries in the body, are attached. The air space inside the balloon corresponds to the pericardial cavity.

The pericardium is lined by a delicate serous membrane that can be subdivided into the visceral pericardium and the parietal pericardium. The **visceral pericardium**, or *epicardium*, covers the outer surface of the heart and adheres closely to it; the **parietal pericardium** lines the inner surface of the **pericardial sac**, which surrounds the heart (Figure 20–2c●). The pericardial sac, or *fibrous pericardium*, which consists of a dense network of collagen fibers, stabilizes the position of the heart and associated vessels within the mediastinum.

The small space between the parietal and visceral surfaces is the pericardial cavity. It normally contains 10–20 ml of **pericardial fluid**, secreted by the pericardial membranes. This fluid acts as a lubricant, reducing friction between the opposing surfaces as the heart beats.

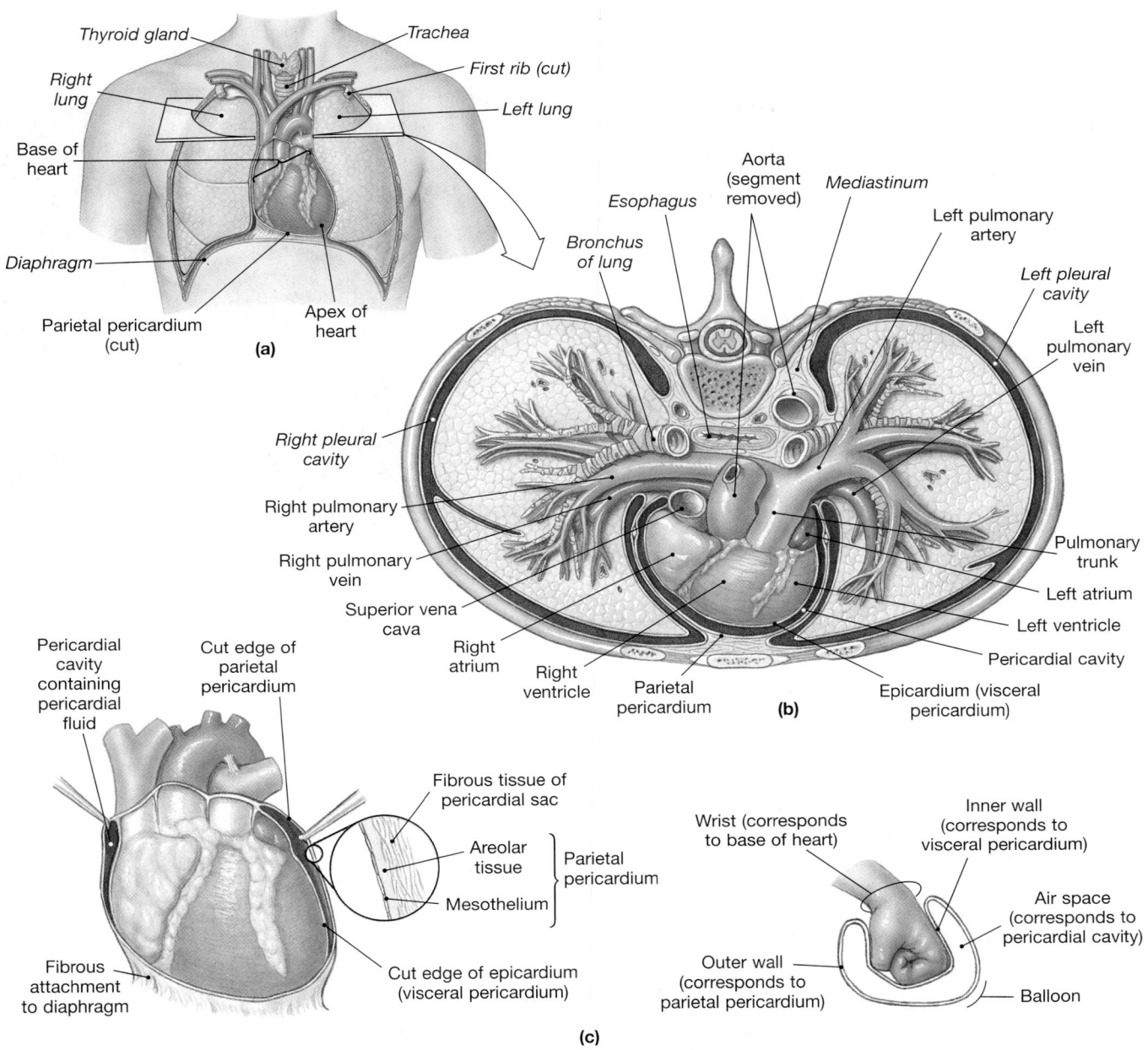

•FIGURE 20–2
The Location of the Heart in the Thoracic Cavity. The heart is situated in the anterior part of the mediastinum, immediately posterior to the sternum. **(a)** An anterior view of the open chest cavity, showing the position of the heart and major vessels relative to the lungs. **(b)** A superior view of the heart and other organs in the mediastinum with the tissues of the lungs removed to reveal the blood vessels and airways. **(c)** The relationship between the heart and the pericardial cavity; compare with the fist-and-balloon example. **ATLAS** Plates 7.4a, 7.6a,b

Pathogens can infect the pericardium, producing the condition **pericarditis**. The inflamed pericardial surfaces rub against one another, producing a distinctive scratching sound. The pericardial irritation and inflammation also commonly result in an increased production of pericardial fluid. Fluid then collects in the pericardial cavity, restricting the movement of the heart. This condition, called **cardiac tamponade** (tam-po-NĀD; *tampon*, plug), can also be caused by traumatic injuries, such as gunshot wounds, that produce bleeding into the pericardial cavity. **AM** Infection and Inflammation of the Heart

☐ SUPERFICIAL ANATOMY OF THE HEART

The four cardiac chambers can easily be identified in a superficial view of the heart (Figure 20–3•). The two atria have relatively thin muscular walls and are highly expandable. When not filled with blood, the outer portion of each atrium deflates and becomes a lumpy, wrinkled flap. This expandable extension of an atrium is called an *atrial appendage*, or an **auricle** (AW-ri-kl; *auris*, ear), because it reminded early anatomists of the external

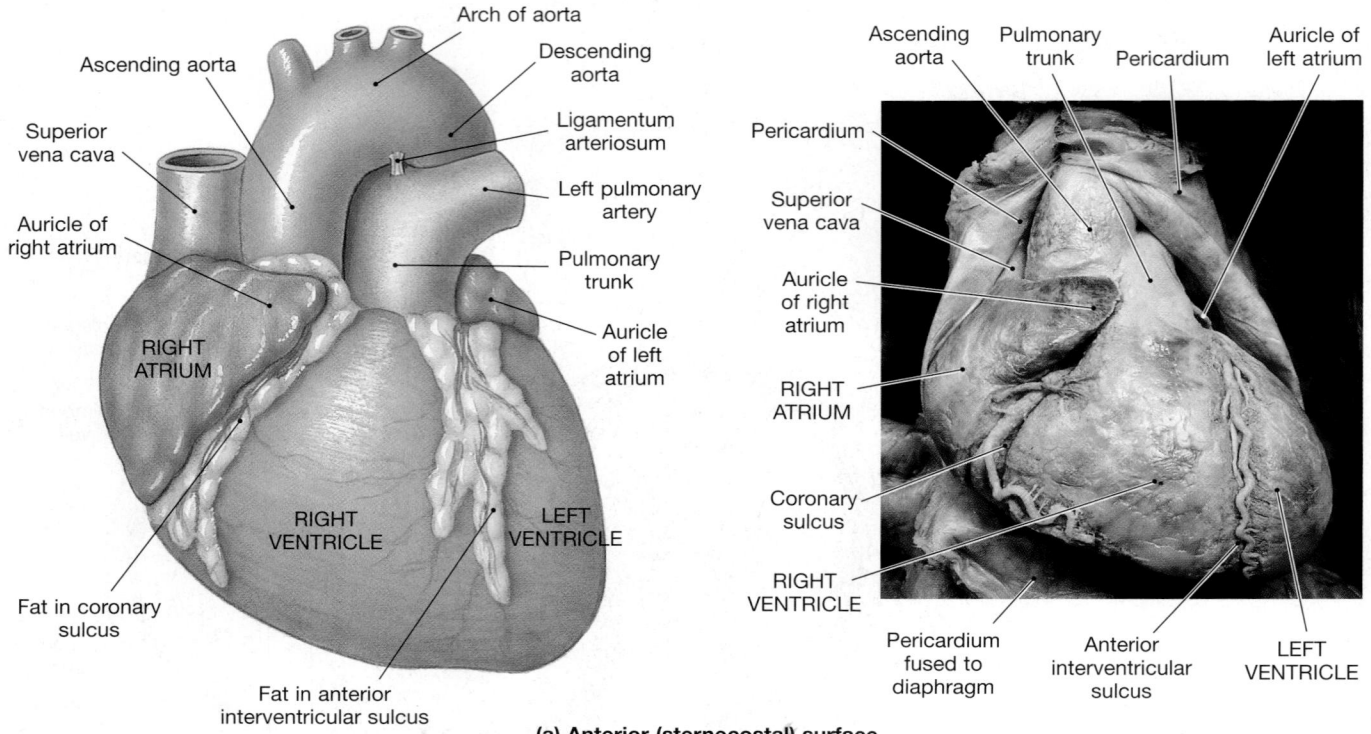

(a) Anterior (sternocostal) surface

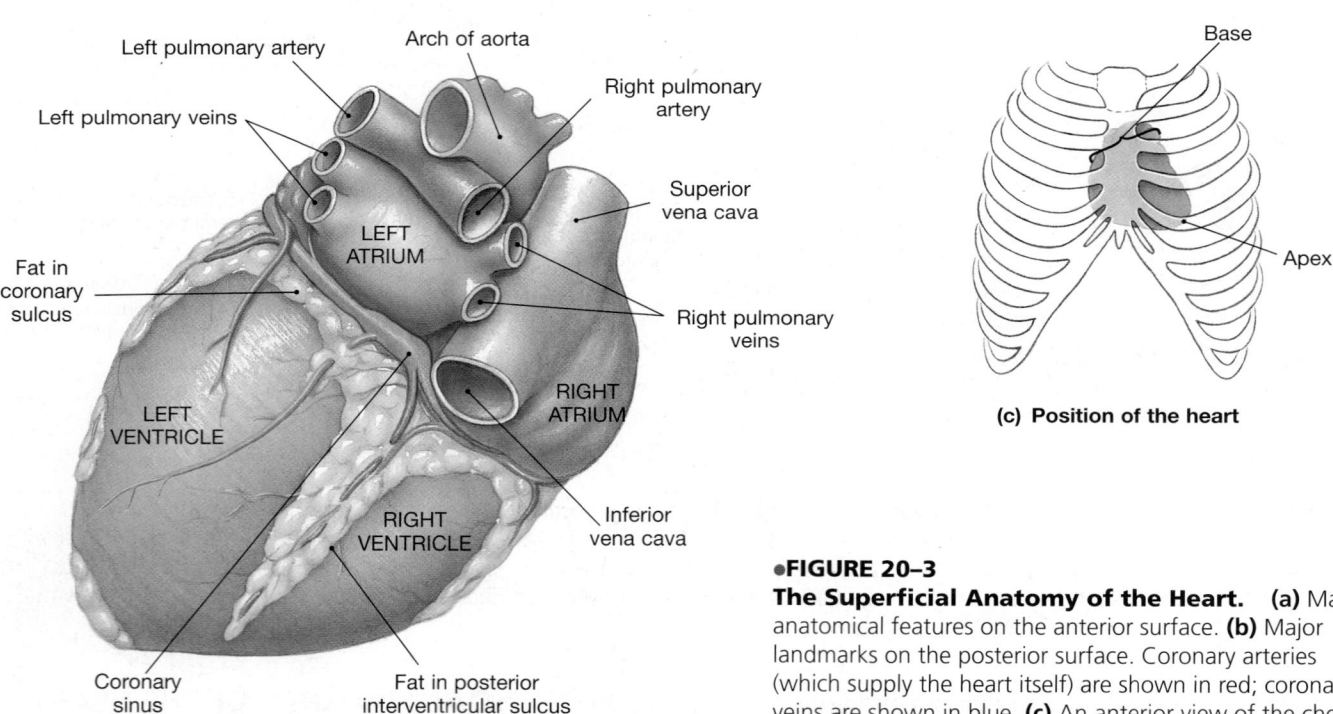

(b) Posterior surface

(c) Position of the heart

•FIGURE 20–3
The Superficial Anatomy of the Heart. **(a)** Major anatomical features on the anterior surface. **(b)** Major landmarks on the posterior surface. Coronary arteries (which supply the heart itself) are shown in red; coronary veins are shown in blue. **(c)** An anterior view of the chest, showing the position of the heart relative to the chest wall.

ear (Figure 20–3a•). The **coronary sulcus**, a deep groove, marks the border between the atria and the ventricles. The **anterior interventricular sulcus** and the **posterior interventricular sulcus**, shallower depressions, mark the boundary between the left and right ventricles (Figure 20–3a,b•).

The connective tissue of the epicardium at the coronary and interventricular sulci generally contains substantial amounts of fat. In fresh or preserved hearts, this fat must be stripped away to expose the underlying grooves. These sulci also contain the arteries and veins that supply blood to the cardiac muscle.

☐ THE HEART WALL

A section through the wall of the heart reveals three distinct layers: an outer epicardium, a middle myocardium, and an inner endocardium. Figures 20–4a● illustrates these three layers:

1. The **epicardium** is the visceral pericardium that covers the outer surface of the heart. This serous membrane consists of an exposed mesothelium and an underlying layer of loose connective tissue that is attached to the myocardium.

2. The **myocardium**, or muscular wall of the heart, forms both atria and ventricles. This layer contains cardiac muscle tissue, blood vessels, and nerves. The myocardium consists of concentric layers of cardiac muscle tissue. The atrial myocardium contains muscle bundles that wrap around the atria and form figure eights that pass through the interatrial septum (Figure 20–4b●). Superficial ventricular muscles wrap around both ventricles; deeper muscle layers spiral around and between the ventricles toward the apex.

3. The inner surfaces of the heart, including those of the heart valves, are covered by the **endocardium**, a simple squamous epithelium that is continuous with the endothelium of the attached blood vessels.

Cardiac Muscle Tissue

Recall from Chapter 10 that **cardiac muscle cells** are interconnected by **intercalated discs** (Figure 20–5a,b●). ⊂⊃ p. 324 At an intercalated disc, the interlocking membranes of adjacent cells are held together by desmosomes and linked by gap junctions (Figure 20–5c●). Intercalated discs convey the force of contraction from cell to cell and propagate action potentials. Table 20–1 (p. 690) provides a quick review of the structural and functional differences between cardiac muscle cells and skeletal muscle fibers. Among the histological characteristics of cardiac muscle cells that differ from those of skeletal muscle fibers are (1) a small size; (2) a single, centrally located nucleus; (3) branching interconnections between cells; and (4) the presence of intercalated discs.

☐ INTERNAL ANATOMY AND ORGANIZATION

This section walks you through the major landmarks and structures visible on the interior surface of the heart. In a sectional view, you can see that the right atrium communicates with the right ventricle and the left atrium with the left ventricle. The atria

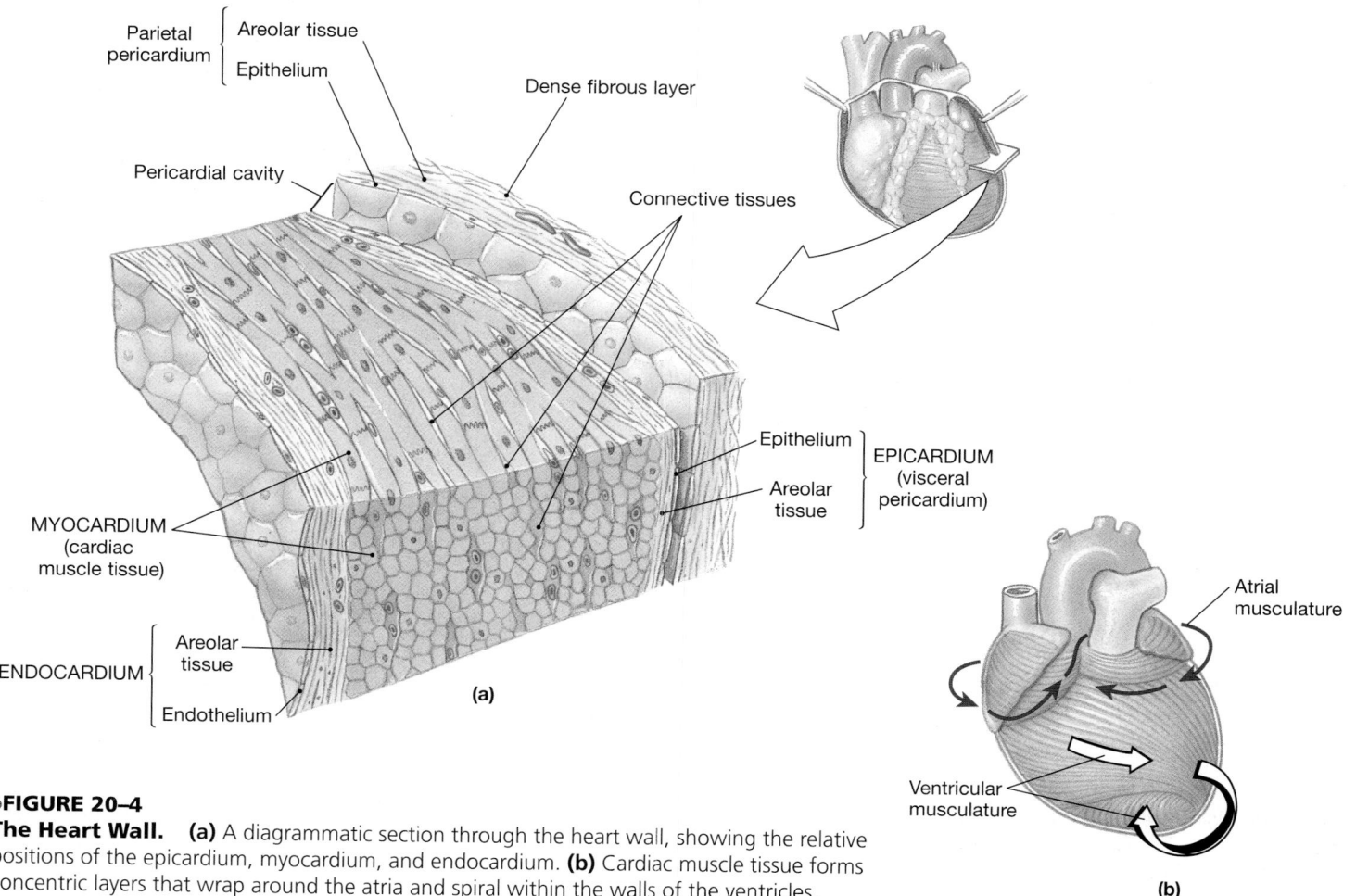

●**FIGURE 20–4**
The Heart Wall. **(a)** A diagrammatic section through the heart wall, showing the relative positions of the epicardium, myocardium, and endocardium. **(b)** Cardiac muscle tissue forms concentric layers that wrap around the atria and spiral within the walls of the ventricles.

are separated by the **interatrial septum** (*septum*, wall); the ventricles are separated by the much thicker **interventricular septum** (Figure 20–6a,c●). Each septum is a muscular partition. **Atrioventricular (AV) valves**, folds of fibrous tissue, extend into the openings between the atria and ventricles. These valves permit blood flow in one direction only: from the atria to the ventricles.

The Right Atrium

The right atrium receives blood from the systemic circuit through the two great veins: the **superior vena cava** (VĒ-na KĀ-vuh; plural, *venae cavae*) and the **inferior vena cava**. The superior vena cava, which opens into the posterior and superior portion of the right atrium, delivers blood to the right atrium from the head, neck, upper limbs, and chest. The inferior vena cava, which opens into the posterior and inferior portion of the right atrium, carries blood to the right atrium from the rest of the trunk, the viscera, and the lower limbs. The *coronary veins* of the heart return blood to the **coronary sinus**, a large, thin-walled vein that opens into the right atrium inferior to the connection with the inferior vena cava.

The opening of the coronary sinus lies near the posterior edge of the interatrial septum. From the fifth week of embryonic development until birth, the **foramen ovale**, an oval opening, penetrates the interatrial septum and connects the two atria. The foramen ovale permits blood flow from the right atrium to the left atrium

while the lungs are developing before birth. At birth, the foramen ovale closes, and the opening will be permanently sealed off within three months of delivery. The **fossa ovalis**, a small, shallow depression, persists at this site in the adult heart (Figure 20–6a,c●). If the foramen ovale does not close, blood will flow from the left atrium into the right atrium rather than the opposite way, because, after birth, blood pressure in the pulmonary circuit is lower than that in the systemic circuit. We will consider the physiological effects of this condition in Chapter 21. **ATLAS** Embryology Summary 15: The Development of the Heart

The posterior wall of the right atrium and the interatrial septum have smooth surfaces. In contrast, the anterior atrial wall and the inner surface of the auricle contain prominent muscular ridges called the **pectinate muscles** (*pectin*, comb), or *musculi pectinati* (Figure 20–6a,c●).

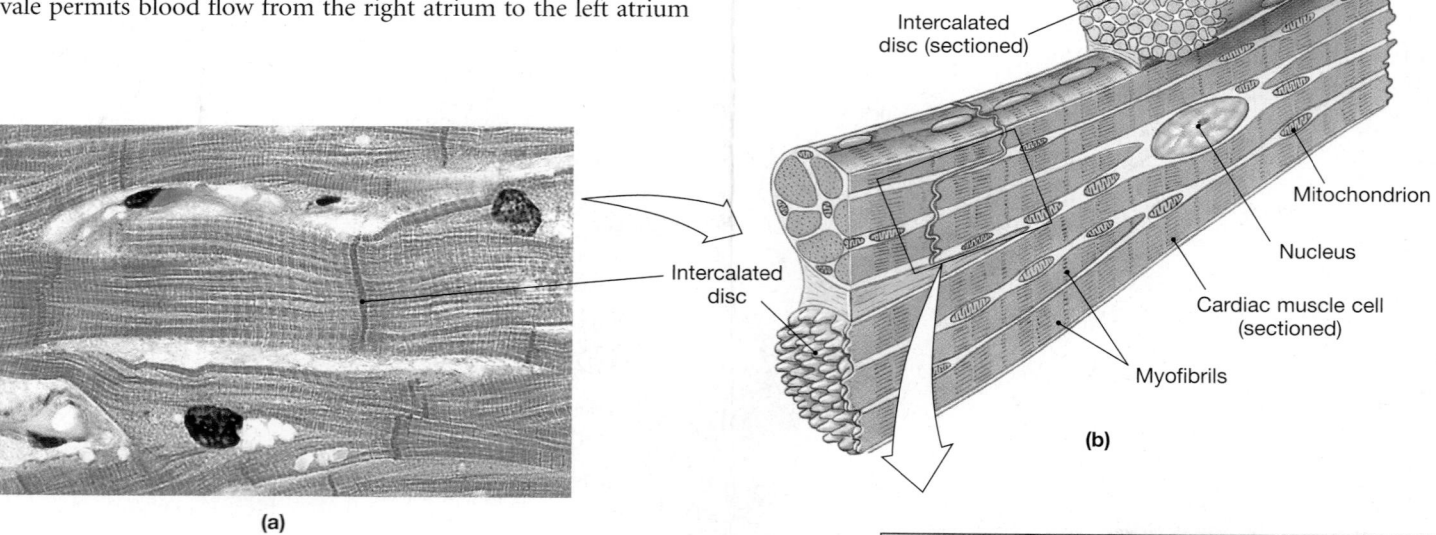

Cardiac muscle cell (intact)

Intercalated disc (sectioned)

Mitochondrion

Nucleus

Cardiac muscle cell (sectioned)

Myofibrils

Intercalated disc

(b)

(a)

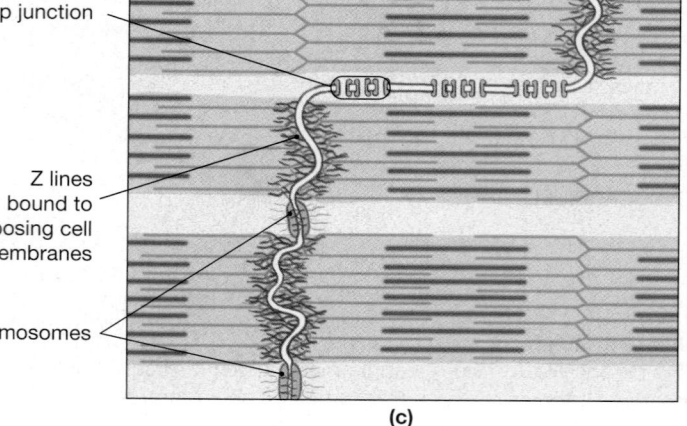

Gap junction

Z lines bound to opposing cell membranes

Intercalated disc

Desmosomes

(c)

●**FIGURE 20–5**
The Heart Wall and Cardiac Muscle Cells.
(a) Sectional and **(b)** diagrammatic views of cardiac muscle tissue. (LM × 575) **(c)** The structure of an intercalated disc.

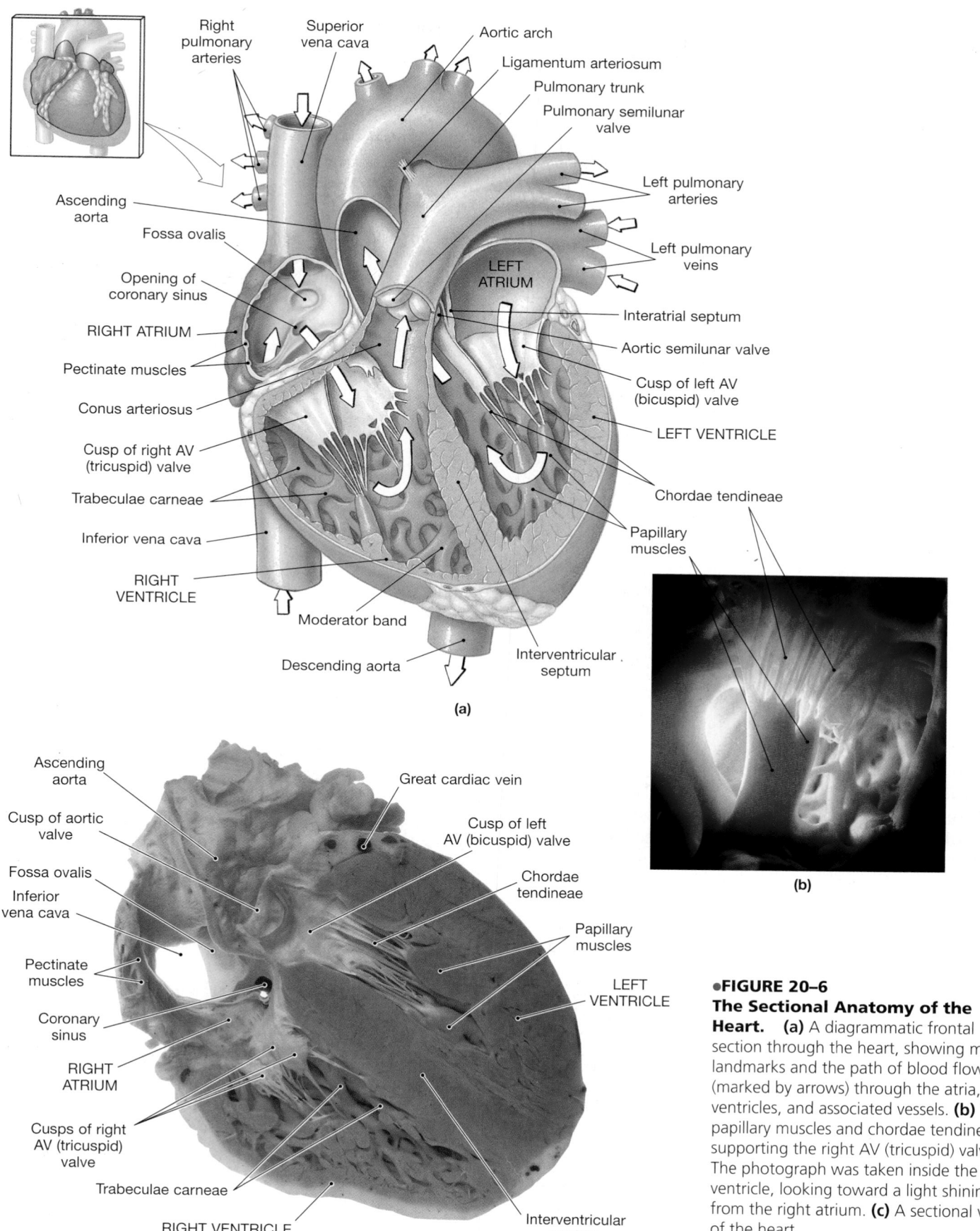

Right pulmonary arteries

Superior vena cava

Aortic arch

Ligamentum arteriosum

Pulmonary trunk

Pulmonary semilunar valve

Ascending aorta

Fossa ovalis

Opening of coronary sinus

RIGHT ATRIUM

Pectinate muscles

Conus arteriosus

Cusp of right AV (tricuspid) valve

Trabeculae carneae

Inferior vena cava

RIGHT VENTRICLE

Moderator band

Descending aorta

LEFT ATRIUM

Left pulmonary arteries

Left pulmonary veins

Interatrial septum

Aortic semilunar valve

Cusp of left AV (bicuspid) valve

LEFT VENTRICLE

Chordae tendineae

Papillary muscles

Interventricular septum

(a)

Ascending aorta

Cusp of aortic valve

Fossa ovalis

Inferior vena cava

Pectinate muscles

Coronary sinus

RIGHT ATRIUM

Cusps of right AV (tricuspid) valve

Trabeculae carneae

RIGHT VENTRICLE

Great cardiac vein

Cusp of left AV (bicuspid) valve

Chordae tendineae

Papillary muscles

LEFT VENTRICLE

Interventricular septum

(c)

(b)

●FIGURE 20–6
The Sectional Anatomy of the Heart. (a) A diagrammatic frontal section through the heart, showing major landmarks and the path of blood flow (marked by arrows) through the atria, ventricles, and associated vessels. (b) The papillary muscles and chordae tendineae supporting the right AV (tricuspid) valve. The photograph was taken inside the right ventricle, looking toward a light shining from the right atrium. (c) A sectional view of the heart.

SUMMARY TABLE 20–1 STRUCTURAL AND FUNCTIONAL DIFFERENCES BETWEEN CARDIAC MUSCLE CELLS AND SKELETAL MUSCLE FIBERS

Feature	Cardiac Muscle Cells	Skeletal Muscle Fibers
Size	10–20 μm × 50–100 μm	100 μm × up to 40 cm
Nuclei	Typically 1 (rarely 2–5)	Multiple (hundreds)
Contractile proteins	Sarcomeres along myofibrils	Sarcomeres along myofibrils
Internal membranes	Short T tubules; no triads formed with sarcoplasmic reticulum	Long T tubules form triads with cisternae of the sarcoplasmic reticulum
Mitochondria	Abundant (25% of cell volume)	Relatively scarce
Inclusions	Myoglobin, lipids, glycogen	Little myoglobin, few lipids, but extensive glycogen reserves
Blood supply	Very extensive	More extensive than in most connective tissues, but sparse compared with supply to cardiac muscle cells
Metabolism (resting)	Not applicable	Aerobic, primarily lipid based
Metabolism (active)	Aerobic, primarily using lipids and carbohydrates	Anaerobic, through breakdown of glycogen reserves
Contractions	Twitches with brief relaxation periods; long refractory period prevents tetanic contractions	Usually sustained tetanic contractions
Stimulus for contraction	Autorhythmicity of pacemaker cells generates action potentials	Activity of somatic motor neuron generates action potentials in sarcolemma
Trigger for contraction	Calcium entry from the ECF and calcium release from the sarcoplasmic reticulum	Calcium release from the sarcoplasmic reticulum
Intercellular connections	Branching network with cell membranes locked together at intercalated discs; connective tissue fibers tie adjacent layers together	Adjacent fibers tied together by connective tissue fibers

The Right Ventricle

Blood travels from the right atrium into the right ventricle through a broad opening bounded by three fibrous flaps. These flaps, or **cusps**, are part of the **right atrioventricular (AV) valve**, also known as the **tricuspid** (trī-KUS-pid; *tri*, three) **valve**. The free edge of each cusp is attached to tendinous connective-tissue fibers called the **chordae tendineae** (KOR-dē TEN-di-nē-ē; tendinous cords). The fibers originate at the **papillary** (PAP-i-ler-ē) **muscles**, conical muscular projections that arise from the inner surface of the right ventricle (Figure 20–6a,b●). The right AV valve closes when the right ventricle contracts, preventing the backflow of blood into the right atrium. Without the chordae tendineae, the cusps would be like swinging doors that permitted blood flow in both directions.

The internal surface of the ventricle also contains a series of muscular ridges: the **trabeculae carneae** (tra-BEK-ū-lē KAR-nē-ē; *carneus*, fleshy). The *moderator band* is a muscular ridge that extends horizontally from the inferior portion of the interventricular septum and connects to the anterior papillary muscle. This ridge contains a portion of the *conducting system*, an internal network that coordinates the contractions of cardiac muscle cells. The moderator band delivers the stimulus for contraction to the papillary muscles, so that they begin tensing the chordae tendineae before the rest of the ventricle contracts.

The superior end of the right ventricle tapers to the **conus arteriosus**, a conical pouch that ends at the **pulmonary valve**, or

pulmonary semilunar valve. The pulmonary valve consists of three semilunar (half-moon–shaped) cusps of thick connective tissue. Blood flowing from the right ventricle passes through this valve to enter the **pulmonary trunk**, the start of the pulmonary circuit. The arrangement of cusps prevents backflow as the right ventricle relaxes. Once in the pulmonary trunk, blood flows into the **left pulmonary arteries** and the **right pulmonary arteries**. These vessels branch repeatedly within the lungs before supplying the capillaries where gas exchange occurs.

The Left Atrium

From the respiratory capillaries, blood collects into small veins that ultimately unite to form the four pulmonary veins. The posterior wall of the left atrium receives blood from two **left** and two **right pulmonary veins**. Like the right atrium, the left atrium has an auricle. A valve, the **left atrioventricular (AV) valve**, or **bicuspid** (bī-KUS-pid) **valve**, guards the entrance to the attached ventricle (Figure 20–6a,c●). As the name *bicuspid* implies, the left AV valve contains a pair, not a trio, of cusps. Clinicians often call this valve the **mitral** (MĪ-tral; *mitre*, a bishop's hat) **valve**. The left AV valve permits the flow of blood from the left atrium into the left ventricle.

The Left Ventricle

The right and left ventricles contain equal amounts of blood. The left ventricle is much larger than the right, however, because it has thicker walls. The thick, muscular wall enables the left ventricle to develop pressure sufficient to push blood through the large systemic circuit, whereas the right ventricle needs to pump blood, at lower pressure, only about 15 cm (6 in.) to and from the lungs. The internal organization of the left ventricle generally resembles that of the right ventricle, except for the absence of a moderator band (Figure 20–6a,c●). The trabeculae carneae are prominent, and a pair of large papillary muscles tense the chordae tendineae that brace the cusps of the AV valve and prevent the backflow of blood into the left atrium.

Blood leaves the left ventricle by passing through the **aortic valve**, or *aortic semilunar valve*, into the **ascending aorta**. The arrangement of cusps in the aortic valve is the same as that in the pulmonary valve. Saclike dilations of the base of the ascending aorta are adjacent to each cusp. These sacs, called aortic sinuses, prevent the individual cusps from sticking to the wall of the aorta when the valve opens. The *right* and *left coronary arteries*, which deliver blood to the myocardium, originate at the **aortic sinuses**. Once the blood has been pumped out of the heart and into the systemic circuit, the aortic valve prevents backflow into the left ventricle. From the ascending aorta, blood flows on through the **aortic arch** and into the **descending aorta** (Figure 20–6a●). The pulmonary trunk is attached to the aortic arch by the *ligamentum arteriosum*, a ligament that is a remnant of an important fetal blood vessel that once linked the pulmonary and systemic circuits.

Structural Differences between the Left and Right Ventricles

The function of an atrium is to collect blood that is returning to the heart and deliver it to the attached ventricle. The functional demands on the right and left atria are similar, and the two chambers look almost identical. The demands on the right and left ventricles, however, are very different, and the two have significant structural differences.

Anatomical differences between the left and right ventricles are best seen in a three-dimensional view (Figure 20–7a●). The lungs are close to the heart, and the pulmonary blood vessels are relatively short and wide. Thus, the right ventricle normally does not need to push very hard to propel blood through the pulmonary circuit. Accordingly, the wall of the right ventricle is relatively thin. In sectional view, it resembles a pouch attached to the massive wall of the left ventricle. When it contracts, the right ventricle acts like a bellows, squeezing the blood against the thick wall of the left ventricle. This mechanism moves blood very efficiently with minimal effort, but it develops relatively low pressures.

A comparable pumping arrangement would not be suitable for the left ventricle, because six to seven times as much force must be exerted to push blood around the systemic circuit as around the pulmonary circuit. The left ventricle has an extremely thick muscular wall and is round in cross section

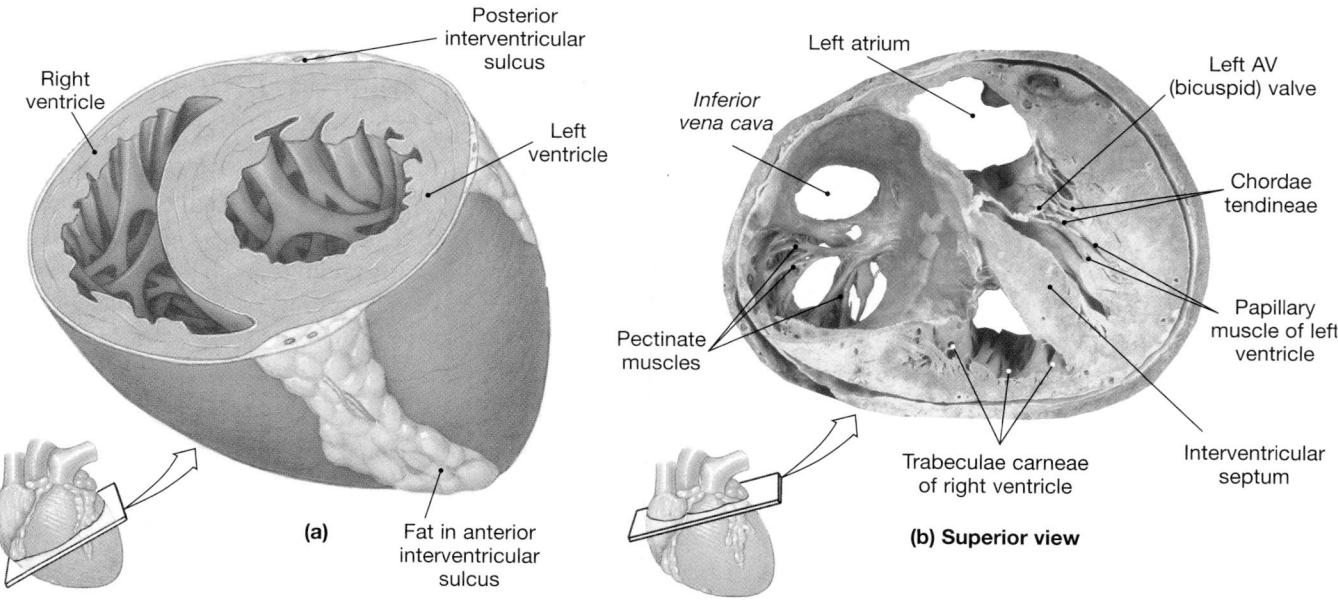

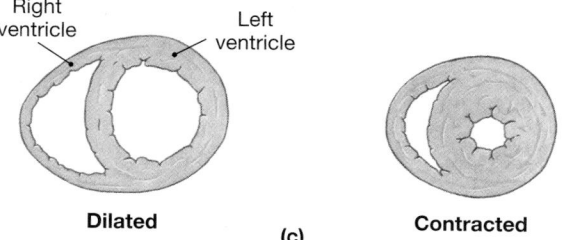

●**FIGURE 20–7**
Structural Differences between the Left and Right Ventricles. **(a)** A diagrammatic sectional view through the heart, showing the relative thicknesses of the two ventricles. Notice the pouchlike shape of the right ventricle and the size of the left ventricle. **(b)** A horizontal section through the heart at the level of vertebra T_8. **(c)** Diagrammatic views of the ventricles just before a contraction (dilated) and just after a contraction (contracted).

(Figure 20–7a,b●). When this ventricle contracts, (1) the distance between the base and apex decreases, and (2) the diameter of the ventricular chamber decreases. If you imagine the effects of simultaneously squeezing and rolling up the end of a toothpaste tube, you will get the idea. The forces generated are quite powerful, more than enough to open the aortic valve and eject blood into the ascending aorta.

As the powerful left ventricle contracts, it also bulges into the right ventricular cavity (Figure 20–7c●). This dual action improves the efficiency of the right ventricle's efforts. Individuals whose right ventricular musculature has been severely damaged may survive, because the contraction of the left ventricle helps push blood into the pulmonary circuit. We shall return to this topic in Chapter 21, where we consider the integrated functioning of the cardiovascular system. **AM** The Cardiomyopathies

The Heart Valves

The heart has a series of one-way valves that prevent the backflow of blood as the chambers contract. We will now describe the structure and function of these heart valves.

THE ATRIOVENTRICULAR VALVES The atrioventricular (AV) valves prevent the backflow of blood from the ventricles to the atria when the ventricles are contracting. The chordae tendineae and papillary muscles play an important role in the normal function of the AV valves. When the ventricles are relaxed, the chordae tendineae are loose and the AV valves offer no resistance to the flow of blood from the atria to the ventricles (Figure 20–8a●). When the ventricles contract, blood moving back toward the atria swings the cusps together, closing the valves (Figure 20–8b●). At the same time, the contraction of the papillary muscles tenses the chordae tendineae and stops the cusps before they swing into the atria. If the chordae tendineae are cut or the papillary muscles are damaged, there is backflow, or **regurgitation**, of blood into the atria each time the ventricles contract.

THE SEMILUNAR VALVES The pulmonary and aortic valves prevent the backflow of blood from the pulmonary trunk and aorta into the right and left ventricles, respectively. Unlike the AV valves, the semilunar valves do not require muscular braces, because the arterial walls do not contract and the relative positions of the cusps are stable. When the semilunar valves close, the three symmetrical cusps support one another like the legs of a tripod (Figure 20–8a,c●).

Serious valve problems can interfere with cardiac function. If valve function deteriorates to the point at which the heart cannot maintain adequate circulatory flow, symptoms of **valvular heart disease (VHD)** appear. Congenital malformations may be responsible, but in many cases the condition develops after *carditis*, an inflammation of the heart, occurs.

One relatively common cause of carditis is **rheumatic (roo-MA-tik) fever**, an acute childhood reaction to infection by streptococcal bacteria. Valve problems serious enough to reduce cardiac function may not appear until 10–20 years after the initial infection. The resulting clinical disorder is known as *rheumatic heart disease (RHD)*. Classic cases of rheumatic fever have become less common in the United States owing to the ready availability of antibiotics. However, failure to complete the full course of antibiotic therapy can result in an increased risk of developing carditis and subsequent rheumatic heart disease. **AM** RHD and Valvular Stenosis

☐ CONNECTIVE TISSUES AND THE FIBROUS SKELETON

The connective tissues of the heart include large numbers of collagen and elastic fibers. Each cardiac muscle cell is wrapped in a strong, but elastic, sheath, and adjacent cells are tied together by fibrous cross-links, or "struts." These fibers are, in turn, interwoven into sheets that separate the superficial and deep muscle layers. The connective-tissue fibers (1) provide physical support for the cardiac muscle fibers, blood vessels, and nerves of the myocardium; (2) help distribute the forces of contraction; (3) add strength and prevent overexpansion of the heart; and (4) provide elasticity that helps return the heart to its original size and shape after a contraction.

The **fibrous skeleton** of the heart consists of four dense bands of tough elastic tissue that encircle the bases of the pulmonary trunk and aorta and the heart valves (Figure 20–8●). These bands stabilize the positions of the heart valves and ventricular muscle cells and physically isolate the ventricular cells from the atrial cells.

☐ THE BLOOD SUPPLY TO THE HEART

The heart works continuously, and cardiac muscle cells require reliable supplies of oxygen and nutrients. The **coronary circulation** supplies blood to the muscle tissue of the heart. During maximum exertion, the demand for oxygen rises considerably. The blood flow to the myocardium may then increase to nine times that of resting levels. The coronary circulation includes an extensive network of coronary blood vessels (Figure 20–9●, p. 694).

The Coronary Arteries

The left and right **coronary arteries** originate at the base of the ascending aorta, at the aortic sinuses (Figure 20–9a●, p. 694). Blood pressure here is the highest in the systemic circuit. Each time the left ventricle contracts, it forces blood into the aorta. The arrival of additional blood at elevated pressures stretches the elastic walls of the aorta, and when the left ventricle relaxes, blood no longer flows into the aorta, pressure declines, and the walls of the aorta recoil. This recoil, called *elastic rebound*, pushes blood both forward, into the systemic circuit, and backward, through the aortic sinus and then into the coronary arteries. Thus, the combination of elevated blood pressure and elastic rebound ensures a continuous flow of blood to meet the demands of active cardiac muscle tissue.

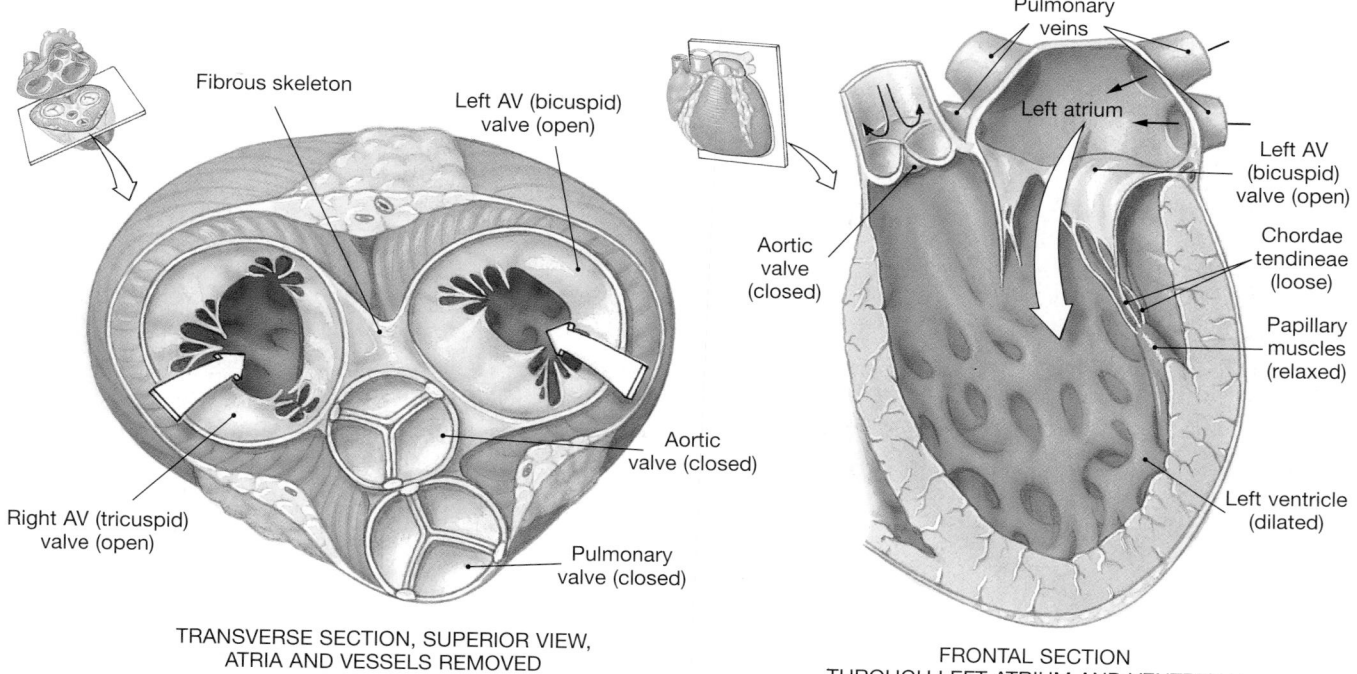

Fibrous skeleton

Left AV (bicuspid) valve (open)

Right AV (tricuspid) valve (open)

Aortic valve (closed)

Pulmonary valve (closed)

TRANSVERSE SECTION, SUPERIOR VIEW, ATRIA AND VESSELS REMOVED

Pulmonary veins

Left atrium

Aortic valve (closed)

Left AV (bicuspid) valve (open)

Chordae tendineae (loose)

Papillary muscles (relaxed)

Left ventricle (dilated)

FRONTAL SECTION THROUGH LEFT ATRIUM AND VENTRICLE

(a) Relaxed ventricles

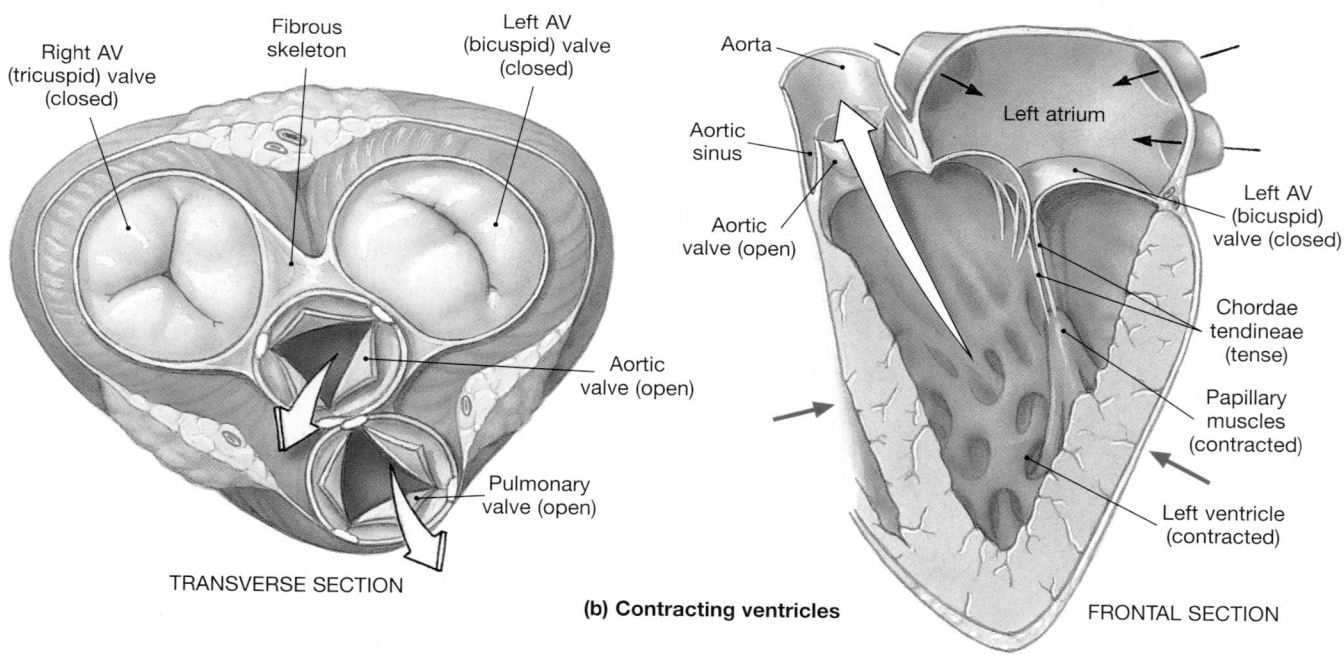

Right AV (tricuspid) valve (closed)

Fibrous skeleton

Left AV (bicuspid) valve (closed)

Aortic valve (open)

Pulmonary valve (open)

TRANSVERSE SECTION

Aorta

Left atrium

Aortic sinus

Aortic valve (open)

Left AV (bicuspid) valve (closed)

Chordae tendineae (tense)

Papillary muscles (contracted)

Left ventricle (contracted)

FRONTAL SECTION

(b) Contracting ventricles

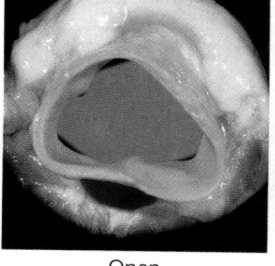

Open

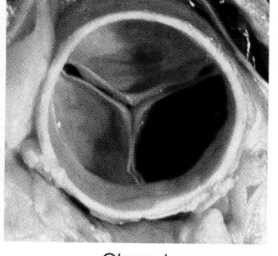

Closed

(c) Semilunar valve function

●**FIGURE 20–8**
Valves of the Heart. White arrows indicate blood flow into or out of a ventricle; black arrows, blood flow into an atrium; and green arrows, ventricular contraction. **(a)** When the ventricles are relaxed, the AV valves are open and the semilunar valves are closed. The chordae tendineae are loose, and the papillary muscles are relaxed. **(b)** When the ventricles are contracting, the AV valves are closed and the semilunar valves are open. In the frontal section, notice the attachment of the left AV valve to the chordae tendineae and papillary muscles. **(c)** The aortic valve in the open (left) and closed (right) positions. The individual cusps brace one another in the closed position.

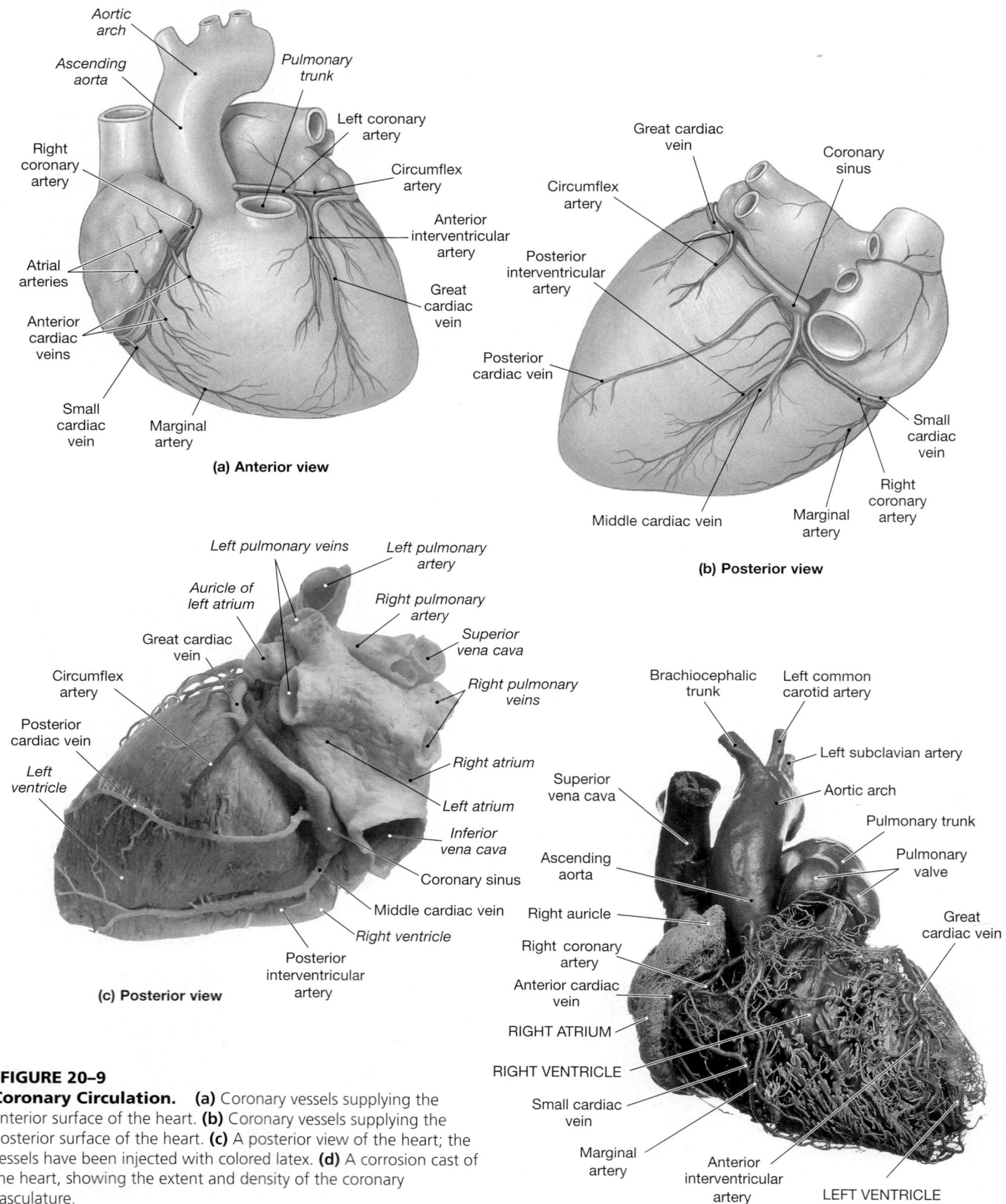

Aortic arch

Ascending aorta

Right coronary artery

Atrial arteries

Anterior cardiac veins

Small cardiac vein

Pulmonary trunk

Left coronary artery

Circumflex artery

Anterior interventricular artery

Great cardiac vein

Marginal artery

(a) Anterior view

Great cardiac vein

Circumflex artery

Posterior interventricular artery

Posterior cardiac vein

Coronary sinus

Middle cardiac vein

Small cardiac vein

Right coronary artery

Marginal artery

(b) Posterior view

Left pulmonary veins

Auricle of left atrium

Great cardiac vein

Circumflex artery

Posterior cardiac vein

Left ventricle

Left pulmonary artery

Right pulmonary artery

Superior vena cava

Right pulmonary veins

Right atrium

Left atrium

Inferior vena cava

Coronary sinus

Middle cardiac vein

Right ventricle

Posterior interventricular artery

(c) Posterior view

Brachiocephalic trunk

Left common carotid artery

Left subclavian artery

Aortic arch

Pulmonary trunk

Pulmonary valve

Great cardiac vein

Superior vena cava

Ascending aorta

Right auricle

Right coronary artery

Anterior cardiac vein

RIGHT ATRIUM

RIGHT VENTRICLE

Small cardiac vein

Marginal artery

Anterior interventricular artery

LEFT VENTRICLE

(d) Coronary circulation and great vessels, anterior view

●**FIGURE 20–9**
Coronary Circulation. **(a)** Coronary vessels supplying the anterior surface of the heart. **(b)** Coronary vessels supplying the posterior surface of the heart. **(c)** A posterior view of the heart; the vessels have been injected with colored latex. **(d)** A corrosion cast of the heart, showing the extent and density of the coronary vasculature.

The **right coronary artery**, which follows the coronary sulcus around the heart, supplies blood to (1) the right atrium, (2) portions of both ventricles, and (3) portions of the conducting system of the heart, including the *sinoatrial (SA)* and *atrioventricular (AV) nodes*. The cells of these nodes are essential to establishing the normal heart rate. We will focus on their functions and their part in regulating the heart rate in a later section.

Inferior to the right atrium, the right coronary artery generally gives rise to one or more **marginal arteries**, which extend across the surface of the right ventricle (Figure 20–9a,b●). The right coronary artery then continues across the posterior surface of the heart, supplying the **posterior interventricular artery**, or *posterior descending artery*, which runs toward the apex within the posterior interventricular sulcus (Figure 20–9b,c●). The posterior interventricular artery supplies blood to the interventricular septum and adjacent portions of the ventricles.

The **left coronary artery** supplies blood to the left ventricle, left atrium, and interventricular septum. As it reaches the anterior surface of the heart, it gives rise to a circumflex branch and an anterior interventricular branch. The **circumflex artery** curves to the left around the coronary sulcus, eventually meeting and fusing with small branches of the right coronary artery (Figure 20–9a,b,c●). The much larger **anterior interventricular artery**, or *left anterior descending artery*, swings around the pulmonary trunk and runs along the surface within the anterior interventricular sulcus (Figure 20–9a●).

The anterior interventricular artery supplies small tributaries continuous with those of the posterior interventricular artery. Such interconnections between arteries are called **arterial anastomoses** (a-nas-tō-MŌ-sēz; *anastomosis*, outlet). Because the arteries are interconnected in this way, the blood supply to the cardiac muscle remains relatively constant despite pressure fluctuations in the left and right coronary arteries as the heart beats.

The Cardiac Veins

The various cardiac veins are shown in Figure 20–9●. The **great cardiac vein** begins on the anterior surface of the ventricles, along the interventricular sulcus. This vein drains blood from the region supplied by the anterior interventricular artery, a branch of the left coronary artery. The great cardiac vein reaches the level of the atria and then curves around the left side of the heart within the coronary sulcus. The vein empties into the coronary sinus, which lies in the posterior portion of the coronary sulcus. The coronary sinus opens into the right atrium near the base of the inferior vena cava.

Other cardiac veins that empty into the great cardiac vein or the coronary sinus include (1) the **posterior cardiac vein**, draining the area served by the circumflex artery, (2) the **middle cardiac vein**, draining the area supplied by the posterior interventricular artery, and (3) the **small cardiac vein**, which receives blood from the posterior surfaces of the right atrium and ventricle. The **anterior cardiac veins**, which drain the anterior surface of the right ventricle, empty directly into the right atrium.

Coronary Artery Disease

The term **coronary artery disease (CAD)** refers to areas of partial or complete blockage of coronary circulation. Cardiac muscle cells need a constant supply of oxygen and nutrients, so any reduction in coronary circulation produces a corresponding reduction in cardiac performance. Such reduced circulatory supply, known as **coronary ischemia** (is-KĒ-mē-uh), generally results from partial or complete blockage of the coronary arteries. The usual cause is the formation of a fatty deposit,

or *plaque*, in the wall of a coronary vessel. The plaque, or an associated *thrombus* (clot), then narrows the passageway and reduces blood flow. Spasms in the smooth muscles of the vessel wall can further decrease or even stop blood flow. Plaques may be visible by *angiography* or high-resolution ultrasound, and the metabolic effects can be detected in digital subtraction angiography (DSA) scans of the heart (Figure 20–10a, b●). We will consider the development and growth of plaques in Chapter 21.

●**FIGURE 20–10**
Coronary Circulation and Clinical Testing. **(a)** A color enhanced DSA image of a healthy heart. The ventricular walls have an extensive circulatory supply. (The atria are not shown.) **(b)** A color enhanced DSA image of a damaged heart. Most of the ventricular myocardium is deprived of circulation.

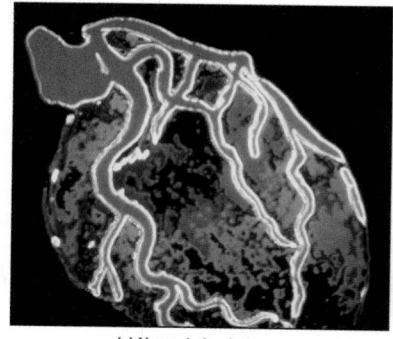

(a) Normal circulation

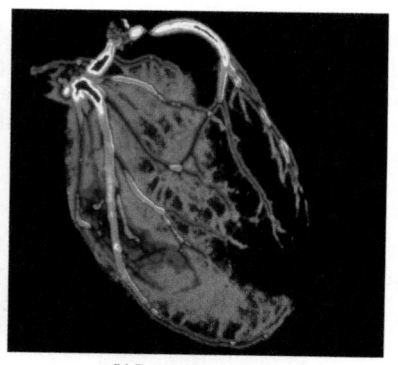

(b) Restricted circulation

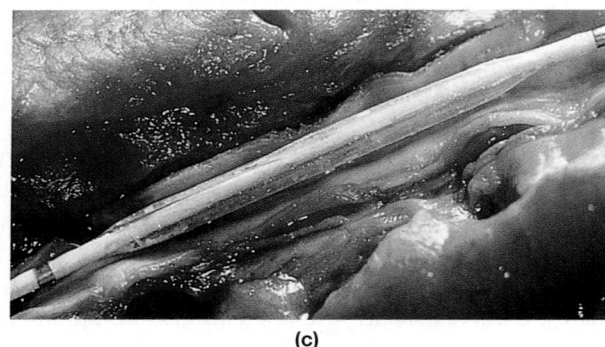
(c)

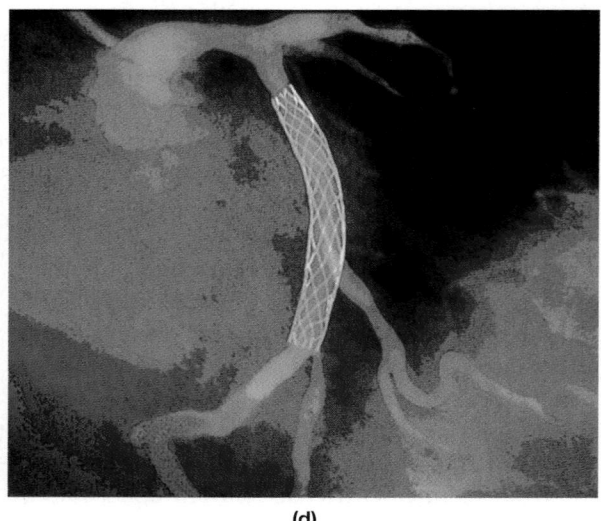
(d)

●**FIGURE 20–10**
Coronary Circulation and Clinical Testing. (*continued*) **(c)** Balloon angioplasty may be used to remove a circulatory blockage. The catheter is guided through the coronary arteries to the site of blockage and inflated to press the soft plaque against the vessel wall. **(d)** A stent is often inserted after balloon angioplasty to help prevent plaques from recurring. This scan shows a stent in the anterior interventricular artery.

One of the first symptoms of CAD is commonly **angina pectoris** (an-JĪ-nuh PEK-tor-is; *angina*, pain spasm + *pectoris*, of the chest). In the most common form, a temporary ischemia develops when the workload of the heart increases. Although the individual may feel comfortable at rest, exertion or emotional stress can produce a sensation of pressure, chest constriction, and pain that may radiate from the sternal area to the arms, back, and neck.

Angina can typically be controlled by a combination of drug treatment and changes in lifestyle, including (1) limiting activities known to trigger angina attacks (e.g., strenuous exercise) and avoiding stressful situations, (2) stopping smoking, and (3) lowering one's fat consumption. Among the medications used to control angina are drugs that block sympathetic stimulation (*propranolol* or *metoprolol*), vasodilators such as *nitroglycerin* (nī-trō-GLIS-er-in), and drugs that block calcium movement into the cardiac muscle cells (*calcium channel blockers*).

Angina can also be treated surgically. A single, soft plaque may be reduced with the aid of a long, slender **catheter** (KATH-e-ter), a small-diameter tube. The catheter is inserted into a large artery and guided into a coronary artery to the plaque. A variety of surgical tools can be slid into the catheter, and the plaque can then be removed with laser beams or chewed to pieces by a device that resembles a miniature version of a Roto-Rooter® drain cleaner. Debris created during the destruction of a plaque is sucked up by the catheter, preventing the blockage of smaller vessels.

In **balloon angioplasty** (AN-jē-ō-plas-tē; *angeion*, vessel), the tip of the catheter contains an inflatable balloon (Figure 20–10c●). Once in position, the balloon is inflated, pressing the plaque against the vessel walls. This procedure works best in small (under 10 mm), soft plaques. Several factors make angioplasty a highly attractive treatment: (1) The mortality rate during surgery is only about 1 percent; (2) the success rate is over 90 percent; and (3) the procedure can be performed on an outpatient basis. Because plaques commonly rebound or redevelop after angioplasty, a fine wire tube called a *stent* may be inserted into the vessel. The stent pushes against the vessel wall, holding it open (Figure 20–10d●). Stents are now part of the standard protocol for many cardiac specialists, as the long-term success rate and incidence of complications are significantly lower than those of balloon angioplasty alone. If the circulatory blockage is too large for a single stent, mutiple stents can be inserted along the length of the vessel.

When angioplasty and stents are not feasible, coronary bypass surgery may be done. In a **coronary artery bypass graft (CABG)**, a small section is removed from either a small artery (commonly the *internal thoracic artery*) or a peripheral vein (such as the *great saphenous vein* of the leg) and is used to create a detour around the obstructed portion of a coronary artery. As many as four coronary arteries can be rerouted this way during a single operation. The procedures are named according to the number of vessels repaired, so we speak of single, double, triple, or quadruple coronary bypasses. The mortality rate during surgery for such operations when they are performed before significant heart damage has occurred is 1–2 percent. Under these conditions, the procedure eliminates angina symptoms in 70 percent of the cases and provides partial relief in another 20 percent.

Although coronary bypass surgery does offer certain advantages, recent studies have shown that for mild angina, it does not yield significantly better results than drug therapy. Current recommendations are that coronary bypass surgery be reserved for cases of severe angina that do not respond to other treatment.

✓ Damage to the semilunar valve on the right side of the heart would affect blood flow to which vessel?

✓ What prevents the AV valves from opening back into the atria?

✓ Why is the left ventricle more muscular than the right ventricle?

Answers start on page Q-1

Review the anatomy of the heart on the **IP CD-ROM**: Cardiovascular System/Anatomy Review: The Heart and on the **Anatomy CD-ROM**:Cardiovascular System/Heart Flythrough.

20–3 THE HEARTBEAT

Objectives

■ Describe the events of an action potential in cardiac muscle, and explain the importance of calcium ions to the contractile process.

■ Discuss the differences between nodal cells and conducting cells, and describe the components and functions of the conducting system of the heart.

■ Identify the electrical events associated with a normal electrocardiogram.

■ Explain the events of the cardiac cycle, including atrial and ventricular systole and diastole, and relate the heart sounds to specific events in the cycle.

☐ CARDIAC PHYSIOLOGY

Figure 20–11● presents an overview of those aspects of cardiac physiology which we will consider in this chapter. In a single heartbeat, the entire heart contracts in series—first the atria and then the ventricles. Two types of cardiac muscle cells are involved in a normal heartbeat: (1) specialized muscle cells of the *conducting system*, which control and coordinate the heartbeat, and (2) *contractile cells*, which produce the powerful contractions that propel blood. Each heartbeat begins with an action potential generated at a pacemaker called the *SA node*, which is part of the conducting system. This electrical impulse is then propagated by the conducting system and distributed so that the stimulated contractile cells will push blood in the right direction at the proper time. The electrical events underway in the conducting system can be monitored from the surface of the body through a procedure known as an *electrocardiogram* (*ECG* or *EKG*).

The arrival of an impulse at a cardiac muscle cell membrane produces an action potential that is comparable to an action potential in a skeletal muscle fiber. As in a skeletal muscle fiber, this action potential triggers the contraction of the cardiac muscle cell. Thanks to the coordination provided by the conducting system, the atria contract first, driving blood into the ventricles through the AV valves, and the ventricles contract next, driving blood out of the heart through the semilunar valves.

The SA node generates impulses at regular intervals, and one heartbeat follows another throughout your life. After each heart-

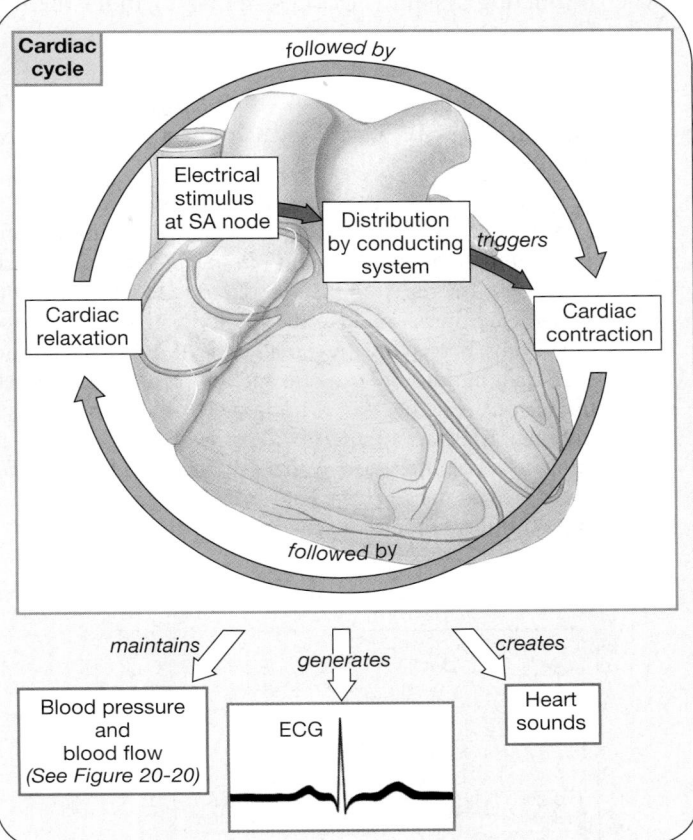

●**FIGURE 20–11**
An Overview of Cardiac Physiology. The major events and relationships are indicated. A simplified version of this figure will appear as a Navigator icon in key figures as we move from one topic to the next.

beat there is a brief pause—less than half a second—before the next heartbeat begins. The period between the start of one heartbeat and the start of the next is called the *cardiac cycle*.

A heartbeat lasts only about 370 msec. Although brief, it is a very busy period! We shall begin our analysis of cardiac function by following the steps that produce a single heartbeat, from the generation of an action potential at the SA node through the contractions of the atria and ventricles.

☐ THE CONDUCTING SYSTEM

In contrast to skeletal muscle, cardiac muscle tissue contracts on its own in the absence of neural or hormonal stimulation. This property is called **automaticity**, or *autorhythmicity*. The cells responsible for initiating and distributing the stimulus to contract are part of the heart's **conducting system**, also known as the *cardiac conduction system* or the *nodal system*. This system is a network of specialized cardiac muscle cells that initiates and distributes electrical impulses. The actual contraction lags behind the passage of an electrical impulse, with the delay representing the time it takes for calcium ions to enter the sarcoplasm and activate the contraction process, as described in Chapter 10. ∞ p. 303

The conducting system (Figure 20–12a●) includes the following elements:

- The *sinoatrial (SA) node*, located in the wall of the right atrium.
- The *atrioventricular (AV) node*, located at the junction between the atria and ventricles.
- *Conducting cells*, which interconnect the two nodes and distribute the contractile stimulus throughout the myocardium. Conducting cells in the atria are found in *internodal pathways*, which distribute the contractile stimulus to atrial muscle cells as the impulse travels from the SA node to the AV node. (The importance of these pathways in relaying the signal to the AV node remains in dispute, because an impulse can also spread from contractile cell to contractile cell, reaching the AV node at about the same time as an impulse that traverses an internodal pathway.) The ventricular conducting cells include those in the *AV bundle* and the *bundle branches*, as well as the *Purkinje* (pur-KIN-jē) *fibers*, which distribute the stimulus to the ventricular myocardium.

Most of the cells of the conducting system are smaller than the contractile cells of the myocardium and contain very few myofibrils. Conducting cells of the SA and AV nodes share another important characteristic: Their excitable membranes cannot maintain a stable resting potential. Each time repolarization occurs, the membrane gradually drifts toward threshold. This gradual depolarization is called a **prepotential** (Figure 20–12b●).

The rate of spontaneous depolarization varies in different portions of the conducting system. It is fastest at the SA node, which in the absence of neural or hormonal stimulation will generate action potentials at a rate of 80–100 per minute. Isolated cells of the AV node depolarize more slowly, generating 40–60 action potentials per minute. Because the SA node reaches threshold first, it establishes the heart rate—the impulse generated by the SA node brings the AV nodal cells to threshold faster than does the prepotential of the AV nodal cells. The normal resting heart rate is somewhat slower than 80–100 per minute, however, due to the effects of parasympathetic innervation. (The influence of autonomic innervation on heart rate is discussed in a later section.)

If any of the atrial pathways or the SA node becomes damaged, the heart will continue to beat, but it will do so at a slower rate, usually 40–60 beats per minute, as dictated by the AV node. Certain cells in the Purkinje fiber network depolarize spontaneously at an even slower rate, and if the rest of the conducting system is damaged, they can stimulate a heart rate of 20–40 beats per minute. Under normal conditions, cells of the AV bundle, the bundle branches, and most Purkinje fibers do not depolarize spontaneously. If, due to damage or disease, these cells *do* begin depolarizing spontaneously, the heart may no longer pump blood effectively, and death can occur if the problem persists.

We will now trace the path of an impulse from its initiation at the SA node, examining its effects on the surrounding myocardium as we proceed.

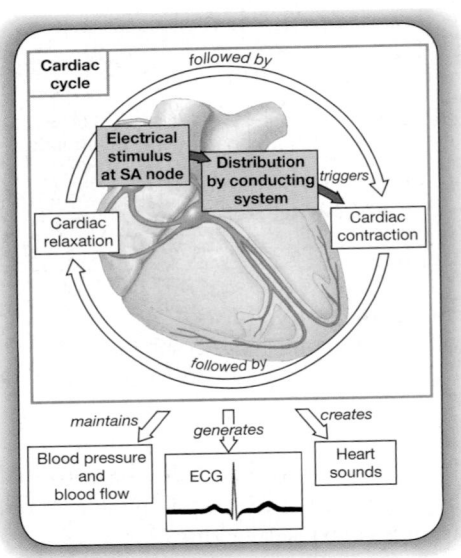

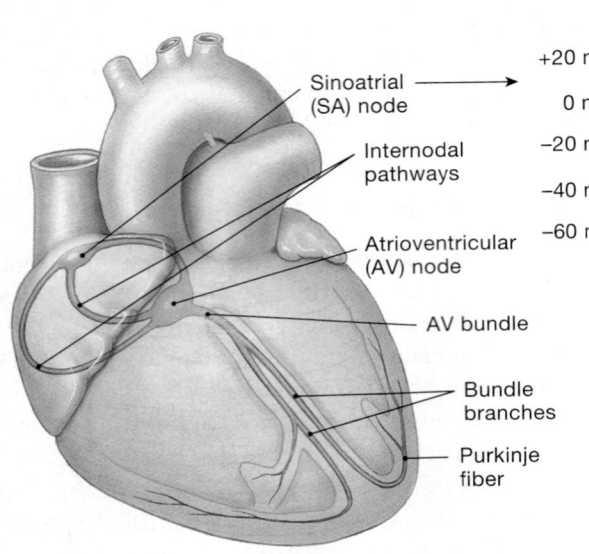

(a) **The conducting system**

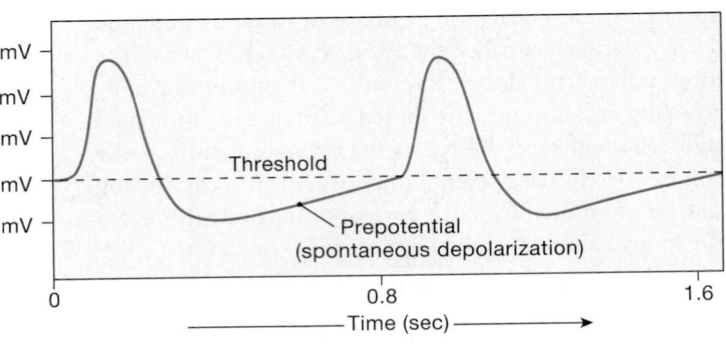

(b) **Depolarization at the SA node**

●**FIGURE 20–12**
The Conducting System of the Heart. The Navigator icon in the shadow box highlights the topics we will consider in this section. **(a)** Components of the conducting system. **(b)** The spontaneous changes in the membrane potential of a pacemaker cell in the SA node that is establishing a heart rate of 72 beats per minute.

The Sinoatrial (SA) Node

The **sinoatrial** (sī-nō-Ā-trē-al) **node** (**SA node**) is embedded in the posterior wall of the right atrium, near the entrance of the superior vena cava (Figures 20–12a and 20–13 [STEP 1]●). The SA node contains **pacemaker cells**, which establish the heart rate. As a result, the SA node is also known as the *cardiac pacemaker* or the *natural pacemaker*. The SA node is connected to the larger AV node by the internodal pathways in the atrial walls. It takes roughly 50 msec for an action potential to travel from the SA node to the AV node along these pathways. Along the way, the conducting cells pass the stimulus to contractile cells of both atria. The action potential then spreads across the atrial surfaces by cell-to-cell contact (Figure 20–13 [STEP 2]●). The stimulus affects only the atria, because the fibrous skeleton isolates the atrial myocardium from the ventricular myocardium.

The Atrioventricular (AV) Node

The relatively large **atrioventricular** (**AV**) **node** (Figures 20–12a and 20–13 [STEP 2]●) sits within the floor of the right atrium near the opening of the coronary sinus. The rate of propagation of the impulse slows as it leaves the internodal pathways and enters the AV node, because the nodal cells are smaller in diameter than the conducting cells. (Chapter 12 discussed the relationship between diameter and propagation speed. ∞ p. 409) In addition, the connections between nodal cells are less efficient than those between conducting cells at relaying the impulse from one cell to another. As a result, it takes about 100 msec for the impulse to pass through the AV node and enter the AV bundle (Figure 20–13 [STEP 3]●).

The delay at the AV node is important, because the atria must contract before the ventricles do. Otherwise, the contraction of the powerful ventricles would close the AV valves and prevent blood flow from the atria into the ventricles. Because there is a delay at the AV node, the atrial myocardium completes its contraction before ventricular contraction begins.

The cells of the AV node can conduct impulses at a maximum rate of 230 per minute. Since each impulse will result in a ventricular contraction, this value is the maximum normal heart rate. Even if the SA node generates impulses at a faster rate, the ventricles will still contract at 230 beats per minute (bpm). This limitation is important, because mechanical factors, discussed later, begin to decrease the pumping efficiency of the heart at rates above approximately 180 bpm. Rates above 230 bpm occur only when the heart or the conducting system has been damaged or stimulated by drugs. As ventricular rates increase toward their theoretical maximum limit of 300–400 bpm, pumping efficiency becomes dangerously, if not fatally, reduced.

A number of clinical problems are the result of abnormal pacemaker function. **Bradycardia** (brā-dē-KAR-dē-uh; *bradys*, slow) is a condition in which the heart rate is slower than normal, whereas **tachycardia** (tak-ē-KAR-dē-uh; *tachys*, swift) indicates a faster-than-normal heart rate. These are relative terms, and in clinical practice the definitions vary with the normal resting heart rate of the individual. **AM** Problems with Pacemaker Function

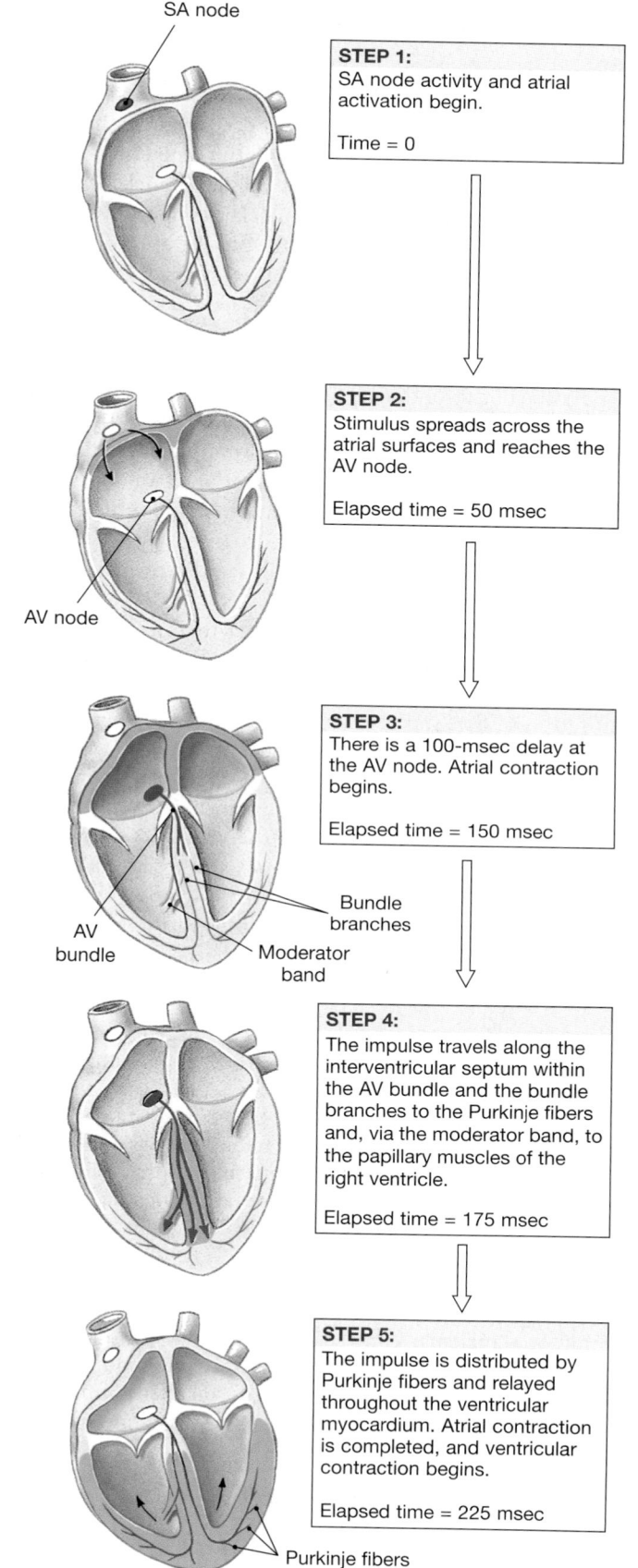

STEP 1: SA node activity and atrial activation begin. Time = 0

STEP 2: Stimulus spreads across the atrial surfaces and reaches the AV node. Elapsed time = 50 msec

STEP 3: There is a 100-msec delay at the AV node. Atrial contraction begins. Elapsed time = 150 msec

STEP 4: The impulse travels along the interventricular septum within the AV bundle and the bundle branches to the Purkinje fibers and, via the moderator band, to the papillary muscles of the right ventricle. Elapsed time = 175 msec

STEP 5: The impulse is distributed by Purkinje fibers and relayed throughout the ventricular myocardium. Atrial contraction is completed, and ventricular contraction begins. Elapsed time = 225 msec

●**FIGURE 20–13**
Impulse Conduction through the Heart

The AV Bundle, Bundle Branches, and Purkinje Fibers

The connection between the AV node and the **AV bundle**, also called the *bundle of His*, is the only electrical connection between the atria and the ventricles. Once an impulse enters the AV bundle, it travels to the interventricular septum and enters the **right** and **left bundle branches**. The left bundle branch, which supplies the massive left ventricle, is much larger than the right bundle branch. Both branches extend toward the apex of the heart, turn, and fan out deep to the endocardial surface. As the branches diverge, they conduct the impulse to **Purkinje fibers** and, through the moderator band, to the papillary muscles of the right ventricle.

Purkinje fibers conduct action potentials very rapidly—as fast as small myelinated axons. Within about 75 msec, the signal to begin a contraction has reached all the ventricular cardiac muscle cells. The entire process, from the generation of an impulse at the SA node to the complete depolarization of the ventricular myocardium, normally takes around 225 msec. By this time, the atria have completed their contractions and ventricular contraction can safely occur.

Because the bundle branches deliver the impulse across the moderator band to the papillary muscles directly rather than by way of Purkinje fibers, the papillary muscles begin contracting before the rest of the ventricular musculature does. Contraction of the papillary muscles applies tension to the chordae tendineae, bracing the AV valves. By limiting the movement of the cusps, tension in the chordae tendineae prevents the backflow of blood into the atria when the ventricles contract. The Purkinje fibers radiate from the apex toward the base of the heart. As a result, ventricular contraction proceeds in a wave that begins at the apex and spreads toward the base. Blood is therefore pushed toward the base of the heart, into the aorta and pulmonary trunk.

If the conducting pathways are damaged, the normal rhythm of the heart will be disturbed. The resulting problems are called *conduction deficits*. If the SA node or internodal pathways are damaged, the AV node will assume command. The heart will continue beating normally, although at a slower rate. If an abnormal conducting cell or ventricular muscle cell begins generating action potentials at a more rapid rate, the impulses can override those of the SA or AV node. The origin of these abnormal signals is called an **ectopic** (ek-TOP-ik; out of place) **pacemaker**. The activity of an ectopic pacemaker partially or completely bypasses the conducting system, disrupting the timing of ventricular contraction. The result is a dangerous reduction in the efficiency of the heart. Such conditions are commonly diagnosed with the aid of an *electrocardiogram*.

AM Diagnosing Abnormal Heartbeats

☐ THE ELECTROCARDIOGRAM

The electrical events occurring in the heart are powerful enough to be detected by electrodes on the surface of the body. A recording of these events is an **electrocardiogram** (ē-lek-trō-KAR-dē-ō-gram), also called an **ECG** or **EKG**. Each time the heart beats, a wave of depolarization radiates through the atria, reaches the AV node, travels down the interventricular septum to the apex, turns, and spreads through the ventricular myocardium toward the base (Figure 20–13●).

By comparing the information obtained from electrodes placed at different locations, a clinician can monitor the electrical activity of the heart, which is directly related to the performance of specific nodal, conducting, and contractile components. For example, when a portion of the heart has been damaged, the affected muscle cells will no longer conduct action potentials. An ECG will reveal an abnormal pattern of impulse conduction.

The appearance of the ECG varies with the placement of the monitoring electrodes, or *leads*. Figure 20–14a● shows the leads in one of the standard configurations. Figure 20–14b● depicts the important features of an ECG obtained with that configuration. Note the following ECG features:

- The small **P wave** accompanies the depolarization of the atria. The atria begin contracting about 100 msec after the start of the P wave.

- The **QRS complex** appears as the ventricles depolarize. This is a relatively strong electrical signal, because the ventricular muscle is much more massive than that of the atria. It is also a complex signal, in part because it incorporates atrial repolarization as well as ventricular depolarization. The ventricles begin contracting shortly after the peak of the **R wave**.

- The smaller **T wave** indicates ventricular repolarization. You do not see a deflection corresponding to atrial repolarization, because it occurs while the ventricles are depolarizing and the electrical events are masked by the QRS complex.

To analyze an ECG, you must measure the size of the voltage changes and determine the durations and temporal relationships of the various components. Attention usually is focused on the amount of depolarization occurring during the P wave and the QRS complex. For example, an excessively large QRS complex often indicates that the heart has become enlarged. A smaller-than-normal electrical signal may mean that the mass of the heart muscle has decreased (although monitoring problems are more often responsible).

The size and shape of the T wave may also be affected by any condition that slows ventricular repolarization. For example, starvation and low cardiac energy reserves, coronary ischemia, or abnormal ion concentrations will reduce the size of the T wave.

You must also measure the time between waves. The values are reported as *segments* or *intervals*. Segments generally extend from the end of one wave to the start of another; intervals are more variable, but always include at least one entire wave. Commonly used segments and intervals are indicated in Figure 20–14b●. The names, however, do not always seem to fit. For example,

- The **P–R interval** extends from the start of atrial depolarization to the start of the QRS complex (ventricular depolarization) rather than to R, because in abnormal ECGs the peak can be difficult to determine. Extension of the P–R interval to more than 0.2 second can indicate damage to the conducting pathways or AV node.

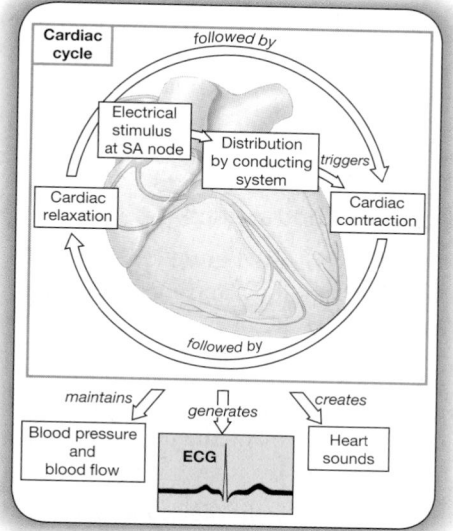

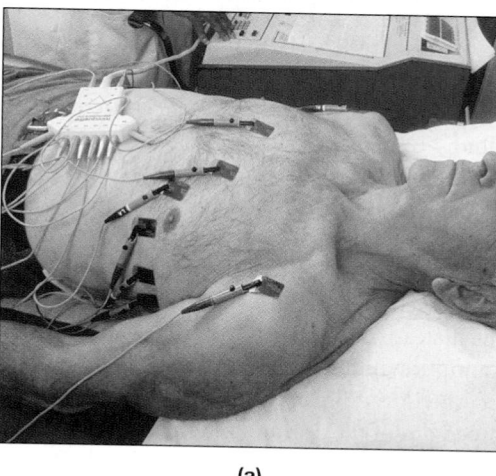

(a)

•FIGURE 20–14
An Electrocardiogram. The Navigator icon in the shadow box highlights the topic we will consider in this section. **(a)** Electrode placement for recording a standard ECG. **(b)** An ECG printout is a strip of graph paper containing a record of the electrical events monitored by the electrodes. The placement of electrodes on the body surface affects the size and shape of the waves recorded. This example is a normal ECG; the enlarged section indicates the major components of the ECG and the measurements most often taken during clinical analysis.

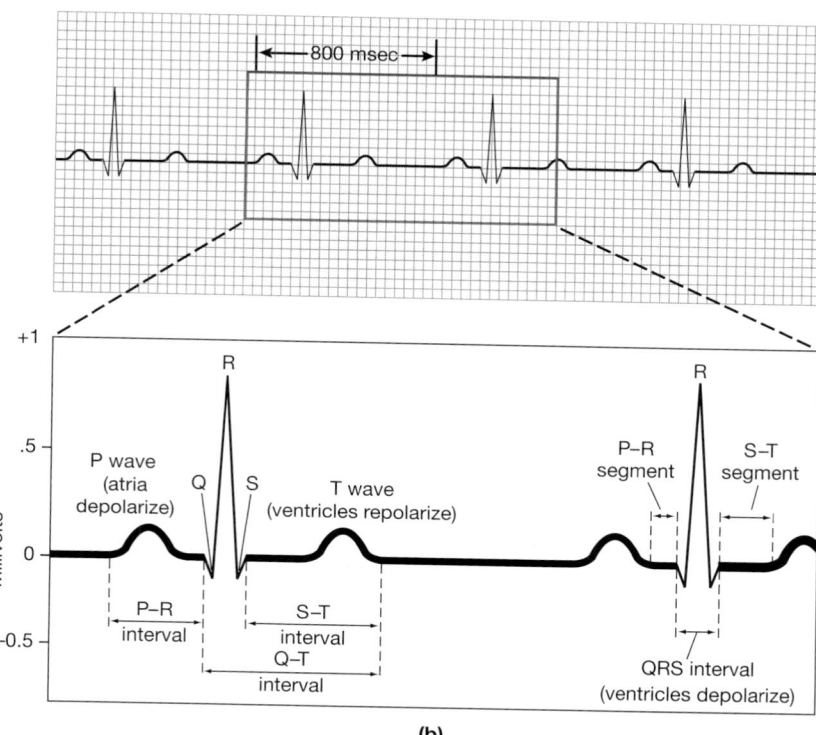

(b)

- The **Q–T interval** indicates the time required for the ventricles to undergo a single cycle of depolarization and repolarization. It is usually measured from the end of the P–R interval rather than from the bottom of the Q wave. The Q–T interval can be lengthened by conduction problems, coronary ischemia, or myocardial damage. A congenital heart defect that can cause sudden death without warning may be detectable as a prolonged Q–T interval.

Despite the variety of sophisticated equipment available to assess or visualize cardiac function, in the majority of cases the ECG provides the most important diagnostic information. ECG analysis is especially useful in detecting and diagnosing **cardiac arrhythmias** (ā-RITH-mē-az)—abnormal patterns of cardiac electrical activity. Momentary arrhythmias are not inherently dangerous; about 5 percent of the healthy population experiences a few abnormal heartbeats each day. Clinical problems appear when the arrhythmias reduce the pumping efficiency of the heart. Serious arrhythmias may indicate damage to the myocardium, injuries to the pacemakers or conduction pathways, exposure to drugs, or variations in the electrolyte composition of extracellular fluids. **AM** Diagnosing Abnormal Heartbeats; Examining the Heart

☐ CONTRACTILE CELLS

The Purkinje fibers distribute the stimulus to the **contractile cells,** which form the bulk of the atrial and ventricular walls. In the discussions of cardiac muscle tissue in earlier chapters, we considered only the structure of contractile cells, which account for roughly 99 percent of the muscle cells in the heart. In both cardiac muscle cells and skeletal muscle fibers, (1) an action potential leads to the appearance of Ca^{2+} among the myofibrils, and (2) the binding of Ca^{2+} to troponin on the thin filaments initiates the contraction. But skeletal and cardiac muscle cells differ in terms of the nature of the action potential, the source of the Ca^{2+}, and the duration of the resulting contraction. ∞ p. 324

The Action Potential in Cardiac Muscle Cells

In Chapter 12, we compared action potentials in skeletal muscle fibers and in cardiac muscle cells. ∞ p. 411 We will now take a closer look at the origin and conduction of an action potential in a contractile cell.

The resting potential of a ventricular contractile cell is approximately −90 mV, comparable to that of a resting skeletal muscle fiber (−85 mV). (The resting potential of an atrial contractile cell is about −80 mV, but the basic principles described here apply to atrial cells as well.) An action potential begins when the membrane of the ventricular muscle cell is brought to threshold, usually at about −75 mV. Threshold is normally reached in a portion of the membrane next to an intercalated disc. The typical stimulus is the excitation of an adjacent muscle cell. Once threshold has been reached, the action potential proceeds in three basic steps (Figure 20–15a●):

STEP 1: *Rapid Depolarization.* The stage of rapid depolarization in a cardiac muscle cell resembles that in a skeletal muscle fiber. At threshold, voltage-regulated sodium channels open and the membrane suddenly becomes permeable to Na^+. The result is a massive influx of sodium ions and the rapid depolarization of the sarcolemma. The channels involved are called **fast channels**, because they open quickly and remain open for only a few milliseconds.

STEP 2: *The Plateau.* As the transmembrane potential approaches +30 mV, the voltage-regulated sodium channels close. They will remain closed and inactivated until the membrane potential reaches −60 mV. The cell now begins actively pumping Na^+ out of the cell. However, a net loss of positive charges does not continue, because as the sodium channels are closing, voltage-regulated calcium channels are opening. These channels are called **slow calcium channels**, because they open slowly and remain open for a relatively long period—roughly 175 msec. While the slow calcium channels are open, calcium ions enter the sarcoplasm. The entry of positive charges through the calcium channels, in the form of Ca^{2+}, roughly balances the loss of positive ions through the active transport of Na^+ and the transmembrane potential remains near 0 mV for an extended period. This portion of the action potential curve is called the plateau. The presence of a plateau is the major difference between action potentials in cardiac muscle cells and in skeletal muscle fibers. In a skeletal muscle fiber, rapid depolarization is immediately followed by rapid repolarization.

STEP 3: *Repolarization.* As the plateau continues, slow calcium channels begin closing and **slow potassium channels** begin opening. As the channels open, potassium ions (K^+) rush out of the cell, and the net result is a period of rapid repolarization that restores the resting potential.

THE REFRACTORY PERIOD As with skeletal muscle contractions, for some time after an action potential begins, the membrane will not respond normally to a second stimulus. This time is called the refractory period. In the absolute refractory period, the membrane cannot respond at all, because the sodium channels either are already open or are closed and inactivated. In a ventricular muscle cell, the absolute refractory period lasts approximately 200 msec, spanning the duration of the plateau and the initial period of rapid repolarization.

The absolute refractory period is followed by a shorter (50 msec) relative refractory period. During this period, the voltage-regulated sodium channels are closed, but can open.

The membrane will respond to a stronger-than-normal stimulus by initiating another action potential. In total, an action potential in a ventricular contractile cell lasts 250–300 msec, roughly 30 times as long as a typical action potential in a skeletal muscle fiber.

Calcium Ions and Cardiac Contractions

The appearance of an action potential in the cardiac muscle cell membrane produces a contraction by causing an increase in the concentration of Ca^{2+} around the myofibrils. This process occurs in two steps:

1. Calcium ions entering the cell membrane during the plateau phase of the action potential provide roughly 20 percent of the Ca^{2+} required for a contraction.

2. The arrival of extracellular Ca^{2+} is the trigger for the release of additional Ca^{2+} from reserves in the sarcoplasmic reticulum (SR).

Extracellular calcium ions thus have both direct and indirect effects on cardiac muscle cell contraction. For this reason, cardiac muscle tissue is highly sensitive to changes in the Ca^{2+} concentration of the extracellular fluid.

In a skeletal muscle fiber, the action potential is relatively brief and ends as the related twitch contraction begins (Figure 20–15b●). The twitch contraction is short and ends as the SR reclaims the Ca^{2+} it released. In a cardiac muscle cell, as we have seen, the action potential is prolonged and calcium ions continue to enter the cell throughout the plateau. As a result, the period of active muscle cell contraction continues until the plateau ends. As the slow calcium channels close, the intracellular calcium ions are absorbed by the SR or are pumped out of the cell, and the muscle cell relaxes.

In skeletal muscle fibers, the refractory period ends before peak tension develops. As a result, twitches can summate and tetanus can occur. In cardiac muscle cells, the absolute refractory period continues until relaxation is under way. Because summation is not possible, tetanic contractions cannot occur in a normal cardiac muscle cell, regardless of the frequency or intensity of stimulation. This feature is vital: A heart in tetany could not pump blood. With a single twitch lasting 250 msec or longer, a normal cardiac muscle cell can reach 300–400 contractions per minute under maximum stimulation. This rate is not reached in a normal heart, due to limitations imposed by the conducting system.

✓ If the cells of the SA node failed to function, how would the heart rate be affected?

✓ In a person with bradycardia, is cardiac output likely to be greater than or less than normal? Explain.

✓ Why is it important for impulses from the atria to be delayed at the AV node before they pass into the ventricles?

Answers start on page Q-1

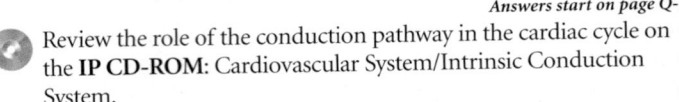 Review the role of the conduction pathway in the cardiac cycle on the **IP CD-ROM**: Cardiovascular System/Intrinsic Conduction System.

•**FIGURE 20–15**
The Action Potential in Skeletal and Cardiac Muscle. The Navigator icon in the shadow box highlights the topic we will consider in this section. **(a)** An action potential in a ventricular muscle cell. **(b)** The relationship between an action potential and a twitch contraction in skeletal muscle and cardiac muscle. The shaded areas indicate the duration of the absolute (green) and relative (beige) refractory periods.

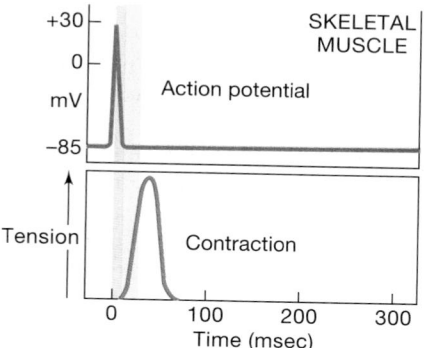

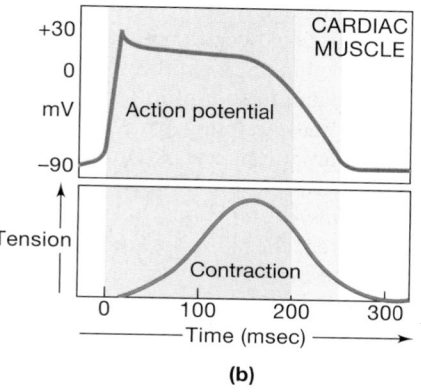

(b)

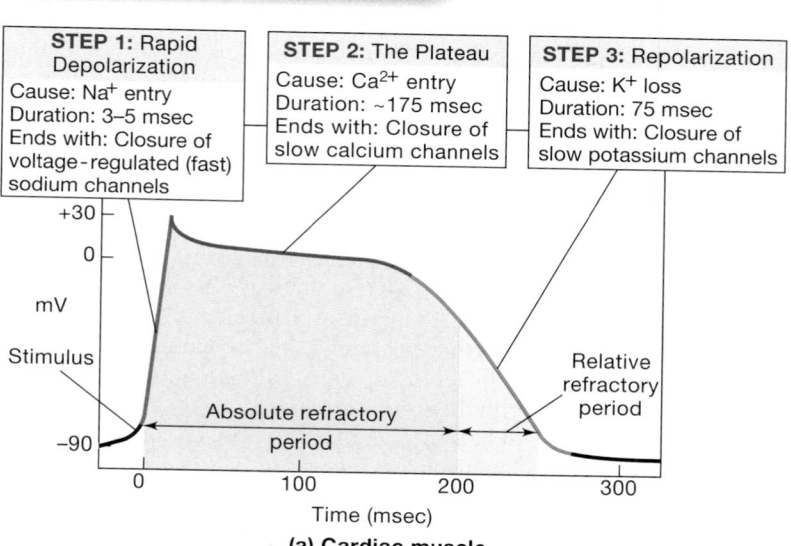

(a) Cardiac muscle

STEP 1: Rapid Depolarization
Cause: Na⁺ entry
Duration: 3–5 msec
Ends with: Closure of voltage-regulated (fast) sodium channels

STEP 2: The Plateau
Cause: Ca²⁺ entry
Duration: ~175 msec
Ends with: Closure of slow calcium channels

STEP 3: Repolarization
Cause: K⁺ loss
Duration: 75 msec
Ends with: Closure of slow potassium channels

◻ THE CARDIAC CYCLE

Each heartbeat is followed by a brief resting phase, allowing time for the chambers to relax and prepare for the next heartbeat. The period between the start of one heartbeat and the beginning of the next is a single **cardiac cycle**. The cardiac cycle, therefore, includes alternating periods of contraction and relaxation. For any one chamber in the heart, the cardiac cycle can be divided into two phases: (1) systole and (2) diastole. During **systole** (SIS-tō-lē), or contraction, the chamber contracts and pushes blood into an adjacent chamber or into an arterial trunk. Systole is followed by **diastole** (dī-AS-tō-lē), or relaxation. During diastole, the chamber fills with blood and prepares for the next cardiac cycle.

Fluids move from an area of high pressure to one of lower pressure. In the course of the cardiac cycle, the pressure within each chamber rises during systole and falls during diastole. Valves between adjacent chambers help ensure that blood flows in the desired direction, but blood will flow from one chamber to another only if the pressure in the first chamber exceeds that in the second. This basic principle governs the movement of blood between atria and ventricles, between ventricles and arterial trunks, and between major veins and atria.

The correct pressure relationships are dependent on the careful timing of contractions. For example, blood could not move in the desired direction if an atrium and its attached ventricle contracted at precisely the same moment. The elaborate pacemaking and conducting systems normally provide the required spacing between atrial and ventricular systoles. At a representative heart rate of 75 bpm, a sequence of systole and diastole in either the atria or the ventricles lasts 800 msec. For convenience, we shall assume that the cardiac cycle is determined by the atria and that it includes one cycle of atrial systole and atrial diastole. This convention follows our description of the conducting system and the propagation of the stimulus for contraction.

Heart Attacks

In a **myocardial** (mī-ō-KAR-dē-al) **infarction** (**MI**), or *heart attack*, part of the coronary circulation becomes blocked and cardiac muscle cells die from lack of oxygen. The affected tissue then degenerates, creating a nonfunctional area known as an *infarct*. Heart attacks most commonly result from severe coronary artery disease (CAD, p. 695). The consequences depend on the site and nature of the circulatory blockage. If it occurs near the start of one of the coronary arteries, the damage will be widespread and the heart may stop beating. If the blockage involves one of the smaller arterial branches, the individual may survive the immediate crisis but may have many complications, all unpleasant. As scar tissue forms in the damaged area, the heartbeat may become irregular and other vessels can become constricted, creating additional circulatory problems.

Myocardial infarctions are generally associated with fixed blockages, such as those seen in CAD. When the crisis develops as a result of thrombus (clot) formation at a plaque, the condition is called **coronary thrombosis**. A vessel already narrowed by plaque formation may also become blocked by a sudden spasm in the smooth muscles of the vascular wall. The individual then may experience intense pain, similar to that felt in an angina attack, but persisting even at rest. However, pain does not always accompany a heart attack, and *silent heart attacks* may be even more dangerous than more apparent, because the condition may go undiagnosed and may not be treated before a fatal MI occurs. Roughly 25 percent of heart attacks are not recognized when they occur.

The cytoplasm of a damaged cardiac muscle cell differs from that of a normal muscle cell. As the supply of oxygen decreases, the cells become more dependent on anaerobic metabolism to meet their energy needs. p. 317 Over time, the cytoplasm accumulates large numbers of enzymes involved with anaerobic energy production. As the membranes of the cardiac muscle cells deteriorate, these enzymes enter the surrounding intercellular fluids. The appearance of such enzymes in the circulation thus indicates that an infarct has occurred. The enzymes that are tested for in a diagnostic blood test include **lactate dehydrogenase** (**LDH**), **serum glutamic oxaloacetic transaminase** (**SGOT**, also called *aspartate aminotransferase*), **creatine phosphokinase** (**CPK** or **CK**), and a special form of creatine phosphokinase that occurs only in cardiac muscle (**CK–MB**). (These enzyme tests are described, and normal and abnormal results are considered, in the cardiovascular section of the *Applications Manual*.)

About 25 percent of MI patients die before obtaining medical assistance, and 65 percent of MI deaths among those under age 50 occur within an hour after the initial infarct. The goals of treatment are to limit the size of the infarct and to avoid additional complications by preventing irregular

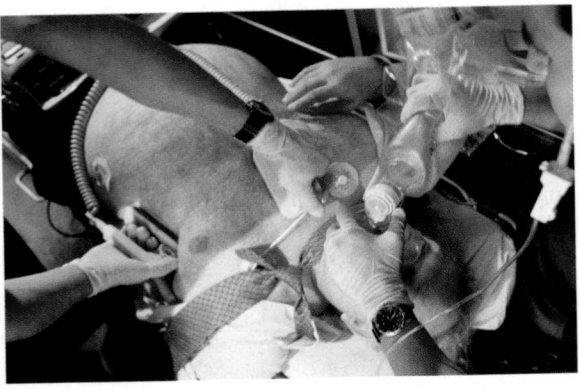

contractions, improving circulation with vasodilators, providing additional oxygen, reducing the cardiac workload, and, if possible, eliminating the cause of the circulatory blockage. Anticoagulants (even aspirin chewed and swallowed at the start of an MI) may help prevent the formation of additional thrombi, and clot-dissolving enzymes may reduce the extent of the damage if they are administered within six hours after the MI occurred. Current evidence suggests that tissue plasminogen activator (t-PA), which is relatively expensive, is more beneficial than other fibrinolytic agents, such as urokinase or streptokinase. p. 676 Follow-up treatment with heparin, aspirin, or both is recommended; without further treatment, the circulatory blockages will reappear in about 20 percent of patients.

Roughly 1.3 million MIs occur in the United States each year, and half the victims die within a year of the incident. The following factors appear to increase the risk of a heart attack: (1) smoking, (2) high blood pressure, (3) high blood cholesterol levels, (4) high circulating levels of low-density lipoproteins (LDLs), (5) diabetes, (6) male gender (below age 70), (7) severe emotional stress, (8) obesity, (9) genetic predisposition, and (10) a sedentary lifestyle. We shall consider the role of lipoproteins and cholesterol in plaque formation and heart disease in Chapter 21. Although the heart attack rate of women under age 70 is lower than that of men, the mortality rate for women is higher—perhaps because heart disease in women is neither diagnosed as early nor treated as aggressively as that in men.

The presence of two risk factors more than doubles the risk of heart attack, so eliminating as many risk factors as possible will improve the chances of preventing or surviving a heart attack. Changing the diet to limit cholesterol, exercising to lower weight, and seeking treatment for high blood pressure are steps in the right direction. It has been estimated that a reduction in coronary risk factors could prevent 150,000 deaths each year in the United States alone.

Phases of the Cardiac Cycle

The phases of atrial systole, atrial diastole, ventricular systole, and ventricular diastole are diagrammed in Figure 20–16● for a heart rate of 75 bpm. As this cardiac cycle begins, all four chambers are relaxed and the ventricles are partially filled with blood. During atrial systole, the atria contract, filling the ventricles completely with blood (Figure 20–16a,b●). Atrial systole lasts 100 msec. Over this period, blood cannot flow into the atria because atrial pressure exceeds venous pressure. Yet there is very little backflow into the veins, even though the connections with the venous system lack valves, because blood takes the path of least resistance. Resistance to blood flow through the broad AV connections and into the ventricles is less than that through the smaller, angled openings of the large veins, which are connected to miles of smaller vessels—many of which do have valves.

The atria next enter atrial diastole, which continues until the start of the next cardiac cycle. Atrial diastole and ventricular systole begin at the same time. Ventricular systole lasts 270 msec. During this period, blood is pushed through the systemic and pulmonary circuits and toward the atria (Figure 20–16c,d●). The heart then enters ventricular diastole (Figure 20–16e,f●), which lasts 530 msec (the 430 msec remaining in this cardiac cycle, plus

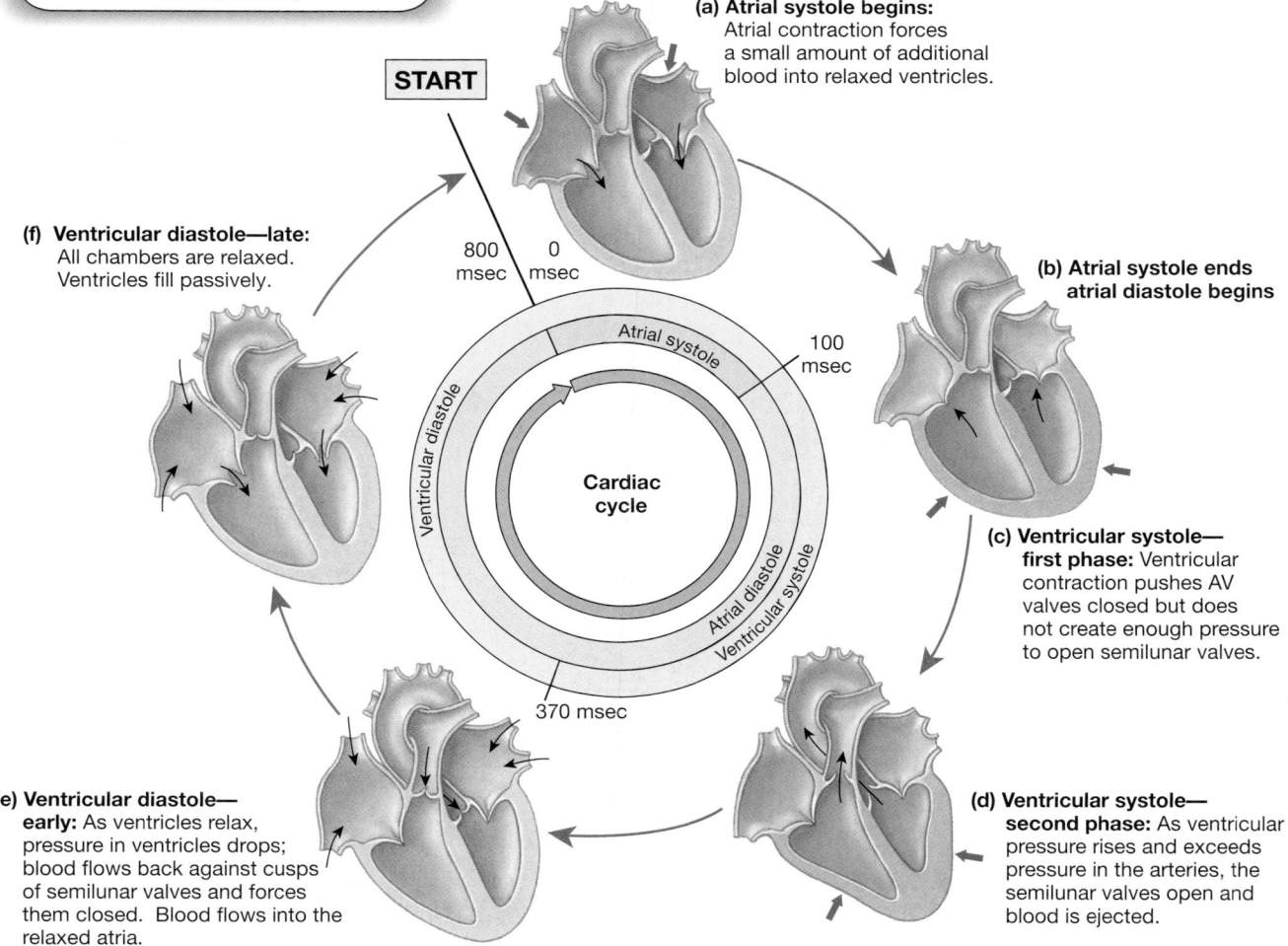

●**FIGURE 20–16**
Phases of the Cardiac Cycle. Thin black arrows indicate blood flow, and green arrows indicate contractions.

(a) Atrial systole begins: Atrial contraction forces a small amount of additional blood into relaxed ventricles.

START

(f) Ventricular diastole—late: All chambers are relaxed. Ventricles fill passively.

800 msec   0 msec

Atrial systole

100 msec

(b) Atrial systole ends atrial diastole begins

Cardiac cycle

Ventricular diastole

Atrial diastole

Ventricular systole

(c) Ventricular systole— first phase: Ventricular contraction pushes AV valves closed but does not create enough pressure to open semilunar valves.

370 msec

(e) Ventricular diastole— early: As ventricles relax, pressure in ventricles drops; blood flows back against cusps of semilunar valves and forces them closed. Blood flows into the relaxed atria.

(d) Ventricular systole— second phase: As ventricular pressure rises and exceeds pressure in the arteries, the semilunar valves open and blood is ejected.

Cardiac cycle
followed by
Electrical stimulus at SA node
Distribution by conducting system
triggers
Cardiac contraction
Cardiac relaxation
followed by
maintains
generates
creates
Blood pressure and blood flow
ECG
Heart sounds

the first 100 msec of the next). For the rest of this cycle, filling occurs passively and both the atria and the ventricles are relaxed. The next cardiac cycle begins with atrial systole and the completion of ventricular filling.

When the heart rate increases, all the phases of the cardiac cycle are shortened. The greatest reduction occurs in the length of time spent in diastole. When the heart rate climbs from 75 bpm to 200 bpm, the time spent in systole drops by less than 40 percent, but the duration of diastole is reduced by almost 75 percent.

Pressure and Volume Changes in the Cardiac Cycle

In considering the pressure and volume changes that occur during the cardiac cycle, we shall use reference numbers that match those in Figure 20–17●, which follow pressures and volumes within the left atrium and left ventricle. The discussion applies to both sides of the heart. Although pressures are lower in the right atrium and right ventricle, both sides of the heart contract at the same time, and they eject equal volumes of blood.

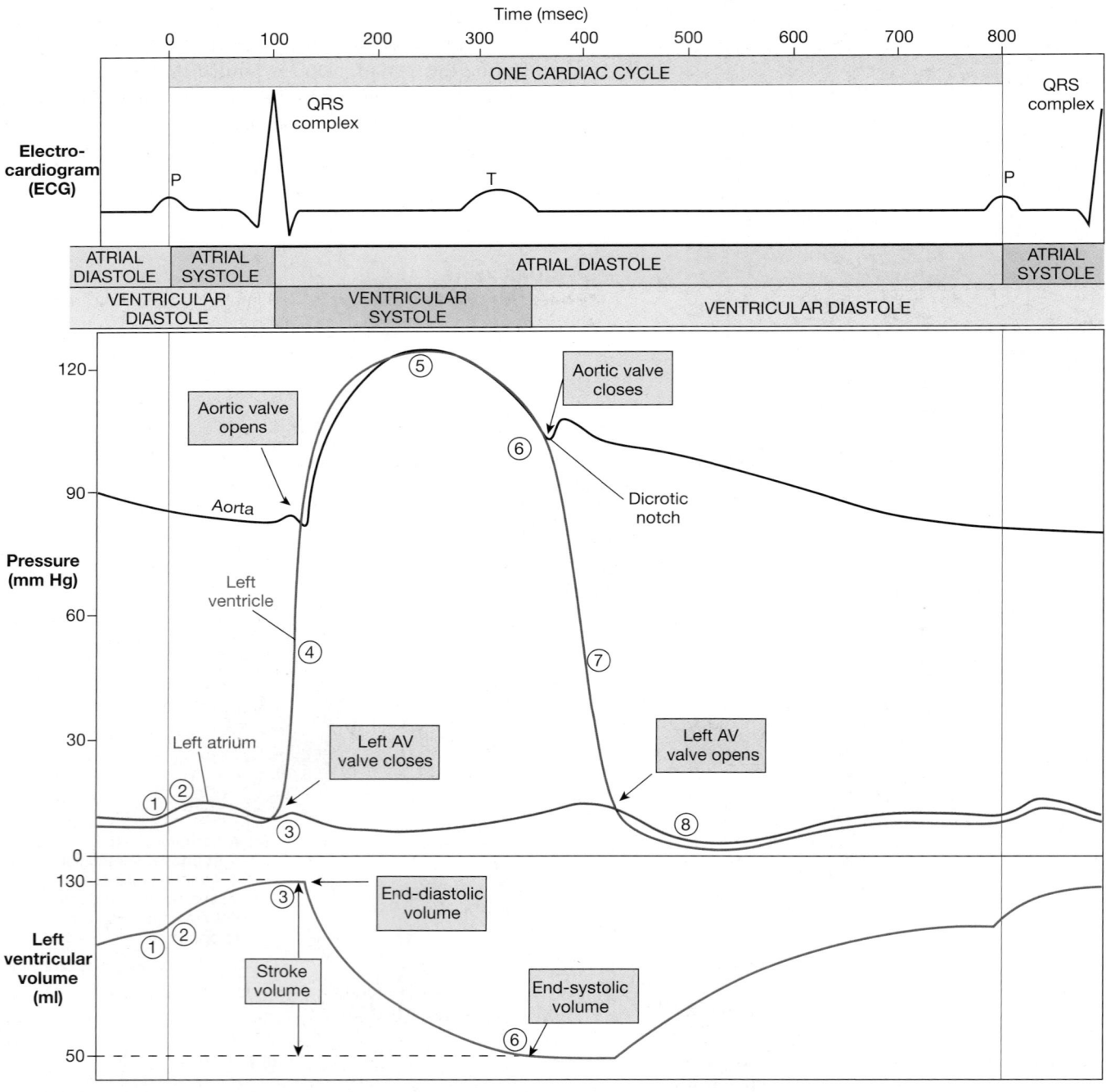

●**FIGURE 20–17**
Pressure and Volume Relationships in the Cardiac Cycle. Major features of the cardiac cycle are shown for a heart rate of 75 bpm. The circled numbers correspond to the associated numbered list in the text.

ATRIAL SYSTOLE The cardiac cycle begins with atrial systole, which lasts about 100 msec in a resting adult:

1. As the atria contract, rising atrial pressures push blood into the ventricles through the open right and left AV valves.

2. At the start of atrial systole, the ventricles are already filled to about 70 percent of their normal capacity, due to passive blood flow toward the end of the previous cardiac cycle. As the atria contract, rising atrial pressures provide the missing 30 percent by pushing blood through the open AV valves. Atrial systole essentially "tops off" the ventricles.

3. At the end of atrial systole, each ventricle contains the maximum amount of blood that it will hold in this cardiac cycle. That quantity is called the **end-diastolic volume** (**EDV**). In an adult who is standing at rest, the end-diastolic volume is typically about 130 ml.

VENTRICULAR SYSTOLE As atrial systole ends, ventricular systole begins. This period lasts approximately 270 msec in a resting adult. As the pressures in the ventricles rise above those in the atria, the AV valves swing shut.

4. During this stage of ventricular systole, the ventricles are contracting. Blood flow has yet to occur, however, because ventricular pressures are not high enough to force open the semilunar valves and push blood into the pulmonary or aortic trunk. Over this period, the ventricles contract isometrically: They generate tension and ventricular pressures rise, but blood flow does not occur. The ventricles are now in the period of **isovolumetric contraction**: All the heart valves are closed, the volumes of the ventricles remain constant, and ventricular pressures rise.

5. Once pressure in the ventricles exceeds that in the arterial trunks, the semilunar valves open and blood flows into the pulmonary and aortic trunks. This point marks the beginning of the period of **ventricular ejection**. The ventricles now contract isotonically: The muscle cells shorten, and tension production remains relatively constant. (To review isotonic versus isometric contractions, see Figure 10–18●, p. 314.)

After reaching a peak, ventricular pressures gradually decline near the end of ventricular systole. Figure 20–17● shows values for the left ventricle and aorta. Although pressures in the right ventricle and pulmonary trunk are much lower, the right ventricle also goes through periods of isovolumetric contraction and ventricular ejection. During ventricular ejection, each ventricle will eject 70–80 ml of blood, the **stroke volume** (**SV**) of the heart. The stroke volume at rest is roughly 60 percent of the end-diastolic volume. This percentage, known as the *ejection fraction*, can vary in response to changing demands on the heart. (We shall discuss the regulatory mechanisms in the next section.)

6. As the end of ventricular systole approaches, ventricular pressures fall rapidly. Blood in the aorta and pulmonary trunk now starts to flow back toward the ventricles, and this movement closes the semilunar valves. As the backflow begins, pressure decreases in the aorta. When the semilunar valves close, pressure rises again as the elastic arterial walls recoil. This small, temporary rise produces a valley in the pressure tracing that is called a *dicrotic* (dī-KRO-tik) *notch* (*dikrotos*, double beating). The amount of

blood remaining in the ventricle when the semilunar valve closes is the **end-systolic volume** (**ESV**). At rest, the end-systolic volume is 50 ml, about 40 percent of the end-diastolic volume.

VENTRICULAR DIASTOLE The period of ventricular diastole lasts for the 430 msec remaining in the current cardiac cycle and continues through atrial systole in the next cycle.

7. All the heart valves are now closed, and the ventricular myocardium is relaxing. Because ventricular pressures are still higher than atrial pressures, blood cannot flow into the ventricles. This is the period of **isovolumetric relaxation**. Ventricular pressures drop rapidly over this period, because the elasticity of the connective tissues of the heart and fibrous skeleton helps re-expand the ventricles toward their resting dimensions.

8. When ventricular pressures fall below those of the atria, the atrial pressures force the AV valves open. Blood now flows from the atria into the ventricles. Both the atria and the ventricles are in diastole, but the ventricular pressures continue to fall as the ventricular chambers expand. Throughout this period, pressures in the ventricles are so far below those in the major veins, that blood pours through the relaxed atria and on through the open AV valves into the ventricles. This passive mechanism is the primary method of ventricular filling. The ventricles will be nearly three-quarters full before the cardiac cycle ends.

The relatively minor contribution that atrial systole makes to ventricular volume explains why individuals can survive quite normally when their atria have been so severely damaged that they can no longer function. In contrast, damage to one or both ventricles can leave the heart unable to maintain adequate blood flow through peripheral tissues and organs. A condition of **heart failure** then exists. **AM** Heart Failure

Heart Sounds

Listening to the heart, a technique called *auscultation,* is a simple and effective method of cardiac diagnosis. Physicians use an instrument called a **stethoscope** to listen to normal and abnormal heart sounds. Where to place the stethoscope depends on which valve is under examination (Figure 20–18a●). Valve sounds must pass through the pericardium, surrounding tissues, and the chest wall, and some tissues muffle sounds more than others. As a result, the placement of the stethoscope does not always correspond to the position of the valve under review.

There are four heart sounds, designated as S_1 through S_4 (Figure 20–18b●). When you listen to your own heart, you usually hear the *first* and *second heart sounds*. These sounds accompany the closing of your heart valves. The first heart sound, known as "lubb" (S_1), lasts a little longer than the second, called "dupp" (S_2). S_1, which marks the start of ventricular contraction, is produced as the AV valves close; S_2 occurs at the beginning of ventricular filling, when the semilunar valves close.

Third and *fourth heart sounds* may be audible, but they are usually very faint and seldom are detectable in healthy adults. These sounds are associated with blood flowing into the ventricles (S_3) and atrial contraction (S_4), rather than with valve action.

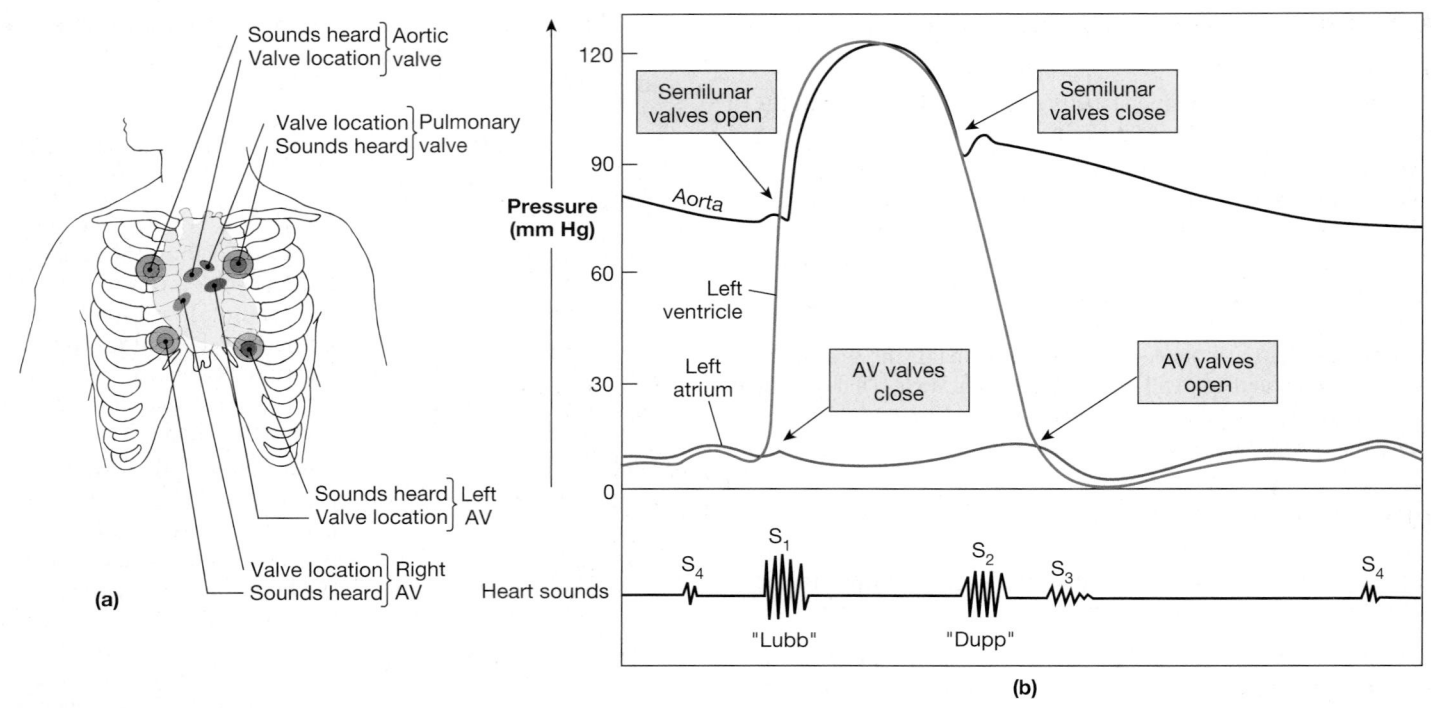

•FIGURE 20–18
Heart Sounds. **(a)** Placements of a stethoscope for listening to the different sounds produced by individual valves.
(b) The timing of heart sounds in relation to key events in the cardiac cycle.

Minor valve abnormalities are relatively common. For example, 5 to 10 percent of healthy individuals have some degree of **mitral valve prolapse**, a condition in which the mitral valve cusps do not close properly. The problem may involve abnormally long (or short) chordae tendineae or malfunctioning papillary muscles. Because the valve does not work perfectly, some regurgitation may occur during left ventricular systole. The surges, swirls, and eddies that accompany regurgitation create a rushing, gurgling sound known as a **heart murmur**.

Minor heart murmurs are very common. Most individuals with this condition are completely asymptomatic and live normal lives, unaware of any circulatory malfunction. Extreme prolapse and valve failure, which may be caused by breakage of the chordae tendineae, can be life threatening. This condition is known as a *mitral valve flail*.

The Energy for Cardiac Contractions

When a normal heart is beating, the energy required is obtained by the mitochondrial breakdown of fatty acids (stored as lipid droplets) and glucose (stored as glycogen). These aerobic reactions can occur only when oxygen is readily available. ∞ p. 316

In addition to obtaining oxygen from the coronary circulation, cardiac muscle cells maintain their own sizable reserves of oxygen. In these cells, oxygen molecules are bound to the heme units of myoglobin molecules. (We discussed this globular protein, which reversibly binds oxygen molecules, and its function in muscle fibers in Chapter 10.) ∞ p. 320 Normally, the combination of circulatory supplies plus myoglobin reserves is enough to meet the oxygen demands of your heart, even when it is working at maximum capacity.

✓ Is the heart always pumping blood when pressure in the left ventricle is rising? Explain.

✓ What factor or factors could cause an increase in the size of the QRS complex of an electrocardiogram recording?

Answers start on page Q-1

 Review the entire cardiac cycle on the **IP CD-ROM**: Cardiovascular System/Cardiac Cycle.

20–4 CARDIODYNAMICS

Objectives

- Define cardiac output, and describe the factors that influence this variable.
- Describe the variables that influence heart rate.
- Describe the variables that influence stroke volume.
- Explain how adjustments in stroke volume and cardiac output are coordinated at different levels of activity.

The term **cardiodynamics** refers to the movements and forces generated during cardiac contractions. Each time the heart beats, the two ventricles eject equal amounts of blood. Earlier we introduced these terms:

- **End-Diastolic Volume (EDV).** The amount of blood in each ventricle at the end of ventricular diastole (the start of ventricular systole).
- **End-Systolic Volume (ESV).** The amount of blood remaining in each ventricle at the end of ventricular systole (the start of ventricular diastole).
- **Stroke Volume (SV).** The amount of blood pumped out of each ventricle during a single beat, which can be expressed as EDV − ESV = SV.
- **Ejection Fraction.** The percentage of the EDV represented by the SV.

Stroke volume is the most important factor in an examination of a single cardiac cycle. If the heart were an old-fashioned bicycle pump, the stroke volume would be the amount of air pumped in one up–down cycle of the handle (Figure 20–19●). Where you stop when you lift the handle determines the end-diastolic volume—how much air the pump contains. How far down you push the handle determines the end-systolic volume—how much air remains in the pump at the end of the cycle. You can increase the amount of air pumped in each cycle by increasing the range of movement of the handle. You pump the maximum amount when the handle moves all the way from the top to the bottom (Figure 20–19b,d●). In other words, you get the largest stroke volume when the EDV is as great as it can be and the ESV is as small as it can be.

When considering cardiac function over time, physicians generally are most interested in the **cardiac output (CO)**, the amount of blood pumped by each ventricle in one minute. In essence, cardiac output is an indication of the blood flow through peripheral tissues—without adequate blood flow, homeostasis cannot be maintained. The cardiac output provides a useful indication of ventricular efficiency over time. We can

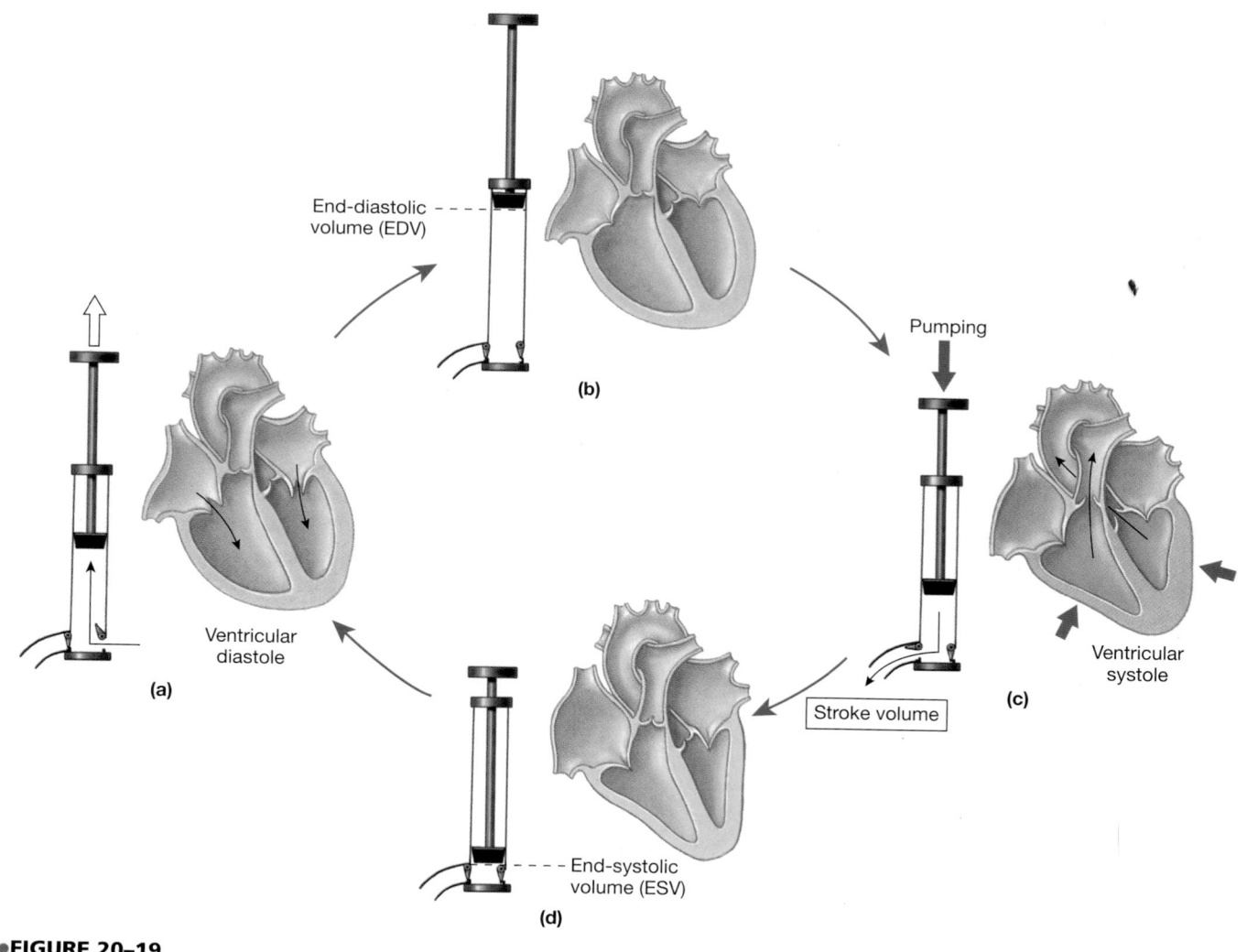

End-diastolic volume (EDV)

Pumping

Ventricular systole

(b)

(c)

Stroke volume

Ventricular diastole

(a)

End-systolic volume (ESV)

(d)

●**FIGURE 20–19**
A Simple Model of Stroke Volume. The stroke volume of the heart can be compared to the amount of air pumped from an old-fashioned bicycle pump, which varies with the amount of movement of the pump handle. The extent of upward movement corresponds to the EDV; the extent of downward movement corresponds to the ESV.

calculate it by multiplying the heart rate (HR) by the average stroke volume (SV):

$$\begin{array}{ccccc} CO & = & Hr & \times & SV \\ \text{cardic} & & \text{heart} & & \text{stroke} \\ \text{output} & & \text{rate} & & \text{volume} \\ \text{(ml/min)} & & \text{(beats/min)} & & \text{(ml/beat)} \end{array}$$

For example, if the heart rate is 75 bpm and the stroke volume is 80 ml per beat, the cardiac output will be

$$CO = 75 \text{ bpm} \times 80 \text{ ml/beat} = 6000 \text{ ml/min} (6 \text{ l/min})$$

Cardiac output is precisely adjusted such that peripheral tissues receive an adequate circulatory supply under a variety of conditions. When necessary, the heart rate can increase by 250 percent, and stroke volume in a normal heart can almost double.

☐ OVERVIEW: THE CONTROL OF CARDIAC OUTPUT

Figure 20–20● summarizes the factors involved in the normal regulation of cardiac output. Cardiac output can be adjusted by changes in either heart rate or stroke volume. For convenience, we can consider these independently as we discuss the individual factors involved. However, changes in cardiac output generally reflect changes in both heart rate and stroke volume.

The heart rate can be adjusted by the activities of the autonomic nervous system or by circulating hormones. The stroke volume can be adjusted by changing the end-diastolic volume (how full the ventricles are when they start to contract), the end-systolic volume (how much blood remains in the ventricle after it contracts), or both. As Figure 20–19● shows, stroke volume peaks when EDV

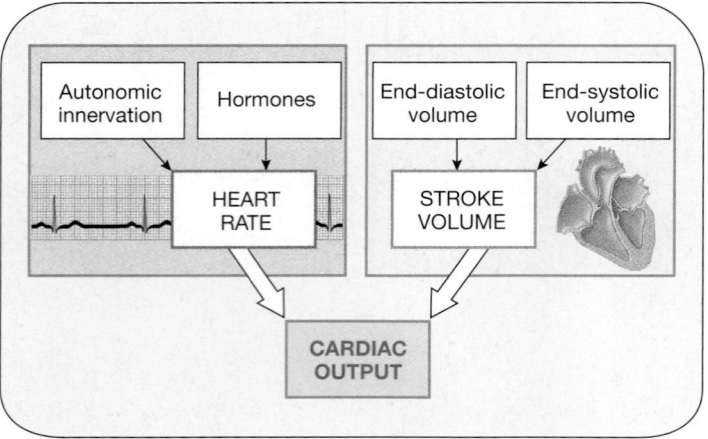

●**FIGURE 20–20**
Factors Affecting Cardiac Output. A simplified version of this figure will appear as a Navigator icon in key figures as we move from one topic to the next.

is high and ESV is low. A variety of other factors can influence cardiac output under abnormal circumstances; we shall consider several examples in a separate section.

☐ FACTORS AFFECTING THE HEART RATE

Under normal circumstances, autonomic activity and circulating hormones are responsible for making delicate adjustments to the heart rate as circulatory demands change. These factors act by modifying the natural rhythm of the heart. Even a heart removed for a heart transplant will continue to beat unless steps are taken to prevent it from doing so.

Autonomic Innervation

The sympathetic and parasympathetic divisions of the autonomic nervous system innervate the heart by means of the *cardiac plexus* (Figures 16–10, p. 547, and 20–21●). Postganglionic sympathetic neurons are located in the cervical and upper thoracic ganglia. The vagus nerves (X) carry parasympathetic preganglionic fibers to small ganglia in the cardiac plexus. Both ANS divisions innervate the SA and AV nodes and the atrial muscle cells. Although ventricular muscle cells are also innervated by both divisions, sympathetic fibers far outnumber parasympathetic fibers there.

The *cardiac centers* of the medulla oblongata contain the autonomic headquarters for cardiac control. ∞ p. 475 The **cardioaccelleratory center** controls sympathetic neurons that increase the heart rate; the adjacent **cardioinhibitory center** controls the parasympathetic neurons that slow the heart rate. The activities of the cardiac centers are regulated by reflex pathways and through input from higher centers, especially from the parasympathetic and sympathetic headquarters in the hypothalamus.

CARDIAC REFLEXES Information about the status of the cardiovascular system arrives over visceral sensory fibers accompanying the vagus nerve and the sympathetic nerves of the cardiac plexus. The cardiac centers monitor baroreceptors and chemoreceptors innervated by the glossopharyngeal (IX) and vagus (X) nerves. ∞ pp. 499, 500 On the basis of the information received, the centers adjust cardiac performance to maintain adequate circulation to vital organs, such as the brain. The centers respond to changes in blood pressure and in arterial concentrations of dissolved oxygen and carbon dioxide. For example, a decline in blood pressure or oxygen concentrations or an increase in carbon dioxide levels generally indicates that the heart must work harder to meet the demands of peripheral tissues. The cardiac centers then call for an increase in cardiac activity. We will detail these reflexes and their effects on the heart and peripheral vessels in Chapter 21.

AUTONOMIC TONE As is the case in other organs with dual innervation, the heart has a resting autonomic tone. Both autonomic divisions are normally active at a steady background level, releasing ACh and NE at the nodes and into the myocardium. Thus, cutting the vagus nerves increases the heart rate, and sympathetic blocking agents slow the heart rate.

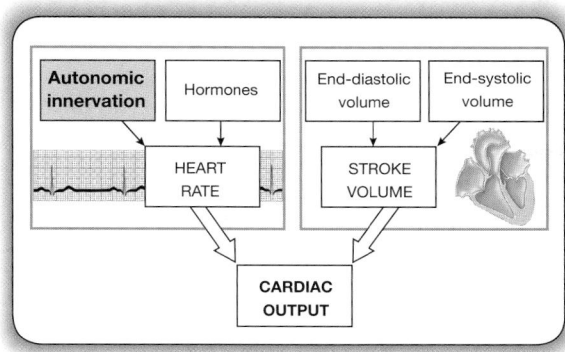

●**FIGURE 20–21**
Autonomic Innervation of the Heart.   The Navigator icon in the shadow box highlights the topic we will consider in this section.

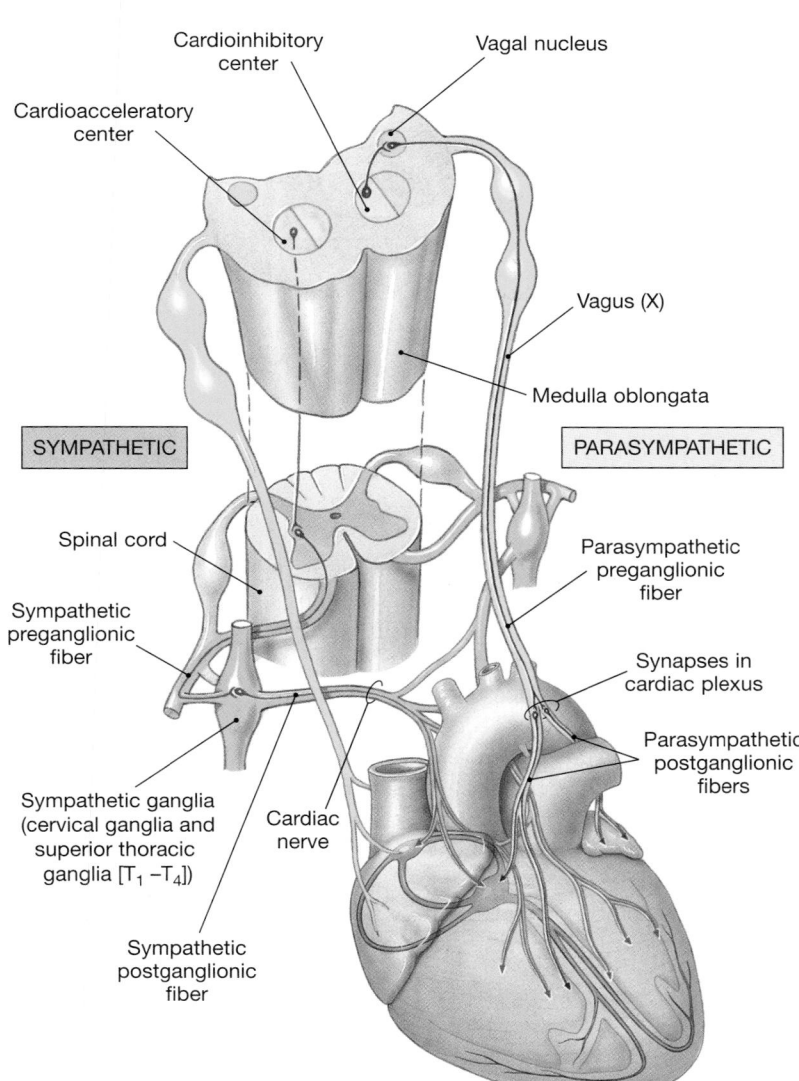

In a healthy, resting individual, parasympathetic effects dominate. In the absence of autonomic innervation, the heart rate is established by the pacemaker cells of the SA node. Such a heart beats at a rate of 80–100 bpm. At rest, a typical adult heart with normal innervation beats at 70–80 bpm due to activity in the parasympathetic nerves innervating the SA node. If parasympathetic activity increases, the heart rate declines further. Conversely, the heart rate will increase if either parasympathetic activity decreases or sympathetic activation occurs. Through dual innervation and adjustments in autonomic tone, the ANS can make very delicate adjustments in cardiovascular function to meet the demands of other systems.

EFFECTS ON THE SA NODE The sympathetic and parasympathetic divisions alter the heart rate by changing the permeabilities of cells in the conducting system. The most dramatic effects are seen at the SA node, where changes in the rate at which impulses are generated affect the heart rate.

Consider the SA node of a resting individual whose heart is beating at 75 bpm (Figure 20–22a●). Any factor that changes the rate of spontaneous depolarization or the duration of repolarization will alter the heart rate by changing the time required to reach threshold. Acetylcholine released by parasympathetic neurons opens chemically regulated K^+ channels in the cell membrane, thereby dramatically slowing the rate of spontaneous depolarization and also slightly extending the duration of repolarization (Figure 20–22b●). The result is a decline in heart rate.

The NE released by sympathetic neurons binds to beta-1 receptors, leading to the opening of calcium ion channels. The subsequent influx of Ca^{2+} increases the rate of depolarization and shortens the period of repolarization. The

nodal cells reach threshold more quickly, and the heart rate increases (Figure 20–22c●).

THE ATRIAL REFLEX The **atrial reflex**, or *Bainbridge reflex*, involves adjustments in heart rate in response to an increase in the venous return. When the walls of the right atrium are stretched, the stimulation of stretch receptors in the atrial walls triggers a reflexive increase in heart rate caused by increased sympathetic activity (Figure 20–22●). Thus, when the rate of venous return to the heart increases, the heart rate, and hence the cardiac output, rises as well.

Hormones

Epinephrine, norepinephrine, and thyroid hormone increase the heartrate by their effect on the SA node. The effects of epinephrine on the SA node are similar to those of norepinephrine. Epinephrine also affects the contractile cells; after massive sympathetic stimulation of the adrenal medullae, the myocardium may become so excitable that abnormal contractions occur.

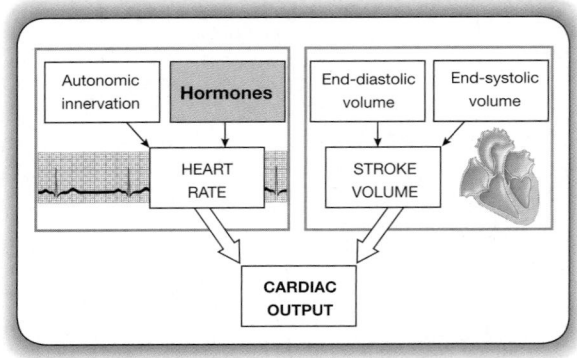

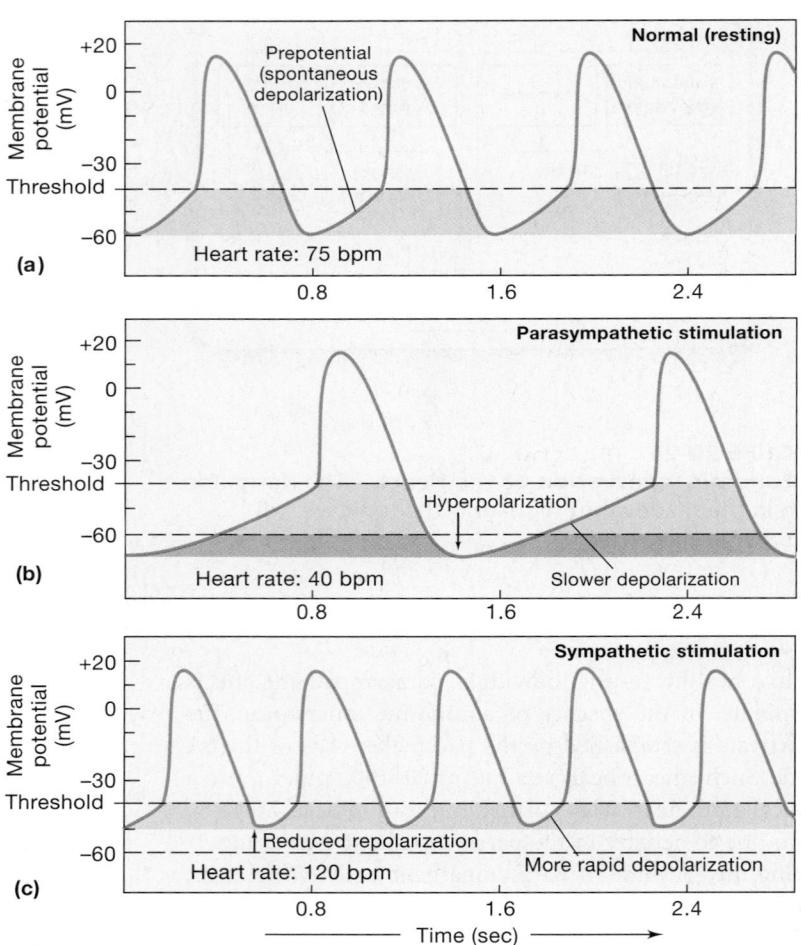

●FIGURE 20–22
Pacemaker Function. **(a)** Pacemaker cells have membrane potentials closer to threshold than those of other cardiac muscle cells (−60 mV versus −90 mV). Their cell membranes undergo spontaneous depolarization to threshold, producing action potentials at a frequency determined by (1) the resting-membrane potential and (2) the rate of depolarization (slope of the prepotential). **(b)** Parasympathetic stimulation releases ACh, which extends repolarization and decreases the rate of spontaneous depolarization. The heart rate slows. **(c)** Sympathetic stimulation releases NE, which shortens repolarization and accelerates the rate of spontaneous depolarization. As a result, the heart rate increases.

Venous Return

Venous return has an indirect effect on heart rate by way of the atrial reflex. It also has direct effects on nodal cells. When venous return increases, the atria receive more blood and the walls are stretched. Stretching of the cells of the SA node leads to more rapid depolarization and an increase in the heart rate.

✓ What effect does drinking large amounts of caffeinated drinks have on the heart?

✓ If the cardioinhibitory center of the medulla oblongata were damaged, which part of the autonomic nervous system would be affected, and how would the heart be influenced?

✓ How does a drug that increases the length of time required for the repolarization of pacemaker cells affect the heart rate?

Answers start on page Q-1

☐ FACTORS AFFECTING THE STROKE VOLUME

The stroke volume is the difference between the end-diastolic volume and the end-systolic volume. Thus, changes in either the EDV or the ESV can change both the stroke volume and cardiac output. The factors involved in the regulation of stroke volume are indicated in Figure 20–23●.

The EDV

The EDV indicates the amount of blood a ventricle contains at the end of diastole, just before a contraction begins. This volume is affected by two factors: the filling time and the venous return. **Filling time** is the duration of ventricular diastole. As such, it depends entirely on the heart rate: The faster the heart rate, the shorter is the available filling time. **Venous return** is the rate of blood flow over this period. Venous return changes in response to alterations in cardiac output, blood volume, patterns of peripheral circulation, skeletal muscle activity, and other factors that affect the rate of blood flow through the venae cavae. (We will explore these factors in Chapter 21.)

The ESV

After the ventricle contracts and the stroke volume has been ejected, the ESV is the amount of blood that remains in the ventricle at the end of ventricular systole. Three factors that influence the ESV are (1) the *preload*, (2) the *contractility* of the ventricle, and (3) the *afterload*.

PRELOAD The degree of stretching experienced by ventricular muscle cells during ventricular diastole is called the **preload**. The preload is directly proportional to the EDV: The greater the EDV, the larger the preload. Preload is significant because it affects the ability of muscle cells to produce tension. In Chapter 10, we considered the length–tension relationship in skeletal muscle fibers; the same principles apply to cardiac muscle cells. ⊙ p. 308 As we noted in Chapter 10, there is a narrow range of optimal resting sarcomere lengths. If the resting length is too short, the sarcomeres cannot contract very far; if the resting length is too great, the zone of overlap disappears and contraction is impossible.

The amount of preload, and hence the degree of myocardial stretching, varies with the demands on the heart. When you are standing at rest, your EDV is low; the ventricular muscle is stretched very little, and the sarcomeres are relatively short. During ventricular systole, the cardiac muscle cells develop little power, and

the ESV is relatively high because the muscle cells contract only a short distance. If you begin exercising, venous return increases and more blood flows into your heart. Your EDV increases, and the myocardium stretches further. As the sarcomeres approach optimal lengths, the ventricular muscle cells can contract more efficiently and produce more forceful contractions. They also shorten more, and more blood is pumped out of your heart.

THE EDV AND STROKE VOLUME In general, the greater the EDV, the larger is the stroke volume. Stretching *past* the optimal length, which would reduce the force of contraction, does not normally occur, because ventricular expansion is limited by myocardial connective tissues, the fibrous skeleton, and the pericardial sac.

The relationship between the amount of ventricular stretching and the contractile force means that, within normal physiological limits, increasing the EDV results in a corresponding increase in the stroke volume. This general rule of "more in = more out" was first proposed by Ernest H. Starling on the basis of an analysis of research performed by Otto Frank. The relationship is therefore known as the **Frank–Starling principle**, or *Starling's law of the heart.*

Autonomic adjustments to cardiac output make the effects of the Frank–Starling principle difficult to see. However, it can be demonstrated effectively in individuals who have received a heart transplant, because the implanted heart is not innervated by the ANS. The most obvious effect of the Frank–Starling principle in these hearts is that the outputs of the left and right ventricles remain balanced under a variety of conditions.

Consider, for example, an individual at rest, with the two ventricles ejecting equal volumes of blood. Although the ventricles

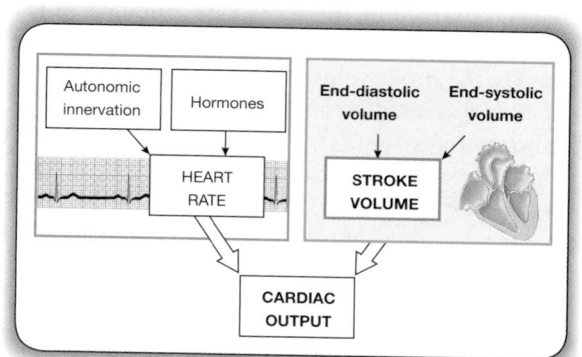

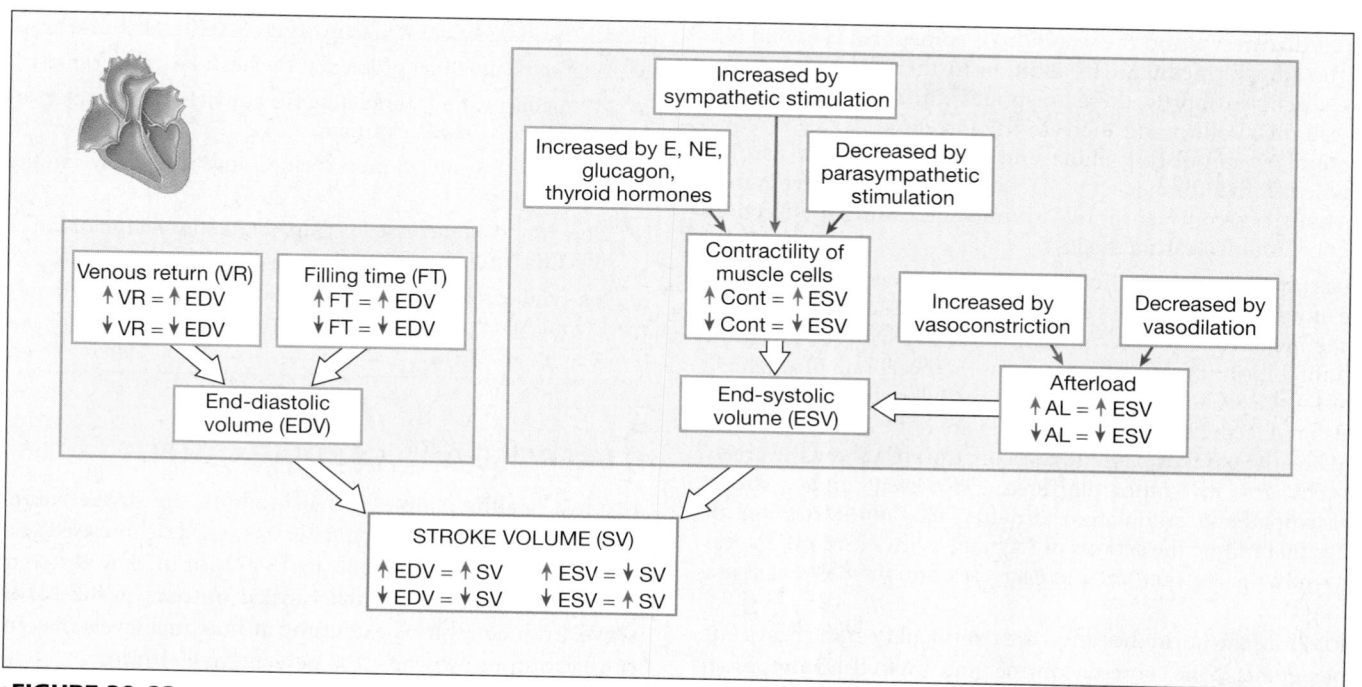

•**FIGURE 20–23**
Factors Affecting Stroke Volume. The arrows indicate the nature of the effects: ↑ = increases, ↓ = decreases.

contract together, they function in series: When the heart contracts, blood leaving the right ventricle heads to the lungs; during the next ventricular diastole, that volume of blood will pass through the left atrium, to be ejected by the left ventricle at the next contraction. If the venous return decreases, the EDV of the right ventricle will decline. During ventricular systole, it will then pump less blood into the pulmonary circuit. In the next cardiac cycle, the EDV of the left ventricle will be reduced, and that ventricle will eject a smaller volume of blood. The output of the two ventricles will again be in balance, but both will have smaller stroke volumes than they did initially.

CONTRACTILITY Contractility is the amount of force produced during a contraction, at a given preload. Under normal circumstances, contractility can be altered by autonomic innervation or circulating hormones. Under special circumstances, contractility can be altered by drugs or as a result of abnormal changes in ion concentrations in the extracellular fluid.

Factors that increase contractility are said to have a *positive inotropic action*; those that decrease contractility have a *negative inotropic action*. Positive inotropic agents typically stimulate Ca^{2+} entry into cardiac muscle cells, thus increasing the force and duration of ventricular contractions. Negative inotropic agents may block Ca^{2+} movement or depress cardiac muscle metabolism. Positive and negative inotropic factors include ANS activity, hormones, and changes in extracelluar ion concentrations.

AUTONOMIC ACTIVITY Autonomic activity alters the degree of contraction and changes the ESV in the following ways:

- Sympathetic stimulation has a positive inotropic effect, causing the release of norepinephrine (NE) by postganglionic fibers of the cardiac nerves and the secretion of epinephrine (E) and NE by the adrenal medullae. In addition to their effects on heart rate, discussed shortly, these hormones stimulate cardiac muscle cell metabolism and increase the force and degree of contraction by stimulating alpha and beta receptors in cardiac muscle cell membranes. ∞ pp. 540–541 The net effect is that the ventricles contract more forcefully, increasing the ejection fraction and decreasing the ESV.

- Parasympathetic stimulation from the vagus nerves has a negative inotropic effect. The primary effect of acetylcholine (ACh) is at the membrane surface, where it produces hyperpolarization and inhibition. The result is a decrease in heart rate through effects on the SA and AV nodes. The force of cardiac contractions is also reduced; because the ventricles are not extensively innervated by the parasympathetic division, the atria show the greatest changes in contractile force. However, under strong parasympathetic stimulation or after the administration of drugs that mimic the actions of ACh, the ventricles contract less forcefully, the ejection fraction decreases, and the ESV enlarges.

HORMONES Many hormones affect the contractility of the heart. For example, epinephrine, norepinephrine, and thyroid hormones all have positive inotropic effects. Glucagon also has a positive inotropic effect. Before synthetic inotropic agents were available, glucagon was widely used to stimulate cardiac function. It is still used in cardiac emergencies and to treat some forms of heart disease.

The drugs *isoproterenol*, *dopamine*, and *dobutamine* mimic the action of E and NE by stimulating beta-1 receptors on cardiac muscle cells. ∞ p. 541 Dopamine (at high doses) and dobutamine also stimulate Ca^{2+} entry through alpha-1 receptor stimulation. *Digitalis* and related drugs elevate intracellular Ca^{2+} concentrations, but by a different mechanism: They interfere with the removal of Ca^{2+} from the sarcoplasm of cardiac muscle cells.

Many of the drugs used to treat hypertension (high blood pressure) have a negative inotropic action. Beta-blocking drugs such as *propranolol*, *timolol*, *metoprolol*, *atenolol*, and *labetalol* block beta receptors, alpha receptors, or both, and prevent sympathetic stimulation of the heart. Calcium channel blockers such as nifedipine or verapamil also have a negative inotropic effect.

AFTERLOAD The afterload is the amount of tension the contracting ventricle must produce to force open the semilunar valve and eject blood. The greater the afterload, the longer is the period of isovolumetric contraction, the shorter the duration of ventricular ejection, and the larger the ESV. In other words, as the afterload increases, the stroke volume decreases.

Afterload is increased by any factor that restricts blood flow through the arterial system. For example, the constriction of peripheral blood vessels or a circulatory blockage will elevate arterial blood pressure and increase the afterload. If the afterload is too great, the ventricle cannot eject blood. Such a high afterload is rare in the normal heart, but damage to the heart muscle can weaken the myocardium enough that even a modest rise in arterial blood pressure can reduce stroke volume to dangerously low levels, producing symptoms of heart failure.

✓ Why is it a potential problem if the heart beats too rapidly?

✓ What effect would stimulating the acetylcholine receptors of the heart have on cardiac output?

✓ What effect would an increase in venous return have on the stroke volume?

✓ How would an increase in sympathetic stimulation of the heart affect the end-systolic volume?

✓ Joe's end-systolic volume is 40 ml, and his end-diastolic volume is 125 ml. What is Joe's stroke volume?

Answers start on page Q-1

EXERCISE AND CARDIAC OUTPUT

In most healthy people, increasing both the stroke volume and the heart rate, such as during heavy exercise, can raise the cardiac output by 300–500 percent, to 18–30 l/min. The difference between resting and maximal cardiac outputs is the **cardiac reserve**. Trained athletes exercising at maximal levels may increase cardiac output by nearly 700 percent, to 40 l/min.

Cardiac output cannot increase indefinitely, primarily because the available filling time shortens as the heart rate increases. At heart rates up to 160–180 bpm, the combination of an increased rate of venous return and increased contractility compensates for the reduction in filling time. Over this range, cardiac output and heart rate increase together. But if the heart rate continues to climb, the stroke volume begins to drop. Cardiac output first plateaus and then declines.

□ SUMMARY: REGULATION OF HEART RATE AND STROKE VOLUME

Figure 20–24● summarizes the factors involved in the regulation of heart rate and stroke volume, which interact to determine cardiac output under normal conditions.

The heart rate is influenced by the autonomic nervous system, circulating hormones, and the venous return.

- Sympathetic stimulation increases the heart rate; parasympathetic stimulation decreases it. Under resting conditions, parasympathetic tone dominates, and the heart rate is slightly slower than the intrinsic heart rate. When activity levels rise, venous return increases and triggers the atrial reflex. The result is an increase in sympathetic tone and an increase in heart rate.

- Circulating hormones, specifically E, NE, and T_3, accelerate the heart rate.

- An increase in venous return stretches the nodal cells and increases the heart rate.

The stroke volume is the difference between the end-diastolic volume (EDV) and the end-systolic volume (ESV).

- The EDV is determined by the available filling time and the rate of venous return.

- The ESV is determined by the amount of preload (the degree of myocardial stretching), the degree of contractility (adjusted by hormones and autonomic innervation), and the afterload (the amount of arterial resistance).

20–5   THE HEART AND THE CARDIOVASCULAR SYSTEM

The goal of cardiovascular regulation is to maintain adequate blood flow to all body tissues. The heart cannot accomplish this goal by itself, and it does not work in isolation. For example, when blood pressure changes, the cardiovascular centers adjust not only the heart rate but also the diameters of peripheral blood vessels. These adjustments work together to keep the blood pressure within normal limits and to maintain circulation to vital tissues and organs. Chapter 21 will complete this story by detailing the cardiovascular responses to changing activity patterns and circulatory emergencies. We will then conclude our discussion of the cardiovascular system by examining the anatomy of the pulmonary and systemic circuits.

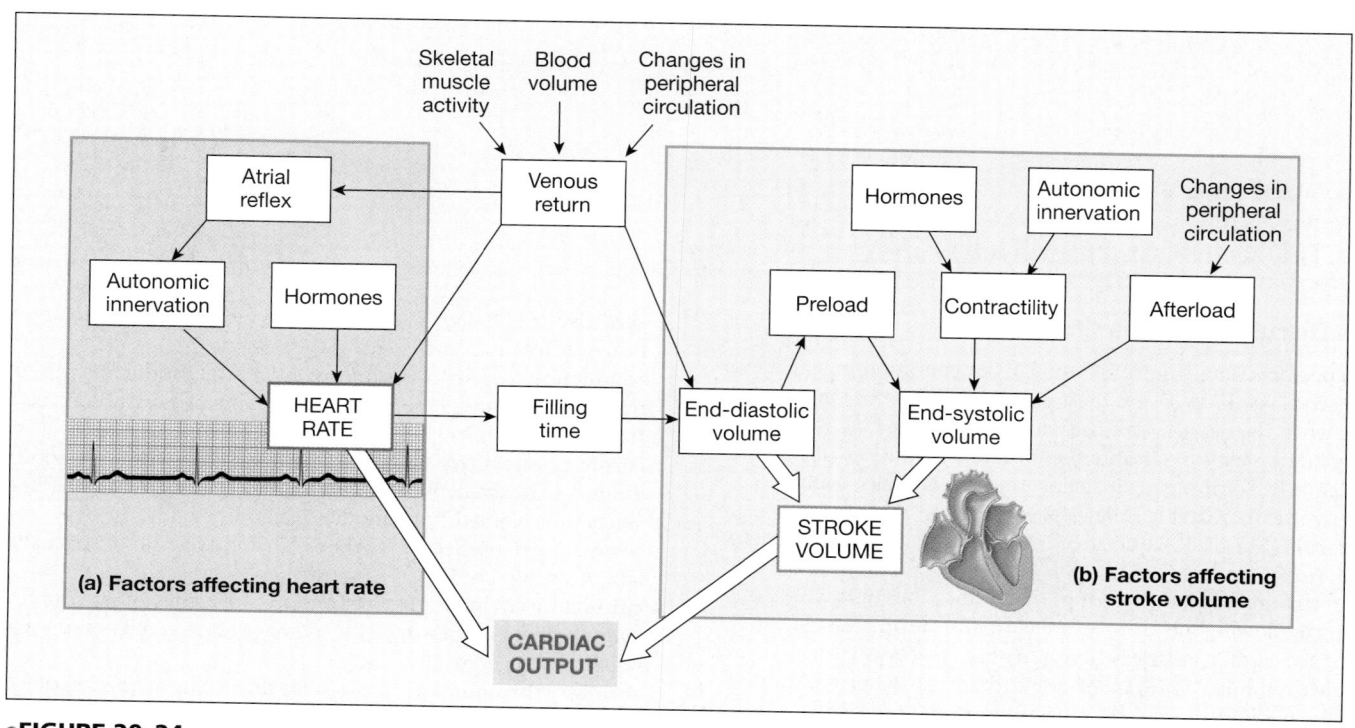

●**FIGURE 20–24**
A Summary of the Factors Affecting Cardiac Output.   (a) Factors affecting heart rate. (b) Factors affecting stroke volume.

Abnormal Conditions Affecting Cardiac Output

Various drugs, abnormal variations in ion concentrations, and changes in body temperature can alter the basic rhythm of contraction established by the SA node. In Chapter 12, we noted that several drugs, including caffeine and nicotine, have a stimulatory effect on excitable membranes in the nervous system. ∞ pp. 415, 420-1 These drugs also cause an increase in heart rate. Caffeine acts directly on the conducting system and increases the rate of depolarization at the SA node. Nicotine acts indirectly by stimulating the activity of sympathetic neurons that innervate the heart.

Disorders affecting ion concentrations or body temperature can have direct effects on cardiac output by changing the stroke volume, the heart rate, or both. Abnormal ion concentrations can change both the contractility of the heart, by affecting the cardiac muscle cells, and the heart rate, by affecting the SA nodal cells. The most obvious and clinically important examples of problems with ion concentrations involve K^+ and Ca^{2+}.

Hyperkalemia and Hypokalemia

Changes in the extracellular K^+ concentration affect the transmembrane potential and the rates of depolarization and repolarization. In **hyperkalemia** (hī-per-ka-LĒ-mē-uh), in which K^+ concentrations are high, the muscle cells depolarize, and repolarization is inhibited. Cardiac contractions become weak and irregular; in severe cases, the heart eventually stops in diastole. In **hypokalemia** (hī-pō-ka-LĒ-mē-uh), in which K^+ concentrations are abnormally low, the membranes of cardiac muscle cells hyperpolarize and become less responsive to stimulation. Meanwhile, the hyperpolarization of nodal cells leads to a reduction in the heart rate. Blood pressure falls, and, in severe cases, the heart eventually stops in systole. Severe hyperkalemia and hypokalemia are life threatening, and require immediate corrective action.

Hypercalcemia and Hypocalcemia

In **hypercalcemia** (hī-per-kal-SĒ-mē-uh), a condition in which the extracellular concentration of Ca^{2+} is elevated, cardiac muscle cells become extremely excitable. Their contractions become powerful and prolonged. In extreme cases, the heart goes into an extended state of contraction that is generally fatal. In **hypocalcemia** (hī-pō-kal-SĒ-mē-uh), in which the Ca^{2+} concentration is abnormally low, the contractions become very weak and may cease altogether.

Abnormal Body Temperatures

Temperature changes affect metabolic operations throughout the body. For example, a reduction in temperature slows the rate of depolarization at the SA node, lowers the heart rate, and reduces the strength of cardiac contractions. (In open-heart surgery, the exposed heart may be deliberately chilled until it stops beating.) An elevated body temperature accelerates the heart rate and the contractile force. That is one reason your heart may seem to race and pound when you have a fever.

Chapter Review

SELECTED CLINICAL TERMINOLOGY

Terms Discussed in This Chapter

angina pectoris (an-JĪ-nuh PEK-tor-is): A condition in which exertion or stress produces severe chest pain, resulting from temporary ischemia when the heart's workload increases. *(p. 696)*
balloon angioplasty: A technique for reducing the size of a coronary plaque by compressing it against the arterial walls, using a catheter with an inflatable collar. *(p. 696)*
bradycardia (brā-dē-KAR-dē-uh): A heart rate that is slower than normal. *(p. 699 and AM)*
calcium channel blockers: Drugs that reduce the contractility of the heart by slowing the influx of calcium ions during the plateau phase of the cardiac muscle action potential. *(p. 696)*
cardiac arrhythmias (ā-RITH-mē-az): Abnormal patterns of cardiac electrical activity, indicating abnormal contractions. *(p. 701 and AM)*
cardiac tamponade: A condition, resulting from pericardial irritation and inflammation, in which fluid collects in the pericardial sac and restricts cardiac output. *(p. 685 and AM)*

carditis (kar-DĪ-tis): A general term indicating inflammation of the heart. *(p. 692 and AM)*
conduction deficit: An abnormality in the conducting system of the heart that affects the timing and pacing of cardiac contractions. *(p. 700 and AM)*
coronary artery bypass graft (CABG): The routing of blood around an obstructed coronary artery (or one of its branches) by a vessel transplanted from another part of the body. *(p. 696)*
coronary artery disease (CAD): The obstruction of coronary circulation. *(p. 695)*
coronary ischemia: The restriction of the circulatory supply to the heart, potentially causing tissue damage and a reduction in cardiac efficiency. *(p. 695)*
coronary thrombosis: A blockage due to the formation of a clot (thrombus) at a plaque in a coronary artery. *(p. 704)*
electrocardiogram (ē-lek-trō-KAR-dē-ō-gram) (**ECG** or **EKG**): A recording of the electrical activities of the heart over time. *(p. 697)*

heart failure: A condition in which the heart weakens and peripheral tissues suffer from oxygen and nutrient deprivation. *(p. 707 and AM)*

heart murmur: The sound produced by regurgitation through an incompletely closed heart valve. *(p. 708)*

mitral valve prolapse: A condition in which the mitral valve cusps do not close properly. *(p. 708)*

myocardial (mī-ō-KAR-dē-al) **infarction (MI):** A condition in which the coronary circulation becomes blocked and cardiac muscle cells die from oxygen starvation; also called a *heart attack*. *(p. 704)*

pericarditis: Inflammation of the pericardium. *(p. 685)*

rheumatic heart disease (RHD): A disorder in which the heart valves become thickened and stiffen into a partially closed position, affecting the efficiency of the heart. *(p. 692 and AM)*

tachycardia ((tak-ē-KAR-dē-uh): A heart rate that is faster than normal. *(p. 699 and AM)*

valvular heart disease (VHD): A condition caused by abnormal functioning of one of the cardiac valves. The severity of the condition depends on the degree of damage and the valve involved. *(p. 692 and AM)*

AM Additional Terms Discussed in the *Applications Manual*

atrial fibrillation
atrial flutter
cardiomyopathies
coronary arteriography
defibrillator
echocardiography
endocarditis
heart block
myocarditis
paroxysmal atrial tachycardia (PAT)
premature atrial contractions (PACs)
premature ventricular contractions (PVCs)
valvular stenosis
ventricular escape
ventricular fibrillation (VF)
ventricular tachycardia (VT)

STUDY OUTLINE

20–1 THE ORGANIZATION OF THE CARDIOVASCULAR SYSTEM p. 683

1. The blood vessels can be subdivided into the **pulmonary circuit** (which carries blood to and from the lungs) and the **systemic circuit** (which transports blood to and from the rest of the body).
2. **Arteries** carry blood away from the **heart; veins** return blood to the heart. **Capillaries**, or *exchange vessels*, are thin-walled, narrow-diameter vessels that connect the smallest arteries and veins. *(Figure 20–1)*
3. The heart has four chambers: the **right atrium** and **right ventricle** and the **left atrium** and **left ventricle**.

20–2 ANATOMY OF THE HEART p. 684

1. The heart is surrounded by the **pericardial cavity** and lies within the anterior portion of the **mediastinum**, which separates the two pleural cavities. *(Figure 20–2)*

THE PERICARDIUM p. 684

2. The pericardial cavity is lined by the **pericardium**. The **visceral pericardium (epicardium)** covers the heart's outer surface, and the **parietal pericardium** lines the inner surface of the **pericardial sac**, which surrounds the heart. *(Figure 20–2)*

SUPERFICIAL ANATOMY OF THE HEART p. 685

3. The **coronary sulcus**, a deep groove, marks the boundary between the atria and the ventricles. Other surface markings also provide useful reference points in describing the heart and associated structures. *(Figure 20–3)*

THE HEART WALL p. 687

4. The bulk of the heart consists of the muscular **myocardium**. The **endocardium** lines the inner surfaces of the heart, and the **epicardium** covers the outer surface. *(Figures 20–4, 20–5)*
5. **Cardiac muscle cells** are interconnected by **intercalated discs**, which convey the force of contraction from cell to cell and conduct action potentials. *(Figure 20–5; Summary Table 20–1)*

INTERNAL ANATOMY AND ORGANIZATION p. 687

6. The atria are separated by the **interatrial septum**, and the ventricles are divided by the **interventricular septum**. The right

atrium receives blood from the systemic circuit via two large veins, the **superior vena cava** and the **inferior vena cava**. (The atrial walls contain the **pectinate muscles**, prominent muscular ridges.) *(Figure 20–6)*
7. Blood flows from the right atrium into the right ventricle via the **right atrioventricular (AV) valve (tricuspid valve)**. This opening is bounded by three **cusps** of fibrous tissue braced by the **chordae tendineae**, which are connected to **papillary muscles**. *(Figure 20–6)*
8. Blood leaving the right ventricle enters the **pulmonary trunk** after passing through the **pulmonary valve**. The pulmonary trunk divides to form the **left** and **right pulmonary arteries**. The **left** and **right pulmonary veins** return blood to the left atrium. Blood leaving the left atrium flows into the left ventricle via the **left atrioventricular (AV) valve (bicuspid**, or **mitral**, **valve)**. Blood leaving the left ventricle passes through the **aortic valve** and into the systemic circuit via the **ascending aorta**. *(Figure 20–6)*
9. Anatomical differences between the ventricles reflect the functional demands placed on them. The wall of the right ventricle is relatively thin, whereas the left ventricle has a massive muscular wall. *(Figure 20–7)*
10. Valves normally permit blood flow in only one direction, preventing the **regurgitation** (backflow) of blood. *(Figure 20–8)*

CONNECTIVE TISSUES AND THE FIBROUS SKELETON p. 692

11. The connective tissues of the heart (mainly collagen and elastic fibers) and the **fibrous skeleton** support the heart's contractile cells and valves. *(Figure 20–8)*

THE BLOOD SUPPLY TO THE HEART p. 692

12. The **coronary circulation** meets the high oxygen and nutrient demands of cardiac muscle cells. The **coronary arteries** originate at the base of the ascending aorta. Interconnections between arteries, called **arterial anastomoses**, ensure a constant blood supply. The **great**, **posterior**, **small**, **anterior**, and **middle cardiac veins** carry blood from the coronary capillaries to the **coronary sinus**. *(Figure 20–9)*
13. In **coronary artery disease (CAD)**, areas of the coronary circulation undergo partial or complete blockage. *(Figure 20–10)*

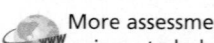 Anatomy of the heart: **IP CD-ROM**: Cardiovascular System/ Anatomy Review: The Heart and on the **Anatomy CD-ROM**: Cardiovascular System/Heart Flythrough.

20–3 THE HEARTBEAT   p. 697

CARDIAC PHYSIOLOGY   p. 697

1. Two general classes of cardiac muscle cells are involved in the normal **heartbeat: contractile cells** and cells of the *conducting system. (Figure 20–11)*

THE CONDUCTING SYSTEM   p. 697

2. The **conducting system** is composed of the *sinoatrial node*, the *atrioventricular node*, and *conducting cells*. The conducting system initiates and distributes electrical impulses within the heart. Nodal cells establish the rate of cardiac contraction, and conducting cells distribute the contractile stimulus from the SA node to the atrial myocardium and the AV node (along *internodal pathways*) and from the AV node to the ventricular myocardium. *(Figure 20–12)*

3. Unlike skeletal muscle, cardiac muscle contracts without neural or hormonal stimulation. **Pacemaker cells** in the **sinoatrial (SA) node** (*cardiac pacemaker*) normally establish the rate of contraction. From the SA node, the stimulus travels to the **atrioventricular (AV) node**, and then to the **AV bundle**, which divides into **bundle branches**. From there, **Purkinje fibers** convey the impulses to the ventricular myocardium. *(Figures 20–12, 20–13)*

THE ELECTROCARDIOGRAM   p. 700

4. A recording of electrical activities in the heart is an **electrocardiogram** (**ECG** or **EKG**). Important landmarks of an ECG include the **P wave** (atrial depolarization), the **QRS complex** (ventricular depolarization), and the **T wave** (ventricular repolarization). *(Figure 20–14)*

CONTRACTILE CELLS   p. 701

5. **Contractile cells** form the bulk of the atrial and ventricular walls. Cardiac muscle cells have a long refractory period, so rapid stimulation produces twitches rather than tetanic contractions. *(Figure 20–15)*

Conduction pathway: **IP CD-ROM**: Cardiovascular System/Intrinsic Conduction System.

THE CARDIAC CYCLE   p. 703

6. The **cardiac cycle** contains periods of **atrial** and **ventricular systole** (contraction) and **atrial** and **ventricular diastole** (relaxation). *(Figure 20–16)*

7. When the heart beats, the two ventricles eject equal volumes of blood. *(Figure 20–17)*

8. The closing of valves and rushing of blood through the heart cause characteristic **heart sounds**, which can be heard during *auscultation. (Figure 20–18)*

Cardiac cycle: **IP CD-ROM**: Cardiovascular System/Cardiac Cycle.

20–4 CARDIODYNAMICS   p. 707

1. The amount of blood ejected by a ventricle during a single beat is the **stroke volume (SV)**. The amount of blood pumped by a ventricle each minute is the **cardiac output (CO)**. *(Figure 20–19)*

OVERVIEW: THE CONTROL OF CARDIAC OUTPUT   p. 710

2. Cardiac output can be adjusted by changes in either stroke volume or heart rate. *(Figure 20–20)*

FACTORS AFFECTING THE HEART RATE   p. 710

3. The **cardioacceleratory center** in the medulla oblongata activates sympathetic neurons; the **cardioinhibitory center** controls the parasympathetic neurons that slow the heart rate. These cardiac centers receive inputs from higher centers and from receptors monitoring blood pressure and the concentrations of dissolved gases. *(Figure 20–21)*

4. The basic heart rate is established by the pacemaker cells of the SA node, but it can be modified by the autonomic nervous system. The **atrial reflex** accelerates the heart rate when the walls of the right atrium are stretched. *(Figure 20–22)*

5. Sympathetic activity produces more powerful contractions that reduce the ESV. Parasympathetic stimulation slows the heart rate, reduces the contractile strength, and raises the ESV.

6. Cardiac output is affected by various factors, including autonomic innervation and hormones. *(Figure 20–22)*

FACTORS AFFECTING THE STROKE VOLUME   p. 712

7. The stroke volume is the difference between the **end-diastolic volume (EDV)** and the **end-systolic volume (ESV)**. The **filling time** and **venous return** interact to determine the EDV. Normally, the greater the EDV, the more powerful is the succeeding contraction (the **Frank-Starling principle**). *(Figure 20–23)*

EXERCISE AND CARDIAC OUTPUT   p. 714

8. The difference between resting and maximal cardiac outputs is the **cardiac reserve**. *(Figure 20–24)*

20–5 THE HEART AND THE CARDIOVASCULAR SYSTEM   p. 715

1. The heart does not work in isolation in maintaining adequate blood flow to all tissues.

REVIEW QUESTIONS

More assessment questions are available to you on the Companion Website. You will find Matching, Multiple Choice, True/False, and other quizzes to help further your understanding of the material covered in this chapter. To access the site, go to www.aw.com/martini.

LEVEL 1 Reviewing Facts and Terms

1. The great cardiac vein drains blood from the heart muscle to the
   (a) left ventricle
   (b) right ventricle
   (c) right atrium
   (d) left atrium

2. The autonomic centers for cardiac function are located in the
   (a) myocardial tissue of the heart
   (b) cardiac centers of the medulla oblongata
   (c) cerebral cortex
   (d) a, b, and c are correct

3. The serous membrane covering the inner surface of the heart is the
   (a) parietal pericardium (b) endocardium
   (c) myocardium (d) visceral pericardium

4. The simple squamous epithelium covering the valves of the heart constitutes the
   (a) epicardium (b) endocardium
   (c) myocardium (d) fibrous skeleton

5. The structure that permits blood flow from the right atrium to the left atrium while the lungs are developing before birth is the
   (a) foramen ovale (b) interatrial septum
   (c) coronary sinus (d) fossa ovalis

6. Blood leaves the right ventricle by passing through the
   (a) aortic valve (b) pulmonary valve
   (c) mitral valve (d) tricuspid valve

7. The P wave of the ECG appears as the
   (a) ventricles depolarize (b) atria depolarize
   (c) ventricles repolarize (d) atria repolarize

8. During diastole, a chamber of the heart
   (a) relaxes and fills with blood
   (b) contracts and pushes blood into an adjacent chamber
   (c) experiences a sharp increase in pressure
   (d) reaches a pressure of approximately 120 mm Hg

9. During the cardiac cycle, the amount of blood ejected from the left ventricle when the semilunar valve opens is the
   (a) stroke volume (SV)
   (b) end-diastolic volume (EDV)
   (c) end-systolic volume (ESV)
   (d) cardiac output (CO)

10. What role do the chordae tendineae and papillary muscles play in the normal function of the AV valves?

11. Describe the three distinct layers that make up the heart wall.

12. What are the principal valves in the heart, and what is the function of each?

13. Trace the normal pathway of an electrical impulse through the conducting system of the heart.

14. What is the cardiac cycle? What phases and events are necessary to complete the cardiac cycle?

15. What three factors regulate stroke volume to ensure that the left and right ventricles pump equal volumes of blood?

LEVEL 2 Reviewing Concepts

16. The cells of the conducting system differ from the contractile cells of the heart in that
   (a) conducting cells are larger and contain more myofibrils
   (b) contractile cells exhibit prepotentials
   (c) contractile cells do not exhibit automaticity
   (d) both a and b are correct

17. Tetanic muscle contractions cannot occur in a normal cardiac muscle cell because
   (a) cardiac muscle tissue contracts on its own
   (b) there is no neural or hormonal stimulation
   (c) the refractory period lasts until the muscle cell relaxes
   (d) the refractory period ends before the muscle cell reaches peak tension

18. The amount of blood that is forced out of the heart depends on
   (a) the degree of stretching at the end of ventricular diastole
   (b) the contractility of the ventricle
   (c) the amount of pressure required to eject blood
   (d) a, b, and c are correct

19. The cardiac output cannot increase indefinitely because
   (a) the available filling time becomes shorter as the heart rate increases
   (b) the cardiovascular centers adjust the heart rate
   (c) the rate of spontaneous depolarization decreases
   (d) the ion concentrations of pacemaker cell membranes decrease

20. Describe the function of the SA node in the cardiac cycle. How does this differ from the function of the AV node?

21. What are the source and significance of heart sounds?

22. Differentiate between stroke volume and cardiac output. How is cardiac output calculated?

23. What factors influence cardiac output?

24. What is the significance of preload to cardiac muscle cells?

25. What effect does sympathetic stimulation have on the heart? What effect does parasympathetic stimulation have on the heart?

26. Describe the effects that epinephrine, norepinephrine, glucagon, and thyroid hormones have on the contractility of the heart.

LEVEL 3 Critical Thinking and Clinical Applications

27. Most of the ATP produced in cardiac muscle is derived from the metabolism of fatty acids. During times of exertion, cardiac muscle cells can use lactic acid as an energy source. Why would this adaptation be advantageous to cardiac function?

28. A patient's ECG recording shows a consistent pattern of two P waves followed by a normal QRS complex and T wave. What is the cause of this abnormal wave pattern?

29. The following measurements were made on two individuals (the values recorded remained stable for one hour):

    Person 1: heart rate, 75 bpm; stroke volume, 60 ml

    Person 2: heart rate, 90 bpm; stroke volume, 95 ml

    Which person has the greater venous return? Which person has the longer ventricular filling time?

30. Karen is taking the medication *verapamil*, a drug that blocks the calcium channels in cardiac muscle cells. What effect should this medication have on Karen's stroke volume?

31. After a myocardial infarction, the cells surrounding the damaged tissue frequently become hyperexcitable and act as ectopic pacemakers. This condition can lead to abnormal ventricular rhythms, with fatal consequences. What would cause the excitability of the uninjured cells?

32. Premature ventricular contractions (PVCs) are triggered when a Purkinje cell or contractile cell in the ventricle depolarizes to threshold. A person experiencing a PVC may feel that his or her heart has "skipped a beat." If the PVC causes an extra contraction, why does the person feel that a beat has been skipped?

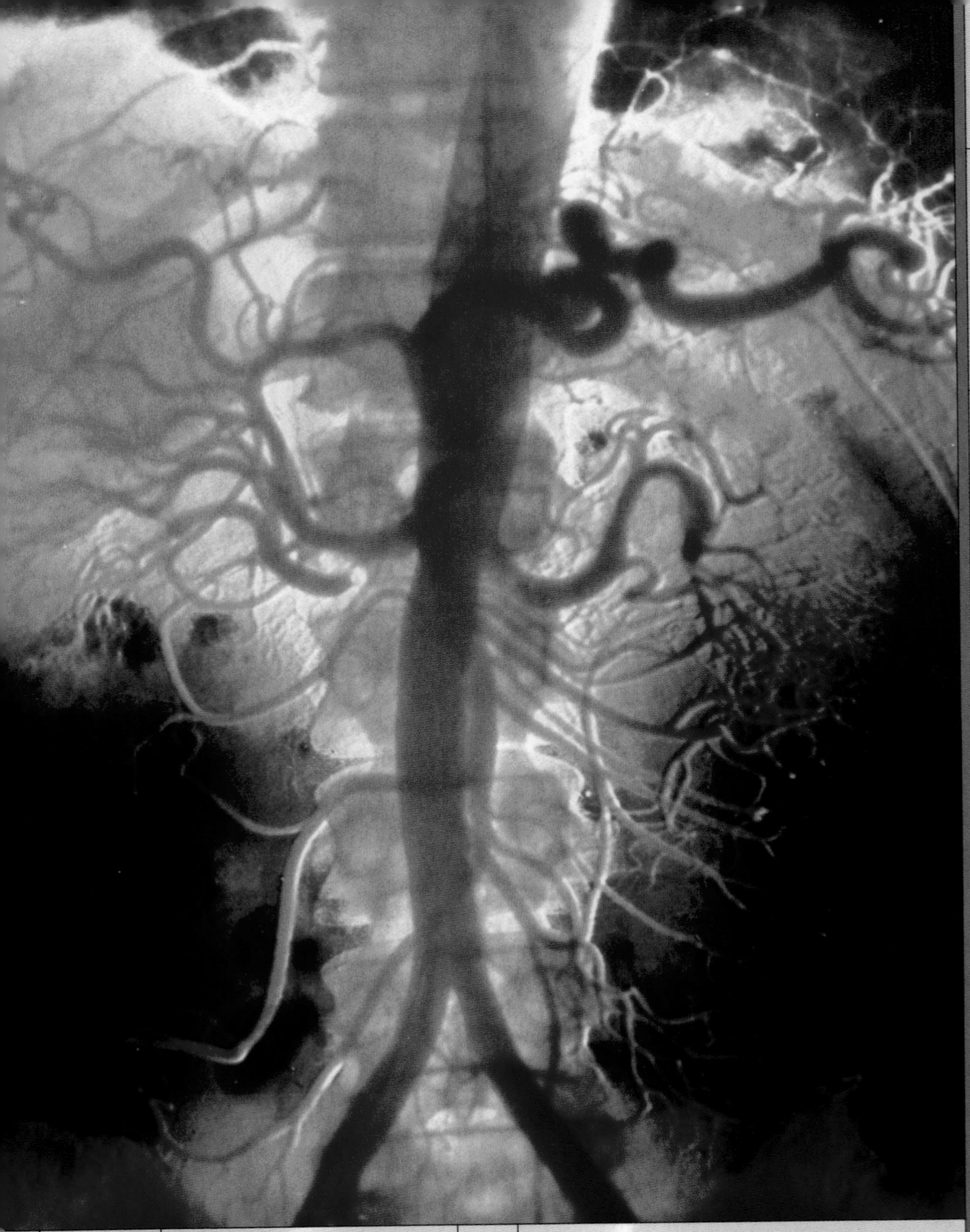

UNIT 4   CHAPTER 19 20 21 22

CLINICAL NOTES
- Aneurysms   726
- Hypotension and Hypertension   735
- Strokes   757

CLINICAL DISCUSSIONS
- Arteriosclerosis   725
- Checking Pulse and Blood
  Pressure   736
- Edema   739
- Circulatory Shock   749
- Congenital Cardiovascular
  Problems   770

BLOOD VESSELS AND CIRCULATION

In the last two chapters, we examined the composition of blood and the structure and function of the heart, whose pumping action keeps the blood in motion. Blood circulates throughout the body, moving from the heart through the tissues and back to the heart, in blood vessels. We will now consider these vessels and the nature of the exchange that occurs between the blood and interstitial fluids of the body.

There are five general classes of blood vessels in the cardiovascular system. **Arteries** carry blood away from the heart. As they enter peripheral tissues, arteries branch repeatedly, and the branches decrease in diameter. The smallest arterial branches are called **arterioles** (ar-TĒ-rē-ōlz). From the arterioles, blood moves into the **capillaries**, where diffusion occurs between blood and interstitial fluid. From the capillaries, blood enters small **venules** (VĒ-nūlz), which unite to form larger **veins** that return blood to the heart.

Two major arteries are connected to the heart, one associated with each ventricle. Blood leaves the heart in the pulmonary trunk, which originates at the right ventricle, and the aorta, which originates at the left ventricle. Each of these arterial trunks has an internal diameter of about 2.5 cm (1 in.). The pulmonary arteries that branch from the pulmonary trunk carry blood to the lungs. The systemic arteries that branch from the aorta distribute blood to all other organs. Within these organs, further branching occurs, creating several hundred million tiny arterioles that provide blood to more than 10 billion capillaries. These capillaries, barely the diameter of a single red blood cell, form extensive, branching networks. If all the capillaries in your body were placed end to end, they would circle the globe with a combined length of over 25,000 miles.

The vital functions of the cardiovascular system depend entirely on events at the capillary level: All chemical and gaseous exchange between blood and interstitial fluid takes place across capillary walls. Tissue cells rely on capillary diffusion to obtain nutrients and oxygen and to remove metabolic wastes, such as carbon dioxide and urea. Diffusion occurs very rapidly, because the distances involved are very small: Few cells lie farther than 125 μm (0.005 in.) from a capillary. Homeostatic mechanisms operating at the local, regional, and systemic levels adjust blood flow through the capillaries to meet the demands of peripheral tissues.

This chapter begins with a description of the histological organization of arteries, capillaries, and veins. We next discuss the functions of these vessels, basic principles of cardiovascular regulation, and the distribution of major blood vessels in the body. We will then be ready to consider the organization and function of the lymphatic system, the focus of Chapter 22.

21–1 THE ANATOMY OF BLOOD VESSELS

Objectives

- Distinguish among the types of blood vessels on the basis of their structure and function.
- Describe how and where fluid and dissolved materials enter and leave the cardiovascular system.

Blood flow requires vessels that are resilient enough to withstand changes in pressure and flexible enough to move with underlying tissues and organs, yet tough enough to remain open and operational under all physiological circumstances. The pressures experienced by vessels vary with distance from the heart, and their structures reflect this fact. Moreover, arteries, veins, and capillaries differ in function, and these functional differences, as always, involve particular structural adaptations. Accordingly, each type of blood vessel exhibits distinctive anatomical features, with modifications based on specific local needs.

☐ THE STRUCTURE OF VESSEL WALLS

The walls of arteries and veins contain three distinct layers—the tunica intima, tunica media, and tunica externa (Figure 21–1●):

1. The **tunica intima** (IN-ti-muh), or *tunica interna*, is the innermost layer of a blood vessel. This layer includes the endothelial lining and an underlying layer of connective tissue with a variable number of elastic fibers. In arteries, the outer margin of the tunica intima contains a thick layer of elastic fibers called the **internal elastic membrane**.

2. The **tunica media**, the middle layer, contains concentric sheets of smooth muscle tissue in a framework of loose connective tissue. The collagen fibers bind the tunica media to the tunica intima and tunica externa. Commonly the thickest layer in the wall of a small artery, the tunica media is separated from the surrounding tunica externa by a thin band of elastic fibers called the **external elastic membrane**. The smooth muscle cells of the tunica media encircle the endothelium lining the lumen of the blood vessel. When these smooth muscles contract, the vessel decreases in diameter; when they relax, the diameter increases. Large arteries also contain layers of longitudinally arranged smooth muscle cells.

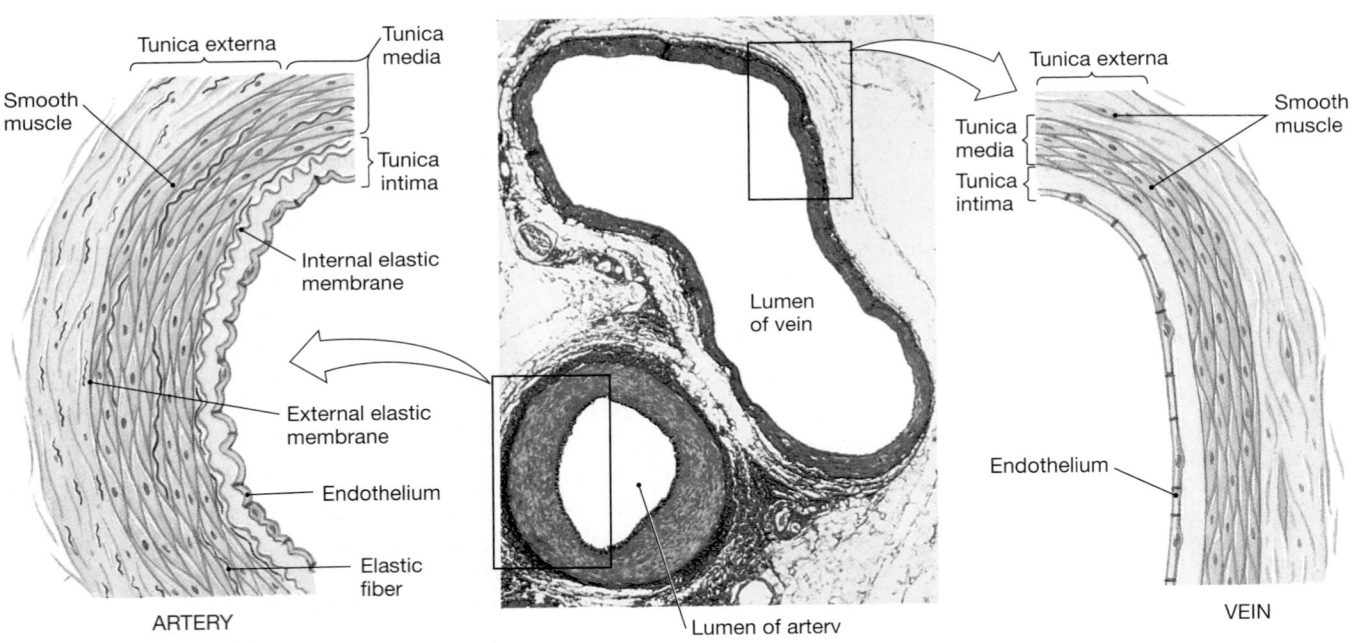

ARTERY

Lumen of artery

VEIN

Feature	Typical Artery	Typical Vein
GENERAL APPEARANCE IN SECTIONAL VIEW	Usually round, with relatively thick wall	Usually flattened or collapsed, with relatively thin wall
TUNICA INTIMA Endothelium	Usually rippled, due to vessel constriction	Often smooth
Internal elastic membrane	Present	Absent
TUNICA MEDIA	Thick, dominated by smooth muscle and elastic fibers	Thin, dominated by smooth muscle and collagen fibers
External elastic membrane	Present	Absent
TUNICA EXTERNA	Collagen and elastic fibers	Collagen, elastic, and smooth muscle fibers

●**FIGURE 21–1
A Comparison of a Typical Artery and a Typical Vein**

3. The **tunica externa** (eks-TER-nuh), or *tunica adventitia* (ad-ven-TISH-ē-uh), is the outermost layer of a blood vessel and forms a connective tissue sheath. In arteries, this layer contains collagen fibers with scattered bands of elastic fibers. In veins, it is generally thicker than the tunica media, and contains networks of elastic fibers and bundles of smooth muscle cells. The connective-tissue fibers of the tunica externa typically blend into those of adjacent tissues, stabilizing and anchoring the blood vessel.

Their layered walls give arteries and veins considerable strength. The muscular and elastic components also permit controlled alterations in diameter as blood pressure or blood volume changes. However, the walls of arteries and veins are too thick to allow diffusion between the bloodstream and surrounding tissues or even between the blood and the tissues of the vessel itself. For this reason, the walls of large vessels contain small arteries and veins that supply the smooth muscle cells and fibroblasts of the tunica media and tunica externa. These blood vessels are called the *vasa vasorum* ("vessels of vessels").

DIFFERENCES BETWEEN ARTERIES AND VEINS

Arteries and veins servicing the same region typically lie side by side (Figure 21–1●). In sectional view, arteries and veins may be distinguished by the following characteristics:

- In general, the walls of arteries are thicker than those of veins. The tunica media of an artery contains more smooth muscle and elastic fibers than does that of a vein. These contractile and elastic components resist the pressure generated by the heart as it forces blood into the circuit.

- When not opposed by blood pressure, the elastic fibers in the arterial walls recoil, constricting the lumen. Thus, seen on dissection or in sectional view, the lumen of an artery often looks smaller than that of the corresponding vein. Because the walls of arteries are relatively thick and strong, they retain their circular shape in section. Cut veins tend to collapse, and in section they often look flattened or grossly distorted.

- The endothelial lining of an artery cannot contract, so when an artery constricts, its endothelium is thrown into folds that give arterial sections a pleated appearance. The lining of a vein lacks these folds.

In gross dissection, arteries and veins can generally be distinguished because:

- The thicker walls of the arteries can be felt if the vessels are compressed.

- Arteries usually retain their cylindrical shape, whereas veins often collapse.

- Arteries are more resilient: When stretched, they keep their shape and elongate, and when released, they snap back. A small vein cannot tolerate as much distortion without collapsing or tearing.

- Veins typically contain *valves*—internal structures that prevent the backflow of blood toward the capillaries. In an intact vein, the location of each valve is marked by a slight distension of the vessel wall. (We shall consider valve structure in a later section.)

ARTERIES

Their relatively thick, muscular walls make arteries elastic and contractile. Elasticity permits passive changes in the diameter of a vessel in response to changes in blood pressure. For example, it allows arteries to absorb the surging pressure waves that accompany the contractions of the ventricles.

The contractility of the arterial walls gives them the ability to change in diameter actively, primarily under the control of the sympathetic division of the autonomic nervous system. When stimulated, arterial smooth muscles contract and thereby constrict the artery—a process called **vasoconstriction**. Relaxation of these smooth muscles increases the diameter of the lumen—a process called **vasodilation**. Vasoconstriction and vasodilation affect (1) the afterload on the heart, (2) peripheral blood pressure, and (3) capillary blood flow. We shall explore these effects in a later section. Contractility is also important during the vascular phase of hemostasis, when the contraction of a damaged vessel wall helps reduce bleeding. ∞ pp. 672-673

In traveling from the heart to peripheral capillaries, blood passes through *elastic arteries, muscular arteries,* and *arterioles* (Figure 21–2●).

Elastic Arteries

Elastic arteries, or *conducting arteries,* are large vessels with a diameter up to 2.5 cm (1 in.). These vessels transport large volumes of blood away from the heart. The pulmonary trunk and aorta, as well as their major arterial branches (the *pulmonary, common carotid, subclavian,* and *common iliac arteries*), are elastic arteries.

The walls of elastic arteries (Figure 21–2●) are extremely resilient because the tunica media contains a high density of elastic fibers and relatively few smooth muscle cells. As a result, elastic arteries can tolerate the pressure changes that occur during the cardiac cycle. We have already mentioned the role played by *elastic rebound* in the aorta in maintaining blood flow in the coronary arteries. ∞ p. 692 However, elastic rebound occurs to some degree in all elastic arteries. During ventricular systole, pressures rise rapidly and the elastic arteries expand. During ventricular diastole, blood pressure within the arterial system falls and the elastic fibers recoil to their original dimensions. Their expansion cushions the sudden rise in pressure during ventricular systole, and their recoil slows the drop in pressure during ventricular diastole. This feature is important because blood pressure is the driving force behind blood flow: The greater the pressure oscillations, the greater are the changes in blood flow. The elasticity of the arterial system dampens the pressure peaks and valleys that accompany the heartbeat. By the time blood reaches the arterioles, the pressure oscillations have disappeared and blood flow is continuous.

Muscular Arteries

Muscular arteries, also known as *medium-sized arteries* or *distribution arteries,* distribute blood to the body's skeletal muscle and internal organs. Most of the vessels of the arterial system are

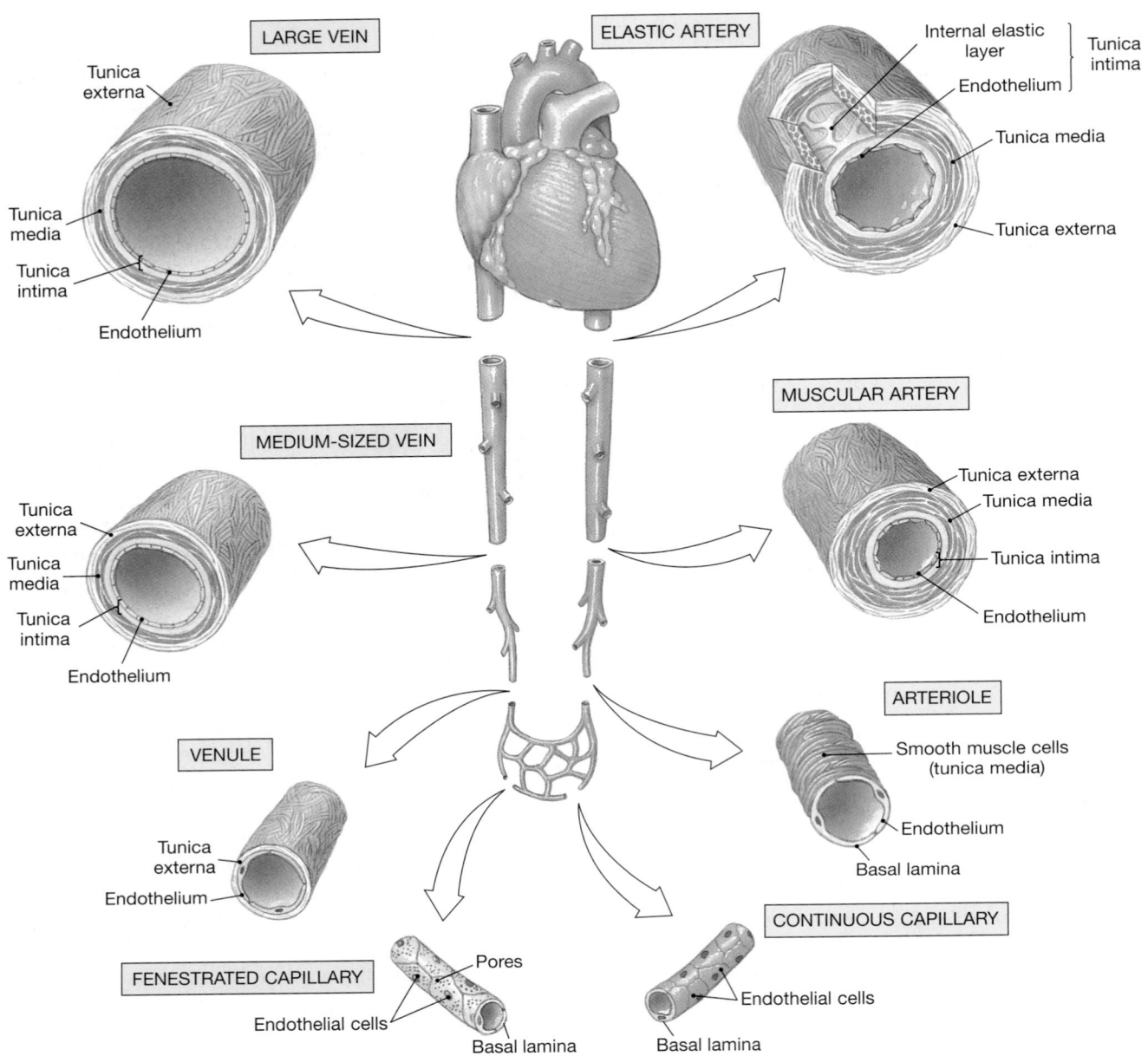

●**FIGURE 21–2**
Histological Structure of Blood Vessels. Representative diagrammatic cross-sectional views of the walls of arteries, veins, and capillaries. Notice the relative sizes of the layers in these vessels.

muscular arteries. These arteries are characterized by a thick tunica media that contains more smooth muscle cells than does the tunica media of elastic arteries (Figures 21–1 and 21–2●). A typical muscular artery has a lumen diameter of approximately 0.4 cm (0.15 in.), but some have diameters as small as 0.5 mm. The *external carotid arteries* of the neck, the *brachial arteries* of the arms, the *mesenteric arteries* of the abdomen, and the *femoral arteries* of the thighs are examples of muscular arteries. You will be familiar with these vessels if you have taken a first-aid class, as these classes usually identify major *pressure points*— places in the body where muscular arteries can be forced against deeper bones to reduce blood flow and control severe bleeding. (See the box on p. 736.)

Arterioles
Arterioles, with an internal diameter of 30 μm or less, are considerably smaller than muscular arteries. Arterioles have a poorly defined tunica externa, and the tunica media in the larger arterioles consists of one or two layers of smooth muscle cells (Figure 21–2●). The tunica media of the smallest arterioles contains scattered smooth muscle cells that do not form a complete layer.

The diameters of smaller muscular arteries and arterioles change in response to local conditions or to sympathetic or endocrine stimulation. For example, arterioles in most tissues vasodilate when oxygen levels are low and, as we saw in Chapter 16, vasoconstrict under sympathetic stimulation. ∞ p. 541 Changes

Arteriosclerosis

Arteriosclerosis (ar-tē-rē-ō-skle-RŌ-sis) is a thickening and toughening of arterial walls. This condition may not sound life threatening, but complications related to arteriosclerosis account for roughly half of all deaths in the United States. The effects of arteriosclerosis are varied; for example, arteriosclerosis of coronary vessels is responsible for *coronary artery disease* (*CAD*), and arteriosclerosis of arteries supplying the brain can lead to strokes. ∞ p. 695 Arteriosclerosis takes two major forms:

1. **Focal calcification** is the gradual degeneration of smooth muscle in the tunica media and the subsequent deposition of calcium salts. Typically, the process involves arteries of the limbs and genital organs. Some focal calcification occurs as part of the aging process, and it may develop in association with atherosclerosis (see below). Rapid and severe calcification may occur as a complication of diabetes mellitus, an endocrine disorder. ∞ p. 633

2. **Atherosclerosis** (ath-er-ō-skle-RŌ-sis) is associated with damage to the endothelial lining and the formation of lipid deposits in the tunica media. Atherosclerosis is the most common form of arteriosclerosis.

Many factors may be involved in the development of atherosclerosis. One major factor is lipid levels in the blood. Atherosclerosis tends to develop in persons whose blood contains elevated levels of plasma lipids—specifically, cholesterol. Circulating cholesterol is transported to peripheral tissues in *lipoproteins*, which are protein–lipid complexes. (We will discuss the various types of lipoproteins in Chapter 25.) Recent evidence indicates that many forms of atherosclerosis are associated with either (1) low levels of *apolipoprotein-E* (*ApoE*), a transport protein whose lipids are quickly absorbed by body tissues, or (2) high levels of *lipoprotein(a)*, a *low-density lipoprotein* (*LDL*) that is absorbed at a much slower rate.

When ApoE levels are low or lipoprotein(a) levels are high, cholesterol-rich lipoproteins remain in circulation for an extended period. Circulating monocytes then begin removing them from the bloodstream. Eventually, the monocytes become filled with lipid droplets. Now called *foam cells*, they attach themselves to the endothelial walls of blood vessels, where they release growth factors. These cytokines stimulate the divisions of smooth muscle cells near the tunica intima, thickening the vessel wall.

Other monocytes then invade the area, migrating between the endothelial cells. As these changes occur, the monocytes, smooth muscle cells, and endothelial cells begin phagocytizing lipids as well. The result is a **plaque**, a fatty mass of tissue that projects into the lumen of the vessel. At this point, the plaque has a relatively simple structure, and evidence suggests that the process can be reversed if appropriate dietary adjustments are made.

If the conditions persist, the endothelial cells become swollen with lipids, and gaps appear in the endothelial lining. Platelets now begin sticking to the exposed collagen fibers. The combination of platelet adhesion and aggregation leads to the formation of a localized blood clot, which will further restrict blood flow through the artery. The structure of the plaque is now relatively complex. Plaque growth can be halted, but the structural changes are generally permanent.

A typical plaque is shown in Figure 21–3●. Elderly individuals—especially elderly men—are most likely to develop atherosclerotic plaques. Evidence suggests that estrogens may slow plaque formation; this may account for the lower incidence of CAD, myocardial infarctions (MIs), and strokes in women. After menopause, when estrogen production declines, the risk of CAD, MIs, and strokes in women increases markedly.

In addition to advanced age and male gender, other important risk factors for atherosclerosis include high blood cholesterol levels, high blood pressure, and cigarette smoking.

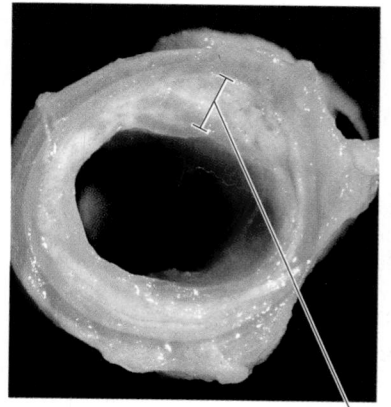
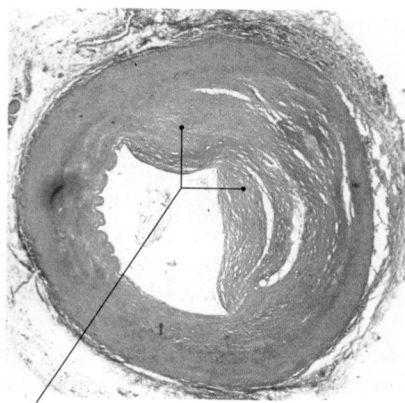

●**FIGURE 21–3**
A Plaque Blocking an Artery. **(a)** A section of a coronary artery narrowed by plaque formation. **(b)** A cross-sectional view of a large plaque.

Plaque deposit in vessel wall

(a)   **(b)**

Roughly 20 percent of middle-aged men have all three of these risk factors; these individuals are four times more likely to experience an MI or a cardiac arrest than are other men in their age group. Although fewer women develop atherosclerotic plaques, elderly women smokers with high blood cholesterol and high blood pressure are at much greater risk than are other women. Factors that can promote the development of atherosclerosis in both men and women include diabetes mellitus, obesity, and stress. Evidence also indicates that at least some forms of atherosclerosis may be linked to chronic infection with *Chlamydia pneumoniae*, a bacterium responsible for several types of respiratory infections, including some forms of pneumonia.

We discussed potential treatments for atherosclerotic plaques, such as catheterization, balloon angioplasty, and bypass surgery, in Chapter 20. ∞ p. 696 In cases in which dietary modifications do not lower circulating LDL levels sufficiently,

drug therapies can bring them under control. Genetic-engineering techniques have recently been used to treat an inherited form of *hypercholesterolemia* (high blood cholesterol) linked to extensive plaque formation. (Individuals with this condition are unable to absorb and recycle cholesterol in the liver.) In this experimental procedure, circulating cholesterol levels declined after copies of appropriate genes were inserted into some of the individual's liver cells.

Without question, the best approach to atherosclerosis is to try to avoid it by eliminating or reducing associated risk factors. Suggestions include (1) reducing the amount of dietary cholesterol, saturated fats, and trans-fatty acids by restricting consumption of fatty meats (such as beef, lamb, and pork), egg yolks, and cream; ∞ p. 50 (2) not smoking; (3) checking your blood pressure and taking steps to lower it if necessary; (4) having your blood cholesterol levels checked annually; (5) controlling your weight; and (6) exercising regularly.

---

in their diameter affect the amount of force required to push blood around the cardiovascular system: More pressure is required to push blood through a constricted vessel than through a dilated one. The force opposing blood flow is called **resistance (R)**, and arterioles are therefore called **resistance vessels**.

Vessel characteristics change gradually as we travel away from the heart. Each type of vessel defined here actually represents the midpoint of a continuum. For example, elastic and muscular arteries are interconnected, so the largest muscular arteries contain a considerable amount of elastic tissue, whereas the smallest resemble heavily muscled arterioles.

Arteries carry blood under great pressure, with the walls of the artery adapted to handle that pressure. Occasionally, local arterial pressure exceeds the capacity of the elastic components of the tunics, producing an **aneurysm** (AN-ū-rizm), or bulge in the weakened wall of an artery. The bulge resembles a bubble in the wall of a tire—and like a bad tire, the affected artery can suffer a catastrophic blowout. The most dangerous aneurysms are those involving arteries of the brain, where they cause strokes, or of the aorta, where a ruptured aneurysm will cause fatal bleeding in a matter of minutes.

Aneurysms most commonly occur in individuals with *arteriosclerosis*. Over time, arteriosclerosis causes vessel walls to become less elastic, and a weak spot can develop. Aneurysms can also be associated with other conditions that weaken arterial walls, such as arterial inflammation or infection, and with *Marfan's syndrome*, a connective-tissue disorder. ∞ p. 124 As you might expect, individuals with high blood pressure are most likely to develop dangerous aneurysms, because the elevated arterial pressures place great stresses on the vessel walls. Unfortunately, because they are often painless, aneurysms are likely to go undetected until catastrophic rupture occurs. **AM** Aneurysms

CAPILLARIES

When we think of the cardiovascular system, we think first of the heart or of the great blood vessels that service it. But the real work of the cardiovascular system is done in the microscopic capillaries that permeate most tissues. These delicate vessels weave throughout active tissues, forming intricate networks that surround muscle fibers, radiate through connective tissues, and branch beneath the basal laminae of epithelia.

Capillaries are the *only* blood vessels whose walls permit exchange between the blood and the surrounding interstitial fluids. Because capillary walls are thin, the diffusion distances are small, so exchange can occur quickly. In addition, blood flows through capillaries relatively slowly, allowing sufficient time for the diffusion or active transport of materials across the capillary walls. Thus, the histological structure of capillaries permits a two-way exchange of substances between blood and interstitial fluid.

A typical capillary consists of an endothelial tube inside a delicate basal lamina. Neither a tunica media nor a tunica externa is present. The average diameter of a capillary is a mere 8 μm, very close to that of a single red blood cell. There are two major types of capillaries: (1) *continuous capillaries* and (2) *fenestrated capillaries*.

Continuous Capillaries

Most regions of the body are supplied by continuous capillaries. In a **continuous capillary**, the endothelium is a complete lining. A cross section through a large continuous capillary cuts across several endothelial cells (Figure 21–4a●). In a small continuous capillary, a single endothelial cell may wrap all the way around the lumen, just as your hand wraps around a small glass.

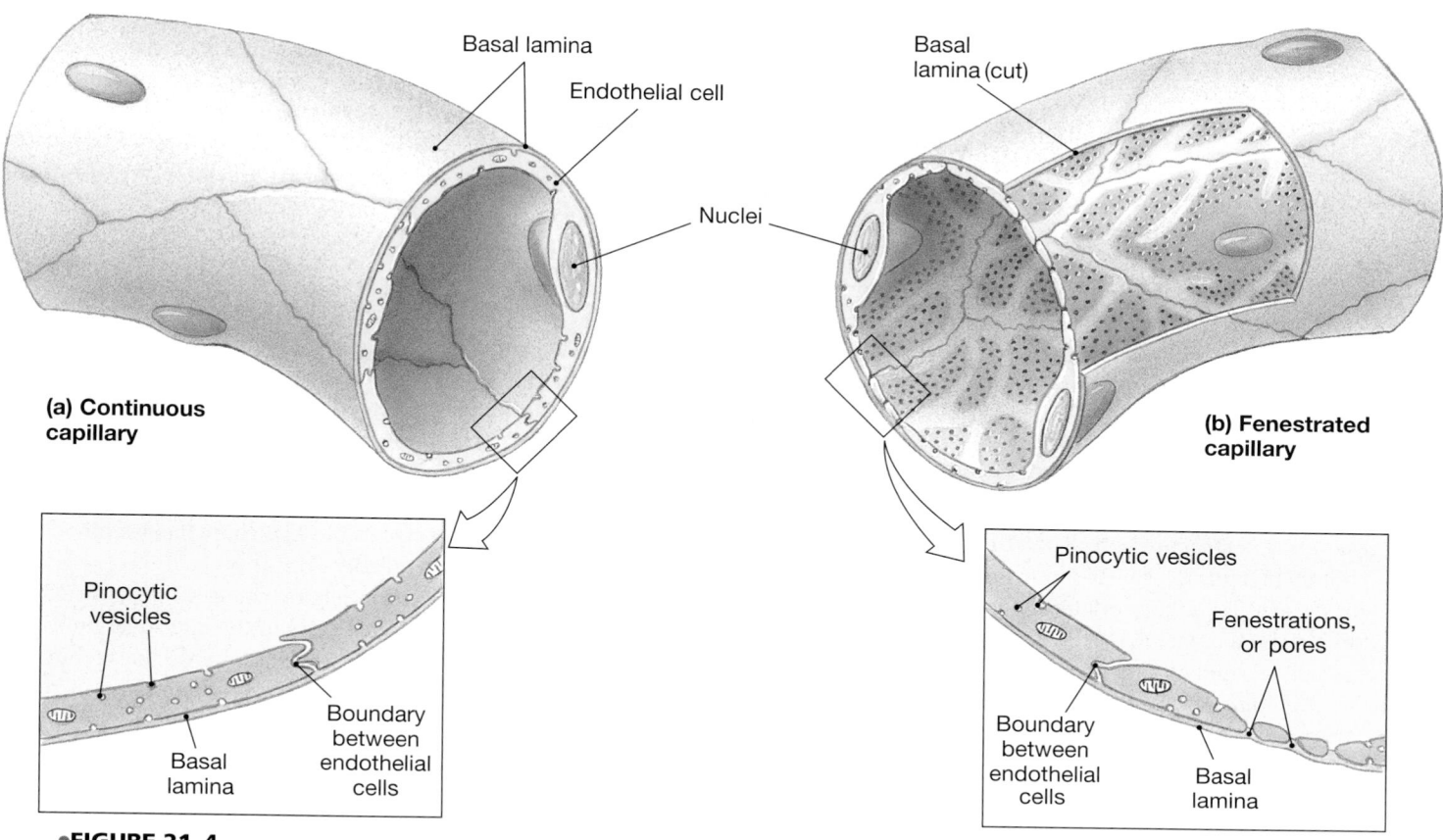

•**FIGURE 21–4**
Capillary Structure. **(a)** A continuous capillary, showing routes for the diffusion of water and solutes. **(b)** A fenestrated capillary, showing the pores that facilitate diffusion across the endothelial lining.

Continuous capillaries are located in all tissues except epithelia and cartilage. Continuous capillaries permit the diffusion of water, small solutes, and lipid-soluble materials into the surrounding interstitial fluid, but prevent the loss of blood cells and plasma proteins. In addition, some exchange may occur between blood and interstitial fluid by vesicular transport—the movement of vesicles that form at the inner endothelial surface. ∞ p. 97

The endothelial cells in specialized continuous capillaries throughout most of the central nervous system and in the thymus are bound together by tight junctions. These capillaries have very restricted permeability characteristics. We discussed one example, the capillaries responsible for the *blood–brain barrier*, in Chapters 12 and 14. ∞ pp. 394, 472

Fenestrated Capillaries

Fenestrated (FEN-es-trā-ted) **capillaries** (*fenestra*, window) are capillaries that contain "windows," or pores, that span the endothelial lining (Figure 21–4b•). The pores permit the rapid exchange of water and solutes as large as small peptides between plasma and interstitial fluid. Examples of fenestrated

capillaries noted in earlier chapters include the *choroid plexus* of the brain and the blood vessels in a variety of endocrine organs, such as the hypothalamus and the pituitary, pineal, and thyroid glands. Fenestrated capillaries are also located along absorptive areas of the intestinal tract and at filtration sites in the kidneys. Both the number of pores and their permeability characteristics may vary from one region of the capillary to another.

Sinusoids (SĪ-nus-oydz) resemble fenestrated capillaries that are flattened and irregular. In contrast to fenestrated capillaries, sinusoids commonly have gaps between adjacent endothelial cells, and the basal lamina is either thinner or absent. As a result, sinusoids permit the free exchange of water and solutes as large as plasma proteins between blood and interstitial fluid.

Blood moves through sinusoids relatively slowly, maximizing the time available for exchange across the sinusoidal walls. Sinusoids occur in the liver, bone marrow, spleen, and many endocrine organs, such as the pituitary and adrenal glands. At liver sinusoids, plasma proteins secreted by the liver cells enter the bloodstream. Along sinusoids of the liver, spleen, and bone marrow, phagocytic cells monitor the passing blood, engulfing damaged red blood cells, pathogens, and cellular debris.

Capillary Beds

Capillaries do not function as individual units but as part of an interconnected network called a **capillary bed**, or **capillary plexus** (Figure 21–5●). A single arteriole generally gives rise to dozens of capillaries that empty into several *venules*, the smallest vessels of the venous system. The entrance to each capillary is guarded by a band of smooth muscle called a **precapillary sphincter**. Contraction of the smooth muscle cells constricts and narrows the diameter of the capillary entrance, thereby reducing the flow of blood. Relaxation of the sphincter dilates the opening, allowing blood to enter the capillary faster.

The capillary bed contains several relatively direct connections between arterioles and venules. The wall in the initial part of such a passageway possesses smooth muscles capable of changing its diameter. This segment is called a **metarteriole** (met-ar-TĒ-rē-ōl). The rest of the passageway, which resembles a typical capillary in structure, is called a **thoroughfare channel**.

A capillary bed may receive blood from more than one artery. The multiple arteries, called **collaterals**, enter the region and fuse before giving rise to arterioles. The fusion of two collateral arteries that supply a capillary bed is an example of an **arterial anastomosis**. (An *anastomosis* is the joining of two tubes.) The interconnections between the *anterior* and *posterior interventricular arteries* of the heart are arterial anastomoses. ∞ p. 695 An arterial anastomosis acts like an insurance policy: If one artery is compressed or blocked, capillary circulation will continue.

Arteriovenous (ar-tē-rē-ō-VĒ-nus) **anastomoses** are direct connections between arterioles and venules. When an arteriovenous anastomosis is dilated, blood will bypass the capillary bed and flow directly into the venous circulation. The pattern of blood flow through an arteriovenous anastomosis is regulated primarily by sympathetic innervation under the control of the cardiovascular centers of the medulla oblongata.

Vasomotion

Although blood normally flows from arterioles to venules at a constant rate, the flow within each capillary is quite variable. Each precapillary sphincter alternately contracts and relaxes, perhaps a dozen times per minute. As a result, the blood flow within any capillary occurs in pulses rather than as a steady and constant stream. The net effect is that blood may reach the venules by one route now and by a different route later. The cycling of contraction and relaxation of smooth muscles that changes blood flow through capillary beds is called **vasomotion**.

Vasomotion is controlled locally by changes in the concentrations of chemicals and dissolved gases in the interstitial fluids. For example, when dissolved oxygen concentrations decline within a tissue, the capillary sphincters relax, so blood flow to the area increases. This process, an example of capillary *autoregulation*, will be the focus of a later section.

When you are at rest, blood flows through roughly 25 percent of the vessels within a typical capillary bed in your body. Your cardio-

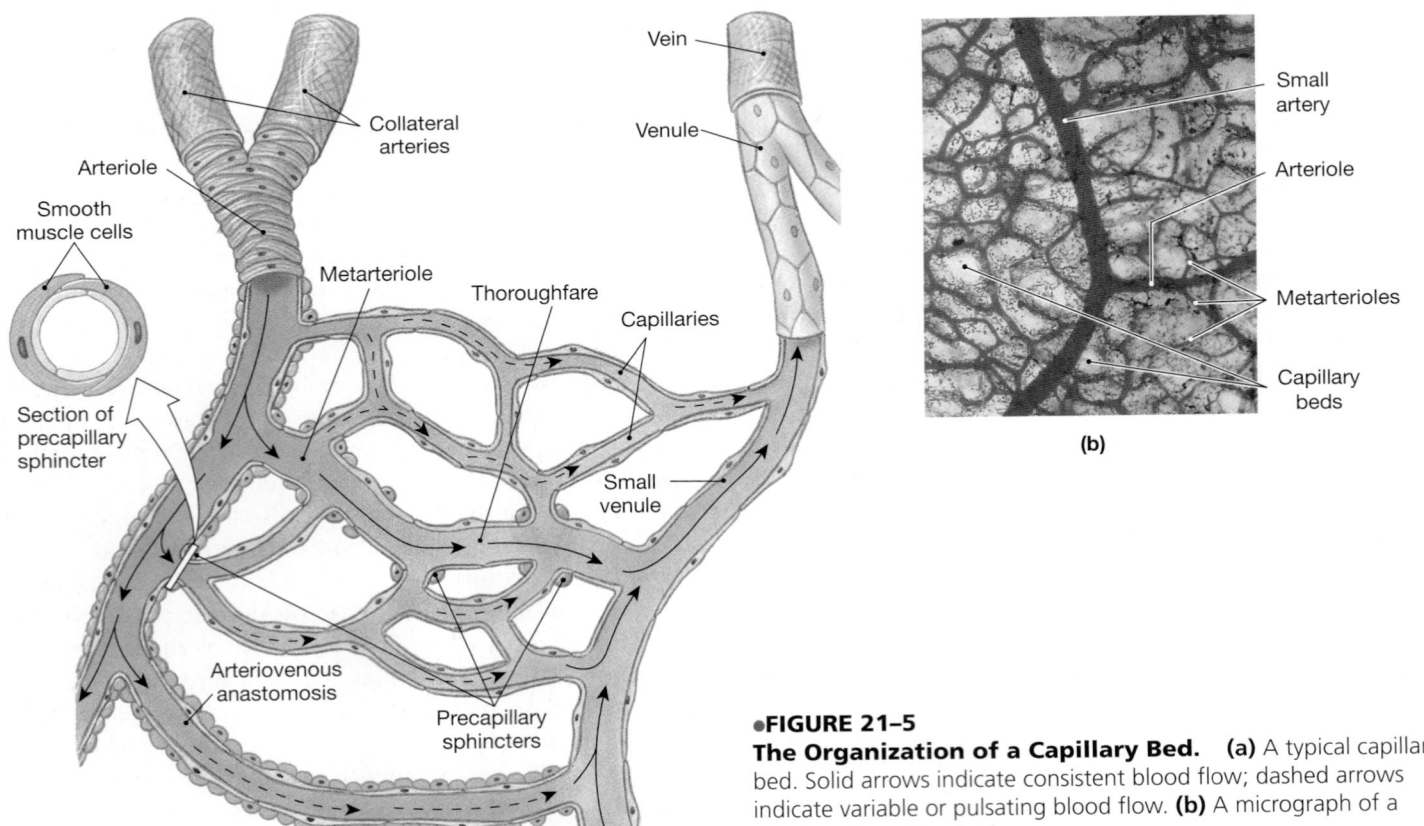

●**FIGURE 21–5**
The Organization of a Capillary Bed. (a) A typical capillary bed. Solid arrows indicate consistent blood flow; dashed arrows indicate variable or pulsating blood flow. (b) A micrograph of a number of capillary beds.

vascular system does not contain enough blood to maintain adequate blood flow to all the capillaries in all the capillary beds in your body at the same time. As a result, when many tissues become active, the blood flow through capillary beds must be coordinated. We shall describe the mechanisms by which the cardiovascular centers perform this coordination later in the chapter.

□ VEINS

Veins collect blood from all tissues and organs and return it to the heart. The walls of veins are thinner than those of corresponding arteries. Venous walls need not be as thick as arterial walls because the blood pressure in veins is lower than that in the arteries. Veins are classified on the basis of their size. Even though their walls are thinner, veins are in general, larger in diameter than their corresponding arteries. (Review Figures 21–1, p. 722, and 21–2•, p. 724, to compare typical arteries and veins.)

Venules

Venules, which collect blood from capillary beds, are the smallest venous vessels. They vary widely in size and character. An average venule has an internal diameter of roughly 20 μm. Venules smaller than 50 μm lack a tunica media, and the smallest venules resemble expanded capillaries.

Medium-Sized Veins

Medium-sized veins range from 2 to 9 mm in internal diameter, comparable in size to muscular arteries. In these veins, the tunica media is thin and contains relatively few smooth muscle cells. The thickest layer of a medium-sized vein is the tunica externa, which contains longitudinal bundles of elastic and collagen fibers.

Large Veins

Large veins include the superior and inferior venae cavae and their tributaries within the abdominopelvic and thoracic cavities. All the tunica layers are present in large veins. The slender tunica media is surrounded by a thick tunica externa composed of a mixture of elastic and collagen fibers.

Venous Valves

The arterial system is a high-pressure system: It takes almost all the force developed by the heart to push blood through the network of arteries and across miles of capillaries. Blood pressure in a peripheral venule is only about 10 percent of that in the ascending aorta, and pressures continue to fall along the venous system.

The blood pressure in venules and medium-sized veins is so low that it cannot oppose the force of gravity. In the limbs, veins of this size contain **valves** (Figure 21–6•)—folds of the tunica intima that project from the vessel wall and point in the direction of blood flow. These valves act like the valves in the heart in that they permit blood flow in one direction only. Venous valves prevent the backflow of blood toward the capillaries.

As long as the valves function normally, any movement that distorts or compresses a vein will push blood toward the heart. This effect improves *venous return*, the rate of blood flow to the heart. ∞ p. 712 The mechanism is particularly important when you are standing, because blood returning from your feet must overcome the pull of gravity to ascend to the heart. Valves compartmentalize the blood within the veins, thereby dividing the weight of the blood between the compartments. Any movement in the surrounding skeletal muscles squeezes the blood toward the heart. Although you are probably not aware of it, when you stand, rapid cycles of contraction and relaxation are occurring within your leg muscles, helping to push blood toward the trunk. When you lie down, venous valves have much less impact on venous return, because your heart and major vessels are at the same level.

If the walls of the veins near the valves weaken or become stretched and distorted, the valves may not work properly. Blood then pools in the veins, and the vessels become grossly distended. The effects range from mild discomfort and a cosmetic problem, as in superficial *varicose veins* in the thighs and legs, to painful distortion of adjacent tissues, as in *hemorrhoids*. **AM** Problems with Venous Valve Function

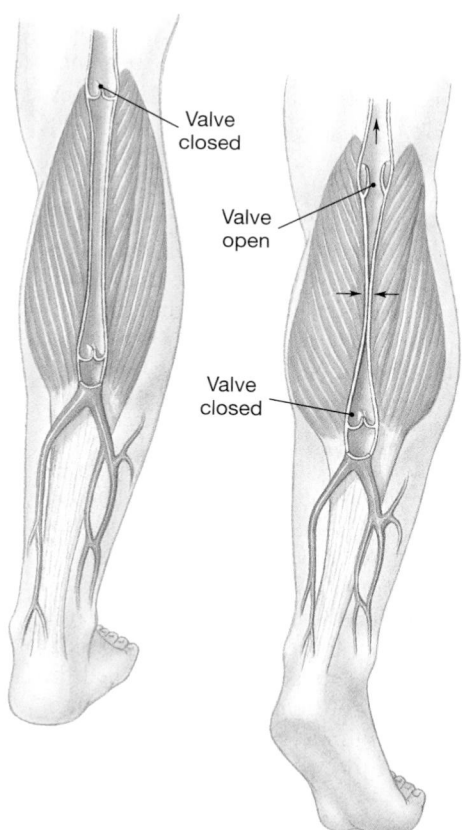

•**FIGURE 21–6**
The Function of Valves in the Venous System. Valves in the walls of medium-sized veins prevent the backflow of blood. Venous compression caused by the contraction of adjacent skeletal muscles assists in maintaining venous blood flow.

Valve closed

Valve open

Valve closed

☐ THE DISTRIBUTION OF BLOOD

The total blood volume is unevenly distributed among arteries, veins, and capillaries (Figure 21–7●). The heart, arteries, and capillaries normally contain 30–35 percent of the blood volume (roughly 1.5 liters of whole blood), and the venous system contains the rest (65–70 percent, or about 3.5 liters). Of the blood in the venous system, roughly one-third (about a liter) is circulating within the liver, bone marrow, and skin. These organs have extensive venous networks that at any moment contain large volumes of blood.

Because their walls are thinner, with a lower proportion of smooth muscle, veins are much more distensible (expandable) than arteries. For a given rise in blood pressure, a typical vein will stretch about eight times as much as a corresponding artery. The *capacitance* of a blood vessel is the relationship between the volume of blood it contains and the blood pressure. If the vessel behaves like a child's balloon, expanding easily with little pressure, it has high capacitance. If it behaves more like a truck tire, expanding only when large pressures are applied, it has low capacitance. Veins expand easily, so they are called **capacitance vessels**. Because veins have high capacitance, large changes in blood volume have little effect on arterial blood pressure. If the blood volume rises or falls, the elastic walls stretch or recoil, changing the volume of blood in the venous system.

If serious hemorrhaging occurs, the *vasomotor centers* of the medulla oblongata stimulate sympathetic nerves that innervate smooth muscle cells in the walls of medium-sized veins. This activity has two major effects:

1. *Systemic Veins Constrict.* This process, called **venoconstriction** (vē-nō-kon-STRIK-shun), reduces the amount of blood within the venous system and increases the volume within the arterial system and capillaries. Venoconstriction can keep the blood volume within the arteries and capillaries at near-normal levels despite a significant blood loss.

2. *The Constriction of Veins in the Liver, Skin, and Lungs Redistributes a Significant Proportion of the Total Blood Volume.* As a result, blood flow to delicate organs, such as the brain, and to active skeletal muscles can be increased or maintained after a blood loss. The amount of blood that can be shifted from veins in the liver, skin, and lungs to the general circulation, called the **venous reserve**, is normally about 20 percent of the total blood volume.

●FIGURE 21–7
The Distribution of Blood in the Cardiovascular System

✓ A cross section of tissue shows several small, thin-walled vessels with very little smooth muscle tissue in the tunica media. Which type of vessels are these?

✓ Why are valves located in veins, but not in arteries?

✓ Where in the body would you find fenestrated capillaries?

Answers start on page Q-1

 To review this material, visit the **IP CD-ROM**: Cardiovascular System/Anatomy Review: Blood Vessel Structure & Function.

21–2 CARDIOVASCULAR PHYSIOLOGY

Objectives

■ Explain the mechanisms that regulate blood flow through arteries, capillaries, and veins.

■ Describe the factors that influence blood pressure and how blood pressure is regulated.

■ Discuss the mechanisms and various pressures involved in the movement of fluids between capillaries and interstitial spaces.

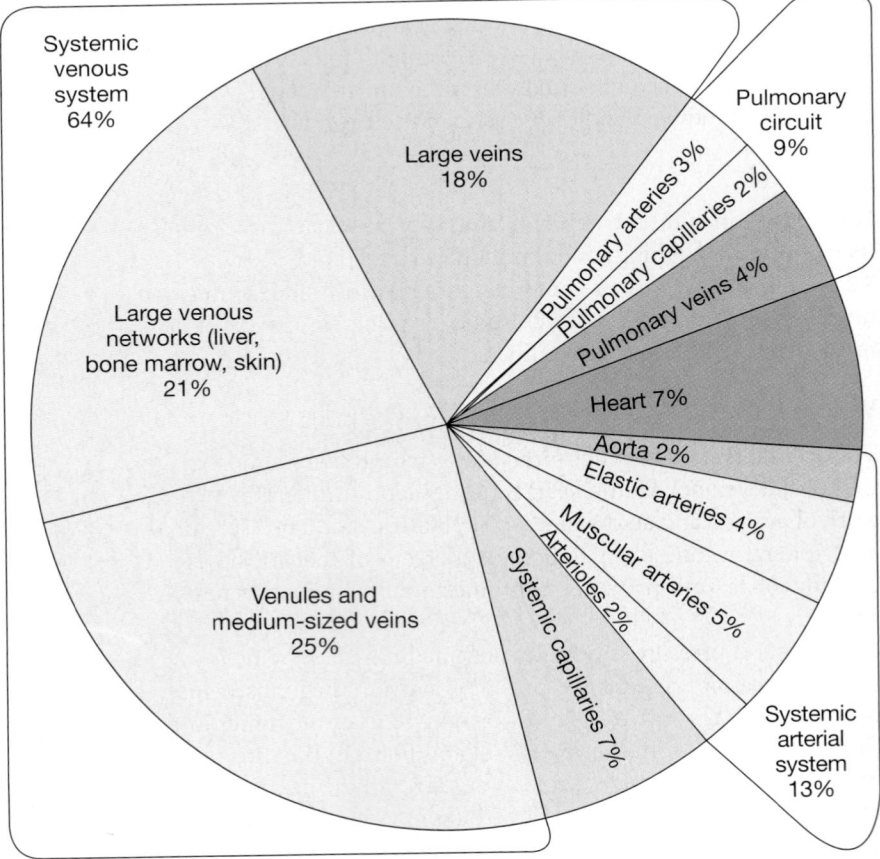

The goal of cardiovascular regulation is the maintenance of adequate blood flow through peripheral tissues and organs. Under normal circumstances, blood flow is equal to cardiac output. When cardiac output goes up, so does the blood flow through capillary beds; when cardiac output declines, capillary blood flow is reduced. We considered the primary factors involved in the regulation of cardiac output in Chapter 20. p. 710 The afterload of the heart is determined by the interplay between *pressure* and *resistance* in the cardiovascular network. If there were no resistance to blood flow in the cardiovascular system, the heart would not have to generate pressure to force blood around the pulmonary and systemic circuits.

Figure 21–8● provides an overview of our discussion of cardiovascular physiology. We shall begin by considering the nature of blood pressure and resistance. We shall pay particular attention to arterial blood pressure, which is responsible for maintaining blood flow within capillaries. Diffusion and osmosis occur within the capillaries, between the blood and the surrounding interstitial fluid. This *capillary exchange* performs vital functions: providing tissues with oxygen and nutrients and removing the carbon dioxide and waste products generated by active cells. Blood then leaves the capillaries and enters the venous system. There, the low venous blood pressure, aided by valves, skeletal muscle contraction, and other factors, returns the blood to the heart. The factors regulating this venous return were considered in Chapter 20. pp. 712, 715

The cardiovascular system must be continuously adjusted to maintain homeostasis. Active tissues require more blood flow than inactive ones; even something as simple as a change in position—going from sitting to standing, for instance—triggers a number of cardiovascular changes. We will end this section with a discussion of what those changes are and how they are coordinated. Some occur automatically, at the local level, but many others are coordinated by the nervous and endocrine systems.

□ PRESSURE

Liquids, including blood, cannot be compressed. A force exerted against a liquid generates **hydrostatic pressure (HP)**, a fluid pressure that is conducted in all directions. p. 93 If a pressure gradient exists, hydrostatic pressure will push a liquid from an area of higher pressure toward an area of lower pressure. The water in the pipes of your home is under hydrostatic pressure generated by some combination of pressure pumps and gravity. That hydrostatic pressure is higher than atmospheric pressure; hence, when you open a faucet, water flows out. The greater the water pressure, the faster is the flow. In other words, the flow rate is directly proportional to the pressure gradient.

In the systemic circuit of the cardiovascular system, the pressure gradient is the **circulatory pressure**—the pressure difference between the base of the ascending aorta and the entrance to the right atrium. Circulatory pressures average about 100 mm Hg. This relatively high pressure is needed primarily to force blood through the arterioles—resistance vessels—and into peripheral capillaries.

For convenience, the circulatory pressure is often divided into three components:

1. *Blood Pressure.* When referring to arterial pressure, we shall use the term **blood pressure (BP)** to distinguish it from the total circulatory pressure. Capillary blood flow is directly proportional to

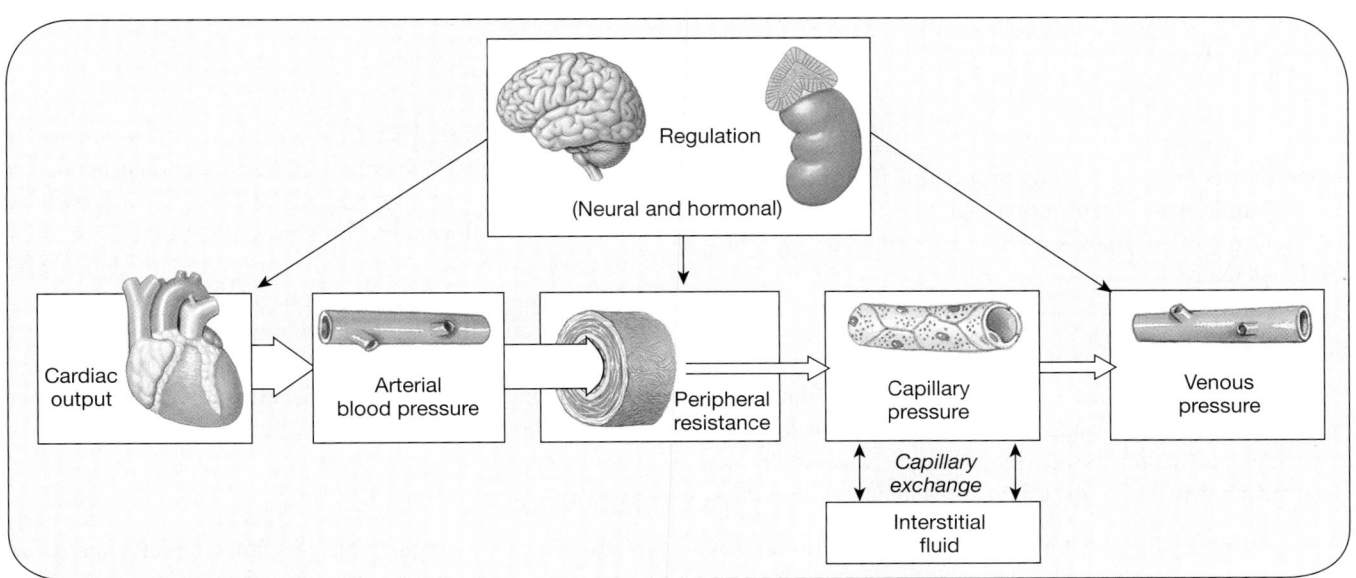

●**FIGURE 21–8**
An Overview of Cardiovascular Physiology. Neural and hormonal activities influence cardiac output, blood pressure, peripheral resistance, and venoconstriction. Peripheral resistance controls blood flow and capillary pressure, which drives the exchange between blood and interstitial fluid. This figure will be repeated, in reduced and simplified form, as a Navigator icon when we change topics.

blood pressure, which is closely regulated by a combination of neural and hormonal mechanisms. Blood pressure in the systemic arterial system ranges from an average of 100 mm Hg to roughly 35 mm Hg at the start of a capillary network.

2. *Capillary Hydrostatic Pressure.* **Capillary hydrostatic pressure (CHP)**, or *capillary pressure*, is the pressure within capillary beds. Along the length of a typical capillary, pressures decline from roughly 35 mm Hg to about 18 mm Hg.

3. *Venous Pressure.* **Venous pressure** is the pressure within the venous system. Venous pressure is quite low: The pressure gradient from the venules to the right atrium is only about 18 mm Hg.

☐ RESISTANCE

A resistance is any force that opposes movement. If you put a kink in a garden hose or cover the mouth of the hose with your finger, you will both increase the resistance of the hose and decrease the flow of water through it. With the tap closed at the kitchen sink, the resistance is sufficient to stop the flow of water. The more open the tap, the lower the resistance and the faster the water flows.

The resistance of the cardiovascular system opposes the movement of blood. The greater the resistance, the slower the blood flow. For circulation to occur, the pressure gradient must be great enough to overcome the **total peripheral resistance**—the resistance of the entire cardiovascular system. Because the resistance of the venous system is very low, for reasons we will describe shortly, attention focuses on the **peripheral resistance (PR)**—the resistance of the arterial system. For blood to flow into peripheral capillaries, the pressure gradient must be great enough to overcome the peripheral resistance. The higher the peripheral resistance, the lower the rate of blood flow. This relationship can be summarized as

$$F \propto \frac{\Delta P}{R}$$

In other words, the flow F is directly proportional to the pressure gradient ΔP and inversely proportional to the resistance R. Sources of peripheral resistance include *vascular resistance, viscosity,* and *turbulence.*

Vascular Resistance

Vascular resistance, the resistance of the blood vessels, is the largest component of peripheral resistance. The most important factor in vascular resistance is friction between blood and the vessel walls. The amount of friction depends on the length and diameter of the vessel.

VESSEL LENGTH Increasing the length of a blood vessel increases friction: The longer the vessel, the larger is the surface area in contact with blood. For example, you can easily blow the water out of a snorkel that is 2.5 cm (1 in.) in diameter and 25 cm (10 in.) long, but you cannot blow the water out of a 15-m-long garden hose, because the total friction is too great. The most dra-

matic changes in blood vessel length occur between birth and maturity, as the individual grows to adult size. In adults, vessel length can increase or decrease gradually when the individual gains or loses weight, but on a day-to-day basis this component of vascular resistance can be considered constant.

VESSEL DIAMETER The effects of friction on blood act primarily in a narrow zone closest to the vessel wall. In a small-diameter vessel, nearly all the blood is slowed down by friction with the walls. Resistance is therefore relatively high. Blood near the center of a large-diameter vessel does not encounter resistance from friction with the walls, so the resistance is relatively low.

Differences in diameter have much more significant effects on resistance than do differences in length. If two vessels are equal in diameter but one is twice as long as the other, the longer vessel will offer twice as much resistance to blood flow. But with two vessels of equal length, one twice the diameter of the other, the narrower one will offer 16 times as much resistance to blood flow. This relationship, expressed in terms of the vessel radius r and resistance R, can be summarized as $R \propto 1/r^4$.

More significantly, there is no way to control vessel length, but vessel diameter can change quickly through vasoconstriction or vasodilation. Most of the peripheral resistance occurs in arterioles, the smallest vessels of the arterial system. As we noted earlier in the chapter, arterioles are extremely muscular: The wall of an arteriole with a lumen diameter of 30 μm can have a 20-μm-thick layer of smooth muscle. When these smooth muscles contract or relax, peripheral resistance increases or decreases. Because a small change in diameter produces a large change in resistance, mechanisms that alter the diameters of arterioles provide control over peripheral resistance and blood flow.

Viscosity

Viscosity, introduced in Chapter 19, is the resistance to flow caused by interactions among molecules and suspended materials in a liquid. ∞ p. 654 Liquids of low viscosity, such as water (viscosity 1.0), flow at low pressures; thick, syrupy fluids, such as molasses (viscosity 300), flow only under higher pressures. Whole blood has a viscosity about five times that of water, owing to the presence of plasma proteins and blood cells. Under normal conditions, the viscosity of blood remains stable, but anemia, polycythemia, and other disorders that affect the hematocrit also change blood viscosity and peripheral resistance.

Turbulence

High flow rates, irregular surfaces, and sudden changes in vessel diameter upset the smooth flow of blood, creating eddies and swirls. This phenomenon, called **turbulence**, increases resistance and slows blood flow.

Turbulence normally occurs when blood flows between the atria and the ventricles and between the ventricles and the aortic and pulmonary trunks. It also develops in large arteries, such as the aorta, when cardiac output and arterial flow rates are very high. However, turbulence seldom occurs in smaller vessels unless their walls are damaged. For example, the formation of scar tissue at an injury site or the development of an atherosclerotic plaque creates abnormal turbulence and restricts blood flow. Because of the distinctive sound, or *bruit* (broo-Ē), produced by turbulence, the presence of plaques in large blood vessels can often be detected with a stethoscope.

Table 21–1 provides a quick review of the terms and relationships discussed in this section.

SUMMARY TABLE 21–1   KEY TERMS AND RELATIONSHIPS PERTAINING TO BLOOD CIRCULATION

Blood Flow (*F*): The volume of blood flowing per unit of time through a vessel or a group of vessels; may refer to circulation through a capillary, a tissue, an organ, or the entire vascular network. Total blood flow is equal to cardiac output.

Blood Pressure (BP): The hydrostatic pressure in the arterial system that pushes blood through capillary beds.

Circulatory Pressure: The pressure difference between the base of the ascending aorta and the entrance to the right atrium.

Hydrostatic Pressure: A pressure exerted by a liquid in response to an applied force.

Peripheral Resistance (PR): The resistance of the arterial system; affected by such factors as vascular resistance, viscosity, and turbulence.

Resistance (*R*): A force that opposes movement (in this case, blood flow).

Total Peripheral Resistance: The resistance of the entire cardiovascular system.

Turbulence: A resistance due to the irregular, swirling movement of blood at high flow rates or to exposure to irregular surfaces.

Vascular Resistance: A resistance due to friction within a blood vessel, primarily between the blood and the vessel walls. Increases with increasing length or decreasing diameter; vessel length is constant, but vessel diameter can change.

Venous Pressure: The hydrostatic pressure in the venous system.

Viscosity: A resistance to flow due to interactions among molecules within a liquid.

RELATIONSHIPS AMONG THE PRECEDING TERMS:

$F \propto \Delta P$ (flow is proportional to the pressure gradient)

$F \propto 1/R$ (flow is inversely proportional to resistance)

$F \propto \Delta P/R$ (flow is directly proportional to the pressure gradient and inversely proportional to resistance)

$F \propto BP/PR$ (flow is directly proportional to blood pressure and inversely proportional to peripheral resistance)

$R \propto 1/r^4$ (resistance is inversely proportional to the fourth power of the vessel radius)

AN OVERVIEW OF CARDIOVASCULAR PRESSURES

The graphs in Figure 21–9● provide an overview of the vessel diameters, areas, pressures, and velocity of blood flow in the systemic circuit.

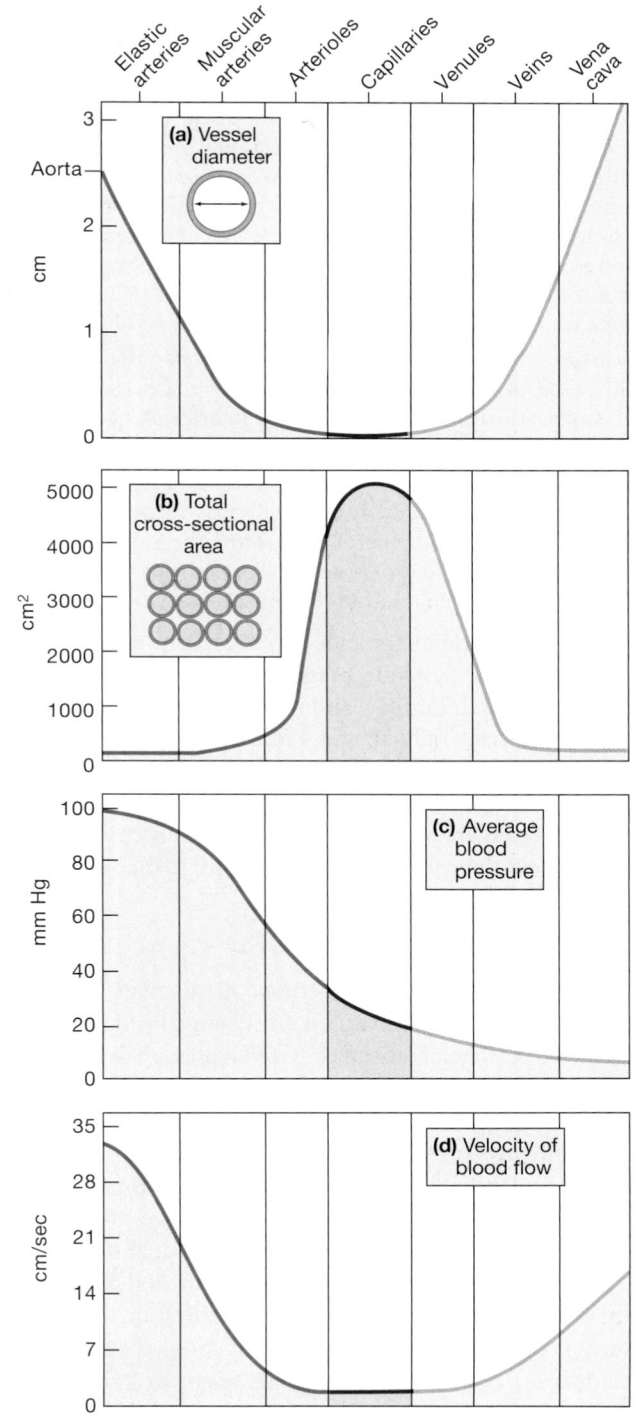

●FIGURE 21–9
Relationships among Vessel Diameter, Cross-Sectional Area, Blood Pressure, and Blood Velocity

- As you proceed from the aorta toward the capillaries, divergence occurs and the arteries branch repeatedly; each branch is smaller in diameter than the preceding one (Figure 21–9a•). As you proceed from the capillaries toward the venae cavae, convergence occurs and the diameters increase as venules combine to form small and medium-sized veins.

- Although the arterioles, capillaries, and venules are small in diameter, the body has a great many of them (Figure 21–9b•). All the blood flowing through the aorta also flows through peripheral capillaries. In effect, your blood moves from one big pipe into countless tiny ones. The blood pressure and the speed of blood flow are proportional to the cross-sectional area of the vessels involved. What is important is not the cross-sectional area of each individual vessel, but the *combined* cross-sectional area of *all* the vessels. The aorta has a cross-sectional area of 4.5 cm², and all the blood entering the systemic circuit flows through it. That systemic blood also flows through the peripheral capillaries. Collectively, these capillaries have a cross-sectional area of 5000 cm².

- As arterial branching occurs, the cross-sectional area increases and blood pressure falls rapidly. Most of the decline occurs in the small arteries and arterioles of the arterial system; venous pressures are relatively low (Figure 21–9c•).

- Like a fast-flowing river delivering water to a floodplain, blood flow decreases in velocity as the total cross-sectional area of the vessels increases from the aorta toward the capillaries. Blood flow then rises as the cross-sectional area drops from the capillaries toward the venae cavae (Figure 21–9d•).

Figure 21–10• graphs the blood pressure throughout the cardiovascular system. Systemic pressures are highest in the aorta, peaking at about 120 mm Hg, and reach a minimum of 2 mm Hg at the entrance to the right atrium. Pressures in the pulmonary circuit are much lower than those in the systemic circuit. The right ventricle does not ordinarily develop high pressures because the pulmonary vessels are much shorter and more distensible than the systemic vessels, thus providing less resistance to blood flow.

☐ ARTERIAL BLOOD PRESSURE

Arterial pressure is important, because it maintains blood flow through capillary beds. To do this, it must always be high enough to overcome the peripheral resistance. Arterial pressure is not stable; it rises during ventricular systole and falls during ventricular diastole. The peak blood pressure measured during ventricular systole is called **systolic pressure**, and the minimum blood pressure at the end of ventricular diastole is called **diastolic pressure**. In recording blood pressure, we separate systolic and diastolic pressures by a slash, as in "120/80" ("one-twenty over eighty") or "110/75".

A *pulse* is a rhythmic pressure oscillation that accompanies each heartbeat. The difference between the systolic and diastolic pressures is the **pulse pressure** (Figure 21–10•). To report a single blood pressure value, we use the **mean arterial pressure** (**MAP**), which is calculated by adding one-third of the pulse pressure to the diastolic pressure:

$$\text{MAP} = \frac{\text{diastolic pressure} + \text{pulse pressure}}{3}$$

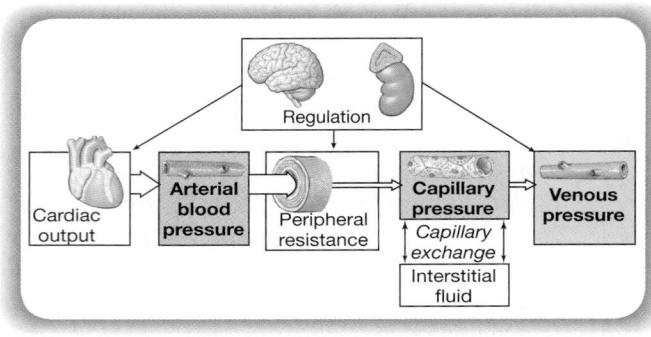

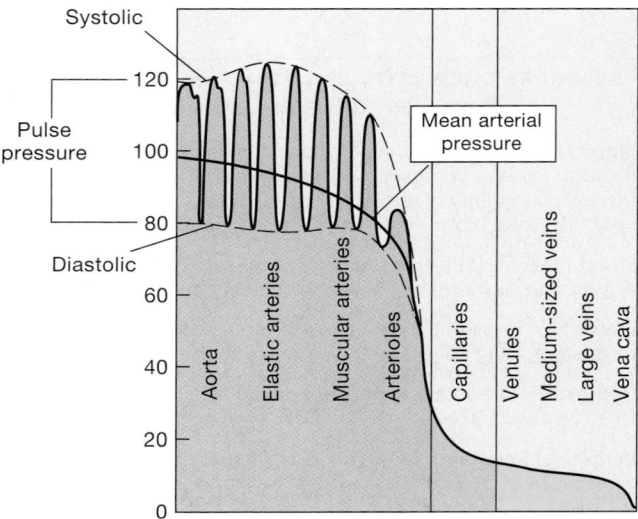

•**FIGURE 21–10**
Pressures within the Cardiovascular System. Notice the general reduction in circulatory pressure within the systemic circuit and the elimination of the pulse pressure within the arterioles. The Navigator icon in the shadow box highlights the importance of this figure to the discussion as a whole.

For a systolic pressure of 120 mm Hg and a diastolic pressure of 90 mm Hg, the MAP is 100 mm Hg:

$$\text{MAP} = 90 + \frac{(120 - 90)}{3} = 90 + 10 = 100 \text{ mm Hg}$$

Elastic Rebound

As systolic pressure climbs, the arterial walls stretch, just as an extra puff of air expands a partially inflated balloon. This expansion allows the arterial system to accommodate some of the blood provided by ventricular systole. When diastole begins and blood pressures fall, the arteries recoil to their original dimensions. Some blood is forced back toward the left ventricle, closing the aortic valve and helping to drive additional blood into the coronary arteries. However, most of the push generated by arterial recoil forces blood toward the capillaries. This phenomenon, called **elastic rebound**, helps to maintain blood flow along the arterial network while the left ventricle is in diastole.

Pressures in Small Arteries and Arterioles

The mean arterial pressure and the pulse pressure become smaller as the distance from the heart increases (Figure 21–10●):

- The mean arterial pressure declines as the arterial branches become smaller and more numerous. In essence, the blood pressure decreases as it overcomes friction and produces blood flow.

- The pulse pressure fades as a result of the cumulative effects of elastic rebound. Each arterial segment reduces the magnitude of the pressure change that is experienced by its downstream neighbors. The effect is like that of a loud shout creating a series of ever-softer echoes. Each time an echo is produced, the reflecting surface absorbs some of the sound energy. Eventually, the echo disappears. The pressure surge accompanying ventricular ejection is analogous to the shout, and it is reflected by the wall of the aorta, echoing down the arterial system until it finally disappears at the level of the small arterioles. By the time blood reaches a precapillary sphincter, no pressure oscillations remain, and the blood pressure remains steady at approximately 35 mm Hg.

There is a normal range of systolic and diastolic pressures found in healthy individuals. When pressures shift outside of the normal range, clinical problems develop. Abnormally high blood pressure is termed **hypertension**, abnormally low blood pressure, **hypotension**. Hypertension is much more common, and in fact many cases of hypotension result from overly aggressive drug treatment for hypertension.

The usual criterion for hypertension in adults is a blood pressure greater than 150/90. One study estimated that 20 percent of the white U.S. population has blood pressures greater than 160/95 and that another 25 percent is on the borderline, with pressures above 140/90. The figures for other racial groups vary; the incidence of hypertension among black Americans is roughly twice that of Caucasian Americans. The extent to which these data reflect genetic rather than environmental factors is not known.

Hypertension significantly increases the workload on the heart, and the left ventricle gradually enlarges. More muscle mass means a greater oxygen demand. When the coronary circulation cannot keep pace, symptoms of coronary ischemia appear. ∞ p. 695 Increased arterial pressures also place a physical stress on the walls of blood vessels throughout the body. This stress promotes or accelerates the development of arteriosclerosis and increases the risk of aneurysms, heart attacks, and strokes. **AM** Hypertension and Hypotension

☐ CAPILLARY EXCHANGE

As blood flows through peripheral tissues, blood pressure forces water and solutes out of the plasma, across capillary walls. Most of this material is reabsorbed by the capillaries, but about 3.6 liters of water and solutes flows through peripheral tissues each day and enters the *lymphatic system*, which then returns the fluid to the bloodstream. The continuous movement of water out of the capillaries, through peripheral tissues, and then back to the bloodstream by way of the lymphatic system has four important functions:

1. It ensures that plasma and interstitial fluid, two major components of extracellular fluid, are in constant communication.

2. It accelerates the distribution of nutrients, hormones, and dissolved gases throughout tissues.

3. It assists in the transport of insoluble lipids and tissue proteins that cannot enter the bloodstream by crossing the capillary walls.

4. It has a flushing action that carries bacterial toxins and other chemical stimuli to lymphoid tissues and organs responsible for providing immunity from disease.

Because capillary exchange plays such an important role in homeostasis, we shall now consider the factors and mechanisms involved. The most important processes that move materials across typical capillary walls are *diffusion, filtration,* and *reabsorption.*

Diffusion

As we saw in Chapter 3, *diffusion* is the net movement of ions or molecules from an area where their concentration is higher to an area where their concentration is lower. ∞ p. 90 The difference between the high and low concentrations represents a *concentration gradient*, and diffusion tends to eliminate that gradient. Diffusion occurs most rapidly when (1) the distances involved are small, (2) the concentration gradient is large, and (3) the ions or molecules involved are small.

Diffusion across capillary walls can occur by five routes:

1. Ions and small organic molecules, such as glucose, amino acids, and urea, can usually enter or leave the bloodstream by diffusion between adjacent endothelial cells or through the pores of fenestrated capillaries. The same routes are involved in the osmotic flow of water.

2. Many ions, including sodium, potassium, calcium, and chloride, can diffuse across endothelial cells by passing through channels in cell membranes.

3. Large water-soluble compounds are unable to enter or leave the bloodstream except at fenestrated capillaries, such as those of the hypothalamus, the kidneys, many endocrine organs, and the intestinal tract.

4. Lipids, such as fatty acids and steroids, and lipid-soluble materials, including soluble gases such as oxygen and carbon dioxide, can cross capillary walls by diffusion through the endothelial cell membranes.

5. Plasma proteins are normally unable to cross the endothelial lining anywhere except in sinusoids, such as those of the liver, where plasma proteins enter the bloodstream.

Filtration

The driving force for filtration is hydrostatic pressure, which, as we saw earlier, pushes water from an area of higher pressure to an area of lower pressure. Blood pressure in a capillary is **capillary hydrostatic pressure (CHP)**.

In *capillary filtration*, water is forced across a capillary wall, and small solute molecules travel with the water (Figure 21–12●). The solute molecules must be small enough to pass between adjacent endothelial cells or through the pores in a fenestrated capillary;

Checking the Pulse and Blood Pressure

You can feel your pulse within any of the large or medi-um-sized arteries. The usual procedure involves using your fingertips to squeeze an artery against a relatively solid mass, preferably a bone. When the vessel is compressed, you feel your pulse as a pressure against your fingertips. The inside of the wrist is commonly used, because the *radial artery* can easily be pressed against the distal portion of the radius (Figure 21–11a●). Other accessible arteries include the *external carotid, brachial, temporal, facial, femoral, popliteal, posterior tibial, and dorsalis pedis arteries*. Firm pressure exerted at these locations, called **pressure points**, can reduce arterial bleeding distal to the site.

Blood pressure not only forces blood through the circulato-ry system, but also pushes outward against the walls of the containing vessels, just as air pushes against the walls of an in-flated balloon. As a result, we can measure blood pressure indi-rectly by determining how forcefully the blood presses against the vessel walls.

The instrument used to measure blood pressure is called a **sphygmomanometer** (sfig-mō-ma-NOM-e-ter; *sphygmos*, pulse + *manometer*, device for measuring pressure). An in-flatable cuff is placed around the arm in such a position that its inflation compresses the brachial artery (Figure 21–11b●). A stethoscope is placed over the artery distal to the cuff, and the cuff is then inflated. A tube connects the cuff to a pressure gauge that reports the cuff pressure in millimeters of mercury (mm Hg). Inflation continues until the cuff pressure is roughly 30 mm Hg above the pressure sufficient to collapse the brachial artery completely, stop the flow of blood, and elimi-nate the sound of the pulse.

The air is then slowly let out of the cuff. When the pressure in the cuff falls below systolic pressure, blood can again enter the artery. At first, blood enters only at peak systolic pressures, and the stethoscope picks up the sound of blood pulsing through the artery. As the pressure falls further, the sound changes, because the artery is remaining open for longer and longer periods. When the cuff pressure falls below diastolic pressure, blood flow becomes continuous and the sound of the pulse becomes muffled or disappears. Thus, the pressure at which the pulse appears corresponds to the peak systolic pres-sure; when the pulse fades, the pressure has reached diastolic levels. The distinctive sounds heard during this test, called *sounds of Korotkoff* (sometimes spelled *Korotkov* or *Korotkow*), are produced by turbulence as blood flows past the constricted portion of the artery.

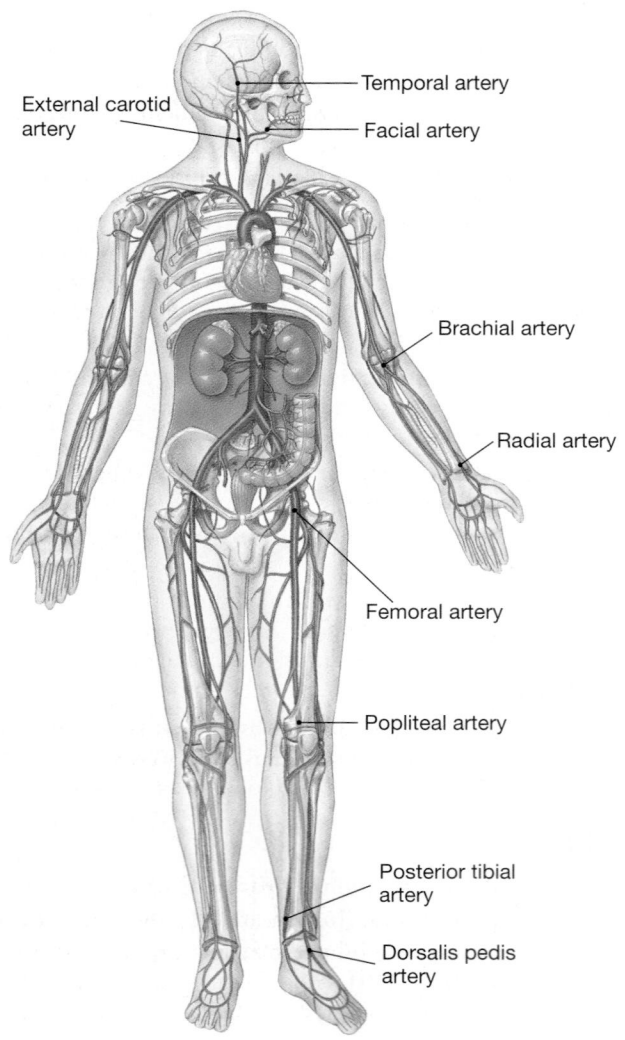

(a) Location of pressure points

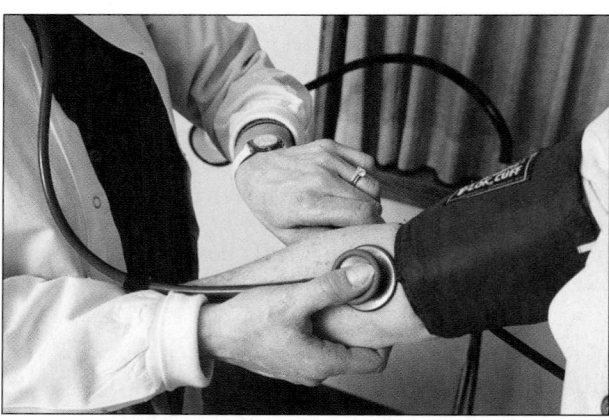

(b) Checking blood pressure using a sphygmomanometer

●**FIGURE 21–11**
Pressure Points and Blood Pressure Measurement.
(a) Pressure points used to monitor the pulse or control peripheral bleeding. **(b)** Using a sphygmomanometer to measure blood pressure.

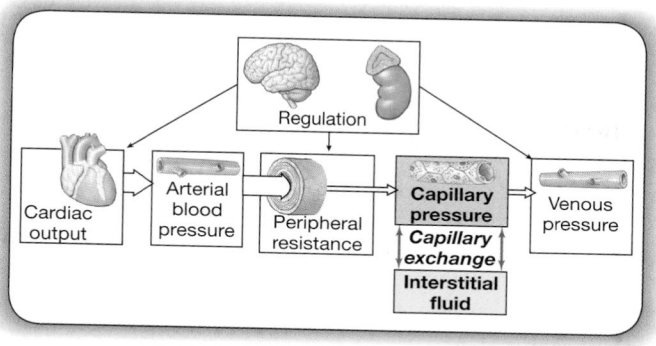

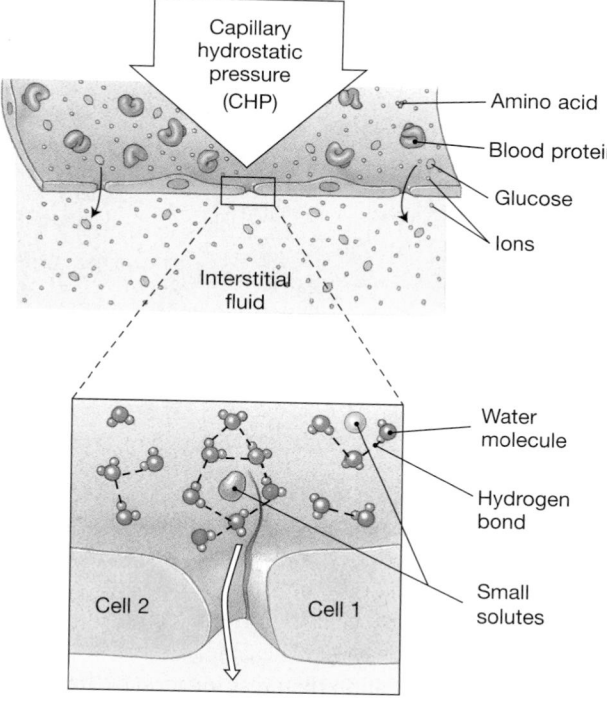

•FIGURE 21–12
Capillary Filtration. Capillary hydrostatic pressure forces water and solutes through the gaps between adjacent endothelial cells in continuous capillaries. The sizes of solutes that move across the capillary wall are determined primarily by the dimensions of the gaps.

larger solutes and suspended proteins are filtered out and remain in the bloodstream.

Along the length of a typical capillary, blood pressure gradually falls from about 35 mm Hg to roughly 18 mm Hg, the pressure at the start of the venous system. Filtration occurs primarily at the arterial end of a capillary, where CHP is highest.

Reabsorption

Reabsorption occurs as the result of osmosis. *Osmosis* is a special term used to refer to the diffusion of water across a selectively permeable membrane separating two solutions of differing solute concentrations. Water molecules tend to diffuse across a

membrane toward the solution containing the higher solute concentration (Figure 3–20•, p. 93).

The **osmotic pressure (OP)** of a solution is an indication of the force of osmotic water movement—in other words, the pressure that must be applied to prevent osmotic movement across a membrane. The higher the solute concentration of a solution, the greater the solution's osmotic pressure. The osmotic pressure of the blood is also called *blood colloid osmotic pressure (BCOP)*, because only the suspended proteins are unable to cross the capillary walls. Clinicians often use the term *oncotic pressure (onkos, a swelling)* when referring to the colloid osmotic pressure of body fluids. The two terms are equivalent. Osmotic water movement will continue until either the solute concentrations are equalized or the movement is prevented by an opposing hydrostatic pressure.

We shall now consider the interplay between filtration and reabsorption along the length of a typical capillary. As the discussion proceeds, remember that hydrostatic pressure forces water *out of* a solution, whereas osmotic pressure draws water *into* a solution.

The Interplay between Filtration and Reabsorption

The rates of filtration and reabsorption gradually change as blood passes along the length of a capillary. The factors involved are diagrammed in Figure 21–13•.

The *net hydrostatic pressure* tends to push water and solutes out of capillaries and into the interstitial fluid. The net hydrostatic pressure is the difference between

1. the *capillary hydrostatic pressure (CHP)*, which ranges from 35 mm Hg at the arterial end of a capillary to 18 mm Hg at the venous end, and

2. the *hydrostatic pressure of the interstitial fluid (IHP)*. Measurements of IHP have yielded very small values that differ from tissue to tissue—from +6 mm Hg in the brain to −6 mm Hg in subcutaneous tissues. A positive IHP opposes CHP, and the tissue hydrostatic pressure must be overcome before fluid can move out of a capillary. A negative IHP assists CHP, and additional fluid will be pulled out of the capillary.

The *net colloid osmotic pressure* tends to pull water and solutes into a capillary from the interstitial fluid. The net colloid osmotic pressure is the difference between

1. the *blood colloid osmotic pressure (BCOP)*, which is roughly 25 mm Hg, and

2. the *interstitial fluid colloid osmotic pressure (ICOP)*. The ICOP is as variable and low as the IHP, because the interstitial fluid in most tissues contains negligible quantities of suspended proteins. Reported values of ICOP are from 0 to 5 mm Hg, within the range of pressures recorded for the IHP.

Because (1) the ICOP and the IHP oppose one another, (2) the two pressures are of comparable size, and (3) they are difficult to measure accurately, both values have been set to 0 mm Hg in our model in Figure 21–13•. This value is acceptable because CHP and BCOP are the primary forces acting across a capillary wall under normal circumstances. However, the method of calculation described here would still apply, regardless of the values selected for IHP and ICOP.

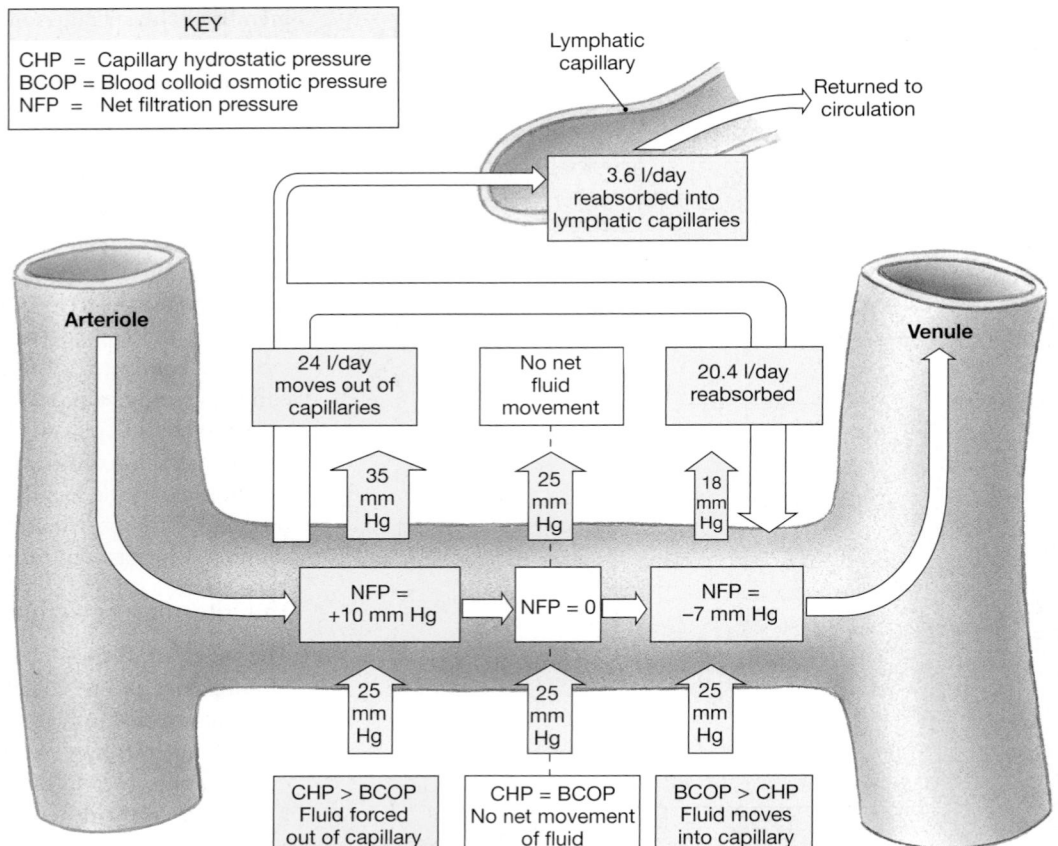

KEY

CHP = Capillary hydrostatic pressure
BCOP = Blood colloid osmotic pressure
NFP = Net filtration pressure

●**FIGURE 21–13**
Forces Acting across Capillary Walls. At the arterial end of the capillary, capillary hydrostatic pressure (CHP) is stronger than blood colloid osmotic pressure (BCOP), so fluid moves out of the capillary. Near the venule, CHP is lower than BCOP, so fluid moves into the capillary. In this model, interstitial fluid colloid osmotic pressure (ICOP) and interstitial fluid hydrostatic pressure (IHP) are assumed to be 0 mm Hg and so are not shown.

The **net filtration pressure (NFP)** is the difference between the net hydrostatic pressure and the net osmotic pressure. In terms of the factors just listed, this means that

$$\begin{array}{ccccc} \text{net filtration} & = & \text{net hydrostatic} & - & \text{net colloid} \\ \text{pressure} & & \text{pressure} & & \text{osmotic pressure} \\ \text{NFP} & = & (\text{CHP} - \text{IHP}) & - & (\text{BCOP} - \text{ICOP}) \end{array}$$

At the arterial end of a capillary, the net filtration pressure is +10 mm Hg:

$$\text{NFP} = (35 - 0) - (25 - 0) = 35 - 25 = 10 \text{ mm Hg}$$

Because this value is positive, it indicates that fluid will tend to move *out of* the capillary and into the interstitial fluid. At the venous end of a capillary, the net filtration pressure is −7 mm Hg:

$$\text{NFP} = (18 - 0) - (25 - 0) = 18 - 25 = -7 \text{ mm Hg}$$

The minus sign indicates that fluid tends to move *into* the capillary; that is, reabsorption is occurring.

The transition between filtration and reabsorption occurs where the CHP is 25 mm Hg, because at that point the hydrostatic and osmotic forces are equal—that is, the NFP is 0 mm Hg. If the maximum filtration pressure at the arterial end of the capillary were equal to the maximum reabsorption pressure at the venous end, this transition point would lie midway along the length of the capillary. Under these circumstances, filtration would occur along the

first half of the capillary, and an identical amount of reabsorption would occur along the second half. However, the maximum filtration pressure is higher than the maximum reabsorption pressure, so the transition point between filtration and reabsorption normally lies closer to the venous end of the capillary than to the arterial end. As a result, more filtration than reabsorption occurs along the capillary. Of the roughly 24 liters of fluid that moves out of the plasma and into the interstitial fluid each day, 85 percent is reabsorbed. The remainder (3.6 liters) flows through the tissues and into lymphatic vessels, for eventual return to the venous system.

Any condition that affects hydrostatic or osmotic pressures in the blood or tissues will shift the balance between hydrostatic and osmotic forces. We can then predict the effects on the basis of an understanding of capillary dynamics. For example,

• If hemorrhaging occurs, both blood volume and blood pressure decline. This reduction in CHP lowers the NFP and increases the amount of reabsorption. The result is a reduction in the volume of interstitial fluid and an increase in the circulating plasma volume. This process is known as a *recall of fluids*.

• If dehydration occurs, the plasma volume decreases owing to water loss, and the concentration of plasma proteins increases. The increase in BCOP accelerates reabsorption and a recall of fluids that delays the onset and severity of clinical symptoms.

• If the CHP rises or the BCOP declines, fluid moves out of the blood and builds up in peripheral tissues, a condition called *edema*.

Edema

✚ **Edema** (e-DĒ-muh) is an abnormal accumulation of interstitial fluid. Edema has many causes, and we shall encounter specific examples in later chapters. The underlying problem in all types of edema is a disturbance in the normal balance between hydrostatic and osmotic forces at the capillary level. For instance,

- When a capillary is damaged, plasma proteins can cross the capillary wall and enter the interstitial fluid. The resulting elevation of the ICOP will reduce the rate of capillary reabsorption and produce a localized edema. This is why you usually have swelling at a bruise.

- In starvation, the liver cannot synthesize enough plasma proteins to maintain normal concentrations in the blood. BCOP declines, and fluids begin moving from the blood into peripheral tissues. In children, fluid accumulates in the abdominopelvic cavity, producing the swollen bellies typical of starvation victims. A reduction in BCOP is also seen after severe burns and in several types of liver or kidney diseases.

- In the U.S. population, most serious cases of edema result from an increase in arterial blood pressure, in venous pressure, or in total circulatory pressure. The increase may result from heart problems, such as heart failure, venous blood clots that elevate venous pressures, or other circulatory abnormalities. The net result is an increase in CHP that accelerates fluid movement into the tissues.

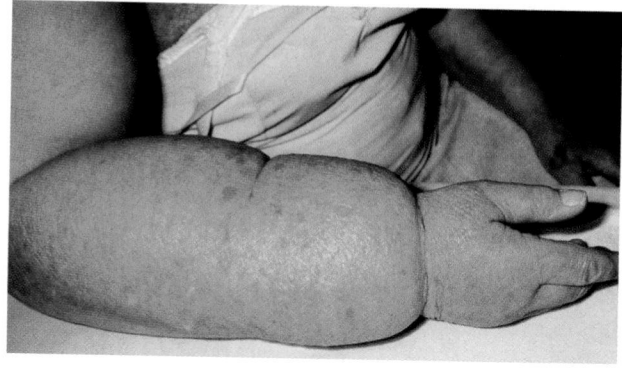

Edema can also result from problems with other systems, such as a blockage of lymphatic vessels or impaired urine formation:

- If the lymphatic vessels in a region become blocked, the volume of interstitial fluid will continue to rise, and the IHP will gradually increase until capillary filtration ceases. In *filariasis*, a condition we shall consider in Chapter 22, parasites can block lymphatic vessels and cause a massive regional edema known as *elephantiasis*.

- If the kidneys are unable to produce urine but the individual continues to drink liquids, the blood volume will rise. This situation ultimately leads to elevated CHP and enhances fluid movement into the peripheral tissues.

☐ VENOUS PRESSURE AND VENOUS RETURN

Venous pressure, although low, determines venous return—the amount of blood arriving at the right atrium each minute. Venous return has a direct impact on cardiac output. ∞ p. 712 Although blood pressure at the start of the venous system is only about one-tenth that at the start of the arterial system, the blood must still travel through a vascular network as complex as the arterial system before returning to the heart.

Pressures at the entrance to the right atrium fluctuate, but they average about 2 mm Hg. Thus, the effective pressure in the venous system is roughly 16 mm Hg (from 18 mm Hg in the venules to 2 mm Hg in the venae cavae), compared with 65 mm Hg in the arterial system (from 100 mm Hg at the aorta to 35 mm Hg at the capillaries). Yet, although venous pressures are low, veins offer comparatively little resistance, so pressure declines very slowly as blood moves through the venous system. As blood continues toward the heart, the veins become larger, resistance drops, and the velocity of blood flow increases (Figure 21–9●, p. 733).

When you stand, the venous blood returning from your body inferior to the heart must overcome gravity as it ascends within the inferior vena cava. Two factors assist the low venous pressures in propelling blood toward your heart: (1) *muscular compression* of peripheral veins and (2) the *respiratory pump*.

Muscular Compression

The contractions of skeletal muscles near a vein compress it, helping push blood toward the heart. The valves in small and medium-sized veins ensure that blood flows in one direction only (Figure 21–6●, p. 729). During normal standing and walking, the cycles of contraction and relaxation that accompany normal movements assist venous return. If you stand at attention, with knees locked and leg muscles immobilized, that assistance is lost. The reduction in venous return then leads to a fall in cardiac output, which reduces the blood supply to the brain. This decline is sometimes enough to cause **fainting**, a temporary loss of consciousness. You would then collapse, but while you were in the horizontal position, both venous return and cardiac output would return to normal.

The Respiratory Pump

As you inhale, your thoracic cavity expands and pressure within the pleural cavities declines. This drop in pressure pulls air into your lungs. At the same time, blood is pulled into the inferior vena cava and right atrium from the smaller veins of your abdominal cavity and lower body. The effect on venous return from the superior vena cava is less pronounced, as blood in that vessel is normally assisted by gravity.

As you exhale, your thoracic cavity decreases in size. Internal pressure then rises, forcing air out of your lungs and pushing venous blood into the right atrium. This mechanism is called the **respiratory pump**, or *thoracoabdominal pump*. The importance of such pumping action increases during heavy exercise, when respirations are deep and frequent.

✓ In a healthy individual, where would the blood pressure be greater, at the aorta or at the inferior vena cava? Explain.

✓ While standing in the hot sun, Sally begins to feel light headed and faints. Explain.

✓ Terry's blood pressure is 125/70. At what pressure did the nurse taking Terry's blood pressure first hear the sounds of Korotkoff?

Answers start on page Q-1

 Review blood pressure regulation on the **IP CD-ROM:** Cardiovascular System/Factors That Affect Blood Pressure.

21-3 CARDIOVASCULAR REGULATION

Objectives

- Describe how central and local control mechanisms interact to regulate blood flow and pressure in tissues.
- Explain how the activities of the cardiac, vasomotor, and respiratory centers are coordinated to control blood flow through the tissues.

Homeostatic mechanisms regulate cardiovascular activity to ensure that **tissue perfusion**, or blood flow through tissues, meets the demand for oxygen and nutrients. The three variable factors are (1) cardiac output, (2) peripheral resistance, and (3) blood pressure. We discussed cardiac output in Chapter 20 (∞ p. 710) and considered peripheral resistance and blood pressure earlier in this chapter.

Most cells are relatively close to capillaries. When a group of cells becomes active, the circulation to that region must increase to deliver the necessary oxygen and nutrients and to carry away the waste products and carbon dioxide they generate. The goal of cardiovascular regulation is to ensure that these blood flow changes occur (1) at an appropriate time, (2) in the right area, and (3) without drastically changing blood pressure and blood flow to vital organs.

Factors involved in the regulation of cardiovascular function include autoregulation, neural mechanisms, and endocrine mechanisms (Figure 21-14●):

- **Autoregulation.** Local factors change the pattern of blood flow within capillary beds in response to chemical changes in interstitial fluids. This is an example of autoregulation at the tissue level. Autoregulation causes immediate, localized homeostatic adjustments. If autoregulation fails to normalize conditions at the tissue level, neural mechanisms and endocrine factors are activated.

- **Neural Mechanisms.** Neural mechanisms respond to changes in arterial pressure or blood gas levels at specific sites. When those changes occur, the cardiovascular centers of the autonomic nervous system adjust cardiac output and peripheral resistance to maintain blood pressure and ensure adequate blood flow.

- **Endocrine Mechanisms.** The endocrine system releases hormones that enhance short-term adjustments and that direct long-term changes in cardiovascular performance.

We will next consider each of these regulatory mechanisms individually.

◻ AUTOREGULATION OF BLOOD FLOW WITHIN TISSUES

Under normal resting conditions, cardiac output remains stable and peripheral resistance within individual tissues is adjusted to control local blood flow.

Factors that promote the dilation of precapillary sphincters are called **vasodilators**. **Local vasodilators** act at the tissue level and accelerate blood flow through the tissue of origin. Examples of local vasodilators include:

- Decreased tissue oxygen levels or increased CO_2 levels.

- The generation of lactic acid or other acids by tissue cells.

- The release of nitric oxide (NO), formerly known as *endothelium-derived relaxation factor* (*EDRF*), from endothelial cells.

- Rising concentrations of potassium ions or hydrogen ions in the interstitial fluid.

- Chemicals released during local inflammation, including histamine and NO. ∞ p. 139

- Elevated local temperatures

These factors work by relaxing the smooth muscle cells of the precapillary sphincters. All of them indicate that conditions in the tissue are abnormal in one way or another. An improvement in blood flow, which will bring oxygen, nutrients, and buffers, may be sufficient to restore homeostasis.

As we noted in Chapter 19, aggregating platelets and damaged tissues produce compounds that stimulate the constriction of precapillary sphincters. These compounds are **local vasoconstrictors**. Examples include prostaglandins and thromboxanes released by activated platelets and white blood cells and the endothelins released by damaged endothelial cells. ∞ p. 673

Local vasodilators and vasoconstrictors control blood flow within a single capillary bed (Figure 21-5●, p 728). In high concentrations, these factors also affect arterioles, increasing or decreasing blood flow to all the capillary beds in a given area.

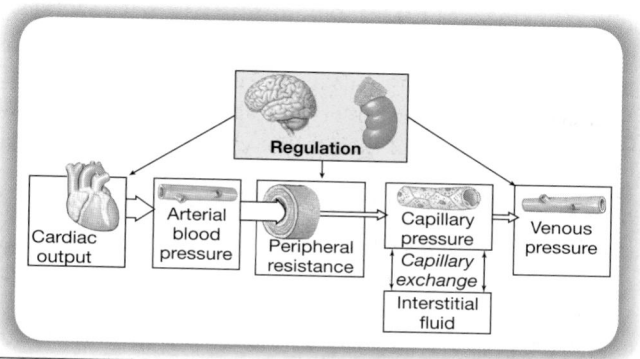

•FIGURE 21–14
Homeostatic Adjustments That Compensate for a Reduction in Blood Pressure and Blood Flow

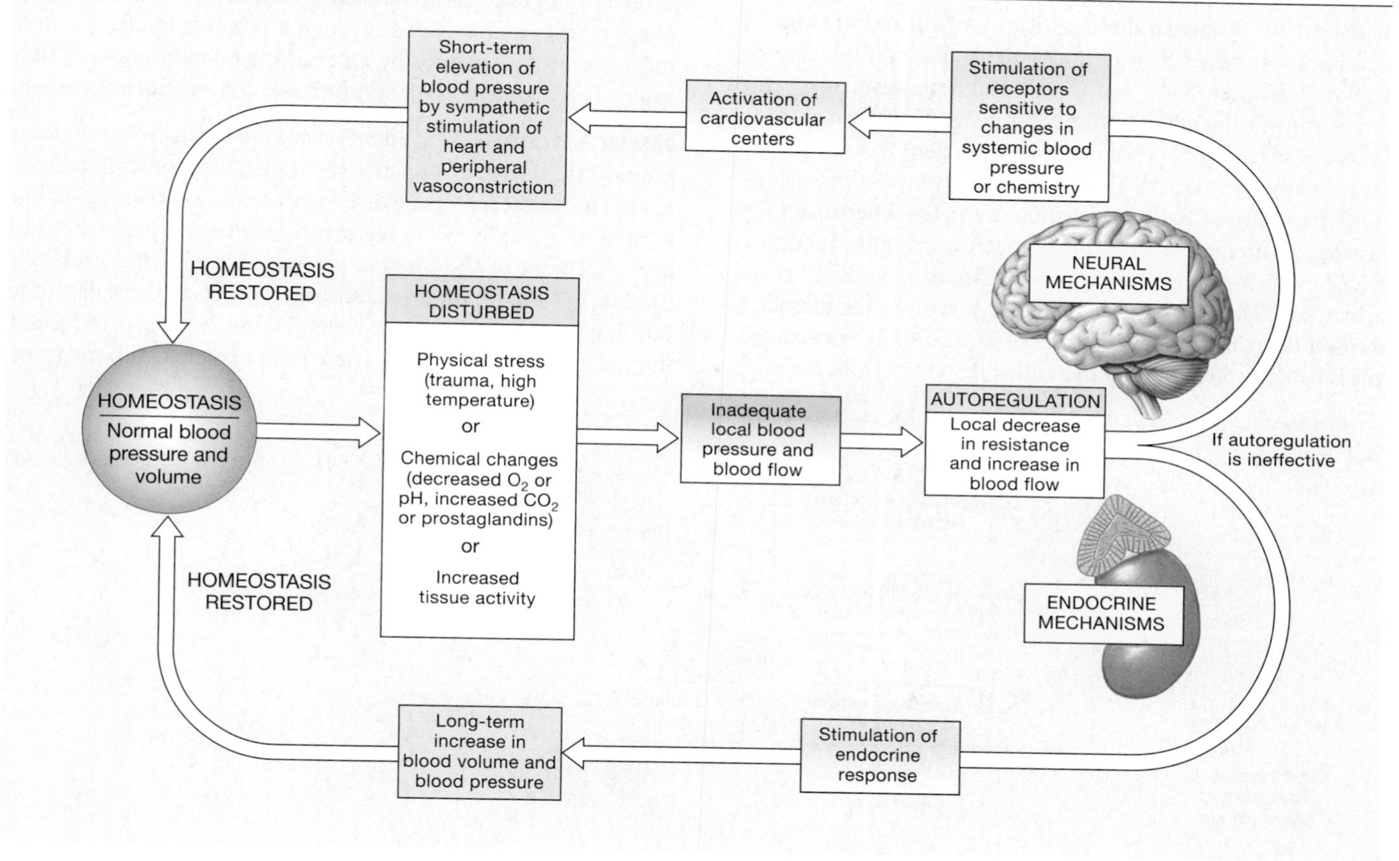

☐ NEURAL MECHANISMS

The nervous system is responsible for adjusting cardiac output and peripheral resistance in order to maintain adequate blood flow to vital tissues and organs. Centers responsible for these regulatory activities include the *cardiac centers* and the *vasomotor centers* of the medulla oblongata. ∞ p. 475 It is difficult to distinguish the cardiac and vasomotor centers anatomically, and they are often considered to form complex **cardiovascular (CV) centers**. In functional terms, however, the cardiac and vasomotor centers often act independently. As we noted in Chapter 20, each cardiac center consists of a *cardioacceleratory center*, which increases cardiac output through sympathetic innervation, and a *cardioinhibitory center*, which reduces cardiac output through parasympathetic innervation. ∞ p. 710

The vasomotor centers contain two populations of neurons: (1) a very large group responsible for widespread vasoconstriction and (2) a smaller group responsible for the vasodilation of arterioles in skeletal muscles and the brain. The vasomotor centers exert their effects by controlling the activity of sympathetic motor neurons:

1. *Control of Vasoconstriction.* The neurons innervating peripheral blood vessels in most tissues are *adrenergic*; that is, they release the neurotransmitter norepinephrine (NE). The response to NE release is the stimulation of smooth muscle in the walls of arterioles, producing vasoconstriction.

2. *Control of Vasodilation.* Vasodilator neurons innervate blood vessels in skeletal muscles and in the brain. The stimulation of these neurons will relax smooth muscle cells in the walls of arterioles, producing vasodilation. The relaxation of smooth

muscle cells is triggered by the appearance of NO in their surroundings. The vasomotor centers may control NO release indirectly or directly. The most common vasodilator synapses are *cholinergic*—their synaptic knobs release ACh. In turn, ACh stimulates endothelial cells in the area to release NO, which causes local vasodilation. Another population of vasodilator synapses is *nitroxidergic*—the synaptic knobs release NO as a neurotransmitter. Nitric oxide has an immediate and direct relaxing effect on the vascular smooth muscle cells in the area.

Vasomotor Tone

In Chapter 16, we discussed the significance of autonomic tone in setting a background level of neural activity that can increase or decrease on demand. ∞ p. 546 The sympathetic vasoconstrictor nerves are chronically active, producing a significant **vasomotor tone**. Vasoconstrictor activity is normally sufficient to keep the arterioles partially constricted. Under maximal stimulation, arterioles constrict to about half their resting diameter, whereas a fully dilated arteriole increases its resting diameter by roughly 1.5 times. Constriction has a significant effect on resistance, because, as we saw earlier (p. 732), the resistance increases sharply as the diameter of the vessel decreases. The resistance of a maximally constricted arteriole is roughly *80 times* that of a fully dilated arteriole. Because blood pressure varies directly with peripheral resistance, the vasomotor centers can control arterial blood pressure very effectively by making modest adjustments in vessel diameters. Extreme stimulation of the vasomotor centers will also produce venoconstriction and a mobilization of the venous reserve.

Reflex Control of Cardiovascular Function

The cardiovascular centers detect changes in tissue demand by monitoring arterial blood, with particular attention to blood pressure, pH, and the concentrations of dissolved gases. The *baroreceptor reflexes* respond to changes in blood pressure, and the *chemoreceptor reflexes* monitor changes in the chemical composition of arterial blood. These reflexes are regulated through a negative feedback loop: The stimulation of a receptor by an abnormal condition leads to a response that counteracts the stimulus and restores normal conditions.

BARORECEPTOR REFLEXES *Baroreceptors* are specialized receptors that monitor the degree of stretch in the walls of expandable organs. ∞ p. 516 The baroreceptors involved with cardiovascular regulation are located in the walls of (1) the **carotid sinuses**, expanded chambers near the bases of the *internal carotid arteries* of the neck (Figure 21–24●, p. 755), (2) the **aortic sinuses**, pockets in the walls of the ascending aorta adjacent to the heart (Figure 20–8b●, p. 693), and (3) the wall of the right atrium. These receptors are components of the

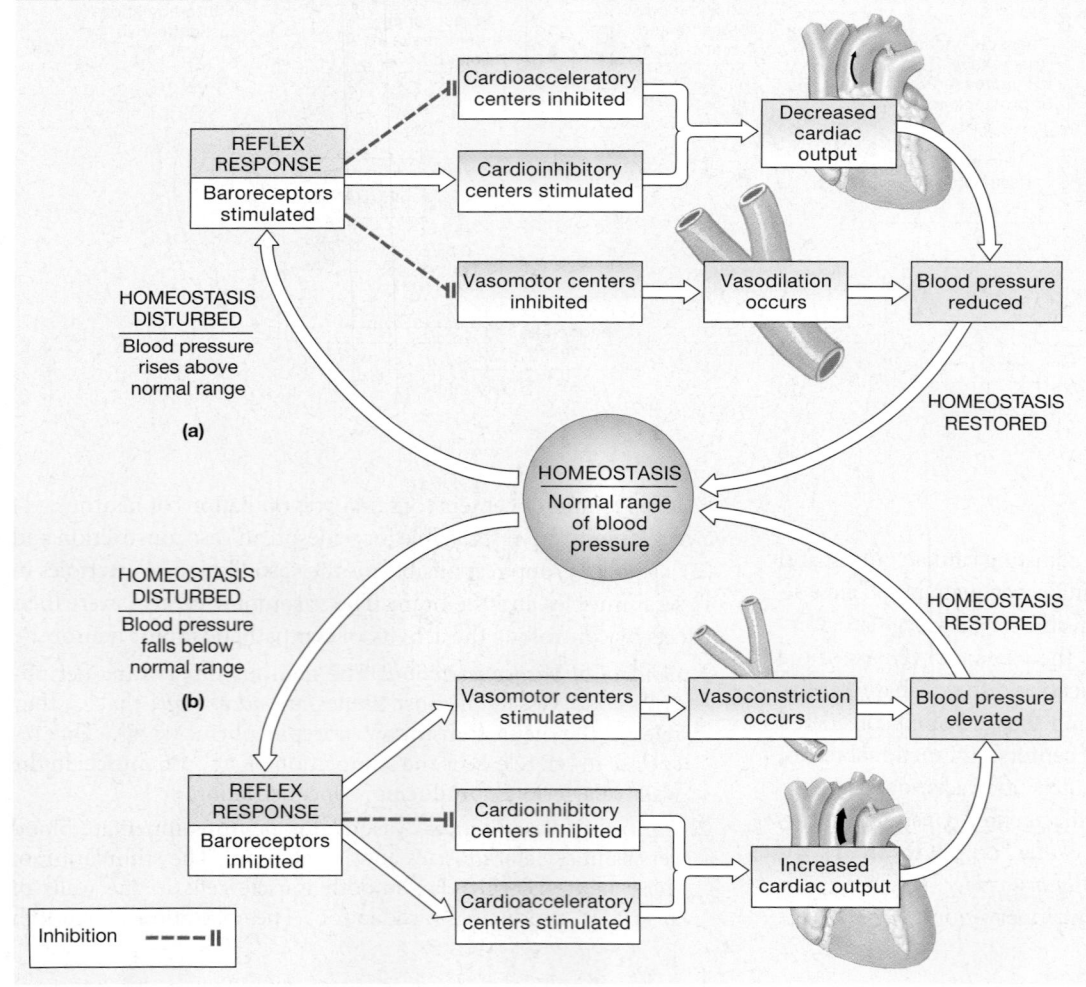

●**FIGURE 21–15**
Baroreceptor Reflexes of the Carotid and Aortic Sinuses

baroreceptor reflexes, which adjust cardiac output and peripheral resistance to maintain normal arterial pressures.

Aortic baroreceptors monitor blood pressure within the ascending aorta. Any changes trigger the **aortic reflex**, which adjusts blood pressure in so as to maintain adequate blood pressure and blood flow through the systemic circuit. In response to changes in blood pressure at the carotid sinus, carotid sinus baroreceptors trigger reflexes that maintain adequate blood flow to the brain. Because blood flow to the brain must remain constant, the carotid sinus receptors are extremely sensitive. Figure 21–15• presents the basic organization of the baroreceptor reflexes triggered by changes in blood pressure at the carotid and aortic sinuses.

When blood pressure climbs, the increased output from the baroreceptors alters activity in the CV centers and produces two major effects (Figure 21–15a•):

1. *A decrease in cardiac output*, due to parasympathetic stimulation and the inhibition of sympathetic activity.
2. *Widespread peripheral vasodilation*, due to the inhibition of excitatory neurons in the vasomotor centers.

The decrease in cardiac output reflects primarily a reduction in heart rate due to the release of acetylcholines at the sinoatrial (SA) node. ∞ p. 711 The widespread vasodilation lowers peripheral resistance, and this effect, combined with a reduction in cardiac output, leads to a decline in blood pressure to normal levels.

When blood pressure falls below normal, baroreceptor output is reduced accordingly (Figure 21–15b•). This change has two major effects:

1. *An increase in cardiac output*, through the stimulation of sympathetic innervation to the heart. This results from the stimula-tion of the cardioaccelerator centers and is accompanied by an inhibition of the cardioinhibitory centers.
2. *Widespread peripheral vasoconstriction*, caused by the stimulation of sympathetic vasoconstrictor neurons by the vasomotor centers.

The effects on the heart result from the release of NE by sympathetic neurons innervating the SA node, the atrioventricular (AV) node, and the general myocardium. In a crisis, sympathetic activation occurs, and its effects will be enhanced by the release of both NE and epinephrine (E) from the adrenal medullae. The net effect is an immediate increase in heart rate and stroke volume and a corresponding rise in cardiac output. The vasoconstriction, which also results from the release of NE by sympathetic neurons, increases peripheral resistance. These adjustments— increased cardiac output and increased peripheral resistance—work together to elevate blood pressure.

Atrial baroreceptors are receptors that monitor blood pressure at the end of the systemic circuit—at the venae cavae and the right atrium. The **atrial reflex** responds to a stretching of the wall of the right atrium. ∞ p. 711

Under normal circumstances, the heart pumps blood into the aorta at the same rate at which blood arrives at the right atrium. When blood pressure rises at the right atrium, blood is arriving at the heart faster than it is being pumped out. The atrial baroreceptors solve the problem by stimulating the CV centers and increasing cardiac output until the backlog of venous blood is removed. Atrial pressure then returns to normal.

CHEMORECEPTOR REFLEXES The **chemoreceptor reflexes** respond to changes in carbon dioxide, oxygen, or pH levels in blood and cerebrospinal fluid (CSF) (Figure 21–16•). The chemoreceptors

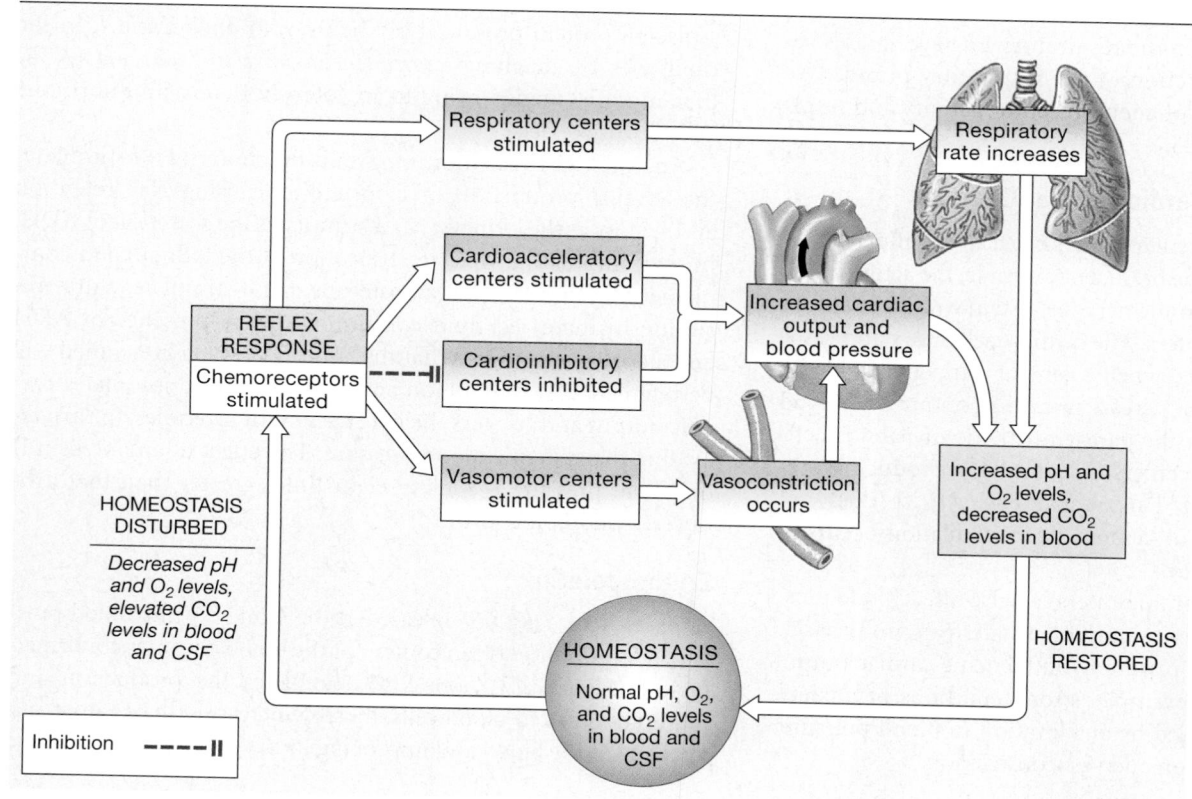

•**FIGURE 21-16**
The Chemoreceptor Reflexes

involved are sensory neurons located in the **carotid bodies**, situated in the neck near the carotid sinus, and the **aortic bodies**, near the arch of the aorta. ∞ p. 516 These receptors monitor the composition of the arterial blood. Additional chemoreceptors located on the ventrolateral surfaces of the medulla oblongata monitor the composition of CSF.

When chemoreceptors in the carotid bodies or aortic bodies detect either a rise in the carbon dioxide content or a fall in the pH of the arterial blood, the cardioacceleratory and vasomotor centers are stimulated and the cardioinhibitory centers are inhibited. This dual effect causes an increase in cardiac output, peripheral vasoconstriction, and an elevation in arterial blood pressure. A drop in the oxygen level at the aortic bodies will have the same effects. Strong stimulation of the carotid or aortic chemoreceptors causes widespread sympathetic activation, with more dramatic increases in heart rate and cardiac output.

The chemoreceptors of the medulla oblongata are involved primarily with the control of respiratory function and secondarily with regulating blood flow to the brain. For example, a steep rise in CSF carbon dioxide levels will trigger the vasodilation of cerebral vessels, but will produce vasoconstriction in most other organs. The result is increased blood flow—and hence increased oxygen delivery—to the brain.

Arterial CO_2 levels can be reduced and O_2 levels increased most effectively by coordinating cardiovascular and respiratory activities. Chemoreceptor stimulation also stimulates the respiratory centers, and the rise in cardiac output and blood pressure is associated with an increased respiratory rate. Coordination of cardiovascular and respiratory activities is vital, because accelerating blood flow in the tissues is useful only if the circulating blood contains an adequate amount of oxygen. In addition, a rise in the respiratory rate accelerates venous return through the action of the respiratory pump. (We will consider other aspects of chemoreceptor activity and respiratory control in Chapter 23.)

CNS Activities and the Cardiovascular Centers

The output of the cardiovascular centers can also be influenced by activities in other areas of the brain. For example, the activation of either division of the autonomic nervous system will affect output from the cardiovascular centers. The cardioacceleratory and vasomotor centers are stimulated when a general sympathetic activation occurs. The result is an increase in cardiac output and blood pressure. In contrast, when the parasympathetic division is activated, the cardioinhibitory centers are stimulated, producing a reduction in cardiac output. Parasympathetic activity does not directly affect the vasomotor centers, but vasodilation occurs as sympathetic activity declines.

The activities of higher brain centers can also affect blood pressure. Our thought processes or emotional states can produce significant changes in blood pressure by influencing cardiac output and vasomotor tone. For example, strong emotions of anxiety, fear, or rage are accompanied by an elevation in blood pressure, caused by cardiac stimulation and vasoconstriction.

☐ HORMONES AND CARDIOVASCULAR REGULATION

The endocrine system provides both short-term and long-term regulation of cardiovascular performance. As we have seen, E and NE from the adrenal medullae stimulate cardiac output and peripheral vasoconstriction. Other hormones important in regulating cardiovascular function include (1) antidiuretic hormone (ADH), (2) angiotensin II, (3) erythropoietin (EPO), and (4) atrial natriuretic peptide (ANP). We introduced these hormones and their functions in Chapter 18. ∞ pp. 618, 636 Although ADH and angiotensin II affect blood pressure, all four hormones are concerned primarily with the long-term regulation of blood volume (Figure 21–17•).

Antidiuretic Hormone

Antidiuretic hormone (ADH) is released at the posterior lobe of the pituitary gland in response to a decrease in blood volume, to an increase in the osmotic concentration of the plasma, or, secondarily, to circulating angiotensin II. The immediate result is a peripheral vasoconstriction that elevates blood pressure. This hormone also stimulates the conservation of water at the kidneys, thus preventing a reduction in blood volume that would further reduce blood pressure (Figure 21–17a•).

Angiotensin II

Angiotensin II appears in the blood after the release of the enzyme renin by specialized kidney cells in response to a fall in renal blood pressure (Figure 21–17a•). Once in the bloodstream, renin functions as an enzyme that starts an enzymatic chain reaction. The initial step occurs when renin converts *angiotensinogen*, a plasma protein produced by the liver, to *angiotensin I*. In the capillaries of the lungs, *angiotensin-converting enzyme (ACE)* then modifies angiotensin I to angiotensin II, an active hormone with diverse effects.

Angiotensin II has four important functions: (1) It stimulates the adrenal production of aldosterone, causing Na^+ retention and K^+ loss at the kidneys; (2) it stimulates the secretion of ADH, in turn stimulating water reabsorption at the kidneys and complementing the effects of aldosterone; (3) it stimulates thirst, resulting in increased fluid consumption (the presence of ADH and aldosterone ensures that the additional water consumed will be retained, elevating blood volume); and (4) it stimulates cardiac output and triggers the constriction of arterioles, in turn elevating the systemic blood pressure. The effect of angiotensin II on blood pressure is four to eight times greater than that produced by norepinephrine.

Erythropoietin

Erythropoietin (EPO) is released at the kidneys if the blood pressure falls or if the oxygen content of the blood becomes abnormally low (Figure 21–17a•). EPO stimulates the production and maturation of red blood cells, thereby increasing the volume and viscosity of the blood and improving its oxygen-carrying capacity.

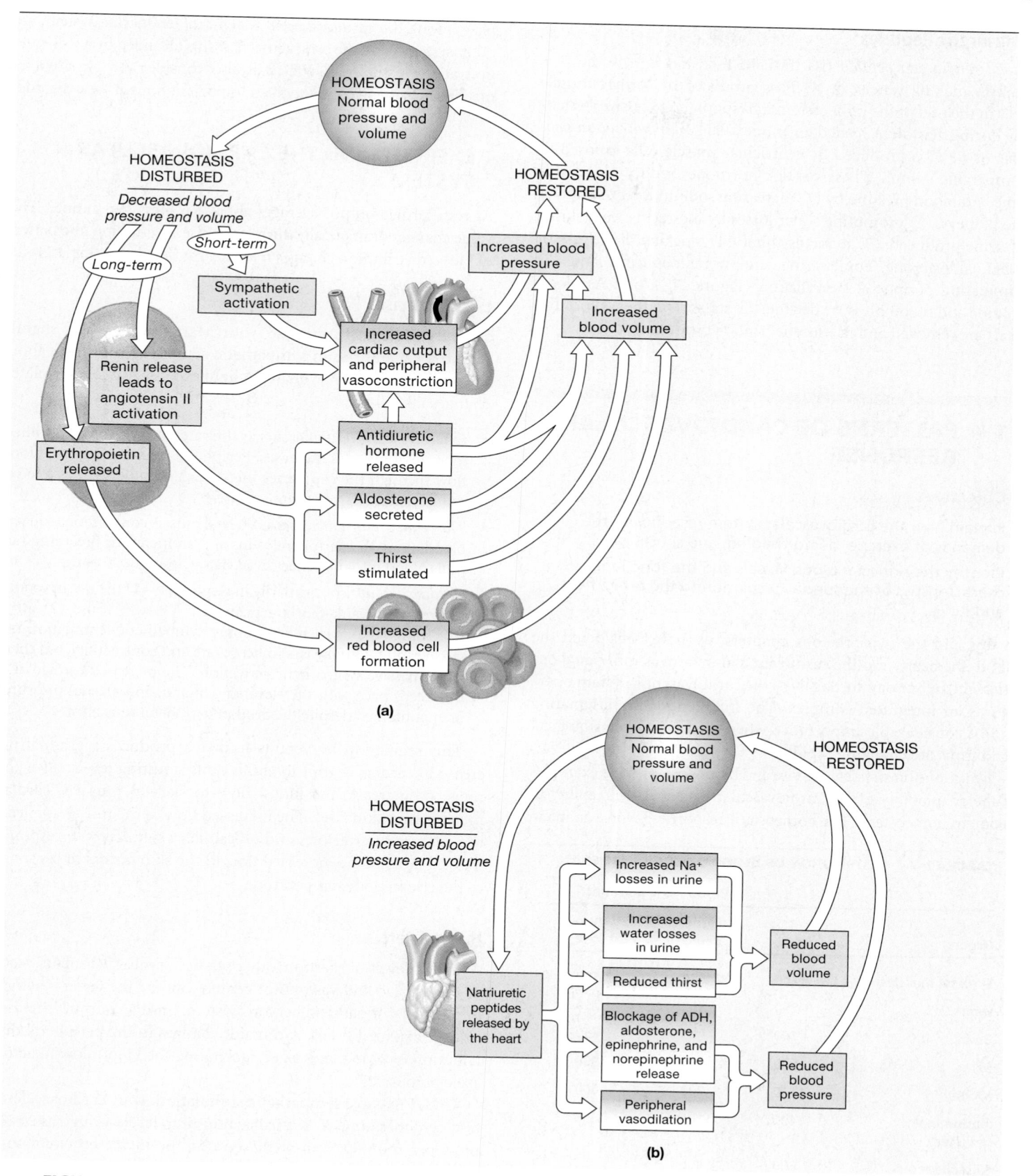

●**FIGURE 21–17**
The Regulation of Blood Pressure and Blood Volume. Factors that compensate for **(a)** decreased blood pressure and volume and for **(b)** increased blood pressure and volume.

Natriuretic Peptides

Atrial natriuretic peptide (nā-trē-ū-RET-ik; *natrium*, sodium + *ouresis*, making water), or *ANP*, is produced by cardiac muscle cells in the wall of the right atrium in response to excessive stretching during diastole. A related hormone called *brain natriuretic peptide*, or *BNP*, is produced by ventricular muscle cells exposed to comparable stimuli. These peptide hormones reduce blood volume and blood pressure by (1) increasing sodium ion excretion at the kidneys, (2) promoting water losses by increasing the volume of urine produced; (3) reducing thirst; (4) blocking the release of ADH, aldosterone, epinephrine, and norepinephrine; and (5) stimulating peripheral vasodilation (Figure 21–17b●). As blood volume and blood pressure decline, the stresses on the walls of the heart are removed and natriuretic peptide production ceases.

21–4 PATTERNS OF CARDIOVASCULAR RESPONSE

Objectives

■ Explain how the cardiovascular system responds to the demands of exercise, hemorrhaging, and shock.

■ Identify the principal blood vessels and the functional characteristics of the special circulation to the brain, heart, and lungs.

In this and the two previous chapters, we have considered the blood, the heart, and the cardiovascular system as individual entities. Yet, in our day-to-day lives, the cardiovascular system operates as an integrated complex. The interactions are fascinating and of considerable importance when physical or physiological conditions are changing rapidly.

Two common stresses, exercise and blood loss, provide examples of the adaptability of the cardiovascular system and its ability to maintain homeostasis. The homeostatic responses involve an interplay among the cardiovascular system, the endocrine system, and other systems, and the central mechanisms are aided by autoregulation at the tissue level. We shall also consider the physiological mechanisms involved in *shock*, an important homeostatic disorder.

☐ EXERCISE AND THE CARDIOVASCULAR SYSTEM

At rest, cardiac output averages about 5.6 liters per minute. That value changes dramatically during exercise. In addition, the pattern of blood distribution changes markedly, as detailed in Table 21–2.

Light Exercise

Before you begin to exercise, your heart rate increases slightly due to a general rise in sympathetic activity as you think about the workout ahead. As you begin light exercise, three interrelated changes take place:

* *Extensive vasodilation occurs* as the rate of oxygen consumption in skeletal muscles increases. Peripheral resistance drops, blood flow through the capillaries increases, and blood enters the venous system at an accelerated rate.

* *The venous return increases* as skeletal muscle contractions squeeze blood along the peripheral veins and an increased breathing rate pulls blood into the venae cavae via the respiratory pump.

* *Cardiac output rises*, primarily in response to (1) the rise in venous return (the Frank–Starling principle ∞ p. 713) and (2) atrial stretching (the atrial reflex). Some sympathetic stimulation occurs, leading to increases in heart rate and contractility, but there is no massive sympathetic activation. The increased cardiac output keeps pace with the elevated demand, and arterial pressures are maintained despite the drop in peripheral resistance.

This regulation by venous feedback produces a gradual increase in cardiac output to about double resting levels. Over the range of increase, the blood flow to skeletal muscles, cardiac muscles, and skin rises. The increased flow to the muscles reflects the dilation of arterioles and precapillary sphincters in response to local factors; the increased flow to the skin occurs in response to the rise in body temperature.

Heavy Exercise

At higher levels of exertion, other physiological adjustments occur as the cardiac and vasomotor centers call for the general activation of the sympathetic nervous system. Cardiac output increases toward maximal levels, and major changes in the peripheral distribution of blood take place, facilitating the blood flow to active skeletal muscles.

Under massive sympathetic stimulation, your cardioacceleratory centers can increase cardiac output to levels as high as 20–25 liters per minute. Even at these rates, the increased circulatory demands of the skeletal muscles can be met only if the vasomotor centers severely restrict the blood flow to "nonessential" organs, such as those of your digestive system. During exercise at maxi-

TABLE 21–2 DISTRIBUTION OF BLOOD DURING EXERCISE

Organ	Rest	Light Exercise	Strenuous Exercise
		Tissue Blood Flow (ml/min)	
Skeletal muscles	1200	4500	12,500
Heart	250	350	750
Brain	750	750	750
Skin	500	1500	1900
Kidney	1100	900	600
Abdominal viscera	1400	1100	600
Miscellaneous	600	400	400
Total cardiac output	5800	9500	17,500

Subject	Heart Weight (g)	Stroke Volume (ml)	Heart Rate (bpm)	Cardiac Output (l/min)	Blood Pressure (systolic/diastolic)
Nonathlete (rest)	300	60	83	5.0	120/80
Nonathlete (maximum)		104	192	19.9	187/75
Trained athlete (rest)	500	100	53	5.3	120/80
Trained athlete (maximum)		167	182	30.4	200/90*

TABLE 21–3 EFFECTS OF TRAINING ON CARDIOVASCULAR PERFORMANCE

*Diastolic pressures in athletes during maximal activity have not been accurately measured.

mal levels, your blood essentially races between the skeletal muscles and the lungs and heart. Although blood flow to most tissues is diminished, skin perfusion increases further, because the body temperature continues to climb. Only the blood supply to the brain remains unaffected.

Exercise, Cardiovascular Fitness, and Health

Cardiovascular performance improves significantly with training. Table 21–3 compares the cardiac performance of athletes with that of nonathletes. Trained athletes have bigger hearts and larger stroke volumes than do nonathletes, and these are important functional differences.

Recall that the cardiac output is equal to the stroke volume times the heart rate; thus, for the same cardiac output, the person with a larger stroke volume has a slower heart rate. An athlete at rest can maintain normal blood flow to peripheral tissues at a heart rate as low as 32 bpm (beats per minute), and, when necessary, the athlete's cardiac output can increase to levels 50 percent higher than those of nonathletes. Thus, a trained athlete can tolerate sustained levels of activity that are well outside the capabilities of nonathletes.

Exercise and Cardiovascular Disease

Regular exercise has several beneficial effects. Even a moderate exercise routine (jogging 5 miles a week, for example) can lower total blood cholesterol levels. A high cholesterol level is one of the major risk factors for atherosclerosis, which leads to cardiovascular disease and strokes. In addition, a healthy lifestyle—regular exercise, a balanced diet, weight control, and not smoking—reduces stress, lowers blood pressure, and slows the formation of plaque.

Regular moderate exercise may cut the incidence of heart attacks almost in half. However, at present only an estimated 8 percent of adults in the United States exercise at recommended levels. Exercise is also beneficial in accelerating one's recovery after a heart attack. Regular light-to-moderate exercise, such as walking, jogging, or bicycling, coupled with a low-fat diet and a low-stress lifestyle, not only reduces symptoms of coronary artery disease (CAD), such as angina, but also improves one's mood and overall quality of life. However, exercise does not remove any underlying medical problem, and atherosclerotic plaques, described on p. 725, do not disappear and seldom grow smaller with exercise.

There is no evidence that *intense* athletic training lowers the incidence of cardiovascular disease. On the contrary, the strains placed on all physiological systems, including the cardiovascular system, during an ultramarathon, iron-man triathlon, or other athletic extreme can be severe. Individuals with congenital aneurysms, cardiomyopathy, or cardiovascular disease risk fatal circulatory problems, such as an arrhythmia or heart attack, during severe exercise. Even healthy individuals can develop acute physiological disorders, such as kidney failure, after extreme exercise. We shall discuss the effects of exercise on other systems in later chapters.

☐ CARDIOVASCULAR RESPONSE TO HEMORRHAGING

In Chapter 19, we considered the local circulatory reaction to a break in the wall of a blood vessel. ∞ p. 672 When hemostasis fails to prevent a significant blood loss, the entire cardiovascular system makes adjustments to maintain blood pressure and restore blood volume (Figure 21–18•). The immediate problem is the maintenance of adequate blood pressure and peripheral blood flow. The long-term problem is the restoration of normal blood volume.

Short-Term Elevation of Blood Pressure

Almost as soon as the pressures start to decline, short-term responses appear. The steps include the following:

- The initial neural response occurs as carotid and aortic reflexes increase cardiac output and cause peripheral vasoconstriction (pp. 742-743). With the blood volume reduced, cardiac output is maintained by increasing the heart rate, typically to 180–200 bpm.

- The combination of stress and anxiety stimulates the sympathetic nervous system headquarters in the hypothalamus, which in turn triggers a further increase in vasomotor tone, constricting the arterioles and elevating blood pressure. At the same time, venoconstriction mobilizes the venous reserve and quickly improves venous return (p. 739).

- Short-term hormonal effects also occur. For instance, sympathetic activation causes the secretion of E and NE by the adrenal medulla, increasing cardiac output and extending peripheral vasoconstriction. In addition, the release of ADH by the posterior lobe of the pituitary gland and the production of angiotensin II enhance vasoconstriction while participating in the long-term response.

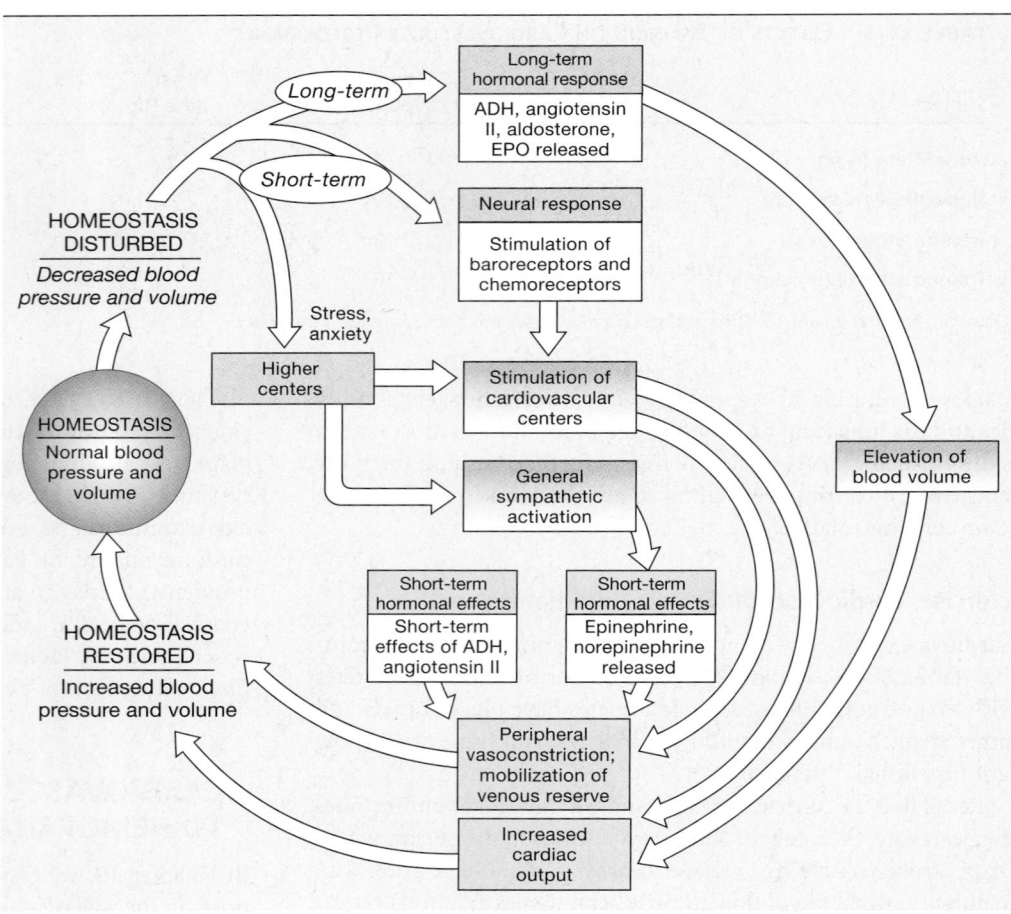

●**FIGURE 21–18**
Cardiovascular Responses to Hemorrhaging and Blood Loss

This combination of short-term responses elevates blood pressure and improves peripheral blood flow, often restoring normal arterial pressures and peripheral circulation after blood losses of up to 20 percent of the total blood volume. Such adjustments are more than sufficient to compensate for the blood loss experienced when you donate blood. (Most blood banks collect 500 ml of whole blood, roughly 10 percent of your total blood volume.)

Long-Term Restoration of Blood Volume

Short-term responses temporarily compensate for a reduction in blood volume. Long-term responses are geared to restoring normal blood volume, a process that can take several days after a serious hemorrhage. The steps include the following:

- The decline in capillary blood pressure triggers a recall of fluids from the interstitial spaces (p. 738).

- Aldosterone and ADH promote fluid retention and reabsorption at the kidneys, preventing further reductions in blood volume.

- Thirst increases, and additional water is obtained by absorption across the digestive tract. This intake of fluid elevates the plasma volume and ultimately replaces the interstitial fluids "borrowed" at the capillaries.

- Erythropoietin targets the bone marrow, stimulating the maturation of red blood cells, which increase blood volume and improve oxygen delivery to peripheral tissues.

◼ SPECIAL CIRCULATION

The vasoconstriction that occurs in response to a fall in blood pressure or a rise in CO_2 levels affects multiple tissues and organs simultaneously. The term *special circulation* refers to the circulation through organs in which blood flow is controlled by separate mechanisms. We shall note three important examples: the circulation to the brain, the heart, and the lungs.

The Brain

In Chapter 14, we noted the existence of the blood–brain barrier, which isolates most CNS tissue from the general circulation. ∞ p. 472 The brain has a very high demand for oxygen and receives a substantial supply of blood. Under a variety of conditions, blood flow to the brain remains steady at about 750 ml/min—roughly 12 percent of the cardiac output delivered to an organ that represents less than 2 percent of the body weight. Neurons do not maintain significant energy reserves, and in functional terms most of the adjustments made by the cardiovascular system treat blood flow to the brain as the number one priority. Even during a cardiovascular crisis, blood flow through the brain remains as near normal as possible: While the cardiovascular centers are calling for widespread peripheral vasoconstriction, the cerebral vessels are instructed to dilate.

Although total blood flow to the brain remains relatively constant, blood flow to specific regions of the brain changes from moment to moment. These changes occur in response to local changes in the composition of interstitial fluid that accompany neural activity. When you read, write, speak, or walk, specific regions of your brain become active. Blood flow to those regions increases almost instantaneously, ensuring that the active neurons will continue to receive the oxygen and nutrients they require.

The brain receives arterial blood from four arteries. Because these arteries form anastomoses inside the cranium, an interruption of any one vessel will not compromise the circulatory supply to the brain. If a plaque or a blood clot blocks an artery or if an artery ruptures, dependent tissues will be injured or killed. Symptoms of a *stroke*, or *cerebrovascular accident (CVA)*, then appear. **AM** The Causes and Treatment of Cerebrovascular Disease

The Heart

We described the anatomy of the coronary circulation in Chapter 20. ∞ p. 692 The coronary arteries arise at the base of the ascending aorta, where systemic pressures are highest. Each time the heart contracts, it squeezes the coronary vessels, so blood flow is reduced. In the left ventricle, systolic pressures are high enough that blood can flow into the myocardium only during diastole; over this period, elastic rebound helps drive blood through the coronary vessels. Normal cardiac muscle cells can tolerate these brief circulatory interruptions because they have substantial oxygen reserves.

When you are at rest, your coronary blood flow is about 250 ml/min. When the workload on your heart increases, local factors, such as reduced O_2 levels and lactic acid production, dilate the coronary vessels and increase blood flow. Epinephrine released during sympathetic stimulation promotes the vasodilation

Circulatory Shock

Shock is an acute circulatory crisis marked by low blood pressure (hypotension) and inadequate peripheral blood flow. Severe and potentially fatal symptoms develop as vital tissues become starved for oxygen and nutrients. Common causes of shock are (1) a drop in cardiac output after hemorrhaging or other fluid losses, (2) damage to the heart, (3) external pressure on the heart, and (4) extensive peripheral vasodilation. We shall focus on the cause, symptoms, and treatment of circulatory shock; the *Applications Manual* discusses other types. **AM** Other Types of Shock

Symptoms of **circulatory shock** appear after fluid losses of about 30 percent of the total blood volume. The cause can be hemorrhaging or fluid losses to the environment, as in dehydration or after severe burns. All cases of circulatory shock share the same six basic symptoms:

1. Hypotension, with systolic pressures below 90 mm Hg.

2. Pale, cool, and moist ("clammy") skin. The skin is pale and cool due to peripheral vasoconstriction; the moisture reflects the sympathetic activation of sweat glands.

3. Frequent confusion and disorientation, due to a drop in blood pressure at the brain.

4. A rise in heart rate and a rapid, weak pulse.

5. A cessation of urination, because the reduced blood flow to the kidneys slows or stops urine production.

6. A drop in blood pH (*acidosis*), due to lactic acid generation in oxygen-deprived tissues.

In mild forms of circulatory shock, homeostatic adjustments can cope with the situation. The short-term and long-term responses detailed in Figure 21–18● are part of the adjustment process. During this period, peripheral blood flow is reduced, but remains adequate to meet tissue demands.

When blood volume declines by more than 35 percent, homeostatic mechanisms are no longer able to cope with the situation. Despite sustained vasoconstriction and the mobiliza-

tion of the venous reserve, blood pressure remains abnormally low, venous return is reduced, and cardiac output is inadequate (Figure 21–19a●). A vicious cycle begins when the low cardiac output damages the myocardium. This damage leads to a further reduction in cardiac output and subsequent reductions in blood pressure and venous return. As blood flow to the brain decreases, the individual becomes disoriented and confused.

When the mean arterial blood pressure (MAP) falls to about 50 mm Hg, carotid sinus baroreceptors trigger a massive activation of the vasomotor centers. In effect, the goal now is to preserve the circulation to the brain at any cost. The sympathetic output causes a sustained and maximal vasoconstriction. This reflex, called the *central ischemic response*, reduces peripheral circulation to an absolute minimum, but it elevates blood pressure to about 70 mm Hg and improves blood flow to cerebral vessels.

The central ischemic response is a last-ditch effort that maintains adequate blood flow to the brain at the expense of other tissues. Unless prompt treatment is provided, the condition will soon prove fatal. Treatment must concentrate on (1) preventing further fluid losses and (2) giving a transfusion of whole blood, plasma expanders, or blood substitutes.

In the absence of treatment, the cardiovascular damage will soon become irreversible (Figure 21–19b●). Conditions in the heart, liver, kidneys, and CNS rapidly deteriorate to the point at which death will occur, even *with* medical treatment.

Irreversible shock begins when conditions in the tissues become so abnormal that the arteriolar smooth muscles and precapillary sphincters become unable to contract, despite the commands of the vasomotor centers. The result is a widespread peripheral vasodilation and an immediate and fatal decline in blood pressure. This event is called *circulatory collapse*. The blood pressure in many tissues then falls so low that the capillaries collapse like deflating balloons. Blood flow through these capillary beds stops completely, and the cells in the affected tissues die. The dying cells release more abnormal chemicals, and the effect quickly spreads throughout the body.

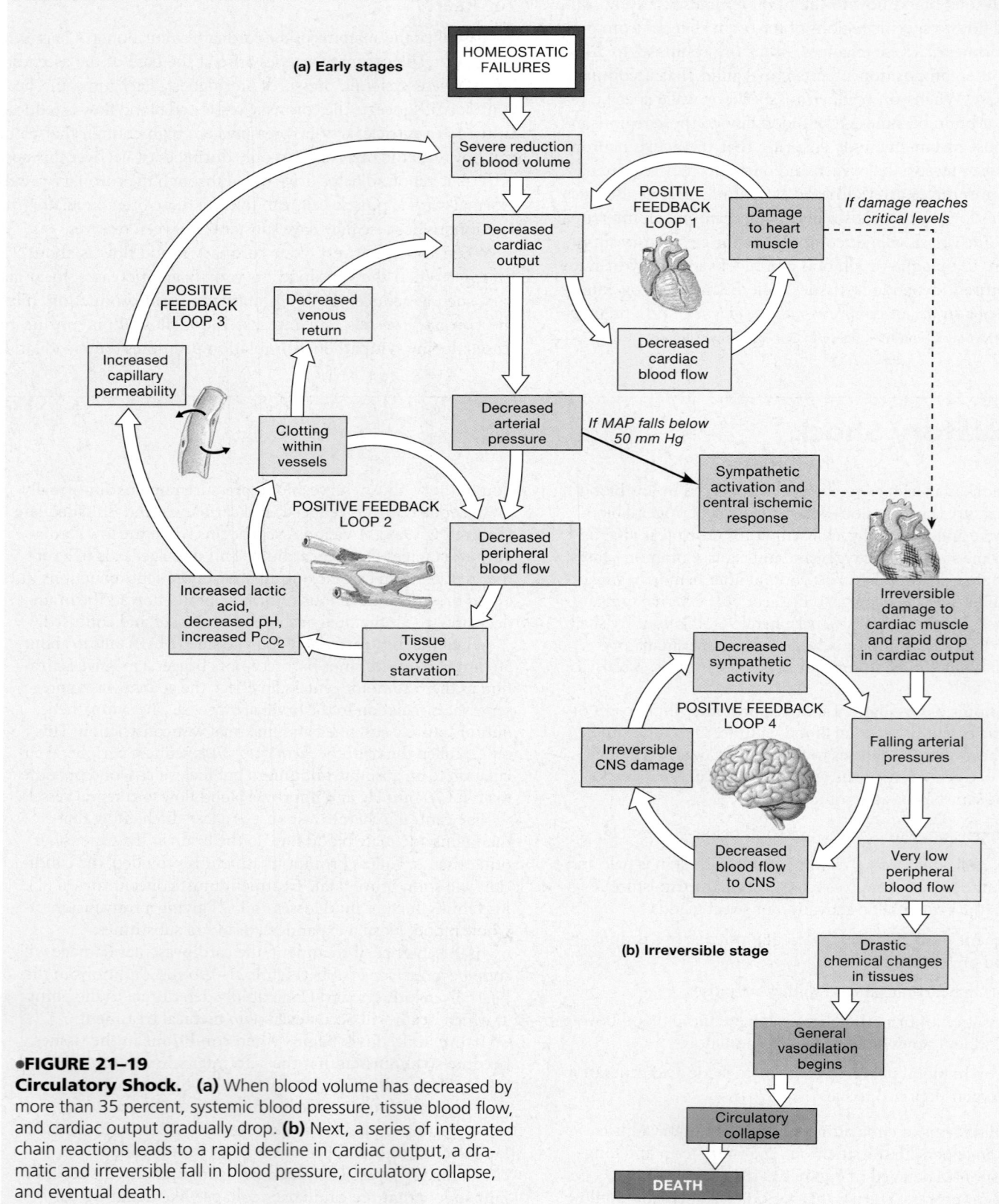

•FIGURE 21–19
Circulatory Shock. **(a)** When blood volume has decreased by
more than 35 percent, systemic blood pressure, tissue blood flow,
and cardiac output gradually drop. **(b)** Next, a series of integrated
chain reactions leads to a rapid decline in cardiac output, a dramatic and irreversible fall in blood pressure, circulatory collapse,
and eventual death.

of coronary vessels while increasing your heart rate and the strength of cardiac contractions. As a result, coronary blood flow increases while vasoconstriction occurs in other tissues.

For uncertain reasons, some individuals experience *coronary spasms*, which can temporarily restrict coronary circulation and produce symptoms of angina. A permanent restriction or blockage of coronary vessels, as in coronary artery disease (CAD), and tissue damage, as caused by a myocardial infarction (MI), can limit the heart's ability to increase its output, even under maximal stimulation. Individuals with these conditions experience symptoms of heart failure when the cardiac workload increases much above resting levels. **AM** Heart Failure

The Lungs

The lungs contain roughly 300 million *alveoli* (al-VĒ-ō-lī; *alveolus*, sac), delicate epithelial pockets where gas exchange occurs. Each alveolus is surrounded by an extensive capillary network. Blood flow through the lungs is regulated primarily by local responses to the levels of oxygen within individual alveoli. When an alveolus contains oxygen in abundance, the associated vessels dilate, so blood flow increases, promoting the absorption of oxygen from the alveolar air. When the oxygen content of the air is very low, the vessels constrict, so blood is shunted to alveoli that still contain significant levels of oxygen. This mechanism maximizes the efficiency of the respiratory system, because the circulation of blood through the capillaries of an alveolus has no benefit unless that alveolus contains oxygen.

The mechanism is precisely the opposite of that in other tissues, where a decline in oxygen levels causes local vasodilation rather than vasoconstriction. The difference makes functional sense, but its physiological basis remains a mystery.

Blood pressure in pulmonary capillaries, averaging 10 mm Hg, is lower than the pressure in systemic capillaries. The BCOP (25 mm Hg) is the same as elsewhere in the bloodstream. As a result, reabsorption exceeds filtration in pulmonary capillaries. Fluid moves continuously into the pulmonary capillaries across the alveolar surfaces, thereby preventing a buildup of fluid in the alveoli that could interfere with the diffusion of respiratory gases. If the blood pressure in pulmonary capillaries rises above 25 mm Hg, fluid will enter the alveoli, causing *pulmonary edema*.

✓ Why does blood pressure increase during exercise?

✓ How would applying a small pressure to the common carotid artery affect your heart rate?

✓ What effect would the vasoconstriction of the renal artery have on blood pressure and blood volume?

✓ Why does a person in circulatory shock have a rapid and weak pulse?

Answers start on page Q-1

Review and explore cardiovascular regulation using the **IP CD-ROM**: Cardiovascular System/Blood Pressure Regulation and Autoregulation and Capillary Dynamics.

21–5 THE DISTRIBUTION OF BLOOD VESSELS: AN OVERVIEW

Objective

- Describe three general functional patterns seen in the pulmonary and systemic circuits of the cardiovascular system.

You already know that the cardiovascular system is divided into the *pulmonary circuit* and the *systemic circuit*. The pulmonary circuit is composed of arteries and veins that transport blood between the heart and the lungs. This circuit begins at the right ventricle and ends at the left atrium. From the left ventricle, the arteries of the systemic circuit transport oxygenated blood and nutrients to all organs and tissues, ultimately returning deoxygenated blood to the right atrium. Figure 21–20● summarizes the primary distribution routes within the pulmonary and systemic circuits.

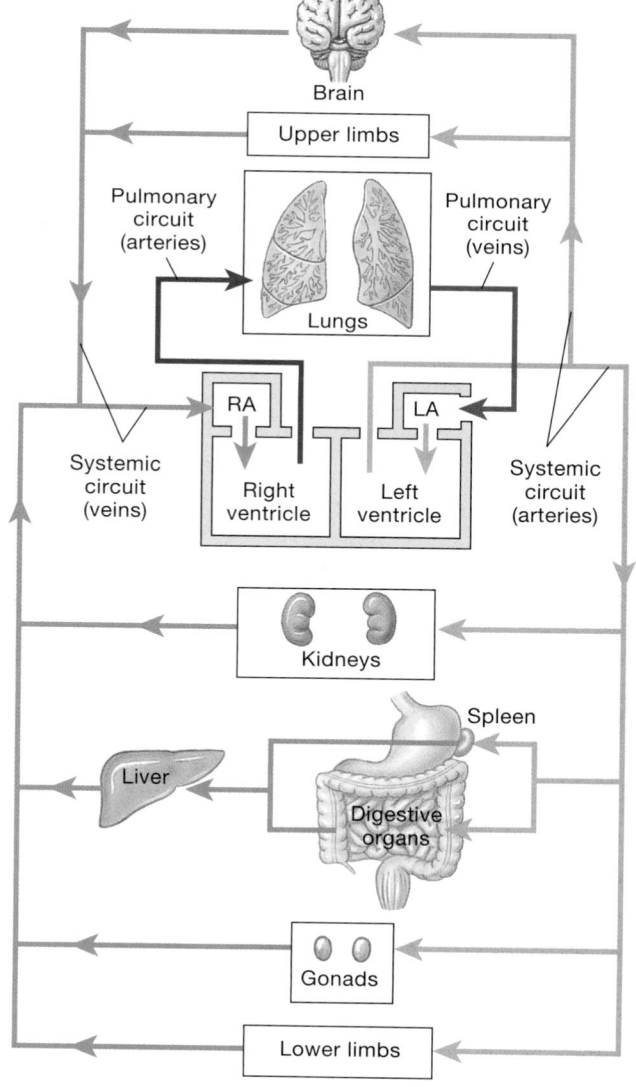

●**FIGURE 21–20**
An Overview of the Pattern of Circulation. Here, *RA* stands for right atrium, *LA* for left atrium.

In the pages that follow, we shall examine the vessels of the pulmonary and systemic circuits further. Three general functional patterns are worth noting at the outset:

1. The peripheral distributions of arteries and veins on the left and right sides are generally identical, except near the heart, where the largest vessels connect to the atria or ventricles. For example, the distributions of the *left* and *right subclavian, axillary, brachial,* and *radial arteries* parallel those of the *left* and *right subclavian, axillary, brachial,* and *radial veins,* respectively.

2. A single vessel may have several names as it crosses specific anatomical boundaries, making accurate anatomical descriptions possible when the vessel extends far into the periphery. For example, the *external iliac artery* becomes the *femoral artery* as it leaves the trunk and enters the lower limb.

3. Arteries and their corresponding veins usually travel together to reach their target organs. Often, anastomoses between adjacent arteries or veins reduce the impact of a temporary or even permanent occlusion (blockage) of a single blood vessel.

21–6 THE PULMONARY CIRCUIT

Objective

■ Identify the major arteries and veins of the pulmonary circuit and the areas they serve.

Blood entering the right atrium has just returned from the peripheral capillary beds, where oxygen was released and carbon dioxide absorbed. After traveling through the right atrium and ventricle, this deoxygenated blood enters the pulmonary trunk, the start of the pulmonary circuit (Figure 21–21●). At the lungs, oxygen is replenished, carbon dioxide is released, and the oxygenated blood is returned to the heart for distribution via the systemic circuit. Compared with the systemic circuit, the pulmonary circuit is short: The base of the pulmonary trunk and the lungs are only about 15 cm (6 in.) apart.

The arteries of the pulmonary circuit differ from those of the systemic circuit in that they carry deoxygenated blood. (For this reason,

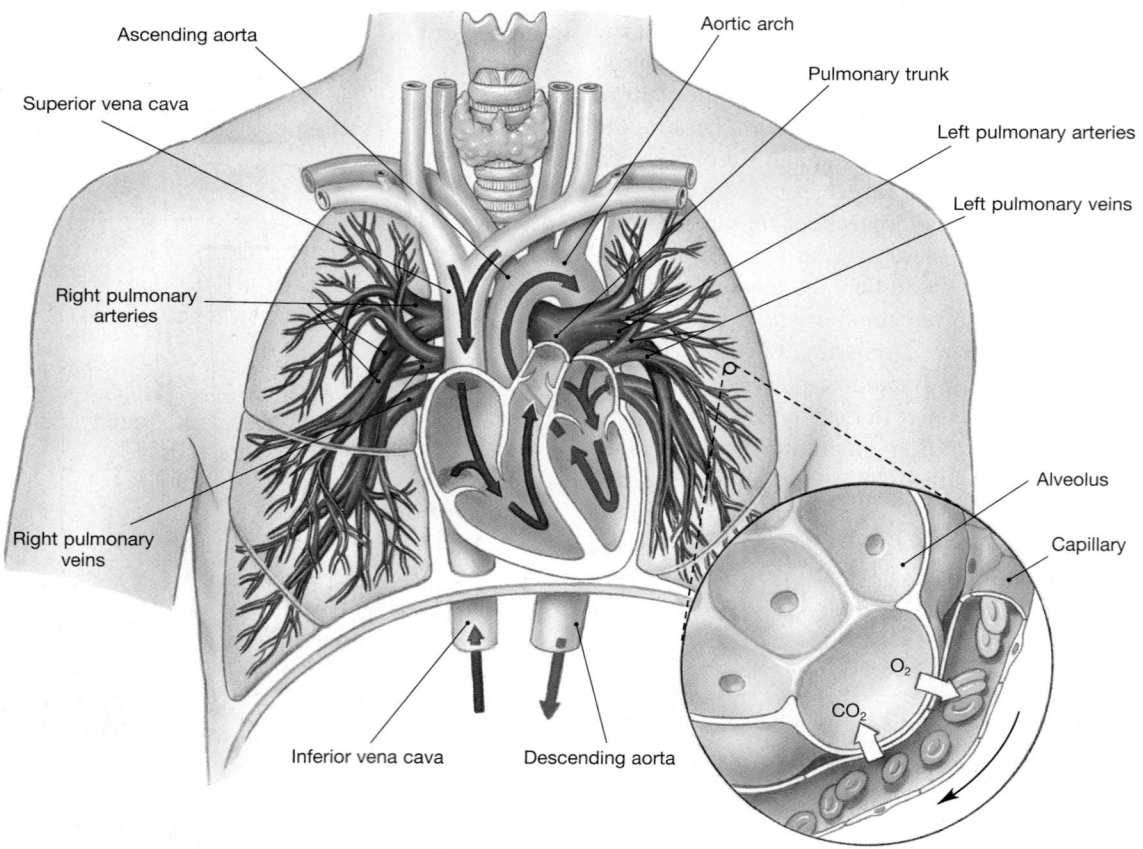

●FIGURE 21–21
The Pulmonary Circuit. The pulmonary circuit consists of pulmonary arteries, which deliver blood from the right ventricle to the lungs; pulmonary capillaries, where gas exchange occurs; and pulmonary veins, which deliver blood to the left atrium. The pulmonary capillaries receive blood containing low levels of oxygen and higher levels of carbon dioxide. As the enlarged view shows, diffusion across the capillary walls at alveoli removes carbon dioxide and provides oxygen to the blood. **ATLAS** Plates 7.4a, 7.6b; Scans 10, 11a

most color-coded diagrams show the pulmonary arteries in blue, the same color as systemic veins.) As it curves over the superior border of the heart, the pulmonary trunk gives rise to the **left** and **right pulmonary arteries**. These large arteries enter the lungs before branching repeatedly, giving rise to smaller and smaller arteries. The smallest branches, the *pulmonary arterioles*, provide blood to capillary networks that surround *alveoli*. The walls of these small air pockets are thin enough for gas to be exchanged between the capillary blood and inspired air. As it leaves the alveolar capillaries, oxygenated blood enters venules that in turn unite to form larger vessels carrying blood toward the **pulmonary veins**. These four veins, two from each lung, empty into the left atrium, completing the pulmonary circuit.

21–7 THE SYSTEMIC CIRCUIT

Objective

- Identify the major arteries and veins of the systemic circuit and the areas they serve.

The systemic circuit supplies the capillary beds in all parts of the body not serviced by the pulmonary circuit. The systemic circuit, which at any moment contains about 84 percent of the total blood volume, begins at the left ventricle and ends at the right atrium.

☐ SYSTEMIC ARTERIES

Figure 21–22● is an overview of the systemic arterial system, indicating the relative locations of major systemic arteries. Figures 21–23 to 21–28● show the detailed distribution of these vessels and their branches. By convention, several large arteries are called *trunks*; examples are the *pulmonary, brachiocephalic, thyrocervical*, and *celiac trunks*. Because most of the major arteries are paired, with one artery of each pair on either side of the body, the terms *right* and *left* will not be used unless the arteries to both sides are labeled.

The Ascending Aorta

The **ascending aorta** begins at the aortic valve of the left ventricle (Figure 21–23●). The left and right coronary arteries originate in the aortic sinus at the base of the ascending aorta, just superior to the aortic valve. We explained the distribution of coronary vessels in Chapter 20 and illustrated them in Figure 20–9●, p. 694.

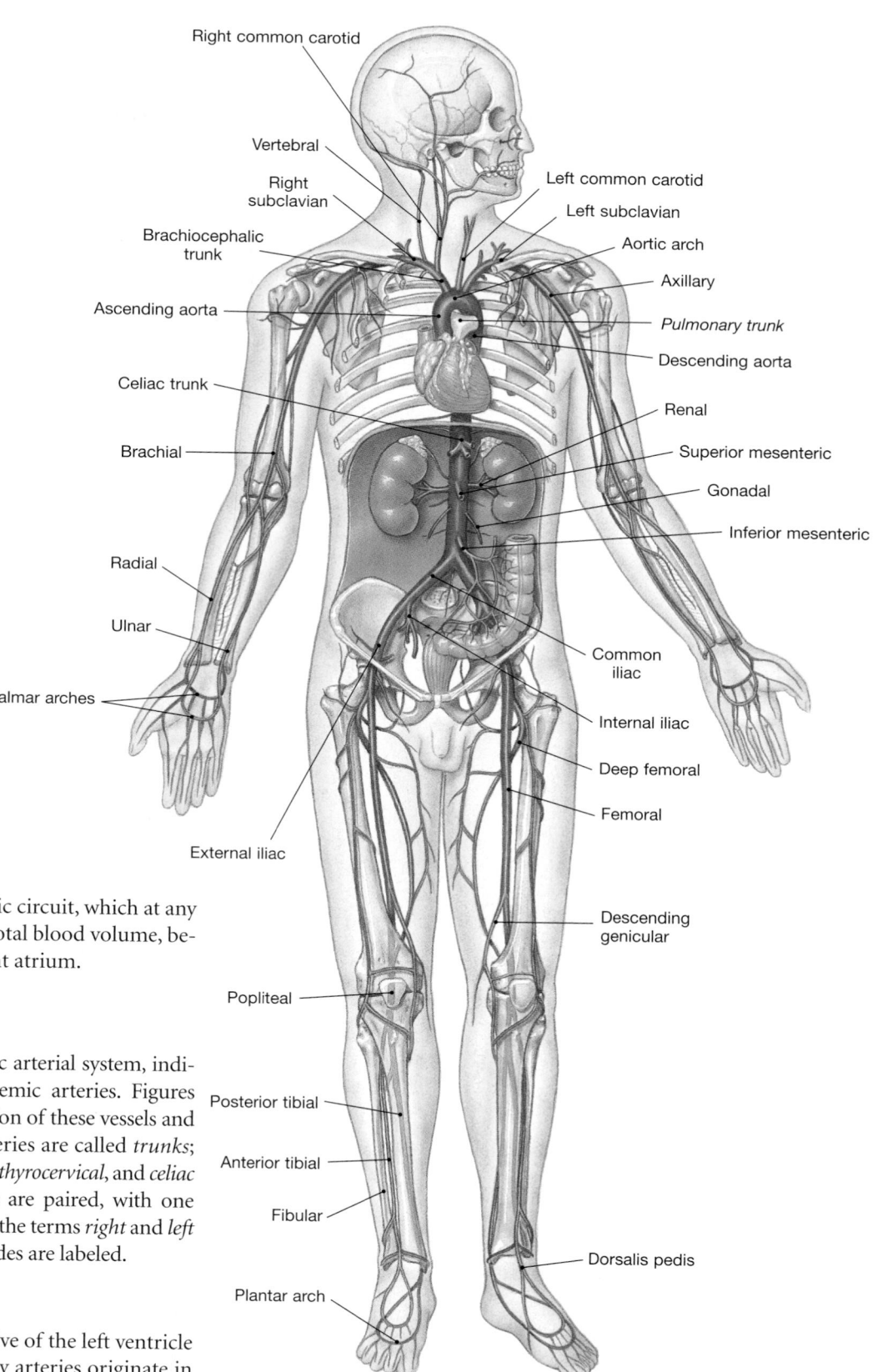

●**FIGURE 21–22**
An Overview of the Major Systemic Arteries

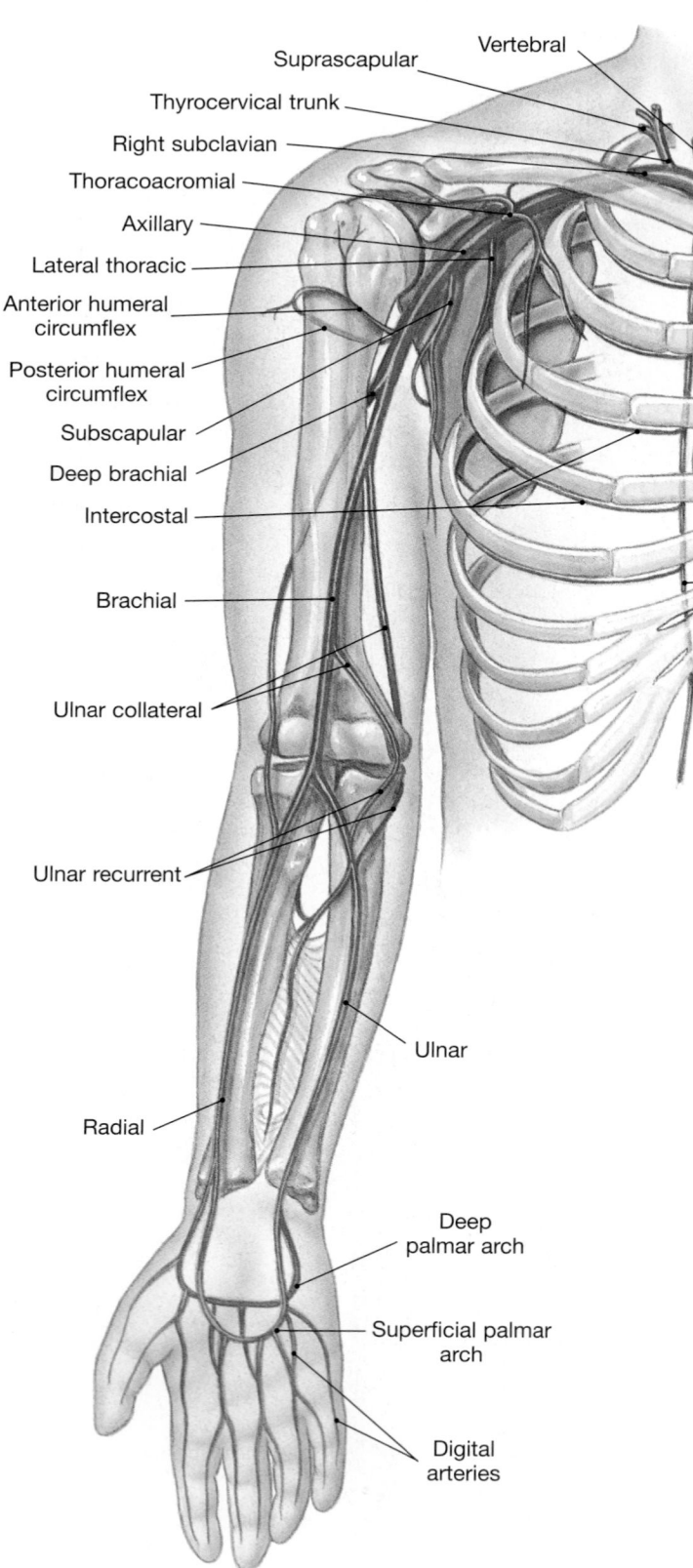

(a)

●**FIGURE 21–23**
Arteries of the Chest and Upper Limb. **(a)** A diagrammatic
view. **ATLAS** Plates 6.1a–c, 6.2a,b; Scan 11b

The Aortic Arch

The **aortic arch** curves like the handle of a cane across the superi-
or surface of the heart, connecting the ascending aorta with the
descending aorta. Three elastic arteries originate along the aortic
arch (Figures 21–22, 21–23, and 21–24●). These arteries, (1) the
brachiocephalic (brā-kē-ō-se-FAL-ik) **trunk**, (2) the **left common
carotid artery**, and (3) the **left subclavian artery**, deliver blood to
the head, neck, shoulders, and upper limbs. The brachiocephalic
trunk, also called the *innominate artery* (in-NOM-i-nāt; un-
named), ascends for a short distance before branching to form the
right subclavian artery and the **right common carotid artery**.

We have only one brachiocephalic trunk, with the left common carotid and left subclavian arteries arising separately from
the aortic arch. However, in terms of their peripheral distribution, the vessels on the left side are mirror images of those on the
right side. Figures 21–23 and 21–24● illustrate the major branches of these arteries.

THE SUBCLAVIAN ARTERIES The subclavian arteries supply blood to
the arms, chest wall, shoulders, back, and CNS (Figures 21–22 and
21–23●). Three major branches arise before a subclavian artery
leaves the thoracic cavity: (1) the **internal thoracic artery**, supplying
the pericardium and anterior wall of the chest, (2) the **vertebral
artery**, which provides blood to the brain and spinal cord, and (3)
the **thyrocervical trunk**, which provides blood to muscles and other
tissues of the neck, shoulder, and upper back.

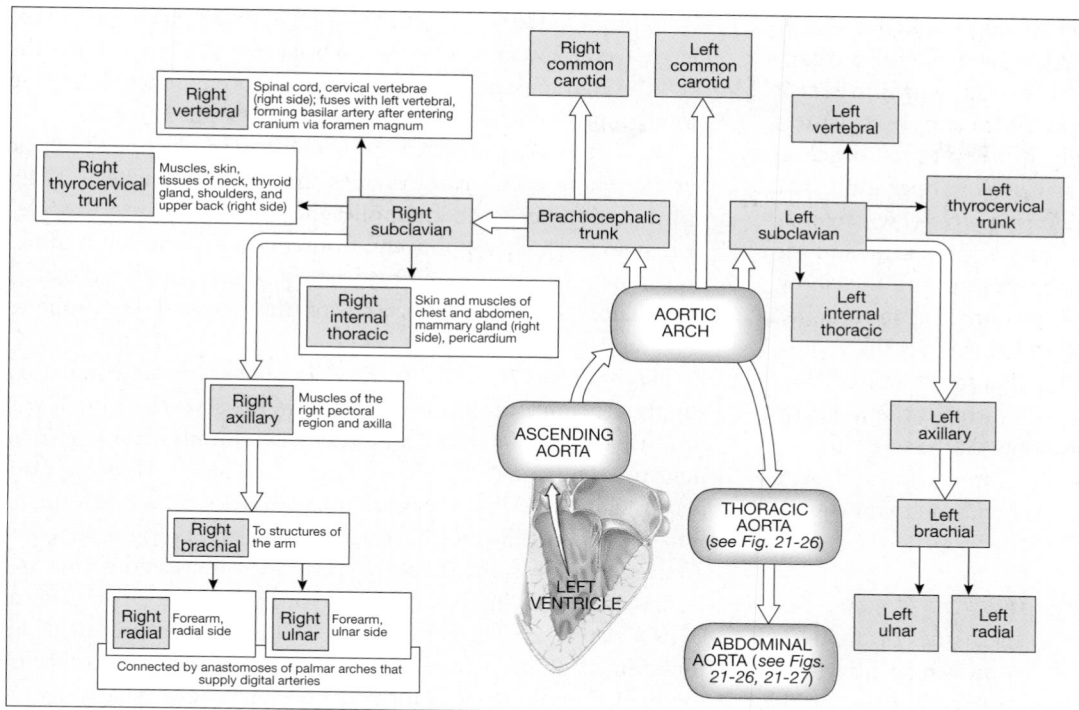

Arteries of the Chest and Upper Limb (continued). **(b)** A flowchart.

(b)

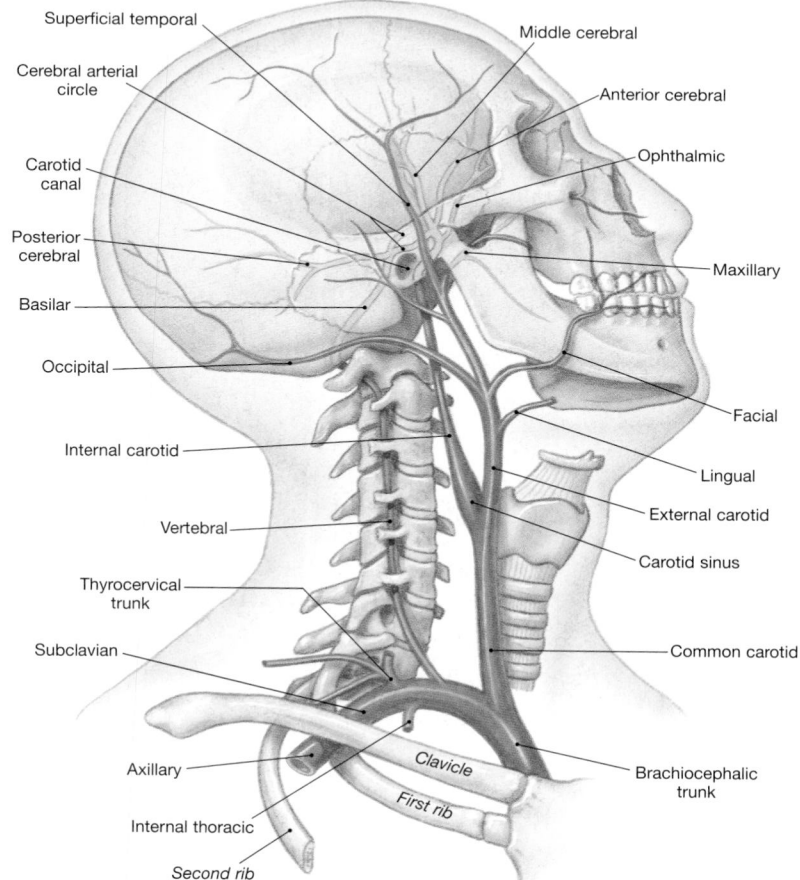

●**FIGURE 21–24**
Arteries of the Neck and Head. Shown as seen from the right side. **ATLAS** Plates 5.2, 5.3a–c; Scan 11c

After leaving the thoracic cavity and passing across the superior border of the first rib, the subclavian is called the **axillary artery**. This artery crosses the axilla to enter the arm, where it gives rise to *humeral circumflex arteries*, which supply structures near the head of the humerus. Distally, it becomes the **brachial artery**, which supplies blood to the rest of the upper limb. The brachial artery gives rise to the *deep brachial artery*, which supplies deep structures on the posterior aspect of the arm, and the *ulnar collateral arteries*, which supply the area around the elbow. As it approaches the coronoid fossa of the humerus, the brachial artery divides into the **radial artery**, which follows the radius, and the **ulnar artery**, which follows the ulna to the wrist. These arteries supply blood to the forearm and, through the *ulnar recurrent arteries*, the region around the elbow. At the wrist, the radial and ulnar arteries fuse to form the **superficial** and **deep palmar arches**, which supply blood to the hand and to the **digital arteries** of the thumb and fingers.

THE CAROTID ARTERY AND THE BLOOD SUPPLY TO THE BRAIN The common carotid arteries ascend deep in the tissues of the neck. You can usually locate the carotid artery by pressing gently along either side of the windpipe (trachea) until you feel a strong pulse.

Each common carotid artery divides into an **external carotid artery** and an **internal carotid artery** (Figure 21–24●). The carotid sinus, located at the base of the internal carotid artery, may extend along a portion of the common carotid. The external carotid arter-

ies supply blood to the structures of the neck, esophagus, pharynx, larynx, lower jaw, and face. The internal carotid arteries enter the skull through the carotid canals of the temporal bones, delivering blood to the brain. (See Figures 7–3 and 7–4●, pp. 213–215.)

The internal carotid arteries ascend to the level of the optic nerves, where each artery divides into three branches: (1) an **ophthalmic artery**, which supplies the eyes, (2) an **anterior cerebral artery**, which supplies the frontal and parietal lobes of the brain, and (3) a **middle cerebral artery**, which supplies the mesencephalon and lateral surfaces of the cerebral hemispheres (Figures 21–24 and 21–25●).

The brain is extremely sensitive to changes in its circulatory supply. An interruption of blood flow for several seconds will produce unconsciousness, and after four minutes some permanent neural damage can occur. Such circulatory crises are rare, because blood reaches the brain through the vertebral arteries as well as by way of the internal carotids. The left and right vertebral arteries arise from the subclavian arteries and ascend within the transverse foramina of the cervical vertebrae. (See Figure 7–19b,c●, p. 232.) The vertebral arteries enter the cranium at the foramen magnum, where they fuse along the ventral surface of the medulla oblongata to form the **basilar artery**. The vertebral arteries and the basilar artery supply blood to the spinal cord, medulla oblongata, pons, and cerebellum before dividing into the **posterior cerebral arteries**, which in turn branch off into the **posterior communicating arteries** (Figure 21–25●).

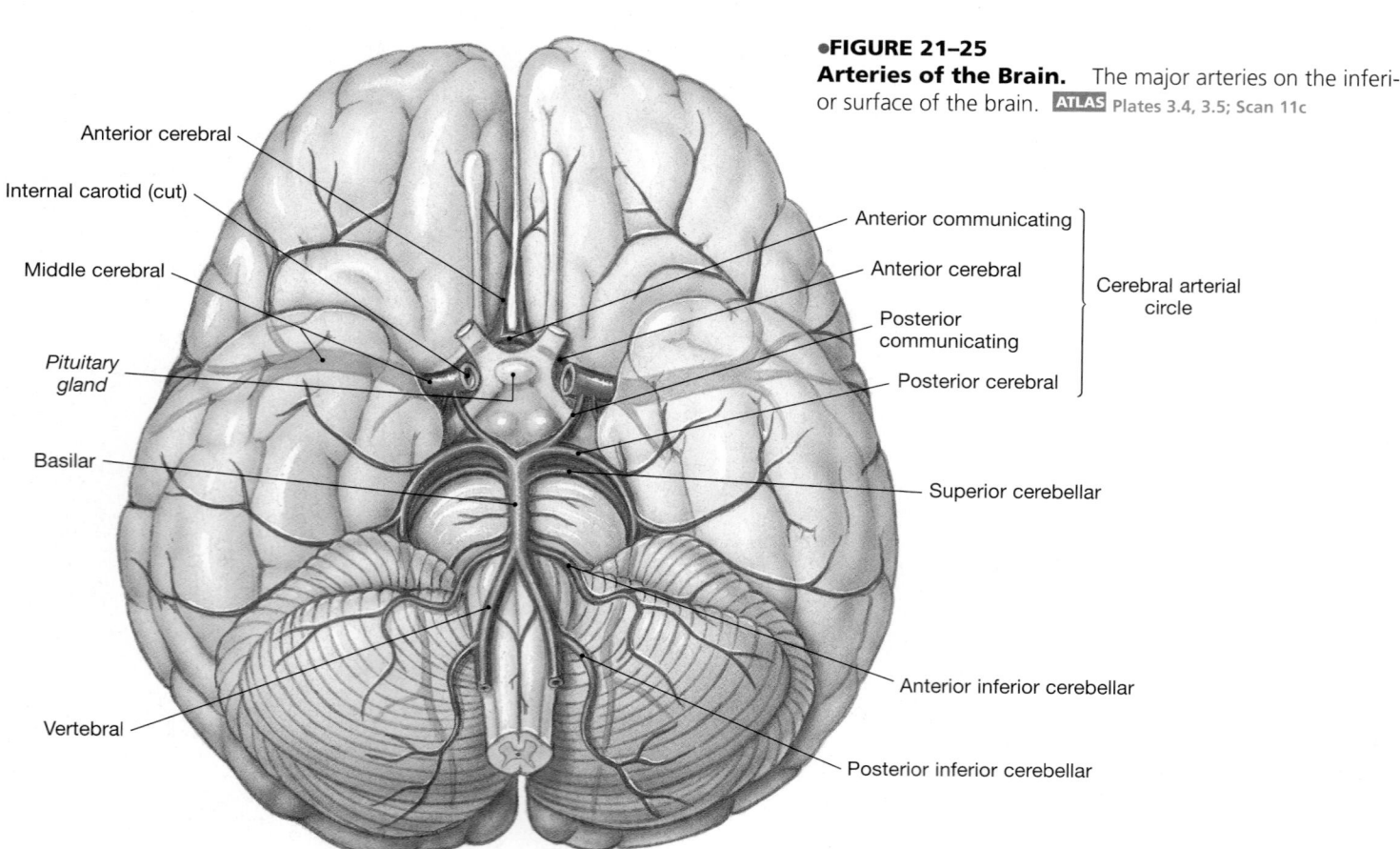

●FIGURE 21–25
Arteries of the Brain. The major arteries on the inferior surface of the brain. **ATLAS** Plates 3.4, 3.5; Scan 11c

Anterior cerebral

Internal carotid (cut)

Middle cerebral

Pituitary gland

Basilar

Vertebral

Anterior communicating

Anterior cerebral

Posterior communicating

Posterior cerebral

Cerebral arterial circle

Superior cerebellar

Anterior inferior cerebellar

Posterior inferior cerebellar

The internal carotids normally supply the arteries of the anterior half of the cerebrum, and the rest of the brain receives blood from the vertebral arteries. But this circulatory pattern can easily change, because the internal carotid arteries and the basilar artery are interconnected in a ring-shaped anastomosis called the **cerebral arterial circle**, or *circle of Willis*, which encircles the infundibulum of the pituitary gland (Figure 21–25●). With this arrangement, the brain can receive blood from either the carotid or the vertebral arteries, and the chances for a serious interruption of circulation are reduced.

◆ **Strokes**, or *cerebrovascular accidents (CVAs)*, are interruptions of the vascular supply to a portion of the brain. The *middle cerebral artery*, a major branch of the cerebral arterial circle, is the most common site of a stroke. Superficial branches deliver blood to the temporal lobe and large portions of the frontal and parietal lobes; deep branches supply the basal nuclei and portions of the thalamus. If a stroke blocks the middle cerebral artery on the left side of the brain, aphasia and a sensory and motor paralysis of the right side of the body result. In a stroke affecting the middle cerebral artery on the right side, the individual experiences a loss of sensation and motor control over the left side of the body and has difficulty drawing or interpreting spatial relationships. Strokes affecting vessels that supply the brain stem also produce distinctive symptoms; those affecting the lower brain stem are commonly fatal.

The Descending Aorta

The **descending aorta** is continuous with the aortic arch. The diaphragm divides the descending aorta into a superior **thoracic aorta** and an inferior **abdominal aorta** (Figures 21–26 and 21–27●, p. 759).

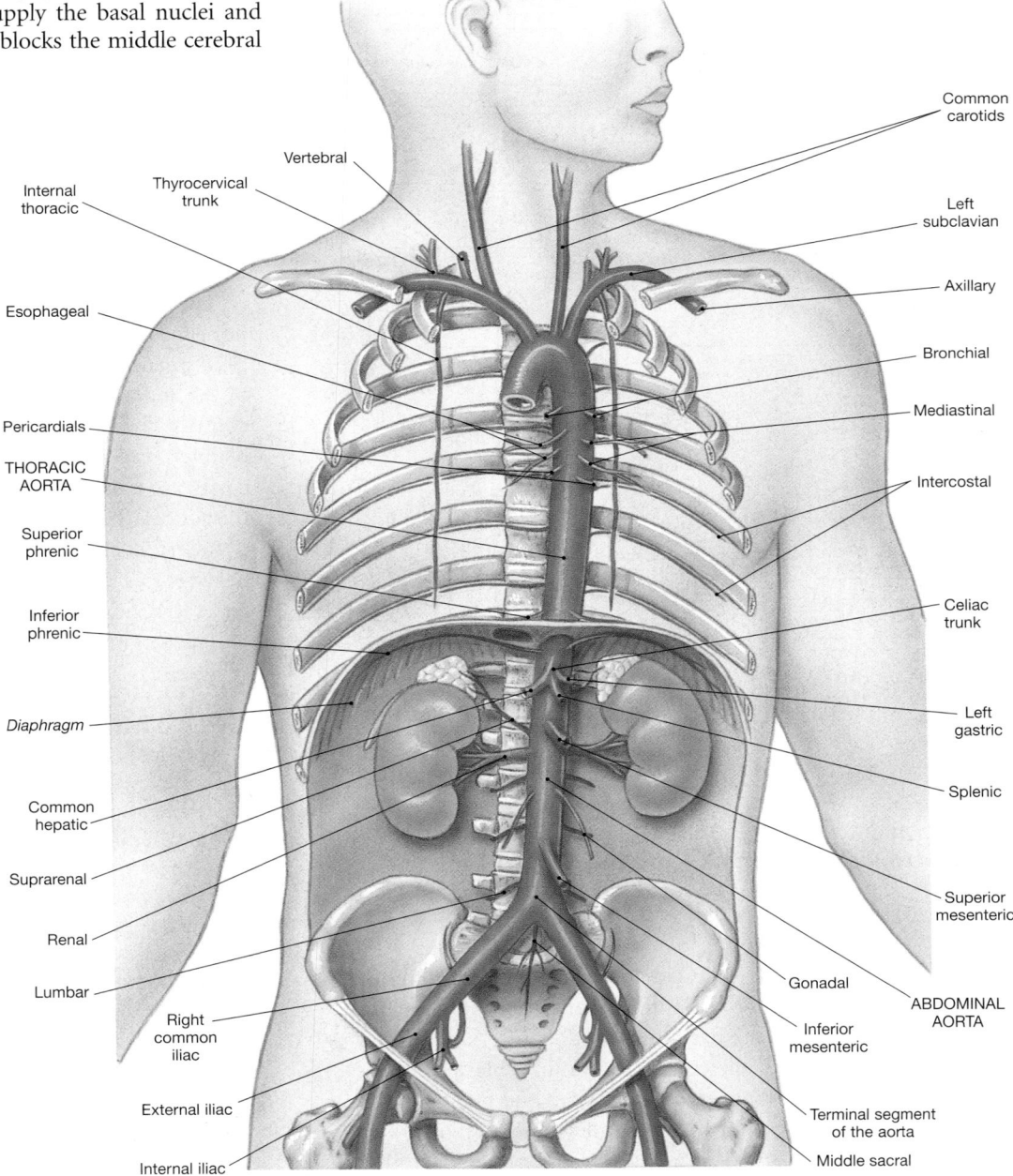

●**FIGURE 21-26**
Major Arteries of the Trunk.
(a) A diagrammatic view, with most of the thoracic and abdominal organs removed. ATLAS Plates 7.6d, 7.10d,e; Scan 11e,f

(a)

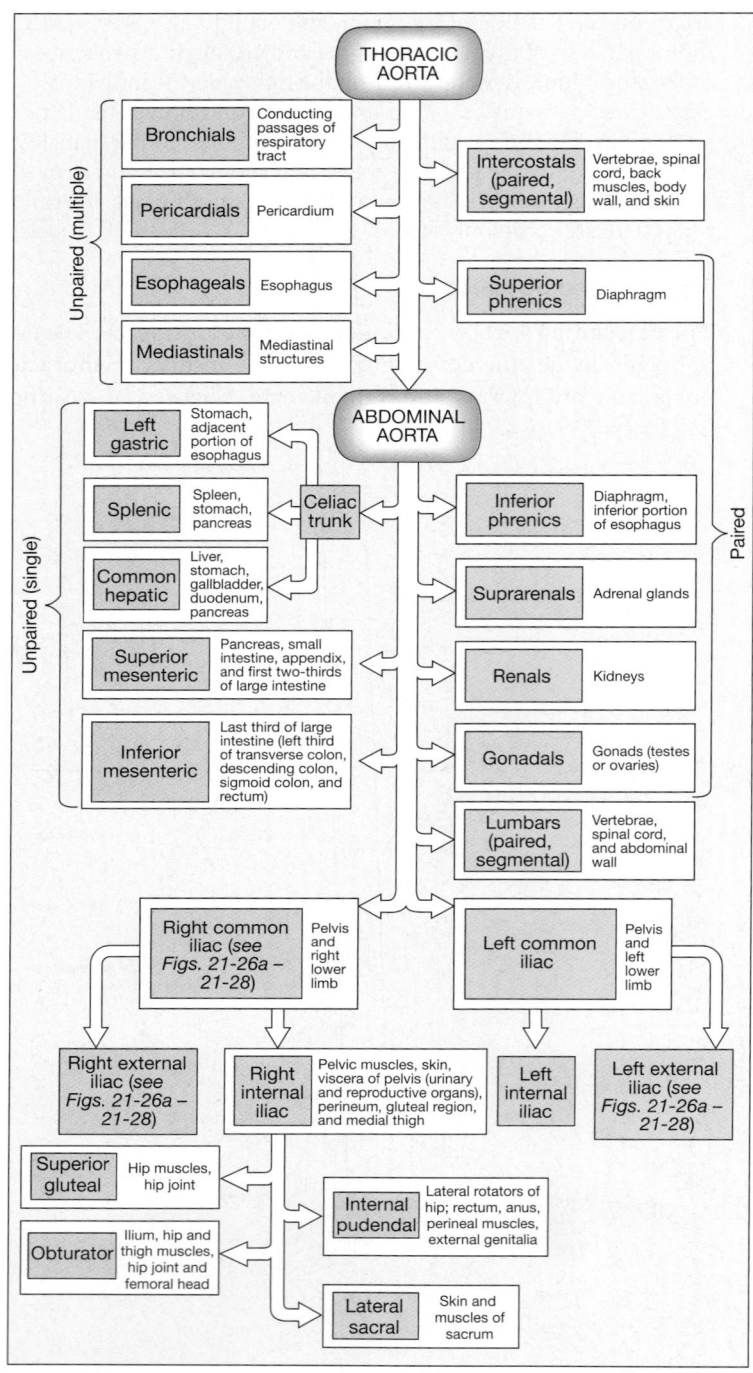

(b)

●**FIGURE 21–26**
Major Arteries of the Trunk (continued). **(b)** A flowchart.

THE THORACIC AORTA The thoracic aorta begins at the level of vertebra T_5 and penetrates the diaphragm at the level of vertebra T_{12}. It travels within the mediastinum, on the posterior thoracic wall, slightly to the left of the vertebral column. This vessel supplies blood to branches that service the tissues and organs of the mediastinum, the muscles of the chest and the diaphragm, and the thoracic spinal cord.

The branches of the thoracic aorta are anatomically grouped as either visceral or parietal:

- *Visceral branches* supply the organs of the chest: The **bronchial arteries** supply the nonrespiratory tissues of the lungs, the **pericardial arteries** supply the pericardium, the **esophageal arteries** supply the esophagus, and the **mediastinal arteries** supply the tissues of the mediastinum.
- *Parietal branches* supply the chest wall: The **intercostal arteries** supply the chest muscles and the vertebral column area, and the **superior phrenic** (FREN-ik) **arteries** deliver blood to the superior surface of the diaphragm, which separates the thoracic and abdominopelvic cavities.

The branches of the thoracic aorta are diagrammed in Figure 21–26●.

THE ABDOMINAL AORTA The abdominal aorta, which begins immediately inferior to the diaphragm, is a continuation of the thoracic aorta (Figure 21–26a●). Descending slightly to the left of the vertebral column but posterior to the peritoneal cavity, the abdominal aorta is commonly surrounded by a cushion of adipose tissue. At the level of vertebra L_4, it splits into two major arteries—the *left* and *right common iliac arteries*—that supply deep pelvic structures and the lower limbs. The region where the abdominal aorta splits is called the *terminal segment of the aorta*.

The abdominal aorta delivers blood to all the abdominopelvic organs and structures. The major branches to visceral organs are unpaired; they arise on the anterior surface of the abdominal aorta and extend into the mesenteries. By contrast, branches to the body wall, the kidneys, the urinary bladder, and other structures outside the abdominopelvic cavity are paired, and originate along the lateral surfaces of the abdominal aorta. Figure 21–26● shows the major arteries of the trunk after the removal of most thoracic and abdominal organs. Figure 21–27● gives the distribution of those arteries to abdominopelvic organs.

The abdominal aorta gives rise to three unpaired arteries (Figures 21–26 and 21–27●).

1. The **celiac** (SĒ-lē-ak) **trunk** delivers blood to the liver, stomach, and spleen. The celiac trunk divides into three branches: (a) the **left gastric artery**, which supplies the stomach and the inferior portion of the esophagus, (b) the **splenic artery**, which supplies the spleen and arteries to the stomach (*left gastroepiploic artery*) and pancreas (*pancreatic arteries*), and (c) the **common hepatic artery**, which supplies arteries to the liver (*hepatic artery proper*), stomach (*right gastric artery*), gallbladder (*cystic artery*), and duodenal area (*gastroduodenal, right gastroepiploic, and superior pancreaticoduodenal arteries*).

2. The **superior mesenteric** (mez-en-TER-ik) **artery** arises about 2.5 cm inferior to the celiac trunk to supply arteries to the pancreas and duodenum (*inferior pancreaticoduodenal artery*), small intestine (*intestinal arteries*), and most of the large intestine (*right and middle colic and the ileocolic arteries*).

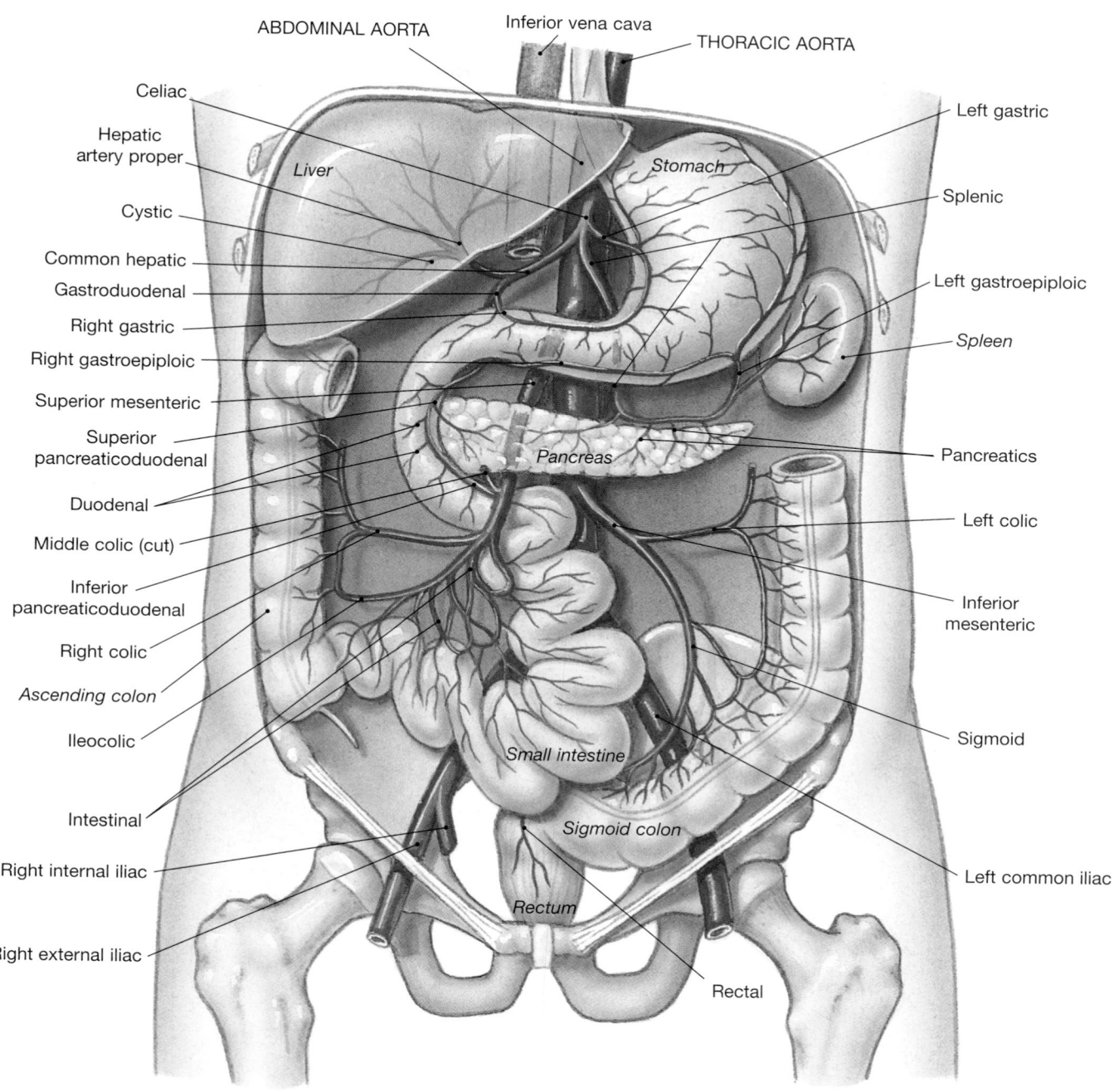

●**FIGURE 21-27**
Arteries Supplying the Abdominopelvic Organs. *(See also Figure 24-23, p. 911.)* ATLAS Plates 7.7g–i, 7.8c,e, 8.2; Scan 11e,f

3. The **inferior mesenteric artery** arises about 5 cm superior to the terminal aorta and delivers blood to the terminal portions of the colon (*left colic* and *sigmoid arteries*) and the rectum (*rectal arteries*).

The abdominal aorta also gives rise to five paired arteries (Figure 21–26●):

1. The **inferior phrenic arteries**, which supply the inferior surface of the diaphragm and the inferior portion of the esophagus.

2. The **suprarenal arteries**, which originate on either side of the aorta near the base of the superior mesenteric artery. Each

suprarenal artery supplies one adrenal gland, which caps the superior part of a kidney.

3. The short (about 7.5 cm) **renal arteries**, which arise along the posterolateral surface of the abdominal aorta, about 2.5 cm (1 in.) inferior to the superior mesenteric artery, and travel posterior to the peritoneal lining to reach the adrenal glands and kidneys. We shall consider the branches of the renal arteries in Chapter 26.

4. The **gonadal** (gō-NAD-al) **arteries**, which originate between the superior and inferior mesenteric arteries. In males, they are called *testicular arteries* and are long, thin arteries that supply blood to the testes and scrotum. In females, they are termed

ovarian arteries and supply blood to the ovaries, uterine tubes, and uterus. The distribution of gonadal vessels (both arteries and veins) differs by gender; we shall describe the differences in Chapter 28.

5. Small **lumbar arteries**, which arise on the posterior surface of the aorta and supply the vertebrae, spinal cord, and abdominal wall.

Arteries of the Pelvis and Lower Limbs

Near the level of vertebra L_4, the terminal segment of the abdominal aorta divides to form a pair of elastic arteries: the **right** and **left common iliac** (IL-ē-ak) **arteries** and the small **middle sacral artery** (Figure 21–26●). These arteries, which carry blood to the pelvis and lower limbs (Figure 21–28●), descend posterior to the cecum and sigmoid colon along the inner surface of the ilium. At the level of the lumbosacral joint, each common iliac divides to form an **internal iliac artery** and an **external iliac artery** (Figure 21–27●). The internal iliac arteries enter the pelvic cavity to supply the urinary bladder, the internal and external walls of the pelvis, the external genitalia, the medial side of the thigh, and, in females, the uterus and vagina. The major tributaries of the internal iliac artery are the *superior gluteal*, *internal pudendal*, *obturator*, and *lateral sacral arteries*. The external iliac arteries supply blood to the lower limbs and are much larger in diameter than the internal iliac arteries.

ARTERIES OF THE THIGH AND LEG Each external iliac artery crosses the surface of an iliopsoas muscle and penetrates the abdominal wall midway between the anterior superior iliac spine and the

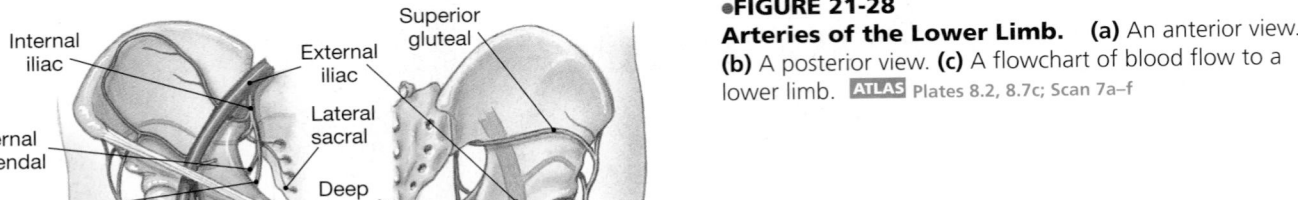

●**FIGURE 21-28**
Arteries of the Lower Limb. **(a)** An anterior view. **(b)** A posterior view. **(c)** A flowchart of blood flow to a lower limb. **ATLAS** Plates 8.2, 8.7c; Scan 7a–f

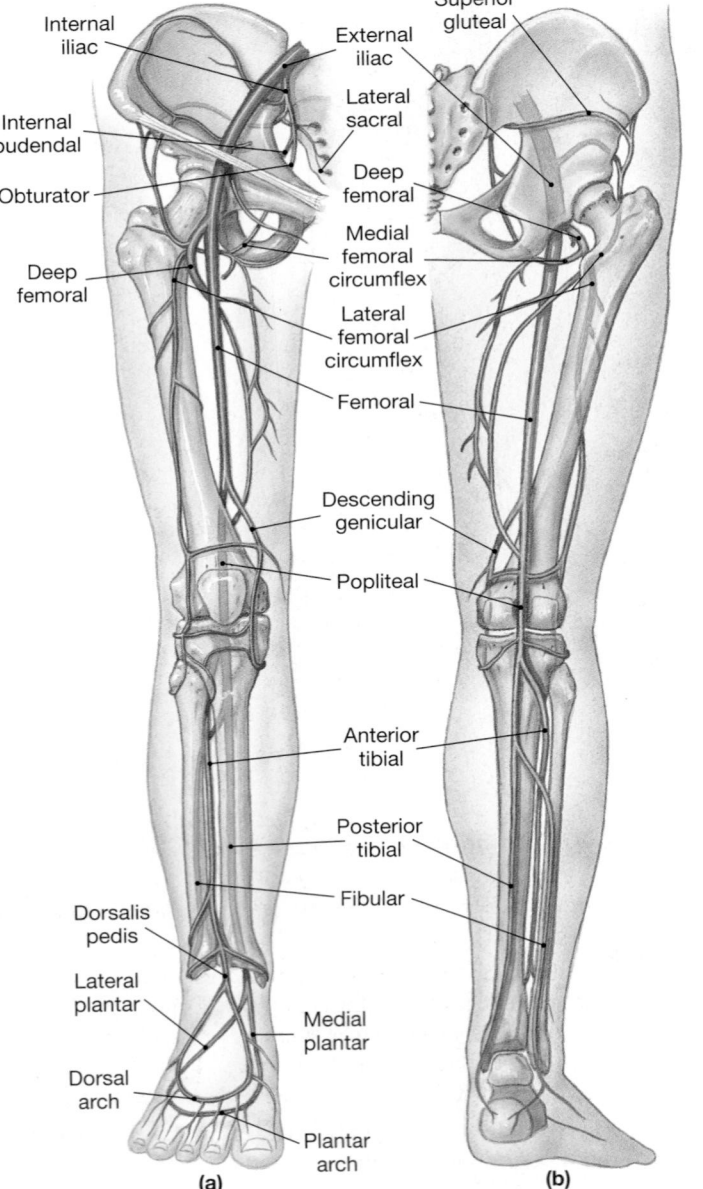

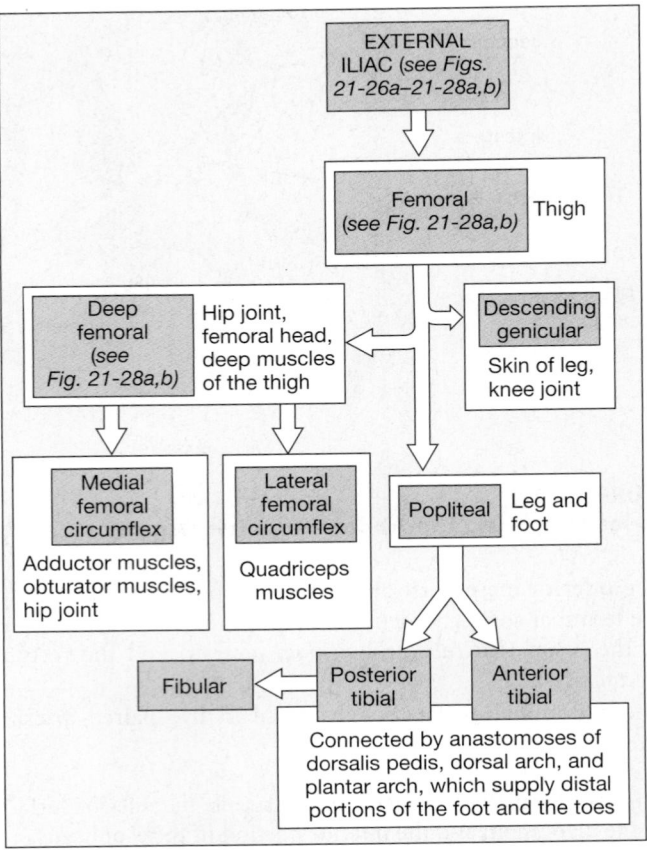

(a)　　　　(b)　　　　(c)

pubic symphysis on that side. It emerges on the anterior, medial surface of the thigh as the **femoral artery**. Roughly 5 cm distal to the emergence of the femoral artery, the **deep femoral artery** branches off its lateral surface (Figure 21–28a,b●). The deep femoral artery, which gives rise to the *femoral circumflex arteries*, supplies blood to the ventral and lateral regions of the skin and deep muscles of the thigh.

The femoral artery continues inferiorly and posterior to the femur. As it approaches the knee, it gives rise to the *descending genicular artery*, which supplies the area around the knee. At the popliteal fossa, posterior to the knee joint, the femoral artery becomes the **popliteal** (pop-LIT-ē-al) **artery**, which then branches to form the **posterior** and **anterior tibial arteries**. The posterior tibial artery gives rise to the **fibular artery**, or *peroneal artery*, before continuing inferiorly along the posterior surface of the tibia. The anterior tibial artery passes between the tibia and fibula, emerging on the anterior surface of the tibia. As it descends toward the foot, the anterior tibial artery provides blood to the skin and muscles of the anterior portion of the leg.

ARTERIES OF THE FOOT When it reaches the ankle, the anterior tibial artery becomes the **dorsalis pedis artery**, which then branches repeatedly, supplying the ankle and dorsal portion of the foot (Figure 21–28a,b●).

As it reaches the ankle, the posterior tibial artery divides to form the **medial** and **lateral plantar arteries**, which supply blood to the plantar surface of the foot. These arteries are connected to the dorsalis pedis artery through a pair of anastomoses. The arrangement produces a **dorsal arch** (*arcuate arch*) and a **plantar arch**; small arteries branching off these arches supply the distal portions of the foot and the toes.

✓ A blockage of which branch from the aortic arch would interfere with blood flow to the left arm?

✓ Why would a compression of the common carotid arteries cause a person to lose consciousness?

✓ Grace is in an automobile accident, and her celiac trunk is ruptured. Which organs will be affected most directly by this injury?

Answers start on page Q-1

 Review the arteries of the body by viewing the dissections on the **Anatomy CD-ROM**: Cardiovascular System.

☐ SYSTEMIC VEINS

Veins collect blood from each of the tissues and organs of the body by means of an elaborate venous network that drains into the right atrium of the heart via the superior and inferior venae cavae (Figure 21–29●). The branching pattern of peripheral veins is much more variable than is the branching pattern of arteries. The discussion that follows is based on the most common arrangement of veins. Complementary arteries and veins commonly run side by side, and in many cases they have comparable names.

One significant difference between the arterial and venous systems concerns the distribution of major veins in the neck and limbs. Arteries in these areas are located deep beneath the skin, protected by bones and surrounding soft tissues. In contrast, the neck and limbs generally have two sets of peripheral veins, one superficial and the other deep. This dual venous drainage is important for controlling body temperature. In hot weather, venous blood flows in superficial veins, where heat loss can occur; in cold weather, blood is routed to the deep veins to minimize heat loss.

The Superior Vena Cava

All the body's systemic veins (except the cardiac veins) drain into either the superior vena cava or the inferior vena cava. The **superior vena cava** (**SVC**) receives blood from the tissues and organs of the head, neck, chest, shoulders, and upper limbs.

VENOUS RETURN FROM THE CRANIUM Numerous veins drain the cerebral hemispheres. The *superficial cerebral veins* and small veins of the brain stem empty into a network of dural sinuses (Figure 21–30a●, p. 763), including the *superior* and *inferior sagittal sinuses*, the *petrosal sinuses*, the *occipital sinus*, the *left* and *right transverse sinuses*, and the *straight sinus* (Figure 21–30b●). The largest, the **superior sagittal sinus**, is in the falx cerebri (see Figure 14–4●, p. 469). Most of the *inferior cerebral veins* converge within the brain to form the **great cerebral vein**, which delivers blood from the interior of the cerebral hemispheres and the choroid plexus to the **straight sinus**. Other cerebral veins drain into the **cavernous sinus** with numerous small veins from the orbit. Blood from the cavernous sinus reaches the internal jugular vein through the petrosal sinuses.

The venous sinuses converge within the dura mater in the region of the lambdoid suture. The left and right transverse sinuses converge at the base of the petrous part of the temporal bone. There, the transverse sinuses form the **sigmoid sinus**, which penetrates the jugular foramen and leaves the skull as the **internal jugular vein**, descending parallel to the common carotid artery in the neck (p. 754).

Vertebral veins drain the cervical spinal cord and the posterior surface of the skull. These vessels descend within the transverse foramina of the cervical vertebrae, in company with the vertebral arteries. The vertebral veins empty into the *brachiocephalic veins* of the chest (discussed later in the chapter).

SUPERFICIAL VEINS OF THE HEAD AND NECK The superficial veins of the head collect to form the **temporal**, **facial**, and **maxillary veins** (Figure 21–30b●). The temporal vein and the maxillary vein drain into the **external jugular vein**. The facial vein drains into the internal jugular vein. A broad anastomosis between the external and internal jugular veins at the angle of the mandible

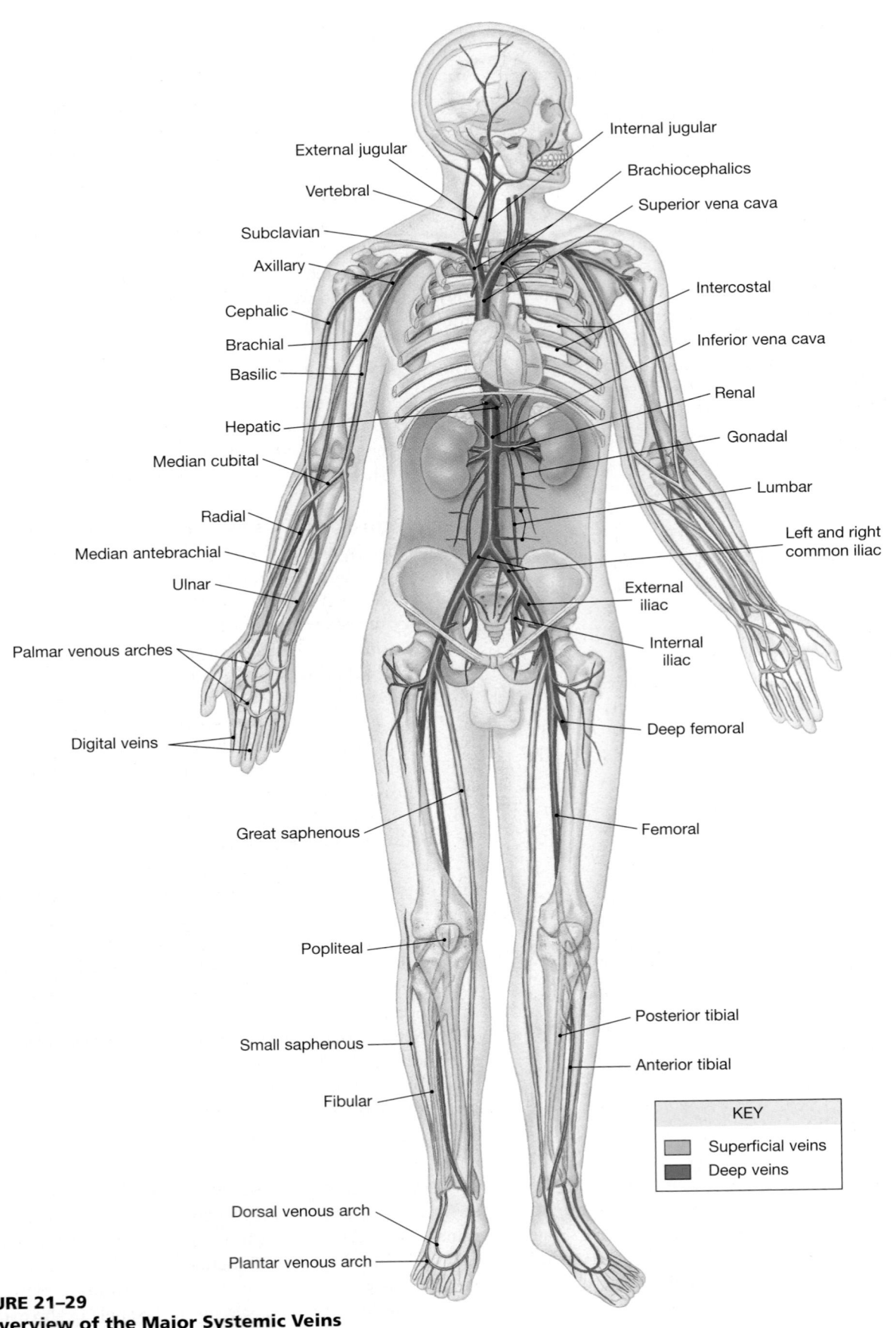

External jugular

Vertebral

Subclavian

Axillary

Cephalic

Brachial

Basilic

Hepatic

Median cubital

Radial

Median antebrachial

Ulnar

Palmar venous arches

Digital veins

Great saphenous

Popliteal

Small saphenous

Fibular

Dorsal venous arch

Plantar venous arch

Internal jugular

Brachiocephalics

Superior vena cava

Intercostal

Inferior vena cava

Renal

Gonadal

Lumbar

Left and right common iliac

External iliac

Internal iliac

Deep femoral

Femoral

Posterior tibial

Anterior tibial

KEY
Superficial veins
Deep veins

●FIGURE 21–29
An Overview of the Major Systemic Veins

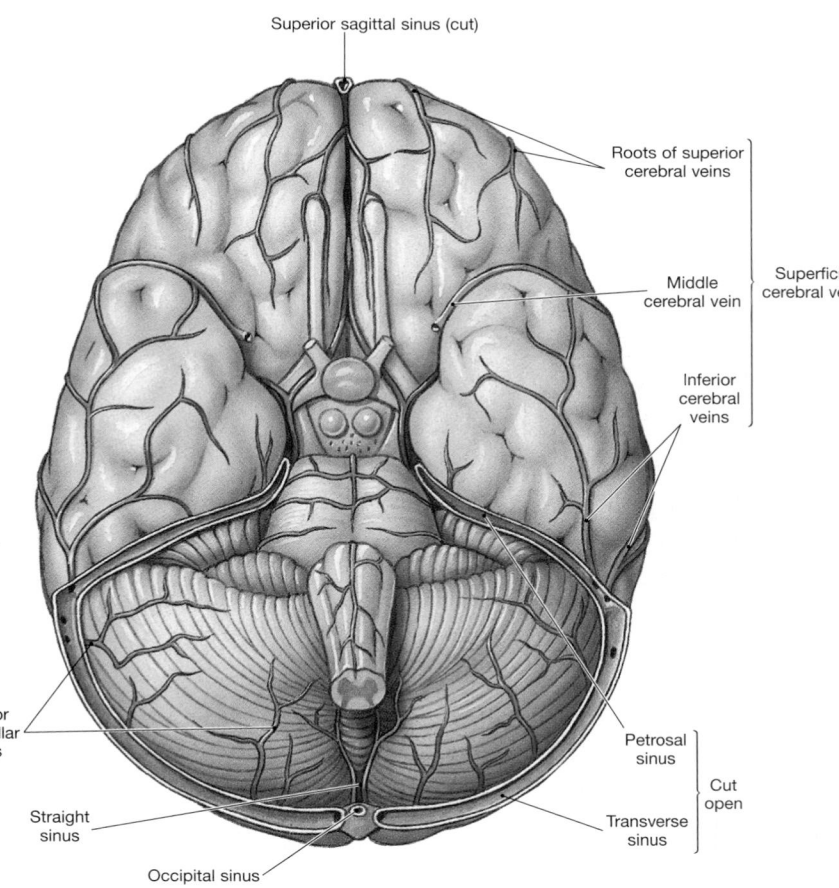

Superior sagittal sinus (cut)

Roots of superior cerebral veins

Middle cerebral vein

Superficial cerebral veins

Inferior cerebral veins

Inferior cerebellar veins

Straight sinus

Occipital sinus

Petrosal sinus

Cut open

Transverse sinus

(a)

provides dual venous drainage of the face, scalp, and cranium. The external jugular vein descends toward the chest just deep to the skin on the anterior surface of the sternocleidomastoid muscle. Posterior to the clavicle, the external jugular vein empties into the *subclavian vein*. In healthy individuals, the external jugular vein is easily palpable, and a *jugular venous pulse* (*JVP*) is sometimes detectable at the base of the neck.

VENOUS RETURN FROM THE UPPER LIMBS The **digital veins** empty into the **superficial** and **deep palmar veins** of the hand, which are interconnected to form the **palmar venous arches**. The superficial arch empties into the **cephalic vein**, which ascends along the radial side of the forearm; the **median antebrachial vein**; and the **basilic vein**, which ascends on the ulnar side (Figure 21–31●). Anterior to the elbow is the superficial **median cubital vein**, which passes from the cephalic vein, medially and at an oblique angle, to connect to the basilic vein. (The median cubital is the vein from which venous blood samples are typically collected.) From the elbow, the basilic vein passes superiorly along the medial surface of the biceps brachii muscle.

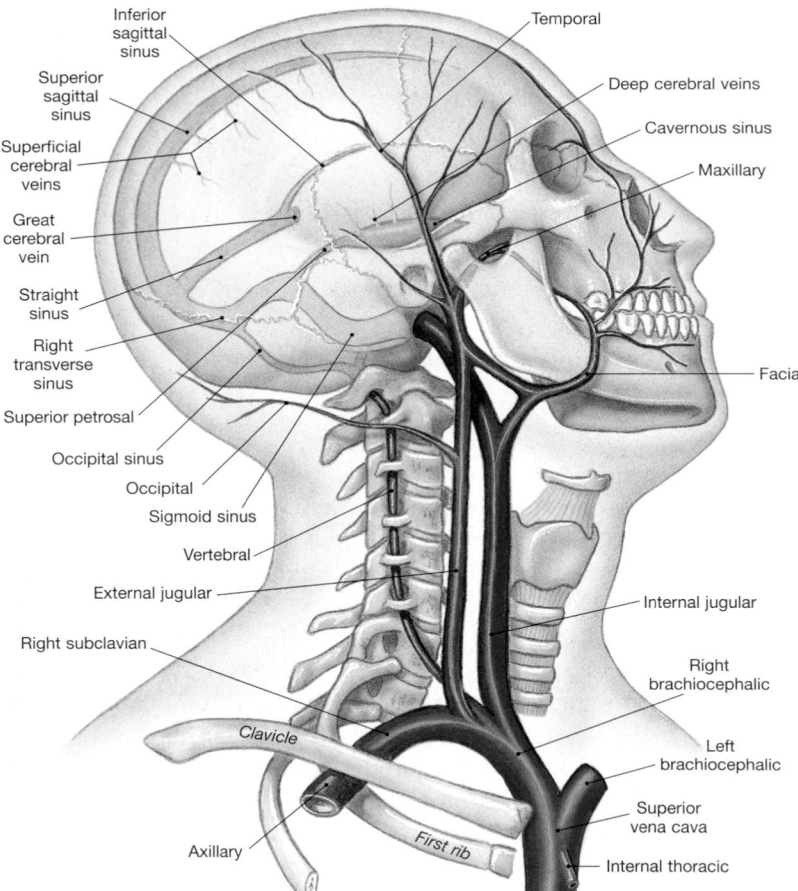

Inferior sagittal sinus

Superior sagittal sinus

Superficial cerebral veins

Great cerebral vein

Straight sinus

Right transverse sinus

Superior petrosal

Occipital sinus

Occipital

Sigmoid sinus

Vertebral

External jugular

Right subclavian

Clavicle

Axillary

First rib

Temporal

Deep cerebral veins

Cavernous sinus

Maxillary

Facial

Internal jugular

Right brachiocephalic

Left brachiocephalic

Superior vena cava

Internal thoracic

(b)

●**FIGURE 21–30**
Major Veins of the Head, Neck, and Brain.
(a) An inferior view of the brain, showing the venous distribution. For the relationship of these veins to meningeal layers, *see Figure 14–4, p. 469.* **(b)** Veins draining the brain and the superficial and deep portions of the head and neck.
ATLAS Plates 5.2, 5.3a–c

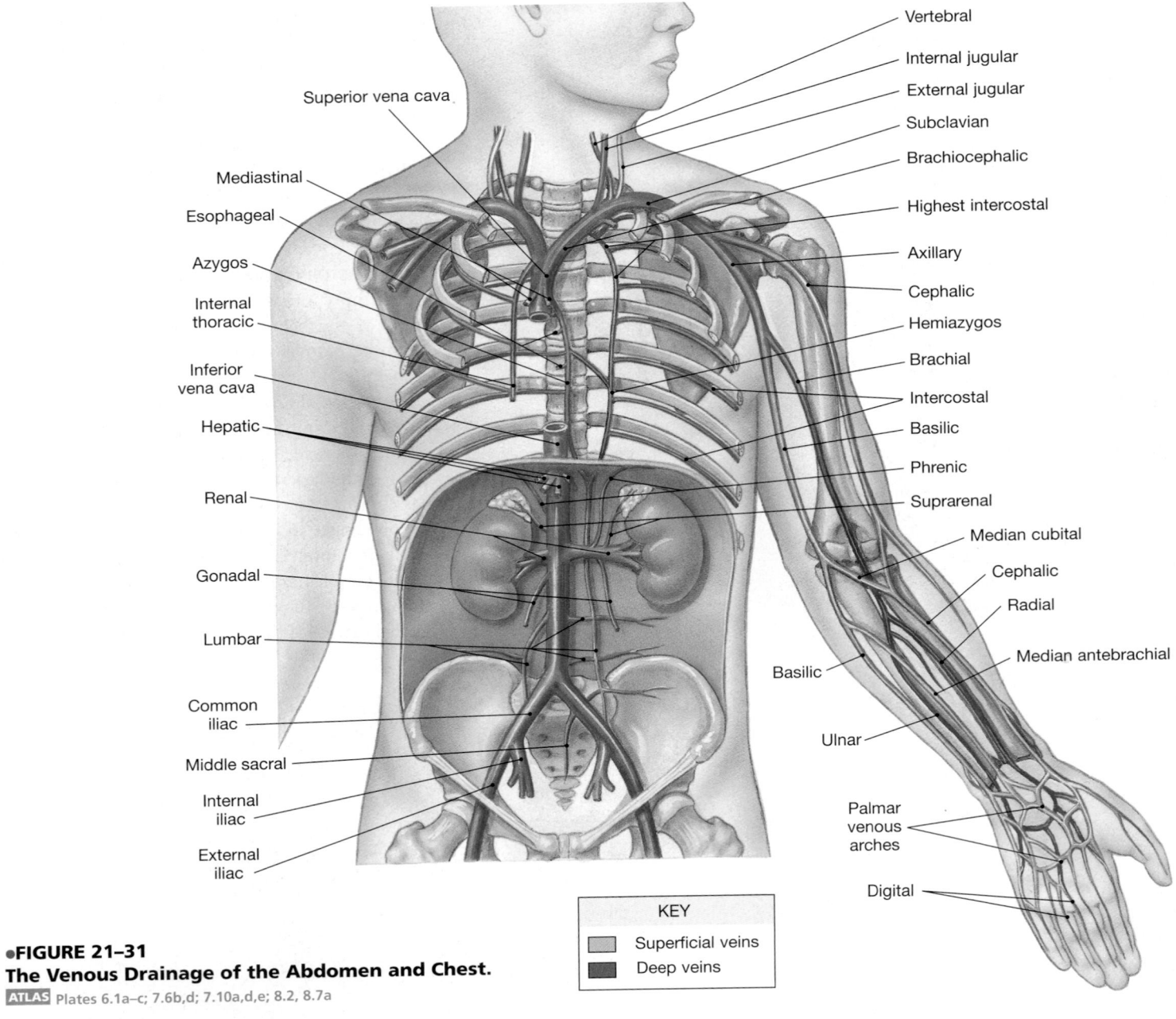

Vertebral

Internal jugular

External jugular

Subclavian

Brachiocephalic

Highest intercostal

Axillary

Cephalic

Hemiazygos

Brachial

Intercostal

Basilic

Phrenic

Suprarenal

Median cubital

Cephalic

Radial

Median antebrachial

Basilic

Ulnar

Palmar venous arches

Digital

Superior vena cava

Mediastinal

Esophageal

Azygos

Internal thoracic

Inferior vena cava

Hepatic

Renal

Gonadal

Lumbar

Common iliac

Middle sacral

Internal iliac

External iliac

KEY	
	Superficial veins
	Deep veins

•FIGURE 21–31
The Venous Drainage of the Abdomen and Chest.
ATLAS Plates 6.1a–c; 7.6b,d; 7.10a,d,e; 8.2, 8.7a

The deep palmar veins drain into the **radial vein** and the **ulnar vein**. After crossing the elbow, these veins fuse to form the **brachial vein**, running parallel to the brachial artery. As the brachial vein continues toward the trunk, it merges with the basilic vein and becomes the **axillary vein**, which enters the axilla.

FORMATION OF THE SUPERIOR VENA CAVA The cephalic vein joins the axillary vein on the lateral surface of the first rib, forming the **subclavian vein**, which continues into the chest. The subclavian vein passes superior to the first rib and along the superior margin of the clavicle, to merge with the external and internal jugular veins of that side. This fusion creates the **brachiocephalic vein**,

or *innominate vein*, which penetrates the body wall and enters the thoracic cavity.

Each brachiocephalic vein receives blood from the **vertebral vein** of the same side, which drains the back of the skull and spinal cord. Near the heart, at the level of the first and second ribs, the left and right brachiocephalic veins combine, creating the superior vena cava. Close to the point of fusion, the **internal thoracic vein** empties into the brachiocephalic vein.

The **azygos** (AZ-i-gos) **vein** is the major tributary of the superior vena cava. This vein ascends from the lumbar region over the right side of the vertebral column to invade the thoracic cavity through the diaphragm. The azygos vein joins the superior vena cava at the level of vertebra T_2. On the left side, the azygos re-

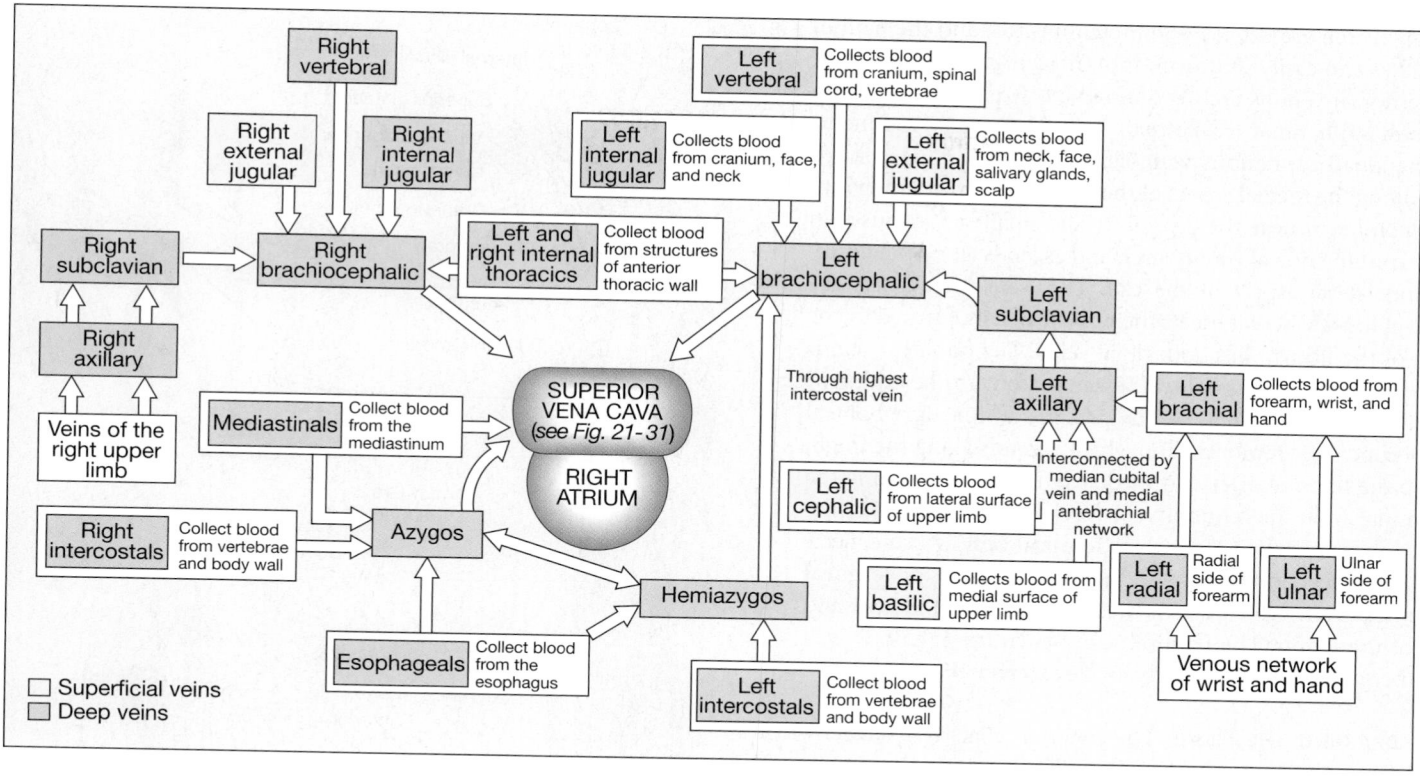

(a) Tributaries of the SVC

●**FIGURE 21–32**
A Flowchart of Circulation to the Superior and Inferior Venae Cavae

ceives blood from the smaller **hemiazygos vein**, which in many people also drains into the left brachiocephalic vein through the *highest intercostal vein.*

The azygos and hemiazygos veins are the chief collecting vessels of the thorax. They receive blood from (1) **intercostal veins**, which in turn receive blood from the chest muscles; (2) **esophageal veins**, which drain blood from the inferior portion of the esophagus; and (3) smaller veins draining other mediastinal structures.

Figure 21–32a● diagrams the venous tributaries of the superior vena cava.

The Inferior Vena Cava

The **inferior vena cava** (**IVC**) collects most of the venous blood from organs inferior to the diaphragm. (A small amount reaches the superior vena cava via the azygos and hemiazygos veins.)

VEINS DRAINING THE LOWER LIMBS Blood leaving capillaries in the sole of each foot collects into a network of **plantar veins**, which supply the **plantar venous arch**. The plantar network provides blood to the deep veins of the leg: the **anterior tibial vein**, the **posterior tibial vein**, and the **fibular vein** (Figure 21–33a●). The **dorsal venous arch** collects blood from capillaries on the superior surface of the foot and the **digital veins** of the toes. The plantar arch and

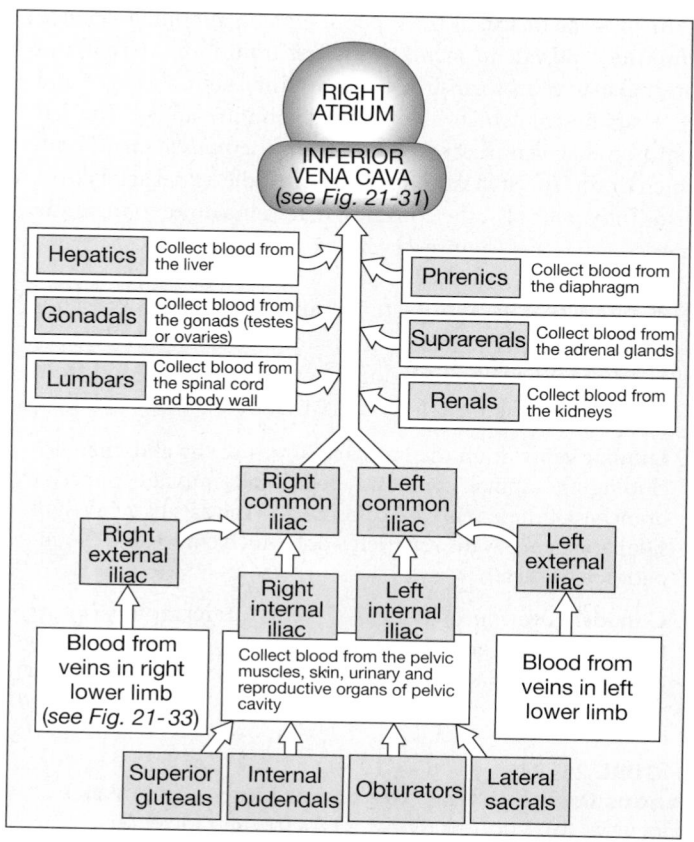

(b) Tributaries of the IVC

the dorsal arch are extensively interconnected, and the path of blood flow can easily shift from superficial to deep veins.

The dorsal venous arch is drained by two superficial veins: the **great saphenous** (sa-Fē-nus) **vein** (*saphenes*, prominent) and the **small saphenous vein**. The great saphenous vein ascends along the medial aspect of the leg and thigh, draining into the *femoral vein* near the hip joint. The small saphenous vein arises from the dorsal venous arch and ascends along the posterior and lateral aspect of the calf. This vein then enters the popliteal fossa, where it meets the **popliteal vein**, formed by the union of the fibular and both tibial veins. The popliteal vein is easily palpated in the popliteal fossa adjacent to the adductor magnus muscle (Figure 21–33b●). At the femur, the popliteal vein becomes the **femoral vein**, which ascends along the thigh, next to the femoral artery. Immediately before penetrating the abdominal wall, the femoral vein receives blood from (1) the great saphenous vein, (2) the **deep femoral vein**, which collects blood from deeper structures in the thigh, and (3) the **femoral circumflex vein**, which drains the region around the neck and head of the femur. The femoral vein penetrates the body wall and emerges in the pelvic cavity as the **external iliac vein**.

VEINS DRAINING THE PELVIS The external iliac veins receive blood from the lower limbs, the pelvis, and the lower abdomen. As the left and right external iliac veins cross the inner surface of the ilium, they are joined by the **internal iliac veins**, which drain the pelvic organs. The internal iliac veins are formed by the fusion of the *superior gluteal, internal pudendal, obturator*, and *lateral sacral veins*. The union of external and internal iliac veins forms the **common iliac vein**, the right and left branches of which ascend at an oblique angle. The left common iliac vein receives blood from the *middle sacral vein*, which drains the area supplied by the middle sacral artery. Anterior to vertebra L_5, the common iliac veins unite to form the inferior vena cava (Figure 21–31●).

VEINS DRAINING THE ABDOMEN The inferior vena cava ascends posterior to the peritoneal cavity, parallel to the aorta. The abdominal portion of the inferior vena cava collects blood from six major veins (Figures 21–31 and 21–32b●):

1. **Lumbar veins** drain the lumbar portion of the abdomen, including the spinal cord and body wall muscles. Superior branches of these veins are connected to the azygos vein (right side) and hemiazygos vein (left side), which empty into the superior vena cava.

2. **Gonadal** (*ovarian* or *testicular*) **veins** drain the ovaries or testes. The right gonadal vein empties into the inferior vena cava; the left gonadal vein generally drains into the left renal vein.

●**FIGURE 21–33**
Venous Drainage from the Lower Limb. **(a)** An anterior view. **(b)** A posterior view. **(c)** A flowchart of venous circulation to a lower limb. ATLAS Plate 8.7a–c; Scan 7a–f

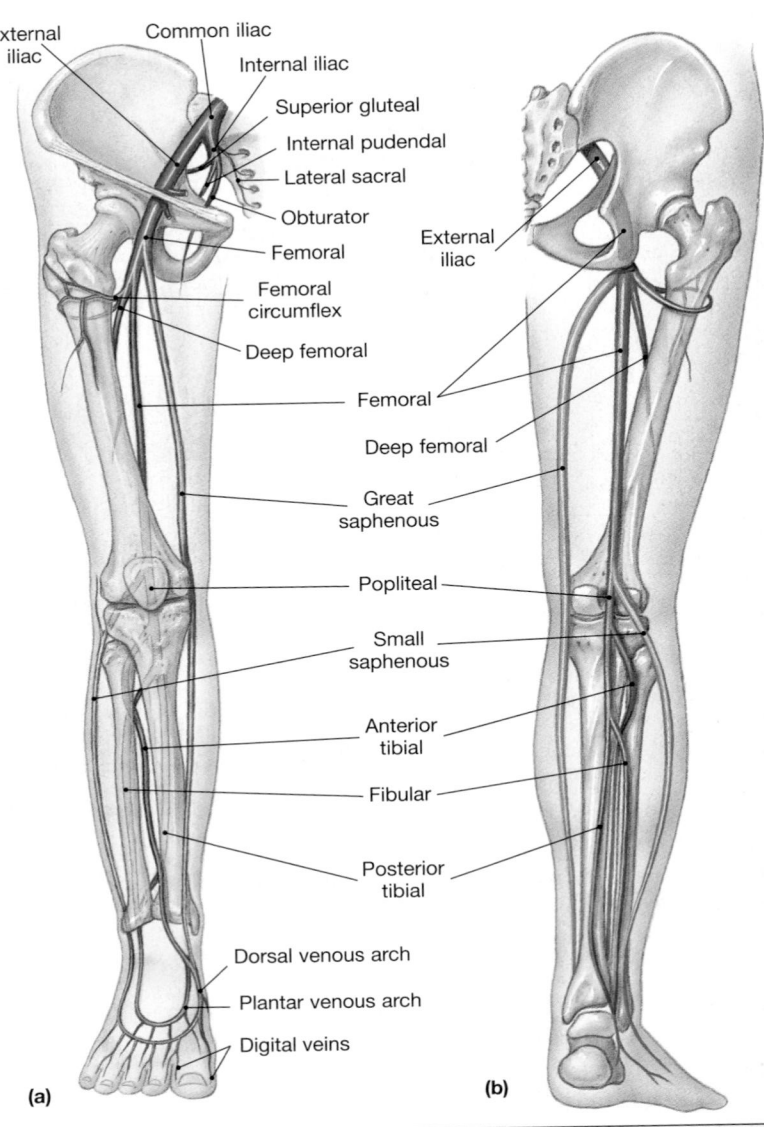

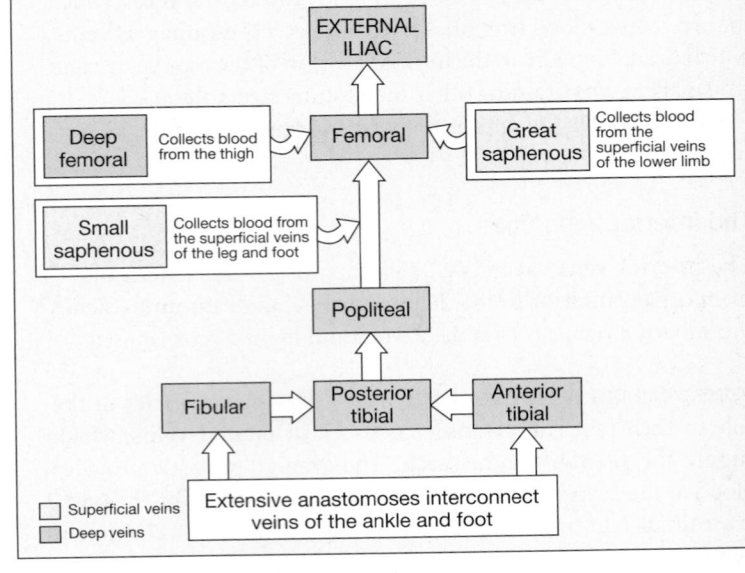

3. **Hepatic veins** from the liver empty into the inferior vena cava at the level of vertebra T_{10}.

4. **Renal veins**, the largest tributaries of the inferior vena cava, collect blood from the kidneys.

5. **Suprarenal veins** drain the adrenal glands. In most people, only the right suprarenal vein drains into the inferior vena cava; the left suprarenal vein drains into the left renal vein.

6. **Phrenic veins** drain the diaphragm. Only the right phrenic vein drains into the inferior vena cava; the left drains into the left renal vein.

Figure 21–32b• diagrams the tributaries of the inferior vena cava.

The Hepatic Portal System

The liver is the only digestive organ that drains directly into the inferior vena cava. Blood leaving the capillaries supplied by the celiac, superior, and inferior mesenteric arteries flows into the **hepatic portal system** (Figure 21–34•). A blood vessel connecting two capillary beds is called a *portal vessel*; the network is a *portal system*. p. 615

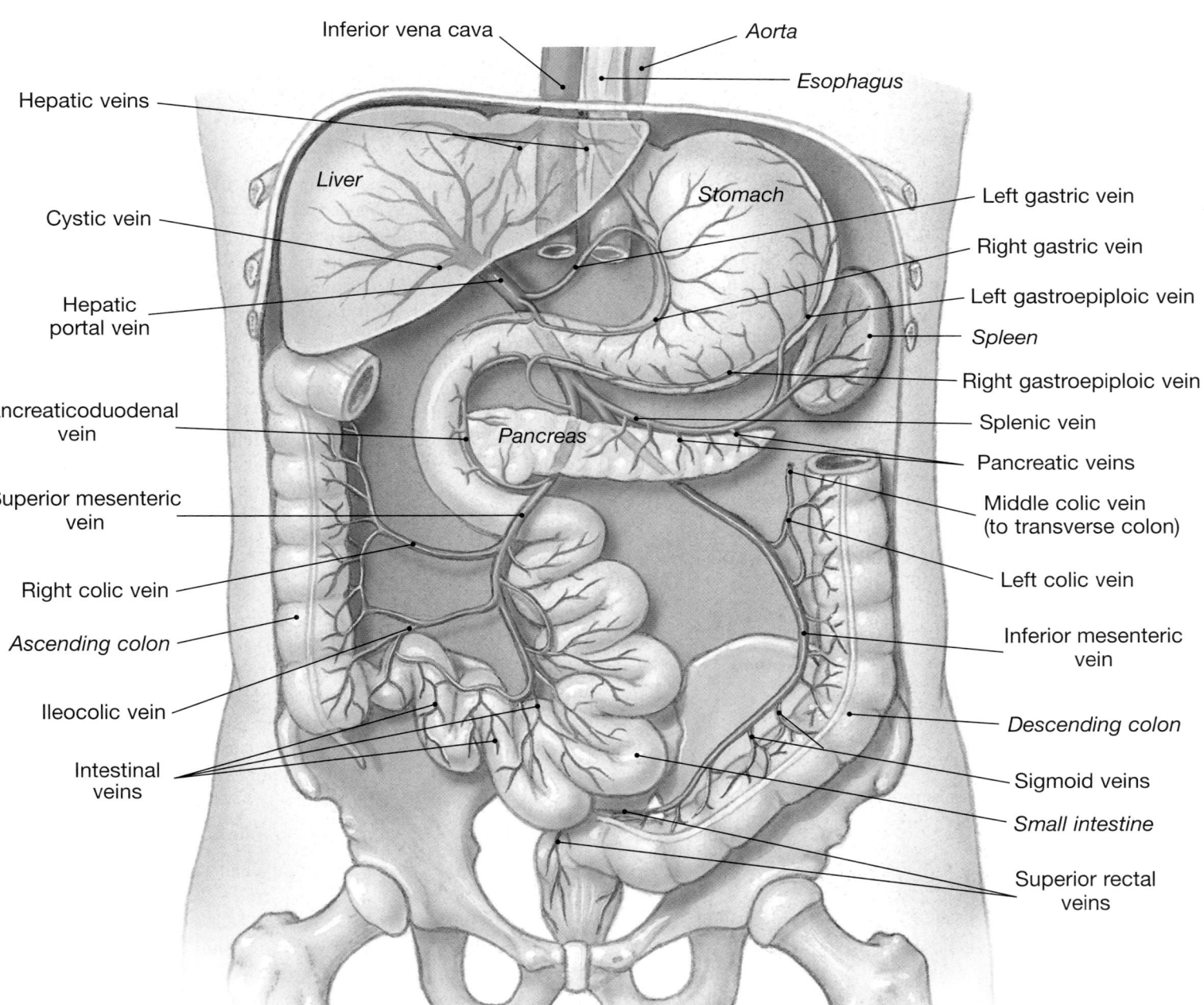

•**FIGURE 21–34**
The Hepatic Portal System. See also *Figure 24–23, p. 911.* **ATLAS** Plates 7.8a–c,e; 7.9a,b

Blood flowing in the hepatic portal system is quite different from that in other systemic veins, because hepatic portal vessels contain substances absorbed by the stomach and intestines. For example, levels of blood glucose and amino acids in the hepatic portal vein often exceed those found anywhere else in the cardiovascular system. The hepatic portal system delivers these and other absorbed compounds directly to the liver for storage, metabolic conversion, or excretion. After passing through liver sinusoids, blood collects in the hepatic veins, which empty into the inferior vena cava. Blood from the digestive organs goes to the liver first, so the composition of the blood in the systemic circuit is fairly stable, regardless of any digestive activities.

The hepatic portal system begins in the capillaries of the digestive organs and ends as the hepatic portal vein discharges blood into liver sinusoids. Tributaries of the hepatic portal vein include (Figure 21–34●):

- The **inferior mesenteric vein**, which collects blood from capillaries along the inferior portion of the large intestine. Its tributaries include the *left colic vein* and the *superior rectal veins*, which drain the descending colon, sigmoid colon, and rectum.
- The **splenic vein**, formed by the union of the inferior mesenteric vein and veins from the spleen, the lateral border of the stomach (*left gastroepiploic vein*), and the pancreas (*pancreatic veins*).
- The **superior mesenteric vein**, which collects blood from veins draining the stomach (*right gastroepiploic vein*), the small intestine (*intestinal* and *pancreaticoduodenal veins*), and two-thirds of the large intestine (*ileocolic, right colic,* and *middle colic veins*).

The hepatic portal vein forms through the fusion of the superior mesenteric and splenic veins. The superior mesenteric vein normally contributes the greater volume of blood and most of the nutrients. As it proceeds, the hepatic portal vein receives blood from the left and right **gastric veins**, which drain the medial border of the stomach, and from the **cystic vein**, emanating from the gallbladder.

21–8 FETAL CIRCULATION

Objectives

- Identify the differences between fetal and adult circulation patterns.
- Describe the changes that occur in the patterns of blood flow at birth.

The fetal and adult cardiovascular systems exhibit significant differences, reflecting different sources of respiratory and nutritional support. Most strikingly, the embryonic lungs are collapsed and nonfunctional, and the digestive tract has nothing to digest. The nutritional and respiratory needs of the fetus are provided by diffusion across the placenta.

☐ PLACENTAL BLOOD SUPPLY

Fetal patterns of blood flow are diagrammed in Figure 21–35●. Blood flow to the placenta is provided by a pair of **umbilical arteries**, which arise from the internal iliac arteries and enter the umbilical cord. Blood returns from the placenta in the single **umbilical vein**, bringing oxygen and nutrients to the developing fetus. The umbilical vein drains into the **ductus venosus**, a vascular connection to an intricate network of veins within the developing liver. The ductus venosus collects blood from the veins of the liver and from the umbilical vein and empties into the inferior vena cava. When the placental connection is broken at birth, blood flow ceases along the umbilical vessels, and they soon degenerate. However, remnants of these vessels persist throughout life as fibrous cords.

☐ CIRCULATION IN THE HEART AND GREAT VESSELS

One of the most interesting aspects of circulatory development reflects the differences between the life of an embryo or fetus and that of an infant. Throughout embryonic and fetal life, the lungs are collapsed; yet, after delivery, the newborn infant must be able to extract oxygen from inspired air rather than across the placenta.

AM Embryology Summary 16: The Development of the Cardiovascular System

Although the interatrial and interventricular septa develop early in fetal life, the interatrial partition remains functionally incomplete until birth. The **foramen ovale**, or *interatrial opening*, is associated with a long flap that acts as a valve. Blood can flow freely from the right atrium to the left atrium, but any backflow will close the valve and isolate the two chambers. Thus, blood can enter the heart at the right atrium and bypass the pulmonary circuit. A second short circuit exists between the pulmonary and aortic trunks. This connection, the **ductus arteriosus**, consists of a short, muscular vessel.

With the lungs collapsed, the capillaries are compressed and little blood flows through the lungs. During diastole, blood enters the right atrium and flows into the right ventricle, but it also passes into the left atrium through the foramen ovale. About 25 percent of the blood arriving at the right atrium bypasses the pulmonary circuit in this way. In addition, over 90 percent of the blood leaving the right ventricle passes through the ductus arteriosus and enters the systemic circuit rather than continuing to the lungs.

☐ CARDIOVASCULAR CHANGES AT BIRTH

At birth, dramatic changes occur. When the infant takes the first breath, the lungs expand, and so do the pulmonary vessels. The resistance in the pulmonary circuit declines suddenly, and blood rushes into the pulmonary vessels. Within a few seconds, rising O_2 levels stimulate the constriction of the ductus arteriosus, isolating the pulmonary and aortic trunks. As pressures rise in the left atrium, the valvular flap closes the foramen ovale. In adults, the interatrial septum bears the *fossa ovalis*, a shallow de-

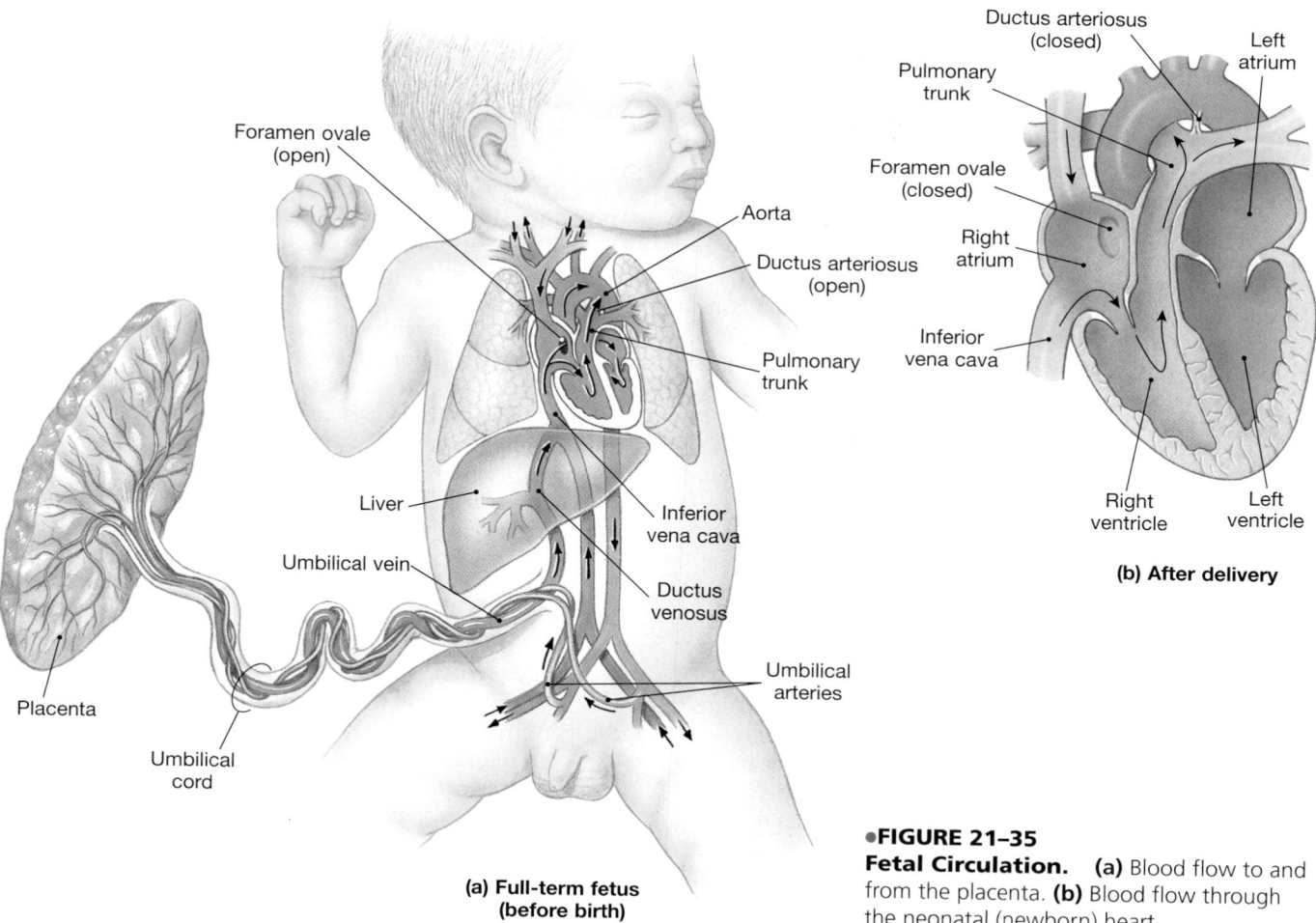

●**FIGURE 21–35**
Fetal Circulation. **(a)** Blood flow to and from the placenta. **(b)** Blood flow through the neonatal (newborn) heart.

pression that marks the site of the foramen ovale. (See Figure 20–6a,c●, p. 689.) The remnants of the ductus arteriosus persist throughout life as the *ligamentum arteriosum*, a fibrous cord.

If the proper circulatory changes do not occur at birth or shortly thereafter, problems will eventually develop. The severity of the problems depends on which connection remains open and on the size of the opening. Treatment may involve the surgical closure of the foramen ovale, the ductus arteriosus, or both. Other forms of congenital heart defects result from abnormal cardiac development or inappropriate connections between the heart and major arteries and veins.

21–9 AGING AND THE CARDIOVASCULAR SYSTEM

Objective

■ Discuss the effects of aging on the cardiovascular system.

The capabilities of your cardiovascular system gradually decline as you age. The major changes are as follows:

- **Age-Related Changes in Blood.** Age-related changes in blood may include (1) a decreased hematocrit; (2) the constriction or blockage of peripheral veins by a *thrombus* (stationary blood clot), which can become detached, pass through the heart, and become wedged in a small artery, commonly in the lungs, causing a *pulmonary embolism*; and (3) a pooling of blood in the veins of the legs because valves are not working effectively.

- **The Aging Heart.** Age-related changes in the heart include (1) a reduction in the maximum cardiac output, (2) changes in the activities of the nodal and conducting cells, (3) a reduction in the elasticity of the fibrous skeleton, (4) a progressive atherosclerosis that can restrict coronary circulation, and (5) the replacement of damaged cardiac muscle cells by scar tissue.

- **Aging and Blood Vessels.** Age-related changes in blood vessels may be linked to arteriosclerosis: (1) The inelastic walls of arteries become less tolerant of sudden pressure increases, which can lead to an *aneurysm*, whose rupture may cause a stroke, myocardial infarction, or massive blood loss (depending on the vessel); (2) calcium salts can be deposited on weakened vascular walls, increasing the risk of a stroke or myocardial infarction; and (3) thrombi can form at atherosclerotic plaques.

Congenital Cardiovascular Problems

Minor individual variations in the vascular network are quite common. For example, very few individuals have identical patterns of venous distribution. Congenital cardiovascular problems serious enough to threaten homeostasis are relatively rare. They generally reflect an abnormal formation of the heart or problems with the interconnections between the heart and the great vessels. Several examples of congenital cardiovascular defects are illustrated in Figure 21–36●. All these conditions can be surgically corrected, although multiple surgeries may be required.

The incomplete closure of the foramen ovale or ductus arteriosus results in similar types of problems. If the foramen ovale remains open, or *patent*, blood recirculates into the pulmonary circuit without entering the left ventricle (Figure 21–36a●). The movement, driven by the relatively high systemic pressure, is called a "left-to-right shunt." Arterial oxyen content is normal, but the left ventricle must work much harder than usual to provide adequate blood flow through the systemic circuit. Hence, pressures rise in the pulmonary circuit. The abnormality may not be immediately apparent, but pulmonary hypertension, pulmonary edema, and cardiac enlargement are eventual results. If the ductus arteriosus remains open, the same basic problems develop as blood ejected by the left ventricle reenters the pulmonary circuit. If valve defects, constricted pulmonary vessels, or other abnormalities occur as well, pulmonary pressures can rise enough to force blood into the systemic circuit through the ductus arteriosus. This movement is called a "right-to-left shunt." Because normal blood oxygenation does not occur, the circulating blood develops a deep red color. The skin then develops the blue tones typical of *cyanosis*, (⚭ p. 161), and the infant is known as a "blue baby."

Ventricular septal defects are openings in the interventricular septum (Figure 21–36b●). These are the most common congenital heart problems, affecting 0.12 percent of newborn infants. The opening between the left and right ventricles has the reverse effect of a connection between the atria: When the more powerful left ventricle beats, it ejects blood into the right ventricle and pulmonary circuit. The end results are the same as for a patent foramen ovale: a left-to-right shunt, with eventual pulmonary hypertension, pulmonary edema, and cardiac enlargement.

The *tetralogy of Fallot* (fa-LŌ) is a complex group of heart and circulatory defects that affect 0.10 percent of newborn infants. In this condition, (1) the pulmonary trunk is abnormally narrow, (2) the interventricular septum is incomplete, (3) the aorta originates where the interventricular septum normally ends, and (4) the right ventricle is enlarged and both ventricles are thickened owing to increased workloads (Figure 21–36c●).

In the *transposition of great vessels*, the aorta is connected to the right ventricle and the pulmonary artery is connected to the left ventricle (Figure 21–36d●). This malformation affects 0.05 percent of newborn infants.

In an *atrioventricular septal defect*, the atria and ventricles are incompletely separated (Figure 21–36e●). The results are quite variable, depending on the extent of the defect and the effects on the atrioventricular valves. This type of defect most commonly affects infants with *Down syndrome*, a disorder caused by the presence of an extra copy of chromosome 21.

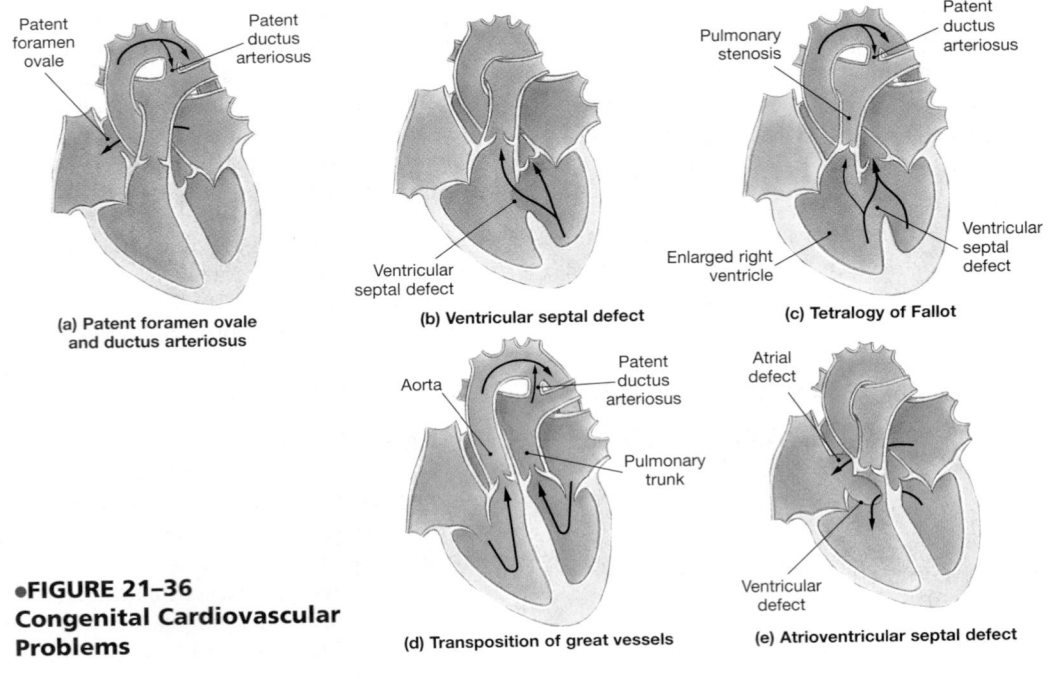

(a) Patent foramen ovale and ductus arteriosus

(b) Ventricular septal defect

(c) Tetralogy of Fallot

(d) Transposition of great vessels

(e) Atrioventricular septal defect

●**FIGURE 21–36**
Congenital Cardiovascular Problems

✓ Whenever Tim gets angry, a large vein bulges in the lateral region of his neck. Which vein is this?

✓ A thrombus that blocks the popliteal vein would interfere with blood flow in which other veins?

✓ A blood sample taken from the umbilical cord contains a high concentration of oxygen and nutrients and a low concentration of carbon dioxide and waste products. Is this sample from an umbilical artery or from the umbilical vein? Explain.

Answers start on page Q-1

Chapter Review

SELECTED CLINICAL TERMINOLOGY

Terms Discussed in This Chapter

aneurysm (AN-ū-rizm):   A bulge in the weakened wall of a blood vessel, generally an artery. *(p. 726 and AM)*

arteriosclerosis (ar-tē-rē-ō-skle-RŌ-sis):   A thickening and toughening of arterial walls. *(p. 725)*

atherosclerosis (ath-er-ō-skle-RŌ-sis):   A type of arteriosclerosis characterized by changes in the endothelial lining and by the formation of a plaque. *(p. 725)*

edema (e-DĒ-muh):   An abnormal accumulation of fluid in peripheral tissues. *(p. 739)*

hemorrhoids (HEM-o-roydz):   Varicose veins in the walls of the rectum, the anus, or both; commonly associated with frequent straining to force bowel movements. *(p. 729 and AM)*

hypertension: Abnormally high blood pressure; usually defined in adults as blood pressure higher than 150/90. *(p. 735 and AM)*

hypotension: Blood pressure so low that circulation to vital organs may be impaired. *(p. 735 and AM)*

pressure points: Locations where muscular arteries can be compressed against skeletal elements to restrict or prevent the flow of blood in an emergency. *(p. 736)*

pulmonary embolism: A vascular blockage caused by the trapping of a detached thrombus in a pulmonary artery. *(p. 769)*

shock: An acute cardiovascular crisis marked by hypotension and inadequate peripheral blood flow. *(p. 749 and AM)*

sounds of Korotkoff: Distinctive sounds, caused by turbulent arterial blood flow, heard while measuring the blood pressure. *(p. 736)*

sphygmomanometer: A device that measures blood pressure by using an inflatable cuff placed around a limb. *(p. 736)*

stroke, or *cerebrovascular accident* (CVA):   An interruption of the vascular supply to a portion of the brain. *(p. 757)*

thrombus: A stationary blood clot within a blood vessel. *(p. 769)*

varicose (VAR-i-kōs) **veins:**   Sagging, swollen veins distorted by gravity and by the failure of the venous valves. *(p. 729)*

AM Additional Terms Discussed in the *Applications Manual*

angiotensin-converting enzyme (ACE) inhibitors
cardiogenic shock
cerebral hemorrhage
cerebral thrombosis
congestive heart failure (CHF)
distributive shock
obstructive shock
orthostatic hypotension
phlebitis
primary hypertension (essential hypertension)
secondary hypertension
transient ischemic attack (TIA)

STUDY OUTLINE

1. Blood flows through a network of arteries, veins, and capillaries. All chemical and gaseous exchange between blood and interstitial fluid takes place across capillary walls.

2. **Arteries** and **veins** form an internal distribution system through which the heart propels blood. Arteries branch repeatedly, decreasing in size until they become **arterioles**. From the arterioles, blood enters the **capillary** networks. Blood flowing from the capillaries enters small **venules** before entering larger veins.

21-1 THE ANATOMY OF BLOOD VESSELS   p. 722

THE STRUCTURE OF VESSEL WALLS   p. 722

1. The walls of arteries and veins contain three layers: the innermost **tunica intima**, the **tunica media**, and the outermost **tunica externa**. *(Figure 21-1)*

DIFFERENCES BETWEEN ARTERIES AND VEINS   p. 723

2. In general, the walls of arteries are thicker than those of veins. Arteries constrict when blood pressure does not distend them; veins con-

strict very little. The endothelial lining cannot contract, so when constriction occurs, the lining of an artery is thrown into folds. *(Figure 21-1)*

ARTERIES   p. 723

3. The arterial system includes the large **elastic arteries**, medium-sized **muscular arteries**, and smaller arterioles. As we proceed toward the capillaries, the number of vessels increases, but the diameter of the individual vessels decreases and the walls become thinner. *(Figure 21-2)*

4. **Atherosclerosis**, a type of **arteriosclerosis**, is associated with changes in the endothelial lining of arteries. Fatty masses of tissue called **plaques** typically develop during atherosclerosis. *(Figure 21-4)*

CAPILLARIES   p. 726

5. Capillaries are the only blood vessels whose walls permit an exchange between blood and interstitial fluid. Capillaries are **continuous** or **fenestrated**. **Sinusoids** have fenestrated walls and form elaborate networks that allow very slow blood flow. Sinusoids are located in the liver and in various endocrine organs. *(Figure 21-3)*

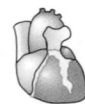

The **CARDIOVASCULAR SYSTEM** in Perspective

Organ/Component	Primary Functions
Heart	Propels blood; maintains blood pressure
Blood Vessels	Distribute blood around the body
Arteries	Carry blood from heart to capillaries
Capillaries	Permit diffusion between blood and interstitial fluids
Veins	Return blood from capillaries to the heart
Blood	Transports oxygen, carbon dioxide, and blood cells; delivers nutrients and hormones; removes waste products; assists in temperature regulation and defense against disease

INTEGRATION WITH OTHER SYSTEMS

The cardiovascular system is both anatomically and functionally linked to all other systems. In the section on vessel distribution, we demonstrated the extent of the anatomical connections. The figure on the opposite page summarizes the physiological relationships between the cardiovascular system and other organ systems.

The most extensive communication occurs between the cardiovascular and lymphatic systems. Not only are the two systems physically interconnected, but cell populations of the lymphatic system use the cardiovascular system as a highway to move from one part of the body to another. In Chapter 22, we shall examine the lymphatic system in detail and consider its role in the immune response.

CLINICAL PATTERNS

Because the cardiovascular system plays a key role in supporting all other systems, cardiovascular disorders will affect virtually every cell in the body. One method of organizing the many potential disorders involving the cardiovascular system is by the nature of the primary problem and whether it affects the blood, the heart, or the vascular network. Some disorders are structural, such as congenital disorders, which may affect blood formation, the structure of the heart, or the arrangement of vessels. Others are primarily functional disorders, such as heart failure (p. 707), hypertension (p. 735), or shock (p. 749). There are also cardiovascular disorders that result from pathogenic infection, tumors, trauma, and degenerative disorders. For a review of the major categories of clinical disorders affecting the cardiovascular system, see the related discussions in the *Applications Manual*.

MEDIA CONNECTIONS

I. Objective: To explore the interdependence of the cardiovascular, nervous, and endocrine systems in the regulation of blood pressure.

Completion time: 15 minutes

The cardiovascular system does not operate in a vacuum; it is both regulated by and responsive to the needs of the rest of the body. This interaction can be studied on the **InterActive Physiology CD-ROM:** Cardiovascular System/Blood Pressure Regulation. Begin with screens 2 through 12, paying attention to which other systems are regulating the return of blood pressure to normal values. In one paragraph per system, summarize the interactions presented thus far. Predict which systems will be involved in long-term regulation of high and low blood pressure. Write out your predictions, indicating how these systems might interact. Continue with the Blood Pressure Regulation tutorial by working through screens 13 through 30. Modify your predictions as necessary to reflect the interactions discussed.

II. Objective: To relate heart physiology to your personal risk factors for heart disease.

Completion time = 10 minutes

Heart disease is still the number one killer of Americans aged 25 and older. What factors increase the risk of heart disease? List as many factors as you can, briefly describing how each could affect the ability of the heart to pump blood efficiently. Indicate which of these risk factors could affect you. To check your list, go to the **Companion Website**, Chapter 21 Media Connections, and click on the keyword **Factors** to read an American Heart Association discussion of this topic. Then, click on the keyword **personal risk** to calculate your coronary heart disease risk.

INTEGUMENTARY SYSTEM

- Stimulation of mast cells produces localized changes in blood flow and capillary permeability

- Delivers immune system cells to injury sites
- Clotting response seals breaks in skin surface
- Carries away toxins from sites of infection
- Provides heat

SKELETAL SYSTEM

- Provides calcium needed for normal cardiac muscle contraction
- Protects blood cells developing in bone marrow

- Provides calcium and phosphate for bone deposition
- Delivers EPO to bone marrow, parathyroid hormone and calcitonin to osteoblasts and osteoclasts

MUSCULAR SYSTEM

- Skeletal muscle contractions assist in moving blood through veins
- Protects superficial blood vessels, especially in neck and limbs

- Delivers oxygen and nutrients, removes carbon dioxide, lactic acid, and heat during skeletal muscle activity

NERVOUS SYSTEM

- Controls patterns of circulation in peripheral tissues
- Modifies heart rate and regulates blood pressure
- Releases ADH

- Endothelial cells maintain blood-brain barrier
- Helps generate CSF

ENDOCRINE SYSTEM

- Erythropoietin regulates production of RBCs
- Several hormones elevate blood pressure
- Epinephrine stimulates cardiac muscle, elevating heart rate and contractile force

- Distributes hormones throughout the body
- Heart secretes ANP and BNP

CARDIOVASCULAR SYSTEM

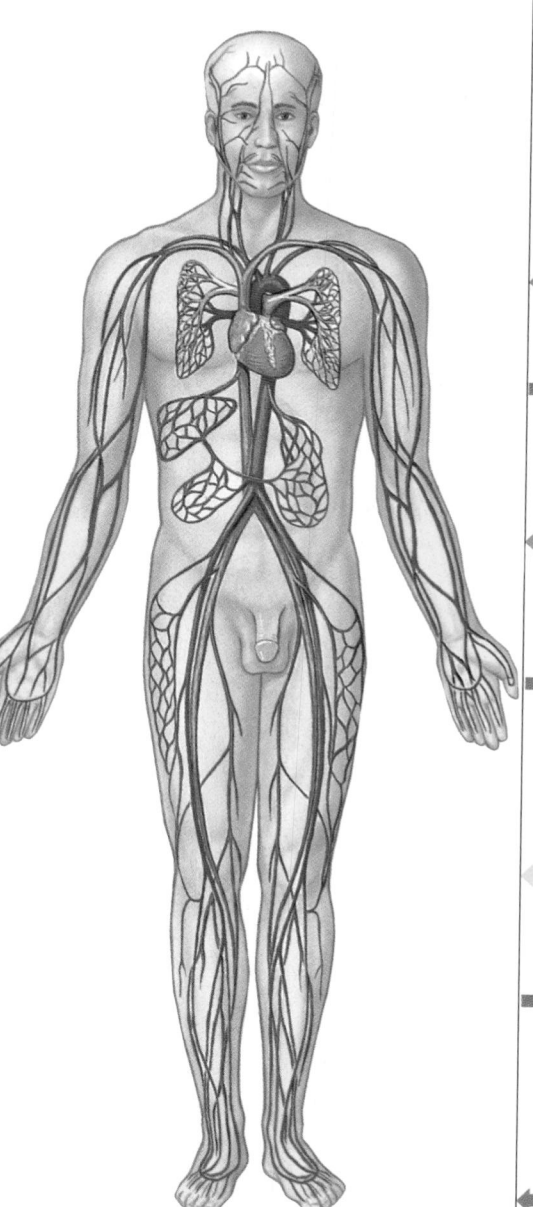

FOR ALL SYSTEMS

Delivers oxygen, hormones, nutrients, and WBCs; removes carbon dioxide and metabolic wastes; transfers heat

LYMPHATIC SYSTEM

- Defends against pathogens or toxins in blood
- Fights infections of cardiovascular organs
- Returns tissue fluid to bloodstream

- Distributes WBCs
- Carries antibodies that attack pathogens
- Clotting response assists in restricting spread of pathogens
- Granulocytes and lymphocytes produced in bone marrow

RESPIRATORY SYSTEM

- Provides oxygen to cardiovascular organs and removes carbon dioxide

- RBCs transport oxygen and carbon dioxide between lungs and peripheral tissues

DIGESTIVE SYSTEM

- Provides nutrients to cardiovascular organs
- Absorbs water and ions essential to maintenance of normal blood volume

- Distributes digestive tract hormones
- Carries nutrients, water, and ions away from sites of absorption
- Delivers nutrients and toxins to liver

URINARY SYSTEM

- Releases renin to elevate blood pressure and erythropoietin to accelerate red blood cell production

- Delivers blood to capillaries, where filtration occurs
- Accepts fluids and solutes reabsorbed during urine production

REPRODUCTIVE SYSTEM

- Sexual hormones maintain healthy vessels
- Estrogen slows development of atherosclerosis

- Distributes reproductive hormones
- Provides nutrients, oxygen, and waste removal for developing fetus
- Local blood pressure changes responsible for physical changes during sexual arousal

6. Capillaries form interconnected networks called **capillary beds** (**capillary plexuses**). A **precapillary sphincter** (a band of smooth muscle) adjusts the blood flow into each capillary. Blood flow in a capillary changes as **vasomotion** occurs. The entire capillary bed may be bypassed by blood flow through **arteriovenous anastomoses**. (*Figure 21–5*)

VEINS   p. 729

7. Venules collect blood from the capillaries and merge into **medium-sized veins** and then **large veins**. The arterial system is a high-pressure system; blood pressure in veins is much lower. **Valves** in veins prevent the backflow of blood. (*Figures 21–1, 21–2, 21–6*)

THE DISTRIBUTION OF BLOOD   p. 730

8. Peripheral **venoconstriction** helps maintain adequate blood volume in the arterial system after a hemorrhage. The **venous reserve** normally accounts for about 20 percent of the total blood volume. (*Figure 21–7*)

Anatomy review: **IP CD-ROM**: Cardiovascular System/Anatomy Review: Blood Vessel Structure & Function.

21–2 CARDIOVASCULAR PHYSIOLOGY   p. 730

1. Cardiovascular regulation involves the manipulation of blood pressure and resistance to control the rates of blood flow and capillary exchange. (*Figure 21–8*)

PRESSURE   p. 731

2. Flow is proportional to pressure difference; blood will flow from an area of higher pressure to one of lower pressure. The **circulatory pressure** is the pressure gradient across the systemic circuit. It is divided into three components: **blood pressure (BP)**, **capillary hydrostatic pressure (CHP)**, and **venous pressure**.

RESISTANCE   p. 732

3. The **resistance (R)** determines the rate of blood flow through the systemic circuit. The major determinant of blood flow rate is the **peripheral resistance (PR)**—the resistance of the arterial system. Neural and hormonal control mechanisms regulate blood pressure and peripheral resistance.

4. **Vascular resistance** is the resistance of the blood vessels. It is the largest component of peripheral resistance and depends on vessel length and vessel diameter.

5. **Viscosity** and **turbulence** also contribute to peripheral resistance. (*Summary Table 21–1*)

AN OVERVIEW OF CARDIOVASCULAR PRESSURES   p. 733

6. The high arterial pressures overcome peripheral resistance and maintain blood flow through peripheral tissues. Capillary pressures are normally low; small changes in capillary pressure determine the rate of movement of fluid into or out of the bloodstream. Venous pressure, normally low, determines *venous return* and affects cardiac output and peripheral blood flow. (*Figures 21–9, 21–10; Summary Table 21–1*)

ARTERIAL BLOOD PRESSURE   p. 734

7. Arterial blood pressure rises during ventricular systole and falls during ventricular diastole. The difference between these two blood pressures is the **pulse pressure**. Blood pressure is measured at the brachial artery with the use of a **sphygmomanometer**. (*Figures 21–10, 21–11*)

CAPILLARY EXCHANGE   p. 735

8. At the capillaries, blood pressure forces water and solutes out of the plasma, across capillary walls. Water moves out of the capillaries, through the peripheral tissues, and back to the bloodstream by way of the lymphatic system. Water movement across capillary walls is determined by the interplay between osmotic pressures and hydrostatic pressures. (*Figure 21–12*)

9. **Osmotic pressure (OP)** is a measure of the pressure that must be applied to prevent osmotic movement across a membrane. Osmotic water movement continues until either solute concentrations are equalized or the movement is prevented by an opposing hydrostatic pressure.

10. The rates of filtration and reabsorption gradually change as blood passes along the length of a capillary, as determined by the **net filtration pressure** (the difference between the net hydrostatic pressure and the net osmotic pressure). (*Figure 21–13*)

VENOUS PRESSURE AND VENOUS RETURN   p. 739

11. Valves, muscular compression, and the **respiratory pump** (*thoracoabdominal pump*) help the relatively low venous pressures propel blood toward the heart. (*Figures 21–6, 21–9*)

Blood pressure regulation: **IP CD-ROM**: Cardiovascular System/Factors That Affect Blood Pressure.

21–3 CARDIOVASCULAR REGULATION   p. 740

1. Homeostatic mechanisms ensure that **tissue perfusion** (blood flow) delivers adequate oxygen and nutrients.

2. Autoregulation, neural mechanisms, and endocrine mechanisms influence the coordinated regulation of cardiovascular function. Autoregulation involves local factors changing the pattern of blood flow within capillary beds in response to chemical changes in interstitial fluids. Neural mechanisms respond to changes in arterial pressure or blood gas levels. Hormones can assist in short-term adjustments (changes in cardiac output and peripheral resistance) and long-term adjustments (changes in blood volume that affect cardiac output and gas transport). (*Figure 21–14*)

AUTOREGULATION OF BLOOD FLOW WITHIN TISSUES   p. 740

3. Peripheral resistance is adjusted at the tissues by local factors that result in the dilation or constriction of precapillary sphincters. (*Figure 21–5*)

NEURAL MECHANISMS   p. 741

4. **Cardiovascular (CV) centers** of the medulla oblongata are responsible for adjusting cardiac output and peripheral resistance to maintain adequate blood flow. The vasomotor centers contain a group of neurons responsible for controlling vasoconstriction and another group responsible for controlling vasodilation.

5. **Baroreceptor reflexes** monitor the degree of stretch within expandable organs. Baroreceptors are located in the **aortic sinuses**, the **carotid sinuses**, and the right atrium. (*Figure 21–15*)

6. **Chemoreceptor reflexes** respond to changes in the oxygen or CO_2 levels in the blood. They are triggered by sensory neurons located in the **carotid bodies** and the **aortic bodies**. (*Figure 21–16*)

HORMONES AND CARDIOVASCULAR REGULATION   p. 744

7. The endocrine system provides short-term regulation of cardiac output and peripheral resistance with epinephrine and norepinephrine from the adrenal medullae. Hormones involved in the long-term regulation of blood pressure and volume are *antidiuretic hormone (ADH), angiotensin II, erythropoietin (EPO),* and *natriuretic peptides (ANP and BNP). (Figure 21–17)*

21-4 PATTERNS OF CARDIOVASCULAR RESPONSE   p. 746

EXERCISE AND THE CARDIOVASCULAR SYSTEM   p. 746

1. During exercise, blood flow to skeletal muscles increases at the expense of blood flow to nonessential organs, so cardiac output rises. Cardiovascular performance improves with training. Athletes have larger stroke volumes, slower resting heart rates, and larger cardiac reserves than do nonathletes. *(Tables 21–2, 21–3)*

CARDIOVASCULAR RESPONSE TO HEMORRHAGING   p. 747

2. Blood loss lowers blood volume and venous return and decreases cardiac output. Compensatory mechanisms include an increase in cardiac output, a mobilization of venous reserves, peripheral vasoconstriction, and the liberation of hormones that promote the retention of fluids and the manufacture of erythrocytes. *(Figure 21–18)*

3. **Shock** is an acute circulatory crisis marked by hypotension and inadequate peripheral blood flow. A severe drop in blood volume produces symptoms of **circulatory shock**. Causes of fluid loss may include hemorrhaging, dehydration, and severe burns. *(Figure 21–19)*

SPECIAL CIRCULATION   p. 748

4. The blood–brain barrier, the coronary circulation, and the circulation to alveolar capillaries in the lungs are examples of special circulation in which cardiovascular dynamics and regulatory mechanisms differ from those in other tissues.

   Cardiovascular regulation: **IP CD-ROM**: Cardiovascular System/ Blood Pressure Regulation and Autoregulation and Capillary Dynamics.

21-5 THE DISTRIBUTION OF BLOOD VESSELS: AN OVERVIEW   p. 751

1. The peripheral distributions of arteries and veins are generally identical on both sides of the body, except near the heart. *(Figure 21–20)*

21-6 THE PULMONARY CIRCUIT   p. 752

1. The pulmonary circuit includes the pulmonary trunk, the **left** and **right pulmonary arteries**, and the **pulmonary veins**, which empty into the left atrium. *(Figure 21–21)*

21-7 THE SYSTEMIC CIRCUIT   p. 753

SYSTEMIC ARTERIES   p. 753

1. The **ascending aorta** gives rise to the coronary circulation. The **aortic arch** communicates with the **descending aorta**. *(Figures 21–22 to 21–28)*

2. Three elastic arteries originate along the aortic arch: the **left common carotid**, the **left subclavian artery**, and the **brachiocephalic trunk**. *(Figures 21–22, 21–23, 21–24)*

3. The remaining major arteries of the body originate from the **descending aorta**. *(Figures 21–26, 21–27, 21–28)*

   Arteries: **Anatomy CD-ROM**: Cardiovascular System.

SYSTEMIC VEINS   p. 761

4. Arteries in the neck and limbs are deep beneath the skin; in contrast, there are generally two sets of peripheral veins, one superficial and one deep. This dual venous drainage is important for controlling body temperature. *(Figure 21–29)*

5. The **superior vena cava** receives blood from the head, neck, chest, shoulders, and arms. *(Figures 21–29 to 21–32)*

6. The **inferior vena cava** collects most of the venous blood from organs inferior to the diaphragm. *(Figures 21–31 to 21–33)*

7. The **hepatic portal system** directs blood from the other digestive organs to the liver before the blood returns to the heart. *(Figure 21–34)*

21-8 FETAL CIRCULATION   p. 768

PLACENTAL BLOOD SUPPLY   p. 768

1. Blood flow to the placenta is provided by a pair of **umbilical arteries** and is drained by a single **umbilical vein**. *(Figure 21–35)*

CIRCULATION IN THE HEART AND GREAT VESSELS   p. 768

2. The interatrial partition remains functionally incomplete until birth. The **foramen ovale** allows blood to flow freely from the right to the left atrium, and the **ductus arteriosus** short-circuits the pulmonary trunk.

CARDIOVASCULAR CHANGES AT BIRTH   p. 768

3. The foramen ovale closes, leaving the fossa ovalis. The ductus arteriosus constricts, leaving the ligamentum arteriosum. *(Figure 21–36)*

4. Congenital cardiovascular problems generally reflect abnormalities of the heart or of interconnections between the heart and great vessels. *(Figure 21–36)*

21-9 AGING AND THE CARDIOVASCULAR SYSTEM   p. 769

1. Age-related changes in the blood include (1) a decreased hematocrit, (2) the constriction or blockage of peripheral veins by a *thrombus* (stationary blood clot), and (3) the pooling of blood in the veins of the legs because valves are not working effectively.

2. Age-related changes in the heart include (1) a reduction in the maximum cardiac output, (2) changes in the activities of the nodal and conducting cells, (3) a reduction in the elasticity of the fibrous skeleton, (4) a progressive atherosclerosis that can restrict coronary circulation, and (5) the replacement of damaged cardiac muscle cells by scar tissue.

3. Age-related changes in blood vessels, commonly related to arteriosclerosis, include (1) a weakening in the walls of arteries, potentially leading to the formation of an *aneurysm*, (2) deposition of calcium salts on weakened vascular walls, increasing the risk of a stroke or myocardial infarction, and (3) the formation of a thrombus at atherosclerotic plaques.

REVIEW QUESTIONS

More assessment questions are available to you on the Companion Website. You will find Matching, Multiple Choice, True/False, and other quizzes to help further your understanding of the material covered in this chapter. To access the site, go to www.aw.com/martini.

LEVEL 1 Reviewing Facts and Terms

Match each numbered item with the most closely related lettered item. Use letters for answers in the spaces provided.

_____ 1. veins
_____ 2. vasa vasorum
_____ 3. conducting arteries
_____ 4. muscular arteries
_____ 5. sinusoids
_____ 6. precapillary sphincter
_____ 7. medulla oblongata
_____ 8. vascular resistance
_____ 9. systolic pressure
_____ 10. diastolic pressure
_____ 11. blood pressure
_____ 12. arterioles
_____ 13. cholinergic
_____ 14. adrenergic
_____ 15. baroreceptors
_____ 16. Frank–Starling principle
_____ 17. aorta
_____ 18. internal iliac artery
_____ 19. renal vein
_____ 20. hepatic vein
_____ 21. great saphenous
_____ 22. interatrial opening
_____ 23. thrombus
_____ 24. embolus

a. specialized fenestrated capillaries
b. minimum blood pressure
c. vasomotor centers
d. drains the kidney
e. vasoconstrictor fibers
f. ventricular stretching
g. elastic arteries
h. blood supply to pelvis
i. vasodilator fibers
j. largest superficial vein
k. capacitance vessels
l. drains the liver
m. largest artery in body
n. stationary blood clot
o. distribution arteries
p. foramen ovale
q. vasomotion
r. resistance vessels
s. migrating blood clot
t. carotid sinus
u. friction
v. sounds of Korotkoff
w. "vessels of vessels"
x. peak blood pressure

25. Thin-walled medium-sized vessels that appear collapsed and that may tear easily in gross dissection are
(a) veins
(b) arterioles
(c) arteries
(d) venules

26. The layer of the arteriole wall that provides the properties of contractility and elasticity is the
(a) tunica adventitia
(b) tunica media
(c) tunica intima
(d) tunica externa

27. Blood vessels that supply the walls of arteries and veins with blood are
(a) coronary vessels
(b) capillaries
(c) vasa vasorum
(d) metarterioles

28. Of the following arteries, the one that is an elastic artery is the
(a) subclavian artery
(b) external carotid artery
(c) brachial artery
(d) femoral artery

29. The two-way exchange of substances between blood and body cells occurs only through
(a) arterioles
(b) capillaries
(c) venules
(d) a, b, and c are correct

30. Large molecules such as peptides and proteins move into and out of the bloodstream by way of
(a) continuous capillaries
(b) fenestrated capillaries
(c) thoroughfare channels
(d) metarterioles

31. The alteration of blood flow due to the action of precapillary sphincters is
(a) vasomotion
(b) autoregulation
(c) selective resistance
(d) turbulence

32. The blood vessels that collect blood from all tissues and organs and return it to the heart are the
(a) veins
(b) arteries
(c) capillaries
(d) arterioles

33. Blood is transported through the venous system by means of
(a) muscular contractions
(b) increasing blood pressure
(c) the respiratory pump
(d) a and c are correct

34. The most important factor in vascular resistance is
(a) the viscosity of the blood
(b) the diameter of blood vessel walls
(c) turbulence due to irregular surfaces of blood vessels
(d) the length of the blood vessels

35. The specialized exchange surfaces of the liver, bone marrow, and pituitary gland are the
(a) continuous capillaries
(b) fenestrated capillaries
(c) sinusoids
(d) arteriovenous anastomoses

36. Net hydrostatic pressure forces water _____ a capillary; net osmotic pressure forces water _____ a capillary.
(a) into, out of
(b) out of, into
(c) out of, out of
(d) a, b, and c are incorrect

37. The two arteries formed by the division of the brachiocephalic trunk are the
(a) aorta and internal carotid
(b) axillary and brachial
(c) external and internal carotid
(d) common carotid and subclavian

38. The unpaired arteries supplying blood to the visceral organs include the
(a) suprarenal, renal, lumbar
(b) iliac, gonadal, femoral
(c) celiac, superior and inferior mesenteric
(d) a, b, and c are correct

39. The paired arteries supplying blood to the body wall and other structures outside the abdominopelvic cavity include the
(a) left gastric, hepatic, splenic, phrenic
(b) suprarenals, renals, lumbars, gonadals
(c) iliacs, femorals, gonadals, ileocecals
(d) celiac, left gastric, superior and inferior mesenteric

40. The vein that drains the dural sinuses of the brain is the
(a) cephalic
(b) great saphenous
(c) internal jugular
(d) superior vena cava

41. The vein that drains the thorax and empties into the superior vena cava is the
(a) azygos
(b) basilic
(c) cardiac
(d) cephalic

42. The vein that collects most of the venous blood from below the diaphragm is the
    (**a**) superior vena cava   (**b**) great saphenous
    (**c**) inferior vena cava   (**d**) azygos

43. The tributaries of the hepatic portal vein include the
    (**a**) lumbar, gonadal, renal, and suprarenal veins
    (**b**) left colic, splenic, inferior and superior mesenteric veins
    (**c**) phrenic, hepatic, renal, and suprarenal veins
    (**d**) peroneal, iliac, saphenous, and femoral veins

44. What are the primary forces that cause fluid to move
    (**a**) out of a capillary at its arterial end and into the interstitial fluid?
    (**b**) into a capillary at its venous end from the interstitial fluid?

45. What two factors assist relatively low venous pressures in propelling blood toward the heart?

46. Give the formula for calculating net filtration pressure, defining each variable.

47. What two effects occur when the baroreceptor response to elevated blood pressure is triggered?

48. What factors affect the activity of chemoreceptors in the carotid and aortic bodies?

49. List six hormones that are important in the regulation of cardiovascular function.

50. What interrelated cardiovascular changes occur before and during light exercise?

51. What criteria are used to identify the basic symptoms of all forms of shock?

52. What circulatory changes occur at birth?

53. What age-related changes take place in the blood, heart, and blood vessels?

LEVEL 2 Reviewing Concepts

54. Differentiate among the following types of pressure:
    (**a**) systolic   (**b**) diastolic
    (**c**) pulse   (**d**) mean arterial

55. When dehydration occurs, there is
    (**a**) accelerated reabsorption of water at the kidneys
    (**b**) a recall of fluids
    (**c**) an increase in the blood colloidal osmotic pressure
    (**d**) a, b, and c are correct

56. Increased CO_2 levels in tissues would promote
    (**a**) the constriction of precapillary sphincters
    (**b**) an increase in the pH of the blood
    (**c**) the dilation of precapillary sphincters
    (**d**) a decrease in blood flow to tissues

57. Elevated levels of the hormones ANP and BNP will produce
    (**a**) increased fluid loss through the kidneys
    (**b**) increased blood volume
    (**c**) increased sodium ion levels in blood
    (**d**) increased venous return

58. The secretion of ADH and aldosterone are typical of the body's long-term compensation following
    (**a**) a heart attack
    (**b**) hypertension
    (**c**) a serious hemorrhage
    (**d**) heavy exercise

59. Relate the anatomical differences between arteries and veins to their functions.

60. Which types of cardiovascular resistance are under autonomic, nervous, or hormonal control? Why is it important that the body have the ability to alter peripheral resistance to blood flow?

61. Why do capillaries permit the diffusion of materials whereas arteries and veins do not?

62. How is blood pressure maintained in veins to cope with the force of gravity?

63. How do pressure and resistance affect cardiac output and peripheral blood flow?

64. Why is blood flow to the brain relatively continuous and constant?

65. Compare the effects of the cardioacceleratory and cardioinhibitory centers on cardiac output and blood pressure.

66. A nurse practitioner tells Mrs. B. that her blood pressure is 150/90. Explain what these numbers represent, and calculate Mrs. B.'s mean arterial pressure.

67. An accident victim displays the following symptoms: hypotension; pale, cool, moist skin; confusion and disorientation. Identify her condition, and explain why these symptoms occur. If you took her pulse, what would you probably find?

68. Emilio is in congestive heart failure. Because of this condition, his ankles and feet appear to be swollen. What is the relationship between congestive heart failure and the accumulation of fluid in the feet and ankles?

LEVEL 3 Critical Thinking and Clinical Applications

69. Bob is sitting outside on a warm day and is sweating profusely. Mary wants to practice taking blood pressures, and he agrees to play patient. Mary finds that Bob's blood pressure is elevated, even though he is resting and has lost fluid from sweating. (She reasons that fluid loss should lower blood volume and thus blood pressure.) Why is Bob's blood pressure high instead of low?

70. The most common site of varicose veins is the greater saphenous vein of the leg. Why?

71. Who would have a higher pulse pressure, a resting athlete or a resting person who never exercises and has a sedentary job? Why?

72. People with allergies commonly take antihistamines with decongestants to relieve their symptoms. The container warns that the medication should not be taken by individuals who are being treated for high blood pressure. Why not?

73. Jolene awakens suddenly to the sound of her alarm clock. Realizing that she is late for class, she jumps to her feet, feels light headed, and falls back on her bed. What probably caused this reaction? Why doesn't this happen all the time?

UNIT 4   CHAPTER   19   20   21   **22**

CLINICAL NOTES
- Infected Lymphoid Nodules 787
- Cancer and the Lymphatic System 788
- Injury to the Spleen 791
- NK Cells, Cancer, and Viral Infections 794
- Necrosis 797
- Drug Reactions 808
- Immunochemistry and Clinical Therapy 814
- Treatment of Anaphylaxis 817

CLINICAL DISCUSSIONS
- Lymphedema 784
- Graft Rejection and Immunosuppression 805
- AIDS 820

22

THE LYMPHATIC SYSTEM AND IMMUNITY

The world is not always kind to the human body. Accidental bumps, cuts, and scrapes, chemical and thermal burns, extreme cold, and ultraviolet radiation are just a few of the hazards in our physical environment. Making matters worse, the world around us contains an assortment of viruses, bacteria, fungi, and parasites capable of not only surviving but thriving inside our bodies—and potentially causing us great harm. These organisms, called **pathogens**, are responsible for many human diseases. Each pathogen has a different lifestyle and attacks the body in a specific way. For example, viruses spend most of their time hiding within cells, which they often eventually destroy, whereas, the largest parasites can actually burrow through internal organs. Many bacteria multiply in interstitial fluids, where they release foreign proteins: enzymes or toxins that can damage cells, tissues, even entire organ systems. And as if that were not enough, we are constantly at risk from renegade body cells that have the potential to produce lethal cancers. ∞ p. 105

AM The Nature of Pathogens: A Closer Look

Many organs and systems work together in an effort to keep us alive and healthy. In this ongoing struggle, the *lymphatic system* plays a central role.

22-1 AN OVERVIEW OF THE LYMPHATIC SYSTEM AND IMMUNITY

Objective

■ Explain the difference between nonspecific and specific defense, and the role of lymphocytes in the immune response.

The **lymphatic system** includes the cells, tissues, and organs responsible for defending the body against both environmental hazards, such as various pathogens, and internal threats, such as cancer cells. We introduced *lymphocytes*, the primary cells of the lymphatic system, in Chapters 4 and 19. ∞ pp. 124, 669 These cells are vital to our ability to resist or overcome infection and disease. Lymphocytes respond to the presence of invading pathogens (such as bacteria or viruses), abnormal body cells (such as virus-infected cells or cancer cells), and foreign proteins (such as the toxins released by some bacteria). They attempt to eliminate these threats or render them harmless by a combination of physical and chemical attacks.

The body has several anatomical barriers and defense mechanisms that either prevent or slow the entry of infectious organisms or attack them if they do succeed in gaining entry. These agencies are known as *nonspecific defenses*, because they do not distinguish one potential threat from another. In contrast, lymphocytes respond to specific threats. If bacteria invade peripheral tissues, lymphocytes organize a defense against that particular type of bacterium. For this reason, lymphocytes are said to provide a *specific defense*, known as the **immune response**. The ability to resist infection and disease through the activation of specific defenses constitutes **immunity**.

All the cells and tissues involved with the production of immunity are sometimes considered to be part of an *immune system*—a physiological system that includes not only the lymphatic system, but also components of the integumentary, cardiovascular, respiratory, digestive, and other systems. For example, interactions between lymphocytes and Langerhans cells of the skin are important in mobilizing specific defenses against skin infections.

We begin this chapter by examining the organization of the lymphatic system. We will then consider the body's nonspecific defenses. Finally, we will see how the lymphatic system interacts with cells and tissues of other systems to defend the body against infection and disease.

22–2 ORGANIZATION OF THE LYMPHATIC SYSTEM

Objectives

- Identify the major components of the lymphatic system, and explain their functions.
- Discuss the importance of lymphocytes, and describe their distribution in the body.
- Describe the structure of lymphoid tissues and organs, and explain their functions.

The lymphatic system consists of (1) **lymph**, a fluid that resembles plasma, but contains a much lower concentration of suspended proteins; (2) a network of **lymphatic vessels**, often called **lymphatics**, which begin in peripheral tissues and end at connections to veins; (3) an array of **lymphoid tissues** and **lymphoid organs** scattered throughout the body; and (4) lymphocytes and smaller numbers of supporting and phagocytic cells. Figure 22–1● provides a general overview of the components of this system. **AM** Embryology Summary 17: The Development of the Lymphatic System

☐ FUNCTIONS OF THE LYMPHATIC SYSTEM

The primary function of the lymphatic system is the production, maintenance, and distribution of lymphocytes that provide de-

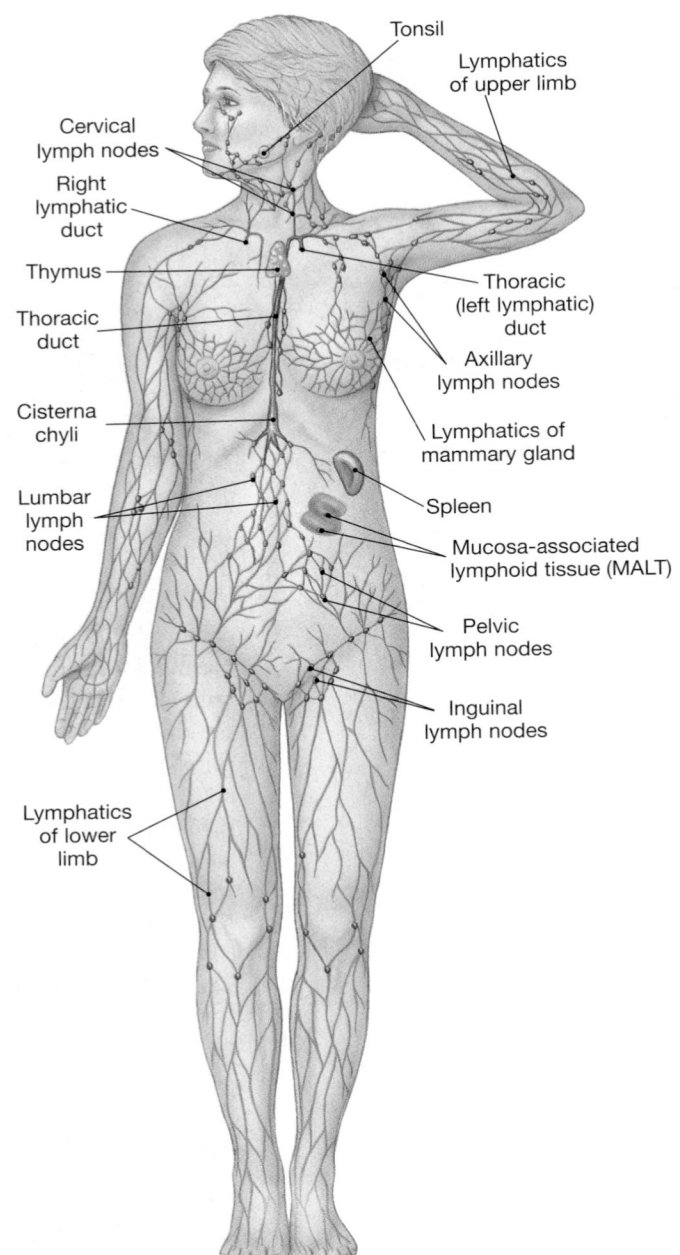

●FIGURE 22–1
The Components of the Lymphatic System

fense against infections and other environmental hazards. Most of the body's lymphocytes are produced and stored within lymphoid tissues, such as the tonsils, and lymphoid organs, such as the spleen and thymus. However, lymphocytes are also produced in areas of red bone marrow, along with various supporting cells, such as monocytes and macrophages.

To provide an effective defense, lymphocytes must detect where problems exist, and they must be able to reach the site of injury or infection. Lymphocytes and other immune cells circulate within the blood and are able to enter or leave the capillaries that supply most of the tissues of the body. As we noted in Chapter 21, capillaries normally deliver more fluid to peripheral tissues than they carry away. ∞ p. 738 The excess fluid returns to the bloodstream through lymphatic vessels. This continuous circulation of extracellular fluid helps transport lymphocytes and other cellular defenders from one organ to another. In the process, it maintains normal blood volume and eliminates local variations in the composition of the interstitial fluid by distributing hormones, nutrients, and waste products from their tissues of origin to the general circulation.

☐ LYMPHATIC VESSELS

Lymphatic vessels carry lymph from peripheral tissues to the venous system. The smallest lymphatic vessels are called *lymphatic capillaries*.

Lymphatic Capillaries

The lymphatic network begins with the **lymphatic capillaries**, or *terminal lymphatics*, which branch through peripheral tissues. They differ from blood capillaries in that lymphatic capillaries (1) originate as blind pockets, (2) are larger in diameter, (3) have thinner walls, and (4) typically have a flattened or irregular outline in sectional view (Figure 22–2●). Although lymphatic capillaries are lined by endothelial cells, the basement membrane is incomplete or absent. The endothelial cells of a lymphatic capillary are not tightly bound together, but they do overlap. The region of overlap acts as a one-way valve, permitting the entry of fluids and solutes (even those as large as proteins), as well as viruses, bacteria, and cell debris, but preventing their return to the intercellular spaces.

Lymphatic capillaries are present in almost every tissue and organ in the body. Prominent lymphatic capillaries in the small intestine called *lacteals* are important in the transport of lipids absorbed by the digestive tract. Lymphatic capillaries are absent in areas that lack a blood supply, such as the cornea of the eye. The bone marrow and the central nervous system also lack lymphatic vessels.

Small Lymphatic Vessels

From the lymphatic capillaries, lymph flows into larger lymphatic vessels that lead toward the trunk. The walls of these vessels contain layers comparable to those of veins, and, like veins, the larger lymphatic vessels contain valves (Figure 22–3●). The valves are quite close together, and at each the lymphatic vessel bulges noticeably. As a result, large lymphatic vessels have a beaded appearance (Figure 22–3a●). The valves prevent the backflow of lymph within lymph vessels, especially those of the limbs. Pressures within the lymphatic system are minimal, and the valves are essential to maintaining normal lymph flow toward the thoracic cavity.

Lymphatic vessels commonly occur in association with blood vessels (Figure 22–3a●). Notice the differences in relative size,

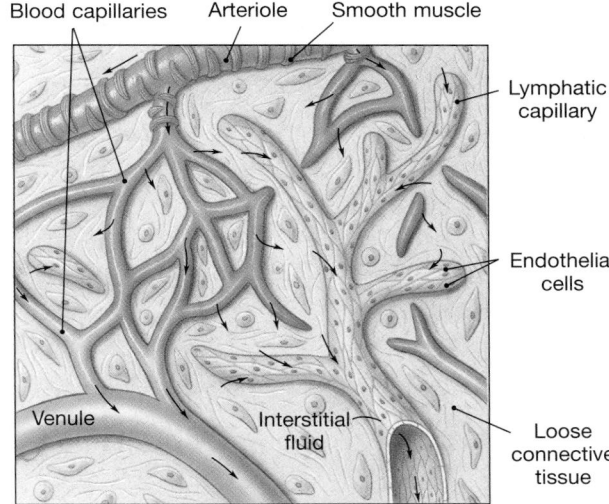

(a) Association of blood capillaries, tissue, and lymphatic capillaries

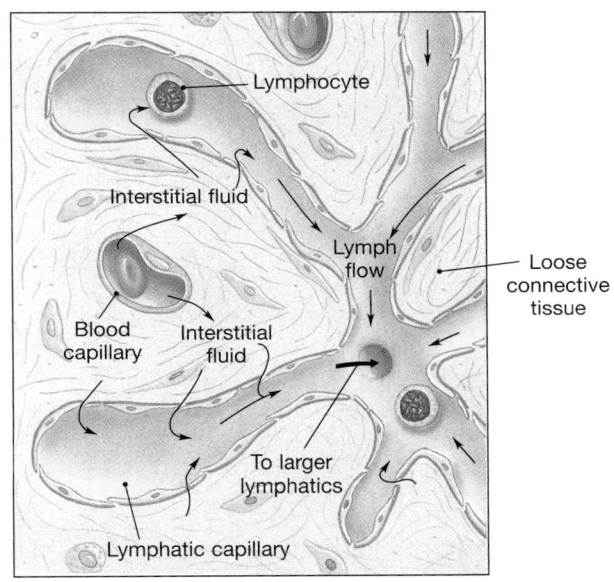

(b) Sectional view

●**FIGURE 22–2**
Lymphatic Capillaries. **(a)** The interwoven network formed by blood capillaries and lymphatic capillaries. Arrows show the movement of fluid out of blood vessels and the net flow of interstitial fluid and lymph. **(b)** A sectional view showing movement of fluid from the plasma through the interstitial fluid and into the lymphatic system.

general appearance, and branching pattern that distinguish the lymphatic vessels from arteries and veins. Characteristic color differences are also apparent on examining living tissues. Most arteries are bright red, veins are dark red, and lymphatic vessels are a pale golden color. In general, a tissue will contain many more lymphatic vessels than veins, but the lymphatic vessels are much smaller.

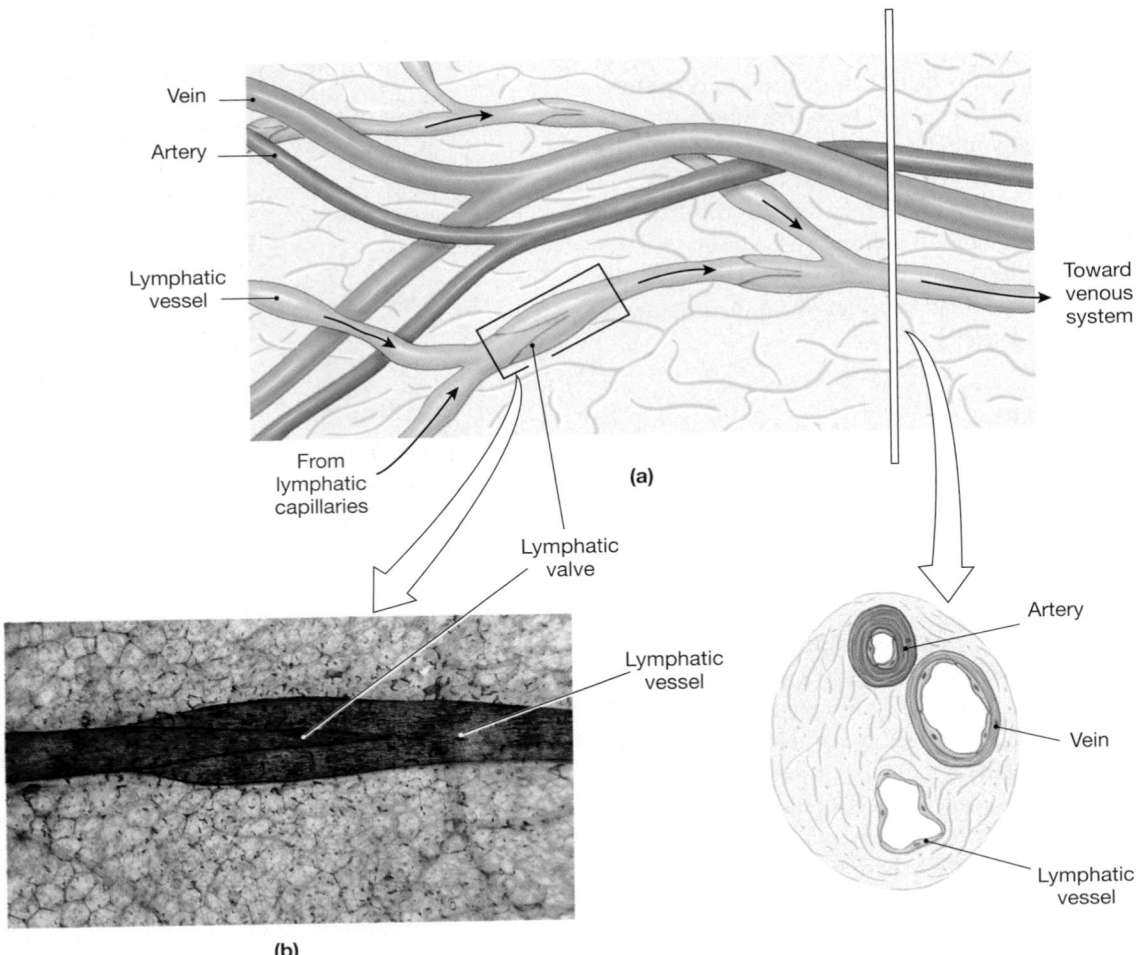

●FIGURE 22–3
Lymphatic Vessels and Valves. **(a)** A diagrammatic view of loose connective tissue, showing small blood vessels and a lymphatic vessel. The cross-sectional view emphasizes the structural differences among these structures. **(b)** A lymphatic valve. Lymphatic valves resemble valves in veins. Each valve consists of a pair of flaps that permit movement of fluid in only one direction. (LM × 43)

Major Lymph-Collecting Vessels

Two sets of lymphatic vessels collect lymph from the lymphatic capillaries: superficial lymphatics and deep lymphatics. **Superficial lymphatics** are located in the subcutaneous layer deep to the skin; in the areolar tissues of the mucous membranes lining the digestive, respiratory, urinary, and reproductive tracts; and in the areolar tissues of the serous membranes lining the pleural, pericardial, and peritoneal cavities. **Deep lymphatics** are larger lymphatic vessels that accompany deep arteries and veins supplying skeletal muscles and other organs of the neck, limbs, and trunk and the walls of visceral organs.

Superficial and deep lymphatics converge to form even larger vessels called *lymphatic trunks*, which in turn empty into two large collecting vessels: the thoracic duct and the right lymphatic duct. The *thoracic duct* collects lymph from the body inferior to the diaphragm and from the left side of the body superior to the diaphragm. The smaller *right lymphatic duct* collects lymph from the right side of the body superior to the diaphragm.

The **thoracic duct** begins inferior to the diaphragm at the level of vertebra L_2. The base of the thoracic duct is an expanded, saclike chamber called the **cisterna chyli** (KĪ-lē) (Figure 22–4●). The cisterna chyli receives lymph from the inferior part of the abdomen, the pelvis, and the lower limbs by way of the *right* and *left lumbar trunks* and the *intestinal trunk*.

The inferior segment of the thoracic duct lies anterior to the vertebral column. From the second lumbar vertebra, it passes posterior to the diaphragm alongside the aorta and ascends along the left side of the vertebral column to the level of the left clavicle. After collecting lymph from the *left bronchomediastinal trunk*, the *left subclavian trunk*, and the *left jugular trunk*, it empties into the left subclavian vein near the left internal jugular vein (Figure 22–4b●). Lymph collected from the left side of the head, neck, and thorax, as well as lymph from the entire body inferior to the diaphragm, reenters the venous circulation in this way.

The **right lymphatic duct** is formed by the merging of the *right jugular*, *right subclavian*, and *right bronchomediastinal*

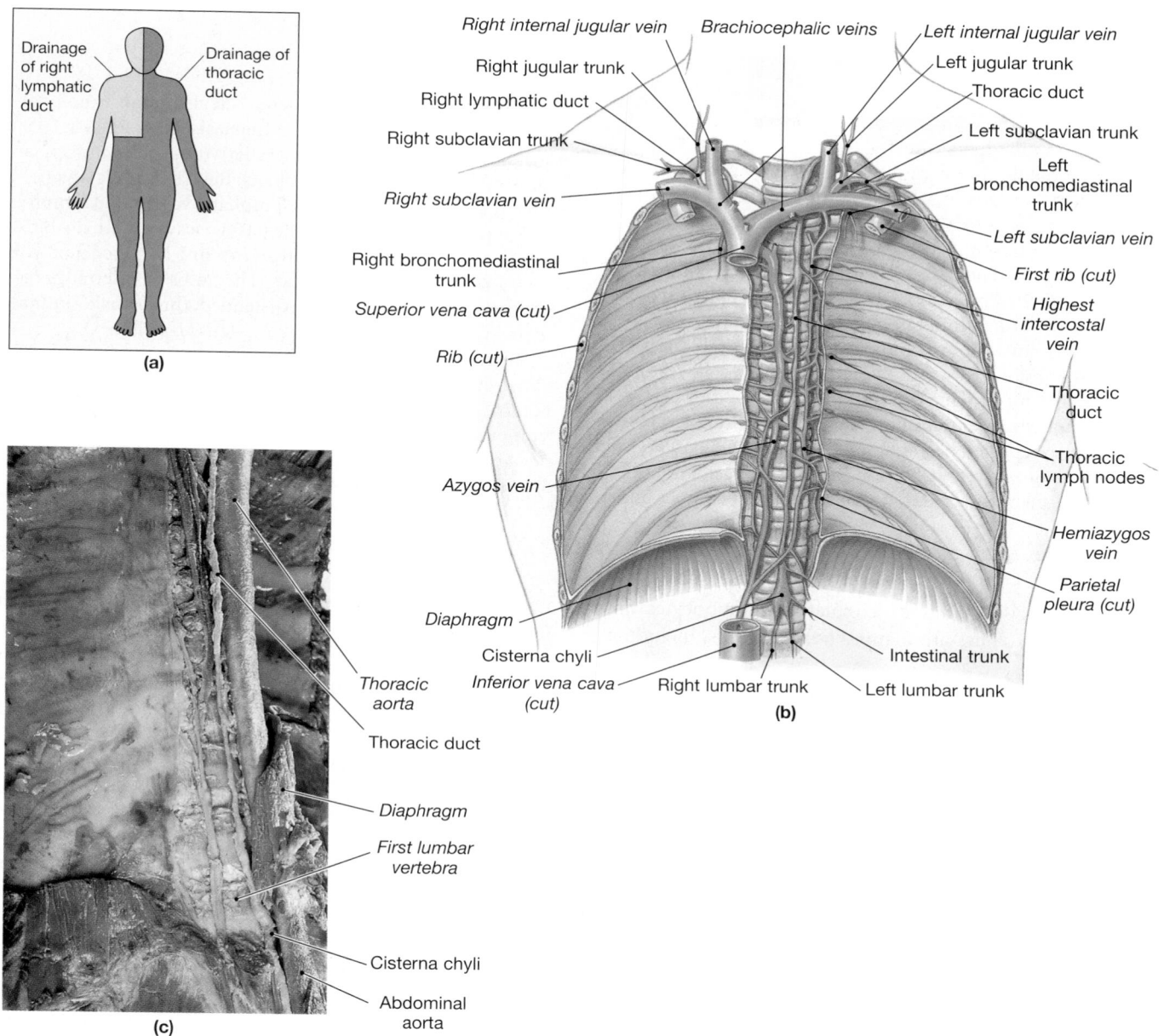

(a)

Drainage of right lymphatic duct

Drainage of thoracic duct

Right internal jugular vein
Brachiocephalic veins
Left internal jugular vein
Right jugular trunk
Left jugular trunk
Right lymphatic duct
Thoracic duct
Right subclavian trunk
Left subclavian trunk
Right subclavian vein
Left bronchomediastinal trunk
Right bronchomediastinal trunk
Left subclavian vein
Superior vena cava (cut)
First rib (cut)
Rib (cut)
Highest intercostal vein
Azygos vein
Thoracic duct
Thoracic lymph nodes
Hemiazygos vein
Diaphragm
Parietal pleura (cut)
Cisterna chyli
Intestinal trunk
Thoracic aorta
Inferior vena cava (cut)
Right lumbar trunk
Left lumbar trunk

(b)

Thoracic duct
Diaphragm
First lumbar vertebra
Cisterna chyli
Abdominal aorta

(c)

●**FIGURE 22–4**
The Relationship between the Lymphatic Ducts and the Venous System. **(a)** The thoracic duct carries lymph originating in tissues inferior to the diaphragm and from the left side of the upper body. The smaller right lymphatic duct services the rest of the body. **(b)** The thoracic duct empties into the left subclavian vein. The right lymphatic duct drains into the right subclavian vein. **(c)** The lymphatic vessels of the trunk. **ATLAS** Scans 13a,b

trunks in the area near the right clavicle. This duct empties into the right subclavian vein, delivering lymph from the right side of the body superior to the diaphragm (Figure 22–4b●).

◻ LYMPHOCYTES

Lymphocytes account for 20–30 percent of the circulating white blood cell population. However, circulating lymphocytes are only a small fraction of the total lymphocyte population. The body contains some 10^{12} lymphocytes, with a combined weight of over a kilogram.

Types of Lymphocytes

Three classes of lymphocytes are in blood: (1) **T** (**t**hymus-dependent) **cells**, (2) **B** (**b**one marrow–derived) **cells**, and (3) **NK** (**n**atural **k**iller) **cells**. Each type has distinctive biochemical and functional characteristics.

Lymphedema

Blockage of the lymphatic drainage from a limb produces **lymphedema** (lim-fe-DĒ-muh), a condition in which interstitial fluids accumulate and the limb gradually becomes swollen and grossly distended. If the condition persists, the connective tissues lose their elasticity and the swelling becomes permanent. Lymphedema by itself does not pose a major threat to life. The danger comes from the constant risk that an uncontrolled infection will develop in the affected area. Because the interstitial fluids are essentially stagnant, toxins and pathogens can accumulate and overwhelm the local defenses without fully activating the immune system.

Temporary lymphedema can result from tight clothing. Chronic lymphedema can result from the formation of scar tissue owing to repeated bacterial infections or from surgery that cuts or removes lymphatic vessels. Breast cancer surgery or radiation treatment can lead to lymphedema of the upper limb on the affected side. Lymphedema can also result from parasitic infections. For example, in **filariasis** (fil-a-RĪ-a-sis), larvae of a parasitic roundworm, generally *Wuchereria bancrofti*, are transmitted by mosquitoes or black flies. The adult worms form massive colonies within lymphatic vessels and lymph nodes. Repeated scarring of the passageways eventually blocks lymphatic drainage and produces extreme lymphedema with permanent distension of tissues. The limbs or external genitalia typically become grossly distended; this parasite-induced edema is called **elephantiasis** (el-e-fan-TĪ-a-sis).

Therapy for chronic lymphedema consists of treating infections with antibiotics and (when possible) reducing the swelling. One treatment involves the application of elastic wrappings that squeeze the tissue. This external compression elevates the hydrostatic pressure of the interstitial fluids and opposes the entry of additional fluid from the capillaries.

Approximately 80 percent of circulating lymphocytes are classified as T cells. There are many types of T cells, including the following:

- **Cytotoxic T cells**, which attack foreign cells or body cells infected by viruses. Their attack commonly involves direct contact. These lymphocytes are the primary cells involved in the production of *cell-mediated immunity*, or *cellular immunity*.

- **Helper T cells**, which stimulate the activation and function of both T cells and B cells.

- **Suppressor T cells**, which inhibit the activation and function of both T cells and B cells.

The interplay between suppressor and helper T cells helps establish and control the sensitivity of the immune response. For this reason, these cells are also known as *regulatory T cells*.

We will examine cytotoxic and regulatory T cells in the course of this chapter. The preceding is not a complete list, however; other types of T cells participate in the immune response. For example, *inflammatory T cells* stimulate regional inflammation and local defenses in an injured tissue, and *suppressor/inducer T cells* suppress B cell activity, but stimulate other T cells.

B cells account for 10–15 percent of circulating lymphocytes. When stimulated, B cells can differentiate into **plasma cells**, which are responsible for the production and secretion of *antibodies*—soluble proteins also known as *immunoglobulins*. ∞ p. 124 These proteins bind to specific chemical targets called **antigens**. Most antigens are pathogens, parts or products of pathogens, or other foreign compounds. Most antigens are proteins, but some lipids, polysaccharides, and nucleic acids can also stimulate antibody production. When an antibody binds to its target antigen, a chain of events begins that leads to the destruction of the target compound or organism. B cells are responsible for *antibody-mediated immunity*, which is also known as *humoral* ("liquid") *immunity* because antibodies occur in body fluids.

The remaining 5–10 percent of circulating lymphocytes are NK cells, also known as **large granular lymphocytes**. These lymphocytes attack foreign cells, normal cells infected with viruses, and cancer cells that appear in normal tissues. Their continuous policing of peripheral tissues has been called *immunological surveillance*.

Life Span and Circulation of Lymphocytes

Lymphocytes are not evenly distributed in the blood, bone marrow, spleen, thymus, and peripheral lymphoid tissues. The ratio of B cells to T cells varies with the tissue or organ in question. For example, B cells are seldom found in the thymus, but in blood, T cells outnumber B cells by a ratio of 8:1. The ratio changes to 1:1 in the spleen and 1:3 in bone marrow.

The lymphocytes in these organs are visitors, not residents. All types of lymphocytes move throughout the body, wandering through tissues and then entering blood vessels or lymphatic vessels for transport.

T cells move relatively quickly. For example, a wandering T cell may spend about 30 minutes in the blood, 5–6 hours in the spleen, and 15–20 hours in a lymph node. B cells, which are responsible for antibody production, move more slowly. A typical B cell spends about 30 hours in a lymph node before moving on.

Lymphocytes have relatively long life spans. Roughly 80 percent survive for 4 years, and some last 20 years or more. Throughout your life, you maintain normal lymphocyte populations by producing new lymphocytes in your bone marrow and lymphoid tissues.

Lymphocyte Production

In Chapter 19, we discussed *hemopoiesis*—the formation of the cellular elements of blood. ⟳ pp. 659, 669 *Erythropoiesis* (red blood cell formation) in adults is normally confined to bone marrow, but lymphocyte production, or **lymphopoiesis** (lim-fō-poy-Ē-sis), involves the bone marrow, thymus, and peripheral lymphoid tissues (Figure 22–5●).

Bone marrow plays the primary role in the maintenance of normal lymphocyte populations. Hemocytoblast divisions in the bone marrow of adults generate the lymphoid stem cells that produce all types of lymphocytes. Two distinct populations of lymphoid stem cells are produced in the bone marrow.

One group of lymphoid stem cells remains in the bone marrow (Figure 22–5a●). Divisions of these cells produce immature B cells and NK cells. B cell development involves intimate contact with large **stromal cells** (*stroma*, a bed) in the bone marrow. The cytoplasmic extensions of stromal cells contact or even wrap around the developing B cells. Stromal cells produce an immune system hormone, or *cytokine*, called *interleukin-7*, that promotes the differentiation of B cells. (We will consider cytokines and their varied effects in a later section.)

As they mature, B cells and NK cells enter the bloodstream and migrate to peripheral tissues (Figure 22–5c●). Most of the B cells move into lymph nodes, the spleen, or other lymphoid tissues. The NK cells migrate throughout the body, moving through peripheral tissues in search of abnormal cells.

The second group of lymphoid stem cells migrates to the thymus (Figure 22–5b●). There, these cells and their descendants develop further in an environment that is isolated from the general circulation by the **blood–thymus barrier**. Under the influence of thymic hormones, the lymphoid stem cells divide repeatedly, producing the various kinds of T cells. At least seven thymic hormones have been identified, although their precise functions and interactions have yet to be determined.

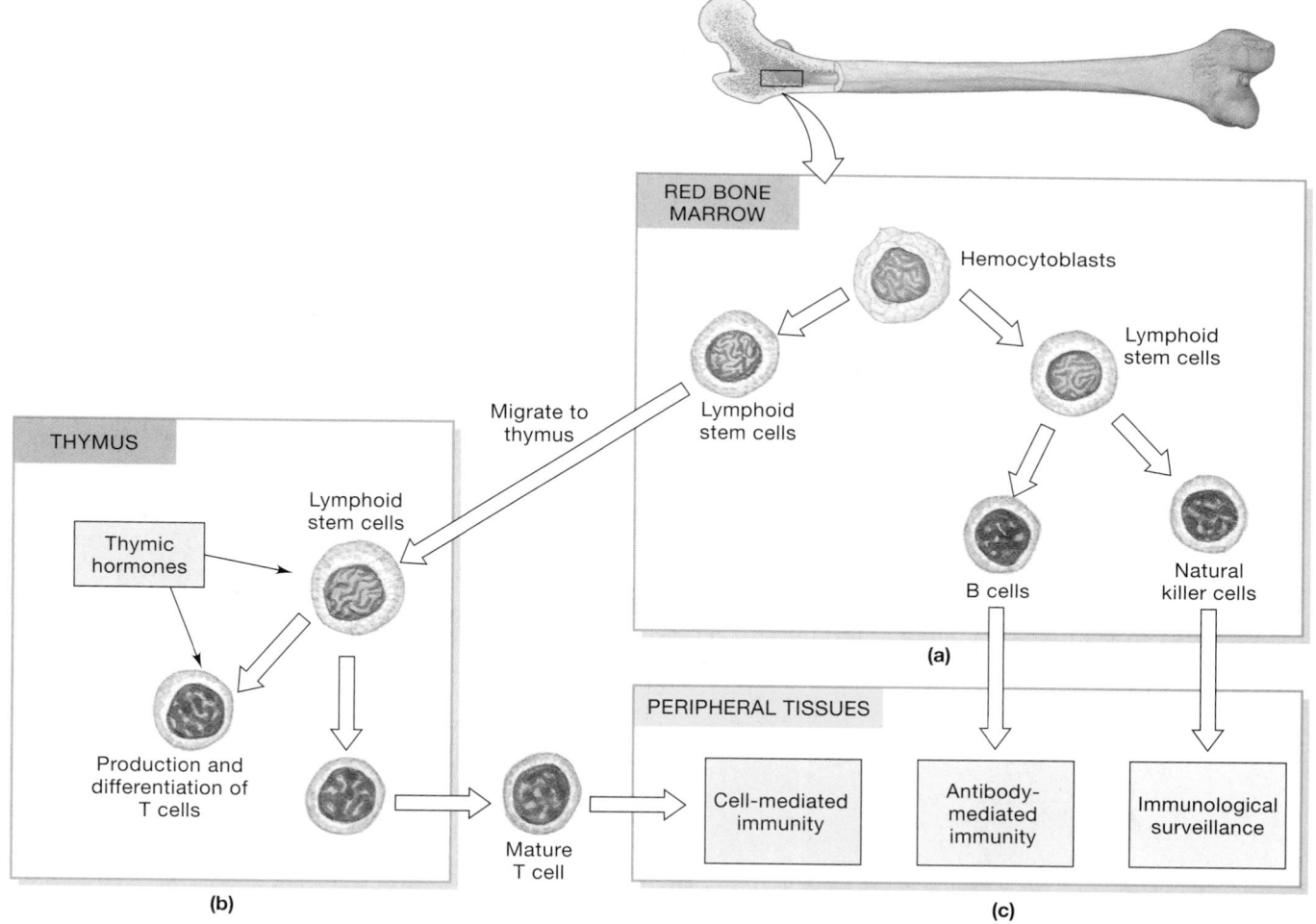

●**FIGURE 22–5**
The Derivation and Distribution of Lymphocytes. Hemocytoblast divisions produce stem cells with two fates. **(a)** One group remains in the bone marrow, producing daughter cells that mature into B cells or NK cells. **(b)** The other group migrates to the thymus, where subsequent divisions produce daughter cells that mature into T cells. The mature B cells, NK cells, and T cells circulate throughout the body in the bloodstream and then **(c)** temporarily reside in peripheral tissues.

When their development is nearing completion, T cells reenter the bloodstream and return to the bone marrow. They also travel to lymphoid tissues and organs, such as the spleen (Figure 22–5c●).

The T cells and B cells that migrate from their sites of origin retain the ability to divide. Their divisions produce daughter cells of the same type; for example, a dividing B cell produces other B cells, not T cells or NK cells. As we shall see, the ability to increase the number of lymphocytes of a specific type is important to the success of the immune response.

☐ LYMPHOID TISSUES

Lymphoid tissues are connective tissues dominated by lymphocytes. In a **lymphoid nodule**, or *lymphatic nodule*, the lymphocytes are densely packed in an area of areolar tissue (Figure 22–6●).

Lymphoid nodules occur in the connective tissue deep to the epithelia lining the respiratory, digestive, and urinary tracts. Typical nodules average about a millimeter in diameter, but the boundaries are not distinct, because no fibrous capsule surrounds them. They commonly have a central zone called a **germinal center**, which contains dividing lymphocytes (Figure 22–6b●).

MALT

The collection of lymphoid tissues linked with the digestive system is called the **mucosa-associated lymphoid tissue (MALT)**. Clusters of lymphoid nodules deep to the epithelial lining of the intestine are known as **aggregated lymphoid nodules**, or *Peyer's patches* (Figure 22–6a●). In addition, the walls of the *appendix*, or *vermiform* ("worm-shaped") *appendix*, a blind pouch that originates near the junction between the small and large intestines, contain a mass of fused lymphoid nodules.

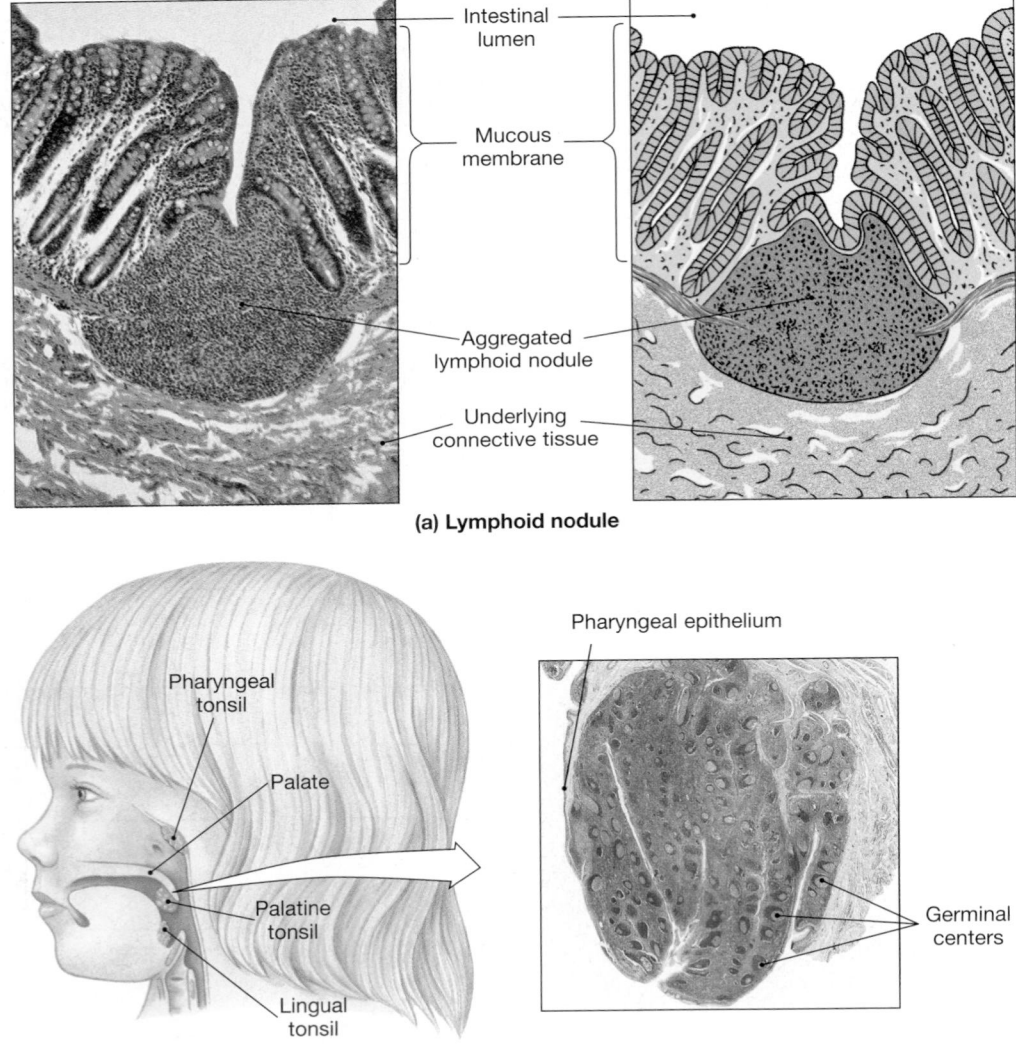

(a) Lymphoid nodule

(b) Pharyngeal tonsil

●**FIGURE 22–6**
Lymphoid Nodules. **(a)** A typical nodule in section. (LM × 17) **(b)** The positions of the tonsils and a tonsil in section. Notice the relatively pale germinal centers, where lymphocyte cell divisions occur.

Tonsils

The **tonsils** are large nodules in the walls of the pharynx (Figure 22–6b●). Most people have five tonsils. Left and right **palatine tonsils** are located at the posterior, inferior margin of the oral cavity, along the boundary with the pharynx. A single **pharyngeal tonsil**, often called the *adenoids*, lies in the posterior superior wall of the nasopharynx, and a pair of **lingual tonsils** lie deep to the mucous epithelium covering the base (pharyngeal portion) of the tongue. Because of their location, the latter are usually not visible unless they become infected and swollen.

The lymphocytes in a lymphoid nodule are not always able to destroy bacterial or viral invaders that have crossed the adjacent epithelium. If pathogens become established in a lymphoid nodule, an infection develops. Two examples are probably familiar to you: *tonsillitis*, an infection of one of the tonsils (generally the pharyngeal or palatine), and *appendicitis*, an infection of the appendix that begins in the lymphoid nodules. Treatment commonly consists of antibiotic therapy, sometimes combined with surgical removal of the infected tissues. **AM** Infected Lymphoid Nodules

☐ LYMPHOID ORGANS

A fibrous connective-tissue capsule separates the lymphoid organs—the *lymph nodes*, the *thymus*, and the *spleen*—from surrounding tissues.

Lymph Nodes

Lymph nodes are small, oval lymphoid organs ranging in diameter from 1 to 25 mm (to about 1 in.). Figure 22–1●, p. 780, shows the general pattern of lymph node distribution in the body. Each lymph node is covered by a capsule of dense connective tissue (Figure 22–7●). Bundles of collagen fibers extend from the capsule into the interior of the node. These fibrous partitions are called **trabeculae** (*trabecula*, a wall).

The shape of a typical lymph node resembles that of a kidney bean (Figure 22–7●). Blood vessels and nerves attach to the lymph node at the **hilus**, a shallow indentation. Two sets of lymphatic vessels, afferent lymphatics and efferent lymphatics, are connected to each lymph node.

1. **Afferent lymphatics** carry lymph to the lymph node from peripheral tissues. The afferent lymphatics penetrate the capsule of the lymph node on the side opposite the hilus.

2. **Efferent lymphatics** are attached to the lymph node at the hilus. These vessels carry lymph away from the lymph node and toward the venous circulation.

LYMPH FLOW Lymph delivered by the afferent lymphatics flows through the lymph node within a network of sinuses, open passageways with incomplete walls (Figure 22–7●). Lymph first enters a *subcapsular sinus*, which contains a meshwork of branching

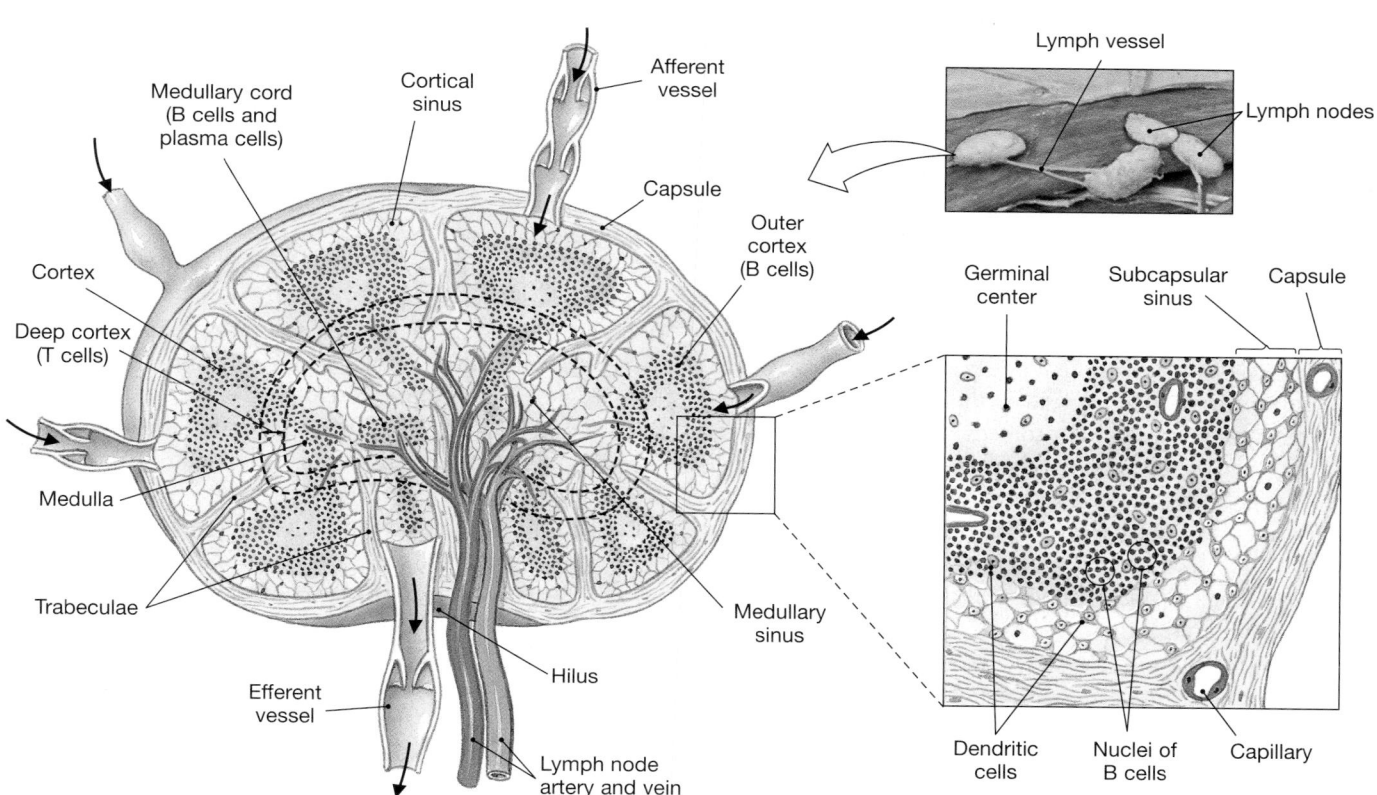

●**FIGURE 22–7**
The Structure of a Lymph Node ATLAS Plate 8.5a

reticular fibers, macrophages, and dendritic cells. **Dendritic cells** are involved in the initiation of the immune response; we will consider their role in a later section. After passing through the subcapsular sinus, lymph flows through the **outer cortex** of the node. The outer cortex contains B cells within germinal centers that resemble those of lymphoid nodules.

Lymph then continues through lymph sinuses in the **deep cortex** (*paracortical area*). Lymphocytes leave the bloodstream and enter the lymph node by crossing the walls of blood vessels within the deep cortex, which is dominated by T cells.

After flowing through the sinuses of the deep cortex, lymph continues into the core, or **medulla**, of the lymph node. The medulla contains B cells and plasma cells organized into elongate masses known as **medullary cords**. After passing through a network of sinuses in the medulla, lymph enters the efferent lymphatics at the hilus.

LYMPH NODE FUNCTION A lymph node functions like a kitchen water filter, purifying lymph before it reaches the venous circulation. As lymph flows through a lymph node, at least 99 percent of the antigens in the lymph are removed. Fixed macrophages in the walls of the lymphatic sinuses engulf debris or pathogens in lymph as it flows past. Antigens removed in this way are then processed by the macrophages and "presented" to nearby lymphocytes. Other antigens bind to receptors on the surfaces of dendritic cells, where they can stimulate lymphocyte activity. This process, *antigen presentation*, is generally the first step in the activation of the immune response.

In addition to filtering, lymph nodes provide an early-warning system. Any infection or other abnormality in a peripheral tissue will introduce abnormal antigens into the interstitial fluid and thus into the lymph leaving the area. These antigens will then stimulate macrophages and lymphocytes in nearby lymph nodes.

To protect a house against intrusion, you might guard all the entrances and exits or place traps by the windows and doors. The distribution of lymphoid tissues and lymph nodes follows such a pattern. The largest lymph nodes are located where peripheral lymphatics connect with the trunk, such as in the groin, the axillae, and the base of the neck. These nodes are often called *lymph glands*. Because lymph is monitored in the cervical, inguinal, or axillary lymph nodes, potential problems can be detected and dealt with before they affect the vital organs of the trunk. Aggregations of lymph nodes also exist in the mesenteries of the gut, near the trachea and passageways leading to the lungs, and in association with the thoracic duct (Figure 22–4•, p. 783). These lymph nodes protect against pathogens and other antigens within the digestive and respiratory systems.

A minor injury commonly produces a slight enlargement of the nodes along the lymphatic vessels draining the region. This symptom, often called "swollen glands," typically indicates inflammation or infection of peripheral structures. The enlargement generally results from an increase in the number of lymphocytes and phagocytes in the node in response to a minor, localized infection. Chronic or excessive enlargement of lymph nodes constitutes **lymphadenopathy** (lim-fad-e-NOP-a-thē), a condition that may occur in response to bacterial or viral infections, endocrine disorders, or cancer.

Lymphatic vessels are located in almost all portions of the body except the central nervous system, and lymphatic capillaries offer little resistance to the passage of cancer cells. As a result, metastasizing cancer cells commonly spread along lymphatic vessels. Under these circumstances, the lymph nodes serve as way stations for migrating cancer cells. Thus, an analysis of lymph nodes can provide information on the spread of the cancer cells, and such information helps determine the appropriate therapies. In Chapter 29, we will discuss one example: identifying the stages of breast cancer by the degree of nodal involvement. *Lymphomas*, one group of cancers originating in the lymphatic system, are discussed in the *Applications Manual*.

AM Lymphomas

The Thymus

The **thymus** is located in the mediastinum, generally just posterior to the sternum. It is pink and has a grainy consistency. In newborn infants and young children, the thymus is relatively large, commonly extending from the base of the neck to the superior border of the heart. The thymus reaches its greatest size (relative to body size) in the first year or two after birth. Although the organ continues to increase in size throughout childhood, the body as a whole grows even faster, so the size of the thymus relative to that of the other organs in the mediastinum gradually decreases. The thymus reaches its maximum absolute size, at a weight of about 40 g (1.4 oz), just before puberty. After puberty, it gradually diminishes in size and becomes increasingly fibrous, a process called *involution*. By the time an individual reaches age 50, the thymus may weigh less than 12 g (0.3 oz). It has been suggested that the gradual decrease in the size and secretory abilities of the thymus may make the elderly more susceptible to disease.

The capsule that covers the thymus divides it into two **thymic lobes** (Figure 22–8a,b•). **Septa**—fibrous partitions (singular, *septum*) that originate at the capsule—divide the lobes into **lobules**, which average 2 mm in width (Figure 22–8b,c•). Each lobule consists of a densely packed outer **cortex** and a paler, central **medulla**. Lymphocytes in the cortex are dividing; as the T cells mature, they migrate into the medulla. After roughly three weeks, these T cells leave the thymus by entering one of the medullary blood vessels.

Lymphocytes in the cortex are arranged in clusters that are completely surrounded by **reticular epithelial cells**. These cells, which developed from epithelial cells of the embryo, also encircle the blood vessels of the cortex. The reticular epithelial cells (1) maintain the blood–thymus barrier and (2) secrete the thymic hormones that stimulate stem cell divisions and T cell differentiation.

As they mature, T cells leave the cortex and enter the medulla of the thymus. The medulla has no blood–thymus barrier. The reticular epithelial cells in the medulla cluster together in concentric layers, forming distinctive structures known as **Hassall's**

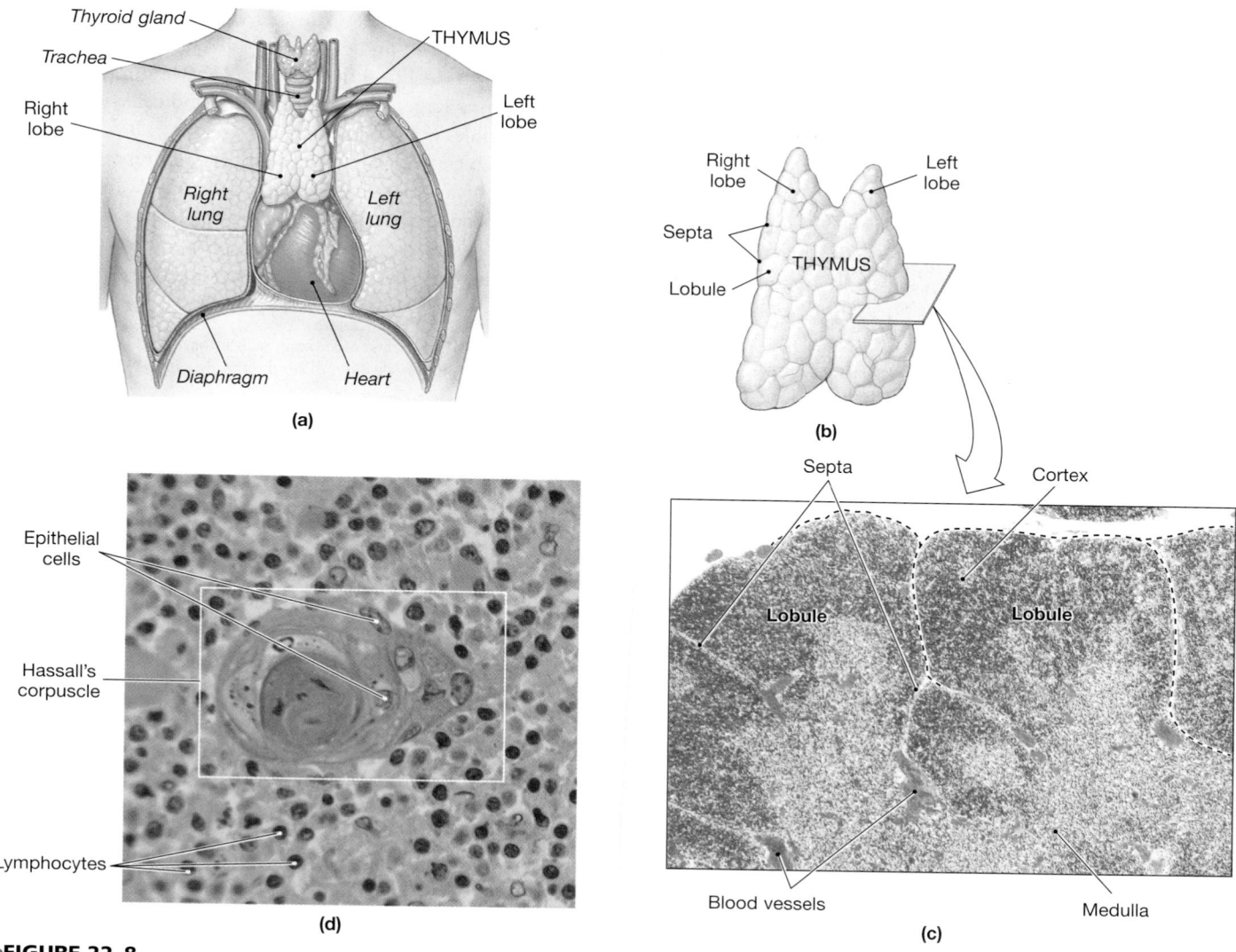

●**FIGURE 22–8**
The Thymus. **(a)** The appearance and position of the thymus in relation to other organs in the chest. **(b)** Anatomical landmarks on the thymus. **(c)** Fibrous septa divide the tissue of the thymus into lobules resembling interconnected lymphatic nodules. (LM × 40) **(d)** At higher magnification, we can see the unusual structure of Hassall's corpuscles. The small cells are lymphocytes in various stages of development. (LM × 532) **ATLAS** Plate 7.6a

corpuscles (Figure 22–8d●). Despite their imposing appearance, the function of Hassall's corpuscles remains unknown. T cells in the medulla can enter or leave the bloodstream across the walls of blood vessels in this region or within one of the efferent lymphatics that collect lymph from the thymus.

HORMONES OF THE THYMUS The thymus produces several hormones that are important to the development and maintenance of normal immunological defenses. *Thymosin* (THĪ-mō-sin) is the name originally given to an extract from the thymus that promotes the development and maturation of lymphocytes. This thymic extract actually contains a blend of several complementary hormones: *thymosin-a, thymosin-b, thymosin V, thymopoietin, thymulin,* and others. The term *thymosins* is now sometimes used to refer to all thymic hormones.

The Spleen

The adult **spleen** contains the largest collection of lymphoid tissue in the body. In essence, the spleen performs the same functions for blood that lymph nodes perform for lymph. Functions of the spleen can be summarized as (1) the removal of abnormal blood cells and other blood components by phagocytosis, (2) the storage of iron from recycled red blood cells, and (3) the initiation of immune responses by B cells and T cells in response to antigens in circulating blood.

ANATOMY OF THE SPLEEN The spleen is about 12 cm (5 in.) long and weighs, on average, nearly 160 g (5.6 oz). In gross dissection, the spleen is deep red, owing to the blood it contains. The spleen lies along the curving lateral border of the stomach, extending between the 9th and 11th ribs on the left side. It is attached to the lateral border of the stomach by the **gastrosplenic ligament**, a broad band of mesentery (Figure 22–9a●).

The spleen has a soft consistency, so its shape primarily reflects its association with the structures around it. The spleen is in contact with the stomach, the left kidney, and the muscular diaphragm. The *diaphragmatic surface* is smooth and convex, conforming to the shape of the diaphragm and body wall. The *visceral surface* contains indentations that conform to the shape of the stomach (the *gastric area*) and the kidney (the *renal area*) (Figure 22–9b●). Splenic blood vessels and lymphatic vessels communicate with the spleen on the visceral surface at the **hilus**, a groove marking the border between the gastric and renal areas. The **splenic artery**, the **splenic vein**, and the lymphatic vessels draining the spleen are attached at the hilus.

HISTOLOGY OF THE SPLEEN The spleen is surrounded by a capsule containing collagen and elastic fibers.[1] The cellular components within constitute the **pulp** of the spleen (Figure 22–9c●). **Red pulp** contains large quantities of red blood cells, whereas **white pulp** resembles lymphoid nodules.

The splenic artery enters at the hilus and branches to produce a number of arteries that radiate outward toward the capsule. These **trabecular arteries** in turn branch extensively, and their finer branches are surrounded by areas of white pulp. Capillaries then discharge the blood into the red pulp.

---

[1]The spleens of dogs, cats, and other mammals of the order *Carnivora* have extensive layers of smooth muscle that can contract to eject blood into the bloodstream. The human spleen lacks those muscle layers and cannot contract.

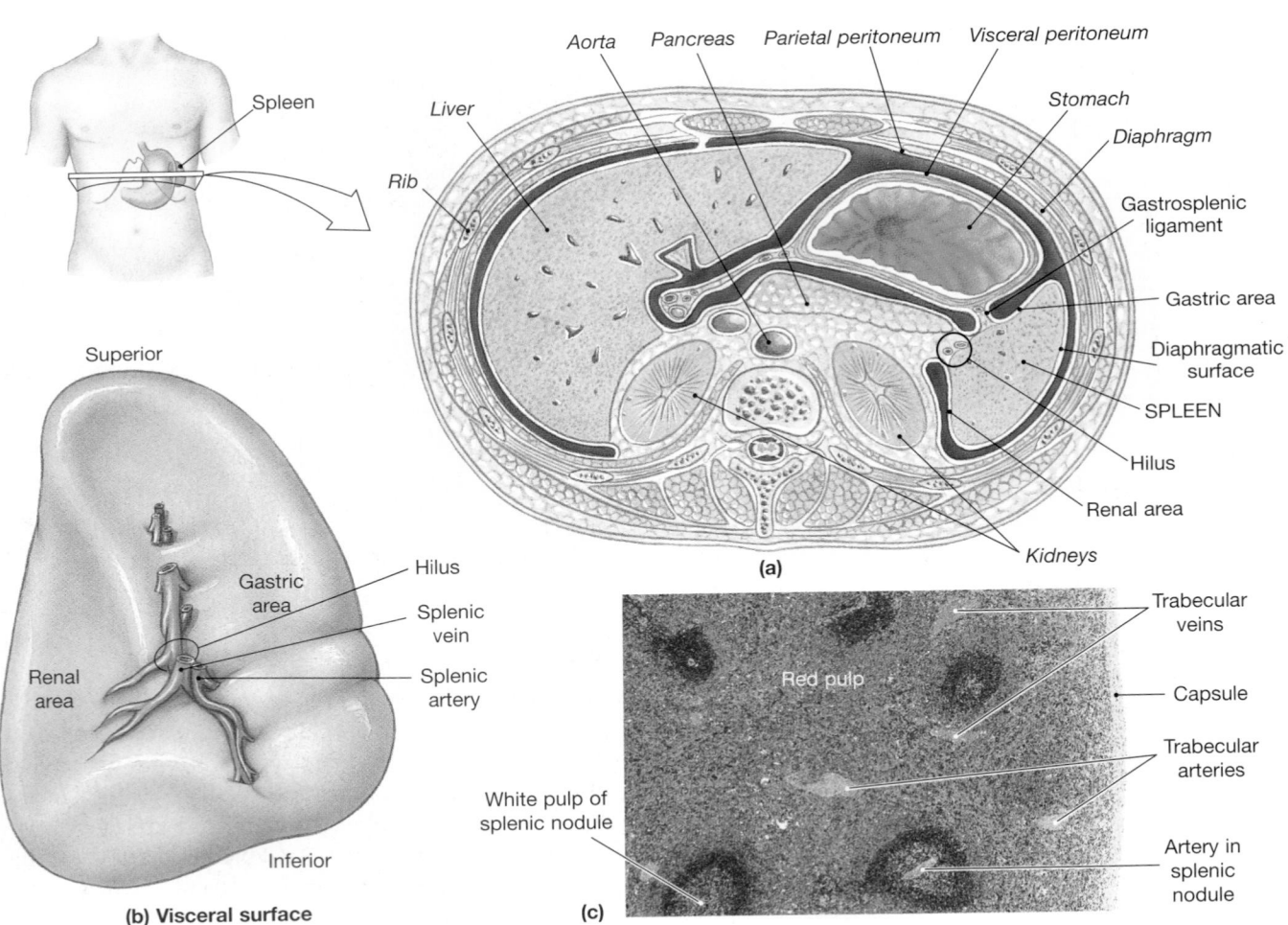

●**FIGURE 22–9**
The Spleen. **(a)** A transverse section through the trunk, showing the typical position of the spleen within the abdominopelvic cavity. The shape of the spleen roughly conforms to the shapes of adjacent organs. **(b)** The external appearance of the intact spleen, showing major anatomical landmarks. Compare this view with that of part (a). **(c)** The histological appearance of the spleen. White pulp is dominated by lymphocytes; it appears blue because the nuclei of lymphocytes stain very darkly. Red pulp contains a preponderance of red blood cells. (LM × 38) **ATLAS** Plates 7.7d, 7.8d,e, 7.9b; Scan 9c

The cell population of the red pulp includes all the normal components of circulating blood, plus fixed and free macrophages. The structural framework of the red pulp consists of a network of reticular fibers. The blood passes through this meshwork and enters large sinusoids, also lined by fixed macrophages. The sinusoids empty into small veins, which ultimately collect into **trabecular veins** that continue toward the hilus.

This circulatory arrangement gives the phagocytes of the spleen an opportunity to identify and engulf any damaged or infected cells in circulating blood. Lymphocytes are scattered throughout the red pulp, and the area surrounding the white pulp has a high concentration of macrophages and dendritic cells. Thus, any microorganism or other antigen in the blood will quickly come to the attention of the splenic lymphocytes.

An impact to the left side of the abdomen can distort or damage the spleen. Such injuries are known risks of contact sports, such as football and hockey, and of more solitary athletic activities, such as skiing and sledding. However, the spleen tears so easily that a seemingly minor blow to the side can rupture the capsule. The result is serious internal bleeding and eventual circulatory shock. The spleen can also be damaged by infection, inflammation, or invasion by cancer cells.

Because the spleen is relatively fragile, it is very difficult to repair surgically. (Sutures typically tear out before they have been tensed enough to stop the bleeding.) A severely ruptured spleen is removed, a process called a **splenectomy** (sple-NEK-to-mē). A person without a spleen survives, but has a greater risk of bacterial infection, particularly involving pneumococcal bacteria than do individuals with a functional spleen. **AM** Disorders of the Spleen

☐ THE LYMPHATIC SYSTEM AND BODY DEFENSES

The human body has multiple defense mechanisms, which can be sorted into two general categories:

1. **Nonspecific defenses** do not distinguish one type of threat from another. Their response is the same, regardless of the type of invading agent. These defenses, which are present at birth, include *physical barriers, phagocytic cells, immunological surveillance, interferons, complement, inflammation,* and *fever.* They provide a defensive capability known as **nonspecific resistance**.

2. **Specific defenses** protect against particular threats. For example, a specific defense may protect against infection by one type of bacterium, but be ineffective against other bacteria and viruses. Many specific defenses develop after birth as a result of accidental or deliberate exposure to environmental hazards. *Specific defenses depend on the activities of lymphocytes.* The body's specific defenses produce a state of protection known as immunity, or **specific resistance**.

Nonspecific and specific resistances are complementary. Both must function normally to provide adequate resistance to infection and disease.

✓ How would blockage of the thoracic duct affect the circulation of lymph?

✓ If the thymus failed to produce thymic hormones, which population of lymphocytes would be affected?

✓ Why do lymph nodes enlarge during some infections?

Answers start on page Q-1

22–3 NONSPECIFIC DEFENSES

Objectives

- List the body's nonspecific defenses and explain the function of each.
- Describe the components and mechanisms of each nonspecific defense.

Nonspecific defenses prevent the approach, deny the entrance, or limit the spread of microorganisms or other environmental hazards. Seven major categories of nonspecific defenses are summarized in Figure 22–10●.

1. *Physical barriers* keep hazardous organisms and materials outside the body. For example, a mosquito that lands on your head may be unable to reach the surface of the scalp if you have a full head of hair.

2. *Phagocytes* are cells that engulf pathogens and cell debris. Examples of phagocytes are the macrophages of peripheral tissues and the microphages of blood.

3. *Immunological surveillance* is the destruction of abnormal cells by NK cells in peripheral tissues.

4. *Interferons* are chemical messengers that coordinate the defenses against viral infection.

5. *Complement* is a system of circulating proteins that assist antibodies in the destruction of pathogens.

6. The *inflammatory response* is a local response to injury or infection that is directed at the tissue level. Inflammation tends to restrict the spread of an injury as well as combat an infection.

7. *Fever* is an elevation of body temperature that accelerates tissue metabolism and defenses.

☐ PHYSICAL BARRIERS

To cause trouble, an antigenic compound or pathogen must enter the body tissues, and so must cross an epithelium. The epithelial covering of the skin has multiple layers, a keratin coating, and a network of desmosomes that lock adjacent cells together. ∞ pp. 113, 157 These barriers provide very effective protection for underlying tissues. Even along the more delicate internal passageways of the respiratory, digestive, and urinary tracts, the epithelial cells are tied together by tight junctions and generally are supported by a dense and fibrous basal lamina.

●**FIGURE 22–10**
Nonspecific Defenses. Nonspecific defenses deny pathogens access to the body or destroy them without distinguishing among specific types.

In addition to the barriers posed by the epithelial cells, most epithelia are protected by specialized accessory structures and secretions. The hairs on most areas of your body surface provide some protection against mechanical abrasion (especially on the scalp), and they often prevent hazardous materials or insects from contacting your skin surface. The epidermal surface also receives the secretions of sebaceous and sweat glands. These secretions, which flush the surface to wash away microorganisms and chemical agents, may also contain bactericidal chemicals, destructive enzymes (*lysozymes*), and antibodies.

The epithelia lining the digestive, respiratory, urinary, and reproductive tracts are more delicate, but they are equally well defended. Mucus bathes most surfaces of your digestive tract, and your stomach contains a powerful acid that can destroy many pathogens. Mucus moves across the respiratory tract lining, urine flushes the urinary passageways, and glandular secretions do the same for the reproductive tract. Special enzymes, antibodies, and an acidic pH add to the effectiveness of these secretions.

☐ PHAGOCYTES

Phagocytes perform janitorial and police services in peripheral tissues, removing cellular debris and responding to invasion by foreign compounds or pathogens. Phagocytes represent the "first line" of cellular defense against pathogenic invasion. Many phagocytes attack and remove microorganisms even before lymphocytes detect their presence. The human body has two general classes of phagocytic cells: *microphages* and *macrophages*.

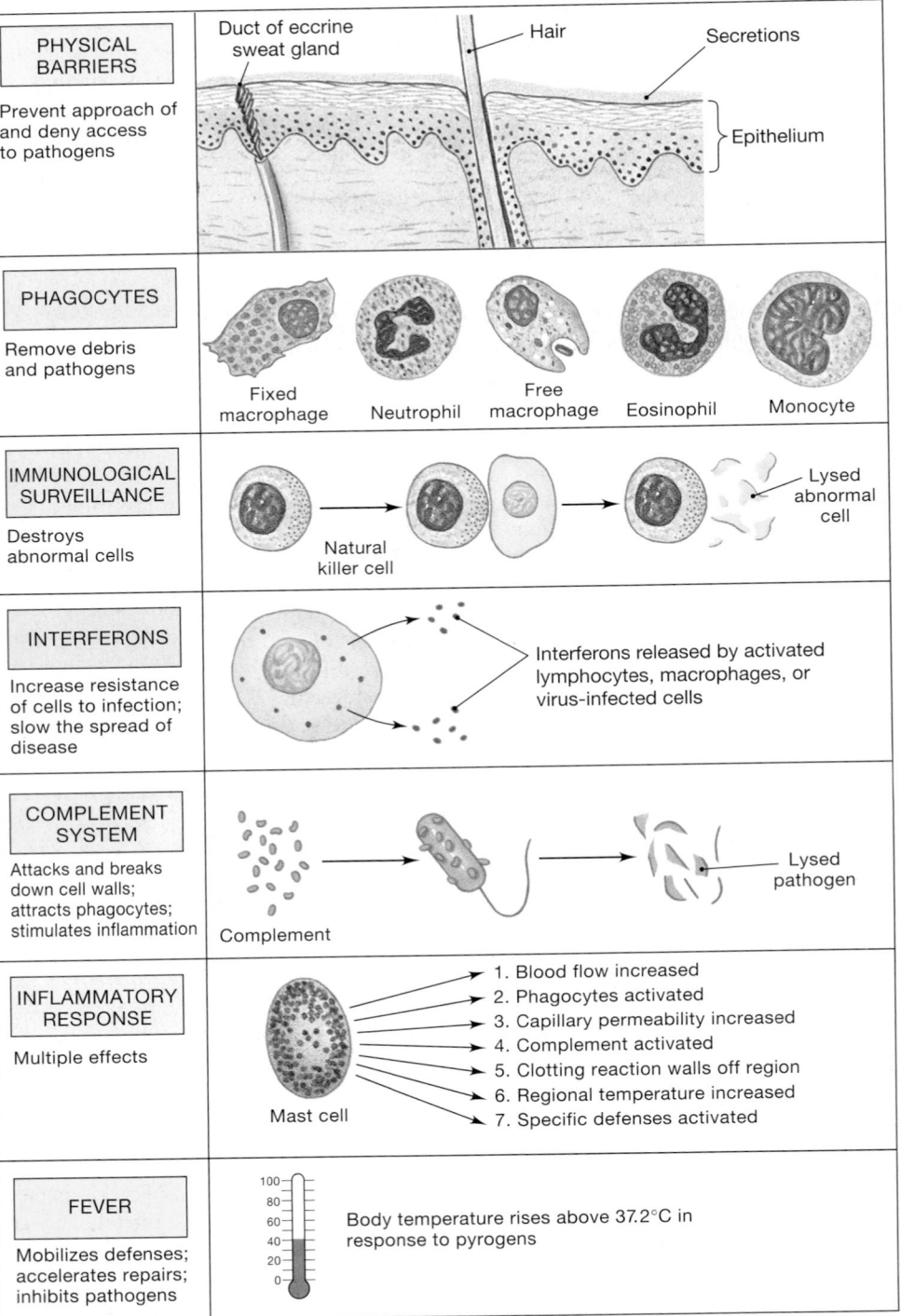

Microphages

Microphages are the neutrophils and eosinophils that normally circulate in the blood. These phagocytic cells leave the bloodstream and enter peripheral tissues that have been subjected to injury or infection. As we noted in Chapter 19, neutrophils are abundant, mobile, and quick to phagocytize cellular debris or invading bacteria.

⊂⊃ p. 667 Eosinophils, which are less abundant, target foreign compounds or pathogens that have been coated with antibodies.

Macrophages

Macrophages are large, actively phagocytic cells. Your body contains several types of macrophages, and most are derived from the monocytes of the circulating blood. Macrophages either are fixed permanently or move freely. Although no organs or tissues are purely phagocytic, almost every tissue in the body shelters resident or visiting macrophages. This relatively diffuse collection of phagocytic cells has been called the **monocyte–macrophage system**, or the *reticuloendothelial system*.

An activated macrophage may respond to a pathogen in several ways. For example,

- It may engulf a pathogen or other foreign object and destroy it with lysosomal enzymes.

- It may bind to or remove a pathogen from the interstitial fluid, but be unable to destroy the invader until assisted by other cells.

- It may destroy its target by releasing toxic chemicals, such as *tumor necrosis factor*, nitric oxide, or hydrogen peroxide, into the interstitial fluid.

We will consider those responses further in a later section.

FIXED MACROPHAGES

Fixed macrophages, or *histiocytes*, are permanent residents of specific tissues and organs. These cells are normally incapable of movement, so the objects of their phagocytic attention must diffuse or otherwise move through the surrounding tissue until they are within range. Fixed macrophages are scattered among connective tissues, usually in close association with collagen or reticular fibers. Their presence has been noted in the papillary and reticular layers of the dermis, in the subarachnoid space of the meninges, and in bone marrow. In some organs, the fixed macrophages have special names: **Microglia** are macrophages in the central nervous system, and **Kupffer cells** are macrophages located in and around the liver sinusoids.

FREE MACROPHAGES

Free macrophages, or *mobile macrophages*, travel throughout the body, arriving at the site of an injury by migration through adjacent tissues or by recruitment from the circulating blood. Some tissues contain free macrophages with distinctive characteristics; for example, the exchange surfaces of the lungs are patrolled by **alveolar macrophages**, also known as *phagocytic dust cells*.

Both classes of macrophages are derived from the monocytes of the blood. The primary difference is that fixed macrophages are always located within a given tissue, but free macrophages come and go. During an infection, this distinction commonly breaks down, because the fixed macrophages may lose their attachments and begin roaming around the damaged tissue.

Movement and Phagocytosis

Free macrophages and microphages share a number of functional characteristics:

- Both can move through capillary walls by squeezing between adjacent endothelial cells, a process known as *emigration*, or *diapedesis*. ⚬⚬ p. 667 The endothelial cells in an injured area develop membrane "markers" that let passing blood cells know that something is wrong. The cells then attach to the endothelial lining and migrate into the surrounding tissues.

- Both may be attracted to or repelled by chemicals in the surrounding fluids, a phenomenon called **chemotaxis**. They are

particularly sensitive to cytokines released by other body cells and to the chemicals released by pathogens.

- For both, phagocytosis begins with **adhesion**, the attachment of the phagocyte to its target. In adhesion, receptors on the cell membrane of the phagocyte bind to the surface of the target. Adhesion is followed by the formation of a vesicle containing the bound target, as detailed in Figure 3–25●, p. 97. The contents of the vesicle are then digested when the vesicle fuses with lysosomes or peroxisomes.

All phagocytic cells function in much the same way, although the target of phagocytosis may differ from one type of cell to another. The life span of an actively phagocytic cell can be rather brief. For example, most neutrophils expire before they have engulfed more than 25 bacteria, and in an infection a neutrophil may attack that many in an hour.

☐ IMMUNOLOGICAL SURVEILLANCE

The immune system generally ignores the body's own cells unless they become abnormal in some way. Natural killer (NK) cells are responsible for recognizing and destroying abnormal cells when they appear in peripheral tissues. The constant monitoring of normal tissues by NK cells is called **immunological surveillance**.

The cell membrane of an abnormal cell generally contains antigens that are not found on the membranes of normal cells. NK cells recognize an abnormal cell by detecting the presence of those antigens. The recognition mechanism differs from that used by T cells or B cells, which are activated only by exposure to a *specific* antigen at a *specific* site on a cell membrane. An NK cell responds to a variety of abnormal antigens that may appear anywhere on a cell membrane. NK cells are therefore much less selective about their targets than are other lymphocytes; if a membrane contains abnormal antigens, it will be attacked. As a result, NK cells are highly versatile: A single NK cell can attack bacteria in the interstitial fluid, body cells infected with virus, or cancer cells.

NK cells also respond much more rapidly than T cells or B cells. The activation of T cells and B cells involves a relatively complex and time-consuming sequence of events; NK cells respond immediately on contact with an abnormal cell.

NK Cell Activation

Activated NK cells react in a predictable way:

STEP 1: If a cell has unusual surface proteins or other components in its cell membrane, an NK cell recognizes that other cell as abnormal. Such recognition activates the NK cell, which then adheres to its target cell.

STEP 2: The Golgi apparatus moves around the nucleus until the maturing face points directly toward the abnormal cell. The process might be compared to the rotation of a tank turret to point the cannon toward the enemy. A flood of secretory vesicles is then produced at the Golgi apparatus. These vesicles, which contain proteins called **perforins**, travel through the cytoplasm toward the cell surface.

STEP 3: The perforins are released at the cell surface by exocytosis and diffuse across the gap separating the NK cell from its target.

STEP 4: On reaching the opposing cell membrane, perforin molecules interact with one another and with the membrane to create a network of pores in it (Figure 22–11●). These pores are large enough to permit the free passage of ions, proteins, and other intracellular materials. As a result, the target cell can no longer maintain its internal environment, and it quickly disintegrates.

It is not clear why perforin does not affect the membrane of the NK cell itself. NK cell membranes contain a second protein, called *protectin*, which may be responsible for binding and inactivating perforin.

NK cells attack cancer cells and cells infected with viruses. Cancer cells probably appear throughout life, but their cell membranes generally contain unusual proteins called **tumor-specific antigens**, which NK cells recognize as abnormal. The affected cells are then destroyed, preserving tissue integrity. Unfortunately, some cancer cells avoid detection, perhaps because they lack tumor-specific antigens or because these antigens are covered in some way. Other cancer cells are able to destroy the NK cells that detect them. This process of avoiding detection or neutralizing body defenses is called **immunological escape**. Once immunological escape has occurred, cancer cells can multiply and spread without interference by NK cells.

NK cells are also important in fighting viral infections. Viruses reproduce inside cells, beyond the reach of circulating antibodies. However, infected cells incorporate viral antigens into their cell membranes, and NK cells recognize these infected cells as abnormal. By destroying them, NK cells can slow or prevent the spread of a viral infection.

□ INTERFERONS

Interferons (in-ter-FĒR-onz) are small proteins released by activated lymphocytes and macrophages and by tissue cells infected with viruses. On reaching the membrane of a normal cell, an interferon binds to surface receptors on the cell and, via second messengers, triggers the production of **antiviral proteins** in the cytoplasm. Antiviral proteins do not interfere with the entry of viruses, but they do interfere with viral replication inside the cell. In addition to their role in slowing the spread of viral infections, interferons stimulate the activities of macrophages and NK cells.

At least three types of interferon exist, each of which has additional specialized functions: (1) **Alpha- (α) interferons**, produced by several types of leukocytes, attract and stimulate NK cells; (2) **beta- (β) interferons**, secreted by fibroblasts, slow inflammation in a damaged area; and (3) **gamma- (γ) interferons**, secreted by T cells and NK cells, stimulate macrophage activity. Most cells other than lymphocytes and macrophages respond to viral infection by secreting beta-interferon.

Interferons are examples of **cytokines** (SĪ-tō-kīnz)—chemical messengers released by tissue cells to coordinate local activities. ∞ p. 605 Cytokines produced by most cells are used only for paracrine communication—that is, cell-to-cell communication within one tissue. But the cytokines released by cellular defenders also act as hormones, affecting cells and tissues throughout the body. We will discuss their role in the regulation of specific defenses in a later section.

□ COMPLEMENT

Your plasma contains 11 special **complement proteins (C)**, which form the **complement system**. The term *complement* refers to the fact that this system complements the action of antibodies.

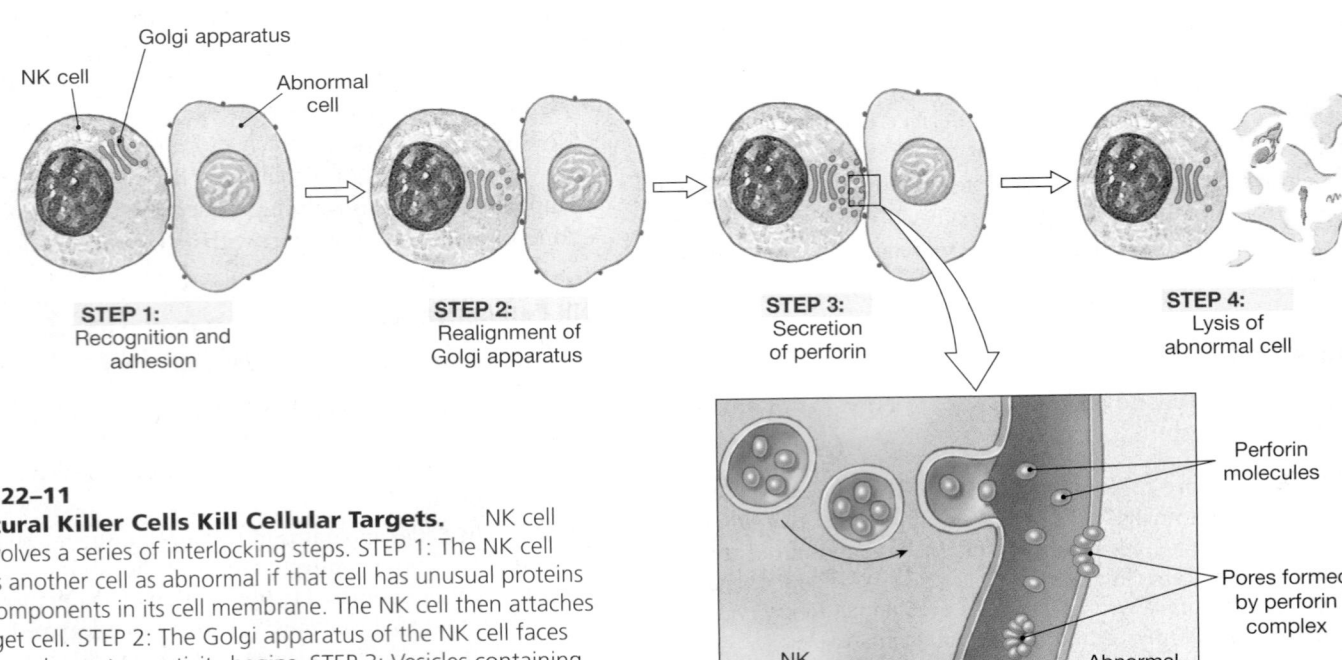

●FIGURE 22–11
How Natural Killer Cells Kill Cellular Targets. NK cell activity involves a series of interlocking steps. STEP 1: The NK cell recognizes another cell as abnormal if that cell has unusual proteins or other components in its cell membrane. The NK cell then attaches to the target cell. STEP 2: The Golgi apparatus of the NK cell faces the target, and secretory activity begins. STEP 3: Vesicles containing perforin are released by exocytosis. STEP 4: Perforin lyses the target cell by creating large pores through the cell membrane.

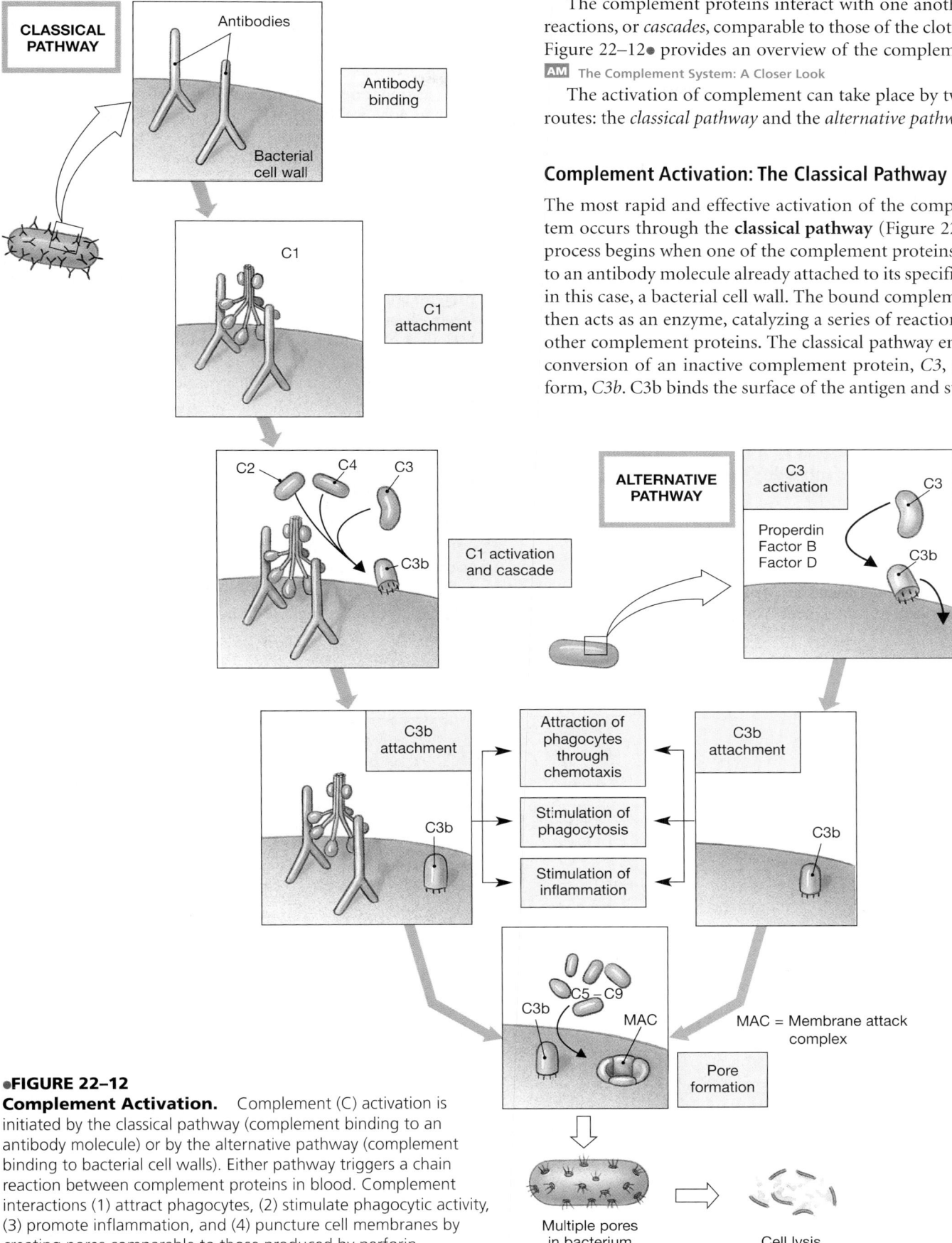

The complement proteins interact with one another in chain reactions, or *cascades*, comparable to those of the clotting system. Figure 22–12● provides an overview of the complement system.

AM The Complement System: A Closer Look

The activation of complement can take place by two different routes: the *classical pathway* and the *alternative pathway*.

Complement Activation: The Classical Pathway

The most rapid and effective activation of the complement system occurs through the **classical pathway** (Figure 22–12●). The process begins when one of the complement proteins (*C1*) binds to an antibody molecule already attached to its specific antigen— in this case, a bacterial cell wall. The bound complement protein then acts as an enzyme, catalyzing a series of reactions involving other complement proteins. The classical pathway ends with the conversion of an inactive complement protein, *C3*, to an active form, *C3b*. C3b binds the surface of the antigen and subsequently

●**FIGURE 22–12**
Complement Activation.   Complement (C) activation is initiated by the classical pathway (complement binding to an antibody molecule) or by the alternative pathway (complement binding to bacterial cell walls). Either pathway triggers a chain reaction between complement proteins in blood. Complement interactions (1) attract phagocytes, (2) stimulate phagocytic activity, (3) promote inflammation, and (4) puncture cell membranes by creating pores comparable to those produced by perforin.

stimulates phagocytosis and promotes inflammation. It also triggers additional reactions that end with the creation of the membrane attack complex from complement factors *C5–C9*.

Complement Activation: The Alternative Pathway

A less effective, slower activation of the complement system occurs in the absence of antibody molecules. This **alternative pathway**, or *properdin pathway*, is important in the defense against bacteria, some parasites, and virus-infected cells. The pathway begins when several complement proteins—including **properdin**, or *factor P, factor B*, and *factor D*—interact in the plasma (Figure 22–12●). The interaction can be triggered by exposure to foreign materials, such as the capsule of a bacterium. As does the classical pathway, the alternative pathway ends with the conversion of C3 to C3b, the stimulation of phagocytosis, and the subsequent formation of the membrane attack complex.

Some bacteria are unaffected by complement, because the complement proteins cannot interact with their cell membranes. These bacteria coat themselves in a protective capsule of carbohydrates. However, the classical pathway still provides a defense against such bacteria, once antibodies have bound to the capsule.

Effects of Complement Activation

Known effects of complement activation include:

- **The Destruction of Target Cell Membranes**. Five of the interacting complement proteins bind to the cell membrane, forming a functional unit called the **membrane attack complex (MAC)**. The MACs create pores in the membrane that are comparable to those produced by perforin, and that have the same effect: The target cell is soon destroyed.

- **The Stimulation of Inflammation**. Activated complement proteins enhance the release of histamine by mast cells and basophils. Histamine increases the degree of local inflammation and accelerates blood flow to the region.

- **The Attraction of Phagocytes**. Activated complement proteins attract neutrophils and macrophages to the area, improving the chances that phagocytic cells will be able to cope with the injury or infection.

- **The Enhancement of Phagocytosis**. A coating of complement proteins and antibodies both attracts phagocytes and makes the target easier to engulf. Macrophage membranes contain receptors that can detect and bind to complement proteins and bound antibodies. After binding, the pathogens are easily engulfed. The antibodies involved are called **opsonins**, and the effect is called **opsonization**.

☐ INFLAMMATION

Inflammation, or the *inflammatory response*, is a localized tissue response to injury. ∞ p. 139 Inflammation produces local sensations of swelling, redness, heat, and pain. Figure 22–13● summarizes the events that occur during inflammation of the skin. Comparable events take place in almost any tissue subjected to

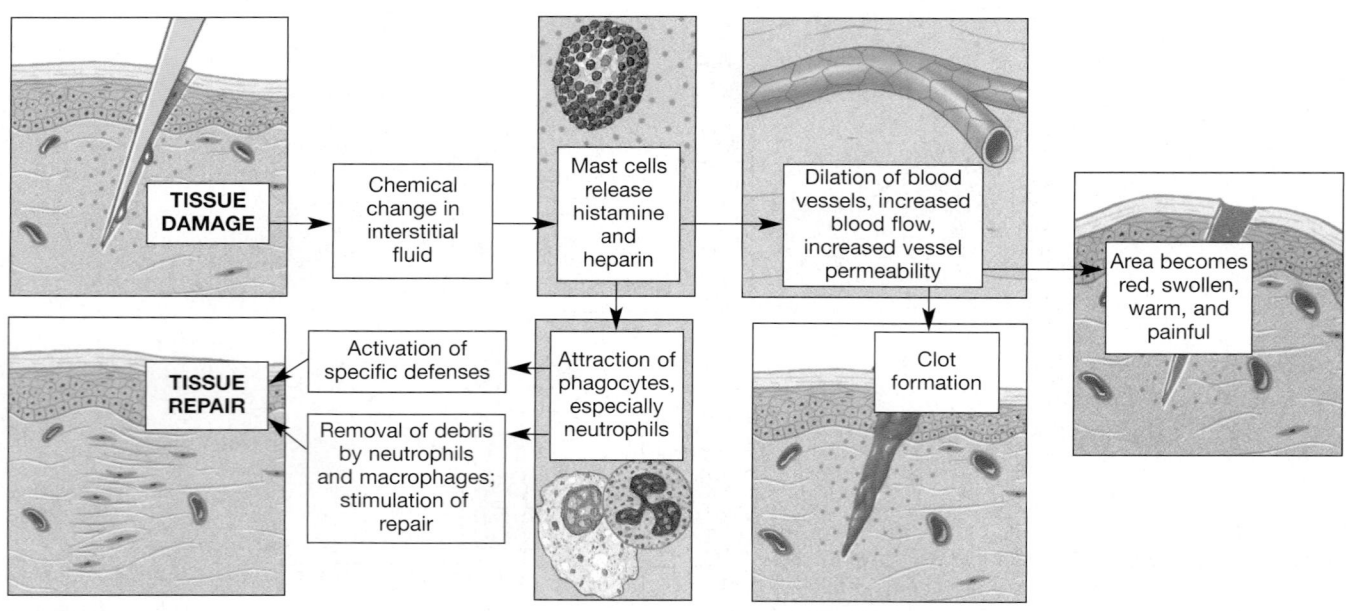

●FIGURE 22–13
Inflammation

physical damage or infection. Many stimuli, including impact, abrasion, distortion, chemical irritation, infection by pathogens, and extreme temperatures (hot or cold), can produce inflammation. Each of these stimuli kills cells, damages connective-tissue fibers, or injures the tissue in some other way. The changes alter the chemical composition of the interstitial fluid. Damaged cells release prostaglandins, proteins, and potassium ions, and the injury itself may have introduced foreign proteins or pathogens. The changes in the interstitial environment trigger the complex process of inflammation.

Inflammation has a number of effects:

- The injury is temporarily repaired, and additional pathogens are prevented from entering the wound.
- The spread of pathogens away from the injury is slowed.
- Local, regional, and systemic defenses are mobilized to overcome the pathogens and facilitate permanent repairs. This repair process is called *regeneration*.

The Response to Injury

Mast cells play a pivotal role in the inflammatory response. When stimulated by mechanical stress or chemical changes in the local environment, these cells release histamine, heparin, prostaglandins, and other chemicals into interstitial fluid. Events then proceed in a series of integrated steps:

1. The histamine that is released increases capillary permeability and accelerates blood flow through the area. The increased blood flow brings more cellular defenders to the site and carries away toxins and debris, diluting them and reducing their local impact.

2. Clotting factors and complement proteins leave the bloodstream and enter the injured or infected area. Clotting does not occur at the actual site of injury, due to the presence of heparin. However, a clot soon forms around the damaged area, and this clot both isolates the region and slows the spread of the chemical or pathogen into healthy tissues. Meanwhile, complement activation through the alternative pathway breaks down bacterial cell walls and attracts phagocytes.

3. The increased blood flow elevates the local temperature, increasing the rate of enzymatic reactions and accelerating the activity of phagocytes. The rise in temperature may also denature foreign proteins or vital enzymes of invading microorganisms.

4. Debris and bacteria are attacked by neutrophils drawn to the area by chemotaxis. As they circulate through a blood vessel in an injured area, neutrophils undergo *activation*, a process in which (1) they stick to the side of the vessel and move into the tissue by diapedesis; (2) their metabolic rate goes up dramatically, and while this *respiratory burst* continues, they generate reactive compounds, such as nitric oxide and hydrogen peroxide, that can destroy engulfed pathogens; and (3) they secrete cytokines that attract other neutrophils and macrophages to the area.

5. Fixed and free macrophages engulf pathogens and cell debris. At first, these cells are outnumbered by neutrophils, but as the macrophages and neutrophils continue to secrete cytokines, the number of macrophages increases rapidly. Eosinophils may get involved if the foreign materials become coated with antibodies.

6. Other cytokines released by active phagocytes stimulate fibroblasts to begin barricading the area with scar tissue, reinforcing the clot and further slowing the invasion of adjacent tissues.

7. The combination of abnormal tissue conditions and chemicals released by mast cells stimulates local sensory neurons, producing sensations of pain. The individual becomes consciously aware of these sensations and may take steps to limit the damage they signal, such as removing a splinter or cleaning a wound.

8. As inflammation is under way, the foreign proteins, toxins, microorganisms, and active phagocytes in the area activate the body's specific defenses.

After an injury, tissue conditions generally become even more abnormal before they begin to improve. The tissue degeneration that occurs after cells have been injured or destroyed is called **necrosis** (ne-KRŌ-sis). The process begins several hours after the initial event, and the damage is caused by lysosomal enzymes. Lysosomes break down by autolysis, releasing digestive enzymes that first destroy the injured cells and then attack surrounding tissues. ∞ p. 80 As local inflammation continues, debris, fluid, dead and dying cells, and necrotic tissue components accumulate at the injury site. This viscous fluid mixture is known as **pus**. An accumulation of pus in an enclosed tissue space is called an **abscess**.

AM Complications of Inflammation

☐ FEVER

Fever is the maintenance of a body temperature greater than 37.2°C (99°F). In Chapter 14, we noted the presence of a temperature-regulating center in the preoptic area of the hypothalamus. ∞ p. 482 Circulating proteins called **pyrogens** (PĪ-rō-jenz) can reset this thermostat and raise body temperature. A variety of stimuli, including pathogens, bacterial toxins, and antigen-antibody complexes, either act as pyrogens themselves or stimulate the release of pyrogens by macrophages. The pyrogen released by active macrophages is a cytokine called **endogenous pyrogen**, or **interleukin-1** (in-ter-LOO-kin), abbreviated **IL-1**.

Within limits, a fever can be beneficial. High body temperatures may inhibit some viruses and bacteria, but the most likely beneficial effect is on body metabolism. For each 1°C rise in temperature, your metabolic rate jumps by 10 percent. Your cells can move faster, so enzymatic reactions occur faster. The net results may be the quicker mobilization of tissue defenses and an accelerated repair process.

✓ What types of cells would be affected by a decrease in the number of monocyte-forming cells in bone marrow?

✓ A rise in the level of interferon in the body would suggest what kind of infection?

✓ What effects do pyrogens have in the body?

Answers start on page Q-1

 The types of immunity are presented on the **Companion Website:** Chapter 22/Tutorials.

22–4 SPECIFIC DEFENSES: AN OVERVIEW OF THE IMMUNE RESPONSE

Objectives

■ Define specific resistance, and identify the forms and properties of immunity.

■ Distinguish between cell-mediated (cellular) immunity and antibody-mediated (humoral) immunity, and identify the cells responsible for each.

Specific resistance, or immunity, is provided by the coordinated activities of T cells and B cells, which respond to the presence of specific antigens. The basic functional relationship can be summarized as follows:

1. *T cells* are responsible for **cell-mediated immunity** (or **cellular immunity**), our defense against abnormal cells and pathogens inside cells.

2. *B cells* provide **antibody-mediated immunity** (or **humoral immunity**), our defense against antigens and pathogens in body fluids.

Both mechanisms are important, because they come into play under different circumstances. Activated T cells do not respond to antigenic materials in solution, and the antibodies produced by activated B cells cannot cross cell membranes. Moreover, helper T cells play a crucial role in antibody-mediated immunity by stimulating the activity of B cells.

Our understanding of immunity has greatly improved in the past two decades, and a comprehensive discussion would involve hundreds of pages and thousands of details. The discussion that follows emphasizes important patterns and introduces general principles that will provide a foundation for future courses in microbiology and immunology.

□ FORMS OF IMMUNITY

Immunity is either *innate* or *acquired* (Figure 22–14●).

Innate immunity is genetically determined; it is present at birth and has no relationship to previous exposure to the antigen in-

volved. For example, people do not get the same diseases that goldfish do. Innate immunity breaks down only in the case of *AIDS* or other conditions that depress all aspects of specific resistance.

Acquired immunity is not present at birth; you acquire immunity to a specific antigen only when you have been exposed to that antigen. Acquired immunity can be *active* or *passive*.

Active immunity appears after exposure to an antigen, as a consequence of the immune response. The immune system is *capable* of defending against a large number of antigens. However, the appropriate defenses are mobilized only after you encounter a particular antigen. Active acquired immunity can develop as a result of natural exposure to an antigen in the environment (*naturally acquired immunity*) or from deliberate exposure to an antigen (*induced active immunity*).

- **Naturally acquired immunity** normally begins to develop after birth, and it is continually enhanced as you encounter "new" pathogens or other antigens. You might compare this process to the development of a child's vocabulary: The child begins with a few basic common words and learns new ones as needed.

- The purpose of **induced active immunity** is to stimulate the production of antibodies under controlled conditions so that you will be able to overcome natural exposure to the pathogen some time in the future. This is the basic principle behind *immunization*, or *vaccination*, to prevent disease. A **vaccine** is a preparation designed to induce an immune response. It contains either a dead or an inactive pathogen, or antigens derived from that pathogen.

Passive immunity is produced by the transfer of antibodies from another source.

- **Natural passive immunity** results when a mother's antibodies protect her baby against infections during gestation (across the placenta) or in early infancy (through breast milk).

- In **induced passive immunity**, antibodies are administered to fight infection or prevent disease. For example, antibodies that attack the rabies virus are injected into a person bitten by a rabid animal.

□ PROPERTIES OF IMMUNITY

Regardless of the form, immunity exhibits four general properties: (1) *specificity*, (2) *versatility*, (3) *memory*, and (4) *tolerance*.

Specificity

A specific defense is activated by a specific antigen, and the immune response targets that particular antigen and no others. **Specificity** results from the activation of appropriate lymphocytes and the production of antibodies with targeted effects. Specificity occurs because T cells and B cells respond to the molecular structure of an antigen. The shape and size of the antigen determine which lymphocytes will respond to its presence. Each T cell or B cell has receptors that will bind to one specific antigen, ignoring all others. The response of an activated T cell or B cell is

•**FIGURE 22–14**
Types of Immunity

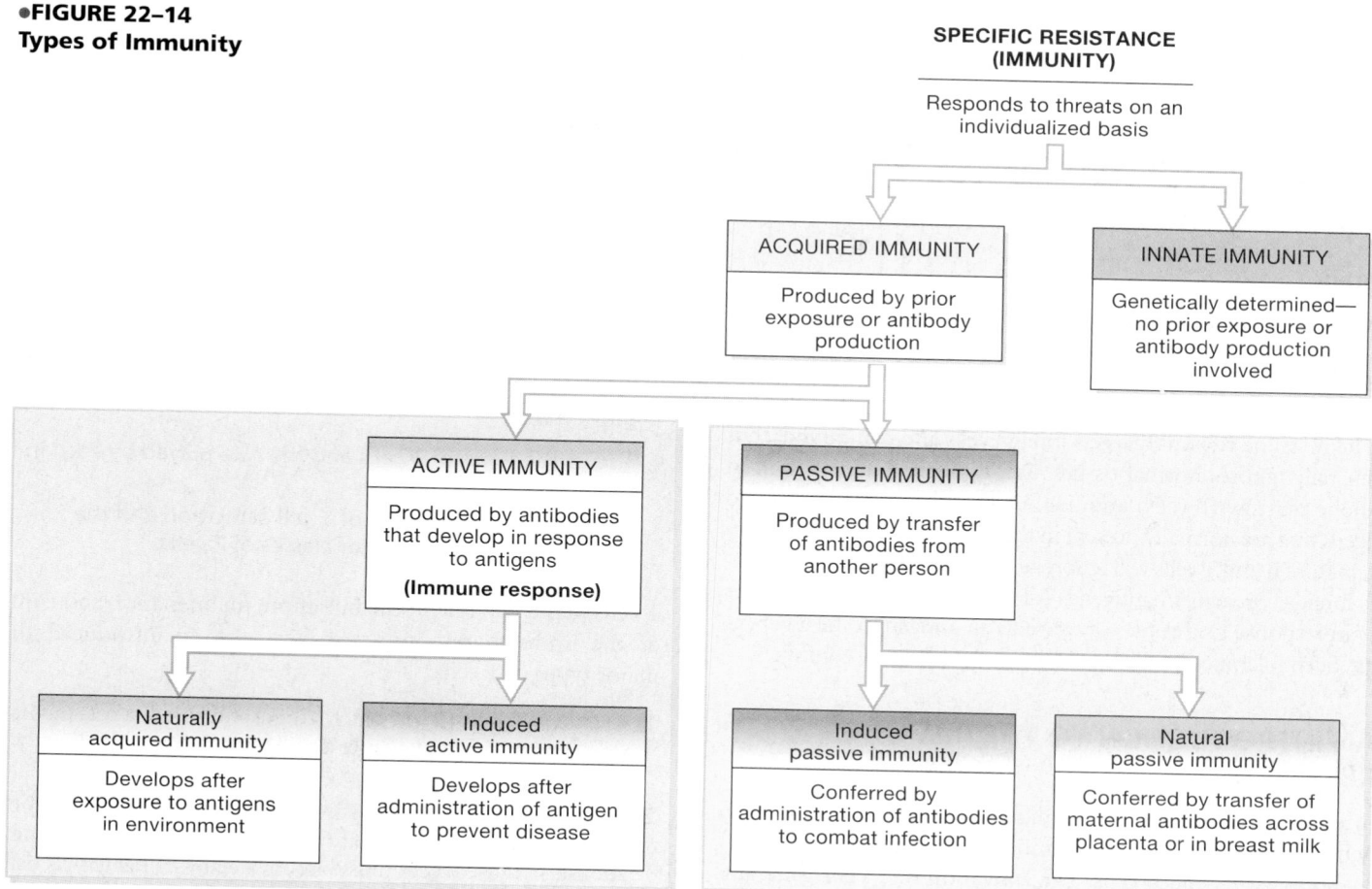

equally specific. Either lymphocyte will destroy or inactivate that antigen without affecting other antigens or normal tissues.

Versatility

Millions of antigens in the environment can pose a threat to health. Over a normal lifetime, an individual encounters only a fraction of that number—perhaps tens of thousands of antigens. Your immune system, however, has no way of anticipating which antigens it will encounter. It must be ready to confront *any* antigen at *any* time. **Versatility** results in part from the large diversity of lymphocytes present in the body and in part from variability in the structure of synthesized antibodies.

During development, differentiation of cells in the lymphatic system produces an enormous number of lymphocytes with varied antigen sensitivities. The trillion or more T cells and B cells in the human body include millions of different lymphocyte populations, distributed throughout the body. Each population consists of several thousand cells with receptors in their membranes that differ from those of other lymphocyte populations. As a result, each group of lymphocytes will respond to the presence of a different antigen.

Several thousand lymphocytes are not enough to overcome a pathogenic invasion. When activated in the presence of an ap-

propriate antigen, however, a lymphocyte begins to divide, producing more lymphocytes with the same specificity. Thus, whenever an antigen is encountered, more lymphocytes are produced. All the cells produced by the division of an activated lymphocyte constitute a **clone**. *All the members of a clone are sensitive to the same specific antigen.*

To understand how this system works, think about running a snack shop with only samples on display. When customers come in and make a selection, you prepare the food on the spot. You can offer a wide selection, because the samples don't take up much space and you do not have to expend energy preparing food that will not be eaten. The same principle applies to the lymphatic system: Your body contains a small number of many different kinds of lymphocytes. When an antigen arrives, the appropriate lymphocytes respond to its presence and start generating additional lymphocytes capable of dealing with that particular antigen.

Memory

The immune system "remembers" antigens that it encounters. As a result of this **memory**, the immune response that occurs after a second exposure to an antigen is stronger and lasts longer than the response to the first exposure. During the initial response to an antigen, lymphocytes that are sensitive to its presence undergo repeated cycles of cell division. Two kinds of cells are produced: some that attack the invader and others that remain inactive unless they are exposed to the same antigen again. These inactive *memory cells* enable your immune system to "remember" antigens it has previously encountered and to launch a faster, stronger counterattack if such an antigen reappears.

Tolerance

The immune system does not respond to all antigens. For example, all cells and tissues in the body contain some antigens that normally fail to stimulate an immune response. The immune system is said to exhibit **tolerance** toward such antigens.

The immune response targets foreign cells and compounds, but it generally ignores normal tissues. During their differentiation in the bone marrow (B cells) and thymus (T cells), cells that react to antigens that are normally present in the body are destroyed. As a result, mature B and T cells will ignore normal, or *self*, antigens and attack foreign, or *nonself*, antigens. Tolerance can also develop over time in response to chronic exposure to an antigen in the environment. Such tolerance lasts only as long as the exposure continues.

☐ AN INTRODUCTION TO THE IMMUNE RESPONSE

Figure 22–15● introduces the highlights of the immune response. When an antigen triggers an immune response, it usually activates both T cells and B cells. The activation of T cells generally occurs first, but only after phagocytes have been exposed to the

antigen. Once activated, T cells attack the antigen and stimulate the activation of B cells. Activated B cells mature into cells that produce antibodies; antibodies distributed in the bloodstream bind to and attack the antigen. We will examine each of these processes more closely in the sections that follow.

22–5 T CELLS AND CELL-MEDIATED IMMUNITY

Objectives

■ Discuss the types of T cells and the role played by each in the immune response.
■ Describe the mechanisms of T cell activation and the differentiation of the major classes of T cells.

T cells play a key role in the initiation, maintenance, and control of the immune response. We have already introduced three major types of T cells:

1. *Cytotoxic T cells* (T_C), which are responsible for cell-mediated immunity. These cells enter peripheral tissues and directly attack antigens physically and chemically.

2. *Helper T cells* (T_H), which stimulate the responses of both T cells and B cells. Helper T cells are absolutely vital to the immune response, because B cells must be activated by helper T cells before the B cells can produce antibodies. The reduction in the helper T

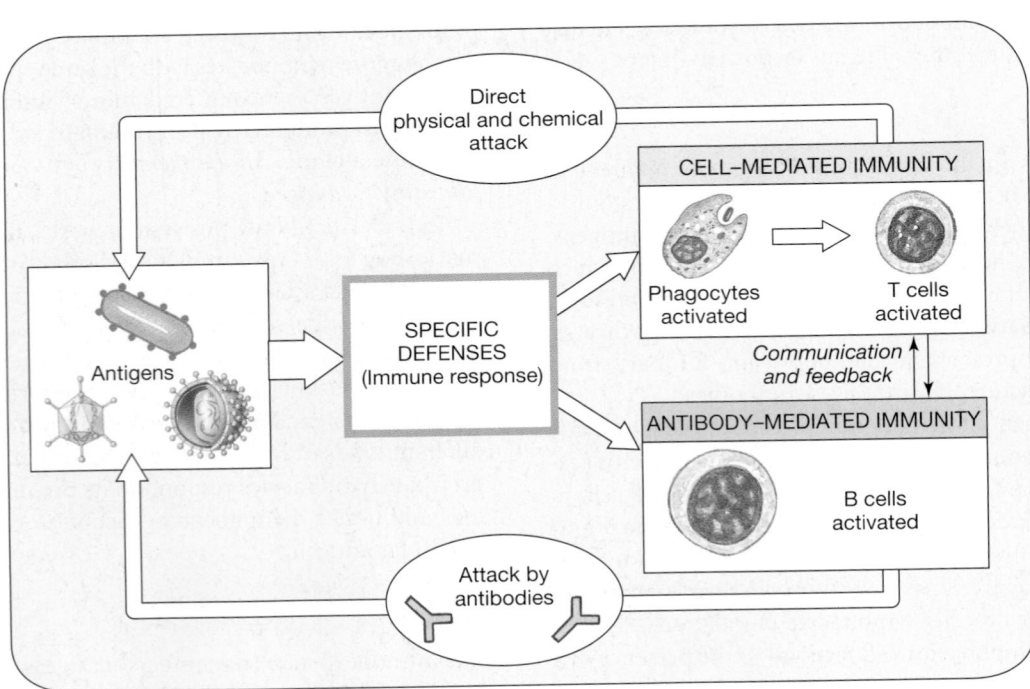

●**FIGURE 22–15**
An Overview of the Immune Response. This figure will be repeated, in reduced and simplified form as Navigator icons, in key figures throughout the chapter as we change topics.

cell population that occurs in AIDS is largely responsible for the loss of immunity. (We will discuss AIDS on p. 820.)

3. *Suppressor T cells* (T_S), which inhibit T cell and B cell activities and moderate the immune response.

Before an immune response can begin, T cells must be activated by exposure to an antigen. This activation seldom happens by direct lymphocyte–antigen interaction, and foreign compounds or pathogens entering a tissue commonly fail to stimulate an immediate immune response.

□ ANTIGEN PRESENTATION

T cells recognize antigens when the antigens are bound to glycoproteins in cell membranes. Glycoproteins are integral membrane components. ○○ p. 72 **Antigen presentation** occurs when an antigen–glycoprotein combination capable of activating T cells appears in a cell membrane. The structure of these glycoproteins is genetically determined. The genes controlling their synthesis are located along one portion of chromosome 6, in a region called the **major histocompatibility complex (MHC)**. These membrane glycoproteins are called **MHC proteins**, or *human leukocyte antigens (HLAs)*.

The amino acid sequences and the shapes of MHC proteins differ among individuals. Each MHC molecule has a distinct three-dimensional shape with a relatively narrow central groove. An antigen that fits into this groove can be held in position by hydrogen bonding.

Two major classes of MHC proteins are known: *Class I* and *Class II*. An antigen bound to a Class I MHC protein acts like a red flag that tells the immune system "Hey, I'm an abnormal cell—please kill me!" An antigen bound to a Class II MHC protein tells the immune system "Hey, this antigen is dangerous—get rid of it!"

Class I MHC proteins are in the membranes of all nucleated cells. These proteins are continuously being synthesized and exported to the cell membrane in vesicles created at the Golgi apparatus. As they form, Class I proteins pick up small peptides from the surrounding cytoplasm and carry them to the cell surface. If the cell is healthy and the peptides are normal, T cells will ignore them. If the cytoplasm contains abnormal (nonself) peptides or viral proteins (Figure 22–16a●), they will soon appear in the cell membrane, and T cells will be activated. Ultimately, their activation will lead to the destruction of the abnormal cells. This is the primary reason that donated organs are commonly rejected by the recipient; despite preliminary cross-match testing, the recipient's T cells recognize the transplanted tissue as foreign.

Class II MHC proteins are present only in the membranes of antigen-presenting cells and lymphocytes. **Antigen-presenting cells (APCs)** are specialized cells responsible for activating T cell defenses against foreign cells (including bacteria) and foreign proteins. Antigen-presenting cells include all the phagocytic cells of the monocyte–macrophage group discussed in other chapters, including (1) the free and fixed macrophages in connective tissues, (2) the Kupffer cells of the liver, and (3) the microglia in the central nervous system (Chapter 12). ○○ pp. 123, 395 The Langerhans cells of the skin and the dendritic cells of the lymph nodes and spleen are APCs that are not phagocytic. ○○ p. 158

Phagocytic APCs engulf and break down pathogens or foreign antigens. This **antigen processing** creates antigenic fragments, which are then bound to Class II MHC proteins and inserted into the cell membrane (Figure 22–16b●). *Class II MHC proteins appear in the cell membrane only when the cell is processing antigens.* Exposure to an APC membrane containing processed antigen can stimulate appropriate T cells.

The Langerhans cells and dendritic cells remove antigenic materials from their surroundings by pinocytosis rather than

●**FIGURE 22–16**

Antigens and MHC Proteins. The Navigator icon in the shadow box highlights how phagocyte activation relates to the rest of the immune response. **(a)** Viral or other foreign antigens appear in cell membranes bound to Class I MHC proteins.

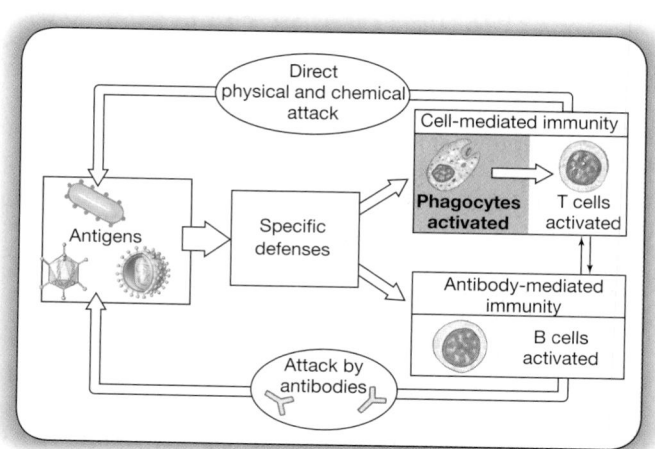

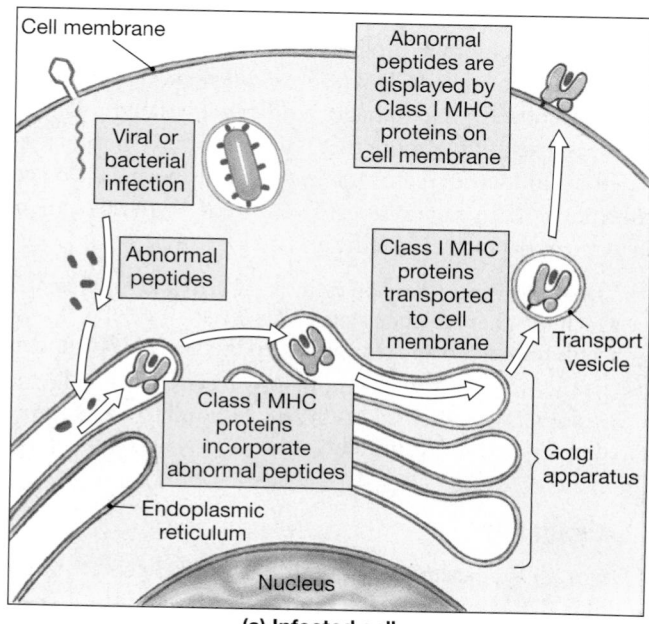

(a) Infected cell

•FIGURE 22–16
Antigens and MHC Proteins *(continued).*
(b) Processed antigens appear on the surfaces of antigen-presenting cells bound to Class II MHC proteins.

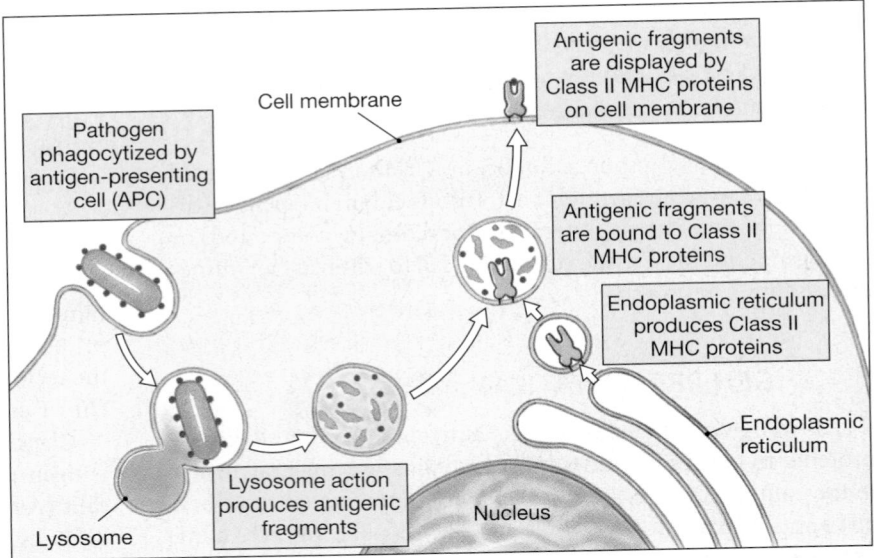

(b) Phagocytic antigen-presenting cell

phagocytosis. However, their cell membranes still present antigens bound to Class II MHC proteins.

☐ ANTIGEN RECOGNITION

Inactive T cells have receptors that recognize Class I or Class II MHC proteins. The receptors also have binding sites that detect the presence of specific bound antigens. If an MHC protein contains any antigen other than the specific target of a particular kind of T cell, the T cell remains inactive. If the MHC protein contains the antigen that the T cell is programmed to detect, binding will occur. This process is called **antigen recognition**, because the T cell recognizes that it has found an appropriate target.

Some T cells can recognize antigens bound to Class I MHC proteins, whereas others can recognize antigens bound to Class II MHC proteins. Whether a T cell responds to antigens held by Class I or Class II proteins depends on the structure of the T cell membrane. The membrane proteins involved are members of a larger class of proteins called **CD** (*cluster of differentiation*) **markers**.

Lymphocytes, macrophages, and other, related cells have CD markers. Each of the more than 70 types of CD markers is designated by an identifying number. All T cells have a **CD3 receptor complex** in their membranes. Two other CD markers are of particular importance in specific groups of T cells:

1. **CD8** markers are found on cytotoxic T cells and suppressor T cells, which together are often called *CD8 T cells* or *CD8 + T cells*. CD8 T cells respond to antigens presented by Class I MHC proteins.

2. **CD4** markers are found on helper T cells, often called *CD4 T cells* or *CD4 + T cells*. CD4 T cells respond to antigens presented by Class II MHC proteins.

Costimulation

CD8 or CD4 markers are bound to the CD3 receptor complex, and it is that complex which ultimately activates the T cell. How-

ever, such activation usually does not occur at the first encounter with the antigen. Antigen recognition simply prepares the cell for activation. Before activation can occur, however, a T cell must bind to the stimulating cell at a second site. This vital secondary binding process, called *costimulation*, essentially confirms the "OK to activate" signal. Appropriate costimulation proteins appear in the presenting cell only if that cell has engulfed antigens or is infected by viruses. Many costimulation proteins are structurally related to the cytokines released by activated lymphocytes. The effects these proteins have on the exposed T cell vary, but typically include the stimulation of transcription at the nucleus and the promotion of cell division and differentiation.

Costimulation is like the safety on a gun: It helps prevent T cells from mistakenly attacking normal (self) tissues. If a cell displays an unusual antigen, but does not display the "I am an active phagocyte" or "I am infected" signal, T cell activation will not occur. Costimulation is important only in determining whether a T cell will become activated. Once activation has occurred, the safety is off and the T cell will attack any cells that carry the target antigens.

☐ ACTIVATION OF CD8 T CELLS

Two different classes of CD8 T cells are activated by exposure to antigens bound to Class I MHC proteins. One type of CD8 T cell responds quickly, giving rise to large numbers of *cytotoxic T cells* and *memory T cells*. The other type of CD8 T cell responds more slowly and produces relatively small numbers of *suppressor T cells*.

Cytotoxic T Cells

Cytotoxic T cells, also called T_c *cells* or *killer T cells*, seek out and destroy abnormal and infected cells. Killer T cells are highly mobile cells that roam throughout injured tissues. When a cytotoxic T cell encounters its target antigens bound to Class I MHC proteins of another cell, it immediately destroys that cell. Three different methods may be used to destroy the target cell (Figure 22–17●). The T cell may (1) destroy the antigenic cell membrane through the release of perforin, (2) kill the target cell by secreting a poisonous **lymphotoxin** (lim-fō-TOK-sin), or (3) activate genes in the target cell's nucleus that tell that cell to die. (We introduced genetically programmed cell death, called *apoptosis*, in Chapter 3.) ⊂⊃ p. 100

The entire sequence of events, from the appearance of the antigen in a tissue to cell destruction by cytotoxic T cells, takes a significant amount of time. After the first exposure to an antigen, two days or more may pass before the concentration of cytotoxic T cells reaches effective levels at the site of injury or infection. Over this period, the damage or infection may spread, making it more difficult to control.

Memory T_c Cells

Memory T_c cells are produced by the same cell divisions that produce cytotoxic T cells. Memory T_c cells ensure that there will be no delay in the response if the antigen reappears. These cells do not differentiate further the first time the antigen triggers an immune response, although thousands of them are produced. Instead, they remain in reserve. If the same antigen appears a second time, memory T cells will *immediately* differentiate into cytotoxic T cells, producing a swift, effective cellular response that can overwhelm an invading organism before it becomes well established in the tissues.

Suppressor T Cells

Suppressor T cells (T_S *cells*) suppress the responses of other T cells and of B cells by secreting *suppression factors*—inhibitory cytokines of unknown structure. The suppression does not occur immediately: Suppressor T cell activation takes much longer than the activation of other types of T cells. In addition, on activation, most of the CD8 T cells in the bloodstream will produce cytotoxic T cells rather than suppressor T cells. As a result, suppressor T cells act *after* the initial immune response. In effect, these cells put on the brakes and limit the degree of immune system activation from a single stimulus.

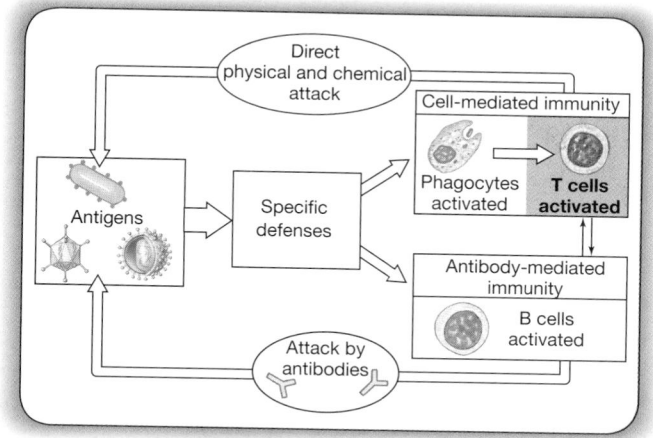

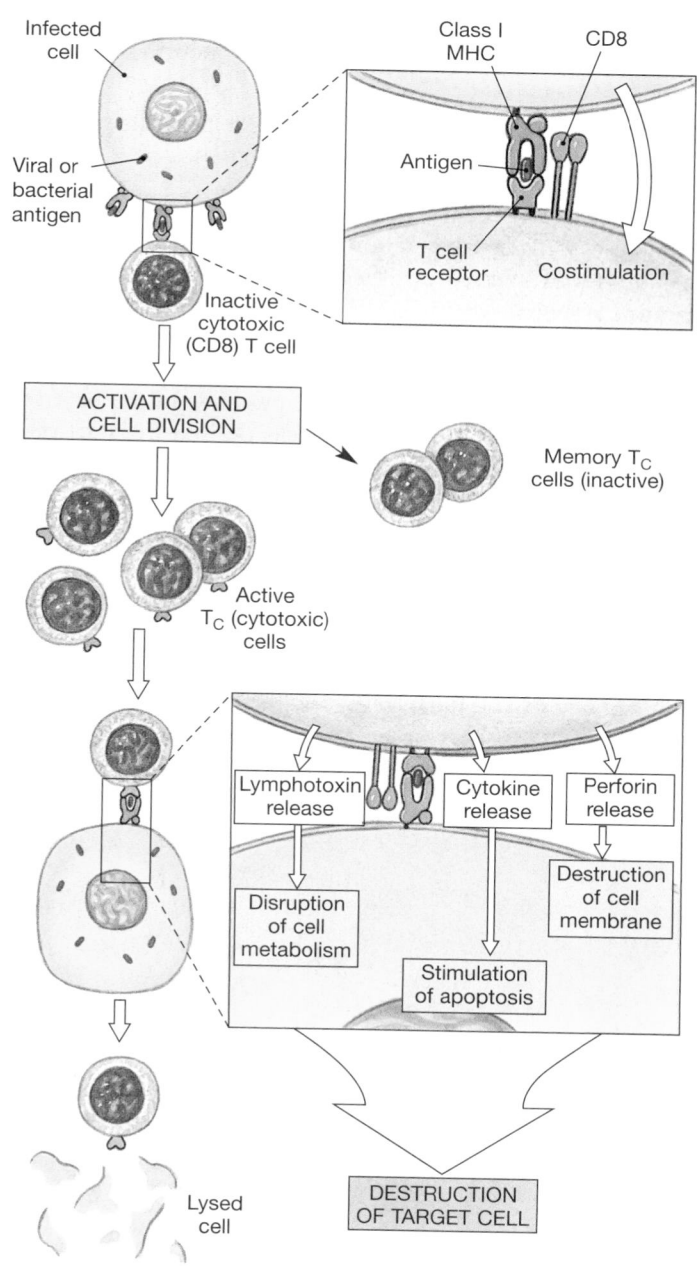

●**FIGURE 22–17**
Antigen Recognition and the Activation of Cytotoxic T Cells. An inactive cytotoxic T cell not only must encounter an appropriate antigen bound to Class I MHC proteins, but also must receive costimulation from the membrane it contacts. It is then activated and undergoes divisions that produce memory T_C cells and active T_C cells. When one of the active T_C cells encounters a membrane displaying the target antigen, it will use one of several methods to destroy the cell. Notice that the dashed box now highlights the activation of T cells.

☐ ACTIVATION OF CD4 T CELLS

Upon activation, CD4 T cells undergo a series of divisions that produce helper T cells (*T_H cells*) and **memory T_H cells** (Figure 22–18●). The memory T_H cells remain in reserve, whereas the helper T cells secrete a variety of cytokines that coordinate specific and nonspecific defenses and stimulate cell-mediated and antibody-mediated immunities. For example, activated helper T cells secrete cytokines that

1. Stimulate the T cell divisions that produce memory T cells and accelerate the maturation of cytotoxic T cells;

2. Enhance nonspecific defenses by attracting macrophages to the affected area, preventing their departure, and stimulating their phagocytic activity and effectiveness;

3. Attract and stimulate the activity of NK cells, providing another mechanism for the destruction of abnormal cells and pathogens; and

4. Promote the activation of B cells, leading to B cell division, plasma cell maturation, and antibody production.

Figure 22–19● concludes this section with a review of the methods of antigen presentation and T cell stimulation. The cell membranes of infected or otherwise abnormal cells trigger an immune response when CD8 T cells recognize antigens bound to Class I MHC proteins. Extracellular pathogens or foreign proteins trigger an immune response when CD4 T cells recognize antigens displayed by Class II MHC proteins. In the next section, we will consider how the T_H cells derived from activated CD4 T cells in turn activate B cells that are sensitive to the antigen involved.

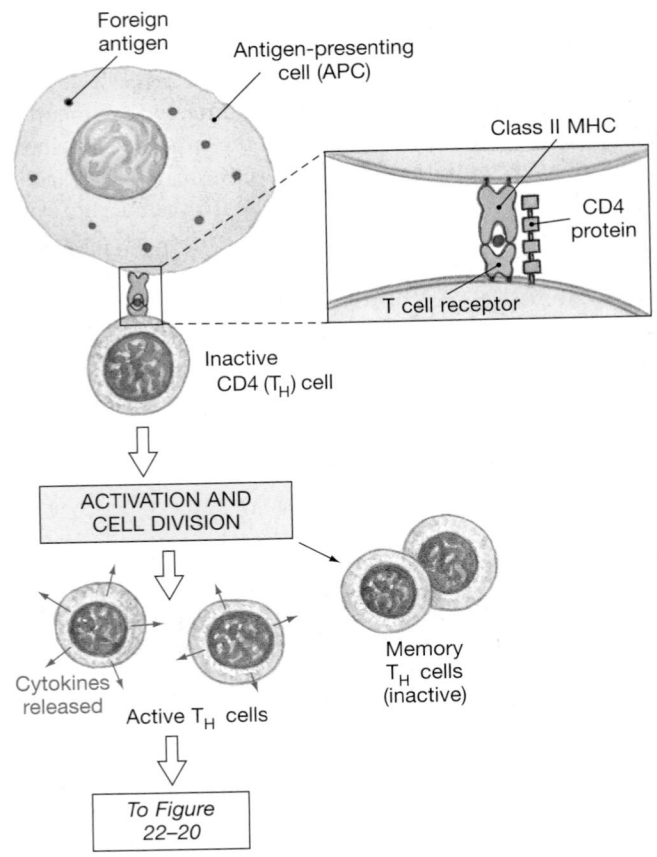

●**FIGURE 22–18**
Antigen Recognition and Activation of Helper T Cells.
Inactive CD4 T cells (T_H cells) must be exposed to appropriate antigens bound to Class II MHC proteins. The T_H cells then undergo activation, dividing to produce active T_H cells and memory T_H cells. Active T_H cells secrete cytokines that stimulate cell-mediated and antibody-mediated immunities. They also interact with sensitized B cells, as Figure 22–20 shows.

●**FIGURE 22–19**
A Summary of the Pathways of T Cell Activation

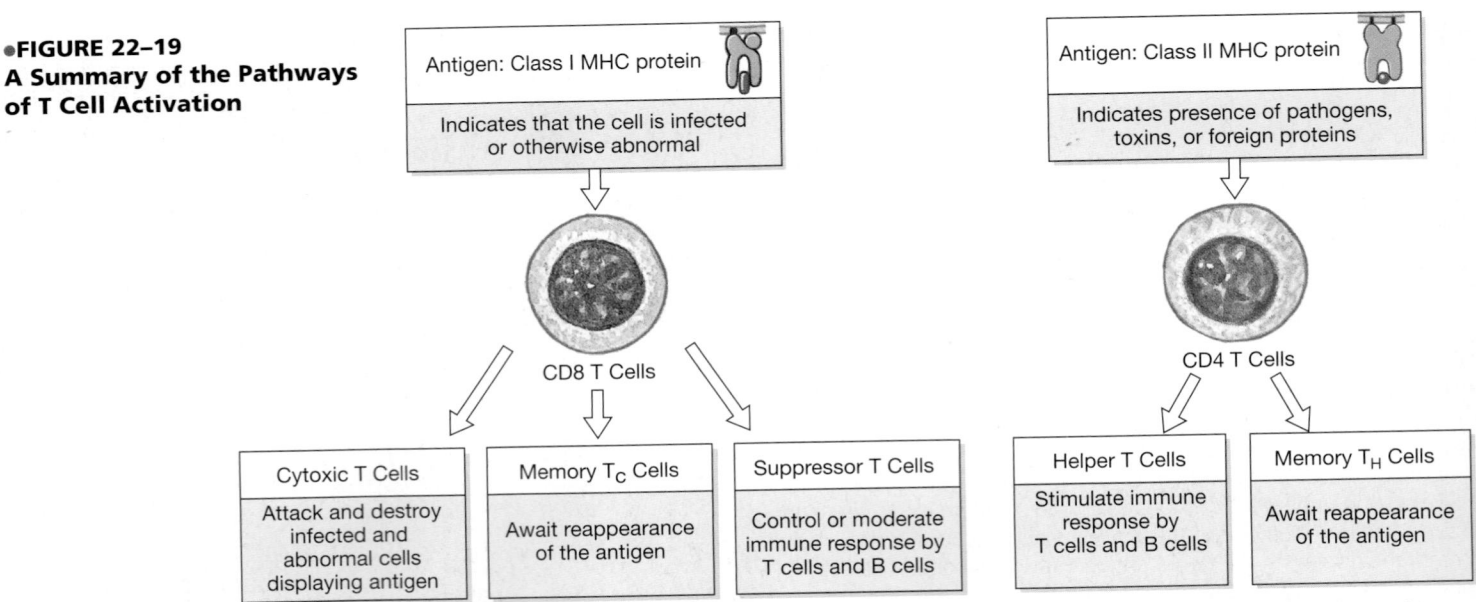

(a) **Activation by Class I MHC proteins**

(b) **Activation by Class II MHC proteins**

Graft Rejection and Immunosuppression

Organ transplantation is a treatment option for patients with severe disorders of the kidneys, liver, heart, lungs, or pancreas. Finding a suitable donor is the first major problem. In the United States, many people die each day while awaiting an organ transplant, and dozens are added to the transplant waiting list. After surgery has been performed, the major problem is **graft rejection**. In graft rejection, T cells are activated by contact with MHC proteins on cell membranes in the donated tissues. The cytotoxic T cells that develop then attack and destroy the foreign cells.

Significant improvements in transplant success can be made by **immunosuppression**, a reduction in the sensitivity of the immune system. Until recently, the drugs used to produce immunosuppression did not selectively target the immune response. For example, **prednisone** (*PRED-ni-sōn*), a corticosteroid, was used because it has anti-inflammatory effects that reduce the number of circulating white blood cells and depress the immune response. However, corticosteroid use also causes undesirable side effects in systems other than the immune system. ⇔ p. 630

An understanding of the communication among T cells, macrophages, and B cells has now led to the development of drugs with more selective effects. **Cyclosporin A (CsA)**, a compound derived from a fungus, was the most important *immunosuppressive drug* developed in the 1980s. This compound suppresses all aspects of the immune response primarily by suppressing helper T cell activity, while leaving suppressor T cells relatively unaffected. **AM** Transplants and Immunosuppressive Drugs

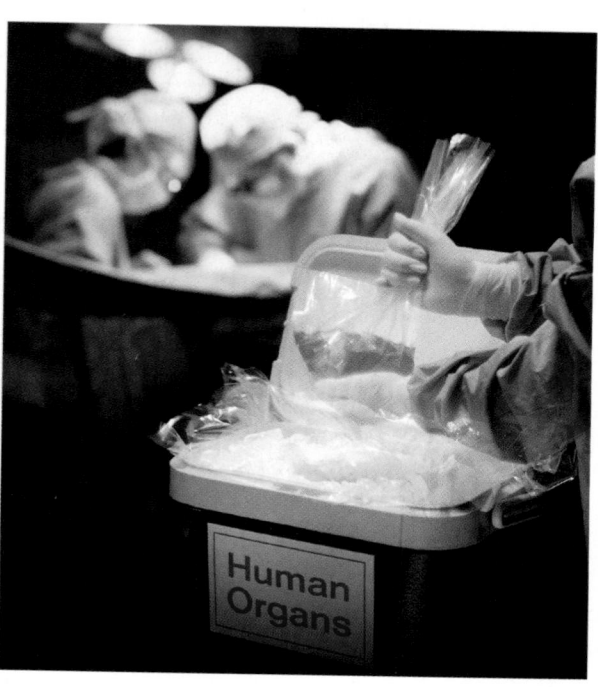

22–6 B CELLS AND ANTIBODY-MEDIATED IMMUNITY

Objectives

- Describe the mechanisms of B cell activation and the differentiation of plasma cells and memory B cells.
- Describe the structure of an antibody, and discuss the types of antibodies in body fluids and secretions.
- Explain the functions of antibodies and how they perform those functions.
- Discuss the primary and secondary responses to antigen exposure.

B cells are responsible for launching a chemical attack on antigens by producing appropriate *antibodies*.

☐ B CELL SENSITIZATION AND ACTIVATION

As we noted earlier, the body has millions of populations of B cells. Each kind of B cell carries its own particular antibody molecules in its cell membrane. If corresponding antigens appear in the interstitial fluid, they will bind to these superficial antibodies. When binding occurs, the B cell prepares to undergo activation. This preparatory process is called **sensitization**. Because B cells migrate throughout the body, pausing briefly in one lymphoid tissue or another, sensitization typically occurs at the lymph node closest to the site of infection or injury.

As noted earlier, B cell membranes contain Class II MHC proteins. During sensitization, antigens are brought into the cell by endocytosis. The antigens subsequently appear on the surface of the B cell, bound to Class II MHC proteins. (The mechanism is comparable to that shown in Figure 22–16b●, p. 802). Once this happens, the sensitized B cell is ready to go, but it generally will not undergo activation unless it receives the "OK" from a helper T cell. The need for activation by a helper T cell helps prevent inappropriate activation, the same way that costimulation acts as a "safety" for cell-mediated immunity.

Next, a sensitized B cell encounters a helper T cell already activated by antigen presentation. The helper T cell binds to the MHC complex, recognizes the presence of an antigen, and begins secreting cytokines that promote B cell activation. After

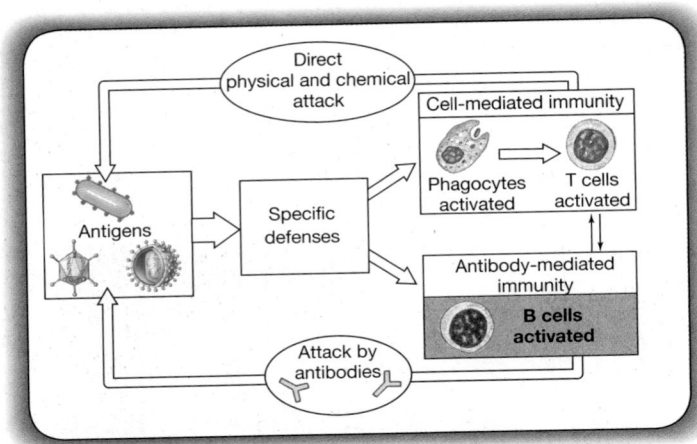

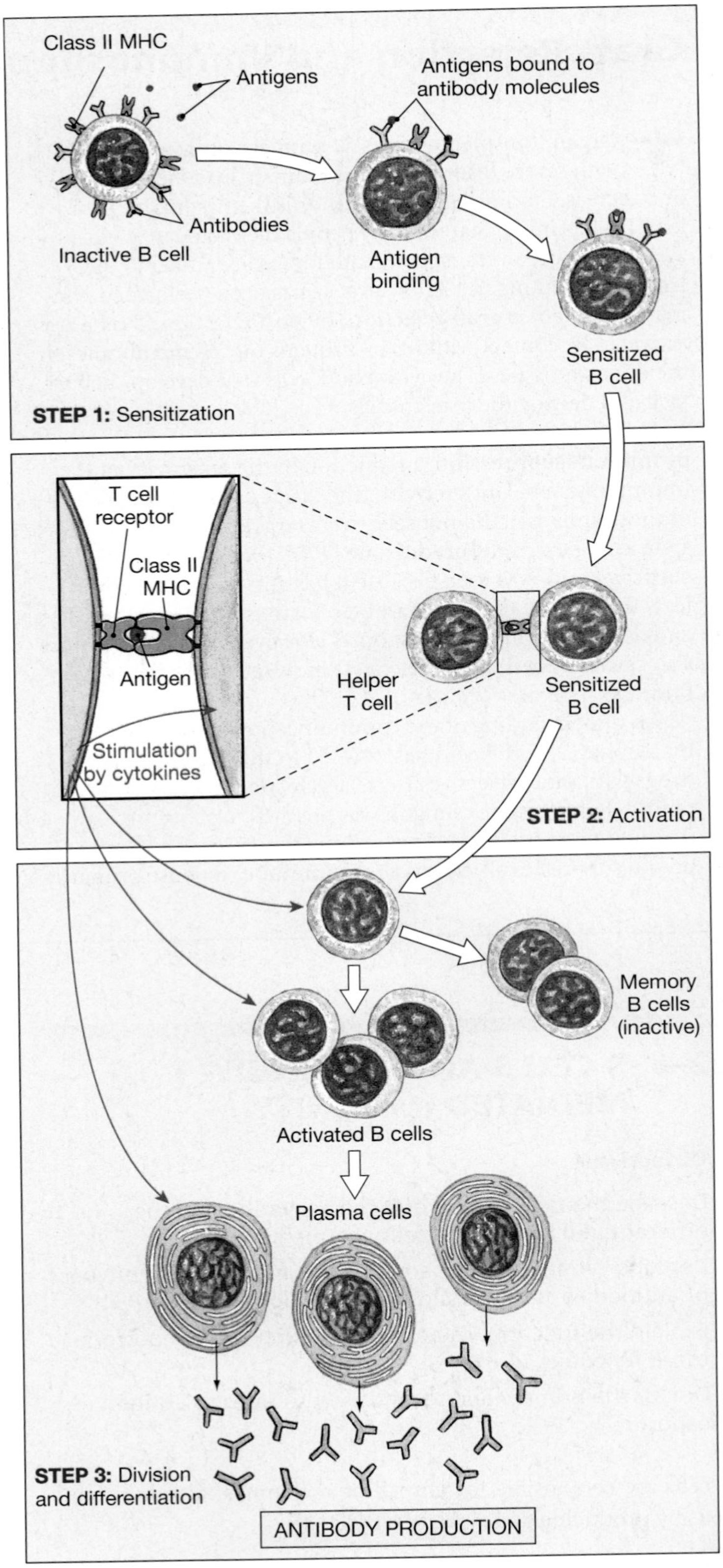

•FIGURE 22–20
The Sensitization and Activation of B Cells. A B cell is
sensitized by exposure to antigens that bind to antibodies in the B
cell membrane. The B cell then displays those antigens in its cell
membrane. Activated helper T cells encountering the antigens
release cytokines that trigger the activation of the B cell. The
activated B cell then divides, producing memory B cells and plasma
cells that secrete antibodies.

activation has occurred, these same cytokines stimulate B cell
division, accelerate plasma cell formation, and enhance antibody production.

Figure 22–20• diagrams the steps involved in activation of a
B cell by a helper T cell. The activated B cell typically divides several times, producing daughter cells that differentiate into plasma cells and *memory B cells*. The plasma cells begin synthesizing
and secreting large numbers of antibodies into the interstitial
fluid. These antibodies have the same target as the antibodies on
the surface of the sensitized B cell. When stimulated by cytokines from helper T cells, a plasma cell can secrete up to 100
million antibody molecules each hour.

Memory B cells perform the same role for antibody-mediated
immunity that memory T cells perform for cell-mediated immunity. Memory B cells do not respond to a threat on first exposure. Instead, they remain in reserve to deal with subsequent
injuries or infections that involve the same antigens. On subsequent exposure, the memory B cells respond by dividing and
differentiating into plasma cells that secrete antibodies in massive quantities.

☐ ANTIBODY STRUCTURE

Figure 22–21• presents different views of a single antibody
molecule. An antibody consists of two parallel pairs of polypeptide chains: one pair of **heavy chains** and one pair of **light**

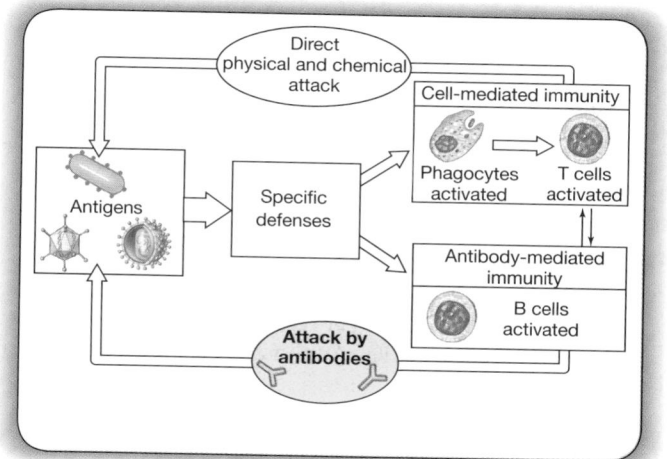

chains. Each chain contains both *constant segments* and *variable segments*.

The constant segments of the heavy chains form the base of the antibody molecule (Figure 22–21a,b●). B cells produce only five types of constant segments. These form the basis of a classification scheme that identifies antibodies as *IgG*, *IgE*, *IgD*, *IgM*, or *IgA*, as we shall discuss in the next section. The structure of the constant segments of the heavy chains determines the way the antibody is secreted and how it is distributed within the body. For example, antibodies in one class circulate in body fluids, whereas those of another class bind to the membranes of basophils and mast cells.

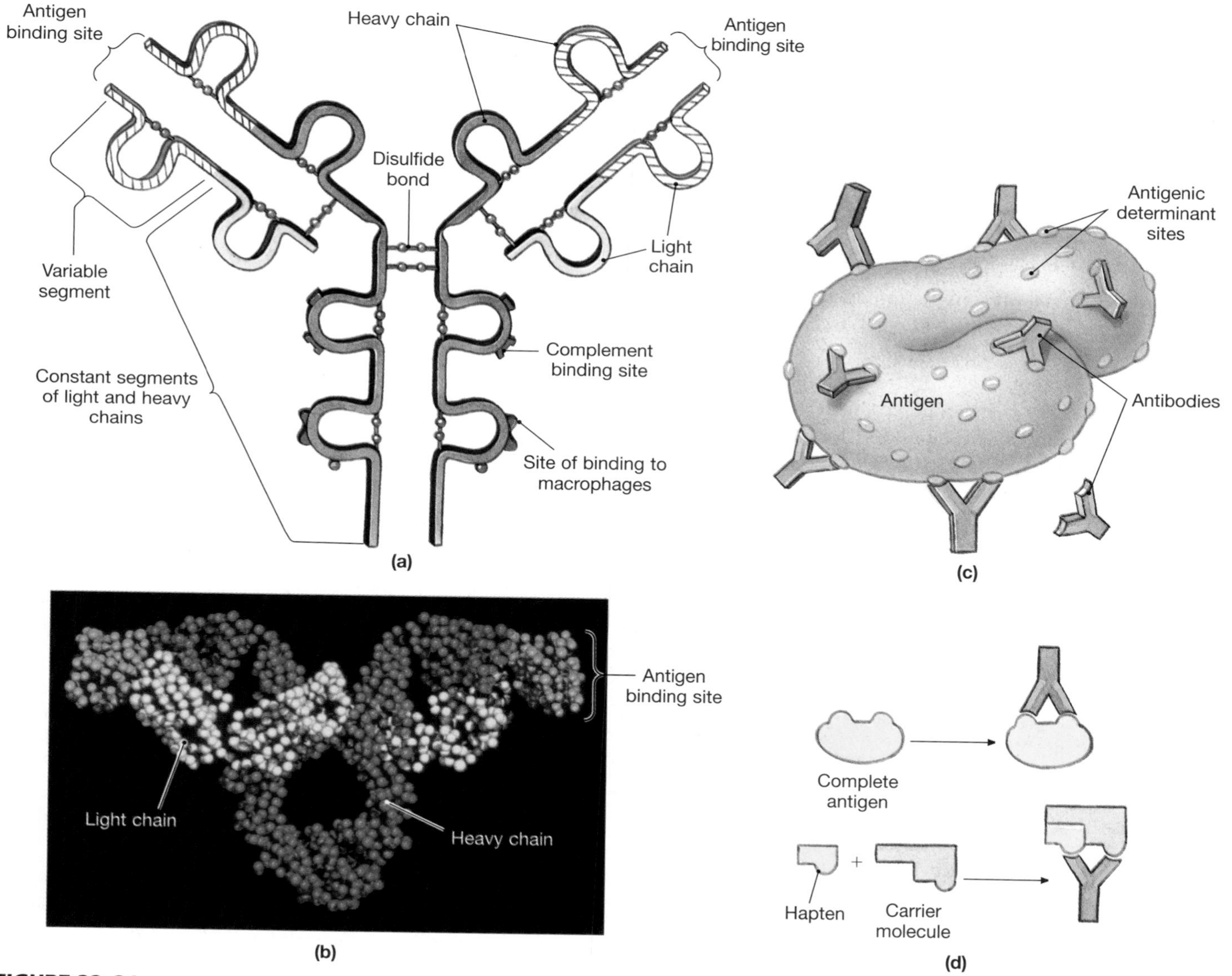

Antigen binding site

Heavy chain

Antigen binding site

Variable segment

Disulfide bond

Light chain

Constant segments of light and heavy chains

Complement binding site

Site of binding to macrophages

Antigenic determinant sites

Antigen

Antibodies

(a)

(c)

Antigen binding site

Light chain

Heavy chain

Complete antigen

Hapten   Carrier molecule

(b)

(d)

●**FIGURE 22–21**
Antibody Structure.   **(a)** A diagrammatic view of the structure of an antibody. **(b)** A computer-generated image of a typical antibody. **(c)** Antigen–antibody binding. **(d)** A hapten, or partial antigen, can become a complete antigen by binding to a carrier molecule.

The heavy-chain constant segments, which are bound to constant segments of the light chains, also contain binding sites that can activate the complement system. These binding sites are covered when the antibody is secreted, but they become exposed when the antibody binds to an antigen.

The specificity of the antibody molecule depends on the structure of the variable segments of the light and heavy chains. The free tips of the two variable segments contain the **antigen binding sites** of the antibody molecule (Figure 22–21a●). These sites can interact with an antigen in the same way that the active site of an enzyme interacts with a substrate molecule. ∞ p. 56

Small differences in the amino acid sequence of the variable segments affect the precise shape of the antigen binding site. These differences account for differences in specificity between the antibodies produced by different B cells. The distinctions are the result of minor genetic variations that occur during the production, division, and differentiation of B cells. The normal adult body contains roughly 10 trillion B cells, which can produce an estimated 100 million types of antibodies, each with a different specificity.

The Antigen–Antibody Complex

When an antibody molecule binds to its corresponding antigen molecule, an **antigen–antibody complex** is formed. Once the two molecules are in position, hydrogen bonding and other weak chemical forces lock them together.

Antibodies do not bind to the entire antigen; rather, they bind to certain portions of its exposed surface, regions called **antigenic determinant sites** (Figure 22–21c●). The specificity of the binding depends initially on the three-dimensional "fit" between the variable segments of the antibody molecule and the corresponding antigenic determinant sites. A **complete antigen** is an antigen with at least two antigenic determinant sites, one for each of the antigen binding sites on an antibody molecule. Exposure to a complete antigen can lead to B cell sensitization and a subsequent immune response. Most environmental antigens have multiple antigenic determinant sites; entire microorganisms may have thousands.

Haptens, or *partial antigens*, do not ordinarily cause B cell activation and antibody production. Haptens include short peptide chains, steroids and other lipids, and several drugs, including antibiotics such as *penicillin*. Haptens may, however, become attached to carrier molecules, forming combinations that can function as complete antigens (Figure 22–21d●). In some cases, the carrier contributes an antigenic determinant site. The antibodies produced will attack both the hapten and the carrier molecule. If the carrier molecule is normally present in the tissues, the antibodies may begin attacking and destroying normal cells. This process is the basis for several drug reactions, including allergies to penicillin.

Classes and Actions of Antibodies

Body fluids have five classes of antibodies, or **immunoglobulins (Igs):** *IgG, IgE, IgD, IgM,* and *IgA* (Table 22–1). The classes are determined by variations in the structure of the heavy-chain constant segments and so have no effect on the antibody's specificity, which is determined by the antigen binding sites. The formation of an antigen–antibody complex may cause the elimination of the antigen in seven ways:

1. *Neutralization.* Both viruses and bacterial toxins have specific sites that must bind to target regions on body cells in order to enter or injure those cells. Antibodies may bind to those sites, making the virus or toxin incapable of attaching itself to a cell. This mechanism is known as **neutralization**.

2. *Agglutination and Precipitation.* Each antibody molecule has two antigen binding sites, and most antigens have many antigenic determinant sites. If separate antigens (such as macromolecules or bacterial cells) are far apart, an antibody molecule will necessarily bind to two antigenic sites on the same antigen. If antigens are close together, however, an antibody can bind to antigenic determinant sites on two different antigens. In this way, antibodies can form "bridges" that tie large numbers of antigens together. The three-dimensional structure created by such binding is known as an **immune complex**. When the antigen is a soluble molecule, such as a toxin, this process may create complexes that are too large to remain in solution. The formation of insoluble immune complexes is called **precipitation**. When the target antigen is on the surface of a cell or virus, the formation of large complexes is called **agglutination**. The clumping of red blood cells that occurs when incompatible blood types are mixed is an agglutination reaction. ∞ p. 663

3. *The Activation of Complement.* On binding to an antigen, portions of the antibody molecule change shape, exposing areas that bind complement proteins. The bound complement molecules then activate the complement system, destroying the antigen (as discussed previously).

4. *The Attraction of Phagocytes.* Antigens covered with antibodies attract eosinophils, neutrophils, and macrophages—cells that phagocytize pathogens and destroy foreign or abnormal cell membranes.

5. *Opsonization.* A coating of antibodies and complement proteins increases the effectiveness of phagocytosis. Recall from page 796 that this effect is called *opsonization*. Microorganisms such as bacteria have slick cell membranes or capsules, and a phagocyte must be able to hang onto its prey before it can engulf the prey. Phagocytes can bind more easily to antibodies and complement proteins on the surface of a pathogen than they can to the bare surface.

6. *The Stimulation of Inflammation.* Antibodies may promote inflammation through the stimulation of basophils and mast cells.

7. *The Prevention of Bacterial and Viral Adhesion.* Antibodies dissolved in saliva, mucus, and perspiration coat epithelia and provide them with an additional layer of defense. A covering of antibodies makes it difficult for pathogens to attach to body surfaces and penetrate their defenses.

TABLE 22–1 CLASSES OF ANTIBODIES

STRUCTURE	DESCRIPTION
IgG 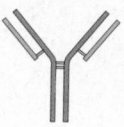	**IgG** is the largest and most diverse class of antibodies. There are several types of IgG, but each type occurs as an individual molecule. Together, they account for 80 percent of all antibodies. The IgG antibodies are responsible for resistance against many viruses, bacteria, and bacterial toxins. These antibodies can cross the placenta, and during embryological development, maternal IgG provides passive immunity to the fetus. However, the anti-Rh (anti-D) antibodies produced by Rh-negative mothers sensitized to Rh surface antigens are also IgG antibodies. These antibodies can cross the placenta and attack fetal Rh-positive red blood cells, producing the *hemolytic disease of the newborn*. ∞ p. 665
IgE 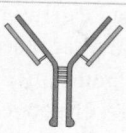	**IgE** attaches as an individual molecule to the exposed surfaces of basophils and mast cells. When a suitable antigen appears and binds to IgE molecules, the cell is stimulated to release histamine and other chemicals that accelerate inflammation in the immediate area. IgE is also important in the allergic response.
IgD 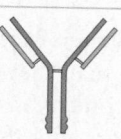	**IgD** is an individual molecule on the surfaces of B cells, where it can bind antigens in the extracellular fluid. This binding can play a role in the activation of the B cell involved.
IgM 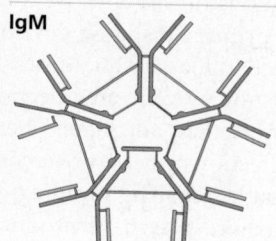	**IgM** is the first type of antibody secreted after an antigen arrives. The concentration of IgM then declines as IgG production accelerates. Although plasma cells secrete individual IgM molecules, IgM circulates as a five-antibody starburst. This configuration makes these antibodies particularly effective in forming immune complexes. The anti-A and anti-B antibodies responsible for the agglutination of cross-matched blood are IgM antibodies. ∞ p. 664 IgM antibodies may also attack bacteria that are insensitive to IgG.
IgA 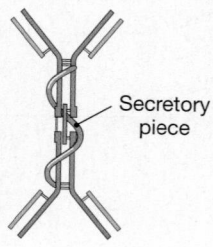 Secretory piece	**IgA** is found primarily in glandular secretions such as mucus, tears, and saliva. These antibodies attack pathogens before they gain access to internal tissues. IgA antibodies circulate in blood as individual molecules or in pairs. Epithelial cells absorb them from the blood and attach a *secretory piece*, which confers solubility, before secreting the IgA molecules onto the epithelial surface.

✓ How can the presence of an abnormal peptide in the cytoplasm of a cell initiate an immune response?

✓ A decrease in the number of cytotoxic T cells would affect which type of immunity?

✓ How would a lack of helper T cells affect the antibody-mediated immune response?

✓ A sample of lymph contains an elevated number of plasma cells. Would you expect the number of antibodies in the blood to be increasing or decreasing? Why?

Answers start on page Q-1

 The processes involved in specific immunity and lymphocyte activation can be reviewed on the **Companion Website:** Chapter 22/Tutorials.

☐ PRIMARY AND SECONDARY RESPONSES TO ANTIGEN EXPOSURE

The initial response to exposure to an antigen is called the **primary response**. When the antigen appears again, it triggers a more extensive and prolonged **secondary response**. This response reflects the presence of large numbers of memory cells that are primed for the arrival of the antigen. Primary and secondary responses are characteristic of both cell-mediated and antibody-mediated immunities. The differences between the responses are most easily demonstrated by following the production of antibodies over time.

The Primary Response

Because the antigen must activate the appropriate B cells, which must then differentiate into plasma cells, the primary response does not appear immediately (Figure 22–22a●). As the plasma cells differentiate and begin secreting, the concentration of circulating antibodies undergoes a gradual, sustained rise.

During the primary response, the **antibody titer** ("standard"), or level of antibody activity, in the plasma does not peak until one to two weeks after the initial exposure. If the individual is no longer exposed to the antigen, the antibody concentration declines thereafter. This reduction in antibody production occurs because (1) plasma cells have very high metabolic rates and survive for only a short time, and (2) further production of plasma cells is inhibited by suppression factors released by suppressor T cells. However, suppressor T cell activity does not begin immediately after antigen exposure, and under normal conditions helper cells outnumber suppressors by more than three to one. As a result, many B cells are activated before suppressor T cell activity has a noticeable effect.

Two types of antibodies are involved in the primary response. Molecules of *immunoglobulin M*, or *IgM*, are the first to appear in the bloodstream. IgM is secreted by the plasma cells that form immediately after B cell activation. These lymphocytes do not

pause to produce memory cells. Levels of *immunoglobulin G*, or *IgG*, rise more slowly, because the stimulated lymphocytes undergo repeated cell divisions and generate large numbers of memory cells as well as plasma cells. In effect, IgM provides an immediate but limited defense that fights the infection until massive quantities of IgG can be produced.

The Secondary Response

Unless they are exposed to the same antigen a second time, memory B cells do not differentiate into plasma cells. If and when that exposure occurs, the memory B cells respond immediately—faster than the B cells stimulated during the initial exposure. This response is immediate in part because memory B cells are activated at relatively low antigen concentrations and in part because they synthesize more effective and destructive antibodies. Activated memory B cells divide and differentiate into plasma cells that secrete these antibodies in massive quantities. This secretion is the secondary response to antigen exposure.

During the secondary response, antibody titers increase more rapidly and reach levels many times higher than they did in the primary response (Figure 22–22b●). The secondary response appears even if the second exposure occurs years after the first, because memory cells may survive for 20 years or more.

The primary response to an antigen is not as effective a defense as the secondary response. The primary response develops slowly, and antibodies are not produced in massive quantities. As a result, the primary response may not prevent an infection, whereas the secondary response is more than adequate. Therefore, a person who survives the first infection can easily overcome a subsequent invasion by the same pathogen. The invading organisms will typically be destroyed before they have produced any disease symptoms. The effectiveness of the secondary response is one of the basic principles behind the use of immunization to prevent disease. **AM** Immunization

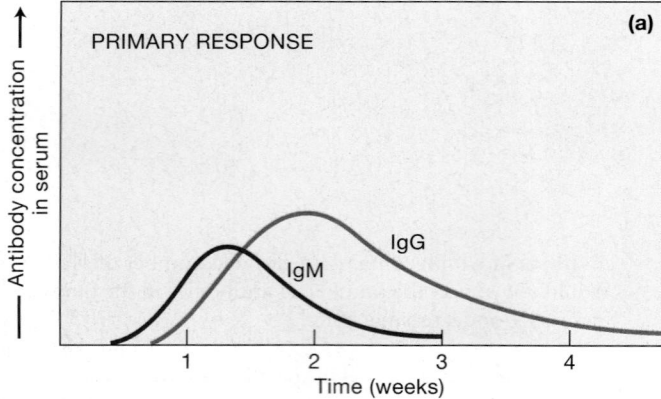
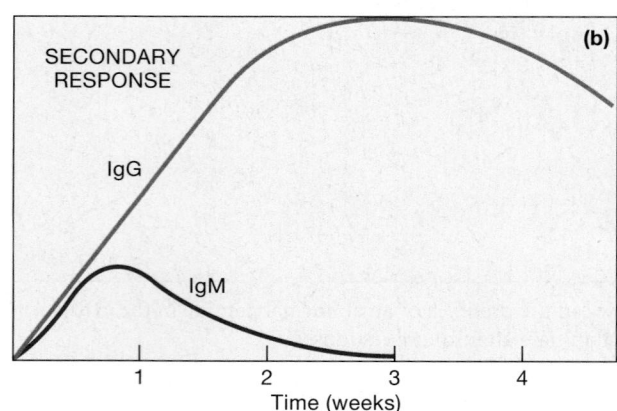

●**FIGURE 22–22**
The Primary and Secondary Immune Responses. **(a)** The primary response takes about two weeks to develop peak antibody titers, and IgM and IgG antibody concentrations do not remain elevated. **(b)** The secondary response is characterized by a very rapid increase in IgG antibody titer, to levels much higher than those of the primary response. Antibody activity remains elevated for an extended period after the second exposure to the antigen.

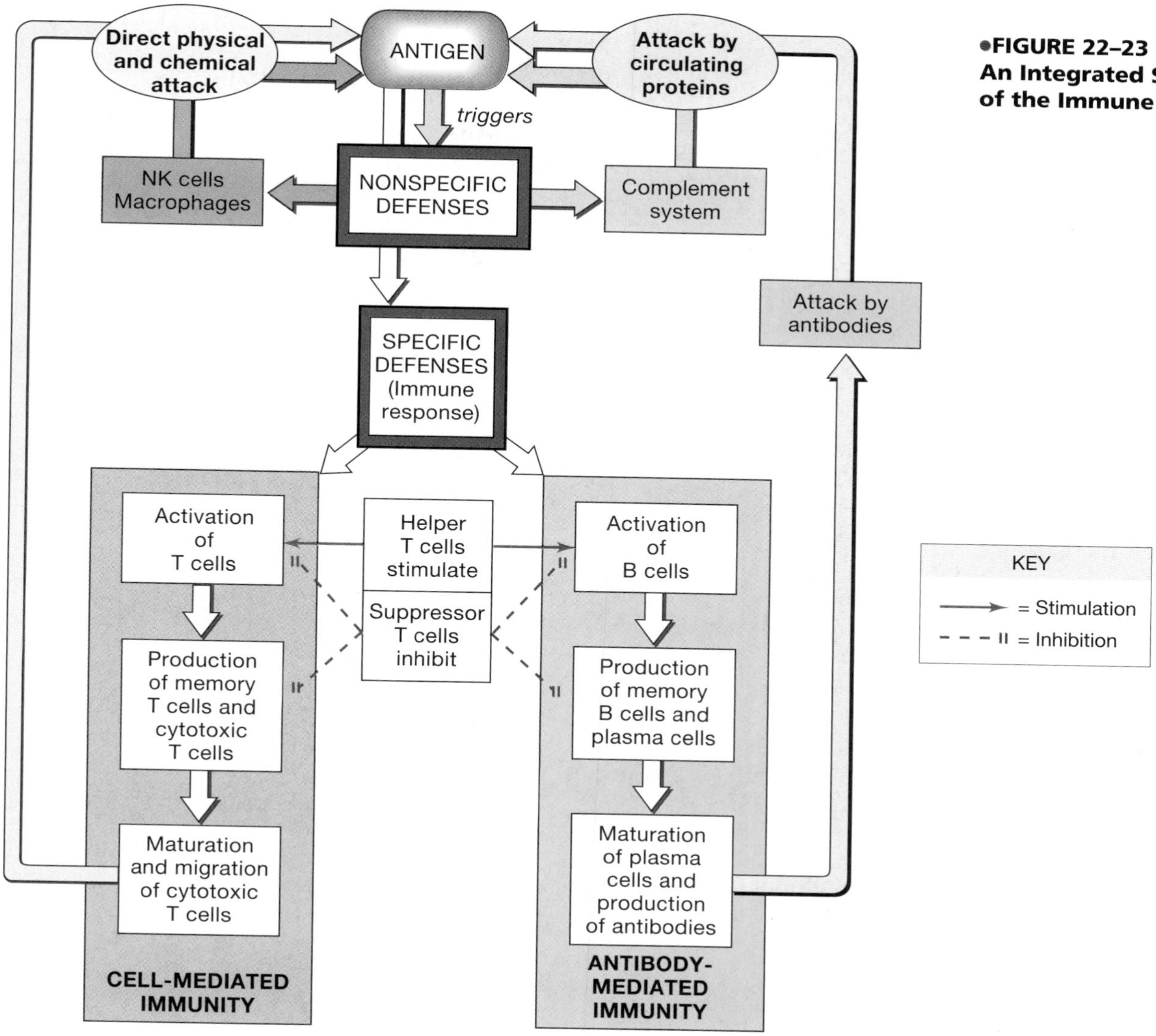

FIGURE 22–23
An Integrated Summary of the Immune Response

- Direct physical and chemical attack
- ANTIGEN
- Attack by circulating proteins
- *triggers*
- NK cells Macrophages
- NONSPECIFIC DEFENSES
- Complement system
- Attack by antibodies
- SPECIFIC DEFENSES (Immune response)

KEY
- → = Stimulation
- – – – II = Inhibition

- Activation of T cells
- Helper T cells stimulate
- Activation of B cells
- Production of memory T cells and cytotoxic T cells
- Suppressor T cells inhibit
- Production of memory B cells and plasma cells
- Maturation and migration of cytotoxic T cells
- Maturation of plasma cells and production of antibodies

CELL-MEDIATED IMMUNITY

ANTIBODY-MEDIATED IMMUNITY

SUMMARY: THE IMMUNE RESPONSE

We have now described the basic chemical and cellular interactions that follow the appearance of a foreign antigen in the body. Figure 22–23● presents an integrated view of the immune response and its relationship to nonspecific defenses. Figure 22–24● provides an overview of the course of events responsible for overcoming bacterial infection.

In the early stages of bacterial infection, before antigen processing has occurred, neutrophils and NK cells migrate into the threatened area and destroy bacteria. Over time, cytokines draw increasing numbers of phagocytes to the region. Cytotoxic T cells appear as arriving T cells are activated by antigen presentation (Figure 22–24●). Last of all, the population of plasma cells rises as activated B cells differentiate. This rise is followed by a gradual, sustained increase in the level of circulating antibodies.

The basic sequence of events is similar when a viral infection occurs. The initial steps are different, however, because cytotoxic T cells and NK cells can be activated by contact with infected cells. Figure 22–25● contrasts the steps involved in defending against bacterial in-

fection with those involved in defending against viral infection. Table 22–2 reviews the cells that participate in tissue defenses.

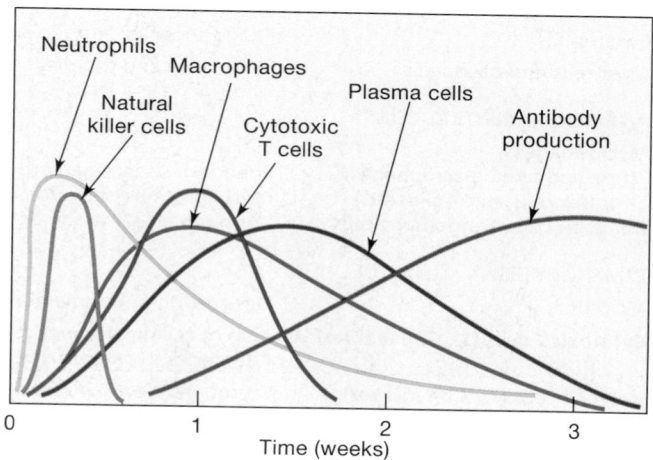

FIGURE 22–24
The Course of the Body's Response to Bacterial Infection. An outline of the basic sequence of events that begins with the appearance of bacteria in peripheral tissues.

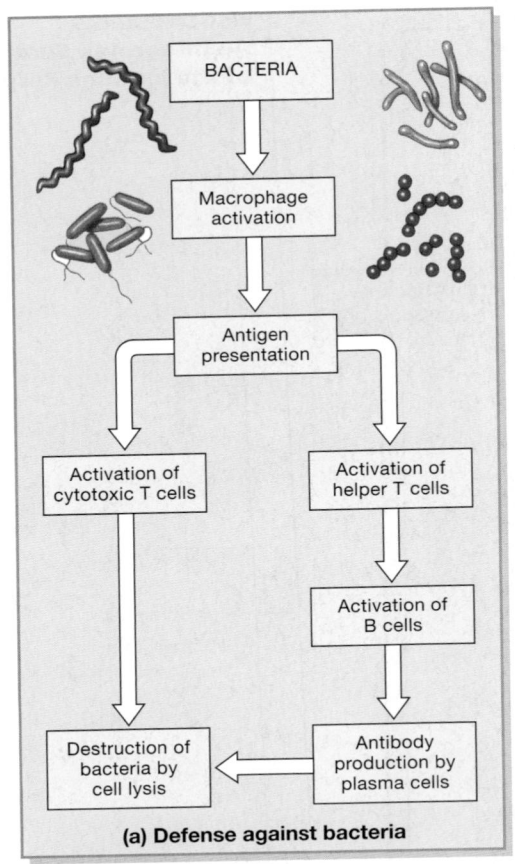

(a) Defense against bacteria

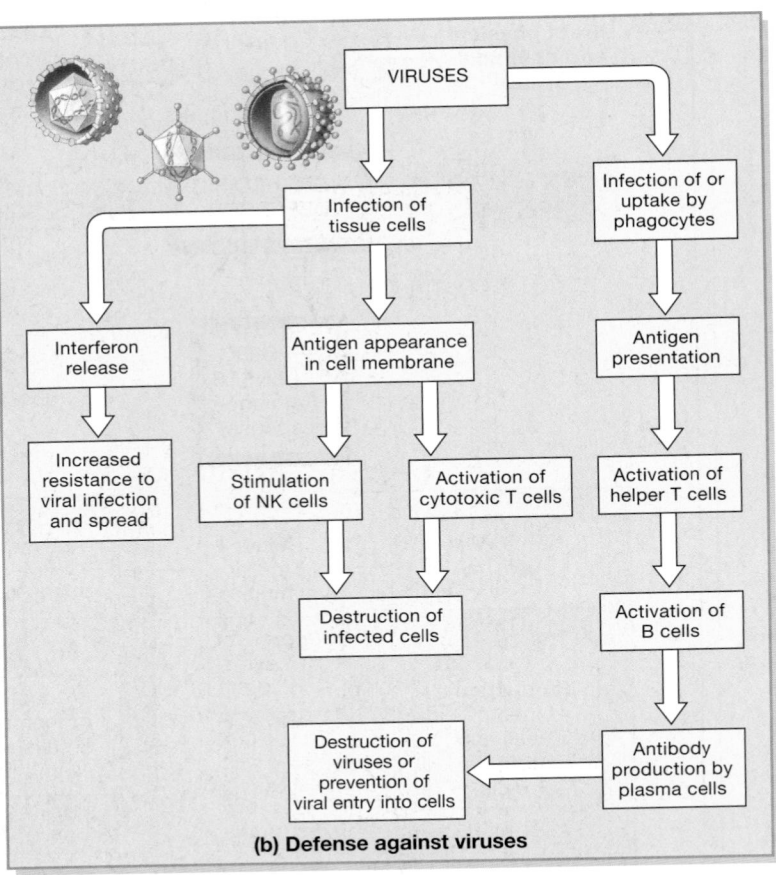

(b) Defense against viruses

•**FIGURE 22–25**
Defenses against Bacterial and Viral Pathogens. **(a)** Defenses against bacteria are usually initiated by active macrophages. **(b)** Defenses against viruses are usually activated after the infection of normal cells.

SUMMARY TABLE 22–2	CELLS THAT PARTICIPATE IN TISSUE DEFENSES
CELL	**FUNCTIONS**
Neutrophils	Phagocytosis; stimulation of inflammation
Eosinophils	Phagocytosis of antigen–antibody complexes; suppression of inflammation; involved in allergic response
Mast cells and basophils	Stimulation and coordination of inflammation by release of histamine, heparin, leukotrienes, prostaglandins
ANTIGEN-PRESENTING CELLS	
Macrophages (free and fixed macrophages, Kupffer cells, microglia, etc.)	Phagocytosis; antigen processing; antigen presentation with Class II MHC proteins; secretion of cytokines, especially interleukins and interferons
Dendritic cells, Langerhans cells	Antigen presentation bound to Class II MHC proteins
LYMPHOCYTES	
NK cells	Destruction of cell membranes containing abnormal antigens
Cytotoxic T cells (T_C, CD8 marker)	Lysis of cell membranes containing antigens bound to Class I MHC proteins; secretion of perforins, defensins, lymphotoxins, and other cytokines
Helper T cells (T_H, CD4 marker)	Secretion of cytokines that stimulate cell-mediated and antibody-mediated immunity; activation of sensitized B cells
B cells	Differentiation into plasma cells that secrete antibodies and provide antibody-mediated immunity
Suppressor T cells (T_S, CD8 marker)	Secretion of suppression factors that inhibit the immune response
Memory cells (T_S, T_H, B)	Produced during the activation of T cells and B cells; remain in tissues awaiting rearrival of antigens

focus: Hormones of the Immune System

- Discuss important hormones of the immune response and explain their significance.

The specific and nonspecific defenses of the body are coordinated by physical interaction and by the release of chemical messengers. One example of physical interaction is antigen presentation by activated macrophages and helper T cells. An example of the release of chemical messengers is the secretion of cytokines by many of the cells involved in the immune response. Cytokines are classified according to their origins: *Lymphokines*

are produced by lymphocytes, *monokines* by active macrophages and other antigen-presenting cells. These terms are misleading, however, because lymphocytes and macrophages may secrete the same cytokines, and cytokines can also be secreted by cells involved with nonspecific defenses and tissue repair.

Table 22–3 summarizes the cytokines that have been identified to date. Five subgroups merit special attention: (1) *interleukins*, (2) interferons, (3) *tumor necrosis factors*, (4) chemicals that regulate phagocytic activities, and (5) *colony-stimulating factors*.

TABLE 22–3 CHEMICAL MEDIATORS (CYTOKINES) OF THE IMMUNE RESPONSE

COMPOUND	FUNCTIONS
Interleukins	
Interleukin-1 (IL-1)	Stimulates T cells to produce IL-2, promotes inflammation; causes fever
IL-2	Stimulates growth and activation of other T cells and NK cells
IL-3	Stimulates production of mast cells and other blood cells
IL-4 (B cell differentiating factor); IL-5 (B cell growth factor); IL-6, IL-7 (B cell stimulating factors); IL-10; IL-11	Promote differentiation and growth of B cells and stimulate plasma cell formation and antibody production; each has somewhat different effects on macrophages and their activities
IL-8	Stimulates blood vessel formation (angiogenesis)
IL-9	Stimulates myeloid cell production (RBCs, platelets, granulocytes, monocytes)
IL-12	Stimulates T cell activity and cell-mediated immunity
IL-13	Suppresses production of several other cytokines (IL-1, IL-8, TNF); stimulates Class II MHC antigen presentation
Interferons (alpha, beta, gamma)	Activate other cells to prevent viral entry; inhibit viral replication; stimulate NK cells and macrophages
Tumor necrosis factors (TNFs)	Kill tumor cells; slow tumor growth; stimulate activities of T cells and eosinophils; inhibit parasites and viruses
Monocyte-chemotactic factor (MCF)	Attracts monocytes; transforms them into macrophages
Migration-inhibitory factor (MIF)	Prevents migration of macrophages from the area
Macrophage-activating factor (MAF)	Makes macrophages more active and aggressive
Microphage-chemotactic factor	Attracts microphages from the blood
Colony-stimulating factors (CSFs)	Stimulate RBC and WBC production
Defensins	Kill cells by piercing cell membrane
Growth-inhibitory factor (GIF)	Reduces or inhibits replication of target cells
Hemopoiesis-stimulating factor	Promotes blood cell production in bone marrow
Leukotrienes	Stimulate regional inflammation
Lymphotoxins	Kill cells; damage tissue; promote inflammation
Perforin	Destroys cell membranes by creating large pores
Transforming growth factor-β (TGF-β)	Stimulates production of IgA and of matrix proteins in connective tissues; inhibits macrophage activation and T_C maturation
Suppression factors	Inhibit T_C cell and B cell activity; depress immune response
Transfer factor	Sensitizes other T cells to same antigen

Interleukins

Interleukins may be the most diverse and important chemical messengers in the immune system. Nearly 20 types of interleukins have been identified; several are given in Table 22–3. Lymphocytes and macrophages are the primary sources of interleukins, but specific interleukins, such as interleukin-1 (IL-1), are also produced by endothelial cells, fibroblasts, and astrocytes. Interleukins have the following general functions:

1. *Increasing T Cell Sensitivity to Antigens Exposed on Macrophage Membranes.* Heightened sensitivity accelerates the production of cytotoxic and regulatory T cells.

2. *Stimulating B Cell Activity, Plasma Cell Formation, and Antibody production.* These events promote the production of antibodies and the development of antibody-mediated immunity.

3. *Enhancing Nonspecific Defenses.* Known effects of interleukin production include (1) the stimulation of inflammation, (2) the formation of scar tissue by fibroblasts, (3) the elevation of body temperature via the preoptic nucleus of the central nervous system, (4) the stimulation of mast cell formation, and (5) the promotion of adrenocorticotropic hormone (ACTH) secretion by the anterior lobe of the pituitary gland.

4. *Moderating the Immune Response.* Some interleukins help suppress immune function and shorten the duration of an immune response.

Two interleukins, IL-1 and IL-2, are important in the stimulation and maintenance of the immune response. When released by activated macrophages and lymphocytes, these cytokines not only stimulate the activities of other immune cells but also further stimulate the secreting cell. The result is a positive feedback loop that promotes the recruitment of additional immune cells. Although mechanisms exist to control the degree of stimulation, the regulatory process sometimes breaks down, and massive production of interleukins can cause problems at least as severe as those of the primary infection. For example, in *Lyme disease* the release of IL-1 by activated macrophages in response to a localized bacterial infection produces symptoms of fever, pain, skin rash, and arthritis that affect the entire body. **AM** Lyme Disease

Some interleukins enhance the immune response, whereas others suppress it. The relative quantities secreted at any moment therefore have a significant effect on the nature and intensity of the response to an antigen. In the course of a typical infection, the pattern of interleukin secretion is constantly changing. Whether the individual succeeds in overcoming the infection is determined in part by whether stimulatory or suppressive interleukins predominate. As a result, interleukins and their interactions are now the focus of an intensive research effort.

Interferons

Interferons make the cell that synthesizes them, and that cell's neighbors, resistant to viral infection, thereby slowing the spread of the virus. These compounds may have other beneficial effects in addition to their antiviral activity. For example, alpha-interferons and gamma-interferons attract and stimulate NK cells, and beta-interferons slow the progress of inflammation associated with viral infection. Gamma-interferons also stimulate macrophages, making them more effective at killing bacterial or fungal pathogens.

Because they stimulate NK cell activity, interferons can be used to fight some cancers. For example, alpha-interferons have been used in the treatment of malignant melanoma, bladder cancer, ovarian cancer, and two forms of leukemia, while alpha- or gamma-interferons may be used to treat Kaposi's sarcoma, a cancer that typically develops in individuals with AIDS. **AM** AIDS

Tumor Necrosis Factors

Tumor necrosis factors (TNFs) slow the growth of a tumor and kill sensitive tumor cells. Activated macrophages secrete one type of TNF and carry the molecules in their cell membranes. Cytotoxic T cells produce a different type of TNF. In addition to their effects on tumor cells, tumor necrosis factors stimulate granular leukocyte production, promote eosinophil activity, cause fever, and increase T cell sensitivity to interleukins.

Chemicals Regulating Phagocytic Activities

Several cytokines coordinate immune defenses by adjusting the activities of phagocytic cells. These cytokines include factors that attract free macrophages and microphages and prevent their premature departure from the site of an injury.

Colony-Stimulating Factors

We introduced **colony-stimulating factors (CSFs)** in Chapter 19. ∞ p. 670 These factors are produced by active T cells, cells of the monocyte–macrophage group, endothelial cells, and fibroblasts. CSFs stimulate the production of blood cells in bone marrow and lymphocytes in lymphoid tissues and organs.

As our understanding of the immune system grows, complex therapies involving combinations of cytokines (including interleukins) and *monoclonal antibodies* are appearing. (Monoclonal antibodies are produced by a single clone of B cells under laboratory conditions.) In one procedure, cytotoxic T cells were removed from patients with *malignant melanoma*, a particularly dangerous type of skin cancer. These lymphocytes were able to recognize tumor cells and had migrated to the tumor, but for some reason they appeared to be unable to kill the tumor cells.

The extracted lymphocytes were cultured, and viruses were used to insert multiple copies of the genes responsible for the production of tumor necrosis factor. The patients were then given periodic infusions of these "supercharged" T cells. To enhance T cell activity further, the researchers also administered doses of interleukin-2. Initial results are promising. It is clear that the ability to manipulate the immune response will revolutionize the treatment of many serious diseases. **AM** Technology, Immunity, and Disease

22-7 NORMAL AND ABNORMAL RESISTANCE

Objectives

- Describe the origin, development, activation, and regulation of normal resistance to disease.
- Explain the origin of autoimmune disorders, immunodeficiency diseases, and allergies, and list important examples of each type of disorder.
- Discuss the effects of stress on the immune function.

The ability to produce an immune response after exposure to an antigen is called **immunological competence**. Cell-mediated immunity can be demonstrated as early as the third month of fetal development, but active antibody-mediated immunity appears later.

☐ THE DEVELOPMENT OF IMMUNOLOGICAL COMPETENCE

The first cells that leave the fetal thymus migrate to the skin and into the epithelia lining the mouth, the digestive tract, and, in females, the uterus and vagina. These cells take up residence in these tissues as antigen-presenting cells, such as the Langerhans cells of the skin, whose primary function will be the activation of T cells. T cells that leave the thymus later in development populate lymphoid organs throughout the body.

The membranes of the first B cells to be produced in the liver and bone marrow carry IgM antibodies. Sometime after the fourth month in utero the fetus may, if exposed to specific pathogens, produce IgM antibodies. Fetal antibody production is uncommon, however, because the developing fetus has natural passive immunity due to the transfer of IgG antibodies from the maternal bloodstream. These are the only antibodies that can cross the placenta, and they include the antibodies responsible for the clinical problems that accompany fetal–maternal Rh incompatibility, discussed in Chapter 19. ∞ p. 665 Because the anti-A and anti-B antibodies are IgM antibodies, and IgM cannot cross the placenta, problems with maternal–fetal incompatibilities involving the ABO blood groups rarely occur.

The natural immunity provided by maternal IgG may not be enough to protect the fetus if the maternal defenses are overrun by a bacterial or viral infection. For example, the microorganisms responsible for syphilis and rubella ("German measles") can cross from the maternal to the fetal bloodstream, producing a congenital infection that leads to the production of fetal antibodies. IgM provides only a partial defense, and these infections can result in severe developmental problems for the fetus. **AM** Fetal Infections

Delivery eliminates the maternal supply of IgG. Although the mother provides IgA antibodies in her breast milk, the infant gradually loses its passive immunity. The amount of maternal IgG in the infant's bloodstream declines rapidly over the first two months after birth. During this period, the infant becomes vulnerable to infection by bacteria or viruses that were previously overcome by maternal antibodies. The infant also begins producing its own IgG, as its immune system begins to respond to infections, environmental antigens, and vaccinations. It has been estimated that, from birth to age 12, children encounter a "new" antigen every six weeks. (This fact explains why most parents, exposed to the same antigens when they were children, remain healthy, while their children develop runny noses and colds.) Over this period, the concentration of circulating antibodies gradually rises toward normal adult levels, and the populations of memory B cells and T cells continue to increase.

Skin tests can be used to determine whether an individual has developed resistance to a particular antigen. In this procedure, small quantities of antigen are injected into the skin, generally on the anterior surface of the forearm. If resistance has developed, the region will become inflamed over the next two to four days. Many states require a tuberculosis test, called a *tuberculin skin test*, when someone applies for a food-service or health-service job. (When preparing or serving food or coming in contact with patients, anyone with tuberculosis can accidentally transmit the bacteria.) If the test is positive, the individual has been exposed to the disease and has developed antibodies. Further tests must then be performed to determine whether an infection is under way. Skin tests are also used to check for allergies to environmental antigens. **AM** Delayed Hypersensitivity and Skin Tests

☐ IMMUNE DISORDERS

Because the immune response is so complex, many opportunities exist for things to go wrong. A variety of clinical conditions result from disorders of the immune function. **Autoimmune disorders** develop when the immune response mistakenly targets normal body cells and tissues. In an **immunodeficiency disease**, either the immune system fails to develop normally or the immune response is blocked in some way. Autoimmune disorders and immunodeficiency diseases are relatively rare—clear evidence of the effectiveness of the immune system's control mechanisms. A far more common, and generally far less dangerous, class of immune disorders is the **allergies**. We next consider examples of each type of disorder. **AM** Immune Complex Disorders; Systemic Lupus Erythematosus

Autoimmune Disorders

Autoimmune disorders affect an estimated 5 percent of adults in North America and Europe. In previous chapters, we noted many examples of autoimmune disorders in discussions of their effects on the function of major systems. **AM** Autoimmune Disorders

The immune system usually recognizes and ignores the antigens normally found in the body—self-antigens. The recognition system can malfunction, however. When it does, the activated B cells make antibodies against other body cells and tissues. These misguided antibodies are called **autoantibodies**. The trigger may be a reduction in suppressor T cell activity, the excessive stimulation of helper T cells, tissue damage that releases large quantities of antigenic fragments, haptens bound to compounds normally ignored, viral or bacterial toxins, or a combination of factors.

The symptoms produced depend on the specific antigen attacked by autoantibodies. For example,

- The inflammation of *thyroiditis* results from the release of autoantibodies against thyroglobulin;
- *Rheumatoid arthritis* occurs when autoantibodies form immune complexes within connective tissues, especially around the joints; and
- *Insulin-dependent diabetes mellitus (IDDM)* is generally caused by autoantibodies that attack cells in the pancreatic islets.

Many autoimmune disorders appear to be cases of mistaken identity. For example, proteins associated with the measles, Epstein–Barr, influenza, and other viruses contain amino acid sequences that are similar to those of myelin proteins. As a result, antibodies that target these viruses may also attack myelin sheaths. This mechanism accounts for the neurological complications that sometimes follow a vaccination or a viral infection. It is also the mechanism that is likely responsible for *multiple sclerosis.*

For unknown reasons, the risk of autoimmune problems increases if an individual has an unusual type of MHC proteins. At least 50 clinical conditions have been linked to specific variations in MHC structure.

Immunodeficiency Diseases

Immunodeficiency diseases result from (1) problems with the embryological development of lymphoid organs and tissues; (2) an infection with a virus, such as HIV, that depresses the immune function; or (3) treatment with, or exposure to, immunosuppressive agents, such as radiation or drugs.

Individuals born with **severe combined immunodeficiency disease (SCID)** fail to develop either cellular or antibody-mediated immunity. Their lymphocyte populations are low, and normal B and T cells are absent. Such infants cannot produce an immune response, and even a mild infection can prove fatal. Total isolation offers protection at great cost, with severe restrictions on lifestyle. Bone marrow transplants from compatible donors, normally a close relative, have been used to colonize lymphoid tissues with functional lymphocytes. Gene-splicing techniques have led to therapies that can treat at least one form of SCID. **AM** *Genetic Engineering and Gene Therapy*

AIDS, an immunodeficiency disease that we consider on page 820, is the result of a viral infection that targets primarily helper T cells. As the number of T cells declines, the normal immune control mechanism breaks down. When an infection occurs, suppressor factors released by suppressor T cells inhibit an immune response before the few surviving helper T cells can stimulate the formation of cytotoxic T cells or plasma cells in adequate numbers.

Immunosuppressive drugs have been used for many years to prevent graft rejection after transplant surgery. But immunosuppressive agents can destroy stem cells and lymphocytes, leading to a complete immunological failure. This outcome is one of the potentially fatal consequences of radiation exposure.

Allergies

Allergies are inappropriate or excessive immune responses to antigens. The sudden increase in cellular activity or antibody titers can have a number of unpleasant side effects. For example, neutrophils or cytotoxic T cells may destroy normal cells while attacking the antigen, or the antigen–antibody complex may trigger a massive inflammatory response. Antigens that trigger allergic reactions are often called **allergens**.

There are several types of allergies. A complete classification recognizes four categories: *immediate hypersensitivity (Type I), cytotoxic reactions (Type II), immune complex disorders (Type III),* and *delayed hypersensitivity (Type IV).* Here we will consider only immediate hypersensitivity, probably the most common form of allergy. One form, *allergic rhinitis,* includes hay fever and environmental allergies that may affect 15 percent of the U.S. population. We discussed one example of a cytotoxic (Type II) reaction, the cross-reaction that follows the transfusion of an incompatible blood type, in Chapter 19. p. 664 Other types of allergies are discussed in the *Applications Manual.* **AM** *Immune Complex Disorders; Delayed Hypersensitivity and Skin Tests*

IMMEDIATE HYPERSENSITIVITY. **Immediate hypersensitivity** is a rapid and unusually severe response to the presence of an antigen. Sensitization to an allergen occurs during the initial exposure, which leads to the production of large quantities of IgE. The tendency to produce IgE antibodies in response to specific allergens may be genetically determined. In drug reactions, such as allergies to penicillin, IgE antibodies are produced in response to a hapten (partial antigen) bound to a larger molecule inside the body.

Due to the lag time needed to activate B cells, produce plasma cells, and synthesize antibodies, the first exposure to an allergen does not produce symptoms, but merely sets the stage for the next encounter. After sensitization, the IgE antibodies become attached to the cell membranes of basophils and mast cells throughout the body. Later, when the individual is exposed to the same allergen, the bound antibodies stimulate these cells to release histamine, heparin, several cytokines, prostaglandins, and other chemicals into the surrounding tissues. A sudden, massive inflammation of the affected tissues results.

The cytokines and other mast cell secretions draw basophils, eosinophils, T cells, and macrophages to the area. These cells release their own chemicals, extending and exaggerating the responses initiated by mast cells.

The severity of the allergic reaction depends on the individual's sensitivity and on the location involved. If allergen exposure occurs at the body surface, the response may be restricted to that area. If the allergen enters the bloodstream, the response could be lethal.

In **anaphylaxis** (a-na-fi-LAK-sis; *ana-*, again + *phylaxis*, protection), a circulating allergen affects mast cells throughout the body (Figure 22–26●). The entire range of symptoms can develop within minutes. Changes in capillary permeabilities produce swelling and edema in the dermis, and raised welts, or *hives,* appear on the surface of the skin. Smooth muscles along the respiratory passageways contract; the narrowed passages make breathing extremely difficult. In severe cases, an extensive peripheral vasodilation occurs, producing a fall in blood pressure that can lead to a circulatory collapse. This response is **anaphylactic shock**.

Many of the symptoms of immediate hypersensitivity can be prevented by the prompt administration of **antihistamines** (an-tē-HIS-ta-mēnz)—drugs that block the action of histamine. *Benadryl®* *(diphenhydramine hydrochloride)* is a popular antihistamine that is available over the counter. The treatment of severe anaphylaxis involves antihistamine, corticosteroid, and epinephrine injections.

☐ STRESS AND THE IMMUNE RESPONSE

One of the normal effects of interleukin-1 secretion is the stimulation of adrenocorticotropic hormone (ACTH) production by the anterior lobe of the pituitary gland. The production of ACTH in turn leads to the secretion of glucocorticoids by the adrenal cortex. ∞ p. 629 The anti-inflammatory effects of the glucocorticoids may help control the size of the immune response. However, the long-term secretion of glucocorticoids, as in the resistance phase of the *general adaptation syndrome*, can inhibit the immune response and lower resistance to disease. ∞ p. 640 Several of the tissue responses to glucocorticoids alter the effectiveness of specific and nonspecific defenses. Examples include the following:

- **The Depression of the Inflammatory Response.** Glucocorticoids inhibit mast cells and reduce the permeability of capillaries. Inflammation is therefore less likely. When it does occur, the reduced permeability of the capillaries slows the entry of fibrinogen, complement proteins, and cellular defenders.

- **A Reduction in the Activities and Numbers of Phagocytes in Peripheral Tissues.** This reduction further impairs nonspecific defense mechanisms and interferes with the processing and presentation of antigens to lymphocytes.

- **The Inhibition of Interleukin Secretion.** A reduction in interleukin production depresses the response of lymphocytes, even to antigens bound to MHC proteins.

The mechanisms responsible for these changes are still under investigation. It is well known, however, that depression of the immune system due to chronic stress can be a serious threat to health.

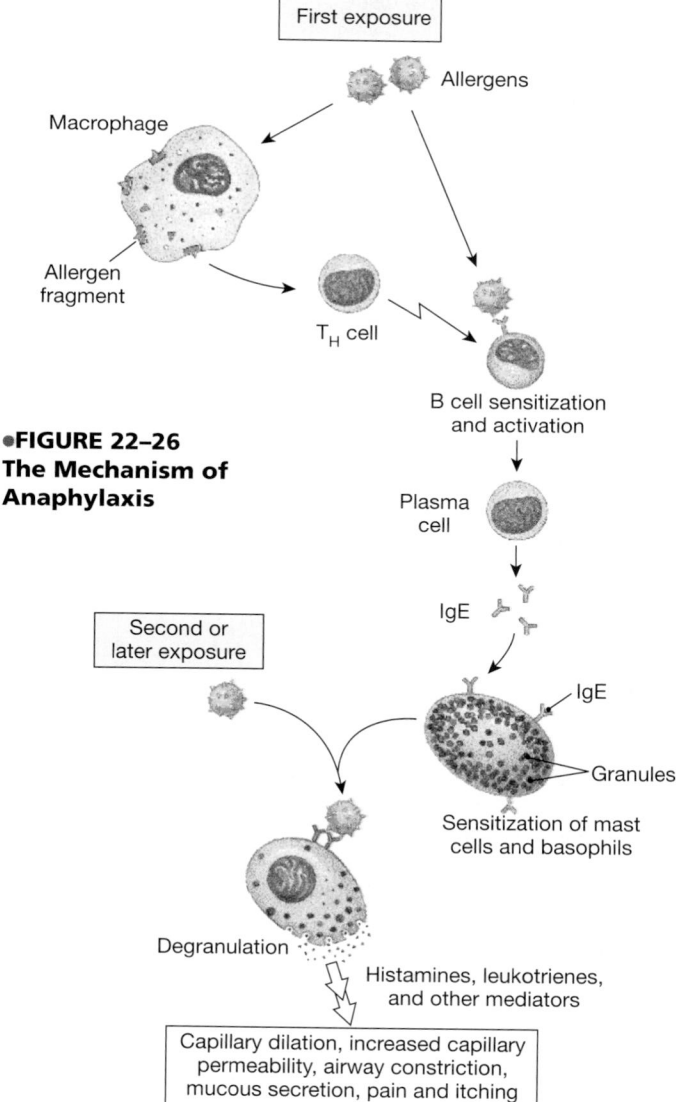

●FIGURE 22–26
The Mechanism of Anaphylaxis

First exposure

Allergens

Macrophage

Allergen fragment

T_H cell

B cell sensitization and activation

Plasma cell

IgE

Second or later exposure

IgE

Granules

Sensitization of mast cells and basophils

Degranulation

Histamines, leukotrienes, and other mediators

Capillary dilation, increased capillary permeability, airway constriction, mucous secretion, pain and itching

22–8 AGING AND THE IMMUNE RESPONSE

Objective

- Describe the effects of aging on the lymphatic system and the immune response.

With advancing age, the immune system becomes less effective at combating disease. T cells become less responsive to antigens; as a result, fewer cytotoxic T cells respond to an infection. This effect may, at least in part, be associated with the gradual involution of the thymus and a reduction in circulating levels of thymic hormones. Because the number of helper T cells is also reduced, B cells are less responsive, so antibody levels do not rise as quickly after antigen exposure. The net result is an increased susceptibility to viral and bacterial infection. For this reason, vaccinations for acute viral diseases, such as the flu (influenza), are strongly recommended for elderly individuals. The increased incidence of cancer in the elderly reflects the fact that immune surveillance declines, so tumor cells are not eliminated as effectively.

✓ Would the primary response or the secondary response be more affected by a lack of memory B cells for a particular antigen?

✓ Which kind of immunity protects a developing fetus, and how is that immunity produced?

✓ How does increased stress decrease the effectiveness of the immune response?

Answers start on page Q-1

 # The **LYMPHATIC SYSTEM** in Perspective

Organ/Component	Primary Functions
Lymphatic Vessels	Carry lymph (water and proteins) and lymphocytes from peripheral tissues to veins of the cardiovascular system
Lymph Nodes	Monitor the composition of lymph; engulf pathogens; stimulate immune response
Spleen	Monitors circulating blood; engulfs pathogens; stimulates immune response
Thymus	Controls development and maintenance of one class of lymphocytes (T cells)

INTEGRATION WITH OTHER SYSTEMS

The figure on the opposite page summarizes the interactions between the lymphatic system and other physiological systems. The relationships among the cells of the immune response and the nervous and endocrine systems are now the focus of intense research.

- The thymus secretes oxytocin, ADH, and endorphins as well as thymic hormones. The effects on the CNS are not known, but removal of the thymus lowers brain endorphin levels.
- Both thymic hormones and cytokines help establish the normal levels of CRH and TRH produced by the hypothalamus.
- Other thymic hormones affect the anterior lobe of the pituitary gland directly, stimulating the secretion of prolactin and GH.

Conversely, the nervous system can apparently adjust the sensitivity of the immune response:

- The CNS innervates dendritic cells in the lymph nodes and spleen, Langerhans cells in the skin, and other antigen-presenting cells. The nerve endings release neurotransmitters that heighten local immune responses. For this reason, some skin conditions, such as *psoriasis*, worsen when a person is under stress.
- Neuroglia in the CNS produce cytokines that promote an immune response.
- A sudden decline in immune function can occur after even a brief period of emotional distress.

CLINICAL PATTERNS

Disorders of the lymphatic system that affect the immune response can be sorted into three general categories:

1. ***Disorders resulting from an insufficient immune response.*** This category includes immunodeficiency disorders, such as AIDS (p. 820) and SCID (p. 816). Individuals with depressed immune defenses can develop life-threatening diseases caused by microorganisms that are harmless to other individuals.

2. ***Disorders resulting from an inappropriate immune response.*** Autoimmune disorders result when normal tissues are mistakenly attacked by T cells or the antibodies produced by activated B cells (p. 815). For instance, in *thrombocytopenic purpura*, the body forms antibodies against its own platelets.

3. ***Disorders resulting from an excessive immune response.*** Conditions such as allergies (p. 816) can result from an immune response that is out of proportion with the size of the stimulus.

The *Applications Manual* discusses representative disorders from each of these categories.

MEDIA CONNECTIONS

I. Objective: To understand the value of monoclonal antibodies in conferring passive immunity and to investigate the techniques used to produce such antibodies.

Completion time = 10 minutes

Prior to visiting certain countries, U.S. citizens are advised to receive gamma globulin shots. Because they contain antibodies against pathogens often encountered in other parts of the world, these shots confer passive immunity to many diseases. One method of obtaining such antibodies is to collect them from the plasma of animals that have been exposed to the pathogens or associated antigens under laboratory conditions. Another method is to extract antibodies from the pooled plasma of people living in places where those pathogens are prevalent. Neither approach is completely satisfactory, and each carries significant risks. Based on your understanding of the immune system, what problems would you expect? To learn about an alternative strategy, go to the **Companion Website**, Chapter 22 Media Connections, and click on the keyword **monoclonals**. Here you will learn how monoclonal antibodies are produced in the lab, using tissue culture techniques. What are the benefits of monoclonal antibody production over other methods?

II. Objective: To explore the coordinated functioning (and malfunctioning) of the immune system by investigating allergic reactions.

Completion time = 12 minutes

Are you one of the millions who suffers from seasonal allergies? Do you truly understand what is happening to you during an allergic reaction? Allergies are a familiar example of the activities associated with specific immunity. For a clear description of the biological interactions that cause the symptoms of allergies, visit the **Companion Website**, Chapter 22 Media Connections, and click on the keyword **allergies**. Then, using Figures 22–17 through 22–19 as additional references, list the cells involved in an allergic response. Define each cell's role in producing the symptoms we have all come to recognize. What causes the runny nose, the itchy eyes, the sneezing? Why are drug companies focused on antihistamines as allergic symptom relief?

INTEGUMENTARY SYSTEM

- Provides physical barriers to pathogen entry
- Langerhans cells in epidermis and macrophages in dermis resist infection and present antigens to trigger immune response
- Mast cells trigger inflammation, mobilize cells of lymphatic system

---

- Provides IgA for secretion onto integumentary surfaces

SKELETAL SYSTEM

- Lymphocytes and other cells involved in the immune response are produced and stored in bone marrow

---

- Assists in repair of bone after injuries
- Macrophages fuse to become osteoclasts

MUSCULAR SYSTEM

- Protects superficial lymph nodes and the lymphatic vessels in the abdominopelvic cavity
- Muscle contractions help propel lymph along lymphatic vessels

---

- Assists in repair after injury

NERVOUS SYSTEM

- Microglia present antigens that stimulate specific defenses
- Neuroglia secrete cytokines
- Innervation stimulates antigen-presenting cells

- Cytokines affect hypothalamic production of CRH and TRH

ENDOCRINE SYSTEM

- Glucocorticoids have anti-inflammatory effects
- Thymic hormones stimulate development and maturation of lymphocytes
- Many hormones affect immune function

- Thymus secretes thymic hormones
- Cytokines affect cells throughout the body

LYMPHATIC SYSTEM

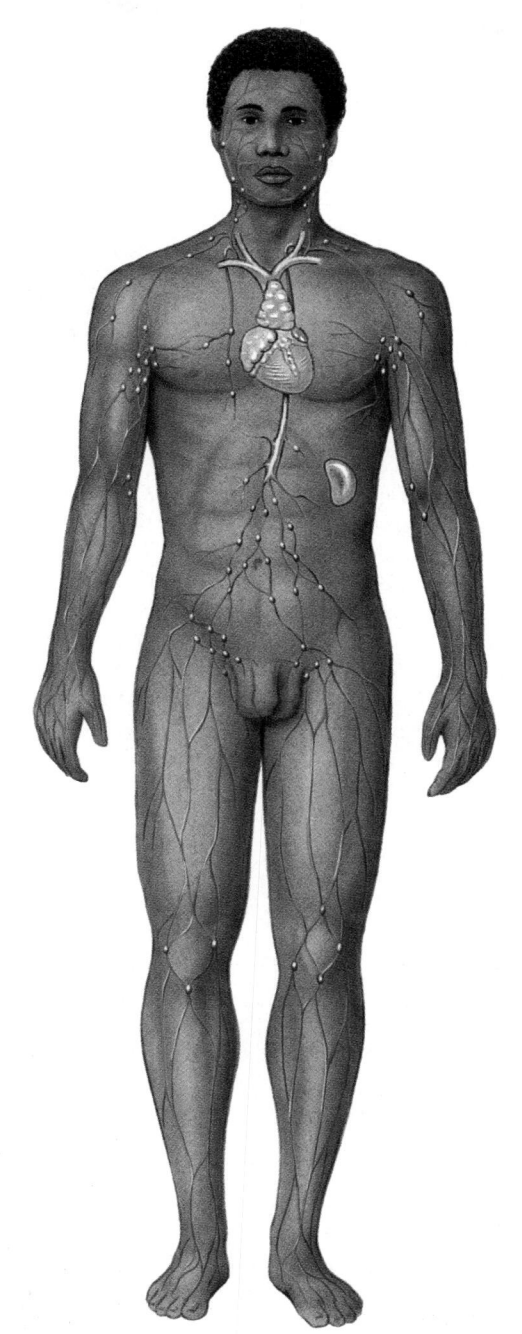

FOR ALL SYSTEMS
Provides specific defenses against infection; immune surveillance eliminates cancer cells; returns tissue fluid to circulation

CARDIOVASCULAR SYSTEM

- Distributes WBCs
- Carries antibodies that attack pathogens
- Clotting response helps restrict spread of pathogens
- Granulocytes and lymphocytes produced in bone marrow

- Fights infections of cardiovascular organs
- Returns tissue fluid to circulation

RESPIRATORY SYSTEM

- Alveolar phagocytes present antigens and trigger specific defenses
- Provides oxygen required by lymphocytes and eliminates carbon dioxide generated during their metabolic activities

---

- Tonsils protect against infection at entrance to respiratory tract

DIGESTIVE SYSTEM

- Provides nutrients required by lymphatic tissues, digestive acids, and enzymes
- Provides nonspecific defense against pathogens

- Tonsils and MALT defend against infection and toxins absorbed from digestive tract
- Lymphatic vessels carry absorbed lipids to venous system

URINARY SYSTEM

- Eliminates metabolic wastes generated by cellular activity

- Acid pH of urine provides nonspecific defense against urinary tract infection
- Provides specific defenses against urinary tract infection

REPRODUCTIVE SYSTEM

- Lysosomes and bactericidal chemicals in secretions provide nonspecific defense against reproductive tract infections

---

- Provides IgA for secretion by epithelial glands

AIDS

Acquired immune deficiency syndrome (AIDS), or *late-stage HIV disease*, is caused by the **human immunodeficiency virus (HIV)**. This virus is a *retrovirus*: It carries its genetic information in RNA rather than in DNA. The virus enters the cell by receptor-mediated endocytosis. ∞ p. 97 In the body, the virus binds to CD4, the membrane protein characteristic of helper T cells. Several types of antigen-presenting cells, including those of the monocyte–macrophage line, also are infected by HIV, but it is the infection of helper T cells that leads to clinical problems.

Once the virus is inside a cell, the viral enzyme *reverse transcriptase* synthesizes a complementary strand of DNA, which is then incorporated into the cell's genetic material. When these inserted viral genes are activated, the infected cell begins synthesizing viral proteins. In effect, the introduced viral genes take over the cell's synthetic machinery and force the cell to produce additional viruses. These new viruses are then shed at the cell surface. **AM** A Closer Look: The Nature of Pathogens

Cells infected with HIV are ultimately killed. Several mechanisms may be responsible for their destruction, including (1) the formation of pores in the cell membrane as the viruses are shed, (2) the cessation of cell maintenance due to the continuing synthesis of viral components, (3) autolysis, and (4) the stimulation of apoptosis.

The gradual destruction of helper T cells impairs the immune response, because these cells play a central role in coordinating cell-mediated and antibody-mediated responses to antigens. To make matters worse, suppressor T cells are relatively unaffected by the virus, and over time the excess of suppressing factors "turns off" the normal immune response. Circulating antibody levels decline, cell-mediated immunity is reduced, and the body is left without defenses against a wide variety of microbial invaders. With the affected person's immune function so reduced, ordinarily harmless pathogens can initiate lethal *opportunistic infections*. Because immune surveillance is also depressed, the risk of cancer increases.

Infection with HIV occurs by intimate contact with the body fluids of infected individuals. Although all body fluids carry the virus, the major routes of transmission involve contact with blood, semen, or vaginal secretions. Most individuals with AIDS become infected through sexual contact with an HIV-infected person (who may *not* necessarily be exhibiting the clinical symptoms of AIDS). The next largest group of infected persons consists of intravenous drug users who shared contaminated needles. Relatively few individuals have become infected with the virus after receiving a transfusion of contaminated blood or blood products. Finally, an increasing number of infants are born with AIDS, having acquired it from infected mothers.

AIDS is a public health problem of massive proportions. Nearly half a million people have already died of AIDS in the

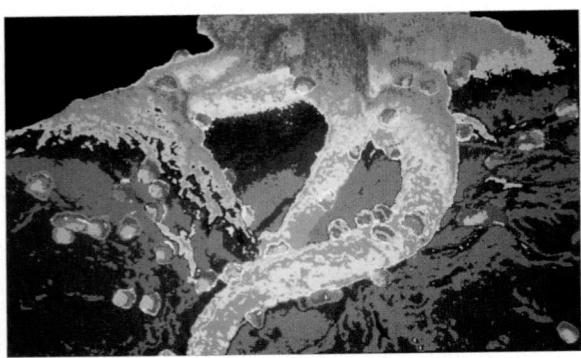

United States alone. Estimates of the total number of individuals who are infected with HIV in the United States range from 650,000 to 1,000,000. In 2002, an estimated 13,000 of these people will die, and 40,000 new cases of AIDS will be diagnosed. The numbers worldwide are even more frightening. The World Health Organization estimates that, as of 2002, 38–40 million individuals were infected. Every 5.4 seconds another person becomes infected with the HIV virus, and every 12 seconds someone dies of AIDS. The total death toll for 2002 is estimated to be 2.5–3 million people.

The best defense against AIDS is to avoid sexual contact or sharing of needles with infected individuals. All forms of sexual intercourse carry the risk of viral transmission. The use of synthetic condoms greatly reduces (but does not eliminate) the chance of infection. Condoms that are not made of synthetic materials prevent pregnancy, but do not block the passage of viruses.

Clinical symptoms of AIDS may not appear until 5–10 years or more after infection. When they do appear, they are commonly mild, consisting of lymphadenopathy and chronic, but nonfatal, infections. So far as is known, however, AIDS is almost always fatal, and most people who carry the virus will eventually die of the disease. (A handful of infected individuals have been able to tolerate the virus without apparent illness for many years; for details, see the *Applications Manual.*)

Despite intensive efforts, a vaccine has yet to be developed that will provide immunity from HIV infection in a noninfected person exposed to the virus. While efforts to prevent the spread of HIV continue, the survival rate for AIDS patients has been steadily increasing, because new drugs are available that slow the progression of the disease and because improved antibiotic therapies help combat secondary infections. This combination is extending the life span of patients while the search for more effective treatment continues. For more information on the distribution of HIV infection, current and future drug therapies, and additional details on HIV disease, consult the *Applications Manual.* **AM** AIDS

Chapter Review

SELECTED CLINICAL TERMINOLOGY

Terms Discussed in This Chapter

acquired immune deficiency syndrome (AIDS): A disorder that develops following HIV infection and that is characterized by reduced circulating antibody levels and depressed cell-mediated immunity. *(p. 820 and AM)*

allergen: An antigen capable of triggering an allergic reaction. *(p. 816)*

allergy: An inappropriate or excessive immune response to antigens, triggered by the stimulation of mast cells bound to IgE. *(p. 816 and AM)*

anaphylactic shock: A drop in blood pressure that may lead to circulatory collapse, resulting from a severe case of anaphylaxis. *(p. 817)*

anaphylaxis (a-na-fi-LAK-sis): A type of allergy in which a circulating allergen affects mast cells throughout the body, producing numerous symptoms very quickly. *(p. 816)*

appendicitis: An infection and inflammation of the aggregated lymphoid nodules in the appendix. *(p. 787 and AM)*

autoimmune disorder: A disorder that develops when the immune response mistakenly targets normal body cells and tissues. *(p. 815)*

bacteria: Prokaryotic cells (cells lacking nuclei and other membranous organelles) that may be extracellular or intracellular pathogens. *(p. 779 and AM)*

filariasis: An infection by larval stages of a parasitic roundworm; the adult worms may scar and block lymphatic vessels, causing acute lymphedema, commonly affecting the external genitalia and lower limbs (**elephantiasis**). *(p. 784)*

fungi (singular, *fungus*): Eukaryotic organisms that absorb organic materials from the remains of dead cells; some fungi are pathogenic. *(p. 784 and AM)*

human immunodeficiency virus (HIV): The virus responsible for AIDS and related immunodeficiency disorders. *(p. 820 and AM)*

immune complex disorder: A disorder caused by the precipitation of immune complexes at sites such as the kidneys, where their presence disrupts normal tissue function. *(p. 816 and AM)*

immunodeficiency disease: A disease in which either the immune system fails to develop normally or the immune response is blocked. *(p. 816)*

immunosuppression: A reduction in the sensitivity of the immune system. *(p. 805)*

immunosuppressive drugs: Drugs administered to inhibit the immune response; examples include prednisone, cyclophosphamide, azathioprine, cyclosporin, and FK506. *(p. 805 and AM)*

lymphadenopathy (lim-fad-e-NOP-a-thē): A chronic or excessive enlargement of lymph nodes. *(p. 788)*

lymphedema: A painless accumulation of lymph in a region whose lymphatic drainage has been blocked. *(p. 784)*

lymphomas: Malignant cancers consisting of abnormal lymphocytes or lymphoid stem cells; examples include *Hodgkin's disease* and *non-Hodgkin's lymphoma*. *(p. 788 and AM)*

severe combined immunodeficiency disease (SCID): A congenital disorder resulting from the failure to develop both cell-mediated and antibody-mediated immunity. *(p. 816 and AM)*

tonsillitis: An infection of one or more tonsils; symptoms include a sore throat, high fever, and leukocytosis (an abnormally high white blood cell count). *(p. 787 and AM)*

vaccine (vak-SĒN): A preparation of antigens derived from a specific pathogen; administered during *immunization*, or *vaccination*. *(p. 798)*

viruses: Noncellular pathogens that replicate by directing the synthesis of virus-specific proteins and nucleic acids inside tissue cells. *(p. 779 and AM)*

AM Additional Terms Discussed in the *Applications Manual*

active immunization
bone marrow transplantation
graft-versus-host (GVH) disease
hybridoma
hypersplenism
hyposplenism
monoclonal antibody
mononucleosis
passive immunization
protozoa
splenomegaly
systemic lupus erythematosus
tonsillectomy

STUDY OUTLINE

22–1 AN OVERVIEW OF THE LYMPHATIC SYSTEM AND IMMUNITY p. 779

1. The cells, tissues, and organs of the **lymphatic system** play a central role in the body's defenses against a variety of **pathogens**, or disease-causing organisms.

2. **Lymphocytes,** the primary cells of the lymphatic system, provide an **immune response** to specific threats to the body. **Immunity** is the ability to resist infection and disease through the activation of specific defenses.

22–2 ORGANIZATION OF THE LYMPHATIC SYSTEM p. 780

1. The lymphatic system includes a network of **lymphatic vessels**, or **lymphatics**, that carries **lymph** (a fluid similar to plasma, but with a lower concentration of proteins). A series of **lymphoid tissues** and **lymphoid organs** is connected to the lymphatic vessels. *(Figure 22–1)*

FUNCTIONS OF THE LYMPHATIC SYSTEM p. 780

2. The lymphatic system produces, maintains, and distributes lymphocytes (which attack invading organisms, abnormal cells, and foreign proteins) and helps maintain blood volume and eliminate local variations in the composition of interstitial fluid.

LYMPHATIC VESSELS p. 781

3. Lymph flows along a network of lymphatic vessels, the smallest of which are the **lymphatic capillaries** (*terminal lymphatics*). The lymphatic vessels empty into the **thoracic duct** and the **right lymphatic duct**. (*Figures 22–2 to 22–4*)

LYMPHOCYTES p. 783

4. The three classes of lymphocytes are **T** (**t**hymus-dependent) **cells**, **B** (**b**one marrow–derived) **cells**, and **NK** (**n**atural **k**iller) **cells**.

5. **Cytotoxic T cells** attack foreign cells or body cells infected by viruses and provide **cell-mediated (cellular) immunity**. *Regulatory T cells* (**helper T cells** and **suppressor T cells**) regulate and coordinate the immune response.

6. B cells can differentiate into **plasma cells**, which produce and secrete *antibodies* that react with specific chemical targets called **antigens**. Antibodies in body fluids are called *immunoglobulins*. B cells are responsible for **antibody-mediated (humoral) immunity**.

7. NK cells (also called **large granular lymphocytes**) attack foreign cells, normal cells infected with viruses, and cancer cells. NK cells provide **immunological surveillance**.

8. Lymphocytes continuously migrate into and out of the blood through the lymphoid tissues and organs. **Lymphopoiesis** (lymphocyte production) involves the bone marrow, thymus, and peripheral lymphoid tissues. (*Figure 22–5*)

LYMPHOID TISSUES p. 786

9. **Lymphoid tissues** are connective tissues dominated by lymphocytes. In a **lymphoid nodule**, the lymphocytes are densely packed in an area of loose connective tissue. The lymphoid tissue embedded within the organs of the digestive system is called **mucosa-associated lymphoid tissue (MALT)**. (*Figure 22–6*)

LYMPHOID ORGANS p. 787

10. Important lymphoid organs include the **lymph nodes**, the **thymus**, and the **spleen**. Lymphoid tissues and organs are distributed in areas that are especially vulnerable to injury or invasion.

11. Lymph nodes are encapsulated masses of lymphoid tissue. The **deep cortex** is dominated by T cells; the **outer cortex** and **medulla** contain B cells. (*Figure 22–7*)

12. The thymus lies behind the sternum, in the anterior mediastinum. **Reticular epithelial cells** scattered among the lymphocytes maintain the blood–thymus barrier and secrete thymic hormones. (*Figure 22–8*)

13. The adult spleen contains the largest mass of lymphoid tissue in the body. The cellular components form the **pulp** of the spleen. **Red pulp** contains large numbers of red blood cells, and **white pulp** resembles lymphoid nodules. (*Figure 22–9*)

THE LYMPHATIC SYSTEM AND BODY DEFENSES p. 791

14. The lymphatic system is a major component of the body's defenses, which are classified as either (1) **nonspecific defenses**, which protect without distinguishing one threat from another, or (2) **specific defenses**, which protect against particular threats only.

22-3 NONSPECIFIC DEFENSES p. 791

1. Nonspecific defenses prevent the approach, deny the entrance, or limit the spread of living or nonliving hazards. (*Figure 22–10*)

PHYSICAL BARRIERS p. 791

2. Physical barriers include hair, epithelia, and various secretions of the integumentary and digestive systems.

PHAGOCYTES p. 792

3. The two types of phagocytic cells are **microphages** and **macrophages** (cells of the **monocyte–macrophage system**). Microphages are the neutrophils and eosinophils in circulating blood.

4. **Phagocytes** move among cells by *emigration*, or *diapedesis* (migration between adjacent endothelial cells), and exhibit **chemotaxis** (sensitivity and orientation to chemical stimuli).

IMMUNOLOGICAL SURVEILLANCE p. 793

5. Immunological surveillance involves constant monitoring of normal tissues by NK cells that are sensitive to abnormal antigens on the surfaces of otherwise normal cells. Cancer cells with **tumor-specific antigens** on their surfaces are killed. (*Figure 22–11*)

INTERFERONS p. 794

6. **Interferons**—small proteins released by cells infected with viruses—trigger the production of **antiviral proteins**, which interfere with viral replication inside the cells. Interferons are **cytokines**—chemical messengers released by tissue cells to coordinate local activities.

COMPLEMENT p. 794

7. At least 11 **complement proteins** make up the **complement system**. These proteins interact with each other in cascades to destroy target cell membranes, stimulate inflammation, attract phagocytes, or enhance phagocytosis. The complement system can be activated by either the **classical pathway** or the **alternative pathway**. (*Figure 22–12*)

INFLAMMATION p. 796

8. **Inflammation** is a localized tissue response to injury. (*Figure 22–13*)

FEVER p. 797

9. A **fever** [body temperature greater than 37.2°C (99°F)] can inhibit pathogens and accelerate metabolic processes. **Pyrogens** can reset the body's thermostat and raise the temperature.

Immunity: **Companion Website**: Chapter 22/Tutorials.

22-4 SPECIFIC DEFENSES: AN OVERVIEW OF THE IMMUNE RESPONSE p. 798

1. T cells are responsible for **cell-mediated (cellular) immunity**. B cells provide **antibody-mediated (humoral) immunity**.

FORMS OF IMMUNITY p. 798

2. **Specific resistance** or immunity involves **innate immunity** (genetically determined and present at birth) or **acquired immunity**. The two types of acquired immunity are **active immunity** (which appears after exposure to an antigen) and **passive immunity** (produced by the transfer of antibodies from another source). (*Figure 22–14*)

PROPERTIES OF IMMUNITY p. 798

3. Immunity exhibits four general properties: **specificity**, **versatility**, **memory**, and **tolerance**. *Memory cells* enable the immune system to "remember" previous target antigens. Tolerance

is the ability of the immune system to ignore some antigens, such as those of body cells.

AN INTRODUCTION TO THE IMMUNE RESPONSE   p. 800

4. The immune response is triggered by the presence of an antigen and includes cell-mediated and antibody-mediated defenses. (*Figure 22–15*)

22–5 T CELLS AND CELL-MEDIATED IMMUNITY   p. 800

ANTIGEN PRESENTATION   p. 801

1. **Antigen presentation** occurs when an antigen–glycoprotein combination appears in a cell membrane (typically, that of a macrophage). T cells sensitive to this antigen will be activated if they contact the membrane of the antigen-presenting cell.
2. All body cells have membrane glycoproteins. The genes controlling their synthesis make up a chromosomal region called the **major histocompatibility complex (MHC).** The membrane glycoproteins are called **MHC proteins.** APCs (**antigen-presenting cells**) are involved in antigen stimulation.

ANTIGEN RECOGNITION   p. 802

3. Lymphocytes are not activated by lone antigens, but will respond to an antigen bound to either a **Class I** or a **Class II** MHC protein in a process called **antigen recognition.**
4. Class I MHC proteins are in all nucleated body cells. Class II MHC proteins are only in **antigen-presenting cells (APCs)** and lymphocytes. (*Figure 22–16*)
5. Whether a T cell responds to antigens held in Class I or Class II MHC proteins depends on the structure of the T cell membrane. T cell membranes contain proteins called **CD** (**c**luster of **d**ifferentiation) **markers. CD3 markers** are present on all **T** cells. **CD8 markers** are on cytotoxic and suppressor T cells. **CD4 markers** are on all helper T cells.

ACTIVATION OF CD8 T CELLS   p. 802

6. One type of CD8 cells responds quickly, giving rise to large numbers of cytotoxic T cells and memory cells. The other type of CD8 cell responds more slowly, giving rise to small numbers of suppressor T cells.
7. **Cytotoxic T cells** seek out and destroy abnormal and infected cells, using three different methods, including the secretion of **lymphotoxin.** (*Figure 22–17*)
8. Cell-mediated immunity (cellular immunity) results from the activation of CD8 T cells by antigens bound to Class I MHCs. When activated, most of these T cells divide to generate cytotoxic T cells and **memory T$_C$ cells,** which remain on reserve to guard against future such attacks. Suppressor T cells depress the responses of other T cells and of B cells. (*Figures 22–17, 22–19*)

ACTIVATION OF CD4 T CELLS   p. 804

9. Helper, or CD4, T cells respond to antigens presented by Class II MHC proteins. When activated, helper T cells secrete lymphokines that aid in coordinating specific and nonspecific defenses and regulate cell-mediated and antibody-mediated immunity. (*Figures 22–18, 22–19*)

22–6 B CELLS AND ANTIBODY-MEDIATED IMMUNITY   p. 805

B CELL SENSITIZATION AND ACTIVATION   p. 805

1. B cells become **sensitized** when antigens bind to their membrane antibody molecules. The antigens are then displayed on the Class II MHC proteins of the B cells, which become activated by helper T cells activated by the same antigen. (*Figure 22–20*)

2. An active B cell may differentiate into a plasma cell or produce daughter cells that differentiate into plasma cells and **memory B cells.** Antibodies are produced by the plasma cells. (*Figure 22–20*)

ANTIBODY STRUCTURE   p. 806

3. An antibody molecule consists of two parallel pairs of polypeptide chains containing *constant* and *variable* segments. (*Figure 22–21*)
4. When antibody molecules bind to an antigen, they form an **antigen–antibody complex.** Effects that appear after binding include **neutralization** (antibody binding such that viruses or bacterial toxins cannot bind); **precipitation** (formation of an insoluble **immune complex**) and **agglutination** (formation of large complexes); **opsonization** (coating of pathogens with antibodies and complement proteins to enhance phagocytosis); stimulation of inflammation; and prevention of bacterial or viral adhesion. (*Figure 22–21*)
5. The five classes of antibodies in body fluids are **immunoglobulin** (1) **G (IgG),** responsible for resistance against many viruses, bacteria, and bacterial toxins; (2) **E (IgE),** which releases chemicals that accelerate local inflammation; (3) **D (IgD),** located on the surfaces of B cells; (4) **M (IgM),** the first type of antibody secreted after an antigen arrives; and (5) **A (IgA),** found in glandular secretions. (*Table 22–1*)

Specific immunity and lymphocyte activation: **Companion Website**: Chapter 22/Tutorials.

PRIMARY AND SECONDARY RESPONSES TO ANTIGEN EXPOSURE   p. 810

6. The antibodies first produced by plasma cells are the agents of the **primary response.** The maximum **antibody titer** appears during the **secondary response** to antigen exposure. (*Figure 22–22*)

FOCUS: HORMONES OF THE IMMUNE SYSTEM   p. 813

7. **Interleukins** increase T cell sensitivity to antigens exposed on macrophage membranes; stimulate B cell activity, plasma cell formation, and antibody production; enhance nonspecific defenses; and moderate the immune response. (*Summary Table 22–2*)
8. Interferons slow the spread of a virus by making the synthesizing cell and its neighbors resistant to viral infections. (*Summary Table 22–2*)
9. **Tumor necrosis factors (TNFs)** slow tumor growth and kill tumor cells. (*Summary Table 22–2*)
10. Several lymphokines adjust the activities of phagocytic cells to coordinate specific and nonspecific defenses. (*Summary Table 22–2*)
11. **Colony-stimulating factors (CSFs)** are factors produced by active T cells, cells of the monocyte-macrophage group, endothelial cells, and fibroblasts. (*Summary Table 22–2*)

SUMMARY: THE IMMUNE RESPONSE   p. 811

1. The initial steps in the immune response to viral and bacterial infections differ. (*Figures 22–23, 22–24, 22–25; Summary Table 22–2*)

22–7 NORMAL AND ABNORMAL RESISTANCE   p. 815

THE DEVELOPMENT OF IMMUNOLOGICAL COMPETENCE   p. 815

1. **Immunological competence** is the ability to demonstrate an immune response after exposure to an antigen. A developing fetus receives passive immunity from the maternal bloodstream.

After delivery, the infant begins developing acquired immunity following exposure to environmental antigens.

IMMUNE DISORDERS   p. 815

2. **Autoimmune disorders** develop when the immune response mistakenly targets normal body cells and tissues.

3. In an **immunodeficiency disease**, either the immune system does not develop normally or the immune response is blocked.

4. **Allergies** are inappropriate or excessive immune responses to **allergens** (antigens that trigger allergic reactions). The four types of allergies are *immediate hypersensitivity (Type I), cytotoxic reactions (Type II), immune complex disorders (Type III)*, and *delayed hypersensitivity (Type IV)*.

5. In **anaphylaxis**, a circulating allergen affects mast cells throughout the body. *(Figure 22–25)*

STRESS AND THE IMMUNE RESPONSE   p. 817

6. **Interleukin-1** released by active macrophages triggers the release of ACTH by the anterior lobe of the pituitary gland. Glucocorticoids produced by the adrenal cortex moderate the immune response, but their long-term secretion can lower a person's resistance to disease.

22–8 AGING AND THE IMMUNE RESPONSE   p. 817

1. With aging, the immune system becomes less effective at combating disease.

REVIEW QUESTIONS

 More assessment questions are available to you on the Companion Website. You will find Matching, Multiple Choice, True/False, and other quizzes to help further your understanding of the material covered in this chapter. To access the site, go to www.aw.com/martini.

LEVEL 1 Reviewing Facts and Terms

1. Lymph from the right arm, the right half of the head, and the right chest is received by the
   (a) cisterna chyli   (b) right lymphatic duct
   (c) right thoracic duct   (d) aorta

2. Lymphoid stem cells that can form all types of lymphocytes occur in the
   (a) bloodstream   (b) thymus
   (c) bone marrow   (d) spleen

3. Lymphatic vessels are located everywhere except the
   (a) lower limbs   (b) central nervous system
   (c) integument   (d) digestive tract

4. The body's largest collection of lymphoid tissue is in the
   (a) adult spleen   (b) adult thymus
   (c) bone marrow   (d) tonsils

5. Red blood cells that are damaged or defective are removed from the bloodstream by the
   (a) thymus   (b) lymph nodes
   (c) spleen   (d) tonsils

6. Phagocytes move through capillary walls by squeezing between adjacent endothelial cells, a process known as
   (a) diapedesis   (b) chemotaxis
   (c) adhesion   (d) perforation

7. Perforins are proteins associated with the activity of
   (a) T cells   (b) B cells
   (c) NK cells   (d) plasma cells

8. Complement activation
   (a) stimulates inflammation   (b) attracts phagocytes
   (c) enhances phagocytosis   (d) a, b, and c are correct

9. The most beneficial effect of fever is that it
   (a) inhibits the spread of some bacteria and viruses
   (b) increases the metabolic rate by up to 10 percent
   (c) stimulates the release of pyrogens
   (d) a and b are correct

10. CD4 markers are associated with
    (a) cytotoxic T cells   (b) suppressor T cells
    (c) helper T cells   (d) a, b, and c are correct

11. List the lymphoid tissues and organs of the body. What are the specific functions of each?

12. Give a function for each of the following:
    (a) cytotoxic T cells   (b) helper T cells
    (c) suppressor T cells   (d) plasma cells
    (e) NK cells   (f) stromal cells
    (g) reticular epithelial cells   (h) interferons
    (i) pyrogens   (j) T cells
    (k) B cells   (l) interleukins
    (m) tumor necrosis factor   (n) colony-stimulating factors

13. What are the three classes of lymphocytes, and where does each class originate?

14. What seven defenses, present at birth, provide the body with the defensive capability known as nonspecific resistance?

LEVEL 2 Reviewing Concepts

15. Compared with nonspecific defenses, specific defenses
    (a) do not distinguish between one threat and another
    (b) are always present at birth
    (c) protect against threats on an individual basis
    (d) deny the entry of pathogens to the body

16. T cells and B cells can be activated only by
    (a) pathogens
    (b) interleukins, interferons, and colony-stimulating factors
    (c) cells infected with viruses, bacteria, or cancer cells
    (d) exposure to a specific antigen at a specific site on a cell membrane

17. Class II MHC proteins appear in the cell membrane only when
    (a) the plasma cells are releasing antibodies
    (b) the cell is processing antigens
    (c) cytotoxic T cells are inhibited
    (d) NK cells are activated

18. List the four general properties of immunity; give an explanation of each.

19. Compare and contrast the effects of complement with those of interferon.

20. How does a cytotoxic T cell destroy another cell displaying antigens bound to Class I MHC proteins?

21. How does the formation of an antigen–antibody complex cause the elimination of an antigen?

22. Give one example of each type of immunity: innate immunity, naturally acquired immunity, induced active immunity, induced passive immunity, and natural passive immunity.

23. An anesthesia technician is advised that she should be vaccinated against hepatitis B, which is caused by a virus. She is given one injection and is told to come back for a second injection in a month and a third injection after six months. Why is this series of injections necessary?

LEVEL 3 Critical Thinking and Clinical Applications

24. An investigator at a crime scene discovers some body fluid on the victim's clothing. The investigator carefully takes a sample and sends it to the crime lab for analysis. On the basis of the analysis of antibodies, could the crime lab determine whether the sample is blood plasma or semen? Explain.

25. Ted finds out that he has been exposed to the measles. He is concerned that he might have contracted the disease, so he goes to see his physician. The physician takes a blood sample and sends it to a lab for antibody titers. The results show an elevated level of IgM antibodies to rubella (measles) virus, but very few IgG antibodies to the virus. Has Ted contracted the disease?

26. While walking along the street, you and your friend see an elderly woman whose left arm appears to be swollen to several times its normal size. Your friend remarks that the woman must have been in the tropics and contracted a form of filariasis that produces elephantiasis. You disagree, saying that it is more likely that the woman had a radical mastectomy (the removal of a breast because of cancer). Explain the rationale behind your answer.

27. Tilly has T cells that can respond to antigen A. Tilly's friend Harry has been exposed to antigen A, and Tilly offers to have some of her T cells transfused to Harry so that he will not come down with an infection. You overhear their conversation and observe that such a transfusion would not work. Why wouldn't it work?

28. You are a researcher interested in studying cells that can respond to the hormone FSH. How can you use antibodies to help you locate FSH-responsive cells?

UNIT 5   CHAPTER   23   24   25   26   27

23

THE RESPIRATORY SYSTEM

When we think of the respiratory system, we generally think of the mechanics of breathing—pulling air into and out of our bodies. However, the requirements for an efficient respiratory system go beyond merely moving air. Cells need energy for maintenance, growth, defense, and replication. Our cells obtain that energy through aerobic mechanisms which require oxygen and produce carbon dioxide. Many aquatic organisms can obtain oxygen and excrete carbon dioxide by diffusion across the surface of the skin or in specialized structures, such as the gills of a fish. Such arrangements are poorly suited for life on land, because the exchange surfaces must be very thin and relatively delicate to permit rapid diffusion. In air, the exposed membranes collapse, evaporation and dehydration reduce blood volume, and the delicate surfaces become vulnerable to attack by pathogens. Our respiratory exchange surfaces are just as delicate as those of an aquatic organism, but they are confined to the inside of the *lungs*—in a warm, moist, protected environment. Under these conditions, diffusion can occur between the air and the blood.

The cardiovascular system provides the link between your interstitial fluids and the exchange surfaces of your lungs. Your circulating blood carries oxygen from the lungs to peripheral tissues; it also accepts and transports the carbon dioxide generated by those tissues, delivering it to the lungs. The respiratory system includes not only the lungs, which are the functional units of this system, but the airways through which air flows as it moves to and from the lungs.

Our discussion of the respiratory system begins by following the path air takes as it travels to the lungs. We will then consider the mechanics of breathing and the physiology of respiration.

23–1 THE RESPIRATORY SYSTEM: AN INTRODUCTION

Objectives

- Describe the primary functions of the respiratory system.
- Explain how the delicate respiratory exchange surfaces are protected from pathogens, debris, and other hazards.

FUNCTIONS OF THE RESPIRATORY SYSTEM

The **respiratory system** has five basic functions:

1. Providing an extensive area for gas exchange between air and circulating blood.
2. Moving air to and from the exchange surfaces of the lungs.
3. Protecting respiratory surfaces from dehydration, temperature changes, or other environmental variations and defending the respiratory system and other tissues from invasion by pathogens.
4. Producing sounds involved in speaking, singing, and nonverbal auditory communication.
5. Providing olfactory sensations to the central nervous system from the olfactory epithelium in the superior portions of the nasal cavity.

In addition, the capillaries of the lungs indirectly assist in the regulation of blood volume and blood pressure, through the conversion of angiotensin I to angiotensin II. ∞ p. 744

ORGANIZATION OF THE RESPIRATORY SYSTEM

We can divide the respiratory system into an upper respiratory system and a lower respiratory system (Figure 23–1●). The **upper respiratory system** consists of the nose, nasal cavity, paranasal sinuses, and pharynx. These passageways filter, warm, and humidify the incoming air—protecting the more delicate surfaces of the lower respiratory system—and cool and dehumidify outgoing air. The **lower respiratory system** includes the larynx (voice box), trachea (windpipe), bronchi, bronchioles, and alveoli of the lungs.

Your **respiratory tract** consists of the airways that carry air to and from the exchange surfaces of your lungs. The respiratory tract consists of a *conducting portion* and a *respiratory portion*. The conducting portion begins at the entrance to the nasal cavity and extends through many passageways (the pharynx and larynx and along the trachea, bronchi, and bronchioles to the *terminal bronchioles*). The respiratory portion of the tract includes the delicate *respiratory bronchioles* and the **alveoli** (al-VĒ-ō-lī), air-filled pockets within the lungs where all gas exchange between air and blood occurs.

●**FIGURE 23–1**
The Components of the Respiratory System. The conducting portion of the respiratory system is shown here. The smaller bronchioles and alveoli are not shown. ATLAS

Plate 7.6a,b

Gas exchange can occur quickly and efficiently because the distance between the blood in an alveolar capillary and the air inside an alveolus is generally less than 1 μm, and in some cases as small as 0.1 μm. To meet the metabolic requirements of peripheral tissues, the exchange surfaces of the lungs must be very large. In fact, the total surface area for gas exchange in the lungs of adults is at least 35 times the surface area of the body. It is difficult to measure the exchange surfaces with precision; estimates of the surface area involved range from 70 m^2 to 140 m^2 (753 ft^2 to 1506 ft^2).

Filtering, warming, and humidification of the inhaled air begin at the entrance to the upper respiratory system and continue throughout the rest of the conducting system. By the time air reaches the alveoli, most foreign particles and pathogens have been removed, and the humidity and temperature are within acceptable limits. The success of this "conditioning process" is due primarily to the properties of the *respiratory mucosa*.

The Respiratory Mucosa

The **respiratory mucosa** (mū-KŌ-suh) lines the conducting portion of the respiratory system. A *mucosa* is a *mucous membrane*, one of the four types of membranes introduced in Chapter 4. It consists of an epithelium and an underlying layer of areolar tissue. ∞ p. 125

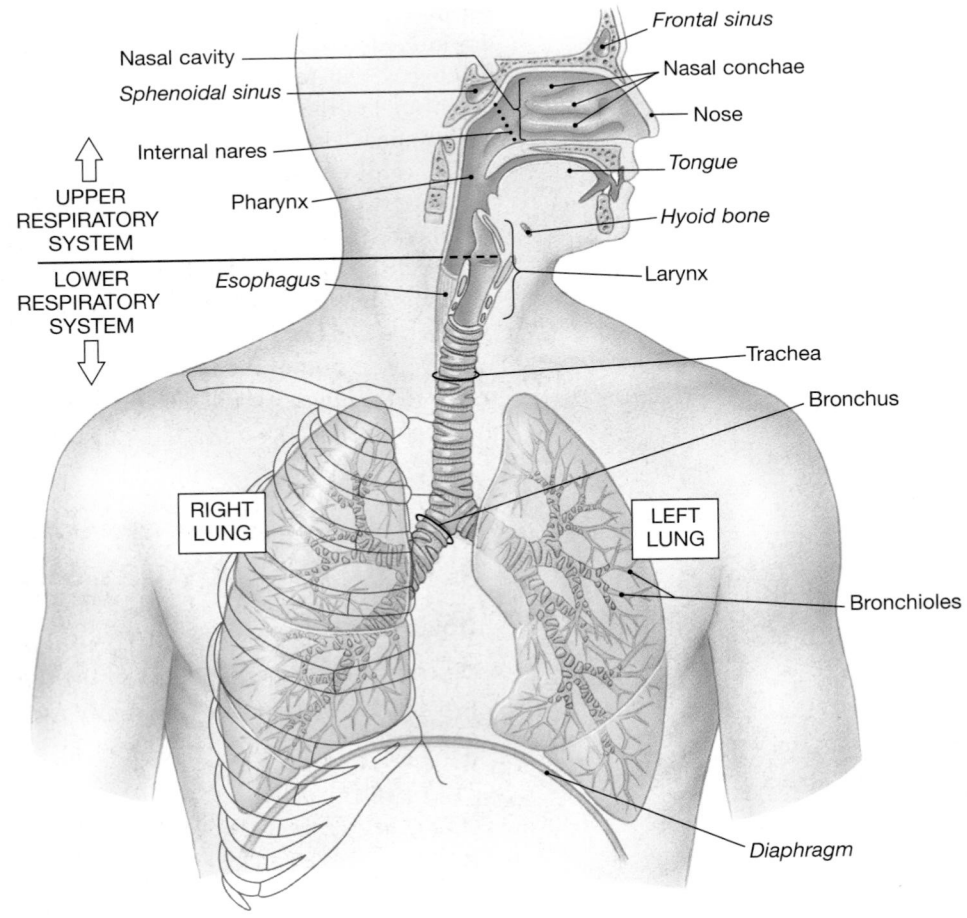

Frontal sinus
Nasal cavity
Sphenoidal sinus
Nasal conchae
Nose
Internal nares
Tongue
UPPER RESPIRATORY SYSTEM
Pharynx
Hyoid bone
LOWER RESPIRATORY SYSTEM
Esophagus
Larynx
Trachea
Bronchus
RIGHT LUNG
LEFT LUNG
Bronchioles
Diaphragm

A pseudostratified ciliated columnar epithelium with numerous goblet cells lines the nasal cavity and the superior portion of the pharynx. ⊂⊃ pp. 118, 121 The structure of the respiratory epithelium changes along the respiratory tract. The epithelium lining inferior portions of the pharynx is a stratified squamous epithelium similar to that of the oral cavity. These portions of the pharynx, which conduct air to the larynx, also convey food to the esophagus. The pharyngeal epithelium must therefore protect against abrasion and chemical attack.

At the beginning of the lower respiratory system is a pseudostratified ciliated columnar epithelium (Figure 23–2a,b●) comparable to that of the nasal cavity. In the smaller bronchioles, this pseudostratified epithelium is replaced by a cuboidal epithelium with scattered cilia. The exchange surfaces of the alveoli are lined by a very delicate simple squamous epithelium. Other, more specialized cells are scattered within that epithelium.

The **lamina propria** (LA-mi-nuh PRŌ-prē-uh) is the underlying layer of areolar tissue that supports the respiratory epithelium. In the upper respiratory system, trachea, and bronchi, the lamina propria contains mucous glands that discharge their secretions onto the epithelial surface. The lamina propria in the conducting portions of the lower respiratory system contains bundles of smooth muscle cells. At the bronchioles, the smooth muscles form relatively thick bands that encircle or spiral around the lumen.

The Respiratory Defense System

The delicate exchange surfaces of the respiratory system can be severely damaged if the inhaled air becomes contaminated with debris or pathogens. Such contamination is prevented by a series of filtration mechanisms that together make up the **respiratory defense system**.

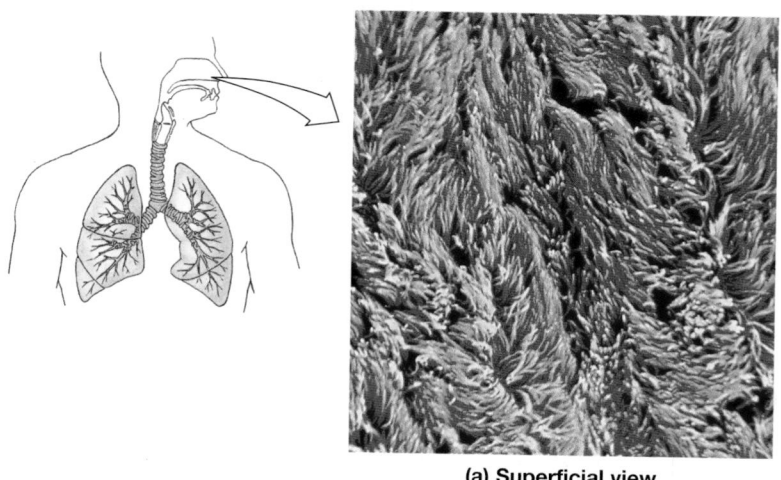

(a) Superficial view

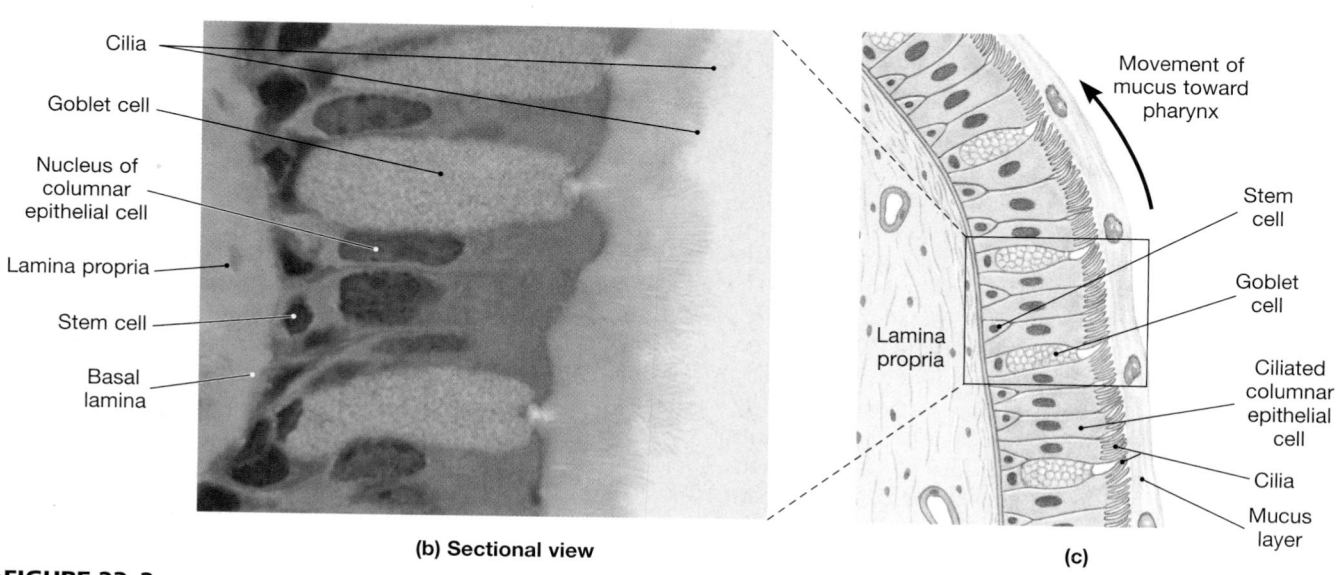

(b) Sectional view

(c)

●**FIGURE 23–2**

The Respiratory Epithelium of the Nasal Cavity and Conducting System. (a) A surface view of the epithelium. The cilia of the epithelial cells form a dense layer that resembles a shag carpet. The movement of these cilia propels mucus across the epithelial surface. (SEM × 1614) (b) The sectional appearance of the respiratory epithelium, a pseudostratified ciliated columnar epithelium. (LM × 1062) (c) A diagrammatic view of the respiratory epithelium of the trachea, showing the mechanism of mucus transport.

Along much of the length of the respiratory tract, goblet cells in the epithelium and mucous glands in the lamina propria produce a sticky mucus that bathes exposed surfaces. In the nasal cavity, cilia sweep that mucus and any trapped debris or microorganisms toward the pharynx, where it will be swallowed and exposed to the acids and enzymes of the stomach. In the lower respiratory system, the cilia also beat toward the pharynx, moving a carpet of mucus in that direction and cleaning the respiratory surfaces. This process is often described as a *mucus escalator* (Figure 23–2c●).

Filtration in the nasal cavity removes virtually all particles larger than about 10 μm from the inhaled air. Smaller particles may be trapped by the mucus of the nasopharynx or by secretions of the pharynx before proceeding along the conducting system. Exposure to unpleasant stimuli, such as noxious vapors, large quantities of dust and debris, allergens, or pathogens, generally causes a rapid increase in the rate of mucus production in the nasal cavity and paranasal sinuses. (The familiar symptoms of the "common cold" result from the invasion of this respiratory epithelium by any of more than 200 types of viruses.)

Most particles 1–5 μm in diameter are trapped in the mucus coating the respiratory bronchioles or in the liquid covering the alveolar surfaces. These areas are outside the boundaries of the mucus escalator, but the foreign particles can be engulfed by alveolar macrophages. Most particles smaller than about 0.5 μm remain suspended in the air.

Large quantities of airborne particles may overload the respiratory defenses and produce a variety of illnesses. For example, the presence of irritants in the lining of the conducting passageways can provoke the formation of abscesses that block airflow and reduce pulmonary function, and damage to the epithelium in the affected area may allow irritants to enter the surrounding tissues of the lung. The irritants then produce local inflammation. Airborne irritants are strongly linked to the development of lung cancer (p. 866). (We noted smoking-induced changes induced in the respiratory epithelium in Chapter 4, Figure 4–22●, p. 141.)

AM Overloading the Respiratory Defenses

Some respiratory illnesses are due to congenital defects. **Cystic fibrosis (CF)** is the most common lethal inherited disease affecting Caucasians of Northern European descent, occurring at a frequency of 1 birth in 2500. It occurs with less frequency in people of Southern European ancestry, in the Ashkenazic Jewish population, and in African-Americans. The condition can result from many different mutations affecting a gene located on chromosome 7. Individuals with CF seldom survive past age 30; death is generally the result of a chronic, massive bacterial infection of the lungs and associated heart failure.

The most serious symptoms appear because the respiratory mucosa in these individuals produces dense, viscous mucus that cannot be transported by the respiratory defense system. The mucus escalator stops working, and mucus blocks the smaller respiratory passageways. This blockage reduces the diameter of the airways, making breathing difficult, and the inactivation of the normal respiratory defenses leads to frequent bacterial infections. **AM** Cystic Fibrosis

23–2 THE UPPER RESPIRATORY SYSTEM

Objective

- Identify the organs of the upper respiratory system, and describe their functions.

As we have seen, the upper respiratory system consists of the nose, nasal cavity, paranasal sinuses, and pharynx (Figures 23–1 and 23–3●).

◻ THE NOSE AND NASAL CAVITY

The nose is the primary passageway for air entering the respiratory system. Air normally enters through the paired **external nares** (NĀ-rēz), or *nostrils* (Figure 23–3a●), which open into the *nasal cavity*. The **vestibule** is the space contained within the flexible tissues of the nose (Figure 23–3c●). The epithelium of the vestibule contains coarse hairs that extend across the external nares. Large airborne particles, such as sand, sawdust, or even insects, are trapped in these hairs and are thereby prevented from entering the nasal cavity.

The **nasal septum** divides the nasal cavity into left and right portions (Figure 23–3b●). The bony portion of the nasal septum is formed by the fusion of the perpendicular plate of the ethmoid bone and the plate of the vomer (Figure 7–3d●, p. 214). The anterior portion of the nasal septum is formed of hyaline cartilage. This cartilaginous plate supports the *dorsum nasi* (DOR-sum NĀ-zī), or *bridge*, and *apex* (tip) of the nose.

The maxillary, nasal, frontal, ethmoid, and sphenoid bones form the lateral and superior walls of the nasal cavity. The mucous secretions produced in the associated *paranasal sinuses* (Figure 7–14●, p. 227), aided by the tears draining through the nasolacrimal ducts, help keep the surfaces of the nasal cavity moist and clean. The *olfactory region*, or superior portion of the nasal cavity, includes the areas lined by olfactory epithelium: (1) the inferior surface of the cribriform plate, (2) the superior portion of the nasal septum, and (3) the superior nasal conchae. Receptors in the olfactory epithelium provide your sense of smell. ∞ p. 566

The *superior, middle,* and *inferior nasal conchae* project toward the nasal septum from the lateral walls of the nasal cavity. ∞ pp. 220, 222 To pass from the vestibule to the internal nares, air tends to flow between adjacent conchae, through the **superior, middle,** and **inferior meatuses** (mē-Ā-tus-ez; *meatus*, a passage) (Figure 23–3b●). These are narrow grooves rather than open passageways; the incoming air bounces off the conchal surfaces and churns like a stream flowing over rapids. This turbulence serves a purpose: As the air eddies and swirls, small airborne particles are likely to come into contact with the mucus that coats the lining of the nasal cavity. In addition to promoting filtration, the turbulence allows extra time for warming and humidifying incoming air. It also creates eddy currents that bring olfactory stimuli to the olfactory receptors.

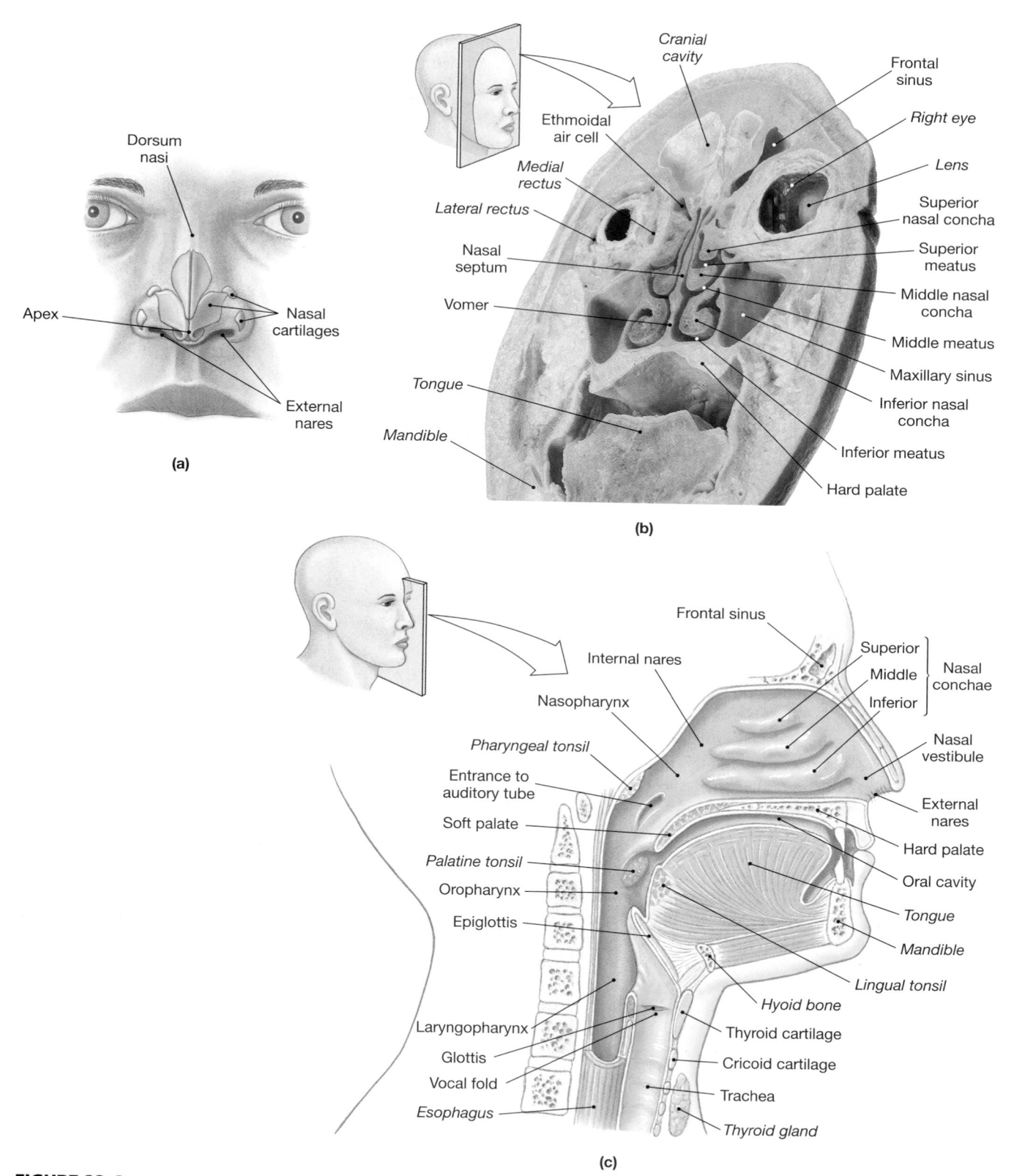

Cranial cavity
Frontal sinus
Ethmoidal air cell
Right eye
Medial rectus
Lateral rectus
Lens
Nasal septum
Superior nasal concha
Vomer
Superior meatus
Tongue
Middle nasal concha
Mandible
Middle meatus
Maxillary sinus
Inferior nasal concha
Inferior meatus
Hard palate

Dorsum nasi
Apex
Nasal cartilages
External nares

(a)

(b)

Frontal sinus
Internal nares
Superior
Middle
Inferior
Nasal conchae
Nasopharynx
Pharyngeal tonsil
Entrance to auditory tube
Nasal vestibule
Soft palate
External nares
Palatine tonsil
Hard palate
Oropharynx
Oral cavity
Epiglottis
Tongue
Mandible
Lingual tonsil
Hyoid bone
Laryngopharynx
Thyroid cartilage
Glottis
Cricoid cartilage
Vocal fold
Trachea
Esophagus
Thyroid gland

(c)

●**FIGURE 23–3**
The Nose, Nasal Cavity, and Pharynx. **(a)** The nasal cartilages and external landmarks on the nose. **(b)** The meatuses and the maxillary sinuses and ethmoidal air cells of the ethmoidal labyrinth. **(c)** The nasal cavity and pharynx, as seen in sagittal section with the nasal septum removed. ATLAS Plate 5.3d

A bony **hard palate**, made up of portions of the maxillary and palatine bones, forms the floor of the nasal cavity and separates it from the oral cavity. A fleshy **soft palate** extends posterior to the hard palate, marking the boundary between the superior **nasopharynx** (nā-zō-FAR-inks) and the rest of the pharynx. The nasal cavity opens into the nasopharynx through a connection known as the **internal nares**.

The Nasal Mucosa

The mucosa of the nasal cavity prepares the air you breathe for arrival at your lower respiratory system. Throughout much of the nasal cavity, the lamina propria contains an abundance of arteries, veins, and capillaries that bring nutrients and water to the secretory cells. The lamina propria of the nasal conchae also contains an extensive network of large and highly expandable veins. This vascularization provides a mechanism for warming and humidifying the incoming air (as well as for cooling and dehumidifying the outgoing air). As cool, dry air passes inward over the exposed surfaces of the nasal cavity, the warm epithelium radiates heat and the water in the mucus evaporates. Air moving from your nasal cavity to your lungs has been heated almost to body temperature, and it is nearly saturated with water vapor. This mechanism protects more delicate respiratory surfaces from chilling or drying out—two potentially disastrous events. Breathing through your mouth eliminates much of the preliminary filtration, heating, and humidifying of the inhaled air. To avoid alveolar damage, patients breathing on a respirator, which utilizes a tube to provide air directly into the trachea, must receive air that has been externally filtered and humidified.

As air moves out of the respiratory tract, it again passes across the epithelium of the nasal cavity. This air is warmer and more humid than the air that enters; it warms the nasal mucosa, and moisture condenses on the epithelial surfaces. Thus, breathing through your nose also helps prevent heat loss and water loss.

The extensive vascularization of the nasal cavity and the relatively vulnerable position of the nose make a nosebleed, or **epistaxis** (ep-i-STAK-sis), a fairly common event. This bleeding generally involves vessels of the mucosa covering the cartilaginous portion of the septum. A variety of factors may be responsible, including trauma, such as a punch in the nose, drying, infections, allergies, or clotting disorders. Hypertension can also provoke a nosebleed by rupturing small vessels of the lamina propria.

◻ THE PHARYNX

The **pharynx** (FAR-inks) is a chamber shared by the digestive and respiratory systems. It extends between the internal nares and the entrances to the larynx and esophagus. The curving superior and posterior walls of the pharynx are closely bound to the axial skeleton, but the lateral walls are flexible and muscular.

The pharynx is divided into the nasopharynx, the oropharynx, and the laryngopharynx (Figure 23–3c●):

1. The **nasopharynx** is the superior portion of the pharynx. It is connected to the posterior portion of the nasal cavity through the internal nares and is separated from the oral cavity by the soft palate. The nasopharynx is lined by the same pseudostratified ciliated columnar epithelium as that in the nasal cavity. The *pharyngeal tonsil* is located on the posterior wall of the nasopharynx; the left and right *auditory tubes* open into the nasopharynx on either side of this tonsil. ∞ pp. 787, 588

2. The **oropharynx** (*oris*, mouth) extends between the soft palate and the base of the tongue at the level of the hyoid bone. The posterior portion of the oral cavity communicates directly with the oropharynx, as does the posterior inferior portion of the nasopharynx. At the boundary between the nasopharynx and the oropharynx, the epithelium changes from pseudostratified columnar to stratified squamous.

3. The narrow **laryngopharynx** (la-rin-gō-FAR-inks), the inferior part of the pharynx, includes that portion of the pharynx between the hyoid bone and the entrance to the larynx and esophagus. Like the oropharynx, the laryngopharynx is lined with a stratified squamous epithelium that resists abrasion, chemical attack, and invasion by pathogens.

23-3 THE LARYNX

Objective

■ Describe the structure of the larynx, and discuss its role in normal breathing and in the production of sound.

Inhaled air leaves the pharynx and enters the larynx through a narrow opening called the **glottis** (GLOT-is). The **larynx** (LAR-inks) is a cartilaginous structure that surrounds and protects the glottis. The larynx begins at the level of vertebra C_4 or C_5 and ends at the level of vertebra C_6. Essentially a cylinder, the larynx has incomplete cartilaginous walls that are stabilized by ligaments and skeletal muscles (Figure 23–4●).

◻ CARTILAGES AND LIGAMENTS OF THE LARYNX

Three large, unpaired cartilages form the larynx: (1) the thyroid cartilage, (2) the cricoid cartilage, and (3) the epiglottis (Figure 23–4●). The **thyroid cartilage** (*thyroid*, shield shaped) is the largest laryngeal cartilage. Consisting of hyaline cartilage, it forms most of the anterior and lateral walls of the larynx. In section, this cartilage is U-shaped; posteriorly, it is incomplete. The prominent anterior surface of the thyroid cartilage, which you can easily see and feel, is called the *laryngeal prominence* or *Adam's apple*. The inferior surface articulates with the cricoid cartilage. The superior surface has ligamentous attachments to the hyoid bone and to the epiglottis and smaller laryngeal cartilages.

The thyroid cartilage sits superior to the **cricoid** (KRĪ-koyd; ring shaped) **cartilage**, another hyaline cartilage. The posterior

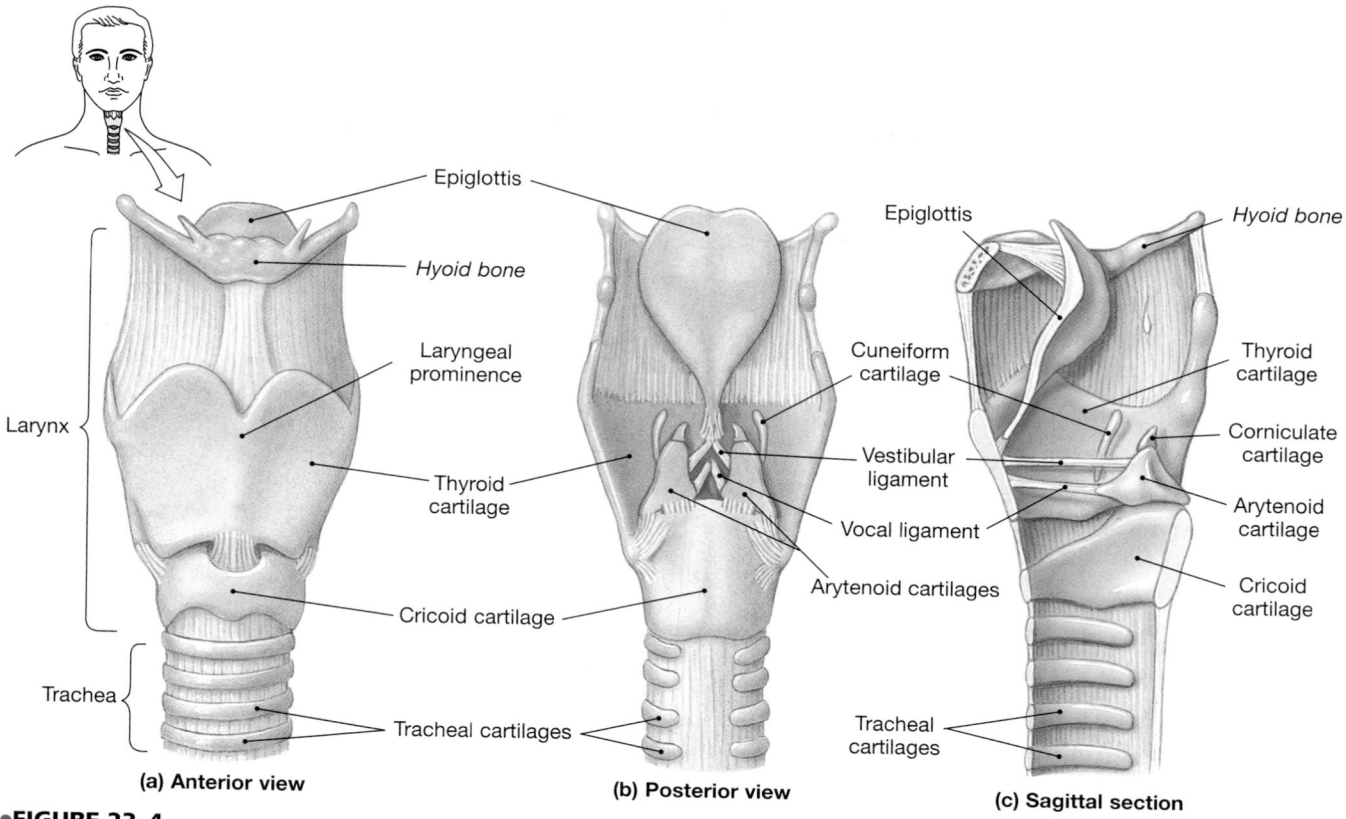

Epiglottis
Hyoid bone
Laryngeal prominence
Larynx
Thyroid cartilage
Cricoid cartilage
Trachea
Tracheal cartilages

(a) Anterior view

Epiglottis
Cuneiform cartilage
Vestibular ligament
Vocal ligament
Arytenoid cartilages
Tracheal cartilages

(b) Posterior view

Epiglottis
Hyoid bone
Thyroid cartilage
Corniculate cartilage
Arytenoid cartilage
Cricoid cartilage
Tracheal cartilages

(c) Sagittal section

●**FIGURE 23–4**
The Anatomy of the Larynx. **(a)** An anterior view. **(b)** A posterior view. **(c)** A sagittal section through the larynx.

portion of the cricoid is greatly expanded, providing support in the absence of the thyroid cartilage. The cricoid and thyroid cartilages protect the glottis and the entrance to the trachea, and their broad surfaces provide sites for the attachment of important laryngeal muscles and ligaments. Ligaments attach the inferior surface of the cricoid cartilage to the first cartilage of the trachea. The superior surface of the cricoid cartilage articulates with the small, paired *arytenoid cartilages.*

The shoehorn-shaped **epiglottis** (ep-i-GLOT-is) projects superior to the glottis and forms a lid over it. Composed of elastic cartilage, the epiglottis has ligamentous attachments to the anterior and superior borders of the thyroid cartilage and the hyoid bone. During swallowing, the larynx is elevated and the epiglottis folds back over the glottis, preventing the entry of both liquids and solid food into the respiratory tract.

The larynx also contains three pairs of smaller hyaline cartilages: (1) The **arytenoid** (ar-i-TĒ-noyd; ladle shaped) **cartilages** articulate with the superior border of the enlarged portion of the cricoid cartilage. (2) The **corniculate** (kor-NIK-ū-lāt; horn shaped) **cartilages** articulate with the arytenoid cartilages. The corniculate and arytenoid cartilages are involved with the opening and closing of the glottis and the production of sound. (3) Elongate, curving **cuneiform** (kū-NĒ-i-form; wedge shaped) **cartilages** lie within folds of tissue (the *aryepiglottic folds*) that

extend between the lateral surface of each arytenoid cartilage and the epiglottis (Figures 23–4c and 23–5●).

The various laryngeal cartilages are bound together by ligaments; additional ligaments attach the thyroid cartilage to the hyoid bone and the cricoid cartilage to the trachea (Figure 23–4a,b●). The **vestibular ligaments** and the **vocal ligaments** extend between the thyroid cartilage and the arytenoid cartilages.

The vestibular and vocal ligaments are covered by folds of laryngeal epithelium that project into the glottis. The vestibular ligaments lie within the superior pair of folds, known as the **vestibular folds**. These folds, which are relatively inelastic, help prevent foreign objects from entering the glottis and protect the more delicate **vocal folds** (Figure 23–5●).

The vocal folds, inferior to the vestibular folds, guard the entrance to the glottis. The vocal folds are highly elastic, because the vocal ligaments consist of elastic tissue. The vocal folds are involved with the production of sound, and for this reason they are known as the **vocal cords**.

□ SOUND PRODUCTION

Air passing through the glottis vibrates the vocal folds and produces sound waves. The pitch of the sound produced depends on the diameter, length, and tension in the vocal folds. The diameter

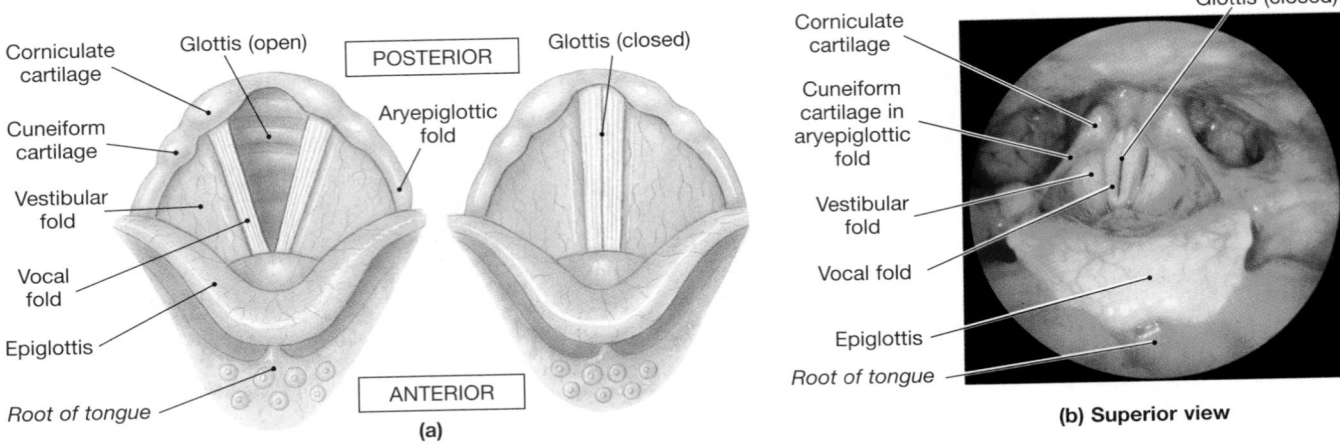

●**FIGURE 23–5**
The Glottis. **(a)** A diagrammatic superior view of the entrance to the larynx, with the glottis open (left) and closed (right). **(b)** A fiber-optic view of the entrance to the larynx, corresponding to the right-hand image in part (a). Note that the glottis is almost completely closed by the vocal folds.

and length are directly related to the size of the larynx. The tension is controlled by the contraction of voluntary muscles that reposition the arytenoid cartilages relative to the thyroid cartilage. When the distance increases, the vocal folds tense and the pitch rises; when the distance decreases, the vocal folds relax and the pitch falls.

Children have slender, short vocal folds; their voices tend to be high pitched. At puberty, the larynx of males enlarges much more than does that of females. The vocal cords of an adult male are thicker and longer, and produce lower tones, than those of an adult female.

Sound production at the larynx is called *phonation* (fō-NĀ-shun; *phone*, voice). Phonation is one component of speech production. However, clear speech also requires *articulation*, the modification of those sounds by other structures. In a stringed instrument, such as a guitar, the quality of the sound produced does not depend solely on the nature of the vibrating string. Rather, the entire instrument becomes involved as the walls vibrate and the composite sound echoes within the hollow body. Similar amplification and resonance occur within your pharynx, oral cavity, nasal cavity, and paranasal sinuses. The combination determines the particular and distinctive sound of your voice. When the nasal cavity and paranasal sinuses are filled with mucus rather than air, as in sinus infections, the sound changes. The final production of distinct words depends further on voluntary movements of the tongue, lips, and cheeks.

☐ THE LARYNGEAL MUSCULATURE

The larynx is associated with (1) muscles of the neck and pharynx, which position and stabilize the larynx (p. 347), and (2) smaller intrinsic muscles that control tension in the vocal folds or open and close the glottis. These latter muscles insert on the thyroid, arytenoid, and corniculate cartilages. The opening or

closing of the glottis involves rotational movements of the arytenoid cartilages that move the vocal folds.

When you swallow, both sets of muscles cooperate to prevent food or drink from entering the glottis. Before the material is swallowed, it is crushed and chewed into a pasty mass known as a *bolus*. Muscles of the neck and pharynx then elevate the larynx, bending the epiglottis over the glottis, so that the bolus can glide across the epiglottis rather than falling into the larynx. While this movement is under way, the glottis is closed.

Food or liquids that touch the vestibular or vocal folds trigger the *coughing reflex*. In a cough, the glottis is kept closed while the chest and abdominal muscles contract, compressing the lungs. When the glottis is opened suddenly, the resulting blast of air from the trachea ejects material that blocks the entrance to the glottis.

An infection or inflammation of the larynx is known as **laryngitis** (lar-in-JĪ-tis). It commonly affects the vibrational qualities of the vocal folds; hoarseness is the most familiar symptom. Mild cases are temporary and seldom serious. However, bacterial or viral infection of the epiglottis can be very dangerous; the resulting swelling may close the glottis and cause suffocation. This condition, acute **epiglottitis** (ep-i-glot-TĪ-tis), can develop rapidly after a bacterial infection of the throat. Young children are most likely to be affected.

✓ Why is the vascularization of the nasal cavity important?

✓ Why is the lining of the nasopharynx different from that of the oropharynx and the laryngopharynx?

✓ When the tension in your vocal folds increases, what happens to the pitch of your voice?

Answers start on page Q-1

23–4 THE TRACHEA AND PRIMARY BRONCHI

Objective

■ Discuss the structure of the extrapulmonary airways.

□ THE TRACHEA

The epithelium of the larynx is continuous with that of the **trachea** (TRĀ-kē-uh), or windpipe, a tough, flexible tube with a diameter of about 2.5 cm (1 in.) and a length of about 11 cm (4.25 in.) (Figure 23–6●). The trachea begins anterior to vertebra C_6 in a ligamentous attachment to the cricoid cartilage. It ends in the mediastinum, at the level of vertebra T_5, where it branches to form the *right* and *left primary bronchi*.

The mucosa of the trachea resembles that of the nasal cavity and nasopharynx (Figure 23–2a●, p. 829). The **submucosa** (sub-mū-KŌ-suh), a thick layer of connective tissue, surrounds the mucosa. The submucosa contains mucous glands that communicate with the epithelial surface through a number of secretory ducts. The trachea contains 15–20 **tracheal cartilages** (Figure 23–6a●), which serve to stiffen the tracheal walls and protect the airway. They also prevent its collapse or overexpansion as pressures change in the respiratory system.

Each tracheal cartilage is C-shaped. The closed portion of the C protects the anterior and lateral surfaces of the trachea. The open portion of the C faces posteriorly, toward the esophagus (Figure 23–6b●). Because these cartilages are not continuous, the posterior tracheal wall can easily distort when you swallow, permitting large masses of food to pass through the esophagus.

An elastic ligament and the **trachealis muscle**, a band of smooth muscle, connect the ends of each tracheal cartilage

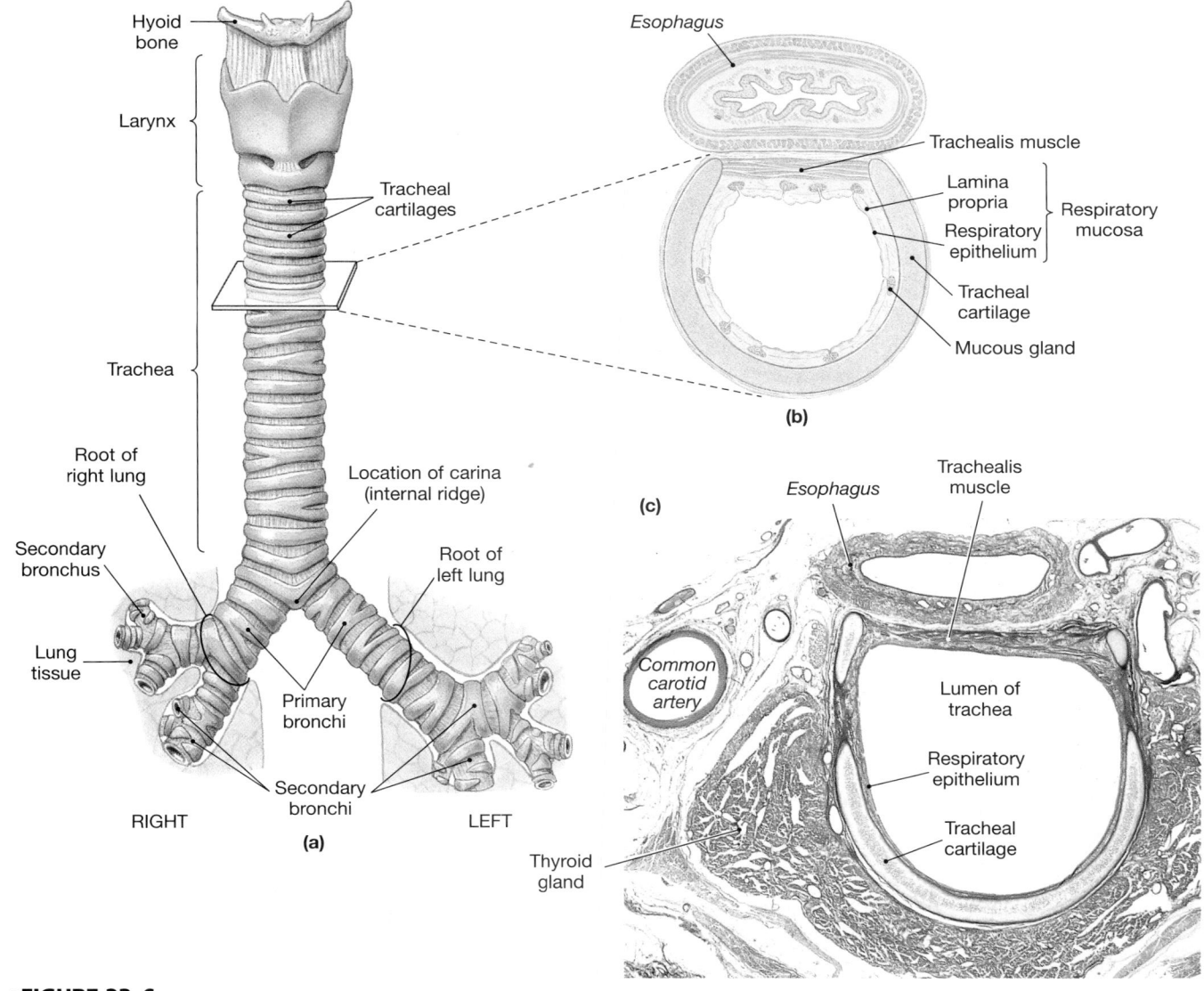

●FIGURE 23–6
The Anatomy of the Trachea.   **(a)** A diagrammatic anterior view. **(b,c)** Cross-sectional views. (LM × 241)   ATLAS Plate 7.4b, Scan 10

Tracheal Blockage

We sometimes inadvertently breathe in foreign objects, a process called *aspiration*. Foreign objects that become lodged in the larynx or trachea are generally expelled by coughing. If the individual can speak or make a sound, the airway is still open, so no emergency measures should be taken. If the person can neither breathe nor speak, an immediate threat to life exists. Unfortunately, many victims become acutely embarrassed by this situation; instead of seeking assistance, they run to the nearest rest room and die there.

In the **Heimlich** (HĪM-lik) **maneuver**, or *abdominal thrust*, a rescuer applies compression to the victim's abdomen just inferior to the diaphragm. This action elevates the diaphragm forcefully and may generate enough pressure to remove the blockage. The maneuver must be performed properly to avoid damage to internal organs. Organizations such as the American Red Cross, local fire departments, and other charitable groups periodically hold brief training sessions in the proper performance of the Heimlich maneuver.

If blockage results from a swelling of the epiglottis or tissues surrounding the glottis, a professionally qualified rescuer may insert a curved tube through the pharynx and glottis to permit airflow. This procedure is called *intubation*. If the blockage is

immovable or the larynx has been crushed, a **tracheostomy** (trā-kē-OS-to-mē) may be performed. In a tracheostomy, an incision is made through the anterior tracheal wall, and a tube is inserted. The tube bypasses the larynx and permits air to flow directly into the trachea.

(Figure 23–6b,c●). Contraction of the trachealis muscle alters the diameter of the trachea, changing the vessel's resistance to airflow. The normal diameter of the trachea changes from moment to moment, primarily under the control of the sympathetic division of the ANS. Sympathetic stimulation increases the diameter of the trachea and makes it easier to move large volumes of air along the respiratory passageways.

☐ THE PRIMARY BRONCHI

The trachea branches within the mediastinum, giving rise to the **right** and **left primary bronchi** (BRONG-kī; singular, *bronchus*). An internal ridge called the **carina** (ka-RĪ-nuh) separates the two bronchi (Figure 23–6a●). Like the trachea, the primary bronchi have cartilaginous C shaped supporting rings. The right primary bronchus supplies the right lung, and the left supplies the left lung. The right primary bronchus is larger in diameter than the left, and descends toward the lung at a steeper angle. Thus, most foreign objects that enter the trachea find their way into the right bronchus rather than the left.

Before branching further, each primary bronchus travels to a groove along the medial surface of its lung. This groove, the **hilus** of the lung, also provides access for entry to pulmonary vessels and nerves (Figure 23–7●). The entire array is firmly anchored in a meshwork of dense connective tissue. This complex, the **root** of the lung (Figure 23–6a●), attaches to the mediastinum and fixes the positions of the major nerves, vessels, and lymphatic vessels. The roots of the lungs are anterior to vertebrae T_5 (right) and T_6 (left).

23–5 THE LUNGS

Objective

■ Describe the superficial anatomy of the lungs, the structure of a pulmonary lobule, and the functional anatomy of the alveoli.

The left and right lungs (Figure 23–7●) are in the left and right pleural cavities, respectively. Each lung is a blunt cone, the tip, or apex, of which points superiorly. The apex on each side extends into the base of the neck superior to the first rib. The broad concave inferior portion, or base, of each lung rests on the superior surface of the diaphragm.

Tuberculosis (too-ber-kū-LŌ-sis), or TB, results from an infection of the lungs by *Mycobacterium tuberculosis*; other organs may be invaded as well. This bacterium may colonize the respiratory passageways, the interstitial spaces, the alveoli, or a combination of the three. Symptoms are variable, but generally include coughing and chest pain, with fever, night sweats, fatigue, and weight loss. In 1900, TB, then known as "consumption," was the leading cause of death. **AM** Tuberculosis

☐ LOBES AND SURFACES OF THE LUNGS

The lungs have distinct **lobes** that are separated by deep fissures (Figure 23–7●). The right lung has three lobes: *superior*, *middle*, and *inferior*, separated by the *horizontal* and *oblique fissures*. The left lung has only two lobes: *superior* and *inferior*, separated by the *oblique*

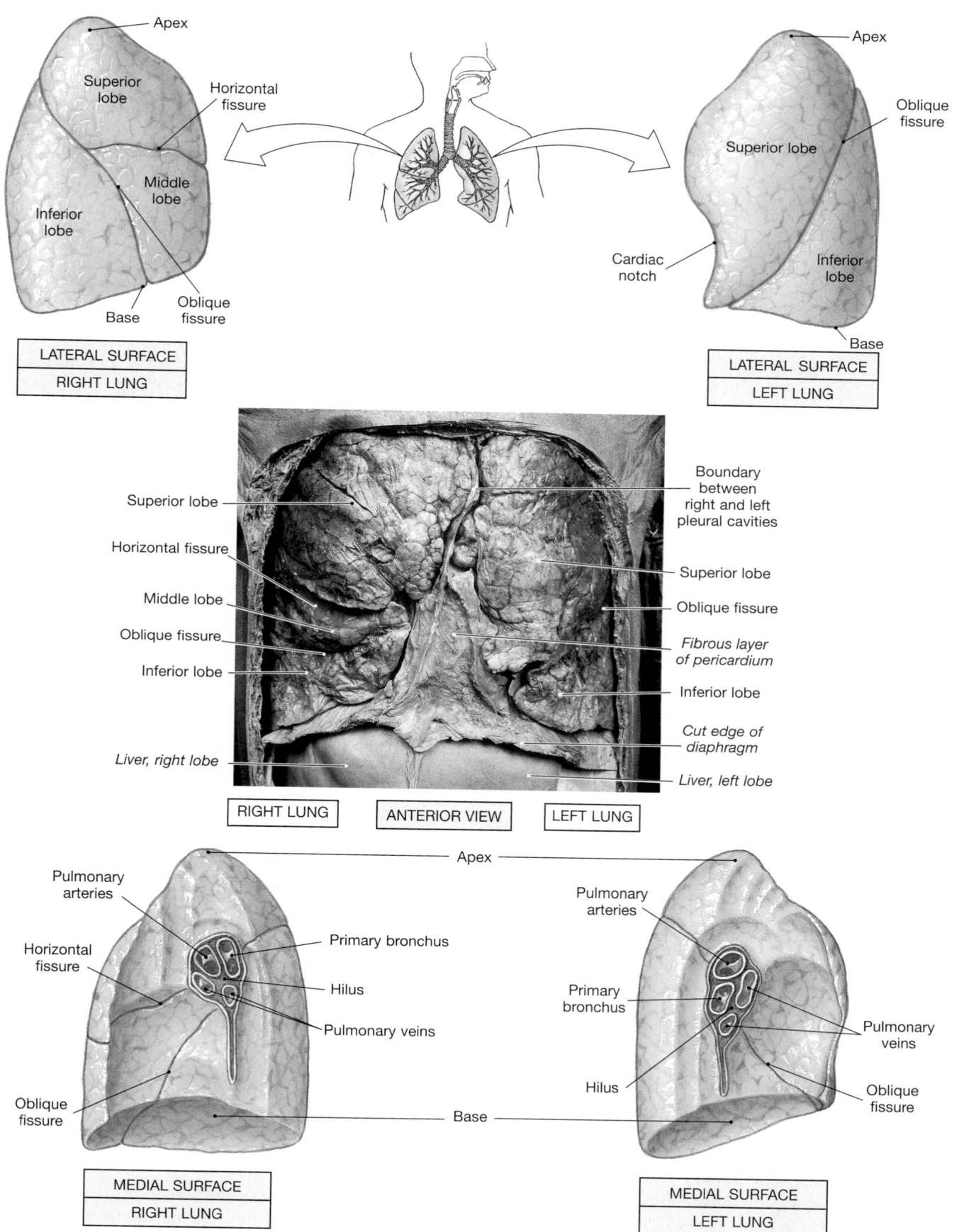

●FIGURE 23–7
The Gross Anatomy of the Lungs. ATLAS Plates 7.4–7.6a–d; Scan 10

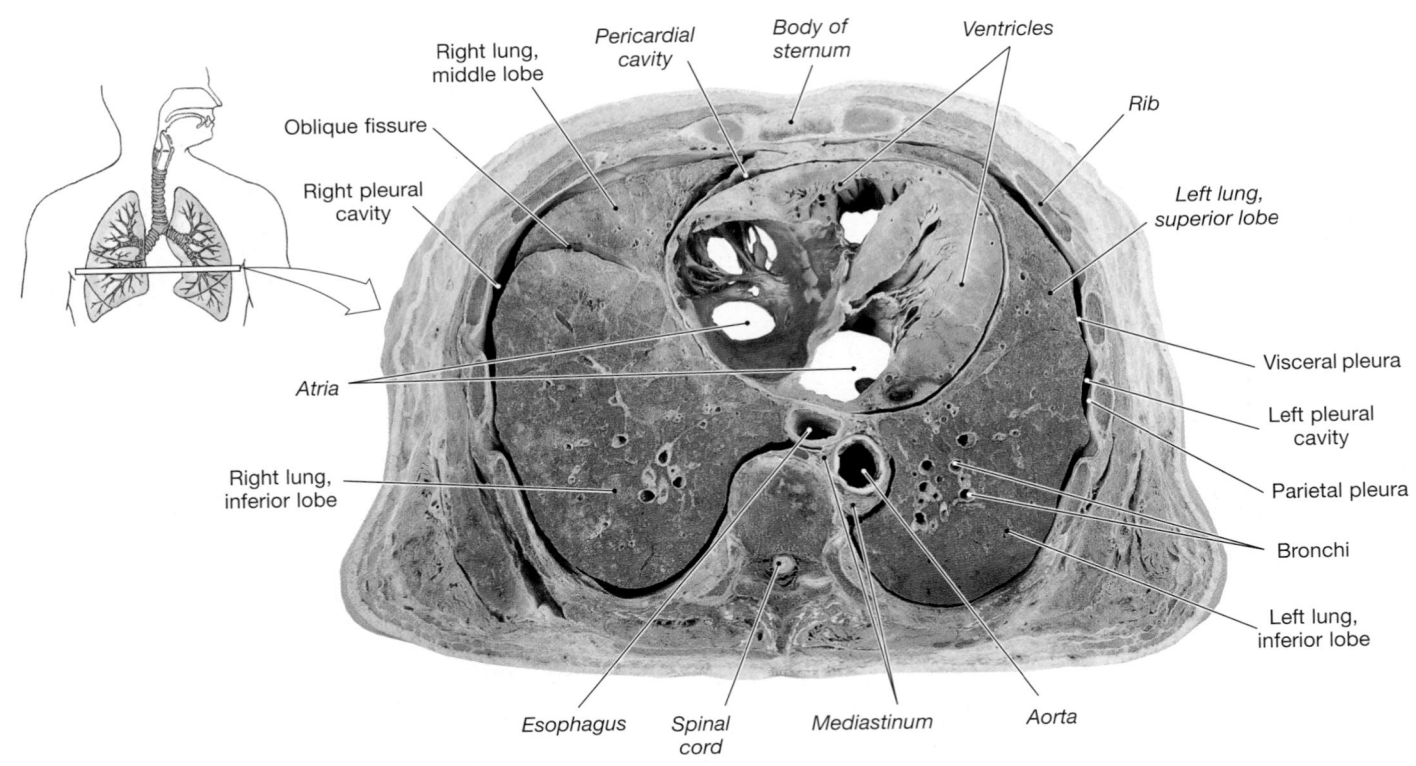

●FIGURE 23–8
The Relationship between the Lungs and Heart. This transverse section was taken at the level of the cardiac notch.

fissure. The right lung is broader than the left, because most of the heart and great vessels project into the left thoracic cavity. However, the left lung is longer than the right lung, because the diaphragm rises on the right side to accommodate the mass of the liver.

The curving anterior and lateral surfaces of each lung follow the inner contours of the rib cage. The medial surface, which contains the hilus, has a more irregular shape. The medial surfaces of both lungs bear grooves that mark the positions of the great vessels and the heart (Figures 23–7 and 23–8●). The heart is located to the left of the midline, and the corresponding impression is larger in the left lung than in the right. In anterior view, the medial edge of the right lung forms a vertical line, whereas the medial margin of the left lung is indented at the **cardiac notch**.

☐ THE BRONCHI

The primary bronchi and their branches form the **bronchial tree** (Figure 23–9●). Because the left and right primary bronchi are outside the lungs, they are called *extrapulmonary bronchi.* As the primary bronchi enter the lungs, they divide to form smaller passageways (Figures 23–6a, and 23–10a●). The branches are collectively called the *intrapulmonary bronchi.*

Each primary bronchus divides to form **secondary bronchi**, also known as *lobar bronchi.* In each lung, one secondary bronchus goes to each lobe, so the right lung has three secondary bronchi and the left lung has two.

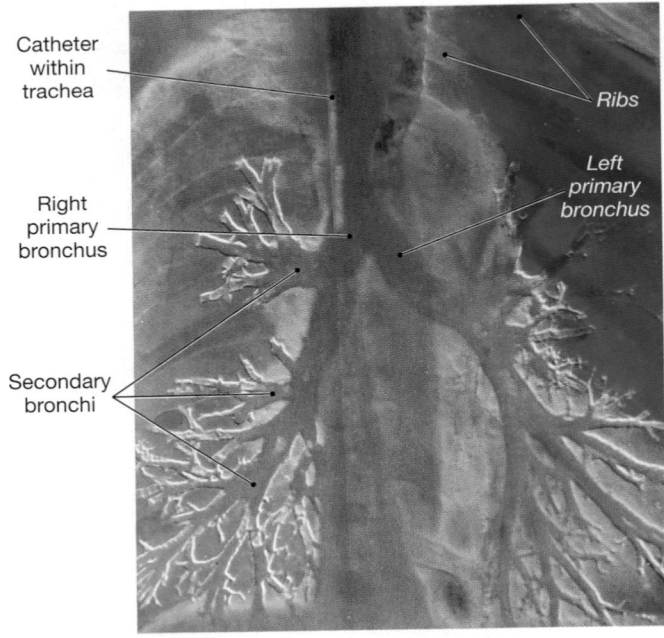

●FIGURE 23–9
The Bronchial Tree. ATLAS Plate 7.4b, Scan 10

Figure 23–10● follows the branching pattern of the left primary bronchus as it enters the lung. (The number of branches has been reduced for clarity.) In each lung, the secondary bronchi branch to form **tertiary bronchi**, or *segmental bronchi.*

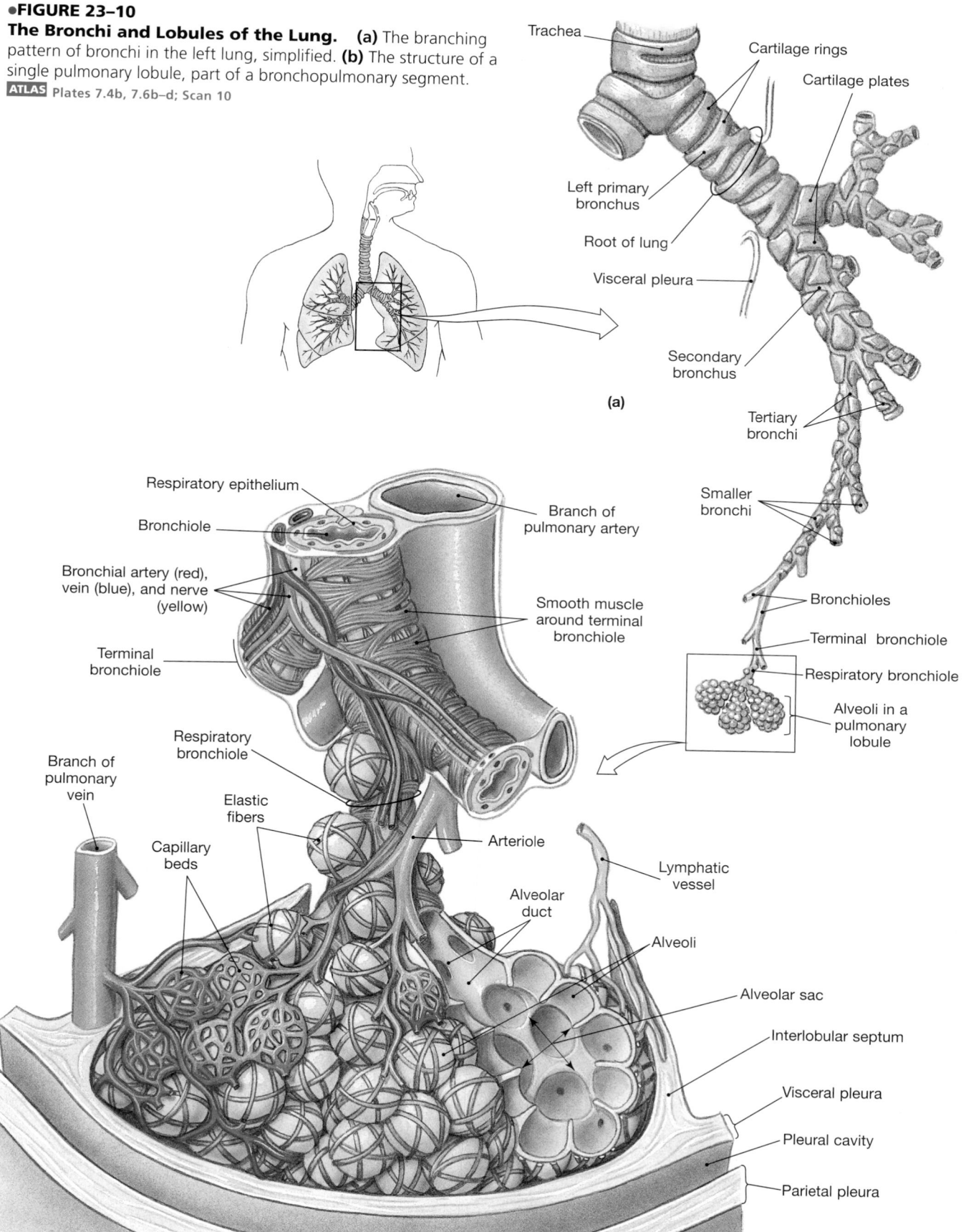

●FIGURE 23–10

The Bronchi and Lobules of the Lung. **(a)** The branching pattern of bronchi in the left lung, simplified. **(b)** The structure of a single pulmonary lobule, part of a bronchopulmonary segment.

ATLAS Plates 7.4b, 7.6b–d; Scan 10

Trachea

Cartilage rings

Cartilage plates

Left primary bronchus

Root of lung

Visceral pleura

Secondary bronchus

Tertiary bronchi

Smaller bronchi

Bronchioles

Terminal bronchiole

Respiratory bronchiole

Alveoli in a pulmonary lobule

(a)

Respiratory epithelium

Bronchiole

Branch of pulmonary artery

Bronchial artery (red), vein (blue), and nerve (yellow)

Smooth muscle around terminal bronchiole

Terminal bronchiole

Respiratory bronchiole

Elastic fibers

Arteriole

Branch of pulmonary vein

Lymphatic vessel

Capillary beds

Alveolar duct

Alveoli

Alveolar sac

Interlobular septum

Visceral pleura

Pleural cavity

Parietal pleura

(b)

The branching pattern differs between the two lungs, but each tertiary bronchus ultimately supplies air to a single **bronchopulmonary segment**, a specific region of one lung (Figure 23–10●). The right lung has 10 bronchopulmonary segments. During development, the left lung also has 10 segments, but subsequent fusion of adjacent tertiary bronchi generally reduces that number to eight or nine.

The walls of the primary, secondary, and tertiary bronchi contain progressively lesser amounts of cartilage. In the secondary and tertiary bronchi, the cartilages form plates arranged around the lumen. These cartilages serve the same purpose as the rings of cartilage in the trachea and primary bronchi. As the amount of cartilage decreases, the relative amount of smooth muscle increases. With less cartilaginous support, the amount of tension in those smooth muscles has a greater effect on bronchial diameter and the resistance to airflow.

During a respiratory infection, the bronchi and bronchioles can become inflamed. In this condition, called **bronchitis**, the smaller passageways may become greatly constricted, leading to difficulties in breathing. One method of investigating the status of the respiratory passageways is the use of a **bronchoscope**, a fiber-optic bundle small enough to be inserted into the trachea and steered along the conducting passageways to the level of the smaller bronchioles. This procedure is called **bronchoscopy** (brong-KOS-ko-pē). In addition to permitting the direct visualization of bronchial structures, the bronchoscope can collect tissue or mucus samples from the respiratory tract. In **bronchography** (brong-KOG-ra-fē), a bronchoscope or catheter introduces a radiopaque material into the bronchi (Figure 23–9●). This technique can permit detailed X-ray analysis of bronchial masses, such as tumors, or other obstructions.

☐ THE BRONCHIOLES

Each tertiary bronchus branches several times within the bronchopulmonary segment, giving rise to multiple **bronchioles**. These passageways branch further into the finest conducting branches, called **terminal bronchioles**. Roughly 6500 terminal bronchioles are supplied by each tertiary bronchus. The lumen of each terminal bronchiole has a diameter of 0.3–0.5 mm.

The walls of bronchioles, which lack cartilaginous supports, are dominated by smooth muscle tissue (Figure 23–10b●). In functional terms, the bronchioles are to the respiratory system what the arterioles are to the cardiovascular system. Varying the diameter of the bronchioles controls the amount of resistance to airflow and the distribution of air in the lungs.

The autonomic nervous system regulates the activity in this smooth muscle layer and thereby controls the diameter of the bronchioles. Sympathetic activation leads to **bronchodilation**, the enlargement of the airway diameter. Parasympathetic stimulation leads to **bronchoconstriction**, a reduction in the diameter of the airway. Bronchoconstriction also occurs during allergic reactions such as anaphylaxis, in response to histamine released by activated mast cells and basophils. ∞ p. 816

Bronchodilation and bronchoconstriction adjust the resistance to airflow toward or away from the respiratory exchange surfaces. Tension in the smooth muscles commonly throws the bronchiolar mucosa into a series of folds, limiting airflow; excessive stimulation, as in *asthma* (AZ-muh), can almost completely prevent airflow along the terminal bronchioles. **AM** Asthma

Pulmonary Lobules

The connective tissues of the root of each lung extend into the lung's parenchyma. These fibrous partitions, or *trabeculae*, contain elastic fibers, smooth muscles, and lymphatic vessels. The trabeculae branch repeatedly, dividing the lobes into ever-smaller compartments. The branches of the conducting passageways, pulmonary vessels, and nerves of the lungs follow these trabeculae. The finest partitions, or **interlobular septa** (*septum*, a wall), divide the lung into **pulmonary lobules** (LOB-ūlz), each of which is supplied by branches of the pulmonary arteries, pulmonary veins, and respiratory passageways (Figure 23–10●). The connective tissues of the septa are, in turn, continuous with those of the *visceral pleura*, the serous membrane covering the lungs.

Each terminal bronchiole delivers air to a single pulmonary lobule. Within the lobule, the terminal bronchiole branches to form several **respiratory bronchioles**. The thinnest and most delicate branches of the bronchial tree, the respiratory bronchioles deliver air to the gas exchange surfaces of the lungs.

The preliminary filtration and humidification of incoming air are completed before incoming air moves beyond the terminal bronchioles. A cuboidal epithelium lines the terminal bronchioles and respiratory bronchioles. There are only scattered cilia, and no goblet cells or underlying mucous glands are present. If particulate matter or pathogens reach this part of the respiratory tract, there is little to prevent them from damaging the delicate exchange surfaces of the lungs.

☐ ALVEOLAR DUCTS AND ALVEOLI

Respiratory bronchioles are connected to individual alveoli and to multiple alveoli along regions called **alveolar ducts** (Figures 23–10b and 23–11●). Alveolar ducts end at **alveolar sacs**, common chambers connected to multiple individual alveoli. Each lung contains about 150 million alveoli, and their abundance gives the lung an open, spongy appearance. An extensive network of capillaries is associated with each alveolus. The capillaries are surrounded by a network of elastic fibers (Figure 23–12a●), which help maintain the relative positions of the alveoli and respiratory bronchioles. Recoil of these fibers during exhalation reduces the size of the alveoli and helps push air out of the lungs.

The alveolar epithelium consists primarily of simple squamous epithelium (Figure 23–12b●). The squamous epithelial cells, called *Type I cells*, are unusually thin and delicate. Roaming **alveolar macrophages** (*dust cells*) patrol the epithelial surface, phagocytizing any particulate matter that has eluded other respiratory defenses and reached the alveolar surfaces. **Septal cells**, also called *Type II cells*, are scattered among the squamous cells.

The large septal cells produce **surfactant** (sur-FAK-tant), an oily secretion containing a mixture of phospholipids and proteins. Surfactant is secreted onto the alveolar surfaces, where it forms a superficial coating over a thin layer of water.

Surfactant reduces surface tension in the liquid coating the alveolar surface. Recall from Chapter 2 that *surface tension* results from the attraction between water molecules at an air–water boundary. ∞ p. 37 Surface tension creates a barrier that keeps small objects from entering the water, but it also tends to collapse small bubbles. Alveolar walls are very delicate—rather like air bubbles; without surfactant, the surface tension would be so high that the alveoli would collapse. Surfactant forms a thin surface layer that interacts with the water molecules, reducing the surface tension and keeping the alveoli open.

If septal cells produce inadequate amounts of surfactant due to injury or genetic abnormalities, the alveoli will collapse after each exhalation and respiration will become difficult. On each breath, the inhalation must then be forceful enough to pop open the alveoli. A person who does not produce enough surfactant is soon exhausted by the effort required to keep inflating and deflating the lungs. This condition is called *respiratory distress syndrome.* **AM** Respiratory Distress Syndrome (RDS)

Gas exchange occurs across the **respiratory membrane** of the alveoli. The respiratory membrane is a composite structure consisting of three parts (Figure 23–12c●):

1. The squamous epithelial cells lining the alveolus.
2. The endothelial cells lining an adjacent capillary.
3. The fused basal laminae that lie between the alveolar and endothelial cells.

At the respiratory membrane, the total distance separating alveolar air from blood can be as little as 0.1 μm; it averages about 0.5 μm. Diffusion across the respiratory membrane proceeds very rapidly, because (1) the distance is small and (2) both

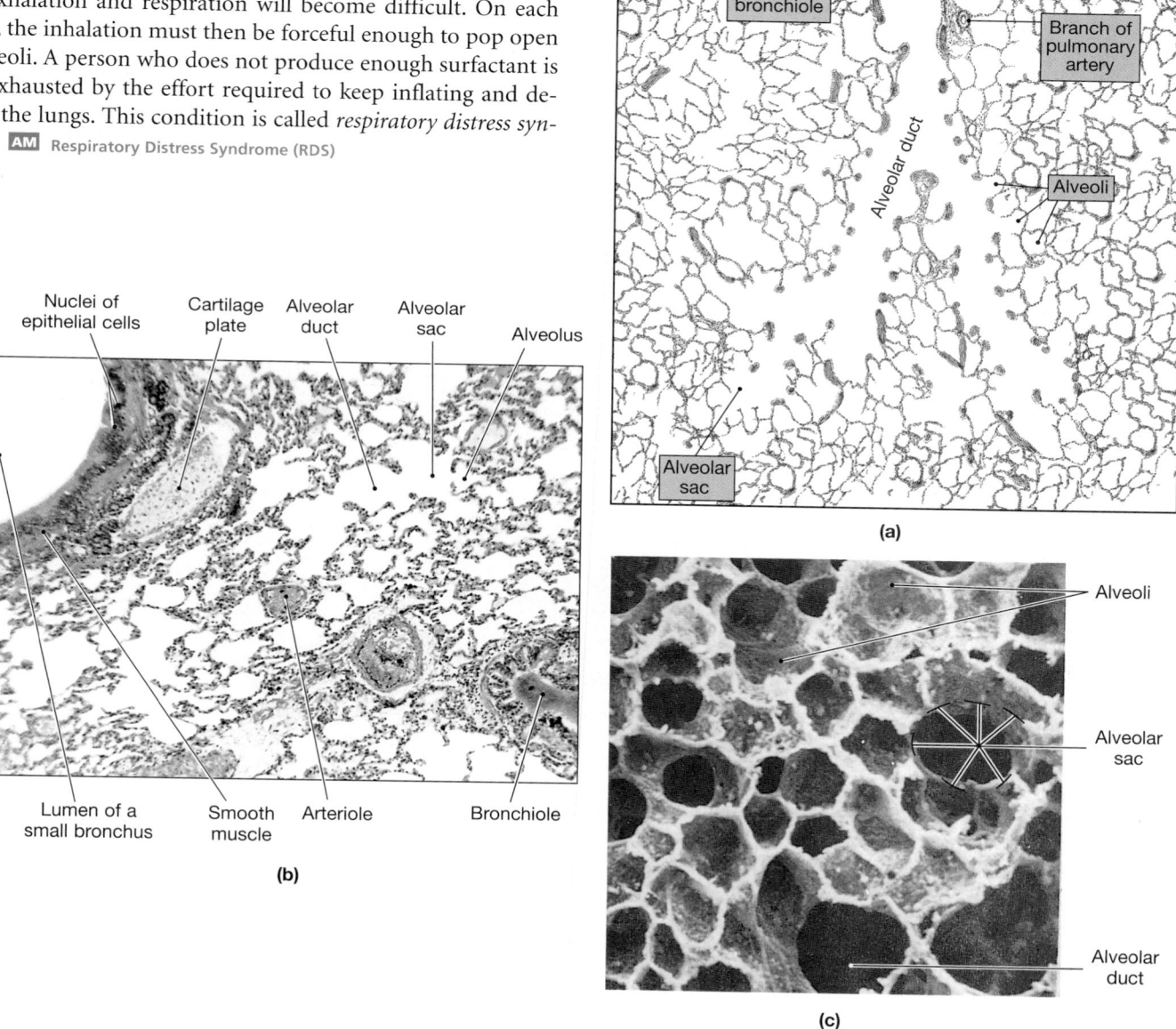

●**FIGURE 23–11**
The Bronchioles. (a) The distribution of a respiratory bronchiole supplying a portion of a lobule. (b) Alveolar sacs and alveoli. (LM × 42) (c) An SEM of the lung. Notice the open, spongy appearance of the lung tissue; compare with Figure 23–10b.

●FIGURE 23–12
Alveolar Organization. **(a)** The basic structure of a portion of a single lobule. A network of capillaries, supported by elastic fibers, surrounds each alveolus. Respiratory bronchioles also contain wrappings of smooth muscle that can change the diameter of these airways. **(b)** A diagrammatic view of alveolar structure. A single capillary may be involved with gas exchange across several alveoli simultaneously. **(c)** The respiratory membrane, which consists of an alveolar epithelial cell, a capillary endothelial cell, and their fused basal laminae.

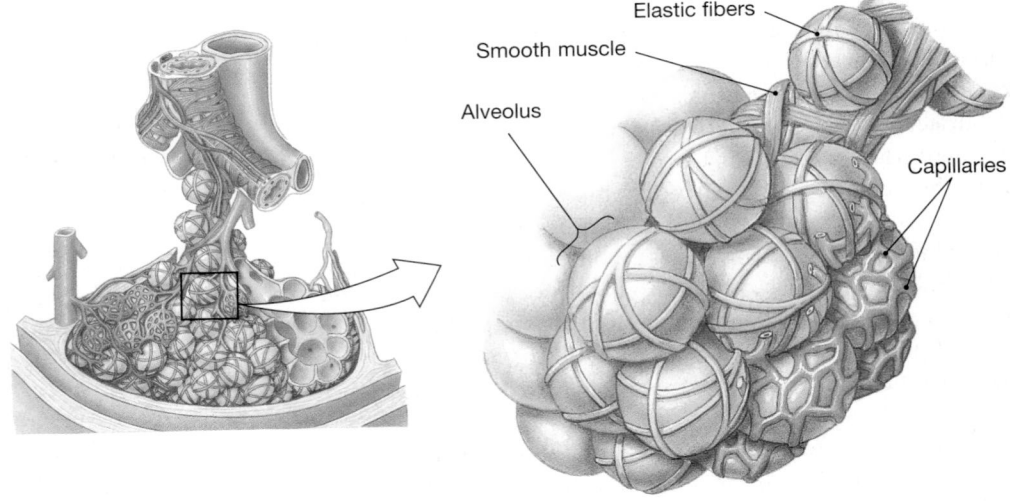

(a)

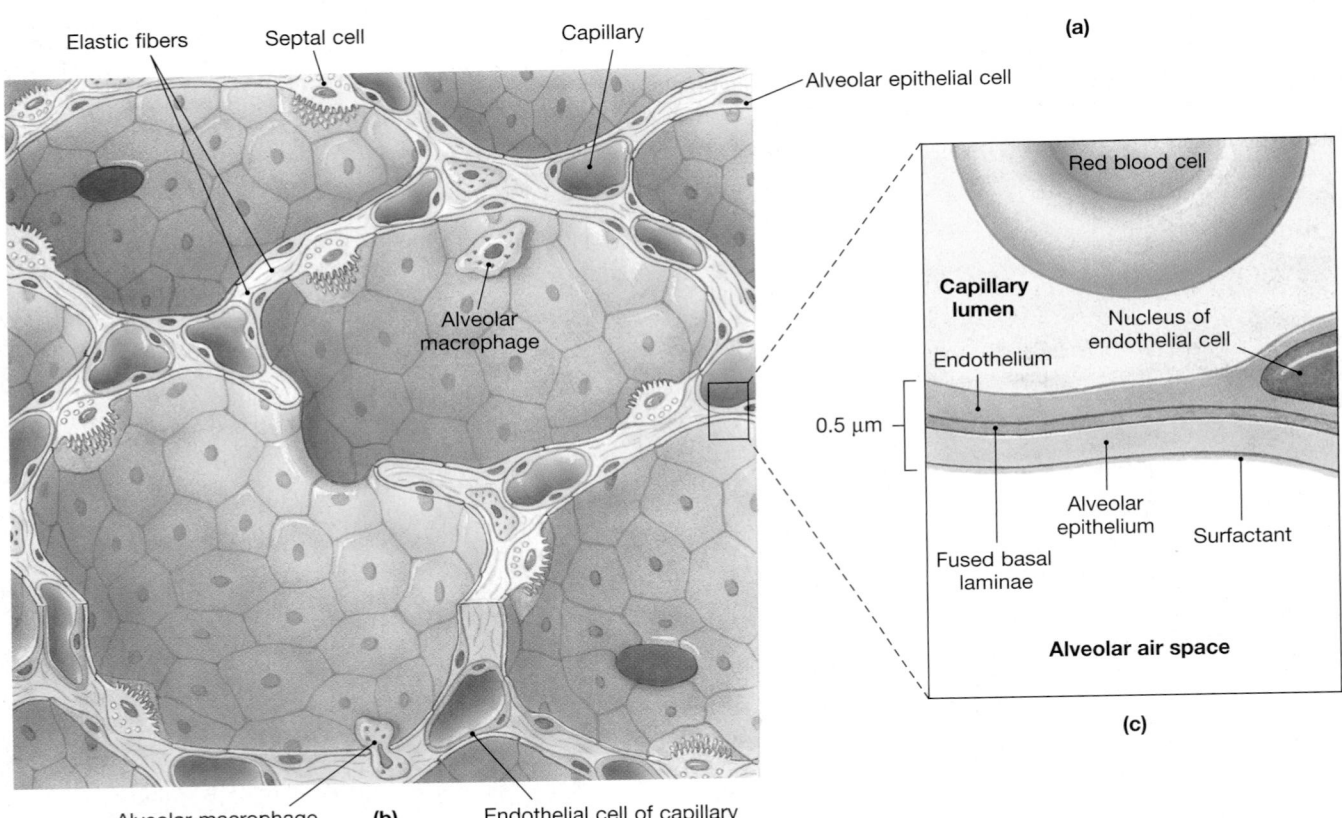

(b)

(c)

oxygen and carbon dioxide are lipid soluble. The membranes of the epithelial and endothelial cells thus do not pose a barrier to the movement of oxygen and carbon dioxide between blood and alveolar air spaces.

In certain disease states, function of the respiratory membrane can be compromised. **Pneumonia** (noo-MŌ-nē-uh) develops from an infection or any other stimulus that causes inflammation of the lobules of the lung. As inflammation occurs, fluids leak into the alveoli and the respiratory bronchioles swell and constrict. Respiratory function deteriorates as

a result. When bacteria are involved, they are generally types that normally inhabit the mouth and pharynx, but have managed to evade the respiratory defenses. Pneumonia becomes more likely when the respiratory defenses have already been compromised by other factors, such as epithelial damage from smoking or the breakdown of the immune system in AIDS. The most common pneumonia that develops in individuals with AIDS results from infection by the fungus *Pneumocystis carinii*. This fungus normally inhabits the alveoli, but the respiratory defenses of healthy individuals are able to prevent infection and tissue damage.

✓ Why are the cartilages that reinforce the trachea C-shaped?

✓ What would happen to the alveoli if surfactant were not produced?

✓ What path does air take in flowing from the glottis to the respiratory membrane?

Answers start on page Q-1

Review the anatomy of the respiratory tract by visiting the **IP CD-ROM**: Respiratory System/Anatomy Review: Respiratory Structures and the **Anatomy CD-ROM**: Respiratory System/Lungs Flythrough.

☐ THE BLOOD SUPPLY TO THE LUNGS

Your respiratory exchange surfaces receive blood from arteries of your pulmonary circuit. The pulmonary arteries enter the lungs at the hilus and branch with the bronchi as they approach the lobules. Each lobule receives an arteriole and a venule, and a network of capillaries surrounds each alveolus as part of the respiratory membrane. In addition to providing a mechanism for gas exchange, the endothelial cells of the alveolar capillaries are the primary source of *angiotensin-converting enzyme (ACE)*, which converts circulating angiotensin I to angiotensin II. This enzyme plays an important role in the regulation of blood volume and blood pressure. ∞ p. 744

Blood from the alveolar capillaries passes through the pulmonary venules and then enters the pulmonary veins, which deliver it to the left atrium. The conducting portions of your respiratory tract receive blood from the *external carotid arteries* (the nasal passages and larynx), the *thyrocervical trunks* (the inferior larynx and trachea), and the *bronchial arteries* (the bronchi and bronchioles). ∞ pp. 756, 753, 758 The capillaries supplied by the bronchial arteries provide oxygen and nutrients to the tissues of conducting passageways of your lungs. The venous blood from these bronchial capillaries ultimately flows into the pulmonary veins, bypassing the rest of the systemic circuit and diluting the oxygenated blood leaving the alveoli.

Blood pressure in the pulmonary circuit is usually relatively low, with systemic pressures of 30 mm Hg or less. With such pressures, pulmonary vessels can easily become blocked by small blood clots, fat masses, or air bubbles in the pulmonary arteries. Because the lungs receive the entire cardiac output, any such objects drifting in blood are likely to cause problems almost at once. The blockage of a branch of a pulmonary artery will stop blood flow to a group of lobules or alveoli. This condition is a **pulmonary embolism**.

Atherosclerosis (damage to arterial walls and plaque formation; and *venous thrombosis* (the blockage of a vein by a blood clot) can promote the development of a pulmonary embolism. In both cases, small blood clots tend to form, break loose, and drift in the bloodstream. If a pulmonary embolism is in place for several hours, the alveoli will permanently collapse. If the blockage occurs in a major pulmonary vessel rather than a minor tributary, pulmonary resistance increases. The resistance places extra strain on the right ventricle, which may be unable to maintain cardiac output, and congestive heart failure can result.

☐ THE PLEURAL CAVITIES AND PLEURAL MEMBRANES

The thoracic cavity has the shape of a broad cone. Its walls are the rib cage, and the muscular diaphragm forms the floor. The two **pleural cavities** are separated by the mediastinum (Figure 23–8●, p. 843). Each lung occupies a single pleural cavity, which is lined by a serous membrane called the **pleura** (PLOO-ra; plural, *pleurae*). The pleura consists of two layers: the parietal pleura and the visceral pleura. The **parietal pleura** covers the inner surface of the thoracic wall and extends over the diaphragm and mediastinum. The **visceral pleura** covers the outer surfaces of the lungs, extending into the fissures between the lobes. Each pleural cavity actually represents a potential space rather than an open chamber, because the parietal and visceral pleurae are usually in close contact.

Both pleurae secrete a small amount of transudate referred to as **pleural fluid**. Pleural fluid forms a moist, slippery coating that provides lubrication, thereby reducing friction between the parietal and visceral surfaces as you breathe. Samples of pleural fluid, obtained by means of a long needle inserted between the ribs, are sometimes obtained for diagnostic purposes. This sampling procedure is called *thoracentesis*. The fluid that is extracted is examined for the presence of bacteria, blood cells, or other abnormal components.

In some disease states, the normal coating of pleural fluid is unable to prevent friction between the opposing pleural surfaces. The result is pain and pleural inflammation, a condition called **pleurisy**. When pleurisy develops, the secretion of pleural fluid may be excessive, or the inflamed pleurae may adhere to one another, limiting relative movement. In either case, breathing becomes difficult and prompt medical attention is required.

✓ Which arteries supply blood to the conducting portions and respiratory exchange surfaces of the lungs?

✓ List the functions of the pleura. What does it secrete?

Answers start on page Q-1

23–6 AN OVERVIEW OF RESPIRATORY PHYSIOLOGY

Objectives

■ Define and compare the processes of external respiration and internal respiration.

■ Describe the major steps involved in external respiration.

The general term *respiration* refers to two integrated processes: *external respiration* and *internal respiration*. The precise definitions of these terms vary from reference to reference. In this discussion, **external respiration** includes all the processes involved

●FIGURE 23–13
An Overview of Key Steps in
Respiration. This figure will be repeated, in reduced and simplified form as Navigator icons in key figures throughout this chapter as we move from one topic to the next.

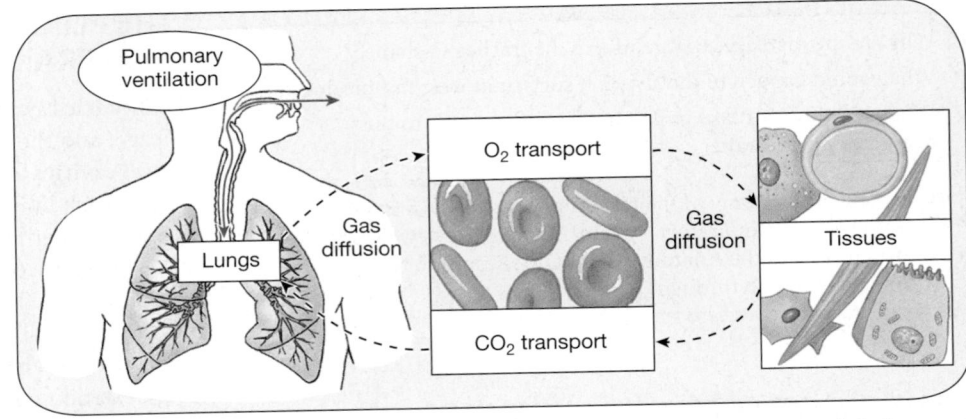

in the exchange of oxygen and carbon dioxide between the body's interstitial fluids and the external environment. The goal of external respiration, and the primary function of the respiratory system, is to meet the respiratory demands of cells. **Internal respiration** is the absorption of oxygen and the release of carbon dioxide by those cells. We shall consider the biochemical pathways responsible for oxygen consumption and for the generation of carbon dioxide by mitochondria, pathways known collectively as *cellular respiration*, in Chapter 25.

Our discussion focuses on three integrated steps involved in external respiration (Figure 23–13●):

1. *Pulmonary ventilation*, or breathing, which involves the physical movement of air into and out of the lungs.
2. *Gas diffusion* across the respiratory membrane between alveolar air spaces and alveolar capillaries and across capillary walls between blood and other tissues.
3. *The transport of oxygen and carbon dioxide* between alveolar capillaries and capillary beds in other tissues.

Abnormalities affecting any one of the steps involved in external respiration will ultimately affect the gas concentrations of interstitial fluids and thereby cellular activities as well. If the oxygen content declines, the affected tissues will become starved for oxygen. **Hypoxia**, or low tissue oxygen levels, places severe limits on the metabolic activities of the affected area. For example, the effects of *coronary ischemia* result from chronic hypoxia affecting cardiac muscle cells. ∞ p. 695 If the supply of oxygen is cut off completely, the condition of **anoxia** (a-NOKS-ē-uh; *a*-, without + *ox*-, oxygen) results. Anoxia kills cells very quickly. Much of the damage caused by strokes and heart attacks is the result of localized anoxia.

23–7 PULMONARY VENTILATION

Objectives

- Summarize the physical principles governing the movement of air into the lungs.
- Describe the origins and actions of the respiratory muscles responsible for respiratory movements.

Pulmonary ventilation is the physical movement of air into and out of the respiratory tract. The primary function of pulmonary ventilation is to maintain adequate *alveolar ventilation*—movement of air into and out of the alveoli. Alveolar ventilation prevents the buildup of carbon dioxide in the alveoli and ensures a continuous supply of oxygen that keeps pace with absorption by the bloodstream.

☐ THE MOVEMENT OF AIR

To understand this mechanical process, we need to look at some basic physical principles governing the movement of air. One of the most basic is that our bodies and everything around us are compressed by the weight of Earth's atmosphere. Although we are seldom reminded of the fact, this **atmospheric pressure** has several important physiological effects. For example, air moves into and out of the respiratory tract as the air pressure in the lungs cycles between below atmospheric pressure and above atmospheric pressure.

Gas Pressure and Volume (Boyle's Law)

The primary differences between liquids and gases reflect the interactions among individual molecules. Although the molecules in a liquid are in constant motion, they are held closely together by weak interactions, such as the hydrogen bonding between adjacent water molecules. ∞ p. 37 Yet, because the electrons of adjacent atoms tend to repel one another, liquids tend to resist compression. If you squeeze a balloon filled with water, it will distort into a different shape, but the volume of the new shape will be the same as that of the original.

In a gas, such as air, the molecules bounce around as independent entities. At normal atmospheric pressures, gas molecules are much farther apart than the molecules in a liquid, so the density of air is relatively low. The forces acting between gas molecules are minimal (the molecules are too far apart for weak interactions to occur), so an applied pressure can push them closer together. Consider a sealed container of air at atmospheric pressure. The pressure exerted by the enclosed gas results from the collision of gas molecules with the walls of the container. The greater the number of collisions, the higher is the pressure.

You can change the gas pressure within a sealed container by changing the volume of the container, thereby giving the gas molecules more or less room in which to bounce around. If you decrease the volume of the container, the collisions will occur more frequently over a given period, elevating the pressure of the gas (Figure 23–14a●). If you increase the volume, fewer collisions will occur per unit time, because it will take longer for a gas molecule to travel from one wall to another. As a result, the gas pressure inside the container will decline (Figure 23–14b●).

An inverse relationship thus exists between the pressure (P) and volume (V) of a gas in a closed container: Gas pressure is inversely proportional to volume. That is, *if you decrease the volume of a gas, its pressure will rise; if you increase the volume of a gas, its pressure will fall*. In particular, the relationship between pressure and volume is reciprocal: If you double the external pressure on a flexible container, its volume will drop by half; if you reduce the external pressure by half, the volume of the container will double. This relationship, $P = 1/V$, first recognized by Robert Boyle in the 1600s, is called **Boyle's law**. [AM] Boyle's Law and Air Overexpansion Syndrome

Pressure and Airflow to the Lungs

Air will flow from an area of higher pressure to an area of lower pressure. This tendency for directed airflow, plus the pressure–volume relationship of Boyle's law, provides the basis for pulmonary ventilation. A single respiratory cycle consists of an *inspiration*, or inhalation, and an *expiration*, or exhalation. Inhalation and exhalation involve changes in the volume of the lungs. These changes create pressure gradients that move air into or out of the respiratory tract.

Each lung lies within a pleural cavity. The parietal and visceral pleurae are separated by only a thin film of pleural fluid. The two membranes can slide across one another, but they are held together by that fluid film. You encounter the same principle when you set a wet glass on a smooth tabletop. You can slide the glass easily, but when you try to lift it, you experience considerable resistance. As you pull the glass away from the tabletop, you create a powerful suction. The only way to overcome it is to tilt the glass so that air is pulled between the glass and the table, breaking the fluid bond.

A comparable fluid bond exists between the parietal pleura and the visceral pleura covering the lungs. As a result, the surface of each lung sticks to the inner wall of the chest and the superior surface of the diaphragm. Movements of the chest wall or diaphragm thus directly affect the volume of the lungs. The basic principle is shown in Figure 23–15a●. The volume of the thoracic cavity changes when the diaphragm changes position or the rib cage moves:

- The diaphragm forms the floor of the thoracic cavity. The relaxed diaphragm has the shape of a dome that projects superiorly into the thoracic cavity. When the diaphragm contracts, it tenses and moves inferiorly. This movement increases the volume of the thoracic cavity and exerts pressure on the contents of the abdominopelvic cavity. When the diaphragm relaxes, it returns to its original position, and the volume of the thoracic cavity decreases.

- Owing to the nature of the articulations between the ribs and the vertebrae, superior movement of the rib cage increases the depth and width of the thoracic cavity. Inferior movement of the rib cage reverses the process and reduces the volume of the thoracic cavity.

At the start of a breath, pressures inside and outside the thoracic cavity are identical, and no air moves into or out of the lungs (Figure 23–15b●). When the thoracic cavity enlarges, the pleural cavities and lungs expand to fill the additional space (Figure 23–15c●). This increase in volume lowers the pressure inside the lungs. Air now enters the respiratory passageways, because the pressure inside the lungs (P_{inside}) is lower than atmospheric pressure ($P_{outside}$). Air continues to enter the lungs until their volume stops increasing and the internal pressure is the same as that outside. When the thoracic cavity decreases in volume, pressures rise inside the lungs, forcing air out of the respiratory tract (Figure 23–15d●).

Compliance

The **compliance** of the lungs is an indication of their expandability. The lower the compliance, the greater is the force required to fill and empty the lungs. Factors affecting compliance include the following:

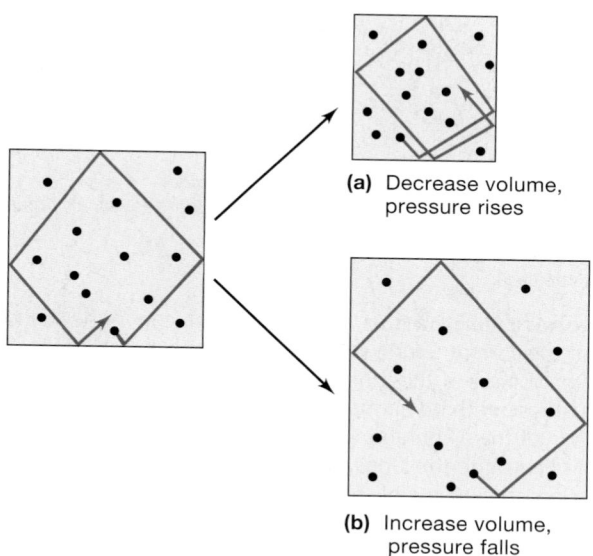

(a) Decrease volume, pressure rises

(b) Increase volume, pressure falls

●**FIGURE 23–14**
Respiratory Pressure and Volume Relationships. Gas molecules in a sealed container bounce off the walls and off one another, traveling the distance indicated in a given period of time. **(a)** If the volume decreases, each molecule will travel the same distance in that same period, but will strike the walls more frequently. The pressure exerted by the gas increases. **(b)** If the volume of the container increases, each molecule will strike the walls less often, lowering the pressure in the container.

- **The Connective-Tissue Structure of the Lungs.** The loss of supporting tissues resulting from alveolar damage, as in *emphysema*, increases compliance (p. 866).
- **The Level of Surfactant Production.** On exhalation, the collapse of alveoli as a result of inadequate surfactant, as in respiratory distress syndrome, reduces the lungs' compliance (p. 841).
- **The Mobility of the Thoracic Cage.** Arthritis or other skeletal disorders that affect the articulations of the ribs or spinal column will also reduce compliance.

At rest, the muscular activity involved in pulmonary ventilation accounts for 3–5 percent of the resting energy demand. If compliance is reduced, that figure climbs dramatically, and an individual may become exhausted simply trying to continue breathing.

☐ PRESSURE CHANGES DURING INHALATION AND EXHALATION

To understand the mechanics of respiration and the principles of gas exchange, we must know the pressures recorded inside and outside the respiratory tract. We can report pressure readings in several ways (Table 23–1); in this text, we shall use millimeters of mercury (mm Hg), as we did for blood pressure. Atmospheric pressure is also measured in *atmospheres*; one atmosphere of pressure (*1 atm*) is equivalent to 760 mm Hg.

The Intrapulmonary Pressure

The direction of airflow is determined by the relationship between atmospheric pressure and intrapulmonary pressure. **Intrapulmonary** (in-tra-PUL-mo-ner-ē) **pressure**, or **intra-alveolar** (in-tra-al-VĒ-ō-lar) **pressure**—pressure measured inside the respiratory tract, at the alveoli.

When you are relaxed and breathing quietly, the pressure difference between atmospheric and intrapulmonary pressures is relatively small. On inhalation, your lungs expand, and the intrapulmonary pressure drops to about 759 mm Hg. Because the intrapulmonary pressure is 1 mm Hg below atmospheric pressure, it is generally reported as −1 mm Hg. On exhalation, your lungs recoil and intrapulmonary pressure rises to 761 mm Hg, or +1 mm Hg (Figure 23–16a●).

The size of the pressure gradient increases when you breathe heavily. When a trained athlete breathes at maximum capacity, the pressure differentials can reach −30 mm Hg during inhala-

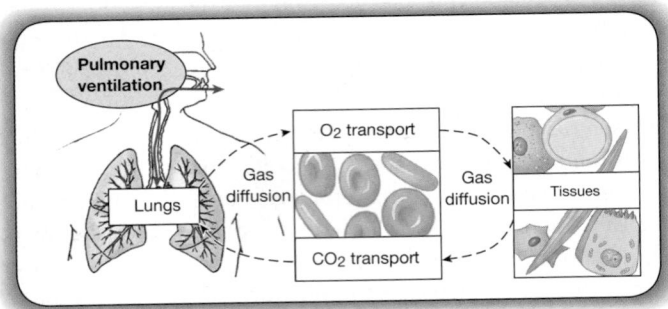

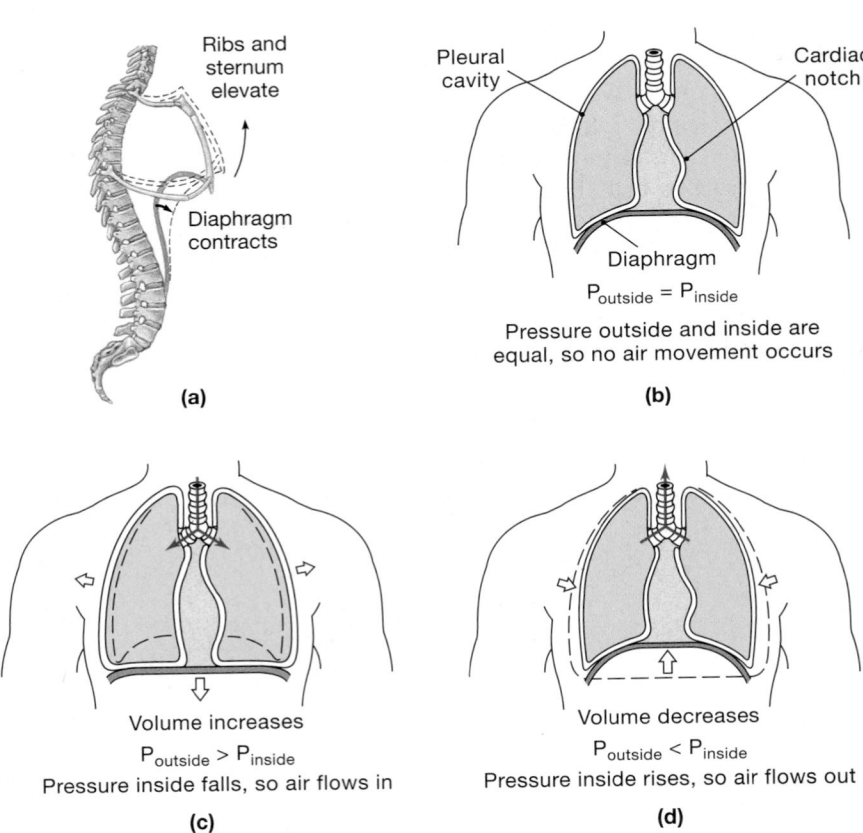

●**FIGURE 23–15**
Mechanisms of Pulmonary Ventilation. The Navigator icon in the shadow box highlights the topic of the current figure. **(a)** As the ribs are elevated or the diaphragm is depressed, the volume of the thoracic cavity increases. **(b)** An anterior view with the diaphragm at rest, with no air movement. **(c)** *Inhalation*: Elevation of the rib cage and contraction of the diaphragm increase the size of the thoracic cavity. Pressure decreases, and air flows into the lungs. **(d)** *Exhalation*: When the rib cage returns to its original position, the volume of the thoracic cavity decreases. Pressure rises, and air moves out of the lungs.

tion and +100 mm Hg if the individual is straining with the glottis kept closed. This is one reason you are told to exhale while lifting weights; exhaling keeps your intrapulmonary pressures and peritoneal pressure from climbing so high that an alveolar rupture or hernia could occur. Exhaling less forcefully against a closed glottis, a procedure known as the *Valsalva maneuver*, caus-

TABLE 23–1 THE FOUR MOST COMMON METHODS OF REPORTING GAS PRESSURES

millimeters of mercury (mm Hg): This is the most common method of reporting blood pressure and gas pressures. Normal atmospheric pressure is approximately 760 mm Hg.

torr: This unit of measurement is preferred by many respiratory therapists; it is also commonly used in Europe and in some technical journals. One torr is equivalent to 1 mm Hg; in other words, normal atmospheric pressure is equal to 760 torr.

centimeters of water (cm H_2O): In a hospital setting, anesthetic gas pressures and oxygen pressures are commonly reported in centimeters of water. One cm H_2O is equivalent to 0.735 mm Hg; normal atmospheric pressure is 1033.6 cm H_2O.

pounds per square inch (psi): Pressures in compressed gas cylinders and other industrial applications are generally reported in psi. Normal atmospheric pressure at sea level is approximately 15 psi.

es reflexive changes in blood pressure and cardiac output due to compression of the aorta and venae cavae. When internal pressures rise, the venae cavae collapse and the venous return decreases. The resulting fall in cardiac output and blood pressure stimulates the aortic and carotid baroreceptors, causing a reflexive increase in heart rate and peripheral vasoconstriction. ∞ pp. 742-743 When the glottis opens and pressures return to normal, venous return and cardiac output rise. Because vasoconstriction has occurred, blood pressure rises sharply, inhibiting the baroreceptors. As a result, cardiac output, heart rate, and blood pressure quickly return to normal levels. The Valsalva maneuver is thus a simple way to check the cardiovascular responses to changes in arterial pressure and venous return.

The Intrapleural Pressure

Intrapleural pressure is the pressure measured in the space between the parietal and visceral pleurae. Intrapleural pressure averages about −4 mm Hg (Figure 23–16b●), but can reach −18 mm Hg during a powerful inhalation. This pressure is below atmospheric pressure, due to the relationship between the lungs and the body wall. Earlier, we noted that the lungs are highly elastic. In fact, they would collapse to about 5 percent of their normal resting volume if the elastic fibers could recoil completely. The elastic fibers cannot recoil significantly, however, because they are not strong enough to overcome the fluid bond between the parietal and visceral pleurae.

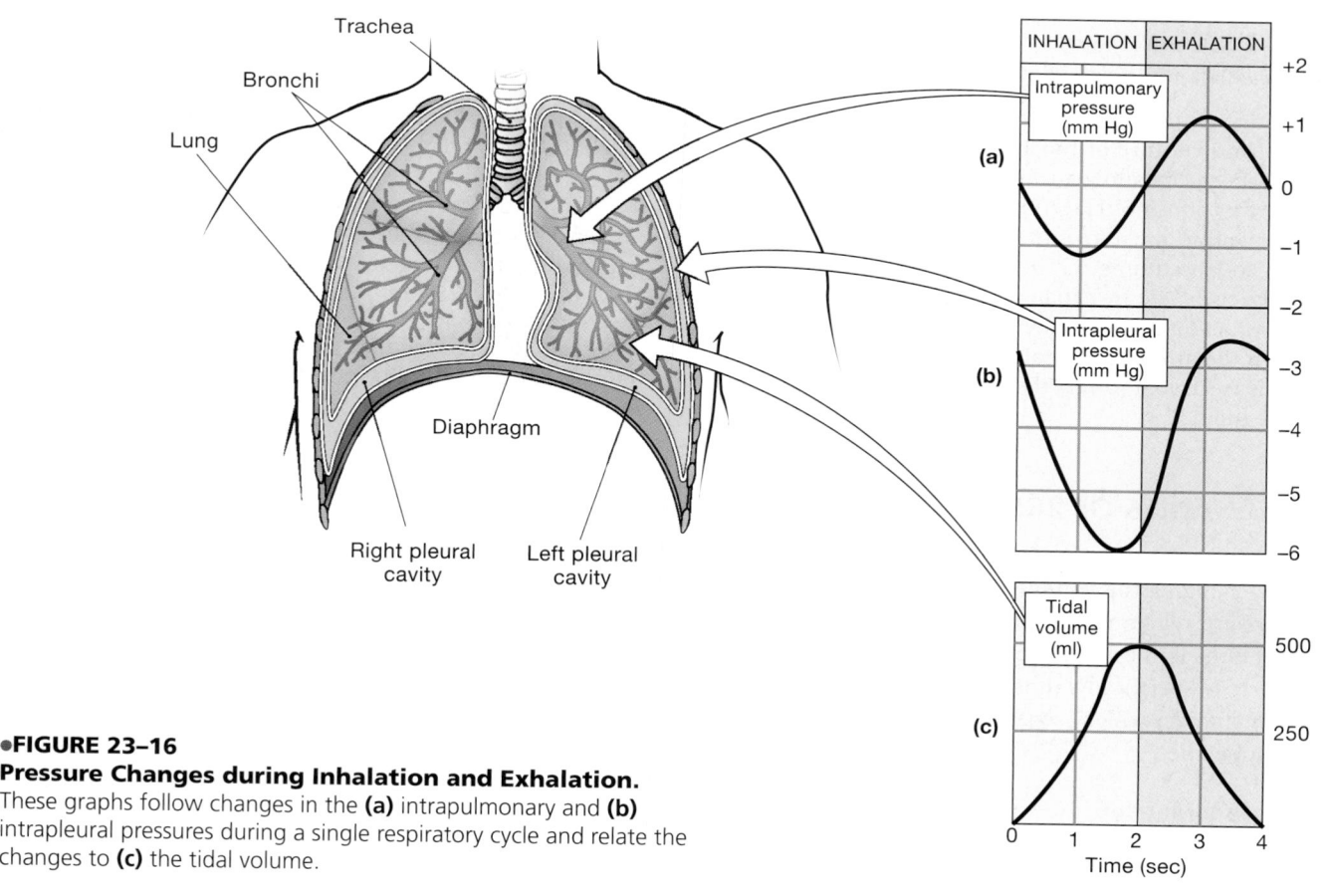

●FIGURE 23–16
Pressure Changes during Inhalation and Exhalation.
These graphs follow changes in the **(a)** intrapulmonary and **(b)** intrapleural pressures during a single respiratory cycle and relate the changes to **(c)** the tidal volume.

The elastic fibers continuously oppose that fluid bond and pull the lungs away from the chest wall and diaphragm, lowering the intrapleural pressure slightly. Because the elastic fibers remain stretched even after a full exhalation, intrapleural pressure remains below atmospheric pressure throughout normal cycles of inhalation and exhalation. The cyclical changes in the intrapleural pressure are responsible for the *respiratory pump*—the mechanism that assists the venous return to the heart. ∞ p. 740

The Respiratory Cycle

A **respiratory cycle** is a single cycle of inhalation and exhalation. The **tidal volume** is the amount of air you move into or out of your lungs during a single respiratory cycle. The graphs in Figure 23–16a,b● follow the intrapulmonary and intrapleural pressures during a single respiratory cycle of an individual at rest. The tidal volume is graphed in Figure 23–16c●.

At the start of the cycle, the intrapulmonary and atmospheric pressures are equal, and no air is moving. Inhalation begins with the fall of intrapleural pressure that accompanies the expansion of the thoracic cavity. This pressure gradually falls to approximately −6 mm Hg. Over the period, intrapulmonary pressure drops to just under −1 mm Hg; it then begins to rise as air flows into the lungs. When exhalation begins, intrapleural and intrapulmonary pressures rise rapidly, forcing air out of the lungs. At the end of exhalation, air movement again ceases when the pressure difference between intrapulmonary and atmospheric pressures has been eliminated. The amount of air moved into the lungs during inhalation is equal to the amount moved out of the lungs during exhalation. That amount is the tidal volume.

An injury to the chest wall that penetrates the parietal pleura or a rupture of the alveoli that breaks through the visceral pleura can allow air into the pleural cavity. This **pneumothorax** (noo-mō-THO-raks) breaks the fluid bond between the pleurae and allows the elastic fibers to recoil, resulting in a "collapsed lung," or **atelectasis** (at-e-LEK-ta-sis; *ateles*, imperfect + *ektasis*, expansion). The treatment for a collapsed lung involves the removal of as much of the air as possible from the affected pleural cavity before the opening is sealed. This treatment lowers the intrapleural pressure and reinflates the lung.

■ THE MECHANICS OF BREATHING

As we have just seen, you move air into and out of the respiratory system by changing the volume of the lungs. Those changes alter the pressure relationships, producing air movement. The changes of volume in the lungs occur through the contraction of skeletal muscles—specifically, those that insert on the rib cage and the *diaphragm*, which separates the thoracic and abdominopelvic cavities.

The Respiratory Muscles

In Chapter 11, we introduced the skeletal muscles involved in respiratory movements. Of those, the most important are the *diaphragm* and the *external intercostal muscles*. ∞ pp. 352, 358, and 359 These muscles are active during normal breathing at rest. The **accessory respiratory muscles** become active when the depth and frequency of respiration must be increased markedly. These muscles include the *internal intercostal, sternocleidomastoid, serratus anterior, pectoralis minor, scalene, transversus thoracis, transversus abdominis, external* and *internal oblique,* and *rectus abdominis muscles* (Figure 23–17●). ∞ pp. 350, 352, and 357-8

MUSCLES USED IN INHALATION Inhalation is an active process involving the contraction of one or more of these muscles:

- The contraction of the diaphragm increases the volume of the thoracic cavity by tensing and flattening its floor, and this increase draws air into the lungs. Diaphragmatic contraction is responsible for roughly 75 percent of the air movement in normal breathing at rest.

- The external intercostal muscles assist in inhalation by elevating the ribs. This action contributes roughly 25 percent to the volume of air in the lungs.

- Accessory muscles, including the sternocleidomastoid, serratus anterior, pectoralis minor, and scalene muscles, can assist the external intercostal muscles in elevating the ribs. These muscles increase the speed and amount of rib movement.

MUSCLES USED IN EXHALATION Exhalation is either passive or active, depending on the level of respiratory activity. When exhalation is active, it may involve one or more of the following muscles:

- The internal intercostal and transversus thoracis muscles depress the ribs and reduce the width and depth of the thoracic cavity.

- The abdominal muscles, including the external and internal oblique, transversus abdominis, and rectus abdominis muscles, can assist the internal intercostal muscles in exhalation by compressing the abdomen and forcing the diaphragm upward.

Artificial respiration is a technique used to provide air to an individual whose respiratory muscles are no longer functioning. In **mouth-to-mouth resuscitation**, a rescuer provides ventilation by exhaling into the mouth, or mouth and nose, of the victim. After each breath, contact is broken to permit passive exhalation by the victim. Air provided in this way contains adequate oxygen to meet the needs of the victim. If the victim's cardiovascular system is nonfunctional as well, a technique called *cardiopulmonary resuscitation (CPR)* is required to maintain adequate blood flow and tissue oxygenation. **AM** CPR

Modes of Breathing

The respiratory muscles are used in various combinations, depending on the volume of air that must be moved into or out of the system. Respiratory movements are usually classified as *quiet breathing* or *forced breathing* by the pattern of muscle activity in the course of a single respiratory cycle.

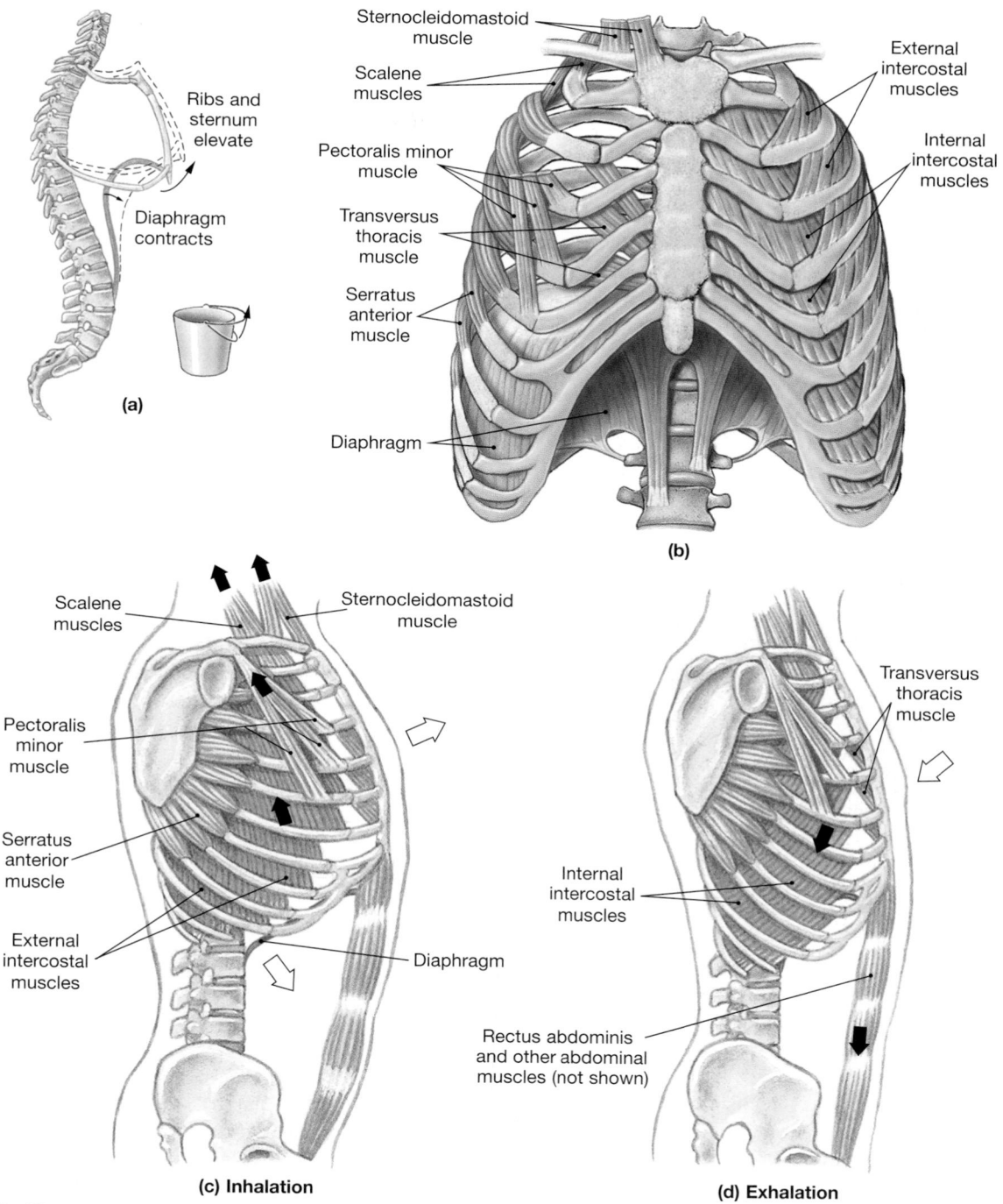

(a) Ribs and sternum elevate

Diaphragm contracts

(b)

Sternocleidomastoid muscle

Scalene muscles

External intercostal muscles

Internal intercostal muscles

Pectoralis minor muscle

Transversus thoracis muscle

Serratus anterior muscle

Diaphragm

Scalene muscles

Sternocleidomastoid muscle

Pectoralis minor muscle

Serratus anterior muscle

External intercostal muscles

Diaphragm

Transversus thoracis muscle

Internal intercostal muscles

Rectus abdominis and other abdominal muscles (not shown)

(c) Inhalation

(d) Exhalation

●**FIGURE 23–17**
The Respiratory Muscles. (a) Movements of the ribs and diaphragm that increase the volume of the thoracic cavity. Diaphragmatic movements were also illustrated in Figure 23–15. (b) Anterior view at rest, with no air movement showing the primary and accessory respiratory muscles. (c) A lateral view during inhalation, showing the muscles that elevate the ribs. (d) A lateral view during exhalation, showing the muscles that depress the ribs. The abdominal muscles that assist in exhalation are represented by a single muscle (the rectus abdominis).

QUIET BREATHING In **quiet breathing**, or **eupnea** (ŪP-nē-uh), inhalation involves muscular contractions, but exhalation is a passive process. Inhalation usually involves the contraction of both the diaphragm and the external intercostal muscles. The relative contributions of these muscles can vary:

• During **diaphragmatic breathing**, or **deep breathing**, contraction of the diaphragm provides the necessary change in thoracic volume. Air is drawn into the lungs as the diaphragm contracts and is exhaled passively when the diaphragm relaxes.

- In **costal breathing**, or **shallow breathing**, the thoracic volume changes because the rib cage alters its shape. Inhalation occurs when contractions of the external intercostal muscles elevate the ribs and enlarge the thoracic cavity. Exhalation occurs passively when these muscles relax.

During quiet breathing, expansion of the lungs stretches their elastic fibers. In addition, elevation of the rib cage stretches opposing skeletal muscles and elastic fibers in the connective tissues of the body wall. When the muscles of inhalation relax, these elastic components recoil, returning the diaphragm, the rib cage, or both to their original positions. This phenomenon is called **elastic rebound**.

Diaphragmatic breathing typically occurs at minimal levels of activity. As increased volumes of air are required, inspiratory movements become larger and the contribution of rib movement increases. Even when you are at rest, costal breathing can predominate when abdominal pressures, fluids, or masses restrict diaphragmatic movements. For example, pregnant women increasingly rely on costal breathing as the uterus enlarges and pushes the abdominal viscera against the diaphragm.

FORCED BREATHING **Forced breathing**, or **hyperpnea** (hī-perp-Nē-uh), involves active inspiratory and expiratory movements. Forced breathing calls on the accessory muscles to assist with inhalation, and exhalation involves contraction of the internal intercostal muscles. At absolute maximum levels of forced breathing, the abdominal muscles are used in exhalation. Their contraction compresses the abdominal contents, pushing them up against the diaphragm and further reducing the volume of the thoracic cavity.

☐ RESPIRATORY RATES AND VOLUMES

Your respiratory system is extremely adaptable. You can be breathing slowly and quietly one moment, rapidly and deeply the next. The respiratory system adapts to meet the oxygen demands of the body by varying the number of breaths per minute and the amount of air moved per breath. When you are exercising at peak levels, the amount of air moving into and out of the respiratory tract can be 50 times the amount moved at rest.

Respiratory Rate

As you read this, you are probably breathing quietly, with a low respiratory rate. Your **respiratory rate** is the number of breaths you take each minute. The normal respiratory rate of a resting adult ranges from 12 to 18 breaths each minute, roughly one for every four heartbeats. Children breathe more rapidly, at rates of about 18–20 breaths per minute.

The Respiratory Minute Volume

We can calculate the amount of air moved each minute, symbolized $\dot{V}_E$, by multiplying the respiratory rate f by the tidal volume V_T. This value is called the **respiratory minute volume**. The tidal

volume at rest varies from individual to individual, but it averages around 500 ml per breath. Therefore, the respiratory minute volume at rest, 12 breaths per minute, is approximately 6 liters per minute:

$$\dot{V}_E = f \times V_T$$

$$\left(\begin{array}{c}\text{volume of air moved}\\\text{each minute}\end{array}\right) = \left(\begin{array}{c}\text{breaths per}\\\text{minute}\end{array}\right) \times \left(\begin{array}{c}\text{tidal}\\\text{volume}\end{array}\right)$$

$$= 12 \times 500 \text{ ml per minute}$$
$$= 6000 \text{ ml per minute}$$
$$= 6.0 \text{ liters per minute}$$

Alveolar Ventilation

The respiratory minute volume measures pulmonary ventilation and provides an indication of how much air is moving into and out of the respiratory tract. However, only some of the inhaled air reaches the alveolar exchange surfaces. A typical inhalation pulls about 500 ml of air into your respiratory system. The first 350 ml inhaled travels along the conducting passageways and enters the alveolar spaces. The last 150 ml of inhaled air never gets farther than the conducting passageways and does not participate in gas exchange with blood. The volume of air in the conducting passages is known as the **anatomic dead space**, denoted V_D. **Alveolar ventilation**, symbolized $\dot{V}_A$, is the amount of air reaching the alveoli each minute. The alveolar ventilation is less than the respiratory minute volume, because some of the air never reaches the alveoli, but remains in the dead space of the lungs. We can calculate alveolar ventilation by subtracting the dead space from the tidal volume:

$$\dot{V}_A = f \times (V_T - V_D)$$

$$\left(\begin{array}{c}\text{alveolar}\\\text{ventilation}\end{array}\right) = \left(\begin{array}{c}\text{breaths}\\\text{per minute}\end{array}\right) \times \left(\begin{array}{cc}\text{tidal} & \text{anatomic}\\\text{volume} - & \text{dead space}\end{array}\right)$$

At rest, alveolar ventilation rates are approximately 4.2 liters per minute (12 × 350 ml). However, the gas arriving in the alveoli is significantly different from that of the surrounding atmosphere, because inhaled air always mixes with "used" air in the conducting passageways (the anatomic dead space) on its way to the exchange surfaces. The air in alveoli thus contains less oxygen and more carbon dioxide than does atmospheric air.

Relationships among V_T, $\dot{V}_E$, and $\dot{V}_A$

The respiratory minute volume can be increased by increasing either the tidal volume or the respiratory rate. Under maximum stimulation, the tidal volume can increase to roughly 4.8 liters. At peak respiratory rates of 40–50 breaths per minute and maximum cycles of inhalation and exhalation, the respiratory minute volume can approach 200 liters (about 55 gal) per minute.

In functional terms, the alveolar ventilation rate is more important than the respiratory minute volume, because it determines the rate of delivery of oxygen to the alveoli. The respiratory rate and the tidal volume together determine the alveolar ventilation rate:

- For a given respiratory rate, increasing the tidal volume will increase the alveolar ventilation rate.
- For a given tidal volume, increasing the respiratory rate will increase the alveolar ventilation rate.

The alveolar ventilation rate can change independently of the respiratory minute volume. In our previous example, the respiratory minute volume at rest was 6 liters and the alveolar ventilation rate was 4.2 l/min. If the respiratory rate rises to 20 breaths per minute, but the tidal volume drops to 300 ml, the respiratory minute volume remains the same ($20 \times 300 = 6000$). However, the alveolar ventilation rate drops to only 3 l/min ($20 \times [300 - 150] = 3000$). Thus, whenever the demand for oxygen increases, both the tidal volume *and* the respiratory rate must be regulated closely. (The mechanisms involved will be the focus of a later section.)

Respiratory Performance and Volume Relationships

Only a small proportion of the air in the lungs is exchanged during a single quiet respiratory cycle; the tidal volume can be increased by inhaling more vigorously and exhaling more completely. We can divide the total volume of the lungs into a series of *volumes* and *capacities* (Figure 23–18●) that can be experimentally determined. The values obtained are useful in diagnosing problems with pulmonary ventilation. Many respiratory volumes and capacities differ by gender, because adult females, on average, have smaller bodies, and thus smaller lung volumes, than do males. Representative values for both genders are indicated in the figure.

Pulmonary volumes include the following:

- The **resting tidal volume** is the amount of air you move into or out of your lungs during a single respiratory cycle under resting conditions. The resting tidal volume averages about 500 ml in both males and females.
- The **expiratory reserve volume (ERV)** is the amount of air that you can voluntarily expel after you have completed a normal, quiet respiratory cycle. As an example, if, with maximum use of the accessory muscles, you can expel an additional 1000 ml of air, your expiratory reserve volume is 1000 ml.
- The **residual volume** is the amount of air that remains in your lungs even after a maximal exhalation—typically, about 1200 ml in males and 1100 ml in females. The **minimal volume**, a component of the residual volume, is the amount of air that would remain in your lungs if they were allowed to collapse. The minimal volume ranges from 30 to 120 ml, but, unlike other volumes, it cannot be measured in a healthy person. The minimal volume and the residual volume are very different, because the fluid bond between the lungs and the chest wall prevents the recoil of the elastic fibers. Some air remains in the lungs, even at minimal volume, because the surfactant coating the alveolar surfaces prevents their collapse.

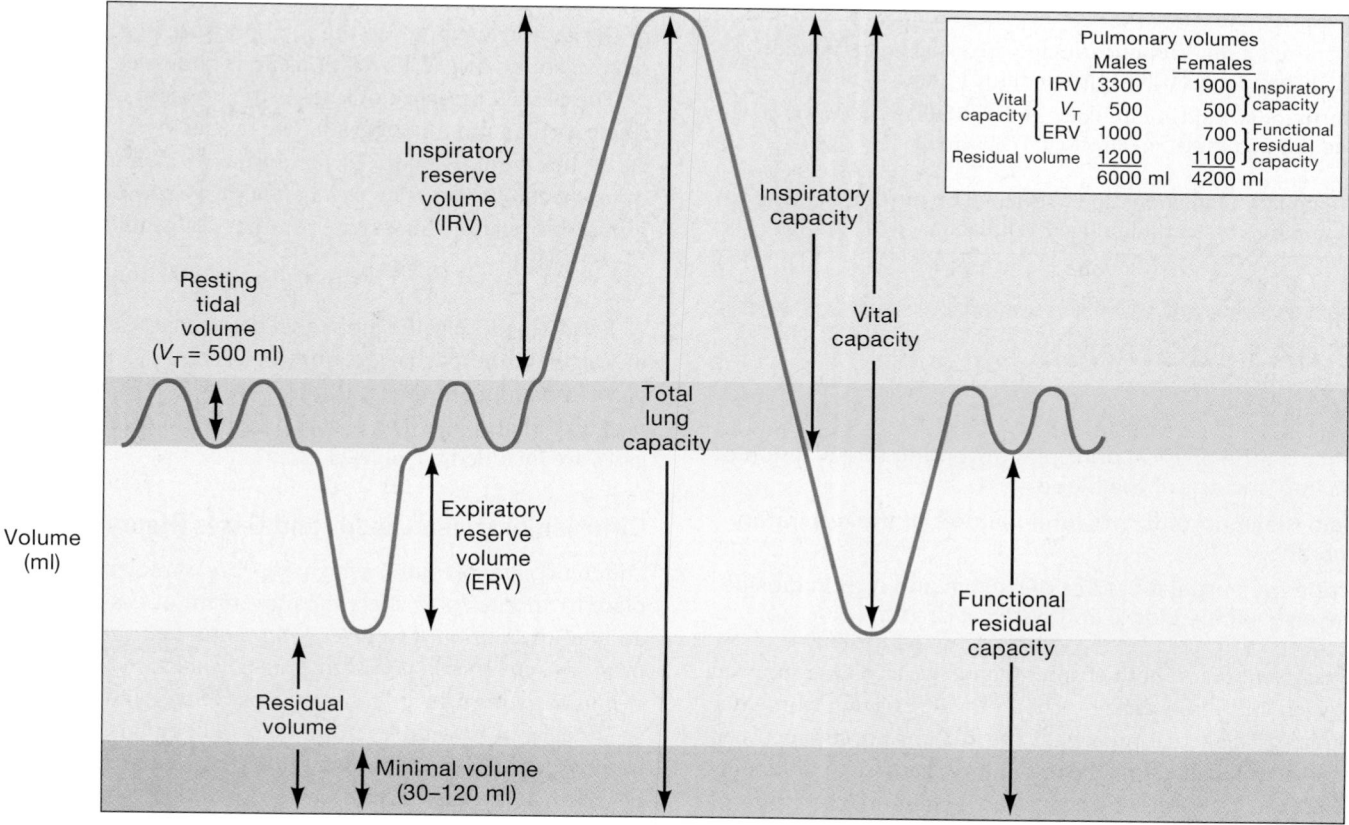

Pulmonary volumes		
	Males	Females
IRV	3300	1900 } Inspiratory capacity
Vital capacity { V_T	500	500 } capacity
ERV	1000	700 } Functional residual
Residual volume	1200	1100 } capacity
	6000 ml	4200 ml

●**FIGURE 23–18**
Respiratory Volumes and Capacities. The graph diagrams the relationships between respiratory volumes and capacities.

- The **inspiratory reserve volume (IRV)** is the amount of air that you can take in over and above the tidal volume. Inspiratory reserve volumes differ significantly by gender, because, on average, the lungs of males are larger than those of females. The inspiratory reserve volume of males averages 3300 ml, compared with 1900 ml in females.

We can determine respiratory capacities by adding the values of various volumes. Examples include the following:

- The **inspiratory capacity** is the amount of air that you can draw into your lungs after you have completed a quiet respiratory cycle. The inspiratory capacity is the sum of the tidal volume and the inspiratory reserve volume.
- The **functional residual capacity (FRC)** is the amount of air remaining in your lungs after you have completed a quiet respiratory cycle. The FRC is the sum of the expiratory reserve volume and the residual volume.
- The **vital capacity** is the maximum amount of air that you can move into or out of your lungs in a single respiratory cycle. The vital capacity is the sum of the expiratory reserve, the tidal volume, and the inspiratory reserve and averages around 4800 ml in males and 3400 ml in females.
- The **total lung capacity** is the total volume of your lungs. The sum of the vital capacity and the residual volume, the total lung capacity averages around 6000 ml in males and 4500 ml in females.

Pulmonary function tests monitor several aspects of respiratory function by measuring rates and volumes of air movement.
AM Pulmonary Function Tests; Asthma

✓ Mark breaks a rib that punctures the chest wall on his left side. What do you expect will happen to his left lung as a result?

✓ In pneumonia, fluid accumulates in the alveoli of the lungs. How would this accumulation affect vital capacity?

Answers start on page Q-1

 Pulmonary ventilation can be reviewed on the **IP CD-ROM:** Respiratory System/Pulmonary Ventilation.

23–8 GAS EXCHANGE

Objectives

- Summarize the physical principles governing the diffusion of gases into and out of the blood.
- Explain the important structural features of the respiratory membrane.
- Describe the partial pressures of oxygen and carbon dioxide in the alveolar air, blood, and systemic circuit.

Pulmonary ventilation both ensures that your alveoli are supplied with oxygen and removes the carbon dioxide arriving from your bloodstream. The actual process of gas exchange occurs between blood and alveolar air across the respiratory membrane. To understand these events, we shall first consider (1) the *partial pressures* of the gases involved and (2) the diffusion of molecules between a gas and a liquid. We can then proceed to discuss the movement of oxygen and carbon dioxide across the respiratory membrane.

☐ THE GAS LAWS

Gases are exchanged between the alveolar air and the blood through diffusion, which occurs in response to concentration gradients. As you saw in Chapter 3, the rate of diffusion varies in response to a variety of factors, such as the size of the concentration gradient and the temperature. ∞ p. 90 The principles that govern the movement and diffusion of gas molecules, such as those in the atmosphere, are relatively straightforward. These principles, known as *gas laws*, have been understood for roughly 250 years. You have already heard about Boyle's law, which determines the direction of air movement. In this section, you will learn about gas laws and other factors that determine the rate of oxygen and carbon dioxide diffusion across the respiratory membrane.

Dalton's Law and Partial Pressures

The air we breathe is not a single gas but a mixture of gases. Nitrogen molecules (N_2) are the most abundant, accounting for about 78.6 percent of the atmospheric gas molecules. Oxygen molecules (O_2), the second most abundant, make up roughly 20.9 percent of the atmospheric content. Most of the remaining 0.5 percent consists of water molecules, with carbon dioxide (CO_2) contributing a mere 0.04 percent.

Atmospheric pressure, 760 mm Hg, represents the combined effects of collisions involving each type of molecule in air. At any moment, 78.6 percent of those collisions will involve nitrogen molecules, 20.9 percent oxygen molecules, and so on. Thus, each of the gases contributes to the total pressure in proportion to its relative abundance. This relationship is known as **Dalton's law**.

The **partial pressure** of a gas is the pressure contributed by a single gas in a mixture of gases. The partial pressure is abbreviated by the prefix P or p. *All the partial pressures added together equal the total pressure exerted by the gas mixture.* For the atmosphere, this relationship can be summarized as follows:

$$P_{N_2} + P_{O_2} + P_{H_2O} + P_{CO_2} = 760 \text{ mm Hg}$$

Because we know the individual percentages in air, we can easily calculate the partial pressure of each gas. For example, the partial pressure of oxygen, P_{O_2}, is 20.9 percent of 760 mm Hg, or roughly 159 mm Hg. The partial pressures for other atmospheric gases are included in Table 23–2.

Diffusion between Liquids and Gases (Henry's Law)

Differences in pressure, which move gas molecules from one place to another, also affect the movement of gas molecules into and out of solution. At a given temperature, *the amount of a particular gas in solution is directly proportional to the partial pressure of that gas*. This principle is known as **Henry's law**.

When a gas under pressure contacts a liquid, the pressure tends to force gas molecules into solution. At a given pressure, the number of dissolved gas molecules will rise until an equilibrium is established. At equilibrium, gas molecules diffuse out of the liquid as quickly as they enter it, so the total number of gas molecules in solution remains constant. If the partial pressure goes up, more gas molecules will go into solution; if the partial pressure goes down, gas molecules will come out of solution (Figure 23–19●).

You see Henry's law in action whenever you open a can of soda. The soda was put into the can under pressure, and the gas (carbon dioxide) is in solution. When you open the can, the pressure falls and the gas molecules begin coming out of solution. The process will theoretically continue until an equilibrium develops between the surrounding air and the gas in solution. In fact, the volume of the can is so small, and the volume of the atmosphere so great, that within a half hour or so, virtually all the carbon dioxide comes out of solution. You are then left with "flat" soda.

The actual *amount* of a gas in solution at a given partial pressure and temperature depends on the solubility of the gas in that particular liquid. Carbon dioxide is highly soluble, oxygen is somewhat less soluble, and nitrogen has very limited solubility in body fluids. The dissolved-gas content is usually reported in milliliters of gas per 100 ml (1 dl) of solution. To see the differ-

ences in relative solubility, we can compare the partial pressure of each gas in the alveoli (Table 23–2) with the amount absorbed by plasma. In a pulmonary vein, plasma generally contains 2.62 ml/dl of dissolved CO_2 (P_{CO_2} = 40 mm Hg), 0.29 ml/dl of dissolved O_2 (P_{O_2} = 100 mm Hg), and 1.25 ml/dl of dissolved N_2 (P_{N_2} = 573 mm Hg).

Decompression sickness is a painful condition that develops when a person is exposed to a sudden drop in atmospheric pressure. Nitrogen is the gas responsible for the problems experienced, owing to its high partial pressure in air. When the pressure drops, nitrogen comes out of solution, forming bubbles like those in a shaken can of soda. The bubbles may form in joint cavities, in the bloodstream, and in the cerebrospinal fluid. Individuals with decompression sickness typically curl up from the pain in affected joints; this reaction accounts for the common name for the condition: *the bends*. Decompression sickness most commonly affects scuba divers, who return to the surface quickly after breathing air under greater-than-normal pressures while submerged. It can also develop in airline passengers subject to sudden losses of cabin pressure. **AM** Decompression Sickness

☐ DIFFUSION AND RESPIRATORY FUNCTION

The gas laws apply to the diffusion of oxygen, carbon dioxide, and nitrogen between a gas and a liquid. We shall now consider how differing partial pressures and solubilities determine the direction and rate of diffusion across the respiratory membrane that separates the air within the alveoli from the blood in alveolar capillaries.

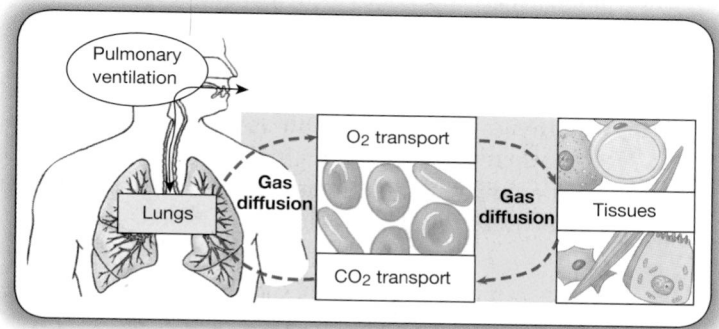

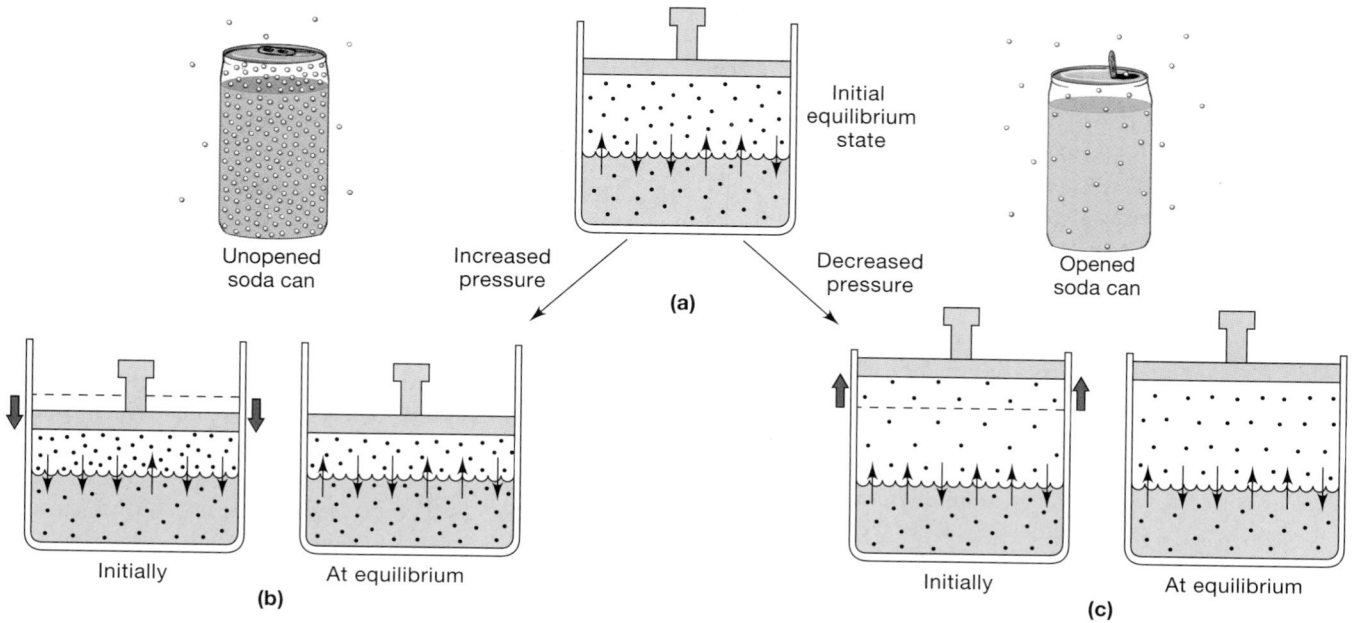

●**FIGURE 23–19**
Henry's Law and the Relationship between Solubility and Pressure. Notice that the dashed box now highlights gas diffusion. **(a)** A solution containing dissolved gas molecules at equilibrium with air under a given pressure. **(b)** Increasing the pressure drives additional gas molecules into solution until a new equilibrium is established. **(c)** When the pressure decreases, some of the dissolved gas molecules leave the solution until a new equilibrium is reached.

TABLE 23–2 PARTIAL PRESSURES (MM HG) AND NORMAL GAS CONCENTRATIONS (%) IN AIR

SOURCE OF SAMPLE	Nitrogen (N_2)	Oxygen (O_2)	Carbon Dioxide (CO_2)	Water Vapor (H_2O)
Inhaled air (dry)	597 (78.6%)	159 (20.9%)	0.3 (0.04%)	3.7 (0.5%)
Alveolar air (saturated)	573 (75.4%)	100 (13.2%)	40 (5.2%)	47 (6.2%)
Exhaled air (saturated)	569 (74.8%)	116 (15.3%)	28 (3.7%)	47 (6.2%)

The Composition of Alveolar Air

As soon as air enters the respiratory tract, its characteristics begin to change. In passing through the nasal cavity, inhaled air becomes warmer and the amount of water vapor increases. Humidification and filtration continue as the air travels through the pharynx, trachea, and bronchial passageways. On reaching the alveoli, the incoming air mixes with air remaining in the alveoli from the previous respiratory cycle. The alveolar air thus contains more carbon dioxide and less oxygen than does atmospheric air.

The last 150 ml of inhaled air never gets farther than the conducting passageways and remains in the anatomic dead space of the lungs. During the subsequent exhalation, the departing alveolar air mixes with air in the dead space, producing yet another mixture that differs from both atmospheric and alveolar samples. The differences in composition between atmospheric (inhaled) and alveolar air are given in Table 23–2.

Efficiency of Diffusion at the Respiratory Membrane

Gas exchange at the respiratory membrane is efficient for the following five reasons:

1. *The Differences in Partial Pressure across the Respiratory Membrane Are Substantial.* This fact is important, because the greater the difference in partial pressure, the faster is the rate of gas diffusion. Conversely, if P_{O_2} in alveoli decreases, the rate of oxygen diffusion into blood will drop. This is why many people feel light headed at altitudes of 3000 m or more; the partial pressure of oxygen in their alveoli has dropped low enough that the rate of oxygen absorption is significantly reduced.

2. *The Distances Involved in Gas Exchange Are Small.* The fusion of capillary and alveolar basal laminae reduces the distance for gas exchange to an average of 0.5 μm. Inflammation of the lung tissue or a buildup of fluid in alveoli increases the diffusion distance and impairs alveolar gas exchange.

3. *The Gases Are Lipid Soluble.* Both oxygen and carbon dioxide diffuse readily through the surfactant layer and the alveolar and endothelial cell membranes.

4. *The Total Surface Area Is Large.* The combined alveolar surface area at peak inhalation may approach 140 m^2 (1506 ft^2). Damage to alveolar surfaces, which occurs in emphysema, reduces the available surface area and the efficiency of gas transfer.

5. *Blood Flow and Airflow Are Coordinated.* This arrangement improves the efficiency of both pulmonary ventilation and pulmonary circulation. For example, blood flow is greatest around alveoli with the highest P_{O_2} values, where oxygen uptake can proceed with maximum efficiency. If the normal blood flow is impaired, as it is in a *pulmonary embolism*, or if the normal air-

flow is interrupted, as it is in various forms of *pulmonary obstruction*, this coordination is lost and respiratory efficiency decreases. **AM** Bronchitis, Emphysema, and COPD

Partial Pressures in Alveolar Air and Alveolar Capillaries

Figure 23–20● illustrates the partial pressures of oxygen and carbon dioxide in the pulmonary and systemic circuits. Blood arriving in the pulmonary arteries has a lower P_{O_2} and a higher P_{CO_2} than does alveolar air. Diffusion between the alveolar mixture and the pulmonary capillaries thus elevates the P_{O_2} of blood while lowering its P_{CO_2}. By the time the blood enters the pulmonary venules, it has reached equilibrium with the alveolar air. Hence, blood departs the alveoli with a P_{O_2} of about 100 mm Hg and a P_{CO_2} of roughly 40 mm Hg (Figure 23–20a●).

Diffusion between alveolar air and blood in the pulmonary capillaries occurs very rapidly. When you are at rest, a red blood cell moves through one of your pulmonary capillaries in about 0.75 second; when you exercise, the passage takes less than 0.3 second. This amount of time is usually sufficient to reach an equilibrium between the alveolar air and the blood.

Partial Pressures in the Systemic Circuit

The oxygenated blood now leaves the alveolar capillaries and returns to the heart, to be discharged into the systemic circuit. As this blood enters the pulmonary veins, it mixes with blood that flowed through capillaries around conducting passageways. Because gas exchange occurs only at alveoli, the blood leaving the conducting passageways carries relatively little oxygen. The partial pressure of oxygen in the pulmonary veins therefore drops to about 95 mm Hg. This is the P_{O_2} in the blood that enters the systemic circuit, and no further changes in partial pressure occur until the blood reaches the peripheral capillaries (Figure 23–20b●).

Normal interstitial fluid has a P_{O_2} of 40 mm Hg. As a result, oxygen diffuses out of the capillaries and carbon dioxide diffuses in, until the capillary partial pressures are the same as those in the adjacent tissues. At a normal tissue P_{O_2}, blood entering the venous system still contains about 75 percent of its total oxygen content.

Blood entering the systemic circuit normally has a P_{CO_2} of 40 mm Hg. Inactive peripheral tissues normally have a P_{CO_2} of about 45 mm Hg. As a result, carbon dioxide diffuses into the blood as oxygen diffuses out (Figure 23–20b●).

Blood samples can be analyzed to determine their concentrations of dissolved gases. The usual tests include the determination of pH, P_{CO_2}, and P_{O_2} in an arterial sample. This analysis can be very helpful in monitoring patients after a heart attack or in chronic respiratory conditions such as obstructive dis-

•FIGURE 23–20
An Overview of Respiratory Processes and Partial Pressures in Respiration. **(a)** Partial pressures and diffusion at the respiratory membrane. **(b)** Partial pressures and diffusion in other tissues.

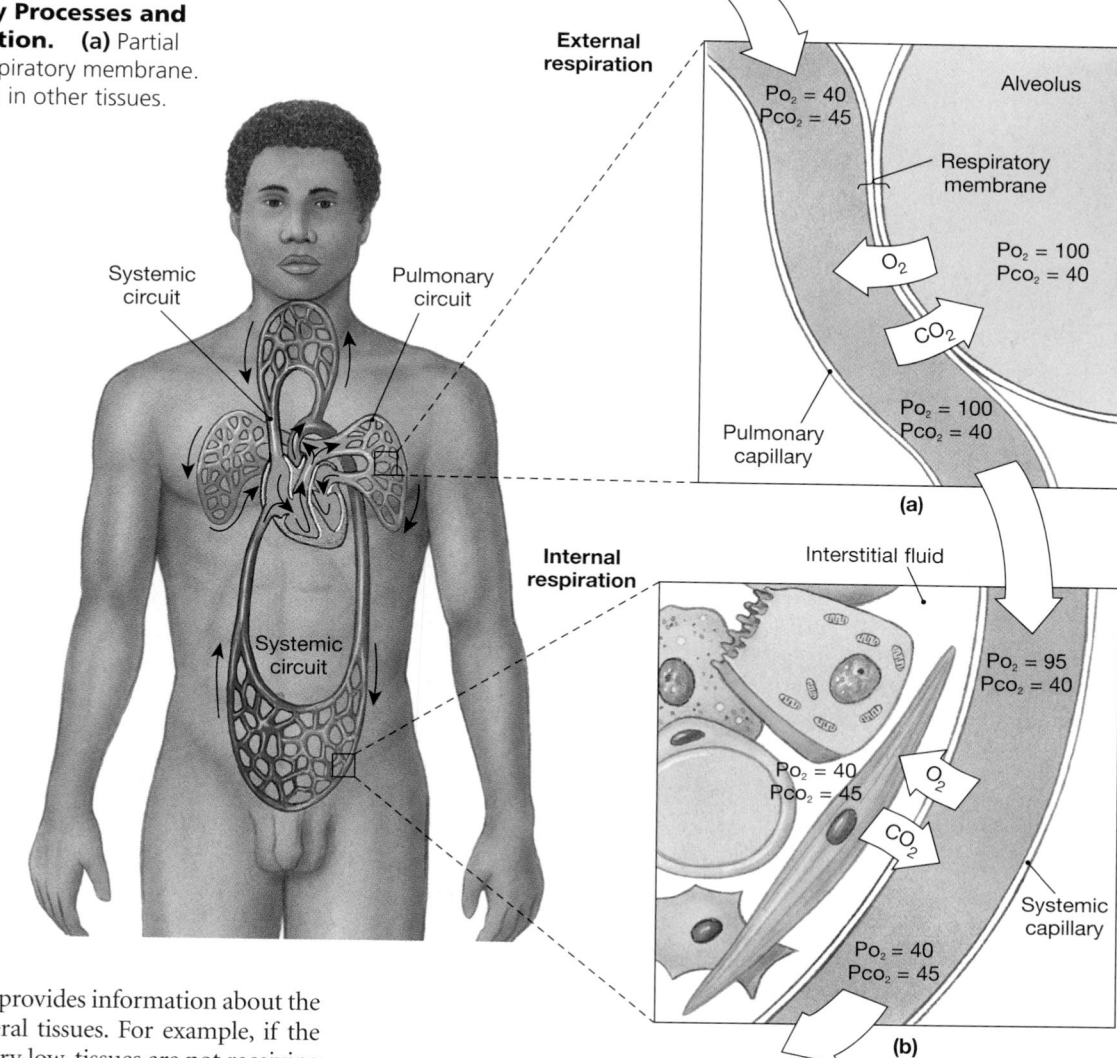

orders or asthma. A blood sample provides information about the degree of oxygenation in peripheral tissues. For example, if the P_{CO_2} is very high and the P_{O_2} is very low, tissues are not receiving adequate oxygen. This condition can be corrected by providing a gas mixture that has a high P_{O_2} (or even pure oxygen, with a P_{O_2} of 760 mm Hg). In addition, blood gas measurements determine the efficiency of gas exchange at the lungs. If the arterial P_{O_2} remains low despite the administration of oxygen, or if the P_{CO_2} is very high, pulmonary exchange problems, such as pulmonary edema, asthma, or pneumonia, must exist.

23–9 GAS PICKUP AND DELIVERY

Objectives

■ Describe how oxygen is picked up, transported, and released in the blood.

■ Discuss the structure and function of hemoglobin.

■ Describe how carbon dioxide is transported in the blood.

Oxygen and carbon dioxide have limited solubilities in blood plasma. For example, at the normal P_{O_2} of alveoli, 100 ml of plasma will absorb only about 0.3 ml of oxygen. The limited solubil-

ities of these gases are a problem, because peripheral tissues need more oxygen and generate more carbon dioxide than the plasma can absorb and transport.

The problem is solved by red blood cells (RBCs), which remove dissolved oxygen and CO_2 molecules from plasma and bind them (in the case of oxygen) or use them to manufacture soluble compounds (in the case of carbon dioxide). Because these reactions remove dissolved gases from blood plasma, gases continue to diffuse into the blood, but never reach equilibrium.

The important thing about these reactions is that they are (1) *temporary* and (2) *completely reversible*. When plasma oxygen or carbon dioxide concentrations are high, the excess molecules are removed by RBCs. When plasma concentrations are falling, the RBCs release their stored reserves.

☐ OXYGEN TRANSPORT

Each 100 ml of blood leaving the alveolar capillaries carries away roughly 20 ml of oxygen. Of this amount, only about 0.3 ml (1.5 percent) consists of oxygen molecules in solution. The rest of the oxygen molecules are bound to *hemoglobin (Hb) molecules*— specifically, to the iron ions in the center of heme units. ∞ p. 656 Recall that the hemoglobin molecule consists of four globular protein subunits, each containing a heme unit. Thus, each hemoglobin molecule can bind four molecules of oxygen, forming oxyhemoglobin. This is a reversible reaction that can be summarized as

$$Hb + O_2 \rightleftharpoons HbO_2.$$

Each red blood cell has approximately 280 million molecules of hemoglobin. Because a hemoglobin molecule contains four heme units, each RBC potentially can carry more than a billion molecules of oxygen.

The percentage of heme units containing bound oxygen at any given moment is called the **hemoglobin saturation**. If all the Hb molecules in the blood are fully loaded with oxygen, saturation is 100 percent. If, on average, each Hb molecule carries two O_2 molecules, saturation is 50 percent.

In Chapter 2, we noted that the shape and functional properties of a protein change in response to changes in its environment. ∞ p. 55 Hemoglobin is no exception: Any changes in shape that occur can affect oxygen binding. Under normal conditions, the most important environmental factors affecting hemoglobin are (1) the P_{O_2} of blood, (2) blood pH, (3) temperature, and (4) ongoing metabolic activity within RBCs.

Hemoglobin and P_{O_2}

An **oxygen–hemoglobin saturation curve**, or *oxygen–hemoglobin dissociation curve*, is a graph that relates the saturation of hemoglobin to the partial pressure of oxygen (Figure 23–21●). The binding and dissociation of oxygen to hemoglobin is a typical reversible reaction. At equilibrium, oxygen molecules bind to heme at the same rate that other oxygen molecules are being released. If you increase the P_{O_2}, you shift the reaction to the right, and more oxygen gets bound to hemoglobin. If you decrease the P_{O_2}, the re-

action shifts to the left, and more oxygen is released by hemoglobin. The graph of this relationship is a curve rather than a straight line, because the shape of the Hb molecule changes slightly each time it binds an oxygen molecule, and this change affects its ability to bind *another* oxygen molecule. In other words, the attachment of the first oxygen molecule makes it easier to bind the second; binding the second promotes binding of the third; and binding of the third enhances binding of the fourth; and so on.

Because each arriving oxygen molecule increases the affinity of hemoglobin for the *next* oxygen molecule, the saturation curve takes the form shown in Figure 23–21●. The slope is gradual until the first oxygen molecule binds to the hemoglobin; then the slope rises rapidly, with a plateau near 100 percent saturation. Over the steep initial slope, a very small decrease in plasma P_{O_2} will result in a large change in the amount of oxygen bound to or released from HbO_2. Because the curve rises quickly, hemoglobin will be more than 90 percent saturated if exposed to an alveolar P_{O_2} above 60 mm Hg. Thus, near-normal oxygen transport can continue despite a decrease in the oxygen content of alveolar air. Without this ability, you could not survive at high altitudes, and conditions significantly reducing pulmonary ventilation would be immediately fatal.

At normal alveolar pressures (P_{O_2} = 100 mm Hg) the hemoglobin saturation is very high (97.5 percent), although complete saturation does not occur until the P_{O_2} reaches excessively high levels (about 250 mm Hg). In functional terms, the maximum saturation is not as important as the ability of hemoglobin to provide oxygen over the normal P_{O_2} range in body tissues. Over that range, from 100 mm Hg at the alveoli to perhaps 15 mm Hg in active tissues, the saturation drops from 97.5 percent to less than 20 percent, and a small change in P_{O_2} makes a big difference in terms of the amount of oxygen bound to hemoglobin.

Note that the relationship between P_{O_2} and hemoglobin saturation remains valid whether the P_{O_2} is rising or falling. *If the* P_{O_2}

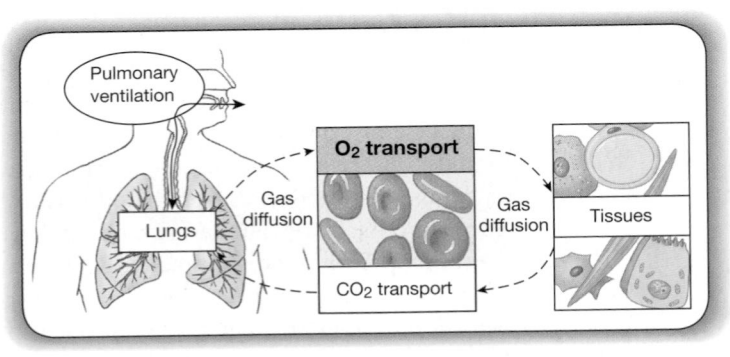

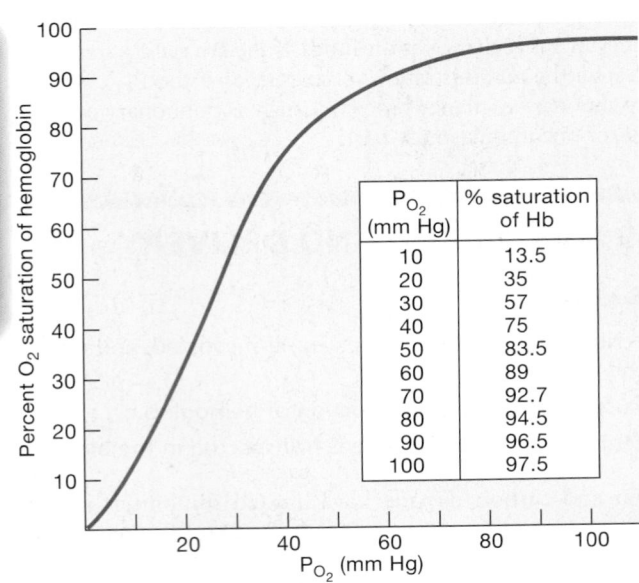

P_{O_2} (mm Hg)	% saturation of Hb
10	13.5
20	35
30	57
40	75
50	83.5
60	89
70	92.7
80	94.5
90	96.5
100	97.5

●**FIGURE 23–21**
The Oxygen–Hemoglobin Saturation Curve. The normal saturation characteristics of hemoglobin at various partial pressures of oxygen.

increases, the saturation goes up and hemoglobin stores oxygen. If the P_{O_2} decreases, hemoglobin releases oxygen into its surroundings. When oxygenated blood arrives in the peripheral capillaries, the blood P_{O_2} declines rapidly as a result of gas exchange with the interstitial fluid. As the P_{O_2} falls, hemoglobin gives up its oxygen.

The relationship between the P_{O_2} and hemoglobin saturation provides a mechanism for automatic regulation of oxygen delivery. Inactive tissues have little demand for oxygen, and the local P_{O_2} is usually about 40 mm Hg. Under these conditions, hemoglobin will not release much oxygen. As it passes through the capillaries, it will go from 97 percent saturation (P_{O_2} = 95 mm Hg) to 75 percent saturation (P_{O_2} = 40 mm Hg). Because it still retains three-quarters of its oxygen, venous blood has a relatively large oxygen reserve. This reserve is important, because it can be mobilized if tissue oxygen demands increase.

Active tissues consume oxygen at an accelerated rate, so the P_{O_2} may drop to 15–20 mm Hg. Hemoglobin passing through these capillaries will then go from 97 percent saturation to about 20 percent saturation. In practical terms, this means that as blood circulates through peripheral capillaries, active tissues will receive 3.5 times as much oxygen as will inactive tissues.

✚ Murder or suicide victims who died in their cars inside a closed garage are popular characters for mystery writers. In real life, entire families are killed each winter by leaky furnaces or space heaters. The cause of death is **carbon monoxide poisoning**. The exhaust of automobiles and other petroleum-burning engines, of oil lamps, and of fuel-fired space heaters contains *carbon monoxide* (CO). Carbon monoxide competes with oxygen molecules for the binding sites on heme units. Unfortunately, the carbon monoxide usually wins, because at very low partial pressures it has a much stronger affinity for hemoglobin than does oxygen. The bond formed between CO and heme is extremely durable, so the attachment of a CO molecule essentially makes that heme unit inactive for respiratory purposes. Carbon monoxide will bind to hemoglobin at very low partial pressures. If CO molecules make up just 0.1 percent of the components of inhaled air, enough hemoglobin will be affected that survival will become impossible without medical assistance. Treatment may include (1) the administration of pure oxygen, because at sufficiently high partial pressures, the oxygen molecules will gradually replace CO at the hemoglobin molecules, and, if necessary, (2) the transfusion of compatible red blood cells.

Hemoglobin and pH

The oxygen–hemoglobin saturation curve in Figure 23–21● was determined in normal blood, with a pH of 7.4 and a temperature of 37°C. In addition to consuming oxygen, active tissues generate acids that lower the pH of the interstitial fluid. When the pH drops, the shape of hemoglobin molecules changes. As a result of this change, the molecules release their oxygen reserves more readily, so the slope of the hemoglobin saturation curve changes (Figure 23–22a●). In other words, the saturation declines. Thus, at a tissue P_{O_2} of 40 mm Hg, hemoglobin molecules release 15 percent more oxygen at a pH of 7.2 than they do at a pH of 7.4. This effect of pH on the hemoglobin saturation curve is called the **Bohr effect**.

Carbon dioxide is the primary compound responsible for the Bohr effect. When CO_2 diffuses into the blood, it rapidly diffuses

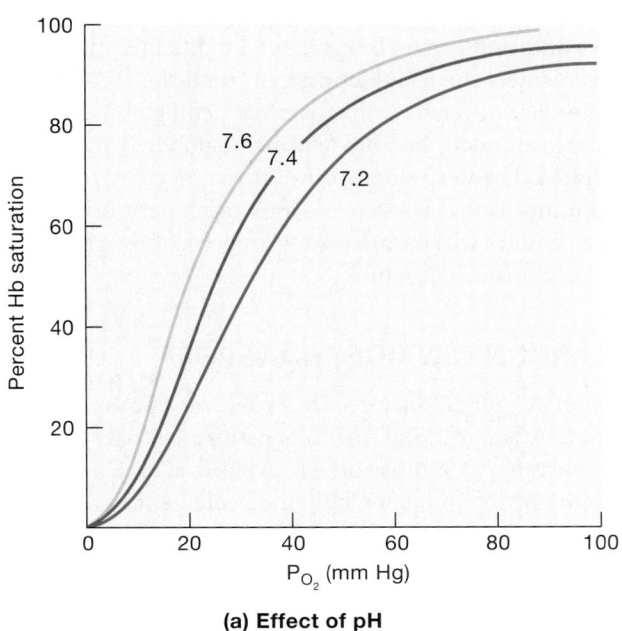

(a) Effect of pH

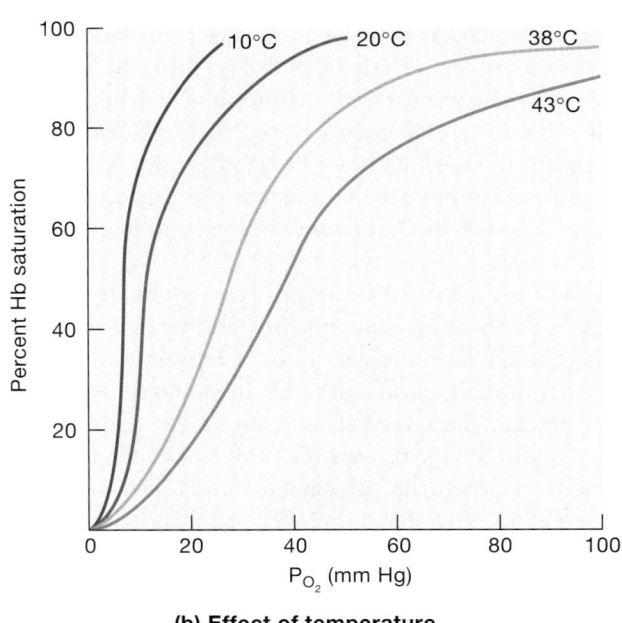

(b) Effect of temperature

●**FIGURE 23–22**
The Effects of pH and Temperature on Hemoglobin Saturation. **(a)** When the pH drops below normal levels, more oxygen is released; the hemoglobin saturation curve shifts to the right. If the pH increases, less oxygen is released; the curve shifts to the left. **(b)** When the temperature rises, the saturation curve shifts to the right.

into red blood cells. There, an enzyme called **carbonic anhydrase** catalyzes the reaction of CO_2 with water molecules:

$$CO_2 + H_2O \underset{\text{carbonic anhydrase}}{\rightleftharpoons} H_2CO_3 \rightleftharpoons H^+ + HCO_3^-$$

The product of this enzymatic reaction, H_2CO_3, is called *carbonic acid*, because it dissociates into a hydrogen ion (H^+) and a bicarbonate ion (HCO_3^-). The rate of carbonic acid formation depends on the amount of carbon dioxide in solution, which, as we noted earlier, depends on the P_{CO_2}. When the P_{CO_2} rises, the reaction proceeds from left to right and the rate of carbonic acid formation accelerates. The hydrogen ions that are generated diffuse out of the RBCs, and the pH of the plasma drops. When the P_{CO_2} declines, the reaction proceeds from right to left; hydrogen ions then diffuse into the RBCs, so the pH of the plasma rises.

Hemoglobin and Temperature

Changes in temperature also affect the slope of the hemoglobin saturation curve (Figure 23–22b●). As the temperature rises, hemoglobin releases more oxygen; as the temperature declines, hemoglobin holds oxygen more tightly. Temperature effects are significant only in active tissues in which large amounts of heat are being generated. For example, active skeletal muscles generate heat, and the heat warms blood that flows through these organs. As the blood warms, the Hb molecules release more oxygen than can be used by the active muscle fibers.

Hemoglobin and BPG

Red blood cells produce adenosine triphosphate (ATP) only by glycolysis, in which, as we saw in Chapter 10, lactic acid is formed. ∞ p. 317 The metabolic pathways involved in glycolysis in an RBC also generate the compound **2,3-bisphosphoglycerate** (biz-fos-fō-GLIS-e-rāt), or **BPG**. Normal RBCs always contain BPG, which has a direct effect on oxygen binding and release. For any partial pressure of oxygen, the higher the concentration of BPG, the more oxygen will be released by the Hb molecules.

The concentration of BPG can be increased by thyroid hormones, growth hormone, epinephrine, androgens, and a high blood pH. These factors improve oxygen delivery to the tissues, because when BPG levels are elevated, hemoglobin will release about 10 percent more oxygen at a given P_{O_2} than it would otherwise. Both BPG synthesis and the Bohr effect improve oxygen delivery when the pH changes: BPG levels rise when the pH increases, and the Bohr effect appears when the pH decreases.

The production of BPG decreases as RBCs age. Thus, the level of BPG can determine how long a blood bank can hold fresh whole blood. When BPG levels get too low, hemoglobin becomes firmly bound to the available oxygen. The blood is then useless for transfusions, because the RBCs will no longer release oxygen to peripheral tissues, even at a disastrously low P_{O_2}.

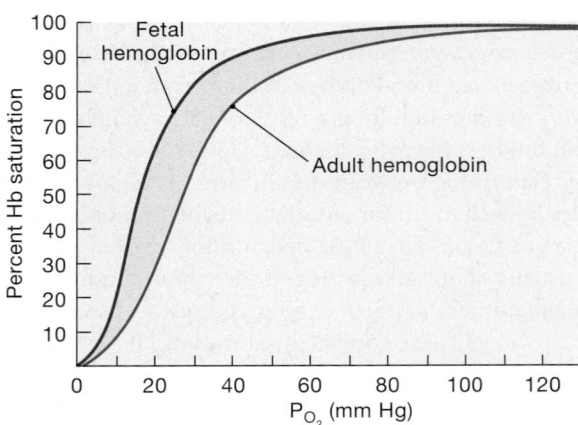

●FIGURE 23–23
A Functional Comparison of Fetal and Adult Hemoglobin

Fetal Hemoglobin

The RBCs of a developing fetus contain **fetal hemoglobin**. The structure of fetal hemoglobin, which differs from that of adult hemoglobin, gives it a much higher affinity for oxygen. At the same P_{O_2}, fetal hemoglobin binds more oxygen than does adult hemoglobin (Figure 23–23●). This trait is important in transferring oxygen across the placenta.

A fetus obtains oxygen from the maternal bloodstream. At the placenta, the maternal blood has a relatively low P_{O_2}, ranging from 35 to 50 mm Hg. If maternal blood arrives at the placenta with a P_{O_2} of 40 mm Hg, hemoglobin saturation will be roughly 75 percent. The fetal blood arriving at the placenta has a P_{O_2} close to 20 mm Hg. However, because fetal hemoglobin has a higher affinity for oxygen, it is still 58 percent saturated.

As diffusion occurs between fetal blood and maternal blood, oxygen enters the fetal bloodstream until the P_{O_2} equilibrates at 30 mm Hg. At this P_{O_2}, the maternal hemoglobin is less than 60 percent saturated, but the fetal hemoglobin is over 80 percent saturated. The steep slope of the saturation curve for fetal hemoglobin means that when fetal RBCs reach peripheral tissues, the Hb molecules will release a large amount of oxygen in response to a very small change in P_{O_2}.

■ CARBON DIOXIDE TRANSPORT

Carbon dioxide is generated by aerobic metabolism in peripheral tissues. After entering the bloodstream, a CO_2 molecule is (1) converted to a molecule of carbonic acid, (2) bound to the protein portion of hemoglobin molecules within red blood cells, or (3) dissolved in plasma. All three are completely reversible reactions. We shall consider the events under way as blood enters peripheral tissues in which the P_{CO_2} is 45 mm Hg.

Carbonic Acid Formation

Most of the carbon dioxide absorbed by blood (roughly 70 percent of the total) is transported as molecules of carbonic acid.

Carbon dioxide is converted to carbonic acid through the activity of the enzyme carbonic anhydrase in RBCs. The carbonic acid molecules immediately dissociate into a hydrogen ion and a bicarbonate ion, as described earlier (p. 858). Hence, we can ignore the intermediate steps in this sequence and summarize the reaction as

$$CO_2 + H_2O \underset{\text{carbonic anhydrase}}{\rightleftharpoons} H^+ + HCO_3^-$$

This reaction is completely reversible. In peripheral capillaries, it proceeds vigorously, tying up large numbers of CO_2 molecules. The reaction continues as carbon dioxide diffuses out of the interstitial fluids.

The hydrogen ions and bicarbonate ions have different fates. Most of the hydrogen ions bind to hemoglobin molecules, forming $Hb \cdot H^+$. The Hb molecules thus function as buffers, tying up the released hydrogen ions before the ions leave the RBCs and affect the plasma pH. The bicarbonate ions move into the surrounding plasma with the aid of a countertransport mechanism that exchanges intracellular bicarbonate ions (HCO_3^-) for extracellular chloride ions (Cl^-). This exchange, which trades one anion for another, does not require ATP. The result is a mass movement of chloride ions into the RBCs, an event known as the **chloride shift**.

Hemoglobin Binding

Roughly 23 percent of the carbon dioxide carried by your blood will be bound to the globular protein portions of the Hb molecules inside RBCs. These CO_2 molecules are attached to exposed amino groups ($-NH_2$) of the Hb molecules. The resulting compound is called **carbaminohemoglobin** (kar-ba-mē-nō-hē-mō-GLŌ-bin), $Hb \cdot CO_2$. The reversible reaction is summarized as follows:

$$CO_2 + HbNH_2 \rightleftharpoons HbNHCOOH$$

This reaction can be abbreviated without the amino groups as

$$CO_2 + Hb \rightleftharpoons Hb \cdot CO_2$$

Plasma Transport

Plasma becomes saturated with carbon dioxide quite rapidly, and only about 7 percent of the carbon dioxide absorbed by peripheral capillaries is transported in the form of dissolved gas molecules. The rest is absorbed by the RBCs for conversion by carbonic anhydrase or storage as carbaminohemoglobin.

Methods of carbon dioxide transport are summarized in Figure 23–24●.

☐ SUMMARY: GAS TRANSPORT

Figure 23–25● summarizes the transportation of oxygen and carbon dioxide in the respiratory and cardiovascular systems. The process is dynamic, capable of varying its responses to meet

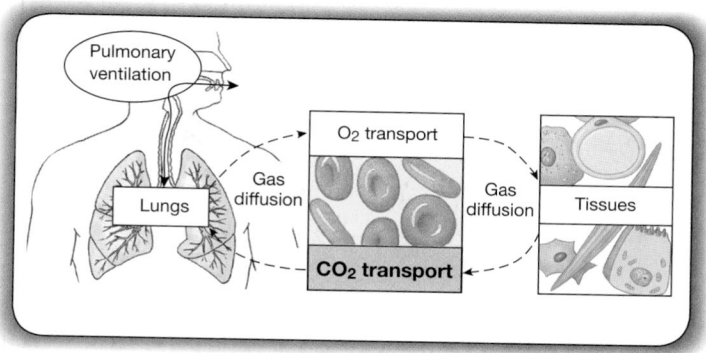

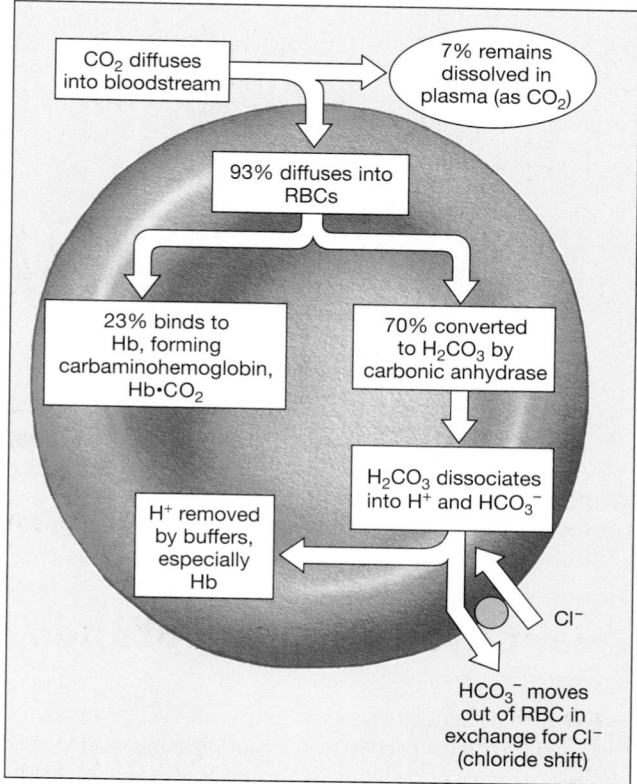

●**FIGURE 23–24**
Carbon Dioxide Transport in Blood

changing circumstances. Some of the responses are automatic and result from the basic chemistry of the transport mechanisms. Other responses require coordinated adjustments in the activities of the cardiovascular and respiratory systems. We will now consider those levels of control and regulation.

✓ As you exercise, hemoglobin releases more oxygen to your active skeletal muscles than it does when the muscles are at rest. Why?

✓ How would blockage of the trachea affect the blood pH?

Answers start on page Q-1

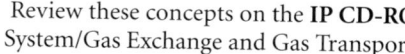 Review these concepts on the **IP CD-ROM**: Respiratory System/Gas Exchange and Gas Transport.

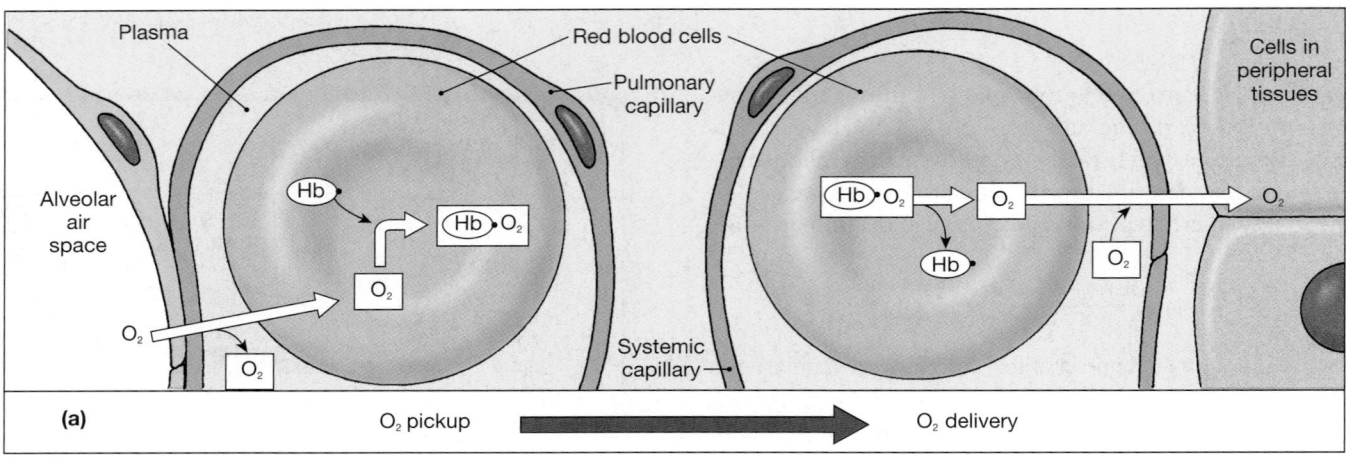

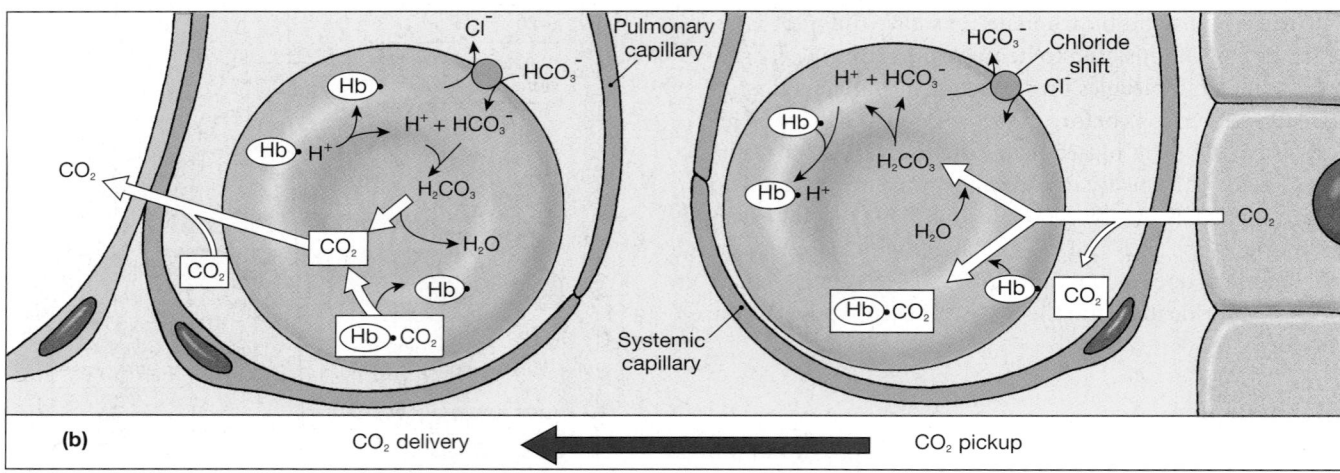

●FIGURE 23–25
A Summary of the Primary Gas Transport Mechanisms. (a) Oxygen transport. (b) Carbon dioxide transport.

Adaptations to High Altitude

Atmospheric pressure decreases with increasing altitude, and so do the partial pressures of the component gases, including oxygen. People living in Denver or Mexico City function normally with alveolar oxygen pressures in the range of 80–90 mm Hg. At higher elevations, the alveolar P_{O_2} continues to decline. At 3300 meters (10,826 ft), an altitude familiar to many hikers and skiers, the alveolar P_{O_2} falls to around 60 mm Hg.

Despite the low alveolar P_{O_2}, millions of people live and work at altitudes this high or higher. Important physiological adjustments include an increased respiratory rate, an increased heart rate, and an elevated hematocrit. Thus, even though the hemoglobin is not fully saturated, the bloodstream holds more of it, and the round-trip between the lungs and the peripheral tissues takes less time. These responses represent an excellent example of the functional interplay between the respiratory and cardiovascular systems. However, most such adaptations take days to weeks to appear. As a result, athletes planning to compete in events held at high altitude must begin training under such conditions well in advance.

Not everyone can tolerate high-altitude conditions. Roughly 20 percent of people who ascend to 2600 meters (8530 ft) or

higher experience *mountain sickness*, or *altitude sickness*. Symptoms may include headache, disorientation, and fatal pulmonary or cerebral edema. **AM** Mountain Sickness

23–10 CONTROL OF RESPIRATION

Objectives

- Describe the factors that influence the respiration rate.
- Identify and discuss reflex respiratory activity and the brain centers involved in the control of respiration.

Peripheral cells continuously absorb oxygen from interstitial fluids and generate carbon dioxide. Under normal conditions, the cellular rates of absorption and generation are matched by the capillary rates of delivery and removal. Both rates are identical to those of oxygen absorption and carbon dioxide excretion at the lungs. If diffusion rates at the peripheral and alveolar capillaries become unbalanced, homeostatic mechanisms intervene to restore equilibrium. Such mechanisms involve (1) changes in blood flow and oxygen delivery that are regulated at the local level and (2) changes in the depth and rate of respiration under the control of the brain's respiratory centers. The activities of the respiratory centers are coordinated with changes in cardiovascular function, such as fluctuations in blood pressure and cardiac output.

▢ LOCAL REGULATION OF GAS TRANSPORT AND ALVEOLAR FUNCTION

The rate of oxygen delivery in each tissue and the efficiency of oxygen pickup at the lungs are largely regulated at the local level. For example, if a peripheral tissue becomes more active, the interstitial P_{O_2} falls and the P_{CO_2} rises. This change increases the difference between the partial pressures in the tissues and in the arriving blood, so more oxygen is delivered and more carbon dioxide is carried away. In addition, the rising P_{CO_2} levels cause the relaxation of smooth muscles in the walls of arterioles and capillaries in the area, increasing local blood flow.

Local factors coordinate (1) *lung perfusion*, or blood flow to the alveoli, with (2) *alveolar ventilation*, or airflow, over a wide range of conditions and activity levels. As blood flows toward the alveolar capillaries, it is directed toward lobules in which the P_{O_2} is relatively high. This movement occurs because alveolar capillaries constrict when the local P_{O_2} is low. (We noted this response, the opposite of that seen in peripheral tissues, in Chapter 21. ⊂⊃ p. 751) Such a shift in blood flow tends to eliminate temporary differences in the oxygen and carbon dioxide contents of alveoli, lobules, or groups of lobules that could otherwise result from minor variations in local blood flow.

Smooth muscles in the walls of bronchioles are sensitive to the P_{CO_2} of the air they contain. When the P_{CO_2} goes up, the bronchioles increase in diameter (bronchodilation). When the P_{CO_2} declines, the bronchioles constrict (bronchoconstriction). Airflow is therefore directed to lobules in which the P_{CO_2} is high. Because their carbon dioxide is obtained from blood, these lobules are actively engaged in gas exchange. This response is especially important, because the improvement of airflow to functional alveoli can at least partially compensate for damage to pulmonary lobules.

By directing blood flow to alveoli with low CO_2 levels and improving airflow to alveoli with high CO_2 levels, local adjustments improve the efficiency of gas transport. When activity levels increase and the demand for oxygen rises, the cardiac output and respiratory rates increase under neural control, but the adjustments in alveolar blood flow and bronchiole diameter occur automatically.

▢ THE RESPIRATORY CENTERS OF THE BRAIN

Respiratory control has both involuntary and voluntary components. Your brain's involuntary centers regulate the activities of the respiratory muscles and control the respiratory minute volume by adjusting the frequency and depth of pulmonary ventilation. They do so in response to sensory information arriving from your lungs and other portions of the respiratory tract, as well as from a variety of other sites.

The voluntary control of respiration reflects activity in the cerebral cortex that affects either the output of the respiratory centers in the medulla oblongata and pons or motor neurons in the spinal cord that control respiratory muscles. The **respiratory centers** are three pairs of nuclei in the reticular formation of the medulla oblongata and pons. The motor neurons in the spinal cord are generally controlled by *respiratory reflexes*, but they can also be controlled voluntarily through commands delivered by the corticospinal pathway. ⊂⊃ p. 524

Respiratory Centers in the Medulla Oblongata

We introduced the *respiratory rhythmicity centers* of the medulla oblongata in Chapter 14. ⊂⊃ p. 475 These paired centers set the pace of respiration. Each center can be subdivided into a **dorsal respiratory group (DRG)** and a **ventral respiratory group (VRG)**. The DRG's *inspiratory center* contains neurons that control lower motor neurons innervating the external intercostal muscles and the diaphragm. The DRG functions in every respiratory cycle, whether quiet or forced. The VRG functions only during forced breathing. It includes neurons that innervate lower motor neurons controlling accessory respiratory muscles involved in active exhalation (an *expiratory center*) and maximal inhalation (an *inspiratory center*).

There is reciprocal inhibition between the neurons involved with inhalation and exhalation. ⊂⊃ p. 453 When the inspiratory neurons are active, the expiratory neurons are inhibited, and vice versa. The pattern of interaction between these groups differs between quiet breathing and forced breathing. During quiet breathing (Figure 23–26a●):

- Activity in the DRG increases over a period of about 2 seconds, providing stimulation to the inspiratory muscles. Over this period, inhalation occurs.
- After 2 seconds, the DRG neurons become inactive. They remain quiet for the next 3 seconds and allow the inspiratory muscles to relax. Over this period, passive exhalation occurs.

During forced breathing (Figure 23–26b●):

- As the level of activity in the DRG increases, it stimulates neurons of the VRG that activate the accessory muscles involved in inhalation.

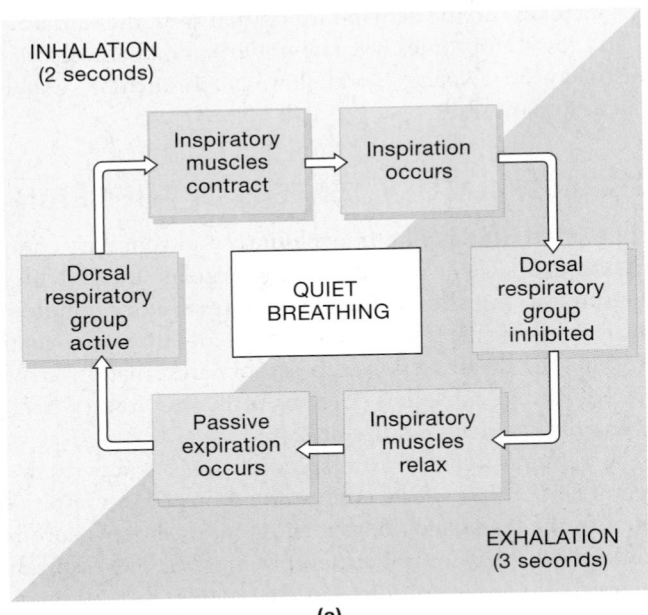

(a)

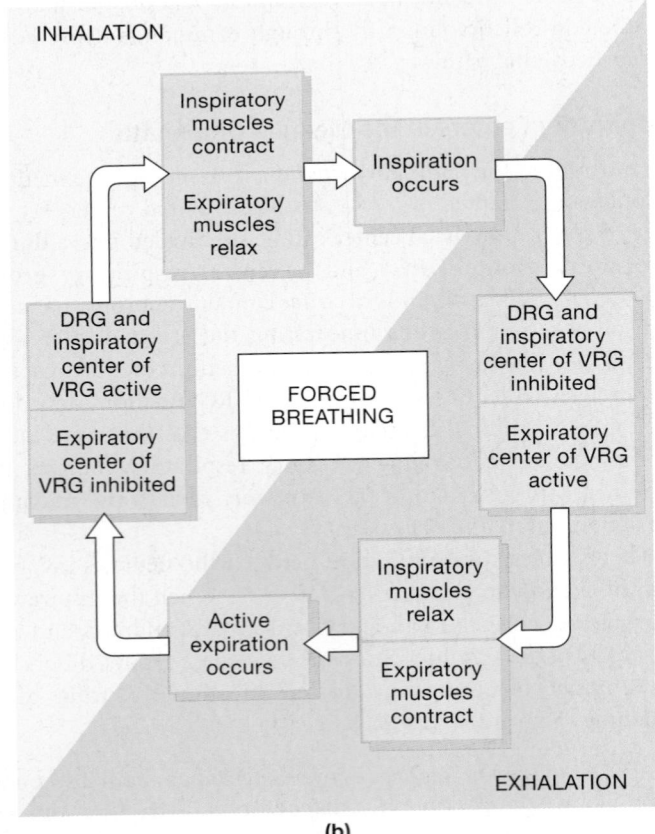

(b)

●**FIGURE 23–26**
Basic Regulatory Patterns of Respiration. (a) Quiet breathing. (b) Forced breathing.

- At the end of each inhalation, active exhalation occurs as the neurons of the expiratory center stimulate the appropriate accessory muscles.

The basic pattern of respiration thus reflects a cyclic interaction between the DRG and the VRG. The pace of this interaction is thought to be established by pacemaker cells that spontaneously undergo rhythmic patterns of activity. Attempts to locate the pacemaker, however, have been unsuccessful.

Central nervous system stimulants, such as amphetamines or even caffeine, increase your respiratory rate by facilitating the respiratory centers. These actions can be opposed by CNS depressants, such as barbiturates or opiates.

The Apneustic and Pneumotaxic Centers

The **apneustic** (ap-NOO-stik) **centers** and the **pneumotaxic** (noo-mō-TAKS-ik) **centers** of the pons are paired nuclei that adjust the output of the respiratory rhythmicity centers. Their activities regulate the respiratory rate and the depth of respiration in response to sensory stimuli or input from other centers in the brain. Each apneustic center provides continuous stimulation to the DRG on that side of the brain stem. During quiet breathing, stimulation from the apneustic center helps increase the intensity of inhalation over the next 2 seconds. Under normal conditions, after 2 seconds the apneustic center is inhibited by signals from the pneumotaxic center on that side. During forced breathing, the apneustic centers also respond to sensory input from the vagus nerves regarding the amount of lung inflation.

The pneumotaxic centers inhibit the apneustic centers and promote passive or active exhalation. Centers in the hypothalamus and cerebrum can alter the activity of the pneumotaxic centers, as well as the respiratory rate and depth. However, essentially normal respiratory cycles continue even if the brain stem superior to the pons has been severely damaged. If the inhibitory output of the pneumotaxic centers is cut off by a stroke or other damage to the brain stem, and if sensory innervation from the lungs is eliminated by cutting the vagus nerves, the person inhales to maximum capacity and maintains that state for 10–20 seconds at a time. Intervening exhalations are brief, and little pulmonary ventilation occurs.

The CNS regions involved with respiratory control are diagrammed in Figure 23–27●. Interactions between the DRG and the VRG establish the basic pace and depth of respiration. The pneumotaxic centers modify that pace: An increase in pneumotaxic output quickens the pace of respiration by shortening the duration of each inhalation; a decrease in pneumotaxic output slows the respiratory pace, but increases the depth of respiration, because the apneustic centers are more active.

☐ RESPIRATORY REFLEXES

The activities of the respiratory centers are modified by sensory information from several sources:

- Chemoreceptors sensitive to the P_{CO_2}, pH, or P_{O_2} of the blood or cerebrospinal fluid.

- Changes in blood pressure in the aortic or carotid sinuses.
- Stretch receptors that respond to changes in the volume of the lungs.
- Irritating physical or chemical stimuli in the nasal cavity, larynx, or bronchial tree.
- Other sensations, including pain, changes in body temperature, and abnormal visceral sensations.

Information from these receptors alters the pattern of respiration. The induced changes have been called *respiratory reflexes*.

⚕ **Sudden infant death syndrome (SIDS)**, also known as *crib death*, kills an estimated 10,000 infants each year in the United States alone. Most crib deaths occur between midnight and 9:00 A.M., in the late fall or winter, and involve infants two to four months old. Eyewitness accounts indicate that the sleeping infant suddenly stops breathing, turns blue, and relaxes. Genetic factors appear to be involved, but controversy remains as to the relative importance of other factors, such as laryngeal spasms, cardiac arrhythmias, upper respiratory system infections, viral infections, and CNS malfunctions. The age at the time of death corresponds with a period when the pacemaker complex and respiratory centers are establishing connections with other portions of the brain. It has recently been proposed that SIDS results from a problem in the interconnection process that disrupts the reflexive respiratory pattern.

The Chemoreceptor Reflexes

The respiratory centers are strongly influenced by chemoreceptor inputs from cranial nerves IX and X and from receptors that monitor the composition of the cerebrospinal fluid (CSF):

- The glossopharyngeal nerve (IX) carries chemoreceptive information from the carotid bodies, adjacent to the carotid sinus. ∞ p. 499 The carotid bodies are stimulated by a decrease in the pH or P_{O_2} of blood. Because changes in P_{CO_2} affect pH, these receptors are indirectly stimulated by a rise in the P_{CO_2}.
- The vagus nerve (X) monitors chemoreceptors in the aortic bodies, near the aortic arch. ∞ p. 500 These receptors are sensitive to the same stimuli as the carotid bodies. Carotid and aortic body receptors are often called *peripheral chemoreceptors*.
- Chemoreceptors are located on the ventrolateral surface of the medulla oblongata in a region known as the *chemosensitive area*. The neurons in that area respond only to the P_{CO_2} and pH of the CSF and are often called *central chemoreceptors*.

We discussed chemoreceptors and their effects on cardiovascular function in Chapters 17 and 21. ∞ pp. 516, 743-4 Stimulation of these chemoreceptors leads to an increase in the depth and rate of respiration. Under normal conditions, a drop in arterial P_{O_2} has little effect on the respiratory centers, until the arterial P_{O_2} drops by about 40 percent, to below 60 mm Hg. If the P_{O_2} of arterial blood drops to 40 mm Hg, the level in peripheral tissues, the respiratory rate will increase by only 50–70 percent. In contrast, a rise of just 10 percent in the arterial P_{CO_2} will cause the respiratory rate to double, even if the P_{O_2} remains completely

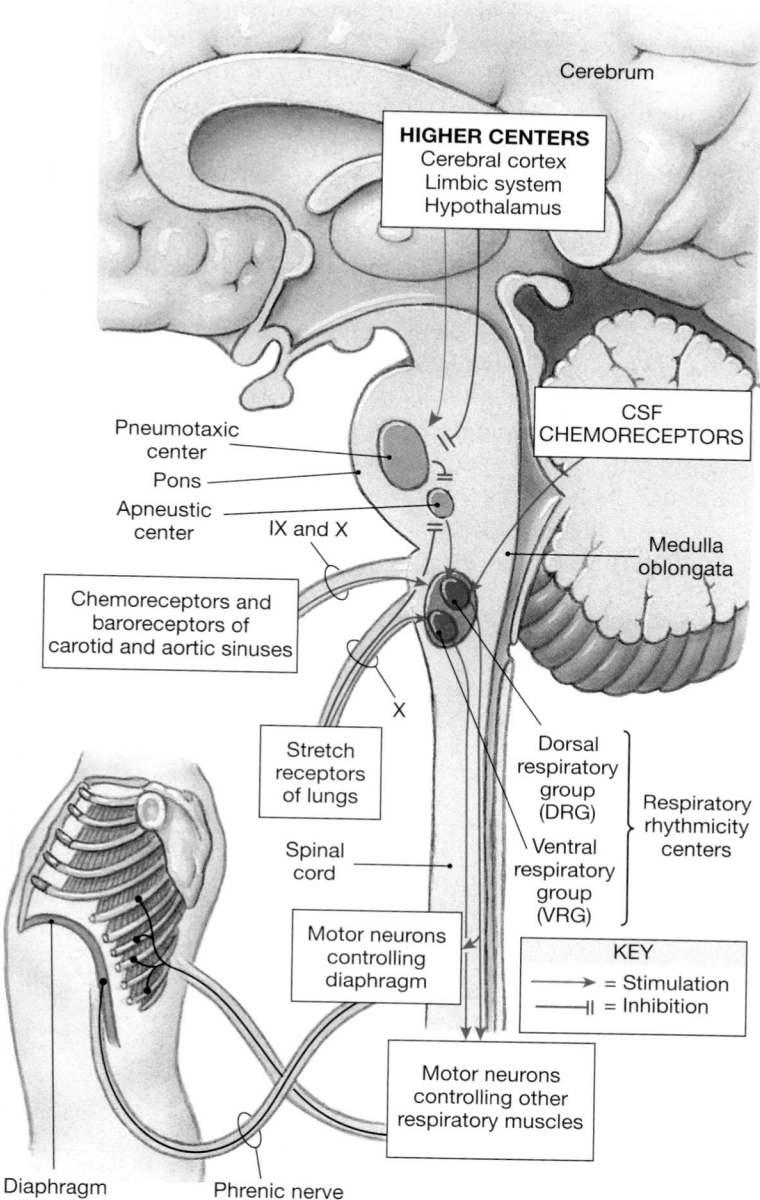

●**FIGURE 23–27**
Respiratory Centers and Reflex Controls. The positions of the major respiratory centers and other factors important to the reflex control of respiration. Pathways for conscious control over respiratory muscles are not shown.

normal. Carbon dioxide levels are therefore responsible for regulating respiratory activity under normal conditions.

Although the receptors monitoring CO_2 levels are more sensitive, oxygen and carbon dioxide receptors work together in a crisis. Carbon dioxide is generated during oxygen consumption, so when oxygen concentrations are falling rapidly, CO_2 levels are usually increasing. This cooperation breaks down only under unusual circumstances. For example, you can hold your breath longer than normal by taking deep, full breaths, but the practice is very dangerous. The danger lies in the fact that the increased ability is due not to extra oxygen, but to the loss of carbon dioxide. If the P_{CO_2} is driven down far enough, your ability to hold your breath can increase to the point at which you become unconscious from oxygen starvation in the brain without ever feeling the urge to breathe. **AM** Shallow-Water Blackout

The chemoreceptors are subject to adaptation—a decrease in sensitivity after chronic stimulation—if the P_{O_2} or P_{CO_2} remains abnormal for an extended period. This adaptation can complicate the treatment of chronic respiratory disorders. For example, if the P_{O_2} remains low for an extended period and the P_{CO_2} remains chronically elevated, the chemoreceptors will reset to those values and will oppose any attempts to return the partial pressures to the proper range. **AM** Chemoreceptor Accommodation and Opposition

Because the chemoreceptors monitoring CO_2 levels are also sensitive to pH, any condition altering the pH of blood or CSF will affect respiratory performance. For example, the rise in lactic acid levels after exercise causes a drop in pH that helps stimulate respiratory activity.

HYPERCAPNIA AND HYPOCAPNIA An increase in the P_{CO_2} of arterial blood constitutes **hypercapnia**. Figure 23–28a● diagrams the central response to hypercapnia, which is triggered by the stimulation of chemoreceptors in the carotid and aortic bodies and is reinforced by the stimulation of CNS chemoreceptors. Carbon

dioxide crosses the blood–brain barrier quite rapidly, so a rise in arterial P_{CO_2} almost immediately elevates CO_2 levels in the CSF, lowering the pH of the CSF and stimulating the chemoreceptive neurons of the medulla oblongata.

These receptors stimulate your respiratory centers to increase the rate and depth of respiration. Your breathing becomes more rapid, and more air moves into and out of your lungs with each breath. Because more air moves into and out of the alveoli each minute, alveolar concentrations of carbon dioxide decline, accelerating the diffusion of carbon dioxide out of alveolar capillaries. Thus, homeostasis is restored.

The most common cause of hypercapnia is hypoventilation. In **hypoventilation**, the respiratory rate remains abnormally low and is insufficient to meet the demands for normal oxygen delivery and carbon dioxide removal. Carbon dioxide then accumulates in the blood.

If the rate and depth of respiration exceed the demands for oxygen delivery and carbon dioxide removal, the condition of **hyperventilation** exists. Hyperventilation will gradually lead to **hypocapnia**, an abnormally low P_{CO_2}. If the arterial P_{CO_2} drops below normal levels, chemoreceptor activity decreases and the respiratory rate falls (Figure 23–28b●). This situation continues until the P_{CO_2} returns to normal and homeostasis is restored.

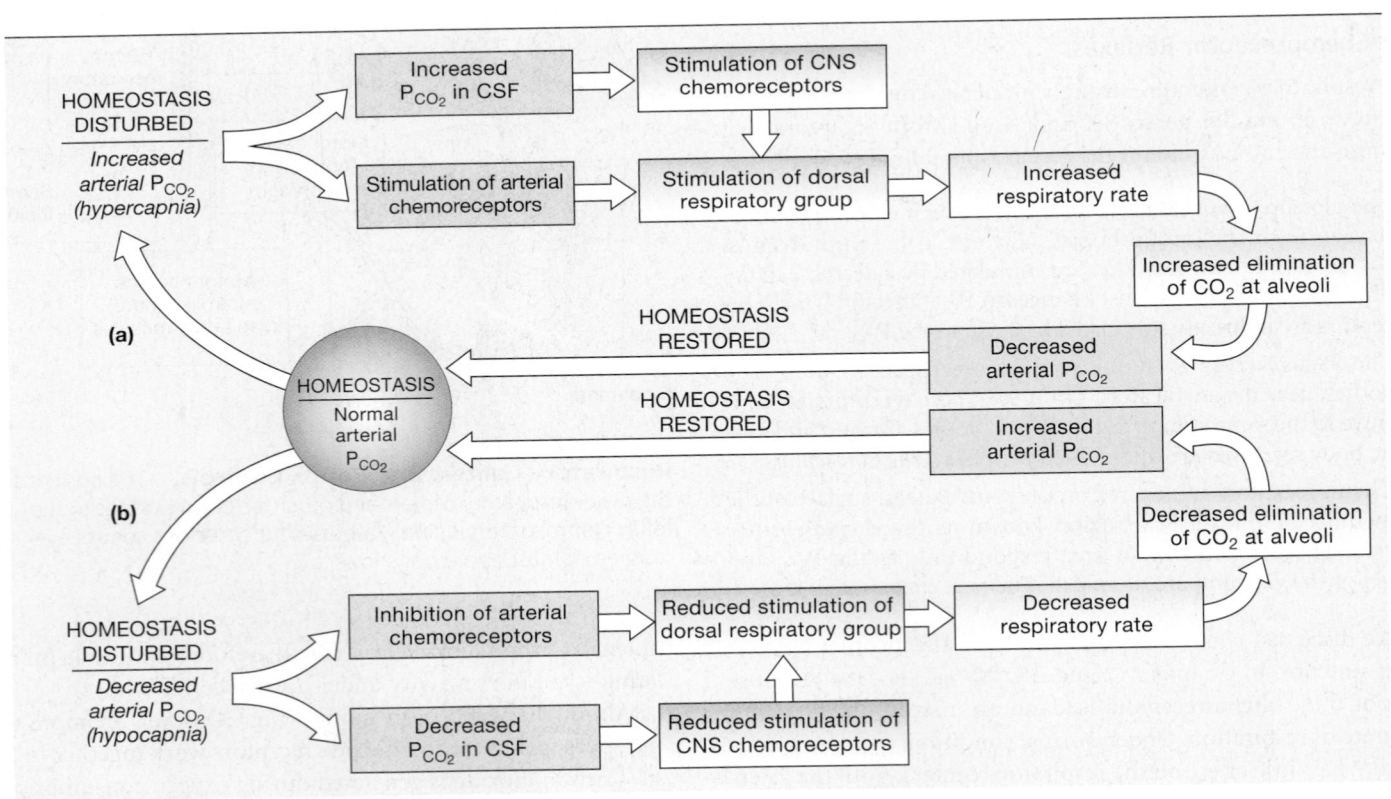

●**FIGURE 23–28**
The Chemoreceptor Response to Changes in P_{CO_2} **(a)** A rise in arterial P_{CO_2} stimulates chemoreceptors that accelerate breathing cycles at the inspiratory center. This change increases the respiratory rate, encourages CO_2 loss at the lungs, and lowers arterial P_{CO_2}. **(b)** A drop in arterial P_{CO_2} inhibits these chemoreceptors. In the absence of stimulation, the rate of respiration decreases, slowing the rate of CO_2 loss and elevating arterial P_{CO_2}.

The Baroreceptor Reflexes

We described the effects of carotid and aortic baroreceptor stimulation on systemic blood pressure in Chapter 21. ∞ pp. 742-744 The output from these baroreceptors also affects the respiratory centers. When blood pressure falls, the respiratory rate increases; when blood pressure rises, the respiratory rate declines. This adjustment results from the stimulation or inhibition of the respiratory centers by sensory fibers in the glossopharyngeal (IX) and vagus (X) nerves.

The Hering–Breuer Reflexes

The **Hering–Breuer reflexes** are named after the physiologists who described them in 1865. The sensory information from these reflexes is distributed to the apneustic centers and the ventral respiratory group. The Hering–Breuer reflexes are not involved in normal quiet breathing or in tidal volumes under 1000 ml. There are two such reflexes:

1. The **inflation reflex** prevents overexpansion of the lungs during forced breathing. Stretch receptors are located in the smooth muscle tissue around bronchioles and are stimulated by lung expansion. Sensory fibers leaving the stretch receptors of each lung reach the respiratory rhythmicity center on the same side via the vagus nerve. As the lung volume increases, the dorsal respiratory group is gradually inhibited, and the expiratory center of the VRG is stimulated. Inhalation stops as the lungs near maximum volume, and active exhalation then begins.

2. The **deflation reflex** inhibits the expiratory centers and stimulates the inspiratory centers when the lungs are deflating. These receptors, which are distinct from those of the inflation reflex, are located in the alveolar wall near the alveolar capillary network. The smaller the volume of the lungs, the greater is the degree of inhibition, until exhalation stops and inhalation begins. This reflex normally functions only during forced exhalation, when both the inspiratory and expiratory centers are active.

Protective Reflexes

Protective reflexes operate when you are exposed to toxic vapors, chemical irritants, or mechanical stimulation of the respiratory tract. The receptors involved are located in the epithelium of the respiratory tract. Examples of protective reflexes include sneezing, coughing, and laryngeal spasms.

Sneezing is triggered by an irritation of the wall of your nasal cavity. Coughing is triggered by an irritation of your larynx, trachea, or bronchi. Both reflexes involve **apnea** (AP-nē-uh), a period in which respiration is suspended. They are usually followed by a forceful expulsion of air intended to remove the offending stimulus. The glottis is forcibly closed while the lungs are still relatively full. The abdominal and internal intercostal muscles then contract suddenly, creating pressures that will blast air out of your respiratory passageways when the glottis reopens. Air leaving the larynx can travel at 160 kph (99 mph), carrying mucus, foreign particles, and irritating gases out of the respiratory tract via the nose or mouth.

Laryngeal spasms result from the entry of chemical irritants, foreign objects, or fluids into the area around the glottis. This reflex generally closes your airway temporarily. A very strong stimulus, such as a toxic gas, could close the glottis so powerfully that you could lose consciousness and die without taking another breath. Fine chicken bones or fish bones that pierce the laryngeal walls can also stimulate laryngeal spasms, swelling, or both, restricting the airway.

Other Sensations That Affect Respiratory Function

Several other sensory stimuli can affect the activities of the respiratory centers. In some cases, the mechanism involved is not known. Examples include the following:

- Sudden pain or immersion in cold water can produce a temporary apnea.

- Chronic pain stimulates the sympathetic division of the autonomic nervous system, leading to an increase in the respiratory rate.

- Both fever and an increase in body temperature due to exertion or overheating cause an increase in the respiratory rate. A reduction in body temperature leads to a decrease in the respiratory rate.

- Curiously, stretching the anal sphincter stimulates the respiratory centers and increases the rate of respiration. Although this reflex is occasionally used to stimulate respiration in an emergency, it is not clear which pathways are involved.

☐ VOLUNTARY CONTROL OF RESPIRATION

Activity of your cerebral cortex has an indirect effect on your respiratory centers, as the following examples show:

- Conscious thought processes tied to strong emotions, such as rage or fear, affect the respiratory rate by stimulating centers in the hypothalamus.

- Emotional states can affect respiration through the activation of the sympathetic or parasympathetic division of the autonomic nervous system. Sympathetic activation causes bronchodilation and increases the respiratory rate; parasympathetic stimulation has the opposite effect.

- An anticipation of strenuous exercise can trigger an automatic increase in the respiratory rate, along with increased cardiac output, by sympathetic stimulation.

Our conscious control over our respiratory activities may bypass the respiratory centers completely, using pyramidal fibers that innervate the same lower motor neurons that are controlled by the DRG and VRG. This control mechanism is an essential part of speaking, singing, and swimming, when respiratory activities must be precisely timed. Higher centers can also have an inhibitory effect on the apneustic centers and on the DRG and VRG; this effect is important when you hold your breath.

Our abilities to override the respiratory centers have limits, however. The chemoreceptor reflexes are extremely powerful respiratory stimulators, and they cannot be consciously suppressed. For example, you cannot kill yourself by holding your breath "till you turn blue." Once the P_{CO_2} rises to critical levels, you will be forced to take a breath.

Emphysema and Lung Cancer

Emphysema and lung cancer are two relatively common disorders that are often associated with cigarette smoking. **Emphysema** (em-fi-SĒ-muh) is a chronic, progressive condition characterized by shortness of breath and an inability to tolerate physical exertion. The underlying problem is the destruction of alveolar surfaces and inadequate surface area for oxygen and carbon dioxide exchange. In essence, respiratory bronchioles and alveoli are functionally eliminated. The alveoli gradually expand, and adjacent alveoli merge to form larger air spaces supported by fibrous tissue without alveolar capillary networks. As connective tissues are eliminated, compliance increases; air moves into and out of the lungs more easily than before. However, the loss of respiratory surface area restricts oxygen absorption, so the individual becomes short of breath.

Emphysema has been linked to the inhalation of air that contains fine particulate matter or toxic vapors, such as those in cigarette smoke. Genetic factors also predispose individuals to the condition. Some degree of emphysema is a normal consequence of aging, however. An estimated 66 percent of adult males and 25 percent of adult females have detectable areas of emphysema in their lungs. **AM** Bronchitis, Emphysema, and COPD

Lung cancer, or *pleuropulmonary neoplasm*, is an aggressive class of malignancies originating in the bronchial passageways or alveoli. These cancers affect the epithelial cells that line conducting passageways, mucous glands, or alveoli. Symptoms generally do not appear until the condition has progressed to the point at which the tumor masses are restricting airflow or compressing adjacent structures. Chest pain, shortness of breath, a cough or a wheeze, and weight loss are common symptoms. Treatment programs vary with the cellular organization of the tumor and whether metastasis (cancer cell migration) has occurred, but surgery, radiation exposure, or chemotherapy may be involved.

Deaths from lung cancer were rare at the turn of the 20th century, but 29,000 such deaths occurred in 1956, 105,000 in 1978, and 154,900 in 2002 in the United States. This rise coincides with an increased rate of smoking in the population. The number of diagnosed cases is also rising, doubling every 15 years. Each year, 22 percent of new cancers detected are lung cancers; in the United States in 2002, an estimated 90,200 men and 79,200 women will be diagnosed with the condition. Lung cancer is increasing markedly among women, but declining among men; in 1989, 101,000 men and 54,000 women were diagnosed with lung cancer in the United States. **AM** Lung Cancer

23–11 CHANGES IN THE RESPIRATORY SYSTEM AT BIRTH

The respiratory system of a fetus and that of a newborn infant differ in several important ways. Before delivery, pulmonary arterial resistance is high, because the pulmonary vessels are collapsed. The rib cage is compressed, and the lungs and conducting passageways contain only small amounts of fluid and no air. During delivery, the lungs are compressed further, and as the placental connection is lost, blood oxygen levels fall and carbon dioxide levels climb rapidly. At birth, the newborn infant takes a truly heroic first breath through powerful contractions of the diaphragmatic and external intercostal muscles. The inhaled air must enter the respiratory passageways with enough force to overcome surface tension and inflate the bronchial tree and most of the alveoli. The same drop in pressure that pulls air into the lungs pulls blood into the pulmonary circulation. The changes in blood flow that occur lead to the closure of the *foramen ovale*, an interatrial connection, and the *ductus arteriosus*, the fetal connection between the pulmonary trunk and the aorta. ∞ p. 768 **ATLAS** Embryology Summary 18: The Development of the Respiratory System

The exhalation that follows fails to empty the lungs completely, because the rib cage does not return to its former, fully compressed state. Cartilages and connective tissues keep the conducting passageways open, and surfactant covering the alveolar surfaces prevents their collapse. Subsequent breaths complete the inflation of the alveoli. Pathologists sometimes use these physical changes to determine whether a newborn infant died before delivery or shortly thereafter. Before the first breath, the lungs are completely filled with amniotic fluids, and they will sink if placed in water. After the infant's first breath, even the collapsed lungs contain enough air to keep them afloat.

23–12 AGING AND THE RESPIRATORY SYSTEM

Many factors interact to reduce the efficiency of the respiratory system in elderly individuals. Three examples are particularly noteworthy:

1. As one's age increases, elastic tissue deteriorates throughout the body, reducing the compliance of the lungs and lowering their vital capacity.

2. Chest movements are restricted by arthritic changes in the rib articulations and by decreased flexibility at the costal cartilages. Along with the changes in item 1, the stiffening and reduction in chest movement effectively limit the respiratory minute volume. This restriction contributes to the reduction in exercise performance and capabilities with increasing age.

3. Some degree of emphysema is normal in individuals over age 50. However, the extent varies widely with the lifetime exposure to cigarette smoke and other respiratory irritants. Figure 23–29● compares the respiratory performance of individuals who have never smoked with individuals who have smoked for varying periods of

time. The message is quite clear: Although some decrease in respiratory performance is inevitable, you can prevent serious respiratory deterioration by stopping smoking or never starting.

✓ What effect would exciting the pneumotaxic centers have on respiration?

✓ Are peripheral chemoreceptors as sensitive to levels of carbon dioxide as they are to levels of oxygen?

✓ Johnny is angry with his mother, so he tells her that he will hold his breath until he turns blue and dies. Should Johnny's mother worry?

Answers start on page Q-1

 Review respiratory controls by visiting the **IP CD-ROM:** Respiratory System/Control of Respiration.

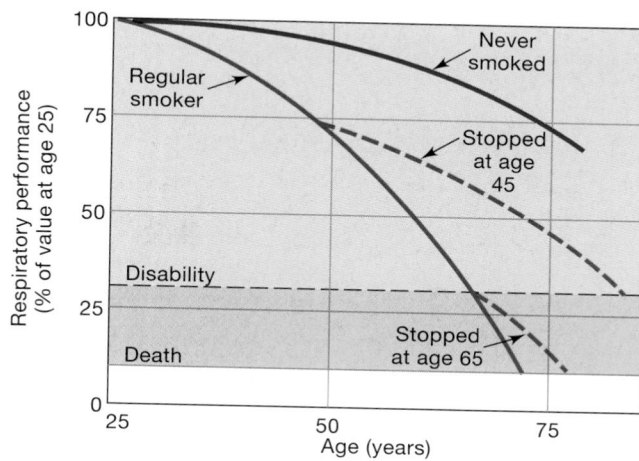

•**FIGURE 23–29**
Aging and the Decline in Respiratory Performance.
The relative respiratory performances of individuals who have never smoked, individuals who quit smoking at age 45, individuals who quit smoking at age 65, and lifelong smokers.

Chapter Review

SELECTED CLINICAL TERMINOLOGY

Terms Discussed in This Chapter

anoxia (a-NOKS-ē-uh): A condition of tissue oxygen starvation caused by (1) circulatory blockage, (2) respiratory problems, or (3) cardiovascular problems. *(p. 844)*

asthma (AZ-muh): An acute respiratory disorder characterized by unusually sensitive, irritated conducting passageways. *(p. 840)*

atelectasis (at-e-LEK-ta-sis): A collapsed lung. *(p. 848)*

bronchitis (brong-KĪ-tis): An inflammation of the bronchial lining. *(p. 840 and AM)*

bronchodilation (brong-kō-dī-LĀ-shun): An enlargement of the respiratory passageways. *(p. 840)*

bronchography (brong-KOG-ra-fē): A procedure in which radiopaque materials are introduced into the airways to improve X-ray imaging of the bronchial tree. *(p. 840)*

bronchoscope: A fiber-optic bundle small enough to be inserted into the trachea and finer airways; the procedure is called *bronchoscopy.* *(p. 840)*

cardiopulmonary resuscitation (CPR): The application of cycles of compression to the rib cage and mouth-to-mouth breathing to maintain cardiovascular and respiratory function. *(p. 848 and AM)*

cystic fibrosis (CF): A relatively common lethal inherited disease caused by an abnormal membrane channel protein; a major symptom is that mucous secretions become too thick to be transported easily, leading to respiratory problems. *(p. 830 and AM)*

decompression sickness, or *the bends:* A condition caused by a drop in atmospheric pressure and the resulting formation of nitrogen gas bubbles in body fluids, tissues, and organs. *(p. 848 and AM)*

emphysema (em-fi-SĒ-muh): A chronic, progressive condition characterized by shortness of breath and an inability to tolerate physical exertion. *(p. 866 and AM)*

epistaxis (ep-i-STAK-sis): A nosebleed. *(p. 832)*

Heimlich (HĪM-lik) **maneuver,** or *abdominal thrust:* Compression applied to the abdomen just inferior to the diaphragm, to force air out of the lungs to clear a blocked trachea or larynx. *(p. 836)*

hypercapnia (hī-per-KAP-nē-uh): An increase in the P_{CO_2} of arterial blood. *(p. 864)*

hypocapnia: An abnormally low arterial P_{CO_2}. *(p. 864)*

hypoxia (hī-POKS-ē-uh): A condition of reduced tissue P_{CO_2}. *(p. 844)*

lung cancer *(pleuropulmonary neoplasm):* A class of aggressive malignancies originating in the bronchial passageways or alveoli. *(p. 866 and AM)*

mountain sickness: An acute disorder resulting from CNS effects of the low gas partial pressures that occur at high altitudes. *(p. 860 and AM)*

pleurisy: An inflammation of the pleurae, accompanied by the secretion of excess amounts of pleural fluid. *(p. 843)*

pneumonia (noo-MŌ-nē-uh): A respiratory disorder characterized by fluid leakage into the alveoli or swelling and constriction of the respiratory bronchioles. *(p. 842)*

pneumothorax (noo-mō-THO-raks): The entry of air into the pleural cavity. *(p. 848)*

pulmonary embolism: A blockage of a branch of a pulmonary artery, with interruption of blood flow to a group of lobules or alveoli. *(p. 843)*

respiratory distress syndrome: A condition resulting from the production of inadequate surfactant and associated alveolar collapse. *(p. 841 and AM)*

sudden infant death syndrome (SIDS), or *crib death:* The death of an infant due to respiratory arrest; the cause remains uncertain. *(p. 863)*

tracheostomy (trā-kē-OS-to-mē): The insertion of a tube directly into the trachea to bypass a blocked or damaged larynx. *(p. 836)*

tuberculosis: A respiratory disorder caused by infection of the lungs by the bacterium *Mycobacterium tuberculosis.* *(p. 836 and AM)*

🔲 Additional Terms Discussed in the *Applications Manual*

chronic obstructive pulmonary disease (COPD)
peak-flow meter
pneumotachometer
shallow-water blackout
spirometer

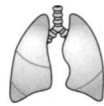

The **RESPIRATORY SYSTEM** in Perspective

Organ/Component	Primary Functions
Nasal Cavities, Paranasal Sinuses	Filter, warm, humidify air; detect smells
Pharynx	Conducts air to larynx; a chamber shared with the digestive tract *(see part i)*
Larynx	Protects opening to trachea and contains vocal cords
Trachea	Filters air, traps particles in mucus; cartilages keep airway open
Bronchi	(Same functions as trachea)
Lungs	Responsible for air movement through volume changes during movements of ribs and diaphragm; include airways and alveoli
Alveoli	Act as sites of gas exchange between air and blood

INTEGRATION WITH OTHER SYSTEMS

The goal of respiratory activity is to maintain homeostatic oxygen and carbon dioxide levels in peripheral tissues. Changes in respiratory activity alone are seldom sufficient to accomplish this; there must also be coordinated changes in cardiovascular activity.

Here are three examples of the integration between the respiratory and cardiovascular systems:

1. At the local level, changes in lung perfusion in response to changes in alveolar P_{O_2} improve the efficiency of gas exchange within or among lobules.

2. Chemoreceptor stimulation not only increases the respiratory drive; it also causes an elevation in blood pressure and increased cardiac output.

3. The stimulation of baroreceptors in the lungs has secondary effects on cardiovascular function. For example, the stimulation of airway stretch receptors not only triggers the inflation reflex, but also increases the heart rate. Thus, as the lungs fill, cardiac output rises and more blood flows through the alveolar capillaries.

The respiratory system is functionally linked to all other systems as well. The figure on the opposite page illustrates these interrelationships.

CLINICAL PATTERNS

Disorders affecting the respiratory system may involve (1) interfering with the movement of air along the respiratory passageways, (2) impeding the diffusion of gases at the respiratory membrane, or (3) reducing the normal circulation of blood through the alveolar capillaries.

These problems can result from trauma, congenital or degenerative problems, tumors, inflammation, or infection of the lungs. Illnesses caused by infections of the upper respiratory tract include some of the most common diseases, such as the "common cold" and influenza. Infections of the lower respiratory tract include two of the most deadly diseases in human history: pneumonia and tuberculosis. Respiratory system disorders also occur secondarily, as a consequence of dysfunctions of other body systems. For instance, asthma is the result of a problem with immune function, and pulmonary emboli result from cardiovascular problems affecting lung perfusion. You will find details on specific disorders and their classification in the *Applications Manual*.

MEDIA CONNECTIONS

I. Objective: To investigate the role of the respiratory control centers in maintaining homeostasis.

Completion time = 15 minutes

The gases brought to the respiratory membrane from the environment diffuse into the bloodstream, and dissolved gases diffusing into the blood from active tissues are eliminated with every breath. This exchange alters the chemical composition of the blood, affecting homeostasis. How is the process monitored so that blood chemistry does not shift too far out of the "normal" range? What other systems are integrally involved in respiratory control? To answer these questions, visit the **InterActive Physiology CD-ROM:** Respiratory System/Control of Respiration, screens 3 through 7. Outline the information presented, including the respiratory centers and the methods utilized by each to maintain homeostasis. For help, review Sections 14–3 and 14–4. Predict what will happen when P_{CO_2} increases, and add the steps of this feedback mechanism to your outline. Complete the tutorial, summarizing each new control pathway. Which systems are central to the control of respira-

tion?

II. Objective: To explore the cooperation of the respiratory system with other body systems in the delivery of oxygen to cells.

Completion time = 7 minutes

According to the game Trivial Pursuit®, the answer to the question, "What causes death in virtually all cases?" is hypoxia. Is this true? What do you know about hypoxia already? Using your text, prepare a simple outline of the steps involved in the delivery of oxygen to the cells of the body. Indicate which body systems are involved in each step. Now visit the **Companion Website**, Chapter 23 Media Connections, and click on the keyword **hypoxia**. After reading the information presented, return to your outline and indicate factors that might interfere with particular steps so as to produce hypoxia. For example, the first step might be pulmonary ventilation. What external conditions could interfere with this process? What physiological problems? Looking at your outline, do you agree with Milton Bradley® that hypoxia is the ONLY true cause of death?

INTEGUMENTARY SYSTEM
- Protects portions of upper respiratory system
- Hairs guard entry to external nares

SKELETAL SYSTEM
- Movements of the ribs important in breathing
- Axial skeleton surrounds and protects lungs

MUSCULAR SYSTEM
- Muscular activity generates carbon dioxide
- Respiratory muscles fill and empty lungs
- Other muscles control entrances to respiratory tract
- Intrinsic laryngeal muscles control airflow through larynx and produce sound

NERVOUS SYSTEM
- Monitors respiratory volume and blood gas levels
- Controls pace and depth of respiration

ENDOCRINE SYSTEM
- Epinephrine and norepinephrine stimulate respiratory activity and dilate respiratory passageways
- Converting enzyme along capillaries of lung converts angiotensin I to angiotensin II

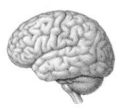

RESPIRATORY SYSTEM

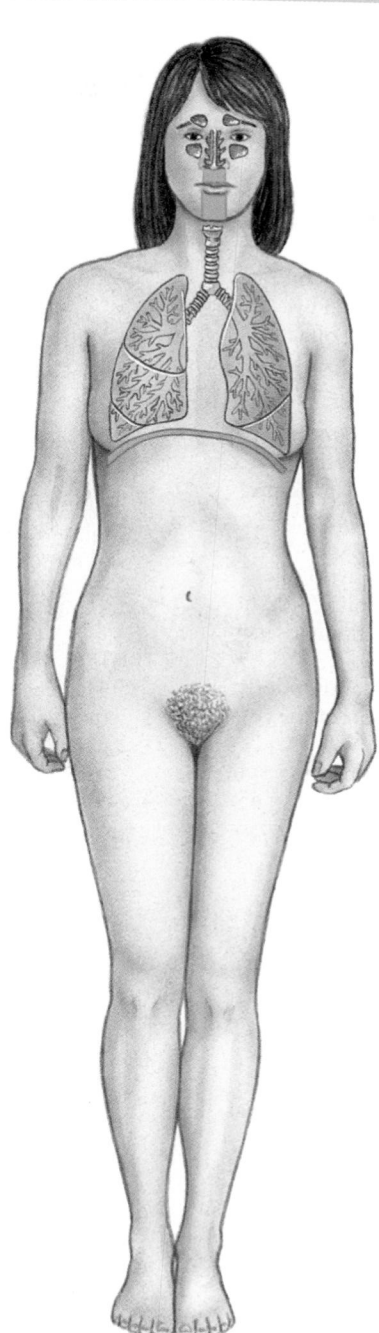

FOR ALL SYSTEMS
Provides oxygen and eliminates carbon dioxide

CARDIOVASCULAR SYSTEM
- Red blood cells transport oxygen and carbon dioxide between the lungs and peripheral tissues
- Activation of angiotensin II by converting enzyme important in regulation of blood pressure and volume
- Bicarbonate ions contribute to buffering capability of blood

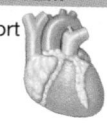

LYMPHATIC SYSTEM
- Tonsils protect against infection at entrance to respiratory tract
- Lymphatic vessels monitor lymph drainage from lungs and mobilize specific defenses when infection occurs
- Alveolar phagocytes present antigens to trigger specific defenses
- Respiratory defense system traps pathogens, protects deeper tissues

DIGESTIVE SYSTEM
- Provides substrates, vitamins, water, and ions that are necessary to all cells of the respiratory system
- Increased thoracic and abdominal pressure through contraction of respiratory muscles can assist in defecation

URINARY SYSTEM
- Eliminates organic wastes generated by cells of the respiratory system
- Maintains normal fluid and ion balance in the blood
- Assists in the regulation of pH by eliminating carbon dioxide

REPRODUCTIVE SYSTEM
- Changes in respiratory rate and depth occur during sexual arousal

STUDY OUTLINE

1. Body cells must obtain oxygen and eliminate carbon dioxide. The respiratory surfaces where gas exchange occurs are inside the lungs.

23-1 THE RESPIRATORY SYSTEM: AN INTRODUCTION   p. 827

FUNCTIONS OF THE RESPIRATORY SYSTEM   p. 828

1. The functions of the **respiratory system** include (1) providing an area for gas exchange between air and circulating blood; (2) moving air to and from exchange surfaces; (3) protecting respiratory surfaces from environmental variations and defending the respiratory system and other tissues from invasion by pathogens; (4) producing sounds; and (5) providing olfactory sensations to the CNS.

ORGANIZATION OF THE RESPIRATORY SYSTEM   p. 828

2. The respiratory system includes the **upper respiratory system**, composed of the nose, nasal cavity, paranasal sinuses, and pharynx, and the **lower respiratory system**, which includes the larynx, trachea, bronchi, bronchioles, and alveoli of the lungs. (*Figure 23–1*)
3. The **respiratory tract** consists of the conducting passageways that carry air to and from the **alveoli**. The passageways of the **upper respiratory system** filter and humidify incoming air. The **lower respiratory system** includes delicate conduction passages and the alveolar exchange surfaces.
4. The **respiratory mucosa** (respiratory epithelium and underlying connective tissue) lines the conducting portion of the respiratory tract.
5. The respiratory epithelium changes in structure along the respiratory tract. It is supported by the **lamina propria**, a layer of areolar tissue. (*Figure 23–2*)
6. Contamination of the respiratory system is prevented by the **respiratory defense system**. (*Figure 23–2*)

23-2 THE UPPER RESPIRATORY SYSTEM   p. 830

1. The components of the upper respiratory system consist of the nose, nasal cavity, paranasal sinuses, and pharynx. (*Figures 23–1, 23–3*)

THE NOSE AND NASAL CAVITY   p. 830

2. Air normally enters the respiratory system through the **external nares**, which open into the **nasal cavity**. The **vestibule** (entryway) is guarded by hairs that screen out large particles. (*Figure 23–3*)
3. Incoming air flows through the **superior**, **middle**, and **inferior meatuses** (narrow grooves) and bounces off the conchal surfaces. (*Figure 23–3*)
4. The **hard palate** separates the oral and nasal cavities. The **soft palate** separates the superior nasopharynx from the rest of the pharynx. The connections between the nasal cavity and nasopharynx are the **internal nares**.
5. The nasal mucosa traps particles, warms and humidifies incoming air, and cools and dehumidifies outgoing air.

THE PHARYNX   p. 832

6. The **pharynx** is a chamber shared by the digestive and respiratory systems. The **nasopharynx** is the superior part of the pharynx. The **oropharynx** is continuous with the oral cavity. The **laryngopharynx** includes the narrow zone between the hyoid bone and the entrance to the esophagus. (*Figure 23–3*)

23-3 THE LARYNX   p. 832

1. Inhaled air passes through the **glottis** en route to the lungs; the **larynx** surrounds and protects the glottis. (*Figure 23–4*)

CARTILAGES AND LIGAMENTS OF THE LARYNX   p. 832

2. The cylindrical larynx is composed of three large cartilages (the **thyroid cartilage**, **cricoid cartilage**, and **epiglottis**) and three smaller pairs of cartilages (the **arytenoid**, **corniculate**, and **cuneiform cartilages**). The epiglottis projects into the pharynx. (*Figures 23–4, 23–5*)
3. Two pairs of folds span the glottis: the inelastic **vestibular folds** and the more delicate **vocal folds**. (*Figure 23–5*)

SOUND PRODUCTION   p. 833

4. Air passing through the glottis vibrates the vocal folds, producing sound. The pitch of the sound depends on the diameter, length, and tension of the vocal folds.

THE LARYNGEAL MUSCULATURE   p. 834

5. The muscles of the neck and pharynx position and stabilize the larynx. The smaller intrinsic muscles regulate tension in the vocal folds or open and close the glottis. During swallowing, both sets of muscles help prevent particles from entering the glottis.

23-4 THE TRACHEA AND PRIMARY BRONCHI   p. 835

THE TRACHEA   p. 835

1. The **trachea** extends from the sixth cervical vertebra to the fifth thoracic vertebra. The **submucosa** contains C-shaped **tracheal cartilages**, which stiffen the tracheal walls and protect the airway. The posterior tracheal wall can distort to permit large masses of food to pass through the esophagus. (*Figure 23–6*)

THE PRIMARY BRONCHI   p. 836

2. The trachea branches within the mediastinum to form the **right** and **left primary bronchi**. Each bronchus enters a lung at the **hilus** (a groove). The **root** is a connective-tissue mass that includes the bronchus, pulmonary vessels, and nerves. (*Figures 23–6, 23–7*)

23-5 THE LUNGS   p. 836

LOBES AND SURFACES OF THE LUNGS   p. 836

1. The **lobes** of the **lungs** are separated by fissures. The right lung has three lobes, the left lung two. (*Figure 23–7*)
2. The anterior and lateral surfaces of the lungs follow the inner contours of the rib cage. The concavity of the medial surface of the left lung is the **cardiac notch**, which conforms to the shape of the pericardium. (*Figure 23–8*)

THE BRONCHI   p. 838

3. The primary bronchi and their branches form the **bronchial tree**. The **secondary** and **tertiary bronchi** are branches within the lungs. As they branch, the amount of cartilage in their walls decreases and the amount of smooth muscle increases. (*Figures 23–9, 23–10*)
4. Each tertiary bronchus supplies air to a single **bronchopulmonary segment**. (*Figure 23–10*)

THE BRONCHIOLES   p. 840

5. **Bronchioles** within the bronchopulmonary segments ultimately branch into **terminal bronchioles**. Each terminal bronchiole delivers air to a single **pulmonary lobule** in which the terminal bronchiole branches into **respiratory bronchioles**. The connective

tissues of the root of the lung extend into the parenchyma of the lung as a series of *trabeculae* (partitions) that branch to form **interlobular septa**, which divide the lung into lobules. (*Figure 23–10*)

ALVEOLAR DUCTS AND ALVEOLI   p. 840

6. The respiratory bronchioles open into **alveolar ducts**, at each of which many alveoli are interconnected. The respiratory exchange surfaces are extensively connected to the circulatory system via the vessels of the pulmonary circuit. (*Figure 23–11*)

7. The **respiratory membrane** consists of a simple squamous epithelium, the endothelial cell lining an adjacent capillary, and the fused basal laminae; **septal cells** scattered in the respiratory membrane produce **surfactant** that keeps the alveoli from collapsing. **Alveolar macrophages** patrol the epithelium and engulf foreign particles. (*Figure 23–12*)

Respiratory tract anatomy: **IP CD-ROM:** Respiratory System/Anatomy Review/Respiratory Structures and **Anatomy CD-ROM:** Respiratory System/Lungs Flythrough.

THE BLOOD SUPPLY TO THE LUNGS   p. 843

8. The conducting portions of the respiratory tract receive blood from the external carotid arteries, the thyrocervical trunks, and the bronchial arteries. Venous blood flows into the pulmonary veins, bypassing the rest of the systemic circuit and diluting the oxygenated blood leaving the alveoli.

THE PLEURAL CAVITIES AND PLEURAL MEMBRANES   p. 843

9. Each lung occupies a single **pleural cavity** lined by a **pleura** (serous membrane). The two types of pleurae are the **parietal pleura**, covering the inner surface of the thoracic wall, and the **visceral pleura**, covering the lungs.

23–6   AN OVERVIEW OF RESPIRATORY PHYSIOLOGY   p. 843

1. Respiratory physiology focuses on a series of integrated processes. **External respiration** (the exchange of oxygen and carbon dioxide between interstitial fluid and the external environment) includes **pulmonary ventilation** (breathing). **Internal respiration** is the exchange of oxygen and carbon dioxide between interstitial fluid and cells. If the oxygen content declines, the affected tissues will suffer from **hypoxia**; if the oxygen supply is completely shut off, **anoxia** and tissue death result. (*Figure 23–13*)

23–7   PULMONARY VENTILATION   p. 844

1. **Pulmonary ventilation** is the physical movement of air into and out of the respiratory tract.

THE MOVEMENT OF AIR   p. 844

2. As pressure on a gas decreases, its volume expands; as pressure increases, gas volume contracts. This inverse relationship is **Boyle's law**. (*Figure 23–14; Table 23–1*)

3. Lung volume is directly affected by movement of the diaphragm and ribs (*Figure 23–15*)846

PRESSURE CHANGES DURING INHALATION AND EXHALATION   p. 846

4. The relationship between **intrapulmonary pressure** (the pressure inside the respiratory tract) and **atmospheric pressure** (**atm**) determines the direction of airflow. **Intrapleural pressure** is the pressure in the space between the parietal and visceral pleurae. (*Figure 23–16*)

5. A **respiratory cycle** is a single cycle of inhalation and exhalation. The amount of air moved in one respiratory cycle is the **tidal volume**. (*Figure 23–16*)

Pulmonary ventilation: **IP CD-ROM:** Respiratory System/Pulmonary Ventilation.

THE MECHANICS OF BREATHING   p. 848

6. The diaphragm and the external and internal intercostal muscles are involved in normal **quiet breathing**, or **eupnea**. Accessory muscles become active during the active inspiratory and expiratory movements of **forced breathing**, or **hyperpnea**. (*Figure 23–17*)

RESPIRATORY RATES AND VOLUMES   p. 850

5. **Alveolar ventilation** is the amount of air reaching the alveoli each minute. The **vital capacity** includes the **tidal volume** plus the **expiratory** and **inspiratory reserve volumes**. The air left in the lungs at the end of maximum exhalation is the **residual volume**. (*Figure 23–18*)

23–8   GAS EXCHANGE   p. 852

THE GAS LAWS   p. 852

1. In a mixed gas, the individual gases exert a pressure proportional to their abundance in the mixture (**Dalton's law**). The pressure contributed by a single gas is its **partial pressure**.

2. The amount of a gas in solution is directly proportional to the partial pressure of that gas (**Henry's law**). (*Figure 23–19*)

DIFFUSION AND RESPIRATORY FUNCTION   p. 853

3. Alveolar air and atmospheric air differ in composition. Gas exchange across the respiratory membrane is efficient due to differences in partial pressures, the small diffusion distance, lipid-soluble gases, the large surface area of all the alveoli combined, and the coordination of blood flow and airflow. (*Table 23–2*)

23–9   GAS PICKUP AND DELIVERY   p. 855

1. Blood entering peripheral capillaries delivers oxygen and absorbs carbon dioxide. The transport of oxygen and carbon dioxide in blood involves reactions that are completely reversible. (*Figure 23–20*)

OXYGEN TRANSPORT   p. 855

2. Oxygen is carried mainly by RBCs, reversibly bound to hemoglobin. At alveolar P_{O_2}, hemoglobin is almost fully saturated; at the P_{O_2} of peripheral tissues, it retains a substantial oxygen reserve. The effect of pH on the hemoglobin saturation curve is called the **Bohr** effect. When low plasma P_{O_2} continues for extended periods, red blood cells generate more **2,3-bisphosphoglycerate** (**BPG**), which reduces hemoglobin's affinity for oxygen. (*Figures 23–21, 23–22*)

3. **Fetal hemoglobin** has a stronger affinity for oxygen than does adult hemoglobin, aiding the removal of oxygen from maternal blood. (*Figure 23–23*)

CARBON DIOXIDE TRANSPORT   p. 858

4. Aerobic metabolism in peripheral tissues generates CO_2. About 7 percent of the CO_2 transported in blood is dissolved in the plasma, 23 percent is bound as **carbaminohemoglobin**, and the rest is converted to carbonic acid, which dissociates into H^+ and HCO_3^-. (*Figure 23–24*)

SUMMARY: GAS TRANSPORT   p. 859

5. Driven by differences in partial pressure, oxygen enters the blood at the lungs and leaves in peripheral tissues; similar forces drive carbon dioxide into the blood at the tissues and into the alveoli at the lungs. (*Figure 23–25*)

Gas transport: **IP CD-ROM:** Respiratory System/Gas Exchange and Gas Transport.

23–10 CONTROL OF RESPIRATION   p. 861

1. Normally, the cellular rates of gas absorption and generation are matched by the capillary rates of delivery and removal and are identical to the rates of oxygen absorption and carbon dioxide excretion at the lungs. When these rates are unbalanced, homeostatic mechanisms restore equilibrium.

LOCAL REGULATION OF GAS TRANSPORT AND ALVEOLAR FUNCTION   p. 861

2. Local factors regulate alveolar blood flow (*lung perfusion*) and airflow (*alveolar ventilation*). Alveolar capillaries constrict under conditions of low oxygen, and bronchioles dilate under conditions of high carbon dioxide.

THE RESPIRATORY CENTERS OF THE BRAIN   p. 861

3. The **respiratory centers** include three pairs of nuclei in the reticular formation of the pons and medulla oblongata. The *respiratory rhythmicity centers* set the pace for respiration, the **apneustic centers** cause strong, sustained inspiratory movements, and the **pneumotaxic centers** inhibit the apneustic centers and promote exhalation. (*Figures 23–26, 23–27*)

RESPIRATORY REFLEXES   p. 862

4. Stimulation of the **chemoreceptor reflexes** is based on the level of carbon dioxide in the blood and CSF. The **inflation reflex** prevents overexpansion of the lungs during forced breathing. The **deflation reflex** stimulates inhalation when the lungs are collapsing. (*Figures 23–27, 23–28*)

VOLUNTARY CONTROL OF RESPIRATION   p. 865

5. Conscious and unconscious thought processes can affect respiration by affecting the respiratory centers.

23–11 CHANGES IN THE RESPIRATORY SYSTEM AT BIRTH   p. 866

1. Before delivery, the fetal lungs are filled with body fluids and collapsed. At the first breath, the lungs inflate and never collapse completely thereafter.

23–12 AGING AND THE RESPIRATORY SYSTEM   p. 866

1. The respiratory system is generally less efficient in the elderly because (1) elastic tissue deteriorates, lowering the vital capacity of the lungs, (2) movements of the chest are restricted by arthritic changes and decreased flexibility of costal cartilages, and (3) some degree of emphysema is generally present. (*Figure 23–29*)

Respiratory controls: **IP CD-ROM:** Respiratory System/Control of Respiration.

REVIEW QUESTIONS

 More assessment questions are available to you on the Companion Website. You will find Matching, Multiple Choice, True/False, and other quizzes to help further your understanding of the material covered in this chapter. To access the site, go to www.aw.com/martini.

LEVEL 1 Reviewing Facts and Terms

1. The C shape of the tracheal cartilages is important because
   (a) large masses of food can pass through the esophagus during swallowing
   (b) large masses of air can pass through the trachea
   (c) it allows greater tracheal elasticity and flexibility
   (d) a, b, and c are correct

2. Control over the amount of resistance to airflow and the distribution of air in the lungs is provided by the
   (a) diaphragm   (b) trachea
   (c) bronchioles   (d) alveoli

3. The presence of an abnormally low carbon dioxide content in blood is
   (a) hypocapnia   (b) anoxia
   (c) hypoxia   (d) hypercapnia

4. During expiration, the lungs contract and the intrapulmonary pressure
   (a) drops to about 759 mm Hg
   (b) remains at 760 mm Hg
   (c) rises to about 761 mm Hg
   (d) does not change

5. During a normal inhalation, the intrapleural pressure is approximately
   (a) −1 mm Hg   (b) +6 mm Hg
   (c) +1 mm Hg   (d) −6 mm Hg

6. According to Boyle's law, if the volume of a gas increases,
   (a) the pressure of the gas will increase
   (b) the pressure of the gas will decrease
   (c) the solubility of the gas will decrease
   (d) the solubility of the gas will increase

7. In tissues at a normal partial pressure of oxygen, blood entering the venous system contains about _____ of its total oxygen content.
   (a) 25 percent   (b) 50 percent
   (c) 75 percent   (d) 90 percent

8. Approximately 70 percent of the carbon dioxide absorbed by blood is transported
   (a) as carbonic acid
   (b) bound to hemoglobin
   (c) in the form of dissolved gas molecules
   (d) bound to oxygen molecules

9. The apneustic centers of the pons
   (a) inhibit the pneumotaxic and inspiratory centers
   (b) provide continuous stimulation to the inspiratory center
   (c) monitor blood gas levels
   (d) alter chemoreceptor sensitivity

10. All of the following provide chemoreceptor input to the respiratory centers of the medulla oblongata, except the
    (a) olfactory epithelium   (b) medullary chemoreceptors
    (c) aortic body   (d) carotid body

11. Sneezing and coughing are classic examples of
    (a) inflation reflexes   (b) deflation reflexes
    (c) protective reflexes   (d) Hering–Breuer reflexes

12. What are the five primary functions of the respiratory system?

13. Distinguish between the structures of the upper respiratory system and those of the lower respiratory system.

14. What defense mechanisms protect the respiratory system from becoming contaminated with debris or pathogens?

15. Name the three regions of the pharynx. Where is each region located?

16. List the cartilages of the larynx. What are the functions of each?

17. How does the parietal pleura differ from the visceral pleura?

18. What four integrated steps are involved in external respiration?

19. What important physiological differences exist between fetal hemoglobin and maternal hemoglobin?

20. By what three mechanisms is carbon dioxide transported in the bloodstream?

21. What effect does the P_{O_2} have on smooth muscles in the walls of bronchioles?

LEVEL 2 Reviewing Concepts

22. Sympathetic stimulation to the smooth muscle tissue layer in the bronchioles causes
    (a) bronchoconstriction
    (b) bronchodilation
    (c) a relaxation of muscle tone
    (d) an increase in tidal volume

23. If you have a respiration rate of 15 breaths per minute and a tidal volume of 500 ml of air, your respiratory minute volume is
    (a) 7.5 l/min   (b) 75 l/min
    (c) 750 l/min   (d) 7500 l/min

24. If an inhalation in one respiratory cycle pulls in 1000 ml of air, the amount of air reaching the alveolar spaces is about
    (a) 300 ml   (b) 850 ml
    (c) 150 ml   (d) 700 ml

25. Gas exchange at the respiratory membrane is efficient because
    (a) the differences in partial pressure are substantial
    (b) the gases are lipid soluble
    (c) the total surface area is large
    (d) a, b, and c are correct

26. For any partial pressure of oxygen, if the concentration of 2,3-bisphosphoglycerate (BPG) increases,
    (a) the amount of oxygen released by hemoglobin will decrease
    (b) the oxygen levels in hemoglobin will be unaffected
    (c) the amount of oxygen released by hemoglobin will increase
    (d) the amount of carbon dioxide carried by hemoglobin will increase

27. The primary physiological adjustment necessary for an athlete to compete at high altitudes is
    (a) an increased respiratory rate
    (b) an increased heart rate
    (c) an elevated hematocrit
    (d) a, b, and c are correct

28. An increase in the partial pressure of carbon dioxide in arterial blood causes chemoreceptors to stimulate the respiratory centers, resulting in
    (a) a decreased respiratory rate
    (b) an increased respiratory rate
    (c) hypocapnia
    (d) hypercapnia

29. What is the functional significance of the decrease in the amount of cartilage and increase in the amount of smooth muscle in the lower respiratory passageways?

30. Why can't you swallow solid food or liquid properly while you are talking?

31. Why is breathing through the nasal cavity more desirable than breathing through the mouth?

32. How would you justify the statement "The bronchioles are to the respiratory system what the arterioles are to the cardiovascular system"?

33. How are septal cells involved with keeping the alveoli from collapsing?

34. Why does oxygen move across the respiratory membrane in one direction while carbon dioxide moves across that same membrane in the opposite direction? Are the two movements functionally related?

35. How does pulmonary ventilation differ from alveolar ventilation, and what is the function of each type of ventilation?

36. What is the significance of (a) Boyle's law, (b) Dalton's law, and (c) Henry's law to the process of respiration?

37. If the pleural cavity is penetrated due to an injury, why does the lung collapse?

38. What happens to the process of respiration when a person is sneezing or coughing?

39. What are the differences between pulmonary volumes and respiratory capacities? How are pulmonary volumes and respiratory capacities determined?

40. What are the primary effects of pH, temperature, and 2,3-bis-phosphoglycerate (BPG) on hemoglobin saturation?

41. What is the functional difference between the dorsal respiratory group (DRG) and the ventral respiratory group (VRG) of the medulla oblongata?

42. What types of sensory information modify the activities of the respiratory centers?

LEVEL 3 Critical Thinking and Clinical Applications

43. Billy's normal alveolar ventilation rate (AVR) during mild exercise is 6.0 l/min. While at the beach on a warm summer day, he goes snorkeling. The snorkel has a volume of 50 ml. Assuming that the water is not too cold and that snorkeling for Billy is mild exercise, what would his respiratory rate have to be for him to maintain an AVR of 6.0 l/min while snorkeling? (Assume a constant tidal volume of 500 ml and an anatomic dead space of 150 ml.)

44. A decrease in blood pressure triggers a baroreceptor reflex that leads to increased ventilation. What is the possible advantage of this reflex?

45. Mr. B. has had chronic advanced emphysema for 15 years. While hospitalized with a respiratory infection, he goes into respiratory distress. Without thinking, his nurse immediately administers pure oxygen, which causes Mr. B. to stop breathing. Why?

46. You spend the night at a friend's house during the winter. Your friend's home is quite old, and the hot-air furnace lacks a humidifier. When you wake up in the morning, you have a fair amount of nasal congestion and decide you might be coming down with a cold. After a steamy shower and some juice for breakfast, the nasal congestion disappears. Explain.

47. Why would a person with kyphosis (see Chapter 7, p. 229) exhibit a lower-than-normal vital capacity?

48. Why do individuals who are anemic generally not exhibit an increase in respiratory rate or tidal volume, even though their blood is not carrying enough oxygen?

49. Doris has an obstruction of her right primary bronchus. As a result, how would you expect the oxygen–hemoglobin saturation curve for her right lung to compare with that for her left?

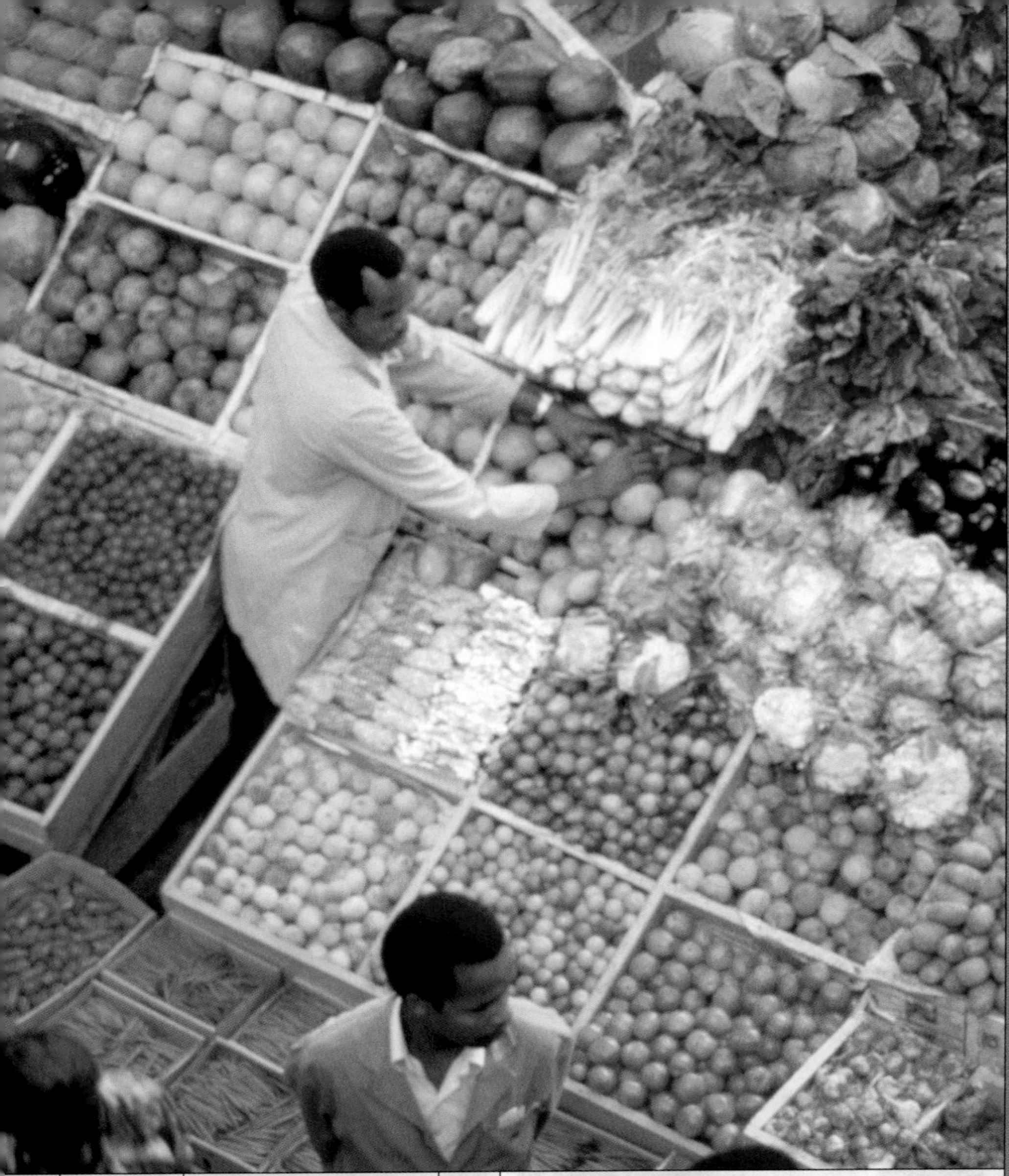

UNIT 5   CHAPTER   23   **24**   25   26   27

THE DIGESTIVE SYSTEM

Although it is important not to take the notion too literally, modern physiology has confirmed that, in a sense, "You are what you eat." We all need a regular supply of nutrients in our diet. Some of those compounds are broken down for energy, but many eventually become part of us—building blocks of our cells, tissues, and organs. Unfortunately, the nutrients in food are not ready for use by our cells. Without processing by the digestive system, food would be of no more use to us than a lump of coal in the gas tank of a car.

While the digestive system is familiar to everyone, few people give it any serious thought unless it malfunctions. Nevertheless, we spend hours of conscious effort filling and emptying it. References to this system are part of our everyday language. Think of how often we use expressions relating to the digestive system: We may "have a gut feeling," "want to chew on" something, or find someone's opinions "hard to swallow." When something does go wrong with the digestive system—even something minor—most of us seek treatment immediately. For this reason, television advertisements promote toothpaste and mouthwash, diet supplements, antacids, and laxatives. This chapter will explain how these commercial products work, as it discusses the organs of the digestive system and their varied functions.

24-1 THE DIGESTIVE SYSTEM: AN OVERVIEW

Objectives

- Identify the organs of the digestive system and their major functions.
- Describe the functional histology of the digestive tract.
- Explain the processes by which materials move through the digestive tract.
- Outline the mechanisms that regulate digestion.

The digestive system may not have the visibility of the integumentary system or the glamour of the reproductive system, but it is certainly just as important. All living organisms must obtain nutrients from their environment to sustain life. These substances are used as raw materials for synthesizing essential compounds (anabolism) or are decomposed to provide the energy that cells need to continue functioning (catabolism). ⚭ pp. 4, 39-40 The catabolic reactions require two essential ingredients: (1) oxygen and (2) organic molecules, such as carbohydrates, fats, or proteins, that can be broken down by intracellular enzymes. Obtaining oxygen and organic molecules can be relatively straightforward for a single-celled organism like an amoeba, but the situation is much more complicated for animals as large and complex as humans. Along with increasing size and complexity comes a division of labor, and the need for the coordination of organ system activities.

In our bodies, the respiratory system works in concert with the cardiovascular system to supply the necessary oxygen. The **digestive system**, working with the cardiovascular and lymphatic systems, provides the organic molecules. In effect, the digestive system provides both the fuel that keeps all the body's cells running and the building blocks needed for cell growth and repair.

The digestive system consists of a muscular tube, the **digestive tract**, also called the *gastrointestinal (GI) tract* or *alimentary canal*, and various **accessory organs**. The *oral cavity* (mouth), *pharynx, esophagus, stomach, small intestine,* and *large intestine* make up the digestive tract. Accessory digestive organs include the teeth, tongue, and various *glandular organs,* such as the salivary glands, liver, and pancreas, which secrete into ducts emptying into the digestive tract. Food enters the digestive tract and

passes along its length. On the way, the secretions of the glandular organs, which contain water, enzymes, buffers, and other components, assist in preparing organic and inorganic nutrients for absorption across the epithelium of the digestive tract.

Figure 24–1● shows the major components of the digestive system. The digestive tract begins at the oral cavity, continues through the pharynx, esophagus, stomach, small intestine, and large intestine, and ends at the anus. These structures have overlapping functions, but each has certain areas of specialization and shows distinctive histological characteristics.

☐ FUNCTIONS OF THE DIGESTIVE SYSTEM

We can regard digestive functions as a series of integrated steps:

1. **Ingestion** occurs when materials enter the digestive tract via the mouth. Ingestion is an active process involving conscious choice and decision making.

●FIGURE 24–1
The Components of the Digestive System.
The major regions and accessory organs of the digestive tract, together with their primary functions.

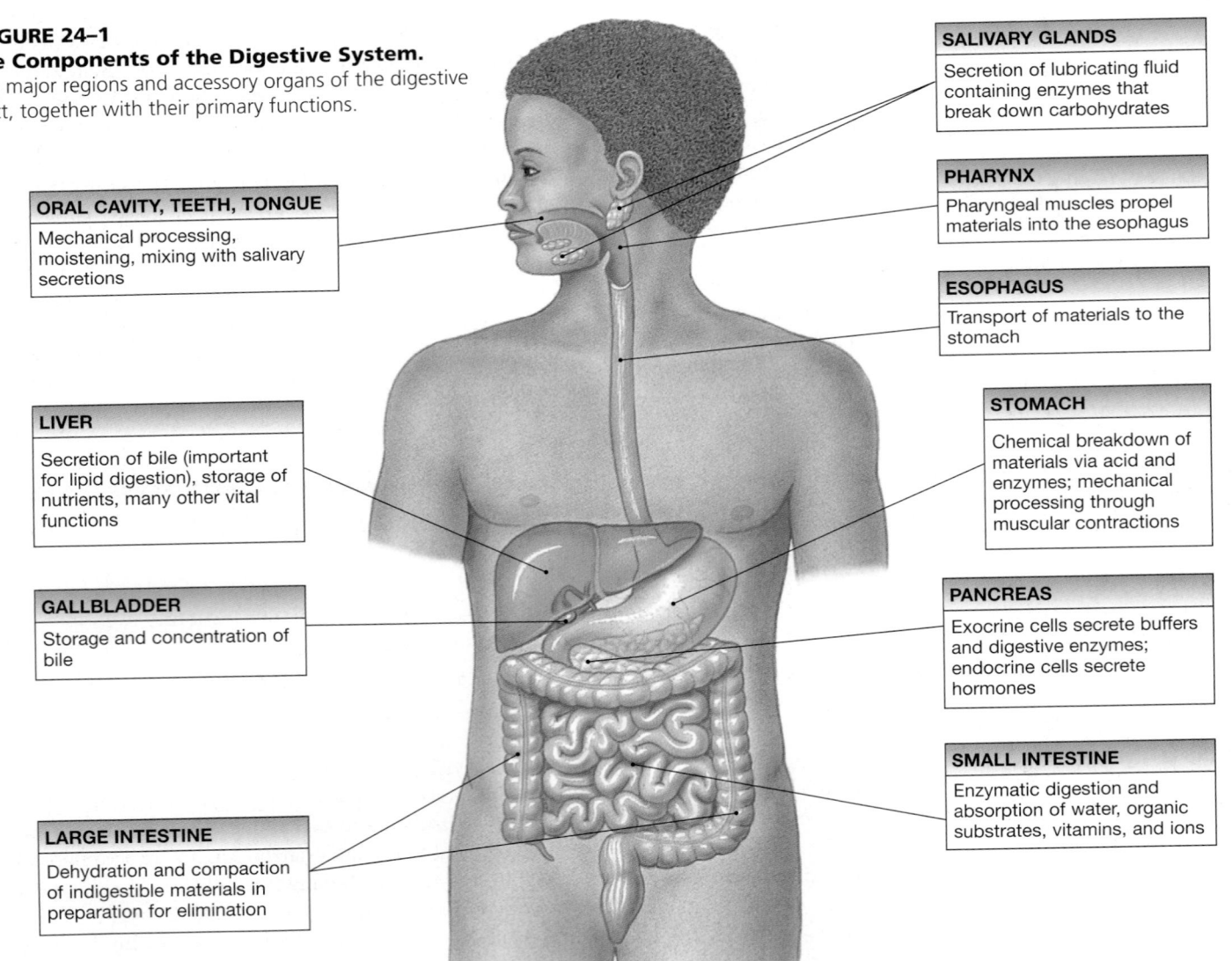

SALIVARY GLANDS
Secretion of lubricating fluid containing enzymes that break down carbohydrates

PHARYNX
Pharyngeal muscles propel materials into the esophagus

ESOPHAGUS
Transport of materials to the stomach

STOMACH
Chemical breakdown of materials via acid and enzymes; mechanical processing through muscular contractions

PANCREAS
Exocrine cells secrete buffers and digestive enzymes; endocrine cells secrete hormones

SMALL INTESTINE
Enzymatic digestion and absorption of water, organic substrates, vitamins, and ions

ORAL CAVITY, TEETH, TONGUE
Mechanical processing, moistening, mixing with salivary secretions

LIVER
Secretion of bile (important for lipid digestion), storage of nutrients, many other vital functions

GALLBLADDER
Storage and concentration of bile

LARGE INTESTINE
Dehydration and compaction of indigestible materials in preparation for elimination

2. **Mechanical processing** is crushing and shearing that makes materials easier to propel along the digestive tract. It also increases their surface area, making them more susceptible to enzymatic attack. Mechanical processing may or may not be required before ingestion; you can swallow liquids immediately, but must process most solids first. Tearing and mashing with the teeth, followed by squashing and compaction by the tongue, are examples of preliminary mechanical processing. Swirling, mixing, and churning motions of the stomach and intestines provide mechanical processing after ingestion.

3. **Digestion** refers to the chemical breakdown of food into small organic fragments suitable for absorption by the digestive epithelium. Simple molecules in food, such as glucose, can be absorbed intact, but epithelial cells have no way to deal with molecules the size and complexity of proteins, polysaccharides, or triglycerides. These molecules must be disassembled by digestive enzymes prior to absorption. For example, the starches in a potato are of no value until enzymes have broken them down to simple sugars that your digestive epithelium can absorb and distribute to your cells.

4. **Secretion** is the release of water, acids, enzymes, buffers, and salts by the epithelium of the digestive tract and by glandular organs.

5. **Absorption** is the movement of organic substrates, electrolytes (inorganic ions), vitamins, and water across the digestive epithelium and into the interstitial fluid of the digestive tract.

6. **Excretion** is the removal of waste products from body fluids. The digestive tract and glandular organs discharge waste products in secretions that enter the lumen of the tract. Most of these waste products, after mixing with the indigestible residue of the digestive process, will leave the body. The ejection of materials from the digestive tract, a process called **defecation** (def-e-KĀ-shun), or *egestion*, eliminates materials as **feces**.

The lining of the digestive tract also plays a protective role by safeguarding surrounding tissues against (1) the corrosive effects of digestive acids and enzymes; (2) mechanical stresses, such as abrasion; and (3) bacteria that either are swallowed with food or reside in the digestive tract. The digestive epithelium and its secretions provide a nonspecific defense against these bacteria. When bacteria reach the underlying layer of areolar tissue, the *lamina propria* ∞ p. 133, they are attacked by macrophages and other cells of the immune system.

We will explore specific functions in more detail as we proceed through the individual regions and components of the system. First, however, we consider several structural and functional characteristics of the system as a whole.

■ THE DIGESTIVE ORGANS AND THE PERITONEUM

The abdominopelvic cavity contains the *peritoneal cavity,* which is lined by a serous membrane that consists of a superficial mesothelium covering a layer of areolar tissue. ∞ pp. 22, 116, 125 We can divide the serous membrane into the serosa, or *visceral peritoneum,* which covers organs within the peritoneal cavity,

and the *parietal peritoneum,* which lines the inner surfaces of the body wall.

The serous membrane lining the peritoneal cavity continuously produces peritoneal fluid, which provides essential lubrication. Because a thin layer of peritoneal fluid separates the parietal and visceral surfaces, relative movement can occur without friction and resulting irritation. About seven liters of fluid is secreted and reabsorbed each day, although the volume within the peritoneal cavity at any one time is very small. Liver disease, kidney disease, and heart failure can cause an increase in the rate at which fluids move into the peritoneal cavity. The accumulation of fluid creates a characteristic abdominal swelling called **ascites** (a-SĪ-tēz). The distortion of internal organs by this fluid can result in symptoms such as heartburn, indigestion, and lower back pain.

Mesenteries

Portions of your digestive tract are suspended within the peritoneal cavity by sheets of serous membrane that connect the parietal peritoneum with the visceral peritoneum. These **mesenteries** (MEZ-en-ter-ēz) are double sheets of peritoneal membrane. The areolar tissue between the mesothelial surfaces provides an access route for the passage of blood vessels, nerves, and lymphatic vessels to and from the digestive tract. Mesenteries also stabilize the positions of the attached organs and prevent your intestines from becoming entangled during digestive movements or sudden changes in body position.

In embryonic development, the digestive tract and accessory organs are suspended within the peritoneal cavity by *dorsal* and *ventral mesenteries* (Figure 24–2a●). The ventral mesentery later disappears along most of the digestive tract, persisting in adults in only two places: on the ventral surface of the stomach, between the stomach and the liver (the lesser omentum); and between the liver and the anterior abdominal wall (the falciform ligament) (Figure 24–2b,c,d●). The **lesser omentum** (ō-MEN-tum; *omentum,* fat skin) stabilizes the position of the stomach and provides an access route for blood vessels and other structures entering or leaving the liver. The **falciform** (FAL-si-form) **ligament** helps stabilize the position of the liver relative to the diaphragm and abdominal wall.

AM Embryology Summary 19: The Development of the Digestive System

As the digestive tract elongates, it twists and turns within the crowded peritoneal cavity. The dorsal mesentery of the stomach becomes greatly enlarged and forms an enormous pouch that extends inferiorly between the body wall and the anterior surface of the small intestine. This pouch, the **greater omentum** (Figure 24–2b,d●), hangs like an apron from the lateral and inferior borders of the stomach. Adipose tissue in the greater omentum conforms to the shapes of the surrounding organs, providing padding and protection across the anterior and lateral surfaces of the abdomen. The lipids in the adipose tissue are an important energy reserve. The greater omentum also provides insulation that reduces heat loss across the anterior abdominal wall.

All but the first 25 cm of the small intestine is suspended by the **mesentery proper**, a thick mesenterial sheet that provides stability,

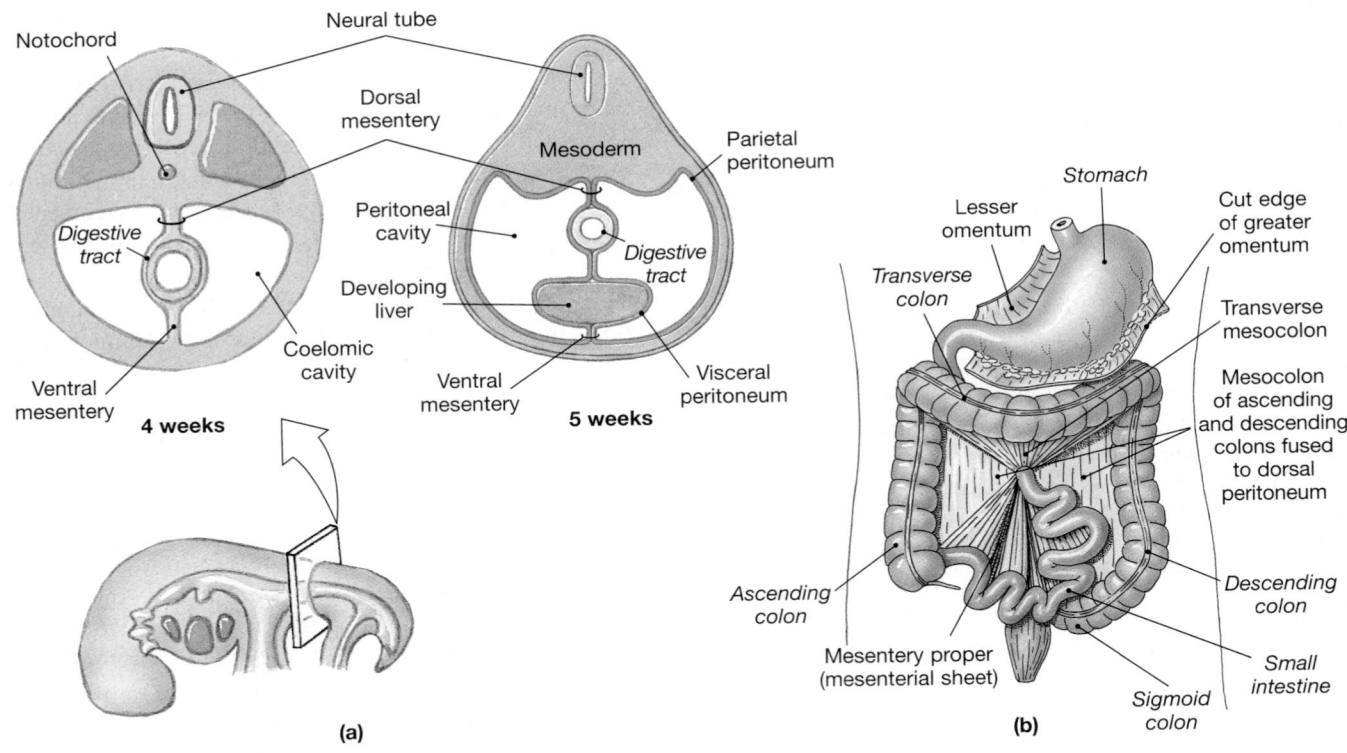

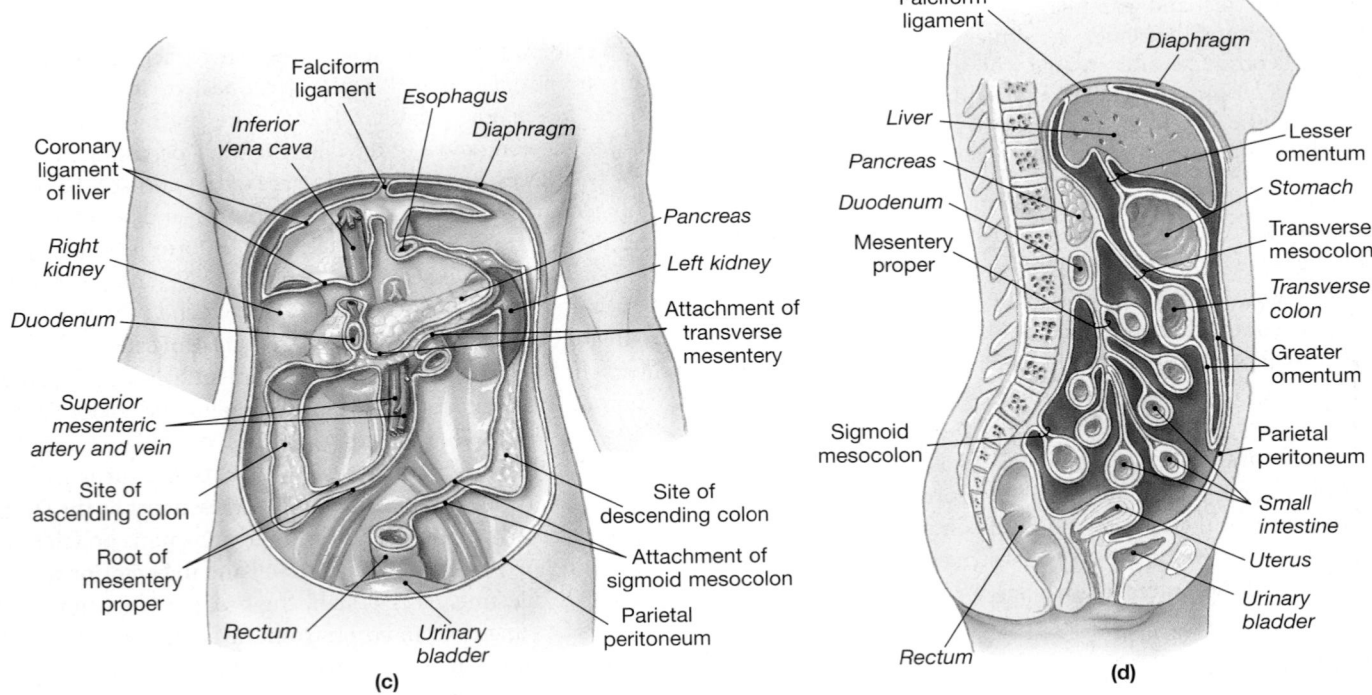

●**FIGURE 24–2**
Mesenteries. **(a)** During embryonic development, the digestive tube is initially suspended by dorsal and ventral mesenteries. In adults, the ventral mesentery is lost, except where it connects the stomach to the liver (at the lesser omentum) and the liver to the anterior body wall and diaphragm (at the falciform ligament). **(b)** A diagrammatic view of the organization of mesenteries in an adult. As the digestive tract enlarges, mesenteries associated with the proximal portion of the small intestine, the pancreas, and the ascending and descending portions of the colon fuse to the body wall. **(c)** An anterior view of the empty peritoneal cavity, showing attachment sites where fusion occurs. **(d)** A sagittal section of the mesenteries of an adult. Notice that the pancreas, duodenum, and rectum are retroperitoneal.

but permits some independent movement. The mesentery associated with the initial portion of the small intestine (the *duodenum*) and the pancreas fuses with the posterior abdominal wall, locking those structures in position. After this fusion is completed, only their anterior surfaces remain covered by peritoneum. Because their mass lies posterior to, rather than within, the peritoneal cavity, these organs are called **retroperitoneal** (*retro*, behind).

A **mesocolon** is a mesentery associated with a portion of the large intestine. During normal development, the mesocolon of the *ascending colon*, the *descending colon*, and the *rectum* of the large intestine fuse to the dorsal body wall. These regions become locked in place. Thereafter, these organs are retroperitoneal, with the visceral peritoneum covering only their anterior surfaces and portions of their lateral surfaces (Figure 24–2b,c,d●). The **transverse mesocolon**, which supports the transverse colon, and the **sigmoid mesocolon**, which supports the sigmoid colon, are all that remains of the original embryonic mesocolon.

An inflammation of the peritoneal membrane produces symptoms of **peritonitis** (per-i-tō-NĪ-tis), a painful condition that interferes with the normal functioning of the affected organs. Physical damage, chemical irritation, and bacterial invasion of the peritoneum can lead to severe and even fatal cases of peritonitis. In untreated appendicitis, peritonitis may be caused by the rupturing of the appendix and the subsequent release of bacteria into the peritoneal cavity. Peritonitis is a potential complication of any surgery in which the peritoneal cavity is opened or of any disease that perforates the walls of the stomach or intestines.

☐ HISTOLOGICAL ORGANIZATION OF THE DIGESTIVE TRACT

The major layers of the digestive tract include (1) the *mucosa*, (2) the *submucosa*, (3) the *muscularis externa*, and (4) the *serosa*. Sectional, diagrammatic views of these layers are presented in Figure 24–3●. The structure of the layers varies by region; the figure is a composite view that most closely resembles the appearance of the small intestine, the longest segment of the digestive tract.

The Mucosa

The inner lining, or **mucosa**, of the digestive tract is a *mucous membrane* consisting of an epithelium, moistened by glandular secretions, and a *lamina propria* of areolar tissue. ◯◯ p. 132-133

THE DIGESTIVE EPITHELIUM The mucosal epithelium is either simple or stratified, depending on its location and the stresses to which it is most often subjected. The oral cavity, pharynx, and esophagus (where mechanical stresses are most severe) are lined by a stratified squamous epithelium, whereas the stomach, the small intestine, and almost the entire length of the large intestine (where absorption occurs) have a simple columnar epithelium that contains goblet cells. Scattered among the columnar cells are **enteroendocrine cells**, which secrete hormones that coordinate the activities of the digestive tract and the accessory glands.

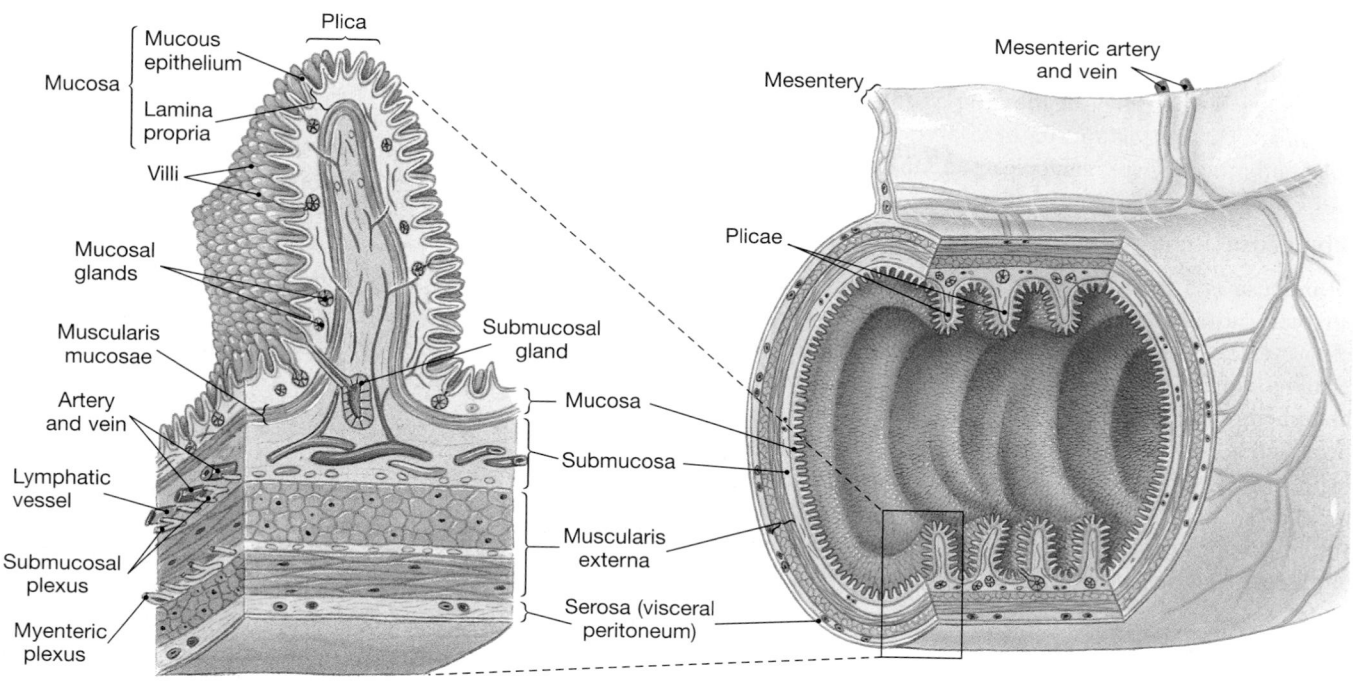

●**FIGURE 24–3**
The Structure of the Digestive Tract. A diagrammatic view of a representative portion of the digestive tract. The features illustrated are those of the small intestine.

The lining of the digestive tract is often thrown into longitudinal folds, which disappear as the tract fills, and permanent transverse folds, or *plicae* (PLĪ-sē; singular, *plica* [PLĪ-ka]) (Figure 24–3●). The folding dramatically increases the surface area available for absorption. The secretions of gland cells located in the mucosa and submucosa—or in accessory glandular organs—are carried to the epithelial surfaces by ducts.

The life span of a typical epithelial cell varies from two to three days in the esophagus to up to six days in the large intestine. The lining of the entire digestive tract is continuously renewed through the divisions of epithelial stem cells, keeping pace with the rate of cell destruction and loss at epithelial surfaces. This high rate of cell division explains why radiation and anticancer drugs that inhibit mitosis have drastic effects on the digestive tract. Lost epithelial cells are no longer replaced, and the cumulative damage to the epithelial lining quickly leads to problems in absorbing nutrients. In addition, the exposure of the lamina propria to digestive enzymes can cause internal bleeding and other serious problems.

THE LAMINA PROPRIA The lamina propria consists of a layer of areolar tissue that also contains blood vessels, sensory nerve endings, lymphatic vessels, smooth muscle cells, and scattered areas of lymphoid tissue. In the oral cavity, pharynx, esophagus, stomach, and *duodenum* (the proximal portion of the small intestine), the lamina propria also contains the secretory cells of mucous glands.

In most areas of the digestive tract, the lamina propria contains a narrow band of smooth muscle and elastic fibers. This band is called the **muscularis** (mus-kū-LAR-is) **mucosae**. The smooth muscle cells in the muscularis mucosae are arranged in two concentric layers (Figure 24–3●). The inner layer encircles the lumen (the *circular muscle*), and the outer layer contains muscle cells oriented parallel to the long axis of the tract (the *longitudinal layer*). Contractions in these layers alter the shape of the lumen and move the epithelial pleats and folds.

The Submucosa

The **submucosa** is a layer of dense irregular connective tissue that surrounds the muscularis mucosae. The submucosa has large blood vessels and lymphatic vessels, and in some regions it also contains exocrine glands that secrete buffers and enzymes into the lumen of the digestive tract. Along its outer margin, the submucosa contains a network of intrinisic nerve fibers and scattered neurons. This **submucosal plexus**, or *plexus of Meissner*, contains sensory neurons, parasympathetic ganglionic neurons, and sympathetic postganglionic fibers that innervate the mucosa and submucosa (Figure 24–3●).

The Muscularis Externa

The submucosal plexus lies along the inner border of the **muscularis externa**, a region dominated by smooth muscle cells. Like the smooth muscle cells in the muscularis mucosae, those in the muscularis externa are arranged in an inner, circular layer and an outer, longitudinal layer. These layers play an essential role in mechanical processing and in the movement of materials along the digestive tract. The movements are coordinated primarily by the sensory neurons, interneurons, and motor neurons of the enteric nervous system (ENS). The ENS is innervated primarily by the parasympathetic division of the ANS. Sympathetic postganglionic fibers also synapse here, although many continue onward to innervate the mucosa and the **myenteric** (mī-en-TER-ik) **plexus** (*mys*, muscle + *enteron*, intestine), or *plexus of Auerbach*. This network of parasympathetic ganglia, sensory neurons, interneurons, and sympathetic postganglionic fibers lies sandwiched between the circular and longitudinal muscle layers. In general, parasympathetic stimulation increases muscle tone and activity; sympathetic stimulation promotes muscular inhibition and relaxation.

The Serosa

Along most portions of the digestive tract inside the peritoneal cavity, the muscularis externa is covered by a serous membrane known as the **serosa** (Figure 24–3●). p. 133 There is no serosa covering the muscularis externa of the oral cavity, pharynx, esophagus, and rectum, where a dense network of collagen fibers firmly attaches the digestive tract to adjacent structures. This fibrous sheath is called an *adventitia*.

☐ THE MOVEMENT OF DIGESTIVE MATERIALS

The muscular layers of the digestive tract consist of *visceral smooth muscle tissue*, a type of smooth muscle introduced in Chapter 10. p. 326 The smooth muscle along the digestive tract shows rhythmic cycles of activity due to the presence of *pacesetter cells*. These smooth muscle cells undergo spontaneous depolarization, and their contraction triggers a wave of contraction that spreads through the entire muscular sheet. Pacesetter cells are located in the muscularis mucosae and muscularis externa, the layers of which surround the lumen of the digestive tract. The coordinated contractions of the muscularis externa play a vital role in the movement of materials along the tract, through *peristalsis*, and in mechanical processing, through *segmentation*.

Peristalsis

The muscularis externa propels materials from one portion of the digestive tract to another by contractions known as **peristalsis** (per-i-STAL-sis). Peristalsis consists of waves of muscular contractions that move a **bolus**, or small oval mass of digestive contents, along the length of the digestive tract. During a peristaltic movement, the circular muscles contract behind the bolus, while circular muscles ahead of the bolus relax. Longitudinal muscles ahead of the bolus contract next, shortening adjacent segments. A wave of contraction in the circular muscles then forces the bolus in the desired direction (Figure 24–4●).

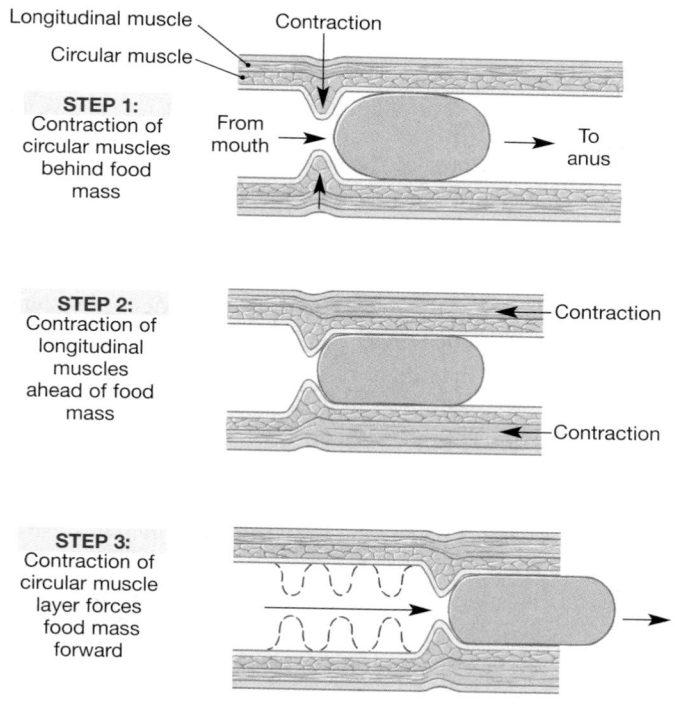

STEP 1:
Contraction of circular muscles behind food mass

STEP 2:
Contraction of longitudinal muscles ahead of food mass

STEP 3:
Contraction of circular muscle layer forces food mass forward

●**FIGURE 24–4**
Peristalsis. Peristalsis propels materials along the length of the digestive tract.

Segmentation

Most areas of the small intestine and some portions of the large intestine undergo cycles of contraction that produce **segmentation**. These movements churn and fragment the bolus, mixing the contents with intestinal secretions. Because they do not follow a set pattern, segmentation movements do not push materials along the tract in any one direction.

□ CONTROL OF THE DIGESTIVE FUNCTION

The activities of the digestive system are regulated by neural, hormonal, and local mechanisms (Figure 24–5●).

Neural Mechanisms

The movement of materials along your digestive tract, as well as many secretory functions, is controlled primarily by neural mechanisms. For example, peristaltic movements limited to a few centimeters of the digestive tract are triggered by sensory receptors in the walls of the digestive tract. The motor neurons that control smooth muscle contraction and glandular secretion are located in the myenteric plexus. These neurons are usually considered to be parasympathetic, because some of them are innervated by parasympathetic preganglionic fibers. However, the plexus also contains sensory neurons, motor neurons, and interneurons responsible for local

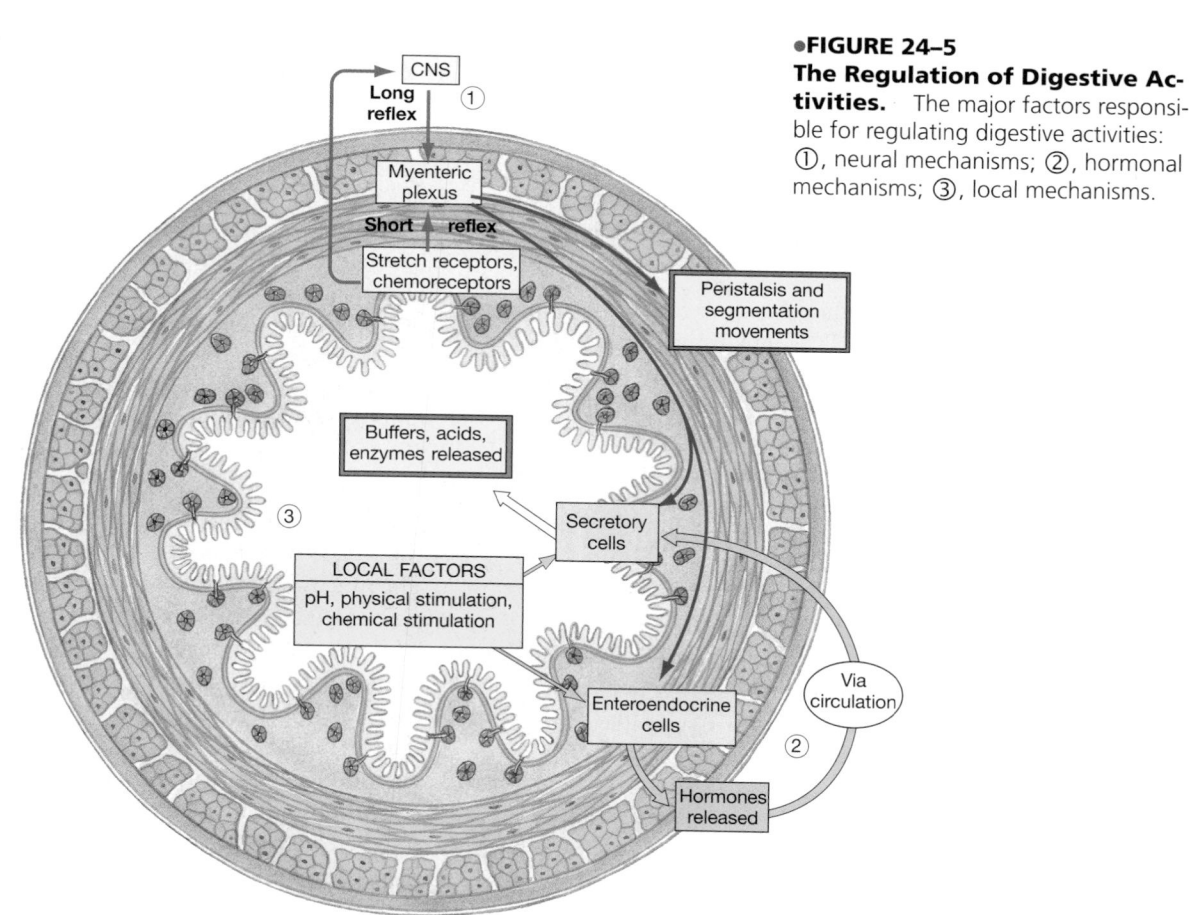

●**FIGURE 24–5**
The Regulation of Digestive Activities. The major factors responsible for regulating digestive activities: ①, neural mechanisms; ②, hormonal mechanisms; ③, local mechanisms.

reflexes that operate entirely outside of the control of the central nervous system. As we noted in Chapter 16, the reflexes controlled by these neurons are called *short reflexes*. ∞ p. 549 These reflexes are also called *myenteric reflexes*, and the term *enteric nervous system* is often used to refer to the neural network that coordinates the myenteric reflexes along the digestive tract.

In general, short reflexes control activities in one region of the digestive tract. The control may involve coordinating local peristalsis and triggering the secretion of digestive glands. Many neurons are involved: The enteric nervous system has roughly as many neurons as the spinal cord, and as many neurotransmitters as the brain. The specific functions and interactions of these neurotransmitters in the enteric nervous system remain largely unknown.

Sensory information from receptors in the digestive tract is also distributed to the CNS, where it can trigger *long reflexes* ∞ p. 549, which involve interneurons and motor neurons in the CNS. Long reflexes provide a higher level of control over digestive and glandular activities, generally controlling large-scale peristaltic waves that move materials from one region of the digestive tract to another. Long reflexes may involve parasympathetic motor fibers in the glossopharyngeal, vagus, or pelvic nerves that synapse in the myenteric plexus.

Hormonal Mechanisms

The sensitivity of the smooth muscle cells to neural commands can be enhanced or inhibited by digestive hormones. Your digestive tract produces at least 18 hormones that affect almost every aspect of digestive function, and some of them also affect the activities of other systems. The hormones (*gastrin, secretin,* and others), which are peptides produced by enteroendocrine cells in the digestive tract, reach their target organs by distribution in the bloodstream. We will consider each of these hormones further as we proceed down the length of the digestive tract.

Local Mechanisms

Prostaglandins, histamine, and other chemicals released into interstitial fluid may affect adjacent cells within a small segment of the digestive tract. These local messengers are important in coordinating a response to conditions (such as a change in the local pH or certain chemical or physical stimuli) that affect only a portion of the tract. For example, the release of histamine in the lamina propria of the stomach stimulates the secretion of acid by cells in the adjacent epithelium.

✓ What is the importance of the mesenteries?

✓ Which would be more efficient in propelling intestinal contents from one place to another—peristalsis or segmentation?

✓ What effect would a drug that blocks the parasympathetic stimulation of the digestive tract have on peristalsis?

Answers start on page Q-1

🔵 The gross anatomy of the digestive system can be reviewed on the **Anatomy CD-ROM**/Digestive system/Digestive/Accessory organs, and /Flythrough.

24–2 THE ORAL CAVITY

Objectives

■ Describe the anatomy of the oral cavity.

■ Discuss the functions of the major structures and regions of the oral cavity.

Our exploration of the digestive tract will follow the path of ingested materials from the mouth to the anus. The mouth opens into the **oral cavity**, or **buccal** (BUK-al) **cavity** (Figure 24–6●). We can summarize the functions of the oral cavity as follows: (1) *analysis* of material before swallowing; (2) *mechanical processing* through the actions of the teeth, tongue, and palatal surfaces; (3) *lubrication* by mixing with mucus and salivary gland secretions; and (4) limited *digestion* of carbohydrates and lipids.

The oral cavity is lined by the **oral mucosa**, which has a stratified squamous epithelium. Only the regions exposed to severe abrasion (such as the superior surface of the tongue and the opposing surfaces of the hard palate, part of the roof of the mouth) are covered by a layer of keratinized cells. The epithelial lining of the cheeks, lips, and inferior surface of the tongue is relatively thin, nonkeratinized, and delicate. Although nutrients are not absorbed in the oral cavity, the mucosa inferior to the tongue is thin enough and vascular enough to permit the rapid absorption of lipid-soluble drugs. *Nitroglycerin* may be administered via this route to treat acute angina attacks. ∞ p. 696

The mucosae of the **cheeks**, or lateral walls of the oral cavity, are supported by pads of fat and the buccinator muscles. ∞ p. 343 Anteriorly, the mucosa of each cheek is continuous with that of the lips, or **labia** (LĀ-bē-uh; singular, *labium*). The **vestibule** is the space between the cheeks (or lips) and the teeth. The **gingivae** (JIN-ji-vē), or *gums*, are ridges of oral mucosa that surround the base of each tooth on the alveolar processes of the maxillary bones and mandible. ∞ pp. 220, 223 In most regions, the gingivae are firmly bound to the periostea of the underlying bones.

The roof of the oral cavity is formed by the hard and soft palates; the tongue dominates its floor (Figure 24–6b●). The floor of the mouth inferior to the tongue receives extra support from the geniohyoid and mylohyoid muscles. ∞ p. 348 The *hard palate* is formed by the palatine processes of the maxillary bones and the horizontal plates of the palatine bones. A prominent central ridge, or *raphe*, extends along the midline of the hard palate. The mucosa lateral and anterior to the raphe is thick, with complex ridges. When your tongue compresses food against your hard palate, these ridges provide traction. The *soft palate* lies posterior to the hard palate. A thinner and more delicate mucosa covers the posterior margin of the hard palate and extends onto the soft palate.

The posterior margin of the soft palate supports the **uvula** (Ū-vū-luh), a dangling process that helps prevent food from entering the pharynx prematurely. On either side of the uvula are two pairs of muscular *pharyngeal arches* (Figure 24–6●). The more anterior **palatoglossal** (pal-a-tō-GLOS-al) **arch** extends between the

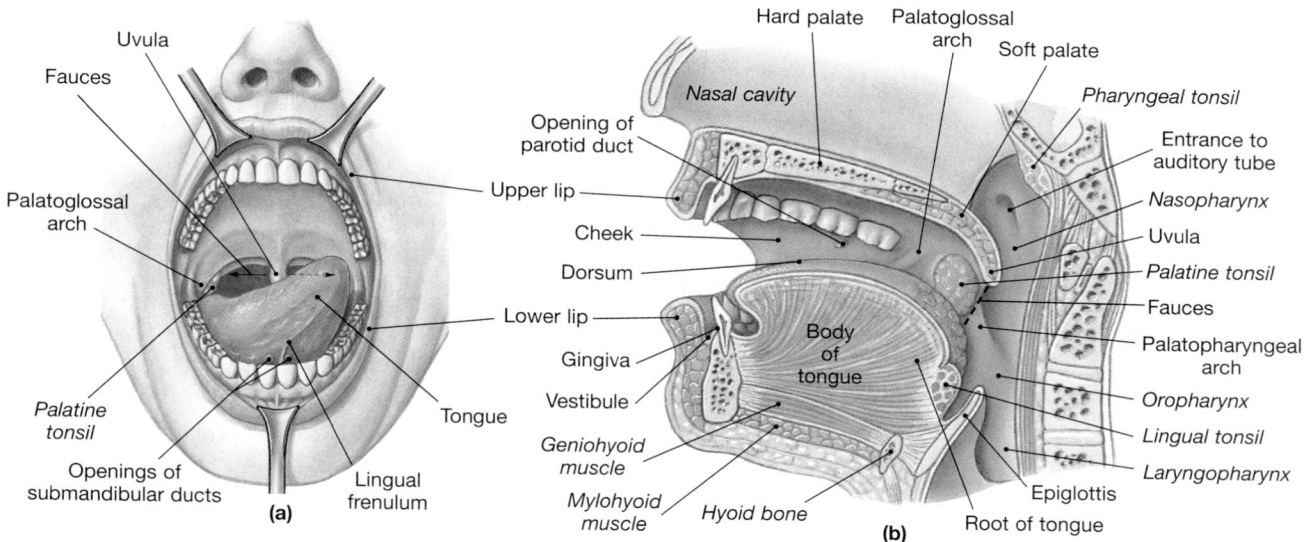

●**FIGURE 24–6**
The Oral Cavity. **(a)** An anterior view of the oral cavity, as seen through the open mouth. **(b)** A sagittal section. ATLAS Plates 3.1, 5.3d

soft palate and the base of the tongue. A curving line that connects the palatoglossal arches and uvula forms the boundaries of the **fauces** (FAW-sēz), the passageway between the oral cavity and the oropharynx. The more posterior **palatopharyngeal** (pal-a-tō-fa-RIN-jē-al) **arch** extends from the soft palate to the pharyngeal wall. A palatine tonsil lies between the palatoglossal and palatopharyngeal arches on either side.

☐ THE TONGUE

The **tongue** (Figure 24–6●) manipulates materials inside the mouth and is occasionally used to bring foods (such as ice cream) into the oral cavity. The primary functions of the tongue are (1) mechanical processing by compression, abrasion, and distortion; (2) manipulation to assist in chewing and to prepare material for swallowing; (3) sensory analysis by touch, temperature, and taste receptors, and (4) secretion of mucins and the enzyme *lingual lipase.*

We can divide the tongue into an anterior **body**, or *oral portion*, and a posterior **root**, or *pharyngeal portion.* The superior surface, or *dorsum*, of the body contains a forest of fine projections, the *lingual papillae.* ∞ p. 567 The thickened epithelium covering each papilla assists the tongue in moving materials. A V-shaped line of circumvallate papillae roughly indicates the boundary between the body and the root of the tongue, which is situated in the oropharynx (Figure 24–6b●).

The epithelium covering the inferior surface of the tongue is thinner and more delicate than that of the dorsum. Along the inferior midline is the **lingual frenulum** (FREN-ū-lum; *frenulum*, a small bridle), a thin fold of mucous membrane that connects the body of the tongue to the mucosa covering the floor of the oral cavity. Ducts from two pairs of salivary glands open on either side of the lingual frenulum, which serves to prevent extreme movements of the tongue. However, if your lingual frenulum is *too* restrictive, you cannot eat or speak normally. Properly diagnosed,

this condition, called **ankyloglossia** (ang-ki-lō-GLOS-ē-uh), can be corrected surgically.

The tongue's epithelium is flushed by the secretions of small glands that extend into the underlying lamina propria. These secretions contain water, mucins, and the enzyme **lingual lipase**, which works over a broad range of pH value (3.0–6.0), so it can start lipid digestion immediately. Because lingual lipase tolerates an acid environment, it can continue to break down lipids—specifically, triglycerides—for a considerable time after the food reaches the stomach.

Your tongue contains two groups of skeletal muscles: (1) **intrinsic tongue muscles** and (2) **extrinsic tongue muscles**. All gross movements of the tongue are performed by the relatively large extrinsic muscles. ∞ p. 346 The smaller intrinsic muscles change the shape of the tongue and assist the extrinsic muscles during precise movements, as in speech. Both intrinsic and extrinsic tongue muscles are under the control of the hypoglossal nerve (XII).

☐ SALIVARY GLANDS

Three pairs of salivary glands secrete into the oral cavity (Figure 24–7a●). Each pair has a distinctive cellular organization and produces *saliva*, a mixture of glandular secretions, with slightly different properties:

1. The large **parotid** (pa-ROT-id) **salivary glands** lie inferior to the zygomatic arch deep to the skin that covers the lateral and posterior surface of the mandible. Each gland has an irregular shape, extending from the mastoid process of the temporal bone across the outer surface of the masseter muscle. ∞ pp. 218, 346 The parotid salivary glands produce a thick, serous secretion containing large amounts of *salivary amylase*, an enzyme that breaks down starches (complex carbohydrates). The secretions of each parotid gland are drained by a **parotid duct** (*Stensen's duct*), which empties into the vestibule at the level of the second upper molar.

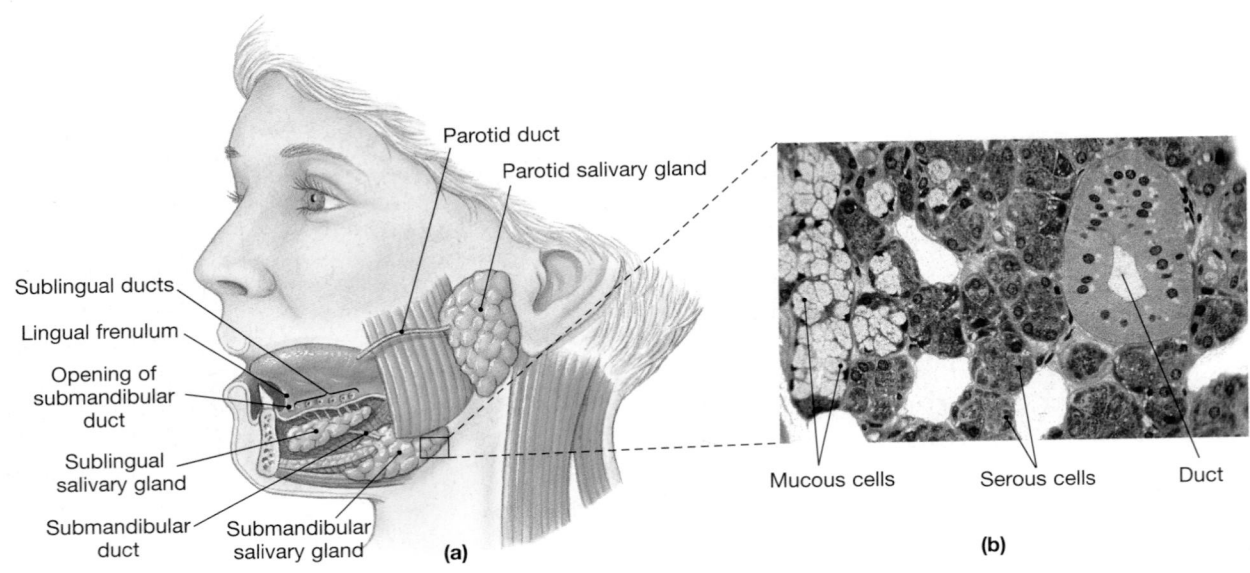

•FIGURE 24–7
The Salivary Glands. **(a)** A lateral view, showing the relative positions of the salivary glands and ducts on the left side of the head. For clarity, the left ramus and body of the mandible have been removed. For the positions of the ducts in the oral cavity, see Figure 24–6. **(b)** The submandibular gland secretes a mixture of mucins, produced by mucous cells, and enzymes, produced by serous cells. (LM × 303)
ATLAS Plates 5.2, 5.3a,b

2. The **sublingual** (sub-LING-gwal) **salivary glands** are covered by the mucous membrane of the floor of the mouth. These glands produce a watery, mucous secretion that acts as a buffer and lubricant. Numerous **sublingual ducts** (*Rivinus' ducts*) open along either side of the lingual frenulum.

3. The **submandibular salivary glands** are situated in the floor of the mouth along the inner surfaces of the mandible within a depression called the *mandibular groove*. The submandibular glands (Figure 24–7a•) secrete a mixture of buffers, glycoproteins called *mucins*, and salivary amylase. The **submandibular ducts** (*Wharton's ducts*) open into the mouth on either side of the lingual frenulum immediately posterior to the teeth (Figure 24–6a•).

Saliva

Your salivary glands produce 1.0–1.5 liters of saliva each day. Saliva is 99.4 percent water, and the remaining 0.6 percent includes an assortment of electrolytes (principally Na^+, Cl^-, and HCO_3^-), buffers, glycoproteins, antibodies, enzymes, and waste products. The glycoproteins, called **mucins**, are primarily responsible for the lubricating action of saliva. ∞ p. 57 About 70 percent of saliva originates in the submandibular salivary glands, 25 percent in the parotids, and the remaining 5 percent in the sublingual salivary glands.

A continuous background level of secretion flushes the oral surfaces, helping keep them clean. Buffers in the saliva keep the pH of your mouth near 7.0 and prevent the buildup of acids produced by bacterial action. In addition, saliva contains antibodies (IgA) and lysozymes that help control populations of oral bacteria. ∞ pp. 807, 571 A reduction in or elimination of salivary secretions, caused by radiation exposure, emotional distress, or other factors, triggers a bacterial population explosion in the oral cavity. This proliferation rapidly leads to recurring infections and the progressive erosion of the teeth and gums.

The saliva produced when you eat has a variety of functions, including:

• Lubricating the mouth.

• Moistening and lubricating materials in the mouth.

• Dissolving chemicals that can stimulate the taste buds and provide sensory information about the material.

• Initiating the digestion of complex carbohydrates before the material is swallowed. The enzyme involved is **salivary amylase**, which is also known as *ptyalin* or *alpha-amylase*. Although the digestive process begins in the oral cavity, it is not completed there, and no absorption of nutrients occurs across the lining of the cavity. Saliva also contains a small amount of lingual lipase that is secreted by the glands of the tongue.

The **mumps virus** most often targets the salivary glands, especially the parotid salivary glands, although other organs can also become infected. Infection typically occurs at 5–9 years of age. The first exposure stimulates the production of antibodies and, in most cases, confers permanent immunity; active immunity can be conferred by immunization. In post-adolescent males, the mumps virus can also infect the testes and cause sterility. Infection of the pancreas by the mumps virus can produce temporary or permanent diabetes; other organ systems, including the central nervous system, are affected in severe cases. An effective mumps vaccine is available. Widespread distribution of that vaccine has almost eliminated the incidence of the disease in the United States.

Control of Salivary Secretions

Salivary secretions are normally controlled by the autonomic nervous system. Each salivary gland receives parasympathetic and sympathetic innervation. The parasympathetic outflow originates in the **salivatory nuclei** of the medulla oblongata and synapses in the submandibular and otic ganglia. ∞ pp. 498, 500, 542 Any object in your mouth can trigger a salivary reflex by stimulating receptors monitored by the trigeminal nerve (V) or by stimulating taste buds innervated by cranial nerves VII, IX, or X. Parasympathetic stimulation accelerates secretion by all the salivary glands, resulting in the production of large amounts of saliva. The role of the sympathetic innervation remains uncertain; evidence suggests that it provokes the secretion of small amounts of very thick saliva.

The salivatory nuclei are also influenced by other brain stem nuclei, as well as by the activities of higher centers. For example, chewing with an empty mouth, the smell of food, or even thinking about food will initiate an increase in salivary secretion rates; that is why chewing gum is so effective at keeping your mouth moist. The presence of irritating stimuli in the esophagus, stomach, or intestines will also accelerate the production of saliva, as will the sensation of nausea. Increased saliva production in response to unpleasant stimuli helps reduce the magnitude of the stimulus by dilution, by a rinsing action, or by buffering strong acids or bases.

☐ THE TEETH

Movements of the tongue are important in passing food across the opposing surfaces of the **teeth**. The **occlusal**, or opposing, **surfaces** of your teeth perform chewing, or **mastication** (mas-ti-KĀ-shun), of food. Mastication breaks down tough connective tissues in meat and the plant fibers in vegetable matter, and it helps saturate the materials with salivary secretions and enzymes.

Figure 24–8a● is a sectional view through an adult tooth. The bulk of each tooth consists of a mineralized matrix similar to that of bone. This material, called **dentin**, differs from bone in that it does not contain cells. Instead, cytoplasmic processes extend into the dentin from cells in the central **pulp cavity**, an interior chamber. The pulp cavity receives blood vessels and nerves from the **root canal**, a narrow tunnel located at the **root**, or base, of the tooth. Blood vessels and nerves enter the root canal through an opening called the **apical foramen** to supply the pulp cavity.

The root of each tooth sits in a bony socket called an *alveolus*. Collagen fibers of the **periodontal ligament** extend from the dentin of the root to the bone of the alveolus, creating a strong articulation known as a gomphosis. ∞ p. 264 A layer of **cementum** (se-MEN-tum) covers the dentin of the root, providing protection and firmly anchoring the periodontal ligament. Cementum is very similar in histological structure to bone and is less resistant to erosion than is dentin.

The **neck** of the tooth marks the boundary between the root and the **crown**, the exposed portion of the tooth that projects above the soft tissue of the gingiva. A shallow groove called the **gingival sulcus** surrounds the neck of each tooth. The mucosa of the gingival sulcus is very thin and is not tightly bound to the periosteum. The epithelium is bound to the tooth over an extensive area. This epithelial attachment prevents bacterial access to the lamina propria of the gingiva and the relatively soft cementum of the root. When you brush and massage your gums, you stimulate the epithelial cells and strengthen the attachment. A condition called *gingivitis*, a bacterial infection of the gingivae, can occur if the attachment breaks down.

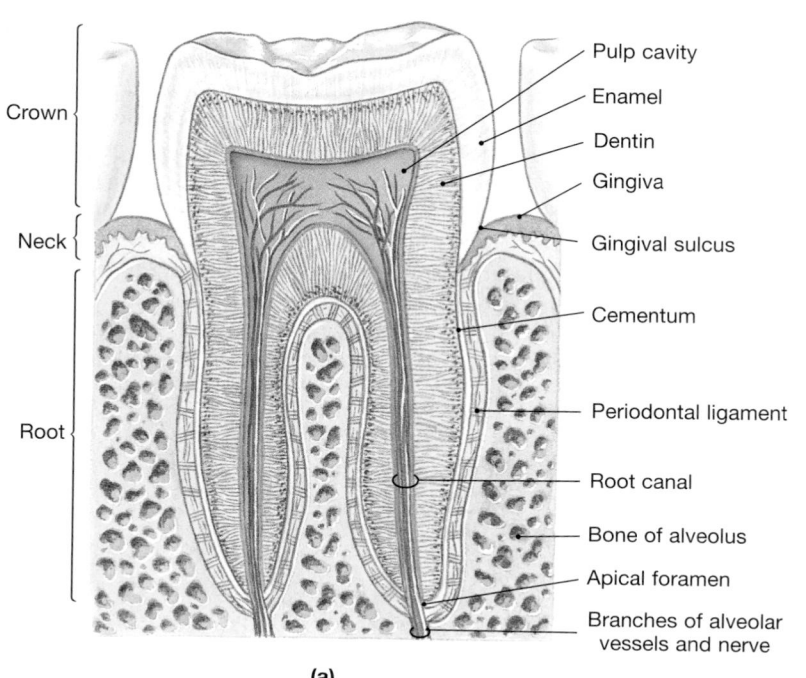

(a)

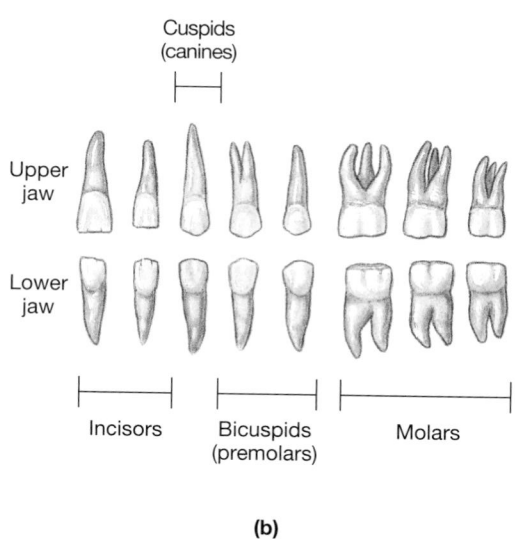

(b)

●**FIGURE 24-8**
Teeth. **(a)** A diagrammatic section through a typical adult tooth. **(b)** The adult teeth.

The dentin of the crown is covered by a layer of **enamel**. Enamel, which contains calcium phosphate in a crystalline form, is the hardest biologically manufactured substance. Adequate amounts of calcium, phosphates, and vitamin D during childhood are essential if the enamel coating is to be complete and resistant to decay.

Tooth decay generally results from the action of bacteria that inhabit your mouth. Bacteria adhering to the surfaces of the teeth produce a sticky matrix that traps food particles and creates deposits known as *dental plaque*. Over time, this organic material can become calcified, forming a hard layer of *tartar*, or *dental calculus*, which can be difficult to remove. Tartar deposits most commonly develop at or near the gingival sulcus, where brushing cannot remove the relatively soft plaque deposits.

Types of Teeth

The alveolar processes of the maxillary bones and the mandible form the *upper* and *lower dental arches*, respectively. These arches contain four types of teeth, each with specific functions (Figure 24–8b●):

1. **Incisors** (in-SĪ-zerz) are blade-shaped teeth located at the front of the mouth. Incisors are useful for clipping or cutting, such as when you nip off the tip of a carrot stick. These teeth have a single root.

2. The **cuspids** (KUS-pidz), or *canines*, are conical, with a sharp ridgeline and a pointed tip. They are used for tearing or slashing. You might weaken a tough piece of celery by the clipping action of the incisors, but then take advantage of the shearing action provided by the cuspids. Cuspids have a single root.

3. **Bicuspids** (bī-KUS-pidz), or *premolars*, have flattened crowns with prominent ridges. They crush, mash, and grind. Bicuspids have one or two roots.

4. **Molars** have very large, flattened crowns with prominent ridges. They excel at crushing and grinding. You can usually shift a tough nut or some spareribs to your bicuspids and molars for successful crunching. Molars typically have three or more roots.

Dental Succession

During embryonic development, two sets of teeth begin to form. The first to appear are the **deciduous teeth** (de-SID-ū-us; *deciduus*, falling off), the temporary teeth of the **primary dentition**. Deciduous teeth are also called *primary teeth, milk teeth,* or *baby teeth*. Most children have 20 deciduous teeth—5 on each side of the upper and lower jaws (Figure 24–9a●). On

Dental Problems and Solutions

The mass of a plaque deposit protects the bacteria that normally reside in the mouth from salivary secretions. As the pathogenic bacteria digest nutrients, they generate acids that erode the enamel and dentin of the teeth. The result is **dental caries**, or "cavities." *Streptococcus mutans* is the most abundant bacterium at these sites, and vaccines are now being developed to promote resistance to it and thereby prevent dental caries. If *S. mutans* (or another bacterium) reaches the pulp and infects it, *pulpitis* (pul-PĪ-tis) results. Treatment generally involves the complete removal of the pulp tissue, especially the sensory innervation and all areas of decay; the pulp cavity is then packed with appropriate materials. This procedure is called a *root canal*.

Brushing the exposed surfaces of your teeth after you eat helps prevent the settling of bacteria and the entrapment of food particles, but bacteria between your teeth, in the region known as the *interproximal space*, and within the gingival sulcus may elude the brush. Dentists therefore recommend the daily use of dental floss to clean these spaces and stimulate the gingival epithelium. If bacteria and plaque remain within the gingival sulcus for extended periods, the acids generated will begin eroding

the connections between the neck of the tooth and the gingiva. The gums appear to recede from the teeth, and **periodontal disease** develops. As this disease progresses, the bacteria attack the cementum, eventually destroying the periodontal ligament and eroding the bone of the alveolus (see photos). This deterioration loosens the tooth, and it falls out or must be pulled. Periodontal disease is the most common cause of the loss of teeth.

Lost or broken teeth have commonly been replaced by "false teeth" attached to a plate or frame inserted into the mouth. In the 1980s, an alternative was developed that uses *dental implants*. A ridged titanium cylinder is inserted into the alveolus, and osteoblasts lock the ridges into the surrounding bone. After 4–6 months, an artificial tooth is screwed into the cylinder.

Dental implants are not suitable for everyone. For example, enough alveolar bone must be present to provide a firm attachment, and complications may occur during or after surgery. Nevertheless, as the technique evolves, dental implants will become increasingly important. Roughly 42 percent of individuals over age 65 have lost all their teeth; the rest have lost an average of 10 teeth.

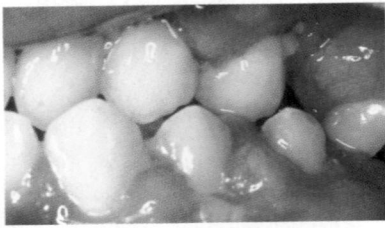

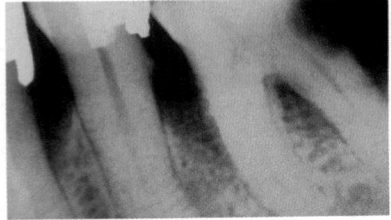

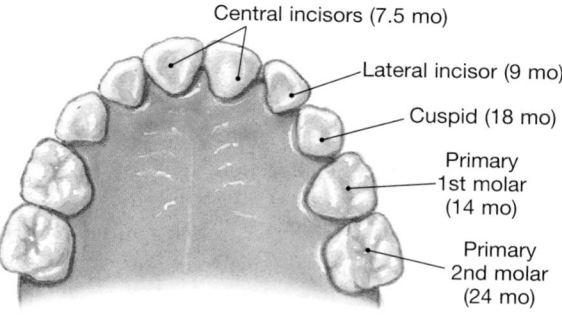

Central incisors (7.5 mo)
Lateral incisor (9 mo)
Cuspid (18 mo)
Primary 1st molar (14 mo)
Primary 2nd molar (24 mo)

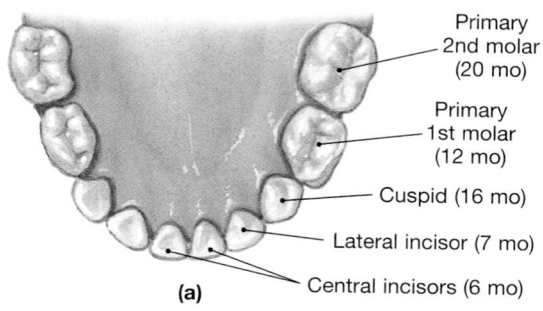

Primary 2nd molar (20 mo)
Primary 1st molar (12 mo)
Cuspid (16 mo)
Lateral incisor (7 mo)
Central incisors (6 mo)

(a)

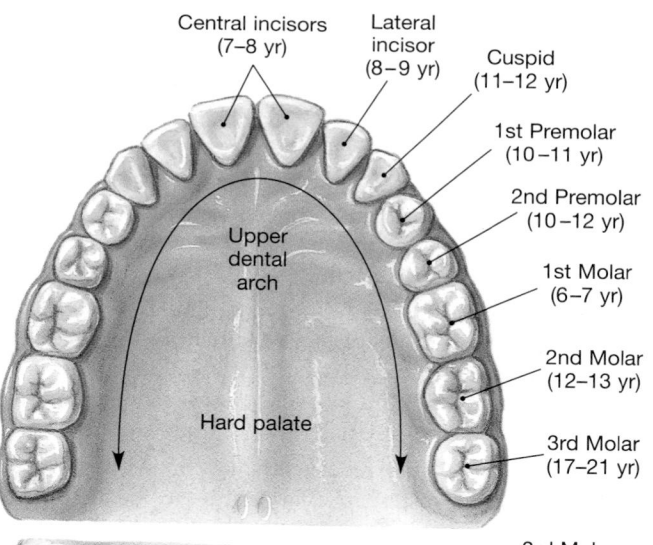

Central incisors (7–8 yr)
Lateral incisor (8–9 yr)
Cuspid (11–12 yr)
1st Premolar (10–11 yr)
2nd Premolar (10–12 yr)
1st Molar (6–7 yr)
2nd Molar (12–13 yr)
3rd Molar (17–21 yr)

Upper dental arch
Hard palate

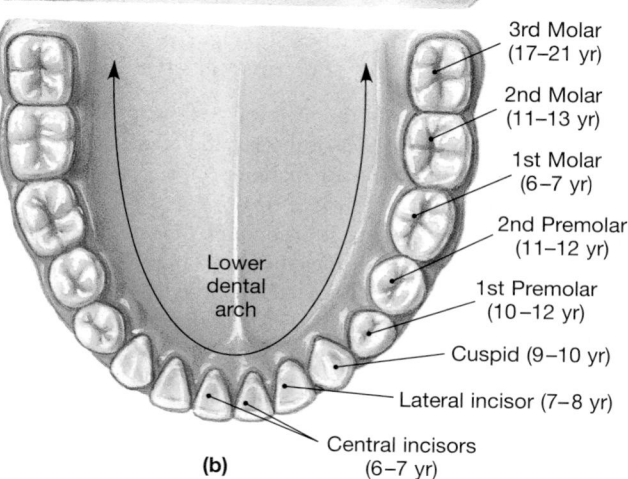

3rd Molar (17–21 yr)
2nd Molar (11–13 yr)
1st Molar (6–7 yr)
2nd Premolar (11–12 yr)
1st Premolar (10–12 yr)
Cuspid (9–10 yr)
Lateral incisor (7–8 yr)
Central incisors (6–7 yr)

Lower dental arch

(b)

•**FIGURE 24–9**
Primary and Secondary Teeth. **(a)** The primary teeth, with the age at eruption given in months. **(b)** The adult teeth, with the age at eruption given in years.

each side of the upper or lower jaw, the primary dentition consists of two incisors, one cuspid, and a pair of deciduous molars. These teeth will later be replaced by the adult **secondary dentition**, or *permanent dentition* (Figure 24–9b•). Adult jaws are larger and can accommodate more than 20 permanent teeth. Three additional molars appear on each side of the upper and lower jaws as the individual ages, extending the length of the tooth rows posteriorly and bring the permanent tooth count to 32.

As replacement proceeds, the periodontal ligaments and roots of the primary teeth erode until the deciduous teeth either fall out or are pushed aside by the **eruption**, or emergence, of the secondary teeth. The adult premolars take the place of the deciduous molars, and the adult molars extend the tooth row as the jaw enlarges. The third molars, or *wisdom teeth*, may not erupt before age 21. When wisdom teeth fail to erupt, it is because they develop in inappropriate positions or because space on the dental arch is inadequate. Any teeth that develop in locations that do not permit their eruption are called *impacted teeth*. Impacted wisdom teeth can be surgically removed to prevent the formation of abscesses.

Mastication

The *muscles of mastication* close your jaws and slide or rock your lower jaw from side to side. ∞ p. 346 Chewing is not a simple process; it can involve any combination of mandibular elevation/depression, protraction/retraction, and medial/lateral movement. (Try classifying the movements involved the next time you eat.)

During mastication, you force food from the oral cavity to the vestibule and back, crossing and recrossing the occlusal surfaces. This movement results in part from the action of the muscles of mastication, but control would be impossible without the aid of the muscles of the cheeks, lips, and tongue. Once you have shredded or torn the material to a satisfactory consistency and have moistened it with salivary secretions, your tongue begins compacting the debris into a bolus. You can swallow a compact, moist, cohesive bolus relatively easily.

✓ Which type of epithelium lines the oral cavity?

✓ The digestion of which nutrient would be affected by damage to the parotid salivary glands?

✓ Which type of tooth is most useful for chopping off bits of relatively rigid foods?

Answers start on page Q-1

24-3 THE PHARYNX

Objective

■ Describe the anatomy and functions of the pharynx.

The **pharynx** serves as a common passageway for solid food, liquids, and air. We described the epithelial lining and divisions of the pharynx—the nasopharynx, the oropharynx, and the laryngopharynx—in Chapter 23. ∞ p. 832 Food normally passes through the oropharynx and laryngopharynx on its way to the esophagus. Both of these divisions have a stratified squamous epithelium similar to that of the oral cavity. The lamina propria contains scattered mucous glands and the lymphoid tissue of the pharyngeal, palatal, and lingual tonsils. Deep to the lamina propria lies a dense layer of elastic fibers, bound to the underlying skeletal muscles.

We described the specific pharyngeal muscles involved in swallowing in Chapter 11. ∞ p. 348

- The *pharyngeal constrictor muscles* push the bolus toward the esophagus.
- The *palatopharyngeus* and *stylopharyngeus muscles* elevate the larynx.
- The *palatal muscles* elevate the soft palate and adjacent portions of the pharyngeal wall.

These muscles cooperate with muscles of the oral cavity and esophagus to initiate swallowing, which pushes the bolus along the esophagus and into the stomach.

24-4 THE ESOPHAGUS

Objective

■ Describe the anatomy and functions of the esophagus.

The **esophagus** (Figure 24–10●) is a hollow muscular tube with a length of approximately 25 cm (1 ft) and a diameter of about 2 cm (0.75 in.) at its widest point. The primary function of the esophagus is to carry solid food and liquids to the stomach.

The esophagus begins posterior to the cricoid cartilage, at the level of vertebra C_6. From this point, where it is at its narrowest, the esophagus descends toward the thoracic cavity posterior to the trachea. It passes inferiorly along the dorsal wall of the mediastinum and enters the abdominopelvic cavity through the **esophageal hiatus** (hī-Ā-tus), an opening in the diaphragm. The esophagus then empties into the stomach anterior to vertebra T_7.

In the neck, the esophagus receives blood from the external carotid arteries and the thyrocervical trunks. In the mediastinum, it is supplied by the esophageal arteries and branches of the bronchial arteries. As it passes through the esophageal hiatus, the esophagus receives blood from the inferior phrenic arteries, and the portion of the esophagus that is adjacent to the stomach is supplied by the left gastric artery. Blood from esophageal capillaries collects into the esophageal, inferior thyroid, azygos, and gastric veins.

The esophagus is innervated by parasympathetic and sympathetic fibers from the esophageal plexus. ∞ p. 546 Resting muscle tone in the circular muscle layer in the superior 3 cm (1 in.) of the esophagus normally prevents air from entering your esophagus. A comparable zone at the inferior end of the esophagus normally remains in a state of active contraction. This condition prevents the backflow of materials from the stomach into the esophagus. Neither region has a well-defined sphincter muscle comparable to those located elsewhere along the digestive tract. Nevertheless, the terms *upper esophageal sphincter* and *lower esophageal sphincter (cardiac sphincter)* are often used to describe these regions at either end of the esophagus, because they are similar in function to other sphincters.

☐ HISTOLOGY OF THE ESOPHAGUS

The wall of the esophagus contains mucosal, submucosal, and muscularis layers comparable to those described in Figure 24–3●. Distinctive features of the esophageal wall (Figure 24–10●) include the following:

- The mucosa of the esophagus contains a nonkeratinized, stratified squamous epithelium similar to that of the pharynx and oral cavity.

- The mucosa and submucosa are thrown into large folds that extend the length of the esophagus. These folds allow for expansion during the passage of a large bolus; except when you swallow, muscle tone in the walls keeps the lumen closed.

- The muscularis mucosae consists of an irregular layer of smooth muscle.

- The submucosa contains scattered *esophageal glands*, which produce a mucous secretion that reduces friction between the bolus and the esophageal lining.

- The muscularis externa has the usual inner circular and outer longitudinal layers. However, in the superior third of the esophagus, these layers contain skeletal muscle fibers; in the middle third, there is a mixture of skeletal and smooth muscle tissue; along the inferior third, only smooth muscle occurs.

- There is no serosa, but an adventitia of connective tissue outside the muscularis externa anchors the esophagus in position against the dorsal body wall. Over the 1–2 cm between the diaphragm and stomach, the esophagus is retroperitoneal, with peritoneum covering the anterior and left lateral surfaces.

The veins draining the inferior portion of the esophagus empty into tributaries of the hepatic portal vein. ∞ p. 767 If the venous pressure in the hepatic portal vein becomes abnormally high, due to liver damage or the constriction of the vessels, blood will pool in the submucosal veins of the esophagus. The veins become grossly distended and create bulges in the esophageal wall. These distorted **esophageal varices** (VAR-i-sēz; *varices*, dilated veins) may constrict the esophageal passageway. They may also rupture, causing massive bleeding into the submucosal tissues or into the lumen of the esophagus and stomach. Esophageal varices commonly develop in individuals with advanced *cirrhosis*, a chronic liver disorder that restricts hepatic blood flow.

●FIGURE 24–10
The Esophagus. **(a)** A transverse section through the esophagus. **(b)** The esophageal mucosa. (LM × 77) **(c)** A color-enhanced SEM of the transition between the esophageal and gastric mucosae at the lower esophageal sphincter.

(a)

(b)

(c)

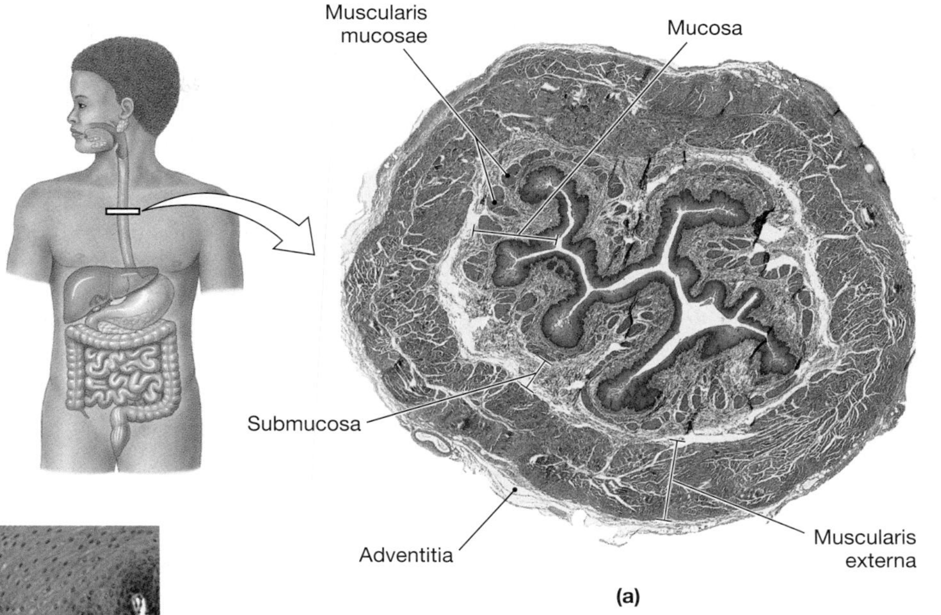

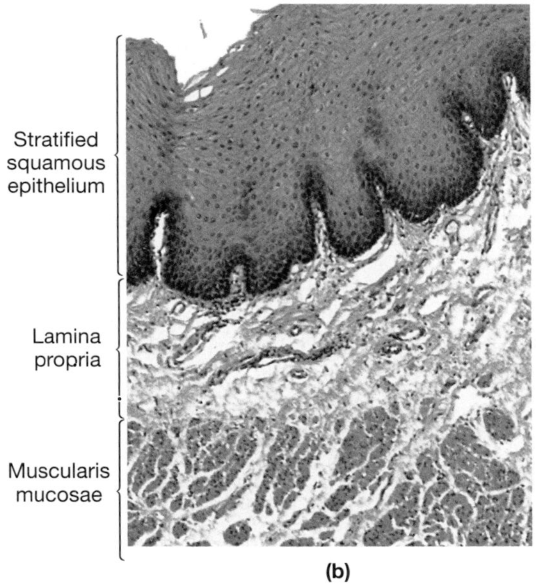

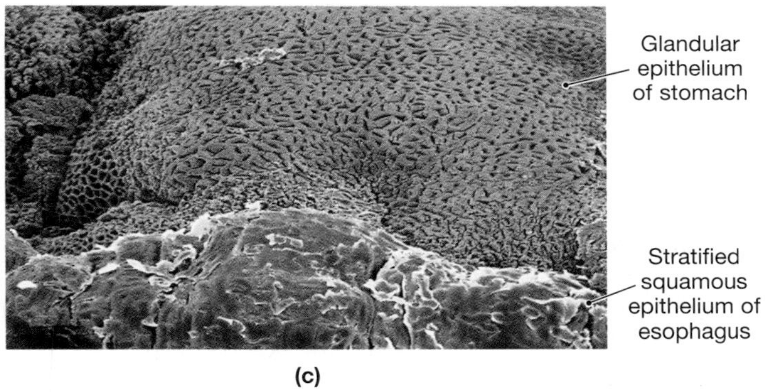

☐ SWALLOWING

Swallowing, or **deglutition**, is a complex process whose initiation can be voluntarily controlled, but that proceeds automatically once it begins. Although you are consciously aware of, and voluntarily control, swallowing when you eat or drink, swallowing can also occur unconsciously, as saliva collects at the back of the mouth. Each day you swallow approximately 2400 times. We can divide swallowing into buccal, pharyngeal, and esophageal phases:

1. The **buccal phase** begins with the compression of the bolus against the hard palate. Subsequent retraction of the tongue then forces the bolus into the oropharynx and assists in the elevation of the soft palate, thereby isolating the nasopharynx (Figure 24–11a,b●). The buccal phase is strictly voluntary. Once the bolus enters the oropharynx, reflex responses are initiated and the bolus is moved toward the stomach.

2. The **pharyngeal phase** begins as the bolus comes into contact with the palatoglossal and palatopharyngeal arches and the pos-

terior pharyngeal wall (Figure 24–11c,d●). The **swallowing reflex** begins when tactile receptors on the palatal arches and uvula are stimulated by the passage of the bolus. The information is relayed to the **swallowing center** of the medulla oblongata over the trigeminal and glossopharyngeal nerves. Motor commands originating at this center then target the pharyngeal musculature, producing a coordinated and stereotyped pattern of muscle contraction. Elevation of the larynx and folding of the epiglottis direct the bolus past the closed glottis, while the uvula and soft palate block passage back to the nasopharynx. It takes less than a second for the pharyngeal muscles to propel the bolus into the esophagus. During this period, the respiratory centers are inhibited and breathing stops.

3. The **esophageal phase** of swallowing begins as the contraction of pharyngeal muscles forces the bolus through the entrance to the esophagus (Figure 24–11e–h●). Once in the esophagus, the bolus is pushed toward the stomach by a peristaltic wave. The approach of the bolus triggers the opening of the lower esophageal sphincter, and the bolus then continues into the stomach.

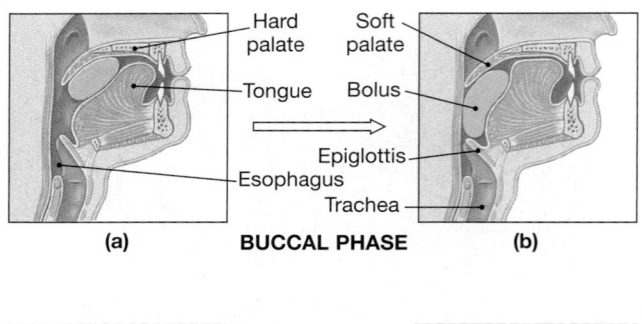

(a)   **BUCCAL PHASE**   (b)

Labels: Hard palate, Soft palate, Tongue, Bolus, Epiglottis, Esophagus, Trachea

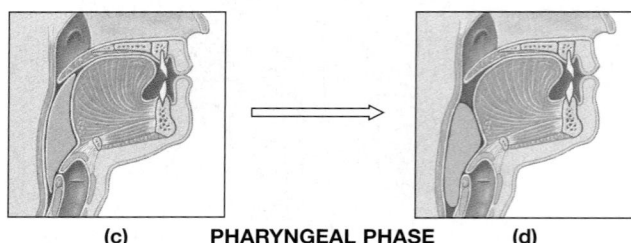

(c)   **PHARYNGEAL PHASE**   (d)

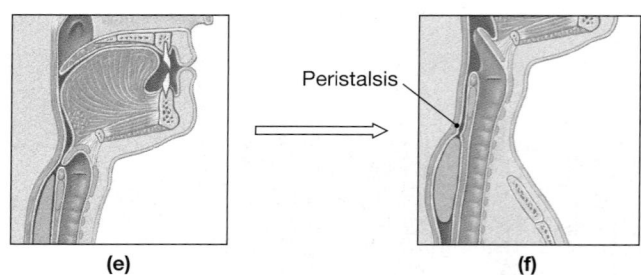

(e)   (f)

Label: Peristalsis

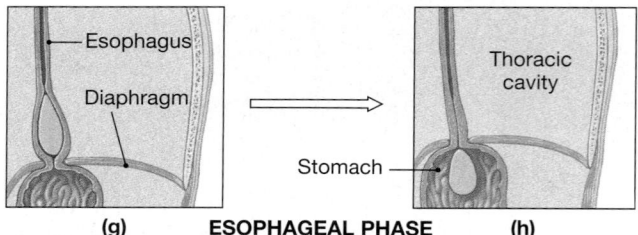

(g)   **ESOPHAGEAL PHASE**   (h)

Labels: Esophagus, Diaphragm, Stomach, Thoracic cavity

•**FIGURE 24-11**
The Swallowing Process. This sequence, based on a series of X rays, shows the stages of swallowing and the movement of materials from the mouth to the stomach.

Primary peristaltic waves are peristaltic movements that are coordinated by afferent and efferent fibers in the glossopharyngeal and vagus nerves. For a typical bolus, the entire trip takes about 9 seconds to complete. Liquids may make the journey in a few seconds, arriving ahead of the peristaltic contractions with the assistance of gravity.

A dry or poorly lubricated bolus travels much more slowly, and a series of *secondary peristaltic waves* may be required to push it all the way to the stomach. Secondary peristaltic waves are local reflexes triggered by the stimulation of sensory receptors in the esophageal walls. These receptors relay information by way of the submucosal and myenteric plexuses, producing peristaltic con-

tractions in the absence of instructions from the CNS. You cannot swallow a completely dry bolus, because friction with the walls of the esophagus will make peristalsis ineffective. (For this reason, it is almost impossible to swallow an entire slice of processed white bread without taking a drink.) Other problems with the movement of a bolus may involve conditions such as *achalasia*, a constriction of the lower esophageal sphincter, and *esophagitis*, an inflammation of the esophagus. **AM** Achalasia and Esophagitis

✓ What is unusual about the muscularis externa of the esophagus?
✓ Where in the human body would you find the fauces?
✓ What is occurring when the soft palate and larynx elevate and the glottis closes?

Answers start on page Q-1

24-5 THE STOMACH

Objective

■ Describe the anatomy of the stomach, its histological features, and its roles in digestion and absorption.

The **stomach** performs four major functions: (1) the bulk storage of ingested food, (2) the mechanical breakdown of ingested food, (3) the disruption of chemical bonds in food material through the action of acids and enzymes, and (4) the production of *intrinsic factor*, a glycoprotein whose presence in the digestive tract is required for the absorption of vitamin B_{12} in the small intestine. Ingested substances combine with the secretions of the glands of the stomach, producing a viscous, highly acidic, soupy mixture of partially digested food called **chyme** (kīm).

☐ ANATOMY OF THE STOMACH

The stomach has the shape of an expanded J (Figure 24–12●). A short **lesser curvature** forms the medial surface of the organ, and a long **greater curvature** forms the lateral surface. The anterior and posterior surfaces are smoothly rounded. The shape and size of the stomach are extremely variable from individual to individual and even from one meal to the next. In an "average" stomach, the lesser curvature has a length of approximately 10 cm (4 in.), and the greater curvature measures about 40 cm (16 in.). The stomach typically extends between the levels of vertebrae T_7 and L_3.

We can divide the stomach into four regions (Figure 24–12●):

1. *The Cardia.* The **cardia** (KAR-dē-uh) is the smallest part of the stomach. It consists of the superior, medial portion of the stomach within 3 cm (1.2 in.) of the junction between the stomach and the esophagus. The cardia contains abundant mucous glands whose secretions coat the connection with the esophagus and help protect that tube from the acids and enzymes of the stomach.

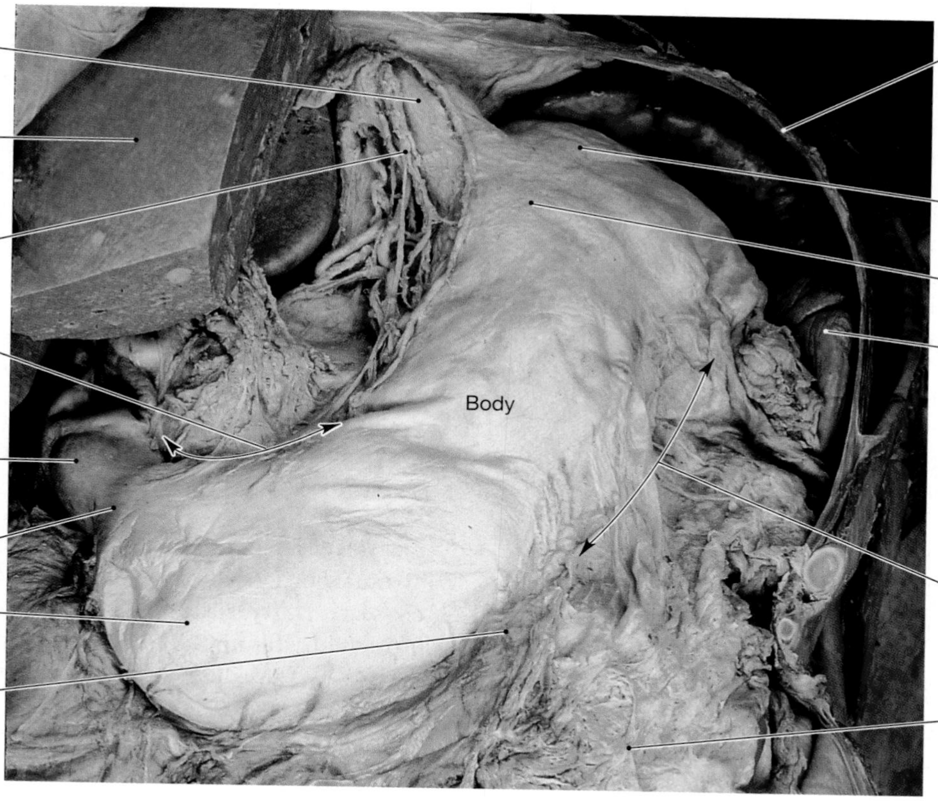

(a)

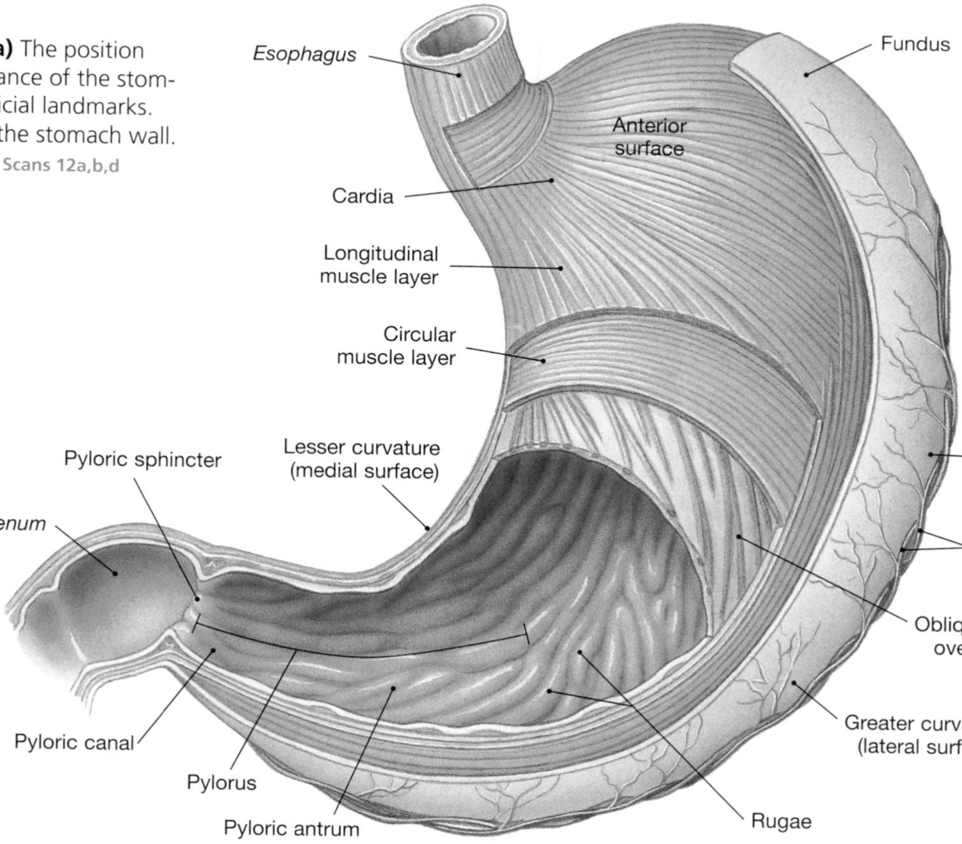

●**FIGURE 24–12**
The Stomach. **(a)** The position and external appearance of the stomach, showing superficial landmarks. **(b)** The structure of the stomach wall.
ATLAS Plates 7.6e, 7.7a; Scans 12a,b,d

(b)

2. *The Fundus.* The **fundus** (FUN-dus) is the portion of the stomach that is superior to the junction between the stomach and the esophagus. The fundus contacts the inferior, posterior surface of the diaphragm (Figure 24–12a●).

3. *The Body.* The area between the fundus and the curve of the J is the **body** of the stomach. The body is the largest region of the stomach, and it functions as a mixing tank for ingested food and secretions produced in the stomach. *Gastric glands* (*gaster*, stomach) in the fundus and body secrete most of the acids and enzymes involved in gastric digestion.

4. *The Pylorus.* The **pylorus** (pī-LOR-us) is the curve of the J. The pylorus is divided into a **pyloric antrum** (*antron*, cavity), which is connected to the body, and a **pyloric canal**, which empties into the *duodenum*, the proximal segment of the small intestine. As mixing movements occur during digestion, the pylorus frequently changes shape. A muscular **pyloric sphincter** regulates the release of chyme into the duodenum. Glands in the pylorus secrete mucus and important digestive hormones, including *gastrin*, a hormone that stimulates the activity of gastric glands.

The stomach's volume increases while you eat and then decreases as chyme enters the small intestine. When your stomach is relaxed (empty), the mucosa is thrown into prominent folds called **rugae** (ROO-gē; wrinkles) (Figure 24–12b●)—temporary features that let the gastric lumen expand. As your stomach fills, the rugae flatten out. When your stomach is full, the rugae almost disappear. When empty, your stomach resembles a muscular tube with a narrow, constricted lumen. When full, it can expand to contain 1–1.5 liters of material.

Musculature of the Stomach

The muscularis mucosae and muscularis externa of the stomach contain extra layers of smooth muscle cells in addition to the usual circular and longitudinal layers. The muscularis mucosae generally contains an outer, circular layer of muscle cells. The muscularis externa has an inner, **oblique layer** of smooth muscle (Figure 24–12b●). The extra layers of smooth muscle strengthen the stomach wall and assist in the mixing and churning activities essential to the formation of chyme.

Blood Supply to the Stomach

The stomach receives blood from (1) the *left gastric artery*, (2) the *splenic artery*, which supplies the *left gastroepiploic artery*, and (3) the *common hepatic artery*, which supplies the *right gastric*, *gastroduodenal*, and *right gastroepiploic arteries*. The stomach is drained by the *gastric* and *gastroepiploic veins*. ∞ pp. 758, 767

Histology of the Stomach

A simple columnar epithelium lines all portions of the stomach (Figure 24–13a●). The epithelium is a *secretory sheet*, which produces a carpet of mucus that covers the interior surfaces of the stomach. The alkaline mucous layer protects epithelial cells against the acids and enzymes in the gastric lumen.

Shallow depressions called **gastric pits** open onto the gastric surface (Figure 24–13b,c,d●). The mucous cells at the base, or *neck*, of each gastric pit actively divide, replacing superficial cells that are shed into the chyme. A typical gastric epithelial cell has a life span of three to seven days, but exposure to strong alcohol or other chemicals that damage or kill epithelial cells will increase the rate of cell turnover.

Gastric Glands

In the fundus and body of the stomach, each gastric pit communicates with several **gastric glands**, which extend deep into the underlying lamina propria (Figure 24–13b,c●). Gastric glands are dominated by two types of secretory cells: (1) *parietal cells* and (2) *chief cells*. Together, they secrete about 1500 ml of **gastric juice** each day.

PARIETAL CELLS **Parietal cells** are especially common along the proximal portions of each gastric gland (Figure 24–13d●). These cells secrete **intrinsic factor**, a glycoprotein that facilitates the absorption of **vitamin B_{12}** across the intestinal lining. (Recall from Chapter 19 that this vitamin is essential for normal erythropoiesis.) ∞ p. 661

Parietal cells also secrete *hydrochloric acid* (HCl). They do not produce HCl in the cytoplasm, however, because it is such a strong acid that it would erode a secretory vesicle and destroy the cell. Instead, H^+ and Cl^-, the two ions that form HCl, are transported independently by different mechanisms (Figure 24–14●). Hydrogen ions are generated inside a parietal cell as the enzyme carbonic anhydrase converts carbon dioxide and water to carbonic acid (H_2CO_3). The carbonic acid promptly dissociates into hydrogen ions and bicarbonate ions (HCO_3^-). The hydrogen ions are actively transported into the lumen of the gastric gland. The bicarbonate ions are ejected into the interstitial fluid by a countertransport mechanism that exchanges intracellular bicarbonate ions for extracellular chloride ions. The chloride ions then diffuse across the cell and through open chloride channels in the cell membrane into the lumen of the gastric gland.

The bicarbonate ions released by the parietal cell diffuse through the interstitial fluid into the bloodstream. When gastric glands are actively secreting, enough bicarbonate ions enter the bloodstream to increase the pH of the blood significantly. This sudden influx of bicarbonate ions has been called the *alkaline tide*.

The secretory activities of the parietal cells can keep the stomach contents at a pH of 1.5–2.0. This highly acidic environment does not by itself digest chyme, but has four important functions:

1. The acidity of gastric juice kills most of the microorganisms ingested with food.

2. The acidity denatures proteins and inactivates most of the enzymes in food.

3. The acidity helps break down plant cell walls and the connective tissues in meat.

4. An acidic environment is essential for the activation and function of *pepsin*, a protein-digesting enzyme secreted by chief cells.

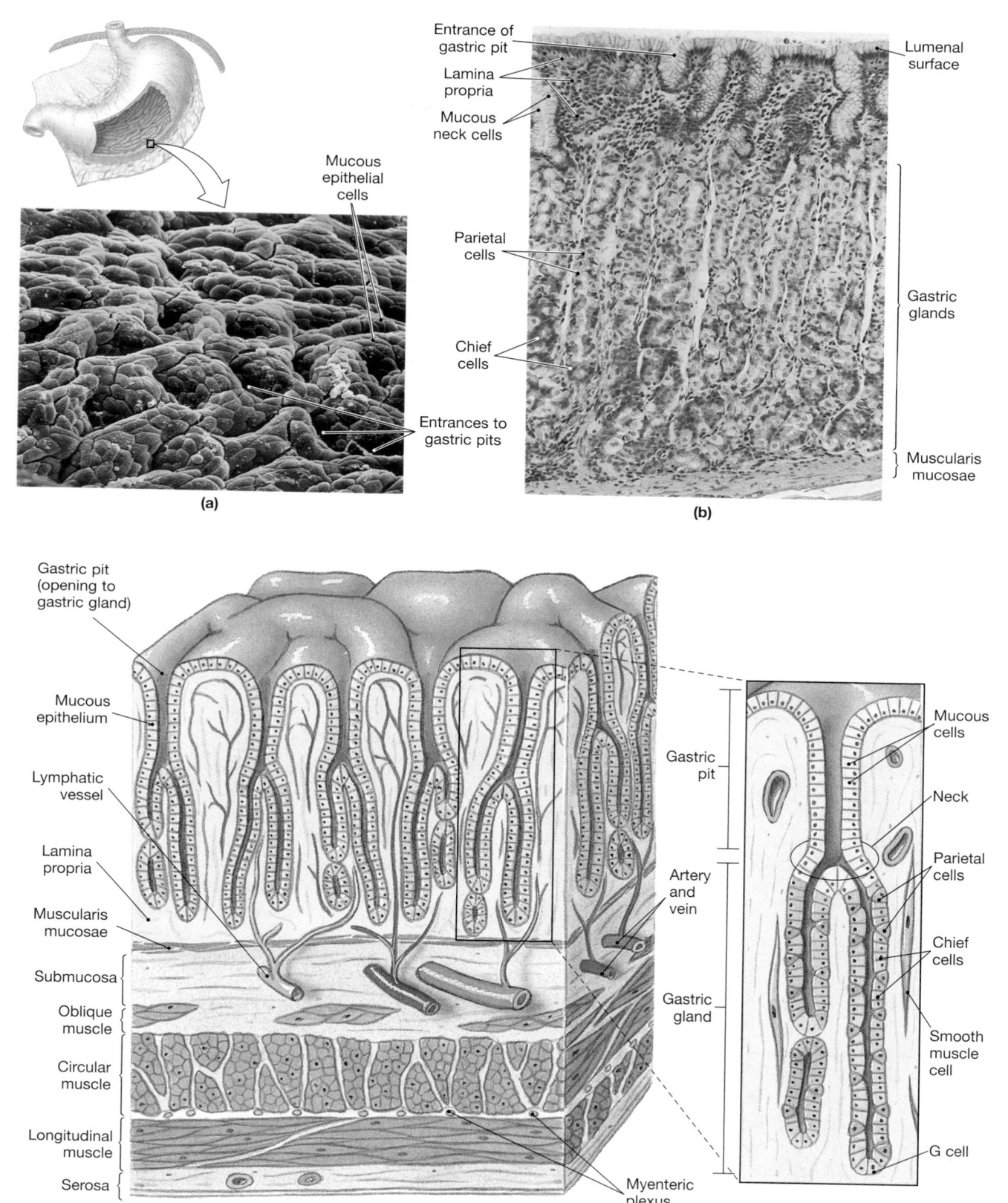

●FIGURE 24–13
The Stomach Lining. **(a)** A surface view of the gastric mucosa of the full stomach, showing the entrances to the gastric pits. (SEM × 35) **(b)** A section through gastric pits and gastric glands. (LM × 300) **(c)** The organization of the stomach wall. **(d)** A gastric gland.

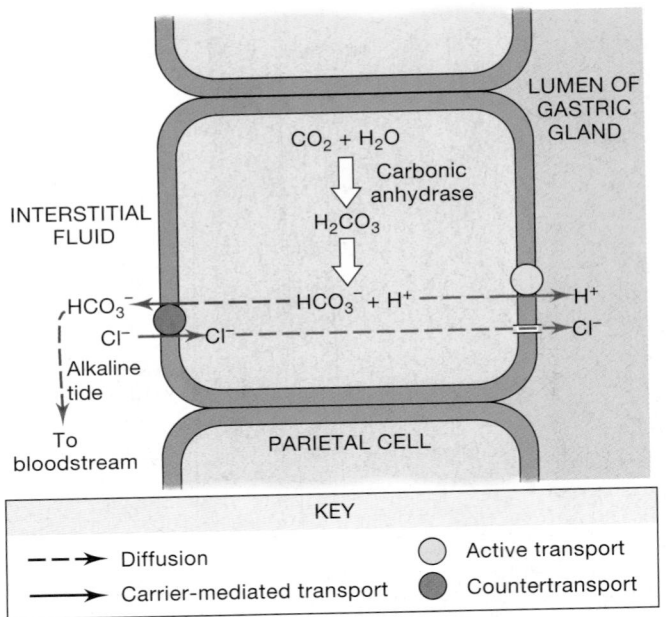

●FIGURE 24–14
The Secretion of Hydrochloric Acid. An active parietal cell generates H^+ by the dissociation of carbonic acid within the cell. The bicarbonate is exchanged for Cl^- in the interstitial fluid; the chloride ions diffuse into the lumen of the gastric gland as the hydrogen ions are transported out of the cell.

CHIEF CELLS **Chief cells** are most abundant near the base of a gastric gland (Figure 24–13d●). These cells secrete **pepsinogen** (pep-SIN-ō-jen), an inactive proenzyme. Pepsinogen is converted by the acid in the gastric lumen to **pepsin**, an active *proteolytic* (protein-digesting) enzyme. Pepsin functions most effectively at a strongly acidic pH of 1.5–2.0. In addition, the stomachs of newborn infants (but not of adults) produce **rennin**, also known as *chymosin*, and **gastric lipase**, enzymes important for the digestion of milk. Rennin coagulates milk proteins; gastric lipase initiates the digestion of milk fats.

Pyloric Glands

Glands in the pylorus produce primarily a mucous secretion, rather than enzymes or acid. In addition, several types of enteroendocrine cells are scattered among the mucus-secreting cells. These enteroendocrine cells produce at least seven hormones, most notably, the hormone **gastrin** (GAS-trin). Gastrin is produced by *G cells*, which are most abundant in the gastric pits of the pyloric antrum. Gastrin stimulates secretion by both parietal and chief cells, as well as contractions of the gastric wall that mix and stir the gastric contents. The pyloric glands also contain *D cells*, which release **somatostatin**, a hormone that inhibits the release of gastrin. D cells continuously release their secretions into the interstitial fluid adjacent to the G cells. This inhibition of gastrin production can be overpowered by neural and hormonal stimuli when the stomach is preparing for digestion or is already engaged in digestion.

A superficial inflammation of the gastric mucosa is called **gastritis** (gas-TRĪ-tis). The condition can develop after a person has swallowed drugs, including alcohol and aspirin. Gastritis can also appear after severe emotional or physical stress, bacterial infection of the gastric wall, or the ingestion of strongly acidic or alkaline chemicals.

A **peptic ulcer** develops when the digestive acids and enzymes manage to erode their way through the defenses of the stomach lining or proximal portions of the small intestine. The specific location of the ulcer is indicated by the terms **gastric ulcer** (stomach) and **duodenal ulcer** (duodenum). Peptic ulcers result from the excessive production of acid or the inadequate production of the alkaline mucus that defends the epithelium against that acid. Since the late 1970s, drugs such as *cimetidine* (*Tagamet®*) have been used to inhibit the production of acid by parietal cells. Gastric mucosal infections involving the bacterium *Helicobacter pylori* are responsible for at least 80 percent of peptic ulcers. Hence, treatment for gastric ulcers commonly involves the administration of antibiotic drugs. **AM** Peptic Ulcers

☐ REGULATION OF GASTRIC ACTIVITY

The production of acid and enzymes by the gastric mucosa can be (1) controlled by the central nervous system, (2) regulated by short reflexes of the enteric nervous system, coordinated in the wall of the stomach, and (3) regulated by hormones of the digestive tract. Three phases of gastric control are identified, although considerable overlap exists among them: the *cephalic phase*, the *gastric phase*, and the *intestinal phase*. These phases, which are named according to the location of the control center, are summarized in Figure 24–15●.

The Cephalic Phase

The **cephalic phase** of gastric secretion begins when you see, smell, taste, or think of food (Figure 24–15a●). This stage, which is directed by the CNS, prepares your stomach to receive food. The neural output proceeds by way of the parasympathetic division of the autonomic nervous system, and the vagus nerves innervate the submucosal plexus of the stomach. Next, postganglionic parasympathetic fibers innervate mucous cells, chief cells, parietal cells, and G cells of the stomach. In response to stimulation, the production of gastric juice accelerates, reaching rates of about 500 ml/h. This phase generally lasts only minutes before the gastric phase commences. Emotional states can exaggerate or inhibit the cephalic phase. For example, anger or hostility leads to excessive gastric secretion, whereas anxiety, stress, or fear decreases gastric secretion and gastric contractions, or *motility*.

The Gastric Phase

The **gastric phase** begins with the arrival of food in the stomach and builds on the stimulation provided during the cephalic phase (Figure 24–15b●). The stimuli that initiate the gastric phase are (1) distension of the stomach, (2) an increase in the pH of the gastric contents, and (3) the presence of undigested materials in the stomach, especially proteins and peptides. The gastric phase consists of the following responses:

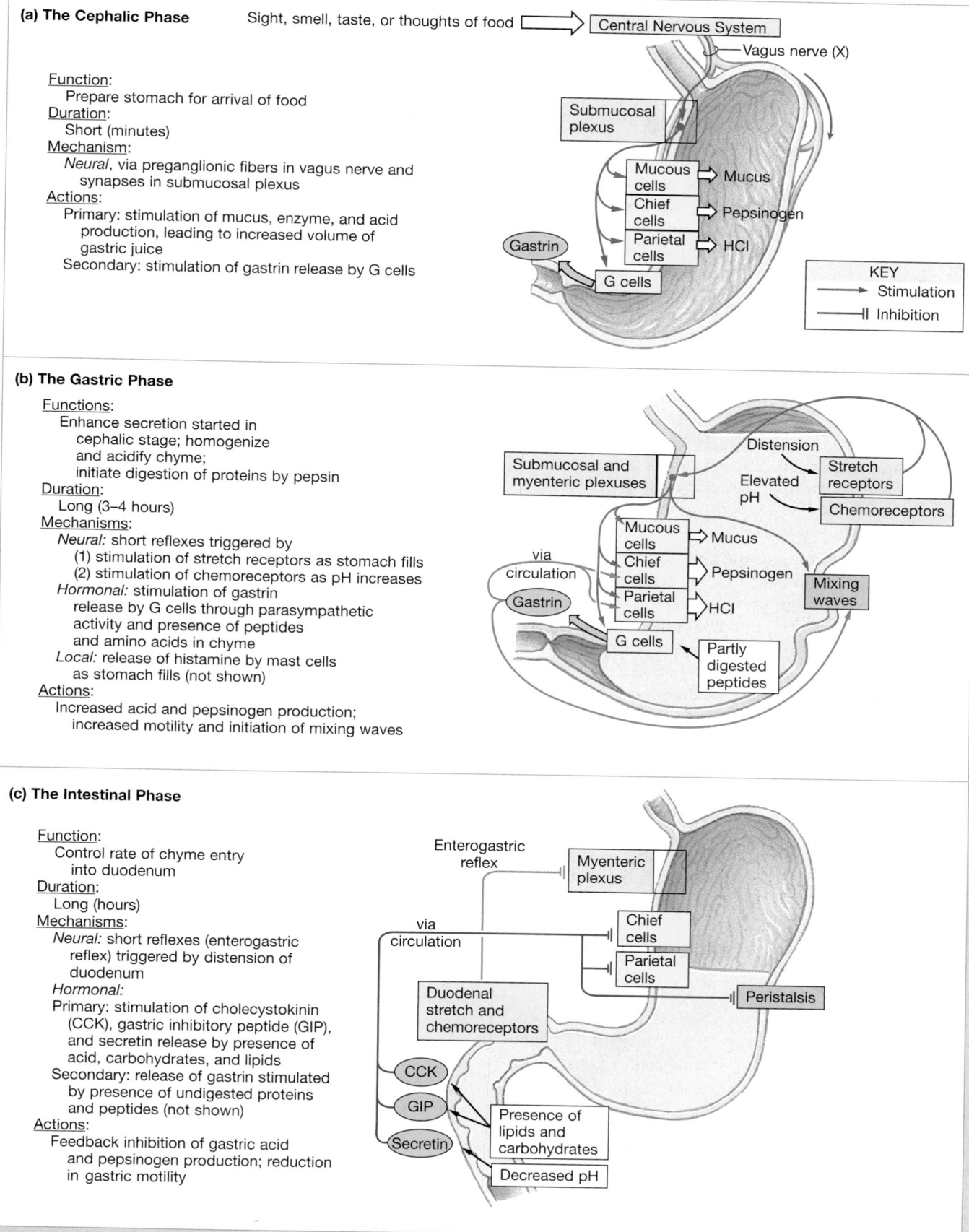

(a) The Cephalic Phase

Sight, smell, taste, or thoughts of food ⟹ Central Nervous System

Vagus nerve (X)

Function:
   Prepare stomach for arrival of food
Duration:
   Short (minutes)
Mechanism:
   Neural, via preganglionic fibers in vagus nerve and
      synapses in submucosal plexus
Actions:
   Primary: stimulation of mucus, enzyme, and acid
      production, leading to increased volume of
      gastric juice
   Secondary: stimulation of gastrin release by G cells

Submucosal plexus

Mucous cells ⟹ Mucus
Chief cells ⟹ Pepsinogen
Parietal cells ⟹ HCl

Gastrin

G cells

KEY
⟶ Stimulation
⊣ Inhibition

(b) The Gastric Phase

Functions:
   Enhance secretion started in
      cephalic stage; homogenize
      and acidify chyme;
      initiate digestion of proteins by pepsin
Duration:
   Long (3–4 hours)
Mechanisms:
   Neural: short reflexes triggered by
      (1) stimulation of stretch receptors as stomach fills
      (2) stimulation of chemoreceptors as pH increases
   Hormonal: stimulation of gastrin
      release by G cells through parasympathetic
      activity and presence of peptides
      and amino acids in chyme
   Local: release of histamine by mast cells
      as stomach fills (not shown)
Actions:
   Increased acid and pepsinogen production;
      increased motility and initiation of mixing waves

Submucosal and myenteric plexuses

Distension
Elevated pH

Stretch receptors
Chemoreceptors

via circulation

Mucous cells ⟹ Mucus
Chief cells ⟹ Pepsinogen
Parietal cells ⟹ HCl

Gastrin

G cells

Mixing waves

Partly digested peptides

(c) The Intestinal Phase

Function:
   Control rate of chyme entry
      into duodenum
Duration:
   Long (hours)
Mechanisms:
   Neural: short reflexes (enterogastric
      reflex) triggered by distension of
      duodenum
   Hormonal:
   Primary: stimulation of cholecystokinin
      (CCK), gastric inhibitory peptide (GIP),
      and secretin release by presence of
      acid, carbohydrates, and lipids
   Secondary: release of gastrin stimulated
      by presence of undigested proteins
      and peptides (not shown)
Actions:
   Feedback inhibition of gastric acid
      and pepsinogen production; reduction
      in gastric motility

Enterogastric reflex

Myenteric plexus

via circulation

Chief cells
Parietal cells

Duodenal stretch and chemoreceptors

Peristalsis

CCK
GIP
Secretin

Presence of lipids and carbohydrates

Decreased pH

●FIGURE 24–15
The Phases of Gastric Secretion. ATLAS Scan 12d

1. *Neural Response.* The stimulation of stretch receptors in the stomach wall and chemoreceptors in the mucosa triggers short reflexes coordinated in the submucosal and myenteric plexuses. The postganglionic fibers leaving the submucosal plexus innervate parietal cells and chief cells, and the release of ACh stimulates their secretion. Proteins, alcohol in small doses, and caffeine enhance gastric secretion markedly by stimulating chemoreceptors in the gastric lining. The stimulation of the myenteric plexus produces mixing waves in the muscularis externa.

2. *Hormonal Response.* Neural stimulation and the presence of peptides and amino acids in chyme stimulate the secretion of gastrin, primarily by G cells of the pyloric antrum. Gastrin entering the interstitial fluid of the stomach must penetrate capillaries and complete a round-trip of the bloodstream before the hormone stimulates parietal and chief cells of the fundus and body. Both parietal and chief cells respond to the presence of gastrin by accelerating their rates of secretion. The effect on the parietal cells is the most pronounced, and the pH of the gastric juice declines as a result. In addition, gastrin stimulates gastric motility.

3. *Local Response.* A distortion of the gastric wall also stimulates the release of histamine in the lamina propria. The source of the histamine is thought to be mast cells in the connective tissue of that layer. Histamine binds to receptors on the parietal cells and stimulates acid secretion.

The gastric phase may continue for three to four hours while the ingested materials are processed by the acids and enzymes. During this period, gastrin stimulates contractions in the muscularis externa of the stomach and intestinal tract. The effects are strongest in the stomach, where stretch receptors are stimulated as well. The initial contractions are weak pulsations in the gastric walls. These *mixing waves* occur several times per minute, and they gradually increase in intensity. After an hour, the material in the stomach is churning like clothing in a washing machine.

When the contractions begin, the pH of the gastric contents is high; only the material in contact with the gastric epithelium is exposed to undiluted digestive acids and enzymes. As mixing occurs, the acid is diluted, but the pH remains elevated until a large volume of gastric juice has been secreted and the contents are thoroughly mixed. This process generally takes several hours. As the pH throughout the chyme reaches 1.5–2.0 and the amount of undigested protein decreases, gastrin production declines, and so do the rates of acid and enzyme secretion by parietal and chief cells.

The Intestinal Phase

The **intestinal phase** of gastric secretion begins when chyme starts to enter the small intestine (Figure 24–15c●). The intestinal phase generally starts after several hours of mixing contractions, when waves of contraction begin sweeping down the length of the stomach. Each time the pylorus contracts, a small quantity of chyme squirts through the pyloric sphincter. The purpose of the intestinal phase is to control the rate of gastric emptying and to ensure that the secretory, digestive, and absorptive functions of the small intestine can proceed with reasonable efficiency. Although we shall consider the intestinal phase now as it affects stomach activity, the arrival of chyme in the small intestine also triggers other neural and hormonal events that coordinate the activities of the intestinal tract and the pancreas, liver, and gallbladder.

The intestinal phase involves a combination of neural and hormonal responses:

1. *Neural Response.* Chyme leaving the stomach relieves some of the distension in the stomach wall, thus reducing the stimulation of stretch receptors. At the same time, the distension of the duodenum by chyme stimulates stretch receptors and chemoreceptors that trigger the **enterogastric reflex**. This reflex temporarily inhibits both central and local stimulation of gastrin production and gastric contractions, as well as stimulating the contraction of the pyloric sphincter. The net result is that immediately after chyme enters the small intestine, gastric contractions decrease in strength and frequency, and further discharge of chyme is prevented, giving the duodenum time to deal with the arriving acids before the next wave of gastric contraction. At the same time, local reflexes at the duodenum stimulate mucus production, which helps protect the intestinal lining from the arriving acids and enzymes.

2. *Hormonal Response.* Several hormonal responses are triggered by the arrival of chyme in the duodenum:

   • The arrival of lipids (especially triglycerides and fatty acids) and carbohydrates in the duodenum stimulates the secretion of the hormones *cholecystokinin* (kō-lē-sis-tō-KĪ-nin), or CCK, and *gastric inhibitory peptide* (GIP). CCK inhibits gastric secretion of acids and enzymes; GIP, which also targets the pancreas, inhibits gastric secretion and reduces the rate and force of gastric contractions. As a result, a meal high in fats stays in your stomach longer, and enters the duodenum at a more leisurely pace, than does a low-fat meal. This delay allows more time for lipids to be digested and absorbed in the small intestine.

   • A drop in pH below 4.5 stimulates the secretion of the hormane *secretin* (sē-KRĒ-tin) by enteroendocrine cells of the duodenum. Secretin inhibits parietal cell and chief cell activity in the stomach. It also targets two accessory organs: the pancreas, where it stimulates the production of buffers that will protect the duodenum by neutralizing the acid in chyme, and the liver, where it stimulates the secretion of bile.

   • The arrival of partially digested proteins in the duodenum stimulates G cells in the duodenal wall. These cells secrete gastrin, which circulates to the stomach and accelerates acid and enzyme production. In effect, this is a feedback mechanism that regulates the amount of gastric processing to meet the requirements of a specific meal.

In general, the rate of movement of chyme into the small intestine is highest when your stomach is greatly distended and the meal contains relatively little protein. A large meal containing small amounts of protein, large amounts of carbohydrates (such as rice or pasta), wine (alcohol), and after-dinner coffee (caffeine) will leave your stomach extremely quickly because both alcohol and caffeine stimulate gastric secretion and motility.

□ DIGESTION AND ABSORPTION IN THE STOMACH

The stomach performs preliminary digestion of proteins by pepsin and, for a variable period, permits the digestion of carbohydrates and lipids by salivary amylase and lingual lipase. Until the pH throughout the contents of the stomach falls below 4.5, salivary amylase and lingual lipase continue to digest carbohydrates and lipids in the meal. These enzymes generally remain active one to two hours after a meal.

As the stomach contents become more fluid and the pH approaches 2.0, pepsin activity increases and protein disassembly begins. Protein digestion is not completed in the stomach, because time is limited and pepsin attacks only specific types of peptide bonds, not all of them. However, pepsin generally has enough time to break down complex proteins into smaller peptide and polypeptide chains before the chyme enters the duodenum.

Although digestion does occur in the stomach, the absorption of nutrients does not take place there, for several reasons: (1) The epithelial cells are covered by a blanket of alkaline mucus and are not directly exposed to chyme, (2) the epithelial cells lack the specialized transport mechanisms of cells that line the small intestine, (3) the gastric lining is relatively impermeable to water, and (4) digestion has not proceeded to completion by the time chyme leaves the stomach. At this stage, most carbohydrates, lipids, and proteins are only partially broken down.

Some drugs can be absorbed in the stomach. For example, ethyl alcohol can diffuse through the mucous barrier and penetrate the lipid membranes of the epithelial cells. As a result, you absorb alcohol in your stomach before any nutrients in a meal reach the bloodstream. Meals containing large amounts of fat will slow the rate of alcohol absorption, because alcohol is lipid soluble, and some of it will be dissolved in fat droplets in the chyme. Aspirin is another lipid-soluble drug that can enter the bloodstream across the gastric mucosa. Such drugs alter the properties of the mucous layer and can promote epithelial damage by stomach acids and enzymes. Prolonged use of aspirin can cause gastric bleeding, so individuals with stomach ulcers usually avoid aspirin.

⚕ **Stomach**, or *gastric*, **cancer** is one of the most common lethal cancers, responsible for roughly 14,000 deaths in the United States each year. The incidence is higher in countries such as Japan or Korea, where the typical diet includes large quantities of pickled, fermented, or smoked foods. Because the symptoms can resemble those of gastric ulcers, the condition may not be recognized in its early stages. Diagnosis generally involves X rays of the stomach at various degrees of distension. The mucosa can also be visually inspected by using a flexible instrument called a *gastroscope*. Attachments permit the collection of tissue samples for analysis. Treatment of stomach cancer involves *gastrectomy* (gas-TREK-to-mē), the surgical removal of part or all of the stomach. People can survive even a total gastrectomy, because the only absolutely vital function of the stomach is the secretion of intrinsic factor. Protein breakdown can still be performed by the small intestine, although at reduced efficiency, and the loss of such functions as food storage and acid production is not life threatening.

✓ How would a large meal affect the pH of blood that leaves the stomach?

✓ When a person suffers from chronic ulcers in the stomach, the branches of the vagus nerve that serve the stomach are sometimes severed. Why?

Answers start on page Q-1

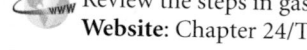 Review the steps in gastric activity on the **Companion Website**: Chapter 24/Tutorials.

24-6 THE SMALL INTESTINE AND ASSOCIATED GLANDULAR ORGANS

Objectives

- Describe the anatomical and histological characteristics of the small intestine.
- Explain the functions of the intestinal secretions, and discuss the regulation of secretory activities.
- Describe the structure, functions, and regulation of the accessory digestive organs.

Your stomach is a holding tank in which food is saturated with gastric juices and exposed to stomach acids and the digestive effects of pepsin. These are preliminary steps; most of the important digestive and absorptive functions occur in your small intestine, where chemical digestion is completed and the products of digestion are absorbed. The mucosa of the small intestine produces only a few of the enzymes involved. The pancreas provides digestive enzymes, as well as buffers that help neutralize chyme. The liver secretes *bile*, a solution stored in the gallbladder for subsequent discharge into the small intestine. Bile contains buffers and *bile salts*, compounds that facilitate the digestion and absorption of lipids.

☐ THE SMALL INTESTINE

The **small intestine** plays the key role in the digestion and absorption of nutrients. Ninety percent of nutrient absorption occurs in the small intestine, and most of the rest occurs in the large intestine. The small intestine averages 6 m (20 ft) in length (range: 4.5–7.5 m; 14.8–24.6 ft) and has a diameter ranging from 4 cm (1.6 in.) at the stomach to about 2.5 cm (1 in.) at the junction with the large intestine. It occupies all abdominal regions except the right and left hypochondriac and epigastric regions (Figure 1–8●, p. 17). The small intestine has three subdivisions: (1) the duodenum, (2) the jejunum, and (3) the ileum.

The **duodenum** (doo-AH-de-num), 25 cm (10 in.) in length, is the section closest to the stomach. This portion of the small intestine is a "mixing bowl" that receives chyme from the stomach and digestive secretions from the pancreas and liver. From its connection with the stomach, the duodenum curves in a C that encloses the pancreas. Except for the proximal 2.5 cm (1 in.), the duodenum is in a retroperitoneal position between vertebrae L_1 and L_4 (Figure 24–2●).

A rather abrupt bend marks the boundary between the duodenum and the **jejunum** (je-JOO-num). At this junction, the small intestine reenters the peritoneal cavity, supported by a sheet of mesentery. The jejunum is about 2.5 meters (8 ft) long. The bulk of chemical digestion and nutrient absorption occurs in the jejunum.

The **ileum** (IL-ē-um) is the final segment of the small intestine. It is also the longest, averaging 3.5 meters (12 ft) in length. The ileum ends at the **ileocecal valve**, a sphincter that controls the flow of materials from the ileum into the *cecum* of the large intestine.

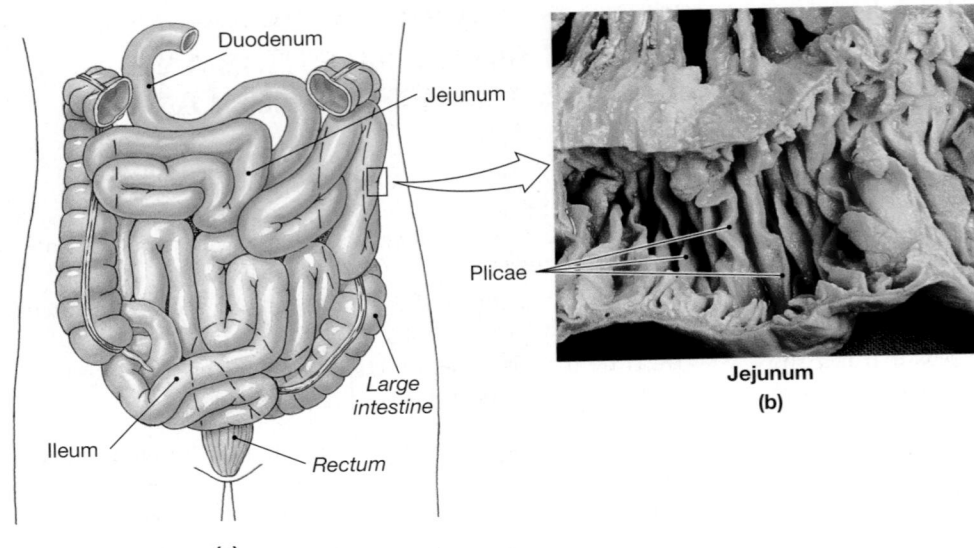

Duodenum

Jejunum

Plicae

Large intestine

Ileum

Rectum

(a)

Jejunum
(b)

•**FIGURE 24–16**
Regions of the Small Intestine.
(a) The positions of the duodenum, jejunum, and ileum in the abdominopelvic cavity. **(b)** A representative view of the jejunum. **ATLAS** Plates 7.6e, 7.7a,c,e,f

The small intestine fills much of the peritoneal cavity, and its position is stabilized by a broad mesentery attached to the dorsal body wall (Figure 24–2•). Movement of the small intestine during digestion is restricted by the stomach, the large intestine, the abdominal wall, and the pelvic girdle. Blood vessels, lymphatic vessels, and nerves reach these segments of the small intestine within the connective tissue of the mesentery. The primary blood vessels involved are branches of the superior mesenteric artery and the superior mesenteric vein. ⚯ pp. 758, 767-8

The subdivisions of the small intestine—the duodenum, jejunum, and ileum—are distinguished by both histological specialization and primary function. The locations of these segments in the peritoneal cavity are indicated in Figure 24–16a•.

☐ HISTOLOGY OF THE SMALL INTESTINE

The intestinal lining bears a series of transverse folds called **plicae**, or *plicae circulares* (Figures 24–16b and 24–17a,b•). Unlike the rugae in the stomach, the plicae are a permanent feature, and do not disappear when the small intestine fills. The small intestine contains roughly 800 plicae; their presence greatly increases the surface area available for absorption.

Figure 24–17c,d,e• presents a more detailed view of the intestinal wall.

INTESTINAL VILLI The mucosa of the small intestine is thrown into a series of fingerlike projections, the **intestinal villi**. These structures are covered by a simple columnar epithelium that is carpeted with microvilli. Because the microvilli project from the epithelium like the bristles on a brush, these cells are said to have a *brush border* (Figure 24–17c,d,e•).

If the small intestine were a simple tube with smooth walls, it would have a total absorptive area of roughly 3300 cm^2 (3.6 ft^2). Instead, the mucosa contains plicae, each plica supports a forest of villi, and each villus is covered by epithelial cells whose exposed surfaces contain microvilli. This arrangement increases the total area for absorption by a factor of more than 600, to approximately 2 million cm^2 (more than 2200 ft^2).

The lamina propria of each villus contains an extensive network of capillaries that originate in a vascular network within the submucosa. These capillaries carry absorbed nutrients to the hepatic portal circulation for delivery to the liver, which adjusts the nutrient concentrations of blood before the blood reaches the general systemic circulation. ⚯ p. 767

In addition to capillaries and nerve endings, each villus contains a lymphatic capillary called a **lacteal** (LAK-tē-al; *lacteus*, milky) (Figure 24–17d,e•). Lacteals transport materials that are unable to enter blood capillaries. For example, absorbed fatty acids are assembled into protein–lipid packages that are too large to diffuse into the bloodstream. These packets, called *chylomicrons*, reach the venous circulation by way of the thoracic duct, which delivers lymph into the left subclavian vein. ⚯ p. 782 The name *lacteal* refers to the pale, cloudy appearance of lymph that contains large quantities of lipids.

Contractions of the muscularis mucosae and smooth muscle cells within the intestinal villi move the villi back and forth, exposing the epithelial surfaces to the liquefied intestinal contents. This movement improves the efficiency of absorption, because local differences in nutrient concentration will be quickly eliminated. Movements of the villi also squeeze the lacteals, thus assisting in the movement of lymph out of the villi.

INTESTINAL GLANDS Between the columnar epithelial cells, goblet cells eject mucins onto the intestinal surfaces. At the bases of the villi are the entrances to the **intestinal glands**, or *crypts of Lieberkuhn*. These glandular pockets extend deep into the underlying lamina propria (Figure 24–17c•). Near the base of each intestinal gland, stem cell divisions produce new generations of

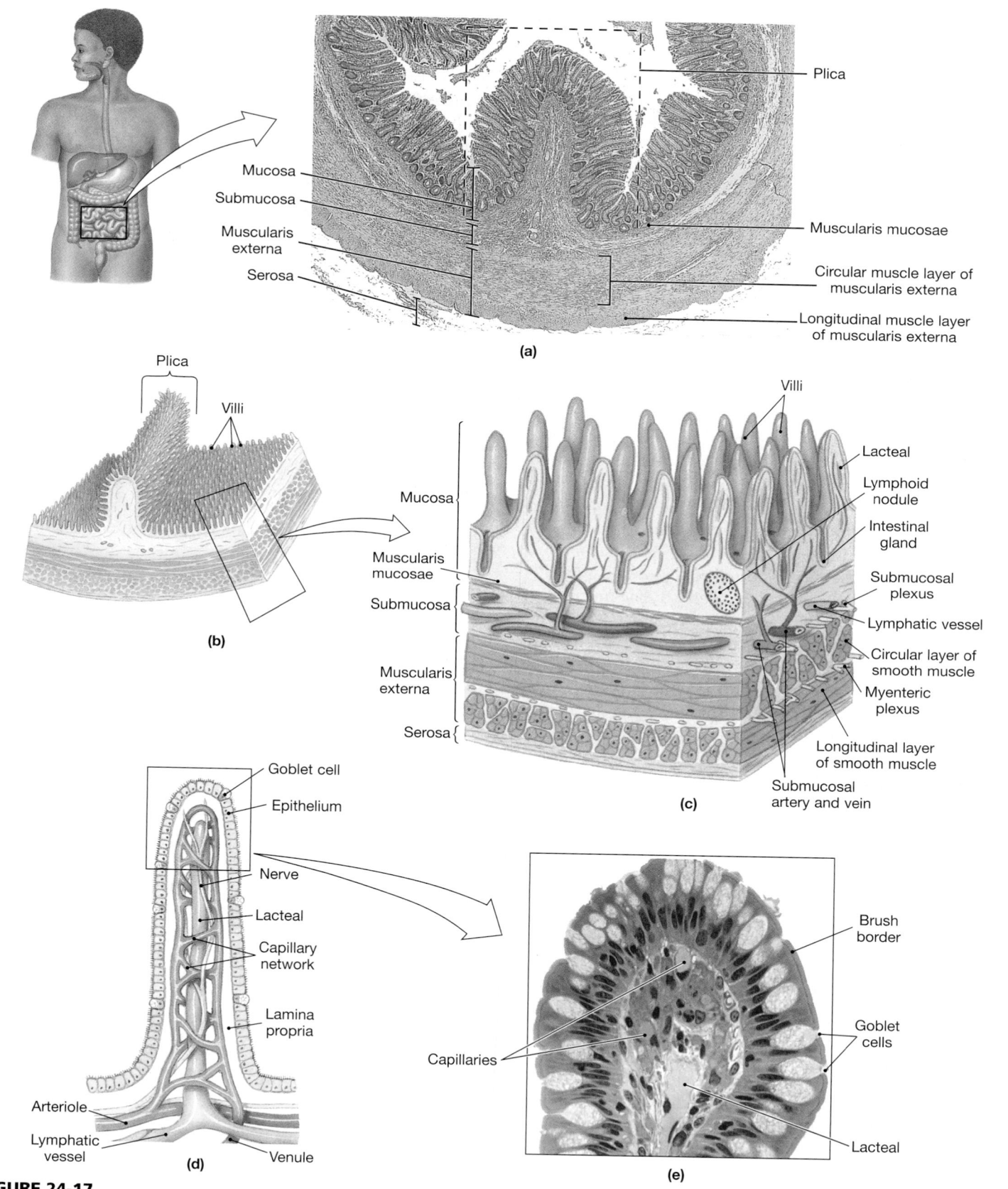

●FIGURE 24-17
The Intestinal Wall. **(a)** A section of the intestine. (LM × 2.5) **(b)** A single plica and multiple villi. **(c)** The organization of the intestinal wall. **(d)** Internal structures in a single villus, showing the capillary and lymphatic supplies. **(e)** A villus in sectional view. (LM × 252)

epithelial cells, which are continuously displaced toward the in-testinal surface. In a few days, the new cells will have reached the tip of a villus, where they are shed into the intestinal lumen. This ongoing process renews the epithelial surface, and the subse-quent disintegration of the shed cells adds enzymes to the lumen.

Several important brush border enzymes enter the intestinal lumen in this way. *Brush border enzymes* are integral membrane proteins located on the surfaces of intestinal microvilli. These enzymes perform the important digestive function of attacking materials that come in contact with the brush border. The breakdown products are then absorbed by the epithelial cells. Once the epithelial cells are shed, they disintegrate within the lumen, and the intracellular and brush border enzymes are re-leased. *Enterokinase* is a brush border enzyme that enters the lumen in this way. Enterokinase does not directly participate in digestion, but it activates a key pancreatic proenzyme, trypsino-gen. (We shall consider the functions of enterokinase and other brush border enzymes in a later section.) Intestinal glands also contain enteroendocrine cells responsible for the production of several intestinal hormones, including gastrin, cholecystokinin, and secretin.

The duodenum has numerous mucous glands, both in the epithelium and deep to it. In addition to intestinal glands, the submucosa contains **duodenal glands**, also called *submucosal glands* or *Brunner's glands*, which produce copious quantities of mucus when chyme arrives from the stomach. The mucus pro-tects the epithelium from the acidity of chyme and also contains buffers that help elevate the pH of the chyme. Along the length of the duodenum, the pH of chyme goes from 1–2 to 7–8. The duodenal glands also secrete the hormone *urogastrone*, which inhibits gastric acid production and stimulates the division of epithelial stem cells along the digestive tract.

REGIONAL SPECIALIZATIONS The duodenum has few plicae and their villi are small. The primary function of the duodenum is to receive chyme from the stomach and neutralize its acids before they can damage the absorptive surfaces of the small intestine. Over the proximal half of the jejunum, however, plicae and villi are very prominent. The plicae and villi thereafter gradually de-crease in size. This reduction parallels a reduction in absorptive activity: Most nutrient absorption has occurred before ingested materials reach the ileum. One rather drastic surgical method of promoting weight loss is the removal of a significant portion of the jejunum. The reduction in absorptive area causes a marked weight loss and may not interfere with adequate nutrition, but the side effects can be very troublesome. **AM** Drastic Weight-Loss Techniques

The distal portions of the ileum lack plicae, and the lamina propria there contains 20–30 masses of lymphoid tissue called *aggregated lymphoid nodules*, or *Peyer's patches*. ∞ p. 786 The lymphocytes in these nodules protect the small intestine from bacteria that are normal inhabitants of the large intestine. Lym-phoid nodules are most abundant in the terminal portion of the ileum, near the entrance to the large intestine.

Intestinal Secretions

Roughly 1.8 liters of watery **intestinal juice** enters your intestinal lumen each day. Intestinal juice moistens chyme, assists in buffering acids, and liquefies both the digestive enzymes provid-ed by the pancreas and the products of digestion. Much of this fluid volume arrives by osmosis, as water flows out of the mucosa and into the relatively concentrated chyme. The rest is secreted by intestinal glands, stimulated by the activation of touch recep-tors and stretch receptors in the intestinal walls.

The duodenal glands help protect the duodenal epithelium from gastric acids and enzymes. These glands increase their secretory ac-tivities in response to (1) local reflexes, (2) the release of the hor-mone *enterocrinin* by enteroendocrine cells of the duodenum, and (3) parasympathetic stimulation via the vagus nerves. Mechanisms 1 and 2 operate only after chyme arrives in the duodenum. Howev-er, because vagus nerve activity triggers their secretion, the duode-nal glands begin secreting during the cephalic phase of gastric secretion, long before the chyme reaches the pyloric sphincter. Thus, the duodenal lining has protection in advance.

Sympathetic stimulation inhibits the activation of the duode-nal glands, leaving the duodenal lining relatively unprepared for the arrival of chyme. This fact probably accounts for the com-mon observation that duodenal ulcers can be caused by chronic stress or other factors that promote sympathetic activation.

☐ INTESTINAL MOVEMENTS

After chyme has arrived in the duodenum, weak peristaltic con-tractions move it slowly toward the jejunum. The contractions are myenteric reflexes that are not under central nervous system control. Their effects are limited to within a few centimeters of the site of the original stimulus. These short reflexes are con-trolled by motor neurons in the submucosal and myenteric plexuses. In addition, some of the smooth muscle cells contract periodically, even without stimulation, establishing a basic con-tractile rhythm that then spreads from cell to cell.

The stimulation of the parasympathetic system increases the sensitivity of the weak myenteric reflexes and accelerates both local peristalsis and segmentation. More elaborate reflexes coordi-nate activities along the entire length of the small intestine. Two examples are triggered by the stimulation of stretch receptors in the stomach as it fills. The **gastroenteric reflex** stimulates motili-ty and secretion along the entire small intestine; the **gastroileal** (gas-trō-IL-ē-al) **reflex** triggers the relaxation of the ileocecal valve. The net result is that materials pass from the small intestine into the large intestine. Thus, the gastroenteric and gastroileal re-flexes accelerate movement along the small intestine—the oppo-site effect of the enterogastric reflex.

Hormones released by the digestive tract can enhance or sup-press reflexes. For example, the gastroileal reflex is triggered by stretch receptor stimulation, but the degree of ileocecal valve re-laxation is enhanced by gastrin, which is secreted in large quanti-ties when food enters the stomach.

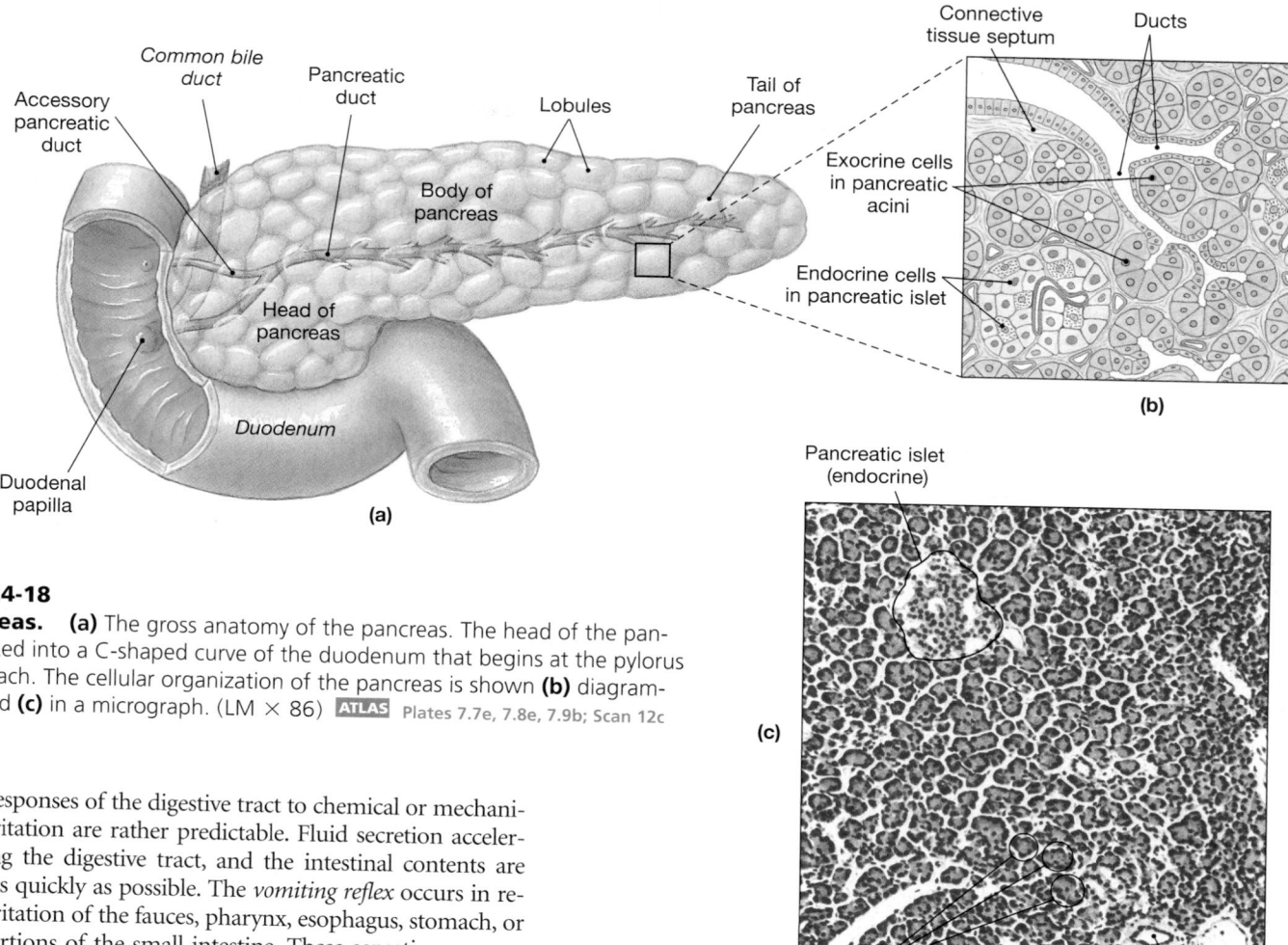

(b)

(c)

●**FIGURE 24-18**
The Pancreas. **(a)** The gross anatomy of the pancreas. The head of the pancreas is tucked into a C-shaped curve of the duodenum that begins at the pylorus of the stomach. The cellular organization of the pancreas is shown **(b)** diagrammatically and **(c)** in a micrograph. (LM × 86) **ATLAS** Plates 7.7e, 7.8e, 7.9b; Scan 12c

The responses of the digestive tract to chemical or mechanical irritation are rather predictable. Fluid secretion accelerates all along the digestive tract, and the intestinal contents are eliminated as quickly as possible. The *vomiting reflex* occurs in response to irritation of the fauces, pharynx, esophagus, stomach, or proximal portions of the small intestine. These sensations are relayed to the *vomiting center* of the medulla oblongata, which coordinates the motor responses. In preparation for vomiting, the pylorus relaxes and the contents of the duodenum and proximal jejunum are discharged into the stomach by strong peristaltic waves that travel toward the stomach rather than toward the ileum. **Vomiting**, or *emesis* (EM-e-sis), then occurs as the stomach regurgitates its contents through the esophagus and pharynx. During regurgitation, the uvula and soft palate block the entrance to the nasopharynx. Increased salivary secretion assists in buffering the stomach acids, thereby preventing erosion of the teeth. In conditions marked by repeated vomiting, severe tooth damage can occur; we shall discuss one example, the eating disorder *bulimia*, in Chapter 25. Most of the force of vomiting comes from expiratory movements that elevate intra-abdominal pressures and force the stomach against the tensed diaphragm.

☐ THE PANCREAS

Your **pancreas** lies posterior to your stomach, extending laterally from the duodenum toward the spleen (Figure 24–18a●). The pancreas is an elongate, pinkish gray organ with a length of about 15 cm (6 in.) and a weight of about 80 g (3 oz). The broad **head** of the pancreas lies within the loop formed by the duodenum as it leaves the pylorus. The slender **body** of the pancreas

extends toward the spleen, and the **tail** is short and bluntly rounded. The pancreas is retroperitoneal and is firmly bound to the posterior wall of the abdominal cavity.

The surface of the pancreas has a lumpy, lobular texture. A thin, transparent capsule of connective tissue wraps the entire organ. Through the anterior capsule and the overlying layer of peritoneum, you can see the pancreatic lobules, associated blood vessels, and excretory ducts. Arterial blood reaches the pancreas by way of branches of the splenic, superior mesenteric, and common hepatic arteries. The pancreatic arteries and pancreaticoduodenal arteries are the major branches from these vessels. The splenic vein and its branches drain the pancreas.

The pancreas is primarily an exocrine organ, producing digestive enzymes and buffers. The large **pancreatic duct** (*duct of Wirsung*) delivers these secretions to the duodenum. A small **accessory pancreatic duct**, or *duct of Santorini*, may branch from the pancreatic duct. The pancreatic duct extends within the attached mesentery to reach the duodenum, where it meets the *common bile duct* from the liver and gallbladder. The two ducts then empty into the *duodenal ampulla*, a chamber located roughly halfway along the length of the duodenum (Figure 24–21b●). When present, the accessory pancreatic duct generally empties into the duodenum independently, outside the duodenal ampulla.

Histological Organization

Partitions of connective tissue divide the interior of the pancreas into distinct lobules. The blood vessels and tributaries of the pancreatic ducts are situated within these connective-tissue septa (Figure 24–18b●). The pancreas is an example of a *compound tubuloalveolar gland*, a structure that we described in Chapter 4. ∞ p. 121 In each lobule, the ducts branch repeatedly before ending in blind pockets called the **pancreatic acini** (AS-i-nī). Each pancreatic acinus is lined with a simple cuboidal epithelium. *Pancreatic islets*, the endocrine tissues of the pancreas, are scattered among the acini (Figure 24–18b,c●). The islets account for only about 1 percent of the cellular population of the pancreas.

The pancreas has two distinct functions, one endocrine and the other exocrine. The endocrine cells of the pancreatic islets secrete insulin and glucagon into the bloodstream. ∞ p. 632 The exocrine cells include the acinar cells and the epithelial cells that line the duct system. Together, the acinar cells and the epithelial cells secrete **pancreatic juice**, an alkaline mixture of digestive enzymes, water, and ions, into the small intestine. Pancreatic enzymes are secreted by the acinar cells. These enzymes do most of the digestive work in the small intestine, breaking down ingested materials into small molecules suitable for absorption. The water and ions, secreted primarily by the cells lining the pancreatic ducts, assist in diluting and buffering the acids in the chyme.

Physiology of the Pancreas

Each day, your pancreas secretes about 1000 ml (1 qt) of pancreatic juice. The secretory activities are controlled primarily by hormones from the duodenum. When chyme arrives in the duodenum, secretin is released. This hormone triggers the pancreatic secretion of a watery buffer solution with a pH of 7.5–8.8. Among its other components, the secretion contains bicarbonate and phosphate buffers that help elevate the pH of the chyme. A different duodenal hormone, cholecystokinin, stimulates the production and secretion of pancreatic enzymes. Pancreatic enzyme secretion also increases under stimulation by the vagus nerves. As we noted earlier, this stimulation occurs during the cephalic phase of gastric regulation, so the pancreas starts to synthesize enzymes before food even reaches the stomach. Such a head start is important, because enzyme synthesis takes much longer than the production of buffers. By starting early, the pancreatic cells are ready to meet the demand when chyme arrives in the duodenum.

The specific pancreatic enzymes involved include the following:

- **Pancreatic alpha-amylase**, a **carbohydrase** (kar-bō-HĪ-drās)— an enzyme that breaks down certain starches. Pancreatic alpha-amylase is almost identical to salivary amylase.

- **Pancreatic lipase**, which breaks down certain complex lipids, releasing products, such as fatty acids, that can be easily absorbed.

- **Nucleases**, which break down nucleic acids.

- **Proteolytic enzymes**, which break certain proteins apart. The proteolytic enzymes of the pancreas include proteases and peptidases. **Proteases** break apart large protein complexes, whereas **peptidases** break small peptide chains into individual amino acids.

Proteolytic enzymes account for about 70 percent of the total pancreatic enzyme production. The enzymes are secreted as inactive proenzymes and are activated only after they reach the small intestine. Proenzymes discussed earlier in the text include pepsinogen, angiotensinogen, plasminogen, fibrinogen, and many of the clotting factors and enzymes of the complement system. ∞ pp. 673, 794 As in the stomach, the release of a proenzyme rather than an active enzyme in the pancreas protects the secretory cells from the destructive effects of their own products. Among the proenzymes secreted by the pancreas are **trypsinogen** (trip-SIN-ō-jen), **chymotrypsinogen** (kī-mō-trip-SIN-ō-jen), **procarboxypeptidase** (prō-kar-bok-sē-PEP-ti-dās), and **proelastase** (pro-ē-LAS-tās).

Once inside the duodenum, enterokinase located on the brush border and in the lumen triggers the conversion of trypsinogen to **trypsin**, an active protease. Trypsin then activates the other proenzymes, producing **chymotrypsin**, **carboxypeptidase**, and **elastase**. Each enzyme attacks peptide bonds linking specific amino acids and ignores others. Together, they break down proteins into a mixture of dipeptides, tripeptides, and amino acids.

✠ **Pancreatitis** (pan-krē-a-TĪ-tis) is an inflammation of the pancreas. A blockage of the excretory ducts, bacterial or viral infections, ischemia, and drug reactions, especially those involving alcohol, are among the factors that may produce this extremely painful condition. These stimuli provoke a crisis by injuring exocrine cells in at least a portion of the organ. Lysosomes in the damaged cells then activate the proenzymes, and autolysis begins. The proteolytic enzymes digest the surrounding, undamaged cells, activating their enzymes and starting a chain reaction. In most cases, only a portion of the pancreas will be affected, and the condition subsides in a few days. In 10–15 percent of pancreatitis cases, the process does not subside; the enzymes can then ultimately destroy the pancreas. If the islet cells are damaged, diabetes mellitus may result. ∞ p. 633

☐ THE LIVER

The **liver**, the largest visceral organ, is one of the most versatile organs in the body. Most of its mass lies in the right hypochondriac and epigastric regions, but it may extend into the left hypochondriac and umbilical regions as well. The liver weighs about 1.5 kg (3.3 lb). This large, firm, reddish-brown organ provides essential metabolic and synthetic services.

Anatomy of the Liver

The liver is wrapped in a tough fibrous capsule and is covered by a layer of visceral peritoneum. On the anterior surface, the **falciform ligament** marks the division between the organ's left lobe and the right lobe (Figure 24–19a,b●). A thickening in the

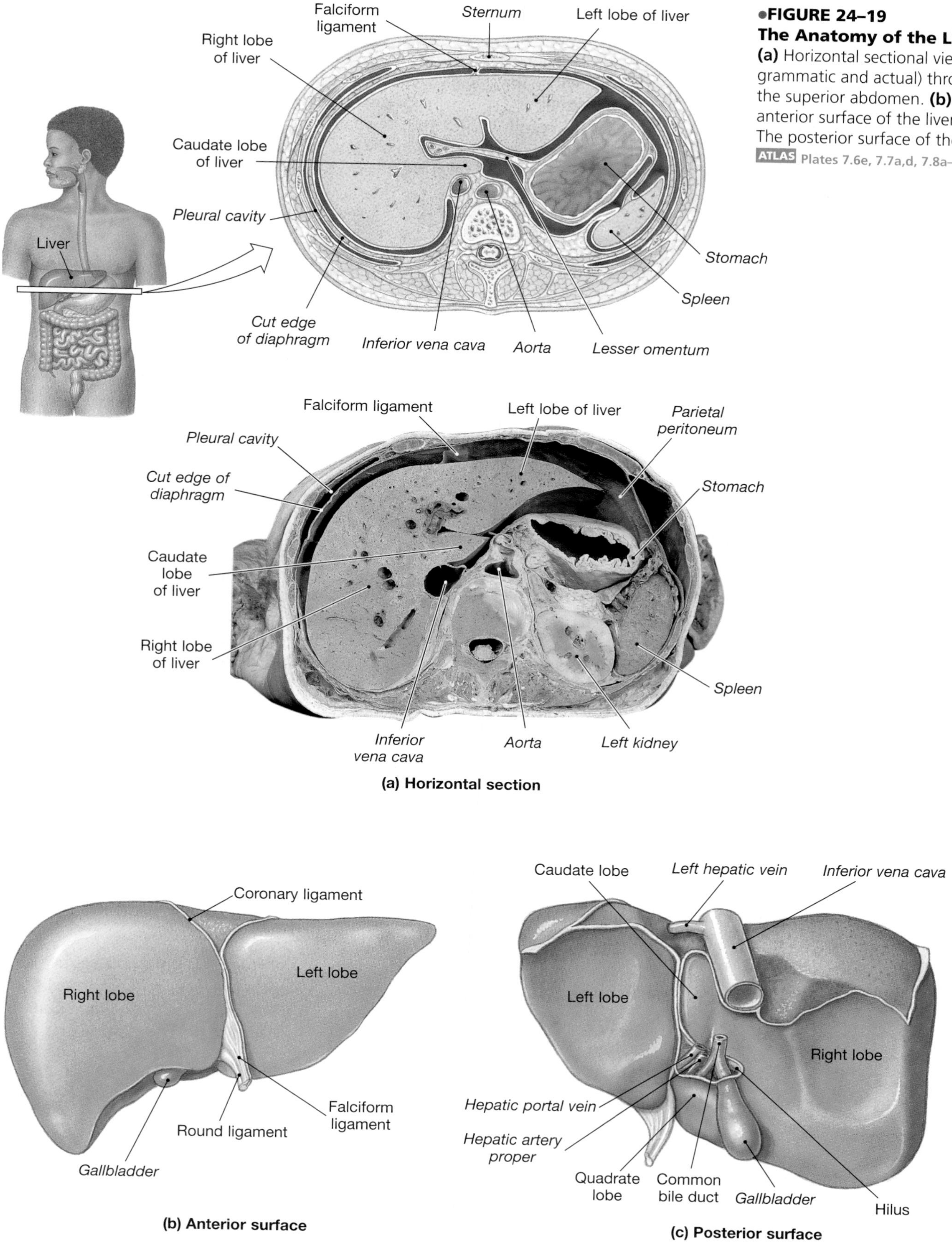

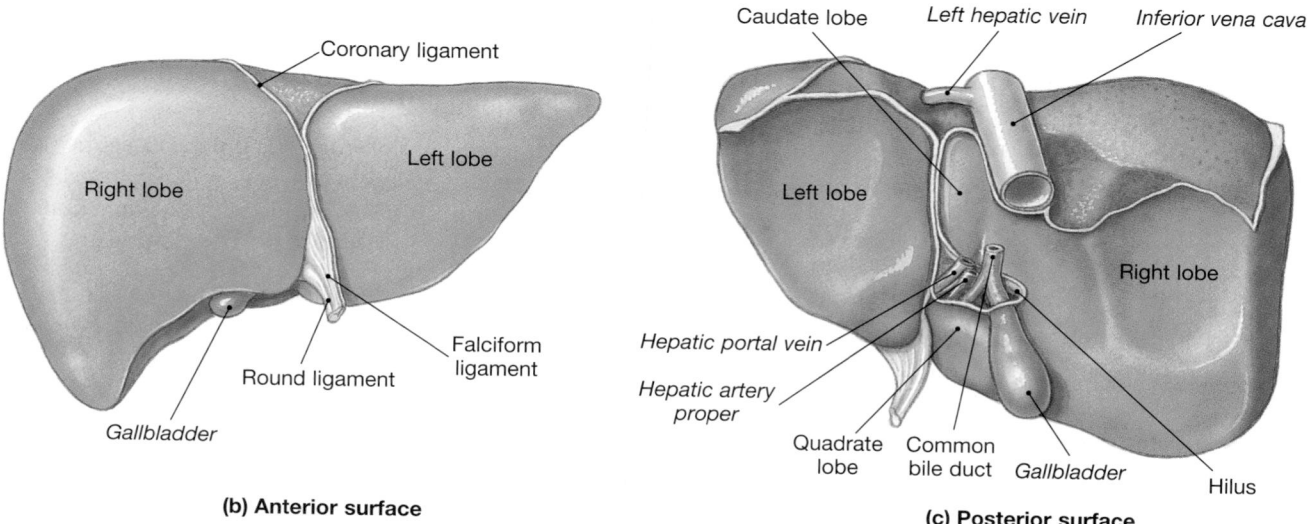

●FIGURE 24–19
The Anatomy of the Liver.
(a) Horizontal sectional views (diagrammatic and actual) through the superior abdomen. **(b)** The anterior surface of the liver. **(c)** The posterior surface of the liver.
ATLAS Plates 7.6e, 7.7a,d, 7.8a–c, 7.9a,b

Falciform ligament
Sternum
Left lobe of liver
Right lobe of liver
Caudate lobe of liver
Pleural cavity
Liver
Cut edge of diaphragm
Inferior vena cava
Aorta
Lesser omentum
Stomach
Spleen

Falciform ligament
Left lobe of liver
Parietal peritoneum
Pleural cavity
Cut edge of diaphragm
Caudate lobe of liver
Right lobe of liver
Stomach
Inferior vena cava
Aorta
Left kidney
Spleen

(a) Horizontal section

Coronary ligament
Left lobe
Right lobe
Gallbladder
Round ligament
Falciform ligament

(b) Anterior surface

Caudate lobe
Left hepatic vein
Inferior vena cava
Left lobe
Right lobe
Hepatic portal vein
Hepatic artery proper
Quadrate lobe
Common bile duct
Gallbladder
Hilus

(c) Posterior surface

posterior margin of the falciform ligament is the **round ligament**, or *ligamentum teres*, a fibrous band that marks the path of the fetal umbilical vein.

On the posterior surface of the liver, the impression left by the inferior vena cava marks the division between the right lobe and the small **caudate** (KAW-dāt) **lobe** (Figure 24–19a,c●). Inferior to the caudate lobe lies the **quadrate lobe**, sandwiched between the left lobe and the gallbladder. Afferent blood vessels and other structures reach the liver by traveling within the connective tissue of the lesser omentum. They converge at a region called the **hilus**, or *porta hepatis* ("doorway to the liver").

We discussed the circulation to the liver in Chapter 21 and summarized that information in Figures 21–27 and 21–34●, pp. 759, 767. Roughly one-third of the blood supply to the liver is arterial blood from the hepatic artery proper. The rest is venous blood from the hepatic portal vein, which begins in the capillaries of the esophagus, stomach, small intestine, and most of the large intestine. Liver cells, called **hepatocytes** (he-PAT-ō-sīts), adjust circulating levels of nutrients by selective absorption and secretion. Blood leaving the liver returns to the systemic circuit through the hepatic veins, which open into the inferior vena cava.

Histological Organization of the Liver

Each lobe of the liver is divided by connective tissue into approximately 100,000 **liver lobules**, the basic functional units of the liver. The histological organization and structure of a typical liver lobule are shown in Figure 24–20●.

Adjacent lobules are separated from each other by an *interlobular septum*. The hepatocytes in a liver lobule form a series of irregular plates arranged like the spokes of a wheel (Figure 24–20a,b●). The plates are only one cell thick, and exposed hepatocyte surfaces are covered with short microvilli. Within a lobule, sinusoids between adjacent plates empty into the **central vein**. (We introduced sinusoids in Chapter 21. ∞ p. 727) The liver sinusoids lack a basement membrane, so large openings between the endothelial cells allow solutes—even those as large as plasma proteins—to pass out of the bloodstream and into the spaces surrounding the hepatocytes.

In addition to containing typical endothelial cells, the sinusoidal lining includes a large number of **Kupffer** (KOOP-fer) **cells**, also known as *stellate reticuloendothelial cells*. ∞ p. 793 These phagocytic cells, part of the monocyte–macrophage system, engulf pathogens, cell debris, and damaged blood cells. Kupffer cells are also responsible for storing (1) iron, (2) some lipids, and (3) heavy metals, such as tin or mercury, that are absorbed by the digestive tract.

Blood enters the liver sinusoids from small branches of the hepatic portal vein and hepatic artery proper. A typical liver lobule has a hexagonal shape in cross section (Figure 24–20b●). There are six **portal areas**, or *hepatic triads*, one at each corner of the lobule. A portal area contains three structures: (1) a branch of the hepatic portal vein, (2) a branch of the hepatic artery proper, and (3) a small branch of the bile duct (Figure 24–20a,b,d●).

Branches from the arteries and veins deliver blood to the sinusoids of adjacent liver lobules (Figure 24–20a,b●). As blood flows through the sinusoids, hepatocytes absorb solutes from the plasma and secrete materials such as plasma proteins. Blood then leaves the sinusoids and enters the central vein of the lobule. The central veins ultimately merge to form the hepatic veins, which then empty into the inferior vena cava. Liver diseases, such as the various forms of *hepatitis*, and conditions such as alcoholism can lead to degenerative changes in the liver tissue and constriction of the circulatory supply. **AM** Liver Disease

Pressures in the hepatic portal system are usually low, averaging 10 mm Hg or less. This pressure can increase markedly, however, if blood flow through the liver becomes restricted as a result of a blood clot or damage to the organ. Such a rise in portal pressure is called **portal hypertension**. As pressures rise, small peripheral veins and capillaries in the portal system become distended and esophageal varices may develop. If these rupture, extensive bleeding can occur. Portal hypertension can also force fluid into the peritoneal cavity across the serosal surfaces of the liver and viscera, producing ascites (p. 134).

The Bile Duct System

The liver secretes a fluid called **bile** into a network of narrow channels between the opposing membranes of adjacent liver cells. These passageways, called **bile canaliculi**, extend outward, away from the central vein. Eventually, they connect with fine **bile ductules** (DUK-tūlz), which carry bile to bile ducts in the nearest portal area. The **right** and **left hepatic ducts** collect bile from all the bile ducts of the liver lobes. These ducts unite to form the **common hepatic duct**, which leaves the liver (Figure 24–21●). The bile in the common hepatic duct either flows into the *common bile duct*, which empties into the duodenal ampulla, or enters the *cystic duct*, which leads to the gallbladder.

The **common bile duct** is formed by the union of the **cystic duct** and the common hepatic duct. The common bile duct passes within the lesser omentum toward the stomach, turns, and penetrates the wall of the duodenum to meet the pancreatic duct at the duodenal ampulla.

The Physiology of the Liver

The liver is responsible for three general categories of functions: (1) *metabolic regulation*, (2) *hematological regulation*, and (3) *bile production*. The liver has more than 200 functions; in this discussion, we shall provide only a general overview.

METABOLIC REGULATION The liver is the primary organ involved in regulating the composition of circulating blood. All blood leaving the absorptive surfaces of the digestive tract enters the hepatic portal system and flows into the liver. Liver cells extract nutrients or toxins absorbed from blood before it reaches the systemic circulation through the hepatic veins. Excess nutrients are removed and stored, and deficiencies are corrected by mobilizing stored reserves or performing synthetic activities such as the following:

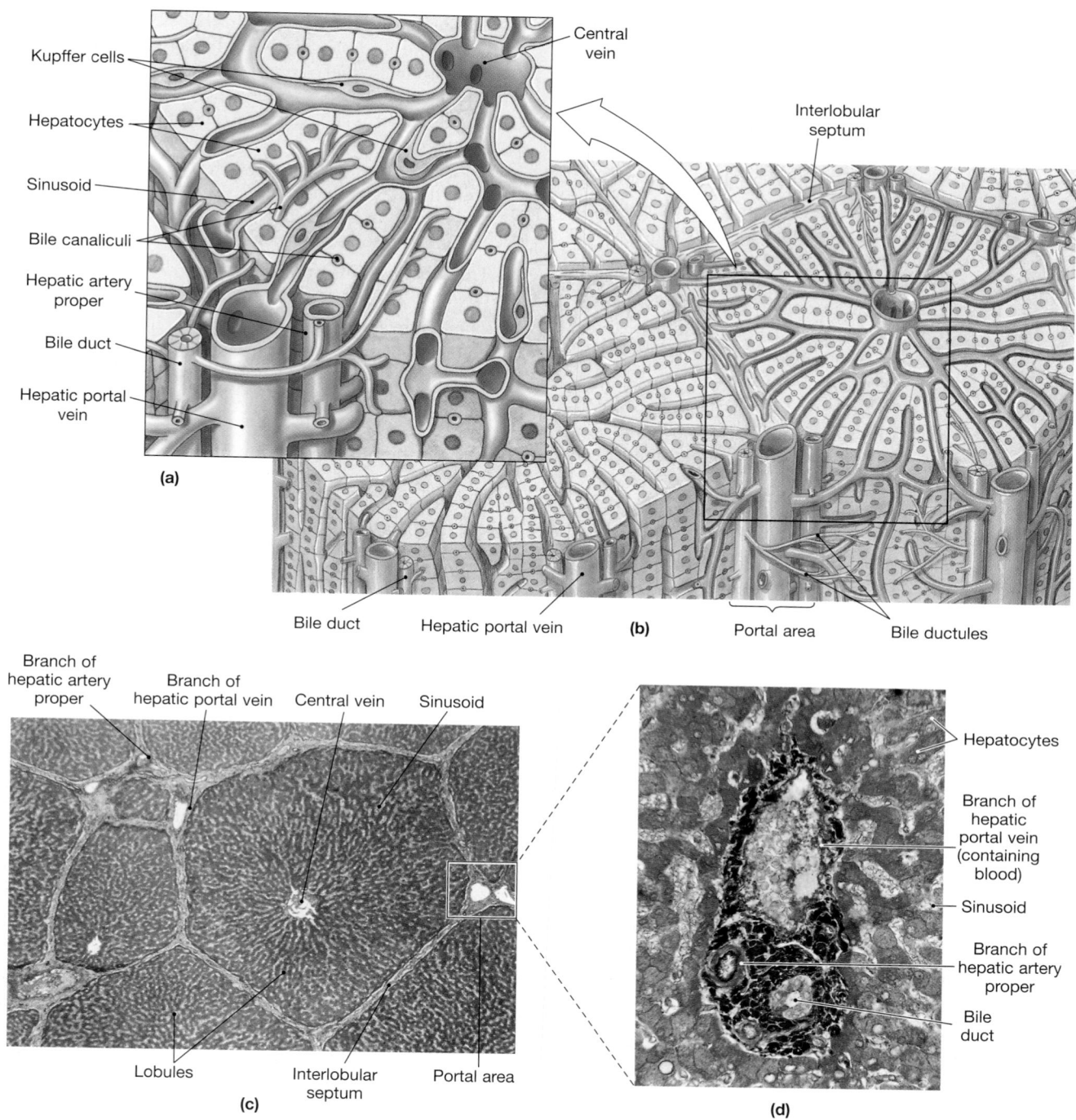

Kupffer cells

Hepatocytes

Sinusoid

Bile canaliculi

Hepatic artery proper

Bile duct

Hepatic portal vein

Central vein

Interlobular septum

(a)

Bile duct

Hepatic portal vein

(b)

Portal area

Bile ductules

Branch of hepatic artery proper

Branch of hepatic portal vein

Central vein

Sinusoid

Lobules

Interlobular septum

Portal area

(c)

Hepatocytes

Branch of hepatic portal vein (containing blood)

Sinusoid

Branch of hepatic artery proper

Bile duct

(d)

●**FIGURE 24-20**
Liver Histology. **(a)** A single liver lobule and its cellular components. **(b)** A diagrammatic view of liver structure, showing relationships among lobules. **(c)** A section through liver lobules from a pig liver. (LM × 38) (The interlobular septa in a human liver are very difficult to see at comparable magnification.) **(d)** A portal area. (LM × 31)

- **Carbohydrate Metabolism.** The liver stabilizes blood glucose levels at about 90 mg/dl. If blood glucose levels drop, hepatocytes break down glycogen reserves and release glucose into the bloodstream. They also synthesize glucose from other carbohydrates or from available amino acids. The synthesis of glucose from other compounds is a process called *gluconeogenesis*. If blood glucose levels climb, hepatocytes remove glucose from

the bloodstream and either store it as glycogen or use it to synthesize lipids that can be stored in the liver or other tissues. These metabolic activities are regulated by circulating hormones, such as insulin and glucagon. ⊂⊃ pp. 632-634

- **Lipid Metabolism.** The liver regulates circulating levels of triglycerides, fatty acids, and cholesterol. When those levels decline, the liver breaks down its lipid reserves and releases them into the bloodstream. When the levels are high, the lipids are removed for storage. However, because most lipids absorbed by the digestive tract bypass the hepatic portal circulation, this regulation occurs only after lipid levels have risen within the general circulation.

- **Amino Acid Metabolism.** The liver removes excess amino acids from the bloodstream. These amino acids can be used to synthesize proteins or can be converted to lipids or glucose for storage.

- **The Removal of Waste Products.** When converting amino acids to lipids or carbohydrates, or when breaking down amino acids to get energy, the liver strips off the amino groups, a process called *deamination*. Ammonia, a toxic waste product, is formed. The liver neutralizes ammonia by converting it to *urea*, a fairly harmless compound excreted at the kidneys. Other waste products, circulating toxins, and drugs are also removed from the blood for inactivation, storage, or excretion.

- **Vitamin Storage.** Fat-soluble vitamins (A, D, E, and K) and vitamin B_{12} are absorbed from the blood and stored in the liver. These reserves are called on when your diet contains inadequate amounts of those vitamins.

- **Mineral Storage.** The liver converts iron reserves to ferritin and stores this protein–iron complex. ∞ **p. 659**

- **Drug Inactivation.** The liver removes and breaks down circulating drugs, thereby limiting the duration of their effects. When they prescribe a particular drug, physicians must take into account the rate at which the liver removes that drug from the bloodstream. For example, a drug that is absorbed relatively quickly must be administered every few hours to keep plasma concentrations at therapeutic levels.

☤ Any condition that severely damages the liver represents a serious threat to life. The liver has a limited ability to regenerate itself after injury, but liver function will not fully recover unless the normal vascular pattern returns. Examples of important types of liver disease include **cirrhosis**, which is characterized by the replacement of lobules by fibrous tissue, and various forms of **hepatitis** caused by viral infections. In some cases, liver transplants are used to treat liver failure, but the supply of suitable donor tissue is limited and the success rate is highest in young, otherwise healthy individuals. Clinical trials are now under way to test an artificial liver known as *ELAD* (*e*xtracorporeal *l*iver *a*ssist *d*evice) that may prove suitable for the long-term support of persons with chronic liver disease. **AM** Liver Disease

HEMATOLOGICAL REGULATION The liver, the largest blood reservoir in your body, receives about 25 percent of the cardiac output. As blood passes through it, the liver performs the following functions:

- **Phagocytosis and Antigen Presentation.** Kupffer cells in the liver sinusoids engulf old or damaged red blood cells, cellular debris, and pathogens from the bloodstream. Kupffer cells are antigen-presenting cells that can stimulate an immune response. ∞ **p. 793**

- **The Synthesis of Plasma Proteins.** Hepatocytes synthesize and release most of the plasma proteins, including the albumins, which contribute to the osmotic concentration of the blood; the various types of transport proteins; clotting proteins; and complement proteins.

- **The Removal of Circulating Hormones.** The liver is the primary site for the absorption and recycling of epinephrine, norepinephrine, insulin, thyroid hormones, and steroid hormones, such as the sex hormones (estrogens and androgens) and corticosteroids. The liver also absorbs cholecalciferol (vitamin D_3) from the blood. Liver cells then convert the cholecalciferol, which may be synthesized in the skin or absorbed in the diet, into an intermediary product, 25-hydroxy-D_3, that is released back into the bloodstream. This intermediary is absorbed by the kidneys and used to generate calcitriol, a hormone important to Ca^{2+} metabolism. ∞ **p. 635**

- **The Removal of Antibodies.** The liver absorbs and breaks down antibodies, releasing amino acids to be recycled.

- **The Removal or Storage of Toxins.** Lipid-soluble toxins in the diet, such as DDT, are absorbed by the liver and stored in lipid deposits, where they do not disrupt cellular functions. Other toxins are removed from the bloodstream and are either broken down or excreted in the bile.

- **The Synthesis and Secretion of Bile.** Bile is synthesized in the liver and excreted into the lumen of the duodenum. Bile consists mostly of water, with minor amounts of ions, *bilirubin* (a pigment derived from hemoglobin), cholesterol, and an assortment of lipids collectively known as the **bile salts**. The water and ions assist in the dilution and buffering of acids in chyme as it enters the small intestine.

Bile salts are synthesized from cholesterol in the liver. Several related compounds are involved; the most abundant are derivatives of the steroids *cholate* and *chenodeoxycholate*.

THE FUNCTIONS OF BILE Most dietary lipids are not water soluble. Mechanical processing in the stomach creates large drops containing a variety of lipids. Pancreatic lipase is not lipid soluble, so the enzymes can interact with lipids only at the surface of a lipid droplet. The larger the droplet, the more lipids are inside, isolated and protected from these enzymes. Bile salts break the droplets apart in a process called **emulsification** (ē-mul-si-fi-KĀ-shun), which dramatically increases the surface area accessible to enzymatic attack.

Emulsification creates tiny *emulsion droplets* with a superficial coating of bile salts. The formation of tiny droplets increases the surface area available for enzymatic attack. In addition, the layer of bile salts facilitates interaction between the lipids and lipid-digesting enzymes supplied by the pancreas. After lipid digestion has been completed, bile salts promote the absorption of lipids by the intestinal epithelium. More than 90 percent of the bile salts are themselves reabsorbed, primarily in the ileum, as lipid digestion is completed. The reabsorbed bile salts enter the hepatic portal circulation and are collected and recycled by the liver. The cycling of bile salts from the liver to the small intestine and back is called the **enterohepatic circulation of bile**.

□ THE GALLBLADDER

The **gallbladder** is a hollow, pear-shaped organ that stores and concentrates bile prior to its excretion into the small intestine. This muscular sac is located in a fossa, or recess, in the posterior

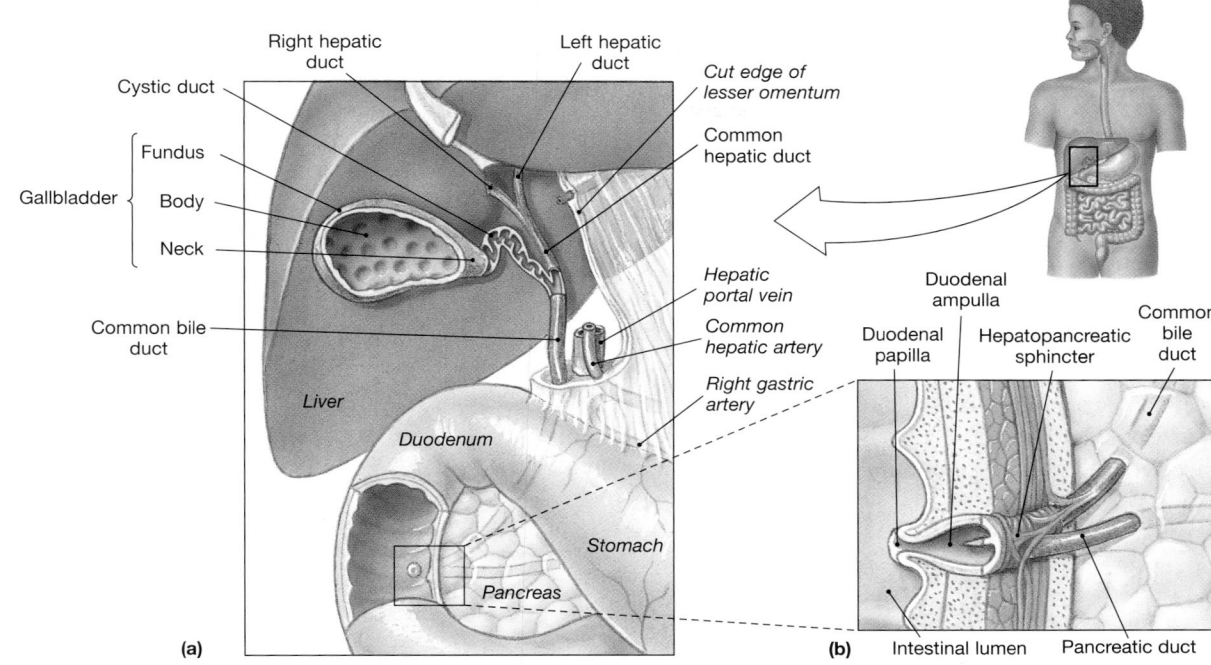

● **FIGURE 24–21 The Gallbladder.**
(a) A view of the inferior surface of the liver, showing the position of the gallbladder and ducts that transport bile from the liver to the gallbladder and duodenum. A portion of the lesser omentum has been cut away. **(b)** An interior view of the duodenum, showing the duodenal ampulla and related structures.
ATLAS Plates 7.7b,d, 7.8b,c; Scan 12c

surface of the liver's right lobe. The gallbladder is divided into three regions: (1) the **fundus**, (2) the **body**, and (3) the **neck** (Figure 24–21a●). The cystic duct leads from the gallbladder toward its union with the common hepatic duct to form the common bile duct. At the duodenum, the common bile duct meets the pancreatic duct before emptying into a chamber called the **duodenal ampulla** (am-PŪL-a) (Figure 24–21b●), which receives buffers and enzymes from the pancreas and bile from the liver and gallbladder. The duodenal ampulla opens into the duodenum at the **duodenal papilla**, a small mound.

The **hepatopancreatic sphincter** (*sphincter of Oddi*), a muscular sphincter, encircles the lumen of the common bile duct and, generally, the pancreatic duct and duodenal ampulla as well. This sphincter remains contracted unless stimulated by the intestinal hormone cholecystokinin (CCK).

Physiology of the Gallbladder

The gallbladder has two major functions: (1) *bile storage* and (2) *bile modification*. Liver cells produce roughly 1 liter of bile each day, but the hepatopancreatic sphincter remains closed until chyme enters the duodenum. In the meantime, when bile cannot flow along the common bile duct, it enters the cystic duct for storage within the expandable gallbladder. When full, the gallbladder contains 40–70 ml of bile. The composition of bile gradually changes as it remains in the gallbladder. Much of the water is absorbed, and the bile salts and other components of bile become increasingly concentrated.

Bile is secreted continuously, but is released into the duodenum only under the stimulation of CCK. In the absence of CCK, the he-

patopancreatic sphincter remains closed, and bile exiting the liver in the common hepatic duct reaches the gallbladder by way of the cystic duct. Cholecystokinin (1) relaxes the hepatopancreatic sphincter and (2) stimulates contractions in the walls of the gallbladder. The contractions push bile into the small intestine. Whenever chyme enters the duodenum, CCK is released, but the amount secreted increases markedly when the chyme contains large amounts of lipids.

If bile becomes too concentrated, crystals of insoluble minerals and salts begin to appear. These deposits are called **gallstones**. Merely having them, a condition termed **cholelithiasis** (kō-lē-li-THĪ-a-sis; *chole*, bile), is not a problem as long as the stones remain small. Small gallstones are normally flushed down the bile duct and excreted. In **cholecystitis**, the gallstones are so large that they can damage the wall of the gallbladder or block the cystic or common bile duct. In severe cases of cholecystitis, the gallbladder can become infected, inflamed, or perforated. Under these conditions, it may be surgically removed in a procedure known as a *cholecystectomy*. This loss does not seriously impair digestion, because bile production continues at normal levels. However, the bile is more dilute, and its entry into the small intestine is not as closely tied to the arrival of food in the duodenum. **AM** Cholecystitis

☐ THE COORDINATION OF SECRETION AND ABSORPTION

A combination of neural and hormonal mechanisms coordinates the activities of the digestive glands. These regulatory mechanisms are centered around the duodenum, where the acids must be neutralized and the appropriate enzymes added.

Neural mechanisms involving the central nervous system (CNS) deal with (1) preparing the digestive tract for activity (parasympathetic innervation) or inhibiting gastrointestinal activity (sympathetic innervation) and (2) coordinating the movement of materials along the length of the digestive tract (the enterogastric, gastroenteric, and gastroileal reflexes).

In addition, motor neurons synapsing in the digestive tract release a variety of neurotransmitters. Many of these chemicals are also released in the CNS, but in general, their functions are poorly understood. Examples of potentially important neurotransmitters include Substance P, enkephalins, and endorphins.

We introduced hormones important to the regulation of intestinal and glandular function in the course of our discussion. We shall now summarize the introductory information and consider some additional details about the regulatory mechanisms involved. Table 24–1 also summarizes this information.

Intestinal Hormones

The intestinal tract secretes a variety of hormones, but it has proved very difficult to determine the primary effects of these hormones, largely because all are peptide hormones with similar chemical structures. Careful analyses have led to a marked increase in the number of intestinal hormones identified, but their specific functions have yet to be sorted out to everyone's satisfaction. Many of these hormones have multiple effects that target several regions of the digestive tract, as well as affecting the accessory glandular organs.

Duodenal enteroendocrine cells produce the following hormones known to coordinate digestive functions:

- **Secretin** is released when chyme arrives in the duodenum. The primary effect of secretin is to cause an increase in the secretion of bile and buffers by the liver and pancreas. Among its secondary effects, secretin reduces gastric motility and secretory rates.

- **Cholecystokinin** (**CCK**) is secreted when chyme arrives in the duodenum, especially when the chyme contains lipids and partially digested proteins. In the pancreas, CCK accelerates the production and secretion of all types of digestive enzymes. It also causes a relaxation of the hepatopancreatic sphincter and contraction of the gallbladder, resulting in the ejection of bile and pancreatic juice into the duodenum. The net effects

SUMMARY TABLE 24–1		IMPORTANT DIGESTIVE HORMONES AND THEIR PRIMARY EFFECTS			
Hormone	Stimulus	Origin	Target	Effects	
Cholecystokinin (CCK)	Arrival of chyme containing lipids and partially digested proteins	Duodenum	Pancreas	Stimulates production of pancreatic enzymes	
			Gallbladder	Stimulates contraction of gallbladder	
			Duodenum	Causes relaxation of hepatopancreatic sphincter	
			Stomach	Inhibits gastric secretion and motion	
			CNS	May reduce hunger	
Enterocrinin	Arrival of chyme in the duodenum	Duodenum	Duodenal glands	Stimulates production of alkaline mucus	
Gastric inhibitory peptide (GIP)	Arrival of chyme containing large quantities of fats and glucose	Duodenum	Pancreas	Stimulates release of insulin by pancreatic islets	
			Stomach	Inhibits gastric secretion and motility	
			Adipose tissue	Stimulates lipid synthesis	
			Skeletal muscle	Stimulates glucose use	
Gastrin	Vagus nerve stimulation or arrival of food in the stomach	Stomach	Stomach	Stimulates production of acids and enzymes; increases motility	
	Arrival of chyme containing large quantities of undigested proteins	Duodenum	Stomach	As above	
Secretin	Arrival of chyme in the duodenum	Duodenum	Pancreas	Stimulates production of alkaline buffers	
			Stomach	Inhibits gastric secretion and motility	
			Liver	Increases rate of bile secretion	
Vasoactive intestinal peptide (VIP)	Arrival of chyme in the duodenum	Duodenum	Duodenal glands, stomach	Stimulates buffer secretion; inhibits acid production; dilates intestinal capillaries	

of CCK are thus to increase the secretion of pancreatic enzymes and to push pancreatic secretions and bile into the duodenum. The presence of CCK in high concentrations has two additional effects: It inhibits gastric activity, and it appears to have CNS effects that reduce the sensation of hunger.

- **Gastric Inhibitory Peptide. Gastric inhibitory peptide (GIP)** is secreted when fats and carbohydrates—especially glucose—enter the small intestine. The inhibition of gastric activity is accompanied by the stimulation of insulin release at the pancreatic islets, so GIP is also known as *glucose-dependent insulinotropic peptide*. This hormone has a number of secondary effects; for instance, it stimulates the activity of the duodenal glands of the duodenum, stimulates lipid synthesis in adipose tissue, and increases glucose use by skeletal muscles.

- **Vasoactive Intestinal Peptide. Vasoactive intestinal peptide (VIP)** stimulates the secretion of intestinal glands, dilates regional capillaries, and inhibits acid production in the stomach. By dilating capillaries in active areas of the intestinal tract, VIP provides an efficient mechanism for removing absorbed nutrients.

- **Gastrin.** Gastrin is secreted by G cells in the duodenum when they are exposed to large quantities of incompletely digested proteins. The functions of gastrin include promoting increased stomach motility and stimulating the production of acids and enzymes.

- **Enterocrinin. Enterocrinin**, a hormone released when chyme enters the small intestine, stimulates mucin production by the submucosal glands of the duodenum.

- **Other Intestinal Hormones.** Several other hormones are produced in relatively small quantities. Examples include *motilin*, which stimulates intestinal contractions; *villikinin*, which promotes the movement of villi and the associated lymph flow; and *somatostatin*, which inhibits gastric secretion.

Functional interactions among gastrin, secretin, CCK, GIP, and VIP are diagrammed in Figure 24–22●.

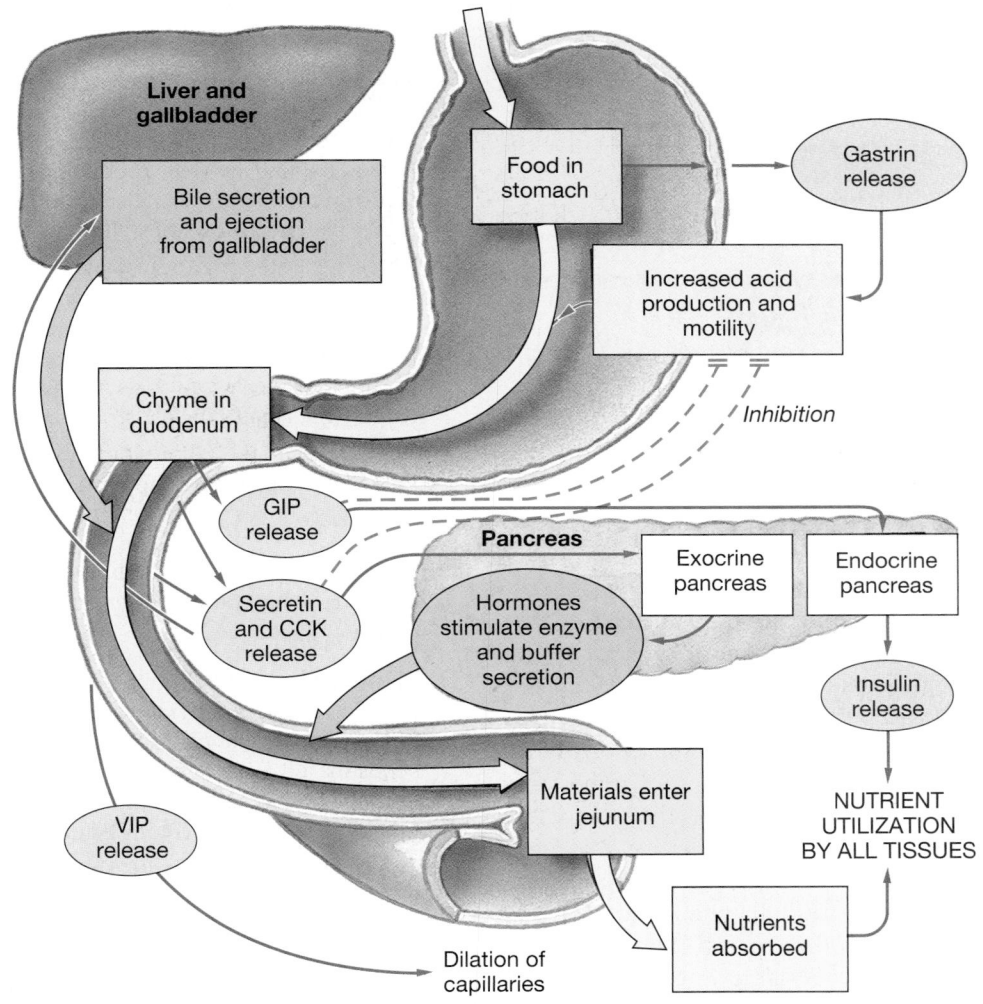

●**FIGURE 24–22**
The Activities of Major Digestive Tract Hormones. The primary actions of gastrin, secretin, CCK, GIP, and VIP.

Intestinal Absorption

On average, it takes about five hours for materials to pass from your duodenum to the end of your ileum, so the first of the materials to enter the duodenum after you eat breakfast may leave the small intestine at lunchtime. Along the way, the organ's absorptive effectiveness is enhanced by the fact that so much of the mucosa is movable. The microvilli can be moved by their supporting microfilaments, the individual villi by smooth muscle cells, groups of villi by the muscularis mucosae, and the plicae by the muscularis mucosae and the muscularis externa. These movements stir and mix the intestinal contents, changing the environment around each epithelial cell from moment to moment.

✔ How is the small intestine adapted for the absorption of nutrients?

✔ How would a meal that is high in fat affect the level of cholecystokinin in the blood?

✔ How would the pH of intestinal contents be affected if the small intestine did not secrete the hormone secretin?

✔ The digestion of which nutrient would be most impaired by damage to the exocrine pancreas?

Answers start on page Q-1

24–7 THE LARGE INTESTINE

Objectives

■ Describe the gross and histological structure of the large intestine.

■ Specify the regional specializations of the large intestine.

■ Explain the significance of the large intestine in the absorption of nutrients.

The horseshoe-shaped **large intestine** begins at the end of the ileum and ends at the anus. The large intestine lies inferior to the stomach and liver and almost completely frames the small intestine (Figure 24–16a●). The major functions of the large intestine include (1) the reabsorption of water and the compaction of the intestinal contents into feces, (2) the absorption of important vitamins liberated by bacterial action, and (3) the storage of fecal material prior to defecation.

The large intestine, also known as the *large bowel*, has an average length of about 1.5 meters (5 ft) and a width of 7.5 cm (3 in.). We can divide it into three parts: (1) the pouch-like *cecum*, the first portion of the large intestine; (2) the *colon*, the largest portion; and (3) the *rectum*, the last 15 cm (6 in.) of the large intestine and the end of the digestive tract (Figure 24–23●).

☐ THE CECUM

Material arriving from the ileum first enters an expanded pouch called the **cecum** (SĒ-kum). The ileum attaches to the medial surface of the cecum and opens into the cecum at the **ileocecal** (il-ē-ō-SĒ-kal) **valve** (Figure 24–23a●). The cecum collects and stores materials from the ileum and begins the process of compaction.

The slender, hollow **appendix**, or *vermiform appendix* (*vermis*, worm), is attached to the posteromedial surface of the cecum (Figure 24–23●). The appendix is generally about 9 cm (3.5 in.) long, but its size and shape are quite variable. A small mesentery called the **mesoappendix** connects the appendix to the ileum and cecum. The mucosa and submucosa of the appendix are dominated by lymphoid nodules, and the primary function of the appendix is as an organ of the lymphatic system. Inflammation of the appendix is known as *appendicitis*. ⟳ p. 787

☐ THE COLON

The **colon** has a larger diameter and a thinner wall than the small intestine. Distinctive features of the colon include the following (Figure 24–23a,c●):

• The wall of the colon forms a series of pouches, or **haustra** (HAWS-truh; singular, *haustrum*). Cutting into the intestinal lumen reveals that the creases between the haustra affect the mucosal lining as well, producing a series of internal folds. Haustra permit the expansion and elongation of the colon, rather like the bellows that allow an accordion to lengthen.

• Three separate longitudinal bands of smooth muscle—called the **taeniae coli** (TĒ-nē-ē KŌ-lē; singular, *taenia*)—are on the outer surfaces of the colon just deep to the serosa. These bands correspond to the outer layer of the muscularis externa in other portions of the digestive tract. Muscle tone within the taeniae coli is what creates the haustra.

• The serosa of the colon contains numerous teardrop-shaped sacs of fat called **fatty appendices**, or *epiploic* (e-pip-LŌ-ik) *appendages*.

We can subdivide the colon into four regions: the ascending colon, transverse colon, descending colon, and sigmoid colon (Figure 24–23a●).

1. The **ascending colon** begins at the superior border of the cecum and ascends along the right lateral and posterior wall of the peritoneal cavity to the inferior surface of the liver. There, the colon bends sharply to the left at the **right colic flexure**, or *hepatic flexure*, which marks the end of the ascending colon and the beginning of the transverse colon (Figure 24–23a●).

2. The **transverse colon** curves anteriorly from the right colic flexure and crosses the abdomen from right to left (Figure 24–23a●). The transverse colon is supported by the transverse mesocolon and is separated from the anterior abdominal wall by the layers of the greater omentum. As the transverse colon reaches the left side of the body, it passes inferior to the greater curvature of the stomach. Near the spleen, the colon makes a 90° turn at the **left colic flexure**, or *splenic flexure*, and becomes the descending colon.

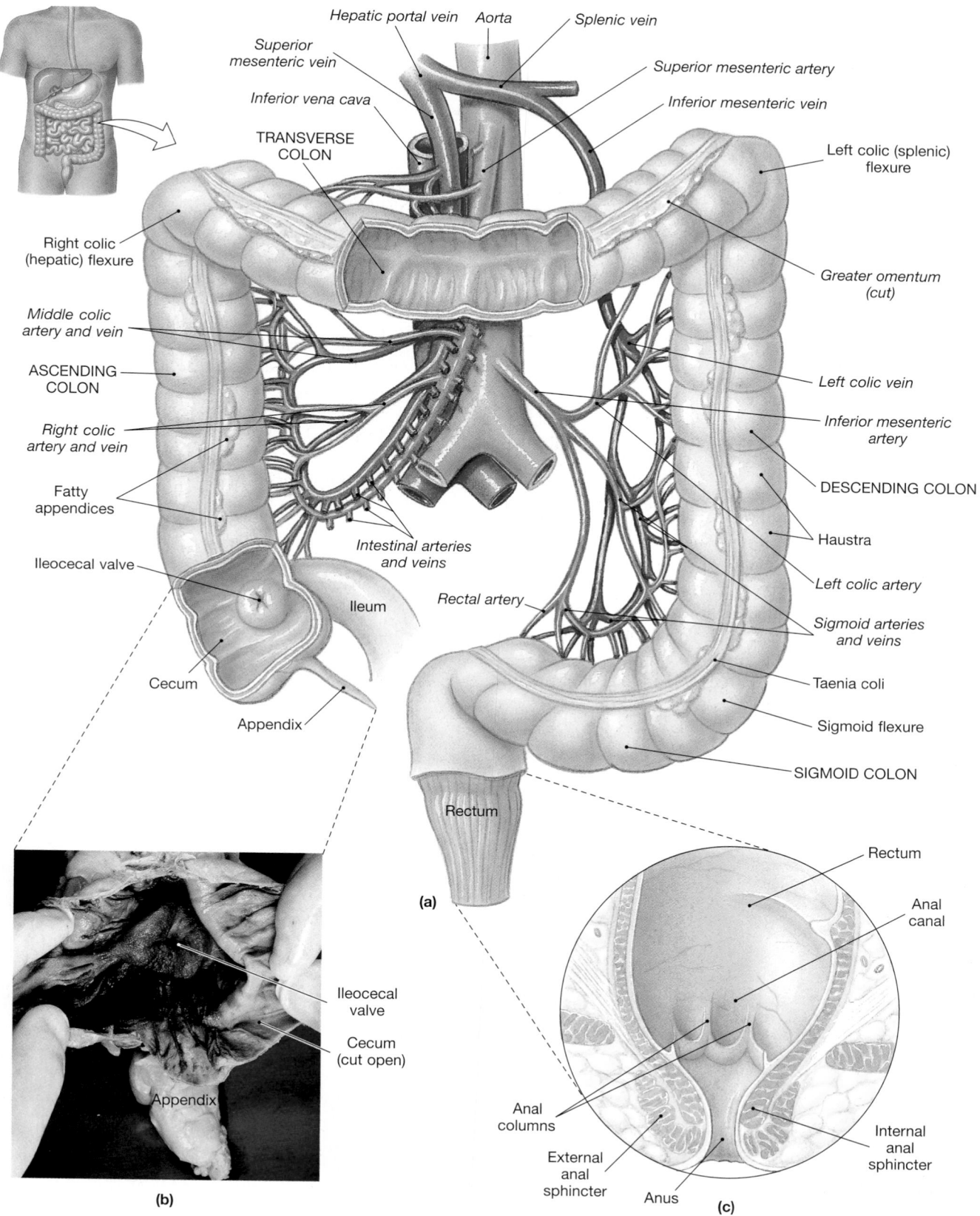

●FIGURE 24–23
The Large Intestine.   (a) The gross anatomy and regions of the large intestine. (b) The appendix. (c) The rectum and anus. **ATLAS** **Plates**
7.6e, 7.7a,b, 7.11; Scans 9d–f, 12e

3. The **descending colon** proceeds inferiorly along the person's left side (Figure 24–23a●) until reaching the iliac fossa. The descending colon is retroperitoneal and firmly attached to the abdominal wall. At the iliac fossa, the descending colon curves at the **sigmoid flexure** and becomes the sigmoid colon.

4. The sigmoid flexure is the start of the **sigmoid** (SIG-moyd) **colon** (*sigmeidos*, the Greek letter *S*), an S-shaped segment that is only about 15 cm (6 in.) long. The sigmoid colon lies posterior to the urinary bladder, suspended from the sigmoid mesocolon (Figure 24–23a). The sigmoid colon empties into the *rectum*.

The large intestine receives blood from tributaries of the superior mesenteric and inferior mesenteric arteries. Venous blood is collected from the large intestine by the superior mesenteric and inferior mesenteric veins. ∞ pp. 758, 768

■ THE RECTUM

The **rectum** (REK-tum), which forms the last 15 cm (6 in.) of the digestive tract (Figure 24–23a,c●), is an expandable organ for the temporary storage of fecal material. The movement of fecal materials into the rectum triggers the urge to defecate.

The last portion of the rectum, the **anal canal**, contains small longitudinal folds called **anal columns**. The distal margins of these columns are joined by transverse folds that mark the boundary between the columnar epithelium of the proximal rectum and a stratified squamous epithelium like that in the oral cavity. The **anus**, or *anal orifice*, is the exit of the anal canal. There, the epidermis becomes keratinized and identical to the surface of the skin. The circular muscle layer of the muscularis externa in this region forms the

internal anal sphincter (Figure 24–23c●), the smooth muscle cells of which are not under voluntary control. The **external anal sphincter**, which guards the anus, consists of a ring of skeletal muscle fibers that encircles the distal portion of the anal canal. This sphincter consists of skeletal muscle and is under voluntary control.

The lamina propria and submucosa of the anorectal canal bear a network of veins. If venous pressures there rise too high due to straining during defecation, the veins can become distended, producing *hemorrhoids*.

Colon cancers are relatively common. Approximately 114,600 cases are diagnosed in the United States each year (in addition to 34,700 cases of rectal cancers). In 2002, an estimated 47,900 deaths occurred from colon cancers (and another 8700 deaths from rectal cancers). The mortality rate for these cancers remains high, and the best defense appears to be early detection and prompt treatment. The standard screening test involves checking the feces for blood. This is a simple procedure that can easily be performed on a stool (fecal) sample as part of a routine physical. For those at increased risk because of family history, associated disease, or older age, visual inspection of the lumen by fiberoptic colonoscopy is recommended. **AM** Colon Inspection and Cancer

■ HISTOLOGY OF THE LARGE INTESTINE

Although the diameter of the colon is roughly three times that of the small intestine, its wall is much thinner. The major characteristics of the colon are the lack of villi, the abundance of goblet cells, and the presence of distinctive intestinal glands (Figure 24–24●). The glands in the large intestine are deeper than those of the small

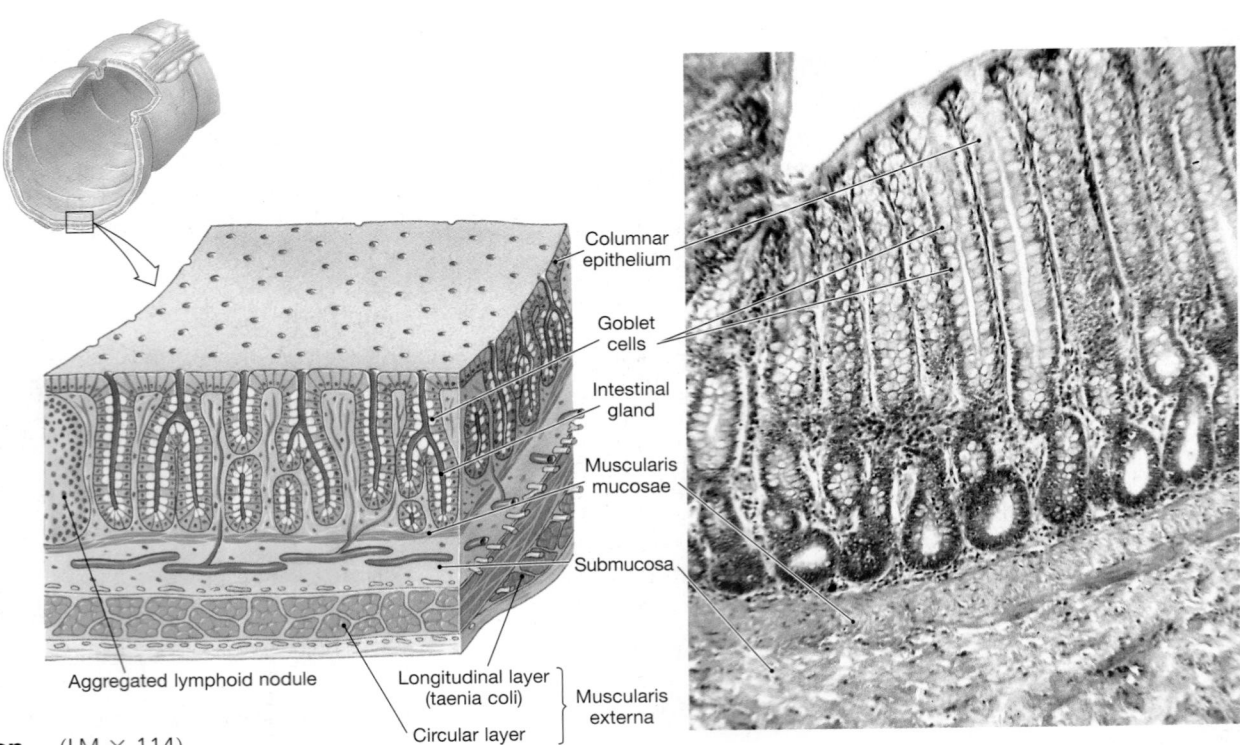

Columnar epithelium

Goblet cells

Intestinal gland

Muscularis mucosae

Submucosa

Aggregated lymphoid nodule

Longitudinal layer (taenia coli)
Circular layer
} Muscularis externa

●FIGURE 24–24
The Mucosa and Glands of the Colon. (LM × 114)

intestine and are dominated by goblet cells. The mucosa of the large intestine does not produce enzymes; any digestion that occurs results from enzymes introduced in the small intestine or from bacterial action. The mucus provides lubrication as the fecal material becomes less moist and more compact. Mucus is secreted as local stimuli, such as friction or exposure to harsh chemicals, trigger short reflexes involving local nerve plexuses. Large lymphoid nodules are scattered throughout the lamina propria and submucosa.

The muscularis externa of the large intestine is unusual, because the longitudinal layer has been reduced to the muscular bands of the taeniae coli. However, the mixing and propulsive contractions of the colon resemble those of the small intestine.

In **diverticulosis** (dī-ver-tik-ū-LŌ-sis), pockets (*diverticula*) form in the mucosa, generally in the sigmoid colon. The pockets get forced outward, probably by the pressures generated during defecation. If they push through weak points in the muscularis externa, the pockets form semi-isolated chambers that are subject to recurrent infection and inflammation. The infections cause pain and occasional bleeding, a condition known as *diverticulitis* (dī-ver-tik-ū-LĪ-tis). Inflammation of other portions of the colon is called *colitis* (kō-LĪ-tis). **AM** Inflammatory Bowel Disease

PHYSIOLOGY OF THE LARGE INTESTINE

Less than 10 percent of the nutrient absorption under way in the digestive tract occurs in the large intestine. Nevertheless, the absorptive operations in this segment of the digestive tract are important. The large intestine also prepares fecal material for ejection from the body.

Absorption in the Large Intestine

The reabsorption of water is an important function of the large intestine. Although roughly 1500 ml of material enters your colon each day, only about 200 ml of feces is ejected. The remarkable efficiency of digestion can best be appreciated by considering the average composition of feces: 75 percent water, 5 percent bacteria, and the rest a mixture of indigestible materials, small quantities of inorganic matter, and the remains of epithelial cells.

In addition to reabsorbing water, the large intestine absorbs a number of other substances that remain in the feces or that were secreted into the digestive tract along its length. Examples include useful compounds, such as bile salts and vitamins, organic waste products, such as urobilinogen, and various toxins generated by bacterial action. Most of the bile salts entering the large intestine are promptly reabsorbed in the cecum and transported in blood to the liver for secretion into bile.

VITAMINS Vitamins are organic molecules that are important as cofactors or coenzymes in many metabolic pathways. Bacteria that reside in the colon generate three vitamins that supplement our dietary supply:

1. *Vitamin K*, a fat-soluble vitamin that the liver needs in order to synthesize four clotting factors, including prothrombin. Intestinal bacteria produce roughly half of our daily vitamin K requirements.

2. *Biotin*, a water-soluble vitamin important to various reactions, notably those of glucose metabolism.

3. *Vitamin B$_5$* (pantothenic acid), a water-soluble vitamin required in the manufacture of steroid hormones and some neurotransmitters.

Vitamin K deficiencies, which lead to impaired blood clotting, result from either (1) a deficiency of lipids in the diet, which impairs the absorption of all fat-soluble vitamins, or (2) problems affecting lipid processing and absorption, such as inadequate bile production or chronic diarrhea. Disorders resulting from deficiencies of biotin or vitamin B$_5$ are extremely rare after infancy, because the intestinal bacteria produce sufficient amounts to supplement any dietary shortage.

ORGANIC WASTES In Chapter 19, we discussed the fate of bilirubin, a breakdown product of heme. ⊂⊃ p. 659 In the large intestine, bacteria convert bilirubin to *urobilinogens* and *stercobilinogens*. Some urobilinogens are absorbed into the bloodstream and then excreted in urine. The urobilinogens and stercobilinogens remaining within the colon are converted to **urobilins** and **stercobilins** by exposure to oxygen. These pigments in various proportions give feces a yellow-brown or brown coloration. Bacterial action breaks down peptides that remain in the feces and generates (1) ammonia, in the form of soluble *ammonium ions* (NH_4^+); (2) *indole* and *skatole*, two nitrogen-containing compounds that are primarily responsible for the odor of feces; and (3) hydrogen sulfide (H_2S), a gas that produces a "rotten egg" odor. Significant amounts of ammonia and smaller amounts of other toxins cross the colonic epithelium and enter the hepatic portal circulation. These toxins are removed by the liver and converted to relatively nontoxic compounds that can be released into the blood and excreted at the kidneys.

Indigestible carbohydrates are not altered by intestinal enzymes, so they arrive in the colon virtually intact. These complex polysaccharides provide a reliable nutrient source for colonic bacteria, whose metabolic activities are responsible for the small quantities of **flatus**, or intestinal gas, in the large intestine. Meals containing large numbers of indigestible carbohydrates (such as frankfurters and beans) stimulate bacterial gas production, leading to distension of the colon, cramps, and the frequent discharge of intestinal gases.

Movements of the Large Intestine

The gastroileal and gastroenteric reflexes move materials into the cecum while you eat. Movement from the cecum to the transverse colon is very slow, allowing hours for water absorption to convert the already thick material into a sludgy paste. Peristaltic waves move material along the length of the colon, and segmentation movements, called *haustral churning*, mix the contents of adjacent haustra. Movement from the transverse colon through the rest of the large intestine results from powerful peristaltic contractions called **mass movements**, which occur a few times each day. The stimulus is distension of the stomach and duodenum; the commands are relayed over the intestinal nerve plexuses. The contractions force feces into the rectum and produce the conscious urge to defecate.

The rectal chamber is usually empty, except when a powerful peristaltic contraction forces feces out of the sigmoid colon. Distension of the rectal wall then triggers the **defecation reflex**, which involves two positive feedback loops (Figure 24–25●). Both loops are triggered by the stimulation of stretch receptors in the walls of the rectum. The first loop is a short reflex that triggers a series of peristaltic contractions in the rectum that move

feces toward the anus (STEPS 1, 2, 3a, and 4a). The second loop is a long reflex coordinated by the sacral parasympathetic system. This reflex stimulates mass movements that push feces toward the rectum from the descending colon and sigmoid colon (STEPS 3b and 4b).

Rectal stretch receptors also trigger two reflexes important to the *voluntary* control of defecation. One is a long reflex mediated by parasympathetic innervation within the pelvic nerves. This reflex causes the relaxation of the internal anal sphincter, the smooth muscle sphincter that controls the movement of feces into the anorectal canal. The second (STEP 3c in Figure 24–25●) is a somatic reflex that stimulates the immediate contraction of the external anal sphincter, a skeletal muscle. ∞ pp. 354-355 The motor commands are carried by the pudendal nerves.

The elimination of feces requires that both the internal and external anal sphincters be relaxed, but the two reflexes just mentioned open the internal sphincter and close the external sphincter. The actual release of feces requires a conscious effort to open the external sphincter. In addition to opening the external sphincter, consciously directed activities such as tensing the abdominal muscles or making expiratory movements while closing the glottis elevate intra-abdominal pressures and help force fecal material out of the rectum.

If the external anal sphincter remains constricted, the peristaltic contractions cease until additional rectal expansion triggers the defecation reflex a second time. The urge to defecate usually develops when rectal pressure reaches about 15 mm Hg. If pressure inside the rectum exceeds 55 mm Hg, the external anal sphincter will involuntarily relax and defecation will occur. This mechanism regulates defecation in infants and in adults with severe spinal cord injuries.

In **diarrhea** (dī-a-RĒ-uh), an individual has frequent, watery bowel movements. The condition results when the mucosa of the colon becomes unable to maintain normal levels of absorption or the rate of fluid entry into the colon exceeds the organ's maximum reabsorptive capacity. Bacterial, viral, or protozoan infection of the colon or small intestine can cause acute bouts of diarrhea lasting several days. Severe diarrhea is life threatening due to cumulative losses of fluids and ions. In *cholera* (KOL-e-ruh), bacteria bound to the intestinal lining release toxins that stimulate a massive secretion of fluids across the intestinal epithelium. Without treatment, a person with cholera can die of acute dehydration in a matter of hours. **AM** Diarrhea

Constipation is infrequent defecation, generally involving dry, hard feces. Constipation occurs when fecal materials move through the colon so slowly that excessive reabsorption of water occurs. The feces become extremely compact, difficult to move, and highly abrasive. Inadequate dietary fiber and fluids, coupled with a lack of exercise, is a common cause of constipation.

Constipation can generally be treated by the oral administration of stool softeners, such as *Colace®*, laxatives, or **cathartics** (ka-THAR- tiks), which promote defecation. These compounds promote movement of water into the feces, increasing fecal mass, or irritating the lining of the colon to stimulate peristalsis. For example, indigestible fiber adds bulk to the feces, retaining moisture and stimulating stretch receptors that promote peristalsis. Thus, the promotion of peristalsis is a benefit of high-fiber cereals. Active movement during exercise also assists in the movement of fecal materials through the colon.

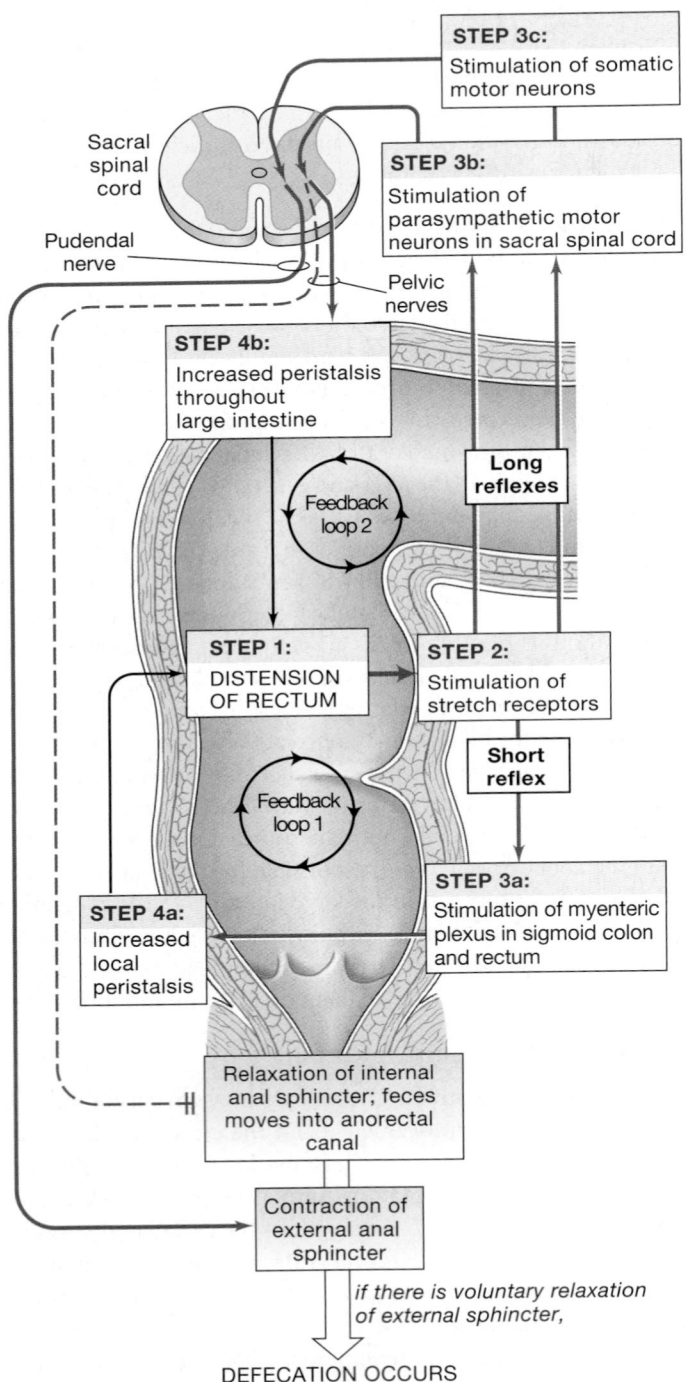

●**FIGURE 24–25**
The Defecation Reflex. Short and long reflexes (STEPS 1, 2, 3a, and 4a and STEPS 3b and 4b, respectively) promote movement of fecal material toward the anus. Another long reflex triggered by rectal stretch receptor stimulation (STEP 3c) prevents involuntary defecation.

24–8 DIGESTION AND ABSORPTION

Objectives

- Specify the nutrients required by the body.
- Describe the chemical events responsible for the digestion of organic nutrients.
- Describe the mechanisms involved in the absorption of organic and inorganic nutrients.

A typical meal contains carbohydrates, proteins, lipids, water, electrolytes, and vitamins. Your digestive system handles each component differently. Large organic molecules must be broken down by digestion before absorption can occur. Water, electrolytes, and vitamins can be absorbed without preliminary processing, but special transport mechanisms may be involved.

☐ SUMMARY: THE PROCESSING AND ABSORPTION OF NUTRIENTS

Food contains large organic molecules, many of them insoluble. The digestive system first breaks down the physical structure of the ingested material and then proceeds to disassemble the component molecules into smaller fragments. This disassembly eliminates any antigenic properties, so that, after absorption, the fragments do not trigger an immune response. The molecules released into the bloodstream will be absorbed by cells and either (1) broken down to provide energy for the synthesis of ATP or (2) used to synthesize carbohydrates, proteins, and lipids. This section will focus on the mechanics of digestion and absorption; the fate of the compounds inside cells will be the focus of Chapter 25.

Most ingested organic materials are complex chains of simpler molecules. In a typical dietary carbohydrate, the basic molecules are simple sugars; in a protein, the building blocks are amino acids; in lipids, they are generally fatty acids; and in nucleic acids, they are nucleotides. Digestive enzymes break the bonds between the component molecules of carbohydrates, proteins, lipids, and nucleic acids in a process called *hydrolysis*. ∞ p. 39

The classes of digestive enzymes differ with respect to their targets. *Carbohydrases* break the bonds between simple sugars, *proteases* split the linkages between amino acids, and *lipases* separate the fatty acids from glycerides. Specific enzymes in each class may be even more selective, breaking bonds between specific molecules. For example, a particular carbohydrase might break bonds between glucose molecules, but not those between glucose and another simple sugar.

Digestive enzymes secreted by the salivary glands, tongue, stomach, and pancreas are mixed into the ingested material as it passes along the digestive tract. These enzymes break down large carbohydrates, proteins, lipids, and nucleic acids into smaller fragments, which in turn must typically be broken down further before absorption can occur. The final enzymatic steps involve brush border enzymes, which are attached to the exposed surfaces of microvilli.

Nucleic acids are broken down into their component nucleotides. Brush border enzymes digest these nucleotides into sugars, phosphates, and nitrogenous bases that are absorbed by active transport. However, nucleic acids represent only a small fraction of the total nutrients absorbed each day. The digestive fates of carbohydrates, lipids, and proteins, the major dietary components, are summarized in Figure 24–26●. Table 24–2 reviews the major digestive enzymes and their functions. We shall now take a closer look at the digestion and absorption of carbohydrates, lipids, and proteins.

☐ CARBOHYDRATE DIGESTION AND ABSORPTION

The digestion of complex carbohydrates (simple polysaccharides and starches) proceeds in two steps. One step involves carbohydrases produced by the salivary glands and pancreas; the other, brush border enzymes.

Salivary and Pancreatic Enzymes

The digestion of complex carbohydrates involves two enzymes— salivary amylase and pancreatic alpha-amylase (Figure 24–26a●)— that function effectively at a pH of 6.7–7.5. Carbohydrate digestion begins in the mouth during mastication, by the action of salivary amylase from the parotid and submandibular salivary glands. Salivary amylase breaks down starches (complex carbohydrates), producing a mixture composed primarily of *disaccharides* (two simple sugars) and *trisaccharides* (three simple sugars). Salivary amylase continues to digest the starches and glycogen in the food for 1–2 hours before stomach acids render the enzyme inactive. Because the enzymatic content of saliva is not high, only a small amount of digestion occurs over this period.

In the duodenum, the remaining complex carbohydrates are broken down by the action of pancreatic alpha-amylase. Any disaccharides or trisaccharides produced, as well as any already present in the food, are ignored by both salivary and pancreatic amylases. Additional hydrolysis does not occur until these molecules contact the intestinal mucosa.

Brush Border Enzymes

Prior to absorption, disaccharides and trisaccharides are fragmented into *monosaccharides* (simple sugars) by brush border enzymes of the intestinal microvilli. The enzyme **maltase** splits bonds between the two glucose molecules of the disaccharide **maltose**. **Sucrase** breaks the disaccharide **sucrose** into glucose and *fructose*, another six-carbon sugar. **Lactase** hydrolyzes the disaccharide **lactose** into a molecule of glucose and one of *galactose*. Lactose is the primary carbohydrate in milk, so by breaking down lactose, lactase provides essential services throughout infancy and early childhood. If the intestinal mucosa stops producing lactase by the time of adolescence, the individual becomes *lactose intolerant*. After ingesting milk and other dairy products, those who are lactose intolerant can have a variety of unpleasant digestive problems. **AM** Lactose Intolerance

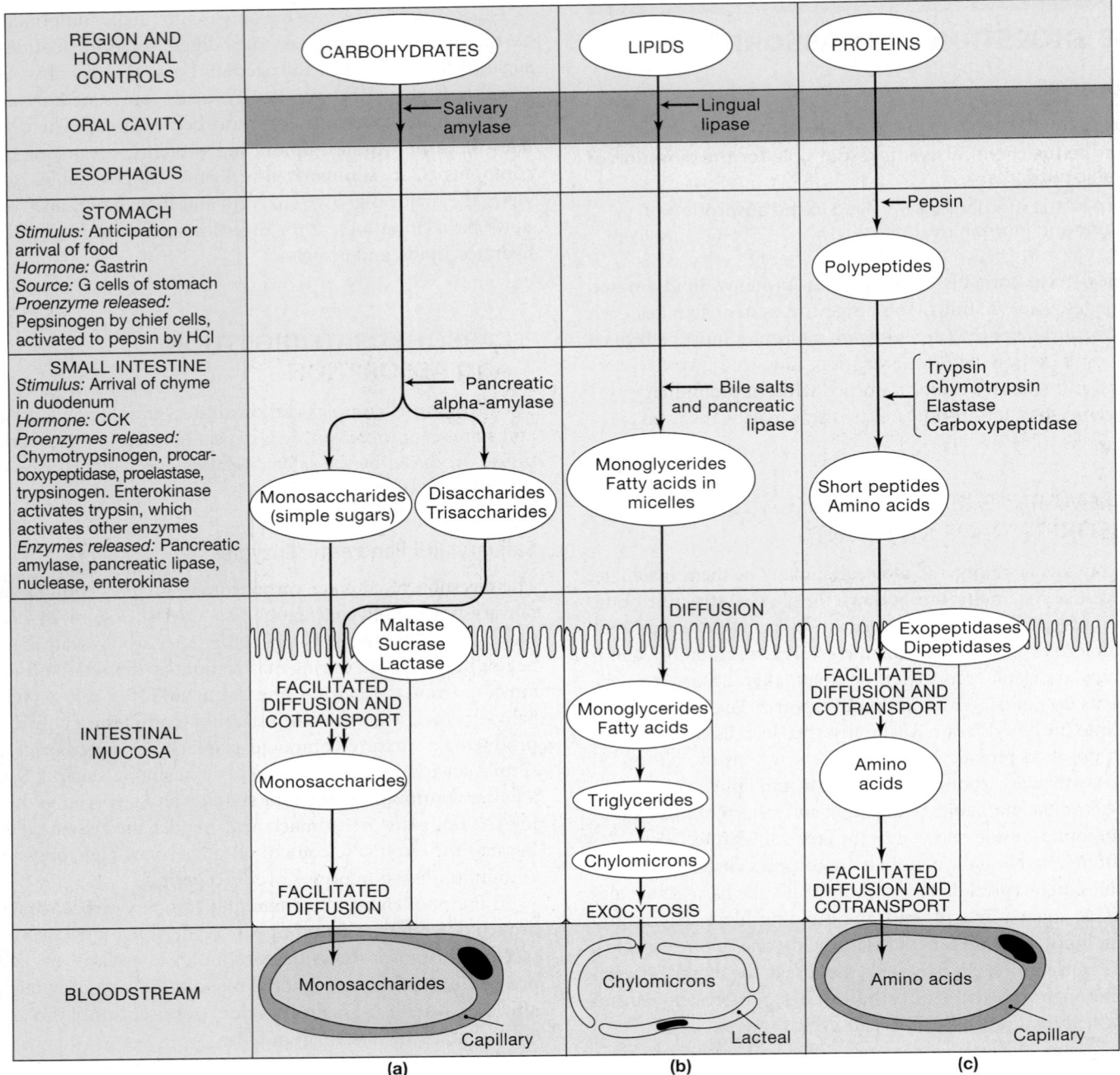

•FIGURE 24–26
A Summary of the Chemical Events in Digestion. For further details on the enzymes involved, see Table 24–2.

Absorption of Monosaccharides

The intestinal epithelium then absorbs the monosaccharides by facilitated diffusion and cotransport mechanisms. (See Figure 3–22•, p. 95). Both methods involve a carrier protein. Facilitated diffusion and cotransport differ in three major ways:

1. *Facilitated Diffusion Moves Only One Molecule or Ion Through the Cell Membrane, Whereas Cotransport Moves More Than One Molecule or Ion Through the Membrane At the Same Time.* In cotransport, the transported materials

move in the same direction: down the concentration gradient for at least one of the transported substances.

2. *Facilitated Diffusion Does Not Require ATP.* Although cotransport by itself does not consume ATP, the cell must often expend ATP to preserve homeostasis. For example, the process may introduce sodium ions that must later be pumped out of the cell.

3. *Facilitated Diffusion Will Not Occur if There is an Opposing Concentration Gradient for the Particular Molecule or Ion.*

SUMMARY TABLE 24–2   DIGESTIVE ENZYMES AND THEIR FUNCTIONS

Enzyme (proenzyme)	Source	Optimal pH	Target	Products	Remarks
CARBOHYDRASES					
Maltase, sucrase, lactase	Brush border of small intestine	7–8	Maltose, sucrose, lactose	Monosaccharides	Found in membrane surface of microvilli
Pancreatic alpha-amylase	Pancreas	6.7–7.5	Complex carbohydrates	Disaccharides and trisaccharides	Breaks bonds between monomers
Salivary amylase	Salivary glands	6.7–7.5	Complex carbohydrates	Disaccharides and trisaccharides	Breaks bonds between monomers
PROTEASES					
Carboxypeptidase (procarboxypeptidase)	Pancreas	7–8	Proteins, polypeptides, amino acids	Short-chain peptides	Activated by trypsin
Chymotrypsin (chymotrypsinogen)	Pancreas	7–8	Proteins, polypeptides	Short-chain peptides	Activated by trypsin
Dipeptidases, peptidases	Brush border of small intestine	7–8	Dipeptides, tripeptides	Amino acids	Found in membrane surface of brush border
Elastase (proelastase)	Pancreas	7–8	Elastin	Short-chain peptides	Activated by trypsin
Enterokinase	Brush border and lumen of small intestine	7–8	Trypsinogen	Trypsin	Reaches lumen through disintegration of shed epithelial cells
Pepsin (pepsinogen)	Chief cells of stomach	1.5–2.0	Proteins, polypeptides	Short-chain polypeptides	Secreted as proenzyme pepsinogen; activated by H^+ in stomach acid
Rennin	Stomach	3.5–4.0	Milk proteins		Secreted only in infants; causes protein coagulation
Trypsin (trypsinogen)	Pancreas	7–8	Proteins, polypeptides	Short-chain peptides	Proenzyme activated by enterokinase; activates other pancreatic proteases
LIPASES					
Lingual lipase	Glands of tongue	3.0–6.0	Triglycerides	Fatty acids and monoglycerides	
Pancreatic lipase	Pancreas	7–8	Triglycerides	Fatty acids and monoglycerides	Bile salts must be present for efficient action
NUCLEASES	Pancreas	7–8	Nucleic acids	Nitrogenous bases and simple sugars	Includes ribonuclease for RNA and deoxyribonuclease for DNA

By contrast, cotransport can occur despite an opposing concentration gradient for one of the transported substances. For example, cells lining the small intestine will continue to absorb glucose when glucose concentrations inside the cells are much higher than they are in the intestinal contents.

The cotransport system responsible for the uptake of glucose also brings sodium ions into the cell. This passive process resembles facilitated diffusion, except that both a sodium ion and a glucose molecule must bind to the carrier protein before they can move into the cell. Glucose cotransport is an example of sodium-linked cotransport. Comparable cotransport mechanisms exist for other simple sugars and for some amino acids. Although these mechanisms deliver valuable nutrients to the cytoplasm, they also bring in sodium ions that must be ejected by the sodium–potassium exchange pump.

The simple sugars that are transported into the cell at the apical surface diffuse through the cytoplasm and reach the interstitial fluid by facilitated diffusion across the basolateral surfaces. These monosaccharides then diffuse into the capillaries of the villus for eventual transport to the liver in the hepatic portal vein.

◻ LIPID DIGESTION AND ABSORPTION

Lipid digestion involves lingual lipase from glands of the tongue and pancreatic lipase from the pancreas (Figure 24–26b●). The most important and abundant dietary lipids are triglycerides, which consist of three fatty acids attached to a single molecule of glycerol. (See Figure 2–15●, p. 51.) The lingual and pancreatic lipases break off two of the fatty acids, leaving monoglycerides.

Lipases are water-soluble enzymes, and lipids tend to form large drops that exclude water molecules. As a result, lipases can attack only the exposed surfaces of the lipid drops. Lingual lipase begins breaking down triglycerides in the mouth and continues for a variable time within the stomach, but the lipid drops are so large and the available time so short, that only about 20 percent of the lipids have been digested by the time the chyme enters the duodenum.

Bile salts improve chemical digestion by emulsifying the lipid drops into tiny emulsion droplets, thus providing better access for pancreatic lipase. The emulsification occurs only after the chyme has been mixed with bile in the duodenum. Pancreatic lipase then breaks apart the triglycerides to form a mixture of fatty acids and monoglycerides. As these molecules are released, they interact with bile salts in the surrounding chyme to form small lipid–bile salt complexes called **micelles** (mī-SELZ). A micelle is only about 2.5 nm ($0.0025\ \mu m$) in diameter.

When a micelle contacts the intestinal epithelium, the lipids diffuse across the cell membrane and enter the cytoplasm. The intestinal cells synthesize new triglycerides from the monoglycerides and fatty acids. These triglycerides, in company with absorbed steroids and phospholipids, are then coated with proteins, creating complexes known as **chylomicrons** (kī-lō-MĪ-kronz; *chylos*, milky lymph + *mikros*, small).

The intestinal cells then secrete the chylomicrons into interstitial fluid by exocytosis. The superficial protein coating of the chylomicrons keeps them suspended in the interstitial fluid, but their size generally prevents them from diffusing into capillaries. Most of the chylomicrons released diffuse into the intestinal lacteals, which lack basement membranes and have large gaps between adjacent endothelial cells. From the lacteals, the chylomicrons proceed along the lymphatic vessels and through the thoracic duct, finally entering the bloodstream at the left subclavian vein.

Most of the bile salts within micelles are reabsorbed by sodium-linked cotransport. Only about 5% of the bile salts secreted by the liver enters the colon, and only about 1% is lost in feces.

◻ PROTEIN DIGESTION AND ABSORPTION

Proteins have very complex structures, so protein digestion is both complex and time consuming. The first problem to overcome is the disruption of the three-dimensional organization of the food so that proteolytic enzymes can attack individual proteins. This step involves mechanical processing in the oral cavity, through mastication, and chemical processing in the stomach, through the action of hydrochloric acid. Exposure of the bolus to a strongly acidic environment breaks down plant cell walls and the connective tissues in animal products, as well as most pathogens.

The acidic contents of the stomach also provide the proper environment for the activity of pepsin, the proteolytic enzyme secreted by chief cells of the stomach (Figure 24–26c●). Pepsin, which works effectively at a pH of 1.5–2.0, breaks the peptide bonds within a polypeptide chain. When chyme enters the duodenum, enterokinase produced in the small intestine triggers the conversion of trypsinogen to **trypsin**, and the pH is adjusted to 7–8. Pancreatic proteases can now begin working. Trypsin, chymotrypsin, and elastase are like pepsin in that they break specific peptide bonds within a polypeptide. Each enzyme has a different specialty. For example, trypsin breaks peptide bonds involving the amino acid *arginine* or *lysine*, while chymotrypsin targets peptide bonds involving *tyrosine* or *phenylalanine*.

Carboxypeptidase also acts in the small intestine. This enzyme chops off the last amino acid of a polypeptide chain, ignoring the identities of the amino acids involved. Thus, while the other peptidases are generating amino acid chains of varying length, carboxypeptidase is producing free amino acids.

Absorption of Amino Acids

The epithelial surfaces of the small intestine contain several peptidases, including **dipeptidases**—enzymes that break short peptide chains into individual amino acids. (Dipeptidases break apart *dipeptides*.) These amino acids, as well as those produced by the pancreatic enzymes, are absorbed through both facilitated diffusion and cotransport mechanisms.

After diffusing to the basal surface of the cell, the amino acids are released into interstitial fluid by facilitated diffusion and cotransport. Once in the interstitial fluid, the amino acids diffuse into intestinal capillaries for transport to the liver by means of the hepatic portal vein.

◻ WATER ABSORPTION

Our cells cannot actively absorb or secrete water. All movement of water across the lining of the digestive tract, as well as the production of glandular secretions, involves passive water flow along osmotic gradients. When two fluids are separated by a selectively permeable membrane, water tends to flow into the solution that has the higher concentration of solutes. ∞ p. 90 Osmotic movements are rapid, so interstitial fluid and the fluids in the intestinal lumen always have the same osmolarity (osmotic concentration of solutes).

Intestinal epithelial cells continuously absorb nutrients and ions, and these activities gradually lower the solute concentration in the lumen. As the solute concentration drops, water moves into the surrounding tissues, maintaining osmotic equilibrium.

Each day, roughly 2000 ml of water enters your digestive tract in the form of food or drink. The salivary, gastric, intestinal, pancreatic, and bile secretions provide an additional 7000 ml. Of that total, only about 150 ml is lost in feces. The sites of secretion and absorption of water are indicated in Figure 24–27●.

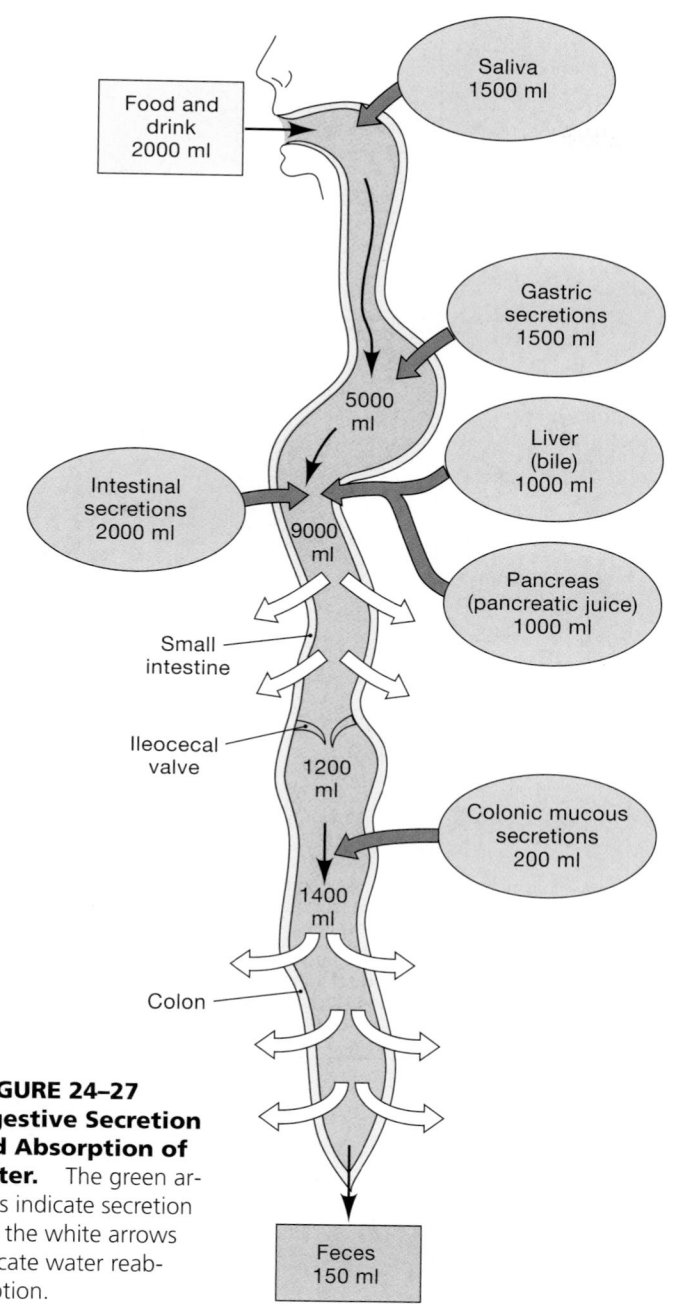

●FIGURE 24–27
Digestive Secretion and Absorption of Water. The green arrows indicate secretion and the white arrows indicate water reabsorption.

pumped into interstitial fluid across the base of the cell. The rate of Na^+ uptake from the lumen is generally proportional to the concentration of Na^+ in the intestinal contents. As a result, eating heavily salted foods leads to increased sodium ion absorption and an associated gain of water through osmosis. The rate of sodium ion absorption by the digestive tract is increased by aldosterone, a steroid hormone from the adrenal cortex. ∞ p. 627

Calcium ion (Ca^{2+}) absorption involves active transport at the epithelial surface. The rate of transport is accelerated by parathyroid hormone (PTH) and calcitriol. ∞ p. 626

As other solutes move out of the lumen, the concentration of potassium ions (K^+) increases. These ions can diffuse into the epithelial cells along the concentration gradient. The absorption of magnesium (Mg^{2+}), iron (Fe^{2+}), and other cations involves specific carrier proteins; the cell must use ATP to obtain and transport these ions to interstitial fluid. Regulatory factors controlling their absorption are poorly understood.

The anions chloride (Cl^-), iodide (I^-), bicarbonate (HCO_3^-), and nitrate (NO_3^-) are absorbed by diffusion or carrier-mediated transport. Phosphate (PO_4^{3-}) and sulfate (SO_4^{2-}) ions enter epithelial cells only by active transport.

☐ VITAMIN ABSORPTION

Vitamins are organic compounds required in very small quantities. There are two major groups of vitamins: fat-soluble vitamins and water-soluble vitamins. Vitamins A, D, E, and K are **fat-soluble vitamins**; their structure allows them to dissolve in lipids. The nine **water-soluble vitamins** include the B vitamins, common in milk and meats, and vitamin C, found in citrus fruits. In Chapter 25, we shall consider the functions of vitamins and associated nutritional problems.

All but one of the water-soluble vitamins are easily absorbed by diffusion across the digestive epithelium. Vitamin B_{12} cannot be absorbed by the intestinal mucosa in normal amounts, unless this vitamin has been bound to intrinsic factor, a glycoprotein secreted by the parietal cells of the stomach (p. 892). The combination is then absorbed via active transport.

Fat-soluble vitamins in the diet enter the duodenum in fat droplets, mixed with triglycerides. The vitamins remain in association with these lipids as they form emulsion droplets and, after further digestion, micelles. The fat-soluble vitamins are then absorbed from the micelles along with the fatty acids and monoglycerides. Vitamin K produced in the colon is absorbed with other lipids released through bacterial action. Taking supplements of fat-soluble vitamins while you have an empty stomach, are fasting, or are on a low-fat diet will be relatively ineffective, because proper absorption of these vitamins requires the presence of other lipids.

Figure 24–28● summarizes the digestive fates of the major electrolytes and vitamins. Review this figure carefully to become familiar with the outlined mechanisms and events. In Chapter 25, we will examine the interplay between the respiratory and digestive systems.

☐ ION ABSORPTION

Osmosis does not distinguish among solutes; all that matters is the total concentration of solutes. To maintain homeostasis, however, the concentrations of specific ions must be closely regulated. Thus, each ion must be handled individually, and the rate of intestinal absorption of each must be tightly controlled (Figure 24–28●). Many of the regulatory mechanisms controlling the rates of absorption are poorly understood.

Sodium ions (Na^+), usually the most abundant cations in food, may enter intestinal cells by diffusion, by cotransport with another nutrient, or by active transport. These ions are then

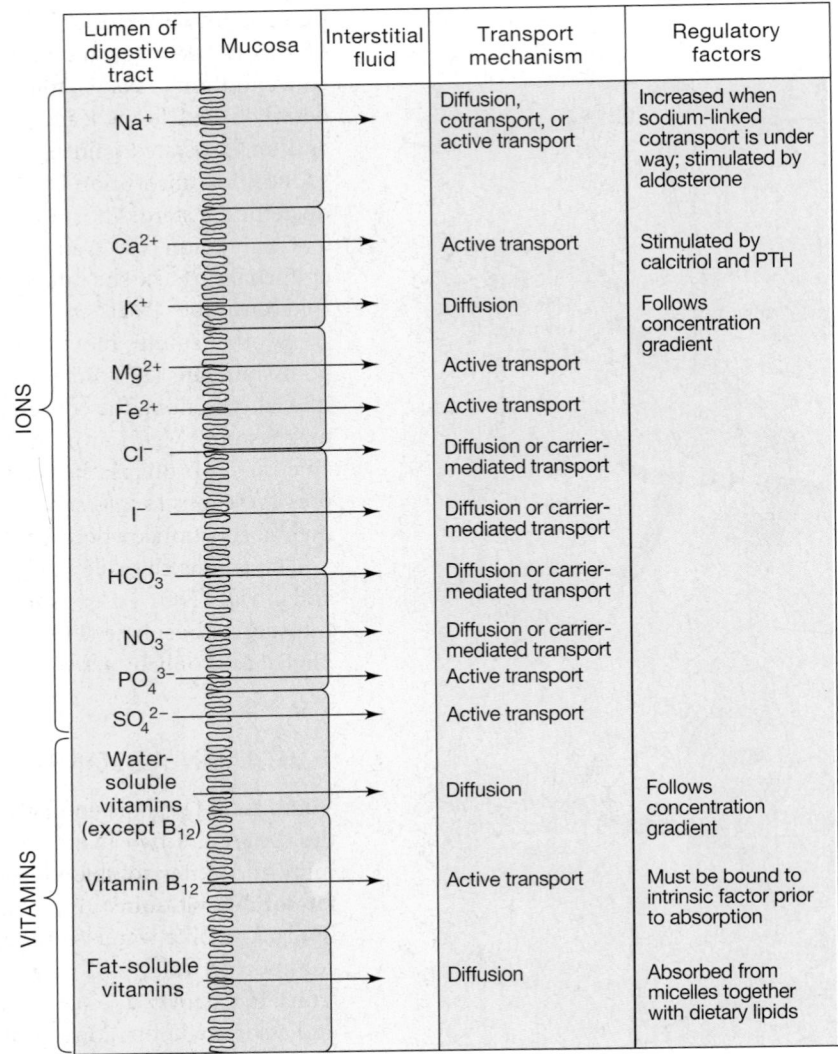

Lumen of digestive tract	Mucosa	Interstitial fluid	Transport mechanism	Regulatory factors
IONS				
Na^+		→	Diffusion, cotransport, or active transport	Increased when sodium-linked cotransport is under way; stimulated by aldosterone
Ca^{2+}		→	Active transport	Stimulated by calcitriol and PTH
K^+		→	Diffusion	Follows concentration gradient
Mg^{2+}		→	Active transport	
Fe^{2+}		→	Active transport	
Cl^-		→	Diffusion or carrier-mediated transport	
I^-		→	Diffusion or carrier-mediated transport	
HCO_3^-		→	Diffusion or carrier-mediated transport	
NO_3^-		→	Diffusion or carrier-mediated transport	
PO_4^{3-}		→	Active transport	
SO_4^{2-}		→	Active transport	
VITAMINS				
Water-soluble vitamins (except B_{12})		→	Diffusion	Follows concentration gradient
Vitamin B_{12}		→	Active transport	Must be bound to intrinsic factor prior to absorption
Fat-soluble vitamins		→	Diffusion	Absorbed from micelles together with dietary lipids

●**FIGURE 24-28**
Ion and Vitamin Absorption by the Digestive Tract

Malabsorption is a disorder characterized by abnormal nutrient absorption. Difficulties in the absorption of all classes of compounds will result from damage to the accessory glands or the intestinal mucosa. If the accessory organs are functioning normally, but their secretions cannot reach the duodenum, the condition is called *biliary obstruction* (bile duct blockage) or *pancreatic obstruction* (pancreatic duct blockage). Alternatively, the ducts may remain open, but the glandular cells may be damaged and unable to continue normal secretory activities. We noted two examples, pancreatitis and cirrhosis, earlier in this chapter.

Even with the normal enzymes present in the lumen, absorption will not occur if the mucosa cannot function properly. A genetic inability to manufacture specific enzymes will result in discrete patterns of malabsorption; lactose intolerance (p. 915) is a good example. Mucosal damage due to ischemia, radiation exposure, toxic compounds, or infection will adversely affect absorption and, as a result, will deplete nutrient and fluid reserves.

24–9 AGING AND THE DIGESTIVE SYSTEM

Objective

■ Summarize the effects of the aging process on the digestive system.

Essentially normal digestion and absorption occur in elderly individuals. However, many changes take place in the digestive system that parallel age-related changes we have already described in connection with other systems:

• **The Division Rate of Epithelial Stem Cells Declines.** The digestive epithelium becomes more susceptible to damage by abrasion, acids, or enzymes. Peptic ulcers therefore become more likely. Stem cells in the epithelium divide less frequently with age, so tissue repair is less efficient. In the mouth, esophagus, and anus, the stratified epithelium becomes thinner and more fragile.

- **Smooth Muscle Tone Decreases.** General motility decreases, and peristaltic contractions are weaker as a result of a decrease in smooth muscle tone. These changes slow the rate of fecal movement and promote constipation. Sagging and inflammation of the haustra in the colon can produce symptoms of diverticulosis (p. 913). Straining to eliminate compacted fecal materials can stress the less resilient walls of blood vessels, producing hemorrhoids. Problems are not restricted to the lower digestive tract; for example, weakening of muscular sphincters can lead to esophageal reflux and frequent bouts of "heartburn."

- **The Effects of Cumulative Damage Become Apparent.** A familiar example is the gradual loss of teeth due to dental caries or gingivitis. Cumulative damage can involve internal organs as well. Toxins such as alcohol and other injurious chemicals that are absorbed by the digestive tract are transported to the liver for processing. The cells of the liver are not immune to these toxic compounds, and chronic exposure can lead to cirrhosis or other types of liver disease.

- **Cancer Rates Increase.** Cancers are most common in organs in which stem cells divide to maintain epithelial cell populations. Rates of colon cancer and stomach cancer rise with age; oral and pharyngeal cancers are particularly common among elderly smokers.

- **Changes in Other Systems Have Direct or Indirect Effects on the Digestive System.** For example, the reduction in bone mass and calcium content in the skeleton is associated with erosion of the tooth sockets and eventual tooth loss. The decline in olfactory and gustatory sensitivities with age can lead to dietary changes that affect the entire body.

✓ What component of food would increase the number of chylomicrons in the lacteals?
✓ The absorption of which vitamin would be impaired by the removal of the stomach?
✓ Why is diarrhea potentially life threatening, but constipation is not?

Answers start on page Q-1

Chapter Review

SELECTED CLINICAL TERMINOLOGY

Terms Discussed in This Chapter

ascites (a-SĪ-tēz):   Fluid leakage into the peritoneal cavity across the serous membranes of the liver and viscera. *(p. 877)*

cathartics (ka-THAR-tiks):   Drugs that promote defecation. *(p. 914)*

cholecystitis (kō-lē-sis-TĪ-tis):   An inflammation of the gallbladder due to a blockage of the cystic or common bile duct by gallstones. *(p. 907 and AM)*

cholelithiasis (kō-lē-li-THĒ-a-sis):   The presence of gallstones in the gallbladder. *(p. 907 and AM)*

cholera (KOL-e-ruh):   A bacterial infection of the digestive tract that causes massive fluid losses through diarrhea. *(p. 914 and AM)*

cirrhosis (sir-Ō-sis):   A disease characterized by the widespread destruction of hepatocytes by exposure to drugs (especially alcohol), by viral infection, by ischemia, or by blockage of the hepatic ducts. *(p. 906 and AM)*

colitis (kō-LĪ-tis):   A general term for a condition characterized by an inflammation of the colon. *(p. 913 and AM)*

constipation: Infrequent defecation, generally involving dry, hard feces. *(p. 914 and AM)*

diarrhea (dī-a-RĒ-uh):   Frequent, watery bowel movements. *(p. 914 and AM)*

diverticulitis (dī-ver-tik-ū-LĪ-tis):   An infection and inflammation of mucosal pockets of the large intestine (diverticula). *(p. 913)*

diverticulosis (dī-ver-tik-ū-LŌ-sis):   The formation of diverticula, generally along the sigmoid colon. *(p. 913)*

esophageal varices (VAR-i-sēz):   Swollen and fragile esophageal veins that result from portal hypertension. *(p. 888)*

gallstones: Deposits of minerals, bile salts, and cholesterol that form if bile becomes too concentrated. *(p. 907 and AM)*

gastrectomy (gas-TREK-to-mē):   The surgical removal of the stomach, generally to treat advanced stomach cancer. *(p. 897)*

gastritis (gas-TRĪ-tis):   An inflammation of the gastric mucosa. *(p. 894)*

gastroscope: A fiber-optic instrument inserted into the mouth and directed along the esophagus and into the stomach; used to examine the interior of the stomach and to perform minor surgical procedures. *(p. 897)*

hepatitis (hep-a-TĪ-tis):   A virus-induced disease of the liver; forms include *hepatitis A, B,* and *C. (p. 906 and AM)*

lactose intolerance: A malabsorption syndrome that results from the lack of the enzyme *lactase* at the brush border of the intestinal epithelium. *(p. 915 and AM)*

mumps: A viral infection in children that tends to focus in the salivary glands, primarily the parotid. *(p. 884)*

pancreatitis (pan-krē-a-TĪ-tis):   An inflammation of the pancreas. *(p. 902)*

peptic ulcer: Erosion of the gastric lining or duodenal lining by stomach acids and enzymes. *(p. 894 and AM)*

periodontal disease: A loosening of the teeth within the alveolar sockets caused by erosion of the periodontal ligaments by acids produced through bacterial action. *(p. 886)*

peritonitis (per-i-tō-NĪ-tis):   An inflammation of the peritoneal membrane. *(p. 879)*

portal hypertension: High venous pressures in the hepatic portal system. *(p. 904)*

pulpitis (pul-PĪ-tis):   An infection of the pulp of a tooth; may be treated by a *root canal* procedure. *(p. 886)*

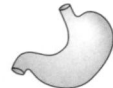

The **DIGESTIVE SYSTEM** in Perspective

Organ/Component	Primary Functions
Salivary Glands	Provide buffers and lubrication; produce enzymes that begin digestion
Pharynx	Conducts solid food and liquids to esophagus; chamber shared with respiratory tract *(see Chapter 23)*
Esophagus	Delivers food to stomach
Stomach	Secretes acids and enzymes
Small Intestine	Secretes digestive enzymes, buffers, and hormones; absorbs nutrients
Liver	Secretes bile; regulates nutrient composition of blood
Gallbladder	Stores bile for release into small intestine
Pancreas	Secretes digestive enzymes and buffers; contains endocrine cells *(see Chapter 18)*
Large Intestine	Removes water from fecal material; stores wastes

INTEGRATION WITH OTHER SYSTEMS

The figure on the opposite page summarizes the physiological relationships between the digestive system and other organ systems. The digestive system has particularly extensive anatomical as well as physiological connections to the nervous, cardiovascular, endocrine, and lymphatic systems. For example, the enteric nervous system contains as many neurons as the spinal cord, and as many neurotransmitters as the brain. As we have seen, the digestive tract is also an endocrine organ that produces a variety of hormones. Many of these hormones, and some of the neurotransmitters produced by the digestive system can enter the circulation, cross the blood-brain barrier, and alter CNS activity. Thus there is a continual exchange of chemical information among these systems.

CLINICAL PATTERNS

As you have seen throughout this chapter, the digestive system has so many components, and those components have so many functions, that digestive system disorders are very diverse, and also relatively common.

The largest category of digestive disorders includes those resulting from inflammation or infection of the digestive tract. This is in part due to the fact that the epithelium lining most of the digestive tract must have two properties that are hard to reconcile: (1) It must be thin and delicate enough to permit the rapid and efficient absorption of nutrients, and (2) it must resist damage by the ingested materials and the enzymes secreted to promote digestion.

The relative delicacy of the epithelium makes it susceptible to damage from chemical attack or abrasion. For example, gastric ulcers (p. 894) develop if acids and enzymes reach and erode the gastric lining. Pathogens in food, such as bacteria, viruses, or multicellular parasites, may also penetrate the epithelial barriers and cause infections. Small battles are continually being fought; the fact that 80% of the plasma cells in the body are normally located within the lamina propria of the digestive tract gives you some indication of how often antigens of one kind or another manage to cross the epithelial barriers.

High rates of cell division and exposure to strong chemical agents are both correlated with an increased risk of cancer. As a result, cancers of the digestive tract are relatively common. As you might predict, most of these are epithelial cancers that develop in the stem cell populations responsible for epithelial cell renewal.

Other classes of digestive system disorders are discussed in the *Applications Manual*.

MEDIA CONNECTIONS

I. Objective: To explore the regulation of the gastrointestinal system by investigating its functioning under various conditions.

Completion time = 12 minutes

The gastrointestinal tract is always active, moving and gurgling (sometimes audibly). GI movements after eating are different from those exhibited in other states, such as during fasting, after a complete night's sleep, while nauseated, or while under stress, indicating that the activities of the digestive system are tightly integrated with those of the body as a whole. Predict how each of the states mentioned above could affect the motility of the GI tract, and what purpose the increase or decrease in motility might serve. For a discussion of the relationship between digestive function and bodily state, visit the **Companion Website**, Chapter 24 Media Connections, and click on the keyword **gut**. Determine whether each of the changes that you predicted is caused by hormones, nervous stimulation, a combination of both, or some other factor. If the changes are due to hormones, identify which hormones are involved.

II. Objective: To investigate the relationship between caloric intake and longevity.

Completion time = 10 minutes

If you are what you eat, what are you if you eat very little? A growing body of evidence suggests that you may live a longer and healthier life if you lower your caloric intake. Experiments have demonstrated that restricted-calorie diets can extend the life-spans of laboratory animals and improve their resistance to disease. Can you speculate as to what mechanisms might underlie such findings? For more information on the scientific study of low-calorie diets, visit the **Companion Website**, Chapter 24 Media Connections, and click on the keyword **diet**. After reading this article, list three alterations in homeostasis that might be expected as a result of a restricted-calorie diet. Is such a diet something you might try in order to increase your fitness? Why or why not?

INTEGUMENTARY SYSTEM
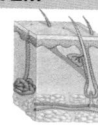
- Provides vitamin D₃ needed for the absorption of calcium and phosphorus
- Provides lipids for storage by adipocytes in subcutaneous layer

SKELETAL SYSTEM
- Skull, ribs, vertebrae, and pelvic girdle support and protect parts of digestive tract
- Teeth important in mechanical processing of food
- Absorbs calcium and phosphate ions for incorporation into bone matrix; provides lipids for storage in yellow marrow

MUSCULAR SYSTEM
- Protects and supports digestive organs in abdominal cavity
- Controls entrances and exits to digestive tract
- Liver regulates blood glucose and fatty acid levels, metabolizes lactic acid from active muscles

NERVOUS SYSTEM
- ANS regulates movement and secretion; reflexes coordinate passage of materials along tract
- Control over skeletal muscles regulates ingestion and defecation
- Hypothalamic centers control hunger, satiation, and feeding behaviors
- Provides substrates essential for neurotransmitter synthesis

ENDOCRINE SYSTEM
- Epinephrine and norepinephrine stimulate constriction of sphincters and depress digestive activity
- Hormones coordinate activity along tract
- Provides nutrients and substrates to endocrine cells
- Endocrine cells of pancreas secrete insulin and glucagon
- Liver produces angiotensinogen

DIGESTIVE SYSTEM
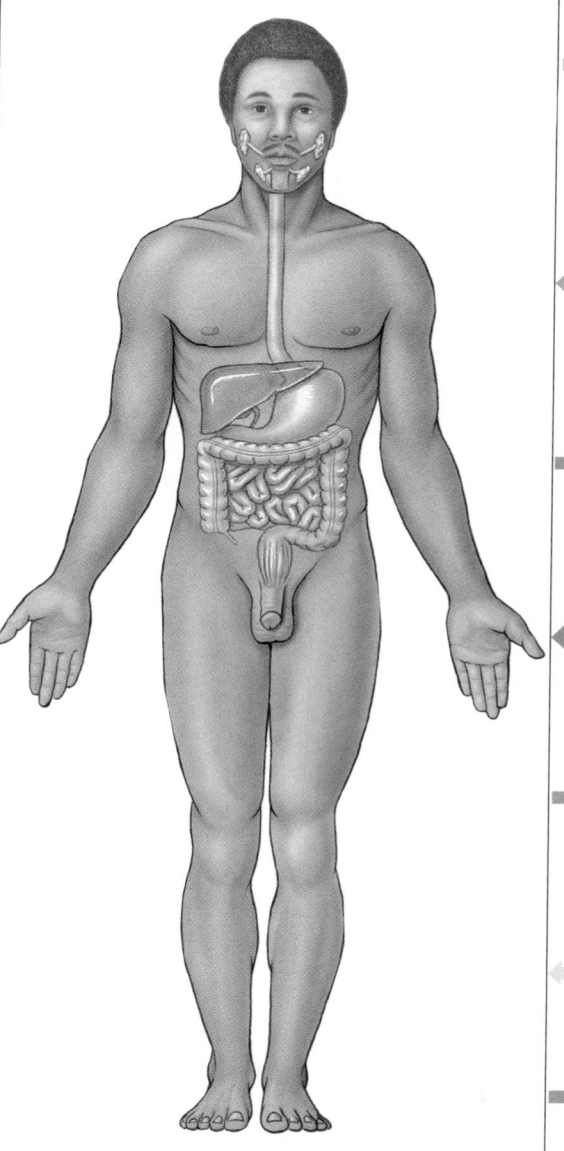

FOR ALL SYSTEMS
Absorbs organic substrates, vitamins, ions, and water required by all cells

CARDIOVASCULAR SYSTEM
- Distributes hormones of the digestive tract
- Carries nutrients, water, and ions from sites of absorption
- Delivers nutrients and toxins to liver
- Absorbs fluid to maintain normal blood volume
- Absorbs vitamin K
- Liver excretes heme (as bilirubin), synthesizes coagulation proteins

LYMPHATIC SYSTEM
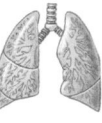
- Tonsils and other lymphoid nodules along digestive tract defend against infection and toxins absorbed from the tract
- Lymphatic vessels carry absorbed lipids to venous system
- Secretions of digestive tract (acids and enzymes) provide nonspecific defense against pathogens

RESPIRATORY SYSTEM
- Increased thoracic and abdominal pressure through contraction of respiratory muscles can assist in defecation
- Pressure of digestive organs against the diaphragm can assist in exhalation and limit inhalation

URINARY SYSTEM
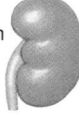
- Excretes toxins absorbed by the digestive epithelium
- Excretes some bilirubin produced by liver
- Absorbs water needed to excrete waste products at the kidneys
- Absorbs ions needed to maintain normal body fluid concentrations

REPRODUCTIVE SYSTEM

- Provides additional nutrients required to support gamete production and (in pregnant women) embryonic and fetal development

AM **Additional Terms Discussed in the *Applications Manual***

achalasia
colectomy
colonoscope
colostomy
esophagitis
gastric stapling
gastroenteritis

giardiasis
ileostomy
inflammatory bowel disease (*ulcerative colitis*)
irritable bowel syndrome
laparoscopy
liver biopsy
liver scans
perforated ulcer
polyps

STUDY OUTLINE

24–1 THE DIGESTIVE SYSTEM: AN OVERVIEW   p. 875

1. The **digestive system** consists of the muscular **digestive tract** and various **accessory organs**. (*Figure 24–1*)

FUNCTIONS OF THE DIGESTIVE SYSTEM   p. 876

2. Digestive functions are **ingestion, mechanical processing, digestion, secretion, absorption,** and **excretion**.

THE DIGESTIVE ORGANS AND THE PERITONEUM   p. 877

3. Double sheets of peritoneal membrane called **mesenteries** suspend the digestive tract. The **greater omentum** lies anterior to the abdominal viscera. Its adipose tissue provides padding, protection, insulation, and an energy reserve. (*Figure 24–2*)

HISTOLOGICAL ORGANIZATION OF THE DIGESTIVE TRACT   p. 879

4. The digestive tract is lined by a mucous epithelium moistened by glandular secretions of the epithelial and accessory organs.

5. The *lamina propria* and epithelium form the **mucosa** (mucous membrane) of the digestive tract. Proceeding outward, we encounter the **submucosa**, the **muscularis externa**, and a layer of areolar tissue called the *adventitia*. Within the peritoneal cavity, the muscularis externa is covered by a serous membrane called the **serosa**. (*Figure 24–3*)

THE MOVEMENT OF DIGESTIVE MATERIALS   p. 880

6. The *visceral smooth muscle tissue* of the digestive tract shows rhythmic cycles of activity due to *pacemaker cells*.

7. The muscularis externa propels materials through the digestive tract by the contractions of **peristalsis. Segmentation** movements in the small intestine churn digestive materials. (*Figure 24–4*)

CONTROL OF THE DIGESTIVE FUNCTION   p. 881

8. Digestive tract activities are controlled by neural, hormonal, and local mechanisms. (*Figure 24–5*)

⊙ Gross anatomy of the digestive system: **Anatomy CD-ROM/** Digestive System/Digestive/Accessory organs, and /Flythrough.

24–2 THE ORAL CAVITY   p. 882

1. The functions of the **oral cavity**, or **buccal cavity,** are (1) *analysis* of foods; (2) *mechanical processing* by the teeth, tongue, and palatal surfaces; (3) *lubrication*, by mixing with mucus and salivary secretions; and (4) initiating the *digestion* of carbohydrates and lipids.

2. The oral cavity is lined by **oral mucosa**. The *hard* and *soft palates* form the roof of the oral cavity, and the *tongue* forms its floor. The **uvula** prevents food from entering the pharynx prematurely. (*Figure 24–6*)

THE TONGUE   p. 883

3. **Intrinsic** and **extrinsic tongue muscles** are controlled by the hypoglossal nerve. (*Figure 24–6*)

SALIVARY GLANDS   p. 883

4. The **parotid, sublingual,** and **submandibular salivary glands** discharge their secretions into the oral cavity. (*Figure 24–7*)

THE TEETH   p. 885

5. **Mastication** (chewing) of the **bolus** occurs through the contact of the **occlusal** (opposing) **surfaces** of the **teeth**. The **periodontal ligament** anchors each tooth in an *alveolus*, or bony socket. **Dentin** forms the basic structure of a tooth. The **crown** is coated with **enamel**, the **root** with **cementum**. (*Figure 24–8*)

6. The 20 primary teeth, or **deciduous teeth**, are replaced by the 32 teeth of the **secondary dentition** during development. (*Figure 24–9*)

24–3 THE PHARYNX   p. 888

1. Propulsion of the bolus through the **pharynx** results from contractions of the *pharyngeal constrictor muscles* and the *palatal muscles* and from elevation of the larynx.

24–4 THE ESOPHAGUS   p. 888

1. The **esophagus** carries solids and liquids from the pharynx to the stomach through the **esophageal hiatus**, an opening in the diaphragm.(*Figure 24–10*)

HISTOLOGY OF THE ESOPHAGUS   p. 888

2. The esophageal mucosa consists of a stratified epithelium. Mucous secretion by esophageal glands of the submucosa reduces friction during the passage of foods. The proportions of skeletal and smooth muscle of the muscularis externa change from the pharynx to the stomach. (*Figure 24–3*)

SWALLOWING   p. 889

3. Swallowing (**deglutition**) can be divided into **buccal, pharyngeal,** and **esophageal phases**. Swallowing begins with the compaction of a bolus and its movement into the pharynx, followed by the elevation of the larynx, reflection of the *epiglottis*, and closure of the *glottis*. After the *upper esophageal sphincter* is opened, peristalsis moves the bolus down the esophagus to the *lower esophageal sphincter*. (*Figure 24–11*)

24–5 THE STOMACH   p. 890

1. The **stomach** has four major functions: (1) bulk storage of ingested food, (2) mechanical breakdown of food, (3) disruption of chemical bonds via acids and enzymes, and (4) production of *intrinsic factor*.

ANATOMY OF THE STOMACH   p. 890

2. The four regions of the stomach are the **cardia, fundus, body,** and **pylorus**. The **pyloric sphincter** guards the exit from the stomach. In a relaxed state, the stomach lining contains numerous **rugae** (ridges and folds). *(Figure 24–12)*

3. Within the **gastric glands, parietal cells** secrete **intrinsic factor** and *hydrochloric acid*. **Chief cells** secrete **pepsinogen**, which is converted by acids in the gastric lumen to the enzyme **pepsin**. **Enteroendocrine** cells of the stomach secrete several compounds, notably the hormone **gastrin**. *(Figures 24–13, 24–14)*

REGULATION OF GASTRIC ACTIVITY   p. 894

4. Gastric secretion involves (1) the **cephalic phase**, which prepares the stomach to receive ingested materials, (2) the **gastric phase**, which begins with the arrival of food in the stomach, and (3) the **intestinal phase**, which controls the rate of gastric emptying. *(Figure 24–15)*

DIGESTION AND ABSORPTION IN THE STOMACH   p. 896

5. The preliminary digestion of proteins by pepsin begins in the stomach. Very little absorption of nutrients occurs in the stomach.

   Gastric activity: **Companion Website**: Chapter 24/Tutorials.

24–6 THE SMALL INTESTINE AND ASSOCIATED GLANDULAR ORGANS   p. 897

1. Most of the important digestive and absorptive functions occur in the **small intestine**. Digestive secretions and buffers are provided by the pancreas, liver, and gallbladder.

THE SMALL INTESTINE   p. 897

2. The small intestine consists of the **duodenum**, the **jejunum**, and the **ileum**. A sphincter, the **ileocecal valve**, marks the transition between the small and large intestines. *(Figures 24–2, 24–16)*

HISTOLOGY OF THE SMALL INTESTINE   p. 898

3. The intestinal mucosa bears transverse folds called **plicae** and small projections called **intestinal villi**. These folds and projections increase the surface area for absorption. Each villus contains a terminal lymphatic called a **lacteal**. Pockets called **intestinal glands** are lined by enteroendocrine, goblet, and stem cells. *(Figures 24–16, 24–17)*

4. **Intestinal juice** moistens chyme, helps buffer acids, and holds digestive enzymes and digestive products in solution.

5. The **duodenal** (*submucosal* or *Brunner's*) **glands** of the duodenum produce mucus, buffers, and the hormone **urogastrone**. The ileum contains masses of lymphoid tissue called *aggregated lymphoid nodules*, or *Peyer's patches*, near the entrance to the large intestine.

INTESTINAL MOVEMENTS   p. 900

6. The **gastroenteric reflex**, initiated by stretch receptors in the stomach, stimulates motility and secretion along the entire small intestine. The **gastroileal reflex** triggers the relaxation of the ileocecal valve.

THE PANCREAS   p. 901

7. The **pancreatic duct** penetrates the wall of the duodenum. Within each lobule of the **pancreas**, ducts branch repeatedly before ending in the **pancreatic acini** (blind pockets). *(Figure 24–18)*

8. The pancreas has two functions: endocrine (secreting insulin and glucagon into the blood) and exocrine (secreting water, ions, and digestive enzymes [**pancreatic juice**] into the small intestine). Pancreatic enzymes include **carbohydrases, lipases, nucleases,** and **proteolytic enzymes**.

THE LIVER   p. 902

9. The **liver** performs metabolic and hematological regulation and produces **bile**. The bile ducts from all the **liver lobules** unite to form the **common hepatic duct**. That duct meets the **cystic duct** to form the **common bile duct**, which empties into the duodenum. *(Figures 24–19, 24–21)*

10. The liver lobule is the organ's basic functional unit. **Hepatocytes** form irregular plates arranged in the form of spokes of a wheel. **Bile canaliculi** carry bile to the **bile ductules**, which lead to **portal areas**. *(Figure 24–20)*

11. In **emulsification, bile salts** break apart large drops of lipids, making them accessible to lipases secreted by the pancreas.

THE GALLBLADDER   p. 906

12. The **gallbladder** stores, modifies, and concentrates bile. *(Figure 24–21)*

THE COORDINATION OF SECRETION AND ABSORPTION   p. 907

13. Neural and hormonal mechanisms coordinate the activities of the digestive glands. Gastrointestinal activity is stimulated by parasympathetic innervation and inhibited by sympathetic innervation. The **enterogastric, gastroenteric,** and **gastroileal reflexes** coordinate movement from the stomach to the large intestine.

14. Intestinal hormones include **enterocrinin, secretin, cholecystokinin (CCK), gastric inhibitory peptide (GIP), vasoactive intestinal peptide (VIP),** and gastrin. *(Figure 24–22; Summary Table 24–1)*

24–7 THE LARGE INTESTINE   p. 910

1. The main functions of the **large intestine** are to (1) reabsorb water and compact materials into feces, (2) absorb vitamins produced by bacteria, and (3) store fecal material prior to defecation. The large intestine consists of the *cecum, colon,* and *rectum*. *(Figure 24–23)*

THE CECUM   p. 910

2. The **cecum** collects and stores material from the ileum and begins the process of compaction. The **appendix** is attached to the cecum. *(Figure 24–23)*

THE COLON   p. 910

3. The **colon** has a larger diameter and a thinner wall than that of the small intestine. The colon bears **haustra** (pouches), **taeniae coli** (longitudinal bands of muscle), and sacs of fat (**fatty appendices**). *(Figure 24–23)*

4. The four regions of the colon are the **ascending colon, transverse colon, descending colon,** and **sigmoid colon**. *(Figure 24–2)*

THE RECTUM   p. 912

5. The **rectum** terminates in the **anal canal**, leading to the **anus**. *(Figure 24–23)*

HISTOLOGY OF THE LARGE INTESTINE   p. 912

6. Characteristics of the colon include the absence of villi and the presence of goblet cells and deep intestinal glands. *(Figure 24–24)*

PHYSIOLOGY OF THE LARGE INTESTINE   p. 913

7. The large intestine reabsorbs water and other substances such as vitamins, urobilinogen, bile salts, and toxins. Bacteria are responsible for intestinal gas, or **flatus**.

8. The gastroileal reflex moves materials from the ileum into the cecum while you eat. Distension of the stomach and duodenum

stimulates **mass movements** of materials from the transverse colon through the rest of the large intestine and into the rectum. Muscular sphincters control the passage of fecal material to the anus. Distension of the rectal wall triggers the **defecation reflex**. *(Figure 24–25)*

24–8 DIGESTION AND ABSORPTION   p. 915

THE PROCESSING AND ABSORPTION OF NUTRIENTS   p. 915

1. The digestive system breaks down the physical structure of the ingested material and then disassembles the component molecules into smaller fragments by *hydrolysis*. *(Figure 24–26; Summary Table 24–2)*

CARBOHYDRATE DIGESTION AND ABSORPTION   p. 915

2. Salivary and pancreatic amylases break down complex carbohydrates into *disaccharides* and *trisaccharides*. These in turn are broken down into *monosaccharides* by enzymes at the epithelial surface. The monosaccharides are then absorbed by the intestinal epithelium by facilitated diffusion or cotransport. *(Figure 24–26)*

LIPID DIGESTION AND ABSORPTION   p. 918

3. *Triglycerides* are emulsified into lipid droplets. The resulting fatty acids and *monoglycerides* interact with bile salts to form **micelles**, from which diffusion occurs across the intestinal epithelium. *(Figure 24–26)*

PROTEIN DIGESTION AND ABSORPTION   p. 918

4. Protein digestion involves a low pH (which destroys tertiary and quaternary structure), the gastric enzyme pepsin, and the various pancreatic proteases. Peptidases liberate amino acids that are absorbed and exported to interstitial fluid. *(Figure 24–26)*

WATER ABSORPTION   p. 918

5. About 2000 ml of water is ingested each day, and digestive secretions provide another 7000 ml. Nearly all is reabsorbed by osmosis. *(Figure 24–27)*

ION ABSORPTION   p. 919

6. Various processes, such as diffusion, cotransport, and carrier-mediated and active transports, are responsible for the movements of cations (sodium, calcium, potassium, and so on) and anions (chloride, iodide, bicarbonate, and so on) into epithelial cells. *(Figure 24–28)*

VITAMIN ABSORPTION   p. 919

7. The **water-soluble vitamins** (except for B_{12}) diffuse easily across the digestive epithelium. **Fat-soluble vitamins** are enclosed within fat droplets and are absorbed with the products of lipid digestion. *(Figure 24–28)*

24-9 AGING AND THE DIGESTIVE SYSTEM   p. 920

1. Age-related changes include a thinner and more fragile epithelium due to a reduction in epithelial stem cell divisions, weaker peristaltic contractions as smooth muscle tone decreases, the effects of cumulative damage, increased cancer rates, and related changes in other systems.

REVIEW QUESTIONS

More assessment questions are available to you on the Companion Website. You will find Matching, Multiple Choice, True/False, and other quizzes to help further your understanding of the material covered in this chapter. To access the site, go to www.aw.com/martini.

LEVEL 1 Reviewing Facts and Terms

1. The enzymatic breakdown of large molecules into their basic building blocks is called
   (a) absorption   (b) secretion
   (c) mechanical digestion (d) chemical digestion

2. The outer layer of the digestive tract is known as the
   (a) serosa   (b) mucosa
   (c) submucosa   (d) muscularis

3. The muscularis externa propels materials from one portion of the digestive tract to another by the contractions of
   (a) segmentation   (b) propulsion
   (c) mass movements (d) peristalsis

4. The activities of the digestive system are regulated by
   (a) hormonal mechanisms (b) local mechanisms
   (c) neural mechanisms   (d) a, b, and c are correct

5. Double sheets of peritoneum that provide support and stability for the organs of the peritoneal cavity are the
   (a) mediastina   (b) mucous membranes
   (c) omenta   (d) mesenteries

6. The peritoneal fold that stabilizes and supports the small intestine is the
   (a) serosa   (b) lesser omentum
   (c) mesentery proper   (d) parietal peritoneum

7. Intrinsic factor and hydrochloric acid are secreted by cells in the stomach wall called
   (a) parietal cells (b) chief cells
   (c) acinar cells   (d) G cells

8. Protein digestion in the stomach results primarily from secretions released by
   (a) G cells   (b) parietal cells
   (c) chief cells   (d) D cells

9. The part of the digestive tract that plays the primary role in the digestion and absorption of nutrients is, or are, the
   (a) large intestine (b) small intestine
   (c) stomach   (d) cecum and colon

10. The hormone that stimulates the secretion of the stomach and contraction of the stomach walls is
    (a) gastrin   (b) enterokinase
    (c) secretin   (d) cholecystokinin

11. The essential metabolic or synthetic service provided by the liver is
    (a) metabolic regulation   (b) hematological regulation
    (c) bile production   (d) a, b, and c are correct

12. Bile release from the gallbladder into the duodenum occurs only under the stimulation of
    (a) cholecystokinin
    (b) secretin
    (c) gastrin
    (d) enterokinase

13. The major function(s) of the large intestine is (are) the
    (a) reabsorption of water and compaction of feces
    (b) absorption of vitamins liberated by bacterial action
    (c) storage of fecal material prior to defecation
    (d) a, b, and c are correct

14. The part of the colon that accepts chyme from the small intestine is the
    (a) sigmoid colon
    (b) transverse colon
    (c) descending colon
    (d) ascending colon

15. The three longitudinal bands of smooth muscle found on the outer surface of the colon are
    (a) fatty appendages
    (b) haustra
    (c) colic flexures
    (d) taeniae coli

16. Three vitamins generated by bacteria in the colon are
    (a) vitamins A, D, and E
    (b) B complex vitamins and vitamin C
    (c) vitamin K, biotin, and pantothenic acid
    (d) niacin, thiamine, and riboflavin

17. The final enzymatic steps in the digestive process are accomplished by
    (a) brush border enzymes of the microvilli
    (b) enzymes secreted by the stomach
    (c) enzymes secreted by the pancreas
    (d) the action of bile from the gallbladder

18. What are the six steps involved in digestion?

19. Name and describe the layers of the digestive tract, proceeding from the innermost to the outermost layer.

20. What three basic mechanisms regulate the activities of the digestive tract?

21. What are the four primary functions of the oral cavity?

22. List the three pairs of salivary glands, and identify the ducts used by their glandular secretions.

23. What are the functions of saliva, and how are salivary secretions controlled?

24. What specific function does each of the four types of teeth perform in the oral cavity?

25. What role do the pharyngeal muscles play in swallowing?

26. What are the three phases of swallowing, and how are they controlled?

27. What contributions do the gastric and pyloric glands of the stomach make to the digestive process?

28. What three subdivisions of the small intestine are involved in the digestion and absorption of food?

29. What are the primary functions of the pancreas, liver, and gallbladder in digestion?

30. Which hormones produced by duodenal enteroendocrine cells effectively coordinate digestive functions?

31. What are the three primary functions of the large intestine?

32. What two positive feedback loops are involved in the defecation reflex?

33. Describe five age-related changes that occur in the digestive system.

LEVEL 2 Reviewing Concepts

34. If the lingual frenulum is too restrictive, an individual
    (a) has difficulty tasting food
    (b) cannot swallow properly
    (c) cannot control movements of the tongue
    (d) cannot eat or speak normally

35. Increased secretion by all the salivary glands results from
    (a) sympathetic stimulation
    (b) hormonal stimulation
    (c) parasympathetic stimulation
    (d) myenteric reflexes

36. The production of acid and enzymes by the gastric mucosa is controlled and regulated by
    (a) the central nervous system
    (b) short reflexes coordinated in the stomach wall
    (c) digestive tract hormones
    (d) a, b, and c are correct

37. During the gastric phase of secretion, the pH of the gastric contents
    (a) decreases continually
    (b) remains the same as the contents become diluted
    (c) increases immediately and remains higher than pH 2.0
    (d) increases slowly with continued mixing

38. Chyme reaches the small intestine most quickly when
    (a) the stomach is greatly distended
    (b) a large amount of gastrin is released
    (c) the meal contains a relatively small quantity of proteins
    (d) a and c are correct

39. A drop in pH below 4.5 in the duodenum stimulates the secretion of
    (a) secretin
    (b) cholecystokinin
    (c) gastrin
    (d) a, b, and c are correct

40. Through which layers of a molar would an orthodontic surgeon drill to perform a root canal (removal of the alveolar nerve in a severely damaged tooth)?

41. What is the purpose of the transverse or longitudinal folds of the epithelium in the digestive tract?

42. Differentiate between peristalsis and segmentation in terms of their actions and their results.

43. How is the epithelium of the stomach protected from digestion?

44. How do the three phases of gastric secretion promote and facilitate gastric control?

45. Why should a person who has stomach ulcers avoid aspirin?

46. Nutritionists have found that after a heavy meal, the pH of blood increases slightly, especially in the veins that carry blood away from the stomach. What causes this "postenteric alkaline tide"?

LEVEL 3 Critical Thinking and Clinical Applications

47. Some persons with gallstones develop pancreatitis. How could this occur?

48. Barb suffers from Crohn's disease, a regional inflammation of the intestine. The disease is thought to have some genetic basis, but the actual cause is as yet unknown. When the disease flares up, she experiences abdominal pain, weight loss, and anemia. Which part(s) of the intestine is (are) probably involved, and what is the cause of Barb's symptoms?

49. What symptoms would you expect to observe in a person whose small intestine is blocked at the level of the jejunum?

50. Recently, more people have turned to surgery to assist them in losing weight. One of the more radical weight control surgeries involves stapling a portion of the stomach shut, creating a smaller volume. What effects might this procedure have on the physiology of the entire digestive system?

UNIT 5   CHAPTER   23   24   25   26   27

METABOLISM AND ENERGETICS

Cells are chemical factories that break down organic molecules to obtain energy, which can then be used to generate ATP. Reactions within mitochondria provide most of the energy needed by a typical cell. ∞ pp. 83, 316 To carry out these metabolic reactions, our cells must have a reliable supply of oxygen and nutrients, including water, vitamins, mineral ions, and organic substrates (the reactants in enzymatic reactions). Oxygen is absorbed at the lungs; the other substances are obtained by absorption at the digestive tract. The cardiovascular system then carries these substances throughout the body. They diffuse from the bloodstream into the tissues, where they can be absorbed and used by our cells.

Mitochondria break down the organic nutrients to provide energy for cell growth, cell division, contraction, secretion, and other functions. Each tissue contains a unique mixture of cells of various kinds. As a result, the energy and nutrient requirements of any two tissues, such as loose connective tissue and cardiac muscle, are quite different. Moreover, activity levels can change rapidly within a tissue, and such changes affect the metabolic requirements of the body. For example, when skeletal muscles start contracting, the tissue demand for oxygen skyrockets. Thus, the energy and nutrient requirements of the body vary from moment to moment (resting versus exercising), hour to hour (asleep versus awake), and year to year (growing child versus adult).

The amount and type of nutrients you obtain in meals can also vary widely. Your body stores energy reserves when nutrients are abundant, and mobilizes them when nutrients are in short supply. Various tissues and organs are specialized to store excess nutrients; adipose tissues, for example, store lipids, whereas muscle and liver tissues store glycogen. The endocrine system, with the assistance of the nervous system, adjusts and coordinates the metabolic activities of the body's tissues and controls the storage and mobilization of these energy reserves.

25–1 AN OVERVIEW OF METABOLISM

Objective

- Define *metabolism*, and explain why cells need to synthesize new organic components.

The term **metabolism** (me-TAB-ō-lizm) refers to all chemical reactions that occur in an organism. Chemical reactions within cells, collectively known as *cellular metabolism*, provide the energy needed to maintain homeostasis and to perform essential functions. Such functions include (1) *metabolic turnover*, the periodic breakdown and replacement of the organic components of a cell; (2) growth and cell division; and (3) special processes, such as secretion, contraction, and the propagation of action potentials.

Figure 25–1● provides a broad overview of the processes involved in cellular metabolism. The cell absorbs organic molecules from the surrounding interstitial fluids. Amino acids, lipids, and simple sugars cross the cell membrane to join nutrients already in the cytoplasm. All the cell's organic building blocks collectively form a *nutrient pool* that the cell relies on to provide energy and to create new intracellular components.

The breakdown of organic substrates is called **catabolism**. This process releases energy that can be used to synthesize ATP or other high-energy compounds. Catabolism proceeds in a series of steps. In general, the initial steps occur in the cytosol, where enzymes break down large organic molecules into smaller fragments. For example, carbohydrates are broken down into short carbon chains, triglycerides are split into fatty acids and glycerol, and proteins are broken down to individual amino acids.

Relatively little ATP is produced during these preparatory steps. However, the simple molecules formed can be absorbed and processed by mitochondria, and the mitochondrial steps release significant amounts of energy. As mitochondrial enzymes break the covalent bonds that hold these molecules together,

they capture roughly 40 percent of the energy released. The captured energy is used to convert ADP to ATP. The other 60 percent escapes as heat that warms the interior of the cell and the surrounding tissues.

The ATP produced by mitochondria provides energy to support **anabolism**—the synthesis of new organic molecules as well as other cell functions. Those additional functions, such as ciliary or cell movement, contraction, active transport, and cell division, vary from one cell to another. For example, muscle fibers need ATP to provide energy for contraction, whereas gland cells need ATP to synthesize and transport their secretions. We have considered such specialized functions in other chapters, so we will restrict our focus here to anabolic processes.

In terms of energy, anabolism is an "uphill" process that involves the formation of new chemical bonds. Cells synthesize new organic components for four basic reasons (Figure 25–2●):

1. *To Perform Structural Maintenance or Repairs.* All cells must expend energy to perform ongoing maintenance and repairs, because most structures in the cell are temporary rather than permanent. Their removal and replacement are part of the process of *metabolic turnover.* ∞ p. 61

2. *To Support Growth.* Cells preparing to divide increase in size and synthesize extra proteins and organelles.

3. *To Produce Secretions.* Secretory cells must synthesize their products and deliver them to the interstitial fluid.

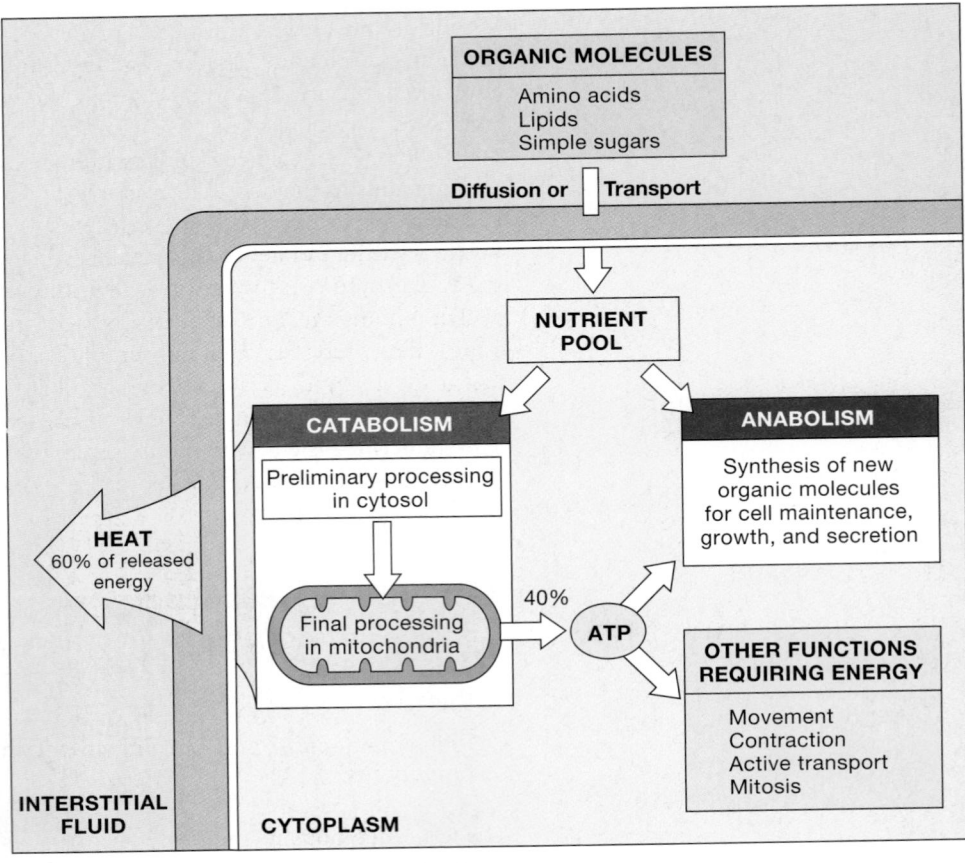

●**FIGURE 25–1**
An Introduction to Cellular Metabolism. The cell obtains organic molecules from the interstitial fluid and breaks them down to ATP. Only about 40 percent of the energy released by catabolism is captured in ATP; the rest is radiated as heat. The ATP generated by catabolism provides energy for all vital cellular activities, including anabolism.

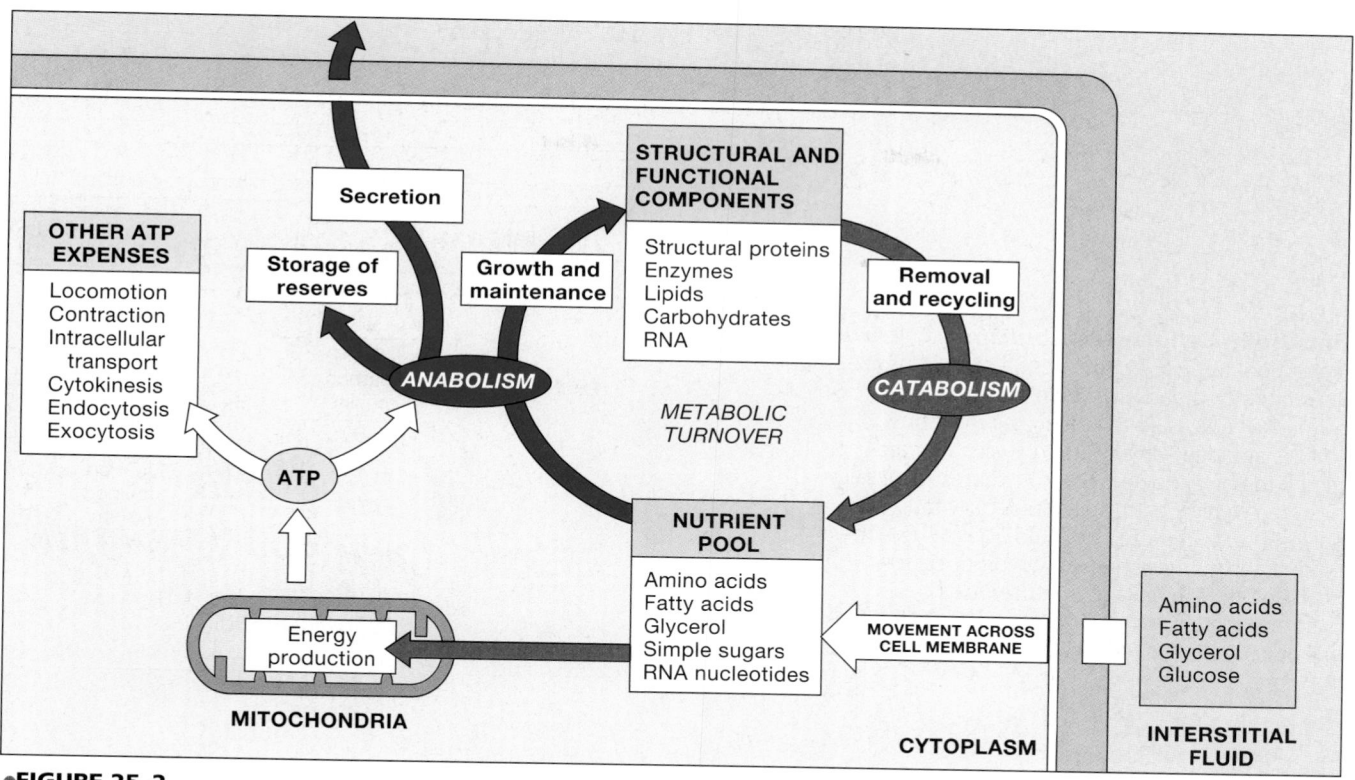

•**FIGURE 25–2**

Metabolic Turnover and Cellular ATP Production. Metabolic turnover is integrated into other cellular functions. Organic molecules released during metabolic turnover can be recycled by means of anabolic pathways and used to build other large molecules. They can also be broken down for energy production. Through catabolism, the cell must provide ATP for anabolism and for other cellular functions.

4. To Build Nutrient Reserves. Most cells prepare for a rainy day—a period of emergency, an interval of extreme activity, or a time when the supply of nutrients in the bloodstream is inadequate. Cells prepare for such times by building up reserves—nutrients stored in a form that can be mobilized as needed. The most abundant storage form of carbohydrate is glycogen, a branched chain of glucose molecules; the most abundant lipids are triglycerides, consisting primarily of fatty acids. Thus, muscle cells and liver cells, for example, store glucose in the form of glycogen, while adipocytes and liver cells store triglycerides. Proteins, the most abundant organic components in the body, perform a variety of vital functions for the cell, and when energy is available, cells synthesize additional proteins. However, when glucose or fatty acids are unavailable, proteins can be broken down into their component amino acids and the amino acids catabolized as an energy source. So, although their primary function is not to serve as an energy source, proteins are so abundant and accessible that they represent an important "last-ditch" nutrient reserve.

The nutrient pool is the source of the substrates for both catabolism and anabolism. As you might expect, the cell tends to conserve the materials needed to build new compounds and

breaks down the rest. The cell continuously replaces membranes, organelles, enzymes, and structural proteins. These anabolic activities require more amino acids than lipids and few carbohydrates. In general, when a cell with excess carbohydrates, lipids, and amino acids needs energy, it will break down carbohydrates first. Lipids are the second choice, and amino acids are seldom broken down if other energy sources are available.

Mitochondria are important because they provide the energy that supports cellular operations. The cell feeds its mitochondria from its nutrient pool, and in return, the cell gets the ATP it needs. However, mitochondria are picky eaters: They will accept only specific organic molecules for processing and energy production. Thus, chemical reactions in the cytoplasm take whichever organic nutrients are available and break them down into smaller fragments that are acceptable to the mitochondria. The mitochondria then break the fragments down further, generating carbon dioxide, water, and ATP (Figure 25–3•). This mitochondrial activity involves two pathways: the *TCA cycle* and the *electron transport system*. In the next section, we shall describe these important catabolic and anabolic reactions that occur in our cells.

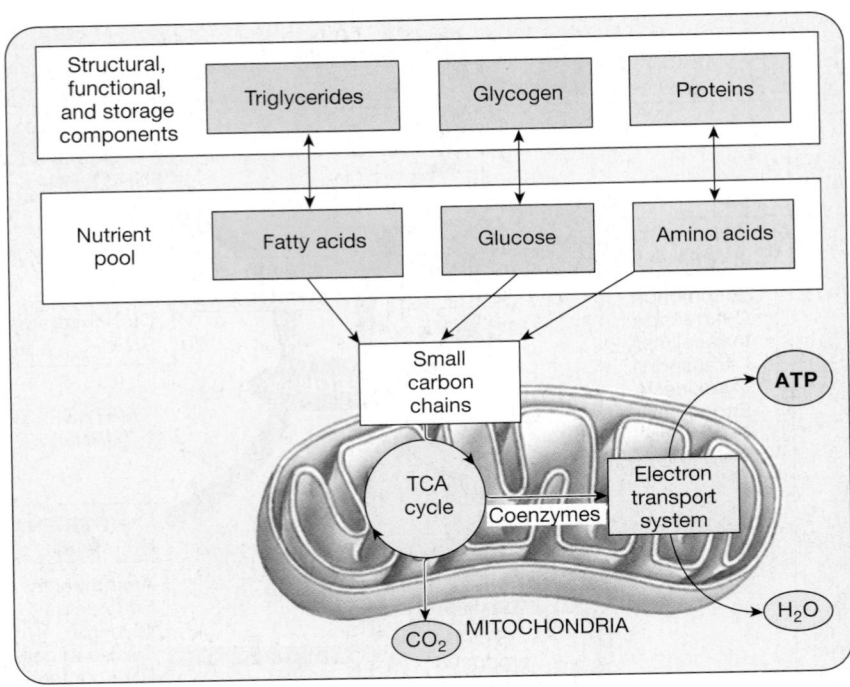

●FIGURE 25–3
Nutrient Use in Cellular Metabolism. Cells use the nutrient pool to build up reserves and to synthesize cellular structures. Catabolism within mitochondria provides the ATP necessary to sustain cell functions. Mitochondria are "fed" small carbon chains that are produced by the breakdown of carbohydrates (especially glucose, stored as glycogen), lipids (especially fatty acids from triglycerides) and proteins (amino acids). The mitochondria absorb these breakdown products for further catabolism by means of the tricarboxylic acid (TCA) cycle and the electron transport system (ETS). This figure will be repeated, in reduced and simplified form as Navigator icons, as we change topics.

25–2 CARBOHYDRATE METABOLISM

Objectives

■ Describe the basic steps in glycolysis, the TCA cycle, and the electron transport chain.

■ Summarize the energy yield of glycolysis and cellular respiration.

Most cells generate ATP and other high-energy compounds by breaking down carbohydrates—especially glucose. The complete reaction sequence can be summarized as

$$C_6H_{12}O_6 \;+\; 6\,O_2 \;\rightarrow\; 6\,CO_2 \;+\; 6\,H_2O$$
$$\text{glucose} \quad\; \text{oxygen} \quad\quad \text{carbon dioxide} \quad\; \text{water}$$

The breakdown occurs in a series of small steps, several of which release sufficient energy to support the conversion of ADP to ATP. The complete catabolism of one molecule of glucose will provide a typical cell in the body with a net gain of 36 molecules of ATP.

Although most of the actual ATP production occurs inside mitochondria, the first steps take place in the cytosol. In Chapter 10 we introduced the process of *glycolysis*, which breaks down glucose in the cytosol and generates smaller molecules that can be absorbed and utilized by mitochondria. Because glycolysis does not require oxygen, the reactions are said to be *anaerobic*. The subsequent reactions, which occur in mitochondria, consume oxygen and are called *aerobic*. The mitochondrial activity responsible for ATP production is called **aerobic metabolism**, or *cellular respiration*.

☐ GLYCOLYSIS

Glycolysis (glī-KOL-i-sis; *glykus*, sweet + *lysis*, dissolution) is the breakdown of glucose to **pyruvic acid**. In this process, a series of enzymatic steps breaks the six-carbon glucose molecule ($C_6H_{12}O_6$) into two three-carbon molecules of pyruvic acid (CH_3–CO—$COOH$). At the normal pH inside the cell, each pyruvic acid molecule loses a hydrogen ion and exists as the negatively charged ion CH_3–CO—COO^-. This ionized form is usually called *pyruvate*, rather than pyruvic acid.

Glycolysis requires (1) glucose molecules, (2) appropriate cytoplasmic enzymes, (3) ATP and ADP, (4) inorganic phosphates, and (5) **NAD** (**n**icotinamide **a**denine **d**inucleotide), a coenzyme that removes hydrogen atoms during one of the enzymatic reactions. (Recall from Chapter 2 that coenzymes are organic molecules which are essential to enzyme function. ∞ p. 57) If any of these participants is missing, glycolysis cannot take place.

The steps in glycolysis are diagrammed in Figure 25–4●. The diagram is an overview, rather than a detailed description, of the pathway involved. Glycolysis begins when an enzyme *phosphorylates*—that is, attaches a phosphate group—to the last (sixth) carbon atom of a glucose molecule, creating **glucose-6-phosphate**. Although this step "costs" the cell one ATP molecule, it has two important results: (1) It traps the glucose molecule within the cell, because phosphorylated glucose cannot cross the cell membrane; and (2) it prepares the glucose molecule for further biochemical reactions.

A second phosphorylation occurs in the cytosol before the six-carbon chain is broken into two three-carbon fragments. Energy benefits begin to appear as these fragments are converted to

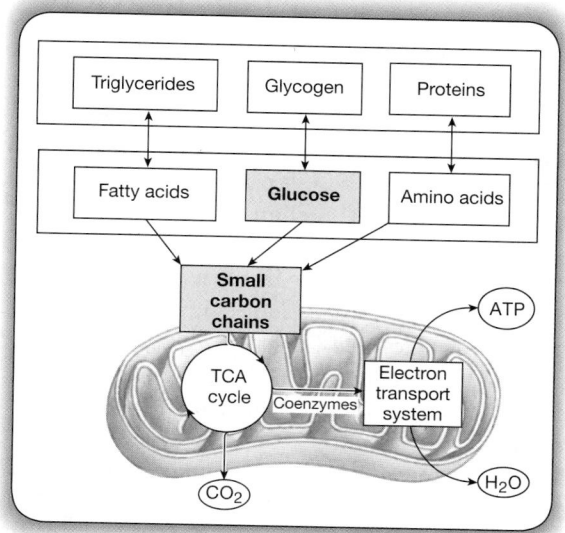

•FIGURE 25–4
Glycolysis. The Navigator icon in the shadow box highlights the topic under discussion. Glycolysis breaks down a six-carbon glucose molecule into two three-carbon molecules of pyruvic acid through a series of enzymatic steps. This diagram follows the fate of the carbon chain. There is a net gain of two ATP molecules for each glucose molecule converted to two pyruvic acid molecules. In addition, two molecules of the coenzyme NAD are converted to NADH. Once transferred to mitochondria, both the pyruvic acid and the NADH can still yield a great deal more energy. The further catabolism of pyruvic acid begins with its entry into a mitochondrion. *(See Figure 25–5.)*

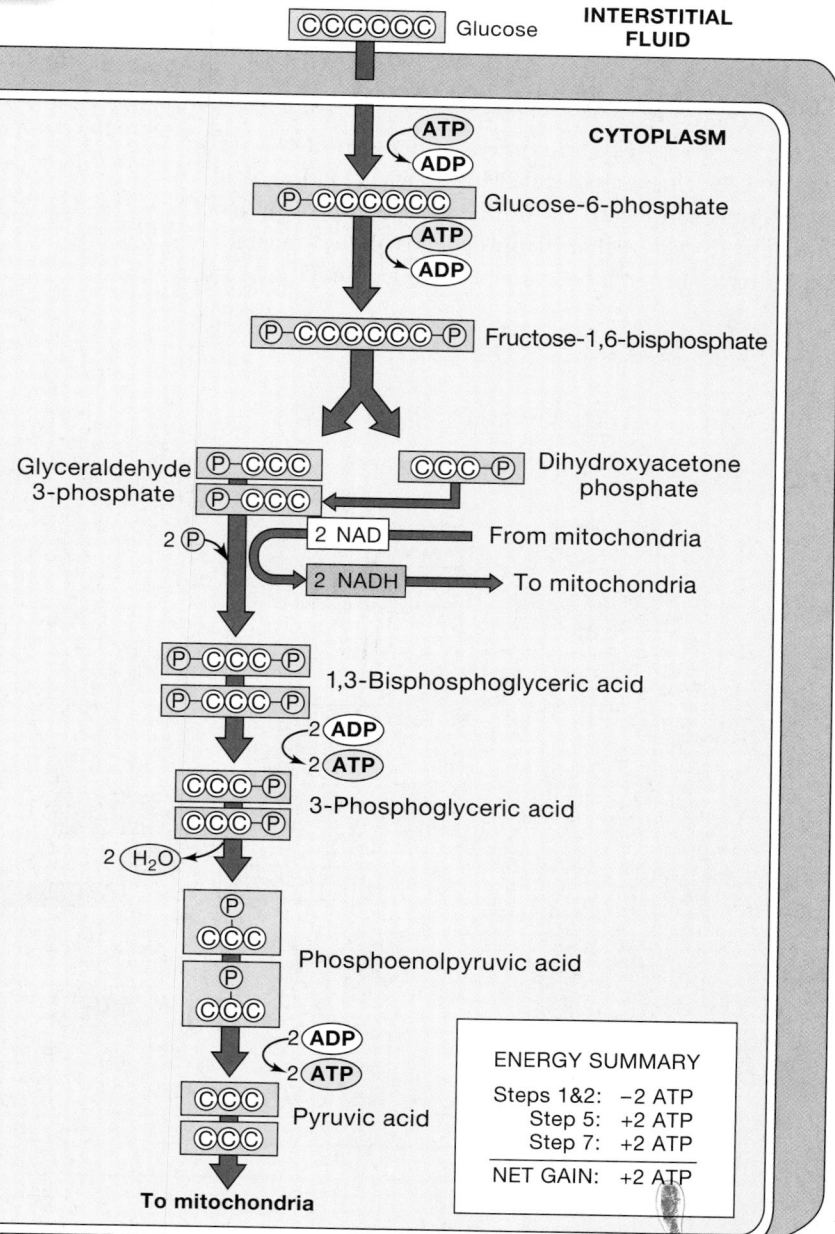

STEPS IN GLYCOLYSIS

STEP 1:
As soon as a glucose molecule enters the cytoplasm, a phosphate group is attached to the molecule.

STEP 2:
A second phosphate group is attached. Together, steps 1 and 2 cost the cell 2 ATP.

STEP 3:
The six-carbon chain is split into two three-carbon molecules, each of which then follows the rest of this pathway.

STEP 4:
Another phosphate group is attached to each molecule, and NADH is generated from NAD.

STEP 5:
One ATP molecule is formed for each molecule processed.

STEP 6:
The atoms in each molecule are rearranged, releasing a molecule of water.

STEP 7:
A second ATP molecule is formed for each molecule processed. Step 7 produces 2 ATP molecules.

INTERSTITIAL FLUID — Glucose

CYTOPLASM

Glucose-6-phosphate

Fructose-1,6-bisphosphate

Glyceraldehyde 3-phosphate — Dihydroxyacetone phosphate

2 NAD — From mitochondria
2 NADH — To mitochondria

1,3-Bisphosphoglyceric acid

3-Phosphoglyceric acid

Phosphoenolpyruvic acid

Pyruvic acid

To mitochondria

ENERGY SUMMARY

Steps 1&2:	−2 ATP
Step 5:	+2 ATP
Step 7:	+2 ATP
NET GAIN:	+2 ATP

pyruvic acid. Two of the steps release enough energy to generate ATP from ADP and inorganic phosphate (PO_4^{3-}, or P_i). In addition, two molecules of NAD are converted to NADH. The net reaction looks like this:

$$Glucose + 2\,NAD + 2\,ADP + 2P_i \rightarrow$$
$$2\,Pyruvic\ acid + 2\,NADH + 2\,ATP$$

This anaerobic reaction sequence provides the cell with a net gain of two molecules of ATP for each glucose molecule converted to two pyruvic acid molecules. A few highly specialized cells, such as red blood cells, lack mitochondria and derive all their ATP through glycolysis. Skeletal muscle fibers rely on glycolysis for ATP production during periods of active contraction; using the ATP provided by glycolysis alone, most cells can survive for brief periods. However, when oxygen is readily available, mitochondrial activity provides most of the ATP required by our cells.

◻ MITOCHONDRIAL ATP PRODUCTION

For the cell, glycolysis yields an immediate net gain of two ATP molecules for each glucose molecule. However, a great deal of additional energy is still locked in the chemical bonds of pyruvic acid. The ability to capture that energy depends on the availability of oxygen. If oxygen supplies are adequate, mitochondria absorb the pyruvic acid molecules and break them down. The hydrogen atoms of each pyruvic acid molecule ($CH_3-CO-COOH$) are removed by coenzymes and will ultimately be the source of most of the energy gain for the cell. The carbon and oxygen atoms are removed and released as carbon dioxide in a process called **decarboxylation** (dē-kar-boks-i-LĀ-shun).

Each mitochondrion has two membranes surrounding it. The *outer membrane* contains large-diameter pores that are permeable to ions and small organic molecules such as pyruvic acid. The ions and molecules easily enter the *intermembrane space* separating the outer membrane from the *inner membrane*. The inner membrane contains a carrier protein that moves pyruvic acid into the mitochondrial matrix.

In the mitochondrion, a pyruvic acid molecule participates in a complex reaction involving NAD and another coenzyme, **coenzyme A**, or **CoA** (Figure 25–5●). This reaction yields one molecule of carbon dioxide, one of NADH, and one of **acetyl-CoA** (as-Ē-til-KŌ-ā)—a two-carbon **acetyl group** (CH_3CO) bound to coenzyme A. Next, the acetyl group is transferred from CoA to a four-carbon molecule of *oxaloacetic acid*, producing *citric acid*.

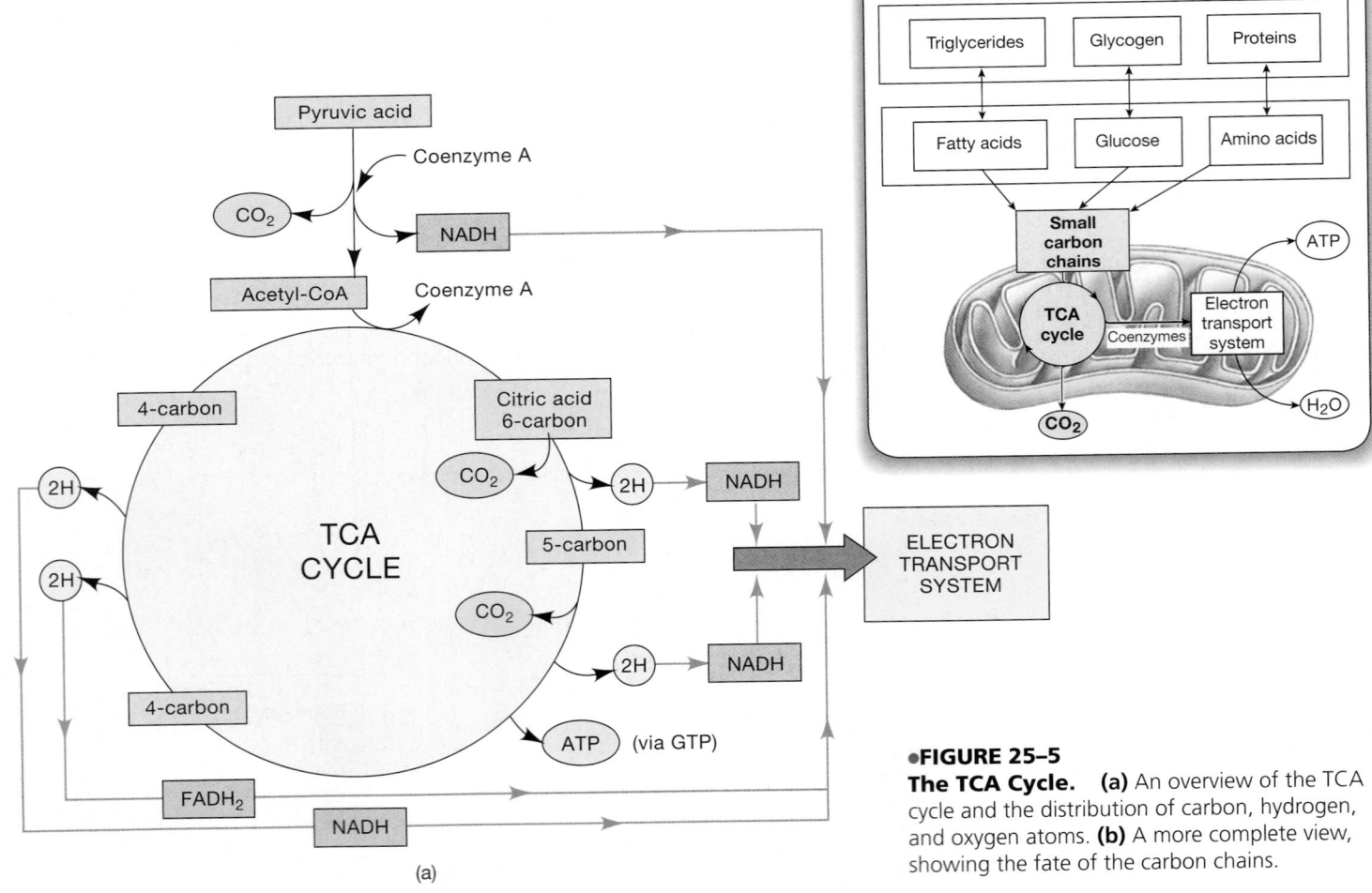

●**FIGURE 25–5**
The TCA Cycle. **(a)** An overview of the TCA cycle and the distribution of carbon, hydrogen, and oxygen atoms. **(b)** A more complete view, showing the fate of the carbon chains.

The TCA Cycle

The formation of citric acid from acetyl-CoA is the first step in a sequence of enzymatic reactions called the **tricarboxylic** (trī-kar-bok-SIL-ik) **acid (TCA) cycle**, or **citric acid cycle**. This reaction sequence is sometimes called the *Krebs cycle* in honor of Hans Krebs, the biochemist who described the various reactions in 1937. The function of the cycle is to remove hydrogen atoms from organic molecules and transfer them to coenzymes. The overall pattern of the TCA cycle is shown in Figure 25–5●. For the biochemical details of glycolysis and the TCA cycle, consult the *Applications Manual*. **AM** Aerobic Metabolism: A Closer Look

At the start of the TCA cycle, the two-carbon acetyl group carried by CoA is attached to a four-carbon oxaloacetic acid mole-

cule to make the six-carbon compound citric acid. Coenzyme A is released intact and can thus bind another acetyl group. A complete revolution of the TCA cycle removes two carbon atoms, regenerating the four-carbon chain. (This is why the reaction sequence is called a *cycle*.) We can summarize the fate of the atoms in the acetyl group as follows:

- The two carbon atoms are removed in enzymatic reactions that incorporate four oxygen atoms and form two molecules of carbon dioxide, a waste product.
- The hydrogen atoms are removed by the coenzyme NAD or a related coenzyme called **FAD** (*flavin adenine dinucleotide*).

Several of the steps involved in a revolution of the TCA cycle involve more than one reaction and require more than one enzyme. Water molecules are tied up in two of those steps. The entire sequence can be summarized as

$$CH_3CO-CoA + 3\,NAD + FAD + GDP + P_i + 2\,H_2O \rightarrow$$
$$CoA + 2\,CO_2 + 3\,NADH + FADH_2 + 2\,H^+ + GTP$$

The only immediate energy benefit of one revolution of the TCA cycle is the formation of a single molecule of GTP (*guanosine triphosphate*) from GDP (*guanosine diphosphate*) and P_i. In practical terms, GTP is the equivalent of ATP, because GTP readily transfers a phosphate group to ADP, producing ATP: GTP + ADP → GDP + ATP.

The formation of GTP from GDP in the TCA cycle is an example of **substrate-level phosphorylation**. In this process, an enzyme uses the energy released by a chemical reaction to transfer a phosphate group to a suitable acceptor molecule. Although GTP is formed in the TCA cycle, many reaction pathways in the cytosol phosphorylate ADP and form ATP directly. For example, the ATP produced during glycolysis is generated through substrate-level phosphorylation. Normally, however, substrate-level phosphorylation provides a relatively small amount of energy compared with *oxidative phosphorylation*.

Oxidative Phosphorylation and the ETS

Oxidative phosphorylation is the generation of ATP within mitochondria in a reaction sequence that requires coenzymes and consumes oxygen. The process produces over 90 percent of the ATP used by our cells. The key reactions take place in the *electron transport system (ETS)*, a series of integral and peripheral proteins in the inner membrane of the

(b)

mitochondrion. The basis of oxidative phosphorylation is very simple:

$$2\,H_2 + O_2 \rightarrow 2\,H_2O.$$

Cells can easily obtain the ingredients for this reaction: Hydrogen is a component of all organic molecules, and oxygen is an atmospheric gas. The only problem is that the reaction releases a tremendous amount of energy all at once. In fact, this reaction releases so much energy that it is used to launch the space shuttle into orbit. Cells cannot handle energy explosions; energy release must be gradual, as it is in oxidative phosphorylation. This powerful reaction proceeds in a series of small, enzymatically controlled steps. Under these conditions, energy can be captured and ATP generated.

OXIDATION, REDUCTION, AND ENERGY TRANSFER The enzymatic steps of oxidative phosphorylation involve oxidation and reduction. (There are different types of oxidation and reduction reactions, but the most important for our purposes are those involving the transfer of electrons.) The loss of electrons is a form of **oxidation**; the acceptance of electrons is a form of **reduction**. The two reactions are always paired. When electrons pass from one molecule to another, the electron donor is oxidized and the electron recipient reduced. Oxidation and reduction are important because electrons carry chemical energy. In a typical oxidation–reduction reaction, *the reduced molecule gains energy at the expense of the oxidized molecule.*

In such an exchange, the reduced molecule does not acquire all the energy released by the oxidized molecule. Some energy is always released as heat, and additional energy may be used to perform physical or chemical work, such as the formation of ATP. By leading the electrons through a series of oxidation–reduction reactions before they ultimately combine with oxygen atoms, cells can capture and use much of the energy released in the formation of water.

Coenzymes play a key role in this process. A coenzyme acts as an intermediary that accepts electrons from one molecule and transfers them to another molecule. In the TCA cycle, NAD and FAD remove hydrogen atoms from organic substrates. Each hydrogen atom consists of an electron (e^-) and a proton (a hydrogen ion, H^+). Thus, when a coenzyme accepts hydrogen atoms, the coenzyme is reduced and gains energy. The donor molecule loses electrons and energy as it gives up its hydrogen atoms.

Next, NADH and $FADH_2$, the reduced forms of NAD and FAD, transfer their hydrogen atoms to other coenzymes. The protons are subsequently released, and the electrons, which carry the chemical energy, enter a sequence of oxidation–reduction reactions. This sequence ends with the electrons' transfer to oxygen and the formation of a water molecule. At several steps along the oxidation–reduction sequence, enough energy is released to support the synthesis of ATP from ADP. We shall now consider that reaction sequence in greater detail.

The coenzyme FAD accepts two hydrogen atoms from the TCA cycle and in doing so gains two electrons, forming $FADH_2$. The ox-

idized form of the coenzyme NAD has a positive charge (NAD^+). This coenzyme also gains two electrons as two hydrogen atoms are removed from the donor molecule, resulting in the formation of NADH and the release of a proton (H^+). For this reason, the reduced form of NAD is often described as "$NADH + H^+$."

Other coenzymes involved in the initial steps of oxidative phosphorylation are **FMN** (*flavin mononucleotide*) and **coenzyme Q** (*ubiquinone*). Coenzymes are either free in the mitochondrial matrix (NAD) or attached to the inner mitochondrial membrane (FAD, FMN, coenzyme Q).

THE ELECTRON TRANSPORT SYSTEM The **electron transport system (ETS)**, or *respiratory chain*, is a sequence of proteins called **cytochromes** (SĪ-tō-krōmz; *cyto-*, cell + *chroma*, color). Each cytochrome has two components: a protein and a pigment. The protein, embedded in the inner membrane of a mitochondrion, surrounds the pigment complex, which contains a metal ion (either iron—Fe^{3+}—or copper—Cu^{2+}). We shall consider four cytochromes: **b**, **c**, **a**, and **a₃**.

Figure 25–6● summarizes the basic steps in oxidative phosphorylation. We shall first consider the path taken by the electrons that are captured and delivered by coenzymes (Figure 25–6a●):

STEP 1: *A Coenzyme Strips a Pair of Hydrogen Atoms from a Substrate Molecule.* As we have seen, different coenzymes are used for different substrate molecules. During glycolysis, which occurs in the cytoplasm, NAD is reduced to NADH. Within mitochondria, both NAD and FAD are reduced through reactions of the TCA cycle, producing NADH and $FADH_2$, respectively.

STEP 2: *NADH and $FADH_2$ Deliver Hydrogen Atoms to Coenzymes Embedded in the Inner Membrane of a Mitochondrion.* The electrons carry the energy, and the protons that accompany them will be released before the electrons are transferred to the ETS. As indicated in Figure 25–6a●, one of two paths is taken to the ETS; which one depends on whether the donor is NADH or $FADH_2$.

STEP 3: *Coenzyme Q Accepts Hydrogen Atoms from $FMNH_2$ and $FADH_2$ and Passes Electrons to Cytochrome b.* The hydrogen atoms are released as hydrogen ions (H^+).

STEP 4: *Electrons Are Passed along the Electron Transport System, Losing Energy in a Series of Small Steps.* The sequence is cytochrome *b* to *c* to *a* to *a₃*.

STEP 5: *At the End of the ETS, an Oxygen Atom Accepts the Electrons, Creating an Oxygen Ion (O^-).* This ion, which has a very strong affinity for hydrogen ions (H^+), quickly combines with two hydrogen ions in the mitochondrial matrix, forming a molecule of water.

Notice that this reaction sequence started with the removal of two hydrogen atoms from a substrate molecule and ended with the formation of water from two hydrogen ions and one oxygen ion. This is the reaction that we described initially as releasing too much energy if performed in a single step. Because the reaction has occurred in a series of small steps, however, the combination of hydrogen and oxygen can take place quietly rather than explosively.

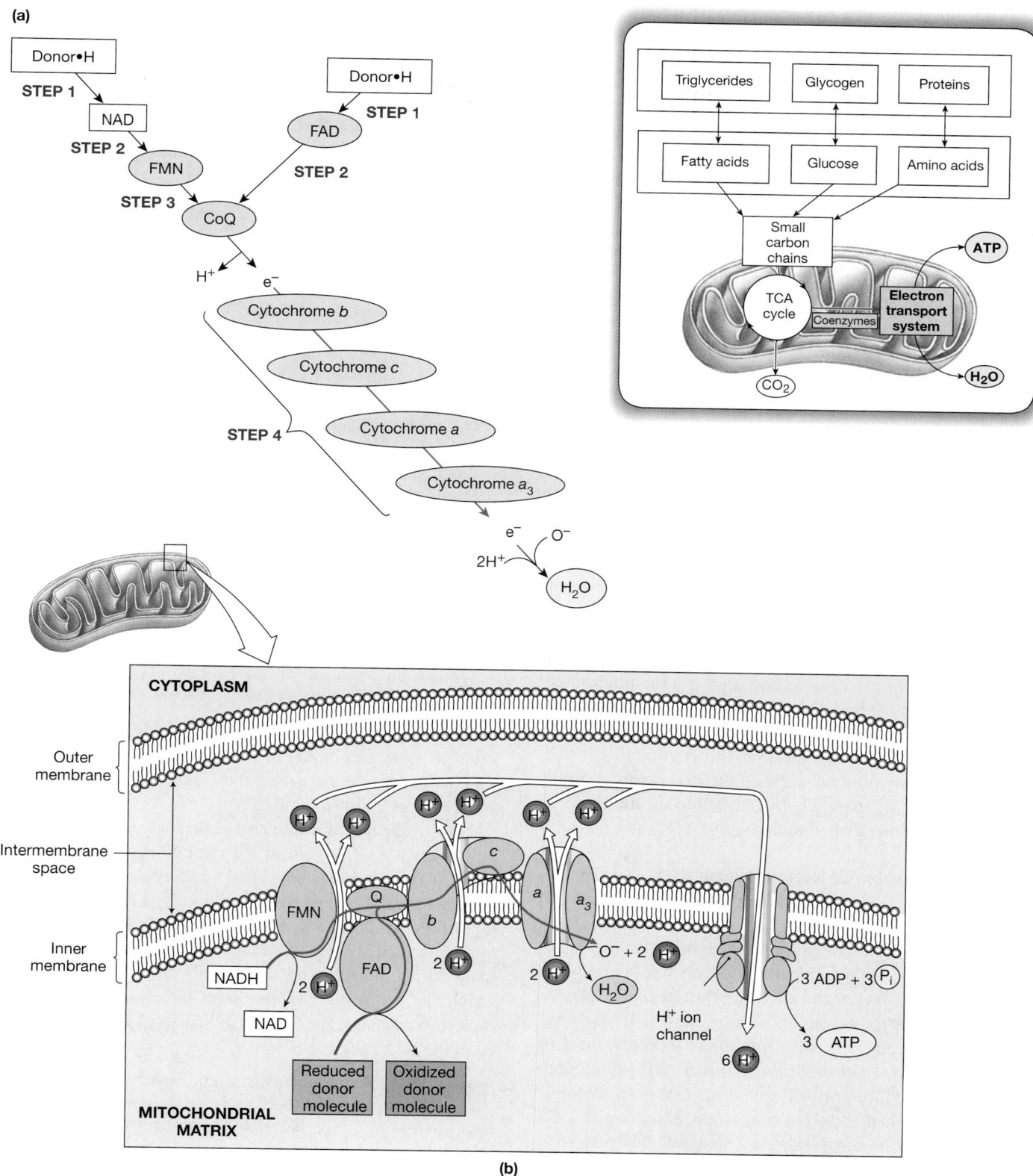

●FIGURE 25–6
Oxidative Phosphorylation. **(a)** The sequence of oxidation–reduction reactions involved. **(b)** The location of the coenzymes and the electron transport system. Notice the sites where hydrogen ions are pumped into the intermembrane space, providing the concentration gradient essential to the generation of ATP.

ATP GENERATION AND THE ETS As we noted in Chapter 3, concentration gradients that exist across membranes represent a form of potential energy, and that energy can be harnessed by the cell. The electron transport system does not produce ATP directly. Instead, it creates the conditions necessary for ATP production by creating a steep concentration gradient across the inner mitochondrial membrane. The electrons that travel along the ETS release energy as they pass from coenzyme to cytochrome and from cytochrome to cytochrome. The energy released at several steps drives hydrogen ion pumps that move hydrogen ions from the mitochondrial matrix into the intermembrane space, between the inner and outer membranes of a mitochondrion. These pumps create a large concentration gradient for hydrogen ions across the inner membrane. It is this concentration gradient that provides the energy to convert ADP to ATP.

Despite the concentration gradient, the hydrogen ions cannot diffuse into the matrix because they are not lipid soluble. However, hydrogen ion channels in the inner membrane permit the diffusion of H^+ into the matrix. These ion channels and their attached *coupling factors* use the kinetic energy of passing hydrogen ions to generate ATP in a process known as *chemiosmosis* (kem-ē-oz-MŌ-sis), or *chemiosmotic phosphorylation*.

Figure 25–6b● diagrams the mechanism of ATP generation. Hydrogen ions are pumped as (1) FMN reduces coenzyme Q, (2) cytochrome *b* reduces cytochrome *c*, and (3) electrons are passed from cytochrome *a* to cytochrome a_3.

For each pair of electrons removed from a substrate in the TCA cycle by NAD, six hydrogen ions are pumped across the inner membrane of the mitochondrion and into the intermembrane space (Figure 25–6b●). Their reentry into the matrix provides the energy to generate three molecules of ATP. For each pair of electrons removed from a substrate in the TCA cycle by FAD, four hydrogen ions are pumped across the inner membrane and into the intermembrane space. Their reentry into the matrix provides the energy to generate two molecules of ATP.

THE IMPORTANCE OF OXIDATIVE PHOSPHORYLATION Oxidative phosphorylation is the most important mechanism for the generation of ATP. In most cases, a chronic suspension, or even a significant reduction, of the rate of oxidative phosphorylation will kill the cell. If many cells are affected, the individual may die. Oxidative phosphorylation requires both oxygen and electrons; the rate of ATP generation is thus limited by the availability of either oxygen or electrons.

Cells obtain oxygen by diffusion from the extracellular fluid. If the supply of oxygen is cut off, mitochondrial ATP production will cease, because reduced cytochrome a_3 will have no acceptor for its electrons. With the last reaction stopped, the entire ETS comes to a halt, like cars at a washed-out bridge. Oxidative phosphorylation can no longer take place, and cells quickly succumb to energy starvation. Because neurons have a high demand for energy, the brain is one of the first organs to be affected.

Hydrogen cyanide gas is sometimes used as a pesticide to kill rats or mice; in some states where capital punishment is legal, it is used to execute criminals. The cyanide ion (CN^-) binds to cy-

tochrome a_3 and prevents the transfer of electrons to oxygen. As a result, cells die from energy starvation.

☐ SUMMARY: ENERGY YIELD OF GLYCOLYSIS AND CELLULAR RESPIRATION

For most cells, the complete reaction pathway that begins with glucose and ends with carbon dioxide and water is the main method of generating ATP. Figure 25–7● reviews the process in terms of energy production:

- **Glycolysis.** During glycolysis, the cell gains two molecules of ATP directly for each glucose molecule broken down anaerobically to pyruvic acid. Two molecules of NADH are also produced. In most cells, electrons are passed from NADH to FAD by means of an intermediate in the intermembrane space and then to CoQ and the electron transport system. This sequence of events ultimately provides an additional four ATP molecules.

- **The Electron Transport System.** The TCA cycle breaks down the 2 pyruvic acid molecules, transferring hydrogen atoms to NADH and $FADH_2$. These coenzymes provide electrons to the ETS; each of the 8 molecules of NADH yields 3 molecules of ATP and 1 water molecule; each of the 2 $FADH_2$ molecules yields 2 ATP molecules and 1 water molecule. Thus, the shuffling from the TCA cycle to the ETS yields 28 molecules of ATP.

- **The TCA Cycle.** Each of the two revolutions of the TCA cycle required to break down both pyruvic acid molecules completely yields one molecule of ATP by way of GTP. This cycling provides an additional gain of two molecules of ATP.

Summing up, for each glucose molecule processed, the cell gains 36 molecules of ATP: 2 from glycolysis, 4 from the NADH generated in glycolysis, 2 from the TCA cycle (by means of GTP), and a total of 28 from the ETS.

Your cardiac muscle cells and liver cells are able to gain an additional two ATP molecules for each glucose molecule broken down. This gain is accomplished by increasing the energy yield from the NADH generated during glycolysis from four to six ATP molecules. In these cells, each NADH molecule passes electrons to an intermediate that generates NADH, not $FADH_2$, in the mitochondrial matrix. The subsequent transfer of electrons to FMN, CoQ, and the ETS results in the production of six ATP molecules, just as if the two NADH molecules had been generated in the TCA cycle.

Although other nutrients can be broken down to provide substrates for the TCA cycle, carbohydrates require the least processing and preparation. It is not surprising, therefore, that athletes have tried to devise ways of exploiting these compounds as ready sources of energy.

Eating carbohydrates *just before* you exercise does not improve your performance and can decrease your endurance by slowing the mobilization of existing energy reserves. Runners or swimmers preparing for lengthy endurance contests, such as a marathon or 5-km swim, do not eat immediately before com-

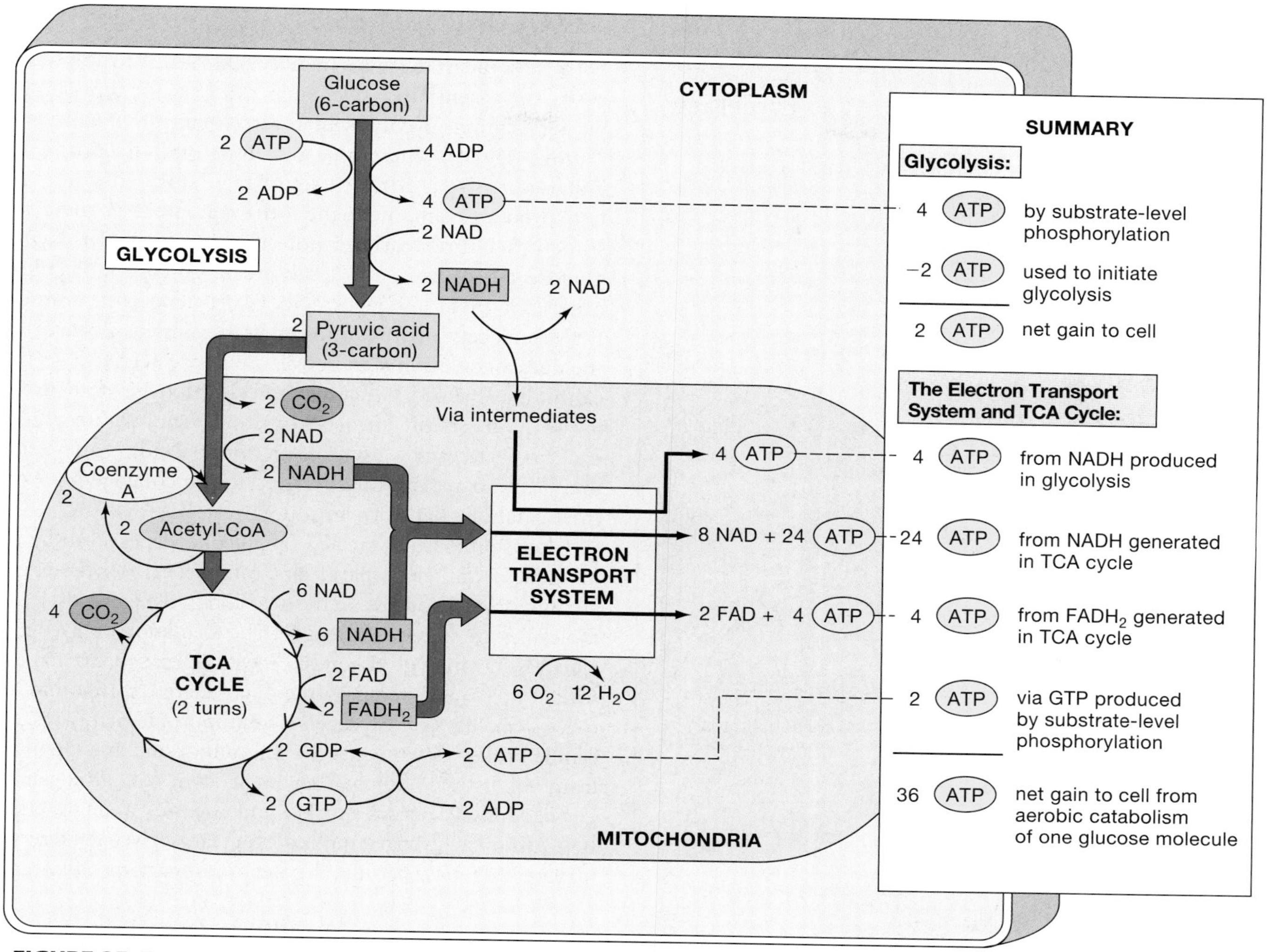

•**FIGURE 25-7**
A Summary of the Energy Yield of Aerobic Metabolism. For each glucose molecule broken down by glycolysis, only two molecules of ATP (net) are produced. However, glycolysis, the formation of acetyl-CoA, and the TCA cycle all yield molecules of reduced coenzymes (NADH or **FADH₂**). Many additional ATP molecules are produced when electrons from these coenzymes pass through the electron transport system. In most cells, each of the two NADH molecules produced in glycolysis provides another two ATP molecules. Each of the eight NADH molecules produced in the mitochondria yields three ATP molecules, for a total of 24. Another two ATP molecules are gained from each of the two **FADH₂** molecules generated in the mitochondria. The TCA cycle generates an additional two ATP molecules in the form of GTP.

peting, and for two hours before the race they limit their intake to drinking water. However, these athletes often eat carbohydrate-rich meals for three days before the event. This **carbohydrate loading** increases the carbohydrate reserves of muscle tissue that will be called on during the competition.

You can obtain maximum effects by exercising to exhaustion for three days before you start a high-carbohydrate diet. This practice, called *carbohydrate depletion/loading*, has a number of potentially unpleasant side effects, including muscle and kidney damage. Sports physiologists recommend that athletes engage in carbohydrate depletion/loading no more than twice a year. Many clinical practitioners advise against the practice altogether, feeling that the potential side effects are too severe to risk for the sake of a competitive advantage.

✓ What is the primary role of the TCA cycle in the production of ATP?

✓ The NADH produced by glycolysis in skeletal muscle fibers leads to the production of two ATP molecules in the mitochondria, but the NADH produced by glycolysis in cardiac muscle cells leads to the production of three ATP molecules. Why?

✓ How would a decrease in the level of cytoplasmic NAD affect ATP production in mitochondria?

Answers start on page Q-1

Review the major metabolic pathways on the **Companion Website:** Chapter 25/Tutorials.

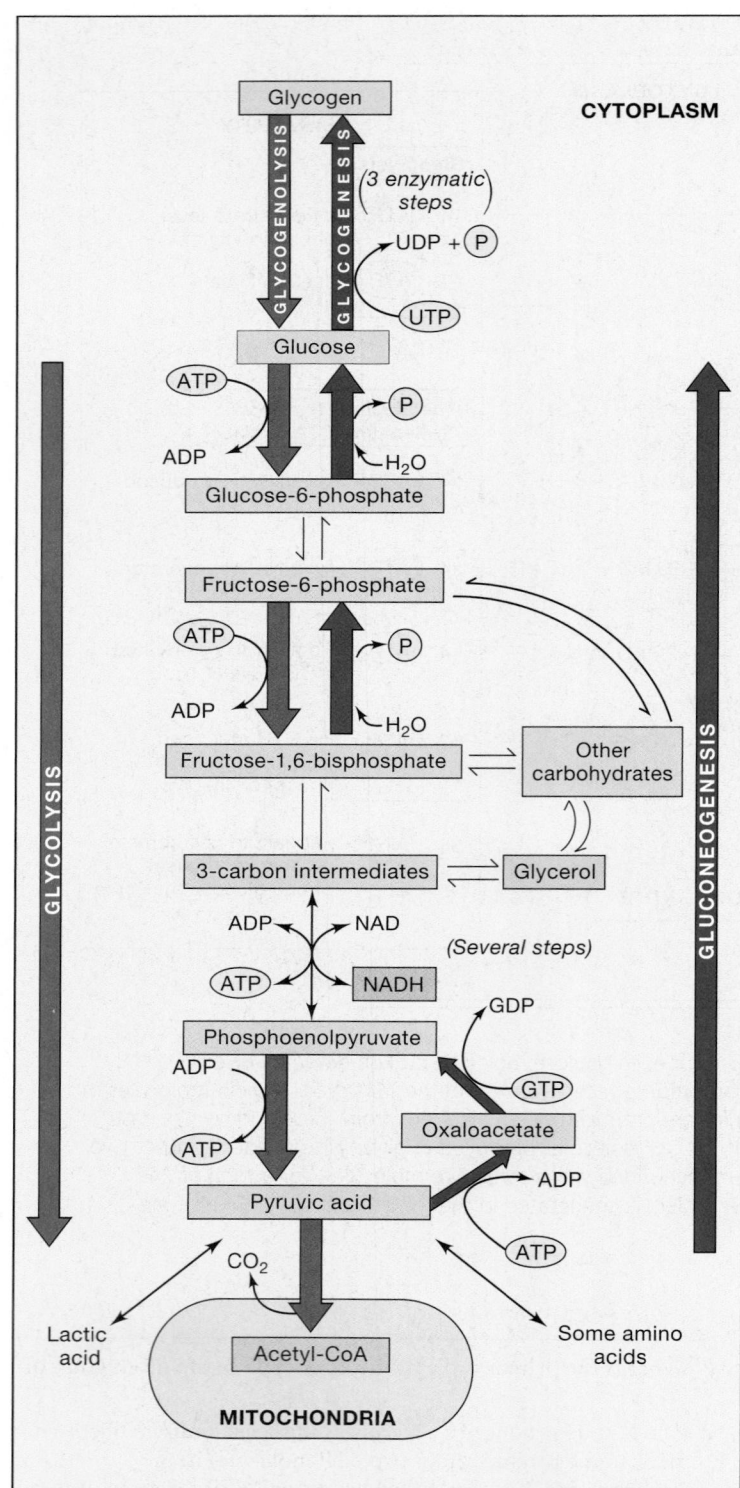

●FIGURE 25–8
Carbohydrate Breakdown and Synthesis.
A flowchart of the pathways for glycolysis and gluconeogenesis.
Many of the reactions are freely reversible, but separate regulatory
enzymes control the key steps, which are indicated by colored
arrows. Some amino acids, carbohydrates other than glucose, and
glycerol can be converted to glucose. The enzymatic reaction that
converts pyruvic acid to acetyl-CoA cannot be reversed.

◻ GLUCONEOGENESIS

Because some of the steps in glycolysis—the breakdown of glucose—are essentially irreversible, cells cannot generate glucose
by performing glycolysis in reverse, using the same enzymes
(Figure 25–8●). Therefore, glycolysis and the production of
glucose involve a different set of regulatory enzymes, and the
two processes are independently regulated. Pyruvic acid or
some other three-carbon molecule can be used to synthesize
glucose. As a result, a cell can create glucose molecules from
other carbohydrates, lactic acid, glycerol, or some amino acids.
However, acetyl-CoA cannot be used to make glucose, because
the decarboxylation step between pyruvic acid and acetyl-CoA
cannot be reversed. **Gluconeogenesis** (gloo-kō-nē-ō-JEN-e-sis)
is the synthesis of glucose from noncarbohydrate precursors,
such as lactic acid, glycerol, or amino acids. Fatty acids and
many amino acids cannot be used for gluconeogenesis, because
their catabolic pathways produce acetyl-CoA.

Glucose molecules created by gluconeogenesis can be used to
manufacture other simple sugars, complex carbohydrates, proteoglycans, or nucleic acids. In the liver and in skeletal muscle, glucose
molecules are stored as glycogen. The formation of glycogen from
glucose, known as **glycogenesis**, is a complex process that involves
several steps and requires the high-energy compound *uridine
triphosphate (UTP)*. Glycogen is an important energy reserve that
can be broken down when the cell cannot obtain enough glucose
from interstitial fluid. The breakdown of glycogen, called
glycogenolysis, occurs quickly and involves a single enzymatic
step. Although glycogen molecules are large, glycogen reserves take
up very little space because they form compact, insoluble granules.

25–3 LIPID METABOLISM

Objectives

- Describe the pathways involved in lipid metabolism.
- Summarize the mechanisms of lipid transport and
  distribution.

Like carbohydrates, lipid molecules contain carbon, hydrogen, and
oxygen, but the atoms are present in different proportions. Triglycerides are the most abundant lipid in the body, so our discussion
will focus on pathways for triglyceride breakdown and synthesis.

◻ LIPID CATABOLISM

During lipid catabolism, or **lipolysis**, lipids are broken down into
pieces that can be either converted to pyruvic acid or channeled
directly into the TCA cycle. A triglyceride is first split into its component parts by hydrolysis. This step yields one molecule of glycerol and three fatty acid molecules. Glycerol enters the TCA cycle
after enzymes in the cytosol convert it to pyruvic acid. The catabolism of fatty acids involves a completely different set of enzymes.

Beta-Oxidation

Fatty acid molecules are broken down into two-carbon fragments in a sequence of reactions known as **beta-oxidation**. This process occurs inside mitochondria, so the carbon chains can enter the TCA cycle immediately. Figure 25–9● diagrams one step in the process of beta-oxidation. Each step generates molecules of acetyl-CoA, NADH, and $FADH_2$, leaving a shorter carbon chain bound to coenzyme A.

Beta-oxidation provides substantial energy benefits. For each two-carbon fragment removed from the fatty acid, the cell gains 12 ATP molecules from the processing of acetyl-CoA in the TCA cycle, plus 5 ATP molecules from the NADH and $FADH_2$. The cell can therefore gain 144 ATP molecules from the breakdown of one 18-carbon fatty acid molecule. This number of ATP molecules yields almost 1.5 times the energy obtained by the complete breakdown of three six-carbon glucose molecules. The catabolism of other lipids follows similar patterns, generally ending with the formation of acetyl-CoA.

Lipids and Energy Production

Lipids are important as an energy reserve because they can provide large amounts of ATP. Since they are insoluble in water, lipids can be stored in compact droplets in the cytosol. This storage method saves space, but when the lipid droplets are large, it is difficult for water-soluble enzymes to get at them. Lipid reserves are therefore more difficult to access than carbohydrate reserves. Also, most lipids are processed inside mitochondria, and mitochondrial activity is limited by the availability of oxygen.

The net result is that lipids cannot provide large amounts of ATP in a short time. However, cells with modest energy demands can shift to lipid-based energy production when glucose supplies are limited. Skeletal muscle fibers normally cycle between lipid metabolism and carbohydrate metabolism. At rest (when energy demands are low), these cells break down fatty acids. During activity (when energy demands are large and immediate), skeletal muscle fibers shift to glucose metabolism.

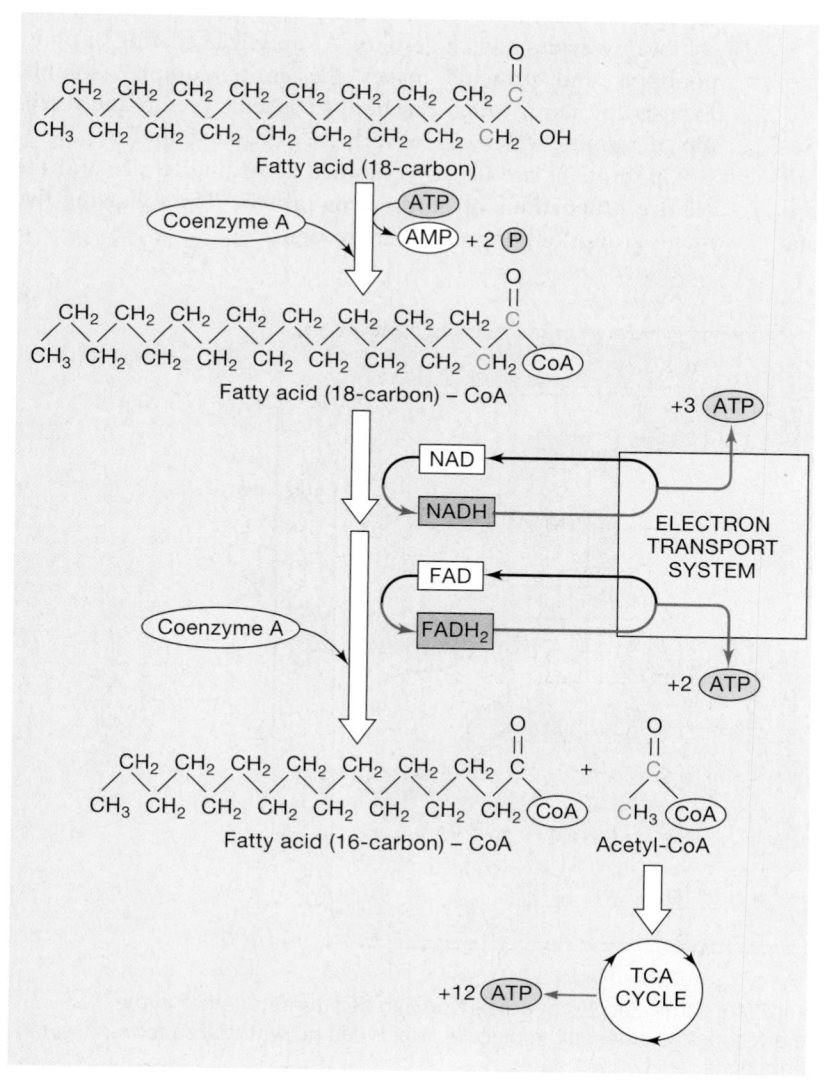

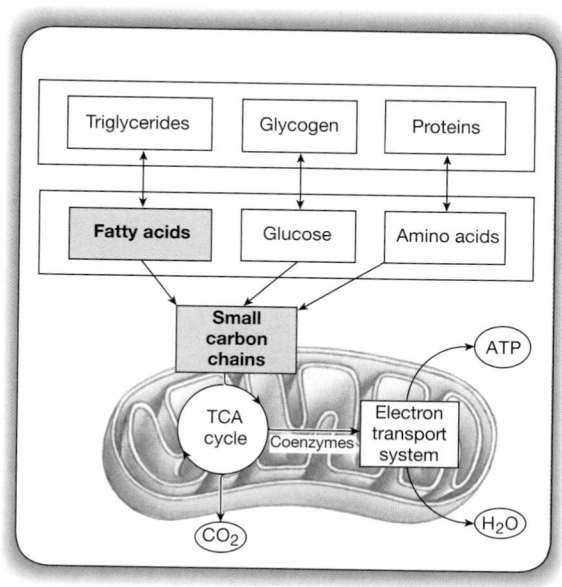

●**FIGURE 25–9**
Beta-Oxidation. During beta-oxidation, the carbon chains of fatty acids are broken down to yield molecules of acetyl-CoA, which can be used in the TCA cycle. The reaction also donates hydrogen atoms to coenzymes, which then deliver them to the electron transport system.

☐ LIPID SYNTHESIS

The synthesis of lipids is known as **lipogenesis** (lip-ō-JEN-e-sis). Figure 25–10● shows the major pathways of this process. Glycerol is synthesized from *dihydroxyacetone phosphate*, one of the intermediate products of glycolysis. The synthesis of most other types of lipids, including nonessential fatty acids and steroids, begins with acetyl-CoA. Lipogenesis can use almost any organic substrate, because lipids, amino acids, and carbohydrates can be converted to acetyl-CoA.

Fatty acid synthesis involves a reaction sequence quite distinct from that of beta-oxidation. Your cells cannot *build* every fatty acid they can break down. **Linoleic acid** and **linolenic acid**, both 18-carbon unsaturated fatty acids, cannot be synthesized. They are called **essential fatty acids**, because they must be included in your diet. Linoleic acid and linolenic acid are synthesized by plants. A deficiency of essential fatty acids generally occurs only among hospitalized individuals receiving nutrients in an intravenous solution. A diet poor in linoleic acid slows growth and alters the appearance of the skin. These fatty acids are also needed to synthesize prostaglandins and some of the phospholipids incorporated into cell membranes throughout the body.

☐ LIPID TRANSPORT AND DISTRIBUTION

Like glucose, lipids are needed throughout the body. All cells require lipids to maintain their cell membranes, and important steroid hormones must reach target cells in many different tissues. Free fatty acids comprise a relatively small percentage of the total circulating lipids. Because most lipids are not soluble in water, special transport mechanisms carry them from one region of the body to another. Most lipids circulate through the bloodstream as lipoproteins.

Free Fatty Acids

Free fatty acids (FFAs) are lipids that can diffuse easily across cell membranes. Free fatty acids in the blood are generally bound to albumin, the most abundant plasma protein. Sources of free fatty acids in the blood include the following:

- Fatty acids that are not used in the synthesis of triglycerides, but that diffuse out of the intestinal epithelium and into the blood.

- Fatty acids that diffuse out of lipid stores, such as the liver and adipose tissue, when triglycerides are broken down. Liver cells, cardiac muscle cells, skeletal muscle fibers, and many other body cells can metabolize free fatty acids, which are an important energy source during periods of starvation, when glucose supplies are limited. **AM** Adaptations to Starvation

Lipoproteins

Lipoproteins are lipid–protein complexes that contain large insoluble glycerides and cholesterol. A superficial coating of phospholipids and proteins makes the entire complex soluble. Exposed proteins, which can bind to specific membrane receptors, determine which cells absorb the associated lipids.

Lipoproteins are usually classified according to size and the relative proportions of lipid versus protein. The following five major groups of lipoproteins are recognized:

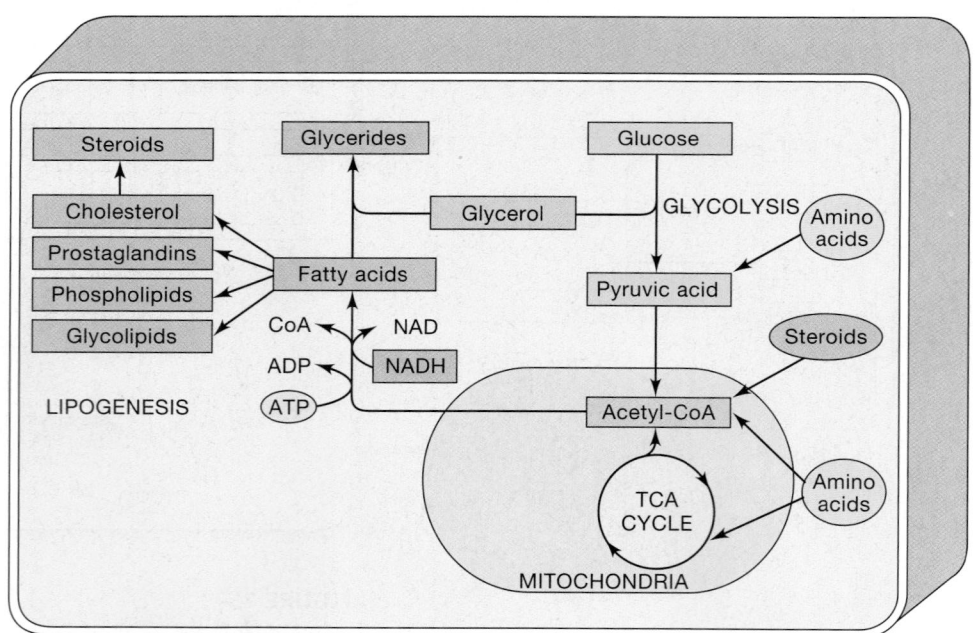

●**FIGURE 25–10**
Lipid Synthesis. Pathways of lipid synthesis begin with acetyl-CoA. Molecules of acetyl-CoA can be strung together in the cytosol, yielding fatty acids, which can be used to synthesize glycerides or other lipid molecules. Lipids can be synthesized from amino acids or carbohydrates by using acetyl-CoA as an intermediate.

1. *Chylomicrons.* Roughly 95 percent of the weight of a chylomicron consists of triglycerides. The largest lipoproteins, ranging in diameter from 0.03 to 0.5 μm, chylomicrons are produced by intestinal epithelial cells. ⟨∞⟩ p. 898 Chylomicrons carry absorbed lipids from the intestinal tract to the bloodstream. The other lipoproteins shuttle lipids among various tissues. The liver is the primary source of all other types of lipoproteins.

2. *Very Low-Density Lipoproteins (VLDLs).* Very low-density lipoproteins contain triglycerides manufactured by the liver, plus small amounts of phospholipids and cholesterol. The primary function of VLDLs is to transport these triglycerides to peripheral tissues. The VLDLs range in diameter from 25 to 75 nm (0.025–0.075 μm).

3. *Intermediate-Density Lipoproteins (IDLs).* Intermediate-density lipoproteins are intermediate in size and lipid composition between VLDLs and low-density lipoproteins (LDLs). IDLs contain smaller amounts of triglycerides than do VLDLs and relatively more phospholipids and cholesterol than do LDLs.

4. *Low-Density Lipoproteins (LDLs).* Low-density lipoproteins contain cholesterol, lesser amounts of phospholipids, and very few triglycerides. These lipoproteins, which are about 25 nm in diameter, deliver cholesterol to peripheral tissues. Because the cholesterol may wind up in arterial plaques, LDL cholesterol is often called "bad cholesterol."

5. *High-Density Lipoproteins (HDLs).* High-density lipoproteins, about 10 nm in diameter, have roughly equal amounts of lipid and protein. The lipids are largely cholesterol and phospholipids. The primary function of HDLs is to transport excess cholesterol from peripheral tissues back to the liver for storage or excretion in the bile. Because HDL cholesterol is returning from peripheral tissues and will not cause circulatory problems, it is called "good cholesterol." Actually, applying the terms *good* and *bad* to cholesterol can be misleading, for cholesterol metabolism is complex and variable. (For more details, see the discussion "Dietary Fats and Cholesterol" on p. 945.)

Figure 25–11● diagrams the probable relationships among these lipoproteins. Chylomicrons produced in the intestinal tract reach the venous circulation by entering lymphatic capillaries and traveling through the thoracic duct (Figure 25–11a●). Although chylomicrons are too large to diffuse across capillary walls, the endothelial lining of capillaries in skeletal muscle, cardiac muscle, adipose tissue, and the liver contains the enzyme **lipoprotein lipase**, which breaks down complex lipids. When chylomicrons contact these endothelial walls, enzymatic activity releases fatty acids and monoglycerides that can diffuse across the endothelium and into interstitial fluid.

The liver controls the distribution of other lipoproteins (Figure 25–11b●):

STEP 1: Liver cells synthesize VLDLs for discharge into the bloodstream.

STEP 2: In peripheral capillaries, lipoprotein lipase removes many of the triglycerides from VLDLs, leaving IDLs; the triglycerides are broken down into fatty acids and monoglycerides.

STEP 3: When IDLs reach the liver, additional triglycerides are removed and the protein content of the lipoprotein is altered. This process creates LDLs, which then return to peripheral tissues to deliver cholesterol.

STEP 4: LDLs leave the bloodstream through capillary pores or cross the endothelium by vesicular transport.

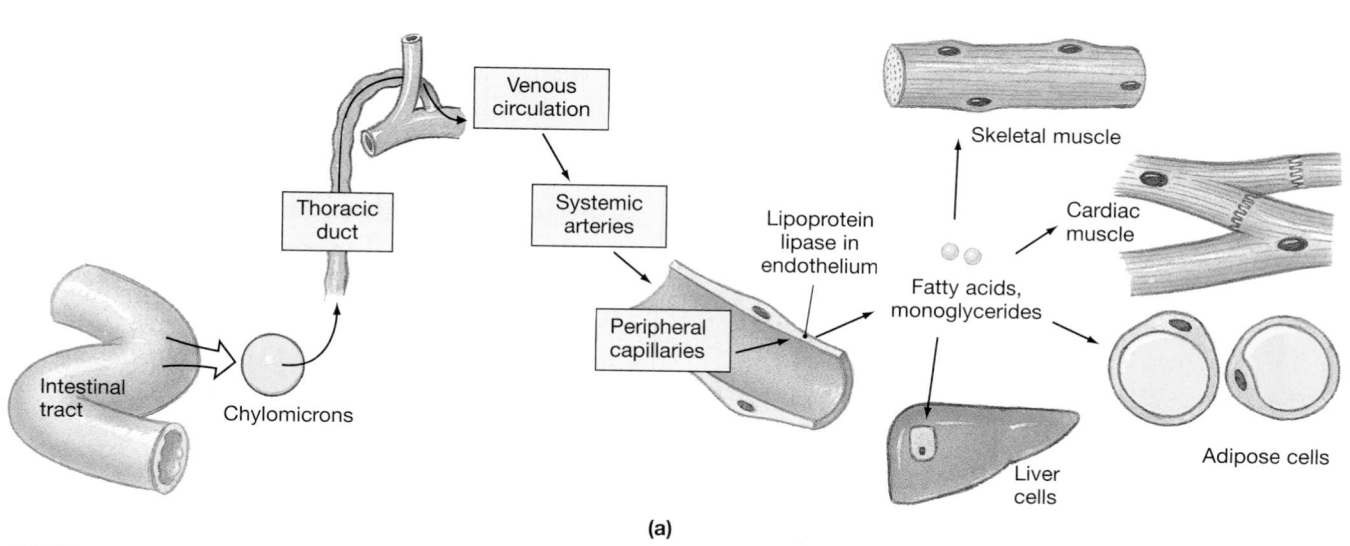

(a)

●**FIGURE 25–11**
Lipid Transport and Utilization. (a) Chylomicrons synthesized at the intestinal epithelium reach the bloodstream by way of the thoracic duct. They are broken down by lipoprotein lipase in capillaries that supply blood to skeletal muscle, cardiac muscle, adipose tissue, and the liver.

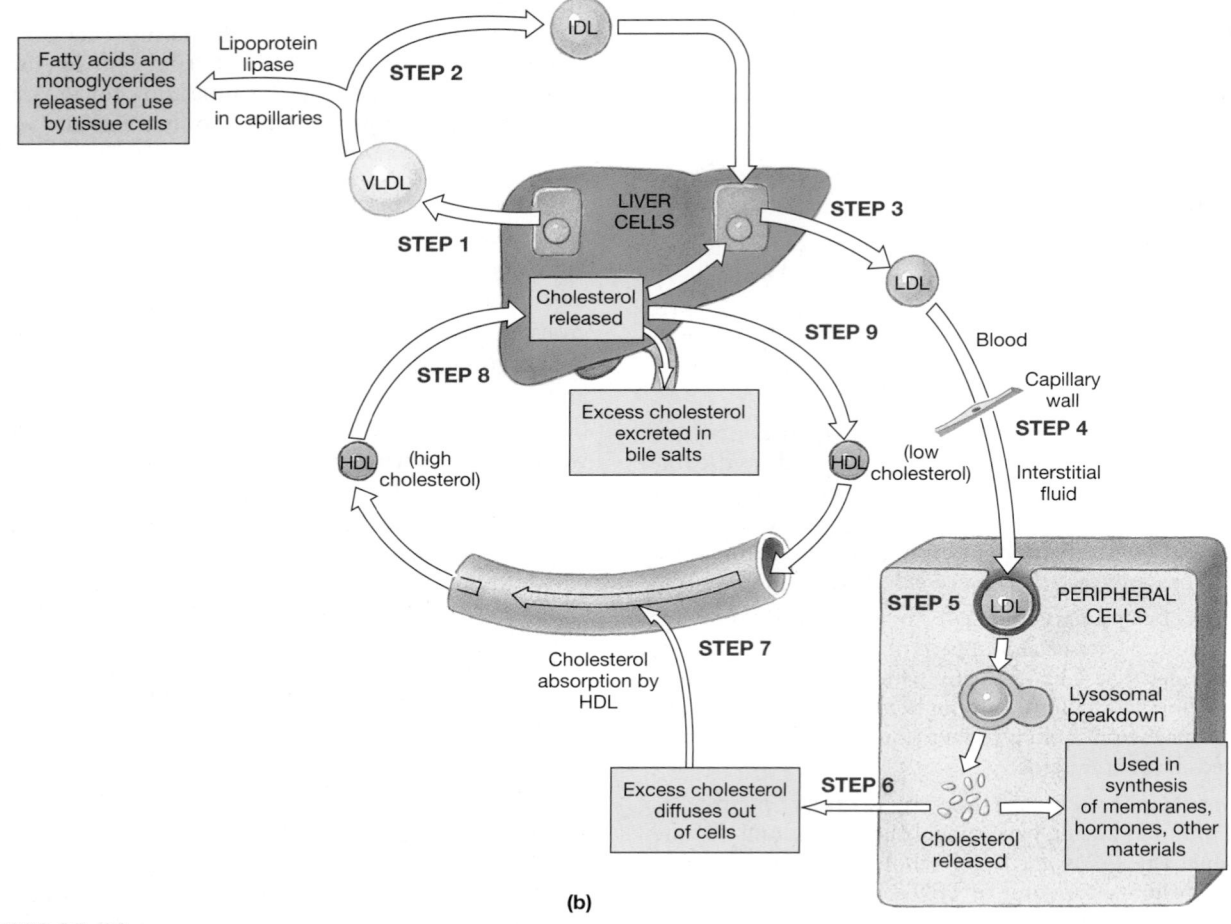

(b)

●**FIGURE 25–11**
Lipid Transport and Utilization. (*continued*) **(b)** (1) Liver cells synthesize VLDLs, which deliver triglycerides to peripheral tissues. (2) Lipoprotein lipase in endothelial cells breaks down these triglycerides and releases fatty acids and monoglycerides, which then diffuse into the surrounding tissues. (3) The IDLs that remain return to the liver, where they are absorbed and converted to LDLs that contain cholesterol. (4) The LDLs circulate to peripheral tissues, cross the endothelium, and are absorbed by cells via endocytosis. (5) Some of the cholesterol is used in cellular processes. (6) The excess cholesterol diffuses back into the bloodstream. (7) In plasma, the cholesterol is absorbed by HDLs produced by the liver. (8) On returning to the liver, the HDLs are absorbed and the cholesterol is extracted. Some of the cholesterol will be exported once again, in LDLs; excess cholesterol is excreted in bile salts. (9) The HDLs stripped of their cholesterol are released into the bloodstream.

STEP 5: Once in peripheral tissues, the LDLs are absorbed by means of receptor-mediated endocytosis. ∞ p. 97 The amino acids and cholesterol then enter the cytoplasm.

STEP 6: The cholesterol not used by the cell in the synthesis of lipid membranes or other products diffuses out of the cell.

STEP 7: The cholesterol then reenters the bloodstream, where it is absorbed by HDLs and returned to the liver.

STEP 8: In the liver, the HDLs are absorbed and their cholesterol is extracted. Some of the cholesterol that is recovered will be used in the synthesis of LDLs, and the rest will be excreted in bile salts.

STEP 9: The HDLs stripped of their cholesterol are released into the bloodstream to travel into peripheral tissues and absorb additional cholesterol.

25–4 PROTEIN METABOLISM

Objective

■ Summarize the main features of protein metabolism and the use of proteins as an energy source.

The body can synthesize 100,000 to 140,000 different proteins with various forms, functions, and structures. Yet, each protein contains some combination of the same 20 amino acids. (Appendix III shows the structures of these amino acids.) Under normal conditions, cellular proteins are continuously recycled. Peptide bonds are broken, and the free amino acids are used in new proteins. This recycling occurs in the cytosol.

Dietary Fats and Cholesterol

Elevated cholesterol levels are associated with the development of atherosclerosis (Chapter 21) and coronary artery disease (Chapter 20). ∞ pp. 725, 695 Current nutritional advice suggests that you reduce cholesterol intake to under 300 mg per day. This amount represents a 40 percent reduction for the average American adult. As a result of rising concerns about cholesterol, such phrases as "low cholesterol," "contains no cholesterol," and "cholesterol free" are now widely used in the advertising and packaging of foods. Cholesterol content alone, however, does not tell the entire story; we must consider some basic information about cholesterol and about lipid metabolism in general:

• **Cholesterol Has Many Vital Functions in the Human Body.** Cholesterol serves as a waterproofing for the epidermis, a lipid component of all cell membranes, a key constituent of bile, and the precursor of several steroid hormones and one vitamin (vitamin D_3). Because cholesterol is so important, dietary restrictions should have the goal of keeping cholesterol levels within acceptable limits. The goal is not to eliminate cholesterol from the diet or from the circulating blood.

• **The Cholesterol Content of the Diet Is Not the Only Source of Circulating Cholesterol.** The human body can manufacture cholesterol from the acetyl-CoA obtained through glycolysis or the beta-oxidation of other lipids. Dietary cholesterol probably accounts for only about 20 percent of the cholesterol in the bloodstream; the rest is the result of metabolism of saturated fats in the diet. If the diet contains an abundance of saturated fats, cholesterol levels in the blood will rise because excess lipids are broken down to acetyl-CoA and used to synthesize cholesterol. Consequently, a person trying to lower his or her serum cholesterol by dietary control must restrict other lipids—especially saturated fats—as well.

• **Genetic Factors Affect Each Individual's Cholesterol Level.** If you reduce your dietary supply of cholesterol, your body will synthesize more to maintain "acceptable" concentrations in the blood. The acceptable level depends on your genetic programming. Because individuals differ in genetic makeup, their cholesterol levels can vary even on similar diets. In virtually all instances, however, dietary restrictions can lower blood cholesterol significantly.

• **Cholesterol Levels Vary with Age and Physical Condition.** At age 19, three out of four males have fasting cholesterol levels (levels measured 8–12 hours after a meal) below 170 mg/dl. Cholesterol levels in females of this age are slightly higher, typically at or below 175 mg/dl. As age increases, the cholesterol values gradually climb; over age 70, the values are 230 mg/dl (males) and 250 mg/dl (females). Cholesterol levels are considered unhealthy if they are higher than those of 90 percent of the population in that age group. For males, this value ranges from 185 mg/dl at age 19 to 250 mg/dl at age 70. For females, the comparable values are 190 mg/dl and 275 mg/dl, respectively.

To determine whether you may need to do anything about your cholesterol level, remember three simple rules:

1. Individuals of any age with total cholesterol values below 200 mg/dl probably do not need to change their lifestyle, unless they have a family history of CAD and atherosclerosis.

2. Those with cholesterol levels between 200 and 239 mg/dl should modify their diet, lose weight (if they are overweight), and have annual checkups.

3. Cholesterol levels over 240 mg/dl warrant drastic changes in dietary lipid consumption, perhaps coupled with drug treatment. Drug therapies are always recommended in cases in which the serum cholesterol level exceeds 240 mg/dl despite dietary modification. Drugs used to lower cholesterol levels include *cholestyramine*, *colestipol*, and *lovastatin*.

When ordering a blood test for cholesterol, most physicians also request information about circulating triglycerides. In fasting individuals, triglycerides are usually present at levels of 40–150 mg/dl. (After a person has consumed a fatty meal, triglyceride levels may be temporarily elevated.)

When cholesterol levels are high, or when an individual has a family history of atherosclerosis or CAD, further tests and calculations may be performed. The HDL level is measured, and the LDL level is calculated as

$$LDL = Cholesterol - HDL - \frac{Triglycerides}{5}$$

A high total cholesterol value linked to a high LDL level spells trouble. In effect, an unusually large amount of cholesterol is being exported to peripheral tissues. Problems can also exist if the individual has high total cholesterol—or even normal total cholesterol—but low HDL levels (below 35 mg/dl). In this case, excess cholesterol delivered to the tissues cannot easily be returned to the liver for excretion. In either event, the amount of cholesterol in peripheral tissues, and especially in arterial walls, is likely to increase. For years, LDL:HDL ratios were taken to be valid predictors of the risk of developing atherosclerosis. Risk-factor analysis and LDL levels are now thought to be more accurate indicators. For males with more than one risk factor, many clinicians recommend dietary and drug therapy if LDL levels exceed 130 mg/dl, regardless of the total cholesterol or HDL levels.

If other energy sources are inadequate, mitochondria can break down amino acids in the TCA cycle to generate ATP. Not all amino acids enter the cycle at the same point, however, so the ATP benefits vary. Nonetheless, the average ATP yield per gram is comparable to that of carbohydrate catabolism.

☐ AMINO ACID CATABOLISM

The first step in amino acid catabolism is the removal of the amino group $(-NH_2)$. This process requires a co-enzyme derivative of **vitamin B_6** (*pyridoxine*). The amino group is removed by *transamination* (tranz-am-i-NĀ-shun) or *deamination* (dē-am-i-NĀ-shun). We shall consider other aspects of amino acid catabolism in a later section.

Transamination

Transamination attaches the amino group of an amino acid to a **keto acid** (Figure 25–12a●), which resembles an amino acid except that the second carbon binds an oxygen atom rather than an amino group. Transamination converts the keto acid into an amino acid that can enter the cytosol and be used for protein synthesis. In the process, the original amino acid becomes a keto acid

that can be broken down in the TCA cycle. Many different tissues perform transaminations, which enable a cell to synthesize many of the amino acids needed for protein synthesis. Cells of the liver, skeletal muscles, heart, lung, kidney, and brain, which are particularly active in protein synthesis, perform many transaminations.

Deamination

Deamination is performed in preparing an amino acid for breakdown in the TCA cycle (Figure 25–12b●). Deamination is the removal of an amino group and a hydrogen atom in a reaction that generates an ammonia (NH_3) molecule or an ammonium ion (NH_4^+). Ammonia molecules are highly toxic, even in low concentrations. Your liver, the primary site of deamination, has the enzymes needed to deal with the problem of ammonia generation. Liver cells convert the ammonia to **urea**, a relatively harmless water-soluble compound excreted in urine. The **urea cycle** is the reaction sequence involved in the production of urea (Figure 25–12c●).

When glucose supplies are low and lipid reserves are inadequate, liver cells break down internal proteins and absorb additional amino acids from the blood. The amino acids are deaminated, and the carbon chains are broken down to provide ATP.

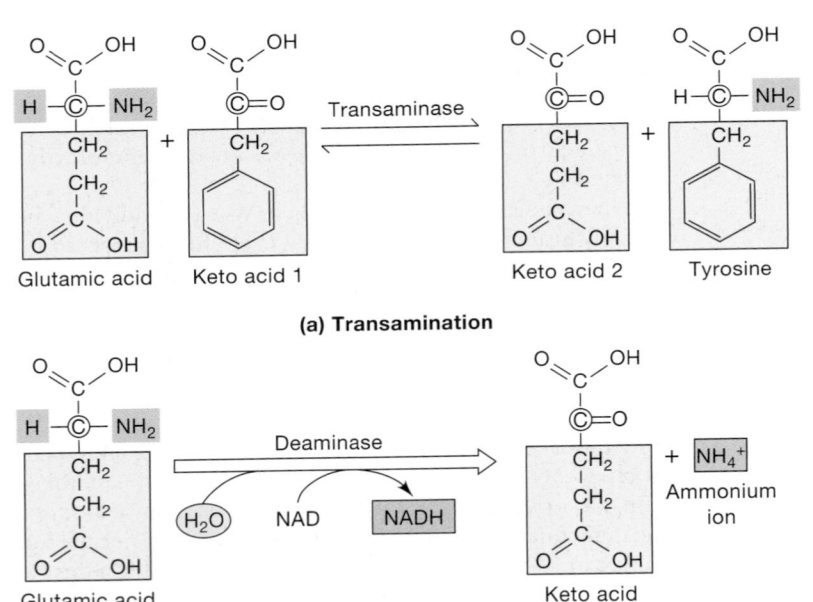

(a) Transamination

(b) Deamination

(c) Urea cycle

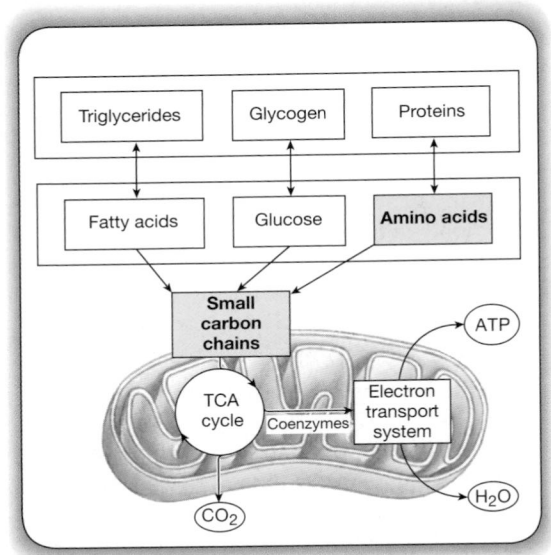

●**FIGURE 25–12**
Amino Acid Catabolism. **(a)** In transamination, an enzyme removes the amino group $(-NH_2)$ from one molecule and attaches it to a keto acid. **(b)** In deamination, an enzyme strips the amino group and a hydrogen atom from an amino acid and produces a keto acid and an ammonium ion. **(c)** The urea cycle takes two metabolic waste products—ammonia and carbon dioxide—and produces a molecule of urea, a relatively harmless, soluble compound that is excreted in the urine.

Proteins and ATP Production

Three factors make protein catabolism an impractical source of quick energy:

1. Proteins are more difficult to break apart than are complex carbohydrates or lipids.

2. One of the by-products, ammonia, is a toxin that can damage cells.

3. Proteins form the most important structural and functional components of any cell. Extensive protein catabolism therefore threatens homeostasis at the cellular and systems levels.

Several inherited metabolic disorders result from an inability to produce specific enzymes involved with amino acid metabolism. *Phenylketonuria* (fen-il-kē-tō-NOO-rē-uh), or PKU, is an example. Individuals with PKU cannot convert phenylalanine to tyrosine, due to a defect in the enzyme *phenylalanine hydroxylase*. This reaction is an essential step in the synthesis of norepinephrine, epinephrine, dopamine, and melanin. If PKU is not detected in infancy, central nervous system development is inhibited and severe brain damage results. **AM** Phenylketonuria

☐ PROTEIN SYNTHESIS

We discussed the basic mechanism for protein synthesis in Chapter 3 (Figures 3–16 and 3–17•, pp. 87, 88–89). Your body can synthesize roughly half of the various amino acids needed to build proteins. There are 10 **essential amino acids**. Your body cannot synthesize eight of them (*isoleucine, leucine, lysine, threonine, tryptophan, phenylalanine, valine,* and *methionine*); the other two (*arginine* and *histidine*) can be synthesized, but in amounts that are insufficient for growing children. Because the body can make other amino acids on demand, they are called the **nonessential amino acids**. Your body can readily synthesize the carbon frameworks of the nonessential amino acids, and a nitrogen group can be added by **amination**—the attachment of an amino group (Figure 25–13•)—or by transamination.

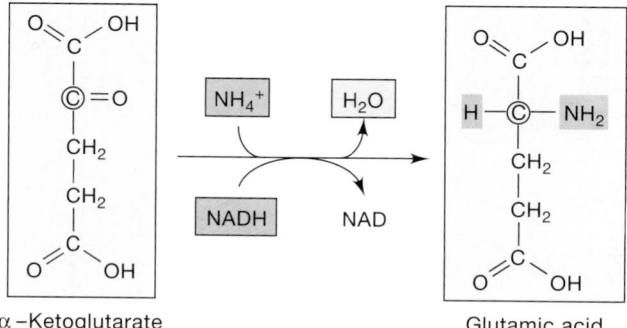

α –Ketoglutarate   Glutamic acid

•**FIGURE 25–13**
Amination. Amination attaches an amino group to a keto acid. This is an important step in the synthesis of nonessential amino acids. Amino groups can also be attached through transamination. (See Figure 25–12a.)

Protein deficiency diseases develop when an individual does not consume adequate amounts of all essential amino acids. All amino acids must be available if protein synthesis is to occur. Every transfer RNA molecule must appear at the active ribosome at the proper time, bearing its individual amino acid. If that does not happen, the entire process comes to a halt. Regardless of its energy content, if the diet is deficient in essential amino acids, the individual will be malnourished to some degree. Examples of protein deficiency diseases include *marasmus, kwashiorkor,* and *pellagra*. More than 100 million children worldwide have symptoms of these disorders, although neither condition is common in the United States today. **AM** Protein Deficiency Diseases

25–5 NUCLEIC ACID METABOLISM

Objective

■ Summarize the main features of nucleic acid metabolism.

Cells contain both DNA and RNA. In Chapter 3, we considered the replication of DNA, the mechanics of cell division, and the importance of DNA in regulating the structural and functional characteristics of the cell. The genetic information contained in the DNA of the nucleus is essential to the cell's long-term survival. As a result, the DNA in the nucleus is never catabolized for energy, even if the cell is dying of starvation. But the RNA in the cell is involved in protein synthesis, and RNA molecules are broken down and replaced regularly.

☐ RNA CATABOLISM

In the breakdown of RNA, the bonds between nucleotides are broken and the molecule is disassembled into individual nucleotides. Although the nucleotides are usually recycled into new nucleic acids, they can be catabolized to simple sugars and nitrogenous bases.

RNA catabolism makes a relatively insignificant contribution to the total energy budget of the cell. Proteins account for 30 percent of the cell's weight, and much more energy can be provided through the catabolism of nonessential proteins. Even when RNA is broken down, only the sugars and pyrimidines provide energy. The sugars can enter the glycolytic pathways. The pyrimidines (cytosine and uracil, in RNA) are converted to acetyl-CoA and metabolized through the TCA cycle. The purines (adenine and guanine) cannot be catabolized. Instead, they are deaminated and excreted as **uric acid**. Like urea, uric acid is a relatively nontoxic waste product, but it is far less soluble than urea. Urea and uric acid are called **nitrogenous wastes**, because they are waste products that contain nitrogen atoms.

Normal uric acid concentrations in plasma average 2.7–7.4 mg/dl, depending on gender and age. When plasma concentrations exceed 7.4 mg/dl, *hyperuricemia* (hī-per-ū-ri-SĒ-mē-uh) exists. This condition may affect 18 percent of the U.S. population. At concentrations over 7.4 mg/dl, body fluids are saturated with uric acid. Although symptoms may not appear at once, uric acid crystals may begin to form in body fluids. The condition that then develops is called *gout*. Most cases of hyperuricemia and gout are linked to problems with the excretion of uric acid by the kidneys. **AM** Gout

■ NUCLEIC ACID SYNTHESIS

Most cells synthesize RNA; DNA synthesis occurs only in cells preparing for mitosis and cell division or for meiosis (nuclear events involved in gamete production). ∞ p. 100 Messenger RNA (mRNA), transfer RNA (tRNA), and ribosomal RNA (rRNA) are transcribed by different forms of RNA polymerase:

- Messenger RNA is manufactured only when specific genes are activated. A strand of mRNA has a life span of minutes to hours.

- Ribosomal RNA and tRNA are more durable than mRNA strands. For example, the average life span of a strand of rRNA is just over five days. However, because a typical cell contains roughly 100,000 ribosomes and many times that number of tRNA molecules, their replacement involves a considerable amount of synthetic activity.

Figure 25–14● summarizes the metabolic pathways for lipids, carbohydrates, and proteins. The diagram follows the reactions in a "typical" cell. Yet no one cell can perform all the anabolic and catabolic operations and interconversions required by the body as a whole. As differentiation proceeds, each type of cell develops

its own complement of enzymes that determines the cell's metabolic capabilities. In the presence of such cellular diversity, homeostasis can be preserved only when the metabolic activities of tissues, organs, and organ systems are coordinated.

✓ How would a diet that is deficient in pyridoxine (vitamin B_6) affect protein metabolism?

✓ Elevated levels of uric acid in the blood can be an indicator of increased metabolism of which organic compound?

✓ Why are high-density lipoproteins (HDLs) considered beneficial?

Answers start on page Q-1

 Reinforce your understanding of the major metabolic pathways by visiting the **Companion Website:** Chapter 25/Reviewing Facts and Terms/Multiple Choice.

25–6 METABOLIC INTERACTIONS

Objective

- Differentiate between the absorptive and postabsorptive metabolic states, and summarize the characteristics of each.

The nutrient requirements of each tissue vary with the types and quantities of enzymes present in the cytoplasm. From a metabolic

●**FIGURE 25–14**
A Summary of the Pathways of Catabolism and Anabolism. An overview of major catabolic and anabolic pathways for lipids, carbohydrates, and proteins.

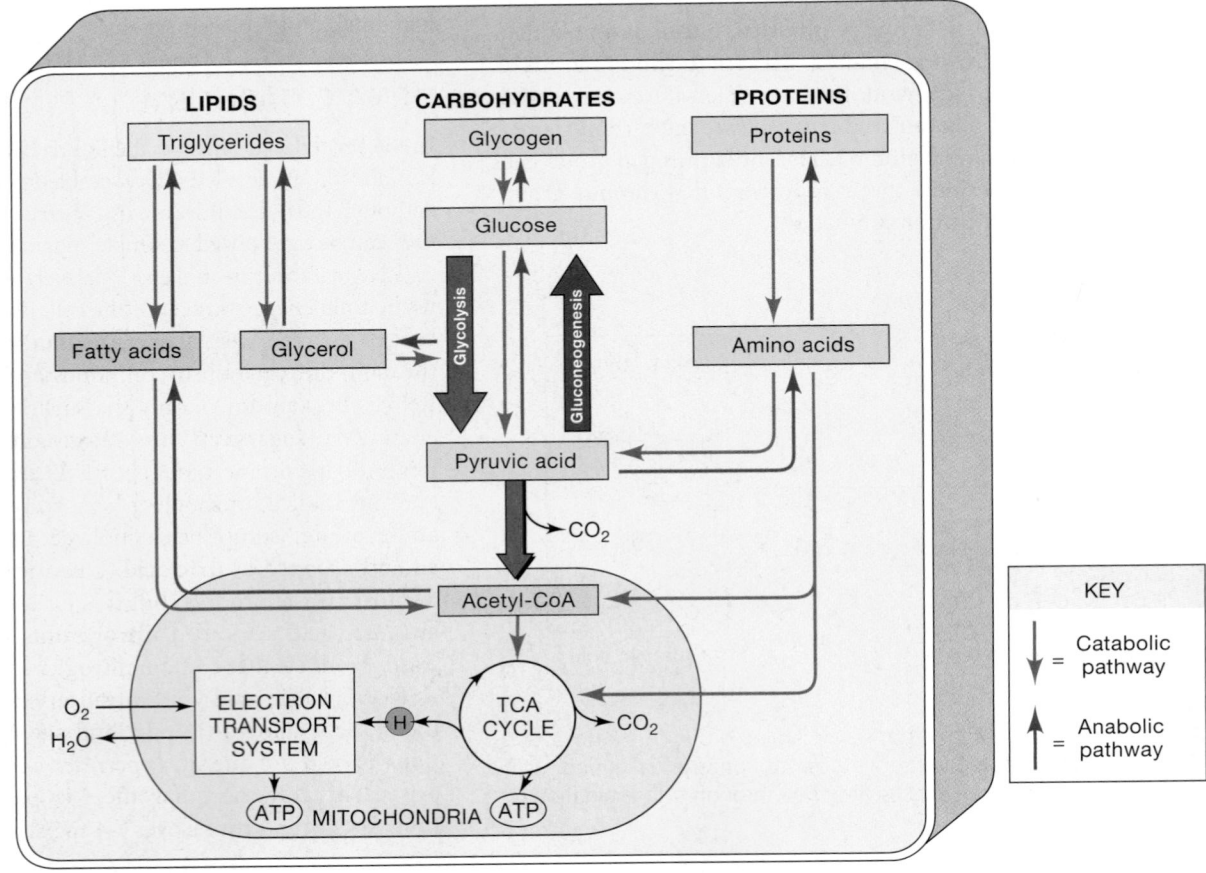

standpoint, we can consider the body in terms of five distinctive components: the liver, adipose tissue, skeletal muscle, neural tissue, and other peripheral tissues:

1. **The Liver.** The liver is the focal point of metabolic regulation and control. Liver cells contain a great diversity of enzymes and so can break down or synthesize most of the carbohydrates, lipids, and amino acids needed by other body cells. Liver cells have an extensive circulatory supply, so they are in an excellent position to monitor and adjust the nutrient composition of circulating blood. The liver also contains significant energy reserves in the form of glycogen deposits.

2. **Adipose Tissue.** Adipose tissue stores lipids, primarily as triglycerides. Adipocytes are located in many areas; in previous chapters, we noted the presence of fat cells in areolar tissue, in mesenteries, within red and yellow marrows, in the epicardium, and around the eyes.

3. **Skeletal Muscle.** Skeletal muscle accounts for almost half of an individual's body weight, and skeletal muscle fibers maintain substantial glycogen reserves. In addition, their contractile proteins can be broken down and the amino acids used as an energy source if other nutrients are unavailable.

4. **Neural Tissue.** Neural tissue has a high demand for energy, but the cells do not maintain reserves of carbohydrates, lipids, or proteins. Neurons must be provided with a reliable supply of glucose, because they are generally unable to metabolize other molecules. If blood glucose levels become too low, neural tissue in the central nervous system cannot continue to function, and the individual falls unconscious.

5. **Other Peripheral Tissues.** Other peripheral tissues do not maintain large metabolic reserves, but they are able to metabolize glucose, fatty acids, or other substrates. Their preferred source of energy varies with the instructions provided by the endocrine system.

The interrelationships among these five components can best be understood by considering the events that occur over a 24-hour period. During this time, the body experiences two broad patterns of metabolic activity: (1) the *absorptive state* and (2) the *postabsorptive state*.

☐ THE ABSORPTIVE STATE

The **absorptive state** is the period following a meal, when nutrient absorption is under way. After a typical meal, the absorptive state continues for about 4 hours. If you are fortunate enough to eat three meals a day, you spend 12 out of every 24 hours in the absorptive state.

A typical meal contains proteins, lipids, and carbohydrates in varying proportions. While you are in the absorptive state, your intestinal mucosa busily absorbs these nutrients. Glucose and amino acids enter the bloodstream, and the hepatic portal vein carries them to the liver. Most of the absorbed fatty acids are packaged in chylomicrons, which enter the lacteals.

Some of the carbohydrates, lipids, and amino acids are broken down at once to provide energy for cellular operations. The remainder is stored, lessening the impact of future shortages. Insulin is the primary hormone of the absorptive state, although various other hormones stimulate amino acid uptake (growth hormone, or GH) and protein synthesis (GH, androgens, and estrogens). We shall now consider the activities under way at specific sites, with particular reference to Table 25–1 and Figure 25–15●.

TABLE 25–1 REGULATORY HORMONES AND THEIR EFFECTS ON PERIPHERAL METABOLISM

Hormone	Effect on General Peripheral Tissues	Selective Effects on Target Tissues
ABSORPTIVE STATE		
Insulin	Increased glucose uptake and utilization	*Liver*: Glycogenesis *Adipose tissue*: Lipogenesis *Skeletal muscle*: Glycogenesis
Insulin and growth hormone	Increased amino acid uptake and protein synthesis	*Skeletal muscle*: Fatty acid catabolism
Androgens, estrogens	Increased amino acid use in protein synthesis	*Skeletal muscle*: Muscle hypertrophy (especially androgens)
POSTABSORPTIVE STATE		
Glucagon		*Liver*: Glycogenolysis
Epinephrine		*Liver*: Glycogenolysis *Adipose tissue*: Lipolysis
Glucocorticoids	Decreased use of glucose; increased reliance on ketone bodies and fatty acids	*Liver*: Glycogenolysis *Adipose tissue*: Lipolysis, gluconeogenesis *Skeletal muscle*: Glycogenolysis, protein breakdown, amino acid release
Growth hormone	Complements effects of glucocorticoids	Acts with glucocorticoids

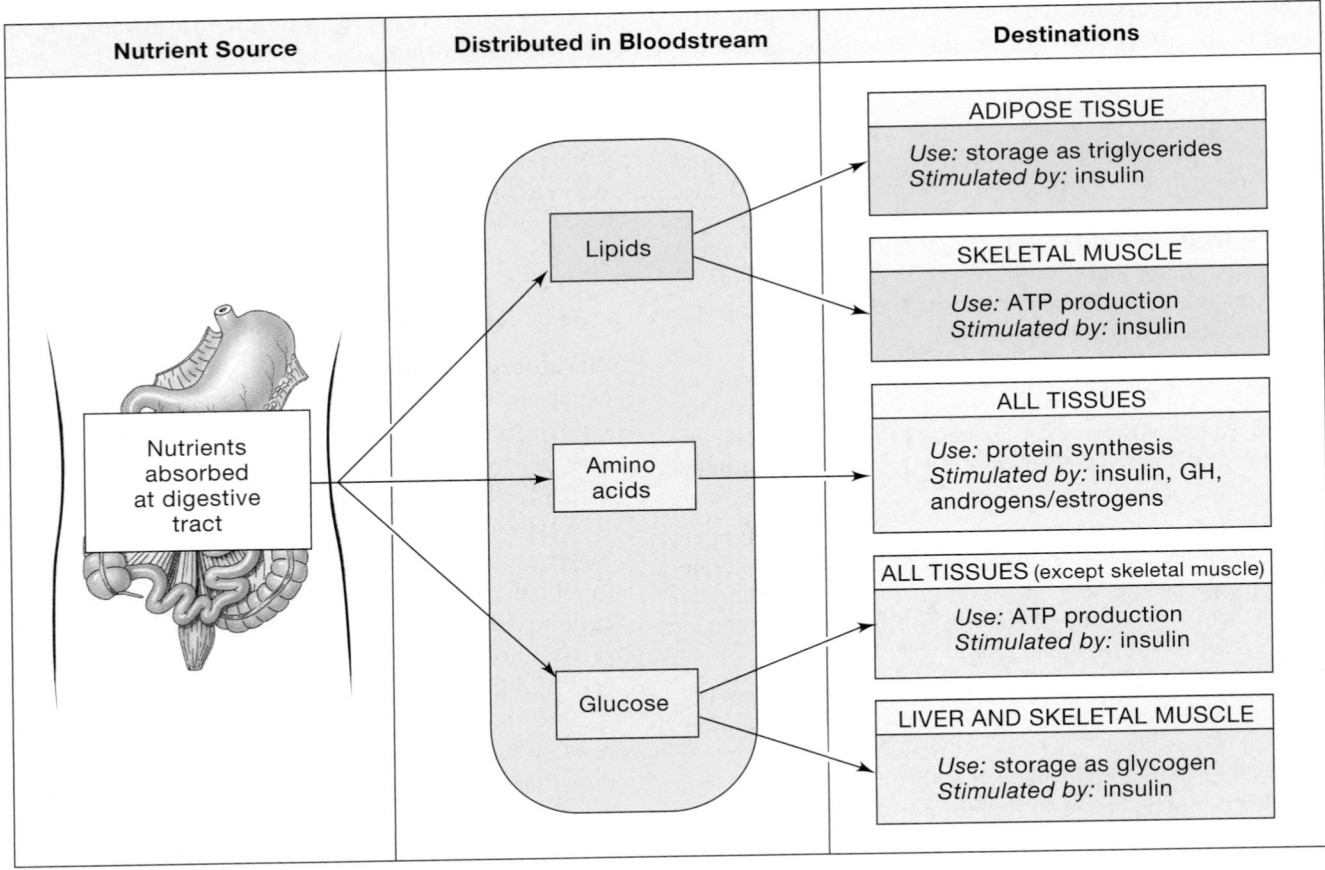

●**FIGURE 25–15**
The Absorptive State. During the absorptive state, the primary metabolic goal is anabolic activity, especially growth and the storage of energy reserves.

The Liver

The liver regulates the levels of glucose and amino acids in the blood arriving in the hepatic portal vein before that blood reaches the inferior vena cava. Despite the continuous absorption of glucose at the intestinal mucosa, blood glucose levels do not skyrocket, because the liver cells, under insulin stimulation, remove glucose from blood delivered by the hepatic portal circulation. Blood glucose levels do rise, but only from about 90 mg/dl to perhaps 150 mg/dl, even after a meal rich in carbohydrates. The liver uses some of the absorbed glucose to generate the ATP required to perform synthetic operations. Glycogenesis (glycogen formation) continues until glycogen accounts for about 5 percent of the total liver weight. If excess glucose remains in the bloodstream, hepatocytes (liver cells) use glucose to synthesize triglycerides. Although small quantities of lipids are normally stored in the liver, most of the synthesized triglycerides are bound to transport proteins and released into the bloodstream as VLDLs. Peripheral tissues—primarily adipose tissues—then absorb these lipids for storage.

The liver does not control circulating levels of amino acids as precisely as it does glucose concentrations. Plasma amino acid levels normally range between 35 and 65 mg/dl, but they may become elevated after a protein-rich meal. The absorbed amino acids are used to support the synthesis of proteins, including plasma proteins and the proenzymes of the clotting system. Liver cells can also synthesize many amino acids, and an amino acid present in abundance may be converted to another, less common type and released into the bloodstream.

Most of the lipids absorbed by the digestive tract do not reach the liver. Triglycerides, cholesterol, and large fatty acids reach the general venous circulation in chylomicrons that are transported in the thoracic duct. Most of these lipids are absorbed by other tissues.

Adipose Tissue

During the absorptive state, adipocytes remove fatty acids and glycerol from the bloodstream. Lipids continue to be removed from the blood for 4–6 hours after you have eaten a fatty meal. Over this period, the presence of chylomicrons may give the plasma a milky appearance, a characteristic called **lipemia** (lip-Ē-mē-uh).

Adipocytes are particularly active in absorbing these lipids and in synthesizing new triglycerides for storage. At normal blood glucose concentrations, any glucose entering adipocytes will be catabolized to provide the energy needed for lipogenesis (lipid synthesis). Adipocytes also absorb amino acids as needed for protein synthesis. Although these cells can use glucose or amino acids to manufacture triglycerides, they do so only if circulating concentrations are unusually high.

If, on a daily basis, you take in more nutrients during the absorptive state than you catabolize during the postabsorptive state, the fat deposits in your adipose tissue will enlarge. Most of the enlargement represents an increase in the size of individual adipocytes. An increase in the total number of adipocytes does not ordinarily occur, except in children before puberty and in extremely obese adults.

A useful definition of **obesity** is "20 percent above ideal weight," because that is the point at which serious health risks appear. On the basis of that criterion, 20–30 percent of men and 30–40 percent of women in the United States can be considered obese. Simply stated, obese individuals are taking in more food energy than they use. Unfortunately, there is very little agreement as to the underlying cause of this situation. The two major categories of obesity—*regulatory obesity* and *metabolic obesity*—are described in the *Applications Manual.* **AM** Body Mass Index and Obesity

Skeletal Muscle, Neural Tissue, and Other Peripheral Tissues

When blood glucose and amino acid concentrations are elevated, insulin is released from the pancreatic islets, and all tissues increase their rates of absorption and utilization. Glucose is catabolized for energy, and the amino acids are used to build proteins.

Glucose is normally retained in the body, because the kidneys prevent the loss of glucose molecules in urine. The kidneys' ability to conserve glucose breaks down only when blood glucose concentrations are extraordinarily high, somewhere in excess of 180 mg/dl. The amino acid content of urine is not as carefully regulated; amino acids commonly appear in the urine after a protein-rich meal.

When blood glucose levels are elevated, most cells ignore the circulating lipids, so the adipocytes have little competition. In resting skeletal muscles, a significant portion of the metabolic demand is met through the catabolism of fatty acids. Glucose molecules are used to build glycogen reserves, which may account for 0.5–1 percent of the weight of each muscle fiber.

☐ THE POSTABSORPTIVE STATE

The **postabsorptive state** is the period when nutrient absorption is not under way and your body must rely on internal energy reserves to continue meeting its energy demands. You spend roughly 12 hours each day in the postabsorptive state, although a person who is skipping meals can extend that time considerably. Metabolic activity in the postabsorptive state is focused on the mobilization of energy reserves and the maintenance of normal blood glucose levels. These activities are coordinated by several hormones, including glucagon, epinephrine, glucocorticoids, and growth hormone (Table 25–1).

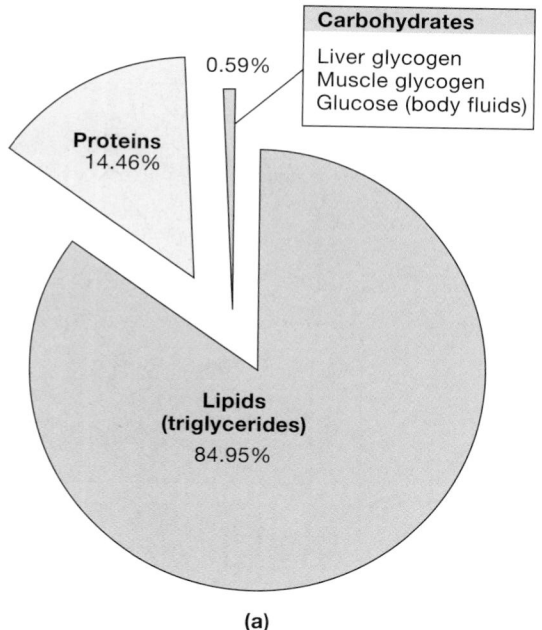

(a)

●**FIGURE 25–16**
Metabolic Reserves. The distribution of the estimated metabolic reserves of a 70-kg individual.

The metabolic reserves of a typical 70-kg (154-lb) individual include carbohydrates, lipids, and proteins (Figure 25–16●). Due to its high energy content, the adipose tissue represents a disproportionate percentage of the total reserve in the form of triglycerides. Most of the available protein reserve is located in the contractile proteins of skeletal muscle. Carbohydrate reserves are relatively small and sufficient for only a few hours or, at most, overnight.

We shall now examine the events under way in specific tissues and organs during the postabsorptive state, as diagrammed in Figure 25–17●.

The Liver

As the absorptive state ends, your intestinal cells stop providing glucose to the portal circulation. At first, the peripheral tissues continue to remove glucose from the blood, and blood glucose levels begin to decline. The liver responds by reducing its synthetic activities. When plasma concentrations fall below 80 mg/dl, liver cells begin breaking down glycogen reserves and releasing glucose into the bloodstream. This glycogenolysis occurs in response to a rise in circulating levels of glucagon and epinephrine. The liver contains 75–100 g of glycogen that is readily available, a reserve that is adequate to maintain blood glucose levels for about four hours.

As glycogen reserves decline and plasma glucose levels fall to about 70 mg/dl, liver cells begin to make glucose in an attempt to stabilize blood glucose levels. The shift from glycogenolysis to gluconeogenesis occurs under stimulation by *glucocorticoids,* steroid hormones from the adrenal cortex. p. 629

●**FIGURE 25–17 The Postabsorptive State.** In the postabsorptive state, energy reserves are mobilized, and peripheral tissues (except neural tissues) shift from glucose catabolism to fatty acid or ketone body catabolism to obtain energy.

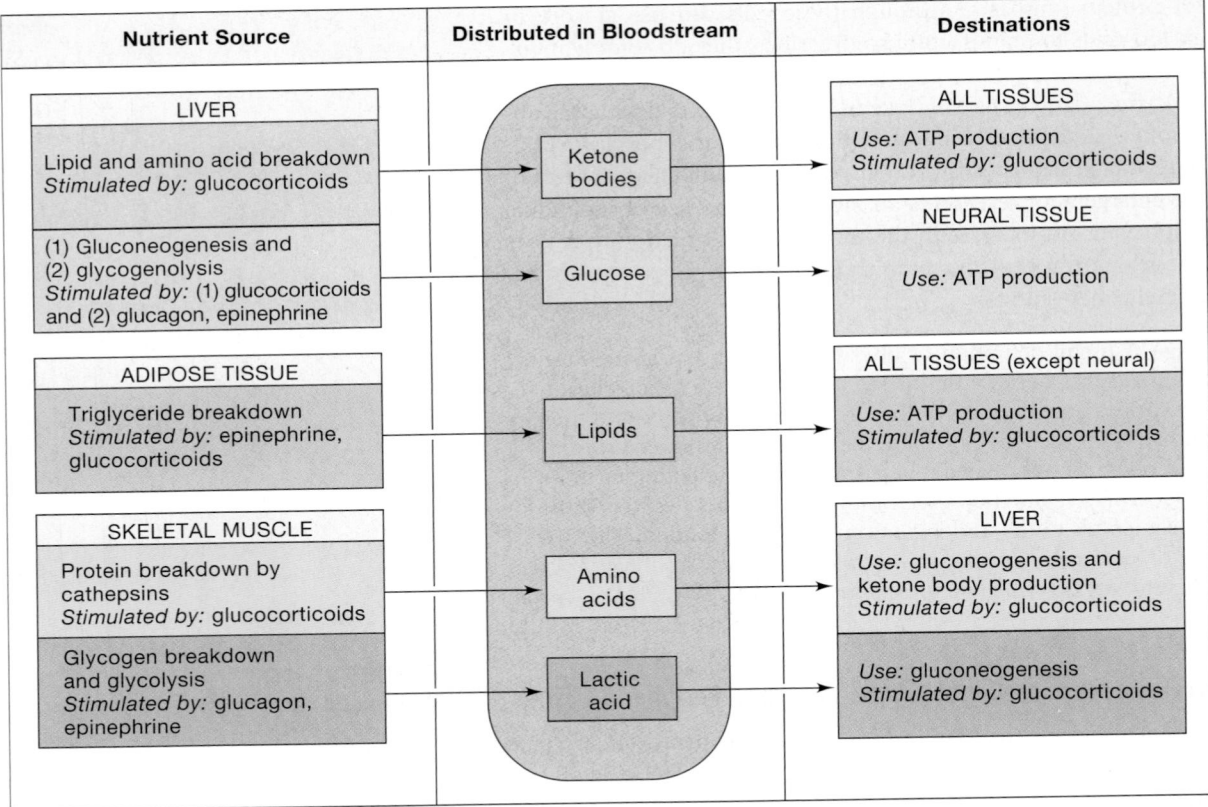

GLUCONEOGENESIS By means of gluconeogenesis, liver cells synthesize glucose molecules from smaller carbon fragments. In effect, any carbon fragment that can be converted to pyruvic acid or one of the three-carbon compounds involved in glycolysis in the cytoplasm can be used to synthesize glucose. (We discussed the conversion of lactic acid to glucose in the liver in Chapter 10. ∞ p. 319) With glucose already in short supply, lipids and amino acids must be catabolized to provide the ATP molecules needed for these syntheses.

UTILIZATION OF LIPIDS In the postabsorptive state, your liver absorbs fatty acids and glycerol from the blood. The glycerol molecules are converted to glucose. Fatty acids are broken down through beta-oxidation to produce large quantities of acetyl-CoA. However, because the enzymatic reaction that converts pyruvic acid to acetyl-CoA cannot be reversed, acetyl-CoA cannot be used to synthesize glucose. Instead,

- Some of the acetyl-CoA molecules deliver their two-carbon acetyl fragments to the TCA cycle, during which they are broken down. The ATP generated can then be used to support gluconeogenesis.

- In addition, some of the molecules of acetyl-CoA are converted to special compounds that can be utilized by peripheral tissues. These compounds, called **ketone bodies**, are organic acids that are also produced during the catabolism of amino acids.

UTILIZATION OF AMINO ACIDS Before an amino acid can be used for either gluconeogenesis or energy production by means of break-

down in the TCA cycle, the amino group ($-NH_2$) must be removed. The structure of the remaining carbon chain determines its subsequent fate. After deamination, some amino acids can be converted to molecules of pyruvic acid or to one of the intermediary molecules of the TCA cycle. Other amino acids—including most of the essential amino acids—can be converted only to acetyl-CoA and must be either broken down further or converted to ketone bodies.

The liver is the most active site of amino acid breakdown. There, the ammonia generated by deamination is converted to urea. This relatively harmless, water-soluble compound circulates in the bloodstream until it is excreted in urine. The urea concentration of blood rises during the postabsorptive period, because the rate of amino acid catabolism increases.

KETONE BODIES During the postabsorptive state, liver cells conserve glucose and break down lipids and amino acids. Both lipid catabolism and amino acid catabolism generate acetyl-CoA. As the concentration of acetyl-CoA rises, ketone bodies begin to form. There are three such compounds: (1) **acetoacetate** (as-ē-tō-AS-e-tāt), (2) **acetone** (AS-e-tōn), and (3) **betahydroxybutyrate** (bā-ta-hī-droks-ē-BŪ-te-rāt). Liver cells do not metabolize any of the ketone bodies, and these compounds diffuse through the cytoplasm and into the general circulation. Cells in peripheral tissues then absorb the ketone bodies and reconvert them to acetyl-CoA for introduction into the TCA cycle.

The normal concentration of ketone bodies in the blood is about 30 mg/dl, and very few of these compounds appear in urine. During even a brief period of fasting, the increased production of ketone bodies results in **ketosis** (kē-TŌ-sis), a high concentration of ketone bodies in body fluids. In ketosis, the concentration of ketone bodies is elevated in blood, a condition called **ketonemia** (kē-tō-NĒ-mē-uh), and in urine, a condition called **ketonuria** (kē-tō-NOO-rē-uh). Ketonemia and ketonuria are clear indications that the catabolism of proteins and lipids is under way. Acetone, which diffuses out of the pulmonary capillaries and into the alveoli very readily, can be smelled on the individual's breath.

SUMMARY During the postabsorptive state, your liver attempts to stabilize blood glucose concentrations, first by the breakdown of glycogen reserves and later by gluconeogenesis. Over the remainder of the postabsorptive state, the combination of lipid and amino acid catabolism provides the necessary ATP and generates large quantities of ketone bodies that diffuse into the bloodstream.

A ketone body is an acid that dissociates in solution, releasing a hydrogen ion. As a result, the appearance of ketone bodies in the bloodstream—ketonemia—is a threat to the plasma pH, which must be controlled by buffers. During prolonged starvation, ketone levels continue to rise. Eventually, buffering capacities are exceeded and a dangerous drop in pH occurs. This acidification of the blood is called **ketoacidosis** (kē-tō-as-i-DŌ-sis). In severe ketoacidosis, the circulating concentration of ketone bodies can reach 200 mg/dl, and the pH may fall below 7.05. A pH that low can disrupt tissue activities and cause coma, cardiac arrhythmias, and death.

In *diabetes mellitus*, most peripheral tissues cannot use glucose, because they lack insulin. ⚭ p. 633 Under these circumstances, cells survive by catabolizing lipids and proteins. The result is the production of large numbers of ketone bodies. This condition leads to *diabetic ketoacidosis*, the most common and life-threatening form of ketoacidosis.

Adipose Tissue

Adipose tissue contains a tremendous storehouse of energy in the form of triglycerides. Fat accounts for approximately 15 percent of the body weight of the average individual—enough to provide a 1–2 month reserve of ATP. Although some areas, including the eyelids, the nose, and the backs of the hands and feet, rarely contain adipose tissue, other regions are preferential sites of deposition. Typically, an individual's adipose tissue is distributed in the hypodermis (50 percent), in the greater omentum (10–15 percent), between muscles (5–8 percent), and around the kidneys (12 percent) and reproductive organs (15–20 percent).

As blood glucose levels decline, the rate of triglyceride synthesis falls. Under the stimulation of epinephrine, glucocorticoids, and growth hormone, adipocytes soon begin breaking down their lipid reserves, releasing fatty acids and glycerol into the bloodstream. This process, called *fat mobilization*, continues for the duration of the postabsorptive state. A normal individual retains about a two-months' supply of energy in the triglycerides of adipose tissue. The evolutionary advantages are obvious: The retention of an energy reserve provides a buffer

against daily, monthly, and even seasonal changes in the available food supply.

Skeletal Muscle

At the start of the postabsorptive state, skeletal muscles obtain energy by breaking down their glycogen reserves and catabolizing the glucose that is released. As the concentrations of fatty acids and ketone bodies in the bloodstream increase, these substrates become increasingly important as an energy source.

Skeletal muscle as a whole contains twice as much glycogen as the liver, but it is distributed throughout the muscular system. These glucose reserves are not directly available to other tissues, because the lack of a key enzyme makes skeletal muscle cells unable to release glucose into the bloodstream. Once skeletal muscle fibers metabolize fatty acids as an energy source, they continue to break down their glycogen reserves and convert the pyruvic acid molecules to lactic acid. Lactic acid then diffuses out of the muscle fibers and into the bloodstream. However, even if all of the available glycogen reserves in your body were mobilized as glucose or as lactic acid, the energy provided would be only enough to get you through a good night's sleep. If the postabsorptive state continues for an unusually long period—long enough that lipid reserves are being depleted—muscle proteins will be broken down. The amino acids that are released diffuse into the blood for use by the liver in gluconeogenesis.

Other Peripheral Tissues

With rising plasma concentrations of lipids and ketone bodies and falling blood glucose levels, peripheral tissues gradually decrease their reliance on glucose. Circulating ketone bodies and fatty acids are absorbed and converted to acetyl-CoA for entry into the TCA cycle.

Neural Tissue

Neurons are unusual in that they continue "business as usual" during the postabsorptive state. Neurons are dependent on a reliable supply of glucose, and changes in the activity of the liver, adipose tissue, skeletal muscle, and other peripheral tissues are intended to ensure that the supply of glucose to the nervous system continues unaffected, despite daily or even weekly changes in the availability of nutrients. Only after a prolonged period of starvation will neural tissue begin to metabolize ketone bodies and lactic acid molecules, as well as glucose. **AM** Adaptations to Starvation

✓ What process in the liver would you expect to increase after you have eaten a high-carbohydrate meal?

✓ Why does the amount of urea in blood increase during the postabsorptive state?

✓ If a cell accumulates more acetyl-CoA than it can metabolize by way of the TCA cycle, what products are likely to form?

Answers start on page Q-1

25-7 DIET AND NUTRITION

Objective

■ Explain what constitutes a balanced diet and why such a diet is important.

The postabsorptive state can be maintained for a considerable period, but homeostasis can be maintained indefinitely only if your digestive tract regularly absorbs fluids, organic substrates, minerals, and vitamins, keeping pace with cellular demands. The absorption of nutrients from food is called **nutrition**.

The individual requirement for each nutrient varies from day to day and from person to person. *Nutritionists* attempt to analyze a diet in terms of its ability to meet the needs of a specific individual. A **balanced diet** contains all the ingredients necessary to maintain homeostasis, including adequate substrates for energy generation, essential amino acids and fatty acids, minerals, and vitamins. In addition, the diet must include enough water to replace losses in urine, feces, and evaporation. A balanced diet prevents **malnutrition**, an unhealthy state resulting from inadequate or excessive intake of one or more nutrients.

☐ FOOD GROUPS AND FOOD PYRAMIDS

One method of avoiding malnutrition is to include members of each of the 6 **basic food groups** in the diet: (1) the *milk, yogurt, and cheese group*; (2) the *meat, poultry, fish, dry beans, eggs, and nuts group*; (3) the *vegetable group*; (4) the *fruit group*; (5) the *bread, cereal, rice, and pasta group*; and (6) the *fats, oils, and sweets group*. Each group differs from the others in the typical balance of proteins, carbohydrates, and lipids it contains, as well as in the amount and identity of vitamins and minerals. The 6 groups are arranged in a *food pyramid*, which has the bread, cereal, rice, and pasta group at the bottom (Figure 25–18● and Table 25–2). The aim of this display is to emphasize the need to restrict dietary fats, oils, and sugar and to increase your consumption of breads and the like, which are rich in complex carbohydrates (polysaccharides such as starch).

These are rather artificial groupings at best and downright misleading at worst. For instance, many processed foods belong in a combination of groups. What is important is that you obtain nutrients in sufficient *quantity* (adequate to meet your energy needs) and *quality* (including essential amino acids, fatty acids, vitamins, and minerals). How these nutrients are packaged is a secondary concern. There is nothing magical about the number 6; since 1940, the U.S. government has advocated 11, 7, 4, or 6 food groups at various times. The key is making intelligent choices about what you eat. The wrong choices can lead to malnutrition even if all 6 groups are represented.

For example, consider the case of the essential amino acids. Your liver cannot synthesize any of these amino acids, and you must obtain them from your diet. Some members of the meat and dairy groups, such as beef, fish, poultry, eggs, and milk, provide all the essential amino acids in sufficient quantities. They are said to contain **complete proteins**. Many plants, while they also supply adequate *amounts* of protein, contain **incomplete proteins**, which are deficient in one or more of the essential amino acids. Vegetarians, who restrict themselves to the fruit and vegetable groups (with or without the bread group), must become adept at juggling the constituents of their meals to include a combination of ingredients that will meet all their amino acid requirements. Even with a proper balance of amino acids, vegetarians face a significant problem, because vitamin B_{12} is obtained only from animal products or from fortified cereals or tofu. (Although some health-food products, such as *Spirulina*®, are marketed as sources of this vitamin, the B_{12} that is present is in a form that humans cannot utilize.)

☐ NITROGEN BALANCE

A variety of important compounds in the body contain nitrogen atoms. These **N compounds** include:

- Amino acids, which are part of the framework of all proteins and protein derivatives, such as glycoproteins and lipoproteins.
- Purines and pyrimidines, the nitrogenous bases of RNA and DNA.
- *Creatine*, important in energy storage in muscle tissue (as creatine phosphate).
- *Porphyrins*, complex ring-shaped molecules that bind metal ions and are essential to the function of hemoglobin, myoglobin, and the cytochromes.

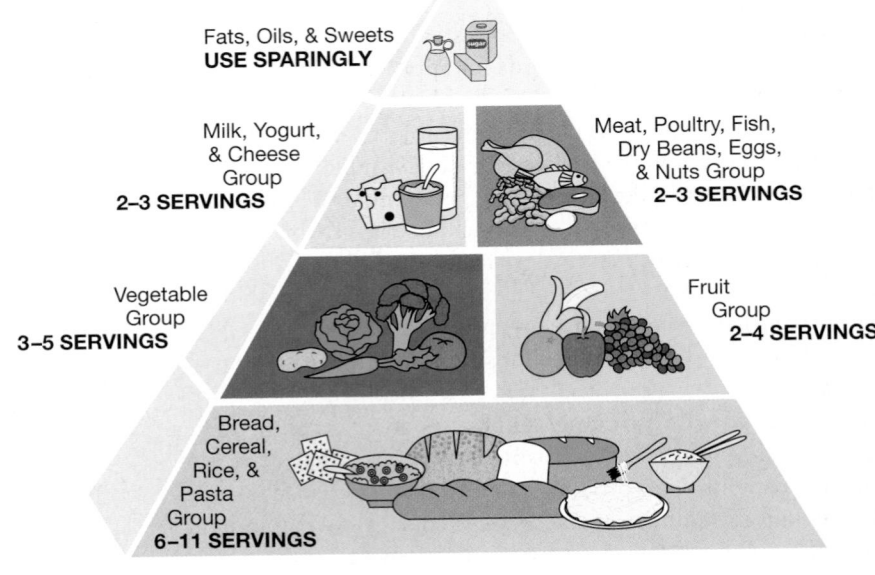

●FIGURE 25–18
The Food Pyramid and Dietary Recommendations

TABLE 25–2 Nutrient Groups

Nutrient Group	Provides	Deficiencies
Fats, oil, sweets	Calories	The majority are deficient in most minerals and vitamins
Milk, yogurt, cheese	Complete proteins; fats; carbohydrates; calcium; potassium; magnesium; sodium; phosphorus; vitamins A, B_{12}, pantothenic acid, thiamine, riboflavin	Dietary fiber, vitamin C
Meat, poultry, fish, dry beans, eggs, nuts	Complete proteins; fats; potassium; phosphorus; iron; zinc; vitamins E, thiamine, B_6	Carbohydrates, dietary fiber, several vitamins
Fruits	Carbohydrates; vitamins A, C, E, folacin; dietary fiber; potassium	Many are low in fats, calories, and protein
Vegetables	Carbohydrates; vitamins A, C, E, folacin; dietary fiber; potassium	Many are low in fats, calories, and protein
Bread, cereal, rice, pasta	Carbohydrates; vitamins E, thiamine, niacin, folacin; calcium; phosphorus; iron; sodium; dietary fiber	Fats

Despite the importance of nitrogen to these compounds, your body neither stores nitrogen nor maintains reserves of N compounds as it does carbohydrates (glycogen) and lipids (triglycerides). Your body can synthesize the carbon chains of the N compounds, but you must obtain nitrogen atoms either by recycling N compounds already in the body or by absorbing nitrogen from your diet. You are in **nitrogen balance** if the amount of nitrogen you absorb from the diet balances the amount you lose in urine and feces. This is the normal condition, and it means that the rates of synthesis and breakdown of N compounds are equivalent.

Growing children, athletes, persons recovering from an illness or injury, and pregnant or lactating women actively synthesize N compounds, so these individuals must absorb more nitrogen than they excrete. Such individuals are in a state of **positive nitrogen balance**. When excretion exceeds ingestion, a **negative nitrogen balance** exists. This is an extremely unsatisfactory situation: The body contains only about a kilogram of nitrogen tied up in N compounds, and a decrease of one-third can be fatal. Even when energy reserves are mobilized, as during starvation, carbohydrates and lipid reserves are broken down first and N compounds are conserved.

Like N compounds, minerals and vitamins are essential components of the diet. Your body cannot synthesize minerals, and your cells can generate only a small quantity of a very few vitamins. We shall consider elements of the diet next.

☐ MINERALS

Minerals are inorganic ions released through the dissociation of electrolytes. Minerals are important for three reasons:

1. *Ions such as Sodium and Chloride Determine the Osmotic Concentrations of Body Fluids.* Potassium is important in maintaining the osmotic concentration of the cytoplasm inside body cells.

2. *Ions in Various Combinations Play Major Roles in Important Physiological Processes.* As we have mentioned, these processes include the maintenance of transmembrane potentials, the construction and maintenance of the skeleton, muscle contraction, the generation of action potentials, the release of neurotransmitters, hormone production, blood clotting, the transport of respiratory gases, buffer systems' fluid absorption, and waste removal, to name but a few examples encountered in earlier chapters.

3. *Ions Are Essential Cofactors in a Variety of Enzymatic Reactions.* For example, the calcium-dependent ATPase in skeletal muscle also requires the presence of magnesium ions, and another ATPase required for the conversion of glucose to pyruvic acid needs both potassium and magnesium ions. Carbonic anhydrase, important in CO_2 transport, buffering systems, and gastric acid secretion, requires the presence of zinc ions. Finally, each component of the electron transport system requires an iron atom, and the final cytochrome (a_3) of the ETS must bind a copper ion as well.

The major minerals and a summary of their functional roles are given in Table 25–3. Your body contains significant reserves of several important minerals; these reserves help reduce the effects of variations in the dietary supply. The reserves are often small, however, and chronic dietary reductions can lead to a variety of clinical problems. Alternatively, because storage capabilities are limited, a dietary excess of mineral ions can be equally dangerous.

Problems involving iron are particularly common. The body of a healthy male contains about 3.5 g of iron in the ionic form Fe^{2+}. Of that amount, 2.5 g is bound to the hemoglobin of circulating red blood cells, and the rest is stored in the liver and bone marrow. In females, the total body iron content averages 2.4 g, with roughly 1.9 g incorporated into red blood cells. Thus, a woman's iron reserves consist of only 0.5 g, half that of a typical man. If the diet contains inadequate amounts of iron, women are therefore more likely to develop signs of iron deficiency than are men.

☐ VITAMINS

Vitamins are assigned to either of two groups, according to their chemical structure and characteristics: (1) *fat-soluble vitamins* or (2) *water-soluble vitamins.*

Fat-Soluble Vitamins

Vitamins A, D, E, and K are the **fat-soluble vitamins**. These vitamins are absorbed primarily from your digestive tract along with the lipid contents of micelles. However, when exposed to sunlight, your skin can synthesize small amounts of vitamin D, and intestinal bacteria produce some vitamin K. The mode of action of these vitamins is unknown. Vitamin A has long been recognized as a structural component of the visual pigment retinal, but its more general metabolic effects are not well understood. Vitamin D is ultimately converted to calcitriol, which binds to cytoplasmic receptors within the intestinal epithelium and promotes an increase in the rate of intestinal calcium and phosphorus absorption. Vitamin E is thought to stabilize intracellular membranes. Vitamin K is a necessary participant in a reaction essential to the synthesis of several proteins, including at least three of the clotting factors. Current information about the fat-soluble vitamins is summarized in Table 25–4.

Alcohol: A Risky Diversion

Alcohol production and sales are big business throughout the Western world. Beer commercials on television, billboards advertising various brands of liquor, and TV or movie characters enjoying a drink—all demonstrate the significance of alcohol in many societies. Many people are unaware of the medical consequences of this cultural fondness for alcohol. Problems with alcohol are usually divided into those stemming from alcohol abuse and those involving alcoholism. The boundary between these conditions is hazy. *Alcohol abuse* is the general term for overuse and the resulting behavioral and physical effects of overindulgence. *Alcoholism* is chronic alcohol abuse, with the physiological changes associated with addiction to other CNS-active drugs. Alcoholism has received the most attention in recent years, although alcohol abuse—especially when combined with driving an automobile—is also in the limelight.

Consider these statistics:

- Alcoholism affects more than 10 million people in the United States alone. The lifetime risk of developing alcoholism for those who drink alcohol is estimated at 10 percent.

- Alcoholism is probably society's most expensive health problem, with an annual estimated direct cost of more than $136 billion. Indirect costs, in terms of damage to automobiles, property, and innocent victims of accidents, are unknown.

- An estimated 25–40 percent of U.S. hospital patients are undergoing treatment related to alcohol consumption. Approximately 200,000 deaths occur annually due to alcohol-related medical conditions. Some major clinical conditions are caused almost entirely by alcohol consumption. For example, alcohol is responsible for 60–90 percent of all liver disease in the United States.

- Alcohol affects all physiological systems. Major clinical symptoms of alcoholism include (1) disorientation and confusion (nervous system); (2) ulcers, diarrhea, and cirrhosis (digestive system); (3) cardiac arrhythmias, cardiomyopathy, and anemia (cardiovascular system); (4) depressed sexual drive and testosterone levels (reproductive system); and (5) itching and angiomas (integumentary system).

- The toll on newborn infants has risen steadily since the 1960s as the number of women drinkers has increased. Women consuming 1 ounce of alcohol per day during

pregnancy have a higher rate of spontaneous abortion and produce children with lower birth weights than do women who consume no alcohol. Women who drink heavily may bear children with *fetal alcohol syndrome (FAS)*, a condition marked by characteristic facial abnormalities, a small head, slow growth, and mental retardation.

- Alcohol abuse is considerably more widespread than alcoholism. Although the medical effects are less well documented, they are certainly significant.

Several factors interact to produce alcoholism. The primary risk factors are gender (males are more likely to become alcoholics than are females) and a family history of alcoholism. There does appear to be a genetic component: A gene on chromosome 11 has been implicated in some inherited forms. The relative importance of genes versus social environment has been difficult to assess. It is likely that alcohol abuse and alcoholism result from a variety of factors. Treatment may consist of counseling and behavior modification. To be successful, treatment must include the total avoidance of alcohol. Supporting groups, such as Alcoholics Anonymous, can be very helpful in providing a social framework for abstinence. The use of the drug *disulfiram (Antabuse®)* has not proved to be as successful as originally anticipated. Antabuse sensitizes the individual to alcohol, so that a drink produces intense nausea; it was anticipated that this would be an effective deterrent. However, clinical tests indicated that it could increase the time between drinks, but could not prevent drinking altogether.

TABLE 25–3 Minerals and Mineral Reserves*

Mineral	Significance	Total Body Content	Primary Route of Excretion	Recommended Daily Intake
BULK MINERALS				
Sodium	Major cation in body fluids; essential for normal membrane function	110 g, primarily in body fluids	Urine, sweat, feces	0.5–1.0 g
Potassium	Major cation in cytoplasm; essential for normal membrane function	140 g, primarily in cytoplasm	Urine	1.9–5.6 g
Chloride	Major anion in body fluids	89 g, primarily in body fluids	Urine, sweat	0.7–1.4 g
Calcium	Essential for normal muscle and neuron function, and normal bone structure	1.36 kg, primarily in skeleton	Urine, feces	0.8–1.2 g
Phosphorus	As phosphate in high-energy compounds, nucleic acids, and bone matrix	744 g, primarily in skeleton	Urine, feces	0.8–1.2 g
Magnesium	Cofactor of enzymes, required for normal membrane functions	29 g (skeleton, 17 g; cytoplasm and body fluids, 12 g)	Urine	0.3–0.4 g
TRACE MINERALS				
Iron	Component of hemoglobin, myoglobin, cytochromes	3.9 g (1.6 g stored as ferritin or hemosiderin)	Urine (traces)	10–18 mg
Zinc	Cofactor of enzyme systems, notably carbonic anhydrase	2 g	Urine, hair (traces)	15 mg
Copper	Required as cofactor for hemoglobin synthesis	127 mg	Urine, feces (traces)	2–3 mg
Manganese	Cofactor for some enzymes	11 mg	Feces, urine (traces)	2.5–5 mg
Cobalt	Cofactor for transaminations	1.1 g	Feces, urine	0.0001 mg
Selenium	Antioxidant	Variable	Feces, urine	0.05–0.2 mg
Chromium	Cofactor for glucose metabolism	0.0006	Feces, urine	0.005–0.2 mg

*For information on the effects of deficiencies and excesses, see Table 27–2, p. 1026.

TABLE 25–4 The Fat-Soluble Vitamins

Vitamin	Significance	Sources	Daily Requirement	Effects of Deficiency	Effects of Excess
A	Maintains epithelia; required for synthesis of visual pigments	Leafy green and yellow vegetables	1 mg	Retarded growth, night blindness, deterioration of epithelial membranes	Liver damage, skin peeling, CNS effects (nausea, anorexia)
D (steroids, including cholecalciferol, or D_3)	Required for normal bone growth, calcium and phosphorus absorption at gut and retention at kidneys	Synthesized in skin exposed to sunlight	None*	Rickets, skeletal deterioration	Calcium deposits in many tissues, disrupting functions
E (tocopherols)	Prevents breakdown of vitamin A and fatty acids	Meat, milk, vegetables	12 mg	Anemia, other problems suspected	None reported
K	Essential for liver synthesis of prothrombin and other clotting factors	Vegetables; production by intestinal bacteria	0.7–0.14 mg	Bleeding disorders	Liver dysfunction, jaundice

*Unless exposure to sunlight is inadequate for extended periods and alternative sources (fortified milk products) are unavailable.

Because they dissolve in lipids, fat-soluble vitamins normally diffuse into cell membranes and other lipids in the body, including the lipid inclusions in the liver and adipose tissue. Your body therefore contains a significant reserve of these vitamins, and normal metabolic operations can continue for several months after dietary sources have been cut off. For this reason, symptoms of **avitaminosis** (ā-vī-ta-min-Ō-sis), or **vitamin deficiency disease**, rarely result from a dietary insufficiency of fat-soluble vitamins. However, avitaminosis involving either fat-soluble or water-soluble vitamins can be caused by a variety of factors other than dietary deficiencies. An inability to absorb a vitamin from the digestive tract, inadequate storage, or excessive demand may each produce the same result.

As Table 25–4 points out, *too much* of a vitamin can produce effects just as unpleasant as too little. **Hypervitaminosis** (hī-per-vī-ta-min-Ō-sis) occurs when the dietary intake exceeds the body's abilities to store, utilize, or excrete a particular vitamin. This condition most commonly involves one of the fat-soluble vitamins, because the excess is retained and stored in body lipids.

"If a little is good, a lot must be better" is a common, but dangerously incorrect, attitude about vitamins. When the dietary supply of fat-soluble vitamins is excessive, the tissue lipids absorb the additional vitamins. Because these vitamins will later diffuse back into the bloodstream, once the symptoms of hypervitaminosis appear, they are likely to persist. When ab-sorbed in massive amounts (from ten to thousands of times the recommended daily allowance), fat-soluble vitamins can produce acute symptoms of *vitamin toxicity*. Vitamin A toxicity is the most common condition, afflicting some children whose parents are overanxious about proper nutrition and vitamins. A single enormous overdose can produce nausea, vomiting, headache, dizziness, lethargy, and even death. Chronic overdose can lead to hair loss, joint pain, hypertension, weight loss, and liver enlargement.

At least 19 cases of vitamin D toxicity were reported in the Boston area during 1992. Symptoms included fatigue, weight loss, and potentially severe damage to the kidneys and cardiovascular system. The problems resulted from drinking milk containing dangerous amounts of vitamin D. Due to problems at one dairy, some of the milk sold had over 230,000 units of vitamin D per quart instead of the usual 400 units per quart. The incident highlighted the need for quality control in the production, and care in the consumption, of vitamin supplements.

Water-Soluble Vitamins

Most of the **water-soluble vitamins** are components of coenzymes (Table 25–5). For example, NAD is derived from niacin, FAD from vitamin B_2 (riboflavin), and coenzyme A from vitamin B_5 (pantothenic acid).

TABLE 25–5 THE WATER-SOLUBLE VITAMINS

Vitamin	Significance	Sources	Daily Requirement	Effects of Deficiency	Effects of Excess
B_1 (thiamine)	Coenzyme in decarboxylations	Milk, meat, bread	1.9 mg	Muscle weakness, CNS and cardiovascular problems, including heart disease; called *beriberi*	Hypotension
B_2 (riboflavin)	Part of FMN and FAD	Milk, meat	1.5 mg	Epithelial and mucosal deterioration	Itching, tingling
Niacin (nicotinic acid)	Part of NAD	Meat, bread, potatoes	14.6 mg	CNS, GI, epithelial, and mucosal deterioration; called *pellagra*	Itching, burning; vasodilation; death after large dose
B_5 (pantothenic acid)	Part of acetyl-CoA	Milk, meat	4.7 mg	Retarded growth, CNS disturbances	None reported
B_6 (pyridoxine)	Coenzyme in amino acid and lipid metabolisms	Meat	1.42 mg	Retarded growth, anemia, convulsions, epithelial changes	CNS alterations, perhaps fatal
Folacin (folic acid)	Coenzyme in amino acid and nucleic acid metabolisms	Vegetables, cereal, bread	0.1 mg	Retarded growth, anemia, gastrointestinal disorders, developmental abormalities	Few noted, except at massive doses
B_{12} (cobalamin)	Coenzyme in nucleic acid metabolism	Milk, meat	4.5 μg	Impaired RBC production, causing *pernicious anemia*	Polycythemia
Biotin	Coenzyme in decarboxylations	Eggs, meat, vegetables	0.1–0.2 mg	Fatigue, muscular pain, nausea, dermatitis	None reported
C (ascorbic acid)	Coenzyme; delivers hydrogen ions, antioxidant	Citrus fruits	60 mg	Epithelial and mucosal deterioration; called *scurvy*	Kidney stones

Eating Disorders

✚ *Eating disorders* are psychological problems that result in either inadequate or excessive food consumption. The most common conditions are **anorexia nervosa**, characterized by self-induced starvation, and **bulimia**, characterized by feeding binges followed by vomiting, the use of laxatives, or both. Adolescent females account for most cases of anorexia nervosa and bulimia; males account for only 5–10 percent. A common thread in the two conditions is an obsessive concern about food and body weight.

According to current estimates, the incidence of anorexia nervosa in the United States ranges from 0.4 to 1.5 per 100,000 population. The incidence among Caucasian women age 12–18 is estimated to be 1 percent. A typical person with this condition is an adolescent Caucasian woman whose weight is roughly 30 percent below normal levels. Although underweight, she is convinced that she is too fat, so she refuses to eat normal amounts of food.

The psychological factors responsible for anorexia are complex. Young women with the condition tend to be high achievers who are attempting to reach an "ideal" weight that will be envied and admired and thereby provide a sense of security and accomplishment. The factors thus tend to be a combination of their view of society ("thin is desirable or demanded"), their view of themselves ("I am not yet thin enough"), and a desire to be able to control their fate ("I can decide when to eat"). Female models, dancers, figure skaters, gymnasts, and theater majors of any age may feel forced to lose weight to remain competitive. The few male anorexics diagnosed typically face comparable stresses. They tend to be athletes, such as jockeys or wrestlers, who need to maintain a minimal weight to succeed in their careers.

Young anorexic women may continue to starve themselves down to a weight of 30–35 kg (66–77 lb). Dry skin, peripheral edema, an abnormally low heart rate and blood pressure, a reduction in bone and muscle mass, and a cessation of menstrual cycles are relatively common symptoms. Some of the changes, especially in bone mass, can be permanent. Treatment is difficult, and only 50–60 percent of anorexics who regain normal weight stay there for five years or more. Death rates from severe anorexia nervosa range from 10 to 15 percent.

Bulimia is more common than anorexia nervosa. In this condition, the individual goes on an "eating binge" that may involve a meal that lasts an hour or two and may include 20,000 or more calories. The meal is followed by induced vomiting, commonly accompanied by the use of laxatives (drugs that pro-

mote the movement of the material through the digestive tract) and diuretics (drugs that promote fluid loss through urination). These often expensive binges may occur several times each week, separated by periods of either normal eating or fasting.

Bulimia generally involves women of the same age group as those who suffer from anorexia nervosa. The actual incidence is difficult to determine; published estimates for young college-age women range from 5 to 18 percent. However, many bulimics are not diagnosed until they are age 30–40. Because they ingest plenty of food, bulimics may have normal body weight, so the condition is harder to diagnose than anorexia nervosa. The health risks of bulimia result from (1) cumulative damage to the stomach, esophagus, oral cavity, and teeth by repeated exposure to stomach acids; (2) electrolyte imbalances resulting from the loss of sodium and potassium ions in the gastric juices, diarrhea, and urine; (3) edema; and (4) cardiac arrhythmias.

The underlying cause of bulimia remains uncertain. Societal factors are certainly involved, but bulimia has also been strongly correlated with depression and with elevated levels of antidiuretic hormone (ADH) in cerebrospinal fluid.

Water-soluble vitamins are rapidly exchanged between the fluid compartments of the digestive tract and the circulating blood, and excessive amounts are readily excreted in urine. For this reason, hypervitaminosis involving water-soluble vitamins is relatively uncommon except among individuals taking large doses of vitamin supplements. However, only vitamins B_{12} and C are stored in significant quantities, so in-

sufficient intake of other water-soluble vitamins can lead to initial symptoms of vitamin deficiency within a period of days to weeks.

The bacterial inhabitants of our intestines help prevent deficiency diseases by producing small amounts of five of the nine water-soluble vitamins, in addition to fat-soluble vitamin K. Your intestinal epithelium can easily absorb all the water-soluble vitamins except B_{12}. The B_{12} molecule is large, and as we discussed in Chapter 24, it must be bound to the *intrinsic factor* from the gastric mucosa before absorption can occur. ∞ p. 892

☐ DIET AND DISEASE

Diet has a profound influence on a person's general health. We have already considered the effects of too many or too few nutrients, above-normal or below-normal concentrations of minerals, and hypervitaminosis or avitaminosis. More subtle long-term problems can occur when the diet includes the wrong proportions or combinations of nutrients. The average diet in the United States contains too many calories, and lipids provide too great a proportion of those calories. This diet increases the incidence of obesity, heart disease, atherosclerosis, hypertension, and diabetes in the U.S. population.

✓ Would an athlete in intensive training try to maintain a positive or a negative nitrogen balance?

✓ How would a decrease in the amount of bile salts in the bile affect the amount of vitamin A in the body?

Answers start on page Q-1

25–8 BIOENERGETICS

Objectives

■ Define *metabolic rate*, and discuss the factors involved in determining an individual's BMR.

■ Discuss the homeostatic mechanisms that maintain a constant body temperature.

The study of **bioenergetics** examines the acquisition and use of energy by organisms. When chemical bonds are broken, energy is released. Inside cells, a significant amount of energy may be used to synthesize ATP, but much of it is lost to the environment as heat. The process of **calorimetry** (kal-o-RIM-e-trē) determines the total amount of energy released when the bonds of organic molecules are broken. The unit of measurement is the **calorie** (KAL-o-rē), defined as the amount of energy required to raise the temperature of 1 g of water 1 degree centigrade. One gram of water is not a very practical measure when you are interested in the metabolic operations that keep a 70-kg human alive, so the **kilocalorie** (KIL-ō-kal-o-rē) (kc), or **Calorie** (with a capital C), also known as "large calorie," is used instead. One Calorie is the amount of energy needed to raise the temperature of 1 *kilo*gram of water 1 degree centigrade. Calorie-counting guides that give the caloric value of various foods indicate Calories, not calories.

☐ FOOD AND ENERGY

In cells, organic molecules are oxidized to carbon dioxide and water. Oxidation also occurs when something burns, and this process can be experimentally controlled. A known amount of food is placed in a chamber called a **calorimeter** (kal-o-RIM-e-ter), which is filled with oxygen and surrounded by a known volume of water. Once the food is inside, the chamber is sealed and the contents are electrically ignited. When the material has completely oxidized and only ash remains in the chamber, the number of Calories released can be determined by comparing the water temperatures before and after the test.

The energy potential of food is usually expressed in Calories per gram (C/g). The catabolism of lipids entails the release of a considerable amount of energy, roughly 9.46 C/g. The catabolism of carbohydrates or proteins is not as rewarding, because many of the carbon and hydrogen atoms are already bound to oxygen. Their average yields are comparable: 4.18 C/g for carbohydrate and 4.32 C/g for protein. Most foods are mixtures of fats, proteins, and carbohydrates, and the values in a "Calorie counter" vary as a result.

☐ METABOLIC RATE

Clinicians can examine your metabolic state and determine how many Calories you are utilizing. The result can be expressed as Calories per hour, Calories per day, or Calories per unit of body weight per day. What is actually measured is the sum of all the varied anabolic and catabolic processes occurring in your body—your **metabolic rate** at that time. The metabolic rate will change according to the activity under way; for instance, measurements taken while one is sprinting are quite different from those taken while one is sleeping. In an attempt to reduce the variations, clinicians standardize the testing conditions so as to determine the **basal metabolic rate (BMR)**. Ideally, the BMR is the minimum resting energy expenditures of an awake, alert person.

A direct method of determining the BMR simply monitors respiratory activity, because in resting persons energy utilization is proportional to oxygen consumption. If we assume that average amounts of carbohydrates, lipids, and proteins are being catabolized, the ratio gives 4.825 Calories per liter of oxygen consumed.

An average individual has a BMR of 70 C per hour, or about 1680 C per day. Although the test conditions are standardized, many uncontrollable factors, including age, gender, physical condition, body weight, and genetic differences (such as ethnic variations), can influence the BMR.

Because the BMR is technically difficult to measure, and because circulating thyroid hormone levels have a profound effect on the BMR, clinicians usually monitor the concentration of thyroid hormones rather than the actual metabolic rate. The results are then compared with normal values, to obtain an index of metabolic activity. One such test, the T_4 **assay**, measures the amount of thyroxine in the blood.

Daily energy expenditures for a given individual vary widely with the activities undertaken. For example, a person leading a sedentary life may have near-basal energy demands, but a single hour of swimming can increase the daily caloric requirements by 500 C or more. If your daily energy intake exceeds your total energy demands, you will store the excess, primarily as triglycerides in adipose tissue. If your daily caloric expenditures exceed your dietary supply, there will be a net reduction in your body's energy reserves and a corresponding loss in weight. This relationship accounts for the significance of Calorie counting and exercise in a weight-control program.

The control of appetite is poorly understood. Stretch receptors along the digestive tract, especially in the stomach, play a role, but other factors are probably more important. Social factors, psychological pressures, and dietary habits are important. Evidence also indicates that complex hormonal stimuli interact to affect appetite. For example, the hormones *cholecystokinin* and *adrenocorticotropic hormone (ACTH)* suppress the appetite. The hormone *leptin*, released by adipose tissues, also plays a role. During the absorptive state, adipose tissues release leptin into the bloodstream as they synthesize triglycerides. Leptin binds to CNS neurons that deal with emotion and the control of appetite. The result is a sense of satiation and the suppression of one's appetite.

☐ THERMOREGULATION

The BMR estimates the rate of energy use by the body. The energy not captured and harnessed by cells is released as heat and serves an important homeostatic purpose. Humans are subject to vast changes in environmental temperatures, but our complex biochemical systems have a major limitation: Enzymes operate over only a relatively narrow range of temperatures. Accordingly, our bodies have anatomical and physiological mechanisms that keep body temperatures within acceptable limits, regardless of the environmental conditions. This homeostatic process is called **thermoregulation**. Failure to control body temperature can result in a series of physiological changes. For example, a body temperature below 36°C (97°F) or above 40°C (104°F) can cause disorientation, and a temperature above 42°C (108°F) can cause convulsions and permanent cell damage.

We continuously produce heat as a by-product of metabolism. When energy use increases, due to physical activity, or when our cells are more active metabolically, as they are during the absorptive state, additional heat is generated. The heat produced by biochemical reactions is retained in water, which accounts for nearly 66 percent of our body weight. Water is an excellent conductor of heat, so the heat produced in one region of the body is rapidly distributed by diffusion, as well as through the bloodstream. If the body temperature is to remain constant, that heat must be lost to the environment at the same rate that it is generated. When the environmental conditions vary from "ideal," becoming too warm or too cold, the body must control the gains or losses to maintain homeostasis.

Mechanisms of Heat Transfer

Heat exchange with the environment involves four basic processes: (1) *radiation*, (2) *conduction*, (3) *convection*, and (4) *evaporation*. These mechanisms are illustrated in Figure 25–19●.

Warm objects lose heat energy as infrared **radiation**. When you feel the heat from the sun, you are experiencing radiant heat. Your body loses heat the same way, but in proportionately smaller amounts. More than 50 percent of the heat you lose is attributable to radiation; the exact amount varies with both body temperature and skin temperature.

Conduction is the direct transfer of energy through physical contact. When you arrive in an air-conditioned classroom and sit on a cold plastic chair, you are immediately aware of this process. Conduction is generally not an effective mechanism for gaining or losing heat. We cannot estimate the value of its contribution, because it varies with the temperature of the object and with the amount of surface area involved. When you are lying on cool sand in the shade, conductive losses can be considerable; when you are standing, conductive losses are negligible.

Convection is the result of conductive heat loss to the air that overlies the surface of the body. Warm air rises, because it is lighter than cool air. As your body conducts heat to the air next to your skin, that air warms and rises, moving away from the surface of the skin. Cooler air replaces it, and as this air in turn becomes warmed, the pattern repeats. Convection accounts for roughly 15 percent of your heat loss and is insignificant as a mechanism of heat gain.

When water evaporates, it changes from a liquid to a vapor. **Evaporation** absorbs energy—roughly 0.58 C per gram of water evaporated. The rate of evaporation occurring at your skin is highly variable. Each hour, 20–25 ml of water crosses epithelia

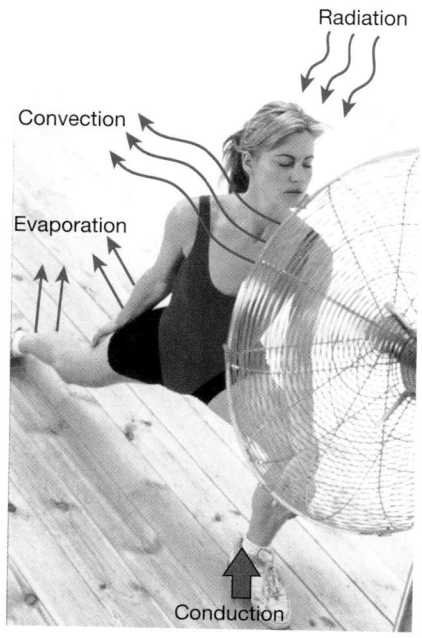

●**FIGURE 25–19**
Routes of Heat Gain and Loss

and evaporates from the alveolar surfaces and the surface of the skin. This insensible perspiration remains relatively constant; at rest, it accounts for roughly 20 percent of your average heat loss. The sweat glands responsible for sensible perspiration have a tremendous scope of activity, ranging from virtual inactivity to secretory rates of 2–4 liters per hour.

The Regulation of Heat Gain and Heat Loss

Heat loss and heat gain involve the activity of many systems. That activity is coordinated by the **heat-loss center** and **heat-gain center**, respectively, in the preoptic area of the anterior hypothalamus. ∞ p. 482 These centers modify the activities of other hypothalamic nuclei. The overall effect is to control temperature by influencing two events: the rate of heat production and the rate of heat loss to the environment. These events may be further supported by behavioral modifications.

MECHANISMS FOR INCREASING HEAT LOSS When the temperature at the preoptic nucleus exceeds its thermostat setting, the heat-loss center is stimulated, producing three major effects:

1. *The Inhibition of the Vasomotor Center Causes Peripheral Vasodilation, and Warm Blood Flows to the Surface of the Body.* The skin takes on a reddish color, skin temperatures rise, and radiational and convective losses increase.

2. *As Blood Flow to the Skin Increases, Sweat Glands Are Stimulated to Increase their Secretory Output.* The perspiration flows across the body surface, and evaporative losses accelerate. A maximal secretion rate would, if it were completely evaporated, remove 2320 C per hour.

3. *The Respiratory Centers Are Stimulated, and the Depth of Respiration Increases.* Often, the individual begins respiring through an open mouth rather than through the nasal passageways, increasing evaporative losses through the lungs.

MECHANISMS FOR PROMOTING HEAT GAIN The function of the heat-gain center of the brain is to prevent **hypothermia** (hī-pō-THER-mē-uh), or below-normal body temperatures. When the temperature at the preoptic nucleus drops below acceptable levels, the heat loss center is inhibited and the heat-gain center is activated.

Heat Conservation The sympathetic vasomotor center decreases blood flow to the dermis of the skin, thus reducing losses by radiation, convection, and conduction. The skin cools, and with blood flow restricted, it may take on a bluish or pale coloration. The epithelial cells are not damaged, because they can tolerate extended periods at temperatures as low as 25°C (77°F) or as high as 49°C (120°F).

In addition, blood returning from the limbs is shunted into a network of deep veins. ∞ p. 761 In warm weather, blood flows in a superficial venous network (Figure 25–20a●). In cold weather, blood is diverted to a network of veins that lie deep to an insulating layer of subcutaneous fat (Figure 25–20b●). This venous network wraps around the deep arteries, and heat is conducted from the warm blood flowing outward to the limbs

into the cooler blood returning from the periphery (Figure 25–20c●). This arrangement traps the heat close to the body core and restricts heat loss. Such exchange between fluids that are moving in opposite directions is called *countercurrent exchange*. (We will return to this topic in Chapter 26.)

Heat Generation The mechanisms available to generate heat can be divided into two broad categories: (1) shivering thermogenesis and (2) nonshivering thermogenesis. In **shivering thermogenesis** (ther-mō-JEN-e-sis), a gradual increase in muscle tone increases the energy consumption of skeletal muscle tissue throughout your body. Both agonists and antagonists are involved, and the degree of stimulation varies with the demand.

If the heat-gain center is extremely active, muscle tone increases to the point at which stretch receptor stimulation will produce brief, oscillatory contractions of antagonistic muscles. In other words, you begin to **shiver**. Shivering increases the workload of the muscles and further elevates oxygen and energy consumption. The heat that is produced warms the deep vessels, to which blood has been shunted by the sympathetic vasomotor center. Shivering can elevate the body temperature quite effectively, increasing the rate of heat generation by as much as 400 percent.

In **nonshivering thermogenesis**, hormones are released that increase the metabolic activity of all tissues:

- The heat-gain center stimulates the adrenal medullae, via the sympathetic division of the autonomic nervous system, and epinephrine is released. Epinephrine increases the rates of glycogenolysis in liver and skeletal muscle and the metabolic rate of most tissues. The effects are immediate.

- The preoptic nucleus regulates the production of thyrotropin-releasing hormone (TRH) by the hypothalamus. In children, when body temperatures are below normal, additional TRH is released, stimulating the release of thyroid-stimulating hormone (TSH) by the anterior lobe of the pituitary gland. The thyroid gland responds to this release of TSH by increasing the rate at which thyroxine is released into the blood. Thyroxine increases not only the rate of carbohydrate catabolism, but also the rate of catabolism of all other nutrients. These effects develop gradually, over a period of days to weeks.

✚ Hypothermia may be intentionally produced during surgery to reduce the metabolic rate of a particular organ or of the patient's entire body. In controlled hypothermia, the individual is first anesthetized to prevent the shivering that would otherwise fight the induction of hypothermia.

During open-heart surgery, the body is typically cooled to 25°–32°C (79°–89°F). This cooling reduces the metabolic demands of the body, which will be receiving blood from an external pump or oxygenator. The heart must be stopped completely during the operation, and it cannot be well supplied with blood over this period. Consequently, the heart is perfused with an *arresting solution* at 0°–4°C (32°–39°F) and maintained at a temperature below 15°C (60°F) for the duration of the opera-

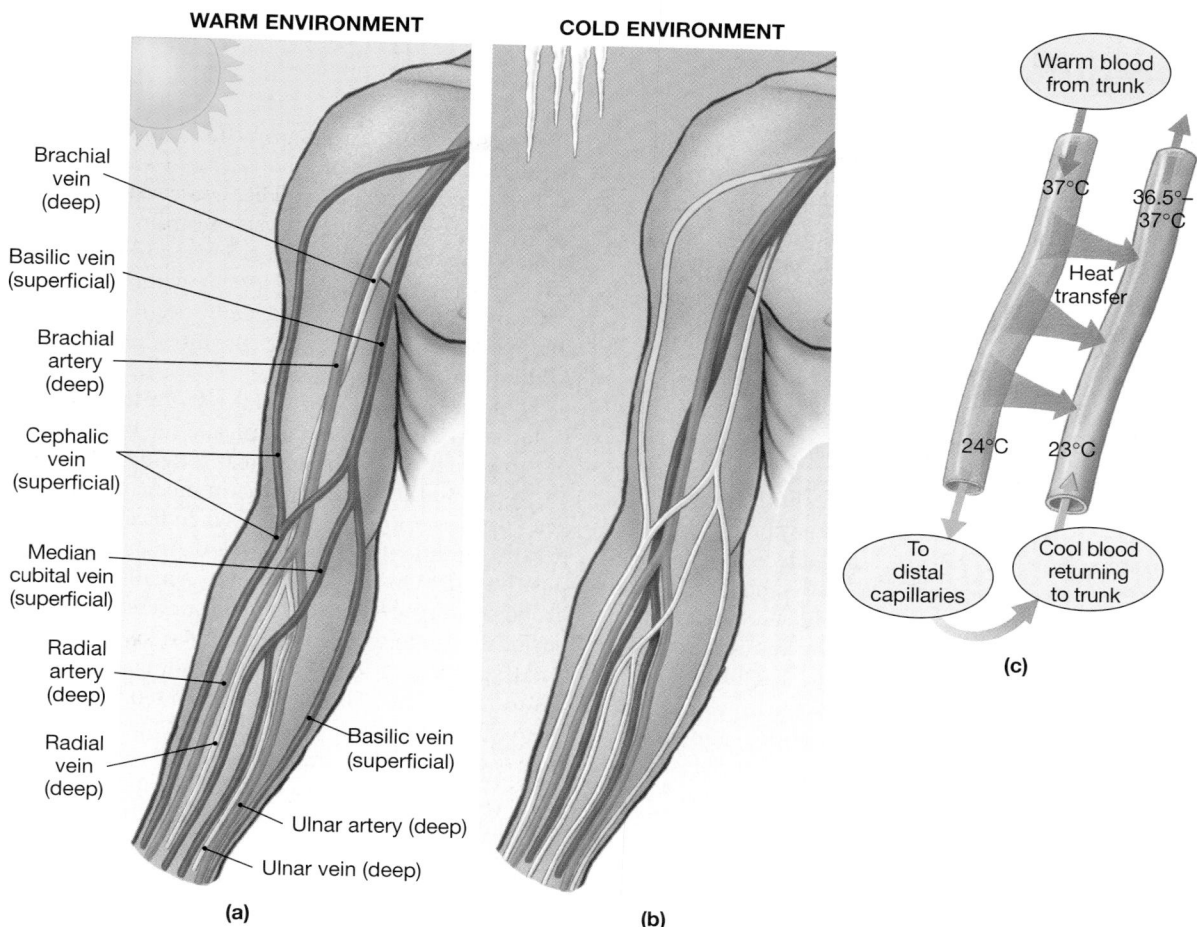

WARM ENVIRONMENT

COLD ENVIRONMENT

Brachial
vein
(deep)

Basilic vein
(superficial)

Brachial
artery
(deep)

Cephalic
vein
(superficial)

Median
cubital vein
(superficial)

Radial
artery
(deep)

Radial
vein
(deep)

Basilic vein
(superficial)

Ulnar artery (deep)

Ulnar vein (deep)

(a)

(b)

Warm blood
from trunk

37°C

36.5°–
37°C

Heat
transfer

24°C

23°C

To
distal
capillaries

Cool blood
returning
to trunk

(c)

●**FIGURE 25–20**
Countercurrent Heat Exchange. (**a**) Circulation through the blood vessels of the forearm in a warm environment. Blood enters the limb in a deep artery and returns to the trunk in a network of superficial veins that radiate heat into the environment through the overlying skin. (**b**) Circulation through the blood vessels of the forearm in a cold environment. Blood now returns to the trunk via a network of deep veins that flow around the artery. The amount of heat loss is reduced, as indicated in (c). (**c**) Countercurrent heat exchange occurs as heat radiates from the warm arterial blood into the cooler venous blood flowing in the opposite direction. By the time the arterial blood reaches distal capillaries, where most of the heat loss to the environment occurs, it is already 13°C cooler than it was when it left the trunk. This mechanism reduces the rate of heat loss while conserving body heat. In effect, the countercurrent exchange traps heat near the trunk.

tion. At these temperatures, the cardiac muscle can tolerate several hours of ischemia without damage.

When cardiac surgery is performed on infants, a deep hypothermia may be produced by cooling the entire body to temperatures as low as 11°C (52°F) for an hour or more. In effect, this procedure duplicates the conditions experienced by accidental drowning victims.

Sources of Individual Variation

The timing of thermoregulatory responses differs from individual to individual. A person may undergo **acclimatization** (a-klī-ma-ti-ZĀ-shun)—the physiological adjustment to a particular environment over time. For example, natives of Tierra del Fuego once lived naked in the snow, but Hawaii residents often unpack their sweaters when the temperature drops below 22°C (72°F).

Another interesting source of variation is body size. Although heat *production* occurs within the mass of the body, heat *loss* must occur across a body surface. As an object (or person) gets larger, its surface area increases at a much slower rate than does its total volume. This relationship affects thermoregulation, because heat is generated by the "volume" (that is, by internal tissues) and is lost at the body surface. Hence, small individuals lose heat more readily than do large individuals.

THERMOREGULATORY PROBLEMS OF INFANTS During embryonic development, the maternal surroundings are at normal body temperature. At birth, the infant's temperature-regulating mechanisms are not fully functional; also, infants lose heat quickly as a result of their small size. Consequently, newborns must be dried promptly and kept bundled up; for those born

Thermoregulatory Disorders

Heat exhaustion and heat stroke are malfunctions of the thermoregulatory system. In **heat exhaustion**, also known as *heat prostration*, the individual has difficulty maintaining blood volume. The heat-loss center stimulates sweat glands, whose secretions moisten the surface of the skin to provide evaporative cooling. As fluid losses mount, blood volume decreases. The resulting decline in blood pressure is not countered by peripheral vasoconstriction, because the heat-loss center is actively stimulating peripheral vasodilation. As blood flow to the brain declines, headache, nausea, and eventual collapse follow. Treatment is straightforward: Provide fluids, salts, and a cooler environment.

Heat stroke is more serious and can follow an untreated case of heat exhaustion. Predisposing factors include any preexisting condition, such as heart disease or diabetes, that affects peripheral circulation. The thermoregulatory center ceases to function, the sweat glands are inactive, and the skin becomes hot and dry. Unless the situation is recognized in time, body temperature may climb to 41°–45°C (106°–113°F). Temperatures in this range quickly disrupt a variety of vital physiological systems and destroy brain, liver, skeletal muscle, and kidney cells. Effective treatment involves lowering the body temperature as rapidly as possible.

If body temperature drops significantly below normal levels, the thermoregulatory system begins to lose sensitivity and effectiveness. Cardiac output and respiratory rate decrease, and if the core temperature falls below 28°C (82°F), cardiac arrest is likely. The individual then has no heartbeat, no respiratory rate, and no response to external stimuli—even painful ones. The body temperature continues to decline, and the skin turns blue or pale and cold.

At this point, we would probably assume that the individual has died. But because metabolic activities have decreased systemwide, the victim may still be saved, even after several hours have elapsed. Treatment consists of cardiopulmonary support and gradual rewarming, both external and internal. The skin can be warmed up to 45°C (110°F) without damage; warm baths or blankets can be used. One effective method of raising internal temperatures involves the introduction of warm saline solution into the peritoneal cavity.

Hypothermia is a significant risk for those engaged in water sports, and it may complicate the treatment of a victim of drowning. Water absorbs heat roughly 27 times as fast as air does, and the body's heat-gain mechanisms are unable to keep pace over long periods or when faced with a large temperature gradient. But hypothermia in cold water does have a positive side: On several occasions, small children who have drowned in cold water have been successfully revived after periods of up to four hours. Children lose body heat quickly, and their systems soon stop functioning as their body temperature declines. This rapid drop in temperature prevents the oxygen starvation and tissue damage that would otherwise occur when breathing ceases.

Resuscitation is not attempted if the individual has actually frozen. Water expands roughly 7 percent during ordinary freezing, and cell membranes throughout the body are destroyed in the process. Very small organisms can be frozen and subsequently thawed without ill effects, because their surface-to-volume ratio is enormous and the freezing process occurs so rapidly that ice crystals never form.

prematurely, a thermally regulated incubator is required. Infants' body temperatures are also less stable than those of adults. Their metabolic rates decline while they sleep and then rise after they awake.

Infants cannot shiver, but they have a different mechanism for raising their body temperature rapidly. In infants, the adipose tissue between the shoulder blades, around the neck, and possibly elsewhere in the upper body is histologically and functionally different from most of the adipose tissue in adults. The tissue is highly vascularized, and the individual adipocytes contain numerous mitochondria. Together, these characteristics give the tissue a deep, rich color that is responsible for the name **brown fat**. ∞ p. 125 The individual adipocytes are innervated by sympathetic autonomic fibers. When these nerves are stimulated, lipolysis accelerates in the adipocytes. The cells do not capture the energy released through fatty acid catabolism, and it radiates into the surrounding tissues as heat. This heat quickly warms the blood that passes through the surrounding network of vessels, and it is then distributed throughout the body. In this way, an infant can accelerate metabolic heat generation by 100 percent very quickly, whereas nonshivering thermogenesis in an adult raises heat production by only 10–15 percent after a period of weeks.

With increasing age and size, body temperature becomes more stable, so the importance of this thermoregulatory mechanism declines. Adults have little if any brown fat; with increased body size, skeletal muscle mass, and insulation, shivering thermogenesis is significantly more effective in elevating body temperature.

THERMOREGULATORY VARIATIONS AMONG ADULTS Adults of the same body weight may differ in their thermal responses if their weight is distributed differently. Which tissues account for their weight is also a factor. Adipose tissue is an excellent insulator, conducting heat at only about one-third the rate of other tissues. As a result, individuals with a more substantial layer of subcutaneous fat may not begin to shiver until long after their more slender companions.

Two otherwise similar individuals may also differ in their response to temperature changes because their hypothalamic thermostats are at different settings. We experience daily oscillations in body temperature, with temperatures falling 1°–2°C (1.8°–3.6°F) at night and peaking during the day or early evening. Individuals vary in terms of their time of maximum temperature setting; some have a series of peaks, with an afternoon low. The origin of these patterns is uncertain. It is not the result of daily activity regimens: People who work at night still show their temperature peaks over the same range of times as the rest of the population.

Fevers

Any elevated body temperature is called **pyrexia** (pī-REK-sē-uh). Pyrexia is usually temporary. A **fever** is a body temperature maintained at greater than 37.2°C (99°F). We discussed fevers when we examined nonspecific defenses. ⚮ p. 797 Fevers occur for a variety of reasons, not all of them pathological. In young children, transient fevers with no ill effects can result from exercise in warm weather. Similar exercise-related elevations were rarely encountered in adults until running marathons became popular. Temperatures from 39° to 41°C (103° to 106°F) may

result. For this reason, competitions are usually held when the air temperature is below 28°C (82°F).

Fevers can also result from such factors as the following:

- Abnormalities affecting the entire thermoregulatory mechanism, such as heat exhaustion or heat stroke.
- Clinical problems that restrict blood flow, such as congestive heart failure.
- Conditions that impair sweat gland activity, such as drug reactions and some skin conditions.
- The resetting of the hypothalamic thermostat by circulating *pyrogens*—most notably, interleukin-1.

We can classify fevers as *chronic* or *acute*. The classification and treatment of fevers is discussed in the *Applications Manual*. **AM** Fevers

---

✓ How would the BMR of a pregnant woman compare with her own BMR before she became pregnant?

✓ What effect would the vasoconstriction of peripheral blood vessels have on an individual's body temperature on a hot day?

✓ Why do infants have greater problems with thermoregulation than adults do?

Answers start on page Q-1

Chapter Review

SELECTED CLINICAL TERMINOLOGY

Terms Discussed in This Chapter

avitaminosis (ā-vī-ta-min-Ō-sis): A vitamin deficiency disease. (*p. 958*)

carbohydrate loading: Eating large quantities of carbohydrates in the days preceding an athletic competition in order to increase one's endurance. (*p. 939*)

eating disorders: Psychological problems that result in inadequate or excessive food consumption. Examples include anorexia nervosa and bulimia. (*p. 959*)

heat exhaustion: A malfunction of the thermoregulatory system, caused by excessive fluid loss in perspiration. (*p. 964*)

heat stroke: A condition in which the thermoregulatory center stops functioning and the body temperature rises uncontrollably. (*p. 964*)

hyperuricemia (hī-per-ū-ri-SĒ-mē-uh): Levels of plasma uric acid above 7.4 mg/dl; can result in the condition called *gout*. (*p. 947*)

hypervitaminosis (hī-per-vī-ta-min-Ō-sis): A disorder caused by the ingestion of excessive quantities of one or more vitamins. (*p. 958*)

hypothermia (hī-pō-THER-mē-uh): Below-normal body temperature. (*p. 962 and AM*)

ketoacidosis (kē-tō-as-i-DŌ-sis): The acidification of blood due to the presence of ketone bodies. (*p. 953 and AM*)

ketonemia (kē-tō-NE-mē-uh): Elevated levels of ketone bodies in blood. (*p. 953*)

ketonuria (kē-tō-NOO-rē-uh): The presence of ketone bodies in urine. (*p. 953*)

ketosis (kē-TŌ-sis): Abnormally high concentration of ketone bodies in body fluids. (*p. 953*)

obesity: A body weight more than 20 percent above the ideal weight for a given individual. (*p. 951 and AM*)

phenylketonuria (fen-il-kē-tō-NOO-rē-uh): An inherited metabolic disorder resulting from an inability to convert phenylalanine to tyrosine. (*p. 947 and AM*)

protein deficiency diseases: Nutritional disorders resulting from a lack of essential amino acids. (*p. 947 and AM*)

AM Additional Terms Discussed in the *Applications Manual*

antipyretic drugs

gout

liposuction

STUDY OUTLINE

25–1 AN OVERVIEW OF METABOLISM   p. 929

1. In general, during *cellular metabolism*, cells break down excess carbohydrates first and then lipids, while conserving amino acids. Only about 40 percent of the energy released through **catabolism** is captured in ATP; the rest is released as heat. *(Figure 25–1)*
2. Cells synthesize new compounds (**anabolism**) (1) to perform structural maintenance or repairs, (2) to support growth, (3) to produce secretions, and (4) to build nutrient reserves. *(Figure 25–2)*
3. Cells feed small organic molecules to their mitochondria; in return, the cells get the ATP they need to perform cellular functions. *(Figure 25–3)*

25–2 CARBOHYDRATE METABOLISM   p. 932

1. Most cells generate ATP and other high-energy compounds through the breakdown of carbohydrates.

GLYCOLYSIS   p. 932

2. **Glycolysis** and **aerobic metabolism**, or *cellular respiration*, provide most of the ATP used by typical cells. Glycogen can be broken down to glucose molecules. In glycolysis, each molecule of glucose yields two molecules of **pyruvic acid** (as pyruvate ions), a net of two molecules of ATP, and two of **NADH**. *(Figure 25–4)*

MITOCHONDRIAL ATP PRODUCTION   p. 934

3. In the presence of oxygen, the pyruvic acid molecules enter mitochondria, where they are broken down completely in the **tricarboxylic acid (TCA) cycle**. Carbon and oxygen atoms are lost as carbon dioxide (**decarboxylation**); hydrogen atoms are passed to coenzymes, which initiate the oxygen-consuming and ATP-generating reaction **oxidative phosphorylation**. *(Figure 25–5)*
4. **Cytochromes** of the **electron transport system (ETS)** pass electrons to oxygen, resulting in the formation of water. As this transfer occurs, the ETS generates ATP. *(Figure 25–6)*

ENERGY YIELD OF GLYCOLYSIS AND CELLULAR RESPIRATION   p. 938

5. For each glucose molecule processed through glycolysis, the TCA cycle, and the ETS, most cells gain 36 molecules of ATP. *(Figure 25–7)*
6. Cells can break down other nutrients to provide substrates for the TCA cycle if supplies of glucose are limited.

 Metabolic pathways: **Companion Website**: Chapter 25/Tutorials.

GLUCONEOGENESIS   p. 940

7. **Gluconeogenesis**, the synthesis of glucose from noncarbohydrate precursors, such as lactic acid, glycerol, or amino acids, enables a liver cell to synthesize glucose molecules when carbohydrate reserves are depleted. **Glycogenesis** is the process of glycogen formation. Glycogen is an important energy reserve when the cell cannot obtain enough glucose from interstitial fluid. *(Figure 25–8)*

25–3 LIPID METABOLISM   p. 940

LIPID CATABOLISM   p. 940

1. During **lipolysis** (lipid catabolism), lipids are broken down into pieces that can be converted into pyruvic acid or channeled into the TCA cycle.
2. Triglycerides, the most abundant lipids in the body, are split into glycerol and fatty acids. The glycerol enters the glycolytic pathways, and the fatty acids enter the mitochondria.

3. **Beta-oxidation** is the breakdown of a fatty acid molecule into two-carbon fragments that can be used in the TCA cycle. The steps of beta-oxidation cannot be reversed, and the body cannot manufacture all the fatty acids needed for normal metabolic operations. *(Figure 25–9)*
4. Lipids cannot provide large amounts of ATP in a short amount of time. However, cells can shift to lipid-based energy production when glucose reserves are limited.

LIPID SYNTHESIS   p. 942

5. In **lipogenesis** (the synthesis of lipids), almost any organic substrate can be used to form glycerol. **Essential fatty acids** cannot be synthesized and must be included in the diet. *(Figure 25–10)*

LIPID TRANSPORT AND DISTRIBUTION   p. 942

6. Most lipids circulate as **lipoproteins** (lipid–protein complexes that contain large glycerides and cholesterol). The largest lipoproteins, **chylomicrons**, carry absorbed lipids from the intestinal tract to the bloodstream. All other lipoproteins are derived from the liver and carry lipids to and from various tissues of the body. *(Figure 25–11)*
7. Capillary walls of adipose tissue, skeletal muscle, cardiac muscle, and the liver contain **lipoprotein lipase**, an enzyme that breaks down complex lipids, releasing a mixture of fatty acids and monoglycerides. *(Figure 25–11)*

25–4 PROTEIN METABOLISM   p. 944

AMINO ACID CATABOLISM   p. 946

1. If other energy sources are inadequate, mitochondria can break down amino acids in the TCA cycle to generate ATP. In the mitochondria, the amino group can be removed by **transamination** or by **deamination**. *(Figure 25–12)*
2. Protein catabolism is impractical as a source of quick energy. *(Figure 25–14)*

PROTEIN SYNTHESIS   p. 947

3. Roughly half the amino acids needed to build proteins can be synthesized. There are 10 **essential amino acids**, which must be acquired through the diet. **Amination**, the attachment of an amino acid group to a carbon framework, is an important step in the synthesis of **nonessential amino acids**. *(Figures 25–13, 25–14)*

25–5 NUCLEIC ACID METABOLISM   p. 947

1. DNA in the nucleus is never catabolized for energy.

RNA CATABOLISM   p. 947

2. RNA molecules are broken down and replaced regularly. They are generally recycled as new nucleic acids, but the nucleotides can be catabolized to simple sugars and nitrogenous bases. In general, nucleic acids do not contribute significantly to the cell's energy reserves.

NUCLEIC ACID SYNTHESIS   p. 948

3. Most cells synthesize RNA, but DNA is synthesized only in cells preparing for mitosis or meiosis. *(Figure 25–14)*

 Metabolic pathways: **Companion Website**: Chapter 25/ Reviewing Facts and Terms/Multiple Choice.

25–6 METABOLIC INTERACTIONS   p. 948

1. No one cell of a human can perform all the anabolic and catabolic operations necessary to support life. Homeostasis can be

preserved only when metabolic activities of different tissues are coordinated. *(Figure 25–14)*

2. The body has five metabolic components: the liver, adipose tissue, skeletal muscle, neural tissue, and other peripheral tissues. The liver is the focal point for metabolic regulation and control. Adipose tissue stores lipids, primarily in the form of triglycerides. Skeletal muscle contains substantial glycogen reserves, and the contractile proteins can be mobilized and the amino acids used as an energy source. Neural tissue does not contain energy reserves; glucose must be supplied to it for energy. Other peripheral tissues are able to metabolize glucose, fatty acids, or other substrates under the direction of the endocrine system.

THE ABSORPTIVE STATE   p. 949

3. For about four hours after a meal, nutrients enter the blood as intestinal absorption proceeds. *(Figure 25–15; Table 25–1)*
4. The liver closely regulates the glucose content of blood and the circulating levels of amino acids.
5. **Lipemia** (a milky appearance of the plasma due to the presence of lipids) commonly marks the **absorptive state**. Adipocytes remove fatty acids and glycerol from the bloodstream and synthesize new triglycerides to be stored for later use.
6. During the absorptive state, glucose molecules are catabolized and amino acids are used to build proteins. Skeletal muscles may also catabolize circulating fatty acids, and the energy obtained therefrom is used to increase glycogen reserves.

THE POSTABSORPTIVE STATE   p. 951

7. The **postabsorptive state** extends from the end of the absorptive state to the next meal. *(Figures 25–16, 25–17)*
8. When blood glucose levels fall, the liver begins breaking down glycogen reserves and releasing the glucose into the bloodstream. As the time between meals increases, liver cells synthesize glucose molecules from smaller carbon fragments and from glycerol molecules. Fatty acids undergo beta-oxidation; the fragments enter the TCA cycle or combine to form **ketone bodies**. *(Table 25–1)*
9. Some amino acids can be converted to pyruvic acid and used for gluconeogenesis; others, including most of the essential amino acids, are converted to acetyl-CoA and are either catabolized or converted to ketone bodies.
10. The average individual carries a one- to two-month energy reserve in adipose tissue. During the postabsorptive state, lipolysis increases and the fatty acids are released into the bloodstream for catabolism.
11. Skeletal muscles metabolize ketone bodies and fatty acids. Their glycogen reserves are broken down to yield lactic acid, which diffuses into the bloodstream. After a prolonged fast, muscle proteins are broken down.
12. Neural tissue continues to be supplied with glucose as an energy source until blood glucose levels become very low.

25–7 DIET AND NUTRITION   p. 954

1. **Nutrition** is the absorption of nutrients from food. A **balanced diet** contains all the ingredients necessary to maintain homeostasis; it prevents **malnutrition**.

FOOD GROUPS AND FOOD PYRAMIDS   p. 954

2. The six **basic food groups** are milk, yogurt, and cheese; meat, poultry, fish, dry beans, eggs, and nuts; vegetables; fruits; bread, cereal, rice, and pasta; and fats, oils, and sweets. These are arranged in a *food pyramid*, with the bread and cereal group forming the base. *(Figure 25–18)*

NITROGEN BALANCE   p. 954

3. Amino acids, purines, pyrimidines, creatine, and porphyrins are **N compounds**, which contain nitrogen atoms. An adequate dietary supply of nitrogen is essential, because the body does not maintain large nitrogen reserves. **Nitrogen balance** is a state in which the amount of absorbed nitrogen equals that lost in urine and feces.

MINERALS   p. 955

4. **Minerals** act as cofactors in a variety of enzymatic reactions. They also contribute to the osmotic concentration of body fluids and play a role in transmembrane potentials, action potentials, the release of neurotransmitters, muscle contraction, skeletal construction and maintenance, gas transport, buffer systems, fluid absorption, and waste removal. *(Table 25–2)*

VITAMINS   p. 955

5. Vitamins are needed in very small amounts. Vitamins A, D, E, and K are the **fat-soluble vitamins**; taken in excess, they can lead to **hypervitaminosis**. **Water-soluble vitamins** are not stored in the body; a lack of adequate dietary supplies may lead to **deficiency disease** (**avitaminosis**). *(Tables 25–3, 25–4)*

DIET AND DISEASE   p. 960

6. A balanced diet can improve one's general health.

25–8 BIOENERGETICS   p. 960

1. Bioenergetics is the study of the acquisition and use of energy by organisms. The energy content of food is usually expressed as **Calories** per gram (C/g).

FOOD AND ENERGY   p. 960

2. The catabolism of lipids releases 9.46 C/g, about twice the amount released by equivalent weights of carbohydrates or proteins.

METABOLIC RATE   p. 960

3. The total of all the anabolic and catabolic processes under way is an individual's **metabolic rate**. The **basal metabolic rate (BMR)** is the rate of energy utilization by a person at rest.

THERMOREGULATION   p. 961

4. The homeostatic regulation of body temperature is **thermoregulation**. Heat exchange with the environment involves four processes: **radiation**, **conduction**, **convection**, and **evaporation**. *(Figure 25–19)*
5. The *preoptic area* of the hypothalamus acts as the body's thermostat, affecting the **heat-loss center** and the **heat-gain center**.
6. Mechanisms for increasing heat loss include both physiological mechanisms (peripheral vasodilation, increased perspiration, and increased respiration) and behavioral modifications.
7. Responses that conserve heat include a decreased blood flow to the dermis and *countercurrent exchange*. *(Figure 25–20)*
8. Heat is generated by **shivering thermogenesis** and **nonshivering thermogenesis**.
9. Thermoregulatory responses differ among individuals. One important source of variation is **acclimatization** (adjusting physiologically to an environment over time).
10. **Pyrexia** is an elevated body temperature. **Fever**, a body temperature above 37.2°C (99°F), can result from problems with the thermoregulatory mechanism, circulation, or sweat gland activity or from the resetting of the hypothalamic thermostat by circulating pyrogens.

REVIEW QUESTIONS

More assessment questions are available to you on the Companion Website. You will find Matching, Multiple Choice, True/False, and other quizzes to help further your understanding of the material covered in this chapter. To access the site, go to www.aw.com/martini.

LEVEL 1 Reviewing Facts and Terms

1. Catabolism refers to
   (a) the creation of a nutrient pool
   (b) the sum total of all chemical reactions in the body
   (c) the production of organic compounds
   (d) the breakdown of organic substrates

2. During the complete catabolism of one molecule of glucose, a typical cell gains
   (a) 4 ATP molecules   (b) 18 ATP molecules
   (c) 36 ATP molecules   (d) 144 ATP molecules

3. The breakdown of glucose to pyruvic acid is an _____ process.
   (a) anaerobic   (b) aerobic
   (c) anabolic   (d) oxidative

4. Glycolysis yields an *immediate* net gain of _____ for the cell.
   (a) 1 ATP molecule   (b) 2 ATP molecules
   (c) 4 ATP molecules   (d) 36 ATP molecules

5. The only *immediate* energy benefit of one turn of the TCA cycle is the formation of a single molecule of
   (a) GTP   (b) NAD
   (c) FAD   (d) CoA

6. The process that produces over 90 percent of the ATP used by our cells is
   (a) glycolysis
   (b) the TCA cycle
   (c) substrate-level phosphorylation
   (d) oxidative phosphorylation

7. When a substrate actively undergoing oxidative phosphorylation loses electrons, it has been
   (a) deaminated   (b) reduced
   (c) oxidized   (d) decarboxylated

8. The synthesis of glucose from nonglucose precursors is
   (a) glycolysis   (b) glycogenesis
   (c) gluconeogenesis   (d) beta-oxidation

9. The sequence of reactions responsible for the breakdown of fatty acid molecules is
   (a) beta-oxidation   (b) the TCA cycle
   (c) lipogenesis   (d) a, b, and c are correct

10. The essential fatty acids that cannot be synthesized by the body, but must be included in the diet, are
    (a) linoleic and linolenic   (b) leucine and lysine
    (c) cholesterol and glycerol (d) HDLs and LDLs

11. The lipoproteins that carry absorbed lipids from the intestinal tract to the bloodstream are the
    (a) HDLs   (b) VLDLs
    (c) LDLs   (d) chylomicrons

12. The removal of an amino group in a reaction that generates an ammonia molecule is called
    (a) ketoacidosis   (b) transamination
    (c) deamination   (d) denaturation

13. The part of the RNA molecule that *cannot* be catabolized to provide energy is the
    (a) phosphate   (b) sugar
    (c) pyrimidine   (d) purine

14. The focal point of metabolic regulation and control is the
    (a) brain   (b) liver
    (c) heart   (d) kidneys

15. When the body is relying on internal energy reserves to continue meeting its energy demands, it is in the
    (a) postabsorptive state (b) absorptive state
    (c) starvation state   (d) deprivation state

16. A complete protein contains
    (a) the proper balance of amino acids
    (b) all the essential amino acids in sufficient quantities
    (c) a combination of nutrients selected from the food pyramid
    (d) N compounds produced by the body

17. All minerals and most vitamins
    (a) are fat soluble
    (b) cannot be stored by the body
    (c) cannot be synthesized by the body
    (d) must be synthesized by the body because they are not present in adequate amounts in the diet

18. The vitamins generally associated with vitamin toxicity are
    (a) fat-soluble vitamins
    (b) water-soluble vitamins
    (c) the B complex vitamins
    (d) vitamins C and B_{12}

19. The basal metabolic rate (BMR) represents the
    (a) maximum energy expenditure when exercising
    (b) minimum resting energy expenditure of an awake, alert person
    (c) minimum energy expenditure during light exercise
    (d) muscular energy expenditure added to the resting energy expenditure

20. The abuse of laxatives and diuretics is commonly associated with
    (a) avitaminosis   (b) kwashiorkor
    (c) bulimia   (d) marasmus

21. Over half the heat loss from our bodies is attributable to
    (a) radiation   (b) conduction
    (c) convection   (d) evaporation

22. The most effective mechanism for elevating body temperature in infants is
    (a) shivering thermogenesis (b) carbohydrate loading
    (c) fatty acid catabolism   (d) lipolysis in brown fat

23. The resetting of the hypothalamic thermostat by circulating pyrogens such as interleukin-1 produces
    (a) hypothermia
    (b) fever
    (c) maintenance of normal body temperature
    (d) shivering

24. Define the terms *metabolism*, *anabolism*, and *catabolism*.

25. Write the complete reaction sequence for carbohydrate metabolism.

26. What is a lipoprotein? What are the major groups of lipoproteins, and how do they differ?

27. What energy yields, in Calories per gram, are associated with the catabolism of carbohydrates, lipids, and proteins?

28. What is the basal metabolic rate (BMR)?

29. What four basic mechanisms are involved in heat exchange between the body and the environment?

30. What hormones are involved in nonshivering thermogenesis?

LEVEL 2 Reviewing Concepts

31. The function of the ETS is to
    (a) produce energy during periods of active muscle contraction
    (b) break six-carbon chains into three-carbon fragments
    (c) prepare the glucose molecule for further reactions
    (d) remove hydrogen atoms from organic molecules and transfer them to coenzymes

32. During periods of fasting or starvation, the presence of ketone bodies in the bloodstream causes
    (a) an increase in blood pH   (b) a decrease in blood pH
    (c) lipemia   (d) diabetes insipidus

33. What happens during protein catabolism? How is this related to nitrogen balance?

34. Why is the TCA cycle called a cycle? What substances enter the cycle, and what substances leave it?

35. What is oxidative phosphorylation? Explain how the electron transport system is involved in this process.

36. How are lipids catabolized in the body? How is beta-oxidation involved with lipid catabolism?

37. How is RNA catabolized to produce energy?

38. How do the absorptive and postabsorptive states maintain normal blood glucose levels?

39. Why is the liver the focal point for metabolic regulation and control?

40. How can the food pyramid be used as a tool to obtain nutrients in sufficient quantity and quality? Why are the dietary fats, oils, and sugars at the top of the pyramid and bread, cereal, rice, and pasta at the bottom?

41. Why are vitamins and minerals essential components of the diet?

42. How is the brain involved in the regulation of body temperature?

43. Some articles in popular magazines refer to "good cholesterol" and "bad cholesterol." To which types and functions of cholesterol might these terms refer? Explain your answer.

LEVEL 3 Critical Thinking and Clinical Applications

44. While resting and alert, Mary has a respiratory rate of 10 breaths per minute and a tidal volume of 300 ml. The air she is breathing contains 18 percent oxygen. Estimate Mary's basal metabolic rate for a 24-hour period.

45. Why is a starving person more susceptible to infectious disease than one who is well nourished?

46. Individuals with anorexia nervosa typically exhibit bradycardia, hypotension, and a decreased heart size. These problems can eventually lead to death from heart failure. How does anorexia cause such symptoms?

47. The drug *colestipol* binds bile salts in the intestine, forming complexes that cannot be absorbed. How would this drug affect cholesterol levels in blood?

48. Bulimia, the binge–purge eating disorder, is more difficult to diagnose than anorexia nervosa because individuals suffering from bulimia are usually of normal weight. What complicating factors of bulimia might be used to diagnose the disease?

49. Ron is a very health-conscious young adult. He eats well and regularly adds vitamin supplements to his diet. Why might he be warned to monitor his blood pressure and report any unusual hair loss?

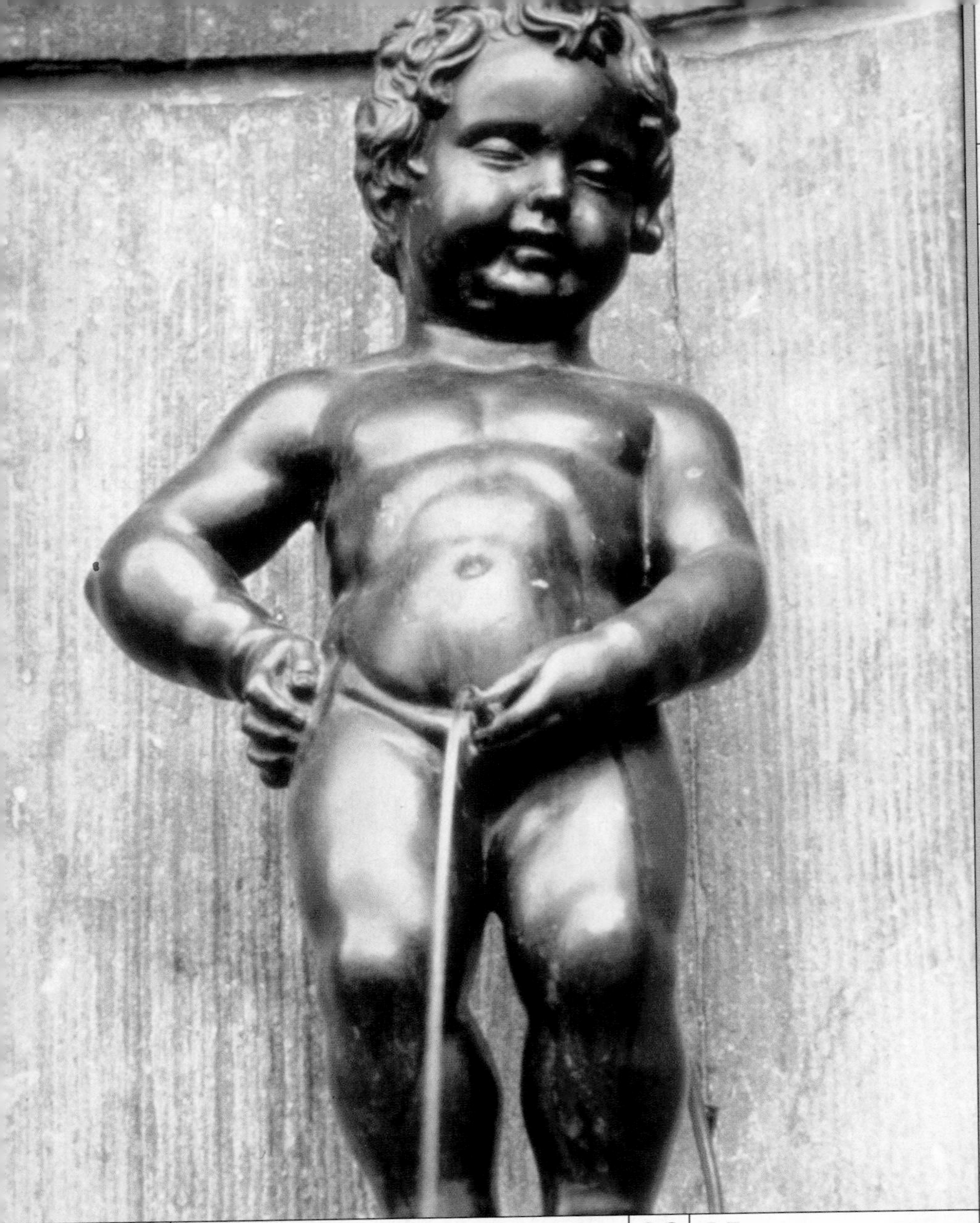

UNIT 5 CHAPTER 23 24 25 **26** 27

CLINICAL NOTES

CLINICAL DISCUSSIONS

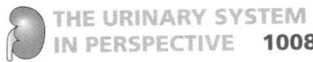

26

THE URINARY SYSTEM

The human body contains trillions of cells bathed in extracellular fluid. In previous chapters, we compared these cells to factories that burn nutrients to obtain energy. Imagine what would happen if *real* factories were built as close together as cells in the body. What a mess they would make! Each would generate piles of garbage, and the smoke they produced would not only deplete the oxygen supply, but also drastically reduce the air quality. In short, a serious pollution problem would result.

The coordinated activities of the digestive, cardiovascular, respiratory, and urinary systems prevent the development of similar pollution problems in the body. The digestive tract absorbs nutrients from food, and the liver adjusts the nutrient concentration of the circulating blood. The cardiovascular system delivers the nutrients and oxygen from the respiratory system to peripheral tissues. As blood leaves these tissues, it carries carbon dioxide and waste products to sites of excretion. The carbon dioxide is removed at the lungs. Most of the physiological waste products are removed by the *urinary system*. In this chapter, we will consider the functional organization of the urinary system and describe the major regulatory mechanisms that control urine production and concentration.

26–1 AN OVERVIEW OF THE URINARY SYSTEM

Objective

- Identify the components of the urinary system, and describe the functions that it performs.

The **urinary system** (Figure 26–1●) has three major functions: (1) *excretion*, or the removal of organic waste products from body fluids, (2) *elimination*, or the discharge of these waste products into the environment, and (3) homeostatic regulation of the volume and solute concentration of blood plasma. The excretory functions of the urinary system are performed by the two **kidneys**—organs that produce **urine**, a fluid containing water, ions, and small soluble compounds. Urine leaving the kidneys travels along the **urinary tract**, which consists of paired tubes called **ureters** (ū-RĒ-terz), to the **urinary bladder**, a muscular sac for temporary storage of urine. On leaving the urinary bladder, urine passes through the **urethra** (ū-RĒ-thra), which conducts the urine to the exterior. The urinary bladder and the urethra are responsible for the elimination of urine, a process called **urination**, or **micturition** (mik-tū-RI-shun). In this process, contraction of the muscular urinary bladder forces urine through the urethra and out of the body.

ATLAS Embryology Summary 20: The Development of the Urinary System

In addition to removing waste products generated by cells throughout the body, the urinary system has several other essential functions that are often overlooked. A more complete list of urinary system functions includes the following:

- *Regulating blood volume and blood pressure*, by adjusting the volume of water lost in urine, releasing erythropoietin, and releasing renin.
- *Regulating plasma concentrations of sodium, potassium, chloride, and other ions*, by controlling the quantities lost in urine and controlling calcium ion levels through the synthesis of calcitriol.
- *Helping to stabilize blood pH*, by controlling the loss of hydrogen ions and bicarbonate ions in urine.
- *Conserving valuable nutrients*, by preventing their excretion in urine while excreting organic waste products—especially nitrogenous wastes, such as *urea* and *uric acid*.
- *Assisting the liver* in detoxifying poisons and, during starvation, deaminating amino acids so that other tissues can break them down.

These activities are carefully regulated to keep the composition of blood within acceptable limits. A disruption of any one of them will have immediate and potentially fatal consequences.

26–2 THE KIDNEYS

Objectives

- Describe the location and structural features of the kidneys.
- Identify the major blood vessels associated with each kidney, and trace the path of blood flow through a kidney.
- Describe the structure of the nephron, and outline the processes involved in the formation of urine.

The kidneys are located on either side of the vertebral column, between vertebrae T_{12} and L_3. The left kidney lies slightly superior to the right kidney (Figures 26–1, anterior view, and 26–2b●, posterior view).

The superior surface of each kidney is capped by an adrenal gland. The kidneys and adrenal glands lie between the muscles of the dorsal body wall and the parietal peritoneum, in a retroperitoneal position (Figure 26–2a●).

The position of the kidneys in the abdominal cavity is maintained by (1) the overlying peritoneum, (2) contact with adjacent visceral organs, and (3) supporting connective tissues. Each kidney is protected and stabilized by three concentric layers of connective tissue (Figure 26–2b●):

1. The **renal capsule**, a layer of collagen fibers that covers the outer surface of the entire organ.
2. The **adipose capsule**, a thick layer of adipose tissue that surrounds the renal capsule.
3. The **renal fascia**, a dense, fibrous outer layer. Collagen fibers extend outward from the renal capsule through the adipose capsule to this layer. The renal fascia anchors the kidney to surrounding structures. Posteriorly, the renal fascia fuses with the deep fascia

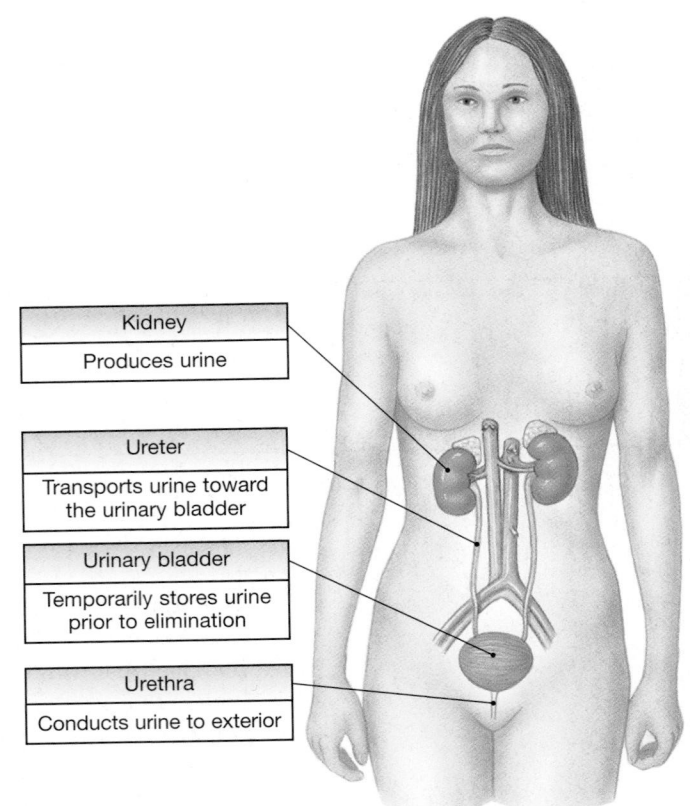

Anterior view

●FIGURE 26–1
An Introduction to the Urinary System. An anterior view of the urinary system, showing the positions of the kidneys and other components.

surrounding the muscles of the body wall. Anteriorly, the renal fascia forms a thick layer that fuses with the peritoneum.

In effect, each kidney hangs suspended by collagen fibers from the renal fascia and is packed in a soft cushion of adipose tissue. This arrangement prevents the jolts and shocks of day-to-day living from disturbing normal kidney function. If the suspensory fibers break or become detached, a slight bump or jar can displace the kidney and stress the attached vessels and ureter. This condition, called a *floating kidney*, is especially dangerous because the ureters or renal blood vessels can become twisted or kinked during movement.

Each reddish-brown kidney is shaped like a kidney bean. A typical adult kidney (Figures 26–3 and 26–4●) is about 10 cm (4 in.) in length, 5.5 cm (2.2 in.) in width, and 3 cm (1.2 in.) in thickness. Each kidney weighs about 150 g (5.25 oz). The **hilus**, a prominent medial indentation, is the point of entry for the *renal artery* and *renal nerves* and the point of exit for the *renal vein* and the ureter.

□ SECTIONAL ANATOMY OF THE KIDNEYS

The fibrous renal capsule covers the outer surface of the kidney and lines the **renal sinus**, an internal cavity within the kidney (Figure 26–4a●). Renal blood vessels and the ureter draining the

●**FIGURE 26–2**
The Position of the Kidneys.   **(a)** A superior view of a section at the level indicated in (b). **(b)** A posterior view of the trunk.
ATLAS Plates 7.9, 7.10a

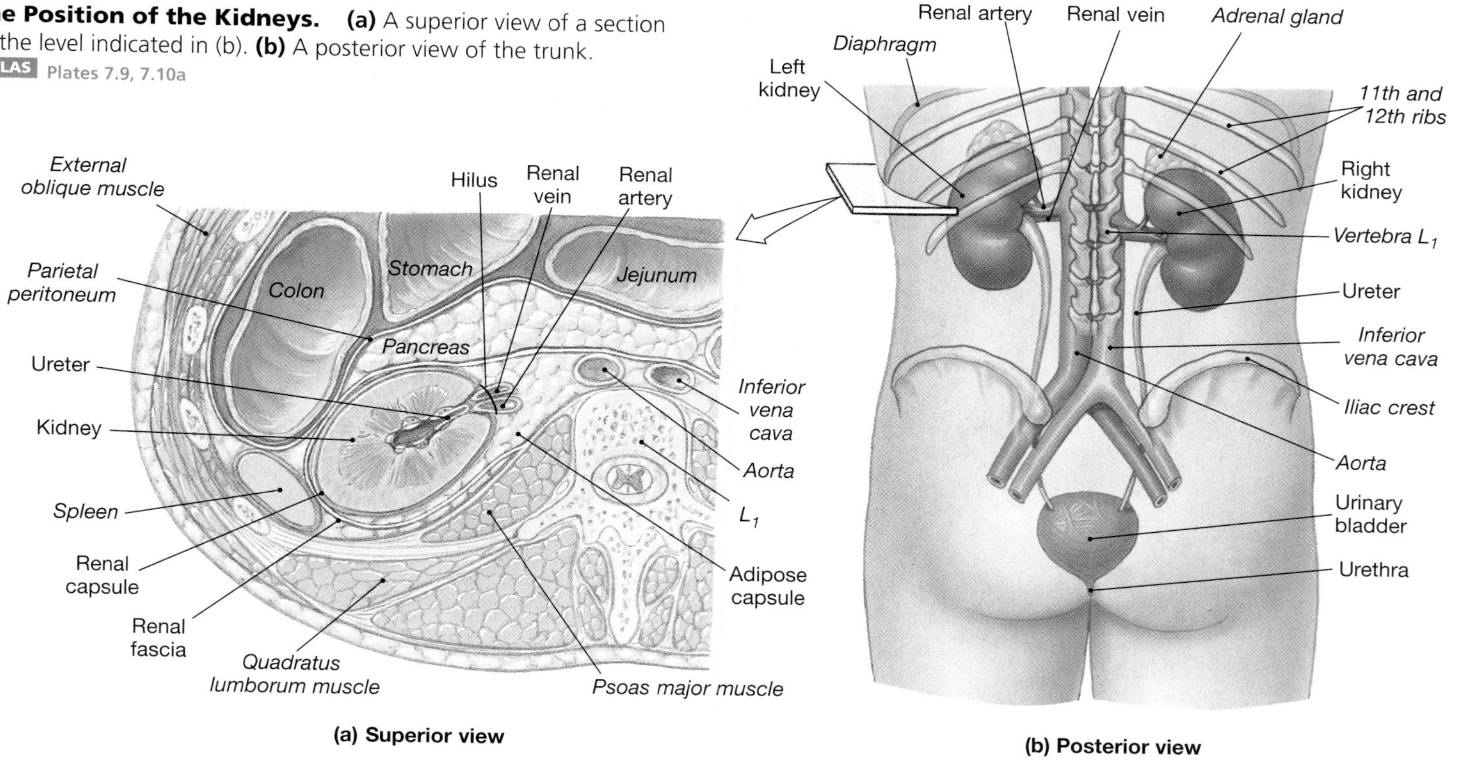

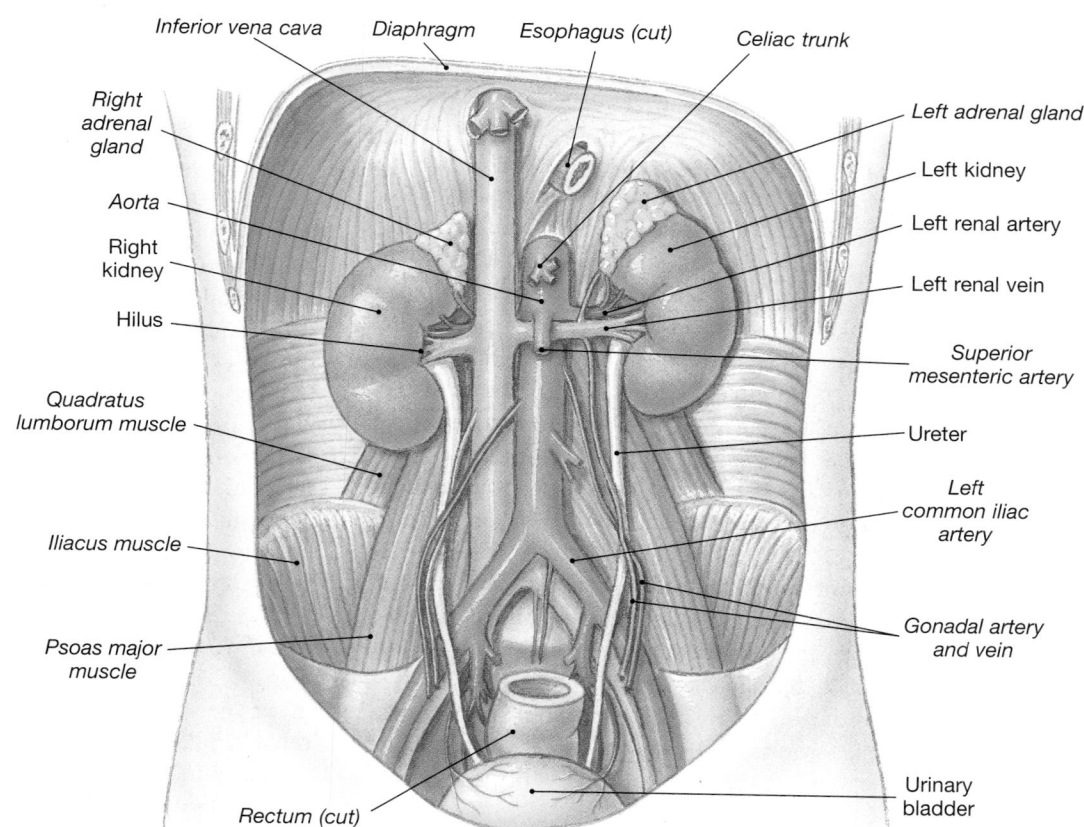

(a) Superior view

Labels (a):
External oblique muscle
Parietal peritoneum
Ureter
Kidney
Spleen
Renal capsule
Renal fascia
Quadratus lumborum muscle
Hilus
Colon
Stomach
Pancreas
Renal vein
Renal artery
Jejunum
Inferior vena cava
Aorta
L_1
Adipose capsule
Psoas major muscle

(b) Posterior view

Labels (b):
Renal artery
Renal vein
Adrenal gland
Diaphragm
Left kidney
11th and 12th ribs
Right kidney
Vertebra L_1
Ureter
Inferior vena cava
Iliac crest
Aorta
Urinary bladder
Urethra

●**FIGURE 26–3**
The Urinary System in Gross Dissection.   The abdominopelvic cavity (with the digestive organs removed), showing the kidneys, ureters, urinary bladder, and blood supply to the urinary structures.
ATLAS Plate 7.10a,d,e

Labels (Figure 26–3):
Inferior vena cava
Diaphragm
Esophagus (cut)
Celiac trunk
Right adrenal gland
Left adrenal gland
Aorta
Left kidney
Right kidney
Left renal artery
Hilus
Left renal vein
Quadratus lumborum muscle
Superior mesenteric artery
Ureter
Iliacus muscle
Left common iliac artery
Psoas major muscle
Gonadal artery and vein
Rectum (cut)
Urinary bladder

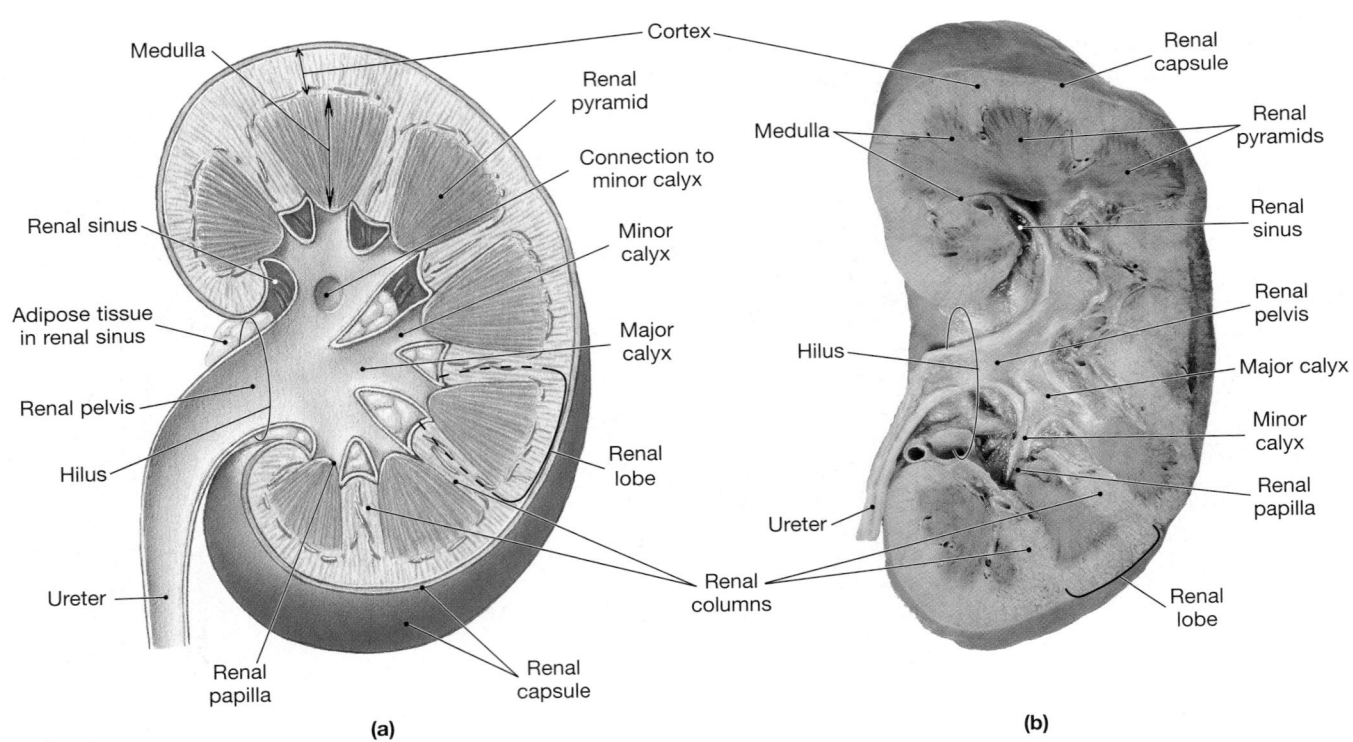

•FIGURE 26–4
The Structure of the Kidney. **(a)** A diagrammatic view of a frontal section through the left kidney. **(b)** A frontal section of the left kidney. **ATLAS** Plate 7.9, 7.10b

kidney pass through the hilus and branch within the renal sinus. The renal capsule is bound to the surfaces of these structures, stabilizing their positions.

The kidney itself has two layers: an outer cortex and an inner medulla. The renal **cortex** is the superficial portion of the kidney, in contact with the renal capsule. The cortex is reddish brown and granular. The renal **medulla** consists of 6 to 18 distinct conical or triangular structures called **renal pyramids**. The base of each pyramid faces the cortex, and the tip of each pyramid, a region known as the **renal papilla**, projects into the renal sinus. Each pyramid has a series of fine grooves that converge at the papilla. Adjacent renal pyramids are separated by bands of cortical tissue called **renal columns**, which extend into the medulla. The columns have a distinctly granular texture, similar to that of the cortex. A **renal lobe** consists of a renal pyramid, the overlying area of renal cortex, and adjacent tissues of the renal columns.

Urine production occurs in the renal lobes. Ducts within each renal papilla discharge urine into a cup-shaped drain called a **minor calyx** (KĀ-liks). Four or five minor calyces (KĀL-i-sēz) merge to form a **major calyx**, and two or three major calyces combine to form the **renal pelvis**, a large, funnel-shaped chamber. The renal pelvis, which fills most of the renal sinus, is connected to the ureter, which drains the kidney.

Urine production begins in microscopic, tubular structures called **nephrons** (NEF-ronz) in the cortex of each renal lobe. Each kidney has roughly 1.25 million nephrons, with a combined length of about 145 km (85 miles).

☐ BLOOD SUPPLY AND INNERVATION OF THE KIDNEYS

Your kidneys receive 20–25 percent of your total cardiac output. In normal, healthy individuals, about 1200 ml of blood flows through the kidneys each minute—a phenomenal amount of blood for organs with a combined weight of less than 300 g (10.5 oz)!

Each kidney receives blood from a **renal artery**, which originates along the lateral surface of the abdominal aorta near the level of the superior mesenteric artery (Figure 21–26a•, p. 757). As it enters the renal sinus, the renal artery provides blood to the **segmental arteries** (Figure 26–5a•). Segmental arteries further divide into a series of **interlobar arteries**, which radiate outward through the renal columns between the renal pyramids. The interlobar arteries supply blood to the **arcuate** (AR-kū-āt) **arteries**, which arch along the boundary between the cortex and medulla of the kidney. Each arcuate artery gives rise to a number of **interlobular arteries**, which supply the cortical portions of the adjacent renal lobes. Branching from each interlobular artery are numerous **afferent arterioles**, which deliver blood to the capillaries supplying individual nephrons (Figure 26–5b•).

From the capillaries of the nephrons, blood enters a network of venules and small veins that converge on the **interlobular veins**. The interlobular veins deliver blood to **arcuate veins**; these in turn empty into **interlobar veins**, which drain directly into the **renal vein** (Figure 26–5a,c•); there are no segmental veins.

The kidneys and ureters are innervated by **renal nerves**. Most of the nerve fibers involved are sympathetic postganglionic

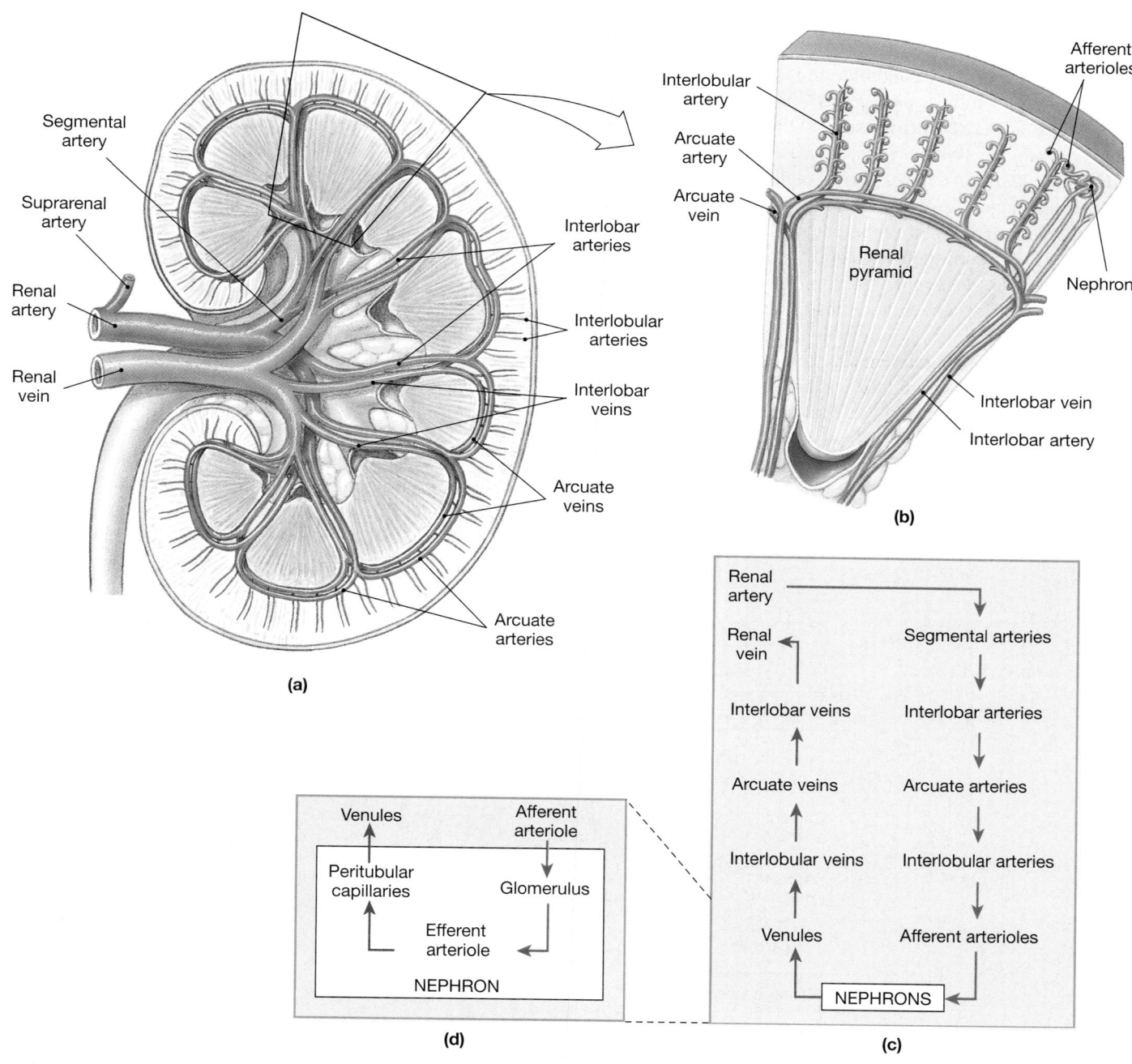

●FIGURE 26–5
The Blood Supply to the Kidneys. **(a)** A sectional view, showing major arteries and veins; compare with Figure 26–4. **(b)** Circulation in the renal cortex. **(c)** A flowchart of renal circulation. **(d)** Blood flow to the nephron. **ATLAS** Plates 7.10a-c; Scans 11e,f

fibers from the celiac plexus and the inferior splanchnic nerves. ∞ p. 538 A renal nerve enters each kidney at the hilus and follows the tributaries of the renal arteries to reach individual nephrons. The sympathetic innervation (1) adjusts rates of urine formation by changing blood flow and blood pressure at the nephron and (2) stimulates the release of renin, which ultimately restricts losses of water and salt in the urine by stimulating reabsorption at the nephron.

The rate of blood flow through the kidneys can be determined by administering the compound *para-aminohippuric acid (PAH)*, which is removed at the nephrons and eliminated in urine. Virtually all the PAH contained in the blood that arrives at the kidneys is removed before the blood departs in the renal veins. Renal blood flow can thus be calculated by comparing plasma concentrations of PAH with the amount secreted in urine. In practice, however, GFR measurement (p. 986) is easier and more widely used. **AM** PAH and the Calculation of Renal Blood Flow

☐ THE NEPHRON

Each nephron consists of a renal tubule and a renal corpuscle. The **renal tubule** is a long tubular passageway which may be 50 mm in length. It begins at the **renal corpuscle** (KOR-pus-ul), a spherical structure consisting of **Bowman's capsule**, a cup-shaped chamber approximately 200 μm in diameter, and a capillary network known as the *glomerulus*. Figure 26–5d● details the pattern of blood flow at the nephron. Blood arrives at the renal corpuscle by way of an afferent arteriole. This arteriole delivers blood to the **glomerulus** (glo-MER-ū-lus; plural, *glomeruli*), which consists of about 50 intertwining capillaries. The glomerulus projects into Bowman's capsule much as the heart projects into the pericardial cavity. Blood leaves the glomerulus in an **efferent arteriole** and flows into a network of capillaries that surround the renal tubule. These capillaries in turn drain into small venules that return the blood to the venous system, as detailed in Figure 26–5c●.

The renal corpuscle is the site where the process of filtration occurs. (We will discuss filtration further in a later section.) In this process, blood pressure forces water and dissolved solutes out of the glomerular capillaries and into a chamber—the *capsular space*—that is continuous with the lumen of the renal tubule (Figure 26–6●). Filtration produces an essentially protein-free solution, known as a **filtrate**, that is otherwise similar to blood plasma.

From the renal corpuscle, filtrate enters the renal tubule, which is responsible for three crucial functions:

- Reabsorbing all the useful organic substrates that enter the filtrate.
- Reabsorbing over 90 percent of the water present in the filtrate.
- Secreting into the tubular fluid any waste products that failed to enter the renal corpuscle through filtration at the glomerulus.

The renal tubule has two convoluted (coiled or twisted) segments—the *proximal convoluted tubule* (PCT) and the *distal convoluted tubule* (DCT)—separated by a simple U-shaped tube, the *loop of Henle* (HEN-lē). The convoluted segments are in the cortex, and the loop of Henle extends partially or completely into the medulla. For clarity, the nephron shown in Figure 26–6● has been shortened and straightened. The regions of the nephron vary by

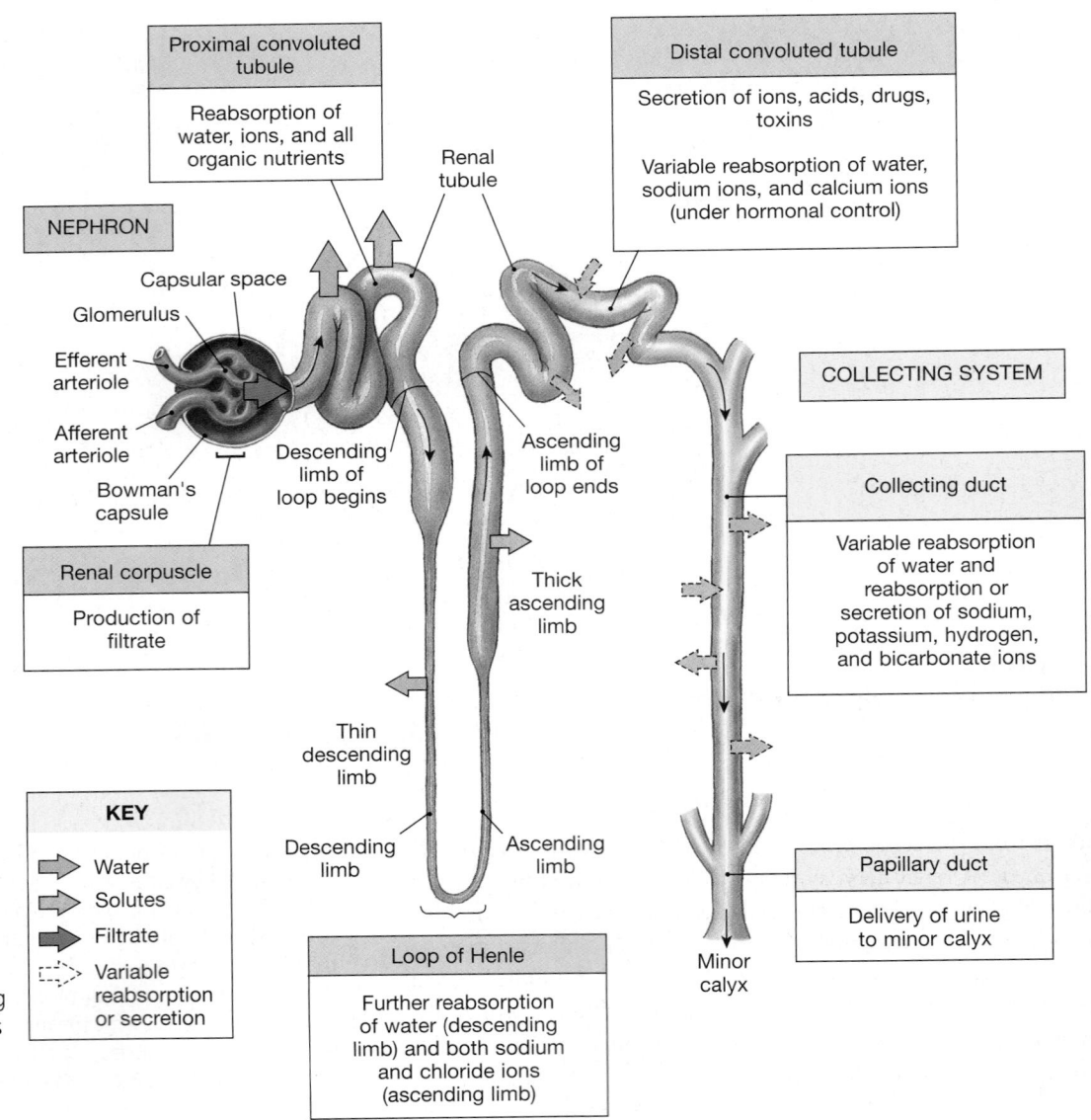

Proximal convoluted tubule

Reabsorption of water, ions, and all organic nutrients

NEPHRON

Capsular space

Glomerulus

Efferent arteriole

Afferent arteriole

Bowman's capsule

Descending limb of loop begins

Renal corpuscle

Production of filtrate

Renal tubule

Distal convoluted tubule

Secretion of ions, acids, drugs, toxins

Variable reabsorption of water, sodium ions, and calcium ions (under hormonal control)

COLLECTING SYSTEM

Ascending limb of loop ends

Thick ascending limb

Thin descending limb

Descending limb

Ascending limb

Collecting duct

Variable reabsorption of water and reabsorption or secretion of sodium, potassium, hydrogen, and bicarbonate ions

Papillary duct

Delivery of urine to minor calyx

Minor calyx

Loop of Henle

Further reabsorption of water (descending limb) and both sodium and chloride ions (ascending limb)

KEY

⇨ Water
⇨ Solutes
⇨ Filtrate
⇢ Variable reabsorption or secretion

●**FIGURE 26–6**
A Representative Nephron.
The major functions of each segment of the nephron and collecting system. For clarity, the nephron has been shortened and straightened considerably.

structure and function. As it travels along the tubule, the filtrate, now called **tubular fluid**, gradually changes in composition. The changes that occur and the characteristics of the urine that results vary with the activities under way in each segment of the nephron. Figure 26–6● and Table 26–1 survey the regional specializations.

Each nephron empties into the **collecting system**, a series of tubes that carry tubular fluid away from the nephron. *Collecting ducts* receive this fluid from many nephrons. Each collecting duct begins in the cortex and descends into the medulla, carrying fluid to a *papillary duct* that drains into a minor calyx.

Nephrons from different locations differ slightly in structure. Roughly 85 percent of all nephrons are **cortical nephrons**, located almost entirely within the superficial cortex of the kidney (Figure 26–7a,b●,). In a cortical nephron, the loop of Henle is relatively short, and the efferent arteriole delivers blood to a network of **peritubular capillaries**, which surround the entire renal tubule. These capillaries drain into small venules that carry blood to the interlobular veins (Figure 26–5c●).

The remaining 15 percent of nephrons, termed **juxtamedullary** (juks-ta-MED-ū-lar-ē) **nephrons** (*juxta*, near), have long loops of Henle that extend deep into the medulla (Figure 26–7a,c●). In these nephrons, the peritubular capillaries are connected to the **vasa recta** (*vasa*, vessel + *recta*, straight)—long, straight capillaries that parallel the loop of Henle.

SUMMARY TABLE 26–1	THE ORGANIZATION OF THE NEPHRON AND COLLECTING SYSTEM IN THE KIDNEY			
Region	**Length**	**Diameter**	**Primary Function**	**Histological Characteristics**
NEPHRON				
Renal corpuscle Glomerulus Parietal epithelium Visceral epithelium	150–250 μm (spherical)	150–250 μm	Filtration of plasma	Glomerulus (capillary knot), supporting cells, and lamina densa, enclosed by Bowman's capsule; visceral epithelium (podocytes) and parietal epithelium separated by capsular space
Renal tubule Proximal convoluted tubule (PCT) Microvilli	14 mm	60 μm	Reabsorption of ions, organic molecules, vitamins, water; secretion of drugs, toxins, acids	Cuboidal cells with microvilli
Loop of Henle Thin segment	30 mm	15 μm	Descending limb: reabsorption of water from tubular fluid	Squamous or low cuboidal cells
Thick segment		30 μm	Ascending limb: reabsorption of ions; assists in creation of a concentration gradient in the medulla	
Distal convoluted tubule (DCT)	5 mm	30–50 μm	Reabsorption of sodium ions and calcium ions; secretion of acids, ammonia, drugs, toxins	Cuboidal cells with few if any microvilli
COLLECTING SYSTEM				
Collecting duct	15 mm	50–100 μm	Reabsorption of water, sodium ions; secretion or reabsorption of bicarbonate ions or hydrogen ions	Cuboidal to columnar cells
Papillary duct	5 mm	100–200 μm	Conduction of tubular fluid to minor calyx; contributes to concentration gradient of the medulla	Columnar cells

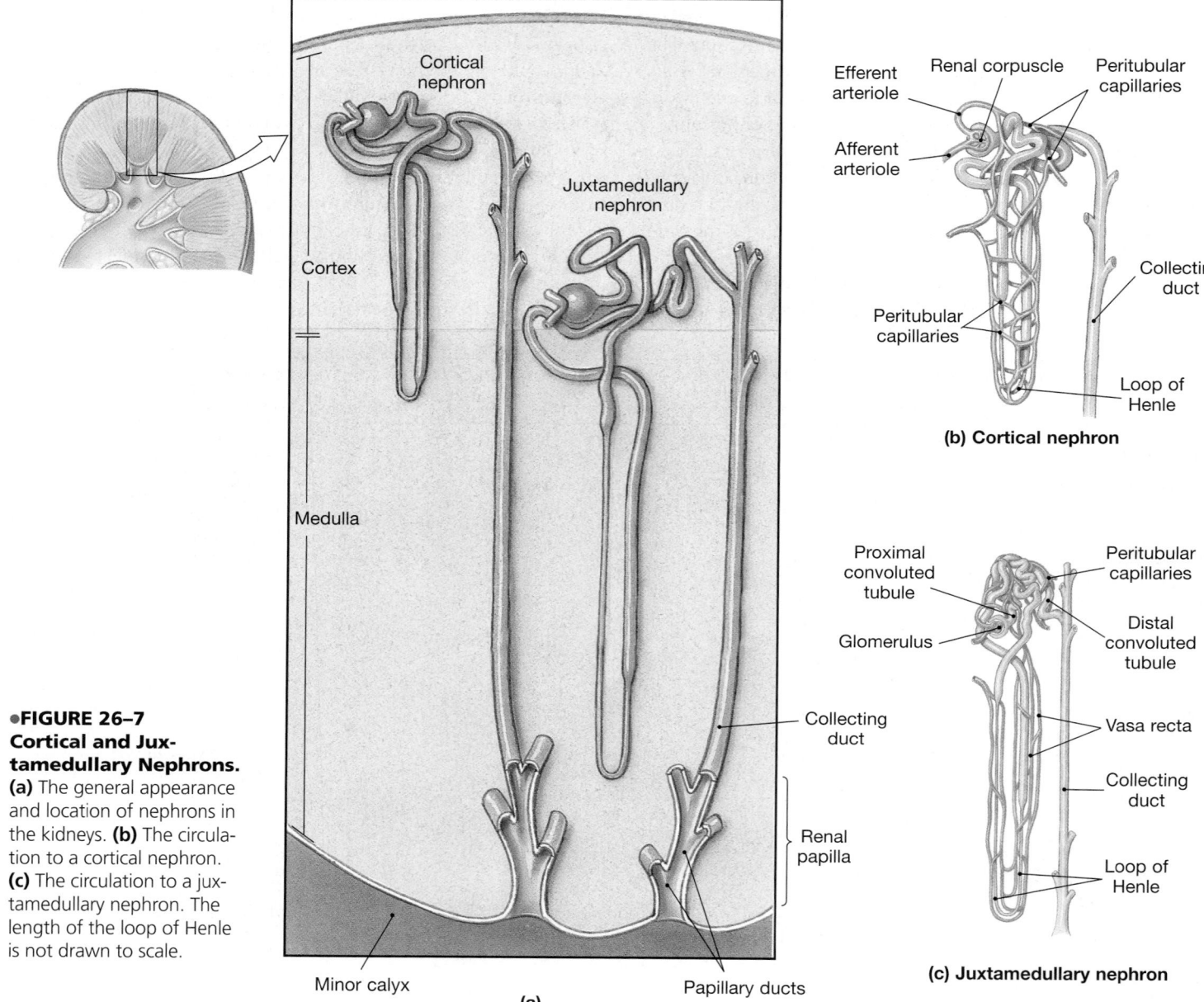

●FIGURE 26–7
**Cortical and Jux-
tamedullary Nephrons.**
(a) The general appearance
and location of nephrons in
the kidneys. **(b)** The circulation to a cortical nephron.
(c) The circulation to a juxtamedullary nephron. The
length of the loop of Henle
is not drawn to scale.

(b) Cortical nephron

(c) Juxtamedullary nephron

(a)

Because they are more numerous than juxtamedullary nephrons, cortical nephrons perform most of the reabsorptive and secretory functions of the kidneys. However, as you will see later in the chapter, it is the juxtamedullary nephrons that give the kidneys the ability to produce concentrated urine.

We shall now examine the structure of each segment of a representative nephron.

The Renal Corpuscle

The renal corpuscle has a diameter of 150–250 μm, on average. It includes (1) a region known as **Bowman's capsule** and (2) the capillary network of the glomerulus (Figure 26–8●). Connected to the initial segment of the renal tubule, Bowman's capsule forms the outer wall of the renal corpuscle and encapsulates the glomerular capillaries.

The glomerulus is surrounded by Bowman's capsule, much as the heart is surrounded by the pericardial cavity. The outer wall of the capsule is lined by a simple squamous **parietal epithelium**. This layer is continuous with the **visceral epithelium**, which covers the glomerular capillaries. The **capsular space** separates the parietal and visceral epithelia. The two epithelial layers are continuous where the glomerular capillaries are connected to the afferent arteriole and efferent arteriole.

The visceral epithelium consists of large cells with complex processes, or "feet," that wrap around the specialized lamina densa of the glomerular capillaries. ∞ p. 114 These unusual cells are called **podocytes** (PŌ-do-sīts; *podos*, foot + -*cyte*, cell). The podocyte feet are known as **pedicels**. Materials passing out of the blood at the glomerulus must be small enough to pass between the narrow gaps, or **filtration slits**, between adjacent pedicels.

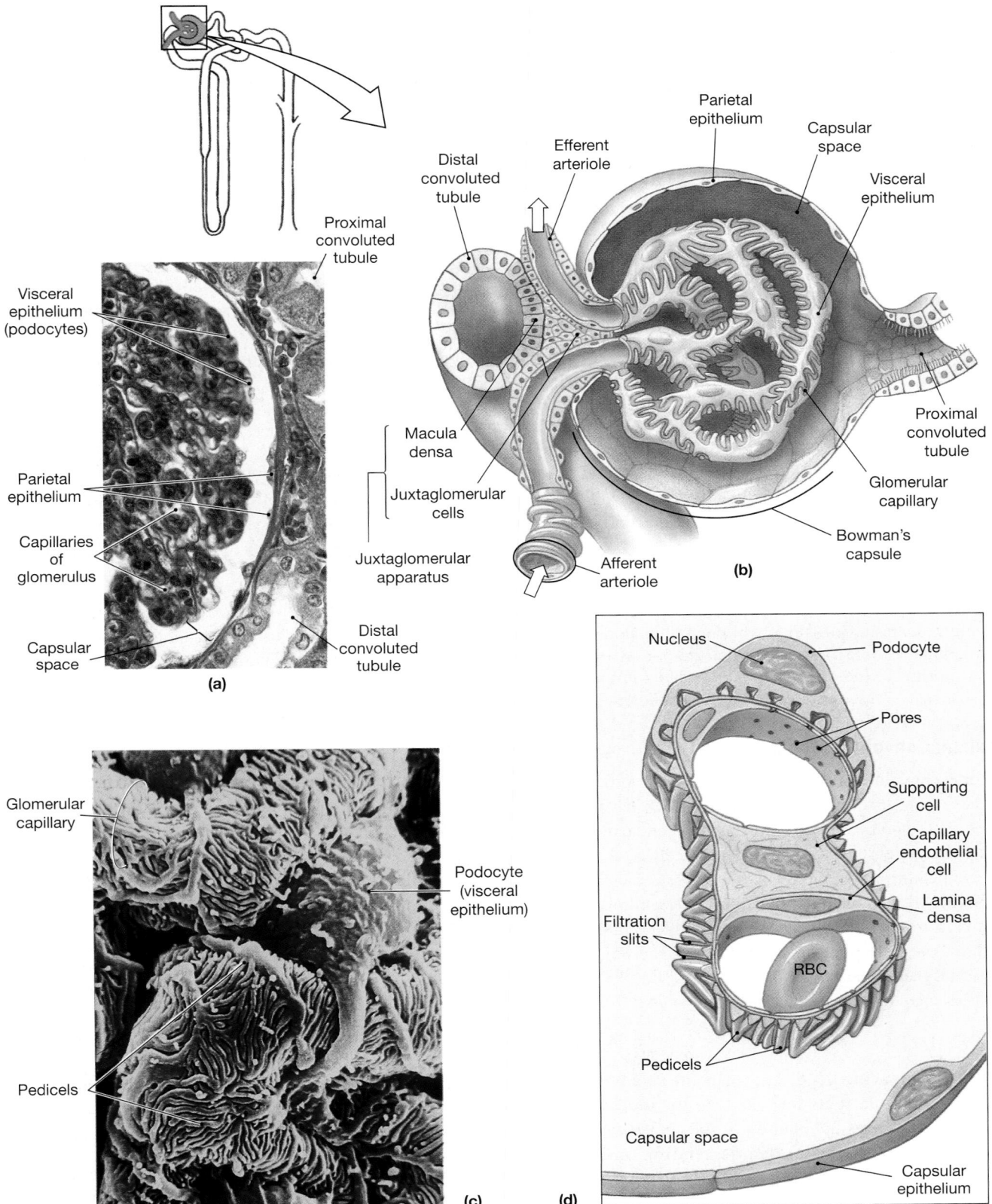

●FIGURE 26–8

The Renal Corpuscle.   **(a)** A renal corpuscle, showing the epithelium of Bowman's capsule and a portion of the glomerular capillary network. (LM × 1120) **(b)** Important structural features of a renal corpuscle. **(c)** The glomerular surface, showing individual podocytes and their processes. (SEM × 27,248) **(d)** A podocyte, with pedicels covering the adjacent surfaces of the lamina densa.

Labels in figure:

(a): Visceral epithelium (podocytes); Proximal convoluted tubule; Parietal epithelium; Capillaries of glomerulus; Capsular space; Distal convoluted tubule

(b): Distal convoluted tubule; Efferent arteriole; Parietal epithelium; Capsular space; Visceral epithelium; Macula densa; Juxtaglomerular cells; Juxtaglomerular apparatus; Afferent arteriole; Proximal convoluted tubule; Glomerular capillary; Bowman's capsule

(c): Glomerular capillary; Podocyte (visceral epithelium); Pedicels

(d): Nucleus; Podocyte; Pores; Supporting cell; Capillary endothelial cell; Lamina densa; Filtration slits; RBC; Pedicels; Capsular space; Capsular epithelium

The glomerular capillaries are fenestrated capillaries—that is, their endothelium contains large-diameter pores (Figure 26–8d●). The lamina densa differs from that found in the basal lamina of other capillary networks in that it may encircle more than one capillary. Special supporting cells that lie between adjacent capillaries play a role in controlling their diameter and thus in the rate of capillary blood flow. Together, the fenestrated endothelium, the lamina densa, and the filtration slits form the *filtration membrane*. During filtration, blood pressure forces water and small solutes across this membrane and into the capsular space. The larger solutes, especially plasma proteins, are excluded. Filtration at the renal corpuscle is both effective and passive, but it has one major limitation: In addition to metabolic wastes and excess ions, compounds such as glucose, free fatty acids, amino acids, vitamins, and other solutes enter the capsular space. These potentially useful materials are recaptured before filtrate leaves the kidneys, with much of the reabsorption occurring in the proximal convoluted tubule.

Glomerulonephritis (glo-mer-ū-lō-nef-RĪ-tis) is an inflammation of the glomeruli that affects the filtration mechanism of the kidneys. The condition, which can develop after an infection involving *Streptococcus* bacteria, is an *immune complex disorder*. The kidneys are not the site of infection, but as the immune system responds, the number of circulating antigen–antibody complexes skyrockets. These complexes are small enough to pass through the lamina densa, but too large to fit between the filtration slits of the filtration membrane. As a result, the filtration mechanism clogs up and filtrate production drops. Any condition that leads to a massive immune response can cause glomerulonephritis, including viral infections and autoimmune disorders. **AM** Immune Complex Disorders; Autoimmune Disorders

The Proximal Convoluted Tubule

The **proximal convoluted tubule (PCT)** is the first segment of the renal tubule (Figure 26–6●). The entrance to the PCT lies almost directly opposite the point where the afferent and efferent arterioles connect to the glomerulus. The lining of the PCT is a simple cuboidal epithelium whose apical surfaces bear microvilli. The tubular cells absorb organic nutrients, ions, water, and plasma proteins (if any) from the tubular fluid and release them into the **peritubular fluid**, the interstitial fluid surrounding the renal tubule. Reabsorption is the primary function of the PCT, but the epithelial cells can also secrete substances into the lumen.

The Loop of Henle

The PCT makes an acute bend that turns the renal tubule toward the renal medulla. This turn leads to the **loop of Henle**, or *nephron loop* (Figures 26–6, and 26–7●, pp. 976, 978). The loop of Henle can be divided into a **descending limb** and an **ascending limb**. Fluid in the descending limb travels toward the renal pelvis, and that in the ascending limb travels toward the renal cortex. Each limb contains a **thick segment** and a **thin segment**. The terms *thick* and *thin* refer to the height of the epithelium, not to the diameter of the lumen: Thick segments have a cuboidal epithelium, whereas a squamous epithelium lines the thin segments.

The thick descending limb has functions similar to those of the PCT: It pumps sodium and chloride ions out of the tubular fluid. The effect of this pumping is most noticeable in the medulla, where the long ascending limbs of juxtamedullary nephrons create unusually high solute concentrations in peritubular fluid. The thin segments are freely permeable to water, but not to solutes; water movement out of these segments helps concentrate the tubular fluid.

The Distal Convoluted Tubule

The thick ascending limb of the loop of Henle ends where it forms a sharp angle near the renal corpuscle. The **distal convoluted tubule (DCT)**, the third segment of the renal tubule, begins there. The initial portion of the DCT passes between the afferent and efferent arterioles (Figure 26–8b●).

In sectional view, the DCT differs from the PCT in that the DCT has a smaller diameter and its epithelial cells lack microvilli (Table 26–1). The DCT is an important site for three important processes:

- The active secretion of ions, acids, and other materials.
- The selective reabsorption of sodium ions and calcium ions from tubular fluid.
- The selective reabsorption of water, which assists in concentrating the tubular fluid.

THE JUXTAGLOMERULAR APPARATUS The epithelial cells of the DCT near the renal corpuscle are taller than those elsewhere along the DCT, and their nuclei are clustered together. This region is called the **macula densa** (MAK-ū-la DEN-sa) (Figure 26–8b●). The cells of the macula densa are closely associated with unusual smooth muscle fibers in the wall of the afferent arteriole. These fibers are known as **juxtaglomerular cells**. Together, the macula densa and juxtaglomerular cells form the **juxtaglomerular apparatus (JGA)**, an endocrine structure that secretes the hormone *erythropoietin* and the enzyme *renin*. ∞ p. 636

The Collecting System

The distal convoluted tubule, the last segment of the nephron, opens into the collecting system, which consists of collecting ducts and papillary ducts (Figure 26–6●). Individual nephrons drain into a nearby **collecting duct**. Several collecting ducts converge to empty into a larger **papillary duct**, which in turn empties into a minor calyx. The epithelium lining the collecting system is typically a columnar epithelium.

In addition to transporting tubular fluid from the nephron to the renal pelvis, the collecting system adjusts its composition and determines the final osmotic concentration and volume of urine.

Polycystic (po-lē-SIS-tik) **kidney disease** is an inherited condition affecting the structure of kidney tubules. Swellings (cysts) develop along the length of the tubules, some growing large enough to compress adjacent nephrons and vessels. Kidney function deteriorates, and the nephrons can eventually become nonfunctional. The process is so gradual that serious problems seldom appear before the individual is 30–40 years of age. Common symptoms include sharp pain in the sides, recurrent urinary infections, and the presence of blood in the urine. Treatment is symptomatic, focusing on the prevention of infection and reduction of pain with analgesics. In severe cases, hemodialysis or kidney transplantation may be required.

✓ Which portions of the nephron are in the renal cortex?

✓ Why don't plasma proteins pass into the capsular space under normal circumstances?

✓ Damage to which part of the nephron would interfere with the control of blood pressure?

Answers start on page Q-1

 The anatomy of the urinary system can be reviewed on the **IP CD-ROM**: Urinary System/Anatomy Review. **Anatomy CD-ROM**: Urinary System/Dissection and Flythrough.

26–3 PRINCIPLES OF RENAL PHYSIOLOGY

Objectives

■ Discuss the major functions of each portion of the nephron and collecting system.

■ Identify and describe the major factors responsible for the production of urine.

■ Describe the normal characteristics, composition, and solute concentrations of a representative urine sample.

The goal of urine production is to maintain homeostasis by regulating the volume and composition of blood. This process involves the excretion of solutes—specifically, metabolic waste products. Three organic waste products are noteworthy:

1. *Urea.* Urea is the most abundant organic waste. You generate roughly 21 g of urea each day, most of it produced during the breakdown of amino acids.

2. *Creatinine.* Creatinine is generated in skeletal muscle tissue by the breakdown of *creatine phosphate*, a high-energy compound that plays an important role in muscle contraction. ∞ p. 316 Your body generates roughly 1.8 g of creatinine each day, and virtually all of it is excreted in urine.

3. *Uric Acid.* Uric acid is formed by the recycling of nitrogenous bases from RNA molecules. ∞ p. 947 You produce approximately 480 mg of uric acid each day.

These waste products are dissolved in the bloodstream and can be eliminated only while dissolved in urine. As a result, their removal is accompanied by an unavoidable water loss. The kidneys are usually capable of producing concentrated urine with an osmotic concentration of 1200–1400 mOsm/l, more than four times that of plasma. (See the box "Expressing Solute Concentrations" on p. 984.) If the kidneys were not able to concentrate the filtrate produced by glomerular filtration, losses of fluid would lead to fatal dehydration in a matter of hours. At the same time, your kidneys ensure that the fluid which *is* lost does not contain potentially useful organic substrates that are present in blood plasma, such as sugars or amino acids. These valuable materials must be reabsorbed and retained for use by other tissues.

□ BASIC PROCESSES OF URINE FORMATION

To perform their functions, your kidneys rely on three distinct processes:

1. *Filtration.* In **filtration**, blood pressure forces water and solutes across the wall of the glomerular capillaries and into the capsular space. Solute molecules small enough to pass through the filtration membrane are carried by the surrounding water molecules.

2. *Reabsorption.* **Reabsorption** is the removal of water and solutes from the filtrate, across the tubular epithelium, and into the peritubular fluid. Reabsorption occurs after filtrate has left the renal corpuscle. Most of the reabsorbed materials are nutrients that your body can use. Whereas filtration occurs solely on the basis of size, reabsorption is a selective process. Solute reabsorption may involve simple diffusion or the activity of carrier proteins in the tubular epithelium. The reabsorbed substances pass into the peritubular fluid and eventually reenter the blood. Water reabsorption occurs passively, through osmosis.

3. *Secretion.* **Secretion** is the transport of solutes from the peritubular fluid, across the tubular epithelium, and into the tubular fluid. Secretion is necessary because filtration does not force all the dissolved materials out of the plasma. Tubular secretion, which provides a backup for the filtration process, can further lower the plasma concentration of undesirable materials. Secretion is often the primary method of excretion for some compounds, including many drugs.

Together, these processes create a fluid that is very different from other body fluids. Table 26–2 provides an indication of the efficiency of the renal system by comparing the concentrations of some substances that are present in both urine and plasma. Before considering the functions of the individual portions of the nephron, we shall briefly examine each of the three major processes involved in urine formation.

Filtration

In filtration, hydrostatic pressure forces water across a membrane, and solute molecules are selected on the basis of size. If the membrane pores are large enough, molecules of solute will be carried along with the water. We can see filtration in action in a coffee machine. Gravity forces hot water through the filter, and the water carries with it a variety of dissolved compounds. The large coffee grounds never reach the pot, because they cannot fit through the pores of the filter. In other words, they are "filtered out" of the solution.

In the body, the heart pushes blood around the cardiovascular system and generates hydrostatic pressure. Filtration occurs across the walls of small blood vessels, pushing water and dissolved materials into the interstitial fluids of the body (Figure 21–10●, p. 734). In some cases, such as the liver, the pores are very large, and even small plasma proteins can be carried into the interstitial fluids. At

TABLE 26–2	SIGNIFICANT DIFFERENCES BETWEEN SOLUTE CONCENTRATIONS OF URINE AND PLASMA	
Component	Urine	Plasma
IONS (mEq/l) [*]		
Sodium (Na^+)	147.5	138.4
Potassium (K^+)	47.5	4.4
Chloride (Cl^-)	153.3	106
Bicarbonate (HCO_3^-)	1.9	27
METABOLITES AND NUTRIENTS (mg/dl)		
Glucose	0.009	90
Lipids	0.002	600
Amino acids	0.188	4.2
Proteins	0.000	7.5 g/dl
NITROGENOUS WASTES (mg/dl)		
Urea	1800	10–20
Creatinine	150	1–1.5
Ammonia	60	<0.1
Uric acid	40	3

* See the box "Expressing Solute Concentrations," on page 984.

the renal corpuscle, however, filtration occurs across a specialized filtration membrane that greatly restricts the passage of even the smallest circulating proteins.

Reabsorption and Secretion

The processes of reabsorption and secretion at the kidneys involve a combination of diffusion, osmosis, and carrier-mediated transport. We considered diffusion and osmosis in Chapter 3, as well as in several other chapters, so we shall pause now only for a brief review of carrier-mediated transport mechanisms.

TYPES OF CARRIER-MEDIATED TRANSPORT In previous chapters, we considered four major types of *carrier-mediated transport:*

- In *facilitated diffusion*, a carrier protein transports a molecule across the cell membrane without expending energy (Figure 3–22●, p. 95). The transport always follows the concentration gradient for the ion or molecule transported.

- *Active transport* is driven by the hydrolysis of ATP to ADP on the inner membrane surface (Figure 3–23●, p. 96). Exchange pumps and other carrier proteins are active along the kidney tubules. Active transport mechanisms can operate despite an opposing concentration gradient.

- In *cotransport*, carrier protein activity is not directly linked to the hydrolysis of ATP (Figure 3–24●, p. 96). Instead, two substrates (ions, molecules, or both) cross the membrane while they are bound to the carrier protein. The movement of the substrates always follows the concentration gradient of at least one of the transported substances. Cotransport mechanisms are responsible for the reabsorption of organic and inorganic compounds from the tubular fluid.

- *Countertransport* resembles cotransport in all respects, except that the two transported ions move in *opposite* directions (Figures 23–24, p. 859, and 24–14●, p. 894). The PCT, DCT, and collecting system contain countertransport mechanisms.

CHARACTERISTICS OF CARRIER-MEDIATED TRANSPORT All of the carrier-mediated processes share five features that are important for an understanding of kidney function:

1. *A Specific Substrate Binds to a Carrier Protein That Facilitates Movement Across the Membrane.*

2. *A Given Carrier Protein Normally Works in One Direction Only.* In facilitated diffusion, that direction is determined by the concentration gradient of the substance being transported. In active transport, cotransport, and countertransport, the location and orientation of the carrier proteins determine whether a particular substance will be reabsorbed or secreted. For example, the carrier protein that transports amino acids from the tubular fluid to the cytoplasm will not carry amino acids back into the tubular fluid.

3. *The Distribution of Carrier Proteins Can Vary from One Portion of the Cell Surface to Another.* Transport between tubular fluid and interstitial fluid involves two steps, because the material must enter the cell at its apical surface and then leave the cell and enter the peritubular fluid at the cell's basolateral surface. Each step involves a different carrier protein. For example, the apical surfaces of cells along the proximal convoluted tubule contain carrier proteins that bring amino acids, glucose, and many other nutrients into these cells by sodium-linked cotransport. In contrast, the basolateral surfaces contain carrier proteins that move those nutrients out of the cell by facilitated diffusion.

4. *The Membrane of a Single Tubular Cell Contains Many Types of Carrier Protein.* Each cell can have multiple functions, and the same cell that reabsorbs one compound can secrete another.

5. *Carrier Proteins, Like Enzymes, Can Be Saturated.* When an enzyme is *saturated*, further increases in substrate concentration have no effect on the rate of reaction ∞ p. 56 When a carrier protein is saturated, further increases in substrate concentration have no effect on the rate of transport across the cell membrane. For any substance, the concentration at saturation is called the **transport maximum** (T_m), or *tubular maximum*. The saturation of carrier proteins involved with tubular secretion seldom occurs in healthy individuals, but carriers involved with tubular reabsorption are often at risk of saturation, especially during the absorptive state that follows a meal. ∞ p. 949

T_m **AND THE RENAL THRESHOLD** Normally, any plasma proteins and nutrients, such as amino acids and glucose, are removed from the tubular fluid by cotransport or facilitated diffusion. If the concentrations of these nutrients rise in the tubular fluid, the rates of reabsorption increase until the carrier proteins are saturated. A concentration higher than the tubular maximum will exceed the reabsorptive abilities of the nephron, so some of the material will remain in the tubular fluid and appear in the urine. The transport maximum thus determines the **renal threshold**—the plasma concentration at which a specific compound or ion will begin appearing in urine.

The renal threshold varies with the substance involved. The renal threshold for glucose is approximately 180 mg/dl. When plasma glucose concentrations remain higher than 180 mg/dl, glucose concentrations in tubular fluid will exceed the T_m of the tubular cells, so glucose will appear in urine. The presence of glucose in urine is a condition called *glycosuria*. After you have eaten a meal rich in carbohydrates, your plasma glucose levels may exceed the T_m for a brief period. Your liver will quickly lower circulating glucose levels, and very little glucose will be lost in your urine. However, chronically elevated plasma and urinary glucose concentrations are highly unusual. (Glycosuria is one of the key symptoms of diabetes mellitus. ∞ p. 633)

The renal threshold for amino acids is lower than that for glucose; amino acids appear in urine when plasma concentrations exceed 65 mg/dl. Your plasma amino acid levels commonly exceed the renal threshold after you have eaten a protein-rich meal, causing some amino acids to appear in your urine. This condition is termed *aminoaciduria*.

T_m values for water-soluble vitamins are relatively low; as a result, excess quantities of these vitamins are excreted in urine. (This is typically the fate of daily water-soluble vitamin supplements.) Cells of the renal tubule ignore a number of other compounds in the tubular fluid. As water and other compounds are removed, the concentration of the ignored materials in the tubular fluid gradually rises. Table 26–3 contains a partial listing of substances that are actively reabsorbed, secreted, or ignored by the renal tubules.

□ AN OVERVIEW OF RENAL FUNCTION

Figure 26–9● reviews the general functions of the various segments of the nephron and collecting system in the formation of urine. Most regions perform a combination of reabsorption and secretion, but the balance between the two shifts from one region to another:

- Filtration occurs exclusively in the renal corpuscle, across the filtration membrane.
- Nutrient reabsorption occurs primarily along the proximal convoluted tubules, but also elsewhere along the renal tubule and within the collecting system.
- Active secretion occurs primarily at the proximal and distal convoluted tubules.
- The loops of Henle—especially the long loops of the juxtamedullary nephrons—and the collecting system interact to regulate the final volume and solute concentration of the urine.

Normal kidney function can continue only as long as filtration, reabsorption, and secretion function within relatively narrow limits. A disruption in kidney function has immediate effects on the composition of the circulating blood. If both kidneys are affected, death will occur within a few days unless medical assistance is provided.

We shall now proceed along the nephron to consider the formation of filtrate and the changes in the composition and concentration that occur as the filtrate passes along the renal tubule. Most of

TABLE 26–3 TUBULAR REABSORPTION AND SECRETION

Reabsorbed	Secreted	No Transport Mechanism
Ions	**Ions**	Urea
Na^+, Cl^-, K^+,	K^+, H^+, Ca^{2+},	Water
Ca^{2+}, Mg^{2+},	PO_4^{3-}	Urobilinogen
SO_4^{2-}, HCO_3^-		Bilirubin
	Wastes	
	Creatinine	
Metabolites	Ammonia	
Glucose	Organic acids and bases	
Amino acids		
Proteins		
Vitamins	**Miscellaneous**	
	Neurotransmitters (ACh, NE, E, dopamine)	
	Histamine	
	Drugs (penicillin, atropine, morphine, many others)	

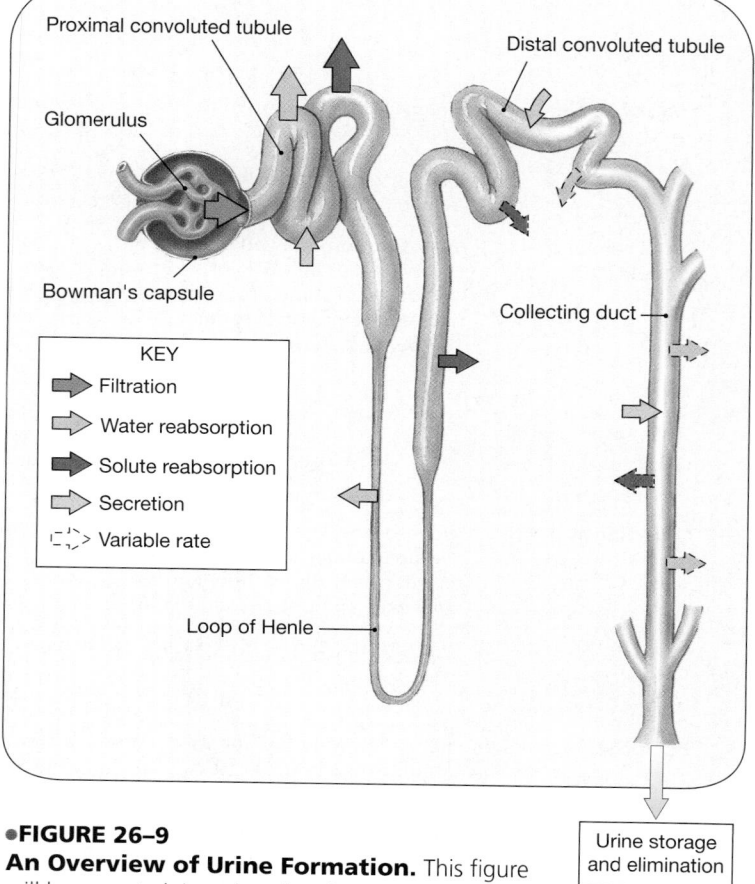

●**FIGURE 26–9**
An Overview of Urine Formation. This figure will be repeated, in reduced and simplified form as Navigator icons, in key figures as we change topics.

what follows applies equally to cortical and juxtamedullary nephrons. The major differences between the two types of nephron are that the loop of Henle of a cortical nephron is shorter and does not extend as far into the medulla as does the loop of Henle of a juxtamedullary nephron (Figure 26–7a●, p. 978). The long loop of Henle in a juxtamedullary nephron extends deep into the renal pyramids, where it plays a vital role in water conservation and the formation of concentrated urine. Because this process is so important, affecting the tubular fluid produced by every nephron in the kidney, and because the functions of the renal corpuscle and of the proximal and distal convoluted tubules are the same in all

nephrons, we shall use a juxtamedullary nephron as our example. Table 26–4 surveys the functions of the various parts of the nephron. Filtration occurs only in the glomerulus, with a combination of reabsorption and secretion occurring along the rest of the nephron and collecting system.

The osmotic concentration, or *osmolarity*, of a solution is the total number of solute particles in each liter. ∞ p. 94 Osmolarity is usually expressed in **osmoles** per liter (Osm/l) or **milliosmoles** per liter (mOsm/l). If each liter of a fluid contains 1 mole of dissolved particles, the solute concentration is 1 Osm/l, or 1000 mOsm/l. Body fluids have an osmotic concentration of

	SUMMARY TABLE 26–4	**RENAL STRUCTURES AND THEIR FUNCTIONS**	
Segment	**General Functions**	**Specific Functions**	**Mechanisms**
Renal corpuscle	*Filtration* of plasma; generates approximately 80 liters/day of filtrate similar in composition to blood plasma without plasma proteins	*Filtration* of water, inorganic and organic solutes from plasma; retention of plasma proteins and blood cells	Glomerular hydrostatic (blood) pressure working across capillary endothelium, lamina densa, and filtration slits
Proximal convoluted tubule (PCT)	*Reabsorption* of 60–70% of the water (108–116 liters/day), 99–100% of the organic substrates, and 60–70% of the sodium and chloride ions in the original filtrate	*Reabsorption:* Active: glucose, other simple sugars, amino acids, vitamins, ions (including sodium, potassium, calcium, magnesium, phosphate, and bicarbonate) Passive: urea, chloride ions, lipid-soluble materials, water *Secretion:* Hydrogen ions, ammonium ions, creatinine, drugs, and toxins (as at DCT)	Carrier-mediated transport, including facilitated transport (glucose, amino acids), cotransport (glucose, ions), and countertransport (with secretion of H^+) Diffusion (solutes) or osmosis (water) Countertransport with sodium ions
Loop of Henle	*Reabsorption* of 25% of the water (45 liters/day) and 20–25% of the sodium and chloride ions present in the original filtrate; creation of the concentration gradient in the medulla	*Reabsorption:* Sodium and chloride ions Water	Active transport via $Na^+ - K^+/2\,Cl^-$ transporter Osmosis
Distal convoluted tubule (DCT)	*Reabsorption* of a variable amount of water (usually 5%, or 9 liters/day), under ADH stimulation, and a variable amount of sodium ions, under aldosterone stimulation	*Reabsorption:* Sodium and chloride ions Sodium ions (variable) Calcium ions (variable) Water (variable) *Secretion:* Hydrogen ions, ammonium ions Creatinine, drugs, toxins	Cotransport Countertransport with potassium ions; aldosterone regulated Carrier-mediated transport stimulated by parathyroid hormone and calcitriol Osmosis; ADH regulated Countertransport with sodium ions Carrier-mediated transport
Collecting system	*Reabsorption* of a variable amount of water (usually 9.3%, or 16.8 liters/day), under ADH stimulation, and a variable amount of sodium ions, under aldosterone stimulation	*Reabsorption:* Sodium ions (variable) Bicarbonate ions (variable) Water (variable) Urea (distal portions only) *Secretion:* Potassium and hydrogen ions (variable)	Countertransport with potassium or hydrogen ions; aldosterone-regulated Diffusion, generated within tubular cells Osmosis; ADH regulated Diffusion Carrier-mediated transport
Peritubular capillaries	*Redistribution* of water and solutes reabsorbed in the cortex	Return of water and solutes to the general circulation	Osmosis and diffusion
Vasa recta	*Redistribution* of water and solutes reabsorbed in the medulla and stabilization of the concentration gradient of the medulla	Return of water and solutes to the general circulation	Osmosis and diffusion

about 300 mOsm/l. In comparison, that of seawater is about 1000 mOsm/l and that of fresh water about 5 mOsm/l. Ion concentrations are often reported in *millequivalents* per liter (mEq/l), whereas the concentrations of large organic molecules are usually reported in grams or milligrams per unit volume of solution (typically, mg or g per dl). **AM** Solutions and Concentrations

26–4 RENAL PHYSIOLOGY: FILTRATION AND THE GLOMERULUS

Objective

- List and describe the factors that influence filtration pressure and the rate of filtrate formation.

Filtration occurs in the renal corpuscle as fluids move across the wall of the glomerulus and into the capsular space. The process of **glomerular filtration** involves passage across the filtration membrane. This membrane has three components: (1) the capillary endothelium, (2) the lamina densa, and (3) the filtration slits (Figures 26–8d and 26–10a●).

Glomerular capillaries are fenestrated capillaries with pores ranging from 60 to 100 nm (0.06 to 0.1 μm) in diameter. These openings are small enough to prevent the passage of blood cells, but they are too large to restrict the diffusion of solutes, even those the size of plasma proteins. The lamina densa is more selective: Only small plasma proteins, nutrients, and ions can cross it. The filtration slits are the finest filters of all. Their gaps are only 6–9 nm wide, which is small enough to block the passage of most small plasma proteins. As a result, under normal circumstances no plasma proteins other than a few albumin molecules (with an average diameter of 7 nm) can cross the filtration membrane and enter the capsular space. However, plasma proteins are all that stay behind, so the filtrate contains dissolved ions and small organic molecules in roughly the same concentrations as in plasma.

□ FILTRATION PRESSURES

We discussed the major forces that act across capillary walls in Chapters 21 and 22. (You may find it helpful to review Figures 21–12 and 21–13● pp. 739, 740, before you proceed.) The primary factor involved in glomerular filtration is basically the same as that governing fluid and solute movement across capillaries throughout the body: the balance between hydrostatic pressure (fluid pressure) and colloid osmotic pressure (pressure due to materials in solution) on either side of the capillary walls.

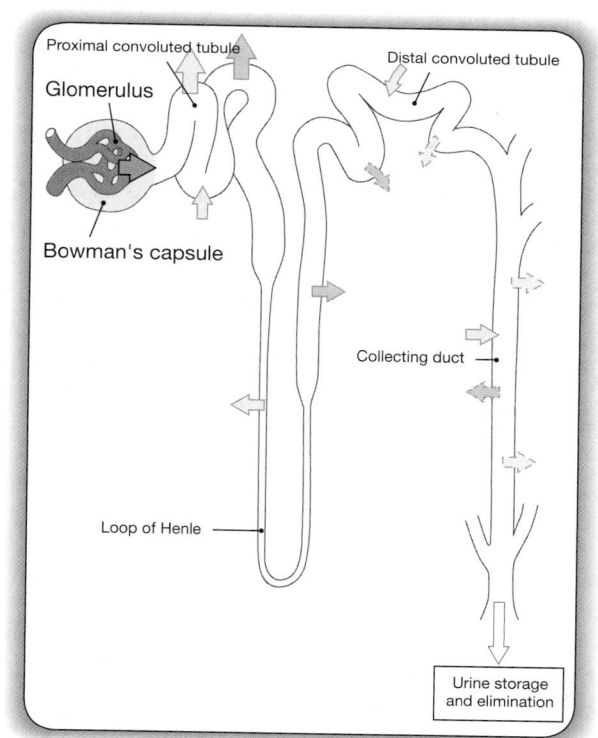

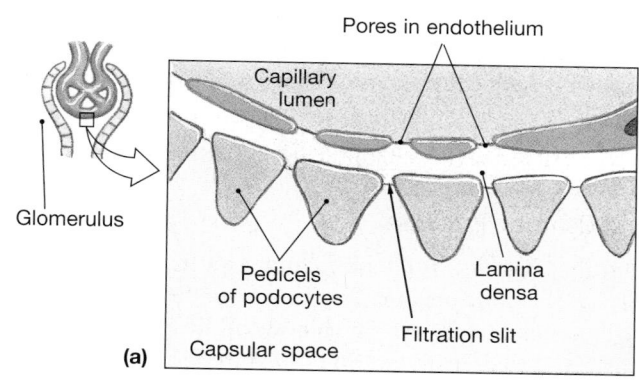

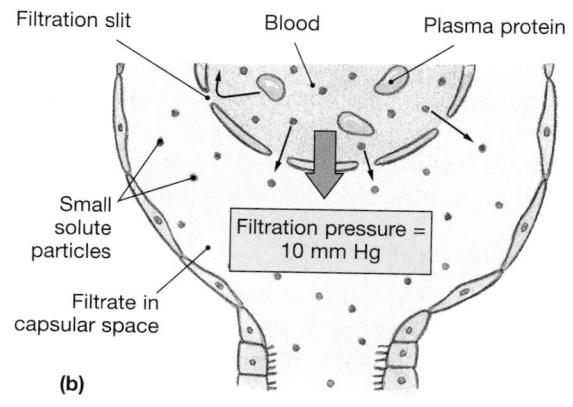

●**FIGURE 26–10**
Glomerular Filtration. **(a)** The filtration membrane. **(b)** Filtration pressure. The Navigator icon in the shadow box repeats Figure 26–9, but highlights the part considered at this point in the discussion.

Hydrostatic Pressure

The **glomerular hydrostatic pressure (GHP)** is the blood pressure in the glomerular capillaries. This pressure tends to push water and solute molecules out of the plasma and into the filtrate. The GHP is significantly higher than capillary pressures elsewhere in the systemic circuit, due to the arrangement of vessels at the glomerulus.

Blood pressure is low in typical systemic capillaries, because capillary blood flows into the venous system, where resistance is relatively low. However, at the glomerulus, blood leaving the glomerular capillaries flows into an efferent arteriole, whose diameter is *smaller* than that of the afferent arteriole. The efferent arteriole offers considerable resistance, so relatively high pressures are needed to force blood into it. As a result, glomerular pressures are similar to those of small arteries, averaging 45–55 mm Hg instead of the 35 mm Hg typical of peripheral capillaries.

Glomerular hydrostatic pressure is opposed by the **capsular hydrostatic pressure (CsHP)**, which tends to push water and solutes out of the filtrate and into the plasma. This pressure results from the resistance to flow along the nephron and the conducting system. (Before additional filtrate can enter the capsule, some of the filtrate already present must be forced into the PCT.) The CsHP averages about 15 mm Hg.

The *net hydrostatic pressure (NHP)* is the difference between the glomerular hydrostatic pressure, which tries to push water and solutes out of the bloodstream, and the capsular hydrostatic pressure, which tries to push water and solutes into the bloodstream. Because glomerular hydrostatic pressure averages 50 mm Hg and capsular hydrostatic pressure averages 15 mm Hg, the net hydrostatic pressure is 35 mm Hg. This relationship can be written as an equation:

$$\text{NHP} = \text{GHP} - \text{CsHP} = 50 \text{ mm Hg} - 15 \text{ mm Hg} = 35 \text{ mm Hg}$$

Colloid Osmotic Pressure

Recall from Chapter 21 that the colloid osmotic pressure of a solution is the osmotic pressure resulting from the presence of suspended proteins. ∞ p. 737 The **blood colloid osmotic pressure (BCOP)** tends to draw water out of the filtrate and into the plasma. It thus opposes filtration. Over the entire length of the glomerular capillary bed, the BCOP averages about 25 mm Hg. Under normal conditions, very few plasma proteins enter the capsular space, so there is no opposing colloid osmotic pressure within the capsule. However, if the glomeruli are damaged by disease or injury and plasma proteins begin passing into the capsular space, a *capsular colloid osmotic pressure* is created that promotes filtration and increases fluid losses in urine.

Filtration Pressure

The **filtration pressure (FP)** at the glomerulus is the difference between the hydrostatic pressure and the colloid osmotic pressure acting across the glomerular capillaries. Under normal circumstances, this relationship can be summarized as

$$\text{FP} = \text{NHP} - \text{BCOP}$$

or

$$\text{filtration pressure} = 35 \text{ mm Hg} - 25 \text{ mm Hg} = 10 \text{ mm Hg.}$$

This number is the average pressure forcing water and dissolved materials out of the glomerular capillaries and into the capsular spaces (Figure 26–10b●). Problems that affect filtration pressure can seriously disrupt kidney function and cause a variety of clinical symptoms. **AM** Conditions Affecting Filtration

☐ THE GLOMERULAR FILTRATION RATE

The **glomerular filtration rate (GFR)** is the amount of filtrate your kidneys produce each minute. Each kidney contains about 6 m^2—some 64 square feet—of filtration surface, and the GFR averages an astounding *125 ml per minute*. This means that roughly 10 percent of the fluid delivered to your kidneys by the renal arteries leaves the bloodstream and enters the capsular spaces.

A *creatinine clearance test* is often used to estimate the GFR. Creatinine, which results from the breakdown of creatine phosphate in muscle tissue, is normally eliminated in urine. Creatinine enters the filtrate at the glomerulus and is not reabsorbed in significant amounts. By monitoring the creatinine concentrations in blood and the amount excreted in urine, a clinician can easily estimate the GFR. Consider, for example, a person who eliminates 84 mg of creatinine each hour and has a plasma creatinine concentration of 1.4 mg/dl. Because the GFR is equal to the amount secreted, divided by the plasma concentration, this person's GFR is

$$\frac{84 \text{ mg/h}}{1.4 \text{ mg/dl}} = 60 \text{ dl/h} = 100 \text{ ml/min.}$$

The GFR is usually reported in milliliters per minute.

The value 100 ml/min is only an approximation of the GFR, because up to 15 percent of creatinine enters the urine by means of active tubular secretion. When necessary, a more accurate GFR determination can be performed by using the complex carbohydrate *inulin*, which is not metabolized in the body and is neither reabsorbed nor secreted by the kidney tubules.

In the course of a single day, your glomeruli generate about 180 liters (50 gal) of filtrate, roughly 70 times the total plasma volume. But as filtrate passes through the renal tubules, about 99 percent of it is reabsorbed. You should now appreciate the significance of tubular reabsorption!

The glomerular filtration rate depends on the filtration pressure across glomerular capillaries. Any factor that alters the filtration pressure will therefore alter the GFR and thereby affect kidney function. One of the most significant factors is a drop in renal blood pressure.

Despite the large volume of filtrate generated, the filtration pressure is relatively low. Whenever the mean arterial pressure falls by 10 percent (from a normal value of about 100 mm Hg to 90 mm Hg), the GFR is severely restricted. If blood pressure at the glomeruli drops by 20 percent, from 50 mm Hg to 40 mm Hg, kidney filtration will cease, because the filtration pressure will be 0 mm Hg. The kidneys are therefore sensitive to changes in blood pressure that have little or no effect on other organs. Hemorrhaging, shock, and dehydration are relatively common clinical conditions that can cause a dangerous decline in the GFR.

☐ CONTROLLING THE GFR

Glomerular filtration is the vital first step essential to all other kidney functions. If filtration does not occur, waste products are not excreted, pH control is jeopardized, and an important mechanism for regulating blood volume is lost. It should be no surprise to find that a variety of regulatory mechanisms exist to ensure that your GFR remains within normal limits.

Filtration depends on adequate blood flow to the glomerulus and on the maintenance of normal filtration pressures. Three interacting levels of control stabilize your GFR: (1) *autoregulation*, at the local level, (2) *hormonal regulation*, initiated by the kidneys, and (3) *autonomic regulation*, primarily by the sympathetic division of the autonomic nervous system.

Autoregulation of the GFR

Autoregulation maintains an adequate GFR despite changes in local blood pressure and blood flow. Maintenance of the GFR is accomplished by changing the diameters of the afferent arterioles, the efferent arterioles, and the glomerular capillaries. The most important regulatory mechanisms stabilize the GFR when the systemic blood pressure declines. A reduction in blood flow and a decline in glomerular blood pressure trigger (1) the dilation of the afferent arteriole, (2) the relaxation of supporting cells and the dilation of the glomerular capillaries, and (3) the constriction of the efferent arteriole. This combination increases blood flow and elevates glomerular blood pressure to normal levels. As a result, filtration rates remain relatively constant. The GFR also remains relatively constant when the systemic blood pressure rises. A rise in renal blood pressure stretches the walls of the afferent arterioles, and the smooth muscle cells respond by contracting. The reduction in the diameter of the afferent arterioles decreases glomerular blood flow and keeps the GFR within normal limits.

Hormonal Regulation of the GFR

The GFR is regulated by the hormones of the renin–angiotensin system and the natriuretic peptides (ANP and BNP). We introduced those hormones and their actions in Chapters 18 and 21. ∞ pp. 636, 746 Renin is an enzyme released by the juxtaglomerular apparatus (JGA) when (1) blood pressure declines at the glomerulus, as the result of a decrease in blood volume, a fall in systemic pressures, or a blockage in the renal artery or its tributaries; (2) juxtaglomerular cells are stimulated by sympathetic innervation; or (3) there is a decline in the osmotic concentration of the tubular fluid at the macula densa. All three of these factors are often involved simultaneously. For example, when the systemic blood pressure declines, it reduces the glomerular filtration rate, and at the same time, baroreceptor reflexes cause sympathetic activation. ∞ p. 742 Meanwhile, a reduction in the GFR slows the movement of tubular fluid along the nephron. Because the tubular fluid then spends more time in the ascending limb of the loop of Henle, the concentration of sodium and chloride ions in the tubular fluid reaching the macula densa and DCT becomes abnormally low.

Figure 26–11a● provides a general overview of the renin–angiotensin system. A fall in GFR leads to the release of renin, which activates angiotensin. Angiotensin II then elevates the blood volume and blood pressure, increasing the GFR. Figure 26–11b● presents a more complete view of the mechanisms involved. Once released into the bloodstream by the juxtaglomerular apparatus, renin converts the inactive protein angiotensinogen to angiotensin I. Angiotensin I, which is also inactive, is then converted to angiotensin II by **angiotensin-converting enzyme (ACE)**. This conversion occurs primarily in the capillaries of the lungs. Angiotensin II is an active hormone whose primary effects include the following:

- *In peripheral capillary beds*, angiotensin II causes a brief but powerful vasoconstriction of arterioles and precapillary sphincters, elevating pressures in the renal artery and their tributaries.
- *At the nephron*, angiotensin II causes the constriction of the efferent arteriole, further elevating glomerular pressures and filtration rates. Angiotensin II also directly stimulates the reabsorption of sodium ions and water at the PCT.
- *At the adrenal glands*, angiotensin II stimulates the secretion of aldosterone by the adrenal cortex. The aldosterone then accelerates sodium reabsorption in the DCT and cortical portion of the collecting system.
- *In the CNS*, angiotensin II (1) causes the sensation of thirst, (2) triggers the release of antidiuretic hormone (ADH), stimulating the reabsorption of water in the distal portion of the DCT and the collecting system, and (3) increases sympathetic motor tone, further stimulating peripheral vasoconstriction.

The combined effect is an increase in systemic blood volume and blood pressure and the restoration of normal GFR.

If the blood volume rises, the GFR increases automatically, and this promotes fluid losses that help return the blood volume to normal levels. If the elevation in blood volume is severe, hormonal factors further increase the GFR and accelerate fluid losses in the urine. As noted in Chapter 18, the *natriuretic peptides* are released in response to the stretching of the walls of the heart by increased blood volume or blood pressure. ∞ p. 636 Atrial natriuretic peptide (ANP) is released by the atria, and brain natriuretic peptide (BNP) is released by the ventricles. Among their other effects, the natriuretic peptides trigger the dilation of the afferent arteriole and constriction of the efferent arteriole. This mechanism elevates glomerular pressures and increases the GFR, leading to increased urine production and decreased blood volume and pressure.

Autonomic Regulation of the GFR

Most of the autonomic innervation of the kidneys consists of sympathetic postganglionic fibers. (The role of the few parasympathetic fibers in regulating kidney function is not known.) Sympathetic activation has one direct effect on the GFR: It produces a powerful vasoconstriction of the afferent arterioles, decreasing the GFR and slowing the production of filtrate. The sympathetic activation triggered by an acute fall in blood pressure or a heart attack will thus override the local regulatory mechanisms that act to stabilize the GFR. As the crisis passes and sympathetic tone decreases, the filtration rate gradually returns to normal levels.

When the sympathetic division alters regional patterns of blood circulation, blood flow to the kidneys is often affected. For example, the dilation of superficial vessels in warm weather

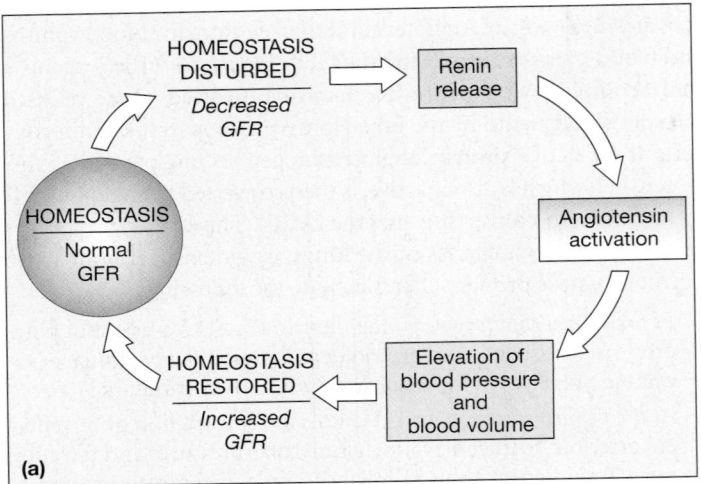

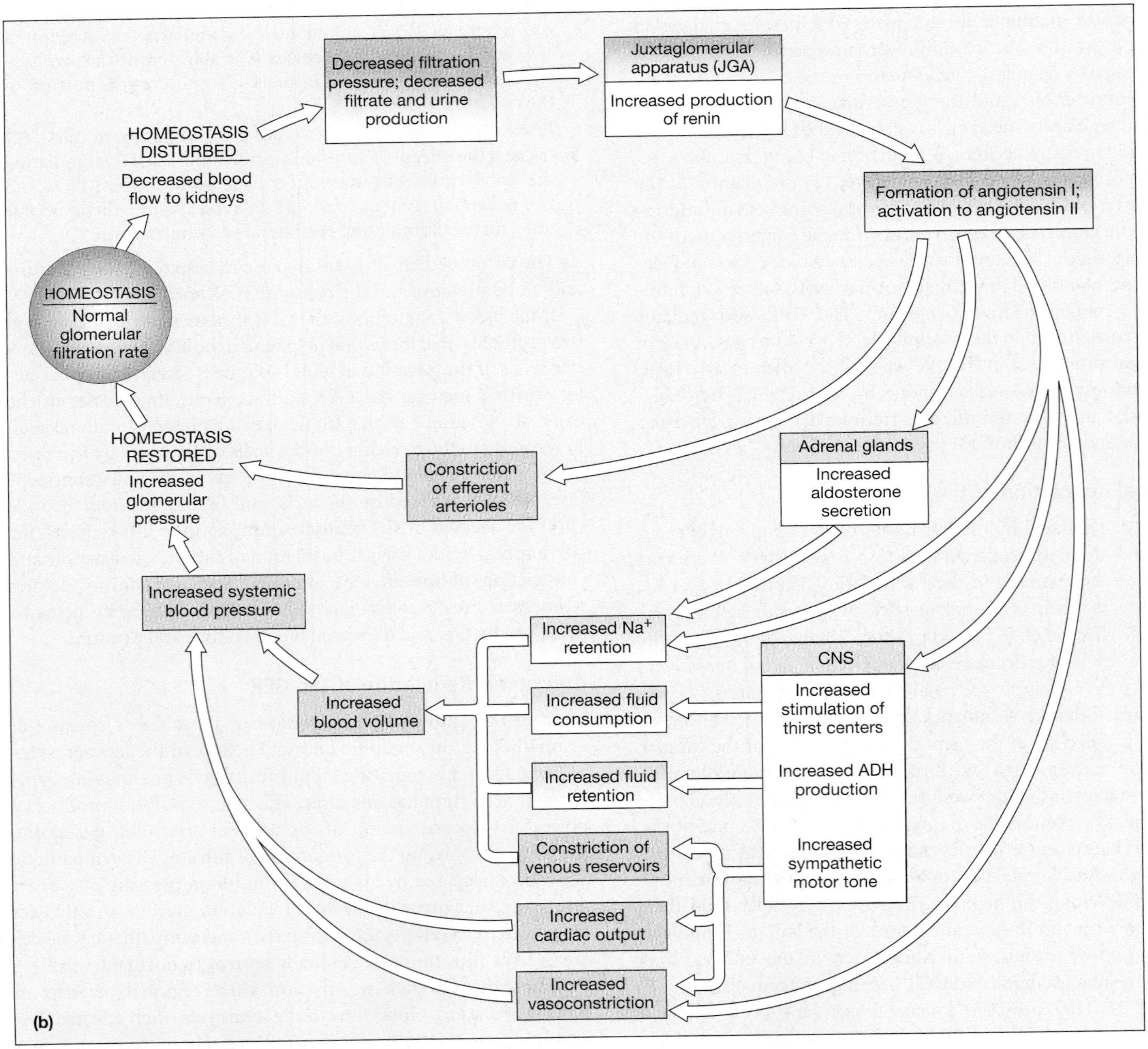

Renal Failure

Renal failure occurs when the kidneys become unable to perform the excretory functions needed to maintain homeostasis. When kidney filtration slows for any reason, urine production declines. As the decline continues, symptoms of renal failure appear because water, ions, and metabolic wastes are retained. Virtually all systems in the body are affected. For example, fluid balance, pH, muscular contraction, metabolism, and digestive function are disturbed. The individual generally becomes hypertensive, anemia develops due to a decline in erythropoietin production, and central nervous system problems can lead to sleeplessness, seizures, delirium, and even coma.

Acute renal failure occurs when exposure to nephrotoxic drugs, renal ischemia, urinary obstruction, or trauma causes filtration to slow suddenly or to stop. The reduction in kidney function occurs over a period of a few days and persists for weeks. Sensitized individuals can also develop acute renal failure after an allergic response to antibiotics or anesthetics. Individuals in acute renal failure may recover if they survive the incident. The kidneys may then regain partial or complete function. (With supportive treatment, the mortality rate is approximately 50 percent.) In *chronic renal failure*, kidney function deteriorates gradually, and the associated problems accumulate over time. The condition generally cannot be reversed, its progression can only be slowed, and symptoms of acute renal failure eventually develop. The symptoms of chronic and acute renal failure can be relieved by dialysis, but this treatment is not a cure. In some cases, kidney transplantation, an option discussed later in the chapter, is the best long-term solution.

---

shunts blood away from your kidneys, so glomerular filtration declines temporarily. The effect becomes especially pronounced during periods of strenuous exercise. As the blood flow to your skin and skeletal muscles increases, kidney perfusion gradually decreases. These changes may be opposed, with variable success, by autoregulation at the local level.

At maximal levels of exertion, renal blood flow may be less than 25 percent of normal resting levels. This reduction can create problems for distance swimmers and marathon runners, because metabolic wastes build up over the course of a long competition. *Proteinuria* (protein loss in urine) commonly occurs after long-distance events because the glomerular cells have been injured by prolonged hypoxia (low oxygen levels). If the damage is substantial, *hematuria* (blood loss in urine) will occur. Hematuria develops in roughly 18 percent of marathon runners. Proteinuria and hematuria generally disappear within 48 hours as the glomerular tissues are repaired. However, a small number of marathon and ultramarathon runners experience *renal failure*, with permanent impairment of their kidney function.

26–5 RENAL PHYSIOLOGY: REABSORPTION AND SECRETION

Objectives

- Identify the types of transport mechanisms found along the nephron, and discuss the reabsorptive or secretory functions of each segment of the nephron and collecting system.

- Explain the role of countercurrent multiplication in the formation of a concentration gradient in the renal medulla.

- Describe how antidiuretic hormone and aldosterone levels influence the volume and concentration of urine.

Reabsorption recovers useful materials that have entered the filtrate, and secretion ejects waste products, toxins, or other undesirable solutes that did not leave the bloodstream at the glomerulus. Both processes occur in every segment of the nephron other than the renal corpuscle, but the relative importance of reabsorption versus secretion changes from segment to segment.

☐ REABSORPTION AND SECRETION AT THE PCT

The cells of the proximal convoluted tubule normally reabsorb 60–70 percent of the volume of the filtrate produced in the renal corpuscle. The reabsorbed materials enter the peritubular fluid and diffuse into peritubular capillaries.

The PCT has five major functions:

1. *Reabsorption of Organic Nutrients.* Under normal circumstances, before the tubular fluid enters the loop of Henle, the PCT reabsorbs more than 99 percent of the glucose, amino acids, and other organic nutrients in the fluid. This reabsorption involves a combination of facilitated transport and cotransport.

2. *Active Reabsorption of Ions.* The PCT actively transports ions, including sodium, potassium, magnesium, bicarbonate, phosphate, and sulfate ions. The ion pumps involved are individually regulated and may be influenced by circulating ion or hormone levels. For example, angiotensin II stimulates Na^+ reabsorption along the PCT. By absorbing carbon dioxide, the PCT indirectly recaptures roughly 90 percent of the bicarbonate ions from tubular fluid (Figure 26–12●). Bicarbonate is important in stabilizing blood pH, a process we shall examine further in Chapter 27.

3. *Reabsorption of Water.* The reabsorptive processes have a direct effect on the solute concentrations inside and outside the tubules. The filtrate entering the PCT has the same osmotic concentration as that of the surrounding peritubular fluid. As transport activities proceed, the solute concentration of tubular fluid decreases and that of peritubular fluid and adjacent capillaries increases. Osmosis then pulls water out of the tubular fluid and into the peritubular fluid. Along the PCT, this mechanism results in the reabsorption of roughly 108 liters of water each day.

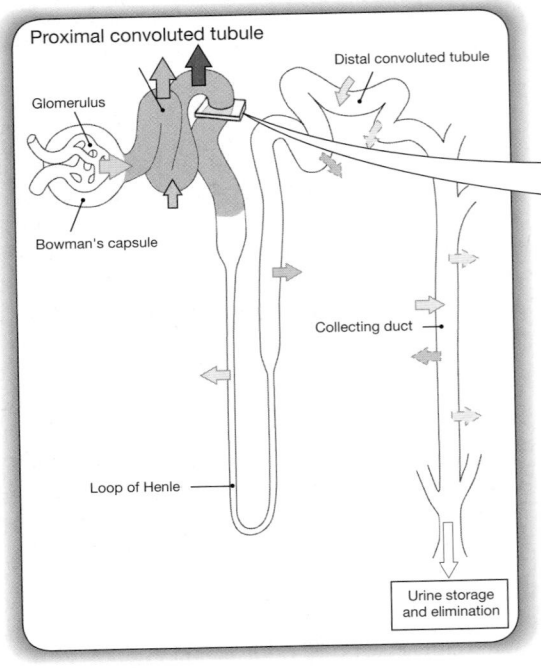

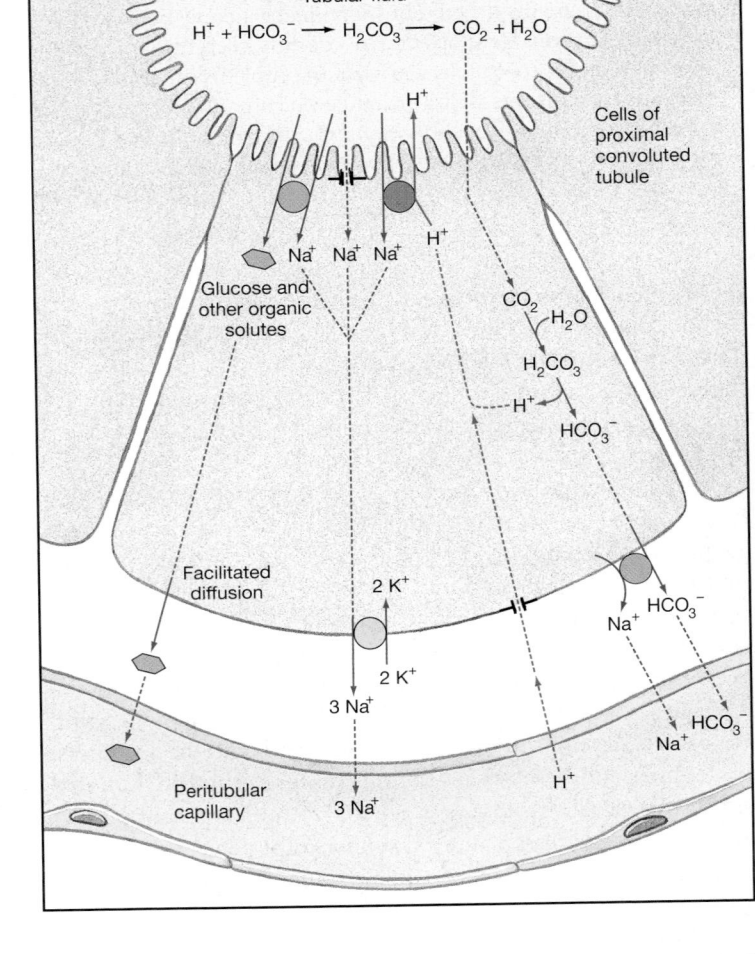

•FIGURE 26–12
Transport Activities at the PCT.
Sodium ions may enter a tubular cell from the filtrate by diffusion, cotransport, or countertransport. The sodium ions are then pumped into the peritubular fluid by the sodium–potassium exchange pump. Other reabsorbed solutes may be ejected into the peritubular fluid by separate active transport mechanisms.

KEY	
⊣⊢ =	Leak channel
⬤ =	Countertransport
◯ =	Exchange pump
⬤ =	Cotransport
---- =	Diffusion
⇢ =	Reabsorption
⇢ =	Secretion

4. *Passive Reabsorption of Ions.* As active reabsorption of ions occurs and water leaves tubular fluid by osmosis, the concentration of other solutes in tubular fluid increases above that in peritubular fluid. If the tubular cells are permeable to them, those solutes will move across the tubular cells and into the peritubular fluid by passive diffusion. Urea, chloride ions, and lipid-soluble materials may diffuse out of the PCT in this way. As this diffusion occurs, it further reduces the solute concentration of the tubular fluid and promotes additional water reabsorption by osmosis.

5. *Secretion.* Active secretion also occurs along the PCT. Because the DCT performs comparatively little reabsorption, we shall consider secretory mechanisms when we discuss the DCT.

Sodium ion reabsorption plays an important role in all of the foregoing processes. Sodium ions may enter the tubular cells by diffusion through Na^+ leak channels; by the sodium-linked cotransport of glucose, amino acids, or other organic solutes; or by countertransport for hydrogen ions (Figure 26–12•). In the tubular cell, the sodium ions diffuse toward the basement membrane. The cell membrane in this area contains sodium–potassium exchange pumps that eject sodium ions in exchange for extracellular potassium ions. Reabsorbed sodium ions then diffuse into the adjacent peritubular capillaries.

The reabsorption of ions and compounds along the PCT involves many different carrier proteins. Some people have an inher-

ited inability to manufacture one or more of these carrier proteins and are therefore unable to recover specific solutes from tubular fluid. For example, in *renal glycosuria* (glī-kō-SOO-rē-uh), a defective carrier protein makes it impossible for the PCT to reabsorb glucose from tubular fluid. **AM** Inherited Problems with Tubular Function

---

✓ What nephron structures are involved in filtration?

✓ What occurs when the plasma concentration of a substance exceeds its tubular maximum?

✓ How would a decrease in blood pressure affect the GFR?

Answers start on page Q-1

🔵 The processes of glomerular filtration and urinary processing can be reviewed on the **IP CD-ROM**: Urinary System/Glomerular Filtration; /Early Filtrate Processing.

☐ THE LOOP OF HENLE AND COUNTERCURRENT MULTIPLICATION

Roughly 60–70 percent of the volume of filtrate produced at the glomerulus has been reabsorbed before the tubular fluid reaches the loop of Henle. In the process, useful organic substrates have been reclaimed, along with many mineral ions. The loop of Henle will reabsorb roughly half of the water, as well as two-thirds of the sodium and chloride ions, remaining in the tubular fluid. This reabsorption is performed efficiently using the principle of countercurrent exchange, which we introduced in Chapter 25 in our discussion of heat conservation techniques. ∞ p. 962

The thin descending limb and the thick ascending limb of the loop of Henle are very close together, separated by peritubular fluid. The exchange that occurs between these segments is called **countercurrent multiplication**. *Countercurrent* refers to the fact that the exchange occurs between fluids moving in opposite directions: Tubular fluid in the descending limb flows toward the renal pelvis, whereas tubular fluid in the ascending limb flows toward the cortex. *Multiplication* refers to the fact that the effect of the exchange increases as movement of the fluid continues.

The two parallel segments of the loop of Henle have very different permeability characteristics. The thin descending limb is permeable to water, but relatively impermeable to solutes. The thick ascending limb, which is relatively impermeable to both water and solutes, contains active transport mechanisms that pump sodium and chloride ions from the tubular fluid into the peritubular fluid of the medulla. Countercurrent multiplication operates as follows:

- Sodium and chloride are pumped out of the thick ascending limb and into the peritubular fluid.

- This pumping action elevates the osmotic concentration in the peritubular fluid around the thin descending limb.

- The result is an osmotic flow of water out of the thin descending limb and into the peritubular fluid, increasing the solute concentration in the thin descending limb.

- The arrival of the highly concentrated solution in the thick ascending limb accelerates the transport of sodium and chloride ions into the peritubular fluid of the medulla.

Notice that this arrangement is a simple positive feedback loop: Solute pumping at the ascending limb leads to higher solute concentrations in the descending limb, which then result in accelerated solute pumping in the ascending limb.

Figure 26–13a● diagrams ion transport across the epithelium of the thick ascending limb. Active transport at the apical surface moves sodium, potassium, and chloride ions out of the tubular fluid. The carrier is called a **Na⁺–K⁺/2 Cl⁻ transporter**, because each cycle of the pump carries a sodium ion, a potassium ion, and two chloride ions into the tubular cell. Potassium and chloride ions are pumped into the peritubular fluid by cotransport carriers. However, potassium ions are removed from the peritubular fluid as the sodium–potassium exchange pump pumps sodium ions out of the tubular cell. The potassium ions then diffuse back into the lumen of the tubule through potassium leak channels. The net result is that Na⁺ and Cl⁻ enter the peritubular fluid of the renal medulla.

The removal of sodium and chloride ions from the tubular fluid in the ascending limb elevates the osmotic concentration of

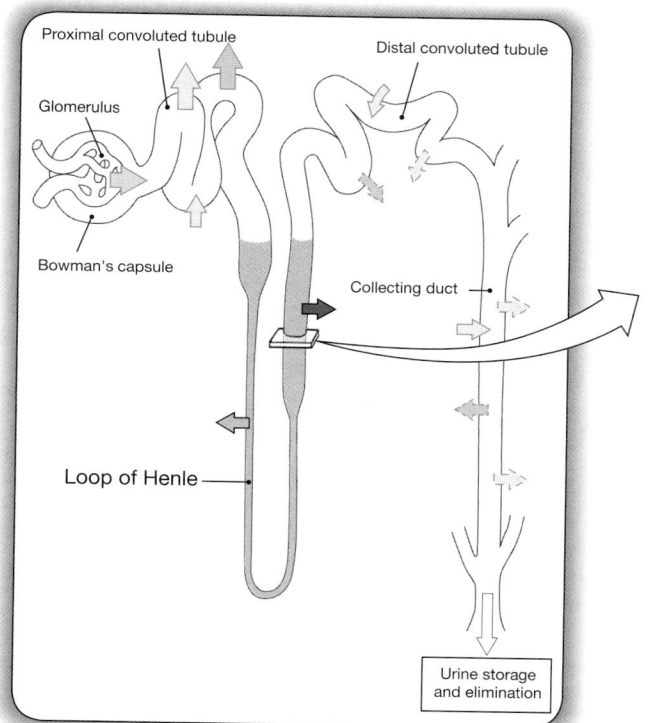

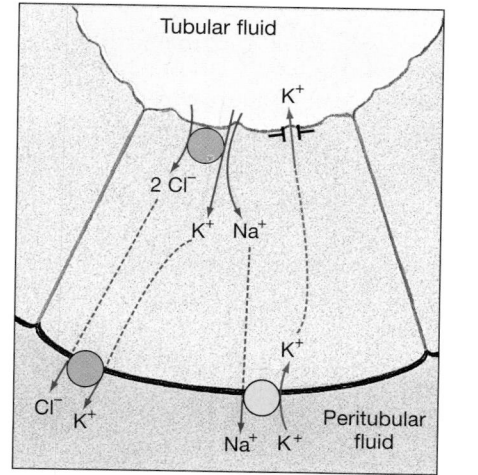

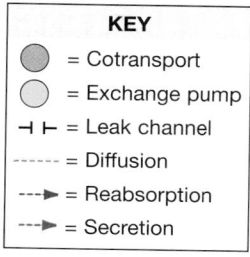

●FIGURE 26–13
Countercurrent Multiplication and Concentration of Urine

KEY
- ● = Cotransport
- ◯ = Exchange pump
- ⊣⊢ = Leak channel
- ----- = Diffusion
- ---▶ = Reabsorption
- ---▶ = Secretion

(a) The mechanism of sodium and chloride ion transport involves Na⁺–K⁺/2 Cl⁻ carrier at the apical surface and two carriers at the basal surface of the tubular cell: a potassium–chloride cotransport pump and a sodium–potassium exchange pump. The net result is the transport of sodium and chloride ions into the peritubular fluid.

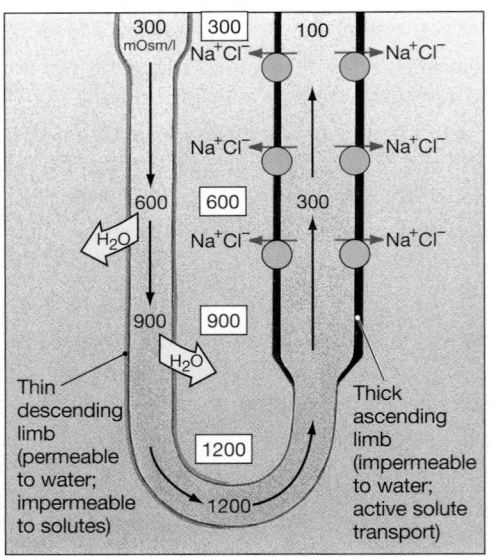

(b) Active transport of NaCl along the ascending thick limb results in the movement of water from the descending limb.

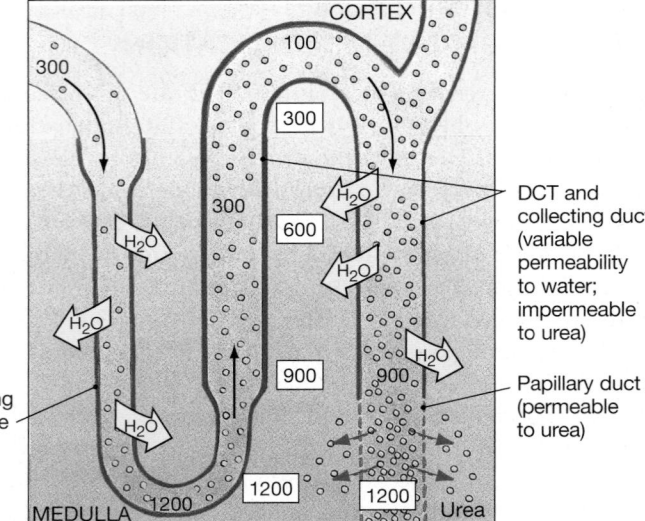

(c) The permeability characteristics of both the loop and the collecting duct tend to concentrate urea in the tubular fluid and in the medulla. The loop of Henle, DCT, and collecting duct are impermeable to urea. As water reabsorption occurs, the urea concentration rises. The papillary ducts' permeability to urea accounts for roughly one-third of the solutes in the deepest portions of the medulla.

●**FIGURE 26–13**
Countercurrent Multiplication and Concentration of Urine   *(continued)*

the peritubular fluid around the thin descending limb (Figure 26–13b●). Because the thin descending limb is permeable to water, but impermeable to solutes, thus, as tubular fluid travels into the medulla along the thin descending limb, osmosis moves water into the peritubular fluid. Solutes remain behind, so the tubular fluid reaching the turn of the loop of Henle has a higher osmotic concentration than it did at the start.

The pumping mechanism of the thick ascending limb is highly effective: Almost two-thirds of the sodium and chloride ions that enter it are pumped out of the tubular fluid before that fluid reaches the DCT. In other tissues, differences in solute concentration are quickly resolved by osmosis. However, osmosis cannot occur across the wall of the thick ascending limb, because the tubular epithelium there is impermeable to water. Thus, as Na^+ and Cl^- are removed, the solute concentration in the tubular fluid declines. Tubular fluid arrives at the DCT with an osmotic concentration of only about 100 mOsm/l, one-third of the concentration of the peritubular fluid of the renal cortex.

The rate of ion transport by the thick ascending limb is proportional to an ion's concentration in tubular fluid. As a result, more sodium and chloride ions are pumped into the medulla at the start of the thick ascending limb, where NaCl concentrations are highest, than near the cortex. This regional difference in the rate of ion transport is the basis of the concentration gradient within the medulla.

The Concentration Gradient of the Medulla

Normally, the maximum solute concentration of the peritubular fluid near the turn of the loop of Henle is about 1200 mOsm/l.

Sodium and chloride ions pumped out of the loop's ascending limb account for roughly two-thirds of that gradient (750 mOsm/l). The rest of the concentration gradient results from the presence of urea. To understand how the urea arrived in the medulla, we must look ahead to events in the last segments of the collecting system (Figure 26–13c●). The thick ascending limb of the loop of Henle, the DCT, and the collecting ducts are impermeable to urea. As water is reabsorbed, the concentration of urea gradually rises. The tubular fluid reaching the papillary duct typically contains urea at a concentration of about 450 mOsm/l. Because the papillary ducts are permeable to urea, the urea concentration in the deepest parts of the medulla also averages 450 mOsm/l.

Benefits of Countercurrent Multiplication

The countercurrent mechanism performs two services:

1. It efficiently reabsorbs solutes and water before the tubular fluid reaches the DCT and collecting system.
2. It establishes a concentration gradient that permits the passive reabsorption of water from the tubular fluid in the collecting system. This reabsorption is regulated by circulating levels of antidiuretic hormone (ADH).

The tubular fluid arriving at the descending limb of the loop of Henle has an osmotic concentration of roughly 300 mOsm/l, due primarily to the presence of ions such as Na^+ and Cl^-. The concentration of organic wastes, such as urea, is low. Roughly half of the tubular fluid entering the loop of Henle is reabsorbed along the thin descending limb, and two-thirds of the

Na$^+$ and Cl$^-$ is reabsorbed along the thick ascending limb. As a result, the DCT receives a reduced volume of tubular fluid with an osmotic concentration of about 100 mOsm/l. Urea and other organic wastes, which were not pumped out of the thick ascending limb, now represent a significant percentage of the dissolved solutes.

◻ REABSORPTION AND SECRETION AT THE DCT

The composition and volume of tubular fluid change dramatically as it travels from the capsular space to the distal convoluted tubule. Only 15–20 percent of the initial filtrate volume reaches the DCT, and the concentrations of electrolytes and organic wastes in the arriving tubular fluid no longer resemble the concentrations in blood plasma. Selective reabsorption or secretion, primarily along the DCT, makes the final adjustments in the solute composition and volume of the tubular fluid.

Reabsorption

Throughout most of the DCT, the tubular cells actively transport Na$^+$ and Cl$^-$ out of the tubular fluid (Figure 26–14a●). Tubular cells along the distal portions of the DCT also contain ion pumps that reabsorb tubular Na$^+$ in exchange for another cation (usually K$^+$) (Figure 26–14b●). The ion pump and the Na$^+$ channels involved are controlled by the hormone *aldosterone*, produced by the adrenal cortex. Aldosterone stimulates the synthesis and incorporation of sodium ion pumps and sodium channels in cell membranes along the DCT and collecting duct. The net result is a reduction in the number of sodium ions lost in urine. However, sodium ion conservation is associated with potassium ion loss. Prolonged aldosterone stimulation can therefore produce *hypokalemia*, a dangerous reduction in the plasma K$^+$ concentration.

The DCT is also the primary site of Ca^{2+} reabsorption, a process regulated by circulating levels of parathyroid hormone and calcitriol. ∞ p. 626

Secretion

Filtration does not force all the dissolved materials out of plasma, and blood entering peritubular capillaries still contains a number of potentially undesirable substances. In most cases, the concentrations of these materials are too low to cause physiological problems. However, any ions or compounds that are present in peritubular capillaries will diffuse into the peritubular fluid. If those concentrations become too high, the tubular cells may absorb these materials from the peritubular fluid and secrete them into the tubular fluid. Table 26–3, p. 983, gives a partial listing of substances secreted into tubular fluid by the distal and proximal convoluted tubules.

The rate of K$^+$ and H$^+$ secretion rises or falls in response to changes in their concentrations in peritubular fluid. The higher the concentration in the peritubular fluid, the higher is the rate of secretion. Potassium and hydrogen ions merit special attention, because their concentrations in body fluids must be maintained within relatively narrow limits.

POTASSIUM ION SECRETION Figure 26–14a,b● diagrams the mechanism of K$^+$ secretion. Potassium ions are removed from the peritubular fluid in exchange for sodium ions obtained from the tubular fluid. These potassium ions diffuse into the lumen through potassium channels at the apical surfaces of the tubular cells. In effect, the tubular cells trade sodium ions in the tubular fluid for excess potassium ions in body fluids.

HYDROGEN ION SECRETION Hydrogen ion secretion is also associated with the reabsorption of sodium. Figure 26–14c● shows two routes of secretion. Both involve the generation of carbonic acid by the enzyme *carbonic anhydrase*. ∞ pp. 858, 892 Hydrogen ions generated by the dissociation of the carbonic acid are secreted by means of sodium-linked countertransport in exchange for Na$^+$ in the tubular fluid. The bicarbonate ions diffuse into the peritubular fluid and then into the bloodstream, where they help prevent changes in plasma pH.

Hydrogen ion secretion acidifies the tubular fluid while elevating the pH of the blood. Hydrogen ion secretion accelerates when the pH of the blood falls, as in *lactic acidosis*, which can develop after exhaustive muscle activity, or *ketoacidosis*, which can develop in starvation or diabetes mellitus. ∞ p. 953 The combination of H$^+$ removal and HCO$_3^-$ production at the kidneys plays an important role in the control of blood pH. Because one of the secretory pathways is aldosterone sensitive, aldosterone stimulates H$^+$ secretion. Prolonged aldosterone stimulation can cause *alkalosis*, or abnormally high blood pH.

In Chapter 25, we noted that the production of lactic acid and ketone bodies during the postabsorptive state can cause acidosis. Under these conditions, the PCT and DCT will deaminate amino acids in reactions that strip off the amino groups (−NH$_2$). The reaction sequence ties up H$^+$ and yields both **ammonium ions** (NH$_4^+$) and HCO$_3^-$. As indicated in Figure 26–14c●, the ammonium ions are then pumped into the tubular fluid by sodium-linked countertransport, and the bicarbonate ions enter the bloodstream by way of the peritubular fluid.

Tubular deamination thus has two major benefits: (1) It provides carbon chains suitable for catabolism, and (2) it generates bicarbonate ions that add to the buffering capabilities of plasma.

◻ REABSORPTION AND SECRETION ALONG THE COLLECTING SYSTEM

The collecting ducts receive tubular fluid from many nephrons and carry it toward the renal sinus along the concentration gradient in the medulla. The normal amount of water and solute loss in the collecting system is regulated in two ways:

- By aldosterone, which controls sodium ion pumps along most of the DCT and the proximal portion of the collecting system.

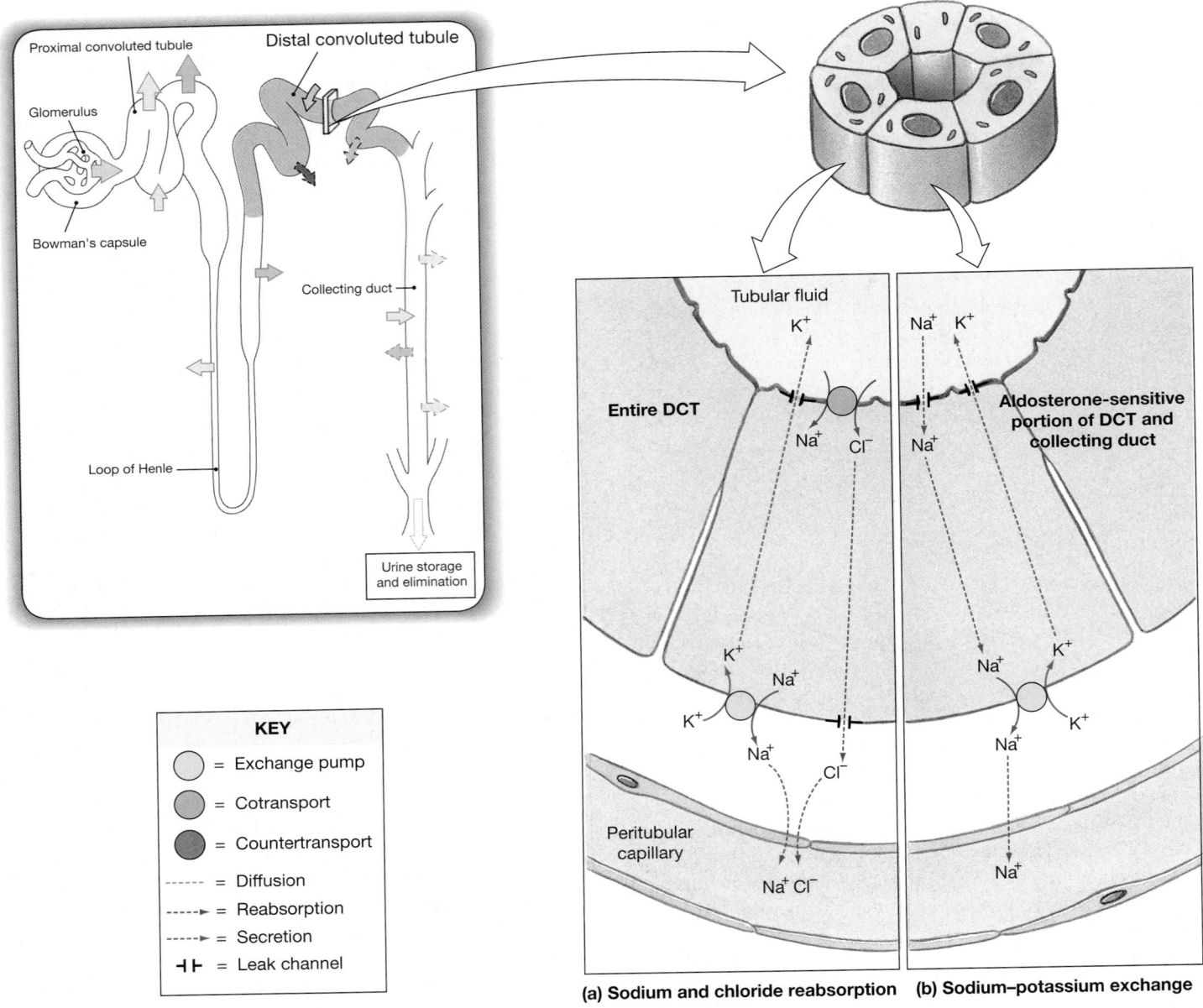

●FIGURE 26–14
Tubular Secretion and Solute Reabsorption at the DCT. **(a)** The basic pattern of the absorption of sodium and chloride ions and the secretion of potassium ions. **(b)** Aldosterone-regulated absorption of sodium ions, linked to the passive loss of potassium ions.

(a) Sodium and chloride reabsorption   (b) Sodium–potassium exchange

KEY

- ○ = Exchange pump
- ○ = Cotransport
- ● = Countertransport
- ----- = Diffusion
- ----▸ = Reabsorption
- ----▸ = Secretion
- ⊣⊢ = Leak channel

- By ADH, which controls the permeability of the DCT and collecting system to water. The collecting system also has other reabsorptive and secretory functions, many of which are important to the control of body fluid pH.

Reabsorption

Important examples of solute reabsorption in the collecting system include the following:

- **Sodium Ion Reabsorption.** The collecting system contains aldosterone-sensitive ion pumps that exchange Na^+ in tubular fluid for K^+ in peritubular fluid (Figure 26–14b●).
- **Bicarbonate Reabsorption.** Bicarbonate ions are reabsorbed in exchange for chloride ions in the peritubular fluid (Figure 26–14c●).

- **Urea Reabsorption.** The concentration of urea in the tubular fluid entering the collecting duct is relatively high. The fluid entering the papillary duct generally has the same osmotic concentration as that of interstitial fluid of the medulla—about 1200 mOsm/l—but contains a much higher concentration of urea. As a result, urea tends to diffuse out of the tubular fluid and into the peritubular fluid of the deepest portion of the medulla.

Secretion

The collecting system is an important site for the control of body fluid pH by means of the secretion of hydrogen or bicarbonate ions. If the pH of the peritubular fluid declines, carrier proteins pump hydrogen ions into the tubular fluid and reabsorb bicarbonate ions that help restore normal pH. If the pH of the per-

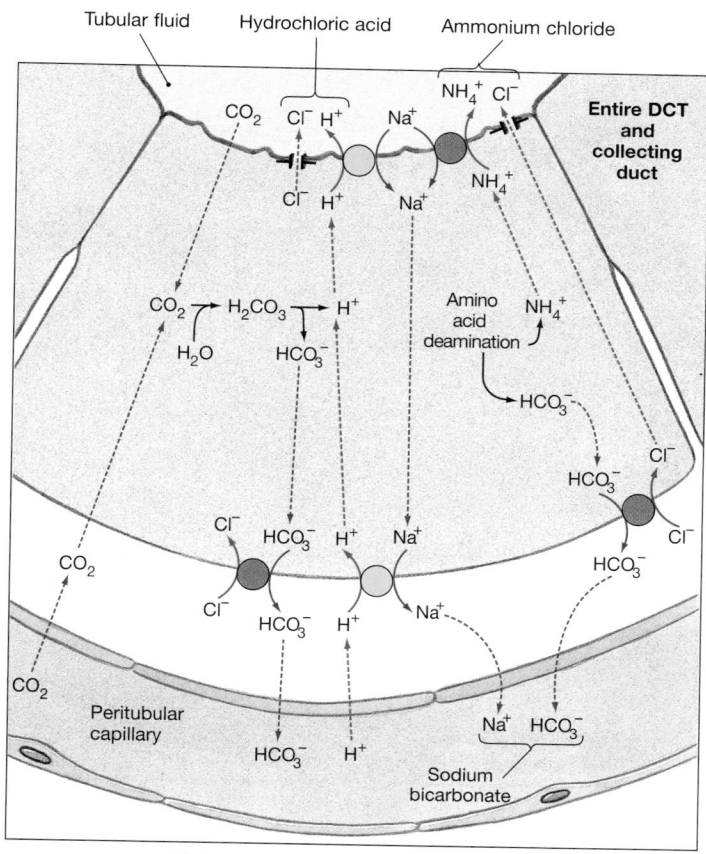

(c) H⁺ secretion and HCO₃⁻ reabsorption

FIGURE 26–14 *(continued)*
Tubular Secretion and Solute Reabsorption at the DCT.
(c) Hydrogen ion secretion and the acidification of urine occur by two routes. The central theme is the exchange of hydrogen ions in the cytoplasm for sodium ions in the tubular fluid and the reabsorption of the bicarbonate ions generated in the process.

itubular fluid rises (a much less common event), the collecting system secretes bicarbonate ions and pumps hydrogen ions into the peritubular fluid. The net result is that the body eliminates a buffer and gains hydrogen ions that will lower the pH. We shall examine these responses in more detail in Chapter 27, when we consider acid–base balance.

◻ THE CONTROL OF URINE VOLUME AND OSMOTIC CONCENTRATION

Urine volume and osmotic concentration are regulated by controlling the reabsorption of water. Water is reabsorbed by osmosis in the proximal convoluted tubule and the descending limb of the loop of Henle. The water permeabilities of these regions cannot be adjusted, and water reabsorption occurs whenever the osmotic concentration of the peritubular fluid exceeds that of the tubular fluid. The ascending limb of the loop of Henle is impermeable to water, but 1–2 percent of the volume of water in the original filtrate is recovered during sodium ion reabsorption in the distal convoluted tubule and collecting system. Because these water

movements cannot be prevented, they represent the *obligatory water reabsorption*, which usually recovers 85 percent of the volume of filtrate produced.

The volume of water lost in urine depends on how much of the water in the remaining tubular fluid (15 percent of the filtrate volume, or roughly 27 liters per day) is reabsorbed along the DCT and collecting system. The amount can be precisely controlled by a process called *facultative water reabsorption*. Precise control is possible because these segments are relatively impermeable to water except in the presence of ADH. This hormone causes the appearance of special *water channels* in the apical cell membranes, dramatically enhancing the rate of osmotic water movement. The higher the circulating levels of ADH, the greater the number of water channels and the greater the water permeability of these segments.

As we noted earlier in this chapter, the tubular fluid arriving at the DCT has an osmotic concentration of only about 100 mOsm/l. In the presence of ADH, osmosis occurs and water moves out of the DCT until the osmotic concentration of the tubular fluid equals that of the surrounding cortex (roughly 300 mOsm/l). The tubular fluid then travels through the collecting duct, which passes along the concentration gradient of the medulla. Additional water is then reabsorbed, and the urine reaching the minor calyx has an osmotic concentration closer to 1200 mOsm/l. Just how closely the osmotic concentration approaches 1200 mOsml/l depends on how much ADH is present.

Figure 26–15● diagrams the effects of ADH on the DCT and collecting system. In the absence of ADH, water is not reabsorbed in these segments, so all the fluid reaching the DCT is lost in the urine. The individual then produces large amounts of very dilute urine (Figure 26–15a●). That is just what happens in cases of *diabetes insipidus* ∞ p. 918, in which urinary water losses may reach 24 liters per day and the urine osmotic concentration is 30–400 mOsm/l. As ADH levels rise, the DCT and collecting system become more permeable to water, the amount of water reabsorbed increases, and the urine osmotic concentration climbs. Under maximum ADH stimulation, the DCT and collecting system become so permeable to water that the osmotic concentration of the urine is equal to that of the deepest portion of the medulla (Figure 26–15b●). Notice that the concentration of urine can never *exceed* that of the medulla, because the concentrating mechanism relies on osmosis.

You continuously secrete ADH at low levels, and your DCT and collecting system always have a significant degree of water permeability. The DCT normally reabsorbs roughly 9 liters of water per day, or about 5 percent of the original volume of filtrate produced by your glomeruli. At normal levels of ADH, the collecting system reabsorbs roughly 16.8 liters per day, or about 9.3 percent of the original volume of filtrate. A healthy adult typically produces 1200 ml of urine per day (about 0.6 percent of the filtrate volume), with an osmotic concentration of 800–1000 mOsm/l.

The effects of ADH are opposed by those of the natriuretic peptides ANP and BNP. ∞ pp. 636, 746 These hormones stimulate the production of a large volume of relatively dilute urine, soon restoring normal plasma volume.

Diuresis (dī-ū-RĒ-sis; *dia*, through + *ouresis*, urination) is the elimination of urine. Whereas *urination* is an equivalent term in a general sense, *diuresis* typically indicates the production of a large volume of urine. **Diuretics** (dī-ū-RET-iks) are drugs that promote the loss of water in urine. The usual goal in diuretic therapy is a reduction in blood volume, blood pressure, extracellular fluid volume, or all three. The ability to control renal water losses with relatively safe and effective diuretics has saved the lives of many individuals, especially those with high blood pressure or congestive heart failure. **AM** Heart Failure

Diuretics have many mechanisms of action, but all such drugs affect transport activities or water reabsorption along the nephron and collecting system. For example, a class of diuretics called *thiazides* (THĪ-a-zīdz) promotes water loss by reducing sodium and chloride ion transport in the proximal and distal convoluted tubules. **AM** Diuretics

Diuretic use for nonclinical reasons is on the rise. For example, large doses of diuretics are taken by some bodybuilders to improve muscle definition temporarily, and by some fashion models or jockeys to reduce body weight for brief periods. This practice of "cosmetic dehydration" is extremely dangerous and has caused several deaths due to electrolyte imbalances and consequent cardiac arrest.

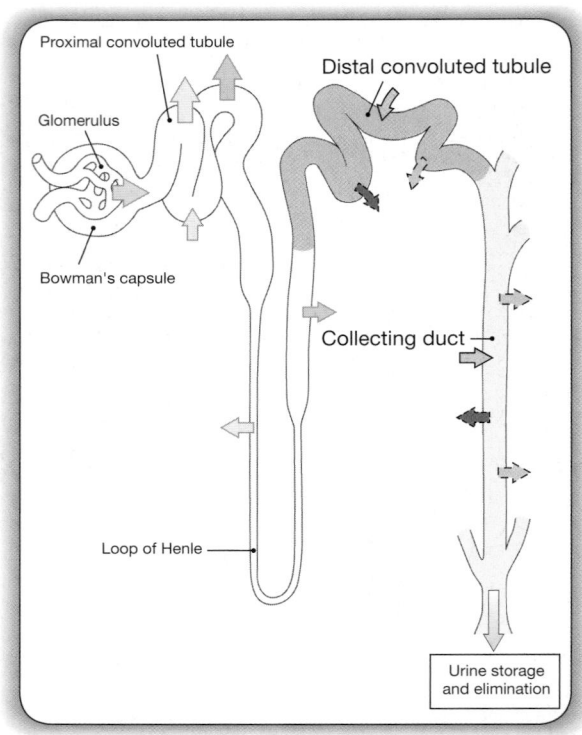

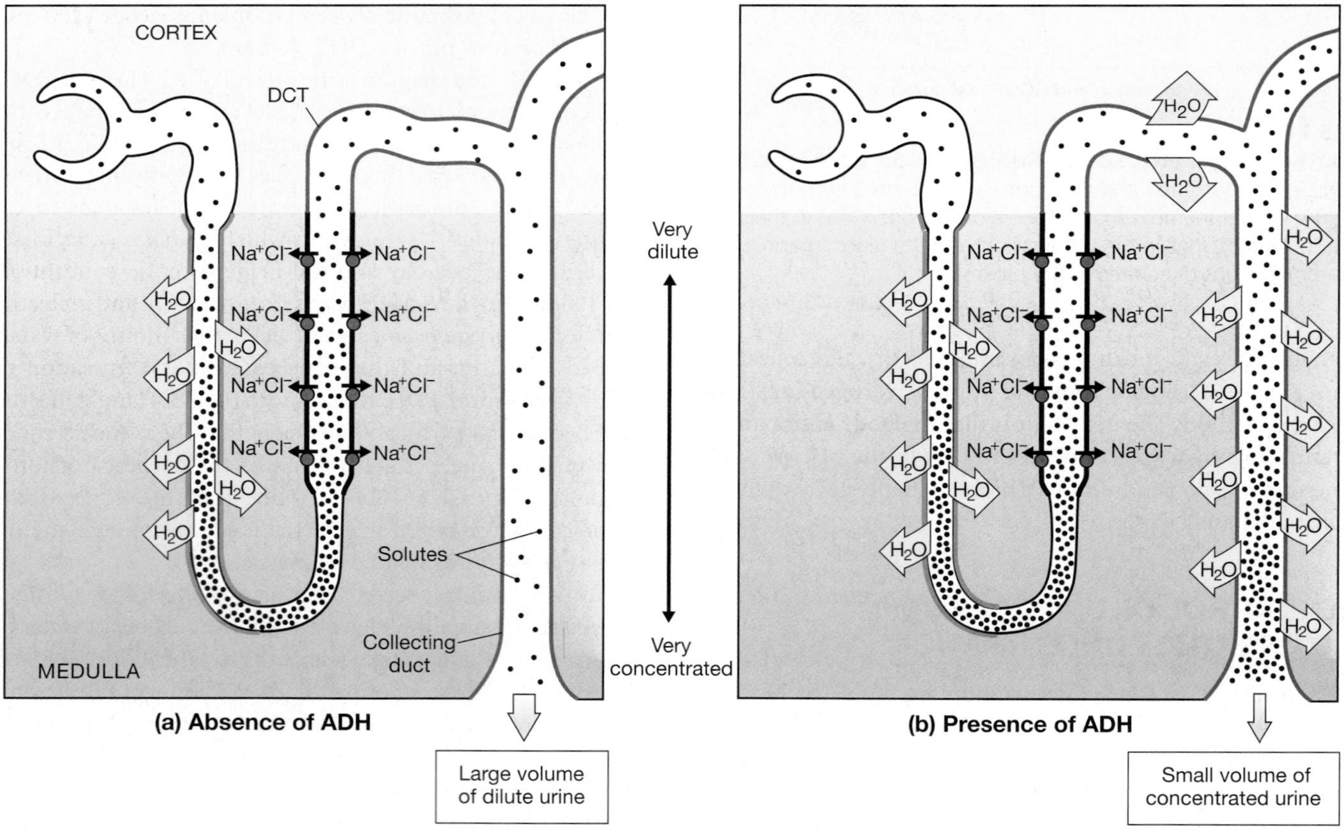

●**FIGURE 26–15**
The Effects of ADH on the DCT and Collecting Duct. **(a)** Tubule permeabilities and the osmotic concentration of urine without ADH.
(b) Tubule permeabilities and the osmotic concentration of urine at maximal ADH levels.

THE FUNCTION OF THE VASA RECTA

The solutes and water reabsorbed in the medulla must be returned to the general circulation without disrupting the concentration gradient. This return is the function of the vasa recta. Blood enters the vasa recta with an osmotic concentration of approximately 300 mOsm/l. The blood descending into the medulla gradually increases in osmotic concentration as the solute concentration in the peritubular fluid rises. This increase in blood osmotic concentration involves both solute absorption and water loss, but solute absorption predominates, because the plasma proteins limit osmotic flow out of the blood. ∞ p. 737 Blood flowing toward the cortex gradually decreases in osmotic concentration as the solute concentration of the peritubular fluid declines. Again, this decrease involves a combination of solute diffusion and osmosis, but in this case osmosis predominates, because the presence of plasma proteins does not oppose the osmotic flow of water into the blood.

The net results are that (1) some of the solutes absorbed in the descending portion of the vasa recta do not diffuse out in the ascending portion and (2) more water moves into the ascending portion of the vasa recta than is moved out in the descending portion. Thus, the vasa recta carries both water and solutes out of the medulla. Under normal conditions, the removal of solutes and water by the vasa recta precisely balances the rates of solute reabsorption and osmosis in the medulla.

THE COMPOSITION OF NORMAL URINE

More than 99 percent of the 180 liters of filtrate produced each day by the glomeruli is reabsorbed and never reaches the renal pelvis. General characteristics of the remaining filtrate, normal urine, are listed in Table 26–5. However, the composition of the 1.2 liters of urine produced varies with the metabolic and hormonal events under way.

The composition and concentration of urine are two integrated but distinct properties. The *composition* of urine reflects the filtration, absorption, and secretion activities of the nephrons. Some compounds, such as urea, are neither actively excreted nor reabsorbed along the nephron. In contrast, organic nutrients are completely reabsorbed, and other compounds, such as creatinine, that are missed by the filtration process are actively secreted into the tubular fluid.

The processes mentioned in the previous paragraph determine the identities and amounts of materials excreted in urine. The *concentration* of these components in a given urine sample depends on the osmotic movement of water across the walls of the tubules and collecting ducts. Because the composition and concentration of urine vary independently, you can produce a small quantity of concentrated urine or a large quantity of dilute urine and still excrete the same amount of dissolved materials. As

TABLE 26–5	GENERAL CHARACTERISTICS OF NORMAL URINE
Characteristic	Normal Range
pH	6.0 (range: 4.5–8)
Specific gravity	1.003–1.030
Osmotic concentration (osmolarity)	855–1335 mOsm/l
Water content	93–97 percent
Volume	1200 ml/day
Color	Clear yellow
Odor	Varies with composition
Bacterial content	Sterile

a result, physicians who are interested in a detailed assessment of renal function commonly rely on a 24-hour urine collection rather than on a single urine sample.

SUMMARY: RENAL FUNCTION

Table 26–4, p. 984, lists the functions of each segment of the nephron and collecting system. Figure 26–16● provides a functional overview that summarizes the major steps involved in the reabsorption of water and the production of concentrated urine:

STEP 1: The filtrate produced at the renal corpuscle has the same osmotic concentration as does plasma—about 300 mOsm/l. It has the composition of blood plasma, without the plasma proteins.

STEP 2: In the PCT, the active removal of ions and organic substrates produces a continuous osmotic flow of water out of the tubular fluid. This process reduces the volume of filtrate, but keeps the solutions inside and outside the tubule isotonic. Between 60 and 70 percent of the filtrate volume has been reabsorbed before the tubular fluid reaches the descending limb of the loop of Henle.

STEP 3: In the PCT and descending limb of the loop of Henle, water moves into the surrounding peritubular fluids, leaving a small volume (15–20 percent of the original filtrate) of highly concentrated tubular fluid. The reduction in volume has occurred by obligatory water reabsorption.

STEP 4: The thick ascending limb is impermeable to water and solutes. The tubular cells actively transport Na^+ and Cl^- out of the tubular fluid, thereby lowering the osmotic concentration of tubular fluid without affecting its volume. The tubular fluid reaching the distal convoluted tubule is hypotonic relative to the peritubular fluid, with an osmotic concentration of only about 100 mOsm/l. Because just Na^+ and Cl^- are removed, urea now accounts for a significantly higher proportion of the total osmotic concentration at the end of the loop than it did at the start.

STEP 5: The final adjustments in the composition of the tubular fluid are made in the DCT and the collecting system. Although the DCT and collecting duct are generally impermeable to solutes, the

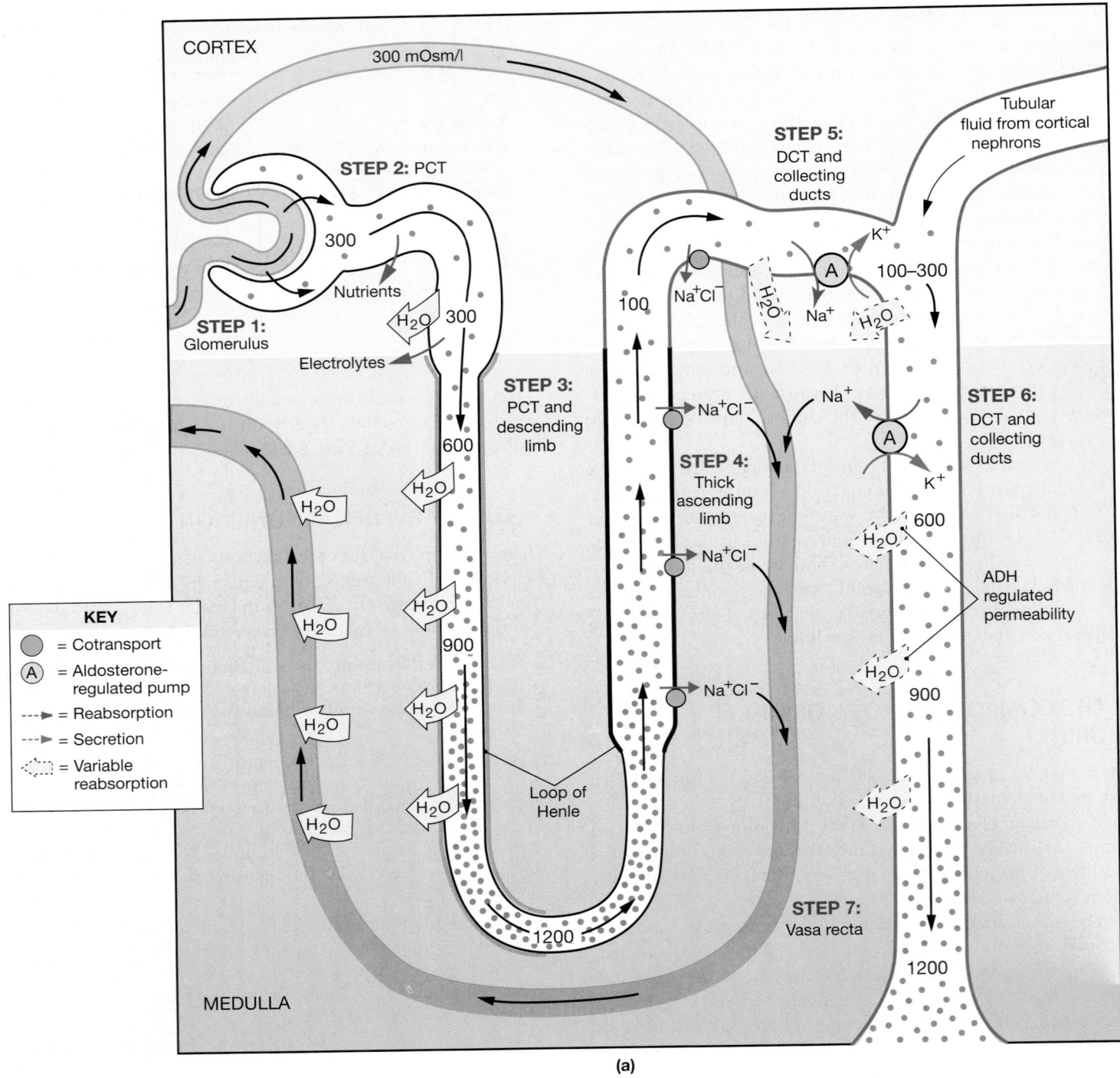

(a)

●**FIGURE 26–16**
A Summary of Renal Function. (a) A general overview of steps and events.

osmotic concentration of tubular fluid can be adjusted through active transport (reabsorption or secretion). Some of these transport activities are hormonally regulated.

STEP 6: The final adjustments in the volume and osmotic concentration of the tubular fluid are made by controlling the water permeabilities of the distal portions of the DCT and the collecting system. These segments are relatively impermeable to water unless

exposed to ADH. Under maximum ADH stimulation, urine volume is at a minimum and urine osmotic concentration is equal to that of the peritubular fluid in the deepest portion of the medulla (roughly 1200 mOsm/l).

STEP 7: The vasa recta absorbs solutes and water reabsorbed by the loop of Henle and the collecting ducts and thereby maintains the concentration gradient of the medulla.

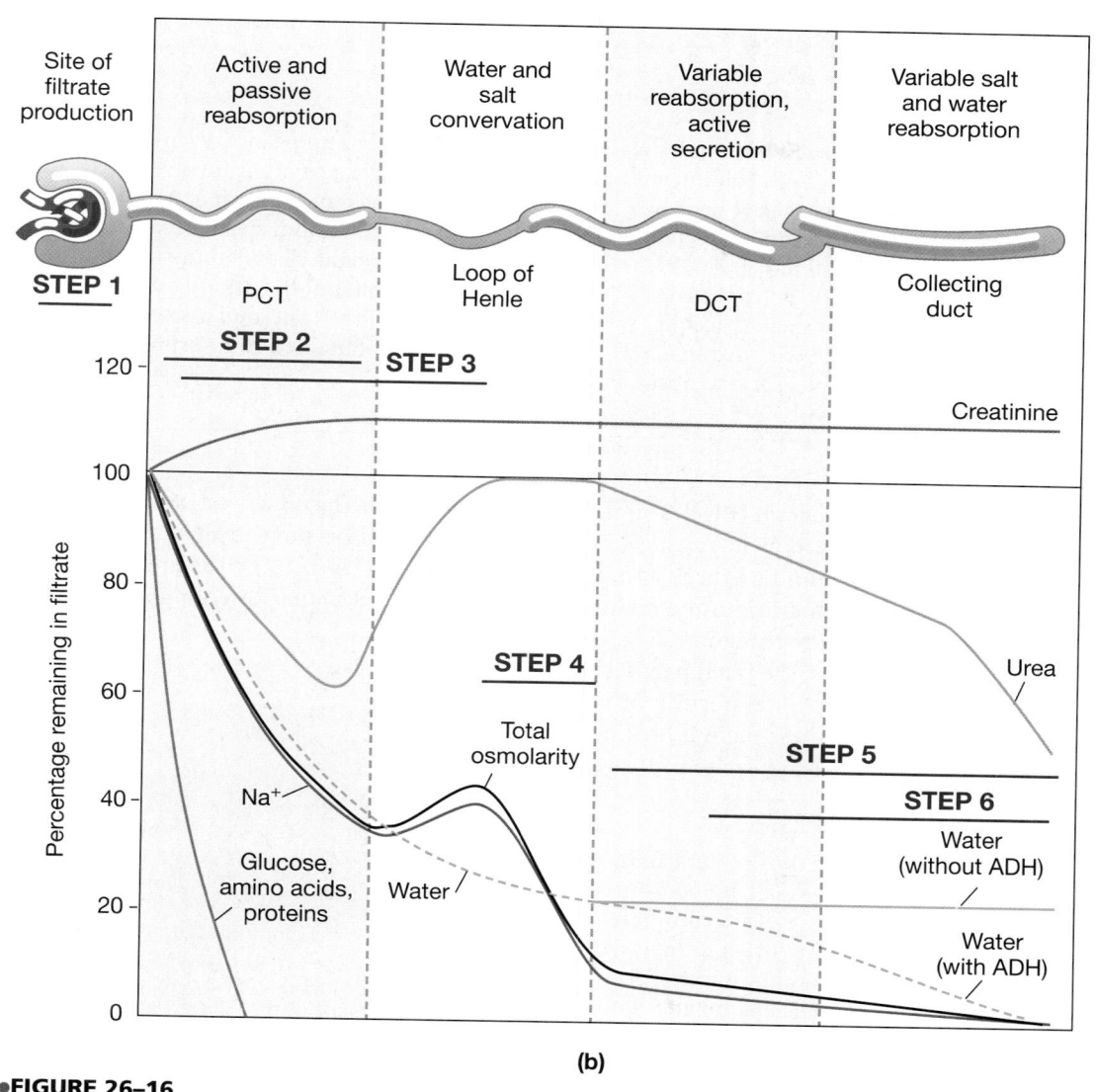

•FIGURE 26–16
A Summary of Renal Function. *(continued).* **(b)** Specific changes in the composition and concentration of the tubular fluid as it flows along the nephron and collecting duct.

✓ What effect would increased amounts of aldosterone have on the K^+ concentration of urine?

✓ What effect would a decrease in the Na^+ concentration of filtrate have on the pH of tubular fluid?

✓ How would the lack of juxtamedullary nephrons affect the volume and osmotic concentration of urine?

✓ Why does a decrease in the amount of Na^+ in the distal convoluted tubule lead to an increase in blood pressure?

Answers start on page Q-1

The processing of urine can be reviewed on the **IP CD-ROM**: Urinary System/Late Filtrate Processing.

26–6 URINE TRANSPORT, STORAGE, AND ELIMINATION

Objectives

■ Describe the structures and functions of the ureters, urinary bladder, and urethra.

■ Discuss the voluntary and involuntary regulation of urination, and describe the micturition reflex.

Filtrate modification and urine production end when the fluid enters the renal pelvis. The urinary tract (the ureters, urinary bladder, and urethra) is responsible for the transport,

Urinalysis

Normal urine is a noncloudy, sterile solution. Its yellow color results from the presence of urobilin, generated in the kidneys from the urobilinogens produced by intestinal bacteria and absorbed in the colon. (See Figure 19–5•, p. 660.) The evaporation of small molecules, such as ammonia, accounts for the characteristic odor of urine. Other substances not normally present, such as acetone or other ketone bodies, can also impart a distinctive smell. **Urinalysis**, the analysis of a urine sample, is a diagnostic tool of considerable importance, even in high-technology medicine. A standard urinalysis involves an assessment of the color and appearance of urine, two characteristics that can be determined without specialized equipment. In the 17th century, physicians classified the taste of the urine as sweet, salty, and so on, but quantitative analytical tests have long since replaced the taste-bud assay. Table 26–6 gives some typical values obtained from urinalysis. **AM** Urinalysis

storage, and elimination of urine. A *pyelogram* (PĪ-el-ō-gram) is an image of the urinary system, obtained by taking an X ray of the kidneys after a radiopaque compound has been administered. Such an image provides an orientation to the relative sizes and positions of the main structures (Figure 26–17•). The sizes of the minor and major calyces, the renal pelvis, the ureters, the urinary bladder, and the proximal portion of the urethra are somewhat variable, because these regions are lined by a *transitional epithelium* that can tolerate cycles of distension and contraction without damage. ⊂⊃ p. 118

Local blockages of the collecting ducts or ureter can result from the formation of *casts*—small blood clots, epithelial cells, lipids, or other materials. Casts are commonly eliminated in urine and are visible in microscopic analysis of urine samples. **Calculi** (KAL-kū-lī), or *kidney stones*, form within the urinary tract from calcium deposits, magnesium salts, or crystals of uric acid. The condition is called *nephrolithiasis* (nef-rō-li-THĪ-a-sis). The blockage of the urinary passage by a stone or by other means, such as external compression, results in **urinary obstruction**, a serious problem because, in addition to causing pain, it reduces or prevents filtration in the affected kidney by elevating the capsular hydrostatic pressure.

Calculi are generally visible on an X ray. If peristalsis and fluid pressures are insufficient to dislodge them, they must be surgically removed or destroyed. One interesting nonsurgical procedure involves breaking the stones apart with a *lithotripter*, a device originally developed from machines used to de-ice airplane wings. Lithotripters focus sound waves on the stones and break them into smaller fragments that can be passed in the urine. Another nonsurgical approach is the insertion of a catheter armed with a laser that can shatter calculi with intense light beams.

☐ THE URETERS

The ureters are a pair of muscular tubes that extend from the kidneys to the urinary bladder—a distance of about 30 cm (12 in.). Each ureter begins at the funnel-shaped renal pelvis (Figure 26–4•, p. 974). The ureters extend inferiorly and medially, passing over the anterior surfaces of the *psoas major muscles* (Figures 26–2b, and 26–3•, p. 973). The ureters are retroperitoneal and are firmly attached to the posterior abdominal wall. The paths taken

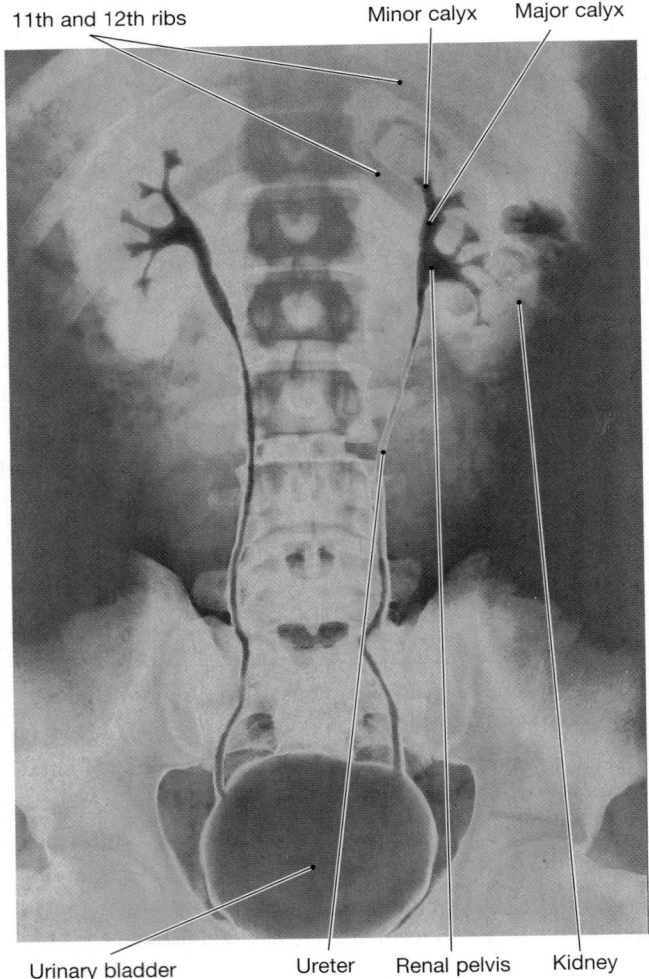

•**FIGURE 26–17**
A Radiographic View of the Urinary System.
A color-enhanced pyelogram (posterior view).

TABLE 26–6 TYPICAL VALUES OBTAINED FROM STANDARD URINE TESTING

Compound	Primary Source	Daily Elimination*	Concentration	Remarks
NITROGENOUS WASTES				
Urea	Deamination of amino acids at liver and kidneys	21 g	1.8 g/dl	Rises if negative nitrogen balance exists
Creatinine	Breakdown of creatine phosphate in skeletal muscle	1.8 g	150 mg/dl	Proportional to muscle mass; decreases during atrophy or muscle disease
Ammonia	Deamination by liver and kidney, absorption from intestinal tract	0.68 g	60 mg/dl	
Uric acid	Breakdown of purines	0.53 g	40 mg/dl	Increases in gout, liver diseases
Hippuric acid	Breakdown of dietary toxins	4.2 mg	350 μg/dl	
Urobilin	Urobilinogens absorbed at colon	1.5 mg	125 μg/dl	Gives urine its yellow color
Bilirubin	Hemoglobin breakdown product	0.3 mg	20 μg/dl	Increase may indicate problem with liver elimination or excess production; causes yellow skin color in jaundice
NUTRIENTS AND METABOLITES				
Carbohydrates		0.11 g	9 μg/dl	Primarily glucose; *glycosuria* develops if T_m is exceeded
Ketone bodies		0.21 g	17 μg/dl	Ketonuria may occur during postabsorptive state
Lipids		0.02 g	1.6 μg/d	May increase in some kidney diseases
Amino acids		2.25 g	287.5 μg/dl	Note relatively high loss compared with other metabolites due to low T_m; excess (*aminoaciduria*) indicates T_m problem
IONS				
Sodium		4.0 g	333 mg/dl	Varies with diet, urine pH, hormones, etc.
Chloride		6.4 g	533 mg/dl	
Potassium		2.0 g	166 m//dl	Varies with diet, urine pH, hormones, etc.
Calcium		0.2 g	17 mg/dl	Hormonally regulated (PTH/CT)
Magnesium		0.15 g	13 mg/dl	
BLOOD CELLS†				
RBCs		130,000/day	100/ml	Excess (*hematuria*) indicates vascular damage
WBCs		650,000/day	500/ml	Excess (*pyuria*) indicates renal infection or inflammation

* Representative values for a 70-kg male.   † Usually estimated by counting the cells in a sample of sediment after urine centrifugation.

Advances in the Treatment of Renal Failure

Many conditions can result in renal failure. The management of chronic renal failure typically involves restricting water and salt intake and minimizing protein intake, with few dietary proteins allowed. This combination reduces strain on the urinary system by (1) minimizing the volume of urine produced and (2) preventing the generation of large quantities of nitrogenous wastes. Acidosis, a common problem in persons with renal failure, can be countered by the ingestion of bicarbonate ions.

If drugs and dietary controls cannot stabilize the composition of blood, more drastic measures are taken. In **hemodialysis** (hē-mō-dī-AL-i-sis), an artificial membrane is used to regulate the composition of blood by means of a *dialysis machine* (Figure 26–18a●). The basic principle involved in this process, called **dialysis**, is passive diffusion across a selectively permeable membrane. The patient's blood flows past an artificial *dialysis membrane*, which contains pores large enough to permit the diffusion of small ions, but small enough to prevent the loss of plasma proteins. On the other side of the membrane flows a special **dialysis fluid**.

The composition of typical dialysis fluid is indicated in Table 26–7. As diffusion takes place across the membrane, the composition of the blood changes. Potassium ions, phosphate ions, sulfate ions, urea, creatinine, and uric acid diffuse across the membrane into the dialysis fluid. Bicarbonate ions and glucose diffuse into the bloodstream. In effect, diffusion across the dialysis membrane takes the place of normal glomerular filtration, and the characteristics of the dialysis fluid ensure that important metabolites remain in the bloodstream rather than diffusing across the membrane.

In practice, silicone rubber tubes called *shunts* are inserted into a medium-sized artery and vein. (The typical location is the forearm, although the lower leg is sometimes used.) The two shunts are then connected as shown in Figure 26–18b●. The connection acts like a short circuit that does not impede the flow of blood, and the shunts can be used like taps in a wine barrel, to draw a blood sample or to connect the individual to a dialysis machine. For long-term dialysis, a surgically created arteriovenous anastomosis provides access.

When connected to the dialysis machine, the individual sits quietly while blood circulates from the arterial shunt, through the machine, and back through the venous shunt. In the machine, the blood flows within a tube composed of dialysis membrane, and diffusion occurs between the blood and the surrounding dialysis fluid.

The use of a dialysis machine is suggested when a patient's *blood urea nitrogen (BUN)* exceeds 100 mg/dl (the normal value is up to 30 mg/dl). Dialysis techniques can maintain patients who are awaiting a transplant, as well as those whose kidney function has been temporarily disrupted. Hemodialysis does have drawbacks, however: (1) The patient must sit by the machine about 15 hours a week, (2) between treatments, the symptoms of uremia gradually appear, (3) hypotension can develop as a result of fluid loss during dialysis, (4) air bubbles in the tubing can cause an embolism to form in the bloodstream, (5) anemia commonly develops, and (6) the shunts can serve as sites of recurring infections.

One alternative to the use of a dialysis machine is **peritoneal dialysis**, in which the peritoneal lining is used as a dialysis membrane. Dialysis fluid is introduced into the peritoneum through a catheter in the abdominal wall, and the fluid is removed and replaced at intervals. For example, one procedure involves cycling 2 liters of fluid in an hour— 15 minutes for infusion, 30 minutes for exchange, and 15 minutes for fluid reclamation. This procedure is usually performed in a hospital. An interesting variation is called **continuous ambulatory peritoneal dialysis (CAPD)**, in which the patient him- or herself administers 2 liters of dialysis fluid through the catheter and then continues with life as usual until 4–6 hours later, when the fluid is removed and replaced.

Probably the most satisfactory solution, in terms of overall quality of life, is *kidney transplantation*. This procedure

TABLE 26–7 THE COMPOSITION OF DIALYSIS FLUID

Component	Plasma	Dialysis Fluid
Electrolytes (mEq/l)		
Potassium	4	3
Bicarbonate	27	36
Phosphate	3	0
Sulfate	0.5	0
Nutrients (mg/dl)		
Glucose	80–100	125
Nitrogenous wastes (mg/dl)		
Urea	20	0
Creatinine	1	0
Uric acid	3	0

Note: Only the significant variations are given; values for other electrolytes are usually similar. Although these values are representative, the precise composition can be tailored to meet specific clinical needs. For example, if plasma potassium levels are too low, the dialysis fluid concentration can be elevated to remedy the situation. Changes in the osmotic concentration of the dialysis fluid can also be used to adjust an individual's blood volume, generally by adjusting the glucose content of the dialysis fluid.

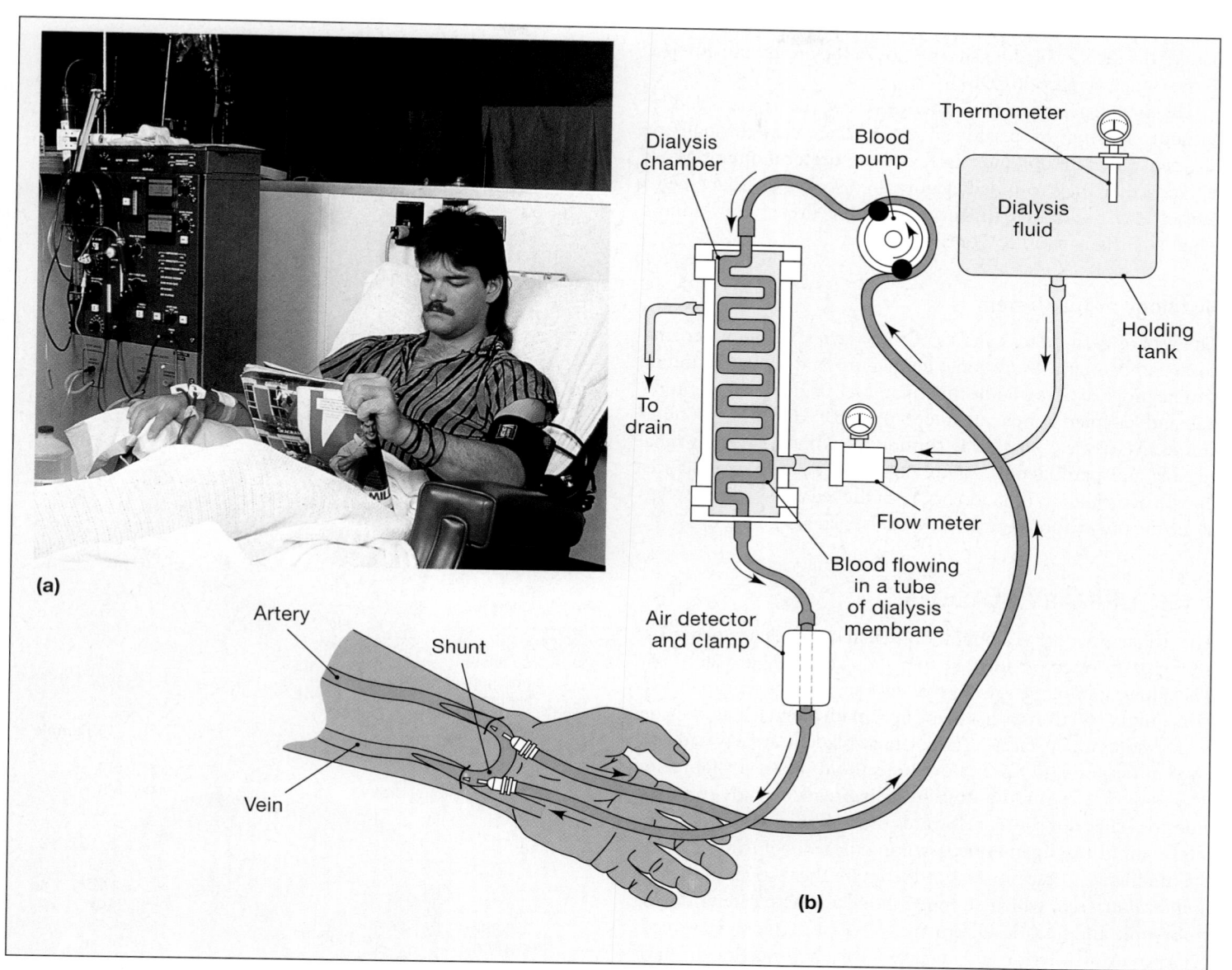

●**FIGURE 26–18**
Hemodialysis. (a) A patient is hooked up to a dialysis machine. (b) The dialysis procedure. Preparation for hemodialysis typically involves the implantation of a pair of shunts connected by a loop that permits normal blood flow when the patient is not hooked up to the machine.

involves the implantation of a new kidney obtained from a living donor or from a cadaver. Roughly half of the 13,372 kidneys transplanted in 2000 were obtained from living, related donors. In most cases, the damaged kidney is removed and its blood supply is connected to the transplant. When the original kidney is left in place, an arterial graft is inserted to carry blood from the iliac artery or the aorta to the transplant, which is placed in the pelvis or lower abdomen.

The success rate for kidney transplantation varies, depending on how aggressively the recipient's T cells attack the donated organ and whether an infection develops. The one-year success rate for transplantation is now 85–95 percent. The use of kidneys taken from close relatives significantly improves the chances that the transplant will succeed. Immunosuppressive drugs are administered to reduce tissue rejection, but unfortunately, this treatment also lowers the individual's resistance to infection.

by the ureters in men and women are different, due to variations in the nature, size, and position of the reproductive organs. As Figure 26–19a• shows, in males, the base of the urinary bladder lies between the rectum and the pubic symphysis; in females, the base of the urinary bladder sits inferior to the uterus and anterior to the vagina (Figure 26–19b•).

The ureters penetrate the posterior wall of the urinary bladder without entering the peritoneal cavity. They pass through the bladder wall at an oblique angle, and the **ureteral openings** are slitlike rather than rounded (Figure 26–19c•). This shape helps prevent the backflow of urine toward the ureter and kidneys when the urinary bladder contracts.

Histology of the Ureters

The wall of each ureter consists of three layers: (1) an inner mucosa, comprising a transitional epithelium and the surrounding lamina propria; (2) a middle muscular layer made up of longitudinal and circular bands of smooth muscle; and (3) an outer connective-tissue layer that is continuous with the fibrous renal capsule and peritoneum (Figure 26–20a•). About every 30 seconds, a peristaltic contraction begins at the renal pelvis and sweeps along the ureter, forcing urine toward the urinary bladder.

☐ THE URINARY BLADDER

The urinary bladder is a hollow, muscular organ that functions as a temporary reservoir for the storage of urine (Figure 26–19c•). The dimensions of the urinary bladder vary with its state of distension, but the full urinary bladder can contain about a liter of urine.

The superior surfaces of the urinary bladder are covered by a layer of peritoneum; several peritoneal folds assist in stabilizing its position. The **middle umbilical ligament** extends from the anterior, superior border toward the umbilicus (navel). The **lateral umbilical ligaments** pass along the sides of the bladder to the umbilicus. These fibrous cords contain the vestiges of the two *umbilical arteries*, which supplied blood to the placenta during embryonic and fetal development. ∞ p. 768 The urinary bladder's posterior, inferior, and anterior surfaces lie outside the peritoneal cavity. In these areas, tough ligamentous bands anchor the urinary bladder to the pelvic and pubic bones.

In sectional view, the mucosa lining the urinary bladder is usually thrown into folds, or **rugae**, that disappear as the bladder fills. The triangular area bounded by the ureteral openings and the entrance to the urethra constitutes a region called the **trigone** (TRĪ-gōn) of the urinary bladder. There, the mucosa is smooth and very thick. The trigone acts as a funnel that channels urine into the urethra when the urinary bladder contracts.

The urethral entrance lies at the apex of the trigone, at the most inferior point in the urinary bladder. The region surrounding the urethral opening, known as the **neck** of the urinary bladder, contains a muscular **internal urethral sphincter**, or *sphincter vesicae*. The smooth muscle fibers of this sphincter provide involuntary control over the discharge of urine from the bladder.

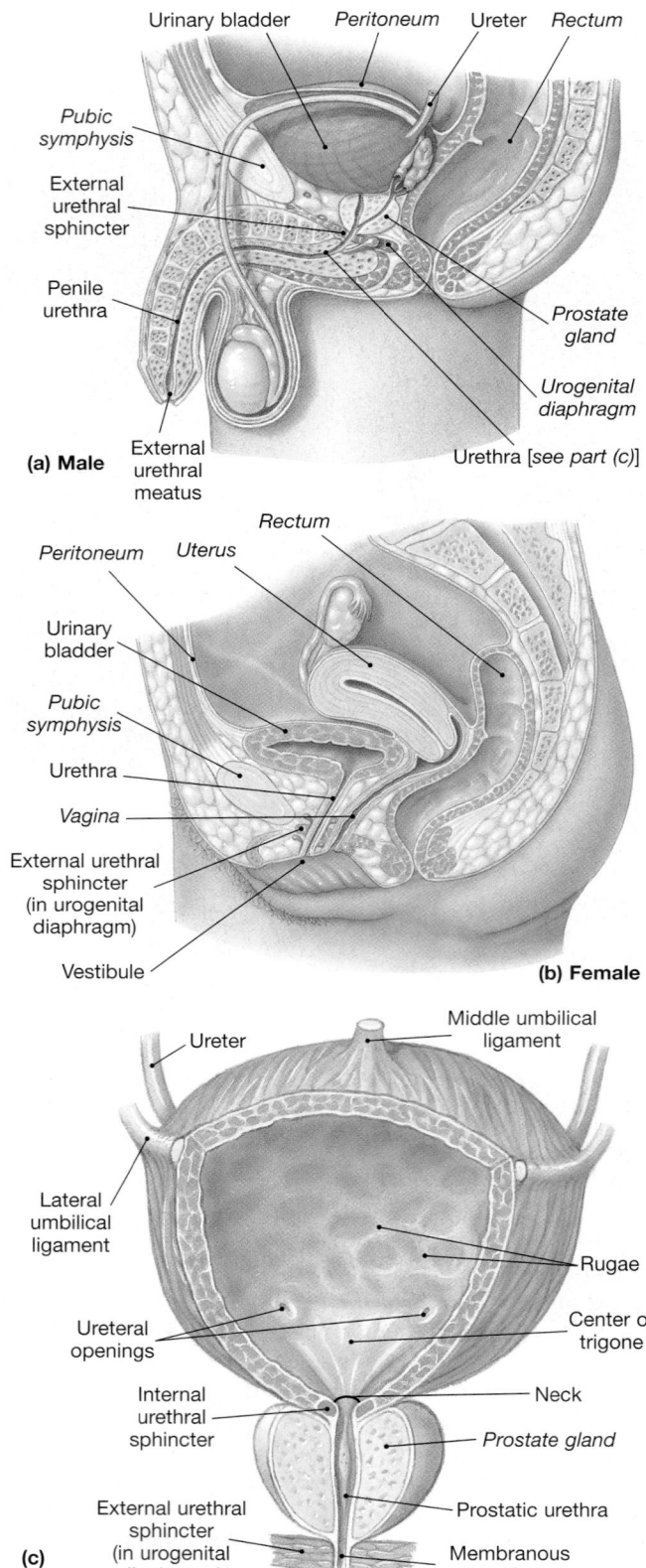

(a) Male

(b) Female

(c)

•FIGURE 26–19
Organs for the Conduction and Storage of Urine. The ureter, urinary bladder, and urethra **(a)** in the male and **(b)** in the female. **(c)** The urinary bladder of a male. **ATLAS** Plates 7.10e, 7.11a,b

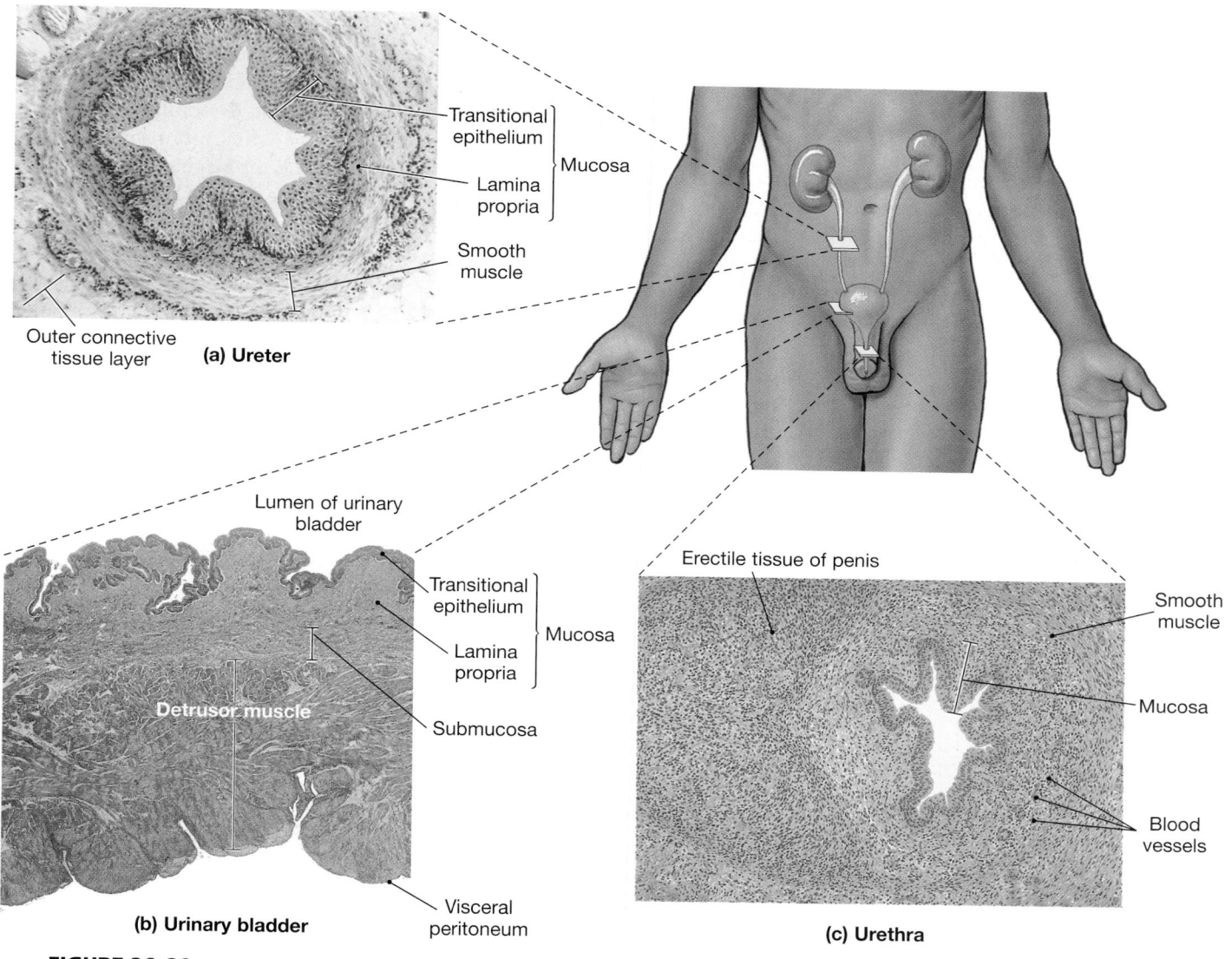

•FIGURE 26–20
The Histology of the Organs that Collect and Transport Urine. **(a)** A transverse section through the ureter. A thick layer of smooth muscle surrounds the lumen. (For a close-up of transitional epithelium, review Figure 4–4c•, p.117.) (LM × 53) **(b)** The wall of the urinary bladder. (LM × 29) **(c)** A transverse section through the male urethra. (LM × 49)

The urinary bladder is innervated by postganglionic fibers from ganglia in the hypogastric plexus and by parasympathetic fibers from intramural ganglia that are controlled by branches of the pelvic nerves.

Histology of the Urinary Bladder

The wall of the urinary bladder contains mucosa, submucosa, and muscularis layers. The muscularis layer consists of inner and outer layers of longitudinal smooth muscle, with a circular layer sandwiched in between. Collectively, these layers form the powerful **detrusor** (de-TROO-sor) **muscle** of the urinary bladder (Figure 26–20b•). Contraction of this muscle compresses the urinary bladder and expels its contents into the urethra.

Each year in the United States, approximately 52,000 new cases of **bladder cancer** are diagnosed, and there are 9500 deaths from this condition. The incidence among males is three times that among females, and most patients are age 60–70. Environmental factors, especially exposure to *2-naphthylamine* or related compounds, are responsible for most bladder cancers. For this reason, the bladder cancer rate is highest among cigarette smokers and employees of chemical and rubber companies. The mechanism appears to involve damage to tumor suppressor genes, such as *p53*, that regulate cell division. The prognosis is reasonably good for localized superficial cancers (88 percent five-year survival), but it is poor for persons with severe metastatic bladder cancer (9 percent five-year survival). Treatment of metastasized bladder cancer is very difficult, because the cancer spreads rapidly through adjacent lymphatic vessels and through the bone marrow of the pelvis.

☐ THE URETHRA

The urethra extends from the neck of the urinary bladder to the exterior of the body. The urethrae of males and females differ in length and in function. In males, the urethra extends from the neck of the urinary bladder to the tip of the penis, a distance that may be 18–20 cm (7–8 in.). The male urethra can be subdivided into three portions (Figure 26–19a,c●): (1) the prostatic urethra, (2) the membranous urethra, and (3) the penile urethra. The **prostatic urethra** passes through the center of the prostate gland. The **membranous urethra** includes the short segment that penetrates the *urogenital diaphragm*, the muscular floor of the pelvic cavity. The **penile** (PĒ-nīl) **urethra** extends from the distal border of the urogenital diaphragm to the external opening, or **external urethral meatus**, at the tip of the penis. In females, the urethra is very short, extending 3–5 cm (1–2 in.) from the bladder to the vestibule (Figure 26–19b●). The external urethral meatus is near the anterior wall of the vagina.

In both genders, as the urethra passes through the urogenital diaphragm, a circular band of skeletal muscle forms the **external urethral sphincter**. This muscular band acts as a valve. The external urethral sphincter, which is under voluntary control via the perineal branch of the pudendal nerve, has a resting muscle tone and must be voluntarily relaxed to permit micturition. ∞ p. 444

Histology of the Urethra

The urethral lining consists of a stratified epithelium that varies from transitional at the neck of the urinary bladder, through stratified columnar at the midpoint, to stratified squamous near the external urethral meatus. The lamina propria is thick and elastic, and the mucous membrane is thrown into longitudinal folds (Figure 26–20c●). Mucin-secreting cells are located in the epithelial pockets. In males, the epithelial mucous glands may form tubules that extend into the lamina propria. Connective tissues of the lamina propria anchor the urethra to surrounding structures. In females, the lamina propria contains an extensive network of veins, and the entire complex is surrounded by concentric layers of smooth muscle. **AM** Urinary Tract Infections

☐ THE MICTURITION REFLEX AND URINATION

Urine reaches the urinary bladder by peristaltic contractions of the ureters. The process of urination is coordinated by the **micturition reflex**, the components of which are diagrammed in Figure 26–21●.

Stretch receptors in the wall of the urinary bladder are stimulated as the bladder fills with urine. Afferent fibers in the pelvic nerves carry the impulses generated to the sacral spinal cord.

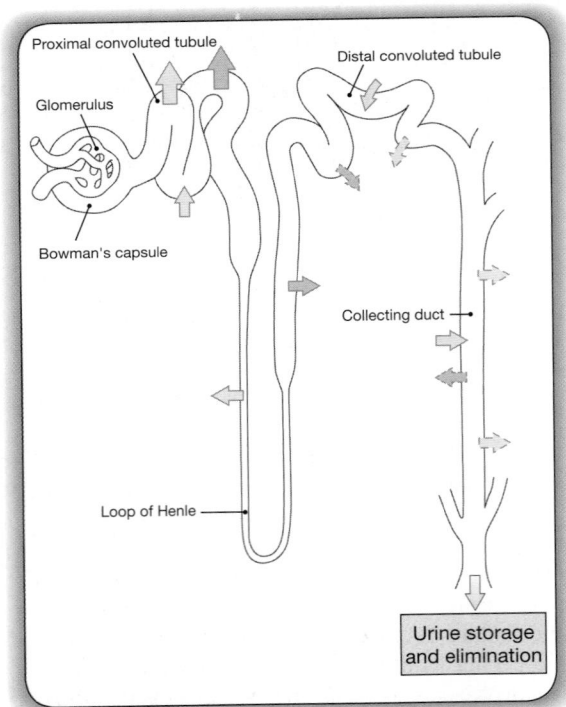

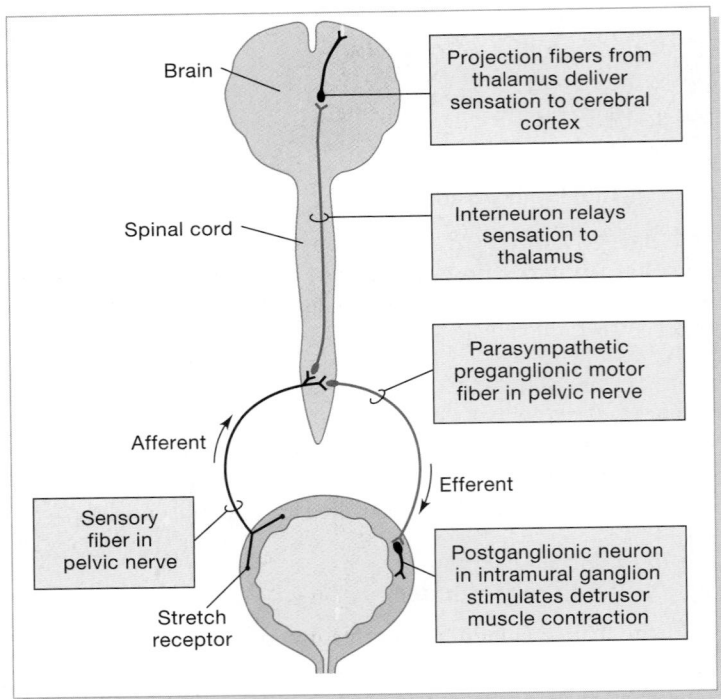

●**FIGURE 26–21**
The Micturition Reflex. The components of the reflex arc that stimulates smooth muscle contractions in the urinary bladder. Micturition occurs after voluntary relaxation of the external urethral sphincter.

The increased level of activity in the fibers (1) facilitates parasympathetic motor neurons in the sacral spinal cord and (2) stimulates interneurons that relay sensations to the thalamus and then, through projection fibers, to the cerebral cortex. As a result, you become consciously aware of the fluid pressure in your urinary bladder.

The urge to urinate generally appears when the bladder contains about 200 ml of urine. The micturition reflex begins to function when the stretch receptors have provided adequate stimulation to parasympathetic preganglionic motor neurons. Action potentials carried by efferent fibers within the pelvic nerves then stimulate ganglionic neurons in the wall of the urinary bladder. These neurons in turn stimulate sustained contraction of the detrusor muscle.

This contraction elevates fluid pressures in the urinary bladder, but urine ejection does not occur unless both the internal and external urethral sphincters are relaxed. The relaxation of the external urethral sphincter occurs under voluntary control. When the external urethral sphincter relaxes, so does the internal urethral sphincter. If the external urethral sphincter does not relax, the internal urethral sphincter remains closed and the urinary bladder gradually relaxes.

A further increase in bladder volume begins the cycle again, usually within an hour. Each increase in urinary volume leads to an increase in stretch receptor stimulation that makes the sensation more acute. Once the volume exceeds 500 ml, the micturition reflex may generate enough pressure to force open the internal urethral sphincter. This opening leads to a reflexive relaxation of the external urethral sphincter, and urination occurs despite voluntary opposition or potential inconvenience. At the end of a normal micturition, less than 10 ml of urine remains in the bladder.

Infants lack voluntary control over urination, because the necessary corticospinal connections have yet to be established. Accordingly, toilet training before age two years typically involves training the parent to anticipate the timing of the reflex rather than training the child to exert conscious control.

Incontinence (in-KON-ti-nens) is the inability to control urination voluntarily. Trauma to the internal or external urethral sphincter can contribute to incontinence in otherwise healthy adults. For example, childbirth can stretch and damage the sphincter muscles, and some mothers then develop *stress incontinence*. In this condition, elevated intra-abdominal pressures, caused, for example, by a cough or sneeze, can overwhelm the sphincter muscles, causing urine to leak out. Incontinence can also develop in older individuals due to a general loss of muscle tone.

Damage to the central nervous system, the spinal cord, or the nerve supply to the urinary bladder or external urethral sphincter can also produce incontinence. For example, incontinence commonly accompanies Alzheimer's disease or spinal cord damage. In most cases, the affected individual develops an *automatic bladder*. The micturition reflex remains intact, but voluntary control of the external urethral sphincter is lost, so the person cannot prevent the reflexive emptying of the urinary bladder. Damage to the pelvic nerves can abolish the micturition reflex entirely, because those nerves carry both afferent and efferent fibers of this reflex arc. The urinary bladder then becomes greatly distended with urine and remains filled to capacity while the excess urine trickles into the urethra in an uncontrolled stream. The insertion of a catheter is commonly required to facilitate the discharge of urine.

26–7 AGING AND THE URINARY SYSTEM

In general, aging is associated with an increased incidence of kidney problems. We described examples, such as *nephrolithiasis*—the formation of calculi, or kidney stones—earlier in the chapter (page 1000). Other age-related changes in the urinary system include:

- **A Decline in the Number of Functional Nephrons.** The total number of kidney nephrons drops by 30–40 percent between ages 25 and 85.
- **A Reduction in the GFR.** This reduction results from decreased numbers of glomeruli, cumulative damage to the filtration apparatus in the remaining glomeruli, and diminished renal blood flow.
- **A Reduced Sensitivity to ADH.** With age, the distal portions of the nephron and collecting system become less responsive to ADH. Reabsorption of water and sodium ions occurs at a reduced rate, and more sodium ions are lost in urine.
- **Problems with the Micturition Reflex.** The following three factors are involved in such problems:
  1. The sphincter muscles lose muscle tone and become less effective at voluntarily retaining urine. *Incontinence* results, commonly involving a slow leakage
  2. The ability to control micturition can be lost after a stroke, Alzheimer's disease, or other CNS problems affecting the cerebral cortex or hypothalamus.
  3. In males, **urinary retention** may develop secondary to enlargement of the prostate gland (*prostatic hypertrophy*). In this condition, swelling and distortion of surrounding prostatic tissues compress the urethra, restricting or preventing the flow of urine.

✓ What effect would a high-protein diet have on the composition of urine?

✓ An obstruction of a ureter by a kidney stone would interfere with the flow of urine between which two points?

✓ The ability to control the micturition reflex depends on your ability to control which muscle?

Answers start on page Q-1

 The **URINARY SYSTEM** in Perspective

Organ/Component	Primary Functions
Kidneys	Form and concentrate urine; regulate blood pH and ion concentrations; perform endocrine functions (*see Chapter 18*)
Ureters	Conduct urine from kidneys to urinary bladder
Urinary Bladder	Stores urine for eventual elimination
Urethra	Conducts urine to exterior

INTEGRATION WITH OTHER SYSTEMS

The urinary system is not the only organ system involved in excretion. Indeed, the urinary, integumentary, respiratory, and digestive systems are together regarded as an anatomically diverse **excretory system** where components perform all the excretory functions of the body that affect the composition of body fluids:

- **Integumentary System.** Water losses and electrolyte losses in sensible perspiration can affect the volume and composi-

tion of the plasma. The effects are most apparent when losses are extreme, such as during peak sweat production. Small amounts of metabolic wastes, including urea, also are eliminated in perspiration.

- **Respiratory System.** The lungs remove the carbon dioxide generated by cells. Small amounts of other compounds, such as acetone and water, evaporate into the alveoli and are eliminated when you exhale.

- **Digestive System.** Your liver excretes small amounts of metabolic waste products in bile, and you lose a variable amount of water in feces.

These excretory activities have an impact on the composition of body fluids. The respiratory system, for example, is the primary site of carbon dioxide excretion. But the excretory functions of these systems are not regulated as closely as are those of the kidneys. Under normal circumstances, the effects of integumentary and digestive excretory activities are minor compared with those of the urinary system.

The figure on the opposite page summarizes the functional relationships between the urinary system and other systems. We will explore many of these relationships further in Chapter 27 when we consider major aspects of fluid, pH, and electrolyte balance.

CLINICAL PATTERNS

Clinical conditions involving the urinary system can be sorted into the same general categories as disorders affecting other systems: (1) degenerative disorders, such as renal failure (p. 989); (2) congenital disorders, such as polycystic kidney disease (p. 980); (3) inflammatory and infectious disorders, such as urinary tract infections (*Applications Manual*); (4) tumors, such as bladder cancer (p. 1005); and (5) immune disorders, such as glomerulonephritis (p. 980). There are also a number of disorders that involve the

urinary system, but that reflect ongoing problems with other systems. Examples include problems with fluid, electrolyte, and acid–base balance that are discussed in Chapter 27. The *Applications Manual* has extended discussions of each of these categories of urinary system disorders. It also discusses common tests of renal function, and gives examples of the calculations involved in estimated renal blood flow, renal clearance, and GFR.

MEDIA CONNECTIONS

I. Objective: To explore the body's dependence on the urinary system in the regulation of the internal fluid environment.

Completion Time = 12 minutes

We know that water makes up a large percentage of the liquid in our body, and that the urinary system is responsible for cleansing that entire volume, but what controls the gain and loss of water in the tissues? Visit the **InterActive Physiology CD-ROM**: Fluids and Electrolytes/Water Homeostasis, and work through screens 3 through 10. How does this information relate to the urinary system? To review the role of the kidneys in water balance, look at Figure 26–9 and then complete screens 11 through 13 of the tutorial. What other systems assist in fluid homeostasis? Continue with screens 14 through 26, noting each new system mentioned and writing down its specific role in water balance. Are there any surprises on your list? Return to the text and review the "System in Perspective" sections for each of these systems. Is water balance discussed in the overview of the system's functions?

II. Objective: To investigate the nature, uses, and potential abuses of the information that can be derived from the analysis of urine.

Completion time = 10 minutes

Have you recently applied for a job? If so, you may have been asked to submit to a urine test, or urinalysis. Recall that urine is merely a by-product of your body's regular fine-tuning of its internal fluids. What can be determined from a chemical and physical study of the urinary system's output? Go to the **Companion Website**, Chapter 26 Media Connections, and click on the keyword **urinalysis**. Read through the information on the tests that comprise a typical urinalysis, paying close attention to the photos and descriptions of what they reveal. List each test and indicate what information it can provide. Place a check mark beside those tests that might be of benefit to a potential employer, and briefly describe why. Is there more information here than an employer needs? Should urinalysis be used as a method of employment screening? Include your thoughts on this at the bottom of your list.

INTEGUMENTARY SYSTEM

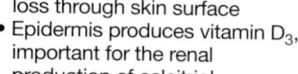

- Sweat glands assist in elimination of water and solutes, especially sodium and chloride ions
- Keratinized epidermis prevents excessive fluid loss through skin surface
- Epidermis produces vitamin D_3, important for the renal production of calcitriol

SKELETAL SYSTEM

- Axial skeleton provides some protection for kidneys and ureters;
- Pelvis protects urinary bladder and proximal portion of urethra

---

- Conserves calcium and phosphate needed for bone growth

MUSCULAR SYSTEM

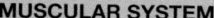

- Sphincter controls urination by closing urethral opening
- Muscle layers of trunk provide some protection for urinary organs

---

- Removes waste products of protein metabolism
- Assists in regulation of calcium and phosphate concentrations

NERVOUS SYSTEM

- Adjusts renal blood pressure
- Monitors distension of urinary bladder and controls urination

ENDOCRINE SYSTEM

- Aldosterone, ADH, ANP, and BNP adjust rates of fluid and electrolyte reabsorption in kidneys

---

- Kidney cells release renin when local blood pressure declines and erythropoietin (EPO) when renal oxygen levels decline

URINARY SYSTEM

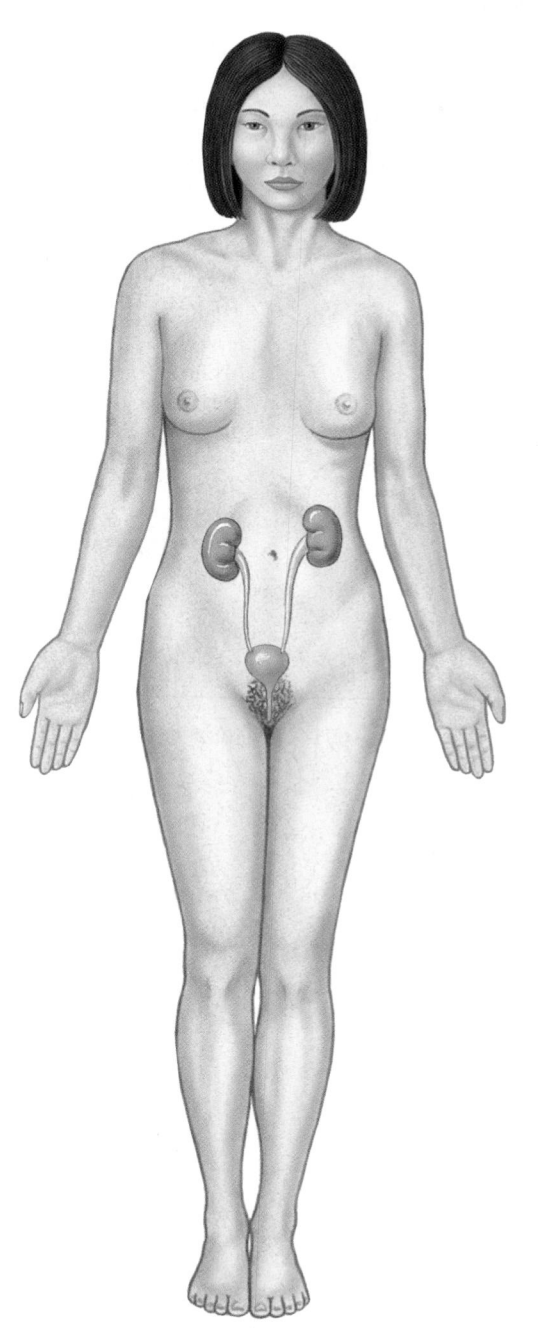

FOR ALL SYSTEMS
Excretes waste products; maintains normal body fluid pH and ion composition

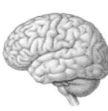

CARDIOVASCULAR SYSTEM

- Delivers blood to capillaries, where filtration occurs;
- Accepts fluids and solutes reabsorbed during urine production

---

- Releases renin to elevate blood pressure and erythropoietin to accelerate red blood cell production

LYMPHATIC SYSTEM

- Provides specific defenses against urinary tract infections

---

- Eliminates toxins and wastes generated by cellular activities
- Acid pH of urine provides nonspecific defense against urinary tract infections

RESPIRATORY SYSTEM

- Assists in the regulation of pH by eliminating carbon dioxide

---

- Assists in the elimination of carbon dioxide
- Provides bicarbonate buffers that assist in pH regulation

DIGESTIVE SYSTEM

- Absorbs water needed to excrete wastes at kidneys
- Absorbs ions needed to maintain normal body fluid concentrations
- Liver removes bilirubin

- Excretes toxins absorbed by the digestive epithelium
- Excretes bilirubin and nitrogenous wastes produced by the liver
- Calcitrol production by kidneys aids calcium and phosphate absorption along digestive tract

REPRODUCTIVE SYSTEM

- Accessory organ secretions may have antibacterial action that helps prevent urethral infections in males

---

- Urethra in males carries semen to the exterior

Chapter Review

SELECTED CLINICAL TERMINOLOGY

Terms Discussed in This Chapter

aminoaciduria: Amino acid loss in urine; the most common form is *cystinuria.* (*p. 983 and AM*)

calculi (KAL-kū-lī): Insoluble deposits that form within the urinary tract from calcium salts, magnesium salts, or uric acid. (*p. 1000*)

clearance test: A test that permits the GFR to be calculated by monitoring plasma and renal concentrations of a specific solute, such as creatinine. (*p. 986 and AM*)

diuretics (dī-ū-RET-iks): Drugs that promote fluid loss in urine. (*p. 999 and AM*)

glomerulonephritis (glo-mer-ū-lō-nef-RĪ-tis): An inflammation of the renal cortex. (*p. 980*)

hematuria: Blood loss in urine. (*p. 989 and AM*)

hemodialysis (hē-mō-dī-AL-i-sis): A technique in which an artificial membrane is used to regulate the composition of blood. (*p. 1002*)

incontinence (in-KON-ti-nens): An inability to control urination voluntarily. (*p. 1007*)

polycystic kidney disease: An inherited abnormality that affects the development and structure of kidney tubules. (*p. 980*)

proteinuria: Protein loss in urine. (*p. 989 and AM*)

pyelogram (PĪ-el-ō-gram): An image obtained by taking an X ray of the kidneys after a radiopaque compound has been administered. (*p. 1000 and AM*)

renal failure: An inability of the kidneys to excrete wastes in sufficient quantities to maintain homeostasis. (*p. 989*)

urinalysis: A physical and chemical assessment of urine. (*p. 1000*)

urinary obstruction: A blockage of the urinary tract. (*p. 1000*)

AM Additional Terms Discussed in the *Applications Manual*

cystitis
dysuria
nephritis
pyelitis
pyelonephritis
uremia
urethritis
urinary tract infection (UTI)

STUDY OUTLINE

26–1 AN OVERVIEW OF THE URINARY SYSTEM   p. 971

1. The two major functions of the **urinary system** are *excretion*—the removal of organic waste products from body fluids—and *elimination*—the discharge of these waste products into the environment. Other functions include regulating blood volume and pressure by adjusting the volume of water lost and releasing hormones; regulating plasma concentrations of ions; helping to stabilize blood pH; conserving nutrients; and assisting the liver in detoxifying poisons and, during starvation, deaminating amino acids so that they can be catabolized by other tissues.

2. The urinary system includes the **kidneys**, the **ureters**, the **urinary bladder**, and the **urethra**. The kidneys produce **urine**, a fluid containing water, ions, and soluble compounds. During **urination** (**micturition**) urine is forced out of the body. (*Figure 26–1*)

26–2 THE KIDNEYS   p. 972

1. The left kidney extends superiorly slightly more than the right kidney. Both kidneys and the adrenal gland that overlies each are retroperitoneal. (*Figures 26–1, 26–2, 26–3*)

2. The **hilus**, a medial indentation, provides entry for the *renal artery* and *renal nerves* and exit for the *renal vein* and the ureter. (*Figures 26–3, 26–4*)

SECTIONAL ANATOMY OF THE KIDNEYS   p. 972

3. The superficial portion of the kidney is the cortex, surrounded by the **medulla**. The ureter communicates with the **renal pelvis**, a chamber that branches into two **major calyces**. Each major calyx is connected to four or five **minor calyces**, which enclose the **renal papillae**. (*Figure 26–4*)

BLOOD SUPPLY AND INNERVATION OF THE KIDNEYS   p. 974

4. The blood supply to the kidneys includes the **renal, segmental, interlobar, arcuate,** and **interlobular arteries**.

5. The **renal nerves**, which innervate the kidneys and ureters, are dominated by sympathetic postganglionic fibers.

THE NEPHRON   p. 976

6. The **nephron** (the basic functional unit in the kidney) consists of the **renal corpuscle** and **renal tubule**. The renal tubule is long and narrow and divided into the *proximal convoluted tubule*, the *loop of Henle*, and the *distal convoluted tubule*. **Filtrate** is produced at the renal corpuscle. The nephron empties **tubular fluid** into the **collecting system**, consisting of **collecting ducts** and **papillary ducts**. (*Figures 26–5, 26–6, 26–7*)

7. Nephrons are responsible for the production of filtrate, the reabsorption of organic nutrients, the reabsorption of water and ions, and the secretion into the tubular fluid of waste products missed by filtration. (*Summary Table 26–1*)

8. Roughly 85 percent of the nephrons are **cortical nephrons**, located in the renal **cortex**. The **juxtamedullary nephrons** are closer to the renal **medulla**, with their *loops of Henle* extending deep into the **renal pyramids**. (*Figure 26–7*)

9. Blood travels from the efferent arteriole to the **peritubular capillaries** and the **vasa recta**. (*Figure 26–8*)

10. The renal tubule begins at the renal corpuscle and includes a knot of intertwined capillaries called the **glomerulus**, surrounded by **Bowman's capsule**. Blood arrives at the glomerulus via the **afferent arteriole** and departs in the **efferent arteriole**. (*Figure 26–8*)

11. At the glomerulus, **podocytes** cover the **lamina densa** of the capillaries that project into the **capsular space**. The **pedicels** of the podocytes are separated by narrow **filtration slits**. (*Figure 26–8*)

12. The **proximal convoluted tubule (PCT)** actively reabsorbs nutrients, plasma proteins, and ions from filtrate. These substances are released into the **peritubular fluid**, which surrounds the nephron. *(Figure 26–6)*
13. The **loop of Henle** includes a **descending limb** and an **ascending limb**; each limb contains a **thick segment** and a **thin segment**. The ascending limb delivers fluid to the **distal convoluted tubule (DCT)**, which actively secretes ions, toxins, and drugs and reabsorbs sodium ions from the tubular fluid. *(Figures 26–6, 26–7, 26–8)*

Urinary system: **IP CD-ROM:** Urinary System/Anatomy Review. **Anatomy CD-ROM:** Urinary System/Dissection and Flythrough

26–3 PRINCIPLES OF RENAL PHYSIOLOGY   p. 981

1. Urine production maintains homeostasis by regulating the blood volume and composition. In the process, organic waste products—notably urea, creatinine, and uric acid—are excreted.

BASIC PROCESSES OF URINE FORMATION   p. 981

2. Urine formation involves **filtration, reabsorption,** and **secretion.** *(Table 26–2)*
3. Four types of *carrier-mediated transport (facilitated diffusion, active transport, cotransport,* and *countertransport)* are involved in modifying filtrate. The saturation limit of a carrier protein is its **transport maximum,** which determines the **renal threshold**—the plasma concentration at which various compounds will appear in urine. *(Table 26–2)*
4. The transport maximum determines the renal threshold for the reabsorption of substances in the tubular fluid.

AN OVERVIEW OF RENAL FUNCTION   p. 981

5. Most regions of the nephron perform a combination of reabsorption and secretion. *(Figure 26–9)*

26–4 RENAL PHYSIOLOGY: FILTRATION AND THE GLOMERULUS   p. 985

FILTRATION PRESSURES   p. 985

1. **Glomerular filtration** occurs as fluids move across the wall of the glomerulus into the capsular space in response to the **glomerular hydrostatic pressure (GHP)**—the hydrostatic (blood) pressure in the glomerular capillaries. This movement is opposed by the **capsular hydrostatic pressure (CsHP)** and by the **blood colloid osmotic pressure (BCOP).** The **filtration pressure (FP)** at the glomerulus is the difference between the blood pressure and the opposing capsular and osmotic pressures. *(Figures 26–8, 26–10)*

THE GLOMERULAR FILTRATION RATE   p. 986

2. The **glomerular filtration rate (GFR)** is the amount of filtrate produced in the kidneys each minute. Any factor that alters the filtration pressure acting across the glomerular capillaries will change the GFR and affect kidney function.

CONTROLLING THE GFR   p. 987

3. A drop in filtration pressures stimulates the **juxtaglomerular apparatus (JGA)** to release renin and erythropoietin. *(Figure 26–11)*
4. Sympathetic activation (1) produces a powerful vasoconstriction of the afferent arterioles, decreasing the GFR and slowing the production of filtrate, (2) alters the GFR by changing the regional pattern of blood circulation, and (3) stimulates the release of renin by the juxtaglomerular apparatus. *(Figure 26–11)*

26–5 RENAL PHYSIOLOGY: REABSORPTION AND SECRETION   p. 989

REABSORPTION AND SECRETION AT THE PCT   p. 989

1. Glomerular filtration produces a filtrate with a composition similar to blood plasma, but with few, if any, plasma proteins.
2. The cells of the PCT normally reabsorb sodium and other ions, water, and almost all the organic nutrients that enter the filtrate. These cells also secrete various substances into the tubular fluid. *(Figure 26–12)*

The processing of urine: **IP CD-ROM:** Urinary System/Early Filtrate Processing; Urinary System/Glomerular Filtration

THE LOOP OF HENLE AND COUNTERCURRENT MULTIPLICATION   p. 991

3. Water and ions are reclaimed from tubular fluid by the loop of Henle. A concentration gradient in the renal medulla encourages the osmotic flow of water out of the tubular fluid. The **countercurrent multiplication** between the ascending and descending limbs of the loop of Henle helps create the osmotic gradient in the medulla. As water is lost by osmosis and the volume of tubular fluid decreases, the urea concentration rises. *(Figure 26–13)*

REABSORPTION AND SECRETION AT THE DCT   p. 993

4. The DCT performs final adjustments by actively secreting or absorbing materials. Sodium ions are actively absorbed, in exchange for potassium or hydrogen ions discharged into tubular fluid. Aldosterone secretion increases the rate of sodium reabsorption and potassium loss. *(Figure 26–14; Table 26–3)*

REABSORPTION AND SECRETION ALONG THE COLLECTING SYSTEM   p. 993

5. The amount of water and solutes in the tubular fluid of the collecting ducts is further regulated by aldosterone and ADH secretions. *(Figure 26–14)*

THE CONTROL OF URINE VOLUME AND OSMOTIC CONCENTRATION   p. 995

6. Urine volume and osmotic concentration are regulated by controlling water reabsorption. Precise control over this is allowed via *facultative water reabsorption.* *(Figure 26–15)*

THE FUNCTION OF THE VASA RECTA   p. 997

7. Normally, the removal of solutes and water by the vasa recta precisely balances the rates of reabsorption and osmosis in the renal medulla.

THE COMPOSITION OF NORMAL URINE   p. 997

8. More than 99 percent of the filtrate produced each day is reabsorbed before reaching the renal pelvis. *(Table 26–5)*
9. **Urinalysis** is the chemical and physical analysis of a urine sample.

SUMMARY: RENAL FUNCTION   p. 997

10. Each segment of the nephron and collecting system contributes to the production of concentrated urine. *(Figure 26–16, Summary Table 26–4)*

The processing of urine: **IP CD-ROM:** Urinary System/Late Filtrate Processing

26–6 URINE TRANSPORT, STORAGE, AND ELIMINATION   p. 999

1. Urine production ends when tubular fluid enters the renal pelvis. The rest of the urinary system is responsible for transporting, storing, and eliminating urine. *(Figure 26–17)*

THE URETERS p. 1000

2. The ureters extend from the renal pelvis to the urinary bladder. Peristaltic contractions by smooth muscles move the urine. (*Figures 26–19, 26–20*)

THE URINARY BLADDER p. 1004

3. The urinary bladder is stabilized by the **middle umbilical ligament** and the **lateral umbilical ligaments**. Internal features include the **trigone**, the **neck**, and the **internal urethral sphincter**. The mucosal lining contains prominent **rugae** (folds). Contraction of the **detrusor muscle** compresses the urinary bladder and expels urine into the urethra. (*Figures 26–19, 26–20*)

THE URETHRA p. 1006

4. In both genders, as the urethra passes through the *urogenital diaphragm*, a circular band of skeletal muscles forms the **external urethral sphincter**, which is under voluntary control. (*Figures 26–19, 26–20*)

THE MICTURITION REFLEX AND URINATION p. 1006

5. Urination is coordinated by the **micturition reflex**, which is initiated by stretch receptors in the wall of the urinary bladder. Voluntary urination involves coupling this reflex with the voluntary relaxation of the external urethral sphincter, which allows the opening of the **internal urethral sphincter**. (*Figure 26–21*)

26-7 AGING AND THE URINARY SYSTEM   p. 1007

1. Aging is generally associated with increased kidney problems. Age-related changes in the urinary system include (1) declining numbers of functional nephrons, (2) reduced GFR, (3) reduced sensitivity to ADH, and (4) problems with the micturition reflex. (**Urinary retention** may develop in men whose prostate gland is enlarged.)

REVIEW QUESTIONS

More assessment questions are available to you on the Companion Website. You will find Matching, Multiple Choice, True/False, and other quizzes to help further your understanding of the material covered in this chapter. To access the site, go to www.aw.com/martini.

LEVEL 1 Reviewing Facts and Terms

1. The layer of collagen fibers covering the outer surface of the entire kidney is the
   - (a) adipose capsule
   - (b) renal cortex
   - (c) renal capsule
   - (d) renal fascia

2. The glomerulus is located within the
   - (a) renal corpuscle
   - (b) renal tubule
   - (c) renal pelvis
   - (d) renal column

3. The capillaries supplying individual nephrons are themselves supplied with blood by the
   - (a) afferent arterioles
   - (b) efferent arterioles
   - (c) peritubular capillaries
   - (d) arcuate veins

4. Large cells with complex processes, or "feet," that wrap around the glomerular capillaries are
   - (a) vasa recta
   - (b) podocytes
   - (c) astrocytes
   - (d) supporting cells

5. The majority of nephrons are located in the _____ of the kidney.
   - (a) vasa recta
   - (b) medulla
   - (c) cortex
   - (d) pelvis

6. The distal convoluted tubule is an important site for
   - (a) active secretion of ions
   - (b) active secretion of acids and other materials
   - (c) selective reabsorption of sodium ions from the tubular fluid
   - (d) a, b, and c are correct

7. The endocrine structure that secretes renin and erythropoietin is
   - (a) the juxtaglomerular apparatus
   - (b) the vasa recta
   - (c) Bowman's capsule
   - (d) the adrenal gland

8. The primary purpose of the collecting system is to
   - (a) transport urine from the urinary bladder to the urethra
   - (b) selectively reabsorb sodium ions from tubular fluid
   - (c) transport urine from the renal pelvis to the ureters
   - (d) make final adjustments to the osmotic concentration and volume of urine

9. Pickup or delivery of substances that are reabsorbed or secreted by the PCT and the DCT is provided by the
   - (a) afferent arteriole
   - (b) peritubular capillaries
   - (c) renal artery
   - (d) efferent arteriole

10. The most abundant organic waste in urine is
    - (a) uric acid
    - (b) creatinine
    - (c) urea
    - (d) creatine phosphate

11. The removal of water and solute molecules from filtrate after it enters the renal tubules is
    - (a) filtration
    - (b) secretion
    - (c) reabsorption
    - (d) elimination

12. Changing the diameter of the afferent and efferent arterioles to alter the GFR can be an example of _____ regulation.
    - (a) hormonal
    - (b) autonomic
    - (c) autoregulation or local
    - (d) all of the above

13. Blood colloid osmotic pressure tends to draw water
    - (a) into Bowman's capsule
    - (b) out of the filtrate and into the plasma
    - (c) across the loop of Henle
    - (d) out of the plasma at the glomerulus

14. Which of the following is not a part of the filtration apparatus?
    - (a) lamina densa
    - (b) parietal epithelium
    - (c) capillary endothelium
    - (d) filtration slits

15. What is the primary function of the urinary system?

16. What structures are components of the urinary system?

17. Trace the pathway of the protein-free filtrate from the time it is produced in the renal corpuscle until it drains into the renal pelvis in the form of urine. (Use arrows to indicate the direction of flow.)

18. Name the segments of the nephron distal to the renal corpuscle, and state the function(s) of each.

19. What role does the lamina densa play in the renal corpuscle?

20. What is the function of the juxtaglomerular apparatus?

21. Using arrows, trace a drop of blood from its entry into the renal artery until its exit via a renal vein.

22. Name and define the three distinct processes involved in the production of urine.
23. What are the primary effects of angiotensin II on kidney function and regulation?
24. Which parts of the urinary system are responsible for the transport, storage, and elimination of urine?

LEVEL 2 Reviewing Concepts

25. The urinary system regulates blood volume and pressure by
    (a) adjusting the volume of water lost in urine
    (b) releasing erythropoietin
    (c) releasing renin
    (d) a, b, and c are correct
26. The balance of solute and water reabsorption in the medulla is maintained by the
    (a) segmental arterioles and veins
    (b) interlobar arteries and veins
    (c) vasa recta
    (d) arcuate arteries
27. Sympathetic activation of the nerve fibers in the nephron causes
    (a) the regulation of glomerular blood flow and pressure
    (b) the stimulation of renin release from the juxtaglomerular apparatus
    (c) the direct stimulation of water and Na^+ reabsorption
    (d) a, b, and c are correct
28. An increase in the colloid osmotic pressure of the filtrate, caused by damage to the glomerulus, would
    (a) decrease the amount of plasma delivered to the kidney
    (b) enhance the movement of water and solutes into the capsular space
    (c) cause a decrease in the renal blood pressure
    (d) decrease movement of water and solutes into the capsular space
29. Sodium reabsorption in the DCT and cortical portion of the collecting system is accelerated by the secretion of
    (a) ADH   (b) renin
    (c) aldosterone   (d) erythropoietin
30. When ADH levels rise,
    (a) the amount of water reabsorbed increases
    (b) the DCT becomes impermeable to water
    (c) the amount of water reabsorbed decreases
    (d) sodium ions are exchanged for potassium ions
31. The control of blood pH by the kidneys during acidosis involves
    (a) the secretion of hydrogen ions and reabsorption of bicarbonate ions from the tubular fluid
    (b) a decrease in the amount of water reabsorbed
    (c) hydrogen ion reabsorption and bicarbonate ion loss
    (d) potassium ion secretion
32. How are proteins excluded from filtrate? Why?
33. How can you determine the filtration pressure (FP) at the glomerulus?
34. What interacting controls stabilize the glomerular filtration rate (GFR)?
35. Differentiate between cotransport and countertransport. What ions are involved in countertransport at the PCT?
36. What primary changes occur in the composition and concentration of filtrate as a result of activity in the proximal convoluted tubule?
37. What two functions does the countercurrent mechanism perform for the kidney?
38. What events in the distal convoluted tubule and collecting system determine the final composition and concentration of filtrate?
39. Describe the micturition reflex.

LEVEL 3 Critical Thinking and Clinical Applications

40. Why might long-haul trailer truck drivers commonly experience kidney problems?
41. Physicians often ask for urine samples collected over a 24-hour period, rather than a single sample. Why?
42. Randy enjoys his four or five cups of coffee a day and his six or seven beers a few nights a week. What are the effects of his caffeine and alcohol consumption on his urinary system?
43. For the past week, Susan has felt a burning sensation in the urethral area when she urinates. She checks her temperature and finds that she has a low-grade fever. What unusual substances are likely to be present in her urine?
44. *Lasix*™ is a diuretic that acts by decreasing the amounts of sodium and chloride ions actively transported by the ascending limb of the loop of Henle. Why would this medication be given to someone with high blood pressure (hypertension)?
45. Carlos has advanced arteriosclerosis. An analysis of his blood indicates elevated levels of aldosterone and decreased levels of ADH. Explain.
46. *Mannitol* is a sugar that is filtered, but not reabsorbed, by the kidneys. What effect would drinking a solution of mannitol have on the volume of urine produced?
47. The drug *Diamox*™ is sometimes used in the treatment of mountain sickness. Diamox inhibits the action of carbonic anhydrase in the proximal convoluted tubule. Polyuria (the elimination of an unusually large volume of urine) is a side effect associated with the medication. Why does this symptom occur?

UNIT 5   CHAPTER   23 24 25 26   27

CLINICAL NOTE
- Hypercalcemia and Hypocalcemia 1025

CLINICAL DISCUSSIONS
- Athletes and Salt Loss 1024
- Hypokalemia and Hyperkalemia 1025
- Fluid, Electrolyte, and Acid–Base Balance in Infants 1040

27

FLUID, ELECTROLYTE, AND ACID–BASE BALANCE

The next time you see a small pond, think about the fish it contains. They live out their lives totally dependent on the quality of that isolated environment. Severe water pollution will kill the fish, but even subtle changes can have equally grave effects. Changes in the volume of the pond, for example, can be quite important. If evaporation removes too much of the water, the fish become overcrowded; oxygen and food supplies run out, and the fish suffocate or starve. The ionic concentration of the water is also crucial. Most of the fish in a freshwater pond will die if the water becomes too salty; those in a saltwater pond will die if their environment becomes too dilute. The pH of the pond water, too, is a vital factor; that is one reason acid rain is such a problem.

Your cells live in a pond whose shores are the exposed surfaces of your skin. Most of your body weight is water. Water accounts for up to 99 percent of the volume of the fluid outside cells, and it is an essential ingredient of cytoplasm. All of a cell's operations rely on water as a diffusion medium for the distribution of gases, nutrients, and waste products. If the water content of the body changes, cellular activities are jeopardized. For example, when the water content reaches very low levels, proteins denature, enzymes cease functioning, and cells die.

27-1 FLUID, ELECTROLYTE, AND ACID–BASE BALANCE: AN OVERVIEW

Objective

- Explain what is meant by the terms "fluid balance," "electrolyte balance," and "acid–base balance," and discuss their importance for homeostasis.

To survive, we must maintain a normal volume and composition of the **extracellular fluid**, or **ECF** (the interstitial fluid, plasma, and other body fluids), and the **intracellular fluid**, or **ICF** (the cytosol). The ionic concentrations and pH, or hydrogen ion concentration, of these fluids are as important as their absolute quantity. If concentrations of calcium or potassium ions in the ECF become too high, cardiac arrhythmias develop and death can result. A pH outside the normal range can also lead to a variety of serious problems. Low pH is especially dangerous, because hydrogen ions break chemical bonds, change the shapes of complex molecules, disrupt cell membranes, and impair tissue functions. ∞ pp. 43-44

1015

In this chapter, we will consider the dynamics of exchange among the various body fluids and between the body and the external environment. Stabilizing the volumes, solute concentrations, and pH of the ECF and the ICF involves three interrelated processes:

1. *Fluid Balance.* You are in **fluid balance** when the amount of water you gain each day is equal to the amount you lose to the environment. The maintenance of normal fluid balance involves regulating content and distribution of body water in the ECF and the ICF. The digestive system is the primary source of water gains, with a small amount of additional water generated by metabolic activity. The urinary system is the primary route for water loss under normal conditions, but as we noted in Chapter 25, sweat gland activity can become important when body temperature is elevated. ∞ p. 962 Because your cells and tissues cannot transport water, fluid balance reflects primarily the control of ionic concentrations inside and outside of the cell; this is known as *electrolyte balance.*

2. *Electrolyte Balance.* **Electrolytes** are ions released through the dissociation of inorganic compounds; they are so named because they will conduct an electrical current in a solution. ∞ p. 42 Each day, your body fluids gain electrolytes from the food and drink you consume and lose electrolytes in urine, sweat, and feces. For each ion, daily gains must balance daily losses. For example, if you lose 500 mg of Na^+ in urine and insensible perspiration, you will need to gain 500 mg of Na^+ from food and drink to remain in sodium balance. If the gains and losses for every electrolyte are in balance, you are said to be in **electrolyte balance**. Electrolyte balance primarily involves balancing the rates of absorption across the digestive tract with rates of loss at the kidneys, although losses at sweat glands and other sites can play a secondary role.

3. *Acid–Base Balance.* You are in **acid–base balance** when the production of hydrogen ions in your body is precisely offset by their loss. When acid–base balance exists, the pH of body fluids remains within normal limits. ∞ p. 44 Preventing a reduction in pH is the primary problem, because your body generates a variety of acids during normal metabolic operations. The kidneys play a major role by secreting hydrogen ions into the urine and generating buffers that enter the bloodstream. Such secretion occurs primarily in the distal segments of the distal convoluted tubule (DCT) and along the collecting system. ∞ p. 993 The lungs also play a key role through the elimination of carbon dioxide.

Much of the material in this chapter was introduced in earlier chapters that focused on aspects of fluid, electrolyte, or acid–base balance that affected specific systems. This chapter provides an overview that integrates those discussions to highlight important functional patterns. Few other chapters have such wide-ranging clinical importance: The treatment of any serious illness affecting the nervous, cardiovascular, respiratory, urinary, or digestive system must include steps to restore normal fluid, electrolyte, and acid–base balances. Because this chapter builds on information provided in earlier chapters, you will encounter many references to relevant discussions and figures that you should use when you need a quick review.

27–2 AN INTRODUCTION TO FLUID AND ELECTROLYTE BALANCE

Objectives

- Compare the composition of intracellular and extracellular fluids.
- Explain the basic concepts involved in the regulation of fluids and electrolytes.
- Identify the hormones that play important roles in regulating fluid balance and electrolyte balance, and describe their effects.

Figure 27–1a● presents an overview of the composition of the body of a 70-kg individual with a minimum of body fat. The distribution was obtained by averaging values for males and females age 18–40 years. Water accounts for roughly 60 percent of the total body weight of an adult male and 50 percent of that of an adult female. This gender difference primarily reflects the proportionately larger mass of adipose tissue in adult females and the greater average muscle mass in adult males. (Adipose tissue is 10 percent water, whereas skeletal muscle is 75 percent water.) In both genders, intracellular fluid contains more of the total body water than does extracellular fluid. Exchange between the ICF and the ECF occurs across cell membranes by osmosis, diffusion, and carrier-mediated transport. (To review the mechanisms involved, see Table 3–3, p. 99.)

◻ THE ECF AND THE ICF

The largest subdivisions of the ECF are the *interstitial fluid* of peripheral tissues and the *plasma* of circulating blood. Minor components of the ECF include lymph, cerebrospinal fluid (CSF), synovial fluid, serous fluids (pleural, pericardial, and peritoneal fluids), aqueous humor, perilymph, and endolymph. Precise measurements of total body water provide additional information on gender differences in the distribution of body water (Figure 27–1b●). The greatest variation is in the ICF, as a result of differences in the intracellular water content of fat versus muscle. Less striking differences occur in the ECF values, due to variations in the interstitial fluid volume of various tissues and the larger blood volume in males versus females.

In clinical situations, it is customary to estimate that two-thirds of the total body water is in the ICF and one-third in the ECF. This ratio underestimates the real volume of the ECF, but that underestimation is appropriate because portions of the ECF, including the water in bone, in many dense connective tissues, and in many of the minor ECF components, are relatively isolated. Exchange between these fluid volumes and the rest of the ECF occurs more slowly than does exchange between plasma and other interstitial fluids. As a result, they can be safely ignored in many cases. Clinical attention is usually focused on the rapid fluid and solute movements associated with the administration of blood, plasma, or saline solutions to counteract blood loss or dehydration.

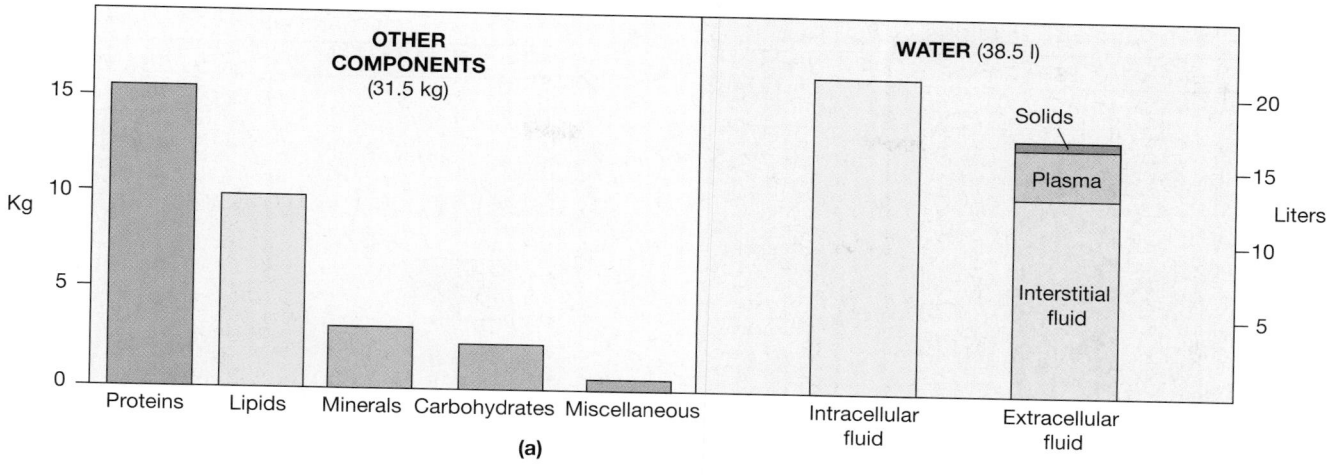

•FIGURE 27–1
The Composition of the Human Body. **(a)** The body composition (averaged for both genders) and major body fluid compartments of a 70-kg individual. For technical reasons, it is extremely difficult to determine the precise volume of any of these compartments; estimates of their relative sizes vary widely. **(b)** A comparison of the body compositions of adult males and females, age 18–40 years.

Body Composition of Adult Males and Females

Material and Compartment	Percentage of Total Body Weight	
	Adult Males	Adult Females
SOLIDS (protein, lipids, minerals, carbohydrates, organic and inorganic materials)	40	50
WATER	60	50
ICF	33	27
ECF	27	23
Interstitial fluid	21.5	18
Plasma	4.5	18
Minor components	1.0	4.0
		1.0

(b)

Exchange among the subdivisions of the ECF occurs primarily across the endothelial lining of capillaries. Fluid may also travel from the interstitial spaces to plasma through lymphatic vessels that drain into the venous system. ∞ pp. 781-783 The identity and quantity of dissolved electrolytes, proteins, nutrients, and waste products in the ECF vary regionally. (For a chemical analysis of the composition of ECF compartments, see Appendix IV.) Yet the variations among the segments of the ECF seem minor compared with the major differences between the ECF and the ICF.

The ECF and ICF are called **fluid compartments**, because they commonly behave as distinct entities. The presence of a cell membrane and active transport at the membrane surface enable cells to maintain internal environments with a composition different from that of their surroundings. The principal ions in the ECF are sodium, chloride, and bicarbonate. The ICF contains an abundance of potassium, magnesium, and phosphate ions, plus large numbers of negatively charged proteins. Figure 27–2• compares the ICF with the two major subdivisions of the ECF.

If the cell membrane were freely permeable, diffusion would continue until these ions were evenly distributed across the membrane. But it does not, because cell membranes are selectively permeable: Ions can enter or leave the cell only via specific membrane channels. In addition, carrier mechanisms move specific ions into or out of the cell.

Despite the differences in the concentration of specific substances, the ICF and ECF osmotic concentrations are identical. Osmosis eliminates minor differences in concentration almost at once, because most cell membranes are freely permeable to water. (The only noteworthy exceptions are the apical surfaces of epithelial cells along the ascending limb of the loop of Henle, the distal convoluted tubule, and the collecting system.) ∞ p. 995 Because changes in solute concentrations lead to immediate changes in water distribution, the regulation of fluid balance and that of electrolyte balance are tightly intertwined.

Physiologists and clinicians pay particular attention to ionic distributions across membranes and to the electrolyte composition of body fluids. Appendix IV reports normal values in the units most often encountered in clinical reports.

☐ BASIC CONCEPTS IN THE REGULATION OF FLUIDS AND ELECTROLYTES

You must understand four basic principles before we can proceed to a discussion of fluid balance and electrolyte balance:

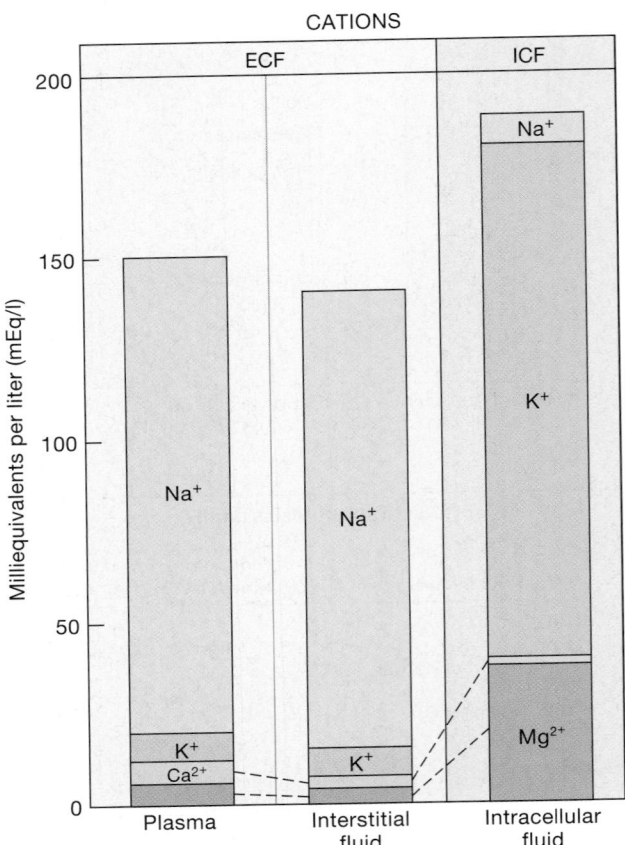

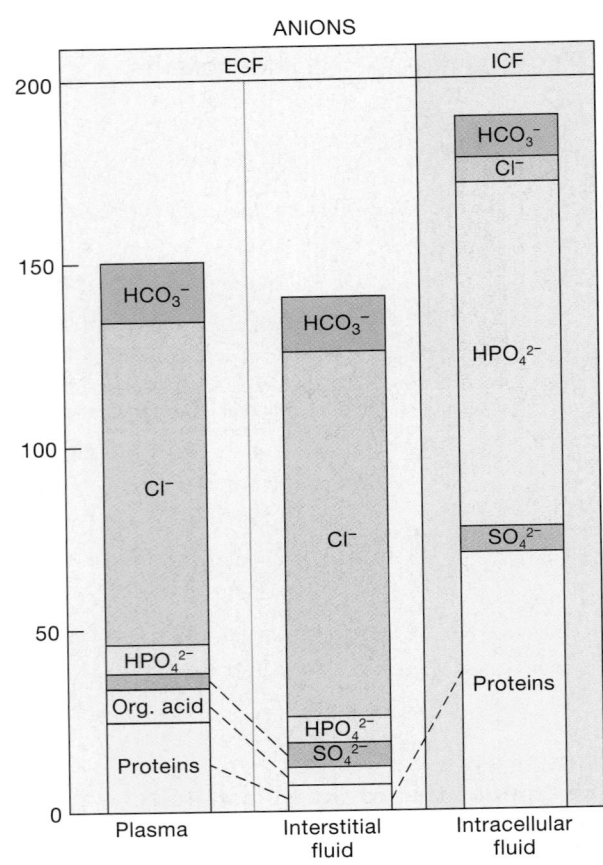

•**FIGURE 27–2**
Cations and Anions in Body Fluids. Notice the differences in cation and anion concentrations in the various body fluid compartments. For information about the composition of other body fluids, see Appendix IV.

1. *All the Homeostatic Mechanisms that Monitor and Adjust the Composition of Body Fluids Respond to Changes in the ECF, Not in the ICF.* Receptors monitoring the composition of two key components of the ECF—plasma and cerebrospinal fluid—detect significant changes in their composition or volume and trigger appropriate neural and endocrine responses. This arrangement makes functional sense, because a change in one ECF component will spread rapidly throughout the extracellular compartment and will affect all the body's cells. In contrast, the ICF is contained within trillions of individual cells that are physically and chemically isolated from one another by their cell membranes. Thus, changes in the ICF in one cell have no direct effect on the composition of the ICF in distant cells and tissues, unless those changes affect the ECF.

2. *No Receptors Directly Monitor Fluid or Electrolyte Balance.* In other words, our receptors cannot detect how many liters of water or grams of sodium, chloride, or potassium the body contains or count how many liters or grams we gain or lose in the course of a day. But our receptors *can* monitor *plasma volume* and *osmotic concentration*. Because fluid continuously circulates between interstitial fluid and plasma, and because exchange occurs between the ECF and the ICF, the plasma volume and osmotic concentration are good indicators of the state of fluid balance and electrolyte balance for the body as a whole.

3. *Our Cells Cannot Move Water Molecules by Active Transport.* All movement of water across cell membranes and epithelia occurs passively, in response to osmotic gradients established by the active transport of specific ions, such as sodium and chloride. You may find it useful to remember the simple phrase "*Water follows salt.*" As we saw in earlier chapters, when sodium and chloride ions (or other solutes) are actively transported across a membrane or epithelium, water follows by osmosis. ∞ p. 991-993 This basic principle accounts for water absorption across the digestive epithelium and water conservation in the kidneys.

4. *The Body's Content of Water or Electrolytes Will Rise if Dietary Gains Exceed Losses to the Environment and Will Fall if Losses Exceed Gains.* This basic rule is important when you consider the mechanics of fluid balance and electrolyte balance. Homeostatic adjustments generally affect the balance between urinary excretion and dietary absorption. As we saw in Chapter 26, the physiological adjustments in renal function are regulated primarily by circulating hormones. These hormones can also produce complementary changes in behavior. For example, the combination of angiotensin II and aldosterone can give you a sensation of thirst—which stimulates you to drink fluids—and a taste for heavily salted foods.

☐ AN OVERVIEW OF THE PRIMARY REGULATORY HORMONES

Major physiological adjustments affecting fluid balance and electrolyte balance are mediated by three hormones: (1) *antidiuretic hormone (ADH)*, (2) *aldosterone*, and (3) *the natriuretic peptides (ANP and BNP)*. These hormones were introduced and discussed in earlier chapters; we shall summarize their effects next. Those interested in a more detailed review should refer to the appropriate sections of Chapters 18, 21, and 26. The interactions among these hormones were shown in Figures 18–20(b), 21–17, 21–18, and 26–11●. ∞ pp. 636, 745, 748, 988

Antidiuretic Hormone

The hypothalamus contains special cells known as **osmoreceptors**, which monitor the osmotic concentration of the ECF. These cells are sensitive to subtle changes: A 2 percent change in osmotic concentration (approximately 6 mOsm/l) is sufficient to alter osmoreceptor activity.

The population of osmoreceptors includes neurons that secrete ADH. These neurons are located in the anterior hypothalamus, and their axons release ADH near fenestrated capillaries in the posterior lobe of the pituitary gland. The rate of ADH release varies directly with the osmotic concentration: The higher the osmotic concentration, the more ADH is released.

Increased release of ADH has two important effects: (1) It stimulates water conservation at the kidneys, reducing urinary water losses and concentrating the urine; and (2) it stimulates the thirst center, promoting the intake of fluids. As we saw in Chapter 21, the combination of decreased water loss and increased water gain gradually restores the normal plasma osmotic concentration. ∞ p. 744

Aldosterone

The secretion of aldosterone by the adrenal cortex plays a major role in determining the rate of Na^+ absorption and K^+ loss along the distal convoluted tubule (DCT) and collecting system of the kidneys. ∞ p. 993 The higher the plasma concentration of aldosterone, the more efficiently the kidneys conserve Na^+. Because "water follows salt," the conservation of Na^+ also stimulates water retention: As Na^+ is reabsorbed, Cl^- follows (Figure 26–14a ●, p. 994), and as sodium and chloride ions move out of the tubular fluid, water follows by osmosis. Aldosterone also increases the sensitivity of salt receptors on the tongue. This effect may increase your interest in, and consumption of, salty foods.

Aldosterone is secreted in response to rising K^+ or falling Na^+ levels in the blood that reaches the adrenal cortex or in response to the activation of the renin–angiotensin system. As we noted in earlier chapters, renin release occurs in response to (1) a drop in plasma volume or blood pressure at the juxtaglomerular apparatus of the nephron, (2) a decline in filtrate osmotic concentration at the DCT, or, as we shall soon see, (3) falling Na^+ or rising K^+ concentrations in the renal circulation.

Natriuretic Peptides

The natriuretic peptides ANP and BNP are released by cardiac muscle cells in response to abnormal stretching of the heart walls, caused by elevated blood pressure or an increase in blood volume. We considered the roles of these peptide hormones in Chapters 18, 21, and 26. ∞ pp. 636, 746, 987 Among their other effects, they reduce thirst and block the release of ADH and aldosterone that might otherwise lead to the conservation of water and salt. The resulting diuresis (fluid loss at the kidneys) lowers both blood pressure and plasma volume, eliminating the source of the stimulation.

☐ THE INTERPLAY BETWEEN FLUID BALANCE AND ELECTROLYTE BALANCE

At first glance, it can be very difficult to distinguish between water balance and electrolyte balance. For example, when you lose body water, your plasma volume decreases and electrolyte concentrations rise. Conversely, when you gain or lose excess electrolytes, there is an associated water gain or loss due to osmosis. However, because the regulatory mechanisms involved are quite different, it is often useful to consider fluid balance and electrolyte balance as distinct entities. This distinction is absolutely vital in a clinical setting, where problems with fluid balance and electrolyte balance must be identified and corrected promptly.

27–3 FLUID BALANCE

Objective

■ Describe the movement of fluid that takes place within the ECF, between the ECF and the ICF, and between the ECF and the environment.

Water circulates freely within the ECF compartment. At capillary beds throughout the body, hydrostatic pressure forces water out of plasma and into interstitial spaces. Some of that water is reabsorbed along the distal portion of the capillary bed, and the rest enters lymphatic vessels for transport to the venous circulation. This circulation is diagrammed in Figure 27–3●, which indicates two additional important relationships between components of the ECF:

1. Water moves back and forth across the mesothelial surfaces that line the peritoneal, pleural, and pericardial cavities and through the synovial membranes that line joint capsules. The flow rate is significant; for example, roughly 7 liters of peritoneal fluid is produced and reabsorbed each day. The actual volume present at any time in the peritoneal cavity, however, is very small—less than 35 ml.

2. Water also moves between blood and cerebrospinal fluid (CSF), the aqueous humor and vitreous humor of the eye, and the perilymph and endolymph of the inner ear. The volumes involved in these water movements are very small, and the volume and composition of the fluids are closely regulated. For those reasons, we shall largely ignore them in the discussion that follows.

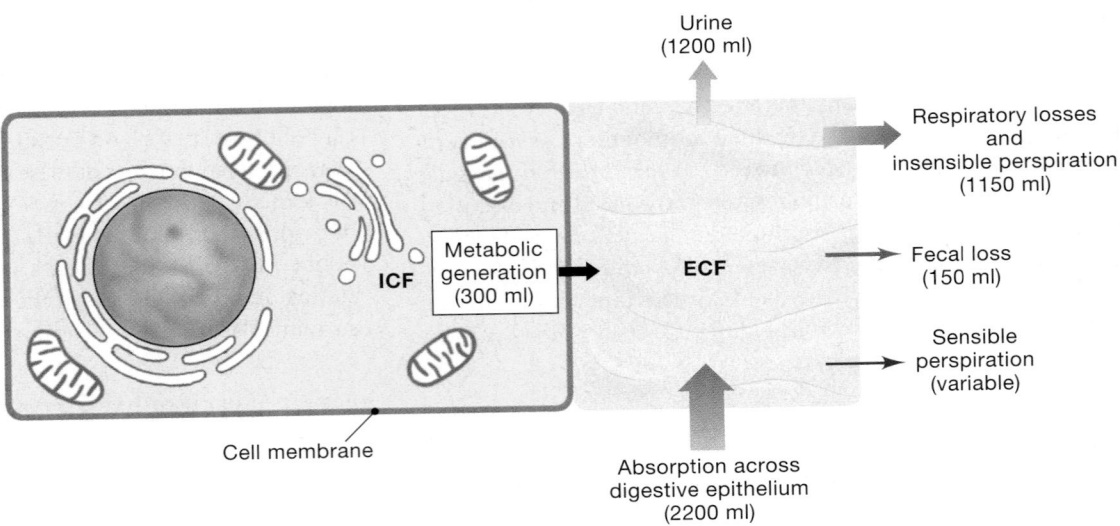

•FIGURE 27–3
Fluid Exchanges. Fluid movement between the ICF and the ECF and between the ECF and the environment. The volumes are not drawn to scale; functionally, the ICF is twice as large as the ECF.

◻ FLUID MOVEMENT WITHIN THE ECF

In the discussion of capillary dynamics in Chapter 21, we considered the basic principles that determine fluid movement among the divisions of the ECF. ∞ p. 735 The exchange between plasma and interstitial fluid, by far the largest components of the ECF, is determined by the relationship between the net hydrostatic pressure, which tends to push water out of the plasma and into the interstitial fluid, and the net colloid osmotic pressure, which tends to draw water out of the interstitial fluid and into the plasma. The interaction between these opposing forces, diagrammed in Figure 21–13• (p. 738), results in the continuous filtration of fluid from the capillaries into the interstitial fluid. This volume of fluid is then redistributed: After passing through the channels of the lymphatic system, the fluid returns to the venous system. At any moment, interstitial fluid and minor fluid compartments contain roughly 80 percent of the ECF volume, and plasma contains the other 20 percent.

Any factor that affects the net hydrostatic pressure or the net colloid osmotic pressure will alter the distribution of fluid within the ECF. The movement of abnormal amounts of water from plasma into interstitial fluid is called **edema**. ∞ p. 739 For example, pulmonary edema can result from an increase in the blood pressure in pulmonary capillaries, and a generalized edema can result from a decrease in blood colloid osmotic pressure, as in advanced starvation, when plasma protein concentrations decline. Localized edema can result from damage to capillary walls, as in bruising, the constriction of regional venous circulation, or a blockage of the lymphatic drainage, as in *lymphedema*, discussed in Chapter 22. ∞ p. 784

◻ FLUID EXCHANGE WITH THE ENVIRONMENT

Figure 27–3• and Table 27–1 indicate the major routes of exchange with the environment:

• **Water Losses.** You lose roughly 2500 ml of water each day through urine, feces, and *insensible perspiration*—the gradual movement of water across the epithelia of the skin and respiratory tract. The

losses due to *sensible perspiration*—the secretory activities of the sweat glands—vary with the activities you undertake. Sensible perspiration can cause significant water deficits, with maximum perspiration rates reaching 4 liters per hour. ∞ p. 962

• **The Temperature Rise Accompanying a Fever.** Fever can also increase water losses. For each degree your temperature rises above normal, your daily insensible water loss increases by 200 ml. The advice "Drink plenty of fluids" for anyone who is sick has a definite physiological basis.

• **Water Gains.** A water gain of roughly 2500 ml/day is required to balance your average water losses. This value amounts to roughly 40 ml/kg of body weight per day. You obtain water through eating (1000 ml), drinking (1200 ml), and *metabolic generation* (300 ml).

Metabolic generation of water is the production of water within cells, primarily as a result of oxidative phosphorylation in mitochondria. (We described the synthesis of water at the end of the electron transport system in Chapter 25.) ∞ p. 936 When a cell breaks down 1 g of lipid, 1.7 ml of water is generated. Breaking down proteins or carbohydrates yields much lower values (0.41 ml/g protein; 0.55 ml/g carbohydrate). A typical mixed diet in the United States contains 46 percent carbohydrates, 40 percent lipids, and 14 percent protein. This diet produces roughly 300 ml of water per day, about 12 percent of your average daily requirement.

TABLE 27–1 WATER BALANCE	
Source	**Daily Input (ml)**
Water content of food	1000
Water consumed as liquid	1200
Metabolic water produced during catabolism	300
Total	**2500**
Method of Elimination	**Daily Output (ml)**
Urination	1200
Evaporation at skin	750
Evaporation at lungs	400
Loss in feces	150
Total	**2500**

Water Excess and Water Depletion

The body's water content cannot easily be determined. However, the concentration of Na^+, the most abundant ion in the ECF, provides useful clues to the state of water balance. When the body water content rises enough to reduce the Na^+ concentration of the ECF below 130 mEq/l, the state of **hyponatremia** (*natrium*, sodium) exists. When the body water content declines, the Na^+ concentration rises; when that concentration exceeds 150 mEq/l, **hypernatremia** exists.

Hyponatremia is a sign of **overhydration**, or *water excess*. Hyponatremia results from (1) the ingestion of a large volume of fresh water or the *infusion* (injection into the bloodstream) of a hypotonic solution; (2) an inability to eliminate excess water in urine, due to chronic renal failure (p. 989), heart failure (p. 707), cirrhosis (p. 906), or some other disorder; and (3) endocrine disorders, such as excessive ADH production. The reduction in Na^+ concentrations leads to a shift of fluid into the ICF. The first signs are the effects on central nervous system function. In the early stages of hyponatremia, the individual behaves as if drunk on alcohol. This condition, *water intoxication*, may sound odd, but is extremely dangerous. Untreated cases can rapidly progress from confusion to hallucinations, convulsions, coma, and then death. Treatment of severe water intoxication generally involves diuretics and infusing a concentrated salt solution that elevates Na^+ levels to near-normal levels.

Dehydration, also known as *volume depletion* or *water depletion*, develops when water losses outpace water gains. The osmotic concentration of plasma gradually increases, so hypernatremia results. The loss of body water is associated with severe thirst, dryness and wrinkling of the skin, and a fall in plasma volume and blood pressure. Eventually, circulatory shock develops, generally with fatal consequences. Treatment for dehydration entails administering hypotonic fluids by mouth or intravenous infusion. This treatment increases ECF volume and restores normal electrolyte concentrations. **AM** Water and Weight Loss

☐ FLUID SHIFTS

Water movement between the ECF and the ICF is called a **fluid shift**. Fluid shifts occur rapidly in response to changes in the osmotic concentration of the ECF and reach equilibrium within minutes to hours. We explored the basic relationships in Chapter 3; see Figure 3–21●, p. 94, which shows the effects of hypertonic and hypotonic solutions on cells:

- **If the Osmotic Concentration of Your ECF Increases, that Fluid Will Become Hypertonic With Respect to Your ICF.** Water will then move from the cells into the ECF until osmotic equilibrium is restored. The osmotic concentration of the ECF will increase if you lose water, but retain electrolytes.

- **If the Osmotic Concentration of Your ECF Decreases, that Fluid Will Become Hypotonic With Respect to Your ICF.** Water will then move from the ECF into the cells, and the ICF volume will increase. The osmotic concentration of the ECF will decrease if you gain water, but do not gain electrolytes.

In sum, if the osmotic concentration of the ECF changes, a fluid shift between the ICF and ECF will tend to oppose the change. Because the volume of the ICF is much greater than that of the ECF, the ICF acts as a water reserve. In effect, instead of a large change in the osmotic concentration of the ECF, smaller changes occur in both the ECF and ICF. Two examples will demonstrate the dynamic exchange of water between the ECF and ICF.

Allocation of Water Losses

When you lose water, but retain electrolytes, the osmotic concentration of the ECF rises. Osmosis then moves water out of the ICF and into the ECF until the two solutions are again isotonic. At that point, both the ECF and ICF will be somewhat more concentrated than normal, and both volumes will be lower than they were before the fluid loss. Because the ICF has roughly twice the functional volume of the ECF, the net change in the ECF is relatively small.

Conditions that cause severe water losses include excessive perspiration (brought about by exercising in hot weather), inadequate water consumption, repeated vomiting, and diarrhea. These conditions promote water losses far in excess of electrolyte losses, so body fluids become increasingly concentrated. Responses that attempt to restore homeostasis include ADH and renin secretion and (as soon as possible) an increase in fluid intake.

Distribution of Water Gains

When you drink a glass of pure water or when you are given hypotonic solutions intravenously, the water content of your body increases without a corresponding increase in the concentration of electrolytes. As a result, the ECF increases in volume, but becomes hypotonic with respect to the ICF. A fluid shift then occurs, and the volume of the ICF increases at the expense of the ECF. Once again, the larger volume of the ICF limits the amount of osmotic change. After the fluid shift, the ECF and ICF have slightly larger volumes and slightly lower osmotic concentrations than they did originally.

Normally, this situation will be promptly corrected. The reduced-plasma osmotic concentration depresses the secretion of ADH, discouraging fluid intake and increasing water losses in urine. If the situation is *not* corrected, a variety of clinical problems will develop as water shifts into the intracellular fluid, distorting cells, changing the solute concentrations around enzymes, and disrupting normal cell functions.

✓ What effect would drinking a pitcher of distilled water have on your level of ADH?

✓ What effect would being in the desert without water for a day have on your plasma osmotic concentration?

Answers start on page Q-1

 Fluid balance review information can be found on the **IP CD-ROM**: Fluids & Electrolytes/Introduction to Body Fluids.

27–4 ELECTROLYTE BALANCE

Objective

■ Discuss the mechanisms by which sodium, potassium, calcium, and chloride ion concentrations are regulated to maintain electrolyte balance.

You are in electrolyte balance when the rates of gain and loss are equal for each electrolyte in your body. Electrolyte balance is important because

• *Total electrolyte concentrations directly affect water balance*, as we described previously, and

• *The concentrations of individual electrolytes can affect cell functions.* We saw many examples in earlier chapters, such as the effect of abnormal Na^+ concentrations on neuron activity and the effects of high or low Ca^{2+} and K^+ concentrations on cardiac muscle tissue.

Two cations, Na^+ and K^+, merit attention, because (1) they are major contributors to the osmotic concentrations of the ECF and the ICF, respectively, and (2) they directly affect the normal functioning of all cells. Sodium is the dominant cation in the ECF. More than 90 percent of the osmotic concentration of the ECF results from the presence of sodium salts, mainly sodium chloride (NaCl) and sodium bicarbonate ($NaHCO_3$), so changes in the osmotic concentration of body fluids generally reflect changes in Na^+ concentration. Normal Na^+ concentrations in the ECF average 136–142 mEq/l, versus 10 mEq/l or less in the ICF. Potassium is the dominant cation in the ICF, where concentrations reach 160 mEq/l. Extracellular K^+ concentrations are generally very low, from 3.8 to 5.0 mEq/l.

Two general rules about sodium balance and potassium balance are worth noting:

1. *The Most Common Problems With Electrolyte Balance Are Caused by an Imbalance Between Gains and Losses of Sodium Ions.*

2. *Problems With Potassium Balance Are Less Common, but Significantly More Dangerous than Are Those Related to Sodium Balance.* (See pp. 1021, 1024, and 1025.)

☐ SODIUM BALANCE

The total amount of sodium in the ECF represents a balance between two factors:

1. *Sodium Ion Uptake Across the Digestive Epithelium.* Sodium ions enter the ECF by crossing the digestive epithelium through diffusion and carrier-mediated transport. The rate of absorption varies directly with the amount of sodium in the diet.

2. *Sodium Ion Excretion at the Kidneys and Other Sites.* Sodium losses occur primarily by excretion in urine and through perspiration. The kidneys are the most important sites of Na^+ regulation. We discussed the mechanisms for sodium reabsorption at the kidneys in Chapter 26. ∞ pp. 990-997

A person in sodium balance typically gains and loses 48–144 mEq (1.1–3.3 g) of Na^+ each day. When sodium gains exceed sodium losses, the total Na^+ content of the ECF goes up; when losses exceed gains, the Na^+ content declines. However, a change in the Na^+ *content* of the ECF does not produce a change in the Na^+ *concentration*. When sodium intake or output changes, a corresponding gain or loss of water tends to keep the Na^+ concentration constant. For example, if you eat a very salty meal, the osmotic concentration of your ECF will not increase. When sodium ions are pumped across your digestive epithelium, the solute concentration in that portion of the ECF increases, whereas that of the intestinal contents decreases. Osmosis then occurs. Additional water enters the ECF from the digestive tract, elevating the blood volume and blood pressure. For this reason, persons with high blood pressure are advised to restrict the amount of salt in their diets.

Sodium Balance and ECF Volume

The sodium regulatory mechanism, diagrammed in Figure 27–4●, changes the ECF volume but keeps the Na^+ concentration relatively stable. If large amounts of salt are consumed *without* adequate fluid, as when you eat salty potato chips without taking a drink, the

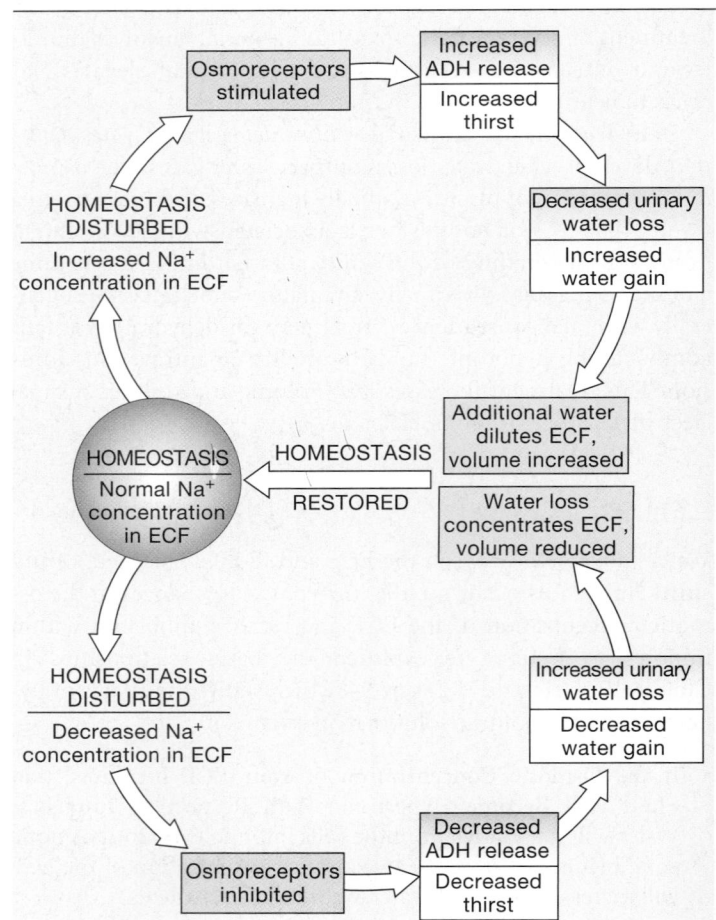

●FIGURE 27–4
The Homeostatic Regulation of Normal Sodium Ion Concentrations in Body Fluids

plasma Na$^+$ concentration will rise temporarily. A change in ECF volume will soon follow, however. Fluid will exit the ICF, increasing ECF volume and lowering Na$^+$ concentrations somewhat. The secretion of ADH restricts water loss and stimulates thirst, promoting additional water consumption. Due to the inhibition of water receptors in the pharynx, ADH secretion begins even before the Na$^+$ absorption occurs; the secretion rate rises further after Na$^+$ absorption, due to osmoreceptor stimulation.

When sodium losses exceed gains, the volume of the ECF decreases. This reduction occurs without a significant change in the osmotic concentration of the ECF. Thus, if you perspire heavily but consume only pure water, you will lose sodium, and the osmotic concentration of your ECF will drop briefly. However, as soon as the osmotic concentration drops by 2 percent or more, ADH secretion will decrease, so water losses at your kidneys will increase. As water leaves the ECF, the osmotic concentration will return to normal.

Minor changes in ECF volume do not matter, because they do not cause adverse physiological effects. If, however, the regulation of Na$^+$ concentrations results in a large change in ECF volume, the situation will be corrected by the same homeostatic mechanisms responsible for regulating blood volume and blood pressure. This is the case because when the ECF volume changes, so does plasma volume and, in turn, blood volume. If the ECF volume rises, blood volume goes up; if the ECF volume drops, blood volume goes down. As we saw in Chapter 21, blood volume has a direct effect on blood pressure. A rise in blood volume elevates blood pressure; a drop lowers blood pressure. The net result is that homeostatic mechanisms can monitor ECF volume indirectly by monitoring blood pressure. The receptors involved are baroreceptors at the carotid sinus, the aortic sinus, and the right atrium. The regulatory steps are reviewed in Figure 27–5●.

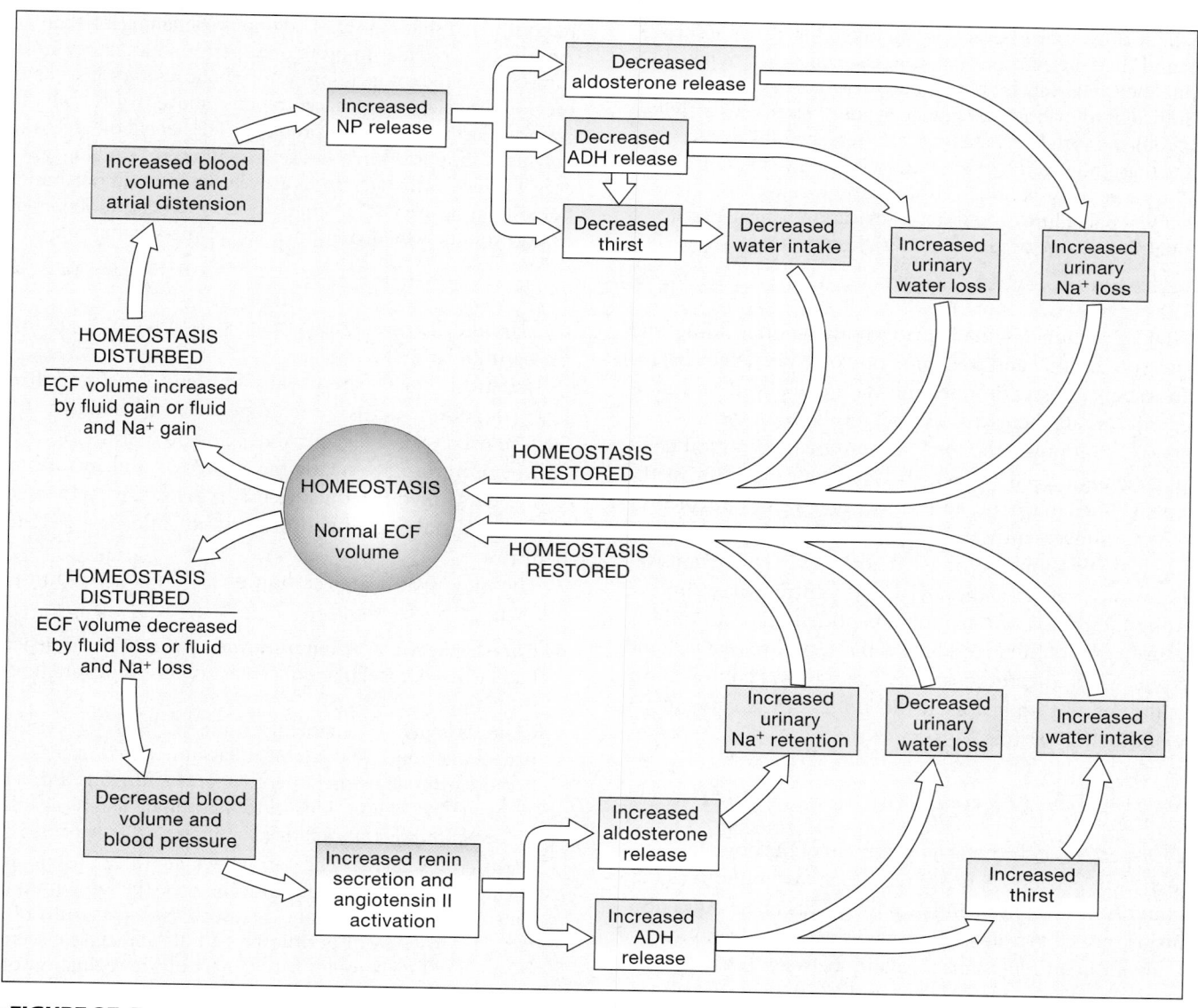

●**FIGURE 27–5**
The Integration of Fluid Volume Regulation and Sodium Ion Concentrations in Body Fluids

Athletes and Salt Loss

Unfounded notions and rumors about water and salt requirements during exercise abound. Sweat is a hypotonic solution that contains Na^+ in lower concentration than the ECF. As a result, a person who is sweating profusely loses more water than salt, and this loss leads to a rise in the Na^+ concentration of the ECF. The water content of the ECF decreases as the water loss occurs, so the blood volume drops. As we noted on page 1021, this condition is called *volume depletion*. Because volume depletion occurs at the same time that blood is being shunted away from the kidneys, waste products accumulate in the blood and kidney function is impaired.

To prevent volume depletion, athletes should pause at intervals to drink liquids. The primary problem in volume depletion is water loss, and research has revealed no basis for the rumor that cramps will result if you drink while exercising. Salt pills and the various sports beverages promoted for "faster absorption" and "better electrolyte balance" have no apparent benefits despite their relatively high cost. Body reserves of electrolytes are sufficient to tolerate extended periods of strenuous activity, and problems with Na^+ balance are extremely unlikely except during marathons or other activities that involve maximal exertion for more than six hours. However, both dehydration (causing acute renal failure) and water intoxication (causing fatal hyponatremia) have occurred in marathon runners.

Some sports beverages contain sugars and vitamins as well as electrolytes. During endurance events (marathons, ultramarathons, and distance cycling), solutions containing less than 10 g/dl of glucose may improve one's performance if they are consumed late in the competition, as metabolic reserves are exhausted. However, high sugar concentrations (above 10 g/dl) can cause cramps, diarrhea, and other problems. The benefit of "glucose polymers" in sports drinks has yet to be proved. Drinking beverages "fortified" with vitamins is actively discouraged: Vitamins are not lost during exercise, and the consumption of these beverages in large volumes could, over time, cause hypervitaminosis.

If the ECF volume is inadequate, both blood volume and blood pressure decline and the renin–angiotensin system is activated. In response, losses of water and Na^+ are reduced, and gains of water and Na^+ are increased. The net result is that the ECF volume increases. Although the total amount of Na^+ in the ECF is increasing (gains exceed losses), the Na^+ concentration in the ECF remains unchanged, because absorption is accompanied by osmotic water movement.

If the plasma volume becomes abnormally large, venous return increases, stretching the atrial and ventricular walls and stimulating the release of natriuretic peptides (ANP and BNP). This in turn reduces thirst and blocks the secretion of ADH and aldosterone, which together promote water or salt conservation. As a result, salt and water loss at the kidneys increases and the volume of the ECF declines.

☐ POTASSIUM BALANCE

Roughly 98 percent of the potassium content of the human body is in the ICF. Cells expend energy to recover potassium ions as they diffuse out of the cytoplasm and into the ECF. The K^+ concentration outside the cell is relatively low, and the concentration in the ECF at any moment represents a balance between (1) the rate of entry across the digestive epithelium and (2) the rate of loss into urine. Potassium loss in urine is regulated by controlling the activities of ion pumps along the distal portions of the nephron and collecting system. Whenever a sodium ion is reabsorbed from the

tubular fluid, it generally is exchanged for a cation (typically K^+) in the peritubular fluid.

Urinary K^+ losses are usually limited to the amount gained by absorption across the digestive epithelium, typically 50–150 mEq (1.9–5.8 g) per day. (The potassium losses in feces and perspiration are negligible.) The K^+ concentration in the ECF is controlled by adjustments in the rate of active secretion along the distal convoluted tubule and collecting system of the nephron.

The rate of tubular secretion of K^+ varies in response to the following three factors:

1. *Changes in the K^+ Concentration of the ECF.* In general, the higher the extracellular concentration of potassium, the higher is the rate of secretion.

2. *Changes in pH.* When the pH of the ECF falls, so does the pH of peritubular fluid. The rate of potassium secretion then declines, because hydrogen ions, rather than potassium ions, are secreted in exchange for sodium ions in tubular fluid. The mechanisms for H^+ secretion were summarized in Figure 26–14c● (p. 995).

3. *Aldosterone Levels.* The rate at which K^+ is lost in urine is strongly affected by aldosterone, because the ion pumps that are sensitive to this hormone reabsorb Na^+ from filtrate in exchange for K^+ from peritubular fluid. Aldosterone secretion is stimulated by angiotensin II as part of the regulation of blood volume. High plasma K^+ concentrations also stimulate aldosterone secretion directly. Either way, under the influence of aldosterone, the amount of sodium conserved and the amount of potassium excreted in urine are directly related.

Hypokalemia and Hyperkalemia

When the plasma concentration of potassium falls below 2 mEq/l, extensive muscular weakness develops, followed by eventual paralysis. We discussed this condition, called **hypokalemia** (*kalium*, potassium), in connection with ion effects on cardiac function. ⊂⊃ p. 716 Causes of hypokalemia include:

- **Inadequate Dietary K⁺ Intake.** If K⁺ gains from the diet do not keep pace with the rate of K⁺ loss in urine, K⁺ concentrations in the ECF will drop.

- **The Administration of Diuretic Drugs.** Several diuretics, including *Lasix*, can produce hypokalemia by increasing the volume of urine produced. Although the concentration of K⁺ in urine is low, the greater the total volume, the more potassium is lost.

- **Excessive Aldosterone Secretion.** The condition of *aldosteronism*, characterized by excessive aldosterone secretion, results in hypokalemia, because the reabsorption of Na⁺ is tied to the secretion of K⁺. ⊂⊃ pp. 629, 993

- **An Increase in the pH of the ECF.** When the H⁺ concentration of the ECF declines, cells exchange intracellular H⁺ for extracellular K⁺. This ion swap helps stabilize the extracellular pH, but gradually lowers the K⁺ concentration of the ECF.

Treatment for hypokalemia generally includes increasing the dietary intake of potassium by salting food with potassium salts (KCl) or by taking potassium liquids or tablets, such as *Slow-K®*. Severe cases are treated by the intravenous infusion of a solution containing K⁺ at a concentration of 40–60 mEq/l.

High K⁺ concentrations in the ECF produce an equally dangerous condition known as **hyperkalemia**. Severe cardiac arrhythmias appear when the K⁺ concentration exceeds 8 mEq/l. ⊂⊃ p. 716 Hyperkalemia results from factors such as:

- **Renal Failure.** Kidney failure due to damage or disease will prevent normal K⁺ secretion and thereby produce hyperkalemia.

- **The Administration of Diuretic Drugs that Block Na⁺ Reabsorption** (also called potassium-sparing diuretics). When sodium reabsorption slows down, so does potassium secretion. Hyperkalemia can result.

- **A Decline in the pH of the ECF.** When the pH of the ECF declines, hydrogen ions move into the ICF in exchange for intracellular potassium ions. In addition, potassium secretion at the kidney tubules slows down, because hydrogen ions are secreted instead of potassium ions. The combination of increased K⁺ entry into the ECF and decreased K⁺ secretion can produce a dangerous hyperkalemia very rapidly.

Treatment for hyperkalemia includes (1) the elevation of ECF volume with a solution low in K⁺; (2) the stimulation of K⁺ loss in urine by using appropriate diuretics, such as Lasix; (3) the cautious administration of buffers (generally sodium bicarbonate) that can control the pH of the ECF; (4) restriction of the dietary K⁺ intake; and (5) the administration of enemas or laxatives which contain compounds, such as *kayexolate*, that promote K⁺ loss across the digestive lining. In cases resulting from renal failure, kidney dialysis may also be required.

☐ OTHER ELECTROLYTES

The ECF concentrations of other electrolytes are regulated as well. We shall note the most important ions involved. Additional information about sodium, potassium, and these other ions is given in Table 27–2.

Calcium Balance

Calcium is the most abundant mineral in the body. A typical individual has 1–2 kg (2.2–4.4 lb) of the element, 99 percent of which is deposited in the skeleton. In addition to forming the crystalline component of bone, calcium ions play key roles in the control of muscular and neural activities, in blood clotting, as cofactors for enzymatic reactions, and as second messengers.

As we noted in Chapters 6 and 18, calcium homeostasis primarily reflects an interplay between the reserves in bone, the rate of absorption across the digestive tract, and the rate of loss at the kidneys. The hormones parathyroid hormone (PTH), calcitriol, and, to a lesser degree, calcitonin maintain calcium homeostasis in the ECF. Parathyroid hormone and calcitriol raise Ca²⁺ concentrations; their actions are opposed by calcitonin. ⊂⊃ p. 626

A small amount of Ca²⁺ is lost in the bile, and under normal circumstances very little Ca²⁺ escapes in urine or feces. To keep pace with biliary, urinary, and fecal Ca²⁺ losses, an adult must absorb only 0.8–1.2 g/day of Ca²⁺. That represents only about 0.03 percent of the calcium reserve in the skeleton. Calcium absorption at the digestive tract and reabsorption along the distal convoluted tubule are stimulated by PTH from the parathyroid glands and calcitriol from the kidneys.

Hypercalcemia exists when the Ca²⁺ concentration of the ECF is above 11 mEq/l. The primary cause of hypercalcemia in adults is *hyperparathyroidism*. ⊂⊃ p. 626 Less common causes include malignant cancers of the breast, lung, kidney, and bone marrow. Severe hypercalcemia (12–13 mEq/l) causes such symptoms as fatigue, confusion, cardiac arrhythmias, and calcification of the kidneys and soft tissues throughout the body.

The condition **hypocalcemia** (a Ca²⁺ concentration under 4 mEq/l) is much less common than hypercalcemia. Hypoparathyroidism, vitamin D deficiency, or chronic renal failure is typically responsible for hypocalcemia. Symptoms include muscle spasms, sometimes with generalized convulsions, weak heartbeats, cardiac arrhythmias, and osteoporosis.

TABLE 27–2 ELECTROLYTE BALANCE

Ion and Normal ECF Range (mEq/l)	Disorder (mEq/l)	Symptoms	Causes	Treatment
Sodium (136–142)	Hypernatremia (>150)	Thirst, dryness and wrinkling of skin, reduced blood volume and pressure, eventual circulatory collapse	Dehydration; loss of hypotonic fluid	Ingestion of water or intravenous infusion of hypotonic solution
	Hyponatremia (<130)	Disturbed CNS function (water intoxication): confusion, hallucinations, convulsions, coma; death in severe cases	Infusion or ingestion of large volumes of hypotonic solution	Diuretic use and infusion of hypertonic salt solution
Potassium (3.8–5.0)	Hyperkalemia (>8)	Severe cardiac arrhythmias	Renal failure; use of diuretics; chronic acidosis	Infusion of hypotonic solution; selection of different diuretics; infusion of buffers; dietary restrictions
	Hypokalemia (<2)	Muscular weakness and paralysis	Low-potassium diet; diuretics; hypersecretion of aldosterone; chronic alkalosis	Increase in dietary K^+ content; ingestion of K^+ tablets or solutions; infusion of potassium solution
Calcium (4.5–5.3)	Hypercalcemia (>11)	Confusion, muscle pain, cardiac arrhythmias, kidney stones, calcification of soft tissues	Hyperparathyroidism; cancer; vitamin D toxicity; calcium supplement overdose	Infusion of hypotonic fluid to lower Ca^{2+} levels; surgery to remove parathyroid gland; administration of calcitonin
	Hypocalcemia (<4)	Muscle spasms, convulsions, intestinal cramps, weak heartbeats, cardiac arrhythmias, osteoporosis	Poor diet; lack of vitamin D; renal failure; hypoparathyroidism; hypomagnesemia	Calcium supplements; administration of vitamin D
Magnesium (1.5–2.5)	Hypermagnesemia (>4)	Confusion, lethargy, respiratory depression, hypotension	Overdose of magnesium supplements or antacids (rare)	Infusion of hypotonic solution to lower plasma concentration
	Hypomagnesemia (<0.8)	Hypocalcemia, muscle weakness, cramps, cardiac arrhythmias, hypertension	Poor diet; alcoholism; severe diarrhea; kidney disease; malabsorption syndrome; ketoacidosis	Intravenous infusion of solution high in Mg^{2+}
Phosphate (1.8–2.6)	Hyperphosphatemia (>6)	No immediate symptoms; chronic elevation leads to calcification of soft tissues	High dietary phosphate intake; hypoparathyroidism	Dietary reduction
	Hypophosphatemia (<1)	Anorexia, dizziness, muscle weakness, cardiomyopathy, osteoporosis	Poor diet; kidney disease; malabsorption syndrome; hyperparathyroidism; vitamin D deficiency	Dietary improvement; hormone or vitamin supplements
Chloride (100–108)	Hyperchloremia (>112)	Acidosis, hyperkalemia	Dietary excess; increased chloride retention	Infusion of hypotonic solution to lower plasma concentration
	Hypochloremia (<95)	Alkalosis, anorexia, muscle cramps, apathy	Vomiting; hypokalemia	Diuretic use and infusion of hypertonic salt solution

Magnesium Balance

The adult body contains about 29 g of magnesium, almost 60 percent of it deposited in the skeleton. The magnesium in body fluids is contained primarily in the ICF, where the concentration of Mg^{2+} averages about 26 mEq/l. Magnesium is required as a cofactor for several important enzymatic reactions, including the phosphorylation of glucose within cells and the use of ATP by contracting muscle fibers. Magnesium is also important as a structural component of bone.

The Mg^{2+} concentration of the ECF averages 1.5–2.5 mEq/l, considerably lower than levels in the ICF. The proximal convoluted tubule reabsorbs magnesium very effectively. Keeping pace with the daily urinary loss requires a minimum dietary intake of only 24–32 mEq (0.3–0.4 g) per day.

Phosphate Balance

Phosphate ions are required for bone mineralization, and roughly 744 g of PO_4^{3-} is bound up in the mineral salts of the skeleton. In body fluids, the most important functions of PO_4^{3-} involve the ICF, where the ions are required for the formation of high-energy compounds, the activation of enzymes, and the synthesis of nucleic acids.

The PO_4^{3-} concentration of the plasma is usually 1.8–2.6 mEq/l. Phosphate ions are reabsorbed from tubular fluid along the proximal convoluted tubule; urinary and fecal losses of PO_4^{3-} amount to 30–45 mEq (0.8–1.2 g) per day. Phosphate ion reabsorption along the PCT is stimulated by calcitriol.

Chloride Balance

Chloride ions are the most abundant anions in the ECF. The plasma concentration ranges from 100–108 mEq/l. In the ICF, Cl^- concentrations are usually low (3 mEq/l). Chloride ions are absorbed across the digestive tract together with sodium ions; several carrier proteins along the renal tubules reabsorb Cl^- with Na^+. ∞ pp. 991-995 The rate of Cl^- loss is small; a gain of 48–146 mEq (1.7–5.1 g) per day will keep pace with losses in urine and perspiration.

✓ Why does prolonged sweating increase plasma sodium ion levels?

✓ Which is more dangerous, disturbances of sodium balance or disturbances of potassium balance?

Answers start on page Q-1

 Ion balance can be investigated on the **IP CD-ROM**: Fluids & Electrolytes/Electrolyte Homeostasis.

27–5 ACID–BASE BALANCE

Objectives

■ Explain the buffering systems that balance the pH of the intracellular and extracellular fluids.

■ Describe the compensatory mechanisms involved in the maintenance of acid–base balance.

In Chapter 2, we introduced the topic of pH and the chemical nature of acids, bases, and buffers. Table 27–3 reviews key terms important to the discussion that follows. If you need a more detailed review, refer to the appropriate sections of Chapter 2 before you proceed. ∞ pp. 43-45

The pH of body fluids can be altered by the introduction of either acids or bases. A general classification of acids and bases sorts them into *strong* versus *weak*:

- *Strong acids* and *strong bases* dissociate completely in solution. For example, hydrochloric acid (HCl), a strong acid, dissociates in solution via the reaction

$$HCl \rightarrow H^+ + Cl^-$$

TABLE 27–3	A REVIEW OF IMPORTANT TERMS RELATING TO ACID–BASE BALANCE
pH	The negative exponent (negative logarithm) of the hydrogen ion concentration
Neutral	A solution with a pH of 7; the solution contains equal numbers of hydrogen ions and hydroxide ions
Acidic	A solution with a pH below 7; in this solution, hydrogen ions predominate
Basic, or alkaline	A solution with a pH above 7; in this solution, hydroxide ions predominate
Acid	A substance that dissociates to release hydrogen ions, decreasing pH
Base	A substance that dissociates to release hydroxide ions or to tie up hydrogen ions, increasing pH
Salt	An ionic compound consisting of a cation other than hydrogen and an anion other than a hydroxide ion
Buffer	A substance that tends to oppose changes in the pH of a solution by removing or replacing hydrogen ions; in body fluids, buffers maintain pH within normal limits (7.35–7.45)

- When *weak acids* or *weak bases* enter a solution, a significant number of molecules remain intact; dissociation is not complete. This means that if you place molecules of a weak acid in one solution and the same number of molecules of a strong acid in another solution, the weak acid will liberate fewer hydrogen ions and have less of an impact on the pH of the solution than the strong acid will. Carbonic acid is a weak acid. At the normal pH of the ECF, an equilibrium state exists, and the reaction can be diagrammed as follows:

$$\underset{\text{carbonic acid}}{H_2CO_3} \leftrightarrow H^+ + \underset{\text{bicarbonate ion}}{HCO_3^-}$$

☐ THE IMPORTANCE OF pH CONTROL

The pH of your body fluids reflects interactions among all the acids, bases, and salts in solution in your body. The pH of the ECF normally remains within relatively narrow limits, usually from 7.35–7.45. Any deviation outside the normal range is extremely dangerous, because changes in H^+ concentrations disrupt the stability of cell membranes, alter the structure of proteins, and change the activities of important enzymes. You could not survive for long with an ECF pH below 6.8 or above 7.7.

When the pH of plasma falls below 7.35, **acidemia** exists. The physiological state that results is called **acidosis**. When the pH of plasma rises above 7.45, **alkalemia** exists. The physiological state that results is called **alkalosis**. Acidosis and alkalosis affect virtually all body systems, but the nervous system and cardiovascular system are particularly sensitive to fluctuations in pH. For example, severe acidosis (pH below 7.0) can be deadly, because (1) the central nervous system function deteriorates, and the individual may become comatose; (2) cardiac contractions grow weak and irregular, and symptoms of heart failure may develop; and (3) peripheral vasodilation produces a dramatic drop in blood pressure, and circulatory collapse can occur.

The control of pH is therefore an extremely important homeostatic process of great physiological and clinical significance. Although both acidosis and alkalosis are dangerous, in practice problems with acidosis are much more common than are problems with alkalosis. This is so because several acids, including carbonic acid, are generated by normal cellular activities.

☐ TYPES OF ACIDS IN THE BODY

The body contains three general categories of acids: (1) *volatile acids*, (2) *fixed acids*, and (3) *organic acids*.

A **volatile acid** is an acid that can leave solution and enter the atmosphere. Carbonic acid (H_2CO_3) is an important volatile acid in body fluids. At the lungs, carbonic acid breaks down into carbon dioxide and water; the carbon dioxide diffuses into the alveoli. In peripheral tissues, carbon dioxide in solution interacts with water to form carbonic acid, which dissociates to release hydrogen ions and bicarbonate ions. The complete reaction sequence is

$$CO_2 + H_2O \leftrightarrow H_2CO_3 \leftrightarrow H^+ + HCO_3^-$$

carbon dioxide   water   carbonic acid   bicarbonate ion

This reaction occurs spontaneously in body fluids, but it takes place much more rapidly in the presence of *carbonic anhydrase (CA)*, an enzyme found in the cytoplasm of red blood cells, liver and kidney cells, parietal cells of the stomach, and many other types of cells.

Because most of the carbon dioxide in solution is converted to carbonic acid and most of the carbonic acid dissociates, the partial pressure of carbon dioxide and the pH are inversely related (Figure 27–6●). When carbon dioxide levels rise, additional hydrogen ions and bicarbonate ions are released, so the pH goes down. (Recall that the pH is a *negative exponent*, so when the concentration of hydrogen ions goes up, the pH goes down.) The P_{CO_2} is the most important factor affecting the pH in body tissues.

At the alveoli, carbon dioxide diffuses into the atmosphere, the number of hydrogen ions and bicarbonate ions in the alveolar capillaries drops, and the blood pH rises. We will consider this process, which effectively removes hydrogen ions from solution, in more detail later in the chapter.

Fixed acids are acids that do not leave solution; once produced, they remain in body fluids until they are eliminated at the kidneys. Sulfuric acid and phosphoric acid are the most important fixed acids. They are generated in small amounts during the catabolism of amino acids and compounds that contain phosphate groups, including phospholipids and nucleic acids.

Organic acids are acid participants in or by-products of aerobic metabolism. Lactic acid, produced by the anaerobic metabolism of pyruvate, and ketone bodies, synthesized from acetyl-CoA, are important organic acids. Under normal conditions, most organic acids are metabolized rapidly, so significant accumulations do not occur. But relatively large amounts of organic acids are produced (1) during periods of anaerobic metabolism, because lactic acid builds up rapidly, and (2) during starvation or excessive lipid catabolism, because ketone bodies accumulate.

☐ MECHANISMS OF pH CONTROL

To maintain acid–base balance over long periods of time, your body must balance gains and losses of hydrogen ions. Hydrogen ions are gained at the digestive tract and through metabolic activities within your cells. You must eliminate these ions at the kidneys, by secreting H^+ into urine, and at the lungs, by forming water and carbon dioxide from H^+ and HCO_3^-. The sites of elimination are far removed from the sites of acid production. As the hydrogen ions travel through the body, they must be neutralized to avoid tissue damage.

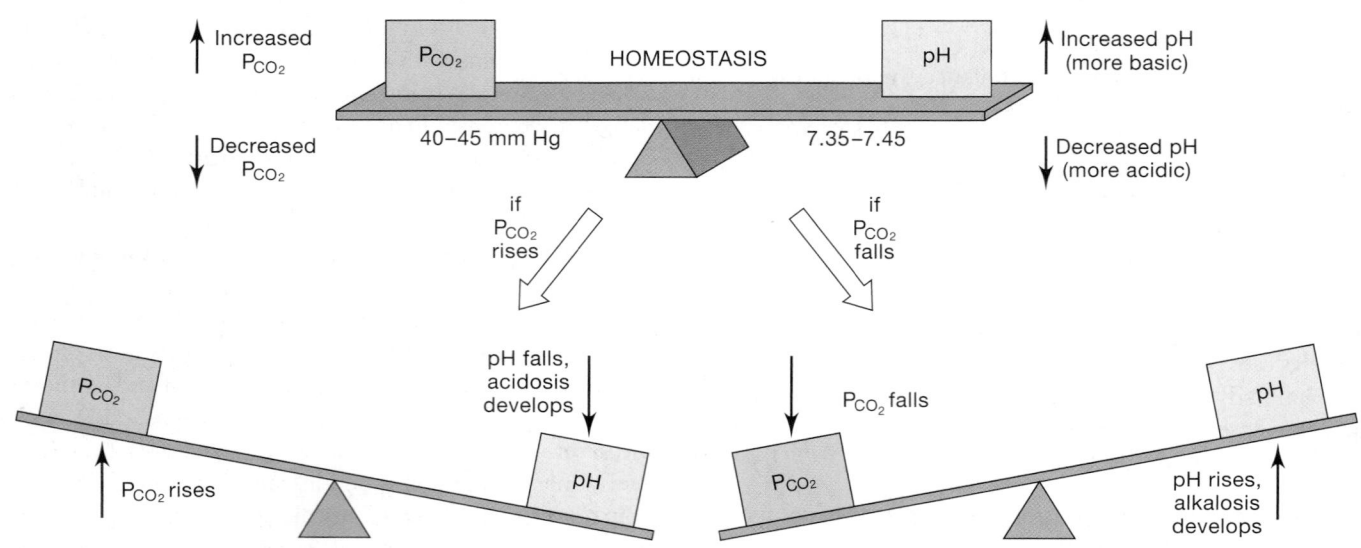

●**FIGURE 27–6**
The Basic Relationship between P_{CO_2} and Plasma pH.   The P_{CO_2} is inversely related to the pH.

The acids produced in the course of normal metabolic operations are temporarily neutralized by buffers and buffer systems in body fluids. *Buffers* are dissolved compounds that can provide or remove H^+ and thereby stabilize the pH of a solution. ∞ p. 45 Buffers include weak acids that can donate H^+ and weak bases that can absorb H^+. A **buffer system** in body fluids generally consists of a combination of a weak acid and the anion released by its dissociation. The anion functions as a weak base. In solution, molecules of the weak acid exists in an equilibrium state with the dissociation products. In chemical notation, this relationship would be represented as

$$HY \leftrightarrow H^+ + Y^-$$

Adding H^+ to the solution will upset the equilibrium, and the resulting formation of additional molecules of the weak acid will remove some of the H^+ from the solution.

The body has three major buffer systems, each with slightly different characteristics and distribution:

1. *Protein buffer systems* contribute to the regulation of pH in the ECF and ICF. These buffer systems interact extensively with the other buffer systems.

2. *The carbonic acid–bicarbonate buffer system* is most important in the ECF.

3. *The phosphate buffer system* has an important role in buffering the pH of the ICF and of urine.

Figure 27–7● indicates the locations of these buffer systems.

Protein Buffer Systems

Protein buffer systems depend on the ability of amino acids to respond to changes in pH by accepting or releasing H^+. The underlying mechanism is shown in Figure 27–8●:

- If the pH climbs, the carboxyl group (−COOH) of the amino acid can dissociate, acting as a weak acid and releasing a hydrogen ion. The carboxyl group then becomes a carboxylate ion (−COO⁻). At the normal pH of body fluids (7.35–7.45), the carboxyl groups of most amino acids have already given up their hydrogen ions. (Proteins carry negative charges primarily for that reason.) However, some amino acids, notably *histidine* and *cysteine*, have R groups (side chains) that will donate hydrogen ions if the pH climbs outside the normal range. Their buffering effects are very important in both the ECF and the ICF.

- If the pH drops, the carboxylate ion and the amino group (−NH₂) can act as weak bases and accept additional hydrogen ions, forming a carboxyl group (−COOH) and an amino ion

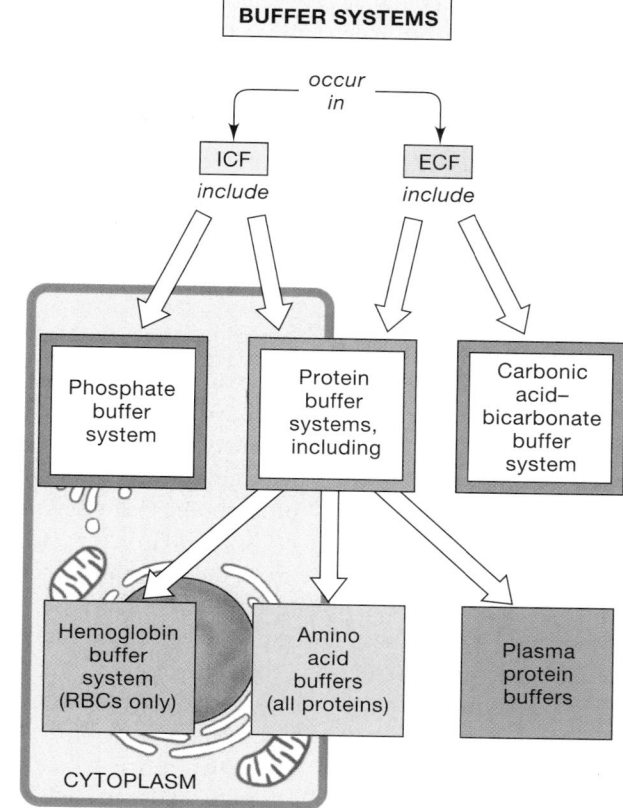

●**FIGURE 27–7**
Buffer Systems in Body Fluids.   Phosphate buffers occur primarily in the ICF, whereas the carbonic acid–bicarbonate buffer system occurs primarily in the ECF. Protein buffer systems are in both the ICF and the ECF. Extensive interactions take place among these systems.

(−NH₃⁺), respectively. This effect is limited primarily to free amino acids and the last amino acid in a polypeptide chain, because the carboxyl and amino groups in peptide bonds cannot function as buffers.

Plasma proteins contribute to the buffering capabilities of blood. Interstitial fluid contains extracellular protein fibers and dissolved amino acids that also assist in the regulation of pH. In the ICF of active cells, structural and other proteins provide an extensive buffering capability that prevents destructive changes in pH when organic acids, such as lactic acid or pyruvic acid, are produced by cellular metabolism.

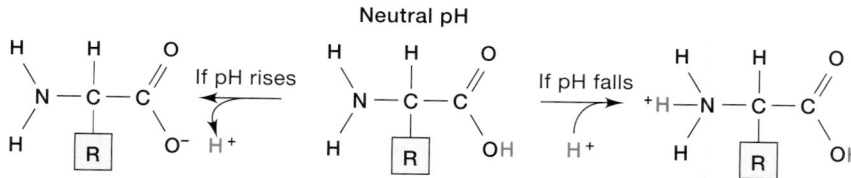

●**FIGURE 27–8**
Amino Acid Buffers.   Amino acids can either donate a hydrogen ion (as at left) or accept one (as at right), depending on the pH of their surroundings.

Because there is an exchange between the ECF and the ICF, the protein buffer system can help stabilize the pH of the ECF. For example, when the pH of the ECF decreases, cells pump H^+ out of the ECF and into the ICF, where they can be buffered by intracellular proteins. When the pH of the ECF rises, pumps in cell membranes exchange H^+ in the ICF for K^+ in the ECF.

These mechanisms can assist in buffering the pH of the ECF. The process is slow, however, because hydrogen ions must be individually transported across the cell membrane. As a result, the protein buffer system in most cells cannot make rapid, large-scale adjustments in the pH of the ECF.

THE HEMOGLOBIN BUFFER SYSTEM The situation is somewhat different for red blood cells. These cells, which contain roughly 5.5 percent of the ICF, are normally suspended in the plasma. They are densely packed with hemoglobin, and their cytoplasm contains large amounts of carbonic anhydrase. Red blood cells have a significant effect on the pH of the ECF, because they absorb carbon dioxide from the plasma and convert it to carbonic acid. Carbon dioxide can diffuse across the RBC membrane very quickly, so no transport mechanism is needed. As the carbonic acid dissociates, the bicarbonate ions move into the plasma in exchange for chloride ions, a swap known as the *chloride shift*. ∞ p. 859 The hydrogen ions are buffered by hemoglobin molecules. At the lungs, the entire reaction sequence, diagrammed in Figure 23–25●, p. 860, proceeds in reverse. This mechanism is known as the **hemoglobin buffer system**.

The hemoglobin buffer system is the only intracellular buffer system that can have an immediate effect on the pH of the ECF. *The hemoglobin buffer system helps prevent drastic changes in pH when the plasma P_{CO_2} is rising or falling.*

The Carbonic Acid–Bicarbonate Buffer System

With the exception of red blood cells, some cancer cells, and tissues temporarily deprived of oxygen, your cells generate carbon dioxide virtually 24 hours a day. As we have described, most of the carbon dioxide is converted to carbonic acid, which then dissociates into a hydrogen ion and a bicarbonate ion. The carbonic acid and its dissociation products form the **carbonic acid–bicarbonate buffer system**.

This buffer system consists of the following reaction, introduced in our discussion of volatile acids (Figure 27–9a●):

$$CO_2 + H_2O \leftrightarrow H_2CO_3 \leftrightarrow H^+ + HCO_3^-$$

| carbon dioxide | water | carbonic acid | bicarbonate ion |

Because the reaction is freely reversible, changing the concentration of any participant will affect the concentrations of all other participants. For example, if hydrogen ions are added, most of them will be removed by interactions with HCO_3^-, forming H_2CO_3 (carbonic acid). In the process, the HCO_3^- acts as a weak base that buffers the excess H^+. The H_2CO_3 formed in this way will in turn dissociate into CO_2 and water (Figure 27–9b●). The extra CO_2 can then be excreted at the lungs. In effect, this reaction takes the H^+ released by a strong organic or fixed acid and generates a volatile acid that can easily be eliminated.

The carbonic acid–bicarbonate buffer system can also protect against increases in pH, although such changes are relatively rare. If hydrogen ions are removed from plasma, the reaction is driven to the right: The P_{CO_2} declines, and the dissociation of H_2CO_3 replaces the missing H^+.

The primary role of the carbonic acid–bicarbonate buffer system is to prevent changes in pH caused by organic acids and fixed acids in the ECF. This buffer system has three important limitations:

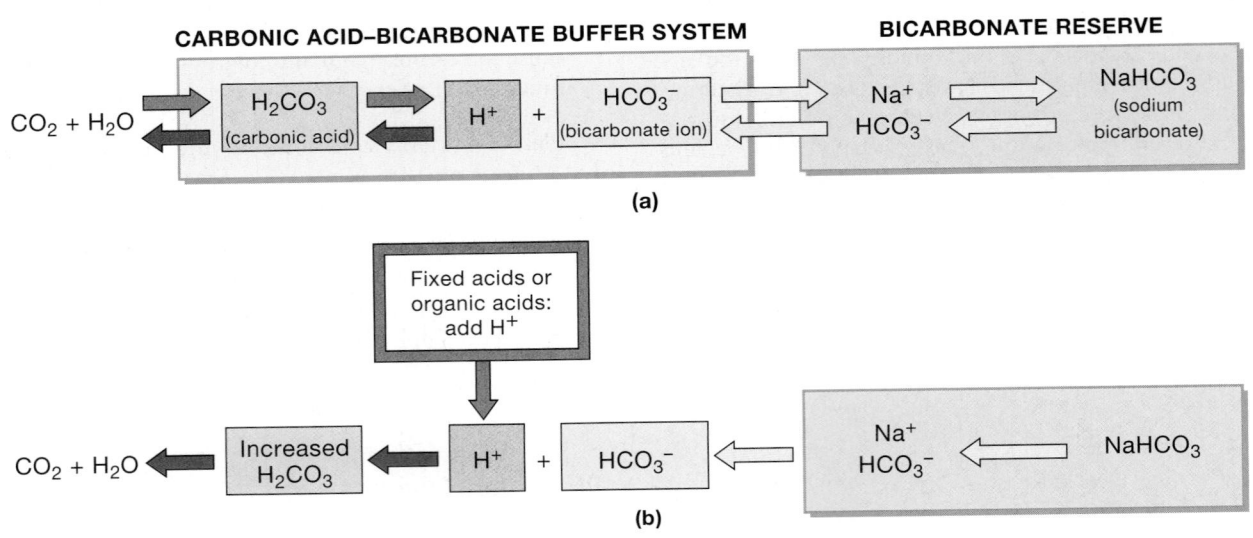

(a)

(b)

●**FIGURE 27–9**
The Carbonic Acid–Bicarbonate Buffer System. **(a)** Basic components of the carbonic acid–bicarbonate buffer system, showing their relationships to carbon dioxide and the bicarbonate reserve. **(b)** The response of the carbonic acid–bicarbonate buffer system to hydrogen ions generated by fixed or organic acids in body fluids.

1. *It Cannot Protect the ECF from Changes in pH that Result From Elevated or Depressed Levels of CO_2.* A buffer system cannot protect against changes in the concentration of its own weak acid. Thus, in this system, the addition of excess H^+ would drive the reaction to the left. But if excess CO_2 were added instead of excess H^+, the elevated CO_2 would drive the reaction to the right. Additional H_2CO_3 would form and dissociate into H^+ and HCO_3^-, reducing the pH of the plasma.

2. *It Can Function Only When the Respiratory System and the Respiratory Control Centers Are Working Normally.* Normally, the elevation in P_{CO_2} that occurs when fixed or organic acids are buffered will stimulate an increase in the respiratory rate. This increase accelerates the removal of CO_2 at the lungs. If the respiratory passageways are blocked, blood flow to the lungs will be impaired, or if the respiratory centers do not respond normally, the efficiency of the buffer system will be reduced. It is important to realize that this buffer system cannot eliminate the H^+ and remove the threat to homeostasis unless the respiratory system is functioning normally.

3. *The Ability to Buffer Acids Is Limited by the Availability of Bicarbonate Ions.* Every time a hydrogen ion is removed from plasma, a bicarbonate ion goes with it. When all the bicarbonate ions have been tied up, buffering capabilities are lost.

Problems due to a lack of bicarbonate ions are rare, for several reasons. First, body fluids contain a large reserve of HCO_3^-, primarily in the form of dissolved molecules of the weak base *sodium bicarbonate* ($NaHCO_3$). This readily available supply of HCO_3^- is known as the **bicarbonate reserve**. The reaction involved (Figure 27–9a●) is

$$Na^+ + HCO_3^- \leftrightarrow NaHCO_3$$
bicarbonate ion   sodium bicarbonate

When hydrogen ions enter the ECF, the bicarbonate ions tied up in H_2CO_3 molecules are replaced by HCO_3^- from the bicarbonate reserve. This relationship is diagrammed in Figure 27–9b●.

Second, additional HCO_3^- can be generated at the kidneys, through mechanisms we introduced in Chapter 26 (Figure 26–14c●, p. 995). In the distal convoluted tubule and collecting system, carbonic anhydrase converts CO_2 within tubular cells into H_2CO_3, which then dissociates. The hydrogen ion is pumped into tubular fluid in exchange for a sodium ion, and the bicarbonate ion is transported into peritubular fluid in exchange for a chloride ion. In effect, tubular cells remove HCl from peritubular fluid in exchange for $NaHCO_3$.

The Phosphate Buffer System

The **phosphate buffer system** consists of the anion $H_2PO_4^-$, which is a weak acid. The operation of the phosphate buffer system resembles that of the carbonic acid–bicarbonate buffer system. The reversible reaction involved is

$$H_2PO_4^- \leftrightarrow H^+ + HPO_4^{2-}$$
dihydrogen phosphate   monohydrogen phosphate

The weak acid is *dihydrogen phosphate* ($H_2PO_4^-$), and the anion released is *monohydrogen phosphate* (HPO_4^{2-}). In the ECF, the phosphate buffer system plays only a supporting role in the regulation of pH, primarily because the concentration of HCO_3^- far exceeds that of HPO_4^{2-}. However, the phosphate buffer system is quite important in buffering the pH of the ICF. In addition, cells contain a *phosphate reserve* in the form of the weak base *sodium monohydrogen phosphate* (Na_2HPO_4). The phosphate buffer system is also important in stabilizing the pH of urine. The dissociation of Na_2HPO_4 provides additional HPO_4^{2-} for use by this buffer system:

$$2Na^+ + HPO_4^{2-} \leftrightarrow Na_2HPO_4$$
monohydrogen phosphate   sodium monohydrogen phosphate

☐ MAINTENANCE OF ACID–BASE BALANCE

Although buffer systems can tie up excess H^+, they provide only a temporary solution to an acid–base imbalance. The hydrogen ions are not eliminated, but merely rendered harmless. For homeostasis to be preserved, the captured H^+ must ultimately either be permanently tied up in water molecules, through the removal of carbon dioxide at the lungs, or be removed from body fluids, through secretion at the kidneys. The underlying problem is that the body's supply of buffer molecules is limited. Suppose that a buffer molecule prevents a change in pH by binding a hydrogen ion that enters the ECF. The buffer is then tied up, reducing the capacity of the ECF to cope with any more H^+. Eventually, all the buffer molecules will be bound to H^+, and pH control will be impossible.

The situation can be resolved only by removing the H^+ from the ECF, thereby freeing the buffer molecules, or by replacing the buffer molecules. Similarly, if a buffer provides a hydrogen ion to maintain normal pH, homeostatic conditions will return only when either another hydrogen ion has been obtained or the buffer has been replaced.

The maintenance of acid–base balance thus includes balancing H^+ gains and losses. This balancing act involves coordinating the actions of buffer systems with respiratory mechanisms and renal mechanisms. These mechanisms support the buffer systems by (1) secreting or absorbing H^+, (2) controlling the excretion of acids and bases, and, when necessary, (3) generating additional buffers. It is the *combination* of buffer systems and these respiratory and renal mechanisms that maintains your pH within narrow limits.

Respiratory Compensation

Respiratory compensation is a change in the respiratory rate that helps stabilize the pH of the ECF. Respiratory compensation occurs whenever your pH strays outside normal limits. Such compensation is effective because respiratory activity has a direct effect on the carbonic acid–bicarbonate buffer system. Increasing or decreasing the rate of respiration alters pH by lowering or raising the P_{CO_2}. When the P_{CO_2} rises, the pH falls, because the addition of CO_2 drives the carbonic acid–bicarbonate buffer system to the right. When the P_{CO_2} falls, the pH rises because the removal of CO_2 drives that buffer system to the left.

We described the mechanisms responsible for the control of respiratory rate in Chapter 23; hence, we will present only a brief summary here. (If necessary, review Figures 23–27 and 23–28●): ∞ pp. 863, 864

- Chemoreceptors of the carotid and aortic bodies are sensitive to the P_{CO_2} of the circulating blood. Other receptors, located on the ventrolateral surfaces of the medulla oblongata, monitor the P_{CO_2} of the CSF. A rise in P_{CO_2} stimulates the receptors.
- The stimulation of the chemoreceptors leads to an increase in the respiratory rate.
- As the rate of respiration increases, more CO_2 is lost at the lungs, so the P_{CO_2} returns to normal levels.

Conversely, when the P_{CO_2} of the blood or CSF declines, the chemoreceptors are inhibited. Respiratory activity becomes depressed and the breathing rate decreases, causing an elevation of the P_{CO_2} in the ECF.

Renal Compensation

Renal compensation is a change in the renal rates of H^+ and HCO_3^- secretion or reabsorption in response to changes in plasma pH. Under normal conditions, the body generates enough organic and fixed acids to add about 100 mEq of H^+ to the ECF each day. An equivalent number of hydrogen ions must therefore be excreted in urine to maintain acid–base balance. In addition, the kidneys assist the lungs by eliminating any CO_2 that enters the renal tubules during filtration or that diffuses into the tubular fluid as it travels toward the renal pelvis.

Hydrogen ions are secreted into the tubular fluid along the proximal convoluted tubule (PCT), the distal convoluted tubule (DCT), and the collecting system. The basic mechanisms involved were detailed in Figures 26–12 and 26–14c● (pp. 990, 995). The ability to eliminate a large number of hydrogen ions in a normal volume of urine depends on the presence of buffers in the urine. The secretion of H^+ can continue only until the pH of the tubular fluid reaches 4.0–4.5. (At that point, the H^+ concentration gradient is so great that hydrogen ions leak out of the tubule as fast as they are pumped in.)

If there were no buffers in the tubular fluid, the kidneys could secrete less than 1 percent of the acid produced each day before the pH reached this limit. To maintain acid balance under these conditions, the kidneys would have to produce about 1000 liters of urine each day just to keep pace with the generation of H^+ in the body. Buffers in tubular fluid are therefore extremely impor-

tant, because they keep the pH high enough for H^+ secretion to continue. Metabolic acids are continuously being generated; without these buffering mechanisms, the kidneys would be unable to maintain homeostasis.

Figure 27–10● diagrams the primary routes of H^+ secretion and shows the buffering mechanisms that stabilize the pH of tubular fluid. The three major buffers involved (Figure 27–10a●) are ① the carbonic acid–bicarbonate buffer system, ② the phosphate buffer system, and ③ ammonia. Glomerular filtration puts components of the carbonic acid–bicarbonate and phosphate buffer systems into the filtrate. The ammonia is generated by tubule cells (primarily those of the PCT).

Figure 27–10a● shows the secretion of H^+, which relies on carbonic anhydrase (CA) activity within the tubular cells. The hydrogen ions generated may be pumped into the lumen in exchange for sodium ions, individually or together with chloride ions. The net result is the secretion of H^+, accompanied by the removal of CO_2 (from the tubular fluid, the tubule cells, and the ECF), and the release of sodium bicarbonate into the ECF.

Figure 27–10b● shows the generation of ammonia within the tubules. As the tubule cells use the enzyme *glutaminase* to break down the amino acid *glutamine*, amino groups are released as ammonium ions (NH_4^+) or as ammonia (NH_3). The ammonium ions are transported into the lumen in exchange for Na^+ in the tubular fluid. The NH_3, which is highly volatile and also toxic to cells, diffuses rapidly into the tubular fluid. There it reacts with a hydrogen ion, forming NH_4^+.

This reaction buffers the tubular fluid and removes a potentially dangerous compound from body fluids. The carbon chains of the glutamine molecules are ultimately converted to HCO_3^-, which is cotransported with Na^+ into the ECF. The generation of ammonia by tubule cells thus ties up H^+ in the tubular fluid and releases sodium bicarbonate into the ECF, where it contributes to the bicarbonate reserve. These mechanisms of H^+ secretion and buffering are always functioning, but their levels of activity vary widely with the pH of the ECF.

THE RENAL RESPONSES TO ACIDOSIS AND ALKALOSIS Acidosis (low body fluid pH) develops when the normal plasma buffer mechanisms are stressed by excessive numbers of hydrogen ions. The kidney tubules do not distinguish among the various acids that may cause acidosis. Whether the fall in pH results from the production of volatile, fixed, or organic acids, the renal contribution remains limited to (1) the secretion of H^+, (2) the activity of buffers in the tubular fluid, (3) the removal of CO_2, and (4) the reabsorption of $NaHCO_3$.

The tubular cells thus bolster the capabilities of the carbonic acid–bicarbonate buffer system. They do so by increasing the concentration of bicarbonate ions in the ECF, replacing those already used to remove hydrogen ions from solution. In a starving individual, the tubular cells break down amino acids, yielding

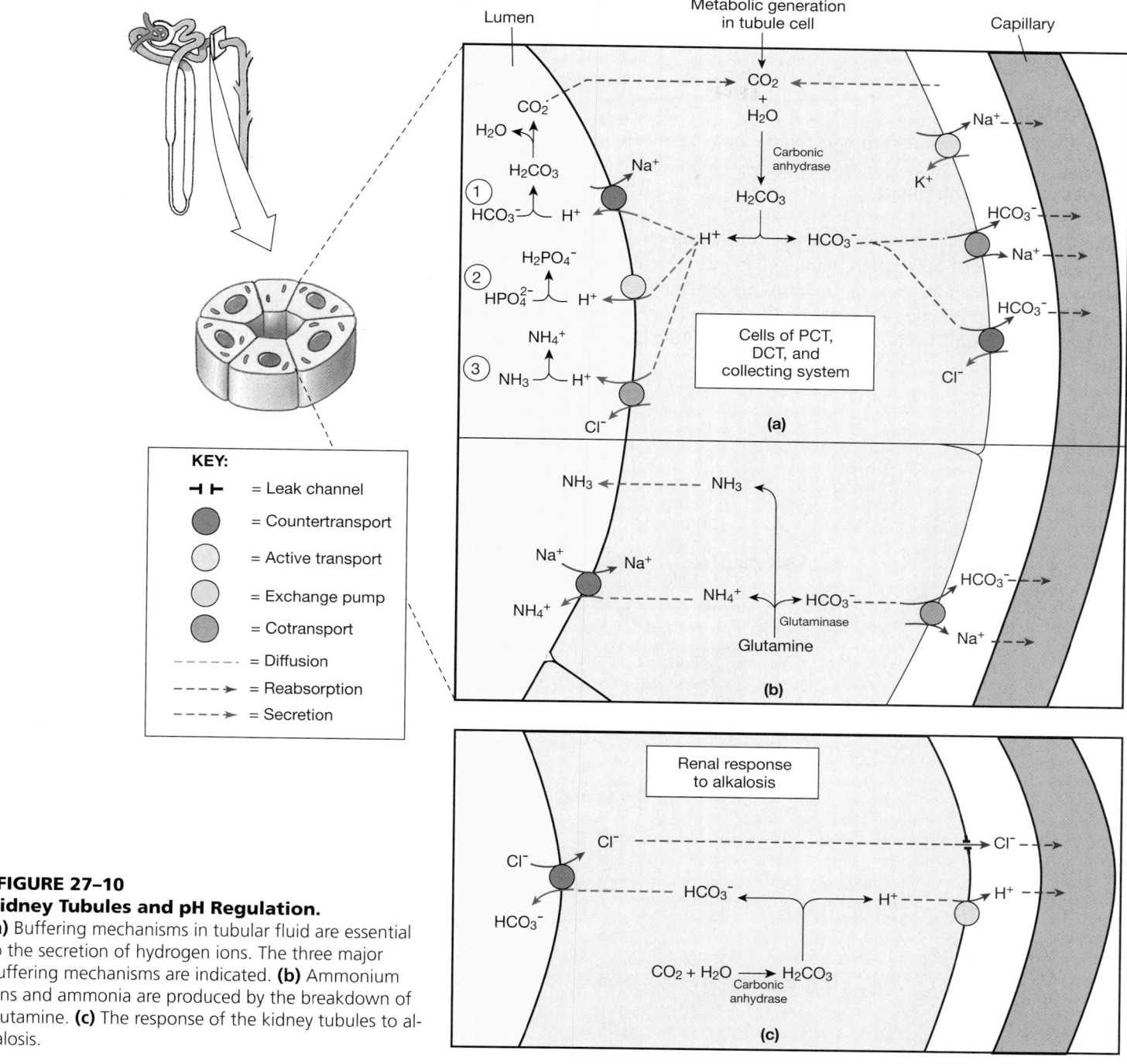

●FIGURE 27–10
Kidney Tubules and pH Regulation.
(a) Buffering mechanisms in tubular fluid are essential to the secretion of hydrogen ions. The three major buffering mechanisms are indicated. (b) Ammonium ions and ammonia are produced by the breakdown of glutamine. (c) The response of the kidney tubules to alkalosis.

ammonium ions that are pumped into the tubular fluid, carbon chains for catabolism, and bicarbonates to help buffer ketone bodies in the blood. (see Figure 26–14c●, p. 995).

When alkalosis (high body fluid pH) develops, (1) the rate of H^+ secretion at the kidneys declines, (2) the tubule cells do not reclaim the bicarbonates in tubular fluid, and (3) the collecting system transports HCO_3^- into tubular fluid while releasing a strong acid (HCl) into peritubular fluid. The concentration of HCO_3^- in plasma decreases, promoting the dissociation of H_2CO_3 and the release of hydrogen ions. The additional H^+ generated at the kidneys helps return the pH to normal levels. The renal response to alkalosis is diagrammed in Figure 27–10c●.

27–6 DISTURBANCES OF ACID–BASE BALANCE

Objectives

■ Identify the most frequent threats to acid–base balance.

■ Explain how the body responds when the pH of body fluids varies outside normal limits.

Figure 27–11● summarizes the interactions among buffer systems, respiration, and renal function in maintaining normal acid–base balance. In combination, these mechanisms are generally able to control pH very precisely, so the pH of the ECF seldom varies more than 0.1 pH unit, from 7.35 to 7.45. When buffering mecha-

nisms are severely stressed, however, the pH wanders outside these limits, producing symptoms of alkalosis or acidosis.

If you are considering a career in a health-related field, an understanding of acid–base dynamics will be essential to you for clinical diagnosis and patient management under a variety of conditions. Temporary shifts in the pH of body fluids occur frequently. Rapid and complete recovery involves a combination of buffer system activity and the respiratory and renal responses. More serious and prolonged disturbances of acid–base balance can result from any factor that affects one of the principal regulatory mechanisms:

• **Any Disorder Affecting Circulating Buffers, Respiratory Performance, or Renal Function.** Several conditions described in earlier chapters, including *emphysema* (Chapter 23) and *renal*

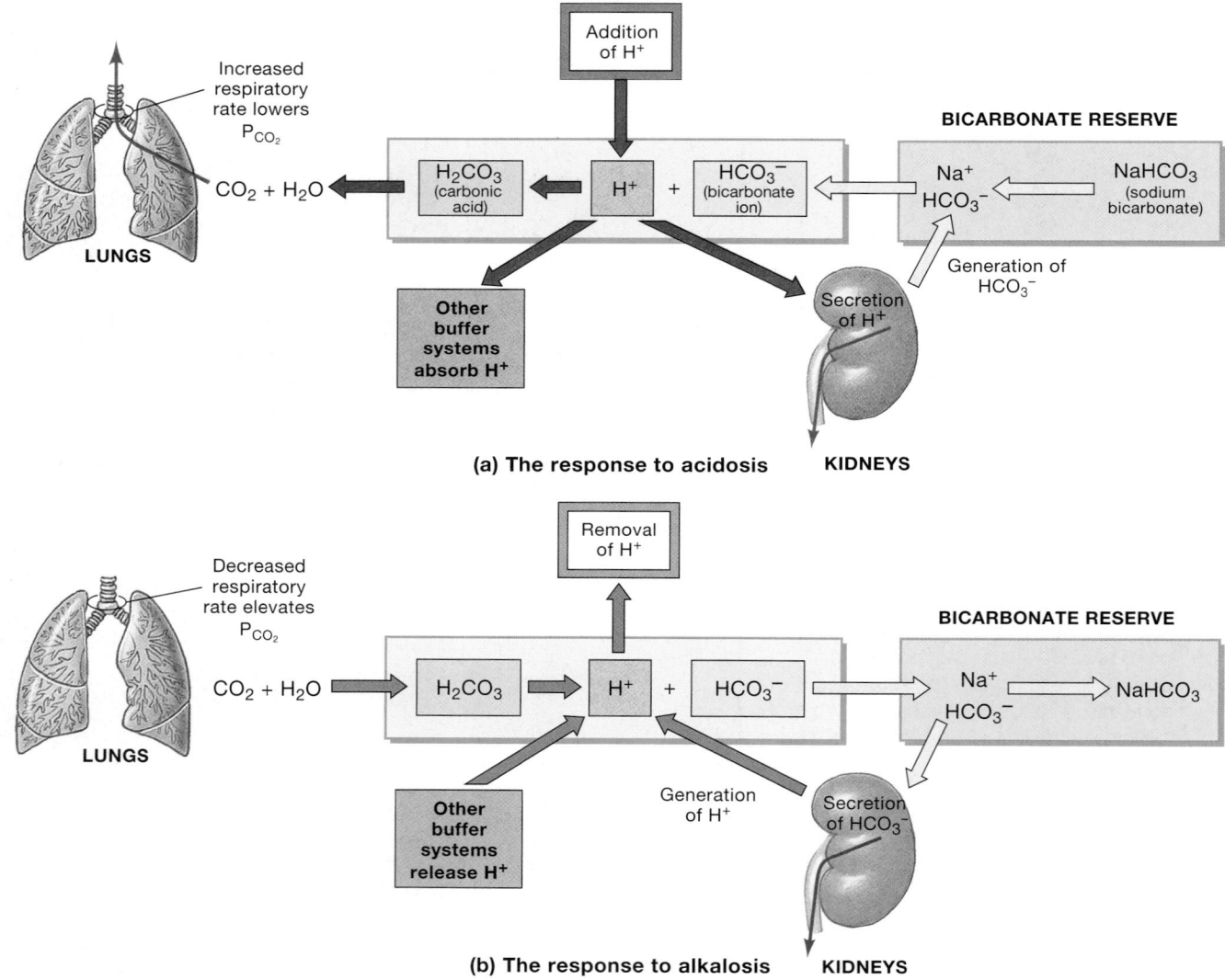

(a) The response to acidosis

(b) The response to alkalosis

●FIGURE 27–11
The Central Role of the Carbonic Acid–Bicarbonate Buffer System in the Regulation of Plasma pH.
Interactions between the carbonic acid–bicarbonate buffer system and other buffer systems and compensation mechanisms. **(a)** The response to a fall in pH caused by the addition of H^+. **(b)** The response to a rise in pH caused by the removal of H^+.

failure (Chapter 26), are associated with dangerous changes in pH. ⬪ pp. 866, 989

- **Cardiovascular Conditions.** Conditions such as *heart failure* or *hypotension* can affect the pH of internal fluids by causing fluid shifts and by changing glomerular filtration rates and respiratory efficiency. ⬪ pp. 707, 735

- **Conditions Affecting the Central Nervous System.** For example, neural damage or disease that affects the CNS will affect the respiratory and cardiovascular reflexes that are essential to normal pH regulation.

Serious abnormalities in acid–base balance generally show an initial *acute phase*, in which the pH moves rapidly away from the normal range. If the condition persists, physiological adjustments occur; the individual then enters the *compensated phase*. Unless the underlying problem is corrected, compensation cannot be completed, and blood chemistry will remain abnormal. The pH typically remains outside normal limits even after compensation has occurred. Even if the pH is within the normal range, the P_{CO_2} or HCO_3^- concentrations can be abnormal.

The primary source of the problem is usually indicated by the name given to the resulting condition. For example,

- *Respiratory acid–base disorders* result from a mismatch between carbon dioxide generation in peripheral tissues and carbon dioxide excretion at the lungs. When a respiratory acid–base disorder is present, the carbon dioxide level of the ECF is abnormal.

- *Metabolic acid–base disorders* are caused by the generation of organic or fixed acids or by conditions affecting the concentration of HCO_3^- in the ECF.

Respiratory compensation alone may restore normal acid–base balance in individuals with respiratory acid–base disorders. In contrast, compensation mechanisms for metabolic acid–base disorders may be able to stabilize pH, but other aspects of acid–base balance (buffer system function, bicarbonate, and P_{CO_2} levels) remain abnormal until the underlying metabolic cause is corrected.

We can subdivide the respiratory and metabolic categories to create four major classes of acid–base disturbances: (1) *respiratory acidosis*, (2) *respiratory alkalosis*, (3) *metabolic acidosis*, and (4) *metabolic alkalosis*.

☐ RESPIRATORY ACIDOSIS

Respiratory acidosis develops when the respiratory system cannot eliminate all the carbon dioxide generated by peripheral tissues. The primary symptom is low plasma pH due to **hypercapnia**, an elevated plasma P_{CO_2}. The usual cause is hypoventilation, an abnormally low respiratory rate. When the P_{CO_2} in the ECF rises, H^+ and HCO_3^- concentrations also begin rising as H_2CO_3 forms and dissociates. Other buffer systems can tie up some of the H^+, but once the combined buffering capacity has

been exceeded, the pH begins to fall rapidly. The effects are diagrammed in Figure 27–12a⬪.

Respiratory acidosis is the most common challenge to acid–base equilibrium. Your tissues generate carbon dioxide rapidly. Even a few minutes of hypoventilation can cause acidosis, reducing the pH of the ECF to as low as 7.0. Under normal circumstances, the chemoreceptors that monitor the P_{CO_2} of plasma and of cerebrospinal fluid (CSF) will eliminate the problem by calling for an increase in breathing (pulmonary ventilation) rates.

If the chemoreceptors fail to respond, if the breathing rate cannot be increased, or if the circulatory supply to the lungs is inadequate, the pH will continue to decline. If the decline is severe, **acute respiratory acidosis** will develop. Acute respiratory acidosis is an immediate, life-threatening condition. It is especially dangerous in persons whose tissues are generating large amounts of carbon dioxide or who are incapable of normal respiratory activity. For this reason, the reversal of acute respiratory acidosis is probably the major goal in the resuscitation of cardiac arrest or drowning victims. Thus first-aid, CPR, and lifesaving courses always stress the "ABCs" of emergency care: *Airway, Breathing,* and *Circulation.*

Chronic respiratory acidosis develops when normal respiratory function has been compromised, but the compensatory mechanisms have not failed completely. Individuals with CNS injuries and those whose respiratory centers have been desensitized by drugs such as alcohol or barbiturates do not respond to warning signals from the chemoreceptors. As a result, these people are prone to developing acidosis due to chronic hypoventilation.

Even when respiratory centers are intact and functional, damage to some components of the respiratory system can prevent increased pulmonary exchange. Examples of conditions fostering chronic respiratory acidosis include emphysema, congestive heart failure, and pneumonia (in which alveolar damage or blockage typically occurs). Pneumothorax and respiratory muscle paralysis have a similar effect, because they, too, limit the ability to maintain adequate breathing rates.

When a normal pulmonary response does not occur, the kidneys respond by increasing the rate of H^+ secretion into tubular fluid. This response slows the rate of change in pH. However, renal mechanisms alone cannot return the pH to normal until the underlying respiratory or circulatory problems are corrected.

The primary problem in respiratory acidosis is that the rate of pulmonary exchange is inadequate to keep the arterial P_{CO_2} within normal limits. The efficiency of breathing can typically be improved temporarily by inducing bronchodilation or by using mechanical aids that provide air under positive pressure. If breathing has ceased, artificial respiration or a mechanical ventilator will be required. These measures may restore normal pH if the respiratory acidosis was neither severe nor prolonged. Treatment of acute respiratory acidosis is complicated by the fact that it causes a complementary *metabolic acidosis*, as we will soon see, due to the generation of lactic acid in oxygen-starved tissues.

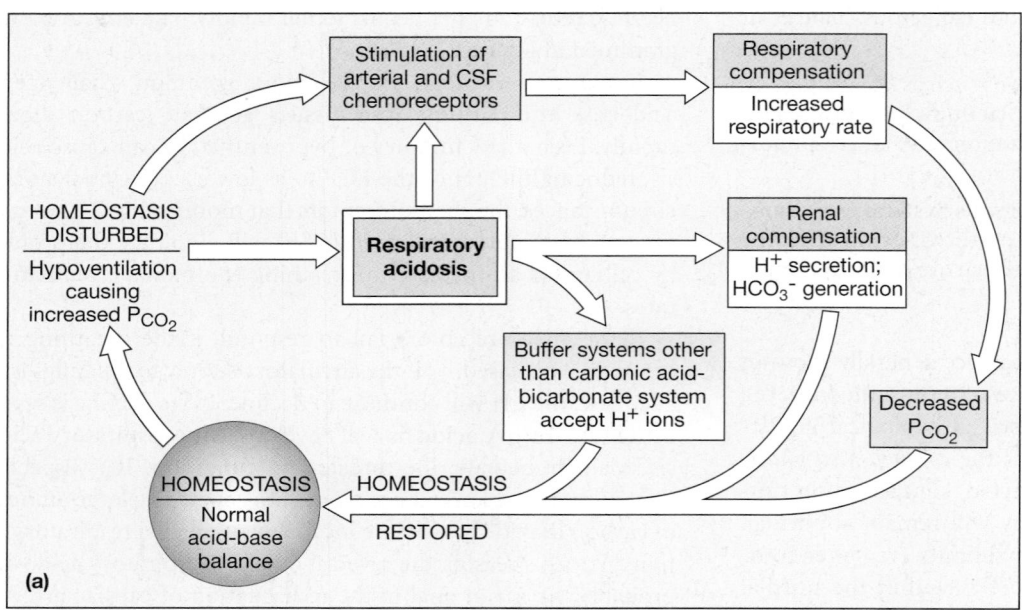

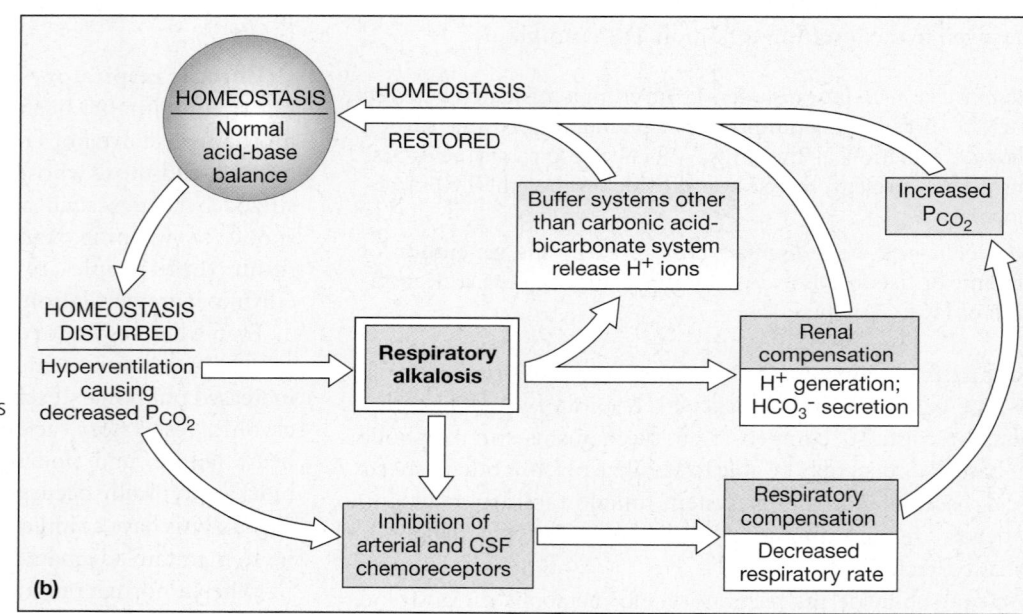

●**FIGURE 27–12**
**Respiratory Acid–Base
Regulation.** **(a)** Respiratory acidosis and **(b)** respiratory alkalosis result from inadequate and excessive breathing, respectively. In healthy individuals, respiratory responses, combined with the responses of the kidneys, are generally able to restore normal acid–base balance.

■ RESPIRATORY ALKALOSIS

Problems with **respiratory alkalosis** (Figure 27–12b●) are relatively uncommon. Respiratory alkalosis develops when respiratory activity lowers plasma to below-normal levels, a condition called **hypocapnia**. A temporary hypocapnia can be produced by *hyperventilation* when increased respiratory activity leads to a reduction in the arterial P_{CO_2}. Continued hyperventilation can elevate the pH to levels as high as 7.8–8.0. This condition generally corrects itself, because the reduction in P_{CO_2} removes the stimulation for the chemoreceptors, so the urge to breathe fades until carbon dioxide levels have returned to normal. *Respiratory alkalosis caused by hyperventilation seldom persists long enough to cause a clinical emergency.*

Common causes of hyperventilation include physical stresses, such as pain, or psychological stresses, such as extreme anxiety. Hyperventilation gradually elevates the pH of the cerebrospinal fluid, and central nervous system function is affected. The initial symptoms involve tingling sensations in the hands, feet, and lips. A light-headed feeling may also be noted. If the hyperventilation continues, the individual may lose consciousness. When unconsciousness occurs, any contributing psychological stimuli are removed, and the breathing rate declines. The P_{CO_2} then rises until the pH returns to normal.

A simple treatment for respiratory alkalosis caused by hyperventilation consists of having the individual breathe and rebreathe the air contained in a small paper bag. As the P_{CO_2} in the bag rises, so do the person's alveolar and arterial CO_2 concentrations. This change eliminates the problem and restores the pH to normal levels. Other problems with respiratory alkalosis are rare and involve primarily (1) individuals adapting to high altitudes, where the low P_{CO_2} promotes hyperventilation, (2) patients on mechanical respirators, or (3) individuals with brain stem injuries who are incapable of responding to shifts in plasma CO_2 concentrations.

☐ METABOLIC ACIDOSIS

Metabolic acidosis is the second most common type of acid–base imbalance. It has three major causes:

1. The most widespread cause of metabolic acidosis is the production of a large number of fixed or organic acids. The hydrogen ions liberated by these acids overload the carbonic acid–bicarbonate buffer system, so the pH begins to decline (Figure 27–13a●). We introduced two examples of metabolic acidosis earlier:

   • **Lactic acidosis** can develop after strenuous exercise or prolonged tissue hypoxia (oxygen starvation) as active cells rely on anaerobic respiration (see Figure 10–20●). ∞ p. 318

   • **Ketoacidosis** results from the generation of large quantities of ketone bodies during the postabsorptive state of metabolism. Ketoacidosis is a problem in starvation and a potentially lethal complication of poorly controlled diabetes mellitus. In either case, peripheral tissues are unable to obtain adequate glucose from the bloodstream and begin metabolizing lipids and ketone bodies. ∞ p. 952

2. A less common cause of metabolic acidosis is an impaired ability to excrete H^+ at the kidneys. For example, conditions marked by severe kidney damage, such as *glomerulonephritis* (Chapter 26), typically result in a severe metabolic acidosis. ∞ p. 980 Metabolic acidosis is also caused by diuretics that turn off the sodium–hydrogen transport system in the kidney tubules. The secretion of H^+ is directly or indirectly linked to the reabsorption of Na^+. When Na^+ reabsorption stops, so does H^+ secretion.

3. Metabolic acidosis occurs after a severe bicarbonate loss (Figure 27–13b●). The carbonic acid–bicarbonate buffer system relies on bicarbonate ions to balance hydrogen ions that threaten pH balance. A drop in the HCO_3^- concentration in the ECF thus reduces the effectiveness of this buffer system, and acidosis soon develops. The most common cause of HCO_3^- depletion is chronic diarrhea. Under normal conditions, most of the bicarbonate ions secreted into the digestive tract in pancreatic, hepatic, and mucous secretions are reabsorbed before the feces are eliminated. In diarrhea, these bicarbonates are lost, and the HCO_3^- concentration of the ECF drops accordingly.

The nature of the problem must be understood before treatment can begin. Potential causes are so varied that a clinician must piece together relevant clues to make a diagnosis. In some cases, the diagnosis is straightforward; for example, a patient with metabolic acidosis after a bicycle race probably has lactic acidosis. In other cases, the clinician must be a detective.

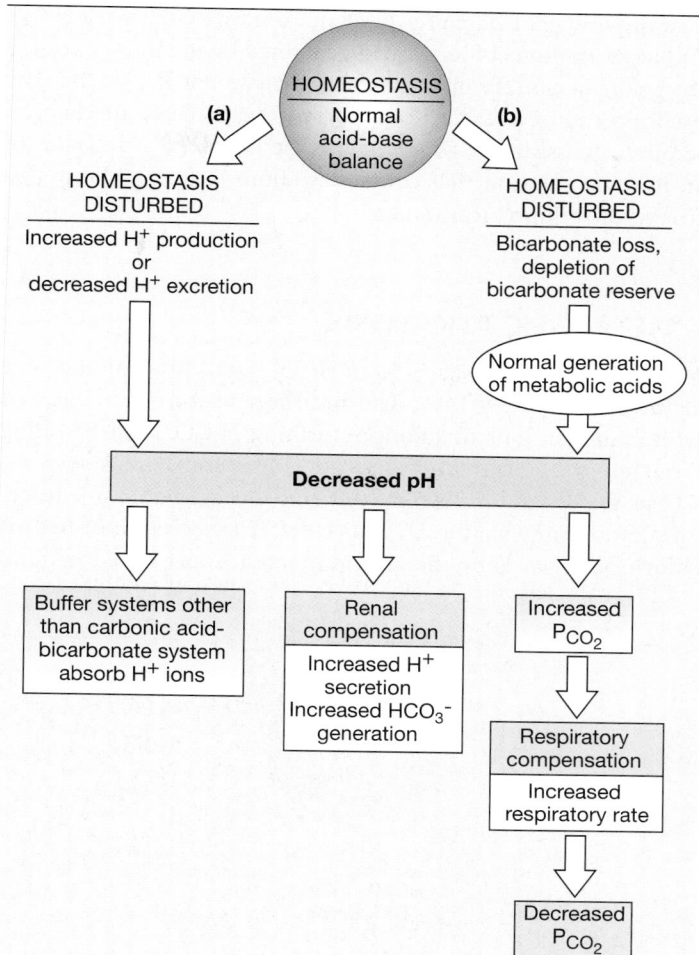

●**FIGURE 27–13**
The Response to Metabolic Acidosis. Metabolic acidosis can result from either **(a)** increased acid production or decreased acid secretion, leading to a buildup of H^+ in body fluids, or **(b)** a loss of bicarbonate ions that makes the carbonic acid–bicarbonate buffer system incapable of preventing a fall in pH. Respiratory and renal compensation mechanisms can stabilize pH, but the blood chemistry remains abnormal until the levels of acid production, acid secretion, and bicarbonate ion concentration return to normal.

Compensation for metabolic acidosis generally involves a combination of respiratory and renal mechanisms. Hydrogen ions interacting with bicarbonate ions form carbon dioxide molecules that are eliminated at the lungs, whereas the kidneys excrete additional hydrogen ions into the urine and generate bicarbonate ions that are released into the ECF.

Combined Respiratory and Metabolic Acidosis

Respiratory acidosis and metabolic acidosis are typically linked, because oxygen-starved tissues generate large quantities of lactic acid and because sustained hypoventilation leads to decreased arterial P_{O_2}. The problem can be particularly serious in cases of near drowning, in which the body fluids have high P_{CO_2}, low P_{O_2}, and large amounts of lactic acid generated by the muscles of the struggling person. Prompt emergency treatment is essential. The usual procedure involves some form of artificial or mechanical respiratory assistance, coupled with the intravenous infusion of an isotonic solution that contains sodium lactate, sodium gluconate, or sodium bicarbonate.

◻ METABOLIC ALKALOSIS

Metabolic alkalosis occurs when HCO_3^- concentrations become elevated (Figure 27–14●). The bicarbonate ions then interact with hydrogen ions in solution, forming H_2CO_3. The resulting reduction in H^+ concentrations causes symptoms of alkalosis.

Metabolic alkalosis is relatively rare, but we noted one interesting cause in Chapter 24. ∞ p. 892 The secretion of hydrochloric acid (HCl) by the gastric mucosa is associated with an influx of large numbers of bicarbonate ions into the ECF. This phenomenon, known as the *alkaline tide*, temporarily elevates the HCO_3^- concentration in the ECF during meals. A person who begins vomiting repeatedly will continue to generate stomach acids to replace those that are lost, so the HCO_3^- concentration of the ECF will continue to rise.

Compensation for metabolic alkalosis involves a reduction in the breathing rate, coupled with the increased loss of HCO_3^- in urine. Treatment of mild cases addresses the primary cause, generally by controlling the vomiting, and may involve the administration of solutions that contain NaCl or KCl.

Treatment of acute cases of metabolic alkalosis may involve the administration of ammonium chloride (NH_4Cl). Metabolism of the ammonium ion in the liver results in the liberation of a hydrogen ion, so, in effect, the introduction of NH_4Cl leads to the internal generation of HCl, a strong acid. As the HCl diffuses into the bloodstream, the pH falls toward normal levels.

◻ THE DETECTION OF ACIDOSIS AND ALKALOSIS

Virtually anyone who has a problem that affects the cardiovascular, respiratory, urinary, digestive, or nervous system may develop potentially dangerous acid–base imbalances. For this reason, most diagnostic blood tests include tests designed to provide information about pH and buffer function. Standard tests include blood pH, P_{CO_2}, and HCO_3^- levels. These measurements make the recognition of acidosis or alkalosis and the classification of a particular condition as respiratory or metabolic in nature relatively straightforward. Figure 27–15● and Table 27–4 indicate the patterns that characterize the four major categories of acid–base disorders. Additional steps, such as determining the *anion gap*, plotting blood test results on a graph or *nomogram*, or using a diagnostic chart can help in identifying possible causes of the problem and in distinguishing compensated from uncompensated conditions. Details are included in the *Applications Manual*. **AM** Diagnostic Classification of Acid–Base Disorders

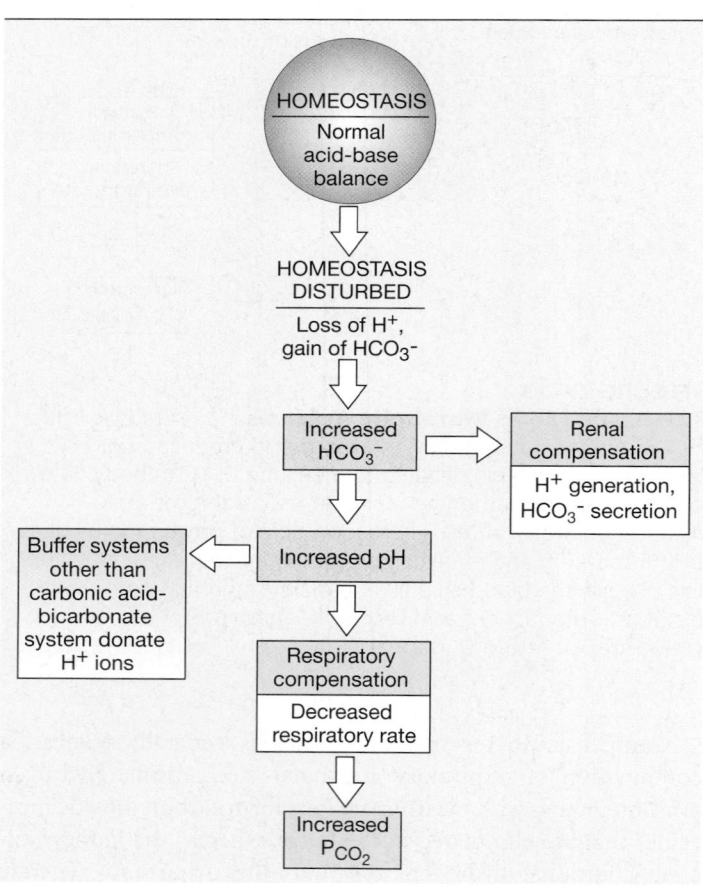

●**FIGURE 27–14**
Metabolic Alkalosis. Metabolic alkalosis most commonly results from the loss of acids, especially stomach acid lost through vomiting. As replacement gastric acids are produced, the alkaline tide introduces a great many bicarbonate ions into the bloodstream, so pH increases.

✔ What effect would a decrease in the pH of the body fluids have on the respiratory rate?

✔ Why does the tubular fluid in nephrons need to be buffered?

✔ How would a prolonged fast affect the body's pH?

✔ Why can prolonged vomiting produce metabolic alkalosis?

Answers start on page Q-1

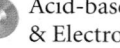 Acid-base balance can be reviewed on the **IP CD-ROM**: Fluids & Electrolytes/Acid-Base Homeostasis.

●FIGURE 27–15
A Diagnostic Chart for Acid–Base Disorders

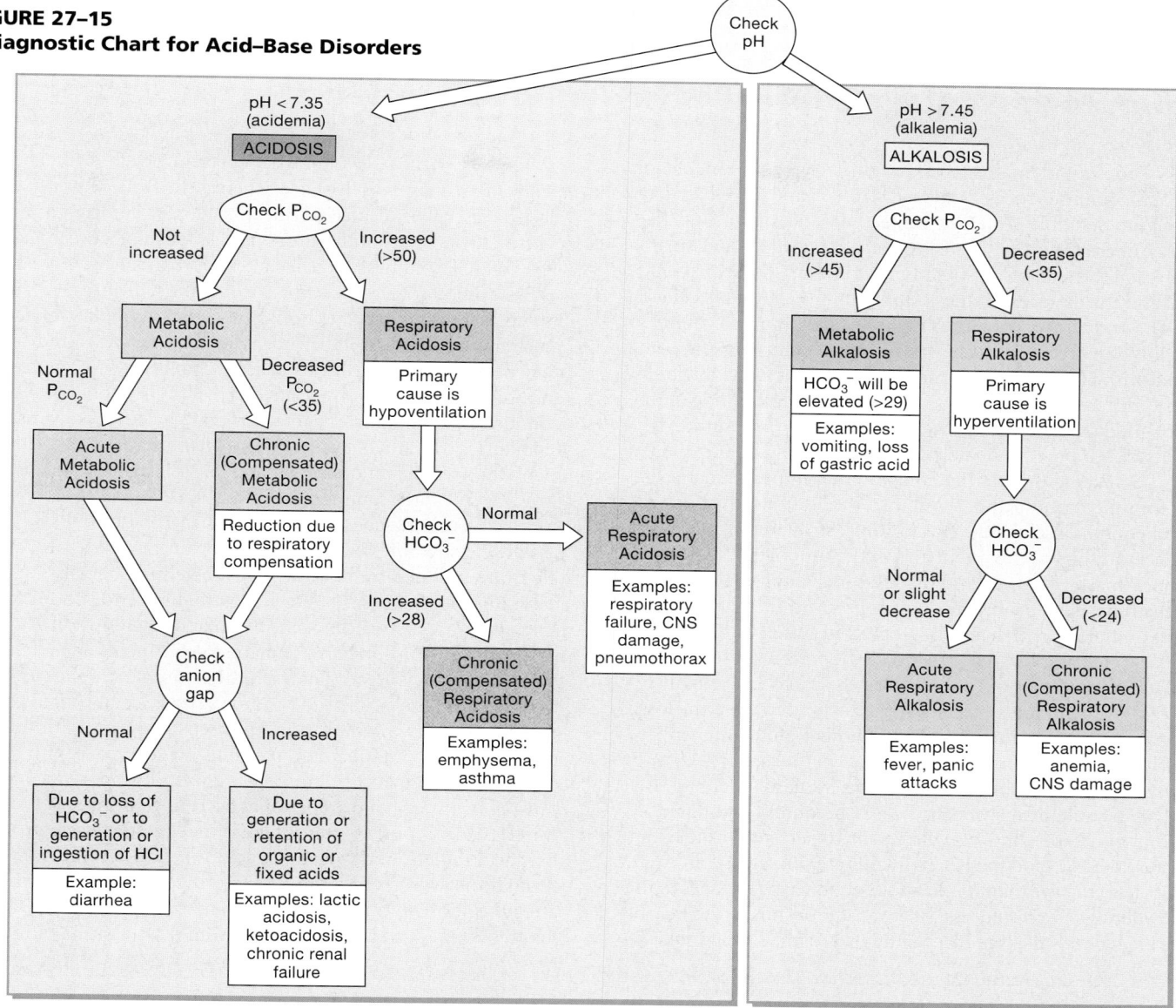

TABLE 27–4   CHANGES IN BLOOD CHEMISTRY ASSOCIATED WITH THE MAJOR CLASSES OF ACID–BASE DISORDERS

Disorder	pH (normal = 7.35–7.45)	HCO₃⁻ (normal = 24–28 mEq/l)	P_{CO_2} (mm Hg) (normal = 35–45)	Remarks	Treatment
Respiratory acidosis	Decreased (below 7.35)	*Acute*: normal *Compensated*: increased (above 28)	Increased (above 50)	Generally caused by hypoventilation and CO_2 buildup in tissues and blood	Improve ventilation; in some cases, with bronchodilation and mechanical assistance
Metabolic acidosis	Decreased (below 7.35)	Decreased (below 24)	*Acute*: normal *Compensated*: decreased (below 35)	Caused by buildup of organic or fixed acid, impaired H^+ elimination at kidneys, or HCO_3^- loss in urine or feces	Administration of bicarbonate (gradual), with other steps as needed to correct primary cause
Respiratory alkalosis	Increased (above 7.45)	*Acute*: normal *Compensated*: decreased (below 24)	Decreased (below 35)	Generally caused by hyperventilation and reduction in plasma CO_2 levels	Reduce respiratory rate, allow rise in P_{CO_2}
Metabolic alkalosis	Increased (above 7.45)	Increased (above 28)	Increased (above 45)	Generally caused by prolonged vomiting and associated acid loss	pH below 7.55: no treatment; pH above 7.55: may require administration of NH_4Cl

Fluid, Electrolyte, and Acid–Base Balance in Infants

Fetuses and infants have very different requirements for the maintenance of fluid, electrolyte, and acid–base balance than do adults. A fetus obtains the water, organic nutrients, and electrolytes it needs from the maternal bloodstream. Buffers in the fetal bloodstream provide short-term pH control, and the maternal kidneys eliminate the H^+ that is generated. The body water content is high: At birth, water accounts for roughly 75 percent of the newborn infant's body weight, compared with 50–60 percent in adults. Several factors contribute to this difference, including (1) the larger infant blood volume (10.0–12.5 percent of body weight, versus 6–7 percent in adults), (2) a proportionately larger volume of cerebrospinal fluid, (3) a larger interstitial fluid volume (in part due to existing for nine months under weightless conditions; the body water content also changes in orbiting astronauts), and (4) differences in the water content and in the proportions, in terms of body mass, of organs and tissues; for example, compared with adults, newborn infants have a proportionately larger heart and brain, which are 75–79 percent water, but less than half as much adipose tissue, which is 10 percent water.

During the first two to three days after delivery, roughly 6 percent of the infant's excess water is lost. Thereafter, the loss is more gradual, with the typical adult body water content appearing after age 2.

At birth, the distribution of body water is also different from that of an adult. In a newborn, the ECF accounts for roughly 35 percent of the total body weight, versus 40 percent for the ICF. Because the ECF and the ICF in infants are similar in volume, the ICF is less effective than it is in adults at buffering changes in the ECF volume. As a result, less water must be lost from the ECF before the ICF volume is reduced enough to damage cells. As

growth occurs, the ICF volume (as a percentage of the total body weight) remains relatively stable, whereas the ECF volume gradually decreases. The relative decline in the ECF volume becomes evident after a few months; the adult relationship between the ECF and ICF volumes is reached after approximately two years.

Basic aspects of electrolyte balance are the same in newborn infants as in adults, but the effects of fluctuations in the diet are much more immediate in newborns because reserves of minerals and energy sources are much smaller. (At any age, the bone mass and adipose tissue mass are relatively small.) The problem is compounded by the fact that the metabolic rate (per unit body mass) of infants is twice that of adults. Thus, infants have an elevated demand for nutrients, and that demand must be met promptly. This is one reason that infants require frequent feedings; another is that they have a much higher demand for water than do adults.

The elevated metabolic rate also means an accelerated production of waste products that must be eliminated at the kidneys. But the kidneys of a newborn are unable to produce urine with an osmotic concentration above 450 mOsm/l. So a newborn must produce more urine to eliminate the metabolic waste products. Water losses at other sites are higher as well. Because the surface-to-volume ratio and the respiratory rate are relatively high, the rate of insensible perspiration is much higher in infants than in adults. To keep pace with rates of water loss in urine and insensible perspiration, newborns must consume, on a proportional basis, seven times as much water as adults do. If water intake is inadequate, waste products accumulate, and metabolic acidosis may develop. Although the kidneys become effective at concentrating urine after about one month, the elevated metabolic rate and water loss remain. As a result, infants continue to consume (and

27–7 AGING AND FLUID, ELECTROLYTE, AND ACID–BASE BALANCE

Objective

- Describe the effects of aging on fluid, electrolyte, and acid–base balance.

Aging affects many aspects of fluid, electrolyte, and acid–base balance, including the following:

- The total body water content gradually decreases with age. Between ages 40 and 60, the total body water content averages 55 percent for males and 47 percent for females. After age 60, the values decline to roughly 50 percent for males and 45 percent for females. Among other effects, this decrease reduces the dilution of waste products, toxins, and any drugs that have been administered.

- A reduction in the glomerular filtration rate and in the number of functional nephrons reduces the ability to regulate pH through renal compensation.

- The ability to concentrate urine declines, so more water is lost in urine. In addition, the rate of insensible perspiration increases as the skin becomes thinner and more delicate. Maintaining fluid balance therefore requires a higher daily water intake. A reduction in ADH and aldosterone sensitivity makes older people less able than younger people to conserve body water when losses exceed gains.

- Many people over age 60 experience a net loss in their body mineral content as muscle mass and skeletal mass decrease. This loss can, at least in part, be prevented by a combination of exercise and an increased dietary mineral supply.

- The reduction in vital capacity that accompanies aging reduces the ability to perform respiratory compensation, increasing the risk of respiratory acidosis. This problem can be compounded by arthritis, which can reduce vital capacity, and by emphysema, another condition that, to some degree, develops with aging.

- Disorders affecting major systems become more common with increasing age. Most, if not all, of these disorders have some impact on either fluid, electrolyte, or acid–base balance.

lose) roughly twice as much water as do adults by body weight. Infants are therefore at greater risk of dehydration; they can survive for only one to three days without water, whereas adults can survive for a week or more.

The increased metabolic rate of infants also means an accelerated demand for oxygen and more rapid generation of CO_2. The respiratory rate in newborns is relatively high—roughly 40 breaths per minute. And the breaths are deep: In proportion to their body weight, newborns must move twice as much air as adults do. As a result, the functional residual capacity of the lungs (the amount of air in the lungs after one quiet respiratory cycle) is about half that of adults. In adults, the functional residual capacity is large relative to the tidal volume (the amount of air moving into or out of the lungs in one respiratory cycle), and this proportion helps prevent rapid changes in the P_{O_2} and P_{CO_2} of alveolar air. In infants, as soon as the respiratory rate changes, the alveolar air composition changes. Because changes in the respiratory rate can cause sudden changes in the P_{O_2} and P_{CO_2} of arterial blood, newborns are at greater risk of developing respiratory acidosis or respiratory alkalosis.

Many newborn infants have some degree of respiratory acidosis caused by the interruption of placental circulation during labor. This condition is known as *fetal stress*. It is therefore important that the newborn begin breathing as soon as possible after delivery so that the excess CO_2 can be eliminated. Most full-term newborns are given oxygen briefly to improve their blood oxygenation and to help them eliminate excess CO_2. (Oxygen administration to premature infants can have unwanted side effects.) The pH of infants in fetal stress typically does not fall below 7.25; in most cases, no treatment other than clearing the airway and administering oxygen is required.

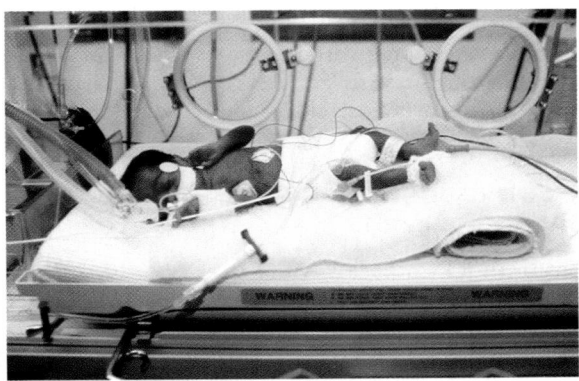

If the fetus struggles during labor or the delivery is prolonged, *fetal distress* can develop. This condition results from the combination of respiratory acidosis, due to decreased placental circulation, and metabolic acidosis, due to lactic acid production. The pH can fall as low as 7.0, with potentially fatal effects on the heart and central nervous system. Because fetal distress is one of the risks of birth, the fetus is commonly monitored during labor and delivery. Early in labor, the heart rate is checked by ultrasound; once delivery is under way, blood samples can be taken from the scalp. If the fetal heart rate becomes too rapid and irregular early in labor, or if the arterial pH drops below 7.2, prompt surgical delivery (by cesarean section) may be recommended. Treatment of a newborn in fetal distress typically includes respiratory assistance, the administration of oxygen, and the intravenous infusion of a sodium bicarbonate solution.

Chapter Review

SELECTED CLINICAL TERMINOLOGY

Terms Discussed in This Chapter

hyperkalemia: Plasma K^+ levels above 8 mEq/l. *(p. 1025)*

hypernatremia: Plasma Na^+ levels above 150 mEq/l. *(p. 1025)*

hypokalemia: Plasma K^+ levels below 2 mEq/l. *(p. 1025)*

hyponatremia: Plasma Na^+ levels below 130 mEq/l. *(p. 1025)*

metabolic acidosis: A type of acidosis caused by the inability to excrete hydrogen ions, the production of numerous fixed or organic acids, or a severe bicarbonate loss. *(p. 1037 and AM)*

metabolic alkalosis: A rare form of alkalosis due to high concentrations of bicarbonate ions in body fluids. *(p. 1038 and AM)*

respiratory acidosis: Acidosis resulting from inadequate respiratory activity, characterized by elevated levels of carbon dioxide (hypercapnia) in body fluids. *(p. 1035 and AM)*

respiratory alkalosis: Alkalosis due to excessive respiratory activity, which depresses carbon dioxide levels and elevates the pH of body fluids. *(p. 1036 and AM)*

AM Additional Terms Discussed in the *Applications Manual*

anion gap
nomogram

STUDY OUTLINE

27–1 FLUID, ELECTROLYTE, AND ACID–BASE BALANCE: AN OVERVIEW p. 1015

1. The maintenance of normal volume and normal composition of extracellular and intracellular fluids is vital to life. Three types of homeostasis are involved: **fluid balance**, **electrolyte balance**, and **acid–base balance**.

27–2 AN INTRODUCTION TO FLUID AND ELECTROLYTE BALANCE p. 1016

THE ECF AND THE ICF p. 1016

1. The **intracellular fluid (ICF)** contains nearly two-thirds of the total body water; the **extracellular fluid (ECF)** contains the rest. Exchange occurs between the ICF and the ECF, but the two **fluid compartments** retain their distinctive characteristics. (*Figures 27–1, 27–2*)

BASIC CONCEPTS IN THE REGULATION OF FLUIDS AND ELECTROLYTES p. 1017

2. Homeostatic mechanisms that monitor and adjust the composition of body fluids respond to changes in the ECF.
3. No receptors directly monitor fluid or electrolyte balance; receptors involved in fluid balance and in electrolyte balance respond to changes in plasma volume or osmotic concentration.
4. Our cells cannot move water molecules by active transport; all movements of water across cell membranes and epithelia occur passively, in response to osmotic gradients.
5. The body content of water or electrolytes will rise if intake exceeds outflow and will fall if losses exceed gains.

AN OVERVIEW OF THE PRIMARY REGULATORY HORMONES p. 1019

6. ADH encourages water reabsorption at the kidneys and stimulates thirst. Aldosterone increases the rate of sodium reabsorption at the kidneys. Natriuretic peptides (ANP and BNP) oppose those actions and promote fluid and electrolyte losses in urine.

THE INTERPLAY BETWEEN FLUID BALANCE AND ELECTROLYTE BALANCE p. 1019

7. The regulatory mechanisms of fluid balance and electrolyte balance are quite different, and the distinction is clinically important.

27–3 FLUID BALANCE p. 1019

FLUID MOVEMENT WITHIN THE ECF p. 1020

1. Water circulates freely within the ECF compartment. (*Figure 27–3*)

FLUID EXCHANGE WITH THE ENVIRONMENT p. 1020

2. Water losses are normally balanced by gains through eating, drinking, and **metabolic generation**. (*Figure 27–3; Table 27–1*)

FLUID SHIFTS p. 1021

3. Water movement between the ECF and ICF is called a **fluid shift**. If the ECF becomes hypertonic relative to the ICF, water will move from the ICF into the ECF until osmotic equilibrium has been restored. If the ECF becomes hypotonic relative to the ICF, water will move from the ECF into the cells, and the volume of the ICF will increase.

Fluid balance: **IP CD-ROM:** Fluids & Electrolytes/Introduction to Body Fluids

27–4 ELECTROLYTE BALANCE p. 1022

1. Problems with electrolyte balance generally result from an imbalance between gains and losses of sodium. Problems with potassium balance are less common, but more dangerous.

SODIUM BALANCE p. 1022

2. The rate of sodium uptake across the digestive epithelium is directly proportional to the amount of sodium in the diet. Sodium losses occur mainly in urine and through perspiration. (*Figure 27–4*)
3. Shifts in sodium balance result in expansion or contraction of the ECF. Large variations in the ECF volume are corrected by the homeostatic mechanisms triggered by changes in blood volume. If the volume becomes too low, ADH and aldosterone are secreted; if the volume becomes too high, ANP is secreted. (*Figure 27–5*)

POTASSIUM BALANCE p. 1024

4. Potassium ion concentrations in the ECF are very low and not as closely regulated as are sodium ion concentrations. Potassium excretion increases as ECF concentrations rise, under aldosterone stimulation, and when the pH rises. Potassium retention occurs when the pH falls.

OTHER ELECTROLYTES p. 1025

5. The ECF concentrations of other electrolytes, such as calcium, magnesium, phosphate, and chloride, are also regulated. (*Table 27–2*)

Ion balance: **IP CD-ROM:** Fluids & Electrolytes/Electrolyte Homeostasis

27–5 ACID–BASE BALANCE p. 1027

1. Acids and bases are either *strong* or *weak*. (*Table 27–3*)

THE IMPORTANCE OF pH CONTROL p. 1027

2. The pH of normal body fluids ranges from 7.35 to 7.45; variations outside this relatively narrow range produce symptoms of **acidosis** or **alkalosis**.

TYPES OF ACIDS IN THE BODY p. 1028

3. **Volatile acids** can leave solution and enter the atmosphere; **fixed acids** remain in body fluids until excreted at the kidneys; **organic acids** are participants in, or by-products of, aerobic metabolism.
4. Carbonic acid is the most important factor affecting the pH of the ECF. In solution, CO_2 reacts with water to form carbonic acid. An inverse relationship exists between pH and the concentration of CO_2. (*Figure 27–6*)
5. Sulfuric acid and phosphoric acid, the most important fixed acids, are generated during the catabolism of amino acids and compounds containing phosphate groups.
6. Organic acids include metabolic products such as lactic acid and ketone bodies.

MECHANISMS OF pH CONTROL p. 1028

7. A **buffer system** typically consists of a weak acid and the anion released by its dissociation. The ion functions as a weak base. The three major buffer systems are (1) **protein buffer systems** in the ECF and ICF; (2) the **carbonic acid–bicarbonate buffer system**, most important in the ECF; and (3) the **phosphate buffer system** in the ICF and urine. (*Figure 27–7*)
8. In protein buffer systems, the component amino acids respond to changes in pH by accepting or releasing hydrogen ions.

The **hemoglobin buffer system** is a protein buffer system that helps prevent drastic changes in pH when the P_{CO_2} is rising or falling. *(Figure 27–8)*

9. The **carbonic acid–bicarbonate buffer system** prevents pH changes caused by organic acids and fixed acids in the ECF. The readily available supply of bicarbonate ions is the **bicarbonate reserve**. *(Figure 27–9)*

10. The phosphate buffer system plays a supporting role in regulating the pH of the ECF, but is important in buffering the pH of the ICF and of urine.

MAINTENANCE OF ACID–BASE BALANCE p. 1031

11. The lungs help regulate pH by affecting the carbonic acid–bicarbonate buffer system. Changing the respiratory rate can raise or lower the P_{CO_2} of body fluids, affecting the body's buffering capacity. This process is called **respiratory compensation**.

12. In **renal compensation**, the kidneys vary their rates of hydrogen ion secretion and bicarbonate ion reabsorption, depending on the pH of the ECF. *(Figure 27–10)*

27–6 DISTURBANCES OF ACID–BASE BALANCE p. 1034

1. Interactions among buffer systems, respiration, and renal function normally maintain tight control of the pH of the ECF, generally within a range of 7.35–7.45. *(Figure 27–11)*

2. **Respiratory acid–base disorders** result when abnormal respiratory function causes an extreme rise or fall in CO_2 levels in the ECF. **Metabolic acid–base disorders** are caused by the generation of organic or fixed acids or by conditions affecting the concentration of bicarbonate ions in the ECF.

RESPIRATORY ACIDOSIS p. 1035

3. **Respiratory acidosis** results from excessive levels of CO_2 in body fluids. *(Figure 27–12)*

RESPIRATORY ALKALOSIS p. 1036

4. **Respiratory alkalosis** is a relatively rare condition associated with hyperventilation. *(Figure 27–12)*

METABOLIC ACIDOSIS p. 1037

5. **Metabolic acidosis** results from the depletion of the bicarbonate reserve, caused by an inability to excrete hydrogen ions at the kidneys, the production of large numbers of fixed and organic acids, or bicarbonate loss that accompanies chronic diarrhea. *(Figure 27–13)*

METABOLIC ALKALOSIS p. 1038

6. **Metabolic alkalosis** occurs when bicarbonate ion concentrations become elevated, as from extended periods of vomiting. *(Figure 27–14)*

THE DETECTION OF ACIDOSIS AND ALKALOSIS p. 1038

7. Standard diagnostic blood tests such as blood pH, P_{CO_2}, and bicarbonate levels are used to recognize and classify acidosis and alkalosis conditions as respiratory or metabolic in nature. *(Figure 27–15; Table 27–4)*

 Acid-base balance: **IP CD-ROM:** Fluids & Electrolytes/Acid-Base Homeostasis

27–7 AGING AND FLUID, ELECTROLYTE, AND ACID–BASE BALANCE p. 1040

1. Changes affecting fluid, electrolyte, and acid–base balance in the elderly include (1) reduced total body water content, (2) impaired ability to perform renal compensation, (3) increased water demands due to reduced ability to concentrate urine and reduced sensitivity to ADH and aldosterone, (4) a net loss of minerals, (5) reductions in respiratory efficiency that affect the ability to perform respiratory compensation, and (6) increased incidence of conditions that secondarily affect fluid, electrolyte, or acid–base balance.

REVIEW QUESTIONS

 More assessment questions are available to you on the Companion Website. You will find Matching, Multiple Choice, True/False, and other quizzes to help further your understanding of the material covered in this chapter. To access the site, go to www.aw.com/martini.

LEVEL 1 Reviewing Facts and Terms

Match each numbered item with the most closely related lettered item. Use letters for answers in the spaces provided.

_____ 1. ECF
_____ 2. ADH
_____ 3. aldosterone
_____ 4. ANP
_____ 5. oxidative phosphorylation
_____ 6. overhydration
_____ 7. dehydration
_____ 8. sodium
_____ 9. potassium
_____ 10. carotid sinus
_____ 11. calcitonin
_____ 12. hydrochloric acid
_____ 13. carbonic acid
_____ 14. carbon dioxide
_____ 15. fixed acid
_____ 16. organic acid
_____ 17. hypoventilation
_____ 18. hyperventilation

a. reduces thirst
b. dominant cation in ECF
c. hyponatremia
d. weak acid
e. baroreceptors

f. promotes calcium loss
g. respiratory alkalosis
h. stimulates water conservation
i. sulfuric acid
j. volatile acid
k. ketone bodies
l. interstitial fluid
m. respiratory acidosis
n. metabolic acidosis
o. hypernatremia
p. metabolic alkalosis
q. dominant cation in ICF
r. conservation of sodium ions

_____ 19. elevated bicarbonate ion concentration
_____ 20. severe bicarbonate loss

s. strong acid

t. metabolic generation of water

21. A person is in acid–base balance when
    (a) the production of hydronium ions is precisely offset by their loss
    (b) the production of hydrogen ions is precisely offset by their loss
    (c) the amount of water gained each day is equal to the amount lost to the environment
    (d) a, b, and c are correct

22. The primary components of the extracellular fluid are
    (a) lymph and cerebrospinal fluid
    (b) plasma and serous fluids
    (c) interstitial fluid and plasma
    (d) a, b, and c are correct

23. The principal anions in the ICF are
    (a) phosphate and chloride
    (b) phosphate and bicarbonate
    (c) sodium and chloride
    (d) sodium and potassium

24. All the homeostatic mechanisms that monitor and adjust the composition of body fluids respond to changes
    (a) in the ICF
    (b) in the ECF
    (c) inside the cell
    (d) a, b, and c are correct

25. Osmoreceptors in the hypothalamus monitor the osmotic concentration of the ECF and secrete _____ in response to higher osmotic concentrations.
    (a) BNP
    (b) ANP
    (c) aldosterone
    (d) ADH

26. When the body water content rises enough to dilute the sodium concentration in the ECF, the body is said to be in a state of
    (a) hypernatremia
    (b) hyponatremia
    (c) osmotic loss
    (d) hypocalcemia

27. The most common problems with electrolyte balance are caused by an imbalance between gains and losses of
    (a) calcium ions
    (b) chloride ions
    (c) potassium ions
    (d) sodium ions

28. The higher the plasma concentration of aldosterone, the more efficiently the kidney will
    (a) conserve sodium ions
    (b) retain potassium ions
    (c) stimulate urinary water loss
    (d) secrete greater amounts of ADH

29. Angiotensin II produces a coordinated elevation in the ECF volume by
    (a) stimulating thirst
    (b) causing the release of ADH
    (c) triggering the production and secretion of aldosterone
    (d) a, b, and c are correct

30. The rate of tubular secretion of potassium ions changes in response to
    (a) changes in pH
    (b) changes in aldosterone levels
    (c) changes in the potassium ion concentration in the ECF
    (d) a, b, and c are correct

31. Respiratory acidosis develops when the plasma pH is
    (a) elevated due to a decreased plasma P_{CO_2} level
    (b) decreased due to an elevated plasma P_{CO_2} level
    (c) elevated due to an elevated plasma P_{CO_2} level
    (d) decreased due to a decreased plasma P_{CO_2} level

32. Metabolic alkalosis occurs when
    (a) bicarbonate ion concentrations become elevated
    (b) there is a severe bicarbonate loss
    (c) the kidneys fail to excrete hydrogen ions
    (d) ketone bodies are generated in abnormally large quantities

33. Identify four hormones which mediate major physiological adjustments that affect fluid and electrolyte balance. What are the primary effects of each hormone?

34. What influence does aldosterone have on the relationship between sodium and potassium?

LEVEL 2 Reviewing Concepts

35. Drinking a hypotonic solution causes the ECF to
    (a) increase in volume and become hypertonic with respect to the ICF
    (b) decrease in volume and become hypertonic with respect to the ICF

    (c) decrease in volume and become hypotonic with respect to the ICF
    (d) increase in volume and become hypotonic with respect to the ICF

36. The osmotic concentration of the ECF decreases if the individual gains water without a corresponding
    (a) gain of electrolytes
    (b) loss of water
    (c) fluid shift from the ECF to the ICF
    (d) a, b, and c are correct

37. Renal failure can result in
    (a) hypernatremia
    (b) hyponatremia
    (c) hyperkalemia
    (d) hypokalemia

38. When the blood pH climbs above 7.35 as a result of increasing bicarbonate ions, the resulting physiological state is
    (a) metabolic acidosis
    (b) metabolic alkalosis
    (c) respiratory acidosis
    (d) respiratory alkalosis

39. When the plasma concentration of potassium decreases to a dangerous level, the result is
    (a) kidney failure
    (b) extensive muscular weakness, followed by paralysis
    (c) decreased blood-clotting capability
    (d) the cessation of enzyme activity in cells

40. In a protein buffer system, if the pH rises,
    (a) the protein acquires a hydrogen ion from carbonic acid
    (b) hydrogen ions are buffered by hemoglobin molecules
    (c) a hydrogen ion is released and a carboxyl ion is formed
    (d) a chloride shift occurs

41. Increasing or decreasing the rate of respiration alters the pH by
    (a) lowering or raising the partial pressure of CO_2
    (b) lowering or raising the partial pressure of O_2
    (c) lowering or raising the partial pressure of N_2
    (d) a, b, and c are correct

42. Differentiate among fluid balance, electrolyte balance, and acid–base balance, and explain why each is important to homeostasis.

43. What are fluid shifts? What is their function, and what factors can cause them?

44. A sample of plasma contains chloride ions at a concentration of 250 mg/dl. How many mmol/l of chloride ions are in the sample?

45. Why should a person with a fever drink plenty of fluids?

46. Describe the effects of losses and gains of sodium on the ECF.

47. Define and give an example of (a) a volatile acid, (b) a fixed acid, and (c) an organic acid. Which represent(s) the greatest threat to acid–base balance? Why?

48. What are the three major buffer systems in body fluids? How does each system work?

49. How do respiratory and renal mechanisms support the buffer systems?

50. Differentiate between respiratory compensation and renal compensation.

51. Distinguish between respiratory and metabolic disorders that disturb acid–base balance.

52. What is the difference between metabolic acidosis and respiratory acidosis? What can cause these conditions?

53. The most recent advice from medical and nutritional experts is to decrease one's intake of salt so that it does not exceed the amount needed to maintain a constant ECF composition. What effect does excessive salt and water ingestion have on (**a**) urine volume, (**b**) urine concentration, and (**c**) blood pressure?

54. Exercise physiologists recommend that adequate amounts of fluid be ingested before, during, and after exercise. Why is fluid replacement during extensive sweating important?

LEVEL 3 Critical Thinking and Clinical Applications

55. After falling into an abandoned stone quarry filled with water and nearly drowning, a young boy is rescued. His rescuers assess his condition. They find that his body fluids have high P_{CO_2} and low P_{O_2} levels and that large amounts of lactic acid were generated by the boy's muscles as he struggled in the water. As a clinician, diagnose the boy's condition and recommend the necessary treatment to restore his body to homeostasis.

56. A person who is hyperventilating and disoriented is brought into the emergency room. Blood analysis indicates that the person has elevated levels of potassium (hyperkalemia). Is the most likely cause acidosis or alkalosis? Explain.

57. Willy has chronic emphysema. A blood analysis indicates that his plasma pH is 7.4, although his plasma bicarbonate level is significantly elevated. How can this be?

58. While visiting a foreign country, Milly inadvertently drinks some water, even though she had been advised not to. She contracts an intestinal disease that causes severe diarrhea. How would you expect her condition to affect her blood pH, urine pH, and pattern of ventilation?

59. Yuka is dehydrated, so her physician prescribes intravenous fluids. The attending nurse becomes distracted and erroneously gives Yuka a hypertonic glucose solution instead of normal saline. What effect will this mistake have on Yuka's plasma levels of ADH and urine volume?

60. Refer to the diagnostic flowchart in Figure 27–15●. Use information from the blood test results in the accompanying table to categorize the acid–base disorders that affect the patients represented in the table.

Results	Patient 1	Patient 2	Patient 3	Patient 4
pH	7.5	7.2	7.0	7.7
P_{CO_2}	32	45	60	50
Na^+	138	140	140	136
HCO_3^-	22	20	28	34
Cl^-	106	102	101	91
Anion gap*	10	18	12	11

* Anion gap = Na^+ concentration − (HCO_3^- concentration + Cl^- concentration).
Refer to the *Applications Manual* for additional information. **AM** Diagnostic Classification of Acid–Base Disorders

UNIT 6   CHAPTER   28   29

28

THE REPRODUCTIVE SYSTEM

The reproductive system is the only system that is not essential to the life of the individual (although its activities do affect other systems). Instead, this system ensures the continued existence of the human species. Many primitive societies failed to discover the basic link between sexual activity and childbirth, and assumed that cosmic forces were responsible for producing new individuals. Although our society has a much clearer view of the reproductive process, a sense of wonder remains. Sexually mature males and females produce individual reproductive cells that are brought together by the sex act. The fusion of these cells starts a chain of events that leads to the appearance of an infant who will mature as part of the next generation.

In this and the next chapter, we shall consider the mechanics of this remarkable process. We begin here by examining the anatomy and physiology of the reproductive system. The human reproductive system produces, stores, nourishes, and transports functional male and female reproductive cells, or **gametes** (GAM-ēts). Chapter 29 starts with **fertilization**, also known as *conception*, in which a male gamete and female gamete unite. All the cells in a human body are the mitotic descendants of a **zygote** (ZĪ-gōt), the single cell created by the fusion of a gamete from the father and a gamete from the mother. The gradual transformation of that single cell into a functional adult, over a span of 15–20 years, is the process of *development*. During this period, hormones produced by the reproductive system direct gender-specific patterns of development.

28–1 INTRODUCTION TO THE REPRODUCTIVE SYSTEM

Objective

- Specify the principal components of the human reproductive system, and summarize their functions.

The **reproductive system** includes the following components:

- **Gonads** (GŌ-nadz), or reproductive organs that produce gametes and hormones.
- Ducts that receive and transport the gametes.
- Accessory glands and organs that secrete fluids into the ducts of the reproductive system or into other excretory ducts.
- Perineal structures that are collectively known as the **external genitalia** (jen-i-TĀ-lē-uh).

In both males and females, the ducts are connected to chambers and passageways that open to the exterior of the body. The structures involved constitute the *reproductive tract*. The male and female reproductive systems are functionally quite different, however. In adult males, the **testes** (TES-tēz; singular, *testis*), or male gonads, secrete sex hormones called *androgens*, (principally *testosterone*, which we introduced, together with other sex hormones, in Chapter 18). ∞ p. 637 The testes also produce the male gametes, called **spermatozoa** (sper-ma-tō-ZŌ-uh; singular, *spermatozoon*), or *sperm*—one-half billion each day. During *emission*, mature spermatozoa travel along a lengthy duct system, where they are mixed with the secretions of accessory glands. The mixture created is known as **semen** (SĒ-men). During *ejaculation*, semen is expelled from the body.

In adult females, the **ovaries**, or female gonads, typically release only one immature gamete, an **oocyte**, per month. This immature gamete travels along short *uterine tubes*, which end in the muscular organ called the *uterus* (Ū-ter-us). If a sperm reaches the oocyte and initiates the process of fertilization, the oocyte matures into an **ovum** (plural, *ova*). A short passageway, the *vagina* (va-JĪ-nuh), connects the uterus with the exterior. During *sexual intercourse*, ejaculation introduces semen into the vagina, and the spermatozoa then ascend the female reproductive tract. If fertilization occurs, the uterus will enclose and support a developing *embryo* as the embryo grows into a *fetus* and prepares for birth.

We shall now examine the anatomy of the male and female reproductive systems further and shall consider the physiological and hormonal mechanisms responsible for the regulation of reproductive function. In earlier chapters, we introduced the anatomical reference points used in the discussions that follow; you may find it helpful to review the figures on the pelvic girdle (Figures 8–8 and 8–10●, pp. 251, 252), perineal musculature (Figure 11–13●, p. 357), pelvic innervation (Figure 13–12●, p. 442), and regional blood supply (Figures 21–28 and 21–33●, pp. 760, 766).

28–2 THE REPRODUCTIVE SYSTEM OF THE MALE

Objectives

- Describe the components of the male reproductive system.
- Outline the processes of meiosis and spermatogenesis in the testes.
- Explain the roles played by the male reproductive tract and accessory glands in the functional maturation, nourishment, storage, and transport of spermatozoa.
- Specify the normal composition of semen.
- Summarize the hormonal mechanisms that regulate male reproductive functions.

The principal structures of the male reproductive system are shown in Figure 28–1●. Proceeding from a testis, the spermatozoa travel within the *epididymis* (ep-i-DID-i-mus); the *ductus deferens* (DUK-tus DEF-e-renz), or *vas deferens*; the *ejaculatory duct*; and the *urethra* before leaving the body. Accessory organs—the *seminal* (SEM-i-nal) *vesicles*, the *prostate* (PROS-tāt) *gland*, and the *bulbourethral* (bul-bō-ū-RĒ-thral) *glands*—secrete various fluids into the ejaculatory ducts and urethra. The external genitalia consist of the *scrotum* (SKRŌ-tum), which encloses the testes, and the *penis* (PĒ-nis), an erectile organ through which the distal portion of the urethra passes.

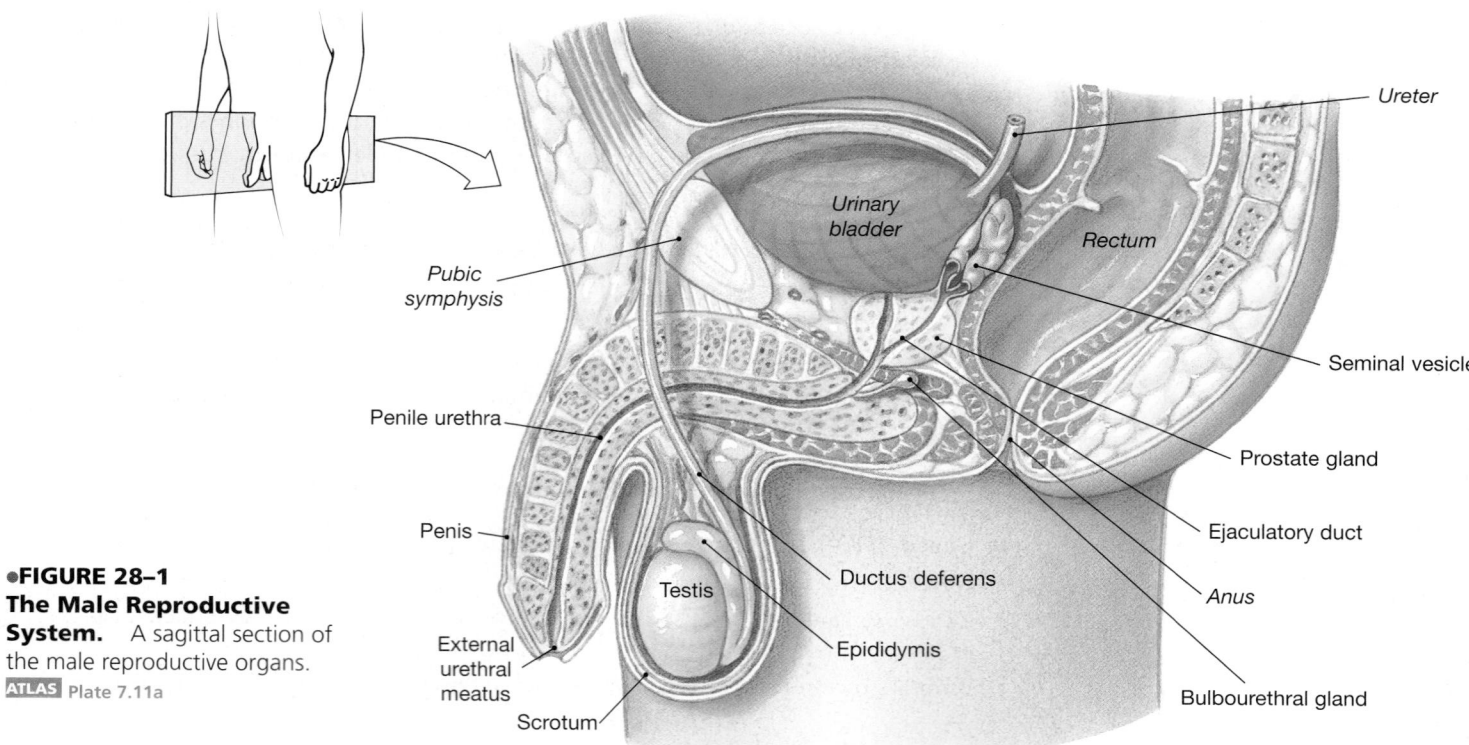

●**FIGURE 28–1**
The Male Reproductive System. A sagittal section of the male reproductive organs.
ATLAS Plate 7.11a

Ureter

Urinary bladder

Rectum

Pubic symphysis

Seminal vesicle

Penile urethra

Prostate gland

Penis

Ejaculatory duct

Ductus deferens

Anus

External urethral meatus

Testis

Epididymis

Scrotum

Bulbourethral gland

●FIGURE 28–2
The Descent of the Testes.
(a) Sagittal sectional views of the positional changes involved in the descent of the testes. Because the size of the gubernaculum testis remains constant (see the scale bar at the left) while the rest of the fetus grows, the relative position of the testis shifts. **(b)** Frontal views showing the descent of the testes and the formation of the spermatic cords.

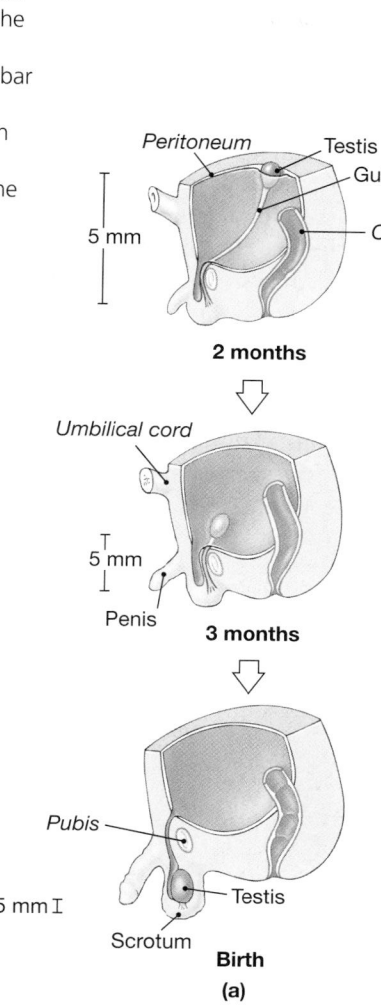

(a)

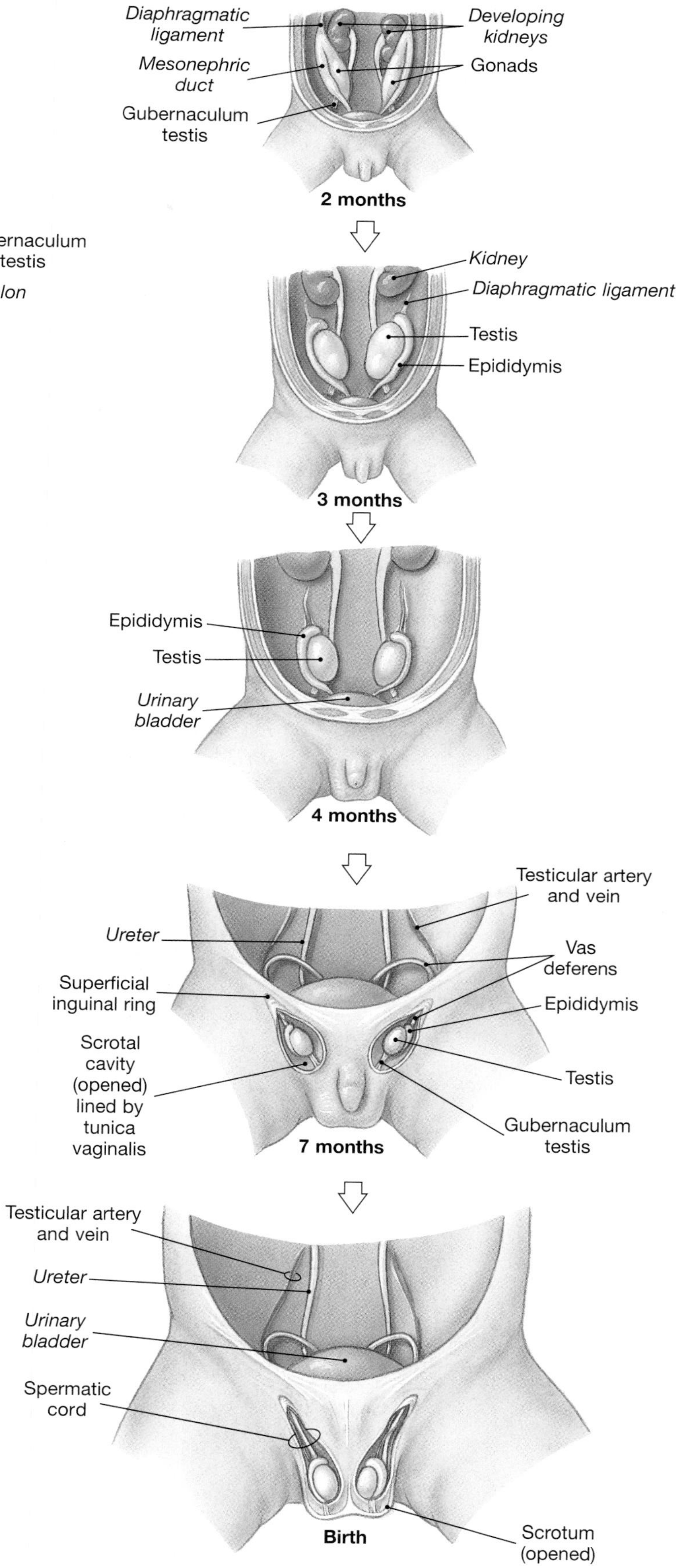

(b)

□ THE TESTES

Each testis has the shape of a flattened egg that is roughly 5 cm (2 in.) long, 3 cm (1.2 in.) wide, and 2.5 cm (1 in.) thick. Each has a weight of 10–15 g (0.35–0.53 oz). The testes hang within the **scrotum**, a fleshy pouch suspended inferior to the perineum, anterior to the anus and posterior to the base of the penis (Figure 28–1●).

AM EMBRYOLOGY SUMMARY 21: The Development of the Reproductive System

Descent of the Testes

During development of the fetus, the testes form inside the body cavity adjacent to the kidneys. A bundle of connective-tissue fibers extends from each testis to the posterior wall of a small anterior and inferior pocket of the peritoneum. These fibers constitute the **gubernaculum testis**. As the fetus grows, the gubernacula do not get any longer, so they lock the testes in position. As a result, the relative position of each testis changes as the rest of the body enlarges: The testis gradually moves inferiorly and anteriorly toward the anterior abdominal wall (Figure 28–2a●). During the seventh developmental month, fetal growth continues at a rapid pace, and circulating hormones stimulate a contraction of the gubernaculum testis. Over this period, each testis moves through the abdominal musculature, accompanied by small pockets of the peritoneal cavity. This process is called the **descent of the testes**.

In **cryptorchidism** (kript-OR-ki-dizm; *crypto*, hidden), one or both of the testes have not descended into the scrotum by the time of birth. Typically, the cryptorchid testes are lodged in the abdominal cavity or within the inguinal canal. Cryptorchidism occurs in about 3 percent of full-term deliveries and in roughly 30 percent of premature births. In most instances, normal descent occurs a few weeks later, but the condition can be surgically corrected if it persists. Corrective measures should be taken before *puberty* (sexual maturation), because cryptorchid (abdominal) testes will not produce spermatozoa. Thus, the individual will be *sterile* (*infertile*), or unable to bear children. If the testes cannot be moved into the scrotum, in most cases they will be removed, because about 10 percent of males with uncorrected cryptorchid testes eventually develop testicular cancer. This surgical procedure is called an *orchiectomy* (or-kē-EK-to-mē; *orchis*, testis).

As it moves through the body wall, each testis is accompanied by the ductus deferens and the testicular blood vessels, nerves, and lymphatic vessels. Together, these structures form the body of the spermatic cord, which we will discuss next.

The Spermatic Cords

The **spermatic cords** are paired structures extending between the abdominopelvic cavity and the testes. Each spermatic cord consists of layers of fascia and muscle enclosing the ductus deferens and the blood vessels, nerves, and lymphatic vessels that supply the testes. The blood vessels include the *deferential artery*, a *testicular artery*, and the **pampiniform** (pam-PIN-i-form; *pampinus*, tendril + *forma*, form) **plexus** of a testicular vein. Branches of the *genitofemoral nerve* from the lumbar plexus provide innervation. Each spermatic cord begins at the *deep inguinal ring*, the entrance to the *inguinal canal* (a passageway through the abdominal musculature). ∞ p. 356 After passing through the inguinal canal, the spermatic cord exits at the *superficial inguinal ring* and descends into the scrotum (Figure 28–3●).

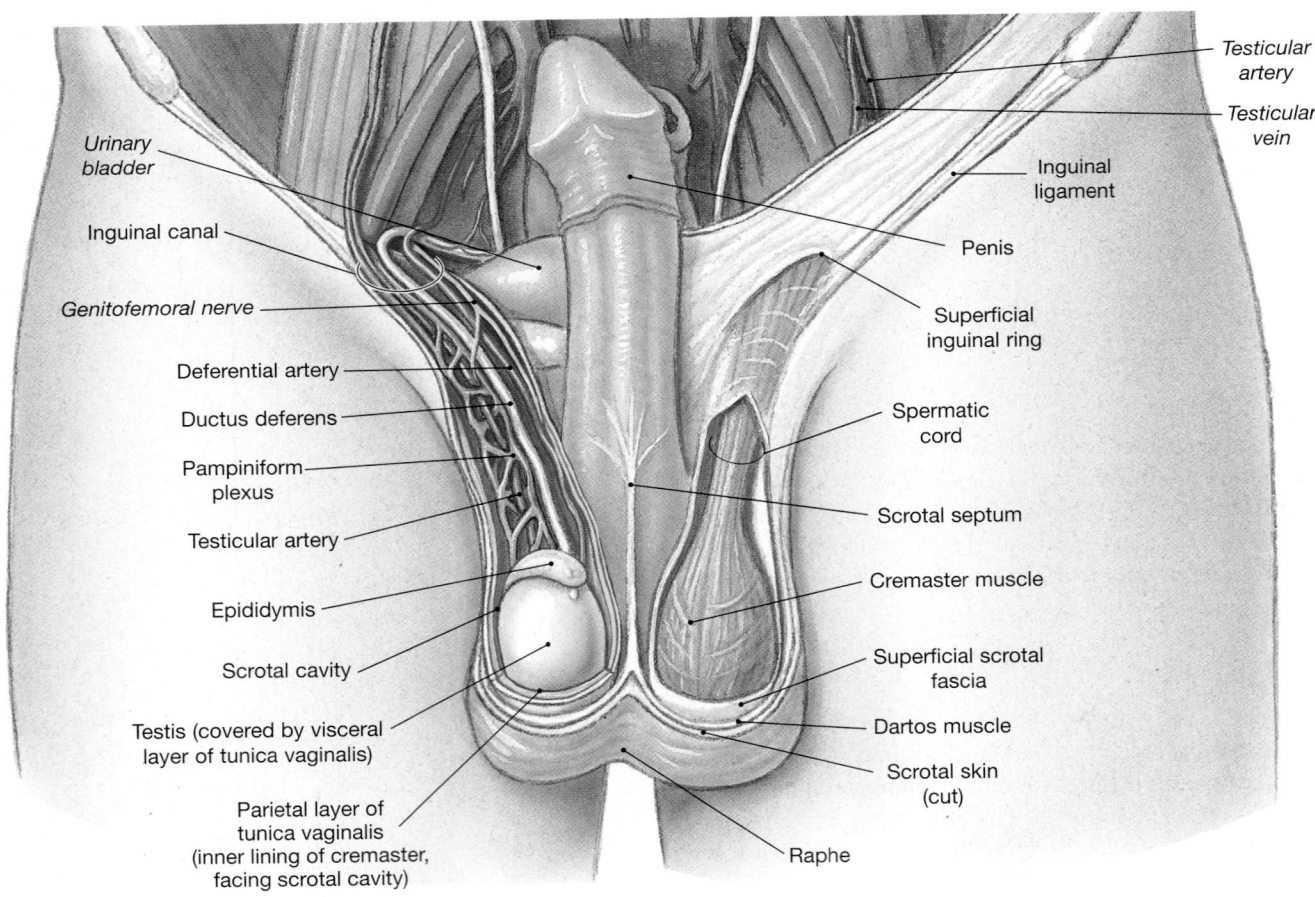

Urinary bladder

Inguinal canal

Genitofemoral nerve

Deferential artery

Ductus deferens

Pampiniform plexus

Testicular artery

Epididymis

Scrotal cavity

Testis (covered by visceral layer of tunica vaginalis)

Parietal layer of tunica vaginalis (inner lining of cremaster, facing scrotal cavity)

Testicular artery

Testicular vein

Inguinal ligament

Penis

Superficial inguinal ring

Spermatic cord

Scrotal septum

Cremaster muscle

Superficial scrotal fascia

Dartos muscle

Scrotal skin (cut)

Raphe

●**FIGURE 28–3**
The Male Reproductive System in Anterior View

The inguinal canals form during development as the testes descend into the scrotum; at that time, these canals link the scrotal cavities with the peritoneal cavity. In normal adult males, the inguinal canals are closed, but the presence of the spermatic cords creates weak points in the abdominal wall that remain throughout life. As a result, *inguinal hernias*—protrusions of a portion of the intestine into the inguinal canal—are relatively common in males. ∞ p. 356 The inguinal canals in females are very small, containing only the *ilioinguinal nerves* and the *round ligaments* of the uterus. The abdominal wall is nearly intact, so inguinal hernias in women are very rare.

The Scrotum and the Position of the Testes

The scrotum is divided internally into two chambers. The partition between the two is marked by a raised thickening in the scrotal surface known as the **raphe** (RĀ-fē) (Figure 28–3●). Each testis lies in a separate chamber, or **scrotal cavity**. Because the scrotal cavities are separated by a partition, infection or inflammation of one testis does not ordinarily spread to the other. A narrow space separates the inner surface of the scrotum from the outer surface of the testis. The **tunica vaginalis** (TOO-ni-ka vaj-i-NAL-is), a serous membrane, lines the scrotal cavity and reduces friction between the opposing parietal (scrotal) and visceral (testicular) surfaces. The tunica vaginalis is an isolated portion of the peritoneum that lost its connection with the peritoneal cavity after the testes descended, when the inguinal canal closed.

The scrotum consists of a thin layer of skin and the underlying superficial fascia. The dermis contains a layer of smooth muscle, the **dartos** (DAR-tōs) **muscle**. Resting muscle tone in the dartos muscle causes the characteristic wrinkling of the scrotal surface. A layer of skeletal muscle, the **cremaster** (krē-MAS-ter) **muscle**, lies deep to the dermis. Contraction of the cremaster during sexual arousal or in response to changes in temperature tenses the scrotum and pulls the testes closer to the body. Normal development of spermatozoa in the testes requires temperatures about 1.1°C (2°F) lower than those elsewhere in the body. The cremaster muscle relaxes or contracts to move the testes away from or toward the body as needed to maintain acceptable testicular temperatures. When air or body temperature rises, the cremaster muscle relaxes and the testes move away from the body. Sudden cooling of the scrotum, as occurs when a man enters a cold swimming pool, results in cremasteric contractions that pull the testes closer to the body and keep testicular temperatures from falling.

Structure of the Testes

Deep to the tunica vaginalis covering the testis is the **tunica albuginea** (al-bū-JIN-ē-uh), a dense layer of connective tissue rich in collagen fibers. These fibers are continuous with those surrounding the adjacent epididymis and extend into the substance of the testis. There they form fibrous partitions, or *septa* (Figure 28–4●),

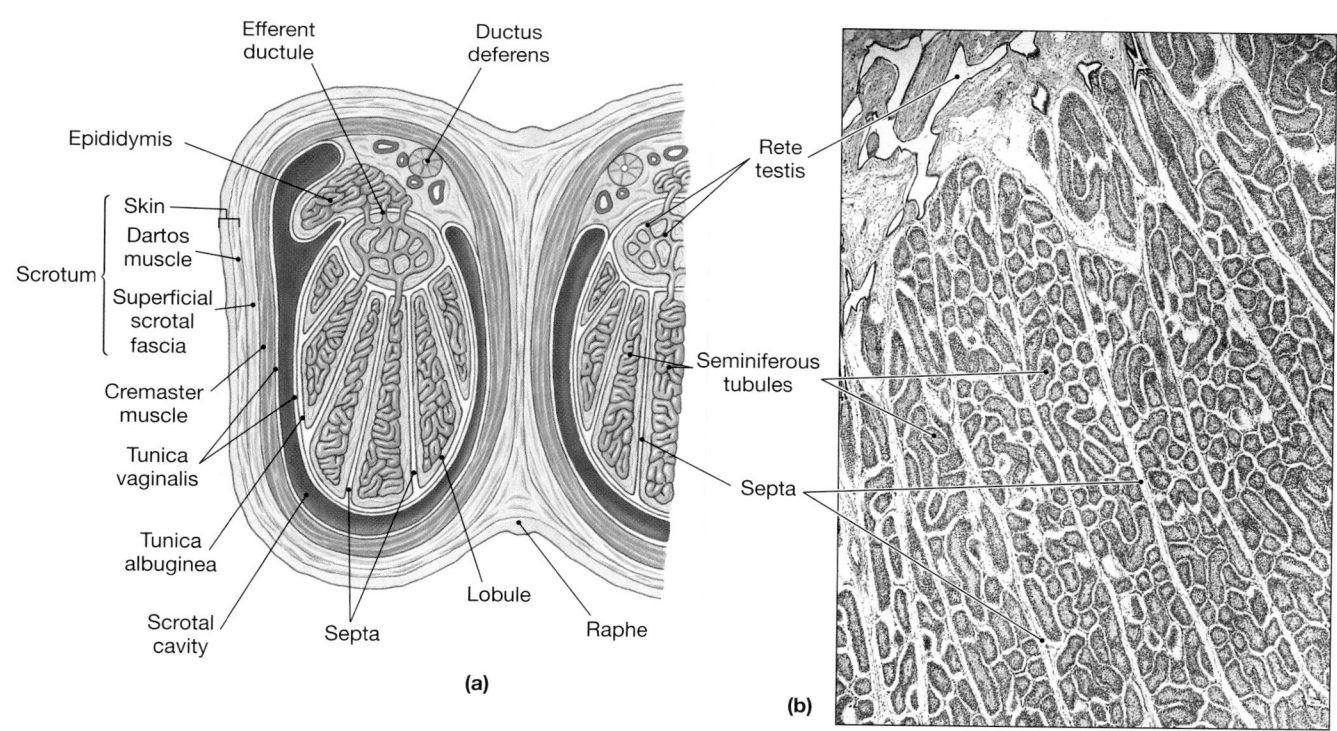

Efferent ductule
Ductus deferens
Epididymis
Rete testis
Skin
Dartos muscle
Scrotum
Superficial scrotal fascia
Cremaster muscle
Seminiferous tubules
Tunica vaginalis
Tunica albuginea
Septa
Lobule
Scrotal cavity
Septa
Raphe

(a)

(b)

●**FIGURE 28–4**
The Structure of the Testes. **(a)** A frontal section. **(b)** A section through a testis. (LM × 26)

that converge toward the region nearest the entrance to the epididymis. The connective tissues in this region support the blood vessels and lymphatic vessels that supply the testis and the *efferent ductules*, which transport spermatozoa to the epididymis.

Histology of the Testes

The septa subdivide the testis into a series of **lobules**. Roughly 800 slender, tightly coiled **seminiferous** (se-mi-NIF-e-rus) **tubules** are distributed among the lobules (Figures 28–4 and 28–5●). Each

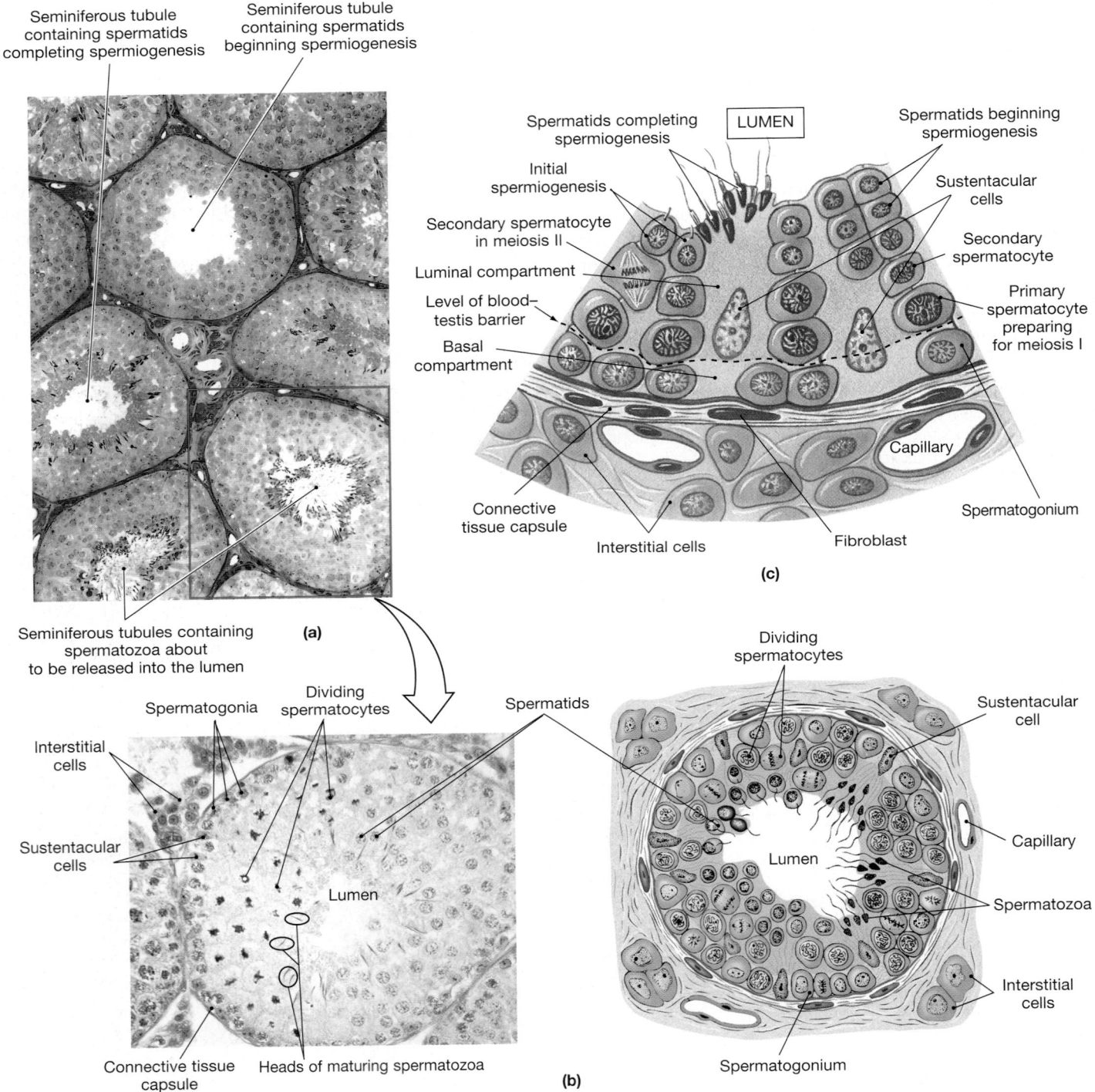

●**FIGURE 28–5**
The Seminiferous Tubules.   (a) A transverse section through a coiled seminiferous tubule. (b) A cross section through a single tubule. (LM × 786) (c) The lining of a tubule. Sustentacular cells surround the stem cells of the tubule and support the developing spermatocytes and spermatids.

tubule averages about 80 cm (31 in.) in length, and a typical testis contains nearly one-half mile of seminiferous tubules. Sperm production occurs within these tubules.

Each seminiferous tubule forms a loop that is connected to a maze of passageways known as the **rete** (RĒ-tē; *rete,* a net) **testis** (Figure 28–4●). Fifteen to 20 large **efferent ductules** connect the rete testis to the epididymis.

Because the seminiferous tubules are tightly coiled, most histological preparations show them in transverse section (Figure 28–5a●). Each tubule is surrounded by a delicate capsule, and areolar tissue fills the spaces between the tubules. Within those spaces are numerous blood vessels and large **interstitial cells** (*cells of Leydig*) (Figure 28–5b●). Interstitial cells are responsible for the production of *androgens,* the dominant sex hormones in males. *Testosterone* is the most important androgen.

Spermatozoa are produced by the process of **spermatogenesis** (sper-ma-tō-JEN-e-sis). Spermatogenesis begins at the outermost layer of cells in the seminiferous tubules and proceeds toward the lumen (Figure 28–5b,c●). At each step in this process, the daughter cells move closer to the lumen. First, stem cells called **spermatogonia** (sper-ma-tō-GŌ-nē-uh) divide by mitosis to produce daughter cells, some of which differentiate into primary spermatocytes. **Primary spermatocytes** (sper-MA-tō-sīts) are the cells that begin *meiosis,* a specialized form of cell division involved only in the production of gametes (spermatozoa in males, ova in females). Spermatocytes give rise to **spermatids** (SPER-ma-tidz)—immature gametes that subsequently differentiate into spermatozoa. The spermatozoa lose contact with the wall of the seminiferous tubule and enter the fluid in the lumen.

Each seminiferous tubule contains spermatogonia, spermatocytes at various stages of meiosis, spermatids, spermatozoa, and large **sustentacular** (sus-ten-TAK-ū-lar) **cells** (or *Sertoli cells*). Sustentacular cells are attached to the tubular capsule and extend toward the lumen between the other types of cells (Figure 28–5b,c●).

☐ SPERMATOGENESIS

Spermatogenesis involves three integrated processes:

1. *Mitosis.* Spermatogonia undergo cell divisions throughout adult life. (You may wish to review the description of mitosis and cell division in Chapter 3. ∞ p. 102) One daughter cell from each division is pushed toward the lumen of the seminiferous tubule. These cells differentiate into primary spermatocytes, which prepare to begin meiosis.

2. *Meiosis.* **Meiosis** (mī-Ō-sis) is a special form of cell division involved in gamete production. Gametes contain 23 chromosomes, half the normal set. As a result, the fusion of the nuclei of a male gamete and a female gamete produces a cell that has the normal number of chromosomes (46), rather than twice that number. In the seminiferous tubules, meiotic divisions that begin with primary spermatocytes produce spermatids, the undifferentiated male gametes.

3. *Spermiogenesis.* Spermatids are small, relatively unspecialized cells. In **spermiogenesis**, spermatids differentiate into physically mature spermatozoa, which are among the most highly specialized cells in the body. Spermiogenesis involves major changes in a spermatid's internal and external structures.

Mitosis and Meiosis

In both males and females, mitosis and meiosis differ significantly in terms of the events that take place in the nucleus. Mitosis is part of the process of somatic cell division, which produces two daughter cells, each containing 23 *pairs* of chromosomes (Figure 28–6a●). Each pair consists of one chromosome provided by the father and another by the mother at the time of fertilization. Because the daughter cells contain both members of each chromosome pair (for a total of 46 chromosomes), they are called **diploid** (DIP-loyd; *diplo,* double) cells. Meiosis involves two cycles of cell division (*meiosis I* and *meiosis II*) and produces four cells, each of which contains 23 individual chromosomes (Figure 28–6b●). Because these cells contain only one member of each pair of chromosomes, they are called **haploid** (HAP-loyd; *haplo,* single) cells. The events in the nucleus shown in Figure 28–6b● are the same whether you consider the formation of spermatozoa or ova.

As a cell prepares to begin meiosis, DNA replication occurs within the nucleus just as it does in a cell preparing to undergo mitosis. This similarity continues as prophase I arrives; the chromosomes condense and become visible with a light microscope. As in mitosis, each chromosome consists of two duplicate *chromatids.*

At this point, the close similarities between meiosis and mitosis end. In meiosis, the corresponding maternal and paternal chromosomes now come together, an event known as **synapsis** (sin-AP-sis). Synapsis involves 23 pairs of chromosomes; each member of each pair consists of two chromatids. A matched set of four chromatids is called a **tetrad** (TET-rad; *tetras,* four). Some exchange of genetic material can occur between the chromatids of a chromosome pair at this stage of meiosis. Such an exchange, called *crossing-over,* increases genetic variation among offspring; we shall discuss it in Chapter 29.

Meiosis includes two division cycles, referred to as meiosis I and meiosis II. The stages within each phase are identified similarly (as prophase I, metaphase II, etc.). The nuclear envelope disappears at the end of prophase I. As metaphase I begins, the tetrads line up along the metaphase plate. As anaphase I begins, the tetrads break up—the maternal and paternal chromosomes separate. This is a major difference between mitosis and meiosis: In mitosis, each daughter cell receives one of the two copies of every chromosome, maternal and paternal; in meiosis I, each daughter cell receives both copies of *either* the maternal chromosome *or* the paternal chromosome from each tetrad. (Compare the two parts of Figure 28–6●.)

As anaphase proceeds, the maternal and paternal components are randomly and independently distributed. That is, as each tetrad splits, you cannot predict which daughter cell will receive copies of the maternal chromosome rather than copies of the

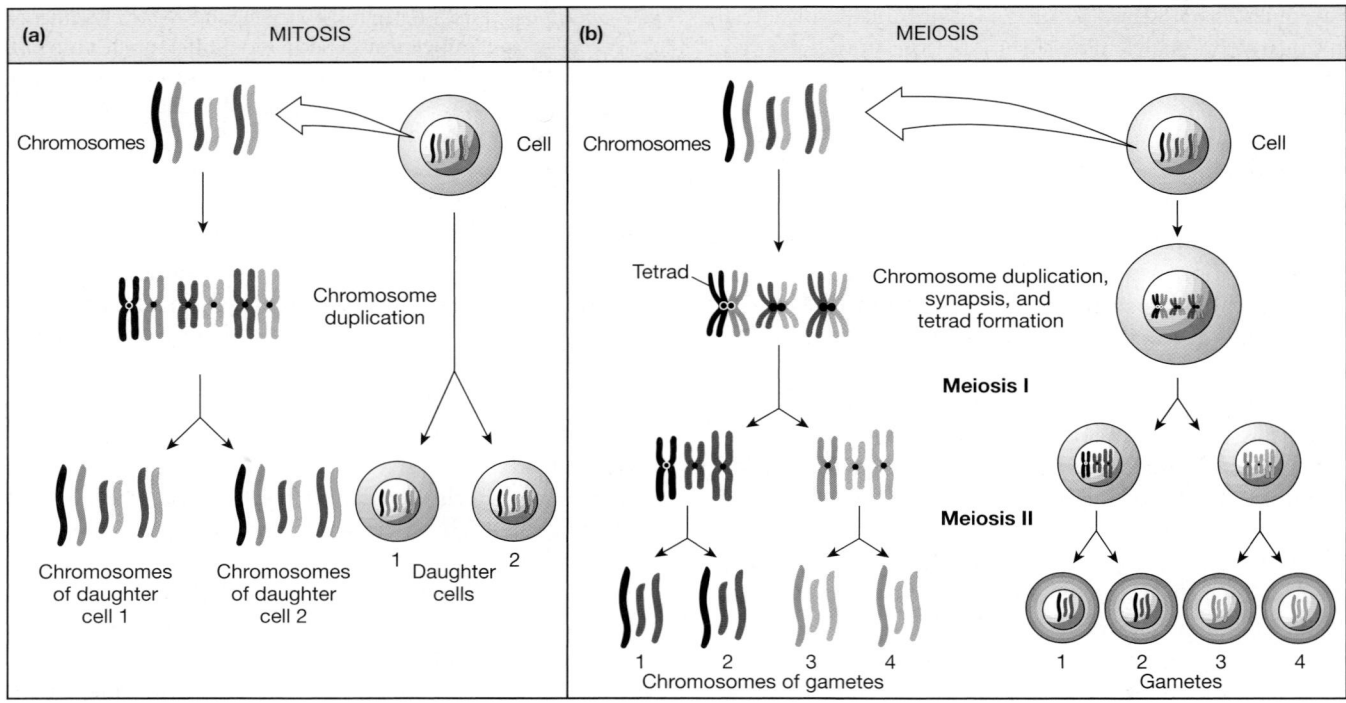

•**FIGURE 28–6**
Chromosomes in Mitosis and Meiosis. (a) Steps in mitosis. (*See Figure 3–29, pp. 102–103.*) (b) Steps in meiosis.

paternal chromosome. As a result, telophase I ends with the formation of two daughter cells containing unique combinations of maternal and paternal chromosomes. Both cells contain 23 chromosomes. Because the first meiotic division reduces the number of chromosomes from 46 to 23, it is called a **reductional division**. Each of these chromosomes still consists of two duplicate chromatids. The duplicates will separate during **meiosis II**.

The interphase separating meiosis I and meiosis II is very brief, and no DNA is replicated during that period. Each cell proceeds through prophase II, metaphase II, and anaphase II. During anaphase II, the duplicate chromatids separate. Telophase II thus yields *four cells*, each containing 23 chromosomes. Because the number of chromosomes has not changed, meiosis II is an **equational division**.

In males, the mitotic division of a spermatogonium produces two daughter cells. One is a spermatogonium that remains in contact with the basement membrane, and the other is a primary spermatocyte that is displaced toward the lumen (Figure 28–7•). As meiosis begins, each primary spermatocyte contains 46 individual chromosomes. At the end of meiosis I, the daughter cells are called **secondary spermatocytes**. Every secondary spermatocyte contains 23 chromosomes, each of which consists of a pair of duplicate chromatids. The secondary spermatocytes soon enter prophase II. The completion of metaphase II, anaphase II, and telophase II yields four spermatids, each containing 23 chromosomes.

For each primary spermatocyte that enters meiosis, four spermatids are produced. Because cytokinesis (cytoplasmic division) is not completed in meiosis I or meiosis II, the four spermatids initially remain interconnected by bridges of cytoplasm. These connections assist in the transfer of nutrients and hormones between the cells, thus helping ensure that the cells develop in synchrony. The bridges are not broken until the last stages of physical maturation.

Spermiogenesis

In **spermiogenesis**, the last step of spermatogenesis, each spermatid matures into a single spermatozoon, or sperm (Figure 28–7•). Developing spermatocytes undergoing meiosis and spermatids undergoing spermiogenesis are not free in the seminiferous tubules. Instead, they are surrounded by the cytoplasm of the sustentacular cells. As spermiogenesis proceeds, the spermatids gradually develop the appearance of mature spermatozoa. At *spermiation*, a spermatozoon loses its attachment to the sustentacular cell and enters the lumen of the seminiferous tubule. The entire process, from spermatogonial division to spermiation, takes approximately nine weeks.

SUSTENTACULAR CELLS Sustentacular cells play a key role in spermatogenesis. These cells have six important functions that directly or indirectly affect mitosis, meiosis, and spermiogenesis within the seminiferous tubules:

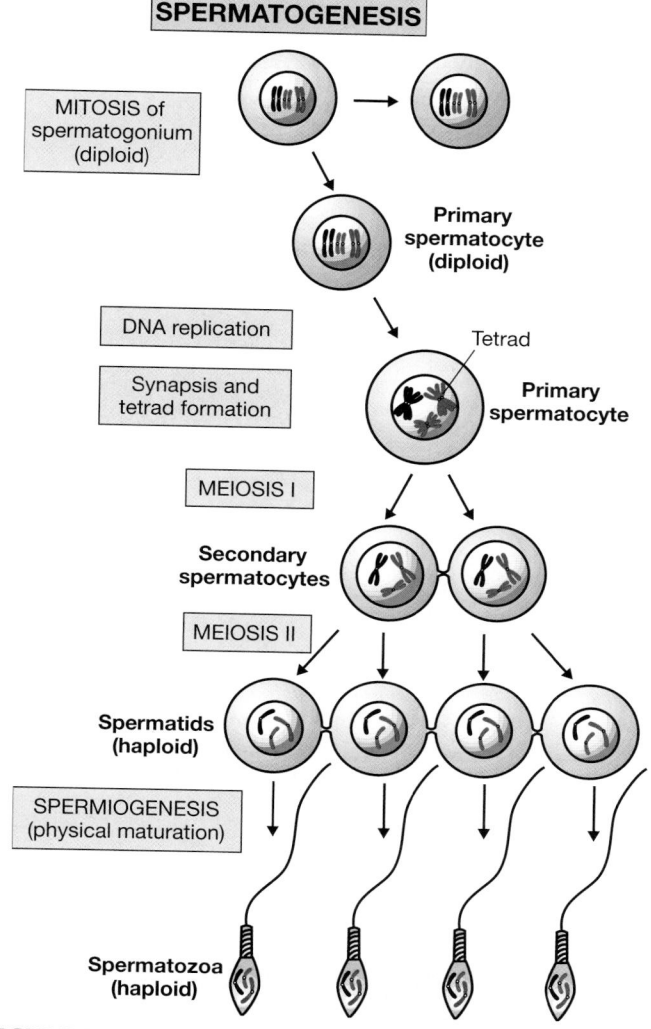

SPERMATOGENESIS

MITOSIS of spermatogonium (diploid)

Primary spermatocyte (diploid)

DNA replication

Synapsis and tetrad formation

Tetrad

Primary spermatocyte

MEIOSIS I

Secondary spermatocytes

MEIOSIS II

Spermatids (haploid)

SPERMIOGENESIS (physical maturation)

Spermatozoa (haploid)

●**FIGURE 28–7**
Spermatogenesis. Meiosis in the seminiferous tubules, showing the distribution of only a few chromosomes.

fluid and the interstitial fluid. In addition, this barrier prevents the cells of the immune system from reaching the developing spermatozoa, which contain sperm-specific antigens in their cell membranes not found in somatic cell membranes. Without some means of isolation from the white blood cells and the immune system, the developing spermatozoa would be detected as foreign and would be attacked and destroyed.

2. *The Support of Mitosis and Meiosis.* Spermatogenesis depends on the stimulation of sustentacular cells by circulating follicle-stimulating hormone (FSH) and testosterone. Stimulated sustentacular cells then promote the division of spermatogonia and the meiotic divisions of spermatocytes.

3. *The Support of Spermiogenesis.* Spermiogenesis requires the presence of sustentacular cells. These cells surround and enfold the spermatids, providing nutrients and chemical stimuli that promote their development.

4. *The Secretion of Inhibin.* Sustentacular cells secrete *inhibin* (in-HIB-in), a peptide hormone, in response to factors released by developing spermatozoa. Inhibin depresses the pituitary production of FSH and perhaps the hypothalamic secretion of gonadotropin-releasing hormone (GnRH). The faster the rate of sperm production, the more inhibin is secreted. By regulating FSH and GnRH secretion, sustentacular cells provide feedback control of spermatogenesis.

5. *The Secretion of Androgen-Binding Protein.* *Androgen-binding protein (ABP)* binds androgens (primarily testosterone) in the fluid contents of the seminiferous tubules. This protein is thought to be important in elevating the concentration of androgens within the seminiferous tubules and in stimulating spermiogenesis. The production of ABP is stimulated by FSH.

6. *The Secretion of Müllerian-Inhibiting Factor.* *Müllerian-inhibiting factor (MIF)* is secreted by sustentacular cells in the developing testes. This hormone causes regression of the fetal *Müllerian ducts*, passageways that participate in the formation of the uterine tubes and the uterus in females. In males, inadequate MIF production during fetal development leads to the retention of these ducts and the failure of the testes to descend into the scrotum.

1. *The Maintenance of the Blood–Testis Barrier.* The seminiferous tubules are isolated from the general circulation by a **blood–testis barrier**, comparable in function to the blood–brain barrier. ∞ p. 472 Sustentacular cells are joined by tight junctions, forming a layer that divides the seminiferous tubule into an outer *basal compartment*, which contains the spermatogonia, and an inner *lumenal compartment* (or *adlumenal compartment*), where meiosis and spermiogenesis occur. Transport across the sustentacular cells is tightly regulated so that conditions in the lumenal compartment remain very stable. The fluid in the lumen of a seminiferous tubule is produced by sustentacular cells, which also regulate the fluid's composition. This fluid is very different from the surrounding interstitial fluid. For example, the fluid in seminiferous tubules is high in androgens, estrogens, potassium, and amino acids. The blood–testis barrier is essential to preserving the differences between the tubular

Testicular cancer occurs at a relatively low rate of about 3 cases per 100,000 males per year. Although only about 7200 new cases occur each year in the United States, testicular cancer is the most common cancer among males age 15–35. The incidence among Caucasian-American males has more than doubled since the 1930s, but the incidence among African-American males has remained unchanged. The reason for this difference is not known.

More than 95 percent of testicular cancers are the result of abnormal spermatogonia or spermatocytes, rather than abnormal sustentacular cells, interstitial cells, or other cells of the testes. Treatment generally consists of a combination of orchiectomy and chemotherapy. The survival rate has increased from about 10 percent in 1970 to about 95 percent in 1999, primarily as a result of earlier diagnosis and improved treatment protocols. Cyclist Lance Armstrong has won the grueling Tour de France race four times after successful treatment for advanced testicular cancer.

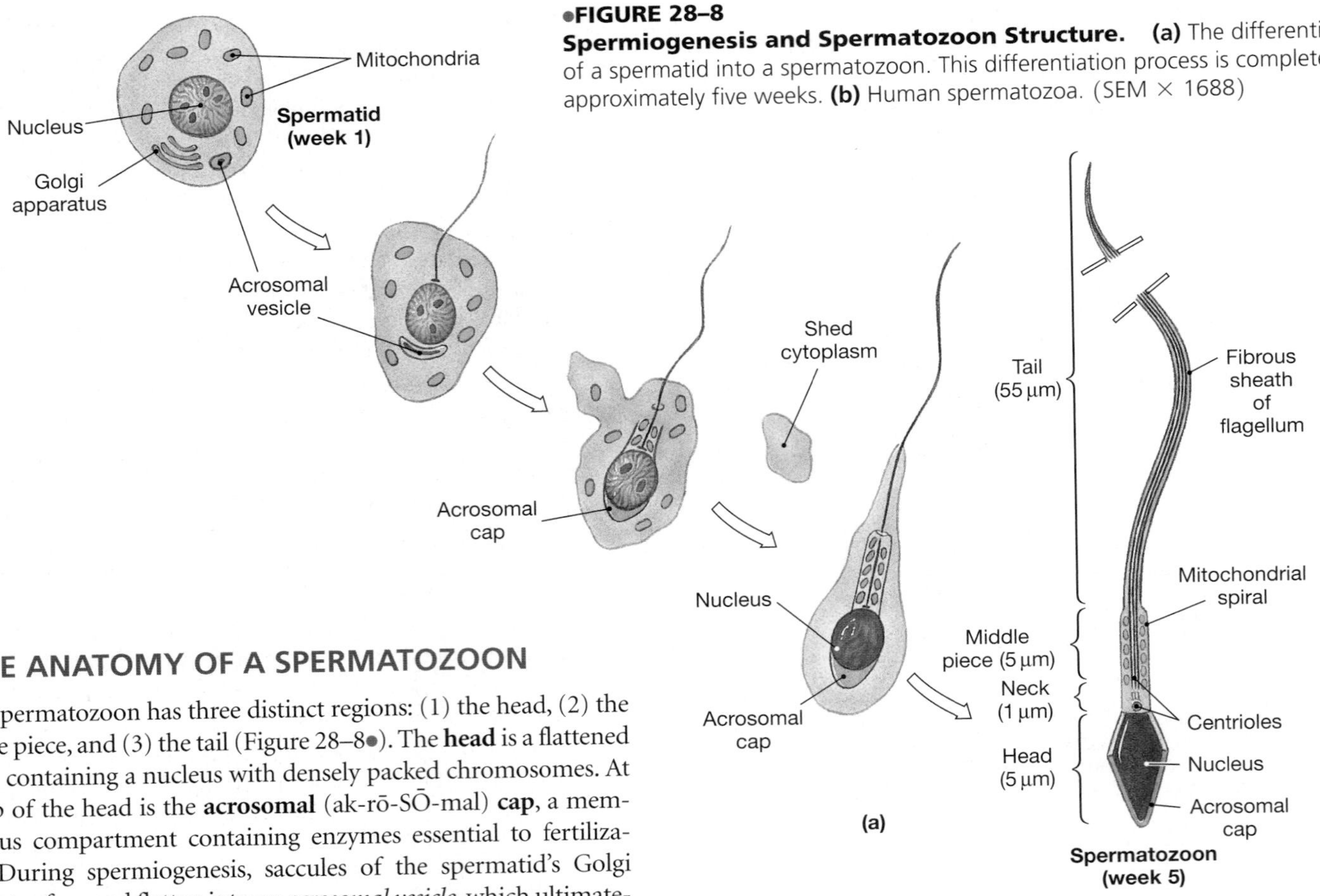

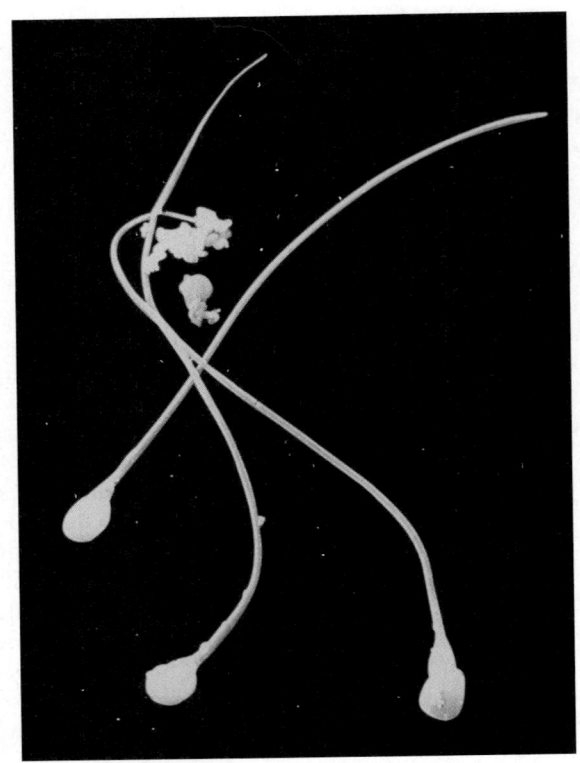

●FIGURE 28–8
Spermiogenesis and Spermatozoon Structure. (a) The differentiation of a spermatid into a spermatozoon. This differentiation process is completed in approximately five weeks. (b) Human spermatozoa. (SEM × 1688)

Mitochondria

Nucleus

Spermatid (week 1)

Golgi apparatus

Acrosomal vesicle

Acrosomal cap

Shed cytoplasm

Nucleus

Acrosomal cap

Tail (55 μm)

Middle piece (5 μm)

Neck (1 μm)

Head (5 μm)

Fibrous sheath of flagellum

Mitochondrial spiral

Centrioles

Nucleus

Acrosomal cap

Spermatozoon (week 5)

(a)

(b)

□ THE ANATOMY OF A SPERMATOZOON

Each spermatozoon has three distinct regions: (1) the head, (2) the middle piece, and (3) the tail (Figure 28–8●). The **head** is a flattened ellipse containing a nucleus with densely packed chromosomes. At the tip of the head is the **acrosomal** (ak-rō-SŌ-mal) **cap**, a membranous compartment containing enzymes essential to fertilization. During spermiogenesis, saccules of the spermatid's Golgi apparatus fuse and flatten into an *acrosomal vesicle*, which ultimately forms the acrosomal cap of the spermatozoon.

A short **neck** attaches the head to the **middle piece**. The neck contains both centrioles of the original spermatid. The microtubules of the distal centriole are continuous with those of the middle piece and tail. Mitochondria in the middle piece are arranged in a spiral around the microtubules. Mitochondrial activity provides the ATP that is needed to move the tail.

The **tail** is the only flagellum in the human body. A *flagellum*, a whiplike organelle, moves a cell from one place to another. Whereas cilia beat in a predictable, waving fashion, the flagellum of a spermatozoon has a complex, corkscrew motion.

Unlike other, less specialized cells, a mature spermatozoon lacks an endoplasmic reticulum, Golgi apparatus, lysosomes, peroxisomes, inclusions, and many other intracellular structures. The loss of these organelles reduces the cell's size and mass; it is essentially a mobile carrier for the enclosed chromosomes, and extra weight would slow it down. Because the cell does not contain glycogen or other energy reserves, however, it must absorb nutrients (primarily fructose) from the surrounding fluid.

□ THE MALE REPRODUCTIVE TRACT

The testes produce physically mature spermatozoa that are incapable of successful fertilization of an oocyte. The other portions of the male reproductive system are responsible for the functional maturation, nourishment, storage, and transport of spermatozoa.

The Epididymis

Late in their development, spermatozoa detach from the sustentacular cells and lie within the lumen of the seminiferous tubule. They have most of the physical characteristics of mature spermatozoa, but are functionally immature and incapable of coordinated locomotion or fertilization. Fluid currents, created by cilia lining the efferent ductules, transport the immobile gametes into the epididymis (Figure 28–4a●). The **epididymis**, the start of the male reproductive tract, is a coiled tube bound to the posterior border of the testis (Figure 28–9a●).

The epididymis can be felt through the skin of the scrotum. A tubule almost 7 meters (23 ft) long, the epididymis is coiled and twisted so as to take up very little space. It has (1) a head, (2) a body, and (3) a tail. The superior **head** is the portion of the epididymis proximal to the testis. The head receives spermatozoa from the efferent ductules.

The **body** begins distal to the last efferent ductule and extends inferiorly along the posterior margin of the testis. Near the inferior border of the testis, the number of coils decreases, marking the start of the **tail**. The tail recurves and ascends to its connection with the ductus deferens. Spermatozoa are stored primarily within the tail of the epididymis.

The epididymis has the following three functions:

1. *It Monitors and Adjusts the Composition of the Fluid Produced By the Seminiferous Tubules.* The pseudostratified columnar epithelial lining of the epididymis bears distinctive stereocilia, which increase the surface area available for absorption from, and secretion into, the fluid in the tubule (Figure 28–9b●).

2. *It Acts as a Recycling Center for Damaged Spermatozoa.* Cellular debris and damaged spermatozoa are absorbed in the epididymis, and the products of enzymatic breakdown are released into the surrounding interstitial fluids for pickup by the epididymal blood vessels.

3. *It Stores and Protects Spermatozoa and Facilitates Their Functional Maturation.* A spermatozoon passes through the epididymis in about two weeks and completes its functional maturation at that time. Over this period, the spermatozoa exist in a sheltered environment that is precisely regulated by the surrounding epithelial cells. Although spermatozoa leaving the epididymis are mature, they remain immobile. To become *motile* (actively swimming) and fully functional, spermatozoa must undergo a process called **capacitation**. Capacitation normally occurs in two steps: (1) Spermatozoa become motile when they are mixed with secretions of the seminal vesicles, and (2) they become capable of successful fertilization when exposed to conditions in the female reproductive tract. The epididymis secretes a substance (as yet unidentified) that prevents premature capacitation.

Transport along the epididymis involves a combination of fluid movement and peristaltic contractions of smooth muscle in the walls of the epididymis. After passing along the tail of the epididymis, the spermatozoa enter the ductus deferens.

The Ductus Deferens

Each **ductus deferens**, or *vas deferens*, is 40–45 cm (16–18 in.) long. It begins at the tail of the epididymis (Figures 28–1, p. 1048, and 28–9a●) and, as part of the spermatic cord, ascends through the inguinal canal (Figure 28–3●, p. 1050). Inside the abdominal

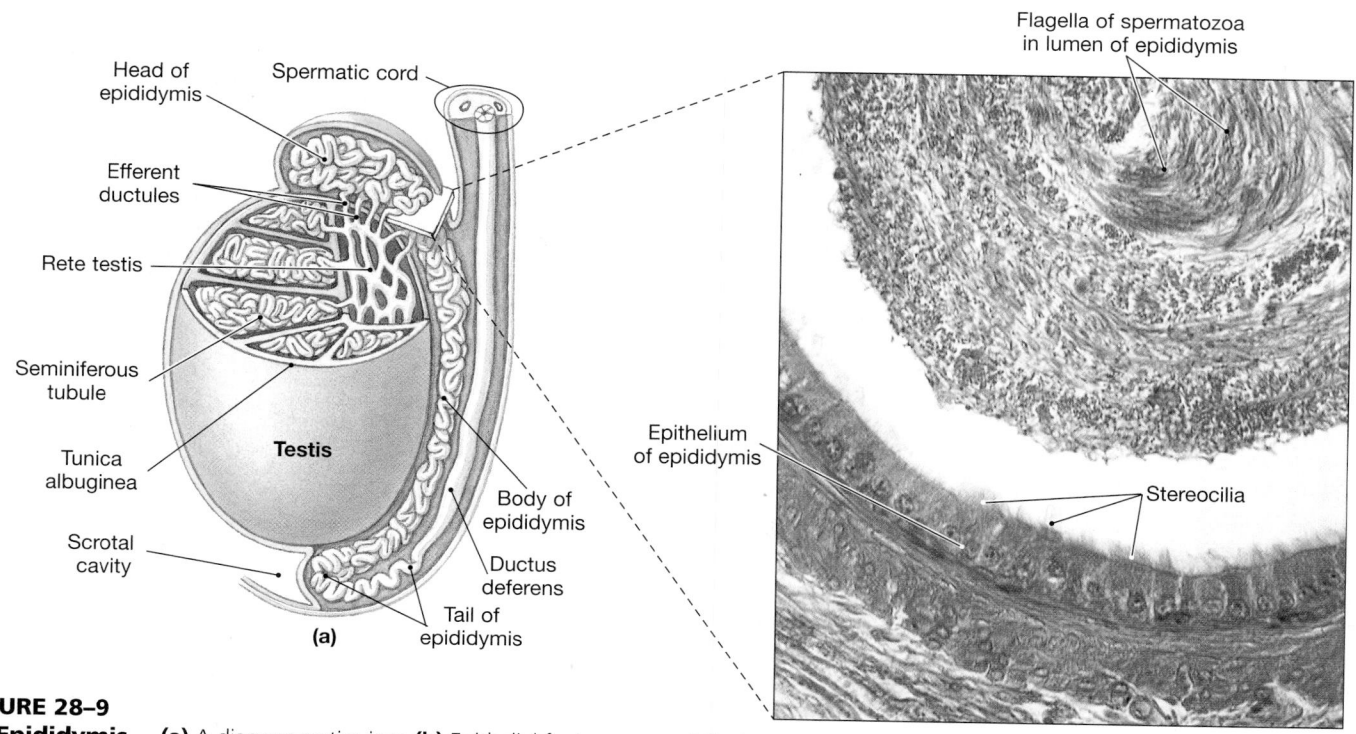

Head of epididymis
Spermatic cord
Efferent ductules
Rete testis
Seminiferous tubule
Tunica albuginea
Testis
Scrotal cavity
Body of epididymis
Ductus deferens
Tail of epididymis
(a)

Flagella of spermatozoa in lumen of epididymis
Epithelium of epididymis
Stereocilia
(b)

●**FIGURE 28–9**
The Epididymis. (a) A diagrammatic view. (b) Epithelial features, especially the elongate stereocilia characteristic of the epididymis. (LM × 1304) **ATLAS** Plate 7.9c

cavity, the ductus deferens passes posteriorly, curving inferiorly along the lateral surface of the urinary bladder toward the superior and posterior margin of the prostate gland. Just before the ductus deferens reaches the prostate gland and seminal vesicles, its lumen enlarges. This expanded portion is known as the **ampulla** (am-PŪL-uh) (Figure 28–10a●).

The wall of the ductus deferens contains a thick layer of smooth muscle (Figure 28–10b●). Peristaltic contractions in this layer propel spermatozoa and fluid along the duct, which is lined by a pseudo-stratified ciliated columnar epithelium. In addition to transporting spermatozoa, the ductus deferens can store spermatozoa for several months. During this time, the spermatozoa remain in a state of suspended animation and have low metabolic rates.

The junction of the ampulla with the duct of the seminal vesicle marks the start of the **ejaculatory duct**. This short passageway (2 cm, or less than 1 in.) penetrates the muscular wall of the prostate

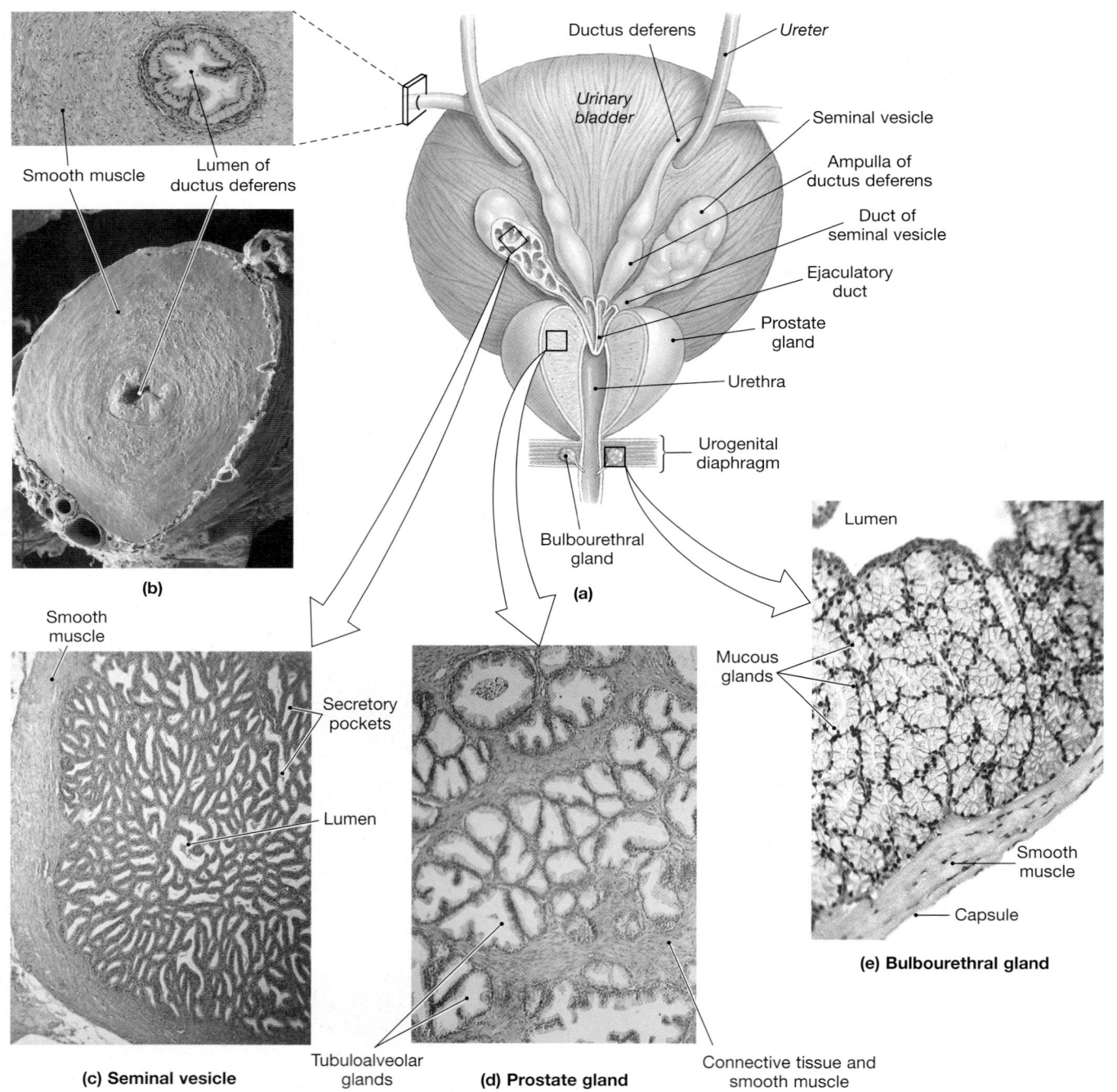

●FIGURE 28–10

The Ductus Deferens and Accessory Glands. **(a)** A posterior view of the prostate gland, showing subdivisions of the ductus deferens in relation to surrounding structures. **(b)** The ductus deferens, showing the smooth muscle around the lumen. (LM [top] × 34, SEM × 42) (Reproduced from R. G. Kessel and R. H. Kardon, *Tissues and Organs: A Text-Atlas of Scanning Electron Microscopy*, W. H. Freeman & Co., 1979.) Sections of **(c)** the seminal vesicle (LM × 44), **(d)** the prostate gland (LM × 49), and **(e)** a bulbourethral gland. (LM × 177)

gland and empties into the urethra near the opening of the ejaculatory duct from the opposite side (Figures 28–1 and 28–10a●).

The Urethra

In males, the **urethra** extends 18–20 cm (7–8 in.) from the urinary bladder to the tip of the penis (Figure 28–1●). It is divided into *prostatic*, *membranous*, and *penile regions*. The male urethra is a passageway used by both the urinary and reproductive systems.

☐ THE ACCESSORY GLANDS

The fluids contributed by the seminiferous tubules and the epididymis account for only about 5 percent of the volume of semen. The fluid component of semen is a mixture of the secretions of many glands, each with distinctive biochemical characteristics. Important glands include the *seminal vesicles*, the *prostate gland*, and the *bulbourethral glands*, all of which occur only in males. Among the major functions of these glands are (1) activating the spermatozoa; (2) providing the nutrients spermatozoa need for motility; (3) propelling spermatozoa and fluids along the reproductive tract, mainly by peristaltic contractions; and (4) producing buffers that counteract the acidity of the urethral and vaginal environments.

The Seminal Vesicles

The ductus deferens on each side ends at the junction between the ampulla and the duct that drains the seminal vesicle (Figure 28–10a●). The **seminal vesicles** are glands embedded in connective tissue on either side of the midline, sandwiched between the posterior wall of the urinary bladder and the rectum. Each seminal vesicle is a tubular gland with a total length of about 15 cm (6 in.). The body of the gland has many short side branches. The entire assemblage is coiled and folded into a compact, tapered mass roughly 5 cm × 2.5 cm (2 in. × 1 in.). The location of the seminal vesicles is shown in Figures 28–1, 28–10a, and 28–11a●.

Seminal vesicles are extremely active secretory glands with an epithelial lining that contains extensive folds (Figure 28–10c●). The seminal vesicles contribute about 60 percent of the volume of semen. Although the vesicular fluid generally has the same osmotic concentration as that of blood plasma, the composition of the two fluids is quite different. In particular, the secretion of the seminal vesicles contains (1) higher concentrations of fructose, which is easily metabolized by spermatozoa; (2) prostaglandins, which can stimulate smooth muscle contractions along the male and female reproductive tracts; and (3) fibrinogen, which after ejaculation forms a temporary clot within the vagina. The secretions of the seminal vesicles are slightly alkaline, helping to neutralize acids in the secretions of the prostate gland and within the vagina. When mixed with the secretions of the seminal vesicles, previously inactive but functional spermatozoa undergo the first step in capacitation and begin beating their flagella, becoming highly motile.

The secretions of the seminal vesicles are discharged into the ejaculatory duct at *emission*, when peristaltic contractions are under way in the ductus deferens, seminal vesicles, and prostate gland. These contractions are under the control of the sympathetic nervous system.

The Prostate Gland

The **prostate gland** is a small, muscular, rounded organ about 4 cm (1.6 in.) in diameter. The prostate gland encircles the proximal portion of the urethra as it leaves the urinary bladder (Figures 28–10a and 28–11a●). The glandular tissue of the prostate consists of a cluster of 30–50 compound tubuloalveolar glands (Figure 28–10d●). ∞ p. 121 These glands are surrounded by and wrapped in a thick blanket of smooth muscle fibers.

The prostate gland produces **prostatic fluid**, a slightly acidic solution that contributes 20–30 percent of the volume of semen. In addition to several other compounds of uncertain significance, prostatic secretions contain **seminalplasmin** (sem-i-nal-PLAZ-min), an antibiotic that may help prevent urinary tract infections in males. These secretions are ejected into the prostatic urethra by peristaltic contractions of the muscular wall.

Prostatic inflammation, or **prostatitis** (pros-ta-TĪ-tis), can occur in males at any age, but it most commonly afflicts older men. Prostatitis can result from bacterial infections, but also occurs in the apparent absence of pathogens. Symptoms can resemble those of prostate cancer. Individuals with prostatitis may complain of pain in the lower back, perineum, or rectum, in some cases accompanied by painful urination and the discharge of mucous secretions from the external urethral meatus. Antibiotic therapy is effective in treating most cases that result from bacterial infection.

The Bulbourethral Glands

The paired **bulbourethral glands**, or *Cowper's glands*, are situated at the base of the penis, covered by the fascia of the urogenital diaphragm (Figures 28–10a and 28–11a●). The bulbourethral glands are round, with diameters approaching 10 mm (less than 0.5 in.). The duct of each gland travels alongside the penile urethra for 3–4 cm (1.2–1.6 in.) before emptying into the urethral lumen. The bulbourethral glands are compound, tubuloalveolar mucous glands (Figure 28–10e●) that secrete a thick, alkaline mucus. The secretion helps neutralize any urinary acids that may remain in the urethra and lubricates the *glans*, or tip of the penis.

☐ SEMEN

A typical ejaculation releases 2–5 ml of semen. This volume of fluid, called **ejaculate**, contains:

- **Spermatozoa.** The normal **sperm count** ranges from 20 million to 100 million spermatozoa per milliliter of sperm.

- **Seminal Fluid.** **Seminal fluid**, the fluid component of semen, is a mixture of glandular secretions with a distinct ionic and nutrient composition. A typical sample of seminal fluid contains the combined secretions of the seminal vesicles (60 percent), the prostate gland (30 percent), the sustentacular cells and epididymis (5 percent), and the bulbourethral glands (less than 5 percent).

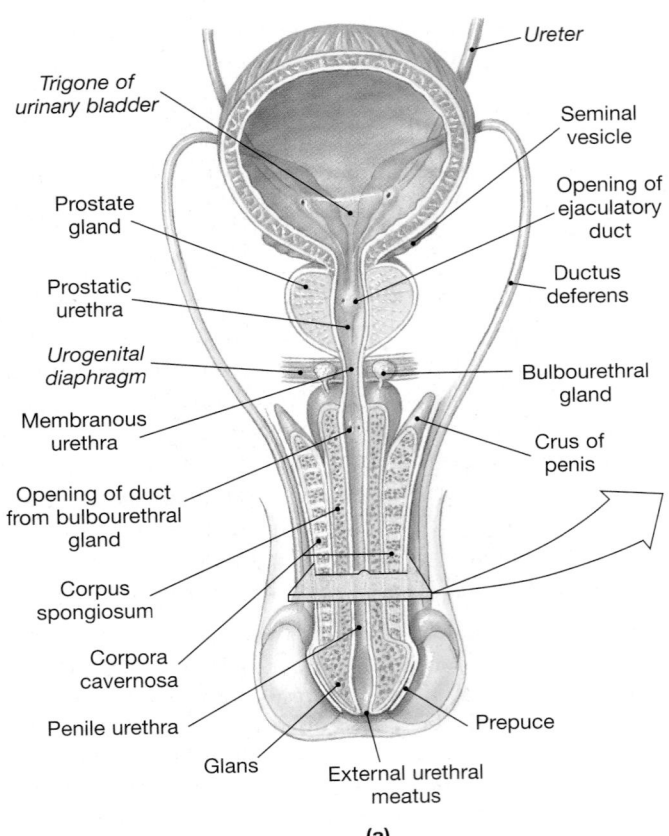

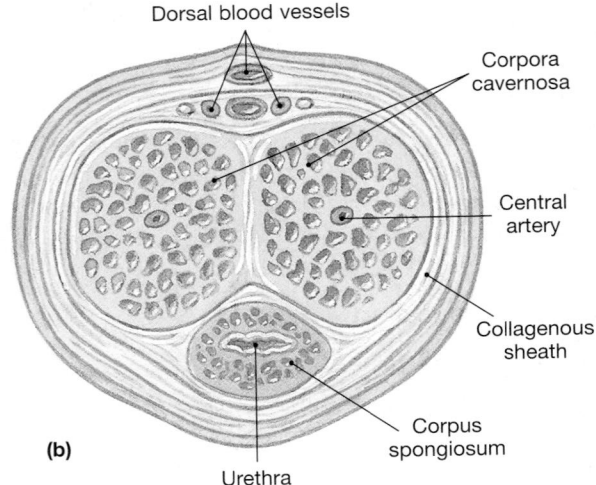

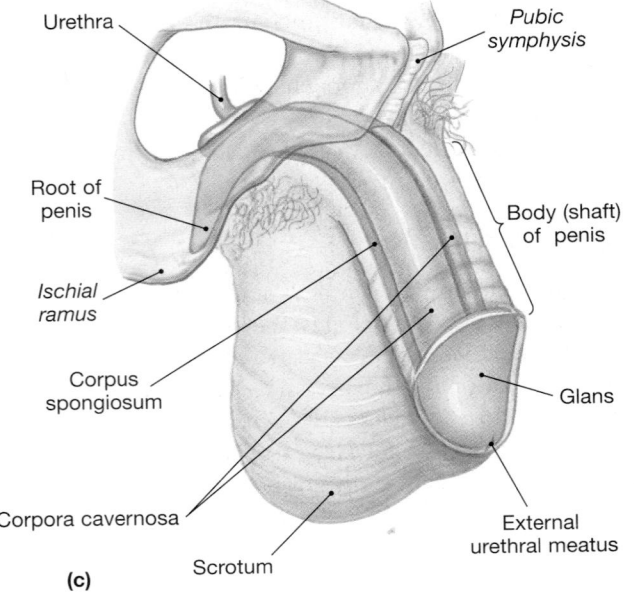

•**FIGURE 28–11**
The Penis. **(a)** A frontal section through the penis and
associated organs. **(b)** A sectional view through the penis.
(c) An anterior and lateral view of the penis, showing
positions of the erectile tissues. **ATLAS** Plate 7.9d

• **Enzymes.** Several important enzymes are present in seminal
fluid, including (1) a protease that may help dissolve mucous
secretions in the vagina; (2) seminalplasmin, an antibiotic enzyme from the prostate gland that kills a variety of bacteria, including *Escherichia coli*; (3) a prostatic enzyme that converts
fibrinogen to fibrin after ejaculation; and (4) *fibrinolysin*, which
subsequently liquefies the clotted semen.

☐ THE EXTERNAL GENITALIA

The male external genitalia consists of the scrotum and penis.
The structure of the scrotum has already been described (p.
1051). The **penis** is a tubular organ through which the distal portion of the urethra passes (Figure 28–11a•). It conducts urine to
the exterior and introduces semen into the female's vagina during sexual intercourse. The penis is divided into three regions: (1)
the root, (2) the body, and (3) the glans (Figure 28–11c•). The
root of the penis is the fixed portion that attaches the penis to the
body wall. This connection occurs within the urogenital triangle
immediately inferior to the pubic symphysis. ∞ pp. 250, 354 The

body (**shaft**) of the penis is the tubular, movable portion of the
organ. The **glans** of the penis is the expanded distal end that surrounds the external urethral meatus. The *neck* is the narrow portion of the penis between the shaft and the glans.

The skin overlying the penis resembles that of the scrotum.
The dermis contains a layer of smooth muscle, and the underlying areolar tissue allows the thin skin to move without distorting
underlying structures. The subcutaneous layer also contains superficial arteries, veins, and lymphatic vessels.

A fold of skin called the **prepuce** (PRĒ-pūs), or *foreskin*, surrounds the tip of the penis. The prepuce attaches to the relatively
narrow neck of the penis and continues over the glans. **Preputial**
(prē-PŪ-shal) **glands** in the skin of the neck and the inner surface of
the prepuce secrete a waxy material known as **smegma** (SMEG-ma).
Unfortunately, smegma can be an excellent nutrient source for bacteria. Mild inflammation and infections in this region are common,

Semen Analysis

Semen analysis is commonly performed to assess male fertility. A sample of semen is donated after a 36-hour period of sexual abstinence. Standard tests include the following:

- **Volume.** The normal ejaculate volume is 2–5 ml. An abnormally low volume may indicate problems with the prostate gland or seminal vesicles.

- **Sperm Count.** Each ejaculate should contain more than 60 million spermatozoa. The concentration of spermatozoa should be over 20 million per milliliter of semen. Most individuals with lower sperm counts are infertile, because too few spermatozoa survive the ascent of the female reproductive tract to perform fertilization. A low sperm count may reflect inflammation of the epididymis, ductus deferens, or prostate gland.

- **Motility.** At least 60 percent of the spermatozoa in the sample should be beating their flagella and swimming actively.

- **Shape.** At least 60 percent of the spermatozoa should have normal shapes. Common abnormalities are malformed heads and "twin" spermatozoa that did not separate at the time of spermiation.

- **Liquefaction.** Within a few minutes after ejaculation, semen coagulates, liquefying again some time later. The function of this clotting is unknown. Normal semen lique-

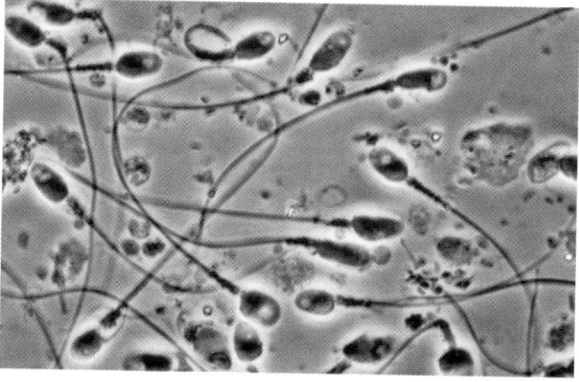

fies after 15–30 minutes. An extended liquefaction time may indicate problems with the secretions of accessory glands.

The determination of male fertility problems in the absence of abnormal semen analysis results may require additional tests. In what is often called the "hamster test," a sample of semen is placed on a slide with the oocyte of a hamster. Normal human spermatozoa will fertilize the oocyte, although further development is impossible. If fertilization does not occur, there may be problems with the enzymes in the acrosomal cap.

A complete chemical analysis of semen appears in Appendix IV.

---

especially if the area is not washed thoroughly and frequently. One way to avoid trouble is to perform a **circumcision** (ser-kum-SIZH-un), the surgical removal of the prepuce. In Western societies (especially the United States), this procedure is generally performed shortly after birth. Although the practice of circumcision remains controversial, strong religious and cultural biases and epidemiological evidence suggest that it will continue.

Deep to the areolar tissue, a dense network of elastic fibers encircles the internal structures of the penis. Most of the body of the penis consists of three cylindrical columns of **erectile tissue**. Erectile tissue consists of a three-dimensional maze of vascular channels, incompletely separated by partitions of elastic connective tissue and smooth muscle fibers. In the resting state, the arterial branches are constricted and the muscular partitions are tense. This combination restricts blood flow into the erectile tissue. The parasympathetic innervation of the penile arteries involves neurons that release nitric oxide (NO) at their synaptic knobs. The smooth muscles in the arterial walls relax when NO is released, at which time the vessels dilate, blood flow increases, the vascular channels become engorged with

blood, and **erection** of the penis occurs. The flaccid (nonerect) penis hangs inferior to the pubic symphysis and anterior to the scrotum, but during erection the penis stiffens and assumes a more upright position.

The anterior surface of the flaccid penis covers two cylindrical masses of erectile tissue: the **corpora cavernosa** (KOR-po-ruh ka-ver-NŌ-suh; singular, *cavernosum*). The two are separated by a thin septum and encircled by a dense collagenous sheath (Figure 28–11a,b●). The corpora cavernosa diverge at their bases, forming the **crura** (*crura*, legs; singular, *crus*) of the penis. Each crus is bound to the ramus of the ischium and pubis by tough connective-tissue ligaments. The corpora cavernosa extend along the length of the penis as far as its neck. The erectile tissue within each corpus cavernosum surrounds a central artery (Figure 28–11b●).

The relatively slender **corpus spongiosum** (spon-jē-Ō-sum) surrounds the penile urethra (Figure 28–11a,b●). This erectile body extends from the superficial fascia of the urogenital diaphragm to the tip of the penis, where it expands to form the glans. The sheath surrounding the corpus spongiosum contains more elastic fibers than does that of the corpora cavernosa, and the erectile tissue contains a pair of small arteries.

As mentioned in the text, circumcision is relatively common in many societies. Worldwide, roughly 25 percent of males are circumcised. However, the practice of neonatal male circumcision continues to be hotly debated. Strong epidemiological evidence favors circumcision: Circumcised males have a lower incidence of neonatal and adult urinary tract infections, genital ulcerative disease, penile carcinoma, and HIV infection (in some groups) than do uncircumcised males. Circumcision also carries serious negative aspects and risks, however, such as pain, bleeding, penile injury, and postoperative infection. Many health professionals, as well as parents, feel that these neonatal problems outweigh the statistical advantages in adulthood. Neither the American Academy of Pediatrics nor the Canadian Paediatric Society advocates routine neonatal circumcision; both support parental choice on the basis of an individual assessment of the potential risks and benefits.

☐ HORMONES AND MALE REPRODUCTIVE FUNCTION

The hormonal interactions in males are diagrammed in Figure 28–12●. We introduced the major reproductive hormones in Chapter 18. ∞ p. 637 The anterior lobe of the pituitary gland releases *follicle-stimulating hormone* (**FSH**) and *luteinizing hormone* (**LH**). The pituitary release of these hormones occurs in the presence of *gonadotropin-releasing hormone* (**GnRH**), a peptide synthesized in the hypothalamus and carried to the anterior lobe by the hypophyseal portal system.

The hormone GnRH is secreted in pulses rather than continuously. In adult males, small pulses occur at 60–90-minute intervals. As levels of GnRH change, so do the rates of secretion of FSH and LH (and testosterone, which is released in response to LH). Unlike the situation in women, which we will consider later in the chapter, the GnRH pulse frequency in adult males remains relatively steady from hour to hour, day to day, and year to year. As a result, plasma levels of FSH, LH, and testosterone remain within a relatively narrow range throughout adult life.

FSH and Spermatogenesis

In males, FSH targets primarily the sustentacular cells of the seminiferous tubules. Under FSH stimulation, and in the presence of testosterone from the interstitial cells, sustentacular cells (1) promote spermatogenesis and spermiogenesis and (2) secrete androgen-binding protein (ABP).

The rate of spermatogenesis is regulated by a negative feedback mechanism involving GnRH, FSH, and inhibin. Under GnRH stimulation, FSH promotes spermatogenesis along the seminiferous tubules. As spermatogenesis accelerates, however, so does the rate of inhibin secretion by the sustentacular cells

of the testes. Inhibin inhibits FSH production in the anterior lobe of the pituitary gland and may also suppress the secretion of GnRH at the hypothalamus.

The net effect is that when FSH levels become elevated, inhibin production increases until FSH levels return to normal. If FSH levels decline, inhibin production falls, so the rate of FSH production accelerates. (Figure 28–12●).

LH and Androgen Production

In males, LH causes the secretion of testosterone and other androgens by the interstitial cells of the testes. Testosterone, the most important androgen, has numerous functions, such as (1) stimulating spermatogenesis and promoting the functional maturation of spermatozoa, through its effects on sustentacular cells; (2) affecting central nervous system (CNS) function, including the libido (sexual drive) and related behaviors; (3) stimulating metabolism throughout the body, especially pathways concerned with protein synthesis and muscle growth; (4) establishing and maintaining the secondary sex characteristics, such as the distribution of facial hair, increased muscle mass and body size, and the quantity and location of characteristic adipose tissue deposits; and (5) maintaining the accessory glands and organs of the male reproductive tract.

Testosterone functions like other steroid hormones, circulating in the bloodstream while bound to one of two types of transport proteins: (1) *gonadal steroid-binding globulin* (GBG), which carries roughly two-thirds of the circulating testosterone, and (2) the albumins, which bind the remaining one-third. ∞ p. 608 Testosterone diffuses across the cell membrane and binds to an intracellular receptor. The hormone–receptor complex then binds to the DNA in the nucleus. In many target tissues, some of the arriving testosterone is converted to **dihydrotestosterone (DHT)**. A small amount of DHT diffuses back out of the cell and into the bloodstream, and DHT levels are usually about 10 percent of circulating testosterone levels. Dihydrotestosterone can also enter peripheral cells and bind to the same hormone receptors targeted by testosterone. In addition, some tissues, notably those of the external genitalia, respond to DHT rather than to testosterone, and other tissues, including the prostate gland, are more sensitive to DHT than to testosterone.

Testosterone production begins around the seventh week of fetal development and reaches a peak after roughly six months. Over this period, the secretion of Müllerian-inhibiting factor by developing sustentacular cells leads to the regression of the Müllerian ducts. The early surge in testosterone levels stimulates the differentiation of the male duct system and accessory organs and affects CNS development. The best-known CNS effects occur in the developing hypothalamus. There, testosterone apparently programs the hypothalamic centers that are involved with (1) GnRH production and the regulation of pituitary FSH and LH secretion, (2) sexual behaviors, and (3)

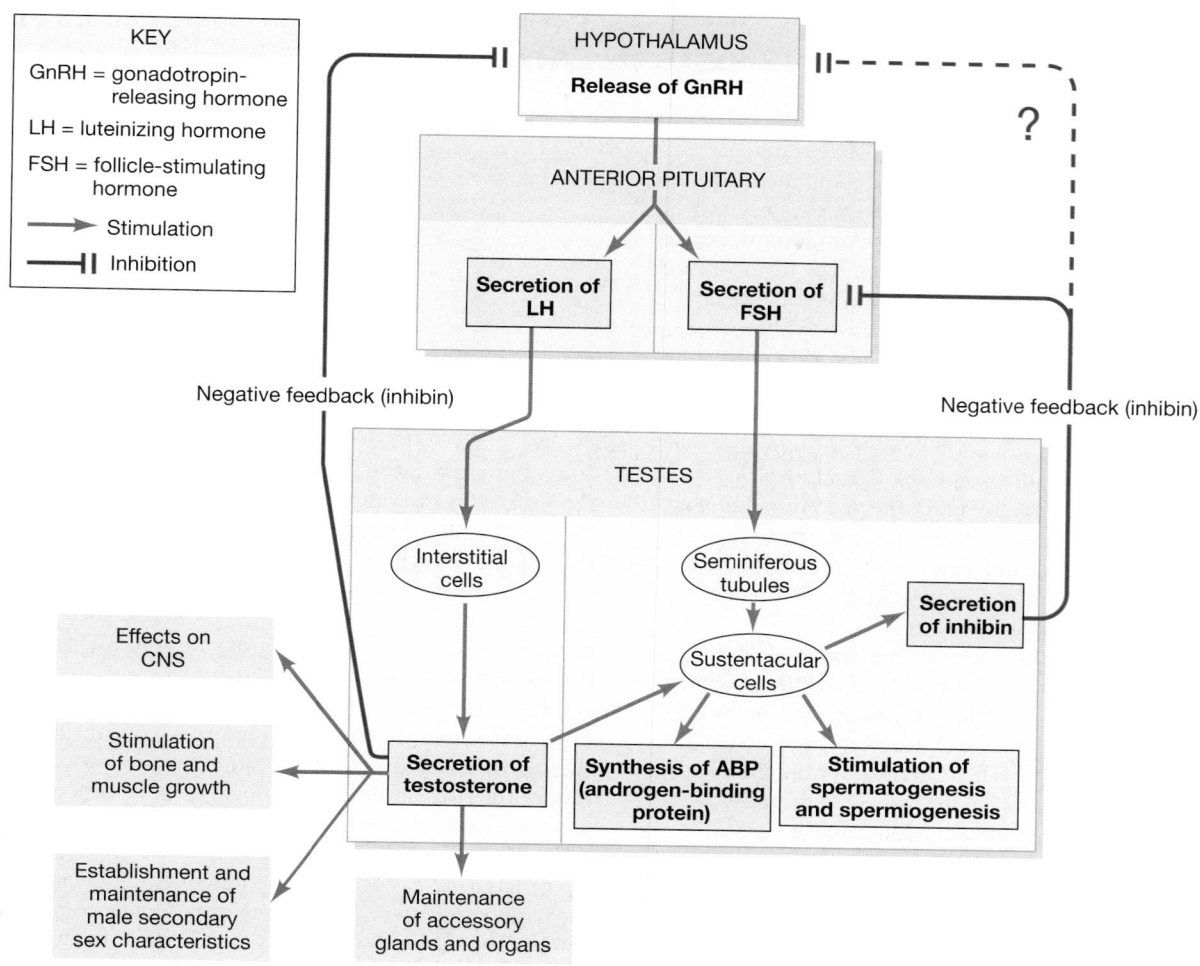

KEY

GnRH = gonadotropin-releasing hormone

LH = luteinizing hormone

FSH = follicle-stimulating hormone

→ Stimulation

⊣ Inhibition

●**FIGURE 28–12**

Hormonal Feedback and the Regulation of the Male Reproductive Function

sexual drive. As a result of this prenatal exposure to testosterone, the hypothalamic centers will respond appropriately when the individual becomes sexually mature. The factors responsible for regulating the fetal production of testosterone are not known.

Testosterone levels are low at birth. Until puberty, background testosterone levels, although still relatively low, are higher in males than in females. Testosterone secretion accelerates markedly at puberty, initiating sexual maturation and the appearance of secondary sex characteristics. In adult males, negative feedback controls the level of testosterone production. Above-normal testosterone levels inhibit the release of GnRH by the hypothalamus, causing a reduction in LH secretion and lowering testosterone levels (Figure 28–12●).

The plasma of adult males also contains relatively small amounts of estradiol (2 ng/dl, versus 525 ng/dl of testosterone). Seventy percent of the estradiol is formed from circulating testosterone. The rest is secreted, primarily by the interstitial and sustentacular cells of the testes. The conversion of testosterone to estradiol is performed by an enzyme called *aromatase.* For unknown reasons, estradiol production increases in older men.

Dehydroepiandrosterone, or **DHEA,** is the primary androgen secreted by the zona reticularis of the adrenal cortex. ⚭ p. 630 As we noted in Chapter 18, these androgens, which are secreted in small amounts, are converted to testosterone (or estrogens) by other tissues. The significance of this small adrenal androgen contribution in both genders remains uncertain, but some people have suggested that DHEA could be used to treat a variety of conditions, including diabetes, heart disease, and depression. The effects of long-term high doses of DHEA remain largely unknown; however, recall from Chapter 18 that the long-term effects of androgen abuse can be quite serious. ⚭ p. 642 High levels of DHEA in women have been linked to an increased risk of ovarian cancer as well as to masculinization, due to the conversion of DHEA to testosterone.

Prostatic Hypertrophy and Prostate Cancer

In most cases, enlargement of the prostate gland, or **benign prostatic hypertrophy**, occurs spontaneously in men over age 50. The increase in size occurs while testosterone production by the interstitial cells decreases. For unknown reasons, small masses, called *prostatic concretions*, may form within the glands (see photo). At the same time, the interstitial cells begin releasing small quantities of estrogens into the bloodstream. The combination of lower testosterone levels and the presence of estrogen probably stimulates prostatic growth. In severe cases, prostatic swelling constricts and blocks the urethra and constricts the rectum. If not corrected, the urinary obstruction can cause permanent kidney damage. Partial surgical removal is the most effective treatment. In the procedure known as a **TURP** (*transurethral prostatectomy*), an instrument pushed along the urethra restores normal function by cutting away the swollen prostatic tissue. Most of the prostate gland remains in place, and no external scars result.

Prostate cancer, a malignant, metastasizing cancer of the prostate gland, is the second most common cancer and the second most common cause of cancer deaths in males. In 2002, approximately 189,000 new cases of prostate cancer were diagnosed in the United States, and there were about 30,200 deaths from the ailment. Most patients are elderly. (The average age at diagnosis is 72.) There are racial differences in susceptibility that are poorly understood. For example, the prostate cancer rates for Asian-American males are relatively low compared with those of either Caucasian-Americans or African-Americans. For all age groups and all races, the rates of prostate cancer are rising sharply. The reason for the increase is not known. Aggressive diagnosis and treatment of localized prostate cancer in elderly patients is controversial because many of these men have non-metastatic tumors, and even if untreated are likely eventually to die of some other disease.

Prostate cancer normally originates in one of the secretory glands. As the cancer progresses, it produces a nodular lump or swelling on the surface of the prostate gland. Palpation of this gland through the rectal wall, a procedure known as a *digital rectal exam* (DRE), is the easiest diagnostic screening procedure. *Transrectal prostatic ultrasound* (TRUS) can be used to obtain more detailed information about the status of the prostate, but at significantly higher cost to the patient. Blood tests are also used for screening purposes. The most sensitive is a blood test for *prostate-specific antigen (PSA)*. Elevated levels of this antigen, normally present in low concentrations, may indicate the presence of prostate cancer. The *serum enzyme assay*, which checks the level of the isozyme *prostatic acid phosphatase*, detects prostate cancer in later stages of development. Screening with periodic PSA tests is now being recommended for men over age 50.

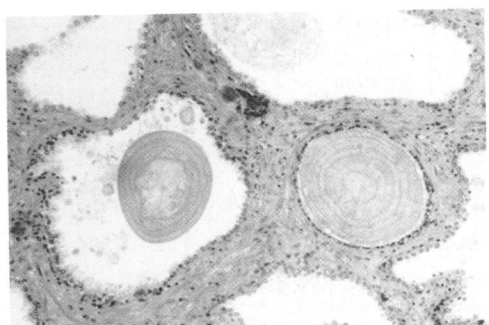

If the condition is detected before the cancer cells have spread to other organs, the usual treatment is localized radiation or surgical removal of the prostate gland. This operation, a **prostatectomy** (pros-ta-TEK-to-mē), can be effective in controlling the condition, but both surgery and radiation can have undesirable side effects, including urinary incontinence and the loss of sexual function. Modified surgical procedures can reduce the risks and maintain normal sexual function in perhaps three out of four patients.

The prognosis is much worse for prostate cancer diagnosed after metastasis has occurred, because metastasis rapidly involves the lymphatic system, lungs, bone marrow, liver, or adrenal glands. The survival rates at this stage become relatively low. Treatments for metastasized prostate cancer include widespread irradiation, hormonal manipulation, lymph node removal, and aggressive chemotherapy. Because the cancer cells are stimulated by testosterone, treatment may involve castration or hormones that depress GnRH or LH production. There are three treatment options. One is an *estrogen*; the usual choice *is diethylstilbestrol (DES)*. A second is *drugs that mimic GnRH*. These drugs are given in high doses, producing a surge in LH production followed by a sharp decline to very low levels, presumably as the endocrine cells adapt to the excessive stimulation. A third is *drugs that block the action of androgens*. Several new drugs of this type, including *flutamide* and *finasateride*, prevent the stimulation of cancer cells by testosterone. Despite these interesting advances in treatment, the average survival time for patients diagnosed with advanced prostatic cancer is only 2.5 years.

The IOC, NCAA, and NFL have banned its use as for muscle enhancement. Nevertheless, DHEA is being promoted as a wonder drug to increase vitality, strength, and muscle mass, and the food supplements prepared from wild Mexican yams are now being advertised as containing "DHEA precursors." These claims are false; the compounds contained in these supplements have no effect on circulating levels of DHEA. The current recommendations are that DHEA use be restricted to controlled, supervised clinical trials and that no one under age 40 use the drug.

✓ On a warm day, would the cremaster muscle be contracted or relaxed? Why?

✓ What will happen if the arteries within the penis dilate?

✓ What effect would low levels of FSH have on sperm production?

Answers start on page Q-1

The anatomy of the male reproductive organs can be reviewed on the **Anatomy CD-ROM**: Reproductive System/Male Flythrough.

28–3 THE REPRODUCTIVE SYSTEM OF THE FEMALE

Objectives

- Describe the components of the female reproductive system.
- Outline the processes of meiosis and oogenesis in the ovaries.
- Identify the phases and events of the ovarian and uterine cycles.
- Describe the structure, histology, and functions of the vagina.
- Summarize the anatomical, physiological, and hormonal aspects of the female reproductive cycle.

A woman's reproductive system produces sex hormones and functional gametes and also must be able to protect and support a developing embryo and nourish the newborn infant. The principal organs of the female reproductive system are the *ovaries*, the *uterine tubes*, the *uterus*, the *vagina*, and the components of the external genitalia (Figure 28–13●). As in males, a variety of accessory glands release their secretions into the female reproductive tract.

The ovaries, uterine tubes, and uterus are enclosed within an extensive mesentery known as the **broad ligament**. The uterine tubes run along the superior border of the broad ligament and open into the pelvic cavity lateral to the ovaries. The **mesovarium** (mes-ō-VAR-ē-um), a thickened fold of mesentery, supports and stabilizes the position of each ovary. The broad ligament attaches to the sides and floor of the pelvic cavity, where it becomes continuous with the parietal peritoneum. The broad ligament thus subdivides this part of the peritoneal cavity. The pocket formed between the posterior wall of the uterus and the anterior surface of the colon is the **rectouterine** (rek-tō-Ū-ter-in) **pouch**; the pocket formed between the uterus and the posterior wall of the bladder is the **vesicouterine** (ves-i-kō-Ū-ter-in) **pouch**. These subdivisions are most apparent in sagittal section (Figure 28–13●).

Several other ligaments assist the broad ligament in supporting and stabilizing the position of the uterus and associated reproductive organs. These ligaments lie within the mesentery sheet of the broad ligament and are connected to the ovaries or uterus. The broad ligament limits side-to-side movement and rotation, and the other ligaments (described in our discussion of the ovaries and uterus) prevent superior–inferior movement.

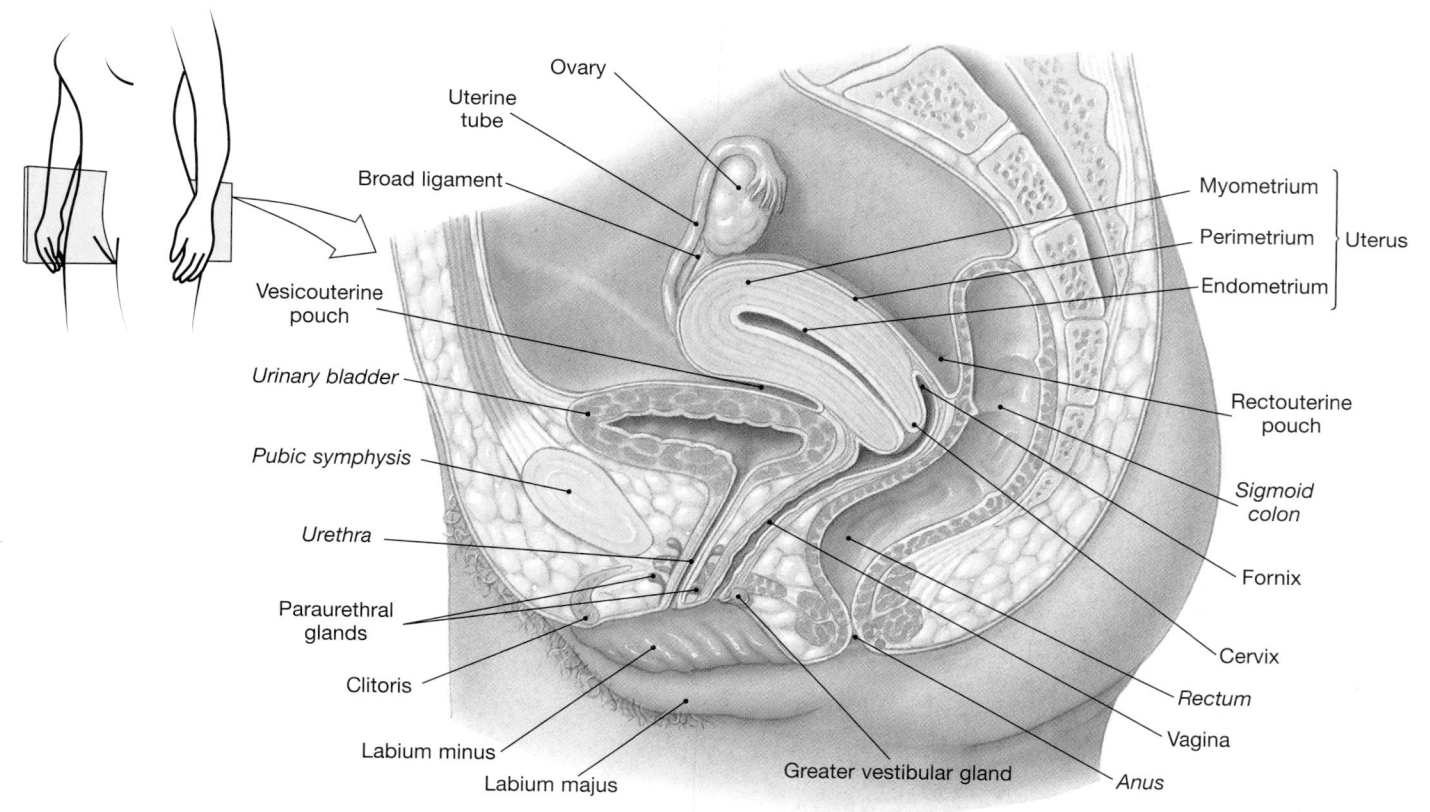

●**FIGURE 28–13**
The Female Reproductive System.   A sagittal section of the female reproductive organs.   ATLAS Plate 7.11b

☐ THE OVARIES

The paired ovaries are small, lumpy, almond-shaped organs near the lateral walls of the pelvic cavity (Figures 28–13 and 28–14●). The ovaries perform three main functions: (1) produce immature female gametes, or oocytes, (2) secrete female sex hormones, including estrogens and progestins, and (3) secrete inhibin, involved in the feedback control of pituitary FSH production.

The position of each ovary is stabilized by the mesovarium and by a pair of supporting ligaments: the ovarian ligament and the suspensory ligament (Figure 28–14a●). The **ovarian ligament** extends from the uterus, near the attachment of the uterine tube, to the medial surface of the ovary. The **suspensory ligament** extends from the lateral surface of the ovary past the open end of the uterine tube to the pelvic wall. The suspensory ligament contains the major blood vessels of the ovary: the **ovarian artery** and **ovarian vein**. These vessels are connected to the ovary at the **ovarian hilum**, where the ovary attaches to the mesovarium.

A typical ovary is a flattened oval about 5 cm in length, 2.5 cm in width, and 8 mm in thickness (2 in. × 1 in. × 0.33 in.) and weighs 6–8 g (roughly 0.25 oz). An ovary is pink or yellowish and has a nodular consistency. The visceral peritoneum, or *germinal epithelium*, covering the surface of each ovary consists of a layer of columnar epithelial cells that overlies a dense connective-tissue layer called the **tunica albuginea**. We can divide the interior tissues, or **stroma**, of the ovary into a superficial *cortex* and a deeper *medulla* (Figure 28–14b●). Gametes are produced in the cortex.

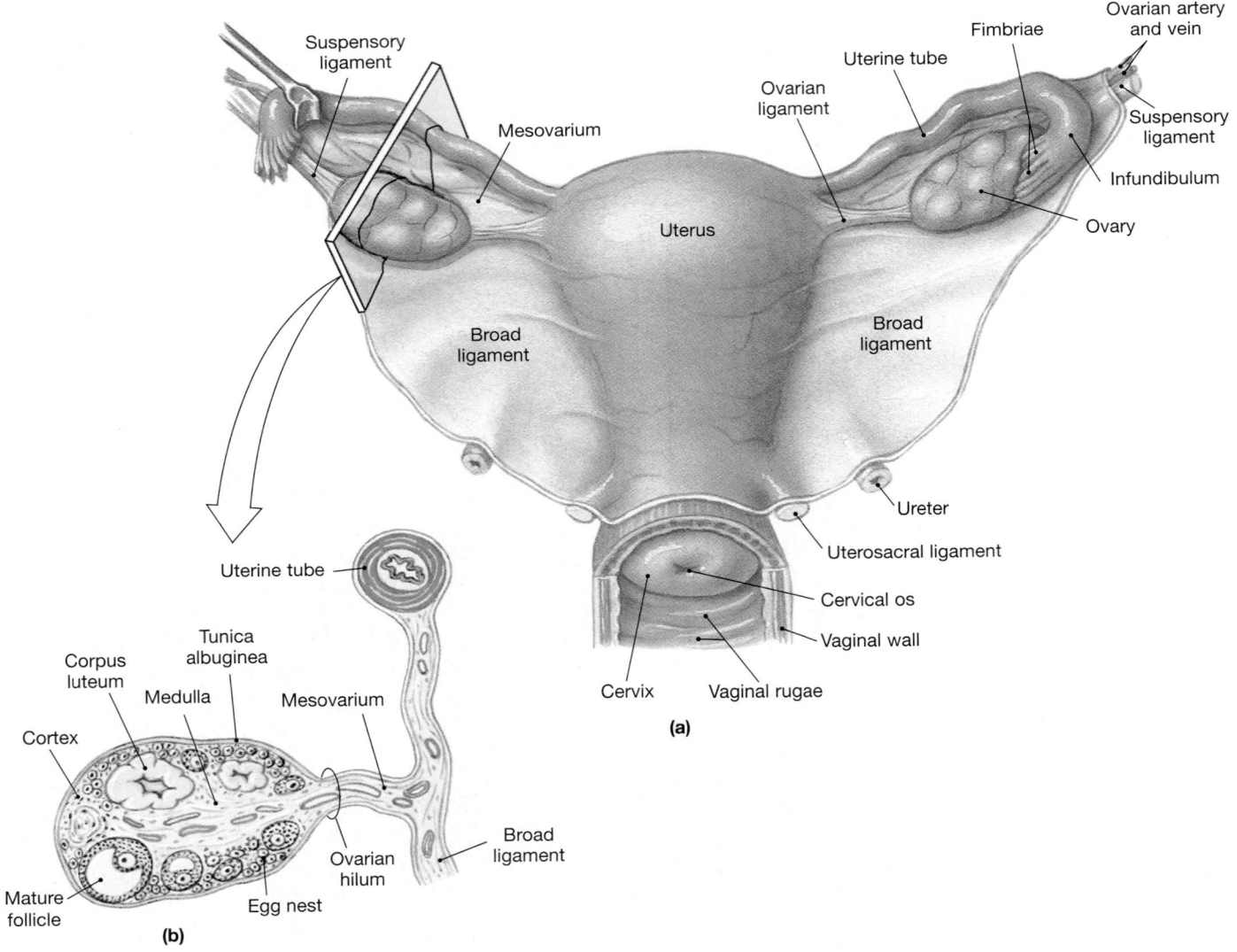

●**FIGURE 28–14**
The Ovaries and Their Relationships to the Uterine Tube and Uterus. **(a)** A posterior view of the uterus, uterine tubes, and ovaries and **(b)** a sectional view of the ovary, uterine tube, and associated mesenteries. **ATLAS** Scan 14

Oogenesis

Ovum production, or **oogenesis** (ō-ō-JEN-e-sis), begins before a woman's birth, accelerates at puberty, and ends at *menopause*. Between puberty and menopause, oogenesis occurs on a monthly basis as part of the *ovarian cycle*.

Unlike spermatogonia, the **oogonia** (ō-ō-GŌ-nē-uh), or stem cells of females, complete their mitotic divisions before birth. Between the third and seventh months of fetal development, the daughter cells, or **primary oocytes** (Ō-ō-sīts), prepare to undergo meiosis. They proceed as far as the prophase of meiosis I, but at that time the process comes to a halt. The primary oocytes then remain in a state of suspended development until the individual reaches puberty, when rising levels of FSH trigger the start of the ovarian cycle. Each month thereafter, some of the primary oocytes will be stimulated to undergo further development. Not all primary oocytes produced during development survive until puberty. The ovaries have roughly 2 million *primordial follicles* at birth, each containing a primary oocyte. By the time of puberty, the number has dropped to about 400,000. The rest of the primordial follicles degenerate in a process called **atresia** (a-TRĒ-zē-uh).

Although the nuclear events under way in the ovaries during meiosis are the same as those in the testes, the process differs in two important details:

1. The cytoplasm of the primary oocyte is unevenly distributed during the two meiotic divisions. Oogenesis produces one functional ovum, which contains most of the original cytoplasm, and two or three **polar bodies**, nonfunctional cells that later disintegrate (Figure 28–15●).
2. The ovary releases a **secondary oocyte** rather than a mature ovum. The secondary oocyte is suspended in metaphase of meiosis II; meiosis will not be completed unless and until fertilization occurs.

The Ovarian Cycle

Ovarian follicles are specialized structures in which both oocyte growth and meiosis I occur. The ovarian follicles are located in the cortex of the ovaries. Primary oocytes are located in the outer portion of the ovarian cortex, near the tunica albuginea in clusters called *egg nests*. Each primary oocyte within an egg nest is surrounded by a single squamous layer of *follicular cells*. The primary oocyte and its follicular cells form a **primordial follicle**. After sexual maturation, a different group of primordial follicles is activated each month. This monthly process is known as the **ovarian cycle**.

The ovarian cycle can be divided into a **follicular phase**, or *preovulatory phase*, and a **luteal phase**, or *postovulatory phase*. Important steps in the ovarian cycle can be summarized as follows (Figure 28–16●):

STEP 1: *The Formation of Primary Follicles.* Follicle formation is stimulated by FSH from the anterior pituitary gland. The ovarian cycle begins as activated primordial follicles develop into **primary follicles**. In a primary follicle, the follicular cells enlarge and undergo re-

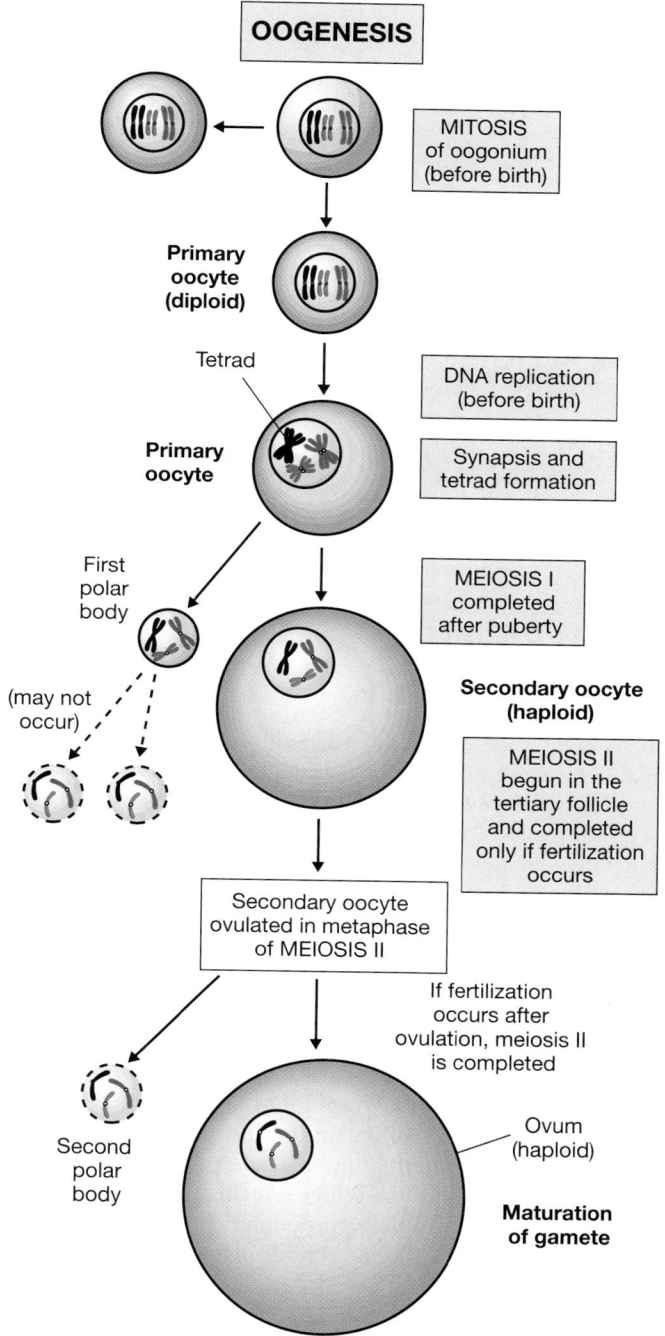

●**FIGURE 28–15**
Oogenesis. In oogenesis, a single primary oocyte produces an ovum and two or three nonfunctional polar bodies. Compare this schematic diagram with Figure 28–7, p. 1055.

peated divisions that create several layers of follicular cells around the oocyte. These follicle cells are now called **granulosa cells**.

As layers of granulosa cells develop around the primary oocyte, microvilli from the surrounding granulosa cells intermingle with those of the primary oocyte. The microvilli are surrounded by a layer of glycoproteins; the entire region is called the **zona pellucida**

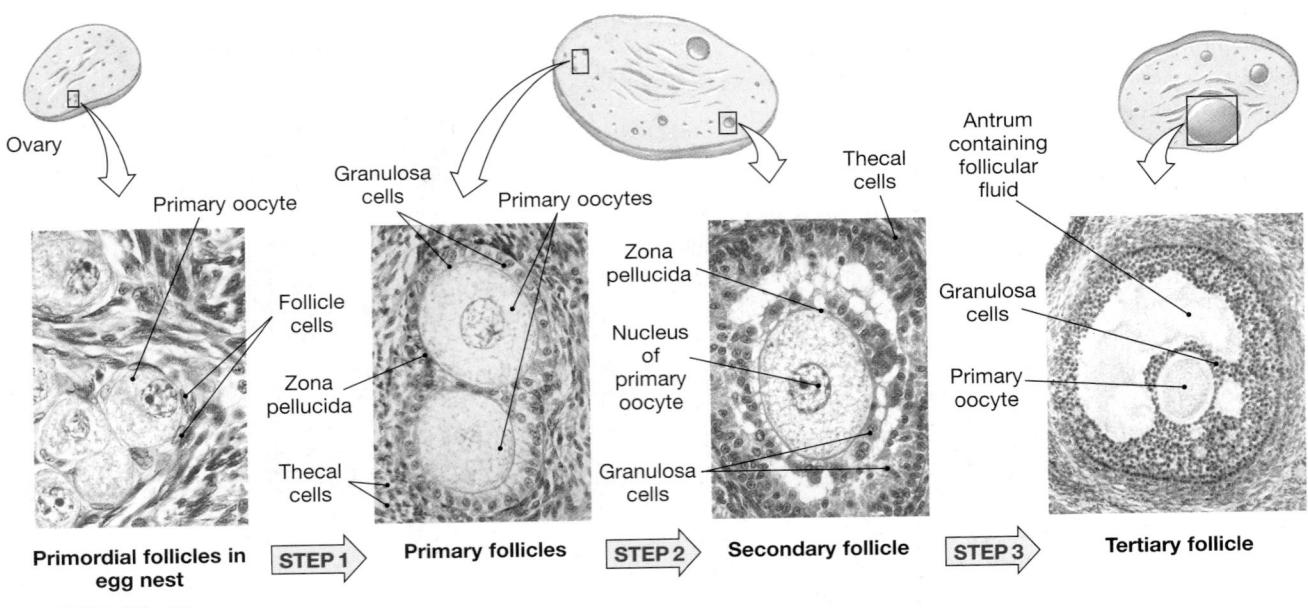

●FIGURE 28–16
The Ovarian Cycle

(ZŌ-na pe-LOO-si-duh). The microvilli increase the surface area available for the transfer of materials from the granulosa cells to the rapidly enlarging oocyte.

The conversion from primordial to primary follicles and subsequent follicular development occurs under FSH stimulation. As the granulosa cells enlarge and multiply, adjacent cells in the ovarian stroma form a layer of **thecal cells** around the follicle. Thecal cells and granulosa cells work together to produce sex hormones called *estrogens.*

STEP 2: *The Formation of Secondary Follicles.* Although many primordial follicles develop into primary follicles, only a few will proceed to the next step. The transformation begins as the wall of the follicle thickens and the granulosa cells begin secreting small amounts of fluid. This **follicular fluid**, or *liquor folliculi*, accumulates in small pockets that gradually expand and separate the inner and outer layers of the follicle. At this stage, the complex is known as a **secondary follicle**. Although the primary oocyte continues to grow at a steady pace, the follicle as a whole now enlarges rapidly because follicular fluid accumulates.

STEP 3: *The Formation of a Tertiary Follicle.* Eight to 10 days after the start of the ovarian cycle, the ovaries generally contain only a single secondary follicle destined for further development. By the 10th to the 14th day of the cycle, that follicle has formed a **tertiary follicle**, or *mature Graafian* (GRAF-ē-an) *follicle*, roughly 15 mm in diameter. This complex spans the entire width of the ovarian cortex and distorts the ovarian capsule, creating a prominent bulge in the surface of the ovary. The oocyte projects into the **antrum** (AN-trum), or expanded central chamber of the follicle. The antrum is surrounded by a mass of granulosa cells.

Until this time, the primary oocyte has been suspended in prophase of meiosis I. As the development of the tertiary follicle ends, LH levels begin rising, prompting the primary oocyte to complete meiosis I. Instead of producing two secondary oocytes, the first meiotic division yields a secondary oocyte and a small, nonfunctional polar body. The secondary oocyte then enters meiosis II, but stops once again on reaching metaphase. Meiosis II will not be completed unless fertilization occurs.

Generally, on day 14 of a 28-day cycle, the secondary oocyte and the surrounding granulosa cells lose their connections with the follicular wall. The granulosa cells immediately surrounding the secondary oocyte now drift free within the antrum and are known as the **corona radiata** (ko-RŌ-nuh rā-dē-A-tuh).

STEP 4: *Ovulation.* At **ovulation**, the tertiary follicle releases the secondary oocyte. The distended follicular wall ruptures, discharging the follicular contents, including the secondary oocyte and corona radiata, into the pelvic cavity. The sticky follicular fluid keeps the corona radiata attached to the surface of the ovary, where direct contact with projections of the uterine tube or with fluid currents established by the ciliated epithelium lining the uterus can transfer the secondary oocyte to the uterine tube.

STEP 5: *The Formation and Degeneration of the Corpus Luteum.* The empty tertiary follicle initially collapses, and ruptured vessels bleed into the antrum. The remaining granulosa cells then invade the area, proliferating to create an endocrine structure known as the **corpus luteum** (LOO-tē-um), named for its yellow color (*lutea*, yellow). This process occurs under LH stimulation.

The lipids contained in the corpus luteum are used to manufacture steroid hormones known as **progestins** (prō-JES-tinz), principally the steroid **progesterone** (prō-JES-ter-ōn). Although moderate amounts of estrogens are also secreted by the corpus luteum, levels are not as high as they were at ovulation, and progesterone is the principal hormone in the interval after ovulation. Its primary function is to prepare the uterus for pregnancy by stimulating the maturation of the uterine lining and the secretions of uterine glands.

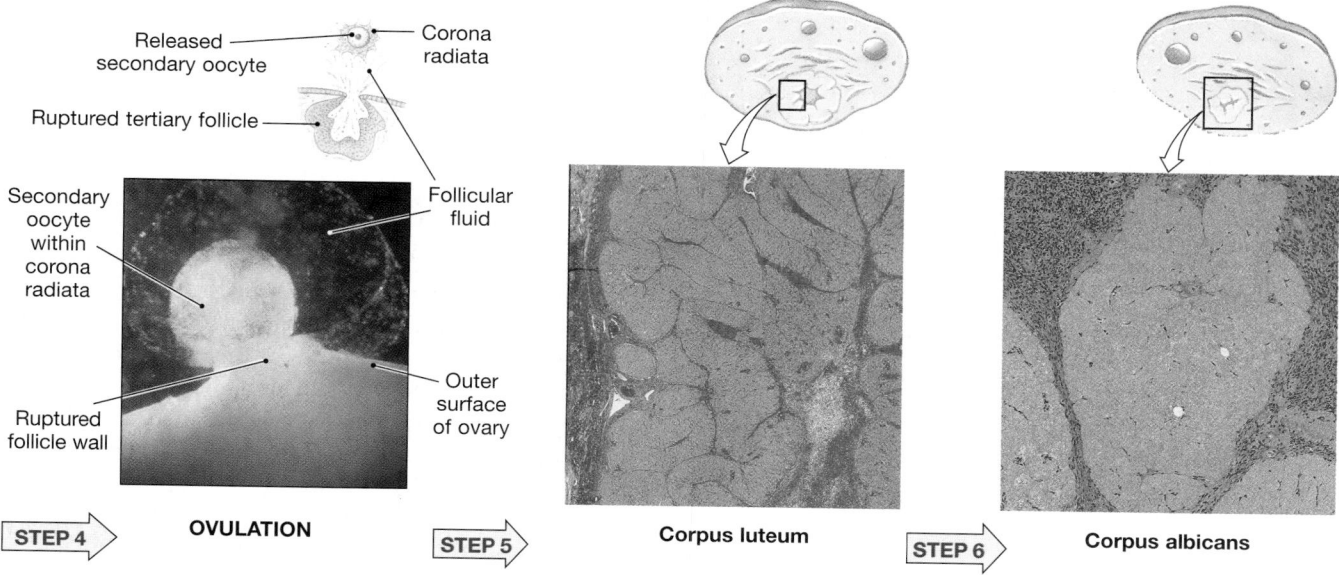

Released secondary oocyte — Corona radiata

Ruptured tertiary follicle

Secondary oocyte within corona radiata

Follicular fluid

Ruptured follicle wall

Outer surface of ovary

STEP 4 **OVULATION** STEP 5 **Corpus luteum** STEP 6 **Corpus albicans**

STEP 6: *Unless Fertilization Occurs, the Corpus Luteum Begins to Degenerate Roughly 12 Days After Ovulation.* Progesterone and estrogen levels then fall markedly. Fibroblasts invade the nonfunctional corpus luteum, producing a knot of pale scar tissue called a **corpus albicans** (AL-bi-kanz). The disintegration, or *involution*, of the corpus luteum marks the end of the ovarian cycle. A new ovarian cycle then begins with the activation of another group of primordial follicles.

Age and Oogenesis

Although many primordial follicles may have developed into primary follicles, and several primary follicles may have been converted to secondary follicles, generally only a single oocyte will be released into the pelvic cavity at ovulation. The rest undergo atresia. At puberty, each ovary contains about 200,000 primordial follicles. Forty years later, few if any follicles remain, although only about 500 will have been ovulated during the interim.

A woman in the United States has a 1 in 70 chance of developing **ovarian cancer** in her lifetime. In 2002, an estimated 23,300 ovarian cancers were diagnosed, and 13,900 women died from this condition. Although ovarian cancer is the third most common reproductive cancer among women, it is the most dangerous because it is seldom diagnosed in its early stages. The prognosis is relatively good for cancers that originate in the general ovarian tissues or from abnormal oocytes. These cancers respond well to some combination of chemotherapy, radiation, and surgery. However, 85 percent of ovarian cancers develop from epithelial cells, and sustained remission can be obtained in only about one-third of the cases of this type. **AM** The Diagnosis and Treatment of Ovarian Cancer

☐ THE UTERINE TUBES

Each **uterine tube** (*Fallopian tube* or *oviduct*) is a hollow, muscular tube measuring roughly 13 cm (5 in.) in length (Figures 28–13, sagittal view; 28–14 and 28–17a●, posterior views). Each uterine tube is divided into three segments (Figure 28–17a●):

1. *The Infundibulum.* The end closest to the ovary forms an expanded funnel, or **infundibulum**, with numerous fingerlike projections that extend into the pelvic cavity. The projections are called **fimbriae** (FIM-brē-ē). The inner surfaces of the infundibulum are lined with cilia that beat toward the middle segment of the uterine tube, called the *ampulla*.

2. *The Ampulla.* The thickness of the smooth muscle layers in the wall of the middle segment, or **ampulla**, of the uterine tube gradually increases as the tube approaches the uterus.

3. *The Isthmus.* The ampulla leads to the **isthmus** (IS-mus), a short segment connected to the uterine wall.

Histology of the Uterine Tube

The epithelium lining the uterine tube is composed of ciliated columnar epithelial cells with scattered mucin-secreting cells (Figure 28–17c●). The mucosa is surrounded by concentric layers of smooth muscle (Figure 28–17b●). Oocyte transport involves a combination of ciliary movement and peristaltic contractions in the walls of the uterine tube. A few hours before ovulation, sympathetic and parasympathetic nerves from the hypogastric plexus "turn on" this beating pattern and initiate peristalsis. The uterine tubes transport a secondary oocyte for final maturation and fertilization. It normally takes three to four days for an oocyte to travel from the infundibulum to the

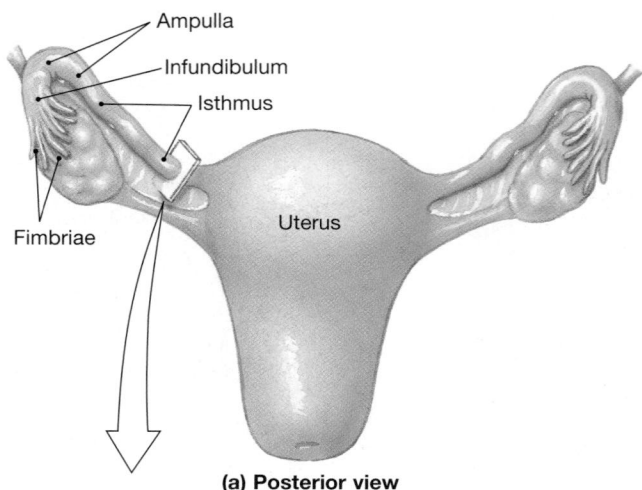

Ampulla
Infundibulum
Isthmus
Fimbriae
Uterus

(a) Posterior view

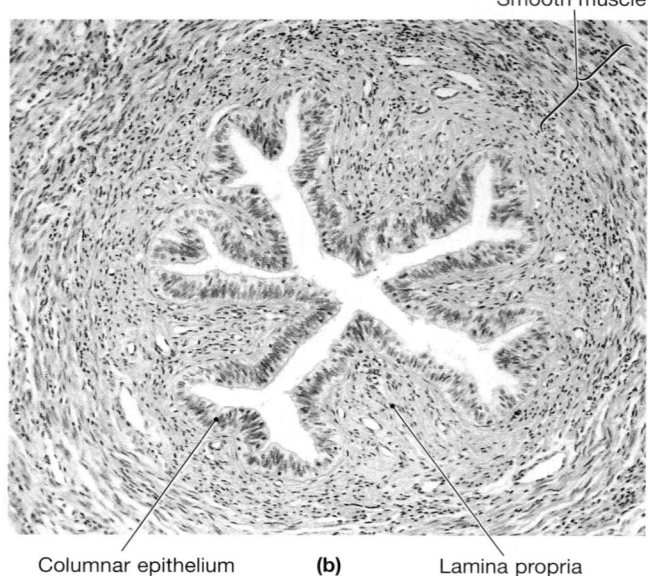

Smooth muscle
Columnar epithelium   **(b)**   Lamina propria

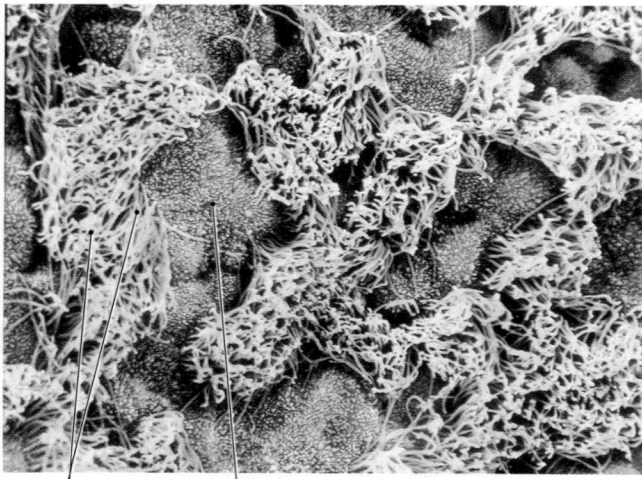

Cilia   Microvilli of mucin-secreting cells
(c)

●**FIGURE 28–17**
The Uterine Tubes. **(a)** Regions of the uterine tubes. **(b)** A sectional view of the isthmus. (LM × 122) **(c)** A colorized SEM of the ciliated lining of the uterine tube. *Professor P. Motta, Dept. of Anatomy, University "La Sapienza," Rome/Science Photo Library/Photo Researchers, Inc.*

uterine cavity. If fertilization is to occur, the secondary oocyte must encounter spermatozoa during the first 12–24 hours of its passage. Fertilization typically occurs near the boundary between the ampulla and isthmus of the uterine tube.

Along with its transport function, the uterine tube provides a nutrient-rich environment that contains lipids and glycogen. This mixture supplies nutrients to both spermatozoa and a developing *pre-embryo* (the cell cluster produced by the initial divisions that follow fertilization). Unfertilized oocytes degenerate in the terminal portions of the uterine tubes or within the uterus.

Pelvic inflammatory disease (PID) in women is a major cause of sterility (infertility). An infection of the uterine tubes, PID affects an estimated 850,000 women each year in the United States. In many cases, sexually transmitted pathogens are involved. As much as 50–80 percent of all first cases may be due to infection by *Neisseria gonorrhoeae*, the organism responsible for symptoms of **gonorrhea** (gon-ō-RĒ-uh), a sexually transmitted disease. Invasion of the region by bacteria that normally reside in the vagina can also cause PID. Symptoms of PID include fever, lower abdominal pain, and elevated white blood cell counts. In severe cases, the infection can spread to other visceral organs or produce a generalized peritonitis.

Sexually active women in the 15–24 age group have the highest incidence of PID. Whereas the use of an oral contraceptive decreases the risk of infection, the presence of an intrauterine device (IUD) may increase the risk by 1.4–7.3 times. Treatment with antibiotics may control the condition, but chronic abdominal pain may persist. In addition, damage and scarring of the uterine tubes can cause infertility by preventing the passage of a zygote to the uterus. Recently, another sexually transmitted bacterium, belonging to the genus *Chlamydia*, has been identified as the probable cause of up to 50 percent of all cases of PID. Despite the fact that women with this infection may develop few, if any, symptoms, scarring of the uterine tubes can still produce infertility.

☐ THE UTERUS

The **uterus** provides mechanical protection, nutritional support, and waste removal for the developing *embryo* (weeks 1–8) and *fetus* (from week 9 to delivery). In addition, contractions in the muscular wall of the uterus are important in ejecting the fetus at the time of birth.

The uterus is a small, pear-shaped organ about 7.5 cm (3 in.) long with a maximum diameter of 5 cm (2 in.). It weighs 30–40 g (1–1.4 oz). In its normal position, the uterus bends anteriorly near its base, a condition known as *anteflexion* (an-tē-FLEK-shun). In this position, the uterus tilts anteriorly, covering the superior and posterior surfaces of the urinary bladder (Figure 28–13•, p. 1065). If the uterus bends backward toward the sacrum, the condition is termed *retroflexion* (re-trō-FLEK-shun). Retroflexion, which occurs in about 20 percent of adult women, has no clinical significance. (A retroflexed uterus generally becomes anteflexed spontaneously during the third month of pregnancy.)

Suspensory Ligaments of the Uterus

In addition to the broad ligament, three pairs of suspensory ligaments stabilize the position of the uterus and limit its range of movement (Figure 28–18a,b•). The **uterosacral** (ū-te-rō-SĀ-kral) **ligaments** extend from the lateral surfaces of the uterus to the anterior face of the sacrum, keeping the body of the uterus from moving inferiorly and anteriorly. The **round ligaments** arise on the lateral margins of the uterus just posterior and inferior to the attachments of the uterine tubes. These ligaments extend through the inguinal canal and end in the connective tissues of the external genitalia. The round ligaments restrict primarily posterior movement of the uterus. The **lateral** (*cardinal*) **ligaments** extend from the base of the uterus and vagina to the lateral walls of the pelvis. These ligaments tend to prevent inferior movement of the uterus. Additional mechanical support is provided by the muscles and fascia of the pelvic floor. ⊂⊃ p. 354

Internal Anatomy of the Uterus

We can divide the uterus into two anatomical regions (Figure 28–18c•, p. 1072): the body and the cervix. The uterine **body**, or *corpus*, is the largest region of the uterus. The **fundus** is the rounded portion of the body superior to the attachment of the uterine tubes. The body ends at a constriction known as the uterine **isthmus**. The **cervix** (SER-viks) is the inferior portion of the uterus that extends from the isthmus to the vagina.

The tubular cervix projects about 1.25 cm (0.5 in.) into the vagina. Within the vagina, the distal end of the cervix forms a curving surface that surrounds the **cervical os** (*os*, an opening or mouth), or *external orifice* of the uterus. The cervical os leads into the **cervical canal**, a constricted passageway that opens into the **uterine cavity** of the body at the **internal os**, or *internal orifice*.

The uterus receives blood from branches of the **uterine arteries**, which arise from branches of the *internal iliac arteries*, and the *ovarian arteries*, which arise from the abdominal aorta inferior to the renal arteries. The arteries to the uterus are extensively interconnected. This arrangement helps ensure a reliable flow of blood to the organ despite changes in its position and the changes in uterine shape that accompany pregnancy. Numerous veins and lymphatic vessels also supply each portion of the uterus. The organ is innervated by autonomic fibers from the hypogastric plexus (sympathetic) and from sacral segments S_3 and S_4 (parasympathetic). Sensory information reaches the central nervous system within the dorsal roots of spinal nerves T_{11} and T_{12}. The most delicate anesthetic procedures used during labor and delivery, known as *segmental blocks*, target only spinal nerves $T_{10}-L_1$.

The Uterine Wall

The dimensions of the uterus are highly variable. In adult women of reproductive age who have not given birth, the uterine wall is about 1.5 cm (0.5 in.) thick. The wall has a thick, outer, muscular **myometrium** (mī-ō-MĒ-trē-um; *myo-*, muscle + *metra*, uterus) and a thin, inner, glandular **endometrium** (en-dō-MĒ-trē-um), or *mucosa*. The fundus and the posterior surface of the uterine body and isthmus are covered by a serous membrane that is continuous with the peritoneal lining. This incomplete serosa is called the **perimetrium** (Figure 28–18c•).

The endometrium contributes about 10 percent to the mass of the uterus. The glandular and vascular tissues of the endometrium support the physiological demands of the growing fetus. Vast numbers of uterine glands open onto the endometrial surface and extend deep into the lamina propria, almost to the myometrium. Under the influence of estrogen, the uterine glands, blood vessels, and epithelium change with the phases of the monthly *uterine cycle.*

The myometrium, the thickest portion of the uterine wall, forms almost 90 percent of the mass of the uterus. Smooth muscle in the myometrium is arranged into longitudinal, circular, and oblique layers. The smooth muscle tissue of the myometrium provides much of the force needed to move a large fetus out of the uterus and into the vagina.

HISTOLOGY OF THE UTERUS We can divide the endometrium into a **functional zone**—the layer closest to the uterine cavity—and an outer **basilar zone**, adjacent to the myometrium (Figure 28–19a•). The functional zone contains most of the uterine glands and contributes most of the endometrial thickness. The basilar zone attaches the endometrium to the myometrium and contains the terminal branches of the tubular endometrial glands.

Within the myometrium, branches of the uterine arteries form **arcuate arteries**, which encircle the endometrium. **Radial arteries** supply **straight arteries**, which deliver blood to the basilar zone of the endometrium, and **spiral arteries**, which supply the functional zone (Figure 28–19b•).

The structure of the basilar zone remains relatively constant over time, but that of the functional zone undergoes cyclical changes in response to sex hormone levels. These cyclical changes produce the characteristic histological features of the uterine cycle.

Cervical cancer is the most common cancer of the reproductive system in women age 15–34. Roughly 12,800 new cases of invasive cervical cancer are diagnosed each year in the United States, and approximately 33 percent of the individuals will eventually die of the condition. Another 34,900 patients are diagnosed with a less aggressive form of cervical cancer. Cervical and other uterine tumors and cancers are discussed in the *Applications Manual*. **AM** Uterine Tumors and Cancers

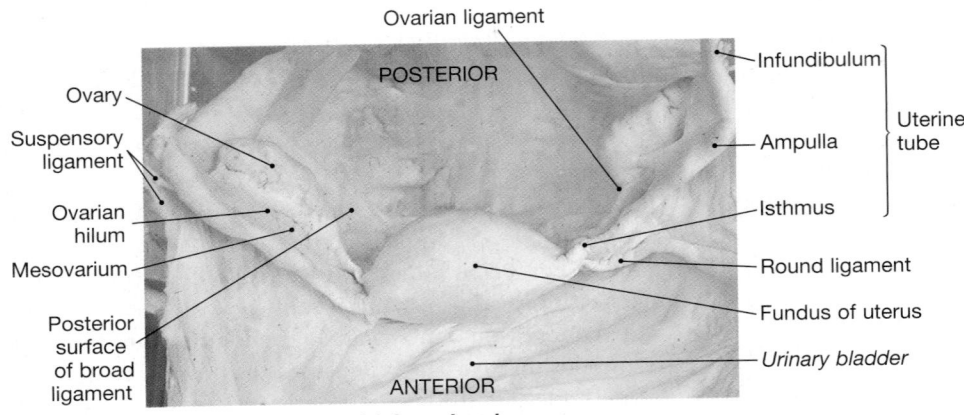

Ovarian ligament

POSTERIOR

Ovary

Suspensory ligament

Ovarian hilum

Mesovarium

Posterior surface of broad ligament

ANTERIOR

Infundibulum

Ampulla } Uterine tube

Isthmus

Round ligament

Fundus of uterus

Urinary bladder

(a) Superior view

●**FIGURE 28–18**
The Uterus. **(a)** The pelvic cavity in superior view, showing the position of the uterus relative to other structures. **(b)** The ligaments that stabilize the position of the uterus in the pelvic cavity. **(c)** A posterior view with the left uterus, uterine tube, and ovary shown in section. **ATLAS** Scan 14

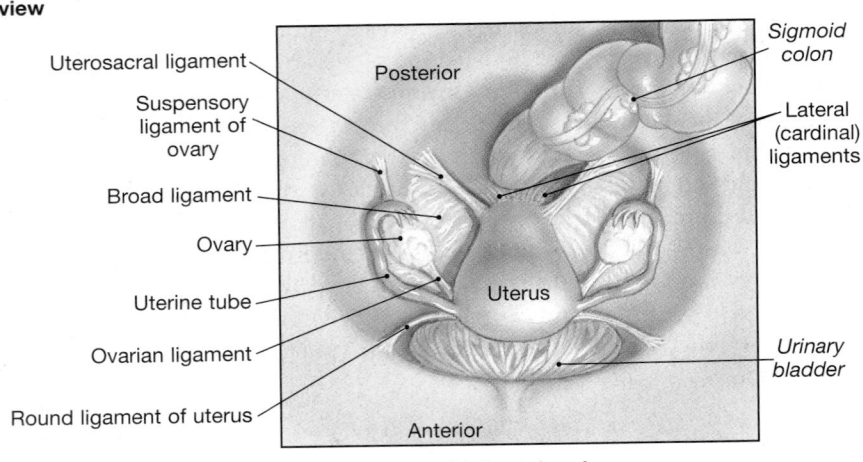

Uterosacral ligament

Suspensory ligament of ovary

Broad ligament

Ovary

Uterine tube

Ovarian ligament

Round ligament of uterus

Posterior

Sigmoid colon

Lateral (cardinal) ligaments

Uterus

Urinary bladder

Anterior

(b) Superior view

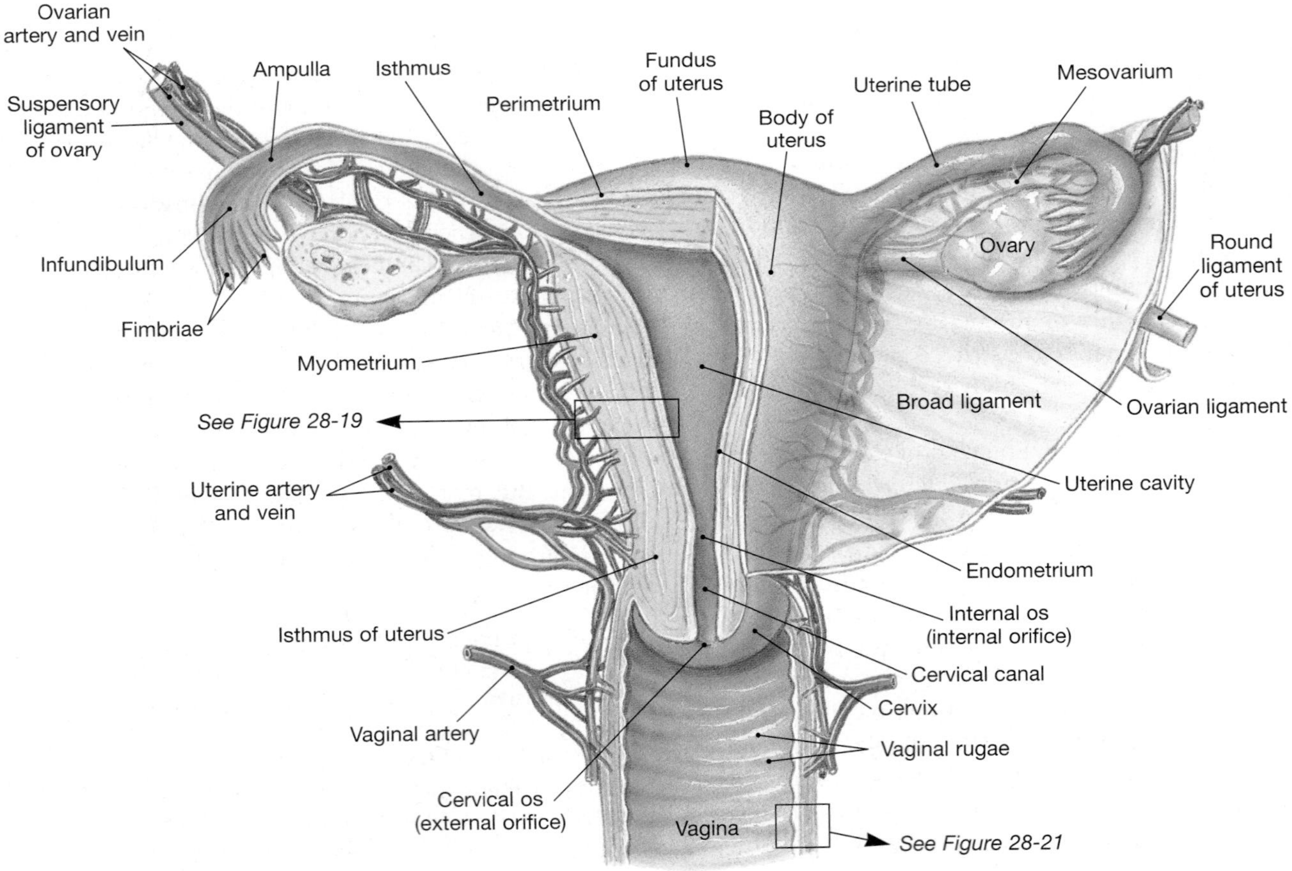

Ovarian artery and vein

Suspensory ligament of ovary

Infundibulum

Fimbriae

Myometrium

See Figure 28-19

Uterine artery and vein

Isthmus of uterus

Vaginal artery

Cervical os (external orifice)

Ampulla

Isthmus

Perimetrium

Fundus of uterus

Body of uterus

Uterine tube

Mesovarium

Ovary

Round ligament of uterus

Broad ligament

Ovarian ligament

Uterine cavity

Endometrium

Internal os (internal orifice)

Cervical canal

Cervix

Vaginal rugae

Vagina

See Figure 28-21

(c) Posterior view

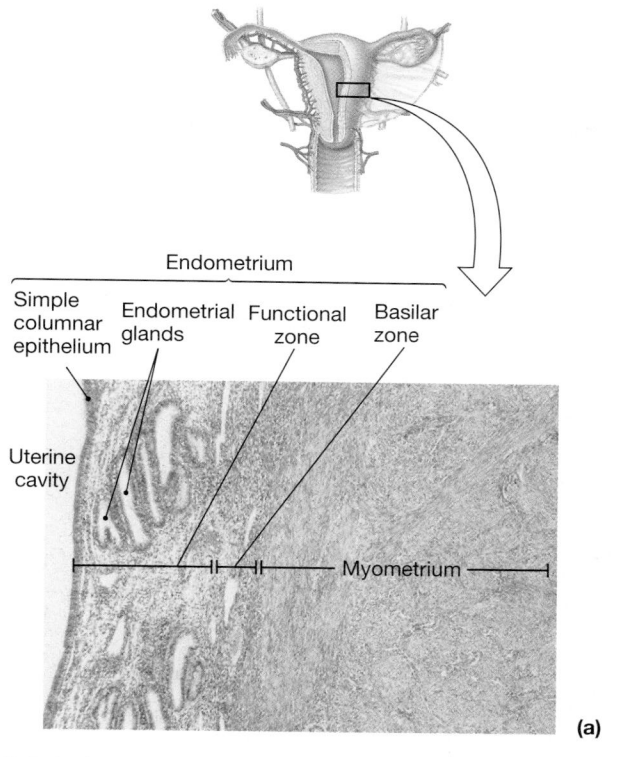

Endometrium

Simple columnar epithelium
Endometrial glands
Functional zone
Basilar zone

Uterine cavity

Myometrium

(a)

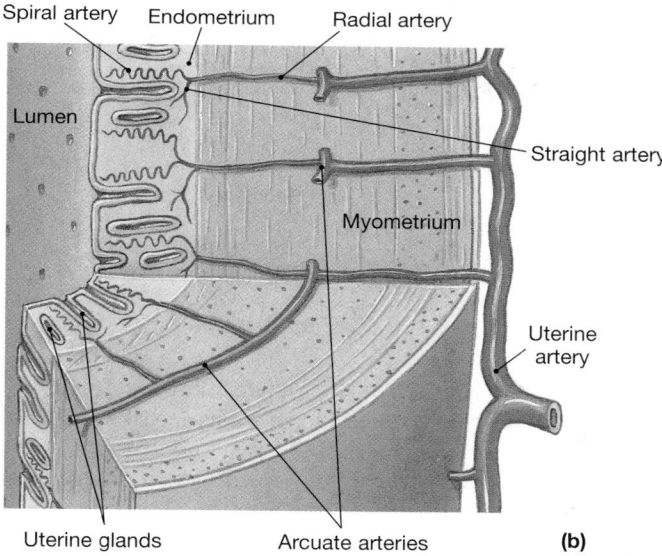

Spiral artery
Endometrium
Radial artery

Lumen

Straight artery

Myometrium

Uterine artery

Uterine glands
Arcuate arteries

(b)

●**FIGURE 28–19**
The Uterine Wall. **(a)** The basic structure of the endometrium. (LM × 32) **(b)** A sectional view of the uterine wall, showing the endometrial regions and the circulatory supply to the endometrium.

The Uterine Cycle

The **uterine cycle**, or *menstrual* (MEN-stroo-al) *cycle*, is a repeating series of changes in the structure of the endometrium. The uterine cycle averages 28 days in length, but it can range from 21 to 35 days in healthy women of reproductive age. We can divide the cycle into three phases: (1) *menses*, (2) the *proliferative phase*,

and (3) the *secretory phase*. The histological appearance of the endometrium during each phase is shown in Figure 28–20●. The phases occur in response to hormones associated with the regulation of the ovarian cycle. Menses and the proliferative phase occur during the follicular phase of the ovarian cycle. The secretory phase corresponds to the luteal phase of the cycle. We shall consider the regulatory mechanism in a later section.

MENSES The uterine cycle begins with the onset of **menses** (MEN-sēz), an interval marked by the degeneration of the functional zone of the endometrium. This degeneration occurs in patches and is caused by the constriction of the spiral arteries, which reduces blood flow to areas of endometrium. Deprived of oxygen and nutrients, the secretory glands and other tissues in the functional zone begin to deteriorate. Eventually, the weakened arterial walls rupture, and blood pours into the connective tissues of the functional zone. Blood cells and degenerating tissues then break away and enter the uterine lumen, to be lost by passage through the cervical os and into the vagina. Only the functional zone is affected, because the deeper, basilar zone is provided with blood from the straight arteries, which remain unconstricted.

The sloughing off of tissue is gradual, and at each site repairs begin almost at once. Nevertheless, before menses has ended, the entire functional zone has been lost (Figure 28–20a●). The process of endometrial sloughing, called **menstruation** (men-stroo-Ā-shun), generally lasts from one to seven days. Over this period roughly 35 to 50 ml of blood is lost. The process can be relatively painless. Painful menstruation, or **dysmenorrhea**, can result from uterine inflammation and contraction or from conditions involving adjacent pelvic structures.

THE PROLIFERATIVE PHASE The basilar zone, including the basal parts of the uterine glands, survives menses intact. In the days after menses, the epithelial cells of these glands multiply and spread across the endometrial surface, restoring the integrity of the uterine epithelium (Figure 28–20b●). Further growth and vascularization result in the complete restoration of the functional zone. As this reorganization proceeds, the endometrium is in the **proliferative phase**. The restoration occurs at the same time as the enlargement of primary and secondary follicles in the ovary. The proliferative phase is stimulated and sustained by estrogens secreted by the developing ovarian follicles.

By the time ovulation occurs, the functional zone is several millimeters thick, and prominent mucous glands extend to the border with the basilar zone. At this time, the endometrial glands are manufacturing a mucus rich in glycogen. The entire functional zone is highly vascularized, with small arteries spiraling toward the inner surface from larger arteries in the myometrium.

THE SECRETORY PHASE During the **secretory phase** of the uterine cycle, the endometrial glands enlarge, accelerating their rates of secretion, and the arteries that supply the uterine wall elongate and spiral through the tissues of the functional zone (Figure 28–20c●). This activity occurs under the combined stimulatory

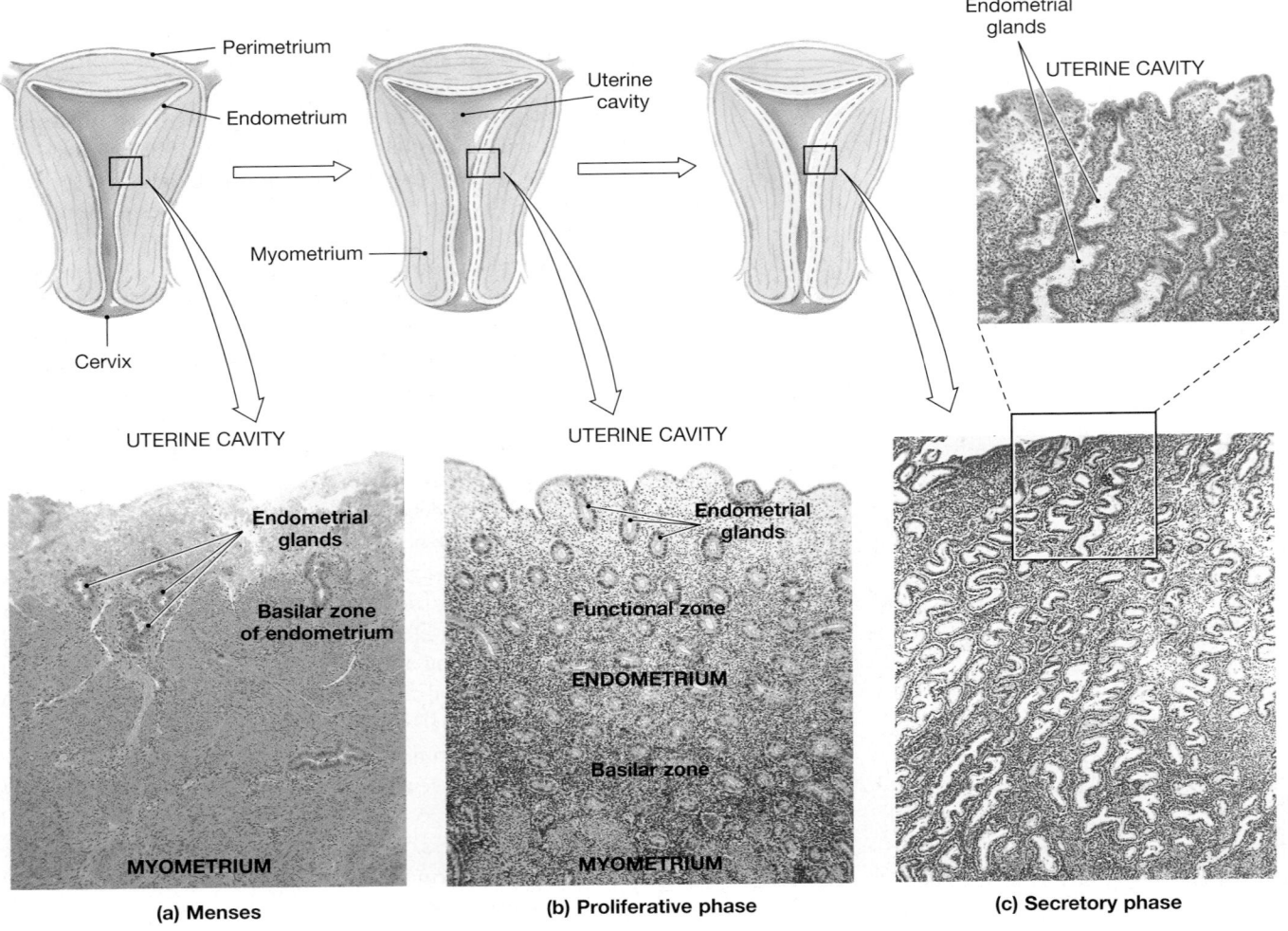

●**FIGURE 28–20**
The Uterine Cycle. The appearance of the endometrium **(a)** at menses (LM × 63) and during the **(b)** proliferative
(LM × 66) and **(c)** secretory phases (LM × 52; blowup, LM × 150) of the uterine cycle.

effects of progestins and estrogens from the corpus luteum. The secretary phase begins at the time of ovulation and persists as long as the corpus luteum remains intact.

Secretory activities peak about 12 days after ovulation. Over the next day or two, the glandular activity declines, and the uterine cycle comes to a close as the corpus luteum stops producing stimulatory hormones. A new cycle then begins with the onset of menses and the disintegration of the functional zone. The secretory phase generally lasts 14 days. As a result, you can determine the date of ovulation by counting backward 14 days from the first day of menses.

Some conditions interfere with the usual menstrual cycle. In **endometriosis** (en-dō-mē-trē-Ō-sis), an area of endometrial tissue begins to grow outside the uterus. The cause is unknown; because this condition is most common in the inferior portion of the peritoneum, one possibility is that pieces of endometrium sloughed off during menstruation are in some way forced through the uterine tubes into the peritoneal cavity, where they reattach. The severity of the condition depends on the size of the abnormal mass and its location. Abdominal pain,

bleeding, pressure on adjacent structures, and infertility are common symptoms. As the island of endometrial tissue enlarges, the symptoms become more severe.

Endometriosis can generally be diagnosed by inserting a laparoscope through a small opening in the abdominal wall. Using this device, a physician can inspect the outer surfaces of the uterus and uterine tubes, the ovaries, and the lining of the pelvic cavity. Treatment of endometriosis may involve hormonal therapy to suppress uterine cycles or surgical removal of the endometrial mass. If the condition is widespread, a *hysterectomy* (removal of the uterus) or *oophorectomy* (removal of the ovaries) may be required.

MENARCHE AND MENOPAUSE The uterine cycle begins at puberty. The first cycle, known as **menarche** (me-NAR-kē; *men*, month + *arche* beginning), typically occurs at age 11–12. The cycles continue until age 45–55, at **menopause** (MEN-ō-pawz), the last uterine cycle. In the interim, the regular appearance of uterine cycles is interrupted only by circumstances such as illness, stress, starvation, or pregnancy.

If menarche does not appear by age 16, or if the normal uterine cycle of an adult woman becomes interrupted for six months or more, the condition of **amenorrhea** (ā-men-ō-RĒ-uh) exists. *Primary amenorrhea* is the failure to initiate menses. This condition may indicate developmental abnormalities, such as nonfunctional ovaries, the absence of a uterus, or an endocrine or genetic disorder. It can also result from malnutrition: Puberty is delayed if leptin levels are too low. ∞ p. 637 Transient *secondary amenorrhea* can be caused by severe physical or emotional stresses. In effect, the reproductive system gets "switched off." Factors that cause either type of amenorrhea include drastic weight loss, anorexia nervosa, and severe depression or grief. Amenorrhea has also been observed in marathon runners and other women engaged in training programs that require sustained high levels of exertion and severely reduce body lipid reserves.

☐ THE VAGINA

The **vagina** is an elastic, muscular tube extending between the cervix and the *vestibule*, a space bounded by the female external genitalia. The vagina has an average length of 7.5–9 cm (3–3.5 in.), but is highly distensible, so its size varies.

At the proximal end of the vagina, the cervix projects into the **vaginal canal**. The shallow recess surrounding the cervical protrusion is known as the **fornix** (FOR-niks). The vagina lies parallel to the rectum, and the two are in close contact posteriorly. Anteriorly, the urethra extends along the superior wall of the vagina from the urinary bladder to the external urethral meatus, which opens into the vestibule. The primary blood supply of the vagina is via the **vaginal branches** of the internal iliac (or uterine) arteries and veins. Innervation is from the hypogastric plexus, sacral nerves S_2–S_4, and branches of the pudendal nerve. ∞ p. 444

The vagina has three major functions:

1. It serves as a passageway for the elimination of menstrual fluids.

2. It receives the penis during sexual intercourse and holds spermatozoa prior to their passage into the uterus.

3. It forms the inferior portion of the *birth canal*, through which the fetus passes during delivery.

Histology of the Vagina

In sectional view, the lumen of the vagina appears constricted, forming a rough H. The vaginal walls contain a network of blood vessels and layers of smooth muscle (Figure 28–21●). The lining is moistened by the secretions of the cervical glands and by the movement of water across the permeable epithelium. The vagina and vestibule are separated by the **hymen** (HĪ-men), an elastic epithelial fold that partially or completely blocks the entrance to the vagina before the initial sexual intercourse. The two *bulbospongiosus muscles* extend along either side of the vaginal entrance, which is constricted by their contractions. ∞ p. 354 These muscles cover the *vestibular bulbs*, masses of erectile tissue

on either side of the vaginal entrance. The vestibular bulbs have the same embryological origins as the corpus spongiosum of the penis in males.

The vaginal lumen is lined by a nonkeratinized stratified squamous epithelium (Figure 28–21●). In the relaxed state, this epithelium is thrown into folds called *rugae*. The underlying lamina propria is thick and elastic, and it contains small blood vessels, nerves, and lymph nodes. The vaginal mucosa is surrounded by an elastic *muscularis* layer, with layers of smooth muscle fibers arranged in circular and longitudinal bundles continuous with the uterine myometrium. The portion of the vagina adjacent to the uterus has a serosal covering that is continuous with the pelvic peritoneum. Along the rest of the vagina, the muscularis layer is surrounded by an *adventitia* of fibrous connective tissue.

The vagina contains a population of resident bacteria, usually harmless, supported by nutrients in the cervical mucus. The metabolic activity of these bacteria creates an acidic environment, which restricts the growth of many pathogens. *Vaginitis* (va-jin-Ī-tis), an inflammation of the vaginal canal, is caused by fungi, bacteria, or parasites. In addition to any discomfort that may result, the condition may affect the survival of spermatozoa and thereby reduce fertility. An acidic environment also inhibits the motility of sperm; for this reason, the buffers in semen are important to successful fertilization. **AM** Vaginitis

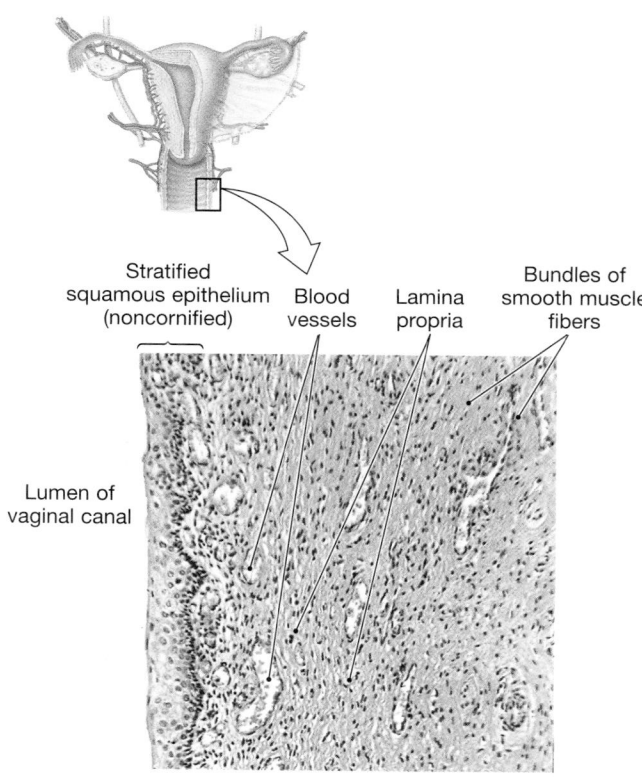

●**FIGURE 28–21**
The Histology of the Vagina. (LM × 27)

The hormonal changes associated with the ovarian cycle also affect the vaginal epithelium. A *vaginal smear* is a sample of epithelial cells shed at the surface of the vagina. By examining these cells, a clinician can estimate the corresponding stage in the ovarian and uterine cycles. This technique is an example of *exfoliative cytology*, a diagnostic procedure. ∞ p. 118

■ THE EXTERNAL GENITALIA

The area containing the female external genitalia is the **vulva** (VUL-vuh), or *pudendum* (Figure 28–22●). The vagina opens into the **vestibule**, a central space bounded by small folds known as the **labia minora** (LĀ-bē-uh mi-NOR-uh; singular, *labium minus*). The labia minora are covered with a smooth, hairless skin. The urethra opens into the vestibule just anterior to the vaginal entrance. The **paraurethral glands**, or *Skene's glands*, discharge into the urethra near the external urethral opening. Anterior to this opening, the **clitoris** (KLI-to-ris) projects into the vestibule. A small, rounded tissue projection, the clitoris is the female equivalent of the penis, derived from the same embryonic structures. Internally, it contains erectile tissue comparable to the corpora cavernosa of the penis. The clitoris engorges with blood during sexual arousal. A small erectile *glans* sits atop it; extensions of the labia minora encircle the body of the clitoris, forming its **prepuce**, or *hood*. **AM**

EMBRYOLOGY SUMMARY 21: The Development of the Reproductive System

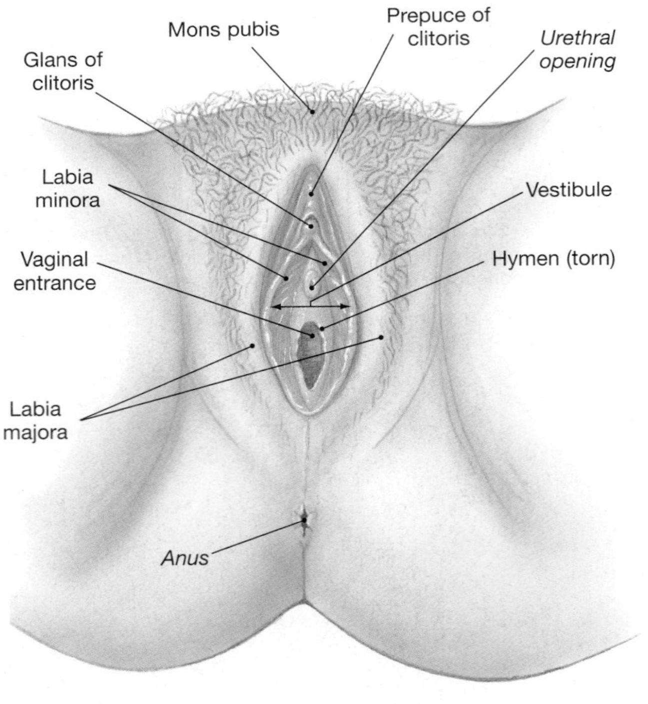

●FIGURE 28–22
The Female External Genitalia

A variable number of small **lesser vestibular glands** discharge their secretions onto the exposed surface of the vestibule, keeping it moist. During sexual arousal, a pair of ducts discharges the secretions of the **greater vestibular glands** (*Bartholin's glands*) into the vestibule near the posterolateral margins of the vaginal entrance. These mucous glands have the same embryological origins as the bulbourethral glands of males.

The outer limits of the vulva are established by the mons pubis and the labia majora. The bulge of the **mons pubis** is created by adipose tissue deep to the skin anterior to the pubic symphysis. Adipose tissue also accumulates within the **labia majora** (singular, *labium majus*), prominent folds of skin that encircle and partially conceal the labia minora and adjacent structures. The outer margins of the labia majora and the mons pubis are covered with coarse hair, but the inner surfaces of the labia majora are relatively hairless. Sebaceous glands and scattered apocrine sweat glands release their secretions onto the inner surface of the labia majora, moistening and lubricating them.

■ THE MAMMARY GLANDS

A newborn infant cannot fend for itself, and several of its key systems have yet to complete their development. Over the initial period of adjustment to an independent existence, the infant gains nourishment from the milk secreted by the maternal **mammary glands**. Milk production, or **lactation** (lak-TĀ-shun), occurs in these glands. In females, mammary glands are specialized organs of the integumentary system that are controlled mainly by hormones of the reproductive system and by the *placenta*, a temporary structure that provides the embryo or fetus with nutrients.

On each side, a mammary gland lies in the subcutaneous tissue of the **pectoral fat pad** deep to the skin of the chest (Figure 28–23a●). Each breast bears a **nipple**, a small conical projection where the ducts of the underlying mammary gland open onto the body surface. The reddish-brown skin around each nipple is the **areola** (a-RĒ-ō-luh). Large sebaceous glands deep to the areolar surface give it a grainy texture.

The glandular tissue of the mammary gland consists of separate lobes, each containing several secretory lobules. Ducts leaving the lobules converge, giving rise to a single **lactiferous** (lak-TIF-e-rus) **duct** in each lobe (Figure 28–23●). Near the nipple, that lactiferous duct enlarges, forming an expanded chamber called a **lactiferous sinus**. Typically, 15–20 lactiferous sinuses open onto the surface of each nipple. Dense connective tissue surrounds the duct system and forms partitions that extend between the lobes and the lobules. These bands of connective tissue, the *suspensory ligaments of the breast*, originate in the dermis of the overlying skin. A layer of areolar tissue separates the mammary gland complex from the underlying pectoralis muscles. Branches of the *internal thoracic artery* supply blood to each mammary gland (Figure 21–23●, p. 754).

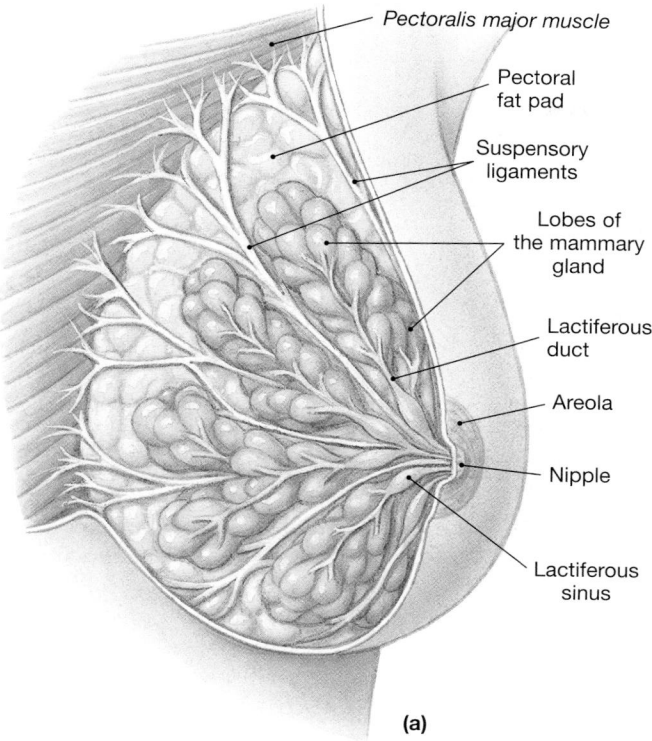

Pectoralis major muscle

Pectoral fat pad

Suspensory ligaments

Lobes of the mammary gland

Lactiferous duct

Areola

Nipple

Lactiferous sinus

(a)

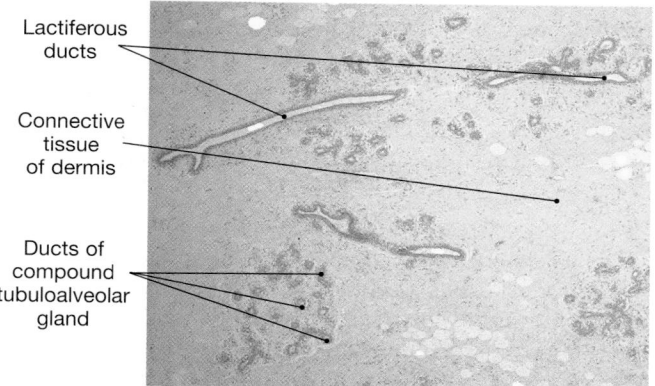

Lactiferous ducts

Connective tissue of dermis

Ducts of compound tubuloalveolar gland

(b) Inactive mammary gland

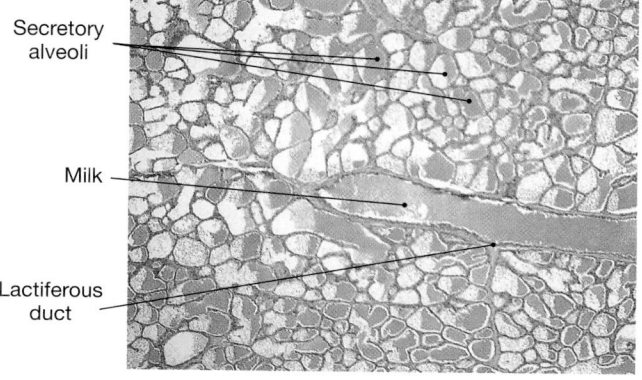

Secretory alveoli

Milk

Lactiferous duct

(c) Active mammary gland

•**FIGURE 28–23**
The Mammary Glands. **(a)** The mammary gland of the left breast.
(b) An inactive mammary gland of a nonpregnant woman. (LM × 53)
(c) An active mammary gland of a nursing woman. (LM × 119)
ATLAS Scan 15

Figure 28–23b,c● compares the histological organizations of inactive and active mammary glands. The inactive, or *resting*, mammary gland is dominated by a duct system rather than by active glandular cells. The size of the mammary glands in a nonpregnant woman reflects primarily the amount of adipose tissue present, not the amount of glandular tissue. The secretory apparatus does not complete its development unless pregnancy occurs. The active mammary gland is a tubuloalveolar gland, consisting of multiple glandular tubes that end in secretory alveoli. In Chapter 29, we will discuss the hormonal mechanisms involved.

✓ As the result of infections such as gonorrhea, scar tissue can block the lumen of each uterine tube. How would this blockage affect a woman's ability to conceive?

✓ What is the advantage of the acidic pH of the vagina?

✓ Which layer of the uterus is sloughed off during menstruation?

✓ Would the blockage of a single lactiferous sinus interfere with the delivery of milk to the nipple? Explain.

Answers start on page Q-1

 The anatomy of the female reproductive organs can be reviewed on the **Anatomy CD-ROM**: Reproductive System/Female Flythrough.

☐ **HORMONES AND THE FEMALE REPRODUCTIVE CYCLE**

The activity of the female reproductive tract is under hormonal control that involves an interplay between secretions of both the pituitary gland and the gonads. But the regulatory pattern in females is much more complicated than in males, because it must coordinate the ovarian and uterine cycles. Circulating hormones control the **female reproductive cycle**, coordinating the ovarian and uterine cycles to ensure proper reproductive function. If the two cycles cannot be coordinated in a normal manner, infertility results. A woman who fails to ovulate cannot conceive, even if her uterus is perfectly normal. A woman who ovulates normally, but whose uterus is not ready to support an embryo, will be just as infertile. Because the processes are complex and difficult to study, many of the biochemical details of the female reproductive cycle still elude us, but the general patterns are reasonably clear.

Breast Cancer

The mammary glands are cyclically stimulated by the changing levels of circulating reproductive hormones that accompany the uterine cycle. The effects usually go unnoticed, but occasional discomfort or even inflammation of mammary gland tissues can occur late in the cycle. If inflamed lobules become walled off with scar tissue, **cysts** are created. Clusters of cysts can be felt in the breast as discrete masses, a condition known as **fibrocystic disease**. Because the symptoms are similar to those of breast cancer, biopsies may be needed to distinguish between this benign condition and breast cancer.

Breast cancer is a malignant, metastasizing cancer of the mammary gland. It is the leading cause of death in women between the ages of 35 and 45, but it is most common in women over age 50. Approximately 39,600 deaths will occur in the United States from breast cancer in 2002, and approximately 203,500 new cases will be reported. An estimated 12 percent of women in the United States will develop breast cancer at some point in their lifetime, and the rate is steadily rising. The incidence is highest among Caucasian-Americans, somewhat lower in African-Americans, and lowest in Asian-Americans and American Indians. Notable risk factors include (1) a family history of breast cancer, (2) a first pregnancy after age 30, and (3) early menarche (first menstrual period) or late menopause (last menstrual period). Breast cancers in males are very rare, but about 400 men die from the disease each year in the United States.

Despite repeated studies, no links have been proven between breast cancer and oral contraceptive use, estrogen therapy, fat consumption, or alcohol use. It appears likely that multiple factors are involved. In some families an inherited genetic variation has been linked to higher than normal risk of developing the disease. However, most women never develop breast cancer—even women in families with a history of the disease. Adequate amounts of nutrients and vitamins and a diet rich in fruits and vegetables appear to offer some protection against the development of breast cancer. Mothers who breast-fed (nursed) their babies have a 20 percent lower incidence of breast cancer after menopause than do mothers who did not breast-feed. The reason is not known. Adding to the mystery, nursing does not appear to affect the incidence of premenopausal breast cancer.

Early detection of breast cancer is the key to reducing mortalities. *Most cases of breast cancer are found through self-examination*, but the use of clinical screening techniques has increased in recent years. **Mammography** involves the use of X rays to examine breast tissues. The radiation dosage can be low, because only soft tissues must be penetrated. This procedure gives the clearest picture of conditions in the breast tissues. Ultrasound can provide some information, but the resulting images lack the detail of standard mammograms.

For treatment to be successful, breast cancer must be identified while it is still relatively small and localized. Once it has grown larger than 2 cm (0.78 in.), the chances of long-term

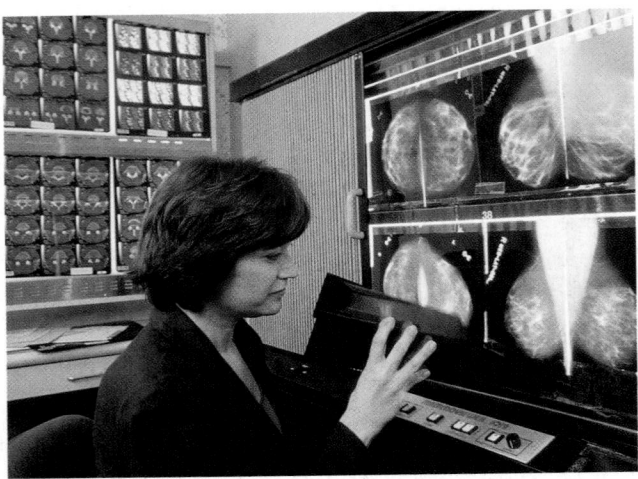

survival worsen. A poor prognosis also follows if the cancer cells have spread through the lymphatic system to the axillary lymph nodes. If the nodes are not yet involved, the chances of five-year survival are about 82 percent, but if four or more nodes are involved, the survival rate drops to 21 percent.

Treatment of breast cancer begins with the removal of the tumor. Because in many cases the cancer cells begin to spread before the condition is diagnosed, part or all of the affected mammary gland is surgically removed:

- In a **segmental mastectomy**, or *lumpectomy*, only a portion of the mammary gland is removed.

- In a **total mastectomy**, the entire mammary gland is removed, but other tissues are left intact.

- In a **radical mastectomy** (rarely done since the 1970s), the pectoralis muscles, the mammary gland, and the axillary lymph nodes are removed. In a *modified radical mastectomy*, the most common operation, the mammary gland and nodes are removed, but the muscular tissue is left intact.

A combination of chemotherapy, radiation treatments, and hormone treatments may be used to supplement the surgical procedures. *Tamoxifen*™ is an antiestrogen drug that is used to treat some cases of breast cancer. It is more effective than conventional chemotherapy for treating breast cancer in women over 50, and it has fewer unpleasant side effects. It can also be used in addition to regular chemotherapy in the treatment of advanced-stage disease. As an added bonus, tamoxifen prevents and even reverses the osteoporosis of aging. This drug has risks, however: When given to premenopausal women, tamoxifen can cause amenorrhea and hot flashes similar to those of menopause. Tamoxifen has also been linked to an increased risk of endometrial cancer, and potentially to liver cancer as well. It has been proposed that this drug and others like it be used to prevent breast cancer in women at risk for the disease as well as to treat it.

As in males, GnRH from the hypothalamus regulates reproductive function in females. However, in females, the GnRH pulse frequency and amplitude (amount secreted per pulse) change throughout the course of the ovarian cycle. If the hypothalamus were a radio, the pulse frequency would correspond to the radio frequency and the amplitude would be the volume control. We will consider changes in pulse frequency, as their effects are both dramatic and reasonably well understood. Changes in GnRH pulse frequency are essential to normal FSH and LH production and thus to normal ovulation. If GnRH is absent or is supplied at a constant rate (without pulses), FSH and LH secretion will stop in a matter of hours.

When you shift from one radio frequency to another, you change stations. You may then hear a very different message—from talk radio to jazz or from hard rock to classical. When the GnRH pulse frequency shifts, the pituitary gland "hears" a different message. The cells responsible for FSH and LH production are called *gonadotropes*. When the GnRH pulse frequency shifts, these cells change their pattern of FSH and LH production. For example, at one pulse frequency, the gonadotropes respond by preferentially secreting FSH, whereas at another frequency, LH is the primary hormone released. FSH and LH production also occurs in pulses that follow the rhythm of GnRH pulses. The precise effects of these pulses on target cells are not known. Changes in GnRH pulse frequency are controlled primarily by circulating levels of estrogens and progestins. Estrogens increase the GnRH pulse frequency, and progestins decrease it.

Hormones and the Follicular Phase

Follicular development begins under FSH stimulation; each month some of the primordial follicles begin to develop into primary follicles. As the follicles enlarge, thecal cells start producing *androstenedione*, a steroid hormone that is a key intermediate in the synthesis of most sex hormones (Figure 28–24●). Androstenedione is absorbed by the granulosa cells and converted to estrogens. In addition, small quantities of estrogens are secreted by interstitial cells scattered throughout the ovarian stroma. Circulating estrogens are bound primarily to albumins, with lesser amounts carried by gonadal steroid-binding globulin (GBG).

Three estrogens circulate in the bloodstream: (1) estradiol, (2) estrone, and (3) estriol. All have similar effects on their target tissues. **Estradiol** (es-tra-DĪ-ol) is the most abundant estrogen, and its effects on target tissues are most pronounced. It is the dominant hormone prior to ovulation. In estradiol synthesis, androstenedione is first converted to testosterone, which the enzyme aromatase converts to estradiol. The synthesis of both *estrone* and *estriol* proceeds directly from androstenedione (Figure 28–24●).

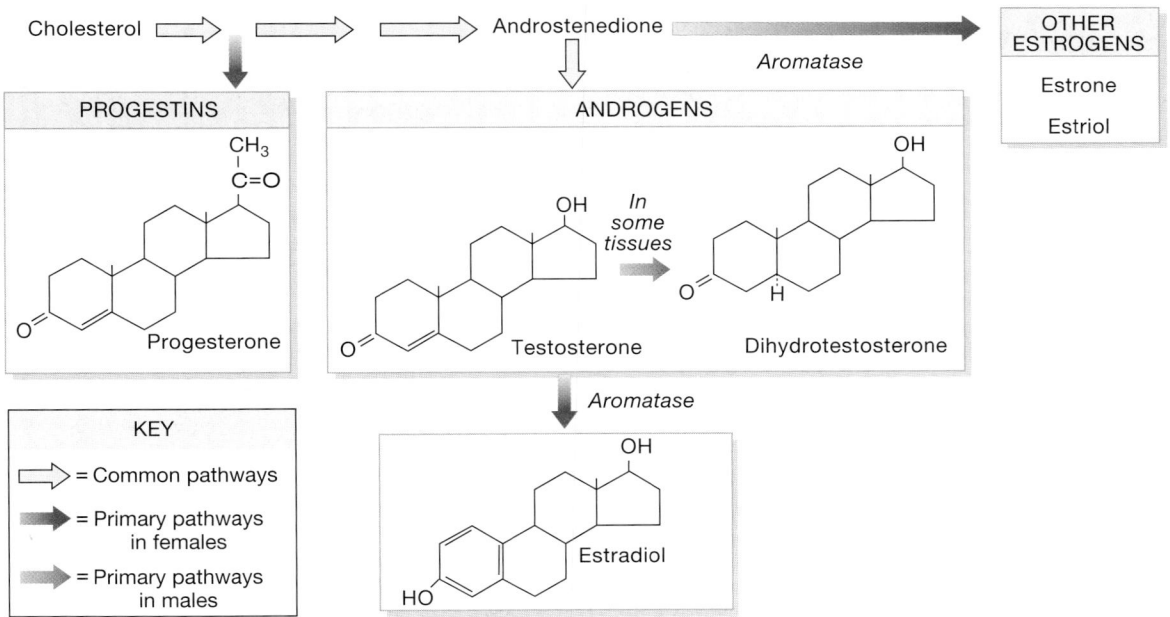

●**FIGURE 28–24**
Pathways of Steroid Hormone Synthesis. All gonadal steroids are derived from cholesterol. In men, the pathway ends with the synthesis of testosterone, which may subsequently be converted to dihydrotestosterone. In women, an additional step past testosterone leads to estradiol synthesis. The synthesis of progesterone and the estrogens other than estradiol involve alternative pathways.

Estrogens have multiple functions that affect the activities of many tissues and organs throughout the body. Among the important general functions of estrogens are (1) stimulating bone and muscle growth, (2) maintaining female secondary sex characteristics, such as body hair distribution and the location of adipose tissue deposits, (3) affecting central nervous system (CNS) activity (especially in the hypothalamus, where estrogens increase the sexual drive), (4) maintaining functional accessory reproductive glands and organs, and (5) initiating the repair and growth of the endometrium. Figure 28–25•, which diagrams the hormonal regulation of ovarian activity, includes an overview of the effects of estrogen on various aspects of reproductive function.

Summary: Hormonal Regulation of the Female Reproductive Cycle

Figure 28–26• shows the changes in circulating hormone levels that accompany the ovarian cycle. Early in the follicular phase, estrogen levels are low and the GnRH pulse frequency is 16–24 per day (one pulse every 60–90 minutes). At this frequency, FSH is the dominant hormone released by the anterior pituitary gland; the estrogen released by developing follicles inhibits LH secretion. As secondary follicles develop, FSH levels decline due to the negative feedback effects of inhibin. Follicular development and maturation continue, however, supported by the combination of estrogens, FSH, and LH.

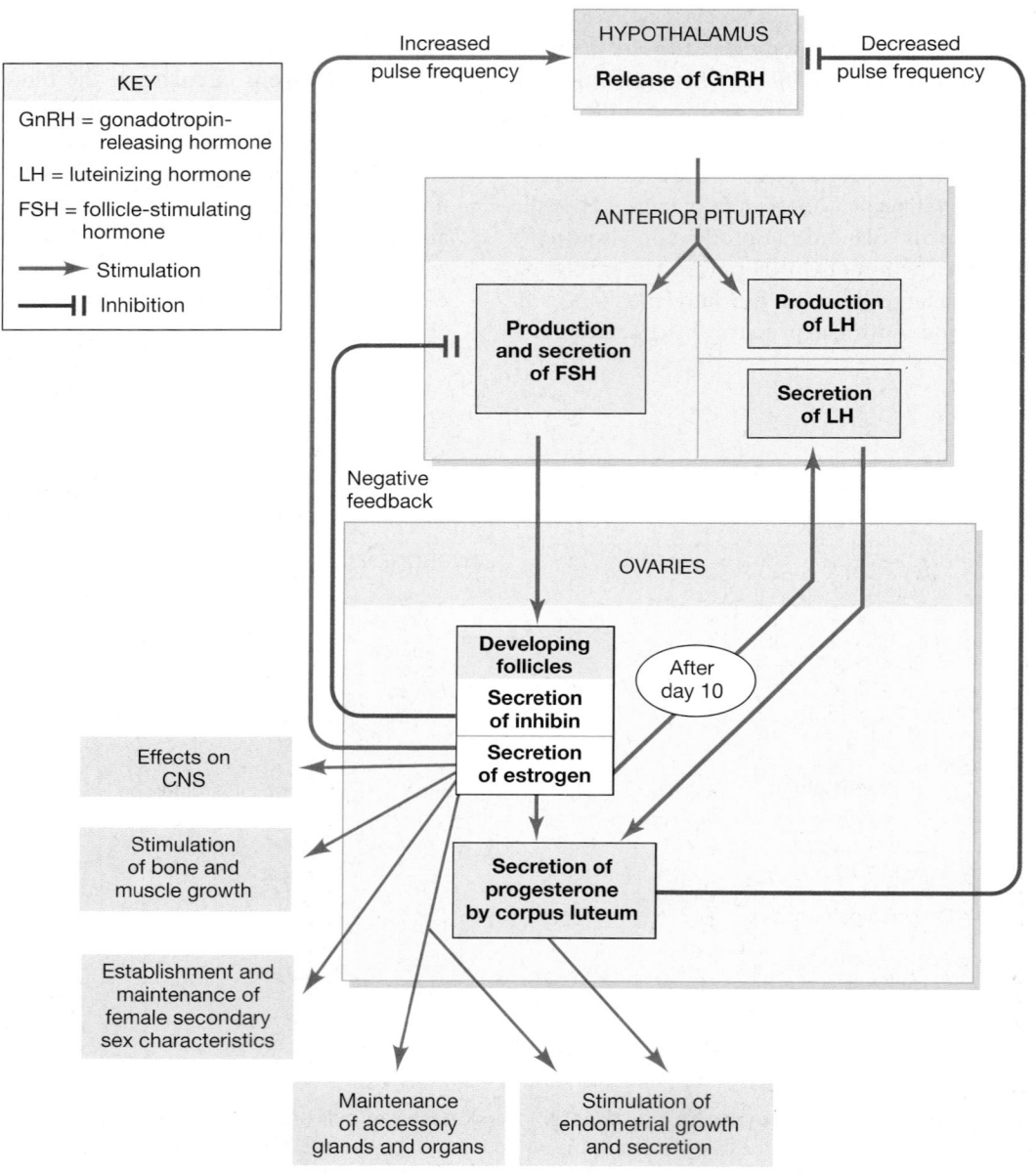

•**FIGURE 28–25**
The Hormonal Regulation of Ovarian Activity

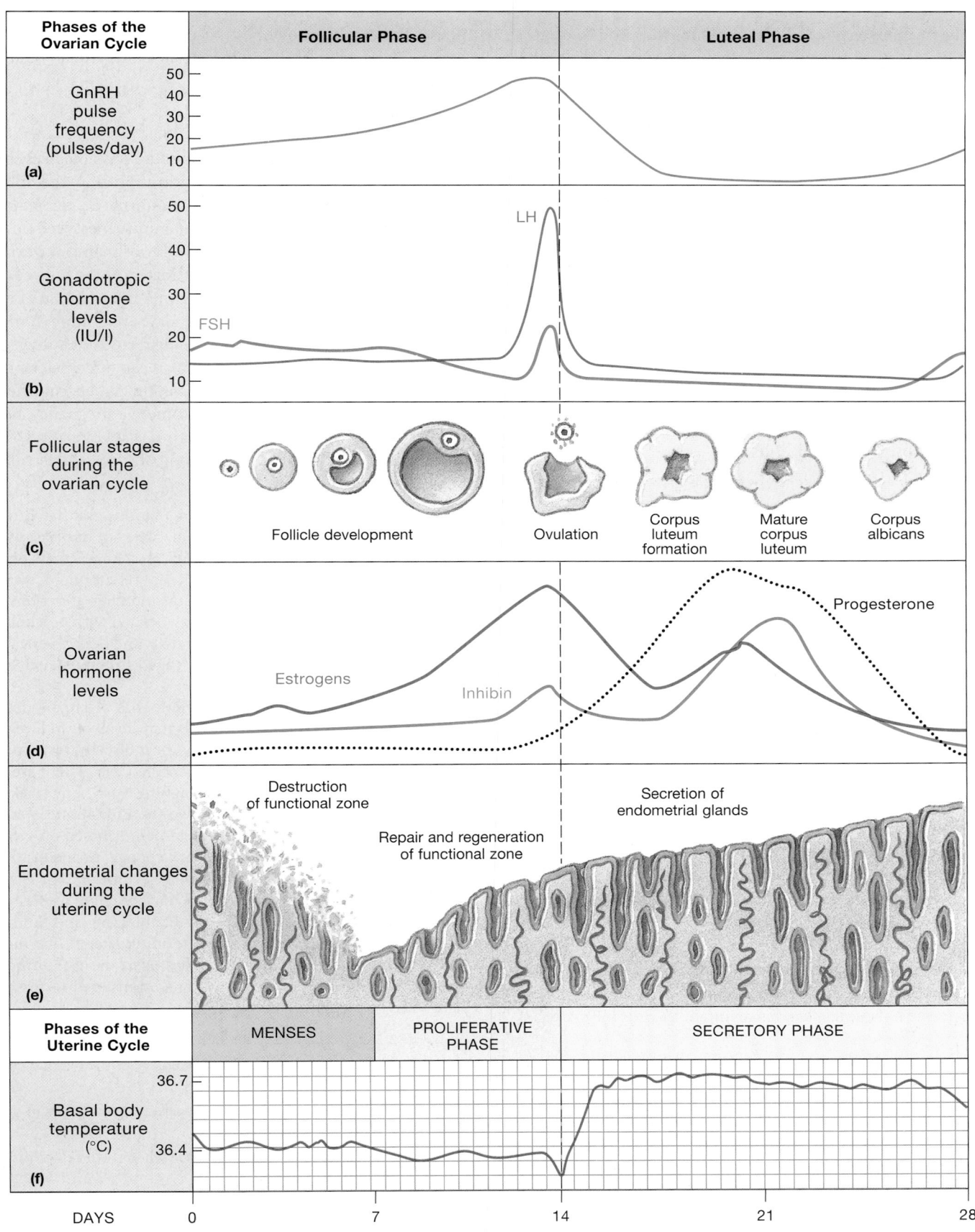

Phases of the Ovarian Cycle	Follicular Phase	Luteal Phase

(a) GnRH pulse frequency (pulses/day) — 50 40 30 20 10

(b) Gonadotropic hormone levels (IU/l) — 50 40 30 20 10 — LH, FSH

(c) Follicular stages during the ovarian cycle — Follicle development, Ovulation, Corpus luteum formation, Mature corpus luteum, Corpus albicans

(d) Ovarian hormone levels — Estrogens, Inhibin, Progesterone

(e) Endometrial changes during the uterine cycle — Destruction of functional zone, Repair and regeneration of functional zone, Secretion of endometrial glands

Phases of the Uterine Cycle	MENSES	PROLIFERATIVE PHASE	SECRETORY PHASE

(f) Basal body temperature (°C) — 36.7 36.4

DAYS 0   7   14   21   28

●FIGURE 28–26
The Hormonal Regulation of the Female Reproductive Cycle

As one or more tertiary follicles begin forming, the concentration of circulating estrogens rises steeply. As a result, the GnRH pulse frequency increases to about 36 per day (one pulse every 30–60 minutes). The increased pulse frequency stimulates LH secretion. In addition, at roughly day 10 of the cycle, the effect of estrogen on LH secretion changes from inhibition to stimulation. The switchover occurs only after rising estrogen levels have exceeded a threshold value for about 36 hours. (The threshold value and the time required vary among individuals.) High estrogen levels also increase gonadotrope sensitivity to GnRH. At about day 14, the estrogen level has peaked, the gonadotropes are at maximum sensitivity, and the GnRH pulses are arriving about every 30 minutes. The result is a massive release of LH from the anterior pituitary gland. This sudden surge in LH concentration triggers (1) the completion of meiosis I by the primary oocyte, (2) the rupture of the follicular wall, and (3) ovulation. Typically, ovulation occurs 34–38 hours after the LH surge begins, roughly 9 hours after the LH peak.

Hormones and the Luteal Phase

The high LH levels that trigger ovulation also promote progesterone secretion and the formation of the corpus luteum. As progesterone levels rise and estrogen levels fall, the GnRH pulse frequency declines sharply, soon reaching 1–4 pulses per day. This frequency of GnRH pulses stimulates LH secretion more than it does FSH secretion, and the LH maintains the structure and secretory function of the corpus luteum.

Although moderate amounts of estrogens are secreted by the corpus luteum, progesterone is the main hormone of the luteal phase. Its primary function is to continue the preparation of the uterus for pregnancy by enhancing the blood supply to the functional zone and stimulating the secretion of the endometrial glands. Progesterone levels remain high for the next week, but unless pregnancy occurs, the corpus luteum begins to degenerate. Roughly 12 days after ovulation, the corpus luteum becomes nonfunctional, and progesterone and estrogen levels fall markedly. The blood supply to the functional zone is restricted, and the endometrial tissues begin to deteriorate. As progesterone and estrogen levels drop, the GnRH pulse frequency increases, stimulating FSH secretion by the anterior lobe of the pituitary gland, and the ovarian cycle begins again.

The hormonal changes involved with the ovarian cycle in turn affect the activities of other reproductive tissues and organs. At the uterus, the hormonal changes maintain the uterine cycle.

Hormones and the Uterine Cycle

Figure 28–26e● follows changes in the endometrium during a single uterine cycle. The declines in progesterone and estrogen levels that accompany the degeneration of the corpus luteum (Figure 28–26c,d●) result in menses. The sloughing off of endometrial tissue continues for several days, until rising estrogen levels stimulate the repair and regeneration of the functional zone of the endometrium. The proliferative phase continues until rising progesterone levels mark the arrival of the secretory phase. The combination of estrogen and progesterone then causes the enlargement of the endometrial glands and an increase in their secretory activities.

Hormones and Body Temperature

The monthly hormonal fluctuations cause physiological changes that affect the core body temperature. During the follicular phase, when estrogen is the dominant hormone, the *basal body temperature*, or the resting body temperature measured on awakening in the morning, is about 0.3°C lower than it is during the luteal phase, when progesterone dominates. At the time of ovulation, the basal body temperature declines noticeably, making the rise in temperature over the next day even more noticeable (Figure 28–26f●). As a result, by keeping records of body temperature over a few uterine cycles, a woman can often determine the precise day of ovulation. This information can be important for individuals who wish to avoid or promote a pregnancy, because fertilization typically occurs within a day of ovulation. Thereafter, oocyte viability and the likelihood of successful fertilization decrease markedly.

In many women, hormonal fluctuations associated with the reproductive cycle produce a variety of unpleasant effects. Several physical and physiological changes may occur 7–10 days before the start of menses. Fluid retention, breast enlargement, headaches, pelvic pain, and bloating are common symptoms. These sensations may be associated with psychological changes producing irritability, anxiety, and depression. This combination of symptoms has been called **premenstrual syndrome (PMS)**.

The mechanism responsible for PMS has yet to be determined. Changes in sex hormone levels may be involved directly, by acting on peripheral organ systems, or indirectly, by modifying the release of neurotransmitters in the CNS. There are no laboratory tests or procedures to diagnose PMS, but tracking the appearance of symptoms over a two- to three-month period can reveal characteristic patterns. Treatment is based on symptoms and may involve exercise, dietary change, or medication, depending on the nature of the primary symptom. For example, if headache is the major problem, analgesics are prescribed, whereas diuretics may be used to combat bloating and fluid retention. For severe PMS, drugs can be administered that block GnRH secretion and stop uterine cycles for six months or more. During the interim, estrogens can be administered to prevent symptoms of premature menopause.

✓ What changes would you expect to observe in the ovarian cycle if the LH surge did not occur?

✓ What effect would a blockage of progesterone receptors in the uterus have on the endometrium?

✓ What event occurs in the uterine cycle when the levels of estrogens and progesterone decline?

Answers start on page Q-1

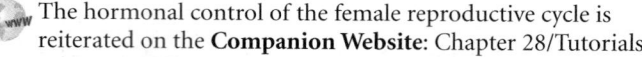 The hormonal control of the female reproductive cycle is reiterated on the **Companion Website**: Chapter 28/Tutorials.

28–4 THE PHYSIOLOGY OF SEXUAL INTERCOURSE

Objective

- Discuss the physiology of sexual intercourse as it affects the reproductive systems of males and females.

Sexual intercourse, also known as *coitus* (KŌ-i-tus) or *copulation*, introduces semen into the female reproductive tract. We shall now consider the process as it affects the reproductive systems of males and females.

☐ MALE SEXUAL FUNCTION

Sexual function in males is coordinated by complex neural reflexes that we do not yet understand completely. The reflex pathways utilize the sympathetic and parasympathetic divisions of the autonomic nervous system. During sexual **arousal**, erotic thoughts, the stimulation of sensory nerves in the genital region, or both lead to an increase in the parasympathetic outflow over the pelvic nerves. This outflow in turn leads to erection of the penis (as discussed on p. 1061). The integument covering the glans of the penis contains numerous sensory receptors, and erection tenses the skin and increases their sensitivity. Subsequent stimulation can initiate the secretion of the bulbourethral glands, lubricating the penile urethra and the surface of the glans.

During intercourse, the sensory receptors of the penis are rhythmically stimulated. This stimulation eventually results in the coordinated processes of emission and ejaculation. **Emission** occurs under sympathetic stimulation. The process begins when the peristaltic contractions of the ampulla push fluid and spermatozoa into the prostatic urethra. The seminal vesicles then begin contracting, and the contractions increase in force and duration over the next few seconds. Peristaltic contractions also appear in the walls of the prostate gland. The combination moves the seminal mixture into the membranous and penile portions of the urethra. While the contractions are proceeding, sympathetic commands also cause the contraction of the urinary bladder and the internal urethral sphincter. The combination of elevated pressure inside the bladder and the contraction of the sphincter effectively prevents the passage of semen into the bladder.

Ejaculation occurs as powerful, rhythmic contractions appear in the *ischiocavernosus* and *bulbospongiosus muscles*, two superficial skeletal muscles of the pelvic floor. The ischiocavernosus muscles insert along the sides of the penis; their contractions serve primarily to stiffen that organ. The bulbospongiosus muscle wraps around the base of the penis; the contraction of this muscle pushes semen toward the external urethral opening. The contractions of both muscles are controlled by somatic motor neurons in the inferior lumbar and superior sacral segments of the spinal cord. (The positions of these muscles are shown in Figure 11–13•, p. 357.)

Ejaculation is associated with intensely pleasurable sensations, an experience known as male **orgasm** (OR-gazm). Several other noteworthy physiological changes occur at this time, including pronounced but temporary increases in heart rate and blood pressure. After ejaculation, blood begins to leave the erectile tissue, and the erection begins to subside. This subsidence, called **detumescence** (de-tū-MES-ens), is mediated by the sympathetic nervous system.

In sum, arousal, erection, emission, and ejaculation are controlled by a complex interplay between the sympathetic and parasympathetic divisions of the autonomic nervous system. Higher centers, including the cerebral cortex, can facilitate or inhibit many of the important reflexes, thereby modifying the patterns of sexual function. Any physical or psychological factor that affects a single component of the system can result in male sexual dysfunction, also called **impotence**.

Impotence is defined as an inability to achieve or maintain an erection. Various physical causes may be responsible for impotence, because erection involves vascular changes as well as neural commands. For example, low blood pressure in the arteries servicing the penis, due to a circulatory blockage such as a plaque, will affect the ability to attain an erection. Drugs, alcohol, trauma, or illnesses that affect the autonomic nervous system or the central nervous system can have the same effect. But male sexual performance can also be strongly affected by the psychological state of the individual. Temporary periods of impotence are relatively common in healthy individuals who are experiencing severe stresses or emotional problems. Depression, anxiety, and fear of impotence are examples of emotional factors that can result in sexual dysfunction. The prescription drug Viagra®, which enhances and prolongs the effects of nitric oxide on the erectile tissue of the penis, has proven useful in treating many cases of impotence.

☐ FEMALE SEXUAL FUNCTION

The phases of female sexual function are comparable to those of male sexual function. During sexual arousal, parasympathetic activation leads to an engorgement of the erectile tissues of the clitoris and increased secretion of cervical mucous glands and the greater vestibular glands. Clitoral erection increases the receptors' sensitivity to stimulation, and the cervical and vestibular glands lubricate the vaginal walls. A network of blood vessels in the vaginal walls becomes filled with blood at this time, and the vaginal surfaces are also moistened by the fluid that moves across the epithelium from underlying connective tissues. (This process accelerates during intercourse as the result of mechanical stimulation.) Parasympathetic stimulation also causes engorgement of blood vessels at the nipples, making them more sensitive to touch and pressure.

Birth Control Strategies

For physiological, logistical, financial, or emotional reasons, most adults practice some form of conception control during their reproductive years. Well over 50 percent of U.S. women age 15–44 employ some method of contraception. When the simplest and most obvious method, sexual abstinence, is unsatisfactory for some reason, another method of contraception must be used to avoid unwanted pregnancies. Many methods are available, so the selection process can be quite involved. Because each method has specific strengths and weaknesses, the potential risks and benefits must be carefully analyzed on an individual basis. We shall consider only a few of the available contraception methods here.

Sterilization is a surgical procedure that makes an individual unable to provide functional gametes for fertilization. Either sexual partner may be sterilized. In a **vasectomy** (vaz-EK-to-mē), a segment of the ductus deferens is removed, making it impossible for spermatozoa to pass from the epididymis to the distal portions of the reproductive tract (Figure 28–27a•). The surgery can be performed in a physician's office in a matter of minutes. The spermatic cords are located as they ascend from the scrotum on either side; after each cord is opened, the ductus deferens is severed. A 1 cm section is removed, and the cut ends are tied shut. The cut ends cannot reconnect, and in time, scar tissue forms a permanent seal. Alternatively, the cut ends of the ductus deferens are blocked with silicone plugs that can later be removed. This more recent vasectomy procedure makes it possible to restore fertility at a later date. After a vasectomy, the man experiences normal sexual function, because the secretions of the epididymis and testes normally account for only about 5 percent of the volume of semen. Spermatozoa continue to develop, but they remain within the epididymis until they degenerate. The failure rate for this procedure is 0.08 percent. (*Failure* for a birth control method is defined as a resulting pregnancy.)

The uterine tubes can be blocked by a surgical procedure known as a **tubal ligation** (Figure 28–27b•). The failure rate for this procedure is estimated at 0.45 percent. Because the surgery requires that the abdominopelvic cavity be opened, complications are more likely than with vasectomy. As in a vasectomy, attempts may be made to restore fertility after a tubal ligation.

A variety of nonsurgical methods are available as well (Figure 28–27c•). **Oral contraceptives** manipulate the female hormonal cycle so that ovulation does not occur. The contraceptive pills produced in the 1950s contained relatively large amounts of estrogen and progestins. These concentrations were adequate to suppress pituitary production of GnRH, so FSH was not released and ovulation did not occur. In most of the oral contraceptive products developed subsequently, smaller amounts of estrogens have been used. Current *combination pills* differ significantly from the earlier products in that the hormonal doses are much lower, with only one-tenth the progestins and less than half the estrogens. The hormones are administered in a cyclic fashion, beginning five days after the start of menses and continuing for three weeks. Then, over the fourth week, the woman takes a placebo or no pills at all. Low-dosage combination pills are sometimes prescribed for women who experience irregular uterine cycles, because these pills create a regular 28-day cycle.

At least 20 brands of combination oral contraceptives are now available, and more than 200 million women are using them worldwide. In the United States, 25 percent of women under age 45 use a combination pill to prevent conception. The failure rate for the combination oral contraceptives, when used as prescribed, is 0.24 percent over a two-year period. Birth control pills are not risk free: Combination pills can worsen problems associated with severe hypertension, diabetes mellitus, epilepsy, gallbladder disease, heart trouble, and acne. Women taking oral contraceptives are also at increased risk of venous thrombosis, strokes, pulmonary embolism, and (for women over 35) heart disease. Pregnancy has similar or higher risks, however.

Three progesterone-only forms of birth control are now available: Depo-provera®, the Norplant® system, and the progesterone-only pill. *Depo-provera* is injected every 3 months. Uterine cycles are initially irregular, and in roughly 50 percent of women using this product they eventually cease. The major problems with this contraceptive method are (1) a tendency to gain weight after three or more years of using it and (2) a slow return to fertility (up to 18 months) after injections are discontinued. The Silastic® (silicone rubber) tubes of the *Norplant system* are saturated with progesterone and inserted under the skin. This method provides birth control for approximately five years, but to date the relatively high cost has limited its use. Because it does not supply estrogen, it produces fewer hormonal side effects than combination pills do. Fertility returns immediately after the removal of a Norplant device. Both Depo-provera and the Norplant system are easy to use and are extremely convenient. The progesterone-only pill is taken daily and may cause irregular uterine cycles. Skipping just one pill may result in pregnancy.

The **condom**, also called a *prophylactic* or "rubber," covers the body of the penis during intercourse and keeps spermatozoa from reaching the female reproductive tract. Condoms are also used to prevent the spread of sexually transmitted diseases, such as syphilis, gonorrhea, and AIDS. The reported condom failure rate varies from 6 to 17 percent, depending on the criteria used in a given study.

Vaginal barriers, such as the *diaphragm* and *cervical cap*, rely on similar principles. A diaphragm, the most popular form of vaginal barrier in use today, consists of a dome of latex rubber with a small metal hoop supporting the rim. Because vaginas vary in size, women choosing this method must be individually fitted. Before intercourse, the diaphragm is inserted so that it covers the cervical os. The diaphragm must be coated with a small amount of spermicidal (sperm-killing) jelly or cream to be an effective contraceptive. The failure rate of a properly fitted and used diaphragm is estimated at 5–6 percent. The cervical cap is smaller and lacks the metal rim. It, too, must be fitted carefully, but unlike the diaphragm, it can be left in place for several days and used without contraceptive chemicals. The failure rate (8 percent) is higher than that for the diaphragm.

An **intrauterine device (IUD)** consists of a small plastic loop or a T that is inserted into the uterine cavity. The mechanism of action remains uncertain, but it is known that IUDs

●FIGURE 28–27
Surgical Sterilization. **(a)** Vasectomy. **(b)** Tubal ligation. **(c)** Contraceptive devices.

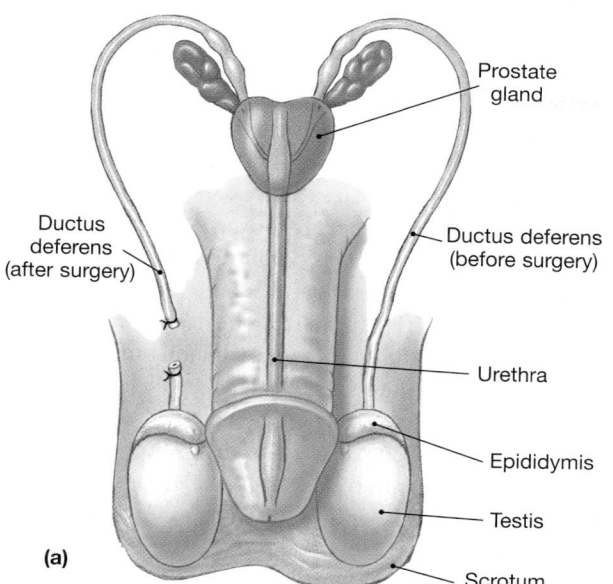

(a)

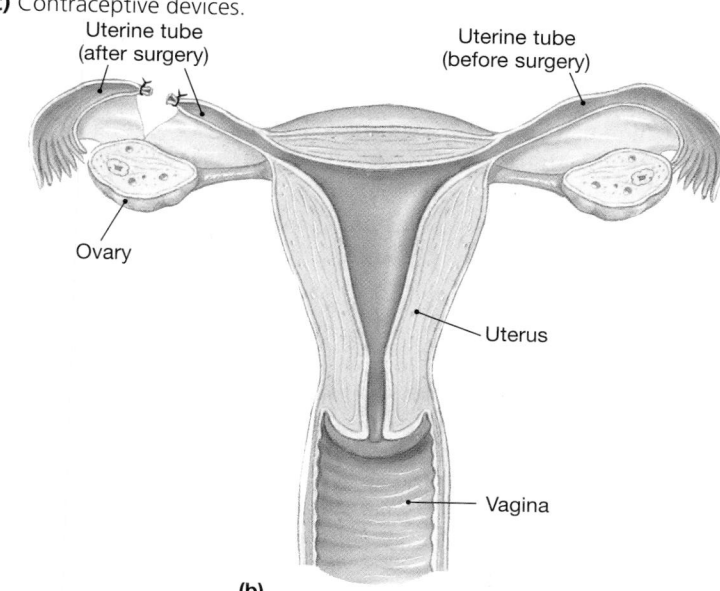

(b)

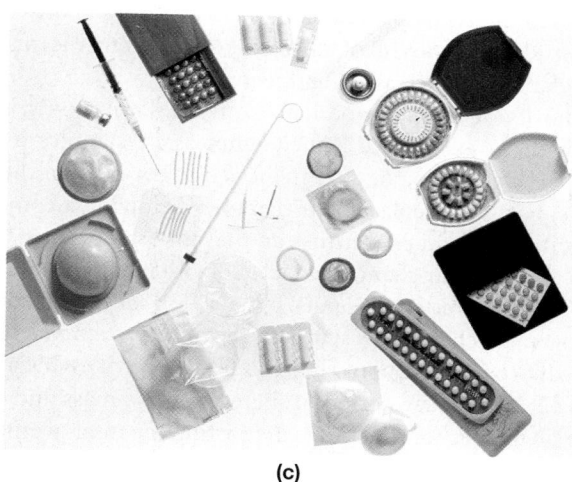

(c)

stimulate prostaglandin production in the uterus, and some are effective for years after insertion. The resulting change in the chemical composition of uterine secretions lowers the chances of fertilization and subsequent *implantation* of the zygote into the uterine lining. (We will discuss implantation in Chapter 29.) In the United States, IUDs are in limited use, but they remain popular in many other countries. The failure rate is estimated at 5–6 percent.

The **rhythm method** involves abstaining from sexual activity on the days ovulation might be occurring. The timing is estimated on the basis of previous patterns of menstruation, and in some cases by monitoring changes in basal body temperature and cervical mucus. The failure rate of the rhythm method is very high—almost 25 percent.

Sterilization, oral contraceptives, condoms, and vaginal barriers are the primary contraception methods for all age groups. But the proportion of the pop-

ulation using a particular method varies by age group. Sterilization is most popular among older women, who may already have had children. Relative availability also plays a role. For example, a sexually active female under age 18 can buy a condom more easily than she can obtain a prescription for an oral contraceptive. But many of the differences are attributable to the relationship between risks and benefits for each age group.

When considering the use and selection of contraceptives, many people simply examine the list of potential complications and make the "safest" choice. For example, media coverage of the risks associated with oral contraceptives made many women reconsider their use. But complex decisions should not be made on such a simplistic basis, and the risks associated with the use of contraceptives must be considered in light of their relative efficiencies. Although pregnancy is a natural phenomenon, it has risks, and the mortality rate for pregnant women in the United States averages about 8 deaths per 100,000 pregnancies. That average incorporates a broad range: The rate is 5.4 per 100,000 women under 20, but 27 per 100,000 among women over 40. Although these risks are small, for pregnant women over age 35 the chances of dying from complications related to pregnancy are almost twice as great as the chances of being killed in an automobile accident and are many times greater than the risks associated with the use of oral contraceptives. For women in developing nations, the comparison is even more striking: The mortality rate for pregnant women in parts of Africa is approximately 1 per 150 pregnancies. In addition to preventing pregnancy, combination birth control pills have been shown to reduce the risks of ovarian and endometrial cancers and fibrocystic breast disease.

Before age 35, *the risks associated with contraceptive use are lower than the risks associated with pregnancy.* The notable exception involves individuals who take the pill, but also smoke cigarettes. Younger women are more fertile, so despite a lower mortality rate for each pregnancy, they are likely to have more pregnancies. As a result, birth control failures imply a higher risk in the younger age groups

After age 35, the risks of complications associated with oral contraceptive use increase, but the risks of using other methods remain relatively stable. Women over age 35 (smokers) or 40 (nonsmokers) are therefore often advised to seek other forms of contraception. Because each contraceptive method has its own advantages and disadvantages, research on contraception control continues. **AM** Experimental Contraceptive Methods

During sexual intercourse, rhythmic contact of the penis with the clitoris and vaginal walls, reinforced by touch sensations from the breasts and other stimuli (visual, olfactory, and auditory), provides stimulation that leads to orgasm. Female orgasm is accompanied by peristaltic contractions of the uterine and vaginal walls and, by means of impulses traveling over the pudendal nerves, rhythmic contractions of the bulbospongiosus and ischiocavernosus muscles. The latter contractions give rise to the sensations of orgasm.

Sexual activity carries with it the risk of infection with a number of microorganisms. The consequences of such an infection may range from rather inconvenient to potentially lethal. **Sexually transmitted diseases (STDs)** are transferred from individual to individual, primarily or exclusively by sexual intercourse. A variety of bacterial, viral, and fungal infections are included in this category. At least two dozen STDs are currently recognized. The bacterium *Chlamydia* can cause PID and infertility. Other types of STDs are also quite dangerous, and a few, including AIDS, are deadly. The incidence of STDs has been increasing in the United States since 1984; an estimated 15 million cases were diagnosed during 1999. Poverty, coupled with drug use, prostitution, and the appearance of drug-resistant pathogens all contribute to the problem. The *Applications Manual* contains a detailed discussion of the most common forms of STD, including *gonorrhea*, *syphilis*, *herpes*, *genital warts*, and *chancroid*. **AM** Sexually Transmitted Diseases

28–5 AGING AND THE REPRODUCTIVE SYSTEM

Objective

- Describe the changes in the reproductive system that occur with aging.

Just as the aging process affects our other systems, it also affects the reproductive systems of men and women. These systems become fully functional at puberty, as noted earlier in the chapter. Thereafter, the most striking age-related changes in the female reproductive system occur at menopause. Comparable age-related changes in the male reproductive system occur more gradually and over a longer period of time.

☐ MENOPAUSE

Menopause is usually defined as the time that ovulation and menstruation cease. Menopause typically occurs at age 45–55, but in the years preceding it, the ovarian and uterine cycles become irregular. This interval is called *perimenopause*. A shortage of primordial follicles is the underlying cause of the irregular cycles. It has been estimated that almost 7 million potential oocytes are in fetal

ovaries after five months of development, but the number drops to about 2 million at birth and to a few hundred thousand at puberty. With the arrival of perimenopause, the number of follicles responding each month begins to drop markedly. As the number of available follicles decreases, levels of estrogen decline and may fail to rise enough to trigger ovulation. By age 50, there are often no primordial follicles left to respond to FSH. In **premature menopause**, this depletion occurs before age 40.

Menopause is accompanied by a decline in circulating concentrations of estrogen and progesterone and a sharp and sustained rise in the production of GnRH, FSH, and LH. The decline in estrogen levels leads to reductions in the size of the uterus and breasts, accompanied by a thinning of the urethral and vaginal epithelia. The reduced estrogen concentrations have also been linked to the development of osteoporosis, presumably because bone deposition proceeds at a slower rate. A variety of neural effects are reported as well, including "hot flashes," anxiety, and depression. Hot flashes typically begin while estrogen levels are declining and cease when estrogen levels reach minimal values. These intervals of elevated body temperature are associated with surges in LH production. The hormonal mechanisms involved in other CNS effects of menopause are poorly understood. In addition, the risk of atherosclerosis and other forms of cardiovascular disease increases after menopause.

The majority of women experience only mild symptoms, but some individuals experience acutely unpleasant symptoms in perimenopause or during or after menopause. For most of those individuals, hormone replacement therapies involving a combination of estrogens and progestins can prevent osteoporosis and the neural and vascular changes associated with menopause. The hormones are administered as pills, by injection, or by transdermal "estrogen patches." The synthetic hormone *etidronate* inhibits osteoporosis by suppressing osteoclast activity. Given at intervals over a two-year period, it increases bone mass and reduces the incidence of fractures in postmenopausal women. However, recent studies suggest that estrogen-replacement therapy should be prescribed with caution and only after a full assessment and discussion of the potential risks and benefits.

☐ THE MALE CLIMACTERIC

Changes in the male reproductive system occur more gradually than do those in the female reproductive system. The period of change is known as the **male climacteric**. Levels of circulating testosterone begin to decline between the ages of 50 and 60, and levels of circulating FSH and LH increase. Although sperm production continues (men well into their eighties can father children), older men experience a gradual reduction in sexual activity. This decrease may be linked to declining testosterone levels. Some clinicians suggest the use of testosterone replacement therapy to enhance the libido (sexual drive) of elderly men.

✔ An inability to contract the ischiocavernosus and bulbospongiosus muscles would interfere with which part(s) of the male sex act?

✔ What changes occur in females during sexual arousal as the result of increased parasympathetic stimulation?

✔ Why does the level of FSH rise and remain high during menopause?

Answer start on page Q-1

Normal human reproduction is a complex process that requires the participation of multiple systems. The hormones discussed in this chapter play a major role in coordinating reproductive events (Table 28–1).

Normal reproduction depends on a combination of hormonal stimuli and physical factors. The man's sperm count must be adequate, the semen must have the correct pH and nutrients, and erection and ejaculation must occur in the proper sequence; the woman's ovarian and uterine cycles must be properly coordinated, ovulation and oocyte transport must occur normally, and her reproductive tract must provide a hospitable environment for the survival, movement, and subsequent fertilization of sperm. For these steps to occur, the reproductive, digestive, endocrine, nervous, cardiovascular, and urinary systems must all be functioning normally.

Even when all else is normal and fertilization occurs at the proper time and place, a healthy infant will not be produced unless the zygote, a single cell the size of a pinhead, manages to develop into a full-term fetus that typically weighs about 3 kg. In Chapter 29, we will consider the process of development, focusing on the mechanisms that determine both the structure of the body and the distinctive characteristics of each individual.

SUMMARY TABLE 28–1 HORMONES OF THE REPRODUCTIVE SYSTEM

Hormone	Source	Regulation of Secretion	Primary Effects
Gonadotropin-releasing hormone (GnRH)	Hypothalamus	*Males*: inhibited by testosterone and possibly by inhibin *Females*: GnRH pulse frequency increased by estrogens, decreased by progestins	Stimulates FSH secretion, LH synthesis As above
Follicle-stimulating hormone (FSH)	Anterior lobe of pituitary gland	*Males*: stimulated by GnRH, inhibited by inhibin *Females*: stimulated by GnRH, inhibited by inhibin	*Males*: stimulates spermatogenesis and spermiogenesis through effects on sustentacular cells *Females*: stimulates follicle development, estrogen production, and oocyte maturation
Luteinizing hormone (LH)	Anterior lobe of pituitary gland	*Males*: stimulated by GnRH *Females*: production stimulated by GnRH, secretion by the combination of high GnRH pulse frequencies and high estrogen levels	*Males*: stimulates interstitial cells to secrete testosterone *Females*: stimulates ovulation, formation of corpus luteum, and progestin secretion
Androgens (primarily testosterone and dihydrotestosterone)	Interstitial cells of testes	Stimulated by LH	Establish and maintain secondary sex characteristics and sexual behavior; promote maturation of spermatozoa; inhibit GnRH secretion
Estrogens (primarily estradiol)	Granulosa and thecal cells of developing follicles; corpus luteum	Stimulated by FSH	Stimulate LH secretion (at high levels); establish and maintain secondary sex characteristics and sexual behavior; stimulate repair and growth of endometrium; increase frequency of GnRH pulses
Progestins (primarily progesterone)	Granulosa cells from mid-cycle through functional life of corpus luteum	Stimulated by LH	Stimulate endometrial growth and glandular secretion; reduce frequency of GnRH pulses
Inhibin	Sustentacular cells of testes and granulosa cells of ovaries	Stimulated by factors released by developing spermatozoa (male) and developing follicles (female)	Inhibits secretion of FSH (and possibly of GnRH)

The **REPRODUCTIVE SYSTEM** in Perspective

Organ/Component	Primary Functions
Testes	Produce sperm and hormones (*see Chapter 18*)
Accessory Organs	
Epididymis	Acts as site of sperm maturation
Ductus Deferens (Sperm Duct)	Conducts sperm between epididymis and prostate gland
Seminal Vesicles	Secrete fluid that makes up much of the volume of semen
Prostate Gland	Secretes fluid and enzymes
Urethra	Conducts semen to exterior
External Genitalia	
Penis	Contains erectile tissue; deposits sperm in vagina of female; produces pleasurable sensations during sexual activities
Scrotum	Surrounds the testes and controls their temperature

Organ/Component	Primary Functions
Ovaries	Produce oocytes and hormones (*see Chapter 18*)
Uterine Tubes	Deliver oocyte or embryo to uterus; normal site of fertilization
Uterus	Site of embryonic development and exchange between maternal and embryonic bloodstreams
Vagina	Site of sperm deposition; acts as birth canal at delivery; provides passageway for fluids during menstruation
External Genitalia	
Clitoris	Contains erectile tissue; produces pleasurable sensations during sexual activities
Labia	Contain glands that lubricate entrance to vagina
Mammary Glands	Produce milk that nourishes newborn infant

INTEGRATION WITH OTHER SYSTEMS

The primary function of the reproductive system—producing children—doesn't play a role in maintaining homeostasis (some would argue that it disrupts homeostasis instead). However, the reproductive process depends on a variety of physical, physiological, and psychological factors, many of which require intersystem cooperation. In addition, the hormones that control and coordinate sexual function have direct effects on the organs and tissues of other systems. For example, testosterone and estradiol affect both muscular development and bone density. The figure on the opposite page summarizes the relationships between the reproductive system and other physiological systems.

CLINICAL PATTERNS

The male and female reproductive systems are complex, and reproductive disorders are many and varied. Major categories of reproductive disorders include:

- Tumors, such as testicular, prostate, ovarian, or uterine cancers.
- Inflammation and infection, such as prostatitis, pelvic inflammatory disease, toxic shock syndrome, and the various sexually transmitted diseases.
- Uterine disorders, such as endometriosis, and hormonally related problems, such as amenorrhea.

- Trauma, such as testicular torsion and inguinal hernias.
- Congenital disorders, such as cryptorchidism.

Most reproductive disorders are primary disorders which reflect problems that originate within the reproductive system. However, amenorrhea and premenstrual syndrome are examples of secondary disorders that can result from problems with the endocrine system, and impotence can result from either neural or hormonal factors. The *Applications Manual* discusses the diagnosis and treatment of major classes of reproductive system disorders.

MEDIA CONNECTION

Objective: Explore the relationship between reproductive cycles and general homeostasis in females through an investigation of PMS.

Completion time = 12 minutes

Have you ever been called "hysterical" or been accused of indulging in "histrionics"? You may know that these terms are derived from the root "hyster," meaning uterus. Since ancient times, women have been stigmatized as exhibiting behavioral peculiarities that were thought to be related in some way to the uterus. Are these behaviors actually due to the uterus, or might they reflect hormonal changes occurring during the reproductive cycle? Referring to Figure 28–26, list the hormonal changes involved in the menstrual cycle. How might these changes be associated with the symptoms usually ascribed to PMS, such as anxiety, depression, cravings, and headaches? Visit the **Companion Website**, Chapter 28 Media Connections, and click on the keyword **PMS** to read a description of this syndrome and current medical treatments. Based on the information in the article, indicate on your list which hormones are thought to be involved in PMS. Knowing what you know about the action of hormones in general, and about PMS in particular, what do you think might help alleviate these symptoms?

INTEGUMENTARY SYSTEM

- Covers external genitalia
- Provides sensations that stimulate sexual behaviors
- Mammary gland secretions provide nourishment for newborn
- Reproductive hormones affect distribution of body hair and subcutaneous fat deposits

SKELETAL SYSTEM

- Pelvis protects reproductive organs of females, portion of ductus deferens and accessory glands in males
- Sex hormones stimulate growth and maintenance of bone
- Sex hormones at puberty accelerate growth and closure of epiphyseal cartilages

MUSCULAR SYSTEM

- Contractions of skeletal muscles eject semen from male reproductive tract
- Muscle contractions during sexual act produce pleasurable sensations in both sexes
- Reproductive hormones, especially testosterone, accelerate skeletal muscle growth

NERVOUS SYSTEM

- Controls sexual behaviors and sexual function
- Sex hormones affect CNS development and sexual behaviors

ENDOCRINE SYSTEM

- Hypothalamic regulatory hormones and pituitary hormones regulate sexual development and function
- Oxytocin stimulates smooth muscle contractions in uterus and mammary glands
- Steroid sex hormones and inhibin inhibit secretory activities of hyopthalamus and pituitary gland

REPRODUCTIVE SYSTEM

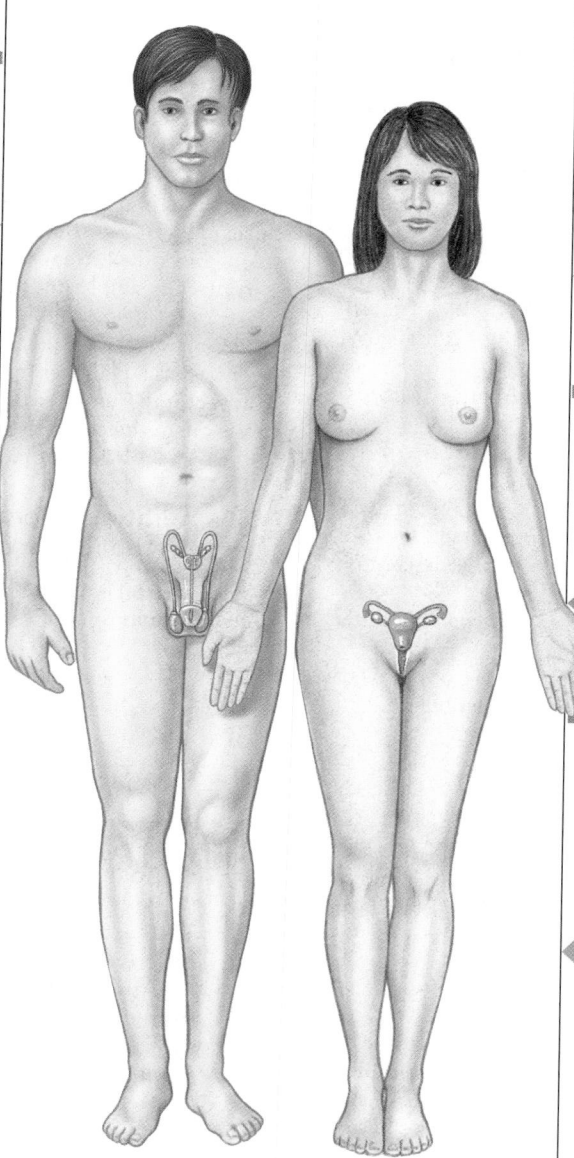

FOR ALL SYSTEMS

Secretion of hormones with effects on growth and metabolism

CARDIOVASCULAR SYSTEM

- Distributes reproductive hormones
- Provides nutrients, oxygen, and waste removal for fetus
- Local blood pressure changes responsible for physical changes during sexual arousal
- Estrogens may help maintain healthy vessels and slow development of atherosclerosis

LYMPHATIC SYSTEM

- Provides IgA for secretions by epithelial glands
- Assists in repairs and defense against infection
- Lysosomes and bacteriocidal chemicals in secretions provide nonspecific defense against reproductive tract infections

RESPIRATORY SYSTEM

- Provides oxygen and removes carbon dioxide generated by tissues of reproductive system
- Changes in respiratory rate and depth occur during sexual arousal, under control of the nervous system

DIGESTIVE SYSTEM

- Provides additional nutrients required to support gamete production and (in pregnant women) embryonic and fetal development

URINARY SYSTEM

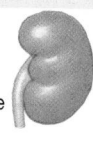

- Urethra in males carries semen to exterior
- Kidneys remove wastes generated by reproductive tissues and (in pregnant women) by growing embryo and fetus
- Accessory organ secretions may have antibacterial action that helps prevent urethral infections in males

Chapter Review

SELECTED CLINICAL TERMINOLOGY

Terms Discussed in This Chapter

amenorrhea: The failure of menarche to appear before age 16, or a cessation of menstruation for six months or more in an adult female of reproductive age. *(p. 1075)*

breast cancer: A malignant, metastasizing cancer of the mammary gland that is the primary cause of death for women age 35–45. *(p. 1078)*

cervical cancer: A malignant, metastasizing cancer of the cervix that is the most common reproductive cancer in women. *(p. 1071)*

cryptorchidism (kript-OR-ki-dizm): The failure of one or both testes to descend into the scrotum by the time of birth. *(p. 1050)*

dysmenorrhea: Painful menstruation. *(p. 1073)*

endometriosis (en-dō-mē-trē-Ō-sis): The growth of endometrial tissue outside the uterus. *(p. 1074)*

fibrocystic disease: Clusters of lobular cysts within the tissues of the mammary gland. *(p. 1078)*

gonorrhea (gon-ō-RĒ-uh): A sexually transmitted disease. *(p. 1070 and AM)*

impotence: An inability to achieve or maintain an erection. *(p. 1083)*

mammography: The use of X rays to examine breast tissue. *(p. 1078)*

mastectomy: The surgical removal of part or all of a cancerous mammary gland. *(p. 1078)*

orchiectomy (or-kē-EK-to-mē): The surgical removal of a testis. *(p. 1050)*

ovarian cancer: A malignant, metastasizing cancer of the ovaries that is the most dangerous reproductive cancer in women. *(p. 1069 and AM)*

pelvic inflammatory disease (PID): An infection of the uterine tubes. *(p. 1070)*

prostate cancer: A malignant, metastasizing cancer of the prostate gland; the second most common cause of cancer deaths in males. *(p. 1064)*

prostatectomy (pros-ta-TEK-to-mē): The surgical removal of the prostate gland. *(p. 1064)*

prostate-specific antigen (PSA): An antigen whose level in blood increases in persons with prostate cancer. *(p. 1064)*

sexually transmitted diseases (STDs): Diseases transferred from one individual to another primarily or exclusively through sexual contact. Examples include gonorrhea, syphilis, herpes genitalis, and AIDS. *(p. 1086 and AM)*

vaginitis (va-jin-Ī-tis): An infection of the vaginal canal by fungal or bacterial pathogens. *(p. 1075 and AM)*

vasectomy (vaz-EK-to-mē): The surgical removal of a segment of the ductus deferens, making it impossible for spermatozoa to reach the distal portions of the male reproductive tract. *(p. 1084)*

AM Additional Terms Discussed in the *Applications Manual*

endometrial polyps
leiomyomas or *fibroids*

STUDY OUTLINE

1. The human reproductive system produces, stores, nourishes, and transports functional **gametes** (reproductive cells). **Fertilization** is the fusion of male and female gametes to create a **zygote**.

28–1 INTRODUCTION TO THE REPRODUCTIVE SYSTEM p. 1047

1. The **reproductive system** includes **gonads** (**testes** or **ovaries**), ducts, accessory glands and organs, and the **external genitalia**.

2. In males, the testes produce **spermatozoa**, which are expelled from the body in **semen** during *ejaculation*. The ovaries of a sexually mature female produce an **oocyte** (immature **ovum**) that travels along *uterine tubes* toward the *uterus*. The *vagina* connects the uterus with the exterior of the body.

28–2 THE REPRODUCTIVE SYSTEM OF THE MALE p. 1048

1. The spermatozoa travel along the *epididymis*, the *ductus deferens*, the *ejaculatory duct*, and the *urethra* before leaving the body. Accessory organs (notably the *seminal vesicles*, *prostate gland*, and *bulbourethral glands*) secrete into the ejaculatory ducts and the urethra. The *scrotum* encloses the testes, and the *penis* is an erectile organ. *(Figure 28–1)*

THE TESTES p. 1049

2. The **descent of the testes** through the *inguinal canals* occurs during fetal development. The testes remain connected to internal structures via the **spermatic cords**. The **raphe** marks the boundary between the two chambers in the **scrotum**. *(Figures 28–2, 28–3)*

3. The **dartos** muscle gives the scrotum a wrinkled appearance; the **cremaster muscle** pulls the testes close to the body.

4. The **tunica albuginea** surrounds each testis. Septa extend from the tunica albuginea to the region of the testis closest to the entrance to the epididymis, creating a series of **lobules**. *(Figure 28–4)*

5. **Seminiferous tubules** within each lobule are the sites of sperm production. From there, spermatozoa pass through the **rete testis**. **Efferent ductules** connect the rete testis to the epididymis. Between the seminiferous tubules are **interstitial cells**, which secrete sex hormones. *(Figures 28–4, 28–5)*

SPERMATOGENESIS p. 1053

6. Seminiferous tubules contain **spermatogonia**, stem cells involved in **spermatogenesis** (the production of spermatozoa), and **sustentacular cells**, which sustain and promote the development of spermatozoa. *(Figures 28–5 to 28–7)*

THE ANATOMY OF A SPERMATOZOON p. 1056

7. Each **spermatozoon** has a **head** tipped by an **acrosomal cap**, a **middle piece**, and a **tail**. *(Figure 28–8)*

THE MALE REPRODUCTIVE TRACT p. 1056

8. From the testis, the spermatozoa enter the **epididymis**, an elongate tubule with **head**, **body**, and **tail** regions. The epididymis

monitors and adjusts the composition of the fluid in the seminiferous tubules, serves as a recycling center for damaged spermatozoa, stores and protects spermatozoa, and facilitates their functional maturation. *(Figure 28–9)*

9. The **ductus deferens**, or *vas deferens*, begins at the epididymis and passes through the inguinal canal as part of the spermatic cord. Near the prostate gland, the ductus deferens enlarges to form the **ampulla**. The junction of the base of the seminal vesicle and the ampulla creates the **ejaculatory duct**, which empties into the urethra. *(Figures 28–9, 28–10)*

10. The **urethra** extends from the urinary bladder to the tip of the penis. The urethra can be divided into *prostatic, membranous,* and *penile regions.*

THE ACCESSORY GLANDS   p. 1059

11. Each **seminal vesicle** is an active secretory gland that contributes about 60 percent of the volume of semen; its secretions contain fructose (which is easily metabolized by spermatozoa), prostaglandins, and fibrinogen. The **prostate gland** secretes slightly acidic **prostatic fluid**. Alkaline mucus secreted by the **bulbourethral glands** has lubricating properties. *(Figures 28–10, 28–11)*

SEMEN   p. 1059

12. A typical ejaculation releases 2–5 ml of semen (an **ejaculate**), which contains 20–100 million spermatozoa per milliliter. The fluid component of semen is **seminal fluid.**

THE EXTERNAL GENITALIA   p. 1060

13. The skin overlying the **penis** resembles that of the scrotum. Most of the **body** of the penis consists of three masses of **erectile tissue**. Beneath the superficial fascia are two **corpora cavernosa** and a single **corpus spongiosum**, which surrounds the urethra. Dilation of the erectile tissue with blood produces an **erection**. *(Figure 28–11)*

HORMONES AND MALE REPRODUCTIVE FUNCTION   p. 1062

14. Important regulatory hormones include **FSH** *(follicle-stimulating hormone)*, **LH** *(luteinizing hormone)*, and **GnRH** *(gonadotropin-releasing hormone)*. *Testosterone* is the most important androgen. *(Figure 28–12)*

Male reproductive organs: **Anatomy CD-ROM:** Reproductive System/Male Flythrough.

28–3 THE REPRODUCTIVE SYSTEM OF THE FEMALE   p. 1065

1. Principal organs of the female reproductive system include the ovaries, *uterine tubes, uterus, vagina,* and external genitalia. *(Figure 28–13)*

2. The ovaries, uterine tubes, and uterus are enclosed within the **broad ligament**. The **mesovarium** supports and stabilizes each ovary. *(Figure 28–13)*

THE OVARIES   p. 1066

3. The ovaries are held in position by the **ovarian ligament** and the **suspensory ligament**. Major blood vessels enter the ovary at the **ovarian hilum**. Each ovary is covered by a **tunica albuginea**. *(Figure 28–14)*

4. **Oogenesis** (ovum production) occurs monthly in **ovarian follicles** as part of the **ovarian cycle**, which is divided into a **follicular** *(preovulatory)* **phase** and a **luteal** *(postovulatory)* **phase.** *(Figures 28–15, 28–16)*

5. As development proceeds, **primordial**, **primary**, **secondary**, and **tertiary follicles** are produced in turn. At **ovulation**, a

secondary oocyte and the surrounding follicular cells of the **corona radiata** are released through the ruptured ovarian wall. The follicular cells remaining within the ovary form the **corpus luteum**, which later degenerates into a **corpus albicans** of scar tissue. *(Figures 28–15, 28–16)*

THE UTERINE TUBES   p. 1069

6. Each **uterine tube** has an **infundibulum** with **fimbriae** (fingerlike projections), an **ampulla**, and an **isthmus**. Each uterine tube opens into the *uterine cavity*. For fertilization to occur, the secondary oocyte must encounter spermatozoa during the first 12–24 hours of its passage from the infundibulum to the uterus. *(Figure 28–17)*

THE UTERUS   p. 1070

7. The **uterus** provides mechanical protection, nutritional support, and waste removal for the developing embryo. Normally, the uterus bends anteriorly near its base *(anteflexion)*. It is stabilized by the broad ligament, **uterosacral ligaments**, **round ligaments**, and **lateral ligaments**. *(Figure 28–18)*

8. Major anatomical landmarks of the uterus include the **body**, **isthmus**, **cervix**, **cervical os** *(external orifice)*, **uterine cavity**, **cervical canal**, and **internal os** *(internal orifice)*. The uterine wall consists of an inner **endometrium**, a muscular **myometrium**, and a superficial **perimetrium** (an incomplete serous layer). *(Figures 28–18, 28–19)*

9. A typical 28-day **uterine**, or *menstrual*, **cycle** begins with the onset of **menses** and the destruction of the **functional zone** of the endometrium. This process of **menstruation** continues from one to seven days. *(Figure 28–20)*

10. After menses, the **proliferative phase** begins, and the functional zone undergoes repair and thickens. The proliferative phase is followed by the **secretory phase**, during which endometrial glands enlarge. Menstrual activity begins at **menarche** and continues until **menopause**. *(Figure 28–20)*

THE VAGINA   p. 1075

11. The **vagina** is a muscular tube extending between the uterus and the external genitalia. Before the first sexual intercourse, a thin epithelial fold, the **hymen**, partially blocks the entrance to the vagina. *(Figure 28–21)*

THE EXTERNAL GENITALIA   p. 1076

12. The components of the **vulva** are the **vestibule**, **labia minora**, **paraurethral glands**, **clitoris**, **labia majora**, and **lesser** and **greater vestibular glands**. *(Figure 28–22)*

THE MAMMARY GLANDS   p. 1076

13. A newborn infant gains nourishment from milk secreted by maternal **mammary glands**. *(Figure 28–23)*

Female reproductive organs: **Anatomy CD-ROM:** Reproductive System/Female Flythrough.

HORMONES AND THE FEMALE REPRODUCTIVE CYCLE   p. 1077

14. Hormonal regulation of the **female reproductive cycle** involves the coordination of the ovarian and uterine cycles.

15. **Estradiol**, the most important *estrogen*, is the dominant hormone of the follicular phase. Ovulation occurs in response to a midcycle surge in LH. *(Figures 28–24, 28–25)*

16. The hypothalamic secretion of GnRH occurs in pulses that trigger the pituitary secretion of FSH and LH. FSH initiates follicular development, and activated follicles and ovarian interstitial cells produce estrogens. High levels of estrogen stimulate LH secretion, increase pituitary sensitivity to GnRH, and increase the GnRH pulse frequency. **Progesterone**, one of the **progestins**, is the principal hormone of the luteal phase. Changes in estrogen

and progesterone levels are responsible for the maintenance of the uterine cycle. (*Figures 28–25, 28–26*)

Hormonal control of the female reproductive cycle: **Companion Website:** Chapter 28/Tutorials.

28–4 THE PHYSIOLOGY OF SEXUAL INTERCOURSE   p. 1083

MALE SEXUAL FUNCTION   p. 1083

1. During sexual **arousal** in males, erotic thoughts, sensory stimulation, or both lead to parasympathetic activity that produces erection. Stimuli accompanying **sexual intercourse** lead to **emission** and **ejaculation**. Contractions of the bulbospongiosus muscles are associated with **orgasm**.

FEMALE SEXUAL FUNCTION   p. 1083

2. The phases of female sexual function resemble those of male sexual function, with parasympathetic arousal and skeletal muscle contractions associated with orgasm.

28–5 AGING AND THE REPRODUCTIVE SYSTEM   p. 1086

MENOPAUSE   p. 1086

1. Menopause (the time that ovulation and menstruation cease) typically occurs at age 45–55. The production of GnRH, FSH, and LH rise, whereas circulating concentrations of estrogen and progesterone decline.

THE MALE CLIMACTERIC   p. 1086

2. During the **male climacteric,** at age 50–60, circulating testosterone levels fall, and levels of FSH and LH rise.

REVIEW QUESTIONS

More assessment questions are available to you on the Companion Website. You will find Matching, Multiple Choice, True/False, and other quizzes to help further your understanding of the material covered in this chapter. To access the site, go to www.aw.com/martini.

LEVEL 1 Reviewing Facts and Terms

1. Perineal structures associated with the reproductive system are collectively known as
   (a) gonads   (b) sex gametes
   (c) external genitalia   (d) accessory glands

2. Sustentacular cells are responsible for the secretion of
   (a) inhibin
   (b) androgen-binding protein
   (c) Müllerian-inhibiting factor
   (d) a, b, and c are correct

3. During meiosis, when synapsis occurs, corresponding maternal and paternal chromosomes come together to produce
   (a) 46 pairs of chromosomes
   (b) 23 chromosomes
   (c) 23 pairs of chromosomes
   (d) the haploid number of chromosomes

4. The completion of meiosis in males produces four spermatids, each containing
   (a) 23 chromosomes
   (b) 23 pairs of chromosomes
   (c) the diploid number of chromosomes
   (d) 46 pairs of chromosomes

5. Erection of the penis occurs when
   (a) there is sympathetic activation of penile arteries
   (b) arterial branches are constricted and muscular partitions are tense
   (c) the vascular channels become engorged with blood
   (d) a, b, and c are correct

6. In males, the primary target of FSH is the
   (a) sustentacular cells of the seminiferous tubules
   (b) interstitial cells of the seminiferous tubules
   (c) cells of Leydig
   (d) epididymis

7. Testosterone and other androgens are secreted by the
   (a) hypothalamus   (b) anterior lobe of the pituitary gland
   (c) sustentacular cells   (d) interstitial cells

8. The ovaries are responsible for
   (a) the production of female gametes
   (b) the secretion of female sex hormones
   (c) the secretion of inhibin
   (d) a, b, and c are correct

9. Gametes are produced in the _____ of the ovary.
   (a) germinal epithelium
   (b) medulla
   (c) cortex
   (d) tunica albuginea

10. In females, meiosis is not completed until
    (a) birth
    (b) puberty
    (c) fertilization occurs
    (d) uterine implantation occurs

11. The part of the endometrium that undergoes cyclical changes in response to levels of sex hormones is the
    (a) serosa
    (b) basilar zone
    (c) muscular myometrium
    (d) functional zone

12. The secretory phase of the uterine cycle is influenced primarily by the stimulatory effects of progestins and estrogens from the
    (a) anterior lobe of the pituitary gland
    (b) hypothalamus
    (c) endometrium
    (d) corpus luteum

13. A sudden surge in LH secretion causes the
    (a) onset of menses
    (b) rupture of the follicular wall and ovulation
    (c) beginning of the proliferative phase
    (d) end of the uterine cycle

14. The principal hormone of the postovulatory phase is
    (a) progesterone   (b) estradiol
    (c) estrogen   (d) luteinizing hormone

15. At ovulation, the basal body temperature
    (a) is not affected
    (b) increases noticeably
    (c) declines sharply
    (d) may increase or decrease a few degrees

16. Impotence is the inability to
    (a) produce sufficient amounts of spermatozoa for fertilization
    (b) achieve or maintain an erection
    (c) achieve orgasm
    (d) ejaculate

17. Erection of the clitoris is caused by the influence of
    (a) parasympathetic activation
    (b) sympathetic activation
    (c) increased output of estrogen
    (d) increased output of progesterone

18. Menopause is accompanied by
    (a) sustained rises in GnRH, FSH, and LH
    (b) drops in circulating levels of estrogen and progesterone
    (c) a thinning of the urethral and vaginal walls
    (d) a, b, and c are correct

19. Which reproductive structures are common to both males and females?

20. Trace the duct system that the sperm traverses from the site of its production to the exterior of the body.

21. Which accessory organs and glands contribute to the composition of semen? What are the functions of each?

22. What are the two primary cell populations in the testes that are responsible for functions related to reproductive activity? What are the functions of these cells?

23. What are the three primary functions of the epididymis?

24. Identify the three regions of the male urethra.

25. Describe the composition of a typical sample of ejaculate.

26. List the functions of testosterone in males.

27. What are the primary functions of the ovaries?

28. List and summarize the important steps in the ovarian cycle.

29. Describe the histological composition of the uterine wall.

30. What are the three major functions of the vagina?

31. What is the role of the clitoris in the female reproductive system?

32. What is the function of the lesser and greater vestibular glands in females?

33. Trace the route that milk takes from its site of production to the outside of the female.

LEVEL 2 Reviewing Concepts

34. Some women experience a sharp pain and spotting (release of a small volume of blood through the vagina) during the middle of their menstrual cycle. These symptoms are most likely due to
    (a) continued buildup of the functional layer of the endometrium
    (b) follicular development in the ovary

    (c) the rupture of a Graafian follicle from the ovary
    (d) PMS

35. Where in the male reproductive system would you expect to find mature spermatozoa?
    (a) in the seminiferous tubules
    (b) in the epididymis
    (c) traveling the ductus deferens and ejaculatory duct
    (d) in the seminal vesicles

36. A sample of female blood is analyzed for reproductive hormone levels. The results indicate a high level of progesterone, relatively high levels of inhibin, and low levels of FSH and LH. The female is most likely experiencing _____ of the uterine cycle.
    (a) the proliferative phase
    (b) menses
    (c) the secretory phase
    (d) menarche

37. In males, the hormone aromatase converts
    (a) testosterone to a form of estrogen
    (b) LH to testosterone
    (c) testosterone to dihydrotestosterone
    (d) estradiol to testosterone

38. How does the human reproductive system differ functionally from all other systems of the body?

39. How are the male and female reproductive systems functionally different?

40. How is the process of meiosis involved in the development of the spermatozoon and the ovum?

41. What are the main differences in gamete production between males and females?

42. Describe the erectile tissues of the penis. How does erection occur?

43. What is the functional significance of the normally acidic pH of the vagina?

44. Using an average cycle of 28 days, describe each of the three phases of the uterine cycle.

45. Describe the hormonal events associated with the ovarian cycle.

46. Describe the hormonal events associated with the uterine cycle.

47. Summarize the steps that occur in sexual arousal and orgasm. Do these processes differ in males and females?

48. How does the aging process affect the reproductive systems of men and women?

49. How do birth control pills prevent conception?

LEVEL 3 Critical Thinking and Clinical Applications

50. Diane has peritonitis (an inflammation of the peritoneum), which her physician says resulted from a urinary tract infection. Why might this condition occur more readily in females than in males?

51. Rod suffers an injury to the sacral region of his spinal cord. Will he still be able to achieve an erection? Explain.

52. A seven-year-old girl develops an ovarian tumor that involves granulosa cells. What symptoms would you expect to observe?

53. Women bodybuilders and women with eating disorders such as anorexia nervosa commonly experience amenorrhea. What does this fact suggest about the relation between body fat and menstruation? What might be the benefit of amenorrhea under such circumstances?

UNIT 6   **CHAPTER**   28   **29**

29

DEVELOPMENT AND INHERITANCE

The physiological processes we have studied thus far are relatively brief in duration. Many last only a fraction of a second; others may take hours at most. But some important processes are measured in months, years, or decades. A human being develops in the womb for nine months, grows to maturity in 15 or 20 years, and may live the better part of a century. During that time, he or she is always changing. Birth, growth, maturation, aging, and death are all parts of a single, continuous process. That process does not end with the individual, because humans can pass at least some of their characteristics on to their offspring. Thus, each generation gives rise to a new generation that will repeat the cycle. In this chapter, we shall explore the continuity of life, from conception to death.

29-1 AN OVERVIEW OF TOPICS IN DEVELOPMENT

Objective

- Explain the relationship between differentiation and development, and specify the various stages of development.

Time refuses to stand still; today's infant will be tomorrow's adult. The gradual modification of anatomical structures during the period from fertilization to maturity is called **development**. The changes that occur during development are truly remarkable. In a mere 40 weeks, all the tissues, organs, and organ systems we have studied thus far take shape and begin to function. What begins as a single cell slightly larger than the period at the end of this sentence becomes an individual whose body contains trillions of cells organized into a complex array of highly specialized structures. The creation of different types of cells required by this process is called **differentiation**. ∞ p. 106 Differentiation occurs through selective changes in genetic activity. As development proceeds, some genes are turned off and others are turned on. The identities of these genes vary from one type of cell to another.

Development begins at **fertilization**, or **conception**. We can divide development into periods characterized by specific anatomical changes. **Embryological development** comprises the events that occur during the first two months after fertilization. The study of these events is called **embryology** (em-brē-OL-o-jē). **Fetal development** begins at the start of the ninth week and continues until birth. Embryological and fetal development are sometimes referred to collectively as **prenatal development**, the primary focus of this chapter. **Postnatal development** commences at birth and continues to **maturity**, when the aging process begins.

A basic understanding of human development provides important insights into anatomical structures. In addition, many of the mechanisms of development and growth are similar to those responsible for the repair of injuries. In this chapter, we shall focus on major aspects of development. We shall consider highlights of the developmental process rather than describe the events in great detail. We shall also consider the regulatory mechanisms and how developmental patterns can be modified—for good or ill. Few topics in the biological sciences hold such fascination, and fewer still confront the investigator with so daunting an array of scientific, technological, and ethical challenges. The ongoing debate over fetal tissue research has brought some of the ethical issues into the public eye. The information presented in this final chapter should help you to formulate your opinions on many difficult moral, legal, and public-policy questions.

Although all humans go through the same developmental stages, differences in their genetic makeup produce distinctive individual characteristics. The term **inheritance** refers to the transfer of genetically determined characteristics from generation to generation. The study of the mechanisms responsible for inheritance is called **genetics**. In this chapter, we shall also consider basic genetics as it applies to inherited characteristics, such as gender, hair color, and various diseases.

29-2 FERTILIZATION

Objectives

- Describe the process of fertilization.
- Explain how developmental processes are regulated.

Fertilization involves the fusion of two haploid gametes, producing a zygote that contains 46 chromosomes, the normal complement in a somatic cell. The functional roles and contributions of the male and female gametes are very different. The spermatozoon simply delivers the paternal chromosomes to the site of fertilization. It must travel a relatively large distance and is small, efficient, and highly streamlined. In contrast, the female gamete must provide all the cellular organelles and inclusions, nourishment, and genetic programming necessary to support development of the embryo for nearly a week after conception. The volume of this gamete is therefore much greater than that of the spermatozoon (Figure 29–1a●). Recall from Chapter 28 that ovulation releases a secondary oocyte suspended in metaphase of meiosis II. At fertilization, the diameter of the secondary oocyte is more than twice the entire length of the spermatozoon. The ratio of their volumes is even more striking—roughly 2000:1.

The spermatozoa arriving in the vagina are already motile, as a result of mixing with secretions of the seminal vesicles—the

first step of *capacitation*. ∞ p. 1057 (A substance secreted by the epididymis appears to prevent premature capacitation.) The spermatozoa, however, cannot perform fertilization until they have been exposed to conditions in the female reproductive tract. The mechanism responsible for this second step of capacitation remains unknown.

Fertilization typically occurs near the junction between the ampulla and isthmus of the uterine tube, generally within a day after ovulation. By this time, a secondary oocyte has traveled a few centimeters, but spermatozoa must cover the distance between the vagina and the ampulla of the uterine tube. A spermatozoon can propel itself at speeds of only about 34 μm per second, roughly equivalent to 12.5 cm (5 in.) per hour, so in theory it should take spermatozoa several hours to reach the upper portions of the uterine tubes. The actual passage time, however, ranges from two hours to as little as 30 minutes. Contractions of the uterine musculature and ciliary currents in the uterine tubes have been suggested as likely mechanisms for accelerating the movement of spermatozoa from the vagina to the site of fertilization.

Even with transport assistance and available nutrients, the passage is not easy. Of the roughly 200 million spermatozoa introduced into the vagina in a typical ejaculation, only about 10,000 enter the uterine tube, and fewer than 100 reach the isthmus. In general, a male with a sperm count below 20 million per milliliter is functionally sterile because too few spermatozoa survive to reach and fertilize an oocyte. While it is true that only one spermatozoon fertilizes an oocyte, dozens of spermatozoa are required for successful fertilization. The additional sperm are essential because one sperm does not contain enough acrosomal enzymes to disrupt the *corona radiata*, the layer of follicle cells that surrounds the oocyte. ∞ p. 1068

☐ THE OOCYTE AT OVULATION

Ovulation occurs before the oocyte is completely mature. The secondary oocyte leaving the follicle is in metaphase of meiosis II. The cell's metabolic operations have been discontinued, and the oocyte drifts in a sort of suspended animation, awaiting the stimulus for further development. If fertilization does not occur, the oocyte disintegrates without completing meiosis.

Fertilization is complicated by the fact that when the secondary oocyte leaves the ovary, it is surrounded by the corona radiata. Fertilization and the events that follow are diagrammed in Figure 29–1b●. The cells of the corona radiata protect the secondary oocyte as it passes through the ruptured follicular wall, across the surface of the ovary, and into the infundibulum of the uterine tube. Although the physical process of fertilization requires only a single spermatozoon in contact with the oocyte membrane, that spermatozoon must first penetrate the corona radiata. The acrosomal cap of the sperm contains several enzymes, including **hyaluronidase** (hī-al-ū-RON-i-dās), which breaks down the bonds between adjacent follicular cells.

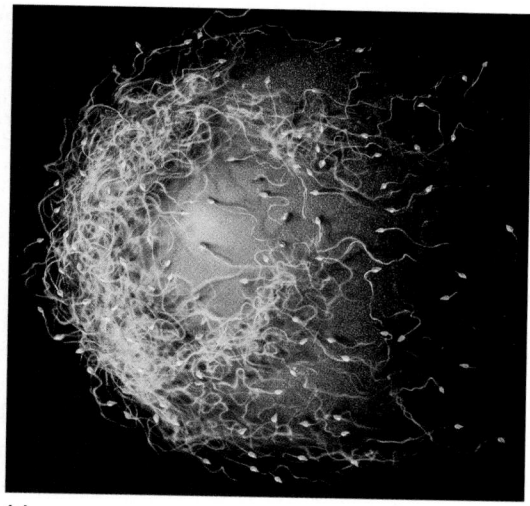

(a)

●**FIGURE 29–1**
Fertilization. **(a)** An oocyte and numerous sperm at the time of fertilization. Notice the difference in size between the gametes. **(b)** Fertilization and the preparations for cleavage.

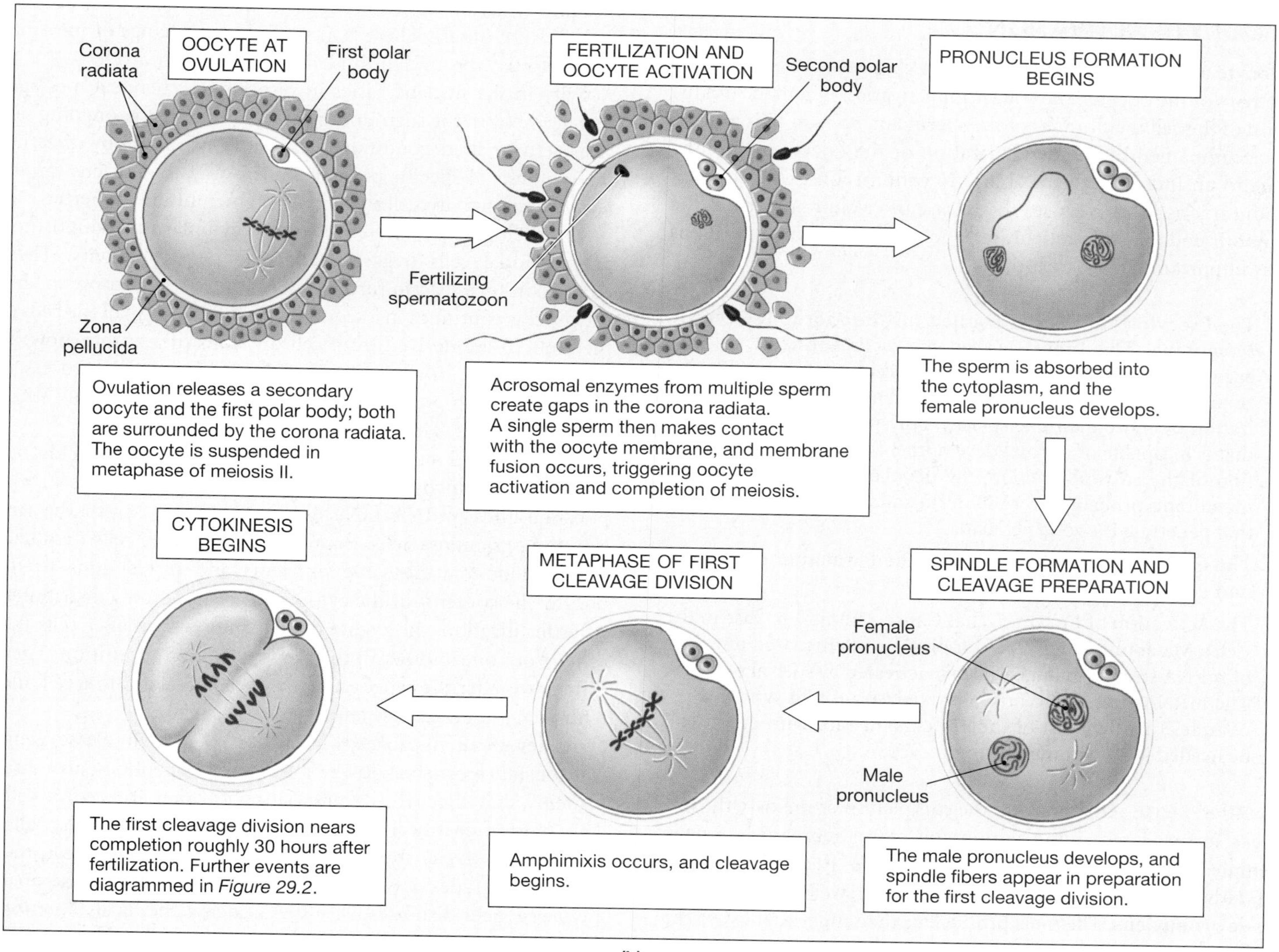

Corona radiata
OOCYTE AT OVULATION
First polar body

FERTILIZATION AND OOCYTE ACTIVATION
Second polar body

PRONUCLEUS FORMATION BEGINS

Fertilizing spermatozoon

Zona pellucida

Ovulation releases a secondary oocyte and the first polar body; both are surrounded by the corona radiata. The oocyte is suspended in metaphase of meiosis II.

Acrosomal enzymes from multiple sperm create gaps in the corona radiata. A single sperm then makes contact with the oocyte membrane, and membrane fusion occurs, triggering oocyte activation and completion of meiosis.

The sperm is absorbed into the cytoplasm, and the female pronucleus develops.

CYTOKINESIS BEGINS

METAPHASE OF FIRST CLEAVAGE DIVISION

SPINDLE FORMATION AND CLEAVAGE PREPARATION

Female pronucleus

Male pronucleus

The first cleavage division nears completion roughly 30 hours after fertilization. Further events are diagrammed in *Figure 29.2*.

Amphimixis occurs, and cleavage begins.

The male pronucleus develops, and spindle fibers appear in preparation for the first cleavage division.

(b)

Dozens of spermatozoa must release hyaluronidase before the connections between the follicular cells break down enough to permit fertilization.

No matter how many spermatozoa slip through the gap in the corona radiata, normally only a single spermatozoon will accomplish fertilization and activate the oocyte. The first step is the binding of the spermatozoon to *sperm receptors* in the zona pellucida. This step triggers the rupture of the acrosomal cap. The hyaluronidase and **acrosin**, another proteolytic enzyme, then digest a path through the zona pellucida toward the surface of the oocyte. When the sperm contacts that surface, the sperm and oocyte membranes begin to fuse. This step is the trigger for *oocyte activation*, a complex process we will discuss in the next section. As the membranes fuse, the entire sperm enters the cytoplasm of the oocyte.

☐ OOCYTE ACTIVATION

Oocyte activation involves a series of changes in the metabolic activity of the oocyte. The trigger for activation is contact and fusion of the cell membranes of the sperm and oocyte. This process is accompanied by the depolarization of the oocyte membrane due to an increased permeability to sodium ions. The entry of sodium ions in turn causes the release of calcium ions from the smooth endoplasmic reticulum. The sudden rise in Ca^{2+} levels has important effects, including:

- **The Exocytosis of Vesicles Located Just Interior to the Oocyte Membrane.** This process, called the *cortical reaction*, releases enzymes that inactivate the sperm receptors and harden the zona pellucida. This combination prevents **polyspermy** (fertilization by more than one sperm). Polyspermy creates a zygote that is incapable of normal development. (Prior to the completion of the cortical reaction, the depolarization of the oocyte membrane probably prevents fertilization by any sperm cells that penetrate the zona pellucida.)

- **The Completion of Meiosis II and the Formation of the Second Polar Body.**

- **The Activation of Enzymes That Cause a Rapid Increase in the Cell's Metabolic Rate.** The cytoplasm contains large numbers of mRNA strands that have been inactivated by special proteins. The mRNA strands are now activated, so protein synthesis accelerates rapidly. Most of the proteins that are synthesized will be needed if development is to proceed.

After oocyte activation and the completion of meiosis, the nuclear material remaining within the ovum reorganizes as the **female pronucleus** (Figure 29–1b●). While these changes are under way, the nucleus of the spermatozoon swells, becoming the **male pronucleus**. The male pronucleus then migrates toward the center of the cell, and spindle fibers form. The two pronuclei fuse in a process called *amphimixis* (am-fi-MIK-sis). Fertilization is now complete, with the formation of a zygote that contains the normal complement of 46 chromosomes.

29–3 THE STAGES OF PRENATAL DEVELOPMENT

Objective

■ List the three prenatal periods, and describe the major events associated with each.

During prenatal development, a single cell ultimately forms a 3–4 kg infant, who in postnatal development will grow through adolescence and maturity toward old age and eventual death. One of the most fascinating aspects of development is its apparent order and simplicity. Continuity exists at all levels and at all times. Nothing leaps into existence unheralded and without apparent precursors; differentiation and increasing structural complexity occur hand in hand.

Differentiation involves changes in the genetic activity of some cells but not others. There is a continuous exchange of information between the nucleus and the cytoplasm in a cell. ∞ p. 83 Activity in the nucleus varies in response to chemical messages that arrive from the surrounding cytoplasm. In turn, ongoing nuclear activity alters conditions within the cytoplasm by directing the synthesis of specific proteins. In this way, the nucleus can affect enzyme activity, cell structure, and membrane properties.

In development, differences in the cytoplasmic composition of individual cells trigger changes in genetic activity. These changes in turn lead to further alterations in the cytoplasm, and the process continues in a sequential fashion. But if all the cells of the embryo are derived from cell divisions of a zygote, how do the cytoplasmic differences originate? What sets the process in motion? The important first step occurs before fertilization, while the oocyte is in the ovary.

Before ovulation, the growing oocyte accepts amino acids, nucleotides, and glucose, as well as more complex materials such as phospholipids, mRNA molecules, and proteins, from the surrounding granulosa cells. Because not all follicle cells manufacture and deliver the same nutrients and instructions to the oocyte, the contents of the cytoplasm are not evenly distributed. After fertilization, the zygote divides into ever smaller cells that differ from one another in their cytoplasmic composition. These differences alter the genetic activity of each cell, creating cell lines with increasingly diverse fates.

As development proceeds, some of the cells will release chemical substances, such as RNA molecules, polypeptides, and small proteins, that affect the differentiation of other embryonic cells. This type of chemical interplay among developing cells, called *induction* (in-DUK-shun), works over very short distances, such as when two types of cells are in direct contact. It may also operate over longer distances, with the inducing chemicals functioning as hormones.

This type of regulation, which involves an integrated series of interacting steps, can control highly complex processes. The mechanism is not risk free, however: The appearance of an abnormal or inappropriate inducer can throw the entire development plan off course. **AM** Teratogens and Abnormal Development

The time spent in prenatal development is known as **gestation** (jes-TĀ-shun). For convenience, we usually think of the gestation period as consisting of three integrated **trimesters**, each three months in duration:

1. The **first trimester** is the period of embryological and early fetal development. During this period, the rudiments of all the major organ systems appear.

2. The **second trimester** is dominated by the development of organs and organ systems, a process that nears completion by the end of the sixth month. Along the way, body proportions change; by the end of this trimester, the fetus looks distinctively human.

3. The **third trimester** is characterized by rapid fetal growth and deposition of adipose tissue. Early in the third trimester, most of the major organ systems become fully functional. An infant born one month or even two months prematurely has a reasonable chance of survival.

The *Atlas* accompanying this text contains "Embryology Summaries," in which we introduce key steps in embryological and fetal development and trace the development of specific organ systems. We will refer to those summaries in the discussions that follow. As we proceed, you should review the material indicated. Doing so will help you understand the "big picture" as well as the specific details.

29–4 THE FIRST TRIMESTER

Objectives

- Explain how the germ layers participate in the formation of extraembryonic membranes.
- Discuss the importance of the placenta as an endocrine organ.

At the moment of conception, the fertilized ovum is a single cell that has a diameter of about 0.135 mm (0.005 in.) and a weight of approximately 150 mg. By the end of the first trimester (the 12th developmental week), the fetus is almost 75 mm (3 in.) long and weighs perhaps 14 g (0.5 oz).

Many important and complex developmental events occur during the first trimester. We shall focus on four general processes: cleavage, implantation, placentation, and embryogenesis:

1. **Cleavage** (KLĒV-ij) is a sequence of cell divisions that begins immediately after fertilization and ends at the first contact with the uterine wall. During cleavage, the zygote becomes a **pre-embryo**, which develops into a multicellular complex known as a *blastocyst*. We introduce cleavage and blastocyst formation in the *Atlas*. **ATLAS** EMBRYOLOGY SUMMARY 1: The Development of Tissues

2. **Implantation** begins with the attachment of the blastocyst to the endometrium of the uterus and continues as the blastocyst invades maternal tissues. During implantation, other important events take place that set the stage for the formation of vital embryonic structures.

3. **Placentation** (plas-en-TĀ-shun) occurs as blood vessels form around the periphery of the blastocyst and the **placenta** develops. The placenta is a complex organ that permits exchange between the maternal and embryonic circulatory systems. The placenta supports the fetus in the second and third trimesters, but it stops functioning and is ejected from the uterus at birth. From that point on, the newborn is physically independent of the mother.

4. **Embryogenesis** (em-brē-ō-JEN-e-sis) is the formation of a viable embryo. This process establishes the foundations for all major organ systems.

The foregoing processes are both complex and vital to the survival of the embryo. Perhaps because the events in the first trimester are so complex, that is the most dangerous period in prenatal life. Only about 40 percent of conceptions produce embryos that survive the first trimester. For that reason, pregnant women are warned to take great care to avoid drugs and other disruptive stresses during the first trimester, in the hope of preventing an error in the delicate processes that are under way.

CLEAVAGE AND BLASTOCYST FORMATION

Cleavage is a series of cell divisions that subdivides the cytoplasm of the zygote (Figure 29–2●). The first cleavage produces a pre-embryo consisting of two identical cells. Identical cells produced by cleavage divisions are called **blastomeres** (BLAS-tō-mērz). The first division is completed roughly 30 hours after fertilization, and subsequent divisions occur at intervals of 10–12 hours. During the initial divisions, all the blastomeres undergo mitosis simultaneously. As the number of blastomeres increases, the timing becomes less predictable.

After three days of cleavage, the pre-embryo is a solid ball of cells resembling a mulberry. This stage is called the **morula** (MOR-ū-la; *morula*, mulberry). The morula typically reaches the uterus on day 4. Over the next two days, the blastomeres form a **blastocyst**, a hollow ball with an inner cavity known as the **blastocoele** (BLAS-tō-sēl). The blastomeres are now no longer identical in size and shape. The outer layer of cells, which separates the outside world from the blastocoele, is called the **trophoblast** (TRŌ-fō-blast). The function is implied by the name: *trophos*, food + *blast*, precursor. These cells are responsible for providing nutrients to the developing embryo. A second group of cells, the **inner cell mass**, lies clustered at one end of the blastocyst. These cells are exposed to the blastocoele, but are insulated from contact with the outside environment by the trophoblast. In time, the inner cell mass will form the embryo.

IMPLANTATION

During blastocyst formation, enzymes released by the trophoblast erode a hole through the zona pellucida, which is then

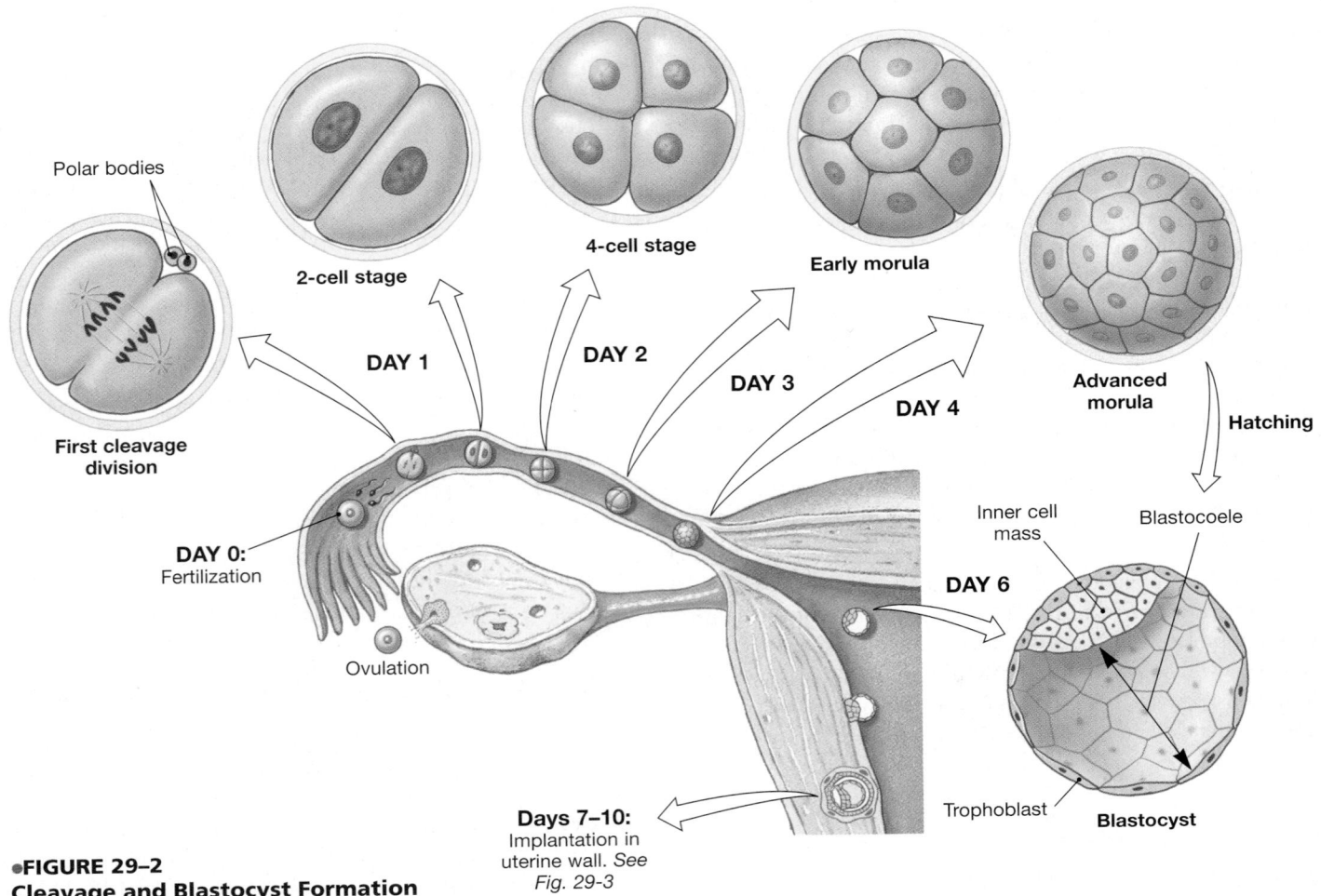

●**FIGURE 29–2**
Cleavage and Blastocyst Formation

shed in a process known as *hatching*. The blastocyst is now freely exposed to the fluid contents of the uterine cavity. This fluid, rich in glycogen, is secreted by the endometrial glands of the uterus. Over the previous few days, the pre-embryo and then early blastocyst had been absorbing fluid and nutrients from its surroundings; the process now accelerates, and the blastocyst enlarges. When fully formed, the blastocyst contacts the endometrium, and implantation occurs. Stages in the implantation process are illustrated in Figure 29–3●.

Implantation begins as the surface of the blastocyst closest to the inner cell mass touches and adheres to the uterine lining. (See Figure 29–3●, day 7.) At the point of contact, the trophoblast cells divide rapidly, making the trophoblast several layers thick. The cells closest to the interior of the blastocyst remain intact, forming a layer of **cellular trophoblast**, or *cytotrophoblast*. Near the endometrial wall, the cell membranes separating the trophoblast cells disappear, creating a layer of cytoplasm containing multiple nuclei (day 8). This outer layer is called the **syncytial** (sin-SISH-al) **trophoblast**, or *syncytiotrophoblast*.

The syncytial trophoblast erodes a path through the uterine epithelium by secreting hyaluronidase. This enzyme dissolves the intercellular cement between adjacent epithelial cells, just as

the hyaluronidase released by spermatozoa dissolves the connections between cells of the corona radiata. At first, the erosion creates a gap in the uterine lining, but migration and divisions of epithelial cells soon repair the surface. When the repairs are completed, the blastocyst loses contact with the uterine cavity. Development occurs entirely within the functional zone of the endometrium. ∞ p. 1071

In most cases, implantation takes place in the fundus or elsewhere in the body of the uterus. In an **ectopic pregnancy**, implantation actually occurs somewhere other than within the uterus, such as in one of the uterine tubes. Approximately 0.6 percent of pregnancies are ectopic pregnancies, which do not produce a viable embryo and can be life threatening. **AM** Ectopic Pregnancies

As implantation proceeds, the syncytial trophoblast continues to enlarge and spread into the surrounding endometrium (day 9). The erosion of uterine glands releases nutrients that are absorbed by the syncytial trophoblast and distributed by diffusion through the underlying cellular trophoblast to the inner cell mass. These nutrients provide the energy needed to support the early stages of embryo formation. Trophoblastic extensions grow around endometrial capillaries. As the capillary walls are de-

larger endometrial veins and arteries, and blood flow through the lacunae accelerates.

The trophoblast undergoes repeated nuclear divisions, shows extensive and rapid growth, has a very high demand for energy, invades and spreads through adjacent tissues, and fails to activate the maternal immune system. In short, the trophoblast has many of the characteristics of cancer cells. In about 0.1 percent of pregnancies, something goes wrong with the regulatory mechanisms, and a normal placenta and embryo do not develop. Instead, the syncytial trophoblast behaves like a tumor. This condition is called **gestational trophoblastic neoplasia**. The least dangerous form, a *hydatidiform* (hī-da-TID-i-form) *mole*, is not malignant. However, about 20 percent of gestational trophoblastic neoplasias will metastasize, invading other tissues, with potentially fatal results. Consequently, prompt surgical removal of the mass is essential, sometimes followed by chemotherapy.

Formation of the Amniotic Cavity

The inner cell mass has little apparent organization early in the blastocyst stage. Yet, by the time of implantation, the inner cell mass has separated from the trophoblast. The separation gradually increases, creating a fluid-filled chamber called the **amniotic** (am-nē-OT-ik) **cavity**. The trophoblast will later be separated from the amniotic cavity by layers of cells that originate at the inner cell mass and line the amniotic cavity. These layers form the *amnion*, a membrane we shall examine later in the chapter. The amniotic cavity can be seen in the depiction of day 9 in Figure 29–3●; additional details from day 10 to day 12 are shown in Figure 29–4●. When the amniotic cavity first appears, the cells of the inner cell mass are organized into an oval sheet that is two layers thick: a superficial layer that faces the amniotic cavity, and a deeper layer that is exposed to the fluid contents of the blastocoele.

Gastrulation and Germ Layer Formation

By day 12, a third layer begins to form through **gastrulation** (gas-troo-LĀ-shun) (Figure 29–4●). During gastrulation, cells in specific areas of the surface move toward a central line known as the **primitive streak**. At the primitive streak, the migrating cells leave the surface and move between the two existing layers. This movement creates three distinct embryonic layers: (1) the **ectoderm**, consisting of superficial cells that did not migrate into the interior of the inner cell mass; (2) the **endoderm**, consisting of the cells that face the blastocoele; and (3) the **mesoderm**, consisting of the poorly organized layer of migrating cells between the ectoderm and the endoderm. Collectively, these three embryonic layers are called **germ layers**. The formation of mesoderm and the fates of each germ layer are summarized in the *Atlas*. **ATLAS** EMBRYOLOGY SUMMARY 4: The Development of Organ Systems Table 29–1 contains a more comprehensive listing of the contributions each germ layer makes to the body systems described in earlier chapters.

Gastrulation produces an oval, three-layered sheet known as the **embryonic disc**. This disc will form the body of the embryo, whereas the rest of the blastocyst will be involved in forming the *extraembryonic membranes*.

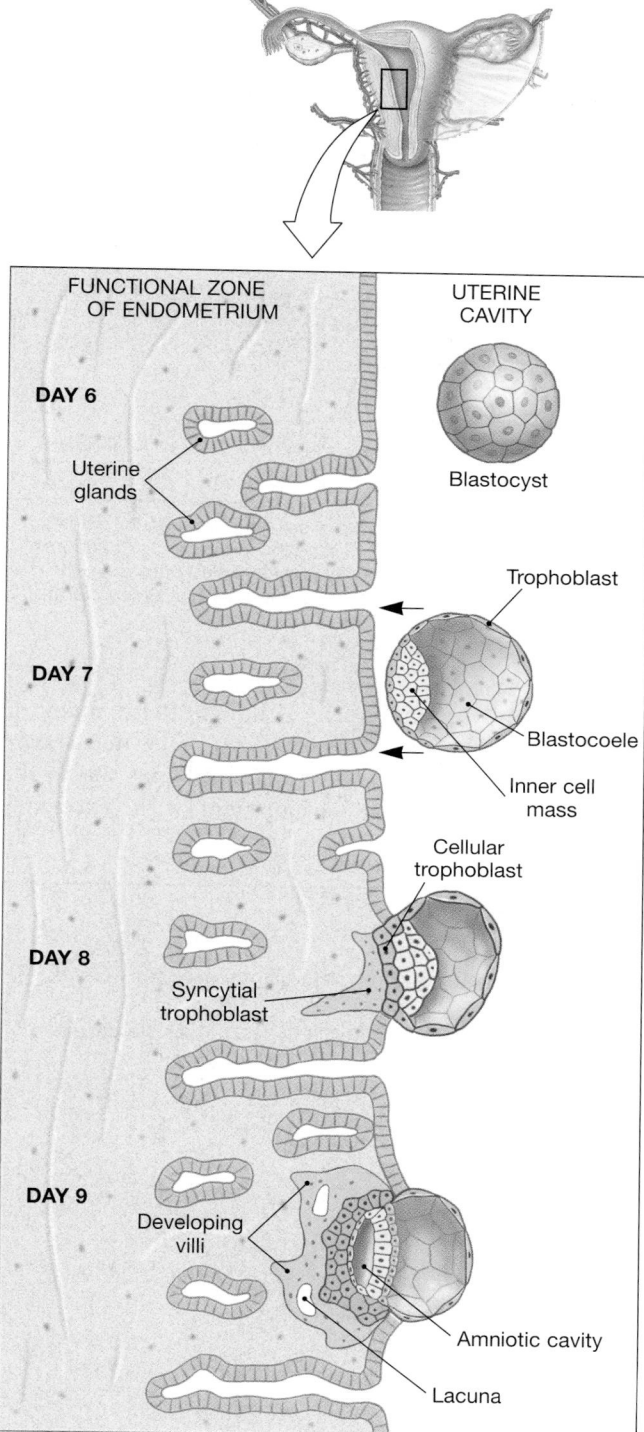

●**FIGURE 29–3**
Stages in Implantation

stroyed, maternal blood begins to percolate through trophoblastic channels known as **lacunae**. Fingerlike **villi** extend away from the trophoblast into the surrounding endometrium and gradually increase in size and complexity until about day 21. By approximately day 9, the syncytial trophoblast begins breaking down

●**FIGURE 29–4**
The Inner Cell Mass and Gastrulation

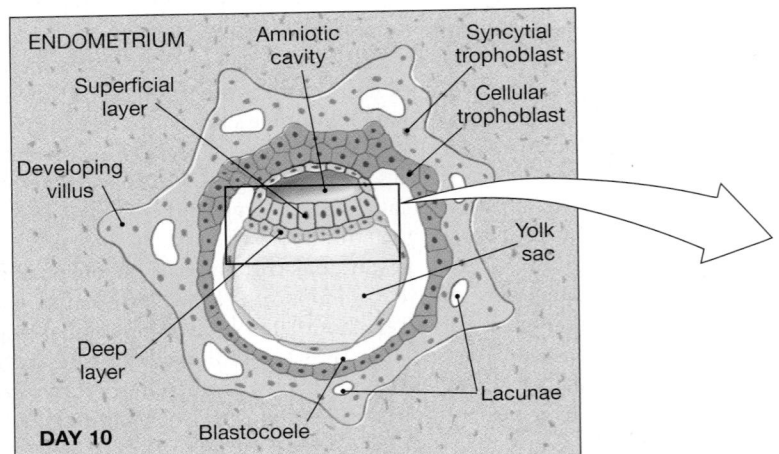

The inner cell mass begins as two layers: a superficial layer, facing the amniotic cavity, and a deep layer, exposed to the blastocoele. Migration of cells around the amniotic cavity is the first step in the formation of the amnion. Migration of cells around the edges of the blastocoele is the first step in yolk sac formation.

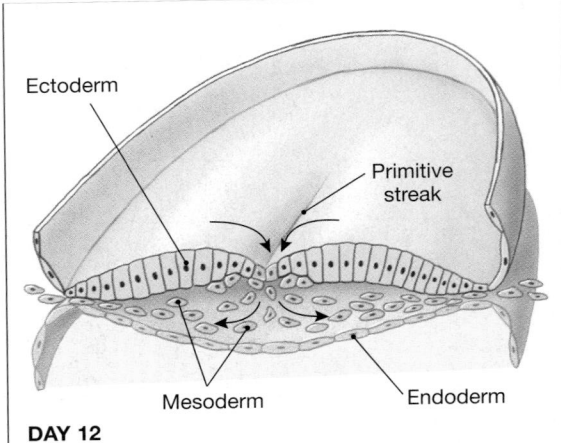

Migration of superficial cells into the interior creates a third layer. From the time this process (gastrulation) begins, the superficial layer is called *ectoderm*, the deep layer *endoderm*, and the migrating cells *mesoderm*.

The Formation of the Extraembryonic Membranes

Germ layers also participate in the formation of four **extra-embryonic membranes**: (1) the *yolk sac* (endoderm and mesoderm), (2) the *amnion* (ectoderm and mesoderm), (3) the *allantois* (endoderm and mesoderm), and (4) the *chorion* (mesoderm and trophoblast). Although these membranes support embryological and fetal development, they leave few traces of their existence in adult systems. Figure 29–5● shows representative stages in the development of the extraembryonic membranes.

SUMMARY TABLE 29–1 THE FATES OF THE GERM LAYERS

ECTODERMAL CONTRIBUTIONS

Integumentary system: epidermis, hair follicles and hairs, nails, and glands communicating with the skin (sweat glands, mammary glands, and sebaceous glands)

Skeletal system: pharyngeal cartilages and their derivatives in adults (portion of sphenoid bone, the auditory ossicles, the styloid processes of the temporal bones, the cornu and superior rim of the hyoid bone)*

Nervous system: all neural tissue, including brain and spinal cord

Endocrine system: pituitary gland and adrenal medullae

Respiratory system: mucous epithelium of nasal passageways

Digestive system: mucous epithelium of mouth and anus, salivary glands

MESODERMAL CONTRIBUTIONS

Skeletal system: all components except some pharyngeal derivatives

Muscular system: all components

Endocrine system: adrenal cortex, endocrine tissues of heart, kidneys, and gonads

Cardiovascular system: all components

Lymphatic system: all components

Urinary system: the kidneys, including the nephrons and the initial portions of the collecting system

Reproductive system: the gonads and the adjacent portions of the duct systems

Miscellaneous: the lining of the body cavities (pleural, pericardial, and peritoneal) and the connective tissues that support all organ systems

ENDODERMAL CONTRIBUTIONS

Endocrine system: thymus, thyroid gland, and pancreas

Respiratory system: respiratory epithelium (except nasal passageways) and associated mucous glands

Digestive system: mucous epithelium (except mouth and anus), exocrine glands (except salivary glands), liver, and pancreas

Urinary system: urinary bladder and distal portions of the duct system

Reproductive system: distal portions of the duct system, stem cells that produce gametes

*The neural crest is derived from ectoderm and contributes to the formation of the skull and the skeletal derivatives of the embryonic pharyngeal arches.

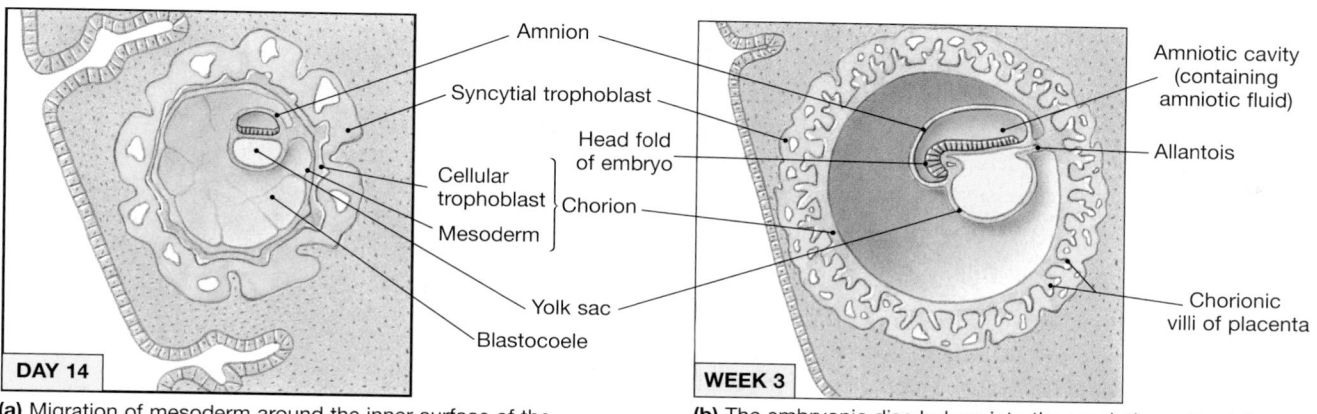

(a) Migration of mesoderm around the inner surface of the trophoblast creates the chorion. Mesodermal migration around the outside of the amniotic cavity, between the ectodermal cells and the trophoblast, forms the amnion. Mesodermal migration around the endodermal pouch creates the yolk sac.

(b) The embryonic disc bulges into the amniotic cavity at the head fold. The allantois, an endodermal extension surrounded by mesoderm, extends toward the trophoblast.

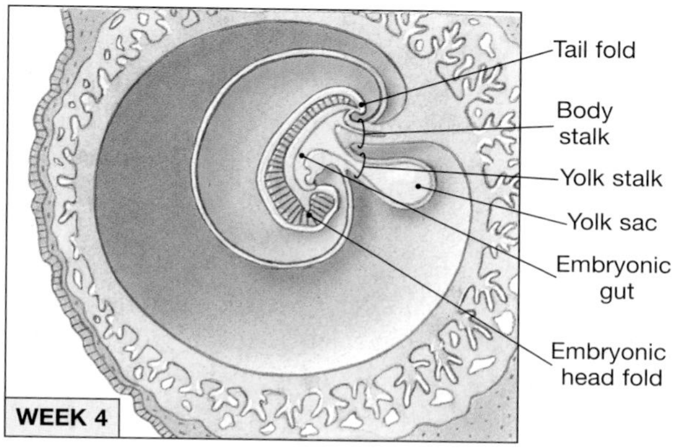

(c) The embryo now has a head fold and a tail fold. Constriction of the connection between the embryo and the surrounding trophoblast narrows the yolk stalk and body stalk.

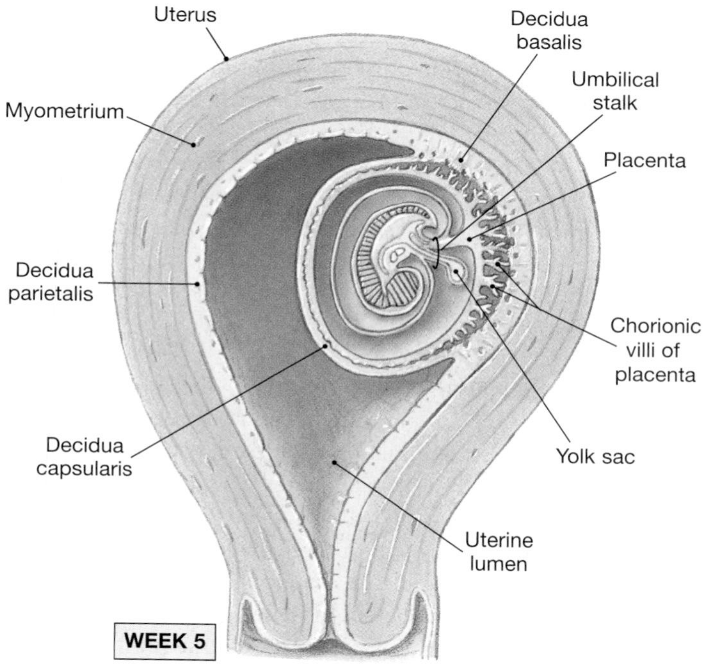

(d) The developing embryo and extraembryonic membranes bulge into the uterine cavity. The trophoblast pushing out into the uterine lumen remains covered by endometrium but no longer participates in nutrient absorption and embryo support. The embryo moves away from the placenta, and the body stalk and yolk stalk fuse to form an umbilical stalk.

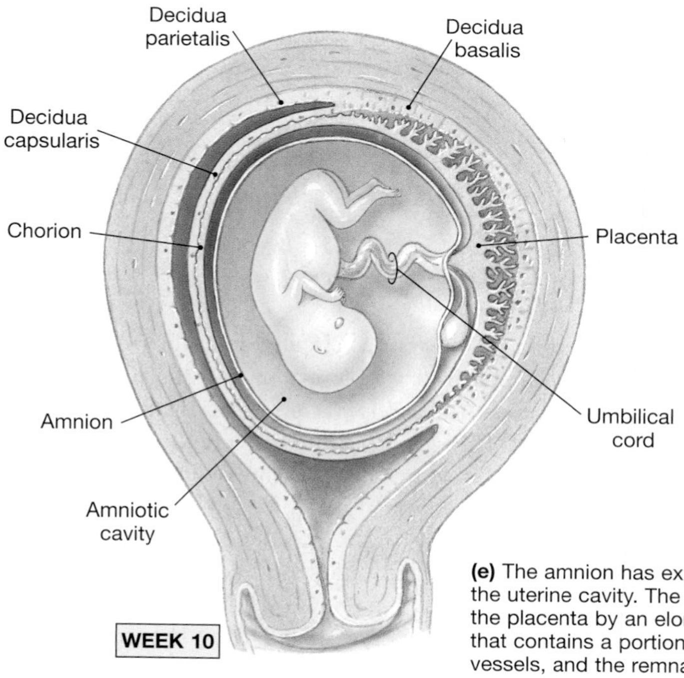

(e) The amnion has expanded greatly, filling the uterine cavity. The fetus is connected to the placenta by an elongated umbilical cord that contains a portion of the allantois, blood vessels, and the remnants of the yolk stalk.

•FIGURE 29–5
Extraembryonic Membranes and Placenta Formation

THE YOLK SAC The **yolk sac** begins as a layer of cells spread out around the outer edges of the blastocoele to form a complete pouch. This pouch is already visible 10 days after fertilization (Figure 29–4•). As gastrulation proceeds, mesodermal cells migrate around the pouch and complete the formation of the yolk sac (Figure 29–5a•). Blood vessels soon appear within the mesoderm, and the yolk sac becomes an important site of blood cell formation.

THE AMNION The ectodermal layer enlarges, and ectodermal cells spread over the inner surface of the amniotic cavity. Mesodermal cells soon follow, creating a second, outer layer (Figure 29–5a•). This combination of mesoderm and ectoderm is the **amnion** (AM-nē-on). As development proceeds, the amnion and the amniotic cavity continue to enlarge. The amniotic cavity contains **amniotic fluid**, which surrounds and cushions the developing embryo or fetus (Figure 29–5b–e•).

THE ALLANTOIS The third extraembryonic membrane begins as an outpocketing of the endoderm near the base of the yolk sac (Figure 29–5b•). The free endodermal tip then grows toward the wall of the blastocyst, surrounded by a mass of mesodermal cells. This sac of endoderm and mesoderm is the **allantois** (a-LAN-tō-is), the base of which later gives rise to the urinary bladder. We illustrate the formation of the allantois and its relationship to the urinary bladder in the *Atlas*. ATLAS EMBRYOLOGY SUMMARY 20: The Development of the Urinary System

THE CHORION The mesoderm associated with the allantois spreads around the blastocyst, separating the cellular trophoblast from the blastocoele. This combination of mesoderm and trophoblast is the **chorion** (KOR-ē-on) (Figure 29–5a,b•).

When implantation first occurs, the nutrients absorbed by the trophoblast can easily reach the inner cell mass by simple diffusion. But as the embryo and the trophoblast enlarge, the distance between them increases, so diffusion alone can no longer keep pace with the demands of the developing embryo. Blood vessels now begin to develop within the mesoderm of the chorion, creating a rapid-transit system that links the embryo with the trophoblast.

The appearance of blood vessels in the chorion is the first step in the creation of a functional placenta. By the third week of development, the mesoderm extends along the core of each trophoblastic villus, forming **chorionic villi** in contact with maternal tissues (Figures 29–5b and 29–6•, p. 1106). These villi continue to enlarge and branch, creating an intricate network within the endometrium. Embryonic blood vessels develop within each villus. Blood flow through those chorionic vessels begins early in the third week of development, when the embryonic heart starts beating. The blood supply to the chorionic villi arises from the allantoic arteries and veins.

As the chorionic villi enlarge, more maternal blood vessels are eroded. Maternal blood now moves slowly through complex lacunae lined by the syncytial trophoblast. Chorionic blood vessels pass close by, and gases and nutrients diffuse between the embryonic and maternal circulations across the layers of the trophoblast. Recall that fetal hemoglobin has a higher affinity for oxygen than does maternal hemoglobin, enabling fetal hemoglobin to strip oxygen from maternal hemoglobin. ∞ p. 658 Maternal blood then reenters the venous system of the mother through the broken walls of small uterine veins. No mixing of maternal and fetal blood occurs, because the two are always separated by layers of trophoblast.

☐ PLACENTATION

At first, the entire blastocyst is surrounded by chorionic villi. The chorion continues to enlarge, expanding like a balloon within the endometrium. By the fourth week, the embryo, amnion, and yolk sac are suspended within an expansive, fluid-filled chamber (Figure 29–5c•). The **body stalk**, the connection between embryo and chorion, contains the distal portions of the allantois and blood vessels that carry blood to and from the placenta. The narrow connection between the endoderm of the embryo and the yolk sac is called the **yolk stalk**. We illustrate the formation of the yolk stalk and body stalk in the *Atlas*. ATLAS EMBRYOLOGY SUMMARY 19: The Development of the Digestive System

The placenta does not continue to enlarge indefinitely. Regional differences in placental organization begin to develop as the expansion of the placenta creates a prominent bulge in the endometrial surface. This relatively thin portion of the endometrium, called the **decidua capsularis** (dē-SID-ū-a kap-sū-LA-ris; *deciduus*, a falling off), no longer participates in nutrient exchange, and the chorionic villi in the region disappear (Figures 29–5d, 29–6a•). Placental functions are now concentrated in a disc-shaped area in the deepest portion of the endometrium, a region called the **decidua basalis** (ba-SA-lis). The rest of the uterine endometrium, which has no contact with the chorion, is called the **decidua parietalis**.

As the end of the first trimester approaches, the fetus moves farther from the placenta (Figure 29–5d,e•). The fetus and placenta remain connected by the **umbilical cord**, or *umbilical stalk*, which contains the allantois, the placental blood vessels, and the yolk stalk.

Placental Circulation

Figure 29–6a• diagrams circulation at the placenta near the end of the first trimester. Blood flows to the placenta through the paired **umbilical arteries** and returns in a single **umbilical vein**. ∞ p. 768 The chorionic villi provide the surface area for active and passive exchanges of gases, nutrients, and waste products between the fetal and maternal bloodstreams (Figure 29–6b•).

The placenta places a considerable demand on the maternal cardiovascular system, and blood flow to the uterus and placenta is extensive. If the placenta is torn or otherwise damaged, the consequences may prove fatal to both fetus and mother. AM Problems with Placentation

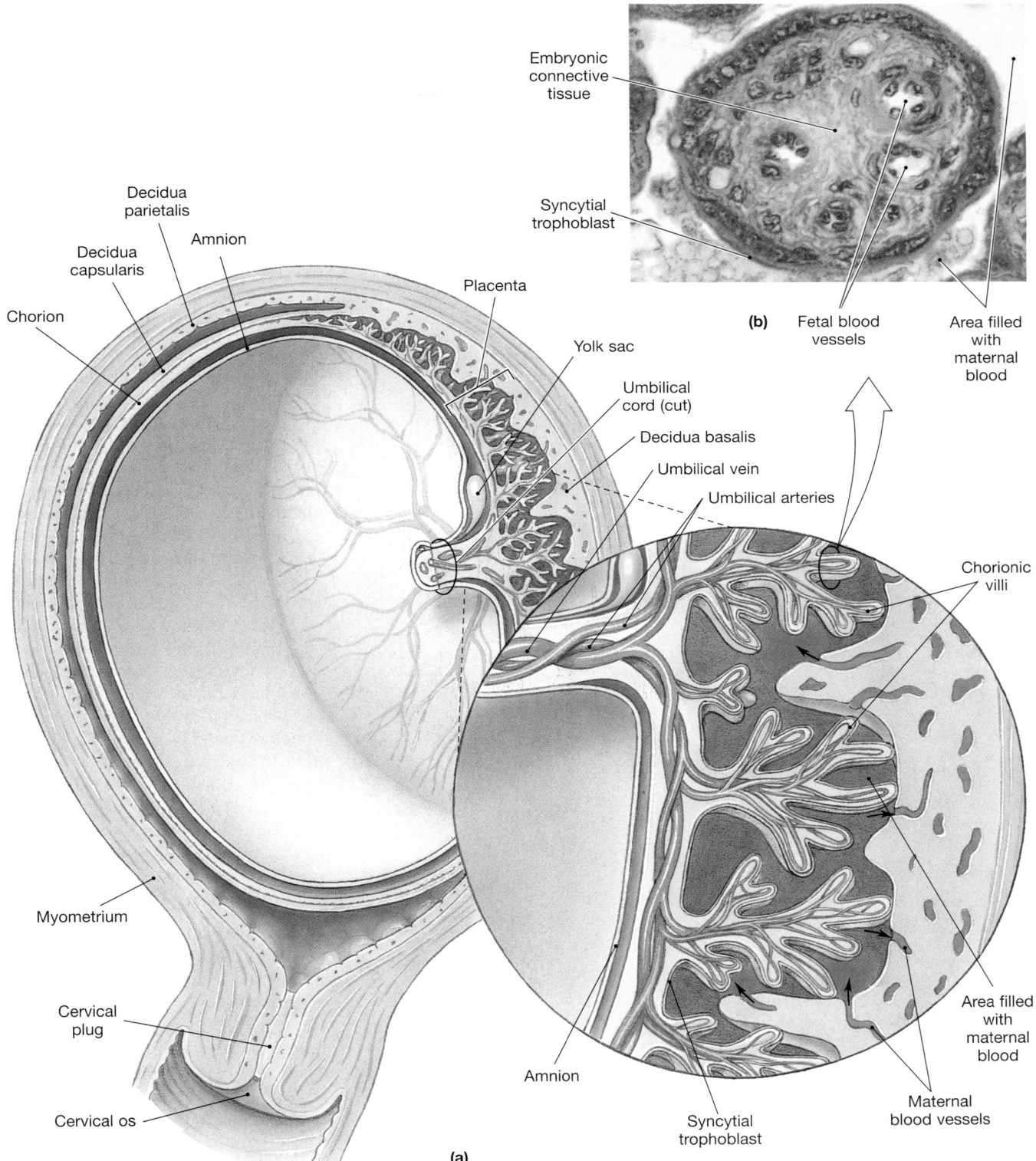

Embryonic connective tissue

Syncytial trophoblast

(b) Fetal blood vessels

Area filled with maternal blood

Decidua parietalis

Decidua capsularis

Amnion

Chorion

Placenta

Yolk sac

Umbilical cord (cut)

Decidua basalis

Umbilical vein

Umbilical arteries

Chorionic villi

Area filled with maternal blood

Maternal blood vessels

Syncytial trophoblast

Amnion

Myometrium

Cervical plug

Cervical os

(a)

●**FIGURE 29–6**
A Three-Dimensional View of Placental Structure. **(a)** For clarity, the uterus is shown after the embryo has been removed and the umbilical cord cut. Arrows indicate the direction of blood flow. Blood flows into the placenta through ruptured maternal arteries and then flows around chorionic villi, which contain fetal blood vessels. **(b)** A cross section through a chorionic villus, showing the syncytial trophoblast exposed to the maternal blood space. (LM × 2045)

Technology and the Treatment of Infertility

Infertility (*sterility*) is usually defined as an inability to achieve pregnancy after one year of appropriately timed intercourse. Problems with fertility are relatively common: An estimated 10–15 percent of U.S. married couples are infertile, and another 10 percent are unable to have as many children as they desire. It is thus not surprising that reproductive physiology has become a popular field and that the treatment of infertility has become a major medical industry.

An infertile woman is unable to produce functional oocytes or to support a developing embryo. An infertile man is incapable of providing a sufficient number of motile sperm capable of successful fertilization. Because the infertility of either sexual partner has the same result, the diagnosis and treatment of infertility must involve the evaluation of both partners. Approximately 40 percent of infertility cases are attributed to the female partner, 40 percent to the male partner, and 20 percent to both partners.

Recent advances in our understanding of reproductive physiology are providing new solutions to fertility problems. These approaches, called **assisted reproductive technologies (ARTs)**, are diagrammed in Figure 29–7●:

- **Low Sperm Count.** In cases of male infertility due to low sperm counts, semen from several ejaculates can be pooled, concentrated, and introduced into the female reproductive tract. This technique, *artificial insemination*, may lead to normal fertilization and pregnancy. In cases in which a male's spermatozoa are unable to penetrate the oocyte, single-sperm fertilization has been accomplished by microscopically manipulating the oocyte and the corona radiata.

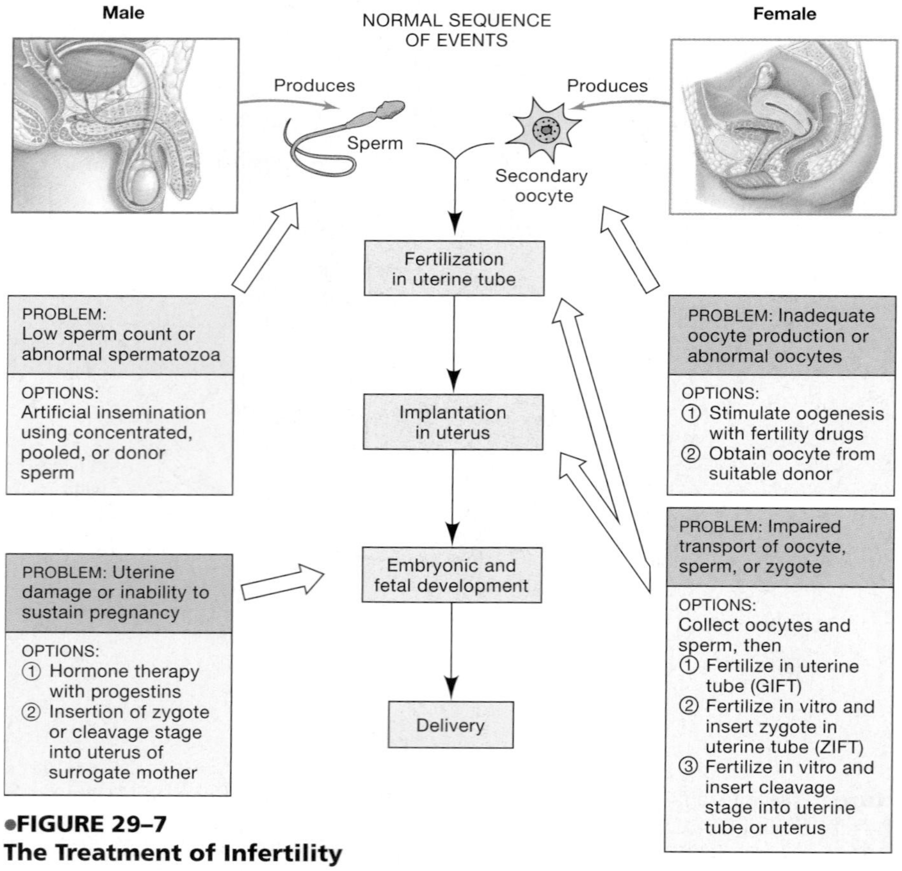

- **Abnormal Spermatozoa.** If the male cannot produce functional spermatozoa, they can be obtained from a *sperm bank*, a storage facility for donor sperm used in artificial insemination.

- **Hormonal Problems.** If the problem involves the woman's inability to ovulate because her gonadotropin or estrogen levels are low or she is unable to maintain adequate progesterone levels after ovulation, these hormones can be provided.

- **Problems with Oocyte Production.** *Fertility drugs*, such as clomiphene (*Clomid®*), stimulate ovarian oocyte production. Clomiphene blocks the feedback inhibition of the hypothalamus and pituitary gland by estrogens. As a result, circulating FSH levels rise, so more follicles are stimulated to complete their development. Injected purified gonadotropins, such as *Pergonal®* (FSH and LH) and *Metrodine®* (FSH), are also used to accelerate the ovum development. The chance that a single oocyte will be fertilized through sexual intercourse is about 1 in 3. Increasing the number of oocytes released raises the odds of fertilization and therefore the odds of a pregnancy. It is not easy to determine just how much ovarian stimulation will be needed, however, so treatment with fertility drugs commonly results in multiple births. Careful monitoring of follicle development reduces the chances of a multiple birth to some extent.

- **Problems with Oocyte Transport From the Ovary to the Uterine Tube.** When problems occur with the transport of the oocyte from the ovary to the uterine tube (for example, due to scarring of the fimbriae), a procedure called *gamete intrafallopian tube transfer* (**GIFT**) can be used. (*Fallopian tube* is another name for the uterine tube.) In this procedure, the ovaries are stimulated with injected hormones, and a large "crop" of mature oocytes is removed from tertiary follicles. The individual oocytes are examined for defects, inserted into the uterine tubes, and exposed to high concentrations of spermatozoa from the husband or donor. The success rate of the procedure is less

●**FIGURE 29–7**
The Treatment of Infertility

than that of natural fertilization (4 percent versus 33 percent), and not every pregnancy produces an infant. The cost of a single procedure (successful or not) averages $5000.

One variation on the GIFT procedure, called *zygote intrafallopian tube transfer* (**ZIFT**), exposes selected oocytes to spermatozoa outside the body and inserts zygotes or pre-embryos, rather than oocytes, into the uterine tubes. If multiple zygotes are available, some can be frozen and stored for later insertion if the initial attempt fails. The cost for a single ZIFT procedure, which has a 4 percent success rate, ranges between $8000 and $10,000.

- **Blocked Uterine Tubes.** Blockage of the uterine tubes or damage to the uterine lining can interfere with oocyte, sperm, and zygote transport. In the GIFT procedure, fertilization occurs in its normal location, within the uterine tube. This site, however, is not essential, and fertilization can also take place in a test tube or petri dish. If a carefully controlled fluid environment is provided, early development will proceed normally. Alternatively, the zygote can be maintained in an artificial environment through the first two to three days of development. This procedure is commonly selected if the uterine tubes are damaged or blocked. The cleavage-stage embryo is then placed directly into the uterus rather than into one of the uterine tubes. The process, called **in vitro fertilization**, or **IVF** (*vitro*, glass), is involved in roughly three-quarters of ART procedures in the United States. The cost of this procedure is comparable to that of ZIFT.

- **Abnormal Oocytes.** If the oocytes released by the ovaries are abnormal in some way, or if menopause has already occurred, viable oocytes can be obtained from a suitable donor. Through treatment with fertility drugs, the donor's ovaries are stimulated to produce a large crop of oocytes, which are then collected and fertilized in vitro, generally by the spermatozoa of the recipient's husband. After cleavage has begun, the pre-embryo is placed in the recipient's uterus, which has been "primed" by progesterone therapy. The pregnancy rate of this procedure is roughly 33 percent for women over age 40, using oocytes donated by women in their early twenties. Oocyte donation has a much higher success rate for these women than does ZIFT or GIFT. This difference suggests that age-related changes in the characteristics and quality of the oocytes, rather than changes in hormone levels or uterine responsiveness, are typically the primary cause of infertility in older women. Despite the fact that the mother has no genetic relationship to the embryo, pregnancy proceeds normally.

- **An Abnormal Uterine Environment.** If fertilization and transport occur normally, but the uterus cannot maintain a pregnancy, the problem may involve low levels of progesterone secretion by the corpus luteum. Hormone therapy may solve this problem. If the maternal uterus simply cannot support development, the zygote or cleavage-stage embryo can be introduced into the uterus of a *surrogate mother*. If the embryo survives and makes contact with the endometrium, development will proceed normally despite the fact that the mother has no genetic relationship to the embryo.

Surrogate motherhood, which sounds relatively simple and straightforward, has proven to be one of the most explosive solutions in terms of ethics and legality. Since 1990, several court cases have resulted from disputes over surrogate motherhood and who merits legal custody of the infant. Legal battles have also broken out over a variety of other complex questions, and some of them will take years to sort out. To understand the problem, consider the following questions:

- Do parents share property rights over frozen and stored zygotes? Can a husband have any of the stored zygotes implanted into the uterus of his second wife without the consent of his first wife, who provided the oocytes? Can a wife use her husband's stored spermatozoa to become pregnant after his death?

- If both donor oocyte and donor sperm are used, do adoption laws apply?

- If the husband provided the spermatozoa that fertilized the oocyte of a donor who is not his wife, for implantation into a surrogate mother, should the wife, the surrogate mother, or the oocyte donor have custody of the child if the father dies?

- If a hospital stores frozen pre-embryos or spermatozoa, but the freezer breaks down, what is the hospital's liability? What is the monetary value of a frozen pre-embryo?

If you use your imagination, you can probably think of even more complex problems, many of which will probably be debated in courtrooms over the next decade.

The Endocrine Placenta

In addition to its role in the nutrition of the fetus, the placenta acts as an endocrine organ. Hormones are synthesized by the syncytial trophoblast and released into the maternal bloodstream. The hormones produced include *human chorionic gonadotropin, human placental lactogen, placental prolactin, relaxin, progesterone,* and *estrogens.*

HUMAN CHORIONIC GONADOTROPIN The hormone **human chorionic gonadotropin (hCG)** appears in the maternal bloodstream soon after implantation has occurred. The presence of hCG in blood or urine samples provides a reliable indication of pregnancy. Kits sold for the early detection of pregnancy are sensitive to the presence of this hormone.

In function, hCG resembles luteinizing hormone (LH), because it maintains the integrity of the corpus luteum and promotes the continued secretion of progesterone. ∞ p. 1082 As a result, the endometrial lining remains perfectly functional, and menses does not occur. In the absence of hCG, the pregnancy would end, because another uterine cycle would begin and the functional zone of the endometrium would disintegrate.

In the presence of hCG, the corpus luteum persists for three to four months before gradually decreasing in size and secretory function. The decline in luteal function does not trigger the return of uterine cycles, because by the end of the first trimester, the placenta actively secretes both estrogens and progesterone.

HUMAN PLACENTAL LACTOGEN AND PLACENTAL PROLACTIN **Human placental lactogen (hPL)**, or *human chorionic somatomammotropin (hCS)*, helps prepare the mammary glands for milk production. It also has a stimulatory effect on other tissues comparable to that of growth hormone (GH). At the mammary glands, the conversion from inactive to active status requires the presence of placental hormones (hPL, **placental prolactin**, estrogen, and progesterone) as well as several maternal hormones (GH, prolactin [PRL], and thyroid hormones). We shall consider the hormonal control of the mammary gland function in a later section.

RELAXIN **Relaxin** is a peptide hormone that is secreted by the placenta, as well as by the corpus luteum, during pregnancy. Relaxin (1) increases the flexibility of the pubic symphysis, permitting the pelvis to expand during delivery, (2) causes the dilation of the cervix, making it easier for the fetus to enter the vaginal canal, and (3) suppresses the release of oxytocin by the hypothalamus and delays the onset of labor contractions.

PROGESTERONE AND ESTROGENS After the first trimester, the placenta produces sufficient amounts of progesterone to maintain the endometrial lining and continue the pregnancy. As the end of the third trimester approaches, estrogen production by the placenta accelerates. As we will see in a later section, the rising estrogen levels play a role in stimulating labor and delivery.

▢ EMBRYOGENESIS

Shortly after gastrulation begins, the body of the embryo begins to separate itself from the rest of the embryonic disc. The body of the embryo and its internal organs now start to form. This process, called **embryogenesis**, begins as folding and differential growth of the embryonic disc produce a bulge that projects into the amniotic cavity (Figure 29–5b●). This projection is known as the **head fold**; similar movements lead to the formation of a **tail fold** (Figure 29–5c●).

The embryo is now physically as well as developmentally distinct from the embryonic disc and the extraembryonic membranes. The definitive orientation of the embryo can now be

seen, complete with dorsal and ventral surfaces and left and right sides. Table 29–2 provides an overview of the subsequent development of the major organs and body systems. The changes in proportions and appearance that occur between the second developmental week and the end of the first trimester are summarized in Figure 29–8●.

The first trimester is a critical period for development, because events in the first 12 weeks establish the basis for **organogenesis**, the process of organ formation. In Embryology Summaries 6–21 in the *Atlas*, we describe major features of organogenesis in each organ system. Important developmental milestones are indicated in Table 29–2, pp. 1112–1113.

✓ What is the developmental fate of the inner cell mass of the blastocyst?

✓ Improper development of which of the extraembryonic membranes would affect the cardiovascular system?

✓ Sue's pregnancy test indicates elevated levels of the hormone hCG (human chorionic gonadotropin). Is she pregnant?

✓ What are two important functions of the placenta?

Answers start on page Q-1

 Review the processes of fertilization and implantation on the **Companion Website**: Chapter 29/Reviewing Concepts/Labeling.

29–5 THE SECOND AND THIRD TRIMESTERS

Objectives

▪ Describe the interplay between the maternal organ systems and the developing fetus.

▪ Discuss the structural and functional changes in the uterus during gestation.

By the end of the first trimester (Figure 29–8d●), the rudiments of all the major organ systems have formed. Over the next three months, the fetus will grow to a weight of about 0.64 kg (1.4 lb). Encircled by the amnion, the fetus grows faster than the surrounding placenta during this second trimester. When the mesoderm on the outer surface of the amnion contacts the mesoderm on the inner surface of the chorion, these layers fuse, creating a compound *amniochorionic membrane*. Figure 29–9a● shows a four-month-old fetus; Figure 29–9b● shows a six-month-old fetus.

During the third trimester, most of the organ systems become ready to fulfill their normal functions without medical assistance. The rate of growth starts to decrease, but in absolute terms this trimester sees the largest weight gain. In the last three months of gestation, the fetus gains about 2.6 kg (5.7 lb), reaching a full-term weight of approximately 3.2 kg (7 lb). The Embryology Summaries in the *Atlas* illustrate organ system development in the second and third trimesters, and highlights are noted in Table 29–2.

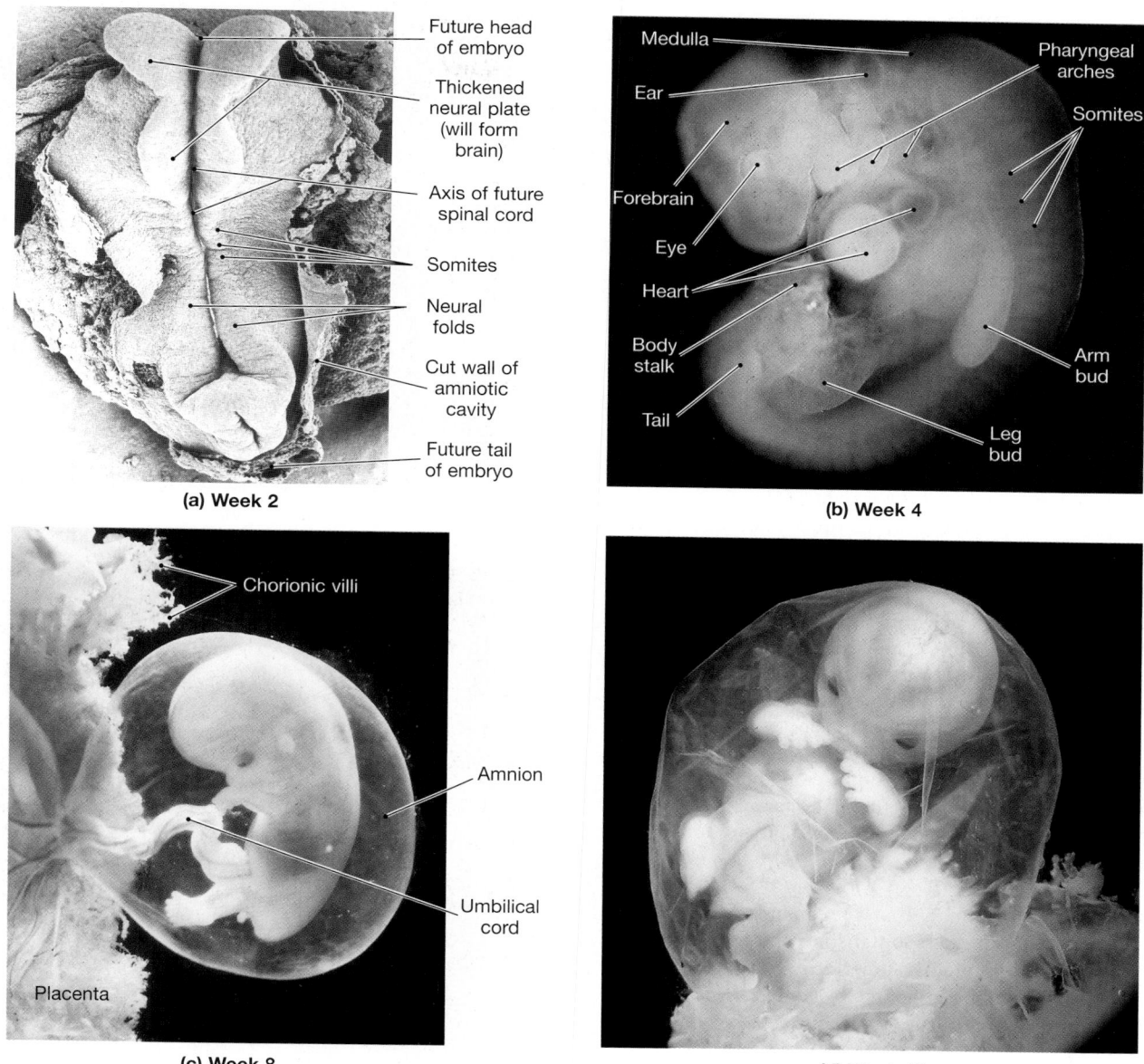

(a) Week 2

Future head of embryo
Thickened neural plate (will form brain)
Axis of future spinal cord
Somites
Neural folds
Cut wall of amniotic cavity
Future tail of embryo

(b) Week 4

Medulla
Ear
Forebrain
Eye
Heart
Body stalk
Tail
Pharyngeal arches
Somites
Arm bud
Leg bud

(c) Week 8

Chorionic villi
Amnion
Umbilical cord
Placenta

(d) Week 12

●**FIGURE 29–8**
The First Trimester. **(a)** An SEM of the superior surface of a monkey embryo after 2 weeks of development. A human embryo at this stage would look essentially the same. **(b–d)** Fiber-optic views of human embryos at 4, 8, and 12 weeks. For actual sizes, see Figure 29–14, p. 1119. **ATLAS** Plate 8.17a

At the end of gestation, a typical uterus will have undergone a tremendous increase in size. Figure 29–10a–c● shows the positions of the uterus, fetus, and placenta from 16 weeks to *full term* (nine months). When the pregnancy is at full term, the uterus and fetus push many of the maternal abdominal organs out of their normal positions (Figure 29–10c,d●).

□ **PREGNANCY AND MATERNAL SYSTEMS**

The developing fetus is totally dependent on maternal organ systems for nourishment, respiration, and waste removal. These functions must be performed by maternal systems in addition to their normal operations. For example, the mother must absorb enough oxygen, nutrients, and vitamins for herself *and* for her fetus, and she must eliminate all the wastes that are generated. Although this is not a burden over the initial weeks of gestation, the demands placed on the mother become significant as the fetus grows. For the mother to survive under these conditions, the maternal systems must compensate for changes introduced by the fetus. In practical terms, the mother must breathe, eat, and excrete for two. Among the major changes that occur in maternal systems are the following:

• **The Maternal Respiratory Rate Goes Up and the Tidal Volume Increases.** As a result, the mother's lungs deliver the extra oxygen required, and remove the excess carbon dioxide generated, by the fetus.

- **The Maternal Blood Volume Increases.** This increase occurs because blood flowing into the placenta reduces the volume in the rest of the systemic circuit, and fetal metabolic activity both lowers the blood P_{O_2} and elevates the P_{CO_2}. The latter combination stimulates the production of renin and erythropoietin, leading to an increase in maternal blood volume through mechanisms detailed in Chapter 21 (see Figures 21–16 and 21–17a●). ∞ pp. 743, 745 By the end of gestation, the maternal blood volume has increased by almost 50 percent.

- **The Maternal Requirements for Nutrients and Vitamins Climb 10–30 Percent.** Pregnant women must nourish both themselves and their fetus and so tend to have increased sensations of hunger.

- **The Maternal Glomerular Filtration Rate Increases by Roughly 50 Percent.** This increase, which corresponds to the increase in blood volume, accelerates the excretion of metabolic wastes generated by the fetus. Because the volume of urine produced increases and the weight of the uterus presses down on the urinary bladder, pregnant women need to urinate frequently.

- **The Uterus Undergoes a Tremendous Increase in Size.** Structural and functional changes in the expanding uterus are so important that we shall discuss them in a separate section.

- **The Mammary Glands Increase in Size, and Secretory Activity Begins.** Mammary gland development requires a combination of hormones, including human placental lactogen and placental prolactin from the placenta, and PRL, estrogens, progesterone, GH, and thyroxine from maternal endocrine organs. By the end of the sixth month of pregnancy, the mammary glands are fully developed and begin to produce secretions that are stored in the duct system of those glands.

⚕ **Abortion** is the termination of a pregnancy. Most references distinguish among spontaneous, therapeutic, and induced abortions. **Spontaneous abortions**, or *miscarriages*, occur as a result of developmental or physiological problems. For example, spontaneous abortions can result from chromosomal defects in the embryo or from hormonal problems, such as inadequate LH production by the maternal pituitary gland (or reduced LH sensitivity at the corpus luteum), inadequate progesterone sensitivity in the endometrium, or placental failure to produce adequate levels of hCG. Spontaneous abortions occur in at least 15 percent of recognized pregnancies. **Therapeutic abortions** are performed when continuing the pregnancy represents a threat to the life of the mother. **AM** Problems with the Maintenance of a Pregnancy

Induced abortions, or *elective abortions*, are performed at the woman's request. Induced abortions remain the focus of considerable controversy. Most induced abortions involve unmarried or adolescent women. The ratio of abortions to deliveries for married women is 1:10, whereas it is nearly 2:1 for unmarried women and adolescents. In most states, induced abortions are legal during the first three months after conception; with restrictions, induced abortions may be permitted until the fifth or sixth month.

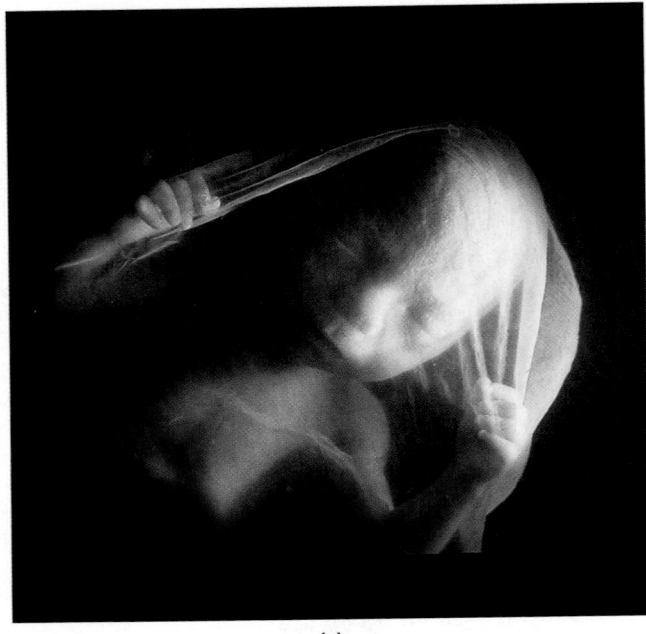

(a)

(b)

●**FIGURE 29–9**
The Second and Third Trimesters. **(a)** A 4-month-old fetus, seen through a fiber-optic endoscope. **(b)** Head of a six-month-old fetus, seen through ultrasound. **ATLAS** Plate 8.17b

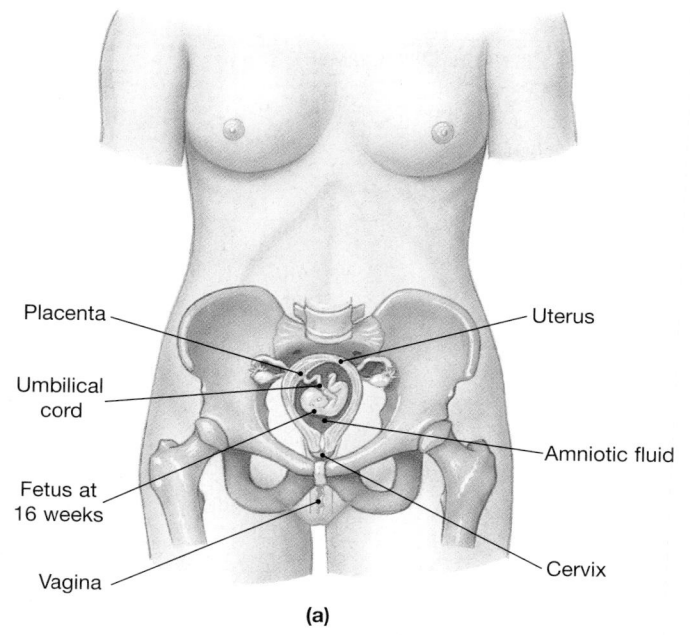

Placenta

Umbilical cord

Fetus at 16 weeks

Vagina

Uterus

Amniotic fluid

Cervix

(a)

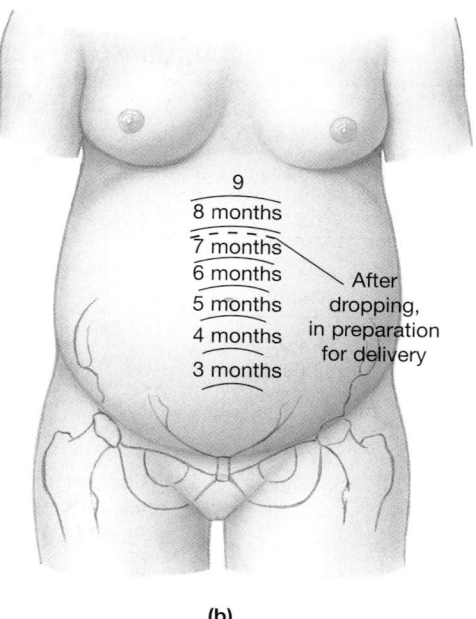

9
8 months
7 months
6 months
5 months
4 months
3 months

After dropping, in preparation for delivery

(b)

●**FIGURE 29–10**
Growth of the Uterus and Fetus. **(a)** Pregnancy at 16 weeks, showing the positions of the uterus, fetus, and placenta. **(b)** Pregnancy at three months to nine months (full term), showing the position of the uterus within the abdomen. **(c)** Pregnancy at full term. Note the positions of the uterus and full-term fetus within the abdomen and the displacement of abdominal organs, compared with **(d)**, a sectional view through the abdominopelvic cavity of a woman who is not pregnant.

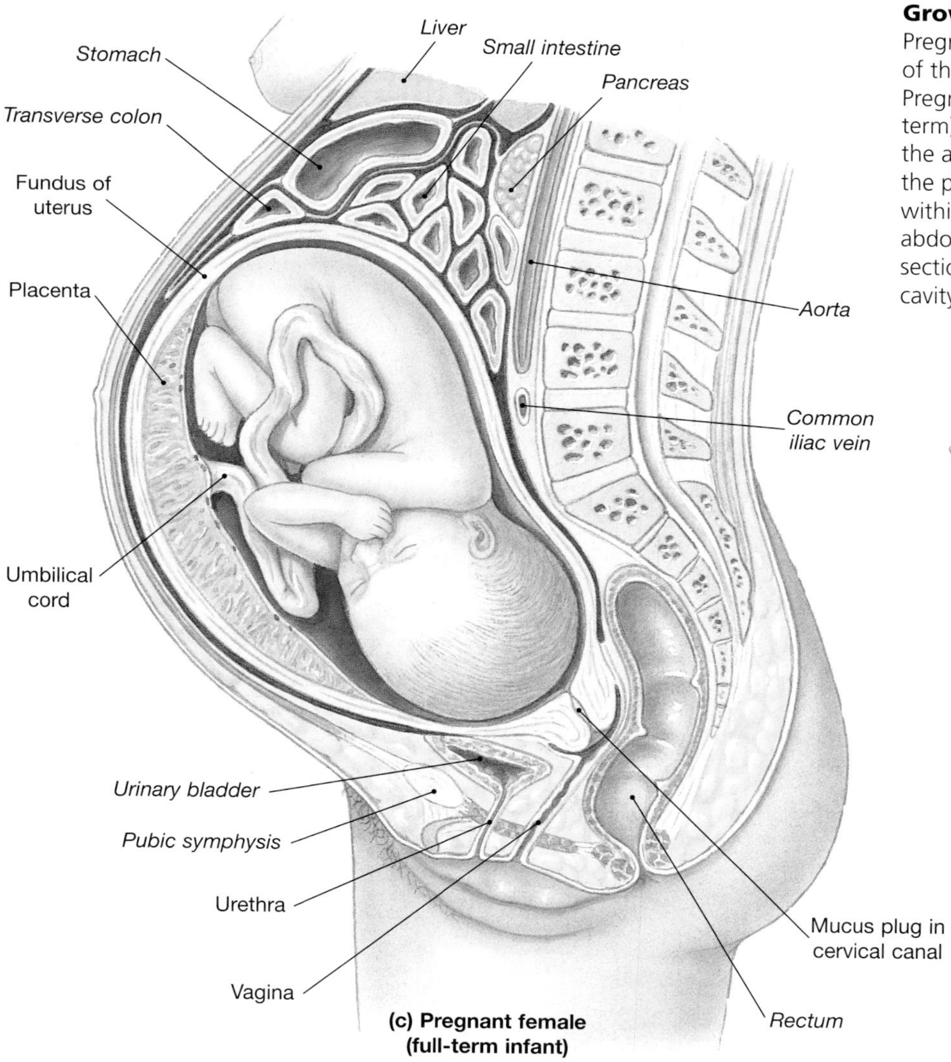

Stomach

Transverse colon

Liver

Small intestine

Pancreas

Fundus of uterus

Placenta

Aorta

Common iliac vein

Umbilical cord

Urinary bladder

Pubic symphysis

Urethra

Vagina

Mucus plug in cervical canal

Rectum

(c) Pregnant female (full-term infant)

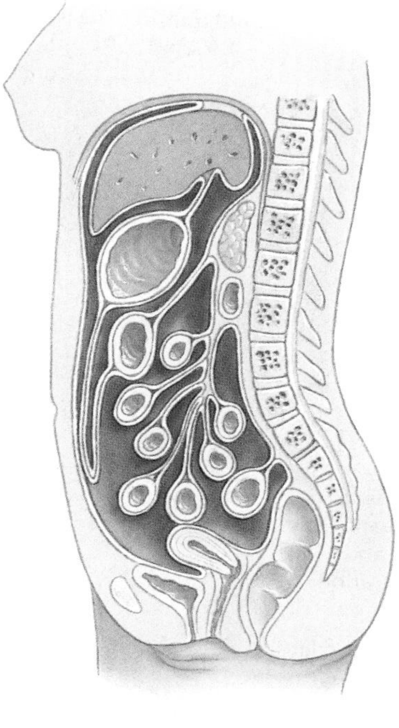

(d) Nonpregnant female

SUMMARY TABLE 29–2 An Overview of Prenatal Development

Background Material

ATLAS EMBRYOLOGY SUMMARIES 1–4:
The Development of Tissues
The Development of Epithelia
The Development of Connective Tissues
The Development of Organ Systems

   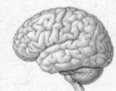 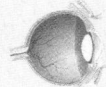

Gestational Age (Months)	Size and Weight	Integumentary System	Skeletal System	Muscular System	Nervous System	Special Sense Organs
1	5 mm, 0.02 g		(b) Formation of somites	(b) Formation of somites	(b) Formation of neural tube	(b) Formation of eyes and ears
2	28 mm, 2.7 g	(b) Formation of nail beds, hair follicles, sweat glands	(b) Formation of axial and appendicular cartilages	(c) Rudiments of axial musculature	(b) CNS, PNS organization, growth of cerebrum	(b) Formation of taste buds, olfactory epithelium
3	78 mm, 26 g	(b) Epidermal layers appear	(b) Spreading of ossification centers	(c) Rudiments of appendicular musculature	(c) Basic spinal cord and brain structure	
4	133 mm, 0.15 kg	(b) Formation of hair, sebaceous glands (c) Sweat glands	(b) Articulations (c) Facial and palatal organization	Fetus starts moving	(b) Rapid expansion of cerebrum	(c) Basic eye and ear structure (b) Formation of peripheral receptors
5	185 mm, 0.46 kg	(b) Keratin production, nail production			(b) Myelination of spinal cord	
6	230 mm, 0.64 kg			(c) Perineal muscles	(b) Formation of CNS tract (c) Layering of cortex	
7	270 mm, 1.492 kg	(b) Keratinization, formation of nails, hair				(c) Eyelids open, retinae sensitive to light
8	310 mm, 2.274 kg		(b) Formation of epiphyseal cartilages			(c) Taste receptors functional
9	346 mm, 3.2 kg					
Early postnatal development		Hair changes in consistency and distribution	Formation and growth of epiphyseal cartilages continue	Muscle mass and control increase	Myelination, layering, CNS tract formation continue	
Location of relevant text and illustrations		ATLAS Embryology Summary 5	Ch. 6: pp. 191–195 ATLAS Embryology Summaries 6, 7, 8	ATLAS Embryology Summary 9	Ch. 13: p. 433 Ch. 14: pp. 464–466 ATLAS Embryology Summaries 10, 11, 12	ATLAS Embryology Summary 13

Note: (b) = beginning; (c) = completion.

Gestational Age (Months)	Endocrine System	Cardiovascular and Lymphatic Systems	Respiratory System	Digestive System	Urinary System	Reproductive System
1		(b) Heartbeat	(b) Formation of trachea and lungs	(b) Formation of intestinal tract, liver, pancreas (c) Yolk sac	(c) Allantois	
2	(b) Formation of thymus, thyroid, pituitary, adrenal glands	(c) Basic heart structure, major blood vessels, lymph nodes and ducts (b) Blood formation in liver	(b) Extensive bronchial branching into mediastinum (c) Diaphragm	(b) Formation of intestinal subdivisions, into villi, salivary glands	(b) Formation of kidneys (metanephros)	(b) Formation of mammary glands
3	(c) Thymus, thyroid gland	(b) Tonsils, blood formation in bone marrow		(c) Gallbladder, pancreas		(b) Formation of gonads, ducts, genitalia
4		(b) Migration of lymphocytes to lymphoid organs; blood formation in spleen			(b) Degeneration of embryonic kidneys (mesonephros)	
5		(c) Tonsils	(c) Nostrils open	(c) Intestinal subdivisions		
6	(c) Adrenal glands	(c) Spleen, liver, bone marrow	(b) Formation of alveoli	(c) Epithelial organization, glands		
7	(c) Pituitary gland			(c) Intestinal plicae	(b) Descent of testes	
8			Complete pulmonary branching and alveolar structure		(c) Nephron formation	Descent of testes complete at or near time of birth
9						
Postnatal Development		Cardiovascular changes at birth; immune response gradually becomes fully operational				
Location of relevant text and illustrations	ATLAS Embryology Summary 14	Ch. 19: pp. 659, 669 Ch. 21: pp. 768–769 Ch. 22: p. 815 ATLAS Embryology Summaries 15, 16, 17	Ch. 23: pp. 858, 866 ATLAS Embryology Summary 18	Ch. 25: pp. 963–964 ATLAS Embryology Summary 19	Ch. 27: p. 1040 ATLAS Embryology Summary 20	Ch. 28: pp. 1049–1051 ATLAS Embryology Summary 21

☐ STRUCTURAL AND FUNCTIONAL CHANGES IN THE UTERUS

At the end of gestation, a typical uterus has grown from 7.5 cm (3 in.) in length and 30–40 g (1–1.4 oz) in weight to 30 cm (12 in.) in length and 1100 g (2.4 lb) in weight. Because the uterus may then contain almost 5 liters of fluid, the organ, together with its contents has a total weight of roughly 10 kg (22 lb). This remarkable expansion occurs through the enlargement (hypertrophy) and elongation of existing cells, especially smooth muscle fibers, rather than by an increase in the total number of cells.

The tremendous stretching of the uterus is associated with a gradual increase in the rate of spontaneous smooth muscle contractions in the myometrium. In the early stages of pregnancy, the contractions are weak, painless, and brief. Evidence indicates that the progesterone released by the placenta has an inhibitory effect on the uterine smooth muscle, preventing more extensive and more powerful contractions.

Three major factors oppose the calming action of progesterone:

1. *Rising Estrogen Levels.* Estrogens produced by the placenta increase the sensitivity of the uterine smooth muscles and make contractions more likely. Throughout pregnancy, progesterone exerts the dominant effect, but as delivery approaches, the production of estrogen accelerates and the myometrium becomes more sensitive to stimulation. Estrogens also increase the sensitivity of smooth muscle fibers to oxytocin.

2. *Rising Oxytocin Levels.* Rising oxytocin levels lead to an increase in the force and frequency of uterine contractions. Oxytocin release is stimulated by high estrogen levels and by the distortion of the cervix. Uterine distortion, especially in the region of the cervix, occurs as the weight of the fetus increases.

3. *Prostaglandin Production.* Estrogens and oxytocin stimulate the production of prostaglandins in the endometrium. These prostaglandins further stimulate smooth muscle contractions.

Late in pregnancy, some women experience occasional spasms in the uterine musculature, but these contractions are neither regular nor persistent. The contractions are called **false labor**. **True labor** begins when the biochemical and mechanical factors reach the point of no return. After nine months of gestation, multiple factors interact to initiate true labor. Once the **labor contractions** have begun in the myometrium, positive feedback ensures that they will continue until delivery has been completed.

Figure 29–11● diagrams important factors that stimulate and sustain labor. The actual trigger for the onset of true labor may be events in the fetus rather than in the mother. When labor commences, the fetal pituitary gland secretes oxytocin, which is then released into the maternal bloodstream at the placenta. The resulting increase in myometrial contractions and prostaglandin production, on top of the priming effects of estrogens and maternal oxytocin, may be the "last straw."

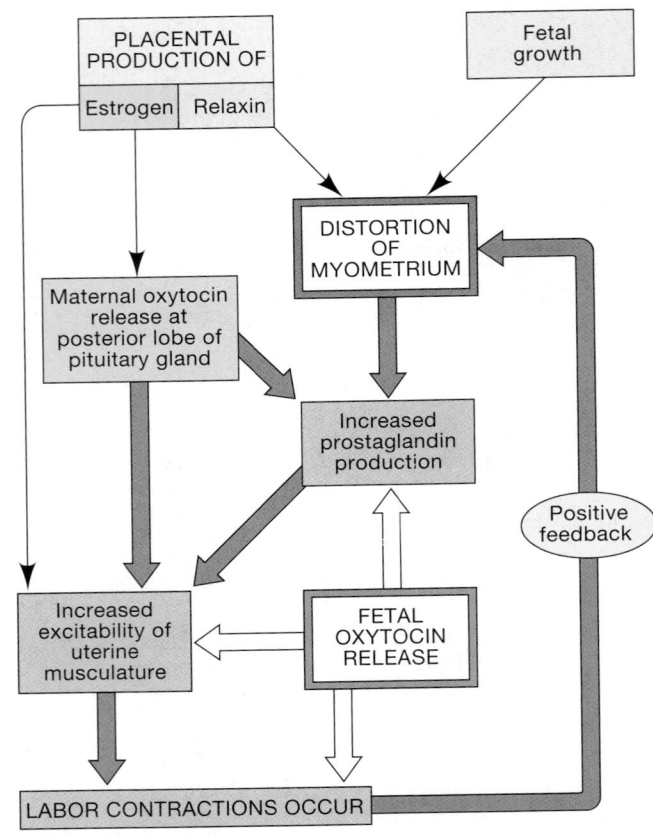

●**FIGURE 29–11**
Factors Involved in the Initiation of Labor and Delivery

29–6 LABOR AND DELIVERY

Objective

■ List and discuss the events that occur during labor and delivery.

The goal of labor is **parturition** (par-tū-RISH-un), the forcible expulsion of the fetus. During true labor, each contraction begins near the top of the uterus and sweeps in a wave toward the cervix. The contractions are strong and occur at regular intervals. As parturition approaches, the contractions increase in force and frequency, changing the position of the fetus and moving it toward the cervical canal.

☐ STAGES OF LABOR

Labor has traditionally been divided into three stages: (1) the *dilation stage*, (2) the *expulsion stage*, and (3) the *placental stage* (Figure 29–12●).

The Dilation Stage

The **dilation stage** begins with the onset of true labor, as the cervix dilates and the fetus begins to move toward the cervical canal (Figure 29–12a●). This stage is highly variable in length,

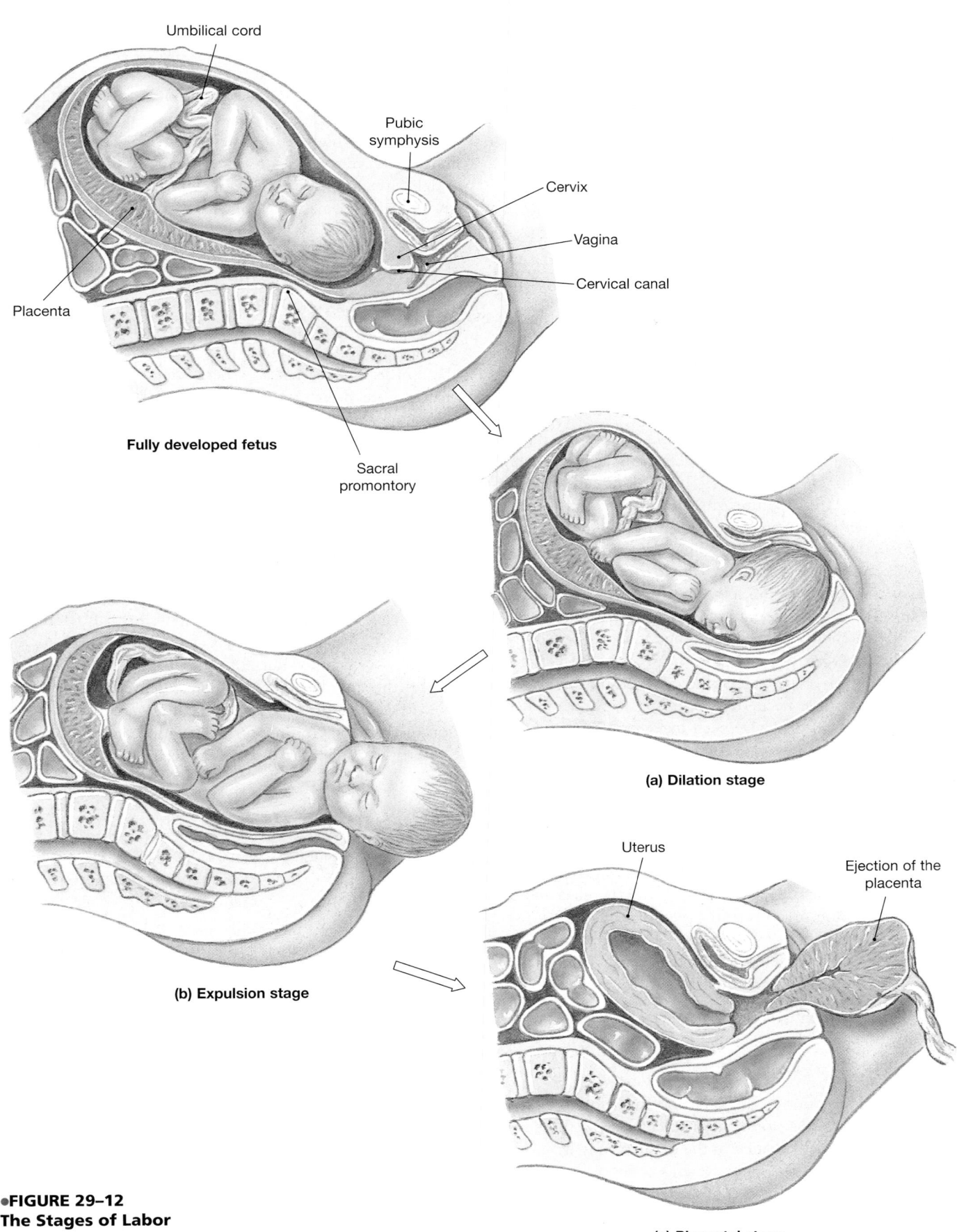

Umbilical cord

Pubic
symphysis

Cervix

Vagina

Cervical canal

Placenta

Fully developed fetus

Sacral
promontory

(a) Dilation stage

(b) Expulsion stage

Uterus

Ejection of the
placenta

(c) Placental stage

●FIGURE 29–12
The Stages of Labor

but typically lasts eight or more hours. At the start of the dilation stage, labor contractions last up to half a minute and occur at intervals of once every 10–30 minutes; their frequency increases steadily. Late in the process, the amniochorionic membrane ruptures, an event sometimes referred to as having one's "water break." If this event occurs before other events of the dilation stage, the life of the fetus is at risk from infection; medical steps may then be taken to accelerate labor.

The Expulsion Stage

The **expulsion stage** begins as the cervix, pushed open by the approaching fetus, completes its dilation (Figure 29–12b●). In this stage, contractions reach maximum intensity, occurring at perhaps two- or three-minute intervals and lasting a full minute. Expulsion continues until the fetus has emerged from the vagina; in most cases, the expulsion stage lasts less than two hours. The arrival of the newborn infant into the outside world is **delivery**, or birth.

If the vaginal canal is too small to permit the passage of the fetus and there is acute danger of perineal tearing, a clinician may temporarily enlarge the passageway by making an incision through the perineal musculature. After delivery, this **episiotomy** (e-pēz-ē-OT-o-mē) can be repaired with sutures, a much simpler procedure than dealing with the bleeding and tissue damage associated with an extensive perineal tear. If complications arise during the dilation or expulsion stage, the infant can be removed by **cesarean section**, or "*C-section*." In such cases, an incision is made through the abdominal wall, and the uterus is opened just enough to allow passage of the infant's head. This procedure is performed during 15–25 percent of the deliveries in the United States—more often than necessary, according to some studies. Over the last decade, efforts have been made to reduce the frequency of both episiotomies and cesarean sections.

The Placental Stage

During the **placental stage** of labor, muscle tension builds in the walls of the partially empty uterus, and the organ gradually decreases in size (Figure 29–12c●). This uterine contraction tears the connections between the endometrium and the placenta. In general, within an hour of delivery, the placental stage ends with the ejection of the placenta, or *afterbirth*. The disruption of the placenta is accompanied by a loss of blood. Because the maternal blood volume has increased greatly during pregnancy, this loss can be tolerated without difficulty.

☐ PREMATURE LABOR

Premature labor occurs when true labor begins before the fetus has completed normal development. The chances of the newborn surviving are directly related to the infant's body weight at delivery. Even with massive supportive efforts, infants weighing less than 400 g (14 oz) at birth will not survive, primarily because the respiratory, cardiovascular, and urinary systems are unable to support life without the aid of maternal systems. As a result, the

dividing line between spontaneous abortion and **immature delivery** is usually set at 500 g (17.6 oz), the normal weight near the end of the second trimester.

Most fetuses born at 25–27 weeks of gestation (a birth weight under 600 g) die despite intensive neonatal care; moreover, survivors have a high risk of developmental abnormalities. **Premature delivery** usually refers to birth at 28–36 weeks (a birth weight over 1 kg). With care, these infants have a good chance of survival and normal development. **AM** Complexity and Perfection

☐ DIFFICULT DELIVERIES

In most pregnancies, by the end of gestation, the fetus has rotated within the uterus to transit the birth canal headfirst, facing the mother's sacrum. In about 6 percent of deliveries, the fetus faces the mother's pubis instead. These infants can be delivered normally, given enough time, but risks to infant and mother are reduced by a *forceps delivery*. Forceps resemble large, curved salad tongs that can be separated for insertion into the vaginal canal, one side at a time. Once in place, they are reunited and used to grasp the head of the fetus. An intermittent pull is applied, so that the forces on the head resemble those of normal delivery.

In 3–4 percent of deliveries, the legs or buttocks of the fetus enter the vaginal canal first. Such deliveries are **breech births**. Risks to the infant are higher in breech births than in normal deliveries, because the umbilical cord can become constricted and cut off placental blood flow. The head is normally the widest part of the fetus; the mother's cervix may dilate enough to pass the baby's legs and body, but not the head. This entrapment compresses the umbilical cord, prolongs delivery, and subjects the fetus to severe distress and potential injury. If attempts to reposition the fetus or promote further dilation are unsuccessful over the short term, delivery by cesarean section may be required.

☐ MULTIPLE BIRTHS

Multiple births (twins, triplets, quadruplets, and so forth) can occur for several reasons. The ratio of twin births to single births in the U.S. population is roughly 1:89. "Fraternal," or **dizygotic** (dī-zī-GOT-ik), twins develop when two separate oocytes were ovulated and subsequently fertilized. Because chromosomes shuffle during meiosis, the odds against any two zygotes from the same parents having identical genes exceeds 1:8.4 million. Seventy percent of twins are dizygotic.

"Identical," or **monozygotic**, twins result either from the separation of blastomeres early in cleavage or by the splitting of the inner cell mass before gastrulation. In either event, the genetic makeup of the twins is identical because both formed from the same pair of gametes. Triplets, quadruplets, and larger multiples can result from multiple ovulations, blastomere splitting, or some combination of the two. For unknown reasons, the statistics for naturally occurring multiple births fall into a pattern: Twins occur at a rate of 1:89, triplets at a rate of $1:89^2$, quadruplets at $1:89^3$, and so forth. The incidence of multiple births can be increased by exposure to fertility drugs that stimulate the maturation of abnormally large numbers of oocytes. (See the discussion "Technology and the Treatment of Infertility" on pp. 1106–1107.)

Multiple pregnancies pose special problems because the strains on the mother are multiplied proportionately. The chances of premature labor are increased, and the risks to the mother are higher than for single births. The risks to the fetuses are also increased, both during gestation and after delivery, because even at full term, the newborn infants have a lower average birth weight. They are also more likely to have problems during delivery. For example, in more than half of twin deliveries, one or both fetuses enter the vaginal canal in an abnormal position.

Complete splitting of the blastomeres or the embryonic disc can produce identical twins. If the separation is not complete, **conjoined** (*Siamese*) **twins** may develop. These infants typically share some skin, a portion of the liver, and perhaps other internal organs as well. When the fusion is minor, the infants can be surgically separated with some success. Most conjoined twins with more extensive fusions fail to survive delivery.

29–7 POSTNATAL DEVELOPMENT

Objective

■ Identify the features and functions associated with the various life stages.

Developmental processes do not cease at delivery, because newborn infants have few of the anatomical, functional, or physiological characteristics of mature adults. In the course of postnatal development, every individual passes through five **life stages**: (1) the *neonatal period*, (2) *infancy*, (3) *childhood*, (4) *adolescence*, and (5) *maturity*. Each stage is typified by a distinctive combination of characteristics and abilities. These stages are a familiar part of human experience. Although each stage has distinctive features, the transitions between them are gradual and the boundaries indistinct. At maturity, development ends and the process of aging, or *senescence*, begins.

☐ THE NEONATAL PERIOD, INFANCY, AND CHILDHOOD

The **neonatal period** extends from birth to one month thereafter. **Infancy** then continues to two years of age, and **childhood** lasts until **adolescence**, the period of sexual and physical maturation. Two major events are under way during these developmental stages:

1. The organ systems other than those associated with reproduction become fully operational and gradually acquire the functional characteristics of adult structures.

2. The individual grows rapidly, and body proportions change significantly.

Pediatrics is a medical specialty focusing on postnatal development from infancy through adolescence. Infants and young children cannot clearly describe the problems they are experiencing, so pediatricians and parents must be skilled observers. Standardized tests are used to assess developmental progress relative to average values. **AM** Monitoring Postnatal Development

The Neonatal Period

Physiological and anatomical changes occur as the fetus completes the transition to the status of newborn infant, or **neonate**. Before delivery, dissolved gases, nutrients, wastes, hormones, and antibodies were transferred across the placenta. At birth, the neonate must become relatively self-sufficient, with respiration, digestion, and excretion performed by its own specialized organs and organ systems. The transition from fetus to neonate can be summarized as follows:

- At birth, the lungs are collapsed and filled with fluid. Filling them with air involves a massive and powerful inhalation. ∞ p. 866

- When the lungs expand, the pattern of cardiovascular circulation changes due to alterations in blood pressure and flow rates. The ductus arteriosus closes, isolating the pulmonary and systemic trunks. Closure of the foramen ovale separates the atria of the heart, completing the separation of the pulmonary and systemic circuits. ∞ p. 768

- The typical heart rate (120–140 beats per minute) and respiratory rate (30 breaths per minute) of neonates are considerably higher than their adult counterparts. In addition, a neonate's metabolic rate is roughly twice that of an adult for equivalent body weights.

- Before birth, the digestive system remains relatively inactive, although it does accumulate a mixture of bile secretions, mucus, and epithelial cells. This collection of debris is excreted during the first few days of life. Over that period, the newborn infant begins to nurse.

- As waste products build up in the arterial blood, they are excreted at the kidneys. Glomerular filtration is normal, but the neonate cannot concentrate urine to any significant degree. As a result, urinary water losses are high, and neonatal fluid requirements are much greater than those of adults. ∞ p. 1040

- The neonate has little ability to control its body temperature, particularly in the first few days after delivery. As the infant grows larger and its insulating subcutaneous adipose "blanket" gets thicker, its metabolic rate also rises. Daily and even hourly shifts in body temperature continue throughout childhood. ∞ p. 963

Over the entire neonatal period, the newborn is dependent on nutrients contained in the milk secreted by the maternal mammary glands.

LACTATION AND THE MAMMARY GLANDS By the end of the sixth month of pregnancy, the mammary glands are fully developed, and the gland cells begin to produce a secretion known as **colostrum** (ko-LOS-trum). Ingested by the infant during the first two or three days of life, colostrum contains more proteins and far less fat than breast milk contains. Many of the proteins are antibodies that may help the infant ward off infections until its own immune system becomes fully functional. In addition, the mucins present in both colostrum and milk can inhibit the replication of a family of viruses (*rotaviruses*) that can cause dangerous forms of gastroenteritis and infant diarrhea.

As colostrum production drops, the mammary glands convert to milk production. Breast milk consists of water, proteins, amino acids, lipids, sugars, and salts. It also contains large quantities of *lysozymes*—enzymes with antibiotic properties. Human milk provides roughly 750 Calories per liter. The secretory rate varies with the demand, but a 5–6-kg (11–13-lb) infant usually requires about 850 ml of milk per day.

Mammary gland secretion is triggered when the infant sucks on the nipple (STEP 1, Figure 29–13●). The stimulation of tactile receptors there leads to the stimulation of secretory neurons in the paraventricular nucleus of the mother's hypothalamus (STEPS 2 and 3). These neurons release oxytocin at the posterior lobe of the pituitary gland (STEP 4). When oxytocin reaches the mammary gland, this hormone causes the contraction of *myoepithelial cells*, contractile cells in the walls of the lactiferous ducts and sinuses. ∞ p. 170 The result is milk ejection (STEP 5), or the **milk let-down reflex**.

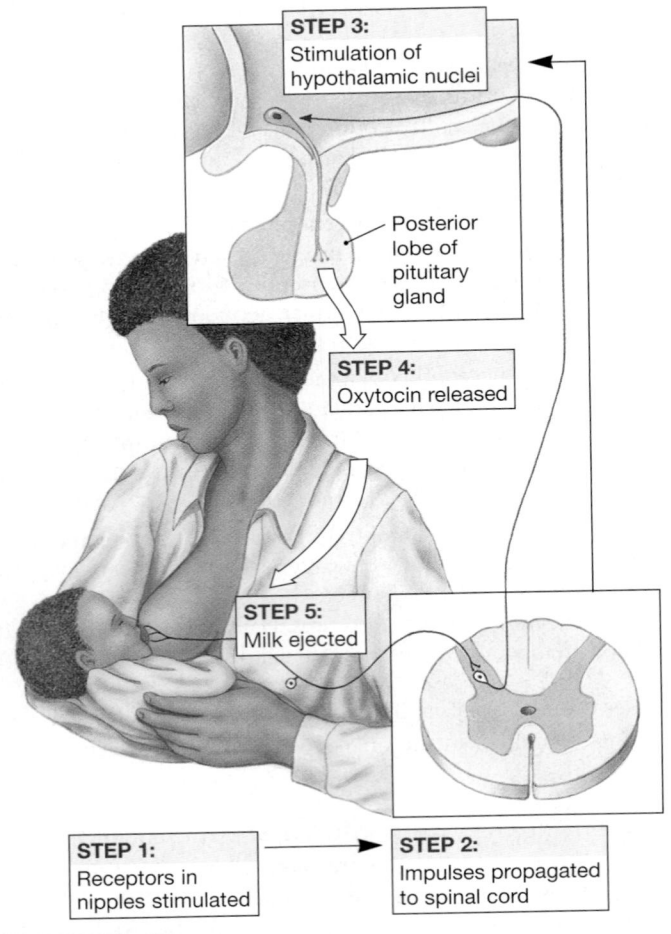

STEP 3:
Stimulation of hypothalamic nuclei

Posterior lobe of pituitary gland

STEP 4:
Oxytocin released

STEP 5:
Milk ejected

STEP 1:
Receptors in nipples stimulated

STEP 2:
Impulses propagated to spinal cord

●**FIGURE 29–13**
The Milk Let-Down Reflex

The milk let-down reflex continues to function until *weaning*, typically one to two years after birth. Milk production ceases soon after, and the mammary glands gradually return to a resting state. Earlier weaning is a common practice in the United States, where women take advantage of commercially prepared milk- or soy-based infant formulas that closely approximate natural breast milk. The major difference between such substitutes and natural milk is that the substitutes lack antibodies and lysozymes, which play an important role in maintaining the health of the infant. Consequently, early weaning is associated with an increased risk of infections and allergies in the infant.

Infancy and Childhood

The most rapid growth occurs during prenatal development, and the rate of growth declines after delivery. Postnatal growth during infancy and childhood takes place under the direction of circulating hormones, notably growth hormone, adrenal steroids, and thyroid hormones. These hormones affect each tissue and organ in specific ways, depending on the sensitivities of the individual cells. As a result, growth does not occur uniformly, so the body proportions gradually change. The head, for example, is relatively large at birth, but decreases in proportion with the rest of the body as the child grows to adulthood (Figure 29–14●).

☐ ADOLESCENCE AND MATURITY

Adolescence begins at **puberty**, the period of sexual maturation, and ends when growth is completed. Three major hormonal events interact at the onset of puberty:

1. The hypothalamus increases its production of gonadotropin-releasing hormone (GnRH). Evidence indicates that this increase is dependent on adequate levels of *leptin*, a hormone released by adipose tissues. ∞ pp. 637–638

2. Endocrine cells in the anterior lobe of the pituitary gland become more sensitive to the presence of GnRH, and circulating levels of FSH and LH rise rapidly.

3. Ovarian or testicular cells become more sensitive to FSH and LH, initiating (1) gamete formation, (2) the production of sex hormones that stimulate the appearance of secondary sex characteristics and behaviors, and (3) a sudden acceleration in the growth rate, culminating in the closure of the epiphyseal cartilages.

The age at which puberty begins varies. In the United States today, puberty generally occurs at about age 14 in boys and 13 in girls, but the normal ranges are broad (9–14 in boys, 8–13 in girls). Many body systems alter their activities in response to circulating sex hormones and to the presence of growth hormone, thyroid hormones, PRL, and adrenocortical hormones, so gender-specific differences in structure and function develop. The following are some of the changes induced by the endocrine system at puberty:

- **Integumentary System.** Testosterone stimulates the development of terminal hairs on the face and chest, whereas those follicles continue to produce fine hairs under estrogen stimula-

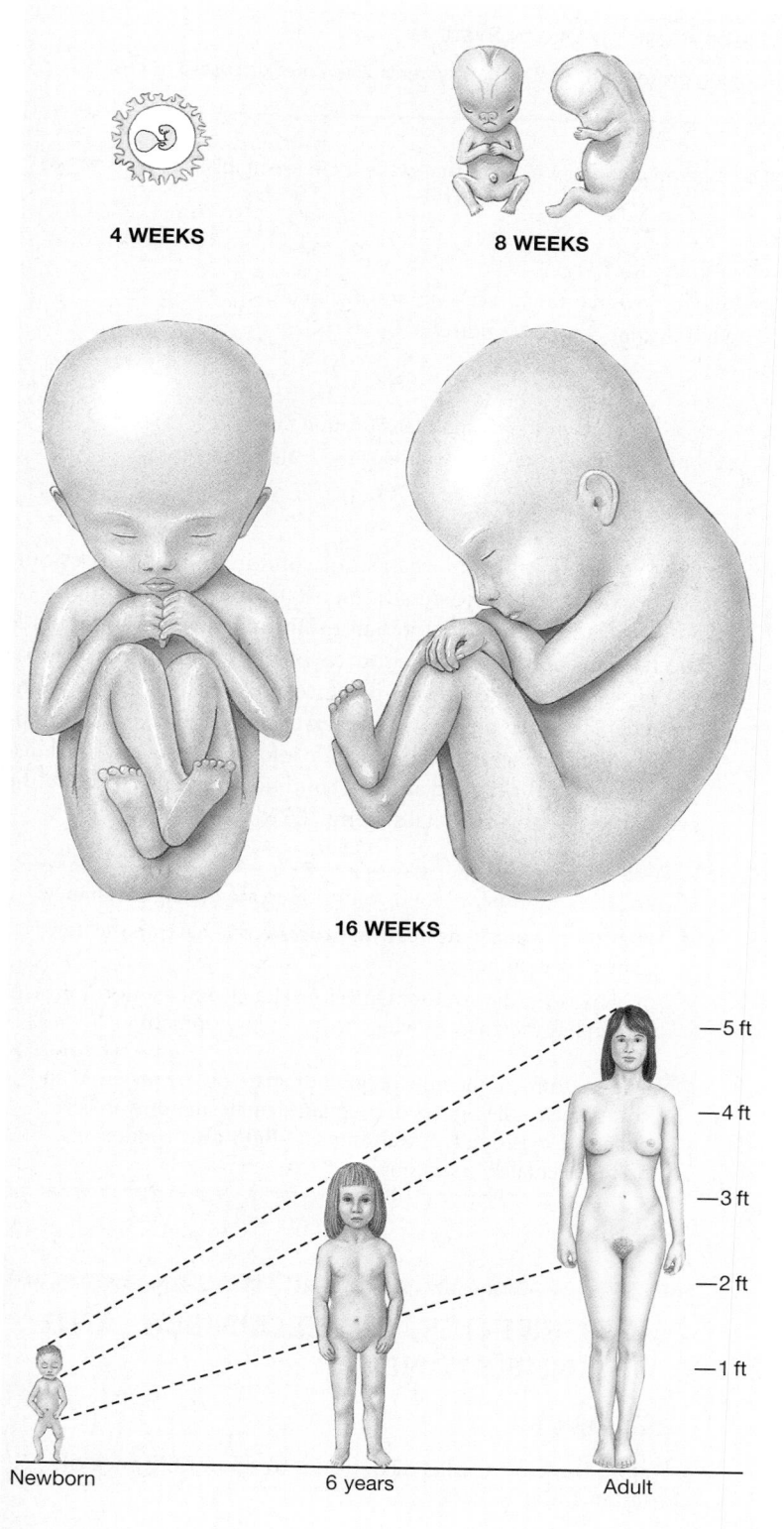

4 WEEKS

8 WEEKS

16 WEEKS

— 5 ft

— 4 ft

— 3 ft

— 2 ft

— 1 ft

Newborn   6 years   Adult

●**FIGURE 29–14**
Growth and Changes in Body Form.   The views at 4, 8, and 16 weeks' gestation are presented (actual size). Notice the changes in body form and proportions as development proceeds. These changes do not stop at birth. For example, the head, which contains the brain and sense organs, is relatively large at birth.

tion. The hairline recedes under testosterone stimulation. Both testosterone and estrogen stimulate terminal hair growth in the axillae and in the genital area. These hormones also stimulate sebaceous gland secretion and may cause acne. Adipose tissues respond differently to testosterone than to estrogens, and this difference produces changes in the subcutaneous distribution of body fat. In women, the combination of estrogens, PRL, growth hormone, and thyroid hormones promotes the initial development of the mammary glands. Although the duct system becomes more elaborate, true secretory alveoli do not develop, and much of the growth of the breasts during this period reflects increased deposition of fat rather than glandular tissue.

- **Skeletal System.** Both testosterone and estrogen accelerate bone deposition and skeletal growth. In the process, they promote the closure of the epiphyses and thus place a limit on growth in height. Estrogens cause more rapid epiphyseal closure than does testosterone. In addition, the period of skeletal growth is shorter in girls than in boys, and girls generally do not grow as tall as boys. Girls grow most rapidly between ages 10 and 13, whereas boys grow most rapidly between ages 12 and 15.

- **Muscular System.** Sex hormones stimulate the growth of skeletal muscle fibers, increasing strength and endurance. The effects of testosterone greatly exceed those of the estrogens, and the increased muscle mass accounts for significant gender differences in body mass, even for males and females of the same height. The stimulatory effects of testosterone on muscle mass have produced an interest in anabolic steroids among competitive athletes of both genders.

- **Nervous System.** Sex hormones affect central nervous system centers concerned with sexual drive and sexual behaviors. These centers differentiated in gender-specific ways during the second and third trimesters, when the fetal gonads secreted either testosterone (in males) or estrogens (in females). The surge in sex hormone secretion at puberty activates the CNS centers.

- **Cardiovascular System.** Testosterone stimulates erythropoiesis and thereby increases the blood volume and the hematocrit. In females, once uterine cycles have begun, the iron loss associated with menses increases the risk of developing iron-deficiency anemia. Late in each uterine cycle, estrogens and progesterone promote the movement of water from plasma into interstitial fluid, leading to an increase in tissue water content. Estrogens decrease plasma cholesterol levels and slow the formation of plaque. As a result, premenopausal women have a lower risk of atherosclerosis than do adult men.

- **Respiratory System.** Testosterone stimulates disproportionate growth of the larynx and a thickening and lengthening of the vocal cords. These changes cause a gradual deepening of the voice of males compared with that of females.

- **Reproductive System.** In males, testosterone stimulates the functional development of the accessory reproductive glands, such as the prostate gland and seminal vesicles, and

SUMMARY TABLE 29-3 EFFECTS OF AGING ON ORGAN SYSTEMS

The characteristic physical and functional changes that are part of the aging process affect all organ systems. Examples discussed in previous chapters include the following:

- A loss of elasticity in the skin that produces sagging and wrinkling. ∞ p. 175
- A decline in the rate of bone deposition, leading to weak bones, and degenerative changes in joints that make them less mobile. ∞ pp. 204, 283
- Reductions in muscular strength and ability. ∞ p. 377
- Impairment of coordination, memory, and intellectual function. ∞ p. 556
- Reductions in the production of, and sensitivity to, circulating hormones. ∞ p. 643
- Appearance of cardiovascular problems and a reduction in peripheral blood flow that can affect a variety of vital organs. ∞ p. 769
- Reduced sensitivity and responsiveness of the immune system, leading to infection, cancer, or both. ∞ p. 817
- Reduced elasticity in the lungs, leading to decreased respiratory function. ∞ p. 866
- Decreased peristalsis and muscle tone along the digestive tract. ∞ p. 920
- Decreased peristalsis and muscle tone in the urinary system, coupled with a reduction in the glomerular filtration rate. ∞ p. 1007
- Functional impairment of the reproductive system, which eventually becomes inactive when menopause or the male climacteric occurs. ∞ p. 1086

helps promote spermatogenesis. In females, estrogens target the uterus, promoting a thickening of the myometrium, increasing blood flow to the endometrium, and stimulating cervical mucus production. Estrogens also promote the functional development of the female accessory reproductive organs. The first few uterine cycles may not be accompanied by ovulation. After the initial stage, the woman will be fertile, even though growth and physical maturation will continue for several years.

After puberty, the continued background secretion of estrogens or androgens maintains the foregoing gender-specific differences. In both genders, growth continues at a slower pace until age 18–21, by which time most of the epiphyseal cartilages have closed. The boundary between adolescence and maturity is hazy, because it has physical, emotional, behavioral, and legal aspects. Adolescence is often said to be over when growth stops, in the late teens or early twenties. The individual is then considered mature.

☐ SENESCENCE

Although growth may cease at maturity, physiological changes continue. The gender-specific differences produced at puberty are retained, but further changes occur when sex hormone levels decline at menopause or the male climacteric. ∞ p. 1086 All these changes are part of the process of **senescence**, or aging, which reduces the efficiency and capabilities of the individual. Even in the absence of other factors, such as disease or injury, senescence-related changes at the molecular level ultimately lead to death.

Table 29–3 summarizes the age-related changes in physiological systems that we have discussed in earlier chapters. Taken together, these changes both reduce the functional abilities of the individual and affect homeostatic mechanisms. As a result, the elderly are less able to make homeostatic adjustments in response to internal or environmental stresses. The risks of contracting a variety of bacterial or viral diseases are proportionately increased as immune function deteriorates. This deterioration leads to drastic physiological changes that affect all internal systems.

Death ultimately occurs when some combination of stresses cannot be countered by existing homeostatic mechanisms.

Physicians attempt to forestall death by adjusting homeostatic mechanisms or removing the sources of stress. **Geriatrics** is a medical specialty that deals with the mechanics of aging; physicians trained in geriatrics are known as **geriatricians**. Problems commonly encountered by geriatricians include infections, cancers, heart disease, strokes, arthritis, and anemia, conditions directly related to the age-induced changes in vital systems. **AM** Death and Dying

---

✓ Why does a mother's blood volume increase during pregnancy?

✓ What effect would a decrease in progesterone have on the uterus during late pregnancy?

✓ An increase in the levels of GnRH, FSH, LH, and sex hormones in blood mark the onset of which stage of development?

Answers start on page Q-1

The anatomy of the non-pregnant female can be reviewed and compared to the figures of pregnant female anatomy in this chapter by visiting the **Anatomy CD-ROM**: Reproductive System/Female Dissections.

29–8 GENETICS, DEVELOPMENT, AND INHERITANCE

Objective

- Relate basic principles of genetics to the inheritance of human traits.

☐ GENES AND CHROMOSOMES

We introduced chromosome structure and the functions of genes in Chapter 3. ∞ pp. 84–90 Chromosomes contain DNA, and genes are functional segments of DNA. Each gene carries the information needed to direct the synthesis of a specific polypeptide. Every nucleated somatic cell in your body carries copies of the original 46 chromosomes present when you were a zygote. Those chromosomes and their component genes constitute your **genotype** (JĒN-ō-tīp).

Through development and differentiation, the instructions contained in the genotype are expressed in many ways. No single cell or tissue uses all the information and instructions contained in the genotype. For example, in muscle fibers, the genes that are important for the formation of excitable membranes and contractile proteins are active, whereas in cells of the pancreatic islets, a different set of genes operates. Collectively, however, the instructions contained in your genotype determine the anatomical and physiological characteristics that make you a unique individual. Those anatomical and physiological characteristics constitute your **phenotype** (FĒN-ō-tīp). In architectural terms, the genotype is a set of plans and the phenotype is the finished building. Specific elements in your phenotype, such as your hair color and eye color, skin tone, and foot size, are called phenotypic *characters*, or *traits*.

Your genotype is derived from the genotypes of your parents. Yet you are not an exact copy of either parent; nor are you an easily identifiable mixture of their characteristics. Our discussion of genetics will begin with the basic patterns of inheritance and their implications. We shall then examine the mechanisms responsible for regulating the activities of the genotype during prenatal development.

☐ PATTERNS OF INHERITANCE

The 46 chromosomes carried by each somatic cell occur in pairs: Every somatic cell contains 23 pairs of chromosomes. At amphimixis, one member of each pair was contributed by the spermatozoon and the other by the ovum. The two members of each pair are known as **homologous** (hō-MOL-o-gus) **chromosomes**. Twenty-two of those pairs are called **autosomal** (aw-tō-SŌ-mal) **chromosomes**. Most of the genes of the autosomal chromosomes affect somatic characteristics, such as hair color and skin pigmentation. The chromosomes of the twenty-third pair are called the **sex chromosomes**; one of their functions is to determine whether the individual will be genetically male or female. Figure 29–15● shows the **karyotype**, or entire set of chromosomes, of a normal male. The discussion that follows will deal first with the inheritance of traits carried on the autosomal chromosomes; we will examine the patterns of inheritance via the sex chromosomes in a later section.

The two chromosomes in a homologous autosomal pair have the same structure and carry genes that affect the same traits. Suppose that one member of the pair contains three genes in a row, with the first determining hair color, the second eye color, and the third skin pigmentation. Then the other chromosome will carry genes that affect the same traits, and the genes will be in the same sequence. The genes will also be located at equivalent positions on their respective chromo-

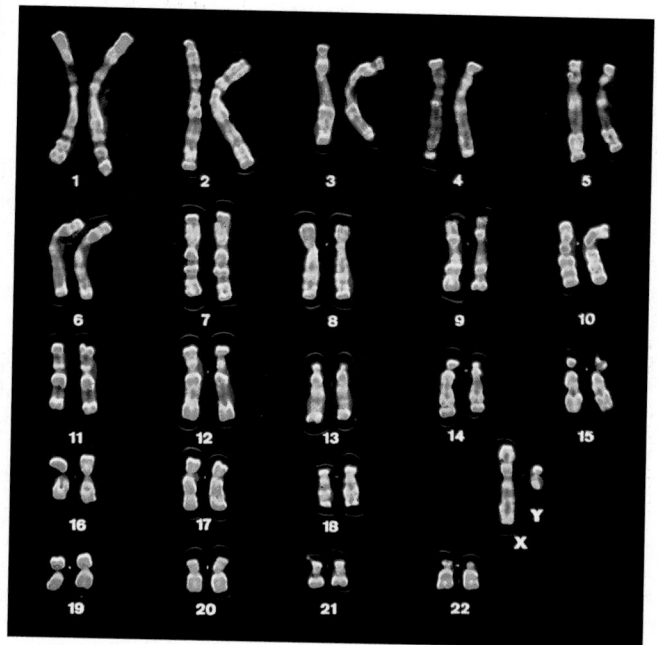

●**FIGURE 29–15**
Human Chromosomes. The 23 pairs of somatic-cell chromosomes of a normal male.

somes. A gene's position on a chromosome is called a **locus** (LŌ-kus; plural, *loci*).

The two chromosomes in a pair may not carry the same *form* of each gene, however. The various forms of any gene are called **alleles** (a-LĒLZ). These *alternate forms* determine the precise effect of the gene on your phenotype. If the two chromosomes of a homologous pair carry the same allele of a particular gene, you are **homozygous** (hō-mō-ZĪ-gus) for the trait affected by that gene. That allele will then indeed be expressed in your phenotype. For example, if you receive a gene for curly hair from your father and a gene for curly hair from your mother, you will be homozygous for curly hair—and you will have curly hair. About 80 percent of an individual's genome consists of homozygous alleles.

Interactions between Alleles

Because the chromosomes of a homologous pair have different origins, one paternal and the other maternal, they do not necessarily carry the same alleles. When you have two different alleles of the same gene, you are **heterozygous** (het-er-ō-ZĪ-gus) for the trait determined by that gene. The phenotype that results from a heterozygous genotype depends on the nature of the interaction between the corresponding alleles. For example, if you received a gene for curly hair from your father, but a gene for straight hair from your mother, whether *you* will have curly hair, straight hair, or even wavy hair depends on the relationship between the alleles for those traits:

- An allele that is **dominant** will be expressed in the phenotype, *regardless of any conflicting instructions carried by the other allele.* For instance, an individual with only one allele for freckles will have freckles, because that allele is dominant over the "non-freckle" allele.

- An allele that is **recessive** will be expressed in the phenotype only if that same allele is present on *both chromosomes* of a homologous pair. For example, in Chapter 5 we described the albino condition, characterized by an inability to synthesize the yellow-brown pigment *melanin*. ∞ p. 160 The presence of one allele that allows for melanin production will result in normal coloration. Two recessive alleles must be present to produce an albino individual. A single gene can have many different alleles, some dominant and others recessive.

- In **incomplete dominance**, heterozygous alleles produce a phenotype that is distinct from the phenotypes of individuals who are homozygous for one allele or the other. A good example is a gene that affects the shape of red blood cells. Individuals with homozygous alleles that carry instructions for normal adult hemoglobin A have red blood cells of normal shape. Individuals with homozygous alleles for hemoglobin S, an abnormal form, have red blood cells that will sickle in peripheral capillaries when the P_{O_2} decreases. These individuals develop *sickle-cell anemia*. ∞ p. 658 Individuals who are heterozygous for this trait do not develop the anemia; instead, they have red blood cells that will sickle only when tissue oxygen levels are extremely low.

- In **codominance**, an individual with a heterozygous allele for a given trait exhibits both the dominant and recessive phenotypes for that trait. Blood type in humans is determined by codominance. ∞ p. 662 The alleles for types A and B blood are dominant over the allele for type O blood, but a person with one type A allele and one type B allele will have type AB blood, not A or B. Type AB blood has *both* type A and type B antigens. The distinction between incomplete dominance and codominance is not always clearcut. For example, a person who is heterozygous for hemoglobin A and hemoglobin S shows incomplete dominance for RBC shape, but codominance for RBC hemoglobin; each red blood cell contains a mixture of hemoglobin A and hemoglobin S.

Penetrance and Expressivity

Differences in genotype lead to distinct variations in phenotype, but the relationships are not always easily predictable. The presence of a particular pair of alleles does not affect the phenotype in the same way in every individual. **Penetrance** is the percentage of individuals with a particular genotype that show the "expected" phenotype. The effects of the genotype in the other individuals may be overridden by the activity of other genes or by environmental factors. For example, *emphysema*, a respiratory disorder discussed in Chapter 23, has been linked to a specific abnormal genotype. ∞ p. 866 Roughly 20 percent of the individuals with this genotype do not develop emphysema. The penetrance of this genotype is therefore approximately 80 percent. The effects of environmental factors are apparent: Most people who develop emphysema are cigarette smokers.

If a given genotype *does* affect the phenotype, it can do so to varying degrees, again depending on the activity of other genes or environmental stimuli. For example, identical twins do not have exactly the same fingerprints, even though they have the same genotype. The extent to which a particular allele is expressed when it is present is termed its **expressivity**.

Environmental effects on genetic expression are particularly evident during embryological and fetal development. Drugs, including certain antibiotics, alcohol, and the nicotine in cigarette smoke, can disrupt fetal development. Stimuli that result in abnormal development are called **teratogens** (TER-a-tō-jenz). AM

Teratogens and Abnormal Development

Predicting Inheritance

Not every allele can be neatly characterized as dominant or recessive. Several that can be are included in Table 29–4. If you consider the traits listed there, you can predict the characteristics of individuals on the basis of their parents' alleles.

In such calculations, dominant alleles are traditionally indicated by capitalized abbreviations, and recessive alleles are abbreviated in lowercase letters. For a given trait, the possibilities are indicated by *AA* (homozygous dominant), *Aa* (heterozygous), or *aa* (homozygous recessive). Each gamete involved in fertilization contributes a single allele for a given trait. That allele must be one of the two contained by all cells in the parent's body. Consider, for example, the offspring of an albino mother and a father with normal skin pigmentation. Because albinism is a recessive trait, the maternal alleles are abbreviated *aa*. No matter which of her oocytes is fertilized, it will carry the recessive *a* allele. The father has normal pigmentation, a dominant trait. He is therefore either homozygous *or* heterozygous for this trait, because both *AA* and *Aa* will produce the same phenotype for normal skin pigmentation. Every sperm produced by a homozygous father will carry the *A* allele. In contrast, half the sperm produced by a heterozygous father will carry the dominant allele *A*, and the other half will carry the recessive allele *a*.

A simple box diagram known as a **Punnett square** lets us predict the probabilities that children will have particular characteristics by showing the various combinations of parental alleles they can inherit. In the Punnett squares shown in Figure 29–16●, the maternal alleles for normal versus albino skin pigmentation are listed along the horizontal axis and the paternal ones along the vertical axis. The combinations are indicated in the small boxes. Figure 29–16a● shows the possible offspring of an *aa* mother and an *AA* father. All the children must have the genotype *Aa*, so all will have normal skin pigmentation. Compare these results with those of Figure 29–16b●, for a heterozygous father (*Aa*). The heterozygous male produces two types of gametes, *A* and *a*, and the secondary oocyte may be fertilized by either one. As a result, the probability is 50 percent that a child of such a father will inherit the genotype *Aa* and so have normal skin pigmentation. The probability of inheriting the genotype *aa*, and thus having the albino phenotype, is also 50 percent.

The Punnett square can also be used to draw conclusions about the identity and genotype of a parent. For example, a man with the genotype *AA* cannot be the father of an albino child (*aa*).

Simple Inheritance

In **simple inheritance**, phenotypes are determined by interactions between a single pair of alleles. (Our examples thus far involve simple inheritance.) We can predict the frequency of

TABLE 29–4 THE INHERITANCE OF SELECTED TRAITS

AUTOSOMAL CHROMOSOMES

Simple Inheritance: The phenotype is determined by a single pair of alleles.

Dominant/Recessive: One allele determines phenotype, the other is suppressed

Examples	Dominant trait	Recessive trait
Skin pigmentation	Normal Freckles	Albino No freckles
Vision	Nearsightedness, farsightedness, astigmatism	Normal vision
Hairline	Widow's peak	Straight hairline
Hair texture	Curly	Straight
Hair color		Blond hair, red hair*
Fingers/toes	Brachydactyly (short digits) Syndactyly (webbed digits) Polydactyly (extra digits) Double-jointed (hyperextended) thumb	Long digits Normal digits Normal number of digits Normal thumb
Earlobes	Free	Attached
Tongue control	Ability to form a U-shape	Inability to form a U-shape
Red blood cells	Rh factor present Type A or B	Rh factor absent Type O
Ability to taste PTC**	Present	Absent
Disorders	Huntington's disease Hypercholesterolemia Achondroplasia Marfan's syndrome	Sickle-cell anemia Cystic fibrosis Tay-Sachs disease Phenylketonuria

Codominance: Both alleles are expressed
Examples: Type AB blood; Structure of serum proteins (albumins, transferrins)

Incomplete Dominance: Two alleles produce intermediate traits
Examples: Hemoglobin A and hemoglobin S production

Polygenic Inheritance: The phenotype is determined by interactions among the alleles of several genes.
Examples: Eye color; Hair color (other than pure blond or red)

SEX CHROMOSOMES

X-linked: The allele on the X chromosome, whether dominant or recessive, determines the phenotype in the absence of a corresponding allele on the Y chromosome. The majority are caused by alleles that are recessive in females.
Examples: Red-green color blindness; Hemophilia (some forms); Duchenne's muscular dystrophy

* Expressed only if individual is also homozygous for blond hair.
** Phenothiocarbamide, a bitter but harmless chemical compound.

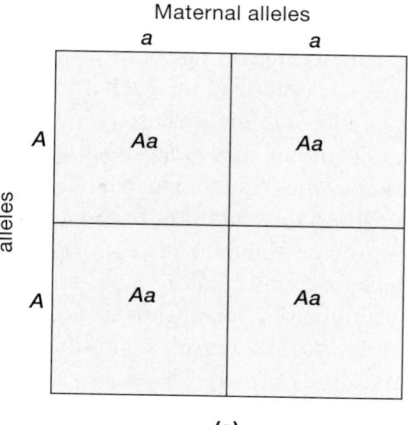

(a)

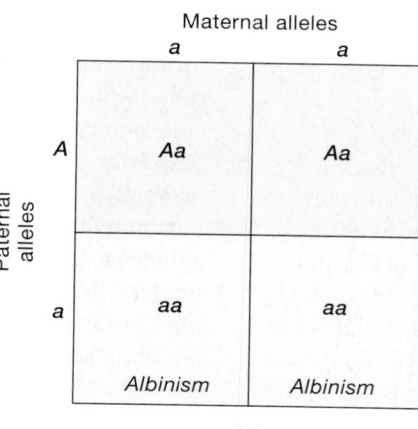

(b)

●**FIGURE 29–16**
Predicting Phenotypic Characteristics by Using Punnett Squares. **(a)** All the offspring of a homozygous dominant father (*AA*) and a homozygous recessive mother (*aa*) will be heterozygous (*Aa*) for that trait. Their phenotype for the trait will be the same as that of the father. **(b)** The offspring of a heterozygous father (*Aa*) and a homozygous recessive mother (*aa*) will be either heterozygous or homozygous for the recessive trait. In this example, half the offspring will have normal skin coloration, and the other half will be albinos.

TABLE 29–5 FAIRLY COMMON INHERITED DISORDERS

Disorder	Text Page or Applications Manual
AUTOSOMAL DOMINANTS	
Polycystic kidney disease	p. 980 and *AM*
Marfan's syndrome	pp. 124, 198 and *AM*
Huntington's disease	p. 556 and *AM*
AUTOSOMAL RECESSIVES	
Deafness	p. 597 and *AM*
Albinism	pp. 160, 1122
Sickle-cell anemia	p. 658 and *AM*
Cystic fibrosis	p. 830 and *AM*
Phenylketonuria	p. 947 and *AM*
Tay–Sachs disease	p. 423 and *AM*
SEX-LINKED	
Duchenne's muscular dystrophy	p. 300 and *AM*
Myotonic muscular dystrophy	*AM*
Hemophilia (A and B)	p. 676
Testicular feminization syndrome	*AM*
Color blindness	p. 585

Note: The genetic bases have been identified for at least some forms of the diseases in blue.

appearance of an inherited disorder that results from simple inheritance by using a Punnett square. Although they are rare in terms of overall numbers, more than 1200 inherited conditions have been identified that reflect the presence of one or two abnormal alleles for a single gene. A partial listing is given in Table 29–5, along with places where additional information can be found.

Polygenic Inheritance

Many phenotypic characters are determined by interactions among several genes. Such interactions constitute **polygenic inheritance**. The resulting phenotype depends on how the genes interact. Because multiple alleles are involved, their frequency of occurrence cannot easily be predicted by using a simple Punnett square. In *suppression*, one gene suppresses the other, so that the second gene has no effect on the phenotype. In *complementary gene action*, dominant alleles on two genes interact to produce a phenotype different from that seen when one gene contains recessive alleles. The risks of developing several important adult disorders, including hypertension and coronary artery disease, are linked to polygenic inheritance.

Many of the developmental disorders responsible for fetal deaths and congenital malformations result from polygenic inheritance. In these cases, the genetic composition of the individual does not by itself determine the onset of the disease. Instead, the conditions regulated by these genes establish a susceptibility to particular environmental influences. Thus, not every individual with the genetic tendency for a certain condition will develop that condition. It is therefore difficult to track polygenic conditions through successive generations. However, because many in-

herited polygenic conditions are *likely* (but not *guaranteed*) to occur, steps can be taken to prevent a crisis. For example, you can reduce hypertension by controlling your diet and fluid volume, and you can prevent coronary artery disease by lowering your serum cholesterol levels.

☐ SOURCES OF INDIVIDUAL VARIATION

Just as you are not a copy of either of your parents, neither are you a 50:50 mixture of their characteristics. We noted one reason for this in Chapter 28: During meiosis, maternal and paternal chromosomes are randomly distributed, so each gamete has a unique combination of maternal and paternal chromosomes. ∞ p. 1053 Thus, you may have an allele for curly hair from your father and an allele for straight hair from your mother, even though your sister received an allele for straight hair from each of your parents. Only in very rare cases will an individual receive both alleles from one parent. The few documented cases appear to have resulted when duplicate maternal chromatids failed to separate during meiosis II and the corresponding chromosome provided by the sperm did not participate in amphimixis. This condition, called *uniparental disomy*, generally remains undetected, because the individuals are phenotypically normal.

Genetic Recombination

During meiosis, various changes can occur in chromosome structure, producing gametes with chromosomes that differ from those of each parent. This phenomenon, called **genetic recombination**, greatly increases the range of possible variation among gametes and thus among members of successive generations, whose genotypes are formed by the combination of gametes in fertilization. Genetic recombination can also complicate the tracing of the inheritance of genetic disorders.

In one common form of recombination, parts of chromosomes become rearranged during synapsis (Figure 29–17●). When tetrads form, adjacent chromatids may overlap, an event called **crossing-over**. The chromatids may then break, and the overlapping segments trade places. This reshuffling process is known as **translocation**.

During recombination, portions of chromosomes may also break away and be lost, or *deleted*. The effects of a deletion on a zygote depend on the nature of the lost genes. In some cases, the effects depend on whether the abnormal gamete is produced through oogenesis or spermatogenesis. This phenomenon is called **genomic imprinting**. For example, the deletion of a specific segment of chromosome 15 in humans causes two very different disorders: *Angelman syndrome* and *Prader–Willi syndrome*. Symptoms of Angelman syndrome include hyperactivity, severe mental retardation, and seizures; the condition results when the abnormal chromosome is provided by the oocyte. Symptoms of Prader–Willi syndrome include a short stature, reduced muscle tone and skin pigmentation, underdeveloped gonads, and mental retardation varying from slight to severe; this condition results when the abnormal chromosome is delivered by the sperm.

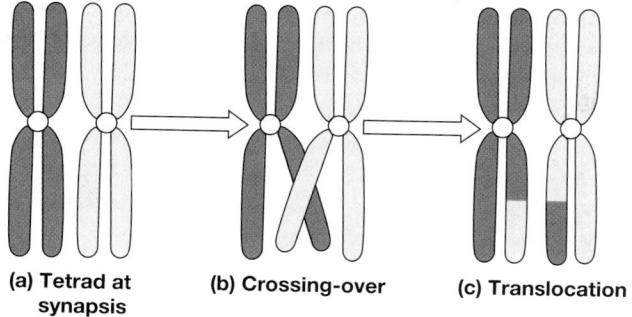

(a) Tetrad at synapsis (b) Crossing-over (c) Translocation

●**FIGURE 29–17**
Crossing-over and Translocation. **(a)** Synapsis, with the formation of a tetrad during meiosis. **(b)** Crossing-over of homologous portions of two chromosomes. **(c)** The breakage and exchange of corresponding sections on the chromosomes in (b).

Recombination that produces abnormal chromosome shapes or numbers is lethal for the zygote in almost all cases. Roughly 10 percent of zygotes have **chromosomal abnormalities**—that is, damaged, broken, missing, or extra copies of chromosomes—but only about 0.5 percent of newborn infants have such abnormalities. Chromosomal abnormalities produce a variety of serious clinical conditions, in addition to contributing to prenatal mortality. *Few individuals with chromosomal abnormalities survive to full term.* The high mortality rate and the severity of the problems reflect the fact that large numbers of genes have been added or deleted. Women who become pregnant later in life run a higher risk of birth defects and miscarriage due to chromosomal abnormalities in the oocyte. It seems that the longer the oocyte remains suspended in meiosis II, the more likely are recombination errors when meiosis is completed.

Mutation

Variations at the level of the individual gene can result from mutations that affect the nucleotide sequence of one allele. **Spontaneous mutations** are the result of random errors in DNA replication. Such errors are relatively common, but in most cases the error is detected and repaired by enzymes in the nucleus. Those errors that go undetected and unrepaired have the potential to change the phenotype in some way.

Mutations occurring during meiosis can produce gametes that contain abnormal alleles. These alleles may be dominant or recessive, and they may occur on autosomal chromosomes or on sex chromosomes. The vast majority of mutations make the zygote incapable of completing normal development. Mutation, rather than chromosomal abnormalities, is probably the primary cause of the high mortality rate among pre-embryos and embryos. (Roughly 50 percent of all zygotes fail to complete cleavage, and another 10 percent fail to reach the fifth month of gestation.) **AM** Complexity and Perfection

If the abnormal allele is dominant, but does not affect gestational survival, the individual's phenotype will show the effects of the mutation. If the abnormal allele is recessive and is on an autosomal chromosome, it will not affect the individual's phenotype as long as the zygote contains a normal allele contributed by the other parent at fertilization. Over generations, a recessive autosomal allele can spread through the population, remaining undetected until a fertilization occurs in which the two gametes contribute identical recessive alleles. This individual, who will be homozygous for the abnormal allele, will be the first to show the phenotypic effects of the original mutation. Individuals who are heterozygous for the abnormal allele, but who do not show the effects of the mutation, are called **carriers**. Genetic tests are available to determine whether an individual is a carrier for one of several autosomal recessive disorders, including Tay–Sachs disease and Huntington's disease. The information obtained from these tests can be useful in counseling prospective parents. For example, if both parents are carriers of the same disorder, they have a 25 percent probability of producing a child with the disease. This information may affect their decision to conceive.

☐ SEX-LINKED INHERITANCE

Unlike the other 22 chromosomal pairs, the sex chromosomes may not be identical in appearance and gene content. There are two types of sex chromosomes: an **X chromosome** and a **Y chromosome**. X chromosomes are considerably larger and have more genes than do Y chromosomes. The Y chromosome includes dominant alleles which specify that an individual with that chromosome will be male. The normal pair of sex chromosomes in males is XY. Females do not have a Y chromosome; their sex chromosome pair is XX.

All oocytes carry an X chromosome, because the only sex chromosomes females have are X chromosomes. But each sperm carries either an X or a Y chromosome, because males have one of each and can pass along either one. Thus, with a Punnett square, you can show that the ratio of males to females in offspring should be 1:1. The birth statistics differ slightly from that prediction, with 106 males born for every 100 females. It has been suggested that more males are born because the sperm that carries the Y chromosome can reach the oocyte first, since that sperm does not have to carry the extra weight of the larger X chromosome.

The X chromosome also carries genes that affect somatic structures. These characteristics are called **X linked** (or *sex linked*), because in most cases there are no corresponding alleles on the Y chromosome. The inheritance of characteristics regulated by these genes does not follow the pattern of alleles on autosomal chromosomes. The best known single-allele characters are those associated with noticeable diseases or functional deficits.

The inheritance of color blindness exemplifies the differences between sex-linked inheritance and autosomal inheritance. Normal color vision is determined by the presence of a dominant allele, *C*, on the X chromosome, whereas red–green color blindness results from the absence of *C* and the presence of a recessive allele, *c*, on the X chromosome. A woman, with her two

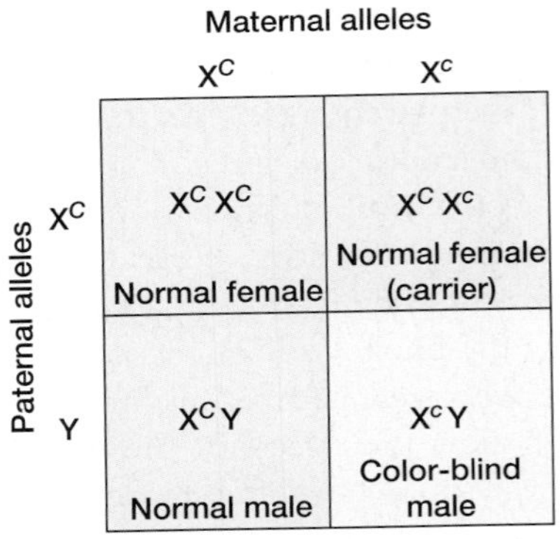

Maternal alleles

●**FIGURE 29–18**
X-Linked Inheritance. Recessive alleles on X, but not on Y, chromosomes produce genetic disorders in males at a higher frequency than in females. Females have two X chromosomes, so a female with a recessive allele will not develop such a disorder if she has a normal allele on the other X chromosome. By contrast, a male with an X chromosome bearing the recessive allele will develop the disorder.

X chromosomes, can be either homozygous dominant (*CC*) or heterozygous (*Cc*) and still have normal color vision. She will be unable to distinguish reds from greens only if she carries two recessive alleles, *cc*. But a male has only one X chromosome, so whichever allele that chromosome carries will determine whether he has normal color vision or is red–green color blind. A Punnett square for an X-linked trait, as in Figure 29–18●, reveals that the sons produced by a father with normal vision and a heterozygous mother will have a 50 percent chance of being red–green color blind, whereas any daughters will have normal color vision.

A number of other clinical disorders noted earlier in the text are X-linked traits, including certain forms of hemophilia, diabetes insipidus, and muscular dystrophy. In several instances, advances in techniques of molecular genetics have enabled geneticists to localize the specific genes on the X chromosome. These techniques provide a relatively direct method of screening for the presence of a particular condition before the symptoms appear and even before birth.

■ THE HUMAN GENOME PROJECT

Few of the genes responsible for inherited disorders have been identified or even associated with a specific chromosome. That situation is changing rapidly, however, due to the **Human Genome Project.** Funded by the National Institutes of Health and the Department of Energy, the project is attempting to transcribe the entire human **genome**—that is, the full complement of genetic material—chromosome by chromosome and gene by

gene. The work began in October 1990 and was expected to take 15 years. Progress has been more rapid than expected, and it is now anticipated that the project will be completed by the end of 2003. A working draft of the entire genome was published in 2001, and 63% of the entire genome was listed as finished, high-quality sequence as of May 2002. High-quality sequence is defined as a complete sequence with no gaps or ambiguities and an error rate of no more than one base per 10,000.

The first step in understanding the human genome is to prepare a map of the individual chromosomes. **Karyotyping** (KAR-ē-ō-tīp-ing; *karyon*, nucleus + *typos*, mark) is the determination of an individual's chromosomal complement. Figure 29–15● shows a set of normal human chromosomes. Each chromosome has characteristic banding patterns, and segments can be stained with special dyes. The patterns are useful as reference points for the preparation of more detailed genetic maps. The banding patterns themselves can be useful, as abnormal patterns are characteristic of some genetic disorders and several cancers.

As of 2002, a progress report included the following:

- Nine chromosomes—chromosomes 3, 5, 11, 12, 16, 19, 21, 22, and the Y chromosome—have been mapped completely (with chromosomes 5 and 21 qualifying as high quality), and preliminary maps have been made of all the other chromosomes.

- Chromosomes 5, 22, 21, and 20 have been completely sequenced, and 16 and 19 are nearing completion.

- More than 38,000 genes have been tentatively identified, and over 12,000 have been mapped. The first number may represent most of the genes in the human genome.

- The genes responsible for more than 60 inherited disorders, including several listed in Table 29–5 and those shown in Figure 29–19●, have been identified. Genetic screenings are now performed for many of these disorders.

It was originally thought that the relationship between genes and proteins was 1:1, and as a result, investigators were anticipating the discovery of as many as 140,000 genes in the human genome. Instead, the number appears to be far fewer—perhaps 40,000 genes in total. This stands in stark contrast with the estimated 2 million different proteins found in the human body. The realization that one gene can carry instructions for more than one protein has revolutionized thinking about genetic diseases and potential therapies. A whole new set of questions and problems has arisen as a result. For example, what factors and enzymes control mRNA processing? How are these factors regulated? Although we may be close to unraveling the human genome, we are still many years from the answers to these questions—and they must be answered if we are to manipulate genes effectively to treat many congenital diseases.

Of course, controversy remains over the advisability of tinkering with our genetic makeup. The Human Genome Project is attempting to determine the normal genetic composition of a "typical" human. Yet we all are variations on a basic theme.

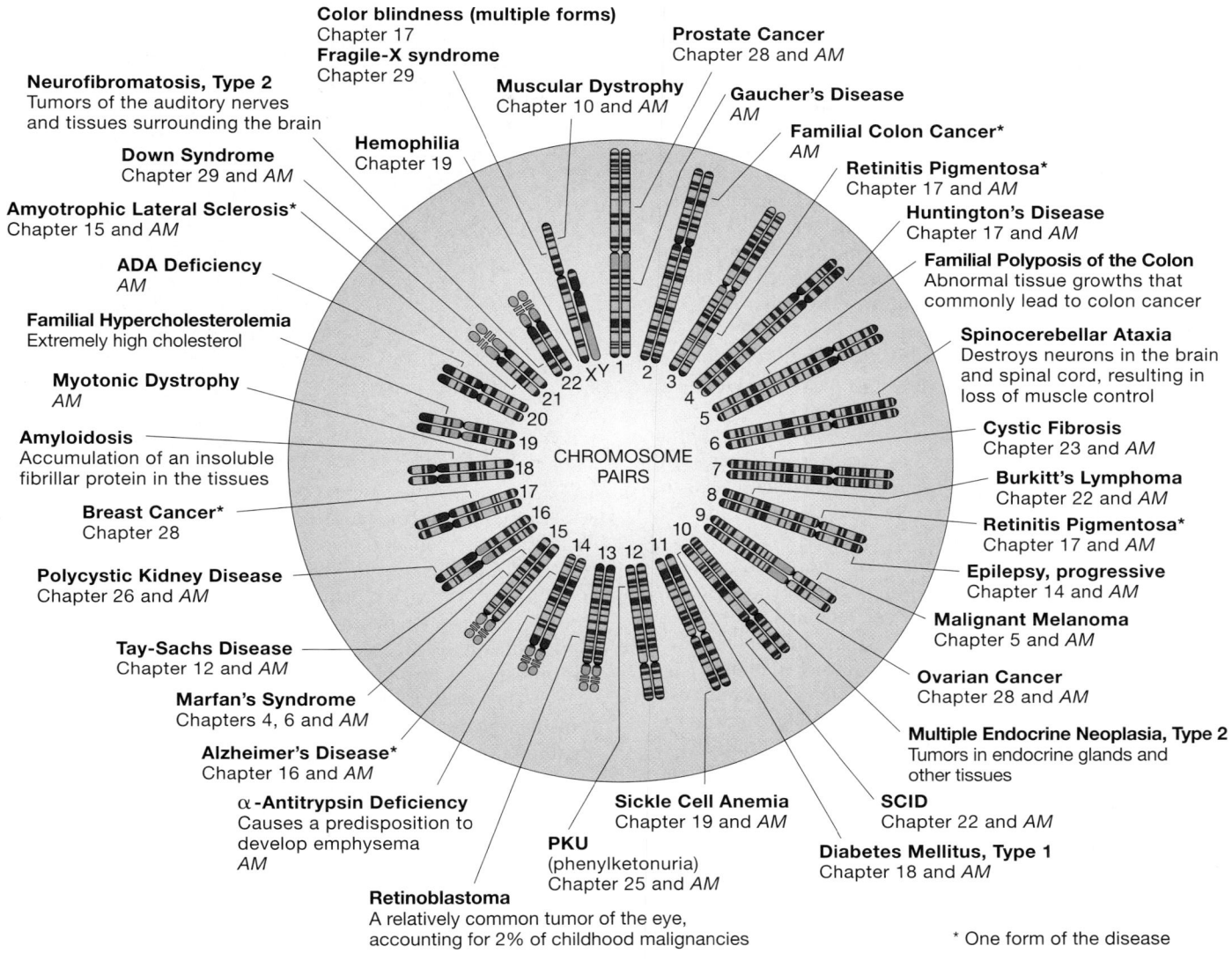

Color blindness (multiple forms)
Chapter 17
Fragile-X syndrome
Chapter 29

Neurofibromatosis, Type 2
Tumors of the auditory nerves
and tissues surrounding the brain

Down Syndrome
Chapter 29 and *AM*

Amyotrophic Lateral Sclerosis*
Chapter 15 and *AM*

ADA Deficiency
AM

Familial Hypercholesterolemia
Extremely high cholesterol

Myotonic Dystrophy
AM

Amyloidosis
Accumulation of an insoluble
fibrillar protein in the tissues

Breast Cancer*
Chapter 28

Polycystic Kidney Disease
Chapter 26 and *AM*

Tay-Sachs Disease
Chapter 12 and *AM*

Marfan's Syndrome
Chapters 4, 6 and *AM*

Alzheimer's Disease*
Chapter 16 and *AM*

α-Antitrypsin Deficiency
Causes a predisposition to
develop emphysema
AM

Retinoblastoma
A relatively common tumor of the eye,
accounting for 2% of childhood malignancies

Hemophilia
Chapter 19

Muscular Dystrophy
Chapter 10 and *AM*

Prostate Cancer
Chapter 28 and *AM*

Gaucher's Disease
AM

Familial Colon Cancer*
AM

Retinitis Pigmentosa*
Chapter 17 and *AM*

Huntington's Disease
Chapter 17 and *AM*

Familial Polyposis of the Colon
Abnormal tissue growths that
commonly lead to colon cancer

Spinocerebellar Ataxia
Destroys neurons in the brain
and spinal cord, resulting in
loss of muscle control

Cystic Fibrosis
Chapter 23 and *AM*

Burkitt's Lymphoma
Chapter 22 and *AM*

Retinitis Pigmentosa*
Chapter 17 and *AM*

Epilepsy, progressive
Chapter 14 and *AM*

Malignant Melanoma
Chapter 5 and *AM*

Ovarian Cancer
Chapter 28 and *AM*

Multiple Endocrine Neoplasia, Type 2
Tumors in endocrine glands and
other tissues

SCID
Chapter 22 and *AM*

Diabetes Mellitus, Type 1
Chapter 18 and *AM*

PKU
(phenylketonuria)
Chapter 25 and *AM*

Sickle Cell Anemia
Chapter 19 and *AM*

CHROMOSOME PAIRS

* One form of the disease

●**FIGURE 29–19**
A Map of Human Chromosomes. The banding patterns of typical chromosomes in a male individual and the locations of the genes responsible for specific inherited disorders. The chromosomes are not drawn to scale.

How do we decide what set of genes to accept as "normal?" Moreover, as we improve our abilities to manipulate our own genetic foundations, we will face troubling ethical and legal decisions. For example, few people object to the insertion of a "correct" gene into somatic cells to cure a specific disease. (See "Genetic Engineering and Gene Therapy" in the *Applications Manual.*) But what if we could insert that modified gene into a gamete and change not only that individual, but his or her descendants as well? And what if the function of the gene was not to correct or prevent any disorder, but rather to "improve" the individual by increasing his or her intelligence, height, or vision

or by altering some other phenotypic characteristic? Such difficult questions will not go away. In the years to come, we will have to find answers that are acceptable to us all.

✓ Curly hair is an autosomal dominant trait. What would be the phenotype of a person who is heterozygous for this trait?

✓ Why are children not identical copies of their parents?

Answers start on page Q-1

Chromosomal Abnormalities and Genetic Analysis

Embryos that have abnormal autosomal chromosomes rarely survive. Two types of abnormalities in autosomal chromosomes, however—translocation defects and trisomy—do not invariably kill the individual before birth.

In a **translocation defect**, crossing-over occurs between different chromosome pairs. For example, a piece of chromosome 8 may become attached to chromosome 14 rather than to the other chromatid of chromosome 8. In their new position, the genes that have been moved may function abnormally, becoming inactive or overactive. (A translocation between chromosomes 8 and 14 is responsible for *Burkitt's lymphoma*, a type of lymphatic system cancer.)

In **trisomy**, something goes wrong in meiosis: At fertilization, one of the gametes contributes an extra copy of one chromosome, so the zygote then has three copies of this chromosome rather than two. The nature of the trisomy is indicated by the number of the chromosome involved. Zygotes with extra copies of chromosomes seldom survive. Individuals with trisomy 13 and trisomy 18 may survive until delivery, but rarely live longer than a year. The notable exception is trisomy 21.

Trisomy 21, or **Down syndrome**, is the most common viable chromosomal abnormality. Estimates of the incidence of this syndrome range from 1.5 to 1.9 per 1000 births for the U.S. population. Affected individuals exhibit mental retardation and characteristic physical malformations, including a facial appearance that gave rise to the term *mongolism*, once used to describe this condition. The degree of mental retardation ranges from moderate to severe. Few individuals with Down syndrome lead independent lives. Anatomical problems affecting the cardiovascular system commonly prove fatal during childhood or early adulthood. Although some individuals survive to moderate old age, many develop Alzheimer's disease while still relatively young (before age 40).

For unknown reasons, maternal age is linked to the risk of having a child with trisomy 21. For a maternal age below 25, the incidence of Down syndrome approaches 1 in 2000 births, or 0.05 percent. For maternal ages 30–34, the odds increase to 1:900, and for ages 35–44, they go from 1:290 to 1:46, or more than 2 percent. These statistics are becoming increasingly significant, because many women have delayed childbearing until their midthirties or later.

Abnormal numbers of sex chromosomes do not produce effects as severe as those induced by extra or missing autosomal chromosomes. In **Klinefelter syndrome**, the individual carries the sex chromosome pattern XXY. The phenotype is male, but the extra X chromosome causes reduced androgen production. As a result, the breasts are slightly enlarged, the testes fail to mature, and affected individuals are sterile. The incidence of this condition among newborn males averages 1:750.

Individuals with **Turner syndrome** have only a single, female sex chromosome; their sex chromosome complement is abbreviated XO. This kind of chromosomal deletion is known as **monosomy**: its incidence at delivery has been estimated as 1:10,000. At birth, the condition may not be recognized, because the phenotype is normal female. But maturational changes do not appear at puberty. The ovaries are nonfunctional, and estrogen is produced only at negligible levels.

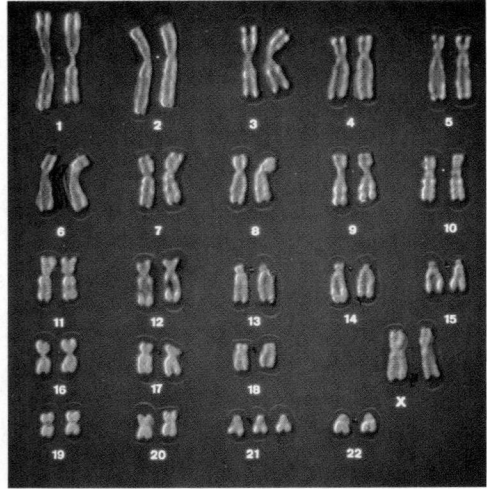

Fragile-X syndrome causes mental retardation, abnormal facial development, and increased testicular size in affected males. The origin is an X chromosome that contains a *genetic stutter*—an abnormal repetition of a single triplet. In this case, the problem is the repetition of the nucleotide sequence CCG at one site on the X chromosome. (A genetic stutter on another chromosome is responsible for Huntington's disease. ∞ p. 556) The presence of the stutter in some way disrupts the normal functioning of adjacent genes and so produces the symptoms of the disorder.

Many of these conditions can be detected prior to birth by the analysis of fetal cells. In **amniocentesis**, a sample of amniotic fluid is removed and the fetal cells that it contains are analyzed. This procedure permits the identification of over 20 congenital conditions, including Down syndrome. The needle inserted to obtain a sample of fluid is guided into position by using ultrasound. Unfortunately, amniocentesis has two major drawbacks:

1. Because the sampling procedure represents a potential threat to the health of the fetus and mother, amniocentesis is performed only when known risk factors are present. Examples of risk factors would include a family history of specific conditions or, in the case of Down syndrome, a maternal age over 35.

2. Sampling cannot safely be performed until the volume of amniotic fluid is large enough that the fetus will not be injured during the sampling process. The usual time for amniocentesis is at a gestational age of 14–15 weeks. It can then take several weeks to obtain results once samples have been collected; by the time the results are received, the option of therapeutic abortion may no longer be available.

In an alternative procedure known as **chorionic villus sampling**, cells collected from the chorionic villi during the first trimester are analyzed. Although it can be performed at an earlier gestational age than can amniocentesis, chorionic villus sampling has largely been abandoned because the risks of misdiagnosis are high and placental damage can cause spontaneous abortion.

CHAPTER REVIEW

SELECTED CLINICAL TERMINOLOGY

Terms Discussed in This Chapter

amniocentesis: An analysis of fetal cells taken from a sample of amniotic fluid. *(p. 1128)*

breech birth: A delivery during which the legs or buttocks of the fetus enter the vaginal canal first. *(p. 1116)*

chorionic villus sampling: An analysis of cells collected from the chorionic villi during the first trimester. *(p. 1128)*

ectopic pregnancy: A pregnancy in which implantation occurs somewhere other than the uterus. *(p. 1100 and AM)*

gestational trophoblastic neoplasia: A tumor formed by undifferentiated, rapid growth of the syncytial trophoblast; if untreated, the neoplasm may become malignant. *(p. 1101)*

infertility: The inability to achieve pregnancy after one year of appropriately timed intercourse. *(p. 1106)*

in vitro fertilization: Fertilization outside the body, generally in a petri dish. *(p. 1107)*

AM Additional Terms Discussed in the *Applications Manual*

abruptio placentae	placenta previa
Apgar rating	preeclampsia
DDST	pseudohermaphrodite
eclampsia	teratogens
fetal alcohol syndrome (FAS)	toxemia of pregnancy

STUDY OUTLINE

29–1 AN OVERVIEW OF TOPICS IN DEVELOPMENT   p. 1095

1. **Development** is the gradual modification of physical and physiological characteristics from **conception** to maturity. The creation of different types of cells is **differentiation**.

2. **Prenatal development** occurs before birth; **postnatal development** begins at birth and continues to **maturity**, when aging begins. **Inheritance** is the transfer of genetically determined characteristics from generation to generation. **Genetics** is the study of the mechanisms of inheritance.

29–2 FERTILIZATION   p. 1096

1. **Fertilization**, or *conception*, normally occurs in the uterine tube within a day after ovulation. Spermatozoa cannot fertilize a secondary oocyte until they have undergone *capacitation*. *(Figure 29–1)*

THE OOCYTE AT OVULATION   p. 1096

2. The acrosomal caps of the spermatozoa release **hyaluronidase** and **acrosin**, enzymes required to penetrate the corona radiata and zona pellucida of the oocyte. When a single spermatozoon contacts the oocyte membrane, fertilization begins and **oocyte activation** follows. *(Figure 29–1)*

OOCYTE ACTIVATION   p. 1098

3. During activation, the oocyte completes meiosis II and thus becomes a functionally mature ovum. **Polyspermy** is prevented by membrane depolarization and the *cortical reaction*.

4. After activation, the **female pronucleus** and the **male pronucleus** fuse in a process called *amphimixis*. *(Figure 29–1)*

29–3 THE STAGES OF PRENATAL DEVELOPMENT   p. 1098

1. During prenatal development, differences in the cytoplasmic composition of individual cells trigger changes in genetic activity. The chemical interplay among developing cells is **induction**.

2. The nine-month **gestation** period can be divided into three **trimesters**.

29–4 THE FIRST TRIMESTER   p. 1099

1. In the **first trimester**, **cleavage** subdivides the cytoplasm of the zygote in a series of mitotic divisions; the zygote becomes a **pre-embryo** and then a **blastocyst**. During **implantation**, the blastocyst burrows into the uterine endometrium. **Placentation** occurs as blood vessels form around the blastocyst and the **placenta** develops. **Embryogenesis** is the formation of a viable embryo.

CLEAVAGE AND BLASTOCYST FORMATION   p. 1099

2. The blastocyst consists of an outer **trophoblast** and an **inner cell mass**. *(Figure 29–2)*

IMPLANTATION   p. 1099

3. Implantation occurs about seven days after fertilization as the blastocyst adheres to the uterine lining. *(Figure 29–3)*

4. As the trophoblast enlarges and spreads, maternal blood flows through open **lacunae**. After **gastrulation**, there is an **embryonic disc** composed of **endoderm**, **ectoderm**, and an intervening **mesoderm**. It is from these **germ layers** that the body systems differentiate. *(Figure 29–4; Summary Table 29–1)*

5. Germ layers help form four **extraembryonic membranes**: the yolk sac, amnion, allantois, and chorion. *(Figure 29–5)*

6. The **yolk sac** is an important site of blood cell formation. The **amnion** encloses fluid that surrounds and cushions the developing embryo. The base of the **allantois** later gives rise to the urinary bladder. Circulation within the vessels of the **chorion** provides a rapid-transit system that links the embryo with the trophoblast. *(Figures 29–5, 29–6)*

PLACENTATION   p. 1104

7. **Chorionic villi** extend outward into the maternal tissues, forming an intricate, branching network through which maternal blood flows. As development proceeds, the **umbilical cord** connects the fetus to the placenta. The trophoblast synthesizes **human chorionic gonadotropin (hCG)**, estrogens, progesterone, **human placental lactogen (hPL)**, **placental prolactin**, and **relaxin**. *(Figure 29–6)*

SEX-LINKED INHERITANCE   p. 1125

9. The two types of sex chromosomes are an **X chromosome** and a **Y chromosome**. The normal sex chromosome complement of males is XY; that of females is XX. The X chromosome carries **X-linked** (*sex-linked*) **genes**, which affect somatic structures, but have no corresponding alleles on the Y chromosome. (*Figure 29–18*)

THE HUMAN GENOME PROJECT   p. 1126

10. The **Human Genome Project** has mapped more than 38,000 of our genes, including some of those responsible for inherited disorders. (*Figure 29–19; Table 29–5*)

REVIEW QUESTIONS

 More assessment questions are available to you on the Companion Website. You will find Matching, Multiple Choice, True/False, and other quizzes to help further your understanding of the material covered in this chapter. To access the site, go to www.aw.com/martini.

LEVEL 1 Reviewing Facts and Terms

1. The gradual modification of anatomical structures during the period from conception to maturity is
   (a) development   (b) differentiation
   (c) embryogenesis   (d) capacitation

2. The enzymes of the acrosomal cap that facilitate fertilization by breaking bonds between adjacent follicular cells include
   (a) hyaluronidase and hCG   (b) hyaluronidase and acrosin
   (c) acrosin and hCS   (d) progesterone and hCG

3. The secondary oocyte leaving the follicle is in
   (a) interphase
   (b) metaphase of the first meiotic division
   (c) telophase of the second meiotic division
   (d) metaphase of the second meiotic division

4. Chemical interplay among developing cells is
   (a) gestation   (b) induction
   (c) capacitation   (d) placentation

5. The stage of development that follows cleavage is the
   (a) blastocyst   (b) morula
   (c) trophoblast   (d) blastocoele

6. The process that secures the developing embryo to the uterine lining is
   (a) cleavage   (b) embryogenesis
   (c) placentation   (d) implantation

7. What begins as a zygote arrives in the uterine cavity as a
   (a) blastocyst   (b) trophoblast
   (c) lacuna   (d) blastomere

8. The first extraembryonic membrane to appear is the
   (a) allantois   (b) yolk sac
   (c) amnion   (d) chorion

9. The primitive streak appears during
   (a) cleavage   (b) implantation
   (c) gastrulation   (d) placentation

10. The part of the uterine endometrium that has no contact with the chorion is the
    (a) decidua capsularis   (b) decidua basalis
    (c) decidua parietalis   (d) allantois

11. The surface that provides for active and passive exchange between the fetal and maternal bloodstreams is the
    (a) yolk stalk   (b) chorionic villi
    (c) umbilical veins   (d) umbilical arteries

12. The placental hormone that appears in the maternal bloodstream soon after implantation is
    (a) human chorionic gonadotropin
    (b) human placental lactogen
    (c) placental prolactin
    (d) relaxin

13. _____ is a critical period for development, as the events that take place at this time establish the basis for organogenesis.
    (a) Gastrulation
    (b) Placentation
    (c) The first trimester
    (d) The second trimester

14. Milk let-down is associated with
    (a) events occurring in the uterus
    (b) placental hormonal influences
    (c) circadian rhythms
    (d) reflex action triggered by nursing

15. An individual who has two different alleles for the same gene is _____ for that trait.
    (a) heterozygous   (b) homozygous
    (c) dominant   (d) recessive

16. If an allele must be present on both the maternal and paternal chromosomes to affect the phenotype, the allele is said to be
    (a) dominant   (b) recessive
    (c) complementary   (d) heterozygous

17. The percentage of individuals with a particular genotype who show the expected phenotype reflects
    (a) dominance
    (b) polygenic inheritance
    (c) spontaneous mutation
    (d) penetrance

18. For a trait *A*, the genotype of a homozygous recessive individual is represented as
    (a) *AA*   (b) *Aa*
    (c) *aa*   (d) *Ab*

19. The chromosome that determines the sex of the individual is
    (a) chromosome 21
    (b) a combination of all 46 chromosomes
    (c) the X chromosome
    (d) the Y chromosome

20. Describe the changes that occur in the oocyte immediately after fertilization.

21. Summarize the developmental changes that occur during the first, second, and third trimesters.

22. (a) What are the four extraembryonic membranes? (b) How do these membranes form, and what are their functions?

23. Indicate when each of the following appears during development, and describe each: embryonic disc, blastocyst, morula, zygote.

24. Identify the three stages of labor, and describe the events that characterize each stage.

25. List the factors involved in initiating labor contractions.

26. Identify the three life stages that occur between birth and approximately age 10. Describe the characteristics of each stage and when it occurs.

27. What hormonal events are responsible for puberty? Which life stage does puberty initiate?

28. What occurs during the life stage of maturity?

29. What is senescence? Give some examples of how it affects organ systems throughout the human body.

LEVEL 2 Reviewing Concepts

30. Human chorionic somatomammotropin will not be converted to active status without
    (a) estrogen
    (b) progesterone
    (c) placental prolactin
    (d) a, b, and c are correct

31. The production of prostaglandins in the endometrium
    (a) initiates the release of oxytocin for parturition
    (b) stimulates smooth muscle contractions
    (c) initiates secretory activity in the mammary glands
    (d) a, b, and c are correct

32. During adolescence, the events that interact to promote increased hormone production and sexual maturation result from activity of the
    (a) hypothalamus
    (b) anterior lobe of the pituitary gland
    (c) ovaries and testicular cells
    (d) a, b, and c are correct

33. What activity during oocyte activation prevents penetration by additional sperm?

34. After implantation, how does the developing embryo obtain nutrients? What structures and processes are involved?

35. What is induction? Explain its significance.

36. In addition to its role in the nutrition of the fetus, what are the primary endocrine functions of the placenta?

37. Discuss the changes that occur in maternal systems during pregnancy. Why are these changes functionally significant?

38. Discuss the changes that occur in the uterus during pregnancy. How do these changes affect uterine tissues, and which hormones are involved?

39. Why are uterine contractions in the early stages of pregnancy weak, painless, and brief?

40. During true labor, what physiological mechanisms ensure that uterine contractions continue until delivery has been completed?

41. What is an episiotomy? When and why is it performed?

42. What physiological adjustments must an infant make during the neonatal period in order to survive?

43. Distinguish between the following paired terms:
    (a) monozygotic and dizygotic
    (b) genotype and phenotype
    (c) heterozygous and homozygous
    (d) simple inheritance and polygenic inheritance

44. What would you conclude about a trait in each of the following situations?
    (a) Children who exhibit the trait have at least one parent who exhibits it also.
    (b) Children exhibit the trait even though neither parent exhibits it.
    (c) The trait is expressed more commonly in sons than in daughters.
    (d) The trait is expressed equally in daughters and sons.

45. How is it possible to produce a red–green color-blind female? What percentage of her sons will also be color blind?

46. Explain the impact of genetic recombination on the production of gametes and on the traits of individuals in future generations.

47. Explain the goals and possible benefits of the Human Genome Project.

LEVEL 3 Critical Thinking and Clinical Applications

48. Hemophilia A, a condition in which blood does not clot properly, is a recessive trait located on the X chromosome (X^h). Suppose that a woman who is heterozygous for this trait (XX^h) mates with a normal male (XY). What is the probability that the couple will have hemophiliac daughters? What is the probability that the couple will have hemophiliac sons?

49. Joan is a 27-year-old nurse who is in labor with her first child. She remembers from her anatomy and physiology class that calcium ions can increase the force of smooth muscle contractions, and because the labor is prolonged, she asks her physician for a calcium injection. The surprised physician informs Joan that such an injection is definitely out of the question. Why?

50. A new mother tells you that when she nurses her baby, she feels as if she is having menstrual cramps. How would you explain this phenomenon?

51. Explain why the normal heart and respiratory rates of neonates are so much higher than those of adults, even though adults are much larger.

52. Pregnant women are advised against taking aspirin during their third trimester. Why?

53. Sally gives birth to a baby with a congenital deformity of the stomach. Sally believes that her baby's affliction is the result of a viral infection that she suffered during her third trimester. Is this a possibility? Explain.

ANSWERS TO CONCEPT CHECK AND END OF CHAPTER QUESTIONS

CHAPTER 1

Page 11
1. White blood cells, or any cells or organisms, that respond to changes in its environment are exhibiting responsiveness, or irritability.
2. A histologist investigates the structure and properties of the tissue level of organization.
3. The study of the physiology of specific organs is called special physiology. In this particular case, the field of study is cardiac physiology (the study of heart function). Because heart failure is typically caused by disease, this specialty would overlap or be closely related to pathological physiology.

Page 15
1. Physiological systems can function normally only under carefully controlled conditions. Homeostatic regulation prevents potentially disruptive changes in the body's internal environment.
2. When homeostasis fails, organ systems function less efficiently or malfunction. The result is the state that we call disease. If the situation is not corrected, death can result.
3. Positive feedback is useful in processes, such as blood clotting, that must move quickly to completion once they have begun. It is harmful in situations in which a stable condition must be maintained, because it tends to increase any departure from the desired condition. For example, positive feedback in the regulation of body temperature would cause a slight fever to spiral out of control, with fatal results. For this reason, physiological systems normally exhibit negative feedback, which tends to oppose any departure from the norm.

Page 22
1. The two eyes would be separated by a midsagittal section.
2. The body cavity inferior to the diaphragm is the abdominopelvic cavity.

Page 27: Level 1 Reviewing Facts and Terms
1. g 2. d 3. a 4. j 5. b 6. l 7. n 8. f 9. h 10. e
11. c 12. o 13. k 14. I 15. m 16. c 17. b 18. b 19. a
20. d 21. a 22. b 23. d 24. a 25. d 26. c 27. d 28. b
29. c 30. c 31. b

Level 2 Reviewing Concepts
32. responsiveness, adaptability, growth, differentiation, reproduction, movement, metabolism, absorption, respiration, and excretion
33. (a) Anatomy is the study of internal and external structures and the physical relationships among body parts. (b) Physiology is the study of how organisms perform their vital functions.
34. The organization at each level determines the characteristics and functions of higher levels.
35. Serous membranes secrete a watery fluid that coats the opposing surfaces and reduces friction.
36. Homeostatic regulation refers to adjustments in physiological systems that are responsible for the preservation of homeostasis.
37. Autoregulation occurs when the activities of a cell, tissue, organ, or organ system change automatically (that is, without neural or endocrine input) when faced with some environmental change. Extrinsic regulation results from the activities of the nervous or endocrine system. It causes more-extensive and potentially more-effective adjustments in activities.

38. Receptor: receives the stimulus; control or integration center: determines whether the stimulus has upset homeostasis, and also initiates an appropriate response; effector: responds to the stimulus according to instructions from the control center.
39. In negative feedback, a variation outside the normal range triggers an automatic response that corrects the situation. In positive feedback, the initial stimulus produces a response that exaggerates the stimulus.
40. The body is erect, and the hands are at the sides with the palms facing forward.
41. Right Upper Quadrant (RUQ) right lobe of liver, gallbladder, right kidney, portions of stomach, large and small intestines; Left Upper Quadrant (LUQ) left lobe of liver, stomach, pancreas, left kidney, spleen, portions of large intestine
42. thoracic cavity; lungs
43. (a) dorsal cavity (b) ventral cavity (c) thoracic cavity (d) abdominopelvic cavity
44. Sectional anatomy deals with the relationship among parts in a three dimensional object, and involves cutting through the object to view internal structures. Superficial anatomy deals with the surface appearance of organs and the visible relationships between structures.

Level 3 Critical Thinking and Clinical Applications
45. Calcitonin is released when calcium levels are elevated. This hormone should bring about a decrease in blood calcium levels, thus decreasing the stimulus for its own release.
46. The initial increase in blood flow to active muscles is an example of autoregulation. The increased demands for oxygen and for waste removal cause a local increase in blood flow even before responses from the nervous or endocrine system take place. Autoregulation is an immediate and automatic response to environmental change (in this case, low oxygen levels and increased wastes) that does not require the nervous or endocrine system.
47. b

CHAPTER 2

Page 38
1. Atoms combine with each other so as to gain a complete set of eight electrons in their outer energy level. Oxygen atoms do not have a full outer energy level, so they readily react with many other elements to attain this stable arrangement. Neon already has a full outer energy level and thus has little tendency to combine with other elements.
2. Hydrogen has three isotopes: hydrogen-1, with a mass of 1; deuterium, with a mass of 2; and tritium, with a mass of 3. The heavier sample must contain a higher proportion of one or both of the heavier isotopes.
3. The atoms in a water molecule are held together by polar covalent bonds. Water molecules are attracted to one another by hydrogen bonds.

Page 41
1. Because this reaction involves a large molecule being broken down into two smaller ones, it is a decomposition reaction. Because energy is released in the process, the reaction can also be classified as exergonic.

2. Enzymes in our cells promote chemical reactions by lowering the activation energy requirements. Enzymes make it possible for chemical reactions to proceed under the conditions compatible with life.

Page 45

1. When it dissolves in water, salt dissociates into ions, charged particles that are capable of conducting an electrical current. Sugar molecules are held together by covalent bonds and so do not dissociate in solution; thus there are no ions to carry a current.

2. Stomach discomfort is commonly the result of excess stomach acidity ("acid indigestion"). Antacids contain a weak base that neutralizes the excess acid.

Page 51

1. A compound with a C:H:O ratio of 1:2:1 is a carbohydrate. The body uses carbohydrates chiefly as an energy source.

2. When two monosaccharides undergo a dehydration synthesis reaction, they form a disaccharide.

3. The most abundant lipid in a sample taken from beneath the skin would be a triglyceride.

4. An analysis of the lipid content of human cell membranes would indicate the presence of mostly phospholipids and small amounts of cholesterol and glycolipids.

Page 57

1. Proteins are chains of small organic molecules called amino acids.

2. An agent that breaks hydrogen bonds would affect the secondary level of protein structure.

3. The heat of boiling will break bonds that maintain the protein's tertiary structure, quaternary structure, or both. The resulting change in shape will affect the ability of the protein molecule to perform its normal biological functions. These changes are known as denaturation.

4. If the active site of an enzyme changes so that the site better fits its substrate, the level of enzyme activity will increase. But if the change alters the active site to the extent that the enzyme's substrate can no longer bind or binds poorly, the enzyme's activity will decrease or be inhibited.

Page 61

1. The nucleic acid RNA (ribonucleic acid) contains the sugar ribose. The nucleic acid DNA (deoxyribonucleic acid) contains the sugar deoxyribose instead; both contain nitrogenous bases and phosphate groups.

2. Phosphorylation of an ADP molecule would yield a molecule of ATP.

Page 63: Level 1 Reviewing Facts and Terms

1. a   2. b   3. d   4. a   5. h   6. n   7. e   8. k   9. a   10. i
11. c   12. o   13. f   14. b   15. m   16. d   17. g   18. j   19. l
20. b   21. c   22. d   23. d   24. b   25. d   26. c   27. a   28. b
29. protons, neutrons, and electrons
30. carbohydrates, lipids, proteins, and nucleic acids
31. (1) They provide a significant energy reserve. (2) They serve as insulation and thus act in heat preservation. (3) They protect organs by cushioning them.
32. (1) support (structural proteins); (2) movement (contractile proteins); (3) transport (transport proteins); (4) buffering; (5) metabolic regulation; (6) coordination and control (hormones and neurotransmitters); and (7) defense (antibodies)
33. (a) DNA: deoxyribose, phosphate, and nitrogenous bases (A, T, C, G); (b) RNA: ribose, phosphate, and nitrogenous bases (A, U, C, G)
34. (1) adenosine, (2) phosphate groups, and (3) appropriate enzymes

Level 2 Reviewing Concepts

35. c   36. a
37. Potential energy is stored energy. Kinetic energy is the energy of motion.

38. Enzymes are specialized protein catalysts that lower the activation energy for chemical reactions. Enzymes speed up chemical reactions but are not used up or changed in the process.

39. A salt is an ionic compound consisting of any cations other than hydrogen ions and any anions other than hydroxide ions. Acids dissociate and release hydrogen ions, while bases remove hydrogen ions from solution (usually by releasing hydroxide ions)

40. Nonpolar covalent bonds have an equal sharing of electrons. Polar covalent bonds have an unequal sharing of electrons. Ionic bonds lose and/or gain electrons.

41. Water molecules are released in dehydration synthesis of polysaccharides.

42. Pure water, with a pH of 7, is neutral because it contains equal numbers of hydrogen ions and hydroxyl ions.

43. Buffer systems maintain pH within normal limits by removing or replacing hydrogen ions as needed.

44. The molecule is a nucleic acid. Carbohydrates and lipids do not contain nitrogen. Although both proteins and nucleic acids contain nitrogen, only nucleic acids normally contain phosphorus.

Level 3 Critical Thinking and Clinical Applications

45. calcium protons = 20; electrons = 20; neutrons = 20; atomic number = 20; atomic weight = 40.08; 2 electrons in shell 1, 8 in shell 2; 8 in shell three and 2 in shell 4

46. The number of neutrons in an atom is equal to the atomic weight minus the atomic number. Thus sulfur has $32 - 16 = 16$ neutrons. The atomic number indicates the number of protons, so a neutral sulfur atom contains 16 protons—plus 16 electrons to balance the protons electrically. The electrons would be distributed as follows: 2 in the first level, 8 in the second level, and the remaining 6 in the third level. To achieve a full 8 electrons in the third (outermost) level, the sulfur atom can accept 2 electrons in an ionic bond or can share 2 electrons in a covalent bond. Because hydrogen atoms can share 1 electron in a covalent bond, the sulfur atom can form two covalent bonds with hydrogen, one with each of two hydrogen atoms.

47. If a person exhales large amounts of CO_2, the equilibrium will shift to the left and the level of H^+ in the blood will decrease. A decrease in the amount of H^+ will cause the pH to rise.

48. Because NaCl is an ionic compound, it will dissociate in water, producing equal numbers of sodium ions and chloride ions. Thus each mole of NaCl will produce 2 moles of ions. To prepare a solution containing 1.2 moles of ions, the student should dissolve 0.6 moles of NaCl. The molecular weight of NaCl is the sum of its atomic weights: 23 g/mol + 35.5 g/mol + 58.5 g/mol. Therefore, 0.6 mol × 58.5 g/mol = 35 g of NaCl should be added to the solution.

49. 0.9% = 0.9 grams in 100 ml. 1 Liter = 1000 ml, therefore you will need to add 9 grams of NaCl to a liter.

CHAPTER 3

Page 74

1. The phospholipid bilayer of the cell membrane form a physical barrier between the internal environment of the cell and the external environment.

2. Channel proteins are integral proteins that allow water and small ions to pass through the cell membrane.

Page 78

1. The fingerlike projections on the surface of the intestinal cells are microvilli. They increase the cells' surface area, enhancing its ability to absorb nutrients.

2. The cytosol has a higher concentration of potassium ions and suspended proteins and a lower concentration of sodium ions than the extracellular fluid. The cytosol also includes small quantities of carbohydrates, and larger reserves of amino acids and lipids.

Page 83
1. The SER functions in the synthesis of lipids such as steroids. Ovaries and testes produce large amounts of steroid hormones, which are lipids, and thus need large amounts of SER.
2. Mitochondria produce energy, in the form of ATP molecules, for the cell. A large number of mitochondria in a cell indicates a high demand for energy.

Page 90
1. The nucleus of a cell contains DNA, which codes for the production of all the cell's polypeptides and proteins. Some of these proteins are structural proteins, which are responsible for the shape and other physical characteristics of the cell. Other proteins are enzymes, which govern cellular metabolism, direct the production of cell proteins, and control all the cell's activities.
2. If a cell lacked the enzyme RNA polymerase, the cell would not be able to transcribe RNA from DNA.

Page 94
1. Diffusion is driven by a concentration gradient. The larger the concentration gradient, the faster the rate of diffusion. The smaller the concentration gradient, the slower the rate of diffusion. If the concentration of oxygen in the lungs were to decrease, the concentration gradient between oxygen in the lungs and oxygen in the blood would decrease (as long as the oxygen level of the blood remained constant). Thus oxygen would diffuse more slowly into the blood.
2. The 10 percent salt solution would be hypertonic with respect to the cells lining the nasal cavity, because this solution contains a higher concentration of salt than do the cells. The hypertonic solution would draw water out of the cells, causing the cells to shrink and adding water to the mucus, thus relieving the congestion.

Page 100
1. To transport hydrogen ions against their concentration gradient—that is, from a region where they are less concentrated (the cells lining the stomach) to a region where they are more concentrated (the interior of the stomach)—energy must be expended. An active transport process must be involved.
2. If the cell membrane were freely permeable to sodium ions, more of these positively charged ions would move into the cell and the transmembrane potential would move closer to zero.
3. This process is an example of phagocytosis.

Page 106
1. This cell would be in the G_1 phase of its life cycle.
2. The deletion of a single nucleotide from a coding sequence of DNA would alter the entire base sequence after the point of deletion. As a result, different codons would be lined up on the messenger RNA that was transcribed from the affected region. These codons, in turn, would result in the incorporation of a different series of amino acids into the polypeptide. The polypeptide product would definitely be abnormal and may not be functional.
3. If spindle fibers failed to form during mitosis, the cell would not be able to separate the chromosomes into two sets. If cytokinesis occurred, the result would be one cell with two sets of chromosomes and one cell with none.

Page 108: Level 1 Reviewing Facts and Terms
1. c 2. d 3. d 4. a 5. c 6. d 7. a 8. c 9. d 10. c
11. b

12. (1) Cells are the building blocks of all plants and animals. (2) Cells are produced by the division of pre-existing cells. (3) Cells are the smallest units that perform all vital physiological functions. (4) Each cell maintains homeostasis at the cellular level. (5) Homeostasis at the tissue, organ, organ system, and organism levels reflect the combined and coordinated actions of many cells.
13. (1) physical isolation, (2) regulation of exchange with the environment, (3) sensitivity, and (4) structural support
14. Membrane proteins function as receptors, channels, carriers, enzymes, anchors, and identifiers.
15. (1) diffusion, (2) carrier-mediated transport, and (3)) vesicular transport
16. (1) distance (2) size of the gradient, (3) molecule size, and (4) temperature
17. Osmosis is the diffusion of water molecules across a membrane. Osmosis occurs across a selectively permeable membrane that is freely permeable to water but not freely permeable to solutes. In osmosis, water will flow across a membrane toward the solution that has the higher concentration of solutes.
18. (a) nonmembranous: cytoskeleton, microvilli, centrioles, cilia, flagella, ribosomes and proteasomes; (b) membranous: nucleus, mitochondria, endoplasmic reticulum, Golgi apparatus, liposomes, and peroxisomes
19. (1) synthesis of proteins, carbohydrates, and lipids; (2) storage of absorbed or synthesized molecules; (3) transport of materials; and (4) detoxification of drugs or toxins
20. (1) prophase, (2) metaphase, (3) anaphase, and (4) telophase

Level 2 Reviewing Concepts
21. b 22. a 23. b 24. c 25. c
26. specificity, saturation, regulation, type of transport (active or passive), and number of ions, molecules, or solutes transported simultaneously
27. (a) Both processes use carrier proteins and exhibit saturation. (b) Facilitated diffusion is passive, does not expend ATP, and has concentration gradient, whereas active transport is active, expends ATP, and has no concentration gradient.
28. By ejecting sodium ions from the cytosol and reclaiming potassium ions from the extracellular fluid in a ratio of 3:2, the sodium–potassium pump maintains the negative resting membrane potential.
29. Cytosol is a colloid that has a high concentration of K^+, and suspended proteins; it also contains small quantities of carbohydrates and large reserves of amino acids and lipids and may contain insoluble materials known as inclusions. Interstitial fluid has a high concentration of Na^+ and fewer suspended proteins and organic nutrients.
30. In transcription, RNA polymerase uses genetic information to assemble a strand of mRNA. In translation, ribosomes use information carried by the mRNA strand to assemble functional proteins.
31. G_0: normal cell functions; G_1: cell growth, duplication of organelles, and protein synthesis; S: DNA replication and synthesis of histones; G_2: protein synthesis
32. Prophase: Chromatin condenses and chromosomes become visible; centrioles migrate to opposite poles of the cell and spindle fibers develop; and the nuclear membrane disintegrates. Metaphase: Chromatids attach to spindle fibers and line up along the metaphase plate. Anaphase: Chromatids separate and migrate toward opposite poles of the cell. Telophase: The nuclear membrane forms; chromosomes disappear as the chromatin relaxes; and nucleoli appear.
33. (a) Cytokinesis is the cytoplasmic movement that separates two daughter cells. (b) It completes mitosis.

Level 3 Critical Thinking and Clinical Applications

34. This process is facilitated diffusion, which requires a carrier molecule but not cellular energy. The energy for this process is provided by the concentration gradient of the substance being transported. When all the carriers are actively involved in transport, the rate of transport reaches a saturation point.

35. Solution A must have initially had more solutes than solution B. As a result, water moved by osmosis across the selectively permeable membrane from side B to side A, increasing the fluid level on side A.

36. For the dialysis fluid to remove urea without removing water, it should not contain urea. Because urea is a small molecule, it will diffuse through the dialysis membrane from an area of high concentration (the blood) to an area of low concentration (the dialysis fluid). To prevent an associated osmotic water movement, the dialysis fluid should have an osmotic concentration similar to that of blood plasma but with higher concentrations of solutes such as bicarbonate ions or glucose. As urea diffuses into the dialysis fluid, glucose and bicarbonate diffuse into the blood; as a result, the solute concentrations remain in balance and no osmotic water movement occurs.

37. All membranous organelles are involved in membrane flow. Those most directly associated are the ER, Golgi, secretory vesicles and cell membrane. A membrane protein would be manufactured in the RER, then flow through the cisternae to a transport vesicle. There the protein will be moved to the cis face of the Golgi apparatus, where it will slowly travel upward toward the trans face, perhaps modified along the way. Once reaching the end of the Golgi apparatus, the protein would be embedded in the membrane of a secretory vesicle, and transported to the cell membrane. There the vesicle will fuse with the membrane, inserting the protein in the cell membrane.

CHAPTER 4

Page 115

1. Epithelial tissue is composed entirely of polar cells bound to a thin basal lamina. Epithelia are avascular and capable of regeneration.

2. The presence of many microvilli on the free surface of epithelial cells greatly increases the cell's surface area, allowing for increased absorption.

3. Gap junctions allow small molecules and ions to pass from cell to cell. Among epithelial cells, they help coordinate functions such as the beating of cilia. In cardiac and smooth muscle tissues, they are essential to the coordination of muscle cell contractions.

Page 122

1. No. A simple squamous epithelium does not provide enough protection against infection, abrasion, or dehydration and so is not located on the skin surface.

2. All these regions are subject to mechanical trauma and abrasion—by food (pharynx and esophagus), feces (anus), and intercourse or childbirth (vagina).

3. This process is holocrine secretion.

4. The gland is an endocrine gland.

Page 128

1. Collagen fibers add strength to connective tissue. A vitamin C deficiency thus might result in the production of connective tissue that is weak and prone to damage.

2. Antihistamines act against the molecule histamine, which is produced by mast cells and basophils that leave the bloodstream.

3. The tissue is adipose (fat) tissue.

Page 132

1. Cartilage lacks a direct blood supply, which is necessary for proper healing to occur. Instead of having chondrocytes repair the injury site with new cartilage, fibroblasts migrate into the area and replace the damaged cartilage with fibrous scar tissue.

2. Intervertebral discs are composed of fibrocartilage.

3. The two connective tissues that contain a fluid matrix are blood and lymph.

Page 135

1. The pleural, peritoneal, and pericardial cavities are all lined by serous membranes.

2. This is an example of a mucous membrane.

3. This tissue is probably fascia, a type of dense connective tissue that attaches muscles to skin and bones.

Page 138

1. This muscle tissue is smooth muscle tissue; both cardiac and skeletal muscle tissues are striated.

2. The cells are probably neurons.

3. New skeletal muscle is produced by the division and fusion of satellite cells, mesenchymal cells that persist in adult skeletal muscle tissue.

Page 144: Level 1 Reviewing Facts and Terms

1. j **2.** g **3.** o **4.** m **5.** a **6.** t **7.** c **8.** p **9.** b **10.** f **11.** i **12.** e **13.** k **14.** h **15.** s **16.** n **17.** d **18.** l **19.** r **20.** q **21.** a **22.** c **23.** d **24.** c **25.** c **26.** b **27.** d **28.** d **29.** b **30.** c **31.** c **32.** a **33.** b **34.** b **35.** a **36.** b

37. (1) It consists mainly of cells rather than extracellular materials; (2) it has a free surface exposed to the external environment or to some internal chamber or passageway; (3) it is avascular—that is, it has no blood vessels; and (4) damaged or lost epithelial cells are continuously being replaced.

38. (1) It provides physical protection; (2) it controls permeability; (3) it provides sensations; and (4) it produces specialized secretions.

39. endocrine glands secrete hormones onto the surface of the gland or directly into the surrounding fluid; exocrine glands secrete via ducts.

40. (1) squamous; (2) cuboidal; and (3) columnar

41. (1) merocrine secretion; (2) apocrine secretion; (3) holocrine secretion

42. (1) specialized cells; (2) extracellular protein fibers; and (3) a fluid ground substance

43. The six basic functions of connective tissue are: 1) establishing a structural framework for the body, 2) transporting fluids, 3) protecting delicate organs, 4) supporting surrounding and interconnecting other tissue types, 5) storing energy reserves, and 6) defending the body from invading microorganisms.

44. (1) collagen fibers; (2) reticular fibers; and (3) elastic fibers

45. (1) tendons; (2) aponeuroses; (3) elastic tissue; and (4) ligaments

46. The four membranes in the body are 1) serous membranes, 2) mucous membranes, 3) cutaneous membrane and 4) synovial membranes.

47. (1) smooth muscle tissue; (2) skeletal muscle tissue; and (3) cardiac muscle tissue

48. (1) neurons, which transmit electrical impulses in the form of changes in the transmembrane potential; and (2) neuroglia, which comprise several kinds of supporting cells and play a role in providing nutrients to neurons

Level 2 Reviewing Concepts

49. Stratified epithelium. The layers provide protection from abrasion and mechanical stresses.

50. Transitional epithelium lines areas where volume changes occur, such as the urinary bladder, the renal pelvis and the ureters.

51. Substances are moved over the epithelial surface by the synchronized beating of cilia, like a continuously moving escalator. The ciliated epithelium that lines the respiratory tract moves mucus, which has trapped dust particles from inhaled air, up from the lungs and toward the throat, where it can then be swallowed and eventually eliminated from the body.

52. Holocrine secretion destroys the gland cell. The entire cell becomes packed with secretory products and then bursts, releasing the secretion but killing the cell. Gland cells must be replaced by the division of stem cells.

53. Exocrine secretions are secreted onto a surface or outward through a duct. Endocrine secretions are secreted by ductless glands into surrounding tissues. The secretions are called hormones, which usually diffuse into the bloodstream for distribution to other parts of the body.

54. Tight junctions block the passage of water or solutes between the cells. In the digestive system, these junctions keep enzymes, acids, and waste products from damaging delicate underlying tissues.

55. Intercellular connections called desmosomes are most abundant in superficial layers of skin. Desmosomes create links that lock cells together, causing dead skin cells to be shed in thick sheets.

56. The fluid connective tissues have a liquid, watery matrix. They differ from supporting connective tissues in that they have many soluble proteins in the matrix and they do not include insoluble fibers.

57. The extensive connections between cells formed by cell junctions, intercellular cement, and physical interlocking hold skin cells together and can deny access to chemicals or pathogens that cover their free surfaces. If the skin is damaged and the connections are broken, infection can easily occur.

58. Similarities: all have contractile proteins, skeletal and cardiac have striations, cardiac and smooth are involuntary; cardiac and smooth are uni-nucleate. Differences: Skeletal is voluntary; cardiac has intercalated discs; smooth is nonstriated; skeletal is multi-nucleate.

59. Adipocytes (fat cells) are metabolically active cells. Their lipids are continuously being broken down and replaced. During a weight-loss program, nutrients are scarce, causing the fat cells to decrease in size. The cells are not killed but merely reduced in size. Once the individual is off the weight-loss program, these cells increase in size and the lost weight will be regained in the same locations.

60. Hyaline cartilage is the most common type of cartilage, and consists of a matrix of closely packed collagen fibers enclosed within a perichondrium. Elastic cartilage has numerous elastic fibers, providing elasticity. Fibrocartilage has little ground substance and the matrix is dominated by densely packed bundles of collagen fibers.

61. Cartilage is avascular, because chondrocytes produce a chemical that discourages the formation of blood vessels. This property makes nutrient and oxygen delivery difficult. In adults, interstitial and appositional growth do not occur, so there is no division of chondrocytes or differentiation of the inner cells of the perichondrium into chondrocytes. Both the avascularity and the lack of division and differentiation contribute to the poor healing of cartilage.

62. Cutaneous membranes are thick, relatively waterproof, and usually dry.

63. Transudate is the fluid that forms between adjacent serous membranes, lubricating the surfaces.

64. Cardiac muscle fibers are incapable of dividing, because they do not contain satellite cells. Therefore, the regeneration of damaged or diseased cardiac muscle tissue is not possible.

65. Elastic ligaments are found along the spinal column, where they are important in stabilizing the position of the vertebrae. It is the elasticity of the elastic ligaments that allows the spine to bend and straighten easily.

Level 3 Critical Thinking and Clinical Applications

66. Because apocrine secretions are released by pinching off a portion of the secreting cell, you could test for the presence of cell membranes, specifically for the phospholipids in cell membranes. Merocrine secretions do not contain a portion of the secreting cell, so they would lack membrane constituents.

67. The respiratory passages are lined by ciliated pseudostratified columnar epithelium. This type of epithelium contains many goblet cells that produce mucus. The mucus traps debris and foreign material that is moved by the beating cilia to the pharynx to be swallowed. Chronic smoking initially paralyzes the cilia, resulting in a buildup of mucus in the airways. The hot air and particulates sear and burn off the cilia. The epithelium is also damaged by the heat and responds by producing even more mucus; the mucus cannot be moved due to the lack of cilia. The only way that the respiratory system can clear the debris and mucus is with forceful bursts of air: coughing.

68. Because animal intestine is modified for absorption, you would look for a slide that shows a single layer of epithelium lining the cavity. The cells would be cuboidal or columnar and would probably have microvilli on the surface to increase surface area. Because the esophagus receives undigested food, it would have a stratified epithelium consisting of squamous cells to protect it.

69. Step 1: Check for striations. (If striations are present, the choices are skeletal muscle or cardiac muscle. If striations are absent, the tissue is smooth muscle.) Step 2: Check for the presence of intercalated discs. (If these discs are present, the tissue is cardiac muscle. If they are absent, it is skeletal muscle.)

70. Periosteum of bone, synovial membrane of hip joint, bone, connective tissue of ligaments and tendons in the area, articular cartilage of joint.

CHAPTER 5

Page 159

1. This outer layer of skin is the stratum corneum.

2. The splinter is lodged in the stratum granulosum.

3. Fresh water is hypotonic with respect to skin cells, so water will move into the cells by osmosis, causing them to swell.

4. Sanding the tips of the fingers will not permanently remove fingerprints. The ridges of the fingerprints are formed in layers of the skin that are constantly regenerated, so these ridges will eventually reappear. The pattern of the ridges is determined by the arrangement of tissue in the dermis, which is not affected by sanding.

Page 163

1. When exposed to the ultraviolet radiation in sunlight or sunlamps, melanocytes in the epidermis and dermis synthesize the pigment melanin, darkening the color of the skin.

2. When skin is warm, blood is diverted to the superficial dermis for the purpose of eliminating heat. Because blood is red, it imparts a red cast to the skin.

3. The hormone cholecalciferol (vitamin D_3) is needed to form strong bones and teeth. The first step in this hormone's production involves the exposure of the skin to specific wavelengths of UV light. When the body surface is covered, UV light cannot penetrate to the blood in the skin to begin vitamin D_3 production, resulting in fragile bones.

Page 166

1. The capillaries and sensory neurons that supply the epidermis are located in the papillary layer of the dermis.

2. The presence of elastic fibers and the resilience of skin turgor allow the dermis to undergo repeated cycles of stretching and recoil.

Page 169

1. The contraction of the arrector pili muscles pulls the hair follicles erect, depressing the area at the base of the hair and making the surrounding skin appear higher. The result is known as "goose bumps."
2. Hair is a derivative of the epidermis, but the follicles are in the dermis. Where the epidermis and deep dermis are destroyed, new hair will not grow.

Page 171

1. Sebaceous glands produce a secretion called sebum. Sebum lubricates and protects the keratin of the hair shaft, lubricates and conditions the surrounding skin, and inhibits the growth of bacteria.
2. Apocrine sweat glands produce a secretion that contains several kinds of organic compounds. Some of these compounds have an odor, and others produce an odor when metabolized by skin bacteria. Deodorants are used to mask the odor of such secretions.
3. Apocrine sweat glands enlarge and increase their secretion production in response to the increase in sex hormones that occurs at puberty.

Page 176

1. The combination of fibrin clot, fibroblasts, and the extensive network of capillaries in tissue that is healing is called granulation tissue.
2. Skin can regenerate effectively even after considerable damage has occurred because stem cells persist in both the epithelial and connective tissue components of skin. When injury occurs, cells of the stratum germinativum replace epithelial cells while mesenchymal cells replace cells lost from the dermis.
3. As a person ages, the blood supply to the dermis decreases and merocrine sweat glands become less active. These changes make it more difficult for the elderly to cool themselves in hot weather.

Page 180: Level 1 Reviewing Facts and Terms

1. a   2. c   3. d   4. b   5. c   6. a   7. d   8. b   9. b   10. d
11. d   12. b   13. a   14. a
15. Epidermal cell division occurs in the stratum germinativum.
16. Keratin is formed from keratin fibers and keratohyalin. Keratin production occurs in keratinocytes of the stratum granulosum.
17. This smooth muscle causes hairs to stand erect when stimulated.
18. Epidermal growth factor promotes the divisions of germinal cells in the stratum germinativum and stratum spinosum. It also accelerates the production of keratin in differentiating epidermal cells and stimulates both epidermal development and epidermal repair after injury and synthetic activity and secretion by epithelial cells.
19. (1) papillary layer, which consists of loose connective tissue and contains capillaries and sensory neurons; (2) reticular layer, which consists of dense irregular connective tissue and bundles of collagen fibers. Both layers contain networks of blood vessels, lymphatic vessels, and nerve fibers.
20. (1) internal root sheath, which surrounds the root; (2) external root sheath, which extends from the skin surface to hair matrix; (3) glassy membrane, the thickened basal lamina
21. (1) bleeding; (2) scab formation; (3) granulation tissue formation; (4) scar

Level 2 Reviewing Concepts

22. Insensible perspiration is water loss via evaporation through the stratum corneum of the skin. Sensible perspiration is produced by active sweat glands.
23. Substances that are fat soluble pass through the permeability barrier easily, because the barrier is composed primarily of lipids surrounding the epidermal cells. Water-soluble drugs are hydrophobic and thus don't penetrate the permeability barrier easily.
24. A tan is a result of the synthesis of melanin in the skin. Melanin helps prevent skin damage by absorbing UV radiation before it reaches the deep layers of the epidermis and dermis. Within the epidermal cells, melanin concentrates around the outer wall of the nucleus, so it absorbs the UV light before it can damage nuclear DNA.
25. Lines of cleavage represent the orientation of the collagen and elastin fibers of the dermis, an orientation that resists normal stresses on the skin. Incisions along the lines of cleavage are more likely to remain closed, and thus will heal more quickly, than would incisions not along lines of cleavage.
26. Inflammation of the skin is painful because sensory receptors are abundant in the skin. Changes associated with swelling stimulate the sensory receptors and bare nerve endings, resulting in a painful sensation.
27. The subcutaneous layer is not highly vascular, and it does not contain major organs; thus the potential for tissue damage is reduced.
28. The cells that produce the nails can be affected by conditions that alter body metabolism; thus the nails hint at possible disease conditions.
29. Scabs temporarily restore epidermal integrity, restricting further entry of microorganisms. They also prevent the loss of fluids, maintaining internal fluid balance.
30. Creams and oils mimic the natural secretions of the sebaceous glands. Applying them to the skin assists the sebaceous glands in maintaining a continuous, flexible, and protective integument.

Level 3 Critical Thinking and Clinical Applications

31. The child probably has a fondness for vegetables that are high in carotene, such as sweet potatoes, squash, and carrots. It is not uncommon for parents to feed babies what they will eat best. If the child consumes large amounts of carotene, the yellow-orange pigment will be stored in the skin, producing a yellow-orange skin color.
32. Most elderly people have poor blood flow to the skin. Thus, temperature receptors in the skin do not sense as much warmth as when there is a rich blood supply. The sensory information is relayed to the brain, and the brain interprets the temperature as cool or cold.
33. (a) Ultraviolet radiation in sunlight converts a cholesterol-related steroid into vitamin D_3, or cholecalciferol. This compound is then converted to calcitriol, which is essential for normal calcium and phosphorus absorption by the small intestine. Calcium and phosphorus are necessary for normal bone maintenance and growth. (b) Milk is routinely fortified with cholecalciferol, normally identified as "vitamin D," which is easily absorbed by the intestines.
34. Sweating from merocrine glands is precisely regulated, and one influencing factor is emotional state. Presumably, a person who is lying is nervous and sweats noticeably; this sweating is detected by the lie detector machine.
35. The chemicals in hair dyes break the protective covering of the cortex allowing the dyes themselves to stain the medulla of the shaft. This is not permanent because the cortex remains damaged, allowing shampoo and UV rays from the sun to enter the medulla and affect the color. Also, the viable portion of the hair remains unaffected, so that when the shaft is replaced the color will be lost.

CHAPTER 6

Page 190

1. If the ratio of collagen to hydroxyapatite in a bone increased, the bone would become less strong (as well as more flexible).
2. The presence of concentric layers of bone around a central canal is indicative of an osteon. Osteons make up compact bone. Because the ends (epiphyses) of long bones are primarily cancellous (spongy) bone, this sample is most likely from the marrow cavity of the shaft (diaphysis) of a long bone.

3. Because osteoclasts break down or demineralize bone, the bone would have a reduced mineral content (less mass); as a result, it would also be weaker.

Page 194
1. During intramembranous ossification, fibrous connective tissue is replaced by bone.
2. In endochondral ossification, cells of the inner layer of the perichondrium differentiate into osteoblasts, and a cartilage model is gradually replaced by bone.
3. Long bones of the body, such as the femur, have an epiphyseal cartilage, a plate of cartilage that separates the epiphysis from the diaphysis as long as the bone is still growing lengthwise. An X ray would indicate whether the epiphyseal cartilage is still present. If it is, growth is still occurring; if it is not, the bone has reached its adult length.

Page 198
1. The larger arm muscles of the weight lifter would apply more mechanical stress to the bones of the upper limbs. In response to that stress, the bones would grow thicker. For similar reasons, we would expect the jogger to have heavier thigh bones.
2. Growth continues throughout childhood. At puberty, a growth spurt occurs and is followed by the closure of the epiphyseal cartilages. The later puberty begins, the taller the child will be when the growth spurt begins, so the taller the individual will be when growth is completed.
3. Increased levels of growth hormone prior to puberty will result in excessive bone growth, making the individual taller.

Page 199
1. The bones of children who have rickets are poorly mineralized and as a result are quite flexible. Under the weight of the body, the leg bones bend. The instability makes walking difficult and can lead to other problems of the legs and feet.
2. Parathyroid hormone (PTH) stimulates osteoclasts to release calcium ions from bone. Increased PTH secretion would result in an increase in the level of calcium ions in the blood.
3. Calcitonin lowers blood calcium levels by (1) inhibiting osteoclast activity and (2) increasing the rate of calcium excretion at the kidneys.

Page 204
1. An external callus forms early in the healing process, when cells from the endosteum and periosteum migrate to the area of the fracture. These cells form an enlarged collar (external callus) that circles the bone in the area of the fracture.
2. The sex hormones known as estrogens play an important role in moving calcium into bones. After menopause, the level of these hormones decreases dramatically; as a result, older women have difficulty replacing the calcium in bones that is being lost due to normal aging. In males, the level of sex hormones (androgens) does not decrease until much later in life.

Page 206: Level 1 Reviewing Facts and Terms
1. b **2.** a **3.** a **4.** c **5.** d **6.** b **7.** b **8.** c **9.** a **10.** d
11. (1) support; (2) storage of minerals and lipids; (3) blood cell production; (4) protection; and (5) leverage
12. (1) osteocytes; (2) osteoblasts; (3) osteoclasts; and (4) osteoprogenitor cells
13. Osteoblasts are responsible for the production of new bone (osteogenesis). Osteoclasts contain lysosomes that secrete enzymes and acids that dissolve the bony matrix and release the stored minerals of calcium and phosphate (osteolysis).
14. (1) diaphysis (shaft); (2) epiphysis; (3) epiphyseal cartilages/line; (4) articular cartilage; (5) medullary canal; (6) periosteum; (7) endosteum

15. In intramembranous ossification, bone develops from mesenchyme or fibrous connective tissue. In endochondral ossification, bone develops from a cartilage model.
16. (1) organic = collagen; (2) inorganic = hydroxyapatite crystals
17. (a) calcium salts and phosphate salts and vitamins A, C, and D_3; (b) calcitriol, growth hormone, thyroxine, estrogens (in females) or androgens (in males,) calcitonin, and parathyroid hormone
18. (1) the bones; (2) the intestinal tract; and (3) the kidneys
19. It stimulates osteoclast activity, increases the rate of intestinal absorption, and decreases the rate of excretion of calcium ions.
20. It inhibits osteoclast activity, decreases the rate of calcium absorption, and increases the rate of excretion of calcium ions.

Level 2 Reviewing Concepts
21. Nutrients reach the osteocytes by diffusion along canaliculi that open onto the surface of the trabeculae.
22. The osteons are parallel to the long axis of the shaft, which does not bend when forces are applied to either end. Stresses or impacts to the side of the shaft can lead to a fracture.
23. Inactivity in unstressed bones leads to the removal of calcium salts. Up to one-third of the bone mass can be lost in this manner, causing the bones to become thin and brittle.
24. The digestive and urinary systems (kidneys) play important roles in providing the calcium and phosphate minerals needed for bone growth. In return, the skeleton provides protection and acts as a reserve of calcium, phosphate, and other minerals that can compensate for changes in the dietary supplies of these ions.
25. The chondrocytes of the epiphyseal cartilage enlarge and divide, increasing the thickness of the cartilage. On the shaft side, the chondrocytes become ossified, "chasing" the expanding epiphyseal cartilage away from the shaft.
26. There are many long bones in the hand, each of which has an epiphyseal cartilage. Measuring the width of these cartilages will provide clues to the hormonal control of growth in the child.
27. The thyroid gland secretes calcitonin, which stimulates calcium excretion and inhibits calcium absorption and osteoclast activity. Without calcitonin, blood calcium levels might become abnormally high.
28. When a bone fracture is repaired, the bone tends to be stronger and thicker than normal at the fracture site.
29. appositional growth.
30. Bone markings give clues as to the size, age, gender, and general appearance of an individual.
31. Elevations and projections on bones are sites of attachment for tendons and ligaments.

Level 3 Critical Thinking and Clinical Applications
32. The fracture might have damaged the epiphyseal cartilage in Sally's right leg. Even though the bone healed properly, the damaged leg did not produce as much cartilage as did the undamaged leg. The result would be a shorter bone on the side of the injury.
33. Ossification can begin in the fetus as early as 6 weeks. By 8–10 weeks, the process is well under way and requires large amounts of calcium and phosphate to make the new bone. If the mother's diet does not supply sufficient amounts of calcium for the fetus, then hormonal regulation will cause minerals to be taken from the mother's bones for use by her fetus. Because Sherry has a diet deficient in calcium, the demineralization of her bone is supplying the calcium needs of her fetus. As a result, Sherry has weak bones that fractured easily.
34. A person deficient in vitamin D_3 would not be able to absorb calcium effectively from the digestive tract, leading to a shortage of calcium in the blood. To maintain homeostasis, the decrease in blood calcium would trigger the release of PTH. The PTH, in turn, would

stimulate osteoclasts to release enough calcium from the bone to maintain proper calcium levels in the blood. Levels of calcitonin would probably decrease, because this hormone lowers blood calcium levels and would aggravate the situation caused by the vitamin D$_3$ deficiency.

35. In kidney failure, the kidneys are not able to retrieve important ions that have been filtered from the blood, including calcium. As a result, calcium would be lost from the body and blood levels of calcium would drop. The reduced levels of calcium in the blood would trigger higher levels of PTH, which, in turn, would trigger an increase in the release of minerals from bone by the osteoclasts. Over time, this mineral release would result in demineralized bones similar to those of an individual with osteoporosis.

36. The matrix of bone will absorb traces of minerals from the diet. These minerals can be identified hundreds of years later. A diet rich in calcium and vitamin B will produce denser bones than will a diet lacking these. Cultural practices such as binding of appendages, or wrapping of infant heads will manifest in misshapen bones. Heavy muscular activity will result in larger bone markings, indicating an athletic lifestyle.

CHAPTER 7

Page 224
1. The foramen magnum is located in the base of the occipital bone.
2. Tomás has fractured his right parietal bone.
3. The sphenoid bone contains the sella turcica, which in turn contains the pituitary gland.

Page 231
1. The adult vertebral column has fewer vertebrae because the five sacral vertebrae fuse to form a single sacrum and the four coccygeal vertebrae fuse to form a single coccyx.
2. The secondary curves of the spine allow us to balance our body weight on our lower limbs with minimal muscular effort. Without the secondary curves, we would not be able to stand upright for extended periods.
3. When you run your finger along a person's spine, you can feel the spinous processes of the vertebrae.

Page 236
1. The dens is located on the axis, or second cervical vertebra, which is in the neck.
2. The presence of transverse foramina indicate that this vertebra is a cervical vertebra.
3. The lumbar vertebrae must support a great deal more weight than do vertebrae that are more superior in the spinal column. The large vertebral bodies allow the weight to be distributed over a larger area.

Page 239
1. True ribs are attached directly to the sternum by their own costal cartilage. False ribs either do not attach to the sternum (as in the floating ribs) or attach by means of a common costal cartilage (as in the vertebrochondral ribs).
2. Improper compression of the chest during CPR can—and commonly does—result in a fracture of the sternum or ribs.
3. Vertebrosternal ribs, or true ribs, attach directly to the sternum whereas vertebrochondral ribs fuse together and merge with the costal cartilages of ribs 8–10 and and then with the cartilages of rib pair 7 before they reach the sternum.

Page 241: Level 1 Reviewing Facts and Terms
1. i **2.** e **3.** l **4.** a **5.** g **6.** j **7.** b **8.** k **9.** d **10.** h **11.** c **12.** f **13.** a **14.** b **15.** d **16.** d **17.** a **18.** d **19.** a **20.** a **21.** d

22. (1) lacrimal bone; (2) nasal bone; (3) maxillary bone; (4) zygomatic bone; and (5) mandible
23. (1) sphenoid; (2) frontal bone; (3) ethmoid; (4) lacrimal bone; (5) maxillary bone; (6) palatine bone; (7) zygomatic bone
24. The vomer forms a bony nasal septum that separates the right and left nasal cavities and separates the external nares anteriorly in the fleshy portion of the nose.
25. the frontal bone, sphenoid, ethmoid, palatine bones, and maxillary bones.

Level 2 Reviewing Concepts
26. d
27. The petrous part of the temporal bone encloses the structures of the inner ear. The middle ear is located in the tympanic cavity within the petrous part. The external acoustic canal ends at the tympanic membrane, which leads to the inner ear. Mastoid air cells within the mastoid process are connected to the tympanic cavity.
28. The ethmoid forms the superior surface of the nasal cavity. The olfactory foramina within the cribriform plate of the ethmoid allows neurons associated with the sense of smell to extend into the nasal cavity.
29. Fontanels are fibrous connections between cranial bones. They permit distortion of the skull without damaging the structure during delivery, helping to ease the child through the birth canal.
30. Shortening of the face and balancing of the weight of the head over the cervical vertebrae.
31. Because it is accessible, the body of the sternum is often used as a site for taking red bone marrow samples. The xiphoid process is used as a landmark in CPR; this part of the sternum is easily broken, because its only attachment is superior, to the sternum. A broken xiphoid process can cause injury to nearby organs.
32. The lumbar vertebrae have massive bodies and carry a lot of weight; these factors contribute to the rupturing of an intervertebral disc. The cervical vertebrae are more delicate and have small bodies; these factors increase the possibility of dislocations and fractures in the cervical vertebrae.
33. The upper seven ribs attach directly to the sternum by separate cartilagenous extensions. The false ribs do not attach directly to the sternum.
34. The ribs raise and lower to increase and decrease the volume of the chest cavity. They move similar to the handle of a bucket. When they rise, the chest cavity expands and we breathe in. When the ribs are lowered to their original position, the volume of the chest cavity decreases and we breathe out.
35. Keeping your back straight keeps the weight aligned along the axis of your vertebral column, where it can be transferred to your lower limbs. Bending your back would strain the muscles and ligaments of the back, increasing the risk of injury.

Level 3 Critical Thinking and Clinical Applications
36. The virus could have been inhaled through the nose and passed into the cranium by way of the olfactory foramina in the cribriform plate of the ethmoid.
37. Joe has lordosis. The extra weight that he is carrying in his abdomen shifts his center of gravity anteriorly. To compensate for this shift, the muscles of the spine exaggerate the lumbar curvature to prevent Joe from falling forward. The situation is similar to that occurring in women in late pregnancy.
38. The large bones of a child's cranium are not yet fused; they are connected by fontanels, areas of connective tissue. By examining the bones, the archaeologist could readily see if sutures had formed. By knowing approximately how long it takes for the various fontanels to close and by determining their sizes, she could estimate the ages of the bones.

CHAPTER 8

Page 246

1. The clavicle attaches the scapula to the sternum and thus restricts the scapula's range of movement. When the clavicle is broken, the scapula has a greater range of movement and is less stable.
2. The head of the humerus articulates with the scapula at the glenoid cavity.

Page 249

1. The two rounded projections on either side of the elbow are the lateral and medial epicondyles of the humerus.
2. The radius is lateral when the forearm is pronated, as in the anatomical position.
3. The first distal phalanx is located at the tip of the thumb; Bill's thumb is broken.

Page 253

1. The three bones that make up the os coxae are the ilium, ischium, and pubis.
2. The pelvis of females is generally smoother and lighter than that of males and has less-prominent markings. The pelvic outlet is enlarged, and there is less curvature on the sacrum and coccyx. The pelvic inlet is wider and more circular. The pelvis as a whole is relatively broad and low. The ilia project farther laterally. The inferior angle between the pubic bones is greater than 100°, as opposed to 90° or less for the pelvis of males. These differences adapt the pelvis of females for supporting the weight of the developing fetus and enable the newborn to pass through the pelvic outlet during delivery.
3. When you are seated, your body weight is borne by the ischial tuberosities of the pelvis.

Page 257

1. Although the fibula is not part of the knee joint and does not bear weight, it is an important point of attachment for many leg muscles. When the fibula is fractured, these muscles cannot function properly to move the leg, and walking is difficult and painful. The fibula also helps stabilize the ankle joint.
2. Joey has most likely fractured the calcaneus (heel bone).
3. The talus transmits the weight of the body from the tibia toward the toes.

Page 260: Level 1 Reviewing Facts and Terms

1. b **2.** b **3.** a **4.** a **5.** d **6.** b **7.** d **8.** b **9.** d **10.** d
11. b **12.** a **13.** c **14.** d
15. the clavicle (collar bone) and the scapula (shoulder blade)
16. The appendicular skeleton includes the bones of the pectoral and pelvic girdle and the associated limbs.
17. the humerus (arm) and the ulna and radius (forearm)
18. extension and flexion
19. the carpal bones and the flexor retinaculum prevent swelling of the connective tissues held between them.
20. femur (thigh), tibia and fibula (leg), tarsal bones (ankle), metatarsal bones (foot), and phalanges (toes)
21. ischium, ilium, and pubis
22. (1) talus; (2) calcaneus; (3) cuboid bone; (4) navicular bone; and (5–7) three cuneiform bones
23. The pollex is the thumb, whereas the hallux is the great toe.

Level 2 Reviewing Concepts

24. The clavicles are small and fragile, so fractures are quite common. The position of the clavicles also makes them vulnerable to injury and damage.
25. The pelvic girdle consists of the ossa coxae. The pelvis is a composite structure; it consists of the ossa coxae of the appendicular skeleton and the sacrum and coccyx of the axial skeleton.

26. The superior limit of the true pelvis is a line that extends from either side of the base of the sacrum, along the arcuate line and pectineal line to the pubis symphysis. The false pelvis includes the bladelike portions of the ilium above the pelvic brim.
27. The slender fibula parallels the tibia of the lower leg and provides an important site for muscle attachment. It does not help in transferring weight to the ankle and foot, however, because it is excluded from the knee joint.
28. The arches absorb shocks as weight distribution shifts during movements. The longitudinal arch absorbs most of the shock of steps, while the transverse arch distributes the weight evenly.
29. The clavicles are small and fragile, so they are easy to break. Once this part of the pectoral girdle is broken, the assailant would no longer have efficient use of the arms.
30. The pelvic girdle transmits the weight of the body to the legs. The pectoral girdle doesn't support the body's weight and so its bones needn't be as massive as those of the pelvic girdle.

Level 3 Critical Thinking and Clinical Applications

31. You could feel the inferior iliac notch, which is at the same level as the head of the femur. You could feel the medial malleolus of the tibia at the ankle.
32. Fred probably dislocated his shoulder, which is quite a common injury due to the weak nature of the glenohumeral joint.
33. Cindy's pelvis might not be broad enough, or the angle inferior to the pubic symphysis might not be large enough, for Cindy to deliver the baby by natural childbirth.
34. Metacarpal bones, carpal bones, radius and ulna, humerus, clavicle, perhaps scapula and sternum. The stresses will follow this pathway, with the largest possibility of break occurring at the clavicle. This is the least supported of the bones, and the stress is not directed along the long axis of this bone as it has been along the previous bones.

CHAPTER 9

Page 267

1. All these joints (other than synostoses) consist of bony regions separated by fibrous or cartilaginous connective tissue.
2. Originally, each of these joints is a syndesmosis. As the bones interlock, they form sutural joints.
3. Articular cartilages lack a blood supply; they rely on synovial fluid to supply nutrients and eliminate wastes. Impairing the circulation of synovial fluid would have the same effect as impairing a tissue's blood supply. Nutrients would not be delivered to meet the tissue's needs, and wastes would accumulate. Damage to and ultimately the death of the cells in the tissue would result.

Page 272

1. When you do jumping jacks and move your lower limbs away from the midline of the body, the movement is abduction. When you bring the lower limbs back together, the movement is adduction.
2. Flexion and extension are the movements associated with hinge joints.

Page 275

1. Intervertebral discs are not found between the first and second cervical vertebrae, between sacral vertebrae in the sacrum, or between coccygeal vertebrae in the coccyx. An intervertebral disc between the first and second cervical vertebrae would prohibit rotation. The vertebrae in the sacrum and coccyx are fused.
2. (a) flexion; (b) lateral flexion; (c) rotation

Page 278

1. Ligaments and muscles provide most of the stability for the shoulder joint.

2. Because the subscapular bursa is located in the shoulder joint, the tennis player would be more likely to develop inflammation of this structure (bursitis). The condition is associated with repetitive motion that occurs at the shoulder, such as swinging a tennis racket. The jogger would be more at risk for injuries to the knee joint.

3. A shoulder separation is an injury involving partial or complete dislocation of the acromioclavicular joint.

4. Terry has most likely damaged his annular ligament.

Page 283

1. The iliofemoral, pubofemoral, and ischiofemoral ligaments are at the hip joint.

2. Damage to the menisci of the knee joint would result in a decrease in the joint's stability. The individual would have a hard time locking the knee in place while standing and would have to use muscle contractions to stabilize the joint. If the person had to stand for a long period, the muscles would fatigue and the knee would "give out." We would also expect the individual to feel pain.

3. "Clergyman's knee" is a bursitis that commonly affects members of the clergy, who spend a great deal of time kneeling. The work of carpet layers and roofers necessitates kneeling and sliding along on their knees, causing a similar inflammation of the bursae in the knee joint.

Page 288: Level 1 Reviewing Facts and Terms

1. d **2.** b **3.** c **4.** c **5.** d **6.** a **7.** b **8.** c **9.** b **10.** b
11. d **12.** c **13.** c **14.** a **15.** b **16.** a **17.** d

18. linear, motion (gliding), angular motion (flexion, extension, adduction, abduction, circumduction), rotation, and a series of special movements (eversion, inversion, protraction, retraction, elevation, depression, etc.)

Level 2 Reviewing Concepts

19. d **20.** c

21. Menisci may subdivide a synovial cavity, channel the flow of synovial fluid, and allow variations in the shape of the articular surfaces.

22. A joint cannot be both highly mobile and very strong. The shoulder (glenohumeral) joint demonstrates the principle that strength and stability must be sacrificed to obtain mobility; it permits the greatest range of motion of any joint in the body. The relatively loose, oversized, poorly reinforced articular capsule lessens the stability of the joint but allows a great range of motion.

23. Articular cartilages do not have perichondrium, and their matrix contains more water than do other cartilages.

24. The nucleus pulposus does not extrude in a slipped disc. In a herniated disc, the nucleus pulposus breaks through the annulus fibrosus.

25. Height decreases during adulthood in part as a result of osteoporosis in the vertebrae and in part as a result of the decline in water content of the nucleus pulposus region of the intervertebral discs.

26. Contraction of the triceps brachii muscle causes extension at the elbow joint.

27. (1) gliding, clavicle and sternum; (2) hinge, elbow; (3) pivot, atlas and axis; (4) ellipsoid, radius and carpal bones; (5) saddle, thumb; (6) ball and socket, shoulder.

Level 3 Critical Thinking and Clinical Applications

28. Dave sprained his ankle. This condition occurs when ligaments are stretched to the point at which some of the collagen fibers are torn. Stretched ligaments in joints can cause the release of synovial fluid, which results in swelling and pain.

29. Bob has probably damaged his menisci, which act as cushions to absorb the force of movements such as jumping. When Bob landed off-balance on his right knee, the excessive force could have pushed the menisci out of place. In turn, the synovial membranes and/or the ligaments were damaged, resulting in swelling.

30. Her pain is probably caused by bursitis, which can result from repetitive motion, infection, trauma, chemical irritation, or pressure over the joint. Given the location of the pain, her pain probably results from the repetitive motion of practicing pitches.

CHAPTER 10

Page 300

1. Because tendons attach muscles to bones, severing the tendon would disconnect the muscle from the bone. If the muscle were to contract, nothing would happen.

2. Skeletal muscle appears striated when viewed through a microscope because the Z lines and thick filaments of the myofibrils within the muscle fibers are aligned.

3. You would expect the greatest concentration of calcium ions to be in the cisternae of the sarcoplasmic reticulum in resting skeletal muscle.

Page 306

1. A muscle's ability to contract depends on the formation of cross-bridges between the myosin and actin myofilaments in the muscle. A drug that interferes with cross-bridge formation would prevent muscle contraction.

2. Because the amount of cross-bridge formation is proportional to the amount of available calcium ions, increased permeability of the sarcolemma to Ca^{2+} would lead to an increased intracellular concentration of Ca^{2+} and a greater degree of contraction. In addition, because the amount of calcium ions in the sarcoplasm must be reduced for relaxation to occur, an increase in the permeability of the sarcolemma to Ca^{2+} could make the muscle unable to relax completely.

3. Without acetylcholinesterase, the motor end plate would be continuously stimulated by acetylcholine; the muscle would lock in a state of contraction.

Page 316

1. A muscle's ability to contract depends on the formation of cross-bridges between the myosin and actin myofilaments in the muscle. In a muscle that is overstretched, the myofilaments would overlap very little. Very few cross-bridges would form, and the contraction would be weak. If the myofilaments did not overlap at all, then no cross-bridges would form and the muscle could not contract.

2. During treppe, there is not enough time between successive contractions to reabsorb all the calcium ions that were released during the prior contraction event. As a result, calcium ions accumulate in the sarcoplasm at higher than normal levels, allowing more cross-bridges to form and tension to increase.

3. Yes. Contraction occurs as thick and thin filaments interact. The entire muscle can shorten (isotonic, concentric), elongate (isotonic, eccentric), or remain the same length (isometric), depending on the relationship between the resistance and the tension produced by actin and myosin interactions.

Page 323

1. A sprinter requires large amounts of energy for a short burst of activity. To supply this demand for energy, the sprinter's muscles switch to anaerobic metabolism. Anaerobic metabolism is not as efficient in producing energy as aerobic metabolism is, and the process also produces acidic waste products; this combination contributes to fatigue. Conversely, marathon runners derive most of their energy from aerobic metabolism, which is more efficient and does not produce the amount of waste products that anaerobic metabolism does.

2. Activities that require short periods of strenuous activity produce a greater oxygen debt, because this type of activity relies heavily on energy production by anaerobic metabolism. Because lifting weights is more strenuous over the short term than swimming is, weight lifting would likely produce a greater oxygen debt than would swimming laps, which is an aerobic activity.

3. Individuals who excel at endurance activities have a higher than normal percentage of slow fibers. Those fibers are physiologically better adapted to this type of activity than are fast fibers, which are less vascular and fatigue faster.

Page 327

1. Cardiac muscle cells are joined by gap junctions, which allow ions and small molecules to flow directly from one cell to another. This type of junction allows for action potentials that are generated in one cell to spread rapidly to adjacent cells. Thus all the cells contract simultaneously as if they were a single unit (a syncytium).

2. Cardiac and smooth muscle contractions are more affected by changes in the concentration of Ca^{2+} in the extracellular fluid than are skeletal muscle contractions because in cardiac and smooth muscles, most of the calcium ions that trigger a contraction come from the extracellular fluid. In skeletal muscle, most of the calcium ions come from the sarcoplasmic reticulum.

3. The actin and myosin filaments of smooth muscle are not as rigidly organized as those of skeletal muscle. This loose organization allows smooth muscle to contract over a wider range of resting lengths.

Page 330: Level 1 Reviewing Facts and Terms

1. d   **2.** c   **3.** b   **4.** c   **5.** d   **6.** d   **7.** c   **8.** d

9. (1) skeletal muscle; (2) cardiac muscle; and (3) smooth muscle

10. (1) producing skeletal movement; (2) maintaining posture and body position; (3) supporting soft tissue; (4) guarding entrances and exits; and (5) maintaining body temperature

11. (1) epimysium: surrounds entire muscle; (2) perimysium: surrounds muscle bundles (fascicles); and (3) endomysium: surrounds skeletal muscle fibers

12. all the muscle fibers controlled by a single motor neuron

13. transverse (T) tubules

14. tropomyosin and troponin

15. (1) exposure of active sites; (2) attachment of cross-bridges; (3) pivoting of myosin head (power stroke); (4) detachment of cross-bridges; and (5) activation of myosin heads (cocking)

16. (1) the frequency of motor unit stimulation and (2) the number of motor units involved

17. The neurotransmitter acetylcholine is responsible for muscle contraction. It is broken down and removed from the synaptic space by acetylcholinesterase.

18. ATP, creatine phosphate, and glycogen

19. (1) aerobic metabolism and (2) glycolysis

20. Hypertrophy is an increase in the size of the tissue without an increase in number of cells.

21. calmodulin

Level 2 Reviewing Concepts

22. d   **23.** b   **24.** d   **25.** a

26. Acetylcholine released by the motor neuron at the neuromuscular junction changes the permeability of the cell membrane at the motor end plate. The permeability change allows the influx of positive charge, which triggers an action potential. The action potential spreads across the entire surface of the muscle fiber and into the interior via T tubules. The cytoplasmic concentration of calcium ions (released from the sarcoplasmic reticulum) increases, triggering the start of a contraction. The contraction ends when AChE removes the ACh from the synaptic cleft and motor end plate.

27. Fatigue at the level of the individual muscle fiber is defined as the period when it can no longer contract despite continued neural stimulation. It may result from the drop in pH that accompanies a build-up of lactic acid.

28. In an initial latent period (after the stimulus arrives and before the tension begins to increase), the action potential in the muscle is generated and triggers the release of calcium ions from the SR. In the contraction phase, the calcium binds to troponin (cross-bridges form) and the tension begins to increase. In the relaxation phase, tension drops because cross-bridges have detached and because calcium levels have fallen; the active sites are once again covered by the troponin–tropomyosin complex.

29. (1) O_2 for aerobic respiration is consumed by liver cells, which need to make a great deal of ATP to convert lactic acid to glucose. (2) O_2 for aerobic respiration is consumed by skeletal muscle fibers as they restore ATP, creatine phosphate, and glycogen concentrations to their former levels. (3) The normal O_2 concentration in blood and peripheral tissues is replenished.

30. Cardiac muscle fibers are mechanically, chemically, and electrically connected to one another, causing the entire tissue to resemble a single, enormous muscle fiber that performs as a functional syncytium.

31. Cardiac muscle tissue has the property of automaticity. The contractions are timed by pacemaker cells, specialized cardiac muscle fibers.

32. At the motor end plates of neuromuscular junctions, the blocking of the binding process by atracurium would inhibit the ability of the muscle to contract. Thus the muscle would remain relaxed. Atracurium could be useful if administered before surgery.

33. They will atrophy. Even a temporary reduction in muscle use, such as when you wear a cast, can cause some degree of atrophy. Physical therapy can offset this effect in cases of temporary inactivity.

34. In rigor mortis, the membranes of the dead cells are no longer selectively permeable; the SR is no longer able to retain calcium ions. As calcium ions enter the sarcoplasm, a sustained contraction develops. The muscles lock in the contracted position, making the body extremely stiff. Contraction persists because the dead muscle cells can no longer make ATP, which is necessary for cross-bridge detachment from the active sites. Rigor mortis begins a few hours after death and lasts 15–25 hours, until the lysosomal enzymes released by autolysis break down the myofilaments.

35. Visceral smooth muscle fibers are interconnected by gap junctions. When one fiber fires an action potential, it is conducted quickly to neighboring cells. The action potential in uninnervated smooth muscle fibers may be the result of exposure to chemicals, hormones, oxygen, carbon dioxide, stretching, or irritation.

36. Powerful, gross movements; each time the motor unit is activated, all the fibers will contract maximally. More-powerful contractions result from the activity of large motor units than from the activity of small motor units. Fine movements require numerous small motor units that can be individually activated.

Level 3 Critical Thinking and Clinical Applications

37. Because organophosphates block the action of acetylcholinesterase, the acetylcholine released into the synaptic cleft would not be removed. It would continue to stimulate the motor end plate, causing a state of persistent contraction (spastic paralysis). If the muscles of respiration were affected (which is likely), Ivan would die of suffocation. Prior to death, the most obvious symptoms would be uncontrolled tetanic contractions of the skeletal muscles.

38. This is a case of competitive inhibition; you would want to increase the level of acetylcholine at the synapse. You would want to use the drug that inhibits acetylcholinesterase, allowing the acetylcholine concentration at the synapse to increase. In turn, the likelihood that the acetylcholine would bind its receptor and lead to muscle stimulation would increase, thus alleviating the symptoms. Blocking acetylcholine would aggravate the symptoms and make the condition worse.

39. If muscle damage had occurred, the doctor would find enzymes and other molecules that normally are present only inside the skeletal muscle cells. These molecules would be in Mary's blood if the cell membranes were broken, allowing the cell contents to escape. Particularly significant would be enzymes associated with energy metabolism, such as CPK.

CHAPTER 11

Page 336

1. A pennate muscle contains more muscle fibers than does a parallel muscle of the same size. A muscle that has more muscle fibers has more myofibrils and sarcomeres. As a result, the contraction of a pennate muscle generates more tension than would the contraction of a parallel muscle of the same size.

2. The opening between the stomach and the small intestine would be guarded by a circular muscle, or sphincter. The concentric circles of muscle fibers found in sphincters are ideally suited for opening and closing openings and for acting as valves in the body.

3. The joint between the occipital bone and the first cervical vertebra is part of a first-class lever system. The joint between the two bones, the fulcrum, lies between the skull, which provides the resistance, and the neck muscles, which provide the applied force.

Page 339

1. The origin of a muscle is the end that remains stationary during an action. Because the gracilis muscle moves the tibia, the origin of this muscle must be on the pelvis (pubis and ischium).

2. Muscles A and B are antagonists, because they perform opposite actions.

3. The name *flexor carpi radialis longus* tells you that this muscle is a long muscle that lies next to the radius and flexes the wrist.

Page 348

1. Contraction of the masseter muscle elevates the mandible; relaxation of this muscle depresses the mandible. You would probably be chewing something.

2. You would expect the buccinator muscle, which forms the mouth for blowing, to be well developed in a trumpet player.

3. Swallowing involves contractions of the palatal muscles, which elevate the soft palate as well as portions of the superior pharyngeal wall. Elevation of the superior portion of the pharynx enlarges the opening to the auditory tube, permitting airflow to the middle ear and the inside of the eardrum. Making this opening larger by swallowing facilitates airflow into or out of the middle ear cavity.

Page 356

1. Damage to the external intercostal muscles would interfere with breathing.

2. A blow to the rectus abdominis muscle would cause that muscle to contract forcefully, resulting in flexion of the vertebral column. In other words, you would "double over."

3. The sore muscles are most likely the erector spinae muscles, especially the longissimus and the iliocostalis muscles of the lumbar region. These muscles would have to contract harder to counterbalance the increased anterior weight you would bear when carrying heavy boxes.

Page 365

1. When you shrug your shoulders, you are contracting your levator scapulae muscles.

2. The rotator cuff muscles include the supraspinatus, infraspinatus, subscapularis, and teres minor muscles. The tendons of these muscles help enclose and stabilize the shoulder joint.

3. Injury to the flexor carpi ulnaris muscle would impair the ability to perform flexion and adduction at the wrist.

Page 376

1. Injury to the obturator muscle would impair your ability to perform lateral rotation at the hip.

2. "Hamstring" refers to a group of muscles that collectively flex the knee. These muscles are the biceps femoris, semimembranosus, and semitendinosus muscles.

3. The calcaneal tendon attaches the soleus and gastrocnemius muscles to the calcaneus (heel bone). When these muscles contract, they produce extension (plantar flexion) at the ankle. A torn calcaneal tendon would make ankle extension difficult.

Page 381: Level 1 Reviewing Facts and Terms

1. b **2.** d **3.** d **4.** d **5.** c **6.** d **7.** a **8.** b **9.** a **10.** c **11.** b **12.** b **13.** c **14.** a **15.** b **16.** d **17.** a

18. (1) the anatomical arrangement of the muscle fibers; and (2) the type of muscle attachment to the skeleton

19. (1) parallel; (2) convergent; (3) pennate; and (4) circular

20. An aponeurosis is a collagenous sheet connecting two muscles. The epicraneal aponeurosis and the linea alba are examples.

21. (1) Muscles of the head and neck, (2) muscles of the vertebral column, (3) oblique and rectus muscles, and (4) muscles of the pelvic floor

22. The masseter, the temporalis and the pterygoid muscles are all used in mastication

23. (1) supporting the organs of the pelvic cavity; (2) flexion at joints of the sacrum and coccyx; and (3) controlling movement of materials through the urethra and anus

24. (1) supraspinatus; (2) infraspinatus; (3) subscapularis; and (4) teres minor muscles

25. (1) muscles that move the thigh, (2) muscles that move the leg, and (3) muscles that move the foot and toes

Level 2 Reviewing Concepts

26. b **27.** b **28.** b

29. Levers differ in the position of their fulcrum in relation to the applied force and the resistance. A first class lever has the fulcrum positioned between the applied force and the resistance. Second class levers have the fulcrum at one end, adjacent to the resistance, with the applied force at the other end. Third class levers have the fulcrum at one end, the resistance at the other, and the applied force in the middle.

30. Agonist, antagonist or synergist. Agonists are prime movers. Antagonists oppose the prime mover, keeping the movements smooth. Synergists provide additional pull or help stabilize the origin.

31. The axial musculature positions the head and vertebral column, moves the rib cage and assists in breathing, and supports the abdominopelvic viscera. The appendicular muscles stabilize or move components of the appendicular (and/or axial) skeleton.

32. The vertebral column does not need a massive series of flexors, because many of the large trunk muscles flex the vertebral column when they contract. In addition, most of the body weight lies anterior to the vertebral column and gravity tends to flex the intervertebral joints.

33. cervical spinal nerves and intercostal nerves.

34. In a convergent muscle, the direction of pull can be changed by stimulating only one group of muscle cells at any one time. When all the fibers contract at once, they do not pull as hard on the tendon as would a parallel muscle of the same size. The reason is that the muscle fibers on opposite sides of the tendon are pulling in different directions rather than working together.

35. A pennate muscle contains more muscle fibers than does a parallel muscle of the same size. A muscle that has more muscle fibers also has more myofibrils and sarcomeres, resulting in a contraction that generates more tension.

36. Lifting heavy objects becomes easier as the elbow approaches a 90° angle. As you decrease the angle at or near full flexion, tension production declines, so movement becomes more difficult.

37. The muscles of mastication will be affected; therefore, chewing will be inhibited.

38. flexion at the knee and extension at the hip

39. The gluteus medius muscle, the deltoid muscle or the vastus lateralis muscle. These are all bulky muscles with few blood vessels or nerves.

Level 3 Critical Thinking and Clinical Applications

40. Contraction of the frontalis muscle would wrinkle Mary's brow; contraction of the procerus muscle would flare her nostrils. Contraction of the levator labii muscle on the right side would raise the right side of her lip, as in sneering. Mary is not happy to see Jill.

41. Jeff should do exercises that flex and twist the trunk. The muscles he should target are the rectus abdominis muscles, the external and internal oblique muscles, and the latissimus dorsi muscle. Sit ups and twisting exercises will effectively tone these areas. If he places a weight on his chest as he does sit ups, he will notice better results, because the rectus abdominis muscle would be working against a greater resistance.

42. As the child grasps the ice cream cone, he uses flexor muscles, principally the flexor digitorum superficialis, flexor digitorum profundus, and flexor pollicis longus muscles. Opening his mouth requires the contraction of the platysma and the relaxation of the masseter, temporalis, and pterygoid muscles. The licking action would involve the genioglossus muscle (to depress and protract the tongue) and the palatoglossus muscle (to elevate the tongue).

CHAPTER 12

Page 395

1. The afferent division of the PNS is composed of nerves that carry sensory information to the brain and spinal cord. Damage to this division would interfere with a person's ability to experience a variety of sensory stimuli.

2. Most sensory neurons of the PNS are unipolar. Thus these neurons are most likely sensory neurons.

3. Microglial cells are small phagocytic cells that occur in increased number in damaged and diseased areas of the CNS.

Page 411

1. The depolarization of the neuron cell membrane involves the opening of gated sodium channels and the rapid influx of sodium ions into the cell. If the sodium channels were blocked, a neuron would not be able to depolarize.

2. If the extracellular concentration of potassium ions were to decrease, more potassium ions would leave the cell; hence the electrical gradient across the membrane (transmembrane potential) would increase. This condition is called hyperpolarization.

3. Action potentials are propagated along myelinated axons by saltatory propagation at speeds much higher than those along unmyelinated axons. The axon with a propagation speed of 50 meters per second must be the myelinated axon.

Page 419

1. When an action potential reaches the presynaptic terminal of a cholinergic synapse, voltage-regulated calcium ion channels open and the influx of Ca^{2+} triggers the release of ACh into the synapse to stimulate the next neuron. If these calcium channels were blocked, the ACh would not be released and transmission across the synapse would cease.

2. Because of synaptic delay, the pathway with fewer neurons (three) will transmit impulses more rapidly.

Page 426: Level 1 Reviewing Facts and Terms

1. c **2.** a **3.** a **4.** d **5.** c **6.** c **7.** d **8.** c **9.** a **10.** b **11.** a **12.** b **13.** a **14.** b **15.** b **16.** a **17.** d **18.** c **19.** a **20.** **(a)** brain and spinal cord **(b)** all other nerve fibers, divided between the efferent division (which consists of the somatic nervous system and the autonomic nervous system) and the afferent division (which consists of receptors and sensory neurons)

21. (1) neurons: transmit nerve impulses; and (2) neuroglia: act as supporting cells

22. (1) satellite cells and (2) Schwann cells

23. (1) sensory neurons: transmit impulses from the PNS to the CNS; (2) motor neurons: transmit impulses from the CNS to peripheral effectors; and (3) interneurons: analyze sensory inputs and coordination of motor outputs

Level 2 Reviewing Concepts

24. b **25.** d **26.** c **27.** b **28.** a **29.** c **30.** a

31. Neurons lack centrioles and therefore cannot divide and replace themselves.

32. Collaterals enable a single neuron to innervate several other cells.

33. Axoplasmic transport is the movement of products that are synthesized in the cell body out to the synaptic knobs. Retrograde flow is the movement of materials toward the cell body.

34. The loss of positive ions is referred to as hyperpolarization. This increases the negativity of the resting membrane potential

35. Voltage-regulated channels open or close in response to changes in the transmembrane potential. Chemically regulated channels open or close when they bind specific extracellular chemicals. Mechanically regulated channels open or close in response to physical distortion of the membrane surface.

36. (1) The change in the transmembrane potential decreases with distance. (2) The graded potential spreads passively due to local currents. (3) The graded potential may involve either depolarization or hyperpolarization. (4) The stronger the stimulus, the greater the change in the transmembrane potential and the larger the area affected.

37. The all-or-none principle of action potentials states that if a depolarization event is sufficient to reach threshold, it will cause an action potential in the cell. This action potential will be of the same strength regardless of the degree of stimulation above threshold.

38. The membrane depolarizes to threshold. Next, voltage-regulated sodium channels are activated, and the membrane rapidly depolarizes. These sodium channels are then inactivated, and potassium channels are activated. Finally, normal permeability returns. The voltage-regulated sodium channels become activated once the repolarization is complete; the voltage-regulated potassium channels begin closing as the transmembrane potential reaches the normal resting potential.

39. In continuous conduction, which occurs in unmyelinated axons, an action potential appears to move across the membrane surface in a series of tiny steps. In saltatory conduction, which occurs in myelinated axons, only the nodes along the axon can respond to a depolarizing stimulus.

40. The larger the diameter of the neuron, the more quickly the action potential propagates.

41. Type A fibers are myelinated and carry action potentials very quickly (140 m/sec). Type B are also myelinated, but carry action potentials more slowly due to their smaller diameter. Type C fibers are extremely slow due to small diameter and lack of myelination.

42. Action potentials last longer in muscle fibers than in nerve fibers. Muscle fibers conduct action potentials at a slower speed.

43. Electrical synapses occur between neighboring cells that are connected by gap junctions. Gap junctions allow for the direct passage of electrical current from one cell to the next, so activity in electrically coupled cells is nearly simultaneous. Chemical synapses involve the release, diffusion, and binding of a neurotransmitter in response to an action potential in the presynaptic cell, before electrical activity can be generated in the postsynaptic cell. Chemical synaptic transmission thus takes longer than electrical synaptic transmission.

44. (1) the action potential arrives at the synaptic knob, depolarizing it; (2) extracellular calcium enters the synaptic knob triggering the exocytosis of ACh: (3) ACh binds to the postsynaptic membrane and depolarizes the next neuron in the chain, (4) ACh is removed by AChE.

45. Neurotransmitters are chemicals that are often classified as excitatory or inhibitory on the basis of their effects on postsynaptic membranes. Neuromodulators are compounds that influence neurotransmitter release or the postsynaptic cell's response to the neurotransmitter.

46. Temporal summation is the addition of stimuli that arrive in rapid succession. It occurs at a single synapse and is active repeatedly. Spatial summation occurs when simultaneous stimuli have a cumulative effect on the transmembrane potential. It involves multiple synapses that are active simultaneously.

47. The synthesis, release, and recycling of neurotransmitter molecules; the movement of materials to and from the cell body via axoplasmic transport; and the recovery from action potentials

48. During long-term strenuous physical activity, endorphins are released in the brain. Endorphins are structurally similar to morphine and relieve pain by suppressing the release of the neurotransmitter Substance P from pain neurons.

49. In MS, the myelin sheath is destroyed. Action potentials travel slower in unmyelinated fibers than in myelinated fibers. The destruction of myelin slows the time it takes for motor neurons to communicate with their effector muscles. This delay results in varying degrees of uncoordinated muscle activity and paralysis. Cumulative sensory and motor losses may eventually lead to generalized sensory deficiencies and muscular paralysis.

Level 3 Critical Thinking and Clinical Applications

50. Brain tumors result from uncontrolled division of neuroglia. Unlike neurons, neuroglia are still capable of cell division. In addition, cells of the meninges can give rise to tumors.

51. The kidney condition is causing Harry to retain potassium ions. As a result, the K^+ concentration of the extracellular fluid is higher than normal. Under these conditions, less potassium diffuses from heart muscle cells than normal, resulting in a resting potential that is less negative (more positive). This change in resting potential moves the transmembrane potential closer to threshold, so it is easier to stimulate the muscle. The ease of stimulation accounts for the increased number of contractions that produces the rapid heart rate.

52. To reach threshold, the postsynaptic membrane must receive enough neurotransmitter to produce an EPSP of 20 mV (10 mV to reach threshold and 10 mV to cancel the IPSPs produced by the inhibitory neurons). Each neuron releases enough neurotransmitter to produce a change of 2 mV, so 10 of the 15 excitatory neurons must be stimulated to produce this effect by spatial summation.

53. Without myelination, information about limb movement and body position moves slowly to the brain, and motor responses move slowly to the muscles. By the time the brain is aware of limb movement or position and issues a motor command, the limb has already moved. When the motor response reaches the skeletal muscle, the response is no longer appropriate. As the peripheral neurons gradually become myelinated, information flow and processing speed up, so we see improved balance, coordination, and capabilities.

54. Initially, the recording would show normal action potentials in response to the constant stimulation. After 50,000–100,000 action potentials have passed, the recording would start to indicate less responsiveness from the neuron; ultimately, the neuron would fail to respond. Because very few ions actually move across the membrane during an action potential, neurons can carry thousands of action potentials before the concentrations of sodium and potassium ions must be reestablished. The activity of the sodium–potassium exchange pump is required. Blocking ATPase activity would shut down the pump, and the cell could not maintain the necessary gradients of sodium and potassium ions. As a result, the membrane could not respond to stimulation until the ion concentration gradients were reestablished.

CHAPTER 13

Page 436

1. The ventral root of a spinal nerve is composed of both visceral and somatic motor fibers. Damage to this root would interfere with motor function.

2. The cerebrospinal fluid that surrounds the spinal cord is located in the subarachnoid space, which lies beneath the epithelium of the arachnoid layer and superficial to the pia mater.

Page 438

1. Because the virus that causes polio would be located in somatic motor neurons, it would be in the anterior gray horns of the spinal cord, where the cell bodies of these neurons are located.

2. A disease that damages myelin sheaths would affect the columns of the spinal cord, because that part of the cord is composed of bundles of myelinated axons.

Page 444

1. The dorsal rami of spinal nerves innervate the skin and muscles of the back. The skin and muscles of the back of the neck and of the shoulders will be affected by such an anesthetic.

2. The phrenic nerves that innervate the diaphragm originate in the cervical plexus. Damage to this plexus, or more specifically to the phrenic nerves, would greatly interfere with the ability to breathe and might result in death.

3. Compression of the sciatic nerve produces the characteristic sensation that you perceive when your leg has "fallen asleep."

Page 453

1. The minimum number of neurons required for a reflex arc is two. One must be a sensory neuron, to bring impulses to the CNS, and the other a motor neuron, to bring about a response to the sensory input.

2. The suckling reflex is an innate reflex.

3. When stretch receptors are stimulated by gamma motor neurons, the muscle spindles become more sensitive. As a result, little if any stretching stimulus would be needed to stimulate the contraction of the quadriceps muscles in the patellar reflex. Thus the reflex response would appear more quickly.

Page 456

1. This response is the tendon reflex.

2. During a withdrawal reflex, the limb on the opposite side is extended. This response is called a crossed extensor reflex.

3. A positive Babinski reflex is abnormal for an adult; it indicates possible damage of descending tracts in the spinal cord.

Page 460: Level 1 Reviewing Facts and Terms
1. d **2.** d **3.** a **4.** c **5.** c **6.** d **7.** b **8.** c **9.** c **10.** a
11. d **12.** a **13.** b **14.** c **15.** c **16.** d
17. (a) 1 (b) 7 (c) 3 (d) 5 (e) 4 (f) 6 (g) 2 (h) 8
18. the epineurium, which surrounds the spinal nerve; the perineurium, which surrounds fascicles; and the endoneurium, which surrounds individual axons

Level 2 Reviewing Concepts
19. The vertebral column continues to grow, extending beyond the cord. The end of the cord is visible as the conus medularis near L_1 and the cauda equina extends the remainder of the column.
20. (1) arrival of stimulus and activation of receptor; (2) activation of sensory neuron; (3) information processing; (4) activation of a motor neuron; and (5) response by an effector (muscle or gland)
21. d
22. The first cervical nerve exits superior to vertebra C_1 (between the skull and vertebra); the last cervical nerve exits inferior to vertebra C_7 (between the last cervical vertebra and the first thoracic vertebra). There are thus 8 cervical nerves but only 7 cervical vertebrae.
23. The cell bodies of spinal motor neurons are located in the anterior gray horns, so damage to these horns would result in a loss of motor control.
24. Inside the CNS, cerebrospinal fluid fills the central canal and the ventricles. Inside the CNS, cerebrospinal fluid fills the subarachnoid space. The CSF acts as a shock absorber as well as a diffusion medium for dissolved gases, nutrients, chemical messengers, and waste products.
25. (1) involvement of pools of interneurons; (2) intersegmental distribution; (3) involvement of reciprocal innervation; (4) motor response prolonged by reverberating circuits; and (5) cooperation of reflexes to produce a coordinated, controlled response
26. Transection of the spinal cord at C_7 would most likely result in paralysis from the neck down. A transection at T_{10} would paralyze the lower half of the body only. Sensory input and motor control of the body from the waist down would be lost.
27. (a) ventral ramus (b) ventral ramus (c) dorsal ramus (d) ventral ramus
28. (a) cervical plexus (b) lumbosacral plexus (c) brachial plexus (d) lumbosacral plexus
29. Transmission across a chemical synapse always involves a synaptic delay, but with only one synapse (a monosynaptic reflex), the delay between stimulus and response is minimized. In a polysynaptic reflex, the length of delay is proportional to the number of synapses involved.
30. The dorsal roots contain axons of sensory neurons, so sensory input from the neurons of the damaged root would not be transmitted to the CNS.
31. Stimulation will increase muscle tone.
32. The spinal cord ends at the level of L_2, so there is little chance of damaging it below that level. A needle can be inserted through the meninges inferior to the conus medullaris into the subarachnoid space with minimal risk to the cauda equina.

Level 3 Critical Thinking and Clinical Applications
33. the median nerve
34. the radial nerve
35. the genitofemoral nerve

36. Ramon has a damaged femoral nerve. This nerve also supplies the sensory innervation of the skin on the anteromedial surface of the thigh and medial surfaces of the leg and foot, so he may also experience numbness in those regions.

CHAPTER 14

Page 467
1. The mesencephalon, the pons and the medulla oblongata make up the brain stem.
2. The three primary brain vesicles are the prosencephalon, the mesencephalon, and the rhombencephalon. The prosencephalon gives rise to the cerebrum and diencephalon; the mesencephalon does not subdivide further; and the rhombencephalon develops into the cerebellum, the pons, and the medulla oblongata.

Page 472
1. If an interventricular foramen became blocked, cerebrospinal fluid could not flow from the lateral ventricles into the third ventricle. Cerebrospinal fluid would continue to form within that ventricle, so the blocked ventricle would swell with fluid—a condition known as hydrocephalus.
2. Diffusion across the arachnoid granulations is the means by which cerebrospinal fluid reenters the bloodstream. If this process decreased, excess fluid would accumulate in the ventricles; the volume of fluid in the ventricles would increase, damaging the brain.
3. The blood–brain barrier restricts and regulates the movement of water-soluble molecules from the blood to the extracellular fluid of the brain.

Page 477
1. The vermis and arbor vitae are part of the cerebellum.
2. Although the medulla oblongata is small, it contains many vital reflex centers, including those that control breathing and regulate the heart rate and blood pressure. Damage to the medulla oblongata can result in a cessation of breathing or in lethal changes in heart rate and blood pressure.
3. Damage to the respiratory centers of the pons could result in loss of ability to modify the rhythmicity center of the medulla oblongata during prolonged inhalation or extensive exhalation.

Page 483
1. Reflexive movements of the eyes, head, and neck are controlled by the superior colliculi of the mesencephalon.
2. The lateral geniculate nuclei are involved with processing visual information. Damage to these nuclei would interfere with the sense of sight.
3. Changes in body temperature would stimulate the preoptic area of the hypothalamus, a division of the diencephalon.
4. The amygdaloid regulates the "fight or flight" response of the sympathetic division of the autonomic nervous system.

Page 487
1. Projection fibers link the cerebral cortex to the spinal cord, passing through the diencephalon, brain stem and cerebellum.
2. The basal nuclei are involved in the subconscious control of skeletal muscle tone and the coordination of learned movement patterns. Damage to the basal nuclei would result in loss of these functions and decreased muscle tone (see Parkinson's disease, p. 487).

Page 492
1. The primary motor cortex is located in the precentral gyrus of the frontal lobe of the cerebrum.
2. Damage to the temporal lobes of the cerebrum would interfere with the processing of olfactory (smell) and auditory (sound) impulses.

3. The stroke has damaged Jake's speech center, located in the frontal lobe.

4. The temporal lobe of Paul's cerebrum is probably involved, specifically the hippocampus and the amygdaloid body. His problems may also involve other parts of the limbic system that act as a gate for loading and retrieving long-term memories.

Page 506: Level 1 Reviewing Facts and Terms

1. d 2. d 3. a 4. d 5. b 6. c 7. a 8. b 9. b 10. a
11. a 12. a

13. (1) cushioning delicate neural structures; (2) supporting the brain; and (3) transporting nutrients, chemical messengers, and waste products

14. (1) portions of the hypothalamus where the capillary endothelium is extremely permeable; (2) capillaries in the pineal gland; and (3) capillaries at the choroid plexus

15. I: olfactory nerve; II: optic nerve; III: oculomotor nerve; IV: trochlear nerve; V: trigeminal nerve; VI: abducens nerve; VII: facial nerve; VIII: vestibulocochlear nerve; IX: glossopharyngeal nerve; X: vagus nerve; XI: accessory nerve; and XII: hypoglossal nerve

Level 2 Reviewing Concepts

16. The brain includes many more interneurons, pathways, and connections than the tracts of the spinal cord, allowing for greater integration of impulses and versatility of response.

17. The functions of the cerebellum include adjusting voluntary and involuntary motor activities on the basis of sensory information and stored memories of previous experiences.

18. c

19. the substantia nigra, which releases the neurotransmitter dopamine at the basal nuclei

20. (1) subconscious control of skeletal muscle contractions, (2) control of autonomic functions, (3) coordination of nervous and endocrine systems, (4) secretion of hormones, (5) production of emotions and drives, (6) coordination of autonomic and voluntary functions, (7) regulation of body temperature, (8) control of circadian rhythms.

21. Stimulation of the feeding and thirst centers of the hypothalamus would produce these sensations.

22. hippocampus, which is part of the limbic system

23. The left hemisphere contains the general interpretive and speech centers and is responsible for language-based skills. Reading, writing, speaking, performing analytical tasks, and logical decision making are dependent on processing done in the left hemisphere. The right hemisphere analyzes sensory information and relates the body to the sensory environment. Interpretive centers in this hemisphere permit the identification of familiar objects by touch, smell, sight, taste, or feel. It is important in understanding three-dimensional relationships and in analyzing the emotional context of a conversation.

24. a 25. d

26. The general intrepretive area (Wernicke's area, sensory) and the speech center (Broca's area, motor) are involved in speech. Aphasia results in the absence of or defects in speech and the inability to comprehend language. Lesions in the general interpretive area produce defective visual and auditory comprehension of language, repetition of spoken sentences, and defective naming of objects. Lesions in the speech center result in hesitant and distorted speech.

Level 3 Critical Thinking and Clinical Applications

27. The sensory innervation of the nasal mucosa is by way of the maxillary branch of the trigeminal nerve (V). Irritation of the nasal lining increases the frequency of action potentials along the maxillary branch of the trigeminal nerve through the semilunar ganglion to reach centers in the mesencephalon, which in turn excite the neurons

of the reticular activating system (RAS). Increased activity by the RAS can raise the cerebrum back to consciousness.

28. The officer is testing the function of Bill's cerebellum. Many drugs, including alcohol, have pronounced effects on the function of the cerebellum. A person who is under the influence of alcohol cannot properly anticipate the range and speed of limb movement, because processing and correction by the cerebellum are slow. As a result, Bill might have a difficult time walking a straight line or touching his finger to his nose.

29. The local anesthetic temporarily numbs portions of the hypoglossal nerve (XII), which controls movements of the tongue. If Tyler tries to eat before the anesthetic wears off, he could bite his tongue severely. Movements of the tongue manipulate food during chewing and help initiate the swallowing reflex, so the inability to move food within the mouth could result in choking.

30. She could have problems with control over her left eye, difficulties with touch perception, and changes in intellectual function.

31. Most of the functional problems are the result of trauma to the cerebral hemispheres due to contact between the brain and the skull. Distortion and damage to the brain stem and medulla can cause death.

CHAPTER 15

Page 518

1. The receptor with the smaller receptor field will provide more-precise sensory information—thus, receptor A.

2. Nociceptors are pain receptors. When they are stimulated, you perceive a painful sensation in the affected hand.

3. Proprioceptors relay information about limb position and movement to the CNS, especially to the cerebellum. If this information were blocked, your movements would be uncoordinated and you would probably not be able to walk.

Page 523

1. The fasciculus gracilis in the posterior column of the spinal cord carries information about touch and pressure from the lower limbs to the brain. It is this tract that is being compressed.

2. Nociceptors are stimulated by pain. The action potentials generated by these receptors are carried by the lateral spinothalamic tracts.

3. Impulses carried along the right fasciculus gracilis are destined for the primary sensory cortex of the left cerebral hemisphere.

Page 527

1. The anatomical basis for opposite-side motor control is that crossing-over (decussation) of axons occurs, so the motor fibers of the corticospinal pathway innervate lower motor neurons on the opposite side of the body.

2. The superior portion of the motor cortex controls the upper limb and upper portion of the lower limb. An injury to this area would affect the ability to control the muscles in those regions of the body.

3. Motor neurons of the red nucleus help control the muscle tone of skeletal muscles in the upper limbs. Increased stimulation of these neurons would increase stimulation of the skeletal muscles, producing increased muscle tone.

Page 530: Level 1 Reviewing Facts and Terms

1. d 2. d 3. c 4. c

5. (1) An arriving stimulus alters the transmembrane potential of the receptor membrane. (2) The receptor potential directly or indirectly affects a sensory neuron. (3) Action potentials travel to the CNS along an afferent fiber.

6. 1) nociceptors: a variety of stimuli associated with tissue damage; (2) thermoreceptors: respond to changes in temperature; (3) mechanoreceptors: stimulated or inhibited by physical distortion, contact, or

pressure; and (4) chemoreceptors: monitor the chemical composition of body fluids and respond to the presence of specific molecules

7. (1) free nerve endings: sensitive to touch and pressure; (2) root hair plexus: monitors distortions and movements across the body surface; (3) tactile discs: detect fine touch and pressure; (4) tactile corpuscles: detect fine touch and pressure; (5) lamellated corpuscles: sensitive to pulsing or vibrating stimuli (deep pressure); and (6) Ruffini corpuscles: sensitive to pressure and distortion of the skin

8. (1) tactile receptors; (2) baroreceptors; and (3) proprioceptors

9. (1) posterior column pathway: provides conscious sensations of highly localized ("fine") touch, pressure, vibration, and proprioception; (2) spinothalamic pathway: provides conscious sensations of poorly localized ("crude") touch, pressure, pain, and temperature; and (3) spinocerebellar pathway: carries proprioceptive information about the position of skeletal muscles, tendons, and joints to the cerebellum

10. (1) corticobulbar tracts; (2) lateral corticospinal tracts; and (3) anterior corticospinal tracts

11. (1) vestibulospinal pathway; (2) tectospinal pathway; and (3) reticulospinal pathway

12. (1) It integrates proprioceptive sensations with visual information from the eyes and equilibrium-related sensations from the inner ear; and (2) it adjusts the activities of the voluntary and involuntary motor centers on the basis of sensory information and the stored memories of previous experiences.

13. (1) The number of neurons in the cerebral cortex continues to increase until at least age 1; (2) the brain as a whole grows in size and complexity until at least age 4; and (3) myelination of CNS neurons continues at least until puberty, reducing the delay between stimulus and response and improving motor control.

Level 2 Reviewing Concepts

14. The sensory neuron that delivers the sensations to the CNS is a first-order neuron. Within the CNS, the axon of the first-order neuron synapses on a second-order neuron, an interneuron located in the spinal cord or brain stem. The second-order neuron synapses on a third-order neuron in the thalamus. The axons of third-order neurons synapse on neurons of the primary sensory cortex of the cerebral hemispheres.

15. d

16. A tonic receptor is always active, while a phasic receptor is normally inactive. Phasic receptors are active only when a change occurs in the condition being monitored.

17. The autonomic nervous system controls visceral effectors, such as smooth and cardiac muscles, glands, and fat cells. The somatic nervous system, which is under voluntary control, controls the contractions of skeletal muscles.

18. A motor homunculus, a mapped-out area of the primary motor cortex, provides an indication of the degree of fine motor control available. A sensory homunculus indicates the degree of sensitivity of peripheral sensory receptors.

19. Injury to the primary motor cortex affects the ability to exert fine control over motor units. Gross movements are still possible, however, because they are controlled by the basal nuclei that use the reticulospinal or rubrospinal tracts. Thus, walking and other voluntary and involuntary movements can be performed with difficulty, and the movements are imprecise and awkward.

20. Muscle tone is controlled by the basal nuclei, cerebellum, and red nuclei through commands distributed by the reticulospinal and rubrospinal tracts.

21. Strong pain sensations arriving at a particular segment of the spinal cord can cause stimulation of the interneurons of the spinothalamic pathway. This stimulation is interpreted by the sensory

cortex as originating in the region of the body surface associated with the origin of that same pathway.

Level 3 Critical Thinking and Clinical Applications

22. The tumor is most likely adjacent to the corticobulbar tracts. The axons of those tracts carry action potentials to motor nuclei of the cranial nerves, which control eye muscles and facial expression.

23. You would expect to observe some degree of paralysis on the left side of Doris' body, relative to the amount of motor cortex deprived of blood. You might also observe some degree of memory loss, mood changes, and impairment of planning abilities.

24. The drug is mimicking the effects of the neurotransmitter serotonin. The areas of the brain that are being affected include the occipital lobe (visual hallucinations, color enhancement), the temporal lobe (auditory hallucinations), and the limbic system (increased sexual appetite).

CHAPTER 16

Page 536

1. Two neurons are required to conduct an action potential from the spinal cord to smooth muscles in the wall of the intestine. One neuron is required to carry the action potential from the spinal cord to the autonomic ganglion, and a second to carry the action potential from the autonomic ganglion to the smooth muscle.

2. The sympathetic division of the autonomic nervous system is responsible for the physiological changes that occur in response to stress and increased activity.

3. The sympathetic division of the autonomic nervous system includes preganglionic fibers from the lumbar and thoracic portions of the spinal cord, whereas the parasympathetic division includes preganglionic fibers from the cranial and sacral portions.

Page 542

1. The nerves that synapse in collateral ganglia originate in the inferior thoracic and superior lumbar portion of the spinal cord; they pass through the chain ganglia to the collateral ganglia.

2. Acetylcholine is the neurotransmitter released by the preganglionic fibers of the sympathetic nervous system. A drug that stimulates ACh receptors would stimulate the postganglionic fibers of the sympathetic nerves, resulting in increased sympathetic activity.

3. Blocking the beta receptors on cells would decrease or prevent sympathetic stimulation of the tissues that contain those cells. The heart rate, the force of contraction of cardiac muscle, and the contraction of smooth muscle in the walls of blood vessels would decrease. These changes would contribute to a lowering of the blood pressure.

Page 545

1. The vagus nerve (X) carries preganglionic parasympathetic fibers that innervate the lungs, heart, stomach, liver, pancreas, and parts of the small and large intestines (as well as several other visceral organs).

2. Muscarinic receptors are a type of acetylcholine receptor located in postganglionic synapses of the parasympathetic nervous system. The stimulation of these receptors at the heart would cause potassium ion channels to open, resulting in hyperpolarization of the cell membrane and a decreased heart rate.

3. The parasympathetic division is sometimes referred to as the anabolic system because parasympathetic stimulation leads to a general increase in the nutrient content of the blood. Cells throughout the body respond to the increase by absorbing the nutrients and using them to support growth and other anabolic activities.

Page 552

1. Most blood vessels receive sympathetic stimulation, so a decrease in sympathetic tone would lead to a relaxation of the smooth muscles in the walls of these vessels and hence to vasodilation. Blood flow to the tissue would, in turn, increase.

2. A patient who is anxious about an impending root canal would probably exhibit some or all of the following changes: a dry mouth, an increased heart rate, an increase in blood pressure, an increased rate of breathing, cold sweats, an urge to urinate or defecate, a change in the motility of the digestive tract (that is, "butterflies in the stomach"), and dilated pupils. These changes would be the result of anxiety or stress causing an increase in sympathetic stimulation.

3. A brain tumor that interferes with hypothalamic function would interfere with autonomic function as well. Centers in the posterior and lateral hypothalamus coordinate and regulate sympathetic function, whereas centers in the anterior and medial hypothalamus control parasympathetic function.

Page 554

1. Higher order functions require the cerebral cortex, involve both conscious and unconscious information processing, and are subject to modification and adjustment over time.

2. An inability to comprehend the spoken or written word indicates a problem with the general interpretive area of the brain, which in most individuals is located in the left temporal lobe of the cerebrum.

3. You are using short term memory, although your teacher would like it if this information were transferred to long term memory.

Page 557

1. The reticular activating system is responsible for rousing the cerebrum to a state of consciousness. If your RAS were suddenly stimulated, you would wake up.

2. A drug that increases the amount of serotonin released in the brain would produce a heightened perception of certain sensory stimuli, such as auditory or visual stimuli, and hallucinations.

3. Some possible reasons for slower recall and for loss of memory in the elderly include a loss of neurons (possibly those involved in specific memories), changes in synaptic organization of the brain, changes in the neurons themselves, and decreased blood flow, which would affect the metabolic rate of neurons and perhaps slow the retrieval of information from memory.

Page 562: Level 1 Reviewing Facts and Terms

1. c **2.** b **3.** c **4.** a **5.** d **6.** d

7. Sympathetic preganglionic fibers emerge from the thoracolumbar area (T_1 through L_2) of the spinal cord. Parasympathetic fibers emerge from the brain stem and the sacral region of the spinal cord (craniosacral).

8. Preganglionic neuron (T_5-L_2) $\rightarrow$ collateral ganglia $\rightarrow$ postganglionic fibers $\rightarrow$ visceral effector in abdominopelvic cavity

9. (1) celiac ganglion; (2) superior mesenteric ganglion; and (3) inferior mesenteric ganglion

10. (1) the release of norepinephrine at specific locations and (2) the secretion of epinephrine (and modest amounts of norepinephrine) into the bloodstream

11. When stimulated, the presynaptic membrane at a cholinergic synapse releases acetylcholine, whereas an adrenergic synapse releases norepinephrine.

12. III, VII, IX, and X

13. (1) ciliary ganglion; (2) sphenopalatine ganglion; (3) submandibular ganglion; and (4) otic ganglion

14. (1) cardiac plexus: heart rate–increase (sympathetic)/decrease (parasympathetic); heart strength–increase (sympathetic)/decrease (parasympathetic); blood pressure–increase (sympathetic)/decrease (parasympathetic); (2) pulmonary plexus: respiratory passageways dilate/constrict; (3) esophageal plexus: respiratory rate increase/decrease; (4) celiac plexus: digestion inhibited/stimulated; (5) inferior mesenteric plexus: digestion inhibited/stimulated; and (6) hypogastric plexus: defecation inhibited/stimulated; urination inhibited/stimulated; sexual organs: stimulate secretion/erection

15. receptor, sensory neuron, interneuron (may or may not be present), and two visceral motor neurons

16. Processing centers in the medulla oblongata coordinate complex sympathetic and parasympathetic reflexes. These medullary centers are regulated by the hypothalamus. Centers in the posterior and lateral hypothalamus coordinate and regulate sympathetic function, and portions of the anterior and medial hypothalamus control the parasympathetic division.

17. (1) They are performed by neurons of the cerebral cortex and involve complex interactions between areas of the cortex and between the cerebral cortex and other parts of the brain; (2) they involve both conscious and unconscious information processing; and (3) their functions are subject to modification and adjustment over time.

18. increased neurotransmitter release, facilitation of synapses, and the formation of additional synaptic connections

19. During non-REM sleep, the entire body relaxes and activity at the cerebral cortex is at a minimum. The heart rate, blood pressure, respiratory rate, and energy utilization decline. During REM sleep, active dreaming occurs, accompanied by alterations in blood pressure and respiratory rates. Muscle tone decreases markedly, and there is less response to outside stimuli.

20. a reduction in brain volume and weight, a reduction in the number of neurons, a decrease in blood flow to the brain, changes in the synaptic organization, and intracellular and extracellular changes in CNS neurons

Level 2 Reviewing Concepts

21. a **22.** a **23.** c

24. The preganglionic fibers innervating the cervical ganglia originate in the white rami of thoracic segments, which are undamaged.

25. Preganglionic fibers entering the adrenal gland proceed to the adrenal medulla, where they synapse on neuroendocrine cells that release the neurotransmitters norepinephrine and epinephrine into the bloodstream.

26. The stimulation of alpha receptors activates enzymes on the inside of the cell membrane; the stimulation of beta receptors in organs and tissues triggers changes in the metabolic activity of the target cell. Beta-receptor stimulation results in the formation of a second messenger that activates or deactivates key enzymes.

27. (1) There is much less divergence within the parasympathetic division, so the effects are more limited in scope. (2) Most of the ACh released is inactivated by AChE within the synapse. ACh diffusing outside of the synapse will be inactivated by other cholinesterases in the tissues.

28. If autonomic motor neurons maintain a background level of activity at all times, they can either increase or decrease their activity, providing a range of control options.

Level 3 Critical Thinking and Clinical Applications

29. Due to the stimulation of the sympathetic division, you would experience increased respiratory rate, increased peripheral vasoconstriction and elevation of blood pressure, increased heart rate and force of contraction, and an increased rate of glucose release into the bloodstream.

30. A CVA is more likely to be devastaing if it occurs in the left hemisphere, which in most people is the site of the general interpretive and speech centers. Specific functional losses would depend on the precise site of the CVA, but in general, damage to the left cerebral hemisphere would affect motor control of the right side, speech, language-based skills, logical decision making, and analytical functions. Damage to the right hemisphere could affect abilities to analyze sensory information and relate the body to the sensory environment.

31. Stress-induced stomach ulcers are due to excessive sympathetic stimulation. The sympathetic division causes vasoconstriction of blood vessels that supply the digestive organs, reducing blood flow to the stomach. The reduction in blood flow slows mucus production and reduces mucosal defenses against stomach acids.

32. You would want to use a beta-blocker, because the stimulation of beta-1 receptors increases heart rate. By blocking these receptors, parasympathetic influence will predominate and the heart rate will slow down.

33. epinephrine, because it would reduce inflammation and relax the smooth muscle of the airways, making it easier for Phil to breathe

34. It is probably mimicking NE and binding to alpha-1 receptors.

35. serotonin, affecting the brain stem, hypothalamus, limbic system, and spinal cord.

CHAPTER 17

Page 569

1. By the end of the lab period, adaptation has occurred. In response to the constant level of stimulation, your receptor neurons have become less active, partially as the result of synaptic fatigue.

2. Your taste receptors (taste buds) are sensitive only to molecules and ions that are in solution. If you dry the surface of your tongue, the salt ions or sugar molecules have no moisture in which to dissolve, so they will not stimulate the taste receptors.

3. The difference in the taste of your grandfather's food is the result of several age-related factors. The number of taste buds declines dramatically after age 50, and those that remain are not as sensitive as they once were. In addition, the loss of olfactory receptors contributes to the perception of less flavor in foods.

Page 580

1. The first layer of the eye that would be affected by inadequate tear production would be the conjunctiva. Drying of this layer would produce an irritated, scratchy feeling.

2. When the lens is round, you are looking at an object that is close to you.

3. Renee will probably not be able to see at all. The fovea of the eye contains cones but no rods. Rods respond to light of low intensity, but cones need high light intensity to be stimulated. In a dimly lit room, the light would not be strong enough to stimulate these photoreceptors.

4. If the canal of Schlemm were blocked, the aqueous humor would not be able to drain; glaucoma would develop. As the quantity of fluid increased, the pressure within the eye would increase, distorting soft tissues and interfering with vision. If untreated, the condition would ultimately cause blindness.

Page 587

1. Even with a congenital lack of cones in your eyes, you would still be able to see—as long as you had functioning rods. But because cone cells function in color vision, you would see only black and white.

2. A deficiency of vitamin A in the diet would affect the quantity of retinal the body could produce and thus would interfere with night vision, which is more sensitive to deficiencies since it is working at the threshold of the body's ability to respond to light.

3. A decrease in phosphodiesterase activity would lead to a higher level of intracellular cGMP. This rise would, in turn, keep the gated sodium channels open and decrease the ability of receptor neurons to respond to photons.

Page 600

1. Without the movement of the round window, the perilymph would not be moved by the vibration of the stapes at the oval window. There would be little or no perception of sound.

2. The loss of stereocilia (as a result of constant exposure to loud noises, for instance) would reduce hearing sensitivity and could lead to deafness.

3. The auditory tube allows pressure to equalize on both sides of the tympanic membrane. If this tube were blocked and external pressures then declined, the pressure on the inside of the tympanic membrane would be greater than that on the outside, forcing the membrane outward and producing pain.

Page 603: Level 1 Reviewing Facts and Terms

1. d **2.** a **3.** a **4.** c **5.** c **6.** b **7.** d **8.** b **9.** d **10.** d
11. c

12. (1) filiform papillae; (2) fungiform papillae; and (3) circumvallate papillae

13. (**a**) the sclera and the cornea (**b**) It (1) provides mechanical support and some physical protection, (2) serves as an attachment site for the extrinsic eye muscles, and (3) contains structures that assist in the focusing process.

14. iris, ciliary body, and choroid

15. (1) malleus, (2) incus, and (3) stapes, which transmit a mechanical vibration (amplified along the way) from the tympanic membrane to the oval window

Level 2 Reviewing Concepts

16. Axons leaving the olfactory epithelium collect into twenty or more bundles that penetrate the cribriform plate of the ethmoid bone to reach the olfactory bulbs of the cerebrum. Axons leaving the olfactory bulb travel along the olfactory tract to reach the olfactory cortex, the hypothalamus, and portions of the limbic system.

17. Olfactory sensations are long lasting and important to memories because the sensory information they provide goes directly to the cerebral cortex via the hypothalamus and the limbic system. These sensations are not first filtered through the thalamus

18. a sty, a painful swelling

19. A visual acuity of 20/15 means that Jane can discriminate images at a distance of 20 feet, whereas someone with "normal" vision must be 5 feet closer (15 feet away) to see the same details. Jane's visual acuity is better than that of someone with 20/20 vision.

20. optic nerve through optic chiasm to lateral geniculate nucleus to visual cortex

21. (1) Sound waves arrive at the tympanic membrane. (2) Movement of the tympanic membrane causes displacement of the auditory ossicles. (3) Movement of the stapes at the oval window establishes pressure waves in the perilymph of the vestibular duct. (4) The pressure waves distort the basilar membrane on their way to the round window of the tympanic duct. (5) Vibration of the basilar membrane causes vibration of hair cells against the tectorial membrane. (6) Information about the region and intensity of stimulation is relayed to the CNS over the cochlear branch of cranial nerve VIII.

Level 3 Critical Thinking and Clinical Applications

22. Your medial rectus muscles would contract, directing your gaze more medially. In addition, your pupils would constrict and the lenses would become more spherical.

23. Sally is apparently nearsighted (myopic) and needs convex lenses to correct the condition.

24. The noise from the fireworks has transferred so much energy to the endolymph in the cochlea of her ears that this fluid continues to move for a long time. The vibrations associated with normal conversation are not strong enough to overcome the currents already moving through the endolymph, making these vibrations difficult to discern against the background "noise." And as long as the endolymph is moving, it will vibrate the tectorial membrane and will stimulate the hair cells, producing a "ringing" sensation.

25. The rapid descent in the elevator causes the maculae in the saccule of your vestibule to slide upward, producing the sensation of downward vertical motion. When the elevator abruptly stops, the maculae do not. It takes a few seconds for them to come to rest in the normal position. As long as the maculae are displaced, you will perceive movement.

26. When Juan closes his eyes, the visual cues are gone and his brain must rely solely on proprioceptive information and information from the static equilibrium centers (saccule and utricle). As a result of his problem with the saccules or utricles, his brain does not receive sufficient information to maintain balance. The movement of the arms toward the side of the impaired receptors is due to the deficit of information arriving from that side of the body.

CHAPTER 18

Page 614

1. Neural responses occur within fractions of a second and do not last long (short duration). Conversely, endocrine responses may be slow to appear but will last for minutes to days (long duration).
2. Adenylate cyclase is the enzyme that converts ATP to cAMP. A molecule that blocks this enzyme would block the action of any hormone that required cAMP as a second messenger.
3. A cell's hormonal sensitivities are determined by the presence or absence of the necessary receptor complex for a given hormone.

Page 619

1. Dehydration increases the osmotic pressure of the blood. The increase in blood osmotic pressure would stimulate the posterior pituitary to release more ADH.
2. Somatomedins mediate the action of growth hormone. When the levels of somatomedins are elevated, the level of growth hormone would be elevated as well.
3. Elevated circulating levels of cortisol would inhibit the cells that control the release of ACTH from the pituitary gland; therefore, the level of ACTH would decrease. This is an example of a negative feedback mechanism.

Page 630

1. An individual whose diet lacks iodine would not be able to form the hormone thyroxine. As a result, we would expect to see the symptoms associated with thyroxine deficiency, such as a decreased rate of metabolism, a decreased body temperature, a poor response to physiological stress, and an increase in the size of the thyroid gland (goiter).
2. Most of the thyroid hormone in the blood is bound to proteins called thyroid-binding globulins. These compounds represent a large reservoir of the thyroid hormone thyroxine that guards against rapid fluctuations in the level of this important hormone. Because such a large amount is stored in this way, it takes several days to deplete the supply of thyroxine, even after the thyroid gland has been removed.
3. The removal of the parathyroid glands would result in a decrease in the blood concentration of calcium ions. This decrease could be counteracted by increasing the amounts of vitamin D and calcium in the diet.
4. One of the functions of cortisol is to decrease the cellular use of glucose while increasing the available glucose by promoting the breakdown of glycogen and the conversion of amino acids to carbohydrates. The net result of elevated cortisol levels would be an elevation in the level of glucose in the blood.

Page 639

1. An individual with Type 1 or Type 2 diabetes has such elevated levels of glucose in the blood that the kidneys cannot reabsorb all the glucose; some glucose is lost in urine. Because the urine contains high concentrations of glucose, less water can be reclaimed by osmosis; the volume of urine production increases. The water losses reduce blood volume and elevate blood osmotic pressure, promoting thirst and triggering the secretion of ADH.
2. Glucagon stimulates the conversion of glycogen to glucose in the liver. Increased levels of glucagon would lead to decreased amounts of glycogen in the liver.
3. Increased amounts of light would inhibit the production (and release) of melatonin from the pineal gland, which receives neural input from the optic tracts. The secretion of melatonin by the pineal gland is influenced by light–dark cycles.

Page 643

1. The type of hormonal interaction exemplified by the effects of insulin and glucagon is antagonism. In this type of hormonal interaction, two hormones have opposite effects on their target tissues.
2. The hormones GH, thyroid hormone, PTH, and the gonadal hormones all play a role in formation and development of the skeletal system.
3. During the resistance phase of the general adaptation syndrome, there is a high demand for glucose, especially by the nervous system. The hormones GH-RH and CRH increase the levels of GH and ACTH, respectively. Growth hormone mobilizes fat reserves and promotes the catabolism of protein; ACTH increases cortisol, which stimulates the conversion of glycogen to glucose as well as the catabolism of fat and protein.

Page 648: Level 1 Reviewing Facts and Terms

1. b **2.** c **3.** d **4.** a **5.** c **6.** d **7.** b **8.** d **9.** c **10.** b
11. a
12. (1) The hypothalamus produces regulatory hormones that control secretion by endocrine cells in the anterior lobe of the pituitary gland, (2) the hypothalamus contains autonomic centers that exert direct neural control over the endocrine cells of the adrenal medulla, and (3) the hypothalamus releases ADH and oxytocin into the bloodstream at the posterior lobe of the pituitary gland. These mechanisms are adjusted through negative feedback loops involving the hormones released by peripheral endocrine tissues and organs.
13. (1) thyroid-stimulating hormone (TSH); (2) adrenocorticotropic hormone (ACTH); (3) follicle-stimulating hormone (FSH); (4) luteinizing hormone (LH); (5) prolactin (PRL); (6) growth hormone (GH); and (7) melanocyte-stimulating hormone (MSH)
14. (1) growth hormone, (2) thyroid hormones, (3) insulin, (4) parathyroid hormone, (5) calcitriol, and (6) the reproductive hormones
15. (1) an increase in the rate of energy consumption and utilization in cells; (2) an acceleration in the production of sodium–potassium ATPase; (3) the activation of genes coding for the synthesis of enzymes involved in glycolysis and energy production; (4) the acceleration of ATP production by mitochondria; and (5) in growing children, the normal development of the skeletal, muscular, and nervous systems
16. calcitonin: decreases the concentration of calcium ions in body fluids; parathyroid hormone: causes an increase in the concentration of calcium ions in body fluids
17. (1) zona glomerulosa: mineralocorticoids; (2) zona fasciculata: glucocorticoids; and (3) zona reticularis: androgens
18. Erythropoietin, which stimulates the production of RBCs by the bone marrow; and calcitriol, responsible for stimulating calcium and phosphate absorption along the digestive tract.

19. (1) It promotes the loss of sodium ions and water at the kidneys; (2) inhibits the secretion of water-conserving hormones, such as ADH and aldosterone; (3) suppresses thirst; and (4) blocks the effects of angiotensin II and norepinephrine on arterioles

20. (1) alpha cells: glucagon; (2) beta cells: insulin; (3) delta cells: somatostatin; and (4) F cells: pancreatic polypeptide

Level 2 Reviewing Concepts

21. In the nervous system, the source and destination of communication are quite specific and the effects are short lived. In endocrine communication, the effects are slow to appear and commonly persist for days. A single hormone can alter the metabolic activities of multiple tissues and organs simultaneously.

22. Hormones direct the synthesis of an enzyme (or other protein) not already present in the cytoplasm. They turn an existing enzyme "on" or "off" and increase the rate of synthesis of a particular enzyme or other protein.

23. The two hormones may have opposing, or antagonistic, effects; they may have additive or synergistic effects; one hormone may have a permissive effect on another (that is, the first hormone is needed for the second to produce its effect); or they may produce different but complementary results in specific tissues and organs.

24. In endocrine reflexes, the functional counterpart of neural reflexes, a stimulus triggers the production of a hormone. Both neural and endocrine reflexes are controlled by negative feedback mechanisms in most cases.

25. Because phosphodiesterase converts cAMP to AMP, the inactivation of this enzyme would prolong the effect of the hormone.

26. The adrenal medulla is controlled by the sympathetic nervous system, while the cortex is stimulated by the release of ACTH from the anterior pituitary gland.

27. The adrenal medulla secretes epinephrine and norepinephrine, both of which stimulate alpha and beta receptors.

28. Hormones can interact in one of four ways: antagonistic interactions oppose one another, synergistic interactions enhance the effects of each hormone, permissive effects where one hormone is required for the other to function, or integrative effects where the action of a hormone differs with differing target organs.

29. **(a)** (1) alarm phase, in which energy reserves are mobilized and the body prepares itself with "fight or flight" responses; (2) resistance phase, in which glucocorticoids are released to mobilize lipid and protein reserves, conserve glucose for neural tissues, elevate and stabilize blood glucose concentrations, and conserve salts and water and lose K^+ and H^+; and (3) exhaustion phase, in which homeostatic regulation breaks down **(b)** epinephrine in the alarm phase and glucocorticoids in the resistance phase

Level 3 Critical Thinking and Clinical Applications

30. Extreme thirst and frequent urination are characteristics of both diabetes insipidus and diabetes mellitus. To distinguish between the two, glucose levels in the blood and urine could be measured. A high glucose concentration would indicate diabetes mellitus.

31. Julie's poor diet would not supply enough Ca^{2+} for her developing fetus, which would remove large amounts of Ca^{2+} from the maternal blood. A lowering of the mother's blood Ca^{2+} would lead to an increase in parathyroid hormone levels and increased mobilization of stored Ca^{2+} from maternal skeletal reserves.

32. Sherry's symptoms suggest hyperthyroidism. Blood tests could be performed to assay the levels of TSH, T_3, and T_4. From these tests, the physician could make a positive diagnosis (hormone levels would be elevated in hyperthyroidism) and also determine whether the condition is primary (a problem with the thyroid gland) or secondary (a problem with hypothalamo-pituitary control of the thyroid gland).

33. All steroid hormones are quite similar in structure. When there is an excess of one steroid, some is converted to another form, commonly a form similar to aldosterone. Increased levels of aldosterone lead to the retention of sodium and water and can produce bloating, as in patients receiving steroid therapy.

34. Anabolic steroids are derivatives of testosterone. The natural effects of this are to increase muscle mass, increase endurance, and enhance the "competitive spirit." Additional side effects in women include hirsutism, enlargement of the laryngeal cartilages, premature closure of the epiphyseal cartilages, and liver dysfunction.

CHAPTER 19

Page 655

1. Venipuncture is a common sampling technique because (1) superficial veins are easy to locate; (2) the walls of veins are thinner than those of arteries; and (3) blood pressure in veins is relatively low, so the puncture wound seals quickly.

2. A decrease in the amount of plasma proteins in the blood would cause a decrease in (1) plasma osmotic pressure; (2) the ability to fight infection; and (3) the transport and binding of some ions, hormones, and other molecules.

3. During a viral infection, you would expect the level of immunoglobulins (antibodies) in the blood to be elevated.

Page 662

1. The hematocrit measures the amount of formed elements (mostly red blood cells) as a percentage of the total blood. In hemorrhage, the loss of blood, especially of red blood cells, would lower the hematocrit.

2. A decreased blood flow to the kidneys will trigger the release of erythropoietin. The elevated erythropoietin will lead to an increase in erythropoiesis (red blood cell formation). Thus, Dave's hematocrit will increase.

3. The liver processes bilirubin prior to excretion in the bile. Diseases that cause damage to the liver, such as hepatitis or cirrhosis, would impair the liver's ability to perform this function. As a result, bilirubin would accumulate in the blood, producing jaundice.

Page 666

1. Surface antigens on RBCs are glycolipids in the cell membrane.

2. A person with Type O blood can accept only Type O blood.

3. If a person with Type A blood receives a transfusion of Type B blood, the red blood cells will clump, or agglutinate, potentially blocking blood flow to various organs and tissues.

Page 672

1. In an infected cut, we would find a large number of neutrophils, phagocytic white blood cells that are generally the first to arrive at the site of an injury. They specialize in dealing with infectious bacteria.

2. The cells that produce circulating antibodies are B lymphocytes; these white blood cells would be found in elevated numbers.

3. During inflammation, basophils release a variety of chemicals, such as histamine and heparin, that exaggerate the inflammation and attract other types of white blood cells.

Page 677

1. Megakaryocytes are the precursors of platelets, which play an important role in hemostasis and the clotting process. A decreased number of megakaryocytes would result in fewer platelets, which in turn would interfere with the ability to clot properly.

2. Fruit juice and water do not contain fats, which are required for vitamin K absorption, leading to a vitamin K deficiency. This would lead to a decrease in the production of several clotting factors—most notably, prothrombin. As a result, clotting time would increase.

3. The activation of Factor XII initiates the intrinsic pathway.

Page 680: Level 1 Reviewing Facts and Terms

1. c **2.** c **3.** a **4.** c **5.** d **6.** a **7.** b **8.** d **9.** d **10.** b
11. (1) the transportation of dissolved gases, nutrients, hormones, and metabolic wastes; (2) the regulation of pH and electrolyte composition of interstitial fluids throughout the body; (3) the restriction of fluid losses through damaged vessels or at other injury sites; (4) defense against toxins and pathogens; and (5) the stabilization of body temperature
12. albumins, which maintain the osmotic pressure of plasma and are important in the transport of fatty acids; globulins, which bind small ions, hormones, or compounds that might otherwise be filtered out of the blood at the kidneys or have very low solubility in water (transport globulins) or attack foreign proteins and pathogens (immunoglobulins); and fibrinogen, which functions in blood clotting
13. (**a**) anti-B antibodies (**b**) anti-A antibodies (**c**) neither anti-A nor anti-B antibodies (**d**) anti-A and anti-B antibodies
14. (1) ameboid movement, the extension of a cellular process; (2) emigration, squeezing between adjacent endothelial cells in the capillary wall; (3) positive chemotaxis, the attraction to specific chemical stimuli, and (4) neutrophils, eosinophils, and monocytes are capable of phagocytosis
15. neutrophils, eosinophils, basophils, and monocytes
16. (1) T cells, which are responsible for cell-mediated immunity; (2) B cells, which are responsible for humoral immunity; and (3) NK cells, which are responsible for immune surveillance
17. (1) M-CSF: monocytes and macrophages; (2) G-CSF: granulocytes; (3) GM-CSF: granulocytes and monocytes; and (4) multi-CSF: granulocytes, monocytes, and megakaryocytes
18. (1) the transport of chemicals important to clotting; (2) the formation of a temporary patch in the walls of damaged blood vessels; and (3) active contraction after the clot has formed
19. Erythropoeitin is released (1) during anemia, (2) when blood flow to the kidneys declines, (3) when oxygen content of the air in the lungs declines, and (4) when the respiratory surfaces of the lungs are damaged
20. (1) anticoagulants in the plasma, (2) heparin, (3) thrombomodulin, (4) Protacyclin and (5) other plasma proteins that inhibit thrombin.
21. The common pathway begins when thromboplastin from either the extrinsic or intrinsic pathway appears in the plasma.
22. An embolus is a drifting blood clot. A thrombus is a blood clot that sticks to the wall of an intact blood vessel.

Level 2 Reviewing Concepts

23. a **24.** d **25.** b **26.** d
27. Red blood cells are biconcave discs that lack mitochondria, ribosomes, and nuclei, and they contain a large amount of hemoglobin. RBC's transport oxygen, while WBC's are involved in immunity.
28. White blood cells defend against toxins and pathogens. Neutrophils, eosinophils, and monocytes engulf and digest bacteria, protozoa, fungi, viruses, and cellular debris. Lymphocytes specialize to attack and destroy specific foreign cells, proteins, and cancerous cells, directly or through the production of antibodies.
29. Blood stabilizes and maintains body temperature by absorbing and redistributing the heat of the body. Heat is absorbed from active skeletal muscles. Dermal capillaries dilate when body temperature rises, thereby increasing blood flow to the skin and dissipating the excess heat to the air. Dermal capillaries constrict when body temperature falls, thereby decreasing blood flow to the skin and thus conserving heat for organs that are more temperature sensitive.
30. Hemoglobin is a protein that demonstrates quaternary structure. An iron is found in the center of each of the four proteins that make up one molecule of hemoglobin. This central iron atom is what actually picks up and releases the oxygen molecules.
31. Arterial puncture is required for checking the efficiency of gas exchange at the lungs. Samples are usually drawn from the radial artery at the wrist or from the brachial artery at the elbow.
32. RhoGam is given to prevent an Rh-negative mother from producing anti-Rh antibodies, which could attack the blood of an Rh-positive fetus. It can be administered soon after the delivery of the first Rh-positive baby. RhoGam contains anti-Rh antibodies that remove fetal Rh-positive antigens from the mother's bloodstream before the mother's immune system recognizes their presence. Thus, the mother's immune system is not induced to produce anti-Rh antibodies.
33. Aspirin inactivates platelet enzymes involved in the production of thromboxanes and prostaglandins, and it inhibits endothelial cell production of prostacyclin. Thus, aspirin inhibits clotting.

Level 3 Critical Thinking and Clinical Applications

34. A prolonged prothrombin time and a normal partial thromboplastin time indicate a deficiency in the extrinsic system but not in the intrinsic system or common pathway. Factor VII would be deficient.
35. Reticulocytes would increase in number as additional numbers of immature RBCs move into the bloodstream to compensate for the reduction in blood volume.
36. Mary would probably exhibit multiple bruises throughout the skin, prolonged bleeding time, and prolonged clotting time (as a result of antibiotic-induced thrombocytopenia).
37. In many cases of kidney disease, the cells responsible for producing erythropoietin are damaged or destroyed. The reduction in erythropoietin levels leads to reduced erythropoiesis and fewer red blood cells, resulting in anemia.
38. Intrinsic factor, an essential part of the transport system for the absorption of vitamin B_{12} by the intestinal cells, is produced by specialized stomach cells. When most of the stomach was removed, intrinsic factor was no longer available to facilitate the absorption of vitamin B_{12}, so injection directly into the bloodstream was necessary. (If Randy did not take the B_{12} shots, he would develop pernicious anemia.)

CHAPTER 20

Page 697

1. The semilunar valves on the right side of the heart guard the opening to the pulmonary artery. Damage to these valves would interfere with the blood flow to this vessel.
2. When the ventricles begin to contract, they force the AV valves to close, tensing the chordae tendineae. The chordae tendineae are attached to the papillary muscles, which begin contracting just before the rest of the ventricular myocardium does. Their contraction prevents the AV valves from opening back into the atria.
3. The left ventricle is more muscular than the right ventricle because the left ventricle must generate enough force to propel blood throughout all the body except the alveoli of the lungs. The right ventricle must generate only enough force to propel blood a few centimeters to the lungs.

Page 702

1. If these cells failed to function, the heart would still continue to beat but at a slower rate; the AV node would act as pacemaker.
2. Cardiac output will be lessened since CO = heart rate times stroke volume. In bradycardia the heart slows down, decreasing rate and therefore decreasing CO.
3. If the impulses from the atria were not delayed at the AV node, they would be conducted through the ventricles so quickly by the

bundle branches and Purkinje cells that the ventricles would begin contracting immediately, before the atria had finished their contraction. As a result, the ventricles would not be as full of blood as they could be, and the pumping of the heart would not be as efficient, especially during activity.

Page 708

1. No. When pressure in the left ventricle first rises, the heart is contracting but no blood is leaving the heart. During this initial phase of contraction, both the AV valves and the semilunar valves are closed. The increase in pressure is the result of increased tension as the cardiac muscle contracts. When the pressure in the ventricle exceeds the pressure in the aorta, the aortic semilunar valves are forced open and blood is rapidly ejected from the ventricle.

2. An increase in the size of the QRS complex indicates a larger-than-normal amount of electrical activity during ventricular depolarization. One possible cause is an increase in the size of the heart. Because more cardiac muscle is depolarizing, the magnitude of the electrical event would be greater.

Page 712

1. Caffeine acts directly on the conducting system and contractile cells of the heart, increasing the rate at which they depolarize. Drinking large amounts of caffeinated coffee would increase the heart rate.

2. The cardioinhibitory center of the medulla oblongata is part of the parasympathetic division of the autonomic nervous system. Damage to this center would result in fewer parasympathetic action potentials to the heart and an increase in heart rate due to sympathetic dominance.

3. A drug that increases the length of time required for the repolarization of pacemaker cells would decrease the heart rate, because the pacemaker cells would generate fewer action potentials per minute.

Page 714

1. If the heart beats too rapidly, it does not have sufficient time to fill completely between beats. The heart pumps in proportion to the amount of blood that enters: The less blood that enters, the less the heart can pump. If it beats too fast, very little blood will enter the bloodstream; tissues will suffer damage from the lack of blood supply.

2. Stimulating the acetylcholine receptors of the heart would cause the heart to slow down. Since the cardiac output is the product of stroke volume and heart rate, if the heart rate decreases, so will the cardiac output (assuming that the stroke volume doesn't change).

3. The venous return fills the heart with blood, stretching the heart muscle. According to the Frank–Starling principle, the more the heart muscle is stretched, the more forcefully it will contract (to a point). The more forceful the contraction, the more blood the heart will eject with each beat (stroke volume). Therefore, increased venous return would increase the stroke volume (if all other factors are constant).

4. An increase in sympathetic stimulation of the heart would result in an increased heart rate and increased force of contraction. The end-systolic volume (ESV) is the amount of blood that remains in a ventricle after a contraction (systole). The more forcefully the heart contracts, the more blood it ejects and the lower the ESV is. Therefore, increased sympathetic stimulation should result in a lower ESV.

5. SV = EDV − ESV, so SV = 125 ml − 40 ml = 85 ml

Page 718: Level 1 Reviewing Facts and Terms

1. c 2. b 3. b 4. b 5. a 6. b 7. b 8. a 9. a

10. During ventricular contraction, tension in the papillary muscles pulls against the chordae tendineae, which keep the cusps of the AV valve from swinging into the atrium. This action prevents the backflow, or regurgitation, of blood into the atrium as the ventricle contracts.

11. (1) The epicardium is the visceral pericardium, which covers the outer surface of the heart. (2) The myocardium is the muscular wall of the heart, which forms both atria and ventricles. It contains cardiac muscle tissue and associated connective tissues, blood vessels, and nerves. (3) The endocardium is a squamous epithelium that covers the inner surfaces of the heart, including the valves.

12. The atrioventricular (AV) valves prevent the backflow of blood from the ventricles into the atria. The right AV valve is the tricuspid valve; the left AV valve is the bicuspid valve. The pulmonary and aortic semilunar valves prevent the backflow of blood from the pulmonary trunk and aorta into the right and left ventricles.

13. SA node → AV node → AV bundle → right and left bundle branches → Purkinje fibers (into the mass of ventricular muscle tissue)

14. The cardiac cycle is the period between the end of one heartbeat and the beginning of the next heartbeat. The cycle begins with atrial systole as the atria contract and push blood into the relaxed ventricles. As the atria relax (atrial diastole), the ventricles contract (ventricular systole), forcing blood through the semilunar valves into the pulmonary trunk and aorta. The ventricles then relax (ventricular diastole). For the rest of the cardiac cycle, both the atria and ventricles are in diastole; passive filling occurs.

15. (1) preload, the stretch on the heart before it contracts; (2) contractility, the forcefulness of contraction of individual ventricular muscle fibers; and (3) afterload, the pressure that must be exceeded before blood can be ejected from the ventricles

Level 2 Reviewing Concepts

16. c 17. c 18. d 19. a

20. The SA node is the pacemaker of the heart, composed of cells that exhibit rapid prepotential. The AV nodal cells are smaller than those of the conduction pathway and the impulse that signals contraction is slowed.

21. The first sound ("lubb"), which marks the start of ventricular contraction, is produced as the AV valves close and the semilunar valves open. The second sound ("dupp") occurs when the semilunar valves close, marking the start of ventricular diastole. The third heart sound is associated with blood flow into the atria, and the fourth with atrial contraction. Listening to the heart sounds (auscultation) is a simple, effective method of cardiac diagnosis.

22. Stroke volume (SV) is the volume of blood ejected by a ventricle in a single contraction. Cardiac output (CO) is the amount of blood pumped by a ventricle in 1 minute. CO (in ml/min) = SV (in ml) × HR (in beats/min)

23. stroke volume and heart rate

24. Preload is the degree of stretching of the ventricular muscle cells during ventricular diastole. There are no antagonistic muscle groups to extend the cardiac muscle tissue after each contraction. The stretching force is provided by the pouring of blood into the heart and by the elasticity of the fibrous skeleton.

25. Sympathetic activation causes the release of NE by postganglionic fibers and the secretion of NE and E by the adrenal medullae. These hormones stimulate the metabolism of cardiac muscle cells and increase the force and degree of contraction. They also depolarize nodal cells toward threshold, which in turn increases the heart rate. Parasympathetic stimulation causes the release of ACh at membrane surfaces, where it produces hyperpolarization and inhibition. The result is a decrease in the heart rate and the force of cardiac contractions.

26. All these hormones have positive inotropic effects, which means that they increase the strength of contraction of the heart.

Level 3 Critical Thinking and Clinical Applications

27. In skeletal muscles, increased accumulation of lactic acid during exercise decreases the efficiency of contraction and can ultimately

prevent contraction due to muscle fatigue. If this situation occurred in the heart, the results could be fatal. Fortunately, the heart can metabolize lactic acid, which is usually broken down before it can interfere with cardiac function. Thus, the heart can continue to beat normally despite a temporary reduction in the local oxygen supply.

28. This patient has a second-degree heart block, a condition in which some signals from the SA node do not reach the ventricular muscle. In this case, for every two action potentials generated by the SA node, only one is reaching the ventricles. This accounts for the two P waves for each QRS complex.

29. Person 1 has a cardiac output of 4500 ml (CO = HR × SV). Person 2 has a cardiac output of 8550 ml. According to Starling's law, in a normal heart the cardiac output is directly proportional to the venous return. Thus, person 2 has the greater venous return. Ventricular filling decreases with increased heart rate; person 1 has the lower heart rate and therefore the longer ventricular filling time.

30. It will decrease the force of cardiac contraction, which directly affects stroke volume. Verpamil should lower Karen's stroke volume.

31. As a result of a myocardial infarction, cardiac muscle cells die. As they die, their cell membranes break down and their cytoplasmic contents leak into the surrounding interstitial fluid. Because the cells contain relatively high concentrations of K^+, the concentration of K^+ in the surrounding fluid would increase as K^+ leaks from the dead and dying cells. The interstitial fluid would contain more K^+, so less K^+ would diffuse through the membranes of living cells, decreasing the potential difference across the membrane and making them more excitable.

32. The extra contraction that is generated by the ectopic pacemaker is accompanied by an extra refractory period. The next normal action potential from the AV node usually arrives while the ventricles are still in the refractory period that accompanies the action potential generated by the ectopic pacemaker. As a result, the action potential does not trigger a contraction and the ventricles are not stimulated again until the next normal signal along the conducting pathways. This longer-than-normal period between subsequent contractions produces the feeling of the heart "skipping" a beat.

CHAPTER 21

Page 730

1. These blood vessels are veins. Arteries and arterioles have a large amount of smooth muscle tissue in a thick, well-developed tunica media.

2. Blood pressure in the arterial system pushes blood into the capillaries. Blood pressure in the venous system is very low, and other forces help keep the blood moving. Valves in veins prevent blood from flowing backward whenever the venous pressure drops; this is not a problem in arteries.

3. You would find fenestrated capillaries in organs and tissues where small peptides move freely into and out of the blood: for instance, endocrine glands, the choroid plexus of the brain, absorptive areas of the intestine, and filtration areas of the kidneys.

Page 740

1. In a healthy individual, the blood pressure should be greater at the aorta than at the inferior vena cava. Blood, like other fluids, moves along a pressure gradient from areas of high pressure to areas of low pressure. If the pressure were higher in the inferior vena cava, the blood would flow backward.

2. While a person stands for periods of time, blood pools in the lower limbs. This pooling decreases the venous return to the heart. In turn, the cardiac output decreases, sending less blood to the brain, causing light-headedness and fainting. A hot day adds to this effect, because body water is lost and blood volume is reduced, through sweating.

3. The nurse first heard the sounds of Korotkoff when the pressure in the cuff reached 125 mm Hg. At this point, the pressure in the vessel during systole is just enough to overcome the pressure in the cuff. Turbulent flow produced in the constricted vessel then produces the audible sounds.

Page 751

1. During exercise, blood pressure increases despite the increased blood flow to skeletal muscles because (1) cardiac output increases and (2) resistance in visceral tissues increases.

2. Pressure at this site would decrease blood pressure at the carotid sinus, where the carotid baroreceptors are located. This decrease would cause a decreased frequency of action potentials along the glossopharyngeal nerve (IX) to the medulla oblongata, and more sympathetic impulses would be sent to the heart. The net result would be an increase in the heart rate.

3. The vasoconstriction of the renal artery would decrease both blood flow and blood pressure at the kidney. In response, the kidney would increase the amount of renin that it releases, which in turn would lead to an increase in the level of angiotensin II. The angiotensin II would bring about increased blood pressure and increased blood volume.

4. A person in circulatory shock has a decreased venous return to the heart and, in turn, a decreased cardiac output. The decreased cardiac output accounts for the weak pulse. Because the cardiac output is decreased, the baroreceptors are stimulated. As a result, sympathetic stimulation to the heart increases, causing the rapid heart rate. Although the heart is beating faster, there is less blood to pump and pulse pressure remains low.

Page 761

1. The left subclavian artery is the branch of the aorta that sends blood to the left arm.

2. The common carotid arteries carry blood to the head. A compression of the common carotid arteries would result in decreased blood pressure at the carotid sinus and a rapid fall in blood flow to the brain, resulting in unconsciousness. An immediate reflexive increase in heart rate and blood pressure would follow.

3. Organs served by the celiac trunk include the stomach, spleen, liver, and pancreas.

Page 771

1. The vein that is bulging is the external jugular vein.

2. A blockage of the popliteal vein would interfere with blood flow in the tibial and peroneal veins (which form the popliteal vein) and the small saphenous vein (which joins the popliteal vein).

3. This blood sample must have come from the umbilical vein, which carries oxygenated, nutrient-rich blood from the placenta to the fetus.

Page 776: Level 1 Reviewing Facts and Terms

1. k **2.** w **3.** g **4.** o **5.** a **6.** q **7.** c **8.** u **9.** x **10.** b **11.** v **12.** r **13.** e **14.** i **15.** t **16.** f **17.** m **18.** h **19.** d **20.** l **21.** j **22.** p **23.** n **24.** s **25.** d **26.** b **27.** c **28.** a **29.** b **30.** b **31.** a **32.** a **33.** d **34.** b **35.** c **36.** b **37.** d **38.** c **39.** b **40.** c **41.** a **42.** c **43.** b

44. Capillary hydrostatic pressure forces fluid out of the capillary at the arteriole end. The blood colloid osmotic pressure causes the movement of fluid back into the capillary at its venous end.

45. (a) muscle pump (b) respiratory pump

46. Net Filtration Pressure = net hydrostatic pressure minus net colloid osmotic pressure. NFP = (CHP−IHP) − (BCOP−ICOP)

where CHP = capillary hydrostatic pressure; IHP = hydrostatic pressure of the interstitial fluid; BCOP = blood colloid osmotic pressure; ICOP = interstitial fluid colloid osmotic pressure

47. Decreased cardiac output and vasodilation occur during baroreceptor response to elevated blood pressure.

48. changes in the carbon dioxide, oxygen, or pH levels in the blood

49. (1) epinephrine/norepinephrine; (2) aldosterone; (3) antidiuretic hormone; (4) angiotensin II; (5) erythropoietin; and (6) atrial natriuretic peptide

50. Before exercise begins, the heart rate increases slightly due to a general rise in sympathetic activity as you think about the workout ahead. Extensive vasodilation occurs; venous return increases; and cardiac output rises.

51. hypotension: systolic pressures below 90 mm Hg; pale, cool, and moist ("clammy") skin; confusion and disorientation; a rapid, weak pulse; the cessation of urination; and a drop in blood pH (acidosis)

52. When an infant takes its first breath, the lungs expand and pulmonary vessels dilate. The smooth muscles in the ductus arteriosus contract, isolating the pulmonary and aortic trunks, and blood begins flowing through the pulmonary circuit. As pressure rises in the left atrium, due to the increased venous return from the lungs, the valvular flap closes the foramen ovale, completing the circulatory remodeling.

53. blood: decreased hematocrit, the formation of thrombin, and valvular malfunction; heart: a reduction in maximum cardiac output, changes in the activities of the nodal and conducting fibers, a reduction in the elasticity of the fibrous skeleton, progressive atherosclerosis, and the replacement of damaged cardiac muscle fibers by scar tissue; blood vessels: a progressive inelasticity in arterial walls, the deposition of calcium salts on weakened vascular walls, and the formation of thrombi at atherosclerotic plaques

Level 2 Reviewing Concepts

54. (**a**) the peak blood pressure measured during ventricular systole (**b**) the minimum blood pressure at the end of ventricular diastole (**c**) the difference between the systolic and diastolic pressures (**d**) a single value for blood pressure, calculated by adding one-third of the pulse pressure to the diastolic pressure

55. d **56.** c **57.** a **58.** c

59. Artery walls are generally thicker, and they contain more smooth muscle and elastic fibers, enabling them to resist and adjust to the pressure generated by the heart. Venous walls are thinner; the pressure in veins is less than that in arteries. Arteries constrict more than veins do when not expanded by blood pressure, due to a greater degree of elastic tissue. Finally, the endothelial lining of an artery has a pleated appearance because it cannot contract and so forms folds. The lining of a vein looks like a typical endothelial layer.

60. Vessel diameter is the only portion of cardiovascular resistance that is under autonomic, nervous, and hormonal control. Peripheral resistance is altered to permit variations in blood flow to meet regional demands for oxygen and nutrients.

61. Capillary walls are thin, so distances for diffusion are small. Continuous capillaries have small gaps between adjacent endothelial cells that permit the diffusion of water and small solutes into the surrounding interstitial fluid but prevent the loss of blood cells and plasma proteins. Fenestrated capillaries contain pores that permit very rapid exchange of fluids and solutes between interstitial fluid and plasma. The walls of arteries and veins are several cell layers thick and are not specialized for diffusion.

62. Movement in the surrounding skeletal muscles squeezes venous blood toward the heart. This mechanism, the muscular pump, is assisted by the presence of valves in the veins, which prevent backflow

of the blood. The respiratory pump results from the increase in internal pressure of the thoracic cavity on exhalation, which pushes venous blood into the right atrium.

63. Cardiac output and peripheral blood flow are directly proportional to blood pressure. Blood pressure is closely regulated by a combination of neural and hormonal mechanisms. The resistance of the circulatory system opposes the movement of blood, so blood flow is inversely proportional to the resistance. Sources of peripheral resistance include vascular resistance, viscosity, and turbulence.

64. The brain receives arterial blood via four arteries that form anastomoses within the cranium. An interruption of any one vessel will not compromise the circulatory supply to the brain.

65. The cardioacceleratory and vasomotor centers are stimulated when a general sympathetic activation occurs. The result is an increase in cardiac output and blood pressure. When the parasympathetic division is activated, the cardioinhibitory center is stimulated, reducing cardiac output.

66. The top number (150) is the systolic pressure, the peak blood pressure measured during ventricular systole. The bottom number (90) is the diastolic pressure, the minimum blood pressure at the end of ventricular diastole. The mean arterial pressure (MAP) is calculated by adding one-third of the pulse pressure to the diastolic pressure: $(150 - 90)/3 + 90 = 110$. (Note: pulse pressure = systolic pressure − diastolic pressure.)

67. The accident victim is in circulatory shock. The hypotension results from a loss of blood volume and decreased cardiac output. The skin is pale and cool due to peripheral vasoconstriction; the moisture results from the sympathetic activation of sweat glands. Falling blood pressure to the brain causes confusion and disorientation. If you took the victim's pulse, you would find it to be rapid and weak, reflecting the heart's response to reduced blood flow and volume.

68. In congestive heart failure, the heart cannot produce enough force to circulate the blood properly. As fluid accumulates in the capillaries, the blood hydrostatic pressure increases and the blood osmotic pressure decreases. The result is a fluid shift from the blood to the interstitial space. The fluid accumulation exceeds the ability of the lymphatic system to drain it, resulting in peripheral edema (swelling in the limbs).

Level 3 Critical Thinking and Clinical Applications

69. There are three contributing factors to Bob's elevated blood pressure: (1) The loss of water by sweating increases blood viscosity. The number of red blood cells remains about the same, but because there is less plasma volume, the concentration of red cells is increased, thus increasing the blood viscosity. Increased viscosity increases peripheral resistance and contributes to increased blood pressure. (2) To cool Bob's body, there is increased blood flow to the skin. This in turn increases venous return, which increases stroke volume and cardiac output (Starling's law of the heart). The increased cardiac output can also contribute to increased blood pressure. (3) The heat stress that Bob is experiencing leads to increased sympathetic stimulation (the reason for the sweating). Increased sympathetic stimulation of the heart will increase heart rate and stroke volume, thus increasing cardiac output and blood pressure.

70. The greater saphenous vein of the leg is a large vein that lies just beneath the skin, with no muscle to support it as with the deeper leg veins. Also, due to its location, the vessel is more susceptible to the effects of gravity than are vessels in the arms. When there is any impediment to blood flow from the legs, blood tends to collect in the greater saphenous vein and to stretch it; the skin does very little to resist the stretching. Even if the valves in the vein are closed, their edges won't meet due to the stretching, allowing more blood to collect inferior to the valves and causing even more stretching. This positive feedback mechanism leads to the black-and-blue nodules of varicose veins.

71. Pulse pressure is the difference between the systolic and diastolic pressures. The heart rate of a resting athlete is slower than that of a nonathlete, due to the size and condition of the heart muscle. The cardiac output of an athlete would be the same or even slightly higher than that of a nonathlete, however, because the athlete's stroke volume is greater. The larger the stroke volume, the greater the difference between the systolic and diastolic pressures. (The systolic pressure is proportional to the volume of blood ejected per beat.) Thus, since the athlete has the larger stroke volume, he or she would have the larger pulse pressure.

72. Antihistamines and decongestants are sympathomimetic drugs; they have the same effects on the body as does stimulation of the sympathetic nervous system. In addition to the desired effects of counteracting the symptoms of the allergy, these medications can produce an increased heart rate, increased stroke volume, and increased peripheral resistance, all of which will contribute to elevating blood pressure. In a person with hypertension (high blood pressure), these drugs would aggravate this condition, with potentially hazardous consequences.

73. When Jolene rapidly moved from a lying position to a standing position, gravity caused her blood volume to move to the lower parts of her body away from the heart, decreasing venous return. The decreased venous return resulted in a decreased EDV, leading to a decreased stroke volume and cardiac output. In turn, blood flow to the brain decreased, so the diminished oxygen supply caused her to be light-headed and feel faint. This reaction doesn't happen all the time because as soon as the pressure drops due to inferior movement of blood, the baroreceptor reflex should be triggered. Normally, a rapid change in blood pressure is sensed by baroreceptors in the aortic arch and carotid sinus. Action potentials from these areas are carried to the medulla oblongata, where appropriate responses are integrated. In this case, we would expect an increase in peripheral resistance to compensate for the decreased blood pressure. If this doesn't compensate enough for the drop, then an increase in heart rate and force of contraction would occur. Normally, these responses occur so quickly that changes in pressure following changes in body position go unnoticed.

CHAPTER 22

Page 791

1. The thoracic duct drains lymph from the area beneath the diaphragm and from the left side of the head and thorax. Most of the lymph enters the venous blood by way of this duct. A blockage of this duct would impair the circulation of lymph through most of the body and would promote the accumulation of fluid in the limbs (lymphedema).

2. The thymic hormones play a role in the differentiation of lymphoid stem cells into T lymphocytes. A lack of these hormones would result in an absence of T lymphocytes.

3. During an infection, the lymphocytes and phagocytes in the lymph nodes in the affected region undergo cell division to deal with the infectious agent more effectively. This increase in the number of cells in the node causes the node to become enlarged or swollen.

Page 798

1. A decrease in the number of monocyte-forming cells in the bone marrow would result in a decreased number of macrophages in the body, because all types of macrophages are derived from monocytes. These include the Kupffer's cells of the liver, Langerhans cells in the skin and digestive tract, and alveolar macrophages.

2. A rise in the level of interferon would suggest a viral infection. Interferon is released from cells that are infected with viruses. It does not help an infected cell but "interferes" with the virus's ability to infect other cells.

3. Pyrogens stimulate the temperature control area of the preoptic nucleus of the hypothalamus. The result is an increase in body temperature, or fever.

Page 809

1. Abnormal peptides in the cytoplasm of a cell can become attached to MHC (major histocompatibility complex) proteins and are then displayed on the surface of the cell's membrane. Peptides presented in this manner are then recognized by T cells, which can initiate an immune response.

2. Cytotoxic T cells function in cell-mediated immunity. A decrease in the number of cytotoxic T cells would interfere with the ability to kill foreign cells and tissues as well as cells infected by viruses.

3. Helper T cells promote B cell division, the maturation of plasma cells, and the production of antibodies by the plasma cells. Without helper T cells, the antibody-mediated immune response would probably not occur.

4. Plasma cells produce and secrete antibodies. In a sample containing an elevated number of plasma cells, we would expect the amount of antibodies in the blood to increase.

Page 817

1. The secondary response would be more affected by the lack of memory B cells for a particular antigen. The ability to produce a secondary response depends on the presence of memory B cells and T cells that are formed during the primary response to an antigen. These cells are not involved in the primary response but are held in reserve against future contact with the same antigen.

2. The developing fetus is protected primarily by natural passive immunity, the product of IgG antibodies that cross the placenta from the mother's bloodstream.

3. Stress can interfere with the immune response by depressing the inflammatory response, reducing the number and activity of phagocytes, and inhibiting interleukin secretion.

Page 824: Level 1 Reviewing Facts and Terms

1. b **2.** c **3.** b **4.** a **5.** c **6.** a **7.** c **8.** d
9. d **10.** c

11. Lymph nodes: filtration of lymph, detection of pathogens, initiation of immune response; tonsils, lymphoid nodules, and aggregated lymphoid nodules: defense of entrance and passageways of digestive tract against pathogens and foreign proteins or toxins; spleen: filtration of blood, recycling of red blood cells, detection of blood-borne pathogens or toxins; thymus: production of mature T cells and hormones that promote immune function; lymphatics: movement of lymph from interstitial spaces to the venous system.

12. (a) lymphocytes responsible for cell-mediated immunity **(b)** stimulate the activation and function of T cells and B cells **(c)** inhibit the activation and function of both T cells and B cells **(d)** produce and secrete antibodies **(e)** recognize and destroy abnormal cells **(f)** produce interleukin-7, which promotes the differentiation of B cells **(g)** maintain the blood–thymus barrier and secrete the thymic hormones that stimulate stem cell division and T cell differentiation **(h)** interfere with viral replication inside the cell and stimulate the activities of macrophages and NK cells **(i)** reset the body's thermostat, causing a rise in body temperature (fever) **(j)** provide cell-mediated immunity, which defends against abnormal cells and pathogens inside cells **(k)** provide humoral immunity, which defends against antigens and pathogens in the body (but not inside cells) **(l)** enhance nonspecific defenses and increase T cell sensitivity and stimulate

B cell activity (**m**) slow tumor growth and kill sensitive tumor cells (**n**) stimulate the production of blood cells in the bone marrow and lymphocytes in lymphoid tissues and organs

13. (1) T cells, derived from the thymus; (2) B cells, derived from bone marrow; and (3) NK cells, derived from bone marrow

14. (1) physical barriers; (2) phagocytic cells; (3) immunological surveillance; (4) interferons; (5) complement; (6) inflammation; and (7) fever

Level 2 Reviewing Concepts

15. c **16.** d **17.** b

18. (1) specificity: The immune response is triggered by a specific antigen and defends against only that antigen; (2) versatility: The immune system can differentiate among tens of thousands of antigens it may encounter during a normal lifetime; (3) memory: The immune response following the second exposure to a particular antigen is stronger and lasts longer than before; and (4) tolerance: Some antigens, such as those on your own cells, don't elicit an immune response.

19. Complement may rupture the cell membrane through the release of perforin, kill the target cell by secreting a poisonous lymphotoxin, or activate genes within the nucleus of the cell that tell the cell to die. Interferon interferes with viral replication inside infected cells by triggering the production of antiviral proteins.

20. by rupturing the antigenic cell membrane; by killing the target cell through lymphotoxin secretions, or by activating genes in the nucleus that program cell death.

21. by neutralization, agglutination and precipitation, the activation of complement, the attraction of phagocytes, opsonization, the stimulation of inflammation, or the prevention of bacterial and viral adhesion

22. innate immunity = genetically programmed immunity; naturally acquired immunity = immunity that develops after birth, due to contact with pathogens. Exposure to chicken pox in grade school is an example of naturally acquired immunity; induced active immunity is immunity that develops after purposeful contact with a pathogen, such as vaccinations; induced passive immunity is temporary immunity provided by injection with antibodies produced in another organism; natural passive immunity is immunity gained through acquiring antibodies from either mother's milk or placental exchange

23. The injections are timed to trigger the primary and secondary responses of the immune system. When the technician is first exposed to the hepatitis antigens, B cells produce daughter cells that differentiate into plasma cells and memory B cells. The plasma cells begin producing antibodies, which represent the primary response to exposure. However, the primary response does not maintain elevated antibody levels for long periods, so the second and third injections are necessary to trigger the secondary (anamnestic) response, when memory B cells differentiate into plasma cells and produce antibody concentrations that remain high much longer.

Level 3 Critical Thinking and Clinical Applications

24. IgA antibodies are found in body secretions such as tears, saliva, semen, vaginal secretions but not in blood plasma. Plasma contains IgM, IgG, IgD, and IgE antibodies. By testing for the presence or absence of IgA and IgG, the lab could determine whether the sample is blood plasma or semen.

25. Yes. On initial contact with a virus, the immune system first produces IgM antibodies. The response is fairly rapid but short-lived. About the time that IgM peaks, IgG levels are beginning to rise. IgG plays the more important role in eventually controlling the disease. The fact that Ted's blood sample has an elevated level of IgM antibodies indicates that he is in the early stages of a primary response to the measles virus.

26. In a radical mastectomy, the neighboring lymph glands in the axilla and surrounding region are removed along with the cancerous breast. (The lymph nodes are removed as a precautionary measure to try to prevent the spread of cancer cells spread by way of the lymphatic system.) The lymphatic vessels from the limb on the affected side are tied off, because there is no place for the lymph to drain to. Over time, lymphedema occurs, causing swelling of the limb.

27. For Harry's body to produce an immune response to antigen A, the antigen would have to be presented on his MHC proteins on the surface of his cells. For T cells to respond to an antigen, those antigens must be bound to MHC proteins of antigen-presenting cells. Tilly and Harry almost certainly have different MHC proteins, so introducing Tilly's cells would not help Harry. Her cells would not recognize the MHC proteins on the antigen-presenting cells and so would not respond to the antigen. In addition, her cells would be recognized as foreign and would be attacked and destroyed by Harry's macrophages.

28. Because FSH is a protein hormone, it binds to receptors on the surface of its target cells. If you could isolate these receptors, you could inject them into an animal, such as a rabbit, that would give an immune response to the human protein. By isolating from the rabbit the plasma cells that produced antibodies to the FSH receptor, you could clone them to produce large amounts of antibodies. Each plasma cell produces only one type of antibody, so this process would produce monoclonal antibodies. You could then tag the antibodies with a marker, such as the fluorescent molecule fluorescein, to make them visible when viewed under a fluorescent microscope. You could then introduce the antibodies into cell populations that you wished to test. If the receptors for FSH were present, the antibodies would bind to them and you could locate them by locating the fluorescent marker.

CHAPTER 23

Page 834

1. The rich blood supply to the nose delivers body heat to the nasal cavity, so the inhaled air is warmed before it leaves the nasal cavity. The heat also evaporates moisture from the epithelium to humidify the incoming air. The moisture is derived from the blood supply as well. The blood supply also brings nutrients and water to the secretory cells of the nasal mucosa.

2. The nasopharynx receives only air from the nasal cavity. The oropharynx and laryngopharynx receive both air from the nasal cavity and food from the oral cavity. Ingested solids and liquids can damage delicate cells; thus, the areas that are in contact with food have a highly protective stratified squamous epithelium, like that of the exterior skin. The lining of the nasopharynx is the same as that of the nasal cavity, a pseudostratified ciliated columnar epithelium.

3. Increased tension in the vocal folds will raise the pitch of your voice.

Page 843

1. The tracheal cartilages are C-shaped to allow room for esophageal expansion when large pieces of food or volumes of liquid are swallowed.

2. Without surfactant, the alveoli would collapse as a result of surface tension in the thin layer of water that moistens the alveolar surfaces.

3. Air that passes through the glottis flows into the larynx and through the trachea. From there, the air flows into a primary bronchus, which supplies the lungs. In the lungs, the air passes to bronchi, bronchioles, a terminal bronchiole, a respiratory bronchiole, an alveolar duct, an alveolar sac, an alveolus, and ultimately to the respiratory membrane.

Page 843
1. The pulmonary arteries supply the exchange surfaces; the external carotid arteries, the thyrocervical trunks, and the bronchial arteries supply the conducting portions of the respiratory system.
2. The pleura is a serous membrane. Pleural surfaces secrete pleural fluid, which lubricates the opposing parietal and visceral surfaces to prevent friction during breathing.

Page 852
1. Because the rib penetrates Mark's chest wall, atmospheric air will enter his thoracic cavity. This condition is a pneumothorax. Pressure within the pleural cavity is normally lower than atmospheric pressure, but when air enters the pleural cavity, the natural elasticity of the lung may cause the lung to collapse. The resulting condition is atelectasis, or a collapsed lung.
2. Because the fluid produced in pneumonia takes up space that would normally be occupied by air, the vital capacity would decrease.

Page 859
1. As skeletal muscles become more active, they generate more heat and more acidic waste products, which lower the pH of the surrounding fluid. The combination of increased temperature and reduced pH causes the hemoglobin to release more oxygen than it would otherwise.
2. Blockage of the trachea would interfere with the body's ability to gain oxygen and to eliminate carbon dioxide. Because most carbon dioxide is carried in blood as bicarbonate ion that is formed from the dissociation of carbonic acid, an inability to eliminate carbon dioxide would result in an excess of hydrogen ions, thus lowering the body's pH.

Page 867
1. The pneumotaxic centers inhibit the inspiratory center and the apneustic center. Exciting the pneumotaxic centers would result in shorter breaths and a more rapid rate of breathing.
2. Peripheral chemoreceptors are more sensitive to carbon dioxide levels than to oxygen levels. When carbon dioxide dissolves, it produces hydrogen ions, thereby lowering pH and altering cell or tissue activity.
3. Johnny's mother shouldn't worry. When Johnny holds his breath, the level of carbon dioxide in his blood will increase. This increase will in turn lead to increased stimulation of the inspiratory center, forcing Johnny to breathe again.

Page 872: Level 1 Reviewing Facts and Terms
1. a 2. c 3. a 4. c 5. d 6. b 7. c 8. a 9. a 10. a
11. c
12. (1) providing an extensive area for gas exchange between inhaled air and circulating blood; (2) moving air to and from the exchange surfaces of the lungs; (3) protecting the respiratory surfaces from dehydration, temperature changes, and other environmental variations and defending the respiratory system and other tissues from invasion by pathogens; (4) producing sounds involved in speaking, singing, or nonverbal communication; and (5) providing olfactory sensations to the CNS from the olfactory epithelium in the superior portions of the nasal cavity (The respiratory system also indirectly assists in the regulation of blood volume, blood pressure, and the control of body fluid pH.)
13. The upper respiratory system consists of the nose, nasal cavity, paranasal sinuses, and pharynx. The lower respiratory system consists of the larynx, trachea, bronchi, bronchioles, and alveoli of the lungs.
14. Goblet cells in the epithelium and mucous glands in the lamina propria of the respiratory tract produce a sticky mucus that bathes exposed surfaces. In the nasal cavity, cilia sweep mucus and any trapped debris or microorganisms toward the pharynx, where it will

be swallowed and exposed to the acids and enzymes of the stomach. In the lower respiratory system, the cilia beat toward the pharynx, creating a mucus escalator that cleans the respiratory passageways. Filtration occurs in the nasal cavity. Particles may become trapped in the liquid covering the alveolar surfaces. Finally, foreign particles can be engulfed by alveolar macrophages.
15. (1) nasopharynx, the superior portion, where the nasal cavity opens into the pharynx; (2) oropharynx, the middle portion, posterior to the oral cavity; and (3) laryngopharynx, the inferior portion that is posterior to the hyoid bone and glottis
16. Thyroid cartilage forms the anterior walls of the larynx; the cricoid cartilage protects the glottis and the entrance to the trachea; the epiglottis forms a lid over the glottis; the arytenoid cartilages and the corniculate cartilages are involved in the formation of sound; the cuneiform cartilages are found in the folds of the larynx.
17. The parietal pleura covers the inner surface of the thoracic wall and extends over the diaphragm and mediastinum. The visceral pleura covers the outer surfaces of the lungs, extending into the fissures between the lobes.
18. (1) pulmonary ventilation (breathing); (2) gas diffusion across the respiratory membrane; (3) the storage and transport of oxygen and carbon dioxide; and (4) the exchange of dissolved gases between blood and interstitial fluids
19. Fetal hemoglobin has a higher affinity for oxygen, enabling it to "steal" oxygen from the maternal hemoglobin. Fetal hemoglobin also binds more oxygen than adult hemoglobin.
20. (1) conversion to carbonic acid; (2) binding to the hemoglobin of red blood cells; and (3) dissolution in plasma
21. When the P_{O_2} goes up, the bronchioles increase in diameter (bronchodilation). When the P_{O_2} declines, the bronchioles constrict (bronchoconstriction). Airflow is therefore directed to lobules where the is high, which helps improve gas exchange, because these will also be the lobules where the is very low.

Level 2 Reviewing Concepts
22. b 23. a 24. d 25. d 26. c 27. d 28. b
29. With less cartilaginous support, the amount of tension in the smooth muscles has a greater effect on bronchial diameter and the resistance to air flow.
30. During talking, the airways are open and the food passages are closed. Trying to swallow while talking is therefore likely to lead to inadvertent entry of food particles into the airways, which will then trigger coughing to expel the food.
31. The nasal cavity is designed to cleanse, moisten, and warm inhaled air, whereas the mouth is not. Air that has entered through the mouth is drier than air that has entered through the nose and as a result can irritate the trachea, causing soreness of the throat.
32. The walls of bronchioles, like the walls of arterioles, are dominated by smooth muscle tissue. Varying the diameter of the bronchioles (bronchodilation or bronchoconstriction) provides control over the amount of resistance to airflow and the distribution of air within the lungs, just as vasodilation and vasoconstriction of the arterioles regulate blood flow and blood distribution.
33. Septal cells produce surfactant, which reduces surface tension in the fluid coating the alveolar surface. The alveolar walls are so delicate that without surfactant, the surface tension would be so high that the alveoli would collapse.
34. The differences in partial pressure are substantial; the diffusion distances involved are small; the gases are lipid soluble; the total surface area is large; and blood flow and airflow are coordinated.
35. Pulmonary ventilation is the physical movement of air into and out of the respiratory tract. Its primary function is to maintain ade-

quate alveolar ventilation. Alveolar ventilation is air movement into and out of the alveoli. This process prevents the buildup of carbon dioxide in the alveoli and ensures a continuous supply of oxygen that keeps pace with absorption by the bloodstream.

36. (a) Boyle's law describes the inverse relationship between pressure and volume: If volume decreases, pressure rises; if volume increases, pressure falls. (b) Dalton's law states that each of the gases that make up a mixture of gases contributes to the total pressure in proportion to its relative abundance; that is, all the partial pressures added together equal the total pressure exerted by the gas mixture. (c) Henry's law states that, at a given temperature, the amount of a particular gas that dissolves in a liquid is directly proportional to the partial pressure of that gas.

37. Such an injury can allow air into the pleural cavity. This pneumothorax breaks the fluid bond between the parietal pleura and the visceral pleura and allows the elastic fibers to recoil, causing the lung to collapse (atelectasis).

38. Both sneezing and coughing involve a temporary cessation of respiration (apnea).

39. Pulmonary volumes include resting tidal volume (averaging 500 ml), expiratory reserve volume (approximately 1200 ml), residual volume (averaging 1200 ml), minimal volume (30–120 ml), and inspiratory reserve volume (approximately 3600 ml). These values are determined experimentally. Respiratory capacities include inspiratory capacity, functional residual capacity, vital capacity, and total lung capacity. Respiratory capacities are determined by adding the values of various volumes.

40. The effect of pH on the hemoglobin saturation curve is called the Bohr effect. When the partial pressure of CO_2 rises, the pH drops; when the partial pressure of CO_2 declines, the pH rises. Hemoglobin molecules release more oxygen when the pH drops than they would when the pH rises. Hydrogen ions tend to "bump" oxygen molecules from the heme units. As the temperature rises, hemoglobin releases more oxygen; as the temperature declines, hemoglobin holds oxygen more tightly. For any partial pressure of oxygen, the higher the concentration of BPG, the more oxygen will be released by the hemoglobin molecules.

41. The DRG is the inspiratory center that contains neurons that control lower motor neurons innervating the external intercostal muscles and the diaphragm. The DRG functions in every respiratory cycle, whether quiet or forced. The VRG functions only during forced respiration—active exhalation and maximal inhalation. The neurons involved with active exhalation are sometimes said to form an expiratory center.

42. chemoreceptors sensitive to the P_{CO_2}, pH, and/or P_{O_2} of the blood or cerebrospinal fluid; changes in blood pressure in the aorta or carotid sinuses; stretch receptors that respond to changes in the volume of the lungs; irritating physical or chemical stimuli in the nasal cavity, larynx, or bronchial tree; and pain, changes in body temperature, and abnormal visceral sensations

Level 3 Critical Thinking and Clinical Applications

43. AVR = respiratory rate × (tidal volume − dead space). In this case, the dead air space is 200 ml (the anatomical dead air space plus the volume of the snorkel), therefore AVR = respiratory rate × (500 − 200). To maintain an AVR of 6 l/min, or 6000 ml/minute, the respiratory rate must be 6,000/(500-200), or 20 breaths per minute.

44. An increase in ventilation will increase the movement of venous blood back to the heart (recall the respiratory pump). This movement would increase venous return and thus help to increase blood pressure (Starling's law of the heart).

45. A person with chronic emphysema has constantly elevated levels of P_{CO_2} in the blood, due to an inability to eliminate CO_2 efficiently

as a result of the physical damage to the lungs. Over time, the brain ignores the stimulatory signals produced by the increased CO_2 and begins to rely on information from the peripheral chemoreceptors to set the pace of breathing (in other words, accommodation has occurred). The peripheral chemoreceptors also accommodate to the elevated CO_2 and respond primarily to the level of O_2 in the blood, increasing breathing when O_2 levels are low and decreasing breathing when O_2 levels are high. When pure O_2 was administered, chemoreceptors responded with fewer action potentials to the medulla oblongata, so Mr. B. stopped breathing.

46. While you were sleeping, the air that you were breathing was so dry that it absorbed more than the normal amount of moisture as it passed thorugh the nasal cavity. The loss of moisture made the mucous secretions quite viscous and harder for the cilia to move. Your nasal epithelia continued to secrete mucus, but very little of it moved; ultimately, the nasal congestion resulted. After the shower and juice, more moisture was transferred to the mucus, loosening it and making it easier to move, thus eliminating the problem.

47. A person with kyphosis has an abnormal thoracic curvature of the spine. The abnormal curvature interferes with rib movement during ventilation as well as lung expansion. As a result, the lungs cannot expand to their full potential. This restriction would decrease the inspiratory reserve and possibly the tidal volume, thus leading to a decreased vital capacity.

48. In anemia, the decreased ability of blood to carry oxygen is due to the lack of functional hemoglobin, red blood cells, or both. The disease does not interfere with the exchange of carbon dioxide within the alveoli nor with the amount of oxygen that will dissolve in the plasma. Because the chemoreceptors respond to dissolved gases and pH, as long as the concentrations of dissolved carbon dioxide, oxygen, and pH are normal, ventilation patterns should not change significantly.

49. The obstruction in Doris' right lung would not allow for gas exchange. Thus, the blood moving through the right lung would not oxygenate and would retain carbon dioxide. The retention of carbon dioxide in the blood would lead to a lower pH than the blood leaving the left lung. The lower pH would shift the oxygen–hemoglobin saturation curve to the left (the Bohr effect) for the right lung, compared with that for the left lung.

CHAPTER 24

Page 882

1. The mesenteries are double layers of serous membrane that support and stabilize the positions of organs in the abdominopelvic cavity and provide a route for the associated blood vessels, nerves, and lymphatic vessels.

2. Peristalsis would be more efficient in propelling intestinal contents. Segmentation is essentially a churning action that mixes intestinal contents with digestive fluids. Peristalsis consists of waves of contractions that not only mix the contents but also propel them along the digestive tract.

3. Parasympathetic stimulation increases muscle tone and activity in the digestive tract. A drug that blocks this activity would decrease the rate of peristalsis.

Page 887

1. The oral cavity is lined by a stratified squamous epithelium. This type of lining is very protective and so is located in areas of the body that receive a great deal of friction or abrasion.

2. Because the parotid salivary glands secrete salivary amylase, an enzyme that digests complex carbohydrates, damage to these glands would interfere with the digestion of carbohydrates.

3. The incisors are the type of tooth best suited for chopping (or cutting or shearing) pieces of relatively rigid food, such as raw vegetables.

Page 890

1. The muscularis externa of the esophagus is unusual because (1) it contains skeletal muscle cells along most of the length of the esophagus and (2) it is surrounded by an adventitia rather than a serosa.

2. The fauces is the opening between the oral cavity and the pharynx.

3. Swallowing (deglutition) is occurring.

Page 897

1. The larger the meal (especially in terms of protein), the more stomach acid is secreted. The hydrogen ions for the acid come from the blood that enters the stomach; therefore, the blood leaving the stomach will have fewer than normal hydrogen ions and will be decidedly alkaline—that is, have a higher pH. This phenomenon is referred to as the alkaline tide.

2. The vagus nerve contains parasympathetic motor fibers that can stimulate gastric secretions. This stimulation can occur even if food is not present in the stomach (the cephalic phase of gastric digestion). Cutting the branches of the vagus that supply the stomach would prevent this type of secretion from occurring and thereby decrease the chance of ulcer formation.

Page 910

1. The small intestine has several adaptations that increase surface area to increase its absorptive capacity. The walls of the small intestine are thrown into folds, the plicae circulares. The tissue that covers the plicae forms fingerlike projections, the villi. The cells that cover the villi have an exposed surface covered by small fingerlike projections, the microvilli. In addition, the small intestine has a very rich supply of blood vessels and lymphatic vessels, which transport the nutrients that are absorbed.

2. The cholecystokinin level in the blood would increase.

3. The hormone secretin, among other things, stimulates the pancreas to release fluid high in buffers to neutralize the chyme that enters the duodenum from the stomach. If the small intestine did not secrete secretin, the pH of the intestinal contents would be lower than normal.

4. Damage to the exocrine pancreas would most affect the digestion of fats (lipids). Enzymes for carbohydrate digestion are produced by salivary glands and the small intestine as well as by the pancreas. Enzymes for protein digestion are produced by the stomach and the small intestine as well as by the pancreas. Even though the digestion of carbohydrates and proteins would not be as complete as it is when the pancreas is functioning, some digestion would still take place. Because the pancreas is the primary source of lipases, lipid digestion would be most impaired.

Page 921

1. Chylomicrons are formed from the fats digested in a meal. A meal that is high in fat would increase the number of chylomicrons in the lacteals.

2. The removal of the stomach would interfere with the absorption of vitamin B_{12}. This vitamin requires intrinsic factor, which is produced by the parietal cells of the stomach.

3. A person can lose fluid and electrolytes faster than these substances can be replaced. This loss would result in dehydration and possibly death. Although it can be quite uncomfortable, constipation is not potentially life-threatening, because it does not interfere with any major body process that supports life. The few toxic waste products that are normally eliminated by way of the digestive system can move into the blood and can ultimately be eliminated by the kidneys.

Page 926: Level 1 Reviewing Facts and Terms

1. d **2.** a **3.** d **4.** d **5.** d **6.** c **7.** a **8.** c **9.** b **10.** a

11. d **12.** a **13.** d **14.** d **15.** d **16.** c **17.** a

18. (1) ingestion; (2) mechanical processing; (3) secretion; (4) digestion; (5) absorption; and (6) excretion

19. mucosa: the epithelial layer that performs chemical digestion and absorption of nutrients; submucosa: the connective tissue layer containing lymphatic and blood vessels and the submucosal nerve plexus; muscularis externa: the smooth muscle layer containing the myenteric nerve plexus; and serosa: the outermost layer epithelium and connective tissue that forms the visceral peritoneum (or connective tissue that forms the adventitia)

20. (1) neural mechanisms; (2) hormonal mechanisms; and (3) local mechanisms

21. (1) sensory analysis of material before swallowing; (2) mechanical processing through the actions of the teeth, tongue, and palatal surfaces; (3) lubrication by mixing with mucus and salivary secretions; and (4) limited digestion of carbohydrates and lipids

22. (1) parotid salivary glands: Stensen's duct; (2) sublingual salivary glands: Rivunus' ducts; and (3) submandibular salivary glands: Wharton's ducts

23. lubricating the mouth; moistening and lubricating materials in the mouth; dissolving chemicals that can stimulate the taste buds and provide sensory information about the material in the mouth; and initiating the digestion of complex carbohydrates before the material is swallowed. Salivary secretions are usually controlled by the ANS. Parasympathetic stimulation accelerates salivary secretion, whereas sympathetic innervation is believed to promote the secretion of small amounts of very thick saliva.

24. (1) incisors: clipping or cutting; (2) cuspids: tearing or slashing; (3) bicuspids: crushing, mashing, grinding; and (4) molars: crushing and grinding

25. The pharyngeal constrictors push the bolus toward the esophagus. The palatopharyngeus and stylopharyngeus muscles elevate the larynx. The palatal muscles raise the soft palate and adjacent portions of the pharyngeal wall.

26. (1) buccal phase; (2) pharyngeal phase; and (3) esophageal phase. Swallowing is controlled by the swallowing center of the medulla oblongata via the trigeminal and glossopharyngeal nerves. The motor commands originating at the swallowing center are distributed by cranial nerves V, IX, X, and XII. Along the esophagus, primary peristaltic contractions are coordinated by afferent and efferent fibers within the glossopharyngeal and vagus nerves, but secondary peristaltic contractions occur in the absence of CNS instructions.

27. The gastric glands are dominated by two types of secretory cells: parietal and chief cells. Parietal cells secrete intrinsic factor and hydrochloric acid. Intrinsic factor facilitates the absorption of vitamin B_{12} across the intestinal lining. The low pH of gastric juice kills most of the microorganisms ingested with food. The acid helps break down plant cell walls and the connective tissues in meat. The acidic environment is essential for the activation and function of pepsin, a protein-digesting enzyme. Chief cells secrete pepsinogen, an inactive proenzyme that is converted by the acid in the lumen to its active form, pepsin. The pyloric glands consist of G cells, which produce gastrin, and D cells, which release somatostatin, a hormone that inhibits gastrin release. Gastrin stimulates the secretion of both parietal and chief cells as well as contractions of the gastric wall that mix and stir the gastric contents.

28. (1) duodenum; (2) jejunum; and (3) ileum

29. The pancreas provides digestive enzymes as well as buffers that assist in the neutralization of chyme. The liver and gallbladder provide bile, which contains additional buffers and bile salts that facilitate the digestion and absorption of lipids. The liver is responsible for metabolic regulation, hematological regulation, and bile production. It is the primary organ involved with regulating the composition of circulating blood.

30. enterocrinin, which stimulates the submucosal glands of the duodenum; secretin, which stimulates the pancreas and liver to increase the secretion of water and buffers; cholecystokinin (CCK), which causes an increase in the release of pancreatic secretions and bile into the duodenum, inhibits gastric activity, and appears to have CNS effects that reduce the sensation of hunger; gastric inhibitory peptide (GIP), which stimulates insulin release at pancreatic islets and the activity of the duodenal submucosal glands; vasoactive intestinal peptide (VIP), which stimulates the secretion of intestinal glands, dilates regional capillaries, and inhibits acid production in the stomach; gastrin, which is secreted by G cells in the duodenum when they are exposed to large quantities of incompletely digested proteins; and, in small quantities, motilin, which stimulates intestinal contractions, villikinin, which promotes the movement of villi and associated lymph flow, and somatostatin, which inhibits gastric secretion

31. (1) resorption of water and compaction of the intestinal contents into feces; (2) absorption of important vitamins liberated by bacterial action; and (3) storage of fecal material prior to defecation

32. (1) Stretch receptors in the rectal walls promote a series of peristaltic contractions in the colon and rectum, moving feces toward the anus; and (2) the sacral parasympathetic system, also activated by the stretch receptors, stimulates peristalsis via motor commands distributed by the pelvic nerves.

33. (1) The rate of epithelial stem cell division declines; (2) smooth muscle tone decreases; (3) the effects of cumulative damage become apparent; (4) cancer rates increase; and (5) changes in other systems have direct or indirect effects on the digestive system.

Level 2 Reviewing Concepts
34. d **35.** c **36.** d **37.** d **38.** d **39.** a
40. through the enamel and the dentin
41. The folds of the epithelium greatly increase surface area for absorption.
42. Peristalsis consists of waves of muscular contractions that move along the length of the digestive tract. During a peristaltic movement, the circular muscles contract behind the digestive contents. Longitudinal muscles contract next, shortening adjacent segments. A wave of contraction in the circular muscles then forces the materials in the desired direction. Segmentation movements churn and fragment the digestive materials, mixing the contents with intestinal secretions. Because they do not follow a set pattern, segmentation movements do not produce directional movement of materials along the tract.
43. The stomach is protected by the mucous secretions of its epithelial lining, and by neural and hormonal control over the times and rates of acid secretion.
44. (1) The cephalic phase begins with the sight or thought of food. Directed by the CNS, this phase prepares the stomach to receive food. (2) The gastric phase begins with the arrival of food in the stomach; this phase is initiated by distension of the stomach, an increase in the pH of the gastric contents, and the presence of undigested materials in the stomach. (3) The intestinal phase begins when chyme starts to enter the small intestine. This phase controls the rate of gastric emptying and ensures that the secretory, digestive, and absorptive functions of the small intestine can proceed at reasonable efficiency.
45. Aspirin is acidic, and acids aggravate stomach ulcers. Also, aspirin promotes bleeding, interfering with the healing of ulcers.
46. After a heavy meal, bicarbonate ions pass from the parietal cells of the stomach into the extracellular fluid, causing the pH of the ex-

tracellular fluid to rise. As the extracellular fluid exchanges ions with the blood, the blood pH also increases.

Level 3 Critical Thinking and Clinical Applications
47. If a gallstone is small enough, it can pass through the common bile duct and block the pancreatic duct. Enzymes from the pancreas then cannot reach the small intestine. As they accumulate, the enzymes irritate the duct and ultimately the exocrine pancreas, producing pancreatitis.
48. The small intestine, especially the jejunum and ileum, are probably involved. Regional inflammation is the cause of Barb's pain. The inflamed tissue will not absorb nutrients; hence she is losing weight. Among the nutrients that are not absorbed are iron and vitamin B_{12}, which are necessary for formation of hemoglobin and red blood cells; hence, she has developed anemia.
49. You would expect to observe pain as the intestinal contents continued to accumulate and stretch the intestinal wall. The abdomen would be distended in the region proximal to the obstruction. Because material cannot leave the small intestine, increased distension and irritation would lead to vomiting, further increasing pressure on the intestine. The vomited material would contain bile, because little of the bile would be absorbed by the blocked intestine. No feces or intestinal gas would be produced because the blockage would prevent movement into the large intestine. Because the blockage is in the region of the jejunum, some absorption of nutrients would take place, but not as much as normal due to the distention and the irritation of the intestinal lining. Various symptoms of malnutrition would result.
50. The primary impact would be a reduction in the volume consumed because the person feels full after eating a small amount. This can result in significant weight loss.

CHAPTER 25

Page 939
1. The primary role of the TCA cycle in ATP production is to transfer electrons from substrates to coenzymes. These electrons carry energy that can then be used as an energy source for the production of ATP by the electron transport system.
2. The NADH produced by glycolysis cannot enter the mitochondria, where the enzymes of the electron transport chain are located. An intermediary in the mitochondrial membrane can, however, transfer the electrons from the NADH to a coenzyme within the mitochondria. In skeletal muscle cells, the intermediary transfers the electrons to FAD, whereas in cardiac muscle cells, a different intermediary is used that transfers the electrons to another NAD. In mitochondria, each NADH yields 3 molecules of ATP, whereas each $FADH_2$ yields just 2 molecules of ATP. The different intermediaries account for the difference in ATP yield.
3. A decrease in the level of cytoplasmic NAD would lead to a decrease in the amount of ATP production in mitochondria. The mitochondria depend on a supply of pyruvic acid from glycolysis. Glycolysis, in turn, requires NAD. A decrease in NAD would decrease the available pyruvic acid for the TCA cycle and thus would decrease overall ATP production.

Page 948
1. Pyridoxine (vitamin B_6) is an important coenzyme in deamination and transamination, the first steps in processing amino acids in the cell. A diet deficient in this vitamin would interfere with the ability to metabolize proteins.
2. Uric acid is the product of purine degradation in the body. The macromolecules that contain purines are the nucleic acids. Elevated levels of uric acid can indicate increased breakdown of nucleic acids.

3. High-density lipoproteins are considered to be beneficial because they reduce the amount of fat (including cholesterol) in the bloodstream by transporting fat back to the liver for storage or excretion in the bile.

Page 953

1. After you have eaten a high-carbohydrate meal, you would expect increased glycogenesis (the formation of glycogen) to occur in the liver.

2. Urea is formed from by-products of protein metabolism. During the postabsorptive state, many amino acids are being metabolized and the ammonia produced by deamination is converted to urea in the liver. Thus, the amount of urea in the blood increases.

3. Excess acetyl-CoA is likely converted into ketone bodies.

Page 960

1. An athlete in extensive training—adding muscle mass—would try to maintain a positive nitrogen balance.

2. Bile salts are necessary for the digestion and absorption of fats and fat-soluble vitamins. Vitamin A is a fat-soluble vitamin. A decrease in the amount of bile salts in the bile would result in a decreased ability to absorb vitamin A from food and over time could result in symptoms of a vitamin A deficiency.

Page 965

1. The BMR of a pregnant woman would be higher than her own BMR when she is not pregnant, due to the increased metabolism associated with support of the fetus as well as the added effect of fetal metabolism.

2. The vasoconstriction of peripheral vessels would decrease both blood flow to the skin and the amount of heat that the body can lose. As a result, the body temperature would increase.

3. Infants have higher surface-to-volume ratios than do adults, and the temperature-regulating mechanisms of the body are not fully functional at birth. As a result, infants must expend more energy to maintain body temperature, and they get cold more easily than do healthy adults.

Page 968: Level 1 Reviewing Facts and Terms

1. d 2. c 3. a 4. c 5. a 6. d 7. c 8. c 9. a 10. a
11. d 12. c 13. d 14. b 15. a 16. b 17. c 18. a 19. b
20. c 21. a 22. d 23. b

24. Metabolism is all of the chemical reactions occurring in the cells of the body. Anabolism is those chemical reactions that result in the synthesis of complex molecules from simpler reactants; products of anabolism are used for maintenance/repair, growth, and secretion. Catabolism is the breakdown of complex molecules into their building block molecules, resulting in the release of energy for the synthesis of ATP and related molecules.

25. $C_6H_{12}O_6 + O_2 \rightarrow CO_2 + 6 H_2O$

26. Lipoproteins are lipid–protein complexes that contain large insoluble glycerides and cholesterol, with a superficial coating of phospholipids and proteins. The major groups are chylomicrons, which consist of 95 percent triglyceride, are the largest lipoproteins, and carry absorbed lipids from the intestinal tract to the bloodstream; very low-density lipoproteins (VLDLs), which consist of triglyceride, phospholipid, and cholesterol and transport triglycerides to peripheral tissues; intermediate-density lipoproteins (IDLs), which are intermediate in size and composition between VLDLs and LDLs; low-density lipoproteins (LDLs, or "bad cholesterols"), which are mostly cholesterol and deliver cholesterol to peripheral tissues; and high-density lipoproteins (HDLs, or "good cholesterols"), which are equal parts protein and lipid (cholesterol and phospholipids) and transport excess cholesterol back to the liver for storage or excretion in bile.

27. carbohydrates: 4.18 C/g; lipids: 9.46 C/g; and proteins: 4.32 C/g

28. The BMR is the minimum, resting energy expenditure of an awake, alert person.

29. (1) radiation (heat loss as infrared waves); (2) conduction (heat loss to surfaces in physical contact); (3) convection (heat loss to the air); and (4) evaporation (heat loss with water being vaporized on the skin)

30. Hormones that increase the metabolic activities of all tissues are involved, including epinephrine, thyrotropin releasing hormone which in turn causes the release of thyroid stimulating hormone and thyroxine.

Level 2 Reviewing Concepts

31. d 32. b

33. Protein catabolism is the breakdown of proteins into constituent amino acids. These amino acids can be either reused or broken down via deamination. Deamination removes the nitrogen group from the amino acid, which is then excreted from the body (nitrogen loss)

34. The TCA reaction sequence is a cycle because the four-carbon starting compound (oxaloacetic acid) is regenerated at the end. Acetyl-CoA and oxaloacetic acid enter the cycle, and CO_2, NADH, ATP, $FADH_2$, and oxaloacetic acid leave the cycle.

35. Oxidative phosphorylation is the generation of ATP within mitochondria, through a reaction sequence that requires coenzymes and consumes oxygen. The electron transport system consists of a sequence of metalloproteins called cytochromes, which pass electrons (from H atoms) along in small steps, gradually releasing energy for the formation of ATP and producing water as a by-product.

36. A triglyceride is hydrolyzed, yielding glycerol and fatty acids. Glycerol is converted to pyruvic acid and enters the TCA cycle. Fatty acids are broken into two-carbon fragments by beta-oxidation inside mitochondria. The two-carbon compounds then enter the TCA cycle.

37. RNA is broken into its nucleotide building blocks, which are either recycled into new nucleic acids or catabolized to sugars (which enter glycolysis) and nitrogenous bases. The pyrimidines are converted into acetyl-CoA for the TCA cycle, and the purines are excreted.

38. The primary hormone of the absorptive state is insulin, which prevents a large surge in blood glucose after a meal. Insulin causes the liver to remove glucose from the hepatic portal circulation. During the postabsorptive state, blood glucose begins to decline, triggering the liver to release glucose via glycogenolysis and gluconeogenesis.

39. Liver cells can break down or synthesize most carbohydrates, lipids, and amino acids. The liver has an extensive blood supply and thus can easily monitor blood composition of these nutrients and regulate accordingly. The liver also stores energy in the form of glycogen.

40. The food pyramid indicates how much of each food group an individual should consume per day to ensure adequate intake of nutrients and calories. The placement of fats, oils, and sugars at the top of the food pyramid indicates that such foods are to be consumed very sparingly, whereas carbohydrates, placed at the bottom of the pyramid, are to be consumed in largest relative quantities.

41. The body cannot synthesize most of its required vitamins and minerals.

42. The brain contains the hypothalamus, the thermostat of the body. The hypothalamus regulates the ANS control of such mechanisms as sweating and shivering thermogenesis via negative feedback homeostatic mechanisms.

43. These terms refer to the high-density lipoproteins (HDL) and low-density lipoproteins (LDL), lipoproteins in the blood that transport cholesterol. HDL ("good cholesterol") transports excess cholesterol to the liver for storage or breakdown, whereas LDL ("bad cholesterol") transports cholesterol to peripheral tissues, which un-

fortunately may include the arteries. The buildup of cholesterol in the arteries is linked to cardiovascular disease.

Level 3 Critical Thinking and Clinical Applications

44. Mary's AVR = 10 breaths/min × (300 ml/breath − 150 ml/breath).
10 breaths/min × 150 ml/breath = 1500 ml/min = 1.5 l/min
daily gas consumption = 1.5 l/min × 60 min/h × 24 h/day = 2160 l/day
daily oxygen consumption = 2160 l air/day × 0.18 O_2 = 388.8 l O_2/day
BMR = 388.8 l O_2/day × 4.825 kcal/l O_2 = 1875 kcal (Calories)/day

45. During starvation, the body must use fat and protein reserves to supply the energy necessary to sustain life. Some of the protein that is metabolized for energy is the gamma globulin fraction of the blood, which is mostly composed of antibodies. This loss of antibodies, coupled with a lack of amino acids to synthesize new ones (as well as protective molecules such as interferon and complement proteins), renders an individual more susceptible than usual to contracting a disease and less likely to recover from it.

46. The response to anorexia is the same as in starvation. To supply the needed energy, the body utilizes reserves of fat and protein. Once the heart has exhausted its available fat reserves, it begins to break down contractile fibers to use the amino acids from the protein as an energy source. Eventually, the size of the heart and the force of cardiac contraction decrease. The weakened heart cannot contract as quickly as before, and bradycardia results. The combination of bradycardia and a decreased force of contraction contribute to hypotension. Ultimately, the heart will pump so inefficiently that it will not be able to serve the needs of the body's tissues; death will occur as the result of heart failure.

47. Colestipol would lead to a decrease in the plasma levels of cholesterol. Bile salts are necessary for the absorption of fats. If bile salts could not be absorbed, the absorption of fats—namely, cholesterol and triglycerides—would decrease. The result would be a decrease in cholesterol from a dietary source and a decrease in fatty acids that could be used to synthesize new cholesterol. In addition, the body would have to replace the bile salts that are being lost with the feces. The drop in cholesterol levels would be further enhanced because bile salts are formed from cholesterol.

48. Complicating factors of bulimia include weakening of the esophageal walls, the stomach and the structures of the oral cavity, electrolyte imbalances, edema and cardiac arrhythmias. These factors can cause esophageal ulcers or bleeding, rotting teeth, bleeding gums, general lethargy, tissue edema, and poor circulation. A patient who presents with any combination of these symptoms may be suffering from bulimia.

49. Ron's physician is concerned that he might overdo his lipid-soluble vitamin supplements. The lipid-soluble vitamins can be stored in the body if taken in excessive amounts. Hypervitaminosis is most commonly an overdose of Vitamin A, leading to hypertension, hair loss, joint pain, and liver enlargement.

CHAPTER 26

Page 981

1. The renal corpuscle, proximal convoluted tubule, distal convoluted tubule, and the proximal portions of the loop of Henle and collecting duct are all in the renal cortex. (In cortical nephrons most of the loops of Henle are in the cortex; in juxtamedullary nephrons, most of the loops of Henle are in the medulla.)

2. The pores of the glomerular capillaries will not allow substances the size of plasma proteins to pass into the capsular space, but the filtration slits of the podocyte will allow only the smallest plasma proteins to pass.

3. Damage to the juxtaglomerular apparatus of the nephron would interfere with the hormonal control of blood pressure.

Page 990

1. The primary components involved are the glomerular capillaries, the lamina densa, and the filtration slits of the podocytes.

2. When the plasma concentration of a substance exceeds its tubular maximum, the excess is not reabsorbed but is excreted in urine.

3. A decrease in blood pressure would reduce the blood hydrostatic pressure within the glomerulus and hence decrease the GFR.

Page 999

1. Aldosterone promotes Na^+ retention and K^+ secretion at the kidneys. In response to increased amounts of aldosterone, the K^+ concentration of urine would increase.

2. The secretion of H^+ by the nephron involves a countertransport mechanism with Na^+. If the concentration of Na^+ in the filtrate decreased, fewer hydrogen ions could be secreted. The result would be a tubular fluid with a higher pH.

3. With no juxtamedullary nephrons, there would not be a large osmotic gradient in the medulla. The kidneys would not be able to form concentrated urine.

4. When the amount of Na^+ in the tubular fluid passing through the distal convoluted tubule decreases, the cells of the macula densa are stimulated to release renin. Renin activates angiotensin, and this activation brings about an increase in blood pressure.

Page 1007

1. Urea, a nitrogenous waste, is formed during the metabolism of amino acids, which are obtained by the breakdown of proteins. Thus, a high-protein diet would lead to increased urea production and an increased amount of urea in the urine. Fluid volume might also increase as a result of the need to flush the excess urea.

2. An obstruction of a ureter would interfere with the passage of urine between the renal pelvis and the urinary bladder.

3. To control the micturition reflex, you must be able to control the external urinary sphincter, a ring of skeletal muscle formed by the urogenital diaphragm, that acts as a valve.

Page 1012: Level 1 Reviewing Facts and Terms

1. c **2.** a **3.** b **4.** b **5.** c **6.** d **7.** a **8.** d **9.** b **10.** c **11.** c **12.** d **13.** b **14.** b

15. The urinary system performs vital excretory functions and eliminates the organic waste products generated by cells throughout the body. It also regulates the volume and solute concentration of body fluids.

16. the kidneys, ureters, urinary bladder, and urethra

17. renal corpuscle (glomerulus/Bowman's capsule) → proximal convoluted tubule → loop of Henle → distal convoluted tubule → collecting duct → papillary duct → renal pelvis

18. proximal convoluted tubule: reabsorbs all the useful organic substrates from the filtrate; loop of Henle: reabsorbs over 90 percent of the water in the filtrate; and distal convoluted tubule: secretes into the filtrate waste products that were missed by filtration

19. The lamina densa is a specialized component of the basal lamina that surrounds the capillaries of the glomerulus. During filtration, it restricts the passage of large plasma proteins but permits the movement of smaller molecules, including albumin, many organic nutrients, and ions.

20. The juxtaglomerular apparatus secretes the enzyme renin and the hormone erythropoietin.

21. renal artery → segmental arteries → interlobar arteries → arcuate arteries → interlobular arteries → afferent arterioles → nephrons → interlobular veins → arcuate veins → interlobar veins → renal vein

22. (1) filtration: the selective removal of large solutes and suspended materials from a solution on the basis of size; requires a filtration membrane and hydrostatic pressure, as provided by gravity or by blood pressure; (2) reabsorption: the removal of water and solute molecules from the filtrate after it enters the renal tubules; and (3) secretion: the transport of solutes from the peritubular fluid, across the tubular epithelium, and into the tubular fluid

23. In peripheral capillary beds, angiotensin II causes powerful vasoconstriction of precapillary sphincters, elevating pressures in the renal arteries and their tributaries. At the nephron, angiotensin II causes the efferent arteriole to constrict, elevating glomerular pressures and filtration rates. At the PCT, it stimulates the reabsorption of sodium ions and water. In the CNS, angiotensin II triggers the release of ADH, stimulating the reabsorption of water in the distal portion of the DCT and the collecting system, and it causes the sensation of thirst. At the adrenal gland, angiotensin II stimulates the secretion of aldosterone by the cortex and epinephrine by the medulla oblongata. The aldosterone accelerates sodium reabsorption in the DCT and the cortical portion of the collection system. Epinephrine causes the heart rate and force of contraction to increase, elevating renal blood pressure.

24. ureters, urinary bladder, and urethra

Level 2 Reviewing Concepts
25. d 26. c 27. d 28. b 29. c 30. a 31. a
32. Proteins are too large to fit through the slit pores. Maintaining proteins in the plasma ensures the blood colloid osmotic pressure will oppose filtration and return water to the plasma.
33. Net filtration pressure = glomerular hydrostatic pressure-capsular hydrostatic pressure-blood colloid osmotic pressure:
NFP = (GHP−CsHP) − BCOP
34. autoregulation at the local level; hormonal regulation initiated by the kidneys; and autonomic regulation (sympathetic division of the ANS)
35. Cotransport occurs when two different ions are brought across a cell membrane together. Countertransport moves one ion into the cell while moving a second out. The ions involved in countertransport at the PCT are sodium and potassium, moving sodium from the tubular fluid to the blood, and potassium from the blood to the cells of the PCT.
36. As a result of facilitated diffusion and cotransport mechanisms, 99 percent of the glucose, amino acids, and other nutrients are reabsorbed before the filtrate leaves the PCT. A reduction of the solute concentration of the tubular fluid occurs due to active ion reabsorption of sodium, potassium, calcium, magnesium bicarbonate, phosphate, and sulfate ions. The passive diffusion of urea, chloride ions, and lipid-soluble materials further reduces the solute concentration of the tubular fluid and promotes additional water reabsorption.
37. (1) It is an efficient way to reabsorb solutes and water before the tubular fluid reaches the DCT and collecting system; and (2) it establishes a concentration gradient that will permit the passive reabsorption of water from urine in the collecting system.
38. In the distal convoluted tubule (DCT) and collecting duct, sodium–chloride cotransport and aldosterone-stimulated reabsorption of sodium ions adjust the osmolarity of the tubular fluid. The osmotic concentration of urine is controlled by variations in the water permeabilities of the distal portion of the DCT and the collecting ducts. The segments are impermeable to water unless they are exposed to antidiuretic hormone (ADH) from the posterior lobe of the pituitary gland. At normal concentrations of ADH, the distal portions of the DCT and the collecting ducts are somewhat permeable to water, and there is an osmotic flow of water out of the urine as it passes along

the collecting ducts. Under these conditions, urine entering the minor calyx has an osmolarity approaching 1200 milliosmoles/liter.
39. The urge to urinate usually appears when the urinary bladder contains about 200 ml of urine. The micturition reflex begins to function when the stretch receptors have provided adequate stimulation to the parasympathetic motor neurons. The activity in the motor neurons generates action potentials that reach the smooth muscle in the wall of the urinary bladder. These efferent impulses travel over the pelvic nerves, producing a sustained contraction of the urinary bladder.

Level 3 Critical Thinking and Clinical Applications
40. Truck drivers may not urinate as frequently as they should. Resisting the urge to urinate can result in urine backing up into the kidney. This backup puts pressure on the kidney tissues, which can lead to tissue death and ultimately to kidney failure.
41. A single sample is useful in detecting the presence or absence of specific solutes (such as glucose), but it is less useful in detecting the amount of fluid lost via urination or the amount of some substance (urea, creatinine, etc.) lost per day. The amount of water reabsorbed from the filtrate can change from moment to moment, making the concentration in any one sample variable. The concentration of a solute in a single sample is much more variable than the amount excreted in a 24-hour period.
42. Caffeine increases the glomerular filtration rate by increasing blood flow into the glomerular capillaries. The increase in filtrate formation and in the rate of flow through the renal tubule reduces the time available for the reabsorption of solutes. The alcohol acts as a diuretic, inhibiting ADH secretion and resulting in excess loss of water by the formation of dilute urine. The resulting dehydration leads to decreased secretion by body glands, including the salivary glands, which makes the mouth dry.
43. Susan may have a urinary tract infection, so her urine is likely to contain blood cells and bacteria.
44. By decreasing the active transport of sodium and chloride ions from the loop of Henle, the drug produces a more concentrated filtrate and a less concentrated interstitial fluid surrounding the loop of Henle. This combination of factors would lead to the production of a large volume of urine for two reasons: First, the more concentrated filtrate would retain water due to the increased osmolarity; second, less water would leave the descending limb because the osmotic gradient would not be as great. The increase in urine volume would lower blood volume. Because blood volume and blood pressure are directly related, if the blood volume decreases, the blood pressure would decrease (assuming no other critical factors change).
45. Arteriosclerosis contributes to hypertension (high blood pressure), which would trigger a baroreceptor reflex that would lead to a decrease in the level of ADH in the blood. Assuming that the arteriosclerosis is affecting all of Carlos' large arteries, including the renal arteries, the stiffening of the vessels would decrease the blood flow to the kidneys, thus triggering the release of renin from the juxtaglomerular apparatus. The renin would catalyze the conversion of angiotensinogen into angiotensin I, and the angiotensin I would be converted into angiotensin II at the lungs. The angiotensin II would stimulate the secretion of aldosterone, which would increase sodium and water reabsorption, thus increasing blood volume. The increase in blood volume would add to the hypertension, further decreasing the levels of ADH.
46. Because mannitol is filtered but not reabsorbed, drinking a mannitol solution would lead to an increase in the osmolarity of the filtrate. Less water would be reabsorbed, and an increased volume of urine would be produced.

47. Carbonic anhydrase catalyzes the reaction that forms carbonic acid, a source of hydrogen ions that are excreted by the kidneys. Hydrogen ion excretion is accomplished by an antiport system in which sodium ions are exchanged for hydrogen ions. Fewer hydrogen ions would be available, so less sodium would be reabsorbed, contributing to an increased osmolarity of the filtrate. In turn, an increased volume of urine and more-frequent urination would result.

CHAPTER 27

Page 1021

1. Drinking a pitcher of distilled water would temporarily lower your blood osmolarity (osmotic concentration). Because ADH release is triggered by increases in osmolarity, a decrease in osmolarity would lead to a decrease in the level of ADH in your blood.
2. Being in the desert without water, you would lose fluid through perspiration, urine formation, and respiration. As a result, the osmotic concentration of your plasma (and other body fluids) would increase.

Page 1027

1. Sweat is a hypotonic solution with lower sodium concentration than the ECF. Sweating causes a greater loss of water than sodium, increasing plasma sodium ion levels.
2. Disturbances in sodium balance are followed by dehydration or edema of tissues. Potassium ion imbalances, on the other hand, can lead to extensive muscle weakness or even paralysis when plasma concentrations are too low, and cardiac arrhythmias when the levels are too high. Potassium ion imbalances are therefore more dangerous than sodium ion imbalances.

Page 1038

1. A decrease in the pH of body fluids would stimulate the respiratory centers of the medulla oblongata. The result would be an increase in the respiratory rate.
2. The kidney tubules modify the pH of the filtrate by secreting H^+ or reabsorbing HCO_3^-. The pH of the tubular fluid must be kept above about 4.5, because H^+ secretion cannot continue against a large concentration gradient. The buffers allow the filtrate to take more H^+ without decreasing the pH below the critical level.
3. In a prolonged fast, fatty acids are mobilized and large numbers of ketone bodies are formed. These molecules are acids that lower the body's pH. (The lowered pH would eventually lead to ketoacidosis.)
4. In vomiting, large amounts of stomach acid are lost from the body. This acid is formed by the parietal cells of the stomach by taking H^+ from blood. Prolonged vomiting would lead to the excessive removal of H^+ from the blood to produce the acid, thus raising the body's pH and leading to metabolic alkalosis.

Page 1043: Level 1 Reviewing Facts and Terms

1. l **2.** h **3.** r **4.** a **5.** t **6.** c **7.** o **8.** b **9.** q **10.** e
11. f **12.** s **13.** d **14.** j **15.** i **16.** k **17.** m **18.** g **19.** p
20. n **21.** b **22.** c **23.** a **24.** b **25.** d **26.** b **27.** d **28.** a
29. d **30.** d **31.** b **32.** a
33. (1) antidiuretic hormone (ADH): stimulates water conservation at the kidneys and stimulates the thirst center; (2) aldosterone: determines the rate of sodium reabsorption and potassium secretion along the DCT and collecting system of the kidney; and (3) natriuretic peptides; reduce thirst and blocks the release of ADH and aldosterone
34. Aldosterone promotes sodium conservation in exchange for potassium excretion in the urine.

Level 2 Reviewing Concepts

35. d **36.** a **37.** c **38.** b **39.** b **40.** c **41.** a
42. Fluid balance is a state in which the amount of water gained each day is equal to the amount lost to the environment. It is vital that the water content of the body remain stable, because water is an essential ingredient of cytoplasm and accounts for about 99 percent of the volume of extracellular fluid. Electrolyte balance exists when there is neither a net gain nor a net loss of any ion in body fluids. It is important that the ionic concentrations in body water remain within normal limits; if levels of calcium or potassium become too high, for instance, cardiac arrhythmias can develop. Acid–base balance exists when the production of hydrogen ions precisely offsets their loss. The pH of body fluids must remain within a relatively narrow range; variations outside this range can be life-threatening.
43. Fluid shifts are rapid water movements between the ECF and the ICF, reaching equilibrium in a matter of minutes. They occur in response to osmotic changes in the ECF. They are caused by increasing or decreasing osmotic concentration of the ECF.
44. To go from mg/dl to mmol/l, multiply by 10 and divide by the atomic weight (35.5 g/mol) of chlorine: $(250 \text{ mg/dl} \times 10 \text{ dl/l})/(35.5 \text{ g/mol}) = 70.42 \text{ mmol/l}$.
45. The temperature rise accompanying a fever can increase water losses. For each degree the temperature rises above normal, the daily water loss increases by 200 ml.
46. Whenever sodium gains exceed sodium losses, the volume of the ECF increases but its osmolarity remains the same. When sodium losses exceed gains, the ECF volume decreases but there is no significant change in the osmolarity of the ECF.
47. A volatile acid is an acid that can leave solution and enter the atmosphere, such as carbonic acid. (b) Sulfuric acid is an example of a fixed acid. These are acids that do not leave solution. (c) Aerobic metabolism produces organic acids, such as lactic acid.
48. (1) protein buffer systems: These depend on the ability of amino acids to respond to changes in pH by accepting or releasing hydrogen ions. If the pH rises, the carboxyl group of the amino acid dissociates to release a hydrogen ion. If the pH drops, the amino group accepts an additional hydrogen ion to form an NH_3^+ group. The plasma proteins contribute to the buffering capabilities of the blood; inside cells, protein buffer systems stabilize the pH of the ECF by absorbing extracellular hydrogen ions or exchanging intracellular hydrogen ions for extracellular potassium. (2) carbonic acid–bicarbonate system: Most carbon dioxide generated in tissues is converted to carbonic acid, which dissociates into a hydrogen ion and a bicarbonate ion. Hydrogen ions released by dissociation of organic or fixed acids combine with bicarbonate ions, elevating the additional CO_2 is lost at the lungs. (3) phosphate buffer systems: This P_{CO_2} buffer system consists of $H_2PO_4^-$, a weak acid that, in solution, reversibly dissociates into a hydrogen ion and HPO_4^{2-}. The phosphate buffer system plays a relatively small role in regulating the pH of the ECF, because the ECF contains far higher concentrations of bicarbonate ions than phosphate ions; however, it is important in buffering the pH of the ICF.
49. Respiratory and renal mechanisms support buffer systems by secreting or absorbing hydrogen ions, controlling the excretion of acids and bases, and generating additional buffers.
50. Respiratory compensation is a change in the respiratory rate that helps stabilize the pH of the ECF. Increasing or decreasing the rate of respiration alters pH by lowering or raising the P_{CO_2}. When the P_{CO_2} goes down, the pH rises; when the P_{CO_2} increases, the pH decreases. Renal compensation is a change in the rates of hydrogen and bicarbonate ion secretion or reabsorption in response to changes in plasma pH. Tubular hydrogen ion secretion results in the diffusion of bicarbonate ions into the ECF.
51. Respiratory disorders result from abnormal carbon dioxide levels in the ECF. An imbalance exists between the rate of CO_2 removal at the lungs and its generation in other tissues. Metabolic disorders are

caused by the generation of organic or fixed acids or by conditions affecting the concentration of bicarbonate ions in the ECF.

52. Acidosis is a decline in the pH of body fluids. Respiratory acidosis (hypercapnia) results from an abnormally high level of carbon dioxide, usually caused by hypoventilation. Metabolic acidosis occurs when bicarbonate ion levels fall, reducing the effectiveness of this buffer system. It can result from over-production of fixed or organic acids, impaired ability to secrete H^+ ions at the kidney, or during severe bicarbonate loss.

53. (a) Excessive salt and water intake causes an increase in total blood volume and blood pressure. Increased blood pressure results in decreased ADH secretion and reduced renin secretion from the kidneys. Decreased renin results in a decreased production of angiotensin II, which reduces the rate at which aldosterone is secreted. Increased sodium ions and increased blood pressure cause the secretion of natriuretic peptides, inhibiting ADH secretion and sodium ion reabsorption in the DCT. These changes cause increased loss of sodium in the urine and an increase in the volume of urine produced. (b) If the amount of salt ingested is excessive, the urine volume will rise and the concentration of salt in the urine will be high. If the amount of water is excessive, the urine volume will increase and the concentration of sodium ions in the urine will decrease. (c) If the amount of salt and water ingested in food exceeds the amount needed to maintain a constant ECF composition, the total blood volume and the blood pressure will increase.

54. Since sweat is usually hypotonic, the loss of a large volume of sweat causes hypertonicity in body fluids. The loss of fluid volume is primarily from the interstitial space, which leads to a reduction in plasma volume and an increase in the hematocrit. Severe dehydration can cause the blood viscosity to increase substantially, resulting in an increased work load on the heart, ultimately increasing the probability of heart failure.

Level 3 Critical Thinking and Clinical Applications

55. The young boy has metabolic and respiratory acidosis. Sustained hypoventilation during drowning leads to decreased arterial and oxygen-starved tissues generate large quantities of lactic acid. Prompt emergency treatment is essential; the usual procedure involves some form of artificial or mechanical respiratory assistance coupled with the intravenous infusion of an isotonic solution containing sodium lactate, sodium gluconate, or sodium bicarbonate.

56. The most likely cause is acidosis. The disorientation could be the result of decreased pH in the ICF of the brain. The hyperventilation would be a response to acidosis, because hyperventilation removes carbon dioxide, the source of a volatile acid, carbonic acid. Finally, in acidosis, hydrogen ions move into the ICF and potassium ions move out in exchange. This exchange, coupled with increased potassium ion retention at the kidney due to increased hydrogen ion excretion, could account for the hyperkalemia.

57. As long as the ratio of bicarbonate ion to carbonic acid in the blood is 20:1, the pH of the blood will be 7.4. If the bicarbonate is significantly elevated without a change in pH, then the concentration of carbon dioxide and carbonic acid in the blood must also be increased. This condition is consistent with emphysema. Persons with emphysema have lost surface area and ventilation capacity and cannot efficiently clear the carbon dioxide from the blood. The result is a chronic elevation in the level of carbon dioxide and carbonic acid. To maintain homeostasis, the body makes adjustments, primarily at the kidneys, to raise the bicarbonate level appropriately to keep the pH in the normal range.

58. Digestive secretions contain high levels of bicarbonate, so persons with diarrhea can lose significant amounts of this important ion, leading to acidosis. We would expect Milly's blood pH to be lower than 7.4 and the pH of her urine to be low (due to increased renal excretion of hydrogen ion). We would also expect an increase in the rate and depth of breathing as the respiratory system tries to compensate by eliminating carbon dioxide.

59. The hypertonic solution will cause fluid to move from the ICF to the ECF, further aggravating Yuka's dehydration. The slight increase in pressure and osmolarity of the blood should lead to an increase in ADH, although ADH levels are probably quite high already. Urine volume would probably increase, because much of the glucose would not be reabsorbed. The osmolarity of the tubular filtrate would increase, decreasing water reabsorption and increasing urine volume.

60. Patient 1 has compensated respiratory alkalosis. Patient 2 has acute metabolic acidosis due to generation or retention of organic or fixed acids. Patient 3 has acute respiratory acidosis. Patient 4 has metabolic alkalosis.

CHAPTER 28

Page 1064

1. On a warm day, the cremaster muscle (as well as the dartos muscle) would be relaxed so that the scrotal sac could descend away from the warmth of the body and cool the testes.

2. The dilation of the arteries within the penis allows blood flow to increase and the vascular chambers to become engorged with blood, resulting in erection.

3. FSH is required for the production of ABP, a protein that binds testosterone and keeps a high level of that hormone available to support spermatogenesis. Low levels of FSH would lead to low levels of testosterone in the seminiferous tubules and thus a lower rate of sperm production and a low sperm count.

Page 1077

1. The blockage of both uterine tubes would cause sterility.

2. The acidic pH of the vagina helps prevent bacterial, fungal, and parasitic infections in this area.

3. The functional layer of the endometrium sloughs off during menstruation.

4. The blockage of a single lactiferous sinus would not interfere with the delivery of milk to the nipple, because each breast generally has 15–20 lactiferous sinuses.

Page 1082

1. If the LH surge did not occur during an ovarian cycle, ovulation and corpus luteum formation would not occur.

2. Progesterone is responsible for the functional maturation and secretion of the endometrium. A blockage of progesterone receptors would inhibit the development of the endometrium (and would make the uterus unprepared for pregnancy).

3. A decline in the levels of estrogens and progesterone signals the beginning of the menses, the end of the uterine cycle.

Page 1087

1. An inability to contract the ischiocavernosus and bulbospongiosus muscles would interfere with a male's ability to ejaculate and to experience orgasm.

2. As the result of parasympathetic stimulation in females during sexual arousal, the erectile tissues of the clitoris engorge with blood, the secretion of cervical and vaginal glands increases, blood flow to the walls of the vagina increases, and the blood vessels in the nipples engorge.

3. At menopause, circulating estrogen levels begin to drop. Estrogen has an inhibitory effect on FSH (and on GnRH). As the level of estrogen declines, the levels of FSH rise and remain high.

Page 1092: Level 1 Reviewing Facts and Terms
1. c   **2.** d   **3.** b   **4.** a   **5.** c   **6.** a   **7.** d   **8.** d
9. c   **10.** c   **11.** d   **12.** d   **13.** b   **14.** a   **15.** c   **16.** b
17. a   **18.** d
19. Structures in common include reproductive organs (gonads) that produce gametes and hormones; ducts that receive and transport gametes; accessory glands and organs that secrete fluids into these or other excretory ducts; and external genitalia.
20. Seminiferous tubules → rete testis → efferent ducts → epididymis → ductus deferens → ejaculatory duct → urethra
21. The accessory organs/glands include the seminal vesicles, prostate gland, and the bulbourethral glands. The major functions of these glands are activating the spermatozoa, providing the nutrients sperm need for motility, propelling sperm and fluids along the reproductive tract, and producing buffers that counteract the acidity of the urethral and vaginal contents.
22. interstitial cells (cells of Leydig): produce male sex hormones (androgens), the most important one being testosterone; sustentacular cells: maintain the blood–testis barrier, support spermatogenesis and spermiogenesis, and secrete inhibin, androgen-binding protein, and Müllerian-inhibiting factor.
23. (1) monitoring and adjusting the composition of tubular fluid; (2) acting as a recycling center for damaged spermatozoa; and (3) storing spermatozoa and facilitating their functional maturation
24. (1) prostatic urethra; (2) membranous urethra; and (3) penile urethra
25. spermatozoa: normal count is 20–100 million/ml; seminal fluid: a mixture of glandular secretions with distinctive ionic and nutrient composition; enzymes: a protease that may help dissolve mucous secretions in the vagina and seminalplasmin, an antibiotic enzyme that kills a variety of bacteria
26. stimulating spermatogenesis and promoting the functional maturation of spermatozoa; maintaining the male reproductive accessory organs; determining male secondary sex characteristics; stimulating metabolic operations, especially those concerned with protein synthesis and muscle growth; and influencing brain development by stimulating sexual behaviors and sexual drive
27. producing and releasing immature female gametes (oocytes); secreting female sex hormones, including estrogens and progestins; and secreting inhibin, involved in the feedback control of pituitary FSH production
28. (1) the formation of primary follicles; (2) the formation of secondary follicles; (3) the formation of a tertiary follicle; (4) ovulation; and (5) the formation and degeneration of the corpus luteum
29. The myometrium is the outer muscular layer; the endometrium is the inner glandular layer; and the perimetrium is an incomplete serosal layer.
30. (1) It serves as a passageway for the elimination of menstrual fluids; (2) it receives the penis during sexual intercourse and holds spermatozoa prior to their passage into the uterus; and (3) in childbirth, it forms the lower portion of the birth canal through which the fetus passes during delivery.
31. The clitoris is a structural component of the external genitalia of the female. It is the female equivalent of the penis; they are derived from the same embryonic structures. Internally, the clitoris contains erectile tissue comparable to the corpus spongiosum of the penis. The clitoris becomes engorged with blood during sexual arousal and provides pleasurable sensations.
32. The lesser vestibular glands discharge their secretions onto the exposed surface of the vestibule, keeping it moistened. During arousal, a pair of ducts discharges the secretions of the greater vestibular glands into the vestibule near the posterolateral margins of the vaginal entrance.
33. secretory lobules of glandular tissue (lobes) → ducts → lactiferous duct → lactiferous sinus → open onto the surface of each nipple

Level 2 Reviewing Concepts
34. c   **35.** c   **36.** c   **37.** a
38. The reproductive system is the only physiological system that isn't required for survival of the individual.
39. In males, the gonads (testes) secrete androgens and produce one-half billion gametes (sperm) each day. The sperm are mixed with secretions, forming semen, and are expelled from the body during ejaculation. In females, the gonads (ovaries) secrete estrogen and progesterone and release one oocyte (gamete) every month. The oocyte may be fertilized within the oviduct.
40. Meiosis is the two-step nuclear division resulting in four haploid cells from one diploid cell. In males, four sperm are produced from each diploid cell. In females, only 1 oocyte (plus up to three polar bodies) is produced from each diploid precursor.
41. Males produce gametes from puberty until death; females produce gametes only from menarche to menopause. Males produce many gametes at a time, females produce one to two per 28 day cycle. Males release mature gametes that have completed meiosis; females release secondary oocytes held in metaphase of meiosis II.
42. The corpora cavernosa extend along the length of the penis as far as the neck of the penis, and the erectile tissue within each corpus cavernosum surrounds a central artery. The slender corpus spongiosum surrounds the urethra. This erectile body extends from the superficial fascia of the urogenital diaphragm to the tip of the penis, where it expands to form the glans. The sheath surrounding the corpus spongiosum contains more elastic fibers than do the corpora cavernosa, and the erectile tissue contains a pair of arteries. Erection occurs when the parasympathetic neurons of the penile arteries release nitric oxide, causing the smooth muscles in the arterial walls to relax. The vessels dilate, blood flow increases, the vascular channels become engorged with blood, and erection of the penis occurs.
43. The normally acidic environment of the vagina serves to prevent the growth of pathogens.
44. (1) menses: the interval marked by the degeneration and loss of the functional zone of the endometrium; lasts from 1–7 days, and 35–50 ml of blood is lost; (2) proliferative phase: growth and vascularization result in the complete restoration of the functional zone; lasts from the end of menses until the beginning of ovulation, around day 14; (3) secretory phase: the endometrial glands enlarge, accelerating their rates of secretion and the arteries elongate and spiral through the tissues of the functional zone; occurs under the combined stimulatory effects of progestins and estrogens from the corpus luteum; begins at ovulation and persists as long as the corpus luteum remains intact
45. As follicular development proceeds, the concentration of circulating estrogen rises. Secondary follicles contain increased numbers of granulosa cells, and the level of circulating inhibin rises. The rising estrogen and inhibin levels inhibit hypothalamic secretion of GnRH and pituitary production and release of FSH. As the follicles develop and estrogen levels rise, the pituitary output of LH gradually increases. Estrogens, FSH, and LH continue to support follicular development and maturation despite a gradual decline in FSH levels. In the second week of the ovarian cycle, estrogen levels sharply increase—and the tertiary follicle enlarges in preparation for ovulation. By day 14, estrogen levels peak, triggering a massive outpouring of LH from the anterior pituitary. The rupture of the follicular wall results in ovulation. Next, LH stimulates the formation of the corpus luteum,

which secretes moderate amounts of estrogens but large amounts of progesterone, the principal hormone of the postovulatory period. About 12 days after ovulation, declining progesterone and estrogen levels stimulate hypothalamic receptors and GnRH production increases, leading to increased FSH and LH production in the anterior pituitary; the cycle begins again.

46. The corpus luteum degenerates and progesterone and estrogen levels drop, resulting in the endometrial breakdown of menses. Next, rising levels of FSH, LH, and estrogen stimulate the repair and regeneration of the functional zone of the endometrium. During the postovulatory phase, the combination of estrogen and progesterone cause the enlargement of the endometrial glands and an increase in their secretory activity.

47. During sexual arousal, erotic thoughts or physical stimulation of sensory nerves in the genital region increases the parasympathetic outflow over the pelvic nerve, leading to erection of the clitoris or penis. Orgasm is the intensely pleasurable sensation associated with perineal muscle contraction and ejaculation in males, and with uterine and vaginal contractions and perineal muscle contraction in females. These processes are comparable in males and females.

48. Women age 45–55 undergo menopause, the time that ovulation and menstruation cease, accompanied by a sharp and sustained rise in the production of GnRH, FSH, and LH and a drop in the concentrations of circulating estrogen and progesterone. The decline in estrogen levels leads to reductions in the size of the uterus and breasts, accompanied by a thinning of the urethral and vaginal walls. In addition to neural and cardiovascular effects, reduced estrogen concentrations have been linked to the development of osteoporosis, presumably because bone deposition proceeds at a slower rate. Men age 50–60+ undergo the male climacteric, a time when circulating testosterone levels begin to decline and circulating levels of FSH and LH rise. Although sperm production continues, there is a gradual reduction in sexual activity in older men.

49. Birth control pills contain progestins and estrogens, which are administered in a cyclic fashion, beginning 5 days after the start of menses and continuing for the next 3 weeks. Over the fourth week, placebo pills or no pills are taken. The slightly elevated levels of progestin and estrogen inhibit GnRH release at the hypothalamus and the release of FSH and LH from the anterior lobe of the pituitary gland. Without FSH, primordial follicles do not initiate development and an LH surge without ovulation cannot occur.

Level 3 Critical Thinking and Clinical Applications

50. There is no direct entry into the abdominopelvic cavity in males as there is in females. In females, the urethral opening is in close proximity to the vaginal orifice, so infectious organisms can exit from the urethral meatus and enter the vagina. They can then proceed through the vagina to the uterus, into the uterine tubes, and finally into the peritoneal cavity.

51. The sacral region of the spinal cord contains the parasympathetic centers that control the external genitalia, so damage to this area of the spinal cord would interfere with the ability to achieve an erection by way of parasympathetic stimuli. However, erection can also occur by way of sympathetic centers in the lower thoracic region of the spinal cord. Visual, auditory, or cerebral stimuli can result in decreased tone in the arteries serving the penis; increased blood flow and erection result. Tactile stimulation of the penis, however, would not generate an erection.

52. Granulosa cells produce estrogens, so a tumor involving these cells would lead to elevated levels of estrogens similar to or higher than those of an adult female. We would expect to observe the development of secondary sex characteristics.

53. It suggests that a certain amount of body fat is necessary for menstrual cycles to occur. The nervous system appears to respond to circulating levels of the hormone leptin; when leptin levels fall below a certain set point, menstruation ceases. Without proper fat reserves, a woman might not be able to have a successful pregnancy. To avoid harm to the mother and the death of a fetus, the body prevents pregnancy by shutting down the ovarian cycle and thus the menstrual cycle. When appropriate energy reserves are available, the cycles begin again.

CHAPTER 29

Page 1108

1. The inner cell mass of the blastocyst eventually develops into the embryo.

2. Cells of the mesoderm migrate to form the yolk sac, in which blood vessels appear. The yolk sac becomes an important site of blood cell formation. Improper development of this extraembryonic membrane would thus affect the development and function of the cardiovascular system.

3. After fertilization, the developing trophoblasts and, later, the placenta produce and release the hormone hCG. Sue is pregnant.

4. Placental functions include (1) supplying the developing fetus with a route for gas exchange, nutrient transfer, and waste product elimination; and (2) producing hormones that affect maternal systems.

Page 1120

1. During pregnancy, blood flow through the placenta reduces the volume of blood in the mother's systemic circuit. This reduction stimulates an increase in maternal blood volume to compensate.

2. Progesterone reduces uterine contractions. A decrease in progesterone at any time during the pregnancy could lead to uterine contractions and, in late pregnancy, labor.

3. An increase in the levels of GnRH, FSH, LH, and sex hormones in blood mark the onset of puberty.

Page 1127

1. A person who is heterozygous for curly hair would have one dominant allele and one recessive allele for that trait. The person's phenotype would be "curly hair."

2. One reason that children are not identical copies of their parents is that during meiosis, parental chromosomes are randomly distributed such that each gamete has a unique set of chromosomes. Also, mutations and the crossing-over that occurs during meiosis introduce new variations.

Page 1131: Level 1 Reviewing Facts and Terms

1. a **2.** b **3.** d **4.** b **5.** b **6.** d **7.** a **8.** b **9.** c **10.** c **11.** b **12.** a **13.** a **14.** d **15.** a **16.** b **17.** d **18.** c **19.** d

20. When a sperm contacts the oocyte, their plasma membranes fuse. The oocyte is then activated: Its metabolic rate rises; it completes meiosis II; and a cortical reaction occurs that prevents additional sperm from entering. (Vesicles beneath the oocyte surface fuse with the cell membrane and discharge their contents.) The male and female pronuclei fuse (amphimixis), and the zygote begins preparing for the first cleavage division.

21. During the first trimester, the rudiments of all major organ systems appear. During the second trimester, the organs and organ systems complete most of their development. During the third trimester, the fetus grows rapidly and most of the major organ systems become fully functional.

22. **(a)** (1) yolk sac; (2) amnion; (3) allantois; and (4) chorion
(b) The yolk sac forms from endoderm and mesoderm; it is an important site of blood cell formation. The amnion forms from ecto-

derm and mesoderm; it encloses the fluid that surrounds and cushions the developing embryo and fetus. The allantois forms from endoderm and mesoderm; its base gives rise to the urinary bladder. The chorion forms from mesoderm and trophoblast; circulation through chorionic vessels provides a "rapid transit system" for blood and nutrients

23. embryonic disc: an oval sheet of cells, develops at days 9–10 from the inner cell mass of the blastocyst; blastocyst: a hollow ball of cells formed after 5 days of cleavage; morula: a solid ball of cells formed during the first 5 days after fertilization; zygote: the fertilized ovum, prior to the start of cleavage. Cleavage begins immediately after fertilization and ends with the formation of the blastocyst

24. (1) dilation stage: It begins with the onset of true labor, as the cervix dilates and the fetus begins to move toward the cervical canal; late in this stage, the amniochorionic membrane ruptures. (2) expulsion stage: It begins as the cervix dilates completely and continues till the fetus has completely emerged from the vagina (delivery). (3) placental stage: The uterus gradually contracts, tearing the connections between the endometrium and the placenta and ejecting the placenta.

25. Relaxin produced by the placenta softens the pubic symphysis, and the weight of the fetus then deforms the cervical orifice. Deformation of the cervix and the rising estrogen levels promote the release of oxytocin, and the already stretched muscles become even more excitable.

26. (1) neonatal period (birth to 1 month): The newborn becomes relatively self-sufficient and begins performing respiration, digestion, and excretion on its own. Heart rates and fluid requirements are higher than those of adults. Neonates have little ability to thermoregulate. (2) infancy (1 month to 2 years): Major organ systems (other than those related to reproduction) become fully operational. (3) Childhood (2 years to puberty): Growth continues; body proportions change significantly.

27. Three events interact to promote increased hormone production and sexual maturation at puberty: (1) The hypothalamus increases its production of GnRH; (2) the anterior pituitary becomes more sensitive to the presence of GnRH and the circulating levels of FSH and LH rise rapidly; and (3) ovarian or testicular cells become more sensitive to FSH and LH. Puberty initiates adolescence, which includes gametogenesis and the production of male or female sex hormones that stimulate the appearance of secondary sexual characteristics and behaviors.

28. Maturity is the cessation of growth, normally occurring in the late teens or early twenties. An individual is reproductively mature immediately after puberty, but skeletal maturation continues until the last epiphyseal cartilages close, sometime after age 20.

29. Senescence is the process of aging, which affects all organ systems. Examples include the loss of elasticity in skin; the declining rate of bone deposition and degenerative changes in joints; reduced muscular strength; impaired coordination, memory, and intellectual function; reduced production of and sensitivity to hormones; cardiovascular problems; reduced immune system function; reduced elasticity in the lungs; decreased muscle tone along the digestive tract; reduced muscle tone in the urinary system; and the functional impairment of the reproductive system.

Level 2 Reviewing Concepts
30. d   31. b   32. d
33. Polyspermy is prevented by the cortical reaction, in which oocyte vesicles release their sperm-inhibiting contents via exocytosis.
34. The post-implantation embryo obtains nutrients through the chorionic villi and later the placenta. The placenta develops during placentation.

35. As cells develop, they communicate with one another via released chemical signals. This communication is referred to as induction.
36. The placenta produces human chorionic gonadotropin, which maintains the integrity of the corpus luteum and promotes the continued secretion of progesterone (keeping the endometrial lining functional) human placental lactogen and placental prolactin, which help prepare the mammary glands for milk production; and relaxin, which increases the flexibility of the pubic symphysis, causes dilation of the cervix, and suppresses the release of oxytocin by the hypothalamus, delaying the onset of labor contractions.
37. The respiratory rate and tidal volume increase, allowing the lungs to obtain the extra oxygen and remove the excess carbon dioxide generated by the fetus. Maternal blood volume increases, compensating for blood that will be lost during delivery. Requirements for nutrients and vitamins climb 10–30 percent, reflecting the fact that some of the mother's nutrients go to nourish the fetus. The glomerular filtration rate increases by about 50 percent, which corresponds to the increased blood volume and accelerates the excretion of metabolic wastes generated by the fetus.
38. By delivery, the uterus has quadrupled in length and grown to almost 20 times its original weight. The myometrium is stretched tremendously, in association with an increase in smooth muscle contractions. Progesterone released by the placenta inhibits the uterine smooth muscle to prevent more-violent contractions. The calming action of progesterone is opposed by estrogens, oxytocin, and prostaglandins; later in pregnancy, the production of estrogens and prostaglandins increases.
39. Progesterone released by the placenta has an inhibitory effect on the uterine smooth muscle.
40. Positive feedback mechanisms ensure that labor contractions continue until delivery has been completed.
41. An episiotomy is a surgical incision in the perineal musculature to prevent tearing of the perineal area during the expulsion phase of labor. It is done to allow passage of the fetus without ripping perineal tissues. A surgical incision is easier to repair than ripped, damaged tissue.
42. The neonate must fill its lungs (which are collapsed and filled with fluid at birth) with air, changing the pattern of cardiovascular circulation due to changes in blood pressure and flow rates. It must excrete the mixture of debris that has collected in the fetal digestive system. The neonate must obtain nourishment from a new source—the mother's mammary glands. Neonatal fluid requirements are high, because the infant cannot concentrate its urine significantly. The infant also has little ability to thermoregulate, although as it grows its insulating adipose tissue increases and its metabolic rate also rises.
43. (a) Monozygotic ("identical") twins result from either the separation of the blastomeres early in cleavage or from the splitting of the inner cell mass prior to gastrulation. The genetic makeup of the pair is almost identical, because both formed from the same gamete. Dizygotic twins result from two separate oocytes that were subsequently fertilized. (b) Genotype refers to all the chromosomes and their genes of the individual. Phenotype refers to the physical and physiological characteristics of the individual. The genotype is a primary determinant of the phenotype, although other factors (injury, disease, environment) can also play a role. (c) If both chromosomes of a homologous pair carry the same allele of a particular gene, the individual is homozygous for that trait. If the two chromosomes carry different alleles for that gene, the individual is heterozygous for that trait. (d) In simple inheritance, phenotypes are determined by interactions between a single pair of alleles. In polygenic inheritance, interactions occur among multiple genes.
44. (a) dominant (b) recessive (c) X-linked (d) autosomal

45. The trait of red–green color blindness is carried on the X chromosome. Women have two X chromosomes, so they are color-blind only if they are homozygous recessive. A color-blind male mating with a heterozygous female has a 50% chance of producing a color-blind female as opposed to a normally sighted female. All the sons produced by a homozygous color-blind female will demonstrate red-green color blindness.

46. Genetic recombination, which occurs at synapsis during meiosis, results in increased variability in traits carried by different gametes and therefore in offspring.

47. The Human Genome Project has transcribed the entire human genome. The project has already helped to identify the genes that are responsible for inherited disorders and will localize the specific chromosomes involved.

Level 3 Critical Thinking and Clinical Applications

48. The probability that this couple's daughters will have hemophilia is 0, because each daughter will receive a normal allele from her father. There is a 50 percent chance that a son will have hemophilia, because each son has a 50 percent chance of receiving the mother's normal allele and a 50 percent chance of receiving the father's recessive allele.

49. Excess calcium could cause tetanic contractions of the smooth muscle of the uterus. These contractions would cut off the blood supply to the placenta and deprive the fetus of oxygen.

50. During nursing, the mechanical stimulus of suckling triggers a neural reflex that leads to the release of oxytocin from the posterior pituitary. Oxytocin enters the bloodstream and stimulates smooth muscle in the ducts of the mammary glands to contract and move milk into the lactiferous sinuses—the milk let-down reflex. Oxytocin receptors are also located on smooth muscle cells of the uterus, so the rise in oxytocin sometimes leads to uterine contractions as well, producing a feeling similar to menstrual cramping.

51. In adults, heat is lost across the skin. In infants, the ratio of surface area to volume is very high, so infants lose heat very quickly. To maintain a constant body temperature despite the heat loss, cellular metabolism must be high. Cellular metabolism requires oxygen, so increased metabolism demands an increased respiratory rate. Cardiac output must increase as well to move the blood from the lungs to the tissues. The range of contraction in the neonate heart is limited, so the greatest increase in cardiac output is achieved by increasing the heart rate.

52. The active ingredient of aspirin blocks enzymes necessary for the production of prostaglandins. Prostaglandins promote uterine contractions during parturition; blocking their synthesis would interfere with normal labor. Prostaglandins are also involved in blood clotting. Taking aspirin could increase the time it would take for blood to clot following delivery, leading to excess blood loss and the possible need for a transfusion.

53. It is very unlikely that the baby's condition is the result of a viral infection contracted during the third trimester. The development of organ systems occurs during the first trimester. By the end of the second trimester, most organ systems are fully formed. During the third trimester, the fetus undergoes tremendous growth but very little new organ formation.

Weights and Measures

Accurate descriptions of physical objects would be impossible without a precise method of reporting the pertinent data. Dimensions such as length and width are reported in standardized units of measurement, such as inches or centimeters. These values can be used to calculate the **volume** of an object, a measurement of the amount of space the object fills. Mass is another important physical property. The **mass** of an object is determined by the amount of matter the object contains; on Earth the mass of an object determines the object's weight.

In the United States, length and width are typically described in inches, feet, or yards; volumes in pints, quarts, or gallons; and weights in ounces, pounds, or tons. These are units of the **U.S. system** of measurement. Table 1 summarizes the familiar and unfamiliar terms used in the U.S. system. For reference purpos-

es, this table also includes a definition of the "household units," popular in recipes. The U.S. system can be very difficult to work with, because there is no logical relationship among the various units. For example, there are 12 inches in a foot, 3 feet in a yard, and 1760 yards in a mile. Without a clear pattern of organization, the conversion of feet to inches or miles to feet can be confusing and time-consuming. The relationships among ounces, pints, quarts, and gallons are no more logical than those among ounces, pounds, and tons.

In contrast, the **metric system** has a logical organization based on powers of 10, as indicated in Table 2. For example, a **meter (m)** is the basic unit for the measurement of size. For measurements of larger objects, data can be reported in **dekameters** (*deka,* ten), **hectometers** (*hekaton,* hundred), or **kilometers (km;** *chilioi,* thousand); for smaller objects, data

TABLE 1   THE U.S. SYSTEM OF MEASUREMENT

Physical Property	Unit	Relationship to Other U.S. Units	Relationship to Household Units
Length	inch (in.)	1 in. = 0.083 ft	
	foot (ft)	1 ft = 12 in.	
		= 0.33 yd	
	yard (yd)	1 yd = 36 in.	
		= 3 ft	
	mile (mi)	1 mi = 5280 ft	
		= 1760 yd	
Volume	fluidram (fl dr)	1 fl dr = 0.125 fl oz	
	fluid ounce (fl oz)	1 fl oz = 8 fl dr	= 6 teaspoons (tsp)
		= 0.0625 pt	= 2 tablespoons (tbsp)
	pint (pt)	1 pt = 128 fl dr	= 32 tbsp
		= 16 fl oz	= 2 cups (c)
		= 0.5 qt	
	quart (qt)	1 qt = 256 fl dr	= 4 c
		= 32 fl oz	
		= 2 pt	
		= 0.25 gal	
	gallon (gal)	1 gal = 128 fl oz	
		= 8 pt	
		= 4 qt	
Mass	grain (gr)	1 gr = 0.002 oz	
	dram (dr)	1 dr = 27.3 gr	
		= 0.063 oz	
	ounce (oz)	1 oz = 437.5 gr	
		= 16 dr	
	pound (lb)	1 lb = 7000 gr	
		= 256 dr	
		= 16 oz	
	ton	1 ton = 2000 lb	

A-1

TABLE 2 THE METRIC SYSTEM OF MEASUREMENT

Physical Property	Unit	Relationship to Standard Metric Units	Conversion to U.S. Units	
Length	nanometer (nm)	1 nm = 0.000000001 m (10^{-9})	= 3.94×10^{-8} in.	25,400,000 nm = 1 in.
	micrometer (μm)	1 μm = 0.000001 m (10^{-6})	= 3.94×10^{-5} in.	25,400 μm = 1 in.
	millimeter (mm)	1 mm = 0.001 m (10^{-3})	= 0.0394 in.	25.4 mm = 1 in.
	centimeter (cm)	1 cm = 0.01 m (10^{-2})	= 0.394 in.	2.54 cm = 1 in.
	decimeter (dm)	1 dm = 0.1 m (10^{-1})	= 3.94 in.	0.254 dm = 1 in.
	meter (m)	standard unit of length	= 39.4 in.	0.0254 m = 1 in.
			= 3.28 ft	0.3048 m = 1 ft
			= 1.093 yd	0.914 m = 1 yd
	kilometer (km)	1 km = 1000 m	= 3280 ft	
			= 1093 yd	
			= 0.62 mi	1.609 km = 1 mi
Volume	microliter (μl)	1 μl = 0.000001 l (10^{-6})		
		= 1 cubic millimeter (mm³)		
	milliliter (ml)	1 ml = 0.001 l (10^{-3})	= 0.0338 fl oz	5 ml = 1 tsp
		= 1 cubic centimeter (cm³ or cc)		15 ml = 1 tbsp
				30 ml = 1 fl oz
	centiliter (cl)	1 cl = 0.01 l (10^{-2})	= 0.338 fl oz	2.95 cl = 1 fl oz
	deciliter (dl)	1 dl = 0.1 l (10^{-1})	= 3.38 fl oz	0.295 dl = 1 fl oz
	liter (l)	standard unit of volume	= 33.8 fl oz	0.0295 l = 1 fl oz
			= 2.11 pt	0.473 l = 1 pt
			= 1.06 qt	0.946 l = 1 qt
Mass	picogram (pg)	1 pg = 0.000000000001 g (10^{-12})		
	nanogram (ng)	1 ng = 0.000000001 g (10^{-9})	= 0.000000015 gr	66,666,666 ng = 1 gr
	microgram (μg)	1 μg = 0.000001 g (10^{-6})	= 0.000015 gr	66,666 μg = 1 gr
	milligram (mg)	1 mg = 0.001 g (10^{-3})	= 0.015 gr	66.7 μg = 1 gr
	centigram (cg)	1 cg = 0.01 g (10^{-2})	= 0.15 gr	6.67 cg = 1 gr
	decigram (dg)	1 dg = 0.1 g (10^{-1})	= 1.5 gr	0.667 dg = 1 gr
	gram (g)	standard unit of mass	= 0.035 oz	28.4 g = 1 oz
			= 0.0022 lb	454 g = 1 lb
	dekagram (dag)	1 dag = 10 g		
	hectogram (hg)	1 hg = 100 g		
	kilogram (kg)	1 kg = 1000 g	= 2.2 lb	0.454 kg = 1 lb
	metric ton (t)	1 t = 1000 kg	= 1.1 ton	
			= 2205 lb	0.907 t = 1 ton

Temperature	Centigrade	Fahrenheit
Freezing point of pure water	0°	32°
Normal body temperature	36.8°	98.6°
Boiling point of pure water	100°	212°
Conversion	°C → °F: °F = (1.8 × °C) + 32	°F → °C: °C = (°F − 32) × 0.56

can be reported in **decimeters** (0.1 m; *decem*, ten), **centimeters** (**cm** = 0.01 m; *centum*, hundred), **millimeters** (**mm** = 0.001 m; *mille*, thousand), and so forth. In the metric system, the same prefixes are used to report weights, based on the **gram (g)**, and volumes, based on the **liter (l).** This text reports data in metric units, in most cases with U.S. system equivalents. Use this opportunity to become familiar with the metric system, because most technical sources report data only in metric units; most of the world outside the United States uses the metric system exclusively. Conversion factors are included in Table 2.

The U.S. and metric systems also differ in their methods of reporting temperatures. In the United States, temperatures are usually reported in degrees Fahrenheit (°F), whereas scientific literature and individuals in most other countries report temperatures in degrees centigrade or Celsius (°C). The relationship between temperatures in degrees Fahrenheit and those in degrees Celsius is indicated in Table 2.

The following illustration spans the entire range of measurements that we will consider in this book. Gross anatomy traditionally deals with structural organization as seen with the naked eye or with a simple hand lens. A microscope can provide higher levels of magnification and can reveal finer details. Before the 1950s, most information was provided by *light microscopy.* A photograph taken through a *light microscope* is called a **light micrograph (LM).** Light microscopy can magnify cellular structures up to about 1000 times and can show details as fine as 0.25 μm. The symbol **μm** stands for **micrometer;** 1 μm = 0.001 mm, or 0.00004 inches. With a light microscope, we can identify cell types, such as muscle fibers or neurons, and can see large structures within a cell. Because individual cells are relatively transparent, thin sections cut through a cell are treated with dyes that stain specific structures to make them easier to see.

Although special staining techniques can show the general distribution of proteins, lipids, carbohydrates, and nucleic acids in the cell, many fine details of intracellular structure remained a mystery until investigators began using *electron microscopy.* This technique uses a focused beam of electrons, rather than a beam of light, to examine cell structure. In *transmission electron microscopy,* electrons pass through an ultrathin section to strike a photographic plate. The result is a **transmission electron micrograph (TEM).** Transmission electron microscopy shows the fine structure of cell membranes and intracellular structures. In *scanning electron microscopy,* electrons bouncing off exposed surfaces create a **scanning electron micrograph (SEM).** Although it cannot achieve as much magnification as transmission microscopy, scanning microscopy provides a three-dimensional perspective of cell structure.

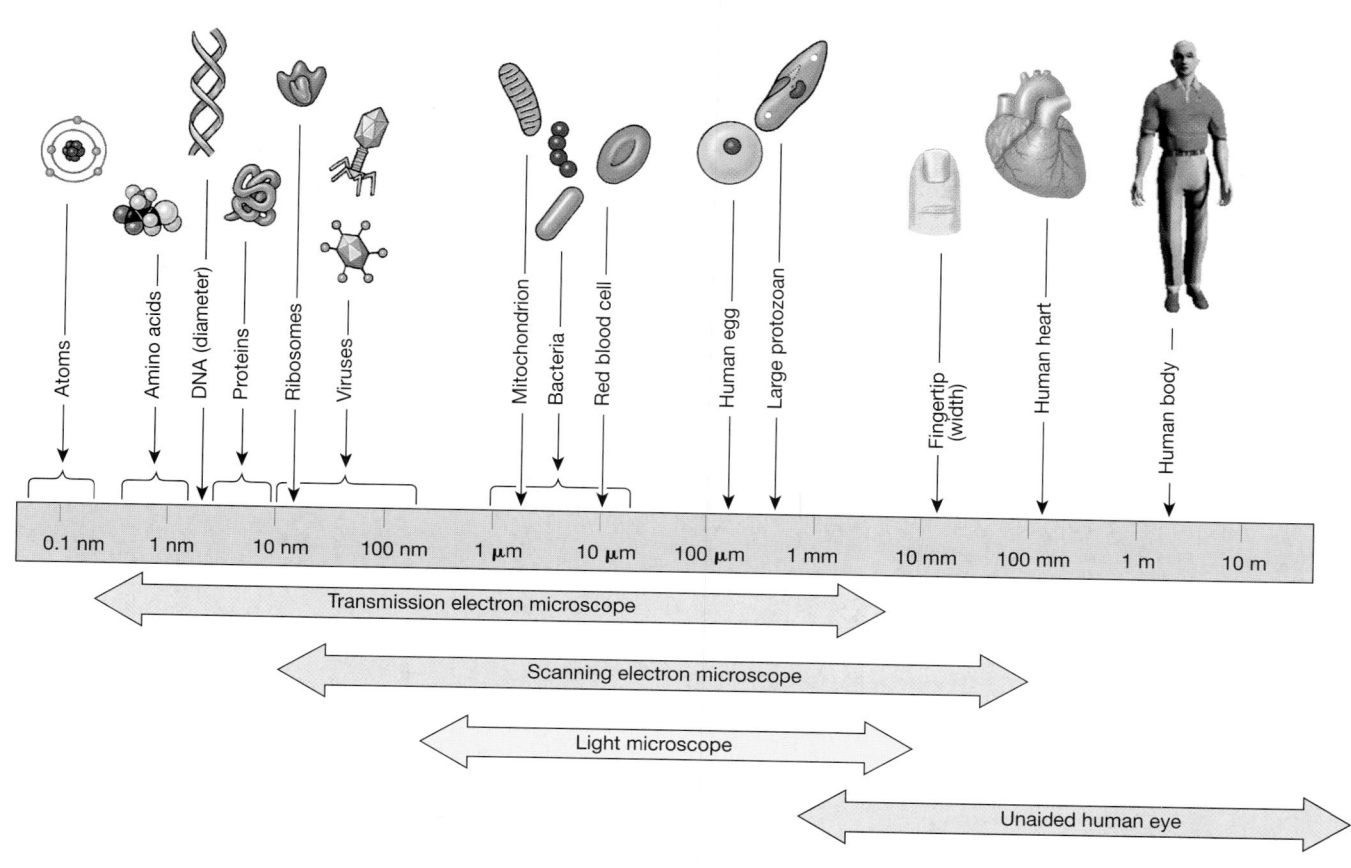

APPENDIX II

Periodic Table

The **periodic table** presents the known elements in order of their atomic weights. Each horizontal row represents a single electron shell. The number of elements in that row is determined by the maximum number of electrons that can be stored at that energy level. The element at the left end of each row contains a single electron in its outermost electron shell; the element at the right end of the row has a filled outer electron shell. Organizing the elements in this fashion highlights similarities that reflect the composition of the outer electron shell. These similarities are evident when you examine the vertical columns. All the gases of the right-most column—helium, neon, argon, krypton, xenon, and radon—have full electron shells; each is a gas at normal atmospheric temperature and pressure, and none reacts readily with other elements. These elements, highlighted in blue, are known as the *noble*, or *inert, gases.* In contrast, the elements of the left-most column below hydrogen—lithium, sodium, potassium, rubidium, cesium, and francium—are silvery, soft metals that are so highly reactive that pure forms cannot be found in nature. The fourth and fifth electron levels can hold up to 18 electrons. Table inserts are used for the so-called *lanthanide* and *actinide series* to save space, as higher levels can store up to 32 electrons. Elements of particular importance to our discussion of human anatomy and physiology are highlighted in pink.

Key:

Atomic number — 1	— Chemical symbol
	H
	Hydrogen — Element name
Atomic weight — 1.01	

1 H Hydrogen 1.01																		2 He Helium 4.00
3 Li Lithium 6.94	4 Be Beryllium 9.01											5 B Boron 10.81	6 C Carbon 12.01	7 N Nitrogen 14.01	8 O Oxygen 16.00	9 F Fluorine 19.00	10 Ne Neon 20.18	
11 Na Sodium 22.99	12 Mg Magnesium 24.31											13 Al Aluminum 26.98	14 Si Silicon 28.09	15 P Phosphorus 30.97	16 S Sulfur 32.07	17 Cl Chlorine 35.45	18 Ar Argon 39.95	
19 K Potassium 39.10	20 Ca Calcium 40.08	21 Sc Scandium 44.96	22 Ti Titanium 47.88	23 V Vanadium 50.94	24 Cr Chromium 52.00	25 Mn Manganese 54.94	26 Fe Iron 55.85	27 Co Cobalt 58.93	28 Ni Nickel 58.69	29 Cu Copper 63.55	30 Zn Zinc 65.39	31 Ga Gallium 69.72	32 Ge Germanium 72.61	33 As Arsenic 74.92	34 Se Selenium 78.96	35 Br Bromine 79.90	36 Kr Krypton 83.80	
37 Rb Rubidium 85.47	38 Sr Strontium 87.62	39 Y Yttrium 88.91	40 Zr Zirconium 91.22	41 Nb Niobium 92.91	42 Mo Molybdenum 95.94	43 Tc Technetium (98)	44 Ru Ruthenium 101.07	45 Rh Rhodium 102.91	46 Pd Palladium 106.42	47 Ag Silver 107.87	48 Cd Cadmium 112.41	49 In Indium 114.82	50 Sn Tin 118.71	51 Sb Antimony 121.76	52 Te Tellurium 127.60	53 I Iodine 126.90	54 Xe Xenon 131.29	
55 Cs Cesium 132.91	56 Ba Barium 137.33	57 La* Lanthanum 138.91	72 Hf Hafnium 178.49	73 Ta Tantalum 180.95	74 W Tungsten 183.85	75 Re Rhenium 186.21	76 Os Osmium 190.2	77 Ir Iridium 192.22	78 Pt Platinum 195.08	79 Au Gold 196.97	80 Hg Mercury 200.59	81 Tl Thallium 204.38	82 Pb Lead 207.2	83 Bi Bismuth 208.98	84 Po Polonium (209)	85 At Astatine (210)	86 Rn Radon (222)	
87 Fr Francium (223)	88 Ra Radium 226.03	89 Ac† Actinium 227.03	104 Rf Rutherfordium (266)	105 Db Dubnium (262)	106 Sg Seaborgium (266)	107 Bh Bohrium (264)	108 Hs Hassium (269)	109 Mt Meitnerium (268)	110 Unnamed (271)	111 Unnamed (272)	112 Unnamed (277)	114 (285)	116 (289)	118 (293)				

*Lanthanide series

58 Ce Cerium 140.12	59 Pr Praseodymium 140.91	60 Nd Neodymium 144.24	61 Pm Promethium (145)	62 Sm Samarium 150.36	63 Eu Europium 151.96	64 Gd Gadolinium 157.25	65 Tb Terbium 158.93	66 Dy Dysprosium 162.50	67 Ho Holmium 164.93	68 Er Erbium 167.26	69 Tm Thulium 168.93	70 Yb Ytterbium 173.04	71 Lu Lutetium 174.97

†Actinide series

90 Th Thorium 232.04	91 Pa Protactinium 231.04	92 U Uranium 238.03	93 Np Neptunium 237.05	94 Pu Plutonium (244)	95 Am Americium (243)	96 Cm Curium (247)	97 Bk Berkelium (247)	98 Cf Californium (251)	99 Es Einsteinium (252)	100 Fm Fermium (257)	101 Md Mendelevium (258)	102 No Nobelium (259)	103 Lr Lawrencium (262)

APPENDIX III

Condensed Structural Formulas of the Amino Acids

Amino acids with acidic or basic side chains (R groups)

Aspartic acid (asp)   Glutamic acid (glu)   Tyrosine (tyr)

Lysine (lys)   Arginine (arg)   Histidine (his)

Amino acids with uncharged but polar side chains

Serine (ser)   Threonine (thr)   Methionine (met)   Cysteine (cys)

Amino acids with hydrocarbon side chains

Glycine (gly)   Alanine (ala)   Valine (val)   Tryptophan (trp)   Asparagine (asn)

Glutamine (gln)   Leucine (leu)   Isoleucine (ile)   Phenylalanine (phe)   Proline (pro)

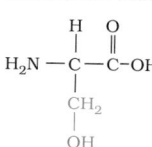
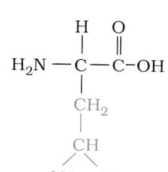
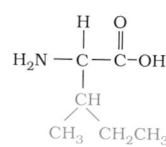
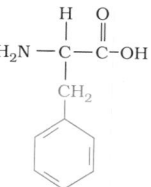
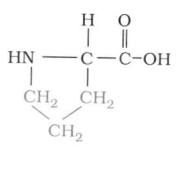

APPENDIX IV

Normal Physiological Values

Tables 3 and 4 present normal averages or ranges for the chemical composition of body fluids. These values are approximations rather than absolute values, because test results vary from laboratory to laboratory owing to differences in procedures, equipment, normal solutions, and so forth. Blanks in the tabular data appear where data are not available; sources used in the preparation of these tables follow. The following locations in the text contain additional information about body fluid analysis:

Table 19–3 (p. 670) presents data on the cellular composition of whole blood.

Table 26–2 (p. 982) compares the average compositions of urine and plasma.

Tables 26–5 (p. 997) and 26–6 (p. 1001) give the general characteristics of normal urine.

SOURCES

Braunwauld, Eugene, Kurt J. Isselbacher, Dennis L. Kasper, Jean D. Wilson, Joseph B. Martin, and Anthony S. Fauci, eds. 1998. *Harrison's Principles of Internal Medicine*, 14th ed. New York: McGraw-Hill.

Lentner, Cornelius, ed. 1981. *Geigy Scientific Tables*, 8th ed. Basel, Switzerland: Ciba–Geigy Limited.

Halsted, James A. 1976. *The Laboratory in Clinical Medicine: Interpretation and Application*. Philadelphia: W.B. Saunders Company.

Wintrobe, Maxwell, G. Richard Lee, Dane R. Boggs, Thomas C. Bitnell, John Foerster, John W. Athens, and John N. Lukens. 1981. *Clinical Hematology*, Philadelphia: Lea and Febiger.

TABLE 3 THE COMPOSITION OF MINOR BODY FLUIDS

Test	Normal Averages or Ranges					
	Perilymph	Endolymph	Synovial Fluid	Sweat	Saliva	Semen
pH			7.4	4–6.8	6.4*	7.19
Specific gravity			1.008–1.015	1.001–1.008	1.007	1.028
Electrolytes (mEq/l)						
Potassium	5.5–6.3	140–160	4.0	4.3–14.2	21	31.3
Sodium	143–150	12–16	136.1	0–104	14*	117
Calcium	1.3–1.6	0.05	2.3–4.7	0.2–6	3	12.4
Magnesium	1.7	0.02		0.03–4	0.6	11.5
Bicarbonate	17.8–18.6	20.4–21.4	19.3–30.6		6*	24
Chloride	121.5	107.1	107.1	34.3	17	42.8
Proteins (total) (mg/dl)	200	150	1.72 g/dl	7.7	386[†]	4.5 g/dl
Metabolites (mg/dl)						
Amino acids				47.6	40	1.26 g/dl
Glucose	104		70–110	3.0	11	224 (fructose)
Urea				26–122	20	72
Lipids (total)	12		20.9	‡	25–500[§]	188

* Increases under salivary stimulation.

† Primarily alpha-amylase, with some lysozymes.

‡ Not present in eccrine secretions.

§ Cholesterol.

TABLE 4   THE CHEMISTRY OF BLOOD, CEREBROSPINAL FLUID, AND URINE

Test	Normal Range		
	Blood*	CSF	Urine
pH	S: 7.38–7.44	7.31–7.34	4.5–8.0
Osmolarity (mOsm/l)	S: 280–295	292–297	855–1335
Electrolytes		(mEq/l unless noted)	(urinary loss per 24-hour period†)
Bicarbonate	P: 21–28	20–24	
Calcium	S: 4.5–5.3	2.1–3.0	6.5–16.5 mEq
Chloride	S: 100–108	116–122	120–240 mEq
Iron	S: 50–150 µg/l	23–52 µg/l	40–150 µg
Magnesium	S: 1.5–2.5	2–2.5	4.9–16.5 mEq
Phosphorus	S: 1.8–2.6	1.2–2.0	0.8–2 g
Potassium	P: 3.8–5.0	2.7–3.9	35–80 mEq
Sodium	P: 136–142	137–145	120–220 mEq
Sulfate	S: 0.2–1.3		1.07–1.3 g
Metabolites		(mg/dl unless noted)	(urinary loss per 24-hour period‡)
Amino acids	P/S: 2.3–5.0	10.0–14.7	41–133 mg
Ammonia	P: 20–150 µg/dl	25–80 µg/dl	340–1200 mg
Bilirubin	S: 0.5–1.0	<0.2	0.02–1.9 mg
Creatinine	P/S: 0.6–1.2	0.5–1.9	1.01–2.5
Glucose	P/S: 70–110	40–70	16–132 mg
Ketone bodies	S: 0.3–2.0	1.3–1.6	10–100 mg
Lactic acid	WB: 5–20§	10–20	100–600 mg
Lipids (total)	S: 400–1000	0.8–1.7	0–31.8 mg
Cholesterol (total)	S: 150–300	0.2–0.8	1.2–3.8 mg
Triglycerides	S: 40–150	0–0.9	
Urea	P/S: 23–43	13.8–36.4	12.6–28.6
Uric acid	S: 2.0–7.0	0.2–0.3	80–976 mg
Proteins	(g/dl)	(mg/dl)	(urinary loss per 24-hour period‡)
Total	S: 6.0–7.8	2.0–4.5	47–76.2 mg
Albumin	S: 3.2–4.5	10.6–32.4	10–100 mg
Globulins (total)	S: 2.3–3.5	2.8–15.5	7.3 mg (average)
Immunoglobulins	S: 1.0–2.2	1.1–1.7	3.1 mg (average)
Fibrinogen	P: 0.2–0.4	0.65 (average)	

* S = serum, P = plasma, WB = whole blood.
† Because urinary output averages just over 1 liter per day, these electrolyte values are comparable to mEq/l.
‡ Because urinary metabolite and protein data approximate mg/l or g/l, these data must be divided by 10 for comparison with CSF or blood concentrations.
§ Venous blood sample.

GLOSSARY

Eponyms in Common Use

Eponym	Equivalent Term(s)	Individual Referenced
The Cellular Level of Organization (*Chapter 3*)		
Golgi apparatus		Camillo Golgi (1844–1926), Italian histologist; shared Nobel Prize in 1906
Krebs cycle	Tricarboxylic, or citric acid, cycle	Hans Adolph Krebs (1900–1981), British biochemist; shared Nobel Prize in 1953
The Skeletal System (*Chapters 6–9*)		
Colles' fracture		Abraham Colles (1773–1843), Irish surgeon
Haversian canals	Central canals	Clopton Havers (1650–1702), English anatomist and microscopist
Haversian systems	Osteons	Clopton Havers
Pott's fracture		Percivall Pott (1713–1788), English surgeon
Sharpey's fibers	Perforating fibers	William Sharpey (1802–1880), Scottish histologist and physiologist
Volkmann's canals	Perforating canals	Alfred Wilhelm Volkmann (1800–1877), German surgeon
Wormian bones	Sutural bones	Olas Worm (1588–1654), Danish anatomist
The Muscular System (*Chapters 10, 11*)		
Achilles' tendon	Calcaneal tendon	Achilles, hero of Greek mythology
Cori cycle		Carl Ferdinand Cori (1896–1984) and Gerty Theresa Cori (1896–1957), American biochemists; shared Nobel Prize in 1947
The Nervous System (*Chapters 12–16*)		
Broca's center	Speech center	Pierre Paul Broca (1824–1880), French surgeon
Foramen of Lushka	Lateral foramina	Hubert von Lushka (1820–1875), German anatomist
Meissner's corpuscles	Tactile corpuscles	Georg Meissner (1829–1905), German physiologist
Merkel's discs	Tactile discs	Friedrich Siegismund Merkel (1845–1919), German anatomist
Foramen of Munro	Interventricular foramen	John Cummings Munro (1858–1910), American surgeon
Nissl bodies		Franz Nissl (1860–1919), German neurologist
Pacinian corpuscles	Lamellated corpuscles	Fillippo Pacini (1812–1883), Italian anatomist
Purkinje cells		Johannes E. Purkinje (1781–1869), Czechoslovakian physiologist
Nodes of Ranvier	Nodes	Louis Antoine Ranvier (1835–1922), French physiologist
Island of Reil	Insula	Johann Christian Reil (1759–1813), German anatomist
Fissure of Rolando	Central sulcus	Luigi Rolando (1773–1831), Italian anatomist
Ruffini's corpuscles		Angelo Ruffini (1864–1929), Italian anatomist
Schwann cells	Neurolemmocytes	Theodor Schwann (1810–1882), German anatomist
Aqueduct of Sylvius	Mesencephalic aqueduct	Jacobus Sylvius (Jacques Dubois, 1478–1555), French anatomist
Sylvian fissure	Lateral sulcus	Franciscus Sylvius (Franz de le Boë, 1614–1672), Dutch anatomist
Pons varolii	Pons	Costanzo Varolio (1543–1575), Italian anatomist
Sensory Function (*Chapter 17*)		
Organ of Corti	Spiral organ	Alfonso Corti (1822–1888), Italian anatomist
Eustachian tube	Auditory tube	Bartolomeo Eustachio (1520–1574), Italian anatomist
Golgi tendon organs	Tendon organs	Camillo Golgi (1844–1926), Italian histologist; shared Nobel Prize in 1906
Hertz (Hz)		Heinrich Hertz (1857–1894), German physicist
Meibomian glands	Tarsal glands	Heinrich Meibom (1638–1700), German anatomist
Canal of Schlemm	Scleral venous sinus	Friedrich S. Schlemm (1795–1858), German anatomist
The Endocrine System (*Chapter 18*)		
Islets of Langerhans	Pancreatic islets	Paul Langerhans (1847–1888), German pathologist
Interstitial cells of Leydig	Interstitial cells	Franz von Leydig (1821–1908), German anatomist

Eponym	Equivalent Term(s)	Individual Referenced
The Cardiovascular System (*Chapters 19–21*)		
Bundle of His		Wilhelm His (1863–1934), German physician
Purkinje fibers		Johannes E. Purkinje (1781–1869), Czechoslovakian
Starling's law		Ernest Henry Starling (1866–1927), English physiologist
Circle of Willis	Cerebral arterial circle	Thomas Willis (1621–1675), English physician
The Lymphatic System (*Chapter 22*)		
Hassall's corpuscles		Arthur Hill Hassall (1817–1894), English physician
Kupffer cells	Stellate reticuloendothelial cells	Karl Wilhelm Kupffer (1829–1902), German anatomist
Langerhans cells		Paul Langerhans (1847–1888), German pathologist
Peyer's patches	Aggregated lymphoid nodules	Johann Conrad Peyer (1653–1712), Swiss anatomist
The Respiratory System (*Chapter 23*)		
Bohr effect		Christian Bohr (1855–1911), Danish physiologist
Boyle's law		Robert Boyle (1621–1691), English physicist
Charles' law		Jacques Alexandre César Charles (1746–1823), French physicist
Dalton's law		John Dalton (1766–1844), English physicist
Henry's law		William Henry (1775–1837), English chemist
The Digestive System (*Chapter 24*)		
Plexus of Auerbach	Myenteric plexus	Leopold Auerbach (1827–1897), German anatomist
Brunner's glands	Duodenal glands	Johann Conrad Brunner (1653–1727), Swiss anatomist
Kupffer cells	Stellate reticuloendothelial cells	Karl Wilhelm Kupffer (1829–1902), German anatomist
Crypts of Lieberkuhn	Intestinal glands	Johann Nathaniel Lieberkuhn (1711–1756), German anatomist
Plexus of Meissner	Submucosal plexus	Georg Meissner (1829–1905), German physiologist
Sphincter of Oddi	Hepatopancreatic sphincter	Ruggero Oddi (1864–1913), Italian physician
Peyer's patches		Johann Conrad Peyer (1653–1712), Swiss anatomist
Duct of Santorini	Accessory pancreatic duct	Giovanni Domenico Santorini (1681–1737), Italian anatomist
Stensen's duct	Parotid duct	Niels Stensen (1638–1686), Danish physician/priest
Ampulla of Vater	Duodenal ampulla	Abraham Vater (1684–1751), German anatomist
Wharton's duct	Submandibular duct	Thomas Wharton (1614–1673), English physician
Duct of Wirsung	Pancreatic duct	Johann Georg Wirsung (1600–1643), German physician
The Urinary System (*Chapter 26*)		
Bowman's capsule	Glomerular capsule	Sir William Bowman (1816–1892), English physician
Loop of Henle	Nephron loop	Friedrich Gustav Jakob Henle (1809–1885), German histologist
The Reproductive System (*Chapters 28, 29*)		
Bartholin's glands	Greater vestibular glands	Casper Bartholin, Jr. (1655–1738), Danish anatomist
Cowper's glands	Bulbourethral glands	William Cowper (1666–1709), English surgeon
Fallopian tube	Uterine tube/oviduct	Gabriele Fallopio (1523–1562), Italian anatomist
Graafian follicle	Tertiary follicle	Reijnier de Graaf (1641–1673), Dutch physician
Interstitial cells of Leydig	Interstitial cells	Franz von Leydig (1821–1908), German anatomist
Glands of Littré	Lesser vestibular glands	Alexis Littré (1658–1726), French surgeon
Sertoli cells	Sustentacular cells	Enrico Sertoli (1842–1910), Italian histologist

Glossary of Key Terms

PRONUNCIATION GUIDE:

The pronunciations given in this text correspond to those given in medical and college dictionaries published in the United States. International pronunciations may vary; some of these pronunciations are included in the Glossary as alternatives. Pronunciation symbols are as follows:

ā *as in* tray	a *as in* track
ē *as in* tree	e *as in* help
ī *as in* spine	i *as in* ink
ō *as in* bone	o *as in* Tom
ū *as in* use	u *as in* run
oo *as in* moon	oy *as in* boy

Emphasized syllables are shown in capital letters, as in united (ū-NĪ-ted) or factor (FAK-tor).

A

abdomen: The region of the trunk bounded by the diaphragm and pelvis.

abdominopelvic cavity: The portion of the ventral body cavity that contains abdominal and pelvic subdivisions; also contains the peritoneal cavity.

abducens (ab-DOO-senz): Cranial nerve VI, which innervates the lateral rectus muscle of the eye.

abduction: Movement away from the midline of the body, as viewed in the anatomical position.

abortion: The premature loss or expulsion of an embryo or fetus.

abruptio placentae (ab-RUP-shē-ō pla-SEN-tē): The premature loss of a placental connection to the uterus, leading to maternal hemorrhaging and shock.

abscess: A localized collection of pus within a damaged tissue.

absorption: The active or passive uptake of gases, fluids, or solutes.

acclimatization: The physical adaptation to a long-term environmental change, such as adaptation that accompany a change in season or in latitude.

accommodation: An alteration in the curvature of the lens of the eye to focus an image on the retina.

acetabulum (a-se-TAB-ū-lum): The fossa on the lateral aspect of the pelvis that accommodates the head of the femur.

acetylcholine (ACh) (as-ē-til-KŌ-lēn): A chemical neurotransmitter in the brain and peripheral nervous system; the dominant neurotransmitter in the peripheral nervous system, released at neuromuscular junctions and synapses of the parasympathetic division.

acetylcholinesterase (AChE): An enzyme found in the synaptic cleft, bound to the postsynaptic membrane, and in tissue fluids; breaks down and inactivates acetylcholine molecules.

acetyl-CoA: An acetyl group bound to coenzyme A, a participant in the anabolic and catabolic pathways for carbohydrates, lipids, and many amino acids.

acetyl group: $-CH_3C = O$.

achalasia (āk-a-LĀ-zē-uh): A condition that develops when the lower esophageal sphincter fails to dilate and ingested materials cannot enter the stomach.

Achilles tendon: See **calcaneal tendon.**

acid: A compound whose dissociation in solution releases a hydrogen ion and an anion; an acidic solution has a pH below 7.0 and contains an excess of hydrogen ions.

acidosis (a-sid-Ō-sis): An abnormal physiological state characterized by a plasma pH below 7.35.

acinus/acini (AS-i-nī): A histological term referring to a blind pocket, pouch, or sac.

acne: A condition characterized by the inflammation of sebaceous glands and follicles; commonly affects adolescents and in most cases involves the face.

acoustic: Pertaining to sound or the sense of hearing.

acquired immune deficiency syndrome (AIDS): A disease caused by the human immunodeficiency virus (HIV); characterized by the destruction of helper T cells and a resulting severe impairment of the immune response.

acromegaly: A condition caused by the overproduction of growth hormone in adults, characterized by a thickening of bones and an enlargement of cartilages and other soft tissues.

acromion (a-KRŌ-mē-on): A continuation of the scapular spine that projects above the capsule of the scapulohumeral joint.

acrosomal cap (ak-rō-SŌ-mal): A membranous sac at the tip of a spermatozoon that contains hyaluronidase.

actin: The protein component of microfilaments that forms thin filaments in skeletal muscles and produces contractions of all muscles through interaction with thick (myosin) filaments; see also **sliding filament theory.**

action potential: A propagated change in the transmembrane potential of excitable cells, initiated by a change in the membrane permeability to sodium ions; *see also* **nerve impulse.**

activation energy: The energy required to initiate a specific chemical reaction.

active transport: The ATP-dependent absorption or secretion of solutes across a cell membrane.

acute: Sudden in onset, severe in intensity, and brief in duration.

adaptation: A change in pupillary size in response to changes in light intensity; a decrease in receptor sensitivity or perception after chronic stimulation; physiological responses that produce acclimatization.

Addison's disease: A condition resulting from the hyposecretion of glucocorticoids; characterized by lethargy, weakness, hypotension, and increased skin pigmentation.

adduction: Movement toward the axis or midline of the body, as viewed in the anatomical position.

adenine: A purine; one of the nitrogenous bases in the nucleic acids RNA and DNA.

adenohypophysis (ad-e-nō-hī-POF-i-sis): The anterior lobe of the pituitary gland.

adenoids: The pharyngeal tonsil.

adenosine: A combination of adenine and ribose.

adenosine diphosphate (ADP): A compound consisting of adenosine with two phosphate groups attached.

adenosine monophosphate (AMP): A nucleotide consisting of adenine plus a phosphate group (PO_4^{3-}); also called *adenosine phosphate.*

adenosine triphosphate (ATP): A high-energy compound consisting of adenosine with three phosphate groups attached; the third is attached by a high-energy bond.

adenylate cyclase: An enzyme bound to the inner surfaces of cell membranes that can convert ATP to cyclic-AMP; also called *adenylyl cyclase.*

adhesion: The fusion of two mesenterial layers after damage or irritation of their opposing surfaces; this process restricts relative movement of the organs involved; the binding of a phagocyte to its target.

adipocyte (AD-i-pō-sīt): A fat cell.

adipose tissue: Loose connective tissue dominated by adipocytes.

adrenal cortex: The superficial portion of the adrenal gland that produces steroid hormones.

adrenal gland: A small endocrine gland that secretes steroids and catecholamines and is located superior to each kidney; also called *suprarenal gland.*

adrenal medulla: The core of the adrenal gland; a modified sympathetic ganglion that secretes catecholamines into the blood during sympathetic activation.

adrenergic (ad-ren-ER-jik): A synaptic terminal that, when stimulated, releases norepinephrine.

adrenocortical hormone: Any steroid produced by the adrenal cortex.

adrenocorticotropic hormone (ACTH): The hormone that stimulates the production and secretion of glucocorticoids by the zona fasciculata of the adrenal cortex; released by the anterior lobe of the pituitary gland in response to corticotropin-releasing hormone.

adventitia (ad-ven-TISH-a): The superficial layer of connective tissue surrounding an internal organ; fibers are continuous with those of surrounding tissues, providing support and stabilization.

aerobic: Requiring the presence of oxygen.

aerobic metabolism: The complete breakdown of organic substrates into carbon dioxide and water, via pyruvic acid; a process that yields large amounts of ATP but requires mitochondria and oxygen.

afferent: Toward.

afferent arteriole: An arteriole that carries blood to a glomerulus of the kidney.

afferent fiber: An axon that carries sensory information to the central nervous system.

afterbirth: The distal portions of the umbilical cord and placenta that are ejected from the uterus during the placental stage of labor.

agglutination (a-gloo-ti-NĀ-shun): The aggregation of red blood cells due to interactions between surface antigens and plasma antibodies.

agglutinins (a-GLOO-ti-ninz): Immunoglobulins in plasma that react with antigens on the surfaces of foreign red blood cells when donor and recipient differ in blood type.

agglutinogens (a-gloo-TIN-ō-jenz): Surface antigens on red blood cells whose presence and structure are genetically determined.

aggregated lymphoid nodules: Lymphoid nodules beneath the epithelium of the small intestine; also called *Peyer's patches.*

agonist: A muscle responsible for a specific movement.

agranular: Without granules; *agranular leukocytes* are monocytes and lymphocytes.

AIDS: See **acquired immune deficiency syndrome.**

alba: White.

albicans (AL-bi-kanz): White.

albinism: The absence of pigment in hair and skin, caused by the inability of melanocytes to produce melanin.

albuginea: (al-bū-JIN-ē-uh): White.

aldosterone: A mineralocorticoid produced by the zona glomerulosa of the adrenal cortex; stimulates sodium and water conservation at the kidneys; secreted in response to the presence of angiotensin II.

aldosteronism: The condition caused by the oversecretion of aldosterone, characterized by fluid retention, edema, and hypertension.

alkalosis (al-ka-LŌ-sis): The condition characterized by a plasma pH greater than 7.45; associated with a relative deficiency of hydrogen ions or an excess of bicarbonate ions.

allantois (a-LAN-tō-is): One of the four extraembryonic membranes; provides vascularity to the chorion and is therefore essential to placenta formation; the proximal portion becomes the urinary bladder.

alleles (a-LĒLZ): Alternate forms of a particular gene.

allergen: An antigenic compound that produces a hypersensitivity response.

alpha-blockers: Drugs that prevent the stimulation of alpha receptors.

alpha cells: Cells in the pancreatic islets that secrete glucagon.

alpha receptors: Membrane receptors sensitive to norepinephrine or epinephrine; stimulation normally results in the excitation of the target cell.

alveolar sac: An air-filled chamber that supplies air to several alveoli.

alveolus/alveoli (al-VĒ-o-lī): Blind pockets at the end of the respiratory tree, lined by a simple squamous epithelium and surrounded by a capillary network; sites of gas exchange with the blood; a bony socket that holds the root of a tooth.

Alzheimer's disease: A disorder resulting from degenerative changes in populations of neurons in the cerebrum, causing dementia characterized by problems with attention, short-term memory, and emotions.

amacrine cells (AM-a-krīn): Modified neurons in the retina that facilitate or inhibit communication between bipolar cells and ganglion cells.

amenorrhea (ā-men-ō-RĒ-uh): A failure to commence menstruation at adolescence or the cessation of menstruation prior to menopause.

amination: The attachment of an amino group to a carbon chain; performed by a variety of cells and important in the synthesis of amino acids.

amino acids: Organic compounds whose chemical structure can be summarized as R — CHNH$_2$ — COOH.

amino group: — NH$_2$.

amnesia: A temporary or permanent memory loss.

amniocentesis: The sampling of amniotic fluid for analytical purposes; used to detect certain genetic abnormalities.

amnion (AM-nē-on): One of the four extraembryonic membranes; surrounds the developing embryo or fetus.

amniotic fluid (am-nē-OT-ik): Fluid that fills the amniotic cavity; cushions and supports the embryo or fetus.

amphiarthrosis (am-fē-ar-THRŌ-sis): An articulation that permits a small degree of independent movement; *see* **interosseous membrane** and **pubic symphysis.**

amphicytes (AM-fi-sīts): Supporting cells that surround neurons in the peripheral nervous system; also called *satellite cells.*

amphimixis (am-fi-MIK-sis): The fusion of male and female pronuclei after fertilization.

ampulla/ampullae (am-PŪL-la): A localized dilation in the lumen of a canal or passageway.

amygdaloid body (ah-MIG-da-loyd): A basal nucleus that is a component of the limbic system and acts as an interface between that system, the cerebrum, and sensory systems.

amylase: An enzyme that breaks down polysaccharides; produced by the salivary glands and pancreas.

anabolism (a-NAB-ō-lizm): The synthesis of complex organic compounds from simpler precursors.

anaerobic: Without oxygen.

anal canal: The distal portion of the rectum that contains the anal columns and ends at the anus.

anal columns: Longitudinal folds in the walls of the anal canal.

analgesic: A substance that relieves pain.

anal triangle: The posterior subdivision of the perineum.

anamnestic response (an-am-NES-tik): The sudden and exaggerated production of antibodies after a second exposure to a specific antigen; due to the activation of memory B cells.

anaphase (AN-a-fāz): The mitotic stage in which the paired chromatids separate and move toward opposite ends of the spindle apparatus.

anaphylaxis (a-na-fi-LAK-sis): A hypersensitivity reaction due to the binding of antigens to immunoglobulins (IgE) on the surfaces of mast cells; the release of histamine, serotonin, and prostaglandins by mast cells then causes widespread inflammation; a sudden decline in blood pressure may occur, producing anaphylactic shock.

anastomosis (a-nas-to-MŌ-sis): The joining of two tubes, usually referring to a connection between two peripheral vessels without an intervening capillary bed.

anatomical position: An anatomical reference position; the body viewed from the anterior surface with the palms facing forward.

anatomy (a-NAT-o-mē): The study of the structure of the body.

anaxonic neuron (an-ak-SON-ik): A central nervous system neuron that has many processes but no apparent axon.

androgen (AN-drō-jen): A steroid sex hormone primarily produced by the interstitial cells of the testis and manufactured in small quantities by the adrenal cortex in either gender.

anemia (a-NĒ-mē-uh): The condition marked by a reduction in the hematocrit, the hemoglobin content of the blood, or both.

anencephaly (an-en-SEF-a-lē): A developmental defect characterized by incomplete development of the cerebral hemispheres and cranium.

anesthesia: The total or partial loss of sensation from a region of the body.

aneurysm (AN-ū-rizm): A weakening and localized dilation in the wall of a blood vessel.

angiogram (AN-jē-ō-gram): An X-ray image of cardiovascular pathways.

angiography: The X-ray examination of vessel distribution after the introduction of radiopaque substances into the bloodstream.

angiotensin I: The hormone produced by the activation of angiotensinogen by renin; angiotensin-converting enzyme converts angiotensin I into angiotensin II in lung capillaries.

angiotensin II: A hormone that causes an elevation in systemic blood pressure, stimulates the secretion of aldosterone, promotes thirst, and causes the release of antidiuretic hormone; angiotensin-converting enzyme in lung capillaries converts angiotensin I into angiotensin II.

angiotensinogen: The blood protein produced by the liver that is converted to angiotensin I by the enzyme renin.

anion (AN-ī-on): An ion bearing a negative charge.

ankyloglossia (ang-ki-lō-GLOS-ē-uh): A condition characterized by an overly robust and restrictive lingual frenulum.

anorexia nervosa: An eating disorder marked by a loss of appetite and pronounced weight loss.

anoxia (a-NOKS-ē-uh): Tissue oxygen deprivation.

antagonist: A muscle that opposes the movement of an agonist.

antebrachium: The forearm.

anteflexion (an-te-FLEK-shun): The normal position of the uterus, with the superior surface bent forward.

anterior: On or near the front, or ventral surface, of the body.

anterograde amnesia: The inability to store memories of events that occur after a specific incident or time.

anthracosis (an-thra-KŌ-sis): "Black lung disease," a deterioration of respiratory exchange efficiency due to the chronic inhalation of coal dust.

antibiotic: A chemical agent that selectively kills pathogens.

antibody (AN-ti-bod-ē): A globular protein produced by plasma cells that will bind to specific antigens and promote their destruction or removal from the body.

antibody-mediated immunity: The form of immunity resulting from the presence of circulating antibodies produced by plasma cells; also called *humoral immunity.*

anticholinesterase: A chemical compound that blocks the action of acetylcholine and causes prolonged and intensive stimulation of post-synaptic membranes.

anticoagulant: A compound that slows or prevents clot formation by interfering with the clotting system.

anticodon: Three nitrogenous bases on a tRNA molecule that interact with an appropriate codon on a strand of mRNA.

antidiuretic hormone (ADH) (an-tī-dī-ū-RET-ik): A hormone synthesized in the hypothalamus and secreted at the posterior lobe of the pituitary gland; causes water retention at the kidneys and an elevation of blood pressure.

antigen: A substance capable of inducing the production of antibodies.

antigen–antibody complex: The combination of an antigen and a specific antibody.

antigenic determinant site: A portion of an antigen that can interact with an antibody molecule.

antigen-presenting cell (APC): A cell that processes antigens and displays them, bound to MHC proteins; essential to the initiation of a normal immune response.

antihistamine (an-tī-HIS-ta-mēn): A chemical agent that blocks the action of histamine on peripheral tissues.

antipyretic agents: Chemicals that reduce fever.

antrum (AN-trum): A chamber or pocket.

anulus (AN-ū-lus): A cartilage or bone shaped like a ring; also spelled *annulus.*

anuria (a-NŪ-rē-uh): A cessation of urine production.

anus: The external opening of the anal canal.

aorta: The large, elastic artery that carries blood away from the left ventricle and into the systemic circuit.

aortic reflex: A baroreceptor reflex triggered by increased aortic pressures; leads to a reduction in cardiac output and a fall in systemic pressure.

Apgar test: A test used to assess the neurological status of a newborn infant.

aphasia: The inability to speak.

apnea (AP-nē-uh): The cessation of breathing.

apneustic center (ap-NOO-stik): A respiratory center whose chronic activation would lead to apnea at full inhalation.

apocrine secretion: A mode of secretion in which the glandular cell sheds portions of its cytoplasm.

aponeurosis/aponeuroses (ap-ō-nū-RŌ-sis): A broad tendinous sheet that may serve as the origin or insertion of a skeletal muscle.

apoplexy: A cerebrovascular accident (stroke).

appendicitis: An inflammation of the appendix.

appendicular: Pertaining to the upper or lower limbs.

appendix: A blind tube connected to the cecum of the large intestine.

appositional growth: The enlargement of a bone by the addition of cartilage or bony matrix at its surface.

aqueduct of the midbrain: The passageway that connects the third ventricle (diencephalon) with the fourth ventricle (metencephalon).

aqueous humor: A fluid similar to perilymph or cerebrospinal fluid that fills the anterior chamber of the eye.

arachidonic acid: One of the essential fatty acids.

arachnoid (a-RAK-noyd): The middle meninx that encloses cerebrospinal fluid and protects the central nervous system.

arachnoid granulations: Processes of the arachnoid that project into the superior sagittal sinus; sites where cerebrospinal fluid enters the venous circulation.

arbor vitae: The central, branching mass of white matter inside the cerebellum.

arcuate (AR-kū-āt): Curving.

areflexia (ā-re-flek-sē-uh): The absence of normal reflex responses to stimulation.

areola (a-RĒ-ō-la): The pigmented area that surrounds the nipple of a breast.

areolar: Containing minute spaces, as in areolar tissue.

areolar tissue: Loose connective tissue with an open framework.

arrector pili (ar-REK-tor PI-lē): Smooth muscles whose contractions force hairs to stand erect.

arrhythmias (ā-RITH-mē-az): Abnormal patterns of cardiac contractions.

arteriole (ar-TĒ-rē-ōl): A small arterial branch that delivers blood to a capillary network.

artery: A blood vessel that carries blood away from the heart and toward a peripheral capillary.

arthritis (ar-THRĪ-tis): An inflammation of a joint.

arthroscope: A fiber-optic device intended for visualizing the interior of joints; also used for certain forms of joint surgery.

articular: Pertaining to a joint.

articular capsule: The dense collagen fiber sleeve that surrounds a joint and provides protection and stabilization.

articular cartilage: The cartilage pad that covers the surface of a bone inside a joint cavity.

articulation (ar-tik-ū-LĀ-shun): A joint; the formation of words.

arytenoid cartilages (ar-i-TĒ-noyd): A pair of small cartilages in the larynx.

ascending tract: A tract carrying information from the spinal cord to the brain.

ascites (a-SĪ-tēz): The overproduction and accumulation of peritoneal fluid.

asphyxia: Unconsciousness due to oxygen deprivation at the central nervous system.

aspirate: To remove or obtain by suction; to inhale.

association areas: Cortical areas of the cerebrum that are responsible for the integration of sensory inputs and/or motor commands.

association neuron: See **interneuron.**

asthma (AZ-ma): A reversible constriction of smooth muscles around respiratory passageways, commonly caused by an allergic response.

astigmatism: A visual disturbance due to an irregularity in the shape of the cornea.

astrocyte (AS-trō-sīt): One of the four types of neuroglia in the central nervous system; responsible for maintaining the blood–brain barrier by the stimulation of endothelial cells.

ataxia: A failure to coordinate muscular activities normally.

atelectasis (at-e-LEK-ta-sis): The collapse of a lung or a portion of a lung.

atherosclerosis (ath-er-ō-skle-RŌ-sis): The formation of fatty plaques in the walls of arteries, restricting blood flow to deep tissues.

atom: The smallest stable unit of matter.

atomic number: The number of protons in the nucleus of an atom.

atomic weight: Roughly, the average total number of protons and neutrons in the atoms of a particular element.

atresia (a-TRĒ-zē-uh): The closing of a cavity, or its incomplete development; refers to the degeneration of developing ovarian follicles.

atria: Thin-walled chambers of the heart that receive venous blood from the pulmonary or systemic circuit atrial natriuretic peptide (ANP): See **natriuretic peptides.**

atrial reflex: The reflexive increase in heart rate after an increase in venous return; due to mechanical and neural factors; also called *Bainbridge reflex.*

atrioventricular (AV) node (ā-trē-ō-ven-TRIK-ū-lar): Specialized cardiocytes that relay the contractile stimulus to the bundle of His, the bundle branches, the Purkinje fibers, and the ventricular myocardium; located at the boundary between the atria and ventricles.

atrioventricular (AV) valve: One of the valves that prevent backflow into the atria during ventricular systole.

atrophy (AT-rō-fē): The wasting away of tissues from a lack of use, ischemia, or nutritional abnormalities.

auditory: Pertaining to the sense of hearing.

auditory ossicles: The bones of the middle ear: malleus, incus, and stapes.

auditory tube: A passageway that connects the nasopharynx with the middle ear cavity; also called *Eustachian tube* or *pharyngotympanic tube.*

auricle: A broad, flattened process that resembles the external ear; in the ear, the expanded, projecting portion that surrounds the external auditory canal, also called *pinna;* in the heart, the externally visible flap formed by the collapse of the outer wall of a relaxed atrium.

autoantibodies: Antibodies that react with antigens on the surfaces of a person's own cells and tissues.

autodigestion: The digestion of tissues by digestive acids or enzymes from the stomach or pancreas.

autoimmunity: The immune system's sensitivity to normal cells and tissues, resulting in the production of autoantibodies.

autolysis: The destruction of a cell due to the rupture of lysosomal membranes in its cytoplasm.

automaticity: The spontaneous depolarization to threshold, characteristic of cardiac pacemaker cells.

autonomic ganglion: A collection of visceral motor neurons outside the central nervous system.

autonomic nerve: A peripheral nerve consisting of preganglionic or postganglionic autonomic fibers.

autonomic nervous system (ANS): Centers, nuclei, tracts, ganglia, and nerves involved in the unconscious regulation of visceral functions; includes components of the central nervous system and the peripheral nervous system.

autopsy: The detailed examination of a body after death.

autoregulation: Changes in activity that maintain homeostasis in direct response to changes in the local environment; does not require neural or endocrine control.

autosomal (aw-to-SŌ-mal): Chromosomes other than the X or Y chromosome.

avascular (ā-VAS-kū-lar): Without blood vessels.

avitaminosis (ā-vī-ta-min-Ō-sis): A condition caused by the inadequate intake of one or more essential vitamins.

avulsion: An injury involving the violent tearing away of body tissues.

axilla: The armpit.

axolemma: The cell membrane of an axon, continuous with the cell membrane of the cell body and dendrites and distinct from any neuroglial coverings.

axon: The elongate extension of a neuron that conducts an action potential.

axon hillock: In a multipolar neuron, the portion of the cell body adjacent to the initial segment.

axoplasm (AK-sō-plazm): The cytoplasm within an axon.

B

Babinski sign: The reflexive extension and abduction of the toes after the medial, plantar surface of the foot is stroked; positive reflex (Babinski sign) is normal up to age 1.5 years; thereafter, a positive reflex indicates damage to descending tracts.

bacteria: Single-celled microorganisms, some pathogenic, that are common in the environment and in and on the body.

Bainbridge reflex: See **atrial reflex.**

baroreception: The ability to detect changes in pressure.

baroreceptor reflex: A reflexive change in cardiac activity in response to changes in blood pressure.

baroreceptors (bar-ō-rē-SEP-torz): The receptors responsible for baroreception.

basal lamina: A layer of filaments and fibers that attach an epithelium to the underlying connective tissue.

basal metabolic rate (BMR): The resting metabolic rate of a normal fasting individual under homeostatic conditions.

basal nuclei: Nuclei of the cerebrum that are important in the subconscious control of skeletal muscle activity.

base: A compound whose dissociation releases a hydroxide ion (OH^-) or removes a hydrogen ion (H^+) from the solution.

basilar membrane: The membrane that supports the organ of Corti and separates the cochlear duct from the scala tympani in the inner ear.

basophils (BĀ-sō-filz): Circulating granulocytes (white blood cells) similar in size and function to tissue mast cells.

B cells: Lymphocytes capable of differentiating into plasma cells, which produce antibodies.

benign: Not malignant.

beta cells: Cells of the pancreatic islets that secrete insulin in response to elevated blood sugar concentrations.

beta oxidation: Fatty acid catabolism that produces molecules of acetyl-CoA.

beta receptors: Membrane receptors sensitive to epinephrine; stimulation may result in the excitation or inhibition of the target cell.

bicarbonate ions: HCO_3^-; anion components of the carbonic acid–bicarbonate buffer system.

bicuspid (bī-KUS-pid): Having two cusps or leafs; refers to a premolar tooth, which has two roots, or to the left AV valve, which has two cusps.

bicuspid valve: The left atrioventricular (AV) valve, also called *mitral valve*.

bifurcate: To branch into two parts.

bile: The exocrine secretion of the liver; stored in the gallbladder and ejected into the duodenum.

bile salts: Steroid derivatives in bile; responsible for the emulsification of ingested lipids.

bilirubin (bil-ē-ROO-bin): A pigment that is the by-product of hemoglobin catabolism.

bioenergetics: The analysis of energy production and utilization by cells.

biofeedback: The use of artificial signals to provide feedback about unconscious, visceral motor activities.

biopsy: The removal of a small sample of tissue for pathological analysis.

bipennate: A muscle whose fibers are arranged on either side of a common tendon.

bladder: A muscular sac that distends as fluid is stored and whose contraction ejects the fluid at an appropriate time; used alone, the term usually refers to the urinary bladder.

blastocoele (BLAS-tō-sēl): A fluid-filled cavity within a blastocyst.

blastocyst (BLAS-tō-sist): An early stage in the developing embryo, consisting of an outer trophoblast and an inner cell mass.

blastomere (BLAS-tō-mēr): One of the cells in the morula; a collection of cells produced by the division of the zygote.

blockers/blocking agents: Drugs that block membrane pores or prevent binding to membrane receptors.

blood–brain barrier: The isolation of the central nervous system from the general circulation; primarily the result of astrocyte regulation of capillary permeabilities.

blood clot: A network of fibrin fibers and trapped blood cells.

blood–CSF barrier: The isolation of the cerebrospinal fluid from the capillaries of the choroid plexus; primarily the result of specialized ependymal cells.

blood pressure: A force exerted against vessel walls by the blood in the vessels, due to the push exerted by cardiac contraction and the elasticity of the vessel walls; usually measured along one of the muscular arteries, with systolic pressure measured during ventricular systole and diastolic pressure during ventricular diastole.

blood–testis barrier: The isolation of the interior of the seminiferous tubules from the general circulation, due to the activities of the sustentacular cells.

Bohr effect: The increased oxygen release by hemoglobin in the presence of elevated carbon dioxide levels.

boil: An abscess of the skin, normally involving a sebaceous gland.

bolus: A compact mass; usually refers to compacted ingested material on its way to the stomach.

bone: See *osseous tissue*.

botulinus toxin (bot-ū-LĪ-nus): A toxin, produced by the anaerobic bacterium *Clostridium botulinum,* that can cause severe food poisoning.

bowel: The intestinal tract.

Bowman's capsule: The cup-shaped initial portion of the renal tubule; surrounds the glomerulus and receives the glomerular filtrate.

Boyle's law: The principle that, in a gas, pressure and volume are inversely related.

brachial: Pertaining to the arm.

brachial plexus: A network formed by branches of spinal nerves C_5–T_1 en route to innervating the upper limb.

brachium: The arm.

bradycardia (brad-ē-KAR-dē-uh): An abnormally slow heart rate, usually below 50 bpm.

brain natural peptide (BNP): see natriuretic peptides

brain stem: The brain minus the cerebrum, diencephalon, and cerebellum.

brevis: Short.

Broca's area: The speech center of the brain, normally located on the neural cortex of the left cerebral hemisphere.

bronchial tree: The trachea, bronchi, and bronchioles.

bronchitis (brong-KĪ-tis): An inflammation of the bronchial passageways.

bronchodilation: The dilation of the bronchial passages; can be caused by sympathetic stimulation.

bronchodilators (brong-kō-dī-LĀ-torz): Drugs that produce bronchodilation; some are used clinically in treating asthma.

bronchoscope: A fiber-optic instrument used to examine the bronchial passageways.

bronchus/bronchi: A branch of the bronchial tree between the trachea and bronchioles.

buccal (BUK-al): Pertaining to the cheeks.

buffer: A compound that stabilizes the pH of a solution by removing or releasing hydrogen ions.

buffer system: Interacting compounds that prevent increases or decreases in the pH of body fluids; includes the carbonic acid–bicarbonate buffer system, the phosphate buffer system, and the protein buffer system.

bulbar: Pertaining to the brain stem.

bulbourethral glands (bul-bō-ū-RĒ-thral): Mucous glands at the base of the penis that secrete into the penile urethra; the equivalent of the greater vestibular glands of females; also called *Cowper's glands.*

bundle branches: Specialized conducting cells in the ventricles that carry the contractile stimulus from the bundle of His to the Purkinje fibers.

bundle of His (hiss): Specialized conducting cells in the interventricular septum that carry the contracting stimulus from the AV node to bundle branches and then to Purkinje fibers.

bursa: A small sac filled with synovial fluid that cushions adjacent structures and reduces friction.

bursitis: A painful inflammation of one or more bursae.

C

calcaneal tendon: The large tendon that inserts on the calcaneus; tension on this tendon produces extension (plantar flexion) of the foot; also called *Achilles tendon.*

calcaneus (kal-KĀ-nē-us or kal-KAN-ē-us): The heelbone, the largest of the tarsal bones.

calcification: The deposition of calcium salts within a tissue.

calcitonin (kal-si-TŌ-nin): The hormone secreted by C cells of the thyroid when calcium ion concentrations are abnormally high; restores homeostasis by increasing the rate of bone deposition and the rate of calcium loss at the kidneys.

calculus/calculi (KAL-kū-lī): Concretions of insoluble materials that form within body fluids, especially the gallbladder, kidneys, or urinary bladder.

callus: A localized thickening of the epidermis due to chronic mechanical stresses; a thickened area that forms at the site of a bone break as part of the repair process.

calorie (c) (KAL-o-rē): The amount of heat that is required to raise the temperature of 1 gram of water 1°C.

Calorie (C): The amount of heat that is required to raise the temperature of 1 kilogram of water 1°C; also called *kilocalorie.*

calorigenic effect: The stimulation of energy production and heat loss by thyroid hormones.

calvaria (kal-VAR-ē-uh or kal-VĀ-rē-ah): The skullcap, consisting of the superior portions of the frontal, parietal, and occipital bones.

calyx/calyces (KĀL-i-sēz): Cup-shaped divisions of the renal pelvis.

canaliculi (kan-a-LIK-ū-lī): Microscopic passageways between cells; bile canaliculi carry bile to bile ducts in the liver; in bone, canaliculi permit the diffusion of nutrients and wastes to and from osteocytes.

cancellous bone (KAN-sel-us): Spongy bone, composed of a network of bony struts.

cancer: An illness caused by mutations leading to the uncontrolled growth and replication of the affected cells.

cannula: A tube that can be inserted into the body; commonly placed in blood vessels prior to transfusion or dialysis.

canthus, medial and lateral (KAN-thus): The angles formed at either corner of the eye between the upper and lower eyelids.

capacitation: The activation process that must occur before a spermatozoon can successfully fertilize an oocyte; occurs in the vagina after ejaculation.

capillary (KAP-i-lār-ē or ka-PIL-a-rē): A small blood vessel, located between an arteriole and a venule, whose thin wall permits the diffusion of gases, nutrients, and wastes between plasma and interstitial fluids.

capitulum (ka-PIT-ū-lum): A general term for a small, elevated articular process; refers to the rounded distal surface of the humerus that articulates with the head of the radius.

caput: The head.

carbaminohemoglobin (kar-bam-ē-nō-HĒ-mō-GLŌ-bin): Hemoglobin bound to carbon dioxide molecules.

carbohydrase (kar-bō-HĪ-drās): An enzyme that breaks down carbohydrate molecules.

carbohydrate (kar-bō-HĪ-drāt): An organic compound containing carbon, hydrogen, and oxygen in a ratio that approximates 1:2:1.

carbon dioxide: CO_2; a compound produced by the decarboxylation reactions of aerobic metabolism.

carbonic anhydrase: An enzyme that catalyzes the reaction $H_2O + CO_2 \rightarrow H_2CO_3$; important in carbon dioxide transport, gastric acid secretion, and renal pH regulation.

carboxypeptidase (kar-bok-sē-PEP-ti-dās): A protease that breaks down proteins and releases amino acids.

carcinogenic (kar-sin-ō-JEN-ik): Stimulating cancer formation in affected tissues.

cardia (KAR-dē-uh): The area of the stomach surrounding its connection with the esophagus.

cardiac: Pertaining to the heart.

cardiac cycle: One complete heartbeat, including atrial and ventricular systole and diastole.

cardiac glands: Mucous glands characteristic of the cardia of the stomach.

cardiac output: The amount of blood ejected by the left ventricle each minute; normally about 5 liters.

cardiac reserve: The potential percentage increase in cardiac output above resting levels.

cardiac tamponade: A compression of the heart due to fluid accumulation in the pericardial cavity.

cardiocyte (KAR-dē-ō-sīt): A cardiac muscle cell.

cardiomyopathy (kar-dē-ō-mī-OP-a-thē): A progressive disease characterized by damage to the cardiac muscle tissue.

cardiopulmonary resuscitation (CPR): A method of artificially maintaining respiratory and circulatory functions.

cardiovascular: Pertaining to the heart, blood, and blood vessels.

cardiovascular centers: Poorly localized centers in the reticular formation of the medulla oblongata of the brain; includes cardioaccelerory, cardioinhibitory, and vasomotor centers.

cardium: The heart.

carina (ka-RĪ-na): A ridge on the inner surface of the base of the trachea that runs anteroposteriorly, between two primary bronchi.

carotene (KAR-ō-tēn): A yellow-orange pigment, found in carrots and in green and orange leafy vegetables, that the body can convert to vitamin A.

carotid artery: The principal artery of the neck, servicing cervical and cranial structures; one branch, the internal carotid, provides a major blood supply to the brain.

carotid body: A group of receptors, adjacent to the carotid sinus, that are sensitive to changes in the carbon dioxide levels, pH, and oxygen concentrations of arterial blood.

carotid sinus: A dilated segment at the base of the internal carotid artery whose walls contain baroreceptors sensitive to changes in blood pressure.

carotid sinus reflex: Reflexive changes in blood pressure that maintain homeostatic pressures at the carotid sinus, stabilizing blood flow to the brain.

carpus/carpal: The wrist.

cartilage: A connective tissue with a gelatinous matrix that contains an abundance of fibers.

castration: The removal of the testes; also called *bilateral orchiectomy.*

catabolism (ka-TAB-ō-lizm): The breakdown of complex organic molecules into simpler components, accompanied by the release of energy.

catalyst (KAT-uh-list): A substance that accelerates a specific chemical reaction but that is not altered by the reaction.

cataract: A reduction in lens transparency that causes visual impairment.

catecholamine (kat-e-KŌL-am-ēn): Epinephrine, norepinephrine, dopamine, and related compounds.

cathepsins (ka-THEP-sinz): Enzymes, present in the sarcoplasm of skeletal muscle cells, that can break down contractile proteins, thereby providing amino acids that can act as a supplemental energy source.

catheter (KATH-e-ter): A tube surgically inserted into a body cavity or along a blood vessel or excretory passageway for the collection of body fluids, monitoring of blood pressure, or introduction of medications or radiographic dyes.

cation (KAT-ī-on): An ion that bears a positive charge.

cauda equina (KAW-da ek-WĪ-na): Spinal nerve roots distal to the tip of the adult spinal cord; they extend caudally inside the vertebral canal en route to lumbar and sacral segments.

caudal/caudally: Closest to or toward the tail (coccyx).

caudate nucleus (KAW-dāt): One of the basal nuclei involved with the subconscious control of skeletal muscular activity.

cavernous tissue: Erectile tissue that can be engorged with blood; located in the penis (males) and clitoris (females).

cecum (SĒ-kum): An expanded pouch at the start of the large intestine.

cell: The smallest living unit in the human body.

cell body: Body; the body of a neuron; also called *soma.*

cell-mediated immunity: Resistance to disease through the activities of sensitized T cells that destroy antigen-bearing cells by direct contact or through the release of lymphotoxins; also called *cellular immunity.*

cellulitis (sel-ū-LĪ-tis): Diffuse inflammation, normally involving areas of loose connective tissue, such as the subcutaneous layer.

cementum (se-MEN-tum): Bony material that covers the root of a tooth and is not shielded by a layer of enamel.

center of ossification: The site in a connective tissue where bone formation begins.

central canal: Longitudinal canal in the center of an osteon that contains blood vessels and nerves; also called *Haversian canal;* a passageway along the longitudinal axis of the spinal cord that contains cerebrospinal fluid.

central nervous system (CNS): The brain and spinal cord.

central sulcus: A groove in the surface of a cerebral hemisphere, between the primary sensory and primary motor areas of the cortex.

centriole: A cylindrical intracellular organelle composed of nine groups of microtubules, three in each group; functions in mitosis or meiosis by organizing the microtubules of the spindle apparatus.

centromere (SEN-trō-mēr): The localized region where two chromatids remain connected after the chromosomes have replicated; site of spindle fiber attachment.

centrosome: A region of cytoplasm that contains a pair of centrioles oriented at right angles to one another.

centrum: The body of a vertebra.

cephalic: Pertaining to the head.

cerebellum (ser-e-BEL-um): The posterior portion of the met-encephalon, containing the cerebellar hemispheres; includes the arbor vitae, cerebellar nuclei, and cerebellar cortex.

cerebral cortex: An extensive area of neural cortex covering the surfaces of the cerebral hemispheres.

cerebral hemispheres: A pair of expanded portions of the cerebrum covered in neural cortex.

cerebral palsy: A chronic condition resulting from damage to motor areas of the brain during development or at delivery.

cerebral peduncle: A mass of nerve fibers on the ventrolateral surface of the mesencephalon; contains ascending tracts that terminate in the thalamus and descending tracts that originate in the cerebral hemispheres.

cerebrospinal fluid (CSF): Fluid bathing the internal and external surfaces of the central nervous system; secreted by the choroid plexus.

cerebrovascular accident (CVA): The occlusion of a blood vessel that supplies a portion of the brain, resulting in damage to the dependent neurons; also called *stroke.*

cerebrum (SER-ē-brum): The largest portion of the brain, composed of the cerebral hemispheres; includes the cerebral cortex, the basal nuclei, and the internal capsule.

cerumen: The waxy secretion of the ceruminous glands along the external acoustic canal.

ceruminous glands (se-ROO-mi-nus): Integumentary glands that secrete cerumen.

cervical enlargement: A thickening of the cervical portion of the spinal cord caused by the abundance of central nervous system neurons involved with motor control of the upper limbs.

cervix: The inferior portion of the uterus.

cesarean section: The surgical delivery of an infant via an incision through the lower abdominal wall and uterus.

chalazion (kah-LĀ-zē-on): An inflammation and distension of a Meibomian gland on the eyelid; also called *sty.*

chancre (SHANG-ker): A skin lesion that develops at the primary site of a syphilis infection.

chemoreception: The detection of changes in the concentrations of dissolved compounds or gases.

chemotaxis (kē-mō-TAK-sis): The attraction of phagocytic cells to the source of abnormal chemicals in tissue fluids.

chemotherapy: The treatment of illness through the administration of specific chemicals.

chloride shift: The movement of plasma chloride ions into red blood cells in exchange for bicarbonate ions generated by the intracellular dissociation of carbonic acid.

cholecystitis (kō-lē-sis-TĪ-tis): An inflammation of the gallbladder.

cholecystokinin (CCK) (kō-lē-sis-tō-KĪ-nin): A duodenal hormone that stimulates the contraction of the gallbladder and the secretion of enzymes by the exocrine pancreas; also called *pancreozymin.*

cholelithiasis (kō-lē-li-THĪ-a-sis): The formation or presence of gallstones.

cholesterol: A steroid component of cell membranes and a substrate for the synthesis of steroid hormones and bile salts.

choline: A breakdown product or precursor of acetylcholine.

cholinergic synapse (kō-lin-ER-jik): A synapse where the presynaptic membrane releases acetylcholine on stimulation.

cholinesterase (kō-li-NES-ter-ās): The enzyme that breaks down and inactivates acetylcholine.

chondrocyte (KON-drō-sīt): A cartilage cell.

chondroitin sulfate (kon-DROY-tin): The predominant proteoglycan in cartilage, responsible for the gelatinous consistency of the matrix.

chordae tendineae (KOR-dē TEN-di-nē-ē): Fibrous cords that stabilize the position of the AV valves in the heart, preventing backflow during ventricular systole.

chorion/chorionic (KOR-ē-on/ko-rē-ON-ik): An extraembryonic membrane, consisting of the trophoblast and underlying mesoderm, that forms the placenta.

choroid: The middle, vascular layer in the wall of the eye.

choroid plexus: The vascular complex in the roof of the third and fourth ventricles of the brain, responsible for the production of cerebrospinal fluid.

chromatid (KRŌ-ma-tid): One complete copy of a DNA strand and its associated nucleoproteins.

chromatin (KRŌ-ma-tin): A histological term referring to the grainy material visible in cell nuclei during interphase; the appearance of the DNA content of the nucleus when the chromosomes are uncoiled.

chromosomes: Dense structures, composed of tightly coiled DNA strands and associated histones, that become visible in the nucleus when a cell prepares to undergo mitosis or meiosis; normal human somatic cells contain 46 chromosomes apiece.

chronic: Habitual or long term.

chylomicrons (kī-lō-MĪ-kronz): Relatively large droplets that may contain triglycerides, phospholipids, and cholesterol in association with proteins; synthesized and released by intestinal cells and transported to the venous blood by the lymphatic system.

chyme (kīm): A semifluid, acidic mixture of ingested food and digestive secretions that forms in the stomach during the early phases of digestion.

chymotrypsin (kī-mō-TRIP-sin): A protease in the small intestine.

chymotrypsinogen: The inactive proenzyme, secreted by the pancreas, that is subsequently converted to chymotrypsin.

ciliary body: A thickened region of the choroid that encircles the lens of the eye; includes the ciliary muscle and the ciliary processes that support the suspensory ligaments of the lens.

cilium/cilia: A slender organelle that extends above the free surface of an epithelial cell and generally undergoes cycles of movement; composed of a basal body and microtubules in a 9×12 array.

circulatory system: The network of blood vessels and lymphatic vessels that facilitate the distribution and circulation of extracellular fluid.

circumduction (sir-kum-DUK-shun): A movement at a synovial joint in which the distal end of the bone describes a circle but the shaft does not rotate.

circumvallate papilla (sir-kum-VAL-āt pa-PIL-la): One of the large, dome-shaped papillae on the superior surface of the tongue that form a V, separating the body of the tongue from the root.

cirrhosis (sir-RŌ-sis): A liver disorder characterized by the degeneration of hepatocytes and their replacement by fibrous connective tissue.

cisterna (sis-TUR-na): An expanded chamber.

citric acid cycle: *See* **TCA cycle.**

cleavage (KLĒ-vij): Mitotic divisions that follow the fertilization of an ovum and lead to the formation of a blastocyst.

cleavage lines: Stress lines in the skin that follow the orientation of major bundles of collagen fibers in the dermis.

clitoris (KLI-to-ris): A small erectile organ of females that is the developmental equivalent of the penis in males.

clone: As a verb, the production of genetically identical cells; as a noun, a genetically identical copy.

clonus (KLŌ-nus): Rapid cycles of muscle contraction and relaxation.

clot: A network of fibrin fibers and trapped blood cells; also called a *thrombus* if it occurs within the circulatory system.

clotting factors: Plasma proteins, synthesized by the liver, that are essential to the clotting response.

clotting response: The series of events that result in the formation of a clot.

coccygeal ligament: The fibrous extension of the dura mater and filum terminale; provides longitudinal stabilization to the spinal cord.

coccyx (KOK-siks): The terminal portion of the spinal column, consisting of relatively tiny, fused vertebrae.

cochlea (KOK-lē-uh): The spiral portion of the bony labyrinth of the inner ear that surrounds the organ of hearing.

cochlear duct (KOK-lē-ar): The central membranous tube within the cochlea that is filled with endolymph and contains the organ of Corti; also called *scala media.*

codon (KŌ-don): A sequence of three nitrogenous bases along an mRNA strand that will specify the location of a single amino acid in a peptide chain.

coelom (SĒ-lōm): The ventral body cavity, lined by a serous membrane and subdivided during fetal development into the pleural, pericardial, and abdominopelvic (peritoneal) cavities.

coenzymes (kō-EN-zīmz): Complex organic cofactors; most are structurally related to vitamins.

cofactor: Ions or molecules that must be attached to the active site before an enzyme can function; examples include mineral ions and several vitamins.

colectomy (kō-LEK-to-mē): The surgical removal of part or all of the colon.

colitis: An inflammation of the colon.

collagen: A strong, insoluble protein fiber common in connective tissues.

collateral ganglion (kō-LAT-er-al): A sympathetic ganglion situated anterior to the spinal column and separate from the sympathetic chain.

Colles' fracture (KOL-lēz): A fracture of the distal end of the radius and possibly the ulna, with posterior and dorsal displacement of the distal bone fragments.

colliculus/colliculi (kol-IK-ū-lus): A little mound; in the brain, refers to one of the thickenings in the roof of the mesencephalon; the superior colliculus is associated with the visual system, and the inferior colliculi with the auditory system.

colloid/colloidal suspension: A solution containing large organic molecules in suspension.

colon: The large intestine.

colonoscope (kō-LON-ō-skōp): A fiber-optic device for examining the interior of the colon.

colostomy (kō-LOS-to-mē): The surgical connection of a portion of the colon to the body wall, sometimes performed after a colectomy to permit the discharge of fecal materials.

colostrum (ko-LOS-trum): The secretion of the mammary glands at the time of childbirth and for a few days thereafter; contains more protein and less fat than does the milk secreted later.

coma (kō-ma): An unconscious state from which an individual cannot be aroused, even by strong stimuli.

comedo (kō-MĒ-dō): An inflamed sebaceous gland.

comminuted: Broken or crushed into small pieces.

commissure: A crossing over from one side to another.

common bile duct: The duct formed by the union of the cystic duct from the gallbladder and the bile ducts from the liver; terminates at the duodenal ampulla, where it meets the pancreatic duct.

common pathway: In the clotting response, the events that begin with the appearance of thromboplastin and end with the formation of a clot.

compact bone: Dense bone that contains parallel osteons.

compensation curves: The cervical and lumbar curves that develop to center the body weight over the lower limbs.

complement: A system of 11 plasma proteins that interact in a chain-reaction after exposure to activated antibodies or the surfaces of certain pathogens; complement proteins promote cell lysis, phagocytosis, and other defense mechanisms.

compliance: Distensibility; the ability of certain organs to tolerate changes in volume; indicates the presence of elastic fibers and smooth muscles.

compound: A molecule containing two or more elements in combination.

concentration: The amount (in grams) or number of atoms, ions, or molecules (in moles) per unit volume.

concentration gradient: Regional differences in the concentration of a particular substance.

conception: Fertilization.

concha/conchae (KONG-kē): Three pairs of thin, scroll-like bones that project into the nasal cavities; the superior and medial conchae are part of the ethmoid, and the inferior conchae are separate bones.

concussion: A violent blow or shock; loss of consciousness due to a violent blow to the head.

condyle: A rounded articular projection on the surface of a bone.

cone: A photoreceptor of the retina, responsible for color vision.

congenital (kon-JEN-i-tal): Present at birth.

congestive heart failure (CHF): The failure to maintain adequate cardiac output due to cardiovascular problems or myocardial damage.

conjunctiva (kon-junk-TĪ-va): A layer of stratified squamous epithelium that covers the inner surfaces of the eyelids and the anterior surface of the eye to the edges of the cornea.

conjunctivitis: An inflammation of the conjunctiva.

connective tissue: One of the four primary tissue types; provides a structural framework that stabilizes the relative positions of the other tissue types; includes connective tissue proper, cartilage, bone, and blood; contains cell products, cells, and ground substance.

continuous propagation: The propagation of an action potential along an unmyelinated axon or a muscle cell membrane, wherein the action potential affects every portion of the membrane surface.

contractility: The ability to contract; possessed by skeletal, smooth, and cardiac muscle cells.

contralateral reflex: A reflex that affects the opposite side of the body from the stimulus.

conus medullaris: The conical tip of the spinal cord that gives rise to the filum terminale.

convergence: In the nervous system, the innervation of a single neuron by axons from several neurons; most common along motor pathways.

coracoid process (KOR-uh-koyd): A hook-shaped process of the scapula that projects above the anterior surface of the capsule of the shoulder joint.

Cori cycle: The metabolic exchange of lactic acid from skeletal muscle for glucose from the liver; performed during the recovery period after muscular exertion.

cornea (KOR-nē-uh): The transparent portion of the fibrous tunic of the anterior surface of the eye.

corniculate cartilages (kor-NIK-ū-lāt): A pair of small laryngeal cartilages.

cornu: Horn-shaped.

corona radiata (ko-RŌ-na rā-dē-A-ta): A layer of follicle cells surrounding a secondary oocyte at ovulation.

coronoid (ko-RŌ-noyd): Hooked or curved.

corpora cavernosa (ka-ver-NŌ-suh): Two parallel masses of erectile tissue within the body of the penis (males) or clitoris (females).

corpora quadrigemina (KOR-po-ra quad-ri-JEM-i-na): The superior and inferior colliculi of the mesencephalic tectum (roof) in the brain.

corpus/corpora: Body.

corpus callosum: The bundle of axons that links centers in the left and right cerebral hemispheres.

corpus luteum (LOO-tē-um): The progestin-secreting mass of follicle cells that develops in the ovary after ovulation.

corpus spongiosum (spon-jē-Ō-sum): The mass of erectile tissue that surrounds the urethra in the penis and expands distally to form the glans.

cortex: The outer layer or portion of an organ.

Corti, organ of: The receptor complex in the cochlear duct that includes the inner and outer hair cells, supporting cells and structures, and the tectorial membrane; provides the sensation of hearing.

corticobulbar tracts (kor-ti-kō-BUL-bar): Descending tracts that carry information or commands from the cerebral cortex to nuclei and centers in the brain stem.

corticospinal tracts: Descending tracts that carry motor commands from the cerebral cortex to the anterior gray horns of the spinal cord.

corticosteroid: A steroid hormone produced by the adrenal cortex.

corticosterone (kor-ti-KOS-te-rōn): A corticosteroid secreted by the zona fasciculata of the adrenal cortex; a glucocorticoid.

corticotropin-releasing hormone (CRH): The releasing hormone, secreted by the hypothalamus, that stimulates secretion of adrenocorticotro-pic hormone by the anterior lobe of the pituitary.

cortisol (KOR-ti-sol): A corticosteroid secreted by the zona fasciculata of the adrenal cortex; a glucocorticoid.

costa/costae: A rib.

cotransport: The membrane transport of a nutrient, such as glucose, in company with the movement of an ion, normally sodium; transport requires a carrier protein but does not involve direct ATP expenditure and can occur regardless of the concentration gradient for the nutrient.

countercurrent exchange: The diffusion between two solutions that travel in opposite directions.

countercurrent multiplication: Active transport between two limbs of a loop that contains a fluid moving in one direction; responsible for the concentration of urine in the kidney tubules.

covalent bond (kō-VĀ-lent): A chemical bond between atoms that involves the sharing of electrons.

cranial: Pertaining to the head.

cranial nerves: Peripheral nerves originating at the brain.

craniosacral division (krā-nē-ō-SAK-ral): See **parasympathetic division.**

craniostenosis (krā-nē-ō-sten-Ō-sis): A skull deformity caused by the premature closure of the cranial sutures.

cranium: The braincase; the skull bones that surround and protect the brain.

creatine: A nitrogenous compound, synthesized in the body, that can form a high-energy bond by connecting to a phosphate group and that serves as an energy reserve.

creatine phosphate: A high-energy compound in muscle cells; during muscle activity, the phosphate group is donated to ADP, regenerating ATP; also called *phosphorylcreatine.*

creatinine: A breakdown product of creatine metabolism.

crenation: Cellular shrinkage due to an osmotic movement of water out of the cytoplasm.

cribriform plate: A portion of the ethmoid that contains the foramina used by the axons of olfactory receptors en route to the olfactory bulbs of the cerebrum.

cricoid cartilage (KRĪ-koyd): A ring-shaped cartilage that forms the inferior margin of the larynx.

crista/cristae: A ridge-shaped collection of hair cells in the ampulla of a semicircular duct; the crista and cupula form a receptor complex sensitive to movement along the plane of the canal.

cross-bridge: A myosin head that projects from the surface of a thick filament and that can bind to an active site of a thin filament in the presence of calcium ions.

cruciate ligaments: A pair of intracapsular ligaments (anterior and posterior) in the knee.

cryosurgery: A surgical technique that involves freezing and killing cells in a localized area.

cryptorchid testis: An undescended testis that is in the abdominopelvic cavity rather than in the scrotum.

cuneiform cartilages (kū-NĒ-i-form): A pair of small cartilages in the larynx.

cupula (KŪ-pū-la): A gelatinous mass that is located in the ampulla of a semicircular duct in the inner ear and whose movement stimulates the hair cells of the crista.

curare: A toxin that prevents neural stimulation of neuromuscular junctions.

Cushing's disease: A condition caused by the oversecretion of adrenal steroids.

cutaneous membrane: The epidermis and papillary layer of the dermis.

cuticle: The layer of dead, keratinized cells that surrounds the shaft of a hair; for nails, *see* **eponychium.**

cyanosis: A bluish coloration of the skin due to the presence of deoxygenated blood in vessels near the body surface.

cyst: A fibrous capsule containing fluid or other material.

cystic duct: A duct that carries bile between the gallbladder and the common bile duct.

cystitis: An inflammation of the urinary bladder.

cytochrome (SĪ-tō-krōm): A pigment component of the electron transport system; a structural relative of heme.

cytokinesis (sī-tō-ki-NĒ-sis): The cytoplasmic movement that separates two daughter cells at the completion of mitosis.

cytology (sī-TOL-o-jē): The study of cells.

cytoplasm: The material between the cell membrane and the nuclear membrane; cell contents.

cytosine: A pyrimidine; one of the nitrogenous bases in the nucleic acids RNA and DNA.

cytoskeleton: A network of microtubules and microfilaments in the cytoplasm.

cytosol: The fluid portion of the cytoplasm.

cytotoxic: Poisonous to cells.

cytotoxic T cells: Lymphocytes involved in cell-mediated immunity that kill target cells by direct contact or by the secretion of lymphotoxins; also called *killer T cells* and T_C *cells.*

D

daughter cells: Genetically identical cells produced by somatic cell division.

deamination (dē-am-i-NĀ-shun): The removal of an amino group from an amino acid.

decarboxylation (dē-kar-boks-i-LĀ-shun): The removal of a molecule of carbon dioxide.

decerebrate: Lacking a cerebrum.

decomposition reaction: A chemical reaction that breaks a molecule into smaller fragments.

decubitis ulcers: Ulcers that form where chronic pressure interrupts blood flow to a portion of the skin.

decussate: To cross over to the opposite side, usually referring to the crossover of the descending tracts of the corticospinal pathway on the ventral surface of the medulla oblongata.

defecation (def-e-KĀ-shun): The elimination of fecal wastes.

deglutition: Swallowing.

degradation: Breakdown, catabolism.

dehydration: A reduction in the water content of the body that threatens homeostasis.

dehydration synthesis: The joining of two molecules associated with the removal of a water molecule.

delta cell: A pancreatic islet cell that secretes growth hormone-inhibiting hormone.

dementia: A loss of mental abilities.

demyelination: The loss of the myelin sheath of an axon, normally due to chemical or physical damage to Schwann cells or oligodendrocytes.

denaturation: A temporary or permanent change in the three-dimensional structure of a protein.

dendrite (DEN-drīt): A sensory process of a neuron.

denticulate ligaments: Supporting fibers that extend laterally from the surface of the spinal cord, tying the pia mater to the dura mater and providing lateral support for the spinal cord.

dentin (DEN-tin): The bonelike material that forms the body of a tooth; differs from bone in that it lacks osteocytes and osteons.

deoxyribonucleic acid (DNA) (dē-ok-sē-rī-bō-noo-KLĀ-ik): A nucleic acid consisting of a chain of nucleotides that contain the sugar deoxyribose and the nitrogenous bases adenine, guanine, cytosine, and thymine.

deoxyribose: A five-carbon sugar resembling ribose but lacking an oxygen atom.

depolarization: A change in the transmembrane potential from a negative value toward 0 mV.

depression: Inferior (downward) movement of a body part.

dermatitis: An inflammation of the skin.

dermatome: A sensory region monitored by the dorsal rami of a single spinal segment.

dermis: The connective tissue layer beneath the epidermis of the skin.

detrusor muscle (de-TROO-sor): A smooth muscle in the wall of the urinary bladder.

detumescence (dē-tū-MES-ens): The loss of a penile erection.

development: Growth and the acquisition of increasing structural and functional complexity; includes the period from conception to maturity.

diabetes insipidus: Polyuria due to inadequate production of antidiuretic hormone.

diabetes mellitus (mel-Ī-tus): Polyuria and glycosuria, most commonly due to the inadequate production of insulin with a resulting elevation of blood glucose levels.

diabetogenic effect: An elevation in blood sugar concentrations after the secretion of growth hormone or glucagon.

dialysis: Diffusion between two solutions of differing solute concentrations across a selectively permeable membrane, which permits the passage of some solutes but not others; regulates the composition of blood.

diapedesis (dī-a-pe-DĒ-sis): The movement of white blood cells through the walls of blood vessels by migration between adjacent endothelial cells.

diaphragm (DĪ-a-fram): Any muscular partition; the respiratory muscle that separates the thoracic cavity from the abdominopelvic cavity.

diaphysis (dī-A-fi-sis): The shaft of a long bone.

diarrhea (dī-a-RĒ-uh): Abnormally frequent defecation, associated with the production of unusually fluid feces.

diarthrosis (dī-ar-THRŌ-sis): A synovial joint.

diastolic pressure: Pressure measured in the walls of a muscular artery when the left ventricle is in diastole.

diencephalon (dī-en-SEF-a-lon): A division of the brain that includes the epithalamus, thalamus, and hypothalamus.

differential count: The determination of the relative abundance of each type of white blood cell on the basis of a random sampling of 100 white blood cells.

differentiation: The gradual appearance of characteristic cellular specializations during development as the result of gene activation or repression.

diffusion: Passive molecular movement from an area of higher concentration to an area of lower concentration.

digestion: The chemical breakdown of ingested materials into simple molecules that can be absorbed by the cells of the digestive tract.

digestive system: The digestive tract and associated glands.

digestive tract: An internal passageway that begins at the mouth, ends at the anus, and is lined by a mucous membrane; also called *gastrointestinal tract*.

dilate: To increase in diameter; to enlarge or expand.

diploid (DIP-loyd): Having a complete somatic complement of chromosomes (23 pairs in human cells).

disaccharide (di-SAK-a-rīd): A compound formed by the joining of two simple sugars by dehydration synthesis.

dislocation: The forceful displacement of an articulating bone to an abnormal position, generally accompanied by damage to tendons, ligaments, the articular capsule, or other structures.

dissociation (di-sō-sē-Ā-shun): *See* **ionization.**

distal: Movement away from the point of attachment or origin; for a limb, away from its attachment to the trunk.

distal convoluted tubule (DCT): The portion of the nephron closest to the connecting tubules and collecting duct; an important site of active secretion.

diuresis (dī-ūr-Ē-sis): Fluid loss at the kidneys; the production of urine.

divergence: In neural tissue, the spread of information from one neuron to many neurons; an organizational pattern common along sensory pathways of the central nervous system.

diverticulitis (dī-ver-tik-ū-LĪ-tis): An inflammation of a diverticulum.

diverticulosis (dī-ver-tik-ū-LŌ-sis): The formation of diverticula in the wall of an organ.

diverticulum: A sac or pouch in the wall of the colon or other organ.

dizygotic twins (dī-zī-GOT-ik): Twins that result from the fertilization of two oocytes.

DNA molecule: Two DNA strands wound in a double helix and held together by weak bonds between complementary nitrogenous base pairs.

dominant trait: An allele whose presence will determine the phenotype, regardless of the nature of its companion allele.

dopamine (DŌ-pa-mēn): An important neurotransmitter in the central nervous system.

dorsal: Toward the back, posterior.

dorsal root ganglion: A peripheral nervous system ganglion containing the cell bodies of sensory neurons.

dorsiflexion: The elevation of the superior surface of the foot through movement at the ankle.

Down syndrome: A genetic abnormality resulting from the presence of three copies of chromosome 21; individuals with this condition have characteristic physical and intellectual deficits.

duct: A passageway that delivers exocrine secretions to an epithelial surface.

ductus arteriosus (ar-tē-rē-Ō-sus): A vascular connection between the pulmonary trunk and the aorta that functions throughout fetal life; normally closes at birth or shortly thereafter and persists as the ligamentum arteriosum.

ductus deferens (DUK-tus DEF-e-renz): A passageway that carries spermatozoa from the epididymis to the ejaculatory duct.

duodenal ampulla: A chamber that receives bile from the common bile duct and pancreatic secretions from the pancreatic duct.

duodenal papilla: A conical projection from the inner surface of the duodenum that contains the opening of the duodenal ampulla.

duodenum (doo-AH-de-num or doo-ō-DĒ-num): The proximal 25 cm of the small intestine that contains short villi and submucosal glands.

dura mater (DOO-ra MĀ-ter or DOO-ra MAH-ter): The outermost component of the cranial and spinal meninges.

dynorphin (dī-NOR-fin): A powerful neuromodulator, produced in the central nervous system, that blocks pain perception by inhibiting pain pathways.

dyslexia: The impaired ability to comprehend written words.

dysmenorrhea: Painful menstruation.

dysmetria (dis-MĒT-rē-uh): Difficulty in performing movements due to problems with the interpretation and anticipation of the distance to be covered.

dysuria (dis-ū-rē-uh): Painful urination.

E

eccrine glands (EK-rin): Sweat glands of the skin that produce a watery secretion.

echocardiography (ek-ō-kar-dē-OG-ra-fē): The examination of the heart by using modified ultrasound techniques.

ectoderm: One of the three primary germ layers; covers the surface of the embryo and gives rise to the nervous system, the epidermis and associated glands, and a variety of other structures.

ectopic (ek-TOP-ik): Outside the normal location.

effector: A peripheral gland or muscle cell innervated by a motor neuron.

efferent: Away from.

efferent arteriole: An arteriole carrying blood away from a glomerulus of the kidney.

efferent fiber: An axon that carries impulses away from the central nervous system.

egestion: *See* **defecation.**

ejaculation (ē-jak-ū-LĀ-shun): The ejection of semen from the penis as the result of muscular contractions of the bulbospongiosus and ischiocavernosus muscles.

ejaculatory ducts (ē-JAK-ū-la-to-rē): Short ducts that pass within the walls of the prostate gland and connect the ductus deferens with the prostatic urethra.

elastase (ē-LAS-tās): A pancreatic enzyme that breaks down elastin fibers.

elastin: Connective tissue fibers that stretch and recoil, providing elasticity to connective tissues.

electrical coupling: A connection between adjacent cells that permits the movement of ions and the transfer of graded or conducted changes in the transmembrane potential from cell to cell.

electrocardiogram (ECG, EKG) (e-lek-trō-KAR-dē-ō-gram): A graphic record of the electrical activities of the heart, as monitored at specific locations on the body surface.

electroencephalogram (EEG): A graphic record of the electrical activities of the brain.

electrolytes (ē-LEK-trō-līts): Soluble inorganic compounds whose ions will conduct an electrical current in solution.

electron: One of the three fundamental subatomic particles; bears a negative charge and normally orbits the protons of the nucleus.

electron transport system (ETS): The cytochrome system responsible for most of the energy production in cells; a complex bound to the inner mitochondrial membrane.

element: All the atoms with the same atomic number.

elephantiasis (el-e-fan-TĪ-a-sis): A lymphedema caused by the infection and blockage of lymphatic vessels by mosquito-borne parasites.

elevation: Movement in a superior, or upward, direction.

elimination: The ejection of wastes from the body through urination or defecation.

embolism (EM-bō-lizm): The obstruction or closure of a vessel by an embolus.

embolus (EM-bō-lus): An air bubble, fat globule, or blood clot drifting in the bloodstream.

embryo (EM-brē-ō): The developmental stage beginning at fertilization and ending at the start of the third developmental month.

embryology (em-brē-OL-o-jē): The study of embryonic development, focusing on the first 2 months after fertilization.

emesis (EM-e-sis): Vomiting.

emmetropia: Normal vision.

emulsification (ē-mul-si-fi-KĀ-shun): The physical breakup of fats in the digestive tract, forming smaller droplets accessible to digestive enzymes; normally the result of mixing with bile salts.

enamel: Crystalline material similar in mineral composition to bone, but harder and without osteocytes, that covers the exposed surfaces of the teeth.

encephalitis: An inflammation of the brain.

endocarditis: An inflammation of the endocardium of the heart.

endocardium (en-dō-KAR-dē-um): The simple squamous epithelium that lines the heart and is continuous with the endothelium of the great vessels.

endochondral ossification (en-dō-KON-dral): The conversion of a cartilaginous model to bone; the characteristic mode of formation for skeletal elements other than the bones of the cranium, the clavicles, and sesamoid bones.

endocrine gland: A gland that secretes hormones into the blood.

endocrine system: The endocrine glands of the body.

endocytosis: The movement of relatively large volumes of extracellular material into the cytoplasm via the formation of a membranous vesicle at the cell surface; includes pinocytosis and phagocytosis.

endoderm: One of the three primary germ layers; the layer on the undersurface of the embryonic disc; gives rise to the epithelia and glands of the digestive system, the respiratory system, and portions of the urinary system.

endogenous: Produced within the body.

endolymph (EN-dō-limf): The fluid contents of the membranous labyrinth (the saccule, utricle, semicircular ducts, and cochlear duct) of the inner ear.

endometrial glands: The secretory glands of the endometrium.

endometrium (en-dō-MĒ-trē-um): The mucous membrane lining the uterus.

endomysium (en-dō-MĪZ-ē-um): A delicate network of connective tissue fibers that surrounds individual muscle cells.

endoneurium: A delicate network of connective tissue fibers that surrounds individual nerve fibers.

endoplasmic reticulum (en-dō-PLAZ-mik re-TIK-ū-lum): A network of membranous channels in the cytoplasm of a cell that function in intracellular transport, synthesis, storage, packaging, and secretion.

endorphins (en-DOR-finz): Neuromodulators, produced in the central nervous system, that inhibit activity along pain pathways.

endosteum: An incomplete cellular lining on the inner (medullary) surfaces of bones.

endothelium (en-dō-THĒ-lē-um): The simple squamous epithelium that lines blood and lymphatic vessels.

enkephalins (en-KEF-a-linz): Neuromodulators, produced in the central nervous system, that inhibit activity along pain pathways.

enteritis (en-ter-Ī-tis): An inflammation of the intestinal tract.

enterocrinin: A hormone secreted by the lining of the duodenum after exposure to chyme; stimulates the secretion of the submucosal glands.

enteroendocrine cells (en-ter-ō-EN-dō-krin): Endocrine cells scattered among the epithelial cells that line the digestive tract.

enterogastric reflex: The reflexive inhibition of gastric secretion; initiated by the arrival of chyme in the small intestine.

enterohepatic circulation: The excretion of bile salts by the liver, followed by the absorption of bile salts by intestinal cells for return to the liver by the hepatic portal vein.

enterokinase: An enzyme in the lumen of the small intestine that activates the proenzymes secreted by the pancreas.

enzyme: A protein that catalyzes a specific biochemical reaction.

eosinophil (ē-ō-SIN-ō-fil): A microphage (white blood cell) with a lobed nucleus and red-staining granules; participates in the immune response and is especially important during allergic reactions.

ependyma (ep-EN-di-mah): The layer of cells lining the ventricles and central canal of the central nervous system.

epicardium: A serous membrane covering the outer surface of the heart; also called *visceral pericardium.*

epidermis: The epithelium covering the surface of the skin.

epididymis (ep-i-DID-i-mus): A coiled duct that connects the rete testis to the ductus deferens; site of functional maturation of spermatozoa.

epidural block: Anesthesia caused by the elimination of sensory inputs from dorsal nerve roots after drugs are introduced into appropriate regions of the epidural space.

epidural space: The space between the spinal dura mater and the walls of the vertebral foramen; contains blood vessels and adipose tissue; a common site of injection for regional anesthesia.

epiglottis (ep-i-GLOT-is): A blade-shaped flap of tissue, reinforced by cartilage, that is attached to the dorsal and superior surface of the thyroid cartilage; folds over the entrance to the larynx during swallowing.

epimysium (ep-i-MĪZ-ē-um): A dense layer of collagen fibers that surrounds a skeletal muscle and is continuous with the tendons/aponeuroses of the muscle and with the perimysium.

epineurium: A dense layer of collagen fibers that surrounds a peripheral nerve.

epiphyseal cartilage (e-pi-FI-zē-al): The cartilaginous region between the epiphysis and diaphysis of a growing bone.

epiphysis (e-PIF-i-sis): The head of a long bone.

epistaxis (ep-i-STAK-sis): A nosebleed.

epithelium (e-pi-THĒ-lē-um): One of the four primary tissue types; a layer of cells that forms a superficial covering or an internal lining of a body cavity or vessel.

eponychium (ep-ō-NIK-ē-um): A narrow zone of stratum corneum that extends across the surface of a nail at its exposed base; also called *cuticle.*

equational division: The second division of meiosis.

equilibrium (ē-kwi-LIB-rē-um): A dynamic state in which two opposing forces or processes are in balance.

erection: The stiffening of the penis due to the engorgement of the erectile tissues of the corpora cavernosa and corpus spongiosum.

erythema (er-i-THĒ-ma): Redness and inflammation at the surface of the skin.

erythrocyte (e-RITH-rō-sīt): A red blood cell; has no nucleus and contains large quantities of hemoglobin.

erythrocytosis (e-rith-rō-sī-TŌ-sis): An abnormally large number of erythrocytes in the circulating blood.

erythropoietin (e-rith-rō-POY-ē-tin): A hormone released by tissues, especially the kidneys, exposed to low oxygen concentrations; stimulates erythropoiesis (red blood cell formation) in bone marrow.

Escherichia coli: A normal bacterial resident of the large intestine.

esophagus: A muscular tube that connects the pharynx to the stomach.

essential amino acids: Amino acids that cannot be synthesized in the body in adequate amounts and must be obtained from the diet.

essential fatty acids: Fatty acids that cannot be synthesized in the body and must be obtained from the diet.

estrogens (ES-trō-jenz): A class of steroid sex hormones that includes estradiol.

eupnea (ŪP-nē-uh): Normal quiet breathing.

evaporation: A movement of molecules from the liquid state to the gaseous state.

eversion (ē-VER-zhun): A turning outward.

excitable membranes: Membranes that propagate action potentials, a characteristic of muscle cells and nerve cells.

excitatory postsynaptic potential (EPSP): The depolarization of a post-synaptic membrane by a chemical neurotransmitter released by the presynaptic cell.

excretion: A removal from body fluids.

exocrine gland: A gland that secretes onto the body surface or into a passageway connected to the exterior.

exocytosis (ek-sō-sī-TŌ-sis): The ejection of cytoplasmic materials by the fusion of a membranous vesicle with the cell membrane.

expiration: Exhalation; breathing out.

expiratory reserve: The amount of additional air that can be voluntarily moved out of the respiratory tract after one normal exhalation.

extension: An increase in the angle between two articulating bones; the opposite of flexion.

external acoustic canal: A passageway in the temporal bone that leads to the tympanic membrane of the inner ear.

external ear: The auricle, external acoustic canal, and tympanic membrane.

external nares (NA-rēz): The nostrils; the external openings into the nasal cavity.

external respiration: The diffusion of gases between the alveolar air and the alveolar capillaries and between the systemic capillaries and peripheral tissues.

exteroceptors: General sensory receptors in the skin, mucous membranes, and special sense organs that provide information about the external environment and about our position within it.

extracellular fluid: All body fluids other than that contained within cells; includes plasma and interstitial fluid.

extraembryonic membranes: The yolk sac, amnion, chorion, and allantois.

extrafusal fibers: Contractile muscle fibers (as opposed to the sensory intrafusal fibers, or muscle spindles).

extrinsic pathway: A clotting pathway that begins with damage to blood vessels or surrounding tissues and ends with the formation of tissue thromboplastin.

F

fabella: A sesamoid bone commonly located in the gastrocnemius muscle.

facilitated: Brought closer to threshold, as in the depolarization of a nerve cell membrane toward threshold; making the cell more sensitive to depolarizing stimuli.

facilitated diffusion: The passive movement of a substance across a cell membrane by means of a protein carrier.

falciform ligament (FAL-si-form): A sheet of mesentery that contains the ligamentum teres, the fibrous remains of the umbilical vein of the fetus.

falx (falks): Sickle-shaped.

falx cerebri (falks ser-Ē-brē): The curving sheet of dura mater that extends between the two cerebral hemispheres; encloses the superior sagittal sinus.

fasciae (FASH-ē-ē): Connective tissue fibers, primarily collagenous, that form sheets or bands beneath the skin to attach, stabilize, enclose, and separate muscles and other internal organs.

fasciculus (fa-SIK-ū-lus): A small bundle; usually refers to a collection of nerve axons or muscle fibers.

fatty acids: Hydrocarbon chains that end in a carboxylic acid group.

fauces (FAW-sēz): The passage from the mouth to the pharynx, bounded by the palatal arches, the soft palate, and the uvula.

febrile: Characterized by or pertaining to a fever.

feces: Waste products eliminated by the digestive tract at the anus; contains indigestible residue, bacteria, mucus, and epithelial cells.

fenestra: An opening.

fertilization: The fusion of a secondary oocyte and a spermatozoon to form a zygote.

fetus: The developmental stage lasting from the start of the third developmental month to delivery.

fibrillation (fi-bri-LĀ-shun): Uncoordinated contractions of individual muscle cells that impair or prevent normal function.

fibrin (FĪ-brin): Insoluble protein fibers that form the basic framework of a blood clot.

fibrinogen (fī-BRIN-ō-jen): A plasma protein that is the soluble precursor of the fibrous protein fibrin.

fibrinolysis (fī-brin-OL-i-sis): The breakdown of the fibrin strands of a blood clot by a proteolytic enzyme.

fibroblasts (FĪ-brō-blasts): Cells of connective tissue proper that are responsible for the production of extracellular fibers and the secretion of the organic compounds of the extracellular matrix.

fibrocartilage: Cartilage containing an abundance of collagen fibers; located around the edges of joints, in the intervertebral discs, the menisci of the knee, and so on.

fibrous tunic: The outermost layer of the eye, composed of the sclera and cornea.

fibula (FIB-ū-luh): The lateral, slender bone of the leg.

filariasis (fil-a-RĪ-a-sis): A condition resulting from infection by mosquito-borne parasites; can cause elephantiasis.

filiform papillae: Slender conical projections from the dorsal surface of the anterior or two-thirds of the tongue.

filtrate: The fluid produced by filtration at a glomerulus in the kidney.

filtration: The movement of a fluid across a membrane whose pores restrict the passage of solutes on the basis of size.

filtration pressure: The hydrostatic pressure responsible for filtration.

filum terminale: A fibrous extension of the spinal cord, from the conus medullaris to the coccygeal ligament.

fimbriae (FIM-brē-ē): Fringes; the fingerlike processes that surround the entrance to the uterine tube.

fissure: An elongate groove or opening.

fistula: An abnormal passageway between two organs or from an internal organ or space to the body surface.

flaccid: Limp, soft, flabby; a muscle without muscle tone.

flagellum/flagella (fla-JEL-uh): An organelle that is structurally similar to a cilium but is used to propel a cell through a fluid.

flatus: Intestinal gas.

flavin adenine dinucleotide (FAD): A coenzyme important in oxidative phosphorylation; cycles between the oxidized (FADH$_2$) and reduced (FAD) states.

flavin adenine mononucleotide (FMN): A coenzyme important in oxidative phosphorylation; cycles between the oxidized (FMNH$_2$) and reduced (FMN) states.

flexion (FLEK-shun): A movement that reduces the angle between two articulating bones; the opposite of extension.

flexor: A muscle that produces flexion.

flexor reflex: A reflex contraction of the flexor muscles of a limb in response to an unpleasant stimulus.

flexure: A bending.

fluoroscope: An instrument that permits the body to be examined in real time with X rays rather than with fixed images on photographic plates.

folia (FŌ-lē-uh): Leaflike folds; the slender folds in the surface of the cerebellar cortex.

follicle (FOL-i-kul): A small secretory sac or gland.

follicle-stimulating hormone (FSH): A hormone secreted by the anterior pituitary; stimulates oogenesis (female) and spermatogenesis (male).

folliculitis (fo-lik-ū-LĪ-tis): An inflammation of a follicle, such as a hair follicle of the skin.

fontanel (fon-tuh-NEL): A relatively soft, flexible, fibrous region between two flat bones in the developing skull; also spelled *fontanelle.*

foramen/foramina (fo-RĀ-men/fo-RAM-in-uh): An opening or passage through a bone.

forearm: The distal portion of the upper limb between the elbow and wrist.

forebrain: The cerebrum.

fornix (FOR-niks): An arch or the space bounded by an arch; in the brain, an arching tract that connects the hippocampus with the mamillary bodies; in the eye, a slender pocket situated where the epithelium of the ocular conjunctiva folds back on itself as the palpebral conjunctiva.

fossa: A shallow depression or furrow in the surface of a bone.

fourth ventricle: An elongate ventricle of the metencephalon (pons and cerebellum) and the myelencephalon (medulla oblongata) of the brain; the roof contains a region of choroid plexus.

fovea (FŌ-vē-uh): The portion of the retina that provides the sharpest vision because it has the highest concentration of cones; also called *macula lutea.*

fracture: A break or crack in a bone.

frenulum (FREN-ū-lum): A bridle; *see* **lingual frenulum.**

frontal plane: A sectional plane that divides the body into an anterior portion and a posterior portion; also called *coronal plane.*

fructose: A hexose (six-carbon simple sugar) in foods and in semen.

fundus (FUN-dus): The base of an organ.

fungiform papillae: The mushroom-shaped papillae on the dorsal and dorsolateral surfaces of the tongue.

furuncle (FUR-ung-kl): A boil, resulting from the invasion and inflammation of a hair follicle or sebaceous gland.

G

gallbladder: The pear-shaped reservoir for bile after it is secreted by the liver.

gametes (GAM-ēts): Reproductive cells (spermatozoa or oocytes) that contain half the normal chromosome complement.

gametogenesis (ga-mē-tō-JEN-e-sis): The formation of gametes.

gamma aminobutyric acid (GABA) (GAM-ma a-MĒ-nō-bū-TIR-ik): A neurotransmitter of the central nervous system whose effects are generally inhibitory.

gamma motor neurons: Motor neurons that adjust the sensitivities of muscle spindles (intrafusal fibers).

ganglion/ganglia: A collection of neuron cell bodies outside the central nervous system.

gangliosides: Glycolipids that are important components of cell membranes in the central nervous system.

gap junctions: Connections between cells that permit electrical coupling.

gaster (GAS-ter): The stomach; the body, or belly, of a skeletal muscle.

gastrectomy (gas-TREK-to-mē): The partial or total surgical removal of the stomach.

gastric: Pertaining to the stomach.

gastric glands: The tubular glands of the stomach whose cells produce acid, enzymes, intrinsic factor, and hormones.

gastric inhibitory peptide (GIP): A duodenal hormone released when the arriving chyme contains large quantities of carbohydrates; triggers the secretion of insulin and a slowdown in gastric activity.

gastrin (GAS-trin): A hormone produced by enteroendocrine cells of the stomach, after exposure to mechanical stimuli or stimulation of the vagus nerve, and of the duodenum, after exposure to chyme that contains undigested proteins.

gastritis (gas-TRĪ-tis): An inflammation of the stomach.

gastroenteric reflex (gas-trō-en-TER-ik): An increase in peristalsis along the small intestine; triggered by the arrival of food in the stomach.

gastroileal reflex (gas-trō-IL-ē-al): Peristaltic movements that shift materials from the ileum to the colon; triggered by the arrival of food in the stomach.

gastrointestinal (GI) tract: See **digestive tract.**

gastroscope: A fiber-optic instrument that permits visual inspection of the stomach lining.

gastrulation (gas-troo-LĀ-shun): The movement of cells of the inner cell mass that creates the three primary germ layers of the embryo.

gene: A portion of a DNA strand that functions as a hereditary unit, is located at a particular site on a specific chromosome, and codes for a specific protein.

genetic engineering: Research and experiments involving the manipulation of the genetic makeup of an organism.

genetics: The study of mechanisms of heredity.

geniculate (je-NIK-ū-lāt): Like a little knee; the medial geniculates and the lateral geniculates are nuclei in the thalamus of the brain.

genitalia (jen-i-TĀ-lē-uh): The reproductive organs.

genotype (JĒN-ō-tīp): An individual's genetic complement, which determines the individual's phenotype.

germinal centers: Pale regions in the interior of lymphoid tissues or nodules, where cell divisions are under way.

gestation (jes-TĀ-shun): The period of intrauterine development.

gingivae (JIN-ji-vē): The gums.

gingivitis: An inflammation of the gums.

gland: Cells that produce exocrine or endocrine secretions.

glans: The expanded tip of the penis that surrounds the external urethral meatus; continuous with the corpus spongiosum.

glaucoma: An eye disorder characterized by rising intraocular pressures due to inadequate drainage of aqueous humor at the canal of Schlemm.

glenoid cavity: A rounded depression that forms the articular surface of the scapula at the shoulder joint.

glial cells (GLĒ-al): *See* **neuroglia**.

globular proteins: Proteins whose tertiary structure makes them rounded and compact.

glomerular capsule: The expanded initial portion of the nephron that surrounds the glomerulus.

glomerular filtration rate: The rate of filtrate formation at the glomerulus.

glomerulonephritis (glo-mer-ū-lō-nef-RĪ-tis): An inflammation of the glomeruli of the kidneys.

glomerulus (glo-MER-ū-lus): A ball or knot; in the kidneys, a knot of capillaries that projects into the enlarged, proximal end of a nephron; the site of filtration, the first step in the production of urine.

glossopharyngeal nerve (glos-ō-fa-RIN-jē-al): Cranial nerve IX.

glottis (GLOT-is): The passageway from the pharynx to the larynx.

glucagon (GLOO-ka-gon): A hormone secreted by the alpha cells of the pancreatic islets; elevates blood glucose concentrations.

glucocorticoids: Hormones secreted by the zona fasciculata of the adrenal cortex to modify glucose metabolism; cortisol and corticosterone are important examples.

glucogenic amino acids: Amino acids that can be broken down, converted to pyruvic acid, and used in gluconeogenesis.

gluconeogenesis (gloo-kō-nē-ō-JEN-e-sis): The synthesis of glucose from protein or lipid precursors.

glucose (GLOO-kōs): A six-carbon sugar, $C_6H_{12}O_6$; the preferred energy source for most cells and normally the only energy source for neurons.

glycerides: Lipids composed of glycerol bound to fatty acids.

glycogen (GLĪ-kō-jen): A polysaccharide that is an important energy reserve; a polymer consisting of a long chain of glucose molecules.

glycogenesis: The synthesis of glycogen from glucose molecules.

glycogenolysis: Glycogen breakdown and the liberation of glucose molecules.

glycolipids (glī-cō-LIP-idz): Compounds created by the combination of carbohydrate and lipid components.

glycolysis (glī-KOL-i-sis): The anaerobic cytoplasmic breakdown of glucose into lactic acid by way of pyruvic acid, with a net gain of two ATP molecules.

glycoprotein (GLĪ-kō-prō-tēn): A compound containing a relatively small carbohydrate group attached to a large protein.

glycosuria (glī-kō-SOO-rē-uh): The presence of glucose in urine.

goblet cell: A goblet-shaped, mucus-producing, unicellular gland in certain epithelia of the digestive and respiratory tracts.

goiter: An enlargement of the thyroid gland.

Golgi apparatus (gol-jē): A cellular organelle consisting of a series of membranous plates that give rise to lysosomes and secretory vesicles.

Golgi tendon organ: A receptor sensitive to tension in a tendon.

gomphosis (gom-FŌ-sis): A fibrous synarthrosis that binds a tooth to the bone of the jaw; *see* **periodontal ligament**.

gonadotropin-releasing hormone (GnRH): A hypothalamic releasing hormone that causes the secretion of both follicle-stimulating hormone and luteinizing hormone by the anterior pituitary gland.

gonadotropins (gō-nad-ō-TRŌ-pinz): Follicle-stimulating hormone and luteinizing hormone, hormones that stimulate gamete development and sex hormone secretion.

gonads (GŌ-nadz): Reproductive organs that produce gametes and hormones.

gout: A condition resulting from elevated uric acid concentrations in blood and in peripheral tissues.

granulocytes (GRAN-ū-lō-sīts): White blood cells containing granules that are visible with the light microscope; includes eosinophils, basophils, and neutrophils; also called *granular leukocytes*.

gray matter: Areas in the central nervous system that are dominated by neuron cell bodies, neuroglia, and unmyelinated axons.

gray ramus: A bundle of postganglionic sympathetic nerve fibers that are distributed to effectors in the body wall, skin, and limbs by way of a spinal nerve.

greater omentum: A large fold of the dorsal mesentery of the stomach; hangs anterior to the intestines.

greater vestibular glands: Mucous glands in the vaginal walls that secrete into the vestibule; the equivalent of the bulbourethral glands of males.

greenstick fracture: A fracture in which a bone cracks and bends, most commonly involving the long bones of young children.

groin: The inguinal region.

gross anatomy: The study of the structural features of the body without the aid of a microscope.

growth hormone (GH): An anterior pituitary hormone that stimulates tissue growth and anabolism when nutrients are abundant and restricts tissue glucose dependence when nutrients are in short supply.

growth hormone–inhibiting hormone (GH-IH): A hypothalamic regulatory hormone that inhibits growth hormone secretion by the anterior pituitary; also called *somatostatin*.

guanine: A purine; one of the nitrogenous bases in the nucleic acids RNA and DNA.

gustation (gus-TĀ-shun): Taste.

gynecologist (gī-ne-KOL-o-jist): A physician specializing in the female reproductive system.

gyrus (JĪ-rus): A prominent fold or ridge of neural cortex on the surfaces of the cerebral hemispheres.

H

hair: A keratinous strand produced by epithelial cells of the hair follicle.

hair cells: Sensory cells of the inner ear.

hair follicle: An accessory structure of the integument; a tube lined by a stratified squamous epithelium that begins at the surface of the skin and ends at the hair papilla.

hair root: A thickened, conical structure consisting of a connective tissue papilla and the overlying matrix, a layer of epithelial cells that produces the hair shaft.

hallux: The big toe.

haploid (HAP-loyd): Possessing half the normal number of chromosomes; a characteristic of gametes.

hapten: A partial antigen that can bind to an antibody but cannot stimulate antibody production; a foreign compound that has only one antigenic determinant site.

hard palate: The bony roof of the oral cavity, formed by the maxillary and palatine bones.

Hassall's corpuscles: Aggregations of epithelial cells in the thymus whose functions are unknown.

haustra (HAWS-truh): Saclike pouches along the length of the large intestine that result from tension in the taenia coli.

heat exhaustion: A condition that is characterized by excessive perspiration and results in dangerous losses of fluids and salts.

heat stroke: A condition resulting from the failure of the body's normal temperature control mechanisms; characterized by a cessation of sweating and a potentially fatal elevation of body temperature.

Heimlich maneuver (HĪM-lik): A technique for removing an airway blockage by external compression of the abdomen and forceful elevation of the diaphragm.

helper T cells: Lymphocytes whose secretions and other activities coordinate cell-mediated and antibody-mediated immunities; also called T_H cells.

hematocrit (he-MAT-ō-krit): The percentage of the volume of whole blood contributed by cells; also called *volume of packed red cells (VPRC)* or *packed cell volume (PCV)*.

hematoma: A tumor or swelling filled with blood.

hematuria (hē-ma-TOO-rē-uh): The presence of abnormal numbers of red blood cells in urine.

heme (hēm): A porphyrin ring containing a central iron atom that can reversibly bind oxygen molecules; a component of the hemoglobin molecule.

hemocytoblasts: Stem cells whose divisions produce each of the various populations of blood cells.

hemodialysis (hē-mō-dī-AL-i-sis): Dialysis of the blood.

hemoglobin (HĒ-mō-glō-bin): A protein composed of four globular subunits, each bound to a heme molecule; gives red blood cells the ability to transport oxygen in the blood.

hemolysis: The breakdown (lysis) of red blood cells.

hemophilia (hē-mō-FĒL-ē-uh): A congenital condition due to the inadequate synthesis of one of the clotting factors.

hemopoiesis (hē-mō-poy-Ē-sis): Blood cell formation and differentiation.

hemorrhage: Blood loss.

hemorrhoids (HEM-o-roydz): Swollen, varicose veins that protrude from the walls of the rectum and/or the anorectal canal.

hemostasis: The cessation of bleeding.

hemothorax: The entry of blood into one of the pleural cavities.

heparin (HEP-a-rin): An anticoagulant released by activated basophils and mast cells.

hepatic duct: The duct that carries bile away from the liver lobes and toward the union with the cystic duct.

hepatic portal vein: The vessel that carries blood between the intestinal capillaries and the sinusoids of the liver.

hepatitis (hep-a-TĪ-tis): An inflammation of the liver, resulting from exposure to toxic chemicals, drugs, or viruses.

hepatocyte (he-PAT-ō-sīt): A liver cell.

hernia: The protrusion of a loop or portion of a visceral organ through the abdominopelvic wall or into the thoracic cavity.

herniated disc: The rupture of the connective tissue sheath of the nucleus pulposus of an intervertebral disc.

heterotopic: Ectopic; outside the normal location.

heterozygous (het-er-ō-ZĪ-gus): Possessing two different alleles at corresponding sites on a chromosome pair; the individual's phenotype is determined by one or both of the alleles.

hexose: A six-carbon simple sugar.

hiatus (hī-Ā-tus): A gap, cleft, or opening.

high-density lipoprotein (HDL): A lipoprotein with a relatively small lipid content; thought to be responsible for the movement of cholesterol from peripheral tissues to the liver.

hilum/hilus (HĪ-lum): A localized region where blood vessels, lymphatic vessels, nerves, and/or other anatomical structures are attached to an organ.

hippocampus: A region, beneath the floor of a lateral ventricle, involved with emotional states and the conversion of short-term to long-term memories.

histamine (HIS-tuh-mēn): The chemical released by stimulated mast cells or basophils to initiate or enhance an inflammatory response.

histology (his-TOL-o-jē): The study of tissues.

histones: Proteins associated with the DNA of the nucleus; the DNA strands are wound around them.

holocrine (HO-lō-krin): A form of exocrine secretion in which the secretory cell becomes swollen with vesicles and then ruptures.

homeostasis (hō-mē-ō-STĀ-sis): The maintenance of a relatively constant internal environment.

homologous chromosomes (hō-MOL-o-gus): The members of a chromosome pair.

homozygous (hō-mō-ZĪ-gus): Having the same allele for a given phenotypic character on two homologous chromosomes.

hormone: A compound that is secreted by one cell and travels through the circulatory system to affect the activities of cells in another portion of the body.

human chorionic gonadotropin (hCG): The placental hormone that maintains the corpus luteum for the first 3 months of pregnancy.

human immunodeficiency virus (HIV): The infectious agent that causes acquired immune deficiency syndrome (AIDS).

human leukocyte antigen (HLA): *See* **MHC protein.**

human placental lactogen (hPL): The placental hormone that stimulates the functional development of the mammary glands.

humoral immunity: *See* **antibody-mediated immunity.**

hyaluronan: A carbohydrate component of proteoglycans in the matrix of many connective tissues.

hyaluronidase: An enzyme that breaks down the bonds between adjacent follicle cells; produced by some bacteria and found in the acrosomal cap of a spermatozoon.

hydrocephalus: A condition resulting from excessive production or inadequate drainage of cerebrospinal fluid.

hydrogen bond: A weak interaction between the hydrogen atom on one molecule and a negatively charged portion of another molecule.

hydrolysis (hī-DROL-i-sis): The breakage of a chemical bond through the addition of a water molecule; the reverse of dehydration synthesis.

hydrophilic (hī-drō-PHIL-ik): Freely associating with water; readily entering into solution.

hydrophobic: Incapable of freely associating with water molecules; insoluble.

hydrostatic pressure: Fluid pressure.

hydroxide ion (hī-DROKS-īd): OH^-.

hypercapnia (hī-per-KAP-nē-uh): High plasma carbon dioxide concentrations, commonly as a result of hypoventilation or inadequate tissue perfusion.

hyperglycemia: Elevated plasma glucose concentrations.

hyperkalemia (hī-per-kā-LĒ-mē-uh): Abnormally high potassium concentrations in extracellular fluid.

hypernatremia: Abnormally high sodium concentrations in extracellular fluid.

hyperopia: Farsightedness, characterized by an inability to focus on nearby objects.

hyperplasia: An abnormal enlargement of an organ due to an increase in the number of cells.

hyperpnea (hī-PERP-nē-uh): Abnormal increases in the rate and depth of respiration.

hyperpolarization: The movement of the transmembrane potential away from the normal resting potential and farther from 0 mV.

hyperreflexia: Abnormally exaggerated reflex responses to stimulation.

hypersecretion: The overactivity of glands that produce exocrine or endocrine secretions.

hypersensitivity: An overreaction to an allergen that results in tissue damage and inflammation.

hypertension: Abnormally high blood pressure.

hyperthermia: Excessively high body temperature.

hyperthyroidism: An excessive production of thyroid hormones.

hypertonic: In comparing two solutions, the solution with the higher osmolarity.

hypertrophy (hī-PER-trō-fē): An increase in the size of tissue without cell division.

hyperventilation (hī-per-ven-ti-LĀ-shun): A rate of respiration sufficient to reduce plasma P_{CO_2} to levels below normal.

hypervitaminosis (hī-per-vī-ta-min-Ō-sis): A clinical condition caused by the excessive ingestion and uptake of vitamins.

hypesthesia: An abnormally decreased sensitivity to stimuli.

hypocapnia: An abnormally low plasma P_{CO_2} commonly as a result of hyperventilation.

hypodermic needle: A needle inserted through the skin to introduce drugs into the subcutaneous layer.

hypodermis: *See* **subcutaneous layer.**

hypokalemia (hī-pō-ka-LĒ-mē-uh): Abnormally low plasma potassium concentrations.

hyponatremia: Abnormally low plasma sodium concentrations.

hyponychium (hī-pō-NIK-ē-um): A thickening in the epidermis beneath the free edge of a nail.

hypophyseal portal system (hī-pō-FI-sē-al): The network of vessels that carry blood from capillaries in the hypothalamus to capillaries in the anterior lobe of the pituitary gland.

hypophysis (hī-POF-i-sis): The pituitary gland.

hyporeflexia: Abnormally depressed reflex responses to stimuli.

hyposecretion: Abnormally low rates of exocrine or endocrine secretion.

hypothalamus: The floor of the diencephalon; the region of the brain containing centers involved with the subconscious regulation of visceral functions, emotions, drives, and the coordination of neural and endocrine functions.

hypothermia (hī-pō-THER-mē-uh): An abnormally low body temperature.

hypothesis: A prediction that can be subjected to scientific analysis and review.

hypotonic: In comparing two solutions, the solution with the lower osmolarity.

hypoventilation: A respiratory rate that is insufficient to keep plasma P_{CO_2} within normal levels.

hypovitaminosis: A clinical condition resulting from inadequate vitamin ingestion and uptake; vitamin deficiency.

hypovolemic (hī-pō-vō-LĒ-mik): An abnormally low blood volume.

hypoxia (hī-POKS-ē-uh): A low tissue oxygen concentration.

I

ileocecal valve (il-ē-ō-SĒ-kal): A fold of mucous membrane that guards the connection between the ileum and the cecum.

ileostomy (il-ē-OS-to-mē): The surgical creation of an opening into the ileum; the opening created when the ileum is surgically attached to the abdominal wall.

ileum (IL-ē-um): The distal 2.5 m of the small intestine.

ilium (IL-ē-um): The largest of the three bones whose fusion creates an os coxae.

immunity: Resistance to injuries and diseases caused by foreign compounds, toxins, or pathogens.

immunization: The production of immunity by the deliberate exposure to antigens under conditions that prevent the development of illness but stimulate the production of memory B cells.

immunodeficiency: An inability to produce normal numbers and types of antibodies and sensitized lymphocytes.

immunoglobulin (i-mū-nō-GLOB-ū-lin): A circulating antibody.

immunosuppression (i-mū-nō-su-PRE-shun): The suppression of immune responses by the administration of drugs or exposure to toxic chemicals, radiation, or infection.

implantation (im-plan-TĀ-shun): The erosion of a blastocyst into the uterine wall.

impotence: The inability to obtain or maintain an erection.

inclusions: Aggregations of insoluble pigments, nutrients, or other materials in cytoplasm.

incontinence (in-KON-ti-nens): The inability to control micturition (or defecation) voluntarily.

incus (IN-kus): The central auditory ossicle, situated between the malleus and the stapes in the middle ear cavity.

inducer: A stimulus that promotes the activity of a specific gene.

inexcitable: Incapable of propagating an action potential.

infarct: An area of dead cells that results from an interruption of blood flow.

infection: The invasion and colonization of body tissues by pathogens.

inferior: Below, in reference to a particular structure, with the body in the anatomical position.

inferior vena cava: The vein that carries blood from the parts of the body inferior to the heart to the right atrium.

infertility: The inability to conceive; also called *sterility.*

inflammation: A nonspecific defense mechanism that operates at the tissue level; characterized by swelling, redness, warmth, pain, and some loss of function.

inflation reflex: A reflex mediated by the vagus nerve (N X) that prevents overexpansion of the lungs.

infundibulum (in-fun-DIB-ū-lum): A tapering, funnel-shaped structure; in the brain, the connection between the pituitary gland and the hypothalamus; in the uterine tube, the entrance bounded by fimbriae that receives the oocytes at ovulation.

ingestion: The introduction of materials into the digestive tract by way of the mouth.

inguinal canal: A passage through the abdominal wall that marks the path of testicular descent and that contains the testicular arteries, veins, and ductus deferens.

inguinal region: The area near the junction of the trunk and the thighs that contains the external genitalia.

inhibin (in-HIB-in): A hormone, produced by sustentacular cells of the testes and follicular cells of the ovaries, that inhibits the secretion of follicle-stimulating hormone by the anterior lobe of the pituitary gland.

inhibitory postsynaptic potential (IPSP): A hyperpolarization of the postsynaptic membrane after the arrival of a neurotransmitter.

initial segment: The proximal portion of the axon where an action potential first appears.

injection: The forcing of fluid into a body part or organ.

inner cell mass: Cells of the blastocyst that will form the body of the embryo.

inner ear: *See* **internal ear.**

innervation: The distribution of sensory and motor nerves to a specific region or organ.

insensible perspiration: Evaporative water loss by diffusion across the epithelium of the skin or evaporation across the alveolar surfaces of the lungs.

insertion: A point of attachment of a muscle; the end that is easily movable.

insoluble: Incapable of dissolving in solution.

insomnia: A sleep disorder characterized by the inability to fall asleep.

inspiration: Inhalation; the movement of air into the respiratory system.

inspiratory reserve: The maximum amount of air that can be drawn into the lungs over and above the normal tidal volume.

insulin (IN-su-lin): A hormone secreted by the beta cells of the pancreatic islets; causes a reduction in plasma glucose concentrations.

integument (in-TEG-ū-ment): The skin.

intercalated discs (in-TER-ka-lā-ted): Regions where adjacent cardiocytes interlock and where gap junctions permit electrical coupling between the cells.

intercellular cement: Proteoglycans situated between adjacent epithelial cells.

intercellular fluid: *See* **interstitial fluid.**

interdigitate: To interlock.

interferons (in-ter-FĒR-onz): Peptides released by virus-infected cells, especially lymphocytes, that slow viral replication and make other cells more resistant to viral infection.

interleukins (in-ter-LOO-kinz): Peptides, released by activated monocytes and lymphocytes, that assist in the coordination of cell-mediated and antibody-mediated immunities.

internal capsule: The collection of afferent and efferent fibers of the white matter of the cerebral hemispheres, visible on gross dissection of the brain.

internal ear: The membranous labyrinth that contains the organs of hearing and equilibrium.

internal nares: The entrance to the nasopharynx from the nasal cavity.

internal respiration: The diffusion of gases between interstitial fluid and cytoplasm.

interneuron: An association neuron; central nervous system neurons that are between sensory and motor neurons.

interoceptors: Sensory receptors monitoring the functions and status of internal organs and systems.

interosseous membrane: The fibrous connective tissue membrane between the shafts of the tibia and fibula and between the radius and ulna; an example of a fibrous amphiarthrosis.

interphase: The stage in the life cycle of a cell during which the chromosomes are uncoiled and all normal cellular functions except mitosis are under way.

intersegmental reflex: A reflex that involves several segments of the spinal cord.

interstitial fluid (in-ter-STISH-al): The fluid in the tissues that fills the spaces between cells.

interstitial growth: A form of cartilage growth through the growth, mitosis, and secretion of chondrocytes in the matrix.

interventricular foramen: The opening that permits fluid movement between the lateral and third ventricles of the brain.

intervertebral disc: A fibrocartilage pad between the bodies of successive vertebrae that absorbs shocks.

intestinal crypt: A tubular epithelial pocket that is lined by secretory cells and opens into the lumen of the digestive tract; also called *intestinal gland.*

intestine: The tubular organ of the digestive tract.

intracellular fluid: The cytosol.

intrafusal fibers: Muscle spindle fibers.

intramembranous ossification (in-tra-MEM-bra-nus): The formation of bone within a connective tissue without the prior development of a cartilaginous model.

intramuscular injection: The injection of medication into the bulk of a skeletal muscle.

intraocular pressure: The hydrostatic pressure exerted by the aqueous humor of the eye.

intrapleural pressure: The pressure measured in a pleural cavity; also called *intrathoracic pressure.*

intrapulmonary pressure (in-tra-PUL-mo-ner-ē): The pressure measured in an alveolus of the lungs; also called *intraalveolar pressure.*

intrauterine: Within the uterus; during prenatal development.

intrinsic factor: A glycoprotein, secreted by the parietal cells of the stomach, that facilitates the intestinal absorption of vitamin B_{12}.

intrinsic pathway: A pathway of the clotting system that begins with the activation of platelets and ends with the formation of platelet thromboplastin.

inversion: A turning inward.

in vitro: Outside the body, in an artificial environment.

in vivo: In the living body.

involuntary: Not under conscious control.

ion: An atom or molecule bearing a positive or negative charge due to the donation or acceptance, respectively, of an electron.

ionic bond (ī-ON-ik): A molecular bond created by the attraction between ions with opposite charges.

ionization (ī-on-i-ZĀ-shun): Dissociation; the breakdown of a molecule in solution to form ions.

ipsilateral: A reflex response that affects the same side as the stimulus.

iris: A contractile structure, made up of smooth muscle, that forms the colored portion of the eye.

ischemia (is-KĒ-mē-uh): An inadequate blood supply to a region of the body.

ischium (IS-kē-um or is-SHĒ-um): One of the three bones whose fusion creates an os coxae.

islets of Langerhans: *See* **pancreatic islets.**

isometric contraction: A muscle contraction characterized by rising tension production but no change in length.

isotonic: A solution with an osmolarity that does not result in water movement across cell membranes.

isotonic contraction: A muscle contraction during which tension climbs and then remains stable as the muscle shortens.

isotopes: Forms of an element whose atoms contain the same number of protons but different numbers of neutrons (and thus differ in atomic weight).

isthmus (IS-mus): A narrow band of tissue connecting two larger masses.

J

jaundice (JAWN-dis): A condition characterized by yellowing of connective tissues due to elevated tissue bilirubin levels; normally associated with damage to the liver or bile ducts.

jejunum (je-JOO-num): The middle part of the small intestine.

joint: An area where adjacent bones interact; also called *articulation.*

juxtaglomerular apparatus: The macula densa and the juxtaglomerular cells; a complex responsible for the release of renin and erythropoietin.

juxtaglomerular cells: Modified smooth muscle cells in the walls of the afferent and efferent arterioles adjacent to the glomerulus and the macula densa.

K

karyotyping (KAR-ē-ō-tī-ping): The determination of the chromosomal characteristics of an individual or cell.

keratin (KER-a-tin): The tough, fibrous protein component of nails, hair, calluses, and the general integumentary surface.

keratinization (KER-a-tin-i-zā-shun): The production of keratin by a stratified squamous epithelium; also called *cornification.*

keto acid: A molecule that ends in — COCOOH; the carbon chain that remains after the deamination or transamination of an amino acid.

ketoacidosis (kē-tō-as-i-DŌ-sis): A reduction in the pH of body fluids due to the presence of large numbers of ketone bodies.

ketogenic amino acids: Amino acids whose catabolism yields ketone bodies rather than pyruvic acid.

ketone bodies: Keto acids produced during the catabolism of lipids and ketogenic amino acids; specifically, acetone, acetoacetate, and beta-hydroxybutyrate.

ketonemia (kē-tō-NĒ-mē-uh): Abnormal concentrations of ketone bodies in blood.

ketonuria (kē-tō-NŪ-rē-uh): Abnormal concentrations of ketone bodies in urine.

ketosis (kē-TŌ-sis): A condition characterized by the abnormal production of ketone bodies.

kidney: A component of the urinary system; an organ functioning in the regulation of plasma composition, including the excretion of wastes and the maintenance of normal fluid and electrolyte balances.

killer T cells: *See* **cytotoxic T cells.**

kilocalorie (KIL-ō-kal-o-rē): *See* **Calorie.**

Krebs cycle: *See* **TCA cycle.**

Kupffer cells (KOOP-fer): Stellate reticular cells of the liver; phagocytic cells of the liver sinusoids.

kyphosis (kī-FŌ-sis): An exaggerated thoracic curvature.

L

labium/labia (LĀ-bē-uh): Lips; the labia majora and labia minora are components of the female external genitalia.

labrum: A lip or rim.

labyrinth: A maze of passageways; the structures of the inner ear.

lacrimal gland (LAK-ri-mal): A tear gland on the dorsolateral surface of the eye.

lactase: An enzyme that breaks down milk proteins.

lactation (lak-TĀ-shun): The production of milk by the mammary glands.

lacteal (LAK-tē-al): A terminal lymphatic within an intestinal villus.

lactic acid: A compound produced from pyruvic acid under anaerobic conditions.

lactiferous duct (lak-TIF-e-rus): A duct draining one lobe of the mammary gland.

lactiferous sinus: An expanded portion of a lactiferous duct adjacent to the nipple of a breast.

lacuna (la-KOO-nuh): A small pit or cavity.

lambdoid suture (lam-DOYD): The synarthrosis between the parietal and occipital bones of the cranium.

lamellae (la-MEL-ē): Concentric layers; the concentric layers of bone within an osteon.

lamellated corpuscle: A receptor sensitive to vibration.

lamina (LA-min-uh): A thin sheet or layer.

lamina propria (LA-min-uh PRŌ-prē-uh): The reticular tissue that underlies a mucous epithelium and forms part of a mucous membrane.

laminectomy: The removal of the spinous processes of a vertebra to gain access and treat a herniated disc.

Langerhans cells (LAN-ger-hanz): Cells in the epithelium of the skin and digestive tract that participate in the immune response by presenting antigens to T cells.

laparoscope (LAP-a-ro-skōp): A fiber-optic instrument used to visualize the contents of the abdominopelvic cavity.

large intestine: The terminal portions of the intestinal tract, consisting of the colon, the rectum, and the anal canal.

laryngopharynx (la-rin-gō-FAR-inks): The division of the pharynx that is inferior to the epiglottis and superior to the esophagus.

larynx (LAR-inks): A complex cartilaginous structure that surrounds and protects the glottis and vocal cords; the superior margin is bound to the hyoid bone, and the inferior margin is bound to the trachea.

latent period: The time between the stimulation of a muscle and the start of the contraction phase.

lateral: Pertaining to the side.

lateral apertures: Openings in the roof of the fourth ventricle that permit the circulation of cerebrospinal fluid into the subarachnoid space.

lateral ventricle: A fluid-filled chamber within a cerebral hemisphere.

laxatives: Compounds that promote defecation through increased peristalsis or an increase in the water content and volume of feces.

lens: The transparent body that is inferior to the iris and pupil and superior to the vitreous humor.

lesion: A localized abnormality in tissue organization.

lesser omentum: A small pocket in the mesentery that connects the lesser curvature of the stomach to the liver.

leukemia (loo-KĒ-mē-uh): A malignant disease of the blood-forming tissues.

leukocyte (LOO-kō-sīt): A white blood cell.

leukocytosis (loo-kō-sī-TŌ-sis): Abnormally high numbers of circulating white blood cells.

leukopenia (loo-kō-PĒ-nē-uh): Abnormally low numbers of circulating white blood cells.

ligament (LI-ga-ment): A dense band of connective tissue fibers that attaches one bone to another.

ligamentum arteriosum: The fibrous strand in adults that is the remnant of the ductus arteriosus of the fetal stage.

ligamentum nuchae (li-guh-MEN-tum NOO-kē or NOO-kā): An elastic ligament between the vertebra prominens and the occipital bone.

ligamentum teres: The fibrous strand in the falciform ligament of adults that is the remnant of the umbilical vein of the fetal stage.

ligate: To tie off.

limbic system (LIM-bik): The group of nuclei and centers in the cerebrum and diencephalon that are involved with emotional states, memories, and behavioral drives.

limbus (LIM-bus): The edge of the cornea, marked by the transition from the corneal epithelium to the ocular conjunctiva.

linea alba: The tendinous band along the midline of the rectus abdominis muscle.

lingual: Pertaining to the tongue.

lingual frenulum: An epithelial fold that attaches the inferior surface of the tongue to the floor of the mouth.

lipase (LĪ-pās): A pancreatic enzyme that breaks down triglycerides.

lipemia (lip-Ē-mē-uh): An elevated concentration of lipids in the blood.

lipid: An organic compound containing carbons, hydrogens, and oxygens in a ratio that does not approximate 1:2:1; includes fats, oils, and waxes.

lipofuscin (li-pō-FŪ-shun): A pigment inclusion in aging nerve cells that is of unknown significance.

lipogenesis (li-pō-JEN-e-sis): The synthesis of lipids from nonlipid precursors.

lipoids: Prostaglandins, steroids, phospholipids, glycolipids, and so on.

lipolysis (lī-POL-i-sis): The catabolism of lipids as a source of energy.

lipoprotein (lī-pō-PRŌ-tēn): A compound containing a relatively small lipid bound to a protein.

liver: An organ of the digestive system that has varied and vital functions, including the production of plasma proteins, the excretion of bile, the storage of energy reserves, the detoxification of poisons, and the interconversion of nutrients.

lobule (LOB-ūl): Histologically, the basic organizational unit of the liver.

local hormone: *See* **prostaglandin.**

long-term memories: Memories that persist for an extended period.

loop of Henle: The portion of the nephron that creates the concentration gradient in the renal medulla.

loose connective tissue: A loosely organized, easily distorted connective tissue that contains several fiber types, a varied population of cells, and a viscous ground substance.

lordosis (lor-DŌ-sis): An exaggeration of the lumbar curvature.

lumbar: Pertaining to the lower back.

lumen: The central space within a duct or other internal passageway.

lungs: The paired organs of respiration, situated in the pleural cavities.

luteinizing hormone (LH) (LOO-tē-in-ī-zing): Also called *lutropin;* a hormone produced by the anterior lobe of the pituitary gland. In females, it assists FSH in follicle stimulation, triggers ovulation, and promotes the maintenance and secretion of endometrial glands. In males, it was formerly called *interstitial cell-stimulating hormone* because it stimulates testosterone secretion by the interstitial cells of the testes.

luxation (luks-Ā-shun): The dislocation of a joint.

lymph: The fluid contents of lymphatic vessels, similar in composition to interstitial fluid.

lymphadenopathy (lim-fad-e-NOP-a-thē): The pathological enlargement of lymph nodes.

lymphatic vessels: The vessels of the lymphatic system; also called *lymphatics.*

lymphedema (lim-fe-DĒ-ma): The swelling of peripheral tissues as a result of excessive lymph production or inadequate drainage.

lymph nodes: Lymphoid organs that monitor the composition of lymph.

lymphocyte (LIM-fō-sīt): A cell of the lymphatic system that participates in the immune response.

lymphokines: Chemicals secreted by activated lymphocytes.

lymphopoiesis (lim-fō-poy-Ē-sis): The production of lymphocytes from lymphoid stem cells.

lymphotoxin (lim-fō-TOK-sin): A secretion of lymphocytes that kills the target cells.

lysis (LĪ-sis): The destruction of a cell through the rupture of its cell membrane.

lysosome (LĪ-sō-sōm): An intracellular vesicle containing digestive enzymes.

lysozyme: An enzyme, present in some exocrine secretions, that has antibiotic properties.

M

macrophage (MAK-rō-fāj): A phagocytic cell of the monocyte–macrophage system.

macula (MAK-ū-la): A receptor complex, located in the saccule or utricle of the inner ear, that responds to linear acceleration or gravity.

macula densa (MAK-ū-la DEN-sa): A group of specialized secretory cells that is located in a portion of the distal convoluted tubule, adjacent to the glomerulus and the juxtaglomerular cells; a component of the juxtaglomerular apparatus.

macula lutea (LOO-tē-uh): *See* **fovea.**

major histocompatibility complex: *See* **MHC protein.**

male climacteric: The age-related cessation of gametogenesis in males as a result of reduced sex hormone production.

malignant tumor: A form of cancer characterized by rapid cell growth and the spread of cancer cells throughout the body.

malleus (MAL-ē-us): The first auditory ossicle, bound to the tympanic membrane and the incus.

malnutrition: An unhealthy state produced by inadequate dietary intake of nutrients, calories, and/or vitamins.

mamillary bodies (MAM-i-lar-ē): Nuclei in the hypothalamus that affect eating reflexes and behaviors; a component of the limbic system.

mammary glands: Milk-producing glands of the female breast.

manus: The hand.

marrow: A tissue that fills the internal cavities in bone; dominated by hemopoietic cells (red bone marrow) or by adipose tissue (yellow bone marrow).

mass peristalsis: A powerful peristaltic contraction that moves fecal materials along the colon and into the rectum.

mass reflex: A hyperreflexia in an area innervated by spinal cord segments that are distal to an area of injury.

mast cell: A connective tissue cell that, when stimulated, releases histamine, serotonin, and heparin, initiating the inflammatory response.

mastectomy: The surgical removal of part or all of a mammary gland.

mastication (mas-ti-KĀ-shun): Chewing.

mastoid sinus: Air-filled spaces in the mastoid process of the temporal bone.

matrix: The extracellular fibers and ground substance of a connective tissue.

maxillary sinus (MAK-si-ler-ē): One of the paranasal sinuses; an air-filled chamber lined by a respiratory epithelium that is located in a maxillary bone and opens into the nasal cavity.

meatus (mē-Ā-tus): An opening or entrance into a passageway.

mechanoreception: The detection of mechanical stimuli, such as touch, pressure, or vibration.

medial: Toward the midline of the body.

mediastinum (mē-dē-as-TĪ-um or mē-dē-AS-ti-num): The central tissue mass that divides the thoracic cavity into two pleural cavities; includes the aorta and other great vessels, the esophagus, trachea, thymus, the pericardial cavity and heart, and a host of nerves, small vessels, and lymphatic vessels; in males, the area of connective tissue attaching a testis to the epididymis, proximal portion of ductus deferens, and associated vessels.

medulla: The inner layer or core of an organ.

medulla oblongata: The most caudal of the brain regions, also called the *myelencephalon.*

medullary cavity: The space within a bone that contains the marrow.

medullary rhythmicity center: The center in the medulla oblongata that sets the background pace of respiration; includes inspiratory and expiratory centers.

megakaryocytes (meg-a-KAR-ē-ō-sīts): Bone marrow cells responsible for the formation of platelets.

meiosis (mī-Ō-sis): Cell division that produces gametes with half the normal somatic chromosome complement.

melanin (ME-la-nin): The yellow-brown pigment produced by the melanocytes of the skin.

melanocyte (me-LAN-ō-sīt): A specialized cell in the deeper layers of the stratified squamous epithelium of the skin; responsible for the production of melanin.

melanocyte-stimulating hormone (MSH): A hormone, produced by the pars intermedia of the anterior lobe of the pituitary gland, that stimulates melanin production.

melanomas (mel-a-NŌ-maz): Dangerous malignant skin cancers that involve melanocytes.

melatonin (mel-a-TŌ-nin): A hormone secreted by the pineal gland; inhibits secretion of MSH and GnRH.

membrane: Any sheet or partition; a layer consisting of an epithelium and the underlying connective tissue.

membrane flow: The movement of sections of membrane surface to and from the cell surface and components of the endoplasmic reticulum, the Golgi apparatus, and vesicles.

membrane potential: *See* **transmembrane potential.**

membranous labyrinth: Endolymph-filled tubes that enclose the receptors of the inner ear.

memory: The ability to recall information or sensations; can be divided into short-term and long-term memories.

menarche (me-NAR-kē): The beginning of menstrual function; the first menstrual period, which normally occurs at puberty.

meninges (men-IN-jēz): Three membranes that surround the surfaces of the central nervous system; the dura mater, the pia mater, and the arachnoid.

meningitis: An inflammation of the spinal or cranial meninges.

meniscectomy: The removal of a meniscus.

meniscus (men-IS-kus): A fibrocartilage pad between opposing surfaces in a joint.

menses (MEN-sēz): The first portion of the uterine cycle, the portion in which the endometrial lining sloughs away.

merocrine (MER-ō-krin): A method of secretion in which the cell ejects materials through exocytosis of the midbrain.

mesencephalon (mez-en-SEF-a-lon): The midbrain; the region between the diencephalon and pons.

mesenchyme: Embryonic or fetal connective tissue.

mesentery (MEZ-en-ter-ē): A double layer of serous membrane that supports and stabilizes the position of an organ in the abdominopelvic cavity and provides a route for the associated blood vessels, nerves, and lymphatic vessels.

mesoderm: The middle germ layer, between the ectoderm and endoderm of the embryo.

mesothelium (mez-ō-THĒ-lē-um): A simple squamous epithelium that lines one of the divisions of the ventral body cavity.

messenger RNA (mRNA): RNA formed at transcription to direct protein synthesis in the cytoplasm.

metabolic turnover: The continuous breakdown and replacement of organic materials within cells.

metabolism (me-TAB-ō-lizm): The sum of all biochemical processes under way within the human body at any moment; includes anabolism and catabolism.

metabolites (me-TAB-ō-līts): Compounds produced in the body as a result of metabolic reactions.

metacarpal bones: The five bones of the palm of the hand.

metalloproteins (met-al-ō-PRŌ-tēnz): Plasma proteins that transport metal ions.

metaphase (MET-a-fāz): The stage of mitosis in which the chromosomes line up along the equatorial plane of the cell.

metaphysis (me-TA-fi-sis): The region of a long bone between the epiphysis and diaphysis, corresponding to the location of the epiphyseal cartilage of the developing bone.

metarteriole (met-ar-TĒ-rē-ōl): A vessel that connects an arteriole to a venule and that provides blood to a capillary plexus.

metastasis (me-TAS-ta-sis): The spread of cancer cells from one organ to another, leading to the establishment of secondary tumors.

metatarsal bone: One of the five bones of the foot that articulate with the tarsal bones (proximally) and the phalanges (distally).

metencephalon (met-en-SEF-a-lon): The pons and cerebellum of the brain.

MHC protein: A surface antigen that is important to the recognition of foreign antigens and that plays a role in the coordination and activation of the immune response; also called *human leukocyte antigen (HLA).*

micelle (mī-SEL): A droplet with hydrophilic portions on the outside; a spherical aggregation of bile salts, monoglycerides, and fatty acids in the lumen of the intestinal tract.

microcephaly (mī-krō-SEF-a-lē): An abnormally small cranium as a result of the premature closure of one or more fontanels.

microfilaments: Fine protein filaments visible with the electron microscope; components of the cytoskeleton.

microglia (mī-KROG-lē-uh): Phagocytic neuroglia in the central nervous system.

microphages: Neutrophils and eosinophils.

microtubules: Microscopic tubules that are part of the cytoskeleton and are a component in cilia, flagella, the centrioles, and spindle fibers.

microvilli: Small, fingerlike extensions of the exposed cell membrane of an epithelial cell.

micturition (mik-tū-RI-shun): Urination.

midbrain: The mesencephalon.

middle ear: The space between the external and internal ears that contains auditory ossicles.

midsagittal plane: A plane passing through the midline of the body that divides it into left and right halves.

mineralocorticoid: Corticosteroids produced by the zona glomerulosa of the adrenal cortex; steroids such as aldosterone that affect mineral metabolism.

miscarriage: A spontaneous abortion.

mitochondrion (mī-tō-KON-drē-on): An intracellular organelle responsible for generating most of the ATP required for cellular operations.

mitosis (mī-TŌ-sis): The division of a single cell nucleus that produces two identical daughter cell nuclei; an essential step in cell division.

mitral valve (MĪ-tral): *See* **bicuspid valve.**

mixed gland: A gland that contains exocrine and endocrine cells, or an exocrine gland that produces serous and mucous secretions.

mixed nerve: A peripheral nerve that contains sensory and motor fibers.

mole: A quantity of an element or compound having a mass in grams equal to the element's atomic weight or to the compound's molecular weight.

molecular weight: The sum of the atomic weights of all the atoms in a molecule.

molecule: A chemical structure containing two or more atoms that are held together by chemical bonds.

monoclonal antibodies (mo-nō-KLŌ-nal): Antibodies produced by genetically identical cells under laboratory conditions.

monocytes (MON-ō-sīts): Phagocytic agranulocytes (white blood cells) in the circulating blood.

monoglyceride (mo-nō-GLI-se-rīd): A lipid consisting of a single fatty acid bound to a molecule of glycerol.

monokines: Secretions released by activated cells of the monocyte–macrophage system to coordinate various aspects of the immune response.

monosaccharide (mon-ō-SAK-uh-rīd): A simple sugar, such as glucose or ribose.

monosynaptic reflex: A reflex in which the sensory afferent neuron synapses directly on the motor efferent neuron.

monozygotic twins: Twins produced by the splitting of a single fertilized egg (zygote).

morula (MOR-ū-la): A mulberry-shaped collection of cells produced by the mitotic divisions of a zygote.

motor unit: All of the muscle cells controlled by a single motor neuron.

mucins (MŪ-sinz): Proteoglycans responsible for the lubricating properties of mucus.

mucosa (mū-KŌ-sa): A mucous membrane; the epithelium plus the lamina propria.

mucosa-associated lymphoid tissue (MALT): The extensive collection of lymphoid tissues linked with the digestive system.

mucous: Indicating the presence or production of mucus.

mucous membrane: *See* **mucosa.**

mucus: A lubricating fluid that is composed of water and mucins and is produced by unicellular and multicellular glands along the digestive, respiratory, urinary, and reproductive tracts.

multipennate: A muscle whose internal fibers are organized around several tendons.

multipolar neuron: A neuron with many dendrites and a single axon; the typical form of a motor neuron.

multiunit smooth muscle: A smooth muscle tissue whose muscle cells are innervated in motor units.

muscarinic receptors (mus-kar-IN-ik): Membrane receptors sensitive to acetylcholine and to muscarine, a toxin produced by certain mushrooms; located at all parasympathetic neuromuscular and neuroglandular junctions and at a few sympathetic neuromuscular and neuroglandular junctions.

muscle: A contractile organ composed of muscle tissue, blood vessels, nerves, connective tissues, and lymphatic vessels.

muscle tissue: A tissue characterized by the presence of cells capable of contraction; includes skeletal, cardiac, and smooth muscle tissues.

muscularis externa (mus-kū-LAR-is): Concentric layers of smooth muscle responsible for peristalsis.

muscularis mucosae: The layer of smooth muscle beneath the lamina propria; responsible for moving the mucosal surface.

mutagens (MŪ-ta-jenz): Chemical agents that induce mutations and may be carcinogenic.

mutation: A change in the nucleotide sequence of the DNA in a cell.

myalgia (mī-AL-jē-uh): Muscle pain.

myasthenia gravis (mī-as-THĒ-nē-a GRA-vis): A muscular weakness due to a reduction in the number of acetylcholine receptor sites on the sarcolemmal surface; suspected to be an autoimmune disorder.

myelencephalon (mī-el-en-SEF-a-lon): *See* **medulla oblongata.**

myelin (MĪ-e-lin): An insulating sheath around an axon; consists of multiple layers of neuroglial membrane; significantly increases the impulse propagation rate along the axon.

myelination: The formation of myelin.

myenteric plexus (mī-en-TER-ik): Parasympathetic motor neurons and sympathetic postganglionic fibers located between the circular and longitudinal layers of the muscularis externa.

myocardial infarction (mī-ō-KAR-dē-al): A heart attack; damage to the heart muscle due to an interruption of regional coronary circulation.

myocarditis: An inflammation of the myocardium.

myocardium: The cardiac muscle tissue of the heart.

myofibril: Organized collections of myofilaments in skeletal and cardiac muscle cells.

myofilaments: Fine protein filaments composed primarily of the proteins actin (thin filaments) and myosin (thick filaments).

myoglobin (MĪ-ō-glō-bin): An oxygen-binding pigment that is especially common in slow skeletal muscle fibers and cardiac muscle cells.

myogram: A recording of the tension produced by muscle fibers on stimulation.

myometrium (mī-ō-MĒ-trē-um): The thick layer of smooth muscle in the wall of the uterus.

myopia: Nearsightedness, an inability to accommodate for distant vision.

myosepta: Connective tissue partitions that separate adjacent skeletal muscles.

myosin: The protein component of thick filaments.

myositis (mī-ō-SĪ-tis): An inflammation of muscle tissue.

N

nail: A keratinous structure produced by epithelial cells of the nail root.

narcolepsy: A sleep disorder characterized by falling asleep at inappropriate moments.

nares, external (NA-rēz): The entrance from the exterior to the nasal cavity.

nares, internal: The entrance from the nasal cavity to the nasopharynx.

nasal cavity: A chamber in the skull that is bounded by the internal and external nares.

nasolacrimal duct: The passageway that transports tears from the nasolacrimal sac to the nasal cavity.

nasolacrimal sac: A chamber that receives tears from the lacrimal ducts.

nasopharynx (nā-zō-FAR-inks): A region that is posterior to the internal nares and superior to the soft palate and ends at the oropharynx.

natriuretic peptides (NP) (nā-trē-ū-RET-ik): Hormones released by specialized cardiocytes when they are stretched by an abnormally large venous return; promotes fluid loss and reductions in blood pressure and in venous return. Includes atrial natural peptide (ANP) and brain natriuretic peptide (BNP).

N compound: An organic compound containing nitrogen atoms.

necrosis (nek-RŌ-sis): The death of cells or tissues from disease or injury.

negative feedback: A corrective mechanism that opposes or negates a variation from normal limits.

neonate: A newborn infant, or baby.

neoplasm: A tumor, or mass of abnormal tissue.

nephritis (nef-RĪ-tis): An inflammation of the kidney.

nephrolithiasis (nef-rō-li-THĪ-a-sis): A condition resulting from the formation of kidney stones.

nephron (NEF-ron): The basic functional unit of the kidney.

nerve impulse: An action potential in a neuron cell membrane.

neural cortex: An area of gray matter at the surface of the central nervous system.

neurilemma (noo-ri-LEM-muh): The outer surface of a neuroglia that encircles an axon.

neurofibrils: Microfibrils in the cytoplasm of a neuron.

neurofilaments: Microfilaments in the cytoplasm of a neuron.

neuroglandular junction: A cell junction at which a neuron controls or regulates the activity of a secretory (gland) cell.

neuroglia (noo-RŌG-lē-ah, noo-RŌ-glī-ah, or noo-rō-GLĪ-ah): Cells of the central nervous system and peripheral nervous system that support and protect neurons; also called *glial cells.*

neurohypophysis (NOO-rō-hī-pof-i-sis): The posterior pituitary, or pars nervosa.

neuromodulator (noo-rō-MOD-ū-la-tor): A compound, released by a neuron, that adjusts the sensitivities of another neuron to specific neurotransmitters.

neuromuscular junction: A synapse between a neuron and muscle cell.

neuron (NOO-ron) **or neurone** (NOO-rōn): A cell in neural tissue that is specialized for intercellular communication through (1) changes in membrane potential and (2) synaptic connections.

neurotransmitter: A chemical compound released by one neuron to affect the transmembrane potential of another.

neurotubules: Microtubules in the cytoplasm of a neuron.

neurulation: The embryological process responsible for the formation of the central nervous system.

neutron: A fundamental particle that does not carry a positive or a negative charge.

neutropenia: An abnormally low number of neutrophils in the circulating blood.

neutrophil (NOO-trō-fil): A microphage that is very numerous and normally the first of the mobile phagocytic cells to arrive at an area of injury or infection.

nicotinic receptors (nik-ō-TIN-ik): Acetylcholine receptors on the surfaces of sympathetic and parasympathetic ganglion cells; respond to the compound nicotine.

nipple: An elevated epithelial projection on the surface of the breast; contains the openings of the lactiferous sinuses.

Nissl bodies: The ribosomes, Golgi apparatus, rough endoplasmic reticulum, and mitochondria of the perikaryon of a typical neuronl.

nitrogenous wastes: Organic waste products of metabolism that contain nitrogen, such as urea, uric acid, and creatinine.

nociception (nō-sē-SEP-shun): Pain perception.

node of Ranvier: The area between adjacent neuroglia where the myelin covering of an axon is incomplete.

nodose ganglion (NŌ-dōs): A sensory ganglion of cranial nerve X.

noradrenaline: *See* **norepinephrine.**

norepinephrine (NE) (nor-ep-i-NEF-rin): A catecholamine neurotransmitter in the peripheral nervous system and central nervous system, released at most sympathetic neuromuscular and neuroglandular junctions and a hormone secreted by the adrenal medulla; also called *noradrenaline.*

nucleic acid (noo-KLĀ-ik): A polymer of nucleotides that contains a pentose sugar, a phosphate group, and one of four nitrogenous bases that regulate the synthesis of proteins and make up the genetic material in cells.

nucleolus (noo-KLĒ-ō-lus): The dense region in the nucleus that is the site of RNA synthesis.

nucleoplasm: The fluid content of the nucleus.

nucleoproteins: Proteins of the nucleus that are generally associated with DNA.

nucleotide: A compound consisting of a nitrogenous base, a simple sugar, and a phosphate group.

nucleus: A cellular organelle that contains DNA, RNA, and proteins; in the central nervous system, a mass of gray matter.

nucleus pulposus (pul-PŌ-sus): The gelatinous central region of an intervertebral disc.

nutrient: An inorganic or organic compound that can be broken down in the body to produce energy.

nystagmus: An unconscious, continuous movement of the eyes as if to adjust to constant motion.

O

obesity: Body weight 10–20 percent above standard values as a result of body fat accumulation.

occlusal surface (o-KLOO-sal): The opposing surfaces of the teeth that come into contact when processing food.

ocular: Pertaining to the eye.

oculomotor nerve (ok-ū-lō-MŌ-ter): Cranial nerve III, which controls the extraocular muscles other than the superior oblique and the lateral rectus muscles.

olecranon (ō-LEK-ruh-non or ō-lah-KRŌ-non): The proximal end of the ulna that forms the prominent point of the elbow.

olfaction: The sense of smell.

olfactory bulb (ōl-FAK-tor-ē): The expanded ends of the olfactory tracts; the sites where the axons of the first cranial nerves (I) synapse on central nervous system interneurons that lie inferior to the frontal lobes of the cerebrum.

oligodendrocytes (o-li-gō-DEN-drō-sīts): Central nervous system neuroglia that maintain cellular organization within gray matter and provide a myelin sheath in areas of white matter.

oligopeptide (ol-i-gō-PEP-tīd): A short chain of amino acids.

oncogene (ON-kō-jēn): A gene that can turn a normal cell into a cancer cell.

oncologists (on-KOL-o-jists): Physicians specializing in the study and treatment of tumors.

oocyte (Ō-ō-sīt): A cell whose meiotic divisions will produce a single ovum and three polar bodies.

oogenesis (ō-ō-JEN-e-sis): Ovum production.

oogonia (ō-ō-GŌ-nē-uh): Stem cells in the ovaries whose divisions give rise to oocytes.

oophorectomy (ō-of-ō-REK-to-mē): The surgical removal of the ovaries.

ooplasm: The cytoplasm of the ovum.

opsin: A protein; one structural component of the visual pigment rhodopsin.

opsonization: An effect of coating an object with antibodies; the attraction and enhancement of phagocytosis.

optic chiasm (KĪ-azm): The crossing point of the optic nerves.

optic nerve: The second cranial nerve (II), which carries signals from the retina of the eye to the optic chiasm.

optic tract: The tract over which nerve impulses from the retina are transmitted between the optic chiasm and the thalamus.

orbit: The bony recess of the skull that contains the eyeball.

orchiectomy (or-kē-EK-to-mē): The surgical removal of one or both testes.

orchitis: An inflammation of the testes.

organelle (or-gan-EL): An intracellular structure that performs a specific function or group of functions.

organic compound: A compound containing carbon, hydrogen, and in most cases oxygen.

organogenesis: The formation of organs during embryological and fetal development.

organs: Combinations of tissues that perform complex functions.

origin: In a skeletal muscle, the point of attachment which does not change position when the muscle contracts; usually defined in terms of movements from the anatomical position.

oropharynx: The middle portion of the pharynx, bounded superiorly by the nasopharynx, anteriorly by the oral cavity, and inferiorly by the laryngopharynx.

os coxae/ossa coxae: The hip bone(s).

osmolarity (oz-mō-LAR-i-tē): The total concentration of dissolved materials in a solution, regardless of their specific identities, expressed in moles; also called *osmotic concentration.*

osmoreceptor: A receptor sensitive to changes in the osmolarity of plasma.

osmosis (oz-MŌ-sis): The movement of water across a selectively permeable membrane from one solution to another solution that contains a higher solute concentration.

osmotic pressure: The force of osmotic water movement; the pressure that must be applied to prevent osmosis across a membrane.

osseous tissue: A strong connective tissue containing specialized cells and a mineralized matrix of crystalline calcium phosphate and calcium carbonate; also called **bone.**

ossicles: Small bones.

ossification: The formation of bone.

osteoblast: (OS-tē-ō-blast): A cell that produces the fibers and matrix of bone.

osteoclast (OS-tē-ō-klast): A cell that dissolves the fibers and matrix of bone.

osteocyte (OS-tē-ō-sīt): A bone cell responsible for the maintenance and turnover of the mineral content of the surrounding bone.

osteogenic layer (os-tē-ō-JEN-ik): The inner, cellular layer of the periosteum that participates in bone growth and repair.

osteolysis (os-tē-OL-i-sis): The breakdown of the mineral matrix of bone.

osteon (OS-tē-on): The basic histological unit of compact bone, consisting of osteocytes organized around a central canal and separated by concentric lamellae.

otic: Pertaining to the ear.

otitis media: An inflammation of the lining of the middle ear cavity.

otolith: A complex formed by the combination of a gelatinous matrix and statoconia, aggregations of calcium carbonate crystals; located above one of the maculae of the vestibule.

oval window: An opening in the bony labyrinth where the stapes attaches to the membranous wall of the vestibular duct.

ovarian cycle (ō-VAR-ē-an): The monthly chain of events that leads to ovulation.

ovary: The female reproductive organ that produces gametes.

ovulation (ov-ū-LĀ-shun): The release of a secondary oocyte, surrounded by cells of the corona radiata, after the rupture of the wall of a tertiary follicle.

ovum/ova (Ō-vum): The functional product of meiosis II, produced after the fertilization of a secondary oocyte.

oxidation: The loss of electrons or hydrogen atoms or the acceptance of an oxygen atom.

oxidative phosphorylation: The capture of energy as ATP during a series of oxidation–reduction reactions; a reaction sequence that occurs in the mitochondria and involves coenzymes and the electron transport system.

oxytocin (oks-i-TŌ-sin): A hormone produced by hypothalamic cells and secreted into capillaries at the posterior lobe of the pituitary gland; stimulates smooth muscle contractions of the uterus or mammary glands in females and the prostate gland in males.

P

pacemaker cells: Cells of the sinoatrial node that set the pace of cardiac contraction.

palate: The horizontal partition separating the oral cavity from the nasal cavity and nasopharynx; divided into an anterior bony (hard) palate and a posterior fleshy (soft) palate.

palatine: Pertaining to the palate.

palpate: To examine by touch.

palpebrae (pal-PĒ-brē): Eyelids.

pancreas: A digestive organ containing exocrine and endocrine tissues; the exocrine portion secretes pancreatic juice, and the endocrine portion secretes hormones, including insulin and glucagon.

pancreatic duct: A tubular duct that carries pancreatic juice from the pancreas to the duodenum.

pancreatic islets: Aggregations of endocrine cells in the pancreas; also called *islets of Langerhans.*

pancreatic juice: A mixture of buffers and digestive enzymes that is discharged into the duodenum under the stimulation of the enzymes secretin and cholecystokinin.

pancreatitis (pan-krē-a-TĪ-tis): An inflammation of the pancreas.

Papanicolaou (Pap) test: A test for the detection of malignancies based on the cytological appearance of epithelial cells, especially those of the cervix and uterus.

papilla (pa-PIL-la): A small, conical projection.

paralysis: The loss of voluntary motor control over a portion of the body.

paranasal sinuses: Bony chambers, lined by respiratory epithelium, that open into the nasal cavity; the frontal, ethmoidal, sphenoidal, and maxillary sinuses.

parasagittal: A section or plane that parallels the midsagittal plane but that does not pass along the midline.

parasympathetic division: One of the two divisions of the autonomic nervous system; also called *craniosacral division;* generally responsible for activities that conserve energy and lower the metabolic rate.

parasympathomimetic drugs: Drugs that mimic the actions of parasympathetic stimulation.

parathyroid glands: Four small glands embedded in the posterior surface of the thyroid gland; secrete parathyroid hormone.

parathyroid hormone (PTH): A hormone secreted by the parathyroid glands when plasma calcium levels fall below the normal range; causes increased osteoclast activity, increased intestinal calcium uptake, and decreased calcium ion loss at the kidneys.

parenchyma (pa-RENG-ki-ma): The cells of a tissue or organ that are responsible for fulfilling its functional role; distinguished from the stroma of that tissue or organ.

paresthesia: A sensory abnormality that produces a tingling sensation.

parietal: Referring to the body wall or outer layer.

parietal cells: Cells of the gastric glands that secrete hydrochloric acid and intrinsic factor.

Parkinson's disease: A progressive motor disorder caused by the degeneration of cerebral nuclei.

parotid salivary glands (pa-ROT-id): Large salivary glands that secrete a saliva containing high concentrations of salivary (alpha) amylase.

pars distalis (dis-TAL-is): The large, anterior portion of the anterior lobe of the pituitary gland.

pars intermedia (in-ter-MĒ-dē-uh): The portion of the anterior lobe of the pituitary gland that is immediately adjacent to the posterior lobe and the infundibulum.

pars nervosa: The posterior lobe of the pituitary gland.

pars tuberalis: The portion of the anterior lobe of the pituitary gland that wraps around the infundibulum superior to the posterior lobe.

parturition (par-tū-RISH-un): Childbirth, delivery.

patella (pa-TEL-uh): The sesamoid bone of the kneecap.

pathogen: A disease-causing organism.

pathogenic: Disease-causing.

pathologist (pa-THO-lo-jist): An M.D. specializing in the identification of diseases on the basis of characteristic structural and functional changes in tissues and organs.

pedicel (PED-i-sel): A slender process of a podocyte that forms part of the filtration apparatus of the kidney glomerulus.

pedicles (PE-di-kulz): Thick, bony struts that connect the vertebral body with the articular and spinous processes.

pelvic cavity: The inferior subdivision of the abdominopelvic cavity; encloses the urinary bladder, the sigmoid colon and rectum, and male or female reproductive organs.

pelvis: A bony complex created by the articulations among the ossa coxae, the sacrum, and the coccyx.

penis (PĒ-nis): A component of the male external genitalia; a copulatory organ that surrounds the urethra and serves to introduce semen into the female vagina; the developmental equivalent of the female clitoris.

pepsin: A proteolytic enzyme secreted by the chief cells of the gastric glands in the stomach.

peptidases: Enzymes that split peptide bonds and release amino acids.

peptide: A chain of amino acids linked by peptide bonds.

peptide bond: A covalent bond between the amino group of one amino acid and the carboxyl group of another.

perforating canal: A passageway in compact bone that is at right angles to the axes of the osteons, between the periosteum and endosteum.

perfusion: The blood flow through a tissue.

pericardial cavity (per-i-KAR-dē-al): The space between the parietal pericardium and the epicardium (visceral pericardium) that covers the outer surface of the heart.

pericarditis: An inflammation of the pericardium.

pericardium (per-i-KAR-dē-um): The fibrous sac that surrounds the heart; its inner, serous lining is continuous with the epicardium.

perichondrium (per-i-KON-drē-um): The layer that surrounds a cartilage, consisting of an outer fibrous region and an inner cellular region.

perikaryon (per-i-KAR-ē-on): The cytoplasm that surrounds the nucleus in the cell body of a neuron.

perilymph (PER-ē-limf): A fluid similar in composition to cerebrospinal fluid; located in the spaces between the bony labyrinth and the membranous labyrinth of the inner ear.

perimysium (pe-ri-MĪZ-ē-um): A connective tissue partition that separates adjacent fasciculi in a skeletal muscle.

perineum (pe-ri-NĒ-um): The pelvic floor and its associated structures.

perineurium: A connective tissue partition that separates adjacent bundles of nerve fibers in a peripheral nerve.

periodontal ligament (per-ē-ō-DON-tal): Collagen fibers that bind the cementum of a tooth to the periosteum of the surrounding alveolus.

periosteum (per-ē-OS-tē-um): The layer that surrounds a bone, consisting of an outer fibrous region and inner cellular region.

peripheral nervous system (PNS): All neural tissue outside the central nervous system.

peripheral resistance: The resistance to blood flow; primarily caused by friction with the vascular walls.

peristalsis (per-i-STAL-sis): A wave of smooth muscle contractions that propels materials along the axis of a tube such as the digestive tract, the ureters, or the ductus deferens.

peritoneal cavity: *See* **abdominopelvic cavity.**

peritoneum (per-i-tō-NĒ-um): The serous membrane that lines the peritoneal cavity.

peritonitis (per-i-tō-NĪ-tis): An inflammation of the peritoneum.

peritubular capillaries: A network of capillaries that surrounds the proximal and distal convoluted tubules of the kidneys.

permeability: The ease with which dissolved materials can cross a membrane; if the membrane is freely permeable, any molecule can cross it; if impermeable, nothing can cross; most biological membranes are selectively permeable.

peroxisome: A membranous vesicle containing enzymes that break down hydrogen peroxide (H_2O_2).

pes: The foot.

petrosal ganglion: A sensory ganglion of the glossopharyngeal nerve (IX).

petrous: Stony; usually refers to the thickened portion of the temporal bone that encloses the inner ear.

pH: The negative exponent of the hydrogen ion concentration, expressed in moles per liter.

phagocyte: A cell that performs phagocytosis.

phagocytosis (fa-gō-sī-TŌ-sis): The engulfing of extracellular materials or pathogens; the movement of extracellular materials into the cytoplasm by enclosure in a membranous vesicle.

phalanx/phalanges (fa-LAN-gēz): Bones of the fingers or toes.

pharmacology: The study of drugs, their physiological effects, and their clinical uses.

pharynx (FAR-inks): The throat; a muscular passageway shared by the digestive and respiratory tracts.

phasic response: A pattern of response to stimulation by sensory neurons that are normally inactive; stimulation causes a burst of neural activity that ends when the stimulus either stops or stops changing in intensity.

phenotype (FĒN-ō-tīp): Physical characteristics that are genetically determined.

phonation (fō-NĀ-shun): Sound production at the larynx.

phosphate group: PO_4^{3-}; a functional group that can be attached to an organic molecule; required for the formation of high-energy bonds.

phospholipid (fos-fō-LIP-id): An important membrane lipid whose structure includes both hydrophilic and hydrophobic regions.

phosphorylation (fos-for-i-LĀ-shun): The addition of a high-energy phosphate group to a molecule.

photoreception: Sensitivity to light.

physiology (fiz-ē-OL-o-jē): The study of function; deals with the ways organisms perform vital activities.

pia mater: The tough, outer meningeal layer that surrounds the central nervous system.

pigment: A compound with a characteristic color.

pineal gland (PIN-ē-ul): Neural tissue in the posterior portion of the roof of the diencephalon; secretes melatonin.

pinna: *See* **auricle.**

pinocytosis (pi-nō-sī-TŌ-sis): The introduction of fluids into the cytoplasm by enclosing them in membranous vesicles at the cell surface.

pituitary gland: An endocrine organ that is situated in the sella turcica of the sphenoid and is connected to the hypothalamus by the infundibulum; includes the posterior lobe (pars nervosa) and the anterior lobe (adenohypophysis).

placenta (pla-SENT-uh): A temporary structure in the uterine wall that permits diffusion between the fetal and maternal circulatory systems; *see* **afterbirth.**

plantar: Referring to the sole of the foot.

plasma (PLAZ-muh): The fluid ground substance of whole blood; what remains after the cells have been removed from a sample of whole blood.

plasma cell: An activated B cell that secretes antibodies.

plasmalemma (plaz-ma-LEM-a): A cell membrane.

platelets (PLĀT-lets): Small packets of cytoplasm that contain enzymes important in the clotting response; manufactured in bone marrow by megakaryocytes.

pleura (PLOO-ra): The serous membrane that lines the pleural cavities.

pleural cavities: Subdivisions of the thoracic cavity that contain the lungs.

pleuritis (ploor-Ī-tis): An inflammation of the pleura; also called *pleurisy.*

plexus (PLEK-sus): A network or braid.

plica (PLĪ-ka): A permanent transverse fold in the wall of the small intestine.

pneumotaxic center (noo-mō-TAKS-ik): A center in the reticular formation of the pons that regulates the activities of the apneustic and respiratory rhythmicity centers to adjust the pace of respiration.

pneumothorax (noo-mō-THO-raks): The introduction of air into the pleural cavity.

podocyte (PŌ-do-sīt): A cell whose processes surround the kidney glomerular capillaries and assist in filtration.

polar body: A nonfunctional packet of cytoplasm that contains chromosomes eliminated from an oocyte during meiosis.

polar bond: A covalent bond in which electrons are shared unequally.

polarized: Referring to cells that have regional differences in organelle distribution or cytoplasmic composition along a specific axis, such as between the basement membrane and free surface of an epithelial cell.

pollex: The thumb.

polymer: A large molecule consisting of a long chain of subunits.

polymorph: A polymorphonuclear leukocyte; a neutrophil.

polypeptide: A chain of amino acids strung together by peptide bonds; those containing more than 100 peptides are called *proteins.*

polyribosome: Several ribosomes linked by their translation of a single mRNA strand.

polysaccharide (pol-ē-SAK-uh-rīd): A complex sugar, such as glycogen or a starch.

polysynaptic reflex: A reflex in which interneurons are interposed between the sensory fiber and the motor neuron(s).

polyunsaturated fats: Fatty acids containing carbon atoms that are linked by double bonds.

polyuria (pol-ē-Ū-rē-uh): Excessive urine production.

pons: The portion of the metencephalon that is anterior to the cerebellum.

popliteal (pop-LIT-ē-al): Pertaining to the back of the knee.

porphyrins (POR-fi-rinz): Ring-shaped molecules that form the basis of important respiratory and metabolic pigments, including heme and the cytochromes.

porta hepatis: A region of mesentery between the duodenum and liver that contains the hepatic artery proper, the hepatic portal vein, and the common bile duct; also called *hilus.*

positive feedback: A mechanism that increases a deviation from normal limits after an initial stimulus.

postabsorptive state: A period that begins 4 hours after a meal; characterized by falling blood glucose concentrations and the mobilization of metabolic reserves.

postcentral gyrus: The primary sensory cortex, where touch, vibration, pain, temperature, and taste sensations arrive and are consciously perceived.

posterior: Toward the back; dorsal.

postganglionic neuron: An autonomic neuron in a peripheral ganglion, whose activities control peripheral effectors.

postovulatory phase: The secretory phase of the menstrual cycle.

postsynaptic membrane: The portion of the cell membrane of a post-synaptic cell that is part of a synapse.

potential difference: The separation of opposite charges; requires a barrier that prevents ion migration.

precentral gyrus: The primary motor cortex of a cerebral hemisphere, located anterior to the central sulcus.

prefrontal cortex: The anterior portion of each cerebral hemisphere; thought to be involved with higher intellectual functions, predictions, calculations, and so forth.

preganglionic neuron: A visceral motor neuron in the central nervous system whose output controls one or more ganglionic motor neurons in the peripheral nervous system.

premolars: Bicuspids; teeth with flattened occlusal surfaces; located anterior to the molar teeth.

premotor cortex: The motor association area between the precentral gyrus and the prefrontal area.

preoptic nucleus: The hypothalamic nucleus that coordinates thermoregulatory activities.

prepuce (PRĒ-pūs): The loose fold of skin that surrounds the glans penis (in males) or the clitoris (in females).

preputial glands (prē-PŪ-shal): Glands on the inner surface of the prepuce that produce a viscous, odorous secretion called *smegma*.

presbyopia: Farsightedness; an inability to accommodate for near vision.

presynaptic membrane: The synaptic surface where neurotransmitter release occurs.

prevertebral ganglion: *See* collateral ganglion.

prime mover: A muscle that performs a specific action.

proenzyme: An inactive enzyme secreted by an epithelial cell.

progesterone (prō-JES-ter-ōn): The most important progestin secreted by the corpus luteum after ovulation.

progestins (prō-JES-tinz): Steroid hormones structurally related to cholesterol; progesterone is an example.

prognosis: A prediction about the possibility or time course of recovery from a specific disease.

projection fibers: Axons carrying information from the thalamus to the cerebral cortex.

prolactin (prō-LAK-tin): The hormone that stimulates functional development of the mammary gland in females; a secretion of the anterior lobe of the pituitary gland.

prolapse: The abnormal descent or protrusion of a portion of an organ, such as the vagina or anal canal.

proliferative phase: A portion of the uterine cycle; the interval of estrogen-induced repair of the functional zone of the endometrium through the growth and proliferation of epithelial cells in the uterine glands.

pronation (prō-NĀ-shun): The rotation of the forearm that makes the palm face posteriorly.

prone: Lying face down with the palms facing the floor.

pronucleus: An enlarged ovum or spermatozoon nucleus that forms after fertilization but before amphimixis.

properdin: The complement factor that prolongs and enhances non-antibody-dependent complement binding to bacterial cell walls.

prophase (PRŌ-fāz): The initial phase of mitosis; characterized by the appearance of chromosomes, the breakdown of the nuclear membrane, and the formation of the spindle apparatus.

proprioception (prō-prē-ō-SEP-shun): The awareness of the positions of bones, joints, and muscles.

prostaglandin (pros-ta-GLAN-din): A fatty acid secreted by one cell that alters the metabolic activities or sensitivities of adjacent cells; also called *local hormone*.

prostatectomy (pros-ta-TEK-to-mē): The surgical removal of the prostate gland.

prostate gland (PROS-tāt): An accessory gland of the male reproductive tract, contributing roughly one-third of the volume of semen.

prostatitis (pros-ta-TĪ-tis): An inflammation of the prostate gland.

prosthesis: An artificial substitute for a body part.

protease: *See* proteinase.

protein: A large polypeptide with a complex structure

proteinase: An enzyme that breaks down proteins into peptides and amino acids.

proteinuria (prō-tēn-ŪR-ē-uh): Abnormal amounts of protein in urine.

proteoglycan (prō-tē-ō-GLĪ-kan): A compound containing a large polysaccharide complex attached to a relatively small protein; examples include hyaluronan and chondroitin sulfate.

prothrombin: A circulating proenzyme of the common pathway of the clotting system; converted to thrombin by the enzyme thromboplastin.

proton: A fundamental particle bearing a positive charge.

protraction: Movement anteriorly in the horizontal plane.

proximal: Toward the attached base of an organ or structure.

proximal convoluted tubule (PCT): The portion of the nephron that is situated between Bowman's capsule and the loop of Henle; the major site of active reabsorption from filtrate.

pruritis (proo-RĪ-tus): Itching.

pseudopodia (soo-dō-PŌ-dē-ah): Temporary cytoplasmic extensions typical of mobile or phagocytic cells.

pseudostratified epithelium: An epithelium that contains several layers of nuclei but whose cells are all in contact with the underlying basement membrane.

psoriasis (sō-RĪ-uh-sis): A skin condition characterized by excessive keratin production and the formation of dry, scaly patches on the body surface.

psychosomatic condition: An abnormal physiological state with a psychological origin.

puberty: A period of rapid growth, sexual maturation, and the appearance of secondary sexual characteristics; normally occurs at ages 10–15 years.

pubic symphysis: The fibrocartilaginous amphiarthrosis between the pubic bones of the ossa coxae.

pubis (PŪ-bis): The anterior, inferior component of the os coxae.

pudendum (pū-DEN-dum): The external genitalia.

pulmonary circuit: Blood vessels between the pulmonary semilunar valve of the right ventricle and the entrance to the left atrium; the blood flow through the lungs.

pulmonary ventilation: The movement of air into and out of the lungs.

pulp cavity: The internal chamber in a tooth, containing blood vessels, lymphatic vessels, nerves, and the cells that maintain the dentin.

pulpitis (pul-PĪ-tis): An inflammation of the tissues of the pulp cavity.

pulvinar: The thalamic nucleus involved in the integration of sensory information prior to projection to the cerebral hemispheres.

pupil: The opening in the center of the iris through which light enters the eye.

purine: A nitrogen compound with a double ring-shaped structure; examples include adenine and guanine, two nitrogenous bases that are common in nucleic acids.

Purkinje cell (pur-KIN-jē): A large, branching neuron of the cerebellar cortex.

Purkinje fibers: Specialized conducting cardiocytes in the ventricles of the heart.

pus: An accumulation of debris, fluid, dead and dying cells, and necrotic tissue.

P wave: A deflection of the ECG corresponding to atrial depolarization.

pyelogram (PĪ-el-ō-gram): A radiographic image of the kidneys and ureters.

pyelonephritis (pī-e-lō-nef-RĪ-tis): An inflammation of the kidneys.

pyloric sphincter (pī-LOR-ik): A sphincter of smooth muscle that regulates the passage of chyme from the stomach to the duodenum.

pylorus (pī-LOR-us): The gastric region between the body of the stomach and the duodenum; includes the pyloric sphincter.

pyrexia (pī-REK-sē-uh): A fever.

pyrimidine: A nitrogen compound with a single ring-shaped structure; examples include cytosine, thymine, and uracil, nitrogenous bases that are common in nucleic acids.

pyrogen (PĪ-rō-jen): A compound that promotes a fever.

pyruvic acid (pī-ROO-vik): A three-carbon compound produced by glycolysis.

Q

quadriplegia: Paralysis of the upper and lower limbs.

quaternary structure: The three-dimensional protein structure produced by interactions between protein subunits.

R

radiodensity: The relative resistance to the passage of X rays.

radiographic techniques: Methods of visualizing internal structures by using various forms of radiational energy.

radiopaque: Having a high radiodensity.

rami communicantes: Axon bundles that link the spinal nerves with the ganglia of the sympathetic chain.

ramus/rami: A branch.

raphe (RĀ-fē): A seam.

receptive field: The area monitored by a single sensory receptor.

recessive trait: An allele that affects the phenotype only when the individual is homozygous for that trait.

recombinant DNA: DNA created by splicing together a specific gene from one organism into the DNA strand of another organism.

rectouterine pouch (rek-tō-Ū-te-rin): The peritoneal pocket between the anterior surface of the colon and the posterior surface of the uterus.

rectum (REK-tum): The inferior 15 cm (6 in.) of the digestive tract.

rectus: Straight.

red blood cell (RBC): *See* **erythrocyte.**

reduction: The gain of hydrogen atoms or electrons or the loss of an oxygen molecule.

reductional division: The first meiotic division, which reduces the chromosome number from 46 to 23.

reflex: A rapid, automatic response to a stimulus.

reflex arc: The receptor, sensory neuron, motor neuron, and effector involved in a particular reflex; interneurons may be present, depending on the reflex considered.

refraction: The bending of light rays as they pass from one medium to another.

refractory period: The period between the initiation of an action potential and the restoration of the normal resting potential; during this period, the membrane will not respond normally to stimulation.

relaxation phase: The period after a contraction when the tension in the muscle fiber returns to resting levels.

relaxin: A hormone that loosens the pubic symphysis; secreted by the placenta.

renal: Pertaining to the kidneys.

renal corpuscle: The initial portion of the nephron, consisting of an expanded chamber that encloses the glomerulus.

renin: The enzyme released by cells of the juxtaglomerular apparatus when renal blood flow declines; converts angiotensinogen to angiotensin I.

rennin: A gastric enzyme that breaks down milk proteins.

replication: Duplication.

repolarization: The movement of the transmembrane potential away from a positive value and toward the resting potential.

residual volume: The amount of air remaining in the lungs after maximum forced exhalation.

respiration: The exchange of gases between cells and the environment; includes pulmonary ventilation, external respiration, internal respiration, and cellular respiration.

respiratory minute volume (V_E): The amount of air moved into and out of the respiratory system each minute.

respiratory pump: A mechanism by which changes in the intrapleural pressures during the respiratory cycle assist the venous return to the heart; also called *thoracoabdominal pump.*

resting potential: The transmembrane potential of a normal cell under homeostatic conditions.

rete (RĒ-tē): An interwoven network of blood vessels or passageways.

reticular activating system (RAS): The mesencephalic portion of the reticular formation; responsible for arousal and the maintenance of consciousness.

reticular formation: A diffuse network of gray matter that extends the entire length of the brain stem.

reticulospinal tracts: Descending tracts of the medial pathway that carry involuntary motor commands issued by neurons of the reticular formation.

retina: The innermost layer of the eye, lining the vitreous chamber; also called *neural tunic.*

retinal (RET-i-nal): A visual pigment derived from vitamin A.

retraction: Movement posteriorly in the horizontal plane.

retroflexion (re-trō-FLEK-shun): A posterior tilting of the uterus that has no clinical significance.

retrograde flow (RET-rō-grād): The transport of materials from the telodendria to the cell body of a neuron.

retroperitoneal (re-trō-per-i-tō-NĒ-al): Behind or outside the peritoneal cavity.

reverberation: A positive feedback along a chain of neurons such that they remain active once stimulated.

rheumatism (ROO-muh-tizm): A condition characterized by pain in muscles, tendons, bones, or joints.

Rh factor: A surface antigen that may be present (Rh-positive) or absent (Rh-negative) from the surfaces of red blood cells.

rhizotomy: The surgical transection of a dorsal root, normally performed to relieve pain.

rhodopsin (rō-DOP-sin): The visual pigment in the membrane disks of the distal segments of rods.

rhythmicity center: A medullary center responsible for the pace of respiration; includes inspiratory and expiratory centers.

ribonucleic acid (rī-bō-noo-KLĀ-ik): A nucleic acid consisting of a chain of nucleotides that contain the sugar ribose and the nitrogenous bases adenine, guanine, cytosine, and uracil.

ribose: A five-carbon sugar that is a structural component of RNA.

ribosome: An organelle that contains rRNA and proteins and is essential to mRNA translation and protein synthesis.

rigor mortis: The extended muscular contraction and rigidity that occurs after death; the result of calcium ion release from the sarcoplasmic reticulum and the exhaustion of cytoplasmic ATP reserves.

rod: A photoreceptor responsible for vision in dim lighting.

rough endoplasmic reticulum (RER): A membranous organelle that is a site of protein synthesis and storage.

round window: An opening in the bony labyrinth of the inner ear that exposes the membranous wall of the tympanic duct to the air of the middle ear cavity.

rubrospinal tracts: Descending tracts of the lateral pathway that carry involuntary motor commands issued by the red nucleus of the mesencephalon.

Ruffini corpuscles (roo-FĒ-nē): Receptors sensitive to tension and stretch in the dermis of the skin.

rugae (ROO-gē): Mucosal folds in the lining of the empty stomach that disappear as gastric distension occurs.

S

saccule (SAK-ūl): A portion of the vestibular apparatus of the inner ear; contains a macula important for static equilibrium.

sagittal plane: A sectional plane that divides the body into left and right portions.

salivary nucleus (SAL-i-va-to-rē): The medullary nucleus that controls the secretory activities of the salivary glands.

salt: An inorganic compound consisting of a cation other than H^+ and an anion other than OH^-.

saltatory propagation: The relatively rapid propagation of an action potential between successive nodes of a myelinated axon.

sarcolemma: The cell membrane of a muscle cell.

sarcoma (sar-KŌ-ma): A tumor of connective tissues.

sarcomere: The smallest contractile unit of a striated muscle cell.

sarcoplasm: The cytoplasm of a muscle cell.

satellite cells: *See* **amphicytes.**

scala media: *See* **cochlear duct.**

scala tympani: *See* **tympanic duct.**

scala vestibuli: *See* **vestibular duct.**

scar tissue: The thick, collagenous tissue that forms at an injury site.

Schlemm, canal of: The passageway that delivers aqueous humor from the anterior chamber of the eye to the venous circulation.

Schwann cells: Neuroglia responsible for the neurilemma that surrounds axons in the peripheral nervous system.

sciatica (sī-AT-i-ka): Pain felt along the peripheral distribution of the sciatic nerve.

sciatic nerve (sī-A-tik): A nerve innervating the posteromedial portions of the thigh and leg.

sclera (SKLER-uh): The fibrous, outer layer of the eye that forms the white area of the anterior surface; a portion of the fibrous tunic of the eye.

sclerosis: A hardening and thickening that commonly occurs secondary to tissue inflammation.

scoliosis (skō-lē-Ō-sis): An abnormal, exaggerated lateral curvature of the spine.

scrotum (SKRŌ-tum): The loose-fitting, fleshy pouch that encloses the testes of the male.

sebaceous glands (se-BĀ-shus): Glands that secrete sebum; normally associated with hair follicles.

sebum (SĒ-bum): A waxy secretion that coats the surfaces of hairs.

secondary sex characteristics: Physical characteristics that appear at puberty in response to sex hormones but are not involved in the production of gametes.

secretin (se-KRĒ-tin): A hormone, secreted by the duodenum, that stimulates the production of buffers by the pancreas and inhibits gastric activity.

semen (SĒ-men): The fluid ejaculate that contains spermatozoa and the secretions of accessory glands of the male reproductive tract.

semicircular ducts: The tubular components of the membranous labyrinth of the inner ear; responsible for dynamic equilibrium.

semilunar valve: A three-cusped valve guarding the exit from one of the cardiac ventricles; the pulmonary and aortic valves.

seminal vesicles (SEM-i-nal): Glands of the male reproductive tract that produce roughly 60 percent of the volume of semen.

seminiferous tubules (se-mi-NIF-e-rus): Coiled tubules where spermatozoon production occurs in the testis.

senescence: Aging.

sensible perspiration: Water loss due to secretion by sweat glands.

septae (SEP-tē): Partitions that subdivide an organ.

serosa: *See* **serous membrane.**

serotonin (ser-ō-TŌ-nin): A neurotransmitter in the central nervous system; a compound that enhances inflammation and is released by activated mast cells and basophils.

serous cell: A cell that produces a serous secretion.

serous membrane: A squamous epithelium and the underlying loose connective tissue; the lining of the pericardial, pleural, and peritoneal cavities.

serous secretion: A watery secretion that contains high concentrations of enzymes; produced by serous cells.

serum: The ground substance of blood plasma from which clotting agents have been removed.

sesamoid bone: A bone that forms within a tendon.

sigmoid colon (SIG-moyd): The S-shaped 18-cm-long portion of the colon between the descending colon and the rectum.

sign: The visible evidence of the presence of a disease.

simple epithelium: An epithelium containing a single layer of cells above the basal lamina.

sinoatrial (SA) node: The natural pacemaker of the heart; situated in the wall of the right atrium.

sinus: A chamber or hollow in a tissue; a large, dilated vein.

sinusitis: An inflammation of a nasal sinus.

sinusoid (SĪ-nus-oyd): An exchange vessel that is similar in general structure to a fenestrated capillary. The two differ in size (sinusoids are larger and more irregular in cross-section), continuity (sinusoids have gaps between endothelial cells), and support (sinusoids have thin basal laminae, if they have them at all).

skeletal muscle: A contractile organ of the muscular system.

skeletal muscle tissue: A contractile tissue dominated by skeletal muscle fibers; characterized as striated, voluntary muscle.

sliding filament theory: The concept that a sarcomere shortens as the thick and thin filaments slide past one another.

small intestine: The duodenum, jejunum, and ileum; the digestive tract between the stomach and the large intestine.

smegma (SMEG-ma): The secretion of the preputial glands of the penis or clitoris.

smooth endoplasmic reticulum (SER): A membranous organelle in which lipid and carbohydrate synthesis and storage occur.

smooth muscle tissue: Muscle tissue in the walls of many visceral organs; characterized as nonstriated, involuntary muscle.

soft palate: The fleshy posterior extension of the hard palate, separating the nasopharynx from the oral cavity.

sole: The inferior surface of the foot.

solute: Any materials dissolved in a solution.

solution: A fluid containing dissolved materials.

solvent: The fluid component of a solution.

soma (SŌ-ma): Body; the body of a neuron; also called *cell body*.

somatic (sō-MAT-ik): Pertaining to the body.

somatic nervous system (SNS): The efferent division of the nervous system that innervates skeletal muscles.

somatomedins: Compounds stimulating tissue growth; released by the liver after the secretion of growth hormone; also called *insulin-like growth factors*.

somatotropin: Growth hormone; produced by the anterior pituitary in response to growth hormone–releasing hormone (GH-RH).

sperm: *See* **spermatozoon.**

spermatic cord: Collectively, the spermatic vessels, nerves, lymphatic vessels, and the ductus deferens, extending between the testes and the proximal end of the inguinal canal.

spermatids (SPER-ma-tidz): The product of meiosis in males; cells that differentiate into spermatozoa.

spermatocyte (sper-MA-tō-sīt): A cell of the seminiferous tubules that is engaged in meiosis.

spermatogenesis (sper-ma-tō-JEN-e-sis): Spermatozoon production.

spermatogonia (sper-ma-tō-GŌ-nē-uh): Stem cells whose mitotic divisions give rise to other stem cells and to primary spermatocytes.

spermatozoon/spermatozoa (sper-ma-to-ZŌ-a): A male gamete; also called *sperm*.

spermicide: A compound toxic to spermatozoa; used as a contraceptive method.

spermiogenesis: The process of spermatid differentiation that leads to the formation of physically mature spermatozoa.

sphincter (SFINK-ter): A muscular ring that contracts to close the entrance or exit of an internal passageway.

spinal nerve: One of 31 pairs of nerves that originate on the spinal cord from anterior and posterior roots.

spindle apparatus: A muscle spindle (intrafusal fibers) and its sensory and motor innervation.

spinocerebellar tracts: Ascending tracts that carry sensory information to the cerebellum.

spinothalamic tracts: Ascending tracts that carry poorly localized touch, pressure, pain, vibration, and temperature sensations to the thalamus.

spinous process: The prominent posterior projection of a vertebra; formed by the fusion of two laminae.

spleen: A lymphoid organ important for the phagocytosis of red blood cells, the immune response, and lymphocyte production.

splenectomy (splē-NEK-to-mē): The surgical removal of the spleen.

sprain: A forceful distortion of an articulation that produces damage to the capsule, ligaments, or tendons but not dislocation of the joint.

sputum (SPŪ-tum): A viscous mucus that is transported to the pharynx by the mucus escalator of the respiratory tract and is ejected from the mouth.

squama: A broad, flat surface.

squamous (SKWĀ-mus): Flattened.

squamous epithelium: An epithelium whose superficial cells are flattened and platelike.

stapedius (sta-Pē-dē-us): A muscle of the middle ear whose contraction tenses the auditory ossicles and reduces the forces transmitted to the oval window.

stapes (STĀ-pēz): The auditory ossicle attached to the tympanic membrane.

statoconia: Densely packed calcium carbonate crystals forming masses that are attached, by a gelatinous matrix, to hair cell receptors of maculae in the vestibule of the inner ear.

stenosis (ste-NŌ-sis): A constriction or narrowing of a passageway.

stereocilia: Elongate microvilli characteristic of the epithelium of the epididymis, portions of the ductus deferens, and the inner ear.

steroid: A ring-shaped lipid structurally related to cholesterol.

stimulus: An environmental change that produces a change in cellular activities; often used to refer to events that alter the transmembrane potentials of excitable cells.

stratified: Containing several layers.

stratum (STRA-tum): A layer.

stretch receptors: Sensory receptors that respond to stretching of the surrounding tissues.

stroma: The connective tissue framework of an organ; distinguished from the functional cells (parenchyma) of that organ.

subarachnoid space: A meningeal space containing cerebrospinal fluid; the area between the arachnoid membrane and the pia mater.

subclavian (sub-KLĀ-vē-an): Pertaining to the region immediately posterior and inferior to the clavicle.

subcutaneous layer: The layer of loose connective tissue below the dermis; also called *hypodermis* or *superficial fascia*.

sublingual salivary glands (sub-LING-gwal): Mucus-secreting salivary glands inferior to the tongue.

submandibular salivary glands: Salivary glands nestled in depressions on the medial surfaces of the mandible; salivary glands that produce a mixture of mucins and enzymes (salivary amylase).

submucosa (sub-mū-KŌ-sa): The region between the muscularis mucosae and the muscularis externa.

submucosal glands: Mucous glands in the submucosa of the duodenum; also called *Brunner's glands* and duodenal glands.

subserous fascia: The loose connective tissue layer beneath the serous membrane that lines the ventral body cavity.

substrate: A participant (product or reactant) in an enzyme-catalyzed reaction.

sulcus (SUL-kus): A groove or furrow.

summation: The temporal or spatial addition of stimuli.

superficial fascia: *See* **subcutaneous layer.**

superior: Above, in reference to a portion of the body in the anatomical position.

superior vena cava (SVC): The vein that carries blood to the right atrium from parts of the body that are superior to the heart.

supination (soo-pi-NĀ-shun): The rotation of the forearm such that the palm faces anteriorly.

supine (SOO-pīn): Lying face up, with palms facing anteriorly.

suppressor T cells: Lymphocytes that inhibit B cell activation and the secretion of antibodies by plasma cells.

suprarenal gland (soo-pra-RĒ-nal): *See* **adrenal gland.**

surfactant (sur-FAK-tant): A lipid secretion that coats the alveolar surfaces of the lungs and prevents their collapse.

sustentacular cells (sus-ten-TAK-ū-lar): Supporting cells of the seminiferous tubules of the testis; responsible for the differentiation of spermatids, the maintenance of the blood–testis barrier, and the secretion of inhibin, androgen-binding protein, and Müllerian-inhibiting factor.

sutural bones: Irregular bones that form in fibrous tissue between the flat bones of the developing cranium; also called *Wormian bones*.

suture: A fibrous joint between flat bones of the skull.

sympathectomy (sim-path-EK-to-mē): The transection of the sympathetic innervation to a region.

sympathetic division: The division of the autonomic nervous system that is responsible for "fight or flight" reactions; primarily concerned with the elevation of metabolic rate and increased alertness.

sympathomimetic drugs: Drugs that mimic the actions of sympathetic stimulation.

symphysis (SIM-fi-sis): A fibrous amphiarthrosis, such as that between adjacent vertebrae or between the pubic bones of the ossa coxae.

symptom: An abnormality of function as a result of disease.

synapse (SIN-aps): The site of communication between a nerve cell and some other cell; if the other cell is not a neuron, the term *neuromuscular* or *neuroglandular junction* is often used.

synaptic delay (sin-AP-tik): The period between the arrival of an impulse at the presynaptic membrane and the initiation of an action potential in the postsynaptic membrane.

synarthrosis (sin-ar-THRŌ-sis): A joint that does not permit relative movement between the articulating elements; *see* **lambdoid suture.**

synchondrosis (sin-kon-DRŌ-sis): A cartilaginous synarthrosis, such as the articulation between the epiphysis and diaphysis of a growing bone.

syncope (SIN-kō-pē): A sudden, transient loss of consciousness; a faint.

syncytial trophoblast (sin-SISH-al): The multinucleate cytoplasmic layer that covers the blastocyst; responsible for uterine erosion and implantation.

syncytium (sin-SI-shē-um): A multinucleate mass of cytoplasm, produced by the fusion of cells or repeated mitoses without cytokinesis.

syndesmosis (sin-dez-MŌ-sis): A fibrous amphiarthrosis.

syndrome: A discrete set of symptoms that occur together.

syneresis (sin-ER-ē-sis): Clot retraction.

synergist (SIN-er-jist): A muscle that assists a prime mover in performing its primary action.

synostosis (sin-os-TŌ-sis): A synarthrosis formed by the fusion of the articulating elements.

synovial cavity (si-NŌ-vē-ul): A fluid-filled chamber in a synovial joint.

synovial fluid: The substance secreted by synovial membranes that lubricates joints.

synovial joint: A freely movable joint where the opposing bone surfaces are separated by synovial fluid; a diarthrosis.

synovial membrane: An incomplete layer of fibroblasts confronting the synovial cavity, plus the underlying loose connective tissue.

synthesis (SIN-the-sis): Manufacture; anabolism.

system: An interacting group of organs that performs one or more specific functions.

systemic circuit: The vessels between the aortic valve and the entrance to the right atrium; the system other than the vessels of the pulmonary circuit.

systole (SIS-tō-lē): A period of contraction in a chamber of the heart, as part of the cardiac cycle.

systolic pressure: The peak arterial pressure measured during ventricular systole.

T

tachycardia (tak-ē-KAR-dē-uh): An abnormally rapid heart rate.

tactile: Pertaining to the sense of touch.

taeniae coli (TĒ-nē-ē KŌ-lī): Three longitudinal bands of smooth muscle in the muscularis externa of the colon.

tarsal bones: The bones of the ankle (the talus, calcaneus, navicular, and cuneiform bones).

tarsus: The ankle.

TCA (tricarboxylic acid) cycle: The aerobic reaction sequence that occurs in the matrix of mitochondria; in the process, organic molecules are broken down, carbon dioxide molecules are released, and hydrogen molecules are transferred to coenzymes that deliver them to the electron transport system; also called *citric acid cycle* or *Krebs cycle.*

T cells: Lymphocytes responsible for cell-mediated immunity and for the coordination and regulation of the immune response; includes regulatory T cells (helpers and suppressors) and cytotoxic (killer) T cells.

tears: The fluid secretion of the lacrimal glands that bathes the anterior surfaces of the eyes.

tectorial membrane (tek-TOR-ē-al): The gelatinous membrane suspended over the hair cells of the organ of Corti.

tectospinal tracts: Descending tracts of the medial pathway that carry involuntary motor commands issued by the colliculi.

tectum: The roof of the mesencephalon of the brain.

telencephalon (tel-en-SEF-a-lon): The forebrain or cerebrum, including the cerebral hemispheres, the internal capsule, and the cerebral nuclei.

telodendria (te-lō-DEN-drē-uh): Terminal axonal branches that end in synaptic knobs.

telophase (TĒL-ō-fāz): The final stage of mitosis, characterized by the disappearance of the spindle apparatus, the reappearance of the nuclear membrane, the disappearance of the chromosomes, and the completion of cytokinesis.

temporal: Pertaining to time (temporal summation) or to the temples (temporal bone).

tendinitis: A painful inflammation of a tendon.

tendon: A collagenous band that connects a skeletal muscle to an element of the skeleton.

tentorium cerebelli (ten-TOR-ē-um ser-e-BEL-ē): A dural partition that separates the cerebral hemispheres from the cerebellum.

teratogen (TER-a-tō-jen): A stimulus that causes developmental defects.

teres: Long and round.

terminal: Toward the end.

tertiary follicle: A mature ovarian follicle, containing a large, fluid-filled chamber.

tertiary structure: The protein structure that results from interactions among distant portions of the same molecule; complex coiling and folding.

testes (TES-tēz): The male gonads, sites of gamete production and hormone secretion.

testosterone (tes-TOS-te-rōn): The principal androgen produced by the interstitial cells of the testes.

tetanic contraction: A sustained skeletal muscle contraction due to repeated stimulation at a frequency that prevents muscle relaxation.

tetanus: A tetanic contraction; also refers to a disease state that results from the stimulation of muscle cells by certain bacterial toxins.

tetany: A tetanic contraction; also refers to abnormally prolonged contractions that result from disturbances in electrolyte balance.

tetrad (TET-rad): Paired, duplicated chromosomes visible at the start of meiosis I.

tetraiodothyronine (tet-ra-ī-ō-dō-THĪ-rō-nēn): T_4, or thyroxine, a thyroid hormone.

thalamus: The walls of the diencephalon.

thalassemia (thal-ah-SĒ-mē-uh): A hereditary disorder that affects hemoglobin synthesis and produces anemia.

theory: A hypothesis that makes valid predictions, as demonstrated by evidence that is testable, unbiased, and repeatable.

therapy: The treatment of disease.

thermogenesis (ther-mō-JEN-e-sis): Heat production.

thermography: A diagnostic procedure involving the production of an infrared image.

thermoreception: Sensitivity to temperature changes.

thermoregulation: Homeostatic maintenance of body temperature.

thick filament: A cytoskeletal filament in a skeletal or cardiac muscle cell; composed of myosin, with a core of titin.

thin filament: A cytoskeletal filament in a skeletal or cardiac muscle cell; consists of actin, troponin, and tropomyosin.

thoracolumbar division (tho-ra-kō-LUM-bar): The sympathetic division of the autonomic nervous system.

thorax: The chest.

threshold: The transmembrane potential at which an action potential begins.

thrombin (THROM-bin): The enzyme that converts fibronogen to fibrin.

thrombocyte (THROM-bō-sīt): See **platelets**; in nonmammalian vertebrates, nucleated cells that are the equivalent of platelets in humans.

thrombocytopenia (throm-bō-sī-tō-PĒ-nē-uh): An abnormally low platelet count in the circulating blood.

thromboembolism (throm-bō-EM-bō-lizm): The occlusion of a blood vessel by a drifting blood clot.

thrombus: A blood clot that develops at the lumenal wall of a blood vessel.

thymine: A pyrimidine; one of the nitrogenous bases in the nucleic acid DNA.

thymosins (THĪ-mō-sinz): Thymic hormones essential to the development and differentiation of T cells.

thymus: A lymphoid organ, the site of T cell formation.

thyroglobulin (thī-rō-GLOB-ū-lin): A circulating transport globulin that binds thyroid hormones.

thyroid gland: An endocrine gland whose lobes are lateral to the thyroid cartilage of the larynx.

thyroid hormones: Thyroxine (T_4) and triiodothyronine (T_3), hormones of the thyroid gland; stimulate tissue metabolism, energy utilization, and growth.

thyroid-stimulating hormone (TSH): The hormone, produced by the anterior lobe of the pituitary gland, that triggers the secretion of thyroid hormones by the thyroid gland.

thyroxine (thī-ROKS-ēn): A thyroid hormone; also called T_4 or *tetraiodothyronine.*

tidal volume: The volume of air moved into and out of the lungs during a normal quiet respiratory cycle.

tissue: A collection of specialized cells and cell products that performs a specific function.

titer: The plasma antibody concentration.

tolerance: The failure to produce antibodies against antigens normally present in the body.

tonic response: An increase or decrease in the frequency of action potentials by sensory receptors that are chronically active.

tonsil: A lymphoid nodule in the wall of the pharynx; the palatine, pharyngeal, and lingual tonsils.

topical: Applied to the body surface.

toxic: Poisonous.

trabecula (tra-BEK-ū-la): A connective tissue partition that subdivides an organ.

trabeculae carneae (tra-BEK-ū-lē KAR-nē-ē): Muscular ridges projecting from the walls of the ventricles of the heart.

trachea (TRĀ-kē-a): The windpipe, an airway extending from the larynx to the primary bronchi.

tracheal ring: A C-shaped supporting cartilage of the trachea.

tracheostomy (trā-kē-OS-to-mē): The surgical opening of the anterior tracheal wall to permit airflow.

trachoma: An infectious disease of the conjunctiva and cornea.

tract: A bundle of axons in the central nervous system.

transamination (tranz-am-i-NĀ-shun): The enzymatic transfer of an amino group from an amino acid to another carbon chain.

transcription: The encoding of genetic instructions on a strand of mRNA.

transdermal medication: The administration of medication by absorption through the skin.

transection: The severing or cutting of an object in the transverse plane.

transfusion: The transfer of blood from a donor directly into the bloodstream of another person.

transient ischemic attack (TIA): A temporary loss of consciousness due to the occlusion of a small blood vessel in the brain.

translation: The process of peptide formation from the instructions carried by an mRNA strand.

transmembrane potential: The potential difference, measured across a cell membrane and expressed in millivolts, that results from the uneven distribution of positive and negative ions across the cell membrane.

transudate (TRAN-sū-dāt): A fluid that diffuses across a serous membrane and lubricates opposing surfaces.

transverse tubules: The transverse, tubular extensions of the sarcolemma that extend deep into the sarcoplasm, contacting cisternae of the sarcoplasmic reticulum; also called *T tubules.*

treppe (TREP-e): A steplike increase in tension production after the repeated stimulation of a muscle, even though the muscle is allowed to complete each relaxation phase.

triad: In a skeletal muscle fiber, the combination of a transverse tubule and two cisternae of the sarcoplasmic reticulum; in the liver, a combination of branches of the hepatic portal vein, hepatic artery proper, and bile duct at a liver lobule.

tricarboxylic acid cycle (trī-kar-bok-SIL-ik): *See* **TCA cycle.**

tricuspid valve (trī-KUS-pid): The right atrioventricular valve, which prevents the backflow of blood into the right atrium during ventricular systole.

trigeminal nerve (trī-JEM-i-nal): Cranial nerve V, which provides sensory information from the lower portions of the face (including the upper and lower jaws) and delivers motor commands to the muscles of mastication.

triglyceride (trī-GLIS-e-rīd): A lipid that is composed of a molecule of glycerol attached to three fatty acids.

trigone (TRĪ-gōn): The triangular region of the urinary bladder that is bounded by the exits of the ureters and the entrance to the urethra.

triiodothyronine: T$_3$, a thyroid hormone.

trisomy (trī-SŌ-mē): The abnormal possession of three copies of a chromosome; trisomy 21 is responsible for Down syndrome.

trochanter (trō-KAN-ter): Large processes near the head of the femur.

trochlea (TRŌK-lē-uh): A pulley; the spool-shaped medial portion of the condyle of the humerus.

trochlear nerve (TRŌK-lē-ar): Cranial nerve IV, controlling the superior oblique muscle of the eye.

trophoblast (TRŌ-fō-blast): The superficial layer of the blastocyst that will be involved in implantation, hormone production, and placenta formation.

tropomyosin (trō-pō-MĪ-ō-sin): A protein on thin filaments that covers the active sites in the absence of free calcium ions.

troponin (TRŌ-pō-nin): A protein on thin filaments that binds to tropomyosin and to G actin.

trunk: The thoracic and abdominopelvic regions; a major arterial branch.

trypsin (TRIP-sin): One of the pancreatic proteases.

trypsinogen (trip-SIN-ō-jen): The inactive proenzyme that is secreted by the pancreas and is converted to trypsin in the duodenum.

T tubules: *See* **transverse tubules.**

tuberculum (too-BER-kū-lum): A small, localized elevation on a bony surface.

tuberosity: A large, roughened elevation on a bony surface.

tubulin: A protein subunit of microtubules.

tumor: A tissue mass formed by the abnormal growth and replication of cells.

tunica (TOO-ni-ka): A layer or covering.

tunica externa: The outermost layer of connective tissue fibers that stabilizes the position of a blood vessel.

tunica intima: The innermost layer of connective tissue fibers in a blood vessel; consists of the endothelium plus an underlying elastic membrane.

tunica media: The middle layer of connective tissue fibers in a blood vessel; contains collagen, elastin, and smooth muscle fibers in varying proportions.

T wave: A deflection of the ECG corresponding to ventricular repolarization.

twitch: A single stimulus–contraction–relaxation cycle in a skeletal muscle.

tympanic duct: The perilymph-filled chamber of the inner ear, adjacent to the basilar membrane; pressure changes there distort the round window; also called *scala tympani.*

tympanic membrane (tim-PAN-ik): The membrane that separates the external acoustic canal from the middle ear; the membrane whose vibrations are trans-

ferred to the auditory ossicles and ultimately to the oval window; also called *eardrum* or *tympanum.*

type A axons: Large myelinated axons.

type B axons: Small myelinated axons.

type C axons: Small unmyelinated axons.

U

ulcer: An area of epithelial sloughing associated with damage to the underlying connective tissues and blood vessels.

ultrasound: A diagnostic visualization procedure that uses high-frequency sound waves.

umbilical cord (um-BIL-i-kal): The connecting stalk between the fetus and the placenta; contains the allantois, the umbilical arteries, and the umbilical vein.

umbilicus: The navel.

unicellular gland: Goblet cells.

unipennate muscle: A muscle whose fibers are on one side of the tendon.

unipolar neuron: A sensory neuron whose cell body is in a dorsal root ganglion or a sensory ganglion of a cranial nerve.

unmyelinated axon: An axon whose neurilemma does not contain myelin and across which continuous propagation occurs.

uracil: A pyrimidine; one of the nitrogenous bases in the nucleic acid RNA.

uremia (ū-RĒ-mē-uh): An abnormal condition caused by impaired kidney function; characterized by the retention of wastes and the disruption of many other organ systems.

ureters (ū-RĒ-terz): Muscular tubes, lined by transitional epithelium, that carry urine from the renal pelvis to the urinary bladder.

urethra (ū-RĒ-thra): A muscular tube that carries urine from the urinary bladder to the exterior.

urethritis: An inflammation of the urethra.

urinalysis: An analysis of the physical and chemical characteristics of urine.

urinary bladder: The muscular, distensible sac that stores urine prior to micturition.

urination: The voiding of urine; micturition.

urobilin: A compound derived from urobilinogen and ultimately from the bilirubin excreted in bile.

uterus (Ū-ter-us): The muscular organ of the female reproductive tract in which implantation, placenta formation, and fetal development occur.

utricle (Ū-tri-kul): The largest chamber of the vestibular apparatus of the inner ear; contains a macula important for static equilibrium.

uvea: The vascular tunic of the eye.

uvula (Ū-vū-luh): A dangling, fleshy extension of the soft palate.

V

vagina (va-JĪ-na): A muscular tube extending between the uterus and the vestibule.

varicose veins (VAR-i-kōs): Distended superficial veins.

vasa recta: Long, straight capillaries that parallel the loop of Henle in juxtamedullary nephrons.

vasa vasorum: Blood vessels that supply the walls of large arteries and veins.

vascular: Pertaining to blood vessels.

vascularity: The blood vessels in a tissue.

vascular spasm: A contraction of the wall of a blood vessel at an injury site; may slow the rate of blood loss.

vasoconstriction: A reduction in the diameter of arterioles due to the contraction of smooth muscles in the tunica media; elevates peripheral resistance; may occur in response to local factors, through the action of hormones, or from the stimulation of the vasomotor center.

vasodilation: An increase in the diameter of arterioles due to the relaxation of smooth muscles in the tunica media; reduces peripheral resistance; may occur in response to local factors, through the action of hormones, or after decreased stimulation of the vasomotor center.

vasomotion: Changes in the pattern of blood flow through a capillary bed in response to changes in the local environment.

vasomotor center: The center in the medulla oblongata whose stimulation produces vasoconstriction and an elevation of peripheral resistance.

vein: A blood vessel carrying blood from a capillary bed toward the heart.

vena cava (VĒ-na KĀ-va): One of the major veins delivering systemic blood to the right atrium; superior and inferior venae cavae.

ventilation: Air movement into and out of the lungs.

ventral: Pertaining to the anterior surface.

ventricle (VEN-tri-kul): A fluid-filled chamber; in the heart, one of the large chambers discharging blood into the pulmonary or systemic circuits; in the brain, one of four fluid-filled interior chambers.

ventricular escape: The initiation of ventricular contractions after a pause caused by impaired conduction of the contractile stimulus from the AV node.

ventricular folds: Mucosal folds in the laryngeal walls that do not play a role in sound production; the false vocal cords.

venule (VEN-ūl): Thin-walled veins that receive blood from capillaries.

vermiform appendix: *See* **appendix**.

vermis (VER-mis): A midsagittal band of neural cortex on the surface of the cerebellum.

vertebral canal: The passageway that encloses the spinal cord; a tunnel bounded by the neural arches of adjacent vertebrae.

vertebral column: The cervical, thoracic, and lumbar vertebrae, the sacrum, and the coccyx.

vertebrochondral ribs: Ribs 8–10; false ribs, connected to the sternum by shared cartilaginous bars.

vertebrosternal ribs: Ribs 1–7; true ribs, connected to the sternum by individual cartilaginous bars.

vertigo: Dizziness.

vesicle: A membranous sac in the cytoplasm of a cell.

vestibular complex: The utricle, saccule, and semicircular canals of the inner ear.

vestibular duct: The perilymph-filled chamber of the inner ear, superior to the vestibular membrane; pressure changes there result from distortions of the oval window; also called *scala vestibuli*.

vestibular membrane: The membrane that separates the cochlear duct from the vestibular duct of the inner ear.

vestibular nucleus: The processing center for sensations that arrive from the vestibular apparatus of the inner ear; located near the border between the pons and the medulla oblongata.

vestibule (VES-ti-būl): A chamber; in the inner ear, the utricle and saccule.

vestibulospinal tracts: Descending tracts of the medial pathway that carry involuntary motor commands issued by the vestibular nucleus to stabilize the position of the head.

villus/villi: A slender projection of the mucous membrane of the small intestine.

virus: A noncellular pathogen.

viscera (VIS-e-rah): Organs in the ventral body cavity.

visceral: Pertaining to viscera or their outer coverings.

visceral smooth muscle: A smooth muscle tissue that forms sheets or layers in the walls of visceral organs; the cells may not be innervated, and the layers often show automaticity (rhythmic contractions).

viscosity: The resistance to flow that a fluid exhibits as a result of molecular interactions within the fluid.

viscous: Thick, syrupy.

vital capacity: The maximum amount of air that can be moved into or out of the respiratory system; the sum of the inspiratory reserve, the expiratory reserve, and the tidal volume.

vitamin: An essential organic nutrient that functions as a coenzyme in vital enzymatic reactions.

vitreous humor: The gelatinous mass in the vitreous chamber of the eye.

vocal folds: Folds in the laryngeal wall that contain elastic ligaments whose tension can be voluntarily adjusted; the true vocal cords, responsible for phonation.

voluntary: Controlled by conscious thought processes.

vulva (VUL-vuh): The female pudendum (external genitalia).

W

Wallerian degeneration: The disintegration of an axon and its myelin sheath distal to an injury site.

white blood cells (WBCs): Leukocytes; the granulocytes and agranulocytes of blood.

white matter: Regions in the central nervous system that are dominated by myelinated axons.

white ramus: A nerve bundle containing the myelinated preganglionic axons of sympathetic motor neurons en route to the sympathetic chain or to a collateral ganglion.

Wormian bones: *See* **sutural bones**.

X

X chromosome: The sex chromosome that is present both in genetic males and in genetic females.

xiphoid process (ZĪ-foyd): The slender, inferior extension of the sternum.

Y

Y chromosome: The sex chromosome whose presence indicates that the individual is a genetic male.

yolk sac: One of the four extraembryonic membranes, composed of an inner layer of endoderm and an outer layer of mesoderm.

Z

zona fasciculata (ZŌ-na fa-sik-ū-LA-ta): The region of the adrenal cortex that secretes glucocorticoids.

zona glomerulosa (glo-mer-ū-LŌ-sa): The region of the adrenal cortex that secretes mineralocorticoids.

zona pellucida (pe-LOO-si-duh): The region between a developing oocyte and the surrounding follicular cells of the ovary.

zona reticularis (re-ti-kū-LAR-is): The region of the adrenal cortex that secretes androgens.

zygote (ZĪ-gōt): The fertilized ovum, prior to the start of cleavage.

ILLUSTRATION CREDITS

Chapter 1 **CO-1** John Paul Endress/Silver Burdett Ginn 1-8a,b Custom Medical Stock Photo, Inc. 1-13aL,aR,bL Science Source/Photo Researchers, Inc. 1-13bR Custom Medical Stock Photo, Inc. 1-14c Ralph Hutchings 1-14b,d Dr. Kathleen Welch 1-15 Monte S. Buchsbaum, M.D. 1-15a Philips Medical Systems 1-15b Alexander Tsiaras/Science Source/Photo Researchers, Inc. 1-15c Photo Researchers, Inc.

Chapter 2 **CO-2** James A. Sugar/CORBIS UN02-1 PhotoEdit

Chapter 3 **CO-3** ©CAMR/A.B. Dowsett/Science Photo Library/Photo Researchers, Inc. 3-5b Fawcett/Hirokawa/Heuser/Science Source/ Photo Researchers, Inc. 3-6a Fawcett/de Harven/Kalnins/Photo Researchers, Inc. 3-8b R. Bollender & Don W. Fawcett/Visuals Unlimited 3-9b Biophoto Associates/Photo Researchers, Inc. 3-10b Dr. Birgit H. Satir 3-12a CNRI/Science Source/Photo Researchers, Inc. 3-13a Don W. Fawcett, M.D., Harvard Medical School 3-13b Biophoto Associates/Photo Researchers, Inc. 3-21a,b,c David M. Philips/Visuals Unlimited 3-26a M. Sahliwa/Visuals Unlimited 3-29a,b,d,e,f Ed Reschke/Peter Arnold, Inc. 3-29c James Solliday/Biological Photo Service

Chapter 4 **CO-4** Photo Researchers, Inc. 4-3a Ward's Natural Science Establishment, Inc. 4-3b Frederic H. Martini 4-4a Pearson Education/PH College 4-4b Courtesy of Gregory N. Fuller, M.D., Ph.D., Section of Neuropathology, M.D. Anderson Cancer Center, Houston, Texas 4-4cL,cR Frederic H. Martini 4-5a,b Frederic H. Martini 4-5c Courtesy of Gregory N. Fuller, M.D., Ph.D., Section of Neuropathology, M.D. Anderson Cancer Center, Houston, Texas 4-6a Phototake/Carolina Biological Supply Company 4-9 Ward's Natural Science Establishment, Inc. 4-10a,b Project Masters, Inc./The Bergman Collection 4-11aL Frederic H. Martini 4-11b Ward's Natural Science Establishment, Inc. 4-12a John D. Cunningham/Visuals Unlimited 4-12b Frederic H. Martini 4-12c Bruce Iverson/ Visuals Unlimited 4-15a Frederic H. Martini 4-15b Robert Brons/Biological Photo Service 4-15c Science Source/Photo Researchers, Inc. 4-15d Ed Reschke/Peter Arnold, Inc. 4-16 Frederic H. Martini 4-19a,b G. W. Willis, MD/Biological Photo Service 4-19c,4-20 Pearson Education/PH College

Chapter 5 **CO-5** Steve Mason/Getty Images, Inc. UN05-1 Doug Martin/Photo Researchers, Inc. 5-2b,c Frederic H. Martini 5-3 ©R.G. Kessel and R.H Kardon, *Tissues and Organs: A Text-Atlas of Scanning Electron Microscopy*, W.H. Freeman & Co., 1979. All Rights Reserved. 5-4 John D. Cunningham/Visuals Unlimited 5-5a Pearson Education/PH College 5-5b Biophoto Associates/Photo Researchers, Inc. 5-6a,b Courtesy of Elizabeth A. Abel, M.D., from Leonard C. Winograd Memorial Slide Collection, Stanford University School of Medicine 5-10c John D. Cunningham/Visuals Unlimited 5-11, 5-12a,b Frederic H. Martini

Chapter 6 **CO-6** Photo Researchers, Inc. 6-2a Ralph T. Hutchings 6-3b Frederic H. Martini 6-3c ©R.G. Kessel and R.H Kardon, *Tissues and Organs: A Text-Atlas of Scanning Electron Microscopy*, W.H. Freeman & Co., 1979. All Rights Reserved. 6-4c Ralph T. Hutchings 6-7L,R Frederic H. Martini 6-8b Pearson Education/PH College 6-9a,b Project Masters, Inc./The Bergman Collection 6-15b,g Southern Illinois University/Visuals Unlimited 6-15c Southern Illinois University/Peter Arnold, Inc. 6-15d Custom Medical Stock Photo, Inc. 6-15e Scott Camazine/Photo Researchers, Inc. 6-15f Frederic H. Martini 6-15h Project Masters, Inc./The Bergman Collection UN06-1 Getty Images Inc.–Hulton Archive Photos 6-16a,b Prof. P. Motta, Dept. of Anatomy, University *La Sapienza*, Rome/Science Photo Library/Photo Researchers, Inc.

Chapter 7 **CO-7** Hendrie, Peter/Getty Images Inc.–Image Bank UN07-1 Will & Deni McIntyre/Photo Researchers, Inc. 7-1aL,aR, 7-5a,b, 7-6a,b, 7-7a,b Ralph T. Hutchings 7-7c Michael J. Timmons 7-8a,b, 7-9a,b, 7-10a, 7-11, 7-12a,b,c, 7-13 Ralph T. Hutchings 7-17a Science Photo Library/Custom Medical Stock Photo, Inc. 7-17b National Medical Slide/Custom Medical Stock Photo, Inc. 7-17c Princess Margaret Rose Orthopaedic Hospital, Edinburgh, Scotland/Science Photo Library/Photo Researchers, Inc. 7-19a,b,c, T7-2L,R,C, 7-20a,b,c, 7-21a,b,c, 7-22a,b,c, 7-23a,b, 7-24a,b Ralph T. Hutchings

Chapter 8 **CO-8** Digital Vision 8-2b,c, 8-3a,b,c, 8-4a,b, 8-5a,b, 8-6a,b, 8-7a,b, 8-8a,b, 8-11a,b, 8-12a,b, 8-13a,b, 8-14a,b Ralph T. Hutchings

Chapter 9 **CO-9** SuperStock, Inc. UN09-1 Tony Neste/Anthony Neste 9-3aT,cL,cR,d, 9-4a,c, 9-5B1,B2,B3,B4, 9-5T1,T2,T3, 9-10a,b, 9-12c,d Ralph T. Hutchings

Chapter 10 **CO-10** Phototake NYC UN10-01 Nishio, Chie/Omni-Photo Communications, Inc. 10-2b Ward's Natural Science Establishment, Inc. 10-4 Photo Researchers, Inc. 10-21a Comack, D. (ed): *Ham's Histology*, 9th ed. Philadelphia: J.B. Lippincott 1987. By Permission. 10-21bB,T Frederic H. 10-22a Phototake NYC 10-23a Frederic H. Martini

Chapter 11 **CO-11** Olivier Rigardiere/Getty Images, Inc. UN11-01 Peter Cade/Getty Images Inc.–Stone

Chapter 12 **CO-12** Hank Morgan/Science Source/Photo Researchers, Inc. 12-3 David Scott/ Phototake NYC 12-7a Frederic H. Martini 12-8a Biophoto Associates/Photo Researchers, Inc. 12-8b Photo Researchers, Inc.

Chapter 13 **CO-13** Frank Pedrick/Index Stock Imagery, Inc. 13-5a Ralph T. Hutchings 13-5b Hinerfeld/Custom Medical Stock Photo, Inc. 13-6 Ralph T. Hutchings 13-7b Michael J. Timmons 13-08a ©R.G. Kessel and R.H Kardon, *Tissues and Organs: A Text-Atlas of Scanning Electron Microscopy*, W.H. Freeman & Co., 1979. All Rights Reserved.

Chapter 14 **CO-14** Margaret Ross/Stock Boston UN14-1 Will Hart 14-1a,b,c,d Ralph T. Hutchings 14-6 Visuals Unlimited 14-9a,b Ralph T. Hutchings 14-10b Daniel P. Perl, M.D., Mount Sinai School of Medicine 14-12b, 14-14a,b Ralph T. Hutchings 14-16b Michael J. Timmons 14-16c Pat Lynch/Photo Researchers, Inc. 14-19e Larry Mulvehill/Photo Researchers, Inc. 14-20a,c Ralph T. Hutchings

Chapter 15 **CO-15** Young-Jin Son, MCP Hahnemann University School of Medicine UN15-01 Elena Rooraid/PhotoEdit 15-3d,e, 15-5 Frederic H. Martini

Chapter 16 **CO-16** Phil Schofield/Getty Images Inc.–Stone

Chapter 17 **CO-17** Blair Seitz/Photo Researchers, Inc. UN17-01 Geoff Tompkins/Science Photo Library/Photo Researchers, Inc. 17-2b Pearson Education/PH College 17-2c G. W. Willis/Terraphotographics/Biological Photo Service 17-3a Ralph T. Hutchings 17-5L Diane Hirsch/Fundamental Photographs 17-5R Diane Schiumo/Fundamental Photographs 17-6a Ed Reschke/Peter Arnold, Inc. 17-6c Custom Medical Stock Photo, Inc. 17-17 Richmond Products, Inc. 17-21b Lennart Nilsson/Bonnier Alba AB, *Behold Man*, pp. 206-207 17-25b Michael J. Timmons 17-26b Ward's Natural Science Establishment, Inc.

Chapter 18 **CO-18** Christopher Bissell/Getty Images Inc.–Stone UN18-01 Russell D. Curtis/Photo Researchers, Inc. UN18-02 AP/Wide World Photos 18-6b Manfred Kage/Peter Arnold, Inc. 18-10 Project Masters, Inc./The Bergman Collection 18-11c Frederic H. Martini 18-13a Project Masters, Inc./The Bergman Collection 18-13b John Paul Kay/Peter Arnold, Inc. 18-14b,c Frederic H. Martini 18-16c Ward's Natural Science Establishment, Inc. 18-17a Custom Medical Stock Photo, Inc. 18-17b Biophoto Associates/Science Source/Photo Researchers, Inc. 18-18b Ward's Natural Science Establishment, Inc.

Chapter 19 **CO-19** Getty Images Inc.-Image Bank 19-1a Martin M. Rotker 19-2a,c David Scharf/Peter Arnold, Inc. 19-2b Ed Reschke/Peter Arnold, Inc. 19-4a,b Stanley Flegler/Visuals Unlimited 19-7 AP/Wide World Photos 19-09 Phototake NYC 19-11a,b,c,d,e Ed Reschke/Peter Arnold, Inc. 19-14b Custom Medical Stock Photo, Inc.

Chapter 20 **CO-20** Darryl Torckler/Getty Images Inc.-Stone UN20-1 Getty Images Inc.-Stone 20-3a Ralph T. Hutchings 20-5a Ed Reschke/Peter Arnold, Inc. 20-6b ©Lennart Nilsson *Behold Man*, Little Brown and Company 20-6c, 20-7b Ralph T. Hutchings 20-8cR Biophoto Associates/Science Source/Photo Researchers, Inc. 20-8cL Science Photo Library/Photo Researchers, Inc. 20-9c,d Ralph T. Hutchings 20-10d Phototake NYC 20-10a,b Howard Sochurek/ Medichrome/The Stock Shop, Inc. 20-10c Peter Arnold, Inc. 20-14a Larry Mulvehill/Photo Researchers, Inc.

Chapter 21 **CO-21** CNRI/Photo Researchers, Inc. UN21-1 Dr. P. Marazzi/Science Photo Library/Photo Researchers, Inc. 21-1 Biophoto Associates/Photo Researchers, Inc. 21-3a B & B Photos/Custom Medical Stock Photo, Inc. 21-3b William Ober/Visuals Unlimited 21-5b Biophoto Associates/Photo Researchers, Inc. 21-11 Jack Star/Getty Images, Inc.-Photodisc

Chapter 22 **CO-22** John Moss/Photo Researchers, Inc. UN22-1 Ben Edwards/Getty Images Inc.–Stone UN22-02 CDC/RG/Peter Arnold, Inc. 22-3b Frederic H. Martini 22-4c Ralph T. Hutchings 22-6a David M. Phillips/Visuals Unlimited 22-6b Biophoto Associates/ Photo Researchers, Inc. 22-8c,d, 22-9c Frederic H. Martini 22-21b Leonard Lessin/Peter Arnold, Inc.

Chapter 23 **CO-23** Derek Berwin/Getty Images Inc.–Image Bank UN23-1 Larry Mulvehill/Photo Researchers, Inc. UN23-2 Nazima Kowall/CORBIS 23-2a Photo Researchers, Inc. 23-2b Frederic H. Martini 23-3b Ralph T. Hutchings 23-5b Phototake NYC 23-6c John D. Cunningham/Visuals Unlimited 23-7, 23-8 Ralph T. Hutchings 23-9 CNRI/Science Photo Library/Photo Researchers, Inc. 23-11b Ward's Natural Science Establishment, Inc. 23-11c Don W. Fawcett, M.D., Micrograph by P. Gehr, from Bloom & Fawcett, *Textbook of Histology*, W.B. Saunders Co.

Chapter 24 **CO-24** Alan Becker/Getty Images Inc.–Image Bank UN24-1L Dr. R. Gottsegen/Peter Arnold, Inc. UN24-1R Science VU/Max Listgarten/Visuals Unlimited 24-7b Frederic H. Martini 24-10a Alfred Pasieka/Peter Arnold, Inc. 24-10b Astrid and Hanns-Frieder Michler/Science Photo Library/Photo Researchers, Inc. 24-10c Prof. P. Motta/Dept. of Anatomy/Univ. *La Sapienza*, Rome/Science Photo Library/Photo Researchers, Inc. 24-12a Ralph T. Hutchings 24-13a P. Motta/Dept. of Anatomy/Univ. *La Sapienza*/SPL/Photo Researchers, Inc. 24-13b John D. Cunningham/Visuals Unlimited 24-16b Ralph T. Hutchings 24-17a Phototake NYC 24-17e M.I. Walker/Photo Researchers, Inc. 24-18c Frederic H. Martini 24-19a Ralph T. Hutchings 24-20c,d Michael J. Timmons 24-23b Ralph T. Hutchings 24-24 Ward's Natural Science Establishment, Inc.

Chapter 25 **CO-25** Tim Bieber/Getty Images Inc.–Image Bank UN25-1 Bill Bachmann/Photri-Microstock, Inc. UN25-2 Express Newspapers/Getty Images, Inc.–Liaison 25-19 Christopher Bissell/Getty Images Inc.-Stone

Chapter 26 **CO-26** Francois Duquesnoy (1594-1643), *Mannaken Pis*, Brussels, Belgium. Kavaler/Art Resource, N.Y. 26-4b Ralph T. Hutchings 26-8c David M. Phillips/Visuals Unlimited 26-17 Photo Researchers, Inc. 26-18 Peter Arnold, Inc. 26-20a Ward's Natural Science Establishment, Inc. 26-20b Frederic H. Martini 26-20c G. W. Willis, MD/Biological Photo Service

Chapter 27 **CO-27** David Young–Wolff/PhotoEdit UN27-1 Robert Harbison UN27-2 SIU/Peter Arnold, Inc.

Chapter 28 **CO-28** Brian Bailey/Getty Images Inc.–Stone UN28-1 Manfred Kage/Peter Arnold, Inc. UN28-2 Frederic H. Martini UN28-3 Spencer Grant/PhotoEdit 28-4b Frederic H. Martini 28-5a Don W. Fawcett, M.D., Harvard Medical School 28-5b Ward's Natural Science Establishment, Inc. 28-8b David M. Phillips/Visuals Unlimited 28-9b Frederic H. Martini 28-10a Ward's Natural Science Establishment, Inc. 28-10b ©R.G. Kessel and R.H Kardon, *Tissues and Organs: A Text-Atlas of Scanning Electron Microscopy*, W.H. Freeman & Co., 1979. All Rights Reserved. 28-10c,d,e, 28-16a,b,c,d Frederic H. Martini 28-16e C. Edelmann/La Villete/Photo Researchers, Inc. 28-16f G.W. Willis, MD/Biological Photo Service 28-16g G. W. Willis/ Terraphotographics/Biological Photo Service 28-17b Frederic H. Martini 28-17c Custom Medical Stock Photo, Inc. 28-18a Ralph T. Hutchings 28-19a Ward's Natural Science Establishment, Inc. 28-20a,b,cT Frederic H. Martini 28-20cB, 28-21 Michael J. Timmons 28-23b Fred E. Hossler/Visuals Unlimited 28-23c Frederic H. Martini 28-27 Charles Thatcher/Getty Images Inc.–Stone

Chapter 29 **CO-29** Charles Gupton/Getty Images Inc.-Stone 29-1a Francis Leroy, Biocosmos/Science Photo Library/Custom Medical Stock Photo, Inc. 29-6b Frederic H. Martini 29-7a Dr. Arnold Tamarin 29-7b,c,d, 29-9a Photo Lennart Nilsson/Albert Bonniers Forlag 29-9b Photo Researchers, Inc. 29-15 Science Photo Library/Photo Researchers, Inc.

INDEX

Selected page references to the Applications Manual appear in blue type.

A band, 296, 298, 299
Abdominal aorta, 758
Abdominal cavity, 20, 21, 22
Abdominal reflex, 456, 457, AM
Abdominal thrust, 836, 867
Abdominopelvic cavity, 20-22
Abdominopelvic quadrants, 17, 26
Abdominopelvic regions, 17, 26
Abducens nerves (VI), 345, 496
Abduction motion, 269, 270
Abductor pollicis brevis muscle, 365
Abortion, 1110
Abruptio placentae, AM
Abscess, 139, 142, 797, 830
Absolute refractory period, 407, 408, 422, 702
Absorptive state
    adipose tissue and, 950-951
    liver functions, 950
    overview, 949, 950
Accessory ligaments, 266
Accessory nerves (XI), 500, 502
Accessory pancreatic duct, 901
Accessory respiratory muscles, 848
Accessory structures (integument)
    glands, 169-171
    hair/hair follicles, 166-169
    nails, 171-172
    overview, 156
Acclimatization and thermoregulation, 963
Accommodation, vision, 578, 579
Accommodation curves, 228
Acetabular labrum, 278
Acetabular notch, 250
Acetabulum, hipbone, 250, 251
Acetate, 413
Acetoacetate, 952
Acetone, 952
Acetyl-CoA, 934-935
Acetyl group, 934
ACh (acetylcholine)
    cholinergic synapses, 412-413, 414
    function mechanism, 418
    heart rate and, 547
    inactivation of, 306
    movement patterns and, 526
    muscle activity and, 301-303, 306
    muscle paralysis and, 306
    as neurotransmitters, 301, 412-413, 414, 416
    parasympathetic division and, 543, 545
    sympathetic division and, 540, 541, 542
    where released, 412, 540
Achalasia, 890, AM
AChE (acetylcholinesterase)
    in cholinergic synapses, 412, 413, 414
    function, 301
    parasympathetic division and, 543
    poisoning, 415
Achilles, 374
Achilles tendon, 255, 374, 376
Aching pain, 513
Achondroplasia, AM
Acid-base balance
    acid types in body, 1028
    deamination, 1032-1033
    disturbances, 1034-1039
    maintenance, 1031-1033
    overview, 1015, 1016
    pH control importance, 1027-1028
    pH control mechanism, 1028-1031, 1034
    respiratory compensation, 1031-1032, 1035
    terminology, 1027
Acidemia, 1027
Acidic solution, 44
Acidophils. *See* Eosinophils

Acidosis
    description, 44, 1002, 1027
    detection, 1038
    ketoacidosis, 993, 1037
    kidneys, 1032-1033
    lactic acidosis, 993, 1037
    metabolic acidosis, 1035, 1037-1038, 1041
    respiratory acidosis, 1035-1036, 1037-1038, 1041
Acids
    description/examples, 44, 45, 60
    fixed acids, 1028
    organic acids, 1028
    strong acids, 44, 1027
    types in body, 1028
    volatile acids, 1028
    weak acids, 45, 1027
Acne, AM
Acoustic canal, external/internal, 218, 588
Acoustic meatus, external/internal, 218, 588
Acquired immune deficiency syndrome. *See* AIDS
Acquired immunity, 798, 799
Acquired reflexes, 449-450
Acromegaly, 198, 205, 621
Acromial end, clavicles, 245
Acromioclavicular ligament, 276
Acromion, 245
Acrosin, 1098
Acrosomal cap, 1056, 1096, 1098
Actin (cytoskeleton), 75
Actin (muscle)
    structure/function, 136, 303
    thin filaments and, 294
Actinic keratosis, 162
Actinins, 296
Action. *See also* Origin/insertion/action/innervation
    methods of describing, 337
    of muscle, 336
    in muscle name, 339
Action potentials
    all-or-none principle, 405
    cardiac muscle cells, 701-703
    definition, 398
    generation, 406-407
    generation summary, 407
    graded potential vs., 409
    in muscle activity, 294, 301, 302, 303
    muscle vs. neurons, 411
    propagation, 408-411
    rate of generation, 422
Activation energy
    description, 40
    enzymes effects, 40
Active immunity, 798, 799
Active immunization, AM
Active site, enzyme, 56
Active transport, 90, 95-96, 982
Acupuncture, 513
Acute pain, 513
Acute phase, acid-base imbalance, 1035
Acute renal failure, 989
Acute respiratory acidosis, 1035
Adam's apple, 832
Adaptation, stimuli, 512, 514
Addison's disease
    cause/effects, 629, 643
    skin effects of, 161
Additive effects, hormones, 639
Adduction motion, 269, 270
Adductor brevis muscle, 368, 370
Adductor longus muscle, 368, 370
Adductor magnus muscles, 368, 370
Adenine (A), structure, 57-58
Adenohypophysis, 615-618
Adenoids, 787, 832

Adenosine, 59
Adenosine diphosphate. *See* ADP
Adenosine monophosphate. *See* AMP
Adenosine triphosphatase. *See* ATPase
Adenosine triphosphate. *See* ATP
ADH. *See* Antidiuretic hormone
Adhesions
    phagocytosis, 793
    serous membranes, 134, 142
Adipocytes, 124
Adipose capsule, 972
Adipose cells. *See* Adipocytes
Adipose tissue
    absorptive state, 950-951
    endocrine activity of, 637, 639
    lipid storage, 949
    postabsorptive state, 953
    structure/function, 125-126
Adolescence, 1118-1120
ADP
    energy and, 59, 60
    in hemostasis, 673
    in muscle contraction, 304
Adrenal glands
    adrenal cortex, 617, 627-630
    adrenal medulla, 537, 539-540, 630
    anatomy, 627, 628
    hormone summary, 628
Adrenergic synapses, 414, 540
Adrenocortical steroids. *See* Corticosteroids
Adrenocorticotropic hormone (ACTH)
    in Addison's disease, 161
    appetite suppression, 961
    functions, 617
Adrenogenital syndrome, 630
Adventitia, 880, 888, 1075
Adventitious bursae, 267
Aerobic endurance, 322
Aerobic metabolism. *See also* Metabolism
    ATP production and, 932
    description, 82, 83, 316
    muscle activity, 316-317, 318, 319
Afferent arterioles, 974
Afferent division, PNS, 386, 387
Afferent fibers, 391
Afferent lymphatics, 787
Afterbirth, 1116
Afterload
    description, 714
    vasoconstriction/vasodilation, 723
Agglutinate/agglutination, 663
Agglutination and antigens, 808
Agglutinins, 662, 663
Agglutinogens, 662
Aggregated lymphoid nodules, 786, 900
Aging. *See also* Wrinkling
    cancer, 140, 921
    cardiovascular system, 769
    digestive system, 920-921
    electrolyte/fluid/acid-base balance, 1040
    endocrine system, 643
    free radicals, 36
    gustation and, 569
    hair color, 167, 175
    hearing loss, 600
    height decrease, 275
    immunity, 817
    integumentary system, 175-176
    intervertebral discs, 275, 276
    joints, 283
    muscular system, 377
    nervous system, 556-557
    olfaction, 567
    oogenesis, 1069
    organ systems, 1120
    platysma changes, 343
    reproductive system, 1086

respiratory system, 866-867
    senile cataracts, 577
    skeletal system, 204, 258
    telomeres/telomerase and, 104, 105
    thymus and, 788
    tissue repair, 140
    urinary system, 1007
    UV radiation, 161, 162
Agonist muscles (prime mover), 337
Agranulocytes/agranular leukocytes, 667
AIDS
    cell death mechanisms of, 820
    defense, 820
    helper T cells and, 800-801, 816, 820
    Pneumocystis carinii, 842
    statistics, 820
    symptoms, 820, 821
    transmission, 820
Ala, sacrum, 236
Alanine, 53
Alarm phase, stress, 640-641
Albinism, AM
Albino individuals, 160, 573
Albumins
    plasma proteins, 654
    as radioactive tracer, 472
    testosterone transport, 1062
Alcohol
    ADH and, 618
    effects on cerebellum, 477
    gastric secretion effects, 896
Alcohol abuse, 956
Alcoholics Anonymous, 956
Alcoholism, 956
Aldosterone
    cardiovascular regulation, 744, 745
    electrolyte/fluid balance, 1019, 1023, 1024
    potassium/sodium ions, 627, 993
    secretion disorders, 629
    in stress response, 641
Aldosteronism, 629, 1025
Alkalemia, 1027
Alkaline solution, 44
Alkaline tide, 892, 1038
Alkalosis
    description, 44, 993, 1027
    detection, 1038-1039
    kidneys, 1032-1033
    metabolic alkalosis, 1038, 1041
    respiratory alkalosis, 1036-1037, 1041
Allantois, 1102, 1103, 1104
Alleles
    definition, 1121
    interactions, 1121-1122
Allergens, 668, 816, 821
Allergic rhinitis, 816
Allergies, 815, 816-817, 818, 821
All-or-none principle
    action potentials, 405
    muscle fiber, 308
Alpha-1 receptors, 541, 630
Alpha-2-macroglobulin, 675
Alpha-2 receptors, 541
Alpha-blockers, 552, 560
Alpha cells, 632, 634
Alpha chains, 656, 657, 658
Alpha-helix, 54
Alpha-interferons, 794, 814
Alpha receptors, 540-541
Alpha waves, 491, 492
Altitude (mountain) sickness, 860, 867
Alveolar (acinar) glands, 121
Alveolar air
    composition, 854
    partial pressures, 854, 855
Alveolar capillaries, 843, 844, 853, 854, 864, 866
Alveolar ducts, 839, 840-842

SOME ABBREVIATIONS USED IN THIS TEXT

ACh — acetylcholine
AChE — acetylcholinesterase
ACTH — adrenocorticotropic hormone
ADH — antidiuretic hormone
ADP — adenosine diphosphate
AIDS — acquired immune deficiency syndrome
ALS — amyotrophic lateral sclerosis
AMP — adenosine monophosphate
ANP — atrial natriuretic peptide
ANS — autonomic nervous system
AP — arterial pressure
ARDS — adult respiratory distress syndrome
atm — atmospheric pressure
ATP — adenosine triphosphate
ATPase — adenosine triphosphatase
AV — atrioventricular
AVP — arginine vasopressin
BMR — basal metabolic rate
BCOP — blood colloid osmotic pressure
BPG — bisphosphoglycerate
bpm — beats per minute
BUN — blood urea nitrogen
C — kilocalorie; centigrade
CABG — coronary artery bypass graft
CAD — coronary artery disease
cAMP — cyclic-AMP
CAPD — continuous ambulatory peritoneal dialysis
CCK — cholecystokinin
CD — cluster of differentiation
CF — cystic fibrosis
CHF — congestive heart failure
CHP — capillary hydrostatic pressure
CsHP — capsular hydrostatic pressure
CNS — central nervous system
CO — cardiac output; carbon monoxide
CoA — coenzyme A
COMT — catecol-O-methyltransferase
COPD — chronic obstructive pulmonary disease
CP — creatine phosphate
CPK, CK — creatine phosphokinase
CPM — continous passive motion
CPR — cardiopulmonary resuscitation
CRF — chronic renal failure
CRH — corticotropin-releasing hormone
CSF — cerebrospinal fluid; colony-stimulating factors
CT — computerized tomography; calcitonin
CVA — cerebrovascular accident
CVS — cardiovascular system
DAG — diacylglycerol
D.C. — Doctor of Chiropractic
DCT — distal convoluted tubule
DDST — Denver Developmental Screening Test
DIC — disseminated intravascular coagulation
DJD — degenerative joint disease
DMD — Duchenne's muscular dystrophy
DNA — deoxyribonucleic acid
D.O. — Doctor of Osteopathy
D.P.M. — Doctor of Podiatric Medicine
DSA — digital subtraction angiography
E — epinephrine
ECF — extracellular fluid
ECG — electrocardiogram
EDV — end-diastolic volume
EEG — electroencephalogram
EKG — electrocardiogram
ELISA — enzyme-linked immunosorbent assay
EPSP — excitatory postsynaptic potential
ERV — expiratory reserve volume
ESV — end-systolic volume
ETS — electron transport system
FAD — flavin adenine dinucleotide

FAS — fetal alcohol syndrome
FES — functional electrical stimulation
FMN — flavin mononucleotide
FRC — functional residual capacity
FSH — follicle-stimulating hormone
GABA — gamma aminobutyric acid
GAS — general adaptation syndrome
GC — glucocorticoid
GFR — glomerular filtration rate
GH — growth hormone
GH-IH — growth hormone–inhibiting hormone
GHP — glomerular hydrostatic pressure
GH-RH — growth hormone–releasing hormone
GIP — gastric inhibitory peptide
GnRH — gonadotropin-releasing hormone
GTP — guanosine triphosphate
Hb — hemoglobin
hCG — human chorionic gonadotropin
HCl — hydrochloric acid
HDL — high-density lipoprotein
HDN — hemolytic disease of the newborn
hGH — human growth hormone
HIV — human immunodeficiency virus
HLA — human leukocyte antigen
HMD — hyaline membrane disease
HP — hydrostatic pressure
hPL — human placental lactogen
HR — heart rate
Hz — Hertz
ICF — intracellular fluid
ICOP — interstitial fluid colloid osmotic pressure
IGF — insulin-like growth factor
IH — inhibiting hormone
IM — intramuscular
IP_3 — inositol triphosphate
IPSP — inhibitory postsynaptic potential
IRV — inspiratory reserve volume
ISF — interstitial fluid
IUD — intrauterine device
IVC — inferior vena cava
IVF — in vitro fertilization
kc — kilocalorie
LDH — lactate dehydrogenase
LDL — low-density lipoprotein
L-DOPA — levodopa
LH — luteinizing hormone
LLQ — left lower quadrant
LM — light micrograph
LSD — lysergic acid diethylamide
LUQ — left upper quadrant
MAO — monoamine-oxidase
MAP — mean arterial pressure
MC — mineralocorticoid
M.D. — Doctor of Medicine
mEq — milliequivalent
MHC — major histocompatibility complex
MI — myocardial infarction
mm Hg — millimeters of mercury
mmol — millimole
mOsm — milliosmole
MRI — magnetic resonance imaging
mRNA — messenger RNA
MS — multiple sclerosis
MSH — melanocyte-stimulating hormone
MSH-IH — melanocyte-stimulating hormone–inhibiting hormone
NAD — nicotinamide adenine dinucleotide
NE — norepinephrine
NFP — net filtration pressure
NHP — net hydrostatic pressure
NO — nitric oxide
NRDS — neonatal respiratory distress syndrome
OP — osmotic pressure
Osm — osmoles
OT — oxytocin
PAC — premature atrial contraction
PAT — paroxysmal atrial tachycardia

PCT — proximal convoluted tubule
PCV — packed cell volume
PEEP — positive end-expiratory pressure
PET — positron emission tomography
PFC — perfluorochemical emulsion
PG — prostaglandin
PID — pelvic inflammatory disease
PIH — prolactin-inhibiting hormone
PIP — phosphatidylinositol
PKC — protein kinase C
PKU — phenylketonuria
PLC — phospholipase C
PMN — polymorphonuclear leukocyte
PNS — peripheral nervous system
PR — peripheral resistance
PRF — prolactin-releasing factor
PRL — prolactin
psi — pounds per square inch
PT — prothrombin time
PTA — post-traumatic amnesia; plasma thromboplastin antecedent
PTC — phenylthiocarbamide
PTH — parathyroid hormone
PVC — premature ventricular contraction
RAS — reticular activating system
RBC — red blood cell
RDA — recommended daily allowance
RDS — respiratory distress syndrome
REM — rapid eye movement
RER — rough endoplasmic reticulum
RH — releasing hormone
RHD — rheumatic heart disease
RLQ — right lower quadrant
RNA — ribonucleic acid
rRNA — ribosomal RNA
RUQ — right upper quadrant
SA — sinoatrial
SCA — sickle cell anemia
SCID — severe combined immunodeficiency disease
SEM — scanning electron micrograph
SER — smooth endoplasmic reticulum
SGOT — serum glutamic oxaloacetic transaminase
SIADH — syndrome of inappropriate ADH secretion
SIDS — sudden infant death syndrome
SLE — systemic lupus erythematosus
SNS — somatic nervous system
STD — sexually transmitted disease
SV — stroke volume
SVC — superior vena cava
T_3 — triiodothyronine
T_4 — tetraiodothyronine, or thyroxine
TB — tuberculosis
TBG — thyroid-binding globulin
TEM — transmission electron micrograph
TIA — transient ischemic attack
T_m — transport (tubular) maximum
TMJ — temporomandibular joint
t-PA — tissue plasminogen activator
TRH — thyrotropin-releasing hormone
tRNA — transfer RNA
TSH — thyroid-stimulating hormone
TSS — toxic shock syndrome
TX — thyroxine
U.S. — United States
UTI — urinary tract infection
UTP — uridine triphosphate
UV — ultraviolet
$\dot{V}_A$ — alveolar ventilation
$\dot{V}_D$ — anatomic dead space
$\dot{V}_E$ — respiratory minute volume
V_T — tidal volume
VF — ventricular fibrillation
VLDL — very low-density lipoprotein
VPRC — volume of packed red cells
VT — ventricular tachycardia
WBC — white blood cell